SOME PHYSICAL PROPERTIES

Air (dry, at 20°C and 1 atm)

Density	1.20 kg/m³
Specific heat capacity at constant pressure	1010 J/kg·K
Ratio of specific heat capacities	1.40
Speed of sound	343 m/s
Electrical breakdown strength	3×10^6 V/m
Effective molecular weight	0.0289 kg/mol

Water

Density	1000 kg/m³
Speed of sound	1460 m/s
Specific heat capacity at constant pressure	4190 J/kg·K
Heat of fusion (0°C)	333 kJ/kg
Heat of vaporization (100°C)	2260 kJ/kg
Index of refraction ($\lambda = 589$ nm)	1.33
Molecular weight	0.0180 kg/mol

Earth

Mass	5.98×10^{24} kg
Mean radius	6.37×10^6 m
Standard gravity	9.81 m/s²
Standard atmosphere	1.01×10^5 Pa
Period of satellite at 100 km altitude	86.3 min
Radius of the geosynchronous orbit	42,200 km
Escape speed	11.2 km/s
Magnetic dipole moment	8.0×10^{22} A·m²
Mean electric field at surface	150 V/m, down

Distance to:

Moon	3.82×10^8 m
Sun	1.50×10^{11} m
Nearest star	4.04×10^{16} m
Galactic center	2.2×10^{20} m
Andromeda galaxy	2.1×10^{22} m
Edge of the observable universe	$\sim 10^{26}$ m

THE GREEK ALPHABET

Alpha	A	α	Iota	I	ι	Rho	P	ρ
Beta	B	β	Kappa	K	κ	Sigma	Σ	σ
Gamma	Γ	γ	Lambda	Λ	λ	Tau	T	τ
Delta	Δ	δ	Mu	M	μ	Upsilon	Y	υ
Epsilon	E	ϵ	Nu	N	ν	Phi	Φ	ϕ, φ
Zeta	Z	ζ	Xi	Ξ	ξ	Chi	X	χ
Eta	H	η	Omicron	O	o	Psi	Ψ	ψ
Theta	Θ	θ	Pi	Π	π	Omega	Ω	ω

FUNDAMENTALS OF PHYSICS

SUPPLEMENTS

FUNDAMENTALS OF PHYSICS, THIRD EDITION is accompanied by a complete supplementary package.

STUDY GUIDE

Stanley Williams, Iowa State University
Kenneth Brownstein, University of Maine
Thomas Marcella, University of Lowell

Chapter introductions outline specific types of problems and tie together work done in previous chapters. Quick reviews of definitions, laws, and concepts are combined with detailed explanations of how to apply them to problems.

SELECTED SOLUTIONS

Edward Derringh, Wentworth Institute of Technology

Provides solutions to selected exercises and problems.

WONDERING ABOUT PHYSICS . . . Using Spreadsheets to Find Out

Dewey I. Dykstra, Jr., Boise State University
Robert G. Fuller, U.S. Air Force Academy and University of Nebraska – Lincoln

Composed of 50 investigations, this specially developed supplement leads students to explore the real world of physical phenomena by using spreadsheet software on a personal computer.

LABORATORY PHYSICS, SECOND EDITION

Harry F. Meiners, Rensselaer Polytechnic Institute
Walter Eppenstein, Rensselaer Polytechnic Institute
Kenneth Moore, Rensselaer Polytechnic Institute
Ralph A. Oliva, Texas Instruments, Inc.

This laboratory manual offers a clear introduction to laboratory procedures and instrumentation, including errors, graphing, apparatus handling, calculators, and computers, in addition to over 70 different experiments grouped by topic.

FOR THE INSTRUCTOR

A complete supplementary package of teaching and learning materials is available for instructors. Contact your local Wiley representative for further information.

FUNDAMENTALS OF PHYSICS

—

Third Edition

David Halliday
University of Pittsburgh

Robert Resnick
Rensselaer Polytechnic Institute

with the assistance of

John Merrill
Brigham Young University

WILEY

JOHN WILEY & SONS   New York · Chichester · Brisbane · Toronto · Singapore

Text and cover design: Karin Gerdes Kincheloe
Production supervised by Lucille Buonocore
Illustration supervised by John Balbalis
Copyediting supervised by Deborah Herbert
Photo research: Safra Nimrod

Library of Congress Cataloging in Publication Data:

Halliday, David
    Fundamentals of physics/David Halliday, Robert Resnick;
  with the assistance of John Merrill. — 3rd ed.
      p. cm.
    Includes index.
    ISBN 0-471-81989-1
    1. Physics.   I. Resnick, Robert
  II. Title.
  QC21.2.H35   1988
  530—dc19   87-31703
                                                CIP

Printed in the United States of America

10 9 8 7 6 5 4 3

PREFACE

This third (1988) edition of *Fundamentals of Physics* is a major revision of both the second (1981) edition of that text and of its revised printing (1986). Although we have retained the basic framework of these earlier versions, we have virtually rewritten the entire book. Users of the earlier editions can appreciate the changes better if we list them in some detail.

(a) In the words of one reviewer, "you have succeeded in maintaining the overall level throughout but have substantially lowered the learning threshold." Many new techniques have been used to achieve this. For example, many hints on problem solving are sprinkled throughout the early chapters, each one focusing on a chronic student hang-up. (A complete list of these Hints appears on page xii). A larger set of worked examples — now called Sample Problems to reflect their consistent focus — is included to provide problem-solving models for all aspects of each chapter; several are put in extended question and answer format to reveal directly the pathways followed by experienced problem solvers. There are more but shorter sections per chapter for easier digestion of the material, and more use of subheads is made within sections for greater clarity and emphasis. In the body of the text, relationships are typically displayed and discussed before they are formally derived, a more inductive procedure that we think will prove effective. Often these formal derivations appear in separate sections or subsections. And, we have greatly expanded the set of confidence-building exercises for homework while increasing the number of problems as well.

(b) Greater clarity has been achieved in many ways. A more student-oriented style is employed than before and a two-column format has been adopted for easier reading. Chapter-head photographs are now included along with captions that make a valid attention-grabbing point about the contents of each chapter. Opening sections discuss the relevance of the topics to be treated in each chapter in order to motivate students from the start. Throughout the chapter, photographs and diagrams are featured in greater numbers, with self-contained captions, to reinforce the text material. In each chapter there are examples that deal with practical and applied situations. Chapters conclude with a detailed Review and Summary section for student reference and study.

(c) The sets of chapter-ending questions, exercises, and problems are by far the largest and most varied of any introductory physics text. We have edited the highly praised sets of the earlier edition to achieve even greater clarity and interest and have added a substantial number of new applied and conceptual ones. A more generous use of figures and photographs serves better to illustrate the questions, exercises, and problems than before.

The thought questions have always been a special feature of our books. They are used as sources of classroom discussion and for clarification of homework concepts. There are nearly 30 per chapter. Their total number, now over 1400 in the entire book, is greater than before and they relate even more to everyday phenomena, serve to arouse curiosity and interest, and stress conceptual aspects of physics.

Exercises typically involve one step or formula or represent a single application and are used for building student confidence. They now constitute about 45 percent of the exercise-problem sets. In preparing the new set of problems, we have been careful not to discard the many tried and true problems that have survived the test of the classroom for many years. Long-time users of our text will not find their favorites missing. Of the substantial number of new problems, many fit the "real world" category of student interest and these and the others serve different pedagogic objectives as well. Amongst the problems are a small number of advanced ones, as well, identified by stars* next to their number. A typical chapter has about 31 exercises and 37 problems, the total number of exercises and problems in the entire book being about 3400.

By labeling exercises "E" and problems "P" and organizing them in order of difficulty for each section of the chapter, we have simplified the selection process for teachers from the voluminous material now made available. The variation of level and the breadth of scope have been enlarged. Hence, teachers can vary the content emphasis and the level of difficulty to suit their tastes and the preparation of the student body while still having a very adequate supply for many years of instruction. Indeed, the book is now somewhat longer principally because of all the self-study and learning features that are now included.

(d) Our treatment of modern physics has been enhanced also. There are now two entirely new modern physics chapters, one on Relativity and the other on Particles and the Cosmos. And, in rewriting the earlier chapters, we have sought to pave the way more effectively than in previous editions for the systematic study of modern physics presented in the later chapters. We have done this in three ways. (i) In appropriate places we have called attention — by specific example — to the impact of relativistic and quantum ideas on our daily lives. (ii) We have stressed those concepts (conservation principles, symmetry arguments, reference frames, role of aesthetics, similarity of methods, use of models, field concepts, wave concepts, etc.) that are common to both classical and modern physics. (iii) Finally, we have included a number of short optional sections in which selected relativistic and quantum ideas are presented in

ways that lay the foundation for the detailed and systematic treatments of relativity, atomic, nuclear, solid state, and particle physics given in later chapters.

(e) To emphasize the relevance of what physicists do, and further motivate the student, we include, within the chapters, numerous applications of physics in engineering, technology, medicine, and familiar everyday phenomena. In addition, we feature 21 separate, self-contained essays, written by distinguished scientists and distributed at appropriate locations in the text, on the application of physics to special topics of student interest such as sports, toys, amusement parks, medicine, lasers, holography, space, superconductivity, concert-hall acoustics, and many more. (See the Table of Contents.)

(f) In the interests of simplification and of greater clarity for students, certain rearrangements of material have been made. For example, motion in one dimension is now treated before vectors. A better balance of the material on rotational motion in mechanics is achieved over two chapters by presenting the simpler concepts in kinematics and dynamics first, and then the more difficult concepts, enabling the instructors to more easily choose the depth desired. Similarly, formerly-scattered material — such as on the Doppler effect or on special relativity — has been drawn together in one place for greater conceptual unity. Material on elasticity, now somewhat longer, fits more naturally into the chapter on equilibrium. There are, of course, many other smaller rearrangements too numerous to mention here.

Like the second edition, this edition is available in a single volume of 42 chapters, ending with relativity, and in an Extended Version of 49 chapters that contains in addition a development of quantum physics and its applications to atoms, solids, nuclei, and particles. The former is meant for introductory courses that treat modern quantum physics in a subsequent separate course or semester. There are also numerous optional sections throughout the text that are of an advanced, historical, general, or specialized nature.

Indeed, just as a textbook alone is not a course, so a course does not include the entire textbook. We have consciously made available much more material than any one course or instructor is expected to "cover." More can be "uncovered" by doing less. The process of physics and its essential unity can be revealed by judicious selective coverage of many fewer chapters than are contained here and by coverage of only portions of many included chapters. Rather than give numerous examples of such coherent selections, we urge the instructor to be guided by his or her own interests and circumstances and to plan ahead so that some topics in modern physics are always included.

A textbook contains far more contributions to the elucidation of a subject than those made by the authors alone. As before, John Merrill (Brigham Young University) has been of special service for all aspects of this work, as has Edward Derringh (Wentworth Institute of Technology). Albert Bartlett (University of Colorado) has been of particular help with the essays and Benjamin Chi (SUNY Albany) with the figures and photographs. At John Wiley, publishers, we have been fortunate to receive strong coordination and support from Robert McConnin and Catherine Faduska, physics editors, with notable contributions from John Balbalis, Lucille Buonocore, Deborah Herbert, Karin Kincheloe, Safra Nimrod, and other members of the production team. We are grateful to all these persons.

Our external reviewers have been outstanding and we acknowledge here our debt to each member of that team, namely, Joseph Buschi (Manhattan College), Philip A. Casabella (Rensselaer Polytechnic Institute), Randall Caton (Christopher Newport College), Roger Clapp (University of South Florida), William P. Crummett (Montana College of Mineral Science and Technology), Robert Endorf (University of Cincinnati), F. Paul Esposito (University of Cincinnati), Andrew L. Gardner (Brigham Young University), John Gieniec (Central Missouri State University), Leonard Kleinman (University of Texas at Austin), Kenneth Krane (Oregon State University), Howard C. McAllister (University of Hawaii at Manoa), Manuel Schwartz (University of Louisville), John Spangler (St. Norbert College), Ross L. Spencer (Brigham Young University), Harold Stokes, (Brigham Young University), David Toot (Alfred University), and George U. Williams (University of Utah). Williams (University of Utah).

We thank all the essayists for their valuable contributions and cooperative spirit. Kathaleen Guyette has been superb in providing the wide range of secretarial services required.

We hope that the final product proves worthy of the effort and that this Third Edition of *Fundamentals of Physics* will contribute to the enhancement of physics education.

DAVID HALLIDAY
5110 Kenilworth Place NE
Seattle, WA 98105

ROBERT RESNICK
Rensselaer Polytechnic Institute
Troy, NY 12180-3590

January, 1988

THE ESSAYISTS

ALBERT A. BARTLETT

Albert A. Bartlett (Essays 4 and 11) is a professor of physics at the University of Colorado, Boulder, where he has been a member of the faculty since 1950. He received his B. A. from Colgate University and his Ph. D. from Harvard in nuclear physics. His interests are centered on teaching physics; he was President of the American Association of Physics Teachers in 1978. He is a founding member of PLAN-Boulder—an environmental organization that is largely responsible for Boulder's Greenbelt and open space land acquisition program.

CHARLES P. BEAN

Charles P. Bean (Essay 14) is Institute Professor of Science at Rensselaer Polytechnic Institute. He received his Ph. D. in physics from the University of Illinois in 1952. For more than 33 years he was a research scientist in the General Electric Research Laboratory and its successor, the General Electric Research and Development Center. While there he made research contributions to the fields of ionic crystals, magnetism, superconductivity, and membrane biophysics. He is a member of the National Academy of Sciences and the American Academy of Arts and Sciences. As an avocation he studies the physics of phenomena in nature.

PETER J. BRANCAZIO

Peter J. Brancazio (Essay 6) is a professor of physics at Brooklyn College, City University of New York. He received his Ph. D. in astrophysics from New York University in 1966. He is the author of two books: *The Nature of Physics* (Macmillan, 1975) and *Sport Science* (Simon & Schuster, 1984). His articles on the physics of baseball, football, and basketball have appeared in *Discover, Physics Today, New Scientist, The Physics Teacher,* and the *American Journal of Physics.* A lifelong athlete and sports fan, he is equally at home on the basketball court and in the classroom.

PATRICIA ELIZABETH CLADIS

Patricia Elizabeth Cladis (Essay 20) was born in Shanghai, China and grew up in Vancouver, British Columbia. She received her Ph. D. in physics from the University of Rochester with a thesis on the dc superconducting transformer. Before joining AT&T Bell Laboratories, she did postdoctoral research at the University of Paris, Orsay, France, where she first learned about liquid crystals and discovered "escape into the third dimension" and point defects in nematics. At Bell Labs she discovered the "reentrant nematic" phase. Currently she uses liquid crystals to study general problems in nonlinear physics. She has published nearly 100 scientific papers and is on the editorial board of the journal *Liquid Crystals.*

ELSA GARMIRE

Elsa Garmire (Essay 19) is professor of electrical engineering and physics, and Director of the Center for Laser Studies at the University of Southern California. Garmire received the A. B. in physics from Harvard University in 1961 and the Ph. D. in physics from M. I. T. in 1965 for research in nonlinear optics under Nobel Prizewinner C. H. Townes. The author of over 120 papers with seven patents, she has been a researcher in quantum electronics and in linear and nonlinear optical devices for 25 years. She is a fellow of the Optical Society of America and of IEEE and has been on the board of both societies. She is associate editor of the journals *Optics Letters* and *Fiber and Integrated Optics,* and was U.S. delegate to the International Commission for Optics.

RUSSELL K. HOBBIE

Russell K. Hobbie (Essays 8 and 21) is a professor of physics at the University of Minnesota. He received his B. S. from M. I. T. and his Ph. D. from Harvard. His research interests include diagnostic radiology, magnetic resonance imaging, impedance cardiography, and computerized medical diagnosis. He is the author of *Intermediate Physics for Medicine and Biology,* published by Wiley, 1988.

TUNG H. JEONG

Tung H. Jeong (Essay 18) received his B. S. from Yale University in 1957, and Ph. D. in nuclear physics from the University of Minnesota in 1963. Presently, he is chairman of the physics department and holder of the Albert Blake Dick endowed chair in Lake Forest College. Besides directing annual summer holography workshops since 1972, he consults and lectures on holography in hundreds of institutions around the world. He is a Fellow of the Optical Society of America and a recipient of the Robert A. Millikan Medal from the American Association of Physics Teachers. His hobbies include skiing, tennis, and playing the violin in the Lake Forest Symphony Orchestra.

KENNETH LAWS

Kenneth Laws (Essay 3) is professor of physics at Dickinson College in Carlisle, Pa., where he has been teaching since 1962. He earned his B. S., M. S., and Ph. D. degrees from Caltech, the University of Pennsylvania, and Bryn Mawr College, respectively. For the last dozen years he has been studying classical ballet at the Central Pennsylvania Youth Ballet, and he has recently been applying the principles of physics to dance movement. This work has lead to numerous lectures and classes around the country, and has culminated in a book, *The Physics of Dance,* published in 1984 (paperback 1986) by Schirmer Books.

PETER LINDENFELD

Peter Lindenfeld (Essay 12) has degrees in electrical engineering and engineering physics from the University of British Columbia and a Ph. D. in physics from Columbia University. Since his graduation he has been at Rutgers University where he is professor of physics. His research and publications are on low temperature physics and superconductivity as well as on activities related to physics teaching. He is a fellow of the American Physical Society and an honorary life member of the New Jersey Section of the American Association of Physics Teachers. He has received awards for his booklet "Radioactive Radiations and their Biological Effects," for his work on a solar calorimeter, and for some of his photographs.

RICHARD L. MORIN

Richard L. Morin (Essays 8 and 21) is an associate professor in the Department of Radiology and Director of the Physics Section in Radiology at the University of Minnesota. He received his undergraduate training in chemistry at Emory University and his Ph. D. in radiological sciences from the University of Oklahoma. His research interests are in the area of computer applications in radiology and nuclear medicine.

SUZANNE R. NAGEL

Suzanne R. Nagel (Essay 17) is head of the Glass Research and Engineering Department at AT&T-Bell Laboratories, Murray Hill, N.J. She received her Ph. D. in ceramic engineering from the University of Illinois in 1972 after her undergraduate studies at Rutgers University. She has authored 30 technical papers in the area of glass science and lightguide technology, and her research has involved the processing and property optimization of optical fibers for communications, as well as close interaction with their manufacture. She is actively involved in promoting careers in science and engineering for women and minorities, and was recently featured in the Chicago Museum of Science and Technology Exhibit "My Daughter the Scientist," which is touring the country.

GERARD K. O'NEILL

Gerard K. O'Neill (Essay 13) holds a bachelor's degree from Swarthmore College, a Ph. D. from Cornell University (1954), and an honorary D. Sc. from Swarthmore. He was a member of the faculty at Princeton University from 1954 to 1985, and was made full professor of physics in 1965. In 1985, he retired early from Princeton, becoming Professor Emeritus of Physics. He is the author of several books and many articles in his field. In March 1985, he was appointed by President Reagan to the National Commission on Space. In 1983 he founded the Geostar Corporation, a communications and navigation satellite company based on patents issued to him. His latest commercial start-up company is O'Neill Communications, Inc.

SALLY K. RIDE

Sally K. Ride (Essay 5) is a NASA Space Shuttle astronaut. She earned a B. S. in physics and a B. A. in English from Stanford University in 1973, and a Ph. D. in physics from Stanford in 1978. After graduate school, she was selected for the Astronaut Corps. She has flown in space twice: on the seventh Space Shuttle mission (STS-7, the second flight of the *Challenger,* launched in June, 1983), and the thirteenth Shuttle mission (STS-41G, launched in October, 1984). In 1986, she was appointed to the Presidential Commission investigating the Space Shuttle *Challenger* accident. Since the completion of the investigation, she has acted as Special Assistant to the Administrator of NASA, helping to develop NASA's long-range plans for human exploration of space. She is currently affiliated with the Center for International Security and Arms Control at Stanford University.

JOHN S. RIGDEN

John S. Rigden (Essay 7) received his Ph. D. from Johns Hopkins University in 1960. After postdoctoral work at Harvard University, he started the physics department at Eastern Nazarene College. After one year at Middlebury College, he moved to St. Louis where he is now professor of physics at the University of Missouri – St. Louis. He is the author of *Physics and the Sound of Music* (Wiley, 1977; Second Edition, 1985). More recently, he has written the definitive biography of the great American physicist, I. I. Rabi: *Rabi: Scientist and Citizen* (Basic Books, 1987). Since 1978 he has been the editor of the *American Journal of Physics.*

JOHN L. ROEDER

John L. Roeder (Essay 2) began investigating the physics of the amusement park with an article in the September 1975 issue of *The Physics Teacher* and now takes his physics classes at The Calhoun School in New York City on field trips to Six Flags Great Adventure, the site of his "original research." In addition to serving as a double Resource Agent — for both the American Association of Physics Teachers and the New York Energy Education Project — John is a cofounder of and the newsletter editor for the Teachers Clearinghouse for Science and Society Education, Inc. He received his A. B. from Washington University and his M. A. and Ph. D. from Princeton University.

WILLIAM A. SHURCLIFF

William A. Shurcliff, Physics Department, Emeritus, Harvard University (Essay 9) received his Ph.D. from Harvard in 1934. He has held positions such as senior scientist and research fellow in numerous government, industrial, and university laboratories, including American Cyanamid Company, Polaroid Corporation, Office of Scientific Research and Development, and the Cambridge Electron Accelerator. He is the author of books on polarized light, solar heated houses, and superinsulated houses.

RAYMOND C. TURNER

Raymond C. Turner (Essay 15) is well known for his work with the physics of toys. He received his B. S. in physics from Carnegie Institute of Technology and his Ph. D. in solid state physics from the University of Pittsburgh in 1966. He is now a professor of physics at Clemson University in South Carolina where he conducts research on electron-spin-resonance studies of polymers. He has presented numerous workshops and lectures at

national teachers' meetings on the use of toys in physics education, and he has served on local and national committees of the American Association of Physics Teachers. He has published articles on physics and toys in the *American Journal of Physics* and *The Physics Teacher.*

JEARL WALKER

Jearl Walker (Essays 1, 2, 10 and 16) is professor of physics at Cleveland State University. He received a B. S. in physics from M. I. T. and a Ph. D. in physics from the University of Maryland. Since 1977 he has conducted "The Amateur Scientist" department of *Scientific American,* where he is read in 10 languages in world-wide publication. His book *The Flying Circus of Physics with Answers* is also published in 10 languages.

LIST OF HINTS

These Hints, which occur in the early chapters and are closely correlated with the Sample Problems, should help in working assigned homework problems and in preparing for exams. Collectively, they represent the stock in trade of experienced problem solvers and of practicing scientists and engineers.

CONTENTS

CHAPTER 9
SYSTEMS OF PARTICLES
180

CHAPTER 10
COLLISIONS
205

CHAPTER 11
ROTATIONAL MOTION
227

CHAPTER 12
ROLLING, TORQUE, AND ANGULAR MOMENTUM
256

CHAPTER 13
EQUILIBRIUM AND ELASTICITY
284

CHAPTER 14
OSCILLATIONS
306

CHAPTER 15
GRAVITY
331

CHAPTER 16
FLUIDS
362

CHAPTER 17
WAVES—I
392

CHAPTER 18
WAVES — II
418

CHAPTER 19
TEMPERATURE
446

CHAPTER 20
HEAT AND THE FIRST LAW OF THERMODYNAMICS
464

CHAPTER 21
THE KINETIC THEORY OF GASES
484

CHAPTER 27

CAPACITANCE
618

CHAPTER 28

CURRENT AND RESISTANCE
640

CHAPTER 29

ELECTROMOTIVE FORCE AND CIRCUITS
662

CHAPTER 30

THE MAGNETIC FIELD
686

CHAPTER 41

DIFFRACTION
922

CHAPTER 42

RELATIVITY
952

APPENDICES
A1

CHAPTER 1

MEASUREMENT

The watches are by Breguet, watchmaker to Marie Antoinette and to Napoleon. Although these watches are beautiful, a cheap modern wristwatch keeps better time. If you want still more accurate time, call the Bureau of Standards at (303) 499-7111. Hydrogen maser clocks, which would take 30,000,000 years to gain or lose a second, are even more accurate. Measuring techniques improve as the years roll by!

1-1 Measuring Things

Physics is based on measurement. What is the time interval between two clicks of this counter? What is the temperature of the liquid helium in this vessel? What is the wavelength of the light from this laser? What is the electric current in this wire? The list goes on.

We start then by learning how to measure the physical quantities in terms of which the laws of physics are expressed. Among these quantities are length, time, mass, temperature, pressure, and electrical resistance. We use many of these words in everyday speech. You might say, for example, "I will go to any *length* to help you as long as you do not *pressure* me." In physics, words like *length* and *pressure* have precise meanings, which

we must not confuse with their everyday meanings. In this example, the scientific meanings of length and pressure have nothing to do with their meanings in the quoted sentence. As Robert Oppenheimer has written, "Often the very fact that the words of science are the same as those of our common life and tongue can be more misleading than enlightening."

We define a physical quantity, such as length, by setting up a *standard* and assigning a *unit*—the meter —to it. We are free to define the standard in any way that we want. The important thing is to do so in such a way that scientists around the world will agree that our definition is both sensible and practical.

Once we have set up a standard, for length, say, we must lay out procedures so that any length whatever, be it the radius of a hydrogen atom, the wheelbase of a car, or the distance to a star, can be expressed in terms of this standard. Many of our comparisons must be indirect. You cannot use a ruler, for example, to measure either the radius of an atom or the distance to a star.

There are so many physical quantities that it becomes a problem to organize them. They are not all independent. Speed, for example, is a ratio of a length to a time. What we do is pick out—by international agreement—a small number of physical quantities and assign standards to them alone. We then define all other physical quantities in terms of these *base standards.*

The base standards must be both accessible and invariable. If we define the length standard as the distance between your nose and your outstretched fingertip, we certainly have an accessible standard but it is not very invariable. The demands of science and engineering push us just the other way. We look first for invariability and we exert great effort, in standardizing laboratories around the world, to make secondary and other auxiliary standards accessible to those who need them.

1–2 The International System of Units*

In 1971, the 14th General Conference on Weights and Measures picked seven quantities as *base units,* forming the basis of the International System of Units, abbreviated SI from its French name and popularly known simply as the *metric system.* The SI base units, particularly those of length, mass, and time, are all on a "human scale." If bacteria have a unit system, they no doubt use smaller base units. Table 1 shows the four base units that we shall deal with in the early chapters of this book.

Many SI *derived units* are defined in terms of these base units. For example, the SI unit for power, called the *watt* (abbr. W), is defined in terms of the base units for mass, length, and time. Thus, as we shall see in Section 7–6,

$$1 \text{ watt} = 1 \text{ W} = 1 \text{ kg} \cdot \text{m}^2/\text{s}^3. \tag{1}$$

To express the very large and the very small numbers that we often run into in physics, we use the so-called scientific notation. For example,

$$3{,}560{,}000{,}000 \text{ m} = 3.56 \times 10^9 \text{ m} \tag{2}$$

and

$$0.000\,000\,492 \text{ s} = 4.92 \times 10^{-7} \text{ s}. \tag{3}$$

* See "Foundations of the International System of Units (SI)," by Robert A. Nelson, *The Physics Teacher,* December 1981.

Table 1   SI Base Units Used in Mechanics

Quantity	Name	Symbol
Length	meter	m
Mass	kilogram	kg
Time	second	s
Amount of substance	mole	mol

You should review this notation and make sure that you can use it with ease, both on paper and on your calculator.

As a further convenience when dealing with very large or very small numbers, we often use the prefixes listed in Table 2. Thus, we can express a particular electric power output as

$$1.27 \times 10^9 \text{ watts} = 1.27 \text{ gigawatts} = 1.27 \text{ GW} \tag{4}$$

and a particular time interval as

$$2.35 \times 10^{-9} \text{ second} = 2.35 \text{ nanoseconds} = 2.35 \text{ ns.} \tag{5}$$

Some of these prefixes, such as in milliliter, centimeter, kilogram, and megabucks, are already familiar to you.

Appendix F gives conversion factors to systems of units other than SI. The United States is the only major country (in fact, almost the only country) that has not officially adopted the International System of Units.

Table 2   SI Prefixes[a]

Factor	Prefix	Symbol	Factor	Prefix	Symbol
10^{18}	exa-	E	10^{-18}	atto-	a
10^{15}	peta-	P	10^{-15}	femto-	f
10^{12}	tera-	T	10^{-12}	**pico-**	**p**
10^9	**giga-**	**G**	**10^{-9}**	**nano-**	**n**
10^6	**mega-**	**M**	**10^{-6}**	**micro-**	**μ**
10^3	**kilo-**	**k**	**10^{-3}**	**milli-**	**m**
10^2	hecto-	h	**10^{-2}**	**centi-**	**c**
10^1	deka-	da	10^{-1}	deci-	d

[a] In all cases, the first syllable is accented, as in na′-no-me′-ter. Prefixes commonly used in this book are shown in boldfaced type.

1–3 Changing Units

We often need to change the units in which a physical quantity is expressed. We do so by what are called *chain-link conversions.* In this method, a conversion factor is written as a ratio that is equal to unity. Thus, 1 min and 60 s are identical physical quantities and we can write

$$\frac{1 \text{ min}}{60 \text{ s}} = 1 \quad \text{and} \quad \frac{60 \text{ s}}{1 \text{ min}} = 1.$$

This is *not* the same as writing $\frac{1}{60} = 1$ or $60 = 1$; the *number* and the *unit* must be treated together.

Because multiplying any quantity by unity leaves it unchanged, we can always introduce such conversion factors where we find it useful. We use the factors in such a way that the unwanted units cancel out. For example,

$$2 \text{ min} = (2 \text{ min})(1) = (2 \text{ min}) \left(\frac{60 \text{ s}}{1 \text{ min}} \right) = 120 \text{ s}. \quad (6)$$

If, by chance, you introduce the conversion factor in a way that the units do *not* cancel, simply invert the factor and try again. Note that units obey the same rules as do algebraic variables and numbers.

Sample Problem 1   The research submersible ALVIN is diving at 36.5 fathoms per minute. (a) Express this speed in meters per second. A *fathom* (abbr. fath) is 6 feet.

To find the result in meters per second, we have

$$36.5 \frac{\text{fath}}{\text{min}} = \left(36.5 \frac{\text{fath}}{\text{min}} \right) \left(\frac{1 \text{ min}}{60 \text{ s}} \right) \left(\frac{6 \text{ ft}}{1 \text{ fath}} \right) \left(\frac{1 \text{ m}}{3.28 \text{ ft}} \right)$$

$$= 1.11 \text{ m/s}. \qquad \text{(Answer)}$$

(b) What is this speed in miles per hour? As above, we have

$$36.5 \frac{\text{fath}}{\text{min}} = \left(36.5 \frac{\text{fath}}{\text{min}} \right) \left(\frac{60 \text{ min}}{1 \text{ h}} \right) \left(\frac{6 \text{ ft}}{1 \text{ fath}} \right) \left(\frac{1 \text{ mi}}{5280 \text{ ft}} \right)$$

$$= 2.49 \text{ mi/h}. \qquad \text{(Answer)}$$

(c) What is this speed in light-years per year? A light-year, defined as the distance that light travels in 1 year, is 9.46×10^{12} km.

We start from the result of (a) above, or

$$1.11 \frac{\text{m}}{\text{s}} = \left(1.11 \frac{\text{m}}{\text{s}} \right) \left(\frac{1 \text{ ly}}{9.46 \times 10^{12} \text{ km}} \right)$$

$$\times \left(\frac{1 \text{ km}}{1000 \text{ m}} \right) \left(\frac{3.16 \times 10^7 \text{ s}}{1 \text{ y}} \right)$$

$$= 3.71 \times 10^{-9} \text{ ly/y}. \qquad \text{(Answer)}$$

We can write this in the even more unlikely form of 3.71 nly/y, where "nly" is an abbreviation for nanolight-year.

If you calculated the answer to part (a) above with your calculator "wide open," you would find 1.112804878 m/s, a result whose implied precision is totally meaningless. We have (properly) rounded this number to 1.11 m/s, which is the preci-

sion that is justified by the precision of the input data. Adjust your calculator so that it does this rounding for you. It will streamline your work and — on examinations — your instructor may well be impressed with your sophistication in handling numbers.*

1-4   Length

In 1792, the newborn French Republic established a metric system of weights and measures. As its cornerstone, the meter was defined to be one ten-millionth of the distance from the north pole to the equator. Eventually, for practical reasons, this earth standard was abandoned and the meter came to be defined as the distance between two fine lines engraved near the ends of a platinum–iridium bar.

This *standard meter bar* was kept at the International Bureau of Weights and Measures near Paris and accurate copies were sent to other standardizing laboratories throughout the world. These secondary standards were used to calibrate other, still more accessible, standards so that ultimately every measuring device derived its authority from the standard meter bar through a complicated chain of comparisons. In 1959, this also became true for the yard, whose legal definition in this country was adopted in that year to be

$$1 \text{ yard} = 0.9144 \text{ meter (exactly)}, \qquad (7)$$

which is equivalent to

$$1 \text{ inch} = 2.54 \text{ centimeters (exactly)}. \qquad (8)$$

Table 3 shows some measured lengths.

Eventually, the accuracy with which length comparisons could be made by aligning fine scratches under a microscope became no longer good enough for modern science and technology. In 1960, an atomic standard for the meter, based on the wavelength of light, was adopted. Specifically, the meter was redefined to be 1,650,763.73 wavelengths of a particular orange-red light emitted by atoms of krypton-86 in a gas discharge tube.† This awkward number of wavelengths was chosen

---

* A fuller discussion of *significant figures* in problem solving appears in Chapter 4. See Hints 1, 2, and 7 of that chapter.

† The number (86) in the notation krypton-86 identifies a particular one of the five stable isotopes of this element. An equivalent notation is ^{86}Kr. The number is called the *mass number* of the isotope in question.

Table 3 Some Lengths

Length	Meters
To the farthest observed quasar (1987)	2×10^{26}
To the Andromeda galaxy	2×10^{22}
To the nearest star (Proxima Centauri)	4×10^{16}
To the farthest planet (Pluto)	6×10^{12}
Radius of the earth	6×10^{6}
Height of Mt. Everest	9×10^{3}
Thickness of this page	1×10^{-4}
Wavelength of light	5×10^{-7}
Size of a typical virus	1×10^{-8}
Radius of the hydrogen atom	5×10^{-11}
Radius of a proton	$\sim 10^{-15}$

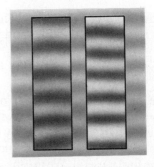

Figure 1 A master machinist's gauge block (left) being compared with a reference standard (right) by means of light waves. If the fringes match, the blocks are of equal length. These two blocks differ in length by about 25 nm.

so that the new standard would be as consistent as possible with the old meter-bar standard.

The krypton-86 atoms of the atomic length standard are available everywhere, are identical, and emit light of precisely the same wavelength. As Philip Morrison has pointed out, every atom is a storehouse of natural units, more secure than the International Bureau of Weights and Measures.

Figure 1 shows how the length of a master machinist's gauge block, used in industry as a precise secondary length standard, is compared with a reference standard at the National Bureau of Standards. The fringes that cross the figure are formed by the interference of light waves. If the fringe patterns on the two rectangular blocks match, the gauge blocks are the same length. If the fringes fail to match by, say, one-tenth of a fringe, the blocks differ in length by one-twentieth of a wavelength, or ~ 30 nm.

By 1983, the demands for higher precision had reached such a point that even the krypton-86 standard could not meet them and in that year a bold step was taken. The meter was redefined as the distance traveled by a light wave in a specified time interval. In the words of the 17th General Conference on Weights and Measures:

The meter is the length of the path traveled by light in vacuum during a time interval of 1/299,792,458 of a second.

This is equivalent to saying that the speed of light c is now *defined as*

$$c = 299{,}792{,}458 \text{ m/s} \quad \text{(exactly).}$$

This new definition of the meter was necessary because measurements of the speed of light had become so precise that the reproducibility of the krypton-86 meter itself became the limiting factor. In view of this, it then made sense to adopt the speed of light as a defined quantity and to use it to redefine the meter.

Sample Problem 2 In track meets, both 100 yd and 100 m are used as distances for dashes. (a) Which is longer?

From Eq. 7, 100 yd is equal to 91.44 m, so that 100 m is longer than 100 yd.

(b) By how many meters is it longer?

We represent the difference by ΔL, where Δ is the capital Greek letter delta. Thus,

$$\Delta L = 100 \text{ m} - 100 \text{ yd} = 100 \text{ m} - 91.44 \text{ m} = 8.56 \text{ m}.$$
(Answer)

(c) By how many feet is it longer?

We can express this difference by the method used in Sample Problem 1,

$$\Delta L = (8.56 \text{ m}) \left(\frac{3.28 \text{ ft}}{1 \text{ m}} \right) = 28.1 \text{ ft.} \quad \text{(Answer)}$$

Thus, 100 m is greater than 100 yd by 28.1 ft.

1-5 Time

Time has two aspects. For civil and some scientific purposes, we want to know the time of day (see Fig. 2) so that we can order events in sequence. In much scientific work, we want to know how long an event lasts. Thus,

Figure 2   When the metric system was proposed in 1792, the hour was redefined to provide a 10-h day. It did not catch on. The maker of this 10-h watch wisely provided a small dial that kept conventional 12-h time. Do the two dials indicate the same time?

Table 4   Some Time Intervals

Time Interval	Seconds
Lifetime of the proton (predicted)	$\sim 10^{39}$
Age of the universe	5×10^{17}
Age of the pyramid of Cheops	1×10^{11}
Human life expectancy (U.S.)	2×10^{9}
Length of a day	9×10^{4}
Time between human heartbeats	8×10^{-1}
Period of concert-A tuning fork	2×10^{-3}
Lifetime of the muon	2×10^{-6}
Shortest laboratory light pulse (1987)	1×10^{-15}
Lifetime of most unstable particle	$\sim 10^{-23}$
The Planck time[a]	$\sim 10^{-43}$

[a] This is the earliest time after the "Big Bang" at which the laws of physics as we know them can be applied.

any time standard must be able to answer the questions: "At what time did it happen?" and "How long did it last?" Table 4 shows some measured time intervals.

Any phenomenon that repeats itself is a possible time standard. The rotating earth, which determines the length of the day, has been used in this way for centuries.

A quartz clock can be calibrated against the rotating earth by astronomical observations and used as a transfer standard to measure time intervals in the laboratory. However, the calibration cannot be carried out with the accuracy called for by modern science and technology.

To meet the need for a better time standard, atomic clocks have been developed in several countries. Figure 3 shows such a clock, based on a characteristic frequency of the isotope cesium-133, at the National Bureau of Standards. It forms the basis in this country for Coordi-

Figure 3   Cesium atomic frequency standard No. NBS-6 at the National Bureau of Standards in Boulder, Colorado. It is the primary standard for the unit of time in the United States. Dial (303) 499-7111 and set your watch by it. Dial (900) 410-8463 for Naval Observatory time si-

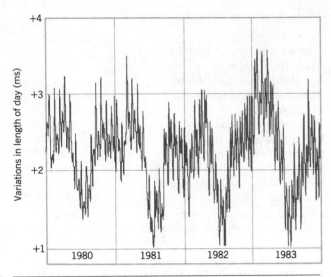

Figure 4 The variation in the length of the day, over a 4-year period. Note that the entire vertical scale amounts to only 3 ms (= 0.003 s).

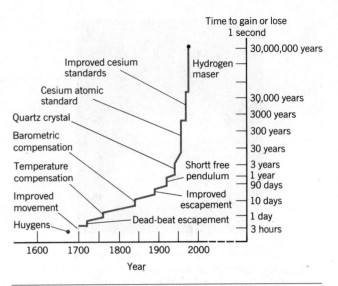

Figure 5 Clocks get better and better. Early pendulum clocks gained or lost 1 s in a few hours. Hydrogen maser clocks would do so only after 30,000,000 years.

nated Universal Time (UTC), for which time signals are available by shortwave radio (stations WWV and WWVH) and by telephone.

Figure 4 shows, by comparison with a cesium clock, variations in the rate of rotation of the earth over a 4-year period.* The relative simplicity of the atom leads us to look to the rotating earth to account for any differences between them as timekeepers. The variations that we see in Fig. 4 can be accounted for as tidal effects caused by the moon and as seasonal variations in the atmospheric winds.

The second based on the cesium clock was adopted as the international standard by the 13th General Conference on Weights and Measures in 1967. The following definition was given:

One second is the time occupied by 9,192,631,770 vibrations of the light (of a specified wavelength) emitted by a cesium-133 atom.

Two cesium clocks could run for 6000 years before their readings would differ by more than 1 s. Hydrogen maser

clocks, not (yet) adopted as time standards, have achieved the incredible precision of 1 s in 30,000,000 years. This is equivalent to 3 μs per century of operation. Figure 5 shows the impressive record of improvements in timekeeping that have occurred over the past 300 years or so, starting with the pendulum clock, invented by Christian Huygens in 1656, and ending with today's hydrogen maser.

Sample Problem 3 Isaac Asimov has proposed a unit of time based on the highest known speed and the smallest measurable distance. It is the *light-fermi,* the time taken by light to travel a distance of 1 fermi (= 1 femtometer = 1 fm = 10^{-15} m). (a) How many seconds are there in a light-fermi?

We find this time by dividing the distance by c, the speed of light (= 3.00×10^8 m/s). Thus,

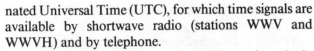

$$1 \text{ light-fermi} = \frac{1 \text{ femtometer}}{\text{speed of light}} = \frac{10^{-15} \text{ m}}{3.00 \times 10^8 \text{ m/s}}$$
$$= 3.33 \times 10^{-24} \text{ s}. \qquad \text{(Answer)}$$

Table 4 tells us that the most unstable elementary particle known to date has an average lifetime of about 10^{-23} s before it disintegrates. We could say that its average lifetime is about 3 light-fermis.

(b) Asimov's proposal finds its symmetrical counterpart in the well-known *light – year,* which is not a time but a dis-

---

* See "The Earth's Inconstant Rotation," by John Wahr, *Sky and Telescope,* June 1986. See also "Studying the Earth by Very-Long-Baseline Interferometry," by William E. Carter and Douglas S. Robertson, *Scientific American,* November 1986.

tance, namely, the distance that light travels in 1 year. How many meters are there in 1 light-year?

This distance is

$$1 \text{ light-year} = ct = (3.00 \times 10^8 \text{ m/s})(3.16 \times 10^7 \text{ s})$$
$$= 9.48 \times 10^{15} \text{ m } (=9.48 \text{ Pm}). \qquad \text{(Answer)}$$

Table 3 tells us that the nearest star to us (Proxima Centauri) is about 4×10^{16} m distant. It is then about 4.2 light-years away. The light from this star that reaches your eye now left it when you were 4.2 years younger.

1-6   Mass

The SI standard of mass is a platinum–iridium cylinder kept at the International Bureau of Weights and Measures near Paris and assigned, by international agreement, a mass of 1 kilogram. Accurate copies have been sent to standardizing laboratories in other countries and the masses of other bodies can be found by the equal-arm balance method to a precision of about two parts in 10^8. Table 5 shows some measured masses expressed in kilograms.

The U.S. copy of the standard kilogram, known as Prototype Kilogram No. 20 (see Fig. 6), is housed in a vault at the National Bureau of Standards. It is removed no more than once a year, for the purpose of checking tertiary standards. Since 1889, it has been taken to

Figure 6   The National Standard Kilogram No. 20, resting in its double bell jar.

Table 5   Some Masses

Object	Kilograms
Known universe	$\sim 10^{53}$
Our galaxy	2×10^{41}
Sun	2×10^{30}
Moon	7×10^{22}
Asteroid Eros	5×10^{15}
Small mountain	1×10^{12}
Ocean liner	7×10^7
Elephant	5×10^3
Grape	3×10^{-3}
Speck of dust	7×10^{-10}
Penicillin molecule	5×10^{-17}
Uranium atom	4×10^{-26}
Proton	2×10^{-27}
Electron	9×10^{-31}

France twice for recomparison with the primary standard.

A Second Mass Standard.   The masses of atoms can be compared with each other more precisely than they can be compared with the standard kilogram. For this reason, we have a second mass standard. It is the carbon-12 atom which, by international agreement, has been assigned a mass of 12 *unified atomic mass units* (abbr. u). The relation between the two standards is

$$1 \text{ u} = 1.6605402 \times 10^{-27} \text{ kg}. \qquad (9)$$

Like all measured quantities, this result carries with it some uncertainty of measurement. In this case, the uncertainty is about ± 10 units in the last two places of decimals.

We can find the masses of other atoms to a considerable accuracy using a mass spectrometer; see Fig. 7. Table 6 shows some atomic masses.

Table 6   Some Atomic Masses

Atom	Mass (u)	Uncertainty (u)
Hydrogen-1	1.00782504	0.00000001
Carbon-12	12.00000000	(exact)
Copper-64	63.9297656	0.0000017
Silver-102	101.91195	0.00012
Cesium-137	136.907073	0.000006
Platinum-190	189.959917	0.000007
Plutonium-238	238.0495546	0.0000024

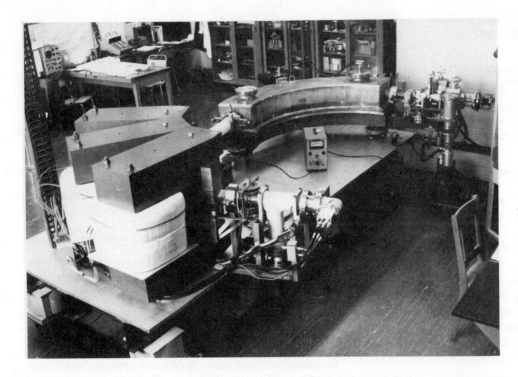

Figure 7 A high-resolution mass spectrometer at the University of Manitoba. It is used for making precise measurements of atomic mass, such as those listed in Table 6. Work in this laboratory is supported by the National Research Council of Canada.

REVIEW AND SUMMARY

Measurement in Physics

Physics is based on the measurement of the physical quantities we use to describe the changes that take place in our universe. Each quantity is measured as a multiple of some unit (for example, meters, seconds, and kilometers per hour). All the units we use may be expressed as combinations of a very few *base* units, which are selected to make their use as convenient as possible.

SI Units

The unit system used in this book is the International System of Units (SI). The four physical quantities displayed in Table 1 serve as a base in the early chapters of this book. *Standards,* which must be both accessible and invariable, define units for these base quantities and are established by international agreement. These standards underlie all physical measurement, for both the base quantities and quantities derived from them, such as force or electric resistivity. The prefixes of Table 2 simplify notation in many cases.

Unit Conversions

Conversion of units from one system to another (for example, from miles per hour to kilometers per second) may be performed by using *chain-link conversions* in which the units are treated as algebraic quantities as the original data are multiplied successively, by conversion factors written as unity, until the desired units are obtained. See Section 1–3 and Sample Problem 1.

The Meter

The unit of length — the meter — is defined as the distance traveled by light during a precisely specified time interval. The yard, together with its multiples and submultiples, is legally defined in this country in terms of the meter.

The Second

The unit of time — the second — was formerly defined in terms of the rotation of the earth. It is now defined in terms of the period of oscillation of the light emitted by an atomic (cesium-133) source. Accurate time signals are disseminated worldwide by radio signals keyed to atomic clocks in various standardizing laboratories.

The Kilogram

The unit of mass — the kilogram — is defined in terms of a particular platinum–iridium prototype kept in Paris, France. For atomic problems, the unified atomic mass unit, defined in terms of the atom carbon-12, is also used.

QUESTIONS

1. How would you criticize this statement: "Once you have picked a standard, by the very meaning of 'standard' it is invariable"?

2. List characteristics other than accessibility and invariability that you would consider desirable for a physical standard.

3. Can you imagine a system of base units (Table 1) in which time was not included?

4. Of the four base units listed in Table 1, only one—the kilogram—has a prefix (see Table 2). Would it be wise to redefine the mass of that platinum–iridium cylinder at the International Bureau of Weights and Measures as 1 g rather than 1 kg?

5. Why are there no SI base units for area or volume?

6. The meter was originally intended to be one ten-millionth of the meridian line from the north pole to the equator that passes through Paris. This definition disagrees with the standard meter bar by 0.023%. Does this mean that the standard meter bar is inaccurate to this extent?

7. In defining the meter bar as the standard of length, the temperature is specified. Can length be called a fundamental quantity if another physical quantity, such as temperature, must be specified in choosing a standard?

8. In redefining the meter in terms of the speed of light, why did not the delegates to the 1983 General Conference on Weights and Measures simplify matters by defining the speed of light to be 3×10^8 m/s exactly? For that matter, why did they not define it to be 1 m/s exactly? Were both of these possibilities open to them? If so, why did they reject them?

9. What does the prefix "micro-" signify in the words "microwave oven"? It has been proposed that food that has been irradiated by gamma rays to lengthen its shelf life be marked "picowaved." What do you suppose that means?

10. Suggest a way to measure (a) the radius of the earth, (b) the distance between the sun and the earth, and (c) the radius of the sun.

11. Suggest a way to measure (a) the thickness of a sheet of paper, (b) the thickness of a soap bubble film, and (c) the diameter of an atom.

12. Name several repetitive phenomena occurring in nature which could serve as reasonable time standards.

13. You could define "1 second" to be one pulse beat of the current president of the American Physical Society. Galileo used his pulse as a timing device in some of his work. Why is a definition based on the atomic clock better?

14. What criteria should be satisfied by a good clock?

15. From what you know about pendulums, cite the drawbacks to using the period of a pendulum as a time standard.

16. On June 30, 1981, the "minute" extending from 10:59 to 11:00 a.m. was arbitrarily lengthened to contain 61 s. This *leap second* was introduced to compensate for the fact that, as measured by our atomic time standard, the earth's rotation rate is slowly decreasing. Why is it desirable to readjust our clocks in this way?

17. A radio station advertises "at 89.5 on your FM dial." What does this number mean?

18. Why do we find it useful to have two standards of mass, the kilogram and the carbon-12 atom?

19. Is the current standard kilogram of mass accessible, invariable, reproducible, and indestructible? Does it have simplicity for comparison purposes? Would an atomic standard be better in any respect? Why don't we adopt an atomic standard, as we do for length and time?

20. Suggest practical ways by which one could determine the masses of the various objects listed in Table 5.

21. Suggest objects whose masses would fall in the wide range in Table 5 between that of an ocean liner and a small mountain and estimate their masses.

22. Critics of the metric system often cloud the issue by saying things such as "Instead of buying one pound of butter you will have to ask for 0.454 kg of butter." The implication is that life would be more complicated. How might you refute this?

EXERCISES AND PROBLEMS

Section 1–2   The International System of Units

1E. Use the prefixes in Table 2 and express (a) 10^6 phones; (b) 10^{-6} phones; (c) 10^1 cards; (d) 10^9 lows; (e) 10^{12} bulls; (f) 10^{-1} mates; (g) 10^{-2} pedes; (h) 10^{-9} Nannettes; (i) 10^{-12} boos; (j) 10^{-18} boys; (k) 2×10^2 withits; (l) 2×10^3 mockingbirds. Now that you have the idea, invent a few more similar expressions. (See, in this connection, p. 61 of *A Random Walk in Science*,

compiled by R. L. Weber; Crane, Russak & Co., New York, 1974.)

2E. Some of the prefixes of the SI units have crept into everyday language. (a) What is the weekly equivalent of an annual salary of 36 K (= 36 kilobucks)? (b) A lottery awards 10 megabucks as the top prize, payable over 20 years. How much is received in each monthly check? (c) The hard disk of a computer has a capacity of 30 MB (= 30 megabytes). At 8 bytes/word, how many words can it store? [In computerese, *kilo* means 1024 (= 2^{10}), not 1000.]

Section 1–4 Length

3E. A space shuttle orbits the earth at an altitude of 300 km. What is this distance in (a) miles and (b) millimeters?

4E. What is your height in meters?

5E. The micrometer (10^{-6} m = 1 μm) is often called the *micron.* (a) How many microns make up 1 km? (b) What fraction of a centimeter equals 1 μm? (c) How many microns are in 1 yd?

6E. The earth is approximately a sphere of radius 6.37×10^6 m. (a) What is its circumference in kilometers? (b) What is its surface area in square kilometers? (c) What is its volume?

7E. Calculate the number of kilometers in 20 mi using only the following conversion factors: 1 mi = 5280 ft, 1 ft = 12 in., 1 in. = 2.54 cm, 1 m = 100 cm, and 1 km = 1000 m.

8E. Give the relation between (a) a square inch and a square centimeter; (b) a square mile and a square kilometer; (c) a cubic meter and a cubic centimeter; and (d) a square foot and a square yard.

9P. A unit of area, often used in expressing areas of land, is the *hectare,* defined as 10^4 m². An open-pit coal mine consumes 75 hectares of land, down to a depth of 26 m, each year. What volume of earth, in cubic kilometers, is removed in this time?

10P. The *cord* is a volume of cut wood equal to a stack 8 ft long, 4 ft wide, and 4 ft high. How many cords of wood are in 1 cubic meter?

11P. A room is 20 ft, 2 in. long and 12 ft, 5 in. wide. What is the floor area in (a) square feet and (b) square meters? If the ceiling is 12 ft, 2½ in. above the floor, what is the volume of the room in (c) cubic feet and (d) cubic meters?

12P. Antarctica is roughly semicircular in shape with a radius of 2000 km. The average thickness of the ice cover is 3000 m. How many cubic centimeters of ice does Antarctica contain? (Ignore the curvature of the earth.)

13P. A typical sugar cube has an edge length of 1 cm. If you had a cubical box that contained a mole of sugar cubes, what would its edge length be? (One mole = 6.02×10^{23} units.)

14P. Hydraulic engineers often use, as a unit of volume of water, the *acre·foot,* defined as the volume of water that will cover 1 acre of land to a depth of 1 ft. A severe thunderstorm dumps 2 in. of rain in 30 min on a town of area 26 km². What volume of water, in acre·feet, fell on the town?

15P. A certain brand of house paint claims a coverage of 460 ft²/gal. (a) Express this quantity in square meters per liter. (b) Express this quantity in SI base units (see Appendix A). (c)

What is the inverse of this quantity and what is its physical significance?

16P. Astronomical distances are so large compared to terrestrial ones that much larger units of length are used for easy comprehension of the relative distances of astronomical objects. An *astronomical unit* (AU) is equal to the average distance from the earth to the sun, about 92.9×10^6 mi. A *parsec* (pc) is the distance at which 1 AU would subtend an angle of 1 second of arc. A *light-year* (ly) is the distance that light, traveling through a vacuum with a speed of 186,000 mi/s, would cover in 1 year. (a) Express the distance from the earth to the sun in parsecs and in light-years. (b) Express a light-year and a parsec in miles. Although the light-year is much used in popular writing, the parsec is the unit used professionally by astronomers.

17P. Assume that the average distance of the sun from the earth is 400 times the average distance of the moon from the earth. Now consider a total eclipse of the sun and state conclusions that can be drawn about (a) the relation between the sun's diameter and the moon's diameter and (b) the relative volumes of the sun and the moon. (c) Hold up a dime (or another small coin) so that it just eclipses the full moon and measure the angle it intercepts at the eye. From this experimental result and the given distance between the moon and the earth (= 3.8×10^5 km), estimate the diameter of the moon.

18P*. The standard kilogram (see Fig. 6) is in the shape of a circular cylinder with its height equal to its diameter. Show that, for a circular cylinder of fixed volume, this equality gives the smallest surface area, thus minimizing the effects of surface contamination and wear.

19P*. The navigator of the oil tanker *Gulf Supernox* uses the satellites of the Global Positioning System (GPS/NAVSTAR) to find latitude and longitude; see Fig. 8. These are 43°36′25.3″ N and 77°31′48.2″ W. If the accuracy of these determinations is ±0.5″, what is the uncertainty in the tanker's position measured along (a) a north–south line and (b) an east–west line? (c) Where is the tanker?

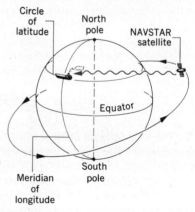

Figure 8 Problem 19.

Section 1–5 Time

20E. Express the speed of light, 3.0×10^8 m/s, in (a) feet per nanosecond and (b) millimeters per picosecond.

21E. Enrico Fermi once pointed out that a standard lecture period (50 min) is close to 1 microcentury. How long is a microcentury in minutes, and what is the percent difference from Fermi's approximation?

22E. There are 365.25 days in 1 year. How many seconds are in 1 year?

23E. A certain pendulum clock (with a 12-h dial) happens to gain 1 min/day. After setting the clock to the correct time, how long must one wait until it again indicates the correct time?

24E. What is the age of the universe (see Table 4) in days?

25E. (a) A unit of time sometimes used in microscopic physics is the *shake*. One shake equals 10^{-8} s. Are there more shakes in a second than there are seconds in a year? (b) Humans have existed for about 10^6 years, whereas the universe is about 10^{10} years old. If the age of the universe is taken to be 1 day, for how many seconds have humans existed?

26E. The maximum speeds of various animals are given roughly as follows in miles per hour: (a) snail, 3×10^{-2}; (b) spider, 1.2; (c) squirrel, 12; (d) human, 23; (e) rabbit, 35; (f) fox, 42; (g) lion, 50; and (h) cheetah, 70. Convert these data to meters per second.

27P. An astronomical unit (AU) is the average distance of the earth from the sun, approximately 150,000,000 km. The speed of light is about 3.0×10^8 m/s. Express the speed of light in terms of astronomical units per minute.

28P. Until 1883, every city and town in the United States kept its own local time. Today, travelers reset their watches only when the time change equals 1 h. How far, on the average, must you travel in degrees of longitude until your watch must be reset by 1 h?

29P. In two *different* track meets, the winners of the mile race ran their races in 3 min 58.05 s and 3 min 58.20 s. In order to conclude that the runner with the shorter time was indeed faster, what is the maximum tolerable error, in feet, in laying out the distances?

30P. Five clocks are being tested in a laboratory. Exactly at noon, as determined by the WWV time signal, on the successive days of a week the clocks read as follows:

Clock	Sun.	Mon.	Tues.	Wed.	Thurs.	Fri.	Sat.
A	12:36:40	13:36:56	12:37:12	12:37:27	12:37:44	12:37:59	12:38:14
B	11:59:59	12:00:02	11:59:57	12:00:07	12:00:02	11:59:56	12:00:03
C	15:50:45	15:51:43	15:52:41	15:53:39	15:54:37	15:55:35	15:56:33
D	12:03:59	12:02:52	12:01:45	12:00:38	11:59:31	11:58:24	11:57:17
E	12:03:59	12:02:49	12:01:54	12:01:52	12:01:32	12:01:22	12:01:12

How would you arrange these five clocks in the order of their relative value as good timekeepers? Justify your choice.

31P. Assuming that the length of the day uniformly increases by 0.001 s in a century, calculate the cumulative effect on the measure of time over 20 centuries. Such a slowing down of the earth's rotation is indicated by observations of the occurrences of solar eclipses during this period.

32P*. The time it takes the moon to return to a given position as seen against the background of the fixed stars is called a *sidereal* month. The time interval between identical phases of the moon is called a *lunar* month. The lunar month is longer than a sidereal month. Why, and by how much?

Section 1–6 Mass

33E. Using conversions and data in the chapter, determine the number of hydrogen atoms required to obtain 1 kg of mass.

34E. One molecule of water (H_2O) contains two atoms of hydrogen and one atom of oxygen. A hydrogen atom has a mass of 1.0 u and an atom of oxygen has a mass of 16 u, approximately. (a) What is the mass in kilograms of one molecule of water? (b) How many molecules of water are in the oceans of the world? The oceans have a total mass of 1.4×10^{21} kg.

35E. The earth has a mass of 5.98×10^{24} kg. The average mass of the atoms that make up the earth is 40 u. How many atoms are in the earth?

36E. In continental Europe, one "pound" is half a kilogram. Which is the better buy: one German pound of coffee for $3.00 or one American pound of coffee for $2.40?

37P. What mass of water fell on the town in Problem 14 during the thunderstorm? One cubic meter of water has a mass of 10^3 kg.

38P. (a) Assuming that the density (mass/volume) of water is exactly 1 g/cm³, express the density of water in kilograms per cubic meter (kg/m³). (b) Suppose that it takes 10 h to drain a container of 5700 m³ of water. What is the mass flow rate, in kilograms per second, of water from the container?

39P. A person on a diet loses 2.3 kg (corresponding to about 5 lb) per week. Express the mass loss rate in milligrams per second.

40P. The grains of fine California beach sand have an average radius of 50 μm. What mass of sand grains would have a total surface area equal to the surface area of a cube 1 m on an edge? Sand is made of silicon dioxide, 1 m³ of which has a mass of 2600 kg.

41P. The distance between neighboring atoms, or molecules, in a solid substance can be estimated by calculating twice the radius of a sphere with volume equal to the volume per atom of the material. Calculate the distance between neighboring atoms in (a) iron and (b) sodium. The densities of iron and sodium are 7.87 g/cm³ and 1.013 g/cm³, respectively; the mass of an iron atom is 9.27×10^{-26} kg, and the mass of a sodium atom is 3.82×10^{-26} kg.

CHAPTER 2

MOTION IN A STRAIGHT LINE

The skyscraper brought with it the elevator and straight-line motion entered its vertical phase. This elevator cab has a total run of 624 ft. It is one of a bank of 12 serving 49 floors of the New York Marriott Hotel. The cab can make the total run, without stops, in 42 s.

2–1 Motion

It is hard to imagine a world in which nothing moves. The study of motion (called *kinematics*) is so important, in fact, that we choose it for the first major topic in this text.* The moving objects that we might examine are — among countless possibilities — a hockey puck sliding on the ice, a baseball flying in a high arc to right field, a car braking to rest, or a space probe on its way to Mars. We shall look at simple cases first, restricting our choices in three ways:

1. We Consider Motion in a Straight Line Only. The line may be vertical (a falling stone), horizontal (a car on

a level highway), or anywhere in between but it must be straight.

2. We Seek Only to Describe the Motion, Not to Account for It. We do not (yet) deal with the mass of the moving object or with the forces that act on it. We simply want to describe motion of the object as it coasts along, speeds up, slows down, or stops and reverses itself.

3. We Consider Only Those Objects that Can Be Represented as Particles. We can represent an object as a particle (that is, as a mass point) if every small part of the object (every atom, say) moves in exactly the same way. A crate sliding down an inclined plane meets this requirement. A rolling wheel does not because a point on the rim of the wheel moves in quite a different way than a point on its axis.

---

* The word *kinematics* is derived from the Greek word for motion. The word *cinema* (moving pictures) has the same root.

Often, an object can be treated as a particle in one situation but not in another. The earth behaves pretty much like a particle if we are interested only in its orbital motion around the sun. To those of us who live on it, however, the earth is no particle. Physicists are particularly interested in the ultimate particles, such as electrons, quarks, and protons; we shall say a little about them in Section 2-8.

2-2 Position

The first question that we are likely to ask about a moving particle is: "Where is it?" Formally, we take its path to be an x axis and we want to know, as a function of time, its *position $x(t)$* on this axis, measured with respect to a selected origin.

Figure 1*a* shows $x(t)$ for the simplest possible kind of motion — no motion at all. It represents a car parked by the side of a highway at a 60-km marker. Figure 1*b* shows a similar plot for a car moving along the highway at a constant speed of 80 km/h ($\approx$ 50 mi/h). Figure 2 shows $x(t)$ plots for several running animals, ranging

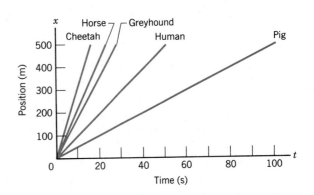

Figure 2 Position–time plots for five animals. We assume that each animal runs for 500 m in a straight line at its top speed.

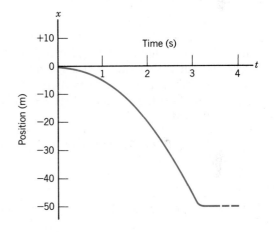

Figure 3 A ball of modeling clay falls 50 m and sticks to the floor. The curve is a position–time plot for the falling object. It is *not* a plot of its path, which is a vertical straight line.

from a cheetah (70 mi/h) to a pig (11 mi/h), with a human (22 mi/h) somewhere in between. Figure 3 is an $x(t)$ plot for a ball of modeling clay, dropped down a stairwell and sticking to the floor 50 m below.

2-3 Average Velocity

Our next question is likely to be: "How fast is the particle moving?" If the $x(t)$ curve is a straight line, as in Figs. 1 and 2, the velocity is constant and is equal to the slope of that line. We see from Fig. 2 that the cheetah travels 500 m in about 16 s. Its velocity is then

$$(500 \text{ m})/(16 \text{ s}) \quad \text{or} \quad 31 \text{ m/s} (= 70 \text{ mi/h}).$$

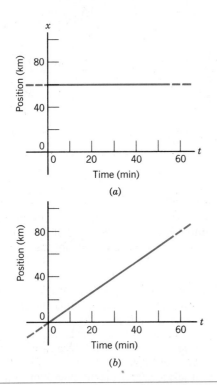

Figure 1 (*a*) A position–time plot for a parked car. (*b*) The same for a car moving at a constant speed of 80 km/h.

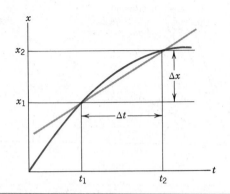

Figure 4 The curve is a position–time plot for a particle moving along the x axis. The average velocity of the particle for the interval specified by the two dots is $\Delta x/\Delta t$, which is the slope of the straight line joining the two dots.

If the $x(t)$ curve is not straight, we can define an *average velocity* $\bar{v}$ of the particle over a specified time interval. Suppose that, as in Fig. 4, the particle is at x_1 at time t_1 and at x_2 at a later time t_2. Its average velocity for this interval is defined as*

$$\bar{v} = \frac{x_2 - x_1}{t_2 - t_1} = \frac{\Delta x}{\Delta t}, \qquad (1)$$

in which

$$\Delta x = x_2 - x_1 \qquad (2)$$

and

$$\Delta t = t_2 - t_1. \qquad (3)$$

Here Δx is the *change in position* or the *displacement* of the particle that occurs during the time interval Δt. We see from the figure that $\bar{v}$ is simply the slope of the straight line that connects the end points of the interval.

The algebraic sign of $\bar{v}$ tells us the direction in which the particle is moving. If $\bar{v}$ is positive, then we must have $x_2 > x_1$ in Eqs. 1 and 2 and the particle is moving so that x increases with time. If $\bar{v}$ is negative, then $x_2 < x_1$ and the particle is moving so that x decreases with time.

Sample Problem 1 You drive your BMW down a straight road for 5.2 mi at 43 mi/h, at which point you run out of gas. You walk 1.2 mi farther, to the nearest gas station, in 27 min.

---

* Throughout this book, a bar over a symbol means an average value of the quantity that the symbol represents.

What is your average velocity from the time that you started your car to the time that you arrived at the gas station?

You can find your average velocity from Eq. 1 if you know both Δx, the net distance that you covered (your displacement) and Δt, the corresponding elapsed time. These quantities are

$$\Delta x = 5.2 \text{ mi} + 1.2 \text{ mi} = 6.4 \text{ mi}$$

and

$$\Delta t = \frac{5.2 \text{ mi}}{43 \text{ mi/h}} + 27 \text{ min}$$
$$= 0.121 \text{ h} + 0.450 \text{ h} = 0.57 \text{ h}.$$

From Eq. 1 we then have

$$\bar{v} = \frac{\Delta x}{\Delta t} = \frac{6.4 \text{ mi}}{0.57 \text{ h}} = 11 \text{ mi/h}. \qquad \text{(Answer)}$$

The $x(t)$ plot of Fig. 5 helps to visualize the problem. Points O and P define the interval for which we want to find the average velocity, this quantity being the slope of the straight line connecting these points.

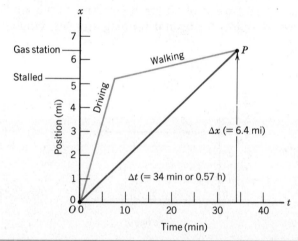

Figure 5 Sample Problem 1. The lines marked "Driving" and "Walking" are the position–time plots for the driver–walker in Sample Problem 1. The slope of the straight line joining O and P is the average velocity for the trip.

Sample Problem 2 Suppose that, in Sample Problem 1, you carry the gas back to your car, covering the return trip in 35 min. What is your average velocity this time, from your starting point at the car to your arrival back at the car on foot?

As before, you must find the net distance covered and the total elapsed time and then divide them to find the average velocity. The displacement Δx is the change in your position between the beginning of the interval (starting your car) and the end of the interval (arriving back at your car on foot). This distance is 5.2 mi (not 7.6 mi). The fact that you walked to the gas station and back does not enter at this point; only your positions at the beginning and at the end of the interval are involved. The elapsed time is

$$\Delta t = \frac{5.2 \text{ mi}}{43 \text{ mi/h}} + 27 \text{ min} + 35 \text{ min}$$

$$= 0.121 \text{ h} + 0.450 \text{ h} + 0.583 \text{ h} = 1.2 \text{ h}.$$

The fact that you went to the gas station and back *does* enter here, in calculating the elapsed time.

The average velocity, from Eq. 1, is then

$$\bar{v} = \frac{\Delta x}{\Delta t} = \frac{5.2 \text{ mi}}{1.2 \text{ h}} = 4.3 \text{ mi/h}. \qquad \text{(Answer)}$$

This is substantially less than the value calculated in Sample Problem 1 (=11 mi/h) because you took a longer time to cover a shorter net distance.

Hint 1: *Read the problem carefully.** For beginning problem solvers, no difficulty is more common than simply not understanding the problem. The best test of understanding that we know of is: Can you explain the problem, in your own words, to a friend? Give it a try.

Hint 2: *Understand what is given and what is requested.* Write down the given data, with units, using the symbols of the chapter. (In Sample Problems 1 and 2, the given data allow you to find your net displacement Δx and the corresponding time interval Δt.) What is the unknown and what is its symbol? (In these Sample Problems, the unknown is your average velocity, symbol $\bar{v}$.) What is the connection between the unknown and the data? (The connection is Eq. 1, the definition of average velocity.)

Hint 3: *Watch the units.* Be sure to use a consistent set of units when putting numbers into the equations. In Sample Problems 1 and 2, which involve a car, the logical units in terms of the given data are miles for distances, hours for time intervals, and miles/hour for velocities. You may need to make conversions, as we did in finding Δt.

Hint 4: *Think about your answer.* Look at your answer and ask yourself whether it makes sense. Is it far too large or far too small? Is the sign correct? Are the units appropriate? In Sample Problem 1, for example, the correct answer is 11 mi/h. If you find 0.00011 mi/h, −11 mi/h, 11 mi/s, or

---

* Collectively, these Hints, which appear in the early chapters and are closely linked to the Sample Problems, represent the stock in trade of experienced problem solvers and of practicing scien- tists and engineers. A List of Hints appears at the front of the book.

11,000 mi/h, you should realize at once that you have done something wrong. The error may lie in your method, in your algebra, or in your arithmetic. Check the problem carefully, being sure to start at the very beginning.

In Sample Problem 1, your answer must be greater than your speed of walking (2–3 mi/h) but less than the speed of the car (=43 mi/h). Finally, the answer to Sample Problem 2 (=4.3 mi/h) must be less than that to Sample Problem 1 (=11 mi/h) for two reasons: you covered a shorter net distance and you took a longer time in which to do so.

Hint 5: *Reading a graph:* Figures 1, 2, 3, and 5 are examples of graphs that you should be able to read easily. For each graph, the variable on the horizontal axis is the time t, the direction of increasing time being to the right. In each graph, the variable on the vertical axis is the position x of the moving object with respect to the origin, the direction of increasing x being upward.

Note the units (seconds or minutes; meters, kilometers, or miles) in which the variables are expressed and note whether the variables are positive or negative (see Fig. 3). In Fig. 2, you should be able to see at once that a cheetah can run faster than a pig (for the same distance, its time is shorter). Each of these graphs describes something that is happening and you should be able to say what it is.

2-4   Instantaneous Velocity

What we usually want to know about a moving particle is not its average velocity over a given time interval but how fast it is moving at a given instant. This *instantaneous velocity* is the value that the average velocity approaches as the interval over which we measure it is made smaller and smaller, approaching zero as a limit. In formal terms,

$$v = \lim_{\Delta t \to 0} \frac{\Delta x}{\Delta t}. \qquad (4)$$

In calculus, the quantity on the right of Eq. 4 is what we mean by the derivative of x with respect to t, which we write as dx/dt. We can then write Eq. 4 as*

$$\boxed{v = \frac{dx}{dt}.} \qquad (5)$$

Thus, for motion in one dimension, the instantaneous velocity of a particle is the rate at which its position is changing with time. From now on, the word *velocity* will

---

* This is the first use of a derivative in this book.

Table 1   The Limiting Process

Beginning Point		End Point		Intervals		Velocity
x_1 (m)	t_1 (s)	x_2 (m)	t_2 (s)	Δx (m)	Δt (s)	$\Delta x / \Delta t$ (m/s)
5.00	1.00	9.00	3.00	4.00	2.00	+2.0
5.00	1.00	8.75	2.50	3.75	1.50	+2.5
5.00	1.00	8.00	2.00	3.00	1.00	+3.0
5.00	1.00	6.75	1.50	1.75	0.50	+3.5
5.00	1.00	5.760	1.200	0.760	0.200	+3.8
5.00	1.00	5.388	1.100	0.388	0.100	+3.9
5.00	1.00	5.196	1.050	0.196	0.050	+3.9
5.00	1.00	5.158	1.040	0.158	0.040	+4.0
5.00	1.00	5.119	1.030	0.119	0.030	+4.0

refer to the instantaneous velocity, unless we say otherwise.

Table 1 will help in understanding the limiting process and the meaning of a derivative. (Hint 7 deals with the same problem graphically.) The first two columns in Table 1 define the beginning of an interval on a particular $x(t)$ curve and the second two columns define its end. The beginning point ($x_1 = 5.00$ m and $t_1 = 1.00$ s) remains fixed and the end point (x_2, t_2) is allowed to approach it.

The last three columns of Table 1 show us that, although the time interval Δt and the corresponding displacement Δx both shrink toward zero, the ratio of these intervals approaches the constant value of $+4.0$ m/s. According to Eq. 4 this is the (instantaneous) velocity of the moving particle at $t = 1.0$ s. The positive sign of the velocity tells us that the particle is moving so that x increases with time.

The magnitude of the velocity of a particle, always a positive quantity, is called its *speed*. For example, suppose that a particle moving along the x axis in the direction of increasing x has a *velocity* of $+5$ m/s. Suppose also that a particle moving along this axis in the opposite direction has a *velocity* of -5 m/s. Both particles have a *speed* of 5 m/s. Appropriately enough, speed is what the speedometer of a car reads. If you are stopped on an interstate highway, the trooper is concerned both with the sign of your velocity (are you going in the right direction?) and with its magnitude (what is your speed?).

Let us now look once more at Fig. 3, the $x(t)$ curve for the falling ball of modeling clay. The initial slope of this curve is zero, which tells us that the ball was released from rest and not thrown. As time goes on, the falling object picks up speed, the slope of the curve becoming steeper and steeper. The slope and thus the velocity are negative while the object is falling, which reminds us that

the object is moving in the negative x direction. The ball strikes the ground at about $t = 3.2$ s and has an abrupt change in speed, sticking to the floor and remaining at rest, as the horizontal slope of the curve beyond $t = 3.2$ s confirms.

Sample Problem 3   Figure 6a is an $x(t)$ plot for an elevator cab that, starting from rest at the ground floor, rises 120 m to the 29th floor and comes to rest. Plot the velocity of the cab as a function of time.

The slope, and hence the velocity, is zero at points a and d of Fig. 6a, indicating that the cab is at rest. For the interval bc during which the cab is moving upward with constant speed, the slope is

$$v_{up} = \frac{\Delta x}{\Delta t} = \frac{90 \text{ m} - 20 \text{ m}}{35 \text{ s} - 21 \text{ s}} = \frac{70 \text{ m}}{14 \text{ s}} = +5.0 \text{ m/s}.$$

The positive sign tells us that the cab is moving toward larger values of x; it is going up. If the cab were descending at the same rate, its velocity would be -5.0 m/s.

The *speed* of the cab, whether it is going up or down, is 5.0 m/s. Figure 6b is a plot of $v(t)$, the velocity of the cab as a function of time. We deal later with Fig. 6c.

Sample Problem 4   In Sample Problem 2, what is your average *speed* for the round trip from the start of your journey to your return to your car?

Your average speed $\bar{s}$ is the total distance covered ($=7.6$ mi) divided by the total elapsed time ($=1.2$ h), or

$$\bar{s} = \frac{7.6 \text{ mi}}{1.2 \text{ h}} = 6.3 \text{ mi/h}. \qquad \text{(Answer)}$$

Note that your average speed, which involves the *total* distance covered, is greater than your average velocity ($=4.3$ mi/h; see Sample Problem 2), which involves only the *net* distance covered.

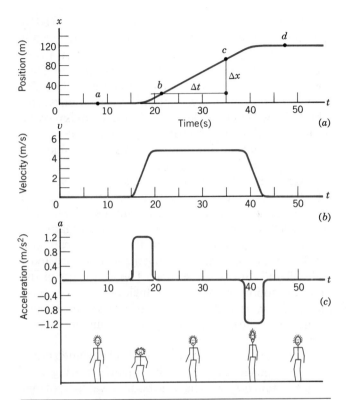

Figure 6   Sample Problem 3. (*a*) An *x*(*t*) curve for a moving elevator cab. (*b*) A *v*(*t*) curve for the moving cab. Check that it is the derivative of the *x*(*t*) curve above ($v = dx/dt$). (*c*) An *a*(*t*) curve for the moving cab. Check that it is the derivative of the *v*(*t*) curve above ($a = dv/dt$). The figures suggest the kinesthetic responses to the accelerations.

Sample Problem 5   A particle moves along the *x* axis in such a way that its position is given by

$$x(t) = 7.8 + 9.2t - 2.1t^3. \qquad (6)$$

The units of the constant coefficients are not shown but they are such that, if *t* is given in seconds, *x* will be in meters.* What is the velocity of the particle at *t* = 3.5 s?

From Eq. 5 we have, substituting *x*(*t*) from Eq. 6,

$$v = \frac{dx}{dt} = \frac{d}{dt}(7.8 + 9.2t - 2.1t^3)$$

or

$$v = 0 + 9.2 - (3 \times 2.1)t^2 = 9.2 - 6.3t^2. \qquad (7)$$

At *t* = 3.5 s, the velocity is

$$v = 9.2 - (6.3)(3.5)^2 = -68 \text{ m/s.} \qquad \text{(Answer)}$$

---

* The coefficients in Eq. 6 are, in sequence and including their units, 7.8 m, 9.2 m/s, and 2.1 m/s³.

The negative sign tells us that the particle is moving in the negative *x* direction.

Hint 6: *Negative numbers.*   The line below is an *x* axis with its origin (*x* = 0) at the center. Using this scale, make sure you understand that, for example, −40 m is less than −10 m and that both are less than 20 m. Note also that 10 m is greater than −30 m.

The four arrows pointing to the right all represent increases in *x*, that is, positive values for Δ*x*, the change in *x*. The four arrows pointing to the left represent decreases in *x*, that is, negative values for Δ*x*, the change in *x*.

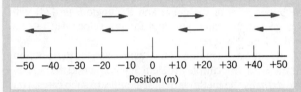

Hint 7: *Derivatives and slopes.*   Every derivative is the slope of a curve. In Sample Problem 3, for example, the velocity of the cab at any instant (a derivative; see Eq. 5) is the slope of the *x*(*t*) curve of Fig. 6*a* at that instant. Let us see how to find a slope (and thus a derivative) graphically.

Figure 7 shows an *x*(*t*) plot for a moving particle. If you want to find the velocity of the particle at *t* = 1 s, put a dot on the curve at that point. Then draw a tangent line (*tangent*

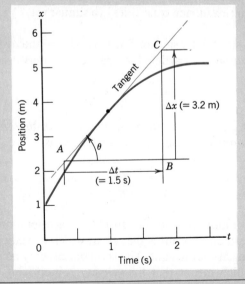

Figure 7   The derivative of a curve at any point is the slope of its tangent line at that point. At *t* = 1.0s, the slope of the tangent line (and thus *dx*/*dt*, the instantaneous velocity) is Δ*x*/Δ*t* = +2.1 m/s.

means *touching;* the tangent line touches the curve at a single point) judging carefully by eye. Construct the right triangle *ABC.* (Although the slope is the same no matter what the size of this triangle, the larger the triangle the more precise will be your graphical measurements.) Find Δx and Δt, using the vertical and horizontal scales to provide the magnitude, the unit, and the sign. Find the slope (derivative) from the following equation:

$$\text{slope} = \frac{\Delta x}{\Delta t} = \frac{5.5 \text{ m} - 2.3 \text{ m}}{1.8 \text{ s} - 0.3 \text{ s}} = \frac{3.2 \text{ m}}{1.5 \text{ s}} = +2.1 \text{ m/s}.$$

As Eq. 5 tells us, this slope is the velocity of the particle at $t = 1$ s. The slope is positive in this case.

If you change the scale on the x or the t axis of Fig. 7, the appearance of the curve and the angle θ ($=47°$) will change but the value you find for the velocity at $t = 1$ s ($= +2.1$ m/s) will not.

If you have a mathematical expression for the function $x(t)$, as we do in Sample Problem 5, you can find the derivative dx/dt by the methods of the calculus. In such cases, you do not need to use the graphical method we have described.

2-5 Acceleration

If the position of a particle changes with time, we have seen how to define an average velocity for a particular interval (Eq. 1) and also an instantaneous velocity (Eq. 5). Suppose now that the velocity changes with time. Can we define useful new quantities in a similar way? Yes, we can.

In strict extension of Eq. 1, we define the *average acceleration* of a particle for a particular interval from

$$\bar{a} = \frac{v_2 - v_1}{t_2 - t_1} = \frac{\Delta v}{\Delta t}. \tag{8}$$

In strict extension of Eq. 5, we define the *instantaneous acceleration* (or simply the *acceleration*) as

$$a = \frac{dv}{dt}. \tag{9}$$

Thus, for motion in one dimension, the acceleration of a particle at any instant is the rate at which its velocity is changing at that instant. The acceleration may be positive, which means that the velocity is increasing with time, or it may be negative, which means that this velocity is decreasing as time goes on.*

---

* The word *deceleration* is often used to mean a decrease in speed.

In Fig. 6c, we show the accelerations experienced by the elevator cab of Sample Problem 3. Compare this figure carefully with Fig. 6b, directly above, and convince yourself that the lower curve is the derivative (slope) of the upper one, as Eq. 9 requires. Where the velocity is constant, for example, the acceleration is zero. This holds whether the elevator cab is at rest or is moving (either up or down) with a constant velocity. The acceleration occurs only during those intervals during which the velocity of the cab is changing.

When the cab starts from rest, its velocity rises from 0 to $+5.0$ m/s and its acceleration is positive. When the cab comes to rest at the top of its run, its velocity falls from $+5.0$ m/s to 0 and the acceleration is negative.

If you were a passenger in the elevator cab, you would feel a little "heavy" during the first interval and a little "light" during the second. The awareness of the body to motion, possessed in a high degree by dancers and athletes, is called the *kinesthetic* sense, a word formed from the Greek words for *motion* and for *perception.*

Although our stomach is a good accelerometer, it is no speedometer. Passengers in a plane moving in a straight line at 600 mi/h have no bodily awareness of that fact. If the velocity of the plane changes, however, as in braking on landing or because of air turbulence, they all know about it. Figure 8 shows Colonel John P. Stapp undergoing an acceleration of about -200 m/s² as the high-speed rocket sled in which he was riding was braked to rest. He seems to know that something is happening.

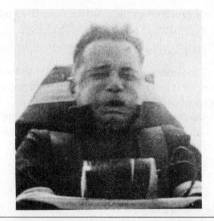

Figure 8 Colonel Stapp, braking to rest in his rocket sled. It is not the fact that he is traveling fast (large velocity) that makes him look this way. It is the fact that his velocity is changing rapidly (large acceleration) that accounts for his distress. His body is an accelerometer, not a speedometer.

Sample Problem 6 (a) Your car, starting from rest, gets up to 55 mi/h in 3.2 s. What is its average acceleration?

Choose the direction in which the car is moving to be an x axis. Your velocity goes from 0 to $+55$ mi/h in 3.2 s. Your average acceleration for this interval is, from Eq. 8,

$$\bar{a} = \frac{\Delta v}{\Delta t} = \frac{55 \text{ mi/h} - 0}{3.2 \text{ s}} = +17 \text{ mi/h} \cdot \text{s.} \quad \text{(Answer)}$$

These mixed units have a certain practical usefulness for problems involving accelerating automobiles. You can also express this acceleration as $+25$ ft/s^2 or as $+7.6$ m/s^2. The positive sign tells us that the velocity is increasing as time goes on.

(b) Later, you brake your car to rest from 55 mi/h in 4.7 s. What is its average acceleration in this case?

From Eq. 8,

$$\bar{a} = \frac{\Delta v}{\Delta t} = \frac{0 - 55 \text{ mi/h}}{4.7 \text{ s}} = -12 \text{ mi/h} \cdot \text{s.} \quad \text{(Answer)}$$

You can also express this acceleration in other units, as -18 ft/s^2 or as -5.5 m/s^2. In this case the minus sign tells us that the car is slowing down.

2-6 Constant Acceleration

An object falling freely near the surface of the earth or a uniformly braked car are familiar examples of the important special case of a particle moving with *constant acceleration;** see Fig. 9c. The distinction between average acceleration and instantaneous acceleration loses its meaning in this case and we can write Eq. 8, with some changes in notation, as

$$a = \frac{v - v_0}{t - 0}.$$

Here v_0 is the velocity at time $t = 0$ and v is the (variable) velocity at any time t. We can recast this as

$$\boxed{v = v_0 + at.} \quad (10)$$

As a check, note that this equation reduces to $v = v_0$ for $t = 0$, as it must. As a further check, take the derivative of Eq. 10. Doing so yields $dv/dt = a$, which is the definition of a. Figure 9b shows a plot of Eq. 10, the $v(t)$ function.

Our next job is to find out how the position of the

---

* In many situations, the acceleration is *not* constant and the results of this section do not apply. It is important to make sure of this point when solving problems.

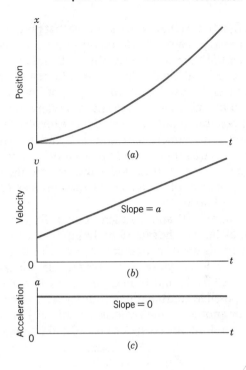

Figure 9 (*a*) The position $x(t)$ of a particle moving with constant acceleration. (*b*) Its velocity $v(t)$, given at each point by the slope of the curve in (*a*) above. (*c*) Its (constant) acceleration, equal to the (constant) slope of the curve $v(t)$ above.

particle varies with time. For the position at any time t, we can write

$$x = x_0 + \bar{v}t, \quad (11)$$

in which x_0 is the position at $t = 0$ and $\bar{v}$ is the average velocity between $t = 0$ and $t = t$.

If you plot v against t using Eq. 10, a straight line results. Under these conditions, the *average* velocity over any time interval (say $t = 0$ to $t = t$) is the average of the velocity at the beginning of the interval ($= v_0$) and the velocity at the end of the interval ($= v$). For the interval $t = 0$ to $t = t$ then, the average velocity is

$$\bar{v} = \tfrac{1}{2}(v_0 + v). \quad (12)$$

Substituting for v from Eq. 10 yields, after a little rearrangement,

$$\bar{v} = v_0 + \tfrac{1}{2}at. \quad (13)$$

Finally, substituting Eq. 13 into Eq. 11 yields

$$\boxed{x - x_0 = v_0t + \tfrac{1}{2}at^2} \quad (14)$$

for the function $x(t)$. As a check, we see that putting $t = 0$ yields $x = x_0$, as it must. The quantity $x - x_0$ is the displacement of the particle between $t = 0$ and $t = t$. As a further check, taking the derivative of Eq. 14 yields Eq. 10, again as it must. Figure 9a is a plot of Eq. 14.

In problems involving constant acceleration, there are five possible quantities that might be involved in any given problem, namely, $x - x_0$, v, t, a, and v_0. Usually, one of these quantities is *not* involved in the problem, *either as a given or as an unknown.* We are then presented with three of the remaining quantities and asked to find the fourth.

Equations 10 and 14 each contain four of these quantities, but not the same four. In Eq. 10, the "missing ingredient" is the displacement, $x - x_0$. In Eq. 14, it is the velocity v. In organizing to solve problems, it helps to recombine Eqs. 10 and 14 in various ways, to make sure that all possible combinations of unknowns are present in one or another of the resulting equations. For example, we can eliminate t between Eqs. 10 and 14, obtaining

$$v^2 = v_0^2 + 2a(x - x_0). \qquad (15)$$

We can eliminate the acceleration a between these same two equations and find

$$x - x_0 = \tfrac{1}{2}(v_0 + v)t. \qquad (16)$$

Finally, we can eliminate v_0, yielding

$$x - x_0 = vt - \tfrac{1}{2}at^2. \qquad (17)$$

Table 2 lists Eqs. 10, 14, 15, 16, and 17 and shows which one of the five possible quantities is the missing ingredient in each equation. To solve a constant acceler-

Table 2 Equations for Motion with Constant Acceleration[a]

Equation Number	Equation	Missing Quantity
10	$v = v_0 + at$	$x - x_0$
14	$x - x_0 = v_0t + \tfrac{1}{2}at^2$	v
15	$v^2 = v_0^2 + 2a(x - x_0)$	t
16	$x - x_0 = \tfrac{1}{2}(v_0 + v)t$	a
17	$x - x_0 = vt - \tfrac{1}{2}at^2$	v_0

[a] Make sure that the acceleration is indeed constant before using the equations in this table. Note that if you differentiate Eq. 14 you get Eq. 10. The other three equations are found by eliminating one or another of the variables between Eqs. 10 and 14.

ation problem, you must decide which of the five quantities is *not* involved in the problem, either as a given or as an unknown, and select the correct equation from Table 2. We shall show how to do this in Sample Problem 7.

Sample Problem 7 You brake your Porsche from 75 km/h to 45 km/h over a displacement of 88 m. (a) What is the acceleration, assumed to be constant?

In this problem the time is not involved, being neither given nor requested. Table 2 then leads us to Eq. 15. Solving this equation for a yields

$$a = \frac{v^2 - v_0^2}{2(x - x_0)} = \frac{(45 \text{ km/h})^2 - (75 \text{ km/h})^2}{(2)(0.088 \text{ km})}$$
$$= -2.05 \times 10^4 \text{ km/h}^2 \approx -1.6 \text{ m/s}^2. \quad \text{(Answer)}$$

The minus sign reminds us that the velocity is decreasing.

(b) What is the elapsed time?

If we use the original data, the missing ingredient is now the acceleration. Table 2 suggests Eq. 16. Solving that equation for t, we obtain

$$t = \frac{2(x - x_0)}{v_0 + v} = \frac{(2)(0.088 \text{ km})}{(75 + 45) \text{ km/h}}$$
$$= 1.5 \times 10^{-3} \text{ h} = 5.4 \text{ s}. \quad \text{(Answer)}$$

(c) If you continue to slow down with the acceleration calculated in (a) above, how much time would elapse in bringing the car to rest from 75 km/h?

The quantity not given or asked for is the displacement, $x - x_0$. Table 2 then suggests that we use Eq. 10. Solving for t gives

$$t = \frac{v - v_0}{a} = \frac{0 - (75 \text{ km/h})}{(-2.05 \times 10^4 \text{ km/h}^2)}$$
$$= 3.7 \times 10^{-3} \text{ h} = 13 \text{ s}. \quad \text{(Answer)}$$

(d) In (c) above, what distance would be covered? From Eq. 14, we have, for the displacement of the car,

$$x - x_0 = v_0t + \tfrac{1}{2}at^2$$
$$= (75 \text{ km/h})(3.7 \times 10^{-3} \text{ h})$$
$$+ \tfrac{1}{2}(-2.05 \times 10^4 \text{ km/h}^2)(3.7 \times 10^{-3} \text{ h})^2$$
$$= 0.137 \text{ km} \approx 140 \text{ m}. \quad \text{(Answer)}$$

(e) Suppose that, on a second trial with the acceleration calculated in (a) above and a different initial velocity, you bring your car to rest after traversing 200 m. What was the total braking time?

The missing quantity here is the initial velocity. We must use Eq. 17. Noting that v (the final velocity) is zero and solving this equation for t gives us

$$t = \left(-\frac{(2)(x - x_0)}{a}\right)^{1/2} = \left(-\frac{(2)(200 \text{ m})}{-1.6 \text{ m/s}^2}\right)^{1/2}$$
$$= 16 \text{ s}. \quad \text{(Answer)}$$

In this Sample Problem, we have used every equation listed in Table 2.

Hint 8: *Check the dimensions.* The dimensions of a velocity are (L/T), that is, length (L) divided by time (T). The dimensions of acceleration are (L/T²) and so on. In any physical equation, the dimensions of all terms must be the same. If you are in doubt about an equation, check its dimensions.

To check the dimensions of Eq. 14 ($x - x_0 = v_0 t + \frac{1}{2} a t^2$), we note that every term must be a length, because that is the dimension of x and of x_0. The dimension of the term $v_0 t$ is (L/T)(T) which is (L). The dimension of $\frac{1}{2} a t^2$ is (L/T²)(T²), which is also (L). This equation checks out. Note that pure numbers such as $\frac{1}{2}$ or π have no dimensions.

You may have remembered this equation incorrectly, perhaps as $x - x_0 = v_0 t^2 + \frac{1}{2} a t$. Checking the dimensions will show that something is wrong.

2-7 Freely-Falling Objects

Figure 10 is a strobe photo of a falling billiard ball. Experiment shows that all objects, no matter what their mass, density, or shape, fall to earth from a given release point with the same *free-fall acceleration* **g**. *Free fall* means falling in a vacuum, so that the frictional resistance and buoyant effect of the air do not affect the motion.

The quantity **g** is often called *the acceleration due to gravity.* However, as we shall see in Chapter 15, the earth's rotation also plays a (small) role in determining this acceleration so that, strictly, this label is a misnomer.*

Figure 11 shows an apparatus at the National Bureau of Standards used to measure *g* by means of light waves reflected from a freely falling mirror. The measurements are so precise that, if the apparatus were raised by even a few centimeters, the (very slightly) reduced value of *g* would be detected.

The direction of **g** at any point determines the direction of the vertical at that point; it defines what we mean by "down." The magnitude of **g** varies from point to point on the earth's surface and also with elevation. Its

---

* The effect of the earth's rotation on the acceleration of a freely falling body amounts—at mid-latitudes—to only about 0.2%.

Figure 10 A strobe photo of a falling ball. The flashes occur 1/30 of a second apart. Note that the ball picks up speed as it falls.

Figure 11 Apparatus at the National Bureau of Standards used to measure the acceleration of gravity to very high precision. It uses light waves to time a falling mirror housed in the central cylinder. The apparatus was designed and built by James E. Faller and his associates.

average value at sea level and mid-latitudes is 9.8 m/s² (≈ 32 ft/s²), a value that we adopt for problems in this chapter.

The equations of Table 2 apply to free fall. We recast them for our purpose, making two small changes. (1) We label the direction of free fall a y axis (rather than an x axis), choosing its positive direction as upward. We do this because, in Chapter 4, we shall be dealing with motion in two dimensions and we want there to reserve the x axis for horizontal motion and the y axis for vertical motion. (2) With this choice, the acceleration will be negative, so we replace a in the equations by $-g$, where g has the value 9.8 m/s².

With these small changes, the equations of Table 2 become

$$v = v_0 - gt, \qquad \{y - y_0\} \quad [10] \quad (18)$$
$$y - y_0 = v_0 t - \tfrac{1}{2}gt^2, \qquad \{v\} \quad [14] \quad (19)$$
$$v^2 = v_0^2 - 2g(y - y_0), \quad \{t\} \quad [15] \quad (20)$$
$$y - y_0 = \tfrac{1}{2}(v_0 + v)t, \qquad \{g\} \quad [16] \quad (21)$$

and

$$y - y_0 = vt + \tfrac{1}{2}gt^2. \qquad \{v_0\} \quad [17] \quad (22)$$

Here the square brackets give the corresponding equation numbers from Table 2. The curly brackets give the missing ingredient.

t	y	v	a
s	m	m/s	m/s²
0	0	0	−9.8
1	−4.9	−9.8	−9.8
2	−19.6	−19.6	−9.8
3	−44.1	−29.4	−9.8
4	−78.4	−39.2	−9.8

Figure 12 Sample Problem 8. The position, velocity, and acceleration for a freely-falling object.

Sample Problem 8   A worker drops a wrench down the elevator shaft of a tall building. (a) Where is the wrench 1.5 s later?

The missing ingredient is the velocity v, which is neither given nor requested. This suggests Eq. 19 above. We choose the release point of the wrench to be the origin of the y axis. Putting $y_0 = 0$, $v_0 = 0$, and $t = 1.5$ s in Eq. 19 gives

$$\begin{aligned} y &= v_0 t - \tfrac{1}{2}gt^2 \\ &= (0)(1.5\text{ s}) - \tfrac{1}{2}(9.8\text{ m/s}^2)(1.5\text{ s})^2 \\ &= -11\text{ m}. \end{aligned} \qquad \text{(Answer)}$$

The minus sign shows us that the wrench is below its release point, which we certainly expect.

(b) How fast is the wrench falling after 1.5 s?

The velocity of the wrench is given by Eq. 18, or

$$v = v_0 - gt = 0 - (9.8\text{ m/s}^2)(1.5\text{ s}) = -15\text{ m/s}. \quad \text{(Answer)}$$

Here the minus sign tells us that the wrench is falling downward. Again, no great surprise. Figure 12 displays the important features of the motion up to $t = 4$ s.

Sample Problem 9   A diver steps off a cliff in Acapulco into the air, as in Fig. 13, falling through the air for 3.1 s before reaching the Pacific Ocean. (a) How high is the cliff?

We take the top of the cliff to be the origin of the y axis and the initial velocity v_0 to be zero. The missing ingredient is the velocity v, so Eq. 19 is required. With $y_0 = 0$, it yields

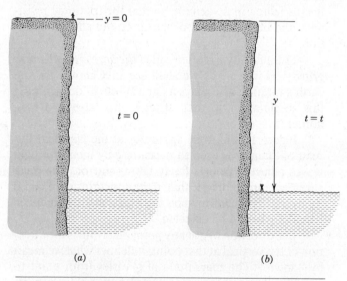

(a)   (b)

Figure 13   Sample Problem 9. (a) A diver about to step off a cliff. (b) Just before hitting the water. If you know the time of fall, you can find the height of the cliff and the speed just before hitting the water.

$$y = v_0t - \tfrac{1}{2}gt^2$$
$$= (0)(3.1 \text{ s}) - (\tfrac{1}{2})(9.8 \text{ m/s}^2)(3.1 \text{ s})^2$$
$$= -47 \text{ m} \ (\approx -150 \text{ ft}). \qquad \text{(Answer)}$$

The minus sign tells us that, at $t = 3.1$ s, the diver is 47 m below his starting point.

(b) How fast is the diver moving when he hits the water? Equation 18 yields

$$v = v_0 - gt$$
$$= 0 - (9.8 \text{ m/s}^2)(3.1 \text{ s})$$
$$= -30 \text{ m/s} \ (\approx -100 \text{ ft/s}). \qquad \text{(Answer)}$$

The minus sign in this case tells us that, when he hits the water, the diver is moving in the direction of decreasing values of y. Again, no surprise.

Sample Problem 10 A pitcher throws a baseball straight up, with an initial speed of 25 m/s (= 82 ft/s). See Figure 14.
(a) How long does it take to reach its highest point?

By *highest point* we mean the point at which its velocity v is zero. Equation 18 is the one we want.

$$t = \frac{v_0 - v}{g} = \frac{25 \text{ m/s} - 0}{9.8 \text{ m/s}^2} = 2.6 \text{ s}. \qquad \text{(Answer)}$$

(b) How high does the ball rise above its release point?

If we stick to the data given in the problem statement, we can use Eq. 20. We take the release point of the ball to be the origin of the y axis. Putting $y_0 = 0$ in Eq. 20 and solving for y, we obtain

$$y = \frac{v_0^2 - v^2}{2g} = \frac{(25 \text{ m/s})^2 - (0)^2}{(2)(9.8 \text{ m/s}^2)}$$
$$= 32 \text{ m} \ (\approx 100 \text{ ft}). \qquad \text{(Answer)}$$

If we wanted to take advantage of the fact that we also know the time of flight, having found it in (a) above, we can also calculate the height of rise from Eq. 22. Check it out.

(c) How long will it take for the ball to reach a point 25 m above its release point?

Inspection of Eqs. 18–22 suggests that we try Eq. 19. With $y_0 = 0$, substitution yields

$$y = v_0t - \tfrac{1}{2}gt^2.$$
$$25 \text{ m} = (25 \text{ m/s})t - (\tfrac{1}{2})(9.8 \text{ m/s}^2)t^2.$$

If we temporarily omit the units (having noted that they form a consistent set), we can rewrite this as

$$4.9t^2 - 25t + 25 = 0.$$

Solving this quadratic equation for t yields*

$$t = 1.4 \text{ s} \quad \text{and} \quad t = 3.7 \text{ s}. \qquad \text{(Answer)}$$

---

* See Appendix G for the formula used to solve a quadratic equation.

There are two such times! This is not really surprising because the ball passes twice through $y = 25$ m, once on the way up and once on the way down; see Hint 10.

We can check our findings because the time at which the ball reaches its maximum height should lie halfway between these two values, or at

$$t = \tfrac{1}{2}(1.4 \text{ s} + 3.7 \text{ s}) = 2.6 \text{ s}.$$

This is exactly what we found in (a) above for the time of flight to maximum height.

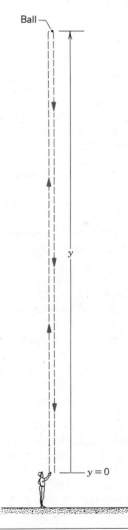

Figure 14 Sample Problem 10. A pitcher throws a baseball straight up in the air. The equations for free fall apply for rising as well as for falling objects. The essential point is that the object is acted on only by the force of gravity during its motion.

Hint 9: Negative signs. In Sample Problems 8, 9, and 10, many answers emerged automatically with negative signs. It is important to know what these signs mean. For falling-body problems, we established a vertical axis (the *y* axis) and we chose—quite arbitrarily—its upward direction to be positive.

We choose the origin of the *y* axis (that is, the *y* = 0 position) to suit the problem. In Sample Problem 8, the origin was the worker's hand; in Sample Problem 9 it was the top of the cliff; in Sample Problem 10 it was the pitcher's hand. A negative value of *y* means that the particle is below the chosen origin; check Sample Problems 8a, 9a, and 10b.

A negative velocity means that the particle is moving in the direction of decreasing *y*, that is, downward. This is true no matter where the particle is located. See Sample Problems 8b and 9b. In Sample Problem 10, the velocity of the ball is positive while it is rising and negative while it is falling.

We have taken the acceleration (= −9.8 m/s²) to be negative in all problems dealing with falling bodies. A negative acceleration means that, as time goes on, the velocity of the particle becomes more and more negative. This is true no matter where the particle is located and no matter how fast or in what direction it is moving. In Sample Problem 10, the acceleration of the ball is negative throughout its flight, whether the ball is rising or falling. The rising ball slows down; the falling ball speeds up. In each case the velocity is becoming more negative.

Hint 10: Unexpected answers—I. Mathematics often generates answers that you might not have thought of as possibilities, as in Sample Problem 10c. If you get more answers than you expect, do not discard out of hand the ones that do not seem to fit. Examine them carefully for physical meaning; it is usually there.

If time is your variable, even a negative value can mean something; negative time simply refers to time before *t* = 0, the (arbitrary) time at which you decided to start your stop watch.

In 1928, the mathematical physicist P. A. M. Dirac found an unexpected negative root to a complicated equation with which he was working. It turned out to be a predictor of the existence of the positive electron. Eugene Wigner has written on the subject, "The Unreasonable Usefulness of Mathematics." In this spirit, pay attention to what the equations cast in your path.

2–8 The Particles of Physics (Optional)

As we progress through the book, we plan to step aside occasionally from the familiar world of large tangible objects and look at nature on a much finer scale. The

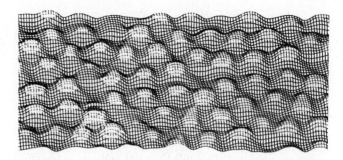

Figure 15 The surface of a specimen of silicon, as revealed by a scanning tunneling microscope. See "The Scanning Tunneling Microscope," by Gerd Binnig and Heinrich Rohrer, *Scientific American,* August 1985. The authors were awarded the 1986 Nobel Prize for developing this instrument.

"particles" that we have dealt with in this chapter, for example, have included baseballs, elevator cabs, and automobiles. In the spirit of our plan, we ask: "How small can a particle be? What are the *ultimate* particles of nature?" *Particle physics*—for so the field that relates to our inquiry is called—attracts the attention of many of the best and the brightest of today's physicists.

The realization that matter, on its finest scale, is not continuous but is made up of atoms was the beginning of understanding for physics and for chemistry. With the scanning tunneling microscope, we can now "see" these atoms, as Fig. 15 makes clear. It is also possible to keep single atoms in tiny electromagnetic "traps" and watch them at leisure. A single electron was kept in such a trap at the University of Washington for 10 months before— by misadventure—it hit the wall and escaped.

We describe the "lumpiness" of matter by saying that matter is *quantized,* the word coming from the German word *quantum,* which means quantity or dose. Quantization is a central feature of nature and we shall point out—as we go along—other physical quantities that are quantized when we look at them on a fine enough scale. This pervasiveness of quantization is reflected in the name we give to physics at the atomic and subatomic level—*quantum physics.*

There is no sharp discontinuity between the quantum world and the world of large-scale objects. The quantum world and the laws that govern it are universal but, as we move from electrons and atoms to baseballs and automobiles, the fact of quantization becomes less noticeable, finally becoming totally undetectable. The "graininess" effectively disappears and the laws of classical physics that govern the motions of large objects are seen as special limiting forms of the more general laws of quantum physics.

The Structure of Atoms.   An atom consists of a central *nucleus* surrounded by one or more *electrons*. The radius of a typical atom is of the order of 10^{-10} m (= 100 pm). That of a typical middle-mass nucleus is some 100,000 times smaller or 10^{-15} m (= 1 fm). The nucleus contains most of the mass of the atom and all of its positive electrical charge.

Consider copper, a typical element. There are 17 kinds of copper atoms or, as we say, 17 *isotopes* of copper. For all electrically neutral copper isotopes, 29 (negatively charged) electrons surround the nucleus, which carries 29 units of positive charge. The isotopes differ in their nuclear mass.

All but two of the 17 copper isotopes are unstable, transforming themselves by radioactive decay into one or the other of the neighboring elements, nickel and zinc, after average lifetimes ranging from a few tenths of a second to a few days. These unstable isotopes clearly cannot exist in nature but must be manufactured in nuclear reactors or in other ways. Ordinary copper is made up of the two remaining (stable) isotopes.

The Structure of Nuclei.   Nuclei are made up of protons and neutrons. The *proton* is the positively charged nucleus of the hydrogen atom, its mass being 1836.15 times that of the electron. At present, no theory can predict the value of the proton:electron mass ratio; if you want a sure-fire Nobel Prize problem, here is a good one. The *neutron* is the uncharged counterpart of the proton, its mass being about 0.14% greater than that of the proton.

The nuclei of all isotopes of copper contain 29 protons, this being the *atomic number Z* of copper. The copper isotopes differ in the numbers of neutrons in their nuclei, these *neutron numbers N* ranging from 28 to 44. The sum of Z and N is the *mass number A* for the isotope, these numbers for the copper isotopes ranging from 57 (= 29 + 28) to 73 (= 29 + 44). The two stable isotopes, with mass numbers 63 and 65, lie roughly in the middle of this range.

Protons and neutrons are not "elementary" particles, since each is made up of other particles called *quarks*. So far, quarks have not been seen in their free state but, because the so-called quark model of the structure of matter is so successful, physicists believe firmly in their reality. Of the six known kinds of quarks,* only two, the so-called *up quark u* and the *down quark d*, are needed to explain "ordinary" matter. As Fig. 16 shows, the proton is made up of two up quarks and one down

---

* The somewhat fanciful names given to the quarks are up, down, charm, strange, top, and bottom.

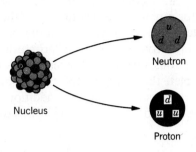

Figure 16   A representation of the nucleus of an atom, showing the neutrons and protons that make it up. These particles, in turn, are composed of "up" and "down" quarks.

quark. The neutron is made up of two down quarks and one up quark.

Electrons.   The electron belongs to a family of particles (called *leptons*) that are unrelated to the quarks. See Table 3 of Chapter 49. Leptons seem to be truly fundamental particles, having no apparent deeper structure and no discernible dimensions.

Particles—the Larger Picture.   So far, we have discussed only those particles needed to explain the structure of ordinary stable matter such as ice cubes or iron nails. However, that is a narrow view of the richness of nature. An impressively large number of particles—all unstable—can be produced by particle accelerators; see Fig. 17. Such particles are also manufactured in our at-

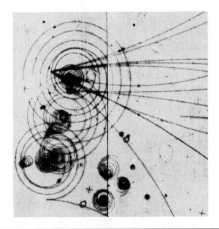

Figure 17   When energetic electrons from the 2-mile-long Stanford Linear Accelerator strike a target, they produce many energetic gamma rays. These gamma rays, which leave no tracks, stream into the bubble chamber from the left. We see the tracks of nine charged particles produced in a collision between a gamma ray and a single proton. The tracks are curved (some are bent into tight spirals) by an external magnet.

mosphere by the impact of cosmic rays from space. Furthermore, for much of the past history of the universe, ordinary stable matter as we know it did not exist, the universe being (at one early stage) a "primordial soup" of protons, neutrons, quarks, and other particles.

A complete theory of the structure of matter must account for the existence of all these additional particles as well. We shall look more closely at this question in

Chapter 49 of the extended version of this book. For now, let us just say that all these particles can be interpreted neatly as combinations of either two or three quarks.*

---

* See James S. Trefil, *From Atoms to Quarks,* Scribner Book Co., New York, 1980, for a readable account of the field of particle physics.

REVIEW AND SUMMARY

Position

We can describe the straight-line motion of a particle by giving its *position x,* which locates it with respect to the origin of a convenient coordinate system. $x(t)$, position x as a function of time t, may be represented in any of the usual ways: an algebraic equation, an x versus t graph, or a table of corresponding values.

During any time interval Δt the position will *change* by Δx. We define the *average velocity $\bar{v}$* of the particle for this time interval as $\Delta x / \Delta t$:*

Average Velocity

$$\bar{v} = \frac{x_2 - x_1}{t_2 - t_1} = \frac{\Delta x}{\Delta t}. \qquad [1]$$

Sample Problems 1 and 2 illustrate the calculation of average velocity. The average velocity $\bar{v}$ is the slope of the straight line connecting the points representing the initial and final positions of the particle on an x versus t graph.

If we allow Δt to approach zero, then Δx will also approach zero; however, their ratio, which is $\bar{v}$, will approach a definite limiting value v, the *instantaneous velocity* (or simply the *velocity*) of the particle at the time in question, or

Velocity

$$v = \lim_{\Delta t \to 0} \frac{\Delta x}{\Delta t} = \frac{dx}{dt}. \qquad [4,5]$$

Instantaneous velocity (at a particular time) may be represented as the slope (at that particular time) of the graph of x versus t. See Sample Problem 3 and Fig. 9. Sample Problem 5 illustrates how we can find velocity by differentiating a function $x(t)$. *Average speed* is defined as the total distance traveled (not the net displacement as in average velocity) divided by the time interval; see Sample Problem 4. *Instantaneous speed* is the magnitude of instantaneous velocity.

Instantaneous acceleration (or simply *acceleration*) is the rate of change of a particle's velocity, or

Acceleration

$$a = \lim_{\Delta t \to 0} \frac{\Delta v}{\Delta t} = \frac{dv}{dt}. \qquad [8,9]$$

Sample Problem 6 shows the calculation of acceleration for an accelerating car.

Figure 9 shows $x(t)$, $v(t)$, and $a(t)$ for the important case in which a is constant. In this circumstance, the five equations in Table 2 describe the motion:

Constant Acceleration

$$v = v_0 + at, \qquad [10]$$
$$x - x_0 = v_0 t + \tfrac{1}{2}at^2, \qquad [14]$$
$$v^2 = v_0^2 + 2a(x - x_0), \qquad [15]$$
$$x - x_0 = \tfrac{1}{2}(v_0 + v)t, \qquad [16]$$

---

* Equation references previously given are enclosed in square brackets.

and

$$x - x_0 = vt - \tfrac{1}{2}at^2. \qquad [17]$$

These equations do *not* hold for more general motion if acceleration is not constant. Sample Problem 7 illustrates the use of these equations.

Falling Objects

An important example of straight-line motion with constant acceleration is that of an object rising or falling freely (in a vacuum) near the earth's surface. The constant acceleration equations describe this motion, but we make two changes in notation: (1) We refer the motion to the vertical y axis with $+y$ vertically *up*; (2) we replace a by $-g$, where g is the magnitude of the free-fall acceleration. Near the earth's surface, $g = 9.8$ m/s^2 ($=32$ ft/s^2). The free-fall equations, with these conventions, are shown as Eqs. 18–22. Sample Problems 8, 9, and 10 show how these equations might be used.

The Structure of Matter

All ordinary matter that we encounter is made of *atoms* which, in turn, are composed of *electrons, protons,* and *neutrons.* The atoms of a particular chemical element all have a certain number (the *atomic number Z* of the element) of protons in their nuclei. An atomic nucleus also has a certain number of neutrons (the *neutron number N*), the sum of N and Z being the *mass number A*. There are several possible mass numbers for each chemical element, each one designating an *isotope* of the element.

Quarks and Leptons

Electrons seem to be elementary particles without internal structure, but protons and neutrons each contain three *quarks*. There appear to be six kinds of quarks in all of nature, two of which (the *up* and *down* quarks) make protons and neutrons. Electrons are part of a family of particles called *leptons.*

QUESTIONS

1. What are some physical phenomena involving the earth in which the earth cannot be treated as a particle?

2. Can the speed of a particle ever be negative? If so, give an example; if not, explain why.

3. Each second a rabbit moves half the remaining distance from its nose to a head of lettuce. Does the rabbit ever get to the lettuce? What is the limiting value of the rabbit's velocity? Draw graphs showing the rabbit's velocity and position as time increases.

4. Average speed can mean the magnitude of the average velocity. Another, more common, meaning given to it is that average speed is the total length of path traveled divided by the elapsed time. Are these meanings different? If so, give an example.

5. A racing car, in a qualifying two-lap heat, covers the first lap with an average speed of 90 mi/h. The driver wants to speed up during the second lap so that the average speed of the two laps together will be 180 mi/h. Show that it cannot be done.

6. Bob beats Judy by 10 m in a 100-m dash. Bob, claiming to give Judy an equal chance, agrees to race her again but to start from 10 m behind the starting line. Does this really give Judy an equal chance?

7. When the velocity is constant, can the average velocity over any time interval differ from the instantaneous velocity at any instant? If so, give an example; if not, explain why.

8. Can the average velocity of a particle moving along the x axis ever be $\tfrac{1}{2}(v_0 + v)$ if the acceleration is not uniform? Prove your answer with the use of graphs.

9. Does the speedometer on an automobile register speed as we defined it?

10. (a) Can an object have zero velocity and still be accelerating? (b) Can an object have a constant velocity and still have a varying speed? In each case, give an example if your answer is yes; explain why if your answer is no.

11. Can the velocity of an object reverse direction when its acceleration is constant? If so, give an example; if not, explain why.

12. Can an object be increasing in speed as its acceleration decreases? If so, give an example; if not, explain why.

13. If a particle is released from rest ($v_0 = 0$) at $x_0 = 0$ at the time $t = 0$, Eq. 14 for constant acceleration says that it is at position x at two different times, namely, $+\sqrt{2x/a}$ and $-\sqrt{2x/a}$. What is the meaning of the negative root of this quadratic equation?

14. What are some examples of falling objects in which it would be unreasonable to neglect air resistance?

15. Figure 18 shows a shot tower in Baltimore, Maryland. It was built in 1829 and used to manufacture lead shot pellets by pouring molten lead through a sieve at the top of the tower. The lead pellets solidify as they fall into a tank of water at the

Figure 18 Question 15.

bottom of the tower, 230 ft below. What are the advantages of manufacturing shot in this way?

16. A person standing on the edge of a cliff at some height above the ground below throws one ball straight up with initial speed u and then throws another ball straight down with the same initial speed. Which ball, if either, has the larger speed when it hits the ground? Neglect air resistance.

17. As an experiment, a skydiver agrees to weigh a pound of butter during the fall using a spring scale. What will the scale read during free fall, just after clearing the plane? What will it read after the parachute has opened and the skydiver is falling at constant terminal speed?

18. What is the downward acceleration of a projectile that is released from a missile accelerating upward at 9.8 m/s²?

19. On another planet, the value of g is one-half the value on earth. How is the time needed for an object to fall to the ground from rest related to the time required to fall the same distance on earth?

20. Consider a ball thrown vertically up. Taking air resistance into account, would you expect the time during which the ball rises to be longer or shorter than the time during which it falls? Why?

21. Make a qualitative graph of speed v versus time t for a falling object (*a*) for which air resistance can be ignored and (*b*) for which air resistance cannot be ignored.

22. A second ball is dropped down an elevator shaft 1 s after the first ball is dropped. (*a*) What happens to the distance between the balls as time goes on? (*b*) How does the ratio v_1/v_2 of the speed of the first ball to the speed of the second ball change as time goes on? Neglect air resistance, and give qualitative answers.

23. Repeat Question 22 taking air resistance into account. Again, give qualitative answers.

EXERCISES AND PROBLEMS

Section 2–3   Average Velocity

1E. A short laboratory light pulse lasts only 10^{-14} s. What is the length of the pulse? (The speed of light is 3.0×10^8 m/s.)

2E. How far does your car, moving at 55 mi/h (=88 km/h) travel forward during the 1 s of time that you take to look at an accident on the side of the road?

3E. Boston Red Sox pitcher Roger Clemens can routinely throw a fastball at a horizontal speed of 160 km/h, as verified by radar gun. How long does it take for the ball to reach homeplate, which is 18.4 m away?

4E. Carl Lewis runs the 100-m dash in about 10 s, and Bill Rodgers runs the marathon (26 mi, 385 yd) in about 2 h 10 min. (*a*) What are their average speeds? (*b*) If Carl Lewis could maintain his sprint speed during a marathon, how long would it taken him to finish?

5E. Figure 19 shows the relation bretween the age of the oldest sediment, in millions of years, plotted against the distance, in kilometers, that the sediment was found from a par-

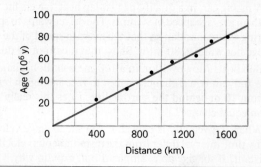

Figure 19 Exercise 5.

ticular ocean ridge. Seafloor material is extruded from this ridge and moves away from it at approximately uniform speed. Find the speed, in centimeters per year, at which this material recedes from the ridge.

6E. The legal speed limit on a thruway is changed from 55 mi/h ($=88.5$ km/h) to 65 mi/h ($=105$ km/h). How much time is thereby saved on a trip from the Buffalo entrance to the New York City exit of the New York Thruway for someone traveling at the legal speed limit over this 435-mi ($=700$-km) stretch of highway?

7E. Using the tables in Appendix F and following the dimensions carefully, find the speed of light ($=3 \times 10^8$ m/s) in miles per hour, feet per second, and light-years per year.

8E. An automobile travels on a straight road for 40 km at 30 km/h. It then continues in the same direction for another 40 km at 60 km/h. What is the average speed of the car during this 80-km trip?

9P. Compute your average speed in the following two cases. (a) You walk 240 ft at a speed of 4.0 ft/s and then run 240 ft at a speed of 10 ft/s along a straight track. (b) You walk for 1.0 min at a speed of 4.0 ft/s and then run for 1.0 min at 10 ft/s along a straight track.

10P. You drive on Interstate 10 from San Antonio to Houston, half the time at 35 mi/h ($=56.3$ km/h) and the other half at 55 mi/h ($=88.5$ km/h). On the way back you travel half the *distance* at 35 mi/h and the other half at 55 mi/h. What is your average speed (a) from San Antonio to Houston, (b) from Houston back to San Antonio, and (c) for the entire trip?

11P. The position of an object moving in a straight line is given by $x = 3t - 4t^2 + t^3$, where x is in meters and t in seconds. (a) What is the position of the object at $t = 1, 2, 3,$ and 4 s? (b) What is the object's displacement between $t = 0$ and $t = 4$ s? (c) What is the average velocity for the time interval from $t = 2$ to $t = 4$ s?

12P. A high-performance jet plane, practicing radar avoidance maneuvers, is in horizontal flight 35 m above the level ground. Suddenly, the plane encounters terrain that slopes gently upward at 4.3°, an amount difficult to detect; see Fig. 20. How much time does the pilot have to make a correction if the plane is to avoid flying into the ground? The airspeed is 1300 km/h.

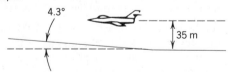

Figure 20 Problem 12.

13P. A car travels up a hill at the constant speed of 40 km/h and returns down the hill at the speed of 60 km/h. Calculate the average speed for the round trip.

14P. Two trains, each having a speed of 30 km/h, are headed at each other on the same straight track. A bird that can fly 60 km/h flies off the front of one train when they are 60 km apart and heads directly for the other train. On reaching the other train it flies directly back to the first train, and so forth. (a) How many trips can the bird make from one train to the other before they crash? (b) What is the total distance the bird travels?

Section 2–5   Acceleration

15E. A car accelerates at 9.2 km/h·s. What is its acceleration in m/s²?

16E. A particle had a velocity of 18 m/s and 2.4 s later its velocity was 30 m/s in the opposite direction. What was the average acceleration of the particle during this 2.4-s interval?

17E. An object moves in a straight line as described by the velocity–time graph (Fig. 21). Sketch a graph that represents the acceleration of the object as a function of time.

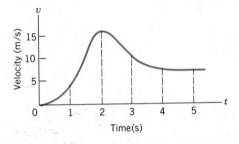

Figure 21   Exercise 17.

18E. The graph of x versus t (see Fig. 22a) is for a particle in straight-line motion. (a) State for each interval whether the velocity v is $+$, $-$, or 0 and whether the acceleration a is $+$, $-$, or 0. The intervals are OA, AB, BC, and CD. (b) From the

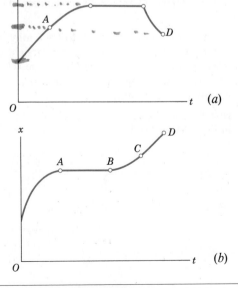

Figure 22   Exercises 18 and 19.

curve, is there any interval over which the acceleration is obviously not constant? (Ignore the behavior at the end points of the intervals.)

19E. Answer the previous questions for the motion described by the graph of Fig. 22b.

20E. A particle moves along the x axis with a displacement versus time as shown in Fig. 23. Sketch roughly curves of velocity versus time and acceleration versus time for this motion.

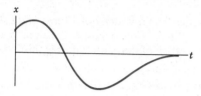

Figure 23 Exercise 20.

21E. For each of the following situations, sketch a graph that is a possible description of position as a function of time for a particle that moves along the x axis. At $t = 1$ s, the particle has (a) zero velocity and positive acceleration; (b) zero velocity and negative acceleration; (c) negative velocity and positive acceleration; (d) negative velocity and negative acceleration. (e) For which of these situations is the speed of the particle increasing at $t = 1$ s?

22E. Consider the two quantities $(dx/dt)^2$ and d^2x/dt^2. (a) Are these merely two equivalent expressions for the same thing? (b) What are the SI units of these two quantities?

23E. A particle moves along the x axis according to the equation $x = 50t + 10t^2$, where x is in meters and t is in seconds. Calculate (a) the average velocity of the particle during the first 3 s of its motion, (b) the instantaneous velocity of the particle at $t = 3.0$ s, and (c) the instantaneous acceleration of the particle at $t = 3.0$ s.

24P. A man stands still from $t = 0$ to $t = 5$ min; from $t = 5$ min to $t = 10$ min he walks briskly in a straight line at a constant speed of 2.2 m/s. What are his average velocity and average acceleration during the time intervals (a) 2 min to 8 min and (b) 3 min to 9 min?

25P. How far does the runner whose velocity–time graph is shown in Fig. 24 travel in 16 s?

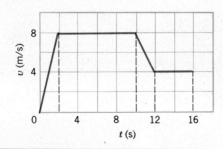

Figure 24 Problem 25.

26P. If the position of an object is given by $x = 2t^3$, where x is measured in meters and t in seconds, find (a) the average velocity and the average acceleration between $t = 1$ s and $t = 2$ s and (b) the instantaneous velocities and the instantaneous accelerations at $t = 1$ s and $t = 2$ s. (c) Compare the average and instantaneous quantities and in each case explain why the larger one is larger.

27P. The position of a particle moving along the x axis is given in centimeters by $x = 9.75 + 1.50t^3$, where t is in seconds. Consider the time interval $t = 2$ s to $t = 3$ s and calculate (a) the average velocity; (b) the instantaneous velocity at $t = 2$ s; (c) the instantaneous velocity at $t = 3$ s; (d) the instantaneous velocity at $t = 2.5$ s; and (e) the instantaneous velocity when the particle is midway between its positions at $t = 2$ s and $t = 3$ s.

28P. In an arcade video game, a spot is programmed to move across the screen according to $x = 9.00t - 0.075t^3$, where x is distance in centimeters measured from the left edge of the screen and t is time in seconds. When the spot reaches a screen edge, at either $x = 0$ or $x = 15$ cm, it starts over. (a) At what time after starting is the spot instantaneously at rest? (b) Where does this occur? (c) What is its acceleration when this occurs? (d) In which direction does it move in the next instant after coming to rest? (e) When does it move off the screen?

29P. The position of a particle moving along the x axis depends on the time according to the equation

$$x = at^2 - bt^3,$$

where x is in feet and t in seconds. (a) What dimensions and units must a and b have? For the following, let their numerical values be 3.0 and 1.0, respectively. (b) At what time does the particle reach its maximum positive x position? (c) What total length of path does the particle cover in the first 4.0 s? (d) What is its displacement during the first 4.0 s? (e) What is the particle's speed at the end of each of the first 4 s? (f) What is the particle's acceleration at the end of each of the first 4 s?

Section 2–6 Constant Acceleration

30E. A rocketship in free space moves with constant acceleration equal to 9.8 m/s². (a) If it starts from rest, how long will it take to acquire a speed one-tenth that of light? (b) How far will it travel in so doing? (The speed of light is 3.0×10^8 m/s.)

31E. A jumbo jet needs to reach a speed of 360 km/h ($=225$ mi/h) on the runway for takeoff. Assuming a constant acceleration and a runway 1.8 km ($=1.1$ mi) long, what minimum acceleration from rest is required?

32E. The head of a rattlesnake can accelerate 50 m/s² in striking a victim. If a car could do as well, how long would it take for it to reach a speed of 100 km/h from rest?

33E. A muon (an elementary particle) is shot with constant speed 5.00×10^6 m/s into a region where an electric field produces an acceleration of the muon of 1.25×10^{14} m/s² directed opposite to the initial velocity. How far does the muon travel before coming to rest?

34E. An electron with initial velocity $v_0 = 1.5 \times 10^5$ m/s enters a region 1.0 cm long where it is electrically accelerated (Fig. 25). It emerges with a velocity $v = 5.7 \times 10^6$ m/s. What was its acceleration, assumed constant? (Such a process occurs in the electron gun in a cathode-ray tube, used in television receivers and oscilloscopes.)

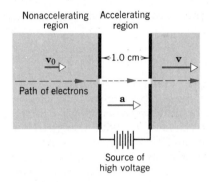

Figure 25 Exercise 34.

35E. An object has a constant acceleration of 3.2 m/s². At a certain instant its velocity is +9.6 m/s. What is its velocity (a) 2.5 s earlier and (b) 2.5 s later?

36E. The Owner's Manual for a 1985 Toyota Cressida says that, if lightly loaded, the car can be braked to a halt from 60 mi/h on dry roads in 43 m. (a) What acceleration does this imply? Express both in SI units and in "g" units. (b) What is the stopping time? If your reaction time T for braking is 400 ms, to how many "reaction times" does the stopping time correspond?

37E. A world's land speed record was set by Colonel John P. Stapp when, on March 19, 1954, he rode a rocket-propelled sled that moved down a track at 1020 km/h. He and the sled were brought to a stop in 1.4 s. See Fig. 8. What acceleration did he experience? Express your answer in terms of g, the acceleration due to gravity.

38E. On a dry road a car with good tires may be able to brake with a deceleration of 11.0 mi/h·s (= 4.92 m/s²). (a) How long does such a car, initially traveling at 55 mi/h (= 24.6 m/s), take to come to rest? (b) How far does it travel in this time?

39E. A rocket-driven sled running on a straight level track is used to investigate the physiological effects of large accelerations on humans. One such sled can attain a speed of 1600 km/h in 1.8 s starting from rest. (a) Assume the acceleration is constant and compare it to g. (b) What is the distance traveled in this time?

40E. The brakes on your automobile are capable of creating a deceleration of 17 ft/s². If you are going 85 mi/h and sudḋ ̣ly see a state trooper, what is the minimum time in which you can get your car under the 55-mi/h speed limit?

41E. An automobile increases its speed uniformly from 25 to 55 km/h in 0.5 min. A bicycle rider uniformly speeds up to 30 km/h from rest in 0.5 min. Calculate the accelerations.

42P. A certain drag racer can accelerate from 0 to 60 km/h in 5.4 s. (a) What is its average acceleration, in m/s², during this time? (b) How far will it travel during the 5.4 s, assuming its acceleration to be constant? (c) How much time would be required to go a distance of 0.25 km if the acceleration could be maintained at the same value?

43P. A train started from rest and moved with constant acceleration. At one time it was traveling 30 m/s, and 160 m farther on it was traveling 50 m/s. Calculate (a) the acceleration, (b) the time required to travel the 160 m mentioned, (c) the time required to attain the speed of 30 m/s, and (d) the distance moved from rest to the time the train had a speed of 30 m/s.

44P. An automobile traveling 35 mi/h (= 56 km/h) is 110 ft (= 34 m) from a barrier when the driver slams on the brakes. Four seconds later the car hits the barrier. (a) What was the automobile's constant deceleration before impact? (b) How fast was the car traveling at impact?

45P. A car moving with constant acceleration covers the distance between two points 60 m apart in 6.0 s. Its speed as it passes the second point is 15 m/s. (a) What is the speed at the first point? (b) What is its acceleration? (c) At what prior distance from the first point was the car at rest?

46P. A subway train accelerates from rest at one station at a rate of 1.2 m/s² for half of the distance to the next station, then decelerates at this same rate for the final half. If the stations are 1100 m apart, find (a) the time of travel between stations and (b) the maximum speed of the train.

47P. A driver's handbook states that an automobile with good brakes and going 50 mi/h can stop in a distance of 186 ft. The corresponding distance for 30 mi/h is 80 ft. Assume that the driver reaction time, during which the acceleration is zero, and the accelerations after the brakes are applied are both the same for the two speeds. Calculate (a) the driver reaction time and (b) the acceleration.

48P. When a driver brings a car to a stop by braking as hard as possible, the stopping distance can be regarded as the sum of a "reaction distance," which is initial speed times reaction time, and "braking distance," which is the distance covered during braking. The following table gives typical values:

Initial Speed (m/s)	Reaction Distance (m)	Braking Distance (m)	Stopping Distance (m)
10	7.5	5.0	12.5
20	15	20	35
30	22.5	45	67.5

(a) What reaction time is the driver assumed to have? (b) What is the car's stopping distance if the initial speed is 25 m/s?

49P. (*a*) If the maximum acceleration that is tolerable for passengers in a subway train is 3.0 mi/h·s and subway stations are located 0.5 mi apart, what is the maximum speed a subway train could attain in this distance? (*b*) What would be the time between stations? (*c*) If the subway train stops for a 20-s interval at each station, what is the maximum average speed of a subway train?

50P. At the instant the traffic light turns green, an automobile starts with a constant acceleration *a* of 2.2 m/s². At the same instant a truck, traveling with a constant speed of 9.5 m/s, overtakes and passes the automobile. (*a*) How far beyond the starting point will the automobile overtake the truck? (*b*) How fast will the car be traveling at that instant? (It is instructive to plot a qualitative graph of *x* versus *t* for each vehicle.)

51P. An elevator cab in the New York Marquis Marriott (see the photo at the start of this chapter) has a total run of 624 ft. Its maximum speed is 1000 ft/min and its (constant) acceleration is 4.0 ft/s². (*a*) How far does it move while accelerating to full speed from rest? (*b*) How long does it take to make the run, starting and ending at rest?

52P. As a high-speed passenger train traveling at 100 mi/h rounds a bend, the engineer is shocked to see that a locomotive has improperly entered onto the track from a siding 0.42 mi ahead; see Fig. 26. The locomotive is moving at 18 mi/h. The

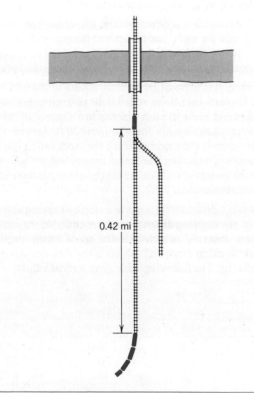

0.42 mi

Figure 26 Problem 52.

engineer of the passenger train immediately applies the brakes. What must be the magnitude of the resulting constant acceleration if a collision is to be just avoided?

53P. Two trains, one traveling at 72 km/h (=45 mi/h) and the other at 144 km/h (=89 mi/h), are headed toward one another along a straight level track. When they are 950 m (=3100 ft) apart, each engineer see's the other's train and applies the brakes. If the brakes decelerate each train at the rate of 1.0 m/s² (=3.3 ft/s²), determine if there is a collision. Assume that neither engineer uses a train's engines to start a train that has stopped.

Section 2–7 Freely-Falling Objects

54E. The single cable supporting an unoccupied construction elevator breaks when the elevator is at rest at the top of a 120-m high building. (*a*) With what speed does the elevator strike the ground? (*b*) For how long was it falling? (*c*) What was its speed when it passed the halfway point on the way down? (*d*) For how long had it been falling when it passed the halfway point?

55E. At a construction site a pipe wrench strikes the ground with a speed of 24 m/s. (*a*) From what height was it inadvertently dropped? (*b*) For how long was it falling?

56E. (*a*) With what speed must a ball be thrown vertically up in order to rise to a height of 50 m? (*b*) How long will it be in the air?

57E. Raindrops fall to earth from a cloud 1700 m above the earth's surface. If they were not slowed by air resistance, how fast would the drops be moving when they strike the ground? Would it be safe to walk outside during a rainstorm?

58E. The Zero Gravity Research Facility at the NASA Lewis Research Center includes a 145-m drop tower. This is an evacuated vertical tower through which, among other possibilities, a 1-m diameter sphere containing an experimental package can be dropped. (*a*) For how long is the experimental package in free fall? (*b*) What is its speed at the bottom of the tower? (*c*) At the bottom of the tower, the sphere experiences an average acceleration of 25*g* as its speed is reduced to zero. Through what distance does it travel in coming to rest?

59E. A shell is fired vertically up from a gun; a rocket, propelled by burning fuel, takes off vertically from a launching area. Plot qualitatively (numbers not required) possible graphs of *a* versus *t*, of *v* versus *t*, and of *y* versus *t* for each. Take *t* = 0 at the instant the shell leaves the gun barrel or the rocket leaves the ground. Continue the plots until the rocket and the shell fall back to earth; neglect air resistance; assume that up is positive and down is negative.

60E. A rock is dropped from a 100-m high cliff. How long does it take to fall (*a*) the first 50 m and (*b*) the second 50 m?

61P. A ball thrown straight up takes 2.25 s to reach a height of 36.8 m. (*a*) What was its initial speed? (*b*) What is its speed at this height? (*c*) How much higher will the ball go?

62P. A rocket is fired vertically and ascends with a constant vertical acceleration of 20 m/s² (= 66 ft/s²) for 1 min. Its fuel is then all used and it continues as a free-fall particle. (*a*) What is the maximum altitude reached? (*b*) What is the total time elapsed from takeoff until the rocket strikes the earth?

63P. An object falls from a bridge that is 45 m above the water. It falls directly into a small boat moving with constant velocity that was 12 m from the point of impact when the object was released. What was the speed of the boat?

64P. A basketball player, standing near the basket to grab a rebound, jumps 76 cm vertically. How much time does the player spend (*a*) in the top 15 cm of this jump and (*b*) in the bottom 15 cm? Does this help explain why such players seem to hang in the air at the tops of their jumps. See Fig. 27.

Figure 27 Problem 64.

65P. At the National Physical Laboratory in England (the British equivalent of our National Bureau of Standards), a measurement of the acceleration *g* was made by throwing a glass ball straight up in an evacuated tube and letting it return, as in Fig. 28. Let ΔT_L be the time interval between the two passages across the lower level, ΔT_U the time interval between the two passages across the upper level, and H the distance between the two levels. Show that

$$g = \frac{8H}{\Delta T_L^2 - \Delta T_U^2}.$$

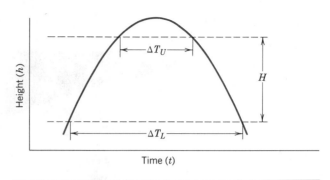

Figure 28 Problem 65.

66P. An arrow is shot straight up in the air with an initial speed of 250 ft/s. If on striking the ground it imbeds itself 6.0 in. into the ground, find (*a*) the acceleration (assumed constant) required to stop the arrow and (*b*) the time required for it to come to rest. Neglect air resistance during the arrow's flight.

67P. A ball of clay falls to the ground from a height of 15 m. It is in contact with the ground for 0.020 s before coming to rest. What is the average accleration of the clay during the time it is in contact with the ground?

68P. A ball is thrown down vertically with an initial speed of 20 m/s from a height of 60 m. (*a*) What will be its speed just before it strikes the ground? (*b*) How long will it take for the ball to reach the ground? (*c*) What would be the answers to (*a*) and (*b*) if the ball were thrown directly up from the same height and with the same initial speed?

69P. Figure 29 shows a simple device for measuring your reaction time. It consists of a strip of cardboard marked with a scale and two large dots. A friend holds the strip with his thumb and forefinger at the upper dot and you position your thumb and forefinger at the lower dot, being careful not to touch the strip. Your friend releases the strip, and you try to pinch it as soon as possible after you see it begin to fall. The mark at the place where you pinch the strip gives your reaction time. How far from the lower dot should you place the 50-ms, 100-ms, 200-ms, and 250-ms marks?

70P. A juggler tosses balls vertically 1 m into the air. How much higher must they be tossed if they are to spend twice as much time in the air?

71P. A stone is thrown vertically upward. On its way up it passes point *A* with speed *v*, and point *B*, 3.0 m higher than *A*, with speed $\frac{1}{2}v$. Calculate (*a*) the speed *v* and (*b*) the maximum height reached by the stone above point *B*.

72P. To test the quality of a tennis ball, you drop it onto the floor from a height of 4.0 ft. It rebounds to a height of 3.0 ft. If the ball was in contact with the floor for 0.010 s, what was its average acceleration during contact?

73P. Water drips from the nozzle of a shower onto the floor

Top

Reaction time, ms

250

200

150

100

50
0

Figure 29 Problem 69.   **Figure 30** Problem 82.

200 cm below. The drops fall at regular intervals of time, the first drop striking the floor at the instant the fourth drop begins to fall. Find the location of the individual drops when a drop strikes the floor.

74P. A lead ball is dropped into a lake from a diving board 5.2 m above the water. It hits the water with a certain velocity and then sinks to the bottom with this same constant velocity. It reaches the bottom 4.8 s after it is dropped. (a) How deep is the lake? (b) What is the average velocity of the ball? (c) Suppose that all the water is drained from the lake. The ball is thrown from the diving board so that it again reaches the bottom in 4.8 s. What is the initial velocity of the ball?

75P. A hoodlum throws a stone vertically down with an initial speed of 12 m/s from the roof of a building, 30 m above the ground. (a) How long does it take the stone to reach the ground? (b) What is the speed of the stone at impact?

76P. If an object travels half its total path in the last second of its fall from rest, find (a) the time and (b) the height of its fall.

Explain the physically unacceptable solution of the quadratic time equation.

77P. A woman fell 144 ft from the top of a building, landing on the top of a metal ventilator box, which she crushed to a depth of 18 in. She survived without serious injury. What acceleration (assumed uniform) did she experience during the collision? Express your answer in terms of g, the acceleration due to gravity.

78P. A stone is dropped into the water from a bridge 144 ft above the water. Another stone is thrown vertically down 1 s after the first is dropped. Both stones strike the water at the same time. (a) What was the initial speed of the second stone? (b) Plot speed versus time on a graph for each stone, taking zero time as the instant the first stone was released.

79P. A parachutist after bailing out falls 50 m without friction. When the parachute opens, he decelerates at 2.0 m/s². He reaches the ground with a speed of 3.0 m/s. (a) How long is the parachutist in the air? (b) At what height did the fall begin?

80P. Two objects begin a free fall from rest from the same height 1 s apart. How long after the first object begins to fall will the two objects be 10 m apart?

81P. An open elevator is ascending with a constant speed of 10 m/s. A ball is thrown straight up by a boy on the elevator when it is a height 30 m above the ground. The initial speed of the ball with respect to the elevator is 20 m/s. (a) What is the maximum height attained by the ball? (b) How long does it take for the ball to return to the elevator?

82P. As Fig. 30 shows, Clara jumps from a bridge, followed closely by Jim. How long did Jim wait after Clara jumped? Assume that Jim is 170 cm tall and that the jumping-off level is at the top of the figure. Make scale measurements directly on the figure.

83P. A balloon is ascending at the rate of 12 m/s at a height 80 m above the ground when a package is dropped. (a) How long does it take the package to reach the ground? (b) With what speed does it hit the ground?

84P*. A steel ball bearing is dropped from the roof of a building (the initial velocity of the ball is zero). An observer standing in front of a window 120 cm high notes that the ball takes 0.125 s to fall from the top to the bottom of the window. The ball bearing continues to fall, makes a completely elastic collision with a horizontal sidewalk, and reappears at the bottom of the window 2.0 s after passing it on the way down. How tall is the building? (The ball will have the same speed at a point going up as it had going down after a completely elastic collision.)

85P*. A dog sees a flowerpot sail up and then back past a window 5.0 ft (= 1.77 m) high. If the total time the pot is in sight is 1 s, find the height above the top of the window to which the pot rises.

CHAPTER 3
VECTORS

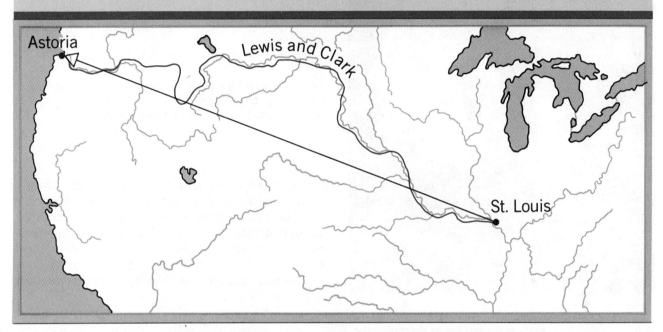

The arrow is a displacement vector *describing the outbound leg of the Lewis and Clark expedition of 1804–1805 that opened up the Northwest. It shows their net change in position but tells nothing about the route they followed or the time that it took them.*

3-1 Vectors and Scalars

A particle confined to a straight line has only two directions in which it can move. We take its velocity to be positive in one of these directions and negative in the other. For a particle moving in three dimensions, however, life is not so simple and a plus or minus sign is no longer enough to define the direction of the velocity. We need an arrow, free to point in any direction whatsoever. We need a *vector*.

A vector is a quantity that has magnitude and direction and that follows certain rules of combination, which we shall spell out below. Some physical quantities that can be represented by vectors are displacement, velocity,

acceleration, force, and magnetic field. Figure 1 shows two of these vectors in a familiar situation. The *position* of the ball with respect to the club head at the moment the photo was taken and its *velocity* at that same instant are shown. We could also have shown, by other vectors, the net force acting on the ball and its acceleration.

Not all physical quantities involve a direction. Some examples are temperature, pressure, energy, mass, and time, none of which "points" in the spatial sense. We call such quantities *scalars* and we deal with them by the rules of ordinary algebra.

The simplest of all vectors is the *displacement vec-*

Figure 1 The arrows show the position, with respect to the club head, and the velocity of the golf ball at the instant the photo was taken. Both quantities are vectors.

tor, a displacement being a change of position.* In Fig. 2*a*, for example, if a particle changes its position by moving from *A* to *B* we say that it undergoes a displacement, which we represent by the arrow pointing from *A* to *B*. To distinguish a vector symbol from other kinds of arrows, we use a closed triangular arrow head.

The arrows from *A* to *B*, from *A'* to *B'*, and from *A"* to *B"* in Fig. 2*a* represent the same *change of position* for the particle and we make no distinction among them. All three arrows have the same magnitude and direction and are identical displacement vectors. If Lewis and Clark had started from Miami, for example, the displacement

---

* The word *vector* comes from the Latin word *to carry*. This seems appropriate for a displacement vector, which carries a particle from one point to another.

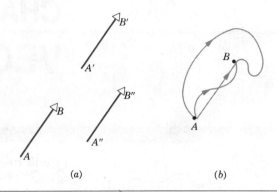

(a)   (b)

Figure 2 (*a*) All three arrows represent the same displacement vector. (*b*) All three paths connecting the two points correspond to the same displacement vector.

vector shown in the map at the head of this chapter would have landed them somewhere in New Mexico.

The displacement vector tells us nothing about the actual path that the particle takes. In Fig. 2*b*, for example, all three paths connecting points *A* and *B* correspond to the same displacement vector, that of Fig. 2*a*. Such vectors represent only the overall effect of the motion, not the motion itself.

Sample Problem 1 Figure 3 shows a trail 5.8 km long, leading from a trailhead (elevation 1770 ft) to the summit of Mt. Lafayette (elevation 5250 ft) in New Hampshire. Sara hikes up this trail. What displacement vector describes her ascent?

The vector we seek is an arrow pointing directly from the trailhead to the summit. Its length is given by

$$L = \sqrt{d^2 + h^2}.$$

Here *d*, the horizontal distance from the trailhead to the summit, can be read from the map scale to be 3.5 km. The symbol *h* is the elevation difference between the trailhead and the summit, which is 3480 ft or 1.1 km. Thus, we have

$$L = \sqrt{d^2 + h^2} = \sqrt{(3.5 \text{ km})^2 + (1.1 \text{ km})^2}$$
$$= 3.7 \text{ km} \qquad \text{(Answer)}$$

for the length of the displacement vector. Note that it has no connection with the actual length of the trail, except that it is certain to be shorter.

For the angle that the displacement vector makes with the horizontal, we have

$$\theta = \tan^{-1} \frac{h}{d} = \tan^{-1} \frac{1.1 \text{ km}}{3.5 \text{ km}} = 17°. \quad \text{(Answer)}$$

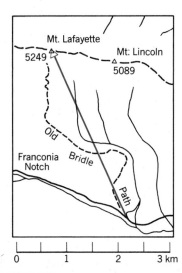

Figure 3   Sample Problem 1. A trail up Mt. Lafayette in New Hampshire.

The *average* angle $\bar{\theta}$ that Sara experiences while hiking on the trail (of length l) will be less than this and is given by

$$\bar{\theta} = \sin^{-1}\frac{h}{l} = \sin^{-1}\frac{1.1 \text{ km}}{5.8 \text{ km}} = 11°.$$

It is in fact the purpose of a trail to be longer and less steep than the displacement vector connecting its end points. If you want to follow a displacement vector, take a chair lift.

3-2   Adding Vectors: Graphical Method

Suppose that, as in Fig. 4*a*, a particle moves from *A* to *B* and then later from *B* to *C*. We can represent its overall displacement (no matter what its actual path) by two successive displacement vectors, *AB* and *BC*. The net effect of these two displacements is a single displacement from *A* to *C*. We call *AC* the *vector sum* of the vectors *AB* and *BC*. This sum is not an algebraic sum and we need more than simple numbers to specify it.

In Fig. 4*b*, we redraw the vectors of Fig. 4*a* and relabel them in the way that we shall use from now on, namely, with boldfaced symbols such as **a**, **b**, or **s**. In handwriting, we often put an arrow over the symbol to represent a vector. If we want to indicate just the magnitude of the vector (a quantity that is always positive) we use the italic symbol, such as *a*, *b*, or *s*. The boldfaced symbol always refers to both properties of the vector, magnitude and direction.

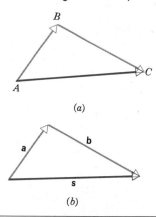

(*a*)

(*b*)

Figure 4   (*a*) *AC* is the vector sum of the vectors *AB* and *BC*. (*b*) The same vectors relabeled.

We can represent the relation among the three vectors in Fig. 4*b* by

$$\mathbf{s} = \mathbf{a} + \mathbf{b}, \tag{1}$$

in which we say that the vector **s** is the *vector sum* of vectors **a** and **b**. The rules for adding vectors in this way are as follows: (1) On a sheet of paper, lay out the vector **a** to some convenient scale. (2) Lay out vector **b** to the same scale, with its tail at the head of vector **a**. (3) Construct the vector sum **s** by drawing a line from the tail of **a** to the head of **b**. You can easily generalize this procedure if you need to add more than two vectors.

Since vectors are new quantities, we must expect some new rules. The symbol "+" in Eq. 1 and the words "add" and "sum" have different meanings than they have in arithmetic or in ordinary algebra. They tell us to carry out quite a different set of operations.

Vector addition, defined in this way, has two important properties. First, the order of addition does not matter. That is,

$$\mathbf{a} + \mathbf{b} = \mathbf{b} + \mathbf{a} \qquad \text{(commutative law)}. \tag{2}$$

Study of Fig. 5 should convince you that this is the case. Second, if there are more than two vectors, it does not matter how we group them as we add them. Thus, if we want to add vectors **a**, **b**, and **c**, we can add **a** and **b** first and then add that sum to **c**. On the other hand, we can add **b** and **c** first, and then add *that* sum to **a**. We get the same result either way. In equation form,

$$(\mathbf{a} + \mathbf{b}) + \mathbf{c} = \mathbf{a} + (\mathbf{b} + \mathbf{c}) \qquad \text{(associative law)}. \tag{3}$$

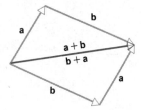

Figure 5 The two vectors **a** and **b** can be added in any order; see Eq. 2.

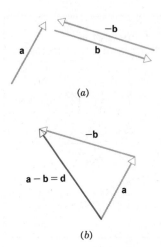

Figure 7 (*a*) Showing vectors **a**, **b**, and −**b**. (*b*) Subtracting vector **b** from vector **a** is the same as adding vector −**b** to vector **a**.

A little quiet contemplation of Fig. 6 will convince you that Eq. 3 is correct. Both the commutative and the associative laws happen to hold also in ordinary arithmetic.

Sometimes, we want to subtract one vector from another. We do this by recognizing that subtraction is a special form of addition. To subtract **b** from **a**,

$$d = a - b = a + (-b) \quad \text{(subtraction)}. \quad (4)$$

That is, we find the difference vector **d** by adding the vector **a** to the vector −**b**, the latter vector being found simply by reversing the direction of the vector **b**. Figure 7 shows how this is done graphically.

Remember, although we have used displacement vectors as a prototype, these rules for addition hold for vectors of all kinds, whether they represent forces, velocities, or anything else. As in ordinary arithmetic, it is still true that we can only add vectors of the same kind. We can add two displacements, for example, or two velocities, but it makes no sense to add a displacement and a velocity. In the world of scalars, that would be like trying to add 21 s and 12 kg.

3-3 Vectors and Their Components

In the vector figures that we have drawn so far, we have not drawn in any coordinate axes. We do so now. For simplicity, we draw the x and the y axes in the plane of the page, the z axis being out of the page at right angles. Furthermore, we consider only vectors that lie in the xy plane.* Figure 8*a* shows such a vector. If we drop perpendicular lines from the ends of the vector **a** to the coordinate axes, the quantities a_x and a_y so formed are called the *components* of the vector **a**. In general, a vector will have three components, although for the case of Fig. 8*a* the component along the z axis happens to be zero. As Fig. 8*b* shows, if you move a vector in such a way that it remains parallel to its original direction, the values of its components remain unchanged.

We can easily find the components of **a** in Fig. 8*a* from

$$a_x = a \cos \theta \quad \text{and} \quad a_y = a \sin \theta, \quad (5)$$

where θ is the angle that the vector **a** makes with the direction of increasing x. Depending on the value of θ,

---

* The extension from the two-dimensional world of the xy plane to the (real) world of three dimensions can easily be made for the situations we consider here. Later, we shall generalize to three dimensions as the physics of the situation requires.

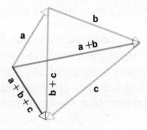

Figure 6 The three vector **a**, **b**, and **c** can be grouped in any way as they are added; see Eq. 3.

the components of a vector may be positive, negative, or zero. In a figure, we use a smaller arrowhead to represent the sign of the component. A component arrow that points in the direction of an increasing coordinate value, as a_x and a_y do in Figs. 8a and 8b, bears a positive sign. Similarly, a component arrow that points in the direction of a decreasing coordinate value bears a negative sign. Figure 9 shows a vector **b** for which b_y is negative and b_x is positive. The components of a vector are scalars because we need only a number, a unit, and an algebraic sign to specify them.

Once a vector has been resolved into its components, the components themselves can be used in place of the vector. Instead of the two numbers, a and θ, we have two other numbers, a_x and a_y. Both sets of numbers contain exactly the same information and we can pass back and forth readily between the two descriptions. To obtain a and θ if we are given a_x and a_y, we note (see Fig. 8a) that

$$a = \sqrt{a_x^2 + a_y^2} \quad \text{and} \quad \tan\theta = \frac{a_y}{a_x}. \tag{6}$$

In solving problems, you may use *either* the a_x, a_y notation *or* the a, θ notation.

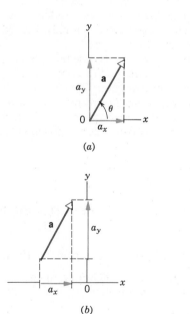

(a)

(b)

Figure 8   (a) The components of vector **a**. (b) The components are unchanged if the vector is translated to another position.

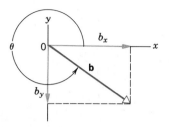

Figure 9   Vector **b** has a positive x component and a negative y component.

3-4   Unit Vectors

It is useful to introduce the notion of a *unit vector,* whose only purpose is to specify a direction. As Fig. 10 shows, we often draw the three unit vectors, labeled **i**, **j**, and **k**, along the x, y, and z axes of a rectangular coordinate system. These three vectors have a magnitude of unity, have no dimensions, and carry no units.

The usefulness of unit vectors is that we can express other vectors in terms of them. Thus, we can write the vectors **a** and **b** of Figs. 8 and 9 as

$$\mathbf{a} = \mathbf{i}a_x + \mathbf{j}a_y \tag{7}$$

and

$$\mathbf{b} = \mathbf{i}b_x + \mathbf{j}b_y; \tag{8}$$

see Fig. 11. The quantities $\mathbf{i}a_x$ and $\mathbf{j}a_y$ are called the *vector components* of the vector **a**, the quantities a_x and a_y alone being the *scalar components* or, more often, simply the *components.* Sample Problem 4 gives one example of the many uses of unit vector notation, namely, in adding vectors.

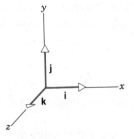

Figure 10   Unit vectors **i**, **j**, and **k** define the directions of a rectangular coordinate system.

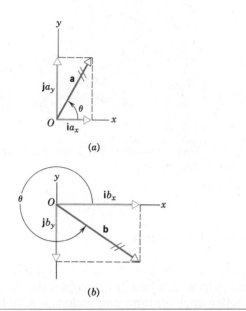

Figure 11 (a) The vector components of vector **a**; see Eq. 7. (b) The vector components of vector **b**; see Eq. 8. The vectors are shown "crossed out," having been replaced by their vector components.

3-5 Adding Vectors: Component Method

The graphical method of adding vectors (see Section 3-2) is cumbersome, of limited accuracy, and hard to carry out in three dimensions. Adding vectors by combining their components analytically is much more practical, straightforward and accurate. Let us see how to do it.

Let **s** be the sum of the vectors **a** and **b**, or

$$s = a + b. \qquad (9)$$

If two vectors, such as **s** and **a** + **b**, are to be equal, they must have the same magnitude and must point in the same direction. This can only happen if their corresponding components are equal. We stress this important conclusion:

Two vectors can only be equal to each other if their corresponding components are equal.

For the vectors of Eq. 9 this requires that

$$s_x = a_x + b_x \qquad (10)$$

and

$$s_y = a_y + b_y. \qquad (11)$$

These two algebraic equations, taken together, are equivalent to the single vector equation 9.*

Here is the rule for adding vectors by this method: (1) Resolve each vector into its components. (2) Add the components for each coordinate axis, taking the algebraic sign into account. (3) The sums so obtained are the components of the sum vector. Once we know the components of the sum vector, we can easily reconstruct that vector in space, using Eq. 6.

When adding vectors by this method, you can often simplify your work by the way you choose the coordinate axes. For example, you can always draw the x axis (or any other axis) so that it lies parallel to one of the vectors. The components of that vector along the other two axes will then be zero. In problems involving motion under gravity, for example, we often choose the y axis to be vertical so as to eliminate the x component of acceleration.

Sample Problem 2 A small plane leaves an airport on an overcast day and later is sighted 215 km away, in a direction making an angle of 22° east of north. How far east and how far north is the plane from its base?

For convenience, we choose the x axis to point east and the y axis to point north. In Fig. 12 we show the position of the plane as a vector **r** that points from the airport to the plane at the instant of sighting. The plane need not have followed a straight line to the point at which it was sighted.

Our problem is to find the components of **r** in Fig. 12. We have

$$\theta = 90° - 22° = 68°,$$

so that (see Eq. 5)

$$r_x = r \cos \theta = (215 \text{ km})(\cos 68°) = 81 \text{ km} \quad \text{(Answer)}$$

and

$$r_y = r \sin \theta = (215 \text{ km})(\sin 68°) = 199 \text{ km}. \quad \text{(Answer)}$$

Thus, the plane is now 199 km north and 81 km east of its base. We have no information about how it got there.

---

* In general, there would be *three* algebraic equations; but since we are dealing here, for simplicity, only with vectors that lie in the xy plane, the z components of such vectors are zero.

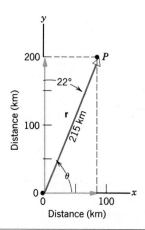

Figure 12 Sample Problem 2. A plane takes off from an airport at the origin and is later sighted at P.

Sample Problem 3 Mort and Melissa, taking part in a rally, drive their Trans Am to checkpoint Able, 36 km due east of their starting point. They then drive to checkpoint Baker, 45 km due north. Finally, they drive to checkpoint Charlie, which is 25 km northwest. Figure 13 shows both the checkpoints and the road network. At checkpoint Charlie, what is their position with respect to their starting point?

This is a matter of reconstructing a vector from its components. We choose a coordinate system with its x axis pointing east and its y axis pointing north. The components of $\mathbf{r}$ are found from

$$r_x = a_x + b_x + c_x = 36 \text{ km} + 0 + (25 \text{ km})(\cos 135°)$$
$$= (36 + 0 - 17.7) \text{ km} = 18.3 \text{ km}$$

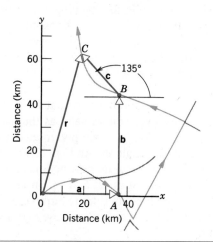

Figure 13 Sample Problem 3. A rally route, showing the origin, checkpoints Able, Baker, and Charlie, and also the road network.

and

$$r_y = a_y + b_y + c_y = 0 + 45 \text{ km} + (25 \text{ km})(\sin 135°)$$
$$= (0 + 45 + 17.7) \text{ km} = 62.7 \text{ km}.$$

The magnitude and direction of $\mathbf{r}$ are then

$$r = \sqrt{r_x^2 + r_y^2} = \sqrt{(18.3 \text{ km})^2 + (62.7 \text{ km})^2}$$
$$= 65 \text{ km} \qquad \text{(Answer)}$$

and

$$\theta = \tan^{-1}\frac{r_y}{r_x} = \tan^{-1}\frac{62.7 \text{ km}}{18.3 \text{ km}} = 74°. \quad \text{(Answer)}$$

Vector $\mathbf{r}$ has a magnitude of 65 km and points 74° north of east.

Note how the work is simplified in this problem by choosing the orientation of the coordinate axes wisely.

Sample Problem 4 Here are three vectors, expressed in unit vector notation:

$$\mathbf{a} = 4.2\mathbf{i} - 1.6\mathbf{j},$$
$$\mathbf{b} = -3.6\mathbf{i} + 2.9\mathbf{j},$$

and

$$\mathbf{c} = -3.7\mathbf{j}$$

All three lie in the xy plane, none of them having a z component. Find the vector $\mathbf{r}$ that is the sum of these three vectors. For convenience, we have suppressed the units in these vector expressions; you may take them to be meters.

From Eqs. 10 and 11 we have

$$r_x = a_x + b_x + c_x = 4.2 - 3.6 + 0 = 0.6$$

and

$$r_y = a_y + b_y + c_y = -1.6 + 2.9 - 3.7 = -2.4.$$

Thus,

$$\mathbf{r} = 0.6\mathbf{i} - 2.4\mathbf{j} \qquad \text{(Answer)}$$

represents the sum vector. Figure 14 shows the three vectors and their sum. Study the figure and convince yourself, by add-

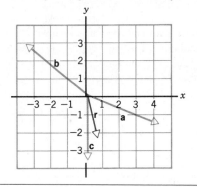

Figure 14 Vector $\mathbf{r}$ is the vector sum of the other three vectors.

ing them graphically in your head or roughly on paper, that **r** really is the sum of the three given vectors.

Hint 1: Angles, degrees, and radians. Angles are usually measured counterclockwise from the positive direction of the x axis. If the angle is greater than 180°, it is often measured clockwise from this axis, with a negative sign attached to show that this has been done. Thus, 210° and −150° are the same angle. Most calculators (try yours) will accept angles in either form when asked to find trig functions.

Angles may be measured in degrees or in radians. One complete traversal of a circle includes 360° or 2π radians. If you know this, you do not need to remember that 1 radian ($=360°/2\pi = 180°/\pi$) is equal to 57.3°. Learn how to change angles from degrees to radians, and conversely, on your calculator. Most calculators, at least in their "wake-up mode," handle trig functions with the angles in degrees.

Hint 2: Trig functions. In this course (and throughout your career if you are going to be an engineer or a scientist), you will need to have the trig functions at your fingertips. If you are a little rusty, review the definitions of the functions in Appendix G, concentrating on the sine, the cosine, and the tangent. In Fig. 15, you should be able to write down without hesitation relations such as $a = c \sin \theta$, $b = c \cos \theta$, and $a = b \tan \theta$.

Be clear about the signs of the functions. Your calculator will show them correctly but you still need to be able to figure them out for yourself, quadrant by quadrant. It helps to keep firmly in mind a plot like Fig. 16, in which the roman numerals indicate the quadrants.

Hint 3: Inverse trig functions. The most important of the inverse trig functions are $\sin^{-1}$, $\cos^{-1}$, and $\tan^{-1}$. Here your calculator may need a little help because, in the range 0–360°, there are always two answers, as Fig. 16 shows. The angle given by your calculator is always the smaller of the two angles. In Sample Problem 1, we have $\theta = \tan^{-1} 0.314$. Your calculator will give you $\theta = 17°$. Although 197° ($=17° + 180°$) has the same tangent, it is clear from the problem that 17° is the angle we want.

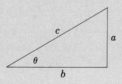

Figure 15 A triangle used to define the trigonometric functions. See Appendix G.

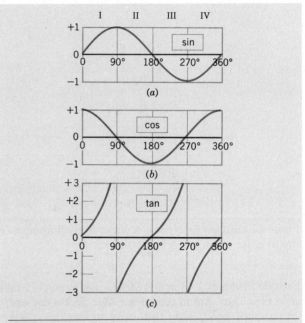

Figure 16 Three useful curves to remember.

3–6 Vectors and the Laws of Physics (Optional)

Figure 17a is a copy of Fig. 8a, showing a vector **a** and its components a_x and a_y. Why did we choose to draw in the coordinate axes as we did? The plain answer is that we drew them so that the two axes would be square with the page and would somehow "look right." There was no deeper reason.

Figure 17b shows another way that we could have drawn in the axes, every bit as valid as the choice we made in Fig. 17a. We have simply rotated the axes through an arbitrary angle ϕ. As Fig. 17b shows, the components of the vector **a** are now no longer a_x and a_y but two different quantities, a'_x and a'_y. In fact, we see that the vector **a** can have an infinite number of component pairs, depending on how we choose to draw in the rotated axes.

Which, then is the "right" component pair? The answer is that it does not matter because all component pairs are equally valid. All can be used to find the magnitude of the vector and its direction in space, using Eq. 6.

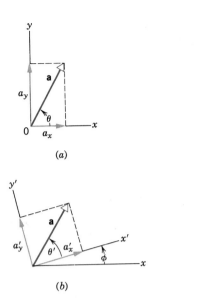

(a)

(b)

Figure 17   (*a*) The vector **a** and its components. (*b*) The same vector, with the axes of the coordinate system rotated through an angle ϕ.

That is,

$$a = \sqrt{a_x^2 + a_y^2} = \sqrt{a_x'^2 + a_y'^2}, \qquad (12)$$

and

$$\theta = \theta' + \phi, \qquad (13)$$

in which

$$\theta = \tan^{-1}\frac{a_y}{a_x} \quad \text{and} \quad \theta' = \tan^{-1}\frac{a_y'}{a_x'}. \qquad (14)$$

We are led to conclude:

> *Relations among vectors (for example, Eq. 1) remain unchanged, no matter where you choose to locate the origin of your coordinate system and no matter how you choose to orient the coordinate axes.*

If this is not true, then the quantities you are dealing with are not vectors.

What has all this to do with the laws of physics? A great deal. The laws of physics (all of them!) are *also* independent of the choice of a coordinate system. Newton's laws of motion, Maxwell's equations of electromagnetism, and Einstein's theory of relativity all hold, no matter how you choose to locate and to orient your coordinate axes. That is why the language of vectors is such a useful and compact way in which to express the laws of physics.

3-7   Multiplying Vectors*

Just as we gave new meaning to the words "adding vectors," we give new meaning to the words "multiplying vectors." There are three ways of multiplying vectors that are especially useful in physics; we discuss each in turn.

Multiplying a Vector by a Scalar.   This is the simplest of the multiplication processes. The product of a scalar *s* and a vector **a** is a new vector, whose direction is the same as that of **a** if *s* is positive but opposite to that direction if *s* is negative. The magnitude of the new vector (which must always be positive) is the magnitude of **a** multiplied by the absolute value of *s*. To divide a vector by a scalar, we simply multiply the vector by the reciprocal of the scalar.

In Fig. 7*a* we saw an example of this kind of multiplication, the scalar in this case being a pure number with the value −1. Often the scalar is not a pure number but a physical quantity, with associated units and dimensions. In such a case, the physical nature of the final vector is different from that of the initial vector.

A Glimpse Ahead.   We pluck an example from Chapter 5, with the promise to look at it in detail later. If the vector **a** represents the acceleration of a particle and if we multiply it by the mass *m* of the particle (a scalar), the resultant vector is the force acting on the particle, which we represent by **F**. Thus,

$$\mathbf{F} = m\mathbf{a}. \qquad (15)$$

This is Newton's famous second law of motion and we show it here — putting off explanation until Section 5-5 — simply to point out that multiplying a vector by a scalar does indeed occur in physics and in very useful contexts.

Note that the vectors **F** and **a** have the same direction, *m* being a positive scalar. They represent different physical quantities, however, with different units and dimensions because *m* is not a pure number but a physical quantity, with units and dimensions of its own. Equation 15, like all vector equations, really represents three separate and independent scalar equations, one for each of the three axes of our coordinate system.

The Scalar Product.   Now let us look at the first of two ways in which we can multiply two vectors together.

---

* The material of this section will be used later throughout the text. The product of a vector and a scalar is first used in Chapter 5. The scalar product is first used in Chapter 7 and the vector product in Chapter 12. The instructor may wish to postpone the assignment of this section accordingly.

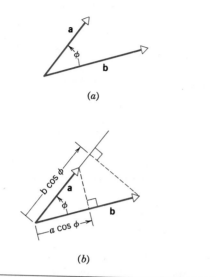

Figure 18 (*a*) Two vectors **a** and **b**, with an angle ϕ between them. (*b*) The vectors **a** and **b**, showing also their components in each other's directions. See Eq. 17.

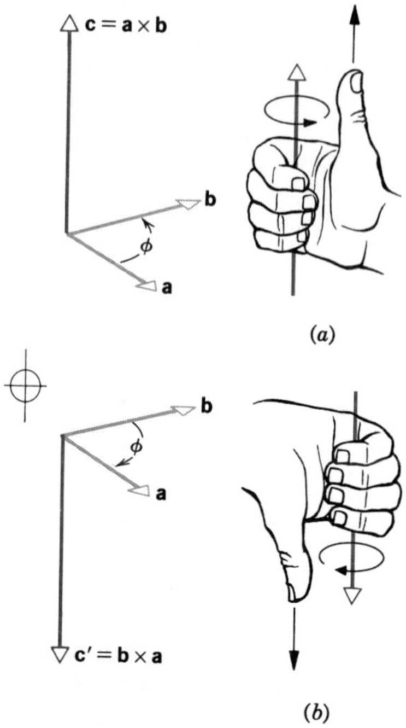

Figure 19 Illustration of the right-hand rule for vector products. (*a*) Swing vector **a** into vector **b** with the fingers of your right hand. Your thumb shows the direction of vector **c**. (*b*) Showing that $(\mathbf{a} \times \mathbf{b}) = -(\mathbf{b} \times \mathbf{a})$

Figure 18*a* shows the vectors **a** and **b**, the angle between them being ϕ. The *scalar product* of these two vectors, represented by **a · b**, is defined to be

$$\mathbf{a} \cdot \mathbf{b} = ab \cos \phi. \tag{16}$$

Here a is the magnitude of **a**, b is the magnitude of **b**, and ϕ is the angle between them.*

Because a and b are scalars and $\cos \phi$ is a pure number, the scalar product of two vectors is a scalar which, of course, is why we call it a scalar product. As Fig. 18*b* shows, we can also regard the scalar product as the product of the magnitude of either vector by the component of the other vector in the direction of the first vector. Thus, we can rewrite Eq. 16 as

$$\mathbf{a} \cdot \mathbf{b} = (a)(b \cos \phi) = (a \cos \phi)(b). \tag{17}$$

Because of the notation, **a · b** is also called the *dot product* of **a** and **b** and is spoken as "a dot b."

A Glimpse Ahead. As one example of a scalar product, we select the definition of the work W (a scalar) that is done by a force **F** as its point of application moves through a displacement **d**. If ϕ is the angle between the

vectors **F** and **d**, the work W is defined from

$$W = \mathbf{F} \cdot \mathbf{d} = Fd \cos \phi. \tag{18}$$

We present this definition here simply to show that scalar products are useful in physics. We shall explore it fully in Chapter 7.

The Vector Product. The *vector product* of vectors **a** and **b**, written as **a × b**, is a third vector **c**. The *magnitude* of **c** is given by

$$c = ab \sin \phi, \tag{19}$$

where ϕ is the (smaller) angle between **a** and **b**.*

The *direction* of **c** is at right angles to the plane formed by the vectors **a** and **b**. Figure 19*a* shows how to establish the *sense* of the vector **c**, that is, the direction in which it points. (1) Place the vectors **a** and **b** tail to tail and set up an imaginary line perpendicular to their plane at the point at which they intersect. (2) Grasp this line

---

* In Fig. 18, there are two angles between these two vectors, ϕ and $360° - \phi$. It does not matter which we choose because the cosine of these two angles is the same, see Fig. 16*b*.

* Note that, in this case, you must pick the smaller of the two angles between the vectors **a** and **b**. This is because $\sin \phi$ and $\sin(360° - \phi)$ differ by an algebraic sign; see Fig. 16*a*.

with your hand, your fingers curling in such a way as to sweep the vector **a** into the vector **b** through the smaller of the two angles between them. (3) Your outstretched thumb then points in the direction of the vector **c**. Because of the notation, **a** × **b** is also called the *cross product* of **a** and **b** and is spoken as "**a** cross **b**."

Note that **b** × **a** is not the same as **a** × **b**, so that the order of the factors in a vector product is important. In a scalar product, on the other hand, you get the same answer no matter what the order of the two factors. Actually, as Fig. 19*b* shows,

$$\mathbf{b} \times \mathbf{a} = -\mathbf{a} \times \mathbf{b}. \tag{20}$$

A Glimpse Ahead. We first run into the vector product in Section 12-4, in which we discuss a force **F** whose point of application is a distance **r** from a certain origin. The torque τ (a turning effect) that this force exerts about the origin is defined to be

$$\tau = \mathbf{r} \times \mathbf{F}. \tag{21}$$

We introduce this equation here, without further explanation at this point, to give a concrete foreshadowing of the use of the vector product in physics.

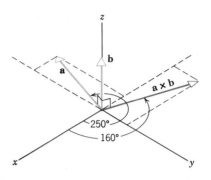

Figure 20 Sample Problem 5. A problem in vector multiplication.

Eq. 16,

$$\mathbf{a} \cdot \mathbf{b} = ab \cos \phi = (18)(12)(\cos 90°) = 0. \quad \text{(Answer)}$$

The scalar product of any two vectors that are at right angles to each other is zero. This is consistent with the fact that neither of these vectors has a component in the direction of the other vector; see Fig. 18*b*.

(b) What is the vector product of the vectors **a** and **b**?

The magnitude of the vector product is, from Eq. 19,

$$ab \sin \phi = (18)(12)(\sin 90°) = 216. \quad \text{(Answer)}$$

The direction of **c** is at right angles to the plane formed by the vectors **a** and **b**. It must then be at right angles to **b**, which means that it is at right angles to the z axis. This in turn means that **c** must lie in the xy plane. The right-hand rule of Fig. 19 shows that **c** points as shown in Fig. 20. Study of Fig. 20 suggests that the direction of **c** makes an angle of 250° − 90° or 160° with the direction of increasing x.

Sample Problem 5 Vector **a** lies in the xy plane. It has a magnitude of 18 units and points in a direction 250° from the direction of increasing x. Vector **b** has a magnitude of 12 units and is pointing along the direction of increasing z. See Fig. 20 (a) What is the scalar product of these two vectors?

The angle ϕ between these two vectors is 90° so that, from

REVIEW AND SUMMARY

Scalars and Vectors

Scalars, such as temperature, have magnitude only. They are specified by a number with a unit (82° F) and obey the rules of ordinary (scalar) algebra. Vectors, such as displacement, have both magnitude and direction (5 m, north) and obey the special rules of vector algebra.

Adding Vectors Geometrically

Two vectors **a** and **b** may be added geometrically by drawing them to a common scale and placing them head to tail. The vector connecting the tail of **a** to the head of **b** is the sum vector **s**, as Fig. 4 shows. To subtract **b** from **a**, reverse the direction of **b** to get − **b**; then add − **b** to **a**: see Fig. 7. Vector addition and subtraction are cummutative and obey the associative law.

The components a_x and a_y of any vector **a** are the perpendicular projections of **a** on the coordinate axes, as Fig. 8 and Sample Problem 2 show. Analytically, the components are given by

Rectangular Components of a Vector

$$a_x = a \cos \theta \quad \text{and} \quad a_y = a \sin \theta. \tag{5}$$

Given the components, we can reconstruct the vector from

$$a = \sqrt{a_x^2 + a_y^2} \quad \text{and} \quad \tan \theta = \frac{a_y}{a_x}. \tag{6}$$

Adding Vectors in
Component Form

To add vectors in component form, we use the rules

$$s_x = a_x + b_x \quad \text{and} \quad s_y = a_y + b_y. \qquad [10,11]$$

See Sample Problem 3.

Often we find it useful to introduce unit vectors **i**, **j**, and **k**, whose magnitudes are unity and whose directions are the x, y, and z axes, respectively; see Fig. 10. We can write any two-dimensional vector **a** in terms of unit vectors as

Unit Vector Notation

$$\mathbf{a} = \mathbf{i}a_x + \mathbf{j}a_y, \qquad [7]$$

in which $\mathbf{i}a_x$ and $\mathbf{j}a_y$ are the vector components of **a**. Sample Problem 4 shows how to add vectors using unit vectors.

Vectors and Physical Laws

Any physical situation involving vectors can be described using several possible coordinate systems (frames of reference). We usually choose the one of these that simplifies our work. However, the relation between the vector quantities does not depend on our choice. The laws of physics, written in vector form, use such relations. We say that the laws of physics are independent of our choice of reference frame.

Scalar Times Vector

The product of a scalar s and a vector **v** is a new vector whose magnitude is sv and whose direction is the same as **v** if s is positive and opposite to **v** if s is negative.

The scalar (or dot) product of two vectors is written as $\mathbf{a} \cdot \mathbf{b}$ and is the scalar quantity given by

The Scalar Product

$$\mathbf{a} \cdot \mathbf{b} = ab \cos \phi, \qquad [16]$$

in which ϕ is the angle between the directions of **a** and **b**; see Fig. 18a. The scalar product may be positive, zero, or negative, depending on the value of ϕ. Being a scalar, it has no directional properties. Figure 18b shows that the scalar product can be viewed as the magnitude of either vector (a, say) multiplied by the component of the other vector ($b \cos \phi$) in the direction of the first.

The vector (or cross) product is written as $\mathbf{a} \times \mathbf{b}$ and is a vector **c** whose magnitude c is given by

The Vector Product

$$c = ab \sin \phi, \qquad [19]$$

in which ϕ is the smaller of the angles between the directions of **a** and **b**. The direction of **c** is at right angles to the plane defined by **a** and **b**. The sense in which **c** points along this direction is given by a right-hand rule described in Fig. 19. Note that (1) $\mathbf{a} \times \mathbf{b} = 0$ if **a** and **b** are either parallel or antiparallel, (2) the magnitude of $\mathbf{a} \times \mathbf{b}$ has its maximum value ($= ab$) if **a** and **b** are at right angles, and (3) $\mathbf{a} \times \mathbf{b} = -\mathbf{b} \times \mathbf{a}$; in contrast, $\mathbf{a} \cdot \mathbf{b} = \mathbf{b} \cdot \mathbf{a}$. Sample Problem 5 illustrates scalar and vector multiplication.

QUESTIONS

1. In 1969, three Apollo astronauts left Cape Canaveral, went to the moon and back, and splashed down in the Pacific Ocean. An admiral bid them good-bye at the Cape and then sailed to the Pacific Ocean in an aircraft carrier where he picked them up. Compare the displacements of the astronauts and the admiral.

2. Can two vectors having different magnitudes be combined to give a zero resultant? Can three vectors?

3. Can a vector have zero magnitude if one of its components is not zero?

4. Can the sum of the magnitudes of two vectors ever be equal to the magnitude of the sum of these two vectors?

5. Can the magnitude of the difference between two vectors ever be greater than the magnitude of either vector? Can it be greater than the magnitude of their sum? Give examples.

6. If three vectors add up to zero, they must all be in the same plane. Make this plausible.

7. Explain in what sense a vector equation contains more information than a scalar equation.

8. Do the unit vectors **i**, **j**, and **k** have units?

9. Name several scalar quantities. Does the value of a scalar quantity depend on the coordinate system you choose?

10. You can order events in time. For example, event b may precede event c but follow event a, giving us a time order of events a, b, c. Hence, there is a sense of time, distinguishing past, present, and future. Is time a vector therefore? If not, why not?

11. Do the commutative and associative laws apply to vector subtraction?

12. Can a scalar product be a negative quantity?

13. (*a*) If **a** · **b** = 0, does it follow that **a** and **b** are perpendicular to one another? (*b*) If **a** · **b** = **a** · **c**, does it follow that **b** equals **c**?

14. If **a** × **b** = 0, must **a** and **b** be parallel to each other? Is the converse true?

15. Must you specify a coordinate system when you (*a*) add two vectors, (*b*) form their scalar product, (*c*) form their vector product, or (*d*) find their components?

EXERCISES AND PROBLEMS

Section 3–2   Adding Vectors: Graphical Method

1E. Consider two displacements, one of magnitude 3 m and another of magnitude 4 m. Show how the displacement vectors may be combined to get a resultant displacement of magnitude (*a*) 7 m, (*b*) 1 m, and (*c*) 5 m.

2E. What are the properties of two vectors **a** and **b** such that
(*a*) **a** + **b** = **c**   and $a + b = c$;
(*b*) **a** + **b** = **a** − **b**;
(*c*) **a** + **b** = **c**   and $a^2 + b^2 = c^2$?

3E. A woman walks 250 m in the direction 30° east of north, then 175 m directly east. (*a*) Using graphical methods, find her final displacement from the starting point. (*b*) Compare the magnitude of her displacement with the distance she walked.

4E. A person walks in the following pattern: 3.1 km north, then 2.4 km west, and finally 5.2 km south. (*a*) Construct the vector diagram that represents this motion. (*b*) How far and in what direction would a bird fly in a straight line to arrive at the same final point?

5E. A car is driven east for a distance of 50 km, then north for 30 km, and then in a direction 30° east of north for 25 km. Draw the vector diagram and determine the total displacement of the car from its starting point.

6P. Vector **a** has a magnitude of 5.0 units and is directed east. Vector **b** is directed 35° west of north and has a magnitude of 4.0 units. Construct vector diagrams for calculating **a** + **b** and **b** − **a**. Estimate the magnitudes and directions of **a** + **b** and **b** − **a** from your diagrams.

7P. Three vectors **a**, **b**, and **c**, each having a magnitude of 50 units, lie in the *xy* plane and make angles of 30°, 195°, and 315° with the positive *x* axis, respectively. Find graphically the magnitudes and directions of the vectors (*a*) **a** + **b** + **c**, (*b*) **a** − **b** + **c**, and (*c*) a vector **d** such that (**a** + **b**) − (**c** + **d**) = 0.

8P. A bank in downtown Boston is robbed (see the map in Fig. 21). To elude police, the thieves escape by helicopter, making three successive flights described by the following displacements: 20 miles, 45° south of east; 33 miles, 26° north of west; 16 miles, 18° east of south. At the end of the third flight they are captured. In what town are they apprehended? (Use the geometric method to add these displacements on the map.)

Section 3–3   Vectors and Their Components

9E. What are the components of a vector **a** in the *xy* plane if

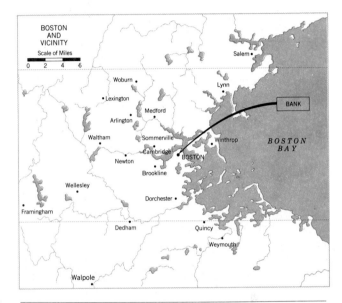

Figure 21   Problem 8.

its direction is 250° counterclockwise from the positive *x* axis and its magnitude is 7.3 units?

10E. The *x* component of a certain vector is −25 units and the *y* component is +40 units. (*a*) What is the magnitude of the vector? (*b*) What is the angle between the direction of the vector and the positive *x* axis?

11E. A displacement vector **r** in the *xy* plane is 15 m long and directed as shown in Fig. 22. Determine the *x* and *y* components of the vector.

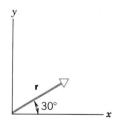

Figure 22   Exercise 11.

12E. A heavy piece of machinery is raised by sliding it 12.5 m along a plank oriented at 20° to the horizontal, as shown in Fig. 23. (*a*) How high above its original position is it raised? (*b*) How far is it moved horizontally?

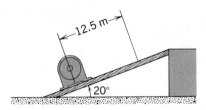

Figure 23 Exercise 12.

13E. The minute hand of a wall clock measures 10 cm from axis to tip. What is the displacement vector of its tip (*a*) from a quarter after the hour to half past, (*b*) in the next half hour, and (*c*) in the next hour?

14E. A ship sets out to sail to a point 120 km due north. An unexpected storm blows the ship to a point 100 km due east of its starting point. How far, and in what direction, must it now sail to reach its original destination?

15P. A person desires to reach a point that is 3.4 km from her present location and in a direction that is 35° north of east. However, she must travel along streets that go either north–south or east–west. What is the minimum distance she could travel to reach her destination?

16P. Rock *faults* are ruptures along which opposite faces of rock have moved past each other, parallel to the fracture surface. Earthquakes often accompany this movement. In Fig. 24, points *A* and *B* coincided before faulting. The component of the net displacement *AB* parallel to the horizontal surface fault line is called the *strike-slip* (*AC*). The component of the net displacement along the steepest line of the fault plane is the *dip-slip* (*AD*). (*a*) What is the net shift if the strike-slip is 22 m and the dip-slip is 17 m? (*b*) If the fault plane is inclined 52° to the horizontal, what is the net *vertical* displacement of *B* as a result of the faulting in (*a*)?

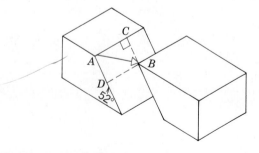

Figure 24 Problem 16.

17P. A wheel with a radius of 45 cm rolls without slipping along a horizontal floor, as shown in Fig. 25. *P* is a dot painted on the rim of the wheel. At time t_1, *P* is at the point of contact between the wheel and the floor. At a later time t_2, the wheel has rolled through one-half of a revolution. What is the displacement of *P* during this interval?

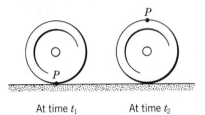

Figure 25 Problem 17.

18P. Two places *A* and *B* on the surface of the earth in South America differ by 1° in latitude and 1° in longitude. Show that the magnitude of the displacement vector from *A* to *B* is approximately $d(1 + \cos^2 \lambda)^{1/2}$, where λ is the latitude of *A* and $d = 111$ km.

19P*. A room has dimensions 10 ft × 12 ft × 14 ft. A fly starting at one corner ends up at a diametrically opposite corner. (*a*) What is the magnitude of its displacement? (*b*) Could the length of its path be less than this distance? Greater than this distance? Equal to this distance? (*c*) Choose a suitable coordinate system and find the components of the displacement vector in this frame. (*d*) If the fly walks rather than flies, what is the length of the shortest path it can take?

Section 3–5 Adding Vectors: Component Method

20E. (*a*) Express the following angles in radians: 20°, 50°, 100°. (*b*) Convert the following angles to degrees: 0.33 rad, 2.1 rad, 7.7 rad.

21E. Find the vector components of the sum **r** of the vector displacements **c** and **d** whose components in meters along three perpendicular directions are $c_x = 7.4$, $c_y = -3.8$, $c_z = -6.1$; $d_x = 4.4$, $d_y = -2.0$, $d_z = 3.3$.

22E. (*a*) What is the sum in unit vector notation of the two vectors $\mathbf{a} = 4\mathbf{i} + 3\mathbf{j}$ and $\mathbf{b} = -3\mathbf{i} + 7\mathbf{j}$? (*b*) What are the magnitude and direction of $\mathbf{a} + \mathbf{b}$?

23E. Calculate the components, magnitudes, and directions of (*a*) $\mathbf{a} + \mathbf{b}$ and (*b*) $\mathbf{b} - \mathbf{a}$ if $\mathbf{a} = 3\mathbf{i} + 4\mathbf{j}$ and $\mathbf{b} = 5\mathbf{i} - 2\mathbf{j}$.

24E. Two vectors are given by $\mathbf{a} = 4\mathbf{i} - 3\mathbf{j} + \mathbf{k}$ and $\mathbf{b} = -\mathbf{i} + \mathbf{j} + 4\mathbf{k}$. Find (*a*) $\mathbf{a} + \mathbf{b}$, (*b*) $\mathbf{a} - \mathbf{b}$, and (*c*) a vector **c** such that $\mathbf{a} - \mathbf{b} + \mathbf{c} = 0$.

25E. Given two vectors, $\mathbf{a} = 4\mathbf{i} - 3\mathbf{j}$ and $\mathbf{b} = 6\mathbf{i} + 8\mathbf{j}$, find the magnitudes and directions of **a**, **b**, $\mathbf{a} + \mathbf{b}$, $\mathbf{b} - \mathbf{a}$, and $\mathbf{a} - \mathbf{b}$.

26P. A flagpole is 15 m high. It is viewed by a person standing in a building that is 21m away. The person's eyes are level with the top of the pole. What is the angle between a horizontal line to the top of the pole and a line from the person's eyes to the bottom of the pole?

27P. Two vectors **a** and **b** have equal magnitudes of 10 units. They are oriented as shown in Fig. 26 and their vector sum is **r**. Find (a) the x and y components of **r**, (b) the magnitude of **r**, and (c) the angle **r** makes with the positive x axis.

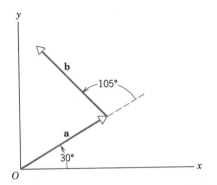

Figure 26   Problem 27.

28P. A golfer takes three putts to get the ball into the hole once it is on the green. The first putt displaces the ball 12 ft north, the second 6.0 ft southeast, and the third 3.0 ft southwest. What displacement was needed to get the ball into the hole on the first putt?

29P. A radar station detects an airplane approaching from the east. At first observation, the range to the plane is 1200 ft at 40° above the horizon. The plane is tracked for another 123° in the east–west plane, the range at final contact being 2580 ft. See Fig. 27. Find the displacement of the plane during the period of observation.

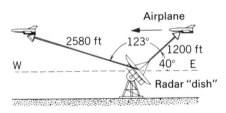

Figure 27   Problem 29.

30P. (a) A man leaves his front door, walks 1000 m east, 2000 m north, and then takes a penny from his pocket and drops it from a cliff 500 m high. Set up a coordinate system and write down an expression, using unit vectors, for the displacement of the penny. (b) The man then returns to his front door, following a different path on the return trip. What is his resultant displacement for the round trip?

31P. A particle undergoes three successive displacements in a plane, as follows: 4.0 m southwest, 5.0 m east, and 6.0 m in a direction 60° north of east. Choose the y axis pointing north

and the x axis pointing east and find (a) the components of each displacement, (b) the components of the resultant displacement, (c) the magnitude and direction of the resultant displacement, and (d) the displacement that would be required to bring the particle back to the starting point.

32P. Prove that two vectors must have equal magnitudes if their sum is perpendicular to their difference.

33P. Two vectors of lengths a and b make an angle θ with each other when placed tail to tail. Prove, by taking components along two perpendicular axes, that the length of their sum is

$$r = \sqrt{a^2 + b^2 + 2ab\cos\theta}.$$

34P. (a) Using unit vectors, express the diagonals (the lines from one corner to another through the center of the cube) of a cube in terms of its edges, which have length a. (b) Determine the angles made by the diagonals with the adjacent edges. (c) Determine the length of the diagonals.

35P*. A person flies from Washington, DC to Manila. (a) Describe the displacement vector. (b) What is its magnitude if the latitude and longitude of the two cities are 39° N, 77° W and 15° N, 121° E?

Section 3–6   Vectors and the Laws of Physics

36E. A vector **a** with a magnitude of 17 m is directed 56° counterclockwise from the +x axis, as shown in Fig. 28. (a) What are the components a_x and a_y of the vector? (b) A second coordinate system is inclined by 18° with respect to the first. What are the components a'_x and a'_y in this primed coordinate system?

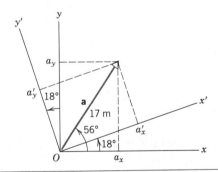

Figure 28   Exercise 36.

*Section 3–7   Multiplying Vectors

37E. A vector **d** has a magnitude of 2.5 m and points north. What are the magnitudes and directions of the vectors (a) **4d** and (b) −**3d**?

38E. Given vector **a** in the +x direction, a vector **b** in the +y direction, and the scalar quantity d: (a) What is the direction of **a** × **b**? (b) What is the direction of **b** × **a**? (c) What is the direction of **b**/d? (d) What is the magnitude of **a** · **b**?

39E. In the coordinate system of Fig. 10, show that

$$\mathbf{i} \cdot \mathbf{i} = \mathbf{j} \cdot \mathbf{j} = \mathbf{k} \cdot \mathbf{k} = 1$$

and

$$\mathbf{i} \cdot \mathbf{j} = \mathbf{j} \cdot \mathbf{k} = \mathbf{k} \cdot \mathbf{i} = 0.$$

40E. In the right-handed coordinate system of Fig. 10, show that

$$\mathbf{i} \times \mathbf{i} = \mathbf{j} \times \mathbf{j} = \mathbf{k} \times \mathbf{k} = 0$$

and

$$\mathbf{i} \times \mathbf{j} = \mathbf{k}; \quad \mathbf{k} \times \mathbf{i} = \mathbf{j}; \quad \mathbf{j} \times \mathbf{k} = \mathbf{i}.$$

41E. Show for any vector $\mathbf{a}$ that $\mathbf{a} \cdot \mathbf{a} = a^2$ and that $\mathbf{a} \times \mathbf{a} = 0$.

42E. Find (a) "north cross west," (b) "down dot south," (c) "east cross up," (d) "west dot west," and (e) "south cross south." Let each vector have unit magnitude.

43E. A vector $\mathbf{a}$ of magnitude 10 units and another vector $\mathbf{b}$ of magnitude 6 units point in directions differing by 60°. Find (a) the scalar product of the two vectors and (b) the vector product $\mathbf{a} \times \mathbf{b}$.

44E. Two vectors, $\mathbf{r}$ and $\mathbf{s}$, lie in the xy plane. Their magnitudes are 4.5 and 7.3 units, respectively, whereas their directions are 320° and 85° measured counterclockwise from the positive x axis. What are the values of (a) $\mathbf{r} \cdot \mathbf{s}$ and (b) $\mathbf{r} \times \mathbf{s}$?

45E. Three vectors add to zero, as in the right triangle of Fig. 29. Calculate (a) $\mathbf{a} \cdot \mathbf{b}$, (b) $\mathbf{a} \cdot \mathbf{c}$, and (c) $\mathbf{b} \cdot \mathbf{c}$.

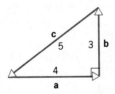

Figure 29 Exercises 45 and 46.

46E. Three vectors add to zero, as in Fig. 29. Calculate (a) $\mathbf{a} \times \mathbf{b}$, (b) $\mathbf{a} \times \mathbf{c}$, and (c) $\mathbf{b} \times \mathbf{c}$.

47P. *Scalar Product in Unit Vector Notation.* Let two vectors be represented in terms of their coordinates as

$$\mathbf{a} = \mathbf{i}a_x + \mathbf{j}a_y + \mathbf{k}a_z$$

and

$$\mathbf{b} = \mathbf{i}b_x + \mathbf{j}b_y + \mathbf{k}b_z.$$

Show analytically that

$$\mathbf{a} \cdot \mathbf{b} = a_x b_x + a_y b_y + a_z b_z.$$

(*Hint:* See Exercise 39.)

48P. Use the definition of scalar product $\mathbf{a} \cdot \mathbf{b} = ab \cos \theta$ and the fact that $\mathbf{a} \cdot \mathbf{b} = a_x b_x + a_y b_y + a_z b_z$ (see Problem 47) to calculate the angle between the two vectors given by $\mathbf{a} = 3\mathbf{i} + 3\mathbf{j} + 3\mathbf{k}$ and $\mathbf{b} = 2\mathbf{i} + \mathbf{j} + 3\mathbf{k}$.

49P. (a) Determine the components and magnitude of $\mathbf{r} = \mathbf{a} - \mathbf{b} + \mathbf{c}$ if $\mathbf{a} = 5\mathbf{i} + 4\mathbf{j} - 6\mathbf{k}$, $\mathbf{b} = -2\mathbf{i} + 2\mathbf{j} + 3\mathbf{k}$, and $\mathbf{c} = 4\mathbf{i} + 3\mathbf{j} + 2\mathbf{k}$. (b) Calculate the angle between $\mathbf{r}$ and the positive z axis.

50P. Two vectors are given by $\mathbf{a} = 3\mathbf{i} + 5\mathbf{j}$ and $\mathbf{b} = 2\mathbf{i} + 4\mathbf{j}$. Find (a) $\mathbf{a} \times \mathbf{b}$, (b) $\mathbf{a} \cdot \mathbf{b}$, and (c) $(\mathbf{a} + \mathbf{b}) \cdot \mathbf{b}$.

51P. *Vector Product in Unit Vector Notation.* Show that $\mathbf{a} \times \mathbf{b} = \mathbf{i}(a_y b_z - a_z b_y) + \mathbf{j}(a_z b_x - a_x b_z) + \mathbf{k}(a_x b_y - a_y b_x)$. (*Hint:* See Exercise 40.)

52P. Two vectors $\mathbf{a}$ and $\mathbf{b}$ have components, in arbitrary units, $a_x = 3.2$, $a_y = 1.6$; $b_x = 0.50$, $b_y = 4.5$. (a) Find the angle between $\mathbf{a}$ and $\mathbf{b}$. (b) Find the components of a vector $\mathbf{c}$ that is perpendicular to $\mathbf{a}$, is in the xy plane, and has a magnitude of 5.0 units.

53P. Vector $\mathbf{a}$ lies in the yz plane 63° from the $+y$ axis with a positive z component and has magnitude 3.2 units. Vector $\mathbf{b}$ lies in the xz plane 48° from the $+x$ axis with a positive z component and has magnitude 1.4 units. Find (a) $\mathbf{a} \cdot \mathbf{b}$, (b) $\mathbf{a} \times \mathbf{b}$, and (c) the angle between $\mathbf{a}$ and $\mathbf{b}$.

54P. Three vectors are given by $\mathbf{a} = 3\mathbf{i} + 3\mathbf{j} - 2\mathbf{k}$, $\mathbf{b} = -\mathbf{i} - 4\mathbf{j} + 2\mathbf{k}$, and $\mathbf{c} = 2\mathbf{i} + 2\mathbf{j} + \mathbf{k}$. Find (a) $\mathbf{a} \cdot (\mathbf{b} \times \mathbf{c})$, (b) $\mathbf{a} \cdot (\mathbf{b} + \mathbf{c})$, and (c) $\mathbf{a} \times (\mathbf{b} + \mathbf{c})$.

55P. Find the angles between the body diagonals of a cube with edge length a. See Problem 34.

56P. (a) We have seen that the commutative law *does not* apply to vector products; that is, $\mathbf{a} \times \mathbf{b}$ does not equal $\mathbf{b} \times \mathbf{a}$. Show that the commutative law *does* apply to scalar products; that is, $\mathbf{a} \cdot \mathbf{b} = \mathbf{b} \cdot \mathbf{a}$. (b) Show that the distributive law applies to both scalar products and vector products; that is, show that

$$\mathbf{a} \cdot (\mathbf{b} + \mathbf{c}) = \mathbf{a} \cdot \mathbf{b} + \mathbf{a} \cdot \mathbf{c}$$

and that

$$\mathbf{a} \times (\mathbf{b} + \mathbf{c}) = \mathbf{a} \times \mathbf{b} + \mathbf{a} \times \mathbf{c}.$$

(c) Does the associative law apply to vector products; that is, does $\mathbf{a} \times (\mathbf{b} \times \mathbf{c})$ equal $(\mathbf{a} \times \mathbf{b}) \times \mathbf{c}$? (d) Does it make any sense to talk about an associative law for scalar products?

57P. Show that the area of the triangle contained between the vectors $\mathbf{a}$ and $\mathbf{b}$ in Fig. 30 is $\frac{1}{2}|\mathbf{a} \times \mathbf{b}|$, where the vertical bars signify magnitude.

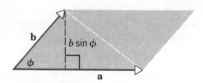

Figure 30 Problem 57.

58P. (*a*) Show that $\mathbf{a} \cdot (\mathbf{b} \times \mathbf{a})$ is zero for all vectors $\mathbf{a}$ and $\mathbf{b}$. (*b*) What is the value of $\mathbf{a} \times (\mathbf{b} \times \mathbf{a})$ if there is an angle ϕ between the directions of $\mathbf{a}$ and $\mathbf{b}$?

59P. Show that $\mathbf{a} \cdot (\mathbf{b} \times \mathbf{c})$ is equal in magnitude to the volume of the parallelepiped formed on the three vectors $\mathbf{a}$, $\mathbf{b}$, and $\mathbf{c}$ as shown in Fig. 31.

60P. The three vectors shown in Fig. 32 have magnitudes $a = 3$, $b = 4$, and $c = 10$. (*a*) Calculate the x and y components of these vectors. (*b*) Find the numbers p and q such that $\mathbf{c} = p\mathbf{a} + q\mathbf{b}$.

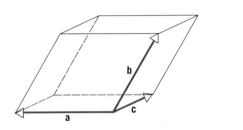

Figure 31   Problem 59.

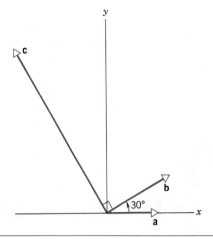

Figure 32   Problem 60.

ESSAY 1
RUSH-HOUR
TRAFFIC FLOW

JEARL WALKER
CLEVELAND STATE
UNIVERSITY

Traffic lights in a small town usually require no special sequencing. The flow of traffic through a system of lights may be haphazard, but the queues at red lights are seldom long. In contrast, traffic flow in a large city, especially during rush hour, requires careful regulation. Without it, the lines of cars lengthen until many intersections are blocked, throwing the whole region into what is called gridlock. Since only the cars on the perimeter of the congested area can then move, hours may be needed to free the cars trapped in the interior.

Suppose you were to engineer the traffic light system for a one-way street that consists of several lanes along which rush-hour traffic flows. The lights are green for 50 s, yellow for 5 s, and red for 25 s—values that are typically employed for a heavily-traveled route through a city. To promote the traffic flow you may be tempted to increase the duration of the green light or decrease the duration of the red light. However, you must remember that the traffic on the perpendicular streets cannot be held in check for too long or lengthy queues will develop there.

How should you time the onsets of green lights at the intersections? If you arrange for all the lights to turn green simultaneously, then the traffic can move for only 50 s. Upon each onset of the green lights, platoons of cars will move along the route until all the lights on the street turn red. To maximize the distance traveled, the drivers will be tempted to race through the system. Large platoons of cars traveling at, say, 55 mi/h along a crowded city street would resemble a "grand prix," presenting an obvious danger.

A better and safer design is one where the light at each intersection does not turn green until the platoon leaders from the previous intersection reach it. For example, suppose the platoon is initially at intersection 1 and headed toward intersection 2. How long after the light at intersection 1 turns green should the light at intersection 2 turn green?

Let d be the distance between the two intersections and v_p be the speed you wish the platoon to have, probably the speed limit. The required time delay on the green light at intersection 2 is d/v_p. If all the subsequent green lights along the street are also appropriately delayed, drivers traveling at v_p should move smoothly through the whole light system. If the intersections are separated by identical distances, then the delay on the green light from one intersection to the next is always d/v_p. The plan works just as well if the distance between intersections varies. You merely use the appropriate value of d to determine the necessary delay needed for the green light at a given intersection. Racing through the system is then futile; eventually the driver is stopped by a red light, whereupon everyone that had been passed catches up before the light turns green.

The plan is slightly modified by two factors. A platoon leader should see the onset of the green light before reaching an intersection. Otherwise, the leader will slow out of fear of entering the intersection on a red light. To adjust for this factor, you must decrease the delay in turning on the green light by a few seconds. However, if the platoon were stopped by a red light at intersection 1, the platoon leader would require a few extra seconds to respond to the onset of green light there and to accelerate to cruising speed. Thus, the leader does not travel the full distance between intersections at v_p. This factor requires an increase in the delay by a few seconds. In

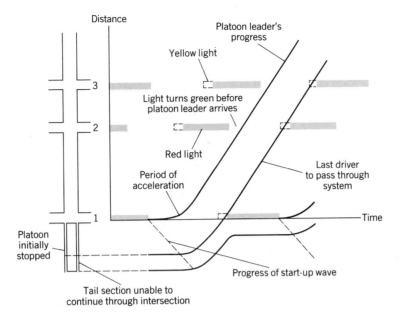

Figure 1 Graphical representation of how a platoon of cars stopped at intersection 1 thereafter proceeds through a system of traffic lights.

some cases, the two factors might offset each other.

All these points are illustrated in Fig. 1. A street system is displayed on the left side. The vertical axis on the graph corresponds to the distance along the street. The platoon, which was initially stopped at intersection 1, travels through the light system. When the platoon leader travels at a constant speed, the slope of the line representing the travel is equal to that speed. The leader's initial period of acceleration is represented by a curved line. Note that the light at intersection 2 is shown turning green a few seconds before the leader reaches it.

With a system of delayed lights, the traffic can still freeze into place. The problem lies in the fact that once a platoon of drivers is stopped and then given another green light, they cannot all accelerate simultaneously. Instead, a "start-up wave" travels from the leader along the length of the platoon. Compared to the leader, the last driver not only has a longer distance to travel to intersection 2 but also must wait for the start-up wave to travel through the platoon. Thus, the last driver in a lengthy platoon is always certain to require much more time to reach the next intersection.

You can measure the speed of a start-up wave by positioning yourself alongside a stationary platoon. Determine the length of the platoon by multiplying the number of cars by an estimate of the average distance between front bumpers. Then measure the time from the onset of the green light until the last car begins to move. The speed of the start-up wave is the length of the platoon divided by the time you measure. A typical value is 5 m/s ($=11$ mi/h).

If the tail of a platoon cannot pass through an intersection with the rest of the platoon, then the subsequent platoons may begin to lengthen. Consider a tail section that has been abandoned at intersection 2 as shown in Fig. 2. Call the section A. As A waits through the red light, the platoon at intersection 1 gets a green light and begins to move forward. Call this platoon B. Since the green light at intersection 2 is delayed to match the progress of the leaders of B, they arrive at the rear of A before the leaders of A begin to move. The problem is compounded by the fact that before the last drivers of A can move, they must wait for the start-up wave to travel to them from their leaders.

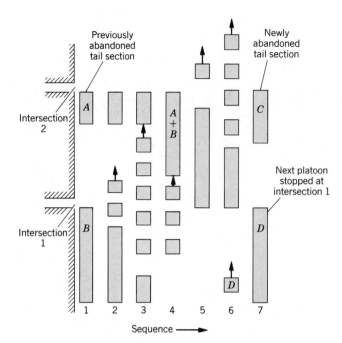

Figure 2 With a delayed light at intersection 2, the platoon length can increase.

The net result is that A and B combine, forming a new platoon at intersection 2. It may be too long to pass entirely through the intersection. Let C be the tail section that is stopped by the next red light at intersection 2. If C is longer than A, the situation has deteriorated. As C waits through the red light, another platoon (call it D) arrives from intersection 1. The combination of C and D may then be longer than the previous combination of A and B, guaranteeing that the next tail section to be abandoned at intersection 2 will be longer than C was. The situation can then quickly become hopeless if with each cycle of the lights the newly abandoned section at intersection 2 is longer than the previous one. Eventually, the abandoned section is so long that it stretches back to intersection 1.

Consider the very next platoon that is waiting at intersection 1. Call the platoon M. When M receives a green light, it moves into intersection 1, blocking it. Since these cars then cannot move until the start-up wave from intersection 2 reaches them, they are still in place when the light at intersection 1 turns red. With intersection 1 blocked, drivers on the perpendicular street cannot move through it during their green light. The queues on the side street lengthen until they block intersections on the streets that run parallel to the route we are considering. The congestion rapidly spreads until the whole region becomes one large parking lot.

A gridlock can develop even if the traffic light system is well designed. I was once trapped in a gridlock when a sudden, heavy snowfall caught the afternoon rush-hour traffic of Cleveland. Since the street I was on was slippery, the platoon leaders proceeded cautiously. The start-up waves also moved slower than normally. Within about 20 min, abandoned sections of platoons stretched back to previously crossed intersections, blocking them. For 2 mi along my route and along five parallel arteries out of the city, traffic came to almost a standstill. I made progress only because cars on the outward end of the route gradually escaped into the suburbs. As they left the pack, a start-up wave moved sedately through the 2-mi platoon, allowing me to creep forward by a few car lengths. The problem worsened as stalled cars blocked lanes. A normal 5 min drive took over 2 h!

A gridlock caused by an unexpected snowfall may be impossible to alleviate. However, a gridlock caused by unexpectedly long platoons may be avoidable if an engineer or a computer intervenes in the normal sequencing of lights. Suppose that through remote sensing you detect the gradual buildup of abandoned sections of platoons at, say, intersection 2. Let A be the section that is abandoned there when you intervene and let B be the platoon that is then stopped at intersection 1. One way to reverse the buildup is to release A early enough that B need not stop as it approaches intersection 2.

If A is already sufficiently long when you intervene, you may need to change the sequencing so that the green lights at intersection 1 and 2 turn on simultaneously. If A is even longer, you will need to turn on the green light at intersection 2 before you turn on the green light at intersection 1.

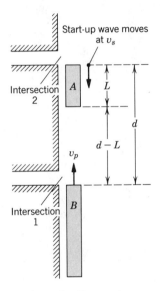

Start-up wave moves at v_s

Intersection 2

A L

d

$d - L$

v_p

Intersection 1

B

Which course of action is best can be stated in terms of the ratio $x = L/d$ of the length L of the abandoned section A to the distance d between intersections 1 and 2. If the leaders of B travel to the rear of A at a constant speed v_p, they require a time of $(d - L)/v_p$ to make the trip (see Fig. 3). Meanwhile, a start-up wave travels through A at a speed of v_s, requiring a time of L/v_s to reach the rear of A. If the leaders of B are to reach the rear of A as the cars there begin to move, then the time t between the onset of the green light at intersection 2 and that at intersection 1 should be

$$t = \frac{d - L}{v_p} - \frac{L}{v_s}.$$

If x is short enough, t is positive, which means that the beginning of the green light at intersection 2 should follow the beginning of the green light at intersection 1 by the time t. If x is long enough, t is negative, which means that the onset of the green light at intersection 2 should precede the onset of the green light at intersection 1 by the time t.

"Switch-over" between the two procedures occurs when t is zero:

$$0 = \frac{d - L}{v_p} - \frac{L}{v_s},$$

Figure 3 Factors important in determining relative onsets of green lights at successive intersections.

which gives x as

$$x = \frac{L}{d} = \frac{v_s}{v_s + v_p}.$$

With typical values of 10 m/s (about 22 mi/h) for the platoon speed and 5 m/s for the speed of the start-up wave, switch-over is required when x is $\frac{1}{3}$. If x is larger than $\frac{1}{3}$, the green light at intersection 2 must turn on before the green light at intersection 1.

You can interpret this result in terms of the speed at which a "wave" of green lights travels along the street. When $x < \frac{1}{3}$, the wave travels in the direction of the traffic and with the same speed. When $x = \frac{1}{3}$, the green lights come on simultaneously. When $x > \frac{1}{3}$, the wave of green lights travels in the opposite direction of the traffic.

This result is only an estimate since the calculation ignores three facts. The leaders of B begin their motion by accelerating from a stop. They must see that the rear of A is moving prior to their arrival or they will slow or stop out of caution. The rear cars in A must, of course, accelerate to the cruising speed prior to the arrival of B. You might enjoy fine tuning the calculation by including these additional factors.

How would you design a light system for optimizing the flow of traffic in two directions on a street? In such a situation is there any way to avoid having cars occasionally stop? Can you simulate traffic platoons moving through a system of street lights on a computer, perhaps making a game out of the simulation? If so, then you could investigate the problems of gridlock without ever having to endure one.

CHAPTER 4
MOTION IN A PLANE

If you move in one dimension, you may or may not be accelerated. If you move in two or more dimensions you cannot escape acceleration. Amusement parks, which sell acceleration to the general public, are based on this principle. The Loch Ness Monster, at Busch Gardens, can change your velocity in many ways.

4-1 Moving in Three Dimensions

If a particle moves in one dimension only, it is easy to answer the question: "Where is the particle?" It is enough to reply: "At $x = +13$ m" or "At $x = -8.5$ m." A single number, with its unit and its algebraic sign, answers the question exactly.

In the three-dimensional world, we need more information than this to tell us where a particle is. We need a *vector*. Let us see how to describe the position, the velocity, and the acceleration of a particle in the language of vectors.

4-2 Where Is the Particle?

In three dimensions, we locate a particle by a vector **r**, extending from the origin of a coordinate system to the particle's position, as in Fig. 1. Thus,

$$\mathbf{r} = \mathbf{i}x + \mathbf{j}y + \mathbf{k}z, \tag{1}$$

in which **i**, **j**, and **k** are *unit vectors* (see Section 3-4) and x, y, and z are the components of the vector **r**. As always, the components can be positive, negative, or zero. For

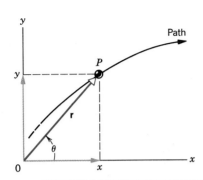

Figure 1   The location of particle *P* at a given instant is specified by a vector **r** from the origin of a coordinate system to the position of the particle. The components of **r** (*x* and *y*) are also shown.

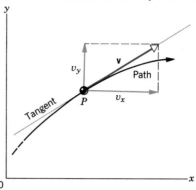

Figure 2   The velocity **v** of particle *P* is shown, along with its components. Note that it lies along the tangent to the path. Compare Fig. 1.

simplicity, we assume $z = 0$ in Eq. 1, so that the motion of the particle in Fig. 1 is confined to the *xy* plane.*

4–3   What Is the Velocity of the Particle?

For one-dimensional motion, the scalar equation $v = dx/dt$ answers this question. In three dimensions, we generalize this definition and write the vector equation

$$\mathbf{v} = \frac{d\mathbf{r}}{dt}. \tag{2}$$

We shall show below that the velocity vector is tangent to the path of the particle.

The Velocity — Magnitude.   Substituting **r** from Eq. 1 into Eq. 2 gives

$$\mathbf{v} = \frac{d}{dt}(\mathbf{i}x + \mathbf{j}y + \mathbf{k}z) = \mathbf{i}\frac{dx}{dt} + \mathbf{j}\frac{dy}{dt} + \mathbf{k}\frac{dz}{dt}$$

or

$$\mathbf{v} = \mathbf{i}v_x + \mathbf{j}v_y + \mathbf{k}v_z. \tag{3}$$

As we see, the three components of the velocity vector are given by

$$v_x = \frac{dx}{dt}, \quad v_y = \frac{dy}{dt}, \quad \text{and} \quad v_z = \frac{dz}{dt}. \tag{4}$$

---

* Although we shall define position, displacement, velocity, and acceleration for the general case of motion in three dimensions, we shall, for reasons of simplicity, illustrate them in the rest of this chapter in two dimensions only. Hence our chapter title.

Note how a single vector equation (Eq. 2) is equivalent to three scalar equations (Eq. 4). Figure 2 shows the velocity vector and its components, again for motion in two dimensions.

The Velocity — Direction.   Figure 3 shows the particle at a time *t*, when its position is given by **r**(*t*), and also at a later time $t + \Delta t$, when its position is given by $\mathbf{r}(t + \Delta t)$. The vector $\Delta\mathbf{r}$ in Fig. 3 represents the *displacement* (that is, the change in position) of the particle during the interval Δt.

The vector $\Delta\mathbf{r}$ lies along a chord of the path of the particle in Fig. 3, pointing in the direction of the average velocity $\bar{\mathbf{v}}$ $(= \Delta\mathbf{r}/\Delta t)$. As we shrink the time interval Δt down to a differential interval dt, however, the vector $\Delta\mathbf{r}$ shrinks to a differential $d\mathbf{r}$ and points, as we see, along the

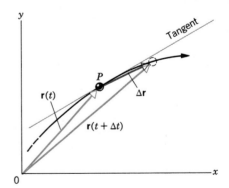

Figure 3   The position of particle *P* along its path is shown, both at time *t* and at a later time $t + \Delta t$. The vector $\Delta\mathbf{r}$ is the displacement of the particle during this interval. As Δt approaches zero, this vector moves closer and closer to the direction of the tangent line. Compare Fig. 2.

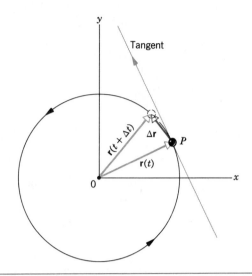

Figure 4 The conditions are the same as in Fig. 3, except that the particle is moving in a circular path. As Δt approaches zero, the vector $\Delta \mathbf{r}$ moves closer and closer to the direction of the tangent line.

tangent to the path, in the direction of the (instantaneous) velocity $\mathbf{v}$ ($= d\mathbf{r}/dt$).

Figure 4 represents a hockey puck held by a string and whirled around on a frictionless air-table. Here $\mathbf{r}$ changes in direction only, its magnitude (the length of the string) remaining constant. Here too the velocity vector at any instant is tangent to the path. If the string were cut, the puck would move at a constant speed along the tangent line. It would *not* fly radially outward, as some may think.

4-4 What Is the Acceleration of the Particle?

For motion in one dimension, the scalar equation $a = dv/dt$ answers this question. In three dimensions, we generalize this definition and write the vector equation

$$\mathbf{a} = \frac{d\mathbf{v}}{dt}. \tag{5}$$

If the velocity changes *either* in magnitude *or* direction (or both), there will be an acceleration.

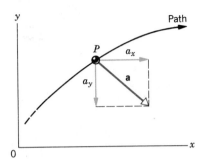

Figure 5 The acceleration $\mathbf{a}$ of particle P is shown, along with its components. Compare Figs. 1 and 2.

Substituting $\mathbf{v}$ from Eq. 3 into Eq. 5 yields

$$\mathbf{a} = \frac{d}{dt}(\mathbf{i}v_x + \mathbf{j}v_y + \mathbf{k}v_z)$$

$$= \mathbf{i}\frac{dv_x}{dt} + \mathbf{j}\frac{dv_y}{dt} + \mathbf{k}\frac{dv_z}{dt}$$

or

$$\mathbf{a} = \mathbf{i}a_x + \mathbf{j}a_y + \mathbf{k}a_z, \tag{6}$$

in which the three components of the acceleration vector are given by

$$a_x = \frac{dv_x}{dt}, \quad a_y = \frac{dv_y}{dt}, \quad \text{and} \quad a_z = \frac{dv_z}{dt}. \tag{7}$$

Figure 5 shows the acceleration vector and its components, again for motion in two dimensions.

Sample Problem 1 A rabbit runs across a parking lot on which a set of coordinate axes has been drawn. The rabbit's path is such that the components of its position with respect to an origin of coordinates are given as functions of time by

$$x(t) = -0.31t^2 + 7.2t + 28$$

and

$$y(t) = 0.22t^2 - 9.1t + 30.$$

The units of the numerical coefficients in these equations have been suppressed but they are such that, if you substitute t in seconds, x and y will emerge in meters. (a) Calculate the rabbit's position (magnitude and direction) at $t = 15$ s.

At $t = 15$ s, the components of the position are

$$x = (-0.31)(15)^2 + (7.2)(15) + (28) = 66 \text{ m}$$

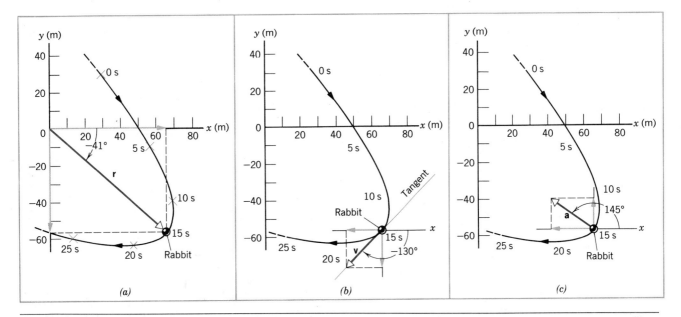

(a)   (b)   (c)

Figure 6   Sample Problems 1, 2 and 3. (a) The path of a rabbit across a parking lot, showing its position at the indicated times. The vector **r** locates the rabbit at $t = 15$ s. (b) The velocity **v** of the rabbit at $t = 15$ s. Note that it is tangent to the path. (c) The acceleration **a** of the rabbit at $t = 15$ s. As it happens, the rabbit has this same acceleration for all points of its path.

and

$$y = (0.22)(15)^2 - (9.1)(15) + (30) = -57 \text{ m}.$$

The components and **r** itself are shown in Fig. 6a.

The magnitude of **r** is given by

$$r = \sqrt{x^2 + y^2} = \sqrt{(66 \text{ m})^2) + (-57 \text{ m})^2} = 87 \text{ m}. \quad \text{(Answer)}$$

The angle θ between **r** and the direction of increasing x is

$$\theta = \tan^{-1} \frac{y}{x} = \tan^{-1} \left(\frac{-57 \text{ m}}{66 \text{ m}} \right) = -41°. \quad \text{(Answer)}$$

Although $\theta = 139°$ has the same tangent as $-41°$, study of the signs of the components of **r** in Fig. 6a rules it out.

(b) Calculate the position of the rabbit at $t = 0, 5, 10, 15,$ 20, and 25 s and sketch the rabbit's path.

Proceeding as in (a) above leads to the following values of r and θ.

$t(s)$	$x(m)$	$y(m)$	$r(m)$	θ
0	28	30	41	$+47°$
5	56	-10	57	$-10°$
10	69	-39	79	$-29°$
15	66	-57	87	$-41°$
20	48	-64	80	$-53°$
25	14	-60	62	$-77°$

Figure 6a shows a plot of the rabbit's path.

Sample Problem 2   In Sample Problem 1, find the magnitude and the direction of the rabbit's velocity vector at $t = 15$ s.

The velocity component in the x direction (see Eq. 4) is

$$v_x(t) = \frac{dx}{dt} = \frac{d}{dt}(-0.31t^2 + 7.2t + 28) = -0.62t + 7.2.$$

At $t = 15$ s, this becomes

$$v_x = (-0.62)(15) + 7.2 = -2.1 \text{ m/s}.$$

Similarly,

$$v_y(t) = \frac{dy}{dt} = \frac{d}{dt}(0.22t^2 - 9.1t + 30) = 0.44t - 9.1.$$

At $t = 15$ s, this becomes

$$v_y = (0.44)(15) - (9.1) = -2.5 \text{ m/s}.$$

The components of **v**, along with the vector itself, are shown in Fig. 6b.

The magnitude and the direction of **v** are given (see Fig. 2) by

$$v = \sqrt{v_x^2 + v_y^2} = \sqrt{(-2.1 \text{ m/s})^2 + (-2.5 \text{ m/s})^2}$$
$$= 3.3 \text{ m/s} \quad \text{(Answer)}$$

and

$$\theta = \tan^{-1} \frac{v_y}{v_x} = \tan^{-1} \left(\frac{-2.5 \text{ m/s}}{-2.1 \text{ m/s}} \right)$$
$$= \tan^{-1} 1.19 = -130°. \qquad \text{(Answer)}$$

Although 50° has the same tangent, inspection of the signs of the velocity components in Fig. 6b indicates that the desired angle is in the third quadrant, given by 50° − 180° = −130°. The velocity vector in Fig. 6b is tangent to the path of the rabbit and points in the direction that the rabbit is running.

Sample Problem 3 In Sample Problem 1, what are the magnitude and the direction of the rabbit's acceleration vector **a** at $t = 15$ s?

The acceleration components (see Eq. 7) are given by

$$a_x = \frac{dv_x}{dt} = \frac{d}{dt}(-0.62t + 7.2) = -0.62 \text{ m/s}^2$$

and

$$a_y = \frac{dv_y}{dt} = \frac{d}{dt}(0.44t - 9.1) = 0.44 \text{ m/s}^2.$$

We see that the acceleration does not vary with time; it is a constant. In fact, we have differentiated the time variable completely away! The components of **a**, along with the vector itself, are shown in Fig. 6c.

The magnitude and direction of **a** are given by

$$a = \sqrt{a_x^2 + a_y^2} = \sqrt{(-0.62 \text{ m/s}^2)^2 + (0.44 \text{ m/s}^2)^2}$$
$$= 0.76 \text{ m/s}^2 \qquad \text{(Answer)}$$

and

$$\theta = \tan^{-1} \frac{a_y}{a_x} = \tan^{-1} \left(\frac{0.44 \text{ m/s}^2}{-0.62 \text{ m/s}^2} \right) = 145°. \quad \text{(Answer)}$$

The acceleration vector has the same magnitude and direction for all parts of the rabbit's path. Perhaps a strong southeast wind was blowing across the parking lot.

Hint 1: Significant figures—I. If you were going to divide 137 jelly beans among 3 people, you would not think of giving each person *exactly* 137/3 or 45.66666666 · · · beans. You would give each person 45 beans and draw straws to see who would not get one of the remaining two. We need to develop that same kind of common sense in dealing with numerical calculations in physics.

In Sample Problem 1, for example, if you calculate the magnitude of **r** with your calculator wide open, you will get $r = 87.20665112$ m. This number has 10 *significant fig-*

ures. Your input data (the values of x and y) have only two significant figures.

> *In general, no calculated result should have more significant figures than the input data from which it was derived.*

Thus, you quickly *round off* that long and misleading number to 87 m, as we did.

It is hard to escape the feeling that you are throwing away good data when you round off in this way but, in fact, you are doing the opposite; you are throwing away useless and misleading numbers. You can set your calculator to do this rounding for you. Note that, regardless of how you set it, your calculator continues to compute wide open internally, displaying only the rounded result that you ask it to show you.

Hint 2: Significant figures and decimal places. Don't confuse these. Consider the following lengths: 35.6 m, 3.56 m, and 0.356 m, and 0.00356 m. They all have three significant figures but, in sequence, they have one, two three, and five decimal places.

Hint 3: Drawing a graph. Figure 6 brings out many of the elements that you need to consider in drawing a graph. In Sample Problem 1b the quantities to be plotted are the position variables x and y, each expressed in meters. In this problem it makes sense to choose the same scale for both axes. You may have to experiment with the scale to get a graph that will fill the space well and be easy to read. If you find that you need more points to get a smooth graph, calculate them from the formulas given in the problem statement. If one of your points falls distinctly off a smooth curve, you have probably made an error in calculating or in plotting.

The crosses on the graph of Fig. 6a show the time t at which the rabbit was at that position. By this device, you can display a third variable on your graph.

Hint 4: Trig functions and angles. In Sample Problem 2, we were given $\theta = \tan^{-1} 1.19$ and asked to find θ. Your calculator will tell you $\theta = 50°$. However, Fig. 16 of Chapter 3 shows that $\theta = 230° (= 50° + 180°)$ has the same tangent. Inspection of the signs of the velocity components v_x and v_y in Fig. 6b tells us that this latter angle is the correct one.

There is still another decision to make. We can stick with 230° or we can relabel it as −130°. They are exactly the same angle, as we pointed out in Hint 1 of Chapter 3. We chose $\theta = -130°$, purely as a matter of taste.

Hint 5: Drawing vectors—direction. The directions of the vectors in Fig. 6 were established in the following way: (1) Choose the point at which you wish the tail of the vector to be. (2) From that point, draw a line in the direction of increasing *x*. (3) Using a protractor, measure the angle θ counterclockwise from this line (if θ is positive) or clockwise (if θ is negative).

Hint 6: Drawing vectors—length. The vector **r** in Fig. 6a should be drawn to the same scale as the two axes because, like them, it is a length. The velocity vector **v** in Fig. 6b and the acceleration vector **a** in Fig. 6c, however, have no established scale in this problem and you may make them as long or as short as you wish.

It makes no sense to ask whether, for example, a velocity vector should be longer or shorter than a displacement vector. They are different physical quantities, expressed in different units, and they have no common scale.

4-5 Projectile Motion

Here we consider a particle—or a *projectile,* as we shall now call it—that moves in two dimensions rather than one. Our projectile might be a golf ball (as in Fig. 7), a

Figure 8 A human projectile. He follows a parabolic path from the mouth of the cannon to (we trust) the net.

baseball, or a cannon ball (human or otherwise; see Fig. 8), among many possibilities. Throughout, we shall assume that atmosphere has no effect on the motion of the projectile. Figure 9, to which we shall return, shows the path followed by a projectile under these *no-atmosphere* conditions.

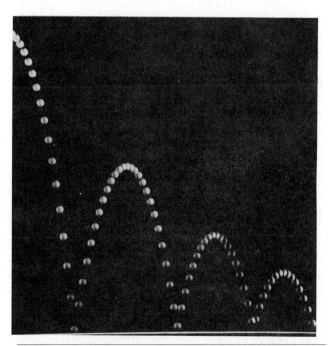

Figure 7 A strobe photo of a golf ball bouncing off a hard surface. Between impacts, the ball exhibits projectile motion.

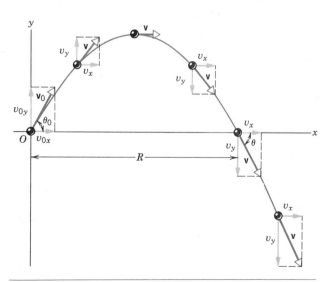

Figure 9 The parabolic path of a projectile. The initial velocity and the velocities at various points along its path are shown, along with their components. Note that the horizontal velocity component remains constant. *R* defines the *range* of the projectile.

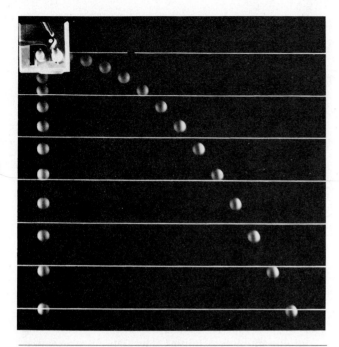

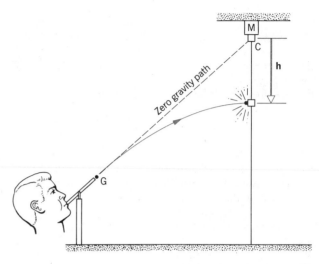

Figure 11 The ball bearing always hits the falling can. The vector **h** shows both the distance of fall of the ball bearing from its zero gravity path and the distance of fall of the can from its support; they are the same.

Figure 10 Ball I is released from rest at the same instant that ball II is fired to the right with initial velocity v_0. Their vertical motions are identical. The exposures in this strobe photo are taken at intervals of $\frac{1}{30}$ s.

Our plan is to resolve the initial velocity vector $\mathbf{v}_0$ of Fig. 9 into its components

$$v_{0x} = v_0 \cos \theta_0 \quad \text{and} \quad v_{0y} = v_0 \sin \theta_0, \quad (8)$$

and to deal with each component separately. In this way, we break up our two-dimensional problem into two separate and easier one-dimensional problems, one for the horizontal motion and one for the vertical motion. It is crucial to our plan that these two motions (except for sharing a common time variable) can be treated independently. Here are two experiments that prove this point.

Two Golf Balls. Figure 10 is a strobe photo of two golf balls, one (ball I) released from rest and the other (ball II) fired horizontally from a spring gun. They both share the same vertical motion, each ball falling through the same vertical distance in the same interval of time. *The fact that ball II is moving horizontally while it is falling has no effect on its vertical motion.* To push the experiment to a limit, we say that, if you were to fire a rifle horizontally and at the same time drop a bullet, in the absence of air resistance the two bullets would reach the level ground at the same time.

A Great Student Rouser. Figure 11 shows a demonstration that has enlivened many a physics lecture. It involves a blow gun G, using a ball bearing as a projectile. The target is a tin can C, suspended from a magnet M, the tube of the blow gun being bore-sighted directly at the can. The experiment is arranged so that the magnet releases the can just as the ball leaves the blow gun.

If g (the free-fall acceleration) were zero, the ball would follow the straight line shown in Fig. 11 and the can would float in place after the magnet released it. The ball would certainly hit the can.

However, g is *not* zero. The ball *still* hits the can! As the vector **h** in Fig. 11 shows, during the time of flight of the ball, both ball and can fall the same distance h. The harder the demonstrator blows, the greater the initial speed of the ball, the shorter the time of flight, and the smaller the value of h.

4-6 Projectile Motion Analyzed

Now we are ready to analyze projectile motion, under several headings.

The Horizontal Motion. Because there is no acceleration in the horizontal direction, the horizontal component of the velocity remains unchanged, as you can check from Fig. 9. The horizontal position is given by

Eq. 14 of Chapter 2, in which we put $a = 0$ and substitute $v_{0x} (= v_0 \cos \theta_0)$ for v_0. Thus,

$$x - x_0 = (v_0 \cos \theta_0)t. \qquad (9)$$

The Vertical Motion. The vertical motion is the same as the motion we discussed in Section 2–7 for a particle in free fall. Equations 18–22 of Chapter 2 apply.

Equation 19 of Chapter 2, for example, becomes

$$y - y_0 = (v_0 \sin \theta_0)t - \tfrac{1}{2}gt^2. \qquad (10)$$

Study of the vertical velocity component in Fig. 9 shows that it behaves just as for a ball thrown vertically upward. It points upward initially, its magnitude steadily decreasing until it drops to zero, marking the maximum height of the path. The component then reverses its direction, its magnitude becoming larger and larger as time goes on.

In addition, we shall find Eqs. 18 and 20 of Chapter 2 useful. Adapted to our purpose, they are

$$v_y = v_0 \sin \theta_0 - gt, \qquad (11)$$

and

$$v_y^2 = (v_0 \sin \theta_0)^2 - 2g(y - y_0). \qquad (12)$$

The Equation of the Path. We can find the equation of the path (the trajectory) of the projectile by eliminating t between Eqs. 9 and 10. Solving Eq. 9 for t and substituting into Eq. 10, we obtain, after a little rearrangement,

$$y = (\tan \theta_0)x - \left(\frac{g}{2(v_0 \cos \theta_0)^2} \right) x^2 \quad \text{(trajectory)}. \quad (13)$$

This is the equation of the path shown in Fig. 9. For simplicity, we have here put $x_0 = 0$ and $y_0 = 0$ in Eqs. 9 and 10, respectively. Because g, θ_0, and v_0 are constants, Eq. 13 is of the form $y = ax + bx^2$, in which a and b are constants. This is the equation of a parabola.

Figure 12 is a strobe photo of a modern reenactment of an experiment done by Galileo in 1608, in which he first proved that a projectile follows a parabolic path.* He did so by rolling a ball down a groove in an inclined plane, starting from various heights and measuring the corresponding positions at which the ball struck the floor.

The Horizontal Range. The horizontal range R of the projectile, as Fig. 9 shows, is the distance of horizontal travel when the projectile has returned to the level at

---

* See "Galileo's Discovery of the Parabolic Trajectory," by Stillman Drake and James MacLachlan, *Scientific American*, March 1975.

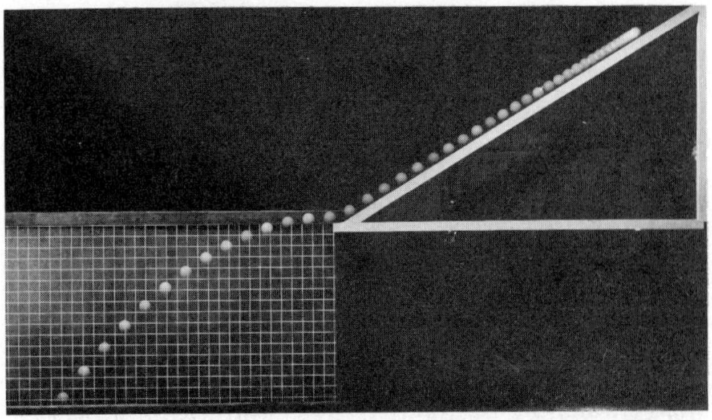

Figure 12 A strobe photo of a ball rolling down an inclined plane and following a parabolic path after it is projected horizontally from the bottom of the plane. This is a modern version of an experiment first done by Galileo in 1608.

which it was launched. To find it, let us put $x - x_0 = R$ in Eq. 9 and $y - y_0 = 0$ in Eq. 10, obtaining

$$x - x_0 = (v_0 \cos \theta)t = R$$

and

$$y - y_0 = (v_0 \sin \theta)t - \tfrac{1}{2}gt^2 = 0.$$

Eliminating t between these two equations yields

$$R = \frac{2v_0^2}{g} \sin \theta_0 \cos \theta_0.$$

Using the identity $\sin 2\theta_0 = 2 \sin \theta_0 \cos \theta_0$ (see Appendix G), we obtain

$$R = \frac{v_0^2}{g} \sin 2\theta_0. \qquad (14)$$

Note that R has maximum value when $\sin 2\theta_0 = 1$, which corresponds to $2\theta_0 = 90°$ or $\theta_0 = 45°$.

The Effects of the Atmosphere. We have assumed that the air through which the projectile moves has no effect on its motion, a reasonable assumption at low speeds. However, as the speed increases, the disagreement between our calculations and the actual motion of the projectile will become large. Figure 13, for example, shows two paths for a fly ball that leaves the bat at an angle of 60° with the horizontal with an initial speed of 100 mi/h.* Path I (the baseball player's fly ball) shows a calculated path that approximates normal conditions of

---

* See "The Trajectory of a Fly Ball," by Peter J. Brancazio, *The Physics Teacher*, January 1985.

Table 1 Two Fly Balls[a]

	Path II (Vacuum)	Path I (Air)
Range	581 ft	323 ft
Maximum height	252 ft	174 ft
Time of flight	7.9 s	6.6 s

[a] See Fig. 13. The launch angle is 60° and the launch speed is 100 mi/h.

play. Path II (the physics professor's fly ball) shows the path that the ball would follow in a vacuum. Table 1 gives some data for the two cases. Our assumption that air resistance can be neglected clearly does not apply to experiments carried out in such laboratories as Shea Stadium or Candlestick Park. We discuss the effect of drag on the motion in Chapter 6 and, in Essay 6 on "The Aerodynamics of Projectiles," Peter Brancazio returns to this topic in more detail.

It may come as a surprise but *no* problem in physics can be solved *exactly,* no matter how many significant figures the calculated answer may contain. Calculating the orbit of the moon, for example, involves hours and hours of computer time and—even though the results are impressive and easily good enough for all practical purposes—they can never be said to be exact. (Has the gravitational attraction of the asteroid Nemausa been taken fully into account?)

It is always necessary to make approximations in solving problems. The physicist P. A. M. Dirac has said that the trick is to divide the problem into two parts, one of which is simple and the other of which is small. You then solve the simple part exactly and do the best you can with the small part. Sometimes, the "small" part is small enough that you can neglect it entirely, as we do for air resistance at low speeds.

Sample Problem 4 A rescue plane is flying at a constant elevation of 1200 m with a speed of 430 km/h toward a point directly over a person struggling in the water (see Fig. 14). At what angle of sight ϕ should the pilot release a rescue capsule if it is to strike (very close to) the person in the water?

The initial velocity of the capsule is the same as that of the plane. That is, the initial velocity $\mathbf{v}_0$ is horizontal and has a magnitude of 430 km/h. We can find the time of flight of the capsule from Eq. 10, or

$$y - y_0 = (v_0 \sin \theta_0)t - \tfrac{1}{2}gt^2.$$

Putting $y - y_0 = -1200$ m (the minus sign because the person is below the origin) and $\theta_0 = 0$, we obtain

$$-1200 \text{ m} = 0 - \tfrac{1}{2}(9.8 \text{ m/s}^2)t^2.$$

Solving for t yields*

$$t = \sqrt{\frac{(2)(1200 \text{ m})}{9.8 \text{ m/s}^2}} = 15.65 \text{ s}.$$

The horizontal distance covered by the capsule (and by the plane) in that time is given by Eq. 9, or

$$
\begin{aligned}
x - x_0 &= (v_0 \cos \theta_0)t \\
&= (430 \text{ km/h})(\cos 0°)(15.65 \text{ s})(1\text{h}/3600 \text{ s}) \\
&= 1.869 \text{ km} = 1869 \text{ m}.
\end{aligned}
$$

The angle of sight is then (see Fig. 14)

$$\phi = \tan^{-1}\frac{x}{h} = \tan^{-1}\frac{1869 \text{ m}}{1200 \text{ m}} = 57°. \quad \text{(Answer)}$$

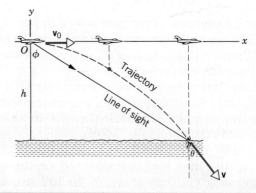

Figure 14 Sample Problem 4. A plane drops a rescue capsule and then continues in level flight. While it is falling, the horizontal velocity component of the capsule remains equal to the velocity of the plane.

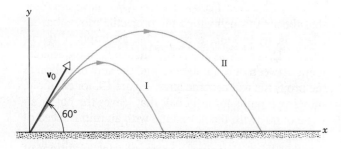

Figure 13 (I) The path of a fly ball, calculated (using a computer) by taking air resistance into account. (II) The path it would follow in a vacuum, calculated by the methods of this chapter. See Table 1.

* We explain in Hint 7 why we keep (temporarily) more significant figures for some quantities than the input data justify.

Because they have the same horizontal velocity, the plane remains vertically over the capsule while the capsule is in flight.

Sample Problem 5   A soccer player (see Fig. 15) kicks a ball at an angle of 38° from the horizontal, with an initial speed of 15 m/s. (a) How long will the ball be in the air?

When the ball hits the ground, it will have returned to its starting level so that $y - y_0 = 0$. Making this substitution in Eq. 10 and solving for t, we obtain

$$y - y_0 = (v_0 \sin \theta)t - \tfrac{1}{2}gt^2 = 0$$

or

$$t = \frac{2 v_0 \sin \theta}{g} = \frac{(2)(15 \text{ m/s})(\sin 38°)}{(9.8 \text{ m/s}^2)}$$
$$= 1.88 \text{ s} \approx 1.9 \text{ s}. \qquad \text{(Answer)}$$

(b) How far down the field did the ball land? From Eq. 9, we have

$$x - x_0 = (v_0 \cos \theta)t$$
$$= (15 \text{ m/s})(\cos 38°)(1.88 \text{ s})$$
$$= 22 \text{ m}. \qquad \text{(Answer)}$$

We could have found this answer from Eq. 14 but we preferred to find it directly from the basic equations describing the motion.

(c) How high did the ball go?

At its maximum height, $v_y = 0$. From Eq. 12,

$$v_y^2 = (v_0 \sin \theta)^2 - 2g(y - y_0) = 0.$$

Figure 15   Sample Problem 5. A soccer ball is kicked down the field.

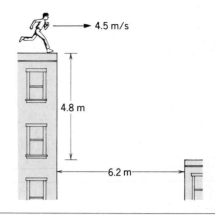

Figure 16   Sample Problem 6. Should the burglar jump? (We think not.)

Solving for $y - y_0$ gives

$$y - y_0 = \frac{v_0^2 \sin^2 \theta}{2g} = \frac{(15 \text{ m/s})^2(\sin 38°)^2}{(2)(9.8 \text{ m/s}^2)}$$
$$= 4.4 \text{ m}. \qquad \text{(Answer)}$$

Sample Problem 6   A police officer is chasing a burglar across a roof top, both are running at 4.5 m/s. Before the burglar reaches the edge of the roof, he has to decide whether or not to try jumping to the roof of the next building, which is 6.2 m away but 4.8 m lower; see Fig. 16. Can he make it? Assume that he jumps horizontally.

He has to fall 4.8 m, which will take him (putting $\theta_0 = 0$ and $y - y_0 = -4.8$ m in Eq. 10 and solving for t)

$$t = \sqrt{-\frac{2(y - y_0)}{g}} = \sqrt{-\frac{(2)(-4.8 \text{ m})}{9.8 \text{ m/s}^2}}$$
$$= 0.990 \text{ s}.$$

We now ask: "How far would the burglar move horizontally in this time?" The answer (see Eq. 9) is

$$x - x_0 = (v_0 \cos \theta_0)t$$
$$= (4.5 \text{ m/s})(\cos 0°)(0.990 \text{ s}) = 4.5 \text{ m}.$$

To reach the next building, the burglar has to move 6.2 m horizontally. Our advice: "Don't jump."

Sample Problem 7   Figure 17 shows a pirate ship, moored 560 m from a fort defending the harbor entrance of an island. The harbor defense cannon, sited at sea level, has a muzzle velocity of 82 m/s.

(a) To what angle must the cannon be elevated to hit the pirate ship?

Solving Eq. 14 for $2\theta_0$ yields

$$2\theta_0 = \sin^{-1} \frac{gR}{v_0^2} = \sin^{-1} \frac{(9.8 \text{ m/s}^2)(560 \text{ m})}{(82 \text{ m/s})^2} = \sin^{-1} 0.816.$$

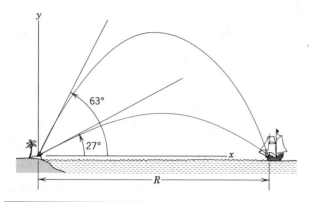

Figure 17 Sample Problem 7. At this range, the harbor defense cannon can hit the pirate ship at two different elevation angles.

There are two angles whose sine is 0.816, namely, 54.7° and 125.3°. Thus, we find

$$\theta_0 = \tfrac{1}{2}(54.7°) \approx 27° \qquad \text{(Answer)}$$

and

$$\theta_0 = \tfrac{1}{2}(125.3°) \approx 63°. \qquad \text{(Answer)}$$

The commandant of the fort can elevate the guns to either of these two angles and (if only there were no intervening air!) hit the pirate ship.

(b) What are the times of flight for the two elevation angles calculated above?

Solving Eq. 9 for t gives, for $\theta_0 = 27°$,

$$t = \frac{x - x_0}{v_0 \cos \theta_0} = \frac{560 \text{ m}}{(82 \text{ m/s})(\cos 27°)} = 7.7 \text{ s.} \quad \text{(Answer)}$$

Repeating the calculation for $\theta_0 = 63°$ yields $t = 15$ s. It is reasonable that the time of flight for the higher elevation angle should be larger.

(c) How far should the pirate ship be from the fort if it is to be beyond range of the cannon balls?

We have seen that maximum range corresponds to an elevation angle θ_0 of 45°. Thus, from Eq. 14,

$$R = \frac{v_0^2}{g} \sin 2\theta_0 = \frac{(82 \text{ m/s})^2}{(9.8 \text{ m/s}^2)} \sin (2 \times 45°)$$

$$= 690 \text{ m.} \qquad \text{(Answer)}$$

As the pirate ship sails away, the two elevation angles at which the ship can be hit draw closer together, eventually merging at $\theta_0 = 45°$ when the ship is 690 m away. Beyond that point the ship is safe.

Hint 7: *Significant figures—II.* In some problems, you calculate a numerical value in one step of the problem and

then use that number in a later step. In such cases, it is proper, strictly for the purposes of calculation, to keep more significant figures than you would accept in the final result.

In Sample Problem 4, for example, we calculated the time of flight of the capsule and found $t = 15.65$ s. If asked what the time of flight was, you would have to round this and say 16 s. However, it is permissible to use $t = 15.65$ in an extended calculation, *provided* that you round the final result. If you round at every step, you will reduce the precision of your answer.

Hint 8: *Numbers versus algebra—I.* One way to avoid numerical rounding errors is to solve the problem algebraically to the very end, substituting numbers only in the final step. That is easy to do in the Sample Problems of this set and is the way experienced problem solvers operate. In these early chapters, however, we prefer to solve most problems in bits and pieces, to give the beginner a firmer numerical grasp of what is going on. As the text progresses, however, we shall stick more and more to the algebra and shall point out the advantages of doing so.

4-7 Uniform Circular Motion

In projectile motion the acceleration, which is that due to gravity, is constant in both magnitude and direction. Here we look at motion in which the acceleration is constant in magnitude but not in direction. We are speaking of *uniform circular motion,* in which a particle moves at uniform speed v in a circle (or a circular arc) of radius r.

A point on a rotating fan blade or on a rotating compact disk (CD) undergoes uniform circular motion. Actually, we all experience uniform circular motion because of the rotation of the earth. The motions of an earth satellite in circular orbit around the earth and of the earth itself in (near circular) orbit around the sun are other examples of this kind of motion.

Figure 18 shows a particle in uniform circular motion with speed v in a circle of radius r. Its velocity vectors at two symmetrical points, marked p and q, are drawn. These vectors have the same magnitude v but—because they point in different directions—they are different vectors. The components of each vector can be seen to be

$$v_{px} = +v \cos \theta, \qquad v_{py} = +v \sin \theta$$

and

$$v_{qx} = +v \cos \theta, \qquad v_{qy} = -v \sin \theta.$$

The time required for the particle to move from p to

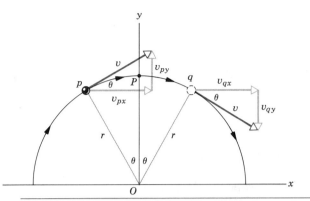

Figure 18   A particle moves in uniform circular motion at constant speed v in a circle of radius r. Its positions at symmetrical points p and q are shown, along with its velocity components at those points. We show that the instantaneous acceleration of the particle at any point is directed toward the center of the circle and has a magnitude v^2/r.

q at constant speed v is

$$\Delta t = \frac{pq}{v} = \frac{r(2\theta)}{v}, \tag{15}$$

in which pq is the length of the arc from p to q.

We now have enough information to calculate the components of the average acceleration experienced by the particle as it moves from p to q in Fig. 18. For the x component, we have

$$\bar{a}_x = \frac{v_{qx} - v_{px}}{\Delta t} = \frac{v\cos\theta - v\cos\theta}{\Delta t} = 0.$$

This conclusion is not surprising because it is clear from symmetry in Fig. 18 that the x component of velocity has the same value at q that it does at p.

For the y component of the average acceleration, we find, making use of Eq. 15,

$$\bar{a}_y = \frac{v_{qy} - v_{py}}{\Delta t} = \frac{-v\sin\theta - v\sin\theta}{\Delta t}$$

$$= -\frac{2v\sin\theta}{2r\theta/v} = -\left(\frac{v^2}{r}\right)\left(\frac{\sin\theta}{\theta}\right).$$

The minus sign tells us that this acceleration component points vertically downward in Fig. 18.

Now let us allow the angle θ in Fig. 18 to become smaller and smaller approaching zero as a limit. This requires that points p and q in that figure approach their midpoint, shown as point P at the top of the circle. The average acceleration, whose components we have just found, then approaches the instantaneous acceleration **a** at point P.

The *direction* of this instantaneous acceleration

vector at point P in Fig. 18 is downward, toward the center of the circle at O, because the direction of the average acceleration does not change as θ becomes smaller. To find the *magnitude* of the instantaneous acceleration vector, we must realize that, as θ becomes smaller and smaller, the ratio $(\sin\theta)/\theta$ approaches unity. In the relation given above for a_y, we then have

$$\boxed{a = \frac{v^2}{r}} \qquad \text{(centripetal acceleration).} \tag{16}$$

Our conclusion: When you see a particle moving at constant speed v in a circle (or a circular arc) of radius r, you may be sure that it has an acceleration, directed toward the center of the circle, of magnitude v^2/r.

Figure 19 shows the relation between the velocity and the acceleration vectors at various stages during uniform circular motion. Both vectors have constant magnitude as the motion progresses but their directions change continuously. The velocity is always tangent to the circle in the direction of motion; the acceleration is always directed radially inward. Because of this, the acceleration associated with uniform circular motion is called a *centripetal* (meaning center seeking) acceleration, a term coined by Isaac Newton.

The acceleration resulting from a change in the direction of a velocity is just as real as one resulting from a change in the magnitude of a velocity. In Fig. 8 of Chapter 2, for example, we showed Colonel John P. Stapp braking his rocket sled to rest. He is experiencing a velocity that is constant in direction but changing rapidly in magnitude. On the other hand, an astronaut whirling in the human centrifuge at the NASA Manned Spacecraft Center in Houston experiences a velocity that is constant in magnitude but changing rapidly in direction.

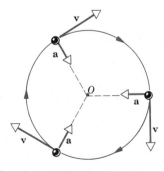

Figure 19   The velocity and the acceleration vectors for a particle in uniform circular motion. They both have constant magnitude but vary continuously in direction.

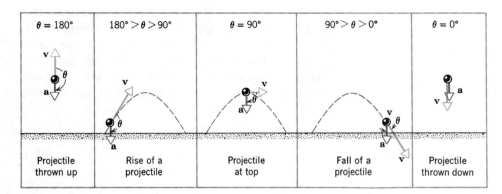

| $\theta = 180°$ | $180° > \theta > 90°$ | $\theta = 90°$ | $90° > \theta > 0°$ | $\theta = 0°$ |
| Projectile thrown up | Rise of a projectile | Projectile at top | Fall of a projectile | Projectile thrown down |

Figure 20 The relation between the velocity and the acceleration of a projectile for various motions. Note that the acceleration and the velocity vectors do not have any fixed directional relation to each other.

The accelerations that each of these people feel are indistinguishable.

There is no fixed relation between the direction of the velocity vector and the acceleration vector for a moving particle. Figure 20 shows examples in which the angle between these two vectors varies from 0° to 180°. In only one case do the two vectors happen to point in the same direction.

Sample Problem 8 A satellite is in circular earth orbit, at an altitude h (= 200 km) above the earth's surface. At that altitude, the free-fall acceleration g is 9.20 m/s². What is the orbital speed of the satellite?

We can recast Eq. 16 ($a = v^2/r$), with $a = g$ and with $r = R + h$, as

$$g = \frac{v^2}{R + h}.$$

Solving for v gives

$$v = \sqrt{g(R + h)}$$
$$= \sqrt{(9.20 \text{ m/s}^2)(6.37 \times 10^6 \text{ m} + 200 \times 10^3 \text{ m})}$$
$$= 7770 \text{ m/s} = 7.77 \text{ km/s}. \qquad \text{(Answer)}$$

You can show that this is equivalent to 17,400 mi/h and that the satellite would take 1.47 h to complete one orbital revolution.

It may puzzle you that, although $g = 9.20$ m/s² at the position of the satellite orbit, an astronaut experiences the familiar weightlessness, as if gravity had somehow been turned off. The explanation is that both the astronaut *and the spacecraft* are accelerating toward the center of the earth at 9.20 m/s². They are both in free fall, just like a passenger in a freely falling elevator cab. In both cases, the person (astronaut or elevator passenger) experiences no relative motion with respect to his or her "vehicle" (spacecraft or elevator cab). See Essay 5, "Physics in Weightlessness," by Sally Ride.

4-8 Relative Motion in One Dimension

You may see a duck flying north at, say, 20 mi/h. To another duck flying alongside, however, your duck is at rest. The velocity of a particle depends on the *reference frame* of the person who is doing the measuring. For our purposes, a reference frame is the physical object to which you attach your coordinate system.

The reference frame that seems most natural to us in our daily comings and goings is the solid earth beneath our feet.* When a traffic officer tells you that you have been driving at 70 mi/h the unspoken qualification, "in a coordinate system attached to the earth," is taken for granted by each of you.

If you are traveling in an automobile, an elevator, an airplane, a free balloon, or a spaceship, the reference frame of the earth may not be the most convenient one. You are free to make any choice that you wish. Once having made it, however, you must always be aware of your choice and you must be careful to make all your measurements with respect to it.

Suppose that Alex (frame A) is parked by the side of a highway, watching car P (the "particle") speed past. Barbara (frame B), driving along the highway at constant speed, is also watching car P. Suppose that, as in Fig. 21, they both measure the position of the car at a given

---

* Even Shakespeare seemed to think so. He had Hamlet say, "this goodly frame, the earth. . . ."

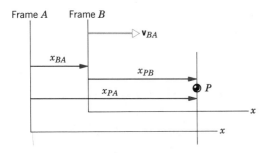

Figure 21 Alex (frame A) and Barbara (frame B) watch car P. All motion is along the common x axis of the two frames. The vector $\mathbf{v}_{BA}$ shows the relative separation velocity of the two frames. The three position measurements shown are all made at the same instant of time.

moment. From the figure we see that

$$x_{PA} = x_{PB} + x_{BA}. \qquad (17)$$

The terms in Eq. 17 are scalars and may be of either sign. The equation is read: "The position of P as measured by A *is equal to* the position of P as measured by B *plus* the position of B as measured by A." Note how this reading is supported by the sequence of the subscripts.

Taking the time derivative of Eq. 17, we obtain

$$\frac{d}{dt}(x_{PA}) = \frac{d}{dt}(x_{PB}) + \frac{d}{dt}(x_{BA}),$$

or (because $v = dx/dt$)

$$v_{PA} = v_{PB} + v_{BA}. \qquad (18)$$

This scalar equation is the relation between the velocities of the same object (car P) as measured in the two frames.* In words, it says: "The velocity of P as measured by A *is equal to* the velocity of P as measured by B *plus* the velocity of B as measured by A." The term v_{BA} is the separation velocity of frame B with respect to frame A. see Fig. 21.

We consider only those frames that move at constant velocity with respect to each other.† In our exam-

---

* It may help to note that, on the right-hand side of this equation, the two inner subscripts (B, B) are the same; the two outer ones (P, A) are the same as those on the left, and appear in the same sequence.

† We call them *inertial reference frames* for reasons that will appear later.

ple, this means that Barbara (frame B) will drive always at constant speed with respect to Alex (frame A). Car P (the moving particle), however, may speed up, slow down, come to rest, or reverse direction.

The time derivative of the velocity equation (Eq. 18) yields the acceleration equation, or

$$a_{PA} = a_{PB}. \qquad (19)$$

Note that, because v_{BA} is a constant, its time derivative is zero. Equation 19 tells us that, for the class of reference frames that we have decided to deal with (that is, those whose velocities of separation are constant), *both observers will measure the same acceleration for the moving particle*.

Sample Problem 9 Alex, parked by the side of an east–west highway, is watching car P, which is moving fast in a westerly direction. Barbara, driving east at a speed v_{BA} ($= 50$ mi/h), watches the same car. Take the easterly direction as positive. (a) If Alex measures a velocity v_{PA} ($= -80$ mi/h) for car P, what velocity will Barbara measure?

From Eq. 18,

$$v_{PB} = v_{PA} - v_{BA}.$$

We have $v_{PA} = -80$ mi/h, the minus sign telling us that car P is moving west, that is, in the negative direction. We also have $v_{BA} = 50$ mi/h so that

$$v_{PB} = (-80 \text{ mi/h}) - (50 \text{ mi/h}) = -130 \text{ mi/h}. \quad \text{(Answer)}$$

If car P were connected to Barbara's car by a string wound up on a spool, the string would be unwinding at this speed as the two cars separated.

(b) If Alex sees car P brake to a halt in 10 s, what acceleration (assumed constant) will he measure for it?

From Eq. 10 of Chapter 2 ($v = v_0 + at$) we have

$$a = \frac{v - v_0}{t} = \frac{(0) - (-80 \text{ mi/h})}{10 \text{ s}}$$

$$= \left(\frac{80 \text{ mi/h}}{10 \text{ s}}\right)\left(\frac{1 \text{ m/s}}{2.24 \text{ mi/h}}\right) = 3.6 \text{ m/s}^2. \quad \text{(Answer)}$$

(c) What acceleration would Barbara measure for the braking car?

Barbara sees the initial speed of the car as -130 mi/h, as we calculated in (b) above. Although the car has braked to rest, it is only at rest in Alex's reference frame. To Barbara, car P is not at rest at all but is receding from her at 50 mi/h so that its final velocity in her reference frame is -50 mi/h. Thus, from the relation $v = v_0 + at$,

$$a = \frac{v - v_0}{t} = \frac{(-50 \text{ mi/h}) - (-130 \text{ mi/h})}{10 \text{ s}}$$

$$= 3.6 \text{ m/s}^2. \qquad \text{(Answer)}$$

This is exactly the same acceleration that Alex measured, thus reassuring us that we made no mistakes.

4–9 Relative Motion at High Speeds (Optional)

An orbiting satellite has a speed of 17,000 mi/h. Before you call that a high speed you must answer this question: "High compared to what?" Nature has given us a standard; it is the speed of light c, where

$$\boxed{c = 299{,}792{,}458 \text{ m/s}} \qquad \text{(speed of light). (20)}$$

As we shall learn later, no entity — be it particle or wave — can move faster than the speed of light, no matter what reference frame is used for observation. By this yardstick, all tangible large-scale objects — no matter how high their speeds seem by ordinary standards — are very slow indeed. The speed of the orbiting satellite, for example, is only 0.0025% of the speed of light. Subatomic particles such as electrons or protons, however, can acquire speeds very close to (but never equal to or greater than) the speed of light. Experiment shows, for example, that an electron accelerated through 10 million volts acquires a speed of 0.9988c; if you accelerate it through 20 million volts, its speed increases, but only to 0.9997c. The speed of light is a barrier that objects can approach but never reach.

We now ask: "How can we be sure that Newtonian mechanics, which was developed by studying very slow objects, also holds for very fast objects, such as energetic electrons or protons?" The answer, which we can only find from experiment, is that Newtonian mechanics does *not* hold at speeds that approach the speed of light. Einstein's theory of special relativity, however, agrees with experiment at *all* speeds. We give here a foretaste of that theory, which we introduce formally in Chapter 42.

In the limit of "low" speeds — and by that we mean all speeds that can be acquired by gross tangible objects — the equations of Einstein's theory reduce to those of Newtonian mechanics. The failure of Newtonian mechanics is gradual, its predictions agreeing less and less well with experiment as the speed increases. Let us give an example.

Equation 18,

$$v_{PA} = v_{PB} + v_{BA} \quad \text{(low speeds; Newton),} \qquad (21)$$

gives the relation of the speed of particle P as seen by observer B to that seen by observer A. We state without proof that Einstein's theory gives for this relation

$$v_{PA} = \frac{v_{PB} + v_{BA}}{1 + v_{PB}v_{BA}/c^2} \quad \text{(all speeds; Einstein).} \quad (22)$$

We see that if $v_{PB} \ll c$ and also $v_{BA} \ll c$ (which will always be the case for ordinary objects), then the denominator in Eq. 22 is very close to unity and Eq. 22 reduces to Eq. 21, as we know it must.

The speed of light c is the central constant of Einstein's theory and every relativistic equation contains it. A device for testing the equations of relativity is to allow c to become infinitely large. Under those conditions, *all* speeds would be "low" and Newtonian mechanics would never fail. Putting $c \to \infty$ in Eq. 22 does indeed reduce that equation to Eq. 21.

Sample Problem 10 (Low speeds) For $v_{PB} = v_{BA} = 0.0001c$ ($= 67{,}000$ mi/h!), what do Eqs. 21 and 22 predict for v_{PA}?

From Eq. 21,

$$v_{PA} = v_{PB} + v_{BA}$$
$$= 0.0001c + 0.0001c = 0.0002c. \quad \text{(Answer)}$$

From Eq. 22,
$$v_{PA} = \frac{v_{PB} + v_{BA}}{1 + v_{PB}v_{BA}/c^2}$$

$$= \frac{0.0001c + 0.0001c}{1 + (0.0001c)^2/c^2} = \frac{0.0002c}{1.00000001}$$

$$\approx 0.0002c. \qquad \text{(Answer)}$$

Conclusion: At any speed acquired by ordinary tangible objects, Eqs. 21 and 22 yield essentially the same answer. We can use Eq. 21 (Newtonian mechanics) without a second thought.

Sample Problem 11 (High speeds) For $v_{PB} = v_{BA} = 0.65c$, what do Eqs. 21 and 22 predict for v_{PA}?

From Eq. 21,
$$v_{PA} = v_{PB} + v_{BA}$$
$$= 0.65c + 0.65c = 1.30c. \qquad \text{(Answer)}$$

From Eq. 22,

$$v_{PA} = \frac{v_{PB} + v_{BA}}{1 + v_{PB}v_{BA}/c^2}$$

$$= \frac{0.65c + 0.65c}{1 + (0.65c)(0.65c)/c^2} = \frac{1.30c}{1.42} = 0.91c. \quad \text{(Answer)}$$

Conclusion: At high speeds, Newtonian mechanics and relativity theory predict very different results. Newtonian mechanics involves no upper limit on speed and can easily (as in this case) predict a speed greater than the speed of light. Relativity theory, on the other hand, *never* predicts a speed that exceeds *c*, no matter how high the combining speeds. Experiment agrees with relativity theory on all counts.

4-10   Relative Motion in Two Dimensions

Now we move from the scalar world of motion in one dimension to the vector world of motion in two (and, by extension, in three) dimensions. Figure 22 shows reference frames *A* and *B*, our two observers again watching the moving particle *P*. We shall once more assume that the two frames are separating at a constant velocity $\mathbf{v}_{BA}$ and, for simplicity, we shall further assume that their *x* and *y* axes remain parallel to each other as they do so.

Let observers *A* and *B* each measure the position of particle *P*. From the vector triangle in Fig. 22, we have the vector equation

$$\mathbf{r}_{PA} = \mathbf{r}_{PB} + \mathbf{r}_{BA}. \qquad (23)$$

This relation is the vector equivalent of Eq. 17, which is a scalar equation holding for one-dimensional motion.

If we take the time derivative of Eq. 23, we find a connection between the (vector) velocities of the particle

as measured by our two observers, namely,

$$\mathbf{v}_{PA} = \mathbf{v}_{PB} + \mathbf{v}_{BA}. \qquad (24)$$

This is the vector equivalent of the scalar relation Eq. 18. Note that the order of the subscripts is the same as in that equation and that again $\mathbf{v}_{BA}$ is the (constant) relative separation velocity of *B* as observed by *A*.

If we take the time derivative of Eq. 24, we obtain a connection between the two measured accelerations, namely,

$$\mathbf{a}_{PA} = \mathbf{a}_{PB}. \qquad (25)$$

It remains true for motion in three dimensions that, if we consider only those reference frames that move with constant velocity with respect to each other, all observers will measure the same (vector) acceleration for a moving particle.

Sample Problem 12   The compass in a plane indicates that it is headed due east; its air speed indicator reads 215 km/h. A steady wind of 65 km/h is blowing due north. (a) What is the velocity of the plane with respect to the ground?

The moving "particle" in this problem is the plane *P*. There are two reference frames, the ground (*G*) and the air mass (*A*). By a simple change of notation, we can rewrite Eq. 24 as

$$\mathbf{v}_{PG} = \mathbf{v}_{PA} + \mathbf{v}_{AG}. \qquad (26)$$

Figure 23*a* shows these vectors, which form a right triangle. The terms in Eq. 26 are, in sequence, the velocity of the plane with respect to the ground, the velocity of the plane with respect to the air, and the velocity of the air with respect to the ground (that is, the wind velocity). Note the orientation of the

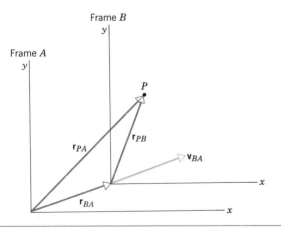

Figure 22   Reference frames in two dimensions. The vectors $\mathbf{r}_{PA}$ and $\mathbf{r}_{PB}$ show the positions of particle *P* in frames *A* and *B*, respectively. Vector $\mathbf{r}_{BA}$ shows the position of frame *B* with respect to frame *A*. Vector $\mathbf{v}_{BA}$ shows the (constant) separation velocity of the two frames.

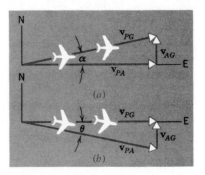

Figure 23   Sample Problem 12. (*a*) A plane, heading due east, is blown to the north. (*b*) To travel due east, the plane must head into the wind.

plane, which is consistent with a due east reading on its compass.

The magnitude of the ground velocity (the ground speed) is found from

$$v_{PG} = \sqrt{v_{PA}^2 + v_{AG}^2}$$

$$= \sqrt{(215 \text{ km/h})^2 + (65 \text{ km/h})^2} = 225 \text{ km/h.} \quad \text{(Answer)}$$

The angle α in Fig. 23a follows from

$$\alpha = \tan^{-1} \frac{v_{AG}}{v_{PA}} = \tan^{-1} \frac{65 \text{ km/h}}{225 \text{ km/h}} = 16.1°. \quad \text{(Answer)}$$

Thus, with respect to the ground, the plane is flying at 225 km/h in a direction 16.1° north of east. Note that its ground speed is greater than its airspeed.

(b) If the pilot wishes to fly due east, what must be the heading? That is, what must the compass read?

In this case the pilot must head into the wind so that the velocity of the plane with respect to the ground points east. The wind remains unchanged and the vector diagram representing Eq. 26 is as shown in Fig. 23b. Note that the three vectors still form a right triangle, as they did in Fig. 23a.

The pilot's ground speed is now

$$v_{PG} = \sqrt{v_{PA}^2 - v_{AG}^2}$$

$$= \sqrt{(215 \text{ km/h})^2 - (65 \text{ km/h})^2} = 205 \text{ km/h.}$$

As the orientation of the plane in Fig. 23b indicates, the pilot must head into the wind by an angle θ given by

$$\theta = \sin^{-1} \frac{v_{AG}}{v_{PA}} = \sin^{-1} \frac{65 \text{ km/h}}{215 \text{ km/h}} = 17.6°. \quad \text{(Answer)}$$

Note that, by heading into the wind as the pilot has done, the ground speed is now less than the airspeed.

REVIEW AND SUMMARY

Position

To describe the motion of a particle in three-dimensional space, we first give its *position* **r**, which locates it with respect to the origin of a convenient coordinate system. Its velocity at any time is then the rate of change of its position with respect to time;

Velocity v

$$\mathbf{v} = \frac{d\mathbf{r}}{dt}. \qquad [2]$$

v is always tangent to the path (the *trajectory*) of the particle.

With $\mathbf{r} = \mathbf{i}x + \mathbf{j}y + \mathbf{k}z$, the components of **v** are

$$v_x = \frac{dx}{dt}, \quad v_y = \frac{dy}{dt}, \quad \text{and} \quad v_z = \frac{dz}{dt}. \qquad [4]$$

See Sample Problems 1 and 2 for examples.

Acceleration a

The acceleration of the particle is defined as the rate at which its velocity is changing;

$$\mathbf{a} = \frac{d\mathbf{v}}{dt}. \qquad [5]$$

Its components are

$$a_x = \frac{dv_x}{dt}, \quad a_y = \frac{dv_y}{dt}, \quad \text{and} \quad a_z = \frac{dv_z}{dt}; \qquad [7]$$

see Sample Problem 3.

Projectile Motion

We analyze projectile motion (or any other motion for which acceleration **a** is constant) by resolving **r**, **v**, and **a** into x and y components, treating each component as an independent one-dimensional motion. Figure 9 illustrates the analysis. In the absence of aerodynamic forces (air friction and lift), the acceleration is constant with magnitude g and is directed in the $-y$ direction. For a launch angle θ_0 and initial position at the origin of coordinates, the motion is described by the equations.

$$x - x_0 = (v_0 \cos \theta_0)t, \qquad [9]$$

$$y - y_0 = (v_0 \sin \theta_0)t - \tfrac{1}{2}gt^2, \qquad [10]$$

$$v_y = v_0 \sin \theta_0 - gt, \qquad [11]$$

and

$$v_y^2 = (v_0 \sin \theta_0)^2 - 2g(y - y_0).$$

The trajectory is a parabola whose equation is

The Trajectory

$$y = (\tan \theta_0)x - \left(\frac{g}{2(v_0 \cos \theta_0)^2}\right) x^2, \qquad [13]$$

and the projectile's range (the horizontal distance between launch and its return to the launching height) is

The Horizontal Range

$$R = \frac{v_0^2}{g} \sin 2\theta_0. \qquad [14]$$

Sample Problems 4 through 7 illustrate the use of these trajectory equations.

Uniform Circular Motion

In *uniform circular motion*, a particle moves with constant speed v in a circle of radius r. This is accelerated motion because the direction of the velocity vector $\mathbf{v}$ is continuously changing. The acceleration $\mathbf{a}$ is always directed toward the center of the circle and has a constant magnitude of

$$a = \frac{v^2}{r}. \qquad [16]$$

Sample Problem 8 shows how this equation might be used.

Motion Viewed from Different Reference Frames

The mathematical description of the motion of a particle depends on the reference frame of the observer. If an object P is observed by two observers, A and B, the velocities measured by them are related by

$$\mathbf{v}_{PA} = \mathbf{v}_{PB} + \mathbf{v}_{BA}, \qquad [18,24]$$

in which $\mathbf{v}_{BA}$ is the velocity of B with respect to A. If neither observer is experiencing accelerated motion, they both measure the same acceleration for P, that is,

$$\mathbf{a}_{PA} = \mathbf{a}_{PB}. \qquad [19,25]$$

Sample Problem 9 illustrates the use of these equations for one-dimensional motion; Sample Problem 12 treats a two-dimensional case.

If speeds near the speed of light are involved, Eq. 18 must be replaced by an equation derived using the *special theory of relativity*. For one-dimensional motion, the correct result is

$$v_{PA} = \frac{v_{PB} + v_{BA}}{1 + v_{PB}v_{BA}/c^2}, \qquad [22]$$

which becomes identical to Eq. 18 if the speeds are all negligible compared to the speed of light c. Sample Problems 10 and 11 illustrate the use of this equation.

QUESTIONS

1. Can the acceleration of a body change its direction (*a*) without its displacement changing direction and (*b*) without its velocity changing direction?

2. If the acceleration of a body is constant in a given reference frame, is it necessarily constant in all other reference frames?

3. In broad jumping, sometimes called long jumping, does it matter how high you jump? What factors determine the span of the jump?

4. Why doesn't the electron in the beam from an electron gun fall as much because of gravity as a water molecule in the

stream from a hose? Assume horizontal motion initially in each case.

5. At what point in its path does a projectile have its minimum speed? Its maximum?

6. Figure 24 shows the path followed by a NASA Learjet in a run designed to simulate low-gravity conditions for a short period of time. Make an argument to show that, if the plane follows a particular parabolic path, the passengers will experience weightlessness.

7. A shot put is thrown from above ground level. The launch angle that will produce the longest range is less than

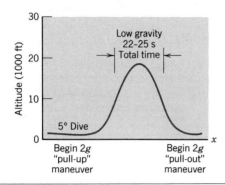

Figure 24   Question 6.

45°; that is, a flatter trajectory has a longer range. Explain why.

8.   You are driving directly behind a pickup truck, going at the same speed as the truck. A crate falls from the bed of the truck to the road. Will your car hit the crate before the crate hits the road if you neither brake nor swerve?

9.   Trajectories are shown in Fig. 25 for three kicked footballs. Pick the trajectory for which (*a*) the time of flight is least; (*b*) the vertical velocity component at launch is greatest; (*c*) the horizontal velocity component at launch is greatest; (*d*) the launch speed is least. Ignore air resistance.

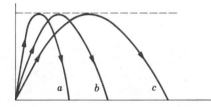

Figure 25   Question 9.

10.   A rifle is "sighted-in" against a target at the same elevation. Show that, at the same range, it will shoot too high when shooting either uphill or downhill. See "A Puzzle in Elementary Ballistics," by Ole Anton Haugland, *The Physics Teacher,* April 1983.

11.   In his book, *Sport Science,* Peter Brancazio, with such projectiles as baseballs and golf balls in mind, writes: "Everything else being equal, a projectile will travel farther on a hot day than on a cold day, farther at high altitude than at sea level, farther in humid than in dry air." How can you explain these claims?

12.   Long-range artillery pieces are not set at the "maximum range" angle of 45° but at larger elevation angles, in the range 55° to 65°. What's wrong with 45°?

13.   In projectile motion when air resistance is negligible, is it ever necessary to consider three-dimensional motion rather than two-dimensional?

14.   Is it possible to be accelerating if you are traveling at constant speed? Is it possible to round a curve with zero acceleration? With constant acceleration?

15.   Show that, taking the earth's rotation and revolution into account, a book resting on your table moves faster at night than it does during the daytime. In what reference frame is this statement true?

16.   An aviator, pulling out of a dive, follows the arc of a circle and is said to have "experienced 3*g*'s" in pulling out of the dive. Explain what this statement means.

17.   A boy sitting in a railroad car moving at constant velocity throws a ball straight up into the air. Will the ball fall behind him? In front of him? Into his hands? What happens if the car accelerates forward or goes around a curve while the ball is in the air?

18.   A woman on the observation platform of a train moving with constant velocity drops a coin while leaning over the rail. Describe the path of the coin as seen by (*a*) the woman on the train, (*b*) a person standing on the ground near the track, and (*c*) a person in a second train moving in the opposite direction to the first train on a parallel track.

19.   The Newtonian velocity transformation, Eq. 21, is so instinctively familiar from everyday experience that it is sometimes claimed to be "obviously correct, requiring no proof." Many so-called refutations of relativity theory turn out to be based on this claim. How would you refute someone who made this claim?

20.   Water is collecting in a bucket during a steady downpour. Will the rate at which the bucket is filling change if a steady horizontal wind starts to blow?

21.   Drops are falling vertically in a steady rain. In order to go through the rain from one place to another in such a way as to encounter the least number of raindrops, should you move with the greatest possible speed, the least possible speed, or some intermediate speed? See "An Optimal Speed for Travers-

Figure 26   Question 22.

ing a Constant Rain," by S. A. Stern, *American Journal of Physics*, September 1983.

22. What's wrong with Fig. 26? The boat is sailing with the wind.

23. An elevator is descending at a constant speed. A passenger drops a coin to the floor. What accelerations would (a) the passenger and (b) a person at rest with respect to the elevator shaft observe for the falling coin?

EXERCISES AND PROBLEMS

Section 4–3 What Is the Velocity of the Particle?

1E. A plane flies 300 mi east from city A to city B in 45 min and then 600 mi south from city B to city C in 1.5 h. (a) What are the magnitude and direction of the displacement vector that represents the total trip? What are (b) the average velocity vector and (c) the average speed for the trip?

2E. A train moving at a constant speed of 60 km/h moves east for 40 min, then in a direction 50° east of north for 20 min, and finally west for 50 min. What is the average velocity of the train during this run?

3E. In 3.5 h, a balloon drifts 21.5 km north, 9.70 km east, and 2.88 km in elevation from its release point on the ground. Find (a) the magnitude of its average velocity and (b) the angle its average velocity makes with the horizontal.

Section 4–4 What Is the Acceleration of the Particle?

4E. A particle moves so that its position as a function of time in SI units is

$$\mathbf{r}(t) = \mathbf{i} + 4t^2\mathbf{j} + t\mathbf{k}.$$

Write expressions for (a) its velocity and (b) its acceleration as functions of time.

5E. The position $\mathbf{r}$ of a particle moving in an xy plane is given by $\mathbf{r} = (2t^3 - 5t)\mathbf{i} + (6 - 7t^4)\mathbf{j}$. Here $\mathbf{r}$ is in meters and t is in seconds. Calculate (a) $\mathbf{r}$, (b) $\mathbf{v}$, and (c) $\mathbf{a}$ when $t = 2$ s.

6E. An ice boat sails across the surface of a frozen lake with constant acceleration produced by the wind. At a certain instant its velocity is $6.30\mathbf{i} - 8.42\mathbf{j}$ in meters per second. Three seconds later the boat is instantaneously at rest. What is its acceleration during this interval?

7P. A particle A moves along the line $y = d$ (30 m) with a constant velocity $\mathbf{v}$ ($v = 3.0$ m/s) directed parallel to the positive x axis (Fig. 27). A second particle B starts at the origin with zero speed and constant acceleration $\mathbf{a}$ ($a = 0.40$ m/s²) at the same instant that particle A passes the y axis. What angle θ between $\mathbf{a}$ and the positive y axis would result in a collision between these two particles?

8P. A particle leaves the origin with an initial velocity $\mathbf{v} = 3.0\mathbf{i}$, in meters per second. It experiences a constant acceleration $\mathbf{a} = -1.0\mathbf{i} - 0.5\mathbf{j}$, in meters per second squared. (a) What is the velocity of the particle when it reaches its maximum x coordinate? (b) Where is the particle at this time?

9P. The velocity $\mathbf{v}$ of a particle moving in the xy plane is

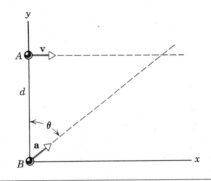

Figure 27 Problem 7.

given by $\mathbf{v} = (6t - 4t^2)\mathbf{i} + 8\mathbf{j}$. Here $\mathbf{v}$ is in meters per second and t (> 0) is in seconds. (a) What is the acceleration when $t = 3$ s? (b) When (if ever) is the acceleration zero? (c) When (if ever) is the velocity zero? (d) When (if ever) does the speed equal 10 m/s?

Section 4–6 Projectile Motion Analyzed

10E. In a cathode-ray tube, a beam of electrons is projected horizontally with a speed of 1.0×10^9 cm/s into the region between a pair of horizontal plates 2.0 cm long. An electric field between the plates exerts a constant downward acceleration on the electrons of magnitude 1.0×10^{17} cm/s². Find (a) the time required for the electrons to pass through the plates, (b) the vertical displacement of the beam in passing through the plates, and (c) the components of the velocity of the beam as it emerges from the plates.

11E. Electrons, like all forms of matter, fall under the influence of gravity. If an electron is projected horizontally with a speed of 3.0×10^7 m/s (one-tenth the speed of light), how far will it fall in traversing 1 m of horizontal distance?

12E. A dart is thrown horizontally toward the bull's eye, point P on the dart board, with an initial speed of 10 m/s. It hits at point Q on the rim, vertically below P, 0.19 s later; see Fig. 28. (a) What is the distance PQ? (b) How far away from the dart board did the player stand?

13E. A rifle is aimed horizontally at a target 100 ft away. The bullet hits the target 0.75 in. below the aiming point. (a) What is the bullet's time of flight? (b) What is the muzzle velocity of the rifle?

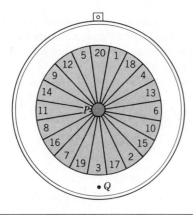

Figure 28 Exercise 12.

14E. A ball rolls off the edge of a horizontal table top 4.0 ft high. It strikes the floor at a point 5.0 ft horizontally away from the edge of the table. (*a*) For how long was the ball in the air? (*b*) What was its speed at the instant it left the table?

15E. A projectile is fired horizontally from a gun located 45 m above a horizontal plane with a muzzle speed of 250 m/s. (*a*) How long does the projectile remain in the air? (*b*) At what horizontal distance does it strike the ground? (*c*) What is the magnitude of the vertical component of its velocity as it strikes the ground?

16E. A baseball leaves the pitcher's hand horizontally at a speed of 100 mi/h. The distance to the batter is 60 ft. (*a*) How long does it take for the ball to travel the first 30 ft horizontally? The second 30 ft? (*b*) How far does the ball fall under gravity during the first 30 ft of its horizontal travel? (*c*) During the second 30 ft? (*d*) Why aren't these quantities equal? Ignore the effect of air resistance.

17E. A projectile is launched with an initial velocity of 30 m/s at an angle of 60° above the horizontal. Calculate the magnitude and direction of its velocity (*a*) 2 s and (*b*) 5 s after launch.

18E. A stone is thrown with an initial velocity of 20 m/s at an angle of 40° above the horizontal. Find its horizontal and vertical displacements (*a*) 1.1 s and (*b*) 5.0 s after launch.

19E. You throw a ball from a cliff with an initial velocity of 15 m/s at an angle of 20° below the horizontal. Find (*a*) its horizontal displacement and (*b*) its vertical displacement 2.3 s later.

20E. You throw a ball with a speed of 25 m/s at an angle of 40° above the horizontal directly toward a wall as shown in Fig. 29. The wall is 22 m from the release point of the ball. (*a*) How long is the ball in the air before it hits the wall? (*b*) How far above the release point does the ball hit the wall? (*c*) What are the horizontal and vertical components of its velocity as it hits the wall? (*d*) Has it passed the highest point on its trajectory when it hits?

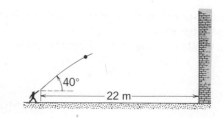

Figure 29 Exercise 20.

21E. Prove that for a projectile fired from the surface of level ground at an angle θ_0 above the horizontal, the ratio of the maximum height H to the range R is given by $H/R = \frac{1}{4} \tan \theta_0$. See Fig. 30.

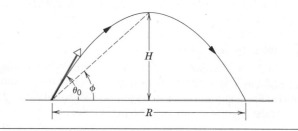

Figure 30 Exercises 21 and 22.

22E. A projectile is fired from the surface of level ground at an angle θ_0 above the horizontal. (*a*) Show that the elevation angle ϕ of the highest point as seen from the launch point is related to θ_0, the elevation angle of projection, by $\tan \phi = \frac{1}{2} \tan \theta_0$. See Fig. 30 and Exercise 21. (*b*) Calculate ϕ for $\theta_0 = 45°$.

23E. A stone is projected at an initial speed of 120 ft/s directed 60° above the horizontal, at a cliff of height h, as shown in Fig. 31. The stone strikes the ground at A 5.5 s after launching. Find (*a*) the height h of the cliff; (*b*) the speed of the stone just before impact at A; and (*c*) the maximum height H reached above the ground.

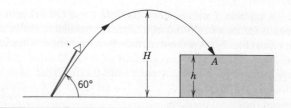

Figure 31 Exercise 23.

24P. In Sample Problem 4, find (*a*) the speed of the capsule when it hits the water and (*b*) the impact angle θ shown in Fig. 14.

25P. In the 1968 Olympics in Mexico City, Bob Beamon shattered the record for the long jump with a jump of 8.90 m.

Assume that his initial speed on takeoff was 9.5 m/s, about equal to that of a sprinter. How close did this world-class athlete come to the maximum possible range in the absence of air resistance? The value of g in Mexico City is 9.78 m/s².

26P. A rifle with a muzzle velocity of 1500 ft/s shoots a bullet at a target 150 ft away. How high above the target must the rifle barrel be pointed so that the bullet will hit the target?

27P. Show that the maximum height reached by a projectile is $y_{max} = (v_0 \sin \theta_0)^2/2g$.

28P. In a detective story, a body is found 15 ft from the base of a building and beneath an open window 80 ft above. Would you guess the death to be accidental or not? Why?

29P. In Galileo's *Two New Sciences,* the author states that "for elevations [angles of projection] which exceed or fall short of 45° by equal amounts, the ranges are equal. . . ." Prove this statement. See Fig. 32.

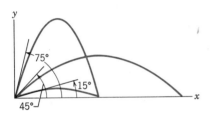

Figure 32 Problem 29.

30P. A ball is thrown from the ground into the air. At a height of 9.1 m, the velocity is observed to be $\mathbf{v} = 7.6\mathbf{i} + 6.1\mathbf{j}$ in meters per second (x axis horizontal, y axis vertical and up). (a) To what maximum height will the ball rise? (b) What will be the total horizontal distance traveled by the ball? (c) What is the velocity of the ball (magnitude and direction) the instant before it hits the ground?

31P. According to Eq. 14, the range of a projectile depends not only on v_0 and θ_0 but also on the value g of the gravitational acceleration, which varies from place to place. In 1936, Jesse Owens established a world's running broad jump record of 8.09 m at the Olympic Games at Berlin (g = 9.8128 m/s²). Assuming the same values of v_0 and θ_0, by how much would his record have differed if he had competed instead in 1956 at Melbourne (g = 9.7999 m/s²)? (In this connection see "The Earth's Gravity," by Weikko A. Heiskanen, *Scientific American,* September 1955.)

32P. A third baseman wishes to throw to first base, 127 ft distant. His best throwing speed is 85 mi/h. (a) If the ball leaves his hand, 3.0 ft above the ground, in a horizontal direction, what will happen to it? (b) At what upward angle must the third baseman launch the ball if the first baseman is to catch it? Assume that the first baseman's glove is also 3.0 ft above the ground. (c) What will be the time of flight?

33P. During volcanic eruptions, chunks of solid rock can be blasted out of the volcano; these projectiles are called *volcanic blocks.* Figure 33 shows a cross section of Mt. Fuji, in Japan. (a) At what initial speed would a block have to be ejected, at 35° to the horizontal, from the vent at A in order to fall at the foot of the volcano at B? (b) What is the time of flight?

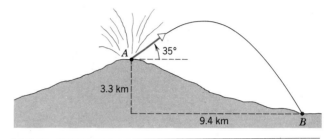

Figure 33 Problem 33.

34P. At what initial speed must the basketball player throw the ball, at 55° above the horizontal, to make the foul shot, as shown in Fig. 34? The basket rim is 18 in. in diameter. See the figure for other dimensions.

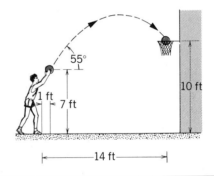

Figure 34 Problem 34.

35P. A football player punts the football so that it will have a "hang time" (time of flight) of 4.5 s and land 50 yd (=45.7 m) away. If the ball leaves the player's foot 5.0 ft (=1.52 m) above the ground, what is its initial velocity (magnitude, and direction)?

36P. The B-52 shown in Fig. 35 is 49 m long and is traveling at an air speed of 820 km/h (=510 mi/h) over a bombing

Figure 35 Problem 36.

range. How far apart will the bomb craters be? Make any measurements you need directly from the figure. Assume that there is no wind and ignore air resistance. How would air resistance affect your answer?

37P. A projectile is fired with an initial speed v_0 from the ground at a target on the ground a distance R away. (*a*) Show that there are two possible paths, a "high" and a "low" trajectory, to the target; see Fig. 36. (*b*) For $v_0 = 30$ m/s and $R = 20$ m, find the two possible elevation angles of fire.

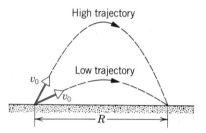

Figure 36   Problem 37.

38P. Find the angle of projection at which the horizontal range and the maximum height of a projectile are equal.

39P. What is the maximum vertical height to which a baseball player can throw a ball if he can throw it a maximum distance of 60 m?

40P. A football is kicked off with an initial speed of 64 ft/s at a projection angle of 45°. A receiver on the goal line 60 yd away in the direction of the kick starts running to meet the ball at that instant. What must be his average speed if he is to catch the ball just before it hits the ground? Neglect air resistance. (In this connection, see "Catching a Baseball," by Seville Chapman, *American Journal of Physics,* October 1968.)

41P. A ball rolls off the top of a stairway with a horizontal velocity of magnitude 5.0 ft/s. The steps are 8.0 in. high and 8.0 in. wide. Which step will the ball hit first?

42P. A certain airplane has a speed of 180 mi/h and is diving at an angle of 30° below the horizontal when a radar decoy is released. The horizontal distance between the release point and the point where the decoy strikes the ground is 2300 ft. (*a*) How high was the plane when the decoy was released? (*b*) How long was the decoy in the air? See Fig. 37.

43P. A plane, diving at an angle of 53° with the vertical, releases a projectile at an altitude of 730 m. The projectile hits the ground 5.0 s after being released. (*a*) What is the speed of the plane? (*b*) How far did the projectile travel horizontally during its flight? (*c*) What were the horizontal and vertical components of its velocity just before striking the ground?

44P. A ball is thrown horizontally from a height of 20 m

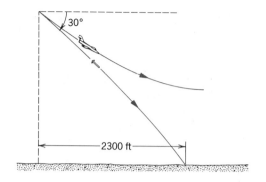

Figure 37   Problem 42.

and hits the ground with a speed that is three times its initial speed. What was the initial speed?

45P. The launching speed of a certain projectile is five times the speed it has at its maximum height. Calculate the elevation angle at launching.

46P. (*a*) During a tennis match, a player serves at 23.6 m/s (as recorded by radar gun), the ball leaving the racquet, 2.37 m above the court surface, horizontally. By how much does the ball clear the net, which is 12 m away and 0.90 m high? (*b*) Suppose the player serves the ball as before except that the ball leaves the racquet at 5.0° below the horizontal. Does the ball clear the net now?

47P. In Sample Problem 7, suppose that a second identical harbor defense cannon is emplaced 30 m above sea level, rather that at sea level. By how much is the range (the horizontal distance from launch to impact) of the second cannon longer than that of the first, which was found to be 690 m in the Sample Problem? Assume a 45° elevation angle of fire.

48P. A batter hits a pitched ball at a height 4.0 ft above ground so that its angle of projection is 45° and its horizontal range is 350 ft. The ball is fair down the left field line where a 24-ft fence is located 320 ft from home plate. Will the ball clear the fence? If so, by how much?

49P. The kicker on a football team can give the ball an initial speed of 25 m/s. Within what angular range must he kick the ball if he is to just score a field goal from a point 50 m in front of the goalposts whose horizontal bar is 3.44 m above the ground?

50P. An antitank gun is located on the edge of a plateau that is 60 m above the surrounding plain. The gun crew sights an enemy tank stationary on the plain at a horizontal distance of 2.2 km from the gun. At the same moment, the tank crew sees the gun and starts to move directly away from it with an acceleration of 0.90 m/s². If the antitank gun fires a shell with a muzzle speed of 240 m/s at an elevation angle of 10° above the horizontal, how long should the gun crew wait before firing if they are to hit the tank? See Fig. 38.

51P*. A rocket is launched from rest and moves in a straight line at 70° above the horizontal with an acceleration of

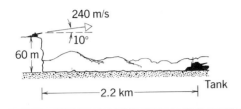

Figure 38 Problem 50.

46 m/s². After 30 s of powered flight, the engines shut off and the rocket follows a parabolic path back to earth; see Fig. 39. (*a*) Find the time of flight from launching to impact. (*b*) What is the maximum altitude reached? (*c*) What is the distance from launch pad to impact point?

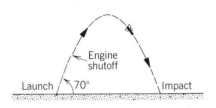

Figure 39 Problem 51.

Section 4–7 Uniform Circular Motion

52E. In Bohr's model of the hydrogen atom, an electron revolves around a proton in a circular orbit of radius 5.28×10^{-11} m with a speed of 2.18×10^6 m/s. What is the acceleration of the electron in this model of the hydrogen atom?

53E. What is the acceleration of a sprinter running at 10 m/s when rounding a bend with a turn radius of 25 m?

54E. A magnetic field can force a charged particle to move in a circular path. An electron experiences a radial acceleration of 3.0×10^{14} m/s² in a particular magnetic field. What is its speed if the radius of its circular path is 15 cm?

55E. A sprinter runs at 9.2 m/s around a circular track with a centripetal acceleration of 3.8 m/s². (*a*) What is the track radius? (*b*) How long does it take to go completely around the track at this speed?

56E. An earth satellite moves in a circular orbit 640 km above the earth's surface. The time for one revolution is 98 min. (*a*) What is the speed of the satellite? (*b*) What is the acceleration of gravity at the orbit?

57E. If a remote-controlled space probe can withstand the stresses of a 20*g* acceleration, (*a*) what is the minimum turning radius of such a craft moving at a speed of one-tenth the speed of light? (*b*) How long would it take to complete a 90° turn at this speed?

58E. A rotating fan completes 1200 revolutions every minute. Consider a point on the tip of the blade, which has a radius of 0.15 m. (*a*) Through what distance does the point move in one revolution? (*b*) What is the speed of the point? (*c*) What is its acceleration?

59E. The fast train known as the TGV (Train Grand Vitesse) that runs south from Paris in France has a scheduled average speed of 216 km/h. (*a*) If the train goes around a curve at this speed and the acceleration experienced by the passengers is to be limited to 0.05*g*, what is the smallest radius of curvature for the track that can be tolerated? (*b*) If there is a curve with a 1.0-km radius, to what speed must the train be slowed?

60E. Certain neutron stars (extremely dense stars) are believed to be rotating at about 1 rev/s. If such a star has a radius of 20 km, (*a*) what is the speed of an object on the equator of the star and (*b*) what is the centripetal acceleration of this point?

61E. An astronaut is rotated in a centrifuge of radius 5 m. (*a*) What is the speed if the acceleration is 7*g*? (*b*) How many revolutions per minute are required to produce this acceleration?

62P. A carnival Ferris wheel has a 15-m radius and completes five turns about its horizontal axis every minute. (*a*) What is the acceleration of a passenger at the highest point? (*b*) What is the acceleration at the lowest point?

63P. Calculate the acceleration of a person at latitude 40° owing to the rotation of the earth.

64P. (*a*) What is the centripetal acceleration of an object on the earth's equator owing to the rotation of the earth? (*b*) What would the period of rotation of the earth have to be in order that objects on the equator have a centripetal acceleration equal to 9.8 m/s²?

65P. A woman 1.6 m tall stands upright at latitude 50° for 24 h. (*a*) During this interval, how much farther does the top of her head move than the soles of her feet? (*b*) How much greater is the acceleration of the top of her head than the acceleration of the soles of her feet? Consider only effects associated with the rotation of the earth.

66P. A particle *P* travels with constant speed on a circle of radius 3.0 m and completes one revolution in 20 s (Fig. 40).

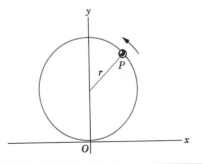

Figure 40 Problem 66.

The particle passes through O at $t = 0$. With respect to the origin O, find (a) the magnitude and direction of the vectors describing its position 5.0 s, 7.5 s, and 10 s later; (b) the magnitude and direction of the displacement in the 5.0-s interval from the fifth to the tenth second; (c) the average velocity vector in this interval; (d) the instantaneous velocity vector at the beginning and at the end of this interval; and (e) the instantaneous acceleration vector at the beginning and at the end of this interval.

67P. (a) Use the data of Appendix C to calculate the ratio of the centripetal accelerations of the Earth and Saturn owing to their revolutions about the sun. Assume that both planets move in circular orbits with constant speed. (b) What is the ratio of the distances of these two planets from the sun? (c) Compare your answers in parts (a) and (b) and suggest a simple relation between centripetal acceleration and distance from the sun. Check your hypothesis by calculating the same ratios for another pair of planets.

68P. A boy whirls a stone in a horizontal circle 2.0 m above the ground by means of a string 1.5 m long. The string breaks, and the stone flies off horizontally striking the ground 10 m away. What was the centripetal acceleration of the stone while in circular motion?

Section 4–8   Relative Motion in One Dimension

69E. A boat is traveling at 14 km/h with respect to the water of a river in the upstream direction. The water itself is flowing at 9 km/h with respect to the ground. (a) What is the velocity of the boat with respect to the ground? (b) A child on the boat walks from bow to stern at 6 km/h with respect to the boat. What is the child's velocity with respect to the ground?

70E. A person walks up a stalled 15-m-long escalator in 90 s. When standing on the same escalator, now moving, the person is carried up in 60 s. How much time would it take that person to walk up the moving escalator? Does the answer depend on the length of the escalator?

71E. A transcontinental flight of 2700 mi is scheduled to take 50 min longer westward than eastward. The airspeed of the jet is 600 mi/h. What assumptions about the jet stream wind velocity, presumed to be east or west, are made in preparing the schedule?

72P. The airport terminal in Geneva, Switzerland has a "moving sidewalk" to speed passengers through a long corridor. Peter, who walks through the corridor but does not use the moving sidewalk, takes 150 s to do so. Paul, who simply stands on the moving sidewalk, covers the same distance in 70 s. Mary not only uses the sidewalk but walks along it. How long does Mary take? Assume that Peter and Mary walk at the same speed.

Section 4–9   Relative Motion at High Speeds

73E. An electron moves at speed $0.42c$ with respect to observer B. Observer B moves at speed $0.63c$ with respect to observer A, in the same direction as the electron. What does observer A measure for the speed of the electron?

74P. Galaxy Alpha is observed to be receding from us with a speed of $0.35c$. Galaxy Beta, located in precisely the opposite direction, is also found to be receding from us at this same speed. What recessional speed would an observer on Galaxy Alpha find (a) for our galaxy and (b) for Galaxy Beta?

Section 4–10   Relative Motion in Two Dimensions

75E. Two highways intersect, as shown in Fig. 41. At the instant shown, a police car P is 800 m from the intersection and moving at 80 km/h. Motorist M is 600 m from the intersection and moving at 60 km/h. At this moment, what is the velocity (magnitude and direction) of the motorist with respect to the police car?

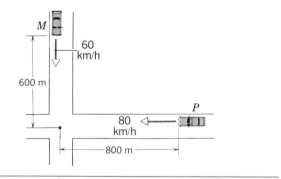

Figure 41   Exercise 75.

76E. Snow is falling vertically at a constant speed of 8.0 m/s. At what angle from the vertical do the snowflakes appear to be falling as viewed by the driver of a car traveling on a straight road with a speed of 50 km/h?

77E. In a large department store, a shopper is standing on the "up" escalator, which is traveling at an angle of 40° above the horizontal and at a speed of 0.75 m/s. He passes his daughter, who is standing on the identical, adjacent "down" escalator. (See Fig. 42.) Find the velocity of the shopper relative to his daughter.

78P. A helicopter is flying in a straight line over a level field at a constant speed of 6.2 m/s and at a constant altitude of 9.5 m. A package is ejected horizontally from the helicopter with an initial velocity of 12 m/s relative to the helicopter, and in a direction opposite to the helicopter's motion. (a) Find the initial speed of the package relative to the ground. (b) What is the horizontal distance between the helicopter and the package at the instant the package strikes the ground? (c) What angle does the velocity vector of the package make with the ground at the instant before impact, as seen from the ground?

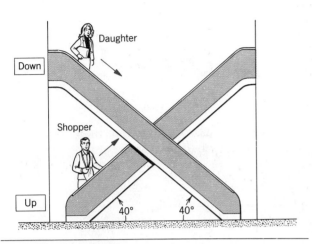

Figure 42 Exercise 77.

79P. A train travels due south at 30 m/s (relative to ground) in a rain that is blown toward the south by the wind. The path of each raindrop makes an angle of 22° with the vertical, as measured by an observer stationary on the earth. An observer seated in the train, however, sees perfectly vertical tracks of rain on the windowpane. Determine the speed of the raindrops relative to the earth.

80P. Two ships, A and B, leave port at the same time. A travels northwest at 24 knots and ship B travels at 28 knots in a direction 40° west of south. (1 knot = 1 nautical mile per hour; see Appendix F.) (*a*) What are the magnitude and direction of the velocity of ship A relative to B? (*b*) After what time will they be 160 nautical miles apart? (*c*) What will be the bearing of B from A at that time?

81P. A light plane attains an airspeed of 500 km/h. The pilot sets out for a destination 800 km to the north, but discovers that the plane must be headed 20° east of north to fly there directly. The plane arrives in 2.0 h. What was the vector wind velocity?

82P. The New Hampshire State Police use aircraft to enforce highway speed limits. Suppose that one of the airplanes has a speed of 135 mi/h in still air. It is flying straight north so that it is at all times directly above a north–south highway. A ground observer tells the pilot by radio that a 70-mi/h wind is blowing, but neglects to give the wind direction. The pilot observes that in spite of the wind the plane can travel 135 mi along the highway in 1 h. In other words, the ground speed is the same as if there were no wind. (*a*) What is the direction of the wind? (*b*) What is the heading of the plane, that is, the angle between its axis and the highway?

83P. A wooden boxcar is moving along a straight railroad track at a speed v_1. A sniper fires a bullet (initial speed v_2) at it from a high-powered rifle. The bullet passes through both walls of the car, its entrance and exit holes being exactly opposite to each other as viewed from within the car. From what direction, relative to the track, was the bullet fired? Assume that the bullet was not deflected upon entering the car, but that its speed decreased by 20%. Take $v_1 = 85$ km/h and $v_2 = 650$ m/s. (Are you surprised that you don't need to know the width of the boxcar?)

84P. A woman can row a boat 4.0 mi/h in still water. (*a*) If she is crossing a river where the current is 2.0 mi/h, in what direction will her boat be headed if she wants to reach a point directly opposite from her starting point? (*b*) If the river is 4.0 mi wide, how long will it take her to cross the river? (*c*) How long will it take her to row 2.0 mi *down* the river and then back to her starting point? (*d*) How long will it take her to row 2.0 mi *up* the river and then back to her starting point? (*e*) In what direction should she head the boat if she wants to cross in the shortest possible time?

85P. A man wants to cross a river 500 m wide. His rowing speed (relative to the water) is 3000 m/h. The river flows at a speed of 2000 m/h. The man's walking speed on shore is 5000 m/h. (*a*) Find the path (combined rowing and walking) he should take to get to the point directly opposite his starting point in the shortest time. (*b*) How long does it take?

86P. A battleship steams due east at 24 km/h. A submarine 4.0 km away fires a torpedo that has a speed of 50 km/h; see Fig. 43. If the bearing of the ship as seen from the submarine is 20° east of north, (*a*) in what direction should the torpedo be fired to hit the ship, and (*b*) what will be the running time for the torpedo to reach the battleship?

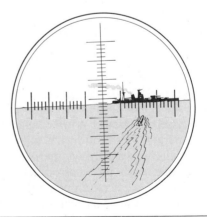

Figure 43 Problem 86.

CHAPTER 5

FORCE AND MOTION—I

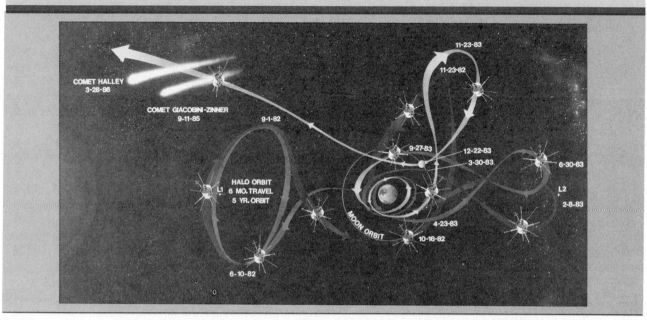

A triumph of Newtonian mechanics. After spending 4 years orbiting a point along the earth–sun axis, spacecraft ISEE 3 went on a complex journey requiring 37 rocket burns and 5 lunar flybys. It intercepted Comet Giacobini-Zinner in 1985, rounded the sun, passed upstream of Comet Halley in 1986, and will return to earth's vicinity in the year 2012. Newton would have been pleased.

5–1 Why Does a Particle Change Its Velocity?

Sometimes an object that we are watching — perhaps an automobile, a baseball, or a cat — will change its velocity. It will accelerate. Observation has taught us that when this happens we can always find one or more nearby objects that seem to be associated with that change. We relate the acceleration of a particle then to some interaction between the particle and its surroundings. We are so used to this that, when we see an object change its velocity without apparent cause, we suspect a

trick. If a rolling ball suddenly changes direction, we look for a hidden magnet or an air jet. Table 1 lists some accelerated motions and the nearby objects that we associate with them.

The central problem of mechanics is this: (1) We are given a particle (usually referred to as a *body,* from now on) whose characteristics (for example, mass, shape, volume, electric charge) we know. (2) We also know the locations and properties of all significant nearby objects. That is, we are fully informed about the body's environment. (3) How will the body move?

Isaac Newton (1642–1727), in putting forward his laws of motion and his theory of gravity, first solved this problem. Here is our plan for learning about what he did:

Table 1   Some Accelerated Motions and Their Causes

The Particle	The Motion	The Cause
Electron	Circulates around the nucleus of a hydrogen atom	The nucleus
Billiard ball	Hits another ball on the table and is deflected	The other ball; the table; the earth
Iron nail	Picked up from the table by a magnet	The magnet
Book	Comes to rest as you slide it across a carpet	The carpet; the earth
Sky diver	Comes to terminal speed after the parachute opens	The atmosphere; the earth
Comet Halley	Comes in from space, rounds the sun, and returns to space	The sun

(1) We introduce the concept of *force,* in terms of the acceleration given to a selected standard body. (2) We show how to assign a *mass* to a body, so that we may understand how it is that different bodies, subject to the same environment, have different accelerations. (3) Finally, we find ways to calculate the force acting on a body from the properties of the body and its environment. That is, we look for *force laws.*

Figure 1 suggests this program. Force appears in both the laws of motion (which tell us what acceleration a body will experience when a force acts on it) and in the force laws (which tell us how to calculate the force that will act on a body in a particular environment.) It is the

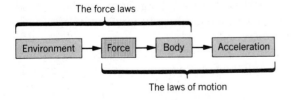

Figure 1   Our program for mechanics. The three left-hand boxes suggest that force is an interaction between a body and its environment. The three right-hand boxes suggest that a force acting on a body will accelerate it.

crowning glory of Newton's mechanics that, for a fantastic range of phenomena, it predicts results that agree with experiment.

There are some important problems to which Newtonian mechanics does not give correct answers. If the speeds of the particles involved are an appreciable fraction of the speed of light, we must replace Newtonian mechanics by Einstein's special theory of relativity. For problems on the scale of atomic structure (for example, the motions of electrons within atoms), we must replace Newtonian mechanics by quantum mechanics. We now view Newton's mechanics as a special case of these two, more comprehensive, theories. It is a very important special case, however, encompassing as it does the motions of objects that range in size from molecules to galaxies. Within this broad range it is highly accurate, as the successful maneuvering of space probes reminds us.

5-2   Newton's First Law

Galileo, who died about a year before Newton was born, laid a firm foundation for Newton's work. Before Galileo's time, it was thought that some influence or "force" was needed to keep a body moving at constant velocity. A body was then thought to be in its "natural state" when it was at rest. For a body to move with constant velocity, it was believed that it had to be propelled in some way. Otherwise, it would "naturally" stop moving.

This is not unreasonable. If you propel a book across a carpet it does indeed come to rest. If you want to make it move across the carpet with constant velocity you might for example, tie a string to it and pull it across.

Slide the book, however, over the ice of a skating rink. It goes a lot farther. You can imagine smoother and longer surfaces, over which the book would slide farther and farther. In the limit you can think of a long, extremely smooth surface, over which the book would show little sign at all of slowing down. We can in fact come close to this in the laboratory, by propelling a book over a horizontal air table, across which it floats almost friction-free on a film of air.

We are led to conclude that you do *not* need a force to keep a body moving with constant velocity. This fits in nicely with what we learned in Section 4-8 about reference frames. A body at rest in one frame moves at constant velocity with respect to an observer in another. *Rest* and *moving with constant velocity* are not all that differ-

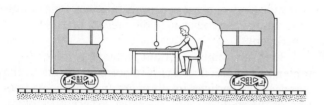

Figure 2 If the hanging ball remains at rest over the marker, your private railroad car is an inertial reference frame; if it doesn't, it isn't. The arrangement is suitable for motion in a horizontal plane only.

ent. This leads us to the first of Newton's three laws of motion:

> Newton's first law: *Consider a body on which no net force acts. If the body is at rest, it will remain at rest. If the body is moving with a constant velocity, it will continue to do so.*

Newton's first law is really a statement about reference frames in that it defines the kinds of reference frames in which the laws of Newtonian mechanics hold. From this point of view the first law is expressed as:

> Newton's first law: *If the net force acting on a body is zero, it is possible to find a set of reference frames in which that body has no acceleration.*

Newton's first law is sometimes called the *law of inertia* and the reference frames that it defines are called *inertial reference frames.*

Figure 2 shows how you can test a reference frame to see whether or not it is an inertial frame. You put a test body at rest in it and arrange things so that no net force acts on it. If the body remains at rest, your frame is inertial; if it does not, your frame is not an inertial frame. Once you have found one inertial frame you can find many more because any frame that moves at constant velocity with respect to an inertial frame is also an inertial frame.

If you put a bowling ball at rest on a rotating merry-go-round, no identifiable force acts on the ball but it does not remain at rest. If you roll the ball out along a radial line it will veer away from that line. Rotating reference frames are *not* inertial frames. Strictly speaking, the earth is not an inertial frame either, because of its rotation. However, it is only if we consider large-scale motions such as wind and ocean currents that we must take into account the noninertial character of the rotating

earth; inept bowlers cannot blame the rotation of the earth for their poor performance. In all that follows, unless we say otherwise, we take the earth to be an inertial reference frame.*

5–3 Force

We now wish to define *force* carefully, in terms of the acceleration that it gives to a standard reference body. As a standard body, we use (or rather we imagine that we use) the standard kilogram of Fig. 6 of Chapter 1. This body has been assigned, exactly and by definition, a mass of 1 kg. Later, we shall show how to assign masses to other bodies.

We put the standard body on a horizontal frictionless table and we fasten a spring to it, as in Fig. 3*a*. We now pull the spring to the right so that, by trial and error, the standard body experiences a measured acceleration of 1 m/s², as in Fig. 3*b*. We then declare, as a matter of definition, that the spring is exerting a force on the standard body whose magnitude is 1 newton (abbr. N).

We can exert a 2-N force on our standard body by pulling it so that its measured acceleration is 2 m/s², and so on. In general, if we see that our standard body has an acceleration a, we know that a force F must be acting on it and that the magnitude of the force (in newtons) is equal to the magnitude of the acceleration (in m/s²).

Acceleration is a vector. Is force a vector? We can easily enough assign a direction to force, namely, that of the acceleration that it produces in our standard body. That, however, is not enough to prove that force is a vector. We must also show that force obeys the laws of vector addition and we can only find out whether or not it does by experiment.

Let us arrange to exert on our standard body either a 4-N force along the x axis or a 3-N force along the y axis. The first of these forces, acting alone, would produce an acceleration of 4 m/s² along the x axis. The second, acting alone, would produce an acceleration of 3 m/s² along the y axis.

What if the forces act simultaneously, as in Fig. 4? We would find by experiment (and only by experiment)

---

* The earth's noninertial character shows up in that a falling object does not fall straight down but veers to the east a little. At latitude 45°, for example, an object dropped from a height of 50 m would (neglecting air resistance) land 5 mm east of the spot it would have hit if the earth were not rotating.

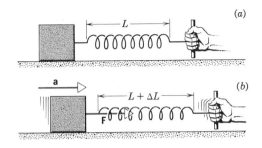

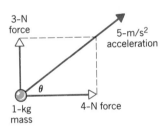

Figure 3 (*a*) A spring is attached to a standard body whose mass is 1 kg. (*b*) The standard body is accelerated by pulling on the spring, which is stretched by an amount ΔL. The surface is frictionless.

Figure 4 A 3-N force and a 4-N force act simultaneously on a body whose mass is 1 kg. The body's acceleration is the same as if it had been acted on by a single force equal to the vector sum of the two actual forces. Forces add like vectors.

that the acceleration of the standard body would be 5 m/s², directed as shown in Fig. 4. This is exactly the same acceleration that you would find if the standard body were acted on by a single force, equal to the vector sum of the two actual forces. Experiments like this show beyond doubt that forces are vectors. They have magnitude; they have direction; they add according to the vector rules.

5-4 Mass

Everyday experience tells us that a given force will produce different accelerations in different bodies. Put a baseball and a bowling ball on the floor and give each one the same sharp kick; the acceleration of the baseball will be much greater.

To get at this quantitatively, let us attach a spring to our standard body as in Fig. 3 and let us give it an acceleration a_0 of, say, 1 m/s². The force that the spring exerts will be 1 N. Note carefully the extension ΔL of the spring

that corresponds to this 1-N force. The mass m_0 of our standard body is, by definition, exactly 1 kg.

Now let us replace the standard body by an arbitrary body (body X) and let us apply the same 1-N force to it. We can do this by pulling on it in such a way that the spring is stretched by the same amount ΔL that it was in the experiment above with the standard body. Suppose that the acceleration a_X of body X turns out to be 0.25 m/s². We can use this experimental result to assign a mass m_X to body X by declaring that if the same force acts on two different bodies, we define the ratio of their masses to be the inverse ratio of their accelerations. Thus,

$$m_X = m_0 \frac{a_0}{a_X} = (1 \text{ kg}) \frac{1 \text{ m/s}^2}{0.25 \text{ m/s}^2} = 4 \text{ kg}.$$

Thus, body X, which receives only one-fourth the acceleration as the standard body when the same force acts on it, has, by this definition, four times the mass.

In this way, we can assign masses to bodies other than the standard body. Before we accept this method, however, let us test it, in two ways.

The First Test. Let us repeat the comparison to the standard body but with a different common force acting. Suppose, for example, that we stretch the spring farther, so that the acceleration a_0' of the standard body is now 5 m/s². That is, we decide to use a 5-N force rather than a 1-N force to make our mass comparisons.

We will find by experiment that, if we let this same 5-N force act on body X, its acceleration a_X' will be 1.25 m/s². The mass that we find for body X will then be

$$m_X = m_0 \frac{a_0'}{a_X'} = (1 \text{ kg}) \frac{5 \text{ m/s}^2}{1.25 \text{ m/s}^2} = 4 \text{ kg},$$

exactly as before.

The Second Test. Suppose that — using the spring method described above — we have compared a second body (body Y) with our standard body and that we have found $m_Y = 6$ kg. Now let us compare body X and body Y directly. That is, exert the same force $\mathbf{F}$ (of any convenient magnitude) on each of them and measure the accelerations a_X'' and a_Y'' that result. Let us say that we find $a_X'' = 2.4$ m/s² and $a_Y'' = 1.6$ m/s².

Now let us find the mass of body Y by comparing it — not with the standard body — but directly with body X, whose mass we already know. We find

$$m_Y = m_X \frac{a_X''}{a_Y''} = (4 \text{ kg}) \frac{2.4 \text{ m/s}^2}{1.6 \text{ m/s}^2} = 6 \text{ kg}.$$

Again we get the same answer.

Thus, our method of assigning masses to arbitrary bodies yields consistent answers, no matter what force we use in making the comparison and no matter what body we use for a comparison standard. Mass seems to be truly an intrinsic characteristic of a body.

5–5 Newton's Second Law

It is a tribute to Newton's genius that all of the detailed definitions, experiments, and observations that we have described so far can be summarized in a simple vector equation, Newton's second law of motion:

$$\boxed{\sum \mathbf{F} = m\mathbf{a}} \quad \text{(Newton's second law).} \quad (1)$$

In using Eq. 1, we must first be quite certain what body we are applying it to. $\sum \mathbf{F}$ in Eq. 1 is the *vector* sum of *all* the forces that act *on* that body. If you miss any forces (or count any of them twice), you will be in trouble. Only forces that act *on* the body are to be included. There may be many forces in a given problem and you must be sure to pick only those that act on the body with which you are dealing. Finally, $\sum \mathbf{F}$ includes only *external* forces, that is, forces exerted on the body by other bodies. We do not count internal forces, in which one part of the body exerts a force on another part.

Like all vector equations, Eq. 1 is equivalent to three scalar equations, or

$$\boxed{\sum F_x = ma_x, \ \sum F_y = ma_y, \ \text{and} \ \sum F_z = ma_z.} \quad (2)$$

These equations relate the three components of the net force acting on a body to the three components of the acceleration of that body.

Finally, we note that Newton's second law includes the formal statement of Newton's first law as a special case. That is, if no force acts on a body, Eq. 1 tells us that the body will not be accelerated. This is not to trivialize Newton's first law; its role in defining the set of reference frames in which Newton's mechanics holds justifies its status as a separate law.

Although we shall use SI units almost exclusively from now on, other systems of units are still in use. Chief among these are the British system and the CGS (centimeter-gram-second) system. Table 2 (see also Appendix F) shows the units in which Eqs. 1 and 2 are expressed in these systems.

Table 2   Units in Newton's Second Law (Eqs. 1 and 2)

System	Force	Mass	Acceleration
SI	newton (N)	kilogram (kg)	m/s^2
CGS	dyne	gram (g)	cm/s^2
British	pound (lb)	slug	ft/s^2

Sample Problem 1   A student pushes a loaded sled whose mass m is 240 kg for a distance d of 2.3 m over the frictionless surface of a frozen lake. He exerts a constant horizontal 130-N force F ($= 29$ lb) as he does so; see Fig. 5a. If the sled starts from rest, what is its final velocity?

As Fig. 5b shows, we lay out a horizontal x axis, we take the direction of increasing x to be to the right, and we treat the sled as a particle. We assume that the force F exerted by the student is the only horizontal force acting on the sled. We can then find the acceleration of the sled from Newton's second law, or

$$a = \frac{F}{m} = \frac{130 \text{ N}}{240 \text{ kg}} = 0.542 \text{ m/s}^2.$$

Because the acceleration is constant, we can use Eq. 15 of Chapter 2 $[v^2 = v_0^2 + 2a(x - x_0)]$ to find the final velocity. Putting $v_0 = 0$ and $x - x_0 = d$ and solving for v, we obtain

$$v = \sqrt{2ad}$$
$$= \sqrt{(2)(0.542 \text{ m/s}^2)(2.3 \text{ m})} = 1.6 \text{ m/s.} \quad \text{(Answer)}$$

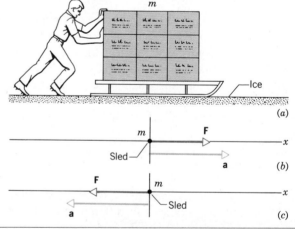

Figure 5   Sample Problem 1. (*a*) A student pushes a loaded sled over a frictionless surface. (*b*) A "free-body diagram", showing the net force acting on the sled and the acceleration it produces. (*c*) A second free-body diagram. The student now pushes in the opposite direction on the sled, reversing its motion.

The force, the acceleration, the displacement, and the final velocity of the sled are all positive, which means that they all point to the right in Fig. 5b.

Sample Problem 2   The student in Sample Problem 1 wants to reverse the direction of the velocity of the sled in 4.5 s. With what constant force must he push on the sled to do so?

Let us find the (constant) acceleration, using Eq. 10 of Chapter 2 ($v = v_0 + at$). Solving for a gives

$$a = \frac{v - v_0}{t} = \frac{(-1.6 \text{ m/s}) - (1.6 \text{ m/s})}{4.5 \text{ s}} = -0.711 \text{ m/s}^2.$$

This is larger in magnitude than the acceleration in Sample Problem 1 (0.542 m/s²) so it stands to reason that the student must push harder this time. We find this (constant) force from

$$F = ma = (240 \text{ kg})(-0.711 \text{ m/s}^2)$$
$$= -170 \text{ N} \ (= -38 \text{ lb}). \quad \text{(Answer)}$$

The negative sign shows that the student is pushing the sled in the direction of decreasing x, that is, to the left in Fig. 5.

Sample Problem 3   A crate whose mass m is 360 kg rests on the bed of a truck that is moving at a speed v_0 of 120 km/h, as in Fig. 6a. The driver applies the brakes and slows to a speed v of 62 km/h in 17 s. What force (assumed constant) acts on the crate during this time? Assume that the crate does not slide on the truck bed.

We first find the (constant) acceleration of the crate. Solving Eq. 10 of Chapter 2 ($v = v_0 + at$) for a yields

$$a = \frac{v - v_0}{t} = \frac{(62 \text{ km/h}) - (120 \text{ km/h})}{17 \text{ s}}$$
$$= \left(-3.41 \frac{\text{km}}{\text{h} \cdot \text{s}}\right)\left(\frac{1 \text{ h}}{3600 \text{ s}}\right)\left(\frac{1000 \text{ m}}{1 \text{ km}}\right) = -0.947 \text{ m/s}^2.$$

As Fig. 6 shows, the velocity vector of the crate points to the right, its acceleration vector to the left.

The force on the crate follows from Newton's second law:

$$F = ma$$
$$= (360 \text{ kg})(-0.947 \text{ m/s}^2) = -340 \text{ N}. \quad \text{(Answer)}$$

This force acts in the same direction as the acceleration, namely, to the left in Fig. 6b. If the crate is not secured by straps, only friction between the truck bed and the crate can provide this force.

If the frictional force is not great enough, the crate will slide forward. Put another way, in the reference frame of the earth, the crate will slow down less rapidly than will the truck. In pop-physics language, the driver may say: "I jammed on the brakes but the crate kept on going and slammed into the back of the cab."

Sample Problem 4   In a two-dimensional tug-of-war, Alex, Betty, and Charles pull on ropes that are tied to an automobile tire. The ropes make angles as shown in Fig. 7a, which is a view from above. Alex pulls with a force F_A (220 N) and Charles with a force F_C (170 N). With what force F_B does Betty pull? The tire is stationary and the orientation of Charles's rope is not given.

Figure 7b, in which the tire is represented by a dot and the forces that act on it are shown, is called a *free-body diagram*. The acceleration of the tire is zero so, from Eq. 1, the net force on the tire must also be zero. That is,

$$\sum \mathbf{F} = \mathbf{F}_A + \mathbf{F}_B + \mathbf{F}_C = 0.$$

This vector equation is equivalent to the two scalar equations

$$\sum F_x = F_C \cos \phi - F_A \cos 47° = 0 \quad (3)$$

and

$$\sum F_y = F_C \sin \phi + F_A \sin 47° - F_B = 0. \quad (4)$$

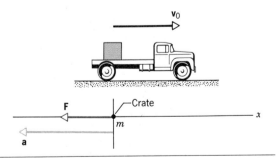

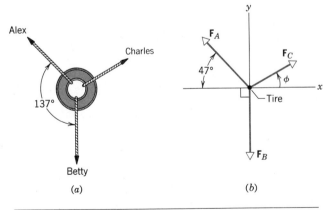

Figure 6   Sample Problem 3. (a) A crate on a truck that is slowing down. (b) The net force on the crate and the acceleration that it produces.

Figure 7   Sample Problem 4. (a) A view from above of three people pulling on a tire. (b) A free-body diagram for the tire.

The signs of the various terms in Eqs. 3 and 4 correspond to the directions of the corresponding force components in Fig. 7b. Substituting known values gives, for Eq. 3,

$$(170 \text{ N})(\cos \phi) = (220 \text{ N})(\cos 47°)$$

or

$$\phi = \cos^{-1} \frac{(220 \text{ N})(0.682)}{170 \text{ N}} = 28.0°.$$

Substituting into Eq. 4 gives us

$$F_B = F_C \sin \phi + F_A \sin 47°$$
$$= (220 \text{ N})(\sin 28.0°) + (170 \text{ N})(\sin 47°)$$
$$= 230 \text{ N}. \qquad \text{(Answer)}$$

Convince yourself that the three force vectors in Fig. 7b, if suitably translated, form a closed triangle. That is, they add up to zero.

Hint 1: Read the problem carefully. Read the problem statement several times until you have a good mental picture of what the setup is, what data are given, and what is requested. In Sample Problems 1 and 2, for example, you should tell yourself: "Someone is pushing a sled. Its speed changes, so acceleration is involved. The motion is along a straight line. A force is given in one problem and asked for in the other. It looks like Newton's second law applied to one-dimensional motion."

Hint 2: Reread the text. If you know what the problem is about but don't know what to do next, put the problem aside and reread the text. If you are hazy about Newton's second law, reread this section. Study the Sample Problems. The one-dimensional motion part of Sample Problems 1 and 2 should send you back to Chapter 2 and especially to Table 2 of that chapter, which displays all the equations you are likely to need.

Hint 3: Draw a figure. You may need two figures. One is a rough sketch of the actual real-world situation, such as Fig. 5a or 6a. The other is a *free-body diagram,* which shows (as a dot) the body and the forces that act on it. In Figs. 5b and 5c the sled is shown as a dot, as is the crate in Fig. 6b and the tire in Fig. 7b. In each case, the forces that act on the body are shown.

Hint 4: What is your system? If you are using Newton's second law, you must know to what body or system you are applying it. In Sample Problems 1 and 2, it is the sled (not the student or the ice). In Sample Problem 3, it is the crate

(not the truck). In Sample Problem 4, it is the tire (not the ropes or the people).

Hint 5: What is your reference frame? Be clear as to what reference frame you are using. We used a reference frame attached to the earth in all the Sample Problems so far. In Sample Problem 3, you have to make sure that you do not think it is the truck. In this problem, the truck is accelerating and is not an inertial reference frame.

Hint 6: Choose your axes wisely. In Sample Problem 4, we saved a lot of work by choosing one of our coordinate axes to coincide with one of the forces (the y axis with $\mathbf{F}_B$). Try working this problem with a coordinate system in which no axis coincides with a force.

5–6 Newton's Third Law

Forces come in pairs. If a hammer exerts a force on a nail, the nail exerts an equal but oppositely directed force on the hammer. If you kick a brick wall, the wall pushes back at you. The situation has been summed up with the gentle words: "You cannot touch without being touched."

Formally (see Fig. 8), let body A exert a force ($\mathbf{F}_{BA}$) on body B; experiment shows that body B then exerts a force ($\mathbf{F}_{AB}$) on body A. These two forces are equal in magnitude and oppositely directed. That is,

$$\boxed{\mathbf{F}_{AB} = -\mathbf{F}_{BA}} \qquad \text{(Newton's third law).} \qquad (5)$$

Note the order of the subscripts. $\mathbf{F}_{AB}$, for example, is the force exerted *on* body A *by* body B.

Equation 5 sums up Newton's third law of motion. Commonly, one of these forces (it does not matter which) is called the *action force.* The other member of the pair is then called the *reaction force.* Every time you

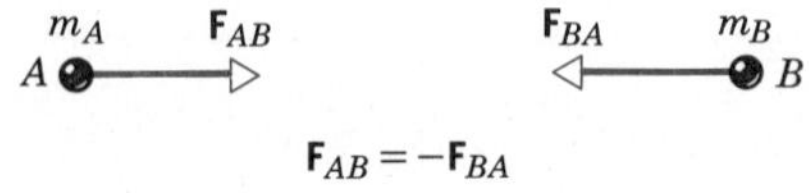

Figure 8 Newton's third law. Body A exerts a force $\mathbf{F}_{BA}$ on body B. Body B must then exert a force $\mathbf{F}_{AB}$ on body A, where $\mathbf{F}_{AB} = -\mathbf{F}_{BA}$.

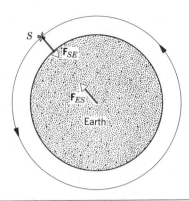

Figure 9   A satellite in earth orbit. The forces shown are an action–reaction pair. Note that they act on different bodies.

see a force, a good question is: "Where is its reaction force?"

The words, "To every action there is always an equal and opposite reaction," have become enshrined in the popular language and mean various things to various speakers. In physics, however, these words mean Eq. 5 and nothing else. In particular, cause and effect are not involved; either force can be the action force.

You may think: "If every force has an equal and opposite force associated with it, why don't they cancel each other? How can anything ever get moving?" The answer is simple. As Fig. 8 shows, the two members of an action–reaction pair *always* act on different bodies so that they cannot possibly cancel each other. If two forces act on the *same* body they are *not* an action–reaction pair, even though they may be equal and opposite. Let's identify the action – reaction pairs in two examples.

An Orbiting Satellite.   Figure 9 shows a satellite orbiting the earth. The only force that acts on it is F_{SE}, the force exerted *on* the satellite *by* the gravitational pull of the earth. Where is the corresponding reaction force? It is F_{ES}, the force acting on the earth owing to the gravita-

tional pull of the satellite; its effective point of application is the center of the earth.

You may think that the tiny satellite cannot exert much of a gravitational pull on the earth but it does, exactly as Newton's third law requires. That is, considering magnitudes only, $F_{ES} = F_{SE}$. The force F_{ES} causes the earth to accelerate, but, because of the earth's large mass, its acceleration is so small that it cannot be detected.

A Book Resting on a Table.*   Figure 10a shows a book resting on a table. The earth pulls downward on the book with a force F_{BE}. The book does not accelerate because this force is canceled by an equal and opposite contact force F_{BT} exerted on the book by the table.

Even though F_{BE} and F_{BT} are equal in magnitude and oppositely directed, they do *not* form an action–reaction pair. Why not? *Because they act on the same body, the book.* They cancel each other (which an action–reaction pair can never do) and thus account for the fact that the book is not accelerating.

Each of these forces must then have a corresponding reaction force somewhere. Where are they? The reaction force to F_{BE} is F_{EB}, the (gravitational) force with which the book attracts the earth. We show this action–reaction pair in Fig. 10b.

Figure 10c shows the reaction force to F_{BT}. It is F_{TB}, the contact force on the table owing to the book. The action–reaction pairs in this problem, and the bodies on which they act, are

First pair: $F_{BE} = -F_{EB}$ (book and earth)

and

Second pair: $F_{BT} = -F_{TB}$ (book and table).

---

* We ignore small complications caused by the rotation of the earth.

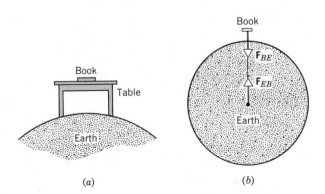

(a)

(b)

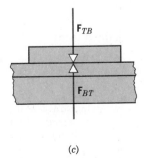

(c)

Figure 10   *(a)* A book rests on a table, which in turn rests on the earth. *(b)* The book and the earth pull on each other gravitationally, forming an action-reaction pair. *(c)* The table and the book exert action-reaction contact forces on each other.

We can use an accelerating elevator cab to sort the four forces shown in Fig. 10 properly into action–reaction pairs. Suppose that the cab is accelerating upward. The book and the table, inside the cab, would then press against each other with a greater force. The contact forces $\mathbf{F}_{TB}$ and $\mathbf{F}_{BT}$ (see Fig. 10c) would increase in magnitude but would remain equal and opposite. However, the gravitational forces $\mathbf{F}_{BE}$ and $\mathbf{F}_{EB}$ (see Fig. 10b) would remain unchanged. Both pairs of forces would continue to obey Newton's third law. The book accelerates because $\mathbf{F}_{BE}$ and $\mathbf{F}_{BT}$ (which are not an action–reaction pair) no longer cancel each other.

Would you have guessed that an orbiting satellite is simpler to analyze—in the context of Newton's third law—than a book resting quietly on a table?

5-7 Mass and Weight

The mass of a body and the weight of a body are totally different properties. They are often confused because, at any point near the earth's surface, they are proportional to each other. For example, if you increase the number of marbles in a sack by 10%, you have increased both the mass and the weight of the marbles by the same factor. This proportionality means that you can measure the mass of a body by measuring its weight and conversely. It also leads to *unit borrowing,* so that mass units are sometimes used to specify weight and the other way around. We need to be very clear about mass and weight.

Mass. Mass is a scalar, its SI unit being the kilogram. The mass of a body—one of its intrinsic properties—may be found by comparing the body with the standard kilogram, using (in principle) the methods of Section 5-4. The mass of a body is the same on the earth's surface, in an orbiting satellite, on Mars, or in interstellar space. If you kick a bowling ball and stub your toe, it is the mass of the ball that accounts for your misery.

Weight.* Weight is a vector (a force), its SI unit being the newton. The weight $\mathbf{W}$ of a body—unlike its mass—is *not* an intrinsic property of the body. If the mass of a body is m, its weight is defined from

$$\mathbf{W} = m\mathbf{g}, \qquad (6)$$

---

*In discussing weight in this and the following section, we assume that the measurements take place in an inertial (that is, a nonaccelerating) reference frame such as a laboratory. Later, in Sample Problem 10, we discuss the case of a noninertial (that is, an accelerating) reference frame such as an accelerating elevator cab. Here we will find that we must distinguish carefully between *weight* and *apparent weight.*

in which $\mathbf{g}$ is the free-fall acceleration at the location of the body. Thus, the weight of a body depends on its location because $\mathbf{g}$ varies from point to point. The weight of a certain bowling ball, for example, is 71 N on the earth's surface, 27 N on Mars, and 12 N on the moon. Its mass is 7.2 kg in all three places. When a child tries to lift a bowling ball from the floor, it is its weight that the child complains about.

5-8 Two Measuring Instruments

It will help to understand mass and weight if we look carefully at two common measuring instruments, the equal arm balance and the spring scale.

The Equal–Arm Balance. Let us put the standard kilogram on the right pan of such a balance (Fig. 11) and balance it by pouring lead shot onto the left pan. At balance, we know that the mass of the pellets in the left pan is 1 kg. The balance compares the *weights* of the two pan loads. As Eq. 6 shows, however, when the weights are equal the masses are also equal, $\mathbf{g}$ having a fixed value at the site of the balance. Thus, a balance determines the relative masses of two bodies—by "weighing" them. The specific value of the free-fall acceleration $\mathbf{g}$ at the balance site does not matter, as long as it is not zero; a balance would work on the moon but not in interstellar space.

The Spring Scale. This device—the bathroom scale being a familiar example—measures the *weight* of a body, by noting how far that force will stretch a spring; see Fig. 12. Metric spring scales, including the digital readout scales in your chemistry laboratory, are almost always marked in *mass* units, such as grams or kilo-

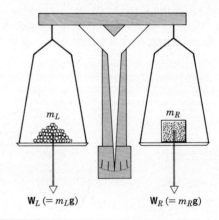

Figure 11   An equal arm balance. When it is in balance, the masses of the objects in the two pans are equal.

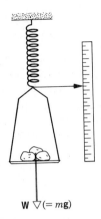

Figure 12   A spring scale. Its reading is proportional to the *weight* of the object placed in its pan. The scale can be marked in mass units if it is both calibrated and used at the same location, so that *g* does not change.

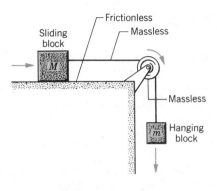

Figure 13   A block of mass *M* on a horizontal frictionless surface is connected to a block of mass *m* by a cord. "Massless" and "frictionless" mean that the values of mass and friction are so small that they would change our answers by less than about 2%. The light arrows show the directions in which the system will move when released from rest.

grams. Spring scales are so marked because it is almost always the mass (and not the weight) that the user is interested in, whether buying potatoes or "weighing out" reagents in a chemistry lab. As Eq. 6 reminds us, spring scales will read incorrectly if they are used at a place where the *g* is different from that at the place at which they were calibrated.

Weighing Atoms.   The equal arm balance and the spring scale are all right for gross objects such as diamonds, apples, or loaded trucks, but they will not do for atoms or subatomic particles. The gravitational forces acting on such tiny particles are far too small to be useful. Atomic masses are measured by the methods of Section 5–4, using electric and magnetic field forces (rather than spring forces) to accelerate the particles. Figure 7 of Chapter 1 shows a *mass spectrometer,* with which the masses of atoms can be determined precisely.

5-9   Applying Newton's Laws of Motion

Special Sample Problem*   Figure 13 shows a block (the *sliding block*) whose mass *M* is 3.3 kg. It is free to move along a horizontal frictionless surface such as an air track. The sliding

---

* We believe that students can best learn to solve problems — not by studying formal rules — but by studying how problems are actually solved. In this Special Sample Problem, the solution is worked out in exhaustive detail, using a question-and-answer format.

block is connected by a cord and a pulley to a second block (the *hanging block*), whose mass *m* is 2.1 kg. The hanging block will fall and the sliding block will accelerate to the right. Find (a) the acceleration of the sliding block, (b) the acceleration of the hanging block, and (c) the tension in the cord.

Q.   *What is this problem all about?*

You are given two massive objects, the sliding block and the hanging block. It might not occur to you but you are also given the earth, which pulls on each of these objects; without the earth, nothing would happen. There are five forces:

1.   The cord pulls to the right on the sliding block with a force of magnitude *T*.

2.   The cord pulls upward on the hanging block with a force of the same magnitude *T*. This keeps the hanging block from falling freely, which it would otherwise do. Note that the cord has the same tension throughout its length; the pulley just serves to change the direction of this force, without changing its magnitude.

3.   The earth pulls down on the sliding block with a force *M***g**, the weight of the sliding block.

4.   The earth pulls down on the hanging block with a force *m***g**, the weight of the hanging block.

5.   The air track pushes up on the sliding block with a vertical force **N**, a contact force called a *normal force* because it acts at right angles to the surface.

There is another thing that you should note. If the hanging block falls 1 mm in a certain time, the sliding block moves 1 mm to the right in that same interval. The blocks move together and their accelerations have the same magnitude a.

Q. *How do I classify this problem? Should it suggest a particular law of physics to me?*

Yes it should. Forces, masses, and accelerations are involved and this should suggest Newton's second law of motion, $\Sigma \mathbf{F} = m\mathbf{a}$.

Q. *If I apply this law to this problem, to what object do I apply it?*

There are two "particles" in this problem, the sliding block and the hanging block. Although they are extended objects, we can treat each block as a particle because every small part of it (every atom, say) moves in exactly the same way. Apply Newton's second law separately to each block.

Q. *What about the pulley?*

We cannot represent the pulley as a particle because different parts of it move in different ways. When we study rotation, we shall see how to deal with pulleys. Meanwhile, we get around the problem in a practical way, by using a pulley whose mass is negligible compared with the masses of the two blocks. The only function of the pulley is to change the direction of the cord that joins the two blocks.

Q. *O.K. Now how do I apply $\Sigma \mathbf{F} = m\mathbf{a}$ to the sliding block?*

Represent the sliding block by a particle of mass M and draw *all* the forces that act *on* it, as in Fig. 14. This is the *free-body diagram*. There are three forces. Next, draw a horizontal axis (an x axis). It makes sense to draw this axis parallel to the air track, in the direction in which the block moves. A y axis is not needed because the block does not move in this direction.

Q. *Thanks, but you still haven't told me how to apply $\Sigma \mathbf{F} = m\mathbf{a}$ to the sliding block. All you have done is to tell me how to draw a free-body diagram.*

Right you are. $\Sigma \mathbf{F} = m\mathbf{a}$ is a vector equation and you can write it as three scalar equations. Thus,

$$\sum F_x = Ma_x, \quad \sum F_y = Ma_y, \quad \text{and} \quad \sum F_z = Ma_z, \quad (7)$$

in which ΣF_x, ΣF_y, and ΣF_z are the components of the net force. There is no net force in the y direction, the weight of the sliding block being balanced by an upward-acting normal force exerted on the sliding block by the air track. No force acts in the z direction, which is at right angles to the page. Thus, only the first of Eqs. 7 is useful.

In the x direction, there is only one force component and $\Sigma F_x = Ma_x$ becomes

$$T = Ma, \quad (8)$$

which contains two unknowns, T and a so we can't yet solve it. Recall, however, that we have not yet said anything about the hanging block.

Q. *I agree. How do I apply $\Sigma \mathbf{F} = m\mathbf{a}$ to the hanging block.*

Draw a free-body diagram as in Fig. 15. This time, Newton's second law yields

$$mg - T = ma. \quad (9)$$

This contains the same two unknowns as Eq. 8 does. If you add these equations, T will cancel out. Solving for a yields

$$a = \frac{m}{M + m}\, g. \quad \text{(Answer)} \quad (10)$$

Substituting this result into Eq. 8 yields

$$T = \frac{Mm}{M + m}\, g. \quad \text{(Answer)} \quad (11)$$

Putting in the numbers gives, for these two quantities,

$$a = \frac{m}{M + m}\, g = \frac{2.1 \text{ kg}}{3.3 \text{ kg} + 2.1 \text{ kg}}\,(9.8 \text{ m/s}^2)$$
$$= 3.8 \text{ m/s}^2 \quad \text{(Answer)}$$

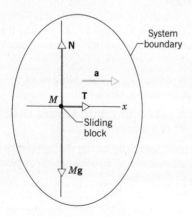

Figure 14 A free-body diagram for the sliding block of Fig. 13.

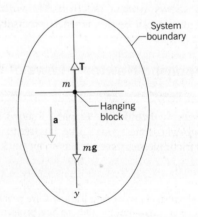

Figure 15 A free-body diagram for the hanging block of Fig. 13.

and

$$T = \frac{Mm}{M+m} g = \frac{(3.3 \text{ kg})(2.1 \text{ kg})}{3.3 \text{ kg} + 2.1 \text{ kg}} (9.8 \text{ m/s}^2)$$
$$= 13 \text{ N}. \qquad \text{(Answer)}$$

Q. The problem is now solved. What is the next problem?

That's a fair question, but we are not here only to solve problems but chiefly to learn physics. This problem is not really finished until we have studied the results to see if they make sense. This is often a much more confidence-building experience than simply getting the right answer.

Look first at Eq. 10. Note that it is dimensionally correct and also that the acceleration a will always be less than g. This is as it must be, because the hanging block is not in free fall. The cord pulls upward on it.

Look now at Eq. 11, which we rewrite in the form

$$T = \frac{M}{M+m} mg. \qquad (12)$$

In this form, it is easier to see that this equation is also dimensionally correct, because both T and mg are forces. Equation 12 also lets us see that the tension in the cord is always less than mg, the weight of the hanging block. That is a comforting thought because, if T were *greater* than mg, the hanging block would accelerate upward!

The numerical values bear out what we have said. The acceleration (3.8 m/s^2) is indeed less than g (9.8 m/s^2). The tension in the cord (13 N) is indeed less than the weight of the hanging block ($=mg=20$ N). However, we learned a lot more from examining the equations in algebraic form than we possibly could from these two numerical examples.

Q. That was worth doing. Are we finished now?

No. There are still other checks that we can make. As Einstein once said, "No number of experiments can prove me right; a single experiment can prove me wrong." We had better make all the checks that we can.

We can learn a lot of physics by studying special cases, in which we know what the answers must be. A simple example is to put $g = 0$, as if the experiment were carried out in interstellar space. We know that, in that case, the blocks would not move from rest and there would be no tension in the cord. Do the formulas predict this? Yes they do. If you put $g = 0$ in Eqs. 10 and 11, you find $a = 0$ and $T = 0$. Other special cases are $M = 0$ and $m \to \infty$. How many other special cases can you identify?

Q. Now we're finished, I assume.

Not really. There is still another way to look at this problem and to learn some more physics. It is to treat both blocks together as a single composite "body." Figure 16 shows the appropriate free-body diagram. Note that three forces only act on this body, the two weights and the normal force. The tension T has now become an internal force, exerted by one

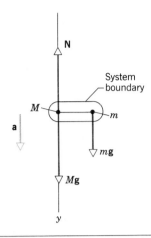

Figure 16   A free-body diagram for a composite system consisting of both blocks shown in Fig. 13.

part of the system on another; we count only *external* forces in applying Newton's second law.

Applying Newton's second law to the "particle" of Fig. 16 gives

$$Mg + mg - N = (M + m)a. \qquad (13)$$

However, we have already seen that $N = Mg$ so that the above equation reduces at once to

$$a = \frac{m}{M+m} g,$$

which is exactly Eq. 10. We can get no information about the tension in the cord from this approach because, as we have said, the cord is now internal to the system.

We can learn an interesting thing by putting $N = 0$ in Eq. 13 above. Doing so leads to the prediction that $a = g$. What does $N = 0$ mean physically? It means removing the air track and the pulley. If we do so, the two blocks will certainly fall freely under gravity, just as our formula predicts. Now we are finished!

Sample Problem 5   A block whose mass M is 33 kg is pushed across a frictionless surface by a stick whose mass m is 3.2 kg, as in Fig. 17a. The block is moved (from rest) a distance d ($=77$ cm) in 1.7 s. (a) Identify all action–reaction force pairs in this problem.

As the exploded view of Fig. 17b shows, there are two action–reaction pairs:

First pair: $\mathbf{F}_{HS} = -\mathbf{F}_{SH}$ (hand and stick)
Second pair: $\mathbf{F}_{SB} = -\mathbf{F}_{BS}$ (stick and block).

$\mathbf{F}_{HS}$, the force exerted on the hand by the stick, is the force that you would feel if the hand in Fig. 17 were yours.

(b) What force must the hand apply to the stick?

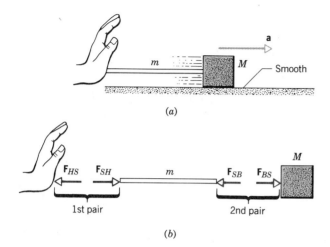

(a)

1st pair   2nd pair

(b)

Figure 17 Sample Problem 5. (*a*) A block of mass *M* is pushed over a smooth horizontal surface by a stick of mass *m*. (*b*) An exploded view, showing the action–reaction pairs between the hand and the stick (first pair) and between the stick and the block (second pair).

We must first find the (constant) acceleration. From Eq. 14 of Chapter 2 ($x - x_0 = v_0 t + \frac{1}{2} at^2$) we have, putting $v_0 = 0$ and $x - x_0 = d$ and solving for a,

$$a = \frac{2d}{t^2} = \frac{(2)(0.77 \text{ m})}{(1.7 \text{ s})^2} = 0.533 \text{ m/s}^2.$$

To find the force that the hand exerts, we apply Newton's second law to a system consisting of the stick and the block taken together. Thus,

$$F_{SH} = (M + m)a = (33 \text{ kg} + 3.2 \text{ kg})(0.533 \text{ m/s}^2)$$
$$= 19.3 \text{ N}. \qquad \text{(Answer)}$$

(c) With what force does the stick push on the block?

To find this force, we apply Newton's second law to the block alone, or

$$F_{BS} = Ma = (33 \text{ kg})(0.533 \text{ m/s}^2)$$
$$= 17.6 \text{ N}. \qquad \text{(Answer)}$$

(d) What is the net force on the stick?

We can find the magnitude F of this force in two ways. First, using results from (b) and (c) above,

$$F = F_{SH} - F_{SB} = 19.3 \text{ N} - 17.6 \text{ N} = 1.7 \text{ N}. \quad \text{(Answer)}$$

Note that we have used Newton's third law here, in assuming that $\mathbf{F}_{SB}$, the force that the block exerts on the stick, has the same magnitude (17.6 N) as the force $\mathbf{F}_{BS}$ that the stick exerts on the block; see (a) above.

The second way to arrive at an answer is to apply Newton's second law to the stick directly. We have

$$F = ma = (3.2 \text{ kg})(0.533 \text{ m/s}^2) = 1.7 \text{ N}, \quad \text{(Answer)}$$

in agreement with our first result. This is as it must be because the two methods are algebraically identical; check it out.

Sample Problem 6 Figure 18*a* shows a block of mass $m = 15$ kg hanging from three cords. What are the tensions in these cords?

The system is at rest so that no accelerations are involved. Applying Newton's second law to the block allows us to conclude that

$$T_C = mg = (15 \text{ kg})(9.8 \text{ m/s}^2) = 147 \text{ N} \approx 150 \text{ N}. \quad \text{(Answer)}$$

In view of the input data, only two significant figures are justified in our final answers so that we have rounded off our computational result (147 N) to 150 N. We retain the computational result, however, for use in later stages of the calculation. See Hint 7 of Chapter 4.

The clue to our next step is to realize that the knot where the three cords join is the only point at which all three forces act and it is this knot to which we should apply Newton's second law. Figure 18*b* shows the free-body diagram. The knot is not accelerated so that the net force acting on it must be zero. Thus,

$$\sum \mathbf{F} = \mathbf{T}_A + \mathbf{T}_B + \mathbf{T}_C = 0.$$

This vector equation is equivalent to the two scalar equations

$$\sum F_y = T_A \sin 28° + T_B \sin 47° - T_C = 0$$

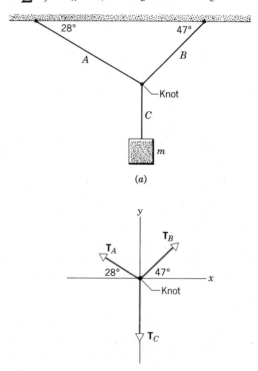

(a)

(b)

Figure 18 Sample Problem 6. (*a*) A block of mass *m* hangs from three strings. (*b*) A free-body diagram for the knot at the intersection of the three strings.

and

$$\sum F_x = T_B \cos 47° - T_A \cos 28° = 0.$$

Substituting numerical values leads to

$$T_A(0.470) + T_B(0.731) = 147 \text{ N} \qquad (14)$$

and

$$T_B(0.682) = T_A(0.883). \qquad (15)$$

From Eq. 15, we have

$$T_B = \frac{0.883}{0.682} T_A = 1.30 \, T_A.$$

Substituting this quantity into Eq. 14 and solving for T_A, we obtain

$$T_A = \frac{147 \text{ N}}{0.470 + (1.30)(0.731)} = 104 \text{ N} \approx 100 \text{ N}. \quad \text{(Answer)}$$

Our last unknown follows from

$$T_B = 1.30 \, T_A = (1.30)(104 \text{ N}) = 135 \text{ N} \approx 140 \text{ N}. \quad \text{(Answer)}$$

Sample Problem 7 Figure 19a shows a block of mass $m = 15$ kg held by a cord on a frictionless 27° plane. What is the tension in the cord? What force does the plane exert on the block?

Let us apply Newton's second law to the block. Figure 19b shows its free-body diagram. The following forces act on the block: (1) a normal force **N**, exerted outward on the block by the plane on which it rests; (2) the tension **T** in the cord; and (3) the weight **W** $(= m\mathbf{g})$ of the block. Because the acceleration of the block is zero, the net force acting on the block must also be zero, or

$$\sum \mathbf{F} = \mathbf{T} + \mathbf{N} + m\mathbf{g} = 0. \qquad (16)$$

We choose a coordinate system with the x axis parallel to the plane. With this choice, not one but two forces (**N** and **T**)

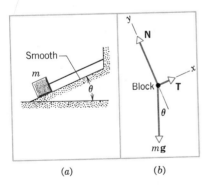

(a)   (b)

Figure 19 Sample Problem 7. (a) A block of mass m rests on a smooth plane, held there by a cord. (b) A free body diagram for the block. Note how the coordinate axes are chosen.

will line up with coordinate axes (a bonus). The scalar equations that are equivalent to Eq. 16 are

$$\sum F_x = T - mg \sin \theta = 0$$

and

$$\sum F_y = N - mg \cos \theta = 0.$$

Thus,

$$\begin{aligned} T &= mg \sin \theta \\ &= (15 \text{ kg})(9.8 \text{ m/s}^2)(\sin 27°) = 67 \text{ N}, \quad \text{(Answer)} \end{aligned}$$

and

$$\begin{aligned} N &= mg \cos \theta \\ &= (15 \text{ kg})(9.8 \text{ m/s}^2)(\cos 27°) = 131 \text{ N}. \quad \text{(Answer)} \end{aligned}$$

Sample Problem 8 Suppose that you cut the cord holding the block on the plane in Fig. 19a. With what acceleration will the block move?

Cutting the cord removes the tension **T** in Fig. 19b. The two remaining forces do not cancel and, indeed, they cannot, because they do not share the same line of action. Applying Newton's second law to the x components of the forces **N** and $m\mathbf{g}$ in Fig. 19b yields

$$\sum F_x = 0 - mg \sin \theta = ma,$$

so that

$$a = -g \sin \theta. \qquad (17)$$

Note that the normal force **N** played no role in finding the acceleration, its x component being zero.

Equation 17 yields

$$a = -(9.8 \text{ m/s}^2)(\sin 27°) = -4.4 \text{ m/s}^2. \quad \text{(Answer)}$$

The minus sign tells us that the acceleration is in the direction of decreasing x, that is, down the plane.

Equation 17 shows us that the acceleration is independent of the mass of the block. This reminds us that the acceleration of a freely falling body is also independent of the mass of the falling body. Indeed, Eq. 17 shows us how an inclined plane can be used to "dilute" gravity, as it were, so that the effects of "falling" can be studied more easily. For $\theta = 90°$, Eq. 17 yields $a = -g$ and for $\theta = 0$, it yields $a = 0$. Both are expected results.

Sample Problem 9 Figure 20a shows two blocks connected by a cord that passes over a (massless) pulley. Let $m = 1.3$ kg and $M = 2.8$ kg. Find the tension in the cord and the (common) acceleration of the two blocks.

Figures 20b and 20c show free-body diagrams for the two blocks. We are given that $M > m$, so we expect M to fall and m to rise. We choose the directions of our y axes with this in mind. That is, we choose up as the direction of increasing y for block m and $down$ for block M.

The tension in the cord must be less than the weight of block M (otherwise, that block would not fall from rest) and greater than the weight of block m (otherwise, that block would

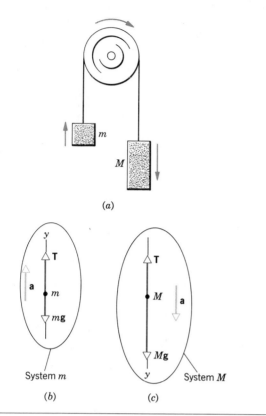

Figure 20 Sample Problem 9. (*a*) A block of mass *M* and one of mass *m* are connected by a cord that passes over a pulley. The directions in which the system will accelerate from rest are shown by the light arrows. (*b*) A free-body diagram for block *m*. The direction of increasing *y* is chosen to be "up." (*c*) A free-body diagram for block *M*. The direction of increasing *y* is chosen to be "down."

not rise). The vectors in the two free-body diagrams of Fig. 20 are drawn to scale to represent these facts.

Applying Newton's second law to block *m* gives

$$T - mg = ma, \qquad (18)$$

and, for block *M*,

$$Mg - T = Ma. \qquad (19)$$

Adding these two equations and solving for *a*, we obtain

$$a = \frac{M - m}{M + m} g. \qquad (20)$$

Substituting this result in either Eq. 18 or Eq. 19 and solving for *T*, we obtain

$$T = \frac{2mM}{M + m} g. \qquad (21)$$

We can write Eq. 21 in two different but totally equivalent ways:

$$T = \frac{M + M}{M + m} mg \quad \text{and} \quad T = \frac{m + m}{M + m} Mg. \qquad (22)$$

The first of these formulations shows that $T > mg$ and the second shows that $T < Mg$. That is, the tension *T* is intermediate between the weights of the two bodies. If $M = m$, Eqs. 20 and 21 give us $a = 0$ and $T = mg = Mg$, as we expect. That is, the acceleration is zero (the blocks do not move from rest) and the tension is equal to the weight of either of the two identical blocks. (Note that it is not *twice* the weight of a block, as some may think.)

Inserting the numbers given in the problem statement gives us, from Eq. 20,

$$a = \frac{M - m}{M + m} g = \frac{2.8 \text{ kg} - 1.3 \text{ kg}}{2.8 \text{ kg} + 1.3 \text{ kg}} (9.8 \text{ m/s}^2)$$

$$= 3.6 \text{ m/s}^2. \qquad \text{(Answer)}$$

Equation 21 gives, for the tension,

$$T = \frac{2Mm}{M + m} g = \frac{(2)(2.8 \text{ kg})(1.3 \text{ kg})}{2.8 \text{ kg} + 1.3 \text{ kg}} (9.8 \text{ m/s}^2)$$

$$= 17 \text{ N}. \qquad \text{(Answer)}$$

You can show that the weights of the two blocks are 13 N ($=mg$) and 27 N ($=Mg$), so that the tension ($=17$ N) does indeed lie between these two values.

Sample Problem 10   A passenger of mass *m* ($=72.2$ kg) stands on a platform scale in an elevator cab, as in Fig. 21. We wish to find the scale reading for various accelerations of the cab.

We consider this problem from the point of view of an observer in an (inertial) reference frame fixed with respect to the earth. Let this observer apply Newton's second law to the accelerating passenger. Figures 21*a*–21*e* are free-body diagrams for the passenger, treated as a particle (some particle!) for various accelerations of the cab.

Regardless of the acceleration of the cab, the earth pulls downward on the passenger with a force *mg*, in which *g* ($=9.80$ m/s^2) is the free-fall acceleration in the reference frame of the earth. The scale platform pushes upward on the passenger with a force whose magnitude W' is equal to the reading of the scale.* Newton's second law yields

$$W' - mg = ma,$$

or

$$W' = m(g + a). \qquad (23)$$

---

* The scale reading, which we have labeled W', is the weight that the accelerating passenger would judge himself to have as he reads the digital scale display. This quantity is often called the *apparent weight,* the term *weight* (or *true weight*) being reserved for the quantity *mg*.

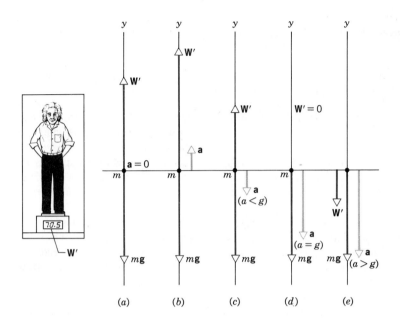

Figure 21 Sample Problem 10. A passenger in an elevator, standing on a spring scale which indicates his apparent weight W'. *(a)* A free-body diagram for the case in which the acceleration of the elevator cab is zero. *(b)* For $a = +3.2$ m/s². *(c)* For $a = -3.2$ m/s². *(d)* For $a = -g = -9.8$ m/s². *(e)* For $a = -12$ m/s².

(a) What does the scale read if the cab is at rest or moving with constant speed (see Fig. 21*a*)?

Here $a = 0$ and we have

$$W' = m(g + a) = (72.2 \text{ kg})(9.80 \text{ m/s}^2 + 0)$$
$$= 708 \text{ N} (= 159 \text{ lb}), \quad \text{(Answer)}$$

just as we expect.

(b) What does the scale read if the cab has an upward acceleration of magnitude 3.20 m/s² (see Fig. 21*b*)?

An upward acceleration means that the cab is either moving up with increasing speed or down with decreasing speed. In either case, Eq. 23 yields

$$W' = m(g + a) = (72.2 \text{ kg})(9.80 \text{ m/s}^2 + 3.20 \text{ m/s}^2)$$
$$= 939 \text{ N} (= 211 \text{ lb}). \quad \text{(Answer)}$$

The passenger presses down on the scale with a greater force than if he were at rest. The passenger—reading the scale—might conclude that he has gained 52 lb!

(c) What does the scale read if the cab has a downward acceleration of magnitude 3.20 m/s² (see Fig. 21*c*)?

A downward acceleration means that the cab is either moving up with decreasing speed or down with increasing speed. Equation 23 yields

$$W' = m(g + a) = (72.2 \text{ kg})(9.80 \text{ m/s}^2 - 3.20 \text{ m/s}^2)$$
$$= 477 \text{ N} (= 107 \text{ lb}). \quad \text{(Answer)}$$

The passenger presses down on the scale with less force than if the cab were at rest. He seems to have lost 52 lb.

(d) What does the scale read if the cable breaks, so that the cab falls freely (see Fig. 21*d*)?

In this case, the passenger (and also the scale) are in free fall, with $a = -g$. Equation 23 then yields

$$W' = m(g + a) = m(g - g) = 0. \quad \text{(Answer)}$$

Thus, in free fall, the scale reads zero and the passenger, in his accelerating frame, concludes that he is weightless. This is the same weightlessness that astronauts in earth orbit experience. In each case (elevator passenger or astronaut), the feeling of weightlessness arises *not* because gravity has ceased to act—it hasn't—but because the vehicle (elevator cab or spacecraft) and its occupant are both accelerating toward the center of the earth *at the same rate.*

In Essay 5 on "Physics in Weightlessness", Sally Ride discusses various aspects of this condition.

(e) What would happen if the cab were pulled (downward) with an acceleration of -12.0 m/s² (see Fig. 21*e*)?

This is an acceleration whose magnitude exceeds that of free fall. Equation 23 yields

$$W' = m(g + a) = (72.2 \text{ kg})(9.80 \text{ m/s}^2 - 12.0 \text{ m/s}^2)$$
$$= -159 \text{ N} (= -36 \text{ lb}). \quad \text{(Answer)}$$

If the scale were screwed to the floor of the cab and the passenger's shoes glued to the scale platform, the scale would read backward, the reading being -36 lb. If the passenger slipped out of his shoes, he would rise with respect to the cab until his head touched the ceiling, pushing against it with a force of 36 lb. As seen from an inertial frame, the passenger would free-fall until his head touched the ceiling.

REVIEW AND SUMMARY

A Program for Mechanics

We may explain the motion of an object by observing that accelerations are caused by forces. Forces, in turn, come about because of interactions between the object and its environment. To predict the motion of the object, we need to know the following: (1) How do we calculate **F** (the sum of the forces acting on the object) from the properties of the particle and its surroundings? The relations that allow us to do this are called the *force laws*. (2) How will the particle move when the net force **F** acts on it? The relations that tell us are called the *laws of motion*.

Newton's First Law

In the simplest case, the environment exerts no force on the object. Newton's first law then tells us that the object moves with zero acceleration; that is, it is either at rest or moving in a straight line with constant speed. Reference frames in which a free particle has no acceleration are *inertial reference frames*.

Measuring Force

If a force **F** *does* act, we must know how to measure it. We define the strength of the force, in newtons, to be numerically equal to the acceleration, in m/s^2, that it would cause if applied by itself to a standard mass (with mass $m_0 = 1$ kg by definition). Experiment shows that force is a vector in that it obeys the laws of vector addition; see Fig. 4. The direction of an individual force vector is the direction of the acceleration it would cause when applied to the standard mass.

Measuring Mass

We must also know how to assign masses to objects other than the standard kilogram. To do so, we apply any force to the object of interest (with unknown mass m_X) and, in a separate experiment, to our standard test body (mass $m_0 = 1$ kg). We measure the accelerations a_X and a_0 that result. We then *define* the mass m_X of the object of interest to be

$$m_X = m_0 \left(\frac{a_0}{a_X} \right) \quad \text{(same force } \mathbf{F} \text{ acting).}$$

Experiment shows that masses assigned to various bodies in this way are internally consistent and add like scalars.

Newton's second law, which successfully predicts the acceleration of an object acted on by one or more external forces, is summarized by the relation

Newton's Second Law

$$\sum \mathbf{F} = m\mathbf{a}, \qquad [1]$$

in which $\Sigma \mathbf{F}$ is the vector sum of *all* the forces that act *on* the object. In scalar component form, this law becomes

$$\sum F_x = ma_x, \quad \sum F_y = ma_y, \quad \text{and} \quad \sum F_z = ma_z. \qquad [2]$$

Sample Problems 1–4 illustrate simple situations analyzed in terms of this law. The Special Sample Problem in Section 5–9 illustrates in considerable detail the steps leading to a complete solution of a two-body acceleration problem.

Units for $\Sigma \mathbf{F} = m\mathbf{a}$

The appropriate SI units for $\Sigma \mathbf{F} = m\mathbf{a}$ are, respectively, the newton (N), the kilogram (kg), and the m/s^2. Table 2 shows the units for SI and for two other unit systems.

Newton's Third Law

Forces come from mutual interactions between pairs of objects. Newton's third law states that, in each interaction, there are always two forces—one on each object—and that the two forces are equal in magnitude and opposite in direction:

$$\mathbf{F}_{AB} = -\mathbf{F}_{BA} \qquad [5]$$

Weight and Mass

An important force acting on any body is its weight **W**, defined as the force

$$\mathbf{W} = m\mathbf{g}, \qquad [6]$$

g being the free-fall acceleration at the body's location. Weight is a vector measured in force units (newtons in SI), whereas mass is a scalar whose SI unit is the kilogram.

Sample Problems 5–10 illustrate the strategies for using Newton's second law in several situations. You should study them carefully, because they illustrate many of the difficulties that are often encountered.

QUESTIONS

1. Why do you fall forward when a moving bus decelerates to a stop and fall backward when it accelerates from rest?

2. Subway standees often find it convenient to face the side of the car when the train is starting or stopping and to face the front or rear when it is running at constant speed? Why?

3. A block with mass m is supported by a cord C from the ceiling, and a similar cord D is attached to the bottom of the block (Fig. 22). Explain this: If you give a sudden jerk to D, it will break; but if you pull on D steadily, C will break.

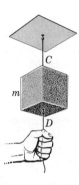

Figure 22 Question 3.

4. Criticize the statement, often made, that the mass of a body is a measure of the "quantity of matter" in it.

5. Suppose that a body that is acted on by exactly two forces is accelerated. Does it then follow that (*a*) the body cannot move with constant speed? (*b*) the velocity can never be zero? (*c*) the sum of the two forces cannot be zero? (*d*) the two forces must act in the same line?

6. You drop two objects of different masses simultaneously from the top of a tower. Show that, if you assume the air resistance to have the same constant value for each object, the one with the larger mass will strike the ground first. How good is this assumption?

7. The owner's manual of a car suggests that your seat belt should be adjusted "to fit snugly" and that the front seat head rest should *not* be adjusted so that it fits comfortably at the back of your neck but so that "the top of the head rest is level with the top of your ears." How do Newton's laws support these good recommendations?

8. A Frenchman, filling out a form, writes "78 kg" in the space marked Poids (weight). However, weight is a force and the kilogram is a mass unit. What do Frenchmen (among others) have in mind when they use a mass unit to report their weight? Why don't they report their weight in newtons? How many newtons does this Frenchman weigh? How many pounds?

9. Figure 23 shows comet Kohoutek as it appeared in 1973. Like all comets, it moves around the sun under the influence of the gravitational pull that the sun exerts on it. The nucleus of the comet is a relatively massive core at a position indicated by P. The tail of a comet is produced by the action of the solar wind, which consists of charged particles streaming outward from the sun. By inspection, what, if anything, can you say about the direction of the force that acts on the nucleus of the comet? What about the direction in which the nucleus is being accelerated? What about the direction in which the comet is moving?

10. In general (see Fig. 23), comets have a dust tail, consisting of dust particles pushed away from the sun by the pressure of sunlight. Why is this tail often curved?

Figure 23 Questions 9 and 10.

11. What is your mass in slugs? Your weight in newtons?

12. Using force, length, and time as fundamental quantities, what are the dimensions of mass?

13. A horse is urged to pull a wagon. The horse refuses to try, citing Newton's third law as a defense: The pull of the horse on the wagon is equal but opposite to the pull of the wagon on the horse. "If I can never exert a greater force on the wagon than it exerts on me, how can I ever start the wagon moving?" asks the horse. How would you reply?

14. Comment on whether the following pairs of forces are examples of action–reaction: (*a*) The earth attracts a brick; the brick attracts the earth. (*b*) A propellered airplane pushes air in toward the tail; the air pushes the plane forward. (*c*) A horse pulls forward on a cart, accelerating it; the cart pulls backward on the horse. (*d*) A horse pulls forward on a cart without moving it; the cart pulls back on the horse. (*e*) A horse pulls forward on a cart without moving it; the earth exerts an equal and opposite force on the cart. (*f*) The earth pulls down on the cart; the ground pushes up on the cart with an equal and opposite force.

15. Comment on the following statements about mass and weight taken from examination papers. (*a*) Mass and weight are the same physical quantities expressed in different units. (*b*) Mass is a property of one object alone, whereas weight results from the interaction of two objects. (*c*) The weight of an object is proportional to its mass. (*d*) The mass of a body varies with changes in its local weight.

16. Describe several ways in which you could, even briefly, experience weightlessness.

17. The mechanical arm on the Space Shuttle can handle a 2200-kg satellite when extended to 12 m. Yet, on the ground, this remote manipulator system (RMS) cannot support its own weight. In the weightlessness of an orbiting Shuttle, why does the RMS have to be able to exert any force at all?

18. In Fig. 24, we show four forces that are equal in magnitude. What combination of three of these, acting together on the same body, might keep that body in translational equilibrium?

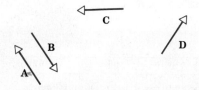

Figure 24 Question 18.

19. An elevator is supported by a single cable. There is no counterweight. The elevator receives passengers at the ground floor and takes them to the top floor, where they disembark. New passengers enter and are taken down to the ground floor. During this round trip, when is the tension in the cable equal to the weight of the elevator plus passengers? Greater? Less?

20. You are on the flight deck of the orbiting space shuttle *Discovery* and someone hands you two wooden balls, outwardly identical. One, however, has a lead core but the other does not. Describe several ways of telling them apart.

21. You are an astronaut in the lounge of an orbiting space station and you remove the cover from a long thin jar containing a single olive. Describe several ways—all taking advantage of the inertia of either the olive or the jar—to remove the olive from the jar?

22. A horizontal force acts on a body that is free to move. Can

it produce an acceleration if the force is less than the weight of that body?

23. Why does the acceleration of a freely-falling object not depend on the weight of the object?

24. What's the relation—if any—between the force acting on an object and the direction in which the object is moving?

25. You shoot an arrow into the air and you keep your eye on it as it follows a parabolic flight path to the ground. You notice that the arrow turns in flight so that it is always tangent to its flight path. What makes it do that?

26. A bird alights on a stretched telegraph wire. Does this change the tension in the wire? If so, by an amount less than, equal to, or greater than the weight of the bird?

27. In November 1984, astronauts Joe Allen and Dale Gardner salvaged a Westar-6 communications satellite from a faulty orbit and placed it into the cargo bay of the space shuttle *Discovery;* see Fig. 25. Describing the experience, Joe Allen said of the satellite, "It's not heavy; it's massive." What did he mean?

Figure 25 Question 27.

28. Why do raindrops fall with constant speed during the latter stages of their descent?

29. In a tug of war, three men pull on a rope to the left at *A* and three men pull to the right at *B* with forces of equal magnitude. Now a 5-lb weight is hung vertically from the center of the rope. (*a*) Can the men get the rope *AB* to be horizontal? (*b*) If not, explain. If so, determine the magnitude of the forces required at *A* and *B* to do this.

30. The following statement is true; explain it. Two teams are having a tug of war; the team that pushes harder (horizontally) against the ground wins.

31. A massless rope is strung over a frictionless pulley. A monkey holds onto one end of the rope and a mirror, having the same weight as the monkey, is attached to the other end of the rope at the monkey's level. Can the monkey get away from its image seen in the mirror (*a*) by climbing up the rope, (*b*) by climbing down the rope, or (*c*) by releasing the rope?

32. You stand on the large platform of a spring scale and note your weight. You then take a step on this platform and note that the scale reads less than your weight at the beginning of the step and more than your weight at the end of the step. Explain.

33. Could you weigh yourself on a scale whose maximum reading is less than your weight? If so, how?

34. A weight is hung by a cord from the ceiling of an elevator. From the following conditions, choose the one in which the tension in the cord will be greatest . . . least: (*a*) elevator at rest; (*b*) elevator rising with uniform speed; (*c*) elevator descending with decreasing speed; (*d*) elevator descending with increasing speed.

35. A woman stands on a spring scale in an elevator. In which of the following cases will the scale record the minimum reading . . . the maximum reading: (*a*) elevator stationary; (*b*) elevator cable breaks, free fall; (*c*) elevator accelerating upward; (*d*) elevator accelerating downward; (*e*) elevator moving at constant velocity?

36. What conclusion might a physicist draw if unequal masses hung over a pulley inside an elevator remain balanced; that is, there is no tendency for the pulley to turn?

37. In Fig. 26, a needle has been placed in each end of a

Figure 26 Question 37.

broomstick, the tips of the needles resting on the edges of filled wine glasses. The experimenter strikes the broomstick a swift and sturdy blow with a stout rod. The broomstick breaks and falls to the floor but the wine glasses remain in place and no wine is spilled. This impressive parlor stunt was popular at the end of the last century. What is the physics behind it? (If you try it, practice first with empty soft drink cans. Come to think of it, you might ask your physics instructor to do it, as a lecture demonstration!)

EXERCISES AND PROBLEMS

Section 5–2 Newton's First Law

1E. An unrestrained child is playing on the front seat of a car traveling in a residential neighborhood at 35 km/h. A small dog runs across the road and the driver applies the brakes, stopping the car and missing the dog. With what speed does the child strike the dashboard, presuming that the car stops before the child does so? Compare this speed with the world-record 100-m dash, about 10 s.

2P. Suppose that gravity were suddenly turned off, so that the earth becomes a free object rather than being confined to orbit the sun. How long would it take for the earth to reach a distance from the sun equal to Pluto's present orbital radius? (*Hint:* You will find some of the data you need in Appendix C.)

Section 5–5 Newton's Second Law

3E. (*a*) Two 10-lb weights are attached to a spring scale as shown in Fig. 27*a*. (*a*) What is the reading of the scale? (*b*) A single 10-lb weight is attached to a spring scale that itself is attached to a wall, as shown in Fig. 27*b*. What is the reading of the scale? (The scale indicates the tension in the string.)

4E. A 5.5-kg block is initially at rest on a frictionless horizontal surface. It is pulled with a constant horizontal force of 3.8 N. (*a*) What is its acceleration? (*b*) How long must it be

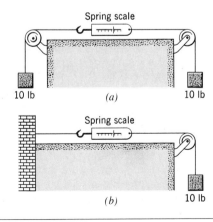

Figure 27 Exercise 3.

pulled before its speed is 5.2 m/s? (c) How far does it move in this time?

5E. An experimental rocket sled can be accelerated from rest to 1600 km/h in 1.8 s. (a) What is the acceleration of the sled? (b) What force would be required if the sled has a mass of 500 kg?

6E. A car traveling at 50 km/h hits a bridge abutment. A passenger in the car moves forward a distance of 65 cm (with respect to the road) while being brought to rest by an inflated air bag. What force (assumed constant) acts on the passenger's upper torso, which has a mass of 40 kg?

7E. A neutron travels at a speed of about 1.4×10^7 m/s. Nuclear forces are of very short range, being essentially zero outside a nucleus but very strong inside. If the neutron is captured and brought to rest by a nucleus whose diameter is 1.0×10^{-14} m, what is the magnitude of the force, presumed to be constant, that acts on this neutron? The neutron's mass is 1.67×10^{-27} kg.

8E. A light beam from a satellite-carried laser strikes an object ejected from an accidentally–launched ballistic missile; see Fig. 28. The beam exerts a force of 2.5×10^{-5} N on the target. If the "dwell time" of the beam on the object is 2.3 s, by how much is the object displaced from its flight path if it is (a) a 280-kg warhead and (b) a 2.1-kg decoy? (These displacements can be measured by observing the reflected beam.)

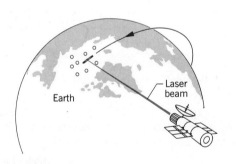

Figure 28 Exercise 8.

9P. A charged sphere of mass 3.0×10^{-4} kg is suspended from a string. An electric force acts horizontally on the sphere so that the string makes an angle of 37° with the vertical when at rest. Find (a) the magnitude of the electric force and (b) the tension in the string.

Section 5–6 Newton's Third Law

10E. Two blocks, with masses $m_1 = 4.6$ kg and $m_2 = 3.8$ kg, are connected by a light spring on a horizontal frictionless table. At a certain instant, when m_2 has an acceleration $a_2 = 2.6$ m/s², (a) what is the force on m_2 and (b) what is the acceleration of m_1?

11P. A 40-kg girl and an 8.4-kg sled are on the surface of a frozen lake, 15 m apart. By means of a rope, the girl exerts a

5.2-N force on the sled, pulling it toward her. (a) What is the acceleration of the sled? (b) What is the acceleration of the girl? (c) How far from the girl's initial position do they meet, presuming the force to remain constant? Assume that no frictional forces act.

12P. Let the only forces acting on two bodies be their mutual interactions. If both bodies start from rest, show that the distance traveled by each is inversely proportional to its mass.

13P. Two blocks are in contact on a frictionless table. A horizontal force is applied to one block, as shown in Fig. 29. (a) If $m_1 = 2.3$ kg, $m_2 = 1.2$ kg, and $F = 3.2$ N, find the force of contact between the two blocks. (b) Show that if the same force F is applied to m_2 rather than to m_1, the force of contact between the blocks is 2.1 N, which is not the same value derived in (a). Explain.

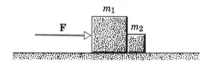

Figure 29 Problem 13.

Section 5–7 Mass and Weight

14E. What are the mass and weight of (a) a 1400-lb snowmobile and (b) a 400-kg heat pump?

15E. What is the weight in newtons and the mass in kilograms of (a) a 5-lb bag of sugar, (b) a 240-lb fullback, and (c) a 1.8-ton automobile? (1 ton = 2000 lb.)

16E. A space traveler whose mass is 75 kg leaves the Earth. Compute his weight (a) on Earth, (b) on Mars, where $g = 3.8$ m/s², and (c) in interplanetary space. (d) What is his mass at each of these locations?

17E. A certain particle has a weight of 20 N at a point where the acceleration due to gravity is 9.8 m/s². (a) What are the weight and mass of the particle at a point where the acceleration due to gravity is 4.9 m/s²? (b) What are the weight and mass of the particle if it is moved to a point in space where the gravitational force is zero?

Section 5–9 Applying Newton's Laws of Motion

18E. A 12,000-kg airplane is in level flight at a speed of 900 km/h. What is the upward–directed lift force exerted by the air on the airplane?

19E. A small 150-g pebble is 3.4 km deep in the ocean and is falling with a constant terminal speed of 25 m/s. What force does the water exert on the falling pebble?

20E. What is the net force acting on a 3800-lb automobile accelerating at 12 ft/s²?

21E. In a modified tug-of-war game, two people pull in opposite directions, not on a rope, but on a 25-kg sled resting on an icy road. If the participants exert forces of 90 N and 92 N, what is the acceleration of the sled?

22E. A 400-lb motorcycle accelerates from rest to 50 mi/h in 6.0 s. (a) What is the acceleration of the motorcycle? (b) What net force (assumed constant) acts on it?

23E. Refer to Fig. 13. Let $m = 2.0$ kg and $M = 4.0$ kg. Find (a) the tension in the string and (b) the acceleration of the two blocks.

24E. A body with mass m is acted on by two forces F_1 and F_2, as shown in Fig. 30. If $m = 5.2$ kg, $F_1 = 3.7$ N, and $F_2 = 4.3$ N, find the vector acceleration of the body.

Figure 30 Exercise 24.

25E. Refer to Fig. 19. Let the mass of the block be 8.5 kg and the angle θ equal 30°. Find (a) the tension in the string and (b) the normal force acting on the block. (c) If the string is cut, find the acceleration of the block.

26E. A jet plane starts from rest on the runway and accelerates for takeoff at 2.3 m/s². It has two jet engines, each of which exerts a thrust of 1.4×10^5 N (≈ 16 tons). What is the weight of the plane?

27E. *Sunjamming.* The sun yacht *Diana,* designed to navigate in the solar system using the pressure of sunlight, has a sail area of 3.0 km² and a mass of 900 kg. Near the earth's orbit, the sun could exert a radiation force of 20 N on its sail. (a) What acceleration would such a force impart to the craft? (b) A small acceleration can produce large effects if it acts steadily for a long enough time. Starting from rest then, how far would the craft have moved after 1 day under these conditions? (c) What would then be its speed? (See "The Wind from the Sun," a fascinating science fiction account by Arthur C. Clarke of a sun yacht race.)

28E. What strength fishing line is needed to stop a 19-lb salmon swimming at 9.2 ft/s in a distance of 4.4 in.?

29E. An electron travels in a straight line from the cathode of a vacuum tube to its anode, which is exactly 1.5 cm away. It starts with zero speed and reaches the anode with a speed of 6.0×10^6 m/s. (a) Assume constant acceleration and compute the force on the electron. The electron's mass is 9.11×10^{-31} kg. This force is electrical in origin. (b) Calculate the gravitational force on the electron.

30E. An electron is projected horizontally at a speed of 1.2×10^7 m/s into an electric field that exerts a constant vertical force of 4.5×10^{-16} N on it. The mass of the electron is 9.11×10^{-31} kg. Determine the vertical distance the electron is deflected during the time it has moved forward 30 mm horizontally.

31E. A car moving initially at a speed of 50 mi/h (≈ 80 km/h) and weighing 3000 lb ($\approx 13,000$ N) is brought to a stop in a distance of 200 ft (≈ 61 m). Find (a) the braking force and (b) the time required to stop. Assuming the same braking force, find (c) the distance and (d) the time required to stop if the car were going 25 mi/h ($= 40$ km/h) initially.

32E. Determine the frictional force of the air on a meteor (a small stone that enters the earth's atmosphere from outer space; see Fig. 31) of mass 0.25 kg falling with an acceleration of 9.2 m/s².

Figure 31 Exercise 32.

33E. Compute the initial upward acceleration of a rocket of mass 1.3×10^4 kg if the initial upward thrust of its engine is 2.6×10^5 N. Do not neglect the weight of the rocket (the downward pull of the earth on it).

34E. A rocket and its payload have a total mass of 50,000 kg. How large is the thrust of the rocket engine when (a) the rocket is "hovering" over the launch pad, just after ignition, and (b) when the rocket is accelerating upward at 20 m/s²?

35P. A 160-lb firefighter slides down a vertical pole with an acceleration of 10 ft/s², directed downward. (a) What upward vertical force does the pole exert on the firefighter? (b) What force is exerted on the pole by the firefighter?

36P. A worker drags a crate across a factory floor by pulling on a rope tied to the crate. The worker exerts a force of 450 N on the rope, which is inclined at 38° to the horizontal. The floor exerts a horizontal resistive force of 125 N, as shown in Fig. 32. Calculate the acceleration of the crate (a) if its mass is 96 kg and (b) if its weight is 96 N.

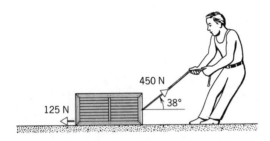

Figure 32   Problem 36.

37P. An 8.5-kg object passes through the origin with a velocity of 30 m/s parallel to the x axis. It experiences a constant 17-N force in the direction of the positive y axis. Calculate (a) the velocity and (b) the position of the particle after 15 s have elapsed.

38P. An elevator and its load have a combined mass of 1600 kg. Find the tension in the supporting cable when the elevator, originally moving downward at 12 m/s, is brought to rest with contant acceleration in a distance of 42 m.

39P. An object is hung from a spring balance attached to the ceiling of an elevator. The balance reads 65 N when the elevator is standing still. (a) What is the reading when the elevator is moving upward with a constant speed of 7.6 m/s? (b) What is the reading of the balance when the elevator is moving upward with a speed of 7.6 m/s and decelerating at a rate of 2.4 m/s²?

40P. A 1400-kg jet engine is fastened to the fuselage of a passenger jet by just three bolts (this is the usual practice). Assume that each bolt supports one-third of the load. (a) Calculate the force on each bolt as the plane waits in line for clearance to take off. (b) During flight, the plane encounters turbulence, which suddenly imparts an upward vertical acceleration of 2.6 m/s² to the plane. Calculate the force on each bolt now.

41P. A 15,000-kg helicopter is lifting a 4500-kg car with an upward acceleration of 1.4 m/s². Calculate (a) the force the air exerts on the helicopter blades and (b) the tension in the upper supporting cable; see Fig. 33.

Figure 33   Problem 41.

42P. A man of mass 80 kg (weight mg = 176 lb) jumps down to a concrete patio from a window ledge only 0.50 m (= 1.6 ft) above the ground. He neglects to bend his knees on landing, so that his motion is arrested in a distance of 2.0 cm (= 0.79 in.). (a) What is the acceleration of the man from the time his feet first touch the patio to the time he is brought fully to rest? (b) With what force does this jump jar his bone structure?

43P. Three blocks are connected, as shown in Fig. 34, on a horizontal frictionless table and pulled to the right with a force T_3 = 65 N. If m_1 = 12 kg, m_2 = 24 kg, and m_3 = 31 kg, calculate (a) the acceleration of the system and (b) the tensions T_1 and T_2. Draw an analogy to bodies being pulled in tandem, such as an engine pulling a train of coupled cars.

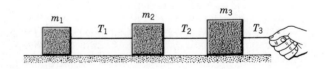

Figure 34   Problem 43.

44P. An elevator weighing 6000 lb is pulled upward by a cable with an acceleration of 4.0 ft/s². (a) Calculate the tension in the cable. (b) What is the tension when the elevator is decelerating at 4.0 ft/s² but is still moving upward?

45P. An 80-kg person is parachuting and experiencing a downward acceleration of 2.5 m/s². The mass of the parachute is 5.0 kg. (a) What is the value of the upward force exerted on the parachute by the air? (*b*) What is the value of the downward force exerted by the person on the parachute?

46P. A 100-kg man lowers himself to the ground from a height of 10 m by means of a rope passed over a frictionless pulley and attached to a 70-kg sandbag. (*a*) With what speed does the man hit the ground? (*b*) Is there anything he could do to reduce the speed with which he hits the ground?

47P. A new 26-ton Navy jet (Fig. 35) requires an airspeed of 280 ft/s for lift-off. Its own engine develops a thrust of 24,000 lb. The jet is to take off from an aircraft carrier with a 300-ft flight deck. What force must be exerted by the catapult of the carrier? Assume that the catapult and the jet's engine each exert a constant force over the 300-ft takeoff distance. (1 ton = 2000 lb.)

Figure 35 Problem 47.

48P. A 60-kg woman stands in an elevator that is accelerating downward at 3.1 m/s². What force does she exert on the elevator floor?

49P. A landing craft approaches the surface of Callisto, one of the satellites (moons) of the planet Jupiter (Fig. 36). If an upward thrust of 3260 N is supplied by the rocket engine, the craft descends with constant speed. Callisto has no atmosphere. If the upward thrust is 2200 N, the craft accelerates downward at 0.39 m/s². (*a*) What is the weight of the landing craft in the vicinity of Callisto's surface? (*b*) What is the mass of the craft? (*c*) What is the acceleration due to gravity near the surface of Callisto?

50P. What is the minimum acceleration with which a 50-kg person can slide down a rope that can withstand a maximum tension of 425 N without breaking?

51P. A chain consisting of five links, each of mass 0.10 kg, is lifted vertically with a constant acceleration of 2.5 m/s², as

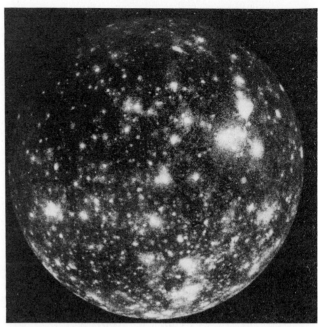

Figure 36 Problem 49.

shown in Fig. 37. Find (*a*) the forces acting between adjacent links, (*b*) the force **F** exerted on the top link by the agent lifting the chain, and (*c*) the *net* force acting on each link.

Figure 37 Problem 51.

52P. A block of mass $m_1 = 3.7$ kg on a smooth inclined plane of angle 30° is connected by a cord over a small frictionless pulley to a second block of mass $m_2 = 2.3$ kg hanging vertically (Fig. 38). (*a*) What is the acceleration of each block? (*b*) What is the tension in the cord?

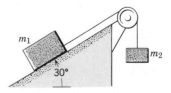

Figure 38 Problem 52.

53P. How could a 100-lb object be lowered from a roof using a cord with a breaking strength of 87 lb without breaking the cord?

54P. A block is projected up a frictionless inclined plane with a speed v_0. The angle of incline is θ. (*a*) How far up the plane does it go? (*b*) How long does it take to get there? (*c*) What is its speed when it gets back to the bottom? Find numerical answers for $\theta = 32°$ and $v_0 = 3.5$ m/s.

55P. A lamp hangs vertically from a cord in a descending elevator. The elevator has a deceleration of 8.0 ft/s² (= 2.4 m/s²) before coming to a stop. (*a*) If the tension in the cord is 20 lb (=89 N), what is the mass of the lamp? (*b*) What is the tension in the cord when the elevator ascends with an acceleration of 8.0 ft/s² (=2.4 m/s²)?

56P. A 100-kg crate is pushed at constant speed up the smooth 30° ramp shown in Fig. 39. (*a*) What horizontal force **F** is required? (*b*) What force is exerted by the ramp on the crate?

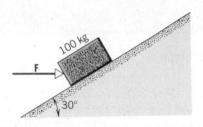

Figure 39 Problem 56.

57P. A 10-kg monkey is climbing a massless rope attached to a 15-kg mass over a (frictionless!) tree limb. (*a*) With what minimum acceleration must the monkey climb up the rope so that it can raise the 15-kg mass off the ground? If, after the mass has been raised off the ground, the monkey stops climbing and holds on to the rope, what will now be (*b*) the monkey's acceleration and (*c*) the tension in the rope?

58P. Figure 40 shows a section of an alpine cable-car system.

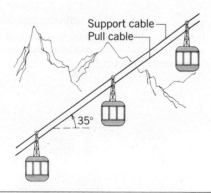

Figure 40 Problem 58.

The maximum permitted mass of each car with occupants is 2800 kg. The cars, riding on a support cable, are pulled by a second cable attached to each pylon. What is the difference in tension between adjacent sections of pull cable if the cars are accelerated up the 35° incline at 0.81 m/s²?

59P. (*a*) Neglecting gravitational forces, what force would be required to accelerate a 1200-metric ton spaceship from rest to one-tenth the speed of light in 3 days? In 2 months? (One metric ton = 1000 kg.) (*b*) Assuming that the engines are shut down when this speed is reached, what would be the time required to complete a 5-light-month journey for each of these two cases?

60P. An elevator consists of the elevator cage (*A*), the counterweight (*B*), the driving mechanism (*C*), and the cable and pulleys as shown in Fig. 41. The mass of the cage is 1100 kg and the mass of the counterweight is 1000 kg. Neglect friction and the mass of the cable and pulleys. The elevator accelerates upward at 2.0 m/s² and the counterweight accelerates downward at the same rate. What are the values of the tensions (*a*) T_1 and (*b*) T_2? (*c*) What force is exerted on the cable by the driving mechanism?

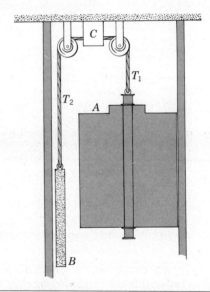

Figure 41 Problem 60.

61P. A 5.0-kg block is pulled along a horizontal frictionless floor by a cord that exerts a force $P = 12$ N at an angle $\theta = 25°$ above the horizontal, as shown in Fig. 42. (*a*) What is the

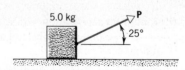

Figure 42 Problem 61.

acceleration of the block? (*b*) The force *P* is slowly increased. What is the value of *P* just before the block is lifted off the floor? (*c*) What is the acceleration of the block just before it is lifted off the floor?

62P. In earlier days, horses pulled barges down canals in the manner shown in Fig. 43. Suppose that the horse is exerting a force of 7900 N at an angle of 18° to the direction of motion of the barge, which is headed straight along the canal. The mass of the barge is 9500 kg and its acceleration is 0.12 m/s². Calculate the force exerted by the water on the barge.

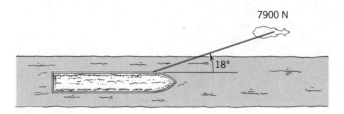

Figure 43 Problem 62.

63P. A certain force gives mass m_1 an acceleration of 12 m/s². The same force gives mass m_2 an acceleration of 3.3 m/s². What acceleration would the force give to an object whose mass is (*a*) the difference between m_1 and m_2, and (*b*) the sum of m_1 and m_2?

64P. A research balloon of total mass *M* is descending vertically with downward acceleration *a* (Fig. 44). How much ballast must be thrown from the car to give the balloon an *upward* acceleration *a*, presuming that the upward lift of the balloon does not change?

Figure 44 Problem 64.

65P. A rocket with mass 3000 kg is fired from rest from the ground at an angle of elevation of 60°. The motor exerts a thrust of 6.0×10^4 N at a constant angle of 60° to the horizontal for 50 s and then cuts out. Ignore the mass of fuel consumed and neglect aerodynamic drag. Calculate (*a*) the altitude of the rocket at motor cutout and (*b*) the total horizontal distance from firing point to impact.

66P. A block of mass *M* is pulled along a horizontal frictionless surface by a rope of mass *m*, as shown in Fig. 45. A horizontal force **P** is applied to one end of the rope. (*a*) Show that the rope *must* sag, even if only by an imperceptible amount. Then, assuming that the sag is negligible, find (*b*) the acceleration of rope and block, (*c*) the force that the rope exerts on the block *M*, and (*d*) the tension in the rope at its midpoint.

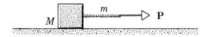

Figure 45 Problem 66.

67P. A man sits in a bosun's chair supported by a light rope passing over a pulley, as shown in Fig. 46. The man pulls on the free end of the rope to lift himself. (*a*) If the mass of the man and chair together is 95 kg, with what force must he pull to raise himself at constant speed? (*b*) With what force must he pull if he desires an upward acceleration of 1.3 m/s²? Ignore friction and the mass of the pulley.

Figure 46 Problem 67.

CHAPTER 6
FORCE AND MOTION—II

This man is pedaling at constant speed and yet he is accelerating. In what direction? Where is the force that provides this acceleration? (It cannot be his weight or the upward force exerted by the ground; both are vertical and he does not accelerate in this direction.) Why doesn't he fall over?

6–1 Friction

The cyclist pictured above is rounding a curved track, its center of curvature being off to his left. As we shall see later, it is a *force of friction,* exerted horizontally by the roadway on his tires and pointing to the right in the figure, that allows him to pedal round the curve.

Although we have ignored its effects up to now, friction is important in our daily lives. Left to act alone, it would stop every rolling wheel, bring to a halt every rotating shaft, and wear away every coin. In an automobile, about 20% of the gasoline is used to counteract friction in the engine and in the drive train. On the other hand, if friction were totally absent, we could not walk or ride a bicycle. We could not hold a pencil and, if we could, it would not write. Nails and screws would be useless, woven cloth would fall apart, and knots would come undone.*

Water exerts a frictional drag force on a swimmer. The atmosphere impedes the motion of an airplane but gives the parachute its legitimacy. Fluids such as lubricating oil can be used to reduce the friction between surfaces in relative motion, as at the axle of a wheel. Air can be used as an almost ideal "lubricant," as in the

---

* It is pointless to think of a totally frictionless world at this level. Friction occurs because atoms attract each other. If they did not, there would be no molecules, no liquids, no solids (and no physics instructors).

almost frictionless air track gliders that you may have seen in your physics lab or lecture room. There is the frictional force exerted by the road on the tires of your car, propelling it forward. There is the (tiny) rolling friction that accounts for the effectiveness of ball bearings.

In this chapter, we deal largely with the frictional forces that exist between dry solid surfaces, moving across each other at relatively slow speeds. Consider two simple experiments:

1. First Experiment. Slide a book across a table top. A frictional force, exerted by the table top on the bottom of the sliding book acts to slow it down and eventually to bring it to rest. If you want to make the book move across the table with constant velocity, you must push or pull it across by applying a steady external force to it, just large enough to counteract the frictional force.

2. Second Experiment. A heavy crate is resting on the floor of a warehouse. You push on it horizontally with a steady force but it does not move. That is because the force that you apply is balanced by a frictional force, exerted horizontally on the bottom of the crate by the floor. Remarkably, this frictional force automatically adjusts itself, both in magnitude and direction, to cancel exactly whatever force you decide to apply. Of course, if you push hard enough, you may succeed in moving the crate (see the first experiment) but we are talking about the case in which you do not.

Figure 1 shows the situation in detail. In Fig. 1*a*, a block rests on a table top, its weight **W** balanced by an equal but opposite normal force **N**. In Fig. 1*b*, you exert an external force **F** on the block, tending to pull it to the left. In response, a frictional force $\mathbf{f}_s$ arises, pointing to the right and exactly compensating for the force that you have applied.

Figures 1*c* and 1*d* show that, as you increase your applied force, the frictional force $\mathbf{f}_s$ increases also and the block remains at rest. At a certain value of the external force, however, the block will "break away" from its intimate contact with the table top and will accelerate, moving to the left as in Fig. 1*e*.

If you wish the block to move across the surface with a constant speed, you must usually decrease the applied force, as in Fig. 1*f*. That is, the *kinetic frictional force* (associated with motion) is usually less than the maximum value of the *static frictional force,* which acts when there is no motion.

Figure 1*g* shows the results of an experiment in which the force on a 400-g block was slowly increased until breakaway occurred. The reduced force needed to

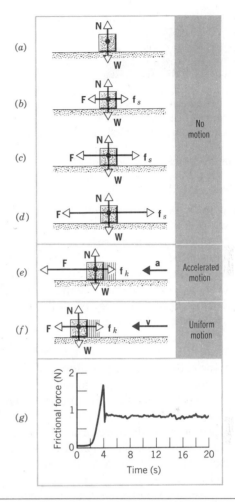

Figure 1 (*a–d*) An external force **F**, applied to a resting block, is counterbalanced by an equal but opposite frictional force $\mathbf{f}_s$. As **F** is increased, $\mathbf{f}_s$ also increases, until $\mathbf{f}_s$ reaches a certain maximum value. (*e*) The block then "breaks away," accelerating suddenly to the left. (*f*) If the block is now to move with constant velocity, the applied force **F** must be reduced from the maximum value it had just before the block broke away. (*g*) Some experimental results.

keep the block moving at constant speed is clearly displayed.*

The transition from the static to the kinetic frictional force, though it may seem abrupt, is nevertheless continuous. In fact, kinetic friction for dry surfaces at low speeds proceeds by a "stick-and-slip" process, the

---

* See "Undergraduate Computer-Interfacing Projects," by Joseph Priest and John Snider, *The Physics Teacher,* May 1987.

Figure 2 A magnified section of a highly polished steel surface. The vertical scale of the surface irregularities is several thousand atomic diameters. The section has been cut at an angle so that the vertical scale is exaggerated with respect to the horizontal scale by a factor of 10.

slip being much like the initial breakaway from the static condition. It is these repetitive stick-and-slip events that cause dry surfaces moving across each other to squeak. Thus, we have the squealing of skidding tires on a dry pavement, the squeaking of rusty hinges, and the painful sound made by chalk dragged across the chalk board by a skillful lecturer. Sometimes, dry friction leads to pleasant sounds, as when a bow sticks and slips its way over a violin string.

Basically, the frictional force is an electromagnetic force, acting between the surface atoms of one body and those of another. If two highly polished metal surfaces are brought together in a vacuum, there is no question of their sliding over each other. They weld together instantly, forming a single piece of metal. If machinist's polished gage blocks are placed flat surface to flat surface in air, they stick firmly to each other and can only be separated by means of a wrenching motion. Under ordinary circumstances, however, such intimate atom-to-atom contact is not possible. Even a highly polished metal surface, as Fig. 2 shows, is far from plane on the atomic scale. Moreover, the surface layers of every object that we come in contact with are formed of oxides and other contaminants.

When two surfaces are placed together, only the high points touch each other. It has been said* that putting two solids together "is rather like turning Switzerland upside down and standing it on Austria." The actual microscopic area of contact is much less than the apparent macroscopic contact area, perhaps by a factor of 10^4 or so. Many contact points actually become *cold welded* together. When the surfaces are pulled across each other, the frictional force that acts on each is associated with the rupturing of thousands of these tiny

* See the informative article on "Friction" in the *Encyclopedia Britannica,* 14th ed.

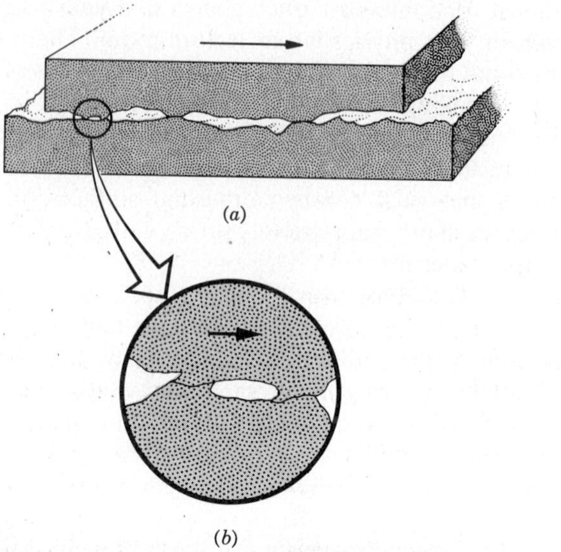

(a)

(b)

Figure 3 The mechanism of sliding friction. (a) The upper surface is sliding to the right over the lower surface in this enlarged view. (b) A detail, showing two spots where cold welding has occurred. Force is required to break these welds and maintain the motion.

welds, which continually reform as additional chance contacts are made; see Fig. 3.

Experiments with radioactive tracers show that, when one dry metal surface is dragged across another, tiny metal fragments are indeed torn from each surface. The wear of piston rings has been tested by using rings that have been made radioactive by exposure in a nuclear reactor. Material torn from the rings is carried away by the lubricating oil, where it can be detected by its radioactivity.

6–2 The Laws of Friction

Experiment shows that, when dry, unlubricated, solid surfaces slide over each other, the magnitude of the frictional force **f** is given by the following laws:

$$\boxed{f \le \mu_s N} \quad \text{(static friction)*} \qquad (1)$$

and

$$\boxed{f = \mu_k N} \quad \text{(kinetic friction).} \qquad (2)$$

Here N is the magnitude of the *normal force,* that is, the perpendicular force with which each surface presses on the other. The (dimensionless) coefficients μ_s and μ_k are, respectively, the coefficient of static friction and the coefficient of kinetic friction. These coefficients are reasonably constant and independent of the area of contact and (for the kinetic case) the speed of the relative motion. For the kinetic case, the direction of the frictional force on a given body is always in the direction of the relative velocity of the opposing surface.

Equations 1 and 2 are relations between the magnitudes only of the normal and the frictional forces; these forces are always directed at right angles to each other. Both equations are examples of the force laws that we discussed in Section 5–1. These particular force laws, which are practical summaries of a variety of data from many sources, are not as firmly based as, for example, Newton's law of gravity (which controls the motions of the solar system) or Coulomb's law of electrostatics (which acts within the atom to establish its structure). Even so, the laws of friction are accurate enough to be useful if they are applied in situations that are not too different from those of the experiments used to measure the value of μ.

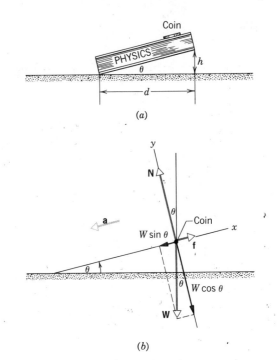

Figure 4   Sample Problem 1. (*a*) A coin is about to slide from rest down the cover of a book. (*b*) A free-body diagram for the coin, showing the three forces (drawn to scale) that act on it. The weight **W** is shown resolved into its components along the x and the y axes.

Sample Problem 1   Figure 4*a* shows a coin resting on a book that has been tilted at an angle θ with the horizontal. By trial and error you find that, when θ is increased to 13°, the coin begins to slide down the book. What is the coefficient of static friction between the coin and the book?

Figure 4*b* shows a free-body diagram for the coin at the moment that sliding is about to take place. The forces that act on the coin are the normal force **N**, pushing outward from the plane of the book, the weight **W** of the coin, and the frictional force **f**, acting on the coin and directed up the plane. Because the coin is in equilibrium, the resultant external force acting on it must be zero, or

$$\sum \mathbf{F} = \mathbf{f} + \mathbf{W} + \mathbf{N} = 0. \tag{3}$$

The x component of this vector equation gives us

$$\sum F_x = f - W \sin \theta = 0 \quad \text{or} \quad f = W \sin \theta. \tag{4}$$

For the y component we have

$$\sum F_y = N - W \cos \theta = 0 \quad \text{or} \quad N = W \cos \theta. \tag{5}$$

When the coin is just on the point of sliding, the static frictional force acting on it has its maximum value. Thus, from Eq. 1, $f = \mu_s N$. Substituting this into Eq. 4 and dividing by Eq. 5, we obtain

$$\frac{f}{N} = \frac{\mu_s N}{N} = \frac{W \sin \theta}{W \cos \theta} = \tan \theta$$

or

$$\mu_s = \tan \theta = \tan 13° = 0.23. \quad \text{(Answer)} \tag{6}$$

Why not try to measure μ_s for a coin and this book? You do not need a protractor. Tan θ is the ratio h/d of the two lengths shown in Fig. 4*a* and you can measure these lengths directly with a ruler.

Sample Problem 2   Renaldo is driving his car at 24 m/s (54 mi/h). What is the shortest distance d in which he can brake to a halt? The coefficient of static friction between the tires and the road surface μ_s is 0.6, a typical value.

Figure 5 shows the car and its free-body diagram. Although the frictional force and the normal force both act on the car at the bottoms of the four tires, we have concentrated each into a single force, as shown.

From Eq. 15 of Chapter 2 [$v^2 = v_0^2 + 2a(x - x_0)$], we have, putting $v = 0$ and $x - x_0 = d$,

$$d = -\frac{v_0^2}{2a}. \tag{7}$$

Applying Newton's second law to the y component of the motion tells us that $W = N$. From the x component, we learn

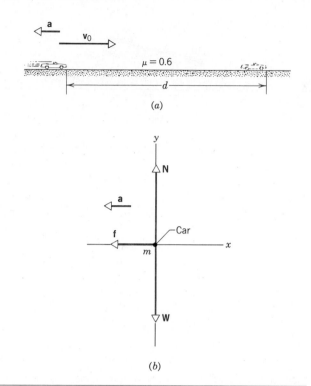

(a)

(b)

Figure 5 Sample Problem 2. (a) A car, moving to the right and braking to a halt after a distance d. (b) A free-body diagram for the car, the three force vectors being drawn to scale. The acceleration of the car points to the left, in the direction of the frictional force **f**.

(because the force acting in this direction is that due to friction) that

$$-f = ma = \frac{W}{g} a = \frac{N}{g} a$$

or

$$a = -g \frac{f}{N} = -g\mu_s. \qquad (8)$$

Substituting this result into Eq. 7 yields

$$d = -\frac{v_0^2}{2a} = \frac{v_0^2}{2g\mu_s} = \frac{(24 \text{ m/s})^2}{(2)(9.8 \text{ m/s}^2)(0.6)}$$
$$\approx 50 \text{ m.} \qquad \text{(Answer)} \quad (9)$$

We should not be too precise about the answer because measured coefficients of friction are rarely reliable to more than a single significant figure. Note that the stopping distance is proportional to the *square* of the speed. If you double your speed, the stopping distance increases by a factor of 4.

You may wonder why we used the static, rather than the kinetic, coefficient of friction in Eq. 9. After all, the car is moving while it is braking to rest. However, if the car does not

skid, there is no relative movement between the bottoms of the tires and the roadway so that the static coefficient is appropriate. We asked for the *minimum* stopping distance and the trick here is to keep the car just on the verge of skidding. If the driver wishes to brake to a halt in a greater distance than this minimum, he can easily do so by exerting a steady light pressure on the brake pedal; the static frictional force exerted by the road on the tires will automatically adjust to an appropriate lower value. If skidding occurs, there *is* relative motion between the road and the tires and we must use the kinetic coefficient of friction (which is less than the static coefficient) in Eq. 9.

Note that the weight of the car does not appear in Eq. 9. It is true that a heavier car requires a larger (frictional) force to stop it in a given distance. On the other hand, the heavier the car the greater the normal force, so this increased frictional force is automatically provided.

Sample Problem 3 A girl pulls a loaded sled, whose mass m is 75 kg, along a horizontal surface at constant speed. The coefficient of kinetic friction μ_k between the runners and the snow is 0.1 and the angle ϕ in Fig. 6 is 42°. (a) What is the tension in the rope?

Figure 6b shows the free-body diagram for the sled. Applying Newton's second law in the horizontal direction yields

$$T \cos \phi - f = ma_x = 0. \qquad (10)$$

For the vertical direction, we have

$$T \sin \phi + N - mg = ma_y = 0, \qquad (11)$$

in which mg is the weight of the sled. From Eq. 2 we have

$$f = \mu_k N. \qquad (12)$$

These three equations contain T, N, and f as unknowns. Eliminating N and f will allow us to find the remaining variable T.

We start by adding Eqs. 10 and 12, obtaining

$$T \cos \phi = \mu_k N,$$

or

$$N = \frac{T \cos \phi}{\mu_k}. \qquad (13)$$

Substituting this value of N into Eq. 11 and solving for T, we obtain

$$T = \frac{\mu_k mg}{\cos \phi + \mu_k \sin \phi} \qquad (14)$$
$$= \frac{(0.1)(75 \text{ kg})(9.8 \text{ m/s}^2)}{\cos 42° + (0.1)(\sin 42°)}$$
$$= 91 \text{ N.} \qquad \text{(Answer)}$$

(b) What is the normal force with which the ground pushes vertically upward on the sled?

As Fig. 7b suggests, the normal force is less than the weight of the sled, part of the weight being compensated by the

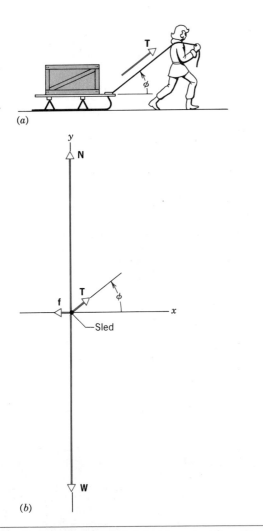

(a)

(b)

Figure 6 Sample Problem 3. (*a*) A girl exerts a force **T** on a sled, pulling it at a constant velocity. (*b*) A free-body diagram for the sled. The four forces are drawn to scale. Note that the frictional force and the tension in the rope are substantially less than the weight of the sled.

upward component of the tension in the rope. Substituting T from Eq. 14 into Eq. 13 yields

$$N = \frac{\cos \phi}{\cos \phi + \mu \sin \phi}\, mg \tag{15}$$

$$= \frac{\cos 42°}{\cos 42° + (0.1)(\sin 42°)}\, mg = 0.92\, mg. \quad \text{(Answer)}$$

Thus, because of the upward component of the pull on the sled, the normal force is reduced to 92% of the weight of the sled.

6-3   The Drag Force and Terminal Speed*

If a body moves through a fluid such as air or water, a frictionlike *drag force* will act on it, tending to retard its motion. We may be dealing with a cannon ball sinking in the ocean, a power boat crossing a lake, a truck driving along a highway, a bumblebee or a jet plane in flight, a baseball rising in the air, or a parachutist drifting down toward the earth. A typical small dust particle is so slowed down as it falls that it may take an hour to fall 1 ft. Since the air in our houses is never still, much dust never settles.

We limit ourselves to the case of bodies falling through air and moving at speeds high enough so that the flow of the air behind the falling body is turbulent. A falling baseball, sky diver, or parachutist meets these conditions. In these circumstances, the drag force acting on the body is given by

$$D = \tfrac{1}{2}C\rho A v^2. \tag{16}$$

Here A is the effective cross-sectional area of the falling body, ρ is the density of air, and v is the speed of fall. C is a dimensionless *drag coefficient* that depends on the shape of the moving object and whose value generally lies in the range 0.5–1.0.

When a body is released from rest, its initial drag force is zero. As it falls through the air, its speed increases from zero and so does its drag force. Eventually, as Fig. 7 suggests, its drag force becomes equal to its weight, at which point the net force acting on the body becomes zero. From Newton's second law, the acceleration of the body also becomes zero and the body falls at its constant *terminal speed* v_t.

At terminal speed, $D = mg$. From Eq. 16,

$$\tfrac{1}{2}C\rho A v_t^2 = mg$$

and

$$v_t = \sqrt{\frac{2mg}{C\rho A}}. \tag{17}$$

Table 1 shows some terminal speeds in air.

Equation 17 predicts that the terminal speed is

---

* See Essay 6, "Physics and Sports: The Aerodynamics of Projectiles," by Peter Brancazio, for a more general and complete treatment.

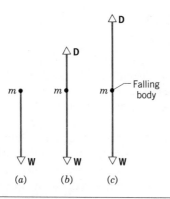

Figure 7 The forces that act on a body falling through air: (*a*) At the instant of release and (*b*) a little later, after a drag force has developed. (*c*) The drag force has increased until it balances the weight of the body. The body now falls at its (constant) terminal speed.

Table 1 Some Terminal Speeds in Air

Object	Terminal Speed (m/s)	95% Distance[a] (m)
16-lb Shot	145	2500
Sky diver (typical)	60	430
Baseball	42	210
Tennis ball	31	115
Basketball	20	47
Ping-Pong ball	9	10
Raindrop (radius = 1.5 mm)	7	6
Parachutist (typical)	5	3

[a] This is the distance through which the body must fall from rest to reach 95% of its terminal speed.
Source: Adapted from Peter J. Brancazio, *Sport Science,* Simon & Schuster, New York, 1984.

greater the smaller the effective area of the falling body. Figure 8 shows how downhill racing skiers assume the "egg position" to reduce their effective area. Figure 9 shows that the cross section of a highly streamlined aircraft wing is closely matched by that of a trout! Sky divers (see Fig. 10) often wish to maximize their effective area, thus reducing their terminal speed.

Equation 17 also shows that, although all bodies fall with the same acceleration in a vacuum, they do not do so in air. It can be shown that, if Galileo had dropped a

Figure 9 The curve is the profile of an efficient, low-drag aircraft wing. The dots suggest the outline of a trout, expanded to the same size. Humans and nature reach the same conclusions about streamlining.

Figure 8 The downhill skier Erwin Stricker in the "egg position" to minimize air resistance. Note also the low-friction clothing and the streamlined helmet.

Figure 10 A sky diver. She can control her terminal speed by adjusting the area she presents to the atmosphere through which she is falling.

lead ball and a wooden ball, each 15 cm in diameter, from the Leaning Tower at Pisa, the lead ball would have struck the ground first, the wooden ball being about 3 m behind it.

Sample Problem 4   A raindrop whose radius R is 1.5 mm falls from a cloud whose height h above the earth is 1200 m. (a) What is the terminal speed of the falling drop?

The mass m and the area A of the drop are

$$m = \tfrac{4}{3}\pi R^3 \rho_w \quad \text{and} \quad A = \pi R^2,$$

in which ρ_w (= 1000 kg/m³) is the density of water. Equation 17 then becomes, with $C = 0.6$ and the density of air written as ρ_a (= 1.2 kg/m³),

$$
\begin{aligned}
v_t &= \sqrt{\frac{8R\rho_w g}{3C\rho_a}} \\
&= \sqrt{\frac{(8)(1.5 \times 10^{-3}\ \text{m})(1000\ \text{kg/m}^3)(9.8\ \text{m/s}^2)}{(3)(0.6)(1.2\ \text{kg/m}^3)}} \\
&= 7.4\ \text{m/s} \ (= 17\ \text{mi/h}). \quad\quad \text{(Answer)}
\end{aligned}
$$

Note that the height of the cloud does not enter. The raindrop (see Table 1) would have reached terminal speed after falling just a few meters.

(b) What would have been the speed just before impact if there had been no drag force?

From Eq. 20 of Chapter 2,

$$
\begin{aligned}
v &= \sqrt{2gh} = \sqrt{(2)(9.8\ \text{m/s}^2)(1200\ \text{m})} \\
&= 153\ \text{m/s} \ (= 340\ \text{mi/h}). \quad\quad \text{(Answer)}
\end{aligned}
$$

Under these conditions, Shakespeare would scarcely have written, "it falleth like the gentle rain from heaven, upon the place beneath."

6–4   Uniform Circular Motion

If a particle moves in a circle (or in a circular arc) of radius r with a uniform* speed v, it is accelerated toward the center of that circle. The (constant) magnitude of this *centripetal acceleration* is given (see Eq. 16 of Chapter 4) by

$$a = \frac{v^2}{r}. \quad\quad (18)$$

---

*We treat the case in which the speed in circular motion is *not* uniform in Chapter 11. In such cases, the acceleration has a tangential, as well as a radial, component.

To account for the centripetal acceleration, a *centripetal force,* directed toward the center of the circle, must act on the body. The (constant) magnitude of this force follows from Newton's second law and is

$$F = ma = \frac{mv^2}{r}. \quad\quad (19)$$

If this force is not provided, there will be no uniform circular motion. Both the centripetal acceleration and the centripetal force are vectors whose magnitudes are constant but whose directions in space are changing continuously.

If the body in uniform circular motion is, say, a hockey puck whirled around on the end of a string as in Fig. 11, the centripetal force is provided by the tension in the string. For the moon in its (nearly) uniform circular motion around the earth, the centripetal force is supplied by the gravitational attraction of the earth. Thus, a centripetal force is not a new kind of force. A force can be a tension force, a gravitational force, or any other kind of force and also be (or not be) centripetal.

Let us compare and contrast two familiar examples of uniform circular motion:

1.   Rounding a Curve in a Car.   You are sitting in the center of the rear seat of a car. The driver suddenly turns left, rounding a corner, with tires squealing, in a wide circular arc. You find yourself sliding across the leather seat to the right and you remain jammed against the right interior wall of the car until the driver straightens out. What is going on?

The car, while it is moving in the circular arc, is in uniform circular motion. What is the source of the cen-

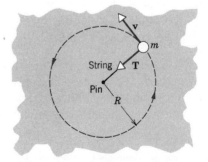

Figure 11   A hockey puck of mass m is whirled with constant speed in a circular path on a horizontal frictionless surface. The resultant external force acting on the puck is the centripetal force **T** with which the string pulls on it.

tripetal force that must act on it to make it move in this way? It is the sideways frictional force exerted by the (unbanked) roadway on the car through the bottoms of its tires. This force points radially inward and—added up over the four tires—must have a magnitude given by Eq. 19.

You yourself would have been in uniform circular motion in the center of the seat if the frictional force exerted on you by the seat had been great enough. However, it was not. Viewed from the reference frame of the ground, you continued moving in a straight line until the wall of the car (which was moving in a circle) reached you and pushed you toward the center of the circle.

2. Orbiting the Earth. This time you are a passenger in the space shuttle *Atlantis,* orbiting the earth and experiencing the familiar weightlessness that we have all seen on television. What is going on in this case?

Both you and the shuttle are in uniform circular motion. There must be a centripetal force acting on each of you to keep you moving in this way. What is its source? In this case it is the gravitational attraction of the earth, which acts directly on both you and the shuttle. This force is directed radially inward, toward the center of the earth, and has a magnitude given by Eq. 19.

In both car and shuttle you are in uniform circular motion, acted on by a centripetal force. Yet your kinesthetic experiences in the two situations are so different. In the car, jammed up against the wall, you are aware of internal uneasiness, of feeling "compressed." In the orbiting shuttle, on the other hand, you are floating around with no sensation of being acted on by a force of any kind. Why this great difference?

The difference is traceable to the nature of the two centripetal forces. In the car, the centripetal force is a contact force, exerted by the wall of the car externally on your body. In the shuttle, the centripetal force is a volume force, exerted by the earth's gravity on every atom of your body and on every atom of the shuttle, in proportion to the mass of that atom. Thus, no internal stresses develop, you experience no discomfort because of centripetal forces, and there is no need (and indeed no possibility) for the wall of the shuttle to press against you.

Sample Problem 5 Igor is a cosmonaut-engineer in the spacecraft *Vostok II,* orbiting above the earth at an altitude h of 520 km with a speed v of 7.6 km/s. Igor's mass is 79 kg. (a) What is his acceleration?

Igor is in uniform circular motion in a circle of radius $R + h$, where R is the radius of the earth. His centripetal acceleration is given by Eq. 18, or

$$a = \frac{v^2}{r} = \frac{v^2}{R + h}$$

$$= \frac{(7.6 \times 10^3 \text{ m/s})^2}{6.37 \times 10^6 \text{ m} + 0.52 \times 10^6 \text{ m}}$$

$$= 8.38 \text{ m/s}^2 \approx 8.4 \text{ m/s}^2. \qquad \text{(Answer)}$$

This is the value of the free-fall acceleration at this altitude. Imagine a tower 520 km high, on a nonrotating earth. The acceleration we have just calculated would be the initial acceleration of a bowling ball dropped from such a tower. Igor and the falling bowling ball *both* experience exactly the same free-fall acceleration toward the center of the earth. The difference between them is that Igor, in addition, is moving in orbit and the bowling ball is not.

(b) What (centripetal) gravitational force does the earth exert on Igor?

The centripetal force is

$$F = ma = (79 \text{ kg})(8.38 \text{ m/s}^2)$$

$$= 660 \text{ N} (= 150 \text{ lb}). \qquad \text{(Answer)}$$

If Igor were to stand on a scale placed on the top of the tower we have described in (a) above, the scale would read 150 lb. In orbit, the scale (assuming that Igor could "stand" on it) would read zero because both he and the scale are in free fall.

Hint 1: Looking things up. In Sample Problem 5, you needed to know the radius of the earth, which was not given in the problem statement. You need to become familiar with sources of this kind of information, starting with this book. Many useful data are given in the inside covers, in the various appendixes, and in the tables that appear throughout the text. The *Handbook of Chemistry and Physics,* published every year by the Chemical Rubber Company (CRC), is an invaluable resource.

For practice, see if you can track down the density of iron, the series expansion of e^x, the number of centimeters in a mile, the mean distance of Saturn from the sun, the mass of the proton, the speed of light, and the atomic number of samarium. You can find them all in this book.

Sample Problem 6 Figure 12*a* shows a *conical pendulum.* Its bob, whose mass m is 1.5 kg, whirls around in a horizontal circle at constant speed v at the end of a cord whose length L, measured to the center of the bob, is 1.7 m. The cord makes an

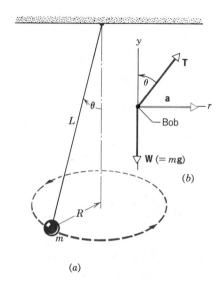

(a)

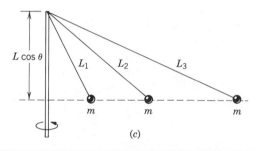

(b)

(c)

Figure 12 Sample Problem 6. (*a*) A conical pendulum, its cord making an angle θ with the vertical. (*b*) A free-body diagram for the pendulum bob. The axes point in the vertical and the radial directions. The resultant force (and thus the acceleration) points radially inward. (*c*) Three pendulums, of different lengths, are whirled around by a rotating vertical shaft; their bobs circulate in the same horizontal plane, as Eq. 24 predicts.

angle θ of 37° with the vertical. As the bob swings around, the cord sweeps out the surface of a cone. Find the period of the pendulum, that is, the time τ for the bob to make one complete revolution.

Figure 12*b* shows the free-body diagram for the bob. The forces that act on it are the tension **T** in the cord and the weight **W**($=m$**g**) of the bob, acting straight down. Applying Newton's second law in the vertical direction yields (bearing in mind that the acceleration a_y in this direction is zero)

$$T \cos \theta - mg = ma_y = 0 \quad \text{or} \quad T \cos \theta = mg. \quad (20)$$

In the radial direction we have, from Eq. 19,

$$T \sin \theta = \frac{mv^2}{R}, \quad (21)$$

where R is the radius of the circular path. Dividing Eq. 21 by Eq. 20 and solving for v, we obtain

$$v = \sqrt{\frac{gR \sin \theta}{\cos \theta}}. \quad (22)$$

We can substitute $2\pi R/\tau$ for the speed v of the bob. Doing so and solving for τ, we obtain

$$\tau = 2\pi \sqrt{\frac{R \cos \theta}{g \sin \theta}}. \quad (23)$$

However, from Fig. 12*a* we see that $R = L \sin \theta$. Making this substitution in Eq. 23 yields

$$\tau = 2\pi \sqrt{\frac{L \cos \theta}{g}} \quad (24)$$

$$= 2\pi \sqrt{\frac{(1.7 \text{ m})(\cos 37°)}{9.8 \text{ m/s}^2}} = 2.3 \text{ s.} \quad \text{(Answer)}$$

From Eq. 24 we see that the period τ does not depend on the mass of the bob, only on $L \cos \theta$, the vertical distance of the bob from the level of its point of support. Thus, if several conical pendulums of different lengths are swung from the same support with the same period, their bobs will all lie in the same horizontal plane, as Fig. 12*c* shows.

Sample Problem 7 Let the block in Fig. 13*a* represent a Cadillac whose mass m is 1610 kg moving at a constant speed v

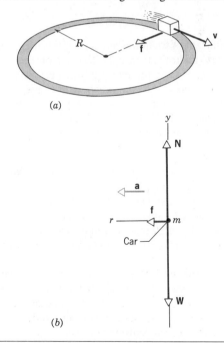

(a)

(b)

Figure 13 Sample Problem 7. (*a*) A car moving around a curved unbanked road at constant speed. Sideways road friction provides the necessary centripetal force. (*b*) A free-body diagram for the car.

of 72 km/h (or 20 m/s) on a curved unbanked roadway whose radius of curvature R is 190 m. What must be the minimum coefficient of friction μ_s between the tires and the roadway?

As Fig. 13b shows, two vertical forces act on the car, its weight $\mathbf{W}$ ($= m\mathbf{g}$) and the normal force $\mathbf{N}$ with which the roadway pushes up on the car. Because there is no acceleration in the vertical direction, these two forces must balance so that we have $N = W = mg$.

In the radial direction, however, the car has a centripetal acceleration given by Eq. 18 ($a = v^2/R$). The centripetal force associated with this acceleration is the frictional force with which the roadway pushes radially inward on the car, acting at the bottoms of its tires. This force is

$$f = \mu_s N = \mu_s mg. \qquad (25)$$

We use the static coefficient here because there is no motion in the radial direction between the roadway and the bottoms of the tires. We use the "=" sign in Eq. 25 because we seek the *minimum* value of the frictional coefficient. That is, we are dealing with the case in which the car is barely able to make the turn.

Newton's second law gives us (see Eq. 19)

$$f = ma = \frac{mv^2}{R}. \qquad (26)$$

Combining Eqs. 25 and 26 leads to

$$\mu_s = \frac{v^2}{gR} = \frac{(20 \text{ m/s})^2}{(9.8 \text{ m/s}^2)(190 \text{ m})} = 0.21. \quad \text{(Answer)} \quad (27)$$

If the coefficient of friction is not *at least* this high, the car will not be able to make the turn and will skid. Note that this critical coefficient depends on the *square* of the speed, so that—as all good drivers know—slowing down a little helps a lot. The mass of the car does not enter into Eq. 27 so that the same critical coefficient holds for a bicycle and a Cadillac provided (of course) that they are moving at the same speed.

Sample Problem 8   You cannot count on a sideways frictional force to get your car around a curve if the road is icy or wet. That is why highways are banked; see Fig. 14a. As in Sample Problem 7, suppose that a car of mass m is moving at a constant speed v of 20 m/s around a curve whose radius R is 190 m. What angle of bank θ would make reliance on friction unnecessary?

The centripetal acceleration and the required centripetal force are the same, whether or not the road is banked. The effect of banking is to tilt the normal force $\mathbf{N}$ toward the center of curvature of the road, so that its inward radial component can supply the needed centripetal force.

There is no acceleration in the vertical direction in Fig. 14 so that

$$N \cos \theta = W = mg. \qquad (28)$$

In the radial direction we have (see Eq. 19)

$$F = N \sin \theta = \frac{mv^2}{R}. \qquad (29)$$

Dividing Eq. 29 by Eq. 28 gives

$$\tan \theta = \frac{v^2}{gR}. \qquad (30)$$

Thus, we have

$$\theta = \tan^{-1} \frac{v^2}{gR}$$

$$= \tan^{-1} \frac{(20 \text{ m/s})^2}{(9.8 \text{ m/s}^2)(190 \text{ m})} \qquad (31)$$

$$= 12°. \qquad \text{(Answer)}$$

Comparing Eqs. 22, 27 and 31 shows that the critical coefficient of friction for an unbanked road is the same as the tangents of the bank angle for a banked road and the vertical angle of a conical pendulum.

In driving a car around a properly banked curve, at its

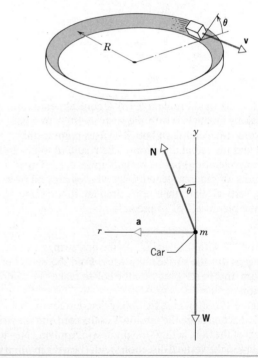

Figure 14   Sample Problem 8. (*a*) A car moving around a banked roadway at constant speed. The radially inward component of the normal force provides the necessary centripetal force. The bank angle is exaggerated for clarity. (*b*) A free-body diagram for the car. The resultant force (and thus the acceleration) point radially inward.

design speed, the driver is unaware of any sideways force, on the car or on himself. On an unbanked curve, where radial frictional forces are necessary, the driver is well aware that he is rounding a curve. The roadway exerts its sideways frictional force on the car and indirectly on him through the fabric of the seat cover.

Sample Problem 9   A Rotor is found in many amusement parks. It is a hollow cylindrical room that can be made to rotate about a central vertical axis, as in Fig. 15. A person enters the Rotor, closes the door, and stands up against the wall. The Rotor starts rotating and gradually increases its speed until, at a certain critical speed, the floor is dropped away, revealing a deep pit. The person does not fall but remains pinned to the wall, supported by an upward-directed frictional force. The static coefficient of friction between the canvas-covered Rotor wall and a person's clothing is 0.4 and the radius R of the Rotor is 2.1 m. (a) At what minimum rotational speed (measured in revolutions per minute) is it safe to drop the floor?

Figure 15 shows the situation just after the critical rotational speed has been reached. There is no acceleration in the vertical direction so that

$$f = \mu_s N = mg. \tag{32}$$

In the radial direction we have (see Eq. 19)

$$N = \frac{mv^2}{R}. \tag{33}$$

Dividing Eq. 33 by Eq. 32 and solving for v, the linear speed of

the Rotor wall, we obtain

$$v = \sqrt{\frac{gR}{\mu_s}}. \tag{34}$$

The circumference of the Rotor wall is $2\pi R$. Dividing this into the speed v of the Rotor wall gives the rotational frequency:

$$frequency = \frac{v}{2\pi R} = \frac{1}{2\pi} \sqrt{\frac{g}{\mu_s R}} \tag{35}$$

$$= \frac{1}{2\pi} \sqrt{\frac{(9.8 \text{ m/s}^2)}{(0.4)(2.1 \text{ m})}}$$

$$= 0.54 \text{ rev/s or } 33 \text{ rev/min.} \quad \text{(Answer)}$$

By chance, this is about the same rotational speed as a phonograph record.

(b) At this rotational speed, what centripetal acceleration does the rider experience?

Combining Eqs. 18 and 34 leads to

$$a = \frac{v^2}{R} = \frac{g}{\mu_s} = \left(\frac{1}{0.4}\right) g = 2.5g. \quad \text{(Answer)} \tag{36}$$

By comparison, a jet plane creates about $1.5g$ on takeoff and a roller coaster can produce an acceleration of about $4g$ on its sharpest turn. Jet fighter pilots routinely experience $6-9g$ turns and, in some cases, the accelerations may go as high as $11g$, equivalent to increasing their weight by this same factor. At higher accelerations, blackout occurs.

6-5   The Forces of Nature*   (Optional)

We have used **F** as a generic symbol for force. We have also used other symbols. Thus, **W** for the weight of a body, **T** for the tension in a cord, **f** for a frictional force, **N** for a normal force (such as keeps us from falling through the floor when we get out of bed in the morning), and **D** for the drag force exerted, for example, by the atmosphere on a skydiver. All these quantities are forces.

At the fundamental level, all the forces mentioned above fall into two categories: (1) the gravitational force, of which weight is our only example, and (2) the electromagnetic force, which includes — without exception — all the others. The force that makes a balloon stick to a wall and the force with which a magnet picks up an iron nail are other examples of the electromagnetic force. In

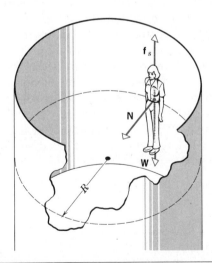

Figure 15   Sample Problem 9. A Rotor in an amusement park. The centripetal force is the normal force with which the wall pushes radially inward on the rider.

---

* See P. C. W. Davies, *The Forces of Nature,* 2nd ed., Cambridge University Press, New York, 1986, for a very readable account.

Table 2 The Quest for the Superforce—a Progress Report

Date	Researcher	Achievement
1687	Newton	Showed that the same laws applied to heavenly bodies and to objects on earth. He unified celestial and terrestrial mechanics.
1820 1830s	Oersted Faraday	They showed, by brilliant experiments, that the then separate sciences of electricity and magnetism were intimately linked.
1873	Maxwell	He showed that the sciences of electricity, magnetism, and optics could be unified into the single subject of electromagnetism, whose governing equations he codified.
1979	Glashow Salam Weinberg	In this year, these workers received the Nobel Prize for showing that the weak force and the electromagnetic force could be viewed as different aspects of a single *electroweak force.* This reduced the number of fundamental forces from four to three.
1984	Rubbia van der Meer	In this year, these workers received the Nobel Prize for verifying experimentally the predictions of the theory of the electroweak force.

———————————————————— Work in Progress (1988) ————————————————————

Grand unification theories: Called GUTs, these theories seek to unify the electroweak force and the strong force.
Supersymmetry theories: These theories seek to unify all forces, including gravity, within a single framework.
Superstring theories: These supersymmetry theories interpret particles as strings and assume a 10-dimensional universe.
Open experimental questions: Do protons decay? Do neutrinos have mass? Do magnetic monopoles exist? Where are the "missing" solar neutrinos? Is there a fifth force? (See "Is There a Fifth Fundamental Force?," by A. P. French, *The Physics Teacher,* May 1986.

fact, aside from the gravitational force, *all* forces that we can experience directly as a push or a pull are electromagnetic when looked at closely. That is, all such forces, including frictional forces, normal forces, contact forces, drag forces, and tension forces involve, at their deepest level, electromagnetic forces exerted by one atom on another. The tension in a rope, for example, is only maintained because the atoms of the rope attract each other.

Beyond the *gravitational force* and the *electromagnetic force,* only two other forces are known to occur in nature. Both have ranges of influence so short that we cannot experience them directly through our sensory perceptions. They are the *weak force,* which is involved in certain kinds of radioactive decay, and the *strong force,* which binds together the constituents of protons and neutrons and is the "glue" that holds the atomic nucleus together.

Physicists believe that nature has an underlying simplicity and that four forces—all so different—seem like too many. Einstein spent most of his working life trying to interpret these forces as different aspects of a single superforce. The quest continues today, at the very forefront of physics. Table 2 lists the progress that has been made in the direction of *unification* (as the goal is called) and gives some hints for the future.

The forces and the particles of nature are intimately connected with the origin and the development of the universe. With a force unification theory in hand, it should become possible—in principle—to answer questions such as, "How did our present universe evolve?" and "What will the universe be like in the future?" In Chapter 49 of the extended version of this book, we shall explore such questions further.

REVIEW AND SUMMARY

Static Friction

Suppose that a horizontal force **F** is applied to a block resting on a rough surface as in Fig. 1. As **F** is slowly increased from zero an opposing force **f** of equal magnitude automatically appears, associated with microscopic interlocking irregularities between the block and the surface at their areas of contact. This is the phenomenon of static friction. When the applied force reaches a certain critical value, the microscopic bonds break and the block begins to move.

Experimentally, the force of static friction turns out to be largely independent of the area of contact and proportional to the normal force N acting between the block and the surface. The law of static friction is

Law of Static Friction

$$f \leq \mu_s N, \qquad [1]$$

where μ_s is the coefficient of static friction, a dimensionless quantity that is approximately constant. See Sample Problems 1 and 2.

Law of Kinetic Friction

Once the block has been set into motion, the force **F** needed to keep it in motion with a constant velocity — that is, to maintain it in equilibrium — is usually less than the critical force needed to get the motion started. The magnitude of the opposing frictional force f is given by the law of kinetic friction:

$$f_k = \mu_k N. \qquad [2]$$

Here μ_k is the coefficient of kinetic friction. See Sample Problem 3.

An object moving through a fluid fast enough to create turbulence experiences a drag force D given by

Fluid Drag

$$D = \tfrac{1}{2} C \rho A v^2, \qquad [16]$$

in which C is a dimensionless *drag coefficient*, A is the effective cross-sectional area of the object, ρ is the density of the fluid, and v is the speed of the object. If the object falls under the influence of gravity, its motion is retarded by this fluid drag. The object initially experiences acceleration g, but its acceleration decreases as its speed increases until it reaches a *terminal speed*, which may be shown to be given by

Terminal Speed

$$v_t = \sqrt{\frac{2mg}{C \rho A}}. \qquad [17]$$

See Table 1 and Sample Problem 4.

Centripetal Force

An object that moves at constant speed v in a circle of radius R is said to be in uniform circular motion. The kinematics of such motion (see Section 4–7) indicate that the object has a centripetal acceleration **a** directed toward the center of the circle. The magnitude of **a** is $a = v^2/R$. Newton's second law yields, for the corresponding centripetal force,

$$F = ma = \frac{mv^2}{R}. \qquad [19]$$

Such a force may be provided, among other ways, by gravity (Sample Problem 5); by a taut string, as in a conical pendulum (Sample Problem 6); by a radially directed frictional force, as when a car drives around a curve on an unbanked roadway (Sample Problem 7), a normal force, as when a car drives around a banked curve (Sample Problem 8), or the wall of an amusement park Rotor (Sample Problem 9).

The Forces of Nature

There seem to be just four *fundamental forces* in nature: gravitational, electromagnetic, weak, and strong. We experience gravitational forces as weight. All the other forces we encounter in ordinary life (pushes and pulls, friction, normal forces, string tension, etc.) are electromagnetic forces arising from the charged particles of which matter is made. The strong interaction holds the constituents of atomic nuclei (protons and neutrons) together. The weak interaction governs radioactive beta decay of nuclei. The electromagnetic and weak forces can be viewed as different aspects of a single *electroweak* force. Current research continues the work of Albert Einstein in search of possible connections between the remaining three fundamental forces.

QUESTIONS

1. There is a limit beyond which further polishing of a surface *increases* rather than decreases frictional resistance. Explain why.

2. Can the coefficient of static friction have a value greater than 1? What about the coefficient of kinetic friction?

3. What are some examples of the beneficial uses of friction?

4. A crate, heavier than you are, rests on a rough floor. The coefficient of static friction between the crate and the floor is the same as that between the soles of your shoes and the floor. Can you push the crate across the floor? (See Fig. 16.)

Figure 16 Question 4.

5. Often a base runner can get to a base quicker by running than by sliding. Explain why this is so. Why slide then?

6. How could a person who is at rest on completely frictionless ice covering a pond reach shore? Could she do this by walking, rolling, swinging her arms, or kicking her feet? How could a person be placed in such a position in the first place?

7. If you want to stop a car in the shortest distance on an icy road, should you (*a*) push hard on the brakes to lock the wheels, (*b*) push as hard as possible without allowing slipping, (*c*) pump the brakes, or (*d*) do something else?

8. Why do tires grip the road better on level ground than they do when going uphill or downhill?

9. What is the purpose of curved surfaces, called spoilers, placed on the rear of sports cars? They are designed so that air flowing past them exerts a downward force.

10. Two surfaces are in contact but are at rest relative to each other. Nevertheless, each exerts a force of friction on the other. Explain how.

11. Which raindrops, if either, fall faster—small ones or large ones?

12. The terminal speed of a baseball is 95 mi/h. However, the measured speeds of pitched balls often exceed this, sometimes exceeding 100 mi/h. How can this be?

13. A log is floating downstream. How would you calculate the drag force acting on it?

14. Estimate the terminal speed of a baseball falling in a vacuum? Of a tennis ball in water? What happens to a baseball that is fired downward at twice its terminal speed?

15. Why are train roadbeds and highways banked on curves?

16. How does the earth's rotation affect the apparent weight of an object at the equator?

17. Why is it that racing drivers actually speed up when traversing a curve?

18. You are flying a plane at constant altitude and wish to make a 90° turn. Why is it useful to bank the plane as you do so?

19. When a wet dog shakes itself, people standing nearby tend to get wet. Why does the water fly outward from the dog in this way?

20. You are driving a station wagon at uniform speed along a straight highway. A beachball rests at the center of the wagon bed and a helium-filled balloon floats above it, touching the roof of the wagon. What happens to each if you (*a*) turn a corner or (*b*) apply the brakes?

21. You must have noticed (Einstein did) that when you stir a cup of tea, the floating tea leaves collect at the center of the cup rather than at the outer rim. Can you explain this? (Einstein could.)

22. Explain why a plumb bob will not hang exactly in the direction of the earth's gravitational attraction at most latitudes.

23. Suppose that you need to measure whether a table top in a train is truly horizontal. If you use a spirit level, can you determine this when the train is moving down or up a grade? When the train is moving along a curve? (*Hint:* There are two horizontal components.)

24. In the conical pendulum of Sample Problem 6, what happens to the period τ and the speed v when $\theta = 90°$? Why is this angle not achievable physically? Discuss the case for $\theta = 0°$.

25. A coin is put on a phonograph turntable. The motor is started but, before the final speed of rotation is reached, the coin flies off. Explain why.

26. A car is riding on a country road that resembles a roller coaster track. If the car travels with uniform speed, compare the force it exerts on a horizontal section of the road to the force it exerts on the road at the top of a hill and at the bottom of a hill. Explain.

27. A passenger in the front seat of a car finds himself sliding toward the door as the driver makes a sudden left turn. Describe the forces on the passenger and on the car at this instant if the motion is viewed from a reference frame (*a*) attached to the earth and (*b*) attached to the car.

28. Your car skids across the center line on an icy highway. Should you turn the front wheels in the direction of skid or in the opposite direction (*a*) when you want to avoid a collision with an oncoming car, (*b*) when no other car is near but you want to regain control of the steering? Assume rear-wheel drive.

EXERCISES AND PROBLEMS

Section 6-2 The Laws of Friction

1E. A bedroom bureau with a mass of 45 kg, including drawers and clothing, rests on the floor. (*a*) If the coefficient of static friction between the bureau and the floor is 0.45, what is the minimum horizontal force a person must apply to start the bureau moving? (*b*) If the drawers and clothing, with 17-kg mass, are removed before the bureau is pushed, what is the minimum horizontal force required to start it moving?

2E. A baseball player with mass $m = 79$ kg, sliding into second base, is retarded by a force of friction $f = 470$ N. What is the value of the coefficient of kinetic friction μ_k between the player and the ground?

3E. A 35-kg crate is at rest on the floor. A man attempts to push it across the floor by applying a 100-N force horizontally. (*a*) Take the coefficient of static friction between the crate and floor to be 0.37 and show that the crate does not move. (*b*) A second man helps by pulling up on the crate. What minimum vertical force must he apply so that the crate starts to move across the floor? (*c*) If the second man applies a horizontal rather than a vertical force, what minimum force, in addition to the 100-N force of the first man, must he exert to get the crate started?

4E. The coefficient of static friction between Teflon and scrambled eggs is about 0.04. What is the smallest angle from the horizontal that will cause the eggs to slide across the bottom of a Teflon-coated skillet?

5E. A 70-kg mountain climber spans a rock chimney as shown in Fig. 17. If the effective coefficient of static friction between his hands and boots and the rock is 3.4, with what minimum force must he press with each hand and foot to avoid falling?

Figure 17   Exercise 5.

6E. A 100-N force is applied at an angle θ above the horizontal to a 25-kg chair sitting on the floor. (*a*) For each of the following angles θ, calculate the normal force of the floor on the chair and the force of friction that the floor must exert on the chair if it is to remain at rest: (*i*) 0°, (*ii*) 30°, (*iii*) 60°. (*b*) Take the coefficient of static friction between the chair and the floor to be 0.42 and, for each of the above directions of the applied force, decide if the chair does indeed remain at rest.

7E. The coefficient of static friction between the tires of a car and a dry road is 0.60. If the mass of the car is 1500 kg, what maximum braking force is obtainable?

8E. Suppose that only the rear wheels of an automobile can accelerate it, and that one-half the total weight of the automobile is supported by those wheels. (*a*) What is the maximum acceleration attainable if the coefficient of static friction between tires and road is μ_s? (*b*) Take $\mu_s = 0.35$ and get a numerical value for this acceleration.

9E. What is the greatest acceleration that can be generated by a runner if the coefficient of static friction between shoes and road is 0.95?

10E. A person applies a 220-N horizontal force to a 55-kg crate to push it across a level floor. The coefficient of kinetic friction is 0.37. (*a*) What is the force of friction? (*b*) What is the acceleration of the crate?

11E. A 50-lb (= 220-N) trunk rests on the floor. The coefficient of static friction between the trunk and the floor is 0.41, while the coefficient of kinetic friction is 0.32. (*a*) What is the minimum horizontal force with which a person must push on the trunk to start it moving? (*b*) Once moving, what horizontal force must the person apply to keep the trunk moving with constant velocity? (*c*) If the person continued to push with the force used to start the motion, what would be the acceleration of the trunk?

12E. A 125-lb (= 556-N) filing cabinet rests on the floor. The coefficient of static friction between it and the floor is 0.68, while the coefficient of kinetic friction is 0.56. In four different attempts to move it, it is pushed with horizontal forces of (*a*) 50 lb (= 222 N), (*b*) 75 lb (= 334 N), (*c*) 100 lb (= 445 N), and (*d*) 125 lb (= 556 N). For each attempt, tell if the cabinet moves and calculate the force of friction the floor exerts on it. The cabinet is initially at rest for each attempt.

13E. A horizontal force F of 12 lb pushes a block weighing 5 lb against a vertical wall (Fig. 18). The coefficient of static

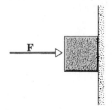

Figure 18   Exercise 13.

friction between the wall and the block is 0.6, and the coefficient of kinetic friction is 0.4. Assume that the block is not moving initially. (a) Will the block start moving? (b) What is the force exerted on the block by the wall?

14E. A house is built on the top of a hill with a 45° slope. Subsequent slumping of material on the slope surface indicates that the slope angle should be reduced. If the coefficient of friction of soil on soil is 0.5, through what additional angle ϕ (see Fig. 19) should the slope be regraded?

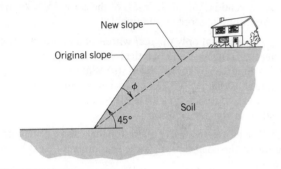

Figure 19 Exercise 14.

15E. The coefficient of kinetic friction in Fig. 20 is 0.2. What is the acceleration of the block?

16E. A 110-g hockey puck slides on the ice for 15 m before it stops. (a) If its initial speed was 6.0 m/s, what is the force of friction between puck and ice? (b) What is the coefficient of kinetic friction?

17E. A student wants to determine the coefficients of static friction and kinetic friction between a box and a plank. She places the box on the plank and gradually raises one end of the plank. When the angle of inclination with the horizontal reaches 30°, the box starts to slip and slides 2.5 m down the plank in 4.0 s. What are the coefficients of friction?

18P. A worker wishes to pile sand onto a circular area in his yard. The radius of the circle is R. No sand is to spill onto the surrounding area; see Fig. 21. Show that the greatest volume of sand that can be stored in this manner is $\pi\mu_s R^3/3$, where μ_s is the static coefficient of friction of sand on sand. (The volume of a cone is $Ah/3$, where A is the base area and h is the cone height.)

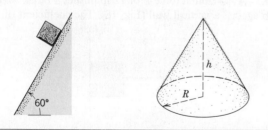

Figure 20 Exercise 15.   **Figure 21** Problem 18.

19P. Frictional heat generated by the moving ski is the chief factor promoting sliding in skiing. The ski sticks at the start but once in motion will melt the snow beneath it. Waxing the ski makes it water repellent and reduces friction with the film of water. A magazine reports that a new type of plastic ski is even more water repellent and that, on a gentle 200-m slope in the Alps, a skier reduced his time from 61 to 42 s with the new skis. (a) Determine the average accelerations for each pair of skis. (b) Assuming a 3° slope, compute the coefficient of kinetic friction for each case.

20P. A 10-kg block of steel is at rest on a horizontal table. The coefficient of static friction between block and table is 0.50. (a) What is the magnitude of the horizontal force that will just start the block moving? (b) What is the magnitude of a force acting upward 60° from the horizontal that will just start the block moving? (c) If the force acts down at 60° from the horizontal, how large can it be without causing the block to move?

21P. A railroad flatcar is loaded with crates having a coefficient of static friction of 0.25 with the floor. If the train is moving at 48 km/h, in how short a distance can the train be stopped without letting the crates slide?

22P. A block slides down an inclined plane of slope angle θ with constant velocity. It is then projected up the same plane with an initial speed v_0. (a) How far up the incline will it move before coming to rest? (b) Will it slide down again?

23P. A 150-lb crate is dragged across a floor by pulling on a rope inclined 15° above the horizontal. (a) If the coefficient of static friction is 0.50, what tension in the rope is required to start the crate moving? (b) If $\mu_k = 0.35$, what is the initial acceleration of the crate?

24P. A piece of ice slides down a 35° incline in twice the time it takes to slide down a frictionless 35° incline. What is the coefficient of kinetic friction between the ice and the incline?

25P. In Fig. 22, A is a 44-N block and B is a 22-N block. (a) Determine the minimum weight (block C) that must be placed on A to keep it from sliding, if μ_s between A and the table is 0.20. (b) Block C suddenly is lifted off A. What is the acceleration of block A, if μ_k between A and the table is 0.15?

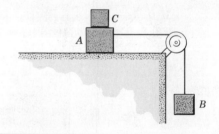

Figure 22 Problem 25.

26P. A 3.5-kg block is pushed along a horizontal floor by a force $P = 15$ N that makes an angle $\theta = 40°$ with the horizon-

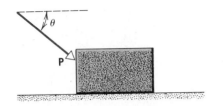

Figure 23 Problem 26.

tal, as shown in Fig. 23. The coefficient of kinetic friction between the block and floor is 0.25. Calculate (*a*) the frictional force exerted on the block and (*b*) the acceleration of the block.

27P. The handle of a floor mop of mass m makes an angle θ with the vertical direction; see Fig. 24. Let μ_k be the coefficient of kinetic friction between mop and floor, and let μ_s be the coefficient of static friction between mop and floor. Neglect the mass of the handle. (*a*) Find the magnitude of the force F directed along the handle required to slide the mop with uniform velocity across the floor. (*b*) Show that if θ is smaller than a certain angle θ_0, the mop cannot be made to slide across the floor no matter how great a force is directed along the handle. What is the angle θ_0?

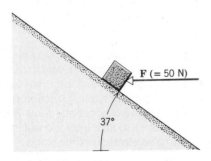

Figure 24 Problem 27.

28P. A 5.0-kg block on an inclined plane is acted on by a horizontal force of 50 N (Fig. 25). The coefficient of friction between block and plane is 0.30. (*a*) What is the acceleration of the block if it is moving up the plane? (*b*) With the horizontal force still acting, how far up the plane will the block go if it has

an initial upward speed of 4.0 m/s? (*c*) What happens to the block after it reaches the highest point?

29P. A wire will break under tensions exceeding 1000 N (=225 lb). (*a*) If the wire, not necessarily horizontal, is used to drag a box across the floor, what is the greatest weight that can be moved if the coefficient of static friction is 0.35? (*b*) If the wire is used to lift a box, what is the greatest weight that can be lifted with an upward acceleration of 0.92 m/s²?

30P. Figure 26 shows the cross section of a road cut into the side of a mountain. The solid line AA' represents a weak bedding plane along which sliding is possible. The block B directly above the highway is separated from uphill rock by a large crack (called a *joint*), so that only the force of friction between the block and the likely surface of failure prevents sliding. The mass of the block is 1.8×10^7 kg, the dip angle of the failure plane is 24°, and the coefficient of static friction between block and plane is 0.63. (*a*) Show that the block will not slide. (*b*) Water seeps into the joint, exerting a hydrostatic force F on the block parallel to the incline. What minimum value of F will trigger a slide?

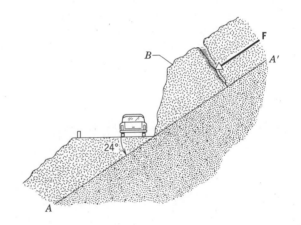

Figure 26 Problem 30.

31P. A block weighing 80 N rests on a plane inclined at 20° to the horizontal, as shown in Fig. 27. The coefficient of static friction is 0.25, while the coefficient of kinetic friction is 0.15. (*a*) What is the minimum force F, parallel to the plane, that will prevent the block slipping down the plane? (*b*) What is the minimum force F that will start the block moving up the

Figure 25 Problem 28.

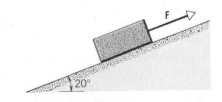

Figure 27 Problem 31.

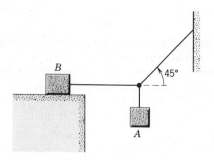

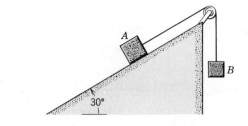

Figure 30 Problem 34.

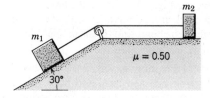

Figure 31 Problem 35.

Figure 28 Problem 32.

plane? (c) What force F is required to move the block up the plane at constant velocity?

32P. Block B in Fig. 28 weighs 710 N. The coefficient of static friction between block and horizontal surface is 0.25. Find the maximum weight of block A for which the system will be in equilibrium.

33P. Body B weighs 100 lb and body A weighs 32 lb (Fig. 29). The coefficients of friction between B and the incline are $\mu_s = 0.56$ and $\mu_k = 0.25$. (a) Find the acceleration of the system if B is initially at rest. (b) Find the acceleration if B is moving up the incline. (c) What is the acceleration if B is moving down the incline?

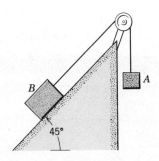

Figure 29 Problem 33.

34P. Two blocks are connected over a massless pulley as shown in Fig. 30. The mass of block A is 10 kg and the coefficient of kinetic friction is 0.20. Block A slides down the incline at constant speed. What is the mass of block B?

35P. Block m_1 in Fig. 31 has a mass of 4.0 kg and m_2 has a mass of 2.0 kg. The coefficient of friction between m_2 and the horizontal plane is 0.50. The inclined plane is smooth. Find (a) the tension in the string and (b) the acceleration of the blocks.

36P. An 8.0-lb block and a 16-lb block connected together by a string slide down a 30° inclined plane. The coefficient of kinetic friction between the 8.0-lb block and the plane is 0.10; between the 16-lb block and the plane it is 0.20. Find (a) the acceleration of the blocks and (b) the tension in the string,

assuming that the 8.0-lb block leads. (c) Describe the motion if the blocks are reversed.

37P. Two masses, $m_1 = 1.65$ kg and $m_2 = 3.30$ kg, attached by a massless rod parallel to the incline on which both slide, as shown in Fig. 32, travel down along the plane with m_1 trailing m_2. The angle of incline is $\theta = 30°$. The coefficient of kinetic friction between m_1 and the incline is $\mu_1 = 0.226$; between m_2 and the incline the corresponding coefficient is $\mu_2 = 0.113$. Compute (a) the tension in the rod linking m_1 and m_2 and (b) the common acceleration of the two masses. (c) How would the answers to (a) and (b) be changed if m_2 trails m_1?

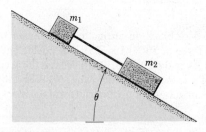

Figure 32 Problem 37.

38P. A 4.0-kg block is put on top of a 5.0-kg block. To cause the top block to slip on the bottom one, held fixed, a horizontal force of 12 N must be applied to the top block. The assembly of blocks is now placed on a horizontal, frictionless table (Fig. 33). Find (a) the maximum horizontal force F that can be applied to the lower block so that the blocks will move together and (b) the resulting acceleration of the blocks.

39P. A 40-kg slab rests on a frictionless floor. A 10-kg block rests on top of the slab (Fig. 34). The coefficient of static fric-

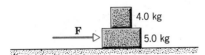

Figure 33 Problem 38.

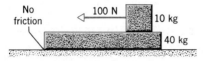

Figure 34 Problem 39.

tion between the block and the slab is 0.60, whereas the kinetic coefficient is 0.40. The 10-kg block is acted on by a horizontal force of 100 N. What are the resulting accelerations of (a) the block and (b) the slab?

40P. The two blocks (with $m = 16$ kg and $M = 88$ kg) shown in Fig. 35 are free to move. The coefficient of static friction between the blocks is $\mu_s = 0.38$ but the surface beneath M is frictionless. What is the minimum horizontal force F required to hold m against M?

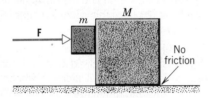

Figure 35 Problem 40.

41P. A crate slides down an inclined right-angled trough as in Fig. 36. The coefficient of kinetic friction between the block and the material composing the trough is μ_k. Find the acceleration of the crate.

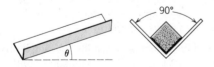

Figure 36 Problem 41.

42P. A locomotive accelerates a 25-car train along a level track. Each car has a mass of 50 metric tons and is subject to a frictional force $f = 250v$, where the speed v is in meters per second and the force f is in newtons. At the instant when the speed of the train is 30 km/h, the acceleration is 0.20 m/s². (a) What is the tension in the coupling between the first car and the locomotive? (b) If this tension is the maximum force the locomotive can exert on the train, what is the steepest grade up

which the locomotive can pull the train at 30 km/h? (One metric ton $= 10^3$ kg.)

43P*. A 1000-kg boat is traveling at 90 km/h when it shuts off its engine. The force of friction f with the water is proportional to the speed v of the boat: $f = 70v$, where v is in meters per second and f is in newtons. Find the time required for the boat to slow down to 45 km/h.

Section 6–3 The Drag Force and Terminal Speed

44E. Calculate the drag force on a missile 53 cm in diameter cruising with a speed of 250 m/s at a low altitude where the density of air is 1.2 kg/m³. Assume $C = 0.75$.

45E. The terminal speed of a sky diver in the spread-eagle position is 160 km/h. In the nose-dive position, the terminal speed is 310 km/h. What is the ratio of the projected areas in these two orientations? Assume that the drag coefficient C does not change.

46E. Calculate the ratio of the drag force on a passenger jet flying with a speed of 1000 km/h at an altitude of 10 km to the drag force on a prop-driven transport flying at half the speed and half the altitude of the jet. At 10 km the density of air is 0.38 kg/m³ and at 5 km the air density is 0.67 kg/m³. Assume that the planes have the same effective cross-sectional area and the same drag coefficient C.

47P. From the data in Table 1, deduce the diameter of the 16-lb shot. Assume $C = 0.49$.

Section 6–4 Uniform Circular Motion

48E. If the coefficient of static friction for tires on a road is 0.25, at what maximum speed can a car round a level 47.5-m radius curve without slipping?

49E. During an Olympic bobsled run, a European team takes a turn of radius 25 ft at a speed of 60 mi/h. How many g's do the riders experience?

50E. What is the smallest radius of an unbanked curve around which a bicyclist can travel if her speed is 18 mi/h and the coefficient of static friction between the tires and the road is 0.32?

51E. A 2400-lb ($= 10.7$ kN) car traveling at 30 mi/h ($= 13.4$ m/s) attempts to round an unbanked curve with a radius of 200 ft ($= 61.0$ m). (a) What force of friction is required to keep the car on its circular path? (b) If the coefficient of static friction for the tires and road is 0.35, is the attempt successful?

52E. A circular curve of highway is designed for traffic moving at 60 km/h ($= 37$ mi/h). (a) If the radius of the curve is 150 m ($= 490$ ft), what is the correct angle of banking of the road? (b) If the curve were not banked, what would be the minimum coefficient of friction between tires and road that would keep traffic from skidding at this speed?

53E. A banked circular highway curve is designed for traffic moving at 60 km/h. The radius of the curve is 200 m. Traffic is moving along the highway at 40 km/h on a stormy day. What is the minimum coefficient of friction between tires and road that will allow cars to negotiate the turn without sliding off the road?

54E. You are driving a car at a speed of 85 km/h (=53 mi/h) when you notice a barrier across the road exactly 60 m (=200 ft) ahead. (*a*) What is the minimum static coefficient of friction between tires and road that will allow you to stop before striking the barrier? (*b*) Suppose that you are driving on a large empty parking lot. What is the minimum coefficient of static friction that would allow you to turn the car in a 60-m radius circle and, in this way, avoid collision with a wall 60 m ahead?

55E. A child places a picnic basket on the outer rim of a merry-go-round that has a radius of 15 ft (=4.6 m) and revolves once every 30 s. (*a*) What is the speed of a point on the rim of the merry-go-round? (*b*) How large must the coefficient of static friction be for the basket to stay on the merry-go-round?

56E. A conical pendulum is formed by attaching a 50-g mass to a 1.2-m string. The mass swings around a circle of radius 25 cm. (*a*) What is the speed of the mass? (*b*) What is the acceleration of the mass? (*c*) What is the tension in the string?

57E. In the Bohr model of the hydrogen atom, the electron revolves in a circular orbit around the nucleus. If the radius is 5.3×10^{-11} m and the electron makes 6.6×10^{15} rev/s, find (*a*) the speed of the electron, (*b*) the acceleration (magnitude and direction) of the electron, and (*c*) the centripetal force acting on the electron. (This force is the result of the attraction between the positively charged nucleus and the negatively charged electron.) The electron's mass is 9.11×10^{-31} kg.

58E. A mass m on a frictionless table is attached to a hanging mass M by a cord through a hole in the table (Fig. 37). Find the speed with which m must move for M to stay at rest.

59E. A stunt man drives a car over the top of a hill, the cross section of which can be approximated by a circle of radius

Figure 38   Exercise 59.

250 m, as in Fig. 38. What is the greatest speed at which he can drive without the car leaving the road at the top of the hill?

60P. A driver's manual states that a driver traveling at 48 km/h and desiring to stop as quickly as possible travels 10 m before the foot reaches the brake. The car travels an additional 21 m before coming to rest. (*a*) What coefficient of friction is assumed in these calculations? (*b*) What is the minimum radius for turning a corner at 48 km/h without skidding?

61P. A small coin is placed on a flat, horizontal turntable. The turntable is observed to make three revolutions in 3.14 s. (*a*) What is the speed of the coin when it rides without slipping at a distance 5.0 cm from the center of the turntable? (*b*) What is the acceleration (magnitude and direction) of the coin in part (*a*)? (*c*) What is the frictional force acting on the coin in part (*a*) if the coin has a mass of 2.0 g? (*d*) What is the coefficient of static friction between the coin and the turntable if the coin is observed to slide off the turntable when it is more than 10 cm from the center of the turntable?

62P. A small object is placed 10 cm from the center of a phonograph turntable. It is observed to remain on the table when it rotates at $33\frac{1}{3}$ rev/min but slides off when it rotates at 45 rev/min. Between what limits must the coefficient of static friction between the object and the surface of the turntable lie?

63P. A bicyclist travels in a circle of radius 25 m at a constant speed of 9.0 m/s. The combined mass of the bicycle and rider is 85 kg. Calculate (*a*) the force of friction exerted by the road and (*b*) the total force exerted by the road. See Fig. 39.

64P. A car is rounding a curve of radius $R = 220$ m (=720 ft) at its design speed $v = 94$ km/h (=58 mi/h). What force does a passenger with mass $m = 85$ kg ($W = mg = 187$ lb) exert on the seat cushion?

65P. A 150-lb student on a steadily rotating Ferris wheel has an apparent weight of 125 lb at the highest point. (*a*) What is the student's apparent weight at the lowest point? (*b*) What would be the student's apparent weight at the highest point if the speed of the Ferris wheel were doubled?

66P. A stone of mass m at the end of a string is whirled around in a vertical circle of radius R. Find the critical speed below which the string would become slack at the highest point.

67P. A certain string can withstand a maximum tension of 9.0 lb without breaking. A child ties a 0.82-lb stone to one end and, holding the other end, whirls the stone in a vertical circle of radius 3.0 ft, slowly increasing the speed until the string

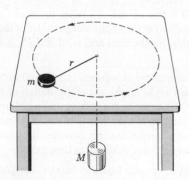

Figure 37   Exercise 58.

Figure 39 Problem 63.

breaks. (*a*) Where is the stone on its path when the string breaks? (*b*) What is the speed of the stone as the string breaks?

68P. An airplane is flying in a horizontal circle at a speed of 480 km/h. If the wings of the plane are tilted 40° to the horizontal, what is the radius of the circle in which the plane is flying? See Fig. 40. Assume that the required force is provided entirely by an aerodynamic lift that is perpendicular to the wing surface.

69P. A frigate bird is soaring in a circular path. Its bank angle is estimated to be 25° and it takes 13 s for the bird to complete one circle. (*a*) How fast is the bird flying? (*b*) What is the radius

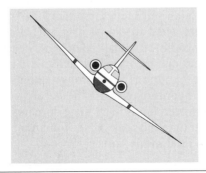

Figure 40 Problem 68.

of the circle? (See "The Amateur Scientist," *Scientific American,* March 1985.)

70P. A model airplane of mass 0.75 kg is flying at constant speed in a horizontal circle at one end of a 30-m cord and at a height of 18 m. The other end of the cord is tethered to the ground. The airplane makes 4.4 rev/min and the lift is perpendicular to the wings. (*a*) What is the acceleration of the plane? (*b*) What is the tension in the cord? (*c*) What is the lift produced by the plane's wings?

71P. An old streetcar rounds a corner on unbanked tracks. If the radius of the tracks is 30 ft and the car's speed is 10 mi/h, what angle with the vertical will be made by the loosely hanging hand straps?

72P. Assume that the standard kilogram would weigh exactly 9.80 N at sea level on the earth's equator if the earth did not rotate. Then take into account the fact that the earth does rotate, so that this mass moves in a circle of radius 6.40×10^6 m (the earth's radius) at a constant speed of 465 m/s. (*a*) Determine the centripetal force needed to keep the standard moving in its circular path. (*b*) Determine the force exerted by the standard kilogram on a spring balance from which it is suspended at the equator (its apparent weight).

73P. A 1.34-kg ball is attached to a rigid vertical rod by means of two massless strings each 1.7 m long. The strings are attached to the rod at points 1.7 m apart. The system is rotating about the axis of the rod, both strings being taut and forming an equilateral triangle with the rod, as shown in Fig. 41. The tension in the upper string is 35 N. (*a*) Draw the free-body diagram for the ball. (*b*) What is the tension in the lower string? (*c*) What is the net force on the ball at the instant shown in the figure? (*d*) What is the speed of the ball?

Figure 41

74P. Because of the rotation of the earth, a plumb bob may not hang exactly along the direction of the earth's gravitational pull on the bob but may deviate slightly from that direction. Calculate the deviation (*a*) at 40° latitude, (*b*) at the poles, and (*c*) at the equator.

CHAPTER 7

WORK AND KINETIC ENERGY

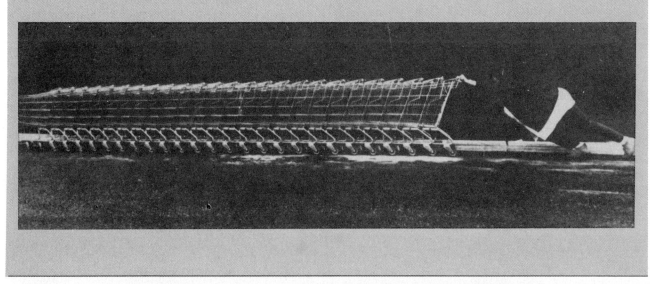

A young man is doing work on a stack of carts at a Toronto supermarket. If he is pushing the carts at a constant velocity, the net force on the carts must be zero. How can work be done if there is no net force? Read this chapter to unravel this puzzler.

7–1 A Walk Around Newtonian Mechanics

Now that we have dealt with each of Newton's three laws of motion, you may feel that our study of Newtonian mechanics is approaching an end. However, mountaineers are never satisfied with one view of a mountain, no matter how impressive that view may be. They always want to walk around it and to study it from as many angles as possible. In that way, they gain new insights and they learn how to do familiar routes more easily. Sometimes, they also see higher peaks, previously hidden from view, and the possibility of new conquests. It is the same with Newtonian mechanics.

In this chapter, we shall introduce two new concepts, *work* and *kinetic energy*. Our rewards will be those of the mountaineer. We shall gain new insights into Newtonian mechanics and learn how to solve some kinds of problems with remarkable ease. Most important of all, we shall catch a glimpse of a higher peak. The concepts of work and of kinetic energy are the first steps on the trail to a universal law that—so far—knows no exceptions, *the law of conservation of energy.*

Newtonian mechanics, grand as its structure may be, does fail when we apply it to particles moving at speeds comparable to the speed of light, yielding there to Einstein's special theory of relativity. It also fails when we apply it to motions of electrons in atoms, yielding in that case to quantum physics. The law of conservation of energy, however, holds in all these domains. It is a peak even higher than the peak of Newtonian mechanics and the trail to its summit starts right here.

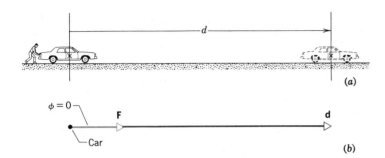

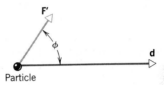

Figure 1   *(a)* You push a stalled car for a distance d, exerting a constant horizontal force. *(b)* A free-body diagram for the car, showing the force **F** that you apply and the displacement **d** of the car. The two vectors are parallel, the angle ϕ between them being zero.

7–2   Work: Motion in One Dimension with a Constant Force

Imagine (see Fig. 1*a*) that you push on a stalled car with a steady horizontal force **F** and that the car moves through a horizontal displacement **d**. We say, as a matter of definition, that you have done an amount of work W on the car given by

$$W = Fd. \tag{1}$$

Here F is the magnitude of the force that you apply and d is the magnitude of the displacement of the point of application of that force. Figure 1*b* shows a free-body diagram, with the car represented as a particle. Note that the angle ϕ between the force vector **F** and the displacement vector **d** is zero.

W in Eq. 1 is the work that *you* do on the car. Alternatively, we can say that W is the work done by the force **F** (that you exert) on the car. Other forces may (and do) act on the car and they may (or may not) also do work. The work done by each force must be calculated separately. To find the *total* work done on the car, we must add up the separate work contributions of all the forces that act on it.

Figure 2 is a generalization of Fig. 1*b*, in which the angle ϕ between the force vector **F** and the displacement vector **d** has a value other than zero. We define the work W done by the force **F** on the particle in this more general case to be

$$\boxed{W = Fd \cos \phi.} \tag{2}$$

Note that if $\phi = 0$, $\cos \phi = 1$; Eq. 2 then reduces to Eq. 1, as it must.

We can write Eq. 2 as

$$W = (d)(F \cos \phi) = (F)(d \cos \phi), \tag{3}$$

which shows that there are two ways to calculate work: Multiply the displacement d by the component of force

Figure 2   A constant force **F** acts on a particle that undergoes a displacement **d**. The two vectors make a constant angle ϕ with each other; compare Fig. 1*b*.

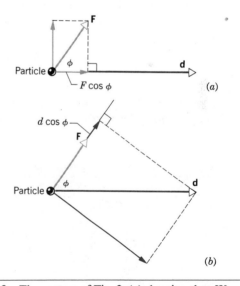

Figure 3   The vectors of Fig. 2: *(a)* showing that $W = (d)(F \cos \phi)$ and *(b)* showing that $W = (F)(d \cos \phi)$.

in the direction of the displacement (Fig. 3*a*) *or* multiply the force F by the component of the displacement in the direction of the force (Fig. 3*b*). The two methods always give the same answer.

If the force and the displacement are at right angles to each other, then $\phi = 90°$ and Eq. 2 tells us that the work done by the force is zero. This means that, if you hold a stack of books in your arms (see Fig. 4*a*) and carry

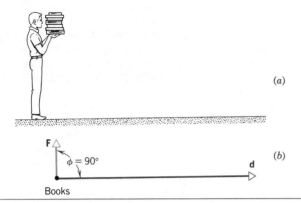

Figure 4 (*a*) You carry a stack of books along a horizontal path. (*b*) The force that you exert on the books is at right angles to their displacement, so that no work is done on the books by that force.

them horizontally from, say, the bookstore to your dorm, no work is done on the books by the upward force that you exert on them. You may dispute this. It is true that forces inside your body are doing work as muscles expand and contract but that is not the work that we are talking about. Our concern in Fig. 4 is the work done by the force **F** on the books, formally defined by Eq. 2. In fact, the athlete in Fig. 5 does no work in *holding* the weights, no matter how much force he must exert; he does work only in *lifting* the weights. The athlete, misled

Figure 5 The weight lifter is exerting a great force on the weights but he is doing no work on them at the instant shown because he is simply holding them in place. There is a force but no displacement.

by the popular meanings attached to the word "work" and perhaps not thinking about Eq. 2 at the moment, may disagree with the first part of this statement.

If the force and the displacement point in opposite directions, so that the angle between them is 180°, then the work done by the force on the particle is negative. This follows from Eq. 2 because cos 180° = −1. Recall that *F* and *d* in that equation, being the magnitudes of vectors, are always positive. When you lift a book vertically, you do positive work on the book because the (upward) force that you exert on the book and the (upward) displacement of the book are in the same direction. If you *lower* a book, however, you do *negative* work on it because, although the force that you exert on the book still points upward, the displacement of the book is now downward.

Work is a *scalar,* although the two quantities involved in its definition, force and displacement, are vectors. We can write Eq. 2 more compactly in vector form, as a scalar (or dot) product.* Thus,

$$W = \mathbf{F} \cdot \mathbf{d} \qquad \text{(work: constant force).} \qquad (4)$$

This equation is identical with Eq. 2 in every respect.

The SI unit of work is the *newton · meter,* a quantity that is used so often it has been given a special name, the *joule* (abbr. J). In the British system of units, the unit of work is the *foot · pound* (abbr. ft · lb). The relations are

$$1 \text{ joule} = 1 \text{ J} = 1 \text{ N} \cdot \text{m} = 1 \text{ kg} \cdot \text{m}^2/\text{s}^2$$
$$= 0.738 \text{ ft} \cdot \text{lb}. \qquad (5)$$

Appendix F gives conversion factors to other units in which work is measured.

A convenient unit of work when dealing with atoms or with subatomic particles is the *electron-volt* (abbr. eV) and its common multiples, the keV ($= 10^3$ eV), the MeV ($= 10^6$ eV), the GeV ($= 10^9$ eV), and the TeV ($= 10^{12}$ eV). We shall define the electron-volt precisely in Section 26–2 but we give here its relation to the joule, namely,

$$1 \text{ electron-volt} = 1 \text{ eV} = 1.60 \times 10^{-19} \text{ J}. \qquad (6)$$

Compared with the joule, the electron-volt is a very small unit.

---

* This is our first application of the dot (or scalar) product notation in this text. You may wish to review Section 3–7, where the dot product of two vectors is defined and discussed and Eq. 4 is given as an example.

Sample Problem 1 A student in a geology lab (see Fig. 6) lifts a rock, whose mass m is 12 kg, from the floor and, raising it (with negligible acceleration) to a height d of 1.8 m, puts it on a shelf. We wish to identify the forces that act on the rock and to calculate the work that they do. (a) Identify these forces and calculate their magnitudes.

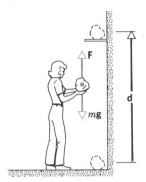

Figure 6 Sample Problem 1. A student lifts a rock specimen from the floor and puts it on a shelf. Two forces act on the rock.

Two forces act on the rock, its weight $m\mathbf{g}$ acting downward and the upward force $\mathbf{F}$ exerted by the student. Because the rock is not accelerated, the resultant force acting on it must be zero. The common magnitude of these forces is

$$F = mg = (12 \text{ kg})(9.8 \text{ m/s}^2) = 118 \text{ N}. \quad \text{(Answer)}$$

(b) How much work does the student do on the rock?

The force $\mathbf{F}$ exerted by the student and the displacement $\mathbf{d}$ point in the same direction, the angle ϕ between them being zero. From Eq. 2 we have

$$\begin{aligned}W_s &= Fd \cos \phi = (mg)(d)(\cos 0°) \\ &= (118 \text{ N})(1.8 \text{ m})(+1) = +212 \text{ J}. \quad \text{(Answer)}\end{aligned}$$

(c) How much work does the earth's gravity do on the rock?

As Fig. 6 shows, the angle ϕ between the weight $m\mathbf{g}$ and the displacement $\mathbf{d}$ is 180°, so that

$$\begin{aligned}W_g &= (mg)(d)(\cos 180°) \\ &= (118 \text{ N})(1.8 \text{ m})(-1) = -212 \text{ J}. \quad \text{(Answer)}\end{aligned}$$

(d) What is the total work done on the rock by all the forces that act on it?

We simply add up the work done by the two forces, or

$$\begin{aligned}W_{tot} &= W_s + W_g \\ &= (+212 \text{ J}) + (-212 \text{ J}) = 0. \quad \text{(Answer)}\end{aligned}$$

Another way to find the total work is to calculate the work done by the *net* force. In this problem, the net force is zero so, of course, it does no work, in agreement with what we have just found.

Do not imagine that we have just said that you can lift a rock without doing any work on it. When we speak of the work to lift a rock, we are referring to the work done *on* the rock *by* the person who is lifting it and *not* to the *total* work done on the rock. The work done by the person is positive and, in this case, is 212 J.

Sample Problem 2 A clerk pushes a filing cabinet, whose mass m is 85 kg, at constant speed across a tiled floor for a distance d of 3.1 m; see Fig. 7a. The coefficient of friction μ_k between the bottom of the cabinet and the tile is 0.22. We wish to identify the forces that act on the cabinet and calculate the work that they do. (a) Identify these forces and calculate their magnitudes.

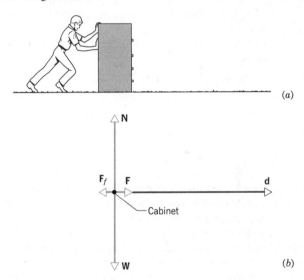

Figure 7 Sample Problem 2. (a) A clerk pushes a filing cabinet across a floor. (b) A free-body diagram of the cabinet, showing the forces that act on it. Its displacement $\mathbf{d}$ is also shown.

As Fig. 7b shows, four forces act on the cabinet—the force $\mathbf{F}$ that the clerk exerts, the frictional force $\mathbf{f}$, the weight $m\mathbf{g}$ of the cabinet, and the normal force $\mathbf{N}$ exerted on the cabinet by the floor. The cabinet is in equilibrium at all times so that

$$\Sigma F_x = F - f = 0$$

and

$$\Sigma F_y = N - mg = 0.$$

The magnitudes of these forces are

$$N = mg = (85 \text{ kg})(9.8 \text{ m/s}^2) = 833 \text{ N} \quad \text{(Answer)}$$

and

$$F = f = \mu_k N = (0.22)(833 \text{ N}) = 183 \text{ N}. \quad \text{(Answer)}$$

(b) How much work does the clerk do on the cabinet? From Eq. 2 we have

$$W_F = Fd \cos \phi$$
$$= (183 \text{ N})(3.1 \text{ m})(\cos 0°) = +567 \text{ J}. \quad \text{(Answer)}$$

(c) How much work does the frictional force **f** do on the cabinet?

This force has the same magnitude as the force exerted by the clerk but is oppositely directed. Thus, the angle ϕ between the vectors **f** and **d** is 180°. From Eq. 2 then

$$W_f = Fd \cos \phi = (180 \text{ N})(3.1 \text{ m})(\cos 180°)$$
$$= -567 \text{ J}. \quad \text{(Answer)}$$

(d) How much work is done on the cabinet by the forces m**g** and **N**?

These forces are at right angles to the displacement so that they do no work on the cabinet, the angle ϕ in Eq. 2 being 90°.

(e) What is the total work done on the cabinet by all the forces that act on it?

We find this by adding up the work done separately by the four forces, or

$$W_{tot} = W_F + W_f + W_{mg} + W_N$$
$$= (+567 \text{ J}) + (-567 \text{ J}) + (0) + (0) = 0. \quad \text{(Answer)}$$

As in Sample Problem 1, this result is not surprising, because the net force on the cabinet is zero.

Sample Problem 3 A crate whose mass m is 15 kg is pulled a distance d ($= 5.7$ m) up a frictionless ramp, to a height h of 2.5 m above its starting point; see Fig. 8a. (a) What force **F** must the winch cable exert?

Figure 8b shows the free-body diagram for the crate. The crate is in equilibrium so that, applying Newton's second law parallel to the ramp, we obtain

$$F = mg \sin \theta$$
$$= (15 \text{ kg})(9.8 \text{ m/s}^2)(2.5 \text{ m/5.7 m}) = 64.5 \text{ N}. \quad \text{(Answer)}$$

Note that we have calculated sin θ directly from the given values of h and d, with no need to find θ itself.

(b) How much work is done by the force **F**?

From Eq. 2, we have

$$W_F = Fd \cos \phi$$
$$= (64.5 \text{ N})(5.7 \text{ m})(\cos 0°) = 368 \text{ J}. \quad \text{(Answer)}$$

Do not confuse the angle ϕ (which is the angle between the vectors **F** and **d** in Fig. 8) with the angle θ (which is the angle of the ramp).

(c) How much work would be required to lift the crate vertically upward, through a height h?

A force equal to the weight of the crate would be needed, so that

$$W_h = Fh \cos \phi = mgh \cos \phi$$
$$= (15 \text{ kg})(9.8 \text{ m/s}^2)(2.5 \text{ m})(\cos 0°) = 368 \text{ J}. \quad \text{(Answer)}$$

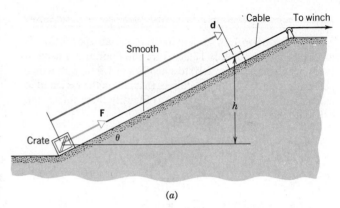

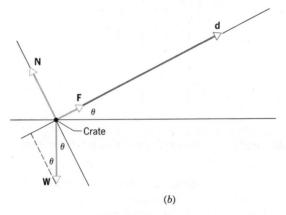

(b)

Figure 8 Sample Problem 3. (a) A crate is pulled up a frictionless ramp by a force parallel to the ramp. (b) A free-body diagram for the crate, showing all the forces that act on it. Its displacement **d** is also shown.

This is the same answer that we found in (b)! The difference is that in (b) we applied a smaller force (65 N versus 150 N) through a larger distance (5.7 m versus 2.5 m). In other words, we used the ramp to lift a crate by applying a force smaller than the weight of the crate. That is what a ramp is for.

Hint 1: Finding the work. In problems in which work is involved, draw a free-body diagram and make sure that you can answer these questions: With what particle am I dealing? What forces act on it? With which of these forces am I concerned? Is it perhaps the *net* force? What is the displacement of the particle? What is the angle between the displacement and the force? What is the sign of the work? Is the sign of the work physically reasonable? Review Sample Problems 1–3 with these questions in mind.

7-3   Work: Motion in One Dimension with a Variable Force

Here we consider that both the force acting on a particle and the displacement of that particle lie along the same straight line, which we may take as an x axis. We assume further that the magnitude of the force is *not constant* but depends on the position of the particle.

Figure 9a shows a plot of such a one-dimensional force. What work is done by this force as the particle on which it acts moves from an initial point x_i to a final point x_f? To find out, let us divide the total displacement of the particle into a number of intervals of width Δx. We choose Δx small enough so that we can take the force $F(x)$ as reasonably constant over that interval.

The work δW done by the force over any particular interval is given by Eq. 2, or

$$\delta W = \overline{F(x)}\Delta x, \qquad (7)$$

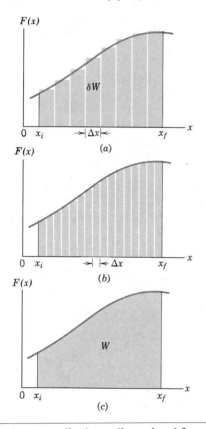

$F(x)$

(a)

$F(x)$

(b)

$F(x)$

(c)

in which $\overline{F(x)}$ is the average value of the force for that interval. On the graph of Fig. 9a, δW is equal in magnitude to the area of the vertical strip to which it refers, $F(x)$ being the height of the strip and Δx its width.

To find the total work done by the force as the particle moves from x_i to x_f, we add up the work increments for the various strips into which the curve of Fig. 9a is divided. That is,

$$W = \sum \delta W = \sum F(x)\Delta x. \qquad (8)$$

Here the Greek capital letter *sigma* stands for the *sum* over all the intervals from x_i to x_f.

Equation 8 is an approximation because the broken "skyline" formed by the tops of the rectangular strips in Fig. 9a only approximates the actual curve in that figure. Put another way, the area defined by the strips in Fig. 9a is only approximately equal to the actual area under the curve.

We can make the approximation better by reducing the strip width Δx and using more strips, as in Fig. 9b. In the limit, we can let the strip width approach zero, the number of strips then becoming infinitely large. We then have, as an exact result,

$$W = \lim_{\Delta x \to 0} \sum F(x)\Delta x. \qquad (9)$$

As you will have learned in your calculus course, this limit is exactly what we mean by the integral of the function $F(x)$ between the limits x_i amd x_f. Thus, Eq. 9 becomes*

$$\boxed{W = \int_{x_i}^{x_f} F(x)dx} \qquad \text{(work: variable force).} \quad (10)$$

If we know the function $F(x)$, we can substitute it into Eq. 10, carry out the integration, introduce the limits of integration, and thus find the work. Geometrically, the work is equal to the area that lies under the $F(x)$ curve between the limits x_i and x_f, as in Fig. 9c.

7-4   Work Done by a Spring

As an important example of a variable force, we consider the force exerted by a spring that is stretched or com-

Figure 9   (a) A generalized one-dimensional force. The particle on which the force acts moves from x_i to x_f. (b) The same as (a) but divided into narrower strips. (c) The limiting case. The work done by the force is given by Eq. 10 and is represented geometrically by the shaded area under the curve.

---

* This is the first use that we have made of an integral in this book. The symbol ∫ is a distorted S (for *sum*) and symbolizes the integration process. See Appendix G for a list of common integrals.

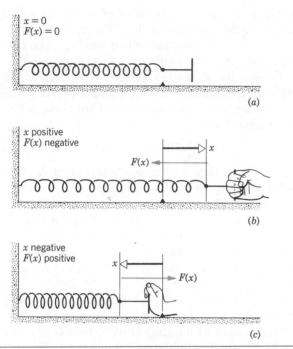

Figure 10 (*a*) A spring, with an attached handle, in its relaxed state. The small triangle marks the end point of the spring. (*b*) The spring is stretched by an amount *x*. Note the restoring force exerted by the spring. (*c*) The spring is compressed by an amount *x*. Again, note the restoring force. $F(x)$ and *x* have opposite signs, as predicted by Eq. 11.

pressed from its relaxed state. It is our aim to calculate the work done by such a spring force, using Eq. 10.

Figure 10*a* shows a spring in its relaxed state. In Fig. 10*b*, the spring has been stretched by pulling the handle to the right. The spring force pulls back on the handle, toward the left. In Fig. 10*c*, the spring has been compressed by pushing the handle to the left. The spring force now acts in the opposite direction, pushing on the handle toward the right.

To a good approximation for many springs, the force $F(x)$ exerted by the spring is proportional to *x*, the extension of the spring. Thus,

$$\boxed{F(x) = -kx} \quad \text{(Hooke's law)}, \quad (11)$$

in which *k* is a positive constant whose value depends on the nature of the spring. Figure 11 is a plot of Eq. 11. Be clear that $F(x)$ in Eq. 11 is the force exerted *by* the spring on the agent (the hand in Fig. 10) that is stretching or compressing it. The force exerted *on* the spring by that agent is $-F(x)$.

In its relaxed state, the extension *x* of a spring is zero, a positive value of *x* corresponding to a stretching of the spring and a negative value to a compression. As Figs. 10 and 11 and Eq. 11 show, $F(x)$ and *x* always have opposite algebraic signs.

The SI unit for the constant *k* in Eq. 11 (called the *spring constant*) is the newton/meter. The larger the value of *k*, the stiffer the spring. As Eq. 11 confirms, the spring constant *k* is the negative of the slope of the straight line of Fig. 11.

We now ask: "If a spring changes from an initial state (in which its extension is x_i) to a final state (in which its extension is x_f), how much work does the spring force do?" Substituting Eq. 11 into Eq. 10 yields

$$W = \int_{x_i}^{x_f} F(x)dx = \int_{x_i}^{x_f} (-kx)dx = -k\int_{x_i}^{x_f} x\, dx. \quad (12)$$

From the list of integrals in Appendix G we note that

$$\int x\, dx = \tfrac{1}{2}x^2.$$

Thus,

$$W = (-\tfrac{1}{2}k)|x^2|_{x_i}^{x_f} = (-\tfrac{1}{2}k)(x_f^2 - x_i^2)$$

or

$$\boxed{W = \tfrac{1}{2}kx_i^2 - \tfrac{1}{2}kx_f^2} \quad \text{(work } by \text{ a spring).} \quad (13)$$

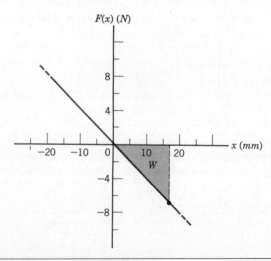

Figure 11 The force–distance plot for the spring of Sample Problems 4–6. It obeys Hooke's law (Eq. 11) and has a spring constant $k = 408$ N/m. See Sample Problem 4 for the significance of the dot and Sample Problem 5 for that of the shaded area marked *W*.

The sign of the work done by the spring force is positive if $x_i^2 > x_f^2$ and negative if $x_i^2 < x_f^2$.

If the spring is initially in its relaxed state ($x_i = 0$) and is stretched (or compressed) to an extension x, the work done by the spring is

$$\boxed{W = -\tfrac{1}{2}kx^2.}$$ (14)

Equation 14 is a special case of Eq. 13, for which $x_i = 0$ and $x_f = x$.

Equation 13 (and its special case, Eq. 14) gives the work done *by* the spring on the agent (the hand in Fig. 10) that is stretching or compressing it. The work done *on* the spring by that external agent is the negative of this quantity.

Note that the length of the spring does not appear explicitly in the expressions for the force exerted by the spring (Eq. 11) or the work done by the spring (Eqs. 13 and 14). The length of the spring is one of a number of factors that contribute to the value of the spring constant k, others being the spring geometry and the elastic properties of the material of which the spring is made.

Sample Problem 4 You apply a 4.9-N force F to the free end of a spring, stretching it from its relaxed state by a distance d of 12 mm, as in Fig. 10b. (a) What is the spring constant of the spring?

From Eq. 11 we have

$$k = -\frac{F}{x} = -\frac{-4.9 \text{ N}}{12 \times 10^{-3} \text{ m}} = 408 \text{ N/m.} \quad \text{(Answer)}$$

Note that, because x is positive (a stretching), the force $F(x)$ exerted *by* the spring is negative. Note that we do not need to know the length of the spring. The plot of Eq. 11 in Fig. 11 refers to this spring.

(b) What force does the spring exert on you if you stretch it by 17 mm?

From Eq. 11 we have

$$F(x) = -kx = -(408 \text{ N/m})(17 \times 10^{-3} \text{ m})$$
$$= -6.9 \text{ N.} \quad \text{(Answer)}$$

The dot on the curve of Fig. 11 represents this force and the corresponding displacement. Note that x is positive and $F(x)$ is negative, as in Fig. 10b.

Sample Problem 5 You stretch the spring of Sample Problem 4 by 17 mm from its relaxed state; see Fig. 10b. How much work does the spring force do on your hand?

Because the spring is initially in its relaxed state, we can

use Eq. 14. Putting $x = d$ in that equation yields

$$W = -\tfrac{1}{2}kd^2 = -(\tfrac{1}{2})(408 \text{ N/m})(17 \times 10^{-3} \text{ m})^2$$
$$= -5.9 \times 10^{-2} \text{ J} = -59 \text{ mJ.} \quad \text{(Answer)}$$

The shaded area in Fig. 11 represents this work. The work done by the spring is negative because the displacement of the end point of the spring and the force exerted by the spring point in opposite directions. Note that the amount of work done by the spring would be the same if it had been compressed (rather than stretched) by 17 mm.

Sample Problem 6 The spring of Fig. 10b is initially stretched by 17 mm. You allow it to return slowly to its relaxed state and then compress it by a further 12 mm. How much work does the spring now do on your hand?

For this situation, we have $x_i = +17$ mm (a stretching) and $x_f = -12$ mm (a compression). Equation 13 becomes

$$W = \tfrac{1}{2}kx_i^2 - \tfrac{1}{2}kx_f^2$$
$$= (\tfrac{1}{2})(408 \text{ N/m})(17 \times 10^{-3} \text{ m})^2$$
$$\quad - (\tfrac{1}{2})(408 \text{ N/m})(12 \times 10^{-3} \text{ m})^2$$
$$= 0.030 \text{ J} = 30 \text{ mJ.} \quad \text{(Answer)}$$

In this case, the spring did more positive work (in moving from its initial stretched to its relaxed state) than negative work (in moving further from its relaxed state to its final compressed state). The net work done by the spring is thus positive.

Hint 2: Derivatives and integrals; slopes and areas. If you know a function $y = F(x)$, you can find its derivative (for any value of x) or its integral (between any two values of x) from the rules of calculus. If you do not know the function analytically but have a numerical plot of it, you can find both its derivative and integral graphically. We showed how to find the derivative graphically in Hint 7 of Chapter 2. Here we show how to find an integral graphically.

Figure 12 is a numerical plot of a particular force function $F(x)$. Let us find graphically the work done by this force as the particle on which it acts moves from $x_i = 2.0$ cm to $x_f = 5.0$ cm. The work is the shaded area under the curve between these two points.

You can approximate this area by an equivalent rectangle, formed by drawing a horizontal line across the figure. Draw it at a level such that the two areas marked "1" and "2" are judged to be equal. $F = 43.5$ N is about right and the area of the equivalent rectangle ($= W$) is then

$$W = \text{height} \times \text{base} = (43.5 \text{ N})(5.0 \text{ cm} - 2.0 \text{ cm})$$
$$= 130 \text{ N·cm} = 1.3 \text{ N·m} = 1.3 \text{ J.}$$

You can find the area more closely by counting the squares underneath the curve. The rectangle on the right of

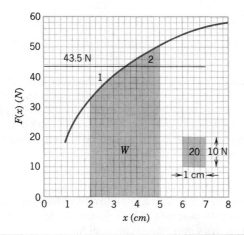

Figure 12 A generalized plot of $F(x)$. The shaded area (the work) is approximated by a rectangle. The small rectangle at the right serves to calibrate the small squares in work units.

Fig. 12 can be used to calibrate a square because we see from it that 20 squares = 10 N·cm. By counting large blocks, you can quickly find that the shaded area contains about 260 squares. The work is then

$$W = (260 \text{ squares}) \left(\frac{10 \text{ N} \cdot \text{m}}{20 \text{ squares}} \right) = 130 \text{ N} \cdot \text{cm} = 1.3 \text{ J},$$

just as above. *Remember:* Every derivative is a slope; every integral is an area.*

7-5 Kinetic Energy

If you see a hockey puck at rest on the ice and—a moment later—you see it hurtling toward the goal, you are entitled to conclude: "Someone hit it with a hockey stick." A pedantic physicist might put it: "Someone did work on the puck, exerting a force on it over a (small) distance."

Indeed, when we see an object moving in our reference frame, its very motion is a signal to us that work has been done on it to establish that motion. The mass of the object must also be involved in this "signal" because a given amount of work will impart different speeds to objects of different mass. What is there about the motion of a particle that we can relate it quantitatively to the work that must be done on the particle?

The property that we are seeking (as we shall see) is

---

* Assuming a single independent variable.

the *kinetic energy* of the particle, which we define as

$$\boxed{K = \tfrac{1}{2}mv^2} \quad \text{(kinetic energy defined).} \quad (15)$$

The kinetic energy of a particle depends on the square of the speed of the particle and is thus always positive. Its SI unit is the joule, the unit in which work is measured. Kinetic energy is a scalar quantity; if a particle of mass m is moving with speed v, its kinetic energy is $\tfrac{1}{2}mv^2$ no matter in what direction the particle is moving. Table 1 lists some kinetic energies.

If the kinetic energy of a particle changes from some initial value K_i to a final value K_f, the work that must be done on the particle (as we shall show below) is given by

$$\boxed{W = K_f - K_i = \Delta K}$$
$$\text{(work-energy theorem).} \quad (16)$$

The work W in Eq. 16 must be the *total work* done on the particle by *all* the forces that act on it. Alternatively, it must be the work done on the particle by the *net force* that acts on it.

The general message of this important *work-energy theorem* is: If you want to change the kinetic energy of a particle, you must do work on it. A more explicit statement is:

> *The change in the kinetic energy of a particle is equal to the total work done on that particle by all the forces that act on it.*

We stress that the work-energy theorem is not a new, independent law of classical mechanics. In this chapter, we have simply defined work (see Eq. 10) and kinetic energy (see Eq. 15) and stated a connection (Eq. 16) between them, which we shall derive directly from Newton's second law. The theorem, as we shall learn, is useful in that it gives us a new way to look at familiar problems and makes the solution of certain kinds of problem much easier. Before proving the work-energy theorem, let us see how to apply it to some familiar situations.

A Particle in Free Fall. If you drop a baseball, as in Fig. 13a, the net force acting on it is its weight mg. This force points in the direction in which the ball is moving so that the work done by the gravitational force on the falling ball is positive. From the work-energy theorem (Eq. 16), the kinetic energy of the falling ball should increase. We know that it does.

If you throw a baseball vertically upward, as in Fig.

Table 1   Some Kinetic Energies

Item	Remarks	Kinetic Energy (J)
Meteor Crater meteor	See Exercise 10-36	1.3×10^{18}
Aircraft carrier *Nimitz*	91,400 tons at 30 knots	9.9×10^{9}
Orbiting satellite	100 kg at 300-km altitude	3.0×10^{9}
Trailer truck	18-wheeler at 60 mi/h	2.2×10^{6}
Football linebacker	110 kg at 9 m/s	4.5×10^{3}
NATO SS 109 bullet	4 g at 950 m/s	1.8×10^{3}
Pitched baseball	100 mi/h	1.5×10^{2}
Falling penny	3.2 g at 50 m	1.6
Bee in flight	1 g at 2 m/s	2×10^{-3}
Snail	5 g at 0.03 mi/h	4.5×10^{-7}
Electron in TV tube	20 keV	3.2×10^{-15}
Electron in copper	At absolute zero	6.7×10^{-19}

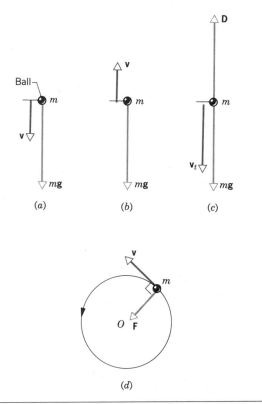

(a)   (b)   (c)

(d)

Figure 13   (*a*) The gravitational force does positive work on a freely falling ball. (*b*) The gravitational force does negative work on a ball thrown vertically upward. (*c*) No work is done on a ball that has reached its terminal speed because the net force acting on the ball is zero, the gravitational force $m\mathbf{g}$ being canceled by the atmospheric drag force **D**. (*d*) A particle in uniform circular motion. The force **F** does no work because it is at right angles to the direction in which the particle is moving.

13*b*, the gravitational force acting on it is now opposite to the motion and the work done by this force on the ball is negative. The work–energy theorem then tells us that the kinetic energy of the ball should decrease. Again, we know that it does.

Suppose that a baseball is dropped from a great height and, eventually, falls through air at its (constant) terminal speed, as in Fig. 13*c*. Its weight is now exactly balanced by an upward-pointing drag force so that the net force acting on the falling ball is zero. The work–energy theorem tells us that the kinetic energy should remain constant as it falls. This is indeed what happens.

A Particle in Uniform Circular Motion.   The speed of a particle in uniform circular motion is constant so that its kinetic energy does not change. The work–energy theorem then tells us that the resultant force acting on the particle does no work on it. We know this to be true because that force points radially inward, at right angles to the direction in which the particle is moving; see Fig. 13*d*. As we pointed out in connection with Fig. 4, no work is done under these circumstances.

Proof of the Work–Energy Theorem.   This theorem, which holds quite generally for a particle moving in three dimensions, is a direct consequence of Newton's second law. We shall prove it, however, only for the special case of a particle moving in one dimension. We assume that the net force acting on that particle may vary in magnitude.

Consider a particle of mass m, moving along the x axis, and acted on by a net force $F(x)$ that points along that axis. The work done on the particle by this force as the particle moves from an initial position x_i to a final position x_f is (see Eq. 10)

$$W = \int_{x_i}^{x_f} F(x)dx = \int_{x_i}^{x_f} ma \, dx, \qquad (17)$$

in which we have used Newton's second law by replacing $F(x)$ by ma. We can recast the quantity $ma \, dx$ in this equation as follows:

$$ma \, dx = m\frac{dv}{dt} dx = m\frac{dx}{dt} dv = mv \, dv. \qquad (18)$$

Substituting Eq. 18 into Eq. 17 yields

$$W = \int_{v_i}^{v_f} mv \, dv = m \int_{v_i}^{v_f} v \, dv = \tfrac{1}{2}mv_f^2 - \tfrac{1}{2}mv_i^2. \qquad (19)$$

Note that when we change the variable from x to v we are required to express the limits on the integral in terms of the new variable. Note also that, because the mass m is a constant, we were able to move it outside the integral.

Recognizing the terms on the right of Eq. 19 (see Eq. 15) as kinetic energies allows us to write this equation as

$$W = K_f - K_i = \Delta K,$$

which (see Eq. 16) is the work–energy theorem that we set out to prove.

Sample Problem 7   What is the kinetic energy of the earth as it moves in its orbit?

The earth completes one orbital revolution in a time T of 1 year. If R is the radius of the earth's orbit, the earth's average orbital speed is

$$v = \frac{2\pi R}{T} = \frac{(2\pi)(1.50 \times 10^{11} \text{ m})}{3.16 \times 10^7 \text{ s}} = 2.98 \times 10^4 \text{ m/s}.$$

We note in passing that the speed of the earth in its orbit is almost exactly 0.01% of the speed of light. From Eq. 15 then, the kinetic energy of the earth is

$$K = \tfrac{1}{2}mv^2 = (\tfrac{1}{2})(5.98 \times 10^{24} \text{ kg})(2.98 \times 10^4 \text{ m/s})^2$$
$$= 2.7 \times 10^{33} \text{ J}. \qquad \text{(Answer)}$$

The U.S. annual energy production, from all sources, is only about 6×10^{19} J. If we could find a way to siphon off and store only one-billionth of the earth's kinetic energy, it would supply our country's energy needs for about 45,000 years.

Sample Problem 8   A neutron in a beam emerging from a nuclear reactor travels a distance d of 6.2 m across a research laboratory in a time t of 180 μs. What is its kinetic energy? The mass of the neutron is 1.67×10^{-27} kg.

We find the speed from

$$v = \frac{d}{t} = \frac{6.2 \text{ m}}{180 \times 10^{-6} \text{ s}} = 3.44 \times 10^4 \text{ m/s}.$$

From Eq. 15, the kinetic energy is

$$K = \tfrac{1}{2}mv^2 = (\tfrac{1}{2})(1.67 \times 10^{-27} \text{ kg})(3.44 \times 10^4 \text{ m/s})^2$$
$$= 9.9 \times 10^{-19} \text{ J} = 6.2 \text{ eV}. \qquad \text{(Answer)}$$

When dealing with subatomic particles such as the neutron, the electron-volt is a much more convenient energy unit than the joule; see Eq. 6.

Sample Problem 9   You drop a penny, whose mass m is 3.2 g, down a stairwell to a floor a distance h (11 m) below. What is the speed v of the coin just before it hits the floor?

We first calculate the work done on the falling coin by the net force that acts on it. In the absence of air resistance, this (constant) force is its weight mg. From Eq. 2 we have

$$W = Fd \cos \phi = (mg)(h)(\cos 0°) = mgh.$$

The initial kinetic energy K_i of the coin is zero and its final kinetic energy K_f is $\tfrac{1}{2}mv^2$. From the work–energy theorem (Eq. 16) we then have

$$W = K_f - K_i$$

or

$$mgh = \tfrac{1}{2}mv^2 - 0.$$

We have at once that

$$v = \sqrt{2gh} = \sqrt{(2)(9.8 \text{ m/s}^2)(11 \text{ m})} = 15 \text{ m/s}. \quad \text{(Answer)}$$

You should also solve this problem in the familiar way, using the equations for constant acceleration listed in Table 2 of Chapter 2, and prove to yourself that you get the same answer. Which method you select is purely a matter of convenience. Both methods are bound to give the same answer because they are both derived from the same definitions and basic assumptions.

Sample Problem 10   A block whose mass m is 5.7 kg slides on a horizontal frictionless table top with a constant speed v of 1.2 m/s. It is brought momentarily to rest by compressing a spring in its path, as in Fig. 14. By what distance d is the spring compressed? The spring constant k is 1500 N/m.

Let us apply the work–energy theorem to the block as it is brought to rest by the spring. From Eq. 14, the work done *by* the spring force *on* the block as the spring is compressed a distance d is given by

$$W = -\tfrac{1}{2}kd^2.$$

The change in kinetic energy is

$$\Delta K = K_f - K_i = 0 - \tfrac{1}{2}mv^2.$$

Figure 14 A block moves toward a spring. It will compress it momentarily by a maximum distance d.

The work–energy theorem requires that these two quantities be equal. Setting them so and solving for d, we obtain

$$d = v\sqrt{\frac{m}{k}} = (1.2 \text{ m/s})\sqrt{\frac{5.7 \text{ kg}}{1500 \text{ N/m}}}$$

$$= 7.4 \times 10^{-2} \text{ m} = 7.4 \text{ cm.} \qquad \text{(Answer)}$$

7-6   Power

A contractor wishes to lift a load of bricks from the sidewalk to the top of a building. It is easy to calculate how much work is required to do this. The contractor, however, is much more interested in the length of time required for the winch to perform this work. Will the job take 1 hour or 1 week?

When we are thinking about buying an outboard motor, we are not concerned at all with how much work the motor can do. We want to know how fast it can do this work. Can it drive our boat at 10 knots?

When a quarryworker uses dynamite for blasting in a marble quarry, he is not interested in how much energy the dynamite can release. This same energy could easily be released by putting an electric heater in the tamp hole and waiting. The quarryworker wants the energy to be released in a very short time, so that the overpressure necessary to blast apart the rock will be developed.

The concept involved here is the rate of doing work, or *power*. If an amount of work W is carried out in a time interval Δt, the average power for that interval is defined to be

$$\boxed{\bar{P} = \frac{W}{\Delta t}} \qquad \text{(average power).} \qquad (20)$$

The *instantaneous power* P is the instantaneous rate of

doing work, which we can write as

$$P = \frac{dW}{dt}. \qquad (21)$$

The SI unit of power is the joule per second. This unit is used so often that it has been given a special name, the *watt* (abbr. W). This unit is named after James Watt, who so greatly improved the rate at which steam engines could do work. In the British system, the unit of power is the foot·pound per second. Often the horsepower (= 550 ft·lb/s) is used. Some relations are

$$1 \text{ watt} = 1 \text{ W} = 1 \text{ J/s} = 0.738 \text{ ft·lb/s} \qquad (22)$$

and

$$1 \text{ horsepower} = 1 \text{ hp} = 550 \text{ ft·lb/s} = 746 \text{ W.} \qquad (23)$$

Inspection of Eq. 20 shows that work can be expressed as power multiplied by time. This is the origin of the common energy unit, the kilowatt·hour. Thus,

$$1 \text{ kilowatt·hour} = 1 \text{ kW·h} = (10^3 \text{ W})(3600 \text{ s})$$
$$= 3.60 \times 10^6 \text{ J} = 3.60 \text{ MJ.} \qquad (24)$$

Perhaps because of our utility bills, the watt and the kilowatt·hour have become identified as electrical units. They can be used equally well to express mechanical power and energy. Thus, if you pick up this book from the floor and put it on a table top, you are free to report the work that you have done as 4×10^{-6} kW·h (or more conveniently as 4 mW·h).

We can also express the power delivered to a body in terms of the force that acts on the body and its velocity. Thus, for a particle moving in one dimension, Eq. 21 becomes

$$P = \frac{dW}{dt} = \frac{F \, dx}{dt} = F\left(\frac{dx}{dt}\right),$$

which we can write simply as

$$P = Fv, \qquad (25)$$

in which P is the instantaneous power. In the more general case of motion in three dimensions, we can extend Eq. 25 to

$$\boxed{P = \mathbf{F} \cdot \mathbf{v}.} \qquad (26)$$

Sample Problem 11 A load of bricks whose mass m is 420 kg is to be lifted through a height h of 120 m in 5.0 min. What must be the minimum power of the winch motor?

The work to be done is

$$W = mgh = (420 \text{ kg})(9.8 \text{ m/s}^2)(120 \text{ m})$$
$$= 4.94 \times 10^5 \text{ J}.$$

From Eq. 20, the average power is

$$\bar{P} = \frac{W}{\Delta t} = \frac{4.94 \times 10^5 \text{ J}}{5 \times 60 \text{ s}}$$
$$= 1650 \text{ W} = 1.65 \text{ kW} = 2.2 \text{ hp}. \qquad \text{(Answer)}$$

The actual power must be greater than this because of frictional power losses and other inefficiencies. We have assumed that the mass of the winch platform and cable were negligible compared with the mass of the bricks and that the load was lifted without appreciable acceleration.

Sample Problem 12 An 80-hp outboard motor, operating flat out (see Fig. 15), can drive a speedboat at 22 knots (= 25 mi/h = 11 m/s). What is the forward thrust (force) of the motor?

Figure 15 Sample Problem 12. A speedboat in equilibrium. The power generated by its engine goes to overcome the drag force exerted on the boat (and on the water skier) by the water. The kinetic energies of the boat and the skier remain constant.

From Eq. 25 we have

$$F = \frac{P}{v} = \frac{(80 \text{ hp})(746 \text{ W/hp})}{11 \text{ m/s}}$$
$$= 5400 \text{ N} (= 1200 \text{ lb}). \qquad \text{(Answer)}$$

Note that, at constant speed, the thrust of the propeller is just balanced by the resistive drag force of the water so that the boat does not accelerate.

7–7 Kinetic Energy at High Speeds (Optional)

For particles moving at speeds that approach the speed of light, Newtonian mechanics fails and must be replaced by Einstein's special theory of relativity.* One consequence is that we can no longer use the expression $K = \frac{1}{2}mv^2$ for the kinetic energy of a particle. Instead, we must use

$$K = mc^2 \left(\frac{1}{\sqrt{1 - (v/c)^2}} - 1 \right), \qquad (27)$$

in which c is the speed of light.†

Figure 16 shows that these two formulas, which look so different, do indeed give widely different results at high speeds. Experiment shows that—beyond any prospect of doubt—the relativistic expression (Eq. 27) is correct and the classical expression (Eq. 15) is not. At low speeds, however, the two formulas merge, yielding the same result. In particular, both formulas yield $K = 0$ for $v = 0$.

All relativistic formulas must reduce to their classical counterparts at low speeds. To see how this comes about for Eq. 27, let us write that equation in the form

$$K = mc^2[(1 - \beta^2)^{-1/2} - 1]. \qquad (28)$$

Here, for convenience, we have substituted the dimensionless *speed parameter* β for the speed ratio v/c.

At very low speeds, $v \ll c$ and therefore $\beta \ll 1$. At low speeds then, we can then expand the quantity $(1 - \beta^2)^{-1/2}$ by the binomial theorem, obtaining (see

---

* Our first encounter with this fact was in Section 4–9 (Relative Motion at High Speeds).

† In relativistic equations as we present them throughout this book, the mass m will always be understood to be the mass measured when the particle is at rest or essentially at rest.

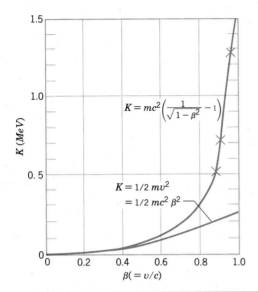

Figure 16 The relativistic (Eq. 27) and the classical (Eq. 15) expressions for the kinetic energy of an electron, plotted as a function of v/c, where v is the speed of the electron and c is the speed of light. Note that at low speeds, the two curves blend into each other. At high speeds, they diverge widely. The crosses are experimental points, showing that, at high speeds, the relativistic curve agrees with experiment but the classical curve does not.

Hint 3)

$$(1 - \beta^2)^{-1/2} = 1 + \tfrac{1}{2}\beta^2 + \cdots. \qquad (29)$$

Substituting Eq. 29 into Eq. 28 leads to

$$K = mc^2[(1 + \tfrac{1}{2}\beta^2 + \cdots) - 1].$$

If β is small enough, successive terms in the sum in the above equation will decrease rapidly in size. We can then, with little error, replace the sum by its first two terms, obtaining

$$K \approx (mc^2)[(1 + \tfrac{1}{2}\beta^2) - 1]$$

or

$$K = (mc^2)(\tfrac{1}{2}\beta^2) = \tfrac{1}{2}mv^2,$$

which is exactly what we hoped to prove. In the last step above, we replaced β by v/c.

Hint 3: Approximations Often we have a quantity written in the form $(a + b)^n$ and we want to find its approximate

value for the case in which $b \ll a$. It is simplest to recast it in the form $(1 + x)^n$, where x is dimensionless and is much less than unity. Thus, we can put

$$(a + b)^n = a^n(1 + b/a)^n = (a^n)(1 + x)^n.$$

This is of the desired form, with $x (= b/a)$ being dimensionless. We can then evaluate $(1 + x)^n$ by the binomial theorem, keeping only as many terms as are appropriate to the problem.

For our purpose, the binomial theorem (see Appendix G) can be written

$$(1 + x)^n = 1 + \frac{n}{1!}x + \frac{n(n-1)}{2!}x^2 + \cdots. \qquad (30)$$

In applying Eq. 30 in connection with Eq. 29, note that $x = -\beta^2$ and $n = -\tfrac{1}{2}$. The exclamation marks in Eq. 30 identify factorials. That is, $4! = 4 \times 3 \times 2 \times 1 (= 24)$; you probably have a key for factorials on your calculator.

For exercise, evaluate $(1 + 0.045)^{-2.3}$ both on your calculator and by expansion, using Eq. 30 with $x = 0.045$ and $n = -2.3$. Check the various terms in the binomial sum and see how rapidly they drop off.

7-8 Reference Frames (Optional)

We have agreed to apply Newton's laws of mechanics only in inertial reference frames. Recall that these are frames that move with respect to each other at a constant velocity.

There are some physical quantities such that, if observers in different inertial reference frames measure them, they will all find the same answer. In Newtonian mechanics, these *invariant* quantities (as they are called) are force, mass, acceleration, and time. That is, if an observer in one frame finds that a certain particle has a mass of 3.15 kg, observers in all other frames will find the same result.

For other measured quantities, such as the displacement of a particle or the velocity of a particle, observers in different frames will find different answers; these quantities are *not* invariant.

If the displacement of a particle depends on the reference frame of the observer, so must the work done on the particle because work ($W = \mathbf{F} \cdot \mathbf{d}$; see Eq. 4) is defined in terms of displacement. If the displacement of a particle during a given interval is measured to be

+2.47 m in one reference frame, it can be measured to be zero in another and −3.64 m in a third. The force **F** does not change (it is invariant) so that work that is positive in one frame can be zero in another and negative in a third. Before we conclude that the work–energy theorem is in trouble, however, let us take a look at the kinetic energy.

If the velocity of a particle depends on the reference frame of the observer, then so must the kinetic energy of a particle because kinetic energy ($K = \frac{1}{2}mv^2$) is defined in terms of velocity. The mass m does not change (it is invariant) so that the kinetic energy of a particle can have a value of 5.62 J in one frame, 8.95 J in another frame, and be zero in a third.

From Galileo to Einstein, physicists have come to believe in the *principle of invariance,* namely:

> *The laws of physics must have the same form in all inertial reference frames.*

That is, it does not matter that some *physical quantities* have different values in different reference frames; nevertheless, the *laws of physics* must hold true in all frames. Behind this formal statement of invariance is a curiosity to see what events look like from different perspectives and a feeling that—in some deep sense—if different observers look at the same event, they must see nature working in the same way.

Among the laws to which this invariance principle applies is the work–energy theorem. Thus, even though different observers, watching the same moving particle, would get different answers for their measurements of work and of kinetic energy, they would each agree that the work–energy theorem holds in their frame. Let us look at a simple case.

In Fig. 17, Sally is riding up in an elevator at constant speed and holding a book. Steve, standing on a facing balcony, observes her as the cab rises through a height h. What is the work–energy situation as viewed from these two reference frames?

1. Sally's Report. "My reference frame is the elevator cab. I am exerting an upward force on the book but this force does no work because the book is not moving in my reference frame. The weight of the book, acting downward, does no work, for the same reason. Thus, the total work done on the book by all the forces that act on it is zero. From the work–energy theorem, the kinetic energy of the book should not change. That is what I ob-

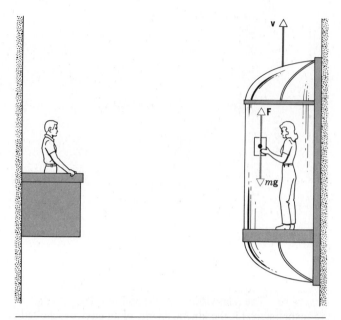

Figure 17 Sally rides in an elevator, holding a book. Steve watches her. Both check out the work–energy theorem in their respective reference frames, as it applies to the motion of the book.

serve; the kinetic energy of the book is zero in my reference frame and remains so. Everything fits together."

2. Steve's Report. "My reference frame is the balcony. I see the force **F** that Sally is exerting on her book. In my frame, its point of application is moving and the work it does as it moves upward through a height h is $+mgh$. I also see the weight vector; it is also doing work on the book, in amount $-mgh$. Thus, the total rate at which work is being done on the book is zero. From the work–energy theorem, the kinetic energy of the book should not change. That is what I observe. In my frame, the kinetic energy is $\frac{1}{2}mv^2$ and remains so. Everything fits together."

Although Steve and Sally do not agree about a number of things (for example, the displacement of the book and its kinetic energy), they both agree that the work–energy theorem works in their reference frame.

It does not matter what (inertial) reference frame you pick in which to solve a problem but (1) make sure you know what that frame is and (2) use that frame consistently throughout the problem.

REVIEW AND SUMMARY

We begin, in this chapter, a study of work and energy, which will culminate with the *law of conservation of energy* in Chapter 8.

Whenever a constant force **F** acts on an object while it experiences a displacement **d**, we say that the force does *work W* on the object. The amount of work done is the scalar quantity calculated from

Work by a Constant Force

$$W = Fd \cos \phi = \mathbf{F} \cdot \mathbf{d} \quad \text{(work: constant force).} \qquad [2,4]$$

where ϕ is the angle between the directions of **F** and **d**. W may be positive, zero, or negative, depending on the value of ϕ; see Sample Problem 1. The SI unit of work is the *joule* (J) with

The Joule

$$1 \text{ J} = 1 \text{ N} \cdot \text{m} = 1 \text{ kg} \cdot \text{m}^2/\text{s}^2 = 0.738 \text{ ft} \cdot \text{lb.} \qquad [5]$$

The *electron-volt* (eV) is an important unit of work used in atomic and nuclear physics. It is defined as

The Electron-Volt

$$1 \text{ eV} = 1.60 \times 10^{-19} \text{ J.} \qquad [6]$$

The *kilowatt · hour* (kW · h) is also commonly used in engineering practice:

The Kilowatt · Hour

$$1 \text{ kilowatt} \cdot \text{hour} = 1 \text{ kW} \cdot \text{h} = 3.6 \times 10^6 \text{ J} = 3.60 \text{ MJ.} \qquad [24]$$

Sample Problems 1–3 show typical work situations involving constant forces.

If the directions of the force and displacement are parallel (and parallel to an x axis), but the magnitude of the force varies with position x, the work done by the force as the object moves from x_i (x_{initial}) to x_f (x_{final}) is given, by extension of Eq. 2, as

Work by a Variable Force

$$W = \int_{x_i}^{x_f} F(x)dx \quad \text{(work: variable force).} \qquad [10]$$

An application is the work done to stretch or compress a spring by a distance x from its relaxed length. If the spring obeys *Hooke's law,* the force exerted *by* the spring is

Hooke's Law

$$F(x) = -kx \quad \text{(Hooke's law),} \qquad [11]$$

in which k is a *spring constant* that is characteristic of the spring; see Sample Problem 4. The work done *by* a spring as it is stretched or compressed from x_i to x_f is

Work to Stretch a Spring

$$W = \tfrac{1}{2}kx_i^2 - \tfrac{1}{2}kx_f^2 \quad \text{(work: } by \text{ a spring).} \qquad [13]$$

Sample Problems 5 and 6 illustrate several applications of this important relation.

When an object with mass m moves with speed v, we assign to it a *kinetic energy K* given by

Kinetic Energy

$$K = \tfrac{1}{2}mv^2 \quad \text{(kinetic energy defined).} \qquad [15]$$

Kinetic energy is a positive scalar quantity and has the same units and dimensions as work. Sample Problems 7 and 8 show examples of calculating the kinetic energies of moving objects.

The *work–energy theorem,*

The Work–Energy Theorem

$$W = K_f - K_i = \Delta K \quad \text{(work-energy theorem),} \qquad [16]$$

states that the *change* in the kinetic energy of a particle ($\Delta K = K_{\text{final}} - K_{\text{initial}}$) during any time interval is equal to the total work done on that particle, during the time interval, by all the forces that act on it. This theorem, which is derived here from Newton's laws of motion, provides an alternate and often simpler way of solving dynamics problems. See Sample Problems 9 and 10.

Power is the *rate* at which work is done. If a force does work W during a time interval Δt, the average *power* is

$$\bar{P} = \frac{W}{\Delta t} \quad \text{(average power).} \qquad [20]$$

Power

Instantaneous power is

$$P = \frac{dW}{dt}.$$ [21]

If force **F** is acting on an object moving along a straight line with velocity **v**, its power is

$$P = \mathbf{F} \cdot \mathbf{v}.$$ [26]

Like work, power is a scalar. Its SI unit is the watt:

The Watt

$$1 \text{ watt} = 1 \text{ W} = 1 \text{ J/s} = 0.738 \text{ ft} \cdot \text{lb/s}.$$ [22]

The unit of power in British units is the ft·lb/s, with the horsepower (hp) being commonly used:

The Horsepower

$$1 \text{ hp} = 550 \text{ ft} \cdot \text{lb/s} = 746 \text{ W}$$ [23]

Sample Problems 11 and 12 illustrate the use of the concept of power.

Kinetic energy for objects moving at speeds comparable to the speed of light must be calculated using the relativistic expression

Relativistic Kinetic Energy

$$K = mc^2 \left(\frac{1}{\sqrt{1 - (v/c)^2}} - 1 \right),$$ [27]

in which c is the speed of light. This formula for kinetic energy, which has been verified experimentally over a wide range of speeds, reduces to Eq. 15 when v is much less than c.

The Principle of Invariance

Some *invariant* quantities (mass, force, acceleration, and time in Newtonian mechanics) have the same numerical values in all inertial reference frames. Others (for example, velocity, kinetic energy, and work) have different values. However, the *laws* of physics must have the same *form* in all inertial reference frames. This is the *principle of invariance.*

QUESTIONS

1. Can you think of other words like "work" whose colloquial meanings are often different from their scientific meanings?

2. Why is it tiring to hold a heavy weight even though no work is done?

3. The inclined plane (Sample Problem 3) is a simple "machine" that enables us to do work with the application of a smaller force than is otherwise necessary. The same statement applies to a wedge, a lever, a screw, a gear wheel, and a pulley combination. Far from saving us work, such machines in practice require that we do a little more work with them than without them. Why is this so? Why do we use such machines?

4. In a tug of war, one team is slowly giving way to the other. What work is being done and by whom?

5. Give an example in which positive work is done by a frictional force.

6. The displacement of a body depends on the reference frame of the observer. It follows that the work done on a body should also depend on the observer's reference frame. You drag a crate across a rough floor by pulling on it with a rope. Identify reference frames in which the work done on the crate by the tension in the rope would be (*a*) positive, (*b*) zero, and (*c*) negative.

7. Why can you much more easily ride a bicycle for a mile on level ground than run the same distance? In each case, you transport your own weight for a mile and in the first you must also transport the bicycle and, moreover, do so in a shorter time! (See *The Physics Teacher,* March 1981, p. 194.)

8. Suppose that the earth revolves around the sun in a perfectly circular orbit. Does the sun do any work on the earth?

9. You slowly lift a bowling ball from the floor and put it on a table. Two forces act on the ball: its weight, *m***g**, and your upward force, − *m***g**. These two forces cancel each other so that it would seem that no work is done. On the other hand, you know that you have done some work. What is wrong?

10. You cut a spring in half. What is the relation of the spring constant k for the original spring to that for either of the half-springs?

11. Springs A and B are identical except that A is stiffer than B; that is, $k_A > k_B$. On which spring is more work expended if they are stretched (*a*) by the same amount and (*b*) by the same force?

12. Does kinetic energy depend on the direction of the motion involved? Can it be negative? Does its value depend on the reference frame of the observer?

13. In picking up a book from the floor and putting it on a table, you do work. However, the kinetic energy of the book does not change. Is there a violation of the work–energy theorem here? Explain why or why not.

14. Does the work–energy theorem hold if friction acts on an object? Explain your answer.

15. You throw a ball vertically in the air and catch it. What does the work–energy theorem say qualitatively about the free flight during this round trip? Answer the question first neglecting air resistance and then taking it into account.

16. Why can a car so easily pass a loaded truck when going uphill? The truck is heavier, of course, but its engine is more powerful in proportion (or is it?). What considerations enter into choosing the design power of a truck engine and of a car engine?

17. Does the power needed to raise a box onto a platform depend on how fast it is raised?

18. You lift some library books from a lower to a higher shelf in time Δt. Does the work that you do depend on (a) the mass of the books, (b) the weight of the books, (c) the height of the upper shelf above the floor, (d) the time Δt, and (e) whether you lift the books sideways or directly upward?

19. We hear a lot about the "energy crisis." Would it be more accurate to speak of a "power crisis"?

20. We say that a 1-keV electron is a "classical" particle, a 1-MeV electron is a "relativistic" particle, and a 1-GeV electron is an "extremely relativistic" particle. What exactly do these terms mean?

21. Sally and Yuri are flying jet planes at the same speed on parallel low-altitude paths. Suddenly, Sally lands at a convenient air strip. Consider how this looks from the reference frame of Yuri, who keeps on flying. (a) Would he say that Sally's plane had gained or lost kinetic energy? (b) Would he say that work was done *on* her plane or *by* her plane as she landed? (c) Would he conclude that the work–energy theorem holds? (d) Answer these same questions from the reference frame of Chang, who is in the control tower.

EXERCISES AND PROBLEMS

Section 7–2   Work: Motion in One Dimension with a Constant Force

1E. To push a 50-kg crate across a floor, a worker applies a force of 200 N, directed 20° above the horizontal. The floor exerts a 175-N force of friction on the crate. As the crate moves 3.0 m, what work is done on the crate by (a) the worker, (b) the force of friction, (c) the force of gravity, and (d) the normal force of the floor on the crate? (e) What is the total work done on the crate?

2E. To push a 25-kg crate up a 25° incline, a worker exerts a force of 200 N, parallel to the incline. The incline exerts a 96-N force of friction on the crate. As the crate slides 1.5 m, what work is done on the crate by (a) the worker, (b) the force of friction, (c) the force of gravity, and (d) the normal force of the incline on the crate? (e) What is the total work done on the crate?

3E. A 100-kg object is initially moving in a straight line with a speed of 50 m/s. (a) If it is brought to a stop with a deceleration of 2.0 m/s², what force is required, what distance does it travel, and what work is done by the force? (b) Answer the same questions if its deceleration is 5.0 m/s².

4E. A 100-lb block of ice ($m = W/g = 45$ kg) slides down an incline 5.0 ft (= 1.5 m) long and 3.0 ft (= 0.91 m) high. A worker pushes up on the ice parallel to the incline so that it slides down at constant speed. The coefficient of friction between the ice and the incline is 0.10. Find (a) the force exerted by the worker, (b) the work done by the worker on the block, (c) the work done by gravity on the block, (d) the work done by the surface of the incline on the block, and (e) the work done by the resultant force on the block.

5P. A cord is used to lower vertically a block of mass M a distance d at a constant downward acceleration of $g/4$. (a) Find the work done by the cord on the block. (b) Find the work done by the force of gravity.

6P. Figure 18 shows an arrangement of pulleys designed to facilitate the lifting of a heavy load L. Assume that friction can be ignored everywhere and that the pulleys to which the load is attached weigh a total of 20 lb. An 840-lb load is to be raised 12 ft. (a) What is the minimum applied force F that can lift the

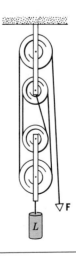

Figure 18   Problem 6.

load? (b) How much work must be done against gravity in lifting the 840-lb load 12 ft? (c) Through what distance must the applied force be exerted to lift the load 12 ft? (d) How much work must be done by F to accomplish this task?

7P. A worker pushed a 60-lb block ($m = W/g = 27$ kg) 30 ft ($= 9.2$ m) along a level floor at constant speed with a force directed 32° below the horizontal. If the coefficient of kinetic friction is 0.20, how much work did the worker do on the block?

8P. A 50-kg trunk is pushed 6.0 m at constant speed up a 30° incline by a constant horizontal force. The coefficient of sliding friction between the trunk and the incline is 0.20. Calculate the work done by (a) the applied force, (b) the frictional force, and (c) the gravitational force.

9P. A block of mass $m = 3.57$ kg is drawn at constant speed a distance $d = 4.06$ m along a horizontal floor by a rope exerting a constant force of magnitude $F = 7.68$ N at an angle $\theta = 15.0°$ above the horizontal. Compute (a) the work done by the rope on the block, (b) the work done by friction on the block, (c) the total work done on the block, and (d) the coefficient of kinetic friction between block and floor.

10P. A crate weighing 500 lb ($m = W/g = 230$ kg) is suspended from the end of a rope 40 ft ($= 12$ m) long. By applying a horizontal force, the crate is then pushed aside 4.0 ft ($= 1.2$ m) from the vertical and held there. (a) What is the force needed to keep the crate in this position? (b) How much work is being done to keep the crate in this position? (c) What work was done by gravity when the crate was moved aside? (d) How much work was done by the horizontal force? (e) Is the answer to (d) equal to the answer to (a) times 4.0 ft ($= 1.2$ m)? Explain why or why not. (f) How much work is done by the tension in the rope? (g) What was the total work done in moving the crate?

Section 7–3 Work: Motion in One Dimension with a Variable Force

11E. A 5.0-kg block moves in a straight line on a horizontal frictionless surface under the influence of a force that varies with position as shown in Fig. 19. How much work is done by the force as the block moves from the origin to $x = 8.0$ m?

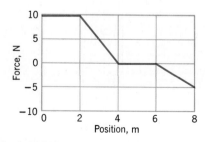

Figure 19 Exercise 11.

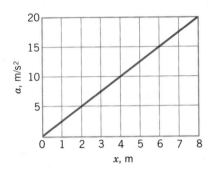

Figure 20 Exercise 12.

12E. A 10-kg mass moves along the x axis. Its acceleration as a function of its position is shown in Fig. 20. What is the net work performed on the mass as it moves from $x = 0$ to $x = 8.0$ m?

13P. (a) Estimate the work done by the force shown on the graph (Fig. 21) in displacing a particle from $x = 1$ m to $x = 3$ m. Refine your method to see how close you can come to the exact answer of 6 J. (b) The curve is given analytically by $F = a/x^2$, where $a = 9$ N·m². Show how to calculate the work by the rules of integration.

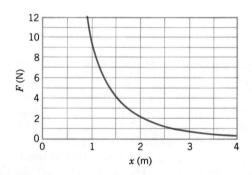

Figure 21 Problem 13.

14P. The force exerted on an object is $F = F_0(x/x_0 - 1)$. Find the work done in moving the object from $x = 0$ to $x = 2x_0$ (a) by plotting $F(x)$ and finding the area under the curve and (b) by evaluating the integral analytically.

Section 7–4 Work Done by a Spring

15E. A spring has a force constant of 15 N/cm. (a) How much work is required to extend the spring 7.6 mm from its equilibrium position? (b) How much work is needed to extend the spring an additional 7.6 mm?

16P. Figure 22 shows a spring with a pointer attached, hanging next to a scale graduated in millimeters. Three different

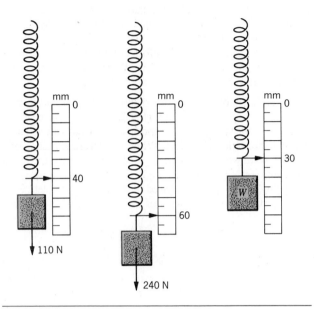

Figure 22   Problem 16.

weights are hung from the spring, in turn, as shown. (*a*) If all weight is removed from the spring, which mark on the scale will the pointer indicate? (*b*) What is the weight *W*?

Section 7–5   Kinetic Energy

17E.   A conduction electron in copper near the absolute zero of temperature has a kinetic energy of 6.7×10^{-19} J. What is the speed of the electron?

18E.   Calculate the kinetic energies of the following objects moving at the given speeds: (*a*) a 110-kg football linebacker running at 8.1 m/s; (*b*) a 4.2-g bullet at 950 m/s; (*c*) the aircraft carrier *Nimitz,* 91,400 tons at 32 knots.

19E.   A proton (nucleus of the hydrogen atom) is being accelerated in a linear accelerator. In each stage of such an accelerator, the proton is accelerated along a straight line at 3.6×10^{15} m/s². If a proton enters such a stage moving initially with a speed of 2.4×10^7 m/s and the stage is 3.5 cm long, compute (*a*) its speed at the end of the stage and (*b*) the gain in kinetic energy resulting from the acceleration. The mass of the proton is 1.67×10^{-27} kg. Express the energy in electron-volts.

20E.   A proton starting from rest is accelerated in a cyclotron to a final speed of 3.0×10^6 m/s (about 1% of the speed of light). How much work, in electron-volts, is done on the proton by the electrical force of the cyclotron that accelerated it?

21E.   An outfielder throws a baseball with an initial speed of 120 ft/s (= 36.6 m/s). Just before an infielder catches the ball at the same level, its speed is reduced to 110 ft/s (= 33.5 m/s). What work was done by the force of the air on the ball? The weight of a baseball is 9.0 oz (*m* = *W/g* = 255 g).

22E.   A 30-g bullet, initially traveling 500 m/s, penetrates 12 cm into a solid wall. (*a*) What work is done by the wall in stopping the bullet? (*b*) Assume that the force of the wall on the bullet is constant and calculate its value.

23E.   A single force acts on a body in rectilinear motion. A plot of velocity versus time for the body is shown in Fig. 23. Find the sign (positive or negative) of the work done *by* the force *on* the body in each of the intervals *AB, BC, CD,* and *DE.*

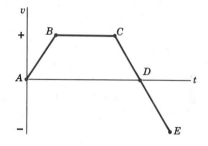

Figure 23   Exercise 23.

24E.   A 75-g frisbee is thrown from a height of 1.1 m above the ground with a speed of 12 m/s. When it has reached a height of 2.1 m, its speed is 10.5 m/s. How much work was done on the frisbee by (*a*) gravity and (*b*) the air?

25E.   A firehose is uncoiled by pulling the end of the hose horizontally along a smooth surface at the steady speed of 2.3 m/s, as shown in Fig. 24. One meter of the hose has a mass of 0.25 kg. How much work is required to uncoil 12 m of hose?

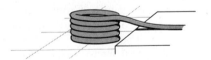

Figure 24   Exercise 25.

26E.   If a 2.9×10^5-kg Saturn V rocket with an Apollo spacecraft attached must achieve an escape velocity of 11.2 km/s (= 25,000 mi/h) near the surface of the earth, how much energy must the fuel contain? Would the system actually need as much or would it need more or less? Why?

27E.   From what height would a 2800-lb (= 12,000 N) automobile have to fall to gain the kinetic energy equivalent to what it would have when going 55 mi/h (= 89 km/h)? Does the answer depend on the weight of the car?

28E.   A 1000-kg car is traveling at 60 km/h on a level road. The brakes are applied long enough to do 50 kJ of work. (*a*)

What is the final speed of the car? (*b*) How much more work must be done by the brakes to stop the car?

29P. A running man has half the kinetic energy that a boy of half his mass has. The man speeds up by 1.0 m/s and then has the same kinetic energy as the boy. What were the original speeds of man and boy?

30P. A force acts on a 3.0-kg particle in such a way that the position of the particle as a function of time is given by $x = 3t - 4t^2 + t^3$, where x is in meters and t is in seconds. Find the work done by the force during the first 4.0 s.

31P. A 0.63-kg ball is thrown straight up into the air with an initial speed of 14 m/s. It reaches a height of 8.1 m, then falls back down. Assume that the only forces acting are those of gravity and air resistance and calculate the work done during the ascent by the force of air resistance.

32P. The earth circles the sun once a year. How much work would have to be done on the earth to bring it to rest relative to the sun? See Appendix C for numerical data and ignore the rotation of the earth about its own axis.

33P. A helicopter is used to lift a 160-lb astronaut ($m = W/g = 72$ kg) 50 ft (= 15 m) vertically from the ocean by means of a cable. The acceleration of the astronaut is $g/10$. (*a*) How much work is done by the helicopter on the astronaut? (*b*) How much work is done by the gravitational force on the astronaut? What are (*c*) the kinetic energy and (*d*) the speed of the astronaut just before he reaches the helicopter?

34P. A 60-lb (= 267-N) girl slides down a 20-ft (= 6.1 m) playground slide that makes an angle of 20° with the horizontal. The coefficient of kinetic friction is 0.10. (*a*) Find the work done by the force of gravity. (*b*) Find the work done by the force of friction. (*c*) If she starts at the top with a speed of 1.5 ft/s (= 0.457 m/s), what is her speed at the bottom?

35P. A 0.55-kg projectile is launched from the edge of a cliff with an initial kinetic energy of 1550 J and at its highest point is 140 m above the launch point. (*a*) What is the horizontal component of its velocity? (*b*) What was the vertical component of its velocity just after launch? (*c*) At one instant during its flight the vertical component of its velocity is found to be 65 m/s. At that time, how far is it above or below the launch point?

36P. A comet having a mass of 8.38×10^{11} kg strikes the earth at a relative speed of 30 km/s. (*a*) Compute the kinetic energy of the comet in megatons of TNT; the detonation of one million tons of TNT releases 4.2×10^{15} J of energy. (*b*) The diameter of the crater blasted by a large explosion is proportional to the one-third power of the explosive energy released, with a 1 megaton of TNT explosion producing a crater about 1 km in diameter. What is the diameter of the crater produced by the impact of the comet? (In the past, impacts by comets may have been the cause of mass extinctions of many species of animals and plants; it is thought by many that dinosaurs became extinct by this mechanism.)

37P. A 2.0-kg block is forced against a horizontal spring of negligible mass, compressing the spring by 15 cm. When the block is released from the compressed spring, it moves 60 cm across a horizontal tabletop before coming to rest. The force constant of the spring is 200 N/m. What is the coefficient of sliding friction between the block and the table?

38P. A 250-g block is dropped onto a vertical spring with spring constant $k = 2.5$ N/cm (Fig. 25). The block becomes attached to the spring and the spring compresses 12 cm before coming momentarily to rest. While the spring is being compressed, what work is done (*a*) by the force of gravity and (*b*) by the spring? (*c*) What was the speed of the block just before it hits the spring? (*d*) If the initial speed of the block is doubled, what is the maximum compression of the spring? Ignore friction.

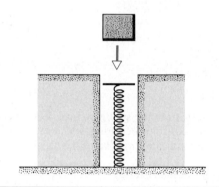

Figure 25 Problem 38.

Section 7–6 Power

39E. A 45,000-kg rocket acquires a speed of 6400 km/h one minute after launch. (*a*) What is its kinetic energy at the end of this first minute? (*b*) What is the average power expended during this first minute, neglecting frictional and gravitational forces?

40E. A boy whose mass is 51 kg climbs, with constant speed, a vertical rope 6.0 m long in 10 s. (*a*) How much work does the boy perform? (*b*) What is the boy's power output during the climb?

41E. A 55-kg woman ($W = mg = 120$ lb) runs up a flight of stairs having a rise of 4.5 m (= 15 ft) in 3.5 s. What average power must she supply?

42E. The loaded cab of an elevator has a mass of 3.0×10^3 kg and moves 200 m up the shaft in 20 s at constant speed. At what average rate does the cable do work on the cab?

43E. In a 100-person ski lift, a machine raises passengers averaging 150 lb (= 667 N) a height of 500 ft (= 152 m) in 60 s, at constant speed. Find the power output of the motor, assuming no frictional losses.

44E. Starting a race, a 150-lb (= 670-N) sprinter runs the first 7.7 yd (= 7.0 m) in 1.6 s, starting from rest and accelerating uniformly. (*a*) What is the sprinter's speed at the end of the 1.6 s? (*b*) What is the sprinter's kinetic energy? (*c*) What average power does the sprinter generate during the 1.6 s interval?

45E. The luxury liner *Queen Elizabeth 2* (see Fig. 26) is powered by a new diesel-electric powerplant, which replaced the original steam engines. The maximum power output is 92 MW at a cruising speed of 32.5 knots. What force is exerted by the propellers on the water at this highest attainable speed? (1 knot = 6076 ft/h.)

Figure 26 Exercise 45.

46E. What power, in horsepower, must be developed by the engine of a 1600-kg car moving at 25 m/s (= 90 km/h) on a level road if the forces of resistance total 700 N?

47E. A swimmer moves through the water at a speed of 0.22 m/s. The drag force opposing this motion is 110 N. What power is developed by the swimmer?

48E. An elevator cab in the New York Marriott Marquis (see the photo at the beginning of Chapter 2) has a mass of 4500 kg and can carry a maximum load of 1800 kg. The cab is moving upward at full load at 3.8 m/s. What power is required to keep the cab in motion?

49E. The energy consumed in running is about 80 kcal/km, regardless of speed. What average power does a runner expend during (*a*) a 100-m dash (time = 10 s) and (*b*) a marathon (distance = 26.2 mi; time = 2 h 10 min)? (1 cal = 4.186 J.)

50E. (*a*) Show that the power output of an airplane cruising at constant speed v in level flight is proportional to v^3. Assume that the aerodynamic drag force is given by Eq. 16 of Chapter 6. (*b*) By what factor must the engines' power be increased to increase the airspeed by 50%?

51P. Each second, 1200 m³ of water passes over a waterfall

100 m high. Assuming that three-fourths of the kinetic energy gained by the water in falling is converted to electrical energy by a hydroelectric generator, what is the power output of the generator? (Note that 1 m³ of water has a mass of 10^3 kg.)

52P. A 1400-kg block of granite is pulled up an incline at a constant speed of 1.34 m/s by a steam winch (Fig. 27). The coefficient of kinetic friction between the block and incline is 0.40. (*a*) How much work is done by each of the forces that act on the block as it moves 9.0 m up the incline? (*b*) How much power must be supplied by the winch?

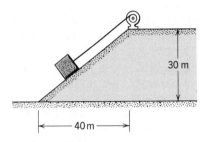

Figure 27 Problem 52.

53P. A 100-kg block is pulled at a constant speed of 5.0 m/s across a horizontal floor by an applied force of 122 N at 37° above the horizontal. (*a*) At what rate is the applied force doing work? (*b*) At what rate is the force of friction doing work?

54P. A horse pulls a cart with a force of 40 lb at an angle of 30° with the horizontal and moves along at a speed of 6.0 mi/h. (*a*) How much work does the horse do in 10 min? (*b*) What is the power output of the horse?

55P. A 3200-lb automobile ($m = W/g = 1500$ kg) starts from rest on a horizontal road and gains a speed of 45 mi/h (= 72 km/h) in 30 s. (*a*) What is the kinetic energy of the auto at the end of the 30 s? (*b*) What is the average net power delivered to the car during the 30-s interval? (*c*) What is the instantaneous power at the end of the 30-s interval assuming that the acceleration was constant?

56P. A 2.0-kg object accelerates uniformly from rest to a speed of 10 m/s in 3.0 s. (*a*) How much work is done on the object? (*b*) What is the instantaneous power delivered to the object 1.5 s after it starts?

57P. A net force of 5.0 N acts on a 15-kg body initially at rest. Compute (*a*) the work done by the force in the first, second, and third seconds and (*b*) the instantaneous power exerted by the force at the end of the third second.

58P. A fully loaded, slow-moving freight elevator has a total mass of 1200 kg. It is required to travel upward 54 m in 3.0 min. The counterweight has a mass of 950 kg. Find the power output, in horsepower, of the elevator motor. Ignore the work required to start and stop the elevator; that is, assume it travels at constant speed.

59P. Your car averages 30 mi/gal of gasoline. (*a*) How far could you travel on 1 kW·h of energy consumed? (*b*) If you are driving at 55 mi/h, at what rate are you expending energy? The heat of combustion of gasoline is 140 MJ/gal.

60P. The force required to tow a boat at constant velocity is proportional to the speed. If it takes 10 hp to tow a certain boat at a speed of 2.5 mi/h, how much horsepower does it take to tow it at a speed of 7.5 mi/h?

61P. What power is developed by a grinding machine whose wheel has a radius of 20 cm and runs at 2.5 rev/s when the tool to be sharpened is held against the wheel with a force of 180 N? The coefficient of friction between the tool and the wheel is 0.32.

62P. A truck can move up a road having a grade of 1.0-ft rise every 50 ft with a speed of 15 mi/h. The resisting force is equal to 1/25 the weight of the truck. If its power output is the same, how fast will the truck move down the hill? Assume that the resisting force remains unchanged.

63P. At full power, a 1.5-MW railroad locomotive accelerates a train from a speed of 10 m/s to 25 m/s in 6.0 min. (*a*) Neglecting friction, calculate the mass of the train. (*b*) Find the speed of the train as a function of time in seconds during the interval. (*c*) Find the force accelerating the train as a function of time during the interval. (*d*) Find the distance moved by the train during the interval.

64P. The resistance to motion of an automobile depends on road friction, which is almost independent of speed, and on aerodynamic drag, which is proportional to speed squared. For a 12,000-N car, the total resistant force F is given by $F = 300 + 1.8v^2$, where F is in newtons and v is in meters per second. Calculate the power required from the motor to accelerate the car at 0.92 m/s^2 when the speed is 80 km/h.

65P*. A *governor* consists of two 200-g masses attached by light, rigid 10-cm rods to a vertical rotating axle. The rods are hinged so that the masses swing out from the axle as they rotate with it. However, when the angle θ is 45°, the masses encounter

the wall of the cylinder in which the governor is rotating; see Fig. 28. (*a*) What is the minimum rate of rotation, in revolutions per minute, required for the masses to touch the wall? (*b*) If the coefficient of kinetic friction between the masses and the wall is 0.35, what power is dissipated as a result of the masses rubbing against the wall when the mechanism is rotating at 300 rev/min?

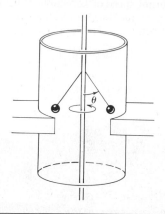

Figure 28 Problem 65.

Section 7–7 Kinetic Energy at High Speeds

66E. An electron is moving at a speed such that it could circumnavigate the earth at the equator in 1 s. (*a*) What is its speed in terms of the speed of light? (*b*) What is its kinetic energy in electron-volts? (*c*) What percent error do you make if you use the classical formula to compute the kinetic energy?

67E. The work–energy theorem applies to particles at all speeds. How much work, in keV, must be done to increase the speed of an electron from rest (*a*) to 0.50*c*, (*b*) to 0.99*c*, and (*c*) to 0.999*c*?

68P. An electron has a speed of 0.999*c*. (*a*) What is its kinetic energy? (*b*) If its speed is increased by 0.05%, by what percentage will its kinetic energy increase?

CHAPTER 8

THE CONSERVATION OF ENERGY

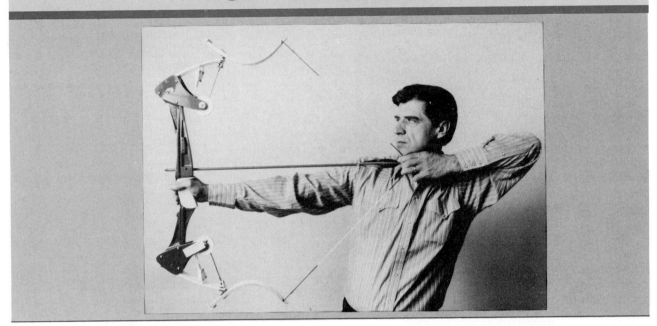

Miroslav Simo, engineer and toxophilite, does work in drawing the string of this high-tech bow of his own design. In so doing, he causes potential energy to be stored in the stressed bow. By releasing the string, he permits that stored potential energy to be converted into the kinetic energy of a moving arrow. Thus, energy is changed from one form to another.

8–1 Conservation Laws

Some of the most powerful laws of physics—called *conservation laws*—are expressed in a very simple form:

Consider a system of particles, completely isolated from outside influences. As the particles move about and interact with each other, there is a certain property of the system that does not change.

In shorthand form, we can express this as

$$X = \text{a constant} \quad \text{(isolated system)}, \quad (1)$$

in which X is the property that remains constant or, as we say, is *conserved*. Confronted with such a law, most phys-

icists feel that they are communicating with nature at a pretty deep level.

The conservation law central to this chapter is the *conservation of energy*. We deal first with frictionless mechanical systems, in which only kinetic and potential energies play a role. Later, we extend the law to include other forms of energy, such as thermal energy. First, we must introduce and define potential energy.

8–2 Potential Energy: Some Insights

In the photo at the head of this chapter, the archer does work in pulling back the bow string but no kinetic energy appears. However, the archer can generate some kinetic

Figure 1 A pole-vaulter does work in bending his fiberglass vaulting pole, thus storing potential energy. Near the top of his path, this potential energy is returned to him, in the form of kinetic energy. The introduction of energy-storing poles revolutionized this sport.

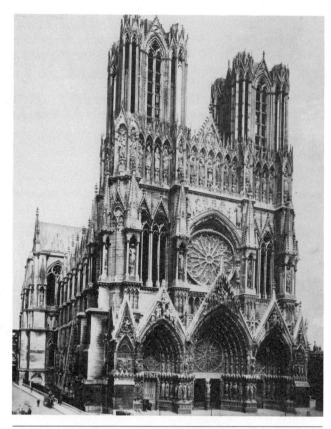

Figure 2 Rheims cathedral, built about A.D. 1240. The work done in lifting each stone into place remains stored in the structure as gravitational potential energy.

energy—in the form of a flying arrow—by simply releasing the bow string.

It makes sense to say that the kinetic energy of the arrow is in some sense "stored" in the stressed bow, put there by the archer as he drew the bow string. Thus appears the notion of *potential energy*, to which we give the symbol U. The name is suggestive: Energy is hidden away but has the potential to reappear in kinetic form. Potential energy could also be called *configuration energy*, because the system on which the work is done—the bow, in our example—stores potential energy by changing its configuration in some sense or other.

As another example, if you lift a book, you do work but no kinetic energy appears. However, you can easily generate some by dropping the book. We cannot speak of lifting a book without taking the earth into account. Lifting means "increasing the distance between the book and the center of the earth" or, equivalently, "changing the *configuration* of the book–earth system." Lifting a book is not unlike drawing a bow. Dropping the book is not unlike releasing the bow string. In each case, we do work and, by a later conscious act, see that work appear

as kinetic energy. Does it not make sense to consider the book–earth system, like a stressed bow, as a repository of potential energy?

Figure 1 shows still another example of potential energy being used. The pole-vaulter stores energy by bending the fiberglass vaulting pole (that is, by changing its configuration) and recovers it as kinetic energy, just when he needs it. In a related application, the indoor running track at Harvard University was computer-designed to incorporate a springiness matched to the requirements of a runner.* Thus, the track surface can deliver stored energy to the runner, much as in the case of pole vaulting. The proof of the track's success is in the record. In the year before the track was built, the at-home speed advantage of Harvard runners was −0.26%. In the first year of the track's use, the at-home advantage improved to +2.9%.

---

* See "Fast Running Tracks," by Thomas A. McMahon and Peter R. Greene, *Scientific American,* December 1978.

In looking at a medieval cathedral (Fig. 2), the discerning physicist or engineer must be impressed by the vast amount of work expended — over centuries — to lift each stone from the level of the quarry to its present elevation. Whatever else they may be, the medieval cathedrals stand with the modern skyscrapers as monuments to (gravitational) potential energy. Similarly, when we consider the damage done by a major earthquake, we are impressed by the enormous amount of (elastic) potential energy released as kinetic energy. Fault lines in the earth's crust are stressed "bows" whose moments of release are vitally important to those who live nearby.

8-3   A Close Look at Three Forces

It is our plan to examine three different forces and to ask of each: "Can we associate a potential energy with this force?" We shall see that the answer to this question for a spring force and for a gravitational force is yes, but for a frictional force the answer is no.

The Spring Force.   Figure 3a shows a spring in its relaxed state, with one end fastened to a wall. A block of mass m, sliding with kinetic energy K, has just made contact with the free end of the spring. We assume that the horizontal plane is frictionless and that the spring obeys Hooke's law (Eq. 11 of Chapter 7), or

$$F(x) = -kx \qquad \text{(Hooke's law).} \qquad (2)$$

Here $F(x)$, the *spring force,* is the force exerted *by* the spring *on* the block and x is the amount by which the spring is stretched (x positive) or compressed (x negative). The *spring constant* k is a measure of the stiffness of the spring: the larger the value of k, the stiffer the spring.

The block of Fig. 3a slows down, losing kinetic energy until it comes momentarily to rest (Fig. 3c) against the compressed spring. Then, as the spring expands, the block reverses its motion, regaining the kinetic energy it had previously lost.

It is useful to say that, as the spring is being compressed, the kinetic energy of the block is gradually transferred to the spring where it is stored as potential energy. As the compressed spring expands, its store of potential energy decreases, being gradually transferred back to the block as kinetic energy. Note that steps (c), (d), and (e) in Fig. 3 could represent an idealized bow and arrow, the stressed "bow" (Fig. 3c) transferring its potential energy to the flying "arrow" (Fig. 3e).

We can represent the energy transfers of Fig. 3 by

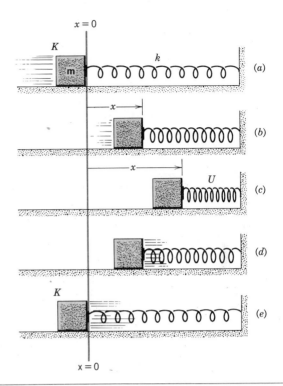

Figure 3   The kinetic energy of a moving block is transferred to a spring, as potential energy; this energy is then returned to the block, as kinetic energy. (a) The block strikes the end of the spring, which is in its relaxed state. (c) The block comes to rest, having transferred all its energy to the (compressed) spring. (e) The block has recovered its initial kinetic energy.

$$\Delta U + \Delta K = 0 \qquad (3)$$

which tells us that every change in the kinetic energy of the block is accompanied by an equal but opposite change in the potential energy of the spring. Equivalently, we can write

$$E = U + K = \text{constant,} \qquad (4)$$

in which E, a constant, is called the *mechanical energy* of the block–spring system. Equations 3 and 4 are equivalent statements of the law of conservation of mechanical energy.

The Gravitational Force.   Figure 4 shows a ball of mass m moving vertically upward near the surface of the earth and acted on only by the gravitational force, whose magnitude is mg. The ball rises, losing kinetic energy and coming momentarily to rest at position c. The ball then reverses its motion, steadily regaining the kinetic energy

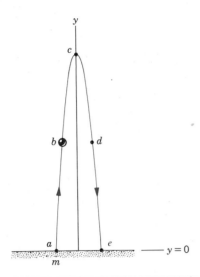

Figure 4 A ball of mass m is thrown upward, transferring its kinetic energy continuously into potential energy of the ball–earth system, until it comes to rest at c. As the ball falls, it regains its kinetic energy, drawing it from the store of potential energy in the ball–earth system. The labeled points correspond to parts (a) through (e) of Fig. 3.

that it lost on the way up. Figures 3 and 4 are similar, equivalent states being labeled with the same letters. Note that, in each figure, a particle (block or ball) makes a "round trip" during which it loses its kinetic energy only to gain it all back by the time it returns to its starting point.

It is again useful to say that, as the ball of Fig. 4 rises, its kinetic energy is gradually transferred to the ball–earth system, where it is stored as potential energy. In the second phase, as the ball falls, the potential energy stored in the ball–earth system decreases, being gradually transferred back to the ball as kinetic energy. Equations 3 and 4 hold for the gravitational force as well as for the spring force.

The Frictional Force. Figure 5 shows a block of mass m projected onto a rough horizontal plane and brought to rest by the frictional force that acts on it. We see at once that this situation is different from that of the spring force and the gravitational force. There is no way that we can get back the original kinetic energy of the block after the frictional force has brought it to rest. If the block's kinetic energy has been stored, it is stored in an inaccessible form, not readily transferable back to the block.

As the block of Fig. 5 comes to rest, the block and the surface over which it slides heat up. As we shall learn in later chapters, what has happened here is that the kinetic energy of the directed, large-scale motion of the

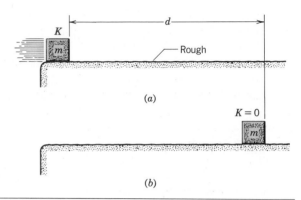

Figure 5 (a) A block is projected, with initial kinetic energy K, onto a rough horizontal surface. (b) The block, having traveled a distance d, is brought to rest by the action of the frictional force.

block has been transformed—not into potential energy—but into the kinetic energy of the randomly directed moving atoms that make up the block and the plane. Reversing this process would be like unscrambling an egg. We cannot associate a potential energy with the frictional force.

8–4 Defining the Potential Energy

Now comes the quantitative part. Suppose that the force we are examining—the gravitational force, say—acts on a particle—perhaps a ball—and does an amount of work W. We suppose further that no other force acts on the particle. The work–energy theorem,

$$W = \Delta K \quad \text{(work–energy theorem)}, \qquad (5)$$

tells us that W is equal to the change in the kinetic energy of the particle. Combining Eqs. 3 and 5 yields

$$\boxed{\Delta U = -W} \quad \text{(definition of } \Delta U). \qquad (6)$$

Thus, if a system changes from one configuration to another because of the action of a force, the change in the potential energy of the system is the negative of the work done by that force. We see also that the SI unit for potential energy is the same as that for work, namely, the joule.

For one-dimensional motion, Eq. 6 becomes

$$\Delta U = -W = -\int_{x_0}^{x} F(x)\, dx. \qquad (7)$$

Here x_0 defines an arbitrary reference configuration of the system and x its general configuration.

If we want the actual potential energy $U(x)$, rather than the change in potential energy ΔU, we can extend Eq. 7 by writing

$$U(x) = U(x_0) + \Delta U = U(x_0) - \int_{x_0}^{x} F(x)\, dx. \qquad (8)$$

Here $U(x_0)$ is the potential energy of the system when it is in its chosen reference configuration.

It turns out that only *changes* in potential energy are physically important. Thus, we are free to assign the arbitrary value of zero to the potential energy of the system when it is in its reference configuration. That is, we are free to put $U(x_0) = 0$ in Eq. 8. Let us clarify these ideas by applying them to two specific cases.

The Spring Force. Let us choose the reference configuration of the spring to be its relaxed state ($x_0 = 0$). Let us further declare that its potential energy in this state is zero [$U(x_0) = 0$]. We also assume that no force other than the spring force is present in the problem. Substituting Hooke's law of force (Eq. 2) for $F(x)$ in Eq. 8

then yields

$$U(x) = 0 - \int_{0}^{x} (-kx)\, dx,$$

or

$$U(x) = \tfrac{1}{2}kx^2. \qquad (9)$$

Because x appears as a square in Eq. 9, the potential energy of a spring has the same (positive) value, whether the spring is stretched or compressed by any given amount.

With the aid of Eq. 9, Eq. 4 becomes, for the block–spring system,

$$\tfrac{1}{2}kx^2 + \tfrac{1}{2}mv^2 = E \qquad \text{(block–spring system),} \quad (10)$$

in which E, a constant, is the mechanical energy of the system. Equation 10 is a statement of the law of conservation of mechanical energy for a block–spring system.

Figure 6 shows how energy shuttles back and forth between the spring and the block as these do work on

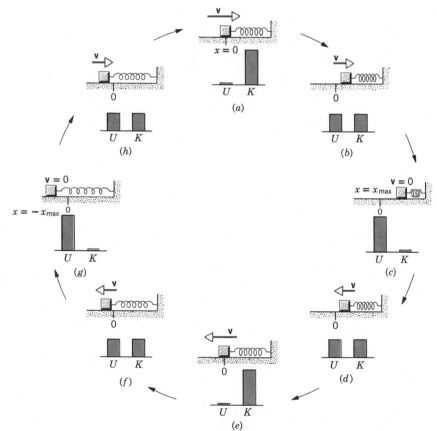

Figure 6 A block attached to a spring oscillates back and forth on a horizontal frictionless surface. The mechanical energy E of the system remains constant but is shunted continuously back and forth between kinetic and potential forms as the cycle repeats itself. Twice per cycle the energy is all kinetic, twice it is all potential, and four times per cycle it is shared equally between these two forms.

each other. At certain stages of the cycle, the energy is all kinetic. At other stages, it is all potential. At intermediate stages, the energy is shared between these two forms, the sum E (in the absence of friction) remaining constant at all times.

The Gravitational Force. Consider a particle moving in a vertical plane near the earth's surface, acted on only by the earth's gravitational force. We construct a y axis in the vertical direction and we put $F(y) = -mg$ for the (constant) force that acts on the particle. The minus sign arises because this force points downward, in the direction of decreasing y. Furthermore, we take $y_0 = 0$ to define the reference configuration in which the potential energy of the ball–earth system is zero.

With these substitutions, Eq. 8 becomes

$$U(y) = 0 - \int_0^y (-mg)dy,$$

or

$$\boxed{U(y) = mgy.} \qquad (11)$$

With the aid of Eq. 11, Eq. 4 becomes, for the ball–earth system,

$$\boxed{mgy + \tfrac{1}{2}mv^2 = E} \qquad \text{(ball–earth system),} \quad (12)$$

in which E is the mechanical energy. Although the altitude y and the speed v of the particle may vary, they will always do so in such a way that E in Eq. 12 remains constant.

Equations 11 and 12 hold not only for a ball thrown vertically but also for a ball thrown at an angle with the vertical and moving in a vertical plane. It also holds for a pendulum swinging in a vertical plane. If a particle moves horizontally near the earth's surface, the force acting on it ($= mg$) is at right angles to this motion and can do no work on the particle. Thus, in calculating changes in the gravitational potential energy of a particle near the surface of the earth, only vertical displacements of the particle need be considered; horizontal displacements do not count.

Figure 7 A frictionless pendulum swings back and forth. The mechanical energy E of the system remains constant but is shunted continuously back and forth between kinetic and potential forms as the cycle repeats itself. Twice per cycle the energy is all kinetic, twice it is all potential, and four times per cycle it is shared equally between these two forms. Check the bar graphs and compare with Fig. 6.

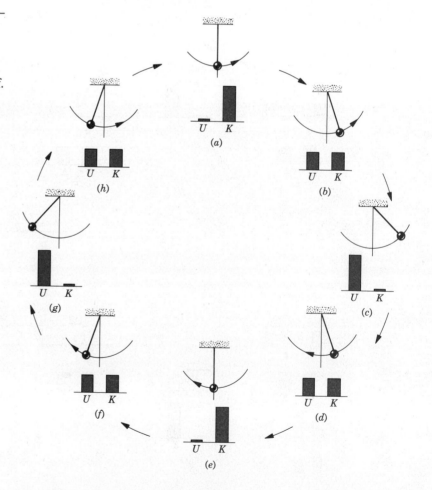

Figure 7 shows the variation with time of the kinetic energy and the gravitational potential energy of a swinging pendulum. We choose the potential energy of the system to be zero when the bob is at the bottom of its swing. In this configuration, the bob is moving with its greatest speed and the system energy is all kinetic. At the ends of its swing, the bob is momentarily at rest and the system energy is all potential. In other configurations, the energy is partly potential and partly kinetic, the sum (in the absence of friction) always remaining constant. (Note that no changes in potential energy are associated with the tension force in the pendulum cord. This force acts at right angles to the motion of the pendulum bob and thus can do no work on it.)

Finding the Force.   Equation 8 shows how to find the potential energy $U(x)$ for a one-dimensional problem if we know the force $F(x)$. Sometimes we want to go the other way. That is, we know the potential energy function $U(x)$ and we want to find the force.

For one-dimensional motion, the work W done by the force that acts on a particle as the particle moves through a distance Δx is $F(x)\,\Delta x$. We can then write Eq. 6 as

$$\Delta U = -W = -F(x)\,\Delta x.$$

Solving for $F(x)$ and passing to the differential limit yields

$$\boxed{F(x) = -\frac{dU}{dx}} \quad \text{(one-dimensional motion)} \quad (13)$$

as the answer we seek.

We can check this result by putting $U(x) = \frac{1}{2}kx^2$, which is the potential energy for the spring force. Equation 13 then yields, as expected, $F(x) = -kx$, which is Hooke's law. Similarly, we can substitute $U(y) = mgy$, which is the potential energy function for a particle of mass m in the gravitational field of the earth, near its surface. Equation 13 then yields $F = -mg$, which is the (constant) gravitational force that acts on the particle.

Sample Problem 1 An elevator cab of mass m (=920 kg) moves from street level to the top of the World Trade Center in New York, 412 m above the ground. What is the change in the gravitational potential energy of the cab?

Strictly, we are talking about the change in potential energy of the cab–earth system. From Eq. 11,

$$\Delta U = mg\,\Delta y = mgh = (920 \text{ kg})(9.8 \text{ m/s}^2)(412 \text{ m})$$

$$= 3.7 \times 10^6 \text{ J} = 3.7 \text{ MJ}. \qquad \text{(Answer)}$$

This is almost exactly 1 kW·h, which costs a few cents at commercial rates.

Sample Problem 2 The spring of a spring gun is compressed a distance d of 3.2 cm from its relaxed state, and a ball of mass m (= 12 g) is put in the barrel. With what speed will the ball leave the barrel once the gun is fired? The spring constant k is 7.5 N/cm. Assume no friction and a horizontal gun barrel.

Let E_i be the mechanical energy in the initial state (before the gun has been fired) and E_f be the mechanical energy in the final state (just after the gun has been fired). Initially, the mechanical energy resides entirely in the spring, as potential energy and in amount $\frac{1}{2}kd^2$. In the final state, the mechanical energy resides entirely in the ball, as kinetic energy and in amount $\frac{1}{2}mv^2$. Because mechanical energy is conserved, we have

$$E_i = E_f,$$
$$U_i + K_i = U_f + K_f,$$
$$\tfrac{1}{2}kd^2 + 0 = 0 + \tfrac{1}{2}mv^2.$$

Solving for v yields

$$v = d\sqrt{\frac{k}{m}} = (0.032 \text{ m})\sqrt{\frac{750 \text{ N/m}}{12 \times 10^{-3} \text{ kg}}}$$

$$= 8.0 \text{ m/s}. \qquad \text{(Answer)}$$

Sample Problem 3 In Fig. 3, the block, whose mass m is 1.7 kg, has an initial speed v of 2.3 m/s. The spring constant k is 320 N/m. (a) By how much is the spring compressed in the configuration of Fig. 3c?

The mechanical energy E of the block–spring system in the configurations of Figs. 3a and 3c must be the same. Thus,

$$E_a = E_c,$$
$$U_a + K_a = U_c + K_c,$$
$$0 + \tfrac{1}{2}mv^2 = \tfrac{1}{2}kx^2 + 0.$$

Solving for x yields

$$x = v\sqrt{\frac{m}{k}} = (2.3 \text{ m/s})\sqrt{\frac{1.7 \text{ kg}}{320 \text{ N/m}}}$$

$$= 0.17 \text{ m} = 17 \text{ cm}. \qquad \text{(Answer)}$$

(b) For what value of x will the energy be equally divided between potential and kinetic forms?

In Fig. 3a, the mechanical energy E is all kinetic and is

$$E = K = \tfrac{1}{2}mv^2 = (\tfrac{1}{2})(1.7 \text{ kg})(2.3 \text{ m/s})^2 = 4.50 \text{ J}.$$

The question is: "For what compression of the spring will its potential energy be half of this value?" We have

$$U(x) = \tfrac{1}{2}kx^2 = \tfrac{1}{2}E.$$

Solving for x yields

$$x = \sqrt{\frac{E}{k}} = \sqrt{\frac{4.50 \text{ J}}{320 \text{ N/m}}} = 0.12 \text{ m} = 12 \text{ cm}. \quad \text{(Answer)}$$

We suggest that, using your best intuition, you estimate the speed of the block under these conditions; then calculate it and see how close you were.

Sample Problem 4   In Fig. 8, a child, whose mass m is 32 kg, is released from rest at the top of a curved water slide, a height h of 8.5 m above the level of the pool. How fast is the child moving when she is projected into the pool? Assume that the slide is essentially frictionless.

At first glance, this problem may seem to be impossible to solve because we are given no information about the shape of the slide. However, it is easy if we apply the conservation of mechanical energy principle. Note that, in the absence of friction, the force exerted on the child by the slide is a normal force, acting at right angles to the surface of the slide. This force then can do no work on the child. The only force that does work is the gravitational force. The mechanical energy E is thus conserved throughout the motion and we can write, from Eqs. 4 and 12,

$$E_b = E_t$$

or

$$\tfrac{1}{2}mv_b^2 + mgy_b = \tfrac{1}{2}mv_t^2 + mgy_t.$$

Here the subscript b refers to the bottom of the slide and t to the top. The mass of the child cancels out and we have

$$v_b^2 = v_t^2 + 2g(y_t - y_b). \tag{14}$$

We recognize this at once as Eq. 15 of Chapter 2. Putting $v_t = 0$ and $y_t - y_b = h$ leads to

$$v_b = \sqrt{2gh} = \sqrt{(2)(9.8 \text{ m/s}^2)(8.5 \text{ m})}$$

$$= 13 \text{ m/s} (= 29 \text{ mi/h}). \tag{Answer}$$

This is the same speed that the child would reach if she were dropped vertically through a height of 8.5 m. In an actual case,

some frictional forces would act and the child would not be moving quite so fast.

This problem would have been hard to solve by straightforward application of Newton's laws and it may seem to be a trick that we are able to solve it so easily by energy methods. However, the law of conservation of mechanical energy follows directly from Newton's laws and the definition of work, so that we were using these laws without realizing it. On the other hand, if you were asked to find the time taken for the child to reach the bottom of the slide, energy methods would be of no use; you would need to know the shape of the slide and you would have a hard problem on your hands.

Hint 1:   *Conservation law problems—I.*   These formal steps should help you to solve problems involving the conservation of mechanical energy. Review Sample Problems 2–4 with these steps in mind.

To what system are you applying the conservation law? In your mind's eye you should be able to draw a closed surface such that whatever is inside is your system and whatever is outside is the environment of that system. In Sample Problem 2, the system is the ball + spring, taken together. In Sample Problem 3, it is the block + spring. In Sample Problem 4, it is the child + earth; the slide is only tangentially involved.

Is friction present? If forces such as friction or air resistance are present, mechanical energy will not be conserved. Make sure that the effects of such forces are small enough to neglect.

Is your system isolated? Conservation laws only apply to isolated systems. That means that no external forces must act on the system and no energy must cross the system boundary. In Sample Problem 2, if you chose the ball alone as your system you would find that it is not isolated; the spring does work on it across the system boundary. Thus, you cannot apply mechanical energy conservation to the ball alone (or to the spring alone).

What are the initial and the final states of your system? The system changes from some appropriate initial state to some appropriate final state. You apply the conservation of mechanical energy by saying that E has the same value in each of these states. That is, $E_i = E_f$. You must be very clear about what these two states are.

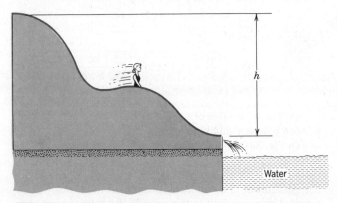

Figure 8   Sample Problem 4. A child slides down a water slide into a pool of water. The three-dimensional slide is shown "rolled out" into a plane. If friction is neglected, the speed of the child at the bottom of the slide does not depend on the shape of the slide.

8-5   Conservative and Nonconservative Forces

If a potential energy can be associated with a force, we call that force *conservative.* If a potential energy cannot be associated with a force, we call that force *nonconservative*. The spring force and the gravitational force are conservative; the frictional force is nonconservative.

There are two tests that we can apply to a force to find out whether it is conservative. The tests are totally equivalent so that, if the force meets either one, it will automatically meet the other.

First Test. Suppose that you throw a ball in the air and then catch it when it returns. If potential energy is to mean anything, the change in the potential energy of the ball–earth system for this round trip must be zero.

From Eq. 6, this requirement that $\Delta U = 0$ for a round trip means that the work W done by the force in question over such a round trip must also be zero. Thus, our first test can be stated as follows:

A force is conservative if the work it does on a particle that moves through a round trip is zero; otherwise, the force is nonconservative.

The gravitational force, by this criterion, is conservative. It did negative work on the ball while the ball was rising and an equal amount of positive work on its return trip. The work for the round trip was zero.

The requirement of zero work for a round trip is not met by the frictional force. If you push an eraser for a certain distance across a chalkboard, the frictional force does negative work on the eraser. However, as you drag the eraser back to its starting point, the direction of the frictional force reverses automatically and that force *still* does negative work on the eraser. The total work done by the frictional force for a round trip is not zero. The force of friction is nonconservative.

Second Test. Suppose that, as in Fig. 9, a particle goes from a to b along path 1 and then back to a along path 2. If the force acting on the particle is conservative,

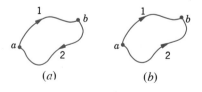

(a)   (b)

Figure 9 (a) A particle, acted on by a conservative force, moves in a round trip, starting from point a, passing through point b, and returning to point a. (b) A particle starts from point a and proceeds to point b, following either of two available paths.

then from our first test above, the work done on the particle for this round trip must be zero. We can write

$$W_{ab,1} + W_{ba,2} = 0$$

or

$$W_{ab,1} = -W_{ba,2}. \tag{15}$$

That is, the work in going from a to b along path 1 is the negative of the work in going from b to a along path 2. Now let us cause the particle to go from a to b along path 2, as in Fig. 9b. This merely reverses the direction of motion along this path, so that

$$W_{ab,2} = -W_{ba,2} \tag{16}$$

From Eqs. 15 and 16, we can then write

$$W_{ab,1} = W_{ab,2}. \tag{17}$$

which tells us that the work done on a particle by a conservative force does not depend on the path followed between those points. Thus, our second (equivalent) test can be stated as follows:

A force is conservative if the work done by it on a particle that moves between two points is the same for all paths connecting those points; otherwise, the force is nonconservative.

Let us test the gravitational force by applying this criterion to it. Pick up a stone at point i in Fig. 10a, move it horizontally to point a, and then raise it through a vertical height h to point f. We label the path iaf as path 1. For the horizontal distance ia, the work done on the stone by the gravitational force is zero because the force and the displacement are at right angles; for the vertical distance af, the work is $-mgh$. Thus, the total work done over path 1 is $-mgh$.

Now consider path 2 in Fig. 10a, a straight line connecting points i and f. To find the work done on this path, we note first that the angle between the force ($=mg$) and the displacement ($=d$) is $180° - \phi$, where the angle ϕ is defined in the figure. Thus, we have

$$W = Fd \cos(180° - \phi) = -Fd \cos \phi.$$

But $d \cos \phi = h$ so that $W = -mgh$, just as for path 1.

We can also show that the work done over a completely arbitrary path, such as path X in Fig. 10b, is also equal to $-mgh$. To do so, let us approximate path X by a series of vertical and horizontal steps. We can have as many steps as we wish so that we can make this stepped path arbitrarily close to path X. If the particle follows

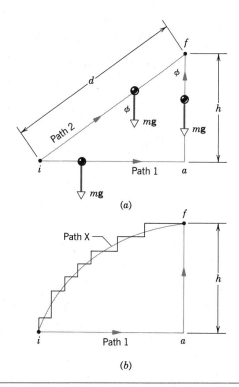

(a)

(b)

Figure 10 (*a*) A particle moves from point *i* to point *f* in the earth's gravitational field. The work done on the particle by the gravitational force is the same for either path. (*b*) The work is also the same for these two paths or for *any* path connecting ponts *i* and *f*.

this stepped path in moving from *i* to *f* in Fig. 10*b*, no work will be done on the horizontal segments. The sum of the vertical segments is just *h*, so that the work on the stepped path is again $-mgh$.

8–6 The Potential Energy Curve

We can learn a lot about the motion of a particle constrained to move along the *x* axis by inspecting a plot of its potential energy function $U(x)$. Figure 11*a* shows such a plot. To start with, we can find the force $F(x)$ that acts on the particle by (graphically) taking the slope of the $U(x)$ curve at various points, as Eq. 13 ($F = -dU/dx$) instructs us to do. Figure 11*b* is a plot of $F(x)$ found in this way.

In the absence of friction, the conservation of mechanical energy *E* holds. That is,

$$U(x) + K = E. \qquad (18)$$

Figure 11*a* displays, at the position of the black triangle on the *x* axis, the values of *K* and *U* that correspond to a particular (constant) value of the total energy *E*. Because *K* cannot be negative, the only regions of the *x* axis available to the particle are those for which $E - U$ is zero or positive, that is, those regions in which the horizontal line corresponding to the mechanical energy *E* lies above the $U(x)$ curve.

Turning Points. In Fig. 11*a*, E_0 is the smallest value of mechanical energy that is possible. For this value, $E = U$ so that $K = 0$ and the particle remains at rest at position x_0. At a slightly higher energy E_1, the particle can move between x_1 and x_2 but cannot stray beyond those limits. As it moves from x_0 in either direction, the speed of the particle decreases, the particle coming momentarily to rest and then reversing its motion at the *turning points* x_1 and x_2.

At an energy like E_2 in Fig. 11*a*, there are four turning points and the particle (constrained as always to the *x* axis) can oscillate back and forth in either one of two *potential valleys*. At an energy such as E_3, there is

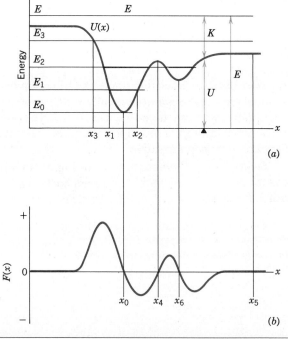

(a)

(b)

Figure 11 (*a*) A plot of $U(x)$, an arbitrary potential energy function for a particle confined to move along the *x* axis. Possible values of the mechanical energy *E* are shown. There is no friction, so that mechanical energy is conserved. (*b*) A plot of the force $F(x)$ acting on the particle, derived from the potential energy plot by taking slopes.

only one turning point, at x_3. If the particle is moving initially in the direction of decreasing x, it will slow down as it approaches x_3, will pause momentarily, and will reverse its motion. For values of E that lie entirely above the potential energy curve, there are no turning points.

Equilibrium Points. At a point such as x_0, where $U(x)$ has a minimum value, the slope of the curve $U(x)$ is zero so that—as Eq. 13 tells us—the force acting on the particle is zero. If the particle is displaced slightly in either direction, a force will appear that tends to push the particle back toward x_0, as study of Fig. 11b will confirm. We say that x_0 is a position of *stable equilibrium.* A particle placed at rest at that position will remain at rest. A marble at the bottom of a tea cup is an example.

At a point, such as x_4, where $U(x)$ has a maximum value, the slope of the curve is zero so that the force acting on the particle is again zero; see Fig. 11b. However, if the particle is displaced slightly from this point, a force appears that tends to push the particle even farther away, as study of Fig. 11b confirms. Points such as x_4 are points of *unstable equilibrium.* A marble balanced on the top of a bowling ball is an example.

At points such as x_5, near which $U(x)$ is constant, the force acting on the particle remains zero for small displacements and the particle is said to be in *neutral equilibrium.* A marble on a flat horizontal table is an example.

8-7 Nonconservative Forces

Suppose that a frictional force acts on the oscillating block of the block–spring system of Fig. 6. We know that the motion decreases continually in amplitude and eventually dies away. Put another way, the mechanical energy E of the system decreases with time. In still other words, because of the frictional force, the mechanical energy of the block–spring system is no longer conserved.

Besides the decrease in mechanical energy, we notice something else: Work is done on the block by the frictional force and this work is negative. We know this because, at any stage, the direction of the frictional force acting on the block of Fig. 6 is always opposite to the direction in which the block is moving. Let us play a hunch and guess that the (negative) work done by the frictional force is equal to the (negative) change in mechanical energy of the block–spring system. Can we prove it?

Consider the general case of a particle acted on by a number of forces, one of them (a frictional force) being

nonconservative and all the others conservative. The total work W done on the particle is the sum of the work done by each of the forces that act on it. We can then write the work–energy theorem as

$$W = \sum W_c + W_f = \Delta K$$
(work–energy theorem),   (19)

in which the first term on the right ($\Sigma\,W_c$) is the work done by all the conservative forces and the second term (W_f) is the work done by the frictional force.

Each conservative force can be identified with a potential energy so that, from Eq. 6, we can write

$$\sum W_c = -\sum U(x). \quad (20)$$

Combining Eqs. 19 and 20 yields

$$W_f = \Delta K + \sum \Delta U(x).$$

The right side of this equation, however, is just the change in the mechanical energy of the system. Thus, we have

$$W_f = \Delta E. \quad (21)$$

Our hunch was correct! (If it had not been, we would not have brought it up here.)

Suppose that we watch the oscillating block–spring system of Fig. 6 for 1.5 min and note that its mechanical energy E decreases by 180 mJ. Equation 21 tells us that the work done by the frictional force during that interval was exactly -180 mJ.

8-8 The Conservation of Energy

We have seen that, if nonconservative forces act, the law of conservation of mechanical energy no longer holds. Before giving up a useful law of physics, we always look for ways to extend or generalize it. We can do so in this case.

What happens to the "missing" energy ΔE when nonconservative forces act? We find a clue in the fact that both the block of Fig. 6 and the surface over which it slides become slightly warmer as the oscillations of the system die away. It is as if the kinetic energy of the direction motion of the sliding block was transformed into kinetic energy of the disordered random motions of the atoms that make up the block and the surface over which it slides. We call such energy *internal energy* (or *thermal energy*) and we represent it by U_{int}, which we define from

$$\Delta U_{int} = -W_f. \quad (22)$$

Thus, if the frictional force acting on the block in Fig. 6 does −180 mJ of work in reducing the mechanical energy of the block–spring system, then, from Eq. 22, 180 mJ of internal energy appear in the block and the surface over which it slides.

We have written the work–energy theorem (Eq. 19) as

$$\sum W_c + W_f = \Delta K \quad \text{(work–energy theorem).} \quad (23)$$

Substituting from Eqs. 20 and 22 yields

$$\boxed{\sum \Delta U + \Delta K + \Delta U_{int} = 0} \quad (24)$$

as a statement of *the law of conservation of energy.* This law is more comprehensive than the law of conservation of *mechanical* energy (Eq. 3), which holds only for cases in which no nonconservative forces act.

In applying this equation, you must choose your system carefully. In Fig. 5, for example, you cannot choose the block as your system but must choose a larger system, made up of the block and the surface over which it slides. The internal energy that is generated as the block comes to rest is shared between these two bodies and there is no simple way to calculate beforehand how the available energy will distribute itself between them.

It turns out that in new situations (involving perhaps electrical or magnetic phenomena) we can *always* identify new quantities like U_{int} that permit us to expand the scope of our definition of energy and to retain, in a more generalized form, the law of the conservation of energy. Thus, we can always write, for an isolated system,

$$\Delta K + \Sigma \Delta U + \Delta U_{int}$$
$$+ \text{(changes in other forms of energy)} = 0. \quad (25)$$

We can express this generalized conservation of energy principle in words as follows:

> *Energy may be transformed from one kind to another in an isolated system but it cannot be created or destroyed; the total energy of the system always remains constant.*

This statement is a generalization of experience, so far not contradicted by any laboratory experiment or observation of nature.

Figure 12, an exhibit at the Museum of Science in Boston, illustrates energy conservation in an instructive and entertaining way. By turning a wheel (doing work), the observer lifts a heavy ball vertically, thus adding

Figure 12 A permanent exhibit at the Museum of Science, in Boston, illustrating the conservation of energy. A ball is given an initial store of potential energy, which it transfers to other forms before returning to its starting point.

some potential energy to the system. The ball then rolls downward through an elaborate caged runway, losing potential energy and gaining kinetic energy. During part of its journey, it rolls around a vertical loop, momentarily "buying back" some of its potential energy at the expense of an equal amount of kinetic energy. It loses energy further by collisions, by friction, by interaction with magnets, and in the form of light and sound, eventually coming to rest near the floor. Energy is conserved in this round trip, although mechanical energy is not.

Figure 13 is another example of energy conservation in which nonconservative forces act. As the climber ascends the rock face in Fig. 13a, he is gaining gravitational potential energy at the expense of his own store of biochemical internal energy. In Fig. 13b, he descends, at a roughly constant speed, sliding down a rope which is fed through a metal brake bar. The gravitational potential energy that he loses in descending appears now as internal or thermal energy in the rope and the brake bar. They get hot. The purpose of the brake bar is to prevent the gravitational potential energy from transforming itself into the kinetic energy of a falling climber!

Often in the history of physics, the conservation of energy principle has seemed to fail. Its apparent failure, however, always stimulated a search for the reason and —so far—the reason has always been found and the law of conservation of energy has been maintained, perhaps in a more generalized form. In later chapters, we shall

(a)   *(b)*

Figure 13 (*a*) Going up! Internal biochemical energy changes into gravitational potential energy. (*b*) Coming down! Gravitational potential energy changes into internal energy of the rope and the brake bar. They get hot.

study a number of new kinds of energy transformations —from mechanical to internal, from mechanical to electrical, from chemical to mechanical, and so on. The energy concept has now become one of the grand unifying ideas of physical science.

Sample Problem 5 A ball bearing whose mass m is 5.2 g is fired vertically downward from a height h of 18 m with an initial speed v_0 of 14 m/s; see Fig. 14. It buries itself in sand to a depth d of 21 cm. What average upward resistive force F does the sand exert on the ball as it comes to rest?

Let us apply energy conservation to the ball bearing as it moves from its initial position at height h above the ground to its final position a distance d below the ground. From Eq. 21, the work done by the frictional force in bringing the ball to rest is equal to the change in the mechanical energy E of the system. These two quantities are

$$W_f = -Fd$$

and

$$\Delta E = \Delta U + \Delta K = -mg(h + d) - \tfrac{1}{2}mv_0^2.$$

Figure 14 Sample Problem 5. A ball bearing is fired downward, coming to rest in sand. Mechanical energy is conserved over path h but not over path d, where a (nonconservative) frictional force acts.

Equating these two quantities yields

$$Fd = mg(h + d) + \tfrac{1}{2}mv_0^2.$$

Solving for F gives

$$\begin{aligned}
F &= \frac{mv_0^2}{2d} + \frac{mgh}{d} + mg \\[4pt]
&= \frac{(5.2 \times 10^{-3}\ \text{kg})(14\ \text{m/s})^2}{(2)(0.21\ \text{m})} \\[4pt]
&\quad + \frac{(5.2 \times 10^{-3}\ \text{kg})(9.8\ \text{m/s}^2)(18\ \text{m})}{0.21\ \text{m}} \\[4pt]
&\quad + (5.2 \times 10^{-3}\ \text{kg})(9.8\ \text{m/s}^2) \\[4pt]
&= 2.43\ \text{N} + 4.37\ \text{N} + 0.051\ \text{N} \\[4pt]
&= 6.8\ \text{N} \ (= 1.5\ \text{lb}) \qquad\qquad \text{(Answer)}
\end{aligned}$$

Sample Problem 6 A block whose mass m is 4.5 kg is fired up a 32° inclined plane (see Fig. 15) with an initial speed v_0 of 5.2 m/s. It travels a distance d of 1.5 m up the plane, comes momentarily to rest, and then slides back down to the bottom of the plane. (*a*) What is the magnitude f of the (constant) frictional force that acts on the block while it is moving?

From Eq. 21, the work W_f done by the frictional force as the block moves up the plane is equal to the change in the

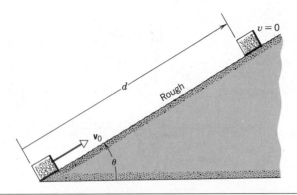

Figure 15 Sample Problem 6. A block is projected up a rough plane, with initial speed v_0. It comes to rest momentarily after a distance d and then slides back down to the bottom of the plane. Because of the (nonconservative) frictional force, mechanical energy is not conserved.

mechanical energy of the block between those two positions. We have

$$W_f = -fd$$

and

$$\Delta E = \Delta U + \Delta K = (+mgd \sin \theta) + (-\tfrac{1}{2}mv_0^2).$$

Equating these two quantities yields

$$-fd = mgd \sin \theta - \tfrac{1}{2}mv_0^2.$$

Solving for f gives

$$f = \frac{mv_0^2}{2d} - mg \sin \theta$$

$$= \frac{(4.5 \text{ kg})(5.2 \text{ m/s})^2}{(2)(1.5 \text{ m})}$$

$$- (4.5 \text{ kg})(9.8 \text{ m/s}^2)(\sin 32°)$$

$$= 17.2 \text{ N} (= 3.9 \text{ lb}). \qquad \text{(Answer)}$$

(b) What is the speed v of the block when it reaches the bottom of the plane?

Consider its round trip journey, from the bottom of the plane, up the plane, and back to the bottom again. From Eq. 21, the work W_f done by the frictional force on this round trip is equal to the change in the mechanical energy of the block. W_f is equal to $-2fd$, the factor 2 arising because the distance covered by the block is $2d$.

The potential energy of the block is unchanged at the end of the block's round trip. Thus, the change in mechanical energy for the block is equal to the change in its kinetic energy, or $\Delta E = \Delta K$. Thus,

$$W_f = \Delta E = \Delta K,$$

or

$$-2fd = \tfrac{1}{2}mv^2 - \tfrac{1}{2}mv_0^2.$$

Solving for v yields

$$v = \sqrt{v_0^2 - \frac{4fd}{m}}$$

$$= \sqrt{(5.2 \text{ m/s})^2 - \frac{(4)(17.2 \text{ N})(1.5 \text{ m})}{4.5 \text{ kg}}}$$

$$= 2.0 \text{ m/s.} \qquad \text{(Answer)}$$

As we expect, this is smaller than the initial speed of the block (= 5.2 m/s) as it started up the plane.

8–9 Mass and Energy (Optional)

The science of chemistry was built up by applying the laws of conservation of energy and conservation of mass to chemical reactions.* In 1905, Einstein showed that, as a consequence of his special theory of relativity, mass is simply another form of energy. These two conservation laws now appear as but different aspects of a single, deeper law, the law of conservation of mass–energy.

In chemical reactions, it turns out that the amount of mass that is turned into other forms of energy (or vice versa) is such a tiny fraction of the total mass involved that there is no hope of detecting the mass change, even with the best laboratory balances. Mass and energy truly *seem* to be separately conserved. In nuclear reactions, however, in which the energy released per event is often about a million times greater than in chemical reactions, the change in mass can easily be measured. Taking mass–energy interchanges into account where nuclear reactions are involved becomes a matter of necessary laboratory routine.

Mass and energy are related by what is certainly the best-known equation in physics (see Fig. 16) namely,

$$\boxed{E = mc^2,} \qquad (26)$$

in which E is the energy equivalent of mass m and c is the speed of light. Table 1 shows the energy equivalence of the mass of a few objects.

The amount of energy lying dormant in ordinary objects is enormous. The energy equivalent of the mass

---

* In this Section, the word *mass* and the symbol m always refer to the mass of an object as ordinarily measured, that is, while the object is at rest. In Chapter 42, which deals with Einstein's theory of relativity, we shall call this the *rest mass* of the object.

Figure 16 Students of Shenandoah Junior High School, in Miami, Florida, honor Einstein on the 100th anniversary of his birth by spelling out his famous formula with their bodies. Courtesy of Rocky Raisen, physics teacher.

Table 1   The Energy Equivalence of a Few Objects

Object	Mass (kg)	Energy Equivalence
Electron	9.1×10^{-31}	8.2×10^{-14} J (= 511 keV)
Proton	1.7×10^{-27}	1.5×10^{-10} J (= 938 MeV)
Uranium atom	4.0×10^{-25}	3.6×10^{-8} J (= 225 GeV)
Dust particle	1×10^{-13}	1×10^{4} J (= 2 kcal)
Penny	3.1×10^{-3}	2.8×10^{14} J (= 78 GW·h)

of a penny, for example, would cost over a million dollars to purchase from your local utility company. The mass equivalents of some energies are equally striking. The annual U.S. electrical energy production, for example, corresponds to a mass equivalence of only a few hundred pounds of matter (stones, potatoes, library books, anything!)

In applying Eq. 26 to reactions between particles, we find it convenient to rewrite it as

$$Q = \Delta m c^2, \tag{27}$$

in which Q (called simply the Q of the reaction) is the energy released (or absorbed) in the reaction and Δm is the decrease (or increase) in the mass of the particles as a result of the reaction. If the reaction is nuclear fission, less than 0.1% of the mass initially present is transformed into other forms of energy. If the reaction is a chemical one, the percentage is much less, typically by a factor of one million. In the matter of extracting energy from matter, there is a long way to go.

In practice, SI units are rarely used in connection with Eq. 27, being too large for convenience. Masses are usually entered in atomic mass units (abbr. u; see Section 1-6), where

$$1 \text{ u} = 1.66 \times 10^{-27} \text{ kg}, \tag{28}$$

and energies in electron-volts or multiples thereof. Equation 6 of Chapter 7 told us that

$$1 \text{ eV} = 1.60 \times 10^{-19} \text{ J}. \tag{29}$$

In the units of Eqs. 28 and 29, the multiplying constant c^2 has the values

$$c^2 = 9.32 \times 10^8 \text{ eV/u}$$
$$= 9.32 \times 10^5 \text{ keV/u} = 932 \text{ MeV/u}. \tag{30}$$

Equation 26 requires that we recast the law of conservation of energy to include the possibility of mass-energy. Thus, Eq. 25 becomes

$$\Delta K + \sum \Delta U + \Delta U_{\text{int}} + \Delta m c^2 + \cdots = 0. \tag{31}$$

We can also write this in the form

$$E_{\text{tot}} = K + \sum U + U_{\text{int}} + mc^2 + \cdots,$$
$$= \text{constant} \tag{32}$$

in which E_{tot} is the total energy.

Sample Problem 7 Suppose that 1 mol of (diatomic) oxygen interacts with 2 mol of (diatomic) hydrogen to produce 2 mol of water vapor, according to the reaction

$$2H_2 + O_2 \rightarrow 2H_2O.$$

The energy Q released is 4.85×10^5 J. What fraction of the mass of the initial reactants vanishes to generate this energy?

The mass decrease needed to supply the released energy follows from Eq. 27 as

$$\Delta m = \frac{Q}{c^2} = \frac{4.85 \times 10^5 \text{ J}}{(3.00 \times 10^8 \text{ m/s})^2} = 5.39 \times 10^{-12} \text{ kg}.$$

The mass of the reactants is twice the molecular weight of hydrogen plus the molecular weight of oxygen, or

$$M = 2(1.01 \text{ g}) + 32.0 \text{ g} = 34.0 \text{ g} = 0.034 \text{ kg}.$$

The fraction desired is

$$f = \frac{\Delta m}{M} = \frac{5.39 \times 10^{-12} \text{ kg}}{0.034 \text{ kg}} = 1.6 \times 10^{-10}. \quad \text{(Answer)}$$

Such a tiny fractional mass loss, typical of chemical reactions, cannot be detected by even the most sophisticated laboratory methods. Thus, although mass is converted into other forms of energy in (exothermic) chemical reactions, the fraction so converted is too small to detect.

Sample Problem 8 A typical nuclear fission reaction is

$$^{235}U + n \rightarrow {}^{140}Ce + {}^{94}Zr + 2n,$$

in which n represents a neutron. The masses involved are

^{235}U	235.04 u	^{94}Zr	93.91 u
^{140}Ce	139.91 u	n	1.00867 u

(a) What fraction of the mass of the two interacting particles is transformed into kinetic energy?

The mass difference between the reacting particles and the product particles is

$$\Delta m = (235.04 + 1.00867)$$
$$- (139.91 + 93.91 + 2 \times 1.00867)$$
$$= 0.211 \text{ u}.$$

The mass of the interacting particles is

$$M = 235.04 + 1.00867 = 236.05 \text{ u},$$

and the fraction f is

$$f = \frac{\Delta m}{M} = \frac{0.211 \text{ u}}{236.05 \text{ u}} = 0.00089, \text{ or about } 0.1\%. \quad \text{(Answer)}$$

Although this is small, it is easily measurable and is, in any case, very much greater than the corresponding fraction calculated for chemical reactions in Sample Problem 7.

(b) How much energy is released in each individual fission reaction?

From Eq. 27 we have

$$Q = \Delta mc^2 = (0.211 \text{ u})(932 \text{ MeV/u}) = 197 \text{ MeV}. \quad \text{(Answer)}$$

Note that we chose our value for c^2 from those displayed in Eq. 30. The fission energy release (≈ 197 MeV) is to be compared with energy releases of a few electron-volts that are typical of chemical reactions.

Sample Problem 9 The nucleus of an atom of deuterium (or heavy hydrogen) is called a *deuteron*. It is composed of a proton and a neutron. How much energy must be expended to tear it apart? The needed masses are

deuteron	m_d =	2.01355 u
proton	m_p =	1.00728 u
neutron	m_n =	1.00867 u

If energy must be added to the deuteron to tear it apart, the mass of its separated constituents must be greater than the deuteron mass, since adding energy means adding mass. That mass difference is

$$\Delta m = m_d - (m_p + m_n)$$
$$= (2.01355) - (1.00728 + 1.00867) = -0.00240 \text{ u}.$$

The corresponding energy is then, from Eq. 27,

$$Q = \Delta mc^2 = (-0.00240 \text{ u})(932 \text{ MeV/u})$$
$$= -2.24 \text{ MeV}. \quad \text{(Answer)}$$

The quantity 2.24 MeV is known as the *binding energy* of the deuteron. By contrast, the energy required to pull the electron out of the hydrogen atom is only about 13.6 eV, smaller by a factor of about 6×10^{-6}

8–10 Energy is Quantized (Optional)

The air, as we wave our hand through it, seems to be perfectly continuous. Yet on a fine enough scale, air is not continuous at all but comes in "lumps," that is, in particles of discrete mass—oxygen and nitrogen molecules in this case. We say that mass is *quantized*. When we step into the atomic and subatomic world, in which we have no direct "fingertip" experience, we find that many other physical quantities are quantized. Energy is one of those quantities.

We would all say that a swinging pendulum can have any reasonable energy that we choose to give it. Things are different in the atomic world, however. The external electrons of an atom can arrange themselves in only certain characteristic patterns—called *quantum*

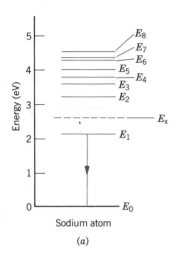

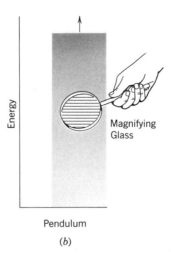

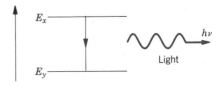

Figure 17   *(a)* Some of the energy levels of a sodium atom, corresponding to the various quantum states in which the atom may exist. The lowest state, marked E_0, is called the ground state. The atom emits characteristic yellow sodium light when it transfers from state E_1 to its ground state, as the vertical arrow suggests. *(b)* The energy of a pendulum is also quantized but the levels are so close together that they cannot be distinguished, even under the closest scrutiny.

states—each with is own discrete energy E. Energies that lie between these values simply do not occur.

Figure 17*a* shows the allowed *energy levels* for an isolated sodium atom, each energy corresponding to a different quantum state. The lowest energy, labeled E_0 in Fig. 17*a* and assigned the arbitrary value of zero, is the *ground state* of the sodium atom. It is the state in which the isolated sodium atom usually exists, just as we normally find a marble in a bowl resting at the bottom of the bowl. To occupy higher states, the sodium atom must absorb energy from some external source, perhaps by colliding with electrons in a sodium vapor lamp.

Actually, the energy of *everything*—including our swinging pendulum—is quantized. As Fig. 17*b* shows, however, for large objects the allowed energy levels are so exceedingly close together that we cannot tell that they are not continuous. For such objects, we can ignore quantization totally because we cannot possibly detect it.

Quantization and the Emission of Light.   Let us see how neatly the concepts of quantization and conservation of energy joint to explain the emission of light by single isolated atoms. Consider the yellow light emitted by sodium atoms, easily seen by throwing salt (sodium chloride) into an open flame.

Light can be analyzed as a wave, with a well-defined characteristic frequency of oscillation, which we represent by the symbol v. The yellow sodium light is emitted by sodium atoms as they transfer from state E_1 in Fig. 17*a* to state E_0, the ground state; the transition is marked in the figure.

In general, radiation such as light is emitted from *any* atom when the atom changes from *any* state E_x to *any* state of lower energy E_y, as Fig. 18 suggests. The conservation of energy principle for light emission then takes the form

$$E_x - E_y = hv, \qquad (33)$$

in which h is a constant and v is the frequency of the light. This equation—first put forward by the Danish physicist Niels Bohr—tells us that $E_x - E_y$, the energy lost by the atom, is carried away by the emitted light.

Equation 33 governs the emission (and also the absorption) of radiation of all kinds—not only from atoms but from nuclei, molecules, or heated solids. It applies not only to visible light but to radiations of all frequencies, ranging from penetrating x rays to radio waves.

The constant h in Eq. 33 is called the *Planck constant* and has the value

$$\boxed{\begin{aligned} h &= 6.63 \times 10^{-34}\ \text{J}\cdot\text{s} \\ &= 4.14 \times 10^{-15}\ \text{eV}\cdot\text{s} \end{aligned}} \qquad \text{(Planck constant).} \quad (34)$$

This constant, named after the German physicist Max Planck who introduced it into physics in 1900, is the central constant of quantum physics. It pops up in almost every quantum equation, just as the speed of light c pops up in every equation of the theory of relativity. We shall meet it again!

Figure 18   An atom goes from one of its excited states to a state of lower energy, emitting light in the process.

REVIEW AND SUMMARY

Conservation Laws

Conservation laws are laws of the form

$$X = \text{a constant} \quad \text{(isolated system)}, \tag{1}$$

in which X is a *conserved* quantity. The *law of conservation of energy,* the subject of this chapter, is an important example.

Whenever *all* the forces acting on a particle are conservative, the total mechanical energy $E = U + K$ remains constant:

Conservation of Mechanical Energy

$$E = U + K = \text{constant}. \tag{4}$$

Said another way, changes in K (kinetic energy) are balanced by changes in potential energy U so that

$$\Delta U + \Delta K = 0. \tag{3}$$

Equations 3 and 4 are equivalent statements of the *law of conservation of mechanical energy.*

Potential energy U is energy "stored" in a system because work has been done against some conservative force. The energy stored in a stretched spring and the energy stored in an elevated mass are common examples. The potential energy associated with any one-dimensional conservative force may be calculated from

Potential Energy

$$U(x) = U(x_0) - W = U(x_0) - \int_{x_0}^{x} F(x)dx, \tag{6,8}$$

in which x_0 is an arbitrary reference position, $U(x_0)$ is the potential energy (usually chosen to be zero) at the reference position, and W is the work done by the conservative force when the particle is moved from the reference point to the point x. If we wish to calculate the force associated with a particular potential energy function, we may do so by writing Eq. 8 in derivative form:

Force from Potential Energy

$$F = -\frac{dU}{dx} \quad \text{(one-dimensional motion).} \tag{13}$$

Using Eq. 8, it is shown that the potential energy associated with any Hooke's law force for which $F = -kx$ as in a block-spring system, is

Hooke's Law Force

$$U(x) = \tfrac{1}{2}kx^2 \tag{9}$$

Gravitational Potential Energy

and the gravitational potential energy associated with an object at a height y above some reference height near the earth's surface is

$$U(y) = mgy. \tag{11}$$

Sample Problems 1–4 illustrate the calculation of potential energy and the use of Eq. 4.

Conservative Forces

A force acting on a particle is conservative if either of the two conditions outlined in Section 8–5 is satisfied; otherwise it is nonconservative. Mechanical energy $(K + U)$ is conserved only if no nonconservative forces are acting.

Turning Points

From the form of $U(x)$ alone, assuming no friction, we can learn a lot about the motion of a particle. Guided by the conservation of mechanical energy principle, we can identify turning points, regions and points of equilibrium, forbidden regions, and speed variations, as Fig. 11 shows.

If some nonconservative force—friction, for example—does work W_f, the conservation of mechanical energy principle takes the form

Nonconservative Work

$$W_f = \Delta E; \tag{21}$$

the change in total mechanical energy of a closed system is exactly equal to the work done by nonconservative forces.

We can *always* associate some kind of energy change with the work done by nonconservative forces. (U_{int}, internal energy associated with temperature and other parameters, is one example.) The sum of the energies so defined always remains constant for a closed system. We write this *conservation of energy principle* (the law of *conservation of total energy*) as

Conservation of Total Energy

$$\Delta K + \sum \Delta U + \Delta U_{int} + \text{(changes in other forms of energy)} = 0 \qquad [25]$$

Energy may be transformed from one kind to another in an isolated system, but it cannot be created or destroyed; the total energy of the system remains constant. Sample Problems 5 and 6 illustrate the use of Eq. 21.

The theory of special relativity allows us to extend this conservation principle to its ultimate form, the conservation of total relativistic energy E_{tot}. We do so by recognizing a new kind of energy, mass energy mc^2, and by putting

Mass Energy

$$E_{tot} = K + \sum U + U_{int} + \cdots + mc^2. \qquad [32]$$

The conservation of energy principle then takes the form

$$\Delta E_{tot} = \Delta K + \sum \Delta U + \Delta U_{int} + \cdots + \Delta mc^2 = 0. \qquad [31]$$

In this view, mass is a form of energy. The mass–energy equivalence formula,

$E = mc^2$

$$E = mc^2, \qquad [26]$$

then shows how to express any mass energy in either energy units (for example, joules, electron-volts) or mass units (for example, kilograms, atomic mass units). In Sample Problems 7–9, we explore some of the implications of the equivalence of mass and energy.

Energy is Quantized

Energy is quantized; systems of particles (for example, a sodium atom) have only certain allowed values of energy that are quite discrete. Intermediate energies do not occur. If an atom changes from one state to another, it must conserve energy by absorbing or emitting the exact amount of energy determined by the difference between its initial and final states. If energy is conserved by the emission of light, the frequency v of the light is determined from the energy change using the Bohr relation

Energy and Light Frequency

$$E_x - E_y = hv, \qquad [33]$$

in which h is the *Planck constant*

$$h = 6.63 \times 10^{-34} \text{ J·s} = 4.14 \times 10^{-15} \text{ eV·s}. \qquad [34]$$

QUESTIONS

1. An automobile is moving along a highway. The driver jams on the brakes and the car skids to a halt. In what form does the lost kinetic energy of the car appear?

2. In the above question, assume that the driver operates the brakes in such a way that there is no skidding or sliding. In this case, in what form does the lost kinetic energy of the car appear?

3. You drop an object and observe that it bounces to one and one-half times its original height. What conclusions can you draw?

4. What happens to the potential energy that an elevator loses in coming down from the top of a building to a stop at the ground floor?

5. Mountain roads rarely go straight up the slope but wind up gradually. Explain why.

6. Air bags greatly reduce the chance of injury in a car accident. Explain how they do so, in terms of energy transfers.

7. Pole vaulting was transformed when the wooden pole was replaced by the fiberglass pole. Explain why.

8. You see a duck flying by and declare it to have a certain amount of kinetic energy. However, another duck, flying alongside the first one, declares it to have no kinetic energy at all. Who is right, you or the second duck? How does the conservation of energy principle fit into this situation?

9. A ball dropped to earth cannot rebound higher than its release point. However, spray from the bottom of a waterfall can sometimes rise higher than the top of the falls. Why is this?

10. An earthquake can release enough energy to devastate a city. Where does this energy reside an instant before the earthquake takes place?

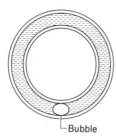

Figure 19   Question 11.

Figure 20   Question 17.

11. Figure 19 shows a circular glass tube fastened to a vertical wall. The tube is filled with water except for an air bubble that is temporarily at rest at the bottom of the tube. Discuss the subsequent motion of the bubble in terms of energy transfers. Do so both neglecting viscous and frictional forces and also taking them fully into account.

12. In Sample Problem 4 (see Fig. 8) we concluded that the speed of the child at the bottom does not depend at all on the shape of the surface. Would this still be true if friction were present?

13. Give physical examples of unstable equilibrium. Of neutral equilibrium. Of stable equilibrium.

14. In an article "Energy and the Automobile," which appeared in the October 1980 issue of *The Physics Teacher,* the author (Gene Waring) states: "It is interesting to note that *all* the fuel input energy is eventually transformed to thermal energy and strung out along the car's path." Analyze the various mechanisms by which this might come about. Consider, for example, road friction, air resistance, braking, the car radio, the headlamps, the battery, internal engine and drive train losses, the horn, and so on. Assume a straight and level roadway.

15. Trace back to the sun as many of our present energy sources as you can. Can you think of any that cannot be so traced?

16. Explain, using work and energy ideas, how a child pumps a swing up to large amplitudes from a rest position. (See "How to Make a Swing Go," by R. V. Hesheth, *Physics Education,* July 1975.)

17. Two disks are connected by a stiff spring. Can you press the upper disk down enough so that when it is released it will spring back and raise the lower disk off the table (see Fig. 20)? Can mechanical energy be conserved in such a case?

18. Discuss the words "energy conservation" as used (*a*) in this chapter and (*b*) in connection with an "energy crisis" (for example, turning off lights). How do these two usages differ?

19. The electric power for a small town is provided by a hydroelectric plant at a nearby river. If you turn off a light bulb in this closed-energy system, conservation of energy requires that an equal amount of energy, perhaps in another form, appears somewhere else in the system. Where and in what form does this energy appear?

20. A spring is compressed by tying its ends together tightly. It is then placed in acid and dissolves. What happens to its stored potential energy?

21. The expression $E = mc^2$ tells us that perfectly ordinary objects such as coins or pebbles contain enormous amounts of energy. Why did these large stores of energy go unnoticed for so long?

22. "Nuclear explosions—weight for weight—release about a million times more energy than do chemical explosions because nuclear explosions are based on Einstein's $E = mc^2$ relation." What do you think of this statement?

23. How can mass and energy be equivalent in view of the fact that they are totally different physical quantities, defined in different ways and measured in different units?

24. A hot metallic sphere cools off as it rests on the pan of a scale. If the scale were sensitive enough, would it indicate a change in mass?

25. Are there quantized quantities in classical (that is, non-quantum) physics? If so, give examples.

EXERCISES AND PROBLEMS

Section 8–4   Defining the Potential Energy

1E. A certain spring stores 25 J of potential energy when it is compressed by 7.5 cm. What is the force constant of the spring?

2E. To disable ballistic missiles during the early boost-phase of their flight, an "electromagnetic rail gun," to be carried in low-orbit earth satellites, is being developed. The gun might fire a 2.4-kg maneuverable projectile at 10 km/s. The kinetic energy carried by the projectile is sufficient on impact to disable a missile even if it carries no explosive. (A weapon of this kind is a "kinetic energy" weapon.) The projectile is accelerated to muzzle speed by electromagnetic forces. Suppose instead that we wish to fire the projectile using a spring (a "spring" weapon). What must the force constant be in order to achieve the desired speed after compressing the spring 1.5 m?

3E. It is claimed that large trees can evaporate as much as 900 kg of water per day. (a) Evaporation takes place from the leaves. To get there the water must be raised from the roots of the tree. Assuming the average rise of water to be 9.0 m from the ground, how much energy must be supplied? (b) What is the average power if the evaporation is assumed to occur during 12 h of the day?

4E. The summit of Mount Everest is 8850 m above sea level. (a) How much energy would a 90-kg climber expend against gravity in climbing to the summit from sea level? (b) How many Mars bars, at 300 kcal per bar, would supply an energy equivalent to this? Your answer should suggest that work done against gravity is a very small part of the energy expended in climbing a mountain.

5E. Approximately 5.5×10^6 kg of water drops 50 m over Niagara Falls every second. (a) How much potential energy is lost every second by the falling water? (b) What would be the power output of an electric generating plant that could convert *all* of the water's potential energy to electrical energy? (c) If the utility company sold this energy at an industrial rate of 1 cent/kW·h, what would be their yearly income from this source?

6E. The potential energy of a diatomic molecule (for example, H_2 or O_2) is given by

$$U = \frac{A}{r^{12}} - \frac{B}{r^6},$$

where r is the separation between the atoms that make up the molecule and A and B are positive constants. Find the equilibrium separation, that is, the distance between the atoms at which the force on each atom is zero.

7E. You drop a physics textbook, whose mass is 2.0 kg, to a friend who is standing on the ground 10 m below, as in Fig. 21. (a) What is the potential energy of the book just before you release it? (b) What is its kinetic energy just before your friend catches it in her outstretched hands, which are 1.5 m above the ground level? (c) How fast is the book moving as it is caught? Take the zero of potential energy to be ground level.

8E. A 200-lb man jumps out a window into a fire net 30 ft below. The net stretches 9.0 ft before bringing him to rest and tossing him back into the air. What is the potential energy of the stretched net if no energy is dissipated by nonconservative forces?

9E. An 8.0-kg mortar shell is fired straight up with an initial speed of 100 m/s. (a) What is the kinetic energy of the shell as it leaves the mortar? (b) What is the shell's potential energy at the top of its trajectory? Take the zero of potential energy to be at the ground level.

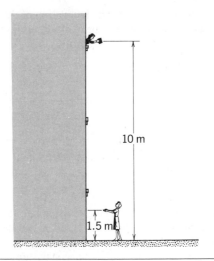

Figure 21 Exercise 7.

10E. A very small ice cube is released from the edge of a hemispherical frictionless bowl whose radius is 20 cm; see Fig. 22. How fast is the cube moving at the bottom of the bowl?

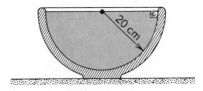

Figure 22 Exercise 10.

11E. A frictionless roller-coaster car starts at point A in Fig. 23 with speed v_0. What will be the speed of the car (a) at point B, (b) at point C, and (c) at point D? Assume that the car can be considered a particle and that it always remains on the track.

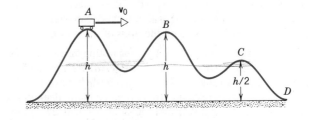

Figure 23 Exercise 11.

12E. A runaway truck with failed brakes is barreling downgrade at 80 mi/h. Fortunately, there is an emergency escape ramp at the bottom of the hill. The inclination of the ramp is 15°; see Fig. 24. How long must it be to make certain of bringing the truck to rest, at least momentarily?

Figure 24 Exercise 12.

13E. A volcanic ash flow is moving across horizontal ground when it encounters a 10° upslope. It is observed to travel 920 m on the upslope before coming to rest. The volcanic ash contains trapped gas, so the force of friction with the ground is very small and can be ignored. At what speed was the ash flow moving just before encountering the upslope?

14E. The magnitude of the gravitational force of attraction between a particle of mass m_1 and one of mass m_2 is given by

$$F(x) = G\,\frac{m_1 m_2}{x^2},$$

where G is a constant and x is the distance between the particles. (a) What is the potential energy function $U(x)$? Assume that $U(x) \rightarrow 0$ as $x \rightarrow \infty$. (b) How much work is required to increase the separation of the particles from $x = x_1$ to $x = x_1 + d$?

15E. A 1.0-kg object is acted on by a net conservative force given by $F = -3.0x - 5.0x^2$, where F is in newtons if x is in meters. (a) What is the potential energy of the object at $x = 2.0$ m? (b) If the object has a speed of 4.0 m/s in the negative x direction when it is at $x = 5.0$ m, what is its speed as it passes through the origin?

16E. A projectile with a mass of 2.4 kg is fired from a cliff 125 m high with an initial velocity of 150 m/s, directed 41° above the horizontal. What is (a) the kinetic energy of the projectile just after firing and (b) its potential energy? Use the ground as $y = 0$. (c) Find the speed of the projectile just before it strikes the ground. As long as air resistance can be ignored, do the answers really depend on the mass of the projectile?

17E. Figure 25 shows an 8.0-kg stone resting on a spring. The spring is compressed 10 cm by the stone. (a) What is the force constant of the spring? (b) The stone is pushed down an additional 30 cm and released. How much potential energy is stored in the spring just before the stone is released? (c) How high above this new (lowest) position will the stone rise?

18E. A 5.0-g marble is fired vertically upward using a spring gun. The spring must be compressed 8.0 cm if the marble is to just reach a target 20 m above it. (a) What is the change in gravitational potential energy of the marble over its upward flight? (b) What is the force constant of the spring?

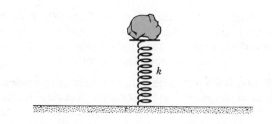

Figure 25 Exercise 17.

19E. A ball with mass m is attached to the end of a very light rod with length L. The other end of the rod is pivoted so that the ball can move in a vertical circle. The rod is pulled aside to the horizontal and given a downward push as shown in Fig. 26 so that the rod swings down and just reaches the vertically upward position. (a) What is the change in potential energy of the ball? (b) What initial speed was imparted to the ball?

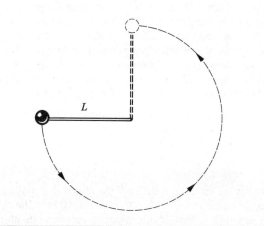

Figure 26 Exercise 19.

20E. A thin rod whose length is $L = 2.0$ m and whose mass is negligible is pivoted at one end so that it can rotate in a vertical circle. The rod is pulled aside through an angle $\theta = 30°$ and then released, as shown in Fig. 27. How fast is the lead ball moving at its lowest point?

21E. Figure 28a shows the force in newtons as a function of stretch or compression in centimeters for the cork gun containing a spring shown in Fig. 28b. The spring is compressed by 5.5 cm and used to propel a cork of mass 3.8 g from the gun. (a) What is the speed of the cork if it is released as the spring passes through its relaxed position? (b) Suppose now that the cork sticks to the spring, causing the spring to extend 1.5 cm beyond its unstretched length before separation occurs: What is the speed of the cork at the time of release in this case?

22P. The area of the continental United States is about 8 × 10^6 km² and the average elevation of its land surface is about

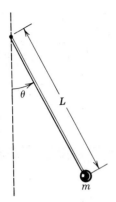

Figure 27   Exercise 20.

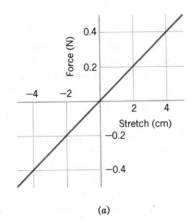

(a)

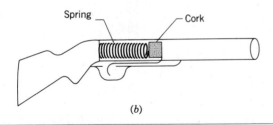

(b)

Figure 28   Exercise 21.

Figure 29   Problem 23.

that "counts" is the vertical distance that her center of gravity rose after her feet left the ground. Estimate that, at the instant her feet lost contact, her center of gravity was 110 cm above ground level. Assume also that, as she clears the bar, her center of gravity is at the same height as the bar.)

24P.   A 2.0-kg block is placed against a compressed spring on a frictionless 30° incline (Fig. 30). The spring, whose force constant is 19.6 N/cm, is compressed 20 cm, after which the block is released. How far up the incline will it go before coming to rest? Measure the final position of the block with respect to its position just before being released.

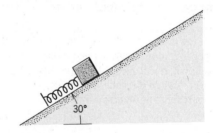

Figure 30   Problem 24.

500 m. The average yearly rainfall is 75 cm. Two-thirds of this rainwater returns to the atmosphere by evaporation, but the rest eventually flows into the oceans. If all this water could be used to generate electricity in hydroelectric power plants, what average power output could be produced?

23P.   In the 1984 Olympic Games, the West German high jumper Ulrike Meyfarth set a women's Olympic record for this event with a jump of 2.02 m; see Fig. 29. Other things being equal, how high might she have jumped on the moon, where the surface gravity is only 0.17 that on earth? (*Hint:* The height

25P.   An ideal massless spring can be compressed 2.0 cm by a force of 270 N. A block whose mass is 12 kg is released from rest at the top of an incline as shown in Fig. 31, the angle of the incline being 30°. The block comes to rest momentarily after it has compressed this spring by 5.5 cm. (*a*) How far has the block moved down the incline at this moment? (*b*) What is the speed of the block just as it touches the spring?

26P.   A 50-g ball is thrown from a window with an initial velocity of 8.0 m/s at an angle of 30° above the horizontal. Using energy methods determine (*a*) the kinetic energy of the

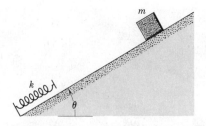

Figure 31 Problem 25.

ball at the top of its flight and (b) its speed when it is 3.0 m below the window.

27P. The spring of a spring gun has a force constant of 4.0 lb/in. When the gun is inclined at an angle of 30°, a 2.0-oz ball is projected to a height of 6.0 ft above the muzzle of the gun. (a) What was the muzzle speed of the ball? (b) By how much must the spring have been compressed initially?

28P. A 5.0-kg mortar shell is fired upward at an angle of 34° with the horizontal and with a muzzle speed of 100 m/s. (a) What is its initial kinetic energy? (b) What is its potential energy at the top of its trajectory? Assume the ground is at $y = 0$. (c) how high does the shell go?

29P. A simple pendulum is made by tying a 2.0-kg stone to a string 4.0 m long. The stone is projected perpendicular to the string, away from the ground, with the string at an angle of 60° with the vertical. It is observed to have a speed of 8.0 m/s when it passes its lowest point. (a) What was the speed of the stone at the moment of release? (b) What is the largest angle with the vertical that the string will reach during the stone's motion? (c) Using the lowest point of the swing as the zero of gravitational potential energy, what is the total mechanical energy of the system?

30P. The string in Fig. 32 is 120 cm long and the distance d to the fixed peg is 75 cm. When the ball is released from rest in the position shown, it will swing along the dotted arc. How fast will it be going (a) when it reaches the lowest point in its swing and (b) when it reaches its highest point, after the string catches on the peg?

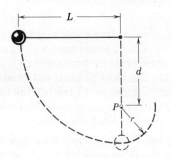

Figure 32 Problems 30 and 38.

31P. A chain is held on a frictionless table with one-fourth of its length hanging over the edge, as shown in Fig. 33. If the chain has a length L and a mass m, how much work is required to pull the hanging part back on the table?

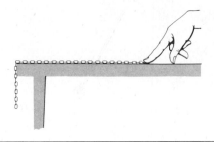

Figure 33 Problem 31.

32P. One end of a vertical spring is fastened to the ceiling. A weight is attached to the other end and slowly lowered to its equilibrium position. Show that the loss of gravitational potential energy of the weight equals one-half the gain in elastic potential energy. (Why are these two quantities not equal?)

33P. A 2.0-kg block is dropped from a height of 40 cm onto a spring of force constant $k = 1960$ N/m; see Fig. 34. Find the maximum distance the spring will be compressed.

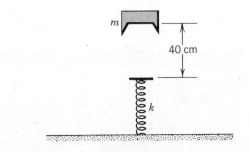

Figure 34 Problem 33.

34P. Figure 35 shows a simple pendulum of length L. Its bob is observed to have a speed v_0 when the cord makes an angle θ_0 with the vertical. (a) Derive an expression for the speed of the bob when it is in its lowest position. (b) What is the least value that v_0 can have if the cord is to swing up to a horizontal position? (c) To a vertical position with the cord remaining straight?

35P. Two children are playing a game in which they try to hit a small box on the floor with a marble fired from a spring-loaded gun that is mounted on a table. The target box is 2.2 m horizontally from the edge of the table; see Fig. 36. Bobby compresses the spring 1.10 cm, but the marble falls 27 cm short. How far should Rhoda compress the spring to score a hit?

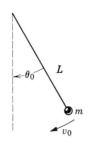

Figure 35 Problem 34.

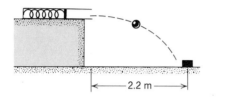

Figure 36 Problem 35.

36P. A small block of mass m slides along the frictionless loop-the-loop track shown in Fig. 37. (a) The block is released from rest at point P. What is the resultant force acting on it at point Q? (b) At what height above the bottom of the loop should the block be released so that it is on the verge of losing contact with the track at the top of the loop?

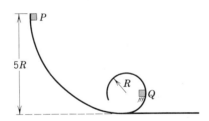

Figure 37 Problem 36.

37P. Tarzan, who weighs 180 lb, swings from a cliff at the end of a convenient 50-ft grapevine; see Fig. 38. From the top of the cliff to the bottom of the swing, Tarzan's center of gravity would fall by 8.5 ft. The grapevine has a breaking strength of 250 lb. Will the grapevine break?

38P. In Fig. 32 show that, if the pendulum bob is to swing completely around the fixed peg, then $d > 3L/5$. (*Hint:* The bob must be moving at the top of its swing; otherwise the string will collapse.)

39P. A light rigid rod of length L has a ball with mass m attached to its end, forming a simple pendulum. It is inverted so that the rod is straight up and then released. What are (a) its speed at the lowest point and (b) the tension in the suspension

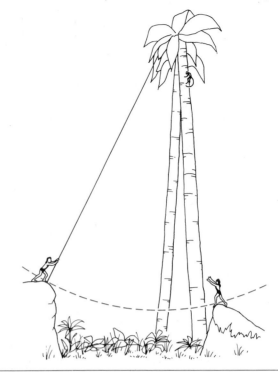

Figure 38 Problem 37.

at that instant? (c) The same pendulum is next put in a horizontal position and released from rest. At what angle from the vertical will the tension in the rod equal the weight in magnitude?

40P. An escalator joins one floor with another one 8.0 m above. The escalator is 12 m long and moves along its length at 60 cm/s. (a) What power must its motor deliver if it is required to carry a maximum of 100 persons per minute, of average mass 75 kg? (b) An 80-kg man walks up the escalator in 10 s. How much work does the motor do on him? (c) If this man turned around at the middle and walked down the escalator so as to stay at the same level in space, would the motor do work on him? If so, what power does it deliver for this purpose? (d) Is there any (other?) way the man could walk along the escalator without consuming power from the motor?

41P*. A 3.2-kg block starts at rest and slides a distance d down a smooth 30° incline where it runs into a spring of negligible mass; see Fig. 39. The block slides an additional 21 cm

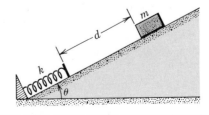

Figure 39 Problem 41.

before it is brought to rest momentarily by compressing the spring, whose force constant is 430 N/m. (a) What is the value of d? (b) The speed of the block continues to increase for a certain interval after the block makes contact with the spring. What additional distance does the block slide before it reaches its maximum speed and begins to slow down?

42P*. A boy is seated on the top of a hemispherical mound of ice (Fig. 40). He is given a very small push and starts sliding down the ice. Show that he leaves the ice at a point whose height is $2R/3$ if the ice is frictionless. (*Hint:* The normal force vanishes as he leaves the ice.)

Figure 40 Problem 42.

Section 8-6 The Potential Energy Curve
43E. A particle moves along the x axis through a region in which its potential energy $U(x)$ varies as in Fig. 41. (a) Make a quantitative plot of the force $F(x)$ that acts on the particle, using the same x axis scale as in Fig. 41. (b) The particle has a (constant) mechanical energy E of 4.0 J. Sketch a plot of its kinetic energy $K(x)$ directly on Fig. 41.

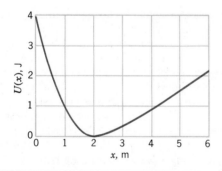

Figure 41 Exercise 43.

44P. A particle of mass 2 kg moves along the x axis through a region in which its potential energy $U(x)$ varies as shown in Fig. 42. When the particle is at $x = 2$ m, its velocity is -2 m/s. (a) What force acts on it at this position? (b) Between what limits does the motion take place? (c) How fast is it moving when it is at $x = 7$ m?

45P. Figure 43a shows an atom of mass m at a distance r from a resting atom of mass M, where $m \ll M$. Figure 43b shows the potential energy function $U(r)$ for various positions

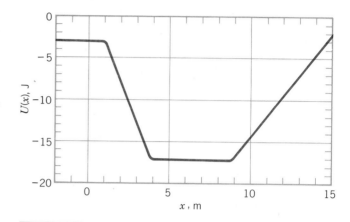

Figure 42 Problem 44.

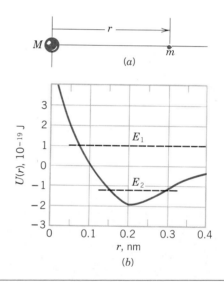

Figure 43 Problem 45.

of the lighter atom. Describe the motion of this atom if (a) the total mechanical energy is greater than zero, as at E_1 and (b) if it is less than zero, as at E_2. For $E_1 = 1 \times 10^{-19}$ J and $r = 0.3$ nm, find (c) the potential energy, (d) the kinetic energy, and (e) the force (magnitude and direction) acting on the moving atom.

Section 8-8 The Conservation of Energy
46E. When a Space Shuttle (mass 79,000 kg) returns to earth from orbit, it enters the atmosphere at a speed of 18,000 mi/h, which it gradually reduces to a touchdown speed of 190 knots (= 220 mi/h). What is its kinetic energy (a) at atmospheric entry and (b) at touchdown? See Fig. 44. (c) What happens to all the "missing" energy?

47E. A 68-kg skydiver falls at a constant terminal speed of

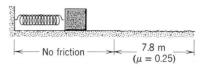

Figure 44   Exercise 46.

59 m/s. At what rate is gravitational potential energy being removed from the earth–sky diver system? What happens to this energy?

48E.   A river descends 15 m in passing through rapids. The speed of the water is 3.2 m/s upon entering the rapids and is 13 m/s as it leaves. What percentage of the potential energy lost by 10 kg of water in traversing the rapids appears as kinetic energy of water downstream? (Does the answer depend on the mass of water considered?) What happens to the rest of the energy?

49E.   A projectile whose mass is 9.4 kg is fired vertically upward. On its upward flight, an energy of 68 kJ is dissipated because of air resistance. How much higher would it have gone if the air resistance had been made negligible (for example, by streamlining the projectile)?

50E.   A 25-kg bear slides, from rest, 12 m down a lodgepole pine tree, moving with a speed of 5.6 m/s at the bottom. (*a*) What is the initial potential energy of the bear? (*b*) What is the kinetic energy of the bear at the bottom? (*c*) What is the average frictional force that acts on the bear? Put $y = 0$ at the bottom of the pole.

51E.   During a rockslide, a 520-kg rock slides from rest down a hillslope that is 500 m long and 300 m high. The coefficient of kinetic friction between the rock and the hill surface is 0.25. (*a*) What is the potential energy of the rock just before the slide? (Take $y = 0$ to be at the bottom of the hill.) (*b*) How much work was done on the rock by frictional forces during the slide? (*c*) What is the kinetic energy of the rock as it reaches the bottom of the hill? (*d*) What is its speed?

52E.   As Fig. 45 shows, a 3.5-kg block is released from a compressed spring whose force constant is 640 N/m. After leaving the spring, the block travels over a horizontal surface, with a coefficient of friction 0.25, for a distance of 7.8 m before coming to rest. (*a*) How much work was done by friction in bringing

the block to rest? (*b*) What was the maximum kinetic energy of the block? (*c*) How far was the spring compressed before the block was released?

53E.   A 2.5-kg block collides with a horizontal massless spring whose force constant is 320 N/m; see Fig. 46. The block compresses the spring a maximum distance of 7.5 cm from its rest position. The coefficient of kinetic friction between the block and the horizontal surface is 0.25. (*a*) How much work was done by the spring in bringing the block to rest? (*b*) How much work was done by the force of friction while the block was being brought to rest by the spring? (*c*) What was the speed of the block when it hit the spring?

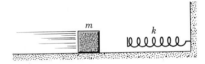

Figure 46   Exercise 53.

54P.   Two snow-covered peaks are at elevations of 850 m and 750 m above the valley between them. A ski run extends from the top of the higher peak to the top of the lower one, with a total length of 3.2 km and an average slope of $30°$; see Fig. 47. (*a*) A skier starts from rest on the higher peak. At what speed will he arrive at the top of the lower peak if he just coasts without using the poles? Ignore friction. (*b*) Approximately how large a coefficient of friction with the snow could be tolerated without preventing him from reaching the lower peak?

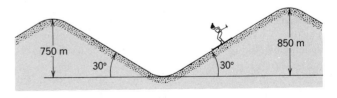

Figure 47   Problem 54.

55P.   A factory worker accidentally releases a 400-lb crate that was being held at rest on a 12-ft-long ramp inclined at $39°$ to the horizontal. The coefficient of kinetic friction between crate and ramp, and also between the crate and the factory floor, is 0.28. (a) How fast is the crate moving as it reaches the bottom of the ramp? (b) How far will it subsequently slide across the factory floor?

56P. Two blocks are connected by a string, as shown in Fig. 48. They are released from rest. Show that, after they have moved a distance L, their common speed is given by

$$v = \sqrt{\frac{2(m_2 - \mu m_1)gL}{m_1 + m_2}},$$

in which μ is the coefficient of kinetic friction between the upper block and the surface.

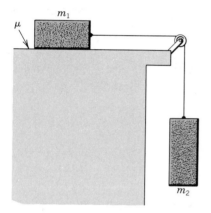

Figure 48 Problem 56.

57P. A 4.0-kg block starts up a 30° incline with 128 J of kinetic energy. How far will it slide up the plane if the coefficient of friction is 0.30?

58P. A block is moving up a 40° incline. At a point 1.8 ft from the bottom of the incline (measured along the incline), it has a speed of 4.5 ft/s. The coefficient of kinetic friction between block and incline is 0.15. (a) How much farther up the incline will the block move? (b) How fast will it be going after it slides back to the bottom of the incline?

59P. A certain spring is found *not* to conform to Hooke's law. The force (in newtons) it exerts when stretched a distance x (in meters) is found to have magnitude $52.8x + 38.4x^2$ in the direction opposing the stretch. (a) Compute the total work required to stretch the spring from $x = 0.50$ m to $x = 1.00$ m. (b) With one end of the spring fixed, a particle of mass 2.17 kg is attached to the other end of the spring when it is extended by an amount $x = 1.00$ m. If the particle is then released from rest, compute its speed at the instant the spring has returned to the configuration in which the extension is $x = 0.50$ m. (c) Is the force exerted by the spring conservative or nonconservative? Explain.

60P. The magnitude of the force of attraction between the positively charged nucleus and the negatively charged electron in the hydrogen atom is given by

$$F = k\frac{e^2}{r^2},$$

where e is the charge of the electron, k is a constant, and r is the separation between electron and nucleus. Assume that the nucleus is fixed. Imagine that the electron is initially moving in a circle of radius r_1 about the nucleus and jumps suddenly into a circular orbit of smaller radius r_2; see Fig. 49. (a) Calculate the change in kinetic energy of the electron, using Newton's second law. (b) Using the relation between force and potential energy, calculate the change in potential energy of the atom. (c) By how much has the total energy of the atom changed in this process? (This energy is often given off in the form of radiation.)

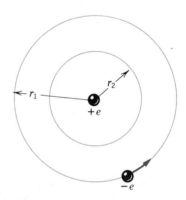

Figure 49 Problem 60.

61P. A stone of weight w is thrown vertically upward into the air with an initial speed v_0. If a constant force f due to air resistance acts on the stone throughout its flight, (a) show that the maximum height reached by the stone is

$$h = \frac{v_0^2}{2g(1 + f/w)}.$$

(b) Show that the speed of the stone upon impact with the ground is

$$u = v_0 \left(\frac{w - f}{w + f}\right)^{1/2}.$$

62P. A child's playground slide is in the form of an arc of a circle with a maximum height of 4.0 m and a radius of curvature of 12 m; see Fig. 50. A 25-kg child starts from rest at the top of the slide and is observed to have a speed of 6.2 m/s at the bottom. (a) What is the length of the slide? (b) What average force of friction acts on the child over this distance?

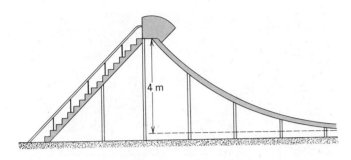

Figure 50 Problem 62.

63P. A small particle slides along a track with elevated ends and a flat central part, as shown in Fig. 51. The flat part has a length $L = 2.0$ m. The curved portions of the track are frictionless. For the flat part, the coefficient of kinetic friction is $\mu_k = 0.20$. The particle is released at point A which is a height $h = 1.0$ m above the flat part of the track. Where does the particle finally come to rest?

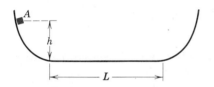

Figure 51 Problem 63.

64P. A very light rigid rod whose length is L has a ball of mass m attached to one end (Fig. 52). The other end is pivoted frictionlessly in such a way that the ball moves in a vertical circle. The system is launched from the horizontal position A with downward initial velocity v_0. The ball just reaches point D

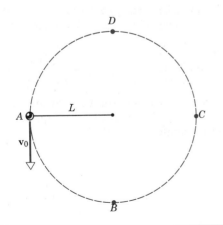

Figure 52 Problem 64.

and then stops. (a) Derive an expression for v_0 in terms of L, m, and g. (b) What is the tension in the rod when the ball is at B? (c) A little grit is placed on the pivot, after which the ball just reaches C when launched from A with the same speed as before. How much work is done by friction during this motion? (d) How much total work is done by friction before the ball finally comes to rest at B after oscillating back and forth several times?

65P. The cable of a 4000-lb elevator in Fig. 53 snaps when the elevator is at rest at the first floor so that the bottom is a distance $d = 12$ ft above a cushioning spring whose spring constant is $k = 10,000$ lb/ft. A safety device clamps the guide rails so that a constant friction force of 1000 lb opposes the motion of the elevator. (a) Find the speed of the elevator just before it hits the spring. (b) Find the distance x that the spring is compressed. (c) Find the distance that the elevator will bounce back up the shaft. (d) Using the conservation of energy principle, find approximately the total distance that the elevator will move before coming to rest. Why is the answer not exact?

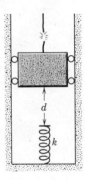

Figure 53 Problem 65.

66P*. While a 1700-kg automobile is moving at a constant speed of 15 m/s, the motor supplies 16 kW of power to overcome friction, wind resistance, and so on. (a) What is the effective retarding force associated with all the frictional forces combined? (b) What power must the motor supply if the car is to move up an 8% grade (8 m vertically for each 100 m horizontally) at 15 m/s? (c) At what downgrade, expressed in percentage terms, would the car coast at 15 m/s?

Section 8–9 Mass and Energy
67E. (a) How much energy in joules is represented by a mass of 100 g? (b) For how many years would this supply the energy needs of a one-family home consuming energy at the average rate of 1 kW?

68E. Verify the energy equivalences listed in Table 1, obtaining the energy both in joules and the other units listed. Use the exact value of c.

69E. The magnitude M of an earthquake on the Richter scale is related to the released energy E in joules by the equation

$$\log E = 4.4 + 1.5\ M.$$

(a) The 1906 San Francisco earthquake, see Fig. 54, was of magnitude 8.2. How much energy was released? (b) How much mass would have to be converted to release this amount of energy?

Figure 54   Exercise 69.

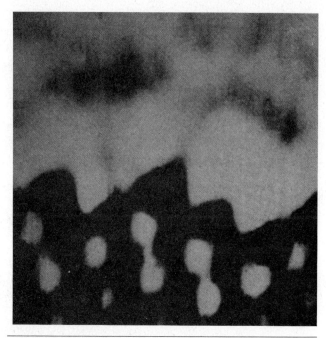

Figure 55   Exercise 70.

70E. The nucleus of an atom of gold (see Fig. 55) contains 79 protons and 118 neutrons and has a mass of 196.9232 u. Calculate the binding energy of the nucleus. See Sample Problem 9 for other needed numerical data.

71P. A nuclear power plant in Oregon supplies 1030 MW of useful power steadily for a year. In addition, 2100 MW of power is discharged as thermal energy to the Columbia River. How much mass is converted to other forms of energy in a year at this plant?

72P. The United States generated about 2.31×10^{12} kW·h of electrical energy in 1983. How many kilograms of mass would have to be completely converted to electrical energy to yield this value?

73P. An aspirin tablet has a mass of 320 mg. For how many miles would the energy equivalent of this mass, in the form of gasoline, power a car? Assume 30 mi/gal and a heat of combustion of gasoline of 130 MJ/gal. Express your answer in terms of the equatorial circumference of the earth.

74P. How much mass must be converted into kinetic energy to accelerate an 1820-ton spaceship from rest to one-tenth the speed of light? Use the nonrelativistic formula for kinetic energy.

75P. The sun radiates energy at the rate of 4×10^{26} W. How many "tons of sunlight" does the earth intercept in 1 day?

Section 8–10   Energy Is Quantized

76E. By how much must the energy of an atom change in order to emit light of frequency 4.3×10^{14} s^{-1}?

77P. (a) A hydrogen atom has an energy of -3.4 eV. If its energy changes to -13.6 eV, what is the frequency of the light? (b) Is the light emitted or absorbed?

ESSAY 2
FEAR AND TREMBLING AT THE AMUSEMENT PARK

JOHN ROEDER
THE CALHOUN SCHOOL

JEARL WALKER
CLEVELAND STATE
UNIVERSITY

The lure of the rides at an amusement park lies in their apparent danger, in illusions of motion and forces—and often in the uncommon experience of rapid rotation. The rides might play such tricks on your common sense that you cherish the experience, thrilling in being hurled toward the ground or whipped around with such abandon that sight of your surroundings blurs. You owe much of the excitement to physics.

Merry-Go-Round

Many of the rides involve rotation because it creates illusions and strange sensations. For example, consider a traditional merry-go-round with imitation horses. When you ride a horse while the ride turns around its center, you feel as though you are being forced radially outward. Is there really such a *centrifugal force* on you? What could generate it? Certainly there is no agent on one side of you pressing you outward. One reason the merry-go-round is fun is that it creates the illusion of a magical centrifugal force. Actually, the only radial force on you is inward as the horse pulls on the lower part of your body, forcing you to continue moving around a circle (Fig. 1). Since the lower part of your body pulls on the rest of you, your body tends to lean outward. You misinterpret the leaning as being due to some unseen agent pushing against you.

Every merry-go-round fan knows that the outer horses are more exciting to ride than the inner horses. The difference is due to the speed the horse gives you. If your speed is small, you require only a small centripetal acceleration to move around a circle. A larger centripetal acceleration is needed when your speed is larger. Since the horses are fastened to the merry-go-round, they each must complete a circle in the same amount of time. The inner horses travel around a small circle with a small speed. When you ride one of them, the acceleration that you undergo is small, as is also the centripetal force you experience. When you ride one of the outer horses, your speed is larger, which means that your acceleration and the centripetal force on you must also be larger. The illusion of a centrifugal force resulting from some unseen agent may then be quite compelling.

Ferris Wheel*

A Ferris wheel also creates illusions. In this ride, you sit in a cage that is suspended on the rim of a vertical structure (Fig. 2). When the structure rotates around its center, you move around a large vertical circle. Throughout the circle your cage is free to rotate around its own axis so that you remain vertical. When you move through the bottom of the circle, you feel especially heavy whereas when you move through the top of the circle, you feel light but not quite weightless. The ride produces two illusions: the sensation that your weight varies and the sensation that you are subject to a

---

* The original Ferris wheel was created for the 1893 World's Columbian Exposition in Chicago by George Washington Gale Ferris, a civil-engineering graduate of Rensselaer Polytechnic Institute. The original wheel was 250 ft in diameter and could carry 2000 passengers at one time in its 36 cars.

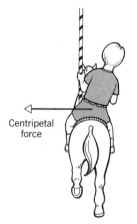

Figure 1 Centripetal force on a merry-go-round.

centrifugal force. The Ferris wheel is scarier than the merry-go-round not only because it generates a fear of falling but also because the forces on you vary with time.

Consider the phase when you move through the bottom of the circle (Fig. 3*a*). The apparent centrifugal force on you seems to press you into the seat more than normal, creating the illusion that your weight is greater than normal. Of course, there is no unseen agent forcing you radially outward from the center of rotation and the earth's gravitational attraction on you is unchanged. You feel especially heavy because the force from the seat is larger than it would be were the Ferris wheel stationary. Let **W** be your weight (a downward vector) and **P** be the force you feel from the seat (an upward vector). Since you accelerate toward the center of the rotation, you are undergoing a positive (upward) acceleration. The magnitude of the centripetal force on you is

$$F_c = \frac{mv^2}{r}.$$

where m is your mass, v is your speed and the speed of the cage you are in, and r is the radius of the Ferris wheel and thus your distance from the ride's axis of rotation. The centripetal force is the vector sum of **W** and **P**:

$$-W + P = \frac{mv^2}{r}.$$

which yields P as

$$P = W + \frac{mv^2}{r}.$$

In addition to supporting you, the seat must also supply the centripetal force necessary for you to continue moving in the circle. You misinterpret the larger than normal force on you from the seat as evidence that your weight has increased.

When you move through the top of the circle, your acceleration is again toward the center of the rotation, but this time the vector is downward and thus negative (Fig. 3*b*):

$$-W + P = -\frac{mv^2}{r}.$$

which yields P as

$$P = W - \frac{mv^2}{r}.$$

Figure 2 Ferris wheel.

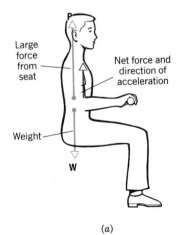

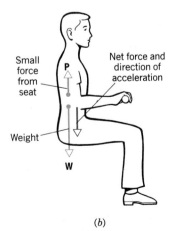

Figure 3 Forces on passenger in rotating Ferris wheel: (*a*) at bottom of circle and (*b*) at top of circle.

(*a*) (*b*)

Figure 4 Tilted Roundup ride.

The force on you from the seat is smaller than it would be were the Ferris wheel stationary. You misinterpret the reduction in the force as evidence that you are lighter than normal.

Roundup

The Roundup is a ride that initially appears to be similar to the Rotor (see Sample Problem 9 in Chapter 6). It consists of a wide horizontal disk that has a sturdy network of metal bars fastened to its perimeter (Fig. 4). You stand on the disk next to a solid wall that is fastened to the bars. A constraining bar lies in front of you, locked into place. When the disk spins around its center, the wall presses against your back, providing centripetal force. As the speed builds, the force grows stronger until it is quite large when the disk reaches its final, constant speed. You feel as though an unseen agent is pushing you radially outward with an irresistible force, pinning you against the wall like some butterfly specimen.

While the disk continues to spin, a machine lifts and tilts it so that you then rotate around an axis that is 60° off the vertical. The ride is then similar to a Ferris wheel, except that a wall and disk push against you instead of a seat. However, it differs from a Ferris wheel in that when you are at the top of the rotation, the centripetal acceleration on you is greater than the acceleration of free fall. Your sensation of weight there, as evidenced by the downward force on you from the wall, may seem normal, even though you are upside down. When you pass through the lowest portion of the circle, you seem to be heavier than normal because the force from the wall is then greatest. The continuous variations in the force you feel and the sensation of your weight make the tilted phase of the ride scarier than when the spinning disk is horizontal and the forces on you are constant.

Roller Coaster

One of the most popular rides at an amusement park is a roller coaster (Fig. 5). The coaster, which consists of several cars with passengers, runs along a track over hills and through valleys. To get it started, a chain driven by a machine pulls the coaster up the first hill, which is the highest of all the hills. The track that descends from the hill on the far side is steep. When the first cars reach the descending track, the chain disengages and the front cars drag the rest of the coaster into the descent. Speed builds, as does fear in the passengers.

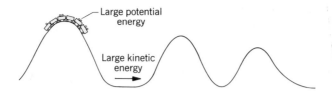

Figure 5 Roller coaster.

In pulling the coaster up the hill, the machine does work, storing gravitational potential energy in the coaster. Just as the coaster begins its descent, nearly all its energy is in that form. With essentially no kinetic energy, the coaster has a small speed at this point. However, as the coaster descends, energy is transferred from potential energy to kinetic energy, increasing the speed. Thereafter, when the coaster climbs a hill, energy is transferred from kinetic energy to potential energy. Whenever it descends, the transfer is reversed.

To have the most fun in a roller coaster, should you sit in the front or rear car? Both positions offer thrills, especially if the hilltop is greatly curved so that when passing over the crest the force on you from the chair is reduced and you feel as though you are being thrown from the chair. However, even when the hilltop is flat, sitting in the front car is frightening because you experience the illusion that you are falling over a cliff when the car begins its descent. Because your view of the track is limited, you do not anticipate the drop. Suddenly, the supporting force you feel from the chair is reduced. The abrupt change creates the illusion that you have fallen out of control. The illusion is weaker when you sit in the rear car because the disappearance of the front cars as they begin to descend alerts you that your descent is imminent.

The rear car offers a different type of thrill when the coaster descends from a flat hill such as the first one. In the rear car, you begin your descent while the rest of the coaster is already headed downward. Passengers in the front car begin their descent with a slow speed. You begin yours with a larger speed because by then an appreciable part of the coaster's energy is in the form of kinetic energy. When your car begins to descend, it falls away from you momentarily as you continue to move horizontally. You have the illusion of being thrown from your chair, prevented from flying out of the car altogether by a lap strap, retaining bar, or some other constraining device (Fig. 6). Some riders enjoy raising their arms as a sign of bravado when beginning their descent at high speed, trusting that the constraining device on them will keep them within the car. At times the devices have failed and riders have been killed.

The rapid pace of the car and the illusion of being thrown from it are two reasons why a ride in the last car is frightening. However, much of the fear comes from a much more subtle factor. When you near the descending track, the force on your back from the chair rapidly increases. You suddenly fear that the chair is going to hurl you free of the car.

Why does the force on you increase in such a dramatic manner? Suppose that the

Figure 6 Passenger's motion when descent begins.

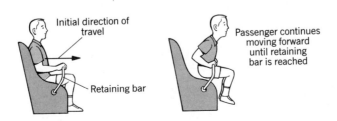

coaster travels from a flat-top hill onto descending track that is at a constant angle θ with the horizontal. Before the first car begins to descend, the coaster's acceleration is zero because there is no net force on it. When the first cars begin to descend, they are pulled along the track by a force that is equal to the product of their mass m and the acceleration of gravity along the track:

$$F = mg \sin \theta.$$

Since the first cars are attached to the rest of the coaster, this force accelerates the entire coaster:

$$F = Ma = mg \sin \theta,$$

where M is the coaster's mass and a is the acceleration of the coaster and you. Rearrange the equation to isolate a:

$$a = \frac{mg \sin \theta}{M}.$$

As more cars begin to descend, m grows larger, increasing your acceleration (Fig. 7).

Assume that the mass of the passengers is uniformly distributed along the length of the coaster. Let x be the length of the coaster on the descending track and L be the total length of the coaster. Then,

$$\frac{m}{M} = \frac{x}{L}.$$

Substitute this ratio into the equation for the acceleration:

$$a = \frac{x(g \sin \theta)}{L}.$$

The acceleration of you and the coaster increases as x increases, until $x = L$.

The equation may appear to be tame but it is responsible for the sudden, wrenching fear you undergo when you ride the last car. You accelerate because the back of your chair pushes on you. The acceleration and the force from the chair depend on the value of x. As x increases, the acceleration increases, but that means that x increases even faster, which makes the acceleration even larger, which makes x increase faster still (Fig. 8). Your mind races with the changes you experience. The force on your back is initially small but as you near the descending track it rapidly increases. You feel as though some unseen agent is behind you, attempting to propel you over the edge of the hill. Suddenly, just when the force on your back has shot up to its peak value, it abruptly disappears as you begin your descent, enhancing the illusion that you have been thrown from the chair.

Some modern roller coaster tracks come equipped with loops and spirals. In the

Figure 7 Force on descending coaster.

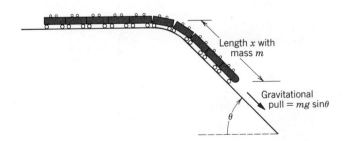

Length x with mass m

Gravitational pull $= mg \sin\theta$

θ

Figure 8 Force on your back when in rear car of coaster.

Double Loop version, the coaster travels through two vertical loops, with the passengers being upside down at the top of each loop. The ride begins in the normal fashion. After the coaster passes over several hills, it enters the first loop, climbing its interior and then coming back down on the opposite side (Fig. 9). The experience is hair-raising. When passing through the top of the loop, the ground seems to be above you, and you fear that you might fall from your car. As you travel around the loop, you undergo the same sensations you experience in a tilted Roundup except that the force on you is from a chair instead of a wall. At the bottom of the loop, you seem to be compressed into the chair while at the top the chair hardly pushes on you. The illusionary decrease of weight at the top of the loop enhances your fear of leaving the car and falling to the ground.

How high should the first hill be in comparison to the top of the loop? For simplicity, suppose the coaster consists of only one car and the energy losses during the ride can be ignored. The problem is then similar to Problem 36 in Chapter 8. The initial hill must provide at least enough potential energy such that when the coaster is at the top of the loop it still makes contact with the track. Any lower starting height means that the coaster will leave the track at the top of the loop or while approaching the top. In practice, the initial hill is considerably higher than the minimum allowed because the coaster experiences energy losses along its route to the loop. For added precaution the coaster is equipped with two sets of wheels, one above the track and the other set below the track. Usually, the weight of the coaster rests on the top set of

Figure 9 Coaster in a vertical loop.

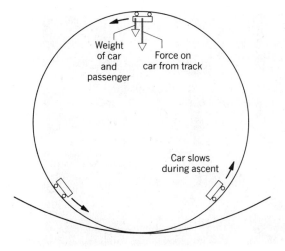

wheels, but when it is at the top of the loop the other set of wheels can prevent the coaster from leaving the track.

Calypso or Scrambler

Another popular type of ride is one where the cage in which you sit is attached to the outer end of a horizontal mechanical arm. Call this arm the secondary arm. Its other end is attached to the outer end of another horizontal arm. Call this arm the primary arm. The other end of the primary arm is attached to the center of the ride. The ride has several primary arms with up to four secondary arms on each. Each secondary arm has a cage with passengers. The primary arms rotate around the center while the secondary arms rotate around the point where they are attached to the primary arms. During the complex rotations you undergo, you are thrown about your seat with abandon even if you hold onto a constraining bar locked into place in front of you.

The manner in which the arms rotate divides this type of ride into two versions. In the Calypso version, the secondary arms turn in the same direction as the primary arms, whereas in the Scrambler version they turn in the opposite direction (Fig. 10). Both versions offer ample thrills and illusions.

Suppose you were to design one of these rides. For simplicity, assume that the ride has only one set of primary and secondary arms. Also, assume that the arms are of equal length. (In practice, the secondary arm is actually shorter than the primary arm to prevent cages from crashing when the ride is in operation.) How fast should the secondary arm rotate in comparison to the primary arm? If you make them rotate at the same speed, the passenger in a Calypso will travel through a large circle while a passenger in a Scrambler will travel back and forth in a straight line through the center of the ride. Both results are boring.

The Calypso is more exciting if you have the secondary arm turn twice as fast as the primary arm. When you pass through the outermost part of the path that you take, your speed is large, as is your acceleration and the force on you. From there you spiral into the center and back out again (Fig. 11). When you pass through the center, your speed is small and the change in your direction of travel is moderate, both factors reducing the force on you. As you again travel toward the outermost part of the path, the force increases. Since you cannot brace for the continuous variation in your direction of travel and for the size and direction of the force on you, you are hurled first one way and then another way, with no letup until the ride stops.

Figure 10 Calypso and/or Scrambler ride.

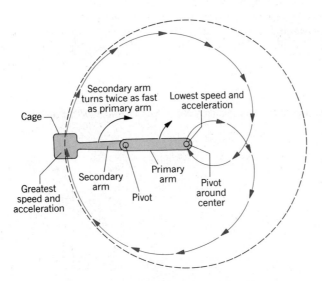

Figure 11 Motion in a simplified Calypso ride.

The same arrangement of relative speeds for the arms also makes the Scrambler stimulating. At the outermost part of your path, your speed is small but the change in your direction of travel is so abrupt that you undergo a large acceleration and feel a large force. From there you move almost directly toward the center of the ride, rapidly picking up speed (Fig. 12). Your speed is greatest when you pass through the center. However, since the change in your direction of travel is only gradual there, your acceleration and the force you feel are both small. When you return to the outermost part of the path, you again undergo rapid acceleration and feel a large force. Bracing yourself in this ride is no more possible than in the Calypso.

When the secondary arm turns twice as fast as the primary arm, the rides give different, bewildering experiences. Are there better choices for the rotation speeds? You might enjoy searching for them by simulating the Calypso and Scrambler on a computer. Although the simulations will not be as thrilling as actually being in one of the rides, they will also not be as taxing on your nerves.

Figure 12 Motion in a simplified Scrambler ride.

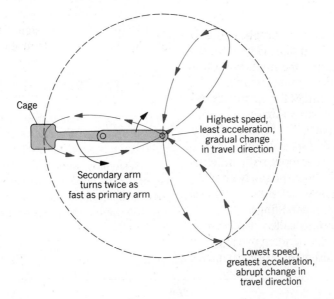

CHAPTER 9

SYSTEMS OF PARTICLES

For some purposes, we can represent systems made up of many particles by a single particle to which we can assign a mass, a position, a velocity, and an acceleration and on which a net force can be said to act. We can represent by a single vector, for example, the velocity of this assembly of skiers at the annual Finlandia cross-country ski run.

9–1 A Special Spot

Physicists love to look at something complicated and find in it something simple and familiar. Here is an example. If you toss a baseball bat into the air, its motion as it twists and turns is clearly more complicated than that of, say, a tossed baseball. Every part of the bat moves in a different way from every other part so that you cannot represent the bat as a single particle. However, if you look closely, you will find that one special spot along the axis of the bat moves in a simple parabolic path, much as a tossed baseball does. This point, surprisingly enough, turns out to be exactly the point at which the bat balances as you support it across your extended finger.

Every body has such a point, called its *center of mass.* Figure 1 shows an ax thrown between two jugglers. The center of mass of the ax (and no other point) moves like a free particle, following a parabolic path. If the jugglers were doing their stunt in a dark room (!) and if a small light bulb were fastened to the center of mass of the ax, the simplicity of its motion would be clear.

9–2 The Center of Mass

In learning about center of mass, our plan is to start with simple systems of particles and then work our way up to solid objects such as baseball bats.

Systems of Particles. Figure 2a shows two particles of masses m_1 and m_2 separated by a distance d. We have arbitrarily chosen the origin of the x axis to coincide with m_1. We *define* the position of the center of mass to be

$$x_{\text{cm}} = \frac{m_2}{m_1 + m_2} d. \tag{1}$$

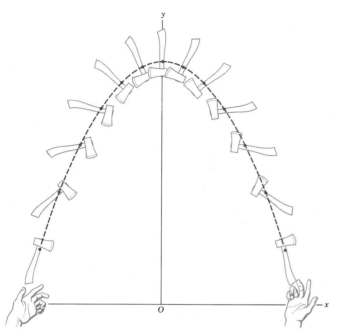

y

O ———————————— x

Figure 1   Two performers toss an ax between them. Note that one special spot on the ax (its center of mass) follows a parabolic path, just as the center of a tossed baseball would. No other point of the ax moves in such a simple way.

This definition is consistent with our interpretation of the center of mass as a balance point. For example, if $m_2 = 0$, there is only one particle (m_1) and the center of mass must lie at the position of that particle; Eq. 1 dutifully reduces to $x_{cm} = 0$. If $m_1 = 0$, there is again only one particle (m_2) and we have, as we expect, $x_{cm} = d$. If $m_1 = m_2$, the masses are equal and the center of mass should be halfway between them; Eq. 1 reduces to $x_{cm} = \frac{1}{2}d$, again as we expect. Finally, Eq. 1 tells us that x_{cm} can only have values that lie between zero and d; that is, the center of mass (balance point) must lie somewhere between the two particles.

Figure 2b shows a more generalized situation, in which the two particles have been moved out along the x axis, away from the origin. The position of the center of mass is now defined from

$$x_{cm} = \frac{m_1 x_1 + m_2 x_2}{m_1 + m_2}.$$ (2)

Note that if we put $x_1 = 0$ (see Fig. 2b), x_2 becomes d and Eq. 2 reduces to Eq. 1, as it must.

We can rewrite Eq. 2 as

$$x_{cm} = \frac{m_1 x_1 + m_2 x_2}{M},$$

in which $M (= m_1 + m_2)$ is the total mass of the system.

We can extend this definition to the case of n particles, strung out along the x axis. The location of the center of mass is

$$x_{cm} = \frac{m_1 x_1 + m_2 x_2 + m_3 x_3 + \cdots + m_n x_n}{M}$$

$$= \frac{1}{M} \sum m_i x_i.$$ (4)

Here the subscript i is a running number that takes on integral values from 1 to n.

If the particles are distributed in three dimensions, the center of mass must be identified by three coordinates. By extension of Eq. 4, they are defined from

$$x_{cm} = \frac{1}{M} \sum m_i x_i,$$

$$y_{cm} = \frac{1}{M} \sum m_i y_i,$$ (5)

$$z_{cm} = \frac{1}{M} \sum m_i z_i.$$

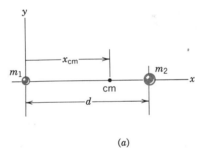

(a)

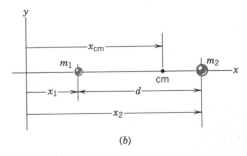

(b)

Figure 2   (a) Two particles of masses m_1 and m_2 are separated by a distance d. The dot shows the position of the center of mass, calculated from Eq. 1. (b) The same as (a) except that the particles are located farther from the origin. The position of the center of mass is calculated from Eq. 3. The relative location of the center of mass (with respect to the particles) is the same in each case.

We can also deal with the center of mass in the language of vectors. The position vector $\mathbf{r}_i$ describing a particle whose coordinates are x_i, y_i, and z_i is

$$\mathbf{r}_i = \mathbf{i}x_i + \mathbf{j}y_i + \mathbf{k}z_i. \qquad (6)$$

Here the subscript i is the running number that identifies the particle and $\mathbf{i}$, $\mathbf{j}$, and $\mathbf{k}$ are unit vectors pointing, respectively, in the direction of the x, y, and z axes.* Similarly, the position of the center of mass of a system of particles can be written

$$\mathbf{r}_{cm} = \mathbf{i}x_{cm} + \mathbf{j}y_{cm} + \mathbf{k}z_{cm}. \qquad (7)$$

The three scalar equations of Eq. 5 can now be replaced by a single vector equation,

$$\mathbf{r}_{cm} = \frac{1}{M} \sum m_i \mathbf{r}_i. \qquad (8)$$

You can check that this equation is correct by substituting Eqs. 6 and 7 into it; the scalar relations of Eq. 5 result.

Center of Mass — Rigid Bodies. An ordinary object, such as an ax or a guitar, contains so many particles (atoms) that we can best treat it as a continuous distribution of matter. The "particles" then become differential mass elements dm, the sums of Eq. 5 become integrals, and the coordinates of the center of mass are defined from

$$x_{cm} = \frac{1}{M} \int x \, dm,$$
$$y_{cm} = \frac{1}{M} \int y \, dm, \qquad (9)$$
$$z_{cm} = \frac{1}{M} \int z \, dm.$$

If we wish to locate the center of mass by means of a vector, we can replace the three scalar relations of Eq. 9 by the single vector relation

$$\mathbf{r}_{cm} = \frac{1}{M} \int \mathbf{r} \, dm. \qquad (10)$$

Many objects have a point, a line, or a plane of symmetry. The center of mass of the object then lies at that point, on that line, or in that plane. For example, the center of mass of a homogeneous sphere (which has a point of symmetry) is at the center of the sphere. The center of mass of a homogeneous cone (which has a line of symmetry) lies on the axis of the cone. The center of mass of a banana (which has a plane of symmetry) lies somewhere in that plane.

The center of mass of an object need not lie within the body of that object. There is no rubber at the center of mass of an automobile tire and no iron at the center of mass of a horse shoe.

Sample Problem 1 Figure 3 shows three particles of masses m_1 (= 1.2 kg), m_2 (= 2.5 kg), and m_3 (= 3.4 kg) located at the corners of an equilateral triangle of edge a (= 140 cm). Where is the center of mass?

We choose our x and y coordinate axes so that one of the particles is located at the origin and so that the x axis coincides with one of the sides of the triangle. The coordinates of the three particles can be shown to have the following values:

Particle	Mass (kg)	x (cm)	y (cm)
m_1	1.2	0	0
m_2	2.5	140	0
m_3	3.4	70	121

Our wise choice of coordinate axes made possible the fact that three of the coordinates in the above table are zero, simplifying the calculations. Note that the total mass M of the system is 7.1 kg.

From Eq. 5, the coordinates of the center of mass are

$$x_{cm} = \frac{1}{M} \sum m_i x_i = \frac{m_1 x_1 + m_2 x_2 + m_3 x_3}{M}$$

$$= \frac{(1.2 \text{ kg})(0) + (2.5 \text{ kg})(140 \text{ cm}) + (3.4 \text{ kg})(70 \text{ cm})}{7.1 \text{ kg}}$$

$$= 83 \text{ cm} \qquad \qquad \text{(Answer)}$$

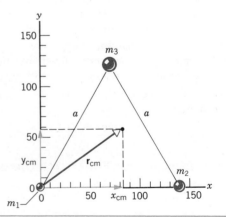

Figure 3 Sample Problem 1. Three particles having different masses form an equilateral triangle of side a. The center of mass is identified by the vector $\mathbf{r}_{cm}$.

---

* You may wish to reread Section 3–4, which deals with unit vectors.

and

$$y_{cm} = \frac{1}{M} \sum m_i y_i = \frac{m_1 y_1 + m_2 y_2 + m_3 y_3}{M}$$

$$= \frac{(1.2 \text{ kg})(0) + (2.5 \text{ kg})(0) + (3.4 \text{ kg})(121 \text{ cm})}{7.1 \text{ kg}}$$

$$= 58 \text{ cm.} \qquad \text{(Answer)}$$

The position of the center of mass is identified by the vector $\mathbf{r}_{cm}$ in Fig. 3.

So far, we have discussed the center of mass as a balance point for a system of particles. We shall show later that the center of mass has more significant properties than this. For example, in the application of Newton's second law, the system of particles may be replaced by a single particle located at the center of mass, the mass of this single equivalent particle being the total mass of the system of particles. To display this equivalence, we repeat the table shown above and we add to it the information about the single particle that—placed at the center of mass—is equivalent to the system of three particles.

Particle	Mass (kg)	x (cm)	y (cm)
m_1	1.2	0	0
m_2	2.5	140	0
m_3	3.4	70	121
M	7.1	83	58

Sample Problem 2   Find the center of mass of the uniform triangular plate of Fig. 4.

Figure 4a shows the plate divided into thin slats, parallel to one side of the triangle. From symmetry, the center of mass of a uniform slat is at its midpoint. The center of mass of the triangular plate must lie somewhere along the line shown in

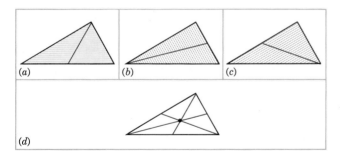

(a)   (b)   (c)

(d)

Figure 4   Sample Problem 2. In (a), (b), and (c), the triangle is divided into thin slats, parallel to each of the three sides. The center of mass must lie along the symmetrical dividing lines shown. (d) The dot, the only point common to all three lines, is the position of the center of mass.

Fig. 4a, connecting one vertex with the midpoint of the opposite side. Surely the plate would balance if it were placed on a knife edge coinciding with this line of symmetry.

In Figs. 4b and 4c, we subdivide the plate into slats parallel to the remaining two sides of the triangle. Again, the center of mass must lie somewhere along each of the bisecting lines shown. Hence, the center of mass of the plate must lie at the intersection of these three symmetry lines, as Fig. 4d shows. It is the only point that the three lines have in common.

You can check this conclusion experimentally by taking advantage of the intuitive notion that an object—suspended from a point—will reorient itself so that its center of mass lies vertically below that point. Suspend the triangle from each vertex in turn, drawing a chalk line vertically downward from the suspension point. The center of mass of the triangle will be the intersection of the three chalk lines.

Sample Problem 3   Figure 5a shows a circular metal plate of radius $2R$ from which a disk of radius R has been removed. Let us call it object X. Its center of mass is shown as a dot on the x axis. Locate this point.

Figure 5b shows object X, its hole filled with a disk of radius R, which we call object D. Let us label as object C the large uniform composite disk so formed. From symmetry, the center of mass of object C is at the origin of the coordinate system, as shown.

In finding the center of mass of a composite object, we can assume that the masses of its components are concentrated at their individual centers of mass. Thus, object C can be treated as equivalent to two mass points, representing objects X and D. Figure 5c shows the positions of the centers of mass of these three objects.

The position of the center of mass of object C is given, from Eq. 5, as

$$x_C = \frac{m_D x_D + m_X x_X}{m_D + m_X},$$

in which x_D and x_X are the positions of the centers of mass of objects D and X, respectively. Noting that $x_C = 0$ and solving for x_X, we obtain

$$x_X = -\frac{x_D m_D}{m_X}. \qquad (11)$$

If ρ is the density of the plate material and t is the thickness of the plate, we have

$$m_D = \pi R^2 \rho t \quad \text{and} \quad m_X = \pi (2R)^2 \rho t - \pi R^2 \rho t.$$

With these substitutions and putting $x_D = -R$, Eq. 11 becomes

$$x_X = -\frac{(-R)(\pi R^2 \rho t)}{\pi (2R)^2 \rho t - \pi R^2 \rho t} = \tfrac{1}{3} R. \qquad \text{(Answer)}$$

Note that the density and the thickness of the plate cancel out.

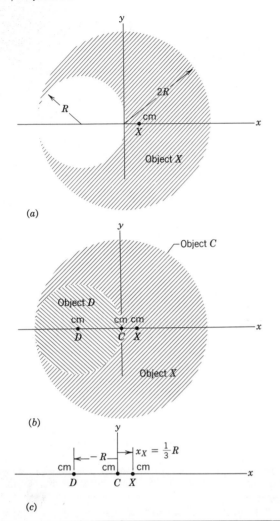

(a)

(b)

(c)

$x_X = \frac{1}{3}R$

Figure 5 Sample Problem 3. (*a*) Object X is a metal disk of radius $2R$ with a hole of radius R cut in it. (*b*) Object D is a metal disk that fills the hole in object X; its center of mass is at $x_D = -R$. Object C is the composite object made up of objects X and D; its center of mass is at the origin of coordinates. (*c*) The centers of mass of the three objects are shown.

Hint 1: *Center of mass problems.* The above Sample Problems have shown us three strategies for simplifying the work in center-of-mass problems. (1) Make full use of the symmetry of the object, be it that of a point, a line, or a plane. (2) If the object can be divided into several parts, treat each of these parts as a particle, located at its own center of mass. (3) Choose your axes wisely. If your system is a group of particles, choose one of the particles as your origin. If your system is a body with a line of symmetry, there is your *x* axis!

9-3 Newton's Second Law for a System of Particles

If the cue ball strikes a resting billiard ball, you expect that the two ball system will continue to have some forward motion after impact. You would be surprised, for example, if both balls came back toward you or if both moved to the right or to the left.

We shall show that the entity that moves forward, its steady motion completely unaffected by the collision, is the center of mass of the two balls. If you focus on this point—which is always halfway between the balls—you can easily convince yourself by trial at a pool table that this is so. No matter whether the collision is glancing, head on, or somewhere in between, the center of mass moves majestically forward, as if the collision had never occurred. Let us look into this more closely.

We replace our pair of billiard balls by an assembly of particles, *n* in number, of (possibly) different masses. We are not interested in the individual motions of these particles but *only* in the motion of their center of mass. Although the center of mass is just a point in space, it moves like a particle whose mass is equal to the total mass of the system; we can assign a position, a velocity, and an acceleration to it. We state (and shall prove below) that the (vector) equation that governs the motion of the center of mass of such a system of particles is

$$\sum \mathbf{F}_{ext} = M\mathbf{a}_{cm} \qquad \text{(system of particles).} \quad (12)$$

Equation 12 is Newton's second law, governing the motion of the center of mass of a system of particles. It is remarkable that it retains the same form ($\Sigma \mathbf{F} = m\mathbf{a}$) that holds for the motion of a single particle. In using Eq. 12, the three quantities that appear in it must be evaluated with some care. We can summarize as follows:

1. $\Sigma \mathbf{F}_{ext}$ is the *vector* sum of *all* the *external* forces that act on the system. Forces exerted by one part of the system on the other are called *internal forces* and we must be careful not to include them in using Eq. 12.

2. *M* is the *total mass* of the system. We assume that no mass enters or leaves the system as it moves, so that *M* remains constant.

3. $\mathbf{a}_{cm}$ is the acceleration of the *center of mass* of the system. Equation 12 gives no information about the acceleration of any other point of the system.

Equation 12, like all vector equations, is equivalent

to three scalar equations, involving the components of $\Sigma\mathbf{F}_{ext}$ and $\mathbf{a}_{cm}$ along the three coordinate axes. These equations are

$$\sum F_{ext,x} = Ma_{cm,x},$$
$$\sum F_{ext,y} = Ma_{cm,y}, \qquad (13)$$
$$\sum F_{ext,z} = Ma_{cm,z}.$$

Now we can understand the behavior of the billiard balls. After the cue ball is in motion, no net external force acts on the (two-ball) system. Because $\Sigma\mathbf{F}_{ext} = 0$, Eq. 12 tells us that $\mathbf{a}_{cm} = 0$ also. Because acceleration is the rate of change of velocity, we conclude that the velocity of the center of mass of the system of two balls does not change. When the two balls collide, the forces that come into play are *internal* forces, exerted by one ball on the other. Such forces do not contribute to $\Sigma\mathbf{F}_{ext}$, which remains zero. Thus, the center of mass of the system, which was moving forward before the collision, must continue to move forward after the collision, with the same speed and in the same direction.

Consider now two other particles that interact with each other, the earth (mass m_e) and the moon (mass m_m). In contrast with the billiard balls of our example above, an external force *does* act on this system, the gravitational attraction of the sun. According to Eq. 12, the center of mass of the earth–moon system moves under the influence of the sun's gravitational force as if a single particle of mass $m_e + m_m$ were concentrated at that point and as if the sun's gravitational force acted there. The earth follows a wavy path around this orbit, as Fig. 6 shows.

Equation 12 applies not only to a system of particles but also to a solid body, such as the ax of Fig. 1. In that case, M in Eq. 12 is the mass of the ax and $\Sigma\mathbf{F}_{ext}$ is $M\mathbf{g}$, the (external) gravitational force acting on the ax. Equation 12 then tells us that $\mathbf{a}_{cm} = \mathbf{g}$. In other words, the center of mass of the ax moves in the (uniform) gravitational field

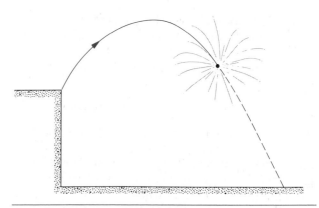

Figure 7   A rocket explodes in flight. The center of mass of the fragments continues to follow the original parabolic path. Effects due to air resistance are neglected.

near the surface of the earth as if it were a single particle of mass M.

Figure 7 shows another interesting case. At a fireworks display, a rocket is launched on a parabolic path. At a certain point, it explodes into fragments. If the explosion had not occurred, the rocket would have continued along the trajectory shown in the figure.

Note that the forces of the explosion are *internal* to the system; that is, they are forces exerted by one part of the system on another part. If we ignore air resistance, the total *external* force $\Sigma\mathbf{F}_{ext}$ acting on the system is $M\mathbf{g}$, the force of gravity, and is unchanged by the explosion. Thus, from Eq. 12, the acceleration of the center of mass of the fragments (while they are in flight) remains equal to $\mathbf{g}$ and the center of mass follows the same parabolic trajectory that the unexploded rocket would have followed.

Proof of Eq. 12.   Now let us prove this important equation, from which we have learned so much. From Eq. 8 we have, for a system of n particles,

$$M\mathbf{r}_{cm} = m_1\mathbf{r}_1 + m_2\mathbf{r}_2 + m_3\mathbf{r}_3 + \cdots + m_n\mathbf{r}_n$$
$$\text{(position)}, \quad (14)$$

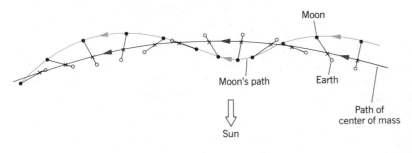

Moon

Moon's path   Earth

⇩
Sun

Path of
center of mass

Figure 6   The sun's gravity acts on the earth —moon system, as an external force. The center of mass of this system orbits the sun as if all the mass of the system were concentrated at that point. The earth and the moon follow wavy paths about this orbit. Not to scale. (The center of mass of the earth–moon system lies within the earth so that the earth's "wobble" is much less than the figure suggests.)

in which M is the total mass of the system and $\mathbf{r}_{cm}$ is the vector that locates the position of the center of mass, just as if the center of mass were a real particle.

Differentiating Eq. 14 with respect to time gives

$$M\mathbf{v}_{cm} = m_1\mathbf{v}_1 + m_2\mathbf{v}_2 + m_3\mathbf{v}_3 + \cdots + m_n\mathbf{v}_n \quad \text{(velocity)}. \quad (15)$$

Here $\mathbf{v}_1 \, (= d\mathbf{r}_1/dt)$ is the velocity of the first particle, and so on, and $\mathbf{v}_{cm} \, (= d\mathbf{r}_{cm}/dt)$ is the velocity at which the center of mass moves. Although the center of mass is just a geometrical point, it has a position, a velocity, and an acceleration, as if it were a particle.

Differentiating Eq. 15 with respect to time leads to

$$M\mathbf{a}_{cm} = m_1\mathbf{a}_1 + m_2\mathbf{a}_2 + m_3\mathbf{a}_3 + \cdots + m_n\mathbf{a}_n \quad \text{(acceleration)}. \quad (16)$$

Here $a_1 \, (= d\mathbf{v}_1/dt)$ is the acceleration of the first particle, and so on, and $\mathbf{a}_{cm} \, (= d\mathbf{v}_{cm}/dt)$ is the acceleration of the center of mass. From Newton's second law, $m_1\mathbf{a}_1$ is the resultant force $\mathbf{F}_1$ that acts on the first particle, and so on. Thus, we can rewrite Eq. 16 as

$$M\mathbf{a}_{cm} = \mathbf{F}_1 + \mathbf{F}_2 + \mathbf{F}_3 + \cdots + \mathbf{F}_n. \quad (17)$$

Among the forces that contribute to the right side of Eq. 17 will be forces that the particles of the system exert on each other (internal forces) and forces exerted on the particles from outside the system (external forces). From Newton's third law, the internal forces form action–reaction pairs and cancel out in the sum that appears on the right side of Eq. 17. This sum then is the same as the vector sum of all the *external* forces that act on the system, internal forces canceling out. Equation 17 then reduces to Eq. 12, the relation that we set out to prove.

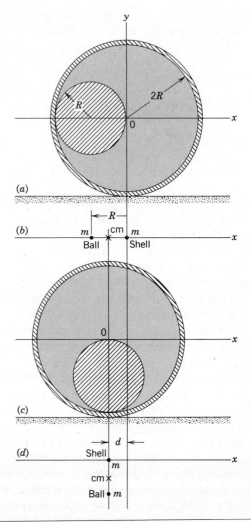

Figure 8 Sample Problem 4. (a) A ball of radius R is released from this initial position, free to roll inside a spherical shell of inner radius $2R$. (b) The centers of mass of the ball, the shell, and the combination of the two. (c) The final state, after the ball has come to rest. The shell has moved so that the center of mass of the system remains in place. (d) The centers of mass of the ball, the shell, and the combination of the two.

Sample Problem 4 A ball of mass m and radius R is placed inside a spherical shell of the same mass m and inner radius $2R$. The combination is at rest on a table top as shown in Fig. 8a. The ball is released, rolls back and forth inside, and finally comes to rest at the bottom, as in Fig. 8c. What will be the displacement d of the shell during this process?

The only external forces acting on the ball–shell system are the downward force of gravity and the normal force exerted vertically upward by the table. Neither force has a horizontal component so that $\Sigma F_{ext,x} = 0$. From Eq. 13, the acceleration component $a_{cm,x}$ of the center of mass must also be zero. Thus, the horizontal position of the center of mass of the system must remain fixed and the shell must move in such a way as to make sure that this happens.

We can represent both ball and shell by single particles of mass m, located at their respective centers. Figure 8b shows the

system before the ball is released and Fig. 8d after the ball has come to rest at the bottom of the shell. We choose our origin to coincide with the initial position of the center of the shell. Figure 8b shows that, with respect to this origin, the center of mass of the ball–shell system is located a distance $\frac{1}{2}R$ to the left, halfway between the two particles. Figure 8d shows that the displacement of the shell is given by

$$d = \tfrac{1}{2}R. \quad \text{(Answer)}$$

The shell must move to the left through this distance as the ball comes to rest.

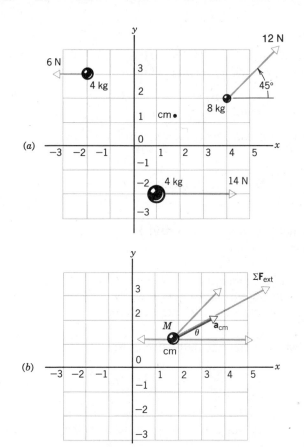

Figure 9   Sample Problem 5. (*a*) Three particles, placed at rest in the positions shown, are acted on by the forces shown. The center of mass of the system is marked. (*b*) The forces are transferred to the center of mass of the system, which behaves like a particle whose mass *M* is equal to the system mass. The net force and the acceleration of the center of mass are shown.

Sample Problem 5   Figure 9*a* shows a system of three resting particles of different masses, each particle acted on by a different external force. What is the acceleration of the center of mass of this system?

The position of the center of mass, calculated by the methods of Sample Problem 1, is marked by a dot in the figure. As Fig. 9*b* suggests, we treat this point as if it were a real particle, assigning to it a mass *M* equal to the system mass of 16 kg and assuming that all external forces are applied at that point.

The *x* component of the net external force acting on the center of mass is (see Fig. 9*b*)

$$F_{ext,x} = 14\ N - 6\ N + (12\ N)(\cos 45°) = 16.5\ N,$$

and the *y* component is

$$F_{ext,y} = (12\ N)(\sin 45°) = 8.49\ N.$$

The net external force thus has a magnitude

$$F_{ext} = \sqrt{(16.5\ N)^2 + (8.49\ N)^2} = 18.6\ N$$

and makes an angle with the *x* axis given by

$$\theta = \tan^{-1}\frac{8.49\ N}{16.5\ N} = \tan^{-1} 0.515 = 27°.\quad \text{(Answer)}$$

This is also the direction of the acceleration vector. From Eq. 12, the magnitude of the acceleration of the center of mass is given by

$$a_{cm} = \frac{F_{ext}}{M} = \frac{18.6\ N}{16\ kg} = 1.16\ m/s^2.\quad \text{(Answer)}$$

All three particles of Fig. 9*a*, and also their center of mass, move with (different) constant accelerations. If the particles start from rest, each will move, with ever-increasing speed, along a straight line in the direction of the force acting on it.

9-4   Momentum

Momentum is another of those words that have a lot of meanings in the common language but only a single precise meaning in physics. The momentum of a particle is a vector **p**, defined from

$$\boxed{\mathbf{p} = m\mathbf{v}}\quad \text{(momentum of a particle),}\quad (18)$$

in which *m* is the mass of the particle and **v** is its velocity.

Newton actually expressed his second law of motion in terms of momentum:

The rate of change of the momentum of a particle is proportional to the net force acting on the particle and is in the direction of that force.

In equation form this becomes

$$\boxed{\sum \mathbf{F} = \frac{d\mathbf{p}}{dt}.}\quad (19)$$

Substituting for **p** from Eq. 18 gives

$$\sum \mathbf{F} = \frac{d\mathbf{p}}{dt} = \frac{d}{dt}(m\mathbf{v}) = m\frac{d\mathbf{v}}{dt} = m\mathbf{a}.$$

Thus, the relations $\sum \mathbf{F} = d\mathbf{p}/dt$ and $\sum \mathbf{F} = m\mathbf{a}$ are completely equivalent as they apply to the motion of single particles in classical mechanics. Both are expressions of Newton's second law of motion.

Momentum at Very High Speeds. (Optional)* For particles moving with speeds that are close to the speed of light, we find that Newtonian mechanics predicts results that do not agree with experiment. In such cases, we must use Einstein's theory of special relativity. In relativity, the formulation $\mathbf{F} = d\mathbf{p}/dt$ holds good, *provided* that we define the momentum of a particle not as $m\mathbf{v}$ but as

$$\mathbf{p} = \frac{m\mathbf{v}}{\sqrt{1 - (v/c)^2}}, \qquad (20)$$

in which c, the speed of light, is that sure marker of a relativistic equation.

The speeds of common macroscopic objects such as baseballs, bullets, or space probes are so much less than the speed of light that the quantity $(v/c)^2$ in Eq. 20 is very much less than unity. Under these conditions, Eq. 20 reduces to Eq. 18 and Einstein's relativity theory reduces to Newtonian mechanics. For electrons and other subatomic particles, however, speeds very close to that of light are easily obtained and the definition in Eq. 20 *must* be used, often as a matter of routine engineering practice.

9-5 The Momentum of a System of Particles

Consider now a system of n particles, each with its own mass, velocity, and linear momentum. The particles may interact with each other and external forces may act on them as well. The system as a whole will have a total

---

* Recall that in Section 7–7 we discussed kinetic energy at very high speeds.

momentum $\mathbf{P}$, defined to be the vector sum of the individual particle momenta. Thus,

$$\mathbf{p} = \mathbf{p}_1 + \mathbf{p}_2 + \mathbf{p}_3 + \cdots + \mathbf{p}_n$$
$$= m_1\mathbf{v}_1 + m_2\mathbf{v}_2 + m_3\mathbf{v}_3 + \cdots + m_n\mathbf{v}_n. \quad (21)$$

If we compare this equation with Eq. 15, we see at once that

$$\boxed{\mathbf{P} = M\mathbf{v}_{cm}} \quad \text{(momentum of a system of particles)}, \quad (22)$$

which is another way to define the momentum of a system of particles:

The momentum of a system of particles is equal to the product of the total mass M of the system and the velocity of the center of mass.

If we take the time derivative of Eq. 22, we find

$$\frac{d\mathbf{P}}{dt} = M\frac{d\mathbf{v}_{cm}}{dt} = M\mathbf{a}_{cm}. \quad (23)$$

Comparing Eqs. 12 and 23 allows us to write Newton's second law for a system of particles in the equivalent form

$$\boxed{\sum \mathbf{F}_{ext} = \frac{d\mathbf{P}}{dt}.} \quad (24)$$

This equation is the generalization of the single-particle equation $\Sigma\mathbf{F} = d\mathbf{p}/dt$ to a system of many particles. Table 1 displays and compares the important relations that we have set out for single particles and for systems of particles.

Table 1 Some Definitions and Laws in Classical Mechanics[a]

Law or Definition	Single Particle	System of Particles
Newton's second law	$\sum \mathbf{F} = m\mathbf{a}$[b]	$\sum \mathbf{F}_{ext} = M\mathbf{a}_{cm}$ (12)
Definition of momentum	$\mathbf{p} = m\mathbf{v}$ (18)	$\mathbf{P} = M\mathbf{v}_{cm}$ (22)
Newton's second law	$\sum \mathbf{F} = d\mathbf{p}/dt$ (19)	$\sum \mathbf{F}_{ext} = d\mathbf{P}/dt$ (24)
Work–energy theorem[c]	$W = \Delta K$[d]	$W_{cm} = \Delta K_{cm}$ (39)

[a] The numbers in parentheses are the equation numbers in the text.
[b] Equation 1 of Chapter 5.
[c] See Section 9–8.
[d] Equation 16 of Chapter 7.

9-6   Conservation of Momentum

Suppose that the sum of the external forces acting on a system of particles is zero. Putting $\Sigma\, \mathbf{F}_{ext} = 0$ in Eq. 24 then yields $d\mathbf{P}/dt = 0$, or

$$\mathbf{P} = \text{a constant}\quad\text{(isolated system).}\qquad(25)$$

This important result is called the law of *conservation of momentum*. It tells us that, if no external forces act on a system of particles, the total momentum of the system remains constant.

Like the law of conservation of energy that we met in Chapter 8, conservation of momentum is a more general law than Newtonian mechanics itself. It continues to hold in the subatomic realm, where Newton's laws fail. It holds for the highest particle speeds, where Einstein's relativity theory prevails; it is only necessary to use Eq. 20, rather than Eq. 18, for the momentum.

From Eq. 22 ($\mathbf{P} = M\mathbf{v}_{cm}$) we see that, if $\mathbf{P}$ is a constant as in Eq. 25, then $\mathbf{v}_{cm}$, the velocity of the center of mass, is also constant. This in turn means that $\mathbf{a}_{cm}$, the acceleration of the center of mass, must be zero, which is consistent with Eq. 12 ($\Sigma\, \mathbf{F}_{ext} = M\mathbf{a}_{cm}$) for the case of $\Sigma\, \mathbf{F}_{ext} = 0$. Thus, the law of conservation of momentum is consistent with both forms of Newton's second law for systems of particles, as displayed in Table 1.

Equation 25 is a vector equation and, as such, is equivalent to three scalar equations corresponding to momentum conservation in three mutually perpendicular directions. It may happen, for example, that there is an external force acting on the system but it acts (say) in only the vertical direction, with no component in any horizontal direction. In such a case, the horizontal components of the total momentum of the system remain constant, even though the vertical component does not.

Sample Problem 6 A stream of bullets whose mass m is 3.8 g is fired horizontally with a speed v of 1100 m/s into a large wooden block of mass M (= 12 kg) that is initially at rest on a horizontal table; see Fig. 10. If the block is free to slide without friction across the table, what speed will it acquire after it has absorbed 8 bullets?

Equation 25 ($\mathbf{P}$ = constant) is generaly valid only for closed systems, in which no particles leave or enter. Thus, our system must include both the block and the 8 bullets, taken together. In Fig. 10, we have identified this system by drawing a closed curve around it.

In the vertical direction, the only net external force is the

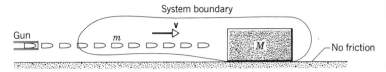

Figure 10 Sample Problem 6. A gun fires a stream of bullets toward a block of wood. The colored line shows the envelope defining the system to which we apply the law of conservation of linear momentum. No net external force pierces the boundary of this system.

force of gravity acting on the bullets while they are in flight. Because this does not affect the horizontal motion, we need not consider it further.

In the horizontal direction, no external force acts on the block + bullets system. The forces called into play when the bullets strike the block are *internal* forces and do not contribute to $\mathbf{F}_{ext}$, which (as we have said) is zero.

Because no (horizontal) external forces act, we can apply the law of conservation of momentum (Eq. 25). The initial (horizontal) momentum, measured while the bullets are still in flight* and the block is at rest, is

$$P_i = n(mv),$$

in which mv is the momentum of an individual bullet and $n = 8$. The final momentum, measured when all the bullets are in the block and the block is sliding over the table with speed V, is

$$P_f = (M + nm)V.$$

Conservation of momentum requires that

$$P_i = P_f$$

or

$$n(mv) = (M + nm)V.$$

Solving for V yields

$$V = \frac{nm}{M + nm}\, v = \frac{(8)(3.8 \times 10^{-3}\ \text{kg})}{12\ \text{kg} + (8)(3.8 \times 10^{-3}\ \text{kg})}\,(1100\ \text{m/s})$$
$$= 2.7\ \text{m/s}\qquad\qquad\text{(Answer)}$$

With the choice of system that we made, we did not have to consider what happened when the bullets hit the block. Those forces are all internal.

Sample Problem 7 As Fig. 11 shows, a cannon whose mass M is 1300 kg fires a 72-kg ball in a horizontal direction with a muzzle speed v of 55 m/s. The cannon is mounted so that it can

---

* For simplicity, we assume a rapid rate of fire, so that all the bullets are in flight before the first bullet strikes. (Actually, it doesn't make any difference.)

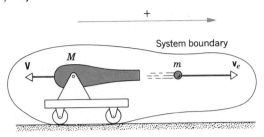

Figure 11 Sample Problem 7. A cannon of mass M fires a ball of mass m. The velocities of the shell and of the recoiling cannon are shown, in a reference frame fixed with respect to the earth. The light arrow shows the direction taken as positive for all velocities.

recoil freely. (a) What is the velocity V of the recoiling cannon with respect to the earth?

We choose the cannon plus the ball as our system. By doing so, the forces called into play by the firing of the cannon are internal to the system and we do not have to deal with them. The external forces acting on the system have no components in the horizontal direction. Thus, the horizontal component of the total linear momentum of the system must remain unchanged as the cannon is fired.

We choose a reference frame fixed with respect to the earth and we assume that all velocities are positive if they point to the right in Fig. 11.

Before the cannon is fired, the system has an initial momentum P_i of zero. After the cannon has fired, the ball has a horizontal velocity v *with respect to the recoiling cannon, v* being the ball's muzzle speed. In the reference frame of the earth, however, the horizontal velocity of the ball is $v + V$. Thus, the total linear momentum of the system after firing is

$$P_f = MV + m(v + V),$$

in which the first term on the right is the momentum of the recoiling cannon and the second term that of the speeding ball.

Conservation of linear momentum in the horizontal direction requires that $P_i = P_f$, or

$$0 = MV + m(v + V).$$

Solving for V yields

$$V = -\frac{mv}{M + m} = -\frac{(72 \text{ kg})(55 \text{ m/s})}{1300 \text{ kg} + 72 \text{ kg}}$$
$$= -2.9 \text{ m/s.} \qquad \text{(Answer)}$$

The minus sign tells us that the cannon recoils to the left in Fig. 11, as we know it does.

(b) What is the initial velocity v_e of the ball with respect to the earth?

The velocity of the ball with respect to the (recoiling) cannon is its muzzle speed v. With respect to the earth, the

velocity of the ball is

$$v_e = v + V$$
$$= 55 \text{ m/s} + (-2.9 \text{ m/s}) = 52 \text{ m/s.} \qquad \text{(Answer)}$$

Because of the recoil, the ball is moving a little slower with respect to the earth than it otherwise would. Note the importance in this problem of choosing the system (cannon + ball) wisely and being absolutely clear about the reference frame (earth or recoiling cannon) to which the various measurements are referred.

Sample Problem 8 Figure 12 shows two blocks, connected by a spring and free to slide on a frictionless horizontal surface. The blocks, whose masses are m_1 and m_2, are pulled apart and then released from rest. What fraction of the total kinetic energy of the system will each block have at any later time?

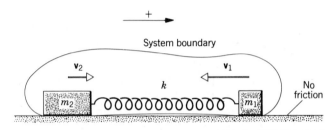

Figure 12 Sample Problem 8. Two blocks, resting on a frictionless surface and connected by a spring, have been pulled apart and then released from rest. The vector sum of their linear momenta remains zero during their subsequent motions. The system boundary is shown. The light arrow shows the direction taken as positive for all velocities.

We take the two blocks and the spring (assumed massless) as our system and the horizontal surface on which they slide as our reference frame. We assume that velocities are positive if they point to the right in Fig. 12.

The initial momentum P_i of the system before the blocks are released is zero. The final momentum, at any time after the blocks are released, is

$$P_f = m_1v_1 + m_2v_2,$$

in which v_1 and v_2 are the velocities of the blocks. Conservation of momentum requires that $P_i = P_f$, or

$$0 = m_1v_1 + m_2v_2.$$

Thus, we have

$$\frac{v_1}{v_2} = -\frac{m_2}{m_1}, \qquad (26)$$

the minus sign telling us that the two velocities always have opposite signs. Equation 26 holds at every instant after release, no matter what the individual speeds of the blocks.

The kinetic energies of the blocks are $K_1 = \frac{1}{2}m_1v_1^2$ and

$K_2 = \frac{1}{2}m_2v_2^2$ and the fraction we seek, for the block of mass m_1, is

$$f_1 = \frac{K_1}{K_1 + K_2} = \frac{\frac{1}{2}m_1v_1^2}{\frac{1}{2}m_1v_1^2 + \frac{1}{2}m_2v_2^2}. \qquad (27)$$

From Eq. 26, $v_2 = -v_1(m_1/m_2)$. Making this substitution into Eq. 27 leads, after a little algebra, to

$$f_1 = \frac{m_2}{m_1 + m_2}. \qquad \text{(Answer)} \quad (28)$$

Similarly, for the block of mass m_2,

$$f_2 = \frac{m_1}{m_1 + m_2}. \qquad \text{(Answer)} \quad (29)$$

Thus, the kinetic energy split is a constant, independent of time, the least massive block receiving the largest share of the available kinetic energy. If, for example, $m_2 = 10m_1$, then

$$f_1 = \frac{10m_1}{m_1 + 10m_1} = 0.91 \quad \text{and} \quad f_2 = \frac{m_1}{m_1 + 10m_1} = 0.09.$$

In this case, the lighter block (m_1) gets 91% of the available kinetic energy and the heavier block (m_2) gets the remaining 9%.

If $m_2 \gg m_1$, we see from Eqs. 28 and 29 that $f_1 \approx 100\%$ and $f_2 \approx$ zero; that is, the lighter block gets essentially all the kinetic energy.

Equations 28 and 29 apply equally well to a stone falling in the gravitational field of the earth. Here m_2 is the mass of the earth and m_1 that of the stone, the spring of Fig. 12 being equivalent to the force of gravitational attraction between the earth and the stone. The logical reference frame to adopt for this problem is a frame in which the center of mass of the stone–earth system is at rest. In this frame, the stone has essentially all the kinetic energy ($f_1 \approx 1$) and the earth very little ($f_2 \approx 0$). From Eq. 26, however, we see that the magnitudes of the linear momentum of the stone and the earth remain equal at all times, the tiny speed v_2 of the earth being compensated for by m_2, its enormous mass.

Hint 2: Conservation law problems—II. This is a good time to reread Hint 1 of Chapter 8, which deals with problems involving the conservation of energy. The same broad suggestions hold here for problems involving the conservation of momentum.

First, make sure that you have chosen a closed, isolated system. *Closed* means that no matter (particles) passes through the system boundary in either direction. *Isolated* means that the resultant external force acting on the system is zero. If the system is *not* isolated, momentum is *not* conserved. That is, Eq. 25 does not hold.

Convince yourself that the systems in Sample Problems 6–8 are both closed and isolated. Note that if you had chosen the block in Sample Problem 6, the cannon ball in Sample Problem 7, and block m_1 in Sample Problem 8 as your system, momentum would not have been conserved. These systems are not isolated; external forces act on them.

Remember that momentum is a vector so that each component can be conserved separately, provided only that the corresponding component of the resultant external force is zero. In Sample Problem 7, the resultant external force acting horizontally on the cannon + ball system is zero, so that the horizontal component of momentum is conserved. However, the resultant external force acting vertically on this system is not zero; gravity acts on the cannon ball while it is in flight. Thus, the vertical component of momentum for this system is not conserved.

Select two appropriate states of the system (which you may choose to call the *initial state* and the *final state*) and write down expressions for the momentum of the system in each of these two states. In writing down these expressions, make sure that you know what inertial reference frame you are using and make sure also that you include the entire system, not missing any part of it and not including objects that do not belong to your system. The caution about reference frames is particularly needed in Sample Problem 7.

Finally, set your expressions for $\mathbf{P}_i$ and for $\mathbf{P}_f$ equal to each other and solve for what is requested.

9-7   Systems with Variable Mass: A Rocket   (Optional)

In the systems we have dealt with so far, we have assumed that the total mass of the system remains constant. Sometimes it does not, the rocket (see Fig. 13) being a familiar example. Most of the mass of a rocket on its launching pad is fuel, all of which will eventually be burned and ejected from the nozzle of the rocket engine.

How do we handle the variation of the mass of the rocket as it accelerates? We solve this puzzle by applying Newton's second law, not to the rocket alone but to the rocket and its ejected combustion products taken together. The mass of *this* system does *not* change as the rocket accelerates.

Finding the Acceleration.   Assume that you are in an inertial reference frame, watching a rocket accelerating through deep space with no gravitational or atmospheric drag forces acting on it. At an arbitrary time t (see Fig. 14a), let M be the mass of the rocket and v its velocity.

Figure 13 An Atlas rocket soon after launch.

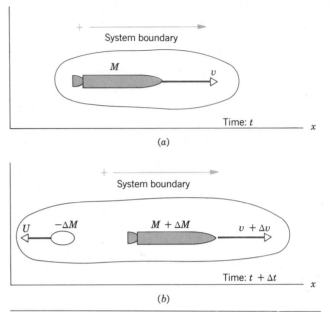

Figure 14 (a) A rocket of mass M, at rest in an inertial reference frame at time t. (b) The rocket at a time $t + \Delta t$, showing its ejected combustion products. The light arrow shows the direction taken as positive for all velocities.

Figure 14b shows how things stand an interval Δt later. The mass of the rocket is now $M + \Delta M$ and the velocity of the rocket has increased to $v + \Delta v$. A mass $-\Delta M$ of combustion products has been ejected from the nozzle, with a velocity U.*

The momentum of the isolated system *rocket + combustion products* is conserved. Thus, the momentum at time t must equal the momentum at time $t + \Delta t$, or (see Fig. 14)

$$Mv = (M + \Delta M)(v + \Delta v) + (-\Delta M)(U). \quad (30)$$

The characteristic exhaust speed of the rocket engine is a necessary part of the problem and we have not yet introduced it. The exhaust velocity U_0 (measured with respect to the rocket) and the exhaust velocity U (measured with respect to the reference frame of Fig. 14b) differ by the velocity of the rocket, or

$$U - U_0 = v + \Delta v. \quad (31)$$

Eliminating U between Eqs. 30 and 31 yields, after a little algebra,

$$\Delta M\, U_0 = M\, \Delta v.$$

Dividing each side Δt yields, in the differential limit,

$$\frac{dM}{dt} U_0 = M \frac{dv}{dt}. \quad (32)$$

Now dM/dt (which describes the loss of mass of the rocket) and U_0 (which points to the left in Fig. 14b) are both negative quantities. We replace dM/dt by $-R$, where R is the (positive) rate of fuel consumption. We also replace U_0 by $-u$, where u is the (positive) speed of the exhaust gases relative to the rocket. Finally, we recognize that dv/dt is the acceleration of the rocket. With these changes, Eq. 32 becomes

$$\boxed{Ru = Ma} \quad \text{(first rocket equation).} \quad (33)$$

Equation 33 holds at any instant, the mass M, the fuel consumption rate R, and the acceleration a being evaluated at that instant.

The left side of Eq. 33 has the dimensions of a force and depends only on design characteristics of the rocket

---

* Note that, because the mass of the rocket decreases as it ejects its burned fuel, ΔM is inherently negative, so that $-\Delta M$ is a positive quantity.

engine, namely, the rate R at which it ejects mass and the speed u with which that mass is ejected relative to the rocket. We call this term the *thrust* of the rocket engine and we represent it by T. Newton's second law emerges clearly if we write Eq. 33 as $T = Ma$, in which a is the acceleration of the rocket at the time that its mass is M.

Finding the Velocity.   How will the velocity of a rocket change as it consumes its fuel? Replacing U_0 by $-u$ in Eq. 32 allows us to write

$$dv = -u \frac{dM}{M}.$$

Integrating leads to

$$\int_{v_i}^{v_f} dv = -u \int_{M_i}^{M_f} \frac{dM}{M},$$

in which M_i is the initial mass of the rocket and M_f its final mass. Evaluating the integral gives

$$\boxed{v_f - v_i = u \ln \frac{M_i}{M_f}} \quad \text{(second rocket equation),} \quad (34)$$

for the increase in the speed of the rocket.* We see the advantage of multistage rockets, in which M_f is reduced by throwing away successive stages as their fuel is used up. An ideal rocket would reach its destination with only its payload surviving.

Sample Problem 9   A rocket whose initial mass M_i is 850 kg ejects mass during a burn at the rate $R = 2.3$ kg/s. The speed u of the exhaust gases relative to the nozzle of the rocket engine is 2800 m/s. (a) What thrust does the rocket engine provide?

From Eq. 33, the thrust is

$$T = Ru = (2.3 \text{ kg/s})(2800 \text{ m/s})$$
$$= 6440 \text{ N} (\approx 1400 \text{ lb}). \quad \text{(Answer)}$$

(b) What is the initial acceleration of the rocket?
From Newton's second law, we have

$$a = \frac{T}{M_i} = \frac{6440 \text{ N}}{850 \text{ kg}} = 7.6 \text{ m/s}^2. \quad \text{(Answer)}$$

To be launched from the earth's surface, a rocket must have an initial acceleration greater than that of the earth's surface gravity. Put another way, the thrust T (= 6400 N) of the rocket

---

* The symbol "ln" in Eq. 34 refers to the *natural logarithm,* which is taken to the base e (= 2.718 . . .); punch "ln", not "log" on your calculator.

---

engine must exceed the initial weight of the rocket (= $M_i g$ = 850 kg × 9.8 m/s² = 8300 N). Because this requirement is not met, our rocket could not be launched from the earth's surface. It must have been launched into space by another, more powerful, rocket.

(c) The mass M_f of the rocket when its fuel is exhausted is 180 kg. What is its speed at that time?
From Eq. 34, with $v_i = 0$,

$$v_f = u \ln \frac{M_i}{M_f}$$
$$= (2800 \text{ m/s}) \ln \frac{850 \text{ kg}}{180 \text{ kg}}$$
$$= (2800 \text{ m/s}) \ln 4.72 \approx 4300 \text{ m/s}. \quad \text{(Answer)}$$

Recall that, in deriving this result, we assumed that the rocket was accelerating in gravity-free space. Note that the ultimate speed of the rocket can easily exceed its exhaust speed u.

9-8   Systems of particles: Work and Energy   (Optional)

As Fig. 15 shows, a skater pushes herself away from a railing, picking up some kinetic energy in the process. Where does this energy come from? If you ask the skater, she will say, "From me!" and she will be right. The skater draws on her internal energy reserves in straightening her arm to push herself away from the railing and would be very aware of muscular exertion.

There is a problem if you try to account for this work by applying the work–energy theorem of Section 7–5. This theorem (see Eq. 16 of Chapter 7) is

$$W = K_f - K_i = \Delta K \quad \text{(single particle).} \quad (35)$$

It tells us that the increase in kinetic energy is equal to the work done by the net force. However, the net force in Fig. 15a is the force $\mathbf{F}_{ext}$ exerted by the railing on the skater and *this force does no work.* We know this because the point of application of the force does not move and the force drops to zero as soon as the skater loses contact with the railing. Yet kinetic energy appears; that is our puzzle.

The solution to our puzzle is that the work–energy theorem of Eq. 35 applies only to a single particle. The criterion for representing a body as a single particle is that every part of it moves in the same way. In the essential act of extending her arm to push herself away from the railing, the skater fails to meet this requirement. She must be treated as a system of particles, to which Eq. 35 simply does not apply.

The remaining question is a formal one. We first met Newton's second law ($\Sigma \mathbf{F} = m\mathbf{a}$) in the context of a

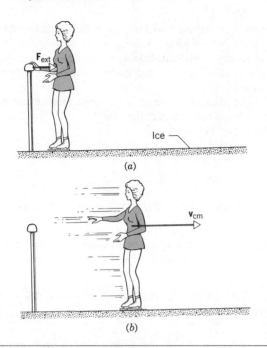

Figure 15 (a) A skater pushes herself away from a railing. The railing exerts a force $\mathbf{F}_{ext}$ on her. (b) The skater glides across the ice after push-off, with velocity $\mathbf{v}_{cm}$.

single particle. In Section 9–3 we saw that, by writing it in the form $\Sigma\mathbf{F}_{ext} = M\mathbf{a}_{cm}$, we could apply this law to a system of particles. The question is: "Can we generalize the work–energy theorem of Eq. 35 so that it too applies to a system of particles?" We can.

Consider a system of particles on which $\mathbf{F}_{ext}$ is the net external force and let us choose an x axis to point in the direction of this force. Suppose that the center of mass of the system moves a distance dx_{cm} along this axis in a time dt. Newton's second law (Eq. 12) applies:

$$F_{ext} = Ma_{cm}.$$

Multiplying each side of this equation by dx_{cm} yields

$$F_{ext}\,dx_{cm} = Ma_{cm}dx_{cm} = M\frac{dv_{cm}}{dt}\,dx_{cm} = M\frac{dx_{cm}}{dt}\,dv_{cm}$$

or

$$F_{ext}\,dx_{cm} = Mv_{cm}dv_{cm}. \qquad (36)$$

As the force acts, let the center of mass of the system move from an initial position x_i to a final position x_f. Integrating Eq. 36 between these limits gives

$$\int_{x_i}^{x_f} F_{ext}\,dx_{cm} = (\tfrac{1}{2}Mv_{cm}^2)_f - (\tfrac{1}{2}Mv_{cm}^2)_i. \qquad (37)$$

The integral in Eq. 37 has the dimensions of force times distance, or work. We represent it by W_{cm}, or

$$W_{cm} = \int_{x_i}^{x_f} F_{ext}dx_{cm} \qquad \text{(center-of-mass work).}$$

$$(38)$$

W_{cm} is called the *center-of-mass work.* Using Eq. 38, we can rewrite Eq. 37 as

$$W_{cm} = K_{cm,f} - K_{cm,i} = \Delta K_{cm}$$

$$\text{(work–energy theorem),} \quad (39)$$

which is the work–energy theorem for a system of particles. Expressed in words, this equation is:

The work done by the net external force acting on a system of particles—assumed to act at its center of mass—is equal to the change in the translational kinetic energy of the system.

To apply the work–energy theorem to a system of particles, we proceed as follows: Treat the system of particles as a single particle of mass M, located at the center of mass. Find the net external force $\mathbf{F}_{ext}$ that acts on the system and apply it at that point. Calculate (using Eq. 38) the work W_{cm} that this force *would do* under these conditions as the center of mass moves. Substitute W_{cm} into Eq. 39 and find ΔK_{cm}.

K_{cm} in Eq. 39 is the kinetic energy associated with the *translational motion only* of the system. Energy associated with rotation of the system about this point, or with internal motions or vibrations, is not included in K_{cm}.

What Makes a Car Go?   Another common situation in which an external force acts on a system but does no work is the acceleration of a car from rest, as in Fig. 16. The external force here is exerted, in a forward direction, by the road on the bottoms of the tires. You may not agree that this force does no work because—after all—

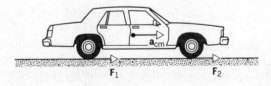

Figure 16 A car, initially at rest, accelerates to the right. The road exerts four frictional forces (two of them shown) on the bottoms of the tires. Taken together, these four forces constitute the net external force $\mathbf{F}_{ext}$ acting on the car.

the car moves. However, what counts is the relative motion between the road and the bottoms of the tires and, in this case, there is none; if this were not so, the tires would not leave tread marks on a snow-covered road.

The kinetic energy acquired by the car (we would all agree) comes from the combustion of gasoline. Because of the (vitally-necessary) internal motions of its engine and its drive train, the car cannot be treated as a single particle, anymore than could the skater of Fig. 15. We must treat it as a system of particles with the work-energy theorem of Eq. 39 applying.

Equation 39 also applies to the braking of a car. In this case, the kinetic energy of the car decreases. The center-of-mass work is now negative and the "lost" kinetic energy appears inside the car, as thermal energy of the braking surfaces. The brakes get hot.

Sample Problem 10   Figure 17 shows a suited astronaut, whose mass M is 110 kg, pushing himself off into space from the wall of a massive spaceship. When he breaks contact with the ship, his center of mass has moved a distance s_{cm} of 18 cm and is moving outward with a speed v_{cm} of 1.1 m/s. What force F_{ext} (assumed constant) does the spaceship exert on the astronaut while he is pushing off?

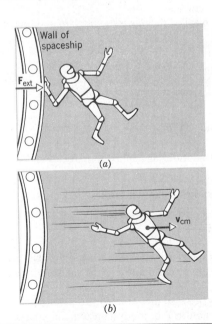

(a)

(b)

Figure 17   Sample Problem 10. (a) An astronaut pushes himself into space from the wall of a spaceship. During this maneuver, the wall exerts a force $\mathbf{F}_{ext}$ on him. (b) After push-off, he coasts away with a velocity $\mathbf{v}_{cm}$.

From Eq. 38 we can write, because F_{ext} is a constant,

$$W_{cm} = \int_{x_i}^{x_f} F_{ext} dx_{cm} = F_{ext} \int_{x_i}^{x_f} dx_{cm} = F_{ext} s_{cm}. \quad (40)$$

From Eq. 39,

$$W_{cm} = K_{cm,f} - K_{cm,i}$$

or

$$F_{ext} s_{cm} = \tfrac{1}{2} M v_{cm}^2 - 0.$$

This yields

$$F_{ext} = \frac{M v_{cm}^2}{2 s_{cm}} = \frac{(110 \text{ kg})(1.1 \text{ m/s})^2}{(2)(0.18 \text{ m})}$$
$$= 370 \text{ N} \ (= 83 \text{ lb}). \qquad \text{(Answer)}$$

We leave to you the problem of getting the astronaut back to the ship.

Sample Problem 11   A car, with its driver, has a mass $M = 1400$ kg. It brakes to rest from a speed $v_{cm} = 24$ m/s (≈ 86 km/h) covering a distance $s_{cm} = 180$ m as it does so. What resultant frictional force $\mathbf{F}_{ext}$, assumed constant, does the road exert on the tires during this process? Assume that the driver manipulates the brakes so that the tires do not skid.

In this case, the force $\mathbf{F}_{ext}$ (which points backward) and the displacement s_{cm} (which points forward) point in opposite directions so that the work W_{cm} is negative. From Eq. 38 we then have, because F_{ext} is a constant,

$$W_{cm} = -F_{ext} s_{cm}.$$

Equation 39 then becomes

$$W_{cm} = K_{cm,f} - K_{cm,i}$$

or

$$-F_{ext} s_{cm} = 0 - \tfrac{1}{2} M v_{cm}^2,$$

so that

$$F_{ext} = \frac{M v_{cm}^2}{2 s_{cm}} = \frac{(1400 \text{ kg})(24 \text{ m/s})^2}{(2)(180 \text{ m})}$$
$$= 2240 \text{ N} \ (\approx 500 \text{ lb}). \qquad \text{(Answer)}$$

Note how different the situation is if the driver simply jams on the brakes and skids to a halt, with the wheels locked. Here the car behaves like a single particle, like a block projected onto a rough horizontal plane and brought to rest by frictional forces. The loss of kinetic energy is now associated with *external* work, done by the frictional force that the road exerts on the (skidding) car. The skid marks on the road and the elevated temperatures of the tires (and of the road) tell us that the initial kinetic energy of the car appears at these sites and not, as previously, as thermal energy in the brake drums.

REVIEW AND SUMMARY

Center of Mass

The *center of mass* of a system of discrete particles is defined to be the point whose coordinates are given by

$$x_{cm} = \frac{1}{M} \sum m_i x_i, \quad y_{cm} = \frac{1}{M} \sum m_i y_i, \quad z_{cm} = \frac{1}{M} \sum m_i z_i \qquad [5]$$

or

$$\mathbf{r}_{cm} = \frac{1}{M} \sum m_i \mathbf{r}_i. \qquad [8]$$

Here M is the total mass of the system. See Sample Problems 1–3. If mass is continuously distributed, these are replaced by the integrals

$$x_{cm} = \frac{1}{M} \int x \, dm, \quad y_{cm} = \frac{1}{M} \int y \, dm, \quad z_{cm} = \frac{1}{M} \int z \, dm \qquad [9]$$

or

$$\mathbf{r}_{cm} = \frac{1}{M} \int \mathbf{r} \, dm. \qquad [10]$$

Newton's Second Law for a System

The motion of the center of mass of any system of particles is governed by *Newton's second law for a system of particles:*

$$\sum \mathbf{F}_{ext} = M \mathbf{a}_{cm} \qquad [12]$$

in which $\sum \mathbf{F}_{ext}$ is the resultant of the *external* forces acting on the system, M is the total mass of the system, and $\mathbf{a}_{cm}$ is the acceleration of the system's center of mass. Sample Problems 4 and 5 illustrate simple applications of this important law.

For a single particle, we define a quantity $\mathbf{p}$ called the *linear momentum* and then we express Newton's second law for linear momentum:

Linear Momentum and Newton's Second Law

$$\mathbf{p} = m\mathbf{v} \quad \text{and} \quad \sum \mathbf{F} = \frac{d\mathbf{p}}{dt}. \qquad [18,19]$$

For a system of particles these relations become

$$\mathbf{P} = M\mathbf{v}_{cm} \quad \text{and} \quad \sum \mathbf{F}_{ext} = \frac{d\mathbf{P}}{dt}. \qquad [22,24]$$

See Table 1 for a summary of these important relations as applied to particles and to systems of particles.

A more complete (relativistic) definition of linear momentum is

Relativistic Momentum

$$\mathbf{p} = \frac{m\mathbf{v}}{\sqrt{1 - (v/c)^2}}, \qquad [20]$$

a form that must be used whenever a particle's speed is a significant fraction of the speed of light. Equation 20 is equivalent to Eq. 18 whenever $v \ll c$. Equation 20 is valid under any circumstances.

Conservation of Linear Momentum

Linear momentum is a conserved quantity; if a system is isolated so that no *external* forces act on any of its particles, the (vector) linear momentum $\mathbf{P}$ of the system remains constant;

$$\mathbf{P} = \text{a constant} \quad \text{(isolated system).} \qquad [25]$$

This is *the law of conservation of linear momentum.* It is thought to be valid under all circumstances, even in situations where Newton's laws of motion are not valid. Sample Problems 6–8 illustrate how momentum is conserved in systems in which there are somewhat violent internal interactions.

Variable Mass Systems

Section 9–7 illustrates a strategy for dealing with systems of variable mass. The trick is to redefine the system, enlarging its boundaries until it encompasses a larger system whose mass *does* remain constant; then apply the laws of mechanics (Eq. 25). For a rocket, this means that the system includes both the rocket and its exhaust gases. The result of an analysis of this kind is that a rocket accelerates, in the absence of external forces, at a rate given by

$$Ru = Ma \quad \text{(first rocket equation)} \qquad [33]$$

in which $M(t)$ is the rocket's mass (including unexpended fuel), R is the fuel consumption rate, and u is the fuel's exhaust speed relative to the rocket. The term Ru is the *thrust* of the engine. For a rocket with constant R and u, the resulting change in rocket speed is

$$v_f - v_i = u \ln \frac{M_i}{M_f} \quad \text{(second rocket equation).} \qquad [34]$$

Sample Problem 9 shows some sample calculations.

Work–Energy Theorem for a System

One part of the kinetic energy of a system of particles is that associated with motion of the center of mass, $K_{cm} = \frac{1}{2}Mv_{cm}^2$. Changes in this center-of-mass kinetic energy are related to the external forces acting on the system by a work–energy theorem

$$W_{cm} = K_{cm,f} - K_{cm,i} = \Delta K_{cm}. \qquad [39]$$

Here W_{cm} is the work that would have been done by an external force if it had been acting on a particle located at the moving center of mass of the system. That is, in one dimension,

Center-of-Mass Work

$$W_{cm} = \int_{x_i}^{x_f} F_{ext} \, dx_{cm}. \qquad [38]$$

This does not necessarily represent an energy exchange with the external agent. The "extra" kinetic energy may be supplied (or absorbed) by internal work (work done on each other by the components of the system). Sample Problems 10 and 11 show examples.

QUESTIONS

1. Does the center of mass of a solid object necessarily lie within the object? If not, give examples.

2. Figure 18 shows (*a*) an isosceles triangle and (*b*) a right circular cone whose diameter is the same length as the base of the triangle. The center of mass of the triangle is one-third of the way up from the base but that of the cone is only one-fourth of the way up. Can you explain this difference?

3. How is the center-of-mass concept related to the concept of geographic center of the country? To the population center of the country? What can you conclude from the fact that the geographic center differs from the population center?

4. Where is the center of mass of the earth's atmosphere?

5. An amateur sculptor decides to portray a bird (Fig. 19). Luckily, the final model is actually able to stand upright. The model is formed of a single thick sheet of metal of uniform thickness. Of the points shown, which is most likely to be the center of mass?

6. Someone claims that when a skillful high jumper clears the bar his center of mass actually goes *under* the bar. Is this possible?

7. A ballet dancer doing a *grand jete* (great leap; see Fig. 20)

Figure 18   Question 2.

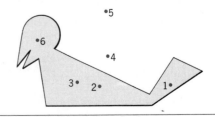

Figure 19   Question 5.

Figure 20 Question 7.

seems to float horizontally in the central portion of her leap. Show how the dancer can maneuver her legs in flight so that, although her center of mass does indeed follow the expected parabolic trajectory, the top of her head moves more or less horizontally. (See "The Physics of Dance," by Kenneth Laws, *Physics Today,* February 1985.)

8. A light and a heavy body have equal kinetic energies of translation. Which one has the larger momentum?

9. A bird is in a wire cage hanging from a spring balance. Is the reading of the balance when the bird is flying about greater than, less than, or the same as that when the bird sits in the cage?

10. Can a sailboat be propelled by air blown at the sails from a fan attached to the boat? Explain your answer.

11. When you run, do you transfer momentum to the earth? Explain your answer.

12. A canoeist in a still pond can reach shore by jerking sharply on the rope attached to the bow of the canoe. How do you explain this? (It really can be done.)

13. How might a person sitting at rest on a frictionless horizontal surface get altogether off it?

14. A man stands still on a large sheet of slick ice; in his hand he holds a lighted firecracker. He throws the firecracker into the air. Describe briefly, but as exactly as you can, the motion of the center of mass of the firecracker and the motion of the center of mass of the system consisting of man and firecracker. It will be most convenient to describe each motion during each of the following periods: (*a*) after he throws the firecracker, but before it explodes; (*b*) between the explosion and the first piece of firecracker hitting the ice; (*c*) between the first fragment hitting the ice and the last fragment landing; and (*d*) during the time when all fragments have landed but none has reached the edge of the ice.

15. Justify the following statement: "The law of conservation of linear momentum, as applied to a single particle, is equivalent to Newton's first law of motion."

16. In 1920, a prominent newspaper editorialized as follows about the pioneering rocket experiments of Robert H. Goddard, dismissing the notion that a rocket could operate in a vacuum: "That Professor Goddard, with his 'chair' in Clark College and the countenancing of the Smithsonian Institution, does not know the relation of action to reaction, and of the need to have something better than a vacuum against which to react—to say that would be absurd. Of course, he only seems to lack the knowledge ladled out daily in high schools." What is wrong with this argument?

17. Are there any possible methods of propulsion in outer space other than rockets? If so, why are they not used?

18. Can a rocket reach a speed greater than the speed of the exhaust gases that propel it? Explain why or why not.

19. If only an external force can change the state of motion of the center of mass of a body, how does it happen that the internal force of the brakes can bring a car to rest?

20. We say that a car is not accelerated by internal forces but rather by external forces exerted on it by the road. Why then do cars need engines?

21. Can the work done by internal forces decrease the kinetic energy of a body? . . . increase it?

22. (*a*) If you do work on a system, does the system necessarily acquire kinetic energy? (*b*) If a system acquires kinetic energy, does it necessarily mean that some external agent did work on it? Give examples. (By "kinetic energy" here we mean kinetic energy associated with the motion of the center of mass.)

23. In Section 9–8, we saw an example (a skater) in which kinetic energy appeared but no external work was done. Consider an opposite case. A screwdriver is held tightly against a rotating grinding wheel. Here external work is done but the kinetic energy of the screwdriver does not change. Explain this apparent violation of the work–energy theorem.

EXERCISES AND PROBLEMS

Section 9–2 The Center of Mass

1E. How far is the center of mass of the earth–moon system from the center of the earth? (From Appendix C, obtain the masses of the earth and moon and the distance between the centers of the earth and moon. It is interesting to compare the answer to the earth's radius.)

2E. Experiments using the diffraction of electrons show that the distance between the centers of the carbon (C) and oxygen (O) atoms in the carbon monoxide gas molecule is 1.131×10^{-10} m. Locate the center of mass of a CO molecule relative to the carbon atom.

3E. Where is the center of mass of the three particles shown in Fig. 21?

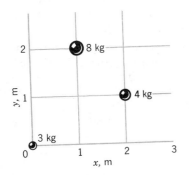

Figure 21 Exercise 3.

4E. Three thin rods each of length L are arranged in an inverted U, as shown in Fig. 22. The two rods on the arms of the U each have mass M; the third rod has mass $3M$. Where is the center of mass of the assembly?

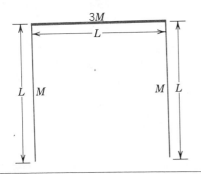

Figure 22 Exercise 4.

5E. A uniform square plate 6 m on a side has a square piece 2 m on a side cut out of it. The center of this notch is at $x = 2$ m, $y = 0$. The center of the square plate is at $x = y = 0$; see Fig. 23. Find the x and y coordinates of the center of mass of the remaining piece.

6P. Show that the ratio of the distances of two particles from their center of mass is the inverse ratio of their masses.

7P. Figure 24 shows a composite slab with dimensions 22 cm $\times$ 13 cm $\times$ 2.8 cm. Half the slab is made of aluminum

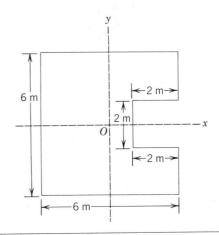

Figure 23 Exercise 5.

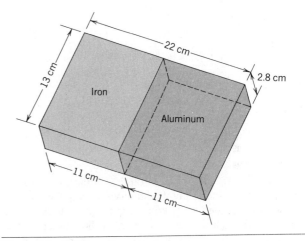

Figure 24 Problem 7.

(density $= 2.70$ g/cm^3) and half of iron (density $= 7.85$ g/cm^3), as shown. Where is the center of mass of the slab?

8P. In the ammonia (NH_3) molecule, the three hydrogen (H) atoms form an equilateral triangle, the distance between centers of the atoms being 16.28×10^{-11} m, so that the center of the triangle is 9.40×10^{-11} m from each hydrogen atom. The nitrogen (N) atom is at the apex of a pyramid, the three hydrogens constituting the base (see Fig. 25). The nitrogen–hydrogen distance is 10.14×10^{-11} m and the nitrogen–hydrogen atomic mass ratio is 13.9. Locate the center of mass relative to the nitrogen atom.

9P. A box, open at the top, in the form of a cube of edge length 40 cm, is constructed from thin metal plate. Find the coordinates of the center of mass of the box with respect to the coordinate system shown in Fig. 26.

10P. A cylindrical storage tank is initially filled with aviation gasoline. The tank is then drained through a valve on the bot-

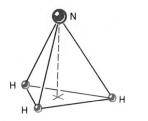

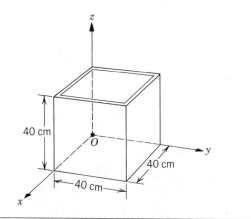

Figure 25 Problem 8.

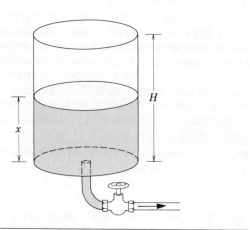

Figure 26 Problem 9.

tom. See Fig. 27. (a) As the gasoline is withdrawn, describe qualitatively the motion of the center of mass of the tank and its remaining contents. (b) What is the depth x to which the tank is filled when the center of mass of the tank and its remaining contents reaches its lowest point? Express your answer in terms of H, the height of the tank; M its mass; and m, the mass of gasoline it can hold.

Figure 27 Problem 10.

Section 9–3 Newton's Second Law for a System of Particles

11E. Two skaters, one with mass 65 kg and the other with mass 40 kg, stand on an ice rink holding a pole with a length of 10 m and a mass that is negligible. Starting from the ends of the pole, the skaters pull themselves along the pole until they meet. How far will the 40-kg skater move?

12E. A Chrysler with a mass of 2400 kg is moving along a straight stretch of road at 80 km/h. It is followed by a Ford with mass 1600 kg moving at 60 km/h. How fast is the center of mass of the two cars moving?

13E. A man of mass m clings to a rope ladder suspended below a balloon of mass M; see Fig. 28. The balloon is stationary with respect to the ground. (a) If the man begins to climb the ladder at a speed v (with respect to the ladder), in what direction and with what speed (with respect to the earth) will the balloon move? (b) What is the state of motion after the man stops climbing?

Figure 28 Exercise 13.

14E. Two particles P and Q are initially at rest 1.0 m apart. P has a mass of 0.10 kg and Q a mass of 0.30 kg. P and Q attract each other with a constant force of 1.0×10^{-2} N. No external forces act on the system. (a) Describe the motion of the center of mass. (b) At what distance from P's original position do the particles collide?

15E. A cannon and a supply of cannon balls are inside a sealed railroad car of length L, as in Fig. 29. The cannon fires to the right; the car recoils to the left. The cannon balls remain in the car after hitting the far wall. (a) After all the cannon balls have been fired, what is the greatest distance the car can have moved from its original position? (b) What is the speed of the car just after all the cannon balls have been fired?

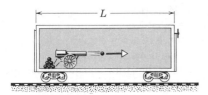

Figure 29 Exercise 15.

16P. A stone is dropped at $t = 0$. A second stone, with twice the mass of the first, is dropped from the same point at $t = 100$ ms. (a) Where is the center of mass of the two stones at $t = 300$ ms? Neither stone has yet reached the ground. (b) How fast is the center of mass of the two-stone system moving at that time?

17P. Ricardo, mass 80 kg, and Carmelita, who is lighter, are enjoying Lake Merced at dusk in a 30-kg canoe. When the canoe is at rest in the placid water, they change seats, which are 3.0 m apart and symmetrically located with respect to the canoe's center. Ricardo notices that the canoe moved 40 cm relative to a submerged log and calculates Carmelita's mass, which she has not told him. What is it?

18P. A shell is fired from a gun with a muzzle velocity of 1500 ft/s, at an angle of 60° with the horizontal. At the top of the trajectory, the shell explodes into two fragments of equal mass. One fragment, whose speed immediately after the explosion is zero, falls vertically. How far from the gun does the other fragment land, assuming level terrain?

19P. Two bodies, each made up of weights from a set, are connected by a light cord that passes over a light, frictionless pulley with a diameter of 50 mm. The two bodies are at the same level. Each originally has a mass of 500 g. (a) Locate their center of mass. (b) Twenty grams are transferred from one body to the other, but the bodies are prevented from moving. Locate the center of mass. (c) The two bodies are now released. Describe the motion of the center of mass and determine its acceleration.

20P. A dog, weighing 10 lb, is standing on a flatboat so that he is 20 ft from the shore. He walks 8.0 ft on the boat toward shore and then halts. The boat weighs 40 lb, and one can assume there is no friction between it and the water. How far is the dog from the shore at the end of this time? The shoreline is also to the left in Fig. 30. (*Hint:* The center of mass of boat + dog does not move. Why?)

Section 9-4 Linear Momentum
21E. What is the linear momentum of a 3000-lb automobile traveling at 55 mi/h?

22E. Suppose that your mass is 80 kg. How fast would you have to run to have the same momentum as a 1600-kg car moving at 1.2 km/h ($\approx$ 0.75 mi/h)?

23E. Find (a) the momentum and (b) the kinetic energy of a 4-g bullet with a speed of 950 m/s. (c) How fast must a 450-kg

Figure 30 Problem 20.

deer move to have the same momentum?

24E. How fast must an 816-kg Volkswagon travel (a) to have the same momentum as a 2650-kg Cadillac going 16 km/h and (b) to have the same kinetic energy? (c) Make the same calculations using a 9080-kg truck instead of a Cadillac.

25E. An electron has a speed of $0.99c$ ($= 2.97 \times 10^8$ m/s). What is its linear momentum?

26E. The momentum of a particle moving at 1.5×10^8 m/s is measured to be 2.9×10^{-19} kg·m/s. By finding its mass, identify the particle.

27P. An object is tracked by a radar station and found to have a position vector given by $\mathbf{r} = (3500 - 160t)\mathbf{i} + 2700\mathbf{j} + 300\mathbf{k}$, with $\mathbf{r}$ in meters and t in seconds. The radar station's x axis points east, its y axis north, and its z axis vertically up. If the object is a 250-kg missile warhead, what are (a) its momentum and (b) its direction of motion?

28P. A 5.0-kg object with a speed of 30 m/s strikes a steel plate at an angle of 40° and rebounds at the same speed and angle (Fig. 31). What is the change (magnitude and direction) of the linear momentum of the object?

Figure 31 Problem 28.

29P. A 2000-kg truck traveling north at 40 km/h turns east and accelerates to 50 km/h. (a) What is the change in kinetic energy of the truck? (b) What is the magnitude and direction of the change of the truck's momentum?

30P. A 50-g ball is thrown from ground level into the air with an initial speed of 16 m/s at an angle of 30° above the horizontal. (a) What are the values of the kinetic energy of the ball initially and just before it hits the ground? (b) Find the corresponding values of the momentum (magnitude and direction). (c) Show that the change in momentum is just equal to the weight of the ball multiplied by the time of flight.

31P. A particle with mass m has linear momentum p equal to mc. What is its speed in terms of c, the speed of light?

Section 9-6 Conservation of Linear Momentum

32E. A 200-lb man standing on a surface of negligible friction kicks forward a 0.15-lb stone lying at his feet so that it acquires a speed of 13 ft/s. What velocity does the man acquire as a result?

33E. Two blocks of masses 1.0 kg (weight $= mg = 2.2$ lb) and 3.0 kg (weight $= mg = 6.6$ lb) connected by a spring rest on a frictionless surface. If the two are given velocities such that the first travels at 1.7 m/s ($= 5.6$ ft/s) toward the center of mass, which remains at rest, what is the velocity of the second?

34E. A space vehicle is traveling at 4000 km/h with respect to the earth when the exhausted rocket motor is disengaged and sent backward with a speed of 80 km/h with respect to the command module. The mass of the motor is four times the mass of the module. What is the speed of the command module after the separation?

35E. A 75-kg man is riding on a 39-kg cart traveling at a speed of 2.3 m/s. He jumps off in such a way as to land on the ground with zero horizontal speed. What is the resulting change in the speed of the cart?

36E. A railroad flatcar of weight W can roll without friction along a straight horizontal track. Initially, a man of weight w is standing on the car, which is moving to the right with speed v_0. What is the change in velocity of the car if the man runs to the left (Fig. 32) so that his speed relative to the car is v_{rel} just before he jumps off at the left end?

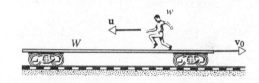

Figure 32 Exercise 36.

37E. Two 2-kg masses, A and B, collide. The velocities before the collision are $v_A = 15i + 30j$ and $v_B = -10i + 5.0j$. After the collision, $v'_A = -5.0i + 20j$. All speeds are given in meters per second. (a) What is the final velocity of B? (b) How much kinetic energy was gained or lost in the collision?

38E. (a) An object with mass m is acted on by a constant force F. As a result, its velocity changes by an amount Δv. Show that

$$F = \frac{m \, \Delta v}{\Delta t},$$

where Δt is the time over which the force acted. (*Hint:* See Eq. 19.) (b) The National Transportation Safety Board is testing the crash-worthiness of a new car. The 2300-kg vehicle, moving at 15 m/s, is allowed to collide with a bridge abutment, being brought to rest in a time of 0.56 s. What force, assumed constant, acted on the car during impact?

39E. A karate expert breaks a pine board, 2.2 cm thick, with a hand chop. Strobe photography shows that his hand, whose mass may be taken as 580 g, strikes the top of the board with a speed of 9.5 m/s and comes to rest 2.8 cm below this level. (a) What is the time duration of the chop (assuming a constant force)? (b) What average force is applied?

40P. The last stage of a rocket is traveling at a speed of 7600 m/s. This last stage is made up of two parts which are clamped together, namely, a rocket case with a mass of 290 kg and a payload capsule with a mass of 150 kg. When the clamp is released, a compressed spring causes the two parts to separate with a relative speed of 910 m/s. (a) What are the speeds of the two parts after they have separated? Assume that all velocities are along the same line. (b) Find the total kinetic energy of the two parts before and after they separate and account for the difference, if any.

41P. A vessel at rest explodes, breaking into three pieces. Two pieces, having equal mass, fly off perpendicular to one another with the same speed of 30 m/s. The third piece has three times the mass of each other piece. What is the direction and magnitude of its velocity immediately after the explosion?

42P. A radioactive nucleus, initially at rest, decays by emitting an electron and a neutrino at right angles to one another. The momentum of the electron is 1.2×10^{-22} kg·m/s and that of the neutrino is 6.4×10^{-23} kg·m/s. (a) Find the direction and magnitude of the momentum of the recoiling nucleus. (b) The mass of the residual nucleus is 5.8×10^{-26} kg. What is its kinetic energy of recoil? A *neutrino* is one of the fundamental particles of physics.

43P. A 2000-kg railroad flatcar, which can move on the tracks with virtually no friction, is sitting motionless next to a station platform. A 100-kg football player is running along the platform parallel to the tracks at 10 m/s. He jumps onto the back of the flatcar. (a) What is the speed of the flatcar after he is aboard and at rest on the flatcar? (b) Now he starts to walk, at 0.5 m/s relative to the flatcar, to the front of the car. What is the speed of the flatcar as he walks?

44P. A rocket sled with a mass of 2900 kg moves at 250 m/s on a set of rails. At a certain point, a scoop on the sled dips into a trough of water located between the tracks and scoops water into an empty tank on the sled. By applying the conservation of linear momentum, determine the speed of the sled after 920 kg of water have been scooped up.

45P. A 3.5-g bullet is fired horizontally at two blocks resting on a smooth tabletop, as shown in Fig. 33a. The bullet passes through the first block, with mass 1.2 kg, and embeds itself in the second, with mass 1.8 kg. Speeds of 0.63 m/s and 1.4 m/s, respectively, are thereby imparted to the blocks, as shown in Fig. 33b. Neglecting the mass removed from the first block by the bullet, find (a) the speed of the bullet immediately after emerging from the first block and (b) the original speed of the bullet.

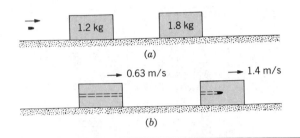

(a)

(b)

Figure 33   Problem 45.

46P.   A pellet gun fires ten 2.0-g pellets per second with a speed of 500 m/s. The pellets are stopped by a rigid wall. (a) What is the momentum of each pellet? (b) What is the kinetic energy of each pellet? (c) What is the average force exerted by the stream of pellets on the wall? (d) If each pellet is in contact with the wall for 0.6 ms, what is the average force exerted on the wall by each pellet while in contact? Why is this so different from (c)?

47P.   A machine gun fires 50-g bullets at a speed of 1000 m/s. The gunner, holding the machine gun in his hands, can exert an average force of 180 N against the gun. Determine the maximum number of bullets he can fire per minute.

48P.   Each minute, a special game warden's machine gun fires 220 10-g rubber bullets with a muzzle velocity of 1200 m/s. How many bullets must be fired at an 85-kg animal charging toward the warden at 4.0 m/s in order to stop the animal in its tracks? (Assume that the bullets travel horizontally and drop to the ground after striking the target.)

49P.   A body of mass 8 kg is traveling at 2 m/s with no external force acting. At a certain instant an internal explosion occurs, splitting the body into two chunks of 4-kg mass each; 16 J of translational kinetic energy are imparted to the two-chunk system by the explosion. Neither chunk leaves the line of the original motion. Determine the speed and direction of motion of each of the chunks after the explosion.

50P.   Assume that the car in Exercise 36 is initially at rest. It holds n men each of weight w. If each man in succession runs with a relative velocity v_{rel} and jumps off the end, do they impart to the car a greater velocity than if they all run and jump at the same time?

51P.   During a violent thunderstorm, hail the size of marbles (diameter = 1.0 cm) falls at a speed of 25 m/s. There are estimated to be 120 hailstones per cubic meter of air. Ignore the bounce of the hail on impact. (a) What is the mass of each hailstone? (b) What force is exerted by hail on a 10 m × 20 m

flat roof at the height of the storm? Assume that, as for ice, 1 cm³ of hail has a mass of 0.92 g.

52P.   A 1400-kg cannon, which fires a 70-kg shell with a muzzle speed of 556 m/s, is set at an elevation angle of 39° above the horizontal. The cannon is mounted on frictionless rails, so that it recoils freely. (a) What is the speed of the shell with respect to the earth? (b) At what angle with the ground is the shell projected? (Hint: The horizontal component of the momentum of the system remains unchanged as the gun is fired.)

53P*.   At a factory, a 300-kg crate is dropped vertically from a packing machine onto a conveyor belt moving at 1.2 m/s; see Fig. 34. The coefficient of kinetic friction between belt and crate is 0.40. After a short time, slipping between the belt and the crate ceases and the crate then moves along with the belt. For the period of time during which the crate is being brought to rest relative to the belt, calculate, for a coordinate system at rest in the factory, (a) the work done by friction, (b) the force of friction, and (c) the work done by the motor driving the belt. (d) Why are the answers to (a) and (c) different and where does the extra energy go?

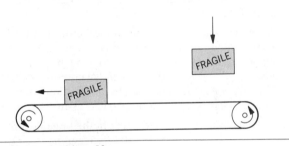

Figure 34   Problem 53.

Section 9–7   Systems with Variable Mass: A Rocket

54E.   A rocket is moving away from the solar system at a speed of 6.0×10^3 m/s. It fires its rocket engine, which ejects exhaust with a relative velocity of 3.0×10^3 m/s. The mass of the rocket at this time is 4.0×10^4 kg, and it experiences an acceleration of 2.0 m/s². (a) What is the thrust of the engine? (b) At what rate, in kilograms per second, was exhaust ejected during the firing?

55E.   A 6000-kg spaceprobe moving nose-first toward Jupiter at 100 m/s relative to the sun fires its rocket engine, ejecting 80 kg of exhaust at a speed of 250 m/s relative to the spaceprobe. What is the final velocity of the probe?

56E.   A rocket at rest in space, where there is virtually no gravity, has a mass of 2.55×10^5 kg, of which 1.81×10^5 kg is fuel. The engine consumes fuel at the rate of 480 kg/s, and the exhaust speed is 3.27 km/s. The engine is fired for 250 s. (a) Find the thrust of the rocket engine. (b) What is the mass of the rocket after the engine burn? (c) What is the final speed attained?

57E.   Consider a rocket at rest in empty space. What must be its *mass ratio* (ratio of initial to final mass) in order that, after firing its engine, the rocket's speed is (a) equal to the exhaust speed and (b) equal to twice the exhaust speed?

58E. A railroad car moves at a constant speed of 3.2 m/s under a grain elevator. Grain drops into it at the rate of 540 kg/min. What force must be applied to the railroad car, in the absence of friction, to keep it moving at constant speed?

59P. During a lunar mission, it is necessary to make a mid-course correction of 2.2 m/s in the speed of the spacecraft, which is moving at 400 m/s. The exhaust speed of the rocket engine is 1000 m/s. What fraction of the initial mass of the spacecraft must be discarded as exhaust?

60P. A single-stage rocket, at rest in a certain inertial reference frame, has a mass M when the rocket engine is ignited. Show that, when the mass has decreased to $0.368M$, the gases streaming out of the rocket engine at that time will be at rest in the original reference frame.

61P. A 5.4-kg toboggan carrying 35 kg of sand slides from rest down an icy slope 90-m long, inclined 30° below the horizontal. The sand leaks from the back of the toboggan at the rate of 2.3 kg/s. How long does it take the toboggan to reach the bottom of the slope?

62P. A 6000-kg rocket is set for vertical firing. If the exhaust speed is 1000 m/s, how much gas must be ejected each second to supply the thrust needed (a) to overcome the weight of the rocket and (b) to give the rocket an initial upward acceleration of 20 m/s²? Note that, in contrast to the situation described in Sample Problem 9, gravity is present here as an external force.

63P. Two long barges are floating in the same direction in still water, one with a speed of 10 km/h and the other with a speed of 20 km/h. While they are passing each other, coal is shoveled from the slower to the faster one at a rate of 1000 kg/min; see Fig. 35. How much additional force must be provided by the driving engines of each barge if neither is to change speed? Assume that the shoveling is always perfectly sideways and that the frictional forces between the barges and the water do not depend on the weight of the barges.

64P. A jet airplane is traveling 180 m/s (≈ 600 ft/s). The engine takes in 68 m³ (= 2400 ft³) of air making a mass of 70 kg (= 4.8 slugs) each second. The air is used to burn 2.9 kg (= 0.20 slugs) of fuel each second. The energy is used to compress the products of combustion and to eject them at the rear of the plane at 490 m/s (≈ 1600 ft/s) relative to the plane. Find (a) the thrust of the jet engine and (b) the delivered power (horsepower).

Section 9–8 Systems of Particles: Work and Energy

65E. An automobile with passengers has weight 3680 lb (= 16,400 N) and is moving at 70 mi/h (= 113 km/h) when the driver brakes to a stop. The road exerts a force of 1850 lb (= 8230 N) on the wheels and there is no skidding. (a) What is

Figure 35 Problem 63.

the stopping distance? (b) Find the coefficient of static friction between tires and road.

66E. You crouch from a standing position, lowering your center of mass 18 cm in the process. Then you jump vertically into the air. The force that the floor exerts on you while you are jumping is three times your weight. What is your upward speed as you pass through your standing position leaving the floor?

67E. A 55-kg woman leaps vertically into the air from a crouching position in which her center of mass is 40 cm above the ground. As her feet leave the floor her center of mass is 90 cm above the ground and rises to 120 cm at the top of her leap. (a) What upward force, assumed constant, does the ground exert on her? (b) What maximum speed does she attain?

68P. A 110-kg ice hockey player skates at 3.0 m/s toward a railing at the edge of the ice and stops himself by grasping the railing with his outstretched arms. During this stopping process his center of mass moves 30 cm toward the rail. (a) What average force must he exert on the rail? (b) How much internal work (magnitude and sign) does he do? List any assumptions that you make.

69P. Many bicycles are braked by pressing a piece of hard rubber against each tire. Suppose that such a bicycle, initially traveling at 40.2 km/h, is stopped in 55 revolutions of its 35.6-cm radius tires. The mass of bicycle plus rider is 51.1 kg. Treat the front and rear brakes as identical and assume that the tires do not skid on the road. (a) What work is done by each brake? (b) Assume that the force of a brake against a tire is constant and calculate its value. (c) What is the acceleration of the bicycle? (d) What is the force of the road on each tire?

CHAPTER 10
COLLISIONS

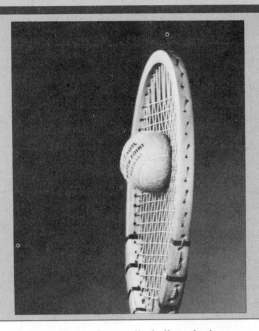

Skill in tennis—as in baseball, bowling, volleyball, and other sports—depends on the proficient manipulation of collisions. The tennis ball in the photo is in contact with the racket for less than about 4 ms (= 0.004 s). During a typical world-class tennis match, a ball is in contact with a racket for a cumulative time of about 1 s per set.

10-1   What Is a Collision?

A layperson might say: "When two things bang into each other, that is a collision." Although we shall refine that definition, it conveys the meaning well enough. Within our daily experience, the things that collide might be billiard balls, a hammer and a nail, a baseball and a bat, and—all too often—automobiles. A stone thrown or dropped toward the earth may be said to collide with it. Figure 1 shows the impressive evidence for one such event that occurred about 20,000 years ago. As Fig. 2 shows, collisions occur that are beyond our direct experience, ranging from the collisions of subatomic particles to the collisions of galaxies.

Figure 1   Meteor Crater in Arizona, the result of a collision about 20,000 years ago. The crater is about 4000 ft in diameter and 600 ft deep. See Exercise 36.

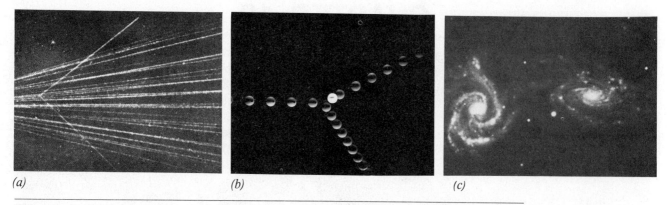

(a)   *(b)*   *(c)*

Figure 2   Collisions over a wide range of masses. *(a)* An alpha particle collides with the nucleus of a helium atom in a cloud chamber ($m \approx 10^{-26}$ kg). *(b)* A billiard ball coming in from the left collides with a resting ball ($m \approx 0.1$ kg). *(c)* Two galaxies colliding ($m \approx 10^{40}$ kg).

Let us formalize our definition: A collision is an isolated event in which a relatively strong force acts on each colliding particle for a relatively short time. It must be possible to make a clean separation between times that are *before the collision* and those that are *after the collision.* Figure 3 suggests the spirit of this definition. The system boundary surrounding the event in that figure reminds us that, in a true collision, external forces play no role.

When a bat strikes a baseball, the beginning and the end of the collision can be determined fairly precisely. The bat–ball contact time (about 1 ms) is short compared with the time that we watch the ball both before and after the collision. As Fig. 4 shows, the force exerted on the ball is strong enough to deform it temporarily.

When a space probe approaches a large planet, swings around it, and then continues its course with increased speed (a *slingshot* encounter), that too is a collision. You may object that the "particles" (probe and planet) do not actually "touch." However, a collision force does not have to be a contact force; it can just as easily be a non-touching gravitational force, as in this case.

Many physicists today spend their time playing what we can call "the collision game." A principal goal of this game is to find out as much as possible about the forces that act during the collision, knowing the state of the particles both before and after the collision. Virtually all of our knowledge of the subatomic world — electrons, protons, neutrons, quarks, and the like — comes from experiments of this kind. The rules of the collision game are the laws of conservation of momentum and of energy. Figure 5*a* shows one of the "playing fields" in which the collision game has been practiced with major success. It is the underground detector for the large, ring-shaped proton accelerator at CERN, the European high-energy physics laboratory near Geneva.

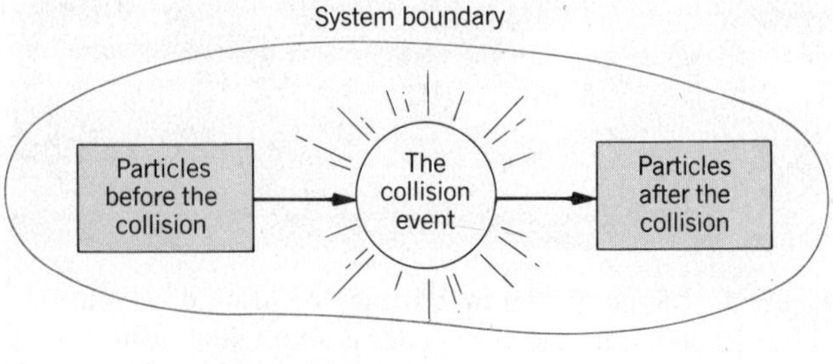

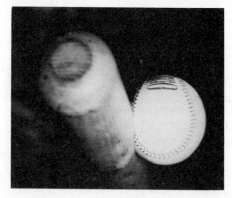

Figure 4   A high-speed flash photo of a baseball bat striking a ball. Note the (temporary) deformation of the ball. The effective duration of the collision is of the order of 1 ms. The average force acting on the ball is more than a ton.

System boundary

```
           Particles        The        Particles
           before the     collision    after the
           collision        event      collision
```

Figure 3   A representation of a collision. Note the sharply defined "before" and "after" states. The "system boundary" suggests that a true collision is isolated from external forces.

(a)

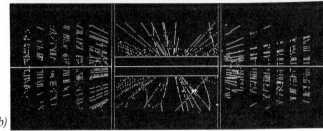

(b)

Figure 5   (a) A view of the underground detector UA1 used in conjunction with the PP Collider accelerator at CERN, the European particle physics laboratory near Geneva, Switzerland. Note the human figures. (b) A computer reconstruction of the particles generated in a single-collision event in this detector. These experiments were used to confirm a crucial theory about the nature of the forces between elementary particles. The 1984 Nobel Prize was awarded for these experiments; see Appendix H.

Figure 5*b* shows one of the events recorded in the CERN detector. The particles whose tracks appear in the figure are produced in a single head-on collision between two high-energy protons.* The colliding protons do not survive the collision—other particles being manufactured from the energy that they carry. It is almost as if a red and a blue billiard ball crashed into each other at high speed, generating an assortment of marbles, ball bearings, and Ping Pong balls!

---

* Actually, between a proton and its antiparticle. The antiproton has the same mass as the proton but carries a negative charge.

10-2   Impulse and Momentum

Figure 6 shows the equal but opposite forces that act in a simple head-on collision between two bodies of different mass. These forces will change the momentum of each body, the amount of the change depending not only on the average value of the force but also on the time during which it acts. To see this quantitatively, let us apply Newton's second law (in the form $\mathbf{F} = d\mathbf{p}/dt$) to the right-hand body in Fig. 6. We have

$$d\mathbf{p} = \mathbf{F}(t)\, dt, \tag{1}$$

in which $\mathbf{F}(t)$ is the time variation of the force, displayed as the curve in Fig. 7. Let us integrate Eq. 1 over the collision interval, that is, from an initial time t_i (before the collision) to a final time t_f (after the collision). Thus,

$$\int_{\mathbf{p}_i}^{\mathbf{p}_f} d\mathbf{p} = \int_{t_i}^{t_f} \mathbf{F}(t)\, dt. \tag{2}$$

The left side of this equation is just $\mathbf{p}_f - \mathbf{p}_i$, the change in momentum of the particle. The right side, which is a measure of both the strength and the duration of the collision force, is called the collision *impulse* **J**. Thus,

$$\boxed{\mathbf{J} = \int_{t_i}^{t_f} \mathbf{F}(t)\, dt} \quad \text{(impulse defined).} \tag{3}$$

Equation 3 tells us that the impulse is equal to the area under the $F(t)$ curve of Fig. 7.

From Eqs. 2 and 3 we see that the change in momentum of each particle in a collision is equal to the impulse that acts on that particle or,

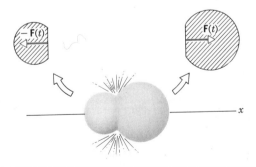

Figure 6   Two "particles" collide elastically with each other. The particles are shown separated in the upper part of the figure to clarify the time-varying force that acts on each of them.

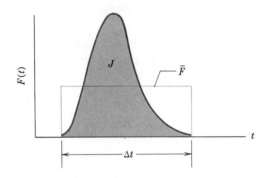

Figure 7 The curve shows the magnitude of the time-varying force $F(t)$ that acts on the colliding particles of Fig. 6. The rectangle shows the average force $\overline{F}$ acting over the time interval Δt. The areas under the curve $F(t)$ and the rectangle are the same and are equal to the magnitude of the impulse J delivered in the collision of Fig. 6.

$$\boxed{\mathbf{p}_f - \mathbf{p}_i = \mathbf{J}}$$

$$(\text{impulse – momentum theorem}). \quad (4)$$

Both impulse and momentum are vectors and have the same units and dimensions. The *impulse – momentum theorem* of Eq. 4 (like the work – energy theorem of Eq. 16, Chapter 7) is not a new and independent theorem but is a direct consequence of Newton's second law. Both theorems are special forms of this law, useful for special purposes.

If $\overline{F}$ is the average value of the force in Fig. 7, we can write the magnitude of the impulse as

$$J = \overline{F}\,\Delta t, \qquad (5)$$

in which Δt is the effective duration of the collision. $\overline{F}$ must be chosen so that the area under the rectangle of Fig. 7 is equal to the area under the $F(t)$ curve.

Sample Problem 1   A 140-g baseball, in horizontal flight with a speed v_i of 39 m/s, is struck by a batter. After leaving the bat, the ball travels in the opposite direction with a speed v_f, also 39 m/s. (a) What impulse J acted on the ball while it was in contact with the bat?

We can calculate the impulse from the effect it produces, using Eq. 4. Let us choose the direction in which the bat is moving to be positive. From Eq. 4 we have

$$\begin{aligned}
J &= p_f - p_i = mv_f - mv_i \\
&= (0.14 \text{ kg})(39 \text{ m/s}) - (0.14 \text{ kg})(-39 \text{ m/s}) \\
&= 10.9 \text{ kg} \cdot \text{m/s}. \qquad \text{(Answer)}
\end{aligned}$$

With our sign convention, the initial velocity of the ball is negative and the final velocity is positive. The impulse turned out to be positive, which tells us that the direction of the impulse vector acting on the ball is the direction in which the bat was swinging, which makes sense.

(b) The impact time Δt for the baseball – bat collision is 1.2 ms, a typical value. What average force acts on the baseball?

From Eq. 5 we have

$$\begin{aligned}
\overline{F} &= \frac{J}{\Delta t} = \frac{10.9 \text{ kg} \cdot \text{m/s}}{0.0012 \text{ s}} \\
&= 9100 \text{ N} \approx 2000 \text{ lb}, \qquad \text{(Answer)}
\end{aligned}$$

which is about a ton. The *maximum* force will be larger than this, as Fig. 7 shows. The sign of the average force exerted on the ball is positive, which means that the direction of the force vector is the same as that of the impulse vector.

(c) What was the average acceleration of the baseball? We find this from

$$\overline{a} = \frac{\overline{F}}{m} = \frac{9100 \text{ N}}{0.14 \text{ kg}} = 6.5 \times 10^4 \text{ m/s}^2. \qquad \text{(Answer)}$$

This is about 6600 times the acceleration of a freely falling body.

In our definition of a collision, we assumed that no external force acts on the colliding particles. That is not quite true in this case, because the ball is always subject to the force of gravity, whether the ball is in free flight or in contact with the bat. However, this force ($= mg = 1.4$ N) is negligible in comparison to the average force exerted by the bat ($= 9100$ N). We are quite safe in treating this collision as "isolated."

Sample Problem 2   (a) In Sample Problem 1, how much work does the bat do on the baseball during the collision?

Let us apply the work – energy theorem (Eq. 16 of Chapter 7), or*

$$W = \Delta K = K_f - K_i = \tfrac{1}{2}mv_f^2 - \tfrac{1}{2}mv_i^2.$$

This tells us (perhaps to our surprise) that — because the final and the initial speeds of the ball are equal — the work done on the ball is zero! How does one explain *this* to the batter?

The answer is that the batter is bodily sensitive to the *force* that he exerts, not to the *work* that he may have done. Although the batter has not changed the kinetic energy of the baseball in this case, he has certainly changed its momentum. Baseball

---

* This theorem applies only to particles. We can treat the baseball as a particle if we assume that it is not permanently deformed by the collision and that it does not rotate.

fans are much more interested in the momentum of the ball (a vector) than in its kinetic energy (a scalar).

(b) The batter now hits a baseball lightly tossed into the air and it leaves his bat with a speed v_f of 55 m/s. How much work is done on the ball in this case?

With $v_i \approx 0$, the work–energy theorem gives

$$W = K_f - K_i = \tfrac{1}{2}mv_f^2 - \tfrac{1}{2}mv_i^2$$
$$= (\tfrac{1}{2})(0.14 \text{ kg})(55 \text{ m/s})^2 - (0)$$
$$= 210 \text{ J.} \qquad \text{(Answer)}$$

(c) What was the average power delivered by the batter during the collision? This work is done during the collision interval of 0.0012 s so that the average power is

$$\bar{P} = \frac{W}{\Delta t} = \frac{210 \text{ J}}{0.0012 \text{ s}} = 180 \text{ kW.} \qquad \text{(Answer)}$$

This is enough power to light up a ball field for a night game. However, we do not have here a solution to the energy crisis; most of us can deliver energy at this rate, but only for about 1 ms.

10-3 Collisions in One Dimension: Elastic

Consider a simple head-on collision of two particles of different mass, perhaps weighted gliders coasting along a virtually frictionless horizontal air track. For convenience, we take one of the particles to be at rest before the collision.*

We assume further that the kinetic energy of the two-particle system is the same both before and after the collision. Collisions in which kinetic energy is conserved form a restricted group, called *elastic collisions.* Momentum is *always* conserved for a collision, whether or not the collision is elastic, because the forces are all internal forces.

Applying the laws of conservation of momentum and of kinetic energy to the collision of Fig. 8 gives us

$$m_1 v_{1i} = m_1 v_{1f} + m_2 v_{2f} \quad \text{(momentum)}, \qquad (6)$$

and

$$\tfrac{1}{2}m_1 v_{1i}^2 = \tfrac{1}{2}m_1 v_{1f}^2 + \tfrac{1}{2}m_2 v_{2f}^2 \quad \text{(kinetic energy).} \qquad (7)$$

In each of these equations, the subscript i identifies the

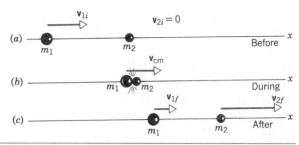

Figure 8   Two particles undergo an elastic collision, one of them (m_2) being initially at rest before the collision. The velocities are shown (*a*) before, (*b*) during, and (*c*) after the collision. (The velocity during the collision is the velocity of the center of mass of the two particles, which are momentarily touching.) The velocities are drawn to scale for the case in which $m_1 = 3m_2$.

initial velocities and the subscript f the final velocities of the particles. If we know the masses of the particles and if we also know v_{1i}, the initial velocity of particle 1, the only unknown quantities are v_{1f} and v_{2f}, the final velocities of the two particles. With two equations at our disposal, we should be able to find these two unknowns. Let us do so.

We can rewrite Eq. 6 as

$$m_1(v_{1i} - v_{1f}) = m_2 v_{2f} \qquad (8)$$

and Eq. 7 as*

$$m_1(v_{1i} - v_{1f})(v_{1i} + v_{1f}) = m_2 v_{2f}^2. \qquad (9)$$

Dividing Eq. 9 by Eq. 8 and carrying out a little more algebra, we obtain

$$v_{1f} = \frac{m_1 - m_2}{m_1 + m_2} v_{1i} \qquad (10)$$

and

$$v_{2f} = \frac{2m_1}{m_1 + m_2} v_{1i}. \qquad (11)$$

We note from Eq. 11 that v_{2f} is always positive (the target particle always moves forward). From Eq. 10 we learn that v_{1f} may be of either sign (the projectile moves forward if $m_1 > m_2$ but rebounds if $m_1 < m_2$). Let us look at a few special situations.

---

* Collisions in which one particle (the target) is at rest in the laboratory reference frame are very common. Even if the target particle is moving with respect to the laboratory, we can always find another inertial reference frame in which the target particle is initially at rest.

* In this step, we use the identity $a^2 - b^2 = (a - b)(a + b)$.

Equal Masses. If $m_1 = m_2$, Eqs. 10 and 11 reduce to

$$v_{1f} = 0 \quad \text{and} \quad v_{2f} = v_{1i},$$

which we might call the pool player's equation. It predicts that particle 1 (initially moving) stops dead in its tracks and particle 2 (initially at rest) takes off with the initial speed of particle 1. In head-on collisions, particles of equal mass simply exchange velocities.*

A Massive Target. In terms of Fig. 8, this requirement is that $m_2 \gg m_1$. We are firing a golf ball at a cannon ball. Equations 10 and 11 then reduce to

$$v_{1f} \approx -v_{1i} \quad \text{and} \quad v_{2f} \approx \left(\frac{2m_1}{m_2}\right) v_{1i}. \tag{12}$$

This tells us that particle 1 (the golf ball) simply bounces back in the same direction from which it came, its speed essentially unchanged. Particle 2 (the cannon ball) moves forward at a low speed; note that the quantity in parentheses in Eq. 12 is much less than unity. All this is what we expect.

A Massive Projectile. This is the opposite case; that is, $m_1 \gg m_2$. This time, we are firing a cannon ball at a golf ball. Equations 10 and 11 reduce to

$$v_{1f} \approx v_{1i} \quad \text{and} \quad v_{2f} \approx 2v_{1i}. \tag{13}$$

Equation 13 tells us that particle 1 (the cannon ball) simply keeps on going, scarcely slowed down by the collision. Particle 2 (the golf ball) leaps ahead, at twice the speed of the cannon ball.

You may wonder: "Why twice?" As a starting point in thinking about the matter, recall the collision described by Eq. 12, in which the velocity of the incident light particle (the golf ball) changed from $+v$ to $-v$, a velocity *change* of $2v$. The same *change* in velocity (from zero to $2v$) occurs in this problem also. The two problems differ only in the reference frame from which they are viewed.

Motion of the Center of Mass. We have pointed out earlier that the center of mass of two colliding particles moves at a steady pace, totally uninfluenced by the collision. This follows from Eq. 22 of Chapter 9,

$$P = Mv_{cm} = (m_1 + m_2)v_{cm}, \tag{14}$$

which relates the momentum P of the system of two particles to v_{cm}, the velocity with which their center of mass moves. If the momentum P is unchanged by the collision (and it is), then v_{cm} must also be unchanged.

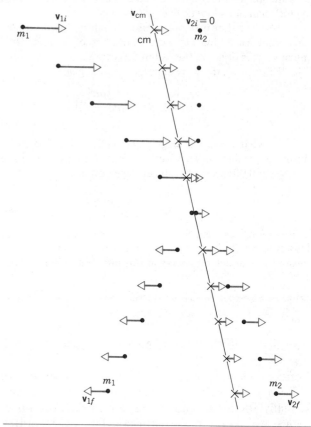

Figure 9 Some "snapshots" of two particles undergoing an elastic collision. We assume that particle 2 is initially at rest and that $m_2 = 3m_1$. The velocity of the center of mass is also shown. Note that it is unaffected by the collision.

From Eq. 14, the velocity of the center of mass for the collision shown in Fig. 8 (target initially at rest) is

$$v_{cm} = \frac{P}{m_1 + m_2} = \frac{m_1}{m_1 + m_2} v_{1i}. \tag{15}$$

Figure 9, a series of "snapshots" of a typical elastic collision, shows that the center of mass does indeed move steadily forward, unaffected by the collision.

Sample Problem 3 In the collision of Fig. 8, let $m_1 = 35$ g, $m_2 = 78$ g, and $v_{1i} = 1.9$ m/s. What are the velocities of the two particles after the collision? What is the velocity of the center of mass?

---

* This is true even though the target particle may not be initially at rest.

From Eq. 10 we have

$$v_{1f} = \frac{35\ g - 78\ g}{35\ g + 78\ g}(1.9\ \text{m/s}) = -0.72\ \text{m/s}.\quad \text{(Answer)}$$

The negative sign tells us that the incoming particle rebounds after the collision. From Eq. 11 we have

$$v_{2f} = \frac{(2)(35\ g)}{35\ g + 78\ g}(1.9\ \text{m/s}) = +1.2\ \text{m/s}.\quad \text{(Answer)}$$

Thus, the (more massive) target particle moves forward, but at a slower speed than that of the incoming particle, which moves at 1.9 m/s. The speed of the center of mass follows from Eq. 15 and is

$$v_{cm} = \frac{35\ g}{35\ g + 78\ g}(1.9\ \text{m/s}) = +0.59\ \text{m/s}.\quad \text{(Answer)}$$

The center of mass moves forward at this speed both before, during, and after the collision.

Sample Problem 4   In a nuclear reactor, newly-produced fast neutrons must be slowed down before they can participate effectively in the chain-reaction process. This is done by allowing them to collide with the nuclei of atoms forming a *moderator*. (a) By what fraction is the kinetic energy of a neutron (mass m_1) reduced in a head-on elastic collision with a nucleus of mass m_2, initially at rest?

The initial and the final kinetic energies of the neutron are

$$K_i = \tfrac{1}{2}m_1 v_{1i}^2 \quad \text{and} \quad K_f = \tfrac{1}{2}m_1 v_{1f}^2.$$

The fraction we seek is then

$$f = \frac{K_i - K_f}{K_i} = \frac{v_{1i}^2 - v_{1f}^2}{v_{1i}^2} = 1 - \frac{v_{1f}^2}{v_{1i}^2}. \quad (16)$$

For such a collision we have, from Eq. 10,

$$\frac{v_{1f}}{v_{1i}} = \frac{m_1 - m_2}{m_1 + m_2}. \quad (17)$$

Substituting Eq. 17 into Eq. 16 yields, after a little algebra,

$$f = \frac{4m_1 m_2}{(m_1 + m_2)^2}. \quad \text{(Answer)}\quad (18)$$

(b) Evaluate f for lead, carbon, and hydrogen. The ratios of nuclear mass to neutron mass ($= m_2/m_1$) for these nuclei are 206 for lead, 12 for carbon, and 1 for hydrogen.

The following values of f are calculated for these nuclei: For lead ($m_2 = 206\ m_1$),

$$f = \frac{(4)(206)}{(1 + 206)^2} = 0.019\ \text{or}\ 1.9\%.\quad \text{(Answer)}$$

For carbon ($m_2 = 12\ m_1$),

$$f = \frac{(4)(12)}{(1 + 12)^2} = 0.28\ \text{or}\ 28\%.\quad \text{(Answer)}$$

For hydrogen ($m_2 = m_1$),

$$f = \frac{(4)(1)}{(1 + 1)^2} = 1\ \text{or}\ 100\%.\quad \text{(Answer)}$$

These results explain why water, which contains lots of hydrogen, is a much better moderator of neutrons than is lead. The collisions that occur in the moderator of a nuclear reactor are random and very few of them will be head-on. The fraction f for an *average* collision will be less than the value calculated above but the preference for hydrogen over lead as a moderator still remains.

Sample Problem 5   A (target) glider, whose mass m_2 is 350 g, rests on an air track, a distance $d = 53$ cm from the end of the track. A second (projectile) glider, whose mass m_1 is 590 g, approaches the target glider with a velocity $v_{1i} = 75$ cm/s and collides elastically with it; see Fig. 10a. The target glider rebounds elastically from the end of the track and meets the projectile glider for a second time, as shown in Fig. 10b. How far from the end of the track will this second collision occur?

Let us use Eqs. 10 and 11 to find the velocities of the two gliders after they collide. From Eq. 10,

$$v_{1f} = v_{1i}\frac{m_1 - m_2}{m_1 + m_2} = (75\ \text{cm/s})\frac{590\ g - 350\ g}{590\ g + 350\ g}$$
$$= 19\ \text{cm/s}. \quad (19)$$

From Eq. 11 we find

$$v_{2f} = v_{1i}\frac{2m_1}{m_1 + m_2} = (75\ \text{cm/s})\frac{(2)(590\ g)}{590\ g + 350\ g}$$
$$= 94\ \text{cm/s}. \quad (20)$$

At the second collision, glider 1 will have traveled a distance $d - x$ and glider 2 a distance $d + x$. Their travel times t

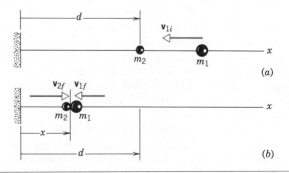

Figure 10   (*a*) Sample Problem 5. Two gliders on an air track are about to collide, glider m_2 being initially at rest. (*b*) Glider m_2 rebounds from a spring at the left end of the track and meets glider m_1 for a second time. Where does this second encounter occur?

for these distances are equal, so that

$$t = \frac{d - x}{v_{1f}} = \frac{d + x}{v_{2f}}.$$

Substituting from Eqs. 19 and 20 and putting $d = 53$ cm, we obtain

$$\frac{53 \text{ cm} - x}{19 \text{ cm/s}} = \frac{53 \text{ cm} + x}{94 \text{ cm/s}}.$$

Solving for x yields, after a little algebra,

$$x = 35 \text{ cm.} \qquad \text{(Answer)}$$

See Hint 1 to learn what is wrong with this method of solving the problem.

Hint 1: Numbers versus algebra. Beginning problem solvers often substitute numbers too soon. This "rush to the numbers" may build a feeling of confidence but it often gives a very restricted view of the overall problem. Sample Problem 5, in which we have deliberately used the numbers too soon, is a good example. If you solve this problem using symbols only, you get, following the steps in the numerical solution above,

$$x = d \frac{m_1 + m_2}{3m_1 - m_2}.$$

We can rewrite this in terms of the dimensionless mass ratio r as

$$x = d \frac{r + 1}{3r - 1} \quad \text{in which} \quad r = \frac{m_1}{m_2}.$$

For this problem, $r = (590 \text{ g})/(350 \text{ g}) = 1.69$ and $x = 35$ cm, in agreement with the numerical solution.

The algebraic solution shows, however, that the result does not depend on the individual masses, *but only on their ratio.* Even more important, *you do not need to know the initial speed of the moving glider;* it has canceled out. In the numerical solution, this cancellation occurred in a hidden way. If you have access to a glider in your physics lab, you can convince yourself by trial that the position of the second collision in Sample Problem 5 is indeed independent of the initial speed of the glider.

The algebraic solution also allows us to check for special cases. For example, $r = 1$ corresponds to $x = d$, which we expect. We can ask questions such as: "For what value of r will $x = \frac{1}{2}d$? or Will $x = 2d$?" (The answers are $r = 3$ and $r = 0.6$.) We can also ask: "What is the smallest value of r for which a second collision will indeed take place?" The answer is $r = \frac{1}{3}$, corresponding to $x \to \infty$.

You may object that if certain input data are not needed, we should not have given them in the problem statement. Our reply: In the real world, problem solvers must supply their own input data and must decide for themselves what is needed and what is not. Our advice: Stick to the algebra as long as you can and study the requirements and implications of the equation with which you end up.

10-4 Collisions in One Dimension: Inelastic

If you drop a Superball onto a hard floor, it loses very little of its kinetic energy on impact and rebounds to essentially its original height. A dropped golf ball will lose *some* of its kinetic energy and will rebound to about 60% of its original height. However, if you drop a ball of putty onto the floor it loses *all* of its kinetic energy and does not rebound at all. Although momentum is conserved in all collisions, kinetic energy is not always conserved.

We restrict ourselves in this section largely to collisions in which the colliding particles stick together after impact, just as the putty ball and the earth did in our example. If particles stick together in this way, they may not have lost *all* of their kinetic energy but they have lost as much of it as they can, consistent with the conservation of momentum. We call such collisions *completely inelastic.* (Although kinetic energy may not be conserved, the "lost" kinetic energy always shows up in some other form — perhaps as thermal energy — so that *total energy* remains conserved.)

Figure 11 shows the situation. The law of conserva-

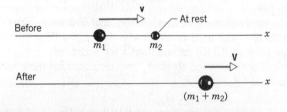

Figure 11 An inelastic collision between two particles, particle m_2 being initially at rest. Afterward, the particles stick together, which is the criterion for a *completely* inelastic collision. The velocities are drawn to scale for the case in which $m_1 = 3m_2$.

tion of momentum holds so that

$$m_1 v = (m_1 + m_2)V, \quad \bullet \qquad (21)$$

or

$$V = v \frac{m_1}{m_1 + m_2}. \qquad (22)$$

We see at once that the final velocity of the combined particle will always be less than that of the incoming particle.

Figure 12 (compare Fig. 9) shows that the center of mass in a completely inelastic collision also moves forward steadily, unaffected by the collision. Figure 12 shows clearly that, even though the collision is completely inelastic, kinetic energy associated with the mo-

tion of the center of mass must still be present. The only way that kinetic energy can vanish *totally* in an inelastic collision is if your reference frame is fixed with respect to the center of mass of the colliding particles. If the target particle m_2 happens to be exceedingly massive, the center of mass of the system essentially coincides with that of the target. This is the situation in which a mud ball is dropped onto the floor, the target particle in this case being the earth.

Sample Problem 6 A flash photo sequence of a golfer driving a ball from a tee (Fig. 13) allows us to measure the following measured speeds:

v_{1i}	Clubhead before impact	45 m/s
v_{1f}	Clubhead after impact	32 m/s
v_{2f}	Ball after impact	62 m/s

The mass m of the ball is 46 g. (a) What is the effective mass M of the clubhead?

The clubhead is not a free particle and it is not certain how much of the shaft or the golfer's hands and arms should be included in its effective mass. We can find the answer from the law of conservation of momentum. In Eq. 6, let m_1 represent the effective mass M of the clubhead and m_2 the mass m of the

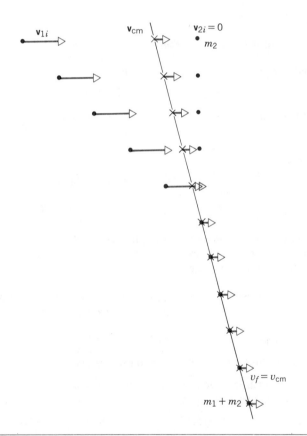

Figure 12 Some "snapshots" of two particles undergoing a completely inelastic collision. We assume that particle 2 is initially at rest and that $m_2 = 3m_1$. Note that the particles stick together after the collision and move forward as a single particle. The velocity of the center of mass is also shown. Note that it is unaffected by the collision.

Figure 13 A high-speed flash photo sequence of a golfer driving a golf ball from the tee. By analyzing such photos, it is possible to measure the speed of the clubhead, both before and after impact, and also the speed of the ball.

ball. Solving this equation for M yields

$$M = \frac{v_{2f}}{v_{1i} - v_{1f}} m$$

$$= \frac{62 \text{ m/s}}{45 \text{ m/s} - 32 \text{ m/s}} (46 \text{ g})$$

$$= 220 \text{ g} = 0.22 \text{ kg}. \qquad \text{(Answer)}$$

The nominal mass of the clubhead of a typical driver is 6 oz or 0.17 kg. Thus, in our example, the effective mass of the driver is about 30% greater than its nominal mass.

(b) What are the initial and the final kinetic energies for this collision?

Before the collision, the kinetic energy is entirely in the clubhead and is

$$K_i = \tfrac{1}{2}Mv_{1i}^2 = (\tfrac{1}{2})(0.22 \text{ kg})(45 \text{ m/s})^2 = 220 \text{ J}. \quad \text{(Answer)}$$

After the collision, the remaining kinetic energy is shared between the ball and the clubhead and is

$$K_f = \tfrac{1}{2}Mv_{1f}^2 + \tfrac{1}{2}mv_{2f}^2$$
$$= (\tfrac{1}{2})(0.22 \text{ kg})(32 \text{ m/s})^2 + (\tfrac{1}{2})(0.046 \text{ kg})(62 \text{ m/s})^2$$
$$= 200 \text{ J}. \qquad \text{(Answer)}$$

Thus, $K_f < K_i$ and kinetic energy is not conserved, the system losing about 9% of its kinetic energy during the collision. Those who hear the thwack of the clubhead striking the ball know where some of this energy goes. Because the ball does not stick to the clubhead, the collision is not *completely* inelastic.

(c) Just as the clubhead hits the ball, somebody sneezes. Where is the golf ball when the golfer reacts to the sneeze? A typical human reaction time τ for such an event is 120 ms.

Traveling at 62 m/s, the ball is a distance from the tee given by

$$d = v_{2f}\tau = (62 \text{ m/s})(0.120 \text{ s}) = 7.4 \text{ m}. \quad \text{(Answer)}$$

The sneeze cannot possibly have any effect on the motion of the ball. In fact, once a golfer has started his downswing, it is already too late to distract him in any way even (some claim) by shooting him!

Sample Problem 7 A *ballistic pendulum* (Fig. 14) is a device that was used to measure the speeds of bullets before electronic timing devices came along. It consists of a large block of wood, whose mass M is 5.4 kg, hanging from two long pairs of cords. A 9.5-g bullet is fired into the block, coming quickly to rest. The *block + bullet* then swing upward, their center of mass rising a vertical distance $h = 6.3$ cm before the pendulum comes momentarily to rest at the end of its arc. (a) What is the speed of the bullet?

As the bullet collides with the block, we have, from the conservation of momentum,

$$mv = (M + m)V.$$

The law of conservation of mechanical energy applies to the

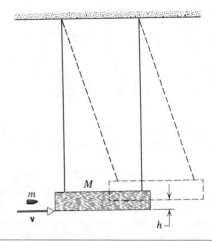

Figure 14 Sample Problem 7. A ballistic pendulum, formerly used to measure the speeds of rifle bullets.

swinging pendulum. The kinetic energy of the system when the block is at the bottom of its arc must then equal the potential energy of the system when the block is at the top, or

$$\tfrac{1}{2}(M + m)V^2 = (M + m)gh.$$

Eliminating V between these two equations leads to

$$v = \frac{M + m}{m} \sqrt{2gh}$$

$$= \left(\frac{5.4 \text{ kg} + 0.0095 \text{ kg}}{0.0095 \text{ kg}} \right) \sqrt{(2)(9.8 \text{ m/s}^2)(0.063 \text{ m})}$$

$$= 630 \text{ m/s}. \qquad \text{(Answer)}$$

We can look at the ballistic pendulum as a kind of transformer, exchanging the high speed of a light object (the bullet) for the low—and thus more easily measurable—speed of a massive object (the block).

(b) What is the initial kinetic energy of the bullet? How much of this energy remains as mechanical energy of the swinging pendulum?

The kinetic energy of the bullet is

$$K_b = \tfrac{1}{2}mv^2 = (\tfrac{1}{2})(0.0095 \text{ kg})(630 \text{ m/s})^2 = 1900 \text{ J}. \quad \text{(Answer)}$$

The mechanical energy of the swinging pendulum is equal to its potential energy when the block is at the top of its swing, or

$$E = (M + m)gh = (5.4 \text{ kg} + 0.0095 \text{ kg})(9.8 \text{ m/s}^2)(0.063 \text{ m})$$
$$= 3.3 \text{ J}. \qquad \text{(Answer)}$$

Thus, only 3.3/1900 or 0.2% of the original kinetic energy of the bullet is transferred to mechanical energy of the pendulum. The rest is dissipated inside the pendulum block as thermal energy.

10-5 Collisions in Two Dimensions

Here we consider a glancing collision (assumed elastic) between a projectile particle and a resting target. Figure 15 shows a typical situation. The distance b by which the extended track of the projectile misses the target is called the *impact parameter*. It is a measure of the directness of the collision, $b = 0$ corresponding to head-on. After the collision, the two particles fly off at angles θ_1 and θ_2, as the figure shows.

From the conservation of momentum (a vector relation), we can write down two scalar equations:

$$m_1 v_{1i} = m_1 v_{1f} \cos \theta_1 + m_2 v_{2f} \cos \theta_2$$
$$(x \text{ component}) \quad (23)$$

and

$$0 = m_1 v_{1f} \sin \theta_1 + m_2 v_{2f} \sin \theta_2 \quad (y \text{ component}). \quad (24)$$

We have assumed that the collision is elastic, so that kinetic energy is conserved. Thus, we have a third equation,

$$\tfrac{1}{2} m_1 v_{1i}^2 = \tfrac{1}{2} m_1 v_{1f}^2 + \tfrac{1}{2} m_2 v_{2f}^2 \quad (\text{kinetic energy}). \quad (25)$$

These three equations contain seven variables: two masses, m_1 and m_2; three speeds, v_{1i}, v_{1f}, and v_{2f}; and two angles, θ_1 and θ_2. If we know any four of these quantities, we can solve the three equations for the remaining three quantities. Often the known quantities are the two masses, the initial speed, and one of the angles. The unknowns to be solved for are then the two final speeds and the remaining angle.

Sample Problem 8 Two particles of equal mass have an elastic collision, the target particle being initially at rest. Show that (unless the collision is head-on) the two particles will always move off at right angles to each other after the collision.

There is a temptation to jump into the problem and solve it in a straightforward way, by applying Eqs. 23, 24, and 25. There is a neater way.

Figure 16a shows the situation both before and after the collision, each particle with its momentum vector attached. Because of conservation of momentum, these vectors must form a closed triangle, as Fig. 16b shows. Because the masses of the particles are equal, the closed momentum triangle of Fig. 16b is also a closed velocity triangle, the mass m canceling. That is,

$$\mathbf{v}_{1i} = \mathbf{v}_{1f} + \mathbf{v}_{2f}. \quad (26)$$

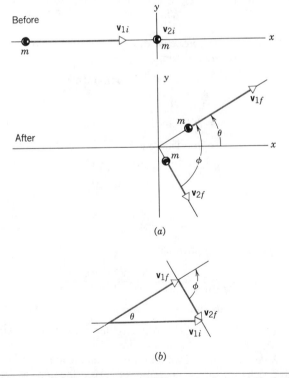

(a)

(b)

Figure 16 Sample Problem 8. Illustrating a neat proof of the fact that, in an elastic collision between two particles of the same mass, the particles fly off at 90° to each other afterward. The target particle must be initially at rest for this to be true.

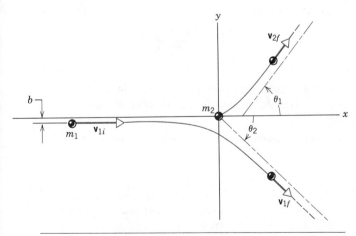

Figure 15 An elastic collision between two particles in which the collision is not head-on. The distance b by which the collision fails to be head-on is called the impact parameter. Particle m_2 (the target) is initially at rest.

Because kinetic energy is conserved, Eq. 25 holds. With the mass m canceled out, this equation becomes

$$v_{1i}^2 = v_{1f}^2 + v_{2f}^2. \tag{27}$$

Equation 27, applied to the triangle of Fig. 16b, is the theorem of Pythagoras (see Appendix G). For it to hold, the triangle of Fig. 16b must be a right triangle and therefore the angle ϕ between the vectors $\mathbf{v}_{1f}$ and $\mathbf{v}_{2f}$ must be 90°, which is what we set out to prove.

Figure 2a illustrates this proposition for colliding alpha particles (the nuclei of helium atoms). The incoming particle enters from the left; the target particle is an atom of the helium gas that fills the chamber in which the collision takes place. Figure 2b shows colliding billiard balls. The target ball stands out as a white disk because, being initially at rest, it is multiply exposed.

Sample Problem 9 Two skaters collide and embrace, in a completely inelastic collision. That is, they stick together after impact, as Fig. 17 suggests. Alfred, whose mass m_A is 83 kg, is originally moving east with a speed $v_A = 6.2$ km/h. Barbara, whose mass m_B is 55 kg, is originally moving north with a speed $v_B = 7.8$ km/h. (a) What is the velocity $\mathbf{V}$ of the couple after impact?

Momentum is conserved during the collision. We can write, for the two momentum components,

$$m_A v_A = MV \cos \theta \quad (x \text{ component}) \tag{28}$$

and

$$m_B v_B = MV \sin \theta, \quad (y \text{ component}) \tag{29}$$

in which $M = m_A + m_B$. Dividing Eq. 29 by Eq. 28 yields

$$\tan \theta = \frac{m_B v_B}{m_A v_A} = \frac{(55 \text{ kg})(7.8 \text{ km/h})}{(83 \text{ kg})(6.2 \text{ km/h})} = 0.834.$$

Thus,

$$\theta = \tan^{-1} 0.834 = 39.8°. \tag{Answer}$$

From Eq. 29 we then have

$$V = \frac{m_B v_B}{M \sin \theta} = \frac{(55 \text{ kg})(7.8 \text{ km/h})}{(83 \text{ kg} + 55 \text{ kg})(\sin 39.8°)}$$

$$= 4.86 \text{ km/h or } 1.35 \text{ m/s.} \tag{Answer}$$

(b) What is the velocity of the center of mass of the two skaters?

We can answer this without further calculation. After the collision, the velocity of the center of mass is the same as the velocity that we calculated in (a) above, namely, 4.8 km/h at 40° north of east. Because the velocity of the center of mass is not changed by the collision, this same value must prevail before the collision, as shown in Fig. 17.

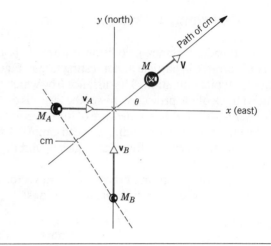

Figure 17 Sample Problem 9. Two skaters, Alfred (A) and Barbara (B), have a completely inelastic collision. Afterward, they move off at angle θ, with combined speed V. The path of their center of mass is shown.

(c) What is the fractional change in the kinetic energy of the skaters because of the collision?

The initial kinetic energy is

$$K_i = \tfrac{1}{2} m_A v_A^2 + \tfrac{1}{2} m_B v_B^2$$
$$= (\tfrac{1}{2})(83 \text{ kg})(6.2 \text{ km/h})^2 + (\tfrac{1}{2})(55 \text{ kg})(7.8 \text{ km/h})^2$$
$$= 3270 \text{ kg} \cdot \text{km}^2/\text{h}^2.$$

The final kinetic energy is

$$K_f = \tfrac{1}{2} MV^2$$
$$= (\tfrac{1}{2})(83 \text{ kg} + 55 \text{ kg})(4.86 \text{ km/h})^2$$
$$= 1630 \text{ kg} \cdot \text{km}^2/\text{h}^2.$$

The fraction we seek is then

$$f = \frac{K_f - K_i}{K_i} = \frac{1630 \text{ kg} \cdot \text{km}^2/\text{h}^2 - 3270 \text{ kg} \cdot \text{km}^2/\text{h}^2}{3270 \text{ kg} \cdot \text{km}^2/\text{h}^2}$$

$$= -0.50. \tag{Answer}$$

Thus, 50% of the initial kinetic energy is lost in the collision. It must be dissipated in some form or other as internal energy of the skating couple.

Hint 2: *More about units.* More often than not, it is wise to express all physical quantities in their basic SI units: thus, all speeds in m/s, all masses in kg, and so on. Sometimes, however, it is not necessary to do this. In Sample Problem 9a, for example, the units cancel out when we calculate the angle θ. In Sample Problem 9c, they cancel when we calculate the dimensionless quantity f. In this latter case, for example, there is no need to change the kinetic energy units to joules, the basic SI energy unit; we can leave them in $\text{kg} \cdot \text{km}^2/\text{h}^2$ because we can look ahead and see that they are going to cancel, because f is a dimensionless ratio.

10-6   Reactions and Decay Processes   (Optional)

Here we consider collisions (called *reactions*) in which the identity and even the number of the interacting particles change because of the collision. We consider too the spontaneous decay of unstable particles. For both kinds of events, there is a clear distinction between times that are "before the event" and times that are "after the event" and the laws of conservation of momentum and of energy apply. In short, we can treat these events by the methods we have already developed for collisions.

Sample Problem 10   A radioactive nucleus of uranium-235 decays spontaneously, emitting an alpha particle* according to the scheme

$$^{235}\text{U} \rightarrow \alpha + {}^{231}\text{Th}.$$

The emitted alpha particle has a kinetic energy K_α of 4.60 MeV. What is the kinetic energy of the recoiling thorium-231 nucleus? Some needed masses are

$$^{231}\text{Th} \quad m_{\text{Th}} = 231 \text{ u}$$
$$^{4}\text{He} \quad m_\alpha = 4.00 \text{ u}.$$

The ^{235}U nucleus is initially at rest in the laboratory. After decay, the two particles fly off in opposite directions, with kinetic energies K_{Th} and K_α, respectively. Applying the law of conservation of momentum leads to

$$0 = m_{\text{Th}} v_{\text{Th}} + m_\alpha v_\alpha,$$

which we can recast as

$$m_{\text{Th}} v_{\text{Th}} = -m_\alpha v_\alpha. \tag{30}$$

Because $K = \frac{1}{2}mv^2$, we can square each side of Eq. 30 and

---

* An alpha particle (symbol α, helium-4, or ^{4}He) is the nucleus of a helium atom.

rewrite it as

$$m_{\text{Th}} K_{\text{Th}} = m_\alpha K_\alpha.$$

Thus,

$$K_{\text{Th}} = K_\alpha \frac{m_\alpha}{m_{\text{Th}}} = (4.60 \text{ MeV}) \frac{4.00 \text{ u}}{231 \text{ u}}$$
$$= 7.97 \times 10^{-2} \text{ MeV} = 79.7 \text{ keV}. \quad \text{(Answer)}$$

We see that, of the total amount of kinetic energy made available during the decay ($= 4.60 + 0.0797 = 4.68$ MeV), the heavy recoiling nucleus receives only about $0.0797/4.68$ or 1.7%.

Sample Problem 11   A nuclear reaction of great importance for the generation of energy by nuclear fusion is the so-called *d-d* reaction, one form of which is

$$d + d \rightarrow t + p. \tag{31}$$

The particles represented by these letters are all isotopes of hydrogen, with the following properties:

Symbols		Name	Mass
p	^{1}H	Proton	$m_p = 1.00783$ u
d	^{2}H	Deuteron	$m_d = 2.01410$ u
t	^{3}H	Triton	$m_t = 3.01605$ u

(a) How much energy is made available because of the mass change that occurs in this reaction?

We represent this energy by the symbol Q, which we recognize as a fundamental property of the reaction. Applying Einstein's $E = mc^2$ relation, we write

$$Q = (2m_d - m_p - m_t)c^2$$
$$= (2 \times 2.01410 \text{ u} - 1.00783 \text{ u} - 3.01605 \text{ u})(932 \text{ MeV/u})$$
$$= (0.00432 \text{ u})(932 \text{ MeV/u}) = 4.03 \text{ MeV}. \quad \text{(Answer)}$$

We have used the value for c^2 ($= 932$ MeV/u) given by Eq. 30 of Chapter 8.

A positive value for Q (as in this case) means that the reaction is *exothermic*, transforming mass energy into kinetic energy. We see that in this reaction only $0.00432/(2 \times 2.01410)$ or about 0.1% of the mass originally present has been so transformed. A negative Q signals an *endothermic* reaction, in which the transfer goes from kinetic energy to mass energy. $Q = 0$ means an elastic encounter, with no mass change and kinetic energy conserved.

(b) A deuteron of kinetic energy $K_d = 1.50$ MeV strikes a resting deuteron, initiating the reaction of Eq. 31. A proton is observed to move off at an angle of 90° with the incident direction, with a kinetic energy of 3.39 MeV; see Fig. 18. What is the kinetic energy of the triton?

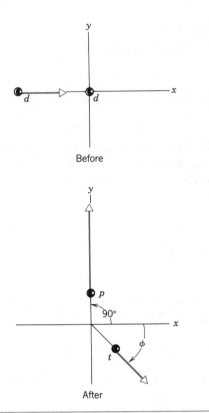

Figure 18 Sample Problem 11. A moving deuteron (d) strikes a resting deuteron. The d-d reaction of Eq. 31 occurs and the product particles (p and t) move off as shown. This nuclear reaction is very important in attempts to build a fusion reactor for generating power.

The energy Q released from the decrease in mass energy must appear as an increase in the kinetic energies of the particles. Thus, we can write

$$Q = K_p + K_t - K_d.$$

Solving for K_t gives us

$$
\begin{aligned}
K_t &= Q + K_d - K_p \\
&= 4.03 \text{ MeV} + 1.50 \text{ MeV} - 3.39 \text{ MeV} \\
&= 2.14 \text{ MeV}. \qquad \text{(Answer)}
\end{aligned}
$$

(c) At what angle ϕ with the incident direction (see Fig. 18) does the triton emerge?

We have not yet made use of the fact that momentum is conserved in the reaction of Eq. 31. The law of the conservation of momentum yields two scalar equations:

$$m_d v_d = m_t v_t \cos \phi \quad (x \text{ component}) \qquad (32)$$

and

$$0 = m_p v_p + m_t v_t \sin \phi \quad (y \text{ component}). \qquad (33)$$

From Eq. 33 we have

$$\sin \phi = -\frac{m_p v_p}{m_t v_t}. \qquad (34)$$

Using the relation $K = \frac{1}{2}mv^2$, we can write the momentum mv as $\sqrt{2mK}$ and thus recast Eq. 34 as

$$
\phi = \sin^{-1}\left(-\sqrt{\frac{m_p K_p}{m_t K_t}}\right) = \sin^{-1}\left(-\sqrt{\frac{(1.01 \text{ u})(3.39 \text{ MeV})}{(3.02 \text{ u})(2.14 \text{ MeV})}}\right)
$$
$$
= \sin^{-1}(-0.728) = -46.7°. \qquad \text{(Answer)}
$$

REVIEW AND SUMMARY

Collisions

In a *collision,* strong mutual forces act between a few particles for a short time, these internal forces being significantly larger than any external forces during the time of the collision. The laws of conservation of momentum and energy, applied to the "before" and "after" situations, often allow us to predict the outcomes of collisions and to learn a great deal about the interactions between the colliding particles.

Applying Newton's second law (in momentum form) to any particle leads to the *impulse–momentum* theorem:

Impulse and Momentum

$$\mathbf{p}_f - \mathbf{p}_i = \mathbf{J}, \quad (\text{impulse-momentum theorem}) \qquad [4]$$

where $\mathbf{p}_f - \mathbf{p}_i = \Delta \mathbf{p}$ is the change in the particle's momentum and $\mathbf{J}$ is the *impulse* due to the force acting on the particle:

$$\mathbf{J} = \int_{t_i}^{t_f} \mathbf{F}(t)\,dt. \quad (\text{impulse defined}) \qquad [3]$$

See Sample Problems 1 and 2.

If kinetic energy is conserved during a collision, we call the collision *elastic*. Equations 10 and 11, which are easily derived from conservation of momentum and kinetic energy, give the final speeds for the two particles after a head-on, one-dimensional collision of this type in which the target particle is initially at rest:

**Elastic Collision
—One Dimension**

$$v_{1f} = \left(\frac{m_1 - m_2}{m_1 + m_2}\right) v_{1i} \quad \text{and} \quad v_{2f} = \left(\frac{2m_1}{m_1 + m_2}\right) v_{1i}. \qquad [10,11]$$

See Sample Problems 3–5 for examples.

Inelastic Collisions

A collision is *inelastic* if the total kinetic energy of the system is changed. It is *completely inelastic* if the colliding particles stick together after the collision. The common final velocity, in this case, is determined by conservation of linear momentum. Kinetic energy is *not* conserved. See Sample Problems 6 and 7 for examples of inelastic collisions.

Collisions in Two Dimensions

Collisions in two dimensions are governed by conservation of vector linear momentum, a condition that leads to two component equations. These determine the final motion if the collision is completely inelastic. Otherwise, conservation of momentum and energy generally leads to equations that cannot be solved completely unless other experimental data, such as the final direction of one of the velocities, is available. Sample Problems 8 and 9 illustrate these ideas.

Reactions and Decay Processes

The identity and perhaps number of the interacting particles change in *reactions* and *decay processes*. Total energy and linear momentum are both conserved, a requirement that often allows calculation of important internal energy changes. An important parameter is the energy balance or Q of the reaction, defined as the change of the system kinetic energy. Positive Q represents an *exothermic* (energy-releasing) reaction while negative Q occurs for *endothermic* (energy-absorbing) reactions. Q always represents changes in internal energy of the reacting system and is always accompanied by a change in rest mass, with

$$Q = \Delta K = -\Delta m\, c^2.$$

Such mass changes are large enough to be detected experimentally in nuclear reactions. Sample Problems 10 and 11 apply these ideas to radioactive decay and to an important nuclear reaction.

QUESTIONS

1. Explain how conservation of momentum applies to a handball bouncing off a wall.

2. Can the impulse of a force be zero, even if the force is not zero? Explain why or why not.

3. Figure 19 shows a popular carnival "strong man" device, in which the contestant tries to see how high he can raise a weighted marker by hitting a target with a sledge hammer. What physical quantity does the device measure? Is it the average force, the maximum force, the work done, the impulse, the energy transferred, the momentum transferred, or something else? Discuss your answer.

4. Although the acceleration of a baseball after it has been hit does not depend on who hit it, something about the baseball's flight must depend on the batter. What is it?

5. Many features of cars, such as collapsible steering wheels and padded dashboards, are meant to change more safely the momentum of passengers during accidents. Explain their usefulness, using the impulse concept.

6. It is said that, during a 30-mi/h collision, a 10-lb child can exert a 300-lb force against a parent's grip. How can such a large force come about?

7. The following statement was taken from an exam paper:

Figure 19 Question 3.

"The collision between two helium atoms is perfectly elastic, so that momentum is conserved." What do you think of this statement?

8. You are driving along a highway at 50 mi/h, followed by another car moving at the same speed. You slow to 40 mi/h but the other driver does not and there is a collision. What are the initial velocities of the colliding cars as seen from the reference frame of (a) yourself, (b) the other driver, and (c) a state trooper, who is in a patrol car parked by the roadside? (d) A judge asks whether you bumped into the other driver or the other driver bumped into you. As a physicist, how would you answer?

9. It is obvious from inspection of Eqs. 8 and 9 that a valid solution to the problem of finding the final velocities of two particles in a one-dimensional elastic collision is $v_{1f} = v_{1i}$ and $v_{2f} = v_{2i} = 0$. What does this mean physically?

10. Two identical cubical blocks, moving in the same direction with a common speed v, strike a third such block initially at rest on a horizontal frictionless surface. What is the motion of the blocks after the collision? Does it matter whether or not the two initially moving blocks were in contact or not? Does it matter whether these two blocks were glued together?

11. C. R. Daish has written that, for professional golfers, the initial speed of the ball off the clubhead is about 140 mi/h. He also says: (a) "if the Empire State Building could be swung at the ball at the same speed as the clubhead, the initial ball velocity would only be increased by about 2%" and (b) that, once the golfer has started his downswing, camera clicking, sneezing, and so on can have no effect on the motion of the ball. Can you give qualitative arguments to support these two statements?

12. Two clay balls of equal mass and speed strike each other head-on, stick together, and come to rest. Kinetic energy is certainly not conserved. What happened to it? How is momentum conserved?

13. A football player, momentarily at rest on the field, catches a football as he is tackled by a running player on the other team. This is certainly a collision (inelastic!) and momentum must be conserved. In the reference frame of the football field, there is momentum before the collision but there seems to be none after the collision. Is linear momentum really conserved? If so, explain how. If not, explain why.

14. Consider a one-dimensional elastic collision between a moving object A and an object B initially at rest. How would you choose the mass of B, in comparison to the mass of A, in order that B should recoil with (a) the greatest speed, (b) the greatest momentum, and (c) the greatest kinetic energy?

15. An hourglass is being weighed on a sensitive balance, first when sand is dropping in a steady stream from the upper to the lower part and then again after the upper part is empty. Are the two weights the same or not? Explain your answer.

16. Give a plausible explanation for the breaking of wooden boards or of bricks by a karate punch (see "Karate Strikes," by Jearl D. Walker, *American Journal of Physics,* October 1975).

17. An evacuated box is at rest on a frictionless table. You punch a small hole in one face so that air can enter. (See Fig. 20.) How will the box move? What argument did you use to arrive at your answer?

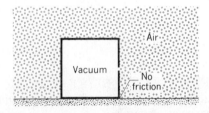

Figure 20 Question 17.

EXERCISES AND PROBLEMS

Section 10–2 Impulse and Momentum

1E. A 95-kg athlete is running at 4.2 m/s. What impulse will stop him?

2E. The momentum of a 1500-kg car increases by 9.0×10^3 kg·m/s in 12 s. (a) What is the magnitude of the force that accelerated the car? (b) By how much did the speed of the car increase?

3E. A cue strikes a pool ball, exerting an average force of 50 N over a time of 10 ms. If the ball has mass 0.20 kg, what speed does it have after impact?

4E. A ball of mass m and speed v strikes a wall perpendicularly and rebounds with undiminished speed. (a) If the time of collision is Δt, what is the average force exerted by the ball on the wall? (b) Evaluate this average force numerically for a rubber ball with mass 140 g moving at 7.8 m/s; the duration of the collision is 3.8 ms.

5E. A stream of water impinges on a stationary "dished" turbine blade, as shown in Fig. 21. The speed of the water is u, both before and after it strikes the curved surface of the blade, and the mass of water striking the blade per unit time is constant at the value μ. Find the force exerted by the water on the blade.

6E. A 150-g (weight ≈ 5.3 oz) baseball pitched at a speed of 40 m/s (≈ 130 ft/s) is hit straight back to the pitcher at a speed of 60 m/s (≈ 200 ft/s). What average force was exerted by the bat if it was in contact with the ball for 5.0 ms?

7E. A force that averages 1000 N is applied to a 0.40-kg steel ball moving at 14 m/s by a collision lasting 27 ms. If the force is

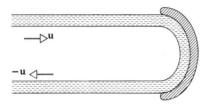

Figure 21   Exercise 5.

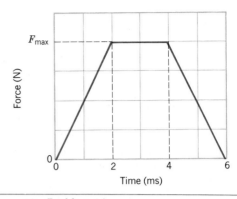

Figure 23   Problem 14.

in a direction opposite to the initial velocity of the ball, find the final speed of the ball.

8E.   A 1.2-kg medicine ball drops vertically onto the floor with a speed of 25 m/s. It rebounds with an initial speed of 10 m/s. (*a*) What impulse acts on the ball during contact? (*b*) If the ball is in contact for 0.020 s, what is the average force exerted on the floor?

9E.   A golfer hits a golf ball, imparting to it an initial velocity of magnitude 50 m/s directed 30° above the horizontal. Assuming that the mass of the ball is 46 g and the club and ball are in contact for 1.0 ms, find (*a*) the impulse imparted to the ball, (*b*) the impulse imparted to the club, (*c*) the average force exerted on the ball by the club, and (*d*) the work done on the ball.

10E.   A 1400-kg car moving at 5.3 m/s is initially traveling north. After completing a 90° right-hand turn in 4.6 s, the inattentive operator drives into a tree, which stops the car in 350 ms. What is the magnitude of the impulse delivered to the car (*a*) during the turn and (*b*) during the collision? What average force acts on the car (*c*) during the turn and (*d*) during the collision?

11E.   Spacecraft *Voyager 2* (mass m and speed v relative to the sun) approaches the planet Jupiter (mass M and speed V relative to the sun) as shown in Fig. 22. The spacecraft rounds the planet and departs in the opposite direction. What is its speed, relative to the sun, after this slingshot encounter? Assume $v = 12$ km/s and $V = 13$ km/s (the orbital speed of Jupiter). The mass of Jupiter is very much greater than the mass of the spacecraft; $M \gg m$. (See "The Slingshot Effect: Explanation and Analogies," by Albert A. Bartlett and Charles W. Hord, *The Physics Teacher,* November 1985.)

12P.   The force on a 10-kg object increases uniformly from

zero to 50 N in 4.0 s. What is the object's final speed if it started from rest?

13P.   A stream of water from a hose is sprayed on a wall. If the speed of the water is 5.0 m/s and the hose sprays 300 cm³/s, what is the average force exerted on the wall by the stream of water? Assume that the water does not spatter back appreciably. Each cubic centimeter of water has a mass of 1.0 g.

14P.   Figure 23 shows an approximate representation of force versus time during the collision of a 58-g tennis ball with a wall. The initial velocity of the ball is 34 m/s perpendicular to the wall; it rebounds with the same speed, also perpendicular to the wall. What is the value of F_{max}, the maximum value of the contact force during the collision?

15P.   A ball having a mass of 150 g strikes a wall with a speed of 5.2 m/s and rebounds with only 50% of its initial kinetic energy. (*a*) What is the final speed of the ball after rebounding? (*b*) What was the impulse delivered by the ball to the wall? (*c*) If the ball was in contact with the wall for 7.6 ms, what was the average force exerted by the wall on the ball during this time interval?

16P.   A 300-g ball with a speed v of 6.0 m/s strikes a wall at an angle θ of 30° and then rebounds with the same speed and angle (Fig. 24). It is in contact with the wall for 10 ms. (*a*) What impulse was experienced by the ball? (*b*) What was the average force exerted by the ball on the wall?

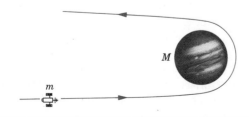

Figure 22   Exercise 11.

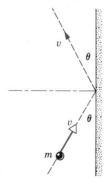

Figure 24   Problem 16.

17P. A 2500-kg unmanned space probe is moving in a straight line at a constant speed of 300 m/s. A rocket engine on the space probe executes a burn in which a thrust of 3000 N acts for 65 s. (a) What is the change in momentum (magnitude only) of the probe if the thrust is backward, forward, or sideways? (b) What is the change in kinetic energy under the same three conditions? Assume that the mass of the ejected fuel is negligible compared to the mass of the space probe.

18P. A force exerts an impulse J on an object of mass m, changing its speed from v to u. The force and the object's motion are along the same straight line. Show that the work done by the force is $\frac{1}{2}J(u + v)$.

19P. Two parts of a spacecraft are separated by detonating the explosive bolts that hold them together. The masses of the parts are 1200 kg and 1800 kg; the magnitude of the impulse delivered to each part is 300 N·s. What is the relative speed of recession of the two parts?

20P. A croquet ball with a mass 0.50 kg is struck by a mallet, receiving the impulse shown in the graph (Fig. 25). What is the ball's velocity just after the force has become zero?

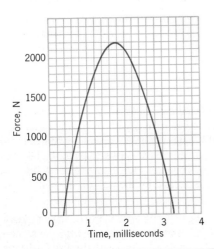

Figure 25   Problem 20.

Section 10-3   Collisions in One Dimension: Elastic

21E. The blocks in Fig. 26 slide without friction. (a) What is the velocity **v** of the 1.6-kg block after the collision? (b) Is the collision elastic?

22E. Refer to Fig. 26. Suppose the initial velocity of the 2.4-kg block is reversed; it is headed directly toward the 1.6-kg block. (a) What would be the velocity **v** of the 1.6-kg block after the collision? (b) Would this collision be elastic?

23E. A hovering fly is approached by an enraged elephant charging at 2.1 m/s. Assuming that the collision is elastic, at

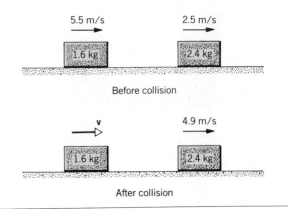

Before collision

After collision

Figure 26   Exercises 21 and 22.

what speed does the fly rebound? Note that the projectile (the elephant) is much more massive than the target (the fly).

24E. An electron collides elastically with a hydrogen atom initially at rest. The initial and final motions are along the same straight line. What fraction of the electron's initial kinetic energy is transferred to the hydrogen atom? The mass of the hydrogen atom is 1840 times the mass of the electron.

25E. An α particle (mass 4 u) experiences an elastic head-on collision with a gold nucleus (mass 197 u) that is originally at rest. What fraction of its original kinetic energy does the α particle lose?

26E. A cart with mass 340 g moving on a frictionless linear air track at an initial speed of 1.2 m/s strikes a second cart of unknown mass at rest. The collision between the carts is elastic. After the collision, the first cart continues in its original direction at 0.66 m/s. (a) What is the mass of the second cart? (b) What is its speed after impact?

27E. The head of a golf club moving at 45 m/s strikes a golf ball (mass = 46 g) resting on a tee. The effective mass of the clubhead (see Sample Problem 6) is 220 g. (a) With what speed does the ball leave the tee? (b) With what speed would it leave the tee if you doubled the mass of the clubhead? If you tripled it? What conclusions can you draw about the use of heavy clubs? Assume that the collisions are perfectly elastic and that the golfer can bring the heavier clubs up to the same speed at impact.

28E. A body of 2.0-kg mass makes an elastic collision with another body at rest and continues to move in the original direction but with one-fourth of its original speed. What is the mass of the struck body?

29P. A steel ball of mass 0.50 kg is fastened to a cord 70 cm long and is released when the cord is horizontal. At the bottom of its path, the ball strikes a 2.5-kg steel block initially at rest on a frictionless surface (Fig. 27). The collision is elastic. Find (a) the speed of the ball and (b) the speed of the block, both just after the collision.

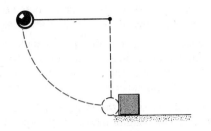

Figure 27 Problem 29.

30P. A platform scale is calibrated to indicate the mass in kilograms of an object placed on it. Particles fall from a height of 3.5 m and collide with the balance pan of the scale. The collisions are elastic; the particles rebound upward with the same speed they had before hitting the pan. If each particle has a mass of 110 g and collisions occur at the rate of 42 s^{-1}, what is the scale reading?

31P. It is well known that bullets and other missiles fired at Superman simply bounce off his chest as in Fig. 28. Suppose that a gangster sprays Superman's chest with 3-g bullets at the rate of 100 bullets/min, the speed of each bullet being 500 m/s. Suppose too that the bullets rebound straight back with no loss in speed. Show that the average force exerted by the stream of bullets on Superman's chest is only 5 N (= 18 oz).

32P. Two titanium spheres approach each other head-on with the same speed and collide elastically. After the collision, one of the spheres, whose mass is 300 g, remains at rest. What is the mass of the other sphere?

Figure 28 Problem 31.

33P. A block of mass m_1 is at rest on a long frictionless table, one end of which is terminated in a wall. Another block of mass m_2 is placed between the first block and the wall and set in motion to the left with constant speed v_{2i}, as in Fig. 29. Assuming that all collisions are completely elastic, find the value of m_2 for which both blocks move with the same velocity after m_2 has collided once with m_1 and once with the wall. Assume the wall to have infinite mass.

Figure 29 Problem 33.

34P. Two 50-lb (mass = 22.7 kg) ice sleds are placed a short distance apart, one directly behind the other, as shown in Fig. 30. An 8-lb (mass = 3.63 kg) cat, standing on one sled, jumps across to the other and immediately back to the first. Both jumps are made at a speed of 10 ft/s (= 3.05 m/s) relative to the ice. Find the final speeds of the two sleds.

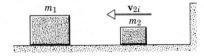

Figure 30 Problem 34.

35P. An elevator is moving up a shaft at 6.0 ft/s. At the instant the elevator is 60 ft from the top, a ball is dropped from the top of the shaft. The ball rebounds elastically from the elevator roof. (*a*) To what height can it rise relative to the top of the shaft? (*b*) Do the same problem assuming the elevator is moving down at 6.0 ft/s.

Section 10–4   Collisions in One Dimension: Inelastic

36E. Meteor Crater in Arizona (see Fig. 1) is thought to have been formed by the impact of a meteor with the earth some 20,000 years ago. The mass of the meteor is estimated to be 5×10^{10} kg and its speed to have been 7200 m/s. What speed would such a meteor impart to the earth in a head-on collision?

37E. A 6.0-kg box sled is coasting across the ice at a speed of 9.0 m/s when a 12-kg package is dropped into it from above. What is the new speed of the sled?

38E. A 5.2-g bullet moving at 672 m/s strikes a 700-g wooden block at rest on a very smooth surface. The bullet emerges with its speed reduced to 428 m/s. Find the resulting speed of the block.

39E. A bullet of mass 10 g strikes a ballistic pendulum of mass 2.0 kg. The center of mass of the pendulum rises a vertical distance of 12 cm. Assuming that the bullet remains embedded in the pendulum, calculate its initial speed.

40E. A 5.0-kg particle with a velocity of 3.0 m/s collides with a 10-kg particle that has a velocity of 2.0 m/s in the same direction. After the collision, the 10-kg particle is observed to be traveling in the original direction with a speed of 4.0 m/s. (a) What is the velocity of the 5.0-kg particle immediately after the collision? (b) By how much does the total kinetic energy of the system of two particles change because of the collision?

41E. A bullet of mass 4.5 g is fired horizontally into a 2.4-kg wooden block at rest on a horizontal surface. The coefficient of kinetic friction between block and surface is 0.20. The bullet comes to rest in the block, which moves 1.8 m. (a) What is the speed of the block immediately after the bullet comes to rest within it? (b) What is the speed of the bullet?

42P. Two cars A and B slide on an icy road as they attempt to stop at a traffic light. The mass of A is 1100 kg and the mass of B is 1400 kg. The coefficient of kinetic friction between the locked wheels of both cars and the road is 0.13. Car A succeeds in coming to rest at the light, but car B cannot stop and rear-ends car A. After the collision, A comes to rest 8.2 m ahead of the impact point and B 6.1 m ahead: see Fig. 31. Both drivers had their brakes locked throughout the incident. (a) From the distances each car moved after the collision, find the speed of each car immediately after impact. (b) Use conservation of momentum to find the speed at which car B struck car A. On what grounds can the use of momentum conservation be criticized here?

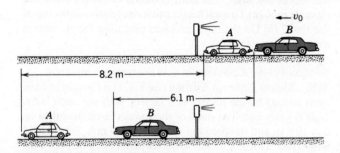

Figure 31 Problem 42.

43P. A 3.0-ton weight falling through a distance of 6.0 ft drives a 0.50-ton pile 1 in. into the ground. Assuming that the weight–pile collision is completely inelastic, find the average force of resistance exerted by the ground.

44P. Two particles, one having twice the mass of the other, are held together with a compressed spring between them. The energy stored in the spring is 60 J. How much kinetic energy does each particle have after they are released? Assume that all the stored energy is transferred to the particles and that neither particle is attached to the spring after they are released.

45P. An object with mass m and speed v explodes into two pieces, one three times as massive as the other; the explosion takes place in gravity-free space. The less-massive piece comes to rest. How much kinetic energy was added to the system?

46P. A box is put on a scale that is adjusted to read zero when the box is empty. A stream of marbles is then poured into the box from a height h above its bottom at a rate of R (marbles per second). Each marble has a mass m. If the collisions between the marbles and the box are completely inelastic, find the scale reading at time t after the marbles begin to fill the box. Determine a numerical answer when $R = 100 \text{ s}^{-1}$, $h = 7.6$ m, $m = 4.5$ g, and $t = 10$ s.

47P. A 35-ton railroad freight car collides with a stationary caboose car. They couple together and 27% of the initial kinetic energy is dissipated as heat, sound, vibrations, and so on. Find the weight of the caboose.

48P. A ball of mass m is projected with speed v_i into the barrel of a spring gun of mass M initially at rest on a frictionless surface; see Fig. 32. The ball sticks in the barrel at the point of maximum compression of the spring. No energy is lost in friction. (a) What is the speed of the spring gun after the ball comes to rest in the barrel? (b) What fraction of the initial kinetic energy of the ball is stored in the spring?

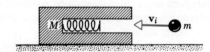

Figure 32 Problem 48.

49P. A block of mass $m_1 = 2.0$ kg slides along a frictionless table with a speed of 10 m/s. Directly in front of it, and moving in the same direction, is a block of mass $m_2 = 5.0$ kg moving at 3.0 m/s. A massless spring with a spring constant $k = 1120$ N/m is attached to the backside of m_2, as shown in Fig. 33. When the blocks collide, what is the maximum compression of the spring? (Hint: At the moment of maximum compression, the two blocks move as one; find the velocity by noting that the collision is completely inelastic to this point.)

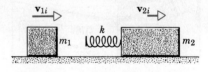

Figure 33 Problem 49.

50P. The bumper of a 1200-kg car is designed so that it can just absorb all the energy when the car runs head-on into a solid stone wall at 5.0 km/h. The car is involved in a collision in which it runs at 70 km/h into the rear of a 900-kg car ahead moving at 60 km/h in the same direction. The 900-kg car is accelerated to 70 km/h as a result of the collision. (a) What is the speed of the 1200-kg car immediately after impact? (b) What is the ratio of the kinetic energy absorbed in the collision to that which can be absorbed by the bumper of the 1200-kg car?

51P. An electron, mass m, collides head-on with an atom, mass M, initially at rest. As a result of the collision, a characteristic amount of energy E is stored internally in the atom. What is the minimum initial speed v_0 that the electron must have? (*Hint:* Conservation principles lead to a quadratic equation for the final elctron speed v and a quadratic equation for the final atom speed V. The minimum value v_0 follows from the requirement that the radical in the solutions for v and V be real.)

52P. A railroad freight car weighing 32 tons and traveling at 5.0 ft/s overtakes one weighing 24 tons and traveling at 3.0 ft/s in the same direction. (a) Find the speeds of the cars after collision and the loss of kinetic energy during collision if the cars couple together. (b) If instead, as is very unlikely, the collision is elastic, find the speeds of the cars after collision.

Section 10-5 Collisions in Two Dimensions

53E. An α particle collides with an oxygen nucleus, initially at rest. The α particle is scattered at an angle of 64° above its initial direction of motion and the oxygen nucleus recoils at an angle of 51° below this initial direction. The final speed of the nucleus is 1.2×10^5 m/s. What is the final speed of the α particle? (The mass of an α particle is 4.0 u and the mass of an oxygen nucleus is 16 u.)

54E. A proton (atomic mass 1 u) with a speed of 500 m/s collides elastically with another proton at rest. The original proton is scattered 60° from its initial direction. (a) What is the direction of the velocity of the target proton after the collision? (b) What are the speeds of the two protons after the collision?

55E. A certain nucleus, at rest, spontaneously disintegrates into three particles. Two of them are detected; their masses and velocities are as shown in Fig. 34. (a) What is the momentum of the third particle, which is known to have a mass of 11.7×10^{-27} kg? (b) How much kinetic energy appears in the disintegration process?

56E. In a game of pool, the cue ball strikes another ball initially at rest. After the collision, the cue ball moves at 3.50 m/s along a line making an angle of 65.0° with its original direction of motion. The second ball acquires a speed of 6.75 m/s. Using momentum conservation, find (a) the angle between the direction of motion of the second ball and the original direction of motion of the cue ball and (b) the original speed of the cue ball.

57E. Two vehicles A and B are traveling west and south,

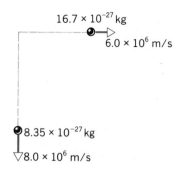

Figure 34 Exercise 55.

respectively, toward the same intersection where they collide and lock together. Before the collision, A (total weight 2700 lb) is moving with a speed of 40 mi/h and B (total weight 3600 lb) has a speed of 60 mi/h. Find the magnitude and direction of the velocity of the (interlocked) vehicles immediately after the collision.

58E. In a game of billiards, the cue ball is given an initial speed V and strikes the pack of 15 stationary balls. All 16 balls then engage in numerous ball–ball and ball–cushion collisions. Some time later, it is observed that (by some accident) all 16 balls have the same speed v. Assuming that all collisions are elastic and ignoring the rotational aspect of the balls' motion, calculate v in terms of V.

59P. A 20-kg body is moving in the direction of the positive x axis with a speed of 200 m/s when, owing to an internal explosion, it breaks into three parts. One part, whose mass is 10 kg, moves away from the point of explosion with a speed of 100 m/s along the positive y axis. A second fragment, with a mass of 4 kg, moves along the negative x axis with a speed of 500 m/s. (a) What is the velocity of the third (6-kg) fragment? (b) How much energy was released in the explosion? Ignore effects due to gravity.

60P. Two balls A and B, having different but unknown masses, collide. A is initially at rest and B has a speed v. After collision, B has a speed $v/2$ and moves at right angles to its original motion. (a) Find the direction in which ball A moves after collision. (b) Can you determine the speed of A from the information given? Explain.

61P. Show that a slow neutron (called a *thermal* neutron) that is scattered through 90° in an elastic collision with a deuteron, that is initially at rest, loses two-thirds of its initial kinetic energy to the deuteron. (The mass of a neutron is 1.0 u; the mass of a deuteron is 2.0 u.)

62P. After a totally inelastic collision, two objects of the same mass and initial speed are found to move away together at half their initial speed. Find the angle between the initial velocities of the objects.

63P. Two pendulums each of length l are initially situated as

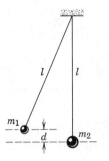

Figure 35 Problem 63.

in Fig. 35. The first pendulum is released and strikes the second. Assume that the collision is completely inelastic and neglect the mass of the strings and any frictional effects. How high does the center of mass rise after the collision?

64P. A billiard ball moving at a speed of 2.2 m/s strikes an identical stationary ball a glancing blow. After the collision, one ball is found to be moving at a speed of 1.1 m/s in a direction making a 60° angle with the original line of motion. (*a*) Find the velocity of the other ball. (*b*) Can the collision be inelastic, given these data?

65P. A ball with an initial speed of 10 m/s collides elastically with two identical balls whose centers are on a line perpendicular to the initial velocity and that are initially in contact with each other (Fig. 36). The first ball is aimed directly at the contact point and all the balls are frictionless. Find the velocities of all three balls after the collision. (*Hint:* With friction absent, each impulse is directed along the line of centers of the balls, normal to the colliding surfaces.)

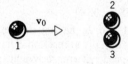

Figure 36 Problem 65.

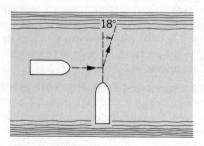

Figure 37 Problem 66.

66P. A barge with mass 1.50×10^5 kg is proceeding downriver at 6.2 m/s in heavy fog when it collides broadside with a barge heading directly across the river; see Fig. 37. The second barge has mass 2.78×10^5 kg and was moving at 4.3 m/s. Immediately after impact, the second barge finds its course deflected by 18° in the downriver direction and its speed increased to 5.1 m/s. The river current was practically zero at the time of the accident. (*a*) What is the speed and direction of motion of the first barge immediately after the collision? (*b*) How much kinetic energy was lost in the collision?

Section 10–6 Reactions and Decay Processes
67E. The precise masses in the reaction

$$p + {}^{19}F \rightarrow \alpha + {}^{16}O$$

have been determined by mass spectrometer measurements and are

$$m_p = 1.007825 \text{ u}, \qquad m_\alpha = 4.002603 \text{ u},$$
$$m_F = 18.998405 \text{ u}, \qquad m_O = 15.994915 \text{ u}.$$

Calculate the Q of the reaction from these data.

68E. A particle called Σ^- (sigma minus), at rest in a certain reference frame, decays spontaneously into two other particles according to

$$\Sigma^- \rightarrow \pi^- + n.$$

The masses are

$$m_\Sigma = 2340.5 m_e,$$
$$m_\pi = 273.2 m_e,$$
$$m_n = 1838.65 m_e,$$

where m_e ($= 9.11 \times 10^{-31}$ kg) is the electron mass. (*a*) How much kinetic energy is generated in this process? (*b*) Which of the decay products (π^- and n) gets the larger share of this kinetic energy? Of the momentum?

69P. An α particle with kinetic energy 7.70 MeV strikes a ^{14}N nucleus at rest. An ^{17}O nucleus and a proton are produced, the proton emitted at 90° to the direction of the incident α particle carrying kinetic energy 4.44 MeV. The masses of the various particles are: α particle, 4.00260 u; ^{14}N, 14.00307 u; proton, 1.007825 u; and ^{17}O, 16.99914 u. (*a*) What is the Q of the reaction? (*b*) What is the kinetic energy of the oxygen nucleus?

70P. Consider the α decay of radium (Ra) to radon (Rn), according to the reaction

$$^{226}Ra \rightarrow \alpha + {}^{222}Rn.$$

The masses of the various nuclei are: ^{226}Ra, 226.0254 u; α, 4.0026 u; ^{222}Rn, 222.0175 u. (*a*) Calculate the Q for the reaction. (*b*) What value of Q would be obtained if the accurate masses given above were rounded off to three significant figures? (*b*) How much kinetic energy is given to the α particle and to the radon nucleus? (For this calculation the rounded-off values of the masses *can* be used; why?)

CHAPTER 11
ROTATION

Around 1640, Edward Somerset, Marquis of Worcester, built a 14-ft diameter perpetual motion machine, in which 40 rolling 50-lb cannon balls were meant to drive the wheel in a steady clockwise rotation. It did not work. (One reason inventors of perpetual motion machines have such rotten luck is that it is not possible to construct a perpetual motion machine.)

11-1 A Skater's Life

A figure skater can illustrate, in an aesthetically pleasing way, two kinds of pure motion. Figure 1*a* shows the skater gliding across the ice in a straight line with constant speed, holding a fixed position. Her motion is one of pure *translation*. Figure 1*b* shows her spinning at a constant rate about a fixed vertical axis, a motion of pure *rotation*. It is this second kind of motion that we deal with in this chapter.

Rotation is the motion of wheels, of gears, of motors, of the hands of clocks, of the rotors of jet engines and of helicopter blades. It is the motion of spinning electrons and atoms, of hurricanes, and of spinning planets, stars, and galaxies. It is the motion of acrobats, of high divers, and of orbiting astronauts. Rotary motion is all around us.

11-2 The Rotational Variables

In this chapter, we deal with the rotation of a *rigid* body about a *fixed* axis. The first of these restrictions means that we shall not examine the rotation of the sun because the sun — a ball of gas — is not a rigid body. Its period of rotation varies from 26 days at its equator to 37 days at its poles. Our second restriction rules out a bowling ball rolling down a bowling lane. Such a ball is in *rolling* motion, rotating about a *moving* axis.

(a)   *(b)*

Figure 1   A skater (Debi Thomas) in motion of (*a*) pure translation and (*b*) pure rotation. In the first case, the motion is along a fixed direction. In the second case, it is about a fixed axis.

Figure 2 shows a rigid body of arbitrary shape in pure *rotation around a fixed axis.* Every point of the body moves in a circle whose center lies on the axis of rotation and every point experiences the same *angular* displacement during a particular time interval. We contrast this with a body moving in pure *translation in a fixed direction.* In pure translation, every point of the body moves in a straight line and every point experiences the same *linear* displacement during a particular time interval. Drawing on the analogy between linear and angular motion will be a constant theme in what follows.

We deal now—one at a time—with the angular equivalents of the linear quantities position (or displacement), velocity, and acceleration.

Angular Position.   Figure 2 shows a reference line, fixed in the body at right angles to the rotation axis and rotating with the body. We can describe the motion of the rotating body by specifying the angular position of this line.

If *s* is an arc length in Fig. 3 and *r* is the radius of that

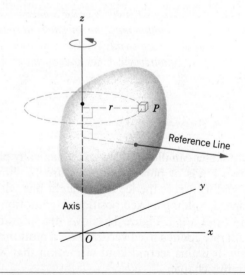

Figure 2   A rigid body of arbitrary shape in pure rotation about the *z* axis of a coordinate system. The *reference line,* whose position with respect to the rigid body is arbitrary, is fixed in the body—at right angles to the rotation axis—and rotates with the body.

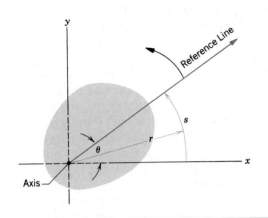

Figure 3   The rotating rigid body of Fig. 2 is shown in cross section. The reference line makes an angle θ with the x axis.

arc, then θ, the angular position of the reference line, is given by

$$\theta = \frac{s}{r} \quad \text{(radian measure)}. \tag{1}$$

An angle defined in this way is given in *radians* rather than in degrees or revolutions. The radian, being the ratio of two lengths, is a pure number and thus has no dimensions. Because the circumference of a circle of radius r is $2\pi r$, there are 2π radians in a complete circle, so that

$$1 \text{ revolution} = 360° = 2\pi \text{ radians}, \tag{2}$$

or

$$1 \text{ radian} = 57.3° = 0.159 \text{ revolution}. \tag{3}$$

In pure translation, we know all that there is to know about a moving particle if we know $x(t)$, its position as a function of time. In pure rotation then, we want to know $\theta(t)$, the angular position of the rotating body as a function of time.

Angular Velocity.   Suppose (see Fig. 4) that our rotating body is at angular position θ_1 at time t_1 and at angular position θ_2 at time t_2. We define the *average angular velocity* of the body to be

$$\overline{\omega} = \frac{\theta_2 - \theta_1}{t_2 - t_1} = \frac{\Delta\theta}{\Delta t}, \tag{4}$$

in which $\Delta\theta$ is the angular displacement that occurs during the time interval Δt. The *instantaneous angular velocity* ω (Greek *omega*), with which we shall be most

concerned, is the limit of this ratio as Δt is made to approach zero. Thus,

$$\omega = \lim_{\Delta t \to 0} \frac{\Delta\theta}{\Delta t} = \frac{d\theta}{dt}. \tag{5}$$

Thus, if we know $\theta(t)$, we can find the angular velocity ω by differentiation. Angular velocity, defined in this way, holds not only for the rotating rigid body as a whole but also for every particle of that body. The units of angular velocity are commonly rad/s or rev/s.

If a particle moves in translation along an x axis, its linear velocity v can be either positive or negative, depending on whether the particle is moving in the direction of increasing or of decreasing x. In the same way, the angular velocity ω of a rotating rigid body can be either positive or negative, depending on whether the body is rotating in the direction of increasing θ (as in Fig. 3) or of decreasing θ.

Angular Acceleration.   If the angular velocity of a rotating body is not constant, then the body has an angular acceleration. Let ω_2 and ω_1 be the angular velocities at times t_2 and t_1, respectively. The *average angular acceleration* of the rotating body is defined from

$$\overline{\alpha} = \frac{\omega_2 - \omega_1}{t_2 - t_1} = \frac{\Delta\omega}{\Delta t}, \tag{6}$$

in which $\Delta\omega$ is the change in the angular velocity that occurs during the time interval Δt. The *instantaneous angular acceleration* α, with which we shall be most concerned, is the limit of this quantity as Δt is made to

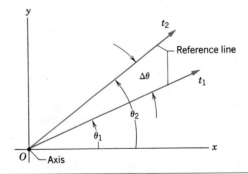

Figure 4   The reference line of the rigid body of Fig. 3 is at angular position θ_1 at time t_1 and at angular position θ_2 at a later time t_2. The quantity $\Delta\theta$ ($= \theta_2 - \theta_1$) is the angular displacement that occurs during the interval Δt ($= t_2 - t_1$). The body itself is not shown.

approach zero. Thus,

$$\alpha = \lim_{\Delta t \to 0} \frac{\Delta \omega}{\Delta t} = \frac{d\omega}{dt}. \tag{7}$$

Angular acceleration, defined in this way, holds not only for the rotating rigid body as a whole but also for every particle of that body. A positive value for α tells us that ω is increasing with time; a negative value means that ω is decreasing with time. The units of angular acceleration are commonly rad/s² or rev/s².

11–3 Angular Quantities as Vectors: An Aside

We can describe the position, velocity, and acceleration of a single particle by means of vectors. If the particle is confined to a straight line, however, we do not need the power of the vector method. Such a particle has only two directions available to it and we can designate these by plus and minus signs.

In the same way, a rigid body rotating about a fixed axis can rotate only clockwise or counterclockwise about this axis, and again we can select between them by means of a plus or a minus sign. The question arises: "If we generalize and consider rotation about a fixed point rather than a fixed axis, can we treat the angular displacement, velocity, and acceleration of a rotating body as vectors?" The answer (with a caution that we clarify below) is: "Yes."

Consider the angular velocity. Figure 5a shows a phonograph record rotating about a fixed spindle. The record has a fixed rotation rate ω (= 33 rev/min) and a fixed direction of rotation (clockwise as viewed from above). We can represent its angular velocity as a vector pointing along the axis of rotation, as in Fig. 5b. We can choose the length of this vector according to some convenient scale, for example, 10 rev/min equals 1 cm.

We can establish a direction for the vector ω by using a right-hand rule, as Fig. 5c shows. Curl your right hand about the rotating disk, your fingers pointing in the direction of rotation. Your extended thumb will then point in the direction of the angular velocity vector. If the disk were to rotate in the opposite sense, the right-hand rule would tell you that the angular velocity vector points in the opposite direction.

It is not easy to get used to representing angular quantities as vectors. We instinctively expect that something should be moving *along* the direction of a vector. That is not the case here. Instead, something (the rigid body) is rotating *around* the direction of the vector. In the world of pure rotation, a vector defines an axis of rotation, not a direction in which something moves.

Now for the caution that we referred to above. Angular *displacements* (unless they are very small) *cannot* be treated as vectors. Why not? We can certainly give them both magnitude and direction, just as we did for the angular velocity vector in Fig. 5. However, that is (as the mathematicians say) necessary but not sufficient. To be a vector, a quantity must *also* obey the rules of vector addition, one of which says that, if you add two vectors, the order in which you add them does not matter. Angular displacements fail this test.

To see this, place a book flat on the floor, as in Fig. 6a. Now give the book two successive 90° rotations, *first* about the (horizontal) x axis and *then* about the (vertical) y axis, using the right-hand rule as a guide in each case.

Now, with a book in the same initial position (Fig. 6b), carry out these two rotations in the reverse order

Figure 5 *(a)* An LP record is shown rotating about a vertical axis that coincides with the axis of the spindle. *(b)* The angular velocity of the rotating record can be represented by the vector ω, lying along the axis and pointing down, as shown. *(c)* We establish the direction of the angular velocity vector as downward by using a right-hand rule. If the fingers of the right hand curl around the way the record is moving, the extended thumb points in the direction of ω.

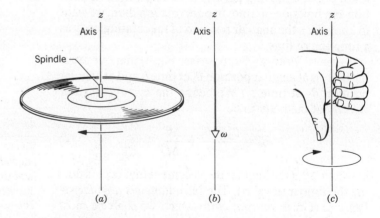

(a)   *(b)*   *(c)*

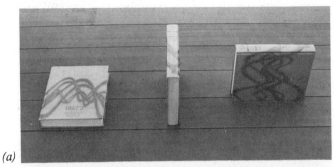

(a)

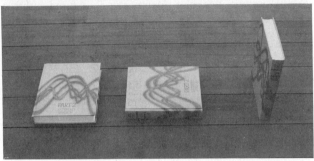

(b)

Figure 6   (a) From its initial position on the left, the book is given two successive 90° rotations, first about the (horizontal) x axis and then about the (vertical) y axis. (b) The book is given the same rotations, but in the reverse order. If angular displacement were truly a vector, the order of these displacements would not matter. It clearly does matter, so that (large) angular displacements are not vectors, even though we can assign magnitude and direction to them.

(that is, *first* about the y axis and *then* about the x axis). As the figure shows, the book ends up in a very different orientation.

Thus, the same two operations have a different result, depending on the order in which you carry them

out. A *quantity that behaves this way cannot be a vector.* With practice, you should be able to show that the final positions of the book are much closer together if you use displacements much smaller than 90°. In the limiting case of differential angular displacements (such as $d\theta$ in Eq. 5), angular displacements *can* be treated as vectors.

11-4   Rotation with Constant Angular Acceleration

In pure translation, motion with a *constant linear acceleration* (for example, a falling body) is an important special case. In Table 2 of Chapter 2, we displayed a series of equations that hold for such motions.

In pure rotation, the case of *constant angular acceleration* is also important and a parallel set of equations holds for this case also. We choose not to derive them here but simply to write them down from the corresponding linear equations, substituting equivalent angular quantities for the linear ones. Table 1 displays both sets of equations. For simplicity, we have chosen both $x_0 = 0$ and $\theta_0 = 0$ in these equations.

Sample Problem 1   Imagine a quarter glued to a table top, with a second quarter touching it, as in Fig. 7. Roll the second quarter around the first, without slippage, until the second quarter returns to its original position. Through what angle θ will the second quarter have turned?

The answer is through 720°, or two revolutions. Try it and see. The second coin will have made one revolution around an axis through the center of the first coin and one about a parallel axis through its own center.

Two angular displacements of the same body about the

Table 1   Equations of Motion for Constant Linear and for Constant Angular Acceleration

Number[a]	Formula	Missing Variables		Formula	Number[b]
10	$v = v_0 + at$	x	θ	$\omega = \omega_0 + \alpha t$	(8)
14	$x = v_0 t + \frac{1}{2}at^2$	v	ω	$\theta = \omega_0 t + \frac{1}{2}\alpha t^2$	(9)
15	$v^2 = v_0^2 + 2ax$	t	t	$\omega^2 = \omega_0^2 + 2\alpha\theta$	(10)
16	$x = \frac{1}{2}(v_0 + v)t$	a	α	$\theta = \frac{1}{2}(\omega_0 + \omega)t$	(11)
17	$x = vt - \frac{1}{2}at^2$	v_0	ω_0	$\theta = \omega t - \frac{1}{2}\alpha t^2$	(12)

[a] Refers to Chapter 2.
[b] Refers to this chapter.

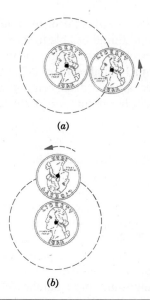

(a)

(b)

Figure 7 Sample Problem 1. (*a*) Roll the right-hand quarter completely around the central one. How many revolutions does it make? (*b*) An intermediate position.

same axis add algebraically. (Think of a merry-go-round at the north pole of the rotating earth.) As this problem demonstrates, the same rule applies if the displacements occur about two different—but parallel—axes. This problem once appeared (with a wrong answer!) as an SAT sample question.

Sample Problem 2 The wheel of a grindstone, as in Fig. 8, has a constant angular acceleration $\alpha = 0.35$ rad/s^2. It starts from rest (that is, $\omega_0 = 0$) with an arbitrary reference line horizontal. (a) What is the angular displacement θ of the reference line (and hence of the wheel) during the interval from $t = 0$ to $t = 18$ s?

We use Eq. 9 of Table 1, or

$$\theta = \omega_0 t + \tfrac{1}{2}\alpha t^2$$
$$= (0)(18 \text{ s}) + (\tfrac{1}{2})(0.35 \text{ rad/s}^2)(18 \text{ s})^2$$
$$= 57 \text{ rad} \approx 3200° = 9.0 \text{ rev.} \qquad \text{(Answer)}$$

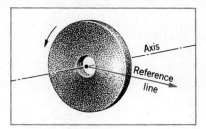

Figure 8 Sample Problems 2 and 3. A grindstone wheel. At $t = 0$ the reference line (which we imagine to be marked on the stone) is horizontal.

(b) What is the angular speed of the wheel at $t = 18$ s? We now use Eq. 8 of Table 1, or

$$\omega = \omega_0 + \alpha t$$
$$= (0) + (0.35 \text{ rad/s}^2)(18 \text{ s})$$
$$= 6.3 \text{ rad/s} = 360°/\text{s} = 1.0 \text{ rev/s.} \qquad \text{(Answer)}$$

Sample Problem 3 For the grindstone of Sample Problem 2, let us assume the same angular acceleration α ($=0.35$ rad/s^2) but let us now assume that the wheel does not start from rest but has an initial angular velocity ω_0 of -4.6 rad/s; that is, the angular acceleration acts initially to slow down the wheel. (a) At what time t will the grindstone come momentarily to rest?

Solving Eq. 8 of Table 1 ($\omega = \omega_0 + \alpha t$) for t yields

$$t = \frac{\omega - \omega_0}{\alpha} = \frac{0 - (-4.6 \text{ rad/s})}{0.35 \text{ rad/s}^2} = 13 \text{ s.} \qquad \text{(Answer)}$$

(b) At what time will the grindstone have turned (in the positive direction of rotation) through five revolutions from its initial angular position?

The wheel, which is initially rotating in the negative direction, must first come momentarily to rest, must then return to its original angular position, and then make five additional revolutions. From Eq. 9 we have

$$\theta = \omega_0 t + \tfrac{1}{2}\alpha t^2.$$

Substitution yields

$$(5 \times 2\pi) \text{ rad} = (-4.6 \text{ rad/s})t + (\tfrac{1}{2})(0.35 \text{ rad/s}^2)t^2.$$

We note that, for t in seconds, the units in this equation form a consistent set. Dropping units (for convenience) and rearranging gives

$$t^2 - 26.3t - 180 = 0. \qquad (13)$$

Solving this quadratic equation for t and discarding the negative root, we obtain

$$t = 32 \text{ s.} \qquad \text{(Answer)}$$

Hint 1: Unexpected answers Do not hasten to throw away one root of a quadratic equation as meaningless. Often, as in Sample Problem 3, a discarded root has physical meaning.

The formal solutions to Eq. 13 in that Sample Problem are $t = 32$ s and $t = -5.6$ s. We chose the first (positive) solution and ignored the second as perhaps meaningless. But is it? A negative time in this problem simply means a time before $t = 0$, that is, a time before you started to pay attention to what was going on.

Figure 9 is a plot of the angular position θ of the grind-

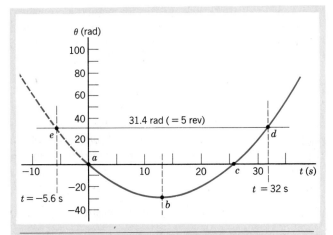

Figure 9   A plot of angular position versus time for the grindstone of Sample Problem 3. Negative times (that is, times before $t = 0$) have been included. The two roots of Eq. 13 are indicated by points d and e.

stone of Sample Problem 3 as a function of time, for both positive and negative times. It is a plot of Eq. 9 of Table 1 ($\theta = \omega_0 t + \frac{1}{2}\alpha t^2$), using as constants $\omega_0 = -4.6$ rad/s and $\alpha = +0.35$ rad/s². Point a corresponds to $t = 0$, at which we arbitrarily took the angular position of the wheel to be zero. The wheel continues to move in the direction of decreasing θ, coming to rest at point $b(t = 13$ s). It then reverses itself, returning to its $\theta = 0$ position at point c and going on for five additional revolutions ($= 31.4$ rad) to point d. This latter point ($t = 32$ s) is the root that we accepted as the answer to our problem.

Note, however, that the wheel occupied this same angular position at $t = -5.6$ s, before the official start of the problem. This root (point e) is just as valid as the root at point d. More important, by asking ourselves if this negative root can have any physical meaning, we have gained new insight into the problem.

The physicist Eugene Wigner once wrote a paper entitled "The Unexpected Usefulness of Mathematics." Our present problem just touches on the deeper issues that he had in mind but his message is valid. An unexpected negative root once led the mathematical physicist P. A. M. Dirac to predict the existence of what turned out to be the positive electron and laid the foundation for the present view that every elementary particle has a symmetrically disposed antiparticle. Always examine, for possible physical meaning, unexpected solutions that the mathematics lays at your feet.

Sample Problem 4   After leaving a helicopter, you notice that the rotor's motion changed from 320 rev/min to 225 rev/

min in 1.5 min. (a) What is the average angular acceleration of the rotor blades during this interval?

From Eq. 6,

$$\bar{\alpha} = \frac{\omega - \omega_0}{t - t_0} = \frac{225 \text{ rev/min} - 320 \text{ rev/min}}{1.5 \text{ min}}$$

$$= -63.3 \text{ rev/min}^2. \qquad \text{(Answer)}$$

The minus sign reminds us that the acceleration is acting to slow down the rotor blades.

(b) If this same acceleration holds, how long will it take for the rotor blades to coast to rest from their initial angular velocity of 320 rev/min?

Solving Eq. 8 of Table 1 ($\omega = \omega_0 + \alpha t$) for t gives

$$t = \frac{\omega - \omega_0}{\alpha} = \frac{0 - 320 \text{ rev/min}}{-63.3 \text{ rev/min}^2} = 5.0 \text{ min}. \quad \text{(Answer)}$$

(c) How many revolutions will the rotor blades make in coming to rest from their initial angular velocity of 320 rev/min?

Solving Eq. 10 of Table 1 ($\omega^2 = \omega_0^2 + 2\alpha\theta$) for θ gives

$$\theta = \frac{\omega^2 - \omega_0^2}{2\alpha} = \frac{0 - (320 \text{ rev/min})^2}{(2)(-63.3 \text{ rev/min}^2)} = 809 \text{ rev.} \quad \text{(Answer)}$$

11–5   The Linear and the Angular Variables

Children on a merry-go-round have different *linear* speeds (in m/s, say) depending on how far from the axis of rotation they happen to be. However, they all have the same *angular* speed (in rev/min, say) no matter where they are located.

We often want to know the relations between the linear variables s, v, and a for a particular point on a rotating body and the angular variables θ, ω, and α for that body. The two sets of variables are connected by r, the perpendicular distance between the point in question and the axis of rotation.

The Position.   From Eq. 1 (see also Fig. 3) we have

$$\boxed{s = \theta r \quad \text{(radian measure)}} \qquad (14)$$

as the relation between the linear position s of a point on a rotating body and the angular position θ of that body. The angle θ must be measured in radians because the equation is itself the definition of angular measure in radians.

The Velocity.   Differentiating Eq. 14 with respect to time — with r held constant — leads to

$$\frac{ds}{dt} = \frac{d\theta}{dt} r.$$

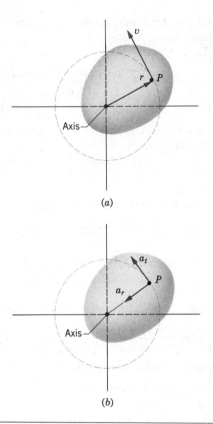

(a)

(b)

Figure 10 The rotating rigid body of Fig. 2 is shown in cross section. Every point of the body (such as P) moves in a circle around the rotation axis. (a) The linear velocity of every point is tangent to the circle in which the point moves. (b) The linear acceleration of the point has (in general) two components. The tangential component a_t is proportional to the angular acceleration of the body (Eq. 17). The radial component a_r is proportional to the square of the angular velocity of the body (Eq. 18).

But ds/dt is the linear speed of the point in question and $d\theta/dt$ is the angular speed of the rotating body, so that

$$v = \omega r \quad \text{(radian measure).} \qquad (15)$$

Again, the angular velocity ω must be expressed in radian measure. Figure 10a shows that the linear velocity is always tangent to the circle described by the point in question as the body rotates.

The Acceleration. Differentiating Eq. 15 with respect to time—again with r held constant—leads to

$$\frac{dv}{dt} = \frac{d\omega}{dt}\, r. \qquad (16)$$

Here, as Fig. 10b shows, we run up against a complication. In Eq. 16, dv/dt is the *tangential component* only of the linear acceleration, representing the changing speed of the particle. We can thus rewrite Eq. 16 as

$$a_t = \alpha r \quad \text{(radian measure).} \qquad (17)$$

However, as Eq. 16 of Chapter 4 tells us, a particle moving in a circular path also has a *radial component* of the linear acceleration, given by v^2/r. By using Eq. 15 above, we can write this as

$$a_r = \frac{v^2}{r} = \omega^2 r \quad \text{(radian measure).} \qquad (18)$$

As Fig. 10b shows, the linear acceleration of a point on a rotating rigid body has, in general, two components. The radial component a_r (given by Eq. 18) is always present as long as the angular velocity of the body is not zero. The tangential component a_t (given by Eq. 17) is present as long as the angular acceleration is not zero.

Figure 11 shows an interesting example of the relation between linear and angular variables. When a tall chimney is toppled by an explosive charge at its base, it will often break as it falls, the rupture starting on the downward side of the falling chimney.

Figure 11 The toppling of a 49-m chimney in Louisville, Kentucky. Note that the top part of the chimney will reach the ground after the bottom part does. See "More on the Falling Chimney," by A. A. Bartlett, *The Physics Teacher,* September 1976, for an account of this phenomenon.

Before rupture, the chimney is a rigid body, rotating about an axis near its base with a certain angular acceleration α. From Eq. 17, the top of the chimney has a tangential acceleration a_t given by αL, where L is the length of the chimney. The vertical component of a_t can easily exceed g, the acceleration of free fall. That is, the top of the chimney is falling downward with a vertical acceleration greater than that of a freely falling brick.

This can only happen as long as the chimney remains a single rigid body. Put simply, the bottom part of the chimney, acting through the mortar that holds the bricks together, must "pull down" on the top part of the chimney to cause it to fall so fast. This shearing force is often more than the mortar can tolerate and the chimney breaks. The chimney has now become *two* rigid bodies, its top part being in free fall and reaching the ground later than it would if the chimney had not broken.

Sample Problem 5 Figure 12 shows the centrifuge used at the Manned Spacecraft Center in Houston to subject astronaut trainees to high accelerations. The effective radius R of the device is 15 m. (a) At what constant angular velocity must the centrifuge rotate if the astronaut is to be subject to a linear acceleration that is 11 times that of gravity? This is about the

maximum acceleration that a highly trained fighter pilot can tolerate—for a short time—without blacking out.

Because the angular velocity is constant, the angular acceleration α ($= d\omega/dt$) is zero and so (see Eq. 17) is the tangential component of the linear acceleration. This leaves only the radial component. From Eq. 18 ($a_r = \omega^2 r$) we have

$$\omega = \sqrt{\frac{a_r}{R}} = \sqrt{\frac{(11)(9.8 \text{ m/s}^2)}{15 \text{ m}}}$$
$$= 2.68 \text{ rad/s} \approx 26 \text{ rev/min.} \qquad \text{(Answer)}$$

(b) What is the linear speed of the astronaut under these conditions?

From Eq. 15,

$$v = \omega R = (2.68 \text{ rad/s})(15 \text{ m}) = 40 \text{ m/s } (= 90 \text{ mi/h.})$$
$$\text{(Answer)}$$

The linear speed of the astronaut is of no concern. It is the acceleration that counts. If you dream of falling from a tall building, it is not the high linear speed before impact that you worry about. It is the large acceleration that occurs when you hit the sidewalk!

Hint 2: Units for angular variables. In writing down Eq. 1 ($\theta = s/r$), we committed ourselves to use radian measure for all angular variables. That is, we committed our-

Figure 12 Sample Problem 5. A centrifuge used at the NASA Manned Spacecraft Center in Houston to subject astronauts to large accelerations. The astronaut is about to enter the gondola.

selves to express angular displacements in radians, angular velocities in rad/s and rad/min, and angular accelerations in rad/s² and rad/min². The only exceptions to this rule are equations that involve *only* angular variables, such as the angular equations listed in Table 1. Here you are free to use any unit you wish for the angular variables. That is, you may use radians, degrees, or revolutions, as long as you use them consistently.

Because s and r in Eq. 1 $(\theta = s/r)$ have the same dimensions, you need not keep track of the radian algebraically, as you must do for other units. You can add or delete it at will, to suit the context. Sample Problem 5a is an example in which the radian was added to the answer; Sample Problem 5b is a case in which it was deleted.

11-6 Kinetic Energy of Rotation

The rapidly rotating blade of a table saw certainly has kinetic energy. How can we express it? We cannot use the familiar formula $K = \frac{1}{2}mv^2$ directly because it applies only to particles and we would not know what to put for v.

Our plan is to treat the table saw (or any other rigid body) as a collection of particles—all with different speeds. We shall then add up the kinetic energies of these particles to find the kinetic energy of the rotating rigid body as a whole.

We then write for the kinetic energy of a rotating body

$$K = \tfrac{1}{2}m_1v_1^2 + \tfrac{1}{2}m_2v_2^2 + \tfrac{1}{2}m_3v_3^2 + \cdots = \sum \tfrac{1}{2}m_iv_i^2, \quad (19)$$

in which m_i is the mass and v_i is the speed of the ith particle. The sum is taken over all the particles that make up the body. The problem with Eq. 19 is that v_i is not the same for all particles. We get around this by using Eq. 15 $(v = \omega r)$ so that we have

$$K = \sum \tfrac{1}{2}m_i(\omega r_i)^2 = \tfrac{1}{2}\Bigl(\sum m_i r_i^2\Bigr)\omega^2, \quad (20)$$

in which ω is the same for all particles.

The quantity in parentheses in Eq. 20 tells us how the mass of the rotating body is distributed about its axis of rotation. We call it the *rotational inertia* * I of the body with respect to that axis. Thus,

$$\boxed{I = \sum m_i r_i^2.} \quad (21)$$

---

* Often called the *moment of inertia*.

Substituting into Eq. 20 yields, as the expression we seek,

$$\boxed{K = \tfrac{1}{2}I\omega^2 \quad \text{(radian measure).}} \quad (22)$$

Because we have used the relation $v = \omega r$ in deriving Eq. 22, ω in that equation must be expressed in radian measure.

Equation 22, which gives the kinetic energy of a rigid body in pure rotation, is the angular equivalent of the formula $K = \frac{1}{2}Mv_{cm}^2$, which gives the kinetic energy of a rigid body in pure translation. In each case, there is a factor of $\frac{1}{2}$. Where mass M (the *translational inertia*) appears in one formula, I (the *rotational inertia*) appears in the other. Finally, each equation contains as a factor the square of a speed, translational or rotational as appropriate. The kinetic energies of translation and of rotation are not different kinds of energy. They are both kinetic energy, expressed in ways that are appropriate to the problem at hand.

The rotational inertia of a rotating body depends not only on its mass but also on how that mass is distributed with respect to the rotation axis. Figure 13 suggests a convincing way to develop a physical feeling for rotational inertia. Figure 13a shows one of two plastic rods that outwardly appear to be identical—each about 1 m long. The dimensions and the weights of the two rods are the same and both rods balance at their midpoints. If you grasp each rod at its center and move them back and forth rapidly in translational motion, you still cannot tell them apart.

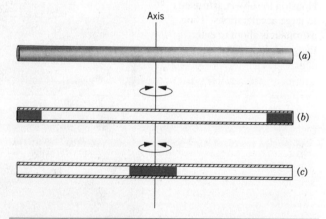

Figure 13 (a) Two plastic rods like this seem identical until you try to twist them rapidly in rotation about their midpoints. Rod (c) wiggles readily; rod (b) does not. The secret is in the internal distribution of the weights about the axis of rotation. Although both have the same weight, the rotational inertia of rod (b) about an axis through its midpoint is considerably greater than that of rod (c).

However, a truly striking difference appears if, with wrist motion, you twist the rods (like a baton) rapidly in back-and-forth angular motion. One rod wiggles quite easily; the other does not. As Figs. 13b and 13c show, the "easy" rod has internal weights concentrated near its center and the "hard" rod has them at its ends. Even though the masses of the rods are equal, their rotational inertias about a central axis are quite different.

11-7   Calculating the Rotational Inertia

If a rigid body is made up of discrete particles, we can calculate its rotational inertia from Eq. 21. If the body is continuous, we must replace the sum in Eq. 21 by an integral and the definition of the rotational inertia becomes

$$I = \int r^2 dm. \tag{23}$$

In the Sample Problems, we show how to calculate I for both kinds of bodies. Table 2 summarizes the rotational inertias of several common bodies, about various axes. Study the table carefully to get a feeling for how the distribution of mass affects the rotational inertia.

The Parallel-Axis Theorem.   If you know the rotational inertia of a body about any axis that passes through its center of mass, you can find its rotational inertia about any other axis parallel to this by the *parallel-axis theorem*, which is

$$I = I_{cm} + Mh^2. \tag{24}$$

Here M is the mass of the body and h is the perpendicular distance between the two (parallel) axes. In words, this theorem can be stated as follows:

The rotational inertia of a body about any axis is equal to the rotational inertia (= Mh²) it would have if all of its mass were concentrated at its center of mass plus its rotational inertia (= I_cm) about a parallel axis through its center of mass.

Proof of the Parallel-Axis Theorem   (Optional).   Let O be the center of mass of an arbitrarily shaped body, shown in cross section in Fig. 14. The center of mass is at the origin of coordinates. Consider an axis through O at right angles to the plane of the figure and another axis parallel to it through point P. Let the coordinates of P be a and b.

Let dm be a mass element, whose coordinates are x and y. The rotational inertia of the body about the axis through P is then, from Eq. 23,

$$I = \int r^2 dm = \int [(x-a)^2 + (y-b)^2] \, dm,$$

which we can rearrange as

$$I = \int (x^2 + y^2) \, dm - 2a \int x \, dm$$
$$- 2b \int y \, dm + \int (a^2 + b^2) \, dm. \tag{25}$$

From the definition of the center of mass, the two integrals in the middle of Eq. 25 are zero. The first integral is simply I_{cm}, the rotational inertia of the body about an axis through its center of mass. Inspection of Fig. 14 shows that the last term in Eq. 25 is Mh^2, where M is the total mass. Thus, Eq. 25 reduces to Eq. 24, which is the relation that we set out to prove.

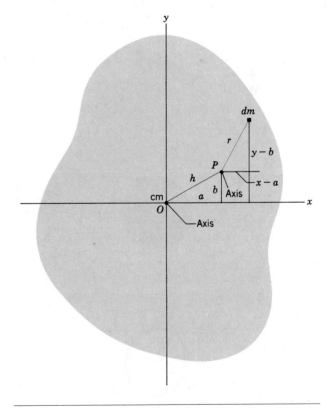

Figure 14   A rigid body is shown in cross section, with its center of mass at O. The parallel-axis theorem (Eq. 24) relates the rotational inertia of the body about an axis through O to that about a parallel axis through a point such as P, a distance h from the center of mass.

Table 2   Some Rotational Inertias

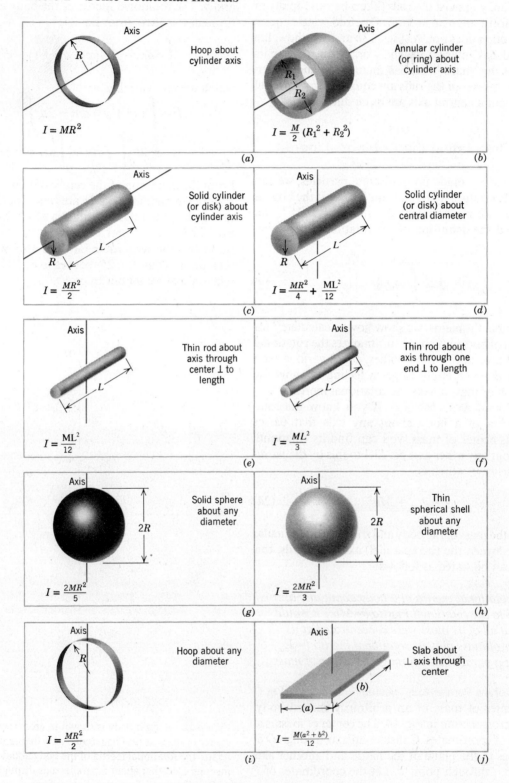

Axis

Hoop about
cylinder axis

$I = MR^2$

(a)

Axis

Annular cylinder
(or ring) about
cylinder axis

R_1

R_2

$I = \frac{M}{2}(R_1^2 + R_2^2)$

(b)

Axis

Solid cylinder
(or disk) about
cylinder axis

L

R

$I = \frac{MR^2}{2}$

(c)

Axis

Solid cylinder
(or disk) about
central diameter

L

R

$I = \frac{MR^2}{4} + \frac{ML^2}{12}$

(d)

Axis

Thin rod about
axis through
center ⊥ to
length

L

$I = \frac{ML^2}{12}$

(e)

Axis

Thin rod about
axis through one
end ⊥ to length

L

$I = \frac{ML^2}{3}$

(f)

Axis

Solid sphere
about any
diameter

$2R$

$I = \frac{2MR^2}{5}$

(g)

Axis

Thin
spherical shell
about any
diameter

$2R$

$I = \frac{2MR^2}{3}$

(h)

Axis

Hoop about any
diameter

R

$I = \frac{MR^2}{2}$

(i)

Axis

Slab about
⊥ axis through
center

(b)

(a)

$I = \frac{M(a^2 + b^2)}{12}$

(j)

Sample Problem 6   Figure 15 shows a rigid body consisting of two particles of mass m connected by a light rod of length L. (a) What is the rotational inertia of this body about an axis through its center, at right angles to the rod (see Fig. 15a)?

From Eq. 21 we have

$$I = \sum m_i r_i^2 = (m)(\tfrac{1}{2}L)^2 + (m)(\tfrac{1}{2}L)^2 = \tfrac{1}{2}mL^2. \quad \text{(Answer)}$$

(b) What is the rotational inertia of the same assembly about a parallel axis through one end of the rod, as in Fig. 15b?

We can use the parallel-axis theorem of Eq. 24. We have just calculated I_{cm} in part (a) and the distance h between the parallel axes is half the length of the rod. Thus, from Eq. 24,

$$I = I_{cm} + Mh^2 = \tfrac{1}{2}mL^2 + (2m)(\tfrac{1}{2}L)^2 = mL^2. \quad \text{(Answer)}$$

We can check this by direct calculation, using Eq. 21. Thus,

$$I = \sum m_i r_i^2 = (m)(0)^2 + (m)(L)^2 = mL^2. \quad \text{(Answer)}$$

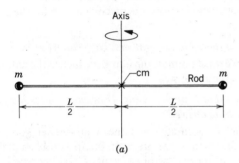

(a)

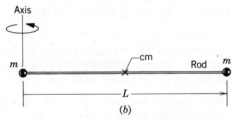

(b)

Figure 15   Sample Problem 6. A rigid body consists of two particles of mass m joined by a rod of negligible mass. Find the rotational inertia about the two axes shown.

Sample Problem 7   Figure 16 shows a uniform rod of mass m and length L. (a) What is its rotational inertia about an axis at right angles to the rod, through its center of mass?

We choose as a mass element a slice of the rod located at position x. The mass per unit length of the rod is m/L so that the mass dm of the element is

$$dm = m\left(\frac{dx}{L}\right).$$

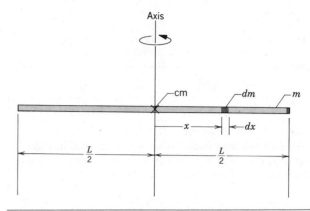

Figure 16   Sample Problem 7. A uniform rod of length L and mass m. Find its rotational inertia about the axes shown.

From Eq. 23 we have

$$I = \int x^2 dm = \int_{x=-L/2}^{x=+L/2} x^2 \left(\frac{m}{L}\right) dx$$
$$= \frac{m}{3L}\left. x^3 \right|_{-L/2}^{+L/2} = \frac{m}{3L}\left[\left(\frac{L}{2}\right)^3 - \left(-\frac{L}{2}\right)^3\right]$$
$$= \tfrac{1}{12} mL^2. \quad \text{(Answer)}$$

This agrees with the result given in Table 2(e).

(b) What is the rotational inertia of the rod about a perpendicular axis through its end point?

We can combine the above result with the parallel-axis theorem (Eq. 24), obtaining

$$I = I_{cm} + Mh^2$$
$$= \tfrac{1}{12} mL^2 + (m)(\tfrac{1}{2}L)^2 = \tfrac{1}{3}mL^2, \quad \text{(Answer)}$$

which agrees with the result given in Table 2(f).

Sample Problem 8   A hydrogen chloride molecule consists of a hydrogen atom whose mass m_H is 1.01 u and a chlorine atom whose mass m_{Cl} is 35.0 u. The centers of the two atoms are a distance d ($= 1.27 \times 10^{-10}$ m $= 127$ pm) apart; see Fig. 17. What is the rotational inertia of the molecule about an axis at right angles to the line joining the two atoms and passing through the center of mass of the molecule?

From Fig. 17 we see that, at the center of mass, we have

$$m_{Cl}x = m_H(d - x),$$

which yields

$$x = d\frac{m_H}{m_{Cl} + m_H}. \quad (26)$$

From Eq. 21, the rotational inertia about an axis through the center of mass is

$$I = \sum mr^2 = m_H(d - x)^2 + m_{Cl}x^2.$$

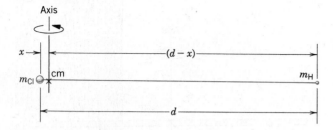

Figure 17 Sample Problem 8. A hydrogen chloride molecule, shown schematically. What is its rotational inertia about an axis through its center of mass and perpendicular to the line joining the two atomic centers?

Substituting Eq. 26 above leads, after some algebra, to

$$I = d^2 \frac{m_H m_{Cl}}{m_{Cl} + m_H} = (127 \text{ pm})^2 \frac{(1.01 \text{ u})(35.0 \text{ u})}{35.0 \text{ u} + 1.01 \text{ u}}$$

$$= 15{,}800 \text{ u} \cdot \text{pm}^2. \qquad \text{(Answer)}$$

These units are convenient enough when dealing with the rotational inertia of molecules. If we use the ångstrom as a length unit (1 Å = 10^{-10} m), the answer above becomes $I = 1.58$ u·Å², an even more convenient unit.

Sample Problem 9 With modern technology, it is possible to construct a flywheel that would store enough energy to run an automobile. Suppose that such a wheel is made in the form of a cylinder whose mass M is 75 kg and whose radius R is 25 cm. If the wheel is spun at 85,000 rev/min, how much rotational kinetic energy can it store?

The rotational inertia of the cylindrical wheel follows from Table 2(c), or

$$I = \tfrac{1}{2}MR^2 = (\tfrac{1}{2})(75 \text{ kg})(0.25 \text{ m})^2 = 2.34 \text{ kg} \cdot \text{m}^2.$$

The angular velocity of the wheel is

$$\omega = (85{,}000 \text{ rev/min})(2\pi \text{ rad/rev})(1 \text{ min/60 s})$$

$$= 8900 \text{ rad/s}.$$

From Eq. 22, the kinetic energy of rotation is then

$$K = \tfrac{1}{2}I\omega^2 = (\tfrac{1}{2})(2.34 \text{ kg} \cdot \text{m}^2)(8900 \text{ rad/s})^2$$

$$= 9.3 \times 10^7 \text{ J} = 26 \text{ kW} \cdot \text{h}. \qquad \text{(Answer)}$$

This amount of energy, used with the expected efficiency, would permit a small car to travel about 200 mi.

Modern flywheels use materials (for example, fused-silica fibers and Kevlar) chosen for their tensile strength rather than for their density. The design aim is to spin the wheel as fast as possible without rupture and, furthermore, to shape the wheel so that the stresses leading to rupture are uniformly distributed throughout the wheel.*

---

* See "Flywheels," by Richard F. Post and Stephen F. Post, *Scientific American*, December 1973, for a fascinating account of the possibilities.

Special Sample Problem. (Optional) There are many interesting problems for which approximate answers are all that are needed. Physicists pride themselves on being able to solve such order-of-magnitude problems. Here is an example:†

> *How many atoms of rubber are worn away from an automobile tire for every revolution of the wheel?*

Problems of this kind are often called *Fermi problems,* after Enrico Fermi, a great practitioner of the craft of proposing them and solving them quickly and cleverly.

Q.   Why would I want to solve such a problem? Does it have any practical significance?

Probably not. However, it is an interesting link between the worlds of the very small (the atom) and the very large (the automobile). The answer will impress upon you how small atoms really are. Most important, the problem should help you to understand how to make estimates and should give you a sense of when an answer is precise enough for the purpose at hand.

Q.   But there are no numbers. How can we even start?

We must estimate the numbers, such as the radius of a tire or the amount of wear.

Q.   But that is just guessing. How can we possibly arrive at an accurate answer?

If by "accurate" you mean an answer good to three significant figures, you are right. But in a problem of this kind, "accurate" means "within a factor of 10 either way." Actually, it is hard to be that far wrong in estimating the input data.

Q.   I understand. Where do we start?

We start with a plan. We shall estimate the volume of rubber worn from the tire and then divide by the volume of an atom. That will give us our answer. Let's deal with the tire first.

Q.   OK. But I don't see any way to guess what volume of rubber is worn from the tire every time the wheel goes round.

We can get an estimate by guessing the volume of rubber worn during the life of the tire and then figure out how many revolutions the wheel makes during that time. Dividing will give the volume of rubber lost per turn.

Let $2\pi R_t$ be the circumference of the tire, W the width of the tread, h the depth of wear, and L the distance traveled during the life of the tire. The number of turns N and the volume of the worn rubber V are

$$N = \frac{L}{2\pi R_t} \quad \text{and} \quad V = (2\pi R_t)Wh.$$

---

† Based on a similar problem suggested by Professor A. A. Bartlett of the University of Colorado.

The volume per turn is then

$$V_t = \frac{V}{N} = \frac{(4\pi^2)R_t^2\,Wh}{L} = \frac{40R_t^2\,Wh}{L}.$$

Note that we have replaced π^2 by 10, which is quite close enough for our purposes.

Q. But there is no need to replace π^2 by 10. My calculator shows 9.87.

You may feel that you are improving the precision of our answer in doing so but you are not. Our other estimates will be so approximate that such precision is misplaced.

Q. I accept that. What do we do next?

We have made great progress. We have reduced part of the problem to quantities that we can estimate. We shall do that soon. Meanwhile, let's think about atoms.

Q. I've been wondering about that. What is a "rubber atom" anyway? I'm sure you won't find it in the periodic table!

You are right of course. Rubber is made up of long-chain molecules formed from carbon, hydrogen, and oxygen atoms. We are interested here only in a sort of generic atom, whose radius we label R_a.

Q. I see. Then the volume V_a of the generic atom would be $(4\pi/3)R_a^3$. Right?

You could say that. It's a little better (and simpler) to put the volume at $(2R_a)^3$, that is, the cube of the diameter. That treats the atoms as little cubes and, in so doing, makes some allowance for the empty space between them.

Q. Now we divide to find our answer. Right?

Right. The number of atoms worn away per turn is

$$n = \frac{V_t}{V_a} = \frac{40R_t^2\,Wh}{L(2R_a)^3} = \frac{5R_t^2\,Wh}{LR_a^3}. \qquad (27)$$

Now we are ready for our estimates. Let's take them one at a time:

R_t About 1 ft or 30 cm or $\frac{3}{10}$ m
W About 4 in. or 10 cm or $\frac{1}{10}$ m
h About 3 mm or $\frac{3}{1000}$ m
L About 75,000 mi or 120,000 km or 12×10^7 m
R_a About 1 Å or 10^{-10} m

Putting these numbers into Eq. 27 gives

$$n = \frac{5 \times 3 \times 3 \times 3}{10 \times 10 \times 1000 \times 12 \times 10^7 \times 10^{-30}}.$$

Q. Shall I work this out on my calculator for you?

No. It's a point of honor not to use a calculator in solving such problems. Let's rewrite this equation by collecting the integers and the powers of 10. Thus,

$$n = \left(\frac{5 \times 3 \times 3 \times 3}{12}\right) \times 10^{18}.$$

You can show in your head that the number in the parentheses is about 10.

Q. So our answer is about 10^{19} atoms per revolution?

Yes. Would you have guessed that it is so large? When someone asks the "tire question" at a party, you can now gaze at the ceiling for a few minutes and say: "about 10^{19} atoms, more or less." That's how quickly Fermi often solved similar problems.

11-8 Torque

A door knob is located as far as possible from the hinge line for a good reason. If you want to open a heavy door you must certainly apply a force; that alone, however, is not enough. Where you apply that force and in what direction you push are also important.

Figure 18a shows a force **F** acting on a body that is free to rotate about an axis.* The force is applied at point P, whose position is defined by the vector **r**. The directions of **F** and **r** make an angle ϕ with each other. As Fig. 18b suggests, only the tangential component of F ($F_t = F \sin \phi$) can have any turning effect on the rigid body. The radial component ($F_r = F \cos \phi$) passes through the axis and cannot cause the body to rotate. With this in mind, we define the *torque* τ acting on the body from†

$$\boxed{\tau = rF \sin \phi.} \qquad (28)$$

The torque, which we may loosely identify as the "turning agent," is what you apply with a screwdriver or a pipe wrench. It plays the same role in generating angular acceleration that force does in generating linear acceleration.

We can write the torque as

$$\tau = (r)(F \sin \phi) = rF_t \qquad (29)$$

or as

$$\tau = (r \sin \phi)(F) = r_\perp F. \qquad (30)$$

Here $r_\perp$, called the *moment arm* of the force, is the perpendicular distance of the line of action of the force

---

* For simplicity, we consider only forces that have no component parallel to the rotation axis.

† The symbol τ is the Greek letter *tau*, pronounced to rhyme with either "saw" or "how;" take your pick. The word *torque* comes from the Latin word meaning "to twist."

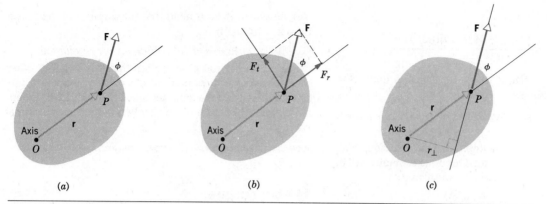

Figure 18 *(a)* A force **F** acts at point P on a rigid body that is free to rotate about an axis through O. The torque exerted by this force is $Fr \sin \phi$. *(b)* The torque can be written as $F_t r$. *(c)* The torque can also be written as $Fr_\perp$.

from the axis of rotation, as Fig. 18c shows. The SI unit of torque is the newton·meter (abbr. N·m).*

We can now explain why the perpetual-motion wheel shown at the head of this chapter does not work. If the wheel is given a powerful initial spin it will coast along as a flywheel until its rotational energy is dissipated by friction. The wheel (14 ft in diameter) must have been an impressive sight in this initial motion, with 40 cannon balls careening back and forth. How easy for the inventor to claim that the eventual stopping of the wheel was due to excessive friction, which he could fix in a later model. However, let the wheel be allowed to come to rest. It will be found, after careful measurement of the moment arms of the gravitional forces acting on the cannon balls, that the resultant torque acting around the axis of the wheel is zero! This may not *seem* to be the case but measurement bears it out.

11–9   Newton's Second Law for Rotation

Figure 19 shows a simple case of rotation about a fixed axis. The rotating rigid body consists of a single particle of mass m fastened to the end of a (massless) rod. A force **F** acts as shown, causing the particle to move in a circle. The particle has a tangential component of acceleration governed by Newton's second law, or

$$F_t = ma_t.$$

The torque acting on the particle is then, from Eq. 29,

$$\tau = F_t\, r = ma_t\, r.$$

From Eq. 17 ($a_t = \alpha r$) we can write this as

$$\tau = m(\alpha r)r = (mr^2)\alpha.$$

The quantity in parentheses (see Eq. 21) is the rotational inertia of the rigid body of Fig. 19 about its axis, so that finally

$$\boxed{\tau = I\alpha \quad \text{(radian measure).}} \tag{31}$$

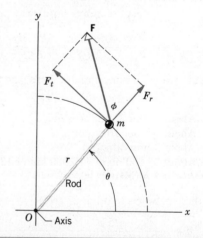

Figure 19   A simple rigid body, free to rotate about an axis through O, consists of a particle of mass m fastened to the end of a rod of length r but negligible mass. An applied force **F** causes the body to turn. The tension in the rod (not shown) also acts on the particle, counteracting the force component F_r and, in addition, providing the necessary centripetal force.

---

* The N · m is also the unit of work. Torque and work, however, are totally different quantities and must not be confused. Work is often expressed in joules (1 J = 1 N · m) but torque never is.

Although we derived Eq. 31 for the special case of a single particle rotating about a fixed axis, it holds for any rigid body rotating about a fixed axis. This follows from the principle of superposition, because any such body can be thought of as an assembly of single particles.

Equation 31 is the angular form of Newton's second law, holding for a rigid body about a fixed axis. It bears the same relation to the *angular* acceleration of a rigid body about a fixed axis that $F = ma$ does to the *linear* acceleration of a rigid body along a fixed direction.

Sample Problem 10   Figure 20a shows a uniform disk whose mass M is 2.5 kg and whose radius R is 20 cm mounted on a fixed horizontal axle. A block whose mass m is 1.2 kg hangs from a light cord that is wrapped around the rim of the disk. Find the acceleration of the falling block, the angular acceleration of the disk, and the tension in the cord.

Figure 20b shows a free-body diagram for the block. The block accelerates downward so that its weight ($= mg$) must exceed the tension T in the cord. We take the downward direction as positive and, from Newton's second law, we have

$$mg - T = ma. \tag{32}$$

Figure 20c shows a free-body diagram for the disk. The torque acting on the disk is TR and the rotational inertia of the disk is $\frac{1}{2}MR^2$. (Two other forces also act on the disk, its weight Mg and the normal force N exerted on the disk by its support. Both of these forces act at the axis of the disk, however, so that they exert no torque on the disk.) Applying Newton's second law in angular form (Eq. 31), we obtain

$$\tau = I\alpha$$

or

$$TR = \frac{1}{2}MR^2 \left(\frac{a}{R}\right).$$

This reduces to

$$T = \frac{1}{2}Ma. \tag{33}$$

In replacing α by a/R, we have assumed—because the cord does not slip—that the linear acceleration of the block is equal to the linear acceleration of the rim of the disk. Combining Eqs. 32 and 33 leads to

$$a = g\,\frac{2m}{M + 2m} = (9.8 \text{ m/s}^2)\,\frac{(2)(1.2 \text{ kg})}{2.5 \text{ kg} + (2)(1.2 \text{ kg})}$$
$$= 4.8 \text{ m/s}^2 \qquad \text{(Answer)}$$

and to

$$T = mg\,\frac{M}{M + 2m} = (1.2 \text{ kg})(9.8 \text{ m/s}^2)\,\frac{2.5 \text{ kg}}{2.5 \text{ kg} + (2)(1.2 \text{ kg})}$$
$$= 6.0 \text{ N}. \qquad \text{(Answer)}$$

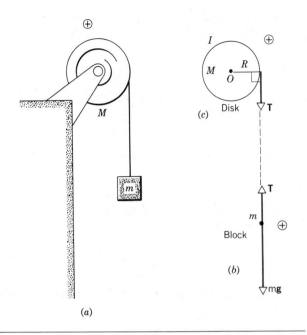

Figure 20   Sample Problems 10 and 11. (a) A falling block causes the disk to rotate. (b) A free-body diagram for the block. (c) A free-body diagram for the disk. The directions taken as positive are shown by the arrows.

As expected, the acceleration of the falling block is less than g and the tension in the cord ($= 6.0$ N) is less than the weight of the hanging block ($= mg = 11.8$ N). We see also that the acceleration of the block and the tension depend on the mass of the disk but not on its radius. As a check, we note that the formulas derived above predict $a = g$ and $T = 0$ for the case of a massless disk ($M = 0$). This is what we expect; the block simply falls as a free body, trailing the string behind it.

The angular acceleration of the disk follows from Eq. 17, or

$$\alpha = \frac{a}{R} = \frac{4.8 \text{ m/s}^2}{0.20 \text{ m}} = 24 \text{ rad/s}^2 = 3.8 \text{ rev/s}^2. \quad \text{(Answer)}$$

11–10   Work, Power, and the Work–Energy Theorem

In Fig. 19, imagine that the rigid body (which consists of a single particle of mass m fastened to the end of a rod of negligible mass) rotates through an angle $d\theta$. The work done by the force **F** is

$$dW = \mathbf{F} \cdot d\mathbf{s} = (F_t)(r\,d\theta),$$

in which ds ($= r\, d\theta$) is the length of arc traversed by the particle. From Eq. 29 we see that $F_t r$ is the torque τ, so that

$$dW = \tau\, d\theta. \qquad (34)$$

The work done during a finite angular displacement, from θ_i to θ_f, is then

$$W = \int_{\theta_i}^{\theta_f} \tau\, d\theta. \qquad (35)$$

Equation 35, which holds for all rigid bodies rotating about a fixed axis, is the rotational equivalent of Eq. 10 of Chapter 7, or

$$W = \int_{x_i}^{x_f} F\, dx.$$

We can find the power for rotational motion from Eq. 34, or

$$P = \frac{dW}{dt} = \tau\, \frac{d\theta}{dt} = \tau\omega. \qquad (36)$$

This is the rotational analog of $P = Fv$, which describes the rate at which a force F does work on a particle moving in a fixed direction with speed v.

To find the angular equivalent of the work–energy theorem, we start from Eq. 35, substituting $I\alpha$ for the torque τ. Thus,

$$W = \int_{\theta_i}^{\theta_f} I\alpha\, d\theta = \int_{\theta_i}^{\theta_f} I\left(\frac{d\omega}{dt}\right) d\theta$$

$$= \int_{\omega_i}^{\omega_f} I\left(\frac{d\theta}{dt}\right) d\omega = \int_{\omega_i}^{\omega_f} I\omega\, d\omega.$$

Carrying out the integration yields, with the help of Eq. 22 ($K = \frac{1}{2}I\omega^2$),

$$W = \tfrac{1}{2}I\omega_f^2 - \tfrac{1}{2}I\omega_i^2 = K_f - K_i = \Delta K. \qquad (37)$$

This *work–energy theorem* tells us that the work done by the resultant torque acting on a rotating rigid body is equal to the change in the rotational kinetic energy of that body. This is the angular equivalent of the work–energy theorem for translational motion, which states that the work done on a rigid body (or on a particle) by the resultant force that acts on it is equal to the change in the translational kinetic energy of that body.

Table 3 summarizes the equations that apply to the rotation of a rigid body about a fixed axis and points out the equivalent relations for translational motion.

Sample Problem 11 (a) In the arrangement of Fig. 20, through what angle does the disk rotate in 2.5 s? Assume that the system starts from rest.

From Eq. 9 of Table 1, we have, putting $\omega_0 = 0$ and using the value of α calculated in Sample Problem 10,

$$\theta = \omega_0 t + \tfrac{1}{2}\alpha t^2$$

$$= (0)(2.5\ \text{s}) + (\tfrac{1}{2})(24\ \text{rad/s}^2)(2.5\ \text{s})^2$$

$$\doteq 75\ \text{rad}. \qquad \text{(Answer)}$$

(b) What is the angular velocity of the disk at $t = 2.5$ s?

We can find this from Eq. 8 of Table 1. Putting $\omega_0 = 0$ and using the value of α calculated in Sample Problem 10, we obtain

$$\omega = \omega_0 + \alpha t$$

$$= 0 + (24\ \text{rad/s}^2)(2.5\ \text{s}) = 60\ \text{rad/s}. \quad \text{(Answer)}$$

Table 3   Some Corresponding Relations for Translational and Rotational Motion

Pure Translation (Fixed Direction)		Pure Rotation (Fixed Axis)	
Position	x	Angular position	θ
Velocity	$v = dx/dt$	Angular velocity	$\omega = d\theta/dt$
Acceleration	$a = dv/dt$	Angular acceleration	$\alpha = d\omega/dt$
Translational inertia	m	Rotational inertia	I
Newton's second law	$F = ma$	Newton's second law	$\tau = I\alpha$
Work	$W = \int F\, dx$	Work	$W = \int \tau\, d\theta$
Kinetic energy	$K = \frac{1}{2}mv^2$	Kinetic energy	$K = \frac{1}{2}I\omega^2$
Power	$P = Fv$	Power	$P = \tau\omega$
Work–energy theorem	$W = \Delta K$	Work–energy theorem	$W = \Delta K$

(c) What is the kinetic energy of the rotating disk at $t = 2.5$ s?

From Eq. 22, the kinetic energy of the disk is $\frac{1}{2}I\omega^2$, in which $I = \frac{1}{2}MR^2$. Thus, using the value of ω found in (b),

$$K_d = \frac{1}{2}I\omega^2 = \frac{1}{2}(\frac{1}{2}MR^2)\omega^2$$
$$= (\frac{1}{4})(2.5 \text{ kg})(0.20 \text{ m})^2(60 \text{ rad/s})^2$$
$$= 90 \text{ J}. \qquad \text{(Answer)}$$

Another approach to this problem is to calculate the change in the rotational kinetic energy from the work–energy theorem. We first calculate the work done on the disk by the resultant torque that acts on it. This torque, exerted by the tension T in the cord, is constant so that, from Eq. 35,

$$W = \int_{\theta_i}^{\theta_f} \tau \, d\theta = \tau \int_{\theta_i}^{\theta_f} d\theta = \tau(\theta_f - \theta_i)$$

For the torque τ we put TR, in which T is the tension in the cord (= 6.0 N; see Sample Problem 10) and R (= 0.20 m) is the radius of the disk. The quantity $\theta_f - \theta_i$ is just the angular displacement that we calculated in (a) above. Thus,

$$W = \tau \, (\theta_f - \theta_i) = TR(\theta_f - \theta_i)$$
$$= (6.0 \text{ N})(0.20 \text{ m})(75 \text{ rad}) = 90 \text{ J}, \quad \text{(Answer)}$$

in full agreement with the value we have already found.

REVIEW AND SUMMARY

The Angular Variables

To describe the rotation of a rigid body about a fixed axis, we define *angular position* θ, *angular displacement* $\Delta\theta$, *angular velocity* ω, and *angular acceleration* α. Angular velocity and angular acceleration are defined as

$$\omega = \frac{d\theta}{dt} \quad \text{and} \quad \alpha = \frac{d\omega}{dt}. \qquad [5,7]$$

If θ is measured in radians, its measure is defined to be

$$\theta = \frac{s}{r} \quad \text{(radian measure).} \qquad [1]$$

Radians are related to other angular measures by

$$1 \text{ revolution} = 360° = 2\pi \text{ radians.} \qquad [2]$$

Angular velocity is a vector quantity whose direction is parallel to the axis of rotation in a sense determined by the *right-hand rule*. However, finite angular displacement is not a vector.

Constant angular acceleration (α = constant) is an important special case. The appropriate kinematic equations, given in Table 1, are

The Kinematic Equations for Constant Angular Acceleration

$$\omega = \omega_0 + \alpha t, \qquad [8]$$
$$\theta = \omega_0 t + \tfrac{1}{2}\alpha t^2, \qquad [9]$$
$$\omega^2 = \omega_0^2 + 2\alpha\theta, \qquad [10]$$
$$\theta = \tfrac{1}{2}(\omega_0 + \omega)t, \qquad [11]$$
$$\theta = \omega t - \tfrac{1}{2}\alpha t^2. \qquad [12]$$

Sample Problems 2–4 illustrate the use of these equations.

We sometimes wish to discuss the motion of a single point P that is part of a rotating rigid body. We can describe the motion of P by three linear variables: the arc distance s (= θr), the velocity $\mathbf{v}$, and the acceleration $\mathbf{a}$. The velocity of any point is tangential to its circular motion and has magnitude

Linear and Angular Variables

$$v = \omega r \quad \text{(radian measure).} \qquad [15]$$

The acceleration $\mathbf{a}$ has radial and tangential components

$$a_t = \alpha r \quad \text{and} \quad a_r = \omega^2 r \quad \text{(radian measure).} \qquad [17,18]$$

The r's in these expressions are the perpendicular distance from the axis of rotation to P. The angular quantities must be expressed in radian measure (rad/s and rad/s²) if v is to be in m/s and the a's are to be in m/s². See Sample Problem 5 and Hint 2.

The kinetic energy K of a rigid body rotating about a fixed axis is given by

Rotational Kinetic Energy

$$K = \tfrac{1}{2}I\omega^2 \quad \text{(radian measure),} \qquad [22]$$

in which I is the *rotational inertia* of the body defined as

$$I = \sum m_i r_i^2 \qquad [21]$$

for a discrete system of particles and as

Rotational Inertia

$$I = \int r^2 \, dm \qquad [23]$$

for a body with continuously distributed mass. The r's in these expressions represent the perpendicular distance from the axis of rotation to each mass element in the body. The rotational inertia of any rigid body depends on (1) the shape of the body, (2) the perpendicular distance from the axis of rotation to the body's center of mass, and (3) the orientation of the body with respect to the axis of rotation. The *parallel-axis theorem* relates the rotational inertia about any axis to that about a parallel axis through the center of mass:

The Parallel-Axis Theorem

$$I = I_{cm} + Mh^2. \qquad [24]$$

Here h is the perpendicular distance between the two axes. Table 2 gives some commonly used expressions for rotational inertia in particular instances. Sample Problems 6–9 illustrate the calculation of rotational inertia and rotational kinetic energy.

Torque τ is a measure of the "turning" action of a force $\mathbf{F}$. For rotation about a fixed axis, and for forces with no components parallel to that axis, torque is given by any one of the three forms

Torque

$$\tau = rF \sin \phi = rF_t = r_\perp F. \qquad [28,29,30]$$

Here r is the perpendicular distance from the axis of rotation to the point P at which $\mathbf{F}$ is applied, ϕ is the angle between the directions of r and $\mathbf{F}$, F_t is the component of $\mathbf{F}$ in the direction of positive translational motion of P, and $r_\perp$ is the perpendicular distance from the axis of rotation to the line of action of $\mathbf{F}$.

The angular acceleration of a rigid body rotating about a fixed axis is given correctly by the rotational analog of Newton's second law:

Newton's Second Law for Rotation

$$\sum \tau = I\alpha. \qquad [31]$$

Sample Problem 10 illustrates its use in a two-body problem.

Rotational equations used for calculating work and power are exact rotational analogs of the corresponding equations governing translational motion, namely,

Rotational Work and Energy

$$W = \int_{\theta_i}^{\theta_f} \tau \, d\theta \qquad [35]$$

and

$$P = \frac{dW}{dt} = \tau \frac{d\theta}{dt} = \tau\omega. \qquad [36]$$

Finally, the form of the work–energy theorem which is useful for rotating objects is

$$W = \tfrac{1}{2}I\omega_f^2 - \tfrac{1}{2}I\omega_i^2 = K_f - K_i = \Delta K. \qquad [37]$$

Sample Problem 9 shows the relations between several of these quantities and applies them to representative problems.

Rotation–Translation Analogs

The expressions describing the kinematics and dynamics of rigid body rotation are similar in form to those describing translational motion of a particle in a straight line. The analogous relations are displayed in Table 3. Careful study of the correspondences shown there can be a real help in understanding the meaning of each of the new equations and can significantly aid problem solving.

QUESTIONS

1. Could the angular quantities θ, ω, and α be expressed in terms of degrees instead of radians in the rotational equations of Table 1?

2. In what sense is the radian a "natural" measure of angle and the degree an "arbitrary" measure of that same quantity?

3. Does the vector representing the angular velocity of a wheel rotating about a fixed axis necessarily have to lie along that axis?

4. Experiment rotating a book after the fashion of Fig. 6, but this time using angular displacements of 180° rather than 90°. What do you conclude about the final positions of the book? Does this change your mind about whether (finite) angular displacements can be treated as vectors?

5. The rotation of the sun can be monitored by tracking sunspots, magnetic storms on the sun that appear dark against the otherwise bright solar disk. Figure 21a shows the initial positions of five spots and Fig. 21b the positions of these same spots one solar rotation later. What can we conclude about the physical nature of the sun from these observations?

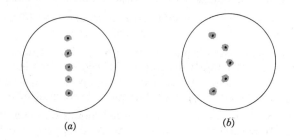

(a)   (b)

Figure 21   Question 5.

6. Why is it suitable to express α in revolutions per second² in Eq. 9 in Table 1 ($\theta = \omega_0 t + \frac{1}{2}\alpha t^2$) but not in Eq. 17 ($a_t = \alpha r$)?

7. A rigid body is free to rotate about a fixed axis. Can the body have nonzero angular acceleration even if the angular velocity of the body is (perhaps instantaneously) zero? What is the linear equivalent of this question? Give physical examples to illustrate both the angular and linear situations.

8. A golfer swings a golf club, making a long drive from the tee. Do all points on the club have the same angular velocity ω at any instant while the club is in motion?

9. When we say that a point on the equator has an angular speed of 2π rad/day, what reference frame do we have in mind?

10. Taking the rotation and the revolution of the earth into account, does a tree move faster during the day or during the night? With respect to what reference frame is your answer given? (The earth's rotation and revolution are both in the same direction.)

11. A wheel is rotating about its axle. Consider a point on the rim. When the wheel rotates with constant angular velocity, does the point have a radial acceleration? A tangential acceleration? When the wheel rotates with constant angular acceleration, does the point have a radial acceleration? A tangential acceleration? Do the magnitudes of these accelerations change with time?

12. Suppose that you were asked to determine the distance traveled by a needle in playing a phonograph record. What information do you need? Discuss from the point of view of reference frames (a) fixed in the room, (b) fixed on the rotating record, and (c) fixed on the arm of the record player.

13. What is the relation between the angular velocities of a pair of coupled gears of different radii?

14. Can the mass of a body be considered as concentrated at its center of mass for purposes of computing its rotational inertia? If yes, explain why. If no, offer a counterexample.

15. About what axis is the rotational inertia of your body the least? About what axis through your center of mass is your rotational inertia the greatest?

16. If two circular disks of the same weight and thickness are made from metals having different densities, which disk, if either, will have the larger rotational inertia about its central axis?

17. The rotational inertia of a body of rather complicated shape is to be determined. The shape makes a mathematical calculation from $\int r^2 dm$ exceedingly difficult. Suggest ways in which the rotational inertia about a particular axis could be measured experimentally.

18. Five solids are shown in cross section (Fig. 22). The cross sections have equal heights and equal maximum widths. The solids have equal masses. Which one has the largest rotational inertia about a perpendicular axis through the center of mass? Which has the smallest?

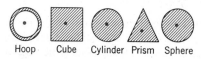

Hoop   Cube   Cylinder   Prism   Sphere

Figure 22   Question 18.

19. Figure 23a shows a meter stick, half of which is wood and half of which is steel, that is pivoted at the wooden end at O. A force is applied to the steel end at a. In Fig. 23b, the stick is pivoted at the steel end at O' and the same force is applied at the wooden end at a'. Does one get the same angular acceleration in each case? If not, in which case is the angular acceleration greater?

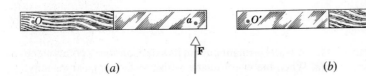

Figure 23   Question 19.

20. You can distinguish between a raw egg and a hardboiled one by spinning each one on the table. Explain how. Also, if you stop a spinning raw egg with your fingers and release it very quickly, it will resume spinning. Why?

21. Comment on each of these assertions about skiing. (a) In downhill racing, one wants skis that do not turn easily. (b) In slalom racing, one wants skis that turn easily. (c) Therefore, the rotational inertia of downhill skis should be larger than that of slalom skis. (d) Considering that there is low friction between skis and snow and that the skier's center of mass is about over the center of the skis, how does a skier exert torques to turn or to stop a turn? (See "The Physics of Ski Turns," by J. I. Shonie and D. L. Mordick, *The Physics Teacher,* December 1972.)

22. Consider a straight stick standing on end on (frictionless) ice. What would be the path of its center of mass if it falls?

EXERCISES AND PROBLEMS

Section 11–2   The Rotational Variables

1E. (a) What angle in radians is subtended from the center of a circle of radius 1.2 m by an arc of length 1.8 m? (b) Express the same angle in degrees. (c) The angle between two radii of a circle is 0.62 rad. What arc length is subtended if the radius is 2.4 m? (d) Repeat (c) if the radius is 61 m.

2E. Show that 1 rev/min = 0.1047 rad/s.

3E. The angle turned through by the flywheel of a generator during a time interval t is given by

$$\theta = at + bt^3 - ct^4,$$

where a, b, and c are constants. What is the expression for its (a) angular velocity and (b) angular acceleration?

4E. Our sun is 2.3×10^4 ly (light-years) from the center of our Milky Way galaxy and is moving in a circle around this center at a speed of 250 km/s. (a) How long does it take the sun to make one revolution about the galactic center? (b) How many revolutions has the sun completed since it was formed about 4.5×10^9 years ago?

5E. The angular position of a point on the rim of a rotating wheel is described by $\theta = 4.0t - 3.0t^2 + t^3$, where θ is in radians if t is given in seconds. (a) What is the angular velocity at $t = 2.0$ s and at $t = 4.0$ s? (b) What is the average angular acceleration for the time interval that begins at $t = 2.0$ s and ends at $t = 4.0$ s? (c) What is the instantaneous angular acceleration at the beginning and end of this time interval?

6E. The angular position of a rotating wheel is given by $\theta = 2 + 4t^2 + 2t^3$, where θ is in radians and t is in seconds. (a) What are the angular position and angular velocity at $t = 0$? (b) What is the angular velocity at $t = 4.0$ s? (c) Calculate the angular acceleration at $t = 2.0$ s. Is the angular acceleration constant?

7P. A wheel rotates with an angular acceleration α given by

$$\alpha = 4at^3 - 3bt^2,$$

where t is the time and a and b are constants. If the wheel has an initial angular speed ω_0, write the equations for (a) the angular speed and (b) the angle turned through as functions of time.

8P. What is the angular speed of (a) the second hand, (b) the minute hand, and (c) the hour hand of a watch?

9P. A good baseball pitcher can throw a baseball toward home plate at 85 mi/h with a spin of 1800 rev/min. How many revolutions does the baseball make on its way to home plate? For simplicity, assume that the 60-ft trajectory is a straight line.

10P. A diver makes 2.5 complete revolutions on the way from a 10-m platform to the water below. Assuming zero initial vertical velocity, calculate the average angular velocity for this dive.

11P. A wheel has eight spokes and a radius of 30 cm. It is mounted on a fixed axle and is spinning at 2.5 rev/s. You want to shoot a 20-cm arrow parallel to this axle and through the wheel without hitting any of the spokes. Assume that the arrow and the spokes are very thin; see Fig. 24. (a) What minimum

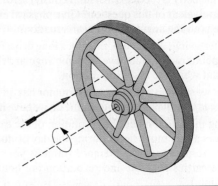

Figure 24   Problem 11.

speed must the arrow have? (*b*) Does it matter where between the axle and rim of the wheel you aim? If so, where is the best location?

Section 11–4 Rotation with Constant Angular Acceleration

12E. A disk, initially rotating at 120 rad/s, is slowed down with a constant angular acceleration of magnitude 4.0 rad/s². (*a*) How much time elapses before the disk stops? (*b*) Through what angle did the disk rotate in coming to rest?

13E. A phonograph turntable rotating at 78 rev/min slows down and stops in 30 s after the motor is turned off. (*a*) Find its (uniform) angular acceleration in rev/min². (*b*) How many revolutions did it make in this time?

14E. The angular speed of an automobile engine is increased from 1200 rev/min to 3000 rev/min in 12 s. (*a*) What is its angular acceleration in rev/min², assuming it to be uniform? (*b*) How many revolutions does the engine make during this time?

15E. A heavy flywheel rotating on its axis is slowing down because of friction in its bearings. At the end of the first minute, its angular velocity is 0.90 of its initial angular velocity of 250 rev/min. Assuming constant frictional forces, find its angular velocity at the end of the second minute.

16E. The flywheel of an engine is rotating at 25 rad/s. When the engine is turned off, the flywheel decelerates at a constant rate and comes to rest after 20 s. Calculate (*a*) the angular acceleration (in rad/s²) of the flywheel, (*b*) the angle (in rad) through which the flywheel rotates in coming to rest, and (*c*) the number of revolutions made by the flywheel in coming to rest.

17E. Starting from rest, a disk rotates about its axis with constant angular acceleration. After 5.0 s, it has rotated through 25 rad. (*a*) What was the angular acceleration during this time? (*b*) What was the average angular velocity? (*c*) What is the instantaneous angular velocity of the disk at the end of the 5.0 s? (*d*) Assuming that the acceleration does not change, through what additional angle will the disk turn during the next 5.0 s?

18E. A pulley wheel 8.0 cm in diameter has a 5.6-m-long cord wrapped around its periphery. Starting from rest, the wheel is given an angular acceleration of 1.5 rad/s². (*a*) Through what angle must the wheel turn for the cord to unwind? (*b*) How long does it take?

19P. A certain wheel turns through 90 rev in 15 s, its angular speed at the end of the period being 10 rev/s. (*a*) What was the angular speed of the wheel at the beginning of the 15-s interval, assuming constant angular acceleration? (*b*) How much time had elapsed between the time the wheel was at rest and the beginning of the 15-s interval?

20P. A wheel has a constant angular acceleration of 3.0 rad/s². In a 4.0-s interval, it turns through an angle of 120 rad.

Assuming the wheel started from rest, how long had it been in motion at the start of this 4.0-s interval?

21P. A wheel, starting from rest, rotates with a constant angular acceleration of 2.0 rad/s². During a certain 3.0-s interval, it turns through 90 rad. (*a*) How long had the wheel been turning before the start of the 3.0-s interval? (*b*) What was the angular velocity of the wheel at the start of the 3.0-s interval?

22P. A flywheel completes 40 rev as it slows from an angular speed of 1.5 rad/s to a complete stop. (*a*) Assuming uniform acceleration, what is the time required for it to come to rest? (*b*) What is the angular acceleration? (*c*) How much time is required for it to complete the first one-half of the 40 rev?

23P. A uniform disk rotates about a fixed axis starting from rest and accelerates with constant angular acceleration. At one time it is rotating at 10 rev/s. After undergoing 60 more complete revolutions its angular speed is 15 rev/s. Calculate (*a*) the angular acceleration, (*b*) the time required to complete the 60 revolutions mentioned, (*c*) the time required to attain the 10-rev/s angular speed, and (*d*) the number of revolutions from rest until the time the disk attained the 10-rev/s angular speed.

24P. Starting from rest at *t* = 0, a wheel undergoes a constant angular acceleration. When *t* = 2.0 s, the angular velocity of the wheel is 5.0 rad/s. The acceleration continues until *t* = 20 s, when it abruptly ceases. Through what angle does the wheel rotate in the interval *t* = 0 to *t* = 40 s?

25P. A pulsar is a rapidly rotating neutron star that emits radio pulses with precise synchronization, there being one such pulse for each rotation of the star. The period *T* of rotation is found by measuring the time between pulses. At present, the pulsar in the central region of the Crab nebula (see Fig. 25) has a period of rotation of *T* = 0.033 s, and this is observed to be increasing at the rate of 1.26×10^{-5} s/y. (*a*) Show that the angular velocity ω of the star is related to the period of rotation by $\omega = 2\pi/T$. (*b*) What is the value of the angular acceleration in rad/s²? (*c*) If its angular acceleration is constant, when will

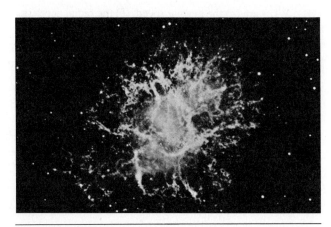

Figure 25 Problem 25.

the pulsar stop rotating? (*d*) The pulsar originated in a super-nova explosion in the year A.D. 1054. What was the period of rotation of the pulsar when it was born? (Assume constant angular acceleration.)

Section 11-5 The Linear and the Angular Variables

26E. What is the acceleration of a point on the rim of a 12-in. ($=30$ cm) diameter record rotating at $33\frac{1}{3}$ rev/min?

27E. A phonograph record on a turntable rotates at $33\frac{1}{3}$ rev/min. (*a*) What is the angular speed in rad/s? What is the linear speed of a point on the record at the needle at (*b*) the beginning and (*c*) the end of the recording? The distances of the needle from the turntable axis are 5.9 in. and 2.9 in., respectively, at these two positions.

28E. What is the angular speed of a car rounding a circular turn of radius 110 m at 50 km/h?

29E. A 1.2-m diameter flywheel is rotating at 200 rev/min. (*a*) What is the angular velocity of the flywheel in rad/s? (*b*) What is the linear velocity of a point on the rim of the flywheel? (*c*) What constant angular acceleration, in rev/min², will cause the wheel to increase its speed to 1000 rev/min in 60 s? (*d*) How many revolutions does the wheel make during this 60-s interval?

30E. A point on the rim of a 0.75-m diameter grinding wheel changes speed uniformly from 12 m/s to 25 m/s in 6.2 s. What is the average angular acceleration of the grinding wheel during this interval?

31E. The earth's orbit about the sun is almost a circle. (*a*) What is the angular velocity of the earth (regarded as a particle) about the sun? (*b*) What is its linear speed in its orbit? (*c*) What is the acceleration of the earth with respect to the sun?

32E. An astronaut is being tested in a centrifuge. The centrifuge has a radius of 10 m and, in starting, rotates according to $\theta = 0.3t^2$, where *t* in seconds gives θ in radians. When $t = 5.0$ s, what are the astronaut's (*a*) angular velocity, (*b*) tangential velocity, (*c*) tangential acceleration, and (*d*) radial acceleration?

33E. What are (*a*) the angular speed, (*b*) the radial acceleration, and (*c*) the tangential acceleration of a spaceship negotiat-

ing a circular turn of radius 2000 mi ($=3220$ km) at a constant speed of 18,000 mi/h (29,000 km/h)?

34E. A coin of mass M is placed a distance R from the center of a phonograph turntable. The coefficient of static friction is μ_s. The angular speed of the turntable is slowly increased to a value ω_0 at which time the coin slides off. (*a*) Find ω_0 in terms of the quantities M, R, g, and μ_s. (*b*) Make a sketch showing the approximate path of the coin as it flies off the turntable.

35P. (*a*) What is the angular speed about the polar axis of a point on the earth's surface at a latitude of 40° N? (*b*) What is the linear speed? (*c*) What are the corresponding values for a point at the equator?

36P. The flywheel of a steam engine runs with a constant angular speed of 150 rev/min. When steam is shut off, the friction of the bearings and of the air brings the wheel to rest in 2.2 h. (*a*) What is the constant angular acceleration, in rev/min², of the wheel? (*b*) How many rotations will the wheel make before coming to rest? (*c*) What is the tangential component of the linear acceleration of a particle that is 50 cm from the axis of rotation when the flywheel is turning at 75 rev/min? (*d*) What is the magnitude of the total linear acceleration of the particle in part (*c*)?

37P. A gyroscope flywheel of radius 2.83 cm is accelerated from rest at 14.2 rad/s² until its angular speed is 2760 rev/min. (*a*) What is the tangential acceleration of a point on the rim of the flywheel? (*b*) What is the radial acceleration of this point when the flywheel is spinning at full speed? (*c*) Through what distance does a point on the rim move during the acceleration?

38P. If an airplane propeller of radius 5.0 ft ($=1.5$ m) rotates at 2000 rev/min and the airplane is propelled at a ground speed of 300 mi/h ($=480$ km/h), what is the speed of a point on the tip of the propeller, as seen by (*a*) the pilot and (*b*) an observer on the ground? Assume that the plane's velocity is parallel to the propeller's axis of rotation.

39P. An early method of measuring the speed of light makes use of a rotating toothed wheel. A beam of light passes through a slot at the outside edge of the wheel, as in Fig. 26, travels to a distant mirror, and returns to the wheel just in time to pass through the next slot in the wheel. One such toothed wheel has

Figure 26 Problem 39.

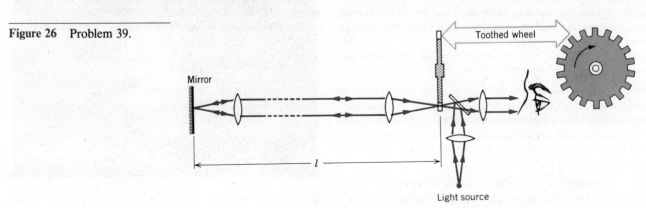

Toothed wheel

Mirror

Light source

a radius of 5.0 cm and 500 teeth at its edge. Measurements taken when the mirror was 500 m from the wheel indicated a speed of light of 3.0×10^5 km/s. (*a*) What was the (constant) angular speed of the wheel? (*b*) What was the linear speed of a point on its edge?

40P. A car starts from rest and moves around a circular track of radius 30 m. Its speed increases at the constant rate of 0.50 m/s². (*a*) What is the magnitude of its linear acceleration 15 s later? (*b*) What angle does the acceleration vector make with the car's velocity at this time?

41P. Wheel *A* of radius $r_A = 10$ cm is coupled by a belt *B* to wheel *C* of radius $r_C = 25$ cm, as shown in Fig. 27. Wheel *A* increases its angular speed from rest at a uniform rate of 1.6 rad/s². Determine the time for wheel *C* to reach a rotational speed of 100 rev/min, assuming the belt does not slip. (*Hint:* If the belt does not slip, the linear speeds at the rims of the two wheels must be equal.)

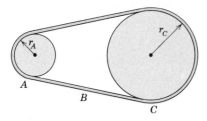

Figure 27   Problem 41.

42P. Four pulleys (*A*, *B*, *B'*, *C*) are connected by two belts (*a*, *b*) as shown in Fig. 28. Pulley *A* (radius 15 cm) is the drive pulley, and it rotates at 10 rad/s. Pulley *B* (radius 10 cm) is connected by belt *a* to pulley *A*. Pulley *B'* (radius 5 cm) is concentric with pulley *B* and is rigidly attached to it. Pulley *C* (radius 25 cm) is connected by belt *b* to pulley *B'*. Calculate (*a*) the linear speed of a point on belt *a*, (*b*) the angular speed of pulley *B*, (*c*) the angular speed of pulley *B'*, (*d*) the linear speed of a point on belt *b*, and (*e*) the angular speed of pulley *C*.

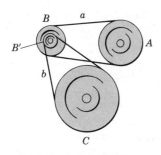

Figure 28   Problem 42.

43P. The disc of a Compact Disc/Digital Audio system has an inner and outer radius for its recorded material (the Tchaikovsky and Mendelssohn violin concertos) of 2.5 cm and 5.8 cm. At playback, the disc is scanned at a constant linear speed of 130 cm/s, starting from its inner edge and moving outward. (*a*) If the initial angular speed of the disc is 50 rad/s, what is its final angular speed? (*b*) The spiral scan lines are 1.6 μm apart; what is the total length of the scan? (*c*) What is the playing time?

44P. A phonograph turntable is rotating at $33\frac{1}{3}$ rev/min. A small object is on the turntable 6.0 cm from the axis of rotation. (*a*) Calculate the acceleration of the object assuming that it does not slip. (*b*) What is the minimum value of the coefficient of static friction between the object and the turntable? (*c*) Suppose that the turntable achieved this angular velocity by starting from rest and undergoing a constant angular acceleration for 0.25 s. Calculate the minimum coefficient of static friction required for the object not to slip during the acceleration period.

Section 11–6   Kinetic Energy of Rotation
45E. Calculate the rotational inertia of a wheel that has a kinetic energy of 24,400 J when it is rotating at 600 rev/min.

46P. The oxygen molecule, O_2, has a total mass of 5.30×10^{-26} kg and a rotational inertia of 1.94×10^{-46} kg·m² about an axis through the center perpendicular to the line joining the atoms. Suppose that such a molecule in a gas has a speed of 500 m/s and that its rotational kinetic energy is two-thirds of its translational kinetic energy. Find its angular velocity.

Section 11–7   Calculating the Rotational Inertia
47E. Calculate the kinetic energies of two uniform cylinders rotating about their axes of symmetry. They have the same mass, 1.25 kg, and rotate with the same angular velocity, 235 rad/s, but the first has a radius of 0.25 m and the second a radius of 0.75 m.

48E. A molecule has a rotational inertia of 14,000 u·pm² and is spinning at an angular speed of 4.3×10^{12} rad/s. (*a*) Express the rotational inertia in kg·m². (*b*) Calculate the rotational kinetic energy in eV.

49E. The masses and coordinates of four particles are as follows: 50 g, $x = 2.0$ cm, $y = 2.0$ cm; 25 g, $x = 0$, $y = 4.0$ cm; 25 g, $x = -3.0$ cm, $y = -3.0$ cm; 30 g, $x = -2.0$ cm, $y = 4.0$ cm. What is the rotational inertia of this collection with respect to the (*a*) *x*, (*b*) *y*, and (*c*) *z* axes?

50E. A communications satellite is a uniform cylinder with mass 1200 kg, diameter 1.2 m, and length 1.7 m. Prior to launching from the shuttle cargo bay, it is set spinning at 1.5 rev/s about the cylinder axis; see Fig. 29. Calculate the satellite's (*a*) rotational inertia about the rotation axis and (*b*) rotational kinetic energy.

51E. Two particles, each with mass *m*, are fastened to each other and to a rotation axis by two light rods, each with length *l*

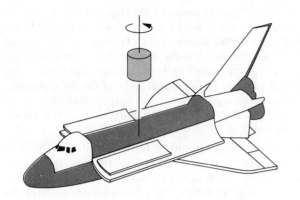

Figure 29 Exercise 50.

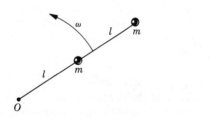

Figure 30 Exercise 51.

and mass M, as shown in Fig. 30. The combination rotates around the rotation axis with angular velocity ω. Obtain algebraic expressions for (a) the rotational inertia of the combination about O and (b) the kinetic energy of rotation about O.

52E. Each of three helicopter rotor blades shown in Fig. 31 is 5.2 m long and has a mass of 240 kg. The rotor is rotating at 350 rev/min. (a) What is the rotational inertia of the rotor assembly about the axis of rotation? (Each blade can be considered a thin rod; why?) (b) What is the kinetic energy of rotation?

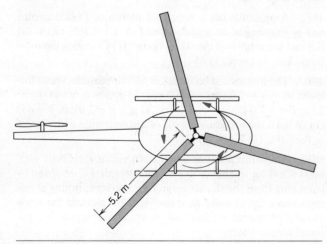

Figure 31 Exercise 52.

53E. Assume the earth to be a sphere of uniform density. Calculate (a) its rotational inertia and (b) its rotational kinetic energy. (c) Suppose that this energy could be harnessed for our use. For how long could the earth supply 1.0 kW of power to each of the 4.2×10^9 persons on earth?

54P. Show that the axis about which a given rigid body has its smallest rotational inertia must pass through its center of mass. (*Hint:* Use the parallel-axis theorem.)

55P. Figure 32 shows a uniform block of mass M and edge lengths a, b, and c. Calculate its rotational inertia about an axis through one corner and perpendicular to the large face of the block. (*Hint:* See Table 2.)

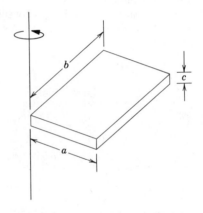

Figure 32 Problem 55.

56P. (a) Show that a solid cylinder of mass M and radius R is equivalent to a thin hoop of mass M and radius $R/\sqrt{2}$, for rotation about a central axis. (b) The radial distance from a given axis at which the mass of a body could be concentrated without altering the rotational inertia of the body about that axis is called the *radius of gyration*. Let k represent the radius of gyration and show that

$$k = \sqrt{\frac{I}{M}}.$$

This gives the radius of the "equivalent hoop" in the general case.

57P. Calculate the rotational inertia of a meter stick, with mass 0.56 kg, about an axis perpendicular to the stick and located at the 20-cm mark.

58P. Delivery trucks that operate by making use of energy stored in a rotating flywheel have been used in Europe. The trucks are charged by using an electric motor to get the flywheel up to its top speed of 200π rad/s. One such flywheel is a solid, homogeneous cylinder with a mass of 500 kg and a radius of 1.0 m. (a) What is the kinetic energy of the flywheel after charging? (b) If the truck operates with an average power requirement of 8.0 kW, for how many minutes can it operate between chargings?

Section 11–8 Torque

59E. A small 0.75-kg ball is attached to a 1.25-m massless rod and hung from a pivot. When the resulting pendulum is 30° from the vertical, what is the magnitude of the torque about the pivot?

60E. If the length of a bicycle pedal arm is 6.0 in. (= 0.152 m) and a downward force of 25 lb (= 111 N) is applied by the foot, what is the torque about the pivot point when the arm makes an angle of (a) 30°, (b) 90°, and (c) 180° with the vertical?

61E. A bicyclist of mass 70 kg puts all his weight on each downward-moving pedal as he climbs up a steep road. Take the diameter of the circle in which the pedals rotate to be 0.40 m and determine the maximum torque he exerts in the process.

62P. The body shown in Fig. 33 is pivoted at O. Three forces act on it in the directions shown on the figure: $F_A = 10$ N at point A, 8.0 m from O; $F_B = 16$ N at point B, 4.0 m from O; and $F_C = 19$ N at point C, 3.0 m from O. What are the magnitude and direction of the resultant torque about O?

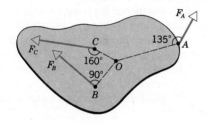

Figure 33 Problem 62.

63P. Figure 34 shows the lines of action and the points of application of two forces about the origin O. Imagine these forces to be acting on a rigid body pivoted at O, all vectors being in the plane of the figure. (a) Find an expression for the magnitude of the resultant torque on the body. (b) If $r_1 = 1.30$ m, $r_2 = 2.15$ m, $F_1 = 4.20$ N, $F_2 = 4.90$ N, $\theta_1 = 75°$, and $\theta_2 = 60°$, what are the magnitude and direction of the resultant torque?

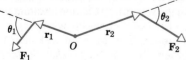

Figure 34 Problem 63.

Section 11–9 Newton's Second Law for Rotation

64E. When a torque of 32 N·m is applied to a certain wheel, it acquires an angular acceleration of 25 rad/s². What is the rotational inertia of the wheel?

65E. In the act of jumping off a diving board, a diver changed his angular velocity from zero to 6.2 rad/s in 220 ms. The diver's rotational inertia is 12 kg·m². (a) What was the angular acceleration during the jump? (b) What external torque acted on the diver during the jump?

66E. A cylinder having a mass of 2.0 kg rotates about an axis through its center. Forces are applied as in Fig. 35. $F_1 = 6.0$ N, $F_2 = 4.0$ N, and $F_3 = 2.0$ N. Also, $R_1 = 5.0$ cm and $R_2 = 12$ cm. Find the magnitude and direction of the angular acceleration of the cylinder.

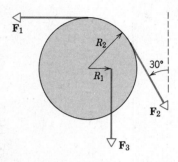

Figure 35 Exercise 66.

67E. A small object with mass 1.3 kg is mounted on one end of a light rod 0.78 m long. The system rotates in a horizontal circle about the other end of the rod at 5000 rev/min. (a) Calculate the rotational inertia of the system about the axis of rotation. (b) Air resistance exerts a force of 2.3×10^{-2} N on the object, directed opposite to its direction of motion. What torque must be applied to the system to keep it rotating at constant speed?

68E. A thin spherical shell has a radius of 1.9 m. An applied torque of 960 N·m imparts an angular acceleration equal to 6.2 rad/s² about an axis through the center of the shell. (a) What is the rotational inertia of the shell about the axis of rotation? (b) Calculate the mass of the shell.

69E. A pulley having a rotational inertia of 1.0×10^{-3} kg·m² and a radius of 10 cm is acted on by a force, applied tangentially at its rim, that varies in time as $F = 0.50t + 0.30\,t^2$, where F is in newtons if t is given in seconds. If the pulley was initially at rest, calculate (a) its angular acceleration and (b) its angular velocity after 3.0 s.

70P. Figure 36 shows the massive shield door at a neutron test facility at Lawrence Livermore Laboratory; this is the world's heaviest hinged door. The door has a mass of 44,000 kg, a rotational inertia about its hinge line of 8.7×10^4 kg·m² and a width of 2.4 m. What steady force, applied at its outer edge at right angles to the door, can move it from rest through an angle of 90° in 30 s?

Figure 36 Problem 70.

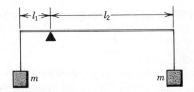

Figure 37 Problem 73.

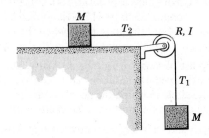

Figure 38 Problem 74.

71P. Two uniform solid spheres have the same mass, 1.65 kg, but one has a radius of 0.226 m while the other has a radius of 0.854 m. (*a*) For each of the spheres, find the torque required to bring the sphere from rest to an angular velocity of 317 rad/s in 15.5 s. Each rotates about an axis through the center of the sphere. (*b*) For each sphere, what force applied tangentially at the equator would provide the needed torque?

72P. In an Atwood's machine (Fig. 20, Chapter 5), one block has a mass of 500 g and the other a mass of 460 g. The pulley, which is mounted in horizontal frictionless bearings, has a radius of 5.0 cm. When released from rest, the heavier block is observed to fall 75 cm in 5.0 s. (*a*) What is the acceleration of each block? (*b*) What is the tension in the part of the cord that supports each block? (*c*) What is the angular acceleration of the pulley? (*d*) Calculate the rotational inertia of the pulley.

73P. Figure 37 shows two blocks each of mass m suspended from the ends of a rigid weightless rod of length $l_1 + l_2$, with $l_1 = 20$ cm and $l_2 = 80$ cm. The rod is held in the horizontal position shown in the figure and then released. Calculate the accelerations of the two blocks as they start to move.

74P. Two identical blocks, each of mass M, are connected by a light string over a frictionless pulley of radius R and rota-

tional inertia I (Fig. 38). The string does not slip on the pulley, and it is not known whether or not there is friction between the plane and the sliding block. When this system is released, it is found that the pulley turns through an angle θ in time t and the acceleration of the blocks is constant. (*a*) What is the angular acceleration of the pulley? (*b*) What is the acceleration of the two blocks? (*c*) What are the tensions in the upper and lower sections of the string? All answers are to be expressed in terms of M, I, R, θ, g, and t.

75P. The length of the day is increasing at the rate of about 1 ms/century. This is primarily due to frictional forces generated by movement of water in the world's shallow seas as a response to the tidal forces exerted by the sun and moon. (*a*) At what rate is the earth losing rotational kinetic energy? (*b*) What is its angular acceleration? (*c*) What tangential force, exerted at (average) latitudes 60° N and 60° S, is applied by the oceans on the near-coastal seabed?

76P. A tall chimney cracks near its base and falls over. Express (*a*) the radial and (*b*) the tangential linear acceleration of the top of the chimney as a function of the angle θ made by the chimney with the vertical. (*c*) Can the resultant linear acceleration exceed g? (*d*) The chimney breaks during the fall. Explain how this can happen. (See "More on the Falling Chimney," by Albert A. Bartlett, *The Physics Teacher*, September 1976.)

Section 11–10   Work, Power, and the Work–Energy Theorem

77E. If $R = 12$ cm, $M = 400$ g, and $m = 50$ g in Fig. 20, find the speed of m after it has descended 50 cm starting from rest. Solve the problem using energy-conservation principles.

78E. An automobile engine develops 100 hp (= 74.6 kW)

when rotating at a speed of 1800 rev/min. What torque does it deliver?

79E. A 32-kg wheel with radius 1.2 m is rotating at 280 rev/min. It must be brought to a stop in 15 s. (*a*) How much work must be done to stop it? (*b*) What is the required power? Assume the wheel to be a thin hoop.

80E. A thin rod of length *l* and mass *m* is suspended freely from one end. It is pulled aside and allowed to swing about a horizontal axis, passing through its lowest position with an angular speed ω. (*a*) Calculate its kinetic energy as it passes through its lowest position. (*b*) How high does its center of mass rise above its lowest position? Neglect friction and air resistance.

81P. A meter stick is held vertically with one end on the floor and is then allowed to fall. Find the speed of the other end when it hits the floor, assuming that the end on the floor does not slip. (*Hint:* Use conservation of energy.)

82P. A rigid body is made of three identical thin rods fastened together in the form of a letter H (Fig. 39). The body is free to rotate about a horizontal axis that passes through one of the legs of the H. The body is allowed to fall from rest from a position in which the plane of the H is horizontal. What is the angular speed of the body when the plane of the H is vertical?

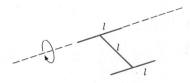

Figure 39 Problem 82.

83P. Calculate (*a*) the torque, (*b*) the energy, and (*c*) the average power required to accelerate the earth from rest to its present angular speed about its axis in 1 day.

84P. A helicopter rotor blade is 7.8 m long and has a mass of 110 kg. (*a*) What force is exerted on the bolt attaching the blade to the rotor axle when the rotor is turning at 320 rev/min? (*Hint:* For this calculation the blade can be considered to be a point mass at the center of mass. Why?) (*b*) Calculate the torque that must be applied to the rotor to bring it to full speed from rest in 6.7 s. Ignore air resistance. (The blade cannot be

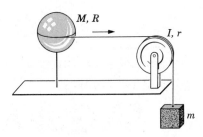

Figure 40 Problem 85.

considered to be a point mass for this calculation. Why not? Assume the distribution of a uniform rod.)

85P. A uniform spherical shell rotates about a vertical axis on frictionless bearings (Fig. 40). A light cord passes around the equator of the shell, over a light, frictionless pulley, and is attached to a small object that is otherwise free to fall under the influence of gravity. What is the speed of the object after it has fallen a distance *h* from rest? Use the work–energy theorem.

86P. A uniform steel rod of length 1.20 m and mass 6.40 kg has attached to each end a small ball of mass 1.06 kg. The rod is constrained to rotate in a horizontal plane about a vertical axis through its midpoint. At a certain instant, it is observed to be rotating with an angular speed of 39.0 rev/s. Because of axle friction, it comes to rest 32.0 s later. Compute, assuming a constant frictional torque, (*a*) the angular acceleration, (*b*) the retarding torque exerted by axle friction, (*c*) the total work done by the axle friction, and (*d*) the number of revolutions executed during the 32.0 s. (*e*) Now suppose that the frictional torque is known not to be constant. Which, if any, of the quantities (*a*), (*b*), (*c*), or (*d*) can still be computed without requiring any additional information? If such exists, give its value.

87P*. A car is fitted with an energy-conserving flywheel, which in operation is geared to the driveshaft so that it rotates at 240 rev/s when the car is traveling at 80 km/h. The total mass of the car is 800 kg, the flywheel weighs 200 N, and it is a uniform disk 1.1 m in diameter. The car descends a 1500-m long, 5° slope, from rest, with the flywheel engaged and no power supplied from the motor. Neglecting friction and the rotational inertia of the wheels, find (*a*) the speed of the car at the bottom of the slope, (*b*) the angular acceleration of the flywheel at the bottom of the slope, and (*c*) the power being absorbed by the rotation of the flywheel at the bottom of the slope.

CHAPTER 12

ROLLING, TORQUE, AND ANGULAR MOMENTUM

The yo-yo was known in ancient Greece as the photo of a bowl figure (~ 450 B.C.) shows. The yo-yo is a physics lab that you can put in your pocket or your purse. This simple device can illustrate many principles of physics, particularly those dealing with rotation, translation, and kinetic energy.

12–1 Discovering the Wheel

The teenager who longs for a set of wheels is in tune with an ancient yearning. Potter's wheels, spinning wheels, and water wheels are found in the earliest records. It is, however, the wheel that rolls that fascinates us all. The novelist Vladimir Nabokov put it well: "The miraculous paradox of smooth round objects conquering space by simply tumbling over and over . . . must have given young mankind a most salutary shock."

Figure 1*a* shows a wheeled cart, requiring a force $\mathbf{F}_c$ to pull it across the ground at constant speed. Compare this with the sledge of Fig. 1*b*, which requires a considerably greater pulling force, $\mathbf{F}_s$. Why is the cart so much

more effective than the sledge? Put another way, how does a wheel work?*

You might suppose that a wheel works because sliding friction is replaced with the much smaller rolling friction. That is part of the answer but note that sliding friction does not disappear; the frictional force merely acts at the axle instead of at the ground. It is true that friction at the axle can be dealt with more readily by lubrication or by the use of ball bearings (more wheels!) but that is not the full answer either. In what follows, we

---

* See "Why Wheels Work," by Richard D. Stepp, *The Physics Teacher*, November 1982, for a full account.

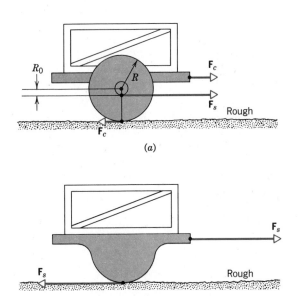

(a)

(b)

Figure 1 (a) A cart, with wheels of radius R and an axle of radius R_0. The frictional force $\mathbf{F}_s$ is exerted by the axle on the wheel as the cart is pulled to the right. (b) A sledge. For the same load, the force $\mathbf{F}_s$ required to pull the sledge is much greater than the force $\mathbf{F}_c$ required to pull the cart.

shall assume (for simplicity) that the frictional force acting at the axle in Fig. 1a has the same effective magnitude as the frictional force acting at the bottom of the sledge in Fig. 1b.

In Fig. 1a, we show the two forces $\mathbf{F}_s$ and $\mathbf{F}_c$ that act on the wheel. As the cart moves forward at constant speed, the wheel has no angular acceleration. Therefore, the net external torque acting on the wheel must be zero. Thus, considering torques about the axle and taking the

clockwise direction as positive,

$$\sum \tau = (F_c)(R) - (F_s)(R_0) = 0,$$

or

$$F_c = F_s(R_0/R). \tag{1}$$

Thus, the magnitude of the cart force is less than that of the sledge force by the factor R_0/R. The secret of the wheel is that, like a lever, it can be used to exchange a greater force (F_s) for a lesser one (F_c). In the language of levers, the ratio R_0/R of Eq. 1 is the *mechanical advantage* of the wheel.

12-2 Rolling

When a bicycle moves along a straight track, the center of each wheel moves forward in pure translation. A point on the rim of the wheel, however, traces out a more complex path, as Fig. 2 shows. In what follows, we analyze the motion of a rolling wheel by viewing it as a combination of pure translation and pure rotation.

Rolling as Rotation and Translation Combined. A ground observer sees the center of mass of a bicycle wheel moving forward at a constant linear speed v_{cm}. The rider, watching the same wheel, sees it in pure rotation, with constant angular speed ω. The rider also sees the road moving backward in his reference frame, at the same linear speed v_{cm}. The relation between these two speeds is

$$\boxed{v_{cm} = \omega R} \quad \text{(pure rolling motion)}, \tag{2}$$

in which R is the radius of the wheel. This relation only

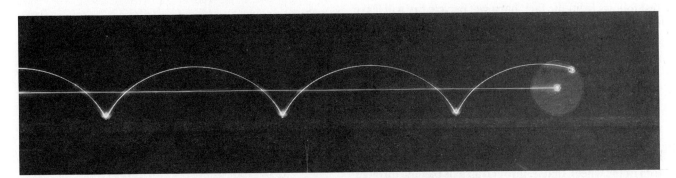

Figure 2 A time exposure photo of a rolling disk. Small lights have been attached to the disk, at its center and at its edge. The latter light traces out a curve called a *cycloid*.

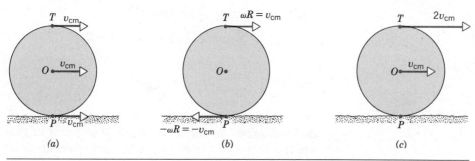

(a)   *(b)*   *(c)*

Figure 3 Rolling as the superposition of translation and rotation about the central axis. (*a*) The translational component: All points move with the same linear velocity (*b*) The rotational component: All points move with the same angular velocity about the central axis. (*c*) The superposition of (*a*) and (*b*). The velocities of the top, the center, and the bottom of the wheel are shown.

holds if the wheel rolls without skidding. For a skidding wheel, ω is zero but v_{cm} is not, so that Eq. 2 cannot possibly hold.

Figure 3 shows how the motion of the rolling wheel can be broken down into its translational and rotational elements. Figure 3*a* shows the translational component, in which every point on the wheel moves forward with the same velocity $\mathbf{v}_{cm}$. Figure 3*b* shows the rotational component, in which every point on the wheel moves with the same angular speed ω about the central axis. Points on the rim all have the same linear speed v_{cm} ($= \omega R$). Figure 3*c* shows the result of superimposing these two motions.

From Fig. 3*c* we conclude that *the top of the wheel moves twice as fast as the center and the bottom does not move at all*. A point on the rim of a flanged railroad wheel, which extends a little below the track level, actually moves backward during a portion of each rotation.

That the bottom of the wheel is instantaneously at rest follows from the fact that a wet bicycle tire will leave tread marks on a dry pavement. Figure 4 shows a time exposure of a rolling bicycle wheel. The blurring of the spokes near the top of the wheel compared with the sharp focus of the spokes near the bottom of the wheel shows that the wheel is moving faster near its top than near its bottom.

Rolling as Pure Rotation. Figure 3*c* suggests another way to look at a rolling wheel, namely, as pure rotation about an instantaneous stationary axis through the bottom of the wheel, that is, through an axis passing through point *P* in Fig. 3*c*, at right angles to the plane of the figure. Figure 5 shows a little more detail, the vectors representing the instantaneous velocities of various points on the rolling wheel.

Figure 4 A photo of a rolling bicycle wheel. Note that the spokes near the top of the wheel are more blurred than those near the bottom of the wheel. This is because they are moving faster, as Fig. 3*c* shows. (The use of a camera with a focal plane shutter complicates the interpretation of the image somewhat but the essential point remains.)

Question. What angular speed about this new axis will a ground observer assign to a rolling bicycle wheel?
Answer. The same angular speed ω that the rider assigns to the wheel as he observes it in pure rotation about an axis through its center of mass.

To gain confidence in this statement, let us use it to calculate the linear speed of the top of the rolling wheel

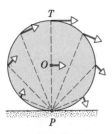

Figure 5 Rolling can be viewed as pure rotation, with angular speed ω, about an instantaneous axis through P. The vectors show the instantaneous linear velocities of selected points on the rolling disk.

from the point of view of the ground observer. The top is distant $2R$ from the instantaneous axis through P in Fig. 3c so that its linear speed should be (using Eq. 2)

$$v_{\text{top}} = (\omega)(2R) = 2(\omega R) = 2v_{\text{cm}},$$

in exact agreement with Fig. 3c.

The Kinetic Energy.   Let us calculate the kinetic energy (as seen by the ground observer) of our rolling wheel. If we view the rolling as pure rotation about an instantaneous axis through P in Fig. 3c, we have

$$K = \tfrac{1}{2}I_P\omega^2, \tag{3}$$

in which ω is the angular speed of rotation and I_P is the rotational inertia of the wheel about the instantaneous axis. From the parallel-axis theorem (see Eq. 24 of Chapter 11) we have

$$I_P = I_{\text{cm}} + MR^2, \tag{4}$$

in which M is the mass of the wheel and I_{cm} is its rotational inertia about a parallel axis through its center of mass. Substituting Eq. 4 into Eq. 3 and using the relation $v_{\text{cm}} = \omega R$ (see Eq. 2), we obtain

$$K = \tfrac{1}{2}I_{\text{cm}}\omega^2 + \tfrac{1}{2}MR^2\omega^2$$

or

$$K = \tfrac{1}{2}I_{\text{cm}}\omega^2 + \tfrac{1}{2}Mv_{\text{cm}}^2. \tag{5}$$

We can interpret the first of these terms ($\tfrac{1}{2}I_{\text{cm}}\omega^2$) as the kinetic energy associated with the rotation of the wheel about an axis through its center of mass (as in Fig. 3b) and the second term ($\tfrac{1}{2}Mv_{\text{cm}}^2$) as the kinetic energy associated with the translational motion of the wheel (as in Fig. 3a).

Sample Problem 1   A solid cylindrical disk, whose mass M is 1.4 kg and whose radius R is 8.5 cm, rolls across a horizontal table at a speed v of 15 cm/s. (a) What is the instantaneous velocity of the top of the rolling cylinder?

When we speak of the speed of a circular rolling object, we always mean the speed of its center. From Fig. 3c we see that the speed of the top of the disk is just twice this, or

$$v_{\text{top}} = 2v_{\text{cm}} = (2)(15 \text{ cm/s}) = 30 \text{ cm/s.} \quad \text{(Answer)}$$

(b) What is the angular speed ω of the rolling disk? From Eq. 2 we have

$$\omega = \frac{v_{\text{cm}}}{R} = \frac{15 \text{ cm/s}}{8.5 \text{ cm}} = 1.8 \text{ rad/s} = 0.28 \text{ rev/s.} \quad \text{(Answer)}$$

The value applies whether the axis of rotation is taken to be the instantaneous axis through point P in Fig. 3 or a parallel axis through the center of mass.

(c) What is the kinetic energy K of the rolling disk?

From Eq. 5 we have, putting $I_{\text{cm}} = \tfrac{1}{2}MR^2$ and using the relation $v_{\text{cm}} = \omega R$,

$$K = \tfrac{1}{2}I_{\text{cm}}\omega^2 + \tfrac{1}{2}Mv_{\text{cm}}^2$$

$$= (\tfrac{1}{2})(\tfrac{1}{2}MR^2)(v_{\text{cm}}/R)^2 + \tfrac{1}{2}Mv_{\text{cm}}^2 = \tfrac{3}{4}Mv_{\text{cm}}^2$$

$$= \tfrac{3}{4}(1.4 \text{ kg})(0.15 \text{ m/s})^2$$

$$= 0.024 \text{ J} = 24 \text{ mJ.} \quad \text{(Answer)}$$

(d) What fraction of the kinetic energy is associated with the motion of translation and what fraction with the motion of rotation about the axis through the center of mass?

The kinetic energy associated with translation is the second term of Eq. 5, or $\tfrac{1}{2}Mv_{\text{cm}}^2$. The fraction we seek is then, using the expression derived in (c) above,

$$f = \frac{\tfrac{1}{2}Mv_{\text{cm}}^2}{\tfrac{3}{4}Mv_{\text{cm}}^2} = \frac{2}{3} \text{ or } 67\%. \quad \text{(Answer)}$$

The remaining 33% is associated with rotation about an axis through the center of mass.

The relative split between translational and rotational energy depends on the rotational inertia of the rolling object. We can summarize as follows:

Object	Rotational Inertia I_{cm}	Percentage of Energy in	
		Translation	Rotation
Hoop	$1MR^2$	50%	50%
Disk	$\tfrac{1}{2}MR^2$	67%	33%
Sphere	$\tfrac{2}{5}MR^2$	71%	29%
General	βMR^2	$100\dfrac{1}{1+\beta}$	$100\dfrac{\beta}{1+\beta}$

We see that the object (hoop) that has its mass the farthest from the central axis of rotation (largest rotational inertia) stores the largest share of its kinetic energy in rotary motion. The sphere, which has its mass closest to the central axis of rotation (smallest rotational inertia), stores the smallest share in that form.

The formulas at the bottom of the table refer to a generic rolling object whose rotational inertia parameter is β. As the table shows, this numerical constant has the value 1 for a hoop, $\frac{1}{2}$ for a disk, and $\frac{2}{3}$ for a sphere.

Sample Problem 2   A bowling ball, whose radius R is 11 cm and whose mass M is 7.2 kg, rolls from rest down a plank whose length L is 2.1 m. The plank is inclined at an angle θ of 34° to the horizontal; see Fig. 6. How fast is the ball moving when it reaches the bottom of the plank?

The center of the ball falls through a vertical distance of L $\sin \theta$, so that the ball loses gravitational energy in amount MgL $\sin \theta$. This loss of potential energy must be equal to the gain in kinetic energy. Thus, we can write (see Eq. 5)

$$MgL \sin \theta = \tfrac{1}{2}Mv_{cm}^2 + \tfrac{1}{2}I_{cm}\omega^2.$$

From Table 2(g) in Chapter 11 we see that, for a solid sphere, $I_{cm} = \frac{2}{3}MR^2$. We can also replace ω by its equal, v_{cm}/R. Substituting both of these quantities into the above equation yields

$$MgL \sin \theta = \tfrac{1}{2}Mv_{cm}^2 + (\tfrac{1}{2})(\tfrac{2}{3})(MR^2)(v_{cm}/R)^2.$$

Solving for v_{cm} yields

$$v_{cm} = \sqrt{(\tfrac{10}{7})gL \sin \theta}$$
$$= \sqrt{(\tfrac{10}{7})(9.8 \text{ m/s}^2)(2.1 \text{ m})(\sin 34°)}$$
$$= 4.1 \text{ m/s.} \qquad \text{(Answer)}$$

Note that the answer does not depend on the mass or the radius of the ball.

Sample Problem 3   Here we generalize the result of Sample Problem 2. As Fig. 6 shows, a hoop, a disk, and a sphere, having the same mass M and the same radius R, are released from rest at the top of a ramp whose length L is 2.5 m and whose ramp angle θ is 12°. (a) Which object will reach the bottom first?

The table in Sample Problem 1 gives us the answer. The sphere puts the largest share of its kinetic energy (71%) into translational motion so it will win the race. Next will come the disk, with the hoop being in third place.

(b) How fast will these objects be moving at the bottom of the ramp?

In rolling down the ramp, the center of mass of each object falls through the same vertical distance h. Like a body in free fall, the object loses potential energy in amount Mgh and thus gains this amount of kinetic energy. At the bottom of the ramp then, the total kinetic energy of each object is the same. How this kinetic energy is divided between translational and rotational forms depends on the object's distribution of mass.

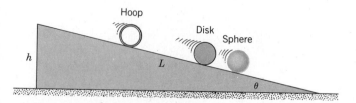

Figure 6   Sample Problem 3. A hoop, a disk, and a sphere roll from rest down a ramp of angle θ. Although released from rest at the same position and time, they arrive at the bottom in the order shown.

From Eq. 5 we can write (putting $\omega = v_{cm}/R$)

$$Mgh = \tfrac{1}{2}I_{cm}\omega^2 + \tfrac{1}{2}Mv_{cm}^2$$
$$= \tfrac{1}{2}I_{cm}(v_{cm}^2/R^2) + \tfrac{1}{2}Mv_{cm}^2$$
$$= \tfrac{1}{2}(I_{cm}/R^2)v_{cm}^2 + \tfrac{1}{2}Mv_{cm}^2. \qquad (6)$$

From the table in Sample Problem 1, we see that $I_{cm}/R^2 = \beta M$, where β is a numerical constant that characterizes the rolling object. Making this substitution into Eq. 6, putting $h = L \sin \theta$, and solving for v_{cm}, we obtain

$$v_{cm} = \sqrt{\frac{2gL \sin \theta}{1 + \beta}}, \qquad \text{(Answer)} \quad (7)$$

which is the answer we seek.

Note that the speed does not depend on the mass or the radius of the rolling object, only on the distribution of its mass about its central axis, which enters through the parameter β. A marble and a bowling ball will have the same speed at the bottom of the ramp and will thus roll down the ramp in the same time. A bowling ball will always beat a disk, of any mass or radius, and anything will beat a hoop.*

For a hoop (see the table in Sample Problem 1) we have $\beta = 1$, so that Eq. 7 yields.

$$v_{cm} = \sqrt{\frac{2gL \sin \theta}{1 + \beta}}$$
$$= \sqrt{\frac{(2)(9.8 \text{ m/s}^2)(2.5 \text{ m})(\sin 12°)}{1 + 1}}$$
$$= 2.26 \text{ m/s.} \qquad \text{(Answer)}$$

A similar calculation yields $v_{cm} = 2.61$ m/s for the disk ($\beta = \frac{1}{2}$) and 2.70 m/s for the sphere ($\beta = \frac{2}{3}$). This supports our prediction of (a) above that the win, place, and show sequence in this race will be sphere, disk, and hoop.

Sample Problem 4   Figure 7 shows a body of mass M and radius R rolling down a ramp that slopes at an angle θ. This

---

* Not quite true; see Question 8.

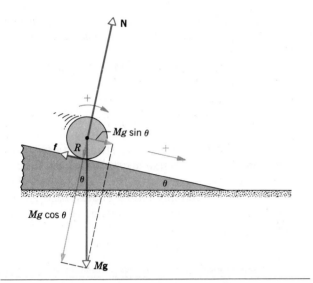

Figure 7   Sample Problem 4. A body of radius R rolls down a ramp. The forces that act on it are its weight $M\mathbf{g}$, an outward normal force $\mathbf{N}$, and a frictional force $\mathbf{f}$ pointing up the ramp.

time let us analyze its motion directly from Newton's laws, rather than by energy methods as we did in Sample Problem 3. (a) What is the linear acceleration of the rolling object?

Figure 7 shows the forces that act on the rolling body. They include its weight $M\mathbf{g}$, a normal force $\mathbf{N}$ exerted outward by the ramp, and a frictional force $\mathbf{f}$ that acts at the contact point and provides the torque that causes the body to rotate. Let us first apply Newton's second law in its linear form, dealing only with components parallel to the ramp. We have (from $\Sigma F = Ma$)

$$\sum F = Mg \sin \theta - f = Ma. \qquad (8)$$

Now we apply Newton's second law in its angular form. We have, applying the relation $\Sigma \tau = I\alpha$ about an axis through the center of mass,*

$$\sum \tau = (f)(R) = I\alpha = I(a/R). \qquad (9)$$

Solving Eq. 9 for the frictional force f gives

$$f = (I/R^2)a = \beta Ma. \qquad (10)$$

Here we have replaced I/R^2 by βM; see the table in Sample Problem 1. Substituting Eq. 10 into Eq. 8 and solving for a

---

* We derived the relation $\Sigma \tau = I\alpha$ for an axis fixed in an inertial reference frame. However, we are applying it here to an axis (that of the rolling body) that is accelerating. It can be shown that this relation still holds provided (1) the axis passes through the center of mass of the accelerating body and (2) the axis does not change its direction as it moves. Both requirements are met in this case.

yields

$$a = \frac{g \sin \theta}{1 + \beta}. \qquad \text{(Answer)} \quad (11)$$

(b) What is the frictional force f?
Substituting Eq. 11 into Eq. 10 yields

$$f = Mg \frac{\beta \sin \theta}{1 + \beta}. \qquad \text{(Answer)} \quad (12)$$

Study of Eq. 12 shows that the frictional force is less than $Mg \sin \theta$, the component of the weight that acts parallel to the ramp. This is necessarily true if the object is to accelerate down the ramp.

If the rolling object is a solid disk, the table in Sample Problem 1 shows that $\beta = \frac{1}{2}$. The acceleration and the frictional force then follow from Eqs. 11 and 12 and are

$$a = \tfrac{2}{3}g \sin \theta \quad \text{and} \quad f = \tfrac{1}{3}Mg \sin \theta.$$

(c) What is the speed of the rolling object at the bottom of the ramp?
The motion is one of constant acceleration so we can use the relation

$$v^2 = v_0^2 + 2a(x - x_0). \qquad (13)$$

Putting $x - x_0 = L$ and $v_0 = 0$ and introducing a from Eq. 11, we obtain Eq. 7—the result that we derived by energy methods. We should not be surprised because, after all, everything that we have said about mechanical energy was derived from Newton's laws in the first place.

12-3   The Yo-yo*

A rolling wheel is not the only way to study rotation and translation. You can use a yo-yo. If the yo-yo rolls down its string for a distance h, it loses potential energy in amount mgh but gains kinetic energy in both translational ($\tfrac{1}{2}mv_{cm}^2$) and rotational ($\tfrac{1}{2}I_{cm}\omega^2$) form. When the string is completely unwound, the yo-yo climbs back up, losing kinetic energy and regaining potential energy.

In a modern yo-yo the string is not tied to the axle but is looped around it. This means that, when the yo-yo reaches the bottom of its string, its kinetic energy can exist only in rotational form. The yo-yo keeps spinning ("sleeping") until you "wake it up" by tugging on the string, causing it to climb back up. The rotational kinetic energy when the yo-yo is at the bottom of its string (and thus the sleeping time) can be considerably increased by

---

* See "The Yo-yo: A Toy Flywheel," by Wolfgang Burger, *American Scientist*, March–April 1984.

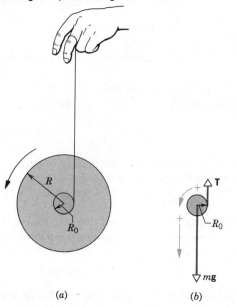

(a)   (b)

Figure 8 (a) A yo-yo, shown in cross section. The string, of assumed negligible thickness, is wound around an axle of radius R_0. (b) A free-body diagram for the falling yo-yo. Only the axle is shown. The directions taken as positive are shown by the light arrows and their associated "+" signs.

throwing the yo-yo down, instead of letting it roll down from rest.

Let us analyze the motion of the yo-yo directly from Newton's second law. Figure 8a shows an idealized yo-yo, in which the thickness of the string can be neglected.* Figure 8b shows a free-body diagram, in which only the yo-yo axle is shown. Applying Newton's second law in its linear form ($\Sigma F = ma$) yields

$$\Sigma F = Mg - T = Ma. \qquad (14)$$

Here M is the mass of the yo-yo and T is the tension in the cord.

Applying Newton's second law in angular form ($\Sigma \tau = I\alpha$) about an axis through the center of mass yields

$$\Sigma \tau = I\alpha,$$

or, using the relation $a = \alpha R_0$,

$$TR_0 = I(a/R_0). \qquad (15)$$

---

* In a real yo-yo, the thickness of the string cannot be neglected. This means that the effective radius of the yo-yo axle is not constant, but varies with the amount of wound-up string.

Here R_0 is the radius of the yo-yo axle, I is the rotational inertia of the yo-yo about its central axis, and α is the angular acceleration of the yo-yo.

Eliminating T between Eqs. 14 and 15 and solving for a, we obtain

$$a = g \frac{1}{1 + I/MR_0^2}. \qquad (16)$$

Thus, an ideal yo-yo rolls down its string with constant acceleration. To have a small acceleration, you need a yo-yo with a large rotational inertia and a small axle radius. Galileo, in studying falling bodies, used an inclined plane to simulate a reduced value of gravity. He could have used a yo-yo.

Sample Problem 5 A yo-yo is constructed of two brass disks whose thickness b is 8.5 mm and whose radius R is 3.5 cm, joined by a short axle whose radius R_0 is 3.2 mm. (a) What is its rotational inertia about its central axis? Neglect the rotational inertia of the axle. The density ρ of brass is 8400 kg/m³.

The rotational inertia I of a disk about its central axis is $\frac{1}{2}MR^2$. In this problem, we can treat the two disks together, as a single disk. We first find its mass from

$$M = V\rho = (2)(\pi R^2)(b)(\rho)$$
$$= (2)(\pi)(0.035 \text{ m})^2(0.0085 \text{ m})(8400 \text{ kg/m}^3)$$
$$= 0.550 \text{ kg}.$$

The rotational inertia is then

$$I = \frac{1}{2}MR^2 = (\frac{1}{2})(0.550 \text{ kg})(0.035 \text{ m})^2$$
$$= 3.37 \times 10^{-4} \text{ kg} \cdot \text{m}^2. \qquad \text{(Answer)}$$

(b) A string of length l (= 1.1 m) and of negligible thickness is wound on the axle. What is the linear acceleration of the yo-yo as it rolls down the string from rest?

From Eq. 16,

$$a = g \frac{1}{1 + I/MR_0^2}$$
$$= \frac{(9.8 \text{ m/s}^2)}{1 + (3.37 \times 10^{-4} \text{ kg} \cdot \text{m}^2)/(0.550 \text{ kg})(0.0032 \text{ m})^2}$$
$$= 0.16 \text{ m/s}^2. \qquad \text{(Answer)}$$

The acceleration points downward and has this same value whether the yo-yo is falling or climbing.

Note that the quantity I/MR_0^2 in the expression for the acceleration (above) is simply the rotational inertia parameter β that we introduced in the table in Sample Problem 1. For this yo-yo, we have $\beta = 60$, a much greater value than that for any of the objects listed in that table. The acceleration of our yo-yo is small, corresponding to that of a hoop rolling down a 1.9° ramp.

(c) What is the tension in the string of the yo-yo?

We can find this by eliminating *a* from Eqs. 15 and 16 and solving for *T*. We find

$$T = \frac{Mg}{1 + MR_0^2/I},$$ (17)

which tells us, as it must, that the tension in the string is less than the weight of the yo-yo. Numerically,

$$T = \frac{(0.550 \text{ kg})(9.8 \text{ m/s}^2)}{1 + (0.550 \text{ kg})(0.0032 \text{ m})^2/(3.37 \times 10^{-4} \text{ kg} \cdot \text{m}^2)}$$

$$= 5.3 \text{ N}.$$ (Answer)

This tension holds whether the yo-yo is falling or climbing.

12-4   Torque Revisited

As a rigid body rotates about a fixed axis, the particles that make up that body move in concentric circles about that axis. We now want to break out of this restriction and consider a particle free to move in three dimensions about the origin of a coordinate system. We start by generalizing our definition of torque.

Figure 9 shows a particle of mass *m* at point *P* in the *xy* plane. Its position with respect to the origin *O* is specified by the vector **r**. A single force **F** acts on the particle, the vectors **r** and **F** making an angle ϕ with each other.

We define the torque exerted on the particle by this force with respect to the origin *O*, as*

$$\boxed{\boldsymbol{\tau} = \mathbf{r} \times \mathbf{F}}$$ (torque defined).   (18)

According to the rules for a vector product (see Fig. 19 of Chapter 3), the vector $\boldsymbol{\tau}$ lies parallel to the *z* axis, in the direction of increasing *z*. In applying the right-hand rule to find this direction, it helps to slide the vector **F** parallel to itself until its tail is at the origin *O*, as the dashed vector in Fig. 9 shows.

The magnitude of the vector $\boldsymbol{\tau}$ is given (see Eq. 19 of Chapter 3) by

$$\tau = rF \sin \phi,$$ (19)

which agrees with our earlier definition, in which we treated only the special case of a torque acting on a rigid

---

* This is the first use of the vector product in this book. You may wish to review Section 3-7.

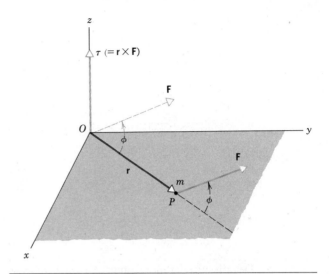

Figure 9 Defining torque. A force **F**, assumed to lie in the *xy* plane, acts on a particle at point *P*. This force exerts a torque $\boldsymbol{\tau}$ (= **r** × **F**) on the particle with respect to the origin *O*. The torque vector points in the direction of increasing *z*.

body confined to rotate about a fixed axis; see Eq. 28 of Chapter 11.

Torque, as defined by Eq. 18, has meaning only with respect to a specified origin. If the particle *P* in Fig. 9 did not lie in the *xy* plane, or if the force **F** did not lie in that plane, the torque $\boldsymbol{\tau}$ would not lie along the *z* axis. The direction of $\boldsymbol{\tau}$ is always at right angles to the plane formed by the vectors **r** and **F**.

If the "particle" at *P* in Fig. 9 is the earth and the sun is located at the origin, the force **F** that acts at *P* would be the gravitational attraction of the sun. This force points directly toward the sun and so can exert no torque about it. This conclusion is consistent with Eq. 19 because the angle ϕ between the vector **r** (which points from sun to earth) and vector **F** (which points from earth to sun) is 180° and sin 180° = 0.

12-5   Angular Momentum

Like all linear quantities, linear momentum has its angular counterpart. Figure 10 shows the linear momentum **p** (= *m***v**) of a particle at point *P* in the *xy* plane. The *angular momentum* **l** of this particle with respect to the origin *O* is defined from

$$\boxed{\mathbf{l} = \mathbf{r} \times \mathbf{p} = m(\mathbf{r} \times \mathbf{v})}$$ (angular momentum defined).   (20)

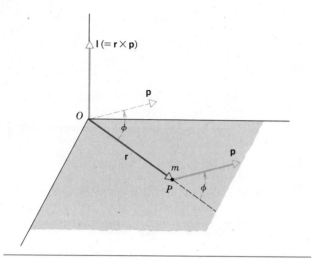

Figure 10 Defining angular momentum. A particle of mass m at point P has a linear momentum $\mathbf{p}$ ($= m\mathbf{v}$), assumed to lie in the xy plane. The particle has an angular momentum l ($= \mathbf{r} \times \mathbf{p}$) with respect to the origin O. The angular momentum vector points in the direction of increasing z.

Comparison of Eqs. 18 and 20 shows that angular momentum bears the same relation to linear momentum that torque does to force. Both have meaning only with respect to a specific origin. The SI unit of angular momentum is the kg·m²/s, equivalent to J·s.

The direction of the angular momentum vector in Fig. 10 is parallel to the z axis, in the direction of increasing z. The magnitude of this vector is given (compare Eq. 19) by

$$l = rmv \sin \phi. \quad (21)$$

If a particle is moving directly away from the origin ($\phi = 0$) or directly toward it ($\phi = 180°$), Eq. 21 tells us that the particle has no angular momentum about that origin.

Just as for torque, angular momentum has meaning only with respect to a specified origin. If the particle P in Fig. 10 did not lie in the xy plane, or if the linear momentum $\mathbf{p}$ of the particle did not also lie in that plane, the angular momentum l would not lie along the z axis. The direction of the angular momentum vector is always at right angles to the plane formed by the vectors $\mathbf{r}$ and $\mathbf{p}$.

12–6 Newton's Second Law in Angular Form

Newton's second law, written in the form

$$\sum \mathbf{F} = \frac{d\mathbf{p}}{dt} \quad \text{(single particle),} \quad (22)$$

expresses the close relation between force and linear momentum for a single particle. We have seen enough of the parallelism between linear and angular quantities to be pretty sure that there is also a close relation between torque and angular momentum. Guided by Eq. 22 we can even guess that it must be

$$\boxed{\sum \tau = \frac{dl}{dt}} \quad \text{(single particle).} \quad (23)$$

Equation 23 is indeed Newton's second law for a single particle, written in its angular form:

The (vector) sum of all the torques acting on a particle is equal to the time rate of change of the angular momentum of that particle.

Equation 23 has no meaning unless the torques τ and the angular momentum l are defined with respect to the same origin.

Proof of Equation 23.   We start with Eq. 20, the definition of angular momentum

$$l = m \, \mathbf{r} \times \mathbf{v}.$$

Differentiating* each side with respect to time t yields

$$\frac{dl}{dt} = m \, \mathbf{r} \times \frac{d\mathbf{v}}{dt} + m \, \frac{d\mathbf{r}}{dt} \times \mathbf{v}. \quad (24)$$

But $d\mathbf{v}/dt$ is the acceleration $\mathbf{a}$ of the particle and $d\mathbf{r}/dt$ is its velocity $\mathbf{v}$. Thus, we can rewrite Eq. 24 as

$$\frac{dl}{dt} = m \, \mathbf{r} \times \mathbf{a} + m \, \mathbf{v} \times \mathbf{v}.$$

Now $\mathbf{v} \times \mathbf{v} = 0$, which leads to†

$$\frac{dl}{dt} = \mathbf{r} \times m\mathbf{a} + 0 = \mathbf{r} \times \left(\sum \mathbf{F} \right) = \sum (\mathbf{r} \times \mathbf{F}). \quad (25)$$

Here we have used Newton's second law ($\sum \mathbf{F} = m\mathbf{a}$) to replace $m\mathbf{a}$ by its equal, the vector sum of the forces that act on the particle. Finally, Eq. 18 shows us that $\mathbf{r} \times \mathbf{F}$ is the torque associated with the force $\mathbf{F}$ so that Eq. 25 becomes

$$\sum \tau = \frac{dl}{dt}.$$

This is Eq. 23, the relation that we set out to prove.

---

* In differentiating a vector product, you must be sure not to change the order of the two quantities ($\mathbf{r}$ and $\mathbf{v}$) that form that product; see Eq. 20 of Chapter 3.

† The vector product of any vector by itself is zero because the angle between the two vectors is necessarily zero.

Sample Problem 6   A particle of mass m is released from rest at point P in Fig. 11. (a) What torque does the gravitational force acting on the particle exert about the origin O?

The torque is given by Eq. 18 ($\boldsymbol{\tau} = \mathbf{r} \times \mathbf{F}$), its magnitude being given by Eq. 19, or

$$\tau = rF\sin\phi.$$

In this example, $r\sin\phi = d$ and $F = mg$. Thus,

$$\tau = mgd = \text{a constant.} \tag{26}$$

Note that the torque is simply the product of the force ($= mg$) and the moment arm ($= d$). The right-hand rule shows that the torque vector $\boldsymbol{\tau}$ is directed into the plane of Fig. 11, at right angles. We represent it by a cross ($\otimes$) at the origin, suggesting the tail of an arrow.

We see that the torque depends very much (through d) on the location of the origin. If the particle is dropped from the origin, we have $d = 0$ and thus no torque.

(b) What is the angular momentum of the falling particle about the origin O?

The angular momentum is given by Eq. 20 ($\boldsymbol{l} = \mathbf{r} \times \mathbf{p}$), its magnitude being given by Eq. 21, or

$$l = rp\sin\phi.$$

In this example, $r\sin\phi = d$ and $p = mv = m(gt)$. Thus,

$$l = mgtd. \tag{27}$$

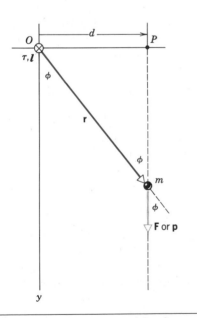

Figure 11   Sample Problem 6. A particle of mass m drops vertically from point P. The torque $\boldsymbol{\tau}$ and the angular momentum $\boldsymbol{l}$ of the falling particle with respect to origin O are directed into the plane of the figure. The cross ($\otimes$) at O (the tail of an arrow) represents these vectors.

The right-hand rule shows that the angular momentum vector $\boldsymbol{l}$ is directed into the plane of Fig. 11, at right angles; the directions of the vectors $\boldsymbol{\tau}$ and $\boldsymbol{l}$ are parallel. The vector $\boldsymbol{l}$ changes with time in magnitude only, its direction remaining unchanged. Simple inspection of Eqs. 26 and 27 shows that the relation $\boldsymbol{\tau} = d\boldsymbol{l}/dt$ is indeed satisfied.

12-7   Systems of Particles

Now we turn our attention from the motion of a single particle with respect to an origin to the motion of a system of particles with respect to an origin. The total angular momentum $\mathbf{L}$ of a system of particles is the (vector) sum of the individual angular momenta $\boldsymbol{l}$ of the particles, or

$$\mathbf{L} = \boldsymbol{l}_1 + \boldsymbol{l}_2 + \boldsymbol{l}_3 + \cdots = \sum \boldsymbol{l}_n, \tag{28}$$

in which n ($= 1, 2, 3, \ldots$) is a *running integer* that identifies the particle.

As time goes on, the angular momenta of individual particles may change, either because of interactions within the system (between the individual particles) or because of influences that may act on the system from the outside. We can find the change with time of the angular momentum $\mathbf{L}$ of the system as a whole as these changes take place by taking the time derivative of Eq. 28. Thus,

$$\frac{d\mathbf{L}}{dt} = \sum \frac{d\boldsymbol{l}_n}{dt}. \tag{29}$$

From Eq. 23, $d\boldsymbol{l}_n/dt$ is just $\sum \boldsymbol{\tau}_n$, the (vector) sum of the torques that act on the nth particle.

Some torques are *internal*, associated with forces that the particles within the system exert on each other; other torques are *external*, associated with forces that act from outside the system. The internal forces, because of Newton's law of action and reaction, cancel in pairs. In adding up the torques then, we need consider only those torques associated with external forces. Equation 29 then becomes

$$\sum \boldsymbol{\tau}_{\text{ext}} = \frac{d\mathbf{L}}{dt} \qquad \text{(system of particles).} \tag{30}$$

Equation 30 is Newton's second law for a system of particles, expressed in terms of angular quantities. In words, Eq. 30 tells us that the (vector) sum of the *external* *torques* acting on a system of particles is equal to the time rate of change of the *angular momentum* of that system. Equation 30 has meaning only if the torque and the angular momentum vectors are referred to the same origin.*

Equation 30 is the angular equivalent of Eq. 24 of Chapter 9, namely,

$$\sum \mathbf{F}_{\text{ext}} = \frac{d\mathbf{P}}{dt} \quad \text{(system of particles).}$$

This tells us that the (vector) sum of the *external forces* acting on a system of particles is equal to the time rate of change of the *linear momentum* of that system.

12–8 The Angular Momentum of a Rigid Body Rotating About a Fixed Axis

Consider a rigid body constrained (perhaps by bearings) to rotate about a fixed axis, which we take to be a *z* axis. As Fig. 12*a* shows, every mass element of the rotating rigid body moves around this axis in a circular path of radius *r*.

As Fig. 12*b* shows, the magnitude of the angular momentum vector associated with this rotating mass element is

$$l = (r)(p)(\sin 90°) = (r)(\Delta m\, v).$$

The *z* component of this angular momentum vector is†

$$l_z = l \sin \theta = (r \sin \theta)(\Delta m\, v) = r_\perp \Delta m\, v.$$

The *z* component of the angular momentum for the rotating rigid body as a whole is found by adding up the contributions of all the mass elements that make up the body. Thus,

$$L_z = \sum l_z = \sum \Delta m\, v r_\perp = \sum \Delta m(\omega r_\perp) r_\perp$$
$$= \omega \left(\sum \Delta m\, r_\perp^2 \right). \tag{31}$$

Here we have used the relation $v = \omega r$, we removed ω

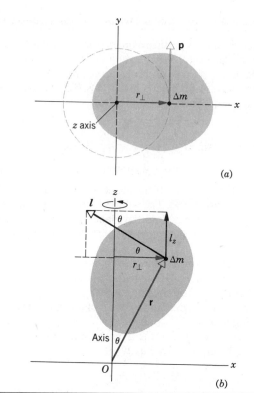

Figure 12 (*a*) The cross section of a rigid body that is rotating about the *z* axis. Every mass element Δm moves in a circle parallel to the *xy* plane, with the same angular velocity $\boldsymbol{\omega}$. (*b*) The same rotating body, showing the mass element in the *xz* plane. The angular momentum $\boldsymbol{l}$ of the rotating element is shown, along with its *z* component l_z.

from the summation because it has the same value for all points of the rotating rigid body.

The quantity $\sum \Delta m\, r_\perp^2$ in Eq. 31 (see Eq. 21 of Chapter 11) is the rotational inertia *I* of the body about the fixed axis. Thus, Eq. 31 reduces to

$$L = I\omega \quad \text{(rigid body, fixed axis).} \tag{32}$$

We have dropped the subscript *z* but must remember that the angular momentum defined by Eq. 32 refers only to the angular momentum component that lies along the rotation axis.† *I* in that equation must also be the rotational inertia about that same axis.

---

* In an inertial reference frame, Eq. 30 can be applied not only with respect to the origin but with respect to *any* other point. In an accelerating frame (such as a wheel rolling down a ramp), Eq. 30 can be applied *only* with respect to the *center of mass* of the system.

† In this section, we are only interested in angular momentum components that lie along the (fixed) rotation axis. Only such components can be related in a simple way to the rotation of the rigid body about this axis.

† If the body has rotational symmetry about an axis passing through its center of mass, then the angular momentum vector of the rotating body lies *entirely* in the *z* direction. That is, the vectors **L** and $\boldsymbol{\omega}$ point in the same direction.

Table 1   Some More Corresponding Relations for Translational and Rotational Motion[a]

Translational		Rotational	
Force	$\mathbf{F}$	Torque	$\boldsymbol{\tau}\,(=\mathbf{r}\times\mathbf{F})$
Linear momentum	$\mathbf{p}$	Angular momentum	$l\,(=\mathbf{r}\times\mathbf{p})$
Linear momentum[b]	$\mathbf{P}\left(=\sum\mathbf{p}_i\right)$	Angular momentum[b]	$\mathbf{L}\left(=\sum l_n\right)$
Linear momentum[b]	$\mathbf{P}=M\mathbf{v}_{\text{cm}}$	Angular momentum[c]	$L=I\omega$
Newton's law[b]	$\sum\mathbf{F}_{\text{ext}}=\dfrac{d\mathbf{P}}{dt}$	Newton's law[b]	$\sum\boldsymbol{\tau}_{\text{ext}}=\dfrac{d\mathbf{L}}{dt}$
Conservation law[d]	$\mathbf{P}=$ a constant	Conservation law[d]	$\mathbf{L}=$ a constant

[a] See also Table 3 of Chapter 11.
[b] For systems of particles.
[c] For a rigid body about a fixed axis.
[d] For an isolated system.

Table 1, which supplements Table 3 of Chapter 11, extends our list of corresponding linear and angular relations.

12-9   Conservation of Angular Momentum

So far we have studied two powerful conservation laws, the conservation of energy and the conservation of linear momentum. Now we meet a third law of this type, the conservation of angular momentum. We start from Eq. 30 ($\sum\boldsymbol{\tau}_{\text{ext}}=d\mathbf{L}/dt$), which is Newton's second law in angular form. If no external torques act on the system, this equation becomes $d\mathbf{L}/dt=0$, or

$$\mathbf{L}=\text{a constant} \quad \text{(isolated system).} \qquad (33)$$

This is the law of conservation of angular momentum:

If no external torques act on a system, the (vector) angular momentum $\mathbf{L}$ *of that system remains constant, no matter what changes take place within the system.*

Like the other two conservation laws that we have studied, this law holds beyond the limitations of Newtonian mechanics. It holds for particles whose speeds approach that of light (where the theory of relativity reigns) and it remains true in the world of subatomic particles (where quantum mechanics reigns). No exceptions to it have ever been found.

12-10   Conservation of Angular Momentum: Some Examples

1. The Spinning Volunteers.   Figure 13 shows a student seated on a stool that can rotate freely about a vertical axis. The student, who has been set into rotation at a modest initial rate ω_i, holds two dumbbells in his outstretched hands. His angular momentum vector $\mathbf{L}$ lies along the vertical axis, pointing upward in the figure.

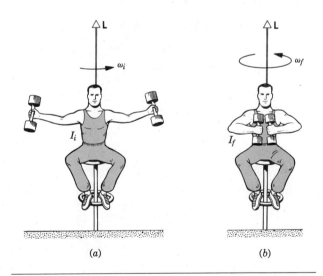

(a)   (b)

Figure 13   (a) The student has a relatively large rotational inertia and a relatively small angular speed. (b) By decreasing his rotational inertia, the student automatically increases his angular speed. The angular momentum $\mathbf{L}$ of the system remains unchanged.

The instructor now asks the student to pull in his arms, thus reducing his rotational inertia from its initial value I_i to a smaller value I_f. His rate of rotation increases markedly, from ω_i to ω_f. If the student wishes to slow down, he has only to extend his arms once more. The same phenomenon is demonstrated by ballet dancers and by figure skaters.

The *student + stool + dumbbells* form an isolated system, on which no external torques act. No matter how the student maneuvers the weights, the angular momentum of that system must remain constant. From Eqs. 32 and 33 we have

$$L = I\omega = \text{a constant,}$$

or

$$I_i\omega_i = I_f\omega_f.$$

In Fig. 13a, the student's angular speed ω_i is relatively low and his rotational inertia I_i relatively high. In Fig. 13b, the reverse is true.

2. The Springboard Diver. Figure 14 shows a diver doing a forward one-and-a-half somersault dive. Note that, as expected, her center of mass follows a parabolic path. She leaves the springboard with a definite angular momentum **L** about an axis through her center of mass,

represented by a vector pointing into the plane of Fig. 14, at right angles. When she is in the air, the diver forms an isolated system and her angular momentum cannot change. By pulling her arms and legs into the *tuck position,* she can considerably reduce her rotational inertia (about an axis at right angles to the plane of Fig. 14 through her center of mass) and thus considerably increase her angular velocity. Pulling out of the tuck position (into the *layout position*) at the end of the dive increases her rotational inertia and thus slows her rotation rate as she enters the water.

No matter how complicated the dive—a forward two-and-a-half somersault with two twists about a head-to-toe axis being an example—angular momentum must be conserved throughout the dive.*

3. Stabilizing a Satellite (or a Frisbee). Before a satellite is launched from the cargo bay of the Space Shuttle, it is made to spin about its central axis; see Fig. 15. Why?

Figure 15 Deployment of the Morelos-D satellite, a communications satellite for Mexico, from the bay of the Space Shuttle on November 17, 1985. The satellite is made to spin about its central axis to stabilize its orientation in space as it makes its way upward to the geosynchronous orbit.

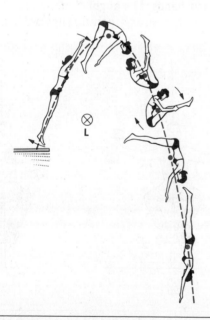

Figure 14 The diver's angular momentum **L** is constant throughout the dive, being represented by the tail of an arrow entering the plane of the figure at right angles. Note also that her center of mass (see the dots) follows a parabolic path.

---

* See "The Physics of Somersaulting and Twisting," by Cliff Frohlich, *Scientific American,* March 1980, for an analysis of such more complicated dives.

The direction of the velocity of a moving particle is harder to change (by a given sidewise impulse) when the velocity is large than when it is small. In the same way, the orientation of a spinning object is harder to change (by a given external torque) when its angular momentum is large than when it is small. The orientation of a satellite that is *not* spinning might be changed by a fairly small external torque, perhaps due to the thin residual atmosphere or to radiation pressure from sunlight. In a more "down-to-earth" example, a Frisbee is stabilized in flight in just the same way.

4. Pumping a Swing.   Impressive results follow if you pump up a swing with careful attention to the conservation of angular momentum.*

The system *swing + child* in Fig. 16 is not isolated; the earth's gravitational attraction exerts a torque on it. Indeed, it is this torque that makes swinging possible. However, near the bottom of the swing's path, the gravitational force passes through or close to the swing support and thus exerts a zero or negligible torque.

The technique is to squat on the seat at the ends of the swings and rise to a standing position as the swing passes through its lowest point. Standing up shifts the center of mass of the child closer to the rotation axis, thus reducing I and increasing ω.

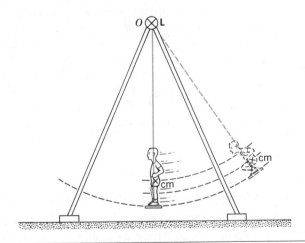

Figure 16   Pumping a swing. The child squats at the end of the swing's arc and stands up as the swing is passing through the bottom of its arc. The child's center of mass is moved closer to the axis of rotation at O as the child stands up. The angular momentum of the swing at the instant shown is represented by the cross at O.

---

* See "How Children Swing," by Stephen M. Curry, *American Journal of Physics,* October 1976, for a fuller account.

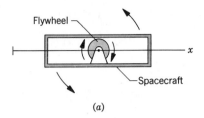

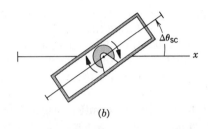

Figure 17   (a) An idealized spacecraft containing a flywheel. If the flywheel is made to rotate clockwise, as shown, the spacecraft itself will rotate counterclockwise because the total angular momentum must remain zero. (b) When the flywheel is braked to rest, the spacecraft will also stop rotating but will have reoriented its axis by the angle $\Delta\theta_{sc}$.

5. Spacecraft Orientation, Acrobatic Twists, and Falling Cats.   An isolated system of particles has no angular momentum. Can its orientation in space be changed by making internal changes in the system? If the system is not a rigid body, the answer is "Yes, under certain conditions."

Figure 17, a spacecraft with a rigidly mounted flywheel, suggests a scheme for orientation control that will be familiar to science fiction fans. The *spacecraft + flywheel* form an isolated system whose total angular momentum **L** is zero and must remain so.

To change the orientation of the spacecraft, start up the flywheel, as in Fig. 17a. To keep the angular momentum equal to zero, the spacecraft will start to rotate in the opposite sense. The next step is to bring the flywheel to rest, at which time the spacecraft will also stop rotating but will have changed its orientation, as in Fig. 17b. At no time during this maneuver does the angular momentum of the system *spacecraft + flywheel* differ from zero.

Conservation of angular momentum requires that

$$I_{sc}\omega_{sc} + I_{fw}\omega_{fw} = 0 \qquad (34)$$

at all times. Here the subscript sc refers to the spacecraft and fw to the flywheel. The two angular velocities have different signs, corresponding to the opposite directions of rotation of the spacecraft and the flywheel.

Because $\omega = \Delta\theta/\Delta t$, we can write Eq. 34 as

$$I_{sc}\Delta\theta_{sc} = -I_{fw}\Delta\theta_{fw}$$

or

$$\Delta\theta_{sc} = -\frac{I_{fw}}{I_{sc}}\Delta\theta_{fw}. \qquad (35)$$

Here $\Delta\theta_{sc}$ is the angle through which the spacecraft rotates in a given time and $\Delta\theta_{fw}$ is the angle through which the flywheel rotates during that same time. The minus sign reminds us that these two angles are oppositely directed. Because $I_{fw} \ll I_{sc}$, the flywheel will have to make many revolutions to rotate the spacecraft through a modest angle.*

Interestingly, the spacecraft *Voyager 2,* on its 1986 flyby of the planet Uranus, was set into unwanted rotation by this mechanism every time its tape recorder was turned on in its high-speed mode. The ground staff at the Jet Propulsion Laboratory had to program the on-board computer to turn on counteracting thruster jets every time the tape recorder was turned on or off.†

This same principle, based (in the simplest case) on generating—for a defined time interval—two equal but opposite internal angular momenta, can be used to change the orientation of other systems. If you hold a cat upside down and drop it, with zero angular momentum, the cat will maneuver its body so as to rotate through 180° and will land on its feet. A dog, being less flexible than a cat, cannot do this. In much the same way, a diver can maneuver her body so as to perform a 180° twist about a head-to-toe axis, initiating the twist *after* leaving the springboard, and a child, sitting on a swing, can pump it up from rest.

6. "Physics from the News: The Airliner Lost an Engine."‡ It may come as a surprise, but the engines on large jet planes are deliberately not bolted to the fuselage as securely as they could be. Why not?

A jet airplane in normal flight has a considerable angular momentum associated with the whirling rotors of its engines. If the pilot shuts one of them down, allowing it to coast to a halt, he can easily compensate for the

"lost" angular momentum by manipulating the control surfaces. Suppose, however, that the rotor of an engine "seizes up" rapidly. To conserve angular momentum, the plane itself would suffer a more or less violent rotation.

To prevent this from happening, the engine mounting bolts are carefully designed to break off on sudden engine seizure, releasing the engine from the plane. Although one can make jokes about an engine dropping off in flight, the passengers should be glad that it did and should thank the engineers who designed the *structural fuses,* as the carefully designed mounting bolts are called.

7. The Incredible Shrinking Star. When the nuclear fire in the core of a star burns low, the star may eventually begin to collapse, building up pressure in its interior. The collapse may go so far as to reduce the radius of the star from something like that of our sun to the incredibly small value of a few kilometers. The star has become a *neutron star,* so called because the material of which it is made has been compressed to an incredibly dense neutron gas.

During this shrinking process, the star forms an isolated system and its angular momentum **L** cannot change. Our sun, a typical star, rotates about one revolution per month. Compared to our sun, the rotational inertia of a neutron star is greatly reduced and its angular velocity correspondingly greatly increased, to as much as 600–800 revolutions per *second.*

Neutron stars can be detected on earth because they may emit continuous jets of radiation, which sweep by the earth like a lighthouse beam, producing periodic pulses of radiation; see Fig. 18. From this aspect, the neutron stars are called *pulsars.* Conservation of angular momentum plays a vital role in all sorts of astrophysical situations.

---

* Actually, engineering analysis favors thruster jets over flywheels as a means of reorienting spacecraft.

† See "Engineering *Voyager 2*'s Encounter with Uranus," by Richard P. Laeser, William I. McLaughlin, and Donna M. Wolff, *Scientific American,* November 1986, for a fascinating account of modifying a spacecraft from a distance of 1.5 billion kilometers.

‡ See the article of this title by Albert A. Bartlett, *The Physics Teacher,* May 1986.

Sample Problem 7 Figure 19 shows the earth rotating on its axis as it orbits the sun. (a) What is the angular momentum of the earth associated with its rotation about its axis?

From Eq. 32 we have

$$L_{rot} = I\omega = \tfrac{2}{5}MR^2\,\frac{2\pi\,\text{rad}}{T},$$

in which $T (= 24\text{ h} = 8.64 \times 10^4\text{ s})$ is the period of rotation of the earth. We find

$$L_{rot} = (\tfrac{2}{5})(5.98 \times 10^{24}\text{ kg})(6.37 \times 10^6\text{ m})^2\,\frac{2\pi}{8.64 \times 10^4\text{ s}}$$

$$= 7.1 \times 10^{33}\text{ kg}\cdot\text{m}^2/\text{s}. \qquad \text{(Answer)}$$

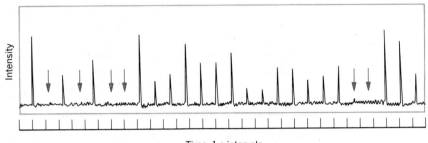

Intensity

Time, 1-s intervals

Figure 18 Electromagnetic pulses received on earth from a rapidly rotating neutron star. The colored vertical arrows suggest pulses too weak to show up in this recording. The pulse interval is remarkably constant, its measured value being 1.187,911,164 s.

This vector is parallel to the earth's axis of rotation, pointing (as the right-hand rule will show) from the south pole to the north pole.

(b) What is the angular momentum of the earth associated with its orbital motion about the sun?

From Eq. 32 we have

$$L_{orb} = I\omega = MR^2 \frac{2\pi \text{ rad}}{T},$$

in which R is now the mean earth–sun distance and T ($= 1$ y $= 3.16 \times 10^7$ s) is now the earth's period of revolution around the sun. We find

$$L_{orb} = (5.98 \times 10^{24} \text{ kg})(1.50 \times 10^{11} \text{ m})^2 \frac{2\pi}{3.16 \times 10^7 \text{ s}}$$

$$= 2.7 \times 10^{40} \text{ kg} \cdot \text{m}^2/\text{s}. \qquad \text{(Answer)}$$

This vector points at right angles to the plane of the earth's orbit. Because of the inclination of the earth's axis, the orbital angular momentum vector and the rotational angular momentum vector make an angle of 23.5° with each other. Both vectors remain constant in magnitude and direction as the earth moves around its orbit.†

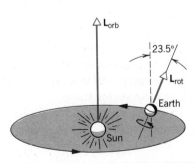

Figure 19 Sample Problem 7. A perspective view of the earth rotating on its axis as it moves in its orbit (assumed circular) around the sun. The angular momentum vectors are not to scale ($\mathbf{L}_{orb}$) is about 4×10^6 times greater than $\mathbf{L}_{rot}$.

---

† This ignores the very slow precession of the earth's rotational axis.

Sample Problem 8 As Fig. 20 shows, a child whirls a ball of mass m in a circle, whose initial radius r_i is 130 cm, at an initial angular speed ω_i of 35 rev/min. The child pulls in the cord, shortening the radius to r_f ($= 85$ cm). What is the angular speed ω_f of the ball in this new orbit?

The net force acting on the ball (that is, the horizontal component of the tension in the cord) is directed toward the center of the circle and hence cannot exert a torque about that point. Thus, no external torque acts on the whirling ball and angular momentum must be conserved.

From Eq. 32,

$$I_i\omega_i = I_f\omega_f,$$

or

$$\omega_f = \frac{I_i}{I_f} \omega_i = \frac{mr_i^2}{mr_f^2} \omega_i = \frac{r_i^2}{r_f^2} \omega_i$$

$$= \frac{(130 \text{ cm})^2}{(85 \text{ cm})^2} (35 \text{ rev/min}) = 82 \text{ rev/min}. \quad \text{(Answer)}$$

The angular speed of the ball increases as the ball is pulled in.

Figure 20 Sample Problem 8. A child whirls a ball around her head. What happens to the angular speed of the ball as she pulls in the string?

Sample Problem 9 Figure 21a shows a student, again sitting on a stool that can rotate freely about a vertical axis. The student, initially at rest, is holding a rim-loaded bicycle wheel whose rotational inertia I about its central axis is 1.2 kg·m². The wheel is rotating at an angular speed ω of 3.9 rev/s. The axis of the wheel is vertical and the angular momentum $\mathbf{L}_i$ of

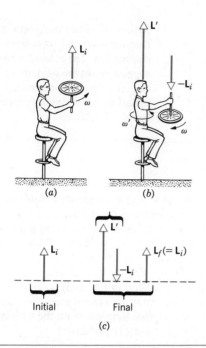

Figure 21 Sample Problem 9. (*a*) A student holds a rotating bicycle wheel. (*b*) The student turns the wheel over, setting himself into rotation. (*c*) The angular momentum must remain $\mathbf{L}_i$ at all times.

the wheel points vertically upward. The student is now asked to turn the wheel end for end. What happens? (The rotational inertia I_0 of *student + stool + wheel* about the stool axis is 6.8 kg·m².)

The *student + stool + wheel* form an isolated system, for which angular momentum must be conserved. The initial angular momentum $\mathbf{L}_i$ is that of the bicycle wheel alone and that must remain the final value after the wheel has been reversed. Figures 21*b* and 21*c* show that, to conserve angular momentum, the student and his stool must acquire an angular momentum $\mathbf{L}'$ in the opposite direction, its magnitude being $2L_i$. Thus,

$$2L_i = I_0\omega',$$

in which ω' is the angular rotational rate of the student on his stool. This yields

$$\omega' = \frac{2L_i}{I_0} = \frac{2I\omega}{I_0} = \frac{(2)(1.2 \text{ kg}\cdot\text{m}^2)(3.9 \text{ rev/s})}{6.8 \text{ kg}\cdot\text{m}^2}$$
$$= 1.4 \text{ rev/s.} \qquad \text{(Answer)}$$

If the student wishes to stop rotating, he has only to reverse the wheel end-for-end once more.

To turn the wheel, the student will be well aware of the need to apply a torque. However, this torque is internal to the *student + stool + wheel* system and so cannot change the total angular momentum of this system.

We can, however, decide to adopt as our system *student + stool* alone, the wheel now being external to this new system. From this point of view, as the student exerts a torque on the wheel, the wheel exerts a reaction torque on him and this torque is now an external torque. It is the action of this external torque that changes the angular momentum of the *student + stool* system, setting it spinning. Whether a torque is internal or external depends on how you choose to define your system.

12–11 The Precessing Top* (Optional)

Every child who owns a top knows that, when it is set spinning, it will rotate slowly about a vertical axis, a motion that we call precession. We analyze this fascinating motion under four headings, using Newton's second law in angular form as our guiding principle.

The Top. As Figure 22*a* shows, a spinning top will precess around the vertical direction. If the top were not spinning, it would simply fall over. In zero gravity, a spinning top would not precess. The precession of the top must then be related to the gravitational force and to the angular momentum of the top. Let us see how.

The Torque. In Fig. 22*b*, we represent the top by a particle of mass M, located at the top's center of mass and traversing a precession circle at an angular rate Ω. The top is not an isolated system. As Fig. 22*b* shows, its weight $M\mathbf{g}$ exerts an external torque τ about O, the point of contact of the top with the ground. The magnitude of this torque is

$$\tau = Mg(r \sin\theta), \qquad (36)$$

in which $r \sin\theta$ is the moment arm of the force $M\mathbf{g}$. The torque, as we see, is tangent to the precession circle.

The Angular Momentum. Figure 22*c* shows the angular momentum $\mathbf{L}$ of the spinning top, represented by a vector lying along the axis of the top. We see that the torque (which we carry over from the previous figure) is perpendicular to the vector $\mathbf{L}$.

Equation 30 ($\Sigma\tau_{\text{ext}} = d\mathbf{L}/dt$) applies. For small changes, we can write it as

$$\Delta\mathbf{L} = \tau\,\Delta t \qquad (37)$$

in which we have dropped the summation symbol and

---

* See "The Amateur Scientist: Tops, Including Some Far-out Ones. The Physics of Spinning," by Jearl Walker, *Scientific American*, March 1981, for a full discussion. Unless a top is started in just the right way, it will "nod" up and down as it precesses; we assume pure precession, without this motion of *nutation*, as it is called.

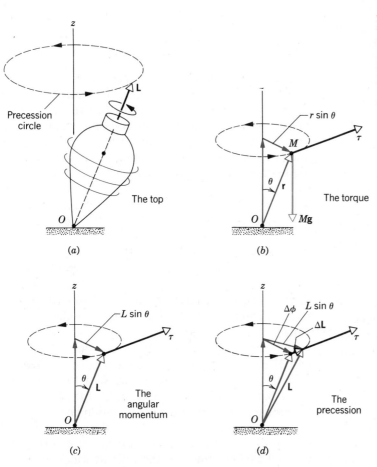

(a)

(b)

(c)

(d)

Figure 22 (*a*) A spinning top precesses about a vertical axis. (*b*) The weight of the top exerts a torque about the point of contact with the floor. (*c*) The torque is at right angles to the angular momentum vector. (*d*) The torque changes the direction of the angular momentum vector, causing precession.

the subscript ext for simplicity. Using Eq. 36, we can write, for the magnitude of ΔL,

$$\Delta L = \tau \, \Delta t = (Mgr \sin \theta)\Delta t. \tag{38}$$

Equation 37 tells us that ΔL, the change in angular momentum that occurs in a time Δt, *must point in the same direction as the torque vector.*

The Precession. Although the torque τ, being at right angles to L, cannot change the magnitude of L, it can change its direction. Figure 22*d* shows how the vector ΔL adds vectorially onto the vector L to bring this about. From the figure we see that

$$\Delta\phi = \frac{\Delta L}{L \sin \theta} = \frac{(Mgr \sin \theta)\Delta t}{L \sin \theta}.$$

Here we have substituted for ΔL from Eq. 38 above. By dividing by Δt, approaching the differential limit, and putting $\Omega = d\phi/dt$, we obtain

$$\boxed{\Omega = \frac{Mgr}{L} = \frac{Mgr}{I\omega}} \tag{39}$$

as the relation between the precession rate Ω (Greek capital *omega*) and the angular momentum L ($= I\omega$) of the spinning top.

Equation 39 tells us, perhaps unexpectedly, that the faster the top spins on its symmetry axis (that is, the greater the value of ω) the slower it precesses. Note too that the precession rate is independent of the angle θ and does indeed go to zero if $g = 0$.

12-12 Angular Momentum Is Quantized (Optional)

A physical quantity is said to be *quantized* if it can exist in nature with only certain discrete values, all intermediate values being prohibited. We have met so far with two examples, the quantization of mass (see Section 2-8) and the quantization of energy (see Section 8-10). Angular momentum is our third example.

Although quantization is universal, it makes itself known most strikingly only at the atomic and subatomic levels and it is there that we must look for our evidence. All of the particles of physics, such as the electron, the proton, the pion, and many others, have intrinsic and characteristic values of angular-momentum, given by the relation

$$l = I \frac{h}{2\pi},\qquad (40)$$

in which I (called the *spin quantum number*) is an integer, a half-integer, or zero.* The quantity h in Eq. 40 is nothing other than the Planck constant, the basic constant of quantum physics.

The electron, for example, has a spin quantum number of $\frac{1}{2}$ so that its intrinsic angular momentum is

$$l = (\tfrac{1}{2})(6.63 \times 10^{-34}\ \text{J}\cdot\text{s})\left(\frac{1}{2\pi}\right) = 5.28 \times 10^{-35}\ \text{J}\cdot\text{s}.$$

The smallness of h means that the quantization of angular momentum is not detectable in even the smallest of macroscopic objects. In the same way, the quantizations of mass and of energy are not directly apparent to our sensory perceptions.

Quantized values of the angular momentum are also associated with the orbital motions of electrons in atoms—and of protons and neutrons in atomic nuclei. Angular momentum considerations lie at the heart of our understanding of the structure of matter at the atomic and subatomic levels. Confronted with a new particle, or with a quantum state of a nucleus, an atom, or a molecule, the first question a physicist is likely to ask is: "What is its angular momentum?" It is no surprise that the Planck constant—the foundation stone of the subatomic enterprise—has units of angular momentum.

---

* The symbol I, a pure number in this case, is also used for the rotational inertia of a body; they should not be confused.

† See Heinz R. Pagels, *The Cosmic Code,* Simon & Schuster, New York, 1982, Part III. See also P. C. W. Davies, *The Forces of Nature,* 2nd ed., Cambridge University Press, Cambridge, 1985, Chapter 5.

12–13 Conservation Laws and Symmetry: An Aside †

We have met—so far—three conservation laws: the conservation of energy, of linear momentum, and of angular momentum. As the course progresses, we shall meet others, the conservation of electric charge being one example. Laws of this type owe their pervasiveness and power to a remarkable principle, first advanced by the mathematician Emmy Noether:

Every conservation law is intimately related to one of the many fundamental symmetries of nature.

Consider this symmetry: Empty space, with all matter removed, is the same in all directions. Surprisingly, the law of conservation of angular momentum flows from this fact. If a totally isolated object—a basketball, say—were suddenly to start spinning, that would certainly violate the law of conservation of angular momentum. It would also violate the symmetry on which this law is based because the spin axis would be a selected direction in space, different from all others. Why that direction more than another? Thus, we begin to see the connection between a symmetry and a conservation law.

For another example, consider the conservation of energy, which is based on this symmetry: In empty space, with all matter removed, the time at which an event occurs is immaterial. One minute is as good as another. One day is as good as another.

To see the connection between this symmetry and the conservation of energy, suppose that (using an example by Heinz Pagels) the world was such that gravity was systematically weaker every Monday. You could then pump water uphill on Monday, at low expenditure of energy. On Tuesday, you could let it flow down again, generating more energy than you expended and thus violating the law of conservation of energy.

Symmetry plays such an important role in music, in art, and in other human endeavors that it is comforting to learn that it plays an equally central role in physics.

REVIEW AND SUMMARY

A wheeled cart has two advantages over a sledge carrying the same load. (1) The force of sliding friction F_s (which acts at the contact between the wheel and its axle) can be greatly reduced for the wheel by using lubricants and ball bearings. (2) F_c, the frictional force between the wheel and the

The Wheel

road, is only a fraction of F_s. In fact,

$$F_c = F_s(R_0/R).$$ [1]

Here R_0 and R are the radii of the axle and wheel, respectively. F_c is the force that must be overcome to pull the cart.

If a wheel rolls without slipping, we have

Rolling Bodies

$$v_{cm} = \omega r.$$ [2]

Here v_{cm} is the speed of the wheel's center and ω is the angular velocity of the wheel about its center. The wheel may also be thought of as rotating instantaneously about a point P of the "road" that is in contact with the wheel. The angular velocity of the wheel about this point is the same as the angular velocity of the wheel about its center. With this in mind, we can show that the rolling wheel has kinetic energy that may be written

Rolling Kinetic Energy

$$K = \tfrac{1}{2}I_{cm}\omega^2 + \tfrac{1}{2}Mv_{cm}^2.$$ [5]

See Sample Problems 1–3 for examples of rolling objects whose motion may be analyzed using energy methods.

Analysis Using Newton's Second Law

Sample Problem 4 illustrates the use of Newton's second law, in forms appropriate for translation ($F = ma_{cm}$) and for rotation ($\tau = I\alpha$), for analyzing accelerated motion of rolling objects. In Section 12–3 and Sample Problem 5, we use these same ideas to analyze the motion of a yo-yo.

In three dimensions, *torque* τ is defined as the vector quantity

Torque as a Vector

$$\boldsymbol{\tau} = \mathbf{r} \times \mathbf{F}.$$ [18]

Angular Momentum of a Particle

The *angular momentum l* of a particle with linear momentum $\mathbf{p}$, mass m, and velocity $\mathbf{v}$ is the vector quantity

$$l = \mathbf{r} \times \mathbf{p} = m(\mathbf{r} \times \mathbf{v}). \quad \text{(angular momentum defined)}$$ [20]

With these definitions of τ and l, *Newton's second law for the angular momentum of a particle* is

Newton's Second Law for Angular Momentum

$$\sum \tau = \frac{dl}{dt} \quad \text{(single particle).}$$ [23]

Sample Problem 6 illustrates the relation between τ, l, and $\mathbf{v}$ in one simple case.

Angular Momentum of a System of Particles

The angular momentum $\mathbf{L}$ of a system of particles is the vector sum of the angular momenta of the individual particles:

$$\mathbf{L} = l_1 + l_2 + \cdots = \sum l_n.$$ [28]

Its time rate of change is equal to the sum of external torques (the torques due to interactions between the particles of the system and particles external to the system). The exact relation is

$$\sum \tau_{ext} = \frac{d\mathbf{L}}{dt} \quad \text{(system of particles).}$$ [30]

Rigid Body Angular Momentum

The component (parallel to the axis of rotation) of the angular momentum of a rigid body rotating about a fixed axis is

$$L = I\omega \quad \text{(rigid body, fixed axis).}$$ [32]

Table 1 summarizes some important analogies between rotational and translational dynamics.

The angular momentum $\mathbf{L}$ of a system remains constant if the net external torque acting on it is zero:

Conservation of Angular Momentum

$$\mathbf{L} = \text{a constant} \quad \text{(isolated system).}$$ [33]

This is the *law of conservation of angular momentum.* It is one of the fundamental conservation laws of nature, having been verified even in situations (high-speed particles or subatomic dimensions) in which Newton's laws are not applicable. Section 12–10 and Sample Problems 7–9 describe several physical situations in which conservation of angular momentum is important.

Precession of a Top

As a final example, the precession of a top is analyzed in terms of Eq. 30. It is shown that the precession rate Ω of the top is given by

$$\Omega = \frac{Mgr}{L} = \frac{Mgr}{I\omega}.$$ [39]

Quantized Angular Momentum

Angular momentum is a *quantized* quantity, occurring in nature only in integral or half-integral multiples of $h/2\pi$, h being the *Planck constant*.

Conservation and Symmetry

Conservation laws are intimately related to the symmetry of the universe. As examples, conservation of energy is required by time symmetry, conservation of angular momentum by direction symmetry, and conservation of momentum by position symmetry.

QUESTIONS

1. A cannon ball and a marble roll from rest down an incline. Which gets to the bottom first?

2. A cylindrical can filled with beef and an identical can filled with apple juice both roll down an incline. Compare their angular and linear accelerations. Explain the difference.

3. A solid wooden cylinder rolls down two different inclined planes of the same height but with different angles of inclination. Will it reach the bottom with the same speed in each case? Will it take longer to roll down one incline than the other? Explain your answers.

4. A solid brass cylinder and a solid wooden cylinder have the same radius and mass, the wooden cylinder being longer. You release them together at the top of an incline. Which will beat the other to the bottom? Suppose that the cylinders are now made to be the same length (and radius) and that the masses are made to be equal by boring a hole along the axis of the brass cylinder. Which cylinder will win the race now? Explain your answers. Assume that the cylinders roll without slipping.

5. Ruth and Roger are cycling along a path at the same speed. The wheels of Ruth's bike are a little larger in diameter than the wheels of Roger's bike. How do the angular speeds of their wheels compare? What about the speeds of the top portions of their wheels?

6. If a car's speedometer is set to read a speed proportional to the rotational speed of its rear wheels, is it necessary to correct the reading when tires with a larger outside diameter (such as snow tires) are used?

7. A cylindrical drum, pushed along by a board from an initial position shown in Fig. 23, rolls forward on the ground a

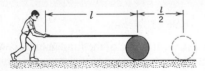

Figure 23 Question 7.

distance $l/2$, equal to half the length of the board. There is no slipping at any contact. Where is the board then? How far has the man walked?

8. Two heavy disks are connected by a short rod of much smaller radius. The system is placed on a ramp so that the disks hang over the sides as in Fig. 24. The system rolls down the ramp without slipping. (*a*) Near the bottom of the ramp the disks touch the horizontal table and the system takes off with greatly increased translational speed. Explain why. (*b*) If this system raced a hoop (of any radius) down the ramp, which would reach the bottom first? (*c*) Show that the system has $\beta > 1$, where β is the rotational inertia parameter that appears in the table of Sample Problem 1.

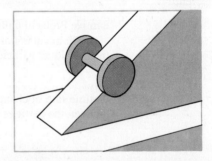

Figure 24 Question 8.

9. A yo-yo falls to the bottom of its cord and then climbs back up. Does it reverse its direction of rotation at the bottom? Explain your answer.

10. A yo-yo is resting on a horizontal table and is free to roll (Fig. 25). If the string is pulled by a horizontal force such as F_1, which way will the yo-yo roll? What happens when the force F_2 is applied (its line of action passes through the point of contact of the yo-yo and table)? If the string is pulled vertically with the force F_3, what happens?

11. A rear-wheel-drive car accelerates quickly from rest. The

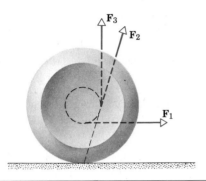

Figure 25   Question 10.

driver observes that the car "noses up." Why does it do that? Would a front-wheel-drive car behave differently?

12.   An arrow turns in flight so as to be tangent to its flight path at all times. However, a football (thrown with considerable spin about its long axis) does not do this. Why this difference in behavior?

13.   The mounting bolts that fasten the engines of jet planes to the structural framework of the plane are arranged to snap apart if the (rapidly rotating) engine suddenly seizes up because of some mishap. Why are such "structural fuses" used?

14.   Is there any advantage in setting the wheels of the landing gear of a plane in rotation just before the plane lands? If so, how would you find the optimum speed and direction of rotation?

15.   A disgruntled hockey player throws a hockey stick along the ice. It rotates about its center of mass as it slides along and is eventually brought to rest by the action of friction. Its motion of rotation stops at the same moment that its center of mass comes to rest, not before and not after. Explain why.

16.   The language we use must be adapted to the case at hand. Ballet instructors, telling a dancer how to do a *pirouette,* may say, "Imagine that your body is being sucked up into a drinking straw." How would you give similar instructions in the language of physics?

17.   When the angular velocity ω of an object increases, its angular momentum may or may not also increase. Give an example in which it does and one in which it does not.

18.   A student stands on a table rotating with an angular speed ω while holding two equal dumbbells at arm's length. Without moving anything else, the two dumbbells are dropped. What change, if any, is there in the student's angular speed? Is angular momentum conserved? Explain your answers.

19.   A helicopter flies off, its propellers rotating. Why doesn't the body of the helicopter rotate in the opposite direction?

20.   If the entire population of the world moved to Antarctica, would it affect the length of the day? If so, in what way?

21.   A circular turntable rotates at constant angular velocity about a vertical axis. There is no friction and no driving torque. A circular pan rests on the turntable and rotates with it; see Fig.

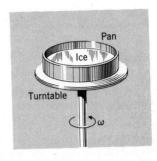

Figure 26   Question 21.

26. The bottom of the pan is covered with a layer of ice of uniform thickness, which is, of course, also rotating with the pan. The ice melts but none of the water escapes from the pan. Is the angular velocity now greater than, the same as, or less than the original velocity? Give reasons for your answer.

22.   Figure 27a shows an acrobat propelled upward by a trampoline with zero angular momentum. Can the acrobat, by maneuvering the body, manage to land on his back as in Fig. 27b? Interestingly, 38% of questioned diving coaches and 34% of a sample of physicists gave the wrong answer. What do *you* think? (See "Do Springboard Divers Violate Angular Momentum Conservation?" by Cliff Frohlich, *American Journal of Physics,* July 1979, for a full discussion.)

(a)   (b)

Figure 27   Question 22.

23.   Explain, in terms of angular momentum and rotational inertia, exactly how one "pumps up" a swing in the sitting position. (See "How Children Swing," by S. M. Curry, *American Journal of Physics,* October 1976.)

24.   Can you "pump" a swing so that it turns in a complete circle, moving completely around its support? Assume (if you wish) that the seat of the swing is connected to its support by a rigid rod rather than by a rope or a chain. Explain your answer.

25.   A massive spinning wheel can be used for a stabilizing effect on a ship. If mounted with its axis of rotation at right angles to the ship deck, what is its effect when the ship tends to roll from side to side?

EXERCISES AND PROBLEMS

Section 12-2 Rolling

1E. A thin-walled pipe rolls along the floor. What is the ratio of its translational kinetic energy to its rotational kinetic energy about an axis through the center of mass?

2E. A uniform sphere rolls down an incline. (*a*) What must be the incline angle if the linear acceleration of the center of the sphere is to be $0.10g$? (*b*) For this angle, what would be the acceleration of a frictionless block sliding down the incline?

3E. A hoop of radius 3.0 m has a mass of 140 kg. It rolls along a horizontal floor so that its center of mass has a speed of 0.15 m/s. How much work must be done on the hoop to stop it?

4E. An automobile traveling 80 km/h (= 50 mi/h) has tires of 75-cm (= 30-in.) diameter. (*a*) What is the angular speed of the tires about the axle? (*b*) If the car is brought to a stop uniformly in 30 turns of the tires (no skidding), what is the angular acceleration of the wheels? (*c*) How far does the car advance during this braking period?

5E. A 1000-kg car has four 10-kg wheels. What fraction of the total kinetic energy of the car is due to rotation of the wheels about their axles? Assume that the wheels have the same rotational inertia as disks of the same mass and size. Explain why you do not need to know the radius of the wheels.

6E. An automobile has a total mass of 1700 kg. It accelerates from rest to 40 km/h in 10 s. Each wheel has a mass of 32 kg and a radius of gyration (see Problem 56 of Chapter 11) of 30 cm. Find, for the end of the 10-s interval, (*a*) the rotational kinetic energy of each wheel about its axle, (*b*) the total kinetic energy of each wheel, and (*c*) the total kinetic energy of the automobile.

7E. A solid 8.0-lb sphere rolls up an incline with an inclination angle of 30°. At the bottom of the incline the center of mass of the sphere has a translational speed of 16 ft/s. (*a*) What is the kinetic energy of the sphere at the bottom of the incline? (*b*) How far does the sphere travel up the plane? Does the answer depend on the weight of the sphere?

8P. (*a*) What are the acceleration and velocity of a point on the top of a 66-cm diameter automobile tire if the automobile is traveling at 80 km/h on a level road? (*b*) What are the acceleration and velocity of a point on the bottom of the tire? (*c*) What are the acceleration and velocity of the center of the wheel? Calculate all quantities first as seen by a passenger in the car and then as seen by an observer standing on the side of the road as the car goes by.

9P. A body of radius R and mass m is rolling horizontally without slipping with speed v. It then rolls up a hill to a maximum height h. (*a*) If $h = 3v^2/4g$, what is the body's rotational inertia? (*b*) What might the body be?

10P. A homogeneous sphere starts from rest at the upper end of the track shown in Fig. 28 and rolls without slipping until it rolls off the right-hand end. If $H = 60$ m and $h = 20$ m and the track is horizontal at the right-hand end, determine the distance to the right of point A at which the ball strikes the horizontal base line.

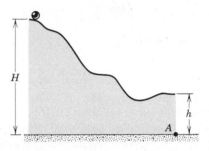

Figure 28 Problem 10.

11P. A small sphere rolls without slipping on the inside of a large fixed hemisphere whose axis of symmetry is vertical. It starts at the top from rest. (*a*) What is its kinetic energy at the bottom? What fraction of the kinetic energy is associated with rotation about an axis through its center of mass? (*b*) What normal force does the small sphere exert on the hemisphere at the bottom? Take the radius of the small sphere to be r, that of the hemisphere to be R, and let m be the mass of the sphere.

12P. A small solid marble of mass m and radius r rolls without slipping along the loop-the-loop track shown in Fig. 29, having been released from rest somewhere on the straight section of track. (*a*) From what minimum height above the bottom of the track must the marble be released in order that it not leave the track at the top of the loop? (The radius of the loop-the-loop is R; assume $R \gg r$.) (*b*) If the marble is released from height $6R$ above the bottom of the track, what is the horizontal component of the force acting on it at point Q?

13P. A solid cylinder of radius 10 cm and mass 12 kg starts from rest and rolls without slipping a distance of 6.0 m down a

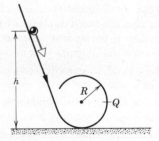

Figure 29 Problem 12.

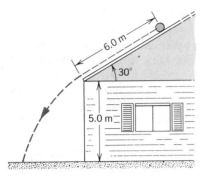

Figure 30   Problem 13.

house roof that is inclined at 30°. (*a*) What is the angular speed of the cylinder about its center as it leaves the house roof? (*b*) The outside wall of the house is 5.0 m high. How far from the wall does the cylinder hit the level ground? See Fig. 30.

14P.   An apparatus for testing the skid resistance of automobile tires is constructed as shown in Fig. 31. The tire is initially motionless and is held in a light framework that is freely pivoted at points *A* and *B*. The rotational inertia of the wheel about its axis is 0.75 kg·m², its mass is 15 kg, and its radius is 0.30 m. The tire is placed on the surface of a conveyor belt that is moving with a surface velocity of 12 m/s, such that *AB* is horizontal. (*a*) If the coefficient of kinetic friction between the tire and the conveyor belt is 0.60, what time will be required for the wheel to achieve its final angular velocity? (*b*) What will be the length of the skid mark on the conveyor surface?

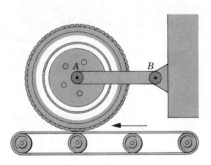

Figure 31   Problem 14.

Section 12-3   The Yo-yo
15E.   A yo-yo has a rotational inertia of 950 g·cm² and a mass of 120 g. Its axle radius is 3.2 mm and its string is 120 cm long. The yo-yo rolls from rest down to the end of the string. (*a*) What is its acceleration? (*b*) How long does it take to reach the end of the string? (*c*) If the yo-yo "sleeps" at the bottom of the string in pure rotary motion, what is its angular speed in rev/s?

16P.   Suppose that the yo-yo in Exercise 15, instead of rolling from rest, is thrown down with an initial speed of 1.3 m/s.

What is its angular speed as it "sleeps" at the bottom of the string?

Section 12-4   Torque Revisited
17E.   Given that $\mathbf{r} = \mathbf{i}x + \mathbf{j}y + \mathbf{k}z$ and $\mathbf{F} = \mathbf{i}F_x + \mathbf{j}F_y + \mathbf{k}F_z$, show that the torque $\boldsymbol{\tau} = \mathbf{r} \times \mathbf{F}$ is given by

$$\boldsymbol{\tau} = \mathbf{i}(yF_z - zF_y) + \mathbf{j}(zF_x - xF_z) + \mathbf{k}(xF_y - yF_x).$$

(*Hint:* See Problem 51 in Chapter 3.)

18P.   Show that, if **r** and **F** lie in a given plane, the torque $\boldsymbol{\tau} = \mathbf{r} \times \mathbf{F}$ has no component in that plane.

Section 12-5   Angular Momentum
19E.   A 1200-kg airplane is flying in a straight line at 80 m/s, 1.3 km above the ground. What is the magnitude of its angular momentum with respect to an observer on the ground directly under the path of the plane?

20E.   Two objects are moving as shown in Fig. 32. What is the total angular momentum about point *O*?

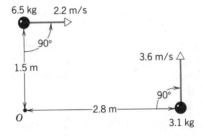

Figure 32   Exercise 20.

21E.   A particle *P* with mass 2.0 kg has position **r** and velocity **v** as shown in Fig. 33. It is acted on by the force **F**. All three vectors lie in a common plane. Presume that $r = 3.0$ m, $v = 4.0$ m/s, and $F = 2.0$ N. Compute (*a*) the angular momentum of the particle and (*b*) the torque, about the origin, acting on the particle. What are the directions of these two vectors?

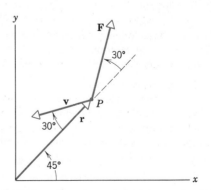

Figure 33   Exercise 21.

22E. If we are given r, p, and ϕ, we can calculate the angular momentum of a particle from Eq. 21. Sometimes, however, we are given the components (x, y, z) of $\mathbf{r}$ and (v_x, v_y, v_z) of $\mathbf{v}$ instead. (a) Show that the components of $\mathbf{l}$ along the x, y, and z axes are then given by

$$l_x = m(yv_z - zv_y),$$
$$l_y = m(zv_x - xv_z),$$
$$l_z = m(xv_y - yv_x).$$

(b) Show that if the particle moves only in the xy plane, the resultant angular momentum vector has only a z component. (*Hint:* See Problem 51 in Chapter 3.)

23E. At a certain time, the position vector in meters of a 0.25-kg object is $\mathbf{r} = 2.0\mathbf{i} - 2.0\mathbf{k}$. At that instant, its velocity in meters per second is $\mathbf{v} = -5.0\mathbf{i} + 5.0\mathbf{k}$, and the force in newtons acting on it is $\mathbf{F} = 4.0\mathbf{j}$. (a) What is the angular momentum of the object about the origin? (b) What torque acts on it? (*Hint:* See Exercises 17 and 22.)

24P. Calculate the angular momentum, about the earth's center, of an 84-kg person on the equator of the rotating earth.

25P. Show that the angular momentum, about any point, of a single particle moving with constant velocity remains constant throughout the motion.

26P. Two particles, each of mass m and speed v, travel in opposite directions along parallel lines separated by a distance d. Find an expression for the total angular momentum of the system about any origin.

27P. A 2.0-kg object moves in a plane with velocity components $v_x = 30$ m/s and $v_y = 60$ m/s as it passes through the point $(x, y) = (3.0, -4.0)$ m. (a) What is its angular momentum relative to the origin at this moment? (b) What is its angular momentum relative to the point $(-2.0, -2.0)$ m at this same moment?

28P. (a) Use the data given in the appendices to compute the total angular momentum of all the planets owing to their revolution about the sun. (b) What fraction of this is associated with the planet Jupiter?

Section 12-6   Newton's Second Law in Angular Form

29E. A sanding disk with rotational inertia 1.2×10^{-3} kg·m² is attached to an electric drill whose motor delivers a torque of 16 N·m. Find (a) the angular momentum and (b) the angular speed of the disk 33 ms after the motor is turned on.

30E. A 3.0-kg particle is at $x = 3.0$ m, $y = 8.0$ m with a velocity of $\mathbf{v} = 5\mathbf{i} - 6\mathbf{j}$ m/s. It is acted on by a 7.0-N force in the negative x direction. (a) What is the angular momentum of the particle? (b) What torque acts on the particle? (c) At what rate is the angular momentum of the particle changing with time?

31E. A wheel of radius 0.25 m, moving initially at 43 m/s, rolls to a stop in 225 m. Calculate (a) its linear acceleration and (b) its angular acceleration. (c) The wheel's rotational inertia is

0.155 kg·m²; calculate the torque exerted by rolling friction on the wheel.

32E. The angular momentum of a flywheel having a rotational inertia of 0.14 kg·m² decreases from 3.0 to 0.80 kg·m²/s in 1.5 s. (a) What is the average torque acting on the flywheel during this period? (b) Assuming a uniform angular acceleration, through what angle will the flywheel have turned? (c) How much work was done on the wheel? (d) What was the average power supplied by the flywheel?

33P. A projectile of mass m is fired from the ground with an initial speed v_0 and an initial angle θ_0 above the horizontal. (a) Find an expression for its angular momentum about the firing point as a function of time. (b) Differentiate (a) to find the rate at which the angular momentum changes with time. (c) Evaluate $\mathbf{r} \times \mathbf{F}$ directly and compare the result with (b). Why should the results be identical?

34P. An impulsive force $F(t)$ acts for a short time Δt on a rotating rigid body with rotational inertia I. Show that

$$\int \tau \, dt = \overline{F}R(\Delta t) = I(\omega_f - \omega_i),$$

where R is the moment arm of the force, $\overline{F}$ is the average value of the force during the time it acts on the object, and ω_i and ω_f are the angular velocities of the object just before and just after the force acts. [The quantity $\int \tau \, dt = \overline{F}R(\Delta t)$ is called the *angular impulse,* in analogy with $\overline{F}(\Delta t)$, the linear impulse.]

35P*. Two cylinders having radii R_1 and R_2 and rotational inertias I_1 and I_2, respectively, are supported by axes perpendicular to the plane of Fig. 34. The large cylinder is initially rotating with angular velocity ω_0. The small cylinder is moved to the right until it touches the large cylinder and is caused to rotate by the frictional force between the two. Eventually, slipping ceases, and the two cylinders rotate at constant rates in opposite directions. Find the final angular velocity ω_2 of the small cylinder in terms of I_1, I_2, R_1, R_2, and ω_0. (*Hint:* Neither angular momentum nor kinetic energy is conserved. Apply the angular impulse equation to each cylinder. See Problem 34.)

36P*. A beginning bowler throws a bowling ball of mass M and radius $R = 11$ cm down the lane with an initial speed

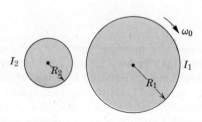

Figure 34   Problem 35.

$v_0 = 8.5$ m/s. The ball is thrown in such a way that it skids for a certain distance before it starts to roll. It is not rotating at all when it first hits the lane, its motion being pure translation. The coefficient of kinetic friction between the ball and the lane is 0.21. (a) For what length of time does the ball skid? (*Hint:* As the ball skids, its speed v decreases and its angular speed ω increases; skidding ceases when $v = R\omega$.) (b) How far down the lane does it skid? (c) How many revolutions does it make before it starts to roll? (d) How fast is it moving when it starts to roll?

Section 12-8 The Angular Momentum of a Rigid Body Rotating About a Fixed Axis

37E. Three particles, each of mass m, are fastened to each other and to a rotation axis by three light strings each with length l as shown in Fig. 35. The combination rotates around the rotational axis with angular velocity ω in such a way that the particles remain in a straight line. (a) Calculate the rotational inertia of the combination about O. (b) What is the angular momentum of the middle particle? (c) What is the total angular momentum of the three particles? Express your answers in terms of m, l, and ω.

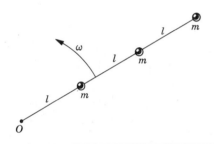

Figure 35 Exercise 37.

38E. A uniform rod rotates in a horizontal plane about a vertical axis through one end. The rod is 6.0 m long, weighs 10 N, and rotates at 240 rev/min clockwise when seen from above. Calculate (a) the rotational inertia of the rod about the axis of rotation and (b) the angular momentum of the rod.

39P. Two wheels, A and B, are connected by a belt as in Fig. 36. The radius of B is three times the radius of A. What would be the ratio of the rotational inertias I_A/I_B if (a) both wheels have the same angular momenta and (b) both wheels have the

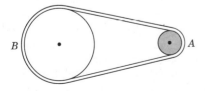

Figure 36 Problem 39.

same rotational kinetic energy? Assume that the belt does not slip.

Section 12-10 Conservation of Angular Momentum: Some Examples

40E. The rotor of an electric motor has a rotational inertia $I_m = 2 \times 10^{-3}$ kg·m² about its central axis. The motor is used to change the orientation of the space probe in which it is mounted. The motor is mounted parallel to the axis of the probe, which has a rotational inertia $I_p = 12$ kg·m² about its axis. Calculate the number of revolutions of the rotor required to turn the probe through 30° about its axis.

41E. A man stands on a frictionless platform that is rotating with an angular speed of 1.2 rev/s; his arms are outstretched and he holds a weight in each hand. With his hands in this position the total rotational inertia of the man, the weights, and the platform is 6.0 kg·m². If by moving the weights the man decreases the rotational inertia to 2.0 kg·m², (a) what is the resulting angular speed of the platform and (b) what is the ratio of the new kinetic energy to the original kinetic energy?

42E. Two disks are mounted on low-friction bearings on the same axle and can be brought together so that they couple and rotate as one unit. (a) The first disk, with rotational inertia 3.3 kg·m², is set spinning at 450 rpm (rev/min). The second disk, with rotational inertia 6.6 kg·m², is set spinning at 900 rpm in the same direction as the first. They then couple together. What is the angular speed after coupling? (b) Now suppose instead that the second disk was set spinning at 900 rpm in the direction opposite to the first disk's rotation. What is the angular speed after coupling in this case?

43E. A wheel is rotating with an angular speed of 800 rev/min on a shaft whose rotational inertia is negligible. A second wheel, initially at rest and with twice the rotational inertia of the first, is suddenly coupled to the same shaft. (a) What is the angular speed of the resultant combination of the shaft and two wheels? (b) What fraction of the original rotational kinetic energy is lost?

44E. The rotational inertia of a collapsing spinning star changes to one-third its initial value. What is the ratio of the new rotational kinetic energy to the initial rotational kinetic energy?

45E. Suppose that the sun runs out of nuclear fuel and suddenly collapses to form a so-called white dwarf star, with a diameter equal to that of the earth. Assuming no mass loss, what would then be the new rotation period of the sun, which currently is about 25 days? Assume that the sun and the white dwarf are uniform spheres.

46E. In a playground, there is a small merry-go-round of radius 1.2 m and mass 180 kg. The radius of gyration (see Problem 56 of Chapter 11) is 91 cm. A child of mass 44 kg runs at a speed of 3.0 m/s tangent to the rim of the merry-go-round when it is at rest and then jumps on. Neglect friction between the bearings and the shaft of the merry-go-round. Calculate (*a*) the rotational inertia of the merry-go-round about its axis of rotation; (*b*) the angular momentum of the child, while running, about the axis of rotation of the merry-go-round; and (*c*) the angular velocity of the merry-go-round and child after the child has jumped on.

47P. With center and spokes of negligible mass, a certain bicycle wheel has a thin rim of radius 1.14 ft and weight 8.36 lb; it can turn on its axle with negligible friction. A man holds the wheel above his head with the axis vertical while he stands on a turntable free to rotate without friction; the wheel rotates clockwise, as seen from above, with an angular speed 57.7 rad/s, and the turntable is initially at rest. The rotational inertia of wheel + man + turntable about the common axis of rotation is 1.54 slug·ft². The man's hand suddenly stops the rotation of the wheel (relative to the turntable). Determine the resulting angular velocity (magnitude and direction) of the system.

48P. Two skaters, each of mass 50 kg, approach each other along parallel paths separated by 3.0 m. They have equal and opposite velocities of 10 m/s. The first skater carries a long light pole, 3.0 m long, and the second skater grabs the end of it as he passes; see Fig. 37. Assume frictionless ice. (*a*) Describe quantitatively the motion of the skaters after they are connected by the pole. (*b*) By pulling on the pole, the skaters reduce their separation to 1.0 m. What is their angular speed then? (*c*) Calculate the kinetic energy of the system in parts (*a*) and (*b*). From where does the change come?

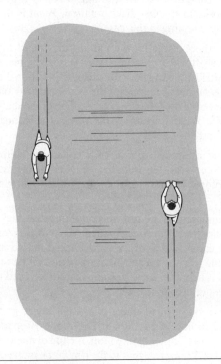

Figure 37 Problem 48.

49P. Two children, each with mass M, sit on opposite ends of a narrow board with length L and mass M (the same as the mass of each child). The board is pivoted at its center and is free to rotate in a horizontal circle without friction. (*a*) What is the rotational inertia of the board plus the children about a vertical axis through the center of the board? (*b*) What is the angular momentum of the system if it is rotating with an angular speed ω_0 in a clockwise direction as seen from above? What is the direction of the angular momentum? (*c*) While the system is rotating, the children pull themselves toward the center of the board until they are half as far from the center as before. What is the resulting angular speed in terms of ω_0? (*d*) What is the change in kinetic energy of the system as a result of the children changing their positions? How is energy conserved during this motion?

50P. In a lecture demonstration, a toy train track is mounted on a large wheel that is free to turn with negligible friction about a vertical axis; see Fig. 38. A toy train of mass m is placed on the track and, with the system initially at rest, the electrical power is turned on. The train reaches a steady speed v with respect to the track. What is the angular velocity ω of the wheel, if its mass is M and its radius R? (Neglect the mass of the spokes of the wheel.)

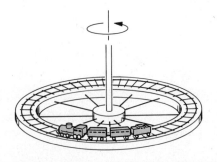

Figure 38 Problem 50.

51P. A cockroach, mass m, runs counterclockwise around the rim of a lazy Susan (a circular dish mounted on a vertical axle) of radius R and rotational inertia I with frictionless bearings. The cockroach's speed (relative to the earth) is v, whereas the lazy Susan turns clockwise with angular speed ω_0. The cockroach finds a bread crumb on the rim and, of course, stops. (*a*) What is the angular speed of the lazy Susan after the cockroach stops? (*b*) Is mechanical energy conserved?

52P. A girl (mass M) stands on the edge of a frictionless merry-go-round (radius R, rotational inertia I) that is not moving. She throws a rock (mass m) in a horizontal direction that is tangent to the outer edge of the merry-go-round. The speed of the rock, relative to the ground, is v. What are (*a*) the angular speed of the merry-go-round and (*b*) the linear speed of the girl after the rock is thrown?

53P. If the polar ice caps of the earth were to melt and the water returned to the oceans, the oceans would be made deeper by about 30 m. What effect would this have on the earth's rotation? Make an estimate of the resulting change in the length of the day. (Concern has been expressed that warming of the atmosphere resulting from industrial pollution could cause the ice caps to melt.)

54P. Particle m in Fig. 39 slides down the frictionless surface and collides with the uniform vertical rod, sticking to it. The rod pivots about O and rotates through the angle θ before coming to rest. Find θ in terms of the other parameters given in the figure.

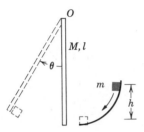

Figure 40 Problem 55.

55P. Two 2.0-kg balls are attached to the ends of a thin rod of negligible mass 50 cm long. The rod is free to rotate without friction about a horizontal axis through its center. A 50-g putty

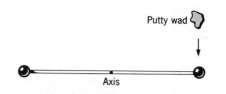

Figure 39 Problem 54.

wad drops onto one of the balls, with a speed of 3.0 m/s, and sticks to it. (See Fig. 40.) (*a*) What is the angular speed of the system just after the putty wad hits? (*b*) What is the ratio of the kinetic energy of the entire system after the collision to that of the putty wad just before? (*c*) Through what angle will the system rotate?

Section 12–11   The Precessing Top
56E. A top is spinning at 30 rev/s about an axis making an angle of 30° with the vertical. Its mass is 0.50 kg and its rotational inertia is 5.0×10^{-4} kg·m². The center of mass is 4.0 cm from the pivot point. If the spin is clockwise as seen from above, what is the magnitude and direction of the angular velocity of precession?

57P. A gyroscope consists of a rotating disk with a 50-cm radius suitably mounted at the center of a 12-cm-long axle so that it can spin and precess freely. Its spin rate is 1000 rev/min. Find the rate of precession (in rev/min) if the axle is supported at one end and is horizontal.

ESSAY 3
BALANCE

KENNETH LAWS
DICKINSON COLLEGE

How do people balance? Suppose you are walking along one rail of a railroad track, striving to maintain your balance as you walk (Fig. 1). You find your body instinctively moving in certain ways that seem to help you perform that feat. These movements may be subtle shifts of body position or may be strange and vigorous gyrations of arms, legs, and other body parts. Do these movements actually help accomplish the purpose, and if so, how and why? How do gymnasts perform on a balance beam, or tightrope walkers avoid falling, or dancers maintain balanced poses for breathtaking moments of timeless suspension?

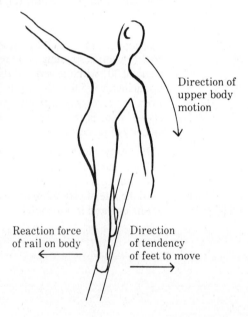

Direction of upper body motion

Reaction force of rail on body

Direction of tendency of feet to move

Figure 1 A person trying to maintain balance while walking on a railroad track instinctively moves in ways that may seem unproductive, but in fact do contribute to balance. (Drawing by Sandra Kopell.)

Conversely, how does one start accelerating away from an initially balanced and static position? For instance, imagine that you are a shortstop waiting for the batter to hit the baseball. You find yourself crouched with feet apart so that you can react rapidly to the direction in which the ball is hit (Fig. 2). Or suppose you are a sprinter ready to begin a race and wish to accelerate as quickly as possible when the starting gun is fired (Fig. 3). In both cases, how does your stance control how quickly you can change your state of motion? Why do swimmers find that the forward acceleration in a standard racing dive is slower than a track sprinter's start, and why are swimming races often plagued by false starts?

Let us first define balance and then analyze the mechanical processes that the body must employ to control that balanced condition.

A body is in balance when, under the influence of gravity and other forces, it is not accelerating away from a static position. In order for this condition to exist for a person on a supporting surface (for example, floor and rail), the body's center of gravity must lie on a vertical line that passes through the area of support. (If a person is standing on one foot, the area of support is defined as the area of contact between

Figure 2 A baseball infielder waiting for the ball to be hit is crouched with feet apart in order to be able to accelerate quickly in the optimum direction for intercepting the ball.

Figure 3 A sprinter starts with both hands and feet on the ground in order to accelerate quickly when the hands are lifted.

the foot and the floor. If more than one distinct contact area is involved, the area of support is defined as the surface circumscribed by the individual contact areas.) If the area of support is large, there is a correspondingly large range of positions of the center of gravity that satisfy the balance condition. Thus, it is easier to maintain balance on a 10-cm-wide balance beam used by gymnasts than on a 7-cm-wide rail or on the small area at the tip of a ballet artist's pointe shoe.

A few simple experiments can be performed by one or two people using a plumb bob or long straight stick. First, try facing a wall with the toes against the wall, then rise onto the toes. Try also standing with the heels against the wall and then lift the toes. Can you maintain balance in either situation? In these cases, the wall imposes a constraint that limits the amount of body lean possible, thereby preventing the center of gravity from lying above the area of support.

Now, moving away from the constraining wall, stand straight with the feet together and parallel, and note that the body can be moved back so that most of the weight is on the heels, or forward so that the weight is on the balls of the feet and toes. Note the range of body orientations that are possible without balance being lost.

The location of the body's center of gravity may be determined crudely by comparing these two balanced configurations (Fig. 4). When the body weight is back on

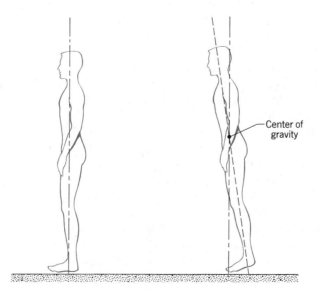

Figure 4 The technique illustrated here, and described in the text, provides a means for locating the center of gravity of the body.

the heels and the fronts of the feet are slightly lifted, hold a plumb line so that the bob is adjacent to the middle of the contact between the heels and the floor. Imagine a projection of that plumb line on the sagittal plane (the vertical plane that bisects the body into left and right halves). That line must pass through the body's center of gravity, somewhere in the hip or abdominal area. Now, keeping the body rigid except for a lean at the ankles, allow the weight to shift forward so that the heels can be lifted slightly from the floor. The plumb line is now held so that the bob is adjacent to the contact between the feet and the floor. A projection of that line on the sagittal plane also passes through the center of gravity. The intersection of those two lines occurs at the center of gravity. Note that the height of the center of gravity relative to total body height depends on the shape of the person. In relation to body height, it is generally lower in women, who have more weight concentrated in the hips, than in men, whose weight is distributed higher, in the chest and shoulders.

Suppose you are standing balanced, but a push from behind starts you falling forward. What must you do with your body to regain balance? Let's try an experiment. Stand balanced with the feet together and imagine that you are on the edge of the roof of a building with a vertical drop immediately in front of you. A "good friend" behind you gives you a push. Note your body's instinctive reaction as it tries to keep from falling over the edge. Let us consider whether instinctive movements accomplish the desired purpose.

Clearly, the center of gravity must be moved backward so that it returns to a position over the area of support. So—we just move the body backward! But wait! According to Newton's second law, the center of gravity of a body cannot be accelerated horizontally unless there is a net horizontal force acting on the body. What can the source of such a force be? A nearby wall or a person standing in front of you perhaps—but in this case all you have to work with is the surface on which you are standing.

One might think that the surface cannot be used as a source of horizontal force for regaining balance, since such a force would occur at the pivot of the toppling motion. But if the body can manipulate itself such that the feet push forward against the surface, then by Newton's third law, the surface will exert a force backward against the feet. If that force is the *only* horizontal force acting on the body, the center of gravity of the body will be accelerated backward. Nothing the body does will contribute to shifting the position of the center of gravity backward unless it produces a forward push of the feet against the surface.

Perhaps you noticed that the body's instinctive reaction to falling forward was to bend the upper body quickly *forward,* the opposite of what one might intuitively expect! Does this really help? If you were standing on a slippery sheet of ice, that movement (a "jackknifing" movement in the extreme) would cause the feet to move forward. But the friction between the feet and the floor prevents that slipping, and the feet just push forward against the floor—exactly the action required for restoring balance! A similar movement is noticed when one walks along one rail of a railroad track, trying to maintain balance. If a fall to the right starts, the body bends suddenly to the right, exerting a force to the right against the rail, causing the rail to push to the left on the body, thereby restoring balance.

You might also notice that the arms may start windmilling when you are falling to the front. Note that the direction—clockwise as viewed from the right—is such as to cause the body to react by trying to rotate in the opposite direction, with the feet moving forward, again producing that forward force of the feet against the surface. So again our instincts are consistent with physical law!

Can this analysis be applied to a tightrope walker who carries a long pole to help maintain balance? Imagine how effective those windmilling arms would be if they were 6 m long, with the enormous associated rotational inertia. Rather than windmilling, they would only have to be rotated by a small amount, with the reaction of the body exerting the necessary force against the tightrope. In this case, of course, the pole is held laterally, since the tightrope walker is concerned about balance in the lateral direction.

Now consider the opposite situation in which a person wishes to *destroy* balance in order to accelerate away from a static position. A body slightly off balance will begin to topple with an angular acceleration that increases as its angle with the vertical increases. That angular acceleration is caused by the torque around the pivot at the floor resulting from the gravitational force acting vertically on the center of gravity, which is now displaced from the vertical line above the area of support.

Toppling can be stopped by means of a horizontal force at the ground that counteracts the toppling torque produced by the gravitational force; that net horizontal force is responsible for the linear acceleration. Thus, a person wanting to initiate a forward acceleration starts to topple forward; the torque due to the gravitational force is clockwise as viewed from the right. A forward force from the surface against the feet produces a counterclockwise torque that can balance the gravitational torque, thereby preventing the angle of the body with the vertical from increasing. But how does one start toppling from an initially balanced condition?

A pertinent question is the following: What do shortstops and sprinters have in common that swimmers preparing for a racing dive don't (other than the fact that they wear more clothing)? The first two employ a stance that has a widely distributed area of support at the ground. The shortstop's feet are apart, and sprinters are supported on both hands and feet. Thus, when baseball players lift one foot, the center of gravity is far from the remaining area of support; when sprinters lift their hands, *their* center of gravity is far in front of the remaining area of support (the feet). In both cases, the body begins to topple rapidly; the force at the feet can then become large, which not only prevents the toppling but allows for a large net horizontal force on the body, causing a rapid acceleration away from the initial position. Note that there is a purpose to the low crouch. If the center of gravity is low, the departure from vertical of the line from the area of support to the center of gravity will be large, with the resulting toppling acceleration correspondingly large.

Our diving swimmers, however, have their two feet at the edge of the pool or a small raised platform. All they can do to shift the location of the area of support is to lift up on the toes so that the center of force (that point where all the vertical force from the floor may be considered to act) is shifted back toward the heels—a small distance. The toppling begins slowly, and the forward linear acceleration is initially small. There is a strong temptation to anticipate the starting gun—hence the tendency toward false starts.

But recall what the body does to *regain* balance when toppling begins. It is possible to manipulate the body in such a way that a net horizontal force is exerted against the supporting surface even if the location of the center of force is unchanged. This technique can be used to help accelerate the swimmer at the beginning of the race. Moving the arms down and forward suddenly will cause the legs to try to rotate in the opposite direction, which will produce a net force backward against the pool edge, thus producing a forward force on the swimmer.

Let us try another experiment. Stand facing a wall about 3 m away, with your feet together. At the "start" signal from a second person with a stopwatch, run forward to touch the wall as quickly as possible. Repeat starting with one foot well in back of the other, but with the front foot in the same position as before. Can the wall be reached more quickly? Speculate on the possible significance for the racing dives of swimmers!

Consider also the effect of the arm movement. With the feet together, time the interval between "start" to the touching of the wall first with the arms initially extended toward the wall, then with the arms initially back and rotated down and forward quickly when starting.

Let us also impersonate a shortstop. Stand halfway between two walls that are about 5 m apart. A second person with a stopwatch will measure the length of time it takes for you to touch the wall to your right or left after the command "right" or "left" is given. In the first case, start with your feet together; then repeat starting with your feet spread apart. Does it take less time to reach the wall when the feet are spread apart than when the initial area of support is small? Also try this experiment first standing erect and then in a low crouch.

A dancer balanced on one foot can use the same process to accelerate away from the balanced position. In a *tombé* movement, shown in Fig. 5, the working leg, initially extended to the rear, swings down and to the front. This causes the supporting foot to exert a backward force against the floor that results in the dancer accelerating forward.

All these analyses and experiments apply to a stationary floor. If the surface is moving, as in a train, balance becomes trickier. Of course, if the velocity of the surface is constant, all of the principles applied above work. But if there is an acceler-

Figure 5 A dancer wishing to accelerate from a balanced position must perform some movement that exerts a horizontal force against the floor. The *tombé* movement performed here by Lisa de Ribère is such a movement.

ation of the surface (a change in magnitude *or* direction of the velocity), balance will be lost unless the torque produced by the horizontal force of the surface against the feet can be balanced by that produced by the gravitational force acting on the center of gravity displaced from the vertical line over the area of support. That is, the mechanism described above for accelerating away from balance is exactly the mechanism used by the body to accelerate *with* the accelerating surface, thus *preventing* a loss of balance! The necessary action requires a lean in the direction of acceleration.

Imagine a ballet artist balanced on pointe. Tiny subtle movements are the only evidence that the artist's body is working to maintain that balance. Or imagine the shortstop who seems to be moving toward a ground ball almost before it is hit. Are these people calculating what they must do using physical analyses based on Newton's laws? These finely tuned movements of the body that seem so impressive are partly instinctive and partly learned by trial and error. It is intriguing for an observer to understand how these actions are accomplished. Is it possible that the performers would also benefit by this understanding?

CHAPTER 13

EQUILIBRIUM AND ELASTICITY

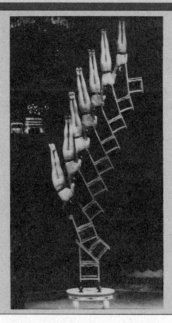

These young Chinese gymnasts seem to know the requirements for static equilibrium. Note that the center of mass of the system falls within the outline of the four legs of the lower chair.

Seven gymnasts, their talents revealing,
Find external torque unappealing.
As for external force—
They avoid it of course
As they clamber right up to the ceiling.

13-1 Equilibrium

Consider these objects: (1) A book resting on a table, (2) a hockey puck sliding across the ice, (3) the rotating blades of a ceiling fan, and (4) the wheel of a bicycle that is traveling along a straight path at constant speed. For each of these four objects:

1. The linear momentum $\mathbf{P}$ of its center of mass is constant.

2. Its angular momentum $\mathbf{L}$ about its center of mass, or about any other point, is also constant.

We say that such objects are in *equilibrium*. The two requirements for equilibrium are then

$$\mathbf{P} = \text{a constant} \quad \text{and} \quad \mathbf{L} = \text{a constant.} \quad (1)$$

Our concern in this chapter is with situations in which the constants in Eq. 1 are in fact zero. That is, we are concerned largely with objects that are not moving in

Figure 1   A balancing rock in the Chiricahua National Monument in Arizona. Although its perch seems precarious, the rock is—for the time being at least—in static equilibrium.

any way—either in translation or in rotation—in the reference frame from which we observe them. Such objects are in *static* equilibrium. Of the four objects mentioned at the beginning of this section, only one—the book resting on the table—is in static equilibrium.

The balancing rock of Fig. 1 is another example of an object that, for the present at least, is in static equilibrium. It shares this property with countless other structures such as bridges, dams, cathedrals, houses, filing cabinets, and tables that neither translate nor rotate as we watch them.

The analysis of static equilibrium is very important in engineering practice. The design engineer must isolate and identify all the external forces and torques that act on the structure. By good design and wise choice of materials, the engineer must make sure that the structure will tolerate the loads that bear on it. Such analyses are necessary to make sure that bridges do not collapse under their traffic and wind loads, that the landing gear of aircraft will survive the shocks of rough landings, and so on.

13-2   The Requirements for Equilibrium

The translational motion of a body is governed by Newton's second law in its linear form, given by Eq. 24 of Chapter 9 as

$$\sum \mathbf{F}_{ext} = \frac{d\mathbf{P}}{dt}.$$ (2)

If the body is in translational equilibrium, that is, if **P** is a constant, then $d\mathbf{P}/dt = 0$ and we must have

$$\boxed{\sum \mathbf{F}_{ext} = 0}$$ (balance of forces).   (3)

The rotational motion of a body is governed by Newton's second law in its angular form, given by Eq. 30 of Chapter 12 as

$$\sum \tau_{ext} = \frac{d\mathbf{L}}{dt}.$$ (4)

If the body is in rotational equilibrium, that is, if **L** is a constant, then $d\mathbf{L}/dt = 0$ and we must have

$$\boxed{\sum \tau_{ext} = 0}$$ (balance of torques).   (5)

Thus, the two requirements for a body to be in equilibrium are as follows:

1.   The vector sum of all the external forces that act on the body must be zero.

2.   The vector sum of all the external torques that act on the body must also be zero.

If you doubt that structures are held in equilibrium by a nice balance of external forces and torques, look at Fig. 2. The weight of the building had been balanced, perhaps for decades, by a force that was exerted upward on the structure by its foundation. If the foundation is blown out, sideways, by explosive charges, this force is suddenly removed and Eq. 3, the first requirement for equilibrium, is suddenly no longer satisfied. At the instant the photograph was taken, the structure was accelerating downward and was certainly not in equilibrium, static or otherwise.

Equations 3 and 5, as vector equations, are each equivalent to three independent scalar equations, one for each direction of the coordinate axes:

Balance of forces	Balance of torques
$\sum F_x = 0$	$\sum \tau_x = 0$
$\sum F_y = 0$	$\sum \tau_y = 0$
$\sum F_z = 0$	$\sum \tau_z = 0.$

(6)

For convenience, we have dropped the subscript ext.

We simplify matters by considering only cases in which the forces that act on the body lie in the *xy* plane. This means that the only torques that can act on the body

Figure 2   A housing project in St. Louis is spectacularly demolished. If the forces that hold a structure in equilibrium are removed, the structure cannot remain in equilibrium.

must be around the z axis or around some axis parallel to it. With this assumption, we eliminate one force equation and two torque equations from Eq. 6, leaving

$$\sum F_x = 0 \quad \text{(balance of forces),}\qquad (7)$$

$$\sum F_y = 0 \quad \text{(balance of forces),}\qquad (8)$$

and

$$\sum \tau_z = 0 \quad \text{(balance of torques).}\qquad (9)$$

Here, F_x and F_y are, respectively, the x and the y components of the external forces that act on the body and τ_z represents the torques that these forces exert about the z axis, or about any axis parallel to it.

A sliding hockey puck satisfies Eqs. 7, 8, and 9 and it is thus in equilibrium but it is not in *static* equilibrium. For that to be true, the linear momentum $\mathbf{P}$ of the puck must not only be constant but the value of the constant must be zero; the puck must be resting on the ice. As the mathematicians say, Eqs. 7, 8, and 9 are *necessary* conditions for static equilibrium but they are not *sufficient*.

13–3   The Force of Gravity

One force that always acts on a body near the surface of the earth is its weight, the force of gravity. Earlier, we did

not hesitate to describe this by a single vector $M\mathbf{g}$ and to attach that vector to the center of mass of the body. Weight, however, is a *body force,* acting separately on every atom of the body. It is not at all obvious that we can replace the vector sum of all these tiny forces by a single force acting at the center of mass. In fact, we can only do so if the gravitational acceleration vector $\mathbf{g}$ has the same magnitude and direction for all points of space occupied by the body. Let us prove that this is so.

A Proof of the Center-of-Mass Theorem.   Figure 3 shows a body of arbitrary shape, with the origin of a

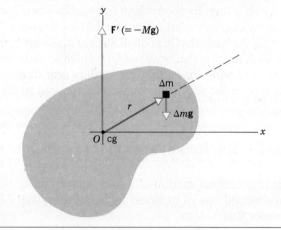

Figure 3   The weight of a body, though distributed throughout its volume, may be balanced by a single force of magnitude Mg acting at the center of gravity. For a uniform gravitational field, the center of gravity coincides with the center of mass.

coordinate system placed at a point O, which we call the *center of gravity* of the body. The body is divided into small mass elements Δm, one of which is shown. The weight $\Delta m \, \mathbf{g}$ of this element acts downward. We put an upward force $\mathbf{F}'$ at O and we hope to show that this single force will hold the body in equilibrium (1) if and only if the magnitude of $\mathbf{F}'$ is Mg and (2) if O, which we have identified as the center of gravity, is in fact the center of mass.

If the body is to be in equilibrium, Eqs. 3 and 5 must be satisfied. We look first at Eq. 3, which deals with the balance of forces. We can write this equation as

$$\sum \mathbf{F} = \mathbf{F}' + \sum \Delta m \, \mathbf{g} = \mathbf{F}' + \mathbf{g} \left(\sum \Delta m \right)$$
$$= \mathbf{F}' + M\mathbf{g} = 0, \qquad (10)$$

in which the sum is taken over all the mass elements that make up the body. Equation 10 tells us that Eq. 3, the balance of forces condition, will be satisfied if $\mathbf{F}'$ is oppositely directed to $M\mathbf{g}$ and has magnitude Mg. Our first point is proved. Note that, in Eq. 10, we were able to remove $\mathbf{g}$ from the sum only because we have assumed that it has the same value (magnitude and direction) for all points of the body.

Now let us look at Eq. 5, the balance of torques. The vector $\mathbf{F}'$ in Fig. 3, which passes through O, has no torque about that point. Thus, Eq. 5 becomes

$$\sum \boldsymbol{\tau} = \sum \mathbf{r} \times (\Delta m \, \mathbf{g}) = \left(\sum \Delta m \, \mathbf{r} \right) \times \mathbf{g} = 0. \quad (11)$$

The quantity $\sum \Delta m \, \mathbf{r}$ in this equation, which tells us how the mass of the body is distributed about point O, is zero if the point O is the center of mass; see Eq. 8 of Chapter 9. Thus, Eq. 11 tells us that Eq. 5, the balance of torques condition, will be satisfied if point O is indeed the center of mass.* Our second point is proved.

Figures 4a and 4b suggest that, if the force $\mathbf{F}'$, which is equal to $-M\mathbf{g}$, is applied to an arbitrary point S of a body, the body will be in equilibrium *only* if its center of gravity lies vertically below its suspension point. Indeed, we saw in Sample Problem 2 of Chapter 9 how we can use this fact to find the center of gravity of a body. However, if $\mathbf{F}'$ is applied at the center of gravity of the body, as in Fig. 4c, the body will be in equilibrium no matter what its orientation. That is, you can turn the body any way

---

* In the cross product of Eq. 11, we were careful not to change the sequences of the vectors $\mathbf{r}$ and $\mathbf{g}$. We were free to change the sequence of Δm however, moving it to the left, because mass is a scalar quantity.

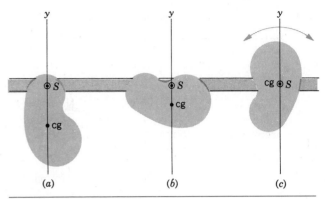

Figure 4   A body suspended from an arbitrary point S, as in (a) and (b), will be in (stable) equilibrium only if its center of gravity hangs vertically below its suspension point. (c) If a body is suspended from its center of gravity, that body is in equilibrium no matter what its orientation.

you wish about that point and it will remain in equilibrium.

If $\mathbf{g}$ does *not* have the same value for all points occupied by the body, we will no longer be able to remove this quantity, as a constant, from the sum in Eq. 10 and our proof fails. The center of gravity and the center of mass are no longer guaranteed to coincide in such cases.

13-4   Some Examples of Static Equilibrium

Sample Problem 1   A uniform beam of length L whose mass m is 1.8 kg rests with its ends on two digital scales, as in Fig. 5a. A block whose mass M is 2.7 kg rests on the beam, its center one-fourth of the way from the beam's left end. What do the scales read?

We choose as our system the beam and the block, taken together. Figure 5b is a free-body diagram for this system, showing all the forces that act. The weight of the beam, $m\mathbf{g}$, acts downward at its center of gravity. Similarly, $M\mathbf{g}$, the weight of the block, acts downward at *its* center of gravity. The scales push upward at the ends of the beam with forces $\mathbf{F}_l$ and $\mathbf{F}_r$. The magnitudes of these latter two forces are the scale readings that we seek.

Our system is in static equilibrium so that the balance of forces equations (Eqs. 7 and 8) and the balance of torques equation (Eq. 9) apply. We shall solve this problem in two equivalent ways.

First solution.   The forces have no x components so that Eq. 7, which is $\sum F_x = 0$, provides no information. Equation 8

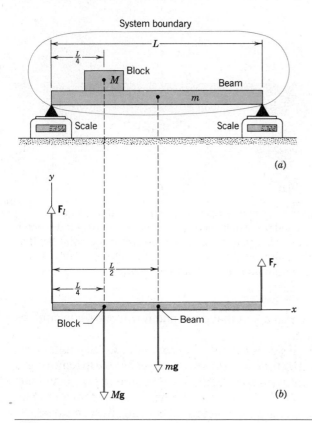

Figure 5   Sample Problem 1. (*a*) A beam of mass *m* supports a block of mass *M*. What are the readings on the digital scales that support the ends of the beam? The system boundary is marked. (*b*) A free-body diagram, showing the forces that act on the system *beam + block*.

gives, for the magnitudes of the *y* components,

$$\sum F_y = F_l + F_r - Mg - mg = 0. \qquad (12)$$

We have two unknown forces (F_l and F_r) but we cannot find them separately because we have only one equation. Fortunately, we have another equation at hand, namely, Eq. 9, the balance of torques equation.

We can apply Eq. 9 to *any* axis at right angles to the plane of Fig. 5. Let us choose an axis through the left end of the beam. Torques that—acting alone—would produce a counterclockwise rotation of the beam about our chosen axis are taken as positive. We then have, from Eq. 9,

$$\sum \tau_z = (F_l)(0) + (F_r)(L) - (mg)(L/2) - (Mg)(L/4) = 0,$$

or

$$F_r = (g/4)(M + 2m).$$
$$= (\tfrac{1}{4})(9.8 \text{ m/s}^2)(2.7 \text{ kg} + 2 \times 1.8 \text{ kg})$$
$$= 15 \text{ N}. \qquad \text{(Answer)} \quad (13)$$

Note how choosing an axis that passes through one of the unknown forces, F_l, eliminates that force from Eq. 9 and allows us to solve directly for the other force.

If we substitute Eq. 13 into Eq. 12 and solve for F_l, we find

$$F_l = (M + m)g - F_r$$
$$= (2.7 \text{ kg} + 1.8 \text{ kg})(9.8 \text{ m/s}^2) - 15 \text{ N}$$
$$= 29 \text{ N}. \qquad \text{(Answer)}$$

Second solution.   As a check, let us solve this problem in a different way, applying the balance of torques equation about two different axes. Choosing first an axis through the left end of the beam, as we did above, we have precisely Eq. 13 and the solution $F_r = 15$ N.

For an axis passing through the right end of the beam, Eq. 9 yields

$$\sum \tau_z = (F_r)(0) - (F_l)(L) + (mg)(L/2) + (Mg)(3L/4) = 0.$$

Solving for F_l, we find

$$F_l = (g/4)(3M + 2m)$$
$$= (\tfrac{1}{4})(9.8 \text{ m/s}^2)(3 \times 2.7 \text{ kg} + 2 \times 1.8 \text{ kg})$$
$$= 29 \text{ N}, \qquad \text{(Answer)}$$

in agreement with our earlier result. Note that the length of the beam does not enter this problem explicitly, but only as it is related to the mass of the beam.

Sample Problem 2   A bowler holds a bowling ball whose mass M is 7.2 kg in the palm of his hand. As Fig. 6*a* shows, his upper arm is vertical and his lower arm is horizontal. What forces must the biceps muscle and the bony structure of the upper arm exert on the lower arm? The forearm and hand together have a mass m of 1.8 kg and the needed dimensions are as shown in Fig. 6*a*.

Our system is the lower arm and the bowling ball, taken together. Figure 6*b* shows a free-body diagram. The unknown forces are **T**, the force exerted by the biceps muscle, and **F**, the force exerted by the upper arm on the lower arm. As in Sample Problem 1, the forces are all vertical.

From Eq. 8, which is $\Sigma F_y = 0$, we find, considering magnitudes only,

$$\sum F_y = T - F - mg - Mg = 0. \qquad (14)$$

Applying Eq. 9 about an axis through O and taking counterclockwise rotations as positive, we obtain

$$\sum \tau_z = (T)(d) + (F)(0) - (mg)(D) - (Mg)(L) = 0. \quad (15)$$

By choosing our axis to pass through point O, we have eliminated the variable F from this equation. Equation 15, solved for T, yields

$$T = g \frac{mD + ML}{d}$$

$$= (9.8 \text{ m/s}^2) \frac{(1.8 \text{ kg})(15 \text{ cm}) + (7.2 \text{ kg})(33 \text{ cm})}{4.0 \text{ cm}}$$

$$= 648 \text{ N, which is 146 lb.} \qquad \text{(Answer)}$$

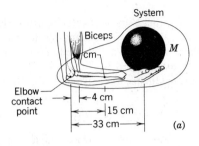

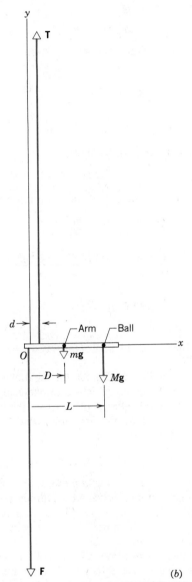

Thus, the biceps muscle must pull up on the forearm with a force that is about nine times larger than the weight of the bowling ball.

If we solve Eq. 14 for F and substitute into it the value of T given above, we find

$$F = T - g(M + m)$$
$$= 648 \text{ N} - (9.8 \text{ m/s}^2)(7.2 \text{ kg} + 1.8 \text{ kg})$$
$$= 560 \text{ N, which is } 126 \text{ lb.} \qquad \text{(Answer)}$$

F is also a strong force, being about eight times the weight of the bowling ball.

Sample Problem 3   A ladder whose length L is 12 m and whose mass m is 45 kg rests against a wall. Its upper end is a distance h of 9.3 m above the ground, as in Fig. 7a. The center of gravity of the ladder is one-third of the way up the ladder. A firefighter whose mass M is 72 kg climbs halfway up the ladder. Assume that the wall, but not the ground, is frictionless. What forces are exerted on the ladder by the wall and by the ground?

Figure 7b shows a free-body diagram. The wall exerts a horizontal force $\mathbf{F}_w$ on the ladder; it can exert no vertical force because the wall–ladder contact is assumed to be frictionless. The ground exerts a force $\mathbf{F}_g$ on the ladder with a horizontal component $\mathbf{F}_{gx}$ and a vertical component $\mathbf{F}_{gy}$. We choose coordinate axes as shown, with the origin O at the point where the wall meets the ground. The distance a from the wall to the foot

Figure 6   Sample Problem 2. (a) A hand holds a bowling ball. The system boundary is marked. (b) A free-body diagram, showing the forces that act. The vectors are to scale, showing the powerful forces exerted by the biceps muscle and the elbow joint.

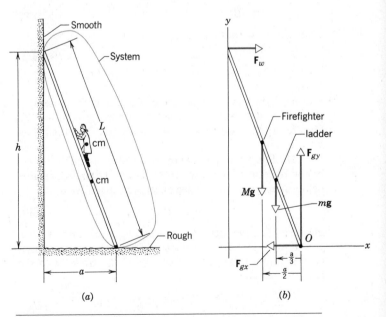

Figure 7   Samle Problems 3 and 4. (a) A firefighter climbs halfway up a ladder that is leaning against a smooth wall. (b) A free-body diagram, showing (to scale) all the forces that act.

of the ladder is readily found from

$$a = \sqrt{L^2 - h^2} = \sqrt{(12\text{ m})^2 - (9.3\text{ m})^2} = 7.6\text{ m}.$$

From Eqs. 7 and 8, the balance of forces equations, we have, respectively,

$$\sum F_x = F_w - F_{gx} = 0 \qquad (16)$$

and

$$\sum F_y = F_{gy} - Mg - mg = 0. \qquad (17)$$

Equation 17 yields

$$F_{gy} = g(M + m) = (9.8\text{ m/s}^2)(72\text{ kg} + 45\text{ kg})$$
$$= 1150\text{ N.} \qquad \text{(Answer)}$$

From Eq. 9, the balance of torques equation, we have, taking an axis through O, the point of contact of the ladder with the ground,

$$\sum \tau_z = (F_w)(h) - (Mg)(a/2) - (mg)(a/3) = 0. \qquad (18)$$

This wise choice of a location for an axis eliminated two variables, F_{gx} and F_{gy}, from the balance of torques equation. Note that when the firefighter has climbed a fraction f of the distance along the ladder, his horizontal distance from O is fa. In our case, $f = \frac{1}{3}$. We find, solving Eq. 18 for F_w,

$$F_w = \frac{ga(M/2 + m/3)}{h}$$
$$= \frac{(9.8\text{ m/s}^2)(7.6\text{ m})(72/2\text{ kg} + 45/3\text{ kg})}{9.3\text{ m}}$$
$$= 410\text{ N.} \qquad \text{(Answer)}$$

From Eq. 16 we have at once

$$F_{gx} = F_w = 410\text{ N.} \qquad \text{(Answer)}$$

Sample Problem 4 In Sample Problem 3, the coefficient of static friction μ_s between the ladder and the ground is 0.53. How far up the ladder can the firefighter go before the ladder starts to slip?

The forces that act have the same labels as in Fig. 7. Let f be the fraction of the way up the ladder the firefighter goes before slippage occurs. At the onset of slippage, we have

$$F_{gx} = \mu_s F_{gy}, \qquad (19)$$

in which F_{gx} is the frictional force and F_{gy} is the normal force.

If we apply Eq. 9, the balance of torques equation, about an axis through the point of contact of the ladder with the ground, we have

$$\sum \tau_z = (F_w)(h) - (mg)(a/3) - (Mg)(f/a) = 0,$$

or

$$F_w = \frac{ga}{h}(Mf + \tfrac{1}{3}m). \qquad (20)$$

This equation shows us that as the firefighter climbs the ladder, that is, as f increases, the force F_w exerted by the wall must increase if equilibrium is to be maintained. To find f at the slippage point, we must first find F_w.

Equation 7, the balance of forces equation for the x direction, gives

$$\sum F_x = F_w - F_{gx} = 0.$$

If we combine this equation with Eq. 19 we find, for the ladder at the point of slipping,

$$F_w = F_{gx} = \mu_s F_{gy}. \qquad (21)$$

From Eq. 8, the balance of forces equation for the y direction, we have

$$\sum F_y = F_{gy} - Mg - mg,$$

or

$$F_{gy} = (M + m)g. \qquad (22)$$

We find, combining Eqs. 21 and 22,

$$F_w = \mu_s g(M + m). \qquad (23)$$

If, finally, we combine Eqs. 20 and 23 and solve for f, we have

$$f = \frac{\mu_s h}{a}\frac{(M + m)}{M} - \frac{m}{3M}$$
$$= \frac{(0.53)(9.3\text{ m})}{7.6\text{ m}}\frac{(72\text{ kg} + 45\text{ kg})}{72\text{ kg}}$$
$$\quad - \frac{45\text{ kg}}{(3)(72\text{ kg})}$$
$$= 0.85. \qquad \text{(Answer)} \quad (24)$$

The firefighter can climb 85% of the way up the ladder before it starts to slip.

You can show from Eq. 24 that the firefighter can climb all the way up the ladder, which corresponds to $f = 1$, without slipping if $\mu_s > 0.61$. On the other hand, the ladder will slip when weight is put on the first rung, which corresponds to $f = 0$, if $\mu_s < 0.11$.

Sample Problem 5 Figure 8a shows a safe, whose mass M is 430 kg, hanging from a boom whose dimensions a and b are 1.9 m and 2.5 m, respectively. The beam has a mass m of 85 kg, the mass of the horizontal cable being negligible. (a) Find the tension T in the cable.

Figure 8b is a free-body diagram of the beam, which we take as our system. The forces acting on it are the tension $\mathbf{T}$ exerted by the cable, the weight $M\mathbf{g}$ of the safe, exerted vertically downward, the weight $m\mathbf{g}$ of the beam itself, and the force components $\mathbf{F}_h$ and $\mathbf{F}_v$ exerted on the beam by the hinge that fastens it to the wall.

Let us apply Eq. 9, the balance of torques equation, to an axis through the hinge at right angles to the plane of the figure.

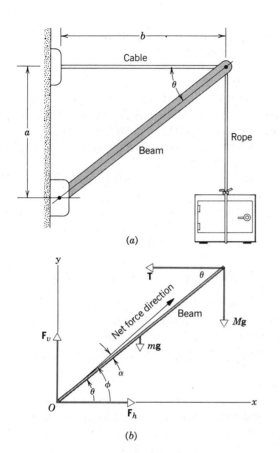

(a)

(b)

Figure 8   Sample Problem 5. (*a*) A heavy safe is hung from a boom consisting of a horizontal steel cable and a wooden beam. (*b*) A free-body diagram for the beam. Note that the net force acting on the beam at its lower end does not point directly along the beam axis.

Taking counterclockwise rotations as positive and dealing with the magnitudes of the forces, we have

$$\sum \tau_z = (T)(a) - (Mg)(b) - (mg)(\tfrac{1}{2}b) = 0.$$

By our wise choice of an axis, we have eliminated the unknown forces F_h and F_v from this equation, leaving only the unknown force T. We find, solving for T,

$$T = \frac{gb(M + \tfrac{1}{2}m)}{a}$$

$$= \frac{(9.8 \text{ m/s}^2)(2.5 \text{ m})(430 \text{ kg} + \tfrac{1}{2} \times 85 \text{ kg})}{1.9 \text{ m}}$$

$$= 6090 \text{ N.} \qquad \text{(Answer)}$$

(b) Find the forces F_h and F_v exerted on the beam by the lower hinge.

We now apply the balance of forces equations. From Eq.

7 we have

$$\sum F_x = T - F_h = 0$$

or

$$F_h = T = 6090 \text{ N.} \qquad \text{(Answer)}$$

From Eq. 8 we have

$$\sum F_y = F_v - mg - Mg = 0$$

or

$$F_v = g(M + m) = (9.8 \text{ m/s}^2)(430 \text{ kg} + 85 \text{ kg})$$
$$= 5050 \text{ N.} \qquad \text{(Answer)}$$

(c) What is the magnitude F of the net force exerted by the hinge on the beam?

From the figure we see that

$$F = \sqrt{F_h^2 + F_v^2}$$
$$= \sqrt{(6090 \text{ N})^2 + (5050 \text{ N})^2} = 7910 \text{ N.}   \text{(Answer)}$$

Note that F, 7910 N, is substantially greater than both the combined weights of the safe and the beam, 5050 N, and the tension in the horizontal wire, 6090 N.

(d) What is the angle α between the direction of the net force $\mathbf{F}$ exerted on the beam by the hinge and the center line of the beam?

From the figure we see that

$$\alpha = \phi - \theta,$$

where

$$\theta = \tan^{-1}\frac{a}{b} = \tan^{-1}\frac{1.9 \text{ m}}{2.5 \text{ m}} = 37.2°,$$

and

$$\phi = \tan^{-1}\frac{F_v}{F_h} = \tan^{-1}\frac{5050 \text{ N}}{6090 \text{ N}} = 39.7°.$$

Thus, we have

$$\alpha = \phi - \theta = 39.7° - 37.2° = 2.5°. \qquad \text{(Answer)}$$

If the weight of the beam were small enough to neglect, you would find $\alpha = 0$; that is, the hinge force would point directly along the beam axis.

Hint 1:   *Static equilibrium problems.*   We have consistently followed these procedures in solving the preceding Sample Problems:

1. Draw a *sketch* of the problem; see Figs. 5–8.

2. Select the *system* to which you will apply the laws

of equilibrium, drawing a closed curve around it on your sketch to fix it clearly in your mind. In Sample Problem 1, the system was the beam and the block, not the scales. In Sample Problem 2, it was the bowling ball and the lower arm, not the upper arm. In Sample Problems 3 and 4, it was the ladder and the firefighter, not the wall or the ground. In Sample Problem 5, it was the beam, not the hanging weight, the wall, or the horizontal cable.

3. Draw a *free-body diagram* of the system. Show all the forces that act on the system, labeling them clearly and making sure that their points of application and lines of action are correctly shown.

4. Draw in the *x and y axes* of a coordinate system. Choose them so that at least one axis is parallel to one or more of the unknown forces. Resolve into components those forces that do not lie along one of the axes. In all of our Sample Problems it made sense to choose the *x* axis horizontal and the *y* axis vertical.

5. Write down the two *balance of force equations,* using symbols throughout.

6. Choose one or more axes at right angles to the plane of the figure and write down the *balance of torques equation* for each axis. If you choose an axis that passes through the line of action of an unknown force, the equation will be simplified because that particular force will not appear in it. In our second solution for Sample Problem 1, we chose axes through $\mathbf{F}_r$ and $\mathbf{F}_l$ in Fig. 5*b*. In Sample Problem 2, our axis passed through $\mathbf{F}$ in Fig. 6*b*. In Sample Problems 3 and 4, our axis passed through $\mathbf{F}_{gx}$ and $\mathbf{F}_{gy}$ in Fig. 7*c*. In Sample Problem 5, our axis passed through point *O*. Any other choice of axes in these problems would have introduced more unknown quantities into the balance of torques equation.

7. *Solve* the equations that you have written down *algebraically* for the unknowns. Beginning students may feel more confident in substituting numbers with units in the independent equations at this stage, especially if the algebra is particularly involved. With experience, however, proceeding algebraically is preferable; see Hint 8 of Chapter 4.

8. Finally, *substitute numbers* with units in your algebraic solutions, obtaining numerical values for the unknowns.

13-5 Indeterminate Structures

For the problems of this chapter, we have only three independent equations at our disposal, two balance of forces equations and one balance of torques equation.

Thus, if a problem has more than three unknowns, we cannot solve it.

It is easy to find such problems. In Sample Problems 3 and 4, for example, we could have assumed that the wall against which the ladder leans is rough. This means that there would be a vertical frictional force acting where the ladder touches the wall, making a total of four unknown forces. With only three equations, we cannot solve this problem.

Consider also an unsymmetrically loaded car. What are the forces—all different—on the four tires? Again, we cannot find them because we have only three independent equations with which to work. Similarly, we can solve the problem of a table with three legs but not one with four legs. Problems like these, in which there are more unknowns than equations, are called *indeterminate.*

And yet, solutions exist to indeterminate problems in the real world. If you rest the tires of a car on four platform scales, each scale will register a definite reading, the sum of the readings being the weight of the car. What is eluding us in our efforts to find the individual scale readings?

The problem is that we have assumed—without making a great point of it—that the bodies to which we apply the equations of static equilibrium are perfectly

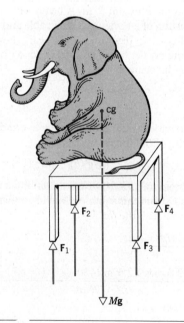

Figure 9 The table is an indeterminate structure. The four forces on the table legs are different in magnitude and cannot be found from the laws of static equilibrium alone.

rigid. By this we mean that they do not deform when forces are applied to them. Strictly, there are no such bodies. The tires of the car, for example, deform easily under load until the car settles down into a position of static equilibrium.

We have all had experience with wobbly restaurant tables, which you have to level by putting folded match covers under one of the legs. If a big enough elephant sat on such a table, however, you may be sure that—if the table did not collapse—its legs, just like the tires of a car, would also deform. The forces acting upward on the legs would settle down to definite values, as in Fig. 9, and the table would no longer wobble.

To solve indeterminate equilibrium problems, the equilibrium equations must be supplemented by some knowledge of *elasticity,* the branch of physics that describes how real bodies deform when forces are applied to them. The following section provides an introduction to this subject.

13-6   Elasticity

When a large number of atoms come together to form a solid, such as an iron nail, they settle down to equilibrium positions in a three-dimensional *lattice* in which each atom has a well-defined equilibrium distance from its nearest neighbors; see Fig. 10.* The lattice is remarkably rigid, which is another way of saying that the "interatomic springs" that hold the lattice together are extremely stiff. It is for this reason that we perceive many ordinary objects such as ladders, tables, and spoons as perfectly rigid. On the other hand, some ordinary objects, such as garden hoses or rubber gloves, do not strike us as rigid at all. The atoms that make up these objects, however, do not form a rigid lattice like that of Fig. 10 but are aligned in long flexible molecular chains, each chain being only loosely bound to its neighbors.

All real "rigid" bodies are to some extent *elastic,* which means that we can change their dimensions, slightly, by pulling, pushing, twisting, or compressing them. To get a feeling for the orders of magnitude involved, consider a steel rod, 1 m long and 1 cm in diame-

---

* Ordinary metal objects, such as an iron nail, are made up of grains of iron, each grain formed as a more or less perfect lattice, such as that of Fig. 10. The forces between the grains are much weaker than the forces that hold the lattice together, so that rupture usually occurs at grain boundaries.

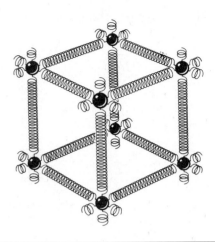

Figure 10   The atoms of a metallic solid are distributed on a repetitive three-dimensional lattice, the interatomic forces being represented here by springs.

ter. If you hang a subcompact car from the end of such a rod, the rod will stretch, but only by about 0.5 mm, or 0.05%. Furthermore, the rod will return to its original length when the car is removed.

If you hang two cars from the rod, the rod will be permanently stretched and will not recover its original length when you remove the load. If you hang three cars from the rod, the rod will break. Just before rupture, the elongation of the rod will be less than 0.2%. Although deformations like this seem small, they are important in engineering practice.

Figure 11 shows three ways that a solid might change its dimensions when forces act on it. In Fig. 11a, a cylinder is stretched. In Fig. 11b, a cylinder is deformed by so-called shearing forces, much as one might deform a pack of cards or a book. In Fig. 11c, a solid object, placed in a fluid under high pressure, is compressed uniformly on all sides. The three modes have in common that there is a *stress,* related to the applied forces, and there is a *strain,* or a deformation of some kind.

The stress and the strain take different forms in the three cases of Fig. 11, but—over the range of useful engineering practice—they are proportional to each other. The constant of proportionality is called a *modulus of elasticity.* Thus,

$$\text{stress} = \text{modulus} \times \text{strain}. \qquad (25)$$

Figure 12 shows the relation between stress and strain for a steel test cylinder such as that of Fig. 13. For a substantial portion of the range of applied stresses, the stress–strain curve is linear and Eq. 25 applies, with a

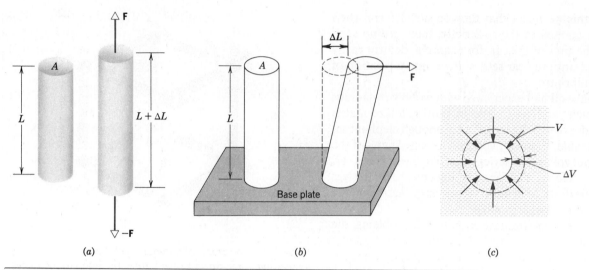

Figure 11 (*a*) A cylinder, subject to tensile stress, stretches by an amount ΔL. (*b*) A cylinder, subject to shearing stress, deforms like a pack of playing cards. (*c*) A solid sphere, subject to uniform hydraulic stress, shrinks in volume. All deformations are greatly exaggerated.

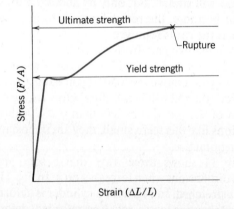

Figure 12 A stress–strain curve for a steel test specimen such as that of Fig. 13. The specimen deforms permanently when the stress is equal to the *yield strength* of the material. It ruptures when the stress is equal to the *ultimate strength* of the material.

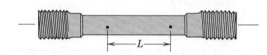

Figure 13 A test specimen, used to determine a stress–strain curve such as that of Fig. 12.

deformation, as the dimensionless quantity $\Delta L/L$, the fractional change in length of the specimen. If the specimen is a long rod, note that not only the entire rod but also any section of it experiences the same strain when a given stress is applied. Because the strain is dimensionless, the modulus in Eq. 25 has the same dimensions as the stress, namely, force per unit area.

The modulus for tensile/compressive stresses is called *Young's modulus,* represented, in engineering practice, by the symbol E. Equation 25 becomes

$$\frac{F}{A} = E\frac{\Delta L}{L}$$

or

$$\Delta L = \frac{FL}{EA}. \qquad (26)$$

The strain in a specimen, $\Delta L/L$, can often be measured conveniently with a *strain gage;* see Fig. 14. These simple and useful devices, which can be attached directly to

constant modulus. If the stress is increased beyond the *yield strength* of the specimen, the specimen becomes permanently changed and does not recover its original dimensions when the stress is removed. Beyond yielding —inevitably—comes rupture, which occurs at a stress called the *ultimate strength.*

Tension and Compression.   For simple stretching or compressing, the stress is defined as F/A, the force divided by the area over which it acts, and the strain, or

Figure 14   A strain gage, of overall dimensions 9.8 mm by 4.6 mm. The gage is fastened with adhesive to the object whose strain is to be measured. The electrical resistance of the gage varies with the strain, permitting strains up to about 3% to be measured.

operating machinery with adhesives, are based on the principle that the electrical resistance of wires made of certain materials is a function of the strain in the wire.

Although the modulus may be almost the same for both tension and compression, the *ultimate strength* may well be different for the two cases. Concrete, for example, is very strong in compression but is so weak in tension that it is almost never used in this way in engineering practice. Table 1 shows the Young's modulus and other elastic properties for some materials of engineering interest.

Shearing.   In the case of shearing, the stress is also a force per unit area but the force vector lies in the plane of the area rather than at right angles to it. The strain is

again the dimensionless ratio $\Delta L/L$, the quantities being defined as shown in Fig. 11*b*. The modulus, which is given the symbol G in engineering practice, is called the *shear modulus.* Equation 26 applies to shearing stresses, with the modulus E replaced by the modulus G.

Shearing stresses play a critical role in shafts that rotate under load, in bone fractures caused by twisting, and in springs.

Hydraulic Compression.   In Fig. 11*c*, the stress is the pressure p in the fluid, once more a force per unit area, and the strain is $\Delta V/V$, where V is the volume of the specimen and ΔV is the change in volume. The modulus, symbol B, is called the *bulk modulus* of the material.

For hydraulic compression, we write Eq. 25 as

$$p = B\,\frac{\Delta V}{V},$$

or

$$\Delta V = \frac{pV}{B}. \tag{27}$$

The bulk modulus of water is 2.2×10^9 N/m² and of steel is 16×10^{10} N/m². The pressure at the bottom of the Pacific Ocean, at its average depth of about 4000 m, is 4.0×10^7 N/m². The fractional compression of water at this depth, caused by pressure alone, is 1.8%; that for a steel object is only about 0.025%. In general, solids—with their rigid atomic lattices—are less compressible than liquids, in which the atoms or molecules are less tightly coupled to their neighbors.

Table 1   Some Elastic Properties of Selected Materials of Engineering Interest

Material	Density (kg/m³)	Young's Modulus (10^9 N/m²)	Ultimate Strength (10^6 N/m²)	Yield Strength (10^6 N/m²)
Steel[a]	7860	200	400	250
Aluminum	2710	70	110	95
Glass	2190	65	50[b]	—
Concrete[c]	2320	30	40[b]	—
Wood[d]	525	13	50[b]	—
Bone	1900	9[b]	170[b]	—
Polystyrene	1050	3	48	—

[a] Structural steel (ASTM-A36).
[b] In compression.
[c] High strength.
[d] Douglas fir.

Sample Problem 6 A structural steel rod has a radius R of 9.5 mm and a length L of 81 cm. A force F of 6.2×10^4 N—which is about 7 tons—stretches it axially. (a) What is the stress in the rod?

The stress is defined from

$$\text{stress} = \frac{F}{A} = \frac{F}{\pi R^2} = \frac{6.2 \times 10^4 \text{ N}}{(\pi)(9.5 \times 10^{-3} \text{ m})^2}$$
$$= 2.2 \times 10^8 \text{ N/m}^2. \qquad \text{(Answer)}$$

The yield strength for structural steel is 2.5×10^8 N/m^2, so that this rod is dangerously close to its yield strength.

(b) What is the elongation of the rod under this load?

From Eq. 26, using the result we have just calculated, we obtain

$$\Delta L = \frac{(F/A)L}{E} = \frac{(2.2 \times 10^8 \text{ N/m}^2)(0.81 \text{ m})}{2.0 \times 10^{11} \text{ N/m}^2}$$
$$= 8.9 \times 10^{-4} \text{ m} = 0.89 \text{ mm}. \qquad \text{(Answer)}$$

Thus, the strain $\Delta L/L$ is $(8.9 \times 10^{-4}$ m$)/(0.81$ m$)$, which is 1.1×10^{-3} or 0.11%.

Sample Problem 7 The femur, which is the principal bone of the thigh, has a minimum diameter in an adult male of about 2.8 cm, corresponding to a cross section A of 6×10^{-4} m^2. At what compressive load would it break?

From Table 1 we see that the ultimate strength S_u for bone in compression is 170×10^6 N/m^2. The compressive force is then

$$F = S_u A = (170 \times 10^6 \text{ N/m}^2)(6 \times 10^{-4} \text{ m}^2)$$
$$= 1.0 \times 10^5 \text{ N}. \qquad \text{(Answer)}$$

This is 23,000 lb or 11 tons. Although this is a large force, it can be encountered during, for example, an unskillful parachute landing on hard ground. The force need not be sustained; a few milliseconds will do it.

Sample Problem 8 A table has three legs 1.00 m in length. One leg is longer by a small distance $d = 0.50$ mm, so that the table wobbles slightly. A heavy steel cylinder whose mass M is 290 kg is placed upright on the table so that all four legs compress and the table no longer wobbles. The legs are wooden cylinders whose cross-sectional area A is 1.0 cm^2. Young's modulus E for the wood is 1.3×10^{10} N/m^2. Assume that the table top remains level and that the legs do not buckle. With what force does the floor push upward on each leg?

We take the table top as our system. If the table top remains level, each of the three short legs must be compressed by the same amount ΔL_3, by the same force F_3. The single long leg must be compressed by a larger amount ΔL_1, by a force F_1, and we must have

$$\Delta L_1 = \Delta L_3 + d.$$

From Eq. 26 ($\Delta L = FL/EA$) we can write this relation as

$$F_1 L = F_2 L + dAE. \qquad (28)$$

From Eq. 8, the balance of forces in the vertical direction, we have

$$\sum F_y = 3F_3 + F_1 - Mg = 0. \qquad (29)$$

If we solve Eqs. 28 and 29 for the unknown forces, we find

$$F_3 = \frac{Mg}{4} - \frac{dAE}{4L}$$
$$= \frac{(290 \text{ kg})(9.8 \text{ m/s}^2)}{4}$$
$$\quad - (5.0 \times 10^{-4} \text{ m})(10^{-4} \text{ m}^2)(1.3 \times 10^{10} \text{ N/m}^2)/4$$
$$= 711 \text{ N} - 163 \text{ N} = 548 \text{ N}. \qquad \text{(Answer)}$$

Similarly

$$F_1 = \frac{Mg}{4} + \frac{3dAE}{4L}$$
$$= 711 \text{ N} + 489 \text{ N} = 1200 \text{ N}. \qquad \text{(Answer)}$$

You can show that, to reach their equilibrium configuration, the three short legs were each compressed by 0.42 mm and the single long leg by 0.92 mm, the difference being 0.50 mm, as expected.

REVIEW AND SUMMARY

Static Equilibrium

A rigid body at rest in an inertial reference frame is said to be in *static equilibrium*. For such a body, (1) the linear acceleration a_{cm} is zero and (2) its angular acceleration α about any point is also.

The first condition requires that the vector sum of all the external forces acting on the body add to zero, or

The Balance of Forces

$$\sum \mathbf{F}_{ext} = 0 \quad \text{(balance of forces)}. \qquad [3]$$

If all the forces lie in the xy plane, this vector equation is equivalent to the two scalar component equations

$$\sum F_x = 0 \quad \text{(balance of forces)} \qquad [7]$$

and

$$\sum F_y = 0 \quad \text{(balance of forces)} \qquad [8]$$

The fact that angular acceleration is zero for a body in static equilibrium implies that the (vector) external torques acting on the body about any point add to zero, or

The Balance of Torques

$$\sum \tau_{\text{ext}} = 0 \quad \text{(balance of torques)}. \qquad [5]$$

If the forces lie in the xy plane, the torques are all parallel to the z axis and this vector equation is equivalent to the single scalar component equation

$$\sum \tau_z = 0 \quad \text{(balance of torques)}. \qquad [9]$$

Center of Gravity

Gravity acts on all the particles of any body and there is no single force that can give the same effect. However, the net force and torque needed for equilibrium calculations may be found by imagining an equivalent total gravitational force $M\mathbf{g}$ acting at a specific point called the *center of gravity*. If the gravitational acceleration $\mathbf{g}$ is the same for all the particles in a body the center of gravity is at the center of mass.

Sample Problems 1–5 show how the equilibrium conditions are applied to rigid bodies.

Elastic Moduli

Three elastic moduli are used to describe the elastic behavior of objects as they respond to the forces that act on them. The strain (fractional deformation) is linearly related to the applied stress (force per unit area) in each case. The general relation is

$$\text{stress} = \text{modulus} \times \text{strain}. \qquad [25]$$

Young's modulus E describes the fractional change in length when a longitudinal tension or compression is applied to a long rod of a particular material (see Fig. 11a). The change ΔL in the rod's length L due to a force F acting on a cross-sectional area A is

Young's Modulus E

$$\Delta L = \frac{FL}{EA}. \qquad [26]$$

Shear Modulus G

The *shear modulus G* is defined in the same way, except that the force F represents shear and the variables A, L, and ΔL are defined as in Fig. 11b. Finally, the *bulk modulus B* describes the volume change that occurs when a sample is compressed by uniform hydrostatic pressure as in Fig. 11c. It is defined by the equation

Bulk Modulus B

$$\Delta V = \frac{pV}{B}, \qquad [27]$$

p being the applied pressure and ΔV being the change in volume V resulting from applied pressure p. Sample Problems 6–8 illustrate some representative situations.

QUESTIONS

1. If a body is not in translational equilibrium, will the torque about any point be zero if the torque about some particular point is zero?

2. Are Eqs. 3 and 5 both necessary and sufficient conditions for static equilibrium?

3. Is a baseball in equilibrium at the instant it comes to rest at the top of a vertical pop fly?

4. In a simple pendulum, is the bob in equilibrium at any point of its swing?

5. If a rigid body is thrown into the air without spinning, it does not begin spinning during its flight, provided that air resistance can be neglected. What does this simple result imply about the location of the center of gravity?

6. The Olympic gymnast Mary Lou Retton did some amazing things on the uneven parallel bars. A friend tells you that careful analysis of films of her exploits shows that, no matter what she does, her center of mass is above her point(s) of support at all times, as required by the laws of physics. Comment on your friend's statement.

7. Do the center of mass and the center of gravity coincide for a building? For a lake? Under what conditions does the difference between the center of mass and the center of gravity of a body become significant?

8. A wheel rotating at constant angular velocity ω about a fixed axis is in mechanical equilibrium because no net external force or torque acts on it. However, the particles that make up the wheel undergo a centripetal acceleration **a** directed toward the axis. Since $\mathbf{a} \neq 0$, how can the wheel be said to be in equilibrium?

9. Give several examples of bodies that are not in equilibrium, even though the resultant of all the forces acting on them is zero.

10. Which is more likely to break in use, a hammock stretched tightly between two trees or one that sags quite a bit? Prove your answer.

11. A ladder is at rest with its upper end against a wall and the lower end on the ground. Is it more likely to slip when a person stands on it at the bottom or at the top? Explain.

12. A picture hangs from a wall by two wires. What orientation should the wires have to be under minimum tension? Explain how equilibrium is possible with any number of orientations and tensions, even though the picture has a definite mass.

13. A book rests on a table. The table pushes up on the book with a force just equal to the weight of the book. Speaking loosely, just how does the table "know" what upward force it must provide? What is the mechanism by which this force comes into play? [See "The Smart Table," by Earl Zwicker (ed.) *The Physics Teacher,* December 1981.]

14. Stand facing the edge of an open door, one foot on each side of the door. You will find that you are not able to stand on your toes. Why?

15. Sit in a straight-backed chair and try to stand up without leaning forward. Why can't you do it?

16. Long balancing poles help a tightrope walker to maintain balance. How?

17. A composite block made up of wood and metal rests on a (rough) table top. In which orientation of the two shown in Fig. 15 can you tip it over with the least force?

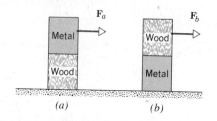

Figure 15 Question 17.

18. Explain, using forces and torques, how a tree can maintain equilibrium in a high wind.

19. You are sitting in the driver's seat of a parked automobile. You are told that the forces exerted upward by the ground on each of the four tires are different. Discuss the factors that enter into a consideration of whether this statement is true or false.

20. In Sample Problem 3, if the wall were rough, would the empirical laws of friction supply us with the extra condition needed to determine the extra (vertical) force exerted by the wall on the ladder?

21. Is Young's modulus for rubber higher or lower than Young's modulus for steel? By this criterion, is rubber more elastic than steel?

22. A beam supported at both ends is loaded in the middle. Show that the upper part of the beam is under compression whereas the lower part is under tension.

23. Why are reinforcing rods used in concrete structures? (Compare the tensile strength of concrete to its compressive strength.)

24. Is there such a thing as a truly rigid body? If so, give an example. If not, explain why.

EXERCISES AND PROBLEMS

Section 13–4   Some Examples of Static Equilibrium

1E. An eight-member family, whose weights in pounds are indicated, is balanced on a see-saw, as shown in Fig. 16. What is the number of the person who causes the largest torque, about the fulcrum, directed (*a*) out of the page and (*b*) into the page?

2E. A rigid square object of negligible weight is acted on by three forces that pull on its corners as shown, to scale, in Fig. 17. (*a*) Is the first condition of equilibrium satisfied? (*b*) Is the second condition of equilibrium satisfied? (*c*) If either of the

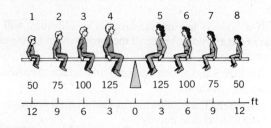

Figure 16 Exercise 1.

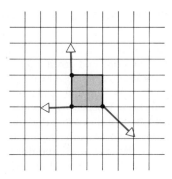

Figure 17   Exercise 2.

preceding answers is no, could a fourth force restore the equilibrium of the object? If so, specify the magnitude, direction, and point of application of the needed force.

3E. A certain nut is known to require forces of 40 N exerted on it from both sides to crack it. What forces *F* will be required when it is placed in the nutcracker shown in Fig. 18.

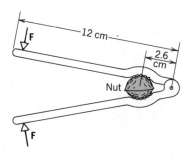

Figure 18   Exercise 3.

4E. The leaning Tower of Pisa (see Fig. 19) is 55 m high and 7.0 m in diameter. The top of the tower is displaced 4.5 m from the vertical. Treating the tower as a uniform, circular cylinder, (*a*) what additional displacement, measured at the top, will bring the tower to the verge of toppling? (*b*) What angle with the vertical will the tower make at that moment?

5E. A particle is acted on by forces given, in newtons, by $F_1 = 10i - 4j$ and $F_2 = 17i + 2j$. (*a*) Find a force F_3 that would keep it in equilibrium. (*b*) What direction does F_3 have relative to the *x* axis?

6E. A bow is drawn until the tension in the string is equal to the force exerted by the archer. What is the angle between the two parts of the string?

7E. A rope, assumed massless, is stretched horizontally between two supports that are 3.44 m apart. When an object of weight 3160 N is hung at the center of the rope, the rope is observed to sag by 35 cm. What is the tension in the rope?

Figure 19   Exercise 4.

8E. In Fig. 20, a man is trying to get his car out of the mud on the shoulder of a road. He ties one end of a rope tightly around the front bumper and the other end tightly around a utility pole 60 ft away. He then pushes sideways on the rope at its midpoint with a force of 125 lb, displacing the center of the rope 1.0 ft from its previous position and the car barely moves. What force does the rope exert on the car? (The rope stretches somewhat under the tension.)

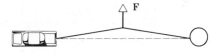

Figure 20   Exercise 8.

9E. The system in Fig. 21 is in equilibrium, but it begins to slip if any additional mass is added to the 5.0-kg object. What is the coefficient of static friction between the 10-kg block and the plane on which it rests?

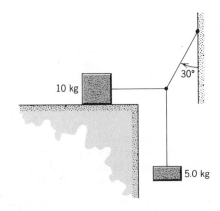

Figure 21   Exercise 9.

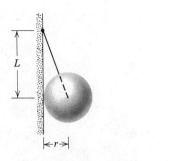

Figure 22 Exercise 10.

10E. A uniform sphere of weight w and radius r is being held by a rope attached to a frictionless wall a distance L above the center of the sphere, as in Fig. 22. Find (*a*) the tension in the rope and (*b*) the force exerted on the sphere by the wall.

11E. An automobile weighing 3000 lb (mass = 1360 kg) has a wheelbase of 120 in. (=305 cm). Its center of gravity is located 70.0 in. (=178 cm) behind the front axle. Determine (*a*) the force exerted on each of the front wheels (assumed the same) and (*b*) the force exerted on each of the back wheels (assumed the same) by the level ground.

12E. A 160-lb man is walking across a level bridge and stops three-fourths of the way from one end. The bridge is uniform and weighs 600 lb. What are the values of the vertical forces exerted on each end of the bridge by its supports?

13E. A diver of weight 580 N stands at the end of a 4.5-m diving board of negligible weight. The board is attached by two pedestals 1.5 m apart, as shown in Fig. 23. Find the tension (or compression) in each of the two pedestals.

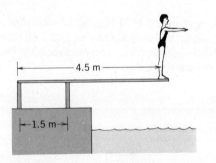

Figure 23 Exercise 13.

14E. A meter stick balances on a knife edge at the 50.0-cm mark. When two nickels are stacked over the 12.0-cm mark, the loaded stick is found to balance at the 45.5-cm mark. A nickel has a mass of 5.0 g. What is the mass of the meter stick?

15E. A beam is carried by three men, one man at one end and the other two supporting the beam between them on a cross piece so placed that the load is equally divided among the three

men. Find where the crosspiece is placed. Neglect the mass of the crosspiece.

16E. A 75-kg window cleaner uses a 10-kg ladder that is 5.0 m long. He places one end down 2.5 m from a wall and rests the upper end against a cracked window and climbs the ladder. He climbs 3.0 m up the ladder when the window breaks. Neglecting friction between the ladder and window and assuming that the base of the ladder did not slip, find (*a*) the force exerted on the window by the ladder just before the window breaks and (*b*) the magnitude and direction of the force exerted on the ladder by the ground just before the window breaks.

17E. Figure 24 shows the anatomical structures in the lower leg and foot that are involved when the heel is raised off the floor so that the foot effectively contacts the floor at only one point, shown as P in the figure. Calculate the forces that must be exerted on the foot by the calf muscle and by the lower-leg bones when a 65.3-kg person stands tip-toe on one foot. Compare these forces to the person's weight. Assume that $a = 5.0$ cm and $b = 15$ cm.

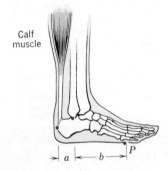

Calf
muscle

Figure 24 Exercise 17.

18E. A uniform cubical crate is 0.75 m on each side and weighs 500 N. It rests on the floor with one edge against a very small, fixed obstruction. At what height above the floor must a horizontal force of 350 N be applied to just tip the crate?

19P. Two identical uniform smooth spheres, each of weight W, rest as shown in Fig. 25 at the bottom of a fixed, rectangular

Figure 25 Problem 19.

container. Find, in terms of W, the forces acting on the spheres by (a) the container surfaces and (b) by one another, if the line of centers of the spheres makes an angle of 45° with the horizontal.

20P. An 1800-lb construction bucket is suspended by a cable that is attached at O to two other cables, these making angles of 66° and 51° with the horizontal. See Fig. 26. Find the tensions in the three cables. (*Hint:* To avoid solving two equations in two unknowns, use the rotated axes shown in the figure.)

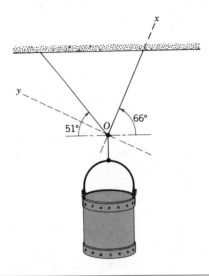

Figure 26 Problem 20.

21P. The system in Fig. 27 is in equilibrium with the string in the center exactly horizontal. Find (a) the angle θ and (b) the tension in each string.

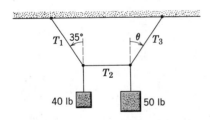

Figure 27 Problem 21.

22P. The force F in Fig. 28 is just sufficient to hold the 14-lb block and weightless pulleys in equilibrium. There is no appreciable friction. Calculate the tension T in the upper cable.

23P. A balance is made up of a rigid rod free to rotate about a point not at the center of the rod. It is balanced by unequal weights placed in the pans at each end of the rod. When an unknown mass m is placed in the left-hand pan, it is balanced

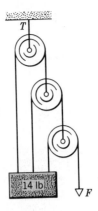

Figure 28 Problem 22.

by a mass m_1 placed in the right-hand pan; and similarly when the mass m is placed in the right-hand pan, it is balanced by a mass m_2 in the left-hand pan. Show that $m = \sqrt{m_1 m_2}$.

24P. A 15-kg weight is being lifted by the pulley system shown in Fig. 29. The upper arm is vertical, whereas the forearm makes an angle of 30° with the horizontal. What forces are being exerted on the forearm by the triceps muscle and by the upper-arm bone (the humerus)? The forearm and hand together have a mass of 2.0 kg with a center of mass 15 cm (measured along the arm) from the point where the two bones are in contact. The triceps muscle pulls vertically upward at a point 2.5 cm behind the contact point.

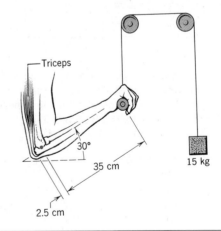

Figure 29 Problem 24.

25P. A 50-kg uniform square sign, 2.0 m on a side, is hung from a 3.0-m rod of negligible mass. A cable is attached to the end of the rod and to a point on the wall 4.0 m above the point where the rod is fixed to the wall, as shown in Fig. 30. (a) What is the tension in the cable? (b) What are the horizontal and vertical components of the force exerted by the wall on the rod?

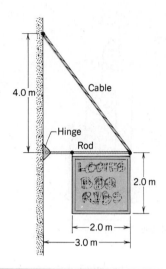

Figure 30 Problem 25.

26P. What force F applied horizontally at the axle of the wheel in Fig. 31 is necessary to raise the wheel over an obstacle of height h? Take r as the radius of the wheel and W as its weight.

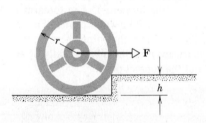

Figure 31 Problem 26.

27P. Forces F_1, F_2, and F_3 act on the structure of Fig. 32 as shown. We wish to put the structure in equilibrium by applying a force, at a point such as P, whose vector components are F_h and F_v. We are given that $a = 2.0$ m, $b = 3.0$ m, $c = 1.0$ m, $F_1 = 20$ N, $F_2 = 10$ N, and $F_3 = 5.0$ N. Find (a) F_h, (b) F_v, and (c) d.

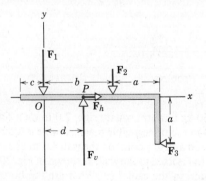

Figure 32 Problem 27.

28P. A trap door in a ceiling is 3.0 ft (=0.91 m) square, weighs 25 lb (mass = 11 kg), and is hinged along one side with a catch at the opposite side. If the center of gravity of the door is 4.0 in. (=10 cm) from the door's center and closer to the hinged side, what forces must (a) the catch and (b) the hinge sustain?

29P. Four identical bricks, each of length L, are put on top of one another (see Fig. 33) in such a way that part of each extends beyond the one beneath. Show that the largest equilibrium extensions are (a) top brick overhanging the one below by $L/2$, (b) second brick from top overhanging the one below by $L/4$, and (c) third brick from top overhanging the bottom one by $L/6$.

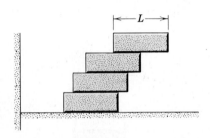

Figure 33 Problem 29.

30P. One end of a uniform beam weighing 50 lb and 3.0 ft long is attached to a wall with a hinge. The other end is supported by a wire (see Fig. 34). (a) Find the tension in the wire. (b) What are the horizontal and vertical components of the force of the hinge?

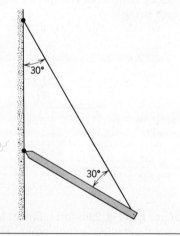

Figure 34 Problem 30.

31P. A door 7.0 ft (=2.1 m) high and 3.0 ft (=0.91 m) wide weighs 60 lb (mass = 27 kg). A hinge 1.0 ft (=0.30 m) from the top and another 1.0 ft from the bottom each support half the door's weight. Assume that the center of gravity is at the geometrical center of the door and determine the horizontal and vertical force components exerted by each hinge on the door.

32P. The system shown in Fig. 35 is in equilibrium. The mass hanging from the end of the strut S weighs 500 lb (mass = 225 kg), and the strut itself weighs 100 lb (mass = 45 kg). Find (a) the tension T in the cable and (b) the horizontal and vertical force components exerted on the strut by the pivot P.

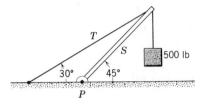

Figure 35 Problem 32.

33P. A nonuniform bar of weight W is suspended at rest in a horizontal position by two light cords as shown in Fig. 36. The angle one cord makes with the vertical is $\theta = 36.9°$; the other makes the angle $\phi = 53.1°$ with the vertical. If the length L of the bar is 6.1 m, compute the distance x from the left-hand end to the center of gravity.

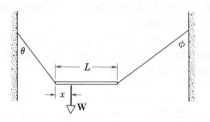

Figure 36 Problem 33.

34P. A thin horizontal bar AB of negligible weight and length L is pinned to a vertical wall at A and supported at B by a thin wire BC that makes an angle θ with the horizontal. A weight W can be moved anywhere along the bar as defined by the distance x from the wall (Fig. 37). (a) Find the tension in the thin wire as a function of x. Find (b) the horizontal and (c) the vertical components of the force exerted on the bar by the pin at A.

35P. In Fig. 37, the length of the bar is 3.0 m and its weight is 200 N. Also, $W = 300$ N and $\theta = 30°$. The wire can withstand a maximum tension of 500 N. (a) What is the maximum distance x possible before the wire breaks? (b) With W placed at this maximum x, what are the horizontal and vertical components of the force exerted on the bar by the pin?

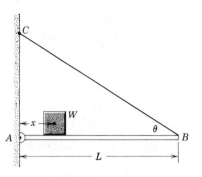

Figure 37 Problems 34 and 35.

36P. Two uniform beams are attached to a wall with hinges and then loosely bolted together as in Fig. 38. Find the horizontal and vertical components of (a) the force on each hinge and (b) the force exerted by the bolt on each beam.

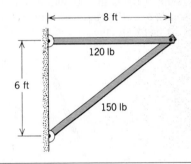

Figure 38 Problem 36.

37P. Four similar, uniform bricks of length L are stacked on a table as shown in Fig. 39 (on this and the next page; compare with Problem 29). We seek to maximize the overhang distance h, measured from the edge of the table. For each of the three arrangements, find the optimum indicated distances and calculate h. [See *Scientific American*, June 1985, for a discussion and an even better version of arrangement (c).]

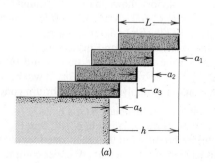

(a)

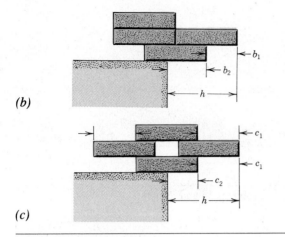

(b)

(c)

Figure 39   Problem 37.

38P. A 100-lb plank, of length $L = 20$ ft, rests on the ground and on a frictionless roller at the top of a wall of height $h = 10$ ft (see Fig. 40). The center of gravity of the plank is at its center. The plank remains in equilibrium for any value of $\theta \geq 70°$ but slips if $\theta < 70°$. Find the coefficient of static friction between the plank and the ground.

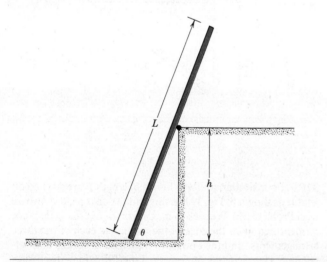

Figure 40   Problem 38.

39P. In the stepladder shown in Fig. 41, AC and CE are 8.0 ft long and hinged at C. BD is a tie rod 2.5 ft long, halfway up. A man weighing 192 lb climbs 6.0 ft along the ladder. Assuming that the floor is frictionless and neglecting the weight of the ladder, find (a) the tension in the tie rod and (b) the forces exerted on the ladder by the floor. (*Hint:* It will help to isolate parts of the ladder in applying the equilibrium conditions.)

40P. By means of a turnbuckle G, a tension force T is produced in bar AB of the square frame $ABCD$ in Fig. 42. Determine the forces produced in the other bars; identify those bars that are in tension and those that are under compression. The

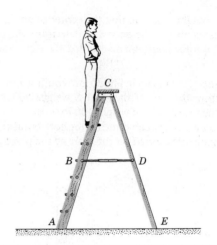

Figure 41   Problem 39.

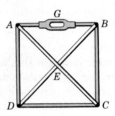

Figure 42   Problem 40.

diagonals AC and BD pass each other freely at E. Symmetry considerations can lead to considerable simplification in this and similar problems.

41P. A uniform cube of side length L rests on a horizontal floor. The coefficient of static friction between cube and floor is μ. A horizontal pull P is applied perpendicular to one of the faces of the cube, at a distance h above the floor on the vertical midline of the cube face. As P is slowly increased, the cube will either (a) begin to slide or (b) begin to tip. What is the condition on μ for (a) to occur? For (b)?

42P. A cubical box is filled with sand and weighs 200 lb ($= 890$ N). It is desired to roll the box by pushing horizontally on one of the upper edges. (a) What minimum force is required? (b) What minimum coefficient of static friction is required? (c) Is there a more efficient way to roll the box? If so, find the smallest possible force that would be required to be applied directly to the box.

43P. A crate in the form of a 4.0-ft cube contains a piece of machinery whose design is such that the center of gravity of the crate and its contents is located 1.0 ft above its geometrical center. The crate rests on a ramp which makes an angle θ with the horizontal. As θ is increased from zero, an angle will be reached at which the crate will either start to slide down the ramp or tip over. Which event will occur if the coefficient of static friction is (a) 0.60 and (b) 0.70? In each case, give the angle at which the event occurs.

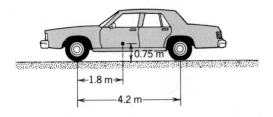

Figure 43   Problem 44.

44P*.   A car on a horizontal road makes an emergency stop by applying the brakes so that all four wheels lock and skid along the road. The coefficient of kinetic friction between tires and road is 0.40. The separation between the front and rear axles is 4.2 m, and the center of mass of the car is located 1.8 m behind the front axle and 0.75 m above the road; see Fig. 43. The car weighs 11 kN. Calculate (a) the braking deceleration of the car, (b) the normal force on each wheel, and (c) the braking force on each wheel. (*Hint:* Although the car is not in translational equilibrium, it *is* in rotational equilibrium.)

Section 13–6   Elasticity

45E.   Figure 44 shows the stress–strain curve for quartzite. Calculate Young's modulus for this material.

46E.   After a fall, a 95-kg rock climber finds himself dangling from the end of a rope 15 m long and 9.6 mm in diameter. The rope stretches by 2.8 cm. Calculate (a) the strain and (b) the stress in the rope. (c) What is Young's modulus of the rope?

47E.   A mine elevator is supported by a single steel cable 2.5 cm in diameter. The total mass of the elevator cage plus occupants is 670 kg. By how much does the cable stretch when the elevator is (a) at the surface, 12 m below the elevator motor, and (b) at the bottom of the 350-m deep shaft? (Neglect the mass of the cable.)

48E.   The (square) beam in Fig. 8a is of Douglas fir. What must be its thickness if a safety factor of 6 is desired? (See Sample Problem 5.)

49E.   A horizontal aluminum pole 4.8 cm in diameter projects 5.3 m from a wall. A 1200-kg object is suspended from

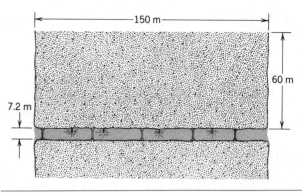

Figure 45   Problem 51.

the end of the pole. The shear modulus of aluminum is 3.0×10^{10} N/m². (a) Calculate the shear stress on the pole. (b) What is the vertical deflection of the end of the pole?

50E.   A solid copper cube has an edge length of 85.5 cm. How much pressure must be applied to the cube to reduce the edge length to 85.0 cm? The bulk modulus of copper is 1.4×10^{11} N/m².

51P.   A 150-m long tunnel 7.2 m high and 5.8 m wide (with a flat roof) is to be constructed 60 m beneath the ground. The tunnel roof is to be supported entirely by square steel columns, each with a cross-sectional area of 960 cm². The density of the ground material is 2.8 g/cm³. (a) What is the weight that the columns must support? (b) How many columns are needed to provide a safety factor of 2? See Fig. 45.

52P.   A rectangular slab of rock rests on a 26° incline; see Fig. 46. The slab has dimensions 43 m long, 2.5 m thick, and 12 m wide. Its density is 3.2 g/cm³. The coefficient of static friction between the slab and the underlying rock is 0.39. (a) Calculate the component of the slab's weight acting parallel to the incline. (b) Calculate the static force of friction. (c) Comparing (a) and (b), convince yourself that the slab is in danger of sliding. Only cohesion between the slab and the incline prevents sliding. It is desired to stabilize the slab with rock bolts driven perpendicular to the incline so that, ignoring cohesion, the slab is stable. If each rock bolt has an area of 6.4 cm² and shear strength 3.6×10^8 N/m², what is the minimum number of bolts needed? Assume that the rock bolts do not affect the normal force.

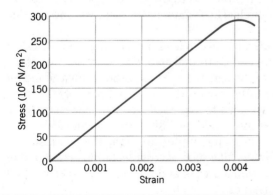

Figure 44   Problem 45.

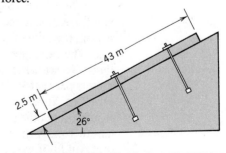

Figure 46   Problem 52.

ESSAY 4
DIMENSIONAL ANALYSIS AND AUTOMOBILE PERFORMANCE: GASOLINE IS STRONGER THAN STEEL

ALBERT A. BARTLETT
UNIVERSITY OF COLORADO, BOULDER

Vehicle Fuel Economy

The conventional measure of automotive fuel economy is the miles per gallon (mpg) that the vehicle will achieve in either city driving or highway driving. The mile is a distance and the gallon is a volume. Thus, mpg has the dimensions of $(\text{area})^{-1}$.

Let us use the definitions given by Hodges[1]

$$\text{fuel economy} = (\text{fuel consumption})^{-1}$$
$$\text{mpg} = (\text{gpm})^{-1}.$$

Thus, the measure of fuel consumption (gpm) of a vehicle is an area.
Question: What area is it?

Let us first estimate this area for a car whose fuel economy is 40 mpg. First, convert everything to SI. To make the conversion, we need the following conversion factors, all of which are exact; that is, all figures to the right of those quoted are zeros:

$$1 \text{ mi} = 5280 \text{ ft}, \qquad 1 \text{ in.} = 2.54 \text{ cm},$$
$$1 \text{ ft} = 12 \text{ in.}, \qquad 1 \text{ gal} = 231 \text{ in}^3.$$

From these we can calculate the following conversion factors, which are given here to only three significant figures:

$$1 \text{ mi} = 1.61 \times 10^3 \text{ m},$$
$$1 \text{ gal} = 3.79 \times 10^{-3} \text{ m}^3,$$
$$1 \text{ gal/mi} = 2.35 \times 10^{-6} \text{ m}^2.$$

Thus, a fuel consumption of $\frac{1}{40}$ gpm is represented by an area of $(\frac{1}{40})(2.35 \times 10^{-6}) = 5.88 \times 10^{-8} \text{ m}^2$. This is the area of a circle whose diameter is just over one-quarter of a millimeter.

When I have asked students to identify this area without telling them its calculated size, I have received answers such as, the frontal area of the car, the area of a piston, the area of the turning circle of the car, or the cross-sectional area of the fuel line. The calculated small magnitude of the area rules out all but possibly the last of these. To see what area is described, we need to express the fuel consumption as cubic meters per meter. For our 40-mpg car this would be $5.88 \times 10^{-8} \text{ m}^3/\text{m}$ (58.8 mm^3/m). Suppose we spread this volume of fuel uniformly along the 1-m distance it would propel the car. The fuel would thus fill a tube of cross-sectional area A and of length 1 m. If the tube had a cross-sectional area A and a length of L meters, the fuel that will fill it would take the car a distance of L meters.

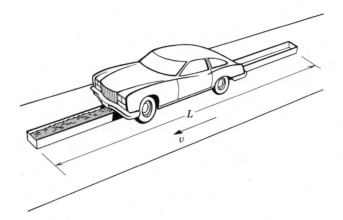

Figure 1 A long tank *full of gasoline in front of the car* and an empty tank behind the car. The car uses all the gas in the tank as it travels.

The area A we calculated above is the cross-sectional area of this long, narrow tank (Fig. 1).

Considerations of Energy and Efficiency

Horton and Compton[2] give the following data for a 1984 Ford Escort. At 80.5 km/h (22.4 m/s), the car requires 4.85×10^3 W (6.5 hp) to overcome aerodynamic drag and 3.1×10^3 W (4.2 hp) to overcome rolling and mechanical losses. Thus, a total of 8.0×10^3 W are required to maintain this car at 22.4 m/s on level pavement in still air.

Let us find the work that is required to move this car 1 m at this speed.

$$\frac{8.0 \times 10^3 \text{ J/s}}{22.4 \text{ m/s}} = 3.6 \times 10^2 \text{ J/m}.$$

Since a joule per meter is a newton, we can say that the total (aerodynamic plus mechanical) frictional drag force on this car at this speed is 3.6×10^2 N. The mass of the car is 1.02×10^3 kg and its weight is 1.0×10^4 N. We can now estimate the coefficient of friction for the car rolling on a level road at 22.4 m/s:

$$\mu = \frac{3.6 \times 10^2}{1.0 \times 10^4} = 0.036.$$

If we assume that this car has a fuel economy of 40 mpg at this speed, we can calculate the chemical energy needed to carry the car 1 m at this speed. The chemical potential energy of gasoline is 3.3×10^{10} J/m³:

$$(3.3 \times 10^{10} \text{ J/m}^3)(5.88 \times 10^{-8} \text{ m}^3/\text{m}) = 1.9 \times 10^3 \text{ J/m}.$$

We now have two measures of the energy required per meter traveled and these may be compared to give an estimate of the thermal efficiency of the car:

$$\frac{3.6 \times 10^2 \text{ J/m}}{1.9 \times 10^3 \text{ J/m}} = 0.18 \text{ or } 18\%.$$

Thus, only 18% of the available chemical energy of the gasoline is used to propel the car at this speed. The other 82% is heat that is exhausted into the atmosphere through the radiator and the exhaust pipe.

Gasoline versus Diesel

The "high-temperature reservoir" of a diesel engine is at a higher temperature than that of a gasoline engine, and hence the diesel engine has the higher thermodynamic efficiency (see Chapter 22). This would make the fuel economy of a diesel car greater than the fuel economy of a gasoline car when they have engines of the same power.

There is a second reason why the fuel economy of a diesel car is greater than that of a gasoline car. This comes from the fact that fuel economy is distance per unit *volume* of fuel rather than per unit *mass* of fuel. Gasoline and diesel fuel have roughly the same chemical potential energy per unit mass but diesel fuel is 10–15% more dense than gasoline. Thus, the energy per unit volume of diesel fuel is 10–15% higher than that for gasoline and so the fuel economy for diesel should be 10–15% higher than for gasoline. This is independent of the higher thermodynamic efficiency of a diesel.

Gasoline Is Stronger than Steel

Imagine the Ford Escort of the earlier discussion being towed along the highway at 22.4 m/s by a fiber whose cross section is the same as that of the tube of fuel. The stress in the fiber is

$$\frac{F}{A} = \frac{3.6 \times 10^2 \text{ N}}{5.88 \times 10^{-8} \text{ m}^2} = 6.1 \times 10^9 \text{ N/m}^2.$$

The ultimate strength of steel is roughly 4×10^8 N/m^2.

We thus have a paradox. A tube of gasoline is able to supply a force that is approximately an order of magnitude larger than the force that would break a steel rod whose diameter was the same as that of the tube!

This has been explained by Purcell as follows. When a steel rod is stretched in the Hooke's law region, all the chemical bonds over a cross section of the rod are stretched simultaneously. If the steel was in the form of a perfect crystalline lattice, the rod would not break until the tension was sufficient to break simultaneously all the chemical bonds over a cross-sectional area of the rod. Ordinary steel is not an ideal crystalline substance and hence it breaks at a much lower force per unit area. Ordinary steel is full of defects. Stresses concentrate at the defects and the bonds break first at these concentrations. This shifts the stresses to other bonds that then break, and so on. A smaller force is required to break bonds sequentially than to break them simultaneously. One can see this by comparing the force necessary to tear apart two postage stamps when all the bonds (between the perforations) are stressed simultaneously and when they are torn apart in the usual way by breaking the bonds one at a time. The burning of gasoline is a process that remakes all the chemical bonds and so it is a process that can generate forces of the size that would be required to break all the chemical bonds simultaneously across a cross section of a sample.

A Dimensionless Measure of Car Performance

The aerodynamic drag force on a car is

$$F_D = C_D(\tfrac{1}{2}\rho v^2)A_F,$$

where ρ is the density of the air, v is the vehicle velocity (in still air), A_F is the frontal area of the car, and C_D is a dimensionless drag coefficient that is small for an aerodynamically efficient car. Since the drag force is proportional to an area and the fuel consumption is an area, one can propose a dimensionless ratio of two areas that would allow one to compare the overall efficiencies of vehicles. One could propose a performance index:

$$PI = \frac{\text{frontal area (in m}^2)}{\text{fuel consumption (in m}^2)} \times 10^{-7}$$

(the factor of 10^{-7} is arbitrary). The larger values of PI would indicate better aerodynamic and engineering design. If one estimates that the frontal area of the Ford Escort is 2.0 m^2, its performance index would be

$$PI = \frac{2.0 \times 10^{-7}}{5.88 \times 10^{-8}} = 3.4.$$

Conclusion

Simple dimensional analysis can yield interesting insights into commonplace things.

REFERENCES   1. L. Hodges, *Am. J. Phys.* **42**, 456 (1974).   2. E. J. Horton and W. D. Compton, *Science* **225**, 587 (1984).

CHAPTER 14
OSCILLATIONS

On July 1, 1940, four months after it was opened to traffic, the Tacoma Narrows Bridge collapsed in a mild gale. The collapse was not due to the brute force of the wind but to a resonance between the natural frequency of oscillation of the bridge and the frequency of wind-generated vortices that pushed and pulled alternately on the bridge structure.

14-1 Oscillations

We are surrounded by motions that repeat themselves. There are swinging chandeliers, boats bobbing at anchor, and the surging pistons in the engines of our cars. There are violin strings, drum heads, pealing bells, the oscillating diaphragm of the telephone and the loudspeaker, and the oscillating quartz crystal in your wristwatch. Less evident are the oscillations of the molecules of the air that transmit the sensation of sound, of the atoms in a solid that convey the sensation of temperature, and of the electrons in the antennas of radio and TV transmitters.

Oscillations are not confined to material objects such as violin strings and electrons. Light, and indeed the whole electromagnetic spectrum from long radio waves to short gamma rays, is an oscillatory motion of electric and magnetic influences. We shall study such oscillations in later chapters, being greatly helped there by the analogy with the mechanical oscillations that we are about to study here.

Oscillations in the real world are usually *damped,* by which we mean that the motion dies out gradually, transferring mechanical energy into thermal energy by the action of frictional forces. Although we cannot eliminate friction from oscillating systems, we can feed energy into the system from an outside source to compensate for the mechanical energy lost to friction. The metronome of Fig. 1 operates on this principle, drawing its maintenance energy from a coiled spring, which you

Figure 1   A metronome. Its frequency of oscillation is adjustable by moving the sliding weight. The oscillations are maintained by means of energy drawn from a wound spring within the case.

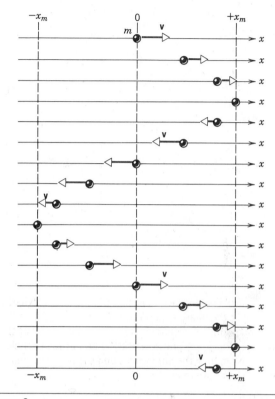

Figure 2   A sequence of "snapshots" (taken at equal time intervals) of a particle oscillating back and forth about the origin on the x axis, between the limits of $+x_m$ and $-x_m$. The arrows show the velocities of the moving particle.

must wind up. Similarly, the falling weight — which you must occasionally lift — in a grandfather's clock supplies energy to maintain the oscillations of the clock pendulum.

14–2   Simple Harmonic Motion

Figure 2 shows a sequence of "snapshots" of a prototype oscillating system, a particle moving back and forth in a periodic way about the origin of the x axis. We take the motion to be undamped and — in this section — we content ourselves with simply describing it. Later, we shall see how to arrange for this motion to happen.

The first question we are likely to ask about any oscillating system is: "How many oscillations does it complete each second?" We can put this more formally as: "What is the *frequency* of oscillation?" The symbol for frequency is the Greek letter v (nu) and its SI unit is the *hertz* (abbr. Hz), where

$$1 \text{ hertz} = 1 \text{ Hz} = 1 \text{ oscillation per second} = 1 \text{ s}^{-1}. \quad (1)$$

Related to the frequency is the *period* T of the motion,

which is the time required to complete one oscillation. That is,

$$T = \frac{1}{v}. \quad (2)$$

Any motion that repeats itself at regular intervals is called *harmonic motion*. We are interested here in motion that repeats itself in a particular way, namely, motion for which the displacement of the particle from the origin is given as a function of time by

$$x(t) = x_m \cos(\omega t + \phi) \quad \text{(displacement),} \quad (3)$$

in which x_m, ω, and ϕ are constants. This motion is not only harmonic, it is *simple* harmonic. Figure 3a is a plot of the *simple harmonic motion* (abbr. SHM) described by Eq. 3.

The quantity x_m in Eq. 3, a positive constant whose value depends on how the motion was started, is called

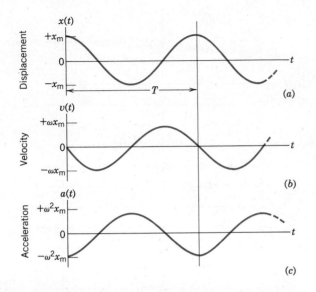

Figure 3 (a) The displacement $x(t)$ of a particle oscillating in SHM with the phase angle ϕ equal to zero; see Eq. 3. T marks one period of oscillation. (b) The velocity $v(t)$ of the particle; see Eq. 5. (c) The acceleration $a(t)$ of the particle; see Eq. 6. Note how the amplitudes and the relative phases of the three curves are related.

the *amplitude* of the motion, the subscript m standing for maximum. The cosine function in Eq. 3 varies between the limits ± 1 so that the displacement $x(t)$ varies between the limits $\pm x_m$, as Figs. 2 and 3a show.

The time-varying quantity $(\omega t + \phi)$ in Eq. 3 is called the *phase* of the motion and the constant ϕ is called the *phase constant*. The value of ϕ also depends on how the motion was started. In particular, the value of ϕ, like that of x_m, is fixed by the displacement and the velocity of the particle at $t = 0$. For the $x(t)$ plot of Fig. 3a, the phase constant ϕ has been taken to be zero.

It remains to interpret the constant ω. The displacement $x(t)$ must return to its initial value after one period T of the motion. That is, $x(t)$ must equal $x(t + T)$. To simplify our analysis, let us put $\phi = 0$ in Eq. 3. From that equation we then have

$$x_m \cos \omega t = x_m \cos \omega (t + T).$$

The cosine function first repeats itself when its angle is increased by 2π radians, so that

$$\omega t + 2\pi = \omega(t + T)$$

or

$$2\pi = \omega T.$$

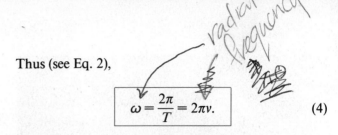

Thus (see Eq. 2),

$$\boxed{\omega = \frac{2\pi}{T} = 2\pi v.} \tag{4}$$

The quantity ω is called the *angular frequency* of the motion, its SI unit being the radian per second. Figure 4 compares $x(t)$ for two simple harmonic motions that differ in various ways as far as their amplitudes, their periods (and thus their angular frequencies), and their phase constants are concerned.

SHM — the Velocity. By differentiating Eq. 3, we can find the velocity of a particle moving with simple harmonic motion. Thus,

$$v(t) = \frac{dx}{dt} = \frac{d}{dt}\left[x_m \cos(\omega t + \phi)\right]$$

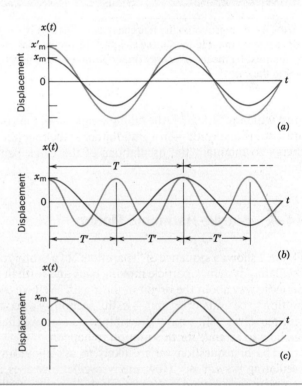

Figure 4 In all three cases, the darker curves are identical with the displacement curve of Fig. 3a. (a) The lighter curve differs from the reference curve *only* in that its amplitude is greater. (b) The lighter curve differs from the reference curve *only* in that its period, shown as T', has been reduced by a factor of 2. (c) The lighter curve differs from the reference curve *only* in that $\phi = -45°$, rather than zero.

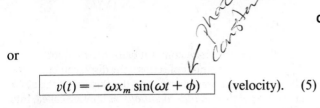

or

$$v(t) = -\omega x_m \sin(\omega t + \phi) \qquad \text{(velocity).} \qquad (5)$$

Figure 3*b* is a plot of Eq. 5 for the case of $\phi = 0$. The positive quantity ωx_m in Eq. 5 is the *velocity amplitude* v_m. That is, the velocity of the oscillating particle varies between the limits $\pm v_m$, as Fig. 3*b* shows. Note in that figure that the phases of the velocity and the displacement differ by 90°. When the magnitude of the displacement is greatest, which occurs at the limits of the motion, the velocity is least, being zero. On the other hand, when the displacement is least, being zero at the midpoint of the motion, the velocity has its greatest magnitude.

SHM — the Acceleration. Knowing the velocity $v(t)$ for simple harmonic motion, we can find the acceleration of the oscillating particle by differentiating once more. Thus, we have, from Eq. 5,

$$a(t) = \frac{dv}{dt} = \frac{d}{dt}[-\omega x_m \sin(\omega t + \phi)]$$

or

$$a(t) = -\omega^2 x_m \cos(\omega t + \phi) \qquad \text{(acceleration).} \qquad (6)$$

Figure 3*c* is a plot of Eq. 6 for the case of $\phi = 0$. The positive quantity $\omega^2 x_m$ in Eq. 6 is the *acceleration amplitude* a_m. That is, the acceleration of the particle varies between the limits $\pm a_m$, as Fig. 3*c* shows.

We can combine Eqs. 3 and 6 to yield

$$a(t) = -\omega^2 x(t), \qquad (7)$$

which shows that, in simple harmonic motion, the acceleration is proportional to the displacement but opposite in sign. That is, when the displacement has its greatest positive value, the acceleration has its greatest negative value and conversely. When the displacement is zero, the acceleration is also zero.

14–3 Simple Harmonic Motion: The Force Law

Once we know how the acceleration of a particle varies with time, we can use Newton's second law to learn what force must act on the particle to make it move in that way. If we combine Newton's second law and Eq. 7, we find, for simple harmonic motion,

$$F = ma = -(m\omega^2)x. \qquad (8)$$

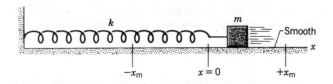

Figure 5   A simple harmonic oscillator. The block — like the particle of Fig. 2 — moves in simple harmonic motion, its displacement being given by Eq. 3.

This result — a force proportional to the displacement but opposite in sign — is familiar. It is Hooke's law

$$F = -kx \qquad \text{(Hooke's law)} \qquad (9)$$

for a spring, the effective spring constant being

$$k = m\omega^2. \qquad (10)$$

We can in fact take Eq. 9 as an alternative definition of simple harmonic motion, which states:

Simple harmonic motion is the motion executed by a particle of mass m subject to a force that is proportional to the displacement of the particle but opposite in sign.

The block–spring system of Fig. 5 forms a *linear simple harmonic oscillator* (linear oscillator, for short), its angular frequency ω being related to the spring constant k and the mass m of the block by Eq. 10, or

$$\omega = \sqrt{\frac{k}{m}} \qquad \text{(angular frequency).} \qquad (11)$$

By combining Eqs. 4 and 11, we can write, for the *period* of the linear oscillator of Fig. 5,

$$T = 2\pi \sqrt{\frac{m}{k}} \qquad \text{(period).} \qquad (12)$$

Equations 11 and 12 tell us that — entirely as we expect on physical grounds — a large angular frequency, and thus a small period, goes with a stiff spring (large k) and a light block (small m).

Every oscillating system, be it the linear oscillator of Fig. 5, a diving board, or a violin string, has some element of "springiness" and some element of "inertia." In the linear oscillator, these elements are located in separate parts of the system, the springiness being entirely in the spring, which we assume to be massless, and the inertia being entirely in the block, which we assume to be rigid. In a violin string, however, the two elements coexist all along the string.

Sample Problem 1 A block whose mass m is 680 g is fastened to a spring whose spring constant k is 65 N/m. The block is pulled a distance x ($= 11$ cm) from its equilibrium position and released from rest. (a) What force does the spring exert on the block just before it is released?

From Hooke's law

$$F(x) = -kx = -(65 \text{ N/m})(0.11 \text{ m}) = -7.2 \text{ N}. \quad \text{(Answer)}$$

The minus sign reminds us that the spring force acting on the block, which points back toward the origin, is opposite in sign to the displacement of the block, which points away from the origin.

(b) What are the angular frequency, the frequency, and the period of oscillation?

From Eq. 11 we have

$$\omega = \sqrt{\frac{k}{m}} = \sqrt{\frac{65 \text{ N/m}}{0.68 \text{ kg}}} = 9.78 \text{ rad/s}. \quad \text{(Answer)}$$

The frequency follows from Eq. 4, or

$$v = \frac{\omega}{2\pi} = \frac{9.78 \text{ rad/s}}{2\pi} = 1.56 \text{ Hz}, \quad \text{(Answer)}$$

and the period from

$$T = \frac{1}{v} = \frac{1}{1.56 \text{ Hz}} = 0.64 \text{ s} = 640 \text{ ms}. \quad \text{(Answer)}$$

(c) What is the amplitude of the oscillation?

The block was released from rest. As it oscillates, the block can never be farther from its equilibrium position than its initial displacement without violating the conservation of energy principle. Therefore,

$$x_m = 11 \text{ cm}. \quad \text{(Answer)}$$

(d) What is the maximum speed of the oscillating block?
From Eq. 5 we see that the velocity amplitude is

$$v_m = \omega x_m = (9.78 \text{ rad/s})(0.11 \text{ m}) = 1.1 \text{ m/s}. \quad \text{(Answer)}$$

This maximum speed occurs when the oscillating block is rushing through the origin; see Fig. 3b.

(e) What is the magnitude of the maximum acceleration of the block?

From Eq. 6 we see that the acceleration amplitude is

$$a_m = \omega^2 x_m = (9.78 \text{ rad/s})^2(0.11 \text{ m}) = 11 \text{ m/s}^2. \quad \text{(Answer)}$$

This maximum acceleration occurs when the block is at the ends of its path; at those points, the force acting on it has its maximum magnitude (see Fig. 3c).

(f) What is the phase constant ϕ for the motion?

At $t = 0$, the moment of release, the displacement of the block has its maximum value x_m and the velocity of the block is zero. If we put these *initial conditions,* as they are called, into Eqs. 3 and 5, we find

$$1 = \cos \phi \quad \text{and} \quad 0 = \sin \phi,$$

respectively. The smallest angle that satisfies both of these requirements is $\phi = 0$.

Sample Problem 2 At $t = 0$, the displacement $x(0)$ of the block in a linear oscillator like that of Fig. 5 is -8.5 cm. Its velocity $v(0)$ is -0.92 m/s and its acceleration $a(0)$ is $+47$ m/s^2. (a) What are the angular frequency ω and the frequency v?

If we put $t = 0$ in Eqs. 3, 5, and 6, we find

$$x(0) = x_m \cos \phi, \tag{13}$$
$$v(0) = -\omega x_m \sin \phi, \tag{14}$$

and

$$a(0) = -\omega^2 x_m \cos \phi. \tag{15}$$

These three equations contain three unknowns, namely, x_m, ϕ, and ω. We should be able to find all three.

If we divide Eq. 15 by Eq. 13, the result is

$$\omega = \sqrt{-\frac{a(0)}{x(0)}} = \sqrt{-\frac{47 \text{ m/s}^2}{-0.085 \text{ m}}} = 23.5 \text{ rad/s}. \quad \text{(Answer)}$$

The frequency v follows from Eq. 4, or

$$v = \frac{\omega}{2\pi} = \frac{23.5 \text{ rad/s}}{2\pi} = 3.74 \text{ Hz}. \quad \text{(Answer)}$$

(b) What is the phase constant ϕ?
If we divide Eq. 14 by Eq. 13, we find

$$\frac{v(0)}{x(0)} = \frac{-\omega x_m \sin \phi}{x_m \cos \phi} = -\omega \tan \phi.$$

Solving for $\tan \phi$, we find

$$\tan \phi = -\frac{v(0)}{\omega x(0)} = -\frac{-0.92 \text{ m/s}}{(23.5 \text{ rad/s})(-0.085 \text{ m})} = -0.461.$$

This equation has two solutions:

$$\phi = -25° \quad \text{and} \quad \phi = 155°.$$

We shall see below how to choose between them.

(c) What is the amplitude x_m of the motion?
From Eq. 13 we have, provisionally putting $\phi = 155°$,

$$x_m = \frac{x(0)}{\cos \phi} = \frac{-0.085 \text{ m}}{\cos 155°} = 0.094 \text{ m} = 9.4 \text{ cm}. \quad \text{(Answer)}$$

If we had chosen $\phi = -25°$, we would have found $x_m = -9.4$ cm. However, the amplitude of the motion must always be a *positive* constant, so $-25°$ cannot be the correct phase constant. We must therefore have

$$\phi = 155°. \qquad \text{(Answer)}$$

14–4 Simple Harmonic Motion: Energy Considerations

Figure 6 of Chapter 8 shows how the energy of a linear oscillator shuttles back and forth between kinetic and potential forms, their sum—the mechanical energy E—remaining constant. We now consider this situation quantitatively.

The potential energy of a linear oscillator like that of Fig. 5 is associated entirely with the spring. Its value depends on how much the spring is stretched or compressed, that is, on $x(t)$. We use Eq. 3 to find

$$U(t) = \tfrac{1}{2}kx^2 = \tfrac{1}{2}kx_m^2 \cos^2(\omega t + \phi). \qquad (16)$$

The kinetic energy of the system is associated entirely with the block. Its value depends on how fast the block is moving, that is, on $v(t)$. We use Eq. 5 to find

$$K(t) = \tfrac{1}{2}mv^2 = \tfrac{1}{2}m(-\omega x_m)^2 \sin^2(\omega t + \phi). \qquad (17)$$

If we substitute for ω from Eq. 11 ($\omega^2 = k/m$), we can write Eq. 17 as

$$K(t) = \tfrac{1}{2}mv^2 = \tfrac{1}{2}kx_m^2 \sin^2(\omega t + \phi). \qquad (18)$$

The mechanical energy follows from Eqs. 16 and 18 and is

$$\begin{aligned}
E &= U + K \\
&= \tfrac{1}{2}kx_m^2 \cos^2(\omega t + \phi) + \tfrac{1}{2}kx_m^2 \sin^2(\omega t + \phi) \\
&= \tfrac{1}{2}kx_m^2 [\cos^2(\omega t + \phi) + \sin^2(\omega t + \phi)].
\end{aligned}$$

However (see Appendix G)

$$\cos^2 \alpha + \sin^2 \alpha = 1,$$

where α is any angle. Thus, the quantity in the square brackets above is unity and we have

$$E = U + K = \tfrac{1}{2}kx_m^2, \qquad (19)$$

which is indeed a constant, independent of the time.

Figure 6a shows the potential energy and the kinetic energy of the linear oscillator as a function of time. Figure 6b shows these two energies as a function of the displacement of the oscillating block.

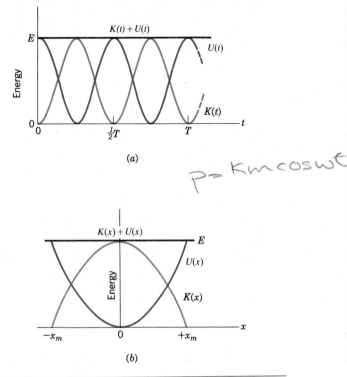

p → Kmcoswt

Figure 6   (a) The potential energy $U(t)$, the kinetic energy $K(t)$, and the mechanical energy E as functions of time, for a particle undergoing simple harmonic motion. Note that all energies are positive and that the potential energy and the kinetic energy peak twice during every period. Compare Fig. 6 of Chapter 8. (b) The potential energy $U(x)$, the kinetic energy $K(x)$, and the mechanical energy E as functions of position, for a particle undergoing simple harmonic motion with amplitude x_m. Note that, for $x = 0$, the energy is all kinetic and for $x = \pm x_m$, it is all potential.

We understand now why an oscillating system must contain an element of springiness and an element of inertia. It needs the former to store its potential energy and the latter to store its kinetic energy. Without either element, the interchange of energy between these two forms would not be possible.

Sample Problem 3 (a) What is the mechanical energy of the linear oscillator of Sample Problem 1?

We find, substituting data from Sample Problem 1 into Eq. 19,

$$\begin{aligned}
E &= \tfrac{1}{2}kx_m^2 = (\tfrac{1}{2})(65 \text{ N/m})(0.11 \text{ m})^2 \\
&= 0.393 \text{ J} = 393 \text{ mJ}. \qquad \text{(Answer)}
\end{aligned}$$

This value remains constant at all phases of the motion.

(b) What is the potential energy of this oscillator when the particle is halfway to its end point, that is, when $x = \pm\frac{1}{2}x_m$?

The potential energy is given by

$$U = \tfrac{1}{2}kx^2 = \tfrac{1}{2}k(\pm\tfrac{1}{2}x_m)^2 = \tfrac{1}{4}(\tfrac{1}{2}kx_m^2).$$

From Eq. 19 then,

$$U = \tfrac{1}{4}E = (\tfrac{1}{4})(393 \text{ mJ}) = 98 \text{ mJ}. \qquad \text{(Answer)}$$

(c) What is the kinetic energy of the oscillator when $x = \frac{1}{2}x_m$?

We find this from

$$K = E - U$$
$$= 393 \text{ mJ} - 98 \text{ mJ} = 295 \text{ mJ}. \qquad \text{(Answer)}$$

Thus, at this phase of the oscillation, 25% of the energy is in potential form and 75% in kinetic form.

14–5   An Angular Simple Harmonic Oscillator

Figure 7 shows an angular form of the linear simple harmonic oscillator of Fig. 5, in which the element of springiness or elasticity is associated with twisting a suspension wire rather than extending or compressing a spring. The device is more commonly known as a *torsion pendulum*.

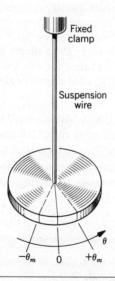
Fixed clamp

Suspension wire

θ

$-\theta_m$ 0 $+\theta_m$

Figure 7   An angular simple harmonic oscillator, or torsional pendulum. It is the angular equivalent of the linear simple harmonic oscillator of Fig. 5. The disk oscillates between limits defined by the angular amplitude θ_m. The torsion in the suspension wire provides the restoring torque.

If we twist the disk in Fig. 7 from its rest position and release it, it will oscillate about that position in *angular simple harmonic motion*. Twisting the disk through an angle θ, in either direction, introduces a restoring torque given by

$$\tau = -\kappa\theta. \qquad (20)$$

Here κ (Greek *kappa*) is a constant that depends on the properties of the suspension wire, that is, on its length, its diameter, and the material of which it is made.

Comparison of Eq. 20 with Eq. 9 shows that Eq. 20 is the angular form of Hooke's law. By the method of analogies, we should be able to transform Eq. 12, which gives the period of linear simple harmonic motion, into an equation for the period of angular simple harmonic motion. Thus, we replace the spring constant k in Eq. 12 by its equivalent, the constant κ of Eq. 20, and we replace the mass m in Eq. 12 by *its* equivalent, the rotational inertia I of the oscillating disk. These replacements lead to

$$\boxed{T = 2\pi \sqrt{\frac{I}{\kappa}}} \qquad \text{(torsional pendulum)} \qquad (21)$$

for the period of the angular simple harmonic oscillator, or torsional pendulum.

Sample Problem 4   As Fig. 8a shows, a thin rod whose length L is 12.4 cm and whose mass m is 135 g is suspended at its midpoint from a long wire. Its period T_a of angular SHM is measured to be 2.53 s. An irregular object, which we call object X, is then hung from the same wire, as in Fig. 8b, and its period T_b found to be 4.76 s. (a) What is the rotational inertia of object X about its suspension axis?

In Table 2(e) of Chapter 11, the rotational inertia of the thin rod about a perpendicular axis through its midpoint is $\frac{1}{12} mL^2$. Thus, we have

$$I_a = \tfrac{1}{12} mL^2 = (\tfrac{1}{12})(0.135 \text{ kg})(0.124 \text{ m})^2$$
$$= 1.73 \times 10^{-4} \text{ kg} \cdot \text{m}^2.$$

Now let us write Eq. 21 twice, once for the rod and once for object X:

$$T_a = 2\pi \sqrt{\frac{I_a}{\kappa}} \quad \text{and} \quad T_b = 2\pi \sqrt{\frac{I_b}{\kappa}}.$$

Here the subscripts refer to Figs. 8a and 8b. Note that the constant κ, which is a property of the wire, is the same for each experiment, only the periods and the rotational inertias being different.

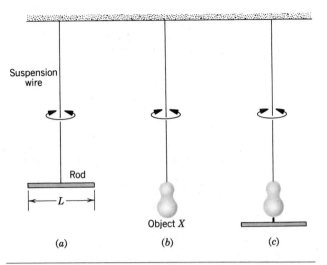

Figure 8 Sample Problem 4. Three torsion pendulums: (a) a rod, (b) an irregular object, and (c) both objects rigidly coupled together.

Let us square each of these two equations, divide them, and solve the resulting equation for I_b. The result is

$$I_b = I_a \frac{T_b^2}{T_a^2} = (1.730 \times 10^{-4} \text{ kg} \cdot \text{m}^2) \frac{(4.76 \text{ s})^2}{(2.53 \text{ s})^2}$$

$$= 6.12 \times 10^{-4} \text{ kg} \cdot \text{m}^2. \quad \text{(Answer)}$$

(b) What would be the period of oscillation if both objects were fastened together and hung from the wire, as in Fig. 8c? We can again write Eq. 21 twice, this time as

$$T_a = 2\pi \sqrt{\frac{I_a}{\kappa}} \quad \text{and} \quad T_c = 2\pi \sqrt{\frac{I_c}{\kappa}}.$$

We find, dividing and putting $I_c = I_a + I_b$,

$$T_c = T_a \sqrt{\frac{I_c}{I_a}} = T_a \sqrt{\frac{I_a + I_b}{I_a}}$$

$$= (2.53 \text{ s}) \sqrt{\frac{1.73 \times 10^{-4} \text{ kg} \cdot \text{m}^2 + 6.12 \times 10^{-4} \text{ kg} \cdot \text{m}^2}{1.73 \times 10^{-4} \text{ kg} \cdot \text{m}^2}}$$

$$= 5.39 \text{ s}. \quad \text{(Answer)}$$

14-6 Pendulums We Have Known

We turn now to a class of simple harmonic oscillators in which the springiness is associated with gravity rather than with the elastic properties of a twisted wire or a compressed spring.

The Simple Pendulum. If you hang an apple down a stairwell at the end of a long thread and set it swinging with a small amplitude, the apple seems to move back and forth in simple harmonic motion. We idealize this problem to that of a *simple pendulum*, which is a particle of mass m suspended from an inextendable, massless string of length L, as in Fig. 9.

In any oscillating system, we look for an inertial element and also for an element of springiness. The inertia is in the mass of the particle but where is the spring? It is in the gravitational attraction between the particle and the earth. Potential energy can be associated with the varying vertical distance of the oscillating particle from the earth, that is, with the varying length of the "gravitational spring." We will not be surprised to find that the expression for the period of oscillation of a simple pendulum contains the free-fall acceleration g.

The forces acting on the particle in Fig. 9 are mg, its weight, and **T**, the tension in the string. We resolve the weight into a radial component $mg \cos \theta$ and a component $mg \sin \theta$ that is tangent to the path. This tangential component—as Fig. 9 shows—changes its direction every time the particle passes through the midpoint of its swing and thus always acts to restore the particle to its central position. Thus, we have, as a restoring force,

$$F = -mg \sin \theta. \quad (22)$$

The minus sign indicates that the force is a restoring force.

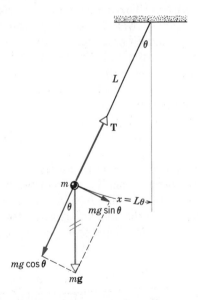

Figure 9 A simple pendulum. The forces acting on the bob are its weight mg and the tension **T** in the cord. The tangential component of the weight provides a restoring force.

If the angle θ in Fig. 9 is small, which we assume, $\sin \theta$ is very nearly equal to θ in radians. For example, if $\theta = 5°$ ($= 0.0873$ rad), $\sin \theta = 0.0872$, a difference of only about 0.1%. The displacement x of the particle measured along its arc is equal to $L\theta$. Thus, if we assume that $\sin \theta \approx \theta$, Eq. 22 becomes

$$F \approx -mg\theta = -mg\frac{x}{L} = -\left(\frac{mg}{L}\right) x. \qquad (23)$$

A glance at Eq. 9 shows that we again have Hooke's law. Thus, a simple pendulum is formally like the linear oscillator of Fig. 5, the effective spring constant k of the gravitational spring being mg/L. We see by inspection that it has the correct dimensions for a spring constant, namely, a force divided by a length.

By making this substitution for k in Eq. 12, we find, for the period of a simple pendulum,

$$T = 2\pi \sqrt{\frac{m}{k}} = 2\pi \sqrt{\frac{m}{mg/L}} \qquad (24)$$

or

$$\boxed{T = 2\pi \sqrt{\frac{L}{g}}} \qquad \text{(simple pendulum)}. \qquad (25)$$

The element of inertia seems to be missing because —as Eq. 25 shows—the period is independent of the mass of the particle. This comes about because the element of springiness, which is the gravitational spring constant mg/L, is itself proportional to the mass of the particle, the two masses canceling in Eq. 24. Figure 7 of Chapter 8 shows how energy shuttles back and forth between potential and kinetic forms during every oscillation of a simple pendulum.

The Physical Pendulum. Most pendulums in the real world are not even approximately "simple." Figure 10 shows a generalized *physical pendulum*, as we call it, its weight $m\mathbf{g}$ acting at its center of gravity C.

When the pendulum of Fig. 10 is displaced through an angle θ, in either direction from its equilibrium position, a restoring torque appears. It acts about an axis through the suspension point O in Fig. 10 and has the value

$$\tau = -(mg \sin \theta)(h).$$

Here $mg \sin \theta$ is the tangential component of the weight $m\mathbf{g}$; h, which is equal to OC, is the moment arm of this force component. The minus sign indicates that the torque is a restoring torque. That is, it is a torque that

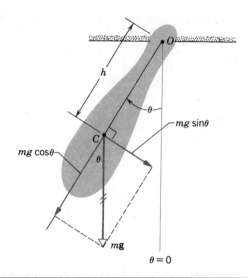

Figure 10 A physical pendulum. The restoring torque is $(mg \sin \theta)(h)$. When $\theta = 0$, the center of gravity C hangs directly below the point of suspension O.

changes sign as the pendulum passes through its equilibrium position, always acting to reduce the angle θ to zero.

We once more decide to limit our interest to small amplitudes, so that $\sin \theta \approx \theta$. The equation above then reduces to

$$\tau \approx -(mgh) \, \theta. \qquad (26)$$

Again we recognize Hooke's law. Thus, a physical pendulum is formally like the linear oscillator of Fig. 5. The spring constant k of the linear oscillator is equivalent to mgh in Eq. 26 and the mass m of the block is equivalent to the rotational inertia I of the pendulum. If we make these replacements in Eq. 12, we find

$$\boxed{T = 2\pi \sqrt{\frac{I}{mgh}}} \qquad \text{(physical pendulum)} \qquad (27)$$

for the period of a physical pendulum. Here I is the rotational inertia of the pendulum—about an axis through its point of support perpendicular to its plane of swing—and h is the distance between the point of support and the center of gravity of the swinging pendulum. We know that a physical pendulum will not swing if we hang it from its center of gravity. Formally, this corresponds to putting $h = 0$ in Eq. 27. That equation then predicts $T \to \infty$, which is the equation's way of telling us that such a pendulum will never complete one swing.

The physical pendulum of Fig. 10 includes the simple pendulum as a special case. Here h would be the length L of the string and I would be mL^2. Making these substitutions in Eq. 27 leads to

$$T = 2\pi \sqrt{\frac{I}{mgh}} = 2\pi \sqrt{\frac{mL^2}{mgL}} = 2\pi \sqrt{\frac{L}{g}},$$

which, as we expect, is exactly Eq. 25, the expression for the period of a simple pendulum.

Measuring g. We can use a physical pendulum to make precise measurements of the free-fall acceleration g. Countless thousands of such measurements have been made in the course of geophysical prospecting; see Fig. 10 of Chapter 15 for some results, which show the high precision with which such measurements can be made.

To analyze a simple case, take the pendulum to be a uniform rod of length L, suspended from one end. For such a pendulum, h in Eq. 27, the distance between the suspension point and the center of gravity, is $\frac{1}{2}L$. Table 2(f) of Chapter 11 tells us that the rotational inertia of this pendulum about a perpendicular axis through one end is $\frac{1}{3}mL^2$. If we put $h = \frac{1}{2}L$ and $I = \frac{1}{3}mL^2$ in Eq. 27 and solve for g, we find

$$g = \frac{8\pi^2 L}{3T^2}. \tag{28}$$

Thus, by measuring L and the period T, we can find the value of g. We can do no more than hint at the refinements that are needed if precise measurements are to be made. They include, for example, swinging the pendulum — which is often made of quartz to minimize thermal expansion — in an evacuated chamber.

The Phantom Pendulum. A source of error in precise pendulum measurements is the fact that, because of earth tremors or other causes, the pendulum support may not be completely rigid. Figure 11 shows a clever way to get around this problem.

Two identical pendulums are used, arranged to swing — with equal amplitudes — exactly 180° out of phase with each other. Thus, when pendulum A in Fig. 11 is moving to the right at the bottom of its swing, pendulum B will be moving to the left with the same speed. If the common support of the two pendulums receives a sudden impulsive jolt, accelerating it (say) to the left, then the motion of one pendulum (A) will be held back a little and that of the other (B) will be advanced by the same amount. The result of such accelerations of the support is that the measured periods of the two (identical) pendulums will not be exactly the same.

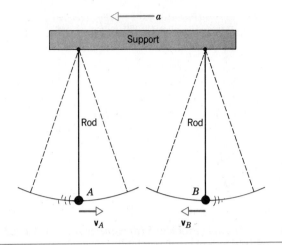

Figure 11 Two pendulums oscillating 180° out of phase. If their common support is given a random jolt, as shown, the apparent period of pendulum A will be shortened and that of pendulum B will be lengthened, by the same small amount. The average of these two quantities represents the period of a similar phantom pendulum whose support is truly stable.

However, the *average* of the two periods will be the value that you would have measured if the suspension support had not been jolted. The average period is that of a *phantom pendulum* that is like the two actual pendulums in every way except two: (1) Its support is immune to random accelerations and (2) it does not actually exist.*

---

* The phantom pendulum technique also deals with another source of error, called *sway*. A swinging pendulum will cause its support to sway a little, no matter how rigid the support may be. However, for the double pendulum arrangement of Fig. 11, the sways are in opposite directions and cancel out.

Sample Problem 5 A meter stick, suspended from one end, swings as a physical pendulum; see Fig. 12a. (a) What is its period of oscillation?

From Table 2(f) of Chapter 11 we see that the rotational inertia of a meter stick of length L_0 about a perpendicular axis through one end is $\frac{1}{3}mL_0^2$. The distance h from the point of suspension to the center of gravity, which is point C in Fig. 12a, is $\frac{1}{2}L_0$. If we substitute these two quantities into Eq. 27, we find

$$T = 2\pi \sqrt{\frac{I}{mgh}} = 2\pi \sqrt{\frac{\frac{1}{3}mL_0^2}{mg(\frac{1}{2}L_0)}} = 2\pi \sqrt{\frac{2L_0}{3g}} \tag{29}$$

$$= 2\pi \sqrt{\frac{(2)(1.00 \text{ m})}{(3)(9.8 \text{ m/s}^2)}} = 1.64 \text{ s.} \tag{Answer}$$

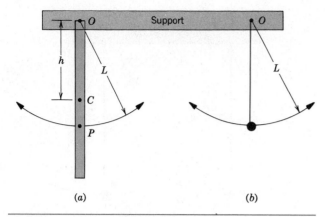

(a)   (b)

Figure 12 Sample Problem 5 (*a*) A meter stick suspended from one end as a physical pendulum. (*b*) A simple pendulum whose length L is chosen so that the periods of the two pendulums are equal. Point P on the pendulum of (*a*) marks this *center of oscillation.*

(b) What would be the length of a simple pendulum that would have the same period; see Fig. 12*b*?

Setting Eqs. 25 and 29 equal yields

$$T = 2\pi \sqrt{\frac{L}{g}} = 2\pi \sqrt{\frac{2L_0}{3g}}.$$

You can see by inspection that

$$L = \tfrac{2}{3}L_0 = (\tfrac{2}{3})(100 \text{ cm}) = 66.7 \text{ cm}. \quad \text{(Answer)}$$

The length of the equivalent simple pendulum is marked as point P in Fig. 12*a*. This point, called the *center of oscillation,* has some interesting properties (see Sample Problem 6 and Problem 72). The center of oscillation is not a fixed point on the pendulum but has meaning only with respect to a specified point of suspension.

Sample Problem 6 Suppose that the pendulum of Fig. 12*a* is inverted and suspended from point P, as in Fig. 13*b*. What will be its period of oscillation?

Let us first find the rotational inertia of the meter stick about an axis through point P. Table 2(*e*) of Chapter 11 tells us that the rotational inertia about point C, the center of gravity, is $\tfrac{1}{12}mL_0^2$. The distance h between points C and P in Fig. 13*b* can be found from Fig. 13*a* to be

$$h = \frac{L_0}{2} - \frac{L_0}{3} = \frac{L_0}{6}.$$

From the parallel-axis theorem we have, for the rotational inertia of the pendulum about a perpendicular axis through P,

$$I = I_{\text{cm}} + mh^2 = \tfrac{1}{12}mL_0^2 + m(L_0/6)^2 = \tfrac{1}{9}mL_0^2.$$

If we substitute these values of h and I into Eq. 27, the result is

$$T = 2\pi \sqrt{\frac{I}{mgh}} = 2\pi \sqrt{\frac{\tfrac{1}{9}mL_0^2}{mg(L_0/6)}} = 2\pi \sqrt{\frac{2L_0}{3g}}. \quad \text{(Answer)}$$

This is exactly Eq. 29. Although we have only shown it for a specific case, the following theorem is true in general:

> *To every point of suspension O of a physical pendulum, there corresponds a* center of oscillation P, *distant from it by the length of the equivalent simple pendulum. The physical pendulum has the same period whether it is suspended from O, as in Fig. 13a, or from P, as in Fig. 13b.*

Sample Problem 7 A disk whose radius R is 12.5 cm is suspended, as a physical pendulum, from a point halfway between its rim and its center; see Fig. 14. Its period T is measured to be 0.871 s. What is the free-fall acceleration g at the location of the pendulum?

The rotational inertia I_{cm} of a disk about its central axis is $\tfrac{1}{2}mR^2$. From the parallel-axis theorem, the rotational inertia about a parallel axis through the point of suspension O, as in Fig. 14, is

$$I = I_{\text{cm}} + mh^2 = \tfrac{1}{2}mR^2 + m(\tfrac{1}{2}R)^2 = \tfrac{3}{4}mR^2.$$

If we put $I = \tfrac{3}{4}mR^2$ and $h = \tfrac{1}{2}R$ in Eq. 27, we find

$$T = 2\pi \sqrt{\frac{I}{mgh}} = 2\pi \sqrt{\frac{\tfrac{3}{4}mR^2}{mg(\tfrac{1}{2}R)}} = 2\pi \sqrt{\frac{3R}{2g}}.$$

We have, solving for g,

$$g = \frac{6\pi^2 R}{T^2} = \frac{(6\pi^2)(0.125 \text{ m})}{(0.871 \text{ s})^2} = 9.76 \text{ m/s}^2. \quad \text{(Answer)}$$

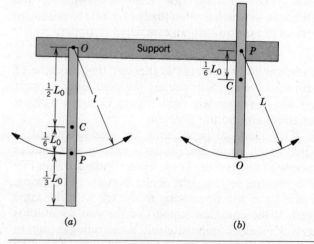

(a)   (b)

Figure 13 Sample Problem 6 (*a*) The pendulum of Fig. 12*a* reproduced. (*b*) The pendulum is reversed, being suspended from its center of oscillation. Reversing the pendulum in this way does not change its period.

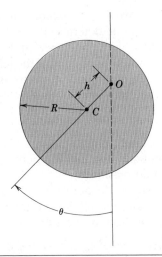

Figure 14   Sample Problem 7 A physical pendulum consisting of a uniform disk suspended from a point (O) that is halfway from the center of the disk to the rim.

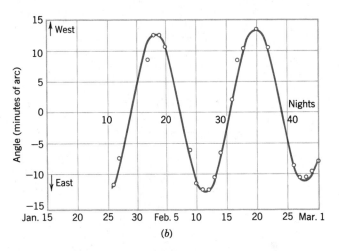

Figure 15   The angular position, as a function of time, of Jupiter's moon Callisto as seen from earth. The circles are based on Galileo's 1610 measurements. The curve is a best fit, strongly suggesting simple harmonic motion. At Jupiter's mean distance, 10 minutes of arc corresponds to about 2×10^6 km. The time axis covers about 6 weeks of observations. For a full discussion, see A. P. French, *Newtonian Mechanics*, W. W. Norton & Company, New York, 1971, p. 288.

14-7   Simple Harmonic Motion and Uniform Circular Motion

In 1610, Galileo, using his newly constructed telescope, discovered the four principal moons of Jupiter. As the nights went by, each moon seemed to him to be moving back and forth in what today we would call simple harmonic motion, the disk of the planet being the midpoint of the motion. The record of Galileo's observations, written in his own hand, are still available. Professor A. P. French of MIT used these data to work out the displacement function for the moon Callisto; see Fig. 15. In that figure, the circles are based on Galileo's observations and the curve is a best fit to the data. The curve strongly suggests Eq. 3, the displacement function for SHM. A period of about 16.8 days can be measured from the plot.

Actually, Callisto moves with essentially constant speed in an essentially circular orbit around Jupiter. Its actual motion—far from being simple harmonic—is uniform circular motion. What Galileo saw—and what you can see with a good pair of binoculars and a little patience—is the projection of this uniform circular motion along a line in the plane of the motion. We are lead by Galileo's remarkable observations to the conclusion

that simple harmonic motion is uniform circular motion viewed edge on. In more formal language:

Simple harmonic motion is the projection of uniform circular motion on a diameter of the circle in which the latter motion occurs.

Let us look at this conclusion more closely. Figure 16a shows a particle P'—the reference particle—in uniform circular motion with angular speed ω in a circle—the *reference circle*—whose radius is x_m. Let us project P' onto the x axis, and imagine a second particle P placed at that point. The location of P as a function of time is given by

$$x(t) = x_m \cos(\omega t + \phi),$$

which is precisely Eq. 3. Our conclusion is correct. If reference particle P' moves in uniform circular motion, its projection particle P will move in simple harmonic motion.

This relation throws new light on the angular frequency ω of simple harmonic motion and we see more clearly how the "angular" originates. The quantity ω is simply the constant angular speed with which the refer-

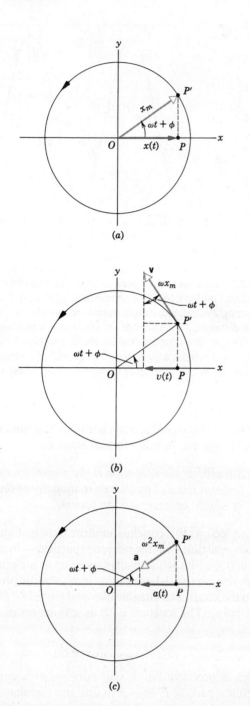

Figure 16 (*a*) A reference particle P' moving with uniform circular motion in a reference circle of radius x_m. Its projection P on the x axis executes simple harmonic motion. (*b*) The projection of the velocity **v** of the reference particle is the velocity of simple harmonic motion. (*c*) The projection of the acceleration **a** of the reference particle is the acceleration of simple harmonic motion.

ence particle P' moves in its reference circle. The phase constant ϕ has a value determined by the position of reference particle P' in its reference circle at $t = 0$.

Figure 16*b* shows the velocity of the reference particle. The magnitude of the velocity vector is ωx_m and its projection on the x axis is

$$v(t) = -\omega x_m \sin(\omega t + \phi),$$

which is exactly Eq. 5. The minus sign appears because the velocity component in Fig. 16*b* points to the left, in the direction of decreasing x.

Figure 16*c* shows the acceleration of the reference particle. The magnitude of the acceleration vector is $\omega^2 x_m$ and its projection on the x axis is

$$a(t) = -\omega^2 x_m \cos(\omega t + \phi),$$

which is exactly Eq. 6. Thus, whether we look at the displacement, the velocity, or the acceleration, the projection of uniform circular motion is indeed simple harmonic motion.

14-8 Damped Simple Harmonic Motion (Optional)

If you try to swing a pendulum under water, you will not have much success. In air, you will do better but the oscillations will gradually die out. In a vacuum, the pendulum will swing for a much longer time but eventually it will also come to rest, by the action of friction at its supports.

Figure 17 shows an idealized damped simple harmonic oscillator. As the oscillations proceed, the mechanical energy of the block–spring system gradually decreases, appearing as thermal energy in the *dash pot*, as the damping device at the bottom part of Fig. 17 is called.

Let us assume that the dash pot provides a damping force that is proportional to the magnitude of the velocity of the oscillating block but opposite to its direction. The total force acting on the block is then

$$F = -kx - b\frac{dx}{dt}. \tag{30}$$

in which $-kx$ is the familiar spring force and b is a damping constant. The value of b depends on the characteristics of the dash pot, such as the viscosity of the oil it contains and the dimensions of the oscillating vane.

We state without proof that, if a particle is acted on

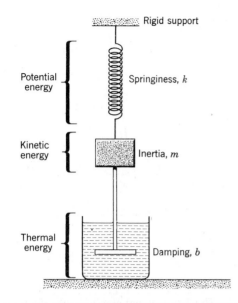

Figure 17   An idealized damped simple harmonic oscillator. A vane immersed in the fluid exerts a damping force on the oscillating block given by $-b(dx/dt)$, where b is the damping constant.

by the force of Eq. 30 and if the damping constant b is reasonably small, its displacement as a function of time will be*

$$x(t) \approx x_m e^{-bt/2m} \cos(\omega t + \phi). \qquad (31)$$

This reduces to the familiar Eq. 3 for $b = 0$, which corresponds to no damping at all. We can regard Eq. 31 as a cosine function whose amplitude, which is $x_m e^{-bt/2m}$, gradually decreases with time, as Fig. 18 suggests. For an undamped oscillator, the mechanical energy is constant and is given by Eq. 19, or $E = \frac{1}{2}kx_m^2$. If the oscillator is damped, the mechanical energy is not constant but decreases with time. If the damping is small enough, we can find $E(t)$ by replacing x_m in Eq. 19 by $x_m e^{-bt/2m}$, the amplitude of the damped oscillations. If we do so, we find

$$E(t) \approx \frac{1}{2}kx_m^2 e^{-bt/m}, \qquad (32)$$

which tells us how the mechanical energy decreases with time.

---

* The angular frequency ω in Eq. 31 is actually a little less than the angular frequency of the undamped motion (see Eq. 11) but, for small enough damping, this difference will be small enough to be neglected.

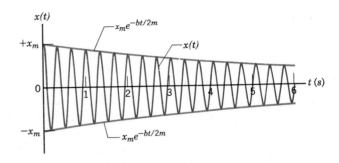

Figure 18   The displacement function $x(t)$ for the damped oscillator of Fig. 17. See Sample Problem 8.

Sample Problem 8   For the damped oscillator of Fig. 17, let us assume that $m = 250$ g, $k = 85$ N/m, and $b = 70$ g/s. (a) What is the period of the motion?

From Eq. 12 we have

$$T = 2\pi \sqrt{\frac{m}{k}} = 2\pi \sqrt{\frac{0.25 \text{ kg}}{85 \text{ N/m}}} = 0.34 \text{ s.} \quad \text{(Answer)}$$

(b) How long does it take for the amplitude of the damped oscillations to drop to half its initial value?

The amplitude, displayed in Eq. 31 as $x_m e^{-bt/2m}$, has the value x_m at $t = 0$. Thus, we must find t in the expression

$$\tfrac{1}{2}x_m = x_m e^{-bt/2m}.$$

Let us cancel x_m and take the natural logarithm of each side of the equation that remains. We find

$$\ln \tfrac{1}{2} = \ln 1 - \ln 2 = 0 - \ln 2 = -bt/2m,$$

or

$$t = \frac{2m \ln 2}{b} = \frac{(2)(0.25 \text{ kg})(\ln 2)}{0.070 \text{ kg/s}} = 5.0 \text{ s.} \quad \text{(Answer)}$$

This is about 15 periods of oscillation.

(c) How long does it take for the mechanical energy to drop to one-half of its initial value?

From Eq. 32 we see that the mechanical energy, which is $\frac{1}{2}kx_m^2 e^{-bt/m}$, has the value $\frac{1}{2}kx_m^2$ at $t = 0$. Thus, we must find t in

$$\tfrac{1}{2}(\tfrac{1}{2}kx_m^2) = \tfrac{1}{2}kx_m^2 e^{-bt/m}.$$

If we divide by $\frac{1}{2}kx_m^2$ and solve for t, as we did above, we find

$$t = \frac{m \ln 2}{b} = \frac{(0.25 \text{ kg})(\ln 2)}{0.070 \text{ kg/s}} = 2.5 \text{ s.} \quad \text{(Answer)}$$

This is exactly half the time calculated in (b), or about 7.5 periods of oscillation. Figure 18 was drawn to illustrate this Sample Problem.

14–9 Forced Oscillations and Resonance (Optional)

A person swinging passively in a swing is an example of a *free oscillation*. If a kind friend pulls or pushes the swing periodically, as in Fig. 19, we have a *forced oscillation*. There are now *two* angular frequencies with which to deal: (1) the "natural" angular frequency of the free oscillations of the system, which we have called ω but which we now relabel as ω_0, and (2) the angular frequency of the external driving force, which we call simply ω.

We can use Fig. 17 to represent an idealized forced simple harmonic oscillator if we allow the structure marked "rigid support" to move up and down at a variable angular frequency ω. A forced oscillator will *always* oscillate at the frequency of the driving force, its displacement $x(t)$ being given by

$$x(t) = x_m \cos(\omega t + \phi), \qquad (33)$$

in which ω is the angular frequency of the driving force and x_m is the amplitude of the oscillations.

The relation between ω, the frequency of the driving force, and ω_0, the natural frequency of free oscillation, determines how large the amplitude x_m of the oscillation will be. The amplitude will be greatest when

$$\boxed{\omega = \omega_0} \qquad \text{(resonance)}, \qquad (34)$$

Figure 19 Two frequencies are suggested in this painting by Nicholas Lancret: (1) the natural frequency at which the lady—left to herself—would swing and (2) the frequency at which her friend is tugging on the rope. If these two frequencies are equal, we have resonance.

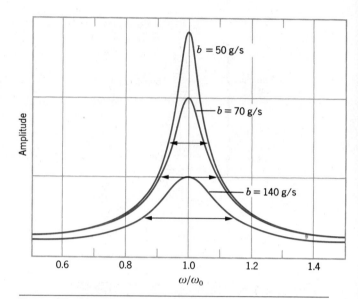

Figure 20 The displacement amplitude x_m of a forced oscillator as the angular frequency ω of the driving force is varied. Note the maximum response at $\omega = \omega_0$, the resonance condition. The three curves correspond to three levels of damping, the smaller the value of b the lighter the damping.

a condition called *resonance*. Thus, if you push a swing at its natural frequency, the amplitude will increase to a large value, a fact that children learn quickly by trial and error. If you push at other frequencies, either higher or lower, the amplitude will be smaller.

Figure 20 shows the resonance phenomenon for three different levels of damping, that is, for three different values of the damping coefficient b. Note that the oscillation amplitude x_m is greatest when $\omega/\omega_0 = 1$, that is, when the resonance condition of Eq. 34 is satisfied. The curves of Fig. 20 show that the smaller the damping, the taller and the narrower is the resonance peak.

All mechanical structures have one or more natural resonant frequencies. It will not do to subject a structure to a strong external driving force that matches one of these frequencies. The image of a soprano shattering a wine glass comes at once to mind. Aircraft designers make sure that none of the natural frequencies at which the wing can vibrate matches the rotational frequency of the engines at cruising speed. It would not do to discover in flight that the wing was flapping violently at certain engine speeds.

The figure at the head of this chapter shows the collapse, in 1940, of the Tacoma Narrows Bridge in Washington state because of resonant vibrations. The

Figure 21 Vortices generated when air moves past a cylinder in a wind tunnel. Note that the vortices occur alternately on each side and can exert an alternating driving force on the structure that generates them. Such vortices caused the bridge collapse shown in the photo at the beginning of this chapter.

driving force in this case was the wind and the collapse occurred, not because of the brute force of the wind, but because the frequency at which vortices were generated as the wind blew through the bridge matched a major natural frequency of the bridge structure. Figure 21, taken in a wind tunnel, shows dramatically how such vortices develop. It is the same phenomenon that causes a flag to flutter in the breeze.

REVIEW AND SUMMARY

Each periodic motion has a *frequency v* (the number of oscillations per second) measured, in SI, in hertz:

Frequency

$$1 \text{ hertz} = 1 \text{ Hz} = 1 \text{ oscillation per second} = 1 \text{ s}^{-1}. \tag{1}$$

The *period T* is the time required to complete one cycle. It is related to frequency as

Period

$$T = \frac{1}{v}. \tag{2}$$

In *simple harmonic motion* (SHM), the displacement $x(t)$ of a particle from its equilibrium position is described by the equation

Simple Harmonic Motion

$$x = x_m \cos(\omega t + \phi) \quad \text{(displacement)}, \tag{3}$$

in which x_m is the *amplitude*. The time-varying quantity $(\omega t + \phi)$ is the *phase* of the motion, with ϕ being the *phase constant*. The *angular frequency* ω is related to the period and frequency of the motion by

$$\omega = \frac{2\pi}{T} = 2\pi v \quad \text{(angular frequency)}. \tag{4}$$

Figure 4 shows how different motions may be represented by appropriate choices for these three constants.

Differentiating Eq. 3 leads to equations for velocity and acceleration as functions of time:

Velocity

$$v = -\omega x_m \sin(\omega t + \phi) \quad \text{(velocity)} \tag{5}$$

and

Acceleration

$$a = -\omega^2 x_m \cos(\omega t + \phi) \quad \text{(acceleration)}. \tag{6}$$

A particle with mass m that moves under the influence of a Hooke's law restoring force given

The Linear Oscillator

by $F = -kx$ exhibits simple harmonic motion with

$$\omega = \sqrt{\frac{k}{m}} \quad \text{(angular frequency)} \quad \text{and} \quad T = 2\pi \sqrt{\frac{m}{k}} \quad \text{(period).} \qquad [11,12]$$

Sample Problems 1 and 2 illustrate the use of these simple harmonic motion relations in describing the behavior of a block–spring oscillator.

Energy

A particle in simple harmonic motion has, at any time, kinetic energy $K = \frac{1}{2}mv^2$ and potential energy $U = \frac{1}{2}kx^2$. If no friction is present, the mechanical energy $E = K + U$ remains constant even though K and U change. Figure 6 and Sample Problem 3 describe the energy changes in a simple harmonic oscillator.

As examples of simple harmonic motion, we analyzed the *torsion pendulum* of Fig. 7, the *simple pendulum* of Fig. 9, and the *physical pendulum* of Fig. 10. Their periods of oscillation are, respectively,

Pendulums

$$T = 2\pi \sqrt{\frac{I}{\kappa}}, \quad T = 2\pi \sqrt{\frac{L}{g}}, \quad \text{and} \quad T = 2\pi \sqrt{\frac{I}{mgh}}. \qquad [21,25,27]$$

In each case, the period is determined by an inertial term divided by an elastic term, which measures the strength of the oscillator's restoring force. Sample Problems 4–7 discuss, by example, some of the features of these systems.

Simple Harmonic Motion and Uniform Circular Motion

Simple harmonic motion is the projection of uniform circular motion onto the diameter of the circle in which the latter motion occurs. Figure 16 shows that all parameters of circular motion (position **r**, velocity **v**, and acceleration **a**) project to the corresponding values for simple harmonic motion.

The mechanical energy E in real oscillating systems decreases steadily, because of friction, and is converted to thermal energy. If the frictional forces causing this decrease are proportional to the oscillator's speed (that is, $F_v = -bv$), as in the case of viscous friction at low enough speeds, the displacement of the oscillator as a function of time is

Damped Harmonic Motion

$$x(t) \approx x_m e^{-bt/2m} \cos(\omega t + \phi), \qquad [31]$$

and the total average energy of the oscillator decreases as

$$E(t) \approx \frac{1}{2}k\, x_m^2 e^{-bt/m}. \qquad [32]$$

If the damping constant b is small, the angular frequency ω is very nearly equal to the natural undamped frequency of the oscillator. Sample Problem 8 describes the decreasing energy in a damped oscillator.

Forced Oscillations and Resonance

It is often necessary to analyze the motion of an oscillator when its motion is "forced" by an external driving mechanism that applies a harmonic force with some angular frequency ω. The response of the oscillator is to oscillate with the frequency of the driving force and with an amplitude that depends on how close the driving angular frequency ω is to the natural angular frequency of the oscillator ω_0. For small damping, the amplitude is largest when the two frequencies are equal, a phenomenon called *resonance*. Figure 20 illustrates the effect.

QUESTIONS

1. Give some examples of motions that are approximately simple harmonic. Why are motions that are exactly simple harmonic rare?

2. Is Hooke's law obeyed, even approximately, by a diving board? A trampoline? A coiled spring made of steel wire?

3. An unstressed spring has a force constant k. It is stretched by a weight hung from it to an equilibrium length well within the elastic limit. Does the spring have the same force constant k for displacements from this new equilibrium position?

4. A spring has a force constant k, and a mass m is suspended from it. The spring is cut in half and the same mass is suspended from one of the halves. How are the frequencies of oscillation, before and after the spring is cut, related?

5. Suppose that we have a block of unknown mass and a spring of unknown force constant. Show how we can predict the period of oscillation of this block–spring system simply by measuring the extension of the spring produced by attaching the block to it.

6. Any real spring has mass. If this mass is taken into account, explain qualitatively how this will affect the period of oscillation of a mass–spring system.

7. Can astronauts in an orbiting space shuttle measure their weight? Their mass? After all, an astronaut is "weightless" in orbit.

8. How are each of the following properties of a simple harmonic oscillator affected by doubling the amplitude: period, force constant, total mechanical energy, maximum velocity, and maximum acceleration?

9. What changes could you make in a harmonic oscillator that would double the maximum speed of the oscillating mass?

10. Could we ever construct a true simple pendulum? Explain your answer.

11. Could standards of mass, length, and time be based on properties of a pendulum? Explain.

12. What would happen to the motion of an oscillating system if the sign of the force term, $-kx$ in Eq. 9, were changed?

13. Will the frequency of oscillation of a torsional pendulum change if you take it to the moon? What about the frequencies of a simple pendulum, a mass–spring oscillator, and a physical pendulum, such as a wooden plank suspended from a nail driven through one end?

14. Predict by qualitative arguments whether a pendulum oscillating with large amplitude will have a period longer or shorter than the period for oscillations with small amplitude. (Consider extreme cases.)

15. How is the period of a pendulum affected when its point of suspension is (a) moved horizontally in the plane of the oscillation with acceleration a; (b) moved vertically upward with acceleration a; (c) moved vertically downward with acceleration $a < g$? Which case, if any, applies to a pendulum mounted on a cart rolling down an inclined plane?

16. How can a pendulum be used so as to trace out a sinusoidal curve?

17. Why are damping devices often used on machinery? Give an example.

18. Give some examples of common phenomena in which resonance plays an important role.

19. In forced damped harmonic motion, is the damping ever useful?

20. In Fig. 20, what value does the amplitude of the forced oscillations approach as the driving frequency ω approaches (a) zero and (b) infinity? (*Hint:* See Exercise 86.)

21. Buildings of different heights sustain different amounts of damage in an earthquake. Explain why.

22. A singer, holding a note of the right frequency, can shatter a glass if the glassware is of high quality. This cannot be done if the glassware quality is low. Explain why.

EXERCISES AND PROBLEMS

Section 14–3 Simple Harmonic Motion: The Force Law

1E. An object undergoing simple harmonic motion takes 0.25 s to travel from one end point of the motion to the other. The distance between the end points is 36 cm. Calculate (a) the period, (b) the frequency, and (c) the amplitude of the motion.

2E. An oscillating mass–spring system takes 0.75 s to begin repeating its motion. Find (a) the period, (b) the frequency in hertz, and (c) the angular frequency in radians per second.

3E. A 4.0-kg block extends a spring 16 cm from its unstretched position. (a) What is the spring constant? (b) The block is removed and a 0.50-kg body is hung from the same spring. If the spring is then stretched and released, what is its period of oscillation?

4E. An oscillator consists of a block of mass 0.50 kg connected to a spring. When set into oscillation with amplitude 35 cm, it is observed to repeat its motion every 0.50 s. Find (a) the period, (b) the frequency, (c) the angular frequency, (d) the spring constant, (e) the maximum speed, and (f) the maximum force exerted on the block.

5E. The vibration frequencies of atoms in solids at normal temperatures are of the order of 10^{13} Hz. Imagine the atoms to be connected to one another by "springs." Suppose that a single silver atom vibrates with this frequency and that all the other atoms are at rest. Compute the effective spring constant. One mole of silver has a mass of 108 g and contains 6.02×10^{23} atoms.

6E. What is the maximum acceleration of a platform that vibrates with an amplitude of 2.2 cm at a frequency of 6.6 Hz?

7E. A loudspeaker produces a musical sound by the oscillation of a diaphragm. If the amplitude of oscillation is limited to 1.0×10^{-3} mm, what frequencies will result in the acceleration of the diaphragm exceeding g?

8E. The scale of a spring balance reading from 0 to 32 lb is 4.0 in. long. A package suspended from the balance is found to oscillate vertically with a frequency of 2.0 Hz. (a) What is the spring constant? (b) How much does the package weigh?

9E. A 20-N weight is hung from the bottom of a vertical

spring, causing the spring to stretch 20 cm. (*a*) What is the spring constant? This spring is now placed horizontally on a frictionless table. One end of it is held fixed and the other end is attached to a 5.0-N weight. The weight is then moved, stretching the spring an additional 10 cm, and released from rest. (*b*) What is the period of oscillation?

10E. A 50-g mass is attached to the bottom of a vertical spring and set vibrating. If the maximum speed of the mass is 15 cm/s and the period is 0.50 s, find (*a*) the spring constant *k* of the spring, (*b*) the amplitude of the motion, and (*c*) the frequency of oscillation.

11E. A 1.0×10^{-20}-kg particle is vibrating with simple harmonic motion with a period of 1.0×10^{-5} s and a maximum speed of 1.0×10^3 m/s. Calculate (*a*) the angular frequency and (*b*) the maximum displacement.

12E. A small body of mass 0.12 kg is undergoing simple harmonic motion of amplitude 8.5 cm and period 0.20 s. (*a*) What is the maximum value of the force acting on it? (*b*) If the oscillations are produced by a spring, what is the force constant of the spring?

13E. In an electric shaver, the blade moves back and forth over a distance of 2.0 mm. The motion is simple harmonic, with frequency 120 Hz. Find (*a*) the amplitude, (*b*) the maximum blade speed, and (*c*) the maximum blade acceleration.

14E. A speaker diaphragm is vibrating in simple harmonic motion with a frequency of 440 Hz and a maximum displacement of 0.75 mm. What are (*a*) the angular frequency, (*b*) the maximum speed, and (*c*) the maximum acceleration of this diaphragm?

15E. An automobile can be considered to be mounted on four springs as far as vertical oscillations are concerned. The springs of a certain car are adjusted so that the vibrations have a frequency of 3.0 Hz. (*a*) What is the force constant of each of the four springs (assumed identical) if the mass of the car is 1450 kg ($W = 3200$ lb)? (*b*) What will the vibration frequency be if five passengers, averaging 73 kg ($W = 160$ lb) each, ride in the car?

16E. A body oscillates with simple harmonic motion according to the equation

$$x = (6.0 \text{ m}) \cos[(3\pi \text{ rad/s})t + \pi/3 \text{ rad}].$$

What are (*a*) the displacement, (*b*) the velocity, (*c*) the acceleration, and (*d*) the phase at the time $t = 2.0$ s? Find also (*e*) the frequency and (*f*) the period of the motion.

17E. A particle executes linear harmonic motion about the point $x = 0$. At $t = 0$, it has displacement $x = 0.37$ cm and zero velocity. The frequency of the motion is 0.25 Hz. Determine (*a*) the period, (*b*) the angular frequency, (*c*) the amplitude, (*d*) the displacement at time *t*, (*e*) the velocity at time *t*, (*f*) the maximum speed, (*g*) the maximum acceleration, (*h*) the displacement at $t = 3.0$ s, and (*i*) the speed at $t = 3.0$ s.

18P. Figure 22 shows an astronaut on a Body Mass Measur-

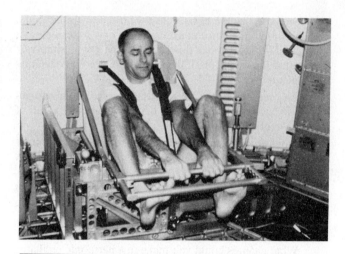

Figure 22 Problem 18.

ing Device (BMMD). Designed for use on orbiting space vehicles, its purpose is to allow astronauts to measure their mass in the weightless conditions in earth orbit. The BMMD is a spring-mounted chair; an astronaut measures his or her period of oscillation in the chair; the mass follows from the formula for the period of an oscillating block–spring system. (*a*) If *M* is the mass of the astronaut and *m* the effective mass of that part of the BMMD that also oscillates, show that

$$M = (k/4\pi^2)T^2 - m,$$

where *T* is the period of oscillation and *k* is the spring constant. (*b*) The spring constant is $k = 605.6$ N/m for the BMMD on Skylab Mission Two; the period of oscillation of the empty chair is 0.90149 s. Calculate the effective mass of the chair. (*c*) With an astronaut in the chair, the period of oscillation becomes 2.08832 s. Calculate the mass of the astronaut.

19P. The piston in the cylinder head of a locomotive has a stroke of 0.76 m (= 2.5 ft). What is the maximum speed of the piston if the drive wheels make 180 rev/min and the piston moves with simple harmonic motion?

20P. A 2.0-kg block hangs from a spring. A 300-g body hung below the block stretches the spring 2.0 cm farther. (*a*) What is the spring constant? (*b*) If the 300-g body is removed and the block is set into oscillation, find the period of motion.

21P. The end point of a spring vibrates with a period of 2.0 s when a mass *m* is attached to it. When this mass is increased by 2.0 kg, the period is found to be 3.0 s. Find the value of *m*.

22P. The end of one of the prongs of a tuning fork that executes simple harmonic motion of frequency 1000 Hz has an amplitude of 0.40 mm. Find (*a*) the maximum acceleration and maximum speed of the end of the prong and (*b*) the acceleration and speed of the end of the prong when it has a displacement 0.20 mm.

23P. A 0.10-kg block slides back and forth along a straight line on a smooth horizontal surface. Its displacement from the origin is given by

$$x = (10 \text{ cm}) \cos[(10 \text{ rad/s})t + \pi/2 \text{ rad})].$$

(a) What is the oscillation frequency? (b) What is the maximum speed acquired by the block? At what value of x does this occur? (c) What is the maximum acceleration of the block? At what value of x does this occur? (d) What force must be applied to the block to give it this motion?

24P. At a certain harbor, the tides cause the ocean surface to rise and fall in simple harmonic motion, with a period of 12.5 h. How long does it take for the water to fall from its maximum height to one-half its maximum height above its average (equilibrium) level?

25P. Two blocks ($m = 1.0$ kg and $M = 10$ kg) and a spring ($k = 200$ N/m) are arranged on a horizontal, frictionless surface as shown in Fig. 23. The coefficient of static friction between the two blocks is 0.40. What is the maximum possible amplitude of the simple harmonic motion if no slippage is to occur between the blocks?

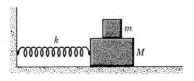

Figure 23 Problem 25.

26P. A block is on a horizontal surface (a shake table) that is moving horizontally with simple harmonic motion of frequency 2.0 Hz. The coefficient of static friction between block and surface is 0.50. How great can the amplitude be if the block does not slip along the surface?

27P. A block is on a piston that is moving vertically with simple harmonic motion of period 1.0 s. (a) At what amplitude of motion will the block and piston separate? (b) If the piston has an amplitude of 5.0 cm, what is the maximum frequency for which the block and piston will be in contact continuously?

28P. An oscillator consists of a block attached to a spring ($k = 400$ N/m). At some time t, the position (measured from the equilibrium location), velocity, and acceleration of the block are $x = 0.10$ m, $v = -13.6$ m/s, $a = -123$ m/s². Calculate (a) the frequency, (b) the mass of the block, and (c) the amplitude of oscillation.

29P. A simple harmonic oscillator consists of a block of mass 2.00 kg attached to a spring of force constant 100 N/m. When $t = 1.00$ s, the position and velocity of the block are $x = 0.129$ m and $v = 3.415$ m/s. (a) What is the amplitude of the oscillation? (b) What were the position and velocity at $t = 0.0$ s?

30P. A massless spring hangs from the ceiling with a small object attached to its lower end. The object is initially held at rest in such a position that the spring is not stretched. The

object is then released from this position and oscillates up and down with its lowest position being 10 cm below the initial position. (a) What is the frequency of the oscillation? (b) What is the speed of the object when it is 8.0 cm below the initial position? (c) An object of mass 300 g is attached to the first object, after which the system oscillates with half the original frequency. What is the mass of the first object? (d) Where is the new equilibrium position with both objects attached to the spring?

31P. Two particles oscillate in simple harmonic motion along a common straight line segment of length A. Each particle has a period of 1.5 s but they differ in phase by 30°. (a) How far apart are they (in terms of A) 0.50 s after the lagging particle leaves one end of the path? (b) Are they moving in the same direction, toward each other, or away from each other at this time?

32P. Two particles execute simple harmonic motion of the same amplitude and frequency along the same straight line. They pass one another when going in opposite directions each time their displacement is half their amplitude. What is the phase difference between them?

33P. Two identical springs are attached to a block of mass m and to fixed supports as shown in Fig. 24. Show that the frequency of oscillation on the frictionless surface is

$$v = \frac{1}{2\pi} \sqrt{\frac{2k}{m}}.$$

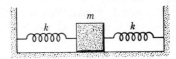

Figure 24 Problems 33 and 34.

34P. Suppose that the two springs in Fig. 24 have different spring constants k_1 and k_2. Show that the frequency v of oscillations of the block is then given by

$$v = \sqrt{v_1^2 + v_2^2}$$

where v_1 and v_2 are the frequencies at which the block would oscillate if connected only to spring 1 or only to spring 2.

35P. Two springs are joined and connected to a mass m as shown in Fig. 25. The surfaces are frictionless. If the springs each have force constant k, show that the frequency of oscilla-

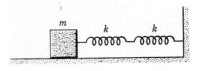

Figure 25 Problem 35.

tion of m is

$$v = \frac{1}{2\pi} \sqrt{\frac{k}{2m}}.$$

36P. A block weighing 14 N, which slides without friction on a 40° incline, is connected to the top of the incline by a light spring of unstretched length 0.45 m and force constant 120 N/m, as shown in Fig. 26. (a) How far from the top of the incline does the block rest in equilibrium? (b) If the block is displaced slightly down the incline, what is the period of the ensuing oscillations?

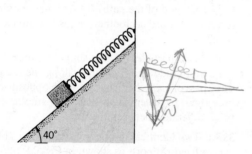

Figure 26 Problem 36.

37P. A uniform spring whose unstressed length is L has a force constant k. The spring is cut into two pieces of unstressed lengths L_1 and L_2, with $L_1 = nL_2$. What are the corresponding force constants k_1 and k_2 in terms of n and k? Does your result seem reasonable for $n = 1$?

38P. Three 10,000-kg ore cars are held at rest on a 30° incline on a mine railway using a cable that is parallel to the incline (Fig. 27). The cable is observed to stretch 15 cm just before a coupling breaks, detaching one of the cars. Find (a) the frequency of the resulting oscillations of the remaining two cars and (b) the amplitude of the oscillations.

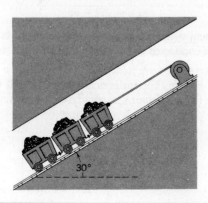

Figure 27 Problem 38.

Section 14–4 Simple Harmonic Motion: Energy Considerations

39E. Find the mechanical energy of a block–spring system having a force constant of 1.3 N/cm and an amplitude of 2.4 cm.

40E. An oscillating block–spring system has a mechanical energy of 1.0 J, an amplitude of 0.10 m, and a maximum speed of 1.2 m/s. Find (a) the force constant of the spring, (b) the mass, and (c) the frequency of oscillation.

41E. A vertical spring stretches 9.6 cm when a 1.3-kg block is hung from its end. (a) Calculate the spring constant. This block is then displaced an additional 5.0 cm downward and released from rest. Find (b) the period, (c) the frequency, (d) the amplitude, (e) the total energy, and (f) the maximum speed.

42E. A 5.0-kg object moves on a horizontal frictionless surface under the influence of a spring with force constant 1.0×10^3 N/m. The object is displaced 50 cm and given an initial velocity of 10 m/s back toward the equilibrium position. (a) What is the frequency of the motion? What are (b) the initial potential energy of the block–spring system, (c) the initial kinetic energy, and (d) the amplitude of the oscillation?

43E. A (hypothetical) large slingshot is stretched 1.5 m to launch a 130-g projectile with speed sufficient to escape from the earth (11.2 km/s). (a) What is the force constant of the device, if all the potential energy is converted to kinetic energy? (b) Assume that an average person can exert a force of 220 N. How many people are required to stretch the slingshot?

44E. (a) When the displacement is one-half the amplitude x_m, what fraction of the total energy is kinetic and what fraction is potential in simple harmonic motion? (b) At what displacement, in terms of the amplitude, is the energy half kinetic and half potential?

45E. A block of mass M, at rest on a horizontal frictionless table, is attached to a rigid support by a spring of constant k. A bullet of mass m and velocity v strikes the block as shown in Fig. 28. The bullet remains embedded in the block. Determine (a) the velocity of the block immediately after the collision and (b) the amplitude of the resulting simple harmonic motion.

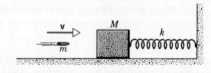

Figure 28 Exercise 45.

46P. A 3.0-kg particle is in simple harmonic motion in one dimension and moves according to the equation

$$x = (5.0 \text{ m}) \cos[(\pi/3 \text{ rad/s})t - \pi/4 \text{ rad}].$$

(a) At what value of x is the potential energy equal to half the total energy? (b) How long does it take the particle to move to this position from the equilibrium position?

47P. A 1.0×10^{-2}-kg particle is undergoing simple harmonic motion with an amplitude of 2.0×10^{-3} m. The maximum acceleration experienced by the particle is 8.0×10^3 m/s². (a) Write an equation for the force on the particle as a function of time. (b) What is the period of the motion? (c) What is the maximum speed of the particle? (d) What is the total mechanical energy of this simple harmonic oscillator?

48P. A massless spring of force constant 19 N/m hangs vertically. A body of mass 0.20 kg is attached to its free end and then released. Assume that the spring was unstretched before the body was released. Find (a) how far below the initial position the body descends, (b) the frequency, and (c) the amplitude of the resulting motion, assumed to be simple harmonic.

49P. Show that the average values, over a complete cycle, of both the kinetic energy and the potential energy of a simple harmonic oscillator equal half the maximum value.

50P. A 4.0-kg block is suspended from a spring with a force constant of 500 N/m. A 50-g bullet is fired into the block from below with a speed of 150 m/s and comes to rest in the block. (a) Find the amplitude of the resulting simple harmonic motion. (b) What fraction of the original kinetic energy of the bullet appears as mechanical energy in the harmonic oscillator?

51P*. A solid cylinder is attached to a horizontal massless spring so that it can roll without slipping along a horizontal surface, as in Fig. 29. The force constant k of the spring is 3.0 N/m. If the system is released from rest at a position in which the spring is stretched by 0.25 m, find (a) the translational kinetic energy and (b) the rotational kinetic energy of the cylinder as it passes through the equilibrium position. (c) Show that under these conditions the center of mass of the cylinder executes simple harmonic motion with a period

$$T = 2\pi \sqrt{\frac{3M}{2k}},$$

where M is the mass of the cylinder.

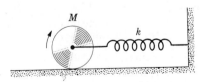

Figure 29   Problem 51.

Section 14–5   An Angular Simple Harmonic Oscillator
52E. A flat uniform circular disk has a mass of 3.0 kg and a radius of 0.70 m. It is suspended in a horizontal plane by a

vertical wire attached to its center. If the disk is rotated 2.5 rad about the wire, a torque of 0.060 N·m is required to maintain the disk in this position. Calculate (a) the rotational inertia of the disk about the wire, (b) the torsional constant κ, and (c) the angular frequency of this torsional pendulum.

53E. An engineer wants to find the rotational inertia of an odd-shaped object of mass 10 kg about an axis through its center of mass. The object is supported with a wire through its center of mass and along the desired axis. The wire has a torsional constant $\kappa = 0.50$ N·m. The engineer observes that this torsional pendulum oscillates through 20 complete cycles in 50 s. What value of rotational inertia is calculated?

54P. The balance wheel of a watch vibrates with an angular amplitude of π rad and a period of 0.50 s. Find (a) the maximum angular speed of the wheel, (b) the angular speed of the wheel when its displacement is $\pi/2$ rad, and (c) the angular acceleration of the wheel when its displacement is $\pi/4$ rad.

55P. A 95-kg solid sphere with a 15-cm radius is suspended by a vertical wire attached to the ceiling of a room. A torque of 0.20 N·m is required to twist the sphere through an angle of 0.85 rad. What is the period of oscillation when the sphere is released from this position?

Section 14–6   Pendulums We Have Known
56E. What is the length of a simple pendulum whose period is 1.00 s at a point where $g = 32.2$ ft/s²?

57E. A 2500-kg demolition ball swings from the end of a crane, as shown in Fig. 30. The length of the swinging segment of cable is 17 m. Find the period of swing, assuming that the system can be treated as a simple pendulum.

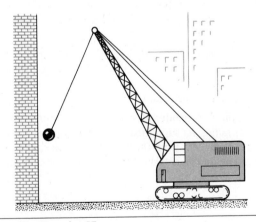

Figure 30   Exercise 57.

58E. What is the length of a simple pendulum that marks seconds by completing a full cycle every 2 s?

59E. If a simple pendulum with length 1.50 m makes 72 oscillations in 3 min, what is the acceleration of gravity at its location?

60E. Two oscillating systems that you have studied are the block–spring and the simple pendulum. There is an interesting relation between them. Suppose that you hang a weight on the end of a spring, and when the weight is in equilibrium, the spring is stretched a distance h. Show that the frequency of this block–spring system is the same as that of a simple pendulum whose length is h. See Fig. 31.

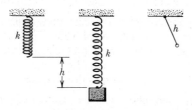

Figure 31 Exercise 60.

61E. A performer, seated on a trapeze, is swinging back and forth with a period of 8.85 s. If she stands up, thus raising her center of gravity by 35 cm, what will be the new period of the swing trapeze? Treat the trapeze + performer as a simple pendulum.

62E. A simple pendulum with length L is swinging freely with small angular amplitude. As the pendulum passes its central or equilibrium position, its cord is suddenly and rigidly clamped at its midpoint. In terms of the original period of the pendulum T, what will the new period be?

63E. A pendulum is formed by pivoting a long thin rod of length L and mass m about a point on the rod that is a distance d above the center of the rod. Find the small amplitude period of this pendulum in terms of d, L, m, and g.

64E. A physical pendulum consists of a meter stick that is pivoted at a small hole drilled through the stick a distance x from the 50-cm mark. The period of oscillation is observed to be 2.5 s. Find the distance x.

65E. A physical pendulum consists of a uniform solid disk (mass M, radius R) supported in a vertical plane by a pivot located a distance d from the center of the disk, as shown in Fig. 32. The disk is displaced by a small angle and released.

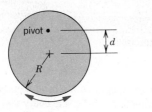

Figure 32 Exercise 65.

Find an expression for the period of the resulting simple harmonic motion.

66E. A uniform circular disk whose radius R is 12.5 cm is suspended, as a physical pendulum, from a point on its rim. (*a*) What is its period of oscillation? (*b*) Show that if suspended from a point halfway from the center to the rim it would have the same period.

67E. A pendulum consists of a uniform disk with radius 10 cm and mass 500 g attached to a 50-cm-long uniform rod with mass 270 g; see Fig. 33. (*a*) Calculate the rotational inertia of the pendulum about the pivot. (*b*) What is the distance between the pivot and the center of mass of the pendulum? (*c*) Calculate the small-angle period of oscillation.

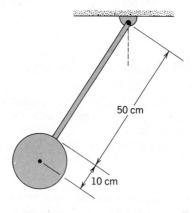

Figure 33 Exercise 67.

68P. For a simple pendulum, find the angular amplitude at which the deviation of the restoring torque from that required for simple harmonic motion is 1.0%.

69P. In Sample Problem 5, we showed that a meter stick, swinging from one end as a physical pendulum, has a center of oscillation located at a distance $2L_0/3$ from its point of suspension. Show that the distance between the point of suspension and the center of oscillation for a physical pendulum of any form is I/mh, where I and h have the meanings assigned to them in Eq. 27.

70P. A meter stick swinging from one end oscillates with a frequency ν_0. What would be the frequency, in terms of ν_0, if the bottom half of the stick were cut off?

71P. A stick with length L oscillates as a physical pendulum, pivoted about point O in Fig. 34. (*a*) Derive an expression for the period of the pendulum in terms of L and x, the distance from the point of support to the center of gravity of the pendulum. (*b*) Show that, if $L = 1.00$ m, the period will have a minimum value for $x = 28.87$ cm. (*c*) Show that, at a site where $g = 9.800$ m/s², this minimum value is 1.525 s.

72P. The center of oscillation of a physical pendulum has this interesting property: If an impulsive force (assumed hori-

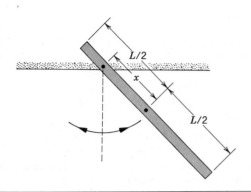

Figure 34 Problem 71.

zontal and in the plane of oscillation) acts at the center of oscillation, no reaction is felt at the point of support. Baseball players (and players of many other sports) know that unless the ball hits the bat at this point (called the "sweetspot" by athletes), the reaction due to impact will sting their hands. To prove this property, let the stick in Fig. 13a simulate a baseball bat. Suppose that a horizontal force F (due to impact with the ball) acts toward the right at P, the center of oscillation. The batter is assumed to hold the bat at O, the point of support of the stick. (a) What acceleration is imparted at O by the translational effect of F? (b) What angular acceleration is produced by F about the center of mass of the stick? (c) As a result of the angular acceleration in (b), what linear acceleration is imparted at O? (d) Considering the magnitudes and directions of the accelerations in (a) and (c), convince yourself that P is indeed the "sweetspot."

73P. (a) What is the frequency of a simple pendulum 2.0 m long? (b) Assuming small amplitudes, what would its frequency be in an elevator accelerating upward at a rate of 2.0 m/s²? (c) What would its frequency be in free fall?

74P. A simple pendulum of length L and mass m is suspended in a car that is traveling with a constant speed v around a circle of radius R. If the pendulum undergoes small oscillations in a radial direction about its equilibrium position, what will its frequency of oscillation be?

75P. The bob on a simple pendulum of length R moves in an arc of a circle. By considering that the acceleration of the bob as it moves through its equilibrium position is that for uniform circular motion (mv^2/R), show that the tension in the string at that position is $mg(1 + \theta_m^2)$ if the angular amplitude θ_m is small. Would the tension at other positions of the bob be larger, smaller, or the same?

76P. A long uniform rod of length L and mass m is free to rotate in a horizontal plane about a vertical axis through its center. A spring with force constant k is connected horizontally between one end of the rod and a fixed wall. What is the period of the small oscillations that result when the rod is pushed slightly to one side and released? See Fig. 35. (When the rod is in equilibrium it is parallel to the wall.)

Figure 35 Problem 76.

77P. A wheel is free to rotate about its fixed axle. A spring is attached to one of its spokes a distance r from the axle, as shown in Fig. 36. Assuming that the wheel is a hoop of mass m and radius R, obtain the angular frequency of small oscillations of this system in terms of m, R, r, and the spring constant k. Discuss the special cases $r = R$ and $r = 0$.

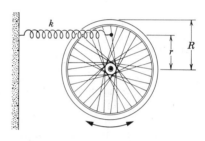

Figure 36 Problem 77.

78P. A 2.5-kg disk, 42 cm in diameter, is supported by a light rod, 76 cm long, which is pivoted at its end, as shown in Fig. 37. (a) The light, torsional spring is initially not connected. What is the period of oscillation? (b) The torsional spring is now connected so that, in equilibrium, the rod hangs vertically. What should be the torsional constant of the spring so that the new period of oscillation is 0.50 s shorter than before?

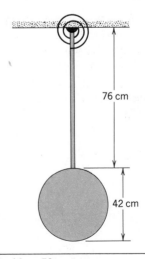

Figure 37 Problem 78.

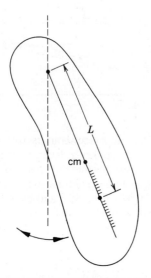

Figure 38 Problem 79.

79P*. A physical pendulum has two possible pivot points; one has a fixed position and the other is adjustable along the length of the pendulum, as shown in Fig. 38. The period of the pendulum when suspended from the fixed pivot is T. The pendulum is then reversed and suspended from the adjustable pivot. The position of this pivot is moved until, by trial and error, the pendulum has the same period as before, namely T. Show that the free-fall acceleration g is given by

$$g = \frac{4\pi^2 L}{T^2},$$

in which L is the distance between the two pivot points. Note that g can be measured in this way without needing to know the rotational inertia of the pendulum or any of its other dimensions except L.

Section 14–8 Damped Simple Harmonic Motion
80E. The amplitude of a lightly damped oscillator decreases by 3.0% during each cycle. What fraction of the energy of the oscillator is lost in each cycle?

81E. In Sample Problem 8, what is the ratio of the amplitude of the damped oscillations after 20 full cycles have elapsed to the initial amplitude?

82E. For the system shown in Fig. 17, the block has a mass of 1.5 kg and the spring constant is 8.0 N/m. The friction force is given by $-b(dx/dt)$, where $b = 230$ g/s. Suppose that the block is pulled down a distance 12 cm and released. (a) Calculate the time interval required for the amplitude to fall to one-third of its initial value. (b) How many oscillations are made by the block in this time?

83P. A damped harmonic oscillator involves a block ($m = 2.0$ kg), a spring ($k = 10$ N/m), and a damping force $F = -bv$. Initially, it oscillates with an amplitude of 25 cm; because of the damping, the amplitude falls to three-fourths of this initial value after four complete cycles. (a) What is the value of b? (b) How much energy has been "lost" during these four cycles?

84P. In Eq. 30, find the ratio of the maximum frictional force ($-b\,dx/dt$) to the maximum spring force ($-kx$) during the first cycle of the damped oscillation. Does this ratio change appreciably during later cycles? Use the data of Sample Problem 8.

85P. Assume that you are examining the characteristics of a suspension system of a 2000-kg automobile. The suspension "sags" 10 cm when the weight of the entire automobile is placed on it. In addition, the amplitude of oscillation decreases by 50% during one complete oscillation. Estimate the values of k and b for the spring and shock absorber system of each wheel. Assume each wheel supports 500 kg.

Section 14–9 Forced Oscillations and Resonance
86E. In Eq. 33, the amplitude x_m is given by

$$x_m = \frac{F_m}{[m^2(\omega^2 - \omega_0^2)^2 + b^2\omega^2]^{1/2}},$$

where F_m is the (constant) amplitude of the externally imposed oscillating force exerted on the spring by the previously rigid support. At resonance, what are (a) the amplitude and (b) the maximum speed of the oscillating object?

87P. A 2200-lb car carrying four 180-lb people is traveling over a rough "washboard" dirt road. The corrugations in the road are 13 ft apart. The car is observed to bounce with maximum amplitude when its speed is 10 mi/h. The car now stops and the four people get out. By how much does the car body rise on its suspension owing to this decrease in weight?

CHAPTER 15
GRAVITY

Gravity, as we all know, deals with falling bodies. Here are several of them. How many features of this photograph can you find that would be different if gravity did not act?

15–1 Gravity and the World Around Us

If you drop a penny, it falls toward the earth. If you lift a suitcase, it may seem "heavy." Facts like these are in our minds when we think about gravity. There is more to it than that! If there were no gravity, there would be no atmosphere. The rivers and oceans would not be confined to depressions in the earth's surface. Trees would not have their upward-reaching forms: There would be no "up"—and we would have little use for roads or bridges. As astronauts quickly learn, legs are not very useful in a weightless environment and we would probably not have evolved them.

At a deeper level, the earth would not have formed, nor would the sun or any other star. There would be no warming solar rays and no sunlight. Because the elements are cooked up in stars from primordial hydrogen

and helium, there would be no carbon, no oxygen, no iron, no uranium, ...

Figure 1 shows the great galaxy in Andromeda, some 2.3 million light-years distant from us. With its thousands of millions of stars it is the near twin of our own Milky Way galaxy and, with it, a prominent member of our Local Group of galaxies. Without gravity, none of this structure would exist. Gravity—we see—is the controlling force in the evolution of the universe; it literally "pulls it all together."

15–2 Newton's Law of Gravity

Physicists like to find areas that seem to have nothing to do with each other and to show that they are related if you look at them closely enough. This search for unifica-

Figure 1   The Andromeda galaxy, faintly visible to the naked eye. Located 2.3×10^6 light-years from us, it is the near twin of our home galaxy. (Isaac Asimov has suggested this comfortable name for what is usually called the Milky Way galaxy.)

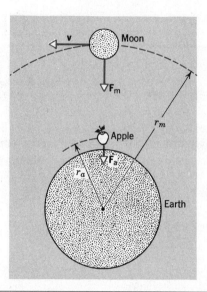

Figure 2   Newton first realized that the gravitational force that causes the apple to fall also keeps the moon in its orbit.

tion has been going on for centuries. In 1665, the 23-year-old Isaac Newton* made a great contribution when he showed that the force that holds the moon in its orbit is the same force that makes an apple fall; see Fig. 2. We take this so much for granted that it is not easy for us to appreciate the ancient view that the motions of objects on the earth and in the heavens were different in kind and were governed by different laws.

Newton concluded that not only does the earth attract an apple and the moon but *every body in the universe attracts every other body*. This thought takes a little getting used to because the familiar gravitational attraction of the earth is so great that it swamps the direct attractions that ordinary bodies have for each other. For example, the earth attracts an apple with a force of a few ounces. Two apples, placed 1 ft apart, also attract each other. However, their mutual force of attraction is less than the weight of a speck of dust. Our sense of gravity as an ever present force in our daily lives comes about simply because the earth is so massive.

Quantitatively, Newton proposed that every particle of matter attracts every other particle with a force

whose magnitude is given by

$$F = G \frac{m_1 m_2}{r^2} \qquad \text{(Newton's law of gravity).} \quad (1)$$

Here m_1 and m_2 are the masses of the particles and r is the distance between them. G is a universal constant,* whose value is

$$G = 6.67 \times 10^{-11} \text{ N} \cdot \text{m}^2/\text{kg}^2$$
$$= 6.67 \times 10^{-11} \text{ m}^3/\text{kg} \cdot \text{s}^2. \quad (2)$$

As Fig. 3 shows, the gravitational forces between two particles act along the line joining them and form an

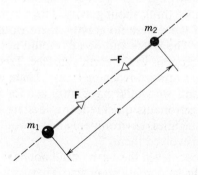

Figure 3   Two particles attract each other according to Newton's law of gravity; see Eq. 1. The two forces, which are equal in magnitude and oppositely directed, form an action–reaction pair.

* See "Isaac Newton," by I. Bernard Cohen, *Scientific American,* December 1955, for a fascinating and authoritative biography.

action–reaction pair. Particle 2 exerts a force **F** on particle 1 and particle 1 exerts a force −**F** on particle 2.

Although Newton's law of gravity applies strictly to particles, we can also apply it to real objects as long as the sizes of the objects are small compared to the distance between them. The moon and the earth in Fig. 2 are far enough apart so that, to a good approximation, we can treat them both as particles. But what about the apple and the earth? From the point of view of the apple, the broad and level earth, stretching out to the horizon beneath it, certainly does not look like a particle.

Newton solved the apple–earth problem by proving an important theorem:

A uniform spherical shell of matter attracts an external particle as if all the shell's mass were concentrated at its center.

The earth can be thought of as a nest of such shells. Thus, from the apple's point of view, the earth *does* behave like a particle, the distance separating them being the radius of the earth. We shall prove this *shell theorem* in Section 15–5.

The magnitude of the gravitational force between two interacting bodies is the same, even though their masses may be quite different. Suppose, as in Fig. 4, that the earth pulls down on an apple with a force of 3 oz. The apple must then pull up on the earth with a force of 3 oz, acting at the center of the earth. What differs for the two bodies are the accelerations that this force produces. For the apple, the acceleration is about 9.8 m/s², the familiar

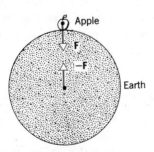

Figure 4   The apple pulls up on the earth just as hard as the earth pulls down on the apple.

---

* Do not confuse G with **g**, the free-fall acceleration of a body near the earth's surface. The first of these quantities is a scalar and a true universal constant of nature. The second, a vector, has different dimensions and is of purely local interest; it is neither universal nor constant.

acceleration of a falling body near the earth's surface. For the earth, the acceleration, measured in a reference frame attached to the center of mass of the apple–earth system, may be shown to be only about 1×10^{-25} m/s².

Sample Problem 1   Two bowling balls whose mass m is 7.3 kg are placed with their centers a distance $r = 50$ cm apart. What gravitational force does each ball exert on the other?

The magnitude of this force follows from Eq. 1, or

$$F = \frac{Gm_1m_2}{r^2} = \frac{(6.67 \times 10^{-11}\ \text{N·m}^2/\text{kg}^2)(7.3\ \text{kg})(7.3\ \text{kg})}{(0.50\ \text{m})^2}$$
$$= 1.4 \times 10^{-8}\ \text{N}. \hspace{2cm} \text{(Answer)}$$

This is about equal to the weight of a grain of very fine beach sand.

As long as the distance between the two balls remains unchanged, the force acting between them does not depend on where they are located or on what other bodies may be nearby. The balls could be resting on the bottom of a swimming pool or drifting in interstellar space. A third bowling ball, or a brick wall, could be placed between them. Our calculation still holds.

15-3   The Gravitational Constant G

Equation 1 gives the *form* of Newton's law of gravitation. The *strength* of the gravitational force, however, depends on the value of the gravitational constant G. If G—by some miracle—were suddenly to increase by a factor of 10, you would be crushed to the floor by your suddenly increased weight. If it were to decrease by this factor, you could jump over a tall building.

G can be found from Eq. 1, by measuring the gravitational force between two spheres of known mass, separated by a known distance. The first person to do this was Henry Cavendish, in 1798, more than a century after Newton advanced his law.

Figure 5 shows Cavendish's apparatus. Two small lead spheres, each of mass m, are fastened to the ends of a rod which is suspended from its midpoint by a fine fiber, forming a *torsion balance;* see Section 14–5. Two large lead spheres, each of mass M, can be placed in either configuration AA or configuration BB, as shown in the figure. In either configuration, the large spheres will attract the small ones, exerting a torque on the rod. The rod will then take up an equilibrium orientation, in which this gravitational torque is just balanced by the torque exerted by the twisted fiber. The two equilibrium

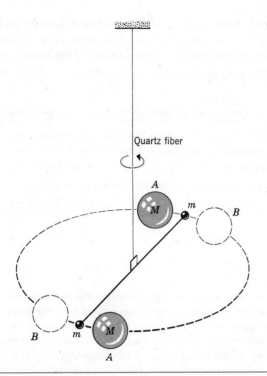

Figure 5 The apparatus used in 1798 by Henry Cavendish to measure the gravitational constant G and thus to "weigh the earth." The large spheres of mass M, shown in configuration AA, can also be moved to configuration BB.

orientations of the rod, corresponding to the two configurations of the large spheres, will differ by an angle 2θ. Sample Problem 2 shows how to calculate G from the measured data.

Knowing G, we can use Newton's law of gravity to find the mass of the earth. The relation between the gravitational force F acting on a particle of mass m at the earth's surface and the acceleration g_0 of that particle is, from Newton's second law,*

$$F = mg_0. \tag{3}$$

The gravitational force is also given, from Newton's law of gravity, by

$$F = \frac{GMm}{R^2}, \tag{4}$$

---

* Here g_0 differs from g, the measured free-fall acceleration, by small effects resulting from the earth's rotation. The correction amounts to less than 0.2% at mid-latitudes.

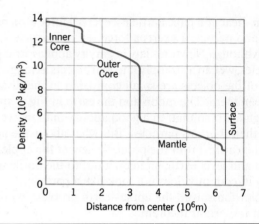

Figure 6 The density of earth as a function of distance from the center. The solid inner core, the largely liquid outer core, and the solid mantle stand out clearly. The crust of the earth is too thin to show up clearly on this plot.

in which M is the mass of the earth and R is its radius. Combining Eqs. 3 and 4 leads to

$$M = \frac{g_0 R^2}{G} = \frac{(9.8 \text{ m/s}^2)(6.37 \times 10^6 \text{ m})^2}{6.67 \times 10^{-11} \text{ m}^3/\text{kg} \cdot \text{s}^2}$$
$$= 6.0 \times 10^{24} \text{ kg}. \tag{5}$$

Knowing the mass of the earth, we can find its average density by dividing by its volume. The value that results is about 5.5 times the density of water. However, the average density of the rocks that form the earth's crust is only about 3 times the density of water. This tells us that the earth's core must be much denser than its crust; see Fig. 6. Who would think that, by measuring the force between two lumps of lead, you could "weigh the earth" and also find out something about its core, far beyond the depth of any drill hole?

Sample Problem 2 In a repetition of the Cavendish experiment of Fig. 5, let us take $M = 12.7$ kg, $m = 9.85$ g, and the length L of the rod connecting the two small spheres to be 52.4 cm. The rod and the fiber form a torsion balance whose rotational inertia I about the central axis is 1.25×10^{-3} kg·m^2 and whose period of oscillation T is 769 s. The angle 2θ between the two equilibrium positions of the rod is 0.516°, the distance R between the centers of the large and the small spheres being 10.8 cm. What value of the gravitational constant G results from these data?

Let us first find κ, the torsional constant of the fiber. The period of oscillation is given by Eq. 21 of Chapter 14, or

$$T = 2\pi \sqrt{\frac{I}{\kappa}}.$$

Solving for κ yields

$$\kappa = \frac{4\pi^2 I}{T^2} = \frac{(4\pi^2)(1.25 \times 10^{-3} \text{ kg} \cdot \text{m}^2)}{(769 \text{ s})^2}$$
$$= 8.34 \times 10^{-8} \text{ N} \cdot \text{m}.$$

The magnitude of the torque exerted by the fiber, with the rod in either of its two equilibrium positions, is related to the angular displacement θ, as Eq. 20 of Chapter 14 shows, by

$$\tau = \kappa\theta.$$

Here θ, which is $0.258°$ or 4.50×10^{-3} rad, is one-half of the angle 2θ given above. Thus,

$$\tau = (8.34 \times 10^{-8} \text{ N} \cdot \text{m})(4.50 \times 10^{-3} \text{ rad})$$
$$= 3.75 \times 10^{-10} \text{ N} \cdot \text{m}.$$

This torque is balanced by the torque associated with the gravitational forces that the large spheres exert on the small ones. The force F on each small sphere is equal to GMm/R^2 and the moment arm for each force is $\frac{1}{2}L$, half the length of the rod. The torque caused by both forces is then

$$\tau = (2F)(\tfrac{1}{2}L) = FL = \frac{GMmL}{R^2}. \tag{6}$$

Solving Eq. 6 for G yields

$$G = \frac{\tau R^2}{MmL} = \frac{(3.75 \times 10^{-10} \text{ N} \cdot \text{m})(0.108 \text{ m})^2}{(12.7 \text{ kg})(9.85 \times 10^{-3} \text{ kg})(0.524 \text{ m})}$$
$$= 6.67 \times 10^{-11} \text{ N} \cdot \text{m}^2/\text{kg}^2, \qquad \text{(Answer)}$$

in agreement with the value reported in Eq. 2.

15-4   Gravity and the Principle of Superposition

Given a group of particles, how do we find the resultant gravitational force exerted on any one of them by all the others? We use the *principal of superposition.* That is, we compute the gravitational force that acts on our selected particle by each of the other particles in turn. We then find the resultant force by adding these forces vectorially. The principle of superposition will pop up several more times throughout this book, in other contexts.

For n interacting particles, we can write the principle of superposition for gravitational forces as

$$\mathbf{F}_1 = \mathbf{F}_{12} + \mathbf{F}_{13} + \mathbf{F}_{14} + \mathbf{F}_{15} + \cdots + \mathbf{F}_{1n}. \tag{7}$$

Here $\mathbf{F}_1$ is the net force on particle 1 and, for example, $\mathbf{F}_{13}$ is the force exerted on particle 1 by particle 3. We can express this equation more compactly as

$$\boxed{\mathbf{F}_1 = \sum \mathbf{F}_{1j}, \qquad j = 2, 3, \ldots, n,} \tag{8}$$

in which the sum is a vector sum and j is a so-called running number.

What about the gravitational force exerted on a particle not by a collection of other particles but by a real extended object? The procedure is to subdivide the extended object into units small enough to treat as particles and to calculate the force on the particle from Eq. 8. In the limiting case, we can subdivide the extended object into differential units of mass dm, each of which will exert only a differential force $d\mathbf{F}$ on the particle in question. In this limit, the sum of Eq. 8 becomes an integral and we have

$$\boxed{\mathbf{F}_1 = \int d\mathbf{F},} \tag{9}$$

in which the integral is taken over the entire extended object.

15-5   Proving the Shell Theorem (Optional)

Let us prove that a uniform spherical shell of matter attracts an external particle as if all the shell's mass were concentrated at its center.

Figure 7 shows a uniform spherical shell of mass M and thickness h. What gravitational force does this shell exert on an external particle of mass m? Note that the line that passes through the particle and the center of the shell is a symmetry axis. Sighting along this line, we see no reason for the gravitational force acting on the particle to lie anywhere but along this axis, pointing toward the shell. Any force component at right angles to this axis will be canceled by a symmetrically located component pointing in the opposite direction.

Let us take as an element of mass the circular hoop labeled dM in Fig. 7. Its circumference is $2\pi(R \sin \theta)$, its width is $R \, d\theta$, and its thickness is h. Its volume is the product of these three quantities, or

$$dV = 2\pi R^2 h \sin \theta \, d\theta.$$

The volume V of the entire shell is $(4\pi R^2)h$, so that the mass of the circular hoop is

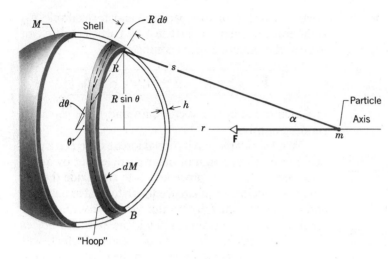

Figure 7 A spherical shell of mass M attracts a particle of mass m. The shell is divided into differential "hoops," of mass dM. The gravitational force on the particle is found by integrating over all the hoops that make up the shell.

$$dM = M\frac{dV}{V} = M\frac{2\pi R^2 h \sin\theta\, d\theta}{4\pi R^2 h} = \frac{M}{2}\sin\theta\, d\theta. \quad (10)$$

Note that all parts of the mass element dM are the same distance s from the particle.

If the hoop of mass dM were shrunk to a point a distance s from the particle, the magnitude of the force that the hoop would exert on the particle would be, from Eq. 1,

$$dF' = G\frac{dM\, m}{s^2}. \quad (11)$$

However, the hoop is not a point mass and, from symmetry, we must add up only the *axial* components of the force exerted on the particle by different segments of the hoop; components at right angles to the axis cancel each other symmetrically.

The magnitude of the force that the hoop of mass dM exerts on the particle of mass m is then

$$dF = dF'\cos\alpha = G\frac{dM\, m}{s^2}\cos\alpha$$
$$= \frac{GMm\cos\alpha\sin\theta\, d\theta}{2s^2}. \quad (12)$$

We have substituted for dM above, using Eq. 10.

Our next task is to find the total force F by integrating over all the mass elements that make up the shell. Substituting Eq. 12 into Eq. 9 leads to

$$F = \int dF = \frac{GMm}{2}\int\frac{\cos\alpha\sin\theta\, d\theta}{s^2} \quad (13)$$

Note that, during this integration process, the quantities

s, α, and θ are variables, with r and R remaining constant.

The three variables in Eq. 13 are not independent. Let us eliminate α and θ, leaving only the single variable s. To do so, we apply the law of cosines twice to the triangle in Fig. 7 whose sides are R, r, and s; see Appendix G. We find

$$s^2 = R^2 + r^2 - 2Rr\cos\theta \quad (14)$$

and

$$R^2 = s^2 + r^2 - 2sr\cos\alpha. \quad (15)$$

By differentiating Eq. 14 we find that

$$\sin\theta\, d\theta = \frac{s\, ds}{Rr} \quad (16)$$

Using Eqs. 15 and 16, we can eliminate α and θ from Eq. 13, leaving only the variable s. After some algebra, we find

$$F = \frac{GMm}{4Rr^2}\int_{r-R}^{r+R}ds + \frac{GMm(r^2 - R^2)}{4Rr^2}\int_{r-R}^{r+R}s^{-2}ds. \quad (17)$$

The limits on the variable s are $r - R$, its smallest value, and $r + R$, its largest value.

The integration of Eq. 17 leads eventually to

$$F = G\frac{Mm}{r^2}, \quad (18)$$

which is exactly Newton's law of gravitation. We have

proved our theorem.*

If you trace out the derivation above for a particle located *inside* a spherical shell, you will be able to prove this related theorem:

> *A uniform shell of matter exerts no gravitational force on a particle located inside it.*

If the density of the earth were uniform—which it is not—the gravitational force acting on a particle would be a maximum at the earth's surface. It would, as we expect, decrease as we move outward. If we were to move inward, perhaps down a deep mine shaft, the gravitational force would change for two reasons. (1) It would tend to increase because we are moving closer to the center of the earth. (2) It would tend to decrease because the shell of the earth's crust that lies outside the particle's radial position would not exert any force on the particle; that is, less of the earth would be gravitationally active. For an earth of uniform density, the second influence would prevail, so that the gravitational force would decrease as we penetrated the earth.

For the real earth, however, the outer crust is so much less dense than the core that the second influence above is greatly weakened and no longer prevails. The gravitational force on a particle actually *increases* slightly as we lower the particle down a deep mine shaft. Eventually, of course, the force must reach a maximum and decrease to zero at the center of the earth.

Sample Problem 3 A tunnel is bored through the earth along a diameter, as in Fig. 8. (a) Show that a package dropped into the tunnel will oscillate back and forth with simple harmonic motion. Assume that the earth is not rotating and is of uniform density throughout.

Figure 8 shows the package when it is a distance r from the center of the earth. The force that acts on the package is associated only with the mass M' of the earth that lies within a sphere of radius r. The shell of the earth that lies outside this sphere exerts no force on the package.

M' is given by

$$M' = \rho V' = \rho \frac{4\pi r^3}{3}, \qquad (19)$$

in which V' is the volume of M' and ρ is the assumed uniform density of the earth.

---

* This is the first serious integration in this book. Be sure to check it out carefully; see Appendix G.

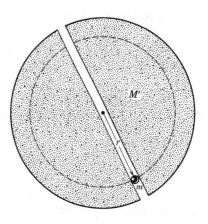

Figure 8 Sample Problem 3. A particle oscillates back and forth along a tunnel drilled through the earth. Mail dropped in at one end emerges at the other about 42 min later. The scheme is not very practical but it's fun to work out.

The force acting on the package is then, using Eq. 19,

$$F = -\frac{GmM'}{r^2} = -\frac{Gm\rho 4\pi r^3}{3r^2} = -\left(\frac{4\pi mG\rho}{3}\right) r = -Kr, \quad (20)$$

in which K, a constant, is equal to $4\pi mG\rho/3$. We have inserted a minus sign to indicate that the force $\mathbf{F}$ and the displacement $\mathbf{r}$ are in opposite directions, the former being toward the center of the earth and the latter away from that point. Thus, Eq. 20 tells us that the force acting on the package is proportional to the displacement of the package but oppositely directed, precisely the criterion for simple harmonic motion.

(b) If this tunnel were used to deliver mail, how long would it take for a letter to travel through the earth?

It would take half the period of the simple harmonic motion. This half period is given by

$$T = \left(\frac{1}{2}\right) 2\pi \sqrt{\frac{m}{K}} = \pi \sqrt{\frac{3m}{4\pi mG\rho}} = \sqrt{\frac{3\pi}{4G\rho}}.$$

For $\rho = 5.5 \times 10^3$ kg/m³, the mean density of the earth, we have

$$T = \sqrt{\frac{3\pi}{4G\rho}}$$

$$= \sqrt{\frac{3\pi}{(4)(6.67 \times 10^{-11} \text{ m}^3/\text{kg} \cdot \text{s}^2)(5.5 \times 10^3 \text{ kg/m}^3)}}$$

$$= 2530 \text{ s} = 42 \text{ min}. \qquad \text{(Answer)}$$

Note that the delivery time is independent of the mass of the package.

15-6 Gravity Near the Earth's Surface

Let us assume, for the time being, that the earth is spherical, that its density depends only on the radial distance from its center, and that it is not rotating. The magnitude of the gravitational force acting on a particle of mass m, located at an external point a distance r from the earth's center, can then be written, from Eq. 1, as

$$F = G\frac{Mm}{r^2},$$

in which M is the mass of the earth. This gravitational force can also be written, from Newton's second law, as

$$F = mg_0.$$

Here g_0 is the free-fall acceleration for an assumed nonrotating earth.* Combining the two equations above gives

$$g_0 = \frac{GM}{r^2}. \tag{21}$$

Table 1 shows some values of g_0 at various altitudes above the surface of the earth, calculated from this equation. Note that, contrary to the impression that gravity drops to zero in an orbiting satellite, we find $g_0 = 8.7$ m/s² at typical space shuttle altitudes. Figure 9 impresses upon us that orbiting astronauts are not much farther from the center of the earth than we are; they are just moving faster.

Table 1   The Variation of g_0 with Altitude

Altitude (km)	g_0 (m/s²)
0	9.83
5	9.81
10	9.80
50	9.68
100	9.53
400[a]	8.70
35,700[b]	0.225
380,000[c]	0.0027

[a] A typical space shuttle altitude.
[b] The altitude of communication satellites
[c] The distance to the moon.

* See the footnote on p. 334.

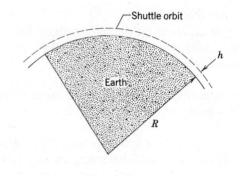

(a)

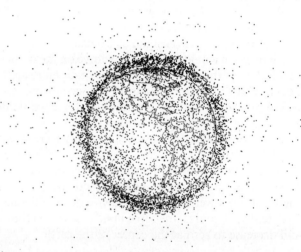

(b)

Figure 9   (a) A sector of the earth, showing (to scale) a typical Space Shuttle orbit at an altitude h above the earth's surface. (b) A computer-generated plot showing man-made objects presently in earth orbit — ranging from functioning satellites to frozen bags of garbage. A plot on a scale smaller than that used here would also show the earth encircled by a faint equatorial ring of objects orbiting in the geosynchronous arc.

The real earth differs from our model earth in three ways.

1. The Earth's Crust Is not Uniform.   There are local density variations everywhere. The precise measurement of local variations in the free-fall acceleration gives information that is useful, for example, for oil prospecting. Figure 10 shows a gravity survey over an underground salt dome. The contours connect points with the same free-fall acceleration, plotted as deviations from a convenient reference value. The unit, named to honor Galileo, is the milligal, where 1 gal = 10^3 mgal = 1 cm/s^2.

2. **The Earth Is not a Sphere.** The earth is approximately an ellipsoid, flattened at the poles and bulging at the equator. The earth's equatorial radius is greater than its polar radius by 21 km. Thus, a point at the poles is closer to the dense core of the earth than is a point on the equator. We would expect that the free-fall acceleration g would increase as one proceeds, at sea level, from the equator toward the poles. Figure 11 shows that this is indeed what happens. The measured values of g in this figure includes both the equatorial bulge effect and effects resulting from the rotation of the earth; see below.

3. **The Earth Is Rotating.** Figure 12a shows the rotating earth from a position in space above the north pole. A crate of mass m rests on a platform scale at the equator. This crate is in uniform circular motion because of the earth's rotation and is accelerated toward the

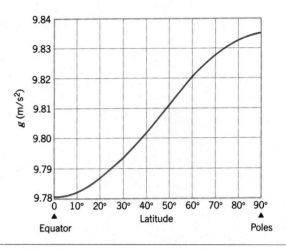

Figure 11 The variation of g with latitude at sea level. About 65% of the effect is due to the rotation of the earth, the remaining 35% resulting from the earth's shape as a flattened sphere.

center of the earth. The resultant force acting on it must then point in that direction.

Figure 12b is a free-body diagram for the crate. The earth exerts a downward gravitational pull of magnitude mg_0. The scale platform also pushes up on the crate with a force mg, the weight of the crate. These two forces do not quite balance and we have, from Newton's second law,

$$F = mg_0 - mg = ma$$

or

$$g_0 - g = a, \qquad (22)$$

in which a is the centripetal acceleration of the crate. For a we can write $\omega^2 R$, where ω is the earth's angular rotation rate and R is its radius. Substituting into Eq. 22 leads to

$$g_0 - g = \omega^2 R = \left(\frac{2\pi}{T}\right)^2 R, \qquad (23)$$

in which $T = 24$ h, the earth's period of rotation. Substituting numerical values in Eq. 23 yields

$$g_0 - g = 0.034 \text{ m/s}^2.$$

We see that g, the measured free-fall acceleration on the equator of the rotating earth, is less than g_0, the expected measurement if the earth were not rotating, by only 0.034/9.8 or 0.35%. The effect decreases as one goes to higher latitudes and vanishes at the poles.

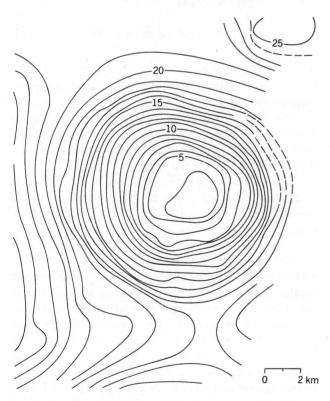

Figure 10 A surface gravity survey over an underground salt dome in Denmark. It is clear that there is something interesting underground. Oil is often found in such formations.

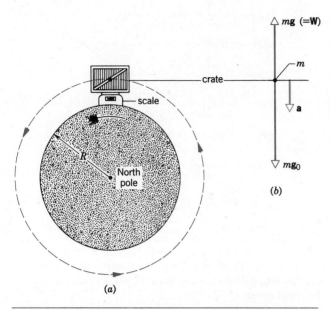

(b)

(a)

Figure 12 (a) A crate on the rotating earth, resting on a platform scale at the equator. The view is along the earth's rotational axis, looking down on the north pole. (b) A free-body diagram for the crate. The crate is in uniform circular motion and is thus accelerated toward the center of the earth.

Sample Problem 4 (a) A neutron star is a collapsed star of extremely high density. The blinking pulsar in the Crab nebula is the best known of many examples. Consider a neutron star with a mass M equal to the mass of the sun, 1.98×10^{30} kg, and a radius R of 12 km. What is the free-fall acceleration at its surface? Ignore rotational effects.

From Eq. 21 we have

$$g_0 = \frac{GM}{R^2} = \frac{(6.67 \times 10^{-11}\ \text{m}^3/\text{kg} \cdot \text{s}^2)(1.98 \times 10^{30}\ \text{kg})}{(12,000\ \text{m})^2}$$

$$= 9.2 \times 10^{11}\ \text{m/s}^2. \qquad \text{(Answer)}$$

Even though pulsars rotate extremely rapidly, rotational effects, because of the small size of pulsars, have only a small influence on the value of g.

(b) The asteroid Ceres has a mass of 1.2×10^{21} kg and a radius of 470 km. What is the free-fall acceleration at its surface?

In this case, we have

$$g_0 = \frac{GM}{R^2} = \frac{(6.67 \times 10^{-11}\ \text{m}^3/\text{kg} \cdot \text{s}^2)(1.2 \times 10^{21}\ \text{kg})}{(4.7 \times 10^5\ \text{m})^2}$$

$$= 0.36\ \text{m/s}^2. \qquad \text{(Answer)}$$

There is quite a contrast between the gravitational forces on the surfaces of these two bodies!

15-7 Gravitational Potential Energy

In Section 8–3, we discussed the gravitational potential energy of a particle, perhaps a baseball, in the gravitational field of the earth. We were careful to keep the particle near the earth's surface, so that we could regard the gravitational force as constant. We arbitrarily defined the potential energy of the particle–earth system to be zero when the particle was on the earth's surface.

Here we broaden our view and consider two particles, of masses m and M, separated by a distance r. For concreteness, we take M to be the earth and m to be a baseball but our conclusions will apply generally, no matter what the relative masses of the particles. We shall find it convenient to change our zero-potential-energy configuration to one in which the two particles are separated by a very large distance.

The gravitational potential energy of the baseball–earth system, as we shall prove below, is

$$U(r) = -\frac{GMm}{r}$$

(gravitational potential energy). (24)

We see that $U(r)$ approaches zero as r approaches infinity. This is as it should be because we made this a requirement at the outset.

We see also that the potential energy $U(r)$ is always negative. If M represents the earth and m a 160-g baseball resting on its surface, we can show from Eq. 24 that the potential energy of the system is -1×10^7 J or -10 MJ. This means that if the baseball is to be pulled away to a very large distance, where the potential energy of the system is zero, some external agent must do 10 MJ of work.

The potential energy given by Eq. 24 is a property of the two particles taken together rather than of either particle alone. There is no way to divide this energy and say that so much belongs to one particle and so much to the other. Nevertheless, if $M \gg m$, as is true for the earth and the baseball, we often speak of "the potential energy of the baseball." We can get away with this because, when a baseball moves in the vicinity of the earth, changes in the potential energy of the baseball–earth system appear almost entirely as changes in the kinetic energy of the baseball, changes in the kinetic energy of the earth being very small by comparison. For bodies of comparable mass, however, we have to be careful to treat them as a system when we speak of their potential energy.

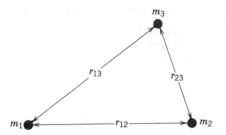

Figure 13   Three particles exert gravitational forces on each other. How can we find the gravitational potential energy of the system?

If the system contains more than two particles, the principle of superposition applies. That is, we consider each pair of particles in turn, calculate the gravitational potential energy of the pair from Eq. 24 as if the other particles were not there, and add up the results. Applying Eq. 24 to each of the three pairs of Fig. 13, for example, gives

$$U = -\left(\frac{Gm_1m_2}{r_{12}} + \frac{Gm_1m_3}{r_{13}} + \frac{Gm_2m_3}{r_{23}}\right). \quad (25)$$

Figure 14, a *globular cluster* in the constellation Sagittarius, is a good example of a naturally occurring system of particles. It contains about 70,000 stars and thus, as you should be able to verify, about 2.5×10^9

Figure 14   A globular cluster in the constellation Sagittarius —an impressive collection of stars locking up a lot of gravitational potential energy. There are many such clusters in our home galaxy. Many are visible with a small telescope.

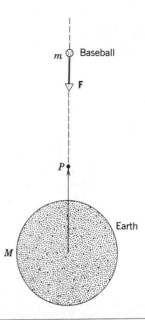

Figure 15   A baseball of mass m falls in from infinity, along a radial line, passing through point P at a distance r from the center of the earth.

pairs of stars. Contemplating this structure suggests the enormous amount of gravitational potential energy stored in the universe.

Proof of Eq. 24.   Let a baseball, starting from rest at a great distance from the earth, fall toward point P, as in Fig. 15. The potential energy of the baseball–earth system when the baseball passes through P is the negative of the work W done by the gravitational force as the baseball moves in from its distant position. Thus, from Eq. 7 of Chapter 8, we have

$$U(r) = -W = -\int_\infty^r \mathbf{F}(r) \cdot d\mathbf{r}. \quad (26)$$

The limits on the integral are the initial position of the baseball, which we take to be infinitely great, and its final position r.

The vector $\mathbf{F}(r)$ in Eq. 26 points radially inward in Fig. 15 and the vector $d\mathbf{r}$ points radially outward, the angle ϕ between them being $180°$. Thus,

$$\mathbf{F}(r) \cdot d\mathbf{r} = F(r)(\cos 180°)(dr) = -F(r)dr. \quad (27)$$

Substituting Newton's law of gravity for $F(r)$ in Eq. 27 and putting that result into Eq. 26, we obtain

$$U(r) = -\int_\infty^r \left(-\frac{GMm}{r^2}\right) dr = -\frac{GMm}{r}\bigg|_\infty^r = -\frac{GMm}{r},$$

which is exactly Eq. 24.

Figure 16 The work done by the gravitational force as a baseball moves from A to E in a gravitational field is independent of the path followed.

Equation 24 holds no matter what path the baseball takes in moving in toward the earth. Consider a path made up of small steps, as in Fig. 16. No work is done along steps like AB, because along them the force is at right angles to the displacement. The total work along all the radial steps, such as BC, is the same as the work done in going along a single radial line, as in Fig. 15. Thus, the work done in moving a particle between two points in the earth's gravitational field is independent of the path that it follows. This is what we mean when we say, as we did in Section 8–5, that the gravitational force is *conservative*. If the work *did* depend on the path, as it does for frictional forces, the whole concept of gravitational potential energy would be meaningless.

The Potential Energy and the Force.   In the proof just given, we derived the potential energy function $U(r)$ from the force function $F(r)$. We should be able to go the other way, that is, to start from the potential energy function and derive the force. Guided by Eq. 13 of Chapter 8, we can write, for spherically symmetric potential energy functions,

$$F(r) = -\frac{dU}{dr} = -\frac{d}{dr}\left(-\frac{GMm}{r}\right) = -\frac{GMm}{r^2}. \quad (28)$$

This is just Newton's law of gravity, the minus sign reminding us that the force is attractive.

Sample Problem 5   What is the gravitational potential energy of the moon–earth system?

The masses of the earth and the moon are 5.98×10^{24} kg and 7.36×10^{22} kg, respectively, and their mean separation distance d is 3.82×10^8 m. From Eq. 24 then,

$$U = -\frac{GMm}{d}$$

$$= -\frac{\left(6.67 \times 10^{-11}\,\frac{\text{N} \cdot \text{m}^2}{\text{kg}^2}\right)(5.98 \times 10^{24}\,\text{kg})(7.36 \times 10^{22}\,\text{kg})}{3.82 \times 10^8\,\text{m}}$$

$$= -7.68 \times 10^{28} \text{ J.} \qquad \text{(Answer)}$$

An energy of this magnitude is about equal to world energy production, at its present rate, for about 10^8 years.

Sample Problem 6   What minimum initial speed must a projectile have at the earth's surface if it is to escape from the earth? Ignore effects caused by atmospheric friction and the earth's rotation.

If you fire a projectile upward, it will usually slow down, come momentarily to rest, and return to earth. There is, however, a certain initial speed at which it will move upward forever, coming to rest only at infinity.

Consider such a projectile, of mass m, leaving the earth's surface with this critical initial speed v. It has a kinetic energy K given by $\frac{1}{2}mv^2$ and a potential energy U given by Eq. 24, or

$$U(R) = -\frac{GMm}{R},$$

in which M is the mass of the earth and R its radius.

When the projectile has reached infinity, it has no kinetic energy—recall that we seek the *minimum* speed for escape—and no potential energy—recall that this is our zero-potential-energy configuration. Its total energy is therefore zero. From the conservation of energy, its total energy at the surface must also be zero, or

$$K + U = 0.$$

This leads to

$$\frac{1}{2}mv^2 + \left(-\frac{GMm}{R}\right) = 0,$$

or

$$v = \sqrt{\frac{2GM}{R}}. \qquad (29)$$

Substituting values into Eq. 29 gives

$$v = \sqrt{\frac{2GM}{R}} = \sqrt{\frac{2(6.67 \times 10^{-11}\,\text{m}^3/\text{kg}\cdot\text{s}^2)(5.98 \times 10^{24}\,\text{kg})}{6.37 \times 10^6\,\text{m}}}$$

$$= 1.12 \times 10^4 \text{ m/s} = 11.2 \text{ km/s or } 25{,}000 \text{ mi/h.} \quad \text{(Answer)}$$

Table 2   Some Escape Speeds

Body	Mass (kg)	Radius (m)	Escape Speed (km/s)
Ceres[a]	1.17×10^{21}	3.8×10^5	0.64
Moon	7.36×10^{22}	1.74×10^6	2.38
Earth	5.98×10^{24}	6.37×10^6	11.2
Jupiter	1.90×10^{27}	7.15×10^7	59.5
Sun	1.99×10^{30}	6.96×10^8	618
Sirius B[b]	2×10^{30}	1×10^7	5200
Neutron star	2×10^{30}	1×10^4	2×10^5

[a] The most massive of the asteroids.
[b] A white dwarf, the companion of the bright star Sirius.

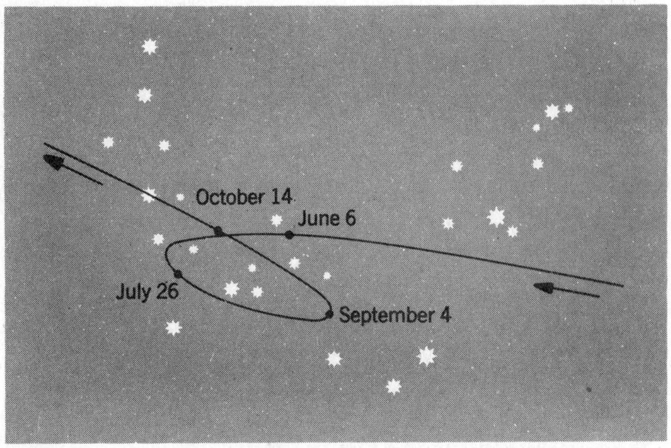

Figure 17   The planet Mars shown moving through the constellation Capricorn during August of 1971. Its position on four selected days is marked. Both Mars and the Earth are moving in orbits around the sun so that we see the position of Mars relative to us, against a background of distant stars.

The escape speed does not depend on the direction in which the projectile is fired. The earth's rotation—which we have ignored so far—does play a role, however. Firing eastward has an advantage in that the earth's tangential surface speed, which is 900 mi/h at Cape Canaveral, can be subtracted from the value calculated from Eq. 29. Table 2 shows some escape speeds.

15-8   Planets and Satellites: Kepler's Laws

The motions of the planets, as they wander* against the background of the stars, have been a puzzle since the dawn of history. The "loop-the-loop" motion of Mars, shown in Fig. 17, was particularly baffling. Johannes Kepler (1571–1630), after a lifetime of study, worked out the empirical laws that govern these motions. Tycho Brahe (1546–1601), the last of the great astronomers to make observations without the help of a telescope (see Fig. 18), compiled the extensive data from which Kepler was able to derive the three laws of planetary motion that now bear his name. Later, Isaac Newton (1642–1727) showed that Kepler's empirical laws followed from his law of universal gravitation.

We discuss each of Kepler's laws in turn. Although we apply the laws here to planets orbiting the sun, they hold equally well for satellites, either natural or artificial, orbiting the earth or any other massive central body.

The Law of Orbits.   *All planets move in elliptical orbits, with the sun at one focus.* Figure 19 shows a planet, of mass m, moving in such an orbit around the sun, whose mass is M. We assume that $M \gg m$, so that

the center of mass of the planet–sun system is virtually at the center of the sun.

The orbit in Fig. 19 is described by giving its *semimajor axis a* and its *eccentricity e*, the latter defined so that *ea* is the distance from the center of the ellipse to either focus. An eccentricity of zero corresponds to a circle, in which the two foci merge to a single central point. The eccentricities of the planets are not large, so that their orbits—sketched on paper—look pretty circular. The eccentricity of the ellipse of Fig. 19, which has been made large for clarity, is 0.74. The eccentricity of the earth's orbit is only 0.0167.

The Law of Areas.   *A line joining the planet to the sun sweeps out equal areas in equal times.* Qualitatively, this law tells us that the planet will move most slowly when it is farthest from the sun and most rapidly when it is nearest to the sun. As it turns out, Kepler's second law is totally equivalent to the law of conservation of angular momentum. Let us prove it.

The long shaded wedge in Fig. 20a shows the area swept out in a time Δt. The area ΔA of this wedge is approximately one-half of its base, $r \, \Delta\theta$, times its altitude r. Thus, $\Delta A = \frac{1}{2} r^2 \Delta\theta$. This expression for ΔA becomes more exact as Δt approaches zero. The instantaneous rate at which area is being swept out is

$$\frac{dA}{dt} = \tfrac{1}{2} r^2 \frac{d\theta}{dt} = \tfrac{1}{2} r^2 \omega, \qquad (30)$$

---

* The word *planet* is derived from the Greek word for *wanderer*.

(a)   (b)

Figure 18 Progress in instrumentation, (a) Tycho Brahe's
Great Quadrant. Much of what
Kepler knew about the solar
system came from observations
made with this and similar
instruments. (b) The Hubbard
space telescope should greatly
extend our knowledge of the
universe.

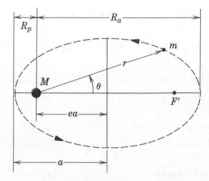

Figure 19 A planet of mass m moving in an elliptical orbit
around the sun. The sun, of mass M, is at one focus of the
ellipse. F' marks the other or "empty" focus. The semimajor
axis a of the ellipse, the perihelion distance R_p, and the
aphelion distance R_a are also shown. The quantity ea locates
the focal points, e being the eccentricity of the orbit.

in which ω is the angular speed of the rotating radius
vector.

Figure 20b shows the linear momentum **p** of the
planet, along with its components. The magnitude of the
angular momentum **L** of the planet about the sun is

$$L = p_\theta r = (mv_\theta)(r) = (m\omega r)(r) = mr^2\omega. \tag{31}$$

Eliminating $r^2\omega$ between Eqs. 30 and 31 leads to

$$\frac{dA}{dt} = \frac{L}{2m}, \tag{32}$$

which tells us that because the angular momentum L for
this isolated planet–sun system remains constant dA/dt
also remains constant, which is just what Kepler discovered.

The Law of Periods. *The square of the period of any
planet is proportional to the cube of the semimajor axis of
its orbit.* Let us consider only circular orbits. Applying
Newton's second law, $F = ma$, to the orbiting planet in
Fig. 21 yields

$$\frac{GMm}{r^2} = (m)(\omega^2 r).$$

Here we have substituted Newton's law of gravity for the
force F and $\omega^2 r$ for the centripetal acceleration. If we
replace ω by its equal, $2\pi/T$, where T is the period of the
motion we find

$$\boxed{T^2 = \left(\frac{4\pi^2}{GM}\right) r^3}$$

(law of periods). (33)

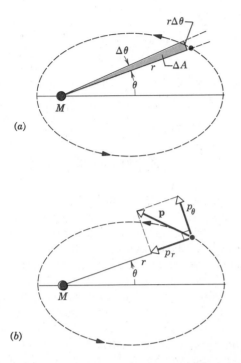

(a)

(b)

Figure 20   (a) In time Δt, the line connecting the planet to the sun sweeps through an angle $\Delta\theta$. (b) The linear momentum of the planet and its components.

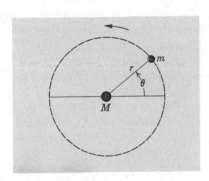

Figure 21   A planet of mass m moving around the sun in a circular orbit of radius a.

The quantity in parentheses is a constant, its value depending only on the mass of the central body.

We state without proof that Eq. 33 holds also for elliptical orbits, provided we replace r by a, the semimajor axis of the ellipse. Table 3 shows how well this law holds for the solar system. As Eq. 33 predicts, the ratio T^2/a^3 has essentially the same value for every planet.

Table 3   Kepler's Law of Periods for the Solar System

Planet	Semimajor Axis a $(10^{10}$ m)	Period T (y)	T^2/a^3 $(10^{-34}$ y^2/m$^3)$
Mercury	5.79	0.241	2.99
Venus	10.8	0.615	3.00
Earth	15.0	1.00	2.96
Mars	22.8	1.88	2.98
Jupiter	77.8	11.9	3.01
Saturn	143	29.5	2.98
Uranus	287	84.0	2.98
Neptune	450	165	2.99
Pluto	590	248	2.99

Sample Problem 7   An earth satellite, in circular orbit at an altitude h of 230 km above the earth's surface, has a period T of 89 min. What mass of the earth follows from these data?

We apply Kepler's law of periods to the satellite–earth system. Solving Eq. 33 for M yields

$$M = \frac{4\pi^2 r^3}{GT^2}. \tag{34}$$

The radius r of the satellite orbit is

$$r = R + h = 6.37 \times 10^6 \text{ m} + 230 \times 10^3 \text{ m} = 6.60 \times 10^6 \text{ m},$$

in which R is the radius of the earth. Substituting this value and the period T into Eq. 34 yields

$$M = \frac{(4\pi^2)(6.60 \times 10^6 \text{ m})^3}{(6.67 \times 10^{-11} \text{ m}^3/\text{kg}\cdot\text{s}^2)(89 \times 60 \text{ s})^2}$$

$$= 6.0 \times 10^{24} \text{ kg}. \tag{Answer}$$

In the same way, we could find the mass of the sun from the period and the radius of the earth's orbit and the mass of Jupiter from the period and the orbital radius of any one of its moons. We do not need to know the mass of the small orbiting body.

Sample Problem 8   Comet Halley has a period of 76 years and, in 1986, had a distance of closest approach to the sun, called the *perihelion distance* R_p, of 8.9×10^{10} m. Table 3 shows that this is between the orbit radii of Mercury and Venus. (a) What is R_a, the comet's farthest distance from the sun, called the *aphelion distance*?

We can find the semimajor axis of the orbit of Comet Halley from Eq. 33. Substituting a for r in that equation and solving for a yields

$$a = \left(\frac{GMT^2}{4\pi^2}\right)^{1/3}. \tag{35}$$

If we substitute the mass M of the sun, 1.99×10^{30} kg, and the period T of the comet, 76 years or 2.4×10^9 s, into Eq. 35 we find that $a = 2.7 \times 10^{12}$ m.

Study of Fig. 19 shows that

$$R_p = a - ea \qquad (36)$$

and

$$R_a = a + ea. \qquad (37)$$

Adding these two equations and solving for R_a yields

$$
\begin{aligned}
R_a &= 2a - R_p \\
&= (2)(2.7 \times 10^{12} \text{ m}) - 8.9 \times 10^{10} \text{ m} \\
&= 5.3 \times 10^{12} \text{ m}. \qquad \text{(Answer)}
\end{aligned}
$$

Table 3 shows that this is just a little less than the semimajor axis of Pluto.

(b) What is the eccentricity of the orbit of Comet Halley?

Subtracting Eq. 36 from Eq. 37 and solving for e, we obtain

$$
\begin{aligned}
e &= \frac{R_a - R_p}{2a} \\
&= \frac{5.3 \times 10^{12} \text{ m} - 8.91 \times 10^{10} \text{ m}}{(2)(2.7 \times 10^{12} \text{ m})} = 0.96. \qquad \text{(Answer)}
\end{aligned}
$$

This cometary orbit, with an eccentricity approaching unity, is a long thin ellipse.

15–9 Satellites: Orbits and Energy (Optional)

As a satellite orbits the earth on its elliptical path, both its speed, which fixes its kinetic energy K, and its distance from the center of the earth, which fixes its potential energy U, fluctuate periodically. Two quantities, however, remain constant: the angular momentum of the satellite and its mechanical energy E.

The potential energy is given by Eq. 24, or

$$U = -\frac{GMm}{r}. \qquad (38)$$

Here r is the radius of the orbit, assumed for the time being to be circular.

To find the kinetic energy, we write down Newton's second law, $F = ma$, for the orbiting satellite,

$$\frac{GMm}{r^2} = m\frac{v^2}{r}.$$

The kinetic energy is then

$$K = \tfrac{1}{2}mv^2 = \frac{GMm}{2r}. \qquad (39)$$

The total energy is

$$E = K + U = \frac{GMm}{2r} - \frac{GMm}{r} = -\frac{GMm}{2r}.$$

Replacing r by a—a useful change of notation—yields

$$\boxed{E = -\frac{GMm}{2a}}$$

(total mechanical energy). (40)

Although we derived Eq. 40 for a circular orbit, we state without proof that this expression holds for the general case of elliptical orbits if a is taken to be the semimajor axis of the ellipse. Put another way, the total energy of an orbiting satellite depends only on the semimajor axis of its orbit and not on its eccentricity e. All four orbits shown in Fig. 22, for example, have the same semimajor axis and thus the same energy. Figure 23 shows the variation of K, U, and E for a satellite moving in a circular orbit about a massive central body.

Suppose that the Space Shuttle *Enterprise* is in a circular orbit of radius a_0 and that, at point P in Fig. 24, the commander orders a short burn from a forward pointing thruster, so that the speed of the ship abruptly decreases. What path will *Enterprise* follow?

The burn reduces the kinetic energy of the ship and thus reduces E; that is, E becomes more negative. This, from Eq. 40, requires a decrease in a, the semimajor axis of the orbit. Equation 33 then requires that T, the orbital period, must also decrease. *Enterprise* will follow the elliptical orbit shown as dashed in Fig. 24. This maneuver is analyzed further in Sample Problem 11.

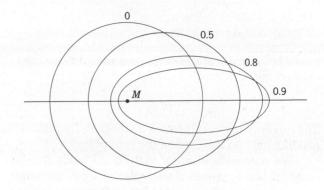

Figure 22 All four orbits have the same semimajor axis a and thus correspond to the same total energy E. Their eccentricities are marked.

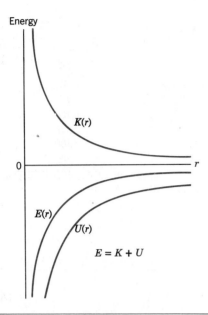

Figure 23   The variations of kinetic energy $K(r)$, potential energy $U(r)$, and total energy $E(r)$ with radius for a satellite in a circular orbit. Check from the curves that $E = K + U$ for all points on the radial axis.

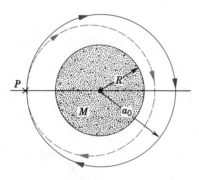

Figure 24   A manned spaceship is in a circular earth orbit of radius a_0. At point P, the commander orders a forward-directed burn, decreasing the speed of the ship. The ship now moves in the elliptical orbit shown dashed.

Sample Problem 9   A communications satellite hovers over a fixed spot on the equator of the rotating earth. What is the altitude of the orbit in which it will do this?

The period T of such a hovering satellite must be equal to the period of rotation of the earth, 8.62×10^4 s as measured with respect to the fixed stars. Solving Eq. 33 for the orbit radius r and substituting for T yields

$$r = \left(\frac{GMT^2}{4\pi^2} \right)^{1/3}$$

$$= \left(\frac{(6.67 \times 10^{-11}\ \text{N·m}^2/\text{kg}^2)\,(5.98 \times 10^{24}\ \text{kg})(86,400\ \text{s})^2}{4\pi^2} \right)^{1/3}$$

$$= 4.22 \times 10^7\ \text{m}.$$

To find the altitude h, we must subtract the radius of the earth. Thus,

$$h = 4.22 \times 10^7\ \text{m} - 6.37 \times 10^6\ \text{m}$$
$$= 3.58 \times 10^7\ \text{m} = 35{,}800\ \text{km or } 22{,}300\ \text{mi.}\quad\text{(Answer)}$$

The official name of the geosynchronous orbit is the Clarke Geosynchronous Orbit, or the Clarke Arc, after Arthur C. Clarke, who first proposed the idea of a communications satellite in 1948. Among many other accomplishments, Clarke is the author of *2001 — A Space Odyssey*.

Sample Problem 10   A playful astronaut puts a bowling ball whose mass m is 7.2 kg into a circular earth orbit at an altitude h of 350 km. (a) What is the kinetic energy of the ball?

The orbit radius r is given by

$$r = R + h = 6370\ \text{km} + 350\ \text{km} = 6.72 \times 10^6\ \text{m},$$

in which R is the radius of the earth. From Eq. 39, the kinetic energy is

$$K = \frac{GMm}{2r}$$

$$= \frac{(6.67 \times 10^{-11}\ \text{N·m}^2/\text{kg}^2)(5.98 \times 10^{24}\ \text{kg})(7.2\ \text{kg})}{(2)(6.72 \times 10^6\ \text{m})}$$

$$= 2.14 \times 10^8\ \text{J} = 214\ \text{MJ}.\quad\text{(Answer)}$$

(b) What is the ball's potential energy?

Equation 38, combined with the result above for K, yields

$$U = -\frac{GMm}{r} = -2K = -4.28 \times 10^8\ \text{J}$$

$$= -428\ \text{MJ}.\quad\text{(Answer)}$$

(c) What is the ball's mechanical energy E?

We have

$$E = K + U = 214\ \text{MJ} + (-428\ \text{MJ}) = -214\ \text{MJ}.\quad\text{(Answer)}$$

(d) What was the mechanical energy E_0 of the ball on the launch pad at Cape Canaveral?

On the launch pad, the ball has some kinetic energy because of the rotation of the earth but we can show that this is small enough to neglect. Thus, the total energy E_0 is equal to the potential energy U_0, or

$$E_0 = -\frac{GMm}{R}$$

$$= -\frac{(6.67 \times 10^{-11}\ \text{N·m}^2/\text{kg}^2)(5.98 \times 10^{24}\ \text{kg})(7.2\ \text{kg})}{6.37 \times 10^6\ \text{m}}$$

$$= -4.51 \times 10^8\ \text{J} = -451\ \text{MJ}.\quad\text{(Answer)}$$

You might be tempted to say that the potential energy of the ball on the earth's surface is zero. Recall, however, that our zero-potential-energy configuration is one in which the ball is removed to a great distance from the earth.

The *increase* in the mechanical energy of the ball as it was launched was

$$\Delta E = E - E_0 = (-214 \text{ MJ}) - (-451 \text{ MJ})$$
$$= 237 \text{ MJ}. \qquad \text{(Answer)}$$

You can buy this amount of energy from your utility company for a few dollars.

Sample Problem 11 Two small spaceships, each with mass $m = 2000$ kg, are in the circular earth orbit of Fig. 24, at an altitude h of 250 km. Igor, the commander of one of the ships, arrives at any fixed point in the orbit 120 s ahead of Sally, the other spaceship commander. (a) What is the radius of the circular orbit?

The radius is

$$a_0 = R + h = 6370 \text{ km} + 250 \text{ km}$$
$$= 6620 \text{ km} = 6.62 \times 10^6 \text{ m}. \qquad \text{(Answer)}$$

(b) What are the period and the speed of the two ships in this circular orbit?

The period follows from Eq. 33, or

$$T_0 = \sqrt{\frac{4\pi^2 a_0^3}{GM}}$$

$$= \sqrt{\frac{(4\pi^2)(6.62 \times 10^6 \text{ m})^3}{(6.67 \times 10^{-11} \text{ m}^3/\text{kg} \cdot \text{s}^2)(5.98 \times 10^{24} \text{ kg})}}$$

$$= 5360 \text{ s} \ (= 89.3 \text{ min}). \qquad \text{(Answer)}$$

The orbital speed is

$$v_0 = \frac{2\pi a_0}{T_0} = \frac{(2\pi)(6.62 \times 10^6 \text{ m})}{5360 \text{ s}}$$
$$= 7760 \text{ m/s}. \qquad \text{(Answer)}$$

(c) At a point such as P in Fig. 24, Sally, wanting to get ahead of Igor, fires a burst in the forward direction, *reducing* her speed by 1.00%. After she executes her burn, Sally will follow the elliptical orbit shown in Fig. 24. What are the speed, kinetic energy, and potential energy of her ship immediately after the burn?

Sally's speed at point P is

$$v = (0.99)v_0 = (0.99)(7760 \text{ m/s}) = 7680 \text{ m/s}. \quad \text{(Answer)}$$

Sally's new kinetic energy at point P is

$$K = \tfrac{1}{2}mv^2 = \tfrac{1}{2}(2000 \text{ kg})(7.68 \times 10^3 \text{ m/s})^2$$
$$= 5.90 \times 10^{10} \text{ J}. \qquad \text{(Answer)}$$

Her potential energy at point P remains unchanged and is

$$U = -\frac{GMm}{a_0}$$

Table 4   The Properties of Two Orbits

Property[a]	Circular Orbit	Elliptical Orbit
Semimajor axis (a, 10^6 m)	6.62	6.43
Closest distance (R_p, 10^6 m)	6.62	6.24
Farthest distance (R_a, 10^6 m)	6.62	6.62
Eccentricity (e)	zero	0.030
Period (T, s)	5360	5130
Energy (E, 10^{10} J)	-6.08	-6.20

[a] The calculation of the values of those properties not included in the Sample Problem is left as an exercise.

$$= -\frac{(6.67 \times 10^{-11} \text{ N} \cdot \text{m}^2/\text{kg}^2)(5.98 \times 10^{24} \text{ kg})(2000 \text{ kg})}{6.62 \times 10^6 \text{ m})}$$

$$= -12.1 \times 10^{10} \text{ J}. \qquad \text{(Answer)}$$

(d) In her new elliptical orbit, what are the total energy, the semimajor axis, and the orbital period?

Sally's new total energy, which is a constant for the new orbit, is

$$E = K + U = 5.90 \times 10^{10} \text{ J} + (-12.1 \times 10^{10} \text{ J})$$
$$= -6.20 \times 10^{10} \text{ J}. \qquad \text{(Answer)}$$

From Eq. 40, the semimajor axis of Sally's new orbit is

$$a = -\frac{GMm}{2E}$$

$$= -\frac{(6.67 \times 10^{-11} \text{ m}^3/\text{kg} \cdot \text{s}^2)(5.98 \times 10^{24} \text{ kg})(2000 \text{ kg})}{(2)(-6.20 \times 10^{10} \text{ J})}$$

$$= 6.43 \times 10^6 \text{ m}. \qquad \text{(Answer)}$$

Note that a is smaller, by 2.9%, than the radius a_0 of the original circular orbit.

The period in Sally's new orbit follows from Eq. 33, or

$$T = \sqrt{\frac{4\pi^2 a^3}{GM}}$$

$$= \sqrt{\frac{(4\pi^2)(6.43 \times 10^6 \text{ m})^3}{(6.67 \times 10^{-11} \text{ m}^3/\text{kg} \cdot \text{s}^2)(5.98 \times 10^{24} \text{ kg})}}$$

$$= 5130 \text{ s} \ (= 85.5 \text{ min}). \qquad \text{(Answer)}$$

This is shorter than the period of Igor's orbit by 230 s. Thus, Sally will arrive back at point P ahead of Igor by $(230 - 120)$ s or 110 s. By slowing down, Sally managed to speed up! To get back into the original circular orbit, still remaining 110 s ahead of Igor, all Sally has to do is to restore her original speed by executing the same burn she did before when she is again at point P but—this time—in a backward direction.

Sally was able to speed up by slowing down for two reasons: (1) As Fig. 24 shows, the length of her new orbit is shorter.

(2) Although, by comparison with her original orbit, she was moving slower at point P, she was moving faster at the orbital position directly opposite to P. Table 4 compares the properties of the original circular orbit and the new elliptical orbit.

15–10   A Closer Look at Gravity   (Optional)

Einstein once said: "I was sitting in a chair in the patent office at Bern when all of a sudden a thought occurred to me: 'If a person falls freely, he will not feel his own weight.' I was startled. This simple thought made a deep impression on me. It impelled me toward a theory of gravitation."

Thus, Einstein tells us how he was led to formulate his general theory of relativity,* which interprets gravitation not as a force but as a curvature in space and time. Although we cannot explore this important theory here, we can at least look at its fundamental postulate, the *principle of equivalence.*

Speaking a little loosely, this principle says that a physicist, locked up in a small box as in Fig. 25, cannot tell the difference between gravity and acceleration. The physicist, standing on a platform scale, observes the acceleration of a falling apple. In Fig. 25a, however, the box is at rest on the earth. In Fig. 25b, it is accelerating through interstellar space at 9.8 m/s². The physicist cannot tell the difference.

Figure 26 shows another aspect. In Fig. 26a, the physicist is floating around in a falling elevator cab. In Fig. 26b, he is floating around in interstellar space. Again, he cannot tell the difference. Figure 27 shows how the falling elevator cab can be realized in practice. From our point of view the astronauts are in uniform circular motion, accelerating toward the center of the earth, as in Fig. 26a. From their point of view, they seem to be in a gravity-free space, as in Fig. 26b. Still again, they cannot tell the difference. Finally, Fig. 28 shows a more "down to earth" example of the equivalence principle. Icicles normally grow downward, following the earth's gravita-

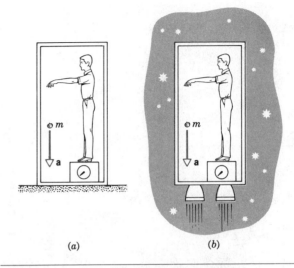

(a)   (b)

Figure 25   (a) A physicist-in-a-box, resting on the earth. (b) The same, accelerating in deep space at 9.8 m/s². It is not possible, by doing experiments within the box, for the physicist to tell which box he is in.

tional field. Here they have grown radially outward on a rotating car wheel. They seem to "think" that there is a radially outward component of gravity.

The principle of equivalence has some practical

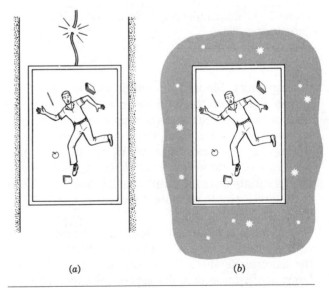

(a)   (b)

Figure 26   (a) A physicist in a falling elevator cab. (b) The elevator cab drifting in deep space. Again, it is not possible, by doing experiments within the cab, for the physicist to tell which box he is in.

---

* Not to be confused with Einstein's *special* theory of relativity, treated in Chapter 42. For an overview of the general theory, which includes the special theory as a limiting case, see Robert Resnick and David Halliday, *Basic Concepts,* 2nd ed., John Wiley & Sons, New York, 1985, Supplementary Topic C.

Figure 27 Astronauts at play in Skylab. Which one is "upside down?"

consequences. Modern navigation systems, for ships, submarines, aircraft, and ballistic missiles, operate by measuring the accelerations encountered along the path of the moving vehicle and, from these inputs, by computing the integrated displacement and thus the position. However, the accelerometers also respond to unexpected variations in the earth's gravitational field. There is no direct way to separate these two effects and thus avoid positional errors.

Another familiar fact follows from the principle of equivalence; namely, in a vacuum, all bodies fall with the same acceleration. If, for example, pears fell with a larger acceleration than did apples, the experiments of Figs. 25 and 26 would fail. In the orbiting Space Shuttle of Fig. 27, pears would fall toward the earth but apples would not.

Figure 28 Icicles grow radially outward on the rim of a car wheel. The icicles cannot "tell" whether the wheel is rotating or whether the wheel is at rest and there is a gravitational field directed radially outward from the axle.

We can express our claim about falling bodies by recalling that there are two different ways that we can assign a mass to a body. We can hang it from a spring scale. We may call the result of this measurement the *gravitational mass* m_g. This is the mass that appears in Newton's law of gravity, or

$$F = G \frac{M m_g}{R^2}. \tag{41}$$

Alternatively, we can measure the mass by the methods of Section 5–4, obtaining what we may call the *inertial mass* m_i. This is the mass that appears in Newton's second law, or

$$F = m_i a. \tag{42}$$

These masses need not be equal to each other. The only way we can find out is to measure them and see. Experiment shows that, to a precision of perhaps 1 part in 10^{12}, *these two masses are identical.*

In Newtonian physics, the experimental fact that $m_g = m_i$ could be regarded as nothing but an astonishing coincidence. In Einstein's general theory of relativity, it enters in a natural way. Thus, we see a direct line of thought about gravity, from Galileo at the tower of Pisa (maybe) through Newton watching the falling apple, to Einstein, daydreaming in his chair in the Swiss patent office. The story of gravity is far from ended.

REVIEW AND SUMMARY

The Law of Gravity

Any pair of particles in the universe attract each other with a force whose magnitude is

$$F = G \frac{m_1 m_2}{r^2} \quad \text{(Newton's law of gravity).} \qquad [1]$$

This is *Newton's law of gravitation.* The *gravitational constant G* is a universal constant whose value is 6.67×10^{-11} N·m²/kg². Sample Problems 1 and 2 show some basic computations using this important law of force.

The Gravitational Behavior of Uniform Spherical Shells

Equation 1 is strictly valid only for mass points. The gravitational force between extended bodies must generally be found by adding (integrating) the individual forces on individual mass points. However, if either of the bodies is a uniform spherical shell or a spherically symmetric solid, the net gravitational force it exerts on an *external* object may be computed as if the mass of the symmetric body were located at its center. The net force it exerts on an *internal* object (one that is inside the spherical shell) is exactly zero. This *shell theorem* is not valid for symmetrical shapes other than spheres. Sample Problem 3 illustrates the use of the complete shell theorem. Most practical problems, such as those illustrated in Sample Problems 1 and 2 (and many others in this chapter), depend on this important theorem.

Gravitational forces obey the *superposition* principle. The total force $\mathbf{F}_1$ on a particle labeled as particle 1 is the sum of the forces exerted on it by all other particles taken one at a time:

Superposition

$$\mathbf{F}_1 = \sum \mathbf{F}_{1j}, \qquad j = 2, 3, \ldots n, \qquad [8]$$

in which the sum is a vector sum of the forces $\mathbf{F}_{1j}$ exerted on particle 1 by particles 2, 3, . . . , *n*.

Using the gravitational force law and the second law of motion, we can predict the value of the *free-fall acceleration* g_0 near the surface of spherical objects (for example, stars, planets, asteroids). The result is that

Free-Fall Acceleration

$$g_0 = \frac{GM}{r^2}. \qquad [21]$$

See Sample Problem 4 for two examples.

Variations of **g** Near the Earth

Although we often take g_0 of the earth as being constant, it varies with altitude, as Eq. 21 and Table 1 show. It also varies with latitude (increasing as latitude increases from the equator to the poles; see Fig. 11) because the earth is slightly nonspherical, rotates, and does not have uniform density. There are also significant local variations in *g* because of nonuniformities in underlying geological structures (see Fig. 10).

The gravitational potential energy $U(r)$ of two objects with masses M and m is the negative of the work done by gravity if the objects were to fall toward each other from infinite (very large) separation. This leads to the expression

Gravitational Potential Energy

$$U(r) = -\frac{GMm}{r} \quad \text{(gravitational potential energy).} \qquad [24]$$

In Sample Problem 5, we estimate the potential energy of the earth–moon system.

If a system contains more than two particles, the total gravitational potential energy is, because of superposition, the sum of terms representing the potential energies of all the pairs; as an example, we have for three particles

Potential Energy of a System

$$U = -\left(\frac{Gm_1 m_2}{r_{12}} + \frac{Gm_1 m_3}{r_{13}} + \frac{Gm_2 m_3}{r_{23}} \right). \qquad [25]$$

A rocket (or any other object) will leave a planet (with mass M and radius R) if its velocity near the planet's surface (but outside its atmosphere) is at least equal to the *escape speed* given by

Escape Speed

$$v = \sqrt{\frac{2GM}{R}}.$$ [29]

We calculate, in Sample Problem 6, the escape speed for the earth to be about 11.2 km/s ($= 25,000$ mi/h).

Planets and Satellites

Gravitational attraction holds the solar system together and makes possible orbiting earth satellites, both natural and artificial. Such motions are governed by *Kepler's three laws of planetary motion*, all of which are direct consequences of Newton's laws of motion and gravitation:

Kepler's Three Laws

1. **The Law of Orbits.**   All planets move in elliptical orbits with the sun at one focus.

2. **The Law of Areas.**   A line joining any planet to the sun sweeps out equal areas in equal times (a statement equivalent to conservation of angular momentum; see Fig. 20).

3. **The Law of Periods.**   The square of the period T of any planet about the sun is proportional to the cube of the semimajor axis a of the orbit. Using Newton's laws we showed, for circular orbits with radius r, that

The Law of Periods

$$T^2 = \left(\frac{4\pi^2}{GM}\right) r^3 \quad \text{(law of periods)},$$ [33]

with M being the mass of the attracting center — the sun in the case of the solar system. This result is generally valid for elliptical planetary orbits, with the semimajor axis a inserted in place of the circular radius r. Sample Problems 7 and 8 use this law to estimate the earth's mass from a satellite's period and to estimate the apogee of Halley's comet.

When a planet or satellite with mass m moves in a circular orbit with radius r, its potential energy U and kinetic energy K are given by

Energy in Planetary Motion

$$U = -\frac{GMm}{r} \quad \text{and} \quad K = +\frac{GMm}{2r}.$$ [38,39]

The total mechanical energy $E = K + U$ is

$$E = -\frac{GMm}{2a},$$ [40]

in which the radius r has been replaced by the semimajor axis a since this expression for total energy is also valid for more general elliptical orbits. Sample Problems 9–11 deal with energies of orbiting satellites.

The Principle of Equivalence

Einstein pointed out that linear acceleration of a frame of reference is indistinguishable from a gravitational field due to masses outside the frame. This *principle of equivalence* led him to a theory of gravitation (the general theory of relativity) that views gravitational effects in terms of distortions of space-time. One of the consequences of the theory is that *gravitational* mass, which determines the strength of an object's participation in the gravitational interaction, is equivalent to *inertial mass,* the mass that appears in Newton's second law of motion. Very careful experiments have shown that these two quantities are identical to a precision of 1 part in 10^{12}.

Questions

1. It is said that Newton was inspired to think of his theory of gravitation by watching the fall of an apple. It would be amusing if an apple, in our modern SI nomenclature, turned out to weigh about 1 N. Is this within the range of the possible weights for apples?

2. If the force of gravity acts on all bodies in proportion to their masses, why doesn't a heavy body fall correspondingly faster than a light body?

3. What approximately is the *gravitational* force of attraction between Rhoda and Ronald if they are 10 m apart? When they are dancing?

4. How does the weight of a space probe vary en route from the earth to the moon? Would its mass change?

5. You can experience weightlessness—however briefly—by jumping from the top of one of the World Trade Center towers. Can you think of any way(s) that you can experience weightlessness while moving *up*?

6. In Sample Problem 3, the tunnel transit time was derived on the assumption of an earth of uniform density. Would this time be larger or smaller if the actual density distribution of the earth, with its dense inner core, were taken into account (see Fig. 6)? Explain your answer.

7. Would we have more sugar to the pound at the pole or the equator? What about sugar to the kilogram?

8. Because the earth bulges near the equator, the source of the Mississippi River, although high above sea level, is nearer to the center of the earth than is its mouth. How can the river flow "uphill"?

9. How could you determine the mass of the moon?

10. One clock is based on an oscillating spring, the other on a pendulum. Both are taken to Mars. Will they keep the same time there that they kept on Earth? Will they agree with each other? Explain. Mars has a mass 0.1 that of Earth and a radius half as great.

11. At the earth's surface, an object resting on a horizontal, frictionless surface is given a horizontal blow by a hammer. The object is then taken to the moon, supported in the same manner, and given an equal blow by the same hammer. To the best of our knowledge, what would be the speed imparted to the object on the moon when compared with the speed resulting from the blow on earth (neglecting any atmospheric effects)?

12. The earth is closer to the sun during winter in the Northern Hemisphere than during summer. Why isn't it colder in summer than in winter?

13. A satellite orbiting the earth around the Arctic circle would be very useful in maintaining surveillance of this strategically important part of the globe. Why don't we put one up?

14. As a car speeds around a curve, the passengers tend to be thrown radially outward. Why are astronauts in a space shuttle not similarly affected as their shuttle speeds in orbit around the earth?

15. The gravitational force exerted by the sun on the moon is about twice as great as the gravitational force exerted by the earth on the moon. Why then doesn't the moon escape from the earth?

16. Explain why the following reasoning is wrong. "The sun attracts all bodies on the earth. At midnight, when the sun is directly below, it pulls on an object in the same direction as the pull of the earth on that object; at noon, when the sun is directly above, it pulls on an object in a direction opposite to the pull of the earth. Hence, all objects should be heavier at midnight (or night) than they are at noon (or day)."

17. The gravitational attraction of the sun and the moon on the earth produces tides. The sun's tidal effect is about half as great as that of the moon's. The direct pull of the sun on the earth, however, is about 175 times that of the moon. Why is it then that the moon causes the larger tides?

18. Particularly large tides, called *spring tides,* occur at full moon and at new moon, when the configurations of the sun, earth, and moon are as shown in Fig. 29. From the figure you might conclude (incorrectly!) that the tidal effects of the sun and of the moon tend to add at new moon but to cancel at full moon. Instead, they add at both of these configurations. Explain why.

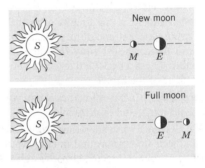

Figure 29 Question 18.

19. The distance between the earth and the sun can be determined by measuring the travel time of radio waves "bounced" off the sun. However, the earth–sun distance was known long before this experiment was performed. How do you suppose the earth–sun distance was first measured?

20. If lunar tides slow down the rotation of the earth (owing to friction), the angular momentum of the earth decreases. What happens to the motion of the moon as a consequence of the conservation of angular momentum? Does the sun (and solar tides) play a role here? (See "Tides and the Earth–Moon System," by Peter Goldreich, *Scientific American,* April 1972.)

21. A satellite in earth orbit experiences a small drag force as it starts to enter the earth's atmosphere. What happens to its speed? (Be careful!)

22. Would you expect the total energy of the solar system to be constant? The total angular momentum? Explain your answers.

23. Does a rocket always need the escape speed of 25,000 mi/h to escape from the earth? What then does "escape speed" really mean?

24. Objects at rest on the earth's surface move in circular paths with a period of 24 h. Are they "in orbit" in the sense that an earth satellite is in orbit? Why not? What would the length of the "day" have to be to put such objects in true orbit?

25. Neglecting air friction and technical difficulties, can a satellite be put into an orbit by being fired from a huge cannon at the earth's surface? Explain your answer.

26. Can a satellite coast in a stable orbit in a plane not passing through the earth's center? Explain your answer.

27. As measured by an observer on earth, would there be any difference in the periods of two satellites, each in a circular orbit near the earth in an equatorial plane, but one moving eastward and the other westward?

28. After *Sputnik I* was put into orbit it was said that it would not return to earth but would burn up in its *descent.* Considering the fact that it did not burn up in its *ascent,* how is this possible?

29. What happens to the angular momentum of an artificial earth satellite as it "speeds down" through the atmosphere?

30. An artificial satellite is in a circular orbit about the earth. How will its orbit change if one of its rockets is momentarily fired (*a*) toward the earth, (*b*) away from the earth, (*c*) in a forward direction, (*d*) in a backward direction, and (*e*) at right angles to the plane of the orbit?

31. Inside a spaceship, what difficulties would you encounter in walking, in jumping, and in drinking?

32. We have all seen TV transmissions from orbiting shuttles and watched objects floating around in effective zero gravity. Suppose that an astronaut, bracing himself against the satellite frame, kicks a floating bowling ball. Will he stub his toe? Explain your answer.

33. If a planet of given density were made larger, its force of attraction for an object on its surface would increase because of the planet's greater mass but would decrease because of the greater distance from the object to the center of the planet. Which effect dominates?

34. The gravitational field associated with the earth is zero both at infinity and at the center of the earth. Is the gravitational potential also zero at each place? Is it indeed the same at each place? *Can* it be zero at either place? *Need* it be zero at either place?

35. Two identical cars travel at the same speed in opposite directions on an east–west highway. Which car presses down harder on the road?

36. What advantage does Florida have over California for launching (nonpolar) U.S. satellites?

37. Does it matter which way a rocket is pointed for it to escape from earth? Assume, of course, that it is pointed above the horizon and neglect air resistance.

38. For a flight to Mars, a rocket is fired in the direction the earth is moving in its orbit. For a flight to Venus, it is fired backward along that orbit. Explain why.

39. Saturn is about six times farther from the sun than Mars. Which planet has (*a*) the greater period of revolution, (*b*) the greater orbital speed, and (*c*) the greater angular speed?

40. See Fig. 30. What is being plotted? Put numbers with units on each axis.

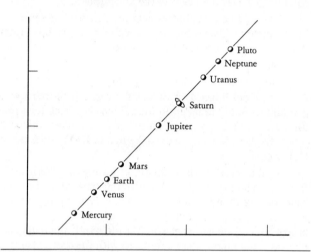

Figure 30 Question 40.

41. How could you determine whether two objects have (*a*) the same gravitational mass, (*b*) the same inertial mass, and (*c*) the same weight?

42. You are a passenger on the *S.S. Arthur C. Clarke,* the first interstellar spaceship. The *Clarke* rotates about a central axis to simulate earth gravity. If you are in an enclosed cabin, how could you tell that you are not on earth?

EXERCISES AND PROBLEMS

Section 15–2 Newton's Law of Gravity

1E. The year 1980 was the 50th anniversary of the discovery of the planet Pluto by the astronomer Clyde Tombaugh. How many revolutions has it made around the sun during that 50-year interval? See Appendix C for needed data.

2E. What must the separation be between a 5.2-kg particle and a 2.4-kg particle for their gravitational attraction to be 2.3×10^{-12} N?

3E. Some believe that the positions of the planets at the time of birth influence the newborn. Others deride this and say that the gravitational force exerted on a baby by the obstetrician is greater than that exerted by the planets. To check this claim, calculate and compare the gravitational force exerted on a 6-kg baby (*a*) by a 70-kg obstetrician who is 1 m away, (*b*) by the massive planet Jupiter ($m = 2 \times 10^{27}$ kg) at its closest approach to earth ($= 6 \times 10^{11}$ m), and (*c*) by Jupiter at its greatest distance from earth ($= 9 \times 10^{11}$ m). (*d*) Is the claim correct?

4E. The sun and earth each exert a gravitational force on the moon. What is the ratio F_{sun}/F_{earth} of these two forces? (The average sun–moon distance is equal to the sun–earth distance.)

5P. How far from the earth must a space probe be along a line toward the sun so that the sun's gravitational pull balances the earth's?

6P. A spaceship is going from the earth to the moon in a trajectory along the line joining the centers of the two bodies. At what distance from the earth will the net gravitational force on the spaceship be zero?

7P. What is the percent change in the acceleration of the earth toward the sun from an eclipse of the sun (with the moon between earth and sun) to an eclipse of the moon (earth between moon and sun)?

Section 15–5   Proving the Shell Theorem

8E. One of the *Echo* satellites consisted of an inflated aluminum balloon 30 m in diameter and of mass 20 kg. A meteor having a mass of 7.0 kg passes within 3.0 m of the surface of the satellite. If the effect of all bodies other than the meteor and satellite are ignored, what gravitational force does the meteor experience at closest approach to the satellite?

9E. Two concentric shells of uniform density having masses M_1 and M_2 are situated as shown in Fig. 31. Find the force on a particle of mass m when the particle is located at (a) $r = a$, (b) $r = b$, and (c) $r = c$. The distance r is measured from the center of the shells.

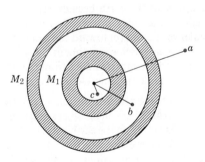

Figure 31 Exercise 9.

10E. With what speed would mail pass through the center of the earth if it were delivered by the chute of Sample Problem 3?

11E. Show that, at the bottom of a vertical mine shaft dug to depth D, the measured value of g will be

$$g = g_s \left(1 - \frac{D}{R} \right),$$

g_s being the surface value. Assume that the earth is a uniform sphere of radius R.

12E. Prove that a uniform shell of matter exerts no gravitational force on a particle located inside it.

13P. Figure 32 shows, not to scale, a cross section through the interior of the earth. Rather than being uniform throughout, the earth is divided into three zones: an outer *crust*, a *mantle,* and an inner *core*. The dimensions of these zones and the mass contained within them are shown on the figure. The earth has total mass 5.98×10^{24} kg and radius 6370 km. Ignore rotation and assume that the earth is spherical. (a) Calculate g at the surface. (b) Suppose that a bore hole is driven to the crust–mantle interface (the *Moho*); what would be the value of g at the bottom of the hole? (c) Suppose that the earth was a uniform sphere with the same total mass and size. What would be the value of g at a depth of 25 km? Use the result of Exercise 11. Precise measurements of g are sensitive probes of the interior structure of the earth, although results can be clouded by local density variations and lack of a precise knowledge of the value of G.

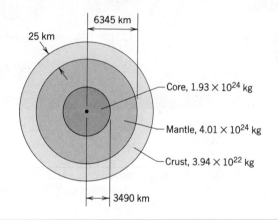

Figure 32 Problem 13.

14P. (a) Figure 33a shows a planetary object of uniform density ρ and radius R. Show that the compressive stress S near the center is given by

$$S = \frac{2}{3} \pi G \rho^2 R^2.$$

(*Hint:* Construct a narrow column of cross-sectional area A extending from the center to the surface. The weight of the material in the column is mg_{av}, where m is the mass of material in the column and g_{av} is the value of g midway between center and surface.) (b) In our solar system, objects (for example, asteroids, small satellites, comets) with "diameters" less than 600 km can be very irregular in shape (see Fig. 33b, which shows Hyperion, a small satellite of Saturn), whereas those with larger diameters are spherical. Only if the rocks have sufficient strength to resist gravity can an object maintain a nonspherical shape. Calculate the ultimate compressive strength of the rocks making up asteroids. Assume a density of 4000

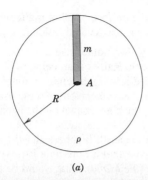

(a)

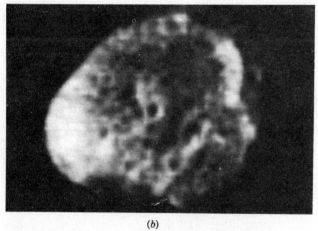

(b)

Figure 33 Problem 14.

kg/m³. (c) What is the largest possible size of a nonspherical self-gravitating satellite made of concrete (see Table 1, Chapter 13); assume $\rho = 3000$ kg/m³.

15P. The following problem is from the 1946 "Olympic" examination of Moscow State University (see Fig. 34): A spherical hollow is made in a lead sphere of radius R, such that its surface touches the outside surface of the lead sphere and passes through its center. The mass of the sphere before hollowing was M. With what force, according to the law of universal gravitation, will the lead sphere attract a small sphere of mass m, which lies at a distance d from the center of the lead sphere on the straight line connecting the centers of the spheres and of the hollow?

16P*. (a) Show that in a chute through the earth along a chord line, rather than along a diameter, the motion of an

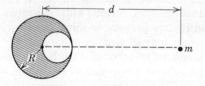

Figure 34 Problem 15.

object will be simple harmonic; assume a uniform earth density and no friction. (b) Find the period. (c) Will the object attain the same maximum speed along a chord as it does along a diameter?

Section 15–6   Gravity Near the Earth's Surface
17E. Verify the values of the free-fall acceleration listed in Table 1.

18E. If a pendulum has a period of exactly 1 s at the equator, what would be its period at the south pole? Use Fig. 11.

19E. (a) Calculate the free-fall acceleration on the surface of the moon from values of the mass and radius of the moon found in Appendix C. (b) What is the period of a "seconds pendulum" (period = 2 s on earth) on the surface of the moon?

20E. At what altitude above the earth's surface would the acceleration of gravity be 4.9 m/s²?

21E. You weigh 120 lb at the sidewalk level outside the World Trade Center in New York City. Suppose that you ride from this level to the top of one of its 1350-ft towers. How much less would you weigh there because you are slightly farther away from the center of the earth?

22E. A typical neutron star may have a mass equal to that of the sun but a radius of only 10 km. (a) What is the acceleration due to gravity at the surface of such a star? (b) How fast would an object be moving if it fell from rest through a distance of 1.0 m on such a star?

23E. An object at rest on the earth's equator is accelerated (a) toward the center of the earth because the earth rotates, (b) toward the sun because the earth revolves around the sun in an almost circular orbit, and (c) toward the center of our galaxy; the period of the sun's revolution about the galactic center is 2.5×10^8 years and the sun's distance from this center is 2.2×10^{20} m. Calculate these three accelerations.

24E. (a) Calculate the acceleration due to gravity on the surface of the moon. (b) What will an object weigh on the moon's surface if it weighs 100 N on the earth's surface? (c) How many earth radii must this same object be from the center of the earth if it is to weigh the same as it does on the moon?

25P. The fact that g varies from place to place over the earth's surface drew attention when Jean Richer in 1672 took a pendulum clock from Paris to Cayenne, French Guiana, and found that it lost 2.5 min/day. If $g = 9.81$ m/s² in Paris, what is g in Cayenne?

26P. (a) If g is to be determined by dropping an object through a distance of (exactly) 10 m, how accurately must the time be measured to obtain a result good to 0.1%? Calculate a percent error and an absolute error, in milliseconds. (b) How accurately (in seconds) would you have to measure the time for 100 oscillations of a 10-m-long pendulum to achieve the same percent error in the measurement of g?

27P. The fastest possible rate of rotation of a planet is that for which the gravitational force on material at the equator barely provides the centripetal force needed for the rotation. (Why?) (*a*) Show then that the corresponding shortest period of rotation is given by

$$T = \sqrt{\frac{3\pi}{G\rho}},$$

where ρ is the density of the planet, assumed to be homogeneous. (*b*) Evaluate the rotation period assuming a density of 3.0 g/cm³, typical of many planets, satellites, and asteroids. No such object is found to be spinning with a period shorter than found by this analysis.

28P. Certain neutron stars (extremely dense stars) are believed to be rotating at about 1 rev/s. If such a star has a radius of 20 km, what must be its minimum mass so that objects on its surface will be attracted to the star and not thrown off by the rapid rotation?

29P. (*a*) Write an expression for the force exerted by the moon, mass M, on a particle of water, mass m, on the earth at A, directly under the moon, as shown in Fig. 35. The radius of the earth is R, and the center-to-center earth–moon distance is r. (*b*) Suppose that the particle of water was at the center of the earth. What force would the moon exert on it then? (*c*) Show that the difference in these forces is given by

$$F_T = \frac{2GMmR}{r^3},$$

and represents the *tidal force,* the force on water relative to the earth. What is the direction of the tidal force? (*d*) Repeat for a particle of water at B, on the far side of the earth from the moon. What is the direction of this tidal force? (*e*) Explain why there are two tidal bulges in the oceans (and solid earth), one pointing toward the moon and the other away from it.

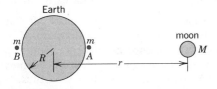

Figure 35 Problem 29.

30P. Masses m, assumed equal, hang from strings of different lengths on a balance at the surface of the earth, as shown in Fig. 36. If the strings have negligible mass and differ in length by h, (*a*) show that the error in weighing, associated with the fact that W' is closer to the earth than W, is $W' - W = 8\pi G\rho mh/3$ in which ρ is the mean density of the earth (5.5 g/cm³). (*b*) Find the difference in length that will give an error of 1 part in a million.

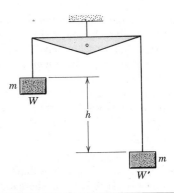

Figure 36 Problem 30.

31P. Sensitive meters that measure the vertical component of the local acceleration due to gravity g can be used to detect the presence of deposits of near-surface rocks of density significantly greater or less than that of the surroundings. Cavities such as caves and abandoned mine shafts can also be located. (*a*) Show that the vertical component of g a distance x from a point directly above the center of a spherical cavern (see Fig. 37) is less than what would be expected assuming a uniform distribution of rock of density ρ, by the amount

$$\Delta g = \frac{4\pi}{3} R^3 G\rho \frac{d}{(d^2 + x^2)^{3/2}},$$

where R is the radius of the cavern and d is the depth of its center. (*b*) These values of Δg, called *anomalies,* are usually very small and expressed in milligals, where 1 gal = 1 cm/s². Oil prospectors doing a gravity survey find Δg varying from 10 milligals to a maximum of 14 milligals over a 150-m distance. Assuming that the larger anomaly was recorded directly over the center of a spherical cavern known to be in the region, find its radius and the depth to the roof of the cave at that point. Nearby rocks have a density of 2.8 g/cm³. (*c*) Suppose that the cavern, instead of being empty, is completely flooded with water. What do the gravity readings in (*b*) now indicate for its radius and depth?

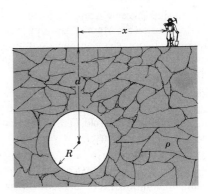

Figure 37 Problem 31.

32P. A scientist is making a precise measurement of g at a certain point in the Indian Ocean (on the equator) by timing the swings of a pendulum of accurately known construction. To provide a stable base the measurements are conducted in a submerged submarine. It is observed that a slightly different result for g is obtained when the submarine is moving eastward than when it is moving westward, the speed in each case being 16 km/h. Account for this difference and calculate the fractional error $\Delta g/g$ in g.

33P. A body is suspended on a spring balance in a ship sailing along the equator with a speed v. (a) Show that the scale reading will be very close to $W_0 (1 \pm 2\omega v/g)$, where ω is the angular speed of the earth and W_0 is the scale reading when the ship is at rest. (b) Explain the plus or minus.

Section 15–7 Gravitational Potential Energy

34E. An 800-kg mass and a 600-kg mass are separated by 25 m. (a) What is the gravitational field due to these masses at a point 20 m from the 800-kg mass and 15 m from the 600-kg mass? (b) What is the gravitational potential energy per kilogram at this point owing to these same masses?

35E. The masses and coordinates of three spheres are as follows: 20 kg, $x = 0.50$ m, $y = 1.0$ m; 40 kg, $x = -1.0$ m, $y = -1.0$ m; 60 kg, $x = 0.0$ m, $y = -0.50$ m. (a) What is the gravitational force on a 20-kg sphere located at the origin? (b) What is the gravitational potential energy of the sphere?

36E. Mars has a mean diameter of 6.9×10^3 km; the Earth's diameter is 1.3×10^4 km. The mass of Mars is $0.11 M_e$. (a) How does the mean density of Mars compare with that of Earth? (b) What is the value of g on Mars? (c) What is the escape velocity on Mars? Compare your answers with values in the appendices.

37E. A spaceship is idling at the fringes of our galaxy, 80,000 light-years from the galactic center. What minimum speed must it have if it is to escape entirely from the gravitational attraction of the galaxy? The mass of the galaxy is 1.4×10^{11} times that of our sun. Assume, for simplicity, that the matter forming the galaxy is distributed with spherical symmetry.

38E. A rocket ship takes off from the Earth on a trip to Jupiter with a stopover on the Moon. Assuming that the mass of the ship does not change, calculate the amount of energy required to escape from these two bodies relative to that required to escape from Earth.

39E. Show that the velocity of escape from the sun at the earth's distance from the sun is $\sqrt{2}$ times the speed of the earth in its orbit, assumed to be a circle. (This is a specific case of a general result for circular orbits: $v_{esc} = \sqrt{2}v_{orb}$.)

40E. The sun, mass 2.0×10^{30} kg, is revolving about the center of the Milky Way galaxy, which is 2.2×10^{20} m away. It completes one revolution every 2.5×10^8 years. Estimate roughly the number of stars in the Milky Way. (*Hint:* Assume for simplicity that the stars are distributed with spherical sym-

metry about the galactic center and that our sun is essentially at the galactic edge.)

41E. It is conjectured that a burned-out star could collapse to a *gravitational radius,* defined as the radius for which the work needed to remove an object of mass m from the star's surface to infinity equals the rest energy mc^2 of the object. Show that the gravitational radius of the sun is GM_s/c^2 and determine its value in terms of the sun's present radius. (For a review of this phenomenon see "Black Holes: New Horizons in Gravitational Theory," by Philip C. Peters, *American Scientist,* Sept.-Oct. 1974.)

42P. Masses of 200 and 800 g are 12 cm apart. (a) Find the gravitational force on a 1.0-g object situated at a point on the line joining the masses 4.0 cm from the 200-g mass. (b) Find the gravitational potential energy per kg at that point. (c) How much work is needed to move this object to a point 4.0 cm from the 800-g mass along the line of centers?

43P. Write an expression for the potential energy of a body of mass m in the gravitational field of the earth and the moon. Let M_e be the earth's mass, M_m the moon's mass, R the distance from the earth's center, and r the distance from the moon's center.

44P. A rocket is accelerated to a speed of $v = 2\sqrt{gR_e}$ near the earth's surface and then coasts upward. (a) Show that it will escape from the earth. (b) Show that very far from the earth its speed is $v = \sqrt{2gR_e}$. (*Hint:* Use conservation of energy.)

45P. (a) What is the escape velocity on a hypothetical planet whose radius is 500 km and whose surface gravity is 3.0 m/s²? (b) How high will a particle rise if it leaves the surface of the planet with a vertical velocity of 1000 m/s? (c) With what speed will an object hit the planet if it is dropped from a height of 1000 km (1500 km from the center of the planet)?

46P. A projectile is fired vertically from the earth's surface with an initial speed of 10 km/s. Neglecting atmospheric friction, how far above the surface of the earth will it go?

47P. In a double star, two stars of mass 3.0×10^{30} kg each rotate about their common center of mass, 1.0×10^{11} m away. (a) What is their common angular speed? (b) Suppose that a meteorite passes through this center of mass moving at right angles to the orbital plane of the stars. What must its speed be if it is to escape from the gravitational field of the double star?

48P. Two neutron stars are separated by a distance of 10^{10} m. They each have a mass of 10^{30} kg and a radius of 10^5 m. They are initially at rest with respect to one another. (a) How fast are they moving when their separation has decreased to one-half its initial value? (b) How fast are they moving just before they collide?

49P. Several planets (Jupiter, Saturn, Uranus) possess nearly circular surrounding rings, perhaps composed of material that failed to form a satellite. In addition, many galaxies contain ringlike structures. Consider a homogeneous ring of mass M

and radius R. (a) What gravitational attraction does it exert on a particle of mass m located a distance x from the center of the ring along its axis? See Fig. 38. (b) Suppose the particle falls from rest as a result of the attraction of the ring of matter. Find an expression for the speed with which it passes through the center of the ring.

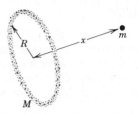

Figure 38 Problem 49.

50P. A sphere of matter, mass M, radius a, has a concentric cavity, radius b, as shown in cross section in Fig. 39. (a) Sketch the gravitational force F exerted by the sphere on a particle of mass m, located a distance r from the center of the sphere, as a function of r in the range $0 \le r \le \infty$. Consider points $r = 0, b, a,$ and ∞ in particular. (b) Sketch the corresponding curve for the potential energy $U(r)$ of the system. (c) From these graphs, how would you obtain graphs of the gravitational field strength and the gravitational potential due to the sphere?

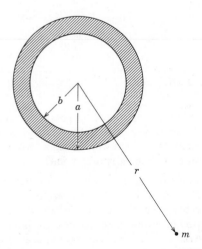

Figure 39 Problem 50.

51P. Two particles with masses m and M are initially at rest an infinite distance apart. Show that at any instant their relative velocity of approach attributable to gravitational attraction is $\sqrt{2G(M + m)/d}$, where d is their separation at that instant. (*Hint:* Use conservation of energy and conservation of linear momentum.)

Section 15-8 Planets and Satellites: Kepler's Laws

52E. How far does the earth travel in its orbit (assumed circular) in 1 min?

53E. The mean distance of Mars from the sun is 1.52 times that of the Earth from the sun. From this, calculate the number of years required for Mars to make one revolution about the sun; compare your answer with the value given in Appendix C.

54E. The planet Mars has a satellite, Phobos, which travels in an orbit of radius 9.4×10^6 m with a period of 7 h 39 min. Calculate the mass of Mars from this information.

55E. Determine the mass of the earth from the period T and the radius r of the moon's orbit about the earth: $T = 27.3$ days and $r = 3.82 \times 10^5$ km.

56E. Calculate the shortest possible period for an earth satellite in circular orbit.

57E. A satellite is placed in a circular orbit with a radius equal to one-half the radius of the moon's orbit. What is its period of revolution in lunar months? (A lunar month is the period of revolution of the moon.)

58E. (a) With what horizontal speed must a satellite be projected at 160 km above the surface of the earth so that it will have a circular orbit about the earth? (b) What will be the period of revolution?

59E. Most asteroids revolve around the sun between Mars and Jupiter. However, several "Apollo asteroids" with diameters of about 30 km move in orbits that cross the orbit of the earth. The orbit of one of these is shown in Fig. 40. By taking measurements directly from the figure, deduce the asteroid's period of revolution in years. (All of these asteroids are expected eventually to collide with the earth.)

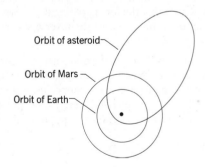

Figure 40 Exercise 59.

60E. A satellite, moving in an elliptic orbit, is 360 km above the earth's surface at its farthest point and 180 km above at its closest point. Calculate (a) the semimajor axis and (b) the eccentricity of the orbit. (*Hint:* See Sample Problem 8.)

61E. The sun's center is at one focus of the earth's orbit. How far from it is the other focus? Express your answer in terms of the solar radius ($= 6.96 \times 10^8$ m). The eccentricity of the earth's orbit is 0.0167 and the semimajor axis may be taken to be 1.5×10^{11} m. See Fig. 19.

62P. (*a*) Express the universal gravitational constant G that appears in Newton's law of gravity in terms of the astronomical unit AU as a length unit, the solar mass M_s as a mass unit, and the year as a time unit. (One astronomical unit $= 1$ AU $= 1.496 \times 10^{11}$ m. One solar mass $= 1M_s = 1.99 \times 10^{30}$ kg. One year $= 1$ y $= 3.156 \times 10^7$ s.) (*b*) What form does Kepler's third law (Eq. 33) take in these units?

63P. In the year 1610, Galileo made a telescope, turned it on Jupiter, and discovered four prominent moons. Their mean orbit radii a and periods T are

Name	a $(10^8$ m)	T (days)
Io	4.22	1.77
Europa	6.71	3.55
Ganymede	10.7	7.16
Callisto	18.8	16.7

(*a*) Plot log a (*y* axis) against log T (*x* axis) and show that you get a straight line. (*b*) Measure its slope and compare it with the value that you expect from Kepler's third law. (*c*) Find the mass of Jupiter from the intercept of this line with the *y* axis. (*Note:* You may also use log–log graph paper.)

64P. A binary star system consists of two stars, each with the same mass as the sun, revolving about their center of mass. The distance between them is the same as the distance between the earth and the sun. What will be the period of revolution in years?

65P. Show how, guided by Kepler's third law (Eq. 33), Newton could deduce that the force holding the moon in its orbit, assumed circular, must vary as the inverse square of the distance from the center of the earth.

66P. A certain triple-star system consists of two stars, each of mass m, revolving about a central star, mass M, in the same circular orbit. The two stars stay at opposite ends of a diameter of the circular orbit; see Fig. 41. Derive an expression for the period of revolution of the stars; the radius of the orbit is r.

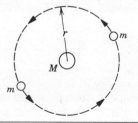

Figure 41 Problem 66.

67P*. What minimum initial speed (measured with respect to the earth) must be imparted to an object resting on the

earth's surface if it is to escape not only from the gravitational field of the earth but also from that of the sun? Ignore the earth's rotation but not its orbital motion around the sun. (*Hint:* Note that for minimum speed the object must be projected in the direction of the earth's orbital motion. Treat the problem in two steps, escape from the sun following that from the earth. The earth's orbital speed, v_0, connects the two reference frames involved.)

68P*. Three identical stars of mass M are located at the vertices of an equilateral triangle with side L. At what speed must they move if they all revolve under the influence of one another's gravity in a circular orbit circumscribing the triangle while still preserving the equilateral triangle?

Section 15-9 Satellites: Orbits and Energy

69E. Spy satellites have been placed in the geosynchronous orbit above the earth's equator. What is the greatest latitude L from which the satellites are visible from the earth's surface? See Fig. 42.

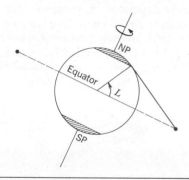

Figure 42 Exercise 69.

70E. A satellite is inserted into the geosynchronous orbit but, by colossal error, it is inserted in the direction *opposite* to the earth's rotation. How often would such a "wrong-way" satellite pass over any given equatorial point?

71E. An asteroid, whose mass is 2.0×10^{-4} times the mass of the earth, revolves in a circular orbit around the sun at a distance that is twice the earth's distance from the sun. (*a*) Calculate the period of revolution of the asteroid in earth years. (*b*) What is the ratio of the kinetic energy of the asteroid to that of the earth?

72E. Consider two satellites, A and B, of equal mass m, moving in the same circular orbit of radius r around the earth but in opposite senses of rotation and therefore on a collision course (see Fig. 43). (*a*) In terms of G, M_e, m, and r, find the total mechanical energy $E_A + E_B$ of the two-satellite-plus-earth system before collision. (*b*) If the collision is completely inelastic so that wreckage remains as one piece of tangled material (mass $= 2m$), find the total mechanical energy immediately after collision. (*c*) Describe the subsequent motion of the wreckage.

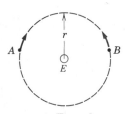

Figure 43 Exercise 72.

73E. Two earth satellites, A and B, each of mass m, are to be launched into (nearly) circular orbits about the earth's center. Satellite A is to orbit at an altitude of 4000 mi. Satellite B is to orbit at an altitude of 12,000 mi. The radius of the earth R_e is 4000 mi (Fig. 44). (a) What is the ratio of the potential energy of satellite B to that of satellite A, in orbit? (Explain the result in terms of the work required to get each satellite from its orbit to infinity.) (b) What is the ratio of the kinetic energy of satellite B to that of satellite A, in orbit? (c) Which satellite has the greater total energy if each has a mass of 1.0 slug? By how much?

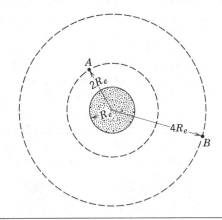

Figure 44 Exercise 73.

74E. Use conservation of energy and Eq. 40 for the total energy to show that the speed v of an object in an elliptical orbit satisfies the relation

$$v^2 = GM \left(\frac{2}{r} - \frac{1}{a} \right).$$

75E. Use the result of Exercise 74 and data contained in Sample Problem 8 to calculate (a) the speed v_p of Halley's comet at perihelion and (b) its speed v_a at aphelion. (c) From these results, show that $r_p v_p = r_a v_a$. Why is this equality expected?

76E. Solve Sample Problem 11 algebraically, inserting numerical values only in the final result. In that way, show that the mass of Sally's spaceship does not enter.

77P. (a) Does it take more energy to get a satellite up to 1000 mi above the earth than to put it in orbit once it is there? (b)

What about 2000 mi? (c) What about 3000 mi? Take the earth's radius to be 4000 miles.

78P. One possibility for damaging a satellite in earth orbit is to launch a swarm of pellets in such a way that they move in the same orbit as the satellite but in the opposite direction. Consider a satellite in a circular orbit whose altitude above the earth's surface is 500 km. An on-board sensor detects a 10-g pellet approaching and determines that a head-on collision is inevitable. (a) What is the kinetic energy of the approaching pellet in the reference frame of the satellite? (b) How does this compare with the kinetic energy of a slug from a modern army rifle? Such a slug has a mass of 4.0 g and a muzzle velocity of 950 m/s.

79P. Consider an artificial satellite in a circular orbit about the earth. State how the following properties of the satellite vary with the radius r of its orbit: (a) period, (b) kinetic energy, (c) angular momentum, and (d) speed.

80P. The asteroid Eros, one of the many minor planets that orbit the sun in the region between Mars and Jupiter, has a radius of 7.0 km and a mass of 5.0×10^{15} kg. (a) If you were standing on Eros, could you lift a 2000-kg pickup truck? (b) Could you run fast enough to put yourself into orbit? Ignore effects due to the rotation of the asteroid. (*Note:* The Olympic records for the 400-m run correspond to speeds of 9.1 m/s for men and 8.2 m/s for women.)

81P. A satellite travels initially in an approximately circular orbit 640 km above the surface of the earth; its mass is 220 kg. (a) Determine its speed. (b) Determine its period. (c) For various reasons, the satellite loses mechanical energy at the (average) rate of 1.4×10^5 J per complete orbital revolution. Adopting the reasonable approximation that the trajectory is a "circle of slowly diminishing radius," determine the distance from the surface of the earth, the speed, and the period of the satellite at the end of its 1500th orbital revolution. (d) What is the magnitude of the average retarding force? (e) Is angular momentum conserved?

82P. The orbit of the earth about the sun is *almost* circular. The closest and farthest distances are 1.47×10^8 km and 1.52×10^8 km, respectively. Determine the maximum variations in (a) potential energy, (b) kinetic energy, (c) total energy, and (d) orbital speed that result from the changing earth–sun distance in the course of 1 year. (*Hint:* Use conservation of energy and angular momentum.)

83P. A weather satellite is in a geosynchronous orbit, hovering over Nairobi, which lies very close to the equator. If its orbit radius is increased by 1.0 km, at what rate and in what direction would its reference spot, which was formerly stationary, move across the earth's surface?

84P. Assume that a geosynchronous communications satellite is in orbit at the longitude of Chicago. You are in Chicago and want to pick up its signals. In what direction should you point the axis of your parabolic antenna? The latitude of Chicago is 47.5°.

ESSAY 5
PHYSICS IN WEIGHT-LESSNESS

SALLY RIDE
CENTER FOR
INTERNATIONAL
SECURITY AND ARMS
CONTROL
STANFORD UNIVERSITY

Newspaper stories describing astronauts "away from the pull of earth's gravity" prompted a high school teacher to complain that "spaceflight sure makes it hard on a science teacher." Her point was that terms like "zero-g" are misleading and that many people interpret that to mean "no gravity." Of course, the Space Shuttle is *not* free of the shackles of gravity. In fact, it is gravity that keeps the spacecraft, and everything inside it, in orbit around earth. The confusion arises because Space Shuttle astronauts *are* "weightless"—they would float above any scales attached to the floor. They aren't weightless because they are "away from gravity," but because the Shuttle and everything in it (astronauts and scales included) are in free-fall. An astronaut could no more stand on scales in the Space Shuttle than an earth-bound scientist could in the classic (though fortunately rare) freely-falling elevator.

An orbiting spacecraft "falls" in that it falls away from the straight line it *would* follow into interplanetary space if there were no forces acting on it. It doesn't come crashing to earth because it has sufficient horizontal velocity to travel "over the horizon:" As it falls, the surface of the earth curves away from it. It is theoretically possible to put a satellite (or a rock) into an orbit only meters above the earth, but its energy would quickly be dissipated by air resistance (and also probably by buildings and hills!). To stay in orbit for more than just a few revolutions, a spacecraft has to be given enough energy to get it into an orbit above most of the earth's atmosphere.

The Space Shuttle is boosted into orbit by the thrust from two solid rockets and three liquid-fueled engines; see Fig. 1. The solid rockets burn for the first 2 min, the launch engines for the first $8\frac{1}{2}$ min. This is enough integrated thrust or impulse to put

Figure 1 The space shuttle is boosted into orbit by five rockets.

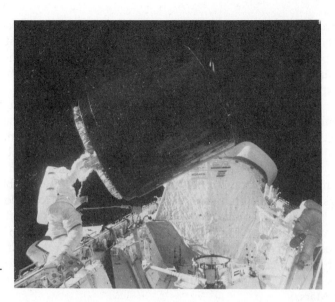

Figure 2 An astronaut manipulating a large satellite.

Space Shuttle into an elliptical low-earth orbit (the precise orbit varies from flight to flight). At the apogee of that orbit (half a world away from the launch pad), the Shuttle's small orbital engines are burned for a couple of minutes to add enough energy to circularize the orbit. These orbital engines shut off when the correct velocity is achieved. No engines are required to keep the Shuttle in orbit. Gravity takes care of that. In a typical circular orbit 400 km above the surface of the earth, the Space Shuttle has a velocity of 8 km/s and takes only 90 min to circle the earth.

Once the Space Shuttle and everything in it have been given the velocity necessary to orbit the earth, gravity does not accelerate objects toward the "floor" of the Shuttle; it accelerates both the objects *and* the floor. All the objects inside are in the same orbit—all falling around the earth together. That they fall together is a result of the equivalence principle, first demonstrated by Galileo. He showed that (neglecting air resistance) if a heavy object and a light object are dropped from the same height, they will hit the ground at the same time. This has been verified many times—often more precisely, but never more dramatically, than by astronaut Dave Scott on *Apollo 15.* He brought a hammer and a feather to the surface of the moon (which has no atmosphere, so no air resistance), stood outside the lunar module, held them at spacesuited arm's length, and let them fall. They hit the lunar surface together. The equivalence principle is demonstrated on every shuttle flight: Since all things inside—astronauts, pencils, satellites, notebooks, extra socks—fall at the same rate, they don't develop motion relative to each other. They "float."

A weightless environment has advantages and disadvantages for astronauts. It's convenient to be able to lift 2000 kg satellites (Fig. 2), float "heavy" equipment across the room, or knock over a carton of milk and not spill it on the floor. On the other hand, it's an annoyance to have to hold a notebook to write in it, or hold a piece of bread to butter it. While it's fun to be able to do a triple somersault and not worry about crashing to the floor, it's distracting to be sent into an unintended somersault by a bump from a friend.

As you would expect, there are also physiological effects associated with weightlessness. The human body evolved on earth; it undergoes changes when it finds itself in this new environment. Perhaps the most visible change is that astronauts' faces become puffy. On earth, gravity is pulling the fluid in the body toward the feet. In orbit, the equilibrium distribution of fluid is different, and it tends to shift toward the upper body. Some astronauts notice this during the flight and are bothered by stuffy sinuses or headaches; some don't notice it until they look in a mirror or at the photographs later.

Another interesting effect is that astronauts grow about an inch in height while in orbit. Since there is no downward force on the spine, the spongy discs in the spinal column are no longer compressed. As the discs relax, astronauts "grow." The effect isn't permanent and astronauts shrink back to their "normal" height on return to earth.

In a weightless environment, the cardiovascular system doesn't need to work very hard to pump blood around the body. It's easier to get the blood back up from the legs, or up to the brain, and the cardiovascular muscles become deconditioned. This isn't a problem as long as astronauts are in orbit, but when they return to earth the cardiovascular system will once again be called on to pump blood against gravity and must be in condition to do so. This isn't a major problem if an astronaut is only in orbit for a week or so, but it is an important consideration for extended flights in a Space Station, and considerable research is required before astronauts can be sent off to Mars.

There is no "preferred direction" in weightlessness, no "upside down" or "rightside up." Physiologically, there's no way to distinguish up from down. The sensors that contribute to our balance and help us determine our orientation are located in the inner ear and are sensitive to gravity on earth. When the head tilts, thin hairlike structures bend and send signals to the brain that the head isn't upright. In weightlessness, these sensors don't register differences in orientation, and there are no other familiar physiological indicators (such as fluid rushing to the head) to provide clues to the brain about the body's orientation. Astronauts feel the same whether their feet are pointed toward the earth or toward the stars. In fact, the Shuttle might be orbiting with its nose toward the Earth, its wing toward the Earth, or its bottom toward the Earth, and astronauts wouldn't be able to tell the difference without looking out a window.

Astronauts have to adapt to an environment that can't be simulated on earth. Things in weightlessness *seem* to be subject to a different set of physical laws. The laws, of course, are the same; but sometimes the implications of those laws are much more apparent. For example, on earth, frictional effects make it difficult to study Newton's laws of motion. Friction is hard to avoid because of gravity. Gravity holds things in contact with the ground, or table, or floor. When gravity no longer holds things in contact with each other, it's easy to avoid frictional effects. Weightlessness is a great improvement on an air-bearing table! In fact, an astronaut may seem to be an involuntary part of an elementary physics lab. Newton's laws of motion become very real . . . and this takes some getting used to. An astronaut put at rest in the middle of the cabin, unable to reach the floor, the ceiling, or any of the walls, will remain at rest —stranded in the middle of the cabin—until a friend comes along to supply an external force. A peanut set in motion will remain in motion until it hits a wall, a ceiling, or somebody's mouth. And a sharp tap on the shoulder can give a sufficient impulse to send an astronaut drifting across the room.

Human beings have learned to deal with the law of action and reaction on earth, where they are anchored to the ground. Anyone who pulls open a drawer unconsciously reacts against the floor. When an unanchored astronaut pulls on a drawer the result is frustrating, but predictable; see Fig. 3. The drawer doesn't open, but the astronaut moves toward the drawer. And if that unanchored astronaut uses a screw-

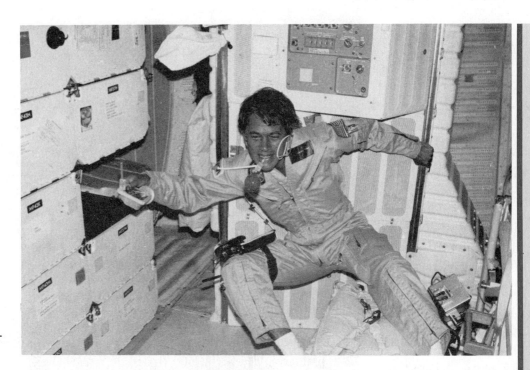

Figure 3 Even simple operations require an astronaut to be anchored.

driver to apply a torque to a screw, the result will be a spinning astronaut, not a turning screw.

The effects of surface tension are very much in evidence on earth: Soap bubbles form, water drops hang on leaky faucets, and a meniscus forms on columns of water that rise up in glass tubes. Surface tension is a result of intermolecular forces. The molecules in a liquid feel some attraction for each other, so those molecules on the surface see a small net force holding them in. Similarly, if a liquid is in contact with a solid, the molecules in the liquid will feel some small attraction to those on the surface of the solid.

As you would expect, surface tension is even more evident in orbit, because its macroscopic effects are not overshadowed by gravitational effects (if the leaky faucet were in the Space Shuttle, the drip would never fall).

Surface tension tends to minimize the surface-to-volume ratio of a liquid. This is evident in weightlessness, where liquids do indeed coalesce into spheres. On earth this isn't as obvious: Spilled milk lies in a puddle on the floor; in weightlessness, the same milk doesn't splatter on the floor but forms a sphere floating in the middle of the room.

The residual forces between molecules at the interface of a solid and a liquid can cause the liquid to "cling" to the solid. It's because of surface tension that near normal (that is, earthlike) dining is possible in space. Astronauts eat out of open cartons and use spoons to get the food to their mouths. The trick, of course, is to have "sticky" foods. Most of the food is dehydrated, and vacuum-packed in plastic cartons with thin plastic tops. It's rehydrated by poking the needle of a water gun through the plastic top, and injecting water. All these foods (e.g., macaroni and cheese, shrimp cocktail, tomato soup) are at least partly liquid once they are rehydrated, and surface tension will hold them inside the container—or on a spoon. Astronauts can snip open a container of soup and eat it with a spoon. The convenient difference is that if the spoon is tilted (or "dropped", or spun), the soup stays on it.

Figure 4 Astronauts at work in the Space Shuttle.

Surface tension keeps rehydrated food on spoons, but it also helps drinks escape from their containers. If a plastic straw is used to drink from a carton, the molecules of the liquid will also feel some attraction to molecules of the straw, including those just above the surface of the liquid. The attraction is just enough that, in weightlessness, the liquid will crawl up a thin straw and collect in a large drop — a sphere — at the opening of the straw. Space Shuttle straws all come with small clamps to pinch them closed and keep the drinks from climbing out.

Some familiar physical effects are a result of the gravitational forces that objects experience on earth; these effects are absent in an orbiting Space Shuttle. Buoyancy is a good example. A stick in a puddle of water will float; mud in that same puddle will sink. The stick floats because it is less dense than the water so buoyant forces hold it up; the mud settles to the bottom because it is denser than the water, and buoyant forces cannot hold it up. On earth, cream rises to the top of milk, an oil slick floats on water, and silt is deposited at the mouths of rivers. These are all familiar phenomena.

In orbit, a column of liquid has no weight; there is no hydrostatic pressure and therefore no buoyant effects and no sedimentation. A cork would not bob, a bubble would not rise to the surface of a liquid (which means that dissolved gas stays in carbonated beverages, so they aren't very good to drink), and there would be no layer of chocolate at the bottom of a glass of chocolate milk.

This same principle — that denser material sinks and less dense material rises — also produces heat convection. On earth, convection occurs when one part of a liquid or gas is heated or cooled. A heated blob expands, becomes less dense, and therefore (on earth) rises; a cooled blob becomes more dense and falls. Convection doesn't occur in weightlessness. Again, since a column of (for example) air has no weight, hot air doesn't rise: It expands when heated but stays where it is.

An interesting demonstration (that for obvious safety reasons hasn't been performed yet) would be to light a match or candle in an orbiting spacecraft. It is convection that keeps a fire going. As oxygen in the air is burned and depleted around a flame, the warmed gas rises and cooler air moves in to replace it and to deliver more oxygen to be consumed. Without convection, a candle would quickly burn itself out.

A weightless world is different from the world with which we are familiar. Some common physical effects are absent, while others are glaringly apparent. As you can imagine, it is an unusual living environment. It's also a unique laboratory environment: one that offers the opportunity to perform fundamental experiments in physics, chemistry, and physiology under new laboratory conditions.

CHAPTER 16

FLUIDS

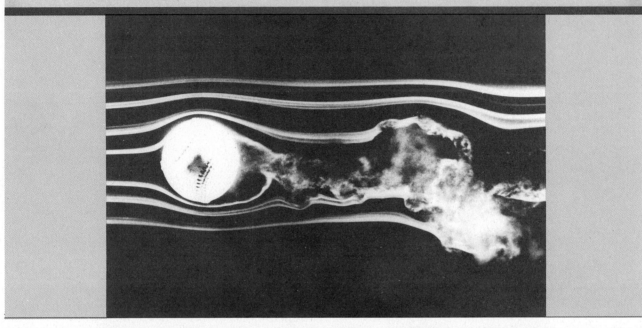

A spinning baseball in a wind tunnel, the flow of the air made visible by smoke streamers.
In play, this ball would take 780 ms to reach the plate and would make about 8 revolutions
on its journey. Which way do you think it would curve?

16-1 Fluids and the World Around Us

Fluids—which include both liquids and gases—play a central role in our daily lives. We breath and drink them and a rather vital fluid circulates in our cardiovascular system. There is the fluid ocean, the fluid atmosphere, and—deep within the earth—its fluid outer core.

In our car, there are fluids in the tires, the gas tank, the radiator, the combustion chambers of the engine, the exhaust manifold, the battery, the air conditioning system, the windshield wiper reservoir, the lubrication system, and the hydraulic system. The next time you see a large piece of earth-moving machinery, count the hydraulic cylinders that permit the machine to do its work. Large jet planes have scores of them.

We use the kinetic energy of a moving fluid in wind-mills and its potential energy in hydraulic power plants. Given time, fluids carve the landscape. We often travel great distances just to watch fluids. As the naturalist-philosopher Loren Eisley has said, "If there is beauty in this world, it is in water."

Perhaps we had better see what physics can tell us about fluids. We shall find that, although it is useful to develop a special formulation of Newton's laws as they apply to fluids, we do not need to introduce any new fundamental principles.

16-2 What Is a Fluid?

A *fluid,* in contrast to a solid, is a substance that can flow. As we know, fluids readily conform to the boundaries of any container in which we put them. They do so because, although a fluid can exert a force at right angles to its surface, it cannot sustain a force that is tangential to that surface.* There are some materials, such as glass or pitch, that take a very long time to conform to the boundaries of a container but they do so eventually so we classify them as fluids.

You may wonder why we lump liquids and gases together and call them fluids. After all (you may say), water is as different from steam as it is from ice. Actually, it is not. Ice, like other crystalline solids, has its constituent atoms organized in a fairly rigid three-dimensional array called a crystalline lattice. In neither water nor steam, however, is there any such orderly long-range arrangement, the intermolecular interactions being restricted to neighboring molecules.

We can demonstrate the close relation between a liquid and a gas experimentally by changing one into the other in a continuous way, without the appearance of a sharp boundary (meniscus) between them and without boiling. By manipulating the pressure and the temperature properly, for example, we can change water into steam in this way. For water—even though we may start and end the process at atmospheric pressure—a pressure exceeding 218 atm (the *critical pressure*) and a temperature exceeding 374°C (the *critical temperature*) must be available during the intermediate steps. Our perception of a sharp distinction (boiling) between water and steam comes about because essentially all our experience with this fluid is at a single pressure, that of the atmosphere.

16-3   Density and Pressure

When we dealt with rigid bodies, we were concerned with particular lumps of matter, such as wooden blocks, baseballs, or metal rods. Physical quantities that we found useful, and in whose terms we expressed Newton's laws, were *mass* and *force.* We typically spoke, for example, of a 3.6-kg block acted on by a 25-N force.

With fluids, we are more interested in properties that vary from point to point in the extended substance than with properties of specific lumps of that substance. It is more useful to speak of *density* and *pressure* than of mass and force.

Table 1   Some Densities

Material or Object		Density (kg/m³)
Interstellar space		10^{-20}
Best laboratory vacuum		10^{-17}
Air:	20°C and 1 atm	1.21
	20°C and 50 atm	60.5
Styrofoam		1×10^2
Water:	20°C and 1 atm	0.998×10^3
	20°C and 50 atm	1.000×10^3
Seawater: 20°C and 1 atm		1.024×10^3
Whole blood		1.060×10^3
Ice		0.917×10^3
Iron		7.8×10^3
Mercury		13.6×10^3
The earth:	average	5.5×10^3
	core	9.5×10^3
	crust	2.8×10^3
The sun:	average	1.4×10^3
	core	1.6×10^5
White dwarf star (core)		10^{10}
Uranium nucleus		3×10^{17}
Neutron star (core)		10^{18}
Black hole (1 solar mass)		10^{19}

Density.   To find the density of a fluid at any point, we construct a small volume element ΔV around that point and measure the mass Δm of the fluid contained within that element. The density is then

$$\rho = \frac{\Delta m}{\Delta V}. \tag{1}$$

Strictly, the density at any point in a fluid is the limit of this ratio as the volume element ΔV at that point is made smaller and smaller. In practice, this requirement causes no difficulty.*

Density is a scalar property, its SI unit being the kilogram per meter³. Table 1 shows the densities of some fluids and some solids and the average densities of some objects. Note that the density of gases (see Air in the table) varies considerably with pressure but that for liquids (see Water) does not. That is, gases are readily compressible; liquids are not.

Pressure.   Let a small pressure-sensing device be

---

* In the more formal language of Section 13-6, a fluid is a substance that cannot support a shearing stress.

---

* In this chapter, we deal only with fluid samples that are large compared with atomic dimensions. This is equivalent to assuming that fluids are perfectly continuous.

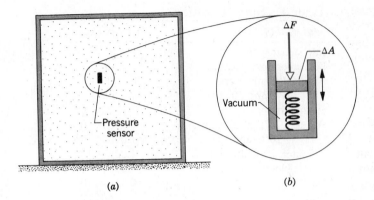

(a)   *(b)*

Figure 1 *(a)* A gas-filled vessel containing a small pressure sensor, the details of which are shown in *(b)*. The pressure is measured by the relative position of the piston in the sensor. At a given location, the pressure is independent of the orientation of the sensor.

suspended inside a fluid-filled vessel, as in Fig. 1*a*. The sensor (see Fig. 1*b*) consists of a piston of area ΔA riding in a close-fitting cylinder and resting against a spring. A readout arrangement allows us to record the amount by which the (calibrated) spring is compressed and thus the magnitude ΔF of the force that acts on the piston. We define the pressure exerted by the fluid on the piston as

$$p = \frac{\Delta F}{\Delta A}. \qquad (2)$$

Strictly, the pressure at any point in the fluid is the limit of this ratio as the area ΔA of the piston, centered on that point, is made smaller and smaller. In practice, this requirement causes no difficulty.

We find by experiment that the pressure p defined by Eq. 2 has the same value at a given point in a fluid at rest, no matter how the pressure sensor is oriented. Pressure is a scalar, having no directional properties. It is true that the force acting on the piston of our pressure sensor is a vector, but Eq. 2 involves only the *magnitude* of that force, a scalar quantity.

The SI unit of pressure is the newton per meter^2, which is given a special name, the *pascal* (abbr. Pa). In metric countries, tire pressure gauges are calibrated in kilopascals. The pascal is related to some other common (non-SI) pressure units by

$$1 \text{ atm} = 1.01 \times 10^5 \text{ Pa} = 760 \text{ torr} = 14.7 \text{ lb/in.}^2.$$

The *atmosphere* (abbr. atm) is, as the name suggests, the approximate average pressure of the atmosphere at sea level. The *torr* (named for Evangelista Torricelli, who invented the mercury barometer in 1674) was formerly called the *millimeter of mercury* (abbr. mm Hg). Table 2 shows some pressures.

Table 2   Some Pressures

	Pressure (Pa)
Center of the sun	2×10^{16}
Center of the earth	4×10^{11}
Highest sustained laboratory pressure	1.5×10^{10}
Deepest ocean trench (bottom)	1.1×10^{8}
Spiked heels on a dance floor	1×10^{6}
Automobile tire (overpressure)	2×10^{5}
Atmosphere at sea level	1.0×10^{5}
Normal blood pressure[a]	1.6×10^{4}
Loudest tolerable sound[b]	30
Faintest detectable sound[b]	3×10^{-5}
Best laboratory vacuum	10^{-12}

[a] The systolic overpressure, corresponding to 120 torr on the physician's pressure gauge.
[b] Overpressure at the ear drum, at 1000 Hz.

Sample Problem 1 A living room has dimensions 3.5 m $\times$ 4.2 m $\times$ 2.4 m. (a) What does the air in the room weigh?

We have, where V is the volume of the room and ρ (see Table 1) is the density of air,

$$\begin{aligned} W = mg &= \rho V g \\ &= (1.21 \text{ kg/m}^3)(3.5 \text{ m} \times 4.2 \text{ m} \times 2.4 \text{ m})(9.8 \text{ m/s}^2) \\ &= 418 \text{ N}. \qquad \text{(Answer)} \end{aligned}$$

This is about 94 lb. Would you have guessed that the air in a room could weigh so much?

(b) What force does the atmosphere exert on the floor of the room?

The force is

$$F = pA = (1.0 \text{ atm}) \left(\frac{1.01 \times 10^5 \text{ N/m}^2}{1 \text{ atm}} \right) (3.5 \text{ m} \times 4.2 \text{ m})$$
$$= 1.5 \times 10^6 \text{ N}. \qquad \text{(Answer)}$$

This force (≈ 170 ton) is the weight of a column of air covering the floor and extending all the way to the top of the atmosphere. It is equal to the force that would be exerted on the floor if (in the absence of the atmosphere) the room were filled with mercury to a depth of 30 in. Why doesn't this enormous force break the floor?

16-4 Fluids at Rest

Figure 2a shows a tank of water—or other liquid—open to the atmosphere. As every diver knows, the pressure increases with depth below the air–water interface. The diver's depth gauge, in fact, is a pressure sensor much like that of Fig. 1b. As every mountaineer and aviator knows, the pressure decreases with altitude as one ascends into the atmosphere above this interface. The aviator's altimeter, in fact, is also an elaborate version of the pressure sensor of Fig. 1b.

Let us look first at the increase in pressure with depth below the water surface. We set up a vertical y axis, its origin being at the air–water interface and the direction of increasing y being up. Consider a water sample contained in a hypothetical cylinder of area A and let y_1 and y_2 (both of which are negative numbers) be the depths of the upper and the lower cylinder faces, respectively, below the surface.

Figure 2b shows a free-body diagram for the water in the cylinder. The water sample is in equilibrium, its weight (downward) being exactly balanced by the difference between the force F_2 ($=p_2A$) acting upward on its lower face and the force F_1 ($=p_1A$) acting downward on

its upper face. Thus,

$$F_2 = F_1 + W. \tag{3}$$

The volume V of the cylinder is $A(y_1 - y_2)$. Thus, the mass m of the water in the cylinder is $\rho A(y_1 - y_2)$ in which ρ is the density of water. The weight W is then $\rho Ag(y_1 - y_2)$. Substitution into Eq. 3 yields

$$p_2A = p_1A + \rho Ag(y_1 - y_2),$$

or

$$\boxed{p_2 = p_1 + \rho g(y_1 - y_2),} \tag{4}$$

in which y_1 and y_2 in Fig. 2 are negative numbers.

If we seek the pressure p at a depth h below the surface we choose level 1 to be the surface and level 2 to be a surface a distance h below, as in Fig. 3. Representing the atmospheric pressure by p_0, we then substitute

$$y_1 = 0, \quad p_1 = p_0 \quad \text{and} \quad y_2 = -h, \quad p_2 = p$$

into Eq. 4, which becomes

$$\boxed{p = p_0 + \rho gh.} \tag{5}$$

As we expect, Eq. 5 reduces to $p = p_0$ for $h = 0$.

The pressure at a given depth depends only on that depth and not on any horizontal dimension. Thus, the pressure on a dam at its base depends only on the depth of the water and is quite independent of the amount of water backed up behind the dam. Lake Mead, for example, extends for many miles behind Hoover dam and is

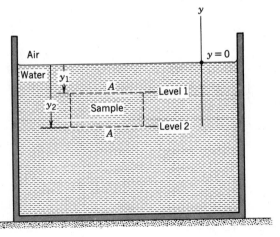

(a)

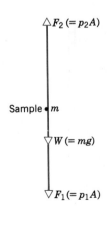

(b)

Figure 2 (a) A sample of water is contained in a hypothetical cylinder of area A. (b) A free-body diagram of the water sample. It is in static equilibrium, its weight being balanced by the net upward buoyant force that acts on it; see Eq. 3.

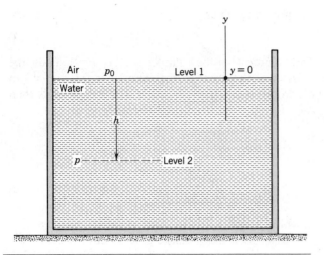

Figure 3 The pressure p increases with depth h below the water surface; see Eq. 5.

700 ft deep at the dam face. This dam would have to be built every bit as strong to hold back just a few thousand gallons of water if the same depth were maintained!

Equation 5 holds no matter what the shape of the containing vessel. Consider Fig. 4a, which shows a tube of irregular shape immersed in a tank of water. The stopcock connected to the tube is open so that the tube can fill freely. You would not doubt that Eq. 5 correctly gives the pressure at all points at any horizontal level such as AA, no matter whether those points are in the tube or in the tank.

Now close the stopcock, an action that causes no pressure changes at any point. With the stopcock closed, lower the tank, leaving the water-filled tube in place as in Fig. 4b. Again, this action causes no pressure changes for the water within the tube, which was removed from contact with the water in the tank when the stopcock was closed. In particular, the pressure of the water in the tube at level AA in Fig. 4b retains its value, namely, that given by Eq. 5. All that counts is the vertical distance h below the free surface.

Pressure decreases as one moves vertically upward in the atmosphere above the liquid surface in Fig. 3. Equation 5 (with h negative, according to the sign convention of Fig. 2) continues to hold. However, because the densities of gases are typically three orders of magnitude less than those of liquids, the magnitude of the pressure change for a given change in elevation in a gas is correspondingly less than in a liquid. Nevertheless, with a pocket aneroid barometer of good quality, you can detect the atmospheric pressure difference (about 0.1%) encountered in climbing a few flights of stairs. Atmospheric pressure drops to half its sea-level value at an altitude of about 18,000 ft.

Sample Problem 2 An enterprising snorkeler reasons that if a 9-in. snorkel tube is good, a 20-ft tube (=6.1 m) would be better; see Fig. 5. What is the pressure at the bottom of the tube? What kind of trouble is the snorkeler in?

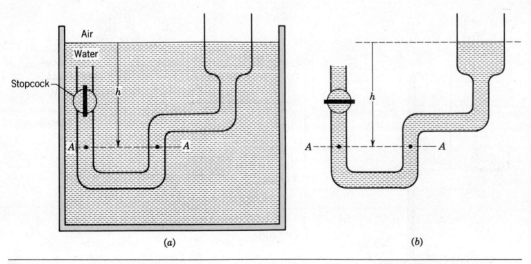

(a)   (b)

Figure 4 (a) A tube is immersed in a tank of water, the stopcock being open. (b) The stopcock is closed and the tank is removed. The pressure at level AA is given by Eq. 5 in each case.

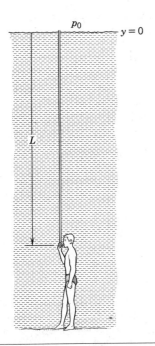

p_0

$y = 0$

L

Figure 5   Sample Problem 2. DON'T TRY THIS. Because the external pressure is so much greater than the internal pressure, you would not be able to expand your lungs to inhale.

His lungs are about to collapse! The pressure increment of the air in his long snorkel tube is negligible so that the air in his lungs is essentially at atmospheric pressure p_0. From Eq. 5, however, the water pressure on the outside of his chest cavity is

$$p = p_0 + \rho g L,$$

where L is the length of his snorkel tube. (If his chest is below the bottom of his snorkel tube, the pressure is actually greater than this.) The pressure difference tending to collapse his lungs is then

$$\Delta p = p - p_0 = \rho g L$$
$$= (1000 \text{ kg/m}^3)(9.8 \text{ m/s}^2)(6.1 \text{ m})$$
$$= 6.0 \times 10^4 \text{ Pa}. \qquad \text{(Answer)}$$

This is about 0.6 atm, a pressure difference certain to cause lung collapse.

Although a snorkeler cannot operate safely at 20-ft depths, a free diver can do so. In this case, the diver's chest cavity contracts, the pressure of the air trapped in his lungs increasing to match the external pressure of the water.

Sample Problem 3   The U-tube in Fig. 6 contains water of density ρ_w in the right arm and oil of unknown density ρ_x in the left. Measurement gives $l = 135$ mm and $d = 12.3$ mm. What is the density of the oil?

We can write down two equivalent expressions for the pressure p_{int} at the oil–water interface in the left arm. By way of the right arm, this interface is a distance l below the free

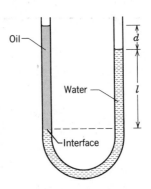

Oil

Water

d

l

Interface

Figure 6   Sample Problem 3. The oil in the left arm stands higher than the water in the right arm because the oil is less dense than the water. Each fluid column produces the same pressure at the interface.

surface of the *water* and we have, from Eq. 5,

$$p_{\text{int}} = p_0 + \rho_w g l \quad \text{(right arm)}.$$

By way of the left arm, the interface is a distance $l + d$ below the free surface of the *oil* and we have, again from Eq. 5,

$$p_{\text{int}} = p_0 + \rho_x g(l + d) \quad \text{(left arm)}.$$

Equating these two quantities and solving for the unknown density yields

$$\rho_x = \rho_w \frac{l}{l + d} = (1000 \text{ kg/m}^3) \frac{135 \text{ mm}}{135 \text{ mm} + 12.3 \text{ mm}}$$
$$= 916 \text{ kg/m}^3. \qquad \text{(Answer)}$$

Note that the answer does not depend on the atmospheric pressure p_0 or the free-fall acceleration g.

16-5   Measuring Pressure

The Pressure of the Atmosphere.   Figure 7a shows the rudiments of a mercury barometer, used to measure the pressure of the atmosphere. To construct it, the long glass tube is filled with mercury and inverted with its open end in a dish of mercury, as the figure shows. The space above the mercury column contains only mercury vapor, whose pressure is so small at ordinary temperatures that it can be neglected.

We can use Eq. 4 to find the atmospheric pressure p_0 in terms of the height h of the mercury column. Choose level 1 in Fig. 2 to be that of the air–mercury interface and level 2 to be that of the top of the mercury column, as in Fig. 7a. We then substitute

$$y_1 = 0, \quad p_1 = p_0 \quad \text{and} \quad y_2 = h, \quad p_2 = 0$$

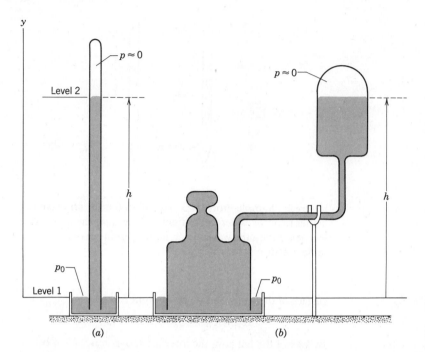

Figure 7 *(a)* A mercury barometer. *(b)* Another mercury barometer. The distance *h* is the same in each case.

into Eq. 4, yielding

$$p_0 = \rho g h. \tag{6}$$

For a given pressure, the height *h* of the mercury column does not depend in any way on the cross-sectional area of the vertical tube. The fanciful barometer of Fig. 7*b* gives the same reading as that of Fig. 7*a*; all that counts is the vertical distance *h* between the mercury levels.

Equation 6 shows that, for a given pressure, the height of the column of mercury depends on the value of *g* at the location of the barometer and also on the density of mercury, which varies with temperature. The column height (in millimeters) is numerically equal to the pressure (in torrs) *only* if the barometer is at a place where *g* has its accepted standard value of 9.80665 m/s² and if the temperature of the mercury is 0°C. If these conditions do not prevail (and they essentially never do), small corrections must be made before the height of the mercury column can be transformed into a pressure.

The Open-Tube Manometer. Many times, as when we inflate the tires of our automobile or measure our blood pressure, we do not want to know the *absolute pressure* but only the so-called overpressure or *gauge pressure* p_g. This is the difference between the absolute pressure and the atmospheric pressure.

An open-tube manometer (Fig. 8) measures the gauge pressure directly. It consists of a U-tube containing a liquid, one end of the tube being connected to the vessel whose gauge pressure we wish to measure, the other end being open to the atmosphere. We can use Eq. 4 to find the gauge pressure in terms of the height *h* of the mercury column. Let us choose levels 1 and 2 as shown

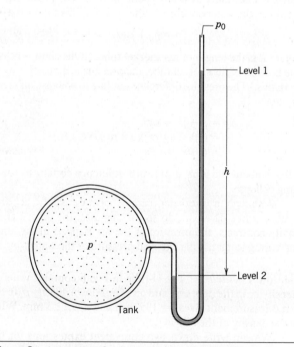

Figure 8 An open-tube manometer, connected so as to read the gauge pressure of the gas in the vessel on the left. The right arm of the U-tube is open to the atmosphere.

in Fig. 8. We then substitute

$$y_1 = 0, \quad p_1 = p_0 \quad \text{and} \quad y_2 = -h, \quad p_2 = p$$

into Eq. 4, yielding

$$p_g = p - p_0 = \rho g h. \tag{7}$$

We see that the gauge pressure p_g is directly proportional to h. The gauge pressure can be positive or negative, depending on whether $p > p_0$ or $p < p_0$.

Sample Problem 4   An open-tube manometer containing oil is connected to a tank containing a gas, as in Fig. 8. The height h of the oil column is 293 mm. The value of the free-fall acceleration g at the location of the manometer is 9.73 m/s^2 and the temperature is 22°C, at which temperature the density ρ of the oil is 835 kg/m^3. What is the gauge pressure of the gas in the tank?

From Eq. 7

$$p_g = \rho g h = (835 \text{ kg/m}^3)(9.73 \text{ m/s}^2)(0.293 \text{ m})$$
$$= 2.38 \times 10^3 \text{ Pa} = 0.0235 \text{ atm.} \qquad \text{(Answer)}$$

Sample Problem 5   The mercury column in a barometer has a measured height h of 740.35 mm. The temperature is -5.0°C, at which temperature the density of mercury is 1.3608×10^4 kg/m^3. The free-fall acceleration g at the site of the barometer is 9.7835 m/s^2. What is the atmospheric pressure?

From Eq. 6 we have

$$p_0 = \rho g h$$
$$= (1.3608 \times 10^4 \text{ kg/m}^3)(9.7835 \text{ m/s}^2)(0.74035 \text{ m})$$
$$= 9.8566 \times 10^4 \text{ Pa.} \qquad \text{(Answer)}$$

Barometer readings are usually expressed in torrs, where 1 torr is the pressure exerted by a column of mercury 1 mm high at a place where g has an accepted standard value of 9.80665 m/s^2 and at a temperature (0.0°C) at which mercury has a density of 1.35955×10^4 kg/m^3. Thus, from Eq. 6,

$$1 \text{ torr} = (1.35955 \times 10^4 \text{ kg/m}^3)(9.80665 \text{ m/s}^2)(1 \times 10^{-3} \text{ m})$$
$$= 133.326 \text{ Pa.}$$

Applying this conversion factor yields, for the atmospheric pressure recorded on the barometer,

$$p_0 = 9.8566 \times 10^4 \text{ Pa} = 739.29 \text{ torr.} \qquad \text{(Answer)}$$

Note that the pressure in torr (739.29 torr) is numerically close to—but otherwise differs significantly from—the height h of the mercury column expressed in mm (740.35 mm). These two quantities will be numerically equal only if the barometer is located at a place where g has its standard value and where the mercury temperature is 0°C.

16-6   Pascal's Principle

When you squeeze one end of a tube of toothpaste, you are watching Pascal's principle in action. This principle is also the basis for the Heimlich maneuver, in which a sharp pressure increase applied to the abdomen is transmitted to the throat, forcefully ejecting a food particle that has become lodged there. The principle was first stated clearly in 1652 by Blaise Pascal (for whom the unit of pressure is named):

> *A change in the pressure applied to an enclosed fluid is transmitted undiminished to every portion of the fluid and to the walls of the containing vessel.*

Demonstrating Pascal's Principle.   Consider the case in which the fluid is an incompressible liquid contained in a tall cylinder, as in Fig. 9. The cylinder is fitted with a piston on which, by the combined action of the atmosphere and of the added lead shot, an external pressure p_{ext} exists. The pressure p at any point P in the liquid is then

$$p = p_{\text{ext}} + \rho g h. \tag{8}$$

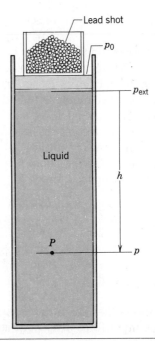

Figure 9   Weights loaded onto the piston create a pressure p_{ext} at the top of the enclosed (incompressible) liquid. If p_{ext} is increased, by adding more weights, the pressure at all points within the liquid increases by the same amount.

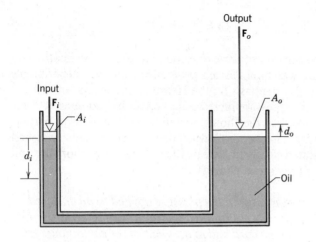

Figure 10 A hydraulic arrangement, used to magnify a force. The work, however, is not magnified, being the same for both the input and the output force.

By adding a little more lead shot to the piston, let us increase p_{ext} by an amount Δp_{ext}. The quantities ρ, g, and h in Eq. 8 remain unchanged so that the pressure change at P is

$$\Delta p = \Delta p_{ext}. \qquad (9)$$

This pressure change is independent of h so that it holds for all points within the closed vessel, as Pascal's principle states.

Pascal's Principle and the Hydraulic Lever. Figure 10 shows how Pascal's principle can be made the basis of a hydraulic lever. In operation, let an external force of magnitude F_i be exerted on the left-hand (or input) piston, whose area is A_i. To keep the system in equilibrium, an external load exerts a compensating force of magnitude F_o on the right-hand (or output) piston, whose area is A_o. We can write, for the change in fluid pressure associated with these applied forces,

$$\Delta p = \frac{F_i}{A_i} = \frac{F_o}{A_o}.$$

Thus,

$$F_o = F_i \frac{A_o}{A_i}. \qquad (10)$$

Equation 10 shows that the output force can be made larger than the input force if $A_o > A_i$, as is the case in Fig. 10.

If we move the input piston downward a distance d_i, the output piston moves upward a distance d_o, where

$$V = A_i d_i = A_o d_o.$$

Here V is the volume of the (incompressible) fluid that is displaced by either piston. We can write this equation as

$$d_o = d_i \frac{A_i}{A_o}. \qquad (11)$$

This shows that, if $A_o > A_i$ (as in Fig. 10), the output piston moves a smaller distance than the input piston does.

From Eqs. 10 and 11 we can write

$$W = F_i d_i = F_o d_o, \qquad (12)$$

which shows that the work W done *on* the input piston by the applied force is equal to the work W done *by* the output piston in lifting the load placed on it.

We see that a given force, exerted over a given distance, can be transformed to a larger force, exerted over a smaller distance. The product of force and distance remains unchanged so that the same work is done. However, there is often tremendous advantage in being able to exert the larger force. Most of us, for example, cannot lift an automobile and welcome the availability of a hydraulic jack, even though we have to pump the handle farther than the automobile rises. Note that, in this device, the displacement d_i in Fig. 10 is not accomplished in a single stroke but over a series of small strokes, each of which pumps more oil into the high-pressure chamber.

The hydraulic brakes on an automobile operate on the principle of Fig. 10, as does the familiar car lift at your local garage. Hydraulic devices are also commonly found on much heavy machinery (see Fig. 11) and on airplanes. In addition to their usefulness as hydraulic

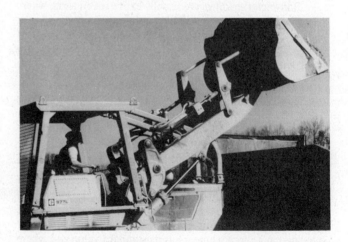

Figure 11 A power shovel. Count the hydraulic cylinders.

levers, such devices are also useful in transmitting forces to remote or inaccessible locations, such as from the cockpit of a jet plane to the tail surfaces.

16–7   Archimedes' Principle

Figure 12 shows a student in a swimming pool, manipulating a very thin plastic sack filled with water. She will find that it is in static equilibrium, tending neither to rise nor to sink. Yet the water in the sack has weight and — for that reason — should sink. The weight must be balanced by an upward force whose magnitude is equal to the weight of the water in the sack.

This upward *buoyant force* is exerted on the water in the sack by the water that surrounds the sack. It arises because — as we have already seen — the pressure in the water increases with depth below the surface so that the pressure near the bottom of the sack is greater than the pressure near the top. Figure 13a shows the forces acting at the hole in the water formerly occupied by the sack of Fig. 12. The buoyant force, which points up, is the vector sum of all these forces.

Let us fill the hole in Fig. 13a with a stone of exactly the same dimensions, as in Fig. 13b. *The same upward buoyant force that acted on the water-filled sack will act on the stone.* However, this force is too small to balance the weight of the stone so the stone will sink. Even though the stone sinks, the water's buoyant force reduces its apparent weight, making the stone easier to lift as long as it is under water.

If we now fill the hole of Fig. 13a with a block of wood of the same dimensions, as in Fig. 13c, the same upward buoyant force acts on the wood. This time, however, the upward force will be *greater* than the weight of

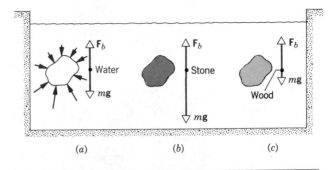

(a)   *(b)*   *(c)*

Figure 13 (*a*) The water surrounding the hole exerts forces on the boundary of the hole, the resultant being an upward buoyant force acting on whatever fills the hole. (*b*) For a stone of the same volume, the weight exceeds the buoyant force. (*c*) For a lump of wood of the same volume, the weight is less than the buoyant force.

the wood so that the wood will rise toward the surface. We summarize these facts by Archimedes' principle:

> *A body wholly or partially immersed in a fluid will be buoyed up by a force equal to the weight of the fluid that it displaces.*

Let us see how this principle can explain floating. As the piece of wood in Fig. 13c rises and breaks through the surface, it displaces less water. According to Archimedes' principle, its buoyant force (which you recall was greater than its weight) decreases. The wood will continue to rise out of the water until the buoyant force acting on it has decreased until it exactly equals its weight. It is then in static equilibrium; it is floating.

The Equilibrium of Floating Objects. Occasionally, sailing vessels or warships are modified, by adding taller masts or heavier guns, in such a way that they become top heavy and tend to capsize in moderately rough seas. Icebergs often tumble as they melt. All this suggests that torques, as well as forces, play a role in the equilibrium of floating objects.

As we have seen, the weight of a floating object (acting downward) is balanced by the buoyant force (acting upward). However, the points of action of these two forces are not the same. The weight acts at the center of mass of the floating object; the buoyant force acts at the center of mass of the hole in the water, a point called the *center of buoyancy.*

If a floating body is tilted by a small angle from its equilibrium position, the shape of the hole in the water changes and so then does the location of the center of

Figure 12 A thin-walled plastic sack of water is in static equilibrium. Its weight must be balanced by an upward force exerted on the sack by the water surrounding it.

Figure 14 A research vessel (FLIP) used to study waves in deep water. It is towed to its work station floating horizontally and, on station, pumps water into its stern tanks and flips to the position shown. It extends 300 ft below the surface.

buoyancy. For a floating object to be in stable equilibrium, its center of buoyancy must shift in such a way that the buoyant force (acting upward) and the weight (acting downward) provide a restoring torque, tending to return the body to its original upright position. If the torque acts in the opposite direction, the floating body will tip over. It is the job of the naval architects to make sure that this does not happen to a ship that they are designing!

Figure 14 shows an unusual research vessel, the Floating Laboratory Instrument Platform (FLIP) that has *two* stable equilibrium positions. It can float "normally" on the surface or, by pumping water into its stern tanks, can "flip" to the position shown. In this position, it extends 55 ft above the surface and 300 ft below, providing a stable instrument platform for the study of ocean waves.

Sample Problem 6 The "tip of the iceberg" in popular speech has come to mean a small visible fraction of something that is mostly hidden. For real icebergs, what is this fraction?

The weight of an iceberg of total volume V_i is

$$W_i = \rho_i V_i g,$$

where ρ_i ($= 917$ kg/m³) is the density of ice.

The weight of the displaced seawater, which is the buoyancy force F_b, is

$$W_w = F_b = \rho_w V_w g,$$

where ρ_w ($= 1024$ kg/m³) is the density of seawater and V_w is the volume of the displaced water, that is, the submerged volume of the iceberg. For the floating iceberg, the two forces are equal, or

$$\rho_i V_i g = \rho_w V_w g.$$

Using this equation, we find that the fraction we seek is

$$f = \frac{V_i - V_w}{V_i} = 1 - \frac{V_w}{V_i} = 1 - \frac{\rho_i}{\rho_w}$$

$$= 1 - \frac{917 \text{ kg/m}^3}{1024 \text{ kg/m}^3} = 0.10 \text{ or } 10\%. \quad \text{(Answer)}$$

Sample Problem 7 A spherical, helium-filled balloon has a radius R of 12 m. The balloon, support cables, and basket have a mass m of 196 kg. What maximum load M can the balloon carry? Take $\rho_{He} = 0.16$ kg/m³ and $\rho_{air} = 1.25$ kg/m³.

The weight of the displaced air, which is the buoyant force, and the weight of the helium in the bag are

$$W_{air} = \rho_{air} Vg \quad \text{and} \quad W_{He} = \rho_{He} Vg,$$

in which V ($= 4\pi R^3/3$) is the volume of the bag.

At balance, from Archimedes' principle,

$$W_{air} = W_{He} + mg + Mg$$

or

$$M = (\tfrac{4}{3}\pi)(R^3)(\rho_{air} - \rho_{He}) - m$$

$$= (\tfrac{4}{3}\pi)(12 \text{ m})^3(1.25 \text{ kg/m}^3 - 0.16 \text{ kg/m}^3) - 196 \text{ kg}$$

$$= 7690 \text{ kg.} \quad \text{(Answer)}$$

A body with this mass would weigh 17,000 lb at sea level.

16-8 Fluids in Motion

In mechanics, you were sometimes assigned a problem and told: "Neglect friction." That was a tacit admission that, if you had included friction, the problem would have been too hard. That is the case here. The motion of *real* fluids is complicated and not yet fully understood. We discuss instead the motion of an *ideal fluid* that is simpler to handle mathematically. Although our results

Figure 15   The smoke from a cigarette gains speed as it rises. At a certain speed, the flow changes from steady to turbulent.

may not agree fully with the behavior of real fluids, they will be close enough to be useful. Here are four assumptions that we make about our ideal fluid:

1.   Steady Flow.   In *steady flow* the velocity of the moving fluid at any given point does not change as time goes on, either in magnitude or in direction. The gentle flow of water near the center of a quiet stream is steady; that in a chain of rapids is not. Figure 15, which shows the smoke rising from a cigarette, shows a transition from steady flow to nonsteady or turbulent flow. The speed of the smoke particles increases as they rise and, at a certain critical speed, the flow changes its character from steady to nonsteady.

2.   Incompressible Flow.   We assume, as we have already done for fluids at rest, that our ideal fluid is incompressible. That is, its density has a constant value.

3.   Nonviscous Flow.   Viscosity in fluids is the analog of friction in solids. Both are mechanisms by which the kinetic energy of moving objects can be converted into thermal energy. In the absence of friction, a block could glide at constant speed along a horizontal surface. In the same way, an object moving through a nonviscous

fluid would experience no viscous drag force. Lord Rayleigh has pointed out that, in an ideal fluid, a ship's propeller would not work but, on the other hand, a ship (once set into motion) would not need a propeller!

4.   Irrotational Flow.   Although it need not concern us further, we also assume that the flow is *irrotational.* To test for this property, let a tiny grain of dust move with the fluid. Although this test body may (or may not) move in a circular path, in irrotational flow the test body will not rotate about an axis through its own center of mass. For a loose analogy, the motion of a Ferris wheel is rotational; that of its passengers is irrotational.

16-9   Streamlines and the Equation of Continuity

Figure 16 shows streamlines traced out by injecting dye into the moving fluid. A *streamline* is the path traced out by a tiny fluid element (which we may call a fluid "particle"). As the fluid particle moves, its velocity may change, both in magnitude and in direction. As Fig. 17 shows, its velocity vector at any point will always be tangent to the streamline at that point. Streamlines never cross because, if they did, a fluid particle arriving at the intersection would have to assume two different velocities simultaneously, an impossibility.

We can build up a *tube of flow* whose boundary is made up of streamlines. Such a tube acts like a pipe because any fluid particle that enters it cannot escape

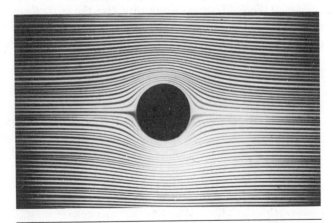

Figure 16   The steady flow of a fluid around a cylinder. An ideal fluid, flowing around a (nonrotating) cylinder in steady flow, exerts no force on it.

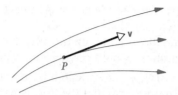

Figure 17 A fluid particle P traces out a streamline. The velocity of the particle is tangent to the streamline at every point.

through its walls; if it did, we would have a case of streamlines crossing each other.

Figure 18 shows two cross sections, of areas A_1 and A_2, along a thin tube of flow. Let us station ourselves at P and watch the fluid for a short time interval Δt. During this interval, a fluid particle will move a small distance $v_1 \Delta t$ and a volume ΔV of fluid, given by

$$\Delta V = A_1 v_1 \Delta t,$$

will pass through the area A_1.

The fluid is incompressible and cannot be created or destroyed. Thus, in this time interval, the same volume of fluid must pass point Q, further down the tube of flow, or

$$\Delta V = A_1 v_1 \Delta t = A_2 v_2 \Delta t.$$

We can write, for any point along the tube of flow,

$$\boxed{R = Av = \text{a constant,}} \qquad (13)$$

in which R, whose SI units are meters³ per second, is the *volume flow rate*. If we multiply R by the (constant) density of the fluid, we get a quantity $Av\rho$, the *mass flow rate* measured in kilograms per second, which also remains constant.

Equation 13 is called the *equation of continuity*. It tells us that, in the narrower parts of a tube of flow, where

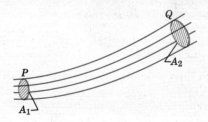

Figure 18 A tube of flow is defined by the streamlines that form its boundary. The flow rate of fluid must be the same for all cross sections of the tube of flow.

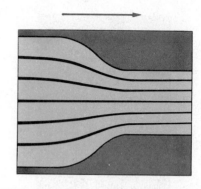

Figure 19 When a channel constricts, the streamlines draw closer together, signaling an increase in the fluid velocity. The arrow shows the direction of flow.

the streamlines are necessarily closer together, the flow will be faster, as Fig. 19 illustrates. Equation 13 is an expression of the law of conservation of mass in a form useful in fluid mechanics.

Sample Problem 8* The cross-sectional area A_0 of the aorta (the major blood vessel emerging from the heart) of a normal resting person is 3 cm² and the speed v_0 of the blood is 30 cm/s. A typical capillary (diameter $\approx$ 6 μm) has a cross-sectional area A of 3×10^{-7} cm² and a flow speed v of 0.05 cm/s. How many capillaries does such a person have?

All the blood that passes through the capillaries must have passed through the aorta so that, from Eq. 13,

$$A_0 v_0 = nAv,$$

where n is the number of capillaries. Solving for n yields

$$n = \frac{A_0 v_0}{Av} = \frac{(3 \text{ cm}^3)(30 \text{ cm/s})}{(3 \times 10^{-7} \text{ cm}^2)(0.05 \text{ cm/s})}$$

$$= 6 \times 10^9 \text{ or 6 billion.} \qquad \text{(Answer)}$$

You can easily show that the combined cross-sectional area of the capillaries is about 600 times the area of the aorta.

Sample Problem 9 Figure 20 shows how the stream of water emerging from a faucet "necks down" as it falls. The cross-sectional area A_0 is 1.2 cm² and that of A is 0.35 cm². The two levels are separated by a vertical distance h (=45 mm). At what rate does water flow from the tap?

---

* See "Life in Moving Fluids," by Steven Vogel, Princeton University Press, 1981, for a fascinating account of the role of fluid flow in biology.

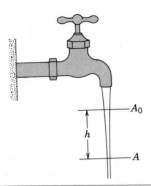

Figure 20 Sample Problem 9. As water falls from a tap, its speed increases. Because the flow rate must be the same at all cross sections, the stream must "neck down." (Effects associated with surface tension are neglected.)

From the equation of continuity (Eq. 13) we have

$$A_0 v_0 = A v, \tag{14}$$

where v_0 and v are the water velocities at the corresponding levels. From Eq. 20 of Chapter 2 we can also write, because the water is falling freely under gravity,

$$v^2 = v_0^2 + 2gh. \tag{15}$$

Eliminating v between Eqs. 14 and 15 and solving for v_0, we obtain

$$v_0 = \sqrt{\frac{2ghA^2}{A_0^2 - A^2}}$$

$$= \sqrt{\frac{(2)(9.8 \text{ m/s}^2)(0.045 \text{ m})(0.35 \text{ cm}^2)^2}{(1.2 \text{ cm}^2)^2 - (0.35 \text{ cm}^2)^2}}$$

$$= 0.286 \text{ m/s} = 28.6 \text{ cm/s}.$$

The volume flow rate R is then

$$R = A_0 v_0 = (1.2 \text{ cm}^2)(28.6 \text{ cm}^2/\text{s})$$
$$= 34 \text{ cm}^3/\text{s}. \qquad \text{(Answer)}$$

At this rate, it would take about 3 s to fill a 100-mL beaker.

16-10   Bernoulli's Equation

Bernoulli's equation, first developed by Daniel Bernoulli* in 1738, is not a new basic principle but is a statement of the conservation of energy in a form that is useful for problems involving fluids.

---

* The Bernoullis were a famous family of scientists and mathematicians, as the cartoon of Fig. 21 suggests.

"WHICH BERNOULLI DO YOU WISH TO SEE— 'HYDRODYNAMICS' BERNOULLI, 'CALCULUS' BERNOULLI, 'GEODESIC' BERNOULLI, 'LARGE NUMBERS' BERNOULLI OR 'PROBABILITY' BERNOULLI ?"

Figure 21 A cartoon by Sidney Harris, a noted chronicler of humor in science. ("Hydrodynamics" Bernoulli is our man.)

Figure 22 represents a tube of flow (or an actual pipe, for that matter) through which an ideal fluid is flowing at a steady rate. In a time interval Δt, suppose that a volume of fluid ΔV, shown colored in Fig. 22a, enters the tube at the left (or input) end and an identical volume, shown colored in Fig. 22b, emerges at the right (or output) end. The emerging volume must be the same as the entering volume because the fluid is incompressible, with an assumed constant density ρ.

Let y_1, v_1, and p_1 be the elevation, the speed, and the pressure of the fluid as it enters at the left and y_2, v_2, and p_2 be the corresponding quantities for the fluid as it emerges at the right. By applying the law of conservation of energy to the fluid, we shall show (see below) that these quantities are related by

$$\boxed{p_1 + \tfrac{1}{2}\rho v_1^2 + \rho g y_1 = p_2 + \tfrac{1}{2}\rho v_2^2 + \rho g y_2.} \tag{16}$$

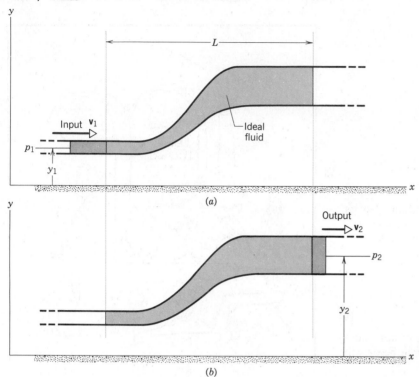

(a)

(b)

Figure 22 Fluid flows through a pipe at a steady rate. During a time interval Δt, the net effect is the transfer from the input end of an amount of fluid shown by the colored area in (a) to the output end, as shown in (b).

We can rewrite this as

$$p + \tfrac{1}{2}\rho v^2 + \rho g y = \text{a constant.} \qquad (17)$$

Equations 16 and 17 are equivalent forms of *Bernoulli's equation.* * Like the equation of continuity (Eq. 13), Bernoulli's equation is not a new principle but simply the reformulation of a familiar principle (the conservation of energy) in a form more suitable to the problem at hand. As a check, let us apply Bernoulli's equation to fluids at rest, by putting $v_1 = v_2 = 0$ in Eq. 16. The result is

$$p_2 = p_1 + \rho g(y_1 - y_2),$$

which is Eq. 4.

A major prediction of Bernoulli's equation emerges if we take y to be a constant ($y = 0$, say) so that the fluid does not change elevation as it flows. Equation 16 then becomes

---

* For irrotational flow (which we assume), the constant in Eq. 17 has the same value for all points within the tube of flow; the points do not have to lie along the same streamline. Similarly, the points 1 and 2 in Eq. 16 can lie anywhere within the tube of flow.

$$p_1 + \tfrac{1}{2}\rho v_1^2 = p_2 + \tfrac{1}{2}\rho v_2^2,$$

which tells us that:

If the speed of a fluid particle increases as it travels along a streamline, the pressure of the fluid must decrease, and conversely.

Put another way, where the streamlines are relatively close together (that is, where the velocity is relatively great), the pressure is relatively low, and conversely.

This result is perhaps the opposite of what you might expect. For example, if you put your hand outside the window of a car, you sense an *increase* of pressure associated with the relative speed of the moving air, not a decrease. The difficulty is that in "sensing" the pressure in this way you interfere with the flow. The pressure must be measured in ways that do not do so. If, for example, you crack the car window a small amount (which will not disturb the flow of the outside air) you will note that smoke generated inside the car drifts outside, in response to the *lower* outside pressure.

To demonstrate the speed–pressure relation in a simple way, hold a strip of paper just below your lower lip and blow gently. You have increased the velocity of the air over the top of the paper and therefore the pressure in that region is reduced. The pressure below the strip (where the velocity is zero) remains unchanged so that the strip rises toward the horizontal.

Bernoulli's equation is strictly valid only to the extent that the fluid is ideal. If viscous forces are present, thermal energy will be involved. We take no account of this in the derivation that follows, so that Bernoulli's equation must be used with great caution in such cases.

Proof of Bernoulli's Equation. Let us take as our system the entire volume of the (ideal) fluid shown in Fig. 22. We now apply the law of energy conservation to this system as it moves from its initial state (Fig. 22a) to its final state (Fig. 22b). We note that the part of the fluid lying between the two vertical planes separated by a distance L in Fig. 22 does not change its properties during this process; we need only be concerned with changes that take place at the input and the output ends.

We apply the law of energy conservation in the form of the work–energy theorem, or

$$W = \Delta K \quad \text{(work–energy theorem)}. \quad (18)$$

This tells us that the change in the kinetic energy of our system must equal the net work done on the system.

The change in kinetic energy occurs only at the ends and is

$$\Delta K = \tfrac{1}{2}\Delta m\, v_2^2 - \tfrac{1}{2}\Delta m\, v_1^2 = \tfrac{1}{2}\rho\, \Delta V(v_2^2 - v_1^2), \quad (19)$$

in which $\Delta m\, (= \rho\, \Delta V)$ is the mass of the fluid that enters at the input end and leaves at the output end.

The work done on the system arises from two sources. The work W_g done by the gravitational force to lift the colored fluid element in Fig. 22 vertically from the input level to the output level is

$$W_g = -\Delta m\, g(y_2 - y_1) = -\rho g\, \Delta V(y_2 - y_1). \quad (20)$$

This work is negative because the displacement (upward) and the gravitational force acting on the particle (downward) point in opposite directions.

Work W_p must also be done *on* the system (at the input end) to push it through the tube and *by* the system (at the output end) to push the fluid ahead of it forward. In general, we can write for work done under these circumstances

$$F\, \Delta x = (pA)(\Delta x) = (p)(A\, \Delta x) = p\, \Delta V,$$

in which a force F, acting on a fluid sample contained in a tube of area A, moves the fluid through a distance Δx.

The net work is then

$$W_p = -p_2\, \Delta V + p_1\, \Delta V = -(p_2 - p_1)\Delta V. \quad (21)$$

The work–energy theorem of Eq. 18 becomes

$$W = W_g + W_p = \Delta K.$$

Substituting from Eqs. 19, 20, and 21 yields

$$-\rho g\, \Delta V(y_2 - y_1) - \Delta V(p_2 - p_1) = \tfrac{1}{2}\rho\, \Delta V(v_2^2 - v_1^2).$$

This, after a slight rearrangement, is exactly Eq. 16, which we set out to prove.

16-11   Some Applications of Bernoulli's Equation

Popping Windows. If a high wind blows past a window, the pressure on the outside of the window is reduced and the window may break outward. This mechanism plays a role when flat roofs are blown off buildings in hurricanes; the roofs are, at least in part, pushed up by pressure from the stagnant air below. Although roofs are designed to withstand a relatively large *downward* pressure difference (due perhaps to snow loading), they are often not designed to withstand a large *upward* pressure difference.

The Venturi Meter. The Venturi meter (see Fig. 23) is a device used to measure the flow speed of a fluid in a pipe. At the throat, the area is reduced from A to a and the velocity is increased from v to V. Note that at the throat, where the velocity is greatest, the pressure is least, as Bernoulli's equation predicts. This seems quite proper, because the pressure difference is in the correct direction to speed up the flow. That is, a particle of fluid entering the throat from the left will be accelerated to the

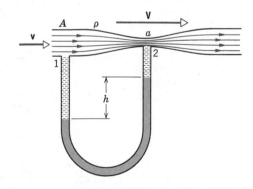

Figure 23 A Venturi meter, used to measure flow speeds in a pipe.

right by the pressure difference between the tube and the throat.

In Problem 80, we ask you to show, using Bernoulli's equation and the equation of continuity (Eq. 13), that

$$v = \sqrt{\frac{2a^2 \, \Delta p}{\rho(A^2 - a^2)}}, \qquad (22)$$

where ρ is the density of the fluid. The volume flow rate R can be found from $R = Av$ and the device can be calibrated to read (and record) this flow rate directly.

The Hole in the Water Tank. In the old West, a desperado fires a bullet into an open water tank, creating a hole a distance h below the water surface, as Fig. 24 shows. How fast is the water moving as it emerges from the hole?

We take the level of the hole as our reference level for measuring elevations and we note that the pressure at the top of the tank, and also at the hole, is atmospheric. Applying Bernoulli's equation (Eq. 16) gives us

$$p_0 + 0 + \rho g h = p_0 + \tfrac{1}{2}\rho v^2 + 0.$$

On the left, the 0 denotes that the velocity of fluid at the top of the tank (that is, the speed with which the level is falling) is negligible. The 0 on the right reminds us that the level of the hole is our reference level for measuring gravitational potential energy. Thus, we find

$$v = \sqrt{2gh}, \qquad (23)$$

which is the same speed that an object would acquire in falling from rest through a distance h.

An Airplane Wing. Figure 25 shows the streamlines about an airplane wing. There is no way to predict, from Bernoulli's equation alone, what the pattern of streamlines will be for a particular wing profile. Given the pat-

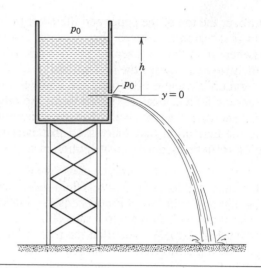

Figure 24 The water flows from the hole with the same speed that it would have acquired had it fallen a vertical distance h.

tern, however, we can note that it is consistent with the existence of an upward force **F** acting on the wing. As the relative spacing of the streamlines shows us, the air speed is greater over the top of the wing than over the bottom. Thus, from Bernoulli's equation, the pressure on the wing must be greater at the bottom than at the top. The force **F** may be resolved into a vertical upward force called the *lift* and a horizontal component, called the induced drag.

We can also interpret the lift force in terms of Newton's third law. As the streamlines show, the effect of the wing is to give the air stream a downward velocity component. The reaction force of the deflected air mass must then act on the wing to give it an equal and opposite

Figure 25 Streamlines showing the flow around an airplane wing. The airspeed below the wing is less than the air speed above the wing so that, from Bernoulli's principle, a dynamic lift force acts upward on the wing.

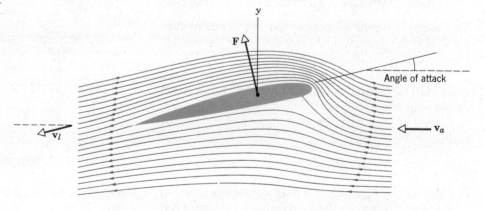

upward component. This interpretation of lift is particularly suitable for understanding the lift on the rotor blades of a helicopter. When the helicopter is hovering near the ground, the downward flow of air is evident to all who stand nearby.

The lift force that acts on an airplane wing (often called *dynamic lift*) must not be confused with the static or buoyant lift exerted, in accord with Archimedes' principle, on balloons or floating icebergs. Dynamic lift acts only when the object and the fluid stream are in relative motion.

16-12   The Flow of "Real" Fluids   (Optional)

The preceding sections dealt with the flow of an *ideal fluid,* its essential property being that it has zero viscosity. All real fluids are viscous and that property has an important influence on their behavior. Let us look at some examples.

The Boundary Layer.   One important effect of viscosity is that, at the surface of a solid object moving through a fluid, the molecules of the fluid tend to be carried along with that surface. Thus, we have a thin *boundary layer* surrounding a moving object, within which the fluid speed drops from its free-stream value to zero at the surface of the solid.

The existence of a boundary layer has many familiar consequences. For example, we might expect that dirt would not stick to a whirling fan blade but it does. The reason is that the air just above the surface of the blade is not moving with respect to the blade so that there is no mechanism for "blowing away" dirt particles. In the same way, you cannot blow fine dust from a table top; you must wipe it off. A dish cloth is far more effective in cleaning dishes than a simple rinse. Flowing water in a mountain stream does not itself carve away the stream bed through which it flows. The boundary layer ensures that the water is stagnant at its points of contact with the stream bed; it is the rock particles carried by the stream that do the carving.

Figure 26 shows how prairie dogs take advantage of the boundary layer to ventilate their burrows. By building a mound around one of the entrances they extend this entrance through the boundary layer so that it opens into the air stream. Other entrances, at ground level, open into stagnant air. The velocity difference between

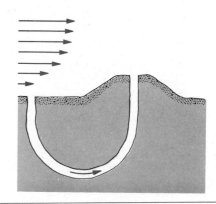

Figure 26   Prairie dogs use the boundary layer phenomenon and Bernoulli's principle to ventilate their burrows.

the two entrances creates a pressure difference (from Bernoulli's principle) and causes air flow through the burrow. If the conical mounds around certain entrances are destroyed, the animals will quickly rebuild them.

Dynamic Lift.   The range of a properly hit golf ball is greatly extended because of the *dynamic lift force* associated with its rotation. The boundary layer is intimately involved.

Figure 27a shows streamlines for the steady flow of air rushing past a stationary, nonrotating ball, at speeds low enough so that turbulence does not occur. Figure 27b shows streamlines for the air carried around by a stationary but rapidly rotating ball. Without viscosity and the boundary layer, the spinning ball could not carry air around in this way and this *circulation* (as it is called) would not exist. Golf balls are systematically roughened by means of dimples to increase this circulation—and the dynamic lift that results from it.

Figure 27c shows the effect of combining the circulation (resulting from the rotation of the ball) and the steady flow (resulting from the translation of the ball through the air). For the case shown, the two velocities add above the ball and subtract below. From the spacing of the resultant streamlines, we see that the velocity of air below the ball is less than that above the ball. From Bernoulli's equation, the pressure of air below the ball must then be greater than that above, so the ball experiences a dynamic lift force.

At the speeds of golf balls and pitched baseballs in normal play, the flow is not steady, as Fig. 27 suggests, but turbulent. The boundary layer "breaks away" from the ball on its back side, forming a turbulent wake; see

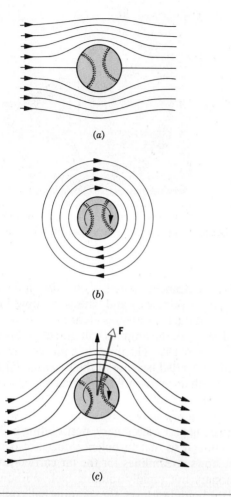

Figure 27 (*a*) Streamline flow around a (nonrotating) ball. (*b*) The circulation of air around a rotating ball, made possible by the boundary layer. (*c*) The combined effects of both motions. From Bernoulli's principle, we see that a dynamic lift force acts upward on the ball.

the photo at the start of this chapter. However, dynamic lift, associated with the circulation, still exists. The effect of the rotation of the ball is to steer the turbulent wake in the direction in which the back of the ball is rotating. The ball deflects in the opposite direction, that is, in the direction in which the front of the ball is rotating. For golf balls, the desired axis of rotation is horizontal, so that the deflection force is upward. If the golfer, by poor technique, partially misaligns the spin axis, the result is a hook or a slice. For baseballs, the desired axis of rotation is usually vertical when throwing a curve, so that the deflection of the ball is sideways.

Flight and Dynamic Lift. Dynamic lift is always associated with *circulation,* that is, with flow such as that suggested by the streamlines of Fig. 27*b*. When an airplane starts from rest and rolls down the runway, such a pattern of circulation quickly develops around its wings. A streamline pattern such as that of Fig. 25 is set up, providing the lift that allows the airplane to rise into the air.

Unlike the ball of Fig. 27, however, the airplane wing does not rotate, nor does any particle of air actually move in a closed loop around the wing. Nevertheless, there is circulation, in this sense: If you draw a closed loop around the wing of Fig. 25 and traverse it in a counterclockwise direction, you will (speaking loosely) be traveling farther with the wind over the top surface than you will against the wind under the bottom surface. Your net experience in traversing the closed loop will be that of a counterclockwise circulation of air.*

---

* For a detailed explanation of how airplanes fly, see "The Science of Flight," by Peter P. Wegener, *American Scientist,* May–June 1986, p. 268.

REVIEW AND SUMMARY

The *density* of any material is defined as its mass per unit volume:

Density

$$\rho = \frac{\Delta m}{\Delta V}.$$

[1]

A fluid is a substance that can flow; it cannot sustain static shear. It can, however, maintain static compression, which is described in terms of *pressure p*:

Fluid Pressure

$$p = \frac{\Delta F}{\Delta A},$$

[2]

in which ΔF is the force acting on a surface element of area ΔA. The force resulting from fluid pressure acts in a direction normal to the fluid surface and, at a particular point in a fluid, its magnitude is independent of the orientation of that surface. *Gauge* pressure is the difference between actual pressure and atmospheric pressure. Many common pressure-measuring devices (pressure gauges) measure gauge pressure directly.

Pressure Units

The SI unit for pressure is the *pascal* ($= 1$ N/m^2). Other units are 1 atm $= 1.01 \times 10^5$ Pa $=$ 760 torr $= 14.7$ lb/in^2. Sample Problem 1 shows sample calculations using pressure and density.

Pressure in a fluid at rest varies with vertical position y. For y measured positive up

Pressure Variation with Height and Depth

$$p_2 = p_1 + \rho g(y_1 - y_2). \tag{4}$$

The pressure is the same for all points at the same level. If h is the *depth* of a fluid sample below some reference level at which the pressure is p_0, Eq. 4 becomes

$$p = p_0 + \rho g h. \tag{5}$$

Sample Problems 2–5 illustrate these ideas.

Pascal's Principle

Pascal's principle, which can be derived from Eq. 4, states that a change in the pressure applied to an enclosed fluid is transmitted undiminished to every portion of the fluid and to the walls of the containing vessel.

Archimedes' Principle

The surface of an immersed object is acted on by forces associated with the fluid pressure. The net force (called the *buoyant* force) acts vertically up through the center of gravity of the displaced fluid (the *center of buoyancy*). Archimedes' principle states that the magnitude of the buoyant force is equal to the weight of the fluid displaced by the object; see Sample Problems 6 and 7.

Fluid Dynamics

Our treatment of *fluid dynamics* is limited to ideal (incompressible with no viscous friction) fluids in steady, irrotational flow. A *streamline* is the path followed by individual fluid particles. A *tube of flow* is a bundle of streamlines. The conservation of mass shows that the flow within any tube of flow obeys the *equation of continuity:*

The Continuity Equation

$$R = Av = \text{a constant}, \tag{13}$$

in which R is the *volume flow rate,* A the cross-sectional area of the tube of flow, and v the speed of the fluid, assumed to be constant across the tube. The *mass flow rate* $Av\rho$ is also constant. See Sample Problems 8 and 9 for examples.

Bernoulli's Equation

The application of Newton's laws to steady, incompressible, nonviscous flow leads to *Bernoulli's equation:*

$$p + \tfrac{1}{2}\rho v^2 + \rho gy = \text{a constant} \tag{17}$$

along any tube of flow. Several applications of this important result are discussed in Section 16–11.

The pressure within real fluids is significantly affected by viscosity and turbulence. These lead to boundary layer phenomena and dynamic lift, both discussed in Section 16–12.

QUESTIONS

1. Explain how it can be that pressure is a scalar quantity when forces, which are vectors, can be produced by the action of pressures.

2. Can you assign a coefficient of static friction between two surfaces, one of which is a fluid surface?

3. Make an estimate of the average density of your body. Explain a way in which you could get an accurate value using ideas in this chapter.

4. In Sample Problem 2 of Chapter 20, we shall learn that an overpressure of only 28 Pa corresponds to the threshold of pain for intense sound. Yet a diver 5 ft below the surface of the

water experiences a much greater pressure than this and feels no pain. Why this difference?

5. Explain the pressure variations in your blood as it circulates through your body.

6. Explain why one could lie on a bed of nails without pain.

7. Explain the statement "water seeks its own level."

8. Water is poured to the same level in each of the vessels shown, all having the same base area (Fig. 28). If the pressure is the same at the bottom of each vessel, the force experienced by the base of each vessel is the same. Why then do the three

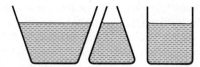

Figure 28 Question 8.

vessels have different weights when put on a scale? This apparently contradictory result is commonly known as the *hydrostatic paradox*.

9. Does Archimedes' principle hold in a vessel in free fall or in a satellite moving in a circular orbit?

10. A spherical bob made of cork floats half submerged in a pot of tea at rest on the earth. Will the cork float or sink aboard a spaceship (*a*) coasting in free space and (*b*) on the surface of Jupiter?

11. How does a suction cup work?

12. Is the buoyant force acting on a submerged submarine the same at all depths?

13. What, if anything, happens to the buoyant force acting on a helium-filled balloon if you replace the helium with the same volume of hydrogen? (Hydrogen is less dense than helium.)

14. A block of wood floats in a pail of water in an elevator. When the elevator starts from rest and accelerates down, does the block float higher above the water surface?

15. Two identical buckets are filled to the brim with water, but one has a block of wood floating in the water. Which bucket, if either, is heavier?

16. "A hot air balloon rises because the heating causes the pressure of the air in the balloon envelope to rise above the external atmospheric pressure, thus providing the necessary lift." What do you think of this sentence?

17. Can you sink an iron ship by siphoning seawater into it? (See Problem 79.)

18. Scuba divers are warned not to hold their breath when swimming upward. Why?

19. A beaker is exactly full of liquid water at its freezing point and has an ice cube floating in it, also at the freezing point. As the cube melts, what happens to the water level in these three cases: (*a*) The cube is solid ice; (*b*) The cube contains some grains of sand; (*c*) The cube contains some bubbles?

20. Although parachutes are supposed to brake your fall, they are often designed with a hole at the top. Explain why.

21. A ball floats on the surface of water in a container exposed to the atmosphere. Will the ball remain immersed at its former depth or will it sink or rise somewhat if (*a*) the container is covered and the air is removed or (*b*) the container is covered and the air is compressed?

22. Explain why an inflated balloon will only rise to a definite height once it starts to rise, whereas a submarine will always

sink to the very bottom of the ocean once it starts to sink, if no changes are made.

23. Why does a balloon weigh the same when empty as when filled with air at atmospheric pressure? Would the weights be the same if measured in a vacuum?

24. During World War II, a damaged freighter that was barely able to float in the North Sea steamed up the Thames estuary toward the London docks. It sank before it arrived. Why?

25. Is it true that a floating object will only be in stable equilibrium if its center of buoyancy lies above its center of gravity? Illustrate with examples.

26. Why will a sinking ship often turn over as it becomes immersed in water?

27. A barge filled with scrap iron is in a canal lock. If the iron is thrown overboard into the water, what happens to the water level in the lock? What if it is thrown onto the land beside the canal?

28. A bucket of water is suspended from a spring balance. Does the balance reading change when a piece of iron suspended from string is immersed in the water? When a piece of cork is put in the water?

29. Why does a uniform wooden stick or log float horizontally? If enough iron is added to one end, it will float vertically. Explain this also.

30. Although there are practical difficulties, it is possible in principle to float an ocean liner in a few barrels of water. How would you go about doing this?

31. Explain why a thin-walled pipe will burst more easily if, when there is a pressure differential between inside and outside, the excess pressure is on the outside.

32. Describe the forces acting on an element of fluid as it flows through a pipe of nonuniform cross section.

33. Explain why the height of the liquid in the standpipes indicates that the pressure drops along the channel, even though the channel has a uniform cross section and the flowing liquid is incompressible (Fig. 29).

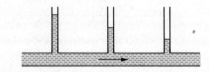

Figure 29 Question 33.

34. Explain why a taller chimney creates a better draft for taking the smoke out of a fireplace. Why doesn't the smoke pour into the room containing the fireplace?

35. In a lecture demonstration a Ping-Pong ball is kept in midair by a vertical jet of air. Is the equilibrium stable, unstable, or neutral? Explain.

36. Two rowboats moving parallel to one another in the same direction are pulled toward one another. Two automobiles moving parallel are also pulled together. Explain such phenomena using Bernoulli's equation.

37. Explain the action of a parachute in retarding free fall using Bernoulli's equation.

38. Why does a stream of water from a faucet become narrower as it falls?

39. Sometimes people remove letters from envelopes by cutting a sliver from a narrow end, holding it firmly, and blowing toward it. Explain, using Bernoulli's equation, why this procedure is successful.

40. On takeoff, would it be better for an airplane to move into the wind or with the wind? On landing?

41. Explain how the difference in pressure between the lower and upper surfaces of an airplane wing depend on the altitude of the moving plane.

42. Explain how accumulation of ice on an airplane wing may change its shape in such a way that its lift is greatly reduced.

43. How is an airplane able to fly upside down?

44. Explain why the destructive effect of a tornado (twister) is greater near the center of the disturbance than near the edge.

45. In steady flow, the velocity vector at any point is constant. Can there then be accelerated motion of the fluid particles? Discuss.

46. Explain why you cannot remove the filter paper from the funnel of Fig. 30 by blowing into the narrow end.

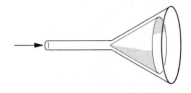

Figure 30 Question 46.

47. According to Bernoulli's equation, an increase in velocity should be associated with a decrease in pressure. Yet, when you put your hand outside the window of a car, increasing the speed at which the air flows by, you sense an *increase* in pressure. Is this a violation of Bernoulli's equation?

48. Why is it that the presence of the atmosphere reduces the maximum range of some objects (for example, tennis balls) but increases the maximum range of others (for example, Frisbees and golf balls)?

49. A discus can be thrown farther *against* a 25-mi/h wind than with it. What is the explanation? (*Hint:* Think about dynamic lift and drag.)

50. Explain why golf balls are dimpled.

EXERCISES AND PROBLEMS

Section 16–3 Density and Pressure

1E. Convert a density of 1.0 g/cm³ to kg/m³.

2E. Three liquids that will not mix are poured into a cylindrical container. The amounts and densities of the liquids are 0.50 L, 2.6 g/cm³; 0.25 L, 1.0 g/cm³; and 0.40 L, 0.80 g/cm³. What is the total force acting on the bottom of the container? One liter = 1L = 1000 cm³. (Ignore the contribution due to the atmosphere.)

3E. Find the pressure increase in the fluid in a syringe when a nurse applies a force of 9.5 lb (= 42 N) to the syringe's piston of radius 1.1 cm.

4E. You inflate the front tires on your car to 28 psi. Later, you measure your blood pressure, obtaining a reading of 120/80, the readings being in mm Hg. In metric countries (which is to say, most of the rest of the world), these quantities are customarily reported in kilopascals (kPa). What are (*a*) your tire pressure reading and (*b*) your blood pressure reading in these units?

5E. An office window is 3.4 m by 2.1 m. As a result of the passage of a storm, the outside air pressure drops to 0.96 atm, but inside the pressure is held at 1.0 atm. What net force pushes out on the window?

6E. A fish maintains its depth in fresh water by adjusting the air content of porous bone or air sacs to make its density the same as that of the water. Suppose that with its air sacs collapsed a fish has a density of 1.08 g/cm³. To what fraction of its expanded body volume must the fish inflate the air sacs to reduce its density to that of water?

7P. An airtight box having a lid with an area of 12 in.² is partially evacuated. If a force of 108 lb is required to pull the lid off the box and the outside atmospheric pressure is 15 lb/in.², what is the pressure in the box?

8P. The total effective blade area of a small windmill is 4.6 m². (*a*) What is the maximum possible power output of the windmill generator when a steady wind is blowing at 6.7 m/s (= 15 mi/h)? (*b*) What does this power become if the windspeed increases by 10%? The density of the air is 1.2 kg/m³.

9P. In 1654 Otto von Guericke, burgomaster of Magdeburg and inventor of the air pump, gave a demonstration before the Imperial Diet in which two teams of eight horses could not pull

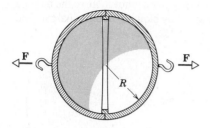

Figure 31   Problem 9.

apart two evacuated brass hemispheres. (*a*) Show that the force *F* required to pull apart the hemispheres is $F = \pi R^2 \Delta p$, where *R* is the (outside) radius of the hemispheres and Δp is the difference in pressure outside and inside the sphere (Fig. 31). (*b*) Taking *R* equal to 1.0 ft and the inside pressure as 0.10 atm, what force would the team of horses have had to exert to pull apart the hemispheres? (*c*) Why are two teams of horses used? Would not one team prove the point just as well?

Section 16–4   Fluids at Rest

10E.   Calculate the hydrostatic difference in blood pressure between the brain and the foot in a person of height 1.83 m. The density of blood is 1.06×10^3 kg/m³.

11E.   Find the total pressure, in pounds per square inch and in pascals, 500 ft (= 150 m) below the surface of the ocean. The density of seawater is 64.3 lb/ft³ (= 1.03 g/cm³) and the atmospheric pressure at sea level is 14.7 lb/in.² (= 1.01×10^5 Pa).

12E.   The sewer outlets of a house constructed on a slope are 8.2 m below street level. If the sewer is 2.1 m below street level, find the minimum pressure differential that must be created by the sewage pump to transfer waste of average density 900 kg/m³.

13E.   Figure 32 displays the phase diagram of carbon, showing the ranges of temperature and pressure in which carbon will crystallize either as diamond or graphite. What is the

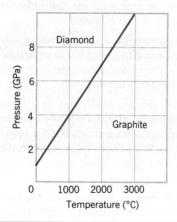

Figure 32   Exercise 13.

minimum depth at which diamonds can form if the local temperature is 1000°C and the subsurface rocks have density 3.1 g/cm³? Assume that, as in a fluid, the pressure is due to the weight of material lying above.

14E.   The human lungs can operate against a pressure differential of up to one-twentieth of an atmosphere. If a diver uses a snorkel (long tube) for breathing, how far below water level can he swim?

15E.   A swimming pool has the dimensions 80 ft × 30 ft × 8.0 ft. (*a*) When it is filled with water, what is the force (resulting from the water alone) on the bottom, on the end, and on the sides? (*b*) If you are concerned with whether or not the concrete walls and floor will collapse, is it appropriate to take the atmospheric pressure into account?

16E.   (*a*) Find the total weight of water on top of a nuclear submarine at a depth of 200 m, assuming that its (cross-sectional) hull area is 3000 m². (*b*) What water pressure would a diver experience at this depth? Express your answer in atmospheres. Do you think that occupants of a damaged submarine at this depth could escape without special equipment? The density of seawater is 1.03 g/cm³.

17E.   Crew members attempt to escape from a damaged submarine 100 m below the surface. What force must they apply to a pop-out hatch, which is 1.2 m by 0.60 m, to push it out? The density of the ocean water is 1025 kg/m³.

18E.   A simple U-tube contains mercury. When 11.2 cm of water is poured into the right arm, how high does the mercury rise in the left arm from its initial level?

19P.   Two identical cylindrical vessels with their bases at the same level each contain a liquid of density ρ. The area of either base is *A*, but in one vessel the liquid height is h_1 and in the other h_2. Find the work done by gravity in equalizing the levels when the two vessels are connected.

20P.   A cylindrical barrel has a narrow tube fixed to the top, as shown with dimensions in Fig. 33. The vessel is filled with water to the top of the tube. Calculate the ratio of the hydrostatic force exerted on the bottom of the barrel to the weight of the water contained inside. Why is the ratio not equal to one? (Ignore the presence of the atmosphere.)

21P.   (*a*) Consider a container of fluid subject to a *vertical upward* acceleration *a*. Show that the pressure variation with depth in the fluid is given by

$$p = \rho h(g + a),$$

where *h* is the depth and ρ is the density. (*b*) Show also that if the fluid as a whole undergoes a *vertical downward* acceleration *a*, the pressure at a depth *h* is given by

$$p = \rho h(g - a).$$

(*c*) What is the state of affairs in free fall?

22P.   In analyzing certain geological features of the earth, it is often appropriate to assume that the pressure at some horizon-

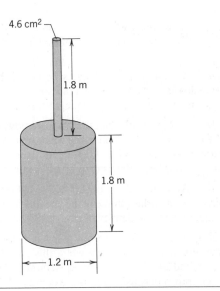

Figure 33   Problem 20.

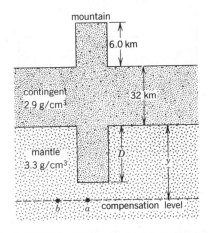

Figure 34   Problem 22.

tal *level of compensation,* deep in the earth, is the same over a large region and is equal to that exerted by the weight of the overlying material. That is, the pressure on the level of compensation is given by the hydrostatic (fluid) pressure formula. This requires, for example, that mountains have low-density *roots;* see Fig. 34. Consider a mountain 6.0 km high. The continental rocks have a density of 2.9 g/cm³; beneath the continent is the mantle, with a density of 3.3 g/cm³. Calculate the depth D of the root. (*Hint:* Set the pressure at points a and b equal; the depth y of the level of compensation will cancel out.)

23P. In Fig. 35, the ocean is about to overrun the continent. Find the depth h of the ocean using the level of compensation method shown in Problem 22.

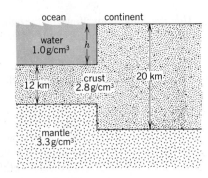

Figure 35   Problem 23.

24P. Water stands at a depth D behind the vertical upstream face of a dam, as shown in Fig. 36. Let W be the width of the dam. (*a*) Find the resultant horizontal force exerted on the dam by the gauge pressure of the water and (*b*) the net torque owing to the gauge pressure of the water exerted about a line through O parallel to the width of the dam. (*c*) Find the moment arm of the resultant hydrostatic force.

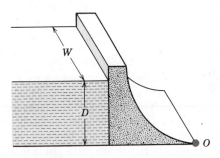

Figure 36   Problem 24.

25P. Figure 37 shows a dam and part of the freshwater reservoir backed up behind it. The dam is made of concrete, density

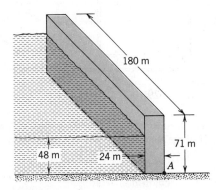

Figure 37   Problem 25.

3.2 g/cm³, and has the dimensions shown on the figure. (a) The force exerted by the water pushes horizontally on the dam face, and this is resisted by the force of static friction between the dam and the bedrock foundation on which it rests. The coefficient of friction is 0.47. Calculate the factor of safety against sliding, that is, the ratio of the maximum possible friction force to the force exerted by the water. (b) The water also tries to rotate the dam about a line running along the base of the dam through point A; see Problem 24. The torque resulting from the weight of the dam acts in the opposite sense. Calculate the factor of safety against rotation, that is, the ratio of the torque owing to the weight of the dam to the torque exerted by the water.

26P. A U-tube is filled with a single homogeneous liquid. The liquid is temporarily depressed in one side by a piston. The piston is removed and the level of the liquid in each side oscillates. Show that the period of oscillation is $\pi\sqrt{2L/g}$, where L is the total length of the liquid in the tube.

Section 16–5 Measuring Pressure

27E. Calculate the height of a column of water that gives a pressure of 1 atm at the bottom. Assume that $g = 9.8$ m/s².

28P. What would be the height of the atmosphere if the air density (a) were constant and (b) decreased linearly to zero with height? Assume a sea-level density of 1.3 kg/m³.

Section 16–6 Pascal's Principle

29E. A piston of small cross-sectional area a is used in a hydraulic press to exert a small force f on the enclosed liquid. A connecting pipe leads to a larger piston of cross-sectional area A (Fig. 38). (a) What force F will the larger piston sustain? (b) If the small piston has a diameter of 1.5 in. and the large piston one of 21 in., what weight on the small piston will support 2.0 tons on the large piston?

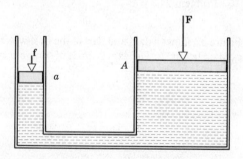

Figure 38 Exercise 29.

30E. In the hydraulic press of Exercise 29, through what distance must the large piston be moved to raise the small piston a distance of 3.5 ft?

Section 16–7 Archimedes' Principle

31E. A tin can has a total volume of 1200 cm³ and a mass of 130 g. How many grams of lead shot could it carry without sinking in water? The density of lead is 11.4 g/cm³.

32E. A boat floating in fresh water displaces 8000 lb of water. (a) How many pounds of water would this boat displace if it were floating in salt water of density 68.6 lb/ft³? (b) Would the volume of water displaced change? If so, by how much?

33E. About one-third of the body of a physicist swimming in the Dead Sea will be above the water line. Assuming that the human body density is 0.98 g/cm³, find the density of the water in the Dead Sea. Why is it so much greater than 1.0 g/cm³?

34E. An iron anchor appears 200 N lighter in water than in air. (a) What is the volume of the anchor? (b) How much does it weigh in air? The density of iron is 7870 kg/m³.

35E. An object hangs from a spring balance. The balance registers 30 N in air, 20 N when this object is immersed in water, and 24 N when the object is immersed in another liquid of unknown density. What is the density of the other liquid?

36E. A cubical object of dimensions $L = 2.0$ ft on a side and weight $W = 1000$ lb in a vacuum is suspended by a rope in an open tank of liquid of density $\rho = 2.0$ slugs/ft³ as in Fig. 39. (a) Find the total downward force exerted by the liquid and the atmosphere on the top of the object. (b) Find the total upward force on the bottom of the object. (c) Find the tension in the rope. (d) Calculate the buoyant force on the object using Archimedes' principle. What relation exists among all these quantities?

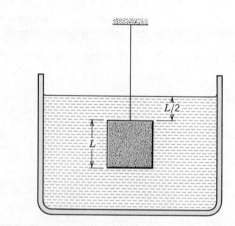

Figure 39 Exercise 36.

37E. A block of wood floats in water with two-thirds of its volume submerged. In oil it has 0.90 of its volume submerged. Find the density of (a) the wood and (b) the oil.

38E. It has been proposed to move natural gas from the North Sea gas fields in huge dirigibles, using the gas itself to provide lift. Calculate the force required to tether such an airship to the ground for off-loading when it is fully loaded with

1.0×10^6 m³ of gas at a density of 0.80 kg/m³. (The weight of the airship is negligible by comparison.)

39E. The Goodyear blimp *Columbia* (see Fig. 40) is cruising slowly at low altitude, filled as usual with helium gas. Its maximum useful payload, including crew and cargo, is 1280 kg. How much more payload could the *Columbia* carry if you replaced the helium with hydrogen? Why not do it? The volume of the helium-filled interior space is 5000 m³. The density of helium gas is 0.16 kg/m³ and the density of hydrogen is 0.081 kg/m³.

Figure 40 Exercise 39.

40E. A helium balloon is used to lift a 40-kg payload to an altitude of 27 km, where the air density is 0.035 kg/m³. The balloon has a mass of 15 kg and the density of the gas in the balloon is 0.0051 kg/m³. What is the volume of the balloon? Neglect the volume of the payload.

41P. A hollow sphere of inner radius 8.0 cm and outer radius 9.0 cm floats half submerged in a liquid of density 800 kg/m³. (a) What is the mass of the sphere? (b) Calculate the density of the material of which the sphere is made.

42P. A hollow spherical iron shell floats almost completely submerged in water. The outer diameter is 60 cm and the density of iron is 7.87 g/cm³. Find the inner diameter.

43P. An iron casting containing a number of cavities weighs 6000 N in air and 4000 N in water. What is the volume of the cavities in the casting? The density of iron is 7.87 g/cm³.

44P. (a) What is the minimum area of a block of ice 0.30 m thick floating on water that will hold up an automobile of mass = 1100 kg? (b) Does it matter where the car is placed on the block of ice?

45P. Three children each of weight 80 lb make a log raft by lashing together logs of diameter 1.0 ft and length 6.0 ft. How many logs will be needed to keep them afloat? Take the density of wood to be 50 lb/ft³.

46P. Assume the density of brass weights to be 8.0 g/cm³ and that of air to be 0.0012 g/cm³. What percent error arises from neglecting the buoyancy of air in weighing an object of mass m and density ρ on a beam balance?

47P. A car has a total mass of 1800 kg. The volume of air space in the passenger compartment is 5.0 m³. The volume of the motor and front wheels is 0.75 m³, and the volume of the rear wheels, gas tank and luggage is 0.80 m³. Water cannot enter these areas. The car is parked on a hill; the handbrake cable snaps and the car rolls down the hill into a lake; see Fig. 41. (a) At first, no water enters the passenger compartment. How much of the car, in cubic meters, is below the water surface with the car floating as shown? (b) As water slowly enters, the car sinks. How many cubic meters of water are in the car as it disappears below the water surface? (The car remains horizontal, owing to a heavy load in the trunk.)

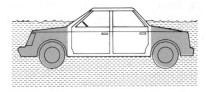

Figure 41 Problem 47.

48P. A block of wood has a mass of 3.67 kg and a density of 600 kg/m³. It is to be loaded with lead so that it will float in water with 0.90 of its volume immersed. What mass of lead is needed (a) if the lead is on top of the wood and (b) if the lead is attached below the wood? The density of lead is 1.13×10^4 kg/m³.

49P. You place a glass beaker, partially filled with water, in a sink (Fig. 42). It has a mass of 390 g and an interior volume of 500 cm³. You now start to fill the sink with water and you find, by experiment, that if the beaker is less than half full, it will float; but if it is more than half full, it remains on the bottom of the sink as the water rises to its rim. What is the density of the material of which the beaker is made?

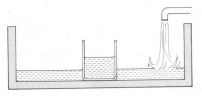

Figure 42 Problem 49.

50P. A cylindrical wooden rod is loaded with lead at one end so that it floats upright in water as in Fig. 43. The length of the submerged portion is $l = 2.5$ m. The rod is set into vertical oscillation. (*a*) Show that the oscillation is simple harmonic. (*b*) Find the period of the oscillation. Neglect the fact that the water has a damping effect of the motion.

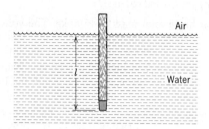

Figure 43   Problem 50.

51P*. The tension in a string holding a solid block below the surface of a liquid (of density greater than the solid) is T_0 when the containing vessel (Fig. 44) is at rest. Show that the tension T, when the vessel has an upward vertical acceleration a, is given by $T_0(1 + a/g)$.

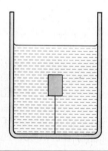

Figure 44   Problem 51.

Section 16–9   Streamlines and the Equation of Continuity

52E. Figure 45 shows the confluence of two streams to form a river. One stream has a width of 8.2 m, depth of 3.4 m, and current speed of 2.3 m/s. The other stream is 6.8 m wide, 3.2 m deep, and flows at 2.6 m/s. The width of the river is 10.5 m and the current speed is 2.9 m/s. What is its depth?

53E. A $\frac{3}{4}$-in. (inside diameter) water pipe is coupled to three $\frac{1}{2}$-in. pipes. (*a*) If the flow rates in the three pipes are 7.0, 5.0, and 3.0 gal/min, what is the flow rate in the $\frac{3}{4}$-in. pipe? (*b*) How will the speed of the water in the $\frac{3}{4}$-in. pipe compare with the speed in the pipe carrying 7.0 gal/min?

54E. A garden hose having an internal diameter of 0.75 in. is connected to a lawn sprinkler that consists merely of an enclosure with 24 holes, each 0.050 in. in diameter. If the water

Figure 45   Exercise 52.

in the hose has a speed of 3.0 ft/s, at what speed does it leave the sprinkler holes?

55P. Water is pumped steadily out of a flooded basement at a speed of 5.0 m/s through a uniform hose of radius 1.0 cm. The hose passes out through a window 3.0 m above the water line. How much power is supplied by the pump?

56P. A river 20 m wide and 4.0 m deep drains a 3000-km² land area in which the average precipitation is 48 cm/y. One-fourth of this subsequently returns to the atmosphere by evaporation, but the remainder ultimately drains into the river. What is the average speed of the river current?

Section 16–11   Some Applications of Bernoulli's Equation

57E. Water is moving with a speed of 5.0 m/s through a pipe with a cross-sectional area of 4.0 cm². The water gradually descends 10 m as the pipe increases in area to 8.0 cm². (*a*) What is the speed of flow at the lower level? (*b*) If the pressure at the upper level is 1.50×10^5 Pa, what is the pressure at the lower level?

58E. Models of torpedoes are sometimes tested in a horizontal pipe of flowing water, much as a wind tunnel is used to test model airplanes. Consider a circular pipe of internal diameter 25 cm and a torpedo model, aligned along the axis of the pipe, with a diameter of 5.0 cm. The torpedo is to be tested with water flowing past it at 2.5 m/s. (*a*) With what speed must the water flow in the unconstricted part of the pipe? (*b*) What will the pressure difference be between the constricted and unconstricted parts of the pipe?

59E. A water intake at a pump storage reservoir (see Fig. 46) has a cross-sectional area of 8.0 ft². The water flows in at a speed of 1.33 ft/s. At the generator building 600 ft below the

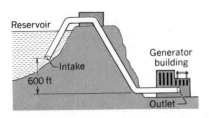

Figure 46   Exercise 59.

intake point, the cross-sectional area is 0.34 ft² and the water flows out at 31.0 ft/s. What is the difference in pressure, in pounds per square inch, between inlet and outlet?

60E. A water pipe having a 1.0-in. inside diameter carries water into the basement of a house at a velocity of 3.0 ft/s at a pressure of 25 lb/in.² If the pipe tapers to $\frac{1}{2}$ in. and rises to the second floor 25 ft above the input point, what are (a) the velocity and (b) the water pressure there?

61E. How much work is done by pressure in forcing 1.4 m³ of water through a 13-mm internal diameter pipe if the difference in pressure at the two ends of the pipe is 1.0 atm?

62E. In a horizontal oil pipeline of constant cross-sectional area, the pressure decrease between two points 1000 ft apart is 5.0 lb/in.² What is the energy loss per cubic foot of oil per foot?

63E. A tank of large area is filled with water to a depth $D = 1.0$ ft. A hole of cross section $A = 1.0$ in.² in the bottom of the tank allows water to drain out. (a) What is the rate at which water flows out in cubic feet per second? (b) At what distance below the bottom of the tank is the cross-sectional area of the stream equal to one-half the area of the hole?

64E. Suppose that two tanks, 1 and 2, each with a large opening at the top, contain different liquids. A small hole is made in the side of each tank at the same depth h below the liquid surface, but the hole in tank 1 has half the cross-sectional area of the hole in tank 2. (a) What is the ratio ρ_1/ρ_2 of the densities of the fluids if it is observed that the mass flux is the same for the two holes? (b) What is the ratio of the flow rates (volume flux) from the two tanks? (c) To what height above the hole in the second tank should fluid be added or drained to equalize the flow rates?

65E. Air flows over the top of an airplane wing, area A, with speed v_t and past the underside of the wing with speed v_u. Show that Bernoulli's equation predicts that the upward lift force L on the wing will be

$$L = \tfrac{1}{2}\rho A(v_t^2 - v_u^2),$$

where ρ is the density of the air.

66E. If the speed of flow past the lower surface of a wing is 110 m/s, what speed of flow over the upper surface will give a lift of 900 Pa? Take the density of air to be 1.3×10^{-3} g/cm³. See Exercise 65.

67E. An airplane has a wing area (each wing) of 10 m². At a certain airspeed, air flows over the upper wing surface at 48 m/s and over the lower wing surface at 40 m/s. What is the mass of the plane? Assume that the plane travels at constant velocity and that lift effects associated with the fuselage and tail assembly are small. Discuss the lift if the airplane, flying at the same airspeed, is (a) in level flight, (b) climbing at 15°, and (c) descending at 15°. See Exercise 65.

68E. A pitot tube, Fig. 47, is used to measure the airspeed of an airplane. It consists of an outer tube with a number of small holes, B, that is connected to one arm of a U-tube. The other arm of the U-tube is connected to a hole, A, at the front end of the device, which points in the direction the plane is headed. At A the air becomes stagnant so that $v_A = 0$. At B, however, the speed of the air presumably equals the airspeed of the aircraft. Use Bernoulli's equation to show that

$$v = \sqrt{\frac{2\rho g h}{\rho_{air}}},$$

where v is the airspeed of the plane and ρ is the density of the liquid in the U-tube. (*Hint:* the gravity terms do not contribute; why?)

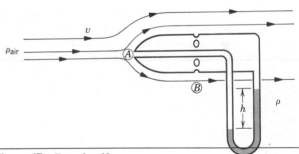

Figure 47   Exercise 68.

69E. A pitot tube (see Exercise 68) is mounted on an airplane wing to determine the speed of the plane relative to the air, which has a density of 1.03 kg/m³. The tube contains alcohol and indicates a level difference of 26 cm. What is the plane's speed relative to the air? The density of alcohol is 0.81×10^3 kg/m³.

70P. A pitot tube (see Exercise 68) on a high-altitude aircraft measures a differential pressure of 180 Pa. What is the airspeed if the density of the air is 0.031 kg/m³?

71P. In a hurricane, the air (density 1.2 kg/m³) is blowing over the roof of a house at a speed of 110 km/h. (a) What is the pressure difference between inside and outside that tends to lift the roof? (b) What would be the lifting force on a roof of area 90 m²?

72P. The windows in an office building are 4.0 m by 5.0 m. On a stormy day, air is blowing at 30 m/s past a window on the 53rd floor. Calculate the net force on the window. The density of air is 1.23 kg/m³.

73P. A sniper fires a rifle bullet into a gasoline tank, making a hole 50 m below the surface of the gasoline. The tank was sealed and is under 3.0-atm absolute pressure, as shown in Fig. 48. The stored gasoline has a density of 660 kg/m³. At what speed does the gasoline begin to shoot out of the hole?

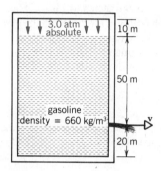

Figure 48   Problem 73.

74P. A hollow tube has a disk *DD* attached to its end (Fig. 49). When air is blown through the tube, the disk attracts the card *CC*. Let the area of the card be *A* and let *v* be the average airspeed between the card and the disk. Calculate the resultant upward force on *CC*. Neglect the card's weight; assume that $v_0 \ll v$, where v_0 is the airspeed in the hollow tube.

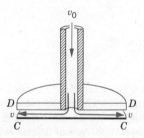

Figure 49   Problem 74.

75P. An 80-cm² plate of 500-g mass is hinged along one side. If air is blown over the upper surface only, what speed must the air have to hold the plate horizontal?

76P. If a person blows air with a speed of 15 m/s across the top of one side of a U-tube containing water, what will be the difference between the water levels on the two sides?

77P. A tank is filled with water to a height *H*. A hole is punched in one of the walls at a depth *h* below the water surface (Fig. 50). (*a*) Show that the distance *x* from the foot of the wall at which the stream strikes the floor is given by $x = 2\sqrt{h(H-h)}$. (*b*) Could a hole be punched at another depth so that this second stream would have the same range? If so, at what depth? (*c*) At what depth should the hole be placed to make the emerging stream strike the ground at the maximum distance from the base of the tank?

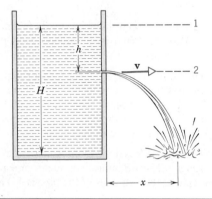

Figure 50   Problem 77.

78P. The fresh water behind a reservoir dam is 15 m deep. A horizontal pipe 4.0 cm in diameter passes through the dam 6.0 m below the water surface, as shown in Fig. 51. A plug secures the pipe opening. (*a*) Find the friction force between plug and pipe wall. (*b*) The plug is removed. What volume of water flows out of the pipe in 3.0 h?

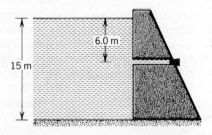

Figure 51   Problem 78.

79P. A *siphon* is a device for removing liquid from a container that cannot be tipped. It operates as shown in Fig. 52. The tube must initially be filled, but once this has been done the liquid will flow until its level drops below the tube opening at *A*. The liquid has density ρ and negligible viscosity. (*a*) With what speed does the liquid emerge from the tube at *C*? (*b*) What is the pressure in the liquid at the topmost point *B*? (*c*) What is the greatest possible height h_1 that a siphon can lift water?

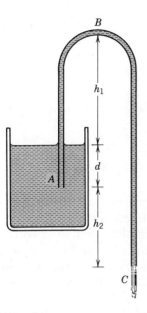

Figure 52 Problem 79.

80P. By applying Bernoulli's equation and the equation of continuity to points 1 and 2 of Fig. 23, show that the speed of flow at the entrance is

$$v = \sqrt{\frac{2a^2 \, \Delta p}{\rho(A^2 - a^2)}}.$$

81P. A Venturi meter has a pipe diameter of 10 in. and a throat diameter of 5.0 in. If the water pressure in the pipe is 8.0 lb/in.^{2} and in the throat is 6.0 lb/in.^{2}, determine the rate of flow of water in cubic feet per second.

82P. Consider the Venturi tube of Fig. 23 without the manometer. Let A equal $5a$. Suppose that the pressure at A is 2.0 atm. (*a*) Compute the values of v at A and v' at a that would make the pressure p' at a equal to zero. (*b*) Compute the corresponding volume flow rate if the diameter at A is 5.0 cm. The phenomenon at a when p' falls to nearly zero is known as cavitation. The water vaporizes into small bubbles.

83P*. A jug contains 15 glasses of orange juice. When you open the tap at the bottom it takes 12 s to fill a glass with juice. If you leave the tap open, how long will it take to fill the remaining 14 glasses and thus empty the jug?

ESSAY 6
PHYSICS AND SPORTS: THE AERODYNAMICS OF PROJECTILES

PETER J. BRANCAZIO

BROOKLYN COLLEGE

An object moving through a fluid always experiences some resistance to its movement. This force, exerted by the fluid on the object, necessarily alters the motion of the object to some degree. We would expect the retarding force to be fairly large if the object is moving through a liquid such as water; but if the fluid is a gas such as air, we might suppose that the force will be so small as to have virtually no effect on the motion of the object. However, as we shall see, the force that we commonly call *air resistance* cannot always be so easily ignored.

Air resistance is one manifestation of *aerodynamic force*—the force exerted by the air on a moving object. (When an object is moving through water, the force is said to be *hydrodynamic.*) Such forces are referred to as *dynamic* forces because they result from motion. Moreover, the force on an object at rest in a moving fluid is the same as when the object is moving in a stationary fluid; that is, the force is created by *relative* motion.

The study of fluid dynamics has considerable practical value in a wide range of areas, ranging from the flow of blood through capillaries to the design of boats, automobiles, and airplanes. It may come as a surprise, however, to learn that these same principles are also used by athletes to improve their performance in a variety of sports, for aerodynamic forces have a substantial effect on the motions of the various projectiles used in sports. These forces make it possible for a baseball pitcher to throw a curve ball and are responsible for the hook or slice of a poorly hit golf drive. They determine the proper technique for throwing a football or a javelin. They also represent the major force of resistance to the movement of a downhill skier or a cyclist.

In general, the fluid-dynamic force on an object will depend on the size, shape, and surface characteristics of the object as well as on its velocity relative to the fluid. The force, of course, will also depend on the properties of the fluid itself. A key property is the *viscosity* of the fluid, which is a measure of internal resistance to flow caused by interactions between the fluid molecules. Where the fluid comes into contact with the surface of an immersed object, its viscosity will create a frictional retarding force parallel to the surface. The viscous forces on an object are larger when the fluid is water rather than air; at room temperature, water is about 40 times more viscous than air— which explains why it takes more effort to wade across a swimming pool than to walk down the street. The nature of the surface of the moving object also plays a role; in general, the smoother the surface, the lower the viscous resistance.

An immersed object necessarily acts as an obstacle to the flow, forcing the fluid to change direction and accelerate around the object. The viscous friction between the fluid and the surface tends to remove energy from the fluid. These energy losses take place in a relatively thin layer of fluid, known as the *boundary layer,* that flows next to the surface. If the fluid is moving slowly, the frictional energy losses will be small; the fluid in the boundary layer will be able to accelerate and remain in contact with

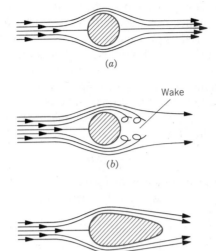

Wake

Figure 1 Fluid flow around a sphere: (*a*) low-velocity flow—no separation; (*b*) high-velocity flow—flow separates from surface, producing a low-pressure wake; and (*c*) flow around streamlined body produces little or no wake.

the surface of the object. However, at higher speeds the energy losses become great enough to prevent the fluid from following the surface contours. The result is that the boundary layer tends to separate from the surface (see Fig. 1), creating a region behind the object known as the *wake* that is characterized by lower pressure and unsteady or turbulent motions. Under these conditions, the fluid pressure on the front or leading surface of the object will exceed the pressure on the rear surface, resulting in a net retarding force. This *form resistance* can be reduced by changing the shape of the object to make it more "streamlined"—that is, by adjusting its contours so that the fluid will not separate from the surface.

The forces created by viscous and form resistance are distributed over the entire surface of an immersed object. However, it is common practice to sum and resolve them into two components: a *drag* force, which acts opposite to the direction of the motion of the object relative to the fluid (that is, antiparallel to the velocity vector), and a *lift* force, which acts at right angles to the direction of motion. Despite its name, the lift force should not be thought of as an upward (gravity-opposing) force, but rather as a lateral or sideways force that can deflect an object in any direction perpendicular to the velocity.

Aerodynamic Drag

The drag force on an object moving through a fluid always acts as a decelerating force, reducing the speed of the object relative to the fluid. In general, drag is produced by both viscous and form resistances. Through a combination of theory and experiment, physicists have discovered that a quantity known as the *Reynolds number* can be used to indicate whether the drag on an immersed object is mainly the result of viscous or form resistance. This dimensionless quantity is defined as

$$R = \frac{\rho v d}{v},$$

where ρ is the fluid density, v is its viscosity, v is the relative velocity of the flow, and d is the cross-sectional diameter of the object at right angles to the flow. In general, viscous drag predominates when the Reynolds number is very small ($R \lesssim 1$)—that is, for small objects moving slowly through viscous fluids. As the Reynolds number becomes larger, form resistance becomes increasingly more significant. For the sizes and speeds of the objects used as projectiles in sports—baseballs, tennis balls, and even the athletes themselves—the Reynolds number is on the order of 10^4–10^5. In this range, viscous effects are relatively small compared to form resistance. In such circumstances, the aerodynamic drag can be represented by a fairly simple equation. From energy and momentum considerations, it can be shown that the force per unit area exerted by a fluid on an immersed object should be proportional to the quantity $\frac{1}{2}\rho v^2$. That is, if D is the drag force,

$$D \propto \tfrac{1}{2}\rho A v^2.$$

This can be made into an equation by introducing as a constant of proportionality a dimensionless quantity known as the *drag coefficient, C_D*:

$$D = \tfrac{1}{2}\rho A C_D v^2,$$

where ρ is the fluid density, A is the effective frontal area of the object (its cross-sectional area perpendicular to the flow), and v is the relative velocity.

The drag coefficient C_D takes into account the relative contributions of viscous and form resistances, and it depends on the nature of the object (size, shape, and irregularity and roughness of its surface) as well as on the characteristics of the flow. In general, the more streamlined an object is, the lower its drag coefficient will be—an important consideration in the design of objects that must move through fluids at high speeds.

As a rule, C_D must be determined by direct measurement. It is common practice to give the drag coefficient for a given object as a function of the Reynolds number. The measurement is usually made by placing the object (or a suitably scaled model) in a wind tunnel, thereby taking advantage of the fact that the drag force depends on the relative velocity of object and fluid. For example, C_D for a baseball moving through still air at 90 mi/h can be measured by mounting a stationary baseball in a 90-mi/h wind.

Figure 2 shows the results of one such set of measurements—namely, the drag coefficient for a smooth sphere. The drag coefficient is rather large at very low Reynolds number (where viscous resistance dominates), but it decreases rapidly as R increases. From $R = 10^3$–10^5, C_D has a constant value of about 0.5. But at about $R = 2 \times 10^5$, the drag coefficient dips sharply and then levels off to a value of about 0.3.

Figure 2 Drag coefficient C_D for a sphere as a function of the Reynolds number R. Note that the Reynolds number is plotted on a logarithmic scale.

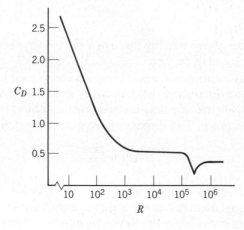

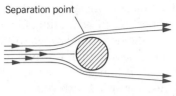

Smooth boundary layer

Figure 3 Transition from smooth to turbulent flow in the boundary layer. The decrease in the size of the wake reduces form resistance and lowers the drag coefficient.

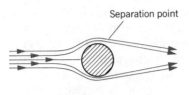

Turbulent boundary layer

This transition takes place because the fluid flow in the boundary layer has changed from a smooth motion to a turbulent one. A turbulent boundary layer carries more energy and thus is able to remain in contact with the surface a bit longer. The effect is to narrow the wake and increase the pressure behind the sphere (see Fig. 3). The resulting reduction in form resistance more than compensates for the increase in viscous resistance and leads to a net reduction in the drag.

Experiments show that roughening the surface of a sphere lowers the speed (and with it the Reynolds number) at which the transition from a smooth to a turbulent boundary layer takes place. Thus, at certain speeds, a roughened sphere may actually experience less drag than a smooth sphere under the same conditions. This unusual phenomenon plays a significant role in the game of golf. At the typical speed of a good golf drive — 120–150 mi/h (190–240 km/h) — a smooth golf ball is below the transition, whereas a ball with a roughened or irregular surface is beyond it. The latter will have a turbulent boundary layer, a smaller wake, and a lower drag coefficient and will travel about 30% farther than a smooth golf ball under the right conditions. This is the reason why there are dimples on a golf ball!

The effect of aerodynamic drag on the motion of a falling body is described in Section 6–3. As a body falls through the air, both the speed and the drag increase until the drag force equals the weight of the body. At this point, the body has achieved its terminal speed (see Eq. 17 of Chapter 6).

One sport that makes use of the fact that falling bodies approach a terminal speed in air is skydiving. A person who jumps out of a plane falls with decreasing acceleration, approaching a terminal speed of as much as 200 mi/h (320 km/h). However, by changing the shape and orientation of their bodies as they fall, sky divers are able to increase or decrease the amount of aerodynamic drag, effectively selecting the terminal speed by changing the drag coefficient and frontal area. When spread-eagled with arms and legs extended, a sky diver feels the greatest drag and has the lowest terminal speed.

The motion of a body falling through the air deviates significantly from that of a freely falling body as its speed approaches its terminal speed. Similarly, the motion of any object launched as a projectile diverges noticeably from the theoretical symmetrical parabolic path if its launching speed is comparable to its terminal speed. When an object is launched upward at an angle to the ground, aerodynamic drag retards both the vertical and horizontal components of its motion. Consequently, the maximum height of the trajectory as well as the horizontal range will be shortened. As the projectile rises, it decelerates more rapidly than it would in a vacuum; as it falls, it accelerates more slowly. The result is that the projectile will take less time to reach its peak than it does to descend, and the speed with which it reaches the ground will be less than the speed at which it was launched. Moreover, since the horizontal velocity component is continually decreasing, the projectile will travel a longer distance horizontally when it is rising than when it is falling. Consequently, the descent of the projectile is both steeper and slower than its rise.

A baseball has a terminal speed of about 95 mi/h (153 km/h). Under game conditions, pitched and batted balls are launched at speeds comparable to or even greater than the terminal speed. (Note that the terminal speed is not the maximum speed that a projectile can travel at in air, but rather the speed that the projectile attains when it is allowed to fall from rest.) Accordingly, it is clear that the trajectory of a baseball in flight must be altered substantially by aerodynamic drag. A typical fly ball has a range in air that is only about 60% of its range in a vacuum (see Fig. 13 of Chapter 4 for a graphic comparison).

Given its large size and asymmetrical shape, a football is particularly susceptible to the influence of aerodynamic drag. When it is traveling nose first, a football presents a smaller cross-sectional area and a more streamlined shape than when it is traveling broadside; the drag is roughly 10 times larger in the broadside orientation. This fact determines the best technique for throwing a football; namely, it should be "spiraled" or thrown nose first with a substantial spin about its long axis. (The spin gives the ball angular momentum, which stabilizes the orientation of the spin axis in flight.) Launching the football with its long axis angled laterally to its path will shorten its range considerably. The effect is especially noticeable when a football is kicked with little or no spin such that it rises in a nose-first orientation and descends broadside. The sudden increase in aerodynamic drag that occurs as the football passes the peak of its trajectory causes it to drop rather steeply and shortens the range of the kick. Good kickers try to get the ball to "turn over" (that is, turn its nose downward on the descent—a result of spiraling the kick) as a way to get more distance.

The reduction of aerodynamic drag is particularly important in achieving more efficient motion, especially when greater speed or economy is a goal. A significant feature of the drag force is that it depends on the square of the speed; thus, for example, the drag on an automobile is four times greater at 60 mi/h than at 30 mi/h, a fact that seriously diminishes the economy of driving at high speeds. At about 30 mi/h (48 km/h), the aerodynamic drag on an automobile exceeds the frictional resistance between the tires and the ground. At 55 mi/h (88 km/h), about half the fuel energy goes to overcoming drag. Automobile designers recently have begun to make use of wind tunnels to find ways to reduce drag; their efforts have led to the gently sloped hoods and windshields found on most economy cars. A streamlined shape and a reduced frontal area are even more essential in the design of high-speed racers. (A typical family automobile has a drag coefficient of about 0.5, whereas for a racer it is around 0.3.)

In sports such as swimming, cycling, or speed skating—where the energy for motion is provided entirely by the athlete—it is particularly important to keep the drag force as low as possible. In downhill skiing, where speeds as high as 80 mi/h (130 km/h) are attained, aerodyanamic drag is virtually the only retarding force. The techniques and equipment used in these sports are designed in large measure to minimize drag. There are three basic approaches to achieve this purpose:

1. Reduce the Effective Frontal Area. Athletes are taught to adopt a stance in which their bodies collide with as little air as possible. By keeping the upper body parallel to the ground instead of upright, the effective frontal area of the body can be reduced by about 25%. Thus, cyclists hunch down over the handlebars; speed skaters bend sharply at the waist and keep one arm tucked behind their backs; skiers crouch down over their skis in the "egg" position.

Swimmers move rather slowly through water (the top racing speeds are less than 5 mi/h), but they nevertheless experience substantial hydrodynamic resistance because of the relatively high density and viscosity of water. Swimmers reduce their effective frontal area in the water by kicking their feet as they swim. The major purpose of kicking the feet is *not* to propel the swimmer (the arm strokes provide approximately 75% of a swimmer's propulsive force) but to reduce drag by keeping the body horizontal in the water.

2. Use Streamlined Equipment. Wind tunnel studies of cyclists and skiers have led to numerous improvements in equipment design. It has been found, for example, that relatively small protrusions such as boot buckles, ski pole baskets, and cables and nuts on bicycles are sources of turbulence and measurable aerodynamic drag. Although these effects are relatively small, they are of some importance to competitors at the world-class level where fractions of a second are often the difference between winning and losing. Accordingly, equipment has been modified so as to remove or reduce protruding parts or else to cover them with smoothly contoured shields. The athlete's head is also a source of some turbulence, so some skiers and cyclists wear tear-drop shaped helmets that produce a more streamlined flow of air around their heads (see Fig. 9 of Chapter 6).

3. Wear Smooth, Skin-Tight Clothing. Loose-fitting clothing increases the athlete's surface area and creates turbulence. "Aerodynamic clothing"—smooth, skin-tight body suits made of specially designed synthetic fabrics—has become increasingly popular with skiers, cyclists, and speed skaters. A smooth body surface is particularly important in competitive swimming; thus, form-fitting swimsuits made of nonabsorbent material have become obligatory. Male swimmers routinely shave off exposed body hair—even on their heads—as a means of reducing hydrodynamic drag.

Aerodynamic Lift

As noted earlier, lift is that component of the interaction between an object and a fluid that is perpendicular to the direction of motion. In general, lift is generated by any effect that causes the fluid to change direction as it flows past the object. If the fluid acquires a component of velocity at right angles to its original direction as a result of its interaction with an immersed object, then the object must have exerted a force on the fluid giving it an acceleration in that direction. According to Newton's third law (the principle of action and reaction), the fluid must apply an equal and opposite force on the object. Thus, if the object diverts the fluid to the left, it will experience a lift force to the right. Among the effects that can create lift are: (1) the object has an asymmetrical shape or orientation with respect to the flow; (2) the object is spinning; and (3) the lateral surface of the object is uneven (that is, one side is rougher than the other).

We have already seen in Section 16–12 that an airplane wing (airfoil) generates lift, diverting the airflow downward as a consequence of its asymmetrical shape and inclination to the flow. Increasing the angle of attack generates more lift, but it also creates more drag by increasing the frontal area of the wing. The best angle of attack is therefore the one that provides the greatest lift-to-drag ratio. This effect can readily be experienced by extending your arm outward, palm down, from the window of a moving automobile. As you tilt your palm into the airflow, you will feel both lift and drag and can observe how each force changes with the angle of tilt.

Just as an airfoil can be used to provide vertical lift to an airplane, a hydrofoil can be used to provide lift to a boat—enough, in fact, to raise its hull entirely out of the water. Since the viscous drag on a submerged hull is normally very large, the use of hydrofoils permits much higher speeds to be attained.

Perhaps the most familiar piece of sports equipment to use lift is the Frisbie. If it is launched with its leading edge tilted slightly upward, a Frisbie will generate enough lift to counteract its own weight, and it will "fly" a considerable distance without falling. The spin given to a Frisbie when it is launched gives it stability much like that of a spiraling football (through the principle of conservation of angular momentum) and serves to maintain a constant angle of attack. If a Frisbie is tilted right or left when it is thrown, then the lift force will have a component in that direction, thereby creating a curved flight path.

Frisbie throwing is actually a more popular variation of a much older sport—the discus throw. A discus is a flat plate that is launched (also with a stabilizing spin) at an angle to the airflow. The lift generated by a properly thrown discus enhances its distance substantially. Discus throwers find that a discus will actually go farther when it is thrown *into* the wind. This is understandable, because when a discus is moving into the wind, its relative speed with respect to the airflow is increased—and the greater the relative speed, the greater the lift. Airplane takeoffs are preferably made into the wind for this same reason.

There are several other sports that depend on the lift force created by launching an object at an angle to the airflow. One of these is ski jumping. When the jumper is airborne, he tilts his skis upward and inclines his body parallel to the skis so as to get the best possible lift-to-drag ratio. Another is javelin throwing. Despite its shape, a javelin can be thrown in such a way as to generate lift. Indeed, javelin designers have used aerodynamic principles to create javelins with enhanced lifting properties and have thus greatly increased the standards of competition in this event.

In all the examples described, the lift force has been used to oppose gravity and increase the time of flight and range of a projectile. There is one application, however, where the lift force is purposely directed *downward*. Auto racing fans are familiar with the "spoiler," which is an airfoil mounted over the rear end of a Formula One racer. Here, however, the airfoil is tilted so as to produce a downward-acting lift force. This force pushes the rear wheels to the track, increases the normal force and hence the frictional force, and thus generates better traction.

The airflow around a sphere cannot generate lift because, by symmetry, it does not deflect the airflow one way or another. However, a *spinning* sphere does cause the airflow to change direction. This effect arises from the frictional losses in the boundary layer that cause the flow to separate from the surface and form a low-pressure wake. On a spinning sphere, one side is turning into the airflow, while the other is turning in the same direction as the airflow. On the latter side, the air is carried along by the surface, causing the separation to occur a little farther downstream. On the side turning into the airflow, the separation point is shifted toward the front. The result is

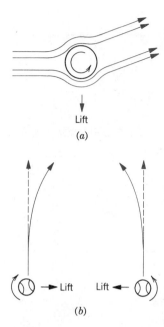

Lift

(a)

Lift   Lift

(b)

Figure 4 Lift on a spinning sphere (the Magnus effect): (a) the airflow around a spinning sphere and (b) paths of a curve ball thrown with different spin directions.

that the wake is not symmetrical; the airflow is deflected to one side, and the sphere experiences a reaction force in the opposite direction (Fig. 4). The direction and strength of this force will depend on the rate and direction of spin. This phenomenon is known as the *Magnus Effect* (after Gustav Magnus, who studied it in the 1850s).

The Magnus effect is familiar to anyone who has watched a ball curve in flight. It plays a key role in such sports as baseball, tennis, golf, and soccer. By applying an appropriate spin, an athlete can make a ball curve in any chosen direction. For example, if a soccer ball is kicked so that the impact point is right of center—giving the ball a counterclockwise spin as seen from above—it will curve from right to left; if the kick is aimed left of center, the ball will curve to the right.

The same principle is used by a baseball pitcher to throw a curve ball. The pitcher imparts spin to the ball by twisting his wrist as the pitch is delivered. However, the anatomy of the human arm is such that it twists more flexibly in one direction than the other. Thus, it is easier for a right-handed pitcher to make a baseball spin counterclockwise than clockwise (as seen from above), whereas the reverse is true for a left-hander. A right-hander's curve ball moves from his right to his left (curving away from a right-handed batter). To make the pitch curve from left to right (a pitch known as a "screwball"), he must twist his wrist in the "wrong" direction. Few right-handed pitchers can throw this pitch effectively. Thus, the Magnus effect, coupled with a quirk of human anatomy, underlie the maneuvering of the various combinations of right-handed and left-handed pitchers and batters that is so central to baseball strategy.

A baseball pitch must necessarily travel on a curved trajectory even if it is not spinning: Because of gravity, it curves downward on a parabolic path. A baseball thrown horizontally at 85 mi/h (137 km/h) will drop about $3\frac{1}{2}$ ft (1.1 m) over the distance from the pitcher's hand to home plate. A pitcher can add to or subtract from this vertical drop by using suitable spin. If the ball is thrown with a topspin (the top of the ball is turning toward the batter) then the lift force will act downward and the pitch will curve more sharply than normal (this pitch is called an overhand curve). If the ball is released with a backspin (by letting the ball roll off the fingertips with a downward snap of the wrist), the lift force will act upward, opposing the force of gravity. This is the so-called "rising" fastball: however, the ball does not actually rise in flight—since it is not humanly possible to impart enough spin to make the lift exceed the weight of the baseball—but because the ball does not drop as much as it does under gravity alone, it gives the illusion of rising.

Topspin and backspin are also important in tennis: Backspin (creating upward lift) tends to prolong the flight of a tennis ball, whereas topspin (producing downward lift) tends to shorten it. It is extremely difficult to hit a hard serve that will clear the net and land in the service box without giving it a topspin. The fact that backspin prolongs the flight of a ball is particularly helpful to golfers. A well-hit drive should have a great deal of backspin on it: The horizontal grooves and sloped face of the clubhead serve to make this possible. Because of the lift produced by backspin, the trajectory of a golf drive is far from parabolic. The optimum launch angle tends to be much lower than 45° (it is generally between 20° and 30°); since gravity is being opposed by lift, the ball can be launched more horizontally for greater distance. However, if the clubface does not hit the ball squarely and is angled right or left, the ball will receive a side spin. This type of spin will make the ball hook (curve left) or slice (curve right) in flight. For the average golfer, the angle of the clubface at the moment of impact can sometimes make all the difference between a birdie and a bogie.

CHAPTER 17
WAVES—I

Hanging ten off the California coast! The energy of this wave comes from work done on the water by the wind, starting hundreds of miles offshore. However, no water travels this great distance and wind may not be blowing at the beach. It is as if the "idea" of wave sweeps in from midocean, giving form and motion to the local water as it passes through.

17-1 Waves and Particles

Two of the ways to get in touch with a friend in a distant city are: You can write a letter or you can pick up the telephone.

The first choice (the letter) is in the spirit of "particle." A material object moves from one point to another, carrying with it information and energy. Most of the previous chapters have dealt with particles or with systems of particles, ranging from electrons to baseballs to automobiles to planets.

The second choice (the telephone) is in the spirit of "wave," the subject of this and the following chapter. In a wave, information and energy move from one point to another but no material object makes that journey. In

your telephone call, a sound wave carries your message from your vocal cords to the telephone. From there, the wave is electromagnetic, passing along a copper wire or an optical fiber or through free space, possibly by way of a communications satellite. At the receiving end there is another short acoustic path to your friend's ear. Although the message is passed, nothing that you have touched reaches your friend. Leonardo da Vinci understood about waves when he wrote of water waves: "it often happens that the wave flees the place of its creation, while the water does not; like the waves made in a field of grain by the wind, where we see the waves running across the field while the grain remains in place."

Particle and *wave* are the two great frameworks available to classical physics and we seem able to asso-

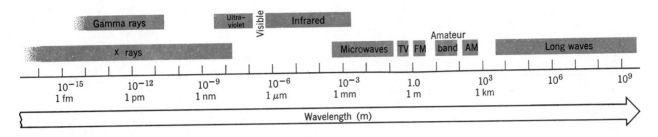

Figure 1   The electromagnetic spectrum. These waves, which differ in wavelength, all travel through free space with the same speed c. Note the small range of wavelengths spanned by visible light.

ciate almost every branch of the subject with one or the other. The two concepts are quite different. The word *particle* suggests a tiny concentration of matter capable of transmitting energy. The word *wave* suggests just the opposite, namely, a broad distribution of energy, filling the space through which it passes. The job at hand is to put aside particles for a while and to learn something about waves.

17-2 Waves*

Mechanical Waves.   A flag waving in the breeze is so familiar that, when the astronauts planted an American flag on the windless moon, they used a flag with built-in ripples so that it would look "natural." There are also water waves, in bodies of water ranging from an ocean to a wash basin. There are sound waves, in air and in water, and seismic waves, in the earth's crust, mantle, and core. The central features of all these *mechanical waves* is that they are governed by Newton's laws and that they require a material medium, such as air, water, a stretched string, or a steel rod, for their propagation.

Electromagnetic Waves.   The most familiar of these waves is visible light. Also part of the *electromagnetic spectrum* to which visible light belongs (see Fig. 1) are x rays, microwaves, and the waves that activate our radios and television sets. Unless you are reading this book in an electrically shielded room (or perhaps underwater!), many such waves will be passing through you at this moment. The x rays, sunlight, and long radio waves behave quite differently when we study them in the laboratory. At root, however, they are very similar, their different behaviors being traceable to their differences in wavelength.

*See John R. Pierce, *Almost All About Waves*, MIT Press, Cambridge, MA, 1974, for a very readable account of the subject.

Electromagnetic waves require no medium for their propagation, traveling freely to us from the distant quasars through the near vacuum of deep space. They all travel through free space with the same speed c, given by

$$c = 299, 792, 458 \text{ m/s} \qquad \text{(speed of light).} \quad (1)$$

We shall return to a study of these waves in Chapter 38.

Matter Waves.   Under certain experimental conditions, a beam of particles—electrons, for example—can exhibit wavelike properties. Such *matter waves* are governed by the laws of quantum physics. We discuss them more fully in Chapter 44 of the extended version of this book and refer to them here only so that our listing of the kinds of waves encountered in nature may be complete.

Much of what we shall say in this chapter will apply to waves of all kinds. When we seek specific illustrations, however, we shall draw them from the family of mechanical waves, focusing especially on waves traveling along a stretched string.

17-3   Waves in a Stretched String

Of all possible mechanical waves, a wave transmitted along a stretched string is perhaps the simplest. If you give one end of a stretched string a single up and down jerk, as in Fig. 2a, an impulse is passed along the string from particle to particle and thus a wave, in the form of a single pulse, travels along the string. If you move your hand up and down in continuous simple harmonic motion (Fig. 2b), an extended traveling sinusoidal wave is generated.

We assume an "ideal" string, in which there are no frictionlike forces to cause the wave to die out as it travels along. We assume further that our string is so long that

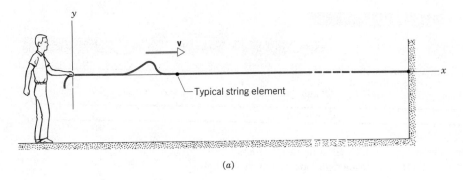

(a)

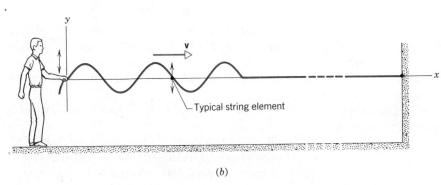

(b)

Figure 2 *(a)* Sending a pulse down a long stretched string. *(b)* Sending a sinusoidal wave down the string. The vibrations of each string element (see the dots) are at right angles to the direction of propagation. The wave is *transverse*.

we do not have to be concerned about any "echos" that might rebound from its far end.

In studying the waves of Fig. 2, we can keep our eye on the wave form as it moves to the right. Alternatively, we can watch the motion of a specified element of the string, of length dx and corresponding mass dm, as it oscillates up and down as the wave passes through it. The displacements of these oscillating string elements are in the y direction, *at right angles* to the direction of travel of the wave, which is the x direction. We call such wave motion *transverse*.

Figure 3 shows a sound wave, set up in a long, air-filled pipe by an oscillating piston. In this kind of mechanical wave, the displacements of small elements of air of mass dm are back and forth in the x direction, *parallel* to the direction in which the wave travels. We call this kind of wave motion *longitudinal;* we shall consider it further in Chapter 18.

17–4 Wavelength and Frequency

A traveling wave on a stretched string can have many shapes, but fundamental to each is a *wavelength* λ and a *frequency* ν. The former—as we shall see in more detail later—is the length of the repeating wave shape and the latter the frequency of the oscillating source that gener-

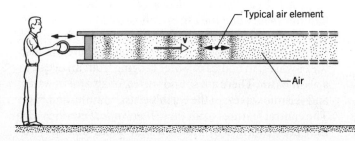

Figure 3 A sound wave is set up in an air-filled pipe by moving a piston back and forth. The vibrations of the elements of the air (see the dot) are parallel to the direction of propagation. The wave is *longitudinal*.

ates the wave. We define a wave shape by giving a relation of the form $y = f(x, t)$, in which y is the transverse displacement—at time t—of a string element as a function of the position x of that element along the string.

As a wave for particular study, we choose the wave of Fig. 2*b*, generated by moving the end of the string transversely in simple harmonic motion. We write for the relation between the displacement y of any string element at position x and at time t

$$y(x, t) = y_m \sin(kx - \omega t), \qquad (2)$$

in which y_m is the *amplitude* of the wave. The quantities k and ω, at this stage, are constants whose physical meaning we are about to discover.

You may wonder why, of the infinite variety of wave forms that are available, we pick the sinusoidal wave of Eq. 2 for detailed study. This is in fact a wise choice because, as we shall learn in Section 17–9, *all* wave forms—including the pulse of Fig. 2a—can be constructed by adding up sinusoidal waves of carefully selected wavelengths and amplitudes. Understanding sinusoidal waves is thus the key to understanding waves of any shape.

There are no limits on the variables x and t in Eq. 2 so that mathematically the equation describes a wave on a string of infinite length, existing for all time, from the remote past to the distant future. As a practical matter, we shall confine our attention to reasonably small ranges of each variable.

Because Eq. 2 has two independent variables (x and t), we cannot represent the dependent variable y on a single two-dimensional plot. We need something like a videotape to show it fully and in real time. We can, however, learn a lot from the two plots of Fig. 4.

Wavelength and Wave Number.   Figure 4a shows how the transverse displacement y of Eq. 2 varies with position x at a fixed time, chosen to be $t = 0$. That is, the figure is a "snapshot" of the wave at that instant. With this restriction, Eq. 2 becomes

$$y(x, 0) = y_m \sin kx \qquad (t = 0). \qquad (3)$$

We define the *wavelength* λ of a wave to be the shortest distance at which the wave pattern (t being held constant) fully repeats itself. A typical wavelength interval is displayed in Fig. 4a. Applying Eq. 3 to each end of this interval yields

$$y = y_m \sin kx = y_m \sin k(x + \lambda). \qquad (4)$$

The sine function first repeats itself when the angle is increased by 2π radians so that Eq. 4 will be true if $k\lambda = 2\pi$, or

$$\boxed{k = \frac{2\pi}{\lambda}} \qquad \text{(angular wave number).} \qquad (5)$$

This relation gives physical meaning to the quantity k that appears in Eq. 2. We see from Eq. 4 that if k is multiplied by a length an angle in radians must result. We call k the *angular wave number* of the wave, its SI unit being the radian per meter.

The *wave number*, symbolized by κ, is defined as $1/\lambda$

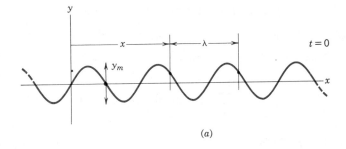

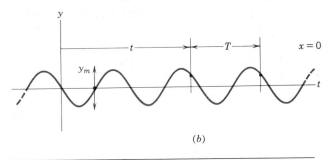

(a)

(b)

Figure 4   (a) A "snapshot" of the sinusoidal curve of Eq. 2 at $t = 0$. A typical wavelength λ is shown. The amplitude y_m is also shown. (b) The displacement as a function of time of an oscillating string element at $x = 0$ as the sinusoidal wave of Eq. 2 passes through it. A typical period T is shown.

and is related to k by

$$\boxed{\kappa = \frac{1}{\lambda} = \frac{k}{2\pi}} \qquad \text{(wave number).} \qquad (6)$$

The wave number κ is the number of waves per unit length of the wave pattern, its SI unit being the reciprocal meter ($= \text{m}^{-1}$).

Frequency and Period.   Figure 4b shows how the displacement y of Eq. 2 varies with time t at a fixed position, taken to be $x = 0$. If you were to station yourself at that position and watch the wiggling string through a narrow vertical slit in a card, the motion of only one "point" of the string would be revealed. Its motion up and down in the y direction would be given by

$$y(0, t) = y_m \sin(-\omega t) = -y_m \sin \omega t \qquad (x = 0). \quad (7)$$

Here we have made use of the fact (see Appendix G) that $\sin(-\alpha) = -\sin \alpha$, where α is any angle.

We define the *period T* of a wave to be the shortest time interval during which the motion of an oscillating string element (at any fixed position x) fully repeats itself. A typical period is displayed in Fig. 4b. Applying

Eq. 7 to each end of this time interval yields

$$y = -y_m \sin \omega t = -y_m \sin \omega(t + T). \tag{8}$$

This can only be true if $\omega T = 2\pi$, or

$$\omega = \frac{2\pi}{T} \qquad \text{(angular frequency).} \tag{9}$$

This gives physical meaning to the quantity ω that appears in Eq. 2. We see from Eq. 8 that if ω is multiplied by a time an angle in radians must result. We call ω the *angular frequency* of the wave, its SI unit being the radian per second.

The *frequency* of the wave, symbolized by ν, is defined simply as $1/T$ and is related to ω by

$$\nu = \frac{1}{T} = \frac{\omega}{2\pi} \qquad \text{(frequency).} \tag{10}$$

The frequency ν is the number of vibrations per unit time made by the wave as it passes a given point. We have already encountered the concepts of frequency and angular frequency in our study of oscillations in Chapter 14. We saw there that the frequency ν is usually reported in units of the hertz or its multiples, where

$$1 \text{ hertz} = 1 \text{ Hz} = 1 \text{ vibration/s.} \tag{11}$$

17–5 The Speed of Traveling Waves

Figure 5 shows two snapshots of the wave of Eq. 2, taken at a small interval Δt apart. The wave is traveling in the direction of increasing x (to the right in Fig. 5), the entire wave pattern moving a distance Δx in that direction during the interval Δt. The ratio $\Delta x/\Delta t$ (or, in the differential limit, dx/dt) is the wave velocity. How can we find its value?

Let us focus on a particular part of the wave pattern —perhaps a point of maximum displacement as suggested in Fig. 5. From Eq. 2, any given displacement y is defined by assigning a fixed value to the quantity in the parentheses, which is called the *phase* of the wave. Thus, putting

$$kx - \omega t = \text{a constant} \tag{12}$$

defines a constant transverse displacement y.

Imagine yourself running along the string at speed v, keeping abreast of the wave form and following your chosen constant displacement. You would have to run

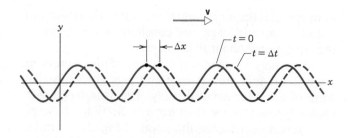

Figure 5 A "snapshot" of the traveling wave of Eq. 2 at $t = 0$ and also at a later time, $t = \Delta t$. During this interval, the entire curve shifts a distance Δx to the right.

to the right (not the left) in Fig. 2*b* to meet the restraint of Eq. 12. That is, as t increases, you would have to increase x to maintain a constant value for $kx - \omega t$.

Taking the derivative of Eq. 12 yields

$$k\frac{dx}{dt} - \omega = 0$$

or

$$\frac{dx}{dt} = v = +\frac{\omega}{k} \tag{13}$$

for the wave velocity. This quantity is positive, telling us that the wave is traveling in the direction of increasing x, that is, to the right in Fig. 5.

Making use of Eq. 5 ($k = 2\pi/\lambda$) and Eq. 9 ($\omega = 2\pi/T$) allows us to write, for the wave speed,

$$v = \frac{\omega}{k} = \frac{\lambda}{T} = \lambda \nu \qquad \text{(wave speed).} \tag{14}$$

Equation 14 seems reasonable; it tells us that the wave moves through one wavelength in one period of oscillation.

Equation 2 describes a wave moving in the direction of increasing x. We can find the equation of a wave traveling in the opposite direction by replacing t in Eq. 2 by $-t$. This corresponds to keeping the quantity $kx + \omega t$ constant, which (compare Eq. 12) requires that x must *decrease* with time. Thus,

$$y(x, t) = y_m \sin (kx - \omega t) \qquad \text{(increasing } x) \tag{15}$$

and

$$y(x, t) = y_m \sin (kx + \omega t) \qquad \text{(decreasing } x). \tag{16}$$

If you analyze the wave of Eq. 16 as we have just done for the wave of Eq. 2 (or Eq. 15) you will find for its velocity

$$\frac{dx}{dt} = -\frac{\omega}{k}.$$ (17)

The minus sign (compare Eq. 13) shows that the wave is indeed moving in the direction of decreasing x as time goes on and justifies our device of switching the sign of the time variable.

Consider now a wave of generalized shape, given by

$$y(x, t) = f(kx + \omega t),$$ (18)

where f represents *any* function, the sine function being only one possibility. Our analysis above shows that all waves in which the variables x and t enter in the combination $kx \pm \omega t$ will be traveling waves. Furthermore, all traveling waves *must* be of this form. Thus, $\sqrt{ax + bt}$ represents a possible (though perhaps physically a little bizarre) traveling wave. The function $\sin(ax^2 - bt)$, on the other hand, is *not* a traveling wave.

Sample Problem 1   A traveling sinusoidal wave is described by

$$y(x, t) = 0.00327 \sin(72.1x - 2.72t),$$ (19)

in which the numerical constants are in SI units (0.00327 m, 72.1 rad/m, and 2.72 rad/s). (a) What is the amplitude of this wave?

By comparison with Eq. 15, we see that

$$y_m = 0.00327 \text{ m} = 3.27 \text{ mm}.$$ (Answer)

(b) What are the wavelength and the period of this wave?

By inspection of Eq. 19, we see that

$$k = 72.1 \text{ rad/m} \quad \text{and} \quad \omega = 2.72 \text{ rad/s}.$$

From Eq. 5 we have

$$\lambda = \frac{2\pi}{k} = \frac{2\pi}{72.1 \text{ rad/m}} = 0.0871 \text{ m} = 8.71 \text{ cm}.$$ (Answer)

You may want to reread Hint 2 of Chapter 11 to justify the disappearance of the rad unit in the above result. From Eq. 9 we have

$$T = \frac{2\pi}{\omega} = \frac{2\pi}{2.72 \text{ rad/s}} = 2.31 \text{ s}.$$ (Answer)

(c) What are the wave number and the frequency of this wave?

From Eq. 6 we have

$$\kappa = \frac{1}{\lambda} = \frac{1}{0.0871 \text{ m}} = 11.5 \text{ m}^{-1}.$$ (Answer)

From Eq. 10 we have

$$v = \frac{1}{T} = \frac{1}{2.31 \text{ s}} = 0.433 \text{ Hz}.$$ (Answer)

(d) What is the speed of this wave?
From Eq. 14 we have

$$v = \frac{\omega}{k} = \frac{2.72 \text{ rad/s}}{72.1 \text{ rad/m}} = 0.0377 \text{ m/s} = 3.77 \text{ cm/s}.$$ (Answer)

Note that all quantities calculated in (b), (c), and (d) are independent of the amplitude of the wave.

Sample Problem 2   For the wave of Eq. 19, what is the displacement y at $x = 22.5$ cm ($= 0.225$ m) and $t = 18.9$ s? Take all numerical values to be exact.

From Eq. 19 we have

$$y = 0.00327 \sin(72.1 \times 0.225 - 2.72 \times 18.9)$$
$$= (0.00327 \text{ m}) \sin(-35.1855 \text{ rad})$$
$$= (0.00327 \text{ m})(0.588)$$
$$= 0.00192 \text{ m} = 1.92 \text{ mm}.$$ (Answer)

Thus, the displacement is positive and is equal to 59% of the amplitude. Be sure to change your calculator mode to radians before evaluating the sine.

We need to assume that the numerical values in this problem are exact because we are asked to specify a displacement in a particular wavelength interval that is far removed from the "home base interval" associated with $x = 0$ and $t = 0$. It is as if you were asked to specify the location of a suitcase in a freight car near the end of a long train. Your measurements backward from the engine had better be precise. If you make what seems like a modest error on a percentage basis, you could even end up in the wrong freight car!

Sample Problem 3   In Sample Problem 2, we showed that the (transverse) displacement of an element of the string for the wave of Eq. 19 at $x = 0.225$ m and $t = 18.9$ s was $y = 1.92$ mm. (a) What is u, the (transverse) velocity of the same element of the string, at that place and at that time? This velocity, which is associated with the transverse oscillation of an element of the string, points in the y direction. Do not confuse it with v, the velocity with which the wave form is propagated and which points in the x direction.

The generalized equation of the wave of Eq. 19 is

$$y(x, t) = y_m \sin(kx - \omega t).$$ (20)

In this expression, let us hold x constant but allow t (for the time being) to be a variable. We then take the derivative of this equation with respect to t, obtaining*

$$u = \frac{\partial y}{\partial t} = -\omega y_m \cos (kx - \omega t).$$ (21)

Substituting numerical values from Sample Problems 1 and 2 and putting $t = 18.9$ s, we obtain

$$u = (-2.72 \text{ rad/s})(3.27 \text{ mm}) \cos (-35.1855 \text{ rad})$$
$$= 7.20 \text{ mm/s}.$$ (Answer)

Thus, the element of the string is moving in the direction of increasing y, with a speed of 7.20 mm/s.

(b) What is the transverse acceleration a_y at that position and at that time?

From Eq. 21, treating x as a constant but allowing t (for the time being) to be a variable, we find

$$a_y = \frac{\partial u}{\partial t} = -\omega^2 y_m \sin (kx - \omega t).$$

Comparison with Eq. 20 shows that we can write this as

$$a_y = -\omega^2 y$$

We see that the (transverse) acceleration of an oscillating string element is proportional to its (transverse) displacement but opposite in sign, completely consistent with the fact that the element is moving (transversely) in simple harmonic motion. Substituting numerical values yields

$$a_y = -(2.72 \text{ rad/s})^2 (1.92 \text{ mm})$$
$$= -14.2 \text{ mm/s}^2.$$ (Answer)

Thus (at $x = 22.5$ cm and $t = 18.9$ s), the displacement of the string is $+1.92$ mm and its acceleration is -14.2 mm/s^2.

Hint 1: Very large angles. Sometimes, as in Sample Problems 2 and 3, an angle much greater than 360° will crop up and you are asked to find its sine or cosine. Adding or subtracting an integral multiple of 360° (or 2π radians) to such an angle does not change the value of any of its trigonometric functions. In Sample Problem 2, for example, the angle was -35.1855 rad, or $-2015.98°$. Adding $6 \times 360°$ to this angle yields

$$-2015.98° + 6 \times 360° = 144.02°,$$

an angle less than 360°; see Fig. 6. The sine of 144.02° is 0.588.

---

* A derivative taken while one (or more) of the variables is treated as a constant is called a *partial derivative* and is represented by the symbol $\partial/\partial x$, rather than d/dx.

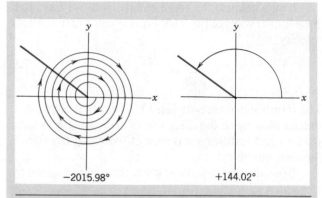

Figure 6   The two angles are different but all their trigonometric functions are identical.

Your calculator will reduce such large angles for you directly, that is,

$$\sin (-2015.98°) = \sin 144.02° = 0.588.$$

Caution: Do not round off such large angles if you intend to take their sines or cosines. Also, if you change a large angle from radians to degrees (as we did above), be sure to use an exact conversion factor ($180° = \pi$ rad) rather than an approximate one ($57.3° \approx 1$ rad). Remember, in taking the sine of a very large angle you are throwing away most of the angle and taking the sine of what is left over. If, for example, you were to round $-2015.98°$ to $-2020°$ (a change of 0.2% and normally a reasonably step), you would find

$$\sin (-2020°) = 0.643,$$

which is badly in error.

17–6   Wave Speed on a Stretched String

If a wave travels through a medium such as water, air, steel, or a stretched string, it must set the particles of that medium into oscillation as it passes. For that to happen, the medium must possess both inertia (so that kinetic energy can be stored) and elasticity (so that potential energy can be stored). It should be possible to calculate the speed of the wave through the medium in terms of these properties. We do so now for a stretched string, in two ways.

Dimensional Analysis.   As its name suggests, the dimensional analysis method is based on analyzing the dimensions of all the physical quantities that enter into the problem. In this case, we seek a speed v, which has the dimensions of length over time, or LT^{-1}.

The inertia characteristic of a stretched string cannot be the mass of the string because the speed is surely the same for a string of any length. The inertia characteristic must be the mass m of the string divided by its length l, a ratio that we call the *linear density* μ of the string. Thus, $\mu = m/l$, its dimensions being ML^{-1}.

You cannot send a wave along a straight stretched string without stretching the string. It is the tension in the string that does the stretching and must therefore represent the elastic characteristic of the string. The tension* τ is a force and has the dimensions (think about $F = ma$) of MLT^{-2}.

The problem is to combine μ (dimensions ML^{-1}) and τ (dimensions MLT^{-2}) in such a way as to generate v (dimensions LT^{-1}). A little juggling of various combinations suggests

$$v = C\sqrt{\frac{\tau}{\mu}}, \tag{22}$$

in which C is a dimensionless constant. The drawback of dimensionless analysis is that it cannot root out such dimensionless factors. In our second approach to the speed problem, we shall see that Eq. 22 is indeed correct and that $C = 1$.

Derivation from Newton's Second Law.   Instead of the sinusoidal wave of Fig. 2b, let us consider a single symmetrical pulse such as that of Fig. 7. For convenience, we choose a reference frame in which the pulse remains stationary. That is, we run along with the pulse, keeping it constantly in view. In this frame, the string will appear to move past us, from right to left in Fig. 7, with speed v.

Consider a small segment of the pulse, of length Δl, forming an arc of a circle of radius R. A force equal in magnitude to the tension τ pulls tangentially on this segment at each end. The horizontal components of these forces cancel but the vertical components add up to form a restoring force **F**. Thus,

$$F = 2\tau \sin\theta \approx \tau(2\theta) = \tau\frac{\Delta l}{R} \quad \text{(force).} \tag{23}$$

We have used here the approximation that $\sin\theta \approx \theta$ for small angles and we note that $2\theta = \Delta l/R$. The mass of the segment is given by

$$\Delta m = \mu\,\Delta l \quad \text{(mass).} \tag{24}$$

---

* We would normally use the symbol T for tension but we want to avoid confusion with our use of this symbol for the period of oscillation.

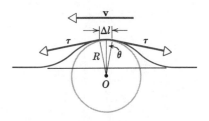

Figure 7   A symmetrical pulse, viewed from a reference frame in which the pulse is stationary. We find the speed by applying Newton's second law to a string element of length Δl, located at the top of the pulse.

At the moment shown in Fig. 7, the string element is moving in an arc of a circle. Thus, it has a centripetal acceleration toward the center of that circle, given by

$$a = \frac{v^2}{R} \quad \text{(acceleration).} \tag{25}$$

Equations 23, 24, and 25 contain the elements of Newton's second law. Combining them gives

$$\text{force} = \text{mass} \times \text{acceleration,}$$

or

$$\frac{\tau\,\Delta l}{R} = (\mu\,\Delta l)\frac{v^2}{R}.$$

Solving this equation for the speed v yields

$$\boxed{v = \sqrt{\frac{\tau}{\mu}}} \quad \text{(speed),} \tag{26}$$

in exact agreement with Eq. 22 if the constant C in that equation is given the value unity.

Equation 26 tells us that the *speed* of a wave along a stretched ideal string depends only on the characteristics of the string and not on the frequency of the wave. The *frequency* of the wave is fixed entirely by the agency that sets up the wave (for example, the person in Fig. 2b). The *wavelength* of the wave is then fixed by Eq. 14 ($v = \lambda\nu$).

Sample Problem 4   A climber, whose mass m is 86 kg, slides down a rope, as in Fig. 8. The leader wishes to signal him by giving the top end of the rope a sharp tap. How long does it take for the signal to travel 32 m down the rope? The linear density μ of the rope is 74 g/m.

From Eq. 26, the speed of the pulse as it travels down the

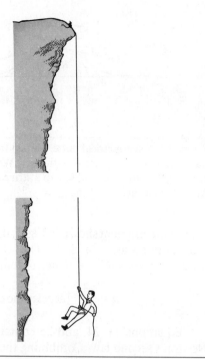

Figure 8 Sample Problem 4. A climber sliding down a rope. How long will it take a pulse to reach him?

rope is

$$v = \sqrt{\frac{\tau}{\mu}} = \sqrt{\frac{mg}{\mu}} = \sqrt{\frac{(86 \text{ kg})(9.8 \text{ m/s}^2)}{0.074 \text{ kg/m}}}$$
$$= 107 \text{ m/s}.$$

Note that, neglecting the weight of the rope itself, we have taken the tension in the rope to be constant along its length and equal to the weight of the climber. The travel time for the pulse is

$$t = \frac{l}{v} = \frac{32 \text{ m}}{107 \text{ m/s}} = 0.30 \text{ s} = 300 \text{ ms}. \quad \text{(Answer)}$$

17-7 The Speed of Light (Optional)

A velocity only has meaning if you specify a reference frame. For the speed of a wave traveling along a stretched string, the logical reference frame is the undisturbed string itself. The situation is the same for sound waves in air, for ripples in water, for seismic waves in the earth, and for all other mechanical waves. When we quote a speed, the implied reference frame is the medium through which the wave travels.

However, visible light (and all other radiations that join with it to form the electromagnetic spectrum) requires no medium for transmission. It travels freely through the vacuum of empty space. When we say, "The speed c of light is 299,792,458 m/s," what reference frame do we have in mind?

The answer to this historic question, which was argued about in one form or another for decades, was given by Einstein in 1905 and it forms one of the two postulates on which his special theory of relativity is based:

The speed of light has the same value c in all directions and in all inertial reference frames.

In measuring the speed of light, the relative velocity of the source of light and the observer simply does not matter.

Think about what this says. Suppose that you have a pulsed laser light source in your laboratory and that you measure the speed of the light in the beam of pulses that it emits, obtaining a value c. Suppose (see Fig. 9) that another experimenter, who is rushing away from you and your light source at very high speed, also measures the speed of the light in that same beam. Einstein's postulate says that you and this second observer will measure the same value c for the speed of light.

Many of us, on first meeting this postulate, tend to reject it on the grounds that it violates "common sense."

Figure 9 A laser in laboratory S emits a beam of light. Its speed is measured by two observers, one in laboratory S and the other in laboratory S′, which is moving at high speed with respect to S. They both find the same result.

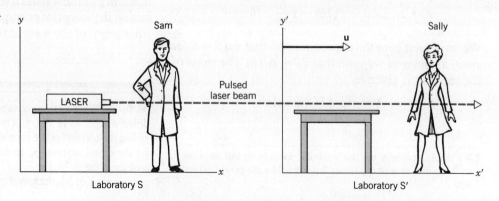

Before doing so, however, consider these facts:

Fact 1.   The postulate does not apply to mechanical waves such as sound waves or to material particles such as baseballs — only to light. Light is different.

Fact 2.   Our perception of "common sense" is developed from watching objects moving at speeds very much less than the speed of light. If the speed of light were very much lower (say 500 mi/h) and thus closer to our daily experiences, none of us would have any conceptual problem with this postulate. It would seem to be common sense of the highest order.

Fact 3.   The postulate itself and the predictions of the theory of relativity for which it forms the foundation have been tested exhaustively in the laboratory; the agreement with experiment is total. No exceptions have ever been confirmed.

We shall return to this postulate and explore its implications fully in Chapter 42, which deals with Einstein's theory of relativity. Meanwhile, we present this foretaste of one of the foundations on which this remarkable theory rests.

17-8   Energy and Power in a Traveling Wave   (Optional)

A traveling wave, passing along a stretched string, transports both kinetic and potential energy. Let us consider each in turn.

Kinetic Energy.   An element of the string of mass dm, oscillating back and forth in simple harmonic motion as the wave passes through it, has kinetic energy associated with its transverse velocity **u**. When the element is rushing through its $y = 0$ position, its transverse velocity — and thus its kinetic energy — is a maximum; see Fig. 10. When the element is in its $y = y_m$ position, its transverse velocity — and thus again its kinetic energy — is zero. An oscillating pendulum bob behaves in a similar way.

Potential Energy.   To send a sinusoidal wave along a previously straight string, you must necessarily stretch the string. As a string element of length dx oscillates back and forth transversely, its length must increase and decrease in a periodic way if the (elastic) string element is to fit the sinusoidal wave form. Potential energy is associated with these length changes, just as for a spring.

When the string element is in its $y = y_m$ position (see Fig. 10), its length has its normal undisturbed value dx so

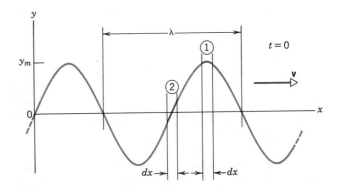

Figure 10   In string element ①, the stored kinetic energy and the stored potential energy are each zero. In string element ②, these stored energies each have their maximum values. The kinetic energy depends on the transverse velocity of the string element. The potential energy depends on the amount by which the string element is stretched as the wave passes through it.

that its stored potential energy is zero. However, when the element is rushing through its $y = 0$ position, it is steeply tilted and thus stretched to its maximum extent. Its stored potential energy then has its maximum value.

The oscillating string element has both its maximum kinetic energy and its maximum potential energy at $y = 0$. An oscillating pendulum bob behaves differently, having its maximum kinetic energy at $y = 0$ but its maximum potential energy at $y = y_m$.

The Transmitted Power.   The kinetic energy dK associated with a string element of mass dm is given by

$$dK = \tfrac{1}{2}\, dm\, u^2. \tag{27}$$

Here u is not the velocity of the wave but the transverse velocity of the oscillating string element, given (see Eq. 21) as

$$u = \frac{\partial y}{\partial t} = -\omega y_m \cos(kx - \omega t). \tag{28}$$

Using this relation and putting $dm = \mu\, dx$ allows us to rewrite Eq. 27 as

$$dK = \tfrac{1}{2}(\mu\, dx)(-\omega y_m)^2 \cos^2(kx - \omega t). \tag{29}$$

Dividing Eq. 29 by dt gives the rate at which kinetic energy is carried along by the wave. The ratio dx/dt is simply the wave velocity so that we have

$$\frac{dK}{dt} = \tfrac{1}{2}\mu v\omega^2 y_m^2 \cos^2(kx - \omega t). \tag{30}$$

The *average* rate at which kinetic energy is transported is

$$\left(\overline{\frac{dK}{dt}}\right) = \tfrac{1}{2}\mu v\omega^2 y_m^2 \,\overline{\cos^2(kx - \omega t)} = \tfrac{1}{4}\mu v\omega^2 y_m^2. \quad (31)$$

We have taken this average over any integral number of wavelengths and have used the fact that the average value of the square of a cosine (or a sine) function over an integral number of wavelengths is $\tfrac{1}{2}$.

Potential energy is also carried along with the wave, and at the same average rate given by Eq. 31. Although we provide no proof, we recall that, in an oscillating system such as a pendulum or a spring–block system, the average kinetic energy and the average potential energy are indeed equal.

The *average power,* which is the average rate at which energy of both kinds is transmitted by the wave, is then

$$\overline{P} = 2\left(\overline{\frac{dK}{dt}}\right) \quad (32)$$

or, from Eq. 31,

$$\boxed{\overline{P} = \tfrac{1}{2}\mu v\omega^2 \, y_m^2.} \quad (33)$$

Two factors in this equation (μ and v) depend on the string alone and two others (ω and y_m) on the wave alone. The fact that the average power transmitted by a wave depends on the square of its amplitude and also on the square of its frequency is a general result, true for waves of all types.

Sample Problem 5 A string has a linear density μ of 525 g/m and is stretched with a tension τ of 45 N ($\approx$ 10 lb). A wave whose frequency ν and amplitude y_m are 120 Hz and 8.5 mm, respectively, is traveling along the string. At what average rate is the wave transporting energy along the string?

Before finding $\overline{P}$ from Eq. 33, we must calculate the angular frequency ω and the wave speed v. From Eq. 10,

$$\omega = 2\pi\nu = (2\pi)(120 \text{ Hz}) = 754 \text{ rad/s.}$$

From Eq. 26 we have

$$v = \sqrt{\frac{\tau}{\mu}} = \sqrt{\frac{45 \text{ N}}{0.525 \text{ kg/m}}} = 9.26 \text{ m/s.}$$

Equation 33 then yields

$$\begin{aligned}
\overline{P} &= \tfrac{1}{2}\mu v\omega^2 y_m^2 \\
&= (\tfrac{1}{2})(0.525 \text{ kg/m})(9.26 \text{ m/s})(754 \text{ rad/s})^2(0.0085 \text{ m})^2 \\
&= 100 \text{ W.} \qquad\qquad\qquad\qquad\qquad\qquad \text{(Answer)}
\end{aligned}$$

17–9 The Principle of Superposition

It often happens that two or more waves pass simultaneously through the same region of space. In listening to a symphony orchestra, for example, sounds from many instruments fall simultaneously on our eardrums. The electrons in the antennas of our radio and TV sets are set in motion by a whole array of signals from different broadcasting centers. The water of a lake or harbor may be churned up by the wakes of many boats.

Suppose that two waves travel simultaneously along the same stretched string. Let $y_1(x, t)$ and $y_2(x, t)$ be the displacements that the string would experience if each wave acted alone. The displacement of the string when

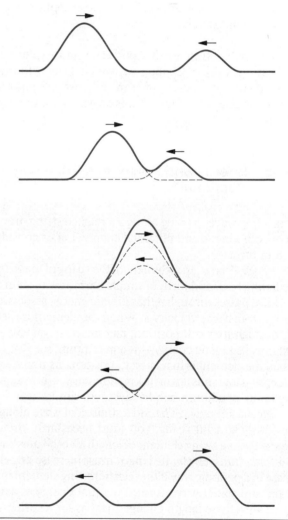

Figure 11 Two pulses travel in opposite directions along a stretched string. The superposition principle applies as they move through each other

both waves act is then

$$y(x, t) = y_1(x, t) + y_2(x, t), \qquad (34)$$

the sum being an algebraic sum. This is an example of the *principle of superposition,* a principle that is broadly useful in physics, cropping up as it does in acoustics, optics, electromagnetism, and quantum physics. For waves in stretched strings, the principle holds as long as the amplitudes of the waves are not too large.

Figure 11 shows a time sequence of "snapshots" of two pulses traveling in opposite directions in the same stretched string. When the pulses overlap, the displacement of the string is the algebraic sum of the individual displacements of the string, as Eq. 34 requires. The pulses simply move through each other, each moving along as if the other were not present.

Fourier Analysis.   The French mathematician Jean Baptiste Fourier (1786–1830) showed how the principle of superposition can be used to analyze nonsinusoidal wave forms. He showed that any wave form can be represented as the sum of a large number of sinusoidal waves, of carefully selected frequencies and amplitudes. The English physicist Sir James Jeans expressed it well:

> [Fourier's] theorem tells us that every curve, no matter what its nature may be, or in what way it was originally obtained, can be exactly reproduced by superposing a sufficient number of simple harmonic curves—in brief, every curve can be built up by piling up waves.

Figure 12 shows an example of Fourier's theorem. The sawtooth wave form in Fig. 12a shows the variation with time (at position $x = 0$) of the wave to which we wish to apply this theorem. We state without proof that the Fou-

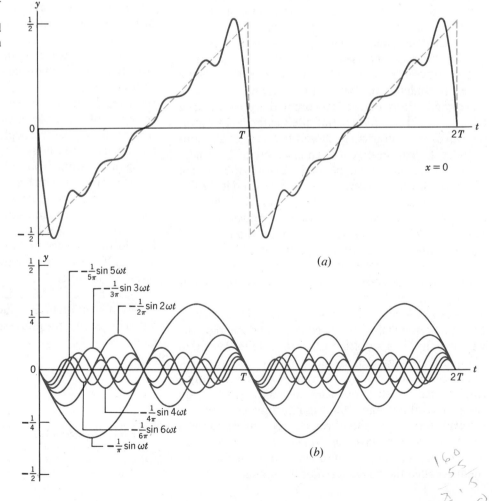

Figure 12   (a) The dashed sawtooth curve is approximated by keeping six terms in the sum of Eq. 35. (b) A separate display of the six terms of that equation.

rier series that represents it is

$$y(t) = -\frac{1}{\pi}\sin \omega t - \frac{1}{2\pi}\sin 2\omega t - \frac{1}{3\pi}\sin 3\omega t \cdots, \quad (35)$$

in which $\omega = 2\pi/T$, where T is the period of the saw-tooth wave. The colored curve of Fig. 12a, which represents the sum of the first six terms of Eq. 35, matches the sawtooth curve rather well. Figure 12b shows these six terms separately. By adding more terms, you could approximate the sawtooth curve as closely as you wish.

We see now why we were justified in spending so much time analyzing the behavior of a sinusoidal wave. Once we understand that, Fourier's theorem opens the door to all other wave shapes for us.

17–10 Signaling with Waves (Optional)

A truly sinusoidal wave, such as that of Fig. 13a, has no beginning and no end, either in space or time, and all its periodic intervals are identical. You cannot use such a wave to send a signal from point to point.

We *can* send a signal with a *pulse,* such as that of Fig. 13b, which might represent the quantity 1 in the binary arithmetic of computers. The pulse of Fig. 13b is *not* a pure sinusoidal wave; that is, it does *not* have a unique wavelength. From Fourier's theorem, this pulse must be built up of a distribution of (infinitely extended) sinusoidal waves of appropriately chosen frequencies and amplitudes. It is impressive that an assembly of sinusoidal waves can be found such that they add up to the pulse of Fig. 13b, canceling to zero everywhere beyond its limits.

You may ask: "Do all the sinusoidal waves that combine to form the pulse of Fig. 13b travel with the same speed?" If the answer is yes, then the pulse will travel at that same common speed and will retain its shape as it does so. If the answer is no, the pulse will travel at a characteristic speed of its own and it will spread out as it travels; it will *disperse.*

For waves on a stretched string, the speed is given by Eq. 26 and we see that this speed does not depend on the wavelength. Thus, a pulse, such as that of Fig. 2a or that of Fig. 13b, will travel along the string at this same speed. Pulses of sound behave in the same way, having only a single speed for all frequencies or wavelengths. If you shout toward a distant cliff, the returning echo retains the wave form of the shout; you can recognize your own words. Waves that have this property are said to be *dispersionless* and the common speed that characterizes them is called the *phase speed* of the waves.

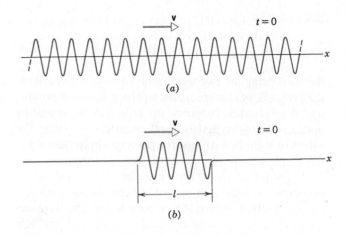

Figure 13 (a) A "snapshot" of the sinusoidal wave of Eq. 2, at $t = 0$. (b) A "snapshot" of a pulse, or wave train, of length l. This can be represented by Fourier methods as the summation of many sinusoidal waves of different wavelengths and amplitudes.

Light waves traveling through free space are also dispersionless. However, when light travels through a transparent medium such as air or glass, the speed *does* depend on the wavelength; such waves exhibit *dispersion.* It is because of dispersion that sunlight spreads out into a spectrum of colors as it passes through a prism or is refracted by drops of water to form a rainbow. The changing of the shape of electromagnetic pulses by dispersion as they pass through the atmosphere is one problem facing those who seek to destroy ballistic missiles by firing pulses from ground-based lasers at them.

The speed at which a pulse travels is called the *group speed* of the pulse. This is the speed at which signals or information can be transmitted. If there is no dispersion, the group speed is the same as the phase speed, given by Eq. 26 for the case of waves traveling along stretched strings. If there *is* dispersion, the group speed and the phase speed differ.*

The motion of waves in water is dispersive. For long waves in deep water, in which the restoring force is provided by gravity, the phase speed and the group speed are given by

$$v_{\text{ph}} = \sqrt{\frac{g\lambda}{2\pi}} \quad \text{and} \quad v_{\text{gr}} = \frac{1}{2}\sqrt{\frac{g\lambda}{2\pi}}. \quad (36)$$

---

* For a more complete discussion of phase and group speed, see Robert Resnick and David Halliday, *Basic Concepts in Relativity and Early Quantum Theory,* 2nd ed., John Wiley & Sons, New York, 1985, Appendix G.

Note that long waves travel faster than short waves and that the group speed is just half of the phase speed. It is not hard to observe this phenomenon. From a boat, focus attention on a particular, noncresting wave (a pulse) as it travels toward the shore. You will see smaller waves superimposed on it and traveling through it, starting at the rear of the pulse and moving through it, to disappear at its leading edge. These smaller waves are traveling at the phase speed, which is twice the group speed.

17-11   Interference of Waves

Suppose that we send two sinusoidal waves of the same wavelength and amplitude in the same direction along a stretched string. The superposition principle applies. What resultant disturbance does it predict for the string?

Everything depends on the extent to which the waves are *in phase* with respect to each other. If they are rigorously in step, they will add up to double the displacement of either wave acting alone. If they are rigorously out of step, they will cancel everywhere, producing no disturbance at all. We call this phenomenon of cancellation and reinforcement *interference;* it applies to waves of all kinds. In Fig. 14, for example, light waves from two tiny adjacent illuminated slits interfere with each other as they fall on a screen, producing a pattern of alternating maximum and minimum light intensity. We shall study the interference of light in detail in Chapter 40.

Let two waves given by

$$y_1(x, t) = y_m \sin (kx - \omega t + \phi) \qquad (37)$$

and

$$y_2(x, t) = y_m \sin (kx - \omega t) \qquad (38)$$

travel along the same stretched string. We see that these waves have the same angular frequency ω, the same angular wave number k, and the same amplitude y_m.

They travel in the same direction, that of increasing x, with the same speed, that given by Eq. 26. They differ only by a constant angle ϕ, called the *phase constant*.

From the principle of superposition (Eq. 34), the combined wave is

$$y(x, t) = y_1(x, t) + y_2(x, t)$$
$$= y_m[\sin (kx - \omega t + \phi) + \sin (kx - \omega t)]. \quad (39)$$

From Appendix G we see that we can write the sum of the sines of two angles as

$$\sin \alpha + \sin \beta = 2 \sin \tfrac{1}{2}(\alpha + \beta) \cos \tfrac{1}{2}(\alpha - \beta). \quad (40)$$

Applying this relation to Eq. 39 yields

$$\boxed{y(x, t) = [2y_m \cos \tfrac{1}{2}\phi] \sin (kx - \omega t + \tfrac{1}{2}\phi).} \quad (41)$$

The resultant wave is thus also a sinusoidal wave, differing from the original waves only in its phase constant, which is $\tfrac{1}{2}\phi$, and in its amplitude, which is the quantity in the square brackets, namely, $2y_m \cos \tfrac{1}{2}\phi$.

If $\phi = 0$ in Eq. 41, this equation reduces to

$$y(x, t) = 2y_m \sin (kx - \omega t) \qquad (\phi = 0). \quad (42)$$

Here the two combining waves are (exactly) in phase and their interference is (fully) constructive. For $\phi = 0$, the two combining waves (see Eqs. 37 and 38) are identical in all respects and the resultant wave differs from either only in having twice the amplitude. Figure 15a (see page 406) shows a case in which, although ϕ is not zero, it is very small.

If $\phi = 180°$, the factor $\cos \tfrac{1}{2}\phi = \cos 90° = 0$, so that the amplitude of the combining wave (Eq. 41) is zero. We have, for all values of x and t,

$$y(x, t) = 0 \qquad (\phi = 180°). \quad (43)$$

Here the combining waves are (exactly) out of phase and their interference is (totally) destructive. Figure 15b shows a case in which, although ϕ is not 180°, it is close to that value. The two combining waves almost — but not quite — cancel.

Figure 14   A pattern of light and dark fringes, caused by the interference of light from two adjacent slitlike sources. The combining waves reinforce each other at the fringe maxima and cancel each other between these maxima.

Sample Problem 6   Two traveling waves, moving in the same direction along a stretched string, interfere with each other. The amplitude y_m of each wave is 9.7 mm and the phase difference ϕ between them is 110°. (a) What is the amplitude y'_m of the wave formed by the interference of these two waves?

From Eq. 41 we have for the amplitude

$$y'_m = 2y_m \cos \tfrac{1}{2}\phi = (2)(9.7 \text{ mm})(\cos 110°/2)$$
$$= 11.1 \text{ mm.} \qquad \text{(Answer)}$$

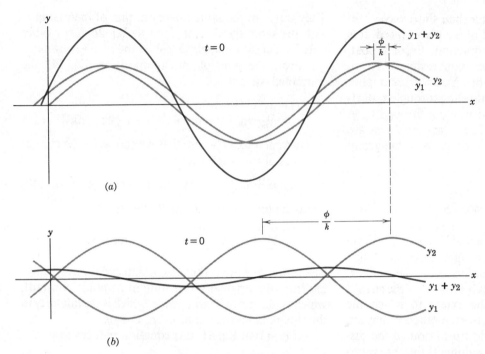

Figure 15 (a) Two waves, whose phases differ by only a small amount, reinforce each other. (b) Two waves, whose phases differ by almost 180°, almost cancel each other.

The amplitude of the combined wave varies (depending on the value of ϕ) between zero and $2y_m$ ($= 19.4$ mm). The amplitude corresponding to $\phi = 110°$ is 57% of the maximum possible value.

(b) What phase difference ϕ between the two combining waves would result in an amplitude for the combined wave that is the same as the (common) amplitudes of the combining waves?

From Eq. 41 we have the requirement

$$2y_m \cos \tfrac{1}{2}\phi = y_m,$$

or

$$\phi = 2 \cos^{-1} (\tfrac{1}{2}) = 120° \text{ or } -120°. \quad \text{(Answer)}$$

Thus, there are two solutions, one corresponding to the first wave leading the second in time and the other to the second wave leading the first.

Hint 2: *Trigonometric identities.* In deriving Eq. 41 above (and in many other places), we have used trigonometric identities such as that of Eq. 40. Practicing engineers and scientists become familiar with these through constant use but, because there are so many of them, the beginner is often at a loss. Appendix G lists all the identities used in this book, plus a few more for completeness. It is helpful to study them, looking for patterns so that, for example, when you see something like $\sin \alpha + \cos \beta$ you will know that this can be expressed in another form.

Try to cast formulas into forms for which identities exist. For example, if you run across $3 \sin \alpha \cos \alpha$, you can write it as $(\tfrac{3}{2})(2 \sin \alpha \cos \alpha)$, which (see Appendix G) is just $\tfrac{3}{2} \sin 2\alpha$. Rejoice when you see $\sin^2 \alpha + \cos^2 \alpha$ because it is equal to unity. Remember that, although the average value of a sine or a cosine function over one wavelength is zero, the average over one full loop (half a wavelength) is $\tfrac{1}{2} \sqrt{2}$ and the average of the *square* of a sine or a cosine function over one wavelength is $\tfrac{1}{2}$.

17–12 Standing Waves

In the preceding section, we treated two sinusoidal waves of the same wavelength and amplitude traveling *in the same direction* along a stretched string. What if they travel in opposite directions? We can learn from the superposition principle.

Figure 16 suggests the situation graphically. We see the two combining waves, one traveling to the left in Fig. 16a, the other to the right in Fig. 16b. Figure 16c shows their sum, obtained by applying the superposition principle graphically. The outstanding feature of the resultant wave is that there are places along the string, called *nodes,* where the string is permanently at rest. Four such

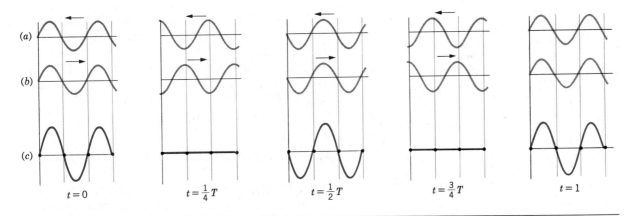

$t = 0$ $t = \frac{1}{4}T$ $t = \frac{1}{2}T$ $t = \frac{3}{4}T$ $t = 1$

Figure 16 How traveling waves produce standing waves. *(a)* and *(b)* represent two waves of the same wavelength and amplitude, traveling in opposite directions. *(c)* Their superposition at four instants during a single period of oscillation. Note the nodes and antinodes in *(c)*, the nodes being represented by dots. There are no nodes or antinodes in the traveling waves *(a)* or *(b)*.

nodes are marked by dots in Fig. 16*c*. Halfway between adjacent nodes are *antinodes*, where the amplitude of the wave disturbance is a maximum. Wave patterns such as that of Fig. 16*c* are called *standing waves*. Let us analyze them mathematically.

We represent the two combining waves by

$$y_1(x, t) = y_m \sin (kx - \omega t) \qquad (44)$$

and

$$y_2(x, t) = y_m \sin (kx + \omega t). \qquad (45)$$

We introduce no phase constants here because it means nothing to speak of a phase difference between waves that travel in opposite directions. In the same spirit, the sweep second hands of two clocks can maintain a constant angular separation if the hands are rotating in the same direction but they cannot do so if the hands are rotating in opposite directions.

The principle of superposition gives, for the combined wave,

$$\begin{aligned} y(x, t) &= y_1(x, t) + y_2(x, t) \\ &= y_m \sin (kx - \omega t) + y_m \sin (kx + \omega t). \end{aligned}$$

Applying the trigonometric relation of Eq. 40 leads to

$$\boxed{y(x, t) = [2y_m \sin kx] \cos \omega t.} \qquad (46)$$

This is *not* a traveling wave because it is not of the form of Eq. 18. Equation 46 describes a *standing wave*.

The quantity in the square brackets of Eq. 46 can be

viewed as the amplitude* of oscillation of the string element located at position *x*. In a traveling sinusoidal wave, the oscillation amplitude is the same for all string elements, regardless of their location. For a standing wave, such as that of Eq. 46, that is not the case. There are certain values of *x* for which the amplitude is zero, namely, those values of *x* for which *kx* has the values $0, \pi, 2\pi, 3\pi$, and so on. Recalling (Eq. 5) that $k = 2\pi/\lambda$, we can write this condition as

$$x = n\frac{\lambda}{2}, \qquad n = 0, 1, 2, 3, \cdots \quad \text{(nodes).} \quad (47)$$

These are the nodes that we mentioned in connection with Fig. 16*c*. Equation 47 tells us that adjacent nodes are one-half wavelength apart.

There are also values of *x* for which the amplitude of vibration has its maximum value, that is, $2y_m$. These occur when $kx = \pi/2, 3\pi/2, 5\pi/2$, and so on. Recalling again that $k = 2\pi/\lambda$, we can write this condition as

$$x = (n + \tfrac{1}{2})\frac{\lambda}{2}, \qquad n = 0, 1, 2, 3, \cdots$$

$$\text{(antinodes).} \quad (48)$$

These are the antinodes of Fig. 16*c*. The antinodes are

---

* The amplitude of oscillation is always positive, so we must identify the amplitude with the *absolute value* of the quantity in square brackets in Eq. 46, regardless of sign.

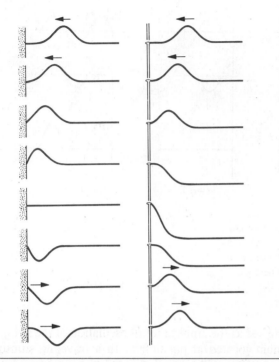

Figure 17 (*a*) A pulse incident from the right is reflected at a rigid wall. Note that the sign of the reflected pulse is reversed. (*b*) Here the termination is flexible, the loop being able to slide without friction up and down the rod. The pulse is reflected without a change of sign.

one-half wavelength apart and are located halfway between pairs of nodes.

Reflections at a Boundary. We can set up a standing wave in a stretched string by allowing a traveling wave to be reflected from the far end of the string. The incident wave and the reflected wave can then be described by Eqs. 44 and 45, respectively, and they can combine to form a pattern of standing waves.

In Fig. 17, we use an incident pulse to show how such reflections take place. In Fig. 17*a*, let the string be fixed at its far end. When the pulse arrives at that end, it exerts an upward force on the support. By Newton's third law, the support exerts an equal but opposite force on the string. This reaction force generates a pulse at the support, which travels back along the string in a direction opposite to that of the incident pulse. In a "hard" reflection of this kind, there must be a node at the support. Thus, the reflected and the incident pulses must have opposite signs, canceling each other at this point.

In Fig. 17*b*, the end of the string is fastened to a light ring that is free to slide without friction along a rod. When the incident pulse arrives, the ring moves up the

rod. Because of its inertia, the ring overshoots and exerts a reaction force on the string. This force sends a reflected pulse back down the string. In a "soft" reflection of this kind, there must be an antinode at the support. Thus, the reflected and the incidence pulses must have the same sign, reinforcing each other at this point.

17–13 Standing Waves and Resonance

Figure 18 shows four standing wave patterns (or *oscillation modes*) that can be set up by wiggling one end of a stretched rubber tube at different frequencies. Note the nodes and antinodes. The important statement about the modes of Fig. 18 is that they occur only at sharply defined frequencies. We say that the system *resonates* at these frequencies.

Consider a string of length *l*, clamped at each end. Because the ends of the string cannot move, a node of the standing wave pattern must exist at each end of the string. The length *l* of the string must then be an integral multiple of a half-wavelength, or $l = n(\frac{1}{2}\lambda)$. The allowed wavelengths are then

$$\lambda = \frac{2l}{n}, \qquad n = 1, 2, 3, \cdots. \qquad (49)$$

The allowed frequencies follow from Eq. 49 and from Eq. 14 ($v = \nu\lambda$). Thus,

$$\boxed{\nu = \frac{v}{\lambda} = \frac{v}{2l}n, \qquad n = 1, 2, 3, \cdots.} \qquad (50)$$

Only if the stretched string is wiggled at one of the frequencies given by Eq. 50 will a pattern of standing waves develop.

If the string is wiggled at a frequency not given by Eq. 50, it will not be possible to transfer energy efficiently from the external wiggling agent to the string. For some time intervals, the external agent (which may be a mechanical vibrator) will do work on the string; at other intervals the string will do work back on the vibrator. At resonance the energy flow is entirely from the vibrator to the string. The vibration amplitude builds up until the vibrating string loses energy by frictionlike losses just as fast as it receives energy from the vibrator.

The phenomenon of resonance is common to all oscillating systems. Figure 19, for example, shows three oscillation modes for a chain hanging under its own weight. Figure 20 shows two of the many (two-dimen-

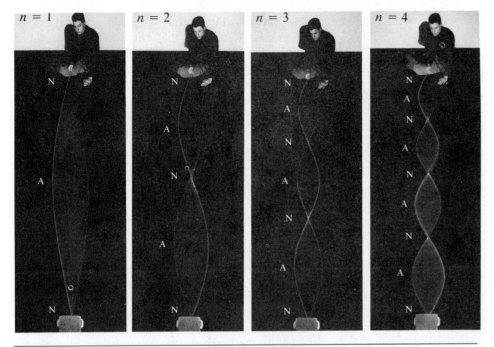

Figure 18   A student wiggles a stretched rubber tube at four critical frequencies, producing four standing wave patterns. The patterns correspond to $n = 1, 2, 3,$ and 4 in Eq. 50.

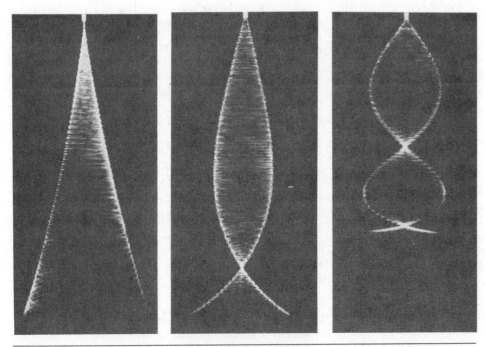

Figure 19   Three modes of oscillation for a freely hanging chain. They occur at different frequencies.

Figure 20 Four of many possible oscillation modes for the head of a kettledrum. They are made visible by sprinkling dark powder on the drumhead. As the drumhead is set into vibration at a single frequency, by a mechanical vibrator, the powder moves toward the nodes, which are lines (rather than points) in this two-dimensional example.

sional) oscillation modes of a kettledrum head.* In quantum physics, the stationary states in which atoms may exist are interpreted as (three-dimensional) oscillation modes of the matter waves that represent the atomic electrons.

Here are two questions that are often asked about oscillating systems:

1. Why Does a Stretched String Resonate Only at Certain Frequencies? Put another way: Why is the oscillation frequency of a stretched string quantized? It is because the string is not infinitely long and thus has two ends. This is what gave rise to the requirement that there must be a node at each end and led directly to Eq. 50. An infinitely long string (no ends) could indeed support a standing wave pattern at *any* frequency.

All oscillating systems that are localized in space—

no matter what their nature—oscillate only at discrete frequencies. Requiring that the wave—whatever its nature—vanish at the boundaries of the system leads directly to frequency quantization. The discrete frequencies of the standing waves in a guitar string and the sharply defined frequencies of the light emitted by atoms are both traceable to this cause. The guitar string has a finite length; the atomic electrons (which can be viewed as a three-dimensional pattern of matter waves) are required, by electrical forces, to stay close to the atomic nucleus. Both systems are localized in space; both systems are frequency quantized.

2. Why Does a Stretched String Have Many Resonant Frequencies but a Mass–Spring System Has Only One? The answer is that, in the mass–spring system, the inertia of the system is concentrated into one system element (the mass) and the elasticity into another (the spring). In a stretched string, on the other hand, the inertia and the elasticity are spread uniformly throughout the system, every element of the string having some of each. In the first case, there is only one possible mechanical arrangement that permits energy to flow back and forth from kinetic to potential forms as the system oscillates. In the second case, there are many such arrangements, each with its own resonant frequency.

Sample Problem 7 In the arrangement of Fig. 21, the frequency v of the vibrator is 120 Hz, the length l of the string is 1.2 m, and the linear density of the string is 1.6 g/m. What must be the tension in the string if it is to vibrate in a mode with one loop?

Combining Eq. 26 ($v = \sqrt{\tau/\mu}$) with Eq. 50 and solving for τ, we obtain

$$\tau = \frac{4l^2v^2\mu}{n^2}, \qquad n = 1, 2, 3, \cdots. \tag{51}$$

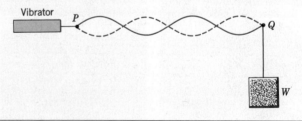

Figure 21 Sample Problem 7. A string under tension connected to a vibrator. For a fixed vibrator frequency, standing wave patterns will occur for discrete values of the tension in the string.

---

* See "The Physics of Kettledrums," by Thomas D. Rossing, *Scientific American*, November 1982.

For one loop (compare Fig. 18) we have $n = 1$, so that

$$\tau = \frac{(4)(1.2 \text{ m})^2(120 \text{ Hz})^2(0.0016 \text{ kg/m})}{1^2}$$

$$= 133 \text{ N.} \qquad \text{(Answer)}$$

This corresponds to a 30-lb hanging weight in Fig. 21. As the demonstrator gradually relaxes the tension in the string, resonance occurs with an increasing number of loops. Although the resonant frequency (which is determined by the vibrator) is always the same in this experiment, the wavelength and the wave speed decrease as the tension is decreased.

REVIEW AND SUMMARY

Transverse and Longitudinal Waves

Waves on a stretched string, the subject of this chapter, are *transverse* mechanical waves governed by Newton's laws. The particles of the medium (the string) move perpendicular to the direction of motion of the wave. Waves in which the medium particles move parallel to the wave propagation are *longitudinal* waves.

The disturbance $y(x, t)$ in a sinusoidal wave moving in the $+x$ direction has the mathematical form

Sinusoidal Waves

$$y(x, t) = y_m \sin(kx - \omega t). \qquad [2]$$

y_m is the *amplitude* of the wave, k the *angular wave number*, ω the *angular frequency*, and $kx - \omega t$ the *phase*. The *wavelength* λ and *wave number* κ (the number of waves per meter) are related to k by

Wave Number and Wavelength

$$\frac{k}{2\pi} = \kappa = \frac{1}{\lambda}. \qquad [5,6]$$

The *period* T and the *frequency* v are related to ω by

Frequency and Period

$$\frac{\omega}{2\pi} = v = \frac{1}{T}. \qquad [9,10]$$

Finally, the wave speed v is related to these other parameters by

Wave Speed

$$v = \frac{\omega}{k} = \frac{\lambda}{T} = \lambda v. \qquad [14]$$

Sample Problems 1–3 illustrate the use of these relations.

In general, any disturbance described by an equation of the form

Traveling Waves

$$y(x, t) = f(kx \pm \omega t) \qquad [18]$$

represents a traveling wave with a wave speed given by Eq. 14 and a wave shape given by the mathematical form of f. The plus (minus) sign denotes a wave traveling in the $-x$ $(+x)$ direction.

The speed of a wave on a stretched string with tension τ and linear density μ is

Wave Speed for Stretched String

$$v = \sqrt{\frac{\tau}{\mu}}. \qquad [26]$$

See Sample Problem 4.

The Speed of Light

The speed of light, in vacuum, has the same value c in all inertial reference frames. This invariance of the speed of light is one of the two postulates of Einstein's special theory of relativity and is now thought to be universally valid.

Power

The *average power*, the average rate at which energy is transmitted by a sinusoidal wave on a stretched string, is given by

$$\bar{P} = \tfrac{1}{2}\mu v \omega^2 y_m^2. \qquad [33]$$

See Sample Problem 5.

Superposition

When two or more waves traverse the same space, the displacement of any particle is the sum of the displacements that the individual waves would give it. This is called *superposition*. See Fig.

Fourier's Theorem

11. *Fourier's theorem* tells us that any wave can be constructed as the superposition of appropriate sinusoidal waves. See Fig. 12.

The transfer of information requires a *wave pulse,* which according to Fourier's theorem can be thought of as a superposition of sinusoidal waves. If the individual waves all have the same wave speed, the pulse shape is transmitted without change at the common *phase* speed of the waves; the wave is *dispersionless.* If the individual waves have different speeds, the wave is said to exhibit *dispersion* and the pulse changes shape as it travels. The pulse shape then travels at the *group* speed. As an example, the group speed v_{gr} for long waves in deep water is just half the phase speed v_{ph} of individual waves:

Signaling with Waves

Group Speed and Phase Speed

$$v_{ph} = \sqrt{\frac{g\lambda}{2\pi}} \quad \text{and} \quad v_{gr} = \frac{1}{2}\sqrt{\frac{g\lambda}{2\pi}}. \tag{36}$$

Interference of Waves

Two sinusoidal waves on the same string exhibit *interference,* the behavior of the string for small amplitudes being governed by superposition. If the two are traveling in the same direction and have the same amplitude y_m and frequency (and hence the same wavelength), but differ in phase by a phase constant ϕ, the result is a single wave with this same frequency:

$$y(x, t) = [2y_m \cos\tfrac{1}{2}\phi] \sin(kx - \omega t + \tfrac{1}{2}\phi). \tag{41}$$

If $\phi = 0°$, the waves are in phase and their interference is (fully) constructive; if $\phi = 180°$, they are out of phase and their interference is destructive. See Fig. 15 and Sample Problem 6.

The interference of two sinusoidal waves having the same frequency and amplitude but moving in opposite directions produces *standing waves* with the equation, for a string with fixed ends,

Standing Waves

$$y = [2y_m \sin kx] \cos \omega t. \tag{46}$$

Standing waves are characterized by fixed positions of zero displacement called *nodes* and positions of maximum displacement called *antinodes;* see Fig. 16 and Eqs. 47 and 48.

Wave Reflection

Standing waves on a string are typically induced by reflection of traveling waves from the ends of the string. If an end is fixed, it must be the position of a node; if it is free, it is the position of an antinode. These *boundary conditions* limit the frequencies of waves for which standing waves will occur on a given string. Each possible frequency is a *resonant* frequency and the corresponding standing wave pattern is an *oscillation mode.* For a stretched string with fixed ends the resonant frequencies are

Resonant Frequencies

$$v = \frac{v}{\lambda} = \frac{v}{2l}\, n, \qquad n = 1, 2, 3, \ldots. \tag{50}$$

The string will readily absorb energy if it is excited at one of its resonant frequencies; excited at other frequencies, it absorbs little energy. This is the phenomenon of *resonance.* Sample Problem 7 analyzes one example.

QUESTIONS

1. How could you prove experimentally that energy can be transported by a wave?

2. Energy can be transferred by particles as well as by waves. How can we experimentally distinguish between these methods of energy transfer?

3. Can a wave motion be generated in which the particles of the medium vibrate with angular simple harmonic motion? If so, explain how and describe the wave.

4. The following functions in which A is a constant are of the form $f(x \pm vt)$:

$$y = A(x - vt), \qquad y = A(x + vt)^2,$$
$$y = A\sqrt{x - vt}, \qquad y = A \ln (x + vt).$$

Explain why these functions are not useful in wave motion.

5. Can one produce on a string a wave form that has a discontinuity in slope at a point, that is, one having a sharp corner? Explain.

6. Compare and contrast the behavior of (*a*) the mass of a mass–spring system, oscillating in simple harmonic motion, and (b) an element of a stretched string through which a traveling sinusoidal wave is passing. Discuss from the point of view of displacement, velocity, acceleration, and energy transfers.

7. How do the amplitude and the intensity of surface water waves vary with the distance from the source?

8. A passing motor boat creates a wake that causes waves to wash ashore. As time goes on, the period of the arriving waves grows shorter and shorter. Why?

9. When two waves interfere, does one alter the progress of the other? Explain.

10. When waves interfere, is there a loss of energy? Explain your answer.

11. Why don't we observe interference effects between the light beams emitted from two flashlights or between the sound waves emitted by two violins?

12. As Fig. 16 shows, twice during a cycle the configuration of standing waves in a stretched string is a straight line, exactly what it would be if the string were not vibrating at all. Where is the energy of the standing wave at these times?

13. If two waves differ only in amplitude and are propagated in opposite directions through a medium, will they produce standing waves? Is energy transported? Are there any nodes? (See Problem 64.)

14. In the discussion of transverse waves in a string, we have dealt only with displacements in a single plane, the xy plane. If all displacements lie in one plane, the wave is said to be *plane polarized.* Can there be displacements in a plane other than the single plane with which we dealt? If so, can two differently plane-polarized waves be combined? What appearance would such a combined wave have?

15. A wave transmits energy. Does it transfer momentum? Can it transfer angular momentum? (See Question 14.) (See "Energy and Momentum Transport in String Waves," by D. W. Juenker, *American Journal of Physics,* January 1976.)

16. In the Mexico City earthquake of 19 September 1985, areas with high damage alternated with areas of low damage. Also, buildings between 5 and 15 stories high sustained the most damage. Discuss these effects in terms of standing waves and resonance.

EXERCISES AND PROBLEMS

Section 17–5 The Speed of Traveling Waves

1E. A wave has a speed of 240 m/s and wavelength 3.2 m. What are (a) the frequency and (b) the period of the wave?

2E. A wave has an angular frequency of 110 rad/s and wavelength of 1.8 m. Calculate (a) the angular wave number and (b) the speed of the wave.

3E. By rocking a boat, a boy produces surface water waves on a previously quiet lake. He observes that the boat performs 12 oscillations in 20 s, each oscillation producing a wave crest 15 cm above the undisturbed surface of the lake. He further observes that a given wave crest reaches shore, 12 m away, in 6.0 s. What are (a) the period, (b) the speed, (c) the wavelength, and (d) the amplitude of this wave?

4E. The speed of electromagnetic waves in vacuum is 3.0×10^8 m/s. (a) Wavelengths in the visible part of the spectrum (light) range from about 400 nm in the violet to about 700 nm in the red. What is the range of frequencies of light waves? (b) The range of frequencies for shortwave radio (for example, FM radio and VHF television) is 1.5–300 MHz. What is the corresponding wavelength range? (c) X rays are also electromagnetic. Their wavelength range extends from about 5.0 nm to about 1.0×10^{-2} nm. What is the frequency range for x rays?

5E. A sinusoidal wave travels along a string. The time for a particular point to move from maximum displacement to zero is 0.17 s; What are the (a) period and (b) frequency? (c) The wavelength is 1.4 m; what is the wave speed?

6E. Write the equation for a wave traveling in the negative direction along the x axis and having an amplitude of 0.010 m, a frequency of 550 Hz, and a speed of 330 m/s.

7E. A traveling wave on a string is described by

$$y = 2 \sin\left[2\pi\left(\frac{t}{0.40} + \frac{x}{80}\right)\right],$$

where x and y are in centimeters and t is in seconds. (a) For $t = 0$, plot y as a function of x for $0 \le x \le 160$ cm. (b) Repeat (a) for $t = 0.05$ s and $t = 0.10$ s. (c) From your graphs, what is the wave speed, and in which direction ($+x$ or $-x$) is the wave traveling?

8E. Show that $y = y_m \sin(kx - \omega t)$ may be written in the alternative forms

$$y = y_m \sin k(x - vt), \qquad y = y_m \sin 2\pi\left(\frac{x}{\lambda} - vt\right),$$

$$y = y_m \sin \omega\left(\frac{x}{v} - t\right), \qquad y = y_m \sin 2\pi\left(\frac{x}{\lambda} - \frac{t}{T}\right).$$

9E. We know (see Exercise 8) that $f(x - vt)$, where f is any function, represents a wave traveling in the positive x direction. To illustrate this, consider the function shown in Fig. 22. (a) What are the values of $f(0), f(1), f(2), f(3), f(4),$ and $f(5)$? (b) Plot $f(x - 5t)$ as a function of x for $0 \le x \le 20$ and $t = 0$. Here x is in centimeters and t is in seconds. (c) Repeat (b) for $t = 1$ s and $t = 2$ s. (d) From your graphs, what is the wave speed? (e) Plot $f(x - 5t)$ as a function of t for $0 \le t \le 2$ s for $x = 10$ cm.

10E. Show (a) that the maximum transverse speed of a particle in a string owing to a traveling wave is given by $u_{max} =$

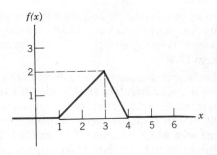

Figure 22 Exercise 9.

$\omega y_m = 2\pi v y_m$ and (b) that the maximum transverse acceleration is $a_{y,max} = \omega^2 y_m = 4\pi^2 v^2 y_m$.

11P. The equation of a transverse wave traveling in a string is given by

$$y = (2.0 \text{ mm}) \sin [(20 \text{ m}^{-1})x - (600 \text{ s}^{-1})t].$$

(a) Find the amplitude, frequency, velocity, and wavelength of the wave. (b) Find the maximum transverse speed of a particle in the string.

12P. (a) Write an expression describing a transverse wave traveling on a cord in the $+y$ direction with a wave number of 60 cm^{-1}, a period of 0.20 s, and an amplitude of 3.0 mm. Take the transverse direction to be the z direction. (b) What is the maximum transverse speed of a point on the cord?

13P. The equation of a transverse wave traveling along a very long string is given by $y = 6.0 \sin (0.020\pi x + 4.0\pi t)$, where x and y are expressed in centimeters and t is in seconds. Calculate (a) the amplitude, (b) the wavelength, (c) the frequency, (d) the speed, (e) the direction of propagation of the wave, and (f) the maximum transverse speed of a particle in the string. (g) What is the transverse displacement at $x = 3.5$ cm when $t = 0.26$ s?

14P. (a) Write an expression describing a transverse wave traveling on a cord in the $+x$ direction with a wavelength of 10 cm, a frequency of 400 Hz, and an amplitude of 2.0 cm. (b) What is the maximum speed of a point on the cord? (c) What is the velocity of the wave?

15P. (a) A continuous sinusoidal longitudinal wave is sent along a coiled spring from a vibrating source attached to it. The frequency of the source is 25 Hz, and the distance between successive rarefactions in the spring is 24 cm. Find the wave speed. (b) If the maximum longitudinal displacement of a particle in the spring is 0.30 cm and the wave moves in the $-x$ direction, write the equation for the wave. Let the source be at $x = 0$ and the displacement at $x = 0$ when $t = 0$ be zero.

16P. For a wave on a stretched cord, find the ratio of the maximum particle speed (the speed with which a single particle in the cord moves transverse to the wave) to the wave speed. If a wave having a certain frequency and amplitude is imposed on a cord, would this speed ratio depend on the material of which the cord is made, such as wire or nylon?

17P. Prove that the slope of a string at any point is numerically equal to the ratio of the particle speed to the wave speed at that point.

18P. A wave of frequency 500 Hz has a velocity of 350 m/s. (a) How far apart are two points 60° out of phase? (b) What is the phase difference between two displacements at a certain point at times 1.0 ms apart?

Section 17–6 Wave Speed on a Stretched String

19E. What is the speed of a transverse wave in a rope of length 2.0 m (= 6.6 ft) and mass 0.060 kg (= 0.0041 slug) under a tension of 500 N (= 110 lb)?

20E. The heaviest and lightest strings in a certain violin have linear densities of 3.0 g/m and 0.29 g/m. What is the ratio, heaviest to lightest, of the diameters of these strings, assuming they are made of the same material?

21E. The speed of a wave on a string is 170 m/s when the tension is 120 N. To what value must the tension be increased in order to raise the wave speed to 180 m/s?

22E. The tension in a wire clamped at both ends is doubled without appreciably changing its length. What is the ratio of the new to the old wave speed for transverse waves in this wire?

23E. Show that, in terms of the tensile stress S and mass density ρ, the speed v of transverse waves in a wire is given by

$$v = \sqrt{\frac{S}{\rho}}.$$

24E. The equation of a transverse wave on a string is

$$y = (2.0 \text{ mm}) \sin [(20 \text{ m}^{-1})x - (600 \text{ s}^{-1})t].$$

The tension in the string is 15 N. (a) What is the wave speed? (b) Find the linear density of this string in grams per meter.

25E. The linear density of a vibrating string is 1.6×10^{-4} kg/m. A transverse wave is propagating on the string and is described by the equation $y = (0.021 \text{ m})\sin[(2.0 \text{ m}^{-1})x + (30 \text{ s}^{-1})t]$. (a) What is the wave speed? (b) What is the tension in the string?

26P. What is the fastest transverse wave that can be sent along a steel wire? Allowing for a reasonable safety factor, the maximum tensile stress to which steel wires should be subject is 7.0×10^8 N/m^2. The density of steel is 7800 kg/m^3. Show that your answer does not depend on the diameter of the wire.

27P. A stretched string has a mass per unit length of 5.0 g/cm and a tension 10 N. A wave on this string has an amplitude of 0.12 mm and a frequency of 100 Hz and is traveling in the negative x direction. Write an equation for this wave.

28P. A simple harmonic transverse wave is propagating along a string toward the left (or $-x$) direction. Figure 23 shows a plot of the displacement as a function of position at time $t = 0$. The string tension is 3.6 N and its linear density is 25 g/m. Calculate (a) the amplitude, (b) the wavelength, (c) the wave speed, (d) the period, and (e) the maximum speed of a

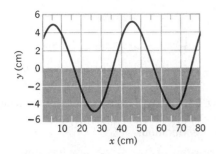

Figure 23 Problem 28.

particle in the string. (*f*) Write an equation describing the traveling wave.

29P. A continuous sinusoidal wave is traveling on a string with velocity 40 cm/s. The displacement of the particles of the string at $x = 10$ cm is found to vary with time according to the equation $y = (5.0 \text{ cm}) \sin[1.0 - (4.0 \text{ s}^{-1})t]$. The linear density of the string is 4.0 g/cm. (*a*) What is the frequency of the wave? (*b*) What is the wavelength of the wave? (*c*) Write the general equation giving the transverse displacement of the particles of the string as a function of position and time. (*d*) Calculate the tension in the string.

30P. In Fig. 24*a*, string #1 has a linear density of 3.0 g/m, and string #2 has a linear density of 5.0 g/m. They are under tension owing to the hanging block of mass $M = 500$ g. (*a*) Calculate the wave speed in each string. (*b*) The block is now divided into two blocks (with $M_1 + M_2 = M$) and the apparatus rearranged as shown in Fig. 24*b*. Find M_1 and M_2 such that the wave speeds in the two strings are equal.

31P. A wire 10 m long and having a mass of 100 g is stretched under a tension of 250 N. If two disturbances, separated in time by 0.030 s, are generated one at each end of the wire, where will the disturbances meet?

32P. The type of rubber band used inside some baseballs and golfballs obeys Hooke's law over a wide range of elongation of

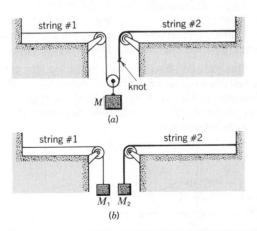

Figure 24 Problem 30.

the band. A segment of this material has an unstretched length *l* and a mass *m*. When a force *F* is applied, the band stretches an additional length Δl. (*a*) What is the speed (in terms of *m*, Δl, and the force constant *k*) of transverse waves on this rubber band? (*b*) Using your answer to (*a*), show that the time required for a transverse pulse to travel the length of the rubber band is proportional to $1/\sqrt{\Delta l}$ if $\Delta l \ll l$ and is constant if $\Delta l \gg l$.

33P*. A uniform rope of mass *m* and length *l* hangs from a ceiling. (*a*) Show that the speed of a transverse wave in the rope is a function of *y*, the distance from the lower end, and is given by $v = \sqrt{gy}$. (*b*) Show that the time it takes a transverse wave to travel the length of the rope is given by $t = 2\sqrt{l/g}$.

Section 17–8 Energy and Power in a Traveling Wave

34E. Show that the average power transmitted by a wave can be written in terms of the frequency *v*, rather than angular frequency ω, in the form

$$\overline{P} = 2\pi^2 \mu v v^2 y_m^2.$$

35E. A string 2.7 m long has a mass of 260 g. The tension in the string is 36 N. What must be the frequency of traveling waves of amplitude 7.7 mm in order that the average transmitted power be 85 W?

36P. A transverse sinusoidal wave is generated at one end of a long, horizontal string by a bar that moves up and down through a distance of 1.0 cm. The motion is continuous and is repeated regularly 120 times per second. The string has linear density 120 g/m and is kept under a tension of 90 N. Find (*a*) the maximum value of the transverse speed *u* and (*b*) the maximum value of the transverse component of the tension. (*c*) Show that the two maximum values calculated above occur at the same phase values for the wave. What is the transverse displacement *y* of the string at these phases? (*d*) What is the maximum power transferred along the string? (*e*) What is the transverse displacement *y* for conditions under which this maximum power transfer occurs? (*f*) What is the minimum power transfer along the string? (*g*) What is the transverse displacement *y* for conditions under which this minimum power transfer occurs?

Section 17–11 Interference of Waves

37E. Two identical traveling waves, moving in the same direction, are out of phase by 90°. What is the amplitude of the combined wave in terms of the common amplitude y_m of the two combining waves?

38E. What phase difference between two otherwise identical traveling waves, moving in the same direction along a stretched string, will result in the combined wave having an amplitude 1.5 times that of the common amplitude of the two combining waves? Express your answer both in degrees and radians.

39P. A source *S* and a detector *D* of radio waves are a distance *d* apart on the ground. The direct wave from *S* is found to be in phase at *D* with the wave from *S* that is reflected from a horizontal layer at an altitude *H* (Fig. 25). The incident and

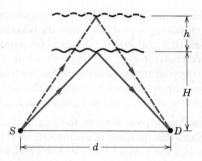

Figure 25 Problem 39.

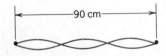

Figure 26 Exercise 45.

reflected rays make the same angle with the reflecting layer. When the layer rises a distance h, no signal is detected at D. Neglect absorption in the atmosphere and find the relation between d, h, H, and the wavelength λ of the waves.

40P. Determine the amplitude of the resultant motion when two sinusoidal motions having the same frequency and traveling in the same direction are combined, if their amplitudes are 3.0 cm and 4.0 cm and they differ in phase by $\pi/2$ rad.

41P. Three sinusoidal waves travel in the positive x direction along the same string. All three waves have the same frequency. Their amplitudes are in the ratio $1 : \frac{1}{2} : \frac{1}{3}$ and their phase angles are 0, $\pi/2$, and π, respectively. Plot the resultant wave form and discuss its behavior as t increases.

42P. Four sinusoidal waves travel in the positive x direction along the same string. Their frequencies are in the ratio $1 : 2 : 3 : 4$ and their amplitudes are in the ratio $1 : \frac{1}{2} : \frac{1}{3} : \frac{1}{4}$, respectively. When $t = 0$, at $x = 0$, the first and third waves are 180° out of phase with the second and fourth. Plot the resultant wave form when $t = 0$ and discuss its behavior as t increases.

Section 17–13 Standing Waves and Resonance

43E. Two waves are propagating on the same very long string. A generator at the left end of the string creates a wave given by

$$y = (6.0 \text{ cm}) \cos \frac{\pi}{2} [(2.0 \text{ m}^{-1})x + (8.0 \text{ s}^{-1})t],$$

and one at the right end of the string creates the wave

$$y = (6.0 \text{ cm}) \cos \frac{\pi}{2} [(2.0 \text{ m}^{-1})x - (8.0 \text{ s}^{-1})t].$$

(a) Calculate the frequency, wavelength, and speed of each wave. (b) Find the points at which there is no motion (the nodes). (c) At which points is the motion a maximum?

44E. A string fixed at both ends is 8.4 m long and has a mass of 0.12 kg. It is subjected to a tension of 96 N and set vibrating. (a) What is the speed of the waves in the string? (b) What is the wavelength of the longest possible standing wave? (c) Give the frequency of that wave.

45E. A nylon guitar string has a linear mass density of 7.2 g/m and is under a tension of 150 N. The fixed supports are 90 cm apart. The string is vibrating in the standing wave pattern shown in Fig. 26. Calculate (a) the speed, (b) wavelength, and (c) frequency of the component waves whose superposition gives rise to this vibration.

46E. The equation of a transverse wave traveling in a string is given by

$$y = 0.15 \sin (0.79x - 13t),$$

in which x and y are expressed in meters and t is in seconds. (a) What is the displacement at $x = 2.3$ m, $t = 0.16$ s? (b) Write down the equation of a wave that, when added to the given one, would produce standing waves on the string. (c) What is the displacement of the resultant standing wave at $x = 2.3$ m, $t = 0.16$ s?

47E. A 15-cm violin string, fixed at both ends, is vibrating in its $n = 1$ mode. The speed of waves in this wire is 250 m/s, and the speed of sound in air is 348 m/s. What are (a) the frequency and (b) the wavelength of the emitted sound wave?

48E. A 120-cm length of string is stretched between fixed supports. What are the three longest possible wavelengths for standing waves in this string? Sketch the corresponding standing waves.

49E. A 125-cm length of string has a mass of 2.0 g. It is stretched with a tension of 7.0 N between fixed supports. (a) What is the wave speed for this string? (b) What is the lowest resonant frequency of this string?

50E. What are the three lowest frequencies for standing waves on a wire 10 m long having a mass of 0.10 kg, which is stretched under a tension of 250 N?

51E. A 1.5-m wire has a mass of 8.7 g and is held under a tension of 120 N. The wire is held rigidly at both ends and set into vibration. Calculate (a) the velocity of waves on the wire, (b) the wavelengths of the waves that produce one- and two-loop standing waves on the string, and (c) the frequencies of the waves that produce one- and two-loop standing waves.

52E. One end of a 120-cm string is held fixed. The other end is attached to a weightless ring that can slide along a frictionless rod as shown in Fig. 27. What are the three longest possible wavelengths for standing waves in this string? Sketch the corresponding standing waves.

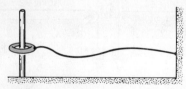

Figure 27 Exercise 52.

53P. A string vibrates according to the equation

$$y = (0.5 \text{ cm}) \left[\sin \left(\frac{\pi}{3} \text{ cm}^{-1} \right) x \right] \cos(40\pi \text{ s}^{-1} t).$$

(a) What are the amplitude and velocity of the component waves whose superposition can give rise to this vibration? (b) What is the distance between nodes? (c) What is the velocity of a particle of the string at the position $x = 1.5$ cm when $t = \frac{2}{8}$ s?

54P. When played in a certain manner, the lowest frequency of vibration of a certain violin string is concert A (440 Hz). What two higher frequencies could also be found on that string if the length is not changed?

55P. A 75-cm string is stretched between fixed supports. It is observed to have resonant frequencies of 420 and 315 Hz, and no other resonant frequencies between these two. (a) What is the lowest resonant frequency for this string? (b) What is the wave speed for this string?

56P. Two transverse sinusoidal waves travel in opposite directions along a string. Each has an amplitude of 0.30 cm and a wavelength of 6.0 cm. The speed of a transverse wave in the string is 1.5 m/s. Plot the shape of the string at times: $t = 0$ (arbitrary), $t = 5.0$, $t = 10$, $t = 15$, and $t = 20$ ms.

57P. Two pulses travel along a string in opposite directions, as in Fig. 28. (a) If the wave velocity is 2.0 m/s and the pulses are 6.0 cm apart, sketch the patterns after 5.0, 10, 15, 20, and 25 ms. (b) What has happened to the energy at $t = 15$ ms?

58P. Two waves on a string are described by the equations

$$y_1 = (0.10 \text{ m}) \sin 2\pi \left[(0.5 \text{ m}^{-1})x + (20 \text{ s}^{-1})t \right]$$

and

$$y_2 = (0.20 \text{ m}) \sin 2\pi \left[(0.5 \text{ m}^{-1})x - (20 \text{ s}^{-1})t \right].$$

Sketch the total response for the point on the string at $x = 3.0$ m; that is, plot y versus t for that value of x.

59P. A 3.0-m long string is vibrating as a three-loop standing wave whose amplitude is 1.0 cm. The wave speed is 100 m/s. (a) What is the frequency? (b) Write equations for two waves that, when combined, will result in this standing wave.

60P. Vibration from a 600-Hz tuning fork sets up standing waves in a string clamped at both ends. The wave speed for the string is 400 m/s. The standing wave has four loops and an amplitude of 2.0 mm. (a) What is the length of the string? (b) Write an equation for the displacement of the string as a function of position and time.

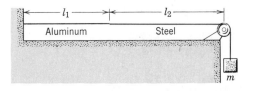

Figure 29 Problem 63.

61P. In an experiment on standing waves, a string 3.0 ft (=0.9 m) long is attached to the prong of an electrically driven tuning fork that vibrates perpendicular to the length of the string at a frequency of 60 vib/s (=60 Hz). The weight of the string is 0.096 lb (mass = 0.044 kg). (a) What tension must the string be under (weights are attached to the other end) if it is to vibrate in four loops? (b) What would happen if the tuning fork were turned so as to vibrate parallel to the length of the string?

62P. Consider a standing wave that is the sum of two waves traveling in opposite directions but otherwise identical. Show that the maximum kinetic energy in each loop of the standing wave is $2\pi^2 \mu y_m^2 v v$.

63P. An aluminum wire of length $l_1 = 60.0$ cm and of cross-sectional area 1.00×10^{-2} cm^2 is connected to a steel wire of the same cross-sectional area. The compound wire, loaded with a block m of mass 10.0 kg, is arranged as shown in Fig. 29 so that the distance l_2 from the joint to the supporting pulley is 86.6 cm. Transverse waves are set up in the wire by using an external source of variable frequency. (a) Find the lowest frequency of excitation for which standing waves are observed such that the joint in the wire is a node. (b) What is the total number of nodes observed at this frequency, excluding the two at the ends of the wire? The density of aluminum is 2.60 g/cm^3, and that of steel is 7.80 g/cm^3.

64P. *Standing wave ratio.* An incident traveling wave, amplitude A_i, is only partially reflected from a boundary, with the amplitude of the reflected wave being A_r. The resulting superposition of two waves with different amplitudes and traveling in opposite directions gives a standing wave pattern of waves whose envelope is shown in Fig. 30. The *standing wave ratio* (SWR) is defined as $(A_i + A_r)/(A_i - A_r) = A_{max}/A_{min}$, and the percent reflection is defined as the ratio of the average power in the reflected wave to the average power in the incident wave, times 100. (a) Show that for 100% reflection SWR $= \infty$ and that for no reflection SWR $= 1$. (b) Show that a measurement of the SWR just before the boundary reveals the percent reflection occurring at the boundary according to the formula

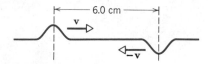

Figure 28 Problem 57.

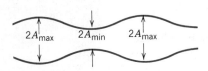

Figure 30 Problem 64.

ESSAY 7
CONCERT HALL ACOUSTICS: SCIENCE OR ART?

JOHN S. RIGDEN
UNIVERSITY OF MISSOURI

On the evening of October 18, 1976, orchestra members of the New York Philharmonic tuned their instruments and readied themselves for a most unusual concert. The setting was Lincoln Center's Avery Fisher Hall, a new name for a concert hall with a troubled past.

Philharmonic Hall had opened in 1962 to high expectations (Fig. 1). One of America's outstanding acousticians, Leo L. Beranek, had worked with the architect who designed the hall. But from the beginning, there were problems: Performers on stage could not hear other parts of the orchestra; the bass sounded weak to people in the audience; echoes from the back wall could be heard; patrons could not hear all that the orchestra was playing. Over the period 1964–1972, many attempts were made to improve the situation, but these attempts failed.

Figure 1 The auditorium of Philharmonic Hall at Lincoln Center for the Performing Arts. This view, from the loge, shows the orchestra level and the stage. Philharmonic Hall was completed in 1962 and at 10:00 a.m. on Monday, 28 May 1962, the young assistant conductor of the New York Philharmonic Orchestra, Seizi Ozawa, commanded the attention of 106 men and women of the orchestra. When his baton came down, the brass and woodwind sections of the orchestra played the opening chords of Brahm's *Third Symphony*.

In December 1974, Cyril M. Harris, an acoustician from Columbia University, was asked to redesign the hall. Harris was most reluctant, but Amyas Ames, chairman of the board of directors of the Lincoln Center for the Performing Arts, was very persuasive. Finally, Harris set down his conditions: First, the inside of the hall would be demolished right back to the girders; second, acoustics would take precedence over aesthetics; third, Philip Johnson would be the architect. These conditions were granted and work began in May 1976, immediately after the Philharmonic's season-ending performance.

Since the hall was to be ready for the start of the fall season, work proceeded through the summer of 1976 at a feverish pace. Harris, wearing a hardhat, watched each step. All was ready for the October 19 beginning of the new concert season, but on October 18 a special concert was performed in the newly designed hall. It was an unusual concert because members of the audience were there by special invitation: They consisted of the construction workers, contractors, subcontractors, and architects who had worked through the summer with such devotion. Music critics were not invited, but they came. When the orchestra finished the fourth movement of Mahler's *Ninth Symphony,* the critics were most favorably impressed; the musicians were delighted; and the construction workers were happy . . . and proud.

Issac Stern is a world-class musician. His instrument is the violin. Yet, when he comes on the stage of a fine concert hall, the concert hall itself becomes an additional instrument he uses to enhance his musical performance. A concert hall is not a passive enclosure in which musical performances occur; rather, it is an active participant in communicating the interpretations of an artist to an attentive patron.

Sound waves, which link a performer and a patron, carry energy. If the source of sound is a violin string oscillating with fundamental and higher harmonic frequencies, pressure undulations with the same frequencies present in the vibrating string are carried by the surrounding air away from the violin. These longitudinal pressure waves transmit acoustic energy away from the source.

A sound wave approaching an open window, reaches the window, passes through, and carries energy from the source-side of the window to the region beyond. An open window is a perfect absorber; it absorbs all the acoustical energy incident on it. The situation is quite different when a sound source is surrounded by reflecting surfaces as is the case in a concert hall: a sound wave strikes a wall and is reflected back into the room; it propagates across the room until it strikes another surface and is again reflected; and so on. If the source of sound delivers energy at a fixed rate into the room, the intensity of the sound builds up rapidly within the enclosed space until it approaches an equilibrium intensity level. A fraction of the incident energy is absorbed by the surface during each reflection; thus, the rate at which the sound level approaches its equilibrium value depends on the nature of the reflecting surfaces. The equilibrium intensity level is reached when the rate of energy absorption by all exposed surfaces equals the rate at which the sound source delivers energy to the enclosure.

A listener in an enclosure hears first the sound that comes directly from the source. Next, after a time interval called the initial-time-delay gap, waves of once-reflected sound reach the listener. Still later, wave after wave of twice-reflected waves pass over the listener. In this fashion, the sound-intensity level increases until the listener is immersed in sound, coming from all directions, at the equilibrium level. The sum of all the reflected sound is called the reverberant sound. At the equilibrium sound level, the ratio of the loudness of the reverberant sound to that of the direct sound determines the fullness of the perceived tone and this fullness of tone is one

Figure 2 Symphony Hall in Boston. This concert hall which opened in 1900, was planned by the Harvard physicist Wallace C. Sabine, a pioneer in the science of acoustics. Sabine developed the quantitative foundations, including an empirical formula for calculating reverberation time, which are today the heart of acoustic design. Note all the irregularities in the walls and ceiling. The conductor Bruno Walter said that Symphony Hall "is the most noble of American concert halls."

characteristic of a fine concert hall. It is reverberant sound, coupled with the absence of interfering noise, that distinguishes musical performances experienced in a fine concert hall from those heard in a town park.

Just as sound *builds up to* an equilibrium level, so it *decays from* an equilibrium level. If the source of sound is terminated, the direct sound is the first to go and a listener perceives an abrupt drop in the sound level. Then the rate of decay slows somewhat until the last once-reflected sound waves reach the listener. The sound level trails off exponentially as weaker and weaker multiply reflected waves come successively to the ears of the listener.

The time required for the sound level to build up to or to decay from its equilibrium value is called the reverberation time and it is the most important acoustical characteristic of a concert hall. Specifically, the reverberation time is defined as the time required for the sound intensity (watts per meter^2) to build up or decrease by a factor of one million. If the reverberation time is too short, musical notes are heard isolated one from the other and the music is perceived as thin. If, on the other hand, the reverberation time is too long, the sounds from earlier notes clash with the notes being played. Typically, the most desirable reverberation time for symphonic music is about 2 s: Symphony Hall in Boston, one of the finest concert halls in the world, has a reverberation time of 1.8 s when it is fully occupied (Fig. 2); the Musikvereinssaal in Vienna, another excellent concert hall, has a reverberation time of 2.05 s (fully occupied).

The reverberation time depends on the volume of the concert hall and the nature of the reflecting surfaces. The larger the volume, the longer it takes sound, traveling at approximately 345 m/s, to traverse the distances between reflecting walls and the longer it takes the reverberant sound field to build up to its equilibrium level. The volume of Symphony Hall is 61,496 m³. The volume of Carnegie Hall in the City of New York is larger, 79,610 m³, yet its reverberation time, 1.7 s, is less than that of Boston's Symphony Hall. The difference is due to the character of the reflecting surfaces. When surfaces exposed to sound waves are highly absorbent, the rate of energy absorption by all surfaces quickly becomes equal to the rate of energy production by all sources; therefore, the smaller is the reverberation time. Thus, we can understand what Isaac Stern means when he says that Carnegie is better in rehearsal than with an audience. Since the absorbing properties of one person are equivalent to 0.5 m² of open window, the reverberation time of a concert hall is longer when it is unoccupied. (This is why reverberation times are measured when the hall is fully occupied. Furthermore, this is why most concert halls have lockers where patrons can hang their winter coats which are very absorbent.)

It all seems quite simple; it is not.

Concert halls are typically located in the middle of large metropolitan areas where sound waves—noise!—are ubiquitous. A concert hall must shield patrons inside from all outside noise: jet planes overhead, city buses and ambulances on adjacent streets, and subways running underground. Sound waves playing over the external structure of a concert hall discover each and every access route to the inside. Even a keyhole can transmit appreciable amounts of sound. The walls themselves transmit sound. Sound waves of all frequencies are incident on the external walls of a concert hall and they drive the walls at the wave frequencies; the vibrating walls, in turn, become the source of sound waves inside the concert hall. The more massive the walls, the greater is the damping of the transmitted wave. As you might predict, massive walls are more likely to vibrate with low frequencies than with high frequencies; thus, as sound waves of all frequencies are incident on the outside walls of a structure, it is the low frequencies to which the walls respond more readily and it is the low tones that are transmitted more readily into the interior of the structure.

Concert halls can be heated and cooled. This means there must be machinery to generate the warm or cool air, fans to push the air, and ducts to transport the air. Mechanical equipment and moving air are sources of noise. The machinery must be placed in a separate building; the ducts must be lined with a sound-absorbing material. Sharp bends, rough joints, or dampers in the ducts must be avoided as they can set the passing air into whirling, turbulent motion. Turbulent air is especially noisy.

One acoustic consultant has told of a potentially disastrous situation involving air-conditioning ducts. When he visited the construction site, he noticed that the ducts had been lowered a few centimeters so that, for support purposes, they rested on the girders. Such contact between the ducts and girders of the concert hall would have coupled the noise of the ducts to the structure itself. Fortunately, the acoustician was able to make a correction before the finished ceiling concealed a problem from view.

Assuming that all ambient sounds, from both external and internal sources, are at a level of a farmhouse bedroom on a quiet night, a conductor can use the full dynamic range of an orchestra, from the loudest *forte fortissimo* to the softest *pianissimo* and even the faintest melodic strain can be heard distinctly. Yet, even if the reverberation time of a quiet concert hall is good, the listening experience of the patron will be marred if the acoustic design of the hall is faulty; furthermore, performers will be unhappy, visiting conductors will fall short of expectations, and orchestra directors will eschew scheduling their orchestras for future performances in a problematic hall.

Concert halls acclaimed by critics, musicians, and audiences can have different shapes (rectangular, fan-shaped, horseshoe-shaped), but they all share certain qualities. Outstanding halls have reverberation times in the 1.7–2.0-s range. Almost as important as reverberation time is a quality, called intimacy, determined by the initial-time-delay gap (the time interval between the receipt of the direct sound by a person in the audience and receipt of the first-reflected sound). The bigger this gap, the less intimate the musical environment. Initial-time-delay gaps range from 10 to 70 ms, but for the best concert halls they are less than 40 ms.

Surfaces in a concert hall not only absorb sound, but they do so selectively; that is, some surfaces absorb high frequencies more readily than low frequencies and vice versa. Suppose, for example, the architectural plans call for the walls and ceiling of a concert hall to have a surface of $\frac{3}{8}$-in. plywood backed by a 4-in. air space. Such surfaces will play havoc with the character of the reverberant sound. For sound waves whose frequencies are below 250 Hz, approximately three-tenths of the intensity of the incident wave will be absorbed *at each reflection.* (Such surfaces can be driven at low frequencies and hence transmit bass sounds to the air space and structure behind.) Since sound waves propagate through air with a speed of 345 m/s, there are about 10 reflections per second in a typical concert hall. Thus, after only 1 s, sound waves with frequencies below 250 Hz are reduced to 1/35 of their original intensity. The result is that the reverberant sound is stripped of its bass component and the concert hall lacks the warmth that comes with a strong low-frequency presence.

In a certain sense, an enclosure is like an organ pipe. Standing waves, with nodes and antinodes, can be set up in an organ pipe whose wavelengths (and frequencies) depend on the length of the pipe. When these frequencies, called resonance frequencies, are excited, the amplitude of the pressure vibrations inside the pipe builds until a strong, dominant sound results. In a similar fashion, standing waves can be established within a concert hall. For example, sound waves can be reflected back and forth between parallel walls. The lowest frequency standing wave, the *fundamental,* has a pressure antinode at each wall and a node in between; thus, the wavelength is equal to twice the distance between the two reflecting walls. Multiples of the fundamental frequency have additional nodes and antinodes. The nodal–antinodal configuration of standing waves can, if a standing wave is prominent, produce both acoustic dead spots and acoustic hot spots—a condition most undesirable for a concert hall where sound uniformity is the objective. For this reason, architects design concert halls so that dimensions are not simple multiples of each other and smooth parallel walls are avoided. Irregularities in the walls and ceiling not only diminish the chance of prominent standing waves being established, they also scatter the sound waves in many directions and contribute to the desired diffuseness of sound (Fig. 3). (Another acoustic hot spot can result from the focusing effect of a curved wall.)

The analogy between an organ pipe and an enclosure is limited by the fact that organ pipes are one dimensional; that is, sound waves propagate back and forth along the length of the pipe. Hence, for the organ pipe, the number of normal modes of oscillation is rather small—on the order of 10. By contrast, concert halls are three-dimensional structures and sound waves can be reflected in all directions; consequently, there are thousands of normal modes. A single tone sounded by a violin may well excite hundreds of the normal modes of the concert hall; vocalists may exploit the normal modes of a concert hall to attain greater dynamic range for certain notes. Thus, standing waves are an important feature of a concert hall; the sound energy briefly captured in the many normal modes of oscillation makes an important contribution to the reverberant sound.

Figure 3 A close-up view of the ceiling of Avery Fisher Hall. The irregularities built into the ceiling reflect sound waves in all directions; thus, they provide an excellent surface for diffusing sound waves. Sound diffusion equalizes the sound intensity throughout a room and eliminates troublesome dead spots, focusing, and echoes.

In spite of our growing body of acoustical knowledge, the first concert in a newly completed concert hall is awaited with both anticipation and uncertainty. This raises the question: Is acoustics a science or is it an art? The many myths about concert hall acoustics lend credence to this question. There is the prevalent idea that the acoustical quality of concert halls improves with age (it does not); there is a tendency to identify prominent features such as decorative gold gild or statuary in great concert halls as the origin of fine acoustics (they are not); there is the mystical idea that the ancients were in possession of secret acoustical principles, principles unknown to moderns, and that if we could rediscover these lost principles, our concert halls would be vastly improved (the ancients had no secret acoustical insights).

The aura of mystery surrounding the subject of acoustics is further enhanced by the fact that acoustics is, in one sense, one of the oldest branches of physics. After all, acoustics played an important role in the design and location of the open-air theaters used by the Greeks and Romans. In another sense, however, and in actual fact, the science of acoustics is young: It was not until around 1900 that acoustical considerations were systematically applied to the design of concert halls. Yet even now, the advice of the acoustician is often ignored by the chief architect. This is what happened in the case of Philharmonic Hall. Over Beranek's reasoned objections, the architect enlarged the seating capacity from 2400 to 2600, not by decreasing the seat dimensions, but by changing the shape and style of the balcony; for economic reasons, both the adjustable ceiling and the irregularities in the side walls (to diffuse the sound) were eliminated. When architects overrule the acoustician, either for economic or for aesthetic reasons, an acoustically inferior concert hall can be the outcome. When this occurs, however, it is not the architect who gets the blame; it is the acoustician and the science of acoustics.

The effect of a concert hall on music goes beyond those moving moments when it effectively links an inspired performer with an eager audience. Enclosures and music have had an effect on each other over centuries. More specifically, the *design* of enclosures and the *composition* of music have had a dramatic influence on each other: Existing auditoriums have affected the type of music composed and new musical compositions have made acoustical demands on enclosed spaces. For example, the long reverberation times of medieval cathedrals — 5 – 10 s — required that music be played or sung very slowly. Because of the long reverberation times in these cathedrals, the spoken word was difficult to understand and chanting replaced speaking. Giovanni Gabrieli, the organist at St. Mark's Basilica in Venice around 1600, wrote music in slow tempo suitable for enclosures with a long reverberation time. Even today, Gregorian chants sound best in acoustical environments similar to medieval cathedrals.

At the time Gabrieli was performing in Venice, new developments were occurring in Italy. Small chapels, rectangular in shape with high ceilings, were built adjacent to large cathedrals. Characteristic of the time, the walls were ornate, rich in sculptural detail. These decorative surfaces were efficient diffusers of sound at all audible frequencies. The small reverberation times — less than 1.5 s — of these small chapels encouraged the development of new musical forms and ushered in the Baroque period. George Frederick Handel wrote much of his music for such environments.

Figure 4 The auditorium of Avery Fisher Hall (the former Philharmonic Hall). During the summer of 1976, the inside of Philharmonic Hall (see Fig. 1) was transformed into the one shown below. Avery Fisher Hall has the classic rectangular shape like Symphony Hall in Boston (Fig. 2). This rectangular design is a direct descendant of the small chapels built during the seventeenth century. Both the walls and the ceiling are amply irregular.

These smaller, more intimate environments considerably enhanced the listening experience (Fig. 4).

With improved acoustical environments, musical performances became more popular. Larger structures were needed to meet the demand of an eager public. Music of the classical period, the symphonies of Josef Haydn, Wolfgang Amadeus Mozart, and Ludwig van Beethoven, are best performed in larger concert halls with reverberation times from 1.5 to 1.7 s. The romantic period followed the classical period and the music of Johannes Brahms, Peter Ilyitch Tchaikovsky, Maurice Ravel, and Richard Strauss thrives in even larger acoustical environments with larger reverberation times—1.8–2.2 s.

The future design of concert halls will exploit not only new materials and methods of construction but also the use of electronic enhancement. If we can use the past as a guide, we can be sure that as concert halls change, composers will create new musical styles that are not only matched to but are also enhanced by the new enclosures.

CHAPTER 18
WAVES—II

A rock group, a symphony orchestra, or a jazz combo are treasure houses of sources of sound waves. There are waves from vibrating strings and wind columns, from drum heads, cymbals, and vocal cords and—as often as not—from a phalanx of loudspeakers. Nobody knows why the overall effect, Fourier-superimposed on the ear drum, often turns out to be pleasant. ("Music has charms to soothe a savage breast,/ To soften rocks, or bend a knotted oak"—Congreve.)

18-1 Sound Waves

Sound waves, broadly defined, are mechanical waves that can travel through a gas, a liquid, or a solid, Seismic prospecting teams use such waves to probe the earth's crust for oil. Ships carry sound ranging gear (sonar) to detect underwater obstacles. Submarines use sound waves to stalk other submarines, largely by listening for the characteristic acoustic signature of propellers or reactor pumps or the occasional dropped mess tray. Figure 1a shows how sonar is used by commercial fishermen to locate schools of fish. Figure 1b, a computer-processed image of a fetal head, shows how sound waves can be used to explore the soft tissues of the human body. Figure 1c shows how sound waves at very high frequencies can be used to create images of very tiny objects, with a resolving power greater than that of any optical microscope.

Suppose that, as in Fig. 2, a seismic exploration crew detonates a test charge at the bottom of a deep well, drilled into granite. Waves of two types travel radially outward—in all directions—from the site of the explosion. There are *transverse* waves, in which the back-and-forth oscillations of small elements of the rock are at right angles to the direction in which the wave is traveling. There are also *longitudinal* waves, in which these oscillations are parallel to the direction of travel.

If the medium through which the waves travel is a

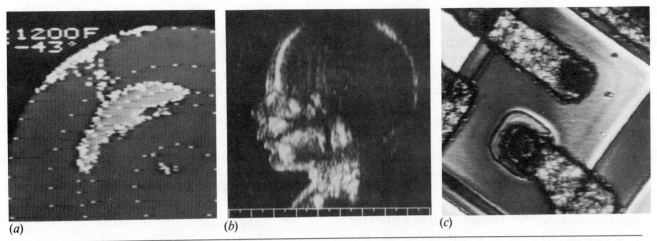

(a) A net full of rock cod, as revealed by the sonar system of a commercial fishing vessel, operating at a frequency of 150 kHz. (b) An ultrasound image of a fetal head at 28 weeks gestation. The frequency of the sound waves used was about 5 MHz. (c) The image of a transistor, viewed in an acoustic microscope at a sound frequency of 4.2 GHz.

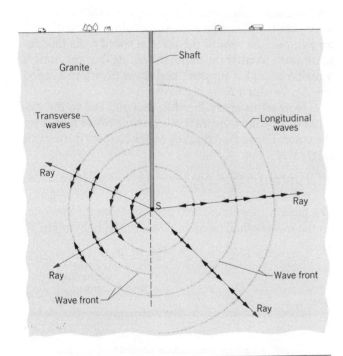

Figure 2   An explosive charge at S sends both longitudinal and (slower) transverse waves traveling throughout the granite in all directions. The waves are shown in separate hemispheres for convenience. The *wave fronts* are surfaces over which the wave disturbance has the same value; *rays* are lines at right angles to the wave fronts, indicating the direction of travel. The short arrows on the rays show the directions of the displacements of small elements of the medium.

fluid (that is, a gas or a liquid), only longitudinal waves can be transmitted. This is because, as we learned in Section 16–2, transverse waves require elasticity of shear to provide a restoring force. The fact that transverse waves from earthquakes are not transmitted through a region near the center of the earth is evidence that part of the earth's central core is liquid.

In this chapter, we shall for the most part be concerned with sound waves defined in the usual sense, that is, as (longitudinal) mechanical waves traveling through air and having frequencies in the audible range.

18–2   The Speed of Sound

The speed of any mechanical wave depends on both an inertial property of the medium (to store kinetic energy) and also on an elastic property (to store potential energy). We can generalize Eq. 26 of Chapter 17, which gives the speed of a transverse wave along a stretched string, by writing

$$v = \sqrt{\frac{\tau}{\mu}} = \sqrt{\frac{\text{elastic property}}{\text{inertial property}}}. \qquad (1)$$

If the medium is air, we can guess that the inertial property, corresponding to the linear density μ for a stretched string, is the volume density ρ of air. What shall we put for the elastic property?

In a stretched string, potential energy is associated with the cyclical stretching of the string elements as the wave passes through. As a sound wave passes through air, potential energy is associated with the periodic compressions and rarefactions of small volume elements of the air. The property that determines the extent to which an element of the medium changes its volume as the pressure applied to it is increased or decreased is the *bulk modulus B. B* is defined (see Eq. 27 of Chapter 13) from*

$$B = -\frac{\Delta p}{\Delta V/V} \quad \text{(definition of } B\text{)}, \qquad (2)$$

in which $\Delta V/V$ is the fractional change in volume produced by a change in pressure Δp. The signs of Δp and ΔV are always opposite. Thus, when we increase the pressure on a fluid element (Δp positive) its volume decreases (ΔV negative). We include a minus sign in the defining equation for B (Eq. 2) so that B will always be a positive quantity. Substituting B for τ and ρ for μ in Eq. 1 yields

$$v = \sqrt{\frac{B}{\rho}} \quad \text{(speed of sound)}. \qquad (3)$$

Table 1 lists the speed of sound in various media.

The density of water is almost 1000 times greater than the density of air. If this were the only relevant factor, we would expect from Eq. 3 that the speed of sound in water would be considerably less than the speed of sound in air. However, Table 1 shows us that the reverse is true. We conclude (again from Eq. 3) that the bulk modulus of water must be greater than that of air by an even larger factor than 1000. This is indeed the case. Water is much more incompressible than air, which (see Eq. 2) is another way of saying that its bulk modulus is much greater.

Formal Derivation of Eq. 3. We now derive Eq. 3 by direct application of Newton's laws. Let a single compressional pulse travel (from left to right) with speed v through the air in a long tube. Let us run along with the pulse at that speed, so that the pulse appears to stand still in our reference frame. Figure 3 shows the situation as it is viewed from that frame. The pulse is standing still and air is moving at speed v through the stationary pulse, from right to left.

Let the pressure of the undisturbed air be p and the

Table 1 The Speed of Sound[a]

Medium	Speed (m/s)
Gases	
Air (0°C)	331
Air (20°C)	343
Helium	965
Hydrogen	1284
Liquids	
Water (0°C)	1402
Water (20°C)	1482
Seawater[b]	1522
Solids	
Aluminum	6420
Steel	5941
Granite	6000

[a] At 0°C and 1 atm pressure.
[b] At 20°C and 3.5% salinity.

pressure inside the pulse be $p + \Delta p$, in which we take Δp to be positive. Consider a slice of air of thickness Δx and area A, moving toward the pulse at speed v. As this element starts to enter the pulse, its leading face encounters a region of higher pressure and slows down, to a speed $v + \Delta v$, in which Δv is negative.

Now let us apply Newton's second law to the element during the time interval Δt required for it to enter the pulse. The net force acting on the element during this interval is

$$F = (pA) - (p + \Delta p)A = -\Delta p\, A \quad \text{(net force)}. \qquad (4)$$

The negative sign indicates (properly) that the net force on the moving fluid element points to the right in Fig. 3.

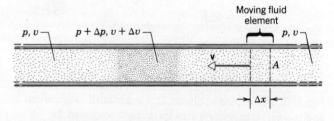

Figure 3 A pulse (compression) is sent down a long air-filled tube. The reference frame of the figure is chosen so that the pulse is at rest and the air is streaming through from left to right. A slice of air of width Δx is shown, moving toward the compressional zone with speed v.

---

* We assume that the compressions and rarefactions of the air are rapid enough so that there is not time for heat transfer to take place from one part of the medium to another. These conditions are met in a sound wave. The bulk modulus for these conditions is called the *adiabatic* bulk modulus. Its value for air at 20°C is 1.4×10^5 Pa.

The volume of the element is $A \, \Delta x$, so that its mass is

$$\Delta m = \rho A \, \Delta x = \rho A v \, \Delta t \quad \text{(mass).} \qquad (5)$$

Finally, the acceleration of the element is

$$a = \frac{\Delta v}{\Delta t} \quad \text{(acceleration).} \qquad (6)$$

Because Δv is negative, the acceleration is also negative and (like the net force) points to the right in Fig. 3.

From Newton's second law ($F = ma$) we have, from Eqs. 4, 5, and 6,

$$(-\Delta p \, A) = (\rho A v \, \Delta t)\frac{\Delta v}{\Delta t},$$

which we can write as

$$\rho v^2 = -\frac{\Delta p}{\Delta v / v}. \qquad (7)$$

Now the air that would occupy a volume $V \, (= Av \, \Delta t)$ outside the pulse is compressed by an amount ΔV ($= A \, \Delta v \, \Delta t$) as it enters the pulse. Thus,

$$\frac{\Delta V}{V} = \frac{A \, \Delta v \, \Delta t}{A v \, \Delta t} = \frac{\Delta v}{v}. \qquad (8)$$

Combining Eqs. 7, 8, and 2 leads to

$$\rho v^2 = -\frac{\Delta p}{\Delta v / v} = -\frac{\Delta p}{\Delta V / V} = B.$$

Solving for v leads at once to Eq. 3, which we set out to prove.

Sample Problem 1   The audible frequency range for normal hearing is from about 20 Hz to 20 kHz. What are the wavelengths of sound waves at these frequencies? Take the speed of sound in air (see Table 1) to be 343 m/s.

From Eq. 14 of Chapter 17 we have, for the lowest frequency,

$$\lambda = \frac{v}{\nu} = \frac{343 \text{ m/s}}{20 \text{ Hz}} = 17 \text{ m} \qquad \text{(Answer)}$$

and, for the highest frequency,

$$\lambda = \frac{v}{\nu} = \frac{343 \text{ m/s}}{20{,}000 \text{ Hz}} = 0.017 \text{ m} = 1.7 \text{ cm.}$$

$$\text{(Answer)}$$

18-3   Traveling Sound Waves

Here we look in some detail at the displacements and pressure variations that we associate with a sound wave passing through air. Figure 4a displays such a wave traveling through a long air-filled tube. Consider a thin slice of air of thickness Δx, located at a position x along the tube. As the wave passes through it, this element oscillates back and forth around its equilibrium position, as the expanded view of Fig. 4b suggests.

The (longitudinal) displacement of the oscillating

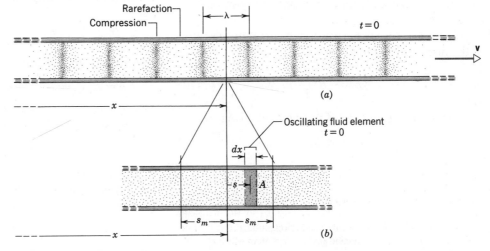

Figure 4   (a) At $t = 0$, a sinusoidal sound wave moves with speed v through a long air-filled tube. (b) An expanded view of a narrow region near position x. A fluid element oscillates about its equilibrium position as the wave passes through. At the moment shown, the central plane of the element is displaced a distance s from this position.

element is given by*

$$s = s_m \cos(kx - \omega t). \qquad (9)$$

The maximum displacement s_m—as we shall show in Sample Problem 2 — is normally very much less than the wavelength of the sound wave. To show the distance s_m in Fig. 4b, its horizontal scale had to be greatly expanded compared with that of Fig. 4a.

As the wave passes by, the pressure at position x in Fig. 4a rises and falls with time, the variation being given (as we shall show below) by

$$\Delta p = \Delta p_m \sin (kx - \omega t). \qquad (10)$$

A negative value of Δp in Eq. 10 corresponds to a rarefaction and a positive value to a compression.

We also show later that the maximum pressure variation in the wave, Δp_m in Eq. 10, is related to the maximum displacement, s_m in Eq. 9, by

$$\Delta p_m = (v\rho\omega)s_m. \qquad (11)$$

The maximum pressure variation Δp_m is normally very much less than the pressure p that prevails when there is no wave.

Figure 5 shows plots of Eqs. 9 and 10 at $t = 0$. Note that the two waves are 90° out of phase, the pressure variation being zero when the displacement is a maximum. Experimentally, it is usually easier to measure the pressure variation than the displacement.

Derivation of Eqs. 10 and 11. Figure 4b shows an oscillating element of air of area A and thickness Δx, its centerline displaced from its equilibrium position by a distance s.

From Eq. 2 we can write, for the pressure variation in the displaced element,

$$\Delta p = -B \frac{\Delta V}{V}. \qquad (12)$$

V in Eq. 12 is the volume of the element, given by

$$V = A \Delta x. \qquad (13)$$

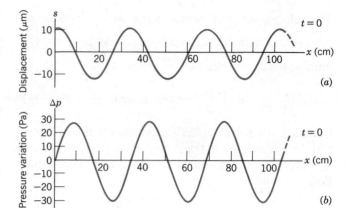

Figure 5 (a) A plot of the displacement function (Eq. 9) for $t = 0$. (b) A similar plot of the pressure variation function (Eq. 10). Both plots are for a 1000-Hz sound wave whose intensity is at the threshold of pain; see Sample Problem 2.

The quantity ΔV in Eq. 12 is the change in volume that occurs because the element is displaced. This volume change comes about because the displacements of the two faces of the element are not quite the same, differing by some amount Δs. Thus,

$$\Delta V = A \, \Delta s. \qquad (14)$$

Substituting Eqs. 13 and 14 into Eq. 12 and passing to the differential limit yields

$$\Delta p = -B \frac{\Delta s}{\Delta x} = -B \frac{\partial s}{\partial x}. \qquad (15)$$

The symbols ∂ shows that the derivative in Eq. 15 is a *partial derivative,** which means that we are interested in how s changes with x at a fixed value of the time variable t. From Eq. 9 we then have, treating t as a constant,

$$\frac{\partial s}{\partial x} = \frac{\partial}{\partial x} [s_m \cos(kx - \omega t)] = -ks_m \sin(kx - \omega t).$$

Substituting this quantity into Eq. 15 yields

$$\Delta p = Bks_m \sin(kx - \omega t) = \Delta p_m \sin(kx - \omega t), \quad (16)$$

which is exactly (see Eq. 10) what we set out to prove.

From Eq. 16 we can write (using Eq. 3)

$$\Delta p_m = (Bk)s_m = (v^2\rho k)s_m.$$

Equation 11, which we also promised to prove, follows at once if we eliminate k by using Eq. 14 of Chapter 17 ($v = \omega/k$).

---

* For the (transverse) displacement of the element of a stretched string (see Eq. 2 of Chapter 17), we used the symbol $y(x, t)$. We change the notation here to $s(x, t)$ because the wave is longitudinal, the displacement being in the same direction as the direction in which the wave travels.

---

* See the footnote on p. 398.

Sample Problem 2   The maximum pressure variation Δp_m that the ear can tolerate in loud sounds is about 28 Pa. What is the displacement amplitude s_m for such a sound in air, at a frequency of 1000 Hz?

From Eq. 11 we have

$$s_m = \frac{\Delta p_m}{v\rho\omega} = \frac{\Delta p_m}{v\rho 2\pi v}$$

$$= \frac{28 \text{ Pa}}{(343 \text{ m/s})(1.21 \text{ kg/m}^3)(2\pi)(1000 \text{ Hz})}$$

$$= 1.1 \times 10^{-5} \text{ m} = 11 \ \mu\text{m}. \qquad \text{(Answer)}$$

We see that the displacement amplitude for even the loudest sound that the ear can tolerate is very small, being about one-seventh the thickness of this page. This amplitude is also very much smaller than the wavelength of a 1000-Hz sound wave in air, which is 34 cm.

The pressure amplitude Δp_m ($= 28$ Pa) is correspondingly small, considering that normal atmospheric pressure is about 10^5 Pa. Figure 5 is plotted for the conditions of this problem.

The pressure amplitude Δp_m for the *faintest* detectable sound at 1000 Hz is 2.8×10^{-5} Pa. Proceeding as above leads to $s_m = 1.1 \times 10^{-11}$ m or 11 pm, which is about 10 times smaller than the radius of a typical atom. The ear is indeed a sensitive detector of sound waves. The ear, in fact, can detect a pulse of sound whose total energy is as small as a few electron-volts, about the same as the energy required to remove an outer electron from a single atom.

18-4   Intensity and Sound Level

If you are trying to sleep when someone is operating a chain saw nearby, you are well aware that there is more to sound than frequency, wavelength, and speed. There is loudness, or *intensity*. The intensity I of a sound wave is defined to be the average rate per unit area at which energy is transmitted by the wave. The SI unit for intensity is thus watts per meter^2 (W/m^2). In a sound wave, the intensity I is related to the displacement amplitude Δs_m by

$$I = \tfrac{1}{2}\rho v\omega^2 s_m^2, \qquad (17)$$

a relation that we shall derive later.

We saw in Sample Problem 2 that the displacement amplitude at the human ear ranged from about 10^{-5} m for the loudest tolerable sound to about 10^{-11} m for the faintest detectable sound, a ratio of 10^6. From Eq. 17 we see that the intensity of a sound varies as the *square* of its amplitude, so that the ratio of intensities between these two limits of sound level would be 10^{12}, an enormous value.

We deal with such an enormous range of variables by using logarithms. Consider the relation

$$y = \log x,$$

in which x and y are general algebraic variables. It is a property of this equation that if we *multiply* x by a factor of 10, y increases by *adding* $\log 10$ ($= 1$) to it. Thus,

$$y' = \log(10x) = \log 10 + \log x = 1 + y.$$

The Richter scale for measuring the energy E of earthquakes is based on a similar formula, namely,

$$\log E = 4.4 + 1.5M,$$

in which M is the Richter magnitude and E is given in joules. You can show that if you *add* unity to a Richter magnitude number, the energy of the corresponding quake increases by a *multiplying* factor of 31.6. The magnitudes of stars and the acidities of solutions (pH values) are also measured on logarithmic scales.

Thus, instead of speaking of the intensity I of a sound wave, we find it much more convenient to speak of a *sound level* β, defined from

$$\beta = (10 \text{ dB}) \log \frac{I}{I_0}, \qquad (18)$$

in which dB is the abbreviation for *decibel* the unit of sound level, the name chosen to recognize the work of Alexander Graham Bell. I_0 in Eq. 18 is a standard reference intensity ($= 10^{-12}$ W/m^2), chosen because it is near the lower limit of human audibility. We see that for $I = I_0$, $\beta = 10 \log 1 = 0$, so that our standard reference level corresponds to zero decibels. Table 2 shows some

Table 2   Equation 18 Explored

β (dB)	I/I_0
0	$10^0 = 1$
10	$10^1 = 10$
20	$10^2 = 100$
30	$10^3 = 1000$
40	$10^4 = 10{,}000$
50	$10^5 = 100{,}000$
.	.
.	.
.	.
120	$10^{12} = 1{,}000{,}000{,}000{,}000$

values of β and the corresponding values of the intensity ratio I/I_0. The table shows us that β increases by the same *constant amount* every time the sound intensity I increases by a given *factor*. As it turns out, that is just the way the human ear operates, so that the ear can be uniformly sensitive over a wide range of sound intensities. The response of the ear is not the same at every frequency. Figure 6 shows how the thresholds of hearing and of pain vary throughout the acoustic spectrum for persons with average hearing. Table 3 shows the intensity levels of some sounds.

Derivation of Eq. 17. Consider, in Fig. 4a, a thin slice of air of thickness dx, area A, and mass dm, oscillating back and forth as the sound wave of Eq. 9 passes through it. Its kinetic energy dK is

$$dK = \tfrac{1}{2}dm\, v_s^2. \tag{19}$$

Here v_s is not the speed of the wave but the speed of the oscillating element of air, given from Eq. 9 by

$$v_s = \frac{\partial s}{\partial t} = -\omega s_m \sin(kx - \omega t).$$

Using this relation and putting $dm = \rho A\, dx$ allows us to rewrite Eq. 19 as

$$dK = \tfrac{1}{2}(\rho A\, dx)(-\omega s_m)^2 \sin^2(kx - \omega t). \tag{20}$$

Dividing Eq. 20 by dt gives the rate at which kinetic

Table 3 Some Sound Levels (dB)

Threshold of hearing	0
Rustle of leaves	10
Whisper (at 1 m)	20
City street, no traffic	30
Office, classroom	50
Normal conversation (at 1 m)	60
Jackhammer (at 1 m)	90
Rock group	110
Threshold of pain	120
Jet engine (at 50 m)	130
Saturn rocket (at 50 m)	200

energy moves along with the wave. The ratio dx/dt is simply the wave velocity so that we have

$$\frac{dK}{dt} = \tfrac{1}{2}\rho A v \omega^2 s_m^2 \sin^2(kx - \omega t). \tag{21}$$

The *average* rate at which kinetic energy is transported is

$$\overline{\left(\frac{dK}{dt}\right)} = \tfrac{1}{2}\rho A v \omega^2 s_m^2 \, \overline{\sin^2(kx - \omega t)} = \tfrac{1}{4}\rho A v \omega^2 s_m^2. \tag{22}$$

We have taken this average over one wavelength* of the wave and have used the fact that the average value of the square of a sine (or a cosine) function over one wavelength is $\tfrac{1}{2}$.

We assume that *potential* energy is carried along with the wave at this same rate. The wave intensity I, which is the average rate per unit area at which energy of both kinds is transmitted by the wave, is then, from Eq. 22,

$$I = \frac{2(\overline{dK/dt})}{A} = \tfrac{1}{2}\rho v \omega^2 s_m^2,$$

which is exactly Eq. 17, the equation we set out to derive.

Sample Problem 3 (a) Two sound waves have intensities I_1 and I_2. How do their sound levels compare?

Let us write the ratio of the two intensities as

$$\frac{I_2}{I_1} = \frac{I_2/I_0}{I_1/I_0}.$$

Taking logarithms of each side and multiplying through by

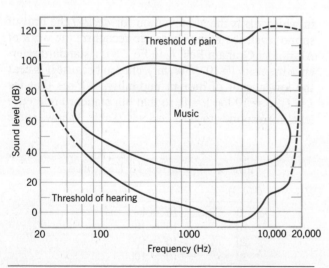

Figure 6 The average range of sound levels for human hearing. Both the threshold of pain and that of hearing depend on the frequency. The approximate ranges of frequencies and sound levels encountered in music are also shown.

---

* Or over any integral number of wavelengths.

10 dB, we have

$$(10 \text{ dB}) \log \frac{I_2}{I_1} = (10 \text{ dB}) \log \frac{I_2}{I_0} - (10 \text{ dB}) \log \frac{I_1}{I_0}.$$

From Eq. 18 we then see that

$$(10 \text{ dB}) \log \frac{I_2}{I_1} = \beta_2 - \beta_1. \quad \text{(Answer)} \quad (23)$$

Thus, the *ratio* of two intensities corresponds to a *difference* in their sound levels.

(b) If you multiply the intensity of a sound wave by 10 you add 10 dB to the sound level, as we have seen. If you multiply the intensity by a factor of 2, by how much do you raise its sound level?

From Eq. 23

$$\beta_2 - \beta_1 = (10 \text{ dB}) \log \frac{I_2}{I_1}$$

$$= (10 \text{ dB}) \log 2 = 3.0 \text{ db.} \quad \text{(Answer)}$$

Sample Problem 4 (a) From Table 3 we see that a rock group ($\beta_2 = 110$ dB) is 20 dB louder than a jackhammer ($\beta_1 = 90$ dB). What is the ratio of their intensities?

From Eq. 23 we have

$$(10 \text{ dB}) \log \frac{I_2}{I_1} = 110 \text{ dB} - 90 \text{ dB} = 20 \text{ dB.}$$

Thus,

$$\log \frac{I_2}{I_1} = \frac{20 \text{ dB}}{10 \text{ dB}} = 2,$$

or

$$\frac{I_2}{I_1} = 100. \quad \text{(Answer)}$$

The rock group makes 100 times as much noise as a iackhammer.

(b) How much more intense is an 80-dB shout than a 20-dB whisper?

From (a) above, we see that every 20 dB increase in sound level corresponds to a factor of 100 in intensity. We are dealing here with an increase in sound level of 60 ($= 20 + 20 + 20$) dB or three factors of 100. Thus, the shout is $100 \times 100 \times 100$ or a million times stronger than the whisper. Would you have imagined that so much additional energy is required to raise your voice?

Sample Problem 5 Spherical sound waves are emitted, uniformly in all directions, from a point source as in Fig. 7, the radiated power P being 25 W. (a) What is the intensity of the sound wave a distance r from the source? Evaluate for $r = 2.5$ m.

All the radiated power' must pass through a sphere of

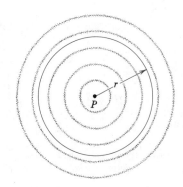

Figure 7 Sample Problem 5. A point source sends out sound waves uniformly in all directions. P is the emitted power, measured in watts. The intensity I (measured in W/m²) at a distance r from the source decreases as $1/r^2$.

radius r centered on the source. Thus,

$$I = \frac{P}{4\pi r^2}. \quad (24)$$

We see that the intensity of the sound drops off as the inverse square of the distance from the source. Numerically, we have

$$I = \frac{25 \text{ W}}{(4\pi)(2.5 \text{ m})^2} = 0.32 \text{ W/m}^2 = 320 \text{ mW/m}^2. \quad \text{(Answer)}$$

(b) What is the corresponding sound level?
From Eq. 18,

$$\beta = (10 \text{ dB}) \log \left(\frac{I}{I_0} \right)$$

$$= (10 \text{ dB}) \log \frac{0.32 \text{ W/m}^2}{10^{-12} \text{ W/m}^2} = 115 \text{ dB.} \quad \text{(Answer)}$$

A comparison of this result with Table 3 raises questions about the wisdom of buying 100-W amplifiers for home use.

18-5 Sources of Musical Sound*

Musical sounds can be set up by vibrating strings (guitar, piano, violin), by vibrating membranes (kettledrum, snare drum), by vibrating air columns (flute, oboe, pipe organ), by vibrating wooden blocks or steel bars (marimba, xylophone), or in still other ways. Most instruments involve more than a single vibrating body. In the

---

* See, for example, John R. Pierce, *The Science of Musical Sound*, W. H. Freeman & Company, New York, 1983.

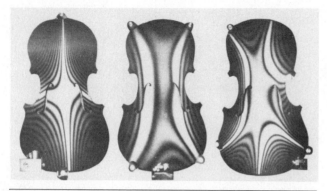

Figure 8 Vibrational patterns for an unattached top plate of a violin, made visible by holographic techniques. See "The Acoustics of Violin Plates," by Carleen Maley Hutchins, *Scientific American,* October 1981.

violin, for example, not only the strings but the body of the instrument participates fully in the sounds that we enjoy; see Fig. 8.

You may recall from Chapter 17 that standing waves can be set up in a stretched string that is fixed at each end. They arise because a wave traveling down the string in one direction will be reflected at the end of the string and return along it in the other direction. If the length of the string is suitably matched (resonance) to the wavelength of the waves, the superposition of these two waves—traveling in opposite directions along the same string—produces a pattern of standing waves in the string.

We can obtain standing sound waves in a fixed length of pipe in the same way, even if one or both ends of the pipe are open to air. Sound waves traveling down the pipe toward its open end partially reflect back into the pipe. (The fact that we can hear the sound outside the pipe is evidence that the reflection is not complete.) If the length of the pipe is suitably matched to the wavelength (or, equivalently, to the frequency) of the sound wave, a pronounced pattern of standing waves can be set up, just as for the resonating string.

The closed end of the pipe, like the fixed end of a string, is a displacement *node,* at which the displacement amplitude of the oscillating elements of air is zero. At the open end of a pipe, however, we find a displacement *antinode* at which the displacement amplitude of the oscillating air elements has its maximum value. (Actually, the antinode is located slightly beyond the end of the pipe.)

Figure 9 shows some of the standing sound waves that can be set up in organ pipes. If the pipe is open at each end (Fig. 9*a*), there must be a displacement antinode at those locations. The wavelengths must be such that an integral number of half-waves will fit into the length L of the pipe. That is, we must have $L = n(\lambda/2)$, where n is a positive integer called the *harmonic number.* The allowed frequencies are then given by

$$\nu = \frac{v}{\lambda} = \frac{vn}{2L}, \qquad n = 1, 2, 3, \ldots \quad \text{(open pipe)}. \quad (25)$$

For a pipe closed at one end, analysis of Fig. 9*b* shows

Figure 9 (*a*) Standing waves in an organ pipe that is open at both ends. The displacement nodes *(N)* and antinodes *(A)* are shown, as is the harmonic number *n*. (*b*) The same for an organ pipe closed at its upper end. See "The Physics of Organ Pipes," by Neville H. Fletcher and Suszanne Thwaites, *Scientific American*, January 1983.

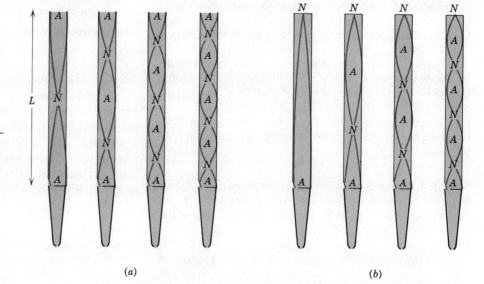

that the allowed frequencies are

$$v = \frac{v}{\lambda} = \frac{vn}{4L}, \qquad n = 1, 3, 5, \ldots \quad \text{(closed pipe).} \quad (26)$$

The lowest frequency that may be excited, corresponding to $n = 1$ in Eqs. 25 and 26, is called the *fundamental* or the *first harmonic,* the remaining frequencies being called the second, third, . . . harmonics. Note that in a closed pipe only odd harmonics are excited.

Many science museums have an exhibit that consists of a collection of tubes of various lengths, open at each end. By putting your ear close to the end of a particular tube, you can hear the fundamental frequency that is characteristic of that tube. This comes about because the room—like most rooms—contains weak background sounds at essentially all frequencies to which the ear is sensitive. Each tube resonates to its own characteristic fundamental frequency, allowing the sound level at that frequency to build up. By moving your ear from tube to tube, you can hear a tune.

The physical size of an instrument reflects the range of frequencies over which the instrument is designed to function. Figure 10, for example, shows the saxophone and the violin families, their frequency ranges being suggested by the piano keyboard. Note that, for every in-

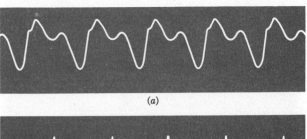

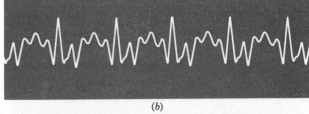

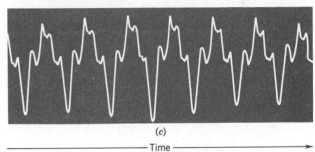

Figure 11   The wave form, as displayed on the screen of a cathode ray tube, of the sound generated by (*a*) a flute, (*b*) an oboe, and (*c*) a saxophone.

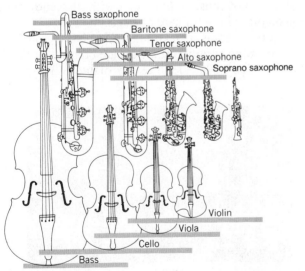

Figure 10   The families of saxophones and of violins, showing the relations between size and frequency range, the latter being shown by the horizontal bars. The frequency scale is suggested by the keyboard at the bottom.

strument, there is considerable overlap with its higher- and lower-frequency neighbors.

In any vibrating system that gives rise to a musical sound, whether it be a violin string or the air in an organ pipe, the fundamental and one or more of the higher harmonics are usually excited at the same time. The resultant sound is the superposition of these components. For different instruments, the components are present in different amounts, which accounts for the fact that a given note sounds characteristically different, depending on the instrument. Figure 11 shows, for example, the wave forms for the same fundamental frequency as sounded on three different instruments.

Sample Problem 6   An open-ended cardboard tube in a science museum exhibit has a length L of 67 cm. (*a*) To what fundamental frequency does it resonate? Assume that your ear is jammed up against one end of the tube.

For the conditions stated, the tube is open at its far end

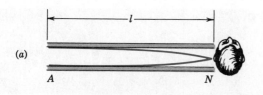

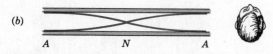

Figure 12 Sample Problem 6. (*a*) A tube closed at one end by the listener's head. (*b*) The head is moved so that the tube is now open at both ends; a different tone is heard.

but closed (by your head) at the other end. The fundamental corresponds to the lowest frequency, which in turn implies the greatest wavelength. As Fig. 12*a* shows, the longest wave that can be matched to the tube has a wavelength four times the length of the tube. The frequency is then given by (see Eq. 26 with $n = 1$)

$$\nu = \frac{v}{\lambda} = \frac{v}{4L} = \frac{343 \text{ m/s}}{(4)(0.67 \text{ m})} = 128 \text{ Hz.} \quad \text{(Answer)}$$

Actually, the frequency will be somewhat lower than this because the antinode in Fig. 12*a* is located somewhat beyond the far end of the tube.

(b) If you move your ear a short distance away from the end of the tube, what fundamental frequency do you now hear?

In so doing, you permit the tube to be open ended at each end. The wavelength of the fundamental is now just twice the length of the tube, as Fig. 12*b* shows. The new fundamental frequency is then (see Eq. 25 with $n = 1$)

$$\nu = \frac{v}{\lambda} = \frac{v}{2L} = \frac{343 \text{ m/s}}{(2)(0.67 \text{ m})} = 256 \text{ Hz.} \quad \text{(Answer)}$$

Again, the frequency actually heard will be a little lower.

18-6 Beats

If we listen, a few minutes apart, to two sounds whose frequencies are 552 and 564 Hz, most of us cannot tell them apart. However, if the sounds reach our ears simultaneously, we hear a sound whose frequency turns out to be 558 Hz, the *average* of the two combining frequencies. Strikingly, the intensity of this sound is modulated by a slow wobbling beat note, whose frequency turns out to be 12 Hz, the *difference* between the two combining frequencies. Figure 13 shows this *beat phenomenon*.

Let the variations with time of the displacements of two sound waves at a particular location be

$$s_1 = s_m \cos \omega_1 t \quad \text{and} \quad s_2 = s_m \cos \omega_2 t. \quad (27)$$

We have assumed, for simplicity, that the waves have the same amplitude. From the superposition principle, the resultant displacement is

$$s = s_1 + s_2 = s_m (\cos \omega_1 t + \cos \omega_2 t).$$

Using the trigonometric identity (see Appendix G)

$$\cos \alpha + \cos \beta = 2 \cos \tfrac{1}{2}(\alpha - \beta) \cos \tfrac{1}{2}(\alpha + \beta)$$

allows us to write the resultant displacement as

$$s = 2s_m \cos \tfrac{1}{2}(\omega_1 - \omega_2)t \cos \tfrac{1}{2}(\omega_1 + \omega_2)t. \quad (28)$$

If we put

$$\omega' = \tfrac{1}{2}(\omega_1 - \omega_2) \quad \text{and} \quad \omega = \tfrac{1}{2}(\omega_1 + \omega_2), \quad (29)$$

we can write Eq. 28 as

$$\boxed{s(t) = [2s_m \cos \omega' t] \cos \omega t.} \quad (30)$$

We now assume that the angular frequencies ω_1 and ω_2 of the combining waves are almost equal to each other, which is equivalent to assuming that $\omega \gg \omega'$ in Eq. 29. We can then regard Eq. 30 as a cosine function whose angular frequency is ω and whose amplitude (which is not constant but varies with frequency ω') is the quantity in the square brackets.

A beat, that is, a maximum of amplitude, will occur whenever the quantity in the square brackets in Eq. 30 has the value $+1$ or -1. Since *each* of these values occurs

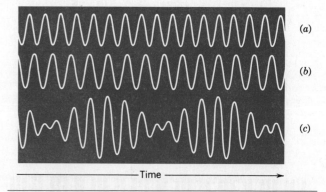

Figure 13 (*a* and *b*) The variation with time of the pressure variations of two sound waves that fall separately on a detector. The frequencies of the waves are nearly equal. (*c*) The resultant pressure variation if the two waves fall simultaneously on the detector.

every cycle, the number of beats per cycle is twice the frequency at which the amplitude varies. Thus (see Eq. 29),

$$\omega_{beat} = 2\omega' = (2)(\tfrac{1}{2})(\omega_1 - \omega_2) = \omega_1 - \omega_2.$$

Because $\omega = 2\pi\nu$, we can recast this as

$$\boxed{\nu_{beat} = \nu_1 - \nu_2}\qquad \text{(beat frequency).}\qquad (31)$$

Musicians use the beat phenomenon in tuning their instruments. If the instrument is sounded against a standard frequency (for example, the lead oboe's reference A) and tuned until the beat disappears, then the instrument is in tune with that standard. In musical Vienna, concert A (440 Hz) is available as a telephone service for the benefit of amateur string quartets and other musicians.

Sample Problem 7 You wish to tune the note A_3 on a piano to its proper frequency of 220 Hz. You have available a tuning fork whose frequency is 440 Hz. How shall you proceed?

These two frequencies are too far apart to expect to hear beats between them. Recall that in our analysis of Eq. 30 we assumed that the two combining frequencies were reasonably close to each other. However, note that the second harmonic of A_3 is 2×220 or 440 Hz.

Let us say that the actual piano string is mistuned and that its fundamental frequency is not exactly 220 Hz. You listen for beats between the fundamental frequency of the tuning fork and the second harmonic of A_3 and you hear a beat frequency of 6 Hz. You then change the tension in the corresponding string until the beat note disappears.

Note that, from the observed beat frequency of 6 Hz, you cannot tell whether the fundamental frequency of the string is 223 Hz or 217 Hz. Both frequencies would produce the same beat frequency. You can easily experiment, however, by tightening the string slightly and noting whether the beat frequency increases or decreases. If it increases, you are going in the wrong direction.

18-7   The Doppler Effect

A police car is parked by the side of the highway, sounding its 1000-Hz siren. If you are also parked by the highway, you will hear that same frequency. If you are driving *toward* the police car at 75 mi/h, you will hear a *higher* frequency (1096 Hz, an *increase* of 96 Hz). If you are driving *away from* the parked police car at that same speed, you will hear a *lower* frequency (904 Hz, a *decrease* of 96 Hz).

These motion-related frequency changes are examples of the *Doppler effect*. The effect was proposed (although not fully worked out) in 1842 by the Austrian physicist Johann Christian Doppler. It was tested experimentally in 1845 by Buys Ballot in Holland, "using a locomotive drawing an open car with several trumpeters."

When the supermarket door opens at your approach, you may be sure that it is operated by a photoelectric cell (quantum physics), by a pressure pad (Pascal's principle), or by a motion-sensitive sound detector (Doppler effect). The Doppler effect holds not only for sound waves but also for all the waves of the electromagnetic spectrum, including microwaves, radio waves, and visible light. Figure 14, for example, shows the Doppler effect for the radio signals from *Sputnik I,* the first artificial earth satellite. The magnitude of a Doppler shift signal such as that of Fig. 14 depends on the *radial component* of the velocity of the satellite at the location of the receiver. At the center of the earth, the satellite (assuming a circular orbit) has no radial velocity component so that a receiver at that location would detect no Doppler shift. A receiver located directly in the satellite orbit would detect a maximum Doppler shift, as the satellite rushed directly toward it and then directly away from it. At intermediate points, the magnitude of the Doppler shift depends on the distance of the receiver from the center of the earth. Using such satellite measurements, astronomer-mountaineer George Wallerstein of the University of Washington was able to measure the altitude of K-2, located on the Pakistan-China border. Although a preliminary measurement suggested the possibility that K-2 might be higher than Mt. Everest, this conclusion was not confirmed by later, more comprehensive satellite measurements.

The earth-satellite link also works in the opposite direction; specialized American and Soviet satellites locate downed fliers by measuring the Doppler effect of signals emitted by the Emergency Locator Transmitters (ELT units) carried by most private planes. There is also a ground-to-ground Doppler link; if you are given a speeding ticket by an officer in a radar-equipped police car, you can blame the Doppler effect. This effect is used universally in astronomy to determine the motions of stars, galaxies, quasars, and other bodies with respect to us. Indeed, it is distressing to imagine how little we would know about the universe without this great experimental tool.

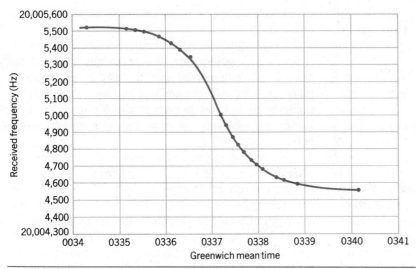

Figure 14 Showing the Doppler effect for the radio signals from *Sputnik I,* the first artificial earth satellite. It was launched from the Soviet Union on October 1, 1957.

In our analyses that follow, we restrict ourselves to sound waves and we take as a reference frame the air mass through which these waves travel. We assume that there is no wind, so that this reference frame is identical to a frame fixed with respect to the earth. Furthermore, we assume that the source S of the sound wave and the detector D of that wave move only along the line that joins them.

Detector Moving: Source at Rest. In Fig. 15, a detector D (represented by an ear) is rushing toward a stationary source at speed V_D. The wave fronts are drawn to be one wavelength apart. Clearly, waves pass the detector at a faster rate when the detector is moving toward the source than they would if it were standing still, which accounts for the higher frequency that the person hears.

Let us watch the detector for time t. If it were at rest, vt/λ waves would pass by in this time. Because of its motion, an additional $V_D t/\lambda$ must be added to this. The rate at which the detector receives waves—which is its perceived frequency—is then

$$\nu' = \frac{vt/\lambda + V_D t/\lambda}{t} = \frac{v + V_D}{\lambda}$$

$$= \frac{v + V_D}{v/\nu} = \nu \frac{v + V_D}{v}. \tag{32}$$

For the general case of a resting source and a moving detector, we have

$$\boxed{\nu' = \nu \frac{v \pm V_D}{v}} \quad \begin{array}{l}\text{(source at rest;}\\ \text{detector moving).}\end{array} \tag{33}$$

Here the plus sign indicates the detector moving toward the source and the minus sign the detector moving away from the source. The general rule for signs in all Doppler effect equations can be stated as follows:

Associate the word toward *with the words* frequency increase. *Conversely, associate the words* away from *with the words* frequency decrease.

Keeping Fig. 15 in mind helps in remembering this simple rule.

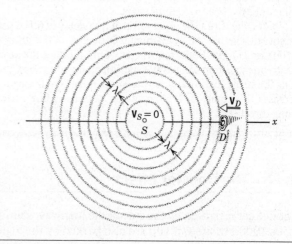

Figure 15 A stationary source of sound S emits spherical wave fronts, shown one wavelength apart. A sound detector D, represented by an ear, moves with speed V_D toward the source. The detector senses a higher frequency because of its motion.

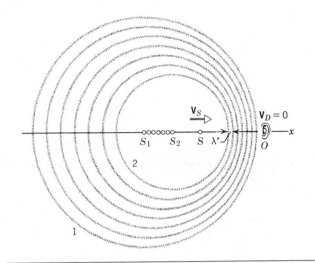

Figure 16   The detector D is at rest, with the source S moving toward it at speed V_S. Wave front 1 was emitted when the source was at S_1, wave front 2 when it was at S_2, and so on. At the moment depicted, the source was at S. The detector perceives a higher frequency because the moving source, chasing its own wave fronts, emits a reduced wavelength in its direction of motion.

Source Moving; Detector at Rest.   Let the detector now be at rest with respect to the air mass and let the source be moving toward it at speed V_S. The effect here (see Fig. 16) is a shortening of the wavelength in the direction in which the source is moving.

Let T ($= 1/v$) be the period of oscillation of the source of sound. During this interval, the source moves toward the detector by a distance $V_S T$ or V_S/v and the wavelength is shortened by that amount. Thus, the wavelength of the sound arriving at the detector is not v/v but $v/v - V_S/v$. The frequency heard by the detector is then

$$v' = \frac{v}{\lambda'} = \frac{v}{v/v - V_S/v} = v\,\frac{v}{v - V_S}. \qquad (34)$$

For the general case of a moving source and a resting detector, we have

$$\boxed{v' = v\,\frac{v}{v \mp V_S}} \quad \begin{array}{l}\text{(source moving;}\\ \text{detector at rest).}\end{array} \qquad (35)$$

Source and Detector Both Moving.   We can combine Eqs. 33 and 35 to produce the general Doppler effect equation, in which both the source and the detector are moving with respect to the air mass. Replacing the v in Eq. 35 (the frequency of the source) by the v' of Eq. 33 (the frequency associated with motion of the detector) leads to

$$\boxed{v' = v\,\frac{v \pm V_D}{v \mp V_S}} \quad \begin{array}{l}\text{(detector moving;}\\ \text{source moving).}\end{array} \qquad (36)$$

Putting $V_S = 0$ in Eq. 36 reduces it to Eq. 33 and putting $V_D = 0$ reduces it to Eq. 35, as we expect. There are four possible combinations of signs in Eq. 36.

The Doppler Effect at Low Speeds.   The Doppler effects for a moving detector (Eq. 33) and for a moving source (Eq. 35) are different, even though the detector and the source may be moving at the same speed. However, if the speeds are low enough (that is, if $V_D \ll v$ and $V_S \ll v$), the frequency changes produced by these two motions are essentially the same.

By using the binomial theorem (see Hint 3 of Chapter 7), you can show that Eq. 36 can be written in the form

$$\boxed{v' \approx v\left(1 \pm \frac{V}{v}\right)} \quad \text{(low speeds only),} \qquad (37)$$

in which V ($= |V_S \pm V_D|$) is the *relative* speed of the source with respect to the detector. The rule for signs remains the same. If the source and the detector are moving *toward* each other, we anticipate a frequency *increase;* this requires that we choose the plus sign in Eq. 37. On the other hand, if the source and the detector are moving away from each other, we anticipate a frequency decrease and must choose the minus sign in Eq. 37.

Supersonic Speeds.   If a source is moving toward a stationary detector at a speed equal to the speed of sound, that is, if $V_S = v$, Eq. 35 predicts that the perceived frequency v' will be infinitely great. This means that the source is moving so fast that it keeps pace with its own wave fronts, as Fig. 17a suggests. What now if the speed of the source *exceeds* the speed of sound?

For such *supersonic* speeds, Eq. 34 no longer applies. Figure 17b depicts the spherical waves that originated at various positions of the source. The radius of each sphere in this figure is vt, where v is the speed of sound and t is the time that has elapsed since the source was at the center of the wave in question. The envelope of these wave fronts is a cone (the *Mach cone*) whose surface makes an angle θ with the direction of motion of the source. From this figure, we see that

$$\sin \theta = \frac{v}{V_S} \quad \text{(Mach cone angle).} \qquad (38)$$

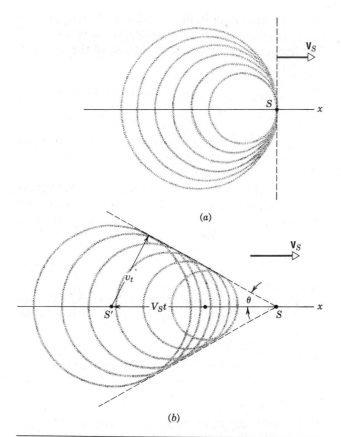

(a)

(b)

Figure 17 (a) A source of sound S moving at a speed V_S very close to the speed of sound. The source moves almost as fast as its own wave fronts. The dashed line represents a plane, tangent to the wave fronts. (b) A source S moving faster than the speed of sound. Note that the plane represented by the dashed line in (a) has folded over to become a conical envelope, of half-angle θ.

The ratio V_S/v is called the *Mach number*. When you hear that a particular plane has flown at Mach 2.3, it means that its speed is 2.3 times greater than the speed of sound in the atmosphere through which the plane is flying.

Some examples of the formation of a Mach cone are the conical shock wave from a supersonic aircraft or projectile; see Fig. 18. When such shock fronts from a supersonic aircraft intercept the earth, they are perceived there as a *sonic boom*. Similar radiation (called Cerenkov radiation) occurs for visible light if electrons travel through water or other transparent substances at speeds that are greater than the speed of light *in that medium*. The blue glow that emanates from the water in which highly radioactive reactor fuel rods are stored is caused by this effect. It is—if you wish—the optical sonic boom

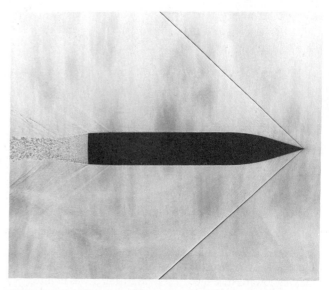

Figure 18 A shadow photograph of a projectile, fired from a gun at Mach 2. Note the Mach cones.

of the speeding electrons that are emitted by the radioactive atoms in the fuel rods.

Sample Problem 8 A police car, parked by the roadside, sounds its siren, which has a frequency v of 1000 Hz. (a) What frequency v' do you hear if you are driving directly toward the police car at 33 m/s ($\approx$ 75 mi/h)?

We put $V_S = 0$ in Eq. 36 because the source (that is, the siren of the police car) is at rest. We choose the plus sign in the numerator of that equation because, according to our rule for signs, if the source and/or the detector move toward each other, the result must be a frequency increase. With these changes, Eq. 36 becomes

$$v' = v\,\frac{v + V_D}{v}$$

$$= (1000 \text{ Hz})\,\frac{343 \text{ m/s} + 33 \text{ m/s}}{343 \text{ m/s}} = 1096 \text{ Hz}, \quad \text{(Answer)}$$

a frequency increase of 96 Hz

(b) If your are driving away from the police car at this same speed, what frequency will you now hear?

The situation is the same as in part (a) except that the sign of V_D must be reversed. Thus,

$$v' = v\,\frac{v - V_D}{v}$$

$$= (1000 \text{ Hz})\,\frac{343 \text{ m/s} - 33 \text{ m/s}}{343 \text{ m/s}} = 904 \text{ Hz}, \quad \text{(Answer)}$$

a frequency decrease of 96 Hz.

(c) Suppose that you are at rest and the police car is coming toward you at 33 m/s. What frequency do you now hear?

This time we put $V_D = 0$ in Eq. 36 because the detector (that is, you) is at rest. We choose the minus sign in the denominator of that equation because only that choice will produce the frequency increase that we know we must have. With these substitutions, Eq. 36 becomes

$$v' = v \frac{v}{v - V_S}$$

$$= (1000 \text{ Hz}) \frac{343 \text{ m/s}}{343 \text{ m/s} - 33 \text{ m/s}} = 1106 \text{ Hz}. \quad \text{(Answer)}$$

This, as expected, is a higher frequency but it is not the *same* higher frequency (= 1096 Hz) that we found for part (a). "Source approaching detector" and "detector approaching source" are physically different situations.

(d) Suppose that the police car is going away from you at this same speed. What frequency do you hear?

The situation is the same as for (c) except that we choose a plus sign for V_S, obtaining

$$v' = v \frac{v}{v + V_S}$$

$$= (1000 \text{ Hz}) \frac{343 \text{ m/s}}{343 \text{ m/s} + 33 \text{ m/s}} = 912 \text{ Hz}. \quad \text{(Answer)}$$

This is a lower frequency, as expected, but it is not the same lower frequency (= 904 Hz) that we found for (b) above.

(e) Suppose that *both* you and the police car are driving toward each other at 33 m/s. What frequency do you hear?

Each of these motions would separately generate a frequency increase. Therefore, we choose the plus sign in the numerator of Eq. 36 and the minus sign in the denominator, obtaining

$$v' = v \frac{v + V_D}{v - V_S}$$

$$= (1000 \text{ Hz}) \frac{343 \text{ m/s} + 33 \text{ m/s}}{343 \text{ m/s} - 33 \text{ m/s}} = 1213 \text{ Hz}. \quad \text{(Answer)}$$

(f) Finally, suppose that both you and the police car are driving away from each other at 33 m/s. What frequency would you hear?

In this case, each motion would separately generate a frequency decrease. We therefore choose the minus sign in the numerator of Eq. 36 and the plus sign in the denominator, obtaining

$$v' = v \frac{v - V_D}{v + V_S}$$

$$= (1000 \text{ Hz}) \frac{343 \text{ m/s} - 33 \text{ m/s}}{343 \text{ m/s} + 33 \text{ m/s}} = 824 \text{ Hz}. \quad \text{(Answer)}$$

For completeness, we should consider the cases in which you and the police car are moving in the same direction, each at 33 m/s. In this case, Eq. 36 predicts that $v' = v$; that is, there is

Table 4   Some Doppler Effect Results[a]

	S away from D	S at rest	S toward D
D toward S	1000	1096	1213
D at rest	912	1000	1106
D away from S	824	904	1000

[a] D = detector; S = source; motion with respect to the air mass.

no Doppler shift. Table 4 summarizes the results of this Sample Problem.

18-8   The Doppler Effect for Light   (Optional)

We are perhaps tempted to apply to light waves the Doppler effect equation (Eq. 36) that we have developed for sound waves in the preceding section, simply substituting c, the speed of light, for v, the speed of sound. We must avoid this temptation.

The reason is that sound waves—like all mechanical waves—require a medium (the air in our examples) for their propagation but light waves, which can travel freely through a vacuum, do not. The speed of sound is always "with respect to the medium" but the speed of light cannot be because there is no medium. As we learned in Section 17-7, the speed of light always has the same value c, in all directions and in all inertial frames.

For sound waves "source in motion" and "detector in motion" refer separately to motion with respect to the medium. They are physically different situations and lead to different equations for the Doppler effect. For light waves, however, all that can possibly matter is the *relative* motion of the source and the detector. "Source moving toward the detector at speed V" and "detector moving toward the source at speed V" are physically identical situations and there should only be one equation (not two) to describe the Doppler effect, an equation that must be derived on the basis of the theory of relativity.

Even though the Doppler equations for light and for sound are necessarily different, at low enough speeds they reduce to the same approximate result. This is not unexpected because *all* the predictions of relativity theory reduce to their classical counterparts at low enough speeds. Thus Eq. 37, with v replaced by c, holds for light waves if $V \ll c$, where V is the relative speed of the source and the detector. That is

$$v' = v(1 \pm V/c) \qquad \text{(light waves; } V \ll c). \qquad (39)$$

If the source and the detector are *approaching* each other, our rule for signs tells us to anticipate a frequency *increase,* which calls for the choice of a *plus* sign in Eq. 39.

In Doppler observations in astronomy, it is the wavelength rather than the frequency that is more readily measured. This suggests that we replace v' and v in Eq. 39 by c/λ' and c/λ respectively. Doing so, we find

$$\lambda' = \lambda(1 \pm V/c)^{-1} \approx \lambda(1 \mp V/c).$$

We can write this as

$$\frac{\lambda' - \lambda}{\lambda} = \mp \frac{V}{c}$$

or

$$V = \frac{\Delta\lambda}{\lambda} c \qquad \text{(light waves; } V \ll c), \qquad (40)$$

in which $\Delta\lambda$ is the magnitude of the Doppler wavelength shift. If the wavelength decreases (called a *blue shift* because the blue portion of the visible spectrum has the shortest wavelengths) the frequency necessarily increases and that—according to our rule for signs—means that the distance between the source and the de-

tector is decreasing. If the wavelength increases (a *red shift*), the distance between the source and the detector is increasing.

Sample Problem 9 Figure 19 shows the light from a distant galaxy known only by its catalog number, NGC 7319.* The intense spectrum line, whose wavelength λ is 513 nm, was emitted by oxygen atoms in the galaxy. The Doppler shift $\Delta\lambda$ shown on the figure (= 12 nm) indicates that this line has a wavelength 12 nm greater than would be measured for similar light emitted by a laboratory source. What is the speed of this galaxy with respect to earth? Is it approaching or receding? Note that *all* lines in the spectrum of Fig. 19 are Doppler shifted; we discuss here only the most intense line.

From Eq. 40 we find

$$V = \frac{\Delta\lambda}{\lambda} c = \frac{(12 \text{ nm})(3.00 \times 10^8 \text{ m/s})}{513 \text{ nm}}$$

$$= 7.0 \times 10^6 \text{ m/s} = 7000 \text{ km/s}. \qquad \text{(Answer)}$$

More careful measurements by astronomers at the Center for Astrophysics, Harvard College Observatory, yield a value of 6764 km/s.

---

* If you want to name a galaxy after a friend, feel free to do so. There are enough of them to allocate about 20 to every living person.

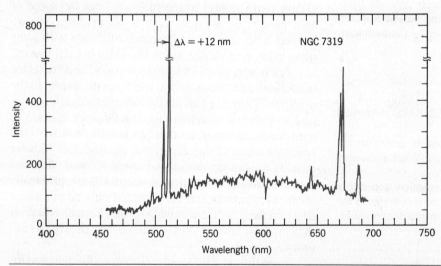

Figure 19 Sample Problem 9. The intensity of the light emitted by a galaxy about 3×10^8 light years from earth is shown here as a function of wavelength. The sharp intensity peaks can be associated with atoms such as hydrogen, nitrogen, and oxygen that are present in the galaxy. All of these lines are Doppler shifted toward longer wavelengths (a *red shift*) because the galaxy is receding from us at a rapid rate. The Doppler shift for the most intense line is shown. Courtesy of J. P. Huchra and his associates.

The observed wavelength is Doppler shifted toward longer wavelengths (a *red shift*) compared with light measured from a resting laboratory source. Thus the observed frequency is *smaller* because of the motion of the galaxy. According to our rule for signs, this means that the galaxy must be *receding* from us. Measurements of this kind are the basis for the belief that the universe is expanding.

REVIEW AND SUMMARY

Sound Waves

Wave Speed

Sound waves are longitudinal mechanical waves propagated in solids, liquids, or gases. The speed v of a longitudinal wave in a medium having a bulk modulus of elasticity B and density ρ is

$$v = \sqrt{\frac{B}{\rho}} \text{ (speed of sound)}. \qquad [3]$$

In air at 20°C, the speed of sound is 343 m/s. The audible range of frequencies is from about 20 Hz ($\lambda = 17$ m) to about 20,000 Hz ($\lambda = 1.7$ cm); see Sample Problem 1.

The equation describing the longitudinal displacements s of a mass element in a sound wave medium is

Wave Equation for Displacement

$$s = s_m \cos(kx - \omega t), \qquad [9]$$

where s_m is the maximum displacement from equilibrium, $k = 2\pi/\lambda$, and $\omega = 2\pi v$, λ and v being the wavelength and frequency, respectively. In terms of the pressure change Δp of the medium from equilibrium, the equation is

Wave Equation for Pressure

$$\Delta p = \Delta p_m \sin(kx - \omega t), \qquad [10]$$

where the pressure amplitude is

$$\Delta p_m = (v\rho\omega)s_m. \qquad [11]$$

Sample Problem 2 illustrates the range of pressures and displacements for sound waves in air.

The average intensity (power per unit area) is

Sound Intensity

$$\bar{I} = \tfrac{1}{2}\rho v \omega^2 s_m^2. \qquad [17]$$

The sound level β in decibels (dB) is defined to correspond to the human sensation of loudness. It is given by

Sound Level in Decibels

$$\beta = (10 \text{ dB}) \log \frac{I}{I_0}. \qquad [18]$$

I_0 ($= 10^{-12}$ W/m²) is the reference intensity level to which I is compared. For every factor of 10 in intensity, 10 dB is added to the sound level. The range of a human ear is approximately 120 dB. Sample Problems 3–5 illustrate calculations using intensity and sound level.

Modes of Vibration

Most sources of sound can vibrate in many standing wave modes, each having a different frequency. The actual motion of the medium in a particular case is represented by the superposition of all the modes, each having a particular amplitude. An organ pipe open at both ends can resonate at frequencies

Organ Pipes

$$v = \frac{v}{\lambda} = \frac{vn}{2L}, \qquad n = 1, 2, 3, \ldots \text{ (open pipe)}, \qquad [25]$$

where v is the speed of the wave in the pipe. For a pipe closed at one end and open at the other, the frequencies are

$$v = \frac{v}{\lambda} = \frac{vn}{4L}, \qquad n = 1, 3, 5, \ldots \text{ (closed pipe)}. \qquad [26]$$

The lowest frequency is the *fundamental* or first harmonic. Frequencies that are integer multiples of the fundamental are called *harmonics,* with, for example, the second harmonic having a frequency twice that of the fundamental. Only odd harmonics are excited in the closed pipe. Sample Problem 6 illustrates a case of resonance in a science museum display.

Beats arise when two waves having slightly different angular frequencies, ω_1 and ω_2, act together. The beat equation giving the resultant displacement as a function of time is

Beats

$$s(t) = [2s_m \cos \omega' t] \cos \omega t. \tag{30}$$

Here $\omega' = \frac{1}{2}(\omega_1 - \omega_2)$ and $\omega = \frac{1}{2}(\omega_1 + \omega_2)$. The beat frequency is

$$\nu_{\text{beat}} = \nu_1 - \nu_2. \tag{31}$$

The Doppler effect describes the change in the observed frequency when the source or the observer moves relative to the medium. The equation for the observed frequency ν' in terms of the source frequency ν is

The Doppler Equation

$$\nu' = \nu \left(\frac{v \pm V_D}{v \mp V_S} \right) \quad \text{(detector and source moving)}, \tag{36}$$

where V_D is the speed of the detector relative to the medium, V_S is that of the source, and v is the speed of sound in the medium; the upper sign on V_S (or V_D) is used when the source (or detector) moves toward the detector (or source), while the lower sign is used when it moves away; there are four different possibilities. When the speeds V_D and V_S are both much smaller than v, all four cases reduce to

$$\nu' \approx \nu \left(1 \pm \frac{V}{v} \right) \quad \text{(low speeds only)}, \tag{37}$$

where V is the relative speed of the source with respect to the detector. Sample Problem 8 illustrates the Doppler effect. The Doppler shift for light follows this same approximate expression, $\nu = \nu'(1 \pm V/c)$, if the relative velocity V is much less than the speed of light c.

If the source speed relative to the medium exceeds the speed of sound in the medium, the Doppler equation no longer applies. In such a case, shock waves result. The half angle θ (see Fig. 17) of the wave front is given by

Shock Wave Half-Angle

$$\sin \theta = \frac{v}{V_S} \quad \text{(Mach cone angle)}. \tag{38}$$

QUESTIONS

1. Why will sound not travel through a vacuum?

2. List some sources of infrasonic waves and of ultrasonic waves. (Infrasonic waves have frequencies below the audible range; ultrasonic waves have frequencies above the audible range.)

3. Ultrasonic waves (frequencies above the audible range) can be used to reveal internal structures of the body. They can, for example, distinguish between liquid and soft human tissues far better than can x rays. Why?

4. What experimental evidence is there for assuming that the speed of sound in air is the same for all wavelengths?

5. How can one create plane waves? Spherical waves?

6. The inverse square law does not apply exactly to the decrease in intensity of sounds with distance. Why not?

7. What is the meaning of zero decibels?

8. Could the reference intensity for audible sound be set so as to permit negative sound levels in decibels? If so, how?

9. How might you go about reducing the noise level in a machine shop?

10. Foghorns emit sounds of very low pitch. For what purpose?

11. Are longitudinal waves in air always audible as sound, regardless of frequency or intensity? What frequencies would give a person the greatest sensitivity, the greatest tolerance, and the greatest range?

12. What is the common purpose of the valves of a cornet and the slide of a trombone?

13. The bugle has no valves. How then can we sound different notes on it? To what notes is the bugler limited? Why?

14. The pitch of the wind instruments rise and that of the string instruments fall as an orchestra warms up. Explain why.

15. Explain how a stringed instrument is tuned.

16. Is resonance a desirable feature of every musical instrument? Give examples.

17. When you strike one prong of a tuning fork, the other prong also vibrates, even if the bottom end of the fork is clamped firmly in a vise. How can this happen? That is, how does the second prong "get the word" that somebody has struck the first prong?

18. How can a sound wave travel down an organ pipe and be reflected at its open end? It would seem that there is nothing there to reflect it.

19. How can we experimentally locate the positions of nodes and antinodes in a string, in an air column, and on a vibrating surface?

20. What physical properties of a sound wave correspond to the human sensation of pitch, of loudness, and of tone quality?

21. What is the difference between a violin note and the same note sung by a human voice that enables us to distinguish between them?

22. Does your singing really sound better in a shower? If so, what are the physical reasons?

23. Explain the audible tone produced by drawing a wet finger around the rim of a wine glass.

24. Would a plucked violin string oscillate for a longer or shorter time if the violin had no sounding board? Explain.

25. A lightning flash dissipates an enormous amount of energy and is essentially instantaneous. How is that energy transformed into the sound waves of thunder? (See "Thunder," by Arthur A. Few, *Scientific American,* July 1975.)

26. Sound waves can be used to measure the speed at which blood flows in arteries and veins. Explain how.

27. Suppose that George blows a whistle and Gloria hears it. She will hear an increased frequency whether she is running toward George or George is running toward her. Are the increases in frequency the same in each case? Assume the same speeds of running.

28. Suppose that, in the Doppler effect for sound, the source and receiver are at rest in some reference frame but the transmitting medium (air) is moving with respect to this frame. Will there be a change in wavelength, or in frequency, received?

29. Jenny, sitting on a bench, sees Lew, also sitting on a bench, across the campus. She blows a whistle to attract his attention. A steady wind is blowing from Jenny to Lew. How does the presence of the wind affect the frequency of the sound that Lew hears? How does the wind affect the sound travel time?

30. You are standing in the middle of the road and a bus is coming toward you at constant speed, with its horn sounding. Because of the Doppler effect is the pitch of the horn (*a*) rising, (*b*) falling, or (*c*) constant?

31. How might the Doppler effect be used in an instrument to detect the fetal heart beat? (Such measurements are routinely made; see "Ultrasound in Medical Diagnosis," by Gilbert B. Devey and Peter N. T. Wells, *Scientific American,* May 1978.)

32. Bats can examine the characteristics of objects—such as size, shape, distance, direction, and motion—by sensing the way the high-frequency sounds they emit are reflected off the objects back to the bat. Discuss qualitatively how each of these features affects the reflected sound waves. (See "Information Content of Bat Sonar Echoes," by J. A. Simmons, D. J. Howell, N. Suga, *American Scientist,* March–April 1975.)

33. Two ships with steam whistles of the same pitch sound off in the harbor. Would you expect this to produce an interference pattern with regions of high and low intensity?

34. A satellite emits radio waves of constant frequency. These waves are picked up on the ground and made to beat against some standard frequency. The beat frequency is then sent through a loudspeaker and one "hears" the satellite signals. Describe how the sound changes as the satellite approaches, passes overhead, and recedes from the detector on the ground.

35. How and why do the Doppler effects for light and for sound differ? In what ways are they the same?

EXERCISES AND PROBLEMS

Where needed in the problems, use

speed of sound in air = 343 m/s = 1125 ft/s;

and

density of air = 1.29 kg/m³

unless otherwise specified.

Section 18-2   The Speed of Sound

1E. Bats emit ultrasonic (high-frequency) waves. The shortest wavelength emitted in air by a bat is about 3.3 mm. What is the highest frequency a bat can emit?

2E. Diagnostic ultrasound of frequency 4.5 MHz is used to examine tumors in soft tissue. (*a*) What is the wavelength in air of such a sound wave? (b) If the speed of sound in tissue is 1500 m/s, what is the wavelength of this wave in tissue?

3E. (a) A conical loudspeaker has a diameter of 15 cm. At what frequency will the wavelength of the sound it emits in air be equal to its diameter? Be ten times its diameter? Be one-tenth its diameter? (b) Make the same calculations for a speaker of diameter 30 cm. (*Note:* If the wavelength is large compared to the diameter of the speaker, the sound waves spread out almost uniformly in all directions from the speaker, but when the wavelength is small compared to the diameter of the speaker, the wave energy is propagated mostly in the forward direction.)

4E. Figure 20 shows a remarkably detailed image, of a transistor in a microelectronic circuit, formed by an acoustic microscope. The sound waves have a frequency of 4.2 GHz. The speed of such waves in the liquid helium in which the specimen is immersed is 240 m/s. (a) What is the wavelength of these ultrahigh-frequency acoustic waves? (b) The ribbonlike conductors in the figure are ~2 μm wide. To how many wavelengths does this correspond?

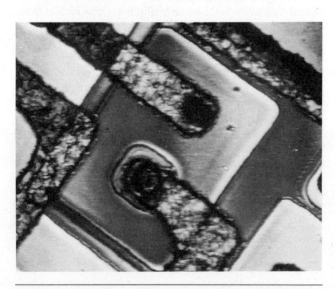

Figure 20 Exercise 4.

5E. (a) A rule for finding your distance from a lightning flash is to count seconds from the time you see the flash until you hear the thunder and then divide the count by 5. The result is supposed to give the distance in miles. Explain this rule and determine the percent error in it at standard conditions. (b) Devise a similar rule for obtaining the distance in kilometers.

6E. A column of soldiers, marching at 120 paces per minute, keep in step with the music of a band at the head of the column. It is observed that the men in the rear end of the column are striding forward with the left foot when those in the band are

advancing with the right. What is the length of the column approximately?

7E. You are at a large outdoor concert, seated 300 m from the stage microphone. The concert is also being broadcast live, in stereo, around the world via satellite. Consider a listener 5000 km away. Who hears the music first and by what time difference?

8E. Two spectators at a soccer game in a large stadium see, and a moment later hear, the ball being kicked on the playing field. The time delay for one spectator is 0.23 s and for the other 0.12 s. The lines through each spectator and the player kicking the ball meet at an angle of 90°. (a) How far is each spectator from the player? (b) How far are the spectators from each other?

9E. The average density of the earth's crust 10 km beneath the continents is 2.7 g/cm³. The speed of longitudinal seismic waves at that depth, found by timing their arrival from distant earthquakes, is 5.4 km/s. Use this information to find the bulk modulus of the earth's crust at that depth. For comparison, the bulk modulus of steel is about 16×10^{10} Pa.

10E. What is the value for the bulk modulus of oxygen at standard temperature and pressure if 1 mol (32 g) of oxygen occupies 22.4 L under these conditions and the speed of sound in oxygen is 317 m/s?

11P. An experimenter wishes to measure the speed of sound in an aluminum rod 10 cm long by measuring the time it takes for a sound pulse to travel the length of the rod. If results good to four significant figures are desired, how precisely must the length of the bar be known and how closely must she be able to resolve time intervals?

12P. The speed of sound in a certain metal is V. One end of a long pipe of that metal of length L is struck a hard blow. A listener at the other end hears two sounds, one from the wave that has traveled along the pipe and the other from the wave that has traveled through the air. (a) If v is the speed of sound in air, what time interval t elapses between the arrival of the two sounds? (b) Suppose that $t = 1.0$ s and the metal is steel. Find the length L.

13P. A man strikes a long aluminum rod at one end. Another man, at the other end with his ear close to the rod, hears the sound of the blow twice, with a 0.12-s interval between. How long is the rod?

14P. Earthquakes generate sound waves in the earth. Unlike in a gas, there are both transverse (S) and longitudinal (P) sound waves in a solid. Typically, the speed of S waves is about 4.5 km/s and that of P waves 8.0 km/s. A seismograph records P and S waves from an earthquake. The first P waves arrive 3 min before the first S waves; see Fig. 21. How far away did the earthquake occur?

15P. A stone is dropped into a well. The sound of the splash is heard 3.0 s later. What is the depth of the well?

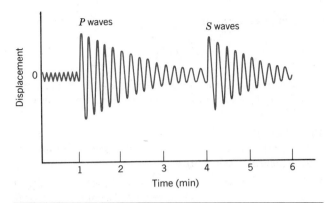

Figure 21   Problem 14.

Section 18-3   Traveling Sound Waves

16E. The pressure in a traveling sound wave is given by the equation

$$p = (1.5 \text{ Pa}) \sin \pi[(1.0 \text{ m}^{-1})x - (330 \text{ s}^{-1})t].$$

Find (a) the pressure amplitude, (b) the frequency, (c) the wavelength, and (d) the speed of the wave.

17P. Two sound waves, from two different sources with the same frequency, 540 Hz, travel at a speed of 330 m/s. What is the phase difference of the waves at a point that is 4.40 m from one source and 4.00 m from the other if the sources are in phase? The waves are traveling in the same direction.

18P. Two waves give rise to pressure variations at a certain point in space given by

$$p_1 = p_m \sin \omega t$$
$$p_2 = p_m \sin(\omega t - \phi).$$

What is the pressure amplitude of the resultant wave at this point when $\phi = 0$, $\phi = \pi/2$, $\phi = \pi/3$, and $\phi = \pi/4$? All ϕ's are measured in radians.

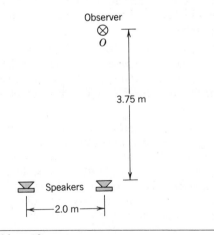

Figure 22   Problem 19.

19P. Two stereo loudspeakers are separated by a distance of 2.0 m. Assume the amplitude of the sound from each speaker is approximately the same at the position of a listener, who is 3.75 m directly in front of one of the speakers; see Fig. 22. (a) For what frequencies in the audible range (20–20,000 Hz) will there be a minimum signal? (b) For what frequencies is the sound a maximum?

Section 18-4   Intensity and Sound Level

20E. Show that the sound wave intensity I can be written in terms of frequency v, rather than angular frequency ω, in the form

$$I = 2\pi^2 \rho v v^2 s_m^2.$$

21E. Spherical waves are emitted from a 1.0-W source in a nonabsorbing medium. What is the wave intensity (a) 1.0 m from the source and (b) 2.5 m from the source?

22E. A source emits spherical waves isotropically (that is, with equal intensity in all directions). The intensity of the wave 2.5 m from the source is 1.91×10^{-4} W/m². What is the power output of this source?

23E. A note of frequency 300 Hz has an intensity of 1.0 μW/m². What is the amplitude of the air vibrations caused by this sound?

24E. What is the intensity ratio of two sounds whose sound levels differ by 1.0 dB?

25E. A certain sound level is increased by an additional 30 dB. Show that (a) its intensity increases by a factor of 1000 and (b) its pressure amplitude increases by a factor of 32.

26E. A salesperson claimed that a stereo system would deliver 120 W of audio power. Testing the system with several speakers set up so as to simulate a point source, the consumer noted that she could get as close as 1.2 m with the volume full on before the sound hurt her ears. Should she report the firm to the Consumer Protection Agency?

27E. A certain loudspeaker produces a sound with a frequency of 2000 Hz and an intensity of 0.96 mW/m² at a distance of 6.1 m. Presume that there are no reflections and that the loudspeaker emits the same in all directions. (a) What would be the intensity at 30 m? (b) What is the displacement amplitude at 6.1 m? (c) What is the pressure amplitude at 6.1 m?

28E. The source of a sound wave delivers 1 μW of power. If it is a point source, (a) what is the intensity 3.0 m away and (b) what is the sound level in decibels at that distance?

29E. (a) If two sound waves, one in air and one in water, are equal in intensity, what is the ratio of the pressure amplitude of the wave in water to that of the wave in air? (b) If the pressure amplitudes are equal instead, what is the ratio of the intensities of the waves? Assume the water is at 20°C.

30P. In Fig. 23, we show an acoustic interferometer, used to demonstrate the interference of sound waves. S is a diaphragm that vibrates under the influence of an electromagnet. D is a sound detector, such as the ear or a microphone. Path SBD can be varied in length, but path SAD is fixed. The interferometer contains air, and it is found that the sound intensity has a minimum value of 100 units at one position of B and continuously climbs to a maximum value of 900 units at a second position 1.65 cm from the first. Find (a) the frequency of the sound emitted from the source and (b) the relative amplitudes of the waves arriving at the detector for each of the two positions of B. (c) How can it happen that these waves have different amplitudes, considering that they originate at the same source?

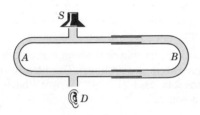

Figure 23 Problem 30.

31P. A sound wave of 40-cm wavelength enters the tube shown in Fig. 24. What must be the smallest radius r such that a minimum will be heard at the detector?

Figure 24 Problem 31.

32P. (a) Show that the intensity I is the product of the energy per unit volume u and the speed of propagation v of a wave disturbance. (b) Radio waves travel at a speed of 3.0×10^8 m/s. Find the energy density in a radio wave 480 km from a 50,000-W source, assuming the waves to be spherical and the propagation to be isotropic.

33P. A line source (for instance, a long freight train on a straight track) emits a cylindrical expanding wave. Assuming that the air absorbs no energy, find how (a) the intensity and (b) the amplitude of the wave depend on the distance from the source. Ignore reflections and consider points near the center of the train.

34P. A wave travels out uniformly in all directions from a point source. (a) Justify the following expression for the displacement y of the medium at any distance r from the source:

$$y = \frac{Y}{r} \sin k(r - vt).$$

Consider the speed, direction of propagation, periodicity, and intensity of the wave. (b) What are the dimensions of the constant Y?

35P. Find the ratios of the intensities, the pressure amplitudes, and the particle displacement amplitudes for two sounds whose intensity levels differ by 37 dB.

36P. Two loudspeakers are located 11 ft apart on the stage of an auditorium. A listener is seated 60 ft from one and 64 ft from the other. A signal generator drives the two speakers in phase with the same amplitude and frequency. The frequency is swept through the audio range. (a) What are the three lowest frequencies for which the listener will hear minimum intensity because of destructive interference? (b) What are the three lowest frequencies for which the listener will hear maximum intensity?

37P. At 1.0 km a 100-Hz horn, assumed to be a point source, is barely audible. At what distance would it begin to cause pain?

38P. You are standing at a distance D from an isotropic source of sound waves. You walk 50 m toward the source and observe that the intensity of these waves has doubled. Calculate the distance D.

39P. A certain loudspeaker (assumed to be a point source) emits 30 W of sound power. A small microphone of effective cross-sectional area 0.75 cm² is located 200 m from the loudspeaker. Calculate (a) the sound intensity at the microphone and (b) the power incident on the microphone.

40P. In a test, a subsonic jet flies overhead at an altitude of 100 m. The sound intensity on the ground as the jet passes overhead is 150 dB. At what altitude should the plane fly so that the ground noise is no greater than 120 dB, the threshold of pain? Ignore the finite time required for the sound to reach the ground.

41P. Two sources of sound are separated by a distance of 5.0 m. They both emit sound at the same amplitude and frequency, 300 Hz, but they are 180° out of phase. At what points along the line between them will the sound intensity be the largest?

42P. A hi-fi engineer has designed a speaker that is spherical in shape and emits sound isotropically (the same intensity in all directions). The speaker emits 10 W of acoustic power into a room with completely absorbent walls, floor, and ceiling (an anechoic chamber). (a) What is the intensity (W/m²) of the sound waves at 3.0 m from the center of the source? (b) How does the amplitude of the waves at 4.0 m compare with that at 3.0 m from the center of the source?

43P. A spherical sound source is placed at P_1 near a reflecting wall AB and a microphone is located at point P_2, as shown in Fig. 25. The frequency of the sound source P_1 is variable. Find two different frequencies for which the sound intensity, as observed at P_2, will be a maximum. Assume the paths of the interfering waves to be parallel. There is no phase change on reflection; the angle of incidence equals the angle of reflection.

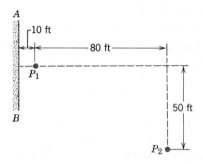

Figure 25   Problem 43.

44P*. Two loudspeakers, S_1 and S_2, are 7.0 m apart. Each emits sound of frequency 200 Hz uniformly in all directions. S_1 has an acoustic output of 1.2×10^{-3} W and S_2 one of 1.8×10^{-3} W. S_1 and S_2 vibrate in phase. Consider a point P that is 4.0 m from S_1 and 3.0 m from S_2. (a) How are the phases of the two waves arriving at P related? (b) What is the intensity of sound at P with both S_1 and S_2 on? (c) What is the intensity of sound at P if S_1 is turned off (S_2 on)? (d) What is the intensity of sound at P if S_2 is turned off (S_1 on)?

45P*. A large parabolic reflector having a circular opening of radius 0.50 m is used to focus sound. If the energy is delivered from the focus to the ear of a listening detective through a tube of diameter 1.0 cm with 12% efficiency, how far away can a whispered conversation be understood? (Assume that the intensity level of a whisper is 20 dB at 1.0 m from the source, considered to be a point, and that the threshold for hearing is 0 dB.)

Section 18–5   Sources of Musical Sound

46E. A sound wave of frequency 1000 Hz propagating through air has a pressure amplitude of 10 Pa. What are the (a) wavelength, (b) particle displacement amplitude, and (c) maximum particle speed? (d) An open-open organ pipe has this frequency as a fundamental. How long is it?

47E. In Fig. 26, a rod R is clamped at its center; a disk D at its end projects into a glass tube that has cork filings spread over its interior. A plunger P is provided at the other end of the tube. The rod is set into longitudinal vibration and the plunger is moved until the filings form a pattern of nodes and antinodes

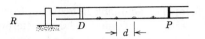

Figure 26   Exercise 47.

(the filings form well-defined ridges at the pressure antinodes). If we know the frequency v of the longitudinal vibrations in the rod, a measurement of the average distance d between successive antinodes determines the speed of sound v in the gas in the tube. Show that

$$v = 2vd.$$

This is Kundt's method for determining the speed of sound in various gases.

48E. A sound wave in a fluid medium is reflected at a barrier so that a standing wave is formed. The distance between nodes is 3.8 cm and the speed of propagation is 1500 m/s. Find the frequency.

49E. (a) Find the speed of waves on a 0.8-g violin string 22 cm long if the frequency of the fundamental is 920 Hz. (b) What is the tension in the string?

50E. If a violin string is tuned to a certain note, by how much must the tension in the string be increased if it is to emit a note of double the original frequency (that is, a note one octave higher in pitch)?

51E. A certain violin string is 30 cm long between its fixed ends and has a mass of 2.0 g. The string sounds an A note (440 Hz) when played without fingering. Where must one put one's finger to play a C (528 Hz)?

52E. An open organ pipe has a fundamental frequency of 300 Hz. The second harmonic of a closed organ pipe has the same frequency as the second harmonic of the open pipe. How long is each pipe?

53E. The water level in a vertical glass tube 1.0 m long can be adjusted to any position in the tube. A tuning fork vibrating at 686 Hz is held just over the open top end of the tube. At what positions of the water level will there be resonance?

54P. The strings of a cello have a length L. (a) By what length l must they be shortened by fingering to change the pitch by a frequency ratio r? (b) Find l, if $L = 0.80$ m and $r = \frac{6}{5}, \frac{5}{4}, \frac{4}{3}$, and $\frac{3}{2}$.

55P. S in Fig. 27 is a small loudspeaker driven by an audio

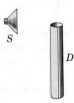

Figure 27   Problem 55.

oscillator and amplifier, adjustable in frequency from 1000 to 2000 Hz only. D is a piece of cylindrical sheetmetal pipe 18 in. long and open at both ends. (a) If the speed of sound in air is 1130 ft/s at the existing temperature, at what frequencies will resonance occur when the frequency emitted by the speaker is varied from 1000 to 2000 Hz? (b) Sketch the displacement nodes for each. Neglect end effects.

56P. A well with vertical sides and water at the bottom resonates at 7.0 Hz and at no lower frequency. The air in the well has a density of 1.1 kg/m³ and a bulk modulus of 1.33×10^5 Pa. How deep is the well?

57P. The width of the terraces in an amphitheater, Fig. 28, is 36 in. A single hand-clap occurring at the center of the stage will reflect back to the stage as a tone of what wavelength?

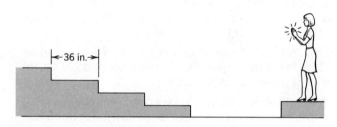

Figure 28 Problem 57.

58P. A tube 1.2 m long is closed at one end. A stretched wire is placed near the open end. The wire is 0.33 m long and has a mass of 9.6 g. It is fixed at both ends and vibrates in its fundamental mode. It sets the air column in the tube into vibration at its fundamental frequency by resonance. Find (a) the frequency of oscillation of the air column and (b) the tension in the wire.

59P. The period of a pulsating variable star may be estimated by considering the star to be executing radial longitudinal pulsations in the fundamental standing wave mode; that is, the radius varies periodically with the time, with a displacement antinode at the surface. (a) Would you expect the center of the star to be a displacement node or antinode? (b) By analogy with the open organ pipe, show that the period of pulsation T is given by

$$T = \frac{4R}{v_s},$$

where R is the equilibrium radius of the star and v_s is the average sound speed. (c) Typical white dwarf stars are composed of material with a bulk modulus of 1.33×10^{22} Pa and a density of 10^{10} kg/m³. They have radii equal to 0.009 solar radii. What is the approximate pulsation period of a white dwarf? (See "Pulsating Stars," by John R. Percy, *Scientific American,* June 1975.)

60P. A 30-cm violin string with linear density 0.65 g/m is

placed near a loudspeaker that is fed by an audio oscillator of variable frequency. It is found that the string is set into oscillation only at the frequencies 880 and 1320 Hz as the frequency of the oscillator is varied continuously over the range 500–1500 Hz. What is the tension in the string?

Section 18–6 Beats

61E. A tuning fork of unknown frequency makes three beats per second with a standard fork of frequency 384 Hz. The beat frequency decreases when a small piece of wax is put on a prong of the first fork. What is the frequency of this fork?

62E. The A string of a violin is a little too taut. Four beats per second are heard when it is sounded together with a tuning fork that is vibrating accurately at the concert pitch of A (440 Hz). What is the period of the violin string vibration?

63E. You are given four tuning forks. The fork with the lowest frequency vibrates at 500 Hz. By using two tuning forks at a time, the following beat frequencies are heard: 1, 2, 3, 5, 7, and 8 Hz. What are the possible frequencies of the other three tuning forks?

64P. Two identical piano wires have a fundamental frequency of 600 Hz when kept under the same tension. What fractional increase in the tension of one wire will lead to the occurrence of 6 beats/s when both wires vibrate simultaneously?

65P. You are given five tuning forks, each of which has a different frequency. By trying every pair of tuning forks, (a) what maximum number of *different* beat frequencies might be obtained? (b) What minimum number of *different* beat frequencies might be obtained?

Section 18–7 The Doppler Effect

66E. A source S generates circular waves on the surface of a lake, the pattern of wave crests being shown in Fig. 29. The speed of the waves is 5.5 m/s and the crest-to-crest separation is 2.3 m. You are in a small boat heading directly toward S at a

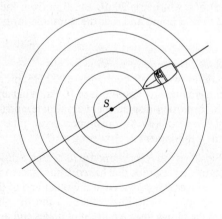

Figure 29 Exercise 66.

constant speed of 3.3 m/s with respect to the shore. What frequency of the waves do you observe?

67E. Trooper *B* is chasing speeder *A* along a straight stretch of road. Both are moving at a top speed of about 100 mi/h. Trooper *B*, failing to catch up, sounds his siren again. Take the speed of sound in air to be 1100 ft/s and the frequency of the source to be 500 Hz. What will be the Doppler shift in the frequency heard by speeder *A*?

68E. A whistle used to call a dog has a frequency of 30 kHz. The dog, however, ignores it. The owner of the dog, who cannot hear sounds above 20 kHz, wants to use the Doppler effect to make certain that the whistle is working. She asks a friend to blow the whistle from a moving car while the owner remains stationary and listens. (*a*) How fast must the car move and in what direction? Is the experiment practical? (*b*) Repeat for a whistle frequency of 22 kHz instead of 30 kHz.

69E. The 16,000-Hz whine of the turbines in the jet engines of an aircraft moving with speed 200 m/s is heard at what frequency by the pilot of a second craft trying to overtake the first at a speed of 250 m/s?

70E. An ambulance emitting a whine at 1600 Hz overtakes and passes a cyclist pedaling a bike at 8.0 ft/s. After being passed, the cyclist hears a frequency of 1590 Hz. How fast is the ambulance moving?

71E. A whistle of frequency 540 Hz moves in a circle of radius 2.0 ft at an angular speed of 15 rad/s. What are (*a*) the lowest and (*b*) the highest frequencies heard by a listener a long distance away at rest with respect to the center of the circle?

72E. In 1845, Buys Ballot first tested the Doppler effect for sound. He put a trumpet player on a flatcar drawn by a locomotive and another player near the tracks. If each player blows a 440-Hz note, and if there are 4.0 beats/s as they approach each other, what is the speed of the flatcar?

73E. Estimate the speed of the projectile illustrated in the photograph in Fig. 18. Assume the speed of sound in the medium through which the projectile is traveling to be 380 m/s.

74E. The speed of light in water is about three-fourths the speed of light in vacuum. A beam of high-speed electrons from a betatron emits Cerenkov radiation in water, the wave front being a cone of angle 60°. Find the speed of the electrons in the water.

75E. A bullet is fired with a speed of 2200 ft/s. Find the angle made by the shock cone with the line of motion of the bullet.

76P. Two identical tuning forks can oscillate at 440 Hz. A person is located somewhere on the line between them. Calculate the beat frequency as measured by this individual if (*a*) she is standing still and the tuning forks both move to the right, say, at 30 m/s, and (*b*) the tuning forks are stationary and the listener moves to the right at 30 m/s.

77P. A plane flies with $\frac{3}{4}$ the speed of sound. The sonic boom reaches a man on the ground exactly 1 min after the plane passed directly overhead. What is the altitude of the plane? Assume the speed of sound to be 330 m/s.

78P. A jet plane passes overhead at a height of 5000 m and a speed of Mach 1.5 (1.5 times the speed of sound). (*a*) Find the angle made by the shock wave with the line of motion of the jet. (*b*) How long after the jet has passed directly overhead will the shock wave reach the ground? Use 331 m/s for the speed of sound.

79P. Figure 30 shows a transmitter and receiver of waves contained in a single instrument. It is used to measure the speed *V* of a target object (idealized as a flat plate) that is moving directly toward the unit, by analyzing the waves reflected from it. (*a*) Apply the Doppler equations twice, first with the target as observer and then with the target as a source, and show that the frequency v_r of the reflected waves at the receiver is related to their source frequency v_s by

$$v_r = v_s \left(\frac{v + V}{v - V} \right),$$

where *v* is the speed of the waves. (*b*) In a great many practical situations, $V \ll v$. In this case, show that the equation above becomes

$$\frac{v_r - v_s}{v_s} \approx \frac{2V}{v}$$

Figure 30 Problem 79.

80P. A sonar device sends sound waves of 0.15 MHz from a hiding police car toward a truck approaching at a speed of 45 m/s. What is the frequency of the reflected waves detected at the police car?

81P. An acoustic burglar alarm consists of a source emitting waves of frequency 28 kHz. What will be the beat frequency of waves reflected from an intruder walking at 0.95 m/s directly away from the alarm?

82P. A siren emitting a sound of frequency 1000 Hz moves away from you toward a cliff at a speed of 10 m/s. (*a*) What is the frequency of the sound you hear coming directly from the siren? (*b*) What is the frequency of the sound you hear reflected off the cliff? (*c*) Could you hear the beat frequency? Take the speed of sound in air as 330 m/s.

83P. A person on a railroad car blows a trumpet sounding at 440 Hz. The car is moving toward a wall at 20 m/s. Calculate (*a*) the frequency of the sound as received at the wall and (*b*) the frequency of the reflected sound arriving back at the source.

84P. Two submarines are on a head-on collision course during maneuvers in the North Atlantic. The first sub is moving at 20 km/h and the second sub at 95 km/h. The first submarine sends out a sonar signal (sound wave in water) at 1000 Hz. Sonar waves travel at 5470 km/h. (*a*) The second sub picks up the signal. What frequency does the second sonar detector hear? (*b*) The first sub picks up the reflected signal. What frequency does the first sonar detector hear? See Fig. 31. The ocean is calm; assume no currents.

20 km/h   95 km/h

Figure 31   Problem 84.

85P. A source of sound waves of frequency 1200 Hz moves to the right with a speed of 98 ft/s relative to the ground. To its right is a reflecting surface moving to the left with a speed of 216 ft/s relative to the ground. Take the speed of sound in air to be 1080 ft/s and find (*a*) the wavelength of the sound emitted in air by the source, (*b*) the number of waves per second arriving at the reflecting surface, (*c*) the speed of the reflected waves, and (*d*) the wavelength of the reflected waves.

86P. In a discussion of Doppler shifts of ultrasonic (high-frequency) waves used in medical diagnosis, the authors remark: "For every millimeter per second that a structure in the body moves, the frequency of the incident ultrasonic wave is shifted approximately 1.3 Hz/MHz." What speed of the ultrasonic waves in tissue do you deduce from this statement?

87P. A bat is flittering about in a cave, navigating very effectively by the use of ultrasonic bleeps (short emissions of high-frequency sound lasting a millisecond or less and repeated several times a second). Assume that the sound emission frequency of the bat is 39,000 Hz. During one fast swoop directly toward a flat wall surface, the bat is moving at $\frac{1}{40}$ of the speed of sound in air. What frequency does the bat hear reflected off the wall?

88P. A submarine moving north with a speed of 75 km/h with respect to the ocean floor emits a sonar signal (sound waves in water used in ways similar to radar) of frequency 1000 Hz. If the ocean at that point has a current moving north at 30 km/h relative to the land, what frequency is observed by a ship north of the submarine that does not have its engine run-

ning? (*Hint:* All speeds in the Doppler equations must be taken with respect to the medium.)

89P. A 2000-Hz siren and a civil defense official are both at rest with respect to the earth. What frequency does the official hear if the wind is blowing at 12 m/s (*a*) from source to observer and (*b*) from observer to source?

90P. Two trains are traveling toward each other at 100 ft/s relative to the ground. One train is blowing a whistle at 500 Hz. (*a*) What frequency will be heard on the other train in still air? (*b*) What frequency will be heard on the other train if the wind is blowing at 100 ft/s toward the whistle and away from the listener? (*c*) What frequency will be heard if the wind direction is reversed?

91P. A girl is sitting near the open window of a train that is moving at a velocity of 10 m/s to the east. The girl's uncle stands near the tracks and watches the train move away. The locomotive whistle vibrates at 500 Hz. The air is still. (*a*) What frequency does the uncle hear? (*b*) What frequency does the girl hear? A wind begins to blow from the east at 10 m/s. (*c*) What frequency does the uncle now hear? (*d*) What frequency does the girl now hear?

Section 18-8   The Doppler Effect For Light

92E. The central spot in Fig. 32*a* is a galaxy in the constellation Corona Borealis; the galaxy is 1.3×10^8 light-years distant. In Fig. 32*b* the central streak is the spectrum of light (distribution in wavelength) from the galaxy. The two vertical dark lines show the presence of calcium. The horizontal arrow shows that these calcium lines occur at longer wavelengths than those for terrestrial light sources containing calcium, the length of the arrow representing the wavelength shift. Previous measurements indicate that this galaxy is receding from us at 2.2×10^4 km/s. Calculate the fractional shift in wavelength of the calcium lines shown.

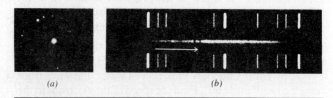

(a)   (b)

Figure 32   Problem 92.

93E. Certain characteristic wavelengths in the light from a galaxy in the constellation Virgo are observed to be increased in wavelength, as compared with terrestrial sources, by about 0.4%. What is the radial speed of this galaxy with respect to the earth? Is it approaching or receding?

94E. The "red shift" of radiation from a distant galaxy consists of the light H$_y$, known to have a wavelength of 434 nm when observed in the laboratory, appearing to have a wavelength of 462 nm. (*a*) What is the speed of the galaxy in the line of sight relative to the earth? (*b*) Is it approaching or receding?

95E. Could you go through a red light fast enough to have it appear green? Take 620 nm as the wavelength of red light and 540 nm as the wavelength for green light.

96P. The period of rotation of the sum at its equator is 24.7 d; its radius is 7.0×10^5 km. What Doppler wavelength shift is expected for light with wavelength 550 nm emitted from the edge of the sun's disk?

97P. An earth satellite, transmitting on a frequency of 40 MHz (exactly), passes directly over a radio receiving station at an altitude of 400 km and at a speed of 3.0×10^4 km/h. Plot the change in frequency, attributable to the Doppler effect, as a function of time, counting $t = 0$ as the instant the satellite is over the station. (*Hint:* The speed V in the Doppler formula is not the actual speed of the satellite but its component in the direction of the station. Neglect the curvature of the earth and of the satellite orbit.)

98P. Microwaves, which travel with the speed of light, are reflected from a distant airplane approaching the wave source. It is found that when the reflected waves are beat against the waves radiating from the source the beat frequency is 990 Hz. If the microwaves are 0.10 m in wavelength, what is the approach speed of the airplane.

ESSAY 8
MEDICAL ULTRASOUND

RUSSELL K. HOBBIE
UNIVERSITY OF
MINNESOTA

RICHARD L. MORIN
UNIVERSITY OF
MINNESOTA

Ultrasound provides a widely used and relatively inexpensive method of examining the shape and movement of organs within the body. Like the sound we hear, ultrasound consists of a longitudinal disturbance that propagates through a medium. The frequencies of ultrasound waves used in medicine range from 1 to 10 MHz, compared to audible frequencies of 30 Hz to 20 kHz. Since typical velocities in tissue are about 1500 m/s, the wavelengths are 0.15–1.5 mm. The images are obtained from waves that have been reflected from the boundaries of organs. The distance to the reflecting surface is obtained from the time delay of the echo. Waves reflected from moving objects, such as red blood cells, exhibit a Doppler shift of frequency that is used to measure the speed of the objects.

The ultrasound signal is produced by a *transducer,* which undergoes mechanical vibration in response to a pulse of electrical voltage. The electronic circuit can be arranged so that the same transducer is used to convert the echo into a weak electric signal. (In continuous wave ultrasound, a steady electrical sine wave is applied to the transducer, and a separate transducer is used to detect the echo.)

The velocity of ultrasound in tissue depends on the bulk modulus and density of the material—$v = \sqrt{B/\rho}$. (This equation is modified slightly if the material is not isotropic.) Soft tissue values range from 1400 to 1600 m/s. A very rigid material such as bone has a velocity of around 4000 m/s. Partial reflections occur whenever the waves strike a boundary where there is a change in the *acoustic impedance* of the material. (This is analogous to the partial reflection of light waves at a boundary where there is a change in the index of refraction.) The acoustic impedance Z is the ratio of the amplitude of the pressure wave p_0 and the amplitude v_0 of the longitudinal velocity with which particles in the material move back and forth: $Z = p_0/v_0$. The impedance is $Z = \rho v = \sqrt{B\rho}$.

Measurements are complicated by the fact that the ultrasound wave loses a considerable amount of energy as it moves through the tissue. This attenuation in soft tissue is typically 100 dB/m for waves at a frequency of 1 MHz. This means that echoes received from distant surfaces are very weak. The electronic amplifiers that are used often produce a signal proportional to the logarithm of the echo, which reduces the difference between strong and weak signals. In pulsed ultrasound, the gain of the amplifier is increased for signals with a long delay to compensate for the attenuation.

Figure 1 shows an ultrasound transducer on the abdomen of a patient. Note that a coupling gel is used to eliminate an air gap between the transducer and the patient's skin. Without this gel, most of the energy would be reflected because of the impedance differences at the transducer–air and air–skin interfaces.

The most common method of making an image is the *B scan.* The transducer is mounted on an arm that records the transducer's position and the direction of the beam. Since the speed of sound in all tissues is approximately the same, the delay t_d of an echo corresponds to the distance L of the interface according to $t_d = 2L/v$. A dot is displayed on the screen; the brightness is corrected for attenuation and is proportional

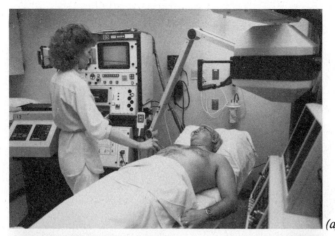

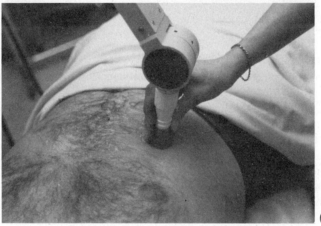

Figure 1 (*a*) An ultrasound transducer is shown on the abdomen of a patient. The arm records the position of the transducer and the direction it points. (*b*) A close-up of the transducer. Note the coupling gel, which eliminates the air gap between the transducer and the patient's skin.

to the echo amplitude, which in turn depends on the impedance ratio at the reflecting surface. Figure 2*a* shows how the transducer can be rocked in place or moved on the surface of the body. Figure 2*b* shows the dots which would be displayed corresponding to the reflections for the rays shown.

It takes a few minutes to move the transducer around to produce the entire image. One can use an array of transducers to collect reflections along many paths simultaneously. It is also possible to "steer" the paths electronically without moving the transducer. As a result, the image can be constructed quite rapidly and the organs can be viewed in "real time."

Figure 2 (*a*) The transducer can be rocked in place or moved about on the surface of the body. (*b*) The dots show the echoes caused by reflctions from the points shown on the surface of an organ. Reflections along many rays traversing the organ constitute the image.

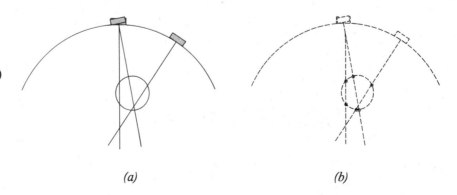

(a)   *(b)*

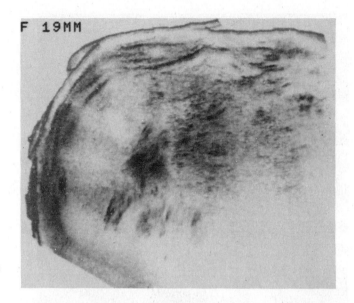

Figure 3 An ultrasound scan of a liver. The dark streaks in the middle of the picture are blood vessels.

Figure 3 shows a cross-section image of a liver. The dark structures in the center are blood vessels. The white areas are fat.

An *M scan* is used to measure the movement of heart valves as a function of time. In this case, the transducer is aimed to record the maximum signal from the desired valve and held there. As the valve moves, the delay of the echo changes, as shown in Fig. 4*a*. The echoes are displayed in Fig. 4*b*. The echoes are recorded versus time on a strip-chart recorder. Figure 5 shows an M scan of a heart valve. The lines marked *W* are from reflections from the moving heart wall; *V* is the reflection from the valve.

Figure 4 A moving reflecting surface ("valve") is enclosed in a stationary cylindrical reflector. (*a*) Echoes from the moving "heart valve" are delayed varying amounts as the "valve" moves. (*b*) The echoes are displayed as shown to make an M scan.

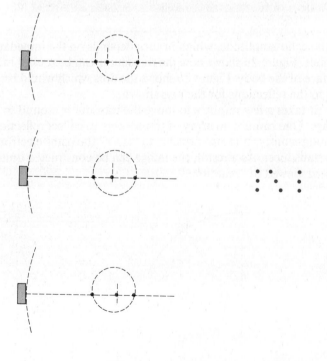

(a)   *(b)*

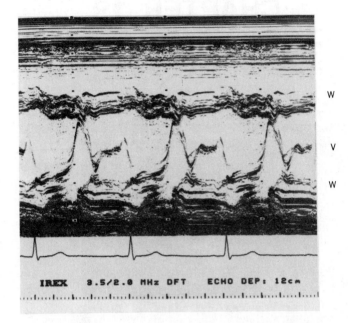

Figure 5 An M scan of a heart valve. The moving lines marked W are reflections from the heart walls; V is a reflection from the valve.

The Doppler effect is used to measure blood flow. If the reflecting particles are moving with velocity v_0 at an angle θ with the ultrasound beam, the frequency v of the beam is shifted by an amount of $\Delta v \approx 2vv_0(\cos\theta)/v$. For a velocity $v_0 = 0.1$ m/s and $v = 2$ MHz, the frequency shift is 260 Hz. This is in the audible range and can be heard by forming a beat of the reflected signal with the sine wave, which was applied briefly to the transducer. The speed of the red cells depends on the average velocity of the blood and is diminished if an artery is obstructed. The carotid artery passes through the neck and supplies blood to the brain. To measure the velocity of the red cells, the physician listens to the beat frequency and adjusts the transducer for the loudest and clearest signal. The signal sounds like a series of "puffs" as the heart beats. Then the apparatus makes a quantitative determination of Δv.

The intensities of the ultrasound beams used in diagnostic studies are typically 10^3 W/m² (100 mW/cm²) or less. Biological effects have not been observed for total energy depositions up to 5×10^5 J/m², and as far as is known, there is no harm from studies at these intensities. However, prudence suggests that no medical study be done unless there is a reason for doing it. At higher intensities, tissue heating is significant. This is sometimes used for physiotherapy of acute back or shoulder pain or muscle injury. At intensities above 7×10^6 W/m², the temperature can rise enough to damage tissue. At 10^7 W/m² or more, the pressure changes in the sound wave are large enough so that voids can suddenly grow and then collapse. This effect is called cavitation. These latter effects have been used experimentally for surgery.

CHAPTER 19
TEMPERATURE

"Some say the world will end in fire,
Some say in ice."
Robert Frost

For the present, it is the sun's fire that keeps us alive. If, in the distant future, the sun
should expand, as a red giant, the fire would increase considerably. Eventually, however,
the ice will come.

19-1 A New Look at Temperature

In physics, we like to take a familiar concept and take a broad look at it, so that we see our daily experience of it as a small part of a grander picture. We understand visible light better, for example, once we see it as one component of an infinite band of radiations—continuous in wavelength—and including radio waves, microwaves, ultraviolet rays, x rays, and so on. If the word *temperature* suggests to you the TV weather report or the dial setting on a microwave oven, perhaps it is time to back off and take a new look at this concept as well.

Temperature—one of the seven SI base standards—is one of the grand variables of physics. Given almost any physical property or measurement, a physicist is likely to ask: "How does it vary with temperature?" Physicists measure temperature on the *Kelvin scale*. Although the temperature of a body can apparently be raised without limit, it cannot be lowered without limit and this limiting low temperature is taken as the zero of the Kelvin scale. Room temperature is about 290 kelvins above this absolute zero, or 290 K as we write it. Figure 1 shows some temperatures of interest.

When the universe began, some 10–20 billion years

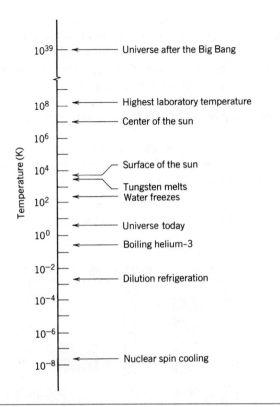

Figure 1   Some temperatures on the Kelvin scale. Note that $T = 0$ corresponds to $10^{-\infty}$ and cannot be plotted on this logarithmic plot.

ago, the temperature was about 10^{39} K. As the universe expanded, it cooled down and has now reached an average temperature of about 3 K. We are a little warmer than this because we happen to live near a star. Without our sun, however, we too would soon be at 3 K.

Physicists around the world are trying to see how close they can come to the absolute zero of temperature. It turns out that the absolute zero is much like the speed c of light in that both are limits that can be approached for material bodies but never attained. As of 1988, physicists have achieved the following in the laboratory:

Fastest electron speed $0.9999999994c$
Lowest temperature   0.00000002 K

You may think that, in each case, that is surely close enough. However, new phenomena keep appearing the closer we get to these unreachable goals. As it turns out, every additional decimal place, either in electron speed or in temperature, must be fought for against ever-increasing experimental difficulties.

Given the broad sweep of possibilities for temperature, the very fact of our existence seems the grandest of miracles. The rates of chemical reactions—including those in our bodies—slow down as the temperature decreases. If the temperature of the earth were just a little lower, we would all freeze to death, as some of our ancestors surely did during the Ice Ages. On the other hand, if the earth's temperature were just a little higher, the atoms of our bodies would be moving around in such an agitated fashion that molecules would break apart. Once more, life would not be possible. As far as temperature is concerned, we hover between fire and ice, in the tiniest of ecological niches.

19-2   Thermodynamics: A New Subject

With this chapter, we leave the subject of mechanics and begin a new subject—*thermodynamics*. Mechanics deals with the mechanical (or external) energies of systems and is governed by Newton's laws. Thermodynamics deals with the internal energy of systems and is governed by a new set of laws, which we shall come to know in this and the next few chapters. To give the flavor of things, some "mechanics words"—as we may call them—are force, kinetic energy, acceleration, Galileo, and Newton's second law. Some "thermodynamics words" are temperature, heat, internal energy, entropy, Kelvin, and the second law of thermodynamics.

The central concept of thermodynamics is temperature. This word is so familiar that most of us—because of our built-in sense of hot and cold—tend to be overconfident in our understanding of it. Our "temperature sense" is in fact not always reliable. On a cold winter day, for example, an iron railing seems much colder to the touch than does a wooden fence post, yet both are at the same temperature. This difference in our sense perception comes about because iron conducts heat away from our fingers much more readily than does wood.

Because of its fundamental importance, we begin our study of thermodynamics by developing the concept of temperature from its foundations, not relying in any way on our temperature sense to do so.

19-3   Temperature: A Fundamental Concept

The properties of many bodies change as you alter their thermal environment, perhaps moving them from a re-

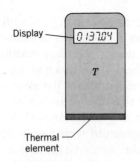

Display
T
Thermal element

Figure 2 A thermoscope. The numbers increase when the device is heated and decrease when it is cooled. The thermally sensitive element could be—among many possibilities—a coil of wire whose electrical resistance is measured and displayed.

frigerator to a warm oven. To give a few examples: The volume of a liquid increases, a metal rod grows a little longer, the electrical resistance of a wire increases, as does the pressure exerted by a confined gas. We can use any one of these properties as the basis of an instrument that will help us to pin down the concept of temperature.

Figure 2 shows such an instrument. Any resourceful engineer could design and construct it, using any one of the properties listed above. The instrument is fitted with a digital readout display and has the following properties: If you heat it with a Bunsen burner, the displayed number starts to increase; if you then put it into a refrigerator, the displayed number starts to decrease. The instrument is not calibrated in any way and the numbers have (as yet) no physical meaning. The device is a *thermoscope* but not (as yet) a *thermometer*.

Suppose that, as in Fig. 3a, you put the thermoscope (which we shall call body T) into intimate contact with another body (body A). The entire system is confined within a thick-walled insulating box. The numbers displayed by the thermoscope roll by until, eventually, they come to rest (let us say the reading is "137.04") and no further change takes place. In fact, every measurable property of body T (the thermoscope) and of body A assumes a stable value and we say that the two bodies are in *thermal equilibrium* with each other.

Now let us put body T in intimate contact with a second body (body B) as in Fig. 3b. Let us say that the two bodies (B and T) come to thermal equilibrium *at the same reading of the thermoscope.*

Finally, as in Fig. 3c, let us bring bodies A and B into intimate contact. Will they be found to be in thermal equilibrium with each other? They will. This answer,

which perhaps seems obvious, is in fact *not* obvious and can only be learned from experiment. Jones and Smith each know Green, but they may or may not know each other. Two pieces of iron each attract a third piece of iron but they may or may not attract each other.

The experimental facts described in Fig. 3 are summed up by the zeroth law of thermodynamics:

> *If bodies A and B are each in thermal equilibrium with a third body T, they are in thermal equilibrium with each other.*

In less formal language, the message of the zeroth law is: "Every body has a property called *temperature*. When

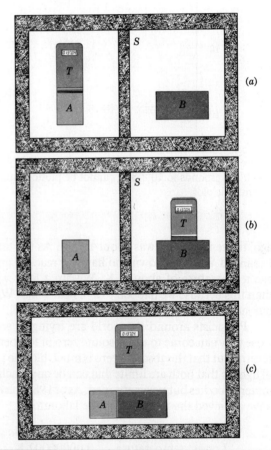

Figure 3 (a) Body T (a thermoscope) and body A are in thermal equilibrium. (b) Body T and body B are also found to be in thermal equilibrium, at the same reading of the thermoscope. (c) If (a) and (b) are true, the zeroth law of thermodynamics states that body A and body B will also be in thermal equilibrium. Body S is a thermally insulating screen.

two bodies are found to be in thermal equilibrium, their temperatures are equal." We can now make our thermoscope (body T) into a thermometer, confident that its readings will have physical meaning. All that we have to do is to calibrate it.

We use the zeroth law constantly in the laboratory. If we want to know whether the liquids in two beakers are at the same temperature, we measure the temperature of each with a thermometer. We do not need to bring the two liquids into intimate contact and observe whether or not they are in thermal equilibrium; we are quite sure that they would be if we were to do so.

The zeroth law, which has been called a logical afterthought, came to light only in the 1930s, long after the first and the second laws of thermodynamics had been discovered and numbered. Because the concept of temperature is fundamental to these two laws, the law that establishes temperature as a valid concept should have the lowest number. Hence the zero.

19-4   Measuring Temperature

Let us see how we define and measure temperatures on the Kelvin scale. Equivalently, let us see how to calibrate a thermoscope so as to make it into a thermometer.

The Triple Point of Water.   The first step in setting up a temperature scale is to pick out some reproducible thermal environment and, quite arbitrarily, assign a certain Kelvin temperature to it. That is, we select a *standard fixed point*. We could, for example, select the freezing point or the boiling point of water but, for various technical reasons, we do not. We select the *triple point of water*.

Liquid water, solid ice, and water vapor can coexist, in thermal equilibrium, at only one set of values of pressure and temperature. Figure 4 shows how this so-called triple point of water can be achieved in the laboratory, in a *triple-point cell*. By international agreement (in 1967), the triple point of water has been assigned a value of 273.16 K, as a standard fixed point for the calibration of thermometers. That is,

$$T_3 = 273.16 \text{ K} \quad \text{(triple-point temperature)}, \quad (1)$$

in which the subscript 3 reminds us of the triple point.

Note that we do not use a degree mark in reporting Kelvin temperatures. It is 300 K (not 300°K) and it is pronounced 300 kelvins (not "300 degrees Kelvin"). The usual SI prefixes apply. Thus, 0.0035 K is 3.5 mK.

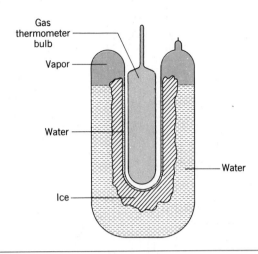

Figure 4   A triple-point cell, in which solid ice, liquid water, and water vapor coexist in thermal equilibrium. By international agreement, the temperature of this mixture has been defined to be 273.16 K. The bulb of a constant volume gas thermometer is shown inserted into the well of the cell.

No distinction in nomenclature is made between temperatures and temperature differences. Thus, we can say, "The boiling point of sulfur is 717.8 K" and "The temperature of this water bath was raised by 8.5 K."

The Constant-Volume Gas Thermometer.   So far, we have not discussed the particular physical property on which, by international agreement, we shall base our thermometer. Shall it be the length of a metal rod, the electrical resistance of a wire, the pressure exerted by a confined gas, or something else? The choice is important because different choices lead to different temperatures for — to give one example — the boiling point of water. For reasons that will emerge below, the standard thermometer, against which all other thermometers are to be calibrated, is based on the pressure exerted by a gas confined to a fixed volume.

Figure 5 shows such a (constant volume) *gas thermometer*. It consists of a bulb made of glass, quartz, or platinum (depending on the temperature range over which the thermometer is to be used) connected by a capillary tube to a manometer. By raising and lowering reservoir R, the mercury meniscus on the left can always be brought to the zero of the manometer scale, thus assuring that the volume of the confined gas remains constant. The temperature of any body in thermal contact with the bulb is then defined to be

$$T = Cp, \quad (2)$$

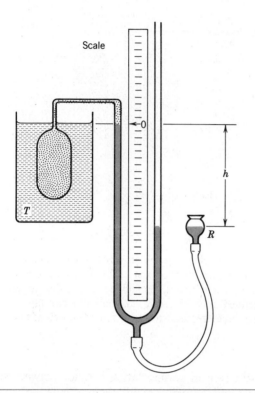

Scale

0

h

R

T

Figure 5 A constant volume gas thermometer, its bulb immersed in a bath whose temperature T is to be measured. The pressure exerted by the gas is $p_{atm} + \rho gh$, where p_{atm} is the atmospheric pressure (read from a barometer) and h is the level difference of the manometer.

in which p is the pressure exerted by the gas and C is a constant. The pressure is calculated from the relation

$$p = p_0 + \rho gh, \tag{3}$$

in which p_0 is the atmospheric pressure, ρ is the density of the mercury in the manometer, and h is the measured level difference.

With the bulb immersed in a triple-point cell, we have

$$T_3 = Cp_3, \tag{4}$$

in which p_3 is the pressure reading under these conditions. Eliminating C between Eqs. 2 and 4 leads to

$$T = T_3 \left(\frac{p}{p_3}\right) = (273.16 \text{ K}) \left(\frac{p}{p_3}\right) \quad \text{(provisional).} \tag{5}$$

Equation 5 is not yet our final definition of temperature. We have said nothing about what gas (or how much gas) we are to use in the thermometer. If our thermometer is used to measure some temperature, such as the boiling point of water, we would find that different

choices lead to slightly different temperatures. However, we find that, as we use smaller and smaller amounts of gas to fill the bulb, the readings converge nicely to a single temperature, no matter what gas we use. Figure 6 shows this comforting convergence.

Thus, we write, as our final recipe for measuring temperature with a gas thermometer,

$$T = (273.16 \text{ K}) \left(\lim_{m \to 0} \frac{p}{p_3}\right). \tag{6}$$

This instructs us to fill the bulb with an arbitrary mass m of *any* gas (for example, nitrogen) and to measure p_3 (using a triple-point cell) and p, the pressure at the point in question. Calculate the ratio p/p_3. Repeat both measurements with a smaller amount of gas in the bulb, again calculating this ratio. Keep on this way, using smaller and smaller amounts of gas, until you can extrapolate to the ratio that you would find if you had essentially no gas in the bulb. Calculate the temperature by substituting that extrapolated ratio into Eq. 6. Temperature defined in this way is the *ideal gas* temperature.

If temperature is to be a truly fundamental physical quantity, one in which the laws of thermodynamics may be expressed, it is absolutely necessary that its definition be independent of the properties of specific materials. It would not do, for example, to have such a basic quantity as temperature depend on the expansivity of mercury, the electrical resistivity of platinum, or any other such "handbook" property. We chose the gas thermometer as our standard instrument precisely because no such specific properties of materials are involved in its operation. You can use *any* gas and you always get the same answer.

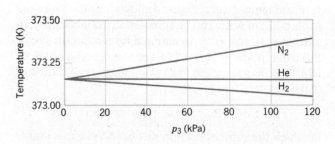

Figure 6 Temperatures calculated from Eq. 5 for a constant volume gas thermometer whose bulb is immersed in condensing steam. Different gases are used, each at a variety of densities. Note that all readings converge in the limit of zero density.

Sample Problem 1 The bulb of a gas thermometer is filled with nitrogen to a pressure of 120 kPa. What provisional value (see Eq. 5 and Fig. 6) would this thermometer yield for the boiling point of water?

From Fig. 6, the curve for nitrogen shows that, at 120 kPa, the provisional temperature for the boiling point of water would be about 373.44 K. The actual boiling point of water (found by extrapolation on Fig. 6; see also Table 1) is 373.125 K. Thus, using the provisional Eq. 5 to determine a temperature (in place of Eq. 6) leads to an error of 0.315 K or 315 mK.

19-5   The International Practical Temperature Scale

Measuring a temperature precisely with a gas thermometer is not an easy task. "The accurate measurement of temperature with a gas thermometer requires months of painstaking laboratory work and mathematical computation and when completed becomes an international event."[*] In practice, the gas thermometer is used only to establish certain fixed points that can then be used to calibrate other more convenient secondary thermometers.

For practical use, as in the calibration of industrial or scientific thermometers, the International Practical Temperature Scale has been adopted. This scale consists of a set of recipes for providing — in practice — the best possible approximations to the Kelvin scale. A set of fixed points (see Table 1) is adopted and a set of instruments is specified to be used in interpolating between these fixed points and in extrapolating beyond the highest fixed point.

19-6   The Celsius and the Fahrenheit Scales

So far, we have only discussed the Kelvin scale, used in basic scientific work. In nearly all countries of the world, the Celsius scale (formerly called the centigrade scale) is the scale of choice for all popular and commercial and much scientific use. The size of the degree is the same on

---

* Quoted from *Heat and Thermodynamics,* 6th ed., by Mark W. Zemansky and Richard H. Dittman, McGraw-Hill Book Company, New York, 1981, p. 21.

Table 1   Primary Fixed Points on the International Practical Temperature Scale

Substance	State	Temperature (K)
Hydrogen	Triple point	13.81
Hydrogen	Boiling point[a]	17.042
Hydrogen	Boiling point	20.28
Neon	Boiling point	27.102
Oxygen	Triple point	54.361
Argon	Triple point	83.798
Oxygen	Boiling point	90.188
Water	Boiling point	373.125
Tin	Melting point	505.074
Zinc	Melting point	692.664
Silver	Melting point	1235.08
Gold	Melting point	1337.58

[a] This boiling point is for a pressure of $\frac{25}{73}$ atm. All other boiling or melting points are for a pressure of 1 atm.

both the Celsius and the Kelvin scales but the zero of the former is shifted to a more convenient value. If T_C represents the Celsius temperature, then

$$T_C = T - 273.15°. \qquad (7)$$

In expressing temperatures on the Celsius scale, the degree symbol is commonly used. Thus, we write 20.00°C, but 293.15 K.

The Fahrenheit scale, used in the United States, employs a smaller degree than the Celsius scale and its zero is set to a different temperature. You can easily verify both statements by comparing an ordinary room thermometer on which both scales are marked. The relation between the Celsius and the Fahrenheit scales is

$$T_F = \tfrac{9}{5}T_C + 32°. \qquad (8)$$

Transferring between these two scales can be done easily by remembering a few corresponding points (such as the freezing and boiling points of water; see Table 2) and by

Table 2   Some Corresponding Temperatures

Temperature	°C	°F
Boiling point of water[a]	100	212
Normal body temperature	37.0	98.6
Accepted comfort level	20	68
Freezing point of water	0	32
Zero of Fahrenheit scale	≈ −18	0
Scales coincide	−40	−40

[a] Strictly, the boiling point of water on the Celsius scale is 99.975°C. See Eq. 7 and Table 1.

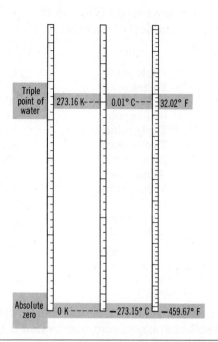

Figure 7 The Kelvin, the Celsius, and the Fahrenheit temperature scales compared.

making use of the fact that 9 degrees on the Fahrenheit scale are equal to 5 degrees on the Celsius scale. Figure 7 compares the Kelvin, the Celsius, and the Fahrenheit scales.

Sample Problem 2 (a) In 1964, the temperature in the Siberian village of Oymyakon reached a value of $-71°C$. What temperature is this on the Fahrenheit scale?

From Eq. 8

$$T_F = \tfrac{9}{5}T_C + 32° = (\tfrac{9}{5})(-71°) + 32°$$
$$= -96°F. \qquad \text{(Answer)}$$

(b) The highest officially recorded temperature in continental United States is 134°F in Death Valley, California. What is this temperature on the Celsius scale?

Solving Eq. 8 for T_C yields

$$T_C = (\tfrac{5}{9})(T_F - 32°) = (\tfrac{5}{9})(134° - 32°)$$
$$= 56.7°C. \qquad \text{(Answer)}$$

19-7 Thermal Expansion

Some Applications. You can often loosen a tight metal jar lid by holding it under a stream of hot water.

The metal lid expands a little relative to the glass jar as its temperature rises. Thermal expansion is not always desirable, as Fig. 8 suggests. We have all seen expansion slots in the roadways of bridges. Pipes at refineries often include an expansion loop, so that the pipe will not buckle as the temperature rises. Dental materials used for fillings must be matched in their expansion properties to those of tooth enamel. In aircraft manufacture, rivets and other fasteners are often designed so that they are to be cooled in dry ice before insertion and then allowed to expand to a tight fit. Thermometers and thermostats may be based on the differences in expansion between the components of a *bimetalic strip;* see Fig. 9. In a thermometer of a familiar type, the bimetallic strip is coiled into a helix that winds and unwinds as the temperature changes; see Fig. 10. Finally, the familiar liquid-in-glass thermometers are based on the fact that liquids such a mercury or alcohol expand to a different (greater) extent than do their glass containers.

Thermal Expansion: Quantitative. If the temperature of a metal rod of length L is raised by an amount ΔT, its length is found to increase by an amount

$$\boxed{\Delta L = L\alpha \, \Delta T,} \qquad (9)$$

Figure 8 Railroad tracks distorted because of thermal expansion on a very hot day.

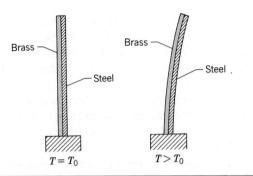

Figure 9   A bimetal strip, consisting of a strip of brass and a strip of steel welded together, at temperature T_0. The strip bends as shown at temperatures above this reference temperature. Below the reference temperature the strip bends the other way. Many thermostats operate on this principle, making and breaking an electrical contact as the temperature rises and falls.

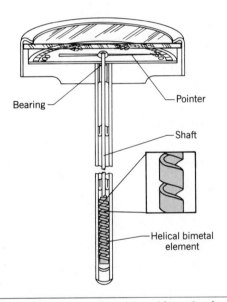

Figure 10   A thermometer based on a bimetal strip. The strip is formed into a helix, which coils or uncoils as the temperature is changed.

Table 3   Some Coefficients of Linear Expansion[a]

Substance	$\alpha \ (10^{-6} \ K^{-1})$
Ice	51
Lead	29
Aluminum	23
Brass	19
Copper	17
Steel	11
Glass (ordinary)	9
Glass (Pyrex)	3.2
Invar[b]	0.7
Fused quartz	0.5

[a] Room temperature values.
[b] This alloy was designed to have a low expansivity. The word is a shortened form of "invariable."

in which α is a constant called the *coefficient of linear expansion*. The value of α depends on the material and on the temperature range of interest. We can rewrite Eq. 9 as

$$\alpha = \frac{\Delta L/L}{\Delta T}, \qquad (10)$$

which shows us that α is the fractional increase in length per unit change in temperature. Although α varies with temperature, for most practical purposes at ordinary temperatures it can be taken as a constant. Table 3 shows some coefficients of expansion.

The thermal expansion of a solid is like a (three-dimensional) photographic enlargement; see Fig. 11. If you have a square metal plate with a hole punched in it, Eq. 9 applies to every linear dimension of the plate, including its edge, its thickness, its face diagonal, and the diameter of the hole. If you scratch your name on the plate in a continuous line, Eq. 9 describes the change in length of that line as the temperature is changed.

There are those who doubt, in spite of the evidence of Fig. 11, that a hole in a flat plate increases in size as the

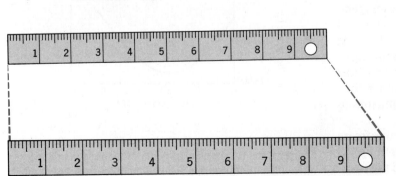

Figure 11   The same steel rule at two different temperatures. On expansion, every dimension is increased in the same proportion. The scale, the numbers, the thickness, and the hole diameter are all increased by the same factor. The expansion has been exaggerated for clarity.

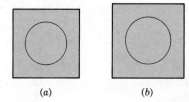

(a)	(b)

Figure 12 (a) A metal plate containing a snuggly fitting disk, at room temperature. (b) The same, at a higher temperature T. The disk (which has expanded) still fits snuggly so that the hole must also have expanded. The expansion has been exaggerated for clarity.

plate is heated. Figure 12 should help to convince such skeptics. Figure 12a shows a metal plate with a snuggly fitting removable central disk, at room temperature. Figure 12b shows the situation at a higher temperature, the plate and the central disk both having expanded uniformly. The disk still fits snugly into its central hole. If the disk expanded, the hole must have done so too.

Thermal Expansion of Liquids. If all dimensions of a solid expand with temperature, the volume of that solid must also expand. For liquids, the volume expansion is the only meaningful parameter. If the temperature of a liquid whose volume is V is increased by an amount ΔT, the increase in volume of the liquid is found to be

$$\Delta V = V\beta \, \Delta T, \qquad (11)$$

where β is the *coefficient of volume expansion* of the liquid. In Problem 36, we ask you to show that the coefficients of volume and of linear expansion for solids are related by

$$\beta = 3\alpha. \qquad (12)$$

The most common liquid, water, does not behave like other liquids. Figure 13a shows its volume expansion curve. Above ~4°C, water expands as the temperature rises. Between 0 and ~4°C, however, water *contracts* with increasing temperature (see Fig. 13b). At about 4°C, the specific volume of water passes through a minimum, which is to say that the density of water passes through a maximum. At all temperatures other than ~4°C, the density of water is less than this maximum value.

This behavior of water is the reason that lakes freeze from the top down rather than from the bottom up. As the surface layers cool from, say, 10°C toward the freezing point, they become denser and sink to the bottom.

Below 4°C, however, further cooling makes the water *less* dense, so that it no longer sinks but stays on the surface until it freezes. If lakes froze from the bottom up, the ice so formed would tend not to melt completely during the summer, being insulated by the water above. After a few years, many bodies of open water in the temperate zones of the earth would essentially be frozen solid all year round. Would you have guessed that so much depends on the (scarcely noticeable) behavior of water in the lower left-hand corner of Fig. 13a?

Thermal Expansion: An Atomic View. Let us see if we can understand why a solid expands if you increase its temperature. The atoms of a crystalline solid are held together in a three-dimensional periodic lattice by springlike interatomic forces. The individual atoms vibrate about these lattice sites with an amplitude that increases with temperature. If the solid as a whole ex-

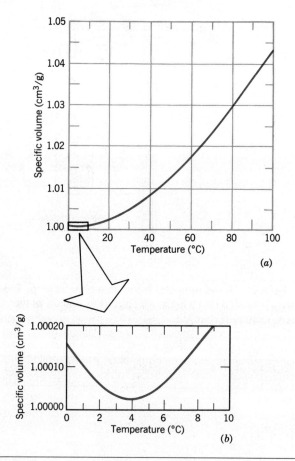

Figure 13 (a) The specific volume of water as a function of temperature. (b) An enlargement of the insert near 4°C, showing a minimum specific volume (that is, a maximum density).

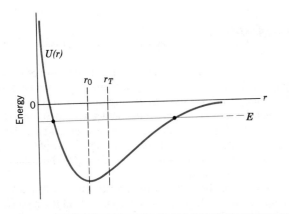

Figure 14   The potential energy $U(r)$ for two atoms separated by a distance r. Note that, as the energy E is increased (corresponding to an increase in temperature), the atoms tend to move farther apart. For a solid with a symmetrical potential energy curve there would be no thermal expansion.

pands, the average distance between neighboring atoms must have increased.

Figure 14 shows the potential energy curve $U(r)$ for a pair of neighboring atoms, r being their separation. The potential energy has a minimum at $r = r_0$, this being the value of lattice spacing that the solid would have at a temperature approaching the absolute zero. We see at once that the curve is not symmetrical, rising more steeply when the atoms are pushed together ($r < r_0$) than when they are pulled apart ($r > r_0$). The interatomic "springs" evidently do not obey Hooke's law.

It is this lack of symmetry of the potential energy function that accounts for the thermal expansion of solids. The solid horizontal line shows the mechanical energy of the pair of atoms at an arbitrary temperature T. At this temperature, the interatomic separation can oscillate between the two limits shown, its average value being marked r_T. Note that r_T is greater than r_0 and also that r_T continues to increase as the energy E (and thus the temperature) is raised. In other words, the lattice spacing—and thus the dimensions of the solid—increases with temperature. A solid with a symmetrical potential energy curve would not expand with temperature.

Sample Problem 3   Steel railroad rails are laid when the temperature is 0°C. What gap should be left between rail sections so that they just touch when the temperature rises to 42°C? The sections are 12.0 m long.

From Table 3, the coefficient of linear expansion for steel is $11 \times 10^{-6}/°C$. From Eq. 9 we then have

$$\Delta L = L\alpha \,\Delta T = (12.0 \text{ m})(11 \times 10^{-6}/°C)(42°C)$$
$$= 5.5 \times 10^{-3} \text{ m} = 5.5 \text{ mm}. \qquad \text{(Answer)}$$

Sample Problem 4   A steel wire whose length L is 130 cm and whose diameter d is 1.1 mm is heated to an average temperature of 830°C and stretched taut between two rigid supports. What tension develops in the wire as it cools to 20°C?

First let us calculate the amount by which the wire would shrink if it were allowed to cool freely. From Eq. 9 we have

$$\Delta L = L\alpha \,\Delta T = (1.3 \text{ m})(11 \times 10^{-6}/°C)(830°C - 20°C)$$
$$= 1.16 \times 10^{-2} \text{ m} = 1.16 \text{ cm}.$$

However, the wire is not permitted to shrink. We therefore calculate what force would be required to stretch the wire by this amount. From Eq. 26 of Chapter 13 ($\Delta L = FL/EA$) we have

$$F = \frac{\Delta L \, EA}{L} = \frac{\Delta L \, E(\pi/4)d^2}{L},$$

in which E is Young's modulus for steel (see Table 1 of Chapter 13) and A is the cross-sectional area of the wire. Substitution yields

$$F = \frac{(1.16 \times 10^{-2} \text{ m})(200 \times 10^9 \text{ N/m}^2)(\pi/4)(1.1 \times 10^{-3} \text{ m})^2}{1.3 \text{ m}}$$
$$= 1700 \text{ N} (= 380 \text{ lb}). \qquad \text{(Answer)}$$

Can you show that this answer is independent of the length of the wire?

Sometimes the bulging brick walls of old buildings are supported by running a steel stress rod from outside wall to outside wall, through the building. The rod is then heated and nuts on the outside walls are tightened. As the rod cools, stress develops in the rod, helping to keep the walls from bulging outward.

Sample Problem 5   On a hot day in Las Vegas, an oil trucker loaded 9785 gal of diesel fuel. He encountered cold weather and, on arriving at Payson, Utah, where the temperature was 41°F lower than in Las Vegas, delivered his entire load. How many gallons did he deliver? The coefficient of volume expansion for diesel fuel is $9.5 \times 10^{-4}/°C$ and the coefficient of linear expansion of his steel truck tank is $11 \times 10^{-6}/°C$.

From Eq. 11,

$$\Delta V = V\beta \,\Delta T = (9785 \text{ gal})(9.5 \times 10^{-4}/°C)(41°F)(5°C/9°F)$$
$$= 212 \text{ gal}.$$

Thus, the amount delivered was

$$V_{\text{del}} = V - \Delta V$$
$$= 9785 \text{ gal} - 212 \text{ gal} = 9573 \text{ gal}. \qquad \text{(Answer)}$$

Note that the thermal expansion of the steel tank has nothing to do with the problem. Question: Who paid for the "missing" diesel fuel?

REVIEW AND SUMMARY

Temperature; Thermometers

Temperature is a quantitative measure of a macroscopic quantity related to our sense of hot and cold. It is measured by a thermometer, which contains a working substance with a measurable property, such as length or pressure, that changes in a regular way as the substance becomes hotter or colder. When a thermometer and some other object are placed in contact with each other, they eventually reach thermal equilibrium. The reading of the thermometer is then taken to be the temperature of the other object. The process is consistent because of the *zeroth law of thermodynamics:* If bodies A and B are each in thermal equilibrium with a third body C (the thermometer), then A and B are in thermal equilibrium with each other.

The Zeroth Law of Thermodynamics

The Kelvin Temperature Scale

Temperature is measured, in the SI system, on the Kelvin scale. The scale is established by first defining the numerical value of the temperature at which water can exist with all three phases in equilibrium (the triple point) to be 273.16 K. Other temperatures are then defined in terms of a constant volume gas thermometer, in which temperature is proportional to the pressure of a constant volume sample of an ideal gas. Since real gases give consistent results, using different gases, only at very low densities, the ideal gas temperature is defined as

$$T = (273.16 \text{ K}) \left(\lim_{m \to 0} \frac{p}{p_3} \right).$$ [6]

Here p_3 and p are the pressures of the gas at the triple point temperature and the measured temperature, respectively, and m is the mass of the gas in the thermometer. See Sample Problem 1.

The International Scale

Several "fixed points" on the Kelvin scale have been measured as a basis for the *International Practical Temperature Scale;* see Table 1.

The Celsius and Fahrenheit Scales

Besides the Kelvin scale, two other temperature scales are in common use: the Celsius scale, defined by

$$T_C = T - 273.15°,$$ [7]

and the Fahrenheit scale, defined by

$$T_F = (\tfrac{9}{5})T_C + 32°.$$ [8]

See Table 2 and Sample Problem 2.

Thermal Expansion

All objects change size with changes in temperature. The change ΔL in any linear dimension L is given by

$$\Delta L = L\alpha \, \Delta T,$$ [9]

in which α is the *coefficient of linear expansion;* see Table 3. The change ΔV in the volume V of a solid, liquid or gas is

$$\Delta V = V\beta \, \Delta T.$$ [11]

Here $\beta = 3\alpha$ is the *coefficient of volume expansion* of the material. See Sample Problems 3 through 5 and Fig. 13 for examples.

QUESTIONS

1. Is temperature a microscopic or macroscopic concept?

2. Are there physical quantities other than temperature that tend to equalize if two different systems are joined?

3. A piece of ice and a warmer thermometer are suspended in an insulated evacuated enclosure so that they are not in contact. Why does the thermometer reading decrease for a time?

4. Let p_3 be the pressure in the bulb of a constant volume gas thermometer when the bulb is at the triple-point temperature of 273.16 K and p the pressure when the bulb is at room temperature. Given three constant volume gas thermometers: for A the gas is oxygen and $p_3 = 20$ cm Hg; for B the gas is also oxygen but $p_3 = 40$ cm Hg; for C the gas is hydrogen and $p_3 = 30$ cm Hg. The measured values of p for the three thermometers are p_A, p_B, and p_C. (a) An approximate value of the room temperature T can be obtained with each of the thermometers using

$$T_A = 273.16 \text{ K} \frac{p_A}{20 \text{ cm Hg}}; \qquad T_B = 273.16 \text{ K} \frac{p_B}{40 \text{ cm Hg}};$$

$$T_C = 273.16 \text{ K} \frac{p_C}{30 \text{ cm Hg}}.$$

Mark each of the following statements true or false: (1) With the method described, all three thermometers will give the same value of T. (2) The two oxygen thermometers will agree with each other but not with the hydrogen thermometer. (3) Each of the three will give a different value of T. (b) In the event that there is disagreement among the three thermometers, explain how you would change the method of using them to cause all three to give the same value of T.

5. A student, when told that the temperature at the center of the sun was thought to be about 1.5×10^7 degrees, asked whether that was on the Celsius or the Kelvin scale. How would you answer? How would you reply if he had asked whether it was on the Celsius or the Fahrenheit scale?

6. The Editor-in-Chief of a well-known business magazine, discussing possible warming effects associated with the increasing concentration of carbon dioxide in the earth's atmosphere, wrote: "The polar regions might be three times warmer than now, ..." What do you suppose he meant? (See "Warmth and Temperature: A Comedy of Errors," by Albert A. Bartlett, *The Physics Teacher,* November 1984.)

7. Although the absolute zero of temperature seems to be experimentally unattainable, temperatures as low as 0.00000002 K have been achieved in the laboratory. Isn't this low enough for all practical purposes? Why would physicists (as indeed they do) strive to obtain still lower temperatures?

8. You put two uncovered pails of water, one containing hot water and one containing cold water, outside in below-freezing weather. The pail with the hot water will usually begin to freeze first. Why? What would happen if you covered the pails?

9. Can a temperature be assigned to a vacuum?

10. Does our "temperature sense" have a built-in sense of direction; that is, does hotter necessarily mean higher temperature, or is this just an arbitrary convention? Celsius, by the way, originally chose the steam point as 0°C and the ice point as 100°C.

11. Many medicine labels inform the user to store below 86°F. Why 86? (*Hint:* Change to Celsius.) (See *The Science Almanac,* 1985–86, p. 430.)

12. How would you suggest measuring the temperature of (*a*) the sun, (*b*) the earth's upper atmosphere, (*c*) an insect, (*d*) the moon, (*e*) the ocean floor, and (*f*) liquid helium?

13. Is one gas any better than another for purposes of a standard constant volume gas thermometer? What properties are desirable in a gas for such purposes?

14. State some objections to using water-in-glass as a thermometer. Is mercury-in-glass an improvement?

15. Explain why the column of mercury first descends and then rises when a mercury-in-glass thermometer is put in a flame.

16. What do the Celsius and Fahrenheit temperature scales have in common?

17. What are the dimensions of α, the coefficient of linear expansion? Does the value of α depend on the unit of length used? When degrees Fahrenheit are used instead of degrees Celsius as the unit of temperature change, does the numerical value of α change? If so, how? If not, prove it.

18. A metal ball can pass through a metal ring. When the ball is heated, however, it gets stuck in the ring. What would happen if the ring, rather than the ball, were heated?

19. A bimetallic strip, consisting of two different metal strips riveted together, is used as a control element in the common thermostat. Explain how it works.

20. Two strips, one of iron and one of zinc, are riveted together side by side to form a straight bar that curves when heated. Why is it that the iron is on the inside of the curve?

21. Explain how the period of a pendulum clock can be kept constant with temperature by attaching vertical tubes of mercury to the bottom of the pendulum.

22. Why should a chimney be free standing, that is, not part of the structural support of the house?

23. Water expands when it freezes. Can we define a coefficient of volume expansion for the freezing process?

24. Explain why the apparent expansion of a liquid in a glass bulb does not give the true expansion of the liquid.

25. Does the change in volume of an object when its temperature is raised depend on whether the object has cavities inside, other things being equal?

26. Why is it much more difficult to make a precise determination of the coefficient of expansion of a liquid than of a solid?

27. A common model of a solid assumes the atoms to be point masses executing simple harmonic motion about mean lattice positions. What would be the coefficient of thermal expansion of such a lattice?

28. Explain the fact that the temperature of the ocean at great depths is very constant the year round, at a temperature of about 4°C.

29. What difficulties would arise if you defined temperature in terms of the density of water?

30. Explain why lakes freeze first at the surface.

31. What is it that causes water pipes to burst in the winter?

32. What can you conclude about how the melting point of ice depends on pressure from the fact that ice floats on water?

EXERCISES AND PROBLEMS

Section 19–4 Measuring Temperature

1E. To measure temperatures, physicists and astronomers often use the variation of intensity of electromagnetic radiation radiated by an object. The wavelength at which the intensity is greatest is given by the equation

$$\lambda_{max}T = 0.2898 \text{ cm} \cdot \text{K},$$

where λ_{max} is the wavelength of greatest intensity and T is the absolute temperature of the object. In 1965, microwave radiation peaked at $\lambda_{max} = 0.107$ cm was discovered coming in all directions from space. To what temperature does this correspond? The interpretation of this background radiation is that it is left over from the Big Bang some 15 billion years ago, when the universe began rapidly expanding and cooling.

2E. A thermocouple is formed from two different metals, joined at two points in such a way that a small voltage is produced when the two junctions are at different temperatures. In a particular iron–constantan thermocouple, with one junction held at $0°C$, the output voltage varies linearly from 0 to 28 mV as the temperature of the other junction is raised from 0 to $510°C$. Find the temperature of the variable junction when the thermocouple output is 10.2 mV.

3E. The amplification or *gain* of a transistor amplifier may depend on the temperature. The gain for a certain amplifier at room temperature ($20°C$) is 30.0, whereas at $55°C$ it is 35.2. What would the gain be at $30°C$ if the gain depends linearly on temperature, over a limited range?

4E. If the gas temperature at the steam point is 373.15 K, what is the limiting value of the ratio of the pressures of a gas at the steam point and at the triple point of water when the gas is kept at constant volume?

5E. A *resistance thermometer* is a thermometer in which the electrical resistance changes with temperature. We are free to define temperatures measured by such a thermometer in kelvins (K) to be directly proportional to the resistance R, measured in ohms (Ω). A certain resistance thermometer is found to have a resistance R of 90.35 Ω when its bulb is placed in water at the triple-point temperature (273.16 K). What temperature is indicated by the thermometer if the bulb is placed in an environment such that its resistance is 96.28 Ω?

6P. Two constant-volume gas thermometers are assembled, one using oxygen as the working gas and the other using hydrogen. Both contain enough gas so that $p_3 = 80$ mm Hg. What is the difference between the pressures in the two thermometers if both are inserted into a water bath at the boiling point? Which pressure is the higher of the two?

7P. A particular gas thermometer is constructed of two gas-containing bulbs, each of which is put into a water bath, as shown in Fig. 15. The pressure difference between the two bulbs is measured by a mercury manometer as shown in the

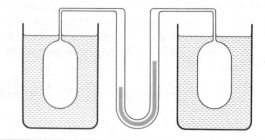

Figure 15 Problem 7.

figure. Appropriate reservoirs, not shown in the diagram, maintain constant gas volume in the two bulbs. There is no difference in pressure when both baths are at the triple point of water. The pressure difference is 120 mm Hg when one bath is at the triple point and the other is at the boiling point of water. Finally, the pressure difference is 90 mm Hg when one bath is at the triple point and the other is at an unknown temperature to be measured. What is the unknown temperature?

8P. A *thermistor* is a semiconductor device with a temperature-dependent electrical resistance. It is commonly used in medical thermometers and to sense overheating in electronic equipment. Over a limited range of temperature, the resistance is given by

$$R = R_a e^{B(1/T - 1/T_a)},$$

where R is the resistance of the thermistor at temperature T and R_a is the resistance at temperature T_a; B is a constant that depends on the particular semiconductor used. For one type of thermistor, $B = 4689$ K and the resistance at 273 K is 1.0×10^4 Ω. What temperature is the thermistor measuring when its resistance is 100 Ω?

9P. In the interval between 0 and $700°C$, a platinum resistance thermometer of definite specifications is used for interpolating temperatures on the International Practical Temperature Scale. The Celsius temperature T_C is given by a formula for the variation of resistance with temperature:

$$R = R_0(1 + AT_C + BT_C^2).$$

R_0, A, and B are constants determined by measurements at the ice point, the steam point, and the zinc point. (*a*) If R equals 10.000 Ω at the ice point, 13.946 Ω at the steam point, and 24.172 Ω at the zinc point, find R_0, A, and B. (*b*) Plot R versus T_C in the temperature range from 0 to $700°C$.

10P. It is an everyday observation that hot and cold objects cool down or warm up to the temperature of their surroundings. If the temperature difference ΔT between an object and its surroundings ($\Delta T = T_{obj} - T_{sur}$) is not too great, the rate of cooling or warming of the object is proportional, approximately, to this temperature difference; that is,

$$\frac{d\,\Delta T}{dt} = -A(\Delta T),$$

where A is a constant. The minus sign appears because ΔT decreases with time if ΔT is positive and increases if ΔT is negative. This is known as *Newton's law of cooling*. (*a*) On what factors does A depend? What are its dimensions? (*b*) If at some instant $t = 0$ the temperature difference is ΔT_0, show that it is

$$\Delta T = \Delta T_0 e^{-At}$$

at a time t later.

11P. The heater of a house breaks down one day when the outside temperature is 7.0°C. As a result, the inside temperature drops from 22 to 18°C in 1.0 h. The owner fixes the heater and adds insulation to the house. Now he finds that, on a similar day, the house takes twice as long to drop from 22 to 18°C when the heater is not operating. What is the ratio of the constant A in Newton's law of cooling (see Problem 10) after the insulation is added to the value before?

12P. A mercury-in-glass thermometer is placed in boiling water for a few minutes and then removed. The temperature readings at various times after removal are as shown in the table. Plot A as a function of time, assuming Newton's law of cooling to apply (see Problem 10). To what extent are you justified in assuming that Newton's law of cooling applies here?

t (s)	T (°C)	t (s)	T (°C)
0	98.4	100	50.3
5	76.1	150	43.7
10	71.1	200	38.8
15	67.7	300	32.7
20	66.4	500	27.8
25	65.1	700	26.5
30	63.9	1000	26.1
40	61.6	1400	26.0
50	59.4	2000	26.0
70	55.4	3000	26.0

Section 19-6   The Celsius and the Fahrenheit Scales

13E. On his 44th birthday celebration, the singer Tom Rush remarked ". . . . or, as I prefer to call it, 5 Celsius." Did Tom work that out correctly? If not, what should his "age in Celsius" be?

14E. At what temperature is the Fahrenheit scale reading equal to (*a*) twice that of the Celsius and (*b*) half that of the Celsius?

15E. If your doctor tells you that your temperature is 310 degrees above absolute zero, should you worry? Explain your answer.

16E. (*a*) The temperature of the surface of the sun is about 6000 K. Express this on the Fahrenheit scale. (*b*) Express normal human body temperature, 98.6°F, on the Celsius scale. (*c*) In the continental United States, the lowest officially recorded temperature is −70°F at Rogers Pass, Montana. Express this on the Celsius scale. (*d*) Express the normal boiling point of oxygen, −183°C, on the Fahrenheit scale. (*e*) At what Celsius temperature would you find a room to be uncomfortably warm?

17E. At what temperature do the following pairs of scales give the same reading: (*a*) Fahrenheit and Celsius, (*b*) Fahrenheit and Kelvin, and (*c*) Celsius and Kelvin?

Section 19-7   Thermal Expansion

18E. A steel rod has a length of exactly 20 cm at 30°C. How much longer is it at 50°C?

19E. An aluminum flagpole is 33 m high. By how much does its length increase as the temperature increases by 15°C?

20E. The Pyrex glass mirror in the telescope at the Mount Palomar Observatory has a diameter of 200 in. The temperature ranges from −10 to 50°C on Mount Palomar. Determine the maximum change in the diameter of the mirror.

21E. A circular hole in an aluminum plate is 2.725 cm in diameter at 0°C. What is its diameter when the temperature of the plate is raised to 100°C?

22E. An aluminum-alloy rod has a length of 10.000 cm at room temperature (20°C) and a length of 10.015 cm at the boiling point of water. (*a*) What is the length of the rod at the freezing point of water? (*b*) What is the temperature if the length of the rod is 10.009 cm?

23E. (*a*) Express the coefficient of linear expansion of aluminum using the Fahrenheit temperature scale. (*b*) Use your answer to calculate the change in length of a 20-ft aluminum rod if it is heated from 40 to 95°F.

24E. Soon after the earth formed, heat released by the decay of radioactive elements raised the average internal temperature from 300 to 3000 K, at about which value it remains today. Assuming an average coefficient of volume expansion of 3×10^{-5} K^{-1}, by how much has the radius of the earth increased since its formation?

25E. A rod is measured to be exactly 20.0 cm long using a steel ruler at a room temperature of 20°C. Both the rod and the ruler are placed in an oven at 270°C, where the rod now measures 20.1 cm using the same ruler. What is the coefficient of thermal expansion for the material of which the rod is made?

26E. The Stanford linear accelerator contains hundreds of brass disks tightly fitted into a steel tube. The system was assembled by cooling the disks in dry ice (−57°C) to enable them to slide into the close-fitting tube. If the diameter of a disk is 80.00 mm at 43°C, what is its diameter in the dry ice?

27E. A glass window is exactly 20 cm (= 7.9 in.) by 30 cm

(= 11.8 in.) at 10°C. By how much has its area increased when its temperature is 40°C?

28E. A brass cube has an edge length of 30 cm. What is the increase in its surface area when it is heated from 20 to 75°C?

29E. Find the change in volume of an aluminum sphere of 10.0-cm radius when it is heated from 0 to 100°C.

30E. What is the volume of a lead ball at 30°C if its volume at 60°C is 50 cm³?

31E. By how much does the volume of an aluminum cube 5.0 cm on an edge increase when it is heated from 10 to 60°C?

32E. Imagine an aluminum cup of 100-cm³ capacity filled with glycerin at 22°C. How much glycerin, if any, will spill out of the cup if the temperature of the cup and glycerin is raised to 28°C? (The coefficient of volume expansion of glycerin is $5.1 \times 10^{-4}/$°C.)

33E. A steel rod at 25°C is bolted securely at both ends and then cooled. At what temperature will it rupture? See Table 1, Chapter 13.

34P. A steel rod is 3.000 cm in diameter at 25°C. A brass ring has an interior diameter of 2.992 cm at 25°C. At what common temperature will the ring just slide onto the rod?

35P. The area A of a rectangular plate is ab. Its coefficient of linear expansion is α. After a temperature rise ΔT, side a is longer by Δa and side b is longer by Δb. Show that if we neglect the small quantity $\Delta a \Delta b/ab$ (see Fig. 16), then $\Delta A = 2\alpha A \, \Delta T$.

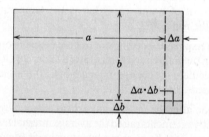

Figure 16 Problem 35.

36P. Prove that, if we neglect extremely small quantities, the change in volume of a solid upon expansion through a temperature rise ΔT is given by $\Delta V = 3\alpha V \, \Delta T$, where α is the coefficient of linear expansion. See Eqs. 11 and 12.

37P. Density is mass divided by volume. If the volume V is temperature dependent, so is the density ρ. Show that the change in density $\Delta\rho$ with change in temperature ΔT is given by

$$\Delta\rho = -\beta\rho \, \Delta T,$$

where β is the volume coefficient of expansion. Explain the minus sign.

38P. When the temperature of a metal cylinder is raised from 0 to 100°C, its length increases by 0.23%. (a) Find the percent change in density. (b) What is the metal?

39P. Show that when the temperature of a liquid in a barometer changes by ΔT, and the pressure is constant, the height h changes by $\Delta h = \beta h \, \Delta T$, where β is the coefficient of volume expansion. Neglect the expansion of the glass tube.

40P. When the temperature of a copper penny is raised by 100°C, its diameter increases by 0.18%. To two significant figures, give the percent increase in (a) the area of a face, (b) the thickness, (c) the volume, and (d) the mass of the penny. (e) Calculate its coefficient of linear expansion.

41P. A clock pendulum made of Invar has a period of 0.500 s and is accurate at 20°C. If the clock is used in a climate where the temperature averages 30°C, what correction (approximately) is necessary at the end of 30 days to the time given by the clock?

42P. A pendulum clock with a pendulum made of brass is designed to keep accurate time at 20°C. What will be the error, in seconds per hour, if the clock operates at 0°C?

43P. The timing of a certain electric watch is governed by a small tuning fork. The frequency of the fork is inversely proportional to the square root of the length of the fork. What is the fractional gain or loss in time for a quartz tuning fork 8.0 mm long at (a) −40°F and (b) +120°F if it keeps perfect time at 25°F?

44P. (a) Show that if the lengths of two rods of different solids are inversely proportional to their respective coefficients of linear expansion at the same initial temperature, the difference in length between them will be constant at all temperatures. (b) What should be the lengths of a steel and a brass rod at 0°C so that at all temperatures their difference in length is 0.30 m?

45P. As a result of a temperature rise of 32°C, a bar with a crack at its center buckles upward, as shown in Fig. 17. If the fixed distance $L_0 = 3.77$ m and the coefficient of linear expansion is $25 \times 10^{-6}/$°C, find x, the distance to which the center rises.

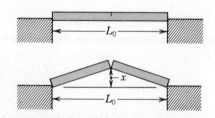

Figure 17 Problem 45.

46P. In a certain experiment, it was necessary to be able to move a small radioactive source at selected, extremely slow speeds. This was accomplished by fastening the source to one

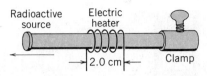

Figure 18 Problem 46.

end of an aluminum rod and heating the central section of the rod in a controlled way. If the effective heated section of the rod in Fig. 18 is 2.0 cm, at what constant rate must the temperature of the rod be made to change if the source is to move at a constant speed of 100 nm/s?

47P. A 1.28-m-long vertical glass tube is half-filled with a liquid at 20°C. How much will the height of the liquid column change when the tube is heated to 30°C? Take $\alpha_{glass} = 1 \times 10^{-5}/°C$ and $\beta_{liquid} = 4 \times 10^{-5}/°C$.

48P. A composite bar of length $L = L_1 + L_2$ is made from a bar of material 1 and length L_1 attached to a bar of material 2 and length L_2, as shown in Fig. 19. (a) Show that the effective coefficient of linear expansion α for this bar is given by $\alpha = (\alpha_1 L_1 + \alpha_2 L_2)/L$. (b) Using steel and brass, design such a composite bar whose length is 52.4 cm and whose effective coefficient of linear expansion is $13 \times 10^{-6}/°C$.

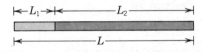

Figure 19 Problem 48.

49P. A thick aluminum rod and a thin steel wire are attached in parallel, as shown in Fig. 20. The temperature is 10°C. Both the rod and the wire are 85 cm long and neither is under stress. The system is heated to 120°C. Calculate the resulting stress in the wire, assuming that the rod expands freely.

50P. A steel boiler in an electric power station is suspended from the ceiling of the boiler room by steel cables; this is to

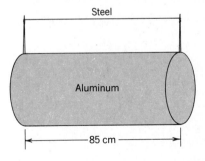

Figure 20 Problem 49.

allow for thermal expansion of the boiler as it heats; see Fig. 21a. At 10°C, a 35.0-m tall boiler is suspended by 2.13-m steel cables and clears the floor by 1.82 m. What are the ceiling and floor clearances when the boiler is at 550°C? Assume that the cables reach 185°C. The variation of the coefficient of linear expansion of steel with temperature is shown in Fig. 21b.

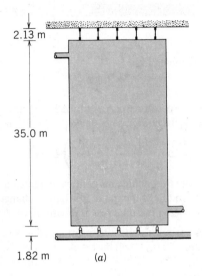

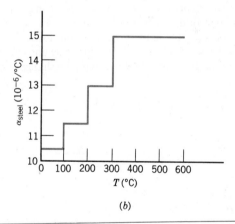

Figure 21 Problem 50.

51P. Three equal-length straight rods, of aluminum, Invar, and steel, all at 20°C, form an equilateral triangle with hinge pins at the vertices. At what temperature will the angle opposite the Invar rod be 59.95°? See Appendix G for needed trigonometric formulas.

52P. Two rods of different materials but having the same lengths L and cross-sectional areas A are arranged end-to-end between fixed, rigid supports, as shown in Fig. 22a. The temperature is T and there is no initial stress. The rods are heated,

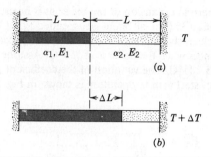

Figure 22 Problem 52.

so that their temperature increases by ΔT. (a) Show that the rod interface is displaced upon heating by an amount given by

$$\Delta L = \left(\frac{\alpha_1 E_1 - \alpha_2 E_2}{E_1 + E_2}\right) L \, \Delta T,$$

where α_1, α_2 are the coefficients of linear expansion and E_1, E_2 are Young's moduli of the materials. Ignore changes in cross-sectional areas. (b) Find the stress at the interface after heating.

53P. An aluminum cube 20 cm on an edge floats on mercury. How much farther will the block sink when the temperature rises from 270 to 320 K? (The coefficient of volume expansion of mercury is $1.8 \times 10^{-4}/°C$.)

54P.* The distance between the towers of the main span of the Golden Gate Bridge near San Francisco is 4200 ft (Fig. 23). The sag of the cable halfway between the towers at 50°F is 470 ft. Take $\alpha = 6.5 \times 10^{-6}/°F$ for the cable and compute (a) the change in length of the cable and (b) the change in sag for a

Figure 23 Problem 54.

temperature change from 10 to 90°F. Assume no bending or separation of the towers and a parabolic shape for the cable.

ESSAY 9
HOW NOT TO BUILD A XYLOPHONE

WILLIAM A SHURCLIFF

CAMBRIDGE,
MASSACHUSETTS

I purchased some steel strips 1 in. wide and $\frac{1}{16}$–$\frac{3}{16}$ in. thick and cut them to various lengths such as 6 in., $6\frac{1}{2}$ in., and 7 in. If I laid such a bar, cushioned by two pieces of thick soft string, on a table and then struck the center of the bar with my knuckle, a pure musical note gladdened the air. If the pieces of string were about one-fifth of the way from the ends of the bar, that is, at the nodes, the note would continue for a long time, such as 10 or 20 s.

Bars of different length produced musical notes of different frequencies, that is, different pitches. I could adjust the pitch of a bar by shortening it (for example, cutting off $\frac{1}{16}$ in. or so) with an ordinary hacksaw, thus raising the pitch, or by filing the central region of the bar to make it thinner, thus lowering the pitch.

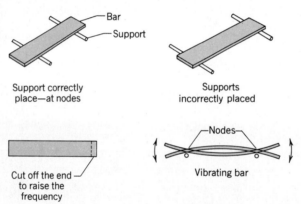

Support correctly place—at nodes

Supports incorrectly placed

Cut off the end to raise the frequency

Nodes

Vibrating bar

After many hours of work, I had about two dozen bars, of varying length and two different thicknesses (thinner bars for the lower notes), adjusted so as to form a two-octave xylophone. I could play almost any tune on it. Using two or more felt-tipped hammers I could play chords.

Now came catch #1: The musical notes were too faint, too feeble. You had to stand close to hear them.

I then obtained a wide-mouthed bottle and blew vigorously across the open top, producing a musical note. I found which bar of the xylophone produced approximately this same note, then placed the bottle beneath one end of the bar. Result: the note was a little louder, but still not loud enough.

Suspecting that the loudness would increase if the natural frequency of the bottle exactly matched the frequency of the bar, I tried adding a little water to the bottle, which slightly increased its natural frequency. After some experimenting to produce a perfect match, I found I had increased the loudness almost 10-fold.

I obtained more bottles, one for each bar—a big bottle for a low-note bar and a

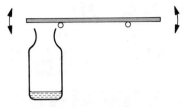

Resonator: a bottle with a small amount of water added to produce frequency match

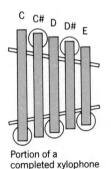

Portion of a completed xylophone

small bottle for a high-note bar—with suitable additions of water to provide frequency matches.

Catch #2 appeared after I returned from a week's trip to the mountains: The xylophone gave only feeble notes. Why? Because much of the water had evaporated, spoiling the frequency matches.

To solve this problem, I again put a carefully tailored amount of liquid in each bottle, but I used a mixture of water and glycerine. Glycerine is hygroscopic; under typical circumstances it does not evaporate. For a while the tuning of the bottles remained excellent.

Catch #3 developed during a long period of rain and high humidity. The amounts of liquid in the bottles increased; the glycerine had "snatched moisture from the air." Again the tuning was spoiled.

I shifted from water and glycerine to oil: ordinary crankcase oil. Oil shows practically no evaporation and is not hygroscopic. I got good results: The system remained in tune for many weeks.

Catch #4 arrived. It seems that the world is full of flies and bugs that love to explore the interiors of bottles. Within a couple of months the bottles contained a disgusting mixture of oil and bugs. The bottle walls were abundantly ornamented with dead, oil-soaked bugs.

I shifted once more—to paraffin. Paraffin is cheap, nontoxic, and easily melted. It does not evaporate, does not absorb moisture, and has no allure for bugs. I cleaned all the bottles and filled each with just enough molten paraffin to provide the correct resonant frequency. Within an hour, all the paraffin had solidified. I seemed, at last, to be all set.

Tragedy struck immediately: when paraffin solidifies, it shrinks. This throws off the tuning.

At this point I quit. But who knows? With a little more experimenting I might have reached the glorious goal of an accurately tuned, stable xylophone that costs less than $20.

CHAPTER 20

HEAT AND THE FIRST LAW OF THERMODYNAMICS

This early steam engine from the Smithsonian Institution is a stalwart reminder of the age of steam. The word 'steam' in its day, like 'electricity' in its day and 'computer' today, was the current buzz word for progress. The scientific underpinnings of these great advances in technology were, in order, the laws of thermodynamics, Maxwell's equations of electromagnetism, and quantum physics.

20-1 Heat

If you take a can of cola from the refrigerator and leave it on the kitchen table, its temperature will rise—rapidly at first and then increasingly slowly—until it equals that of the room. In the same way, the temperature of a cup of hot coffee, left to stand, will fall until it also reaches room temperature.

In generalizing this situation, we describe the cola or the coffee as a *system* (temperature T_S) and the relevant parts of the kitchen as the *environment* (temperature T_E) of that system; see Fig. 1. Our observation is that if T_S is not equal to T_E, then T_S will change until the two temperatures are equal.

In Fig. 1*a*, in which $T_S > T_E$, we say that *heat*

energy—to which we give the symbol Q—flows from the system to the environment. In Fig. 1*c*, the flow is in the other direction. In Fig. 1*b*, where $T_S = T_E$, the net flow is zero. As the figure shows, we (arbitrarily) choose Q to be *positive* when heat flows *into* a system and *negative* when it flows *out of* a system. We are lead to this definition of heat:

Heat is energy that flows between a system and its environment by virtue of a temperature difference that exists between them.

Because heat is energy, its SI unit is the joule. If you suspect that heat is flowing into or out of a system, your key question is: " Where is the temperature difference?"

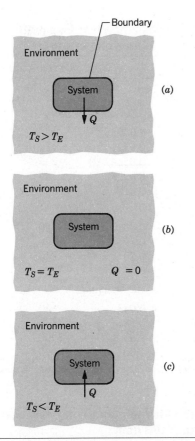

Figure 1   (*a*) If the temperature of the system exceeds that of the environment, heat will flow out of the system until thermal equilibrium is established, as in (*b*). (*c*) If the temperature of the system is less than that of the environment, heat will flow into the system until thermal equilibrium is established.

Energy can also be transferred between a system and its environment by means of *work* (symbol W), which we always associate with a force moving through a distance. If you suspect that work is being done on or by a system, your key question is: " Where is the force and how does its point of application move?"

Both heat and work represent energy-in-transit between a system and its environment. Heat and work, unlike temperature, pressure, and volume, are not intrinsic properties of a system. They have meaning only as they describe energy transfers into or out of the system, adding to or subtracting from the system's store of *internal energy*. Thus, it is proper to say: "During the last 3 min, 15 J of heat flowed from the system to its environment" or "During the last minute, 12 J of work were done on the system by its environment." It is without meaning to say: "This system contains 450 J of heat" or "This system contains 385 J of work." The bookkeeping relation for energy in the form of heat, work, and internal energy is summed up by the first law of thermodynamics, the subject of this chapter.

In popular usage, the word "heat" is often used where "temperature" is intended, as in the cookbook instruction: "Put into an oven at 300 degrees of heat." We also say that when we add heat to something it gets "hotter" by which we mean that its temperature increases. When we say: "It is a hot day," we are referring to temperature and not to heat. Do not confuse these two totally different quantities.

20-2   Measuring Heat: Units

Before it was realized that heat was a form of energy, heat was measured in terms of its ability to raise the temperature of water. Thus, the *calorie* (abbr. cal) was defined as the amount of heat that would raise the temperature of 1 g of water from 14.5 to 15.5°C. In the British system, the corresponding unit of heat was the *British thermal unit* (abbr. Btu), defined as the amount of heat that would raise the temperature of 1 lb of water from 63 to 64°F.

In 1948, it was decided that, since heat (like work) is a form of energy, the SI unit for heat should be the same as for all other forms of energy, namely, the *joule*. The calorie is now defined to be 4.1860 J (exactly) with no reference to the heating of water. The "calorie" used in nutrition, sometimes called a Calorie (abbr. Cal) is really a kilocalorie.

The relations among the various heat units are

$$
\begin{aligned}
1 \text{ J} &= 0.239 \text{ cal} = 9.48 \times 10^{-4} \text{ Btu,} \\
1 \text{ Btu} &= 1055 \text{ J} = 252 \text{ cal,} \\
1 \text{ cal} &= 3.97 \times 10^{-3} \text{ Btu} = 4.19 \text{ J,} \\
1 \text{ Cal} &= 10^{3} \text{ cal} = 3.97 \text{ Btu} = 4190 \text{ J.}
\end{aligned} \tag{1}
$$

In scientific work, heat is increasingly expressed in joules, with the calorie and the Btu being gradually phased out. However, the calorie continues to be used in some areas of chemistry and the Btu in some aspects of engineering practice.

20-3   The Absorption of Heat by Solids and Liquids

Heat Capacity.   The *heat capacity C* of an object (for example, a Pyrex coffee pot, an iron skillet, or a marble slab) is the proportionality constant between the heat added to the object and the change in temperature that results. Thus,

$$Q = C(T_f - T_i), \qquad (2)$$

in which T_i and T_f are the initial and the final temperatures of the object. The heat capacity C of a marble slab used in a bun warmer might, for example, be 179 cal/°C, which we can also write as 179 cal/K or as 747 J/K.

The word "capacity" in this context is really misleading in that it suggests an analogy to the capacity of a bucket to hold water. The analogy is false. Although heat can be transferred to an object, we do not speak of the object as "containing" heat. Indeed, the heat transfer can proceed without limit as long as the necessary temperature difference is maintained. The object may, of course, melt or vaporize during the process.

Specific Heat Capacity. Two objects made of the same material, say marble or iron, will have a heat capacity proportional to their masses. It is therefore convenient to define a "heat capacity per unit mass" or a *specific heat capacity* that refers not to an object but to the material of which the object is made. Thus,

$$Q = cm(T_f - T_i), \qquad (3)$$

in which the constant c is the specific heat capacity of the material. We can say that, although the heat capacity of the marble slab mentioned above is 179 cal/°C (or 747 J/K), the specific heat capacity of marble is 0.21 cal/g·°C (or 880 J/kg·K).

From the way the calorie and the British thermal unit were initially defined, the specific heat capacity of water is

$$c = 1 \text{ cal/g}\cdot°\text{C} = 1 \text{ Btu/lb}\cdot°\text{F} = 4190 \text{ J/kg}\cdot\text{K}. \qquad (4)$$

Table 1 shows the specific heat capacities of some substances at room temperature. Note that the value for water is particularly high.

Molar Heat Capacity. If the mass of a specimen is expressed in moles, the specific heat capacity is called the *molar heat capacity*. Table 1 shows the values for some elemental solids at room temperature.

Note that the molar heat capacities of all the elements listed in Table 1 have about the same value, namely, 25 J/mol·K. As Fig. 2 suggests, the molar heat capacities of all solids vary with temperature, approaching 25 J/mol·K at high enough temperatures. Some substances, such as carbon and beryllium, do not reach this limiting value until temperatures well above room

Table 1   The Specific Heat Capacities of Some Substances at Room Temperature

| Substance | Specific Heat Capacity | | |
	cal/g·K	J/kg·K	J/mol·K[a]
Elemental Solids			
Lead	0.0305	128	26.5
Tungsten	0.0321	134	24.8
Silver	0.0564	236	25.5
Copper	0.0923	386	24.5
Aluminum	0.215	900	24.4
Other Solids			
Brass	0.092	380	
Granite	0.19	790	
Glass	0.20	840	
Ice (−10°C)	0.530	2220	
Liquids			
Mercury	0.033	140	
Ethyl alcohol	0.58	2430	
Seawater	0.93	3900	
Water	1.00	4190	

[a] Called the *molar heat capacity* when expressed in these units.

temperature. Other substances may melt or vaporize before they reach this limit.

When we compare two substances on a molar basis, we are comparing samples that contain the same number of atoms. The fact that, at high enough temperatures,

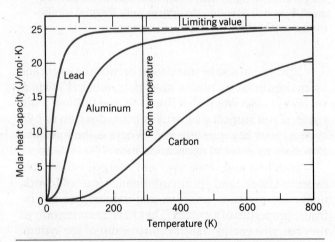

Figure 2   The molar specific heat of three elements as a function of temperature. At high enough temperatures, all solids approach the same limiting value. For lead and aluminum, that value is already essentially reached at room temperature; for carbon it is not.

all solid elements have the same molar heat capacity tells us that all kinds of atoms—whether they be aluminum, copper, uranium, or anything else—absorb energy in the same way.

An Important Point. In assigning a specific heat capacity to any substance, it is important to know not only how much heat was added but the conditions under which that transfer took place. For solids and liquids, we usually assume that the sample was under constant pressure (usually atmospheric) during the heat transfer. It is also conceivable that the sample could have been held at constant volume while the heat was added. This means that the thermal expansion of the sample must be prevented, by applying external pressure. For solids and liquids, this is very hard to arrange experimentally but the effect can be calculated and it amounts to a change of usually no more than a few percent in the specific heat capacity. For gases, as we shall see, adding heat under constant pressure conditions and under constant volume conditions leads to quite different values for the specific heat capacities.

Heats of Transformation. When heat is added to a solid or liquid, the temperature of the sample does not necessarily rise. Instead, the sample may change from one *phase* or *state* (that is, solid, liquid, or gas) to another. Thus, ice melts and water boils, absorbing heat in each case without a temperature change. In the reverse processes (water freezes, steam condenses), heat is released by the sample, again at a constant temperature.

The amount of heat per unit mass transferred during a phase change is called the heat of transformation (symbol L) for the process. The total heat transferred is then

$$Q = Lm, \qquad (5)$$

Table 2 Some Heats of Transformation

| Substance | Melting | | Boiling | |
	Melting Point (K)	Heat of Fusion (kJ/kg)	Boiling Point (K)	Heat of Vaporization (kJ/kg)
Hydrogen	14.0	58.0	20.3	455
Oxygen	54.8	13.9	90.2	213
Mercury	234	11.4	630	296
Water	273	333	373	2256
Lead	601	23.2	2017	858
Silver	1235	105	2323	2336
Copper	1356	207	2868	4730

in which m is the mass of the sample. For water we have, for the *heat of vaporization* at its normal boiling point,

$$L_V = 539 \text{ cal/g} = 40.7 \text{ kJ/mol} = 2260 \text{ kJ/kg}. \quad (6)$$

The heat per unit mass liberated when water freezes is called the *heat of fusion;* for water at its normal freezing point it is

$$L_F = 79.5 \text{ cal/g} = 6.01 \text{ kJ/mol} = 333 \text{ kJ/kg}. \quad (7)$$

Table 2 shows the heats of transformation for some substances.

Sample Problem 1 A candy bar has a marked nutritional value of 350 Cal. How many kilowatt·hours of energy will it deliver to the body as it is digested?

The Calorie in this case is a kilocalorie so that

energy = $(350 \times 10^3 \text{ cal})(4.19 \text{ J/cal})$
= $(1.466 \times 10^6 \text{ J})(1 \text{ W·s/J})(1 \text{ h}/3600 \text{ s})(1 \text{ kW}/1000 \text{ W})$
= 0.407 kW·h.   (Answer)

This amount of energy would keep a 100-W light bulb burning for 4.1 h. To burn up this much energy by exercise, a person would have to jog about 3 or 4 mi.

A generous daily human diet corresponds to about 3.5 kW·h per day which represents the absolute maximum amount of work that a human can do in one day. In an industrialized country, this amount of energy can be purchased for perhaps 25 cents, a sum that represents the value of a day's work if a human is competing with an electric motor.

Sample Problem 2 How much heat is needed (a) to raise the temperature of 725 g of lead from room temperature (20°C or 293 K) to its melting point?

From Eq. 3 we have, using data from Tables 1 and 2,

$$Q = cm(T_f - T_i)$$
$$= (128 \text{ J/kg·K})(0.725 \text{ kg})(602 \text{ K} - 293 \text{ K})$$
$$= 2.87 \times 10^4 \text{ J} = 28.7 \text{ kJ}. \qquad \text{(Answer)}$$

(b) How much additional heat is required to melt the lead at its melting point?

From Eq. 5, again using data from Table 2, we have

$$Q = Lm = (23.2 \text{ kJ/kg})(0.725 \text{ kg})$$
$$= 16.8 \text{ kJ}. \qquad \text{(Answer)}$$

Sample Problem 3 A copper slug whose mass m_c is 75 g is heated in a laboratory oven to a temperature T of 312°C. The slug is then dropped into a glass beaker containing a mass $m_w (=220 \text{ g})$ of water. The effective heat capacity C_b of the beaker is 45 cal/K. The initial temperature T_i of the water and the beaker is 12.0°C. What is the final temperature T_f of the slug, the beaker, and the water?

Let us take as our system the *water* + *beaker* + *copper slug*. No heat enters or leaves this system so that the algebraic sum of the internal heat transfers that occur must be zero. There are three such transfers:

Heat flow into the water: $Q_w = m_w c_w (T_f - T_i)$.
Heat flow into the beaker: $Q_b = C_b (T_f - T_i)$.
Heat flow into the copper: $Q_c = m_c c_c (T_f - T)$.

The temperature difference is written—in all three cases—as the final temperature minus the initial temperature. We see by inspection that Q_w and Q_b are positive (indicating that heat has been added to the water and to the beaker) and that Q_c is negative (indicating that heat has been withdrawn from the copper slug).

From what we have said above, we must have

$$Q_w + Q_b + Q_c = 0. \qquad (8)$$

Substituting the heat transfer expressions above into Eq. 8 yields

$$m_w c_w (T_f - T_i) + C_b (T_f - T_i) + m_c c_c (T_f - T) = 0. \qquad (9)$$

We see that temperatures enter Eq. 9 only as differences. Thus, because the intervals on the Celsius and the Kelvin scale are identical, we can use either of these scales in this equation. Solving Eq. 9 for T_f and substituting, we have

$$T_f = \frac{m_c c_c T + C_b T_i + m_w c_w T_i}{m_w c_w + C_b + m_c c_c}$$

$$= \frac{(75 \text{ g})(0.092 \text{ cal/g·K})(312°C) + (45 \text{ cal/K})(12°C) + (220 \text{ g})(1.00 \text{ cal/g·K})(12°C)}{(220 \text{ g})(1.00 \text{ cal/g·K}) + 45 \text{ cal/K} + (75 \text{ g})(0.092 \text{ cal/g·K})}$$

$$= 19.6°C.$$

From the given data you can show that

$$Q_w \approx 1672 \text{ cal}, \quad Q_b \approx 342 \text{ cal}, \quad \text{and} \quad Q_c \approx -2024 \text{ cal}.$$

The algebraic sum of these three heat transfers is indeed zero, as Eq. 8 requires.

20-4 A Closer Look at Heat and Work

Here we look in some detail at how heat and work are exchanged between a system and its environment. Let us take as our system a gas confined to a cylinder with a movable piston, as in Fig. 3. The pressure of the confined gas is balanced by loading lead shot onto the top of the piston. The walls of the cylinder are insulating but the bottom of the cylinder rests on a heat reservoir (a hot plate, if you will) whose temperature T you can control at will by turning a knob.

The system starts from an *initial state*, described by a pressure p_i, a volume V_i, and a temperature T_i. You

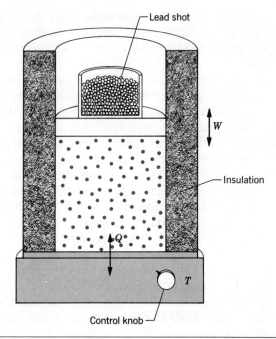

Figure 3 A gas is confined to a cylinder with a movable piston. Heat can be added to, or withdrawn from, the gas by regulating the temperature T of the adjustable heat reservoir.

want to move the system to a *final state*, described by a pressure p_f, a volume V_f, and a temperature T_f. Both states are represented by dots on the pressure–volume diagram of Fig. 4a.

The operation of changing the system from its initial state to its final state is called a *thermodynamic process*. During such processes, heat may be transferred into (or out of) the system from the reservoir. Work may be done *by* the system as the expanding gas lifts the loaded piston. Alternatively, work may be done *on* the system if the loaded piston falls. We assume that all such transfers are carried out very slowly, so that the system remains essentially in thermodynamic equilibrium at all stages.

A Closer Look at Work. Suppose that you remove a little lead shot from the piston of Fig. 3, causing it to rise by a distance ds. The work done by the expanding gas is then

$$dW = \mathbf{F} \cdot d\mathbf{s} = (pA)(ds) = (p)(A \, ds) = p \, dV, \qquad (10)$$

in which A is the area of the piston. The total work done during a finite displacement of the piston follows from

$$\boxed{W = \int dW = \int_{V_i}^{V_f} p \, dV,} \qquad (11)$$

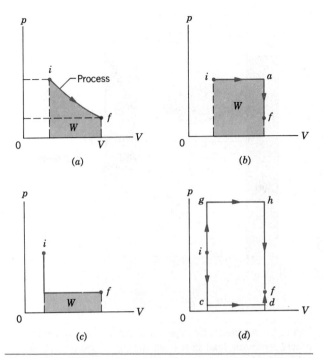

Figure 4 (a) The system of Fig. 3 goes from an *initial state i* to a *final state f* by means of a *thermodynamic process.* The area marked W represents the work done by the system during this process. (b) Another process, connecting the same two states; the work has a different value. (c) Still another process and still another value for the work. (d) The work can be made as small as you like (path *icdf*) or as large as you like (path *ighf*).

in which V_i and V_f are the initial and the final volumes of the gas. To evaluate this integral, you must know how the pressure of the gas varies with its volume for the particular process being carried out. The integral of Eq. 11 (and thus the work W) can be represented graphically as the area under the curve connecting points i and f, as shown by the shaded area in Fig. 4a.

If the work calculated from Eq. 11 is *positive,* that tells you that work was done *by* the system, the expanding gas lifting the loaded piston. Conversely, a *negative* value for W tells you that work was done *on* the system. Recall that the sign for heat is *positive* if heat is *added to* the system and negative if heat is withdrawn from the system. You can remember these sign conventions by recalling that, in a steam engine, both Q and W are positive, heat being *added to* the boiler and work being done *by* the engine.

Consider the path *iaf* shown in Fig. 4b, one of the many different ways in which you can take the system

from its initial to its final state. Step *ia* of this process is carried out at constant pressure, which means that you leave undisturbed the lead shot that rides on top of the piston in Fig. 3. You cause the volume to increase (from V_i to V_f) by slowly turning up the temperature control knob, raising the temperature of the gas to some higher value T_a. During this process, work is done by the expanding gas (to lift the weights on the piston) and heat is added to the system from the heat reservoir (in response to the arbitrarily small temperature differences that you create as you turn up the temperature).

Step *af* of the process of Fig. 4b is carried out at constant volume, so that you must wedge the piston, preventing it from moving. And then use the control knob to reduce the temperature from T_a to its final value T_f, heat being transferred out of the system and into the reservoir during this step.

For the overall process *iaf,* the work W, which is positive and is carried out only during step *ia,* is represented by the shaded area under the curve. Heat is transferred during both steps *ia* and *af,* being positive during the first step and negative (but smaller in magnitude) during the second.

Figure 4c shows a process in which the two steps above are carried out in reverse order. The work W in this case is smaller than for Fig. 4b, as is the net heat transferred. Figure 4d suggests that you can make the work done as small as you want (by following a path like *icdf*) or as large as you want (by following a path like *ighf*).

To sum up: A system can be taken from a given initial state to a given final state by an infinite number of processes. In general, the work W and also the heat Q will have different values for each of these processes. We say that heat and work are *path-dependent* quantities.

20-5 The First Law of Thermodynamics

We have just learned that when a system goes from a given initial to a given final state both W and Q depend on the nature of the process. Experimentally, however, we find a surprising thing. *The quantity $Q - W$ is the same for all processes.* It depends only on the initial and the final states and it does not matter at all how you get from one to the other. All other combinations of Q and W, including Q alone, W alone, $Q + W$, and $Q - 2W$, are *path dependent;* the quantity $Q - W$ alone is not.

$Q - W$ must represent a change in some intrinsic

property of the system. We call it the *internal energy U* and we write

$$\Delta U = U_f - U_i = Q - W \quad \text{(first law).} \quad (12)$$

Equation 12 is the *first law of thermodynamics*. It is an extension of the conservation of energy principle to include thermodynamic systems. If the thermodynamic system changes by only a differential amount, we can write the first law as*

$$dU = dQ - dW \quad \text{(first law).} \quad (13)$$

We saw in Chapter 19 that the essential message of the zeroth law of thermodynamics is: "Every thermodynamic system that is in thermal equilibrium has an important physical property called its *temperature T*." The essential message of the first law is: "Every thermodynamic system that is in thermal equilibrium has an important physical property called its *internal energy U*."

20-6 The first Law of Thermodynamics: Some Special Cases

Here we look at four different thermodynamic processes, in each of which a certain restriction is imposed on the system. We then see what consequences follow when we apply the first law of thermodynamics to the process.

1. Adiabatic Processes. These are processes in which the system is so well insulated that no transfer of heat occurs between it and its environment. Putting $Q = 0$ in the first law (Eq. 12) leads to

$$\Delta U = -W \quad \text{(adiabatic processes).} \quad (14)$$

This tells us that if work is done *by* the system (that is, if W is positive) that must be a decrease in the internal energy of the system. Conversely, if work is done *on* the system (that is, if W is negative) there must be an increase in the internal energy of the system.

For a gas, an increase in internal energy means an increase in temperature, and conversely. The increase in

---

* Here dQ and dW, unlike dU, are not true differentials. That is, there are no such functions as $Q(p, V)$ and $W(p, V)$ that depend only on the state of the system. dQ and dW are called *inexact differentials* and are usually represented by the symbols $đQ$ and $đW$. For our purposes, we can treat them simply as infinitesimally small energy transfers.

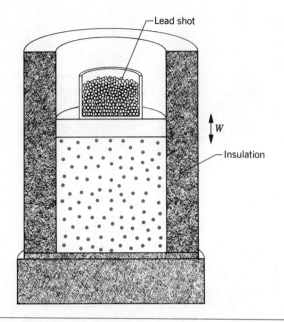

Figure 5 An adiabatic expansion can be carried out by slowly removing lead shot from the top of the piston. Adding lead shot can reverse the process at any stage.

temperature because of the adiabatic compression of air is familiar from the heating of a bicycle pump.

Figure 5 shows an idealized adiabatic process. Heat cannot enter or leave the system because of the insulation. Thus, the only interaction permitted between the system and its environment is the performance of work. We can cause work to be done *on* the system by adding lead shot to the top of the piston, thus compressing the gas. Conversely, we can cause work to be done *by* the system by removing lead shot.

There is a second way of ensuring that heat transfer does not take place during a thermodynamic process: carry it out very rapidly so that there is no time for appreciable heat flow. Thus, the compressions and rarefactions of air as a sound wave passes through are adiabatic. There is simply no time for heat to flow back and forth in synchronism with the rapidly oscillating sound wave. The compressions and expansions of steam in the cylinder of a steam engine, or of the hot gases in the cylinders of an internal combustion engine, are also essentially adiabatic, for the same reason.

2. Constant Volume Processes. If the volume of a system is held constant, that system can do no work. Putting $W = 0$ into the first law (Eq. 12) yields

$$\Delta U = Q \quad \text{(constant volume processes).} \quad (15)$$

Thus, if heat is added to a system (that is, if Q is positive) the internal energy of the system increases. Conversely, if heat is removed during the process (that is, if Q is negative) the internal energy of the system must decrease.

3. Cyclical Processes. There are processes in which, after certain interchanges of heat and work, the system is restored to its initial state. In that case, no intrinsic property of the system — including its internal energy — can possibly change. Putting $\Delta U = 0$ in the first law (Eq. 12) yields

$$Q = W \quad \text{(cyclical processes).} \qquad (16)$$

Thus, the net work done during the process must exactly equal the net amount of heat transferred; the store of internal energy of the system remains unchanged. Such processes form a closed loop on a pressure–volume plot; we shall study cyclical processes in some detail in Chapter 22.

4. Free Expansion. This is an adiabatic process in which no work is done on or by the system. Thus, $Q = W = 0$ and the first law requires that

$$\Delta U = 0 \quad \text{(free expansion).} \qquad (17)$$

Figure 6 shows how such an expansion can be carried out. A gas is confined by a closed stopcock to half of an insulated double chamber, the other half being evacuated. The process consists of opening the stopcock and waiting until equilibrium is established, with the gas now filling both halves of the double chamber. No heat is transferred because of the insulation. No work is done because the expanding gas rushes into an evacuated space, its motion unopposed by any counteracting pressure.

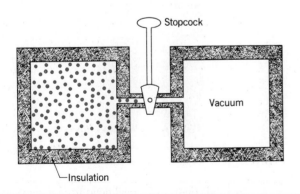

Figure 6   The initial stage of a free-expansion process. After the stopcock is opened, the gas eventually settles down to an equilibrium final state, filling both chambers.

Table 3   The First Law of Thermodynamics: Some Special Cases

The Law: $\Delta U = Q - W$ (Eq. 12)		
Process	Restriction	Consequence
Adiabatic	$Q = 0$	$\Delta U = -W$
Constant volume	$W = 0$	$\Delta U = Q$
Closed cycle	$\Delta U = 0$	$Q = W$
Free expansion	$Q = W = 0$	$\Delta U = 0$

A free expansion differs from all other processes that we have considered so far in that there is no way to carry it out slowly. Thus, although the system is in thermal equilibrium in its initial and its final states, it is *not* in equilibrium during the process. At intermediate states, the temperature, pressure, and volume do not have unique values. Thus, we cannot plot the course of the expansion on a pressure–volume diagram. All that can be done is to plot the initial and the final states.

Table 3 summarizes the characteristics of the processes of this section.

Sample Problem 4   Let 1.00 kg of liquid water be converted to steam by boiling at standard atmospheric pressure; see Fig. 7. The volume changes from an initial value of 1.00×10^{-3} m³ as a liquid to 1.671 m³ as steam. (a) How much work is done by the system during this process?

The work is given from Eq. 11. Because the presssure is constant (at 1.00 atm or 1.01×10^5 Pa) during the boiling process, we can write this equation as

$$W = \int_{V_i}^{V_f} p \, dV = p \int_{V_i}^{V_f} dV = p(V_f - V_i)$$
$$= (1.01 \times 10^5 \text{ Pa})(1.671 \text{ m}^3 - 1 \times 10^{-3} \text{ m}^3)$$
$$= 1.69 \times 10^5 \text{ J} = 169 \text{ kJ}. \qquad \text{(Answer)}$$

The result is positive, indicating that work is done *by* the system on its environment, in lifting the weighted piston of Fig. 7.

(b) How much heat must be added to the system during the process?

From Eqs. 5 and 6 we have

$$Q = Lm = (2260 \text{ kJ/kg})(1.00 \text{ kg})$$
$$= 2260 \text{ kJ}. \qquad \text{(Answer)}$$

The result is positive, which indicates that heat is *added to* the system, as we expect.

(c) What is the change in the internal energy of the system during the boiling process?

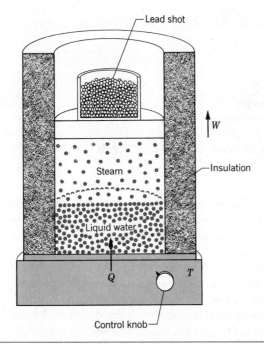

Figure 7 Sample Problem 4. Water boiling at constant pressure. Heat is added from the reservoir until the liquid water has changed completely into steam. Work is done by the expanding gas as it lifts the loaded piston.

We find this from the first law (Eq. 12) or

$$\Delta U = Q - W = 2260 \text{ kJ} - 169 \text{ kJ}$$
$$\approx 2090 \text{ kJ} = 2.09 \text{ MJ}. \qquad \text{(Answer)}$$

This quantity is positive, indicating that the internal energy of the system has increased during the boiling process. This energy represents the internal work done in overcoming the strong attraction that the H_2O molecules have for each other in the liquid state.

We see that, when water is boiled, about 7.5% ($= 169 \text{ kJ}/2260 \text{ kJ}$) of the added heat goes into external work in pushing back the atmosphere. The rest goes into internal energy that is added to the system.

20–7 The Transfer of Heat

We have discussed the transfer of heat between a system and its environment but we have not yet described how that transfer takes place. There are three mechanisms: conduction, convection, and radiation. We discuss each in turn.

Conduction. If you leave a poker in a fire for any length of time, its handle will get hot. Energy is trans-

ferred from the fire to the handle by *conduction* along the length of the metal shaft. The vibration amplitudes of the atoms and electrons of the metal at the hot end of the shaft take on relatively large values, reflecting the elevated temperature of their environment. These increased vibrational amplitudes are passed along the shaft, from atom to atom, during collisions between adjacent atoms. In this way, a region of rising temperature extends itself along the shaft to your hand.

Consider a slab of cross-sectional area A and length L, whose ends are maintained at temperatures T_H and T_C; see Fig. 8. Let Q be the heat that flows through the slab, from its hot face to its cold face, in time t. Experiment shows that the rate of heat flow ($= Q/t$) is given by

$$H(= Q/t) = kA \frac{T_H - T_C}{L}, \qquad (18)$$

in which k, called the *thermal conductivity,* is a constant that depends on the material. Large values of k define good heat conductors, and conversely.

Thermal Resistance (*R*-Value). If you are interested in insulating your house (see Fig. 9) or in keeping cola cans cold on a picnic, you are more concerned with poor heat conductors than with good ones. For this reason, the concept of *thermal resistance R* has been introduced into engineering practice. The *R*-value of a slab of thickness L is defined from

$$R = \frac{L}{k}. \qquad (19)$$

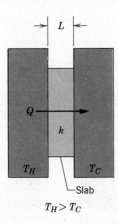

Figure 8 Thermal conduction. Heat flows from a reservoir at temperature T_H to a cooler reservoir at temperature T_C through a conducting slab of thickness L and thermal conductivity k.

Figure 9 The snow reveals an unheated garage. Heat conduction through the roof has melted the snow over the living quarters.

Table 4   Some Thermal Conductivities and R-Values[a]

	Conductivity, k (W/m·K)	R-Value (ft²·°F·h/Btu)
Metals		
Stainless steel	14	0.010
Lead	35	0.0041
Aluminum	235	0.00061
Copper	401	0.00036
Silver	428	0.00034
Gases		
Air (dry)	0.026	5.5
Helium	0.15	0.96
Hydrogen	0.18	0.80
Building Materials		
Polyurethane foam	0.024	5.9
Rock wool	0.043	3.3
Fiber glass	0.048	3.0
White pine	0.11	1.3
Window glass	1.0	0.14

[a] Values are for room temperature. Note that the values of k are given in SI units and those of R in the customary British units. The R-values are for a 1-in. slab.

Thus, the lower the thermal conductivity of the material of which a slab is made, the higher the R-value of the slab. Note that R is a property of a slab of a specified thickness, not of a material. The commonly used units for R (which, in this country at least, are almost never stated) are ft²·°F·h/Btu.

Combining Eqs. 18 and 19 leads to

$$H = A\frac{T_H - T_C}{R}, \qquad (20)$$

which allows one to calculate the rate of heat flow through a slab if its R-value, its area, and the temperature difference between its faces are known.

In reasonably severe climates, it is recommended that the ceilings of single-faimly dwellings be insulated to the level of R-30. From Eq. 20 we see that this means that an average square foot of such a ceiling would lose heat by conduction at a rate of $\frac{1}{30}$ Btu/h for every 1°F difference in temperature.

Table 4 shows the thermal conductivities of various materials and the R-values calculated for 1-in. slabs of that material. The use of R-values is normally restricted to commercial insulating materials, but we show them for a wide range of materials for comparison.

Inspection of the table shows why the Sierra Club recommends using a stainless steel (rather than an aluminum) cup for hot coffee. Stagnant air has an R-value as great as that of any of the commercial building materials shown. In fact, many such materials owe their effectiveness to their ability to entrap isolated pockets of air.

In cold climates, double or triple pane windows are often installed to reduce heat loss. The glass itself is not a particularly good insulator. The air (or other gas) be-

tween the panes would insulate well if it were stagnant —which it is not because of convection currents. The insulating value of a window comes almost entirely from the thin *boundary layer* (see Section 16–12) of stagnant air that clings to each surface of a pane. Doubling the number of panes doubles the number of such insulating boundary layers.

You can further deduce from the table that, to build a slab insulating to R-30, you could use either 5.1 in. of polyurethane foam, 23 in. of white pine, 18 ft of window glass, or 1.4 mi of silver!

A Composite Slab.   Figure 10 shows a composite slab, consisting of two materials having different thicknesses, L_1 and L_2, and different thermal conductivities, k_1 and k_2. The temperatures of their outer surfaces are T_H and T_C. Let us derive an expression for the rate of heat transfer through such a composite slab.

Let T_X be the temperature of the interface between the two slabs. In a steady state, the rate of heat flow through each slab is the same. We can then write, from Eq. 18,

$$H = \frac{k_2 A(T_H - T_X)}{L_2} = \frac{k_1 A(T_X - T_C)}{L_1}. \qquad (21)$$

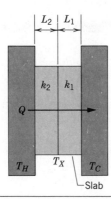

Figure 10 Heat flows through a composite slab made up of materials with different thicknesses and different thermal conductivities.

Solving Eq. 21 for T_X yields, after a little algebra,

$$T_X = \frac{k_1 L_2 T_C + k_2 L_1 T_H}{k_1 L_2 + k_2 L_1}. \tag{22}$$

Substituting this expression for T_X into either form of Eq. 21 yields

$$H = \frac{A(T_H - T_C)}{(L_1/k_1) + (L_2/k_2)}. \tag{23}$$

Equation 19 reminds us that $L/k = R$. We can extend Eq. 23 to any number of slabs, in the form

$$H = \frac{A(T_H - T_C)}{\Sigma (L/k)} = \frac{A(T_H - T_C)}{\Sigma R}, \tag{24}$$

in which the summation sign in the denominator tells us to add up the corresponding values for each slab.

Convection. If you look at the flame of a candle or a match, you are watching heat energy being transported upward by *convection*. Heat transfer by convection occurs when a fluid, such as air or water, is in contact with an object whose temperature is higher than that of its surroundings. The temperature of the fluid that is in contact with the hot object increases and (in most cases) the fluid expands. Being less dense than the surrounding cooler fluid, it rises because of buoyant forces; see Fig. 11. The surrounding cooler fluid falls to take the place of the rising warmer fluid and a convective circulation is set up.

Atmospheric convection plays a fundamental role in determining the global climate patterns and in our daily weather variations. Glider pilots and condors alike seek the convective thermals that, rising from the warmer earth beneath, keep them aloft. Huge energy

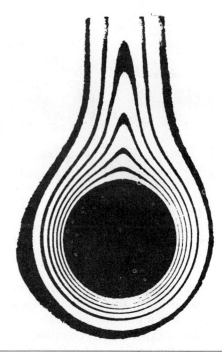

Figure 11 Air rises by convection around a heated cylinder. The fringes are lines of equal temperature.

transfers take place within the oceans by the same process. Finally, energy is transported to the surface of the sun from the nuclear furnace at its core by enormous convection cells, the tops of which can be seen as a granulation of the solar surface.

We have been describing *free,* or *natural* convection. Convection can also be *forced,* as when a furnace blower causes air circulation to heat the rooms of a house.

Radiation. Energy is carried from the sun to us by electromagnetic waves that travel freely through the near vacuum of the intervening space. If you stand near a bonfire or an open fireplace, you are warmed by the same process. All objects emit such electromagnetic radiation because of their temperature and also absorb some of the radiation that falls on them from other objects. The average temperature of our earth, for example, levels off at about 300 K because at that temperature the earth radiates energy into space at the same rate that it receives it from the sun; see Fig. 12. If the temperature of the earth were — by some miracle — to change suddenly from 300 K to 280 K or to 320 K, it would quickly either warm up or cool down to 300 K, restoring its nice thermal balance.

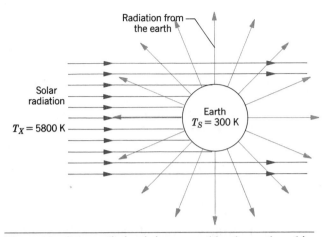

Figure 12   Solar radiation is intercepted by the earth and is (largely) absorbed. The temperature T_E of the earth adjusts itself to a value at which the earth's heat loss by radiation is just equal to the solar heat that it absorbs.

Sample Problem 5   A composite slab (see Fig. 10) whose area A is 26 ft² is made up of 2.0 in. of rock wool and 0.75 in. of white pine. The temperature difference between the faces of the slab is 65°F. What is the rate of heat flow through the slab?

The R-values given in Table 4 area for 1-in. slabs. Thus, the R-value for the rock wool is 3.3×2.0 or 6.6 ft²·°F·h/Btu. For the wood it is 1.3×0.75 or 0.98, in the same units. The composite slab thus has an R-value of $6.6 + 0.98$ or 7.58 ft²·°F·h/Btu. Substitution into Eq. 24 yields

$$H = \frac{A(T_H - T_C)}{\Sigma R} = \frac{(26 \text{ ft}^2)(65 \text{ °F})}{7.58 \text{ ft}^2 \cdot \text{°F} \cdot \text{h/Btu}}$$

$$= 223 \text{ Btu/h} \; (= 65 \text{ W}). \qquad \text{(Answer)}$$

Thus, at this temperature difference, each such insulating slab would transit heat continuously to the outdoors at the rate of 64 W.

REVIEW AND SUMMARY

Heat

Heat Q is the energy that flows between a system and its environment by virtue of a temperature difference between them. It can be measured in *joules* (J), *calories* (cal), *kilocalories* (Cal or kcal), or *British thermal units* (Btu), with

$$1 \text{ Cal} = 10^3 \text{ cal} = 3.97 \text{ Btu} = 4190 \text{ J}. \qquad [1]$$

Heat Capacity

If heat Q is added to an object with mass m, the temperature change $T_f - T_i$ is related to Q by

$$Q = C(T_f - T_i), \qquad [2]$$

in which C is the *heat capacity* of the object. The *specific heat capacity c* (the heat capacity per unit mass; see Table 1) is defined by

Specific Heat Capacity

$$Q = cm(T_f - T_i). \qquad [3]$$

The heat capacity per mole, equal to c times the molecular or atomic weight of the material, shows interesting regularities that help us to understand the mechanisms involved in heat absorption; see Table 1. Measured heat capacities depend on the conditions of measurement (under constant pressure or constant volume, for example) and these must be specified clearly.

Heat supplied to a material may also change its physical state, for example, from solid to liquid or from liquid to gas. The amount required per unit mass for a particular material is its heat of

Heats of Transformation

transformation L; see Table 2. Thus,

$$Q = Lm. \qquad [5]$$

The *heat of vaporization* L_V refers to the amount of energy per unit mass required to vaporize a liquid. For water at atmospheric pressure it has the value 539 cal/g = 40.7 kJ/mol = 2260 kJ/kg (Eq. 6). The *heat of fusion* L_F is the amount of heat energy per unit mass that is liberated when a liquid freezes. Its value for water at atmospheric pressure is 79.5 cal/g = 6.01 kJ/mole = 333 kJ/kg (Eq. 7). Sample Problems 1–3 show examples of calculations using these quantities.

Work is another process by which a system may exchange energy with its surroundings. The amount of energy as work W that a system transmits to its surroundings as it expands (or contracts) from an initial volume V_i to a final volume V_f may be computed from

Work Associated with Volume Change

$$W = \int dW = \int_{V_i}^{V_f} p \, dV. \tag{11}$$

The integral calculation is necessary because the pressure p may vary during the volume change. W may also be computed as the area under the curve of p versus V representing the change (see Fig. 4).

The conservation of energy for a sample of material exchanging energy by work and heat with its surroundings is expressed in the *first law of thermodynamics*. In mathematical form it assumes either of the forms

The First Law of Thermodynamics

$$\Delta U = U_f - U_i = Q - W \quad \text{(first law)} \tag{12}$$

or

$$dU = dQ - dW \quad \text{(first law)}, \tag{13}$$

in which U represents the internal energy of the material, which depends only on its state (temperature, pressure, volume, and constitution). Q represents heat *added* to the material and W represents the work *done by* the material on its surroundings.

Q and W Are Path Dependent; U Is Not

Q and W (as well as dQ and dW), in general, depend on the exact process through which the material goes from the initial to the final state. $U_f - U_i$ (or dU), however, does not; it depends only on the nature of the initial and final states. U represents energy stored within the material in the form of microscopic kinetic energy and potential energy.

Applications of the First Law

Section 20–6 applies the first law of thermodynamics to *adiabatic* processes ($\Delta Q = 0$; $\Delta U = -W$), *constant volume* processes ($W = 0$; $\Delta U = Q$), *cyclical* processes ($\Delta U = 0$; $Q = W$), and *free expansions* ($Q = W = 0$; $\Delta U = 0$). See Table 3. Sample Problem 4 applies the first law to isobaric (constant pressure) processes, using the boiling of water as an example.

Heat Conduction

Heat energy may be transferred between objects with differing temperatures by *conduction, convection, or radiation*. One-dimensional heat flow (conduction) through a slab whose ends are maintained at temperatures T_H and T_C obeys the law of heat conduction:

$$H(=Q/t) = kA \frac{T_H - T_C}{L}, \tag{18}$$

in which H represents the time rate of heat flow, A and L are the cross-sectional area and length of the slab, and k is the thermal conductivity of the material (see Fig. 8). When we deal with commercial building and insulating materials formed into slabs of thickness L, we often use the thermal resistance R (the R-value; $R = L/k$). See Table 4 and Sample Problem 5 for representative values and a sample calculation.

Convection

Convection occurs when temperature differences (or mechanical pumps or blowers) cause motion within a fluid, which carries higher-temperature fluid from the hot object to the cooler one.

Radiation

Radiation refers to the electromagnetic energy emitted by all objects, the amount increasing with increasing temperature.

QUESTIONS

1. Temperature and heat are often confused, as in, "Bake in an oven at moderate heat." By example, distinguish between these two concepts as carefully as you can.

2. Give an example of a process in which no heat is transferred to or from the system but the temperature of the system changes.

3. Can heat be considered a form of stored (or potential) energy? Would such an interpretation contradict the concept of heat as energy in the process of transfer because of a temperature difference?

4. Can heat be added to a substance without causing the temperature of the substance to rise? If so, does this contradict the concept of heat as energy in the process of transfer because of a temperature difference?

5. Why must heat energy be supplied to melt ice—after all, the temperature doesn't change?

6. Explain the fact that the presence of a large body of water nearby, such as a sea or ocean, tends to moderate the temperature extremes of the climate on adjacent land.

7. An electric fan not only does not cool the air but heats it slightly. How then can it cool you?

8. Both heat conduction and wave propagation involve the transfer of energy. Is there any difference in principle between these two phenomena?

9. When a hot object warms a cool one, are their temperature changes equal in magnitude? Give examples. Can one then say that temperature passes from one to the other?

10. A block of wood and a block of metal are at the *same* temperature. When the blocks feel cold, the metal feels colder than the wood; when the blocks feel hot, the metal feels hotter than the wood. Explain. At what temperature will the blocks feel equally cold or hot?

11. How can you best use a spoon to cool a cup of coffee? Stirring—which involves doing work—would seem to heat the coffee rather than to cool it.

12. How does a layer of snow protect plants during cold weather? During freezing spells, citrus growers in Florida often spray their fruit with water, hoping that it will freeze. How does that help?

13. Explain the wind-chill effect.

14. A thick glass is more likely to crack than a thin glass when you pour hot water into it. Why?

15. Why defrost a refrigerator?

16. You put your hand in a hot oven to remove a casserole and burn your fingers on the hot dish. However, the air in the oven is at the same temperature as the casserole dish but it does not burn your fingers. Why not?

17. Metal workers have observed that they can dip a hand very briefly into hot molten metal without ill effect. Explain. (See Essay 10.)

18. Why is thicker insulation used in an attic than in the walls of a house?

19. Is ice always at 0°C? Can it be colder? Can it be warmer? What about an ice–water mixture?

20. Explain why your finger sticks to a metal ice tray just taken from the refrigerator.

21. The water in a kettle makes quite a bubbling noise while it is being heated to boiling. Once it starts boiling, however, it does so quietly. What is the explanation? (*Hint:* Think of the fate of a bubble of vapor rising from the bottom of the kettle before the water is uniformly heated.)

22. On a winter day the temperature of the inside surface of a wall is much lower than room temperature and that of the outside surface is much higher than the outdoor temperature. Explain.

23. The physiological mechanisms that maintain a person's internal temperature operate in a limited range of external temperature. Explain how this range can be extended at each extreme by the use of clothes. (See "Heat, Cold, and Clothing," by James B. Kelley, *Scientific American,* February 1956.)

24. What requirements for thermal conductivity, specific heat capacity, and coefficient of expansion would you want a material to be used as a cooking utensil to satisfy?

25. Suppose that, for some strange reason, the single-pane glass windows of a house are replaced with sheets of aluminum of the same thickness as the glass. How is the rate at which heat is conducted through these window areas affected?

26. Is the temperature of an isolated system (no interaction with the environment) conserved? Explain.

27. Is heat the same as internal energy? If not, give an example in which a system's internal energy changes without a flow of heat across the system's boundary.

28. Can you tell whether the internal energy of a body was acquired by heat transfer or by performance of work?

29. If the pressure and volume of a system are given, is the temperature always uniquely determined?

30. Does a gas do any work when it expands adiabatically? If so, what is the source of the energy needed to do this work?

31. A quantity of gas occupies an initial volume V_0 at a pressure p_0 and a temperature T_0. It expands to a volume V (*a*) at constant temperature and (*b*) at constant pressure. In which case does the gas do more work?

32. Discuss the process of the freezing of water from the point of view of the first law of thermodynamics. Remember that ice occupies a greater volume than an equal mass of water.

33. A thermos bottle contains coffee. The thermos bottle is vigorously shaken. Consider the coffee as the system. (*a*) Does its temperature rise? (*b*) Has heat been added to it? (*c*) Has work been done on it? (*d*) Has its internal energy changed?

34. We have seen that "energy conservation" is a universal law of nature. At the same time national leaders urge "energy conservation" upon us (for example, driving slower). Explain the two quite different meanings of these words.

35. Can heat energy be transferred through matter by radiation? If so, give an example. If not, explain why.

36. Why does stainless steel cookware often have a layer of copper or aluminum on the bottom?

37. Consider that heat can be transferred by convection and radiation, as well as by conduction, and explain why a thermos bottle is double-walled, evacuated, and silvered.

38. A lake freezes first at its upper surface. Is convection involved? What about conduction and radiation?

EXERCISES AND PROBLEMS

Section 20–3 The Absorption of Heat by Solids and Liquids

1E. It is possible to melt ice by rubbing one block of it against another. How much work, in joules, would you have to do to get 1 g of ice to melt?

2E. A certain substance has a molecular weight of 50 g/mol. When 75 cal of heat are added to a 30-g sample of this material, its temperature rises from 25 to 45°C. (*a*) What is the specific heat of this substance? (*b*) How many moles of the substance are present? (*c*) What is the molar heat capacity of the substance?

3E. In a certain solar house, energy from the sun is stored in barrels filled with water. In a particular winter stretch of five cloudy days, 1.0×10^6 kcal are needed to maintain the inside of the house at 22°C. Assuming that the water in the barrels is at 50°C, what volume of water is required?

4E. A diet doctor encourages dieting by drinking ice water. His theory is that the body must burn off enough fat to raise the temperature of the water from 0°C to the body temperature of 37°C. How many liters of ice water would have to be consumed to burn off 454 g ($w = mg = 1$ lb) of fat, assuming that this requires 3500 food calories? (Each food calorie is equal to 1 kcal.) Why is it not advisable to follow this diet? One liter = 10^3 cm³. The density of water is 1.0 g/cm³.

5E. Icebergs in the North Atlantic present hazards to shipping (see Fig. 13), causing the length of shipping routes to increase by about 30% during the iceberg season. Attempts to destroy icebergs include planting explosives, bombing, torpedoing, shelling, ramming, and painting with lampblack. Suppose that direct melting of the iceberg, by placing heat sources in the ice, is tried. How much heat is required to melt 10% of a 200,000-metric-ton iceberg?

Figure 13 Exercise 5.

6E. How much water remains unfrozen after 12 kcal of heat have been extracted from 260 g of liquid water initially at 0°C?

7E. Calculate the minimum amount of heat, in joules, required to completely melt 130 g of silver initially at 15°C. Assume that the specific heat does not change with temperature. See Tables 1 and 2.

8E. A room is lighted by four 100-W incandescent light bulbs. Assuming that 90% of the energy is converted to heat, how much heat is added to the room in 1 h?

9E. What quantity of butter (6.0 Cal/g = 6000 cal/g) would supply the energy needed for a 160-lb man to ascend to the top of Mt. Everest, elevation 29,000 ft, from sea level?

10E. An energetic athlete dissipates all the energy in a diet of 4000 Cal/day. If he were to release this energy at a steady rate, how would this output compare with the energy output of a 100-W bulb?

11E. If the heat energy necessary to raise the temperature of 1.0 lb of water from 68 to 78°F were converted into kinetic energy of translation of the mass of water as a whole, what would the speed of the 1.0-lb "chunk" of water be?

12E. Power is supplied at the rate of 0.40 hp for 2.0 min in drilling a hole in a 1.6-lb copper block. (*a*) How much heat in Btu is generated? (*b*) What is the rise in temperature of the copper if only 75% of the power warms the copper? (1 ft·lb = 1.285×10^{-3} Btu.)

13E. An object of mass 6.0 kg falls through a height of 50 m and, by means of a mechanical linkage, rotates a paddle wheel that stirs 0.60 kg of water. The water is initially at 15°C. What is the maximum possible temperature rise?

14E. (*a*) Compute the possible increase in temperature for water going over Niagara Falls, 162 ft high. (*b*) What factors would tend to prevent this possible rise?

15E. A small electric immersion heater is used to boil 100 g of water for a cup of instant coffee. The heater is labeled 200 watts. Calculate the time required to bring this water from 23°C to the boiling point, ignoring any heat losses.

16E. A pickup truck whose mass is 2200 kg is speeding along the highway at 65 mi/h. (*a*) If you could use all this kinetic energy to vaporize water already at 100°C, how much water could you vaporize? (*b*) If you had to buy this amount of energy from your local utility company at 12¢/kW·h, how much would it cost you? Guess at the answers before you figure them out; you may be surprised.

17E. A 150-g copper bowl contains 220 g of water; both bowl and water are at 20°C. A very hot 300-g copper cylinder is dropped into the water. This causes the water to boil, with 5.0 g being converted to steam, and the final temperature of the entire system is 100°C. (*a*) How much heat was transferred to the water? (*b*) How much to the bowl? (*c*) What was the original temperature of the cylinder?

18P. Calculate the specific heat of a metal from the following data: A container made of the metal has a mass of 3.6 kg and contains 14 kg of water. A 1.8-kg piece of the metal initially at a temperature of 180°C is dropped into the water. The container and water initially have a temperature of 16°C and the final temperature of the entire system is 18°C.

19P. A thermometer of mass 0.055 kg and of specific heat 0.20 cal/g·K reads 15.0°C. It is then completely immersed in 0.300 kg of water and it comes to the same final temperature as the water. If the thermometer reads 44.4°C, what was the temperature of the water before insertion of the thermometer, neglecting other heat losses?

20P. How long does it take a 2.0×10^5-Btu/h water heater to raise the temperature of a 40-gal tub of water from 70 to 100°F?

21P. An athlete needs to lose weight and decides to do it by "pumping iron." (a) How many times must an 80-kg weight be lifted a distance of 1 m in order to burn off 1 lb of fat, assuming that it takes 3500 Cal to do this? (b) If the weight is lifted once every 2 s, how long does it take?

22P. A 1500-kg Mercedes moving at 90 km/h brakes to rest, at uniform acceleration and without skidding, over a distance of 80 m. At what average rate is thermal energy delivered to the brake system?

23P. A chef, upon awaking one morning to find his stove out of order, decides to boil the water for his wife's coffee by shaking it in a thermos flask. Suppose that he uses 500 cm³ of tap water at 59°F, and that the water falls 1.0 ft each shake, the chef making 30 shakes each minute. Neglecting any loss of energy, how long must he shake the flask before the water boils?

24P. A block of ice at 0°C whose mass initially is 50 kg slides along a horizontal surface, starting at a speed of 5.38 m/s and finally coming to rest after traveling 28.3 m. Compute the mass of ice melted as a result of the friction between the block and the surface. (Assume that all the heat generated owing to friction goes into the block of ice.)

25P. The specific heat of a substance varies with temperature according to the formula $c = 0.20 + 0.14T + 0.023T^2$, with T in °C and c in cal/g·K. Find the heat required to raise the temperature of 2.0 g of this substance from 5.0 to 15°C.

26P. In a solar water heater, energy from the sun is gathered by rooftop collectors, which circulate water through tubes in the collector. The solar radiation enters the collector through a transparent cover and warms the water in the tubes; this water is pumped into a holding tank. Assuming that the efficiency of the overall system is 20% (that is, 80% of the incident solar energy is lost from the system), what collector area is necessary to take water from a 200-L tank and raise its temperature from 20 to 40°C in 1.0 h? The intensity of incident sunlight is 700 W/m².

27P. An insulated cup contains 130 cm³ of hot coffee, at a temperature of 80°C. You put in a 12-g ice cube at 0°C to cool it. By how many degrees will your coffee have cooled once the ice has melted? The latent heat of fusion of water is 79.5 cal/g. Treat the coffee as though it were pure water.

28P. What mass of steam at 100°C must be mixed with 150 g of ice at 0°C, in a thermally insulated container, to produce liquid water at 50°C?

29P. A person makes a quantity of iced tea by mixing 500 g of the hot tea (essentially water) with an equal mass of ice at 0°C. What are the final temperature and mass of ice remaining if the initial hot tea is at a temperature of (a) 90°C and (b) 70°C?

30P. (a) Two 50-g ice cubes are dropped into 200 g of water in a glass. If the water were initially at a temperature of 25°C, and if the ice came directly from a freezer at −15°C, what will be the final temperature of the drink? The specific heat of ice is 0.53 cal/g·K and the heat required to melt ice to water is 79.5 cal/g. (b) Suppose that only one ice cube had been used in (a); what would be the final temperature of the drink? Neglect the heat capacity of the glass.

31P. A 20-g copper ring has a diameter of exactly 1.00000 in. at its temperature of 0°C. An aluminum sphere has a diameter of exactly 1.00200 in. at its temperature of 100°C. The sphere is placed on top of the ring (Fig. 14), and the two are allowed to come to thermal equilibrium, no heat being lost to the surroundings. The sphere just passes through the ring at the equilibrium temperature. What is the mass of the sphere?

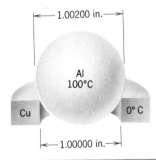

← 1.00200 in. →

Al
100°C

Cu   0° C

← 1.00000 in. →

Figure 14 Problem 31.

32P. A *flow calorimeter* is used to measure the specific heat of a liquid. Heat is added at a known rate to a stream of the liquid as it passes through the calorimeter at a known rate. Then a measurement of the resulting temperature difference between the inflow and the outflow points of the liquid stream enables us to compute the specific heat of the liquid. A liquid of density 0.85 g/cm³ flows through a calorimeter at the rate of 8.0 cm³/s. Heat is added by means of a 250-W electric heating coil, and a temperature difference of 15°C is established in steady-state conditions between the inflow and the outflow points. Find the specific heat of the liquid.

33P. By means of a heating coil, energy is transferred at a constant rate to a substance in a thermally-insulated container. The temperature of the substance is measured as a function of time. (*a*) Show how we can deduce from this information the way in which the heat capacity of the body depends on the temperature. (*b*) Suppose that in a certain temperature range it is found that the temperature T is proportional to t^3, where t is the time. How does the heat capacity depend on T in this range?

34P. Two metal blocks are insulated from their surroundings. The first block, which has a mass $m_1 = 3.16$ kg and is at a temperature of $T_1 = 17°C$, has a specific heat four times that of the second block. This second block is at a temperature $T_2 = 47°C$, and its coefficient of linear expansion is $15 \times 10^{-6}/°C$. When the two blocks are brought together and allowed to come to thermal equilibrium, the area of one face of the second block is found to have decreased by 0.030%. Find the mass of the second block.

Section 20–6   The First Law of Thermodynamics: Some Special Cases

35E. A sample of gas expands from 1.0 to 4.0 m³ while its pressure decreases from 40 to 10 Pa. How much work is done by the gas if its pressure changes with volume according to each of the three processes shown in the p–V diagram in Fig. 15?

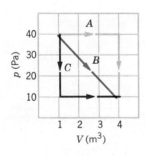

Figure 15 Exercise 35.

36E. Suppose that a sample of gas expands from 1.0 to 4.0 m³ along the path B in the p–V diagram shown in Fig. 16. It is then compressed back to 1.0 m³ along either path A or path C. Compute the net work done by the gas for the complete cycle in each case.

37E. Consider that 200 J of work are done on a system and 70 cal of heat are extracted from the system. In the sense of the first law of thermodynamics, what are the values (including algebraic signs) of (*a*) W, (*b*) Q, and (*c*) ΔU?

38E. A thermodynamic system is taken from an initial state

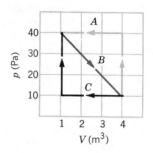

Figure 16 Exercise 36.

A to another B and back again to A, via state C, as shown by the path $ABCA$ in the p–V diagram of Fig. 17*a*. (*a*) Complete the table in Fig. 17*b* by filling in appropriate + or − indications for the signs of the thermodynamic quantities associated with each process. (*b*) Calculate the numerical value of the work done by the system for the complete cycle $ABCA$.

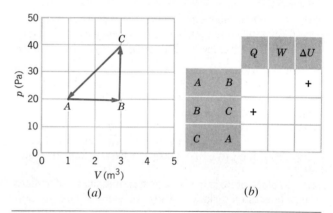

(*a*)   (*b*)

Figure 17 Exercise 38.

39E. Gas within a chamber passes through the cycle shown in Fig. 18. Determine the net heat added to the system during process CA if $Q_{AB} = 4.77$ cal, $Q_{BC} = 0$, and $W_{BCA} = 15.0$ J.

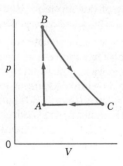

Figure 18 Exercise 39.

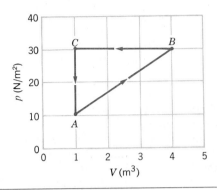

Figure 19   Exercise 40.

40E.   Gas within a chamber undergoes the processes shown in the p–V diagram of Fig. 19. Calculate the net heat added to the system during one complete cycle.

41P.   Figure 20a shows a cylinder containing gas and closed by a movable piston. The cylinder is submerged in an ice–water mixture. The piston is *quickly* pushed down from position 1 to position 2. The piston is held at position 2 until the gas is again at 0°C and then is *slowly* raised back to position 1. Figure 20b is a p–V diagram for the process. If 100 g of ice are melted during the cycle, how much work has been done *on* the gas?

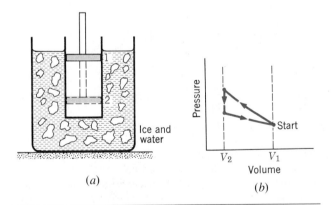

(a)

(b)

Figure 20   Problem 41.

42P.   When a system is taken from state i to state f along the path iaf in Fig. 21, it is found that $Q = 50$ cal and $W = 20$ cal. Along the path ibf, $Q = 36$ cal. (a) What is W along the path ibf? (b) If $W = -13$ cal for the curved return path fi, what is Q for this path? (c) Take $U_i = 10$ cal. What is U_f? (d) If $U_b = 22$ cal, what are the Q values for process ib and process bf?

43P*.   A cylinder has a well-fitted 2.0-kg metal piston whose cross-sectional area is 2.0 cm² (Fig. 22). The cylinder contains water and steam at constant temperature. The piston is ob-

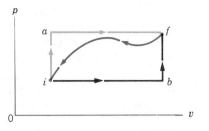

Figure 21   Problem 42.

served to fall slowly at a rate of 0.30 cm/s because heat flows out of the cylinder through the cylinder walls. As this happens, some steam condenses in the chamber. The density of the steam inside the chamber is 6.0×10^{-4} g/cm³ and the atmospheric pressure is 1.0 atm. (a) Calculate the rate of condensation of steam. (b) At what rate is heat leaving the chamber? (c) What is the rate of change of internal energy of the steam and water inside the chamber?

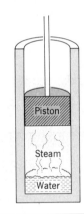

Figure 22   Problem 43.

Section 20–7   The Transfer of Heat

44E.   The average rate at which heat flows out through the surface of the earth in North America is 54 mW/m², and the average thermal conductivity of the near surface rocks is 2.5 W/m·K. Assuming a surface temperature of 10°C, what should be the temperature at a depth of 35 km (near the base of the crust)? Ignore the heat generated by the presence of radioactive elements.

45E.   The thermal conductivity of Pyrex glass at 0°C is 2.9×10^{-3} cal/cm·°C·s. (a) Express this in W/m·K and in Btu/ft·°F·h. (b) What is the R-value for a $\frac{1}{4}$-in. sheet of such glass?

46E.   (a) Calculate the rate at which body heat flows out through the clothing of a skier, given the following data: The body surface area is 1.8 m² and the clothing is 1.0 cm thick; skin surface temperature is 33°C, whereas the outer surface of the clothing is at 1.0°C; the thermal conductivity of the clothing is 0.040 W/m·K. (b) How would the answer change if, after a fall, the skier's clothes become soaked with water? Assume that the thermal conductivity of water is 0.60 W/m·K.

47E. Consider the slab shown in Fig. 8. Suppose that $L = 25$ cm, $A = 90$ cm², and the material is copper. If $T_H = 125$°C, $T_C = 10$°C, and a steady state is reached, find the rate of heat transfer.

48E. A cylindrical copper rod of length 1.2 m and cross-sectional area 4.8 cm² is insulated to prevent heat loss through its surface. The ends are maintained at a temperature difference of 100°C by having one end in a water–ice mixture and the other in boiling water and steam. (a) Find the rate at which heat is transferred along the rod. (b) Find the rate at which ice melts at the cold end.

49E. Show that the temperature T_x at the interface of a compound slab (see Section 7 and Fig. 10) is given by

$$T_x = \frac{R_1 T_H + R_2 T_C}{R_1 + R_2}.$$

50E. Show that in a compound slab such as that in Fig. 23a the temperature change in each portion is inversely proportional to the thermal conductivity.

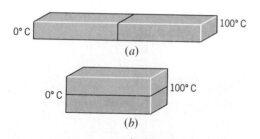

(a)

(b)

Figure 23 Exercise 50 and Problem 52.

51E. Four square pieces of insulation of two different materials, all with the same thickness and area A, are available to cover an opening of area $2A$. This can be done in either of the two ways shown in Fig. 24. Which arrangement, (a) or (b), would give the lower heat flow if $k_2 \neq k_1$?

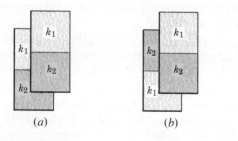

(a)   *(b)*

Figure 24 Exercise 51.

52P. Two identical rectangular rods of metal are welded end to end as shown in Fig. 23a, and 10 cal of heat flows through the rods in 2.0 min. How long would it take for 10 cal to flow through the rods if they are welded as shown in Fig. 23b?

53P. Compute the heat flow through two storm doors 2.0 m high and 0.75 m wide. (a) One door is made with aluminum panels 1.5 mm thick and a 3.0-mm glass pane that covers 75% of its surface (the structural frame is considered to have a negligible area). (b) The second door is made entirely of white pine averaging 2.5 cm in thickness. Take the temperature drop through the doors to be 33°C (=60°F). See Table 4.

54P. An idealized representation of the air temperature as a function of distance from a single-pane window on a calm, winter day is shown in Fig. 25. The window dimensions are 60 cm × 60 cm × 0.50 cm. (a) At what rate does heat flow out through the window area? (*Hint:* The temperature drop across the window glass is very small.) (b) Estimate the difference in temperature between the inner and outer glass surfaces.

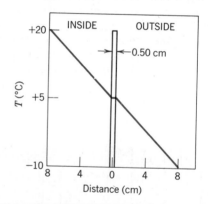

Figure 25 Problem 54.

55P. A large water tank with a bottom 1.7 m in diameter is made of iron boilerplate 5.2 mm thick. As the water is being heated, the gas burner underneath is able to maintain a temperature difference of 2.3°C between the surfaces of the iron. How much heat flows through to the water in 5.0 min? (Iron has a thermal conductivity of 67 W/m·K.)

56P. (a) What is the rate of heat loss in watts per square meter through a glass window 3.0 mm thick if the outside temperature is −20°F and the inside temperature is +72°F? (b) A storm window is installed having the same thickness of glass but with an air gap of 7.5 cm between the two windows. What will be the corresponding rate of heat loss presuming that conduction is the only important heat-loss mechanism?

57P. A container of water has been outdoors in cold weather until a 5.0-cm-thick slab of ice has formed on its surface (Fig. 26). The air above the ice is at −10°C. Calculate the rate of formation of ice (in centimeters per hour) on the bottom sur-

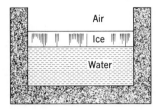

Figure 26 Problem 57.

face of the ice slab. Take the thermal conductivity and density of ice to be 0.0040 cal/s·cm·°C and 0.92 g/cm³. Assume that no heat enters or leaves the water through the walls of the tank.

58P. Ice has formed on a shallow pond and a steady state has been reached with the air above the ice at −5°C and the bottom of the pond at 4°C. If the total depth of ice + water is 1.4 m, how thick is the ice? (Assume that the thermal conductivities of ice and water are 0.40 and 0.12 cal/m·°C·s, respectively.)

59P. Three metal rods, made of copper, aluminum, and brass are each 6.0 cm long and 1.0 cm in diameter. These rods are placed end to end, with the aluminum between the other two. The free ends of the copper and brass rods are maintained at 100°C and 0° C, respectively. Find the steady-state temperatures of the copper–aluminum junction and the aluminum–brass junction. The thermal conductivity of brass is 109 W/m·K. (*Hint:* Equate the rates at which heat flows through each rod.)

60P. A wall assembly consists of a 20 ft × 12 ft frame made of 16 two-by-four vertical studs, each 12 ft long and set with their center lines 16 in. apart. The outside of the wall is faced with ¼-in. plywood sheet (R = 0.30) and ¾-in. white pine siding (R = 0.98). The inside is faced with ¼-in, plasterboard (R = 0.47), and the space between the studs is filled with polyurethane foam (R = 5.9 for a 1-in. layer). A "two-by-four" is actually 1.75 in. × 3.75 in. in size. Assume that they are made of wood for which R = 1.3 for a 1-in. slab. (a) At what rate does heat flow through this wall for a 30°F temperature difference? (b) What is the R-value for the assembled wall? (c) What fraction of the wall area contains studs, as opposed to foam? (d) What fraction of the heat flow is through the studs, as opposed to the foam?

ESSAY 10
BOILING AND THE LEIDENFROST EFFECT

JEARL WALKER
CLEVELAND STATE
UNIVERSITY

How does water boil? As commonplace as the event is, you may not have noticed all its curious features. Some of the features are important in industrial applications, while others appear to be the basis for certain dangerous stunts once performed by daredevils in carnival sideshows.

Arrange for a pan of tap water to be heated from below by a flame or electric heat source. Once the water warms, air molecules are driven out of solution in the water, collecting as tiny bubbles in crevices along the bottom of the pan. These sites are small enough that surface tension prevented water from flooding them when the pan was filled. As air molecules come out of solution, they pass into the crevices, inflating the bubbles and increasing their buoyancy. Eventually the water's surface tension pinches a bubble off from the crevice, allowing the bubble to float to the top surface of the water. Since the crevice is still filled with air, another bubble begins to develop within it. The formation of air bubbles is a sign that the water is heating but has nothing to do with boiling.

Water that is directly exposed to the atmosphere boils at what is sometimes called its normal boiling temperature T_S. For example, T_S is 100°C when the air pressure is 1 atm. Since the water at the bottom of your pan is not directly exposed to atmosphere, it remains liquid even when it superheats above T_S by as much as a few degrees. As it warms and then superheats, it is constantly mixed by convection as hot water rises while cooler water replaces it.

If you continue to increase the pan's temperature, the bottom layer of water begins to vaporize, with water molecules gathering in small vapor bubbles in the dry crevices. This phase of boiling is marked by pops, pings and eventually buzzing. The water almost sings its displeasure at being heated. Every time a vapor bubble expands upward into slightly cooler water, it suddenly collapses because the vapor within it condenses. Each collapse sends out a sound wave, the ping you hear. Once the temperature of the bulk water increases, the bubbles may not collapse until after they pinch off from the crevices and ascend part of the way to the top surface of the water (Fig. 1).

If you still increase the pan's temperature, the clamor of collapsing bubbles first grows louder and then disappears. The noise begins to soften when the bulk liquid is sufficiently hot that the vapor bubbles reach the top surface of the water. There they pop open with a light splash. The water is now in full boil.

If your heat source is a kitchen stove, the story stops at this point. However, with a laboratory burner you can continue to increase the pan's temperature. The vapor bubbles next become so abundant and pinch off from their crevices so frequently that they coalesce, forming columns of vapor that violently and chaoticaly churn upward, sometimes meeting previously detached "slugs" of vapor.

The production of vapor bubbles and columns is called *nucleate boiling* because the formation and growth of the bubbles depends on crevices serving as nucleating

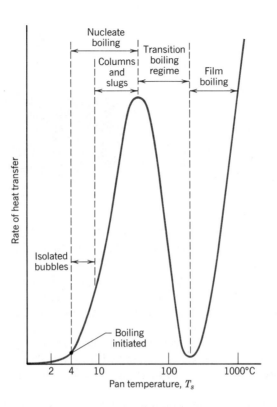

Figure 1 Boiling curve for water.

sites. Whenever you increase the pan's temperature, the rate at which heat is transferred to the water increases. If you continue to raise the pan's temperature past the stage of columns and slugs, the boiling enters a new phase called the *transition regime*. Then each increase in the pan's temperature reduces the rate at which heat is transferred to the water. The decrease is not paradoxical. In the transition regime, much of the bottom of the pan is covered by a layer of vapor. Since water vapor conducts heat about an order of magnitude poorer than does liquid water, the transfer of heat to the water is diminished. The hotter the pan becomes, the less direct contact the water has with it and the worse the transfer of heat becomes. This situation can be dangerous in a heat exchanger whose purpose is to remove heat from a heat source. If the water in the heat exchanger is allowed to enter the transition regime, the source may destructively overheat because of diminished transfer of heat from it.

Such may have caused the explosion of the steam locomotive shown in Fig. 2. The locomotive was driven by steam that was produced by heat from oil burning in the firebox. The boiler that held the water was above the firebox, separated from it by a sturdy metal layer called the "crown sheet." Hot gases from the fire flowed from the firebox and into tubes that extended through the water. Heat was transferred to the water along the length of the tubes and along the surface of the crown sheet. Since the steam was confined, it was under high pressure. Its pressure on the water raised the boiling point of the water.

Normally, this arrangement was safe. However, if the crown sheet became too hot, it entered the transition regime, thereby greatly reducing the transfer of heat through it. Since the crown sheet was directly exposed to the fire, its temperature continued to increase. If the situation were unchecked, the crown sheet softened and sagged until it cracked under the large pressure and the weight of the water in the

Figure 2 A steam locomotive that exploded.

boiler. When the water was dumped into the firebox, the sudden decrease of the pressure dropped the boiling point of the water. Since the water was then hotter than the new boiling point, part of it dramatically flashed to vapor, which then expanded as an explosion. For the locomotive in the figure, the explosion ruptured the boiler, blowing it free of the engine and flipping it over a horizontal axis so that it landed upside down near the track, facing in the opposite direction. Three men died in the accident.

Suppose you continue to increase the temperature of the pan. Eventually, the whole of the bottom surface is covered with vapor. Then heat is slowly transferred to the liquid above the vapor by radiation and gradual conduction. This phase is called *film boiling*.

Although you cannot obtain film boiling in a pan of water on a kitchen stove, it is commonplace in the kitchen. My grandmother once demonstrated how it serves to indicate when her skillet is hot enough for pancake batter. After she heated the empty skillet for a while, she sprinkled a few drops of water into it. The drops sizzled away within seconds. Their rapid disapearance warned her that the skillet was insufficiently hot for the batter. After further heating the skillet, she repeated her test with a few more water drops. This time they beaded up and danced over the metal, lasting well over a minute before they disappeared. The skillet was then hot enough for the batter.

To study her demonstration, I arranged for a flat metal plate to be heated by a laboratory burner. While monitoring the temperature of the plate with a thermocouple, I carefully released a drop of distilled water from a syringe held just above the plate. The drop fell into a dent I had made in the plate with a ball-peen hammer. The syringe allowed me to release drops of uniform size. Once a drop was released, I timed how long it survived on the plate. Afterward, I plotted the survival times of the drops versus the plate temperature. The graph has a curious peak. When the plate temperature is between 100 and about 200°C, each drop spread over the plate in a thin layer and rapidly vaporized. When the plate temperature was about 200°C, a drop deposited on the plate beaded up and survived for over a minute (Fig. 3). At even higher plate temperatures, the water beads did not survive quite as long. Similar experiments with tap water generated a graph with a flatter peak, probably because suspended particles breached the vapor layer, conducting heat into the drops.

The fact that a water drop is long-lived when deposited on metal that is much hotter than the boiling temperature of water was first reported by Hermann Boerhaave in 1732. It was not investigated extensively until 1756 when Johann Gottlieb Leidenfrost published "A Tract about Some Qualities of Common Water." Because Leidenfrost's work was not translated from the Latin until 1965, it was not widely read. Still,

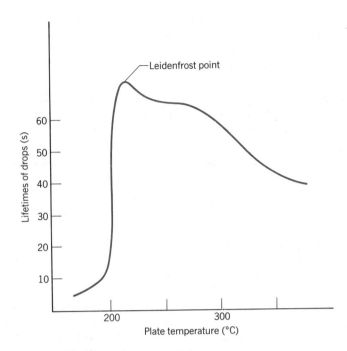

Figure 3 Drop lifetimes on a hot plate.

his name is now associated with the phenomenon. In addition, the temperature corresponding to the peak in a graph such as I made is called the Leidenfrost point.

Leidenfrost conducted his experiments with an iron spoon that was heated red-hot in a fireplace. After placing a drop of water into the spoon, he timed its duration by the swings of a pendulum. He noted that the drop seemed to suck the light and heat from the spoon, leaving a spot duller than the rest of the spoon. The first drop deposited in the spoon lasted 30 s while the next drop lasted only 10 s. Additional drops lasted only a few seconds.

Leidenfrost misunderstood his demonstrations because he did not realize that the longer-lasting drops were actually boiling. Let me explain in terms of my experiments. When the temperature of the plate is less than the Leidenfrost point, the water spreads over the plate and rapidly conducts heat from it, resulting in complete vaporization within seconds. When the temperature is at or above the Leidenfrost point, the bottom surface of a drop deposited on the plate almost immediately vaporizes. The gas pressure from this vapor layer prevents the rest of the drop from touching the plate. The layer thus protects and supports the drop for the next minute or so. The layer is constantly replenished as additional water vaporizes from the bottom surface of the drop because of heat radiated and conducted through the layer from the plate. Although the layer is less than 0.1 mm thick near its outer boundary and only about 0.2 mm thick in its center, it dramatically slows the vaporization of the drop (Fig. 4).

Figure 4 A Leidenfrost drop in cross section.

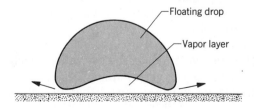

To demonstrate the flow of vapor out from beneath a Leidenfrost drop, I sprinkled fine powder over the plate. As a drop skirted over the plate, the vapor escaping from beneath blew the powder grains out of its way. I also built large amoebalike globs of water by inserting the syringe into the top of a floating drop and adding several squirts of water. The globs were unwieldy, often collapsing the vapor layers beneath them because of their large weight. Each collapse was signaled by a loud sizzle as part of the glob suddenly vaporized. Similar Leidenfrost drops and globs can be made with other liquids. For example, white vinegar has a Leidenfrost point near 250°C, while vodka has one near 150°C.

After reading the translation of Leidenfrost's research, I happened upon a description of a curious stunt that was performed in the sideshows of carnivals around the turn of the century. Reportedly, a performer was able to dip wet fingers into molten lead. Assuming that the stunt involved no trickery, I conjectured that it must depend on the Leidenfrost effect. As soon as the performer's wet flesh touched the hot liquid metal, part of the water vaporized, coating the hand with a vapor layer. If the plunge of fingers was brief, the flesh would not be heated significantly.

I could not resist the temptation to test my explanation. With a laboratory burner I melted down a sizeable slab of lead in a crucible. I heated the lead until its temperature was over 400°C, well above its melting temperature of 328°C. After wetting a finger in tap water, I prepared to touch the top surface of the molten lead. I must confess that I had an assistant standing ready with first-aid materials. I must also confess that my first several attempts failed because my brain refused to allow this ridiculous experiment, always directing my finger to miss the lead.

When I finally overcame my fears and briefly touched the lead, I was amazed. I felt no heat. Just as I had guessed, part of the water on the finger vaporized, forming a protective layer. Since the contact was brief, radiation and conduction of heat through the vapor layer was insufficient to raise perceptibly the temperature of my flesh. I grew braver. After wetting my hand, I dipped all my fingers into the lead, touching the bottom of the container. The contact with the lead was still too brief to result in a burn. Apparently, the Leidenfrost effect, or more exactly, the immediate presence of film boiling, protected my fingers. (Fig. 5)

I still questioned my explanation. Could I possibly touch the lead with a dry finger without suffering a burn? Leaving aside all rational thought, I tried it, immediately realizing my folly when pain raced through the finger. Later, I tested a dry weiner, forcing it into the molten lead for several seconds. The skin of the weiner quickly blackened. It lacked the protection of film boiling just as my dry finger had.

I must caution that dipping fingers into molten lead presents several serious dangers. If the lead is only slightly above its melting point, the loss of heat from it when the water is vaporized may solidify the lead around the fingers. If I were to pull the resulting glove of hot, solid lead up from the container, it will be in contact with my fingers so long that my fingers are certain to be badly burned. I must also contend with the possibility of splashing and spillage. In addition, there is the acute danger of having too much water on the fingers. When the surplus water rapidly vaporizes, it can blow molten lead over the surroundings and, most seriously, into the eyes. I now have scars on my arms and face from such explosive vaporizations. *You should never repeat this demonstration.*

A similar example of the Leidenfrost effect is described in Robert Ruark's best-selling novel *Something of Value.* To determine which of two men was telling the truth, the leader of an African village required each man to lick a hot knife. Presumably, the truthful man would have a tongue that was wet with saliva. When part of the

Figure 5 Walker demonstrating the Leidenfrost effect.

saliva underwent film boiling, his tongue would be spared from burn. On the other hand, the man who was lying would have a dry mouth, thus lacking any such protection from film boiling.

Film boiling can also be seen when liquid nitrogen is spilled. The drops and globs bead up as they skate over the floor. The liquid is at a temperature of about $-200°C$. When the spilled liquid nears the floor, its bottom surface vaporizes. The vapor layer then provides support for the rest of the liquid, allowing the liquid to survive for a surprisingly long time.

I was told of a stunt where a performer poured liquid nitrogen into his mouth without being hurt by its extreme cold. The liquid immediately underwent film boiling on its bottom surface and thus did not directly touch the tongue. Foolishly, I repeated this demonstration. For several dozen times the stunt went smoothly and dramatically. With a large glob of liquid nitrogen in my mouth, I concentrated on not swallowing while I breathed outward. The moisture in my cold breath condensed, creating a terrific plume that extended about a meter from my mouth. However, on my last attempt the liquid thermally contracted two of my front teeth so severely that the enamel ruptured into a "road map" of fissures. My dentist convinced me to drop this demonstration.

The Leidenfrost effect may also play a role in another foolhardy demonstration: walking over hot coals. At times the news media have carried reports of a performer striding over red-hot coals with much hoopla and mystic nonsense, perhaps claiming that protection from a bad burn is afforded by "mind over matter." Actually, physics protects the feet when the walk is successful. Particularly important is the fact that although the surface of the coals is quite hot, it contains surprisingly little energy. If the performer walks at a moderate pace, a footfall is so brief that the foot conducts little energy from the coals. Of course, a slower walk invites a burn because the longer contact allows heat to be conducted to the foot from the interior of the coals.

If the feet are wet prior to the walk, the liquid might also help protect them. To wet the feet a performer might walk over wet grass just before reaching the hot coals. Instead, the feet might just be sweaty because of the heat from the coals or the excitement of the performance. Once the performer is on the coals, some of the heat vaporizes the liquid on the feet, leaving less heat to be conducted to the flesh. In addition, there may be points of contact where the liquid undergoes film boiling, thereby providing brief protection from the coals.

I have walked over hot coals on five occasions. For four of the walks I was fearful enough that my feet were sweaty. However, on the fifth walk I took my safety so much for granted that my feet were dry. The burns I suffered then were extensive and terribly painful. My feet did not heal for weeks.

My failure may have been due to a lack of film boiling on the feet, but I had also neglected an additional safety factor. On the other days I had taken the precaution of clutching an earlier edition of this book to my chest during the walks so as to bolster my belief in physics. Alas, I forgot the book on the day when I was so badly burned.

I have long argued that degree-granting programs should employ "fire-walking" as a last exam. The chairman of the program should wait on the far side of a bed of red-hot coals while a degree candidate is forced to walk over the coals. If the candidate's belief in physics is strong enough that the feet are left undamaged, the chairman hands the candidate a graduation certificate. The test would be far more revealing than traditional final exams.

CHAPTER 21

THE KINETIC THEORY OF GASES

Molecules of the air are bouncing continually from this page. These "molecular tennis balls" are so small that it would take about 10^{13} of them to cover the period that closes this sentence. There are so many of them and they are so fast that they make about 10^{24} collisions with this tiny area every second. It is in this way that we account for the pressure of the atmosphere.

21-1 A New Way to Look at Gases

Classical thermodynamics—the subject of the last two chapters—has nothing to say about atoms. When we apply its laws to a gas, we are concerned only with such macroscopic variables as pressure, volume, and temperature. Although we know that a gas is made up of atoms or molecules, the laws of classical thermodynamics take no account of that fact and give no evidence for it.

However, the pressure exerted by a gas must surely be related to the steady drumbeat of its molecules on the walls of its container. The temperature and the internal energy of a gas must surely be related to the kinetic energy carried by these rapidly moving molecules. Perhaps we can learn something about gases by approaching the subject from this direction. The name we give to this

approach is the *kinetic theory of gases*. It is the subject of this chapter.

21-2 The Avogadro Constant

When our thinking is slanted toward molecules, it makes sense to measure the sizes of our samples in moles. If we do so, we can be certain that we are comparing samples that contain the same number of molecules. The *mole* (abbr. mol) is one of the seven SI base units and is defined as follows:

One mole is the amount of any substance that contains as many elementary entities as there are atoms in a 12-g sample of carbon-12.

The "elementary entities" of the definition are usually atoms or molecules but they need not be. We can speak of a mole of helium, a mole of water, a mole of neutrons, or (for that matter) a mole of tennis balls.

An important question—one that sets the scale for the atomic point of view—is: "Just how many elementary entities are there in a mole?" The answer can only be learned in the laboratory and proves to be

$$N_A = 6.02 \times 10^{23} \text{ mol}^{-1} \quad \text{(the Avogadro constant).} \quad (1)$$

This number is called the *Avogadro constant*, after the Italian scientist Amedeo Avogadro (1776–1856) who first made the important suggestion that all gases—under the same conditions of temperature and pressure—contain the same number of molecules.

When dealing with molecules, the number of moles contained in a sample can be found from

$$n = \frac{\text{number of molecules in the sample}}{\text{the Avogadro constant}} \quad (2)$$

or from

$$n = \frac{\text{mass of sample}}{\text{molecular weight}}$$
$$= \frac{\text{mass of sample}}{(\text{mass of one molecule})(\text{the Avogadro constant})}. \quad (3)$$

The enormously large value of the Avogadro constant suggests how tiny and how numerous atoms must be. A mole of air, for example, can easily fit into a suitcase. Yet, if these molecules were spread uniformly over the surface of the earth, there would be about 120,000 of them in every square centimeter. A second example: One mole of tennis balls would fill a volume equal to that of seven moons!

Measuring the Avogadro Constant.   It is not possible to see individual gas molecules as they move rapidly about, colliding with each other and with the walls of their container. Yet it *is* possible to see the direct effects of such motion and to deduce a value of the Avogadro constant from these observations.

In 1827, the English botanist Robert Brown observed that, when he looked through his microscope at pollen grains and other small objects suspended in water, they "danced around" in a random way. Figure 1 shows an example of such *Brownian motion*.* The suspended particles are continually bombarded on all sides by the

* See "Brownian Motion," by Bernard H. Lavenda, *Scientific American,* February 1985.

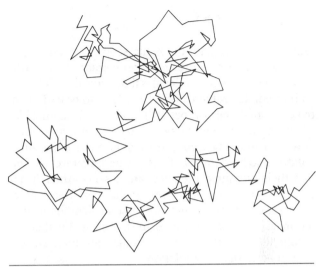

Figure 1   The motion of a tiny particle, suspended in water and viewed through a microscope. The short line segments connect its position at 30-s intervals. The path of the particle is a good example of a *fractal,* which is a curve for which any small section resembles the curve as a whole. For example, if the path of the particle for any of the short line segments shown had been explored at, say, 0.1-s intervals, the resulting plot would be very similar to the present plot.

molecules of the fluid. The numbers of molecules striking opposite sides of the particle in any short time interval—being determined by chance—will not be exactly equal. Because of these fluctuations, a randomly directed unbalanced force will act on the suspended particle, accounting for its "Brownian dance."

It is as if a bowling ball, floating in a gravity-free space, were bombarded from all sides by a swarm of randomly directed, rapidly moving Ping-Pong balls. The bowling ball would jiggle around slightly, and in a random way, exhibiting a kind of Brownian motion. By watching the erratic motion of the bowling ball, you could deduce something about the Ping-Pong balls, even if you could not see them.

It is possible to deduce the value of the Avogadro constant from measurements of the Brownian motion. There are also many other ways to get at this important constant. In the early part of this century, phenomena ranging from radioactivity to the blueness of the sky were employed, Einstein himself proposing four independent methods. It was this firm and broad experimental base for the Avogadro constant that went far to convince doubters—and there were many—of the usefulness of the concept that our familiar tangible world is made up of atoms.

21-3   An Ideal Gas

Our plan is to explain the macroscopic properties of a gas—such as its pressure and its temperature—in terms of the behavior of the molecules that make it up. There is an immediate problem: Which gas? Is it to be hydrogen, oxygen, methane, or perhaps uranium hexafluoride? They are all different. However, if we take 1-mole samples of various gases, confine them in boxes of identical volume and hold them at the same temperature, we find that their measured pressures are nearly—though not exactly—the same. If we repeat the measurements at lower gas densities, we find that these small differences in the measured pressures tend to disappear. Further experiments show that, at low enough densities, all real gases tend to obey the relation

$$\boxed{pV = nRT} \qquad \text{(ideal gas law)}, \qquad (4)$$

in which p is the absolute (not gauge) pressure, n is the number of moles, and R, the *gas constant,* has the same value for all gases, namely,

$$R = 8.31 \text{ J/mol} \cdot \text{K}. \qquad (5)$$

The temperature T in Eq. 4 must be expressed in absolute (Kelvin) units.

You may well ask, "What is so 'ideal' about an ideal gas?"* The answer lies in the simplicity of the law (Eq. 4) that governs its macroscopic properties. Using this law —as we shall see—we can deduce many properties of the ideal gas in a simple way. Although there is no such thing in nature as a truly ideal gas, *all* gases approach the ideal state at low enough densities, that is, under conditions in which their molecules are far enough apart. Thus, the ideal gas concept allows us to gain useful insights into the limiting behavior of real gases.

Work Done by an Ideal Gas.   Suppose that a sample of n moles of an ideal gas, confined to a piston–cylinder arrangement, is allowed to expand from an initial volume V_i to a final volume V_f. Suppose further that the temperature T of the gas is held constant throughout the process. Let us calculate the work done by the (ideal) gas during such an *isothermal expansion.*

Our starting point is Eq. 11 of Chapter 20, or

$$W = \int_{V_i}^{V_f} p \, dV.$$

* Many prefer *perfect gas* to ideal gas; they are the same thing.

Because the gas is ideal, we can substitute for p from Eq. 4, obtaining

$$W = \int_{V_i}^{V_f} \frac{nRT}{V} \, dV.$$

The temperature is required to be constant so that

$$W = nRT \int_{V_i}^{V_f} \frac{dV}{V}$$

or

$$\boxed{W = nRT \ln \frac{V_f}{V_i}} \qquad \text{(ideal gas, isothermal)}. \qquad (6)$$

During an expansion, $V_f > V_i$ by definition so that the ratio V_f/V_i in Eq. 6 is greater than unity. The logarithm of a quantity greater than unity is positive so that the work W done by an ideal gas during an isothermal expansion is positive, as we expect. During a compression, we have $V_f < V_i$ so that the ratio of volumes in Eq. 6 is less than unity. The logarithm in that equation— and hence the work W—is negative, again as we expect. Recall that the symbol ln specifies a *natural* logarithm, that is, a logarithm to base e.

Sample Problem 1   A cylinder contains oxygen at 20°C and a pressure of 15 atm at a volume of 12 L. The temperature is raised to 35°C and the volume reduced to 8.5 L. What is the final pressure of the gas? Assume that the gas is ideal.

From Eq. 4 we have

$$R = \frac{p_i V_i}{T_i} = \frac{p_f V_f}{T_f}.$$

Solving for p_f yields

$$p_f = \frac{p_i T_f V_i}{T_i V_f}. \qquad (7)$$

Before substituting numbers, we must make sure that we express the temperatures on the Kelvin scale. Thus,

$$T_i = (273 + 20) \text{ K} = 293 \text{ K}$$

and

$$T_f = (273 + 35) \text{ K} = 308 \text{ K}.$$

Inserting the given data into Eq. 7 then yields

$$P_f = \frac{(15 \text{ atm})(308 \text{ K})(12 \text{ L})}{(293 \text{ K})(8.5 \text{ L})} = 22 \text{ atm}. \quad \text{(Answer)}$$

Sample Problem 2   One mole of oxygen (assume an ideal gas) expands at a constant temperature T of 310 K from an

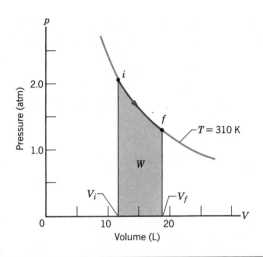

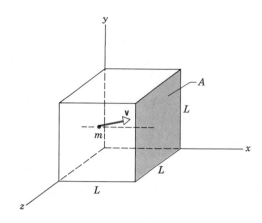

Figure 3 A cubical box of edge L, containing an ideal gas. A molecule of mass m and velocity $\mathbf{v}$ is about to collide with the shaded face of area A.

Figure 2 Sample Problem 2. The shaded area represents the work done by 1 mol of oxygen, assumed to be an ideal gas, in expanding at constant temperature T of 310 K from an initial volume V_i ($= 12$ L) to a final volume V_f($= 19$ L). The curve is an *isotherm*.

21-4 Pressure and Temperature: A Molecular View

initial volume V_i of 12 L to a final volume V_f of 19 L. (a) How much work is done by the expanding gas?

From Eq. 6 we have

$$W = nRT \ln \frac{V_f}{V_i}$$

$$= (1 \text{ mol})(8.31 \text{ J/mol·K})(310 \text{ K}) \ln \frac{19 \text{ L}}{12 \text{ L}}$$

$$= +1180 \text{ J}. \qquad \text{(Answer)}$$

The work done by the expanding gas is positive, consistent with the sign convention for work that we outlined in Section 20-5. The work done is shown by the shaded area in Fig. 2. The curved line is an *isotherm*, that is, a curve giving the relation of pressure to volume for a gas if its temperature is held constant.

(b) How much work is done by the gas during an isothermal *compression*, from $V_i = 19$ L to $V_f = 12$ L?

We proceed as in (a), finding

$$W = nRT \ln \frac{V_f}{V_i}$$

$$= (1 \text{ mol})(8.31 \text{ J/mol·K})(310 \text{ K}) \ln \frac{12 \text{ L}}{19 \text{ L}}$$

$$= -1180 \text{ J}. \qquad \text{(Answer)}$$

This result is equal in magnitude but opposite in sign to the result found in (a) above for an isothermal expansion. The minus sign tells us that an external agent must have done +1180 J of work *on* the gas to compress it.

Here is our first kinetic theory problem: Let n moles (or nN_A molecules; see Eq. 2) of an ideal gas be confined in a cubical box of volume V, as in Fig. 3. The walls of the box are held at temperature T. What is the connection between the pressure p exerted by the gas on the walls and the speeds of the molecules?

The molecules in the box are moving in all directions and with varying speeds, bumping into each other and bouncing from the walls of the box like balls in a racquet court. We ignore (for the time being) collisions of the molecules with each other and consider only collisions with the walls.

Figure 3 shows a typical molecule, of velocity $\mathbf{v}$. We can resolve this velocity into components v_x, v_y, and v_z, parallel to the edges of the box. The molecule will rebound from the shaded face with the x component of its velocity reversed in sign, the other two components being unchanged. The change in the particle's momentum will then be

$$(-mv_x) - (mv_x) = -2mv_x.$$

Hence, the momentum Δp delivered to the face by the colliding molecule will be $+2mv_x$.

The molecule of Fig. 3 will hit the shaded face repeatedly, the time Δt between collisions being the travel time to the opposite face and back again, or $2L/v_x$.*

---

* Convince yourself that collisions with other walls of the box than the two under discussion do not affect v_x.

Thus, the total rate at which momentum is delivered to the shaded face by this single molecule is

$$\frac{\Delta p}{\Delta t} = \frac{2mv_x}{2L/v_x} = \frac{mv_x^2}{L}.$$

From Newton's second law ($\mathbf{F} = d\mathbf{p}/dt$), the total rate at which momentum is delivered to the face is the force acting on that face. To find this force, we must add up the contributions of all the other molecules that strike this face, allowing for the possibility that they all have different speeds. Dividing the total force by the area of the face ($= L^2$) gives the pressure on that face, which is the quantity we seek. Thus,

$$p = \frac{F}{L^2} = \frac{mv_{x1}^2/L + mv_{x2}^2/L + mv_{x3}^3/L + \cdots}{L^2}$$

$$= \left(\frac{m}{L^3}\right)(v_{x1}^2 + v_{x2}^2 + v_{x3}^2 + \cdots). \qquad (8)$$

There are nN_A molecules in the box and thus nN_A terms in the second parentheses of Eq. 8. We can thus replace that quantity by $nN_A \overline{v_x^2}$, where $\overline{v_x^2}$ is the average value of the square of the x component of the molecular speed. Equation 8 becomes

$$p = \frac{nmN_A}{L^3} \overline{v_x^2}.$$

But mN_A is the molecular weight M of the gas and L^3 is the volume of the box, so that

$$p = \frac{nM\overline{v_x^2}}{V}. \qquad (9)$$

For any molecule, $v^2 = v_x^2 + v_y^2 + v_z^2$. Because there are many molecules and because they are all moving in random directions, the average values of the squares of their velocity components are equal, so that $\overline{v_x^2} = \frac{1}{3}\overline{v^2}$. Thus, Eq. 9 becomes

$$p = \frac{nM\overline{v^2}}{3V}. \qquad (10)$$

The square root of $\overline{v^2}$ is a kind of average speed, called the *root-mean-square* speed of the molecules and symbolized by v_{rms}. Its name describes it rather well: You take each speed, you *square* it, you find the *mean* (that is, the average) of all these squared speeds, and then you take the square *root*. We can then write Eq. 10 as

$$p = \frac{nMv_{rms}^2}{3V}. \qquad (11)$$

Equation 11 is very much in the spirit of kinetic theory. It tells us how the pressure of the gas (a purely macroscopic quantity) depends on the speed of the molecules (a purely microscopic quantity).

We can turn Eq. 11 around and use it to calculate v_{rms}. Combining Eq. 11 with the ideal gas law ($pV = nRT$) leads to

$$v_{rms} = \sqrt{\frac{3RT}{M}}. \qquad (12)$$

Table 1 shows some rms speeds calculated from Eq. 12. The speeds are surprisingly high. For hydrogen molecules at room temperature (300 K), the rms speed is 1920 m/s or 4300 mi/h—faster than a speeding bullet! Remember, too, the rms speed is only a kind of average speed; many molecules move significantly faster than this and some are much slower.

The rms speed of the molecules of a gas is closely related to the speed of sound in that gas. For example, at room temperature (300 K), the rms speeds in hydrogen and in nitrogen are 1920 m/s and 517 m/s. The speeds of sound in these two gases at this temperature are 1350 m/s and 350 m/s. In a sound wave, the disturbance is passed on from molecule to molecule by means of collisions. The wave cannot move any faster than the "average" speed of the molecules. In fact, the speed of sound must be somewhat less than this "average" molecular speed because not all molecules are moving in exactly the same direction as the wave.

A question often arises: "If molecules move so fast, why does it take as long as a minute or so before you can smell perfume if someone opens a bottle across a room?

Table 1 Some Molecular Speeds at Room Temperature ($T = 300$ K)[a]

Gas	*Molecular Mass*[b] (g/mol)	v_{rms} (m/s)
Hydrogen	2.02	1920
Helium	4.0	1370
Water vapor	18.0	645
Nitrogen	28.0	517
Oxygen	32.0	483
Carbon dioxide	44.0	412
Sulfur dioxide	64.1	342

[a] For convenience, we often set room temperature = 300 K even though (at 27°C or 81°F) it represents a fairly warm room.
[b] Although molecular masses are most often expressed in g/mol, the proper SI unit is the kg/mol.

A glance ahead at Fig. 5 suggests that—although the molecules move very fast between collisions—a given molecule will wander only very slowly away from its release point. In practical situations, the travel time for an "alien" molecule across the air of a room is controlled by the convection currents that are always present in an ordinary room.

Sample Problem 3 Here are five pure numbers 5, 11, 32, 67, and 89. (a) What is the average value of these numbers?

We find this from

$$\bar{n} = \frac{5 + 11 + 32 + 67 + 89}{5} = 40.8. \quad \text{(Answer)}$$

(b) What is the rms value of these numbers?

We find this from

$$n_{rms} = \sqrt{\frac{5^2 + 11^2 + 32^2 + 67^2 + 89^2}{5}} = 52.1. \quad \text{(Answer)}$$

The rms value is greater than the average value because the larger numbers—being squared—are relatively more important in forming the rms value. To test this, let us replace 89 in our set of five numbers by 300. The average value of the new set of five numbers (as you can easily show) increases by a factor of 2.0. The rms value, however, increases by a factor of 2.7.

The rms values of variables occur in many branches of physics and engineering. The value 120 volts printed on an electric light bulb, for example, is an rms voltage.

Sample Problem 4 What is the rms speed of the molecules of hydrogen gas at room temperature ($T = 300$ K)?

The molecular mass of hydrogen is 2.02 g/mol or (in SI units) 0.00202 kg/mol. Substituting into Eq. 12 yields

$$v_{rms} = \sqrt{\frac{3RT}{M}} = \sqrt{\frac{(3)(8.31 \text{ J/mol}\cdot\text{K})(300 \text{ K})}{0.00202 \text{ kg/mol}}}$$

$$= 1920 \text{ m/s}. \quad \text{(Answer)}$$

As we shall see, many molecules have greater speeds than this and (of course) may have smaller speeds.

21-5 Translational Kinetic Energy

Consider a single molecule as it moves around in the box of Fig. 3, changing its speed from time to time as it collides with other molecules. Its translational kinetic energy at any instant is $\frac{1}{2}mv^2$. Its *average* translational

kinetic energy over the time that we watch it is

$$\bar{K} = \overline{\tfrac{1}{2}mv^2} = \tfrac{1}{2}m\overline{v^2} = \tfrac{1}{2}mv_{rms}^2. \quad \text{(13)}$$

Substituting for v_{rms} from Eq. 12 leads to

$$\bar{K} = (\tfrac{1}{2}m)\frac{3RT}{M}.$$

But M/m, the molecular weight divided by the mass of a molecule, is simply the Avogadro constant, so that

$$\bar{K} = \frac{3RT}{2N_A},$$

which we can write as

$$\boxed{\bar{K} = \tfrac{3}{2}kT.} \quad \text{(14)}$$

The constant k, called the *Boltzmann constant,* is the ratio of the gas constant R to the Avogadro constant N_A. It is sometimes called the gas constant for a single molecule (rather than for a mole) and its value is*

$$k = \frac{R}{N_A} = \frac{8.31 \text{ J/mol}\cdot\text{K}}{6.02 \times 10^{23} \text{ mol}^{-1}}$$

$$= 1.38 \times 10^{-23} \text{ J/K} = 8.62 \times 10^{-5} \text{ eV/K}. \quad \text{(15)}$$

Equation 14 tells us what is perhaps an unexpected thing:

At a given temperature, all gas molecules—no matter what their mass—have the same average translational kinetic energy, namely, $\tfrac{3}{2}kT$. When we measure the temperature of a gas, we are measuring the average translational kinetic energy of its molecules.

Actually, nothing in our derivation has specified that we are dealing with molecules. They could be any objects, such as tiny pollen grains or even tennis balls! We see here the basis for the Brownian motion. A pollen grain, suspended in water and in thermal equilibrium with it, behaves like a very large molecule and has the same translational kinetic energy as do the water molecules that surround it. Because of its very much larger mass, of course, the pollen grain has a correspondingly smaller rms speed, small enough to make its motion observable.

---

* Do not confuse k (the Boltzmann constant) with K (kinetic energy) and K (the Kelvin temperature).

Sample Problem 5 What is the average translational kinetic energy of the oxygen molecules in the air at room temperature (= 300 K)? Of the nitrogen molecules?

The average translational kinetic energy depends only on the temperature and not on the nature of the molecule. It is given from Eq. 14 as

$$\overline{K} = \tfrac{3}{2}kT = (\tfrac{3}{2})(8.62 \times 10^{-5} \text{ eV/K})(300 \text{ K})$$
$$= 0.039 \text{ eV}. \qquad \text{(Answer)}$$

Physicists find it useful to remember that the mean translational kinetic energy of *any* molecule at room temperature is about $\tfrac{1}{25}$ eV, which is essentially the above result.

From Table 1 we see that the rms speed for the molecules of oxygen (for which $M = 32.0$ g/mol) is 483 m/s. That for the molecules of nitrogen ($M = 28.0$ g/mol) is 517 m/s. Thus, the lighter molecule has the larger rms speed, consistent with the fact that the two kinds of molecules have the same average translational kinetic energy.

Sample Problem 6 To enhance the effectiveness of the nuclear fission of uranium it is necessary to separate the (highly fissionable) U-235 isotope from the (less readily) fissionable U-238 isotope. One way of doing this is to form the uranium into a gas (UF_6) and allow it to diffuse repeatedly through a porous barrier. The lighter molecule will diffuse faster, the effectiveness of the barrier being determined by a *separation factor* α, defined as the ratio of the two rms speeds. What is the separation factor for the two kinds of uranium hexafluoride gas molecules?

From Eq. 12 ($v_{\text{rms}} = \sqrt{3RT/M}$) we can write

$$\alpha = \frac{v_{\text{rms},235}}{v_{\text{rms},238}} = \sqrt{\frac{M_{238}}{M_{235}}},$$

in which the M's are the molecular masses of the two gas molecules. We can find these molecular masses by adding six times the atomic mass of fluorine (19.0 g/mol) to the atomic mass of the appropriate uranium atom (235 g/mol or 238 g/mol). Thus,

UF_6 (uranium-238) $M_{238} = 238 + 6 \times 19.0 = 352$ g/mol,
UF_6 (uranium-235) $M_{235} = 235 + 6 \times 19.0 = 349$ g/mol,

so that

$$\alpha = \sqrt{\frac{352 \text{ g/mol}}{349 \text{ g/mol}}} = 1.0043. \qquad \text{(Answer)}$$

As many as 4000 passages through the barrier are required in a practical isotope diffusion plant. Figure 4 shows the huge gaseous diffusion plant at Oak Ridge, covering hundreds of acres, in which this process is carried out.

21-6 Mean Free Path

Figure 5 shows the path of a typical molecule as it moves through a gas, changing both speed and direction abruptly as it collides with other molecules. Between collisions, our typical molecule moves in a straight line

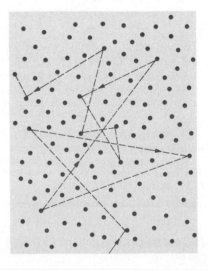

Figure 5 A molecule traveling through a gas, colliding with other molecules in its path. Although the other molecules are shown in stationary positions, they are also moving in a similar fashion.

Figure 4 The TN Gaseous Diffusion Plant at Oak Ridge, used to separate uranium isotopes.

at constant speed. Although the figure shows all the other molecules as stationary, they too are moving in much the same way.

One useful parameter to describe this random motion is the *mean free path* λ. As its name implies, λ is the average distance traversed by a molecule between collisions. We expect λ to vary inversely with N, the number of molecules per unit volume. The larger N, the more collisions there should be and the smaller the mean free path. We also expect λ to vary inversely with the size of the molecules. If the molecules were true points, they would never collide and the mean free path would be infinite. Thus, we expect that the larger the molecules, the smaller the mean free path. We can even predict that λ should vary (inversely) as the *square* of the diameter because it is the cross section of a molecule—not its diameter—that determines its effective target area.

As we shall show, the expression for the mean free path is

$$\lambda = \frac{1}{\sqrt{2}\pi N d^2} \qquad \text{(mean free path)}. \qquad (16)$$

Proof of Eq. 16. We focus attention on a single molecule and assume—as Fig. 5 suggests—that our molecule is traveling with a constant speed v and that all the other molecules are at rest. Later, we shall relax this assumption.

We assume further that the molecules are spheres of diameter d. A collision will then take place if the centers of the molecules come within a distance d of each other, as in Fig. 6a. Another way to look at it is to consider our single molecule to have a *radius* of d and all the other molecules to be *points*, as in Fig. 6b. This does not change our criterion for a collision.

As our single molecule moves through the gas, it sweeps out a jaggedly broken cylinder whose cross-sectional area is πd^2. If we watch this molecule for a time interval Δt, it moves a distance $v\,\Delta t$, where v is its assumed speed. The stretched-out length of our broken cylinder is then $v\,\Delta t$ and the volume of the cylinder is $(\pi d^2)(v\,\Delta t)$. The number of collisions that occur is then equal to the number of (point) molecules that lie within this cylinder; see Fig. 7.

If N is the number of molecules per unit volume, the number we seek is N times the volume of the cylinder, or $(N)(\pi d^2 v\,\Delta t)$. The mean free path is the length of the cylinder divided by this number, or

$$\lambda = \frac{\text{length of path}}{\text{number of collisions}} \approx \frac{v\,\Delta t}{N\pi d^2 v\,\Delta t} = \frac{1}{\pi N d^2}. \qquad (17)$$

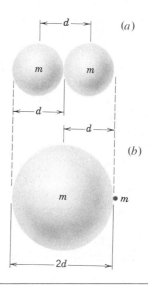

Figure 6 (*a*) A collision occurs when the centers of two molecules come within a distance d of each other, d being the molecular diameter. (*b*) An equivalent but more convenient representation is to think of the moving molecule as having a *radius* d, all other molecules being points.

This equation is approximate because it is based on the assumption that all the molecules except one are at rest. In fact, *all* the molecules are moving; when this is taken properly into account, Eq. 16 results. Note that it differs from the (approximate) Eq. 17 only by a factor of $1/\sqrt{2}$.

We can even get a glimpse of what is "approximate" about Eq. 17. The v in the numerator and that in the denominator are—strictly—not the same. The v in the numerator is $\bar{v}$, the mean speed of the molecule *relative to the container*. The v in the denominator is $\overline{v_{\text{rel}}}$, the mean speed of our single molecule *relative to the other*

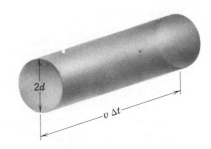

Figure 7 In time Δt the moving molecule sweeps out a broken cylinder of length $v\,\Delta t$ and radius d. The cylinder is shown straightened out for convenience. The number of collisions is equal to the number of molecules whose centers lie within the cylinder.

molecules, which are moving. It is this latter average speed that determines the number of collisions. A detailed calculation, taking into account the actual speed distribution of the molecules, gives $\overline{v_{rel}} = \sqrt{2}\,\overline{v}$, which is where the factor $\sqrt{2}$ originates.

The mean free path of air molecules at sea level is about 0.1 μm. At an altitude of 100 km, the density of air has dropped to such an extent that the mean free path rises to about 16 cm. At 300 km, the mean free path is about 20 km. A problem faced by those who would study the physics and the chemistry of the upper atmosphere in the laboratory is the fact that no available containers are large enough to permit the contained gas samples to simulate upper atmospheric conditions. Studies of the concentrations of Freon, of carbon dioxide, and of ozone in the upper atmosphere are of vital public concern.

Sample Problem 7 The molecular diameters of different kinds of gas molecules can be found experimentally by measuring the rates at which different gases diffuse into each other. For oxygen, $d = 2.9 \times 10^{-10}$ m has been reported. (a) What is the mean free path for oxygen at room temperature ($T = 300$ K) and at atmospheric pressure?

Let us first find N, the number of molecules per unit volume under these conditions. From the ideal gas law, 1 mol of any gas occupies a volume equal to

$$V = \frac{nRT}{p} = \frac{(1\text{ mol})(8.31\text{ J/mol}\cdot\text{K})(300\text{ K})}{(1\text{ atm})(1.01 \times 10^5\text{ Pa/atm})}$$

$$= 2.47 \times 10^{-2}\text{ m}^3.$$

The number of molecules per unit volume is then

$$N = \frac{nN_A}{V} = \frac{(1\text{ mol})(6.02 \times 10^{23}\text{ molecules/mol})}{2.47 \times 10^{-2}\text{ m}^3}$$

$$= 2.44 \times 10^{25}\text{ molecules/m}^3.$$

Equation 16 then gives

$$\lambda = \frac{1}{\sqrt{2}\pi Nd^2} = \frac{1}{(\sqrt{2}\pi)(2.44 \times 10^{25}\text{ m}^{-3})(2.9 \times 10^{-10}\text{ m})^2}$$

$$= 1.1 \times 10^{-7}\text{ m}. \hspace{2cm} \text{(Answer)}$$

This is about 380 molecular diameters. On average, the molecules in such a gas are about 11 molecular diameters apart.

(b) If the average speed of an oxygen molecule is taken to be 450 m/s, what is the average collision rate for a typical molecule?

We find this rate by dividing the average speed by the mean free path, or

$$\text{rate} = \frac{v}{\lambda} = \frac{450\text{ m/s}}{1.1 \times 10^{-7}\text{ m}} = 4.1 \times 10^9\text{ s}^{-1}. \hspace{0.5cm} \text{(Answer)}$$

Thus—on average—every oxygen molecule makes more than 4 billion collisions per second!

21–7 The Distribution of Molecular Speeds (Optional)

The root-mean-square speed v_{rms} gives us a general idea of the molecular speed at a given temperature. We often want to know more. For example, what fraction of the molecules have speeds greater than twice the rms speed? To answer such questions, we need to know the detailed speed distribution among the molecules. Figure 8a shows this distribution for oxygen molecules at room temperature. The root-mean-square speed v_{rms} (= 483 m/s) is marked.

Two other measures of the speed distribution are also shown in Fig. 8a. The *most probable speed* v_P (= 395 m/s) marks the maximum of the speed distribution. As its name indicates, it is the speed most likely to occur. The *average speed* v (= 445 m/s) is—as its name suggests—a simple average of the molecular speeds. A small number of molecules, lying in the extreme tail of the distribution curve, can have speeds much higher than the average speed. This simple fact, as we shall demonstrate, makes possible both rain and sunshine.

Rain. The speed distribution of water molecules at room temperature can be represented by a curve similar to that of Fig. 8a. A water molecule of average speed does not have nearly enough kinetic energy to escape from the water through its surface. Small numbers of very fast molecules far out in the tail can do so, however. Thus, water evaporates, making clouds and rain a possibility.

As the fast water molecules leave the surface, carrying energy with them, the temperature of the remaining water is maintained by heat flow from the surroundings. Other fast molecules—produced in particularly favorable collisions—quickly take the place of those that have left and the speed distribution is maintained.

Sunshine. Let the distribution curve of Fig. 8a refer to protons in the core of the sun. The sun's energy is supplied by a nuclear fusion process that starts with the merging of two protons. However, the protons repel each other because of their electrical charges so that protons of average speed do not have nearly enough kinetic energy to initiate this reaction. Very fast protons in the tail of the distribution curve can do so, however, and thus the sun can shine.

Maxwell's Distribution Law. In 1852, the Scottish

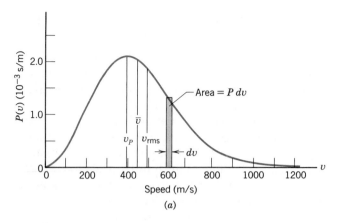

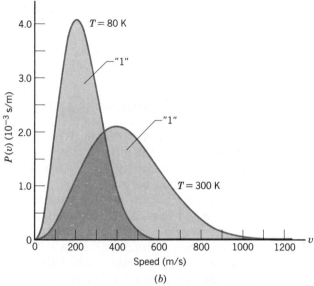

Figure 8   (a) The Maxwell speed distribution curve for oxygen molecules at $T = 300$ K. The three characteristic speeds are marked. (b) The curve for $T = 300$ K is compared with the curve for 80 K. Note that the molecules move slower at the lower temperature. The areas under each curve have a numerical value of unity.

physicist James Clerk Maxwell first solved the problem of finding the speed distribution of gas molecules. This *Maxwell speed distribution* law is

$$P(v) = 4\pi \left(\frac{M}{2\pi RT}\right)^{3/2} v^2 e^{-Mv^2/2RT}. \qquad (18)$$

Here v is the molecular speed, T is the gas temperature, M is the molecular weight of the gas, and R is the gas constant. $P(v)$ in Eq. 18 is a *distribution function*, defined

as follows:

The product P(v) dv (which is a dimensionless quantity) is the fraction of molecules whose speeds lie in the range v to v + dv.

As Fig. 8a shows, this fraction is equal to the area of a strip whose height is $P(v)$ and whose width is dv. The total area under the distribution curve (see Fig. 8b) corresponds to the fraction of the molecules whose speeds lie between zero and infinity. All molecules fall into this category so that the value of this total area is unity.

Sample Problem 8   A container is filled with oxygen gas maintained at 300 K. What fraction of the molecules have speeds in the range 599–601 m/s? The molecular mass M of oxygen is 0.032 kg/mol.

This speed interval Δv (= 2 m/s) is so small that we can treat it as a differential and say that the fraction f that we seek is given very closely by $P(v)\Delta v$, where $P(v)$ is to be evaluated at $v = 600$ m/s, the midpoint of the range; see the shaded strip in Fig. 8a. Thus, using Eq. 18, we find

$$f = P(v)\Delta v = 4\pi \left(\frac{M}{2\pi RT}\right)^{3/2} v^2 e^{-Mv^2/2RT} \Delta v.$$

For convenience in calculating, let us break this down into five factors, as

$$f = (4\pi)(A)(v^2)(e^B)(\Delta v) \qquad (19)$$

in which A and B are

$$A = \left(\frac{M}{2\pi RT}\right)^{3/2} = \left(\frac{0.032 \text{ kg/mol}}{(2\pi)(8.31 \text{ J/mol·K})(300 \text{ K})}\right)^{3/2}$$
$$= 2.92 \times 10^{-9} \text{ s}^3/\text{m}^3$$

and

$$B = -\frac{Mv^2}{2RT} = -\frac{(0.032 \text{ kg/mol})(600 \text{ m/s})^2}{(2)(8.31 \text{ J/mol·K})(300 \text{ K})} = -2.31.$$

Substituting A and B into Eq. 19 yields

$$f = (4\pi)(A)(v^2)(e^B)(\Delta v)$$
$$= (4\pi)(2.92 \times 10^{-9} \text{ s}^3/\text{m}^3)(600 \text{ m/s})^2(e^{-2.31})(2 \text{ m/s})$$
$$= 2.62 \times 10^{-3} \text{ or } 0.262\%. \qquad \text{(Answer)}$$

Thus, at room temperature, 0.217% of the oxygen molecules will have speeds that lie in the narrow range between 599 and 601 m/s. If the shaded strip of Fig. 8a were drawn to the scale of this problem, it would be a very thin strip indeed.

Sample Problem 9   (a) What is the average speed v of the oxygen molecules at $T = 300$ K? The molecular mass M of oxygen is 0.032 kg/mol.

To find the average speed, we weight each speed v by $P(v)dv$, which is the fraction of the molecules whose speeds lie in the interval v to $v + dv$. We then add up (that is, integrate) these fractions over the entire range of speeds. Thus,

$$\bar{v} = \int_0^\infty vP(v)dv. \qquad (20)$$

The next step is to substitute for $P(v)$ from Eq. 18 and evaluate the integral that results. From a table of integrals* we find

$$\bar{v} = \sqrt{\frac{8RT}{\pi M}} \quad \text{(average speed).} \qquad (21)$$

Substituting numerical values yields

$$\bar{v} = \sqrt{\frac{(8)(8.31 \text{ J/mol} \cdot \text{K})(300 \text{ K})}{(\pi)(0.032 \text{ kg/mol})}} = 445 \text{ m/s.} \quad \text{(Answer)}$$

(b) What is the root-mean-square speed v_{rms} of the oxygen molecules in the container?

We proceed as in (a) above except that we multiply v^2 (rather than simply v) by the weighting factor $P(v)dv$. This leads, after another integration, to

$$\overline{v^2} = \int_0^\infty v^2 P(v)dv = \frac{3RT}{M}.$$

The rms speed is the square root of this quantity, or

$$v_{\text{rms}} = \sqrt{\overline{v^2}} = \sqrt{\frac{3RT}{M}} \quad \text{(rms speed).} \qquad (22)$$

Equation 22 is identical to Eq. 12, which we derived earlier. The numerical calculation gives

$$v_{\text{rms}} = \sqrt{\frac{(3)(8.31 \text{ J/mol} \cdot \text{K})(300 \text{ K})}{(0.032 \text{ kg/mol})}} = 483 \text{ m/s.} \quad \text{(Answer)}$$

(c) What is the most probable speed v_P?

The most probable speed is the speed at which $P(v)$ of Eq. 18 has its maximum value. We find it by requiring that $dP/dv = 0$ and solving for v. Doing so yields (as you should show)

$$v_P = \sqrt{\frac{2RT}{M}} \quad \text{(most probable speed).} \qquad (23)$$

Numerically, this yields

$$v_P = \sqrt{\frac{(2)(8.31 \text{ J/mol} \cdot \text{K})(300 \text{ K})}{(0.032 \text{ kg/mol})}} = 395 \text{ m/s.} \quad \text{(Answer)}$$

Table 2 summarizes the three measures of the Maxwell speed distribution.

---

* One such table is to be found in Section A of the familiar *CRC Handbook of Chemistry and Physics;* see integral 667 listed under "Definite Integrals."

21–8 The Heat Capacities of an Ideal Gas

In this section, we want to see how the internal energy U of an ideal gas is related to the kinetic energy of its molecules. We will then use that result to derive an expression for the molar heat capacity of an ideal gas.

The Internal Energy U. Let us first assume that our ideal gas is a monatomic gas, such as helium, neon, or argon. We next assume that the internal energy U of such a gas is simply the sum of the translational kinetic energies of its molecules. The average translational kinetic energy of a single molecule depends only on the gas temperature and is given (see Eq. 14) by $\bar{K} = \frac{3}{2}kT$. A sample of n moles of such a gas contains nN_A molecules. The internal energy U is then

$$U = (nN_A)\bar{K} = (nN_A)(\tfrac{3}{2}kT)$$

or

$$\boxed{U = \tfrac{3}{2}nRT} \quad \text{(monatomic ideal gas).} \qquad (24)$$

Here (see Eq. 15) we have replaced $N_A k$ by its equal, the gas constant R. We see that the internal energy U is a function of the gas temperature only, being independent of other variables such as the pressure or the density.

With Eq. 24 in hand, we are now able to derive an expression for the molar heat capacity of an ideal gas. Actually, we derive two expressions, one for the case in which the volume of the gas remains constant as heat is added to it and the other for the case in which the pressure of the gas remains constant during this process.

The Molar Heat Capacity at Constant Volume. Figure 9a shows n moles of an ideal gas at pressure p and temperature T, confined to a cylinder of fixed volume V. This *initial state* of the gas is marked on the pV curve of Fig. 9b. Suppose now that you add a small amount of heat Q to the gas, by slowly turning up the temperature of the reservoir on which the cylinder rests. The gas temper-

Table 2 Some Parameters for the Maxwell Speed Distribution

Parameter	Symbol	Formula	For Oxygen at 300 K
Most probable speed	v_P	$\sqrt{2RT/M}$	395 m/s
Average speed	$\bar{v}$	$\sqrt{8RT/\pi M}$	445 m/s
Root-mean-square speed	v_{rms}	$\sqrt{3RT/M}$	483 m/s

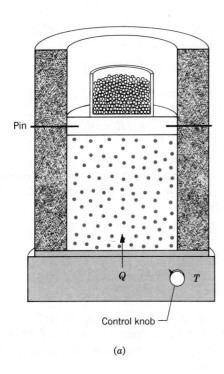

Pin

Q

T

Control knob

(a)

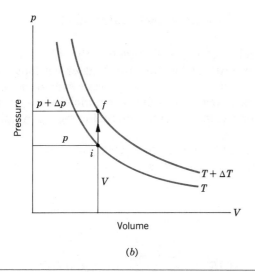

p

$p + \Delta p$

f

p

i

V

$T + \Delta T$

T

Pressure

Volume

V

(b)

Figure 9 (a) The temperature of an ideal gas is raised from T to $T + \Delta T$ in a constant volume process. Heat is added but no work is done. (b) The process is shown on a pV diagram.

ature rises a small amount to $T + \Delta T$ and its pressure to $p + \Delta p$. This *final state* is also marked on Fig. 9b.

The defining equation for C_v, the molar heat capacity at constant volume, is

$$Q = nC_V \, \Delta T \quad \text{(constant volume).} \quad (25)$$

Substituting this expression Q into the first law of thermodynamics, we find

$$\Delta U + p \, \Delta V = nC_V \, \Delta T.$$

Putting $\Delta V = 0$ and solving for C_V, we obtain

$$C_V = \frac{1}{n} \frac{\Delta U}{\Delta T}. \quad (26)$$

From Eq. 24 we see that $\Delta U / \Delta T = \frac{3}{2} nR$. Substituting this result into Eq. 26 yields

$$\boxed{C_V = \tfrac{3}{2}R = 12.5 \text{ J/mol} \cdot \text{K}}$$

$$\text{(monatomic gas).} \quad (27)$$

As Table 3 shows, this prediction of kinetic theory agrees very well with experiment for real monatomic gases, the case that we have assumed. The experimental values of C_V for diatomic and polyatomic gases are substantially higher than those for monatomic gases.

The Molar Heat Capacity at Constant Pressure. We now assume that the temperature of the gas is increased by the same small amount ΔT but that the necessary heat Q is added with the gas under constant pressure. A mechanism for doing this is shown in Fig. 10a, the process being plotted in Fig. 10b. We can guess at once that the molar heat capacity at constant pressure, defined from

$$Q = nC_p \, \Delta T \quad \text{(constant pressure),} \quad (28)$$

will be greater than the molar heat capacity at constant volume because not only must energy be supplied to raise the temperature but it must also be supplied to do external work, that is, to lift the weighted piston of Fig. 10a.

From the first law of thermodynamics we can write Eq. 28 as

$$\Delta U + p \, \Delta V = nC_p \, \Delta T. \quad (29)$$

Table 3   Molar Heat Capacities

Molecule	Example	C_V (J/mol·K)
	Ideal	12.5
Monatomic	He	12.5
	Ar	12.6
	Ideal	20.8
Diatomic	N$_2$	20.7
	O$_2$	20.8
	Ideal	24.9
Polyatomic	NH$_4$	29.0
	CO$_2$	29.7

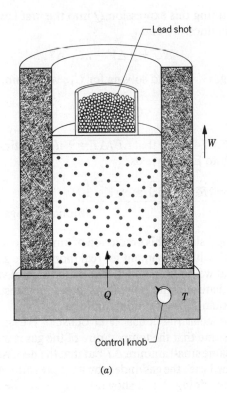

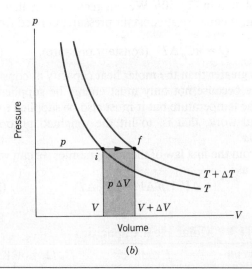

Figure 10 (*a*) The temperature of an ideal gas is raised from T to $T + \Delta T$ by a constant pressure process. Heat is added and work is done in lifting the loaded piston. (*b*) The process is shown on a pV diagram. The work, $p\,\Delta V$, is the shaded area under the line connecting the initial and the final states.

From the ideal gas law ($pV = nRT$) we can write, for constant pressure conditions,

$$p\,\Delta V = nR\,\Delta T,$$

so that Eq. 29 becomes

$$\frac{\Delta U}{\Delta T} + nR = nC_p. \tag{30}$$

But U depends only on the temperature T, no matter what restrictions are placed on the volume or the pressure. Thus, we can use Eq. 26 to replace $\Delta U/\Delta T$ in Eq. 30 by nC_V, obtaining

$$nC_V + nR = nC_p,$$

or

$$\boxed{C_p - C_V = R.} \tag{31}$$

This prediction of kinetic theory agrees well with experiment, not only for monatomic gases, but for gases in general, as long as their density is low enough so that we may treat them as ideal.

21–9 The Equipartition of Energy

As Table 3 shows, the prediction that $C_V = \frac{3}{2}R$ agrees with experiment for monatomic gases but fails for diatomic and polyatomic gases. Let us extend the theory by considering the possibility that molecules with more than one atom can store internal energy in rotational form.

Figure 11 shows kinetic theory models of helium (a monatomic gas), oxygen (diatomic), and ammonia

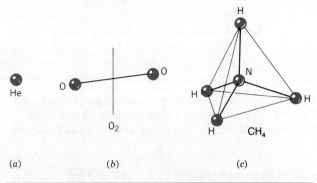

Figure 11 Models of molecules as used in kinetic theory: (*a*) helium, a typical monatomic molecule; (*b*) oxygen, a typical diatomic molecule; and (*c*) methane, a typical polyatomic molecule.

Table 4 Degrees of Freedom for Various Molecules

		Degrees of Freedom		
Molecule	Example	Translational	Rotational	Total (f)
Monatomic	He	3	0	3
Diatomic	O_2	3	2	5
Polyatomic	CH_4	3	3	6

(polyatomic). On the basis of these simplified models, we can believe that monatomic molecules—which are essentially pointlike and have only a very small rotational inertia about any axis—can store energy only in their translational motion. Diatomic and polyatomic molecules, however, should be able to store substantial additional amounts of energy by rotating or by vibrating.

How do we take these possibilities quantitatively into account? James Clerk Maxwell first showed how to do so, by introducing the theorem of the *equipartition of energy:*

Every kind of molecule has a certain number f of degrees of freedom, *which are independent ways in which it can store energy. Each such degree of freedom has associated with it—on average—an energy of $\frac{1}{2}kT$ per molecule (or $\frac{1}{2}RT$ per mole).*

For translational motion, there are three degrees of freedom, corresponding to the three independent axes along which such motion can occur. For rotational motion, a monatomic molecule has no degrees of freedom. A diatomic molecule—see the rigid dumbbell of Fig. 11*b*—has two rotational degrees of freedom, corresponding to the two independent axes about which it can store rotational energy. Such a molecule cannot store rotational energy about the axis connecting the nuclei of its two constituent atoms because its rotational inertia about this axis is essentially zero. A molecule with more than two atoms has six degrees of freedom, three rotational and three translational.

To extend the treatment of Section 21-8 to diatomic and polyatomic gases, it is necessary to retrace the derivations of that section in detail, replacing Eq. 24 ($U = \frac{3}{2}nRT$) by $U = (f/2)nRT$, where f is the number of degrees of freedom listed in the last column of Table 4. Doing so leads to the prediction

$$C_V = (f/2)R = 4.16f \text{ J/mol} \cdot \text{K}, \qquad (32)$$

which agrees—as it must—with Eq. 27 for $f = 3$. As

Table 3 shows, this prediction agrees with experiment for monatomic ($f = 3$) and for diatomic ($f = 5$) but not for polyatomic molecules.

21-10 A Hint of Quantum Theory: An Aside

The next logical step is to see if we cannot improve the agreement of kinetic theory with experiment by taking into account the internal energy stored in the form of molecular vibrations. However, it proves unrewarding to push forward any further with classical kinetic theory, which is based on Newtonian mechanics.

In the middle of the last century, Maxwell (and others) were struggling to build a coherent picture of the structure of an atom, based on kinetic theory and on measurements of the wavelengths of light emitted by atoms. It was to no avail and Maxwell was led to speculate that classical theory—in this context—was in some way flawed.

Maxwell,* baffled in his attempts to understand the atom, wrote that nothing remained but to adopt an attitude of the "thoroughly conscious ignorance that is the prelude to every real advance in knowledge." It was only in the first quarter of the present century that—with the development of modern quantum physics—these real advances in knowledge came to pass.

Figure 12 suggests how the phenomenon of energy quantization (which is central to quantum theory) enters the picture. The ratio C_V/R for hydrogen gas is plotted against temperature, the temperature scale being logarithmic for convenience. At temperatures below about 80 K, this ratio has the value 1.5, characteristic of a monatomic gas. Even though hydrogen is a diatomic

---

* See the biography of Maxwell (by C. W. F. Everitt) in *The Dictionary of Scientific Biography.*

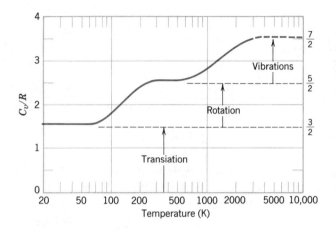

Figure 12 A plot of the ratio C_V/R versus temperature for hydrogen gas. Because the rotational and vibrational motions occur at quantized energies, only translational modes are possible at low temperatures. As the temperature increases, rotational motion can be excited during collisions. At still higher temperatures, vibrational motion can be excited.

molecule, only its translational degrees of freedom seem to be excited. The reason is that rotational motion occurs at quantized energies and, at these low temperatures, the molecules simply do not have enough kinetic energy to set each other rotating when they collide. As the temperature rises, however, rotation becomes possible so that, at "ordinary" temperatures, the hydrogen molecule behaves like the rigid dumbbell that we have assumed in our classical theory. At still higher temperatures, vibrational motions (which also occur only at quantized energies) become possible and the molecule dissociates into two atoms at about 3200 K. Quantum physics accounts in detail for this full range of phenomena. In retrospect, it is not incorrect to say that the seeds of quantum physics lay in the kinetic theory of gases.*

21–11 The Adiabatic Expansion of an Ideal Gas

We learned in Section 18–2 that sound waves are propagated through air and other gases as a series of compressions and rarefactions that take place so rapidly that there is no time for heat to flow from one part of the

---

* See "On Teaching Quantum Phenomena," by Sir N. F. Mott, *Contemporary Physics,* August 1964.

medium to another. Volume changes for which $Q = 0$ are called *adiabatic processes.* We can ensure that $Q = 0$ either by carrying out the volume change very quickly (as in sound waves) or by doing it slowing in a heavily insulated environment. Let us see what kinetic theory has to say about adiabatic processes.

Figure 13a shows an insulated cylinder containing an ideal gas and resting on an insulating stand. By removing weight from the piston, we can allow the gas to expand adiabatically. As the volume increases, both the pressure and the temperature drop. We shall prove below that the relation between the pressure and the volume during an adiabatic process is

$$\boxed{pV^\gamma = \text{a constant}} \quad \text{(adiabatic process)}, \quad (33)$$

in which $\gamma(= C_p/C_V)$ is the ratio of molar heat capacities for the gas. Figure 13b shows a single adiabatic line (or *adiabat*) for the ideal gas sample, cutting across three isotherms.

By using the ideal gas law ($pV = nRT$) to eliminate p from Eq. 33, we can write that equation as

$$\left(\frac{nRT}{V}\right) V^\gamma = \text{a constant}.$$

Because n and R are constants, we can rewrite this, as an alternative expression that holds for adiabatic processes,

$$\boxed{TV^{\gamma-1} = \text{a constant}} \quad \text{(adiabatic process)}. \quad (34)$$

The Heat Capacity Ratio γ. The ratio γ can be measured directly in various ways, without the need to measure separately the two molar heat capacities C_p and C_V. A theoretical expression can also be derived for it from kinetic theory. Thus, by combining Eqs. 32 ($C_V = \frac{1}{2}fR$) and 31 ($C_p - C_V = R$), we find

$$\gamma = \frac{C_p}{C_V} = \frac{R + C_V}{C_V} = \frac{R + \frac{1}{2}fR}{\frac{1}{2}fR} = 1 + \frac{2}{f}. \quad (35)$$

This result agrees well with experiment for monatomic gases (for which $f = 3$ and thus $\gamma = 1.67$) and for diatomic gases (for which $f = 5$ and $\gamma = 1.40$).

Proof of Eq. 33 (Optional). Let the volume of n moles of an ideal gas increase adiabatically (see Fig. 13) by a small amount ΔV. In the first law of thermodynamics,

$$Q = \Delta U + p\,\Delta V, \quad (36)$$

we can put $Q = 0$. Because the gas is ideal, we can also

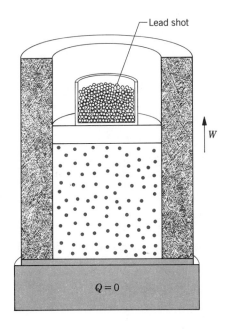

Lead shot

W

$Q = 0$

(a)

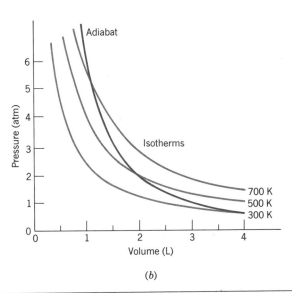

Adiabat

Isotherms

700 K
500 K
300 K

Volume (L)

(b)

Figure 13   (a) The volume of an ideal gas is decreased by adding weight to the piston. The process is adiabatic ($Q = 0$). (b) The process is plotted on a pV diagram.

put (see Eq. 26) $\Delta U = nC_V \Delta T$. With these substitutions, Eq. 36 reduces to

$$n \Delta T = -\left(\frac{p}{C_V}\right)\Delta V. \qquad (37)$$

From the ideal gas law ($pV = nRT$) we have

$$p \Delta V + V \Delta p = nR \Delta T.$$

Replacing R by its equal, $C_p - C_V$ leads to

$$n \Delta T = \frac{p \Delta V + V \Delta p}{C_p - C_V} \qquad (38)$$

Equating Eqs. 37 and 38 and rearranging leads to

$$\frac{\Delta p}{p} + \left(\frac{C_p}{C_V}\right)\frac{\Delta V}{V} = 0.$$

Replacing the molar heat capacity ratio by γ and passing to the differential limit yields

$$\frac{dp}{p} + \gamma \frac{dV}{V} = 0.$$

Integrating (see Appendix G) yields

$$\ln p + \gamma \ln V = \text{a constant},$$

or

$$pV^\gamma = \text{a constant}, \qquad (39)$$

which (compare Eq. 33) is what we set out to prove.

Sample Problem 10   In Sample Problem 2, 1 mol of oxygen (assume an ideal gas) expanded isothermally (at 310 K) from an initial volume of 12 L to a final volume of 19 L. What would be the final temperature if the gas had expanded adiabatically to this same final volume? Oxygen (O_2) is diatomic, so that $\gamma = 1.40$.

From Eq. 33 we can write

$$p_i V_i^\gamma = p_f V_f^\gamma.$$

From the ideal gas law we can write

$$\frac{p_i V_i}{T_i} = \frac{p_f V_f}{T_f}.$$

Dividing these two equations and solving for T_f yields

$$T_f = \frac{T_i V_i^{\gamma-1}}{V_f^{\gamma-1}}$$

$$= \frac{(310 \text{ K})(12 \text{ L})^{1.40-1}}{(19 \text{ L})^{1.40-1}} = 258 \text{ K}. \qquad \text{(Answer)}$$

The fact that the gas has cooled (from 310 to 258 K) means that its internal energy has been correspondingly reduced. This energy has gone to do the work of expansion, lifting the weighted piston in Fig. 13.

REVIEW AND SUMMARY

The Kinetic Theory of Gases

The *kinetic theory of gases* relates the *macroscopic* properties of gases (for example, pressure and temperature) to the *microscopic* properties of the gas molecules (for example, speeds and kinetic energies).

The Avogadro Constant

One mole contains N_A (the *Avogadro constant*) entities; experimentally

$$N_A = 6.02 \times 10^{23} \text{ mol}^{-1} \quad \text{(the Avogadro constant)}. \quad [1]$$

One gram-molecular mass M contains one mole of molecules.

An *ideal gas* is one for which the pressure p, volume V, and temperature T are related by

An Ideal Gas

$$pV = nRT \quad \text{(ideal gas law)}. \quad [4]$$

Here n is the number of moles of the gas and R ($= 8.31$ J/mol·K) is the *gas constant*. See Sample Problem 1.

Work in Isothermal Expansion

The work done *by* an ideal gas during an *isothermal* change from volume V_i to volume V_f is

$$W = nRT \ln \frac{V_f}{V_i} \quad \text{(ideal gas, isothermal)}. \quad [6]$$

See Sample Problem 2.

Pressure and Molecular Speed

The pressure exerted by n moles of ideal gas is

$$p = \frac{nMv_{\text{rms}}^2}{3V}. \quad [11]$$

Temperature and Molecular Speed

Here $v_{\text{rms}} = \sqrt{\overline{v^2}}$ is the *root-mean-square speed* of the molecules. With Eq. 4 this gives

$$v_{\text{rms}} = \sqrt{\frac{3RT}{M}}. \quad [12]$$

See Table 1 and Sample Problems 3 and 4.

Temperature and Kinetic Energy

$\overline{K}$, the average translational kinetic energy per molecule, is

$$\overline{K} = \tfrac{3}{2}kT. \quad [14]$$

Here k ($= R/N_A = 1.38 \times 10^{-23}$ J/K) is the *Boltzmann constant*. See Sample Problems 5 and 6.

Mean Free Path

The *mean free path* λ of a gas molecule (its average path length between collisions; see Fig. 5) is given by

$$\lambda = \frac{1}{\sqrt{2}\pi N d^2}, \quad [16]$$

were N is the number of molecules per unit volume and d is the molecular diameter; see Sample Problem 7.

The *Maxwell speed distribution* $P(v)$ is a function such that $P(v)\,dv$ gives the *fraction* of molecules with speeds between v and $v + dv$:

The Maxwell Speed Distribution

$$P(v) = 4\pi \left(\frac{M}{2\pi RT} \right)^{3/2} v^2 e^{-Mv^2/2RT}. \quad [18]$$

See Sample Problem 8 and Fig. 8. This figure and Sample Problem 9 illustrate the definitions of the *most probable* speed v_P, the *average speed* $\bar{v}$, and the *root-mean-square speed* v_{rms}; see Table 2.

Molar Heats

If the internal energy U of a gas depends only on its temperature, C_V (the *molar heat capacity at constant volume*) is

C_V and U

$$C_V = \frac{1}{n}\frac{\Delta U}{\Delta T}. \quad [26]$$

The first law of thermodynamics then leads to

$C_p - C_V$

$$C_p - C_V = R, \qquad\qquad [31]$$

in which C_p is the molar heat capacity at *constant pressure.*

The Equipartition Theorem

We find C_V itself by using the *equipartition of energy* theorem, which states that every *degree of freedom* of a molecule (that is, every independent way it can store energy) has associated with it—on average—an energy $\frac{1}{2}kT$ per molecule ($= \frac{1}{2}RT$ per mole). If f is the number of degrees of freedom, $U = (f/2)nRT$ and

C_V

$$C_V = (f/2)R = 4.16f \text{ J/mol}\cdot\text{K}. \qquad\qquad [32]$$

For monatomic gases $f = 3$ (three translational degrees); for diatomic gases $f = 5$ (three translational and two rotational degrees). Table 3 shows that these estimates of C_V agree well with experiment.

C_V and Quantum Physics

The temperature dependence of C_V for real gases provides striking evidence for the quantum nature of rotational motion. At high temperatures, C_V is as predicted by the equipartition principle. At low temperatures, however, molecular collisions are not energetic enough to excite the quantized rotational motion. See Fig. 12.

Adiabatic Expansion

When an ideal gas undergoes an adiabatic volume change, its pressure and volume are related by

$$pV^\gamma = \text{a constant} \quad \text{(adiabatic process)}, \qquad\qquad [33]$$

in which $\gamma \ (= C_p/C_V)$ is the ratio of molar heat capacities for the gas. See Fig. 13 and Sample Problem 10.

QUESTIONS

1. In discussing the fact that it is impossible to apply the laws of mechanics individually to atoms in a macroscopic system, Mayer and Mayer state: "The very complexity of the problem [that is, the fact that the number of atoms is large] is the secret of its solution." Discuss this sentence.

2. Is there any such thing as a truly continuous body of matter?

3. In kinetic theory we assume that there are a large number of molecules in a gas. Real gases behave like an ideal gas at low densities. Are these statements contradictory? If not, what conclusion can you draw from them?

4. We have assumed that the walls of the container are elastic for molecular collisions. Actually, the walls may be inelastic. Why does this make no difference as long as the walls are at the same temperature as the gas?

5. On a humid day, some say that the air is "heavy." How does the density of humid air compare with that of dry air at the same temperature and pressure?

6. Where does the average speed of air molecules in still air at room temperature fit into this sequence: 0; 2 m/s (walking speed); 30 m/s (fast car); 500 m/s (supersonic airplane); 1.1×10^4 m/s (escape speed from earth); 3×10^8 m/s (speed of light)?

7. Two equal-size rooms communicate through an open doorway. However, the average temperatures in the two rooms are maintained at different values. In which room is there more air?

8. Molecular motions are maintained by no outside force, yet continue indefinitely with no sign of diminishing speed. What is the reason that friction does not bring these tiny particles to rest, as it does other moving particles?

9. What justification is there in neglecting the changes in gravitational potential energy of molecules in a gas?

10. We have assumed that the force exerted by molecules on the wall of a container is steady in time. How is this justified?

11. The average velocity of the molecules in a gas must be zero if the gas as a whole and the container are not in translational motion. Explain how it can be that the average *speed* is not zero.

12. Consider a hot, stationary golf ball sitting on a tee and a cold golf ball just moving off the tee after being hit. The total kinetic energy of the molecules' motion relative to the tee can be the same in the two cases. Explain how. What is the difference between the two cases?

13. Justify the fact that the pressure of a gas depends on the *square* of the speed of its particles by explaining the dependence of pressure on the collision frequency and the momentum transfer of the particles.

14. Why does the boiling temperature of a liquid increase with pressure?

15. Pails of hot and cold water are set out in freezing weather. Explain (*a*) if the pails have lids, the cold water will freeze first but (*b*) if the pails do not have lids, it is possible for the hot water to freeze first.

16. How is the speed of sound related to the gas variables in the kinetic theory model?

17. Far above the earth's surface the gas kinetic temperature is reported to be on the order of 1000 K. However, a person placed in such an environment would freeze to death rather than vaporize. Explain.

18. Why doesn't the earth's atmosphere leak away? At the top of the atmosphere molecules will occasionally be headed out with a speed exceeding the escape speed. Isn't it just a matter of time?

19. Titan, one of Saturn's many moons, has an atmosphere, but our own moon does not. What is the explanation?

20. As ice is heated it melts and then it boils. However, as solid carbon dioxide is heated it goes directly to the vapor state—we say it *sublimes*—without passing through a liquid state. How could liquid carbon dioxide be produced?

21. Does the concept of temperature apply to a vacuum? Consider interplanetary space, for example.

22. What direct evidence do we have for the existence of atoms? What indirect evidence is there?

23. How, if at all, would you expect the composition of the air to change with altitude?

24. We often say that we see the steam emerging from the spout of a kettle in which water is boiling. However, steam itself is a colorless gas. What is it that you really see?

25. Why does smoke rise, rather than fall, from a lighted candle?

26. Would a gas whose molecules were true geometric points obey the ideal gas law?

27. If you fill a saucer with water at room temperature the water, under normal conditions, will evaporate completely. It is easy to believe that some of the more energetic molecules can escape from the water surface but how can *all* of them eventually escape? Many of them—in fact the vast majority—do not have enough energy to do so.

28. Consider a situation in which the mean free path is greater than the longest straight line in a vessel. Is this a perfect vacuum for a molecule in this vessel?

29. When a can of mixed nuts is shaken, why does the largest nut generally end up on the surface, even if it is denser than the others?

30. Give a qualitative explanation of the connection between the mean free path of ammonia molecules in air and the time it takes to smell the ammonia when a bottle is opened across the room.

31. List effective ways of increasing the number of molecular collisions per unit time in a gas.

32. If molecules are not spherical, what meaning can we give to d in Eq. 16 for the mean free path? In which gases would the molecules act the most nearly as rigid spheres?

33. Suppose that we dispense with the hypothesis of elastic collisions in kinetic theory and consider the molecules as centers of force acting at a distance. What does the concept of mean free path mean under these circumstances?

34. The two opposite walls of a container of gas are kept at different temperatures. The air between the panes of glass in a storm window is a familiar example. Describe in terms of kinetic theory the mechanism of heat conduction through the gas.

35. A gas can transmit only those sound waves whose wavelength is long compared with the mean free path. Can you explain this? Describe a situation for which this limitation would be important.

36. Justify qualitatively the statement that, in a mixture of molecules of different kinds in complete equilibrium, each kind of molecule has the same Maxwellian distribution in speed that it would have if the other kinds were not present.

37. Is it possible for a gas to consist of molecules that all have the same speed?

38. What observation is good evidence that not all molecules of a body are moving with the same speed at a given temperature?

39. The fraction of molecules within a given range Δv of the rms speed decreases as the temperature of a gas rises. Explain.

40. (*a*) Do half the molecules in a gas in thermal equilibrium have speeds greater than v_P? Than $\bar v$? Than v_{rms}? (*b*) Which speed, v_P, $\bar v$, or v_{rms}, corresponds to a molecule having average kinetic energy?

41. Keeping in mind that the internal energy of a body consists of kinetic energy and potential energy of its particles, how would you distinguish between the internal energy of a body and its temperature?

42. The gases in two identical containers are at 1 atm of pressure and at room temperature. One contains helium gas (monatomic, atomic mass 4 g/mol) and the other contains an equal number of moles of argon gas (monatomic, atomic mass 40 g/mol). If 1 cal of heat added to the helium gas increases its temperature a given amount, what amount of heat must be added to the argon gas to increase its temperature by the same amount?

43. Explain how we might keep a gas at a constant temperature during a thermodynamic process.

44. Why is it more common to excite radiation from gaseous atoms by use of electrical discharge than by thermal methods?

45. Explain why the specific heat at constant pressure is greater than the specific heat at constant volume.

46. A certain quantity of an ideal gas is compressed to half its initial volume. The process may be adiabatic, isothermal, or isobaric. For which process is the greatest amount of mechanical work required?

47. Explain why the temperature of a gas drops in an adiabatic expansion.

48. If hot air rises, why is it cooler at the top of a mountain than near sea level? (*Hint:* Air is a poor conductor of heat.)

49. Comment on this statement: "There are two ways to carry out an adiabatic process. One is to do it quickly and the other is to do it in an insulated box."

50. A sealed rubber balloon contains a very light gas. The balloon is released and it rises high into the atmosphere. Describe and explain the thermal and mechanical behavior of the balloon.

EXERCISES AND PROBLEMS

Section 21–2 The Avogadro Constant

1E. Gold has an atomic mass of 197 g/mol. Consider a 2.5-g sample of pure gold. (*a*) Calculate the number of moles of gold present. (*b*) How many atoms of gold are in the sample?

2E. Find the mass in kilograms of 7.5×10^{24} atoms of arsenic, which has an atomic mass of 74.9 g/mol.

3P. If the water molecules in 1.0 g of water were distributed uniformly over the surface of the earth, how many such molecules would there be on 1.0 cm² of the earth's surface?

4P. A _____ of water contains about as many molecules as there are _____ s of water in all the oceans. What single word best fits the two blank spaces: drop, teaspoon, tablespoon, cup, quart, barrel, or ton? The oceans cover 75% of the earth's surface and have an average depth of about 5 km. (After Edward M. Purcell.)

5P. A distinguished scientist has written: "There are enough molecules in the ink that makes one letter of this sentence to provide not only one for every inhabitant of the earth, but one for every creature if each star of our galaxy had a planet as populous as the earth." Check up on this statement. Assume the ink sample (molecular weight = 18 g/mol) to have a mass of 1 μg, the population of the earth to be 5×10^9, and the number of stars in our galaxy to be 10^{11}.

Section 21–3 An Ideal Gas

6E. (*a*) What is the volume occupied by 1.0 mol of an ideal gas at standard conditions, that is, pressure of 1.0 atm (= 1.01×10^5 Pa) and temperature of 0°C (= 273 K)? (*b*) Show that the number of molecules per cubic centimeter (the Loschmidt number) at standard conditions is 2.69×10^{19}.

7E. Compute (*a*) the number of moles and (*b*) the number of molecules in a gas contained in a volume of 1.0 cm³ at a pressure of 100 Pa and a temperature of 220 K.

8E. The best vacuum that can be attained in the laboratory corresponds to a pressure of about 10^{-18} atm, or 1.01×10^{-13} Pa. How many molecules are there per cubic centimeter in such a vacuum at 293 K?

9E. A quantity of ideal gas at 10°C and a pressure of 100 kPa occupies a volume of 2.5 m³. (*a*) How many moles of the gas

are present? (*b*) If the pressure is now raised to 300 kPa and the temperature is raised to 30°C, how much volume will the gas now occupy? Assume no leaks.

10E. Oxygen gas having a volume of 1000 cm³ at 40°C and a pressure of 1.01×10^5 Pa expands until its volume is 1500 cm³ and its pressure is 1.06×10^5 Pa. Find (*a*) the number of moles of oxygen in the system and (*b*) its final temperature.

11E. An automobile tire has a volume of 1000 in.³ and contains air at a gauge pressure of 24 lb/in.² when the temperature is 0°C. What is the gauge pressure of the air in the tires when its temperature rises to 27°C and its volume increases to 1020 in.³? (*Hint:* It is not necessary to convert from British units to SI units; why? Use P_{atm} = 14.7 lb/in.².)

12E. Calculate the work done by an external agent in compressing 1.00 mol of oxygen from a volume of 22.4 L at 0°C and 1.00 atm pressure to 16.8 L at the same temperature.

13P. (*a*) What is the number of molecules per cubic meter in air at 20°C and at a pressure of 1.0 atm (= 1.01×10^5 Pa)? (*b*) What is the mass of this 1 m³ of air? Assume that 75% of the molecules are nitrogen (N_2) and 25% oxygen (O_2).

14P. Pressure *p*, volume *V*, and temperature *T* for a certain material are related by

$$p = \frac{AT - BT^2}{V}.$$

Find an expression for the work done by the material if the temperature changes from T_1 to T_2 while the pressure remains constant.

15P. Air that occupies 5.0 ft³ (= 0.14 m³) at 15 lb/in.² (= 1.03×10^5 Pa) gauge pressure is expanded isothermally to atmospheric pressure and then cooled at constant pressure until it reaches its initial volume. Compute the work done by the gas.

16P. Consider a given mass of an ideal gas. Compare curves representing constant pressure, constant volume, and isothermal processes on (*a*) a *p–V* diagram, (*b*) a *p–T* diagram, and (*c*) a *V–T* diagram. (*d*) How do these curves depend on the mass of gas chosen?

17P. A container encloses two ideal gases. Two moles of the first gas are present, with molecular mass M_1. Molecules of the second gas have a molecular mass $M_2 = 3M_1$, and 0.5 mol of this gas is present. What fraction of the total pressure on the container wall is attributable to the second gas? (The kinetic theory explanation of pressure leads to the experimentally discovered law of partial pressures for a mixture of gases that do not react chemically: *The total pressure exerted by the mixture is equal to the sum of the pressures that the several gases would exert separately if each were to occupy the vessel alone.*)

18P. A weather balloon is loosely inflated with helium at a pressure of 1.0 atm (= 76 cm Hg) and a temperature of 20°C. The gas volume is 2.2 m³. At an elevation of 20,000 ft, the atmospheric pressure is down to 38 cm Hg and the helium has expanded, being under no restraint from the confining bag. At this elevation the gas temperature is −48°C. What is the gas volume (in m³) now?

19P. An air bubble of 20 cm³ volume is at the bottom of a lake 40 m deep where the temperature is 4°C. The bubble rises to the surface, which is at a temperature of 20°C. Take the temperature of the bubble to be the same as that of the surrounding water and find its volume just before it reaches the surface.

20P. An open–closed pipe of length $L = 25$ m contains air at atmospheric pressure. It is thrust vertically into a freshwater lake until the water rises halfway up in the pipe, as shown in Fig. 14. What is the depth h of the lower end of the pipe? Assume that the temperature is the same everywhere and does not change.

21P. The envelope and basket of a hot-air balloon have a combined weight of 550 lb, and the envelope has a capacity of 77,000 ft³. When fully inflated, what should be the temperature of the enclosed air to give the balloon a lifting capacity of 600 lb (in addition to its own weight)? Assume that the surrounding air, at 20°C, has a weight density of 7.56×10^{-2} lb/ft³.

22P. A steel tank contains 300 g of ammonia gas (NH_3) at an absolute pressure of 1.35×10^6 Pa and temperature 77°C. (*a*) What is the volume of the tank? (*b*) The tank is checked later when the temperature has dropped to 22°C and the absolute pressure has fallen to 8.7×10^5 Pa. How many grams of gas leaked out of the tank?

23P. Container A contains an ideal gas at a pressure of 5.0×10^5 Pa and at a temperature of 300 K. It is connected by a thin tube to container B with four times the volume of A; see Fig. 15. B contains the same ideal gas at a pressure of 1.0×10^5 Pa and at a temperature of 400 K. The connecting valve X is opened, and equilibrium is achieved at a common pressure while the temperature of each container is kept constant at its initial value. What is the final pressure in the system?

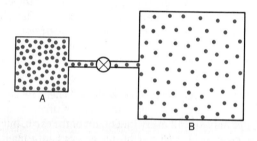

Figure 15 Problem 23.

Section 21–4 Pressure and Temperature: A Molecular View

24E. Calculate the root-mean-square speed of helium atoms at 1000 K. The atomic mass of helium is 4.0 g/mol.

25E. The lowest possible temperature in outer space is 2.7 K. What is the root-mean-square speed of hydrogen molecules at this temperature? (See Table 1.)

26E. Find the rms speed of argon atoms at 40°C (= 313 K). The atomic mass of argon is 39.9 g/mol.

27E. The sun is a huge ball of hot ideal gas. The glow surrounding the sun in the x-ray photo shown in Fig. 16 is the corona—the atmosphere of the sun. Its temperature and pressure are 2.0×10^6 K and 0.030 Pa. Calculate the rms speed of free electrons (mass = 9.11×10^{-31} kg) in the corona.

28E. (*a*) Compute the root-mean-square speed of a nitrogen molecule at 20°C. (*b*) At what temperatures will the root-mean-square speed be half that value and twice that value?

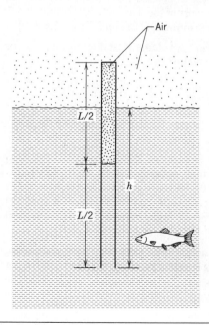

Figure 14 Problem 20.

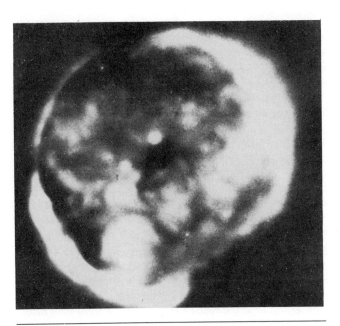

Figure 16 Exercise 27.

29E. At what temperature do the atoms of helium gas have the same rms speed as the molecules of hydrogen gas at 20°C?

30P. At 273 K and 1.00×10^{-2} atm the density of a gas is 1.24×10^{-5} g/cm³. (a) Find v_{rms} for the gas molecules. (b) Find the molecular mass of the gas and identify it.

31P. The mass of the H_2 molecule is 3.3×10^{-24} g. If 10^{23} hydrogen molecules per second strike 2.0 cm² of wall at an angle of 55° with the normal when moving with a speed of 1.0×10^5 cm/s, what pressure do they exert on the wall?

Section 21-5   Translational Kinetic Energy

32E. What is the average translational kinetic energy of individual nitrogen molecules at 1600 K (a) in joules and (b) in electron-volts?

33E. (a) Determine the average value in electron-volts of the kinetic energy of the particles of an ideal gas at 0°C and at 100°C. (b) What is the kinetic energy per mole of an ideal gas at these temperatures, in joules?

34E. At what temperature is the average translational kinetic energy of a molecule equal to 1.0 eV?

35E. Oxygen (O_2) gas at 273 K and 1.0-atm pressure is confined to a cubical container 10 cm on a side. Calculate the ratio of the change in gravitational potential energy of an oxygen molecule falling the height of the box to its mean translational kinetic energy.

36P. Show that the ideal gas equation, Eq. 4, can be written in the alternate forms: (a) $p = \rho RT/M$, where ρ is the mass density of the gas and M the atomic (or molecular) mass per mole; (b) $p = NkT$, where N is the gas particle (atoms or molecules) number density.

37P. Water standing in the open at 32°C evaporates because of the escape of some of the surface molecules. The heat of vaporization (539 cal/g) is approximately equal to ϵn, where ϵ is the average energy of the escaping molecules and n is the number of molecules per gram. (a) Find ϵ. (b) What is the ratio of ϵ to the average kinetic energy of H_2O molecules, assuming that the kinetic energy is related to temperature in the same way as it is for gases?

38P. *Avogadro's law* states that under the same condition of temperature and pressure equal volumes of gas contain equal numbers of molecules. Is this law equivalent to the ideal gas law?

Section 21-6   Mean Free Path

39E. The mean free path of nitrogen molecules at 0°C and 1.0 atm is 0.80×10^{-5} cm. At this temperature and pressure there are 2.7×10^{19} molecules/cm³. What is the molecular diameter?

40E. At 2500 km above the earth's surface the density is about 1 molecule/cm³. (a) What mean free path is predicted by Eq. 16 and (b) what is its significance under these conditions? Assume a molecular diameter of 2.0×10^{-8} cm.

41E. What is the mean free path for 15 spherical jelly beans in a bag that is vigorously shaken? Take the volume of the bag to be 1.0 L and the diameter of a jelly bean to be 1.0 cm.

42E. Derive an expression, in terms of N, $\bar{v}$, and d, for the frequency of collisions suffered by a gas atom or molecule.

43P. In a certain particle accelerator the protons travel around a circular path of diameter 23 m in a chamber of 10^{-6}-mm Hg pressure and 295-K temperature. (a) Calculate the number of gas molecules per cubic centimeter at this pressure. (b) What is the mean free path of the gas molecules under these conditions if the molecular diameter is 2.0×10^{-8} cm?

44P. At what frequency would the wavelength of sound in air be equal to the mean free path in oxygen at 1.0-atm pressure and 0°C? Take the diameter of the oxygen molecule to be 3.0×10^{-8} cm.

45P. (a) What is the molar volume (the volume per mole) of an ideal gas at standard conditions (0°C, 1.0 atm)? (b) Calculate the ratio of the root-mean-square speeds of helium to neon atoms under these conditions. (c) What would be the mean free path of helium atoms under these conditions? Assume the atomic diameter d to be 1.0×10^{-8} cm. (d) What would be the mean free path of neon atoms under these conditions? Assume the same atomic diameter as for helium. (e) Comment on the results of parts (c) and (d) in view of the fact that the helium atoms are traveling faster than the neon atoms.

46P. The mean free path λ of the molecules of a gas may be determined from measurements (for example, from measurement of the viscosity of the gas). At 20°C and 75-cm Hg pressure such measurements yield values of λ_{Ar} (argon) = 9.9 × 10^{-6} cm and λ_{N_2} (nitrogen) = 27.5 × 10^{-6} cm. (a) Find the ratio of the effective cross-section diameters of argon to nitrogen. (b) What would the value be of the mean free path of argon at 20°C and 15 cm Hg? (c) What would the value be of the mean free path of argon at −40°C and 75 cm Hg?

47P. Verify the numbers (10^{13} and 10^{24}) in the caption to the photo at the beginning of this chapter.

Section 21−7   The Distribution of Molecular Speeds

48E. The speeds of a group of ten molecules are 2, 3, 4, . . . , 11 km/s. (a) What is the average speed of the group? (b) What is the root-mean-square speed for the group?

49E. You are given the following group of particles (N_i represents the number of particles that have a speed v_i):

N_i	v_i (cm/s)
2	1
4	2
6	3
8	4
2	5

(a) Compute the average speed $\bar{v}$. (b) Compute the root-mean-square speed v_{rms}. (c) Among the five speeds shown, which is the most probable speed v_P for the entire group?

50E. (a) Ten particles are moving with the following speeds: four at 200 m/s, two at 500 m/s, and four at 600 m/s. Calculate the average and root-mean-square speeds. Is $v_{rms} > \bar{v}$? (b) Make up your own speed distribution for the 10 particles and show that $v_{rms} \geq \bar{v}$ for your distribution. (c) Under what condition (if any) does $v_{rms} = \bar{v}$?

51E. Consider the distribution of speeds shown in Fig. 17. (a) List v_{rms}, $\bar{v}$, and v_P in the order of increasing speed. (b) How does this compare with the Maxwellian distribution?

52E. It is found that the most probable speed of molecules in a gas at equilibrium temperature T_2 is the same as the rms speed of the molecules in this gas when its equilibrium temperature is T_1. Calculate T_2/T_1.

53P. (a) Compute the temperatures at which the rms speed is equal to the speed of escape from the surface of the earth for molecular hydrogen and for molecular oxygen. (b) Do the same for the moon, assuming gravity on its surface to be 0.16g. (c) The temperature high in the earth's upper atmosphere is about 1000 K. Would you expect to find much hydrogen there? Much oxygen?

54P. A molecule of hydrogen (diameter $1.0 × 10^{-8}$ cm) escapes from a furnace ($T = 4000$ K) with the root-mean-square speed into a chamber containing atoms of cold argon (diameter $3.0 × 10^{-8}$ cm) at a density of $4.0 × 10^{19}$ atoms/cm³. (a) What is the speed of the hydrogen molecule? (b) If the molecule and an argon atom collide, what is the closest distance between their centers, considering each as spherical? (c) What is the initial number of collisions per unit time experienced by the hydrogen molecule?

55P. Two containers are at the same temperature. The first contains gas at pressure p_1 whose molecules have mass m_1 with a root-mean-square speed v_{rms1}. The second contains molecules of mass m_2 at pressure $2p_1$ that have an average speed $\bar{v}_2 = 2v_{rms1}$. Find the ratio m_1/m_2 of the masses of their molecules.

56P. For the hypothetical gas speed distribution of N particles shown in Fig. 18 [$P(v) = Cv^2$, $0 < v < v_0$, $P(v) = 0$, $v > v_0$], find (a) an expression for C in terms of N and v_0, (b) the average speed of the particles, and (c) the rms speed of the particles.

57P. A hypothetical gas of N particles has the speed distribution shown in Fig. 19. $P(v) = 0$ for $v > 2v_0$. (a) Express a in

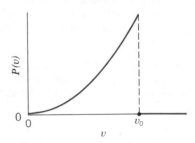

Figure 18   Problem 56.

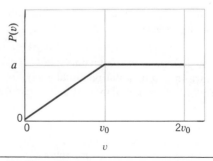

Figure 19   Problem 57.

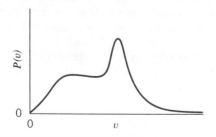

Figure 17   Exercise 51.

terms of N and v_0. (b) How many of the particles have speeds between $1.5v_0$ and $2.0v_0$? (c) Express the average speed of the particles in terms of v_0. (d) Find v_{rms}.

Section 21-8 The Heat Capacities of an Ideal Gas

58E. (a) What is the internal energy of one mol of an ideal gas at 273 K? (b) Does it depend on volume or pressure? Does it depend on the nature of the gas?

59E. One mole of an ideal gas undergoes an isothermal expansion. Find the heat flow into the gas in terms of the initial and final volumes and the temperature. (*Hint:* Use the first law of thermodynamics.)

60E. The mass of a helium atom is 6.66×10^{-27} kg. Compute the specific heat at constant volume for helium gas (in J/kg·K) from the molar heat capacity at constant volume.

61P. Let 5.0 cal of heat be added to a particular ideal gas. As a result, its volume changes from 50 to 100 cm³ while the pressure remains constant at 1.0 atm. (a) By how much did the internal energy of the substance change? (b) If the quantity of gas present is 2.0×10^{-3} mol, find the molar heat capacity at constant pressure. (c) Find the molar heat capacity at constant volume.

62P. A quantity of ideal monatomic gas consists of n moles initially at temperature T_1. The pressure and volume are then slowly doubled in such a manner as to trace out a straight line on the p–V diagram. In terms of n, R, and T_1, what are (a) W, (b) ΔU, and (c) Q? (d) If one were to define an equivalent specific heat for this process, what would be its value?

63P. A container holds a mixture of three nonreacting gases: n_1 moles of the first gas with molar specific heat at constant volume C_1, and so on. Find the molar specific heat at constant volume of the mixture, in terms of the molar specific heats and quantities of the three separate gases.

64P. The mass of a gas molecule can be computed from the specific heat at constant volume. Take $c_V = 0.075$ cal/g·°C for argon and calculate (a) the mass of an argon atom and (b) the atomic weight of argon.

Section 21-9 The Equipartition of Energy

65E. One mole of oxygen (O_2) is heated at constant pressure starting at 0°C. How much heat must be added to the gas to double its volume?

66E. Twelve grams of oxygen (O_2) are heated at constant atmospheric pressure from 25 to 125°C. (a) How many moles of oxygen are present? See Table 1. (b) How much heat is transferred to the oxygen? (c) What fraction of the heat is used to raise the internal energy of the oxygen?

67P. Four moles of an ideal diatomic gas at high temperature experiences a temperature increase of 60 K under constant pressure conditions. (a) How much heat was added to the gas? (b) By how much did the internal energy of the gas increase? (c) How much work was done by the gas? (d) By how much did the internal translational kinetic energy of the gas increase?

68P. The atomic weight of iodine is 127. A standing wave in a tube filled with iodine gas at 400 K has nodes that are 6.77 cm apart when the frequency is 1000 Hz. Is iodine gas monatomic or diatomic?

69P. A room of volume V is filled with diatomic ideal gas (air) at temperature T_1 and pressure p_0. The air is heated to a higher temperature T_2, the pressure remaining constant at p_0 because the walls of the room are not airtight. Show that the internal energy content of the air remaining in the room is the same at T_1 and T_2 and that the energy supplied by the furnace to heat the air has all gone to heat the air *outside* the room. If we add no energy to the air, why bother to light the furnace? (Ignore the furnace energy used to raise the temperature of the walls, and consider only the energy used to raise the air temperature.)

Section 21-11 The Adiabatic Expansion of an Ideal Gas

70E. A mass of gas occupies a volume of 4.3 L at a pressure of 1.2 atm and a temperature of 310 K. It is compressed adiabatically to a volume of 0.76 L. Determine (a) the final pressure and (b) the final temperature, assuming it to be an ideal gas for which $\gamma = 1.4$. (*Hint:* It is not necessary to make any unit conversions.)

71E. (a) One liter of gas with $\gamma = 1.3$ is at 273 K and 1.0-atm pressure. It is suddenly (adiabatically) compressed to half its original volume. Find its final pressure and temperature. (b) The gas is now cooled back to 0°C at constant pressure. What is its final volume?

72E. One mole of an ideal gas expands adiabatically from an initial temperature T_1 to a final temperature T_2. Prove that the work done by the gas is $C_V(T_1 - T_2)$, where C_V is the molar heat capacity. (*Hint:* Use the first law of thermodynamics.)

73E. We know that $pV^\gamma =$ constant for a reversible adiabatic process. Evaluate the "constant" for an adiabatic process involving exactly 2 mol of an ideal gas passing through the state having exactly $p = 1$ atm and $T = 300$ K. Assume a diatomic gas.

74E. For adiabatic processes in an ideal gas, (a) show that the bulk modulus is given by

$$B = -V \frac{dp}{dV} = \gamma p,$$

and therefore (b) the speed of sound is

$$v_s = \sqrt{\frac{\gamma p}{\rho}} = \sqrt{\frac{\gamma RT}{M}}.$$

See Eqs. 2 and 3 in Chapter 18.

75E. Air at 0°C and 1.0-atm pressure has a density of 1.291×10^{-3} g/cm³ and the speed of sound is 331 m/s at that temperature. Compute the ratio of specific heats of air. (*Hint:* See Exercise 74.)

76E. The speed of sound in different gases at the same temperature depends on the molecular weight of the gas. Show that $v_1/v_2 = \sqrt{M_2/M_1}$ (constant T), where v_1 is the speed of sound in the gas of molecular weight M_1 and v_2 is the speed of sound in the gas of molecular weight M_2. (*Hint:* See Exercise 74.)

77P. Use the result of Exercise 74 to show that the speed of sound in air increases about 0.61 m/s for each Celsius degree rise in temperature near 0°C.

78P. From the knowledge that C_V, the molar heat capacity at constant volume, for a gas in a container is $5R$, calculate the ratio of the speed of sound in that gas to the rms speed of its molecules at a temperature T. (*Hint:* See Exercise 74.)

79P. (*a*) An ideal gas initially at pressure p_0 undergoes a free expansion (adiabatic, no external work) until its final volume is 3.0 times its initial volume. What is the pressure of the gas after the free expansion? (*b*) The gas is then slowly and adiabatically compressed back to its original volume. The pressure after compression is $(3.0)^{1/3}p_0$. Determine whether the gas is monatomic, diatomic, or polyatomic. (*c*) How does the average kinetic energy per molecule in this final state compare with that of the initial state?

80P. An ideal gas experiences an adiabatic compression from $p = 1.0$ atm, $V = 1.0 \times 10^6$ L, $T = 0°C$ to $p = 1.0 \times 10^5$ atm, $V = 1.0 \times 10^3$ L. (*a*) Is this a monatomic, a diatomic, or a polyatomic gas? (*b*) What is the final temperature? (*c*) How many moles of the gas are present? (*d*) What is the total translational kinetic energy per mole before and after the compression? (*e*) What is the ratio of the squares of the rms speeds before and after the compression?

81P. A quantity of ideal gas occupies an initial volume V_0 at a pressure p_0 and a temperature T_0. It expands to a volume V_1 (*a*) at constant pressure, (*b*) at constant temperature, (*c*) adiabatically. Graph each case on a p–V diagram. In which case is Q greatest? Least? In which case is W greatest? Least? In which case is ΔU greatest? Least?

82P. C_V for a certain ideal gas is 6.00 cal/mol·K. The temperature of 3.0 mol of the gas is raised 50 K by each of three different processes: at constant volume, at constant pressure, and by an adiabatic compression. Complete the accompanying table, showing for each process the heat added (or subtracted), the work done by the gas, the change in internal energy of the gas, and the change in total translational kinetic energy of the gas.

Process	Heat Added	Work Done by Gas	Change in Internal Energy	Change in Kinetic Energy
Constant volume	____	____	____	____
Constant pressure	____	____	____	____
Adiabatic	____	____	____	____

83P. A reversible heat engine carries 1.00 mol of an ideal monatomic gas around the cycle shown in Fig. 20. Process $1 \rightarrow 2$ takes place at constant volume, process $2 \rightarrow 3$ is adiabatic, and process $3 \rightarrow 1$ takes place at a constant pressure. (*a*) Compute the heat Q, the change in internal energy ΔU, and the work done W, for each of the three processes and for the cycle as a whole. (*b*) If the initial pressure at point 1 is 1.0 atm, find the pressure and the volume at points 2 and 3. Use 1 atm = 1.013×10^5 Pa and $R = 8.314$ J/mol·K.

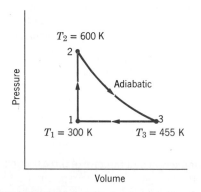

Figure 20 Problem 83.

84P. In a motorcycle engine, after combustion occurs in the top of the cylinder, the piston is forced down as the mixture of gaseous products undergoes an adiabatic expansion. Find the average power involved in this expansion when the engine is running at 4000 rpm, assuming that the gauge pressure immediately after combustion is 15 atm, the initial volume is 50 cm³, and the volume of the mixture at the bottom of the stroke is 250 cm³. Assume that the gases are diatomic, that the time involved in the expansion is one-half that of the total cycle, and that the cycle is reversible. Express your answer in watts and horsepower.

CHAPTER 22

ENTROPY AND THE SECOND LAW OF THERMODYNAMICS

The North Anna nuclear power plant near Charlottesville, Virginia generates electrical energy at the rate of 900 MW. At the same time, by deliberate design, it discards energy into the nearby river at the rate of about 2100 MW. Why does this plant—and all others like it, nuclear or not—throw away more energy than it delivers in useful form to the power grid?

22-1 Some Things that Don't Happen

If you put a quarter on a table top it simply never—entirely by itself—rises into the air, gets so hot that you cannot touch it, or flattens itself out until it is as large as a saucer. We are not surprised at any of these nonevents and we account for them by saying: "It takes *energy* to lift the coin, to heat it, or to hammer it out flat. All these things violate the conservation of energy principle."

Here are three more things that never happen but the difference is that you cannot account for them in this same easy way: (1) Coffee, resting quietly in your cup, spontaneously cools down and starts to swirl around. (2) One end of a spoon resting on a table spontaneously gets hot while the other end cools down. (3) The molecules of

the air in the room all move to one corner and stay there. *These* nonevents do *not* require energy. The coffee could get its energy by cooling down. The hot end of the spoon could get its energy from the cool end. The molecules of air need not change their kinetic energies, just their positions.

Note one thing, however. These last three events occur quite naturally and spontaneously *in the other direction.* Coffee, swirling in your cup, will eventually stop swirling, its rotational energy changing into thermal energy and thus heating the coffee a little. Temperature differences set up between two ends of a spoon tend to equalize. The air molecules rush from their special corner and fill the room uniformly.

The world is full of events that happen in one way

but never in the other. We are so used to this that we take these events for granted if they happen in the "right" direction but we would be astonished beyond all belief if they happened the other way around.

The directions in which natural events happen is governed by *the second law of thermodynamics,* the subject of this chapter.* The second law can be expressed in several equivalent forms, two of which involve simple statements about heat and work. We shall explore these in the next few sections and then consider a third formulation of the law in terms of a new and useful concept — *entropy.*

22–2 Engines

You can change *work into heat* completely and rather easily, by rubbing your hands together briskly. The sensation is as if your hands had absorbed heat from an open fire. For another familiar example, the kinetic energy of a moving car — which is a manifestation of work done on the car — is changed completely into heat in the braking system when you brake the car to a halt.

The reverse process — changing *heat into work* — is quite another matter and we lay down, as a challenge, this formulation of the second law of thermodynamics:

Second law — first form: *It is not possible to change heat completely into work, with no other change taking place.*

Rewards beyond the wildest dreams of avarice await you if you can build a device that violates this law. Although we won't succeed in doing so, let us give it a try.

Figure 1a shows a cylinder containing an ideal gas and resting on a heat reservoir at temperature T. By removing weight gradually from the piston, we can permit the gas to expand. The gas remains at constant temperature while doing so, absorbing heat Q from the reservoir. The system follows the isotherm shown in Fig. 1b and — in lifting the weight — does work W as indicated by the shaded area in that figure. The internal energy U — which depends only on the temperature for an ideal gas — does not change during this isothermal expansion. From the first law of thermodynamics, $Q = \Delta U + W$,

the work W is thus exactly equal to the heat Q extracted from the reservoir. Have we not turned heat completely into work?

We have indeed done so but we have not met the essential requirement *with no other change taking place.*

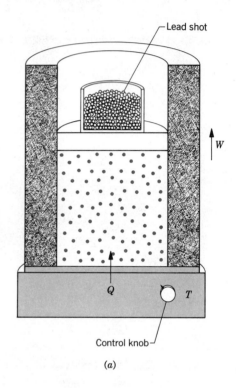

(a)

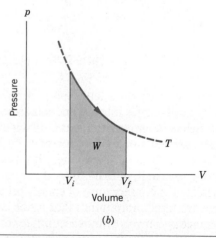

(b)

Figure 1 (a) An ideal gas expands isothermally, absorbing heat Q and doing work W. Although all the heat is transformed into work, there is no violation of the second law of thermodynamics because the system has not been restored to its original state at the end of the process.

---

* See P. W. Atkins, *The Second Law,* W. H. Freeman & Company, New York, 1984, for a very readable account of the second law.

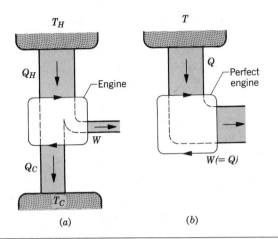

Figure 2 An engine is represented by arrows pointing in a clockwise direction around the central block. (*a*) In a real engine, heat extracted from a reservoir is converted partially into work, the rest being discharged into a reservoir of lower temperature. (*b*) In a perfect engine, all the extracted heat is converted into work. Nobody has ever built such an engine.

The gas in the cylinder is not in the same state as it was when we started; its volume has changed, for example, and so has its pressure. To meet our challenge, we must somehow restore the gas to its original condition. This means that the piston–cylinder arrangement must operate in a cycle. A device that changes heat into work, while operating in a cycle, is called a *heat engine* or, more simply, an *engine.*

Figure 2*a* suggests the overall operation of an engine. During every cycle heat Q_H is extracted from a reservoir at temperature T_H; a portion is diverted to do useful work W and the rest is discharged as heat Q_C to a reservoir at a cooler temperature T_C.

Because an engine operates in a cycle, the internal energy U of the system, that is, of the gas in the cylinder, returns to its original value at the end of the cycle. Thus, $\Delta U = 0$ and from the first law of thermodynamics ($Q = \Delta U + W$), the net work done per cycle must equal the net heat transferred per cycle. We write this as

$$|W| = |Q_H| - |Q_C|. \tag{1}$$

We have chosen here to deal with the (positive) absolute values of Q and W, which we write as $|Q|$ and $|W|$, respectively. Thus, for both $Q = +10$ J and $Q = -10$ J we have $|Q| = 10$ J and similarly for work. We will always make it clear in the context whether heat is being added to or taken from the system, and whether work is being done by the system or on it.

The purpose of an engine is to transform as much of the extracted heat Q_H into work as possible. We measure its success in doing so by its *thermal efficiency e,* defined as the ratio of the work done per cycle — what you get — to the heat absorbed per cycle — what you pay for. Using Eq. 1, we have

$$e = \frac{|W|}{|Q_H|} = \frac{|Q_H| - |Q_C|}{|Q_H|}. \tag{2}$$

Equation 2 shows that the efficiency of an engine can only be unity, or 100%, if $Q_C = 0$, that is, if no heat is delivered to the low-temperature reservoir. Figure 2*b* suggests such a "perfect" engine. From accumulated experience to date, physicists have concluded that it is impossible to build such an engine — which has never stopped hopeful inventors from trying. Another way of expressing the second law of thermodynamics (first form) is: *There are no "perfect" engines.*

The first nonevent that we proposed in Section 22–1, namely, that the coffee in your cup should cool down spontaneously and start swirling around, would be the equivalent of a perfect engine. Heat would be withdrawn from a single reservoir, the coffee, and transformed completely into work, a violation of the second law. If such things were possible, an ocean liner could extract heat from the ocean — an enormous thermal reservoir — and use it to propel itself.

Real Engines. We need to be clear about the relation between the schematic engine of Fig. 2*a* and engines in the real world. Consider, for example, the nuclear power plant shown on the first page of this chapter. The high temperature reservoir, marked T_H in Fig. 2*a*, is the nuclear reactor chamber, through which water circulates to heat water in the steam generator. The low-temperature reservoir, marked T_C in Fig. 2*a*, is the steam condenser, which is cooled by the river water that is pumped through it. The *working substance,* which corresponds to the ideal gas in the cylinder of Fig. 1*a*, is the steam that passes through the turbine, changing its temperature and pressure as it does so. Do not confuse the *working substance* with the *fuel.* The purpose of the latter, which may be uranium fuel pellets, coal, or oil, is to maintain the temperature of the high-temperature reservoir.

Figure 3 is a sketch of a proposed power plant that could operate between the relatively high temperature of surface ocean water in the tropics and the colder water that can be pumped up from deeper in the ocean. A temperature difference of 20 K should, in principle,

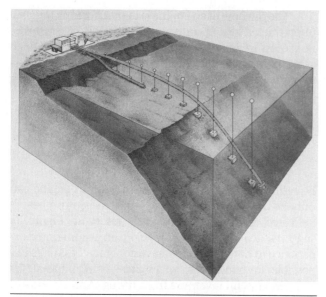

Figure 3 A plan for an electric power generating plant, designed to operate in tropical waters. The short pipe draws warm water from the shallows. The long pipe draws colder water from the deeper ocean. These two sources constitute the thermal reservoirs that drive the plant.

support a practical plant using OTEC technology.* Ammonia has been proposed as a working substance; there is no fuel.

The connection between the engine in your car and the schematic engine of Fig. 2a is complicated by the fact that your car engine is an *internal* combustion engine. The high-temperature reservoir is provided—inside the cylinders—by the combustion of the fuel–air mixture. The low-temperature is that of the exhaust gases as they vent to air. The fuel is the gasoline and the working substance is the mixture of air and the burned fuel.

Sample Problem 1 An automobile engine, whose thermal efficiency e is 22%, operates at 95 cycles per second and does work at the rate of 120 hp. (a) How much work per cycle does the engine do?

The work per cycle is

$$W = \frac{(120 \text{ hp})(746 \text{ W/hp})(1 \text{ J/W·s})}{95 \text{ s}^{-1}} = 942 \text{ J}. \quad \text{(Answer)}$$

---

* The acronym stands for Ocean Thermal Energy Conversion. See "Power from the Sea," by Terry R. Penney and Desikan Bharathan, *Scientific American*, January 1987.

Do not confuse the symbols W for work and W for the watt, a unit of power.

(b) How much heat does the engine absorb per cycle? From Eq. 2, $e = |W|/|Q_H|$, we have

$$Q_H = \frac{W}{e} = \frac{942 \text{ J}}{0.22} = 4282 \text{ J}. \quad \text{(Answer)}$$

(c) How much heat is rejected by the engine per cycle, being discharged to a low-temperature reservoir?

From Eq. 1,

$$|Q_C| = |Q_H| - |W| = 4282 \text{ J} - 942 \text{ J} = 3340 \text{ J}.$$

Heat *rejected by* the engine carries a negative sign so that

$$Q_C = -3340 \text{ J}. \quad \text{(Answer)}$$

We see that this engine extracts 4282 J of heat per cycle, which must be paid for at the gas pump, does 942 J of work, and discards 3340 J of heat to the exhaust. The engine discards 3340/942 or 3.6 times more energy than it converts to useful purposes.

22-3 Refrigerators

Heat flows naturally from a hot place to a cool place, as from the sun to the earth. There is never any "natural" net heat flow in the other direction. We express this observation more precisely in another formulation of the second law of thermodynamics:

> Second law—second form: *It is not possible for heat to flow from one body to another body at a higher temperature, with no other change taking place.*

A device that causes heat to move from a cold place to a warm place is called a *refrigerator*. Figure 4a shows the heat and work transfers that occur. Heat Q_C is extracted from a low-temperature reservoir and some work W is done on the system by an external agent; the heat and the work are combined and discharged as heat Q_H to a high-temperature reservoir. In your household refrigerator, the low-temperature reservoir is the cold chamber in which the food is stored. The high-temperature reservoir is the room in which the unit is housed. Work, which shows up on your utility bill, is done by the motor that drives the unit. In an air conditioner, the low-temperature reservoir is the room to be cooled, the high-temperature reservoir is the outside air, where the condenser coils

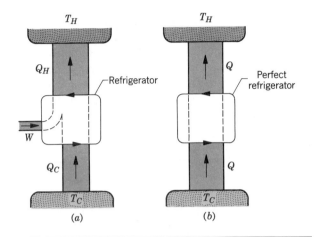

T_H T_H

Q_H —Refrigerator Q Perfect refrigerator

W

Q_C Q

T_C T_C

(a)   (b)

Figure 4 A refrigerator is represented by arrows pointing counterclockwise around the central block. (*a*) In a real refrigerator, heat is extracted from a low-temperature reservoir, some work is done, and the energy equivalent of the two is discharged as heat into a reservoir of higher temperature. (*b*) In a perfect refrigerator, no work is required. Nobody has ever built such a refrigerator.

are located, and again the work is done by the motor that drives the unit.

The purpose of a refrigerator or an air conditioner is to transfer heat from the low-temperature to the high-temperature reservoir, doing as little work on the system as possible. We rate such units by their *coefficient of performance K*, defined from

$$K = \frac{|Q_C|}{|W|} = \frac{|Q_C|}{|Q_H| - |Q_C|}. \qquad (3)$$

Design engineers, and those who pay utility bills, want the coefficient of performance of a refrigerator to be as high as possible. A value of 5 is typical for a household refrigerator and a value in the range 2–3 for a room air conditioner. Figure 4*b* shows a "perfect" refrigerator— one that cools without the expenditure of work; it would have a coefficient of performance of infinity. Long experience has shown that it is impossible to build such a device. Another way to express the second law of thermodynamics (second form) is: *There are no "perfect" refrigerators.*

The second of the nonevents that we proposed in Section 22–1, namely, the spontaneous arising of a temperature difference between the two ends of a spoon, is equivalent to a perfect refrigerator. In the spoon, heat

would flow spontaneously from a cool place to a warm place, a violation of the second form of the second law.

Sample Problem 2 A household refrigerator, whose coefficient of performance K is 4.7, extracts heat from the cooling chamber at the rate of 250 J per cycle. (a) How much work per cycle is required to operate the refrigerator?

From Eq. 3, $K = |Q_C|/|W|$, we have

$$|W| = \frac{|Q_C|}{K} = \frac{250 \text{ J}}{4.7} = 53 \text{ J}.$$

According to our sign conventions, work *done on* a system is negative, so that the work per cycle is

$$W = -53 \text{ J}. \qquad \text{(Answer)}$$

(b) How much heat per cycle is discharged to the room, which forms the high-temperature reservoir of the refrigerator?

Equation 1, which is the first law of thermodynamics for a cyclic device, holds for refrigerators as well as for engines. We then have

$$|Q_H| = |W| + |Q_C| = 53 \text{ J} + 250 \text{ J} = 303 \text{ J}. \quad \text{(Answer)}$$

We see that a refrigerator is also an efficient room heater! By paying for 53 J of work (the motor), you get 303 J of heat delivered to the room from the condenser coils at the back of the unit. If you heated the room with an electric heater, you would get only 53 J of heat for every 53 J of work that you pay for. Think about the wisdom (?) of trying to cool the kitchen on a hot day by leaving the refrigerator door open!

22-4 The Second Law of Thermodynamics

The shorter versions of the two forms of the second law of thermodynamics are:

First form: *There are no perfect engines.*
Second form: *There are no perfect refrigerators.*

Although these statements seem quite different, we want to show that they are exactly equivalent, in the sense that a violation of either implies a violation of the other. That is, if you succeed in building a perfect engine, you can also build a perfect refrigerator and conversely.

Let us assume first that you have built a perfect engine. You can use that engine to provide the work

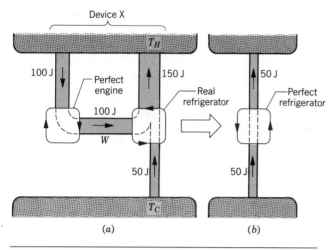

Figure 5 (*a*) The work *W* done by an engine (assumed perfect) is used to drive a real refrigerator. (*b*) The combination of these two devices (called device X) acts like a perfect refrigerator.

input to a (real!) refrigerator and thus transform it into a perfect refrigerator. Figure 5 shows how this is done.

Couple the engine and the refrigerator together as a single unit and adjust things so that the work done per cycle by the engine is just the amount needed per cycle to operate the refrigerator. Thus, no external work is involved in the combined engine + refrigerator, which we call *device X.*

Consider a numerical example. In Fig. 5*a*, the perfect engine extracts 100 J from the high-temperature reservoir and converts it into 100 J of work. The refrigerator part of the combination extracts 50 J from the low-temperature reservoir, combines it with the 100 J of work delivered to it by the engine, and discharges 150 J of heat to the high-temperature reservoir. As Fig. 5*b* shows, the overall effect of device X is to extract 50 J of heat from the low-temperature reservoir and transfer it to the high-temperature reservoir, no work being needed. Device X is a perfect refrigerator! Thus, if you can build a perfect engine, you can also build a perfect refrigerator.

We leave as an exercise to show that, if you can build a perfect refrigerator, you can use it to transform a real engine into a perfect engine. The two formulations of the second law of thermodynamics are indeed identical. If you violate either you automatically violate the other.

22–5 An Ideal Engine

There are no perfect engines. That is, no real engine can have an efficiency of 100%. A question remains: If not 100%, how high can the efficiency of a real engine be? To answer this question, we must probe into the detailed workings of an engine.

In studying gases, we avoided the complexities of real gases by introducing a useful idea: an *ideal gas.* Its usefulness lies in the fact that the ideal gas represents the limiting behavior of real gases at low enough densities. In studying engines, we follow this same path. We avoid the complexities of real engines by introducing another useful idea: an *ideal engine.* The ideal engine—in ways that we shall explore—represents a limiting behavior of real engines.

Our ideal engine consists of a piston–cylinder arrangement containing an ideal gas. A heat reservoir at temperature T_H, another heat reservoir at temperature T_C, and an insulating stand are also provided. The ideal gas constitutes the *system* to which we shall apply the laws of thermodynamics. The cylinder with its weighted piston, the insulating stand, and the two thermal reservoirs constitute the *environment* of that system.

We assume first that our ideal engine has no friction, no fluid turbulence, and no unwanted heat transfers. These are all the obvious things that an engineer would strive to eliminate. Beyond that, however, we assume that all the processes that make up the operating cycle of the engine—all expansions, compressions, and changes in temperature and pressure—are carried out extremely slowly. By doing so we ensure that the system will be essentially in thermal equilibrium at all times and that we can plot the status of the system on a pV diagram.

A process carried out in this way is called a *reversible process,* the test being that the process can be made to go in the opposite direction by making only a *tiny*—strictly, a *differential*—change in the external conditions. Thus, if we are slowly removing weight from a loaded piston, permitting the gas to expand, we can—at any stage—decide to *add* rather than subtract the weight increments, thus turning the expansion into a compression.

Because all its processes are reversible, the cycle as a whole is also reversible. This means that—at will—the engine can be run backward as an ideal refrigerator, the heat and work transfers changing in sign but not in magnitude. Our ideal engine is a *reversible* engine; indeed, that is what is ideal about it.

22-6 The Carnot Cycle

It remains to describe the details of the cycle to which we shall subject the ideal gas that forms the working substance of our ideal, reversible engine. We choose a *Carnot cycle,** which consists of two isothermal and two adiabatic processes. Figure 6 suggests the mechanics of

---

* Named for the French engineer and scientist N. L. Sadi Carnot (pronounced "car-no") who first proposed the concept in 1824.

carrying out this cycle; Fig. 7 shows the course of the cycle on a pV diagram. The following four steps make up the cycle:

Step 1. Put the cylinder on the high-temperature reservoir, with the system, which is an ideal gas, in a state represented by point a in Fig. 7. Gradually, remove some weight from the piston, allowing the system to expand slowly to point b. During this process, heat Q_H is absorbed by the system from the high-temperature reservoir. Because this process is isothermal, the internal energy of the system does not change and all the added heat appears as work. We see from Eq. 2 that we can get an

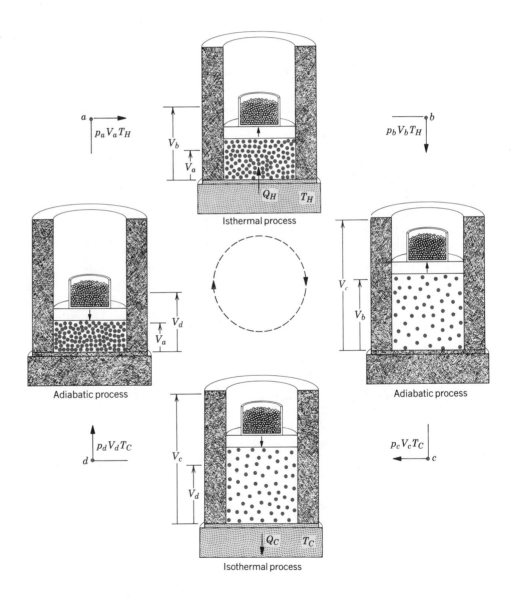

Figure 6 A Carnot cycle. The points a, b, c, and d correspond to the points so labeled in Fig. 7. The cylinder-piston arrangements show intermediate steps in the processes that connect adjacent points of the cycle.

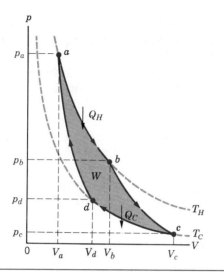

Figure 7 A pV diagram for the Carnot cycle illustrated in Fig. 6. The working substance is taken to be an ideal gas.

expression for e if we can get expressions for $|Q_H|$ and $|Q_C|$.

Step 2. Put the cylinder on the insulating stand and, by removing more weight from the piston, allow the system to further expand slowly to point c in Fig. 7. This expansion is adiabatic because no heat enters or leaves the system. The system does work in lifting the piston further and the temperature of the system drops to T_C, because the energy to do the work must come from the internal energy of the system.

Step 3. Put the cylinder on the colder heat reservoir and, by gradually adding weight to the piston, compress the gas slowly to point d in Fig. 7. During this process, heat Q_C is transferred from the gas to the reservoir. The compression is isothermal at temperature T_C and work is done on the gas by the descending piston and its load.

Step 4. Put the cylinder on the insulating stand and, by adding still more weight, compress the gas slowly back to its initial point a of Fig. 7, thus completing the cycle. The compression is adiabatic because no heat enters or leaves the system. Work is done on the gas and its temperature rises to T_H.

The special property of the Carnot engine is that — as we shall show — its thermal efficiency can be written (compare Eq. 2)

$$e = \frac{T_H - T_C}{T_H} \quad \text{(Carnot engine).} \quad (4)$$

Thus, the efficiency of a Carnot engine depends *only* on the temperatures of the two reservoirs between which it operates. The importance of the Carnot engine is that — as we shall see in the next section — no real engine working between the same two temperatures can have an efficiency greater than a Carnot engine working between those same temperatures. It is in this sense that the Carnot engine represents the limiting behavior of real engines. If, as a practicing engineer, you are striving to increase the efficiency of a (very real!) engine, it is useful to know that there is a fundamental limit, dictated by the laws of thermodynamics, beyond which you cannot possibly go.

A Carnot engine—because it is reversible—can be operated backward as a Carnot refrigerator. As we shall prove below, its coefficient of performance (compare Eq. 3) is given by

$$K = \frac{T_C}{T_H - T_C} \quad \text{(Carnot refrigerator).} \quad (5)$$

Equation 5 tells us that the coefficient of performance of a Carnot refrigerator increases as $T_H \to T_C$. Curiously, the less we need the refrigerator, the better it performs!

Proof of Eqs. 4 and 5.   Along the isothermal path ab in Fig. 7, the temperature remains constant. Because the gas is ideal, its internal energy, which depends only on the temperature, also remains constant. From the first law of thermodynamics then, $\Delta U = 0$ and the heat transferred from the high-temperature reservoir must equal the work done by the expanding gas. From Eq. 5 of Chapter 21 we then have

$$|Q_H| = |W_H| = nRT_H \ln \frac{V_b}{V_a}.$$

Similarly, for the isothermal process cd in Fig. 7, we can write

$$|Q_C| = |W_C| = nRT_C \ln \frac{V_c}{V_d}.$$

Dividing these two equations yields

$$\frac{|Q_H|}{|Q_C|} = \frac{T_H}{T_C} \frac{\ln(V_b/V_a)}{\ln(V_c/V_d)}. \quad (6)$$

Equation 34 of Chapter 21 allows us to write, for the two adiabatic processes bc and da,

$$T_H V_b^{\gamma-1} = T_C V_c^{\gamma-1} \quad \text{and} \quad T_H V_a^{\gamma-1} = T_C V_d^{\gamma-1}.$$

Dividing these two equations results in

$$\frac{V_b^{\gamma-1}}{V_a^{\gamma-1}} = \frac{V_c^{\gamma-1}}{V_d^{\gamma-1}}$$

or

$$\frac{V_b}{V_a} = \frac{V_c}{V_d}. \tag{7}$$

Combining Eqs. 6 and 7 yields

$$\frac{|Q_H|}{|Q_C|} = \frac{T_H}{T_C}. \tag{8}$$

Combining this result with Eq. 2 leads at once to Eq. 4; combining it with Eq. 3 leads to Eq. 5. These are the equations whose proof we sought.

22-7 The Efficiencies of Real Engines

The importance of the Carnot engine is summed up in this theorem:

No real engine operating between two specified temperatures can have a greater efficiency than that of a Carnot engine operating between those same two temperatures.

To prove this statement, let us assume that an inventor, working in his garage, has constructed an engine, engine X, whose efficiency e_x—he claims—is greater than e_c, the efficiency of a Carnot engine. That is,

$$e_x > e_c \quad \text{(a claim).} \tag{9}$$

Let us couple engine X to a Carnot engine operating backward as a Carnot refrigerator, as in Fig. 8a. We adjust the Carnot refrigerator so that the work it requires per cycle is just that provided by engine X.

If Eq. 9 is true then, from the definition of efficiency (see Eq. 2), we must have

$$\frac{|W|}{|Q_H'|} > \frac{|W|}{|Q_H|},$$

which requires that

$$|Q_H| > |Q_H'|. \tag{10}$$

Because of the work equality, we have, from the first law of thermodynamics,

$$|Q_H| - |Q_C| = |Q_H'| - |Q_C'|,$$

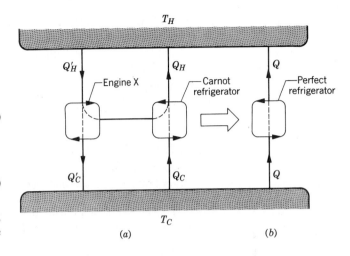

Figure 8 (a) Engine X drives a Carnot refrigerator. (b) If, as claimed, engine X is more efficient than a Carnot engine, then the combination shown in (a) is equivalent to a perfect refrigerator. This violates the second law of thermodynamics, so we conclude that engine X *cannot* be more efficient than a Carnot engine.

which we can write as

$$|Q_H| - |Q_H'| = |Q_C| - |Q_C'| = Q. \tag{11}$$

Because of Eq. 10, the quantity Q in Eq. 11 must be positive.

Comparison of Eq. 11 and Fig. 8 shows that the net effect of engine X and the Carnot refrigerator, working as a combination, is to transfer heat Q from a cold-temperature reservoir to a hot-temperature reservoir, without the requirement of work. Thus, the combination acts like a perfect refrigerator, whose existence is a violation of the second law of thermodynamics.

Something must be wrong with one of our assumptions; it can only be Eq. 9. We conclude that no real engine can have an efficiency greater than that of a Carnot engine working between the same two temperatures. At most, it can have an efficiency equal to this.

A Note on the Proof. The efficiency of a Carnot engine depends only on the temperatures of the two reservoirs and is given by Eq. 4. Possibly, however, some other reversible engine is more efficient than a Carnot engine based on an ideal gas. Perhaps if we used helium, or ammonia, or ethanol?

It is a forlorn hope. It can be shown that *all* reversible engines working between a given pair of temperatures have *exactly* the same efficiency as a Carnot engine employing an ideal gas and working between those same temperatures. Thus, the Carnot efficiency given by Eq. 4 can be extended to all reversible engines; the requirement of an ideal gas as the working substance is not essential. The proof follows along the lines of the proof we have just given above for engine X.

Sample Problem 3 The turbine in a steam power plant takes steam from a boiler at 520°C and exhausts it into a condenser at 100°C. What is its maximum possible efficiency?

Its maximum efficiency is the efficiency of a Carnot engine operating between the same two temperatures. From Eq. 4 then,

$$e_{max} = \frac{T_H - T_C}{T_H} = \frac{793 \text{ K} - 373 \text{ K}}{793 \text{ K}}$$
$$= 0.53 \text{ or } 53\%. \qquad \text{(Answer)}$$

Recall that the temperatures in this equation must be expressed on the Kelvin scale. Because of friction, turbulence, and unwanted thermal losses, actual efficiencies of about 40% may be realized for such a steam engine. Note that the theoretical maximum efficiency depends only on the two temperatures involved, not on the pressures or other factors.

The theoretical efficiency of an ordinary automobile engine is about 56% but practical considerations reduce this to about 25%.

Sample Problem 4 An inventor claims to have developed an engine that, during a certain time interval, takes in 110 MJ of heat at 415 K, rejects 50 MJ of heat at 212 K, and does 16.7 kW·h of work. Would you invest money in this project?

From Eq. 2, the claimed efficiency of this device is

$$e = \frac{|W|}{|Q_H|} = \frac{(16.7 \text{ kW·h})(3.60 \text{ MJ/kW·h})}{110 \text{ MJ}}$$
$$= 0.55 \text{ or } 55\%.$$

From Eq. 4 the maximum theoretical efficiency for the two given temperatures is

$$e = \frac{T_H - T_C}{T_H} = \frac{415 \text{ K} - 212 \text{ K}}{415 \text{ K}} = 0.49 \text{ or } 49\%.$$

The claimed efficiency is greater than the theoretical maximum. Our advice: Don't invest.

Sample Problem 5 A *heat pump* (see Fig. 9) is a device that—acting as a refrigerator—can heat a house by drawing heat from the outside, doing some work, and discharging heat inside the house. The outside temperature is −10°C and the interior is to be kept at 22°C. It is necessary to deliver heat to the interior at the rate of 16 kW to make up for the normal heat

Figure 9 Sample Problem 5. A heat pump.

losses. At what minimum rate must energy be supplied to the heat pump?

From Eq. 5, the maximum coefficient of performance of the heat pump, acting as a refrigerator, is

$$K = \frac{T_C}{T_H - T_C} = \frac{(273 - 10)\text{K}}{(273 + 22)\text{K} - (273 - 10)\text{K}}$$
$$= 8.22.$$

We can recast Eq. 3 as

$$K = \frac{|Q_C|}{|W|} = \frac{|Q_H| - |W|}{|W|}.$$

Solving for $|W|$ and dividing by time to express the result in terms of power, we obtain

$$\left|\frac{W}{t}\right| = \frac{|Q_H/t|}{K + 1} = \frac{16 \text{ kW}}{8.22 + 1} = 1.7 \text{ kW}. \qquad \text{(Answer)}$$

Herein lies the "magic" of the heat pump. By using the heat pump as a refrigerator to cool the great outdoors, you can deliver 16 kW to the interior of the house but you need pay for only the 1.7 kW it takes to run the pump. Actually, the 1.7 kW is a theoretical minimum requirement because it is based on an ideal performance. In practice, a greater power input would be required but there would still be a very considerable saving over, say, heating the house directly with electric heaters. In that case, you would have to pay directly for every kilowatt of heat transfer. When the outside temperature is greater than the inside temperature, the heat pump can be used as an air conditioner. Still operating as a refrigerator, it now pumps heat from inside the house to the great outdoors. Again, work must be done (and paid for) but the energy removed as heat from the house interior exceeds the energy equivalent of the work done. Another thermodynamic bargain!

22-8   Entropy: A New Variable

Each of the three laws of thermodynamics is associated with a specific thermodynamic variable. For the zeroth law (see Chapter 19), the variable is the temperature T. For the first law (see Chapter 20), it is the internal energy U. For the second law, the variable will prove to be one we have not met before; it is the *entropy S*.

It is our plan to define entropy in this section and, in later sections, to express the second law of thermodynamics in terms of that variable. We start by considering the Carnot cycle of Fig. 7, for which (see Eq. 8)

$$\frac{|Q_H|}{T_H} = \frac{|Q_C|}{T_C}.$$

We now discard the absolute value notation, recognizing in the process that whether the Carnot cycle is carried out clockwise, as an engine, or counterclockwise, as a refrigerator, Q_H and Q_C have opposite algebraic signs. With this understanding, we can write the above equation as

$$\frac{Q_H}{T_H} + \frac{Q_C}{T_C} = 0. \qquad (12)$$

Equation 12 tells us that the algebraic sum of the quantity Q/T, taken around the closed cycle of Fig. 7, is zero.

We now wish to generalize Eq. 12, writing it in a form that applies not only to a Carnot cycle but to any reversible cycle. Figure 10a shows such a generalized cycle, superimposed on a family of isotherms. We can approximate this arbitrary cycle as closely as we wish by connecting the isotherms by short, suitably chosen, adiabatic lines, as in Fig. 10b. In this way, we form an assembly of long, thin Carnot cycles. Convince yourself that traversing the individual Carnot cycles in Fig. 10b in sequence is exactly equivalent—in terms of heat transferred and work done—to traversing the jagged series of isotherms and adiabats that approximate the actual cycle. This is so because adjacent Carnot cycles have a common isotherm and the two traversals, in opposite directions, cancel each other in the region of overlap as far as heat transferred and work done are concerned.

We extend Eq. 12 by writing for the isotherm–adiabat sequence of lines in Fig. 10b

$$\sum \frac{Q}{T} = 0.$$

In the limit of infinitesimal temperature differences between the isotherms of Fig. 10b, this becomes

$$\oint \frac{dQ}{T} = 0 \quad \text{(reversible cycle)}. \qquad (13)$$

Here $\oint$ indicates that the integral is evaluated for a com-

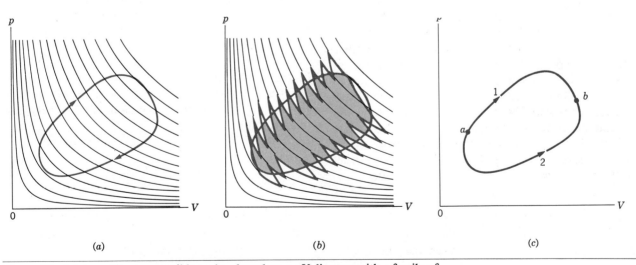

(a)   (b)   (c)

Figure 10   (a) An arbitrary reversible cycle, plotted on a pV diagram with a family of isotherms in the background. (b) The cycle is represented as a family of adjacent Carnot cycles. (c) Paths 1 and 2 are independent paths connecting points a and b.

plete traversal of the reversible cycle, starting and ending at any arbitrarily selected point.

We have established that the temperature T and the internal energy U are intrinsic properties of a system. One test of the validity of such a *state variable* is that, if we take the system around a closed, reversible cycle, the algebraic sum of the changes that occur in the variable must be zero. If this were not so, the variable would not return to its initial value and could not be a meaningful property of the system. Thus, the test of any proposed state variable X is that

$$\oint dX = 0 \quad \text{(reversible cycle).} \quad (14)$$

Comparing Eqs. 13 and 14 shows that dQ/T must represent a differential change in some state variable that we have not previously encountered. We call this new variable the *entropy,* symbol S, of the system and we write, from Eq. 13,

$$dS = \frac{dQ}{T} \quad \text{and} \quad \oint dS = 0. \quad (15)$$

The SI unit for entropy is the unit for heat divided by the unit for temperature, that is, the joule per kelvin.

Note that heat Q and work W are *not* state variables because $\oint dQ \neq 0$ and $\oint dW \neq 0$, as you can easily verify for the special case of a Carnot cycle.

We shall prove below that the property of a state variable X represented by Eq. 14 is exactly equivalent to saying that $\int dX$ between any two states has the same value for all reversible paths connecting those states. Suppose, for example, that a system is changed from an initial state i to a final state f, in which its temperature is higher by 15 K. That same temperature difference would hold no matter which of the infinite number of possible paths the system took in moving from state i to state f. The same is true for any state variable. The pressure difference would be the same for all paths; so would the volume difference. Because entropy is also a state variable, the same must hold true for the entropy difference. Thus, replacing dX in Eq. 14 by dS, we can write, for the entropy difference between any two states,

$$\boxed{S_f - S_i = \int_i^f dS = \int_i^f \frac{dQ}{T}}$$

$$\text{(reversible path),} \quad (16)$$

where the integral is carried out over any reversible path connecting these two states.

We see that Eq. 16 defines the *difference in entropy* between two states, rather than the actual entropy of a state. As for internal energy, however, our concern is always with such differences. An arbitrary constant can be added to the absolute entropy, or to the absolute internal energy, of any state without altering any of our conclusions.

Entropy Changes in Reversible Processes. Equation 16 holds—we stress—for reversible processes only, that is, for processes carried out without friction and so slowly that the process can be reversed at any stage by making an *infinitesimal* change in the environment of the system. If heat is exchanged between the system and its environment in such a process, Eq. 16 tells us that the entropy of the system will change. If heat is *added* to the system, the increments dQ in Eq. 16 (according to our sign convention for heat) are all *positive* so that the resulting entropy change for the system will be an *increase.* At the same time, the entropy of the environment of the system, that is, of the heat reservoir from which the heat was withdrawn and transferred to the system, will decrease, and by the same amount as the entropy of the system increases. This must be so because every time a millijoule of heat enters the system at a given temperature, a millijoule of heat must leave the reservoir, at that same temperature. We conclude the following:

For reversible processes, the entropy of the system *may increase, decrease, or remain unchanged. The entropy change for the* environment *of the system will always be equal in magnitude but opposite in sign. Thus, in a reversible process, the entropy of the* system + environment *remains constant.*

In the next section, we shall look closely at entropy changes in *irreversible* processes and then pull our conclusions together as a new formulation of the second law of thermodynamics.

Proof of Eq. 16. We can write Eq. 15 (see Fig. 10c) as

$$\int_{1a}^b dS + \int_{2b}^a dS = 0, \quad (17)$$

where a and b are arbitrary points and 1 and 2 describe the paths connecting these points. Because the cycle is reversible, we can write Eq. 17 as

$$\int_{1a}^b dS - \int_{2a}^b dS = 0$$

or

$$\int_{1a}^{b} dS = \int_{2a}^{b} dS. \qquad (18)$$

In Eq. 18, we have simply decided to traverse path 2 in the opposite direction, that is, from a to b rather than from b to a. We do this by changing the order of the limits in the second integral of Eq. 17, which requires that we also change the sign of that integral, thus yielding Eq. 18. Therefore, the integral of dS between any two equilibrium states of a system has the same value for all (reversible) paths connecting those states, which is what we set out to prove. If the difference in entropy between two states does not depend on the path, we can calculate that difference by any (reversible) path, which is what Eq. 16 says.

Sample Problem 6   A lump of ice whose mass m is 235 g melts (reversibly) to water, the temperature remaining at 0°C throughout the process. (a) What is the entropy change for the ice? The heat of fusion of ice is 333 kJ/kg.

The requirement that we melt the ice reversibly means that we must put the ice in contact with a heat reservoir whose temperature exceeds 0°C by only a differential amount. (If we lower the reservoir temperature to a differential amount *below* 0°C, the melted ice will start to freeze.) Because the process is reversible, we can use Eq. 16, or

$$S_{\text{water}} - S_{\text{ice}} = \int \frac{dQ}{T} = \frac{1}{T} \int dQ = \frac{Q}{T}.$$

But

$$Q = mL = (0.235 \text{ kg})(333 \text{ kJ/kg}) = 7.83 \times 10^4 \text{ J}.$$

Thus,

$$S_{\text{water}} - S_{\text{ice}} = \frac{Q}{T} = \frac{7.83 \times 10^4 \text{ J}}{273 \text{ K}} = 287 \text{ J/K}. \quad \text{(Answer)}$$

(b) What is the entropy change of the environment?

In this case, the environment is the heat reservoir from which the heat required to melt the ice is drawn. Every unit of heat that *enters* the ice must have *left* the reservoir, the temperature of both ice and reservoir being the same. Therefore, the entropy change of the reservoir is equal in magnitude but opposite in sign to that of the ice, or

$$\Delta S_{\text{reservoir}} = -287 \text{ J/K}. \quad \text{(Answer)}$$

The entropy change for the ice + reservoir, taken together, is thus zero, as it must be for a reversible process.

In practice, the melting of ice is likely to be irreversible, as when you toss an ice cube into a glass of water at room temper-

ature. The temperature difference between the ice and the reservoir, the water, in this case is not a differential amount but is about 20°C. The process proceeds in only one direction — the ice melts — and cannot be reversed at any stage by making only a differential change in the water temperature. You cannot use Eq. 16 in such a case and the calculations we have made above are not valid.

22-9   Entropy and Irreversible Processes

Equation 16 tells us how to calculate the entropy change of a system that undergoes a reversible process. Strictly speaking, there are no such processes in the real world. Friction and unwanted heat transfers are always present and the differences in pressure and temperature between a system and its environment are usually not infinitesimal. Every actual thermodynamic process is — to a greater or lesser extent — irreversible.

How are we to calculate the entropy change between the initial and final states in such cases? We take advantage of the fact that the difference in entropy — or of any other state variable, such as temperature or internal energy — between two equilibrium states does not depend on how the system passes from one state to the other:

To find the entropy change for an irreversible process between two equilibrium states, find a reversible process connecting those same states and calculate the entropy change for that process, using Eq. 16.

Consider two examples.

1.   Free expansion.   As in Section 20-6 (see Fig. 6 of Chapter 20), let an ideal gas increase its volume by expanding into an evacuated space. Because no work is done against the vacuum, $W = 0$. Because the system is enclosed by insulating walls, $Q = 0$. From the first law of thermodynamics, $Q = \Delta U + W$, it follows that $\Delta U = 0$ or

$$U_f = U_i, \qquad (19)$$

where i and f refer to the initial and the final equilibrium states. Because the gas is ideal, the internal energy U depends on temperature only so that Eq. 19 tells us that $T_f = T_i$.

If we try to use Eq. 16 to calculate the entropy difference $S_f - S_i$ for a free expansion, we have an immediate problem. The fact that $Q = 0$ might lead us to predict from Eq. 16 that $\Delta S = 0$, which as we shall see is incorrect. Following our procedure above, we must find a reversible process—*any* reversible process—that connects the initial and the final states and apply Eq. 16—not to the free expansion but to *that* process.

Because $T_f = T_i$ for the free expansion of an ideal gas, a convenient reversible process is an isothermal expansion between the initial and the final states, such as that carried out between points a and b of the Carnot cycle of Fig. 7. The isothermal expansion involves quite a different set of operations from the free expansion, the two processes having in common *only* the fact that they have the same initial and final states. Applying Eq. 16 to the isothermal expansion yields

$$S_f - S_i = \int_{V_i}^{V_f} \frac{dQ}{T} = \frac{1}{T} \int_{V_i}^{V_f} dQ = \frac{Q}{T}. \quad (20)$$

We have here removed the temperature T from the integral because the temperature remains constant in an isothermal process. Q in Eq. 20, the heat transferred during this process, is given by Eq. 6 of Chapter 21 as

$$Q = W = nRT \ln\left(\frac{V_f}{V_i}\right). \quad (21)$$

Substituting Eq. 21 into Eq. 20 leads to

$$\boxed{S_f - S_i = nR \ln \frac{V_f}{V_i}} \quad \text{(free expansion).} \quad (22)$$

Although we calculated this entropy change specifically for a reversible isothermal expansion between two states, it holds for *any* process connecting those same states, including a free expansion.

Because $V_f > V_i$, Eq. 22 tells us that the entropy of the system increases during a free expansion. We note that the entropy of the *environment* of the system does not change during a free expansion because the expansion takes place inside a rigid insulating box; see Fig. 6 of Chapter 20. Thus, the entropy of the *system + environment* increases during a free expansion.

2. Irreversible Heat Transfer. Figure 11a shows two metal blocks, each of mass m and specific heat capacity c, that are thermally insulated from each other within an insulating box. The blocks are alike in every way except that one is at a higher temperature than the other. If we remove the insulating barrier that separates the blocks

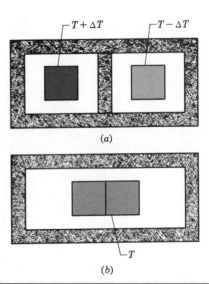

Figure 11 (a) The initial state: Two metal blocks are at different temperatures in individual insulating enclosures. (b) The final state: The insulating wall between the blocks is removed and they are allowed to come to thermal equilibrium at a temperature T.

and put them in thermal contact (see Fig. 11b), they will eventually reach a common temperature T. Like the free expansion, this process is irreversible because we totally lose control of it once we put the two blocks in thermal contact with each other.

To find the entropy change between the initial state of Fig. 11a and the final state of Fig. 11b, we must once more find a *reversible* process connecting these states and calculate the entropy change by applying Eq. 16 to *that* process.

We can carry out a reversible process using a temperature reservoir of large heat capacity whose temperature is under our control, perhaps by turning a knob. We first adjust the reservoir temperature to $T + \Delta T$, the temperature of the hotter block, and we put that block in thermal contact with the reservoir. We then slowly (reversibly) lower the reservoir temperature from $T + \Delta T$ to T, extracting heat from the hot block as we do so. Because heat leaves the hot block, the entropy of that block decreases, the entropy change being

$$\Delta S_H = \int \frac{dQ}{T} = \int_{T+\Delta T}^{T} \frac{mc\, dT}{T} = mc \int_{T+\Delta T}^{T} \frac{dT}{T}$$

$$= mc \ln \frac{T}{T + \Delta T}. \quad (23)$$

Here we have replaced dQ, the heat extracted from the

hot block as its temperature changes by dT, by $mc \, dT$. Because the quantity whose logarithm is to be taken in Eq. 23 is less than unity, the logarithm is a negative number, verifying that the entropy change represents a decrease.

To continue the reversible process, we now adjust the reservoir temperature to $T - \Delta T$, the temperature of the cooler block, and put *that* block in thermal contact with the reservoir. We then slowly (reversibly) raise the temperature of the reservoir from $T - \Delta T$ to T, adding heat to the cooler body as we do so. Because heat is added to it, the entropy of the cooler block increases, the entropy change being

$$\Delta S_C = mc \ln \frac{T}{T - \Delta T}. \qquad (24)$$

The two blocks are now in their final equilibrium state and the reversible process is completed. The entropy change for the system is found by adding Eqs. 23 and 24, or

$$S_f - S_i = \Delta S_H + \Delta S_C$$
$$= mc \ln \frac{T}{T + \Delta T} + mc \ln \frac{T}{T - \Delta T}$$

or

$$\boxed{S_f - S_i = mc \ln \frac{T^2}{T^2 - \Delta T^2}}$$

(irreversible heat transfer).   (25)

In the algebraic steps above, we have made use of the facts that $\ln a + \ln b = \ln ab$ and that $(a + b)(a - b) = a^2 - b^2$.

The quantity whose logarithm is taken in Eq. 25 is greater than unity so that the logarithm is positive. This means that $S_f > S_i$ so that the entropy of the system increases during this irreversible heat transfer. Because the system is thermally isolated from its surroundings, the entropy of the *system + environment* also increases during this irreversible process, just as it did for the (irreversible) free expansion.

Sample Problem 7   One mole of an ideal gas expands to twice its original volume in a free expansion, as in Fig. 6 of Chapter 20. What is the change in entropy of the gas? Of the environment?

The change in entropy of the gas is given by Eq. 22, or

$$S_f - S_i = nR \ln \frac{V_f}{V_i} = (1 \text{ mol})(8.31 \text{ J/mol} \cdot \text{K})(\ln 2)$$
$$= +5.76 \text{ J/K.} \qquad \text{(Answer)}$$

This is an entropy increase. The entropy of the environment of the free expansion does not change because the expanding gas is thermally isolated from its surroundings. Thus, the entropy change for the system + environment is $+5.76$ J/K.

Sample Problem 8   Two blocks of copper, the mass m of each being 850 g, are put into thermal contact in an insulated box, as in Fig. 11b. The initial temperatures of the two blocks are 325 K and 285 K and the constant specific heat capacity c of copper is 0.386 J/g·K. (a) What is the final equilibrium temperature of the two blocks?

(a) The heat lost by the hotter block must be absorbed by the cooler one, or

$$mc(325 \text{ K} - T) = mc(T - 285 \text{ K}).$$

Canceling the factors mc and solving for T yields

$$T = \tfrac{1}{2}(325 \text{ K} + 285 \text{ K}) = 305 \text{ K.} \qquad \text{(Answer)}$$

(b) What is the change in entropy for the two blocks? The quantity ΔT that appears in Eq. 25 is

$$\Delta T = 325 \text{ K} - 305 \text{ K} = 305 \text{ K} - 285 \text{ K} = 20 \text{ K.}$$

From Eq. 25 we then have

$$S_f - S_i = mc \ln \frac{T^2}{T^2 - \Delta T^2}$$
$$= (0.85 \text{ kg})(386 \text{ J/kg} \cdot \text{K}) \ln \frac{(305 \text{ K})^2}{(305 \text{ K})^2 - (20 \text{ K})^2}$$
$$= +1.41 \text{ J/K.} \qquad \text{(Answer)}$$

As for the free expansion, the entropy change for this irreversible process is an increase. Because the process is carried out in an insulated box, the entropy change of the environment is zero so that the net entropy change for the *system + environment* is $+1.41$ J/K, an increase.

22-10   Entropy and the Second Law of Thermodynamics

We are now ready to formulate the second law of thermodynamics in terms of entropy:

Second law—third form: *In any thermodynamic process that proceeds from one equilibrium state to another, the entropy of the* system + environment *either remains unchanged or increases.*

There is no way in which you can make the entropy of the system + environment decrease. It is true that the entropy of a *system* can be made to decrease but that decrease must always be accompanied by an automatic equal or greater increase in the entropy of the environment.

The third nonevent presented in Section 22–1, the spontaneous movement of the air in the room to one corner, is what we may call a *free compression,* the opposite of a free expansion. We saw in the preceding section that a free expansion is always accompanied by an increase of entropy for the system + environment. A free compression would result in an entropy decrease and would thus be a violation of the entropy statement of the second law.

Let us now make sure that the entropy statement of the second law is consistent with the two forms that we have previously presented.

1. There Are no Perfect Engines. This is the short version of the first form of the second law, presented in Section 22–4. Because engines operate in a cycle, the entropy change for the system, that is, for the gas in the cylinder, must be zero for one cycle of operation. We need only be concerned with entropy changes of the environment. For a perfect engine, the environment is the single heat reservoir of Fig. 2b and the entropy change is a decrease, because heat is withdrawn from the reservoir. Thus, a perfect engine generates an entropy decrease, a violation of the entropy statement of the second law.

2. There Are no Perfect Refrigerators. This is the short version of the second form of the second law, presented in Section 22–4. Again, the entropy change for the gas in a perfect refrigerator is zero for one cycle of operation and we need concern ourselves only with the entropy change for the environment. In this case, the environment is the two reservoirs of Fig. 4b and the entropy change is

$$\Delta S = \frac{Q}{T_H} - \frac{Q}{T_C}.$$

Because $T_H > T_C$, this entropy change is negative, again a violation of the entropy statement of the second law.

We see that entropy is the "hidden arrow" of nature. To decide whether or not a proposed event will happen spontaneously, ask: "Will it result in an increase in the entropy of the system and its environment?" If the an-

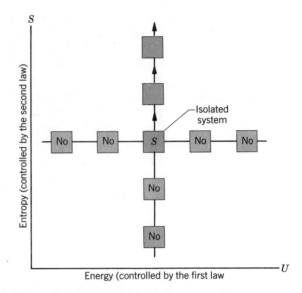

Figure 12 The central block represents the initial state S of an isolated thermodynamic system. The other blocks represent conceivable final states. Only transitions to the states vertically above S are consistent with both the first and the second laws of thermodynamics. Transitions to all other states are forbidden by one or the other of these two laws.

swer is yes the change will occur spontaneously. If it is no, it will not occur.

Figure 12 sums up the processes that can occur spontaneously for an isolated system. The central box represents such a system, in an initial state S with a definite value of energy, measured on the horizontal axis, and of entropy, measured on the vertical axis. The only transitions that can occur spontaneously for this system are to states lying directly above the initial state. For these transitions, the energy remains constant and the entropy increases, as the first and second law of thermodynamics require.

Transitions to states directly below S cannot happen because they represent entropy decreases; such decreases are forbidden — for isolated systems — by the second law. Transitions to states that lie to the right or to the left of S in Fig. 12 cannot happen because they represent changes in energy; such changes are forbidden — for isolated systems — by the first law.

Although a process may be judged to happen spontaneously, entropy and energy considerations do not tell us *when* that spontaneous process will take place. Some such processes, such as the free expansion of a gas into an evacuated chamber or the explosion of a stick of dynamite, take place without delay. Other spontaneous processes, such as the radioactive decay of the long-lived uranium isotopes, take billions of years.

22-11   What Is Entropy All About?

Because our treatment of entropy has been fairly formal, you may not yet have developed a physical feeling for this concept, risen so newly on your horizon. Let us then summarize what we have learned and try to make the concept more physical by extending its meaning in directions we have not yet explored.

Entropy: The Arrow of Time—A Review.   Entropy S, like energy U and temperature T, is simply another of the several physical properties of a system that can be measured in the laboratory and to which a number and a unit can be assigned. (Indeed, there are those who believe that—at the deepest level—entropy is the *simplest* of these three properties to understand in physical terms.)

Consider a system, isolated from its environment, that can exist in two states—call them A and B—that have the same energy. If the system is in state A, will it move spontaneously to state B as time goes on? If it is in state B, will it move spontaneously to state A? Entropy provides the answer. The second law of thermodynamics tells us:

The only changes that are possible for an isolated system are those in which the entropy of the system either increases or remains the same. Changes in which the entropy decreases will not happen.

The entropy remains the same only for reversible processes. No process in nature is truly reversible so that, in essentially all cases, we look for an entropy *increase* when we see a process occur spontaneously.

You may find that the entropy of a particular system *does* decrease in a spontaneous process but you may be sure that a (greater) increase of entropy is going on simultaneously somewhere else.

The first and second laws of thermodynamics have been summed up in this way:

The energy of the universe remains constant; the entropy of the universe always increases.

Energy obeys a conservation law; entropy does not.

Entropy: A Measure of Atomic Disorder.   Entropy is also associated with the *disorder* of a system, a notion that has even captured popular attention. The claim that the disorder of the universe (or of our small isolated corner of it) always increases as time goes on is one that most of us have no trouble in accepting. If disorder and entropy both increase as time goes on, perhaps they are related. They are; the trick is to define *disorder* in a quantitative and useful way.

The formal treatment of this aspect of entropy is the subject of *statistical thermodynamics,* which we do not treat in this book.* We can give a few qualitative examples, however, that show ways in which the disorder of a system—as it is formally defined—can be increased. We hope in this way to convey at least the spirit of the relation between entropy and disorder.

1.   Suddenly double the volume of a thermally insulated container of gas, as in a free expansion. We have seen earlier that such an expansion involves an increase in entropy. The atomic disorder *also* increases because there are now more ways in which positions in space can be assigned to the individual atoms of the gas.

2.   Raise the temperature of a gas in a container of fixed volume by adding heat. If we add heat to the gas in a reversible manner (which we assume), Eq. 16 tells us that the entropy of the gas must increase. The disorder *also* increases because there are now more ways in which velocities can be assigned to the atoms of the gas.

3.   Let the swirling motion of the coffee in a cup gradually dissipate. We have seen that this spontaneous process results in an entropy increase. It also results in an increase in atomic disorder, the initial swirling motion of the coffee being a relatively ordered state.

Examples of this sort could be added without limit. Suffice it to say that disorder can be defined quantitatively and can be related to entropy in a formal manner. The person who pointed the way was Ludwig Boltzmann (1844-1906), whose constant k we have used. The equation

$$S = k \log W,$$

---

* See, for example, *Thermodynamics, Kinetic Theory, and Statistical Thermodynamics,* by Francis W. Sears and Gerhard L. Salinger, Third Edition, Addison-Wesley, Publishing Company, Reading, MA, 1975.

which he first put forward, is the central equation of statistical thermodynamics and relates entropy to disorder quantitatively. It is engraved on his tombstone. Here S is the entropy of the system and W—which is usually a very large number—is a measure of the disorder of the system. We may define W as the number of different ways the atoms of a system can be arranged without changing the external macroscopic properties of the system.

Speaking loosely, W is much greater for a scrambled egg than for a whole egg. That is, there are more ways to construct a scrambled egg from its constituents than there are to construct a whole egg. As we know, if you drop a whole egg, it will scramble itself spontaneously; it never works the other way!

Entropy: A Nice Application.   Entropy considerations lie at the heart of an important method for producing very low temperatures, called *adiabatic demagnetization*. A sample of a solid such as a chrome-alum salt, some of whose atoms are equivalent to tiny magnets, is placed in an insulating enclosure at the lowest attainable temperature (perhaps a few millikelvins). A strong magnetic field is applied by an external magnet so that, as Fig. 13a shows, the tiny atomic magnets line up, forming a very ordered state. The magnet is now wheeled away so that its field is no longer present. By thermal agitation, the atomic magnets now assume random orientations,

as in Fig. 13b. The disorder (and thus the entropy) associated with the atomic alignments has clearly increased.

The system is isolated adiabatically, so that no heat can leave or enter. The process of removing the magnetic field (demagnetization) is very closely reversible. From the second law, then, we expect that the system should exhibit no change in entropy. However, the increase in disorder associated with the randomizing of the directions of the atomic magnets represents an entropy *increase*.

Where is the compensating entropy *decrease?* It can only come from a spontaneous lowering of the temperature of the specimen. This technique has been used with great success to achieve record low temperatures.

22-12   The Nature of Physical Law: An Aside*

So far in this book we have encountered a number of physical laws, among them being: Newton's laws of motion; Newton's law of gravity; the laws of conservation of energy, of linear momentum, and of angular momentum; and the first and second laws of thermodynamics. As we leave the subject of thermodynamics, it is perhaps well to back off and consider the nature of physical laws in general.

A law of physics is simply a statement—in word form, as for the second law of thermodynamics, or in equation form, as for Newton's law of gravity—that summarizes the results of experiment and observation for a certain range of physical phenomena. Because "truth" is such an abstract word, loaded with philosophical and ethical overtones, physicists rarely ask about a law: "Is this law true?" The question almost always is the much more specific and answerable: "Do the results predicted by this law agree with experiment?"

Physicists are continually pushing the limits of the variables in which the law is expressed, probing to see whether the law remains valid. Does the law hold at high temperatures and at high speeds? A law of physics is not an eternal truth but is a claim to be tested and probed. The law may survive these probings unchanged or it may have to be modified in some way. In any case, it must always await the challenge of new experimental data. As Einstein said: "No number of experiments can prove me right; a single experiment can prove me wrong."

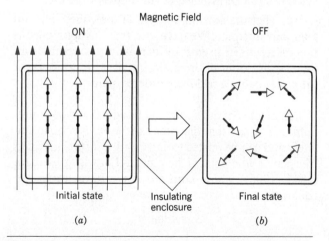

Magnetic Field

ON   OFF

Initial state Insulating Final state
enclosure

(a)   (b)

Figure 13   A method for generating low temperatures. When the magnetic field is removed, the disorder (entropy) associated with the magnetic orientations of the atoms *increases*. Because there can be no change in entropy in this thermally isolated reversible process, the disorder (entropy) associated with the temperature of the specimen must *decrease* by the same amount. Thus, the temperature falls as the magnetic field is removed.

---

* See Richard Feynmann, *The Character of Physical Law*, MIT Press, Cambridge, MA, 1965.

For example, Newton's laws of motion—whose importance should be clear to all—fail to agree with experiment when tested using particles whose speeds are an appreciable fraction of the speed of light. The failure is not sudden but gradual, the discrepancy with experiment growing larger and larger as the particle speeds get closer and closer to the speed of light.

As it happens another law, Einstein's special theory of relativity, turns out to agree with experiment over the full range of observable particle speeds. Where does this leave Newton's laws? It leaves them as an extremely useful special case of the more comprehensive law. The usefulness of Newton's laws in the vast and important region in which they agree with experiment remains undiminished.

What about the second law of thermodynamics? Its present status is that, if we confine ourselves to processes that start and end in states of thermal equilibrium, no

exceptions to it have ever been found. If an inventor claims to have built an engine that violates this law, the crushing weight of decades of experience by gifted scientists and engineers the world around lies heavily against him. The lone inventor may be right but the burden of proof is on him to prove his case, not on the scientific community to disprove it. Still, no physicist would deny that the second law of thermodynamics may one day be reinterpreted and seen as a special case of a more comprehensive law.

Perhaps the most impressive thing about physical laws is the very fact that they exist and are so simple in form. It is a source of wonderment to the discerning thinker that such vast realms of experience can be summarized in a single sentence or a single equation. Einstein put it well when he remarked: "the most incomprehensible thing about the Universe is that it is comprehensible."

REVIEW AND SUMMARY

The Role of the Second Law of Thermodynamics

Many events do not occur even though they would not violate the first law of thermodynamics. The spontaneous flow of heat from a cold to a hot body is an example. The *second law of thermodynamics* sets up a criterion for identifying such "forbidden" processes.

Engines

An *engine* is a device that cyclically accepts heat Q_H from a high-temperature source (at Kelvin temperature T_H) and does work W. There is always a discharge of heat Q_C at a lower temperature T_C. The *efficiency e* of an engine is

$$e = \frac{|W|}{|Q_H|} = \frac{|Q_H| - |Q_C|}{|Q_H|}. \tag{2}$$

The second law of thermodynamics denies the existence of perfect engines: *It is not possible to change heat completely into work with no other changes.* See Fig. 2 and Sample Problem 1.

Refrigerators

A *refrigerator* (or a *heat pump*) is a device that cyclically removes heat from a cold system and, by doing work, delivers it to a warmer one. Its effectiveness is measured by a *coefficient of performance K* defined to be

$$K = \frac{|Q_C|}{|W|} = \frac{|Q_C|}{|Q_H| - |Q_C|}. \tag{3}$$

The second law denies the existence of perfect refrigerators: *It is not possible for heat to flow from one body to another at a higher temperature, with no other changes.* See Fig. 4 and Sample Problem 2.

The Carnot Cycle

The *Carnot cycle* describes the action of an ideal (that is, reversible) heat engine. As Figs. 6 and 7 show, it consists of two *adiabatic processes* (for which $Q = 0$) alternating with two *isothermal processes* (for which T = constant). The efficiency of an engine using a Carnot cycle is

Carnot Engine

$$e = \frac{T_H - T_C}{T_H} \quad \text{(Carnot engine)} \tag{4}$$

and the performance coefficient of a Carnot refrigerator is

Carnot Refrigerator

$$K = \frac{T_C}{T_H - T_C} \quad \text{(Carnot refrigerator).} \tag{5}$$

No real engine (refrigerator) operating between two temperatures can have a greater efficiency (coefficient of performance) than that of a Carnot engine (refrigerator) operating between the same temperatures. See Sample Problems 3–5.

Entropy

Entropy, S, like pressure, volume, and so on, is a property (a *state function*) of a system in equilibrium. The change in S for a system that goes *reversibly* from state i to state f is *defined* to be

$$S_f - S_i = \int_i^f dS = \int_i^f \frac{dQ}{T} \quad \text{(reversible path)}, \qquad [16]$$

where dQ is an increment of heat transferred at temperature T. $S_f - S_i$ depends only on the initial and final states and not in any way on the nature of the reversible path connecting them. The SI unit for entropy is the joule per kelvin.

Entropy Changes for Irreversible Processes

The entropy change for *irreversible* processes may be evaluated by (1) finding a *reversible* path that connects the states i and f and (2) calculating $S_f - S_i$ for *this* path, using Eq. 16. The answer is the entropy change for the irreversible process. Sample Problem 6 applies this strategy to melting ice; Section 22–9 treats the *free expansion* of a gas and the *transfer of heat* between two identical objects. See Sample Problems 7 and 8.

The Entropy Statement of the Second Law

The importance of entropy is that *in any thermodynamic process that proceeds from one equilibrium state to another, the entropy of the* system + environment *either remains unchanged or it increases.* All three statements of the second law of thermodynamics are equivalent.

Physical Law

A law of physics is a statement that summarizes the results of experiment and observation for a certain range of physical phenomena. It is always subject to further observation and, sometimes, modification as a new evidence becomes available. When an older version of a law is replaced by a newer one, the older one often remains as a useful special case of the more general law.

QUESTIONS

1. Is a human being a heat engine? Explain.

2. Couldn't we just as well define the efficiency of an engine as $e = W/Q_C$ rather than as $e = W/Q_H$? Why don't we?

3. The efficiencies of nuclear power plants are less than those of fossil-fuel plants. Why?

4. Can a given amount of mechanical energy be converted completely into heat energy? If so, give an example.

5. An inventor suggested a house might be heated in the following manner: A system resembling a refrigerator draws heat from the earth and rejects heat to the house. He claimed that the heat supplied to the house can exceed the work done by the engine of the system. What is your comment?

6. Give a qualitative explanation of how frictional forces between moving surfaces produce heat energy. Why does the reverse process (heat energy producing relative motion of these surfaces) not occur?

7. A block returns to its initial position after dissipating mechanical energy to heat through friction. Why is this process not thermodynamically reversible?

8. Are any of the following phenomena reversible: (*a*) breaking an empty soda bottle; (*b*) mixing a cocktail; (*c*) melting an ice cube in a glass of iced tea; (*d*) burning a log of firewood; (*e*) puncturing an automobile tire; (*f*) finishing the "Unfinished Symphony"; (*g*) writing this book?

9. Give some examples of irreversible processes in nature.

10. Are there any natural processes that are reversible?

11. Can we calculate the work done during an irreversible process in terms of an area on a p–V diagram? Is any work done?

12. Suggest a reversible process whereby heat can be added to a system. Why would adding heat by means of a Bunsen burner not be a reversible process?

13. To carry out a Carnot cycle, we need not start at point a in Fig. 7 but may equally well start at points b, c, or d, or indeed any intermediate point. Explain.

14. If a Carnot engine is independent of the working substance, then perhaps real engines should be similarly independent, to a certain extent. Why then, for real engines, are we so concerned to find suitable fuels such as coal, gasoline, or fissionable material? Why not use stones as a fuel?

15. Under what conditions would an ideal heat engine be 100% efficient?

16. What factors reduce the efficiency of a heat engine from its ideal value?

17. You wish to increase the efficiency of a Carnot engine as much as possible. You can do this by increasing T_H a certain amount, keeping T_C constant, or by decreasing T_C the same amount, keeping T_H constant. Which would you do?

18. Can a kitchen be cooled by leaving the door of an electric refrigerator open? Explain.

19. Why do you get poorer gasoline mileage from your car in winter than in summer?

20. From time to time inventors will claim to have perfected a device that does useful work but consumes no (or very little) fuel. What do you think is most likely true in such cases: (a) the claimants are right, (b) the claimants are mistaken in their measurements, or (c) the claimants are swindlers? Do you think that such a claim should be examined closely by a panel of scientists and engineers? In your opinion, would the time and effort be justified?

21. We have seen that real engines always discard substantial amounts of heat to their low-temperature reservoirs. It seems a shame to throw this heat energy away. Why not use this heat to run a second engine, the low-temperature reservoir of the first engine serving as the high-temperature reservoir of the second?

22. Give examples in which the entropy of a system decreases and explain why the second law of thermodynamics is not violated.

23. Do living things violate the second law of thermodynamics? As a chicken grows from an egg, for example, it becomes more and more ordered and organized. Increasing entropy, however, calls for disorder and decay. Is the entropy of a chicken actually decreasing as it grows?

24. Two samples of a gas initially at the same temperature and pressure are compressed from a volume V to a volume $V/2$, one isothermally, the other adiabatically. In which sample is the final pressure greater? Does the entropy of the gas change in either process?

25. Suppose that we had chosen to represent the state of a system by its entropy and its absolute temperature rather than by its pressure and volume. (a) What would a Carnot cycle look like on a T–S diagram? (b) What physical significance, if any, can be attached to the area under a curve on a T–S diagram?

26. Is there a change in entropy in purely mechanical motions?

27. Show that the total entropy increases when work is converted into heat by friction between sliding surfaces. Describe the increase in disorder.

28. Two pieces of molding clay of equal mass are moving in opposite directions with equal speed. They strike and stick together. Treat the two pieces as a single system and state whether each of the following quantities is positive, negative, or zero for this process: ΔU, W, Q, and ΔS. Justify your answers.

29. Heat energy flows from the sun to the earth. Show that the entropy of the earth–sun system increases during this process.

30. Is it true that the heat energy of the universe is steadily growing less available? If so, why?

31. Discuss the following comment of Panofsky and Phillips: "From the standpoint of formal physics there is only one concept which is asymmetric in the time, namely, entropy. But this makes it reasonable to assume that the second law of thermodynamics can be used to ascertain the sense of time independently in any frame of reference; that is, we shall take the positive direction of time to be that of statistically increasing disorder, or increasing entropy." (See, in this connection, "The Arrow of Time," by David Layzer, *Scientific American*, December 1975.)

32. Explain the statement "Cosmic rays continually *decrease* the entropy of the earth on which they fall." Why does this not contradict the second law of thermodynamics?

EXERCISES AND PROBLEMS

Section 22–2 Engines

1E. A heat engine absorbs 52 kcal of heat and exhausts 36 kcal of heat each cycle. Calculate (a) the efficiency and (b) the work done in kilocalories per cycle.

2E. A car engine delivers 8.2 kJ of work per cycle. (a) Before a tuneup, the efficiency is 25%. Calculate, per cycle, the heat absorbed from the combustion of fuel and the heat exhausted to the atmosphere. (b) After a tuneup, the efficiency is 31%. What are the new values of the quantities calculated in (a)?

3E. Calculate the efficiency of a fossil-fuel power plant that consumes 380 metric tons of coal each hour to produce useful work at the rate of 750 MW. The heat of combustion of 1 kg of coal is 28 MJ.

4P. One mole of a monatomic ideal gas is caused to go through the cycle shown in Fig. 14. Process bc is a reversible adiabatic expansion. $p_b = 10$ atm, $V_b = 1.0$ m³, and $V_c = 8.0$ m³. Calculate (a) the heat added to the gas, (b) the heat

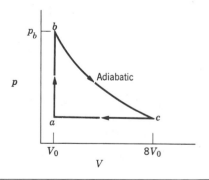

Figure 14 Problem 4.

leaving the gas, (c) the net work done by the gas, and (d) the efficiency of the cycle.

5P. One mole of a monatomic ideal gas initially at a volume of 10 L and a temperature of 300 K is heated at constant volume to a temperature of 600 K, allowed to expand isothermally to its initial pressure, and finally compressed isobarically (that is, at constant pressure) to its original volume, pressure, and temperature. (a) Compute the heat input to the system during one cycle. (b) What is the net work done by the gas during one cycle? (c) What is the efficiency of this cycle?

Section 22-3 Refrigerators

6E. A refrigerator does 150 J of work to remove 560 J of heat from its cold compartment. (a) What is the refrigerator's coefficient of performance? (b) How much heat is exhausted to the kitchen?

7E. To make ice, a freezer extracts 42 kcal of heat at $-12°C$. The freezer has a coefficient of performance of 5.7. The room temperature is $26°C$. (a) How much heat was delivered to the room? (b) How much work was required to run the freezer?

Section 22-7 The Efficiencies of Real Engines

8E. An ideal gas heat engine operates in a Carnot cycle between 235 and $115°C$. It absorbs 6.3×10^4 cal per cycle at the higher temperature. (a) What is the efficiency of the engine? (b) How much work per cycle is this engine capable of performing?

9E. How much work must be done to extract 1.0 J of heat (a) from a reservoir at $7°C$ and transfer it to one at $27°C$ by means of a refrigerator using a Carnot cycle; (b) from one at $-73°C$ to one at $27°C$; (c) from one at $-173°C$ to one at $27°C$; and (d) from one at $-223°C$ to one at $27°C$?

10E. In a Carnot cycle, the isothermal expansion of an ideal gas takes place at 400 K and the isothermal compression at 300 K. During the expansion, 500 cal of heat energy are transferred to the gas. Determine (a) the work performed by the gas during the isothermal expansion, (b) the heat rejected from the gas during the isothermal compression, and (c) the work done on the gas during the isothermal compression.

11E. In a hypothetical nuclear fusion reactor, the fuel is deuterium (D) gas at a temperature of about 7×10^8 K. If this gas could be used to operate an ideal heat engine with $T_C = 100°C$, what would be its efficiency?

12E. A Carnot engine has an efficiency of 22%. It operates between heat reservoirs differing in temperature by $75°C$. What are the temperatures of the reservoirs?

13E. Apparatus that liquefies helium is in a room at 300 K. If the helium in the apparatus is at 4.0 K, what is the minimum ratio of heat energy delivered to the room to the heat energy removed from the helium?

14E. An air conditioner takes heat from a room at $70°F$ and transfers it to the outdoors, which is at $96°F$. For each joule of electrical energy required to operate the air conditioner, how many joules of heat are removed from the room?

15E. For the Carnot cycle illustrated in Fig. 7, show that the

work done by the gas during process bc (step 2) has the same absolute value as the work done on the gas during process da (step 4).

16E. (a) For a Carnot (ideal) refrigerator, show that

$$|W| = |Q_C| \frac{T_H - T_C}{T_C}.$$

(b) In a mechanical refrigerator the low-temperature coils are at a temperature of $-13°C$, and the compressed gas in the condenser has a temperature of $26°C$. What is the theoretical coefficient of performance?

17E. (a) A Carnot engine operates between a hot reservoir at 320 K and a cold reservoir at 260 K. If it absorbs 500 J of heat per cycle at the hot reservoir, how much work per cycle does it deliver? (b) If the same engine, working in reverse, functions as a refrigerator between the same two reservoirs, how much work per cycle must be supplied to remove 1000 J of heat from the cold reservoir?

18E. A combination mercury–steam turbine takes saturated mercury vapor from a boiler at $876°F$ and exhausts it to heat a steam boiler at $460°F$. The steam turbine receives steam at this temperature and exhausts it to a condenser at $100°F$. What is the maximum efficiency of the combination?

19E. In a heat pump, heat from the outdoors at $-5°C$ is transferred to a room at $17°C$, energy being supplied by an electric motor. How many joules of heat will be delivered to the room for each joule of electric energy consumed? Assume an ideal heat pump.

20P. A heat pump is used to heat a building. The outside temperature is $-5.0°C$ and the temperature inside the building is to be maintained at $22°C$. The coefficient of performance is 3.8, and the pump delivers 1.8 Mcal of heat to the building each hour. At what rate must work be done to run the pump?

21P. A Carnot engine produces a power of 500 W. It operates between heat reservoirs at 100 and $60°C$. Calculate (a) the rate of heat input and (b) the rate of exhaust heat output in kilocalories per second.

22P. The motor in a refrigerator has a power output of 200 W. If the freezing compartment is at 270 K and the outside air is at 300 K, assuming ideal efficiency, what is the maximum amount of heat that can be extracted from the freezing compartment in 10 min?

23P. How is the efficiency of a reversible ideal heat engine related to the coefficient of performance of the reversible refrigerator obtained by running the engine backward?

24P. (a) Show that a Carnot cycle, plotted on an absolute temperature versus entropy (T–S) diagram, graphs as a rectangle. For the Carnot cycle shown in Fig. 15, calculate (b) the heat gained and (c) the work done by the system.

25P. In a two-stage Carnot heat engine, a quantity of heat Q_1 is absorbed at a temperature T_1, work W_1 is done, and a quan-

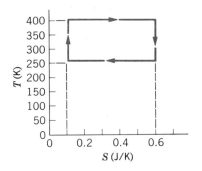

Figure 15   Problem 24.

tity of heat Q_2 is expelled at a lower temperature T_2 by the first stage. The second stage absorbs the heat expelled by the first, does work W_2, and expels a quantity of heat Q_3 at a lower temperature T_3. Prove that the efficiency of the combination engine is $(T_1 - T_3)/T_1$.

26P. (a) Plot accurately a Carnot cycle on a p-V diagram for 1.0 mL of an ideal gas. Let point a correspond to $p = 1.0$ atm, $T = 300$ K, and let b correspond to $p = 0.50$ atm, $T = 300$ K; take the low-temperature reservoir to be at 100 K. Let $\gamma = 1.5$. (b) Compute graphically the work done in this cycle. (c) Compute the work analytically.

27P. A Carnot engine works between temperatures T_1 and T_2. It drives a Carnot refrigerator that works between two different temperatures T_3 and T_4 (see Fig. 16). Find the ratio $|Q_3|/|Q_1|$ in terms of the four temperatures.

28P. An air conditioner operating between 93 and 70°F is rated at 4000 Btu/h cooling capacity. Its coefficient of performance is 27% of that of a Carnot refrigerator operating between the same two temperatures. What is the required horsepower of the motor?

29P. Suppose that a deep shaft were drilled in the earth's crust near one of the poles where the surface temperature is −40°C to a depth where the temperature is 800°C. (a) What is

the theoretical limit to the efficiency of an engine operating between these temperatures? (b) If all of the heat released into the low-temperature reservoir were used to melt ice that was initially at −40°C, at what rate could water at 0°C be produced by a power plant having an output of 100 MW? The specific heat of ice is 0.50 cal/g · °C; its heat of fusion is 80 cal/g. (Note that the engine can operate only between 0 and 800°C in this case. Energy exhausted at −40°C cannot be used to raise the temperature of anything else above −40°C.)

30P. One mole of an ideal monatomic gas is used as the working substance of an engine that operates on the cycle shown in Fig. 17. Calculate (a) the work done per cycle, (b) the heat added per cycle during the expansion stroke abc, and (c) the engine efficiency. (d) What is the Carnot efficiency of an engine operating between the highest and lowest temperatures present in the cycle? How does this compare to the efficiency calculated in (c)? Assume that $p = 2p_0$, $V = 2V_0$, $p_0 = 1.01 \times 10^5$ Pa, and $V_0 = 0.0225$ m^3.

31P. Compute the efficiency of the cycle shown in Problem 83 in Chapter 21. In this case, the heat input will not be at a fixed temperature, as it is in the Carnot cycle.

32P*. A gasoline internal combustion engine can be approximated by the cycle shown in Fig. 18. Assume an ideal gas and use a compression ratio of 4:1 ($V_4 = 4V_1$). Assume that $p_2 = 3p_1$. (a) Determine the pressure and temperature of each

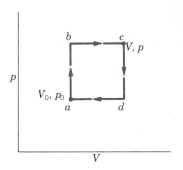

Figure 17   Problem 30.

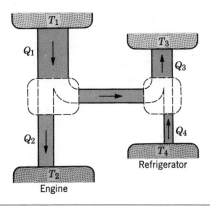

Figure 16   Problem 27.

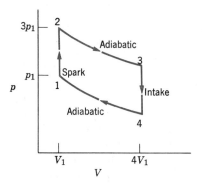

Figure 18   Problem 32.

of the vertex points of the p–V diagram in terms of p_1, T_1, and the ratio of specific heats of the gas. (*b*) What is the efficiency of the cycle?

Section 22–8 Entropy: A New Variable

33E. In Fig. 10*c*, suppose that the change in entropy of the system in passing from state *a* to state *b* along path 1 is +0.602 cal/K. What is the entropy change in passing (*a*) from state *a* to *b* along path 2 and (*b*) from state *b* to *a* along path 2?

34E. Find (*a*) the heat absorbed and (*b*) the change in entropy of a 1.0-kg block of copper whose temperature is increased reversibly from 25 to 100°C. The specific heat of copper is 9.2×10^{-2} cal/g·°C.

35E. Heat can be removed from water at 0°C and atmospheric pressure without causing the water to freeze, if done with little disturbance of the water. Suppose the water is cooled to -5.0°C before ice begins to form. What is the change in entropy per gram occurring during the sudden freezing that then takes place?

36P. An inventor claims to have invented four engines, each of which operates between heat reservoirs at 400 and 300 K. Data on each engine, per cycle of operation, are as follows: Engine (*a*): $Q_H = 200$ J, $Q_C = -175$ J, $W = 40$ J; engine (*b*): $Q_H = 500$ J, $Q_C = -200$ J, $W = 400$ J; engine (*c*): $Q_H = 600$ J, $Q_C = -200$ J, $W = 400$ J; engine (*d*): $Q_H = 100$ J, $Q_C = -90$ J, $W = 10$ J. Which of the first and second laws of thermodynamics (if either) does each engine violate?

37P. At very low temperatures, the molar specific heat of many solids is (approximately) proportional to T^3; that is, $C_V = AT^3$, where A depends on the particular substance. For aluminum, $A = 7.53 \times 10^{-6}$ cal/mol·K⁴. Find the entropy change of 4.0 mol of aluminum when its temperature is raised from 5.0 to 10 K.

Section 22–9 Entropy and Irreversible Processes

38E. An ideal gas undergoes a reversible isothermal expansion at 77°C, increasing its volume from 1.3 to 3.4 L. The entropy change of the gas is 22 J/K. How many moles of gas are present?

39E. An ideal gas undergoes a reversible isothermal expansion at 132°C. The entropy of the gas increases by 46 J/K. How much heat was absorbed?

40E. One mole of an ideal gas expands isothermally at 360 K until its volume is doubled. What is the increase of entropy of the gas?

41E. Suppose that the same amount of heat energy, say 260 J, is transferred by conduction from a heat reservoir at a temperature of 400 K to another reservoir, the temperature of which is (*a*) 100 K, (*b*) 200 K, (*c*) 300 K, and (*d*) 360 K. Calculate the changes in entropy and discuss the trend.

42E. A brass rod is in thermal contact with a heat reservoir at 130°C at one end and a heat reservoir at 24°C at the other end. (*a*) Compute the total change in the entropy arising from the process of conduction of 1200 cal of heat through the rod. (*b*) Does the entropy of the rod change in the process?

43E. In a specific heat experiment, 200 g of aluminum ($c_p = 0.215$ cal/g·°C) at 100°C is mixed with 50 g of water at 20°C. (*a*) Calculate the equilibrium temperature. Find the entropy change (*b*) of the aluminum and (*c*) of the water. (*d*) Calculate the entropy change of the system. (*Hint:* See Eqs. 23 and 24.)

44P. A 10-g ice cube at -10°C is placed in a lake whose temperature is +15°C. Calculate the change in entropy of the system as the ice cube comes to thermal equilibrium with the lake. The specific heat of ice is 0.50 cal/g·°C. (*Hint:* Will the ice cube affect the temperature of the lake?)

45P. An 8.0-g ice cube at -10°C is dropped into a thermos flask containing 100 cm³ of water at 20°C. What is the change in entropy of the system when a final equilibrium state is reached? The specific heat of ice is 0.50 cal/g·°C.

46P. Four moles of an ideal gas are caused to expand from a volume V_1 to a volume $V_2 = 2V_1$. (*a*) If the expansion is isothermal at the temperature $T = 400$ K, find the work done by the expanding gas. (*b*) Find the change in entropy, if any. (*c*) If the expansion were reversibly adiabatic instead of isothermal, what is the entropy change?

47P. A 50-g block of copper having a temperature of 400 K is placed in an insulating box with a 100-g block of lead having a temperature of 200 K. (*a*) What is the equilibrium temperature of this two-block system? (*b*) What is the change in the internal energy of the two-block system as it changes from the initial condition to the equilibrium condition? (*c*) What is the change in the entropy of the two-block system? (See Table 1 in Chapter 20.)

48P. A mole of a monatomic ideal gas is taken from an initial state of pressure p and volume V to a final state of pressure $2p$ and volume $2V$ by two different processes. (I) It expands isothermally until its volume is doubled, and then its pressure is increased at constant volume to the final state. (II) It is compressed isothermally until its pressure is doubled, and then its volume is increased at constant pressure to the final state. Show the path of each process on a p–V diagram. For each process calculate in terms of p and V: (*a*) the heat absorbed by the gas in each part of the process; (*b*) the work done by the gas in each part of the process; (*c*) the change in internal energy of the gas, $U_f - U_i$; and (*d*) the change in entropy of the gas, $S_f - S_i$.

49P. An ideal diatomic gas is caused to pass through the cycle shown on the p–V diagram in Fig. 19, where $V_2 = 3V_1$. Determine, in terms of p_1, V_1, T_1, and R: (*a*) p_2, p_3, and T_3 and (*b*) W, Q, ΔU, and ΔS per mole for all three processes.

50P. An object of constant heat capacity C is heated from an initial temperature T_i to a final temperature T_f, by being placed

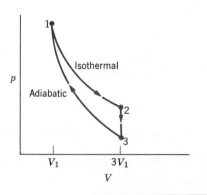

Figure 19 Problem 49.

in contact with a heat reservoir at T_f. Represent the process on a graph of C/T versus T and show graphically that the total change in entropy ΔS (object plus reservoir) is positive and (b) show how the use of heat reservoirs at intermediate temperatures would allow the process to be carried out in a way that makes ΔS as small as desired.

51P. A mixture of 1773 g of water and 227 g of ice at 0°C is, in a reversible process, brought to a final equilibrium state where the water–ice ratio, by mass, is 1 : 1 at 0°C. (a) Calculate the entropy change of the system during this process. (The heat of fusion for water is 79.5 cal/g.) (b) The system is then returned to the first equilibrium state, but in an irreversible way (by using a Bunsen burner, for instance). Calculate the entropy

change of the system during this process. (c) Is your answer consistent with the second law of thermodynamics?

52P. A round silver rod 15 cm long, with a diameter of 1.0 cm, has its ends in contact with heat reservoirs at 60 and 20°C and steady-state heat flow has been established. What will be the initial rate of change of the entropy of the rod if (a) the hot end is suddenly insulated from the 60°C reservoir, or (b) if the entire rod is suddenly insulated?

53P. One mole of an ideal monatomic gas is caused to go through the cycle shown in Fig. 20. (a) How much work is done in expanding the gas from a to c along path abc? (b) What is the change in internal energy and entropy in going from b to c? (c) What is the change in internal energy and entropy in going through one complete cycle? Express all answers in terms of the pressure p_0, volume V_0, and temperature T_0 at point a in the diagram.

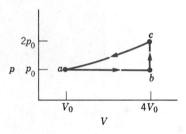

Figure 20 Problem 53.

CHAPTER 23
ELECTRIC CHARGE

Electric charges over Seattle. Perhaps 10 coulombs of charge are exchanged in this lightning flash. By contrast, a quarter contains about 250,000 coulombs of positive charge, neatly balanced by the same amount of negative charge. There is an enormous amount of electric charge locked up in ordinary matter.

23-1 Electromagnetism

The early Greek philosophers knew that if you rubbed a piece of amber it could pick up bits of straw. There is a direct line of development from this ancient observation to the electronic age in which we live. The strength of the connection is indicated by our word "electron," which is derived from the Greek word for amber.

The Greeks also knew that some naturally occurring "stones," which we know today as the mineral magnetite, would attract iron. Such were the modest origins of the sciences of electricity and magnetism. These two sciences developed quite separately for centuries, until 1820 in fact, when Hans Christian Oersted found a connection between them: An electric current in a wire can

deflect a magnetic compass needle. Interestingly enough, Oersted made this discovery while preparing a demonstration lecture for his physics students.

The new science of electromagnetism was developed further by workers in many countries. One of the very best was Michael Faraday,* a truly gifted experimenter with a talent for physical intuition and visualization. His collected laboratory notebooks, for example, do not contain a single equation. James Clerk Maxwell†

---

* See "Michael Faraday," by Herbert Kondo, *Scientific American,* October 1953. For the definitive biography see L. Pearce Williams, *Michael Faraday,* Basic Books, New York, 1964.

† See "James Clerk Maxwell," by James R. Newman, *Scientific American,* June 1955.

put Faraday's ideas into mathematical form, introduced many new ideas of his own, and put electromagnetism on a sound theoretical basis.

Table 2 of Chapter 37 shows the basic laws of electromagnetism, called Maxwell's equations. We plan to work our way through them in the chapters that follow but you might want to glance at them now, just to see where we are headed. These equations play the same role in electromagnetism that Newton's laws of motion do in classical mechanics or that the laws of thermodynamics do in the study of heat.

Maxwell's great discovery in electromagnetism was that light is an electromagnetic wave and that you can measure its speed by making purely electrical and magnetic measurements. With this discovery, Maxwell linked the ancient science of optics to those of electricity and magnetism. Heinrich Hertz* took a giant step forward when he produced electromagnetic "Maxwellian waves" in his laboratory. We now call them short radio waves. It remained for Marconi and others to push forward with the practical applications. Today, Maxwell's equations are used the world over in the solution of a wide variety of practical engineering problems.

23-2   Electric Charge

If you walk across a carpet in dry weather, you can draw a spark by touching a metal door knob. Television advertising has alerted us to problem of "static cling." On a grander scale, lightning is familiar to everyone. All these phenomena represent the merest glimpse of the vast amount of *electric charge* that is stored up in the familiar objects that surround us and—indeed—in our own bodies.

The electrical neutrality of most objects in our visible and tangible world conceals the fact that such objects contain enormous amounts of positive and negative electric charge that largely cancel each other in their external effects. Only when this nice electrical balance is slightly disturbed does nature reveal to us the effects of uncompensated positive or negative charge. When we say that a body is "charged" we mean that it has a slight charge imbalance.

Charged bodies exert forces on each other. To show this, let us charge a glass rod by rubbing it with silk. The

* See "Heinrich Hertz," by Philip and Emily Morrison, *Scientific American,* December 1957.

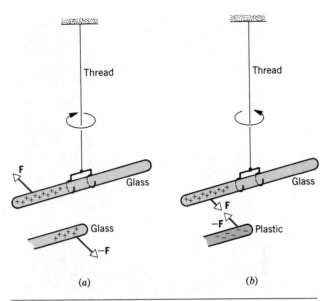

Figure 1 (*a*) Two similarly charged rods repel each other. (*b*) Two oppositely charged rods attract each other.

process of rubbing transfers a tiny amount of charge from one body to the other, thus slightly upsetting the electrical neutrality of each. If you suspend this charged rod from a thread, as in Fig. 1*a*, and if you bring a second charged glass rod nearby, the two rods will repel each other. However, if you rub a plastic rod with fur it will attract the charged end of the hanging glass rod; see Fig. 1*b*.

We explain all this by saying there are two kinds of charge, one of which (the one on the glass rubbed with silk) we have come to call *positive* and the other (the one on the plastic rubbed with fur) we have come to call *negative*. These simple experiments can be summed up by saying:

Like charges repel and unlike charges attract.

In Section 23-4, we put this rule into quantitative form, as Coulomb's law of force. We shall consider only charges that are either at rest with respect to each other or moving very slowly, a restriction that defines the subject of *electrostatics*.

The positive and negative labels for electric charge are due to Benjamin Franklin who, among many other accomplishments, was a scientist of international reputation. It has even been said that Franklin's triumphs in diplomacy in France during the American War of Inde-

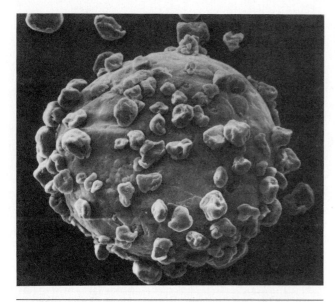

Figure 2 A carrier bead from a Xerox copying machine, covered with toner particles that cling to it by electrostatic attraction.

pendence were facilitated, and perhaps even made possible, because he was so highly regarded as a scientist.

Franklin introduced the words "charge" and "battery" into the language of electricity. When the *battery* pack in your pocket calculator loses its *charge* and, in charging it, you see the *plus* sign marking the *positive* battery terminal, think of Franklin.

Electrical forces between charged bodies have many industrial applications, among them being electrostatic paint spraying and powder coating, fly-ash precipitation, nonimpact ink-jet printing, and photocopying. Figure 2, for example, shows a tiny carrier bead in a Xerox copying machine, covered with particles of black powder, called *toner,* that stick to it by electrostatic forces. These negatively charged toner particles are eventually attracted from their carrier beads to a positively charged latent image of the document to be copied, formed on a rotating drum. A charged sheet of paper then attracts the toner particles from the drum to itself, after which they are heat-fused in place and you have your copy.

23–3 Conductors and Insulators

You cannot seem to charge up a copper rod, no matter how hard you rub it or with what you rub it. However, if you fit the rod with a plastic handle, you will be able to build up a charge. The explanation is that charges placed on some materials—we call them *insulators*—are not free to move around; they stay where you put them. In other materials—we call them *conductors*—charges can move around more or less freely. If you touch a charged isolated copper rod with your finger, the charges will quickly move from the rod through your body to the ground.

Glass, chemically pure water, and plastics are common examples of insulators. Although there are no perfect insulators, fused quartz is pretty good, its insulating ability being about 10^{25} times greater than that of copper.

Copper, metals in general, tap water, and the human body, as Fig. 3a shows, are common examples of conductors, In metals, a fairly subtle experiment called the *Hall effect* shows that it is the negative charges that are free to move; we discuss this effect in Section 30–4. When copper atoms come together to form solid copper, their outer electrons do not remain attached to the individual atoms but become free to wander about within the rigid lattice structure formed by the positively charged ion cores. We call these mobile electrons the *conduction electrons.* The positive charges in a copper rod are just as immobile as they are in a glass rod.

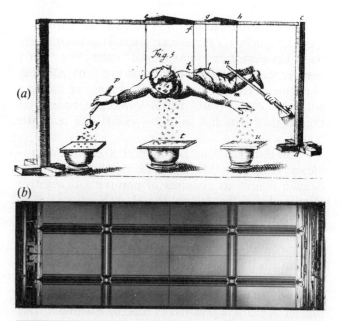

Figure 3 Progress in electricity. (*a*) Not a parlor stunt but a serious experiment carried out in 1774 to prove that the human body is a conductor of electricity. (*b*) A megabyte chip.

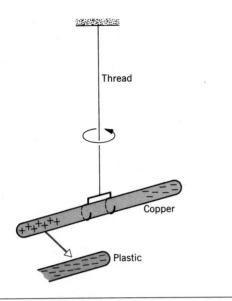

Figure 4   Either end of an isolated uncharged copper rod will be attracted by a charged rod of either sign. In this case, conduction electrons in the copper rod are repelled to the far end of that rod, leaving the near end positive.

The experiment of Fig. 4 demonstrates the mobility of charge in a conductor. A negatively charged plastic rod will attract either end of a suspended but uncharged copper rod. The (mobile) conduction electrons in the copper rod are repelled by the negative charge on the plastic rod to the far end of the copper rod, leaving the near end of the copper rod with a positive charge. An uncharged copper rod will also be attracted by a positively charged glass rod. In this case, the conduction electrons in the copper are attracted by the positively charged glass rod to the near end of the copper rod; the far end of the copper rod is then left with a positive charge.

There are also *semiconductors,* such as silicon and germanium. The microelectronic revolution that has transformed our lives in so many ways rests solidly on semiconducting devices; see Fig. 3*b*. We describe the operation of semiconductors, whose electrical characteristics can often be controlled to suit the requirements of the problem at hand, in Chapter 46 of the extended version of this text.

Finally, there are *superconductors.* The resistance of these materials to the flow of electricity is not just small; it is absolutely zero. If you set up a current in a superconducting ring, it persists without change for as long as you care to watch it, no battery or other source of energy being needed in the circuit.

Superconductivity was discovered in 1911 by the Dutch physicist Kammerlingh Onnes, who observed that solid mercury lost its electrical resistance completely at temperatures below 4.2 K. Until 1986, superconductivity was limited in its usefulness because superconducting materials needed to be cooled to temperatures below about 20 K. For example, the magnet coils on the large particle accelerator at Fermilab are made of a superconducting alloy kept cool with liquid helium. The central feature of the particle accelerator called the Superconducting Supercollider, or SSC, will be a ring of superconducting magnets some 52 miles in circumference.

In recent years, however, alloys have been developed that become superconducting at much higher temperatures so that a new era of useful applications seems to be upon us. Superconductivity at room temperature has not been ruled out as a possibility. See Essay 12.

It is not possible to understand semiconductors or superconductors without some background in quantum physics; we discuss both subjects in greater detail from this point of view in Chapter 46 of the extended version of this book.

23-4   Coulomb's Law

The law that gives the electrostatic force acting between the charges of two particles has exactly the same form as the law that gives the gravitational force acting between the masses of two particles. The two laws are

$$F_{\text{grav}} = G \, \frac{m_1 m_2}{r^2} \quad \text{(Newton's law—gravitation)} \quad (1)$$

and

$$F_{\text{elec}} = C \, \frac{q_1 q_2}{r^2} \quad \text{(Coulomb's law—electrostatics),} \quad (2)$$

in which G and C are constants.

Coulomb's law has survived every experimental test, no exceptions to it having ever been found. It holds deep within the atom, correctly describing the force between the positively charged nucleus and each extranuclear electron. Although classical Newtonian mechanics fails in that realm—where it is replaced by quantum physics—Coulomb's simple law continues to give correct answers. This law also correctly accounts for the forces that bind atoms together to form molecules and the forces that bind atoms or molecules together to form solids or liquids. We ourselves are assemblies of nuclei and electrons held together by forces arising from Eq. 2. There is much more to Coulomb's law than accounting for the forces between two charged rods.

In Eq. 2, F is the magnitude of the force acting on either particle owing to the charge on the other; q_1 and q_2 are the absolute values of the charges of the two particles and r is the distance between them. The constant C, by analogy with the gravitational constant G, may be called the *electrostatic constant*. Both laws are inverse square laws and both involve a property of the interacting particles—the mass in one case and the charge in the other.

The laws differ in that gravitational forces are always attractive but electrostatic forces may be either attractive or repulsive, depending on the signs of the two charges. These differences arise from the fact that, although there are two kinds of charge, there is apparently only one kind of mass.

For practical reasons having to do with the accuracy of measurements, the SI unit of charge is derived from the SI unit of electric current. If you connect the ends of a long wire to the terminals of a battery, a current is set up in the wire. We visualize this current as a flow of charge. The SI unit of current is the *ampere* (abbr. A). It is an SI base unit and in Section 31–4 we shall describe how it is defined in terms of experimental operations.

The SI unit of charge is the *coulomb* (abbr. C). A coulomb is defined as the amount of charge that passes through any cross section of a wire in 1 second if there is a current of 1 ampere in the wire. In general,

$$dq = i\,dt, \qquad (3)$$

where dq (in coulombs) is the charge transferred by a current i (in amperes) during the interval dt (in seconds).

The constant G in Eq. 1 determines the absolute magnitude of the gravitational force. In the same way, the electrostatic constant C in Eq. 2 determines the absolute magnitude of the electrostatic force. For historical reasons, this constant is written, not simply as C, but in a more complex form, so that Coulomb's law appears as

$$\boxed{F = \frac{1}{4\pi\epsilon_0}\frac{q_1 q_2}{r^2}} \quad \text{(Coulomb's law).} \qquad (4)$$

Certain equations that are derived from Eq. 4, but are

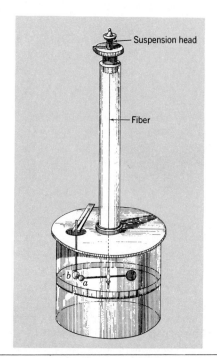

Figure 5 Coulomb's torsion balance, from his 1785 memoir to the Paris Academy of Sciences.

used much more often than it is, are simpler in form if we do it this way. The electrostatic constant in Eq. 4 has the value

$$\frac{1}{4\pi\epsilon_0} = 8.99 \times 10^9 \text{ N}\cdot\text{m}^2/\text{C}^2. \qquad (5)$$

The quantity ϵ_0, called the *permittivity constant,* sometimes appears separately in equations; its value is

$$\epsilon_0 = 8.85 \times 10^{-12} \text{ C}^2/\text{N}\cdot\text{m}^2. \qquad (6)$$

Coulomb's law is named for Charles Augustus Coulomb who, in 1785, measured the electrical forces between small charged spheres, using the apparatus shown in Fig. 5. It is a torsion balance, operating on much the same principle as the torsion balance used by Henry Cavendish in 1798 to measure gravitational forces; see Section 15–3.

Although torsion balance methods are direct, they are not very accurate. We could not be convinced, on the basis of Coulomb's data, that the exponent 2 in Eq. 4 was not really, say, 2.0003. Fortunately, there are indirect methods that convince us that this exponent, if it is not 2 exactly, differs from it by a number that is less than

3×10^{-16}. We shall describe these powerful methods in Section 25-8.

Another parallel between the gravitational and the electrostatic force is that they both obey the principle of superposition. If we have n point charges, they interact independently in pairs and the force on any one of them, let us say q_1, is given by the vector sum

$$\mathbf{F}_1 = \mathbf{F}_{12} + \mathbf{F}_{13} + \mathbf{F}_{14} + \mathbf{F}_{15} + \cdots + \mathbf{F}_{1n}, \quad (7)$$

in which, for example, $\mathbf{F}_{14}$ is the force acting on particle 1 owing to the presence of particle 4. An identical formula holds for the gravitational force.

Finally, the two shell theorems that we found so useful in our study of gravitation hold equally well in electrostatics:

Theorem 1 A uniform spherical shell of charge behaves, for external points, as if all its charge were concentrated at its center.

Theorem 2 A uniform spherical shell of charge exerts no force on a charged particle placed inside the shell.

The proof of these theorms follows exactly the proof for the gravitational case in Section 15-5. All you have to do is replace the mass m in that proof by the charge q, wherever it appears, and replace the gravitational constant G by the corresponding electrostatic constant $1/4\pi\epsilon_0$. The formal identity of the proofs follows from the fact that the two fundamental laws, Eqs. 1 and 2, have exactly the same mathematical form.

Sample Problem 1 Figure 6 shows three charged particles, held in place by forces not shown. What electrostatic force, owing to the other two charges, acts on q_1? Take $q_1 = -1.2\,\mu C$, $q_2 = +3.7\,\mu C$, $q_3 = -2.3\,\mu C$, $r_{12} = 15$ cm, $r_{13} = 10$ cm, and $\theta = 32°$.

This problem calls for the superposition principle. We start by computing the magnitudes of the forces that q_2 and q_3 exert on q_1. We substitute only the absolute values of the charges into Eq. 4, disregarding—for the time being—their signs. We then have

$$F_{12} = \frac{1}{4\pi\epsilon_0} \frac{q_1 q_2}{r_{12}^2}$$

$$= \frac{(8.99 \times 10^9 \text{ N}\cdot\text{m}^2/\text{C}^2)(1.2 \times 10^{-6} \text{ C})(3.7 \times 10^{-6} \text{ C})}{(0.15 \text{ m})^2}$$

$$= 1.77 \text{ N}.$$

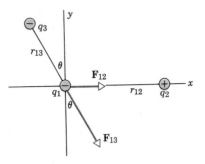

Figure 6 Sample Problem 1. The three charges exert three action–reaction pairs of forces on each other. Only the two forces acting on charge q_1 are shown.

The charges q_1 and q_2 have opposite signs so that the force between them is attractive. Hence, $\mathbf{F}_{12}$ points to the right in Fig. 6.

We also have

$$F_{13} = \frac{(8.99 \times 10^9 \text{ N}\cdot\text{m}^2/\text{C}^2)(1.2 \times 10^{-6} \text{ C})(2.3 \times 10^{-6} \text{ C})}{(0.10 \text{ m})^2}$$

$$= 2.48 \text{ N}.$$

These two charges have the same (negative) sign so that the force between them is repulsive. Thus, $\mathbf{F}_{13}$ points as shown in Fig. 6.

The components of the resultant force $\mathbf{F}_1$ acting on q_1 are

$$F_{1x} = F_{12x} + F_{13x} = F_{12} + F_{13} \sin\theta$$
$$= 1.77 \text{ N} + (2.48 \text{ N})(\sin 32°) = 3.08 \text{ N}$$

and

$$F_{1y} = F_{12y} + F_{13y} = 0 - F_{13} \cos\theta$$
$$= -(2.48 \text{ N})(\cos 32°) = -2.10 \text{ N}.$$

From these components, you can show that the magnitude of $\mathbf{F}_1$ is 3.73 N and that this vector makes an angle of $-34°$ with the x axis.

23-5 Charge is Quantized

In Franklin's day, electric charge was thought to be a continuous fluid, an idea that was useful for many purposes. However, we now know that fluids themselves, such as air or water, are not continuous but are made up of atoms and molecules; matter is discrete. Experiment shows that the "electrical fluid" is not continuous either but that it is made up of multiples of a certain elementary charge. That is, any charge q that can be observed and

Table 1 Some Properties of Three Particles

Particle	Symbol	Charge[a] e	Mass[b] m_e	Angular Momentum[c] $h/2\pi$
Electron	e	-1	1	$\frac{1}{2}$
Proton	p	$+1$	1836.15	$\frac{1}{2}$
Neutron	n	0	1838.68	$\frac{1}{2}$

[a] In units of the elementary charge e.
[b] In units of the electron mass m_e.
[c] The intrinsic spin angular momentum, in units of $h/2\pi$. We introduced this concept in Section 12–12 and give a fuller treatment in Chapter 45 of the extended version of this book.

measured* directly can be written

$$q = ne \qquad n = 0, \pm 1, \pm 2, \pm 3, \cdots, \qquad (8)$$

in which e, the *elementary charge,* has the value

$$e = 1.60 \times 10^{-19} \text{ C}. \qquad (9)$$

The elementary charge is one of the important constants of nature.

When a physical quantity such as charge exists only in discrete "packets" rather than in continuously variable amounts, we say that quantity is *quantized.* We have already seen that matter, energy, and angular momentum are quantized; charge adds one more important physical quantity to the list. Equation 8 tells us that it is possible, for example, to find a particle that carries a charge of zero or of $+10e$ or of $-6e$ but not a particle with a charge of, say, $3.57e$. Table 1 shows the charges and some other properties of the three particles that can be said to make up the material world around us.

The quantum of charge is small. In an ordinary 100-W, 120-V light bulb, for example, about 10^{19} elementary charges enter and leave the bulb every second. The graininess of electricity does not show up in large-scale phenomena, just as you cannot feel the individual molecules of water when you move your hand through it.

Sample Problem 2 A penny, being electrically neutral, contains equal amounts of positive and negative charge. What is the magnitude of these equal charges?

---

* Quarks carry charges of $-\frac{1}{3}e$ or $+\frac{2}{3}e$ but these particles, thought to be the constituent particles of protons and neutrons, have never been detected as free particles; see Section 2–8.

The charge q is given by NZe, in which N is the number of atoms in a penny and Ze is the magnitude of the positive and the negative charges carried by each atom.

The number N of atoms in a penny, assumed for simplicity to be made of copper, is $N_A m/M$, in which N_A is the Avogadro constant. The mass m of the coin is 3.11 g and the atomic weight M of copper is 63.5 g/mol. We find

$$N = \frac{N_A m}{M} = \frac{(6.02 \times 10^{23} \text{ atoms/mol})(3.11 \text{ g})}{63.5 \text{ g/mol}}$$
$$= 2.95 \times 10^{22} \text{ atoms}.$$

Every neutral atom has a negative charge of magnitude Ze associated with its electrons and a positive charge of the same magnitude associated with its nucleus. Here e is the magnitude of the charge on the electron, which is 1.60×10^{-19} C, and Z is the atomic number of the element in question. For copper, Z is 29. The magnitude of the total negative, or positive, charge in a penny is then

$$q = NZe$$
$$= (2.95 \times 10^{22})(29)(1.60 \times 10^{-19} \text{ C})$$
$$= 137,000 \text{ C}. \qquad \text{(Answer)}$$

This is an enormous charge. By comparison, the charge that you might get by rubbing a plastic rod is perhaps 10^{-9} C, smaller by a factor of about 10^{14}. For another comparison, it would take about 38 h for a charge of 137,000 C to flow through the filament of a 100-W, 120-V light bulb. There is a lot of electric charge in ordinary matter.

Sample Problem 3 In Sample Problem 2 we saw that a copper penny contains both positive and negative charges, each of magnitude 1.37×10^5 C. Suppose that these charges could be concentrated into two separate bundles, held 100 m apart. What attractive force would act on each bundle?

From Eq. 4 we have

$$F = \frac{1}{4\pi\epsilon_0} \frac{q^2}{r^2} = \frac{(8.99 \times 10^9 \text{ N} \cdot \text{m}^2/\text{C}^2)(1.37 \times 10^5 \text{ C})^2}{(100 \text{ m})^2}$$
$$= 1.69 \times 10^{16} \text{ N}. \qquad \text{(Answer)}$$

This is about 2×10^{12} tons of force! Even if the charges were separated by one earth diameter, the attractive force would still be about 120 tons. In all of this, we have sidestepped the problem of forming each of the separated charges into a "bundle" whose dimensions are small compared to their separation. Such bundles, if they could ever be formed, would be blasted apart by mutual Coulomb repulsion forces.

The lesson of this sample problem is that you cannot disturb the electrical neutrality of ordinary matter very much. If you try to pull out any sizable fraction of the charge contained in a body, a large Coulomb force appears automatically, tending to pull it back.

Sample Problem 4   The average distance r between the electron and the central proton in the hydrogen atom is 5.3×10^{-11} m. (a) What is the magnitude of the average electrostatic force that acts between these two particles?

From Eq. 4 we have, for the electrostatic force,

$$F_e = \frac{1}{4\pi\epsilon_0} \frac{q_1 q_2}{r^2} = \frac{(8.99 \times 10^9 \text{ N} \cdot \text{m}^2/\text{C}^2)(1.60 \times 10^{-19} \text{ C})^2}{(5.3 \times 10^{-11} \text{ m})^2}$$

$$= 8.2 \times 10^{-8} \text{ N}. \qquad \text{(Answer)}$$

(b) What is the magnitude of the average gravitational force that acts between these particles?

From Eq. 1 we have, for the gravitational force,

$$F_g = G \frac{m_e m_p}{r^2}$$

$$= \frac{(6.67 \times 10^{-11} \text{ m}^3/\text{kg} \cdot \text{s}^2)(9.11 \times 10^{-31} \text{ kg})(1.67 \times 10^{-27} \text{ kg})}{(5.3 \times 10^{-11} \text{ m})^2}$$

$$= 3.6 \times 10^{-47} \text{ N}. \qquad \text{(Answer)}$$

We see that the gravitational force is weaker than the electrostatic force by the enormous factor of about 10^{39}. Although the gravitational force is weak, it is always attractive. Thus, it can act to build up very large masses, as in the formation of stars and planets, so that large gravitational forces can develop. The electrostatic force, on the other hand, is repulsive for like charges so that it is not possible to accumulate large concentrations of either positive or negative charge. We must always have the two together, so that they largely compensate for each other. The charges that we are accustomed to in our daily experiences are slight disturbances of this overriding balance.

Sample Problem 5   The nucleus of an iron atom, which has a radius of about 4×10^{-15} m, contains 26 protons. What repulsive electrostatic force acts between two protons in such a nucleus if they are that far apart?

From Eq. 4 we have

$$F = \frac{1}{4\pi\epsilon_0} \frac{q_p q_p}{r^2}$$

$$= \frac{(8.99 \times 10^9 \text{ N} \cdot \text{m}^2/\text{C}^2)(1.60 \times 10^{-19} \text{ C})^2}{(4 \times 10^{-15} \text{ m})^2}$$

$$= 14 \text{ N}. \qquad \text{(Answer)}$$

This enormous force, more than 3 lb and acting on a single proton, must be more than balanced by the attractive strong nuclear force that binds the nucleus together. The strong force, whose range is so short that its effects cannot be felt very far outside the nucleus, is very well named; see Section 6–5.

23–6   Charge Is Conserved

If you rub a glass rod with silk, a positive charge appears on the rod. Measurement shows that a negative charge of equal magnitude appears on the silk. This suggests that rubbing does not create charge but only transfers it from one body to another, upsetting the electrical neutrality of each during the process. This hypothesis of *conservation of charge,* first put forward by Benjamin Franklin, has stood up under close examination, both for macroscopic charges and at the level of interactions between elementary particles. No exceptions have ever been found. Thus, we add electric charge to our growing list of quantities—including energy and both linear and angular momentum—that obey a conservation law.

Radioactive decay gives us ready examples of charge conservation at the nuclear level. The common uranium isotope uranium-238, or ^{238}U, is typical:

$$^{238}\text{U} \rightarrow {}^{234}\text{Th} + {}^{4}\text{He} \quad \text{(radioactive decay)}. \quad (10)$$

The atomic number of the radioactive *parent* nucleus, ^{238}U, is 92, which tells us that this nucleus contains 92 protons. It decays spontaneously by emitting an alpha particle, ^{4}He, for which $Z = 2$, leaving behind the *daughter* nucleus ^{234}Th, with $Z = 90$. Thus, the amount of charge present before the decay, $92e$, is equal to the amount present after the decay, $90e + 2e$ or $92e$. Charge is conserved.

Another example of charge conservation occurs when an electron, whose charge is $-e$, happens to find itself close to a positive electron, called a *positron,* charge $+e$, both particles being essentially at rest. The two particles may simply disappear, in an *annihilation process,* converting all the energy associated with their masses, which is $2mc^2$, into the radiant energy of two oppositely directed gamma rays. Thus,

$$e^- + e^+ \rightarrow \gamma + \gamma \quad \text{(annihilation)}. \quad (11)$$

In applying the conservation-of-charge principle, we must add the charges algebraically, with due regard for their signs. In the annihilation process of Eq. 11, the net charge of the system is zero, both before and after the event. Charge is conserved.

The converse of annihilation also occurs. Here an energetic gamma ray, passing near a heavy nucleus that serves as a catalyst, simply disappears, converting its radiant energy into the creation of an electron and a positron. Thus,

$$\gamma \rightarrow e^- + e^+ \quad \text{(pair production)}. \quad (12)$$

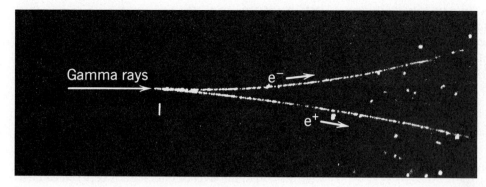

Figure 7 An energetic gamma ray coming in from the left generates an electron-positron pair. The two electrons leave tracks of bubbles in the chamber in which they were created. The tracks curve, in opposite directions, because of the action of a strong external magnetic field.

Again, charge is conserved. The two particles produced in the pair-production process carry off any energy that the incoming gamma ray may have left after providing the energy $2mc^2$ needed to create the pair of electrons.

Figure 7 shows such a pair-production event occurring in a bubble chamber at the Lawrence Berkeley Laboratory. Gamma rays entering the chamber from the left, being uncharged, leave no bubble tracks in the chamber. The bubble tracks left in the wakes of the electron and the positron are curved because of the action of a magnetic field in the chamber. The fact that the particle tracks curve in opposite directions is evidence that the electron and the positron carry charges of opposite sign.

An interesting consequence of the conservation of charge is that it ensures that the electron is an absolutely stable particle. From the point of view of the conservation of energy, there is nothing to stop a resting electron from decaying into two* oppositely-directed gamma rays, in this way

$$e^- \rightarrow \gamma + \gamma \quad \text{(doesn't happen!)}. \quad (13)$$

If such a process occurred, there would be a charge of $-e$ before the event but a charge of zero after the event, which would violate the law of charge conservation. Because charge must be conserved, we conclude that the process does not happen. There is no lighter charged particle into which the electron can decay; we conclude that a single isolated electron must be absolutely stable.

23-7 The Constants of Physics: An Aside

In this chapter we have introduced yet another of the fundamental constants of physics, the elementary charge

e. Perhaps it is time to step aside and to review the role that these constants play in the structure of physics. Table 2 lists four of them that are particularly central.

We note at once how precisely these constants are known. Although we have been using only three significant figures in our illustrative Sample Problems, we see that the constants are typically known to at least seven or eight significant figures. An exception is the gravitational constant, the least well known of all the important physical constants. Experiments seeking improved values of the various constants are going on, in laboratories all over the world, on a continuing basis. A particular constant may be involved, either alone or with other constants, in a wide variety of experiments. Unraveling all these data is no simple task. Every decade or so it seems appropriate to survey the accumulated measurements and, with the help of an elaborate computer program, extract from this vast array of data a set of "best values" of the physical constants. The last such survey was in 1985.*

The improvement in our knowledge of the constants over time is impressive. Table 3, for example, shows how the precision of measurement of the speed of light has improved over the years. Note the variety of methods and the geographical spread of the effort. The measurements finally reached a point at which the precision was limited by the practical reproducibility of the standard of length that was in use at that time. As a result, it was decided to assign a value to the speed of light *by definition* and to redefine the length standard in terms of the speed of light; see Section 1-4.

Each of the constants in Table 2 plays an important role in the structure of physics. We discuss each in turn.

---

* We need *two* gamma rays so that linear momentum can be conserved.

* See Appendix B. For a very readable account of the physical constants, see also B. W. Petley, *The Fundamental Physical Constants and the Frontiers of Measurement,* A. Hilger, Boston, 1985.

Table 2   Four Fundamental Constants of Physics

Constant	Symbol	Value (1985)	Uncertainty[a]	Section
Gravitational constant	G	6.67260×10^{-11} m³/kg · s²	100	15–3
Speed of light	c	2.99792458×10^{8} m/s	Exact	17–7
Planck constant	h	$6.6260754 \times 10^{-34}$ J · s	0.6	8–10
Elementary charge	e	$1.60217733 \times 10^{-19}$ C	0.3	23–5

[a] In parts per million.

The Gravitational Constant G. This constant, which appears in Newton's law of gravity, is the central constant in both Newton's theory of gravity and Einstein's general theory of relativity. Any theory of the large-scale structure and development of the universe must involve this constant at a deep level.

The Speed of Light c. This constant, which appears in all relativistic equations, is the foundation stone of Einstein's special theory of relativity. The speed of light is large by ordinary standards but it is not infinitely great. If it were, the world would be governed by the laws of classical physics and would be an unimaginably different place.

The Planck Constant h. This constant is the central constant of quantum physics. The Planck constant is small but it is not zero. If it were, the world would be governed by the laws of classical physics and would, again, be unimaginably different. We introduced this constant briefly in Section 8–10. In Chapters 43 and 44 of the extended version of this book—in which we develop the concepts of quantum physics from their origins—the Planck constant will play a central role.

The Elementary Charge e. The basic importance of this constant lies in the fact that it can be combined with two other constants to form a dimensionless number, called the *fine structure constant,*[*] α. Thus,

$$\alpha = \frac{e^2}{2\epsilon_0 hc} \approx \frac{1}{137}. \tag{14}$$

This dimensionless constant is central to the theory of quantum electrodynamics, or QED as it is called.[†] This theory, which combines quantum physics with the theory of relativity, is perhaps the most successful theory in physics in terms of predicting results that agree with experiment. The number 137 has fascinated physicists for decades as they sought—and seek—to explore the significance of the fine structure constant. It is an unusual physicist who, coming upon page 137, does not have a fleeting thought of this constant.

* It received its name for historical reasons, having to do with the detailed structure of the spectrum lines. The quantity ϵ_0 that appears in Eq. 14 has a value that is exact by definition and does not play a fundamental role.

† See Richard P. Feynman, *QED—The Strange Theory of Light and Matter,* Princeton University Press, Princeton, NJ, 1985.

Table 3   The Speed of Light: Some Selected Measurements

Date	Experimenter	Country	Method	Speed (10⁸ m/s)	Uncertainty (m/s)
1600	Galileo	Italy	Lanterns and shutters	"Fast"	?
1676	Roemer	France	Moons of Jupiter	2.14	?
1729	Bradley	England	Aberration of light	3.08	?
1849	Fizeau	France	Toothed wheel	3.14	?
1879	Michelson	United States	Rotating mirror	2.99910	75,000
	Michelson	United States	Rotating mirror	2.99798	22,000
1950	Essen	England	Microwave cavity	2.997925	1,000
1958	Froome	England	Interferometer	2.997925	100
1972	Evenson et al.	United States	Laser method	2.997924574	1.1
1974	Blaney et al.	England	Laser method	2.997924590	0.6
1976	Woods et al.	England	Laser method	2.997924588	0.2
1983	Internationally adopted value:			2.99792458	Exact

Sample Problem 6 It is possible to combine the three constants G, h, and c in such a way as to yield a quantity that has the dimensions of a time. This *Planck time* is given by

$$T_P = \sqrt{\frac{hG}{2\pi c^5}}. \tag{15}$$

Show that this quantity does indeed have the dimensions of time and find its value.

From Table 2 we write the three constants as

$$h = (6.63 \times 10^{-34} \text{ J·s}) \left(\frac{1 \text{ kg·m}^2/\text{s}^2}{1 \text{ J}}\right)$$

$$= 6.63 \times 10^{-34} \text{ kg·m}^2/\text{s},$$

$$G = 6.67 \times 10^{-11} \text{ m}^3/\text{kg·s}^2,$$

and

$$c = 3.00 \times 10^8 \text{ m/s}.$$

To find the Planck time we substitute these values into Eq. 15,

obtaining

$$T_P = \sqrt{\frac{(6.63 \times 10^{-34} \text{ kg·m}^2/\text{s})(6.67 \times 10^{-11} \text{ m}^3/\text{kg·s}^2)}{(2\pi)(3.00 \times 10^8 \text{ m/s})^5}}$$

$$= 5.38 \times 10^{-44} \text{ s}. \quad \text{(Answer)}$$

Check this result carefully, to convince yourself that the units do indeed reduce to those of time.

It is perhaps not surprising that the Planck time, which is built up from the fundamental constants of three great theories, should have a fundamental significance. It turns out to be the time following the Big Bang before which we cannot have confidence that our present theories of physics are valid.

The constants h, G, and c can also be arranged to form quantities that have the dimensions of length and of mass. In Exercise 41 and Problem 42 we ask you to evaluate these quantities, which are called the *Planck length* and the *Planck mass*. Like the Planck time, each has physical significance in studies bearing on the origin and evolution of the universe.

REVIEW AND SUMMARY

Electromagnetism

Electromagnetism is a description of the interactions involving electric charge. Classical electromagnetism, summarized by Maxwell's equations, includes the phenomena of electricity, magnetism, electromagnetic induction (electric generators), and electromagnetic radiation (including all of classical optics).

Electric Charge

The strength of a particle's electromagnetic interaction is partly determined by its electric charge, which can be either positive or negative. Like charges repel and unlike charges attract each other. Objects with equal amounts of the two kinds of charge are electrically neutral, whereas those with an imbalance are electrically charged.

Conducting Materials

Conductors are materials in which a significant number of charged particles (electrons in metals) are free to move. The charged particles in insulators are not free. Semiconductors are intermediate between conductors and insulators in this respect. Superconductors offer no resistance to the flow of electrons.

The Coulomb and Ampere

The SI unit of charge is the coulomb (C). It is defined in terms of the unit of current, the ampere (A), as the charge passing a particular point in one second when a current of one ampere is flowing; see Eq. 3. Sample Problem 2 illustrates the large amount of electric charge in ordinary samples of matter.

Coulomb's Law

Coulomb's law describes *electrostatic* forces; the forces between small (point) electric charges at rest. In SI form,

$$F = \frac{1}{4\pi\epsilon_0} \frac{q_1 q_2}{r^2} \quad \text{(Coulomb's law)}. \tag{4}$$

Here $\epsilon_0 = 8.85 \times 10^{-12} \text{ C}^2/\text{N·m}^2$ is the *permittivity constant*; $1/4\pi\epsilon_0 = 8.99 \times 10^9 \text{ N·m}^2/\text{C}^2$. Sample Problem 1 demonstrates a representative calculation.

The force of attraction or repulsion between point charges at rest acts along the line joining the two charges. If more than two charges are present, Eq. 4 holds for the forces between each pair. The resultant force on each charge is then found, using the superposition principle, as the vector sum of the forces exerted on it by each of the others. Sample Problem 1 shows how such forces can be calculated.

| The Shell Theorems | Electrostatic forces obey the same shell theorems proved earlier for gravitational forces: |

Theorem 1. A uniform spherical shell of charge behaves, for external points, as if all its charge were concentrated at its center.

Theorem 2. A uniform spherical shell of charge exerts no force on a charged particle placed inside the shell.

The Elementary Charge *e*

Electric charge is *quantized*. This means that any charge found in nature can be written as *ne*, where *n* is a positive or negative integer and *e* is a constant of nature called the *elementary charge;* its value is approximately 1.60×10^{-19} C.

Electric Charges in Matter

Matter as we ordinarily encounter it can be regarded as composed of protons, neutrons, and electrons whose properties are summarized in Table 1. Atoms contain a small, dense, positively charged nucleus (composed of neutrons and protons) surrounded by a cloud of electrons attracted to the nucleus by the electrical force. Most forces we ordinarily deal with, such as friction, fluid pressure, contact forces between solid surfaces, and elastic forces, are all manifestations of the electrical interactions between the charged particles of which atoms are made. Sample Problems 4 and 5 show that the electrical force is much stronger than gravity, when both interactions occur simultaneously, and that the nuclear force is still stronger.

Conservation of Charge

Electric charge is conserved. This means that the (algebraic) net charge of an isolated system of charges does not change, no matter what interactions occur within the system. Section 23–6 describes three illustrations of this important conservation law.

QUESTIONS

1. You are given two metal spheres mounted on portable insulating supports. Find a way to give them equal and opposite charges. You may use a glass rod rubbed with silk but may not touch it to the spheres. Do the spheres have to be of equal size for your method to work?

2. In Question 1, find a way to give the spheres equal charges of the same sign. Again, do the spheres need to be of equal size for your method to work?

3. A charged rod attracts bits of dry cork dust which, after touching the rod, often jump violently away from it. Explain.

4. The experiments described in Section 23–2 could be explained by postulating four kinds of charge, that is, on glass, silk, plastic, and fur. What is the argument against this?

5. A positive charge is brought very near to an uncharged insulated conductor. The conductor is grounded while the charge is kept near. Is the conductor charged positively or negatively or not at all if (*a*) the charge is taken away and then the ground connection is removed and (*b*) the ground connection is removed and then the charge is taken away?

6. A charged insulator can be discharged by passing it just above a flame. Explain why.

7. If you rub a coin briskly between your fingers, it will not seem to become charged by friction. Why?

8. If you walk briskly across a carpet, you often experience a spark upon touching a door knob. (*a*) What causes this? (*b*) How might it be prevented?

9. Why do electrostatic experiments not work well on humid days?

10. An insulated rod is said to carry an electric charge. How could you verify this and determine the sign of the charge?

11. If a charged glass rod is held near one end of an insulated uncharged metal rod as in Fig. 8, electrons are drawn to one end, as shown. Why does the flow of electrons cease? After all, there is an almost inexhaustible supply of them in the metal rod.

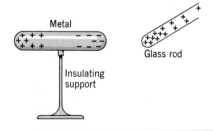

Figure 8 Questions 11 and 12.

12. In Fig. 8, does any electric force act on the metal rod? Explain.

13. A person standing on an insulated stool touches a charged, insulated conductor. Is the conductor discharged completely?

14. (*a*) A positively-charged glass rod attracts a suspended object. Can we conclude that the object is negatively charged? (*b*) A positively charged glass rod repels a suspended object. Can we conclude that the object is positively charged?

15. Is the electric force that one charge exerts on another changed if other charges are brought nearby?

16. A solution of copper sulfate is a conductor. What particles serve as the charge carriers in this case?

17. If the electrons in a metal such as copper are free to move about, they must often find themselves headed toward the metal surface. Why don't they keep on going and leave the metal?

18. Would it have made any important difference if Benjamin Franklin had chosen, in effect, to call electrons positive and protons negative?

19. Coulomb's law predicts that the force exerted by one point charge on another is proportional to the product of the two charges. How might you go about testing this aspect of the law in the laboratory?

20. An electron (charge $= -e$) circulates around a helium nucleus (charge $= +2e$) in a helium atom. Which particle exerts the larger force on the other?

21. The charge of a particle is a true characteristic of the particle, independent of its state of motion. Explain how you can test this statement by making a rigorous experimental check of whether the hydrogen atom is truly electrically neutral.

22. Earnshaw's theorem says that no particle can be in stable equilibrium under the action of electrostatic forces alone. Consider, however, point P at the center of a square of four equal

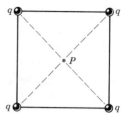

Figure 9 Question 22.

positive charges, as in Fig. 9. If you put a positive test charge there it might seem to be in equilibrium. Every one of the four external charges pushes it toward P. Yet Earnshaw's theorem holds. Can you explain how?

23. The quantum of charge is 1.60×10^{-19} C. Is there a corresponding single quantum of mass?

24. What does it mean to say that a physical quantity is (a) quantized or (b) conserved? Give some examples.

25. In Sample Problem 4 we show that the electrical force is about 10^{39} times stronger than the gravitational force. Can you conclude from this that a galaxy, or a star, or a planet must be essentially neutral electrically?

26. How do we know that electrostatic forces are not the cause of gravitational attraction, between the earth and moon, for example?

EXERCISES AND PROBLEMS

Section 23–4 Coulomb's Law

1E. What would be the force of attraction between two 1.0-C charges separated by a distance of (a) 1.0 m and (b) 1.0 km?

2E. A point charge of $+3.0 \times 10^{-6}$ C is 12 cm distant from a second point charge of -1.5×10^{-6} C. Calculate the magnitude of the force on each charge.

3E. What must be the distance between point charge $q_1 = 26$ μC and point charge $q_2 = -47$ μC in order that the attractive electrical force between them has a magnitude of 5.7 N?

4E. In the return stroke of a typical lightning bolt (see photo at the beginning of this chapter), a current of 2.5×10^4 A flows for 20 μs. How much charge is transferred in this event?

5E. Two equally-charged particles, held 3.2×10^{-3} m apart, are released from rest. The initial acceleration of the first particle is observed to be 7.0 m/s^2 and that of the second to be 9.0 m/s^2. If the mass of the first particle is 6.3×10^{-7} kg, what are (a) the mass of the second particle and (b) the magnitude of the common charge?

6E. Figure 10a shows two charges, q_1 and q_2, held a fixed distance d apart. (a) What is the strength of the electric force that acts on q_1? Assume that $q_1 = q_2 = 20$ μC and $d = 1.5$ m.

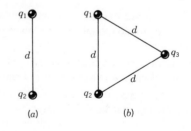

Figure 10 Exercise 6.

(b) A third charge $q_3 = 20$ μC is brought in and placed as shown in Fig. 10b. What now is the strength of the electric force on q_1?

7E. Two identical conducting spheres, 1 and 2, carry equal amounts of charge and are separated by a distance large compared with their diameters. They repel each other with an electrical force $\mathbf{F}$. Suppose now that a third identical sphere 3, having an insulating handle and initially uncharged, is touched first to sphere 1, then to sphere 2, and finally removed. What now is the force $\mathbf{F}'$ between spheres 1 and 2, in terms of $\mathbf{F}$? See Fig. 11.

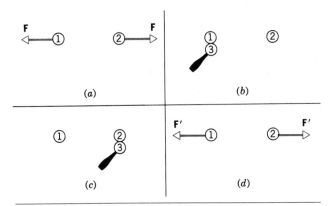

Figure 11 Exercise 7.

8P. Three charged particles lie on a straight line and are separated by a distance d as shown in Fig. 12. Charges q_1 and q_2 are held fixed. Charge q_3, which is free to move, is found to be in equilibrium under the action of the electric forces. Find q_1 in terms of q_2.

Figure 12 Problem 8.

9P. Charges q_1 and q_2 lie on the x axis at points $x = -a$ and $x = +a$, respectively. (a) How must q_1 and q_2 be related for the net force on charge $+Q$, placed at $x = +a/2$, to be zero? (b) Answer the same question if the $+Q$ charge is placed at $x = +3a/2$.

10P. In Fig. 13, what are the horizontal and vertical components of the resultant electric force on the charge in the lower left corner of the square? Assume that $q = 1.0 \times 10^{-7}$ C and $a = 5.0$ cm. The charges are at rest.

11P. Each of two small spheres is charged positively, the combined charge being 5.0×10^{-5} C. If each sphere is repelled from the other by a force of 1.0 N when the spheres are 2.0 m apart, calculate the charge on each sphere.

12P. Two identical conducting spheres, having charges of opposite sign, attract each other with a force of 0.108 N when separated by 50.0 cm. The spheres are connected by a thin conducting wire, which is then removed, and thereafter the spheres repel each other with a force of 0.036 N. What were the initial charges on the spheres?

13P. Two fixed charges, $+1.0$ μC and -3.0 μC, are 10 cm apart. Where may a third charge be located so that no force acts on it?

14P. The charges and coordinates of two charged particles held fixed in the xy plane are: $q_1 = +3.0$ μC, $x = 3.5$ cm, $y = 0.50$ cm, and $q_2 = -4.0$ μC, $x = -2.0$ cm, $y = 1.5$ cm. (a) Find the magnitude and direction of the electrical force on q_2. (b) Where could you locate a third charge $q_3 = +4.0$ μC such that the total electrical force on q_2 is zero?

15P. Two *free* point charges $+q$ and $+4q$ are a distance L apart. A third charge is so placed that the entire system is in equilibrium. (a) Find the location, magnitude, and sign of the third charge. (b) Show that the equilibrium is unstable.

16P. (a) What equal positive charges would have to be placed on the earth and on the moon to neutralize their gravitational attraction? Do you need to know the lunar distance to solve this problem? Why or why not? (b) How many thousand kilograms of hydrogen would be needed to provide the positive charge calculated in part (a)?

17P. A charge Q is fixed at each of two opposite corners of a square. A charge q is placed at each of the other two corners. (a) If the resultant electrical force on Q is zero, how are Q and q related? (b) Could q be chosen to make the resultant electrical force on *every* charge zero? Explain your answer.

18P. A certain charge Q is to be divided into two parts $(Q - q)$ and q. What is the relation of Q to q if the two parts, placed a given distance apart, are to have a maximum Coulomb repulsion?

19P. Two similar tiny conducting balls of mass m are hung from silk threads of length L and carry similar charges q as in

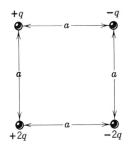

Figure 13 Problem 10.

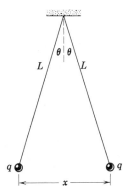

Figure 14 Problems 19 and 20.

Fig. 14. Assume that θ is so small that $\tan\theta$ can be replaced by its approximate equal, $\sin\theta$. (a) To this approximation show that, for equilibrium,

$$x = \left(\frac{q^2 L}{2\pi\epsilon_0 mg}\right)^{1/3},$$

where x is the separation between the balls. (b) If $L = 120$ cm, $m = 10$ g, and $x = 5.0$ cm, what is q?

20P. If the balls of Fig. 14 are conducting, (a) what happens to them after one is discharged? Explain your answer. (b) Find the new equilibrium separation.

21P. Figure 15 shows a long, insulating, massless rod of length L, pivoted at its center and balanced with a weight W at a distance x from the left end. At the left and right ends of the rod are attached positive charges q and $2q$, respectively. A distance h directly beneath each of these charges is a fixed positive charge Q. (a) Find the distance x for the position of the weight when the rod is balanced. (b) What value should h have so that the rod exerts no vertical force on the bearing when balanced? Neglect the interaction between charges at the opposite ends of the rod.

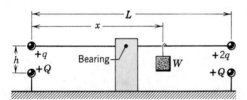

Figure 15 Problem 21.

Section 23–5 Charge Is Quantized

22E. What is the force of attraction between a singly charged sodium ion and an adjacent singly charged chlorine ion in a salt crystal if their separation is 2.82×10^{-10} m?

23E. A neutron is thought to be composed of one "up" quark of charge $+\frac{2}{3}e$ and two "down" quarks each having charge $-\frac{1}{3}e$. If the down quarks are 2.6×10^{-15} m apart inside the neutron, what is the repulsive electrical force between them?

24E. What is the total charge in coulombs of 75 kg of electrons?

25E. How many coulombs of charge, positive and negative, are there in 1 mol of molecular hydrogen gas?

26E. The electrostatic force between two identical ions that are separated by a distance of 5.0×10^{-10} m is 3.7×10^{-9} N. (a) What is the charge on each ion? (b) How many electrons are missing from each ion?

27E. Two small water droplets in air are 1.0 cm apart. Each has acquired a charge of 1.0×10^{-16} C. (a) Find the magnitude of the electric force on each droplet. (b) How many excess electrons are on each droplet?

28E. (a) How many electrons would have to be removed from a penny to leave it with a charge of $+1.0 \times 10^{-7}$ C? (b) To what fraction of the electrons in the penny does this correspond? See Sample Problem 2.

29E. How far apart must two protons be if the electrical repulsive force acting on either one is equal to its weight at the earth's surface?

30E. An electron is in a vacuum near the surface of the earth. Where should a second electron be placed so that the net force on the first electron, owing to the other electron and to gravity, is zero?

31P. A 100-W lamp operated on a 120-V circuit has a current (assumed steady) of 0.83 A in its filament. How long does it take for one mole of electrons to pass through the lamp?

32P. Protons in cosmic rays strike the earth's atmosphere at a rate, averaged over the earth's surface, of 1500 protons/m²·s. What total current does the earth receive from beyond its atmosphere in the form of incident cosmic ray protons?

33P. Calculate the number of coulombs of positive charge in a glass of water. Assume the volume of the water to be 250 cm³.

34P. In the compound CsCl (cesium chloride), the Cs atoms are situated at the corners of a cube with a Cl atom at the cube's center. The edge length of the cube is 0.40 nm; see Fig. 16. The Cs atoms are each deficient in one electron and the Cl atom carries one excess electron. (a) What is the strength of the net electric force on the Cl atom resulting from the eight Cs atoms shown? (b) Suppose that the Cs atom marked with an arrow is missing (crystal defect). What now is the net electric force on the Cl atom resulting from the seven remaining Cs atoms?

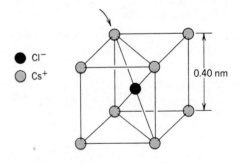

Figure 16 Problem 34.

35P. We know that, within the limits of measurement, the magnitudes of the negative charge on the electron and the positive charge on the proton are equal. Suppose, however, that these magnitudes differed from each other by as little as 0.00010%. With what force would two copper pennies, placed

1.0 m apart, then repel each other? What do you conclude? (*Hint:* See Sample Problems 2 and 3.)

36P. Two engineering students (John at 200 lb and Mary at 100 lb) are 100 ft apart. Let each have a 0.01% imbalance in their amount of positive and negative charge, one student being positive and the other negative. Estimate *roughly* the electrostatic force of attraction between them. (*Hint:* Replace the students by equivalent spheres of water.)

Section 23-6 Charge Is Conserved

37E. In beta decay a heavy fundamental particle changes to another heavy particle and an electron or a positron is emitted. (*a*) If a proton becomes a neutron, which particle is emitted? (*b*) If a neutron becomes a proton, which particle is emitted?

38E. Identify X in the following nuclear reactions:

(*a*) $^1H + {}^9Be \rightarrow X + n$;

(*b*) $^{12}C + {}^1H \rightarrow X$;

(*c*) $^{15}N + {}^1H \rightarrow {}^4He + X$.

(*Hint:* See Appendix D.)

39E. In the radioactive decay of ^{238}U (see Eq. 10), the center of the emerging ^4He particle is, at a certain instant, 9.0×10^{-15} m from the center of the residual nucleus ^{234}Th. At this instant, (*a*) what is the force on the ^4He particle and (*b*) what is its acceleration?

Section 23-7 The Constants of Physics: An Aside

40E. Verify that the fine structure constant is dimensionless and that its numerical value can be expressed as shown in Eq. 14.

41E. (*a*) Arrange the quantities h, G, and c to form a quantity with dimensions of length. (*Hint:* Combine the Planck time with the speed of light; see Sample Problem 6.) (*b*) Evaluate this "Planck length" numerically.

42P. (*a*) Arrange the quantities h, G, and c to form a quantity with dimensions of mass. Do not include any dimensionless factors. (*Hint:* Consider the units of h, G, and c as displayed in Sample Problem 6.) (*b*) Evaluate this "Planck mass" numerically.

CHAPTER 24
THE ELECTRIC FIELD

In a fly ash precipitator, an electric field exerts a force on a charged particle, deflecting it sideways. On the right, the electric field is ON. On the left, it has been turned OFF momentarily. The electric field moves a lot of fly ash.

24–1 Charges and Forces: A Closer Look

During the 1986 flyby of Uranus by the spacecraft *Voyager 2,* the round trip radio communication time from earth was about 5.5 h. What has this to do with charges, forces, and Coulomb's law? As it turns out, a great deal.

Coulomb's law tells us that two point charges exert forces on each other. The law says nothing, however, about how one charge "senses" the distant presence of the other. If the charges are stationary, we do not need to raise this question because we can solve all problems that arise if we know the magnitudes and the positions of the charges. Suppose, however, that one charge suddenly moves a little closer to the second charge. According to

Coulomb's law, the force on the second charge must increase. Speaking loosely, how does the second charge "know" that the first charge has moved?

In our *Voyager 2* example, the first charge might be in the transmitting antenna of the space probe and the second in the antenna of the receiver back on earth. We know that the second charge responds to the motion of the first only after a time L/c, where L is the distance between the charges and c is the speed of light.

The key to understanding this kind of communication between charges is the concept of the *electromagnetic field.* We say that the second charge "learns" that the first charge has moved by means of an electromagnetic field disturbance that travels through the intervening space at the speed of light. This concept leads to the view that light is an electromagnetic wave and that the once separate sciences of electricity, magnetism, and optics can be joined together into a single comprehensive body of knowledge. Among the many practical consequences of the electromagnetic field idea were the invention of radio, the development of microwave radar and television, and a full understanding of a host of electromagnetic devices such as motors, generators, and transformers.

Our plan in these chapters is to establish separately the concepts of the electric field (for stationary charges) and of the magnetic field (for constant currents) and then to use what we learn to develop the concept of light as a traveling electromagnetic wave.

24-2 The Electric Field

The temperature has a definite value at every point in space in the room in which you may be sitting. You can measure it by putting a thermometer at that point. We call such a distribution of temperatures a *temperature field*. In much the same way, you can think of a *pressure field* extending throughout the atmosphere.

These two examples are *scalar fields,* temperature and pressure being scalar quantities. The gravitational field near the earth is a ready example of a *vector field*. To every point of space we can assign a vector **g** that represents the acceleration of a test body—released at that point—resulting from the earth's gravitational attraction. If m is the mass of the test body and **F** the gravitational force that acts on it, **g** is given by

$$\mathbf{g} = \mathbf{F}/m \quad \text{(gravitational field).} \quad (1)$$

If we place a test body carrying a positive electric charge q near a charged rod, an electrostatic force **F** will act on it. We speak of an *electric field* in this region, and we represent it by a vector **E**, defined by

$$\boxed{\mathbf{E} = \mathbf{F}/q} \quad \text{(electric field).} \quad (2)$$

The direction of the vector **E** is that of the vector **F**. That is, it is the direction in which a resting *positive* test charge, placed at the point, would be accelerated.

The SI unit for the electric field is the newton/coulomb (N/C). Although we usually write the unit for the gravitational field as the meter/second², we could just as easily write it as the newton/kilogram. Thus, both E and g are expressed as a force divided by a property—charge or mass—of the test body. Table 1 shows the electric fields that occur in a few situations.

Table 1 Some Electric Fields[a]

Field	Value (N/C)
At the surface of a uranium nucleus	3×10^{21}
Within a hydrogen atom, at the electron orbit	5×10^{11}
Electric breakdown occurs in air	3×10^{6}
At the charged drum of a photocopier	10^{5}
The electron beam accelerator in a TV set	10^{5}
Near a charged plastic comb	10^{3}
In the lower atmosphere	10^{2}
Inside the copper wire of household circuits	10^{-2}

[a] Approximate values.

If the primary charges that set up the electric field that we are examining were fixed in position, we could use a test charge of any size. If the primary charges are not fixed, however, the test charge should be as small as possible. Otherwise, it might cause the primary charges to change their positions. Put another way, we do not want the act of measurement to change the thing being measured. Because charge is quantized, we cannot, of course, use a test charge smaller than the elementary charge e.

The force acting between charged particles was originally thought of as a direct and instantaneous interaction between the charges. We can represent this action-at-a-distance view as

$$\text{charge} \rightleftharpoons \text{charge.} \quad (3)$$

Today, we think of the electric field as an intermediary between the charges. Thus, (1) charge q_1 in Fig. 1a sets up an electric field in the surrounding space, suggested by the dots in the figure and (2) this field acts on charge q_2, the action showing up as the force $\mathbf{F}_2$ that q_2 experiences.

As Fig. 1b suggests, we could just as well say that q_1 is immersed in an electric field set up by q_2 and for that reason experiences a force $\mathbf{F}_1$. Note in Fig. 1 that the fields set up by q_1 and q_2 are different but that the forces acting on the two charges have the same magnitude and form an action–reaction pair. That is, $\mathbf{F}_2 = -\mathbf{F}_1$.

In thinking about how charges exert forces on each other, we see our task divided into two parts: (1) calculating the field set up by a given distribution of charge and (2) calculating the force that a given field will exert on a charge placed in it. We think in terms of

$$\text{charge} \rightleftharpoons \text{field} \rightleftharpoons \text{charge} \quad (4)$$

rather than in terms of direct action between the charges, as Eq. 3 suggests.

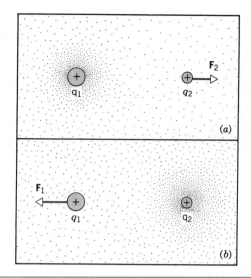

Figure 1 (a) Charge q_1 sets up an electric field that exerts a force $\mathbf{F}_1$ on charge q_2. (b) Charge q_2 sets up a field that exerts a force $\mathbf{F}_2$ on charge q_1. If the charges are different, the fields they set up will be different. The forces, however, are always equal in magnitude and form an action–reaction pair. That is, $\mathbf{F}_1 = -\mathbf{F}_2$.

Sample Problem 1 A proton is placed in a uniform electric field $\mathbf{E}$. What must be the magnitude and direction of this field if the electrostatic force acting on the proton is to just balance its weight?

From Eq. 2, replacing q by e and F by mg we have

$$E = \frac{F}{q} = \frac{mg}{e} = \frac{(1.67 \times 10^{-27} \text{ kg})(9.8 \text{ m/s}^2)}{1.60 \times 10^{-19} \text{ C}}$$

$$= 1.0 \times 10^{-7} \text{ N/C, directed up.} \qquad \text{(Answer)}$$

This is a very weak field indeed. $\mathbf{E}$ must point vertically upward to float the (positively charged) proton.

24–3 Lines of Force

Michael Faraday (1791–1867), who introduced the field concept, thought of the space around a charged body as filled with *lines of force*. Although we no longer attach the same kind of reality to these lines that Faraday did, they are still a nice way to visualize field patterns and that is how we shall use them.

Figure 2 shows the field lines, as we shall call them, for a uniform sphere of negative charge. The lines point radially inward because a positive test charge would be accelerated in that direction. You can tell that the elec-

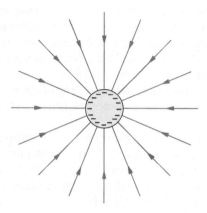

Figure 2 Field lines for a uniform sphere of negative charge. The lines originate on positive charges on distant bodies that are not shown, perhaps the walls of the room. The direction of the electric field at all points near the sphere is radially inward.

tric field becomes weaker as you move away from the central charge because the field lines become progressively farther apart.

In interpreting field line patterns, bear in mind that the field lines in Fig. 2, and in similar figures in this chapter, are representations on the plane of the page of what are really three-dimensional patterns. The relation between the field lines and the electric field vector is this: (1) The tangent to a field line at any point gives the *direction* of $\mathbf{E}$ at that point, and (2) the field lines are drawn so that the number of lines per unit area, in a plane at right angles to the lines, is proportional to the *magnitude* of $\mathbf{E}$. This means that where the field lines are close together E is large and that where they are far apart E is small. Field lines always originate on positive charges and terminate on negative charges. When field lines are shown "hanging in the air," as in Fig. 2 and in other figures of this chapter, we must imagine the lines extended to originate or terminate on charges located on objects not shown, such as the walls of the room.

Figure 3 shows the field lines for a section of an infinitely large, uniform sheet of positive charge. No real sheet can be infinitely large but if we consider only points that are close to a real sheet and not near its edges the pattern of Fig. 3 will hold.

A positive test charge, released in front of such a sheet, would move away from it along a perpendicular line, all other directions being ruled out by considerations of symmetry. Thus, the electric field vector at any point near the sheet must point away from the sheet at right angles. The lines are uniformly spaced, which

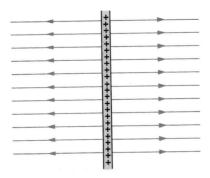

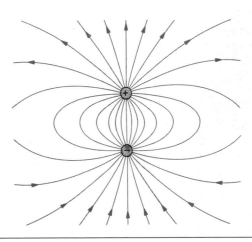

Figure 3 Field lines close to a section of a very large sheet of uniformly distributed positive charge. The lines terminate on negative charges on distant bodies that are not shown. The electric field points away from the sheet and at right angles to it.

Figure 5 Field lines for a positive and a negative point charge of equal magnitude. Two such charges attract each other. The pattern has rotational symmetry about an axis passing through the two charges.

means that **E** has the same magnitude for all points near the sheet.

Figure 4 shows the field lines for two equal positive charges. Figure 5 shows the pattern for two charges that are equal in magnitude but of opposite sign, a configuration that we call an *electric dipole*. Although we do not

often use field lines quantitatively, they are very useful in visualizing what is going on. Can you not almost "see" the charges being pushed apart in Fig. 4 and pulled together in Fig. 5?

Sample Problem 2 In Fig. 2, how does the magnitude of the electric field vary with the distance from the center of the uniformly charged sphere?

Suppose that N field lines terminate on the sphere of Fig. 2. Draw an imaginary concentric sphere of radius r. The number of lines per unit area at any point on this sphere is $N/4\pi r^2$. Because E is proportional to this quantity, we can write $E \propto 1/r^2$. Thus, the electric field set up by a uniform sphere of charge varies as the inverse square of the distance from the center of the sphere. In much the same way, you can show that the electric field set up by an infinitely long uniformly charged rod will vary as $1/r$, where r is the perpendicular distance from the axis of the rod.

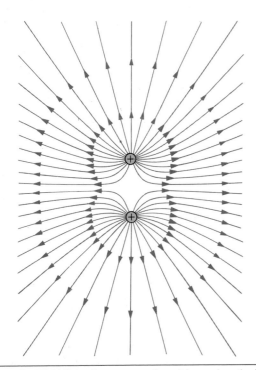

Figure 4 Field lines for two equal positive point charges. Two such charges repel each other. the pattern has rotational symmetry about a vertical axis. The lines terminate on negative charges on distant bodies that are not shown.

24-4 Calculating the Field: A Point Charge

Given a charged object, what electric field does it set up at nearby points? We look first at the case of a single point charge q. Put a test charge q_0 a distance r from this charge. The magnitude of the force acting on q_0, from Coulomb's law (see Eq. 4 of Chapter 23), is

$$F = \frac{1}{4\pi\epsilon_0} \frac{qq_0}{r^2}. \tag{5}$$

The magnitude of the electric field follows from Eq. 2, or

$$E = \frac{F}{q_0} = \frac{1}{4\pi\epsilon_0} \frac{q}{r^2} \quad \text{(point charge).} \quad (6)$$

The direction of **E** is along a radial line from q, pointing outward if q is positive and inward if q is negative.

Sample Problem 3 Figure 6 shows a charge q_1 of $+1.5\ \mu C$ and a charge q_2 of $+2.3\ \mu C$. The first charge is at the origin of an x axis and the second is at a position $x = L$, where $L = 13$ cm. At what point P along the x axis is the electric field zero?

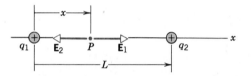

Figure 6 Sample Problem 3. Two positive point charges are separated by a distance L. For what value of x does the electric field vanish?

The point must lie between the charges because only in this region do the forces exerted by q_1 and by q_2 on a test charge oppose each other. If E_1 is the electric field due to q_1 and E_2 is that due to q_2, the magnitudes of these vectors must be equal, or

$$E_1 = E_2.$$

From Eq. 6 we then have

$$\frac{1}{4\pi\epsilon_0} \frac{q_1}{x^2} = \frac{1}{4\pi\epsilon_0} \frac{q_2}{(L-x)^2},$$

where x is the position of point P. Taking the square root of each side and solving for x, we obtain

$$x = \frac{L}{1 + \sqrt{q_2/q_1}} = \frac{13\ \text{cm}}{1 + \sqrt{2.3\ \mu C/1.5\ \mu C}} = 5.8\ \text{cm.} \quad \text{(Answer)}$$

This result is positive and is less than L, confirming that the zero-field point lies between the two charges, as we know it must.

Sample Problem 4 The nucleus of a uranium atom has a radius R of 6.8 fm. What is the magnitude of the electric field at its surface?

The nucleus has a positive charge of Ze where the atomic number $Z (= 92)$ is the number of protons in the nucleus and e ($= 1.60 \times 10^{-19}$ C) is the charge of the proton. This charge, which we assume to be distributed with spherical symmetry,

behaves electrically for external points as if it were concentrated at the nuclear center. From Eq. 6 then,

$$E = \frac{1}{4\pi\epsilon_0} \frac{Ze}{R^2} = \frac{(8.99 \times 10^9\ \text{N}\cdot\text{m}^2/\text{C}^2)(92)(1.60 \times 10^{-19}\ \text{C})}{(6.8 \times 10^{-15}\ \text{m})^2}$$
$$= 2.9 \times 10^{21}\ \text{N/C.} \quad \text{(Answer)}$$

This value appears as an entry in Table 1.

24–5 Calculating the Field: An Electric Dipole

Figure 7 shows two charges of magnitude q but of opposite sign, separated by a distance d. As we indicated in connection with Fig. 5, we call this configuration an *electric dipole*. What is the electric field due to the dipole of Fig. 7 at a point P, a distance z from the midpoint of the dipole on its central axis?

From symmetry, the electric field **E** at point P—and also the fields E_+ and E_- due to the separate charges that make up the dipole—must lie along the dipole axis, which we take to be a z axis. The superposition principle

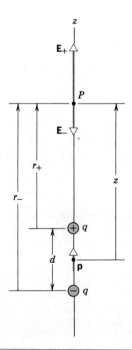

Figure 7 An electric dipole. the electric fields at point P on the dipole axis resulting from the separate charges are shown. How does the resultant field vary with z, the distance from the center of the dipole?

applies to electric fields so that we can write for the magnitude of the field at P,

$$E = E_+ - E_-$$
$$= \frac{1}{4\pi\epsilon_0}\frac{q}{r_+^2} - \frac{1}{4\pi\epsilon_0}\frac{q}{r_-^2}$$
$$= \frac{q}{4\pi\epsilon_0(z - \frac{1}{2}d)^2} - \frac{q}{4\pi\epsilon_0(z + \frac{1}{2}d)^2}. \quad (7)$$

After a little algebra, we can rewrite this equation as

$$E = \frac{q}{4\pi\epsilon_0 z^2}\left(1 - \frac{d}{2z}\right)^{-2} - \frac{q}{4\pi\epsilon_0 z^2}\left(1 + \frac{d}{2z}\right)^{-2}. \quad (8)$$

Physically, we are usually interested in the electrical effect of a dipole only at distances that are large compared with the dimensions of the dipole, that is, at distances such that $z \gg d$. At such large distances, we have $d/2z \ll 1$ in Eq. 8. We can then expand the two quantities in the parentheses in that equation by the binomial theorem, obtaining

$$E = \frac{q}{4\pi\epsilon_0 z^2}\left[\left(1 + \frac{d}{z} - \cdots\right) - \left(1 - \frac{d}{z} + \cdots\right)\right]. \quad (9)$$

As an approximate result that holds at large distances, we then have

$$E \approx \frac{q}{4\pi\epsilon_0 z^2}\frac{2d}{z} = \frac{1}{2\pi\epsilon_0}\frac{qd}{z^3}. \quad (10)$$

The product qd, which involves the two intrinsic properties of the dipole, is called the *electric dipole moment p* of the dipole. Thus, we can write Eq. 10 as

$$\boxed{E = \frac{1}{2\pi\epsilon_0}\frac{p}{z^3}} \quad \text{(electric dipole).} \quad (11)$$

By defining the electric dipole moment as a vector **p**, we can use it to specify the direction of the dipole axis. The magnitude of **p**, as we have seen, is qd and its direction is taken to be from the negative to the positive end of the dipole. The dipole moment vector is indicated in Fig. 7.

Equation 11 shows that, if we measure the electric field of a dipole only at distant points, we can never find q and d separately, only their product. The field at distant points would be unchanged if, for example, q were doubled and d simultaneously cut in half.

Although Eq. 11 holds only for distant points along the dipole axis, it turns out that E for a dipole varies as $1/r^3$ for *all* distant points, whether or not they lie on the z axis; here r is the radial distance of the point in question

from the dipole center. In Problem 31, for example, we ask you to show this by deriving an expression for $E(r)$ at distant points in the equatorial plane of the dipole.

Inspection of Fig. 7 and of the field lines in Fig. 5 shows that the direction of **E** for distant points on the dipole axis is always in the direction of the dipole moment vector **p**. This is true no matter whether point P in Fig. 7 is on the upper or the lower part of the dipole axis.

Inspection of Eq. 11 shows that if you were to double your distance from a dipole, the electric field would drop by a factor of 8. If you were to double your distance from a single point charge, however (see Eq. 6), the electric field would drop only by a factor of 4. It is not hard to see why the electric field of a dipole should decrease more rapidly with distance than does the field of a single charge. For distant points a dipole looks like two equal but opposite charges that are almost — but not quite — on top of each other. We are not surprised then to learn that their electric fields at distant points almost — but not quite — cancel each other.

Sample Problem 5 A molecule of water vapor (H_2O) contains 10 protons and 10 electrons and is thus electrically neutral. However, the centers of positive and of negative charge do not quite coincide so that, for external points, the molecule behaves as if it were an electric dipole like that of Fig. 7, its dipole moment p being 6.2×10^{-30} C·m. What is the electric field at a distance z of 1.1 nm from the molecule on the dipole axis? This distance is about 10 diameters of a hydrogen atom. From Eq. 11

$$E = \frac{1}{2\pi\epsilon_0}\frac{p}{z^3} = \frac{6.2 \times 10^{-30}\text{ C·m}}{(2\pi)(8.85 \times 10^{-12}\text{ C}^2/\text{N·m}^2)(1.1 \times 10^{-9}\text{ m})^3}$$
$$= 8.4 \times 10^7\text{ N/C.} \quad \text{(Answer)}$$

24-6 Calculating the Field: A Ring of Charge

Sometimes we find it desirable — or even necessary — to express the charge on a body in terms other than its total magnitude q. In situations involving charge spread out along a line, for example, we may report the *linear charge density* λ, whose SI unit is coulombs/meter. Table 2, to which we shall refer in future problems, summarizes the possibilities.

Table 2 Some Measures of Electric Charge

Name	Symbol	SI Unit
Charge	q	C
Linear charge density	λ	C/m
Surface charge density	σ	C/m²
Volume charge density	ρ	C/m³

Figure 8 shows a thin ring of radius R, charged to a constant linear charge density λ around its circumference. We may imagine the ring to be made of plastic or some other insulator, so that the charges can be regarded as fixed in place. What is the electric field at a point P, a distance z from the plane of the ring along its central axis?

We cannot apply Coulomb's law directly because the ring is not a point charge. We must, instead, break up the ring into charge elements that are small enough so that we can apply Coulomb's law to *them*. We will then find the electric field due to the ring by adding up the field contributions of all these charge elements.

Consider a differential element of the ring of length ds located at an arbitrary position on the ring in Fig. 8. It contains an element of charge given by

$$dq = \lambda \, ds. \tag{12}$$

This element sets up a differential field $d\mathbf{E}$ at point P. From Eq. 6 we have

$$dE = \frac{1}{4\pi\epsilon_0} \frac{\lambda \, ds}{r^2} = \frac{\lambda \, ds}{4\pi\epsilon_0(z^2 + R^2)}. \tag{13}$$

Figure 8 A uniform ring of charge. What is the electric field at point P, located on the axis of the ring?

Note that all charge elements that make up the ring are the same distance r from point P.

To find the resultant field at P we must add up, vectorially, all the field contributions $d\mathbf{E}$ made by the differential elements of the ring. This may seem like a hard job but, as so often happens, symmetry comes to our rescue. From symmetry, we know that the resultant field $\mathbf{E}$ must lie along the axis of the ring. Thus, only the components of $d\mathbf{E}$ parallel to this axis need be counted. Components of $d\mathbf{E}$ at right angles to the axis will cancel in pairs, the contribution from a charge element at any ring location being canceled by the contribution from the diametrically opposite charge element. Thus, our vector addition becomes a scalar addition of parallel axial components.

The axial component of $d\mathbf{E}$ is $dE \cos \theta$. From Fig. 8 we see that

$$\cos \theta = \frac{z}{r} = \frac{z}{(z^2 + R^2)^{1/2}}. \tag{14}$$

If we combine Eq. 14 and Eq. 13, we find

$$dE \cos \theta = \frac{z\lambda \, ds}{4\pi\epsilon_0(z^2 + R^2)^{3/2}}. \tag{15}$$

To add the various contributions, we need add only the ds elements, because all other quantities in Eq. 15 have the same value for all charge elements. Thus,

$$
\begin{aligned}
E = \int dE \cos \theta &= \frac{z\lambda}{4\pi\epsilon_0(z^2 + R^2)^{3/2}} \int ds \\
&= \frac{z\lambda(2\pi R)}{4\pi\epsilon_0(z^2 + R^2)^{3/2}},
\end{aligned} \tag{16}
$$

in which the integral is simply $2\pi R$, the circumference of the ring. But $\lambda(2\pi R)$ is q, the total charge on the ring, so that we can write Eq. 16 as

$$\boxed{E = \frac{qz}{4\pi\epsilon_0(z^2 + R^2)^{3/2}}} \quad \text{(charged ring).} \tag{17}$$

For points far enough away from the ring so that $z \gg R$, we can put $R = 0$ in Eq. 17. Doing so yields

$$E \approx \frac{1}{4\pi\epsilon_0} \frac{q}{z^2}, \tag{18}$$

which (with z replaced by r) is Eq. 6. We are not surprised because, at large enough distances, the ring behaves electrically like a point charge. We note also from Eq. 17 that $E = 0$ for $z = 0$. This is also not surprising because a test charge at the center of the ring would be pushed or pulled equally in all directions in the plane of the ring and would experience no net force. In general, if symmetry considerations do not permit a (single) preferred direction for a force to act on a test charge, the field at that site must be zero. If there *is* a preferred direction, however, that is the direction of the field.

24-7   Calculating the Field: A Charged Disk

Figure 9 shows a circular plastic disk of radius R, carrying a uniform surface charge of density σ on its upper surface; see Table 2. What is the electric field at point P, a distance z from the disk along its axis?

Our plan is to divide the disk up into concentric rings and then to calculate the electric field by adding up, that is, by integrating, the contributions of the various rings. Figure 9 shows a flat ring with radius s and of width ds, its total charge being

$$dq = \sigma \, dA = \sigma(2\pi s) \, ds, \qquad (19)$$

where dA is the differential area of the ring.

We have already solved the problem of the electric field due to a ring of charge. Substituting dq from Eq. 19 for q in Eq. 17, and replacing R in Eq. 17 by s, we obtain

$$dE = \frac{z\sigma 2\pi s \, ds}{4\pi\epsilon_0(z^2 + s^2)^{3/2}}$$

$$= (\sigma z/4\epsilon_0)(z^2 + s^2)^{-3/2}(2s)ds.$$

We can now find E by integrating over the surface of the disk, that is, by integrating with respect to the vari-

able s. Note that z remains constant during this process. Thus,

$$E = \int dE = \frac{\sigma z}{4\epsilon_0} \int_0^R (z^2 + s^2)^{-3/2} (2s)ds, \qquad (20)$$

the limits on the variable being $s = 0$ and $s = R$. We see that this integral is of the form $\int X^m dX$, in which $X = (z^2 + s^2)$, $m = -3/2$, and $dX = (2s)ds$; see Appendix G. Integrating and substituting the limits, we obtain

$$\boxed{E(r) = \frac{\sigma}{2\epsilon_0}\left(1 - \frac{z}{\sqrt{z^2 + R^2}}\right)}$$

$$\text{(charged disk)} \quad (21)$$

as the final result.

For $R \gg z$, the second term in the parentheses in Eq. 21 approaches zero and this equation reduces to

$$E = \frac{\sigma}{2\epsilon_0} \quad \text{(infinite sheet)}. \qquad (22)$$

This is the electric field set up by a uniform sheet of charge of infinite extent; see Fig. 3. Note that Eq. 22 also follows as $z \to 0$ in Eq. 21; for such nearby points the charged disk does indeed behave as if it were infinite in extent. In Exercise 40 we ask you to show that Eq. 21 reduces to the field of a point charge for $z \gg R$.

Sample Problem 6   The disk of Fig. 9 has a radius R of 2.5 cm and a surface charge density σ of $= +5.3$ μC/m^2 on its upper face. (This, incidentally, is a possible value for the surface charge density on the photosensitive cylinder of a photocopying machine.) (a) What is the electric field at an axial point a distance $z = 12$ cm from the disk?

From Eq. 21 we have

$$E = \frac{\sigma}{2\epsilon_0}\left(1 - \frac{z}{\sqrt{z^2 + R^2}}\right)$$

$$= \frac{5.3 \times 10^{-6} \text{ C/m}^2}{(2)(8.85 \times 10^{-12} \text{ C}^2/\text{N}\cdot\text{m}^2)}\left(1 - \frac{12 \text{ cm}}{\sqrt{(12 \text{ cm})^2 + (2.5 \text{ cm})^2}}\right)$$

$$= 6.3 \times 10^3 \text{ N/C}. \qquad \text{(Answer)}$$

Note that the values of R and r in this problem can be left in centimeters because the units cancel.

(b) What is the electric field at the surface of the disk?
From Eq. 22 we have

$$E = \frac{\sigma}{2\epsilon_0} = \frac{5.3 \times 10^{-6} \text{ C/m}^2}{(2)(8.85 \times 10^{-12} \text{ C}^2/\text{N}\cdot\text{m}^2)}$$

$$= 3.0 \times 10^5 \text{ N/C}. \qquad \text{(Answer)}$$

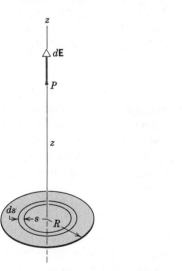

Figure 9   A uniformly charged disk of radius R. What is the electric field at point P, on the axis of the disk?

This value holds for all points close to the surface of the disk and not near its edge. Comparison with Table 1 shows that this value is less than the value at which air breaks down electrically so that the charged disk in this problem will not spark.

24–8   A Point Charge in an Electric Field

In the preceding four sections we looked into the first half of the charge–field–charge interaction: Given a charge distribution, what electric field does it set up in the surrounding space? Here we look into the second half of the interaction: What happens to a charged particle placed in a known external electric field?

The answer is that an electric force given by $\mathbf{F} = \mathbf{E}q$ acts on the particle. Unless this force is counterbalanced by other forces, the particle will accelerate. We look at two situations of special interest, an important example chosen from the history of physics and another example chosen from modern technology.

Measuring the Elementary Charge.*   Figure 10 shows the apparatus used by the American physicist Robert A. Millikan in 1910–1913 to measure the elementary charge e. Oil droplets are introduced into chamber A by an atomizer, some of them becoming charged, either positively or negatively, in the process. Consider a drop that finds its way through a small hole in plate P_1 and drifts into chamber C. Let us assume that this drop carries a charge q, which we take to be negative.

If there is no electric field, two forces act on the drop, its weight mg and an upwardly directed viscous drag force, whose magnitude F is proportional to the speed of the falling drop. The drop quickly comes to a constant terminal speed v at which these two forces are just balanced, as Fig. 11a shows.

An electric field $\mathbf{E}$ is now set up in the chamber, by connecting battery B between plates P_1 and P_2. A third force, $q\mathbf{E}$, now acts on the drop. If q is negative, this force will point upward, and—we assume—the drop will now drift upward, at a new terminal speed v'. Figure 11b shows how the upward electric force $q\mathbf{E}$ is just balanced

---

* For details of Millikan's experiments, see Henry A. Boorse and Lloyd Motz (eds.), *The World of the Atom*, Basic Books, New York, 1966, Chapter 40. To see Millikan from the point of view of two physicists who knew him as graduate students, see "Robert A. Millikan, Physics Teacher," by Alfred Romer, *The Physics Teacher*, February 1978, and "My Work with Millikan on the Oil-Drop Experiment," by Harvey Fletcher, *Physics Today*, June 1982.

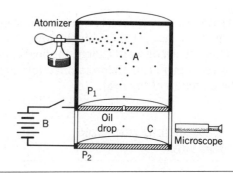

Figure 10   The Millikan oil-drop apparatus for measuring the elementary charge e. In chamber C the drop is acted on by the gravitational force, a force due to the electric field, and, if the drop is moving, a viscous drag force.

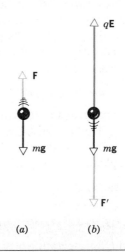

Figure 11   (a) An oil drop falls at terminal speed v in a field-free region. Its weight is balanced by an upward drag force. (b) An electric field force acts upward on the drop, which now rises with a terminal speed v'. The drag force, which always opposes the velocity, now acts downward.

by the weight mg and the new drag force F'. Note that, in each case, the drag force points in the direction opposite to that in which the drop is moving and has a magnitude proportional to the speed of the drop. The charge q on the drop can be found from measurements of v and v'.

Millikan found that the values of q were all consistent with the relation

$$q = ne \qquad n = 0, \pm 1, \pm 2, \pm 3, \cdots, \qquad (23)$$

in which e is a fundamental constant called the *elementary charge* and whose value is 1.60×10^{-19} C. Millikan's experiment is convincing proof that charge is quantized. He earned the 1923 Nobel Prize in physics in

Paper
Input signals
E
Drop generator
Charging unit
Deflecting plates
Gutter

Figure 12 The essential features of an ink-jet printer. An input signal from the computer controls the charge given to the drop and thus the position on the page at which the drop lands. It takes about 100 drops to form a single character.

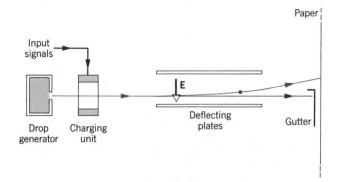

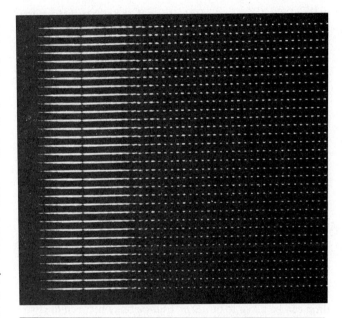

ABCDEFGHIJKLMNOPQRSTUVWXYZ
abcdefghijklmnopqrstuvwxyz
±@#$%¢&*()_+°":?.,1234567890
MNO

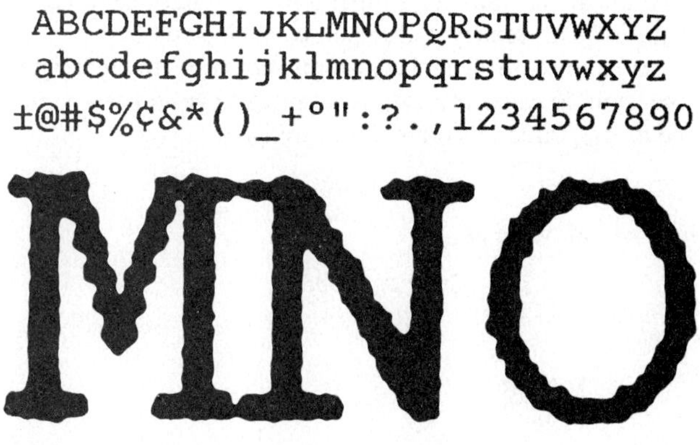

Figure 13 A sample of ink-jet printing, with three letters shown enlarged.

part for this work. Modern measurements of the elementary charge rely on a variety of interlocking experiments, all more precise than the pioneering experiment of Millikan.

Ink-Jet Printing.* The need for high-quality, high-speed printing has caused a search for an alternative to impact printing, such as occurs in a standard typewriter. Building up letters by squirting tiny droplets of ink at the paper has proved to be one way to do it.

Figure 12 shows a charged drop moving between two conducting plates, between which an electric field E has been set up. The drop will be deflected upward and, after passing the edge of the plates, will proceed to the paper and strike it at a position that is determined by the value of E and of the charge q on the drop.

In practice, E is held constant and the position of the drop is determined by the charge q delivered to the drop in the charging unit, through which the drop must pass before entering the deflecting system. The charging unit, in turn, is activated by electronic signals that encode the material to be printed. The printing head moves sideways, from left to right, and each individual character is built up as the motion proceeds.

The drops, which are about 30 μm in diameter, are produced at the rate of about 100,000 per second and move toward the paper at about 18 m/s. Figure 13 shows a sample of ink-jet printing. It takes about 100 drops to form a single character. Using multiple jets, printing speeds exceeding 45,000 lines/min have been achieved. Figure 14 shows the drops issuing from such a multijet drop generator; the height of the nozzle array is only that of a single printed character.

Figure 14 Ink drops issuing from a multijet drop generator in an ink-jet printer. The vertical height of the array of jets is that of a single character. The drops are moving from left to right.

Sample Problem 7 In the Millikan oil-drop apparatus of Fig. 10, a drop of radius $R = 2.76$ μm carries a charge q of three electrons. What electric field (magnitude and direction) is re-

* See "Ink-Jet Printing," by Larry Kuhn and Robert A. Myers, *Scientific American,* April 1979.

quired to balance the drop so that it remains stationary in the apparatus? The density ρ of the oil is 920 kg/m³.

At balance, the field force qE must be equal to the weight of the drop. Because the drop is assumed to be stationary, no viscous drag force acts on it. Thus, we have

$$mg = qE,$$

which we can write as

$$\tfrac{4}{3}\pi R^3 \rho g = (3e)E,$$

so that

$$E = \frac{4\pi R^3 \rho g}{9e}$$

$$= \frac{(4\pi)(2.76 \times 10^{-6} \text{ m})^3 (920 \text{ kg/m}^3)(9.80 \text{ m/s}^2)}{(9)(1.60 \times 10^{-19} \text{ C})}$$

$$= 1.65 \times 10^6 \text{ N/C.} \qquad \text{(Answer)}$$

Because the drop is negatively charged, the electric field must point down. This will give an upward force to balance the weight of the drop.

Sample Problem 8 Figure 15 shows the deflecting electrode system of an ink-jet printer. An ink drop whose mass m is 1.3×10^{-10} kg carries a charge q of 1.5×10^{-13} C and enters the deflecting plate system with a speed $v = 18$ m/s. The length L of these plates is 1.6 cm and the electric field E between the plates is 1.4×10^6 N/C. What will be the vertical deflection of the drop at the far edge of the plates? Ignore the varying electric field at the edges of the plates.

Let t be the time of passage of the drop through the deflecting system. The vertical and the horizontal displacements are given by

$$y = \tfrac{1}{2}at^2 \quad \text{and} \quad L = vt,$$

respectively, in which a is the acceleration of the drop.

The electric force acting on the drop, qE, is much greater than the gravitational force mg so that the acceleration of the drop can be taken to be qE/m. Eliminating t between the two equations above and substituting this value for a leads to

$$y = \frac{qEL^2}{2mv^2}$$

$$= \frac{(1.5 \times 10^{-13} \text{ C})(1.4 \times 10^6 \text{ N/C})(1.6 \times 10^{-2} \text{ m})^2}{(2)(1.3 \times 10^{-10} \text{ kg})(18 \text{ m/s})^2}$$

$$= 6.4 \times 10^{-4} \text{ m} = 0.64 \text{ mm.} \qquad \text{(Answer)}$$

The deflection at the paper will be larger than this. To aim the ink drops so that they form the characters well, it is necessary to control the charge q on the drops—to which the deflection is proportional—to within a few percent. In our treatment, we have neglected the viscous drag forces that act on the drop; they are substantial at these high drop speeds. The analysis is the same as for the deflection of the electron beam in an electrostatic cathode ray tube.

24–9 A Dipole in an Electric Field

We have defined the electric dipole moment **p** of an electric dipole to be a vector whose direction is along the dipole axis, pointing from the negative to the positive charge. We shall see that the behavior of a dipole in a uniform external electric field can be described completely in terms of the two vectors **E** and **p**, with no need to give any details about the structure of the dipole.

As we have seen in Sample Problem 5, a molecule of water vapor (H_2O) has an electric dipole moment of magnitude $p = 6.2 \times 10^{-30}$ C·m. Figure 16 is a representation of this molecule, showing the three nuclei and the surrounding electron cloud. The electric dipole moment **p** is represented by the vector on the axis of symmetry. This moment arises because the effective center

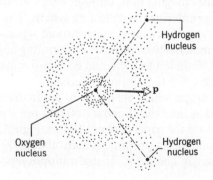

Figure 16 A molecule of H_2O, showing the three nuclei, the electron cloud, and the electric dipole moment vector **p**.

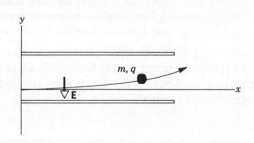

Figure 15 Sample Problem 8. An ink drop of mass m and charge q is deflected in the electric field of an ink-jet printer.

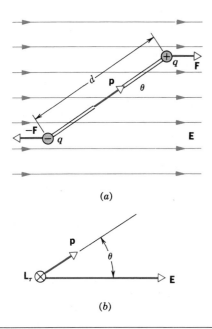

(a)

(b)

Figure 17   (a) An electric dipole in a uniform electric field. (b) Illustrating the relation $\tau = \mathbf{p} \times \mathbf{E}$.

of positive charge does not coincide with the effective center of negative charge.

Figure 17a shows a dipole in a uniform electric field **E**, its dipole moment **p** making an angle θ with the field direction. Two equal and opposite forces **F** and $-\mathbf{F}$ act as shown, where $F = qE$. The net force on the dipole is zero but there is a net torque about the center of mass of the molecule, given by

$$\tau = -Fd \sin \theta.$$

The minus sign indicates that the torque, which is exerted *by* the electric field *on* the dipole, is clockwise in Fig. 17, acting to reduce the value of θ. Substituting qE for F in the above and recalling that $p = qd$, we have

$$\tau = -qdE \sin \theta = -pE \sin \theta. \qquad (24)$$

We can generalize Eq. 24 to vector form as

$$\boxed{\tau = \mathbf{p} \times \mathbf{E}} \quad \text{(torque on a dipole),} \qquad (25)$$

the appropriate vectors being shown in Fig. 17b. This torque tends to orient the dipole in the field direction. In the same way, the torque exerted by a *magnetic* field tends to orient a compass needle (a *magnetic* dipole) in the direction of that field.

An electric dipole in an electric field has its smallest potential energy when it is lined up with the field, just as a pendulum has its smallest gravitational potential energy when it is at the bottom of its swing. To increase its potential energy we must do work on the dipole, the increase being the work that we do.

In any potential energy situation, we are free to define the zero potential energy configuration in a perfectly arbitrary way because only differences in potential energy have physical meaning. It turns out that our final result will be simplest if we choose the potential energy of a dipole in an external field to be zero when the angle θ in Fig. 17b is 90°. The potential energy at any other orientation is then given by

$$U(\theta) = W = \int_{90°}^{\theta} (-\tau)d\theta = \int_{90°}^{\theta} pE \sin \theta \, d\theta, \quad (26)$$

in which W is the work done *by an external agent* to turn the dipole from its reference orientation to angle θ.* Evaluating the integral between the indicated limits leads to

$$U(\theta) = -pE \cos \theta. \qquad (27)$$

We can generalize this to vector form as

$$\boxed{U(\theta) = -\mathbf{p} \cdot \mathbf{E}} \quad \text{(energy of a dipole).} \qquad (28)$$

This equation confirms that U is a minimum (most negative) when the **p** points in the field direction ($\theta = 0$) and is a maximum (most positive) when **p** points opposite to that direction ($\theta = 180°$).

Sample Problem 9   A water molecule (H_2O) in its vapor state has an electric dipole moment of 6.2×10^{-30} C·m. (a) How far apart are the effective centers of positive and negative charge in this molecule?

There are 10 electrons and, correspondingly, 10 positive charges in this molecule; see Fig. 16. We can write, for the magnitude of the dipole moment,

$$p = qd = (10e)(d),$$

in which d is the separation we are seeking and e is the elemen-

---

* The torque applied *by the external agent* must be opposite in sign to the torque applied *by the electric field* (see Eq. 24)—hence the minus sign in Eq. 26.

tary charge. Thus,

$$d = \frac{p}{10e} = \frac{6.2 \times 10^{-30} \text{ C} \cdot \text{m}}{(10)(1.60 \times 10^{-19} \text{ C})}$$
$$= 3.9 \times 10^{-12} \text{ m} = 3.9 \text{ pm.} \qquad \text{(Answer)}$$

This is about 4% of the OH bond distance in this molecule.

(b) If the molecule is placed in an electric field of 1.5×10^4 N/C, what maximum torque can the field exert on it? Such a field can easily be set up in the laboratory.

As Eq. 25 shows, the torque is a maximum when $\theta = 90°$. Substituting this value in that equation yields

$$\tau = pE \sin \theta$$
$$= (6.2 \times 10^{-30} \text{ C} \cdot \text{m})(1.5 \times 10^4 \text{ N/C})(\sin 90°)$$
$$= 9.3 \times 10^{-26} \text{ N} \cdot \text{m.} \qquad \text{(Answer)}$$

(c) How much work must an external agent do to turn this molecule end for end in this field, starting from its fully aligned position, for which $\theta = 0$?

The work is the difference in potential energy between the positions $\theta = 180°$ and $\theta = 0$. Thus, from Eq. 28,

$$W = U(180°) - U(0)$$
$$= (-pE \cos 180°) - (-pE \cos 0)$$
$$= 2pE = (2)(6.2 \times 10^{-30} \text{ C} \cdot \text{m})(1.5 \times 10^4 \text{ N/C})$$
$$= 1.9 \times 10^{-25} \text{ J.} \qquad \text{(Answer)}$$

By comparison, the average translational energy of a molecule at room temperature ($= \frac{3}{2}kT$) is 6.2×10^{-21} J, which is 33,000 times larger. For the conditions of this problem, thermal agitation would swamp the tendency of the dipoles to align themselves with the field. If we wish to align the dipoles, we must use much stronger fields and/or much lower temperatures.

REVIEW AND SUMMARY

The Field Interpretation of Electric Force

One way to explain the electric force between charges is to presume that each charge sets up an electric field in the space surrounding itself. Each charge then interacts with the resulting field of the other charges at its own location.

We define the electric field at a point in space in terms of the electric force exerted on a small test charge placed at that point. The defining equation is

Definition of Electric Field

$$\mathbf{E} = \mathbf{F}/q \quad \text{(electric field).} \qquad [2]$$

See Sample Problem 1. If the value obtained using Eq. 2 is different for different test charges, the field is defined as the limit of this ratio as the test charge q approaches zero.

Lines of Force

Lines of force are a convenient representation of an electric field. They are drawn so that the tangent to a given line is in the direction of the field and so that strength of the field is indicated by the closeness of the lines, closer lines indicating stronger fields. Section 24–3 discusses field lines for a large sheet of charge, a charged sphere (see also Sample Problem 2), two equal charges, and two equal but opposite charges.

The Electric Field Due to a Point Charge

Any electric field can be calculated from Coulomb's law by knowing the locations and charges of the charged particles that cause it. The field due to a point charge has magnitude

$$E = \frac{1}{4\pi\epsilon_0} \frac{q}{r^2} \quad \text{(point charge).} \qquad [6]$$

The Electric Field Due to a Collection of Charges

The field points directly away from a positive charge and directly toward a negative one. The field due to a collection of point charges is obtained by superposition—adding vectorially the contributions from the individual charges. See Sample Problems 3 and 4. We also use superposition to calculate fields due to continuous distributions of charge; the sum of fields due to point charges becomes an integral. We often reduce the complexity of a vector sum by taking advantage of symmetry. We illustrate the procedure in Sections 24–6 (a ring of charge) and 24–7 (a charged disk).

Electric Dipoles

An assembly of two equal but opposite charges $\pm q$ separated by a distance d is called an *electric dipole*. It may be described in terms of a vector dipole moment $\mathbf{p}$ of magnitude qd and direction parallel to a line from the negative to the positive charge. The field due to an electric dipole is described in Section 24–5; see also Problem 31. Many naturally occurring molecules, such as the water molecule discussed in Sample Problem 5, have dipole electrical properties.

Force and Acceleration in an Electric Field

A charged particle in an electric field **E** experiences an electric force **F** = **E**q and an acceleration in accord with Newton's second law. Sample Problems 7 and 8 illustrate the application of these principles to charged particles in uniform electric fields.

A dipole in a uniform electric field experiences no net force but does experience a torque given by

Torque on an Electric Dipole

$$\tau = \mathbf{p} \times \mathbf{E} \quad \text{(torque on a dipole).} \tag{25}$$

The torque tends to align the dipole in a direction parallel to the field. The potential energy of a dipole in an electric field may conveniently be taken to be

Potential Energy of an Electric Dipole

$$U(\theta) = -\mathbf{p} \cdot \mathbf{E} \quad \text{(energy of a dipole),} \tag{28}$$

in which the zero of potential energy occurs when the dipole is perpendicular to the electric field. See Sample Problem 9 for an example.

QUESTIONS

1. Name as many scalar fields and vector fields as you can.

2. (a) In the gravitational attraction between the earth and a stone, can we say that the earth lies in the gravitational field of the stone? (b) How is the gravitational field due to the stone related to that due to the earth?

3. A positively charged ball hangs from a long silk thread. We wish to measure E at a point in the same horizontal plane as that of the hanging charge. To do so, we put a positive test charge q_0 at the point and measure $\mathbf{F}/q_0$. Will F/q_0 be less than, equal to, or greater than E at the point in question?

4. In exploring electric fields with a test charge, we have often assumed, for convenience, that the test charge was positive. Does this really make any difference in determining the field? Illustrate in a simple case of your own devising.

5. Electric lines of force never cross. Why?

6. In Fig. 4, why do the lines of force around the edge of the figure appear, when extended backward, to radiate uniformly from the center of the figure?

7. A point charge q of mass m is released from rest in a nonuniform field. Will it necessarily follow a line of force?

8. A point charge is moving in an electric field at right angles to the lines of force. Does any force act on it?

9. What is the origin of "static cling," a phenomenon that sometimes affects clothes as they are removed from a dryer?

10. Two point charges of unknown magnitude and sign are a distance d apart. The electric field is zero at one point between them, on the line joining them. What can you conclude about the charges?

11. In Sample Problem 3, a charge placed at point P in Fig. 6 is in equilibrium because no force acts on it. Is the equilibrium stable (a) for displacements along the line joining the charges and (b) for displacements at right angles to this axis?

12. Two point charges of unknown sign and magnitude are fixed a distance L apart. Can we have $\mathbf{E} = 0$ for off-axis points (excluding ∞)? Explain.

13. In Fig. 5, the force on the lower charge points up and is finite. The crowding of the lines of force, however, suggests that E is infinitely great at the site of this (point) charge. A charge immersed in an infinitely great field should have an infinitely great force acting on it. What is the solution to this dilemma?

14. Three small spheres x, y, and z carry charges of equal magnitude and with signs as shown in Fig. 18. They are placed at the vertices of an isosceles triangle with the distance between x and y equal to the distance between x and z. Spheres y and z are held in place but sphere x is free to move on a frictionless surface. Which path will sphere x take when released?

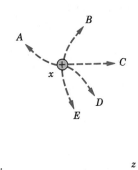

Figure 18 Question 14.

15. A positive and a negative charge of the same magnitude lie on a long straight line. What is the direction of **E** for points on this line that lie (a) between the charges, (b) outside the charges in the direction of the positive charge, (c) outside the charges in the direction of the negative charge, and (d) off the line but in the median plane of the charges?

16. In the median plane of an electric dipole, is the electric field parallel or antiparallel to the electric dipole moment **p**?

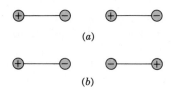

Figure 19 Question 17.

17. (*a*) Two identical electric dipoles are placed in a straight line, as shown in Fig. 19*a*. What is the direction of the electric force on each dipole owing to the presence of the other? (*b*) Suppose that the dipoles are rearranged as in Fig. 19*b*. What now is the direction of the force?

18. Compare the way E varies with r for (*a*) a point charge (Eq. 6), (*b*) a dipole (Eq. 11), and (*c*) a quadrupole (Problem 32).

19. What mathematical difficulties would you encounter if you were to calculate the electric field of a charged ring (or disk) at points *not* on the axis?

20. Figure 3 shows that **E** has the same value for all points in front of an infinite uniformly charged sheet. Is this reasonable? One might think that the field should be stronger near the sheet because the charges are so much closer.

21. Describe, in your own words, the purpose of the Millikan oil-drop experiment.

22. How does the sign of the charge on the oil drop affect the operation of the Millikan experiment?

23. You turn an electric dipole end for end in a uniform electric field. How does the work you do depend on the initial orientation of the dipole with respect to the field?

24. For what orientations of an electric dipole in a uniform electric field is the potential energy of the dipole (*a*) the greatest and (*b*) the least?

25. An electric dipole is placed in a nonuniform electric field. Is there a net force on it?

26. An electric dipole is placed at rest in a uniform external electric field, as in Fig. 17*a*, and released. Discuss its motion.

EXERCISES AND PROBLEMS

Section 24–2 The Electric Field

1E. An electron is released from rest in a uniform electric field of magnitude 2.0×10^4 N/C. Calculate the acceleration of the electron. (Ignore gravity.)

2E. An electron is accelerated eastward at 1.8×10^9 m/s² by an electric field. Determine the magnitude and direction of the electric field.

3E. Humid air breaks down (its molecules become ionized) in an electric field of 3.0×10^6 N/C. What is the magnitude of the electric force on (*a*) an electron and (*b*) an ion (with a single electron missing) in this field?

4E. An α particle, the nucleus of a helium atom, has a mass of 6.7×10^{-27} kg and a charge of $+2e$. What are the magnitude and direction of the electric field that will balance its weight?

5E. In a uniform electric field near the surface of the earth, a particle having a charge of -2.0×10^{-9} C is acted on by a downward electric force of 3.0×10^{-6} N. (*a*) What is the magnitude of the electric field? (*b*) What is the magnitude and direction of the electric force exerted on a proton placed in this field? (*c*) What is the gravitational force on the proton? (*d*) What is the ratio of the electric force to the gravitational force in this case?

6E. An electric field **E** with an average magnitude of about 150 N/C points downward in the earth's atmosphere. We wish to "float" a sulfur sphere weighing 1.0 lb (= 4.4 N) in this field by charging it. (*a*) What charge (sign and magnitude) must be used? (*b*) Why is the experiment not practical?

Section 24–3 Lines of Force

7E. Figure 20 shows field lines of an electric field. (*a*) If the magnitude of the field at A is 40 N/C, what force does an electron at that point experience? (*b*) What is the magnitude of the field at B?

Figure 20 Exercise 7.

8E. Sketch qualitatively the lines of force associated with two separated point charges $+q$ and $-2q$.

9E. Three charges are arranged in an equilateral triangle as in Fig. 21. Consider the lines of force due to $+Q$ and $-Q$, and

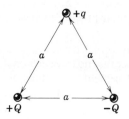

Figure 21 Exercise 9.

from them identify the direction of the force that acts on $+q$ because of the presence of the other two charges. (*Hint:* See Fig. 5.)

10E. Sketch qualitatively the lines of force between two concentric conducting spherical shells, charge $+q_1$ being placed on the inner sphere and $-q_2$ on the outer. Consider the cases $q_1 > q_2$, $q_1 = q_2$, and $q_1 < q_2$.

11E. Sketch qualitatively the lines of force associated with a thin, circular, uniformly-charged disk of radius R. (*Hint:* Consider as limiting cases points very close to the disk, where the electric field is perpendicular to the surface, and points very far from it, where the electric field is like that of a point charge.)

12P. Sketch qualitatively the lines of force associated with three long parallel lines of charge, in a perpendicular plane. Assume that the intersections of the lines of charge with such a plane form an equilateral triangle (Fig. 22) and that each line of charge has the same linear charge density λ.

Figure 22 Problem 12.

13P. Assume that the exponent in Coulomb's law is not 'two' but n. Show that for $n \neq$ 'two' it is impossible to construct lines that will have the properties listed for lines of force in Section 24–3. For simplicity, treat an isolated point charge.

Section 24–4 Calculating the Field: A Point Charge

14E. What is the magnitude of a point charge that would create an electric field of exactly 1 N/C at points exactly 1 m away?

15E. In Fig. 23, charges are placed at the vertices of an equilateral triangle. For what value of Q (both sign and magnitude) does the total electric field vanish at C, the center of the triangle?

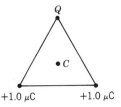

Figure 23 Exercise 15.

16E. What is the magnitude of a point charge chosen so that the electric field 50 cm away has the magnitude 2.0 N/C?

17E. Two point charges of magnitude 2.0×10^{-7} C and 8.5×10^{-8} C are 12 cm apart. (*a*) What electric field does each produce at the site of the other? (*b*) What force acts on each?

18E. Two equal and opposite charges of magnitude 2.0×10^{-7} C are held 15 cm apart. (*a*) What are the magnitude and direction of **E** at a point midway between the charges? (*b*) What force (magnitude and direction) would act on an electron placed there?

19P. The nucleus of an atom of uranium-238 has a radius of 6.8 fm (= 6.8×10^{-15} m) and carries a (positive) charge of Ze in which $Z (= 92)$ is the atomic number of uranium and e is the elementary charge. What is the electric field at the surface of such a nucleus? In what direction does it point? Does its numerical value surprise you?

20P. Two point charges are fixed at a distance d apart (Fig. 24). Plot $E(x)$, assuming $x = 0$ at the left-hand charge. Consider both positive and negative values of x. Plot E as positive if **E** points to the right and negative if **E** points to the left. Assume $q_1 = +1.0 \times 10^{-6}$ C, $q_2 = +3.0 \times 10^{-6}$ C, and $d = 10$ cm.

Figure 24 Problem 20.

21P. (*a*) In Fig. 25, locate the point (or points) at which the electric field is zero. (*b*) Sketch qualitatively the lines of force.

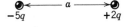

Figure 25 Problem 21.

22P. Charges $+q$ and $-2q$ are fixed a distance d apart as in Fig. 26. (*a*) Find **E** at points A, B, and C. (*b*) Sketch roughly the lines of force.

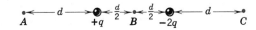

Figure 26 Problem 22.

23P. Two charges $q_1 = 2.1 \times 10^{-8}$ C and $q_2 = -4q_1$ are placed 50 cm apart. Find the point along the straight line passing through the two charges at which the electric field is zero.

24P. A clock face has negative point charges $-q$, $-2q$, $-3q$, . . . , $-12q$ fixed at the positions of the corresponding numerals. The clock hands do not perturb the field. At what time does the hour hand point in the same direction as the electric field at the center of the dial? (*Hint:* Consider diametrically opposite charges.)

25P. An electron is placed at each corner of an equilateral triangle having sides 20 cm long. (*a*) What is the electric field at the midpoint of one of the sides? (*b*) What force would another electron placed there experience?

26P. Calculate **E** (direction and magnitude) at point P in Fig. 27.

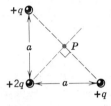

Figure 27   Problem 26.

27P. What is **E** in magnitude and direction at the center of the square of Fig. 28? Assume that $q = 1.0 \times 10^{-8}$ C and $d = 5.0$ cm.

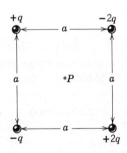

Figure 28   Problem 27.

Section 24-5   Calculating the Field: An Electric Dipole

28E. Calculate the dipole moment of an electron and a proton 4.3 nm apart.

29E. Calculate the magnitude of the force, due to a small electric dipole of dipole moment 3.6×10^{-29} C·m, on an electron 25 nm away along the dipole axis.

30E. In Fig. 7, assume that both charges are positive. Show that E at point P in that figure, assuming $z \gg d$, is given by

$$E = \frac{1}{4\pi\epsilon_0} \frac{2q}{z^2}.$$

31P. Calculate the electric field, magnitude and direction, due to an electric dipole, at a point P located at a distance $r \gg d$ along the perpendicular bisector of the line joining the charges; see Fig. 29. Express your answer in terms of the electric dipole moment **p**.

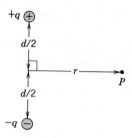

Figure 29   Problem 31.

32P*. *Electric quadrupole.* Figure 30 shows a typical electric quadrupole. It consists of two dipoles whose effects at external points do not quite cancel. Show that the value of E on the axis of the quadrupole for points a distance z from its center (assume $z \gg d$) is given by

$$E = \frac{3Q}{4\pi\epsilon_0 z^4},$$

where Q $(= 2qd^2)$ is called the *quadrupole moment* of the charge distribution.

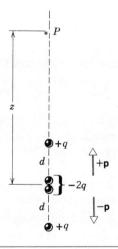

Figure 30   Problem 32.

Section 24-6   Calculating the Field: A Ring of Charge

33E. Make a quantitative plot of the electric field on the axis of a charged ring having a diameter of 6.0 cm and a uniformly distributed charge of 1.0×10^{-8} C.

34P. At what distance along the axis of a charged ring of radius R is the axial electric field strength a maximum?

35P. An electron is constrained to move along the axis of the ring of charge discussed in Section 24–6. Show that the electron can perform small oscillations, through the center of the ring, with a frequency given by

$$\omega = \sqrt{\frac{eq}{4\pi\epsilon_0 mR^3}}.$$

36P. A thin glass rod is bent into a semicircle of radius r. A charge $+Q$ is uniformly distributed along the upper half and a charge $-Q$ is uniformly distributed along the lower half, as shown in Fig. 31. Find the electric field $\mathbf{E}$ at P, the center of the semicircle.

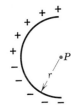

Figure 31   Problem 36.

37P. A thin nonconducting rod of finite length L carries a total charge q, spread uniformly along it. Show that E at point P on the perpendicular bisector in Fig. 32 is given by

$$E = \frac{q}{2\pi\epsilon_0 y} \frac{1}{(L^2 + 4y^2)^{1/2}}.$$

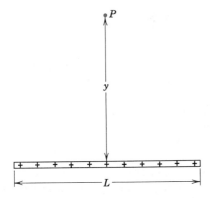

Figure 32   Problem 37.

38P. An insulating rod of length L has charge $-q$ uniformly distributed along its length, as shown in Fig. 33. (a) What is the linear charge density of the rod? (b) What is the electric field at point P a distance a from the end of the rod? (c) If P were very far from the rod compared to L, the rod would look like a point charge. Show that your answer to (b) reduces to the electric field of a point charge for $a \gg L$.

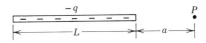

Figure 33   Problem 38.

39.* A "semi-infinite" insulating rod (Fig. 34) carries a constant charge per unit length of λ. Show that the electric field at the point P makes an angle of 45° with the rod and that this result is independent of the distance R.

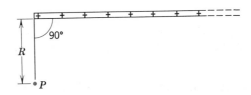

Figure 34   Problem 39.

Section 24–7 Calculating the Field: A Charged Disk

40E. Show that Eq. 21, for the electric field of a charged disk at points on its axis, reduces to the field of a point charge for $z \gg R$.

41P. (a) What total charge q must the disk in Sample Problem 6 carry in order that the electric field on the surface of the disk at its center equals the value at which air breaks down electrically, producing sparks? See Table 1. (b) Suppose that each atom at the surface has an effective cross-sectional area of 0.015 nm². How many atoms are at the disk's surface? (c) The charge in (a) results from some of the surface atoms carrying one excess electron. What fraction of the surface atoms must be so charged?

42P. At what distance along the axis of a charged disk of radius R is the electric field strength equal to one-half the value of the field at the surface of the disk at the center?

Section 24–8 A Point Charge in an Electric Field

43E. (a) What is the acceleration of an electron in a uniform electric field of 1.4×10^6 N/C? (b) How long would it take for the electron, starting from rest, to attain one-tenth the speed of light? (c) How far would it travel? Assume that Newtonian mechanics holds.

44E. One defensive weapon being considered for the Strategic Defense Initiative (Star Wars) uses particle beams. For example, a proton beam striking an enemy missile could render it harmless. Such beams can be produced in "guns" using electric fields to accelerate the charged particles. (a) What acceleration would a proton experience if the electric field is 2.0×10^4 N/C? (b) What speed would the proton attain if the field acts over a distance of one centimeter?

45E. An electron moving with a speed of 5.0×10^8 cm/s is shot parallel to an electric field of strength 1.0×10^3 N/C arranged so as to retard its motion. (*a*) How far will the electron travel in the field before coming (momentarily) to rest and (*b*) how much time will elapse? (*c*) If the electric field ends abruptly after 0.8 cm, what fraction of its initial kinetic energy will the electron lose in traversing it?

46E. A spherical water droplet 1.2 μm in diameter is suspended in calm air owing to a downward-directed atmospheric electric field $E = 462$ N/C. (*a*) What is the weight of the drop? (*b*) How many excess electrons does it carry?

47E. In Millikan's experiment, a drop of radius 1.64 μm and density 0.851 g/cm³ is balanced when an electric field of 1.92×10^5 N/C is applied. Find the charge on the drop, in terms of e.

48P. In a particular early run (1911), Millikan observed that the following measured charges, among others, appeared at different times on a single drop:

6.563×10^{-19} C	13.13×10^{-19} C	19.71×10^{-19} C
8.204×10^{-19} C	16.48×10^{-19} C	22.89×10^{-19} C
11.50×10^{-19} C	18.08×10^{-19} C	26.13×10^{-19} C

What value for the elementary charge e can be deduced from these data?

49P. An object having a mass of 10 g and a charge of $+8.0 \times 10^{-5}$ C is placed in an electric field defined by $E_x = 3.0 \times 10^3$ N/C, $E_y = -600$ N/C, and $E_z = 0$. (*a*) What are the magnitude and direction of the force on the object? (*b*) If the object starts from rest at the origin, what will be its coordinates after 3.0 s?

50P. A uniform electric field exists in a region between two oppositely charged plates. An electron is released from rest at the surface of the negatively charged plate and strikes the surface of the opposite plate, 2.0 cm away, in a time 1.5×10^{-8} s. (*a*) What is the speed of the electron as it strikes the second plate? (*b*) What is the magnitude of the electric field E?

51P. At some instant the velocity components of an electron moving between two charged parallel plates are $v_x = 1.5 \times 10^5$ m/s and $v_y = 3.0 \times 10^3$ m/s. If the electric field between the plates is given by $\mathbf{E} = (120 \text{ N/C})\mathbf{j}$, (*a*) what is the acceleration of the electron? (*b*) What will be the velocity of the electron after its x coordinate has changed by 2.0 cm?

52P. Two large parallel copper plates are 5.0 cm apart and have a uniform electric field between them as depicted in Fig. 35. An electron is released from the negative plate at the same time that a proton is released from the positive plate. Neglect the force of the particles on each other and find their distance from the positive plate when they pass each other. Does it surprise you that you need not know the electric field to solve this problem?

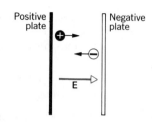

Figure 35 Problem 52.

53P. A uniform vertical field E is established in the space between two large parallel plates. In this field one suspends a small conducting sphere of mass m from a string of length l. Find the period of this pendulum when the sphere is given a charge $+q$ if the lower plate (*a*) is charged positively and (*b*) is charged negatively.

54P. An electron is projected as in Fig. 36 at a speed of 6.0×10^6 m/s and at an angle θ of 45°; $E = 2.0 \times 10^3$ N/C (directed upward), $d = 2.0$ cm, and $L = 10$ cm. (*a*) Will the electron strike either of the plates? (*b*) If it strikes a plate, where does it do so?

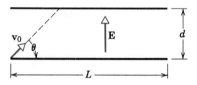

Figure 36 Problem 54.

Section 24–9 A Dipole in an Electric Field

55E. An electric dipole, consisting of charges of magnitude 1.5 nC separated by 6.2 μm, is in an electric field of strength 1100 N/C. (*a*) What is the magnitude of the electric dipole moment? (*b*) What is the difference in potential energy corresponding to a dipole orientation parallel and antiparallel to the field?

56E. An electric dipole consists of charges $+2e$ and $-2e$ separated by 0.78 nm. It is in an electric field of strength 3.4×10^6 N/C. Calculate the magnitude of the torque on the dipole when the dipole moment is (*a*) parallel, (*b*) at a right angle, and (*c*) opposite to the electric field.

57P. Find the work required to turn an electric dipole end for end in a uniform electric field E, in terms of the dipole moment $\mathbf{p}$ and the initial angle θ_0 between $\mathbf{p}$ and $\mathbf{E}$.

58P. Find the frequency of oscillation of an electric dipole, of moment p and rotational inertia I, for small amplitudes of oscillation about its equilibrium position in a uniform electric field E.

CHAPTER 25
GAUSS' LAW

This bronze sculpture by Henry Moore graces the campus of Princeton University. The sculptor's art lies in the shaping of the surfaces of three-dimensional objects. Physicists and mathematicians are also interested in surfaces. In this chapter we introduce Gaussian surfaces, *which have a beauty of their own. It lies in the elegance and simplicity that the use of such constructions brings to the solution of many problems in electrostatics.*

25–1   A New Look at Coulomb's Law

If you want to find the center of mass of a potato, you can do so by experiment or by laborious calculation, involving the numerical evaluation of a triple integral. However, if the potato happens to be a perfect ellipsoid, there is no problem. You know exactly where the center of mass is without calculation. Such are the advantages of symmetry. Symmetrical situations arise in all fields of physics and, when possible, it makes sense to cast the laws of physics in forms that take full advantage of this fact.

Coulomb's law is the governing law in electrostatics, but it is not cast in a form that particularly simplifies the

work in cases of high symmetry. In this chapter we introduce a new formulation of Coulomb's law, called *Gauss' law,*[*] that *can* take easy advantage of such special cases. Gauss' law—for electrostatic problems—is the equivalent of Coulomb's law; which of them we choose depends on the problem at hand. Although both laws are valid for all electrostatic problems, it is one thing for a law to be valid and quite another thing for it to be useful.

---

[*] This law was originally derived for another inverse square force, gravity, by the German mathematician and physicist Carl Friedrich Gauss (1777–1855). For information about the life of this great scientist, see "Gauss," by Ian Stewart, *Scientific American*, July 1977.

We use Coulomb's law, the workhorse of electrostatics, for all problems in which the degree of symmetry is low. Even the most complicated of these can be solved, given enough computer capacity. We use Gauss' law when the symmetry is appropriately high. In such cases, this law not only tremendously simplifies the work but also — because of its simplicity — often provides new insights.

As Table 2 of Chapter 37 shows, Gauss' law is one of Maxwell's four equations. It is the entire purpose of Chapters 23 – 38 of this book to work toward a full understanding of these equations, which govern all classical electromagnetism and optics.

25–2   What Gauss' Law Is All About

Gauss' law is a relation between electric charge and the electric field that it sets up. Coulomb's law, which we write in the form of Eq. 6 of Chapter 24, namely,

$$E = \frac{1}{4\pi\epsilon_0} \frac{q}{r^2} \quad \text{(Coulomb's law)}, \quad (1)$$

also connects these two quantities. Gauss' law is simply a second, equivalent, way of doing so. Before presenting it formally, we want to give some notion of its nature.

Central to Gauss' law is a hypothetical closed surface — called a *Gaussian surface* — that you can set up in space. The Gaussian surface can be of any shape that you wish, but you will find it most useful to draw it in such a way as to conform to the symmetry of the problem you are facing. Thus, the Gaussian surface will often turn out to be a sphere, a cylinder, or some other symmetrical form. It must always be a *closed* surface, so that there can be a clear distinction between points that are inside the surface, on the surface, and outside the surface.

Imagine now that you wander over the Gaussian surface with an electric field meter in hand; you may — or may not — encounter electric fields at various points. You can note how strong they are and in what direction they point. Imagine also that you wander throughout the volume enclosed by this surface with a charge meter in hand; you may — or may not — encounter electric charges. You can note their sign and magnitude.

Gauss' law tells how the fields at the Gaussian surface are related to the charges contained within that surface.

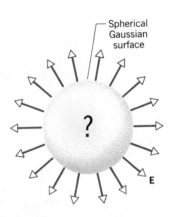

Figure 1   A spherical Gaussian surface. If the electric field vectors are of uniform magnitude and point radially outward as shown, you can conclude that a net positive distribution of charge must lie within the surface and that it must have spherical symmetry.

Figure 1 shows a simple example, in which the Gaussian surface is a sphere, shown in cross section in the figure. Suppose that you find, as you explore this surface, that there is an electric field at every point on the surface, of the same magnitude and pointing radially outward. Without knowing anything about Gauss' law, what are you likely to conclude? That there must be a distribution of net positive charge inside the spherical Gaussian surface, symmetrically distributed about its center. It could be a point charge located at the center. It could be a uniform sphere or shell of charge, centered within the Gaussian sphere. Without further information, you cannot select from these possibilities.

Before we can make these ideas quantitative, we must pause to develop a new concept, that of the *flux* of a vector field.

25–3   Flux

Suppose that you dip a square wire loop into a uniformly flowing stream, the plane of the loop being at right angles to the flow as in Fig. 2a. The rate Φ at which water flows through the loop, measured perhaps in m^3/s, is given by

$$\Phi = Av, \quad (2)$$

in which A is the area of the loop and v is the speed of the water. We call Φ the *flux*. If you turn the loop to the position shown in Fig. 2b, the flux becomes

$$\Phi = Av \cos \theta, \quad (3)$$

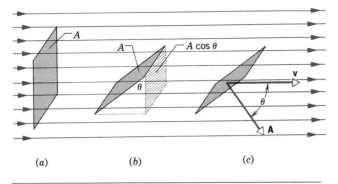

(a)   (b)   (c)

Figure 2 A wire loop of area A is immersed in a flowing stream. (a) The loop is at right angles to the flow. (b) The loop is turned at an angle θ to the flow, its projected area being $A \cos \theta$. (c) The area of the loop is represented by a vector **A** and the stream velocity by **v**, the angle between them being θ.

because only the projected area of the loop counts.

Figure 2c shows how we can express Eq. 3 in vector language. We represent the area of the loop by a vector **A**, at right angles to the plane of the loop. The angle θ is the angle between the vector **A** and the velocity vector **v**. We may write

$$\Phi = Av \cos \theta = \mathbf{A} \cdot \mathbf{v}. \qquad (4)$$

Equation 4 shows that flux, being the scalar product of two vectors, is itself a scalar quantity. The SI unit of flux in this case is m³/s.

The word "flux" comes from the Latin word meaning "to flow," and its use is appropriate in our flowing stream model. However, it is most useful for our purpose not to think about the water—which is what is doing the flowing—but about the array of velocity vectors, one for every point in the stream, that describes the flowing stream. From this point of view we regard the stream velocity as a *vector field*, and we say that Eq. 3 gives the flux of the *velocity field* through the surface of the wire loop.

The flux concept can be applied to *any* vector field, including the electric field, which is our special concern in this chapter. In the case of the electric field, nothing is actually flowing but we retain the useful word flux just the same. One more point: The surfaces shown in Fig. 2 are open surfaces, because they do not define an enclosed volume. Gauss' law deals only with closed surfaces; from now on, unless we specifically say otherwise, we shall be dealing only with the flux of an *electric field* through a *closed* (Gaussian) surface.

25-4 Flux of the Electric Field

To define the flux of the electric field, consider Fig. 3a, which shows an arbitrary Gaussian surface immersed in a nonuniform electric field. Let us divide the surface into small squares, of area ΔA, each square being small enough so that we can consider it to be a plane. We represent each such element of area by a vector $\Delta \mathbf{A}$, whose magnitude is the area ΔA. The direction of $\Delta \mathbf{A}$ is defined to be that of the *outward-drawn normal* to the surface.

At every square in Fig. 3a we can consider the electric field vector **E**. Because the squares have been taken to be arbitrarily small, **E** may be taken as constant for all points on a given square.

The vectors $\Delta \mathbf{A}$ and **E** that characterize each square make an angle θ with each other. Figure 3b shows an enlarged view of the three squares on the Gaussian surface marked x, y, and z. Table 1 summarizes the information about each square needed to calculate the flux.

A provisional definition for the flux of the electric field for the Gaussian surface of Fig. 3 is

$$\Phi = \sum \mathbf{E} \cdot \Delta \mathbf{A}. \qquad (5)$$

Equation 5 instructs us to visit each square on the Gaussian surface, to evaluate the scalar product $\mathbf{E} \cdot \Delta \mathbf{A}$ for the two vectors **E** and $\Delta \mathbf{A}$ that we find there, and to add up the results for all the squares that make up the surface. As Table 1 shows, squares like x in which **E** points inward make a negative contribution to the sum of Eq. 5. Squares like y, in which **E** lies in the surface, make zero contribution and squares like z, in which **E** points outward, make a positive contribution.

The exact definition of the flux of the electric field through a closed surface is found by allowing the area of the squares shown in Fig. 3a to become smaller and smaller, approaching a differential limit. The sum of Eq. 5 then becomes an integral and we have, for the definition of flux,

$$\Phi = \oint \mathbf{E} \cdot d\mathbf{A} \qquad \text{(flux through a Gaussian surface).} \qquad (6)$$

Table 1 Three Squares on a Gaussian Surface

Square	θ	Direction of **E**	Sign of $\mathbf{E} \cdot \Delta \mathbf{A}$
x	$>90°$	Into the surface	Negative
y	$=90°$	Parallel to the surface	Zero
z	$<90°$	Out of the surface	Positive

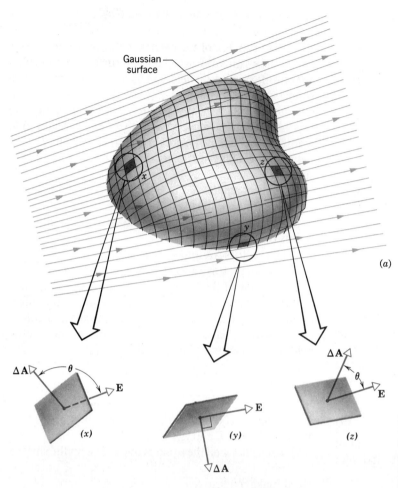

Gaussian surface

(a)

(b)

Figure 3 (a) A Gaussian surface of arbitrary shape immersed in an electric field. Its surface is divided into small squares of area ΔA. (b) The electric field vectors $\mathbf{E}$ and the area vectors $\Delta\mathbf{A}$ for three representative squares, marked x, y, and z.

The circle on the integral sign reminds us that the integration is to be taken over the entire (closed) surface. The flux of the electric field is a scalar, its directly derivable SI unit being $N \cdot m^2/C$.

Sample Problem 1 Figure 4 shows a Gaussian surface in the form of a cylinder of radius R immersed in a uniform electric field $\mathbf{E}$, the cylinder axis being parallel to the field. What is the flux Φ of the electric field through this closed surface?

We can write the flux as the sum of three terms, an integral over (a) the left cylinder cap, (b) the cylindrical surface, and (c) the right cap. Thus, from Eq. 6,

$$\Phi = \oint \mathbf{E} \cdot d\mathbf{A}$$

$$= \int_a \mathbf{E} \cdot d\mathbf{A} + \int_b \mathbf{E} \cdot d\mathbf{A} + \int_c \mathbf{E} \cdot d\mathbf{A}. \quad (7)$$

For the left cap, the angle θ for all points is $180°$, $\mathbf{E}$ is constant, and all the vectors $d\mathbf{A}$ are parallel. Thus,

$$\int \mathbf{E} \cdot d\mathbf{A} = \int E(\cos 180°)\, dA = -E \int dA = -EA,$$

where A, which is πR^2, is the cap area. Similarly, for the right cap,

$$\int \mathbf{E} \cdot d\mathbf{A} = +EA,$$

the angle θ for all points being zero there. Finally, for the cylinder wall,

$$\int \mathbf{E} \cdot d\mathbf{A} = 0,$$

the angle θ being $90°$ for all points on the cylindrical surface. Substituting these terms into Eq. 7 leads us to

$$\Phi = -EA + 0 + EA = 0. \qquad \text{(Answer)}$$

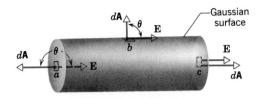

Figure 4   Sample Problem 1. A cylindrical Gaussian surface, closed by end caps, is immersed in a uniform electric field. The cylinder axis is parallel to the field direction.

This result is perhaps not surprising because the lines of force that represent the field pass right through the Gaussian surface, entering through the left end cap and leaving through the right end cap. If the field lines were taken to be stream lines of flowing water, we would say that water flows in at the left cap, out at the right cap, and doesn't flow at all through the cylindrical surface. The total flux, or net flow into the closed surface, is zero.

25-5   Gauss' Law

Now that we have defined the flux of the electric field, we are ready to state Gauss' law, namely,

$$\epsilon_0 \Phi = q \qquad \text{(Gauss' law).} \qquad (8)$$

Equation 8 refers to a Gaussian surface. The quantity Φ is the flux of the electric field over that surface and q is the *net* charge enclosed by that surface. The quantity ϵ_0 is the *permittivity constant,* its value being 8.85×10^{-12} C^2/N·m^2.

By introducing Eq. 6, the definition of flux, we can also write Gauss' law as*

$$\epsilon_0 \oint \mathbf{E} \cdot d\mathbf{A} = q \qquad \text{(Gauss' law).} \qquad (9)$$

Note that q in Eqs. 8 and 9 is the *net* charge within the Gaussian surface, taking its algebraic sign into account. Charge outside the surface, no matter how large or how nearby it may be, is not included in the term q in Gauss' law. The exact form or location of the charges inside the

Gaussian surface is also of no concern; the only thing that matters, on the right side of Eq. 9, is the magnitude and sign of the net enclosed charge. The $\mathbf{E}$ on the left side of Eq. 9, however, is the electric field resulting from *all* charges, both those inside and those outside the Gaussian surface.

Let us apply these ideas to Fig. 5, which shows two charges, equal in magnitude but opposite in sign, and also the lines of force describing the electric fields that they set up in the surrounding space. Four irregularly shaped Gaussian surfaces are also shown, in cross section. Let us consider each in turn.

Surface S_1. The electric field is outward for all points on this surface. Thus, the flux of the electric field is positive. So is the net charge within the surface, as Gauss' law requires.

Surface S_2. The electric field is inward for all points. Thus, the flux of the electric field is negative and so is the enclosed charge, as Gauss' law requires.

Surface S_3. This surface contains no charge. Gauss' law then requires that the flux of the electric field be zero for this surface. This is reasonable because, as we see, the lines of force pass right through the surface, entering it at the top and leaving at the bottom.

Surface S_4. This surface encloses no *net* charge, the positive and the negative charges being of equal magnitude. Gauss' law then requires that the flux for this surface be zero. Again, as for surface S_3, that is reasonable. The lines of force that leave surface S_4 at the top all curve around and reenter it at the bottom.

What would happen if we were to bring an enormous charge Q up close to the surface S_4 in Fig. 5? The pattern of the lines of force would certainly change but the net flux for the four Gaussian surfaces in Fig. 5 would not change. We can understand this because the lines of force associated with the added Q alone would pass right through each of the four Gaussian surfaces they encounter, making no contribution to the total flux through any of them. The value of Q would not enter Gauss' law in any way because Q lies outside all four of the Gaussian surfaces that we are considering.

---

* Gauss' law as presented here is for the important special case in which the charges are in a vacuum or, for most practical purposes, in air. In Section 27-8, we extend Gauss' law to include cases in which other materials, such as mica, oil, or glass, are present.

Sample Problem 2   Figure 6 shows three lumps of plastic, each carrying an electric charge, and a dime, carrying no charge. The cross sections of two Gaussian surfaces are indicated. What is the flux of the electric field for each of these

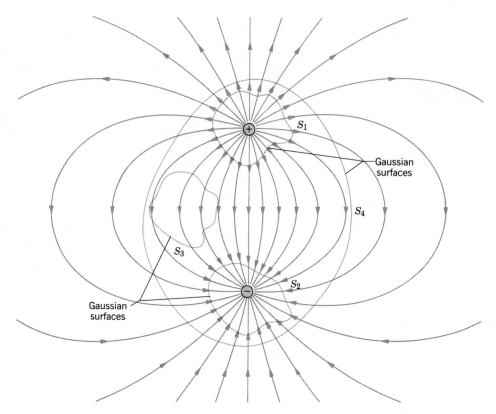

Figure 5 Two point charges, equal in magnitude but opposite in sign, and the lines of force that represent their electric field pattern. Four Gaussian surfaces are shown, in cross section.

surfaces? Assume $q_1 = +3.1$ nC, $q_2 = -5.9$ nC, and $q_3 = -3.1$ nC.

For surface S_1, the net enclosed charge q is q_1. The uncharged coin makes no contribution even though the positive and negative charges it contains may be separated by the action of the field in which the coin is immersed. Charges q_2 and q_3 are outside the surface S_1 and are therefore not included in q. From Eq. 8, we then have

$$\Phi = \frac{q}{\epsilon_0} = \frac{q_1}{\epsilon_0} = \frac{+3.1 \times 10^{-9} \text{ C}}{8.85 \times 10^{-12} \text{ C}^2/\text{N} \cdot \text{m}^2}$$

$$= +350 \text{ N} \cdot \text{m}^2/\text{C}. \qquad \text{(Answer)}$$

The plus sign indicates that the net charge within the surface is positive and also that the net flux through the surface is outward.

For surface S_2, the net charge q is $q_1 + q_2 + q_3$ so that

$$\Phi = \frac{q}{\epsilon_0} = \frac{q_1 + q_2 + q_3}{\epsilon_0}$$

$$= \frac{+3.1 \times 10^{-9} \text{ C} - 5.9 \times 10^{-9} \text{ C} - 3.1 \times 10^{-9} \text{ C}}{8.85 \times 10^{-12} \text{ C}^2/\text{N} \cdot \text{m}^2}$$

$$= -670 \text{ N} \cdot \text{m}^2/\text{C}. \qquad \text{(Answer)}$$

The minus sign shows that the net charge within the surface is negative and that the net flux through the surface is inward.

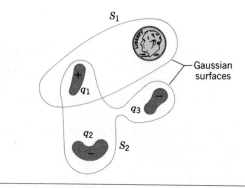

Figure 6 Sample Problem 2. Three plastic objects, each carrying an electric charge, and a coin, which carries no charge. The outlines of two possible Gaussian surfaces are shown.

25–6 Gauss' Law and Coulomb's Law

If Gauss' law and Coulomb's law are equivalent, we should be able to derive each from the other. Here we

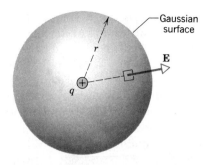

Figure 7   A spherical Gaussian surface centered on a point charge q.

derive Coulomb's law from Gauss' law and from some symmetry considerations.*

Figure 7 shows a point charge q, around which we have drawn a concentric spherical Gaussian surface of radius r. If we divide this surface into differential squares, we know from symmetry that both $\mathbf{E}$ and $d\mathbf{A}$ will be at right angles to the surface, the angle θ between them being zero. Thus, the quantity $\mathbf{E} \cdot d\mathbf{A}$ becomes simply $E\, dA$ and Gauss' law (see Eq. 9) becomes

$$\epsilon_0 \oint \mathbf{E} \cdot d\mathbf{A} = \epsilon_0 \oint E\, dA = q.$$

Because $\mathbf{E}$ has the same magnitude for all points on the sphere, we can factor it out of the integral, leaving

$$\epsilon_0 E \oint dA = q. \tag{10}$$

However, the integral in Eq. 10 is just the area of the spherical surface, or $4\pi r^2$, so that the equation becomes

$$\epsilon_0 E(4\pi r^2) = q$$

or

$$E = \frac{1}{4\pi\epsilon_0} \frac{q}{r^2}, \tag{11}$$

which is Coulomb's law, in the form in which we have written it in Eq. 1.

---

* These two laws are totally equivalent when—as in these Chapters—we apply them to problems involving charges that are either stationary or slowly moving. Gauss' law is more general in that it also covers the case of a rapidly moving charge. For such charges the electric lines of force become compressed in a plane at right angles to the direction of motion, thus losing their spherical symmetry.

25-7   A Charged Isolated Conductor

Gauss' law permits us to prove an important theorem about isolated conductors:

If you put an excess charge on an isolated conductor, that charge will move entirely to the surface of the conductor. None of the excess charge will be found within the body of the conductor.

This might not seem unreasonable considering that like charges repel each other. You might imagine that, by moving to the surface, the added charges are getting as far away from each other as they can. We turn to Gauss' law for a quantitative proof of this qualitative speculation.

Figure 8a shows, in cross section, an isolated lump of copper hanging from a thread and carrying an added charge q. The light gray line shows a Gaussian surface that lies just below the actual surface of the conductor.

The key to our proof is the realization that, under equilibrium conditions, the electric field inside the conductor must be zero. If this were not so, the field would exert a force on the conduction electrons that are ever present in the metal conductor, and internal currents

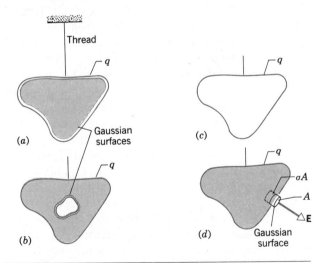

Figure 8   (a) A lump of copper, carrying a charge q, hangs from a thread. A Gaussian surface is drawn within the metal, just below the actual surface. (b) The lump of copper now has an internal cavity. A Gaussian surface lies within the metal, close to the cavity wall. (c) Nothing changes if the cavity is enlarged to include practically the entire lump of metal. (d) A squat cylindrical Gaussian surface pierces the surface of the conductor. It contains a charge σA.

would be set up. However, we know that there are no such enduring currents in an isolated lump of copper. Electric fields will appear inside a conductor during the process of charging it but these fields will not last long. Internal currents will act quickly to redistribute the added charge in such a way that the electric fields inside the conductor vanish, the currents stop, and equilibrium (electrostatic) conditions prevail.

If **E** is zero everywhere inside the conductor, it must be zero for all points on the Gaussian surface because that surface, though close to the surface of the conductor, is definitely inside it. This means that the flux through the Gaussian surface must be zero. Gauss' law then tells us that the charge inside the Gaussian surface must also be zero. If the added charge is not inside the Gaussian surface it can only be outside that surface, which means that it must lie on the actual surface of the conductor.

An Isolated Conductor with a Cavity. Figure 8b shows the same hanging conductor in which a cavity has been scooped out. It is perhaps reasonable to suppose that scooping out the electrically neutral material to form the cavity should not change the distribution of charge or the pattern of the electric field that exists in Fig. 8a. Again, we must turn to Gauss' law for a quantitative proof.

Draw a Gaussian surface surrounding the cavity, close to its walls but inside the conducting body. Because **E** = 0 inside the conductor, there can be no flux through this new Gaussian surface. Therefore, from Gauss' law, that surface can enclose no net charge. We conclude that there is no charge on the cavity walls; it remains on the outer surface of the conductor, as in Fig. 8a.

The Conductor Removed. Suppose that, by some magic, the excess charges could be "frozen" into position on the conductor surface, perhaps by embedding them in a thin plastic coating, and suppose that then the conductor could be removed completely, as in Fig. 8c. This is equivalent to enlarging the cavity of Fig. 8b until it consumes the entire conductor, leaving only the charges. The electric field pattern would not change at all; it would remain zero inside the thin shell of charge and would remain unchanged for all external points. The electric field is set up by the charges and not by the conductor. The conductor simply provides a pathway so that the charges can change their positions.

The External Electric Field. Although the excess charge on an isolated conductor moves entirely to its surface, that charge—except for a spherical conductor—does not distribute itself uniformly over that surface. Put another way, the surface charge density σ, whose SI unit is C/m^2, will vary from point to point over the surface.

We can use Gauss' law to find a relation—at any surface point—between the surface charge density σ at that point and the electric field **E** just outside the surface at that same point. Figure 8d shows a squat cylindrical Gaussian surface, the (small) area of its two end caps being A. The end caps are parallel to the surface, one lying entirely inside the conductor and the other entirely outside. The short cylindrical walls are perpendicular to the surface of the conductor.

The electric field just outside a charged isolated conductor in electrostatic equilibrium must be at right angles to the surface of the conductor. If this were not so, there would be a component of **E** lying in the surface and this component would set up surface currents that would redistribute the surface charges—hence violating our assumption of electrostatic equilibrium. Thus, **E** is perpendicular to the surface and the flux through the exterior end cap of the Gaussian surface of Fig. 8d is EA. The flux through the interior end cap is zero because **E** = 0 for all interior points of the conductor. The flux through the cylindrical walls is also zero because the lines of **E** are parallel to the surface, so they cannot pierce it. The charge enclosed by the Gaussian surface is σA.

To sum up: The flux through the entire Gaussian surface of Fig. 8d is EA and the charge enclosed by that surface is σA. Gauss' law then yields

$$\epsilon_0 \Phi = q$$

or

$$\epsilon_0 EA = \sigma A.$$

Thus, we find

$$\boxed{E = \frac{\sigma}{\epsilon_0}} \quad \text{(conducting surface).} \quad (12)$$

Earlier (see Eq. 22 of Chapter 24), we found that the electric field caused by a charged sheet has a magnitude $\sigma/2\epsilon_0$. Equation 12 shows a value twice as large just outside a charged conductor. Where does the factor of 2 come from? Let us explain.

Consider a point P just outside the surface of the charged conductor and a corresponding point P' just inside that surface. It is convenient to divide the surface charge on the conductor into two parts: (1) the *local charges,* which are those charges on the surface very close to these two points, and (2) the *distant charges,* which are the charges on the rest of the surface of the conductor. The electric fields, both at P and at P', are the superposition of the fields set up by these two charge systems.

The field set up by the local charges is just like that set up by the plane sheet of charge in Fig. 3 of Chapter 24; it has a magnitude of $\sigma/2\epsilon_0$ (see Eq. 22 of Chapter 24) and points *away from* the surface, both at P and at P'.

Because we must have $\mathbf{E} = 0$ inside the conductor, the distant charges must arrange themselves so that they produce a field of magnitude $\sigma/2\epsilon_0$ at P', just canceling the field of the local charges at that point. At P, however, the fields set up by the local charges and the distant charges add, producing a resultant field of magnitude $\sigma/2\epsilon_0 + \sigma/2\epsilon_0$ or σ/ϵ_0, in agreement with Eq. 12.

Sample Problem 3   The electric field just above the surface of the charged drum of a photocopying machine has a magnitude E of 2.3×10^5 N/C. What is the surface charge density on the drum if it is a conductor?

From Eq. 12 we have

$\sigma = \epsilon_0 E = (8.85 \times 10^{-12} \text{ C}^2/\text{N} \cdot \text{m}^2)(2.3 \times 10^5 \text{ N/C})$
$\quad = 2.0 \times 10^{-6} \text{ C/m}^2 = 2.0 \ \mu\text{C/m}^2.$ (Answer)

Sample Problem 4   The magnitude of the average electric field normally present in the earth's atmosphere just above the surface of the earth is about 150 N/C, directed downward. What is the total net surface charge carried by the earth? Assume the earth to be a conductor.

Lines of force end on negative charges so that, if the earth's electric field points downward, the earth's average surface charge density σ must be negative. From Eq. 12 we find

$\sigma = \epsilon_0 E = (8.85 \times 10^{-12} \text{ C}^2/\text{N} \cdot \text{m}^2)(-150 \text{ N/C})$
$\quad = -1.33 \times 10^{-9} \text{ C/m}^2.$

The earth's total charge q is the surface charge density multiplied by $4\pi R^2$, the surface area of the (presumed spherical) earth. Thus,

$q = \sigma 4\pi R^2$
$\quad = (-1.33 \times 10^{-9} \text{ C/m}^2)(4\pi)(6.37 \times 10^6 \text{ m})^2$
$\quad = -6.8 \times 10^5 \text{ C} = -680 \text{ kC.}$ (Answer)

25-8   A Sensitive Test of Coulomb's Law

If an excess charge on an isolated conductor does *not* move entirely to its surface — as we proved it did in the preceding section — then Gauss' law cannot be true because our proof was based on that law. If Gauss' law is not true, then Coulomb's law cannot be true. In particular, the exponent 2 in the inverse square law might not be

exactly 2. Thus, this law might be

$$E = \frac{1}{4\pi\epsilon_0} \frac{q}{r^{2\pm\delta}},\qquad (13)$$

in which δ—if not zero—is a small number.

Coulomb's law is vitally important in physics and if δ in Eq. 13 is not zero, there are serious consequences for our understanding of electromagnetism and quantum physics. The best way to measure δ is to find out *by experiment* whether an excess charge, placed on an isolated conductor, does or does not move *entirely* to its outside surface.

Benjamin Franklin seems to have been the first to carry out experiments along these lines. Figure 9 shows his simple arrangements. Charge a metal ball and lower it by a thread deep inside a metal can. Touch the ball to the inside of the can. When the ball touches the can, the *ball + can* form a single "isolated conductor." If the charge does indeed flow *entirely* to the outside of the can, the ball should be found to be entirely uncharged when removed from the can. Within the accuracy of his experiments, Franklin found it so.

Franklin recommended this "singular fact" to Jo-

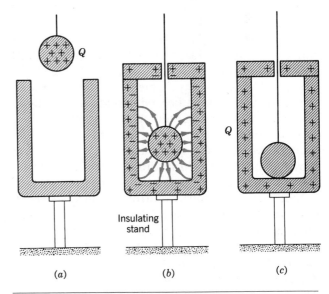

Figure 9   An arrangement conceived by Benjamin Franklin to show that charge placed on a conductor moves to its surface. (*a*) A charged metal ball is lowered into an uncharged metal can. (*b*) The ball is inside the can and an (almost) closed cover is added. (*c*) The ball is touched to the can, thus forming a single conducting object as in Fig. 8*a*. When the ball is removed from the can it is found to be completely uncharged, thus showing that the charge must have been transferred entirely to the can.

Table 2   Tests of Coulomb's Inverse Square Law

Experimenters	*Date*	*δ(Eq. 13)*
Franklin	1755	
Priestley	1767	. . . according to the squares . . .
Robison	1769	<0.06
Cavendish	1773	<0.02
Coulomb	1785	a few percent at most
Maxwell	1873	$<5 \times 10^{-5}$
Plimpton and Lawton	1936	$<2 \times 10^{-9}$
Bartlett, Goldhagen, and Phillips	1970	$<1.3 \times 10^{-16}$
Williams, Faller, and Hill	1971	$<1.0 \times 10^{-16}$

seph Priestly, who checked Franklin's experiments and realized that Coulomb's inverse square law followed from them. Many others, including Cavendish* and Maxwell, repeated the experiments, with ever increasing precision. Modern experiments, carried out with remarkable precision, have shown that if δ in Eq. 13 is not zero it is certainly very very small. Table 2 summarizes the most important of these experiments.

Figure 10 is a sketch of the apparatus used by Plimpton and Lawton to measure δ in Eq. 13 and thus check up on Coulomb's law. The apparatus consists of two concentric metal shells, A and B, the former being 5 ft in diameter. The inner shell contains a sensitive electrometer E connected so that it will indicate whether any charge moves between shells A and B. If the shells are connected by a wire, any charge placed on the shell assembly should reside entirely on the outside of shell A if Gauss' law—and thus Coulomb's law—holds.

By throwing switch S to the left, you can put a substantial charge on the sphere assembly. If any charge moves to shell B, it would have to pass through electrometer E and cause a deflection, which can be observed optically using telescope T, mirror M, and windows W.

However, when the switch S was thrown alternately from left to right, thus connecting the shell assembly alternately to the battery and to the ground, no effect was observed. Knowing the sensitivity of their electrometer, the experimenters concluded that if δ in Eq. 13 is not zero it is no more than $\sim 0.000\ 000\ 002$, a very small number indeed. The inverse square law seems to be on safe ground.

* The same Cavendish who "weighed the earth" with a torsion balance; see Section 15-3.

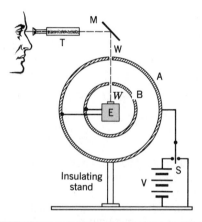

Figure 10   A modern and more precise version of the apparatus of Fig. 9, also designed to verify that charge resides on the outside surface of a metal object. If a charge is placed on sphere A, by throwing switch S to the left, any charge that moved to the inner sphere B would be detected by the sensitive electrometer E. No such charge transfer was found; the charge remained on the outside surface of A.

25-9   Gauss' Law: Linear Symmetry

Figure 11 shows a section of an infinitely long charged plastic rod, its linear charge density, or charge per unit length, being λ. Let us find an expression for the magnitude of the electric field **E** at a distance r from the axis of the rod.

Our Gaussian surface should match the symmetry of the problem, which is cylindrical. We choose a circular cylinder of radius r and length h, coaxial with the rod. The Gaussian surface must be closed so we include two end caps as part of the surface.

Imagine now that, while you are not watching, someone rotates the plastic rod around its cylindrical axis and/or turns it end for end. When you look again at the rod, you will not be able to detect any change. We conclude that, from symmetry, the only uniquely specified direction in this problem is a radial line. Thus, **E** must have a constant magnitude E and (for a positively charged rod) must be directed radially outward at every point on the cylindrical Gaussian surface.

The flux of **E** through this surface is $(E)(2\pi r)(h)$, where $2\pi r$ is the circumference of the cylinder and h is its height, so that $2\pi rh$ is the area of the cylindrical surface. There is no flux through the end caps because **E**, being radially directed, lies parallel to the surface at every point.

The charge enclosed by the surface is λh so that Gauss' law (Eq. 9),

$$\epsilon_0 \oint \mathbf{E} \cdot d\mathbf{A} = q,$$

reduces to

$$\epsilon_0 E(2\pi rh) = \lambda h,$$

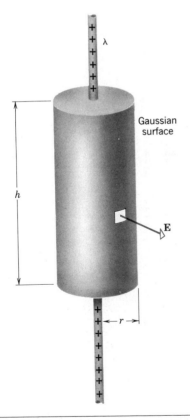

Figure 11 A Gaussian surface in the form of a closed cylinder surrounds a section of a very long, uniformly charged, plastic rod.

yielding

$$\boxed{E = \frac{\lambda}{2\pi\epsilon_0 r}} \quad \text{(line of charge)}. \quad (14)$$

The direction of **E** is radially outward if the line of charge is positive and radially inward if it is negative.

Sample Problem 5 A plastic rod, whose length L is 220 cm and whose radius R is 3.6 mm, carries a negative charge q of magnitude 3.8×10^{-7} C, spread uniformly over its surface. What is the electric field near the midpoint of the rod, at a point on its surface?

Although the rod is not infinitely long, from a point on its surface and near its midpoint it is effectively very long so that we are justified in using Eq. 14. The linear charge density for the rod is

$$\lambda = \frac{q}{L} = \frac{-3.9 \times 10^{-7}\,\text{C}}{2.2\,\text{m}} = -1.73 \times 10^{-7}\,\text{C/m}.$$

From Eq. 14 we then have

$$E = \frac{\lambda}{2\pi\epsilon_0 r}$$
$$= \frac{-1.73 \times 10^{-7}\,\text{C/m}}{(2\pi)(8.85 \times 10^{-12}\,\text{C}^2/\text{N}\cdot\text{m}^2)(0.0036\,\text{m})}$$
$$= -8.6 \times 10^5\,\text{N/C}. \quad \text{(Answer)}$$

The minus sign tells us that, because the rod is negatively charged, the direction of the electric field is radially inward, toward the axis of the rod. Sparking occurs in dry air at atmospheric pressure at an electric field strength of about 3×10^6 N/C. The field strength we calculated above is lower than this by a factor of about 3.4 so that sparking should not occur.

25-10 Gauss' Law: Planar Symmetry

Figure 12 shows a portion of a thin infinite sheet of charge, its surface charge density, or charge per unit area, being σ. A sheet of thin plastic wrap, uniformly charged on one side by friction, can serve as a prototype. Let us find the electric field **E** a distance r in front of the sheet.

A useful Gaussian surface is a closed cylinder of cross-sectional area A and height $2r$, arranged to pierce the sheet at right angles as shown. From symmetry, the only preferred direction for **E** is at right angles to the end cap. Furthermore, **E** must point *away* from the posi-

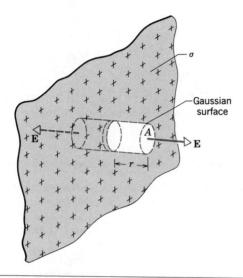

Figure 12 A portion of a very large thin plastic sheet, uniformly charged on one side to surface charge density σ. A Gaussian surface, in the form of a short cylinder closed at each end, pierces the sheet at right angles to it.

tively charged sheet of Fig. 12, thus piercing the Gaussian surface in an outward direction. Because **E** does not pierce the cylinder walls, there is no flux through this portion of the Gaussian surface. Thus, Gauss' law

$$\epsilon_0 \oint \mathbf{E} \cdot d\mathbf{A}$$

becomes

$$\epsilon_0(EA + EA) = \sigma A,$$

where σA is the enclosed charge. This gives

$$\boxed{E = \frac{\sigma}{2\epsilon_0}} \quad \text{(sheet of charge).} \quad (15)$$

This result (Eq. 15) agrees with our earlier result (Eq. 22 of Chapter 24), which we obtained by direct integration.

You may wonder why we introduce such seemingly unrealistic problems as the field set up by an infinite line of charge or by an infinite sheet of charge. It is not enough to say that we do so because it is simple to solve such problems, although that is indeed true. The proper answer is that the solutions for "infinite" problems apply to real world problems to a very good approximation. Thus, Eq. 15 holds quite well for a finite sheet as long as you are close to the sheet and not too near its edges. After all, an ant in the center of a large parking lot probably thinks that it is walking on an infinite sheet, and it might as well be.

Sample Problem 6 Figure 13a shows portions of two large sheets of charge with uniform surface charge densities of $\sigma_+ = +6.8\ \mu C/m^2$ and $\sigma_- = -4.3\ \mu C/m^2$. Find the electric field **E** (a) to the left of the sheets, (b) between the sheets, and (c) to the right of the sheets.

Our strategy is to deal with each sheet separately and then to add the resulting electric fields algebraically, using the superposition principle.

For the positive sheet we have, from Eq. 15,

$$E_+ = \frac{\sigma_+}{2\epsilon_0} = \frac{6.8 \times 10^{-6}\ C/m^2}{(2)(8.85 \times 10^{-12}\ C^2/N \cdot m^2)} = 3.84 \times 10^5\ N/C.$$

Similarly, for the negative sheet

$$E_- = \frac{|\sigma_-|}{2\epsilon_0} = \frac{4.3 \times 10^{-6}\ C/m^2}{(2)(8.85 \times 10^{-12}\ C^2/N \cdot m^2)} = 2.43 \times 10^5\ N/C.$$

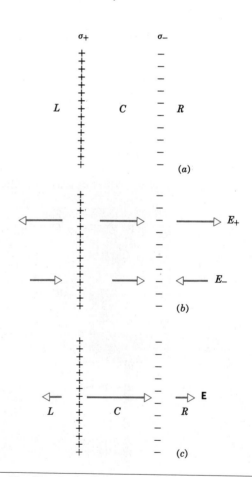

Figure 13 Sample Problem 6. (a) Two large thin sheets of charge face each other. (b) Showing, separately, the electric fields resulting from each charge sheet. (c) The resultant field due to both charge sheets, found from the principle of superposition.

Figure 13*b* shows the two sets of fields calculated above, to the left of the sheets, between them, and to the right of the sheets.

The resultant fields in these three regions follow from the superposition principle. To the left of the sheets, we have

$$E_L = E_+ - E_- = 3.84 \times 10^5 \text{ N/C} - 2.43 \times 10^5 \text{ N/C}$$
$$= 1.4 \times 10^5 \text{ N/C.} \qquad \text{(Answer)}$$

The resultant electric field in this region points to the left, as Fig. 13*c* shows. To the right of the sheets, the electric field has this same magnitude but points to the right in Fig. 13*c*.

Between the sheets, the two fields add and we have

$$E_C = E_+ + E_- = 3.84 \times 10^5 \text{ N/C} + 2.43 \times 10^5 \text{ N/C}$$
$$= 6.3 \times 10^5 \text{ N/C.} \qquad \text{(Answer)}$$

Outside the sheets, the electric field behaves like that from a single sheet whose surface charge density is $\sigma_+ + \sigma_-$ or $+2.5 \times 10^{-6}$ C/m². The field pattern of Fig. 13*c* bears this out. In Exercise 30 and Problem 35 we ask you to investigate the case in which the two surface charge densities are equal in magnitude but opposite in sign and also the case in which they are equal in both magnitude and sign.

25-11   Gauss' Law: Spherical Symmetry

Here we use Gauss' law to prove the two theorems presented without proof in Section 23-4, namely:

A uniform spherical shell of charge behaves, for external points, as if all its charge were concentrated at its center,

and

A uniform spherical shell of charge exerts no electrical force on a charged particle placed inside the shell.

These two shell theorems are the electrostatic analogs of the two gravitational shell theorems presented in Chapter 15. We shall see how much simpler is our Gauss' law proof than the detailed proof of Section 15-5, in which full advantage of the spherical symmetry was not taken.

Figure 14 shows a spherical shell of charge q and radius R and two concentric Gaussian surfaces, S_1 and S_2. Applying Gauss' law to surface S_1, for which $r > R$, leads directly to

$$\boxed{E = \frac{1}{4\pi\epsilon_0} \frac{q}{r^2}} \quad \text{(spherical shell, } r > R\text{),} \quad (16)$$

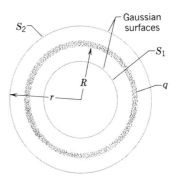

Figure 14   A thin uniformly charged shell. Two Gaussian surfaces are shown in cross section, one enclosing the shell and one enclosed by it.

just as it did in connection with Fig. 7.

Applying Gauss' law to surface S_1, for which $r < R$, leads directly to

$$\boxed{E = 0} \quad \text{(spherical shell, } r < R\text{),} \quad (17)$$

because this Gaussian surface encloses no charge.

Any spherically symmetric charge distribution, such as that of Fig. 15, can be made up of a nest of such shells. The volume charge density ρ, whose SI units are C/m³, must have a constant value for each shell but need not be the same from shell to shell. That is, for the charge distribution as a whole, the volume charge density ρ need not be constant but can vary only with r, the radial distance from the center.

In Fig. 15*a* we show a Gaussian surface with $r > R$ so that the entire charge lies within the surface. In this case the charge behaves as if it were a point charge located at the center and Eq. 16 holds.

Figure 15*b* shows a Gaussian surface of radius r, where $r < R$. Here, by virtue of Eq. 16, that part of the charge distribution that lies outside the Gaussian surface makes no contribution to the field at radius r, which is determined only by the charge q' that lies within that surface. Thus, from Eq. 16

$$E = \frac{1}{4\pi\epsilon_0} \frac{q'}{r^2}, \quad (18)$$

where q' is that portion of the charge within the Gaussian surface at radius r.

Table 3 summarizes the relations between charge and field for the symmetrical charge distributions that we have considered in this chapter.

Table 3 A Summary of Formulas

Situation	Formula for E	Remarks	Equation
Charged conductor	σ/ϵ_0	On the conductor surface	12
	0	Inside the conductor	
Point charge	$q/4\pi\epsilon_0 r^2$		11
Spherical shell	$q/4\pi\epsilon_0 r^2$	Outside the shell	16
	0	Inside the shell	17
Infinite line of charge	$\lambda/2\pi\epsilon_0 r$		14
Infinite sheet of charge	$\sigma/2\epsilon_0$		15

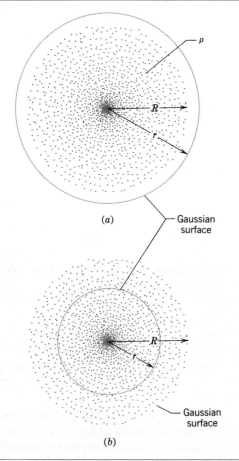

(a)

Gaussian surface

(b)

Gaussian surface

Figure 15   The dots represent a spherically symmetric distribution of charge of radius R, the volume charge density ρ being a function of distance from the center only. The object is not a conductor and the charges are assumed to be held fixed in position. (a) A concentric Gaussian surface with $r > R$. (b) A Gaussian surface with $r < R$.

Sample Problem 7   The nucleus of an atom of gold has a radius $R = 6.2 \times 10^{-15}$ m and carries a positive charge $q = Ze$, where Z (= 79) is the atomic number of gold and e is the elementary charge. Plot the magnitude of the electric field from the center of the gold nucleus outward to a distance of about twice its radius. Assume that the nucleus is spherical and that the charge is distributed uniformly throughout its volume.

The total charge q on the nucleus is

$$q = Ze = (79)(1.60 \times 10^{-19}\ \text{C}) = 1.264 \times 10^{-17}\ \text{C}.$$

Outside the nucleus, the situation is represented by Fig. 15a and by Eq. 16. From this equation we have, for a point on the surface of the nucleus,

$$E = \frac{1}{4\pi\epsilon_0}\frac{q}{r^2}$$

$$= \frac{1.264 \times 10^{-17}\ \text{C}}{(4\pi)(8.85 \times 10^{-12}\ \text{C}^2/\text{N} \cdot \text{m}^2)(6.2 \times 10^{-15}\ \text{m})^2}$$

$$= 3.0 \times 10^{21}\ \text{N/C}. \qquad \text{(Answer)}$$

Inside the nucleus, Fig. 15b and Eq. 18 apply. The charge q' contained within a sphere of radius r, where $r < R$, is proportional to the respective volumes. The volume of a sphere of radius r is $\frac{4}{3}\pi r^3$ so that we have

$$q' = q\frac{\frac{4}{3}\pi r^3}{\frac{4}{3}\pi R^3} = q\frac{r^3}{R^3}.$$

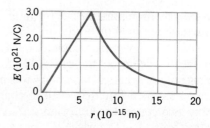

Figure 16   The variation of electric field with distance from the center for the nucleus of a gold atom. The positive charge is assumed to be distributed uniformly throughout the volume of the nucleus.

If we substitute this result into Eq. 18, we find

$$E = \frac{1}{4\pi\epsilon_0}\frac{q'}{r^2} = \left(\frac{q}{4\pi\epsilon_0 R^3}\right)r. \qquad (19)$$

The quantity in parentheses is a constant so that, within the nucleus, E is directly proportional to r, being zero at the nuclear center. Comparison of Eqs. 19 and 16 shows that they give the same result—that calculated above—for $r = R$. This simply tells us that the "inside" equation and the "outside" equation must yield the same value for the electric field at the nuclear surface, where the two equations both apply. Figure 16 shows these results graphically.

REVIEW AND SUMMARY

Gauss' law and Coulomb's law, although expressed in different forms, are equivalent ways of describing the relation between charge and the electric field in static situations. Gauss' law is

Gauss' Law

$$\epsilon_0\Phi = q \quad \text{(Gauss' law)}, \qquad [8]$$

in which q is the net charge inside an imaginary closed surface (a *Gaussian surface*) and Φ is the outward flux of the electric field through the surface:

The Flux of the Electric Field

$$\Phi = \oint \mathbf{E} \cdot d\mathbf{A} \quad \text{(flux through a Gaussian surface)}. \qquad [6]$$

See Sample Problems 1 and 2 and the discussions related to Figs. 3 and 4.

Coulomb's Law and Gauss' Law

Coulomb's law can readily be derived from Gauss' law, the details being given in Section 25–6. Experimental verification of Gauss' law—and thus of Coulomb's law—shows that the exponent of r in Coulomb's law is now known to be exactly 2 within an experimental uncertainty of less than 1×10^{-16}. See Section 25–8 and Table 2.

Using Gauss' law and, in some cases, symmetry arguments, we can derive several important results. Among these are:

Charge on a Conductor

a. An excess charge on an *insulated conductor* is, in equilibrium, entirely on its outer surface (Section 25–7).

Field Near a Charged Conductor

b. The electric field near the *surface of a charged conductor* in equilibrium is perpendicular to the surface and has magnitude

$$E = \frac{\sigma}{\epsilon_0} \quad \text{(conducting surface)} \qquad [12]$$

(Section 25–7 and Sample Problems 3 and 4).

c. The electric field due to an infinite *line of charge* with uniform charge per unit length, λ, is in a direction perpendicular to the line of charge and has magnitude

Field Due to an Infinite Line of Charge

$$E = \frac{\lambda}{2\pi\epsilon_0 r} \quad \text{(line of charge)} \qquad [14]$$

(Section 25–9 and Sample Problem 5).

Field Due to an Infinite Sheet of Charge

d. The electric field due to an infinite sheet of charge is perpendicular to the plane of the sheet and has magnitude

$$E = \frac{\sigma}{2\epsilon_0} \quad \text{(sheet of charge)} \qquad [15]$$

(Section 25–10 and Sample Problem 6).

e. The electric field outside a *spherically symmetrical shell* with radius R and total charge q is directed radially and has magnitude

Field of a Spherical Shell

$$E = \frac{1}{4\pi\epsilon_0} \frac{q}{r^2} \quad \text{(spherical shell, } r > R\text{).} \qquad [16]$$

The charge behaves, for external points, as if it were all at the center of the sphere. The field *inside* a uniform spherical shell is exactly zero:

$$E = 0 \quad \text{(spherical shell, } r < R\text{).} \qquad [17]$$

The electric field *inside a uniform sphere of charge* is directed radially and has magnitude

Field Inside a Uniform
Spherical Charge

$$E = \left(\frac{q}{4\pi\epsilon_0 R^3}\right) r \qquad [19]$$

(Section 25–11 and Sample Problem 7).

QUESTIONS

1. What is the basis for the statement that lines of electric force begin and end only on electric charges?

2. Positive charges are sometimes called "sources" and negative charges "sinks" of electric field. How would you justify this terminology? Are there sources and/or sinks of gravitational field?

3. Does Gauss' law hold for the flow of water? Consider various Gaussian surfaces intersecting a fountain or a waterfall in different ways. What would correspond to positive and negative charges in this case?

4. By analogy with Φ, how would you define the flux Φ_g of a gravitational field? What is the flux of the earth's gravitational field through the boundaries of a room, assumed to contain no matter? Through a sphere closely surrounding the earth? Through a sphere the size of the moon's orbit? See Problem 14.

5. Consider a Gaussian surface that surrounds part of the charge distribution shown in Fig. 17. (*a*) Which of the charges contribute to the electric field at point P? (*b*) Would the value obtained for the flux through the surface, calculated using only the electric field due to q_1 and q_2, be greater than, equal to, or less than that obtained using the total field?

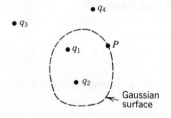

Figure 17 Question 5.

6. Suppose that an electric field in some region is found to have a constant direction but to be decreasing in strength in that direction. What do you conclude about the charge in the region? Sketch the lines of force.

7. A point charge is placed at the center of a spherical Gaussian surface. Is Φ changed (*a*) if the surface is replaced by a cube of the same volume, (*b*) if the sphere is replaced by a cube of one-tenth the volume, (*c*) if the charge is moved off-center in the original sphere, still remaining inside, (*d*) if the charge is moved just outside the original sphere, (*e*) if a second charge is placed near, and outside, the original sphere, and (*f*) if a second charge is placed inside the Gaussian surface?

8. In Gauss' law,

$$\epsilon_0 \oint \mathbf{E} \cdot d\mathbf{A} = q,$$

is $\mathbf{E}$ necessarily the electric field attributable to the charge q?

9. A surface encloses an electric dipole. What can you say about Φ_E for this surface?

10. Suppose that a Gaussian surface encloses no net charge. Does Gauss' law require that $\mathbf{E}$ equal zero for all points on the surface? Is the converse of this statement true; that is, if $\mathbf{E}$ equals zero everywhere on the surface, does Gauss' law require that there be no net charge inside?

11. Is Gauss' law useful in calculating the field due to three equal charges located at the corners of an equilateral triangle? Explain why or why not.

12. A total charge Q is distributed uniformly throughout a cube of edge length a. Is the resulting electric field at an external point P, a distance r from the center of the cube, given by $E = Q/4\pi\epsilon_0 r^2$? See Fig. 18. If not, can E be found by constructing a "concentric" cubical Gaussian surface? If not, explain why not. Can you say anything about E if $r \gg a$?

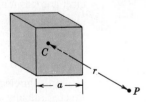

Figure 18 Question 12.

13. Is E necessarily zero inside a charged rubber balloon if the balloon is (a) spherical or (b) sausage shaped? For each shape, assume the charge to be distributed uniformly over the surface. How would the situation change, if at all, if the balloon has a thin layer of conducting paint on its outside surface?

14. A spherical rubber balloon carries a charge that is uniformly distributed over its surface. As the balloon is blown up, how does E vary for points (a) inside the balloon, (b) at the surface of the balloon, and (c) outside the balloon?

15. In Section 25–6 we have seen that Coulomb's law can be derived from Gauss' law. Does this necessarily mean that Gauss' law can be derived from Coulomb's law?

16. A large, insulated, hollow conductor carries a positive charge. A small metal ball carrying a negative charge of the same magnitude is lowered by a thread through a small opening in the top of the conductor, allowed to touch the inner surface, and then withdrawn. What is then the charge on (a) the conductor and (b) the ball?

17. Can we deduce from the argument of Section 25–7 that the electrons in the wires of a house wiring system move along the surfaces of those wires? If not, why not?

18. Does Gauss' law, as applied in Section 25–7, require that all the conduction electrons in an insulated conductor reside on the surface?

19. Suppose that you have a Gaussian surface in the shape of a donut and that there is a single point charge inside. Does Gauss' law hold? If not, why not? If so, is there enough symmetry in the situation to apply it usefully?

20. A positive point charge q is located at the center of a hollow metal sphere. What charges appear on (a) the inner surface and (b) the outer surface of the sphere? (c) If you bring an (uncharged) metal object near the sphere, will it change your answers in (a) or (b) above? Will it change the way charge is distributed over the sphere?

21. Explain why the symmetry of Fig. 11 restricts us to a consideration of E that has only a radial component at any point. Remember, in this case, that the field must not only look the same at any point along the line but must also look the same if the figure is turned end for end.

22. In Section 25–9, the *total* charge on the infinite rod is infinite. Why is not E also infinite? After all, according to Coulomb's law, if q is infinite, so is E.

23. Explain why the symmetry of Fig. 12 restricts us to a consideration of E that has only a component directed away from the sheet. Why, for example, could E not have components parallel to the sheet? Remember, in this case, that the field must not only look the same at any point along the sheet in any direction but must also look the same if the sheet is rotated about any line perpendicular to the sheet.

24. The field due to an infinite sheet of charge is uniform, having the same strength at all points no matter how far from the surface charge. Explain how this can be, given the inverse square nature of Coulomb's law.

25. Explain why the spherical symmetry of Fig. 7 restricts us to a consideration of E that has only a radial component at any point. (*Hint:* Imagine other components, perhaps along the equivalent of longitude or latitude lines on the earth's surface. Spherical symmetry requires that these look the same from any perspective. Can you invent such field lines that satisfy this criterion?)

26. As you penetrate a uniform sphere of charge, E should decrease because less charge lies inside a sphere drawn through the observation point. On the other hand, E should increase because you are closer to the center of this charge. Which effect dominates and why?

27. Given a spherically symmetric charge distribution (not of uniform density of charge radially), is E necessarily a maximum at the surface? Comment on various possibilities.

EXERCISES AND PROBLEMS

Section 25–3   Flux

1E. Water in an irrigation ditch of width w = 3.22 m and depth d = 1.04 m flows with a speed of 0.207 m/s. Find the mass flux through the following surfaces: (a) a surface of area wd, entirely in the water, perpendicular to the flow; (b) a surface with area 3wd/2, of which wd is in the water, perpendicular to the flow; (c) a surface of area wd/2, entirely in the water, perpendicular to the flow; (d) a surface of area wd, half in the water and half out, perpendicular to the flow; (e) a surface of area wd, entirely in the water, with its normal 34° from the direction of flow.

Section 25–4   Flux of the Electric Field

2E. The square surface shown in Fig. 19 measures 3.2 mm on each side. It is immersed in a uniform electric field with

E = 1800 N/C. The field lines make an angle of 35° with the "outward pointing" normal, as shown. Calculate the flux through the surface.

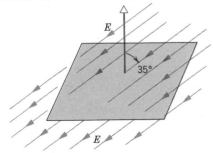

Figure 19 Exercise 2.

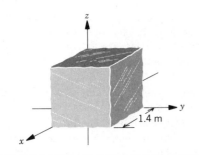

Figure 20 Exercise 3 and Problem 12.

3E. A cube with 1.4-m edges is oriented as shown in Fig. 20 in a region of uniform electric field. Find the electric flux through the right face if the electric field, in newtons per coulomb, is given by (*a*) 6**i**, (*b*) −2**j**, and (*c*) −3**i** + 4**k**. (*d*) What is the total flux through the cube for each of these fields?

4P. Calculate Φ through (*a*) the flat base and (*b*) the curved surface of a hemisphere of radius *R*. The field **E** is uniform, is parallel to the axis of the hemisphere, and the lines of **E** enter through the flat base.

Section 25–5 Gauss' Law

5E. Four charges, 2*q*, *q*, −*q*, and −2*q* are arranged at the corners of a square as shown in Fig. 21. If possible, describe a closed surface that encloses the charge 2*q* and through which the net electric flux is (*a*) 0, (*b*) +3*q*/ϵ_0, and (*c*) −2*q*/ϵ_0.

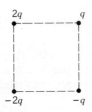

Figure 21 Exercise 5.

6E. Charge on an originally uncharged insulated conductor is separated by holding a positively charged rod very closely nearby, as in Fig. 22. Calculate the flux for the five Gaussian surfaces shown. Assume that the induced negative charge on the conductor is equal to the positive charge *q* on the rod.

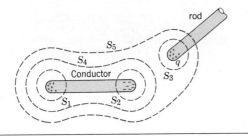

Figure 22 Exercise 6.

7E. A point charge of 1.8 μC is at the center of a cubical Gaussian surface 55 cm on edge. What is Φ_E through the surface?

8E. The net electric flux through each face of a die (singular of dice) has magnitude in units of 10^3 N·m²/C equal to the number of spots on the face (1 through 6). The flux is inward for *N* odd and outward for *N* even. What is the net charge inside the die?

9E. A point charge +*q* is a distance *d*/2 from a square surface of side *d* and is directly above the center of the square as shown in Fig. 23. What is the electric flux through the square? (*Hint:* Think of the square as one face of a cube with edge *d*.)

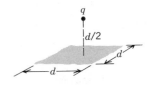

Figure 23 Exercise 9.

10E. A butterfly net is in a uniform electric field as shown in Fig. 24. The rim, a circle of radius *a*, is aligned perpendicular to the field. Find the electric flux through the netting.

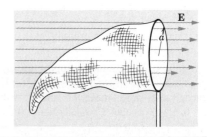

Figure 24 Exercise 10.

11P. It is found experimentally that the electric field in a certain region of the earth's atmosphere is directed vertically down. At an altitude of 300 m the field is 60 N/C and at an altitude of 200 m it is 100 N/C. Find the net amount of charge contained in a cube 100 m on edge located at an altitude between 200 and 300 m. Neglect the curvature of the earth.

12P. Find the net flux through the cube of Exercise 3 and Fig. 20 if the electric field is given by (*a*) **E** = 3*y***j** and (*b*) **E** = −4**i** + (6 + 3*y*)**j**. *E* is in newtons per coulomb if *y* is in meters. (*c*) In each case, how much charge is inside the cube?

13P. A point charge *q* is placed at one corner of a cube of edge *a*. What is the flux through each of the cube faces? (*Hint:* Use Gauss' law and symmetry arguments.)

14P. "Gauss' law for gravitation" is

$$\frac{1}{4\pi G} \Phi_g = \frac{1}{4\pi G} \oint \mathbf{g} \cdot d\mathbf{A} = -m,$$

where m is the enclosed mass and G is the universal gravitation constant (Section 15–6). Derive Newton's law of gravitation from this. What is the significance of the minus sign?

Section 25–7 A Charged Isolated Conductor

15E. A uniformly charged conducting sphere of 1.2-m diameter has a surface charge density of 8.1 μC/m². (a) Find the charge on the sphere. (b) What is the total electric flux leaving the surface of the sphere?

16E. Space vehicles traveling through the earth's radiation belts collide with trapped electrons. Since in space there is no ground, the resulting charge buildup can become significant and can damage electronic components, leading to control-circuit upsets and operational anomalies. A spherical metallic satellite 1.3 m in diameter accumulates 2.4 μC of charge in one orbital revolution. (a) Find the surface charge density. (b) Calculate the resulting electric field just outside the surface of the satellite.

17E. A conducting sphere carrying charge Q is surrounded by a spherical conducting shell. (a) What is the net charge on the inner surface of the shell? (b) Another charge q is placed outside the shell. Now what is the net charge on the inner surface of the shell? (c) If q is moved to a position between the shell and the sphere, what is the net charge on the inner surface of the shell? (d) Are your answers valid if the sphere and shell are not concentric?

18P. An insulated conductor of arbitrary shape carries a net charge of $+10 \times 10^{-6}$ C. Inside the conductor is a hollow cavity within which is a point charge $q = +3.0 \times 10^{-6}$ C. What is the charge (a) on the cavity wall and (b) on the outer surface of the conductor?

19P. An irregularly-shaped conductor has an irregularly-shaped cavity inside. A charge $+q$ is placed on the conductor but there is no charge inside the cavity. Show that (a) $E = 0$ inside the cavity and (b) there is no charge on the cavity wall.

Section 25–9 Gauss' Law: Linear Symmetry

20E. An infinite line of charge produces a field of 4.5×10^4 N/C at a distance of 2.0 m. Calculate the linear charge density.

21E. (a) The drum of the photocopying machine in Sample Problem 3 has a length of 42 cm and a diameter of 12 cm. What is the total charge on the drum? (b) The manufacturer wishes to produce a desktop version of the machine. This requires reducing the size of the drum to a length of 28 cm and a diameter of 8.0 cm. The electric field at the drum surface must remain unchanged. What must be the charge on this new drum?

22E. A very long straight wire carries -3.6 nC/m of fixed negative charge. The wire is to be surrounded by a uniform cylinder of positive charge, radius 1.5 cm, coaxial with the wire. The charge density ρ of the cylinder is to be selected so that the net electric field outside the cylinder is zero. Calculate the required positive charge density ρ.

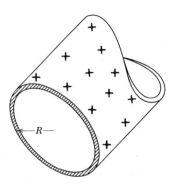

Figure 25 Problem 23.

23P. Figure 25 shows a section through a long, thin-walled metal tube of radius R, carrying a charge per unit length λ on its surface. Derive expressions for E for various distances r from the tube axis, considering both (a) $r > R$ and (b) $r < R$. Plot your results for the range $r = 0$ to $r = 5.0$ cm, assuming that $\lambda = 2.0 \times 10^{-8}$ C/m and $R = 3.0$ cm. (*Hint:* Use cylindrical Gaussian surfaces, coaxial with the metal tube.)

24P. Figure 26 shows a section through two long thin concentric cylinders of radii a and b. The cylinders carry equal and opposite charges per unit length λ. Using Gauss' law, prove (a) that $E = 0$ for $r < a$ and (b) that between the cylinders E is given by

$$E = \frac{1}{2\pi\epsilon_0} \frac{\lambda}{r}.$$

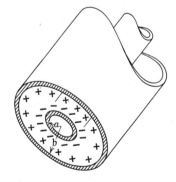

Figure 26 Problem 24.

25P. Figure 27 shows a Geiger counter, used to detect ionizing radiation. The counter consists of a thin central wire, carrying positive charge, surrounded by a concentric circular conducting cylinder, carrying an equal negative charge. Thus, a strong radial electric field is set up inside the cylinder. The cylinder contains a low-pressure inert gas. When a particle of radiation enters the tube through the cylinder walls, it ionizes a few of the gas atoms. The resulting free electrons are drawn to the positive wire. However, the electric field is so intense that,

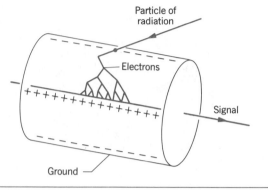

Particle of
radiation

Electrons

+ + + + + + + + + + + + + + +

Signal

Ground

Figure 27 Problem 25.

between collisions with the gas atoms, they have gained energy sufficient to ionize these atoms also. More free electrons are thereby created, and the process is repeated until the electrons reach the wire. The "avalanche" of electrons is collected by the wire, generating a signal recording the passage of the incident particle of radiation. Suppose that the radius of the central wire is 25 μm, the radius of the cylinder 1.4 cm, and the length of the tube 16 cm. The electric field at the cylinder wall is 2.9×10^4 N/C. Calculate the amount of positive charge on the central wire. (*Hint:* See Problem 24.)

26P. A very long conducting cylinder (length L) carrying a total charge $+q$ is surrounded by a conducting cylindrical shell (also of length L) with total charge $-2q$, as shown in cross section in Fig. 28. Use Gauss' law to find (*a*) the electric field at points outside the conducting shell, (*b*) the distribution of the charge on the conducting shell, and (*c*) the electric field in the region between the cylinders.

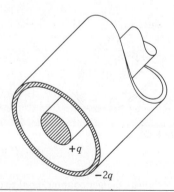

$+q$

$-2q$

Figure 28 Problem 26.

27P. Two long charged concentric cylinders have radii of 3.0 cm and 6.0 cm. The charge per unit length on the inner cylinder is 5.0×10^{-6} C/m and that on the outer cylinder is -7.0×10^{-6} C/m. Find the electric field at (*a*) $r = 4.0$ cm and (*b*) $r = 8.0$ cm.

28P. In Problem 24 a positron revolves in a circular path of radius r, between and concentric with the cylinders. What must be its kinetic energy K in electron-volts? Assume that $a = 2.0$ cm, $b = 3.0$ cm, and $\lambda = 30$ nC/m.

29P. Charge is distributed uniformly throughout an infinitely long cylinder of radius R. (*a*) Show that E at a distance r from the cylinder axis ($r < R$) is given by

$$E = \frac{\rho r}{2\epsilon_0},$$

where ρ is the density of charge. (*b*) What result do you expect for $r > R$?

Section 25–10 Gauss' Law: Planar Symmetry
30E. Two large sheets of positive charge face each other as in Fig. 29. What is **E** at points (*a*) to the left of the sheets, (*b*) between them, and (*c*) to the right of the sheets? Assume the same surface charge density σ for each sheet. Consider only points not near the edges whose distance from the sheets is small compared to the dimensions of the sheet. (*Hint:* See Sample Problem 6.)

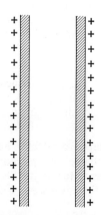

Figure 29 Exercise 30.

31E. A metal plate 8.0 cm on a side carries a total charge of 6.0×10^{-6} C. (*a*) Using the infinite plate approximation, calculate the electric field 0.50 mm above the surface of the plate near the plate's center. (*b*) Estimate the field at a distance of 30 m.

32E. A large flat nonconducting surface carries a uniform charge density σ. A small circular hole of radius R has been cut in the middle of the sheet, as shown in Fig. 30. Ignore fringing of the field lines around all edges and calculate the electric field at point P, a distance z from the center of the hole along its axis. (*Hint:* See Eq. 21 of Chapter 24 and use the principle of superposition.)

33P. A small sphere whose mass m is 1.0 mg carries a charge $q = 2.0 \times 10^{-8}$ C. It hangs in the earth's gravitational field

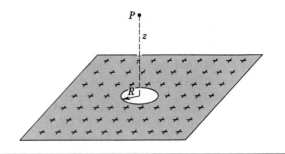

Figure 30   Exercise 32.

from a silk thread that makes an angle $\theta = 30°$ with a large, uniformly-charged nonconducting sheet as in Fig. 31. Calculate the uniform charge density σ for the sheet.

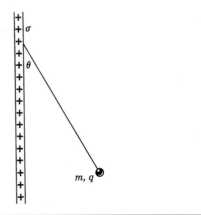

Figure 31   Problem 33.

34P. A 100-eV electron is fired directly toward a large metal plate that has a surface charge density of -2.0×10^{-6} C/m². From what distance must the electron be fired if it is to just fail to strike the plate?

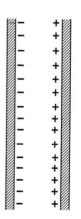

Figure 32   Problem 35.

35P. Two large metal plates face each other as in Fig. 32 and carry charges with surface charge density $+\sigma$ and $-\sigma$, respectively, on their inner surfaces. Find **E** at points (a) to the left of the sheets, (b) between them, and (c) to the right of the sheets. Consider only points not near the edges whose distances from the sheets are small compared to the dimensions of the sheet. (*Hint:* See Sample Problem 6.)

36P. Two large metal plates of area 1.0 m² face each other. They are 5.0 cm apart and carry equal and opposite charges on their inner surfaces. If E between the plates is 55 N/C, what is the charge on the plates? Neglect edge effects.

37P. An electron remains stationary in an electric field directed downward in the earth's gravitational field. If the electric field is due to charge on two large parallel conducting plates, oppositely charged and separated by 2.3 cm, what is the surface charge density, assumed to be uniform, on the plates?

38P.* A plane slab of thickness d has a uniform volume charge density ρ. Find the magnitude of the electric field at all points in space both (a) inside and (b) outside the slab, in terms of x, the distance measured from the median plane of the slab.

Section 25-11   Gauss' Law: Spherical Symmetry

39E. A conducting sphere of radius 10 cm carries an unknown net charge. If the electric field 15 cm from its center is 3.0×10^3 N/C and points radially inward, what is the net charge on the sphere?

40E. A point charge at the origin causes an electric flux of -750 N·m²/C to pass through a spherical Gaussian surface of 10-cm radius centered at the origin. (a) If the radius of the Gaussian surface is doubled, how much flux would then pass through the surface? (b) What is the value of the point charge?

41E. A thin-walled metal sphere has a radius of 25 cm and carries a charge of 2.0×10^{-7} C. Find E for a point (a) inside the sphere, (b) just outside the sphere, and (c) 3.0 m from the center of the sphere.

42E. Two charged concentric spheres have radii of 10 cm and 15 cm. The charge on the inner sphere is 4.0×10^{-8} C and that on the outer sphere is 2.0×10^{-8} C. Find the electric field (a) at $r = 12$ cm and (b) at $r = 20$ cm.

43E. A thin, metallic, spherical shell of radius a carries a charge q_a. Concentric with it is another thin, metallic, spherical shell of radius b ($b > a$) carrying a charge q_b. Find the electric field at radial points r where (a) $r < a$, (b) $a < r < b$, and (c) $r > b$. (d) Discuss the criterion one would use to determine how the charges are distributed on the inner and outer surfaces of each shell.

44E. In a 1911 paper, Ernest Rutherford said, "In order to form some idea of the forces required to deflect an α particle through a large angle, consider an atom containing a point positive charge Ze at its centre and surrounded by a distribution of negative electricity, $-Ze$ uniformly distributed within a sphere of radius R. The electric field E . . . at a distance r

from the center for a point *inside* the atom [is]

$$E = \frac{Ze}{4\pi\epsilon_0}\left(\frac{1}{r^2} - \frac{r}{R^3}\right).$$ ''

Verify this equation.

45P. An uncharged, spherical, thin, metallic shell has a point charge q at its center. Derive expressions for the electric field (a) inside the shell and (b) outside the shell, using Gauss' law. (c) Has the shell any effect on the field due to q? (d) Has the presence of q any effect on the shell? (e) If a second point charge is held outside the shell, does this outside charge experience a force? (f) Does the inside charge experience a force? (g) Is there a contradiction with Newton's third law here? Why or why not?

46P. Equation 12 ($E = \sigma/\epsilon_0$) gives the electric field at points near a charged conducting surface. Apply this equation to a conducting sphere of radius r, carrying a charge q on its surface, and show that the electric field outside the sphere is the same as the field of a point charge at the position of the sphere center.

47P. Figure 33 shows a charge $+q$ arranged as a uniform conducting sphere of radius a and placed at the center of a spherical conducting shell of inner radius b and outer radius c. The outer shell carries a charge of $-q$. Find $E(r)$ (a) within the sphere ($r < a$), (b) between the sphere and the shell ($a < r < b$), (c) inside the shell ($b < r < c$), and (d) outside the shell ($r > c$). (e) What charges appear on the inner and outer surfaces of the shell?

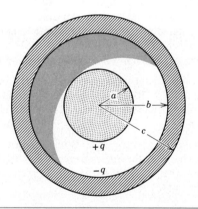

Figure 33   Problem 47.

48P. Figure 34 shows a spherical shell of charge of uniform density ρ. Plot E for distances r from the center of the shell ranging from zero to 30 cm. Assume that $\rho = 1.0 \times 10^{-6}$ C/m³, $a = 10$ cm, and $b = 20$ cm.

49P. Figure 35 shows a point charge of 1.0×10^{-7} C at the center of a spherical cavity of radius 3.0 cm in a piece of metal. Use Gauss' law to find the electric field (a) at point P_1, halfway from the center to the surface and (b) at point P_2.

50P. A proton orbits with a speed $v = 3.0 \times 10^5$ m/s just outside a charged sphere of radius $r = 10^{-2}$ m. What is the charge on the sphere?

51P. A solid nonconducting sphere of radius R carries a non-uniform charge distribution, the charge density being $\rho = \rho_s r/R$, where ρ_s is a constant and r is the distance from the center of the sphere. Show that (a) the total charge on the sphere is $Q = \pi\rho_s R^3$ and (b) the electric field inside the sphere is given by

$$E = \frac{1}{4\pi\epsilon_0}\frac{Q}{R^4}r^2.$$

52P. The spherical region $a < r < b$ carries a charge per unit volume of $\rho = A/r$, where A is constant. At the center ($r = 0$) of the enclosed cavity is a point charge q. What should the value of A be so that the electric field in the region $a < r < b$ has constant magnitude?

53.* A spherical region carries a uniform charge per unit volume ρ. Let $\mathbf{r}$ be the vector from the center of the sphere to a general point P within the sphere. (a) Show that the electric field at P is given by $\mathbf{E} = \rho\mathbf{r}/3\epsilon_0$. (b) A spherical cavity is

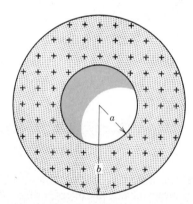

Figure 34   Problem 48.

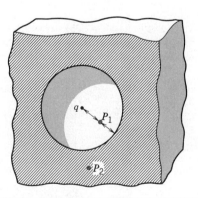

Figure 35   Problem 49.

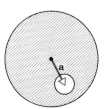

Figure 36   Problem 53.

created in the above sphere, as shown in Fig. 36. Using super-position concepts, show that the electric field at all points within the cavity is $\mathbf{E} = \rho\mathbf{a}/3\epsilon_0$ (uniform field), where $\mathbf{a}$ is the vector connecting the center of the sphere with the center of the cavity. Note that both these results are independent of the radii of the sphere and the cavity.

54P.* Show that stable equilibrium under the action of electrostatic forces alone is impossible. (*Hint:* Assume that at a certain point P in an electric field $\mathbf{E}$ a charge $+q$ would be in stable equilibrium if it were placed there. Draw a spherical Gaussian surface about P, imagine how $\mathbf{E}$ must point on this surface, and apply Gauss' law to show that the assumption leads to a contradiction.) This result is known as Earnshaw's theorem.

CHAPTER 26
ELECTRIC POTENTIAL

This young lady has been raised to an electric potential of possibly 50,000 volts above the potential of her surroundings. The strands of her hair suggest the beginnings of lines of force that reach out to the walls of the room. Although 120 volts can be lethal, she remains unharmed. Can you explain why?

26-1 Gravity, Electrostatics, and Potential Energy

There is little point in solving for a second time a problem that you have already solved. That is why physicists are always on the lookout for areas of physics that—different as they may be in terms of the physical quantities involved—are expressed in the same mathematical framework. As we pointed out in Section 23-4, Newton's law of gravity and Coulomb's law of electrostatics—both inverse square laws—are mathematically identical.* Thus, whatever you can deduce about gravity by

---
* True, the gravitational force is always attractive and the electrostatic force may be either attractive or repulsive. However, we can easily keep this difference straight in our calculations.

---

analyzing its basic law you can carry over with full confidence to electrostatics, and conversely. All that you need to do is change the symbols.

Figure 1a shows a test particle of mass m_0—perhaps a baseball—in free fall in the earth's gravitational field. We can analyze its motion from the point of view of energy and say that, as the baseball falls, its gravitational potential energy is transformed into kinetic energy.

Figure 1b shows the electrostatic parallel, a test charge q_0—perhaps a proton—in "free fall" in an electric field. Drawing on the mathematical identity of the basic laws, we can say with confidence that we must also be able to analyze the motion of the test charge in terms of energy transfers. In particular, we must be able to assign to the test charge an *electric potential energy U*,

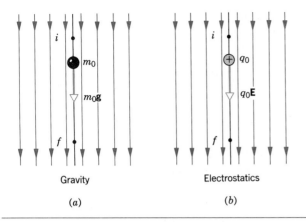

Figure 1   (a) A gravitational force $m_0\mathbf{g}$ acts on a freely falling test particle of mass m_0. (b) An electrostatic force $q_0\mathbf{E}$ acts on a "freely falling" test particle of charge q_0.

whose value for a given test charge depends only on the position of that charge in the electric field.

Let us review the gravitational case. If the test body in Fig. 1a moves from an initial point i to a final point f, we define the difference in its gravitational potential energy (see Section 8–4) to be

$$\Delta U = U_f - U_i = -W_{if}. \qquad (1)$$

Here W_{if} is the work done by the gravitational force on the test body as it moves between these points. Although Fig. 1a shows a uniform gravitational field, the definition of Eq. 1 holds for any field configuration whatever, uniform or not. Other forces may act on the moving test body and may—or may not—do work on it. W_{if} in Eq. 1, however, includes only the work done by the gravitational force.

We proved in Sections 8–5 and 15–7 that the difference in the gravitational potential energy of a test particle as it moves between any two points is independent of the path taken between those points. We summarized this conclusion by labeling the gravitational force as a *conservative* force. If the gravitational force were not conservative, the whole notion of potential energy would fall apart.

We can take over Eq. 1 directly as a definition of the difference in the *electric potential energy* of a test charge q_0 as it moves from an initial point i to a final point f in an electric field:

Let a test charge move from one point to another in an electric field. The difference in the electric

potential energy of the test charge between those points is the negative of the work done by the electric field on *that charge during its motion.*

Again, the difference in the electric potential energy of a test charge between two points is independent of the path taken between those points. That is, the electrostatic force, like the gravitational field force, is a *conservative* force. We do not need to prove this because we have already done so in the gravitational case.

We now move from the definition of the difference in the electric potential energy of a test charge between two points to the definition of the electric potential energy of a test charge at a single point. To do so we make two decisions that are both arbitrary and independent of each other: (1) We take our initial point i to be a standard reference point whose location we specify, and (2) we assign an arbitrary value to the potential energy of the test charge at that point. Specifically, we choose to locate point i at a very large—strictly, an infinite—distance from all relevant charges and we assign the value zero to the potential energy of any test charge at such points. If we put $U_i = 0$ and $U_f = U$ in Eq. 1, we can write that equation as

$$U = -W_{\infty}. \qquad (2)$$

In words, Eq. 2 can be stated as follows:

The potential energy U of a test charge q_0 at any point is equal to the negative of the work W_{∞} done on the test charge by the electric field as that charge moves in from infinity to the point in question.

Keep in mind that *potential energy differences* are fundamental and that *potential energy,* defined by Eq. 2, depends on defining a reference point and assigning a potential energy to it. Instead of choosing $U_i = 0$, we could have chosen $U_i = -137$ J, and it would not have changed the value of the potential energy difference of a given test charge between any pair of field points.

Recall that, in studying the motions of baseballs in the earth's gravitational field, we chose the earth's surface as the zero level of gravitational potential energy. In studying the motions of satellites, however, we found it more convenient to choose an infinite separation as the zero-potential energy configuration.

Sample Problem 1 A child's helium-filled balloon, carrying a charge $q = -5.5 \times 10^{-8}$ C, rises vertically into the air. The electric field that normally exists in the atmosphere near the surface of the earth has a magnitude $E = 150$ N/C and is directed downward. What is the difference in electric potential energy of the balloon between its release position (point i) and its position at an altitude $h = 520$ m (point f)?

The work done on the balloon by the electric field is

$$W_{if} = \mathbf{F} \cdot \mathbf{h} = q\mathbf{E} \cdot \mathbf{h} = qE(\cos 180°)h = -qEh$$
$$= -(-5.5 \times 10^{-8} \text{ C})(150 \text{ N/C})(520 \text{ m})$$
$$= +4.3 \times 10^{-3} \text{ J} = +4.3 \text{ mJ}.$$

Let us review the signs carefully. Although the electric field $\mathbf{E}$ points down, the electric force $q\mathbf{E}$ points up because the charge q on the balloon is negative. Thus, the electric force on the balloon acts in the direction the balloon is moving so that the work done by this force on the balloon is indeed positive, in agreement with our solution.

The difference in the electric potential energy of the balloon follows from Eq. 1, or

$$U_f - U_i = -W_{if} = -4.3 \text{ mJ}. \qquad \text{(Answer)}$$

Can you show that this same answer holds even if the balloon drifts horizontally as it rises to altitude h?

Other forces also do work on the balloon, including the gravitational force, the atmospheric buoyant force, the atmospheric drag force, and possibly the wind force. All these forces are usually much stronger than the electric force but our concern in this problem is only with the electric force. The gravitational force, like the electric force, is conservative so that the balloon has potential energy of two kinds. The minus sign in our calculated result shows that the *electric* potential energy of the balloon decreases as it rises. On the other hand, the *gravitational* potential energy of the balloon increases as it rises.

26-2 The Electric Potential

As we have seen, the potential energy of a point charge in an electric field depends not only on the nature of the field but also on the magnitude of the charge. On the other hand, the potential energy *per unit charge* would have a unique value at any point in the field, independent of the magnitude of the test charge. We call this useful quantity the *electric potential V* (or simply the *potential*) at the point in question.*

---

* In earlier chapters we define *gravitational potential energy.* We could also have defined *gravitational potential* but we did not do so.

We define the difference in potential between any two points as $\Delta U/q_0$ so that, guided by Eq. 1, we can write

$$\boxed{\Delta V = V_f - V_i = -\frac{W_{if}}{q_0}}$$

(potential difference defined).   (3)

The work W_{if} done by the electric field on the positive test charge as it moves from point i to point f may be positive, negative, or zero. Correspondingly, because of the minus sign in Eq. 3, the potential at f will then be less than, greater than, or the same as the potential at i.

We can also regard Eq. 3 from another point of view. If we happen to know the potential difference ΔV between any two points, the work that *we* must do to transport a charge q_0 from one point to the other is given by $\Delta V q_0$.

Guided by Eq. 2, we define the potential at a point from

$$\boxed{V = -\frac{W_\infty}{q_0}} \quad \text{(potential defined)} \qquad (4)$$

in which W_∞ is the work done by the electric field on the test charge as that charge moves in from infinity to the point in question.

Equation 4 tells us that the potential V at a point near an isolated positive charge is positive. To see this, imagine that you push a small positive test charge in from infinity to a point near an isolated positive charge. The electric force acting on the test charge will point away from the central charge, acting to repel the test charge. Thus, the work done *on* the test charge *by* the field force will be negative.* The minus sign in Eq. 4 assures that the potential at the point will be positive. Similarly, the potential for any point near an isolated negative charge will be negative.

The SI unit for potential that follows from Eq. 4 is the joule/coulomb. This combination occurs so often that a special unit, the *volt* (abbr. V) is used to represent it. That is,

$$1 \text{ volt} = 1 \text{ joule/coulomb}. \qquad (5)$$

The word "volt" is familiar, being associated with light

---

* The work that the agent pushing the charge does is positive but the W in Eqs. 2–4 is the work done *by the electric field,* not the work done by the agent.

bulbs, electric appliances, electric outlets, and the batteries in your car or your portable stereo. If you touch the probes of a voltmeter to two points in an electric circuit, you are measuring the potential difference between those points. Potential, which is measured in volts, and potential energy, which is measured in joules, are quite different quantities and must not be confused.

We point out that this new unit allows us to adopt a more conventional unit for the electric field **E**, which we have measured up to now in newtons/coulomb. Thus,

$$1 \text{ N/C} = \left(1 \, \frac{\text{N}}{\cancel{\text{C}}}\right)\left(\frac{1 \, \text{V} \cdot \cancel{\text{C}}}{1 \, \cancel{\text{J}}}\right)\left(\frac{1 \, \cancel{\text{J}}}{1 \, \text{N} \cdot \text{m}}\right) = 1 \text{ V/m.} \quad (6)$$

The conversion factor in the second set of parentheses is derived from Eq. 4; that in the third set of parentheses is derived from the definition of the joule. From now on, we shall report values of the electric field in volts/meter rather than in newtons/coulomb.

Finally, we are now in a position to define the electron-volt, the energy unit that we introduced in Section 7–2 as a convenient one for energy measurements in the atomic and subatomic domain.

One electron-volt *(abbr. eV) is an energy equal to the work required to move a single elementary charge e, such as that carried by the electron or the proton, through a potential difference of 1 volt.*

Guided by Eq. 3, we can write (since 1 V = 1 J/C)

$$1 \text{ eV} = e \, \Delta V$$
$$= (1.60 \times 10^{-19} \text{ C})(1 \text{ J/C}) = 1.60 \times 10^{-19} \text{ J.}$$

As we pointed out earlier, multiples of this unit, such as the keV, MeV, and GeV, are in common use.

26–3 Equipotential Surfaces

The locus of points, all of which have the same potential, is called an *equipotential surface.* A family of equipotential surfaces, each surface corresponding to a different value of the potential, can be used to represent the electric field throughout a certain region. We have seen earlier that electric lines of force can also be used for this purpose. In later sections, we shall look into the intimate connection between these two equivalent ways of describing the electric field.

No net work is done by the electric field as a charge moves between any two points on the same equipoten-

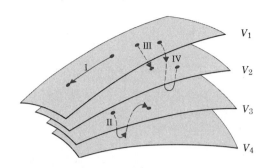

Figure 2 Portions of four equipotential surfaces. Four typical paths along which a test charge may move are also shown.

tial surface. This follows from Eq. 3 ($V_f - V_i = - W_{if}/q_0$) because W_{if} must be zero if $V_f = V_i$. Because of the path independence, $W_{if} = 0$ for *any* path connecting points *i* and *f*, whether or not that path lies entirely on the equipotential surface.

Figure 2 shows a family of equipotential surfaces, associated with a distribution of charges not shown. The work done by the electric field on a test charge as it moves from one end to the other of paths I and II is zero because each of these paths begins and ends on the same equipotential surface. The work done as a test charge moves from one end to the other of paths III and IV is not zero but has the same value for both of these paths because the initial and also the final potentials are identical for the two paths. Put another way, paths III and IV connect the same pair of equipotential surfaces.

From symmetry, the equipotential surfaces for a point charge or a spherically symmetric charge distribution are a family of concentric spheres. For a uniform field, they are a family of planes at right angles to the field. In all cases, including these two examples, the equipotential surfaces are at right angles to the lines of force and thus to **E**, which is tangent to these lines. If **E** were *not* at right angles to the equipotential surface, it would have a component lying in that surface. This component would then do work on a test charge as it moves about on the surface. But work cannot be done if the surface is to be truly an equipotential; the only conclusion is that **E** must be everywhere perpendicular to the surface. Figure 3 shows the lines of force and the equipotential surfaces for a uniform electric field and for the fields associated with a point charge and with an electric dipole.

Finally, we consider the case of the electrified young woman whose photo is shown on the first page of this chapter. Her safety is assured because all parts of her

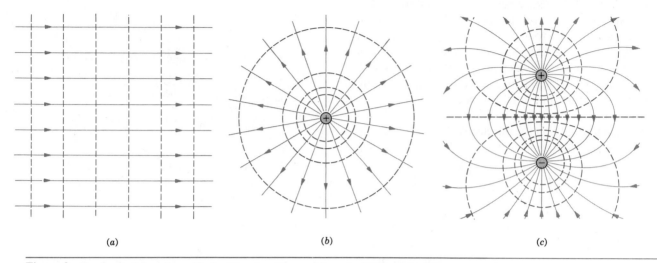

Figure 3   Equipotential surfaces (dashed lines) and lines of force (solid lines) for (*a*) a uniform field, (*b*) the field resulting from a point charge, and (*c*) the field resulting from an electric dipole.

body are raised to the same uniform potential. Damage is only done when a *potential difference* is maintained between two points on the body; in such cases, modest potential differences, such as 120 V or less, can easily be lethal.

26–4   Calculating the Potential from the Field

Here we show how to calculate the potential difference between any two points if you know the electric field **E** at all positions along some path that connects them.

In Fig. 4 a test charge q_0 moves from an initial point *i* to a final point *f* in an electric field, along the path shown. As the charge moves a distance $d\mathbf{s}$ along this path, the electric field does an element of work on it given by $(q_0\mathbf{E}) \cdot d\mathbf{s}$, in which $q_0\mathbf{E}$ is the force exerted by the field on the test charge. To find the total work W_{if} done by the field, we add up—that is, we integrate—the contributions to the work from all the differential segments into which the path is divided. Thus,

$$W_{if} = q_0 \int_i^f \mathbf{E} \cdot d\mathbf{s}. \qquad (7)$$

Such an integral is called a *line integral*. If we substitute

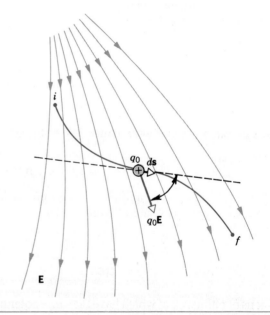

Figure 4   A test charge q_0 moves from point *i* to point *f* along the path shown in a nonuniform electric field.

W_{if} from Eq. 7 into Eq. 3 we find

$$V_f - V_i = -\int_i^f \mathbf{E} \cdot d\mathbf{s}. \qquad (8)$$

If the initial point i is taken to be at infinity and the potential V_i at the point is set equal to zero, Eq. 8 becomes

$$V = -\int_i^f \mathbf{E} \cdot d\mathbf{s}. \qquad (9)$$

If the electric field is known throughout a certain region, Eq. 8 allows us to calculate the difference in potential between any two points in the field. Because the electric force is conservative, all paths will yield the same result. Some paths, of course, are easier to use than others.

Sample Problem 2   Use Eq. 8 to calculate $V_f - V_i$ for the special case of Fig. 5, in which the electric field is uniform and in which the path connecting the initial and final points is a straight line parallel to the field direction.

As the test charge moves from i to f in Fig. 5, its path element $d\mathbf{s}$, which is always in the direction of motion, points down. The electric field $\mathbf{E}$ also points down so that the angle θ between these two vectors is zero. Equation 8 then becomes

$$V_f - V_i = -\int_i^f \mathbf{E} \cdot d\mathbf{s} = -\int_i^f E(\cos 0°)\, ds = -\int_i^f E\, ds.$$

E is constant over the path in this problem and can be removed

from the integral, leaving

$$V_f - V_i = -E \int_i^f ds = -Ed, \qquad \text{(Answer)}$$

in which the integral is simply the length of the path. The minus sign shows that the potential at point f in Fig. 5 is lower than the potential at point i.

Sample Problem 3   In Fig. 6, let a test charge q_0 move from i to f over path icf. Calculate the potential difference $V_f - V_i$ for this path.

At all points along the line ic, $\mathbf{E}$ and $d\mathbf{s}$ are at right angles to each other. Thus, $\mathbf{E} \cdot d\mathbf{s} = 0$ everywhere along this part of the path. Equation 8 then tells us that points i and c are at the same potential. In other words, i and c lie on the same equipotential surface.

For path cf we have $\theta = 45°$ and, from Eq. 8,

$$V_f - V_i = -\int_c^f \mathbf{E} \cdot d\mathbf{s} = -\int_c^f E\,(\cos 45°)\, ds = -\frac{E}{\sqrt{2}} \int_c^f ds.$$

The integral in this equation is the length of the line cf, which is $\sqrt{2}d$. Thus,

$$V_f - V_i = -\frac{E}{\sqrt{2}} \sqrt{2}d = -Ed. \qquad \text{(Answer)}$$

This is the same result that we found in Sample Problem 2, as it must be because the potential difference between two points,

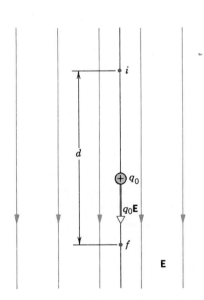

Figure 5   Sample Problem 2. A test charge q_0 moves in a straight line from point i to point f in a uniform electric field.

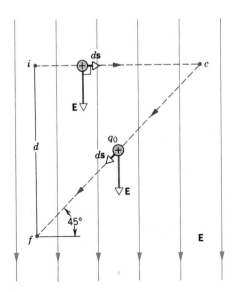

Figure 6   Sample Problem 3. A test charge q_0 moves from point i to point f along path icf in a uniform electric field.

such as i and f in Figs. 5 and 6, does not depend on the path connecting them.

26–5 Calculating the Potential: A Point Charge

Figure 7 shows an isolated positive point charge q. We wish to find the potential at point P, a radial distance r from that charge. Guided by Eq. 9, let us imagine that a test charge q_0 moves from infinity to point P. Because the path followed by the test charge does not matter, we make the simplest choice, a radial line from infinity to q, passing through P.

Let the test charge be at a position r', as Fig. 7 shows. The electric field $\mathbf{E}$ at the site of the test charge points in the direction of increasing r'. The displacement $d\mathbf{s}$ of the test charge is in the direction of decreasing r' (so that $ds = -dr'$) and we have

$$\mathbf{E} \cdot d\mathbf{s} = (E)(\cos 180°)(-dr') = E\, dr'. \qquad (10)$$

Substituting this result into Eq. 9 gives us

$$V = -\int_i^f \mathbf{E} \cdot d\mathbf{s} = -\int_\infty^r E\, dr'. \qquad (11)$$

The magnitude of the electric field at the site of the test charge is given by Eq. 6 of Chapter 24 as

$$E = \frac{1}{4\pi\epsilon_0} \frac{q}{r'^2}.$$

Substituting this result into Eq. 11 leads to

$$V = -\frac{q}{4\pi\epsilon_0} \int_\infty^r \frac{1}{r'^2}\, dr' = -\frac{q}{4\pi\epsilon_0}\left| -\frac{1}{r'} \right|_\infty^r$$

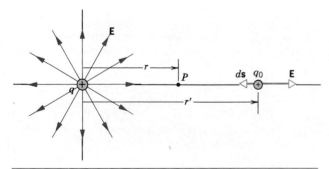

Figure 7 A test charge q_0 moves in from infinity along a radial line to point P. The field is that resulting from a positive point charge q.

or

$$V = \frac{1}{4\pi\epsilon_0} \frac{q}{r}. \qquad (12)$$

We see that the sign of V is the same as the sign of q, confirming our earlier conclusion. Figure 8 shows computer-generated plots of Eq. 12 for a positive and a negative point charge.

To find the potential difference between any two points near an isolated point charge, all that is needed is to apply Eq. 12 to each point and subtract one potential from the other.

Sample Problem 4 What must be the magnitude of an isolated positive charge for which the potential V at a distance of 15 cm is $+120$ V?

Solving Eq. 12 for q yields

$$q = V4\pi\epsilon_0 r = (120\text{ V})(4\pi)(8.85 \times 10^{-12}\text{ C}^2/\text{N} \cdot \text{m}^2)(0.15\text{ m})$$
$$= 2.0 \times 10^{-9}\text{ C}. \qquad \text{(Answer)}$$

A charge of this size can easily be produced on a balloon by rubbing it. In sorting out the units in this problem, it helps to recall from Eq. 5 that $1\text{ V} = 1\text{ J/C} = 1\text{ N} \cdot \text{m/C}$.

Sample Problem 5 What is the potential on the surface of a gold nucleus? The radius R of the nucleus is 6.2 fm and the atomic number Z of gold is 79.

The nucleus, assumed to be spherical, behaves for all outside points as if it were a point charge at the center. The charge q on the nucleus is Ze, where e is the elementary charge. We then have, from Eq. 12,

$$V = \frac{1}{4\pi\epsilon_0} \frac{q}{r} = \frac{1}{4\pi\epsilon_0} \frac{Ze}{R}$$
$$= \frac{(8.99 \times 10^9\text{ N} \cdot \text{m}^2/\text{C}^2)(79)(1.60 \times 10^{-19}\text{ C})}{6.2 \times 10^{-15}\text{ m}}$$
$$= 1.8 \times 10^7\text{ V} = 18\text{ MV}. \qquad \text{(Answer)}$$

This large positive potential cannot be detected outside a gold object such as a coin because it is compensated by an equally large negative potential owing to the electrons in the coin.

26–6 Calculating the Potential: An Electric Dipole

We can find the potential for a group of point charges with the help of the superposition principle. We calcu-

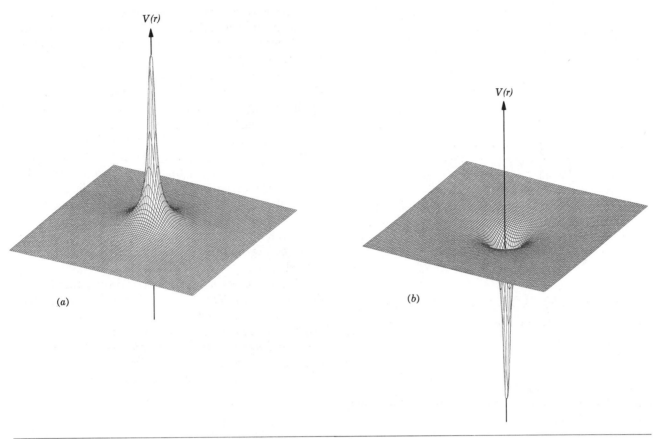

Figure 8   (*a*) A computer-generated plot of the potential $V(r)$ near a positive point charge. (*b*) The same for a negative point charge. The region close to such a charge is often described as a "potential well."

late the potential resulting from each charge at any given point separately, using Eq. 12. Then we add the potentials so calculated algebraically. In equation form, for n charges,

$$V = \sum_n V_n = \frac{1}{4\pi\epsilon_0} \sum \frac{q_n}{r_n}. \tag{13}$$

Here q_n is the value of the nth charge and r_n is the radial distance of the point in question from the nth charge. The sum in Eq. 13 is an algebraic sum and not a vector sum like the sum that would be used to calculate the electric field resulting from a group of point charges. Herein lies an important computational advantage of potential over electric field.

Now let us apply Eq. 13 to two equal charges of opposite sign that constitute an *electric dipole.* We first encountered the electric dipole in Section 24–5, where

we calculated the value of the electric field for points along its axis.

A point P in the field of a dipole can be specified by giving the quantities r and θ in Fig. 9. From symmetry, the potential V_P will not change as the point P rotates about the z axis, r and θ being fixed. Thus, we need only find $V(r, \theta)$ for any plane containing this axis; the plane of Fig. 9 is such a plane. Equations 13 and 12 yield

$$V_P = \sum V_n = V_+ + V_- = \frac{1}{4\pi\epsilon_0} \left(\frac{q}{r_+} + \frac{-q}{r_-} \right)$$

$$= \frac{q}{4\pi\epsilon_0} \frac{r_- - r_+}{r_- r_+},$$

which is an exact relation.

Because naturally occurring dipoles — such as those possessed by many molecules — are small, we are usually interested only in points far from the dipole, such that

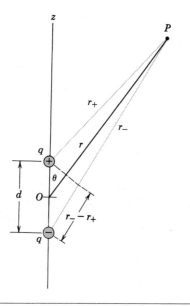

Figure 9 Point P is a distance r from the midpoint O of a dipole, the line OP making an angle θ with the dipole axis.

$r \gg d$, where d is the distance between the charges. Under these conditions, the approximations that follow from a study of Fig. 9 are

$$r_- - r_+ \approx d \cos \theta \quad \text{and} \quad r_- r_+ \approx r^2.$$

If we substitute these quantities into the above expression for V_p, we find

$$V_P \approx \frac{q}{4\pi\epsilon_0} \frac{d \cos \theta}{r^2},$$

or

$$V_P = \frac{1}{4\pi\epsilon_0} \frac{p \cos \theta}{r^2}, \tag{14}$$

in which $p(= qd)$ was defined in Section 24-5 as the electric dipole moment. The quantity p is actually the magnitude of a vector $\mathbf{p}$, the direction of which is along the dipole axis, pointing from the negative to the positive charge.

Equation 14 shows that $V = 0$ everywhere in the equatorial plane of the dipole, defined by $\theta = 90°$. This reflects the fact that a test charge lying in this plane is always equidistant from the positive and the negative charges that make up the dipole so that the (scalar) potentials set up by each charge cancel each other. For a

given distance, V_P has its greatest positive value for $\theta = 0$ and its greatest negative value for $\theta = 180°$. Note that the potential does not depend separately on q and d but only on their product.

As we saw in Section 24–5, many molecules have electric dipole moments. In the absence of a permanent dipole moment, one can always be induced by placing an atom or a molecule in an external electric field. The action of the field, as Fig. 10 shows, is to stretch the atom or molecule, thus separating the centers of positive and negative charge. We say that the atom or molecule becomes *polarized* and acquires an *induced* electric dipole moment. For solids whose properties are the same in all directions, an electric dipole moment $\mathbf{p}$ induced by an external electric field $\mathbf{E}$ points in the direction of that field (see Fig. 10) and disappears when that field is removed.

Electric dipoles are important in situations other than atomic or molecular ones. Radio and TV antennas are often in the form of a metal wire in which electrons surge back and forth periodically. At a certain time, one end of the wire will be negative and the other end positive. Half a cycle later, the polarity of the ends will be reversed. Such an antenna is an oscillating electric di-

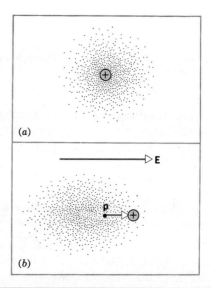

Figure 10 (*a*) An atom, showing the positively charged nucleus and the electron cloud. The centers of positive and of negative charge coincide. (*b*) If the atom is placed in an external electric field, the electron cloud is distorted so that the centers of positive and negative charge no longer coincide. An induced dipole moment appears. The distortion is greatly exaggerated here.

pole, so named because its electric dipole moment changes in a periodic way with time.

Sample Problem 6   What is the potential at point P, located at the center of the square of point charges shown in Fig. 11a? Assume that $d = 1.3$ m and that the charges are

$$q_1 = +12 \text{ nC}, \qquad q_3 = +31 \text{ nC},$$
$$q_2 = -24 \text{ nC}, \qquad q_4 = +17 \text{ nC}.$$

From Eq. 13 we have

$$V_P = \sum_n V_n = \frac{1}{4\pi\epsilon_0} \frac{q_1 + q_2 + q_3 + q_4}{R}.$$

The distance R of each charge from the center of the square is $d/\sqrt{2}$ or 0.919 m, so that

$$V_P = \frac{(8.99 \times 10^9 \text{ N} \cdot \text{m}^2/\text{C}^2)(12 - 24 + 31 + 17) \times 10^{-9} \text{ C}}{0.919 \text{ m}}$$

$$\approx 350 \text{ V.} \qquad\qquad\qquad\qquad \text{(Answer)}$$

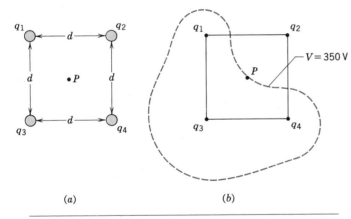

(a)   (b)

Figure 11   Sample Problem 6. (a) Four charges are held fixed at the corners of a square. What is the potential at P, the center of the square? (b) The curve is the intersection with the plane of the figure of the equipotential surface that contains point P.

Close to the three positive charges in Fig. 11a, the potential can assume very large positive values. Close to the single negative charge in that figure, the potential can assume arbitrarily large negative values. There must then be other points within the boundaries of the square that have the same potential as that at point P. Figure 11b shows the trace of a portion of the equipotential surface that contains this point.

26-7   Calculating the Potential: A Charged Disk

If the charge distribution is continuous, rather than an assembly of point charges, the sum in Eq. 13 must be replaced by an integral and we have

$$V = \int dV = \frac{1}{4\pi\epsilon_0} \int \frac{dq}{r}, \qquad (15)$$

where dq is a differential element of the charge distribution and r is its distance from the point at which V is to be calculated. The integral is to be taken over the entire charge distribution.

In Section 24-7, we calculated the magnitude of the electric field for points on the axis of a plastic disk of radius R that is uniformly charged on one face with a surface charge density σ. Here we wish to derive an expression for $V(z)$, the potential for axial points. It will be instructive to compare these two derivations, one involving a vector integration and the other a scalar integration.

In Fig. 12, consider a charge element dq consisting of a flat circular strip of radius s and width ds. We have

$$dq = \sigma(2\pi s)(ds),$$

in which $(2\pi s)(ds)$ is the area of the strip. All parts of this charge element are the same distance r from the axial

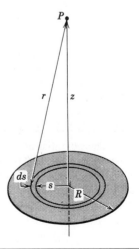

Figure 12   A plastic disk of radius R is charged on its front surface to a surface charge density σ. What is the potential at various points along its axis?

Table 1 Fields and Potentials for Some Charge Configurations

Configuration	Field E	Equation	Potential V	Equation
Point charge	$\dfrac{1}{4\pi\epsilon_0}\dfrac{q}{r^2}$	(24–6)	$\dfrac{1}{4\pi\epsilon_0}\dfrac{q}{r}$	(26–12)
Dipole[a]	$\dfrac{1}{2\pi\epsilon_0}\dfrac{p}{z^3}$	(24–11)	$\dfrac{1}{4\pi\epsilon_0}\dfrac{p\cos\theta}{r^2}$	(26–14)
Charged disk[b]	$\dfrac{\sigma}{2\epsilon_0}\left(1-\dfrac{z}{\sqrt{z^2+R^2}}\right)$	(24–21)	$\dfrac{\sigma}{2\epsilon_0}(\sqrt{z^2+R^2}-z)$	(26–17)
Infinite sheet	$\dfrac{\sigma}{2\epsilon_0}$	(25–15)	$V_0-\left(\dfrac{\sigma}{2\epsilon_0}\right)z$	(Exercise 6)
Insulated conductor	$E=0$, inside $E=\sigma/\epsilon_0$, at surface		$V=$ a constant, inside and on the surface	

[a] The field equation is for distant *axial* points. The potential equation is for *all* distant points.
[b] For *axial* points.

point P so that their contribution to the potential at P is given by Eq. 12, or

$$dV = \frac{1}{4\pi\epsilon_0}\frac{dq}{r} = \frac{1}{4\pi\epsilon_0}\frac{\sigma(2\pi s)(ds)}{\sqrt{z^2+s^2}}. \qquad (16)$$

We find the potential at P by adding up the contributions of all the strips into which the disk can be divided, or

$$V_P = \int dV = \frac{\sigma}{2\epsilon_0}\int_0^R (z^2+s^2)^{-1/2}\, s\, ds$$

$$= \frac{\sigma}{2\epsilon_0}(\sqrt{z^2+R^2}-z). \qquad (17)$$

Be sure to check all the steps in the evaluation of this integral. Note that the variable is s and not z, which remains constant while the integration over the surface of the disk is carried out.

Table 1 summarizes the electric field and the electric potential expressions that we have derived in this and previous chapters for various charge configurations.

Sample Problem 7 The potential at the center of a uniformly charged circular disk of radius R is 550 V. (a) What is the total charge on the disk? Take $R = 3.5$ cm.

At the center of the disk, z in Eq. 17 is zero, so that equation reduces to

$$V_0 = \frac{\sigma R}{2\epsilon_0}. \qquad (18)$$

The total charge q is $\sigma(\pi R^2)$, where πR^2 is the area of the disk. If we solve Eq. 18 for σ and insert it in the expression for q, we

find

$$q = \sigma(\pi R^2) = 2\pi\epsilon_0 R V_0$$
$$= (2\pi)(8.85\times10^{-12}\ \text{C}^2/\text{N}\cdot\text{m}^2)(0.035\ \text{m})(550\ \text{V})$$
$$= 1.1\times10^{-9}\ \text{C} = 1.1\ \text{nC}. \qquad \text{(Answer)}$$

Again, it helps to note from Eq. 5 that $1\ \text{V} = 1\ \text{J/C} = 1\ \text{N}\cdot\text{m/C}$.

(b) What is the potential at a point on the axis of the disk a distance z of 5.0 radii from the center of the disk?

If we put $z = 5R$ in Eq. 17, we find

$$V = \frac{\sigma}{2\epsilon_0}[\sqrt{(5R)^2+R^2}-5R].$$

But, from Eq. 18, $\sigma = 2\epsilon_0 V_0/R$. Substituting this expression in the above yields

$$V = V_0(\sqrt{5^2+1^2}-5)$$
$$= (550\ \text{V})(0.09902) = 54\ \text{V}. \qquad \text{(Answer)}$$

26–8 Calculating the Field from the Potential

In Section 26–4, we showed how to find the potential if you know the electric field. In this section, we propose to go the other way, that is, to show how to find the electric field if you know the potential. As Fig. 3 shows, we have already solved this problem graphically. If you know the potential V for all points near an assembly of charges, you can draw in a family of equipotential surfaces. The lines of force, sketched in at right angles to those surfaces, describe the variation of $\mathbf{E}$. What we seek here is the

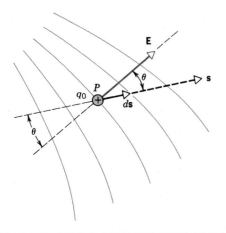

Figure 13   A test charge q_0 moves a distance ds from one equipotential surface to another. The displacement ds makes an angle θ with the direction of the electric field $\mathbf{E}$.

mathematical equivalent of this graphical procedure.

Figure 13 shows the intersection with the plane of the page of a family of closely spaced equipotential surfaces, the potential difference between each pair of adjacent surfaces being dV. The figure shows that $\mathbf{E}$ at a typical point P is at right angles to the equipotential surface through P, as it must be.

Suppose that a test charge q_0 moves through a displacement ds from one equipotential surface to the adjacent surface. From Eq. 3, we see that the work that the electric field does on the test charge as it moves is $-q_0\,dV$. From another point of view, we can say that the work done by the electric field is $(q_0\mathbf{E})\cdot d\mathbf{s}$ or $q_0E(\cos\theta)\,ds$. Equating these two expressions for the work yields

$$-q_0\,dV = q_0E(\cos\theta)ds$$

or

$$E\cos\theta = -\frac{dV}{ds}. \qquad (19)$$

But $E\cos\theta$ is the component of $\mathbf{E}$ in the $\mathbf{s}$ direction. Equation 19 then becomes*

$$E_s = -\frac{\partial V}{\partial s}. \qquad (20)$$

In words, Eq. 20 (which is essentially the inverse of Eq. 8)

___

* The partial derivative symbol indicates that we consider only the variation of V in one specified direction, the $\mathbf{s}$ direction.

states:

The rate of change of the potential with distance in any direction is, when changed in sign, the component of $\mathbf{E}$ in that direction.

If we take the $\mathbf{s}$ direction in Eq. 20 to be, in turn, the directions of the x, y, and z axes, we can find the three components of $\mathbf{E}$ at any point from

$$E_x = -\frac{\partial V}{\partial x}; \quad E_y = -\frac{\partial V}{\partial y}; \quad E_z = -\frac{\partial V}{\partial z}. \qquad (21)$$

Thus, if we know V for all points in the region around a charge distribution, that is, if we know the function $V(x, y, z)$, we can find the components of $\mathbf{E}$—and thus $\mathbf{E}$ itself—at any point by taking partial derivatives.

___

Sample Problem 8   The potential for points on the axis of a charged disk is given by Eq. 17, which we write as

$$V_z = \frac{\sigma}{2\epsilon_0}[(z^2 + R^2)^{1/2} - z].$$

Starting with this expression, derive an expression for the electric field at axial points.

From symmetry, $\mathbf{E}$ must lie along the axis of the disk. If we choose the $\mathbf{s}$ direction in Eq. 20 to be the z direction, we then have

$$E_z = -\frac{\partial V}{\partial z} = -\frac{\sigma}{2\epsilon_0}\frac{d}{dz}[(z^2 + R^2)^{1/2} - z]$$

$$= \frac{\sigma}{2\epsilon_0}\left(1 - \frac{z}{\sqrt{z^2 + R^2}}\right). \qquad \text{(Answer)}$$

This is the same expression that we derived in Section 24-7 by direct integration, using Coulomb's law; compare Eq. 21 of that chapter. Verify that other expressions for E listed in Table 1 can be derived from the corresponding expressions for V that are also listed in that Table.

___

26-9   Electric Potential Energy

In Section 26-1, we discussed the electric potential energy of a test charge as a function of its position in an electric field. In that section, we assumed that the charges that gave rise to the field were fixed in place, so that the field itself could not be influenced by the presence of the test charge. In this section, we take a broader view of electric potential energy, applying it to the energy that a given charge configuration has because of the magnitude and position of its component charges.

For a simple example, if you pull apart two bodies that carry charges of opposite signs, the work that you must do will be stored as electric potential energy in the two-charge system. If you release the charges, you can recover this stored energy, in whole or in part, as kinetic energy of the charged bodies as they rush toward each other. We define the electric potential energy of a system of point charges, held in fixed positions by forces not specified, as follows:

The electric potential energy of a system of fixed point charges is equal to the work that must be done by an external agent to assemble the system, bringing each charge in from an infinite distance.

We assume that the charges are at rest in their initial infinitely distant positions and also in their final configuration.

Figure 14 shows two point charges, separated by a distance r_{12}. Let us imagine q_2 removed to infinity and at rest. The potential V at the original site of q_2, caused by q_1, is given by Eq. 12 as

$$V = \frac{1}{4\pi\epsilon_0} \frac{q_1}{r_{12}}.$$

As q_2 is moved in from infinity to its original position, the work that an external agent must do is Vq_2. Thus, from our definition, the electric potential energy of the two-charge system of Fig. 14 is

$$\boxed{U = W = \frac{1}{4\pi\epsilon_0} \frac{q_1q_2}{r_{12}}.} \qquad (22)$$

If the charges have the same sign, an external agent would have to do positive work to push them together against their mutual repulsion. Hence, as Eq. 22 shows, their mutual potential energy will be positive. If the charges have opposite signs, the external agent will have to do negative work to restrain them against their mutual

attraction or to bring them to rest at their final positions and the mutual potential energy of the two charges will be negative.

Sample Problem 9 Two protons in the nucleus of a uranium-238 atom are 6.0 fm apart. What is the potential energy associated with the repulsive electric force that acts between these two particles?

From Eq. 22, with $q_1 = q_2 = q$,

$$U = \frac{1}{4\pi\epsilon_0} \frac{q^2}{r} = \frac{(8.99 \times 10^9 \text{ N} \cdot \text{m}^2/\text{C}^2)(1.60 \times 10^{-19} \text{ C})^2}{6.0 \times 10^{-15} \text{ m}}$$

$$= 3.8 \times 10^{-14} \text{ J} = 2.4 \times 10^5 \text{ eV} = 240 \text{ keV}. \qquad \text{(Answer)}$$

The two protons do not fly apart because they are held together by the attractive *strong force* that serves to bind the nucleus together. There is a potential energy associated with this force but it is not an electric potential energy.

Sample Problem 10 Figure 15 shows three charges held in fixed positions by forces that are not shown. What is the electric potential energy of this system of charges? Assume that $d = 12$ cm and that

$$q_1 = +q, \quad q_2 = -4q, \quad \text{and} \quad q_3 = +2q,$$

where $q = 150$ nC.

Imagine that q_1 alone is in place and that you bring up q_2 from infinity. From Eq. 22, the work W_{12} that you must do is U_{12}, where

$$W_{12} = U_{12} = \frac{1}{4\pi\epsilon_0} \frac{q_1q_2}{d}.$$

If you then bring up charge q_3 and put it in place, the additional work that you must do is $U_{13} + U_{23}$.

We see that the total work that you must do to assemble the three charges is just the sum of the potential energies of the

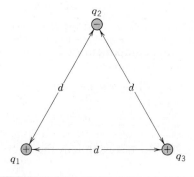

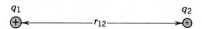

Figure 14 Two charges are held a fixed distance r_{12} apart. What is the electric potential energy of the configuration?

Figure 15 Sample Problem 10. Three charges are held fixed at the vertices of a triangle. What is the electric potential energy of the configuration?

three pairs of charges and is independent of the order in which the charges are brought together. Thus,

$$U = U_{12} + U_{13} + U_{23}$$
$$= \frac{1}{4\pi\epsilon_0}\left(\frac{(+q)(-4q)}{d} + \frac{(+q)(+2q)}{d} + \frac{(-4q)(+2q)}{d}\right)$$
$$= -\frac{10q^2}{4\pi\epsilon_0 d}$$
$$= -\frac{(8.99 \times 10^9 \text{ N}\cdot\text{m}^2/\text{C}^2)(10)(150 \times 10^{-9} \text{ C})^2}{0.12 \text{ m}}$$
$$= -1.7 \times 10^{-2} \text{ J} = -17 \text{ mJ}. \qquad \text{(Answer)}$$

The fact that the potential energy in this case is negative means that negative work would have to be done to assemble this structure, starting with the three charges infinitely separated and at rest. Put another way, an external agent would have to do 17 mJ of work to dismantle the structure completely.

26-10   An Insulated Conductor

In Section 25-7, we concluded that $\mathbf{E} = 0$ for all points inside an insulated conductor and we then used Gauss' law to prove the following:

Once equilibrium has been established, an excess charge placed on an insulated conductor will be found to lie entirely on its surface. This remains true even if the conductor has an empty internal cavity.*

Here we use the fact that $\mathbf{E} = 0$ for all points inside an insulated conductor to prove another fact about such conductors, namely:

An excess charge placed on an insulated conductor will distribute itself on the surface of that conductor so that all points of the conductor—whether on the surface or inside—come to the same potential. This remains true whether or not the conductor has an internal cavity.

Our proof follows directly from Eq. 8, which is

$$V_f - V_i = -\int_i^f \mathbf{E} \cdot d\mathbf{s}.$$

* If the cavity encloses an insulated charge, then some of the conductor's charge will be found on its inner surface as well as on its outer surface.

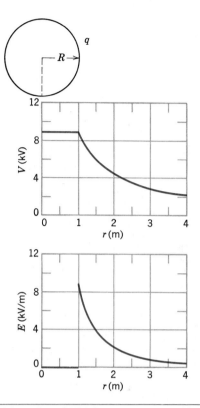

Figure 16   (*a*) A plot of $V(r)$ for a charged spherical shell. (*b*) A plot of $E(r)$ for the same shell.

Since $\mathbf{E} = 0$ for all points within a conductor, it follows directly that $V_f = V_i$ for all possible pairs of points. We might also note that, because the surface of the conductor is an equipotential surface, $\mathbf{E}$ for points on the surface must be at right angles to the surface, as we saw in Chapter 25.

Figure 16*a* is a plot of potential against radial distance for an insulated spherical conducting shell of 1.0-m radius, carrying a charge of 1.0 μC. For points outside the shell, we can calculate $V(r)$ from Eq. 12 because the charge q behaves for such external points as if it were concentrated at the center of the shell. This equation holds right up to the surface of the shell. Now let us push a small test charge right through the shell—assuming a small hole exists—to its center. No extra work is needed because no electric forces act on the test charge once it is inside the shell. Thus, the potential for all points inside the shell has the same value as that on the surface, as Fig. 16*a* shows.

Figure 16*b* shows the variation of electric field with radial distance for the same shell. Note that $E = 0$ every-

Figure 17 A lightning strike has burned the grass on the green of a golf course.

Figure 18 The car is a safe haven in a thunderstorm. Its safety owes more to Gauss' law than to the insulating ability of its rubber tires.

where inside. The lower of these two curves can be derived from the upper by differentiating with respect to r, using Eq. 20; the derivative of a constant, for example, is zero. The upper curve can be derived from the lower by integrating with respect to r, using Eq. 9. The negative of the integral of $1/r^2$, for example, is $1/r$.

Except for spherical conductors, the surface charge does not distribute itself uniformly over the surface of a conductor.* At sharp points or edges, the surface charge density—and thus the external electric field, which is proportional to it—may reach very high values. The air around such sharp points may become ionized, producing the corona discharge that golfers and mountaineers may experience when thunderstorms threaten. Such corona discharges are often the precursors of lightning strikes; see Fig. 17. In such circumstances, it is wise to enclose oneself in a cavity inside a conducting shell, where the electric field is guaranteed to be zero; Fig. 18 shows a suitable mobile arrangement.

If an insulated conductor is placed in an *external electric field,* as in Fig. 19, all points of the conductor still come to a single potential and they do so whether or not the conductor carries an excess charge. The conduction electrons distribute themselves on the surface in such a way that the electric field they produce at interior points cancel the external electric field that would otherwise be

present at those points. If the surface charges in Fig. 19 could be somehow frozen in place and the conductor removed, the pattern of the electric field would remain absolutely unchanged, for both exterior and interior points.

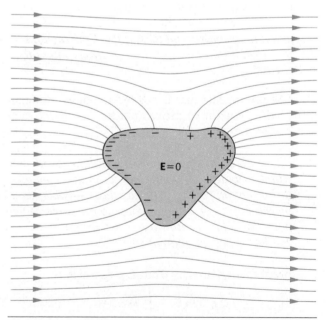

Figure 19 An uncharged conductor is suspended in an external electric field. The conduction electrons distribute themselves on the surface as shown, reducing the electric field inside the conductor to zero.

---

* See "The Lightning-rod Fallacy," by Richard H. Price and Ronald J. Crowley, *American Journal of Physics,* September 1985, p. 843 for a careful discussion.

26-11   The Van de Graaff Accelerator

The heart of a Van de Graaff accelerator* is an arrangement for generating potential differences of the order of several million volts. By allowing charged particles such as electrons or protons to "fall" through this potential difference, a beam of energetic particles can be produced. In medicine, such beams are widely used in the management of malignancies. In physics, accelerated particle beams can be used in a variety of "atom-smashing" experiments. Figure 20 shows such an accelerator in a nuclear physics laboratory at Purdue University.

Figure 20   A view of a nuclear physics laboratory at Purdue University. The particle beam emerges from the Van de Graaff accelerator, which is housed in the large tank at the rear. The experimental area is in the foreground.

Figure 21 suggests how the high potential is generated in a Van de Graaff accelerator. A small conducting shell of radius r is located inside a larger shell of radius R. The two shells carry charges q and Q, respectively. If the inner shell is connected to the outer shell by a conducting path, the two shells then form a single insulated conductor. The charge q then moves *entirely* to the outer surface of the large shell, no matter how much charge there may already be on that shell. Every such charge transfer increases the potential of the outer shell.

---

* So called after Robert J. Van de Graaff, who first put a suggestion by Lord Kelvin into useful practice. See "The Biggest Van de Graaff Machine," by Joe Watson, *The New Scientist,* March 1974. The original Van de Graaff machine, used as a "lightning generator," is on display at the Museum of Science in Boston.

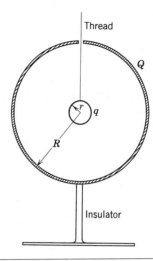

Figure 21   Illustrating the operating principle of a Van de Graaff accelerator.

In practice, charge is carried into the inner shell by a rapidly moving charged belt. Charge is "sprayed" onto the belt outside the machine by a comb of "corona points" and removed from the belt inside the machine in the same way. The motor driving the belt provides the energy needed to move the charge to the higher potential of the shells. The maximum potential that may be achieved with a given accelerator occurs when the rate at which charge is being carried into the inner shell is equal to the rate at which charge leaves the outer shell by leakage along the supports and by corona discharge.

A Clever Idea.   Figure 22 shows an arrangement by which a particle can be accelerated *twice* by the same accelerating potential, thus emerging from the machine with double the kinetic energy that it otherwise would have. Let the high-potential terminal, located in the center of the device, be at a positive potential V. If the

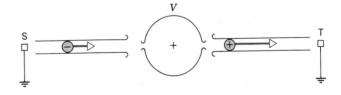

Figure 22   A schematic diagram of an energy-doubling arrangement. Particles leave the source S at low speed, are accelerated twice by the potential difference V maintained in the center of the device, and strike the target T at high speed. The sign of the net charge of the particles changes as they pass through a thin foil in the central terminal.

particle approaching from the left carries a charge of $-e$, it will be *attracted* to the terminal, gaining a kinetic energy Ve. If the particle could somehow change its charge to $+e$ as it passes through the central terminal, it would then be *repelled* by this terminal as it emerged from it on the right, gaining an additional Ve of kinetic energy at the time of its arrival at target T.

This "charge switching" can be done! The particle approaching from the left is a hydrogen atom carrying an excess electron, so that its net charge is indeed $-e$. Inside the terminal, the beam is made to pass through a thin "stripping foil" in which its two outer electrons are stripped off by grazing collisions in the foil. The beam then emerges from the terminal on the right as a single proton, with a charge of $+e$.*

---

* In the early days of its development, an accelerator based on this principle was affectionately called a "Swindletron."

REVIEW AND SUMMARY

The change ΔU in the electric potential energy U of a charged object as it moves from an initial point i to a final point f is

Electric Potential Energy

$$\Delta U = U_f - U_i = -W_{if}.$$ [1]

Work W_{if} is that done by the electric field. For zero potential energy at infinity, the *electric potential energy U* of the body *at a point* is

$$U = -W_\infty.$$ [2]

W_∞ is the work done by the electric field if the object were to move from infinity to the point. See Sample Problem 1.

We define the *potential difference ΔV* between two points in an electric field as

Electric Potential Difference

$$\Delta V = V_f - V_i = -\frac{W_{if}}{q_0} \quad \text{(potential difference defined)},$$ [3]

q_0 being a test charge on which work is done by the field. The *potential* at a point is

Electric Potential V

$$V = -\frac{W_\infty}{q_0} \quad \text{(potential defined).}$$ [4]

The SI unit of potential is the *volt*: 1 volt = 1 joule/coulomb.

Equipotential Surfaces

The points on an *equipotential surface* all have the same potential. The work done on a test charge moving from one such surface to another is independent of the locations of the initial and terminal points on these surfaces and of the path that joins them; see Fig. 2. The electric field **E** is always at right angles to equipotential surfaces; see Fig. 3.

The potential difference between any two points is

Finding V from E

$$V_f - V_i = -\int_i^f \mathbf{E} \cdot d\mathbf{s}.$$ [8]

The line integral is taken over any path connecting the points. For i at infinity we have, for the potential at a particular point,

$$V = -\int_\infty^f \mathbf{E} \cdot d\mathbf{s}.$$ [9]

See Sample Problems 2 and 3.

The potential due to a single point charge is

Potential Due to Point Charges

$$V = \frac{1}{4\pi\epsilon_0} \frac{q}{r}.$$ [12]

The potential due to a collection of point charges is

$$V = \sum_n V_n = \frac{1}{4\pi\epsilon_0} \sum \frac{q_n}{r_n}. \tag{13}$$

(See Sample Problems 4–6.)

The potential due to an electric dipole with dipole moment $p = qd$ is

Electric Dipole Potential

$$V(r, \theta) = \frac{1}{4\pi\epsilon_0} \frac{p \cos \theta}{r^2} \tag{14}$$

for $r \gg d$; r and θ are defined in Fig. 9.

For a continuous distribution of charge, Eq. 13 becomes

Continuous Charge

$$V = \frac{1}{4\pi\epsilon_0} \int \frac{dq}{r}. \tag{15}$$

See Section 26–7, in which we show the potential at a distance z from the center of a charged disk along its axis to be

$$V = \frac{\sigma}{2\epsilon_0} (\sqrt{z^2 + R^2} - z), \tag{17}$$

and Sample Problem 7.

Calculating E from V

Any component of **E** may be found from V by differentiation. The rate of change of potential with distance in any direction is, when changed in sign, the component of **E** in that direction:

$$E_s = -\frac{\partial V}{\partial s}. \tag{20}$$

The Cartesian components of **E** may be found from

$$E_x = -\frac{\partial V}{\partial x}; \quad E_y = -\frac{\partial V}{\partial y}; \quad E_z = -\frac{\partial V}{\partial z}. \tag{21}$$

See Sample Problem 8 and Table 1 for examples.

The electric potential energy of a system of point charges is the work needed to assemble the system with the charges initially at rest and infinitely distant from each other. For two charges,

Electric Potential Energy

$$U = W = \frac{1}{4\pi\epsilon_0} \frac{q_1 q_2}{r_{12}}. \tag{22}$$

Sample Problems 9 and 10 show applications.

A Charged Conductor

An excess charge placed on a conductor will, in equilibrium, be on its outer surface. The charge brings the entire conductor, including both surface and interior points, to a uniform potential. Figure 16 shows $V(r)$ and $E(r)$ for a charged conducting sphere or spherical shell. An electrostatic generator is an important application. Figure 21 illustrates its working principle. If a connection is made between the two conductors in that figure the charge on the inner sphere will move *entirely* to the outer shell, no matter what charge is already on that shell.

QUESTIONS

1. Are we free to call the potential of the earth $+100$ V instead of zero? What effect would such an assumption have on measured values of (*a*) potentials and (*b*) potential differences?

2. What would happen to you if you were on an insulated stand and your potential was increased by 10 kV with respect to the earth?

3. Why is the electron volt often a more convenient unit of energy than the joule?

4. How would a proton-volt compare with an electron-volt? The mass of a proton is 1840 times that of an electron.

5. Do electrons tend to go to regions of high potential or of low potential?

6. Why is it possible to shield a room against electrical forces but not against gravitational forces?

7. Suppose that the earth has a net charge that is not zero. Why is it still possible to adopt the earth as a standard reference point of potential and to assign the potential $V = 0$ to it?

8. Does the potential of a positively-charged insulated conductor have to be positive? Give an example to prove your point.

9. Can two different equipotential surfaces intersect?

10. An electrical worker was accidentally electrocuted and a newspaper account reported: "He accidentally touched a high-voltage cable and 20,000 V of electricity surged through his body." Criticize this statement.

11. Advice to mountaineers caught in lightning and thunderstorms is (a) get rapidly off peaks and ridges and (b) put both feet together and crouch in the open, only the feet touching the ground. What is the basis for this good advice?

12. If **E** equals zero at a given point, must V equal zero for that point? Give some examples to prove your answer.

13. If you know **E** only at a given point, can you calculate V at that point? If not, what further information do you need?

14. In Fig. 2, is the electric field E greater at the left or at the right side of the figure?

15. Is the uniformly-charged, nonconducting disk of Section 26–7 a surface of constant potential? Explain.

16. We have seen that, inside a hollow conductor, you are shielded from the fields of outside charges. If you are *outside* a hollow conductor that contains charges, are you shielded from the fields of these charges? Explain why or why not.

17. Distinguish between potential difference and difference of potential energy. Give examples of statements in which each term is used properly.

18. If the surface of a charged conductor is an equipotential, does that mean that charge is distributed uniformly over that surface? If the electric field is constant in magnitude over the surface of a charged conductor, does *that* mean that the charge is distributed uniformly?

19. In Section 26–10 we learned that charge delivered to the *inside* of an isolated conductor is transferred *entirely* to the outer surface of the conductor, no matter how much charge is already there. Can you keep this up forever? If not, what stops you?

20. Ions and electrons act like condensation centers; water droplets form around them in air. Explain why.

21. If V equals a constant throughout a given region of space, what can you say about **E** in that region?

22. In Chapter 15 we saw that the gravitational field strength is zero inside a spherical shell of matter. The electrical field strength is zero not only inside an isolated charged spherical conductor but inside an isolated conductor of any shape. Is the gravitational field strength inside, say, a cubical shell of matter zero? If not, in what respect is the analogy not complete?

23. How can you ensure that the electric potential in a given region of space will have a constant value?

24. Devise an arrangement of three point charges, separated by finite distances, that has zero electric potential energy.

25. We have seen (Section 26–10) that the potential inside a conductor is the same as that on its surface. (a) What if the conductor is irregularly shaped and has an irregularly shaped cavity inside? (b) What if the cavity has a small "worm hole" connecting it to the outside? (c) What if the cavity is closed but has a point charge suspended within it? Discuss the potential within the conducting material and at different points within the cavities.

26. An isolated conducting spherical shell carries a negative charge. What will happen if a positively charged metal object is placed in contact with the shell interior? Discuss the three cases in which the positive charge is (a) less than, (b) equal to, and (c) greater than the negative charge in magnitude.

EXERCISES AND PROBLEMS

Section 26–2   The Electric Potential

1E. The electric potential difference between discharge points during a particular thunderstorm is 1.2×10^9 V. What is the magnitude of the change in the electrical potential energy of an electron that moves between these points?

2E. A particular 12-V car battery is rated to deliver a charge of 84 A·h (ampere · hours). (a) How many coulombs of charge does this represent? (b) If this entire charge is delivered at 12 V, how much energy is available?

3P. In a typical lightning flash the potential difference between discharge points is about 10^9 V and the quantity of charge transferred is about 30 C. (a) How much energy is released? (b) If all the energy released could be used to accelerate a 1000-kg automobile from rest, what would be its final speed? (c) If it could be used to melt ice, how much ice would it melt at 0°C? The heat of fusion of ice is 3.3×10^5 J/kg.

Section 26–3   Equipotential Surfaces

4E. Two line charges are parallel to the z axis. One, of charge per unit length $+\lambda$, is a distance a to the right of this axis. The other, of charge per unit length $-\lambda$, is a distance a to the left of this axis (the lines and the z axis being in the same plane). Sketch some of the equipotential surfaces.

5E. In moving from A to B along an electric field line, the electric field does 3.94×10^{-19} J of work on an electron in the field illustrated in Fig. 23. What are the differences in the electric potential (a) $V_B - V_A$, (b) $V_C - V_A$, and (c) $V_C - V_B$?

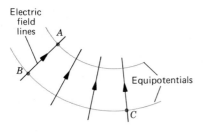

Figure 23 Exercise 5.

6E. Figure 24 shows, edge-on, an "infinite" sheet of positive charge density σ. (a) How much work is done by the electric field of the sheet as a small positive test charge q_0 is moved from an initial position on the sheet to a final position located a perpendicular distance z from the sheet? (b) Use the result from (a) and Eq. 8 to show that, as displayed in Table 1, the electric potential of an infinite sheet of charge can be written

$$V = V_0 - (\sigma/2\epsilon_0)z,$$

where V_0 is the potential at the surface of the sheet.

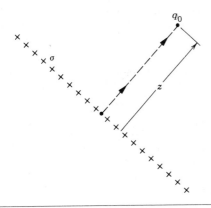

Figure 24 Exercise 6.

7E. In the Millikan oil-drop experiment (see Section 24–8 in Chapter 24), an electric field of 1.92×10^5 N/C is maintained at balance across two plates separated by 1.50 cm. Find the potential difference between the plates.

8E. Two large parallel conducting plates are 12 cm apart and carry equal but opposite charges on their facing surfaces. An electron placed midway between the two plates experiences a force of 3.9×10^{-15} N. (a) Find the electric field at the position of the electron. (b) What is the potential difference between the plates?

9E. An infinite sheet of charge has a charge density $\sigma = 0.10$ μC/m². How far apart are the equipotential surfaces whose potentials differ by 50 V?

10P. Three long parallel lines of charge have the relative linear charge densities shown in Fig. 25. Sketch some lines of force and the intersections of some equipotential surfaces with the plane of this figure.

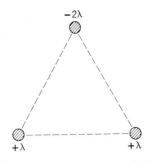

Figure 25 Problem 10.

11P. The electric field inside a nonconducting sphere of radius R, containing uniform charge density, is radially directed and has magnitude

$$E(r) = \frac{qr}{4\pi\epsilon_0 R^3},$$

where q is the total charge in the sphere and r is the distance from the sphere center. (a) Find the potential $V(r)$ inside the sphere, taking $V = 0$ at the center of the sphere. (b) What is the difference in electric potential between a point on the surface and the sphere center? If q is positive, which point is at the higher potential?

12P. A Geiger counter has a metal cylinder 2.0 cm in diameter along whose axis is stretched a wire 1.3×10^{-4} cm in diameter. If 850 V is applied between them, what is the electric field at the surface of (a) the wire and (b) the cylinder? (Hint: Use the result of Problem 24, Chapter 25.)

13P.* A charge q is distributed uniformly throughout a spherical volume of radius R. (a) Show that the potential a distance r from the center, where $r < R$, is given by

$$V = \frac{q(3R^2 - r^2)}{8\pi\epsilon_0 R^3},$$

where the zero of potential is taken at $r = \infty$. (b) Why does this result differ from Problem 11?

Section 26–5 Calculating the Potential: A Point Charge
14E. A point charge has $q = +1.0$ μC. Consider point A, which is 2.0 m distant, and point B, which is 1.0 m distant in a direction diametrically opposite, as in Fig. 26a. (a) What is the potential difference $V_A - V_B$? (b) Repeat if points A and B are located as in Fig. 26b.

15E. Consider a point charge with $q = 1.5 \times 10^{-8}$ C. (a) What is the radius of an equipotential surface having a potential of 30 V? (b) Are surfaces whose potentials differ by a constant amount (1.0 V, say) evenly spaced?

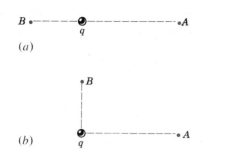

Figure 26   Exercise 14.

16E.   A charge of 1.5×10^{-8} C can be produced by simple rubbing. To what potential would such a charge raise an insulated conducting sphere of 16-cm radius? (See Sample Problem 5.)

17E.   As a Space Shuttle moves through the dilute ionized gas of the earth's ionosphere, its potential is typically changed by -1.0 V before it completes one revolution. By assuming that the Shuttle is a sphere of radius 10 m, estimate the amount of charge it collects.

18E.   Much of the material comprising Saturn's rings (see Fig. 27) is in the form of tiny dust particles having radii on the order of 10^{-6} m. These grains are in a region containing a dilute ionized gas, and they pick up excess electrons. If the electric potential at the surface of a grain is -400 V, how many excess electrons has it picked up?

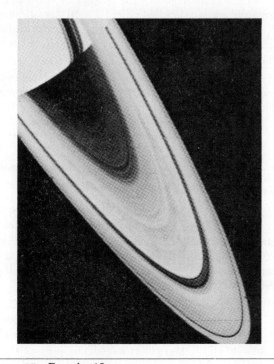

Figure 27   Exercise 18.

19E.   In Fig. 28 sketch qualitatively (a) the lines of force and (b) the intersections of the equipotential surfaces with the plane of the figure. (*Hint:* Consider the behavior close to each point charge and at considerable distances from the pair of charges.)

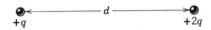

Figure 28   Exercise 19.

20E.   Repeat the procedure explained in Exercise 19 for Fig. 29.

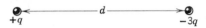

Figure 29   Exercise 20.

21P.   Can a conducting sphere 10 cm in radius hold a charge of 4 μC in air without breakdown? The dielectric strength (minimum field required to produce breakdown) of air at 1 atm is 3 MV/m.

22P.   What are (a) the charge and (b) the charge density on the surface of a conducting sphere of radius 0.15 m whose potential is 200 V?

23P.   An electric field of approximately 100 V/m is often observed near the surface of the earth. If this field were the same over the entire surface, what would be the electric potential of a point on the surface? See Sample Problem 5.

24P.   Suppose that the negative charge in a copper one-cent coin were removed to a very large distance from the earth — perhaps to a distant galaxy — and that the positive charge were distributed uniformly over the earth's surface. By how much would the electric potential at the surface of the earth change? (See Sample Problem 2 in Chapter 23.)

25P.   A spherical drop of water carrying a charge of 30 pC has a potential of 500 V at its surface. (a) What is the radius of the drop? (b) If two such drops of the same charge and radius combine to form a single spherical drop, what is the potential at the surface of the new drop so formed?

26P.   In Fig. 30, locate the points, if any, (a) where $V = 0$ and (b) where $\mathbf{E} = 0$. Consider only points on the axis and choose $d = 1.0$ m.

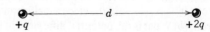

Figure 30   Problem 26.

27P.   A copper sphere whose radius is 1.0 cm has a very thin surface coating of nickel. Some of the nickel atoms are radioactive, each atom emitting an electron as it decays. Half of these electrons enter the copper sphere, each depositing 100 keV of energy there. The other half of the electrons escape, each carrying away a charge of $-e$. The nickel coating has an activity of 10 mCi ($= 10$ millicuries $= 3.70 \times 10^8$ radioactive decays per second). The sphere is hung from a long, nonconducting string and insulated from its surroundings. (*a*) How long will it take for the potential of the sphere to increase by 1000 V? (*b*) How long will it take for the temperature of the sphere to increase by 5.0°C? The heat capacity of the sphere is 14.3 J/°C.

28P. A point charge $q_1 = +6e$ is fixed at the origin of a rectangular coordinate system, and a second point charge $q_2 = -10e$ is fixed at $x = 8.6$ nm, $y = 0$. The locus of all points in the xy plane with $V = 0$ is a circle centered on the x axis, as shown in Fig. 31. Find (*a*) the location x_c of the center of the circle and (*b*) the radius R of the circle. (*c*) Is the $V = 5$ V equipotential also a circle?

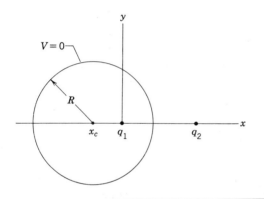

Figure 31 Problem 28.

29P.* A thick spherical shell of charge of uniform charge density is bounded by radii r_1 and r_2, where $r_2 > r_1$. Find the electric potential V as a function of the distance r from the center of the distribution, considering the regions (*a*) $r > r_2$, (*b*) $r_2 > r > r_1$, and (*c*) $r < r_1$. (*d*) Do these solutions agree at $r = r_2$ and at $r = r_1$?

Section 26–6   Calculating the Potential: An Electric Dipole
30E. The ammonia molecule NH_3 has a permanent electric dipole moment equal to 1.47 D, where D = debye unit = 3.34×10^{-30} C·m. Calculate the electric potential due to an ammonia molecule at a point 52 nm away along the axis of the dipole.

31P. For the charge configuration of Fig. 32, show that $V(r)$ for points on the vertical axis, assuming $r \gg d$, is given by

$$V = \frac{1}{4\pi\epsilon_0}\frac{q}{r}\left(1 + \frac{2d}{r}\right).$$

(*Hint:* The charge configuration can be viewed as the sum of an isolated charge and a dipole.)

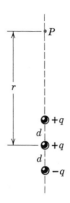

Figure 32 Problem 31.

Section 26–7   Calculating the Potential: A Charged Disk
32P. (*a*) Show that the electric potential at a point on the axis of a ring of charge of radius R, computed directly from Eq. 15, is given by

$$V = \frac{1}{4\pi\epsilon_0}\frac{q}{\sqrt{z^2 + R^2}}.$$

(*b*) From this result derive an expression for E at axial points; compare with the direct calculation of E in Section 24–6 of Chapter 24.

Section 26–8   Calculating the Field from the Potential
33E. Two large parallel metal plates are 1.5 cm apart and carry equal but opposite charges on their facing surfaces. The negative plate is grounded and its potential is taken to be zero. If the potential halfway between the plates is $+5.0$ V, what is the electric field in this region?

34E. The electric potential varies along the x axis as shown in the graph of Fig. 33. For each of the intervals shown (ignore the behavior at the end points of the intervals), determine the x component of the electric field and plot E_x versus x.

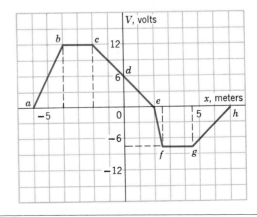

Figure 33 Exercise 34.

35E. Starting from Eq. 14, find E_r, the radial component of the electric field due to a dipole.

36E. In Section 26–7 the potential at an axial point for a charged disk was shown to be

$$V = \frac{\sigma}{2\epsilon_0} (\sqrt{z^2 + R^2} - z).$$

Use Eq. 21 to show that E for axial points is given by

$$E = \frac{\sigma}{2\epsilon_0} \left(1 - \frac{z}{\sqrt{R^2 + z^2}} \right).$$

37E. The electric potential V in the space between the plates of a particular, and now obsolete, vacuum tube is given by $V = 1500x^2$, where V is in volts if x, the distance from one of the plates, is in meters. Calculate the magnitude and direction of the electric field at $x = 1.3$ cm.

38E. Exercise 44 in Chapter 25 deals with Rutherford's calculation of the electric field a distance r from the center of an atom. He also gave the electric potential as

$$V = \frac{Ze}{4\pi\epsilon_0} \left(\frac{1}{r} - \frac{3}{2R} + \frac{r^2}{2R^3} \right).$$

(a) Show how the expression for the electric field given in Exercise 44 of Chapter 25 follows from the above expression for V. (b) Why does this expression for V not go to zero as $r \to \infty$?

39P. A charge per unit length λ is distributed uniformly along a straight-line segment of length L. (a) Determine the potential (chosen to be zero at infinity) at a point P a distance y from one end of the charged segment and in line with it (see Fig. 34). (b) Use the result of (a) to compute the component of the electric field at P in the y direction (along the line). (c) Determine the component of the electric field at P in a direction perpendicular to the straight line.

Figure 34 Problem 39.

40P. On a thin rod of length L lying along the x axis with one end at the origin ($x = 0$), as in Fig. 35, there is distributed a charge per unit length given by $\lambda = kx$, where k is a constant. (a) Taking the electrostatic potential at infinity to be zero, find

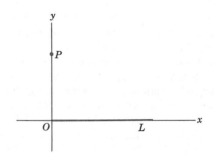

Figure 35 Problem 40.

V at the point P on the y axis. (b) Determine the vertical component, E_y, of the electric field intensity at P from the result of part (a) and also by direct calculation. (c) Why cannot E_x, the horizontal component of the electric field at P, be found using the result of part (a)?

Section 26–9 Electric Potential Energy

41E. (a) For Fig. 36, derive an expression for $V_A - V_B$. (b) Does your result reduce to the expected answer when $d = 0$? When $a = 0$? When $q = 0$?

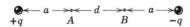

Figure 36 Exercise 41.

42E. Two charges $q = +2.0 \ \mu$C are fixed in space a distance $d = 2.0$ cm apart, as shown in Fig. 37. (a) What is the electric potential at point C? (b) You bring a third charge $q = +2.0 \ \mu$C slowly from infinity to C. How much work must you do? (c) What is the potential energy U of the configuration when the third charge is in place?

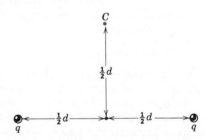

Figure 37 Exercise 42.

43E. The charges and coordinates of two point charges located in the xy plane are: $q_1 = +3.0 \times 10^{-6}$ C, $x = 3.5$ cm, $y = +0.50$ cm; and $q_2 = -4.0 \times 10^{-6}$ C, $x = -2.0$ cm, $y = +1.5$ cm. (a) Find the electric potential at the origin. (b) How much total work must be done to locate these charges at their given positions, starting from infinite separation?

44E. A decade before Einstein published his theory of relativity, J. J. Thomson proposed that the electron might be made up of small parts and that its mass is due to the electrical interaction of the parts. Furthermore, he suggested that the energy equals mc^2. Make a rough estimate of the electron mass in the following way: Assume that the electron is composed of three identical parts that are brought in from infinity and placed at the vertices of an equilateral triangle having sides equal to the classical radius of the electron, 2.82×10^{-15} m. (*a*) Find the total electrical potential energy of this arrangement. (*b*) Divide by c^2 and compare your result to the accepted electron mass (9.11×10^{-31} kg). The result improves if more parts are assumed.

45E. In the quark model of fundamental particles, a proton is composed of three quarks: two "up" quarks, each having charge $+\frac{2}{3}e$, and one "down" quark, having charge $-\frac{1}{3}e$. Suppose that the three quarks are equidistant from each other. Take the distance to be 1.32×10^{-15} m and calculate (*a*) the potential energy of the interaction between the two "up" quarks and (*b*) the total electrical potential energy of the system.

46E. Derive an expression for the work required to put the four charges together as indicated in Fig. 38.

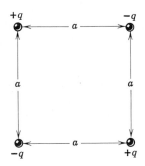

Figure 38 Exercise 46.

47E. What is the electric potential energy of the charge configuration of Fig. 11*a*? Use the numerical values of Sample Problem 6.

48P. Three charges of $+0.12$ C each are placed on the corners of an equilateral triangle, 1.7 m on a side. If energy is supplied at the rate of 0.83 kW, how many days would be required to move one of the charges onto the midpoint of the line joining the other two?

49P. In the rectangle shown in Fig. 39, the sides have lengths 5.0 cm and 15 cm, $q_1 = -5.0 \ \mu C$ and $q_2 = +2.0 \ \mu C$. (*a*) What are the electric potentials at corner B and at corner A? (*b*) How much work is required to move a third charge $q_3 = +3.0 \ \mu C$

Figure 39 Problem 49.

from B to A along a diagonal of the rectangle? (*c*) In this process, is the external work converted into electrostatic potential energy or vice versa? Explain.

50P. A particle of (positive) charge Q is assumed to have a fixed position at P. A second particle of mass m and (negative) charge $-q$ moves at constant speed in a circle of radius r_1, centered at P. Derive an expression for the work W that must be done by an external agent on the second particle in order to increase the radius of the circle of motion, centered at P, to r_2.

51P. Calculate (*a*) the electric potential established by the nucleus of a hydrogen atom at the average distance of the circulating electron ($r = 5.3 \times 10^{-11}$ m), (*b*) the electric potential energy of the atom when the electron is at this radius, and (*c*) the kinetic energy of the electron, assuming it to be moving in a circular orbit of this radius centered on the nucleus. (*d*) How much energy is required to ionize the hydrogen atom? Express all energies in electron-volts.

52P. An electric charge of -9.0 nC is uniformly distributed around a ring of radius 1.5 m that lies in the yz plane with its center at the origin. A point charge of -6.0 pC is located on the x axis at $x = 3.0$ m. Calculate the work done in moving the point charge to the origin.

53P. A particle of charge q is kept in a fixed position at a point P and a second particle of mass m, having the same charge q, is initially held at rest a distance r_1 from P. The second particle is then released and is repelled from the first one. Determine its speed at the instant it is a distance r_2 from P. Let $q = 3.1 \ \mu C$, $m = 20$ mg, $r_1 = 0.90$ mm, and $r_2 = 2.5$ mm.

54P. Two small metal spheres of mass $m_1 = 5.0$ g and mass $m_2 = 10$ g carry equal positive charges $q = 5.0 \ \mu C$. The spheres are connected by a massless string of length $d = 1.0$ m, which is much greater than the sphere radii. (*a*) What is the electrostatic potential energy of the system? (*b*) You cut the string. At that instant what is the acceleration of each of the spheres? (*c*) A long time after you cut the string, what is the speed of each sphere?

55P. Between two parallel, flat, conducting surfaces of spacing $d = 1.0$ cm and potential difference $V = 10$ kV, an electron is projected from one plate directly toward the second. What is the initial velocity of the electron if it comes to rest just at the surface of the second plate?

56P. A gold nucleus contains a positive charge equal to that of 79 protons and has a radius of 6.2 fm; see Sample Problem 5. An α particle (which consists of two protons and two neutrons) has a kinetic energy K at points far from the nucleus and is traveling directly toward it. The α particle just touches the surface of the nucleus where its velocity is reversed in direction. (a) Calculate K. (b) The actual α particle energy used in the experiment of Rutherford and his collaborators that led to the discovery of the concept of the atomic nucleus was 5.0 MeV. What do you conclude?

57P. A particle of mass m, charge $q > 0$, and initial kinetic energy K is projected (from "infinity") toward a heavy nucleus of charge Q, assumed to have negligible size and a fixed position in our reference frame. If the aim is "perfect," how close to the center of the nucleus is the particle when it comes momentarily to rest?

58P. A thin, spherical, conducting shell of radius R is mounted on an insulated support and charged to a potential $-V$. An electron is fired from point P a distance r from the center of the shell ($r \gg R$) with an initial speed v_0, directed radially inward. What value of v_0 is needed for the electron to just reach the shell?

59P. Two electrons are fixed 2.0 cm apart. Another electron is shot from infinity and comes to rest midway between the two. What was its initial speed?

60P. Compute the escape speed for an electron from the surface of a uniformly charged sphere of radius 1.0 cm and total charge 1.6×10^{-15} C. Neglect gravitational forces.

61P. An electron is projected with an initial speed of 3.2×10^5 m/s directly toward a proton that is essentially at rest. If the electron is initially a great distance from the proton, at what distance from the proton is its speed instantaneously equal to twice its initial value?

Section 26–10   An Insulated Conductor

62E. A hollow metal sphere is charged to a potential of $+400$ V with respect to ground and carries a charge of 5.0×10^{-9} C. Find the electric potential at the center of the sphere.

63E. A thin conducting spherical shell of outer radius 20 cm carries a charge of $+3.0 \ \mu$C. Sketch (a) the magnitude of the electric field **E** and (b) the potential V versus the distance r from the center of the shell.

64E. Two identical conducting spheres of radius $r = 0.15$ m are separated by a distance $a = 10$ m. What is the charge on each sphere if the potential of one is $+1500$ V and if the other is -1500 V? What assumptions have you made?

65E. Consider two widely separated conducting spheres, 1, and 2, the second having twice the diameter of the first. The smaller sphere initially has a positive charge q and the larger one is initially uncharged. You now connect the spheres with a long thin wire. (a) How are the final potentials V_1 and V_2 of the spheres related? (b) Find the final charges q_1 and q_2 on the spheres in terms of q.

66P. The metal object in Fig. 40 is a figure of revolution about the horizontal axis. If it is charged negatively, sketch roughly a few equipotentials and lines of force. Use physical reasoning rather than mathematical analysis.

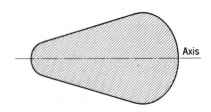

Figure 40   Problem 66.

67P. If the earth had a net charge equivalent to 1 electron/m^2 of surface area (a very artificial assumption), (a) what would be the earth's potential? (b) What would be the electric field due to the earth just outside its surface?

68P. Two metal spheres are 3.0 cm in radius and carry charges of $+1.0 \times 10^{-8}$ C and -3.0×10^{-8} C, respectively, assumed to be uniformly distributed. If their centers are 2.0 m apart, calculate (a) the potential of the point halfway between their centers and (b) the potential of each sphere.

69P. A charged metal sphere of radius 15 cm has a net charge of 3.0×10^{-8} C. (a) What is the electric field at the sphere's surface? (b) What is the electric potential at the sphere's surface? (c) At what distance from the sphere's surface has the electric potential decreased by 500 V?

70P. Two thin, insulated, concentric conducting spheres of radii R_1 and R_2 carry charges q_1 and q_2. Derive expressions for $E(r)$ and $V(r)$, where r is the distance from the center of the spheres. Plot $E(r)$ and $V(r)$ from $r = 0$ to $r = 4.0$ m for $R_1 = 0.50$ m, $R_2 = 1.0$ m, $q_1 = +2.0 \ \mu$C, and $q_2 = +1.0 \ \mu$C. Compare with Fig. 16.

Section 26–11   The Van de Graaff Accelerator

71E. (a) How much charge is required to raise an isolated metallic sphere of 1.0-m radius to a potential of 1.0 MV? Repeat for a sphere of 1.0-cm radius. (b) Why use a large sphere in an electrostatic accelerator when the same potential can be achieved using a smaller charge with a small sphere?

72E. Let the potential difference between the high-potential inner shell of a Van de Graaff accelerator and the point at which charges are sprayed onto the moving belt be 3.4 MV. If the belt transfers charge to the shell at the rate of 2.8 mC/s, what minimum power must be provided to drive the belt?

73E. An α particle (which consists of two protons and two neutrons) is accelerated through a potential difference of 1 million volts in a Van de Graaff accelerator. (a) What kinetic energy does it acquire? (b) What kinetic energy would a proton acquire under these same circumstances? (c) Which particle would acquire the greater speed, starting from rest?

74P. (a) Show, for the electrostatic Van de Graaff accelerator of Fig. 21, that the potential difference between the small sphere and the large sphere is

$$V_r - V_R = \frac{q}{4\pi\epsilon_0}\left(\frac{1}{r} - \frac{1}{R}\right).$$

Note that the potential difference is independent of the charge Q on the outer sphere. (b) Assume q is positive. Show that if the spheres are connected by a fine wire, the charge q will flow entirely to the outer sphere, regardless of the charge Q that may already be present.

75P. The high-voltage electrode of an electrostatic accelerator is a charged spherical metal shell having a potential $V = +9.0$ MV. (a) Electrical breakdown occurs in the gas in this machine at a field $E = 100$ MV/m. To prevent such breakdown, what restriction must be made on the radius r of the shell? (b) A long moving rubber belt transfers charge to the shell at 300 μC/s, the potential of the shell remaining constant because of leakage. What minimum power is required to transfer the charge? (c) The belt is of width $w = 0.50$ m and travels at speed $v = 30$ m/s. What is the surface charge density on the belt?

CHAPTER 27
CAPACITANCE

The NOVA laser at the Lawrence Livermore National Laboratory. This powerful laser, used for research in thermonuclear fusion, generates nanosecond pulses of light at a peak power level of ~ 10^{14} W, about 200 times the present power-generating capacity of the United States. This energy for the pulse is accumulated and stored in the electric fields of a large capacitor bank.

27–1 The Uses of Capacitors

You can store potential energy by pulling a bow, stretching a spring, compressing a gas, or lifting a book. In the latter case, we think of the energy as stored in the *gravitational field* of the book + earth system.

You can also store potential energy in an *electrostatic field*. From this point of view, a *capacitor* is a device designed to "package" an electric field for that purpose. The capacitor in your portable battery-operated photoflash unit does just that, accumulating energy relatively slowly during the charging process and releasing it rapidly during the short duration of the flash. On a

grander scale, we have the enormous capacitor bank powering the NOVA laser shown above.

Capacitors have many uses in our electronic and microelectronic age beyond serving as storehouses for potential energy. For one example, they are vital components of the electromagnetic oscillators that are central components of radio and TV transmitters and receivers. For another example, microscopic capacitors form the memory banks of computers. The electric fields in these tiny devices are significant — not so much for the stored energy they represent as for the ON–OFF information that their presence or absence provides. Figure 1 shows some capacitors of a kind typically used in electronic circuits.

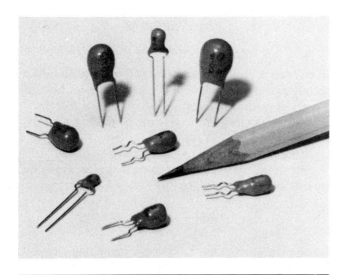

Figure 1   An assortment of capacitors that may be found in electronic circuits. Check the motherboard of your computer.

27-2   Capacitance

Figure 2 shows the elements of a *capacitor*—two isolated conductors of arbitrary shape. No matter what their geometry, we call these conductors *plates*.

Figure 3a shows a less general but more conventional arrangement, a *parallel-plate capacitor* formed of two parallel conducting plates of area A separated by a distance d. The symbol that we use to represent a capacitor (⊣⊢) is based on the structure of a parallel-plate

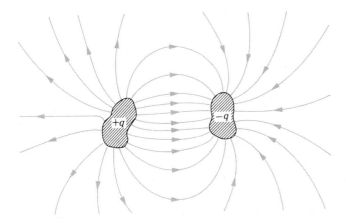

Figure 2   Two conductors, isolated from each other and from their surroundings, form a *capacitor*. When the capacitor is charged, the conductors, or plates as they are called, carry equal but opposite charges of magnitude q.

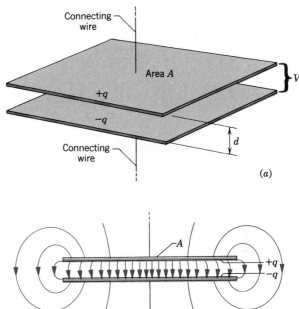

Figure 3   (a) A parallel-plate capacitor, made up of two plates of area A separated by a distance d. (b) As the field lines show, the electric field is uniform in the central region between the plates. The field lines "fringe" at the edges of the plates, showing that the field is not uniform there.

capacitor but is used for capacitors of all geometries. We assume for the time being that no material medium such as glass or plastic is present in the region between the plates. In Section 27-6, we shall remove this restriction.

We describe a capacitor as *charged* if its plates carry equal but opposite charges, of absolute value q.* Note that q is *not* the net charge on the capacitor, which is zero. One way to charge a capacitor is to connect its plates momentarily to the terminals of a battery; equal but opposite charges will then be transferred by the battery to the two plates.

Because the plates are conductors, they are equipotentials. A definite potential difference of absolute value V will exist between the two plates; V is *not* the potential of either conductor but the *potential difference* between

---

* Up to now we have treated q as a scalar, possessing both absolute value and sign. When dealing with capacitors, we find it convenient to treat q as the absolute value only of the charge, its sign to be specified separately.

them. Both q and V, being absolute values, are positive quantities.

The charge q and the potential difference V for a capacitor are proportional to each other. That is,

$$q = CV, \qquad (1)$$

in which the proportionality constant C, whose value is determined by the geometry of the plates, is called the *capacitance* of the capacitor.*

The SI unit of capacitance that follows from Eq. 1 is the coulomb/volt. This unit occurs so often that it is given a special name, the *farad* (abbr. F): That is,

$$1 \text{ farad} = 1 \text{ F} = 1 \text{ coulomb/volt} = 1 \text{ C/V}. \qquad (2)$$

As we shall see, the farad is a very large unit. Submultiples of the farad, such as the microfarad (1 μF = 10^{-6} F) and the picofarad (1 pF = 10^{-12} F) are more convenient units in practice.

27–3 Calculating the Capacitance

Our task here is to calculate the capacitance of a capacitor once we know its geometry. Because we are going to consider a number of different geometries, it seems wise to develop a general plan to simplify the work. In brief our plan is as follows: (1) Assume a charge q on the plates. (2) Calculate the electric field $\mathbf{E}$ between the plates in terms of this charge, using Gauss' law. (3) Knowing $\mathbf{E}$, calculate the potential difference V between the plates from Eq. 8 of Chapter 26. (4) Calculate C from $C = q/V$ (Eq. 1).

Before we start, we can simplify the calculation of both the electric field and the potential difference by making certain simplifying assumptions. We discuss each in turn.

Calculating the Electric Field. The electric field is related to the charge on the plates by Gauss' law, or

$$\epsilon_0 \oint \mathbf{E} \cdot d\mathbf{A} = q. \qquad (3)$$

Here q is the charge contained within the Gaussian surface and the integral is carried out over that surface. In all cases that we shall consider, the Gaussian surface will be such that whenever flux passes through it $\mathbf{E}$ will have a

constant magnitude E and the vectors $\mathbf{E}$ and $d\mathbf{A}$ will be parallel. Equation 3 then reduces to

$$q = \epsilon_0 EA \quad \text{(special case of Eq. 3)}, \qquad (4)$$

in which A is the area of that part of the Gaussian surface through which flux passes. For convenience, we shall always draw the Gaussian surface in such a way that it completely encloses the charge on the positive plate; see Fig. 4 for an example.

Calculating the Potential Difference. The potential difference between the plates is related to the electric field $\mathbf{E}$ by Eq. 8 of Chapter 26, or

$$V_f - V_i = -\int_i^f \mathbf{E} \cdot d\mathbf{s}, \qquad (5)$$

in which the integral is to be evaluated along any path that starts on one plate and ends on the other. We will always choose a path that follows an electric field line from the positive plate to the negative plate. For this path, the vectors $\mathbf{E}$ and $d\mathbf{s}$ will always point in the same direction, so that the quantity $V_f - V_i$ will be negative. Since we are looking for V, the *absolute value* of the potential difference between the plates, we can set $V_f - V_i = -V$. Thus, we can recast Eq. 5 as

$$V = \int_+^- E \, ds \quad \text{(special case of Eq. 5)}, \qquad (6)$$

in which the $+$ and the $-$ signs remind us that our path of integration starts on the positive plate and ends on the negative plate.

We conclude by forming the ratio $C = q/V$, which will always be independent of the value chosen for q. We are now ready to apply Eqs. 4 and 6 to some particular cases.

A Parallel-Plate Capacitor. We assume, as Fig. 4 suggests, that the plates of this capacitor are so large and so close together that we can neglect the "fringing" of the electric field at the edges of the plates, taking $\mathbf{E}$ to be constant throughout the volume between the plates.

Let us draw in a Gaussian surface that includes the charge q on the positive plate, as Fig. 4 shows. From Eq. 4 we can then write

$$q = \epsilon_0 EA, \qquad (7)$$

where A is the area of the plate.

Equation 6 yields

$$V = \int_+^- E \, ds = E \int_0^d ds = Ed. \qquad (8)$$

---

* Equation 1 is a theorem whose proof, which is based on Coulomb's law, is beyond the scope of this text.

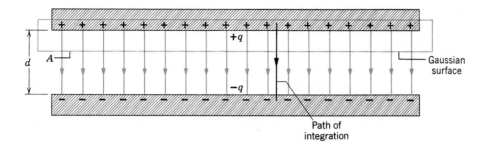

Figure 4   A charged parallel-plate capacitor. The light gray lines represent a Gaussian surface enclosing the charge on the positive plate. The heavy vertical line shows the path of integration along which we apply Eq. 6.

In Eq. 8, E is constant and can be removed from the integral; the second integral above is simply the plate separation d.

If we substitute q from Eq. 7 and V from Eq. 8 into the relation $q = CV$ (Eq. 1), we find

$$\boxed{C = \epsilon_0 \frac{A}{d}} \quad \text{(parallel-plate capacitor).} \quad (9)$$

The capacitance does indeed depend only on geometrical factors, namely, the plate area A and the plate separation d.

As an aside we point out that Eq. 9 suggests one reason why we wrote the electrostatic constant in Coulomb's law in the form $1/4\pi\epsilon_0$. If we had not done so, Eq. 9 — which is used more often in engineering practice than is Coulomb's law — would have been less simple in form. We note further that Eq. 9 permits us to express the permittivity constant ϵ_0 in units more appropriate for use in problems involving capacitors, namely,

$$\epsilon_0 = 8.85 \times 10^{-12} \text{ F/m} = 8.85 \text{ pF/m.} \quad (10)$$

We have previously expressed this constant as

$$\epsilon_0 = 8.85 \times 10^{-12} \text{ C}^2/\text{N} \cdot \text{m}^2, \quad (11)$$

units that prove useful when dealing with problems that involve Coulomb's law; see Section 23–4. The two sets of units are equivalent.

A Cylindrical Capacitor. Figure 5 shows, in cross section, a cylindrical capacitor of length L formed by two coaxial cylinders of radii a and b. We assume that $L \gg b$ so that we can neglect the "fringing" of the electric field that occurs at the ends of the cylinders.

As a Gaussian surface, we choose a cylinder of length L and radius r, closed by end caps. Equation 4 yields

$$q = \epsilon_0 EA = \epsilon_0 E(2\pi rL)$$

in which $2\pi rL$ is the area of the curved part of the Gaussian surface. Solving for E yields

$$E = \frac{q}{2\pi\epsilon_0 Lr}. \quad (12)$$

Substitution of this result into Eq. 6 yields

$$V = \int_+^- E \, ds = \frac{q}{2\pi\epsilon_0 L} \int_a^b \frac{dr}{r} = \frac{q}{2\pi\epsilon_0 L} \ln\left(\frac{b}{a}\right). \quad (13)$$

From the relation $C = q/V$, we then have

$$\boxed{C = 2\pi\epsilon_0 \frac{L}{\ln(b/a)}} \quad \text{(cylindrical capacitor).} \quad (14)$$

We see that the capacitance of the cylindrical capacitor, like that of a parallel-plate capacitor, depends only on geometrical factors, in this case L, b, and a.

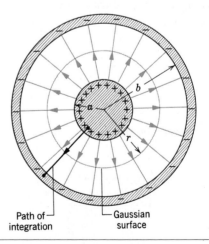

Figure 5   A long cylindrical capacitor shown in cross section. The figure shows a cylindrical Gaussian surface and also the radial path of integration along which Eq. 6 is to be applied. The figure also serves to illustrate a spherical capacitor in central cross section.

A Spherical Capacitor. Figure 4 can also serve as a central cross section of a capacitor that consists of two concentric spherical shells, of radii a and b. As a Gaussian surface we draw a sphere of radius r. Applying Eq. 4 to this surface yields

$$q = \epsilon_0 EA = \epsilon_0 E(4\pi r^2)$$

in which $4\pi r^2$ is the area of the spherical Gaussian surface. We solve this equation for E, obtaining

$$E = \frac{1}{4\pi\epsilon_0} \frac{q}{r^2}, \tag{15}$$

which we recognize as the expression for the electric field due to a uniform spherical charge distribution.

If we substitute this expression into Eq. 6, we find

$$V = \int_+^- E \, ds = \frac{q}{4\pi\epsilon_0} \int_a^b \frac{dr}{r^2} = \frac{q}{4\pi\epsilon_0}\left(\frac{1}{a} - \frac{1}{b}\right)$$
$$= \frac{q}{4\pi\epsilon_0} \frac{b-a}{ab}. \tag{16}$$

If we substitute Eq. 16 into Eq. 1 and solve for C, we find

$$\boxed{C = 4\pi\epsilon_0 \frac{ab}{b-a}} \quad \text{(spherical capacitor).} \tag{17}$$

An Isolated Sphere. We can assign a capacitance to a single isolated conductor by assuming that the "missing plate" is a conducting sphere of infinite radius. After all, the field lines that leave the surface of a charged isolated conductor must end somewhere; the walls of the room in which the conductor is housed can serve effectively as our sphere of infinite radius.

If we let $b \to \infty$ in Eq. 17 and substitute R for a, we find

$$\boxed{C = 4\pi\epsilon_0 R} \quad \text{(isolated sphere).} \tag{18}$$

Table 1 summarizes the various capacitances that we

Table 1 Some Capacitances

Capacitor	Capacitance	Equation
Parallel plate	$\epsilon_0 \dfrac{A}{d}$	9
Cylindrical	$2\pi\epsilon_0 \dfrac{L}{\ln(b/a)}$	14
Spherical	$4\pi\epsilon_0 \dfrac{ab}{b-a}$	17
Isolated sphere	$4\pi\epsilon_0 R$	18

have derived in this section. Note that every formula involves the constant ϵ_0 multiplied by a quantity that has the dimensions of a length.

Sample Problem 1 The plates of a parallel-plate capacitor are separated by a distance $d = 1.0$ mm. What must be the plate area if the capacitance is to be 1.0 F?

From Eq. 9 we have

$$A = \frac{Cd}{\epsilon_0} = \frac{(1.0 \text{ F})(1.0 \times 10^{-3} \text{ m})}{8.85 \times 10^{-12} \text{ F/m}}$$
$$= 1.1 \times 10^8 \text{ m}^2. \quad \text{(Answer)}$$

This is the area of a square more than 10 km on edge. The farad is indeed a large unit. Modern technology, however, has permitted the construction of 1- F capacitors of very modest size. These "Supercaps" are used as backup voltage sources for computers; they can maintain the computer memory for up to 30 days in case of power failure.

Sample Problem 2 The space between the conductors of a long coaxial cable, used to transmit TV signals, has an inner diameter $a = 0.15$ mm and an outer diameter $b = 2.1$ mm. What is the capacitance per unit length of this cable?

From Eq. 14 we have (see Eq. 10)

$$\frac{C}{L} = \frac{2\pi\epsilon_0}{\ln(b/a)} = \frac{(2\pi)(8.85 \text{ pF/m})}{\ln(2.1 \text{ mm}/0.15 \text{ mm})}$$
$$= 21 \text{ pF/m.} \quad \text{(Answer)}$$

Sample Problem 3 A storage capacitor on a random access memory (RAM) chip has a capacitance of 55 fF. If it is charged to 5.3 V, how many excess electrons are there on its negative plate?

We can write, using Eq. 1,

$$n = \frac{q}{e} = \frac{CV}{e} = \frac{(55 \times 10^{-15} \text{ F})(5.3 \text{ V})}{1.60 \times 10^{-19} \text{ C}}$$
$$= 1.8 \times 10^6 \text{ electrons.} \quad \text{(Answer)}$$

For electrons, this is a very small number. A speck of houshold dust, so tiny that it essentially never settles, contains about 10^{17} electrons (and the same number of protons).

Sample Problem 4 What is the capacitance of the earth, viewed as an isolated conducting sphere of radius 6370 km?

From Eq. 18 we have

$$C = 4\pi\epsilon_0 R = (4\pi)(8.85 \times 10^{-12} \text{ F/m})(6.37 \times 10^6 \text{ m})$$
$$= 7.1 \times 10^{-4} \text{ F} = 710 \ \mu\text{F.} \quad \text{(Answer)}$$

A tiny Supercap has a capacitance that is about 1400 times larger than that of the earth.

27-4 Capacitors in Series and in Parallel

We often want to know the equivalent capacitance of two or more capacitors that are connected in a particular way. By "equivalent capacitance" we mean the capacitance of a single capacitor that can be substituted for the actual combination with no change in the operation of the external circuit. We discuss two limiting cases.

Capacitors in Parallel.   Figure 6 shows three capacitors connected *in parallel,* a battery B being connected across the output terminals of the combination.

Capacitors are said to be connected in parallel when the same potential difference is applied to each.

We seek the single capacitance C_{eq} that is "equivalent" to this parallel combination.

For each capacitor we can write, from Eq. 1,

$$q_1 = C_1 V, \quad q_2 = C_2 V, \quad \text{and} \quad q_3 = C_3 V.$$

The total charge on the parallel combination is then

$$q = q_1 + q_2 + q_3 = (C_1 + C_2 + C_3)V.$$

The equivalent capacitance is then

$$C_{eq} = \frac{q}{V} = C_1 + C_2 + C_3,$$

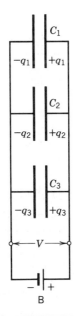

Figure 6   Three capacitors connected in parallel. The criterion for a parallel connection is that each capacitor has the same potential difference V across its plates.

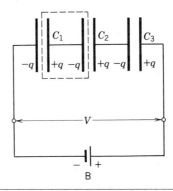

Figure 7   Three capacitors connected in series. The criterion for a series connection is that the sum of the potential differences across each capacitor must equal the potential difference that is applied to the combination.

a result that we can easily extend to any number n of capacitors, as

$$\boxed{C_{eq} = \sum_n C_n} \quad \text{(capacitors in parallel).} \quad (19)$$

Thus, to find the equivalent capacitance of a parallel combination you simply add the individual capacitances.

Capacitors in Series.   Figure 7 shows the three capacitors connected *in series,* a battery B being connected across the output terminals of the combination.

Capacitors are said to be connected in series if the sum of the potential differences across each is equal to the potential difference applied to the combination. *

We seek the single capacitance C_{eq} that is equivalent to this series combination.

We assert that, when the battery is connected, each capacitor in Fig. 7 carries the same charge q. This is true even though the three capacitors may be of different types and may have different capacitances. To understand this, note that the element of the circuit enclosed by the dashed lines in Fig. 7 is "floating;" that is, it is electrically isolated from the rest of the circuit. This element initially carries no net charge and—barring electrical breakdown of the capacitors—there is no way that

---

* We assume further that the combination of capacitors has no side branches.

charge can be moved onto it. Connecting the battery simply has the effect of producing a charge separation in this element, a charge $+q$ moving to the left-hand plate and a charge $-q$ to the right-hand plate; the net charge within the dashed line in Fig. 7 remains zero.

Application of Eq. 1 to each capacitor yields

$$V_1 = \frac{q}{C_1}, \quad V_2 = \frac{q}{C_2}, \quad \text{and} \quad V_3 = \frac{q}{C_3}.$$

The potential difference for the series combination is then

$$V = V_1 + V_2 + V_3$$
$$= q\left(\frac{1}{C_1} + \frac{1}{C_2} + \frac{1}{C_3}\right).$$

The equivalent capacitance is then

$$C_{eq} = \frac{q}{V} = \frac{1}{1/C_1 + 1/C_2 + 1/C_3},$$

which we can easily extend to any number n of capacitors as

$$\boxed{\frac{1}{C_{eq}} = \sum_n \frac{1}{C_n}} \quad \text{(capacitors in series).} \quad (20)$$

From Eq. 20 you can deduce that the equivalent series capacitance is always less than the smallest capacitance in the series of capacitors.

We note that capacitors can be connected in ways that cannot be broken down into series/parallel combinations.

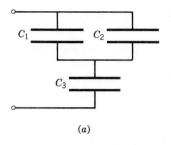

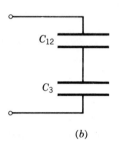

(a)

(b)

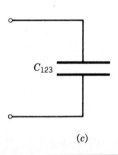

(c)

Figure 8 (a) Three capacitors. What is the equivalent capacitance of the combination? (b) C_1 and C_2, a parallel combination, are replaced by C_{12}. (c) C_{12} and C_3, a series combination, are replaced by the equivalent capacitance C_{123}.

Sample Problem 5 (a) Find the equivalent capacitance of the combination shown in Fig. 8a. Assume

$$C_1 = 12.0 \ \mu F, \quad C_2 = 5.3 \ \mu F, \quad \text{and} \quad C_3 = 4.5 \ \mu F.$$

Capacitors C_1 and C_2 are in parallel. From Eq. 19, their equivalent capacitance is

$$C_{12} = C_1 + C_2 = 12.0 \ \mu F + 5.3 \ \mu F = 17.3 \ \mu F.$$

As Fig. 8b shows, the combination C_{12} and C_3 is in series. From Eq. 20, the final equivalent combination (see Fig. 8c) is found from

$$\frac{1}{C_{123}} = \frac{1}{C_{12}} + \frac{1}{C_3} = \frac{1}{17.3 \ \mu F} + \frac{1}{4.5 \ \mu F} = 0.280 \ \mu F^{-1},$$

or

$$C_{123} = \frac{1}{0.280 \ \mu F^{-1}} = 3.57 \ \mu F. \quad \text{(Answer)}$$

(b) A potential difference $V = 12.5$ V is applied to the input terminals in Fig. 8a. What is the charge on C_1?

We treat the equivalent capacitors C_{12} and C_{123} exactly as we would real capacitors of the same capacitance. For the charge on C_{123} in Fig. 8c we then have

$$q_{123} = C_{123}V = (3.57 \ \mu F)(12.5 \ V) = 44.6 \ \mu C.$$

This same charge exists on each capacitor in the series combination of Fig. 8b. The potential difference across C_{12} in that figure is then

$$V_{12} = \frac{q_{12}}{C_{12}} = \frac{44.6 \ \mu C}{17.3 \ \mu F} = 2.58 \ V.$$

This same potential difference appears across C_1 in Fig. 8a, so that

$$q_1 = C_1 V_1 = (12 \ \mu F)(2.68 \ V)$$
$$= 31 \ \mu C. \quad \text{(Answer)}$$

27-5   Storing Energy in an Electric Field

Work must be done by an external agent to charge a capacitor. Starting with an uncharged capacitor, for example, imagine that—using "magic tweezers"—you remove electrons from one plate and transfer them one at a time to the other plate. The electric field that builds up in the space between the plates will be in a direction that tends to oppose further transfer. Thus, as charge accumulates on the capacitor plates, you will have to do increasingly larger amounts of work to transfer additional electrons. In practice, this work is not done by "magic tweezers" but by a battery, at the expense of its store of chemical energy.

We visualize the work required to charge a capacitor as stored in the form of *electrical potential energy U* in the electric field between the plates. You can recover this energy at will, by discharging the capacitor, just as you can recover the potential energy stored in a bow by releasing the bow string.

Suppose that, at a given instant, a charge q' has been transferred from one plate to the other. The potential difference V' between the plates at that moment will be q'/C. If an extra increment of charge dq' is transferred, the increment of work required will be

$$dW = V'dq' = \frac{q'}{C}\, dq'.$$

The work required to bring the total capacitor charge up to a final value q is

$$W = \int dW = \frac{1}{C} \int_0^q q'\, dq' = \frac{q^2}{2C}.$$

This work is stored as potential energy in the capacitor, so that

$$U = \frac{q^2}{2C} \quad \text{(potential energy).} \qquad (21)$$

From Eq. 1, we can also write this as

$$U = \tfrac{1}{2}CV^2 \quad \text{(potential energy).} \qquad (22)$$

Equations 21 and 22 remain true no matter what the geometry of the capacitor.

To gain some physical insight into energy storage, consider two parallel-plate capacitors C_1 and C_2 that are identical except that C_1 has twice the plate separation of C_2. C_1 will then have twice the volume between its plates and also, from Eq. 9, half the capacitance of C_2. Equa-

tion 4 tells us that, if each capacitor carries the same charge q, the electric fields between their plates will be identical. We then see from Eq. 21 that C_1, which has twice the volume between its plates, also has twice the stored potential energy. It is reasonable to reach the following conclusion from arguments like this:

The potential energy of a charged capacitor may be viewed as stored in the electric field between its plates.

Energy Density.   In a parallel-plate capacitor, neglecting fringing, the electric field has the same value for all points between the plates. Thus, the *energy density u,* that is, the potential energy per unit volume, should also be uniform. We find u by dividing the energy by the volume Ad of the space between the plates. Thus, using Eq. 22, we obtain

$$u = \frac{U}{Ad} = \frac{CV^2}{2Ad}.$$

From Eq. 9, $C = \epsilon_0 A/d$, this result becomes

$$u = \tfrac{1}{2}\epsilon_0 \left(\frac{V}{d}\right)^2.$$

But V/d is the electric field E so that

$$u = \tfrac{1}{2}\epsilon_0 E^2 \quad \text{(energy density).} \qquad (23)$$

Although we derived this result for the special case of a parallel-plate capacitor, it holds generally, whatever may be the source of the electric field. If an electric field $\mathbf{E}$ exists at any point in space, we can think of that point as the site of potential energy in amount, per unit volume, given by Eq. 23.

Sample Problem 6   A 3.55-μF capacitor C_1 is charged to a potential difference $V_0 = 6.30$ V, using a battery. The charging battery is then removed and the capacitor is connected as in Fig. 9 to an uncharged 8.95-μF capacitor C_2. Charge then starts to flow from C_1 to C_2 until an equilibrium is established, with both capacitors at the same potential difference V. (a) What is this common potential difference?

The original charge q_0 is now shared by two capacitors, or

$$q_0 = q_1 + q_2.$$

Applying the relation $q = CV$ to each term yields

$$C_1 V_0 = C_1 V + C_2 V,$$

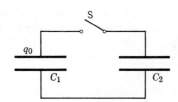

Figure 9 A potential difference V_0 is applied to C_1 and the charging battery is removed. Switch S is then closed so that the charge on C_1 is shared with C_2. What potential difference then appears across the combination?

or

$$V = V_0 \frac{C_1}{C_1 + C_2} = \frac{(6.30 \text{ V})(3.55 \text{ }\mu\text{F})}{3.55 \text{ }\mu\text{F} + 8.95 \text{ }\mu\text{F}}$$

$$= 1.79 \text{ V}. \qquad \text{(Answer)}$$

This suggests a way to measure unknown capacitances.

(b) What is the potential energy before and after the switch S in Fig. 9 is thrown?

The initial potential energy is

$$U_1 = \tfrac{1}{2}C_1 V_0^2 = (\tfrac{1}{2})(3.55 \times 10^{-6} \text{ F})(6.30 \text{ V})^2$$

$$= 7.05 \times 10^{-5} \text{ J} = 70.5 \text{ }\mu\text{J}. \qquad \text{(Answer)}$$

The final potential energy is

$$U_f = \tfrac{1}{2}C_1 V^2 + \tfrac{1}{2}C_2 V^2 = \tfrac{1}{2}(C_1 + C_2)V^2$$

$$= (\tfrac{1}{2})(3.55 \times 10^{-6} \text{ F} + 8.95 \times 10^{-6} \text{ F})(1.79 \text{ V})^2$$

$$= 2.00 \times 10^{-5} \text{ J} = 20.0 \text{ }\mu\text{J}. \qquad \text{(Answer)}$$

Thus, $U_f < U_i$, by about 72%.

This is not a violation of energy conservation. The "missing" energy appears as thermal energy in the connecting wires.*

Sample Problem 7 An isolated conducting sphere whose radius R is 6.85 cm carries a charge $q = 1.25$ nC. (a) How much potential energy is stored in the electric field of this charged conductor?

From Eqs. 21 and 18 we have

$$U = \frac{q^2}{2C} = \frac{q^2}{8\pi\epsilon_0 R} = \frac{(1.25 \times 10^{-9} \text{ C})^2}{(8\pi)(8.85 \times 10^{-12} \text{ F/m})(0.0685 \text{ m})}$$

$$= 1.03 \times 10^{-7} \text{ J} = 103 \text{ nJ}. \qquad \text{(Answer)}$$

(b) What is the energy density at the surface of the sphere?

---

* Some slight amount of energy is also radiated away. For a critical discussion, see "Two-Capacitor Problem: A More Realistic View," by R. A. Powell, *American Journal of Physics,* May 1979.

From Eq. 23

$$u = \tfrac{1}{2}\epsilon_0 E^2,$$

so that we must first find E at the surface of the sphere. This is given by

$$E = \frac{1}{4\pi\epsilon_0} \frac{q}{R^2}.$$

The energy density is then

$$u = \tfrac{1}{2}\epsilon_0 E^2 = \frac{q^2}{32\pi^2\epsilon_0 R^4}$$

$$= \frac{(1.25 \times 10^{-9} \text{ C})^2}{(32\pi^2)(8.85 \times 10^{-12} \text{ C}^2/\text{N}\cdot\text{m}^2)(0.0685 \text{ m})^4}$$

$$= 2.54 \times 10^{-5} \text{ J/m}^3 = 25.4 \text{ }\mu\text{J/m}^3. \qquad \text{(Answer)}$$

(c) What is the radius R_0 of a spherical surface such that one-half of the stored potential energy lies within it?

The problem requires that

$$\int_R^{R_0} dU = \tfrac{1}{2} \int_R^{\infty} dU. \qquad (24)$$

The energy that lies in a spherical shell between radii r and $r + dr$ is

$$dU = (u)(4\pi r^2)(dr),$$

where $(4\pi r^2)(dr)$ is the volume of the spherical shell. The energy density is

$$u = \tfrac{1}{2}\epsilon_0 E^2.$$

At any radius r, the electric field is

$$E = \frac{1}{4\pi\epsilon_0} \frac{q}{r^2}.$$

Therefore,

$$dU = \frac{q^2}{8\pi\epsilon_0} \frac{dr}{r^2}.$$

From Eq. 24 above,

$$\int_R^{R_0} \frac{dr}{r^2} = \tfrac{1}{2} \int_R^{\infty} \frac{dr}{r^2},$$

which becomes

$$\frac{1}{R} - \frac{1}{R_0} = \frac{1}{2R}.$$

Solving for R_0 yields

$$R_0 = 2R = (2)(6.85 \text{ cm}) = 13.7 \text{ cm}. \qquad \text{(Answer)}$$

Thus, half the stored energy is contained within a sphere whose radius is twice the radius of the conducting sphere.

27-6 Capacitor with a Dielectric

If you fill the space between the plates of a capacitor with an insulating material such as mineral oil or plastic, what happens to the capacitance? In 1837, Michael Faraday —to whom the whole concept of capacitance is largely due and for whom the SI unit of capacitance is named— first looked into the matter. Using simple apparatus much like that shown in Fig. 10, he found that the capacitance *increased* by a numerical factor κ, which he called the *dielectric constant* of the introduced material. Table 2 shows some dielectric materials and their dielectric constants. The dielectric constant of a vacuum is unity by definition. Because air is mostly empty space, the measured dielectric constant is only slightly greater than unity; the difference is usually insignificant.

Another effect of the dielectric is to limit the potential difference that can be applied between the plates to a certain value V_{max}. If this value is substantially exceeded, the dielectric material will break down and form a conducting path between the plates. Every dielectric material has a characteristic *dielectric strength,* which is the maximum value of the electric field that it can tolerate without breakdown. A few such values are listed in Table 2.

As Table 1 suggests, the capacitance of any capacitor can be written in the form

$$C = \epsilon_0 L \qquad (25)$$

Figure 10   The simple electrostatic apparatus used by Faraday.

in which L has the dimensions of a length. Table 1 shows, for example, that $L = A/d$ for a parallel-plate capacitor and that $L = 4\pi ab/(b - a)$ for a spherical capacitor. Faraday's discovery was that, with a dielectric *completely* filling the space between the plates, Eq. 25 becomes

$$C = \kappa \epsilon_0 L = \kappa C_{air}, \qquad (26)$$

where C_{air} is the value of the capacitance with air between the plates.

Figure 11 provides some insight into Faraday's experiments. In Fig. 11a the battery ensures that the potential difference V between the plates will remain constant. As a dielectric slab is inserted, we see that the charge q increases by a factor of κ, the additional charge being delivered to the capacitor plates by the battery. In Fig. 11b there is no battery and therefore the charge q must remain constant as the slab is inserted; we see that the potential difference V between the plates decreases by a factor of κ. Both of these observations are consistent— through the relation $q = CV$—with the increase in capacitance caused by the dielectric.

Comparison of Eqs. 25 and 26 suggests that the effect of a dielectric can be summed up in more general terms:

In a region completely filled by a dielectric, all electrostatic equations containing the permittivity constant ϵ_0 are to be modified by replacing that constant by $\kappa \epsilon_0$.

Thus, Coulomb's law (see Eq. 1 of Chapter 25) becomes,

Table 2   Some Properties of Dielectrics[a]

Material	Dielectric Constant κ	Dielectric Strength (kV/mm)
Air (1 atm)	1.00054	3
Polystyrene	2.6	24
Paper	3.5	16
Transformer oil	4.5	
Pyrex	4.7	14
Ruby mica	5.4	
Porcelain	6.5	
Silicon	12	
Germanium	16	
Ethanol	25	
Water (20°C)	80.4	
Water (25°C)	78.5	
Titania ceramic	130	
Strontium titanate	310	8
For a vacuum, $\kappa = $ unity		

[a] Measured at room temperature.

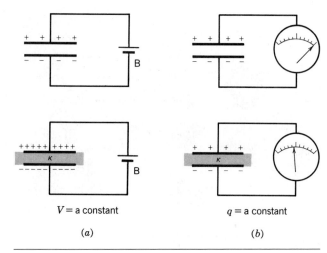

V = a constant

q = a constant

(a)   (b)

Figure 11 (a) If the potential difference between the plates of a capacitor is maintained, as by battery B, the effect of a dielectric is to increase the charge on the plates. (b) If the charge on the capacitor plates is maintained, as in this case, the effect of a dielectric is to reduce the potential difference between the plates.

for a point charge inside a dielectric,

$$E = \frac{1}{4\pi\kappa\epsilon_0} \frac{q}{r^2},$$ (27)

and the expression for the electric field just outside an isolated conductor immersed in a dielectric (see Eq. 12 of Chapter 25) becomes

$$E = \frac{\sigma}{\kappa\epsilon_0}.$$ (28)

Both of these expressions show that, for a fixed distribution of charges, the effect of a dielectric is to weaken the electric field that would otherwise be present.

Sample Problem 8 A parallel-plate capacitor whose capacitance C is 13.5 pF has a potential difference V = 12.5 V between its plates. The charging battery is now disconnected and a porcelain slab (κ = 6.5) is slipped between the plates as in Fig. 11b. What is the potential energy of the unit, both before and after the slab is introduced?

The initial potential energy is given by Eq. 22 as

$$U_i = \tfrac{1}{2}CV^2 = (\tfrac{1}{2})(13.5 \times 10^{-12} \text{ F})(12.5 \text{ V})^2$$

$$= 1.055 \times 10^{-9} \text{ J} = 1055 \text{ pJ}.$$ (Answer)

We can also write the initial potential energy, from Eq. 21, in

the form

$$U_i = \frac{q^2}{2C}.$$

We choose to do so because, from the conditions of the problem statement, q (but not V) remains constant as the slab is introduced. After the slab is in place, C increases to κC so that

$$U_f = \frac{q^2}{2\kappa C} = \left(\frac{1}{\kappa}\right) U_i = \frac{1055 \text{ pJ}}{6.5}$$

$$= 162 \text{ pJ}.$$ (Answer)

The energy after the slab is introduced is smaller by a factor of $1/\kappa$.

The "missing" energy, in principle, would be apparent to the person who introduced the slab. The capacitor would exert a tiny tug on the slab and would do work on it, in amount

$$W = U_i - U_f = (1055 - 162) \text{ pJ} = 893 \text{ pJ}.$$

If the slab were introduced with no restraint and if there were no friction, the slab would oscillate back and forth between the plates with a (constant) mechanical energy of 893 pJ, the system energy shuttling back and forth between kinetic energy of the moving slab and potential energy stored in the electric field.

27–7 Dielectrics: An Atomic View

What happens, in atomic and molecular terms, when we put a dielectric in an external electric field? There are two possibilities.

Polar Dielectrics. The molecules of some dielectrics, like water, have permanent electric dipole moments. In such materials (called *polar dielectrics*), the electric dipoles tend to line up with an external electric field as in Fig. 12. Because the molecules are in constant thermal agitation, the alignment will not be complete but will increase as the strength of the applied field is increased or as the temperature is decreased.

Nonpolar Dielectrics. Whether or not molecules have permanent electric dipole moments, they acquire dipole moments by induction when placed in an external electric field. In Section 26–6 (see Fig. 10 of that chapter), we saw that this external field tends to "stretch" the molecule, separating slightly the centers of negative and positive charge. Figure 13a shows a dielectric slab with no external electric field applied. In Fig. 13b, an external field $\mathbf{E}_0$ is applied, its effect being to separate slightly the centers of the positive and negative distributions. The

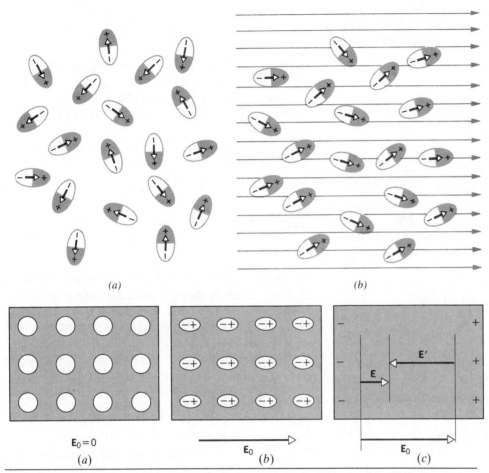

Figure 12   (*a*) Molecules with a permanent electric dipole moment, showing their random orientation in the absence of an external electric field. (*b*) An electric field is applied, producing a partial alignment of the dipoles. Thermal agitation prevents complete alignment.

Figure 13   (*a*) A dielectric slab, suggesting the neutral atoms within the slab. (*b*) An external electric field is applied, stretching out the atoms and separating the centers of positive and negative charge. (*c*) The overall effect is the production of surface charges, as shown. These charges set up a field $\mathbf{E}'$, which opposes the applied external field $\mathbf{E}_0$.

overall effect is a buildup of positive charge on the right face of the slab and of negative charge on the left face. The slab as a whole remains electrically neutral and — within the slab — there is no excess charge in any volume element.

Figure 13*c* shows that the induced surface charges appear in such a way that the electric field $\mathbf{E}'$ set up by them opposes the external electric field $\mathbf{E}_0$. The resultant field $\mathbf{E}$ inside the dielectric, which is the vector sum of $\mathbf{E}_0$ and $\mathbf{E}'$, points in the same direction as $\mathbf{E}_0$ but is smaller in magnitude. Thus, the effect of the dielectric is to weaken the applied field inside the dielectric.

This induced surface charge is the explanation of the most elementary fact of electrostatics, namely, that a charged rod will attract bits of uncharged nonconducting material such as paper. In Fig. 14 we see how surface charges are induced on a bit of paper placed near a charged rod. The attraction of the induced negative charges to the rod exceeds the repulsion of the more distant induced positive charge so that the net effect is an

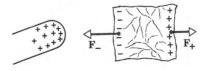

Figure 14   A charged rod attracts an uncharged bit of paper because unbalanced forces act on the induced surface charges.

attraction. If the bit of paper were placed in a *uniform* electric field, induced surface charges would appear but the forces on them would be equal and opposite so that there would be no net attraction.

27–8 Dielectrics and Gauss' Law (Optional)

In our presentation of Gauss' law in Chapter 25, we assumed that the charges existed in a vacuum. Here we shall see how to modify and generalize that law if dielectric materials, such as those listed in Table 2, are present. Figure 15 shows a parallel-plate capacitor both with and without a dielectric. We assume that the charge q on the plates is the same in each case. For the case of Fig. 15a, Gauss' law yields

$$\epsilon_0 \oint \mathbf{E} \cdot d\mathbf{A} = \epsilon_0 E_0 A = q$$

or

$$E_0 = \frac{q}{\epsilon_0 A}. \tag{29}$$

If a dielectric is present, as in Fig. 15b, Gauss' law yields

$$\epsilon_0 \oint \mathbf{E} \cdot d\mathbf{A} = \epsilon_0 E A = q - q' \tag{30}$$

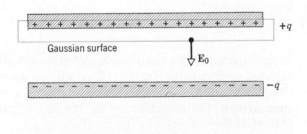

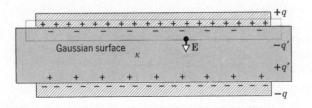

Figure 15 (*a*) A parallel-plate capacitor. (*b*) The same, with a dielectric slab inserted. The charge q on the plates is assumed to be the same in each case.

or

$$E = \frac{q - q'}{\epsilon_0 A}, \tag{31}$$

in which $-q'$, the induced surface charge on the dielectric slab, must be distinguished from q, the *free charge* on the plates. Both of these charges lie within the Gaussian surface of Fig. 15b, the *net* charge within that surface being $q - q'$.

The effect of the dielectric is to weaken the field E_0 by a factor κ, or

$$E = \frac{E_0}{\kappa} = \frac{q}{\kappa \epsilon_0 A}. \tag{32}$$

Comparison of Eqs. 31 and 32 shows that

$$q - q' = \frac{q}{\kappa}, \tag{33}$$

is the *net* charge inside the Gaussian surface. Equation 33 shows correctly that the absolute value q' of the induced surface charge is less than that of the free charge q and is equal to zero if no dielectric is present, that is, if $\kappa = 1$ in Eq. 33.

If we substitute $q - q'$ from Eq. 33 into Eq. 30, we see that we can write Gauss' law in the form

$$\boxed{\epsilon_0 \oint \kappa \mathbf{E} \cdot d\mathbf{A} = q}$$

(Gauss' law with dielectric).   (34)

This important equation, although derived for a parallel-plate capacitor, is true generally and is the most general form in which Gauss' law can be written. Note the following:

1. The flux integral now deals with $\kappa \mathbf{E}$, not with $\mathbf{E}$.*

2. The charge q enclosed by the Gaussian surface is taken to be the *free charge only*. Induced surface charge is deliberately ignored on the right side of this equation, having been taken fully into account by introducing the dielectric constant κ on the left side.

3. Equation 34 differs from Eq. 9 of Chapter 25, our original statement of Gauss' law, only in that ϵ_0 in the latter equation has been replaced by $\kappa \epsilon_0$ and κ has been taken inside the integral to allow for cases in which κ is not constant over the entire Gaussian surface. This is in full accord with our statement on

---

* The vector $\epsilon_0 \kappa \mathbf{E}$ is called the *electric displacement* **D**, which allows us to write Eq. 34 in the simplified form $\oint \mathbf{D} \cdot d\mathbf{A} = q$.

page 627 of the consequences for electrostatics of filling a region of space with a dielectric.

Sample Problem 9 Figure 16 shows a parallel-plate capacitor of plate area A and plate separation d. A potential difference V is applied between the plates. The battery is then disconnected and a dielectric slab of thickness b and dielectric constant κ is placed between the plates as shown. Assume

$$A = 115 \text{ cm}^2, \quad d = 1.24 \text{ cm},$$
$$b = 0.78 \text{ cm}, \quad \kappa = 2.61,$$
$$V_0 = 85.5 \text{ V}.$$

(a) What is the capacitance C_0 before the slab is inserted? From Eq. 9 we have

$$C_0 = \frac{\epsilon_0 A}{d} = \frac{(8.85 \times 10^{-12} \text{ F/m})(115 \times 10^{-4} \text{ m}^2)}{1.24 \times 10^{-2} \text{ m}}$$
$$= 8.21 \times 10^{-12} \text{ F} = 8.21 \text{ pF}. \qquad \text{(Answer)}$$

(b) What free charge appears on the plates? From Eq. 1,

$$q = C_0 V_0 = (8.21 \times 10^{-12} \text{ F})(85.5 \text{ V})$$
$$= 7.02 \times 10^{-10} \text{ C} = 702 \text{ pC}. \qquad \text{(Answer)}$$

Because the charging battery was disconnected before the slab was introduced, the free charge remains unchanged as the slab is put into place.

(c) What is the electric field E_0 in the gaps between the plates and the dielectric slab?

Let us apply Gauss' law in the form given in Eq. 34 to the Gaussian surface in Fig. 16 that includes the free charge on the upper capacitor plate. We have

$$\epsilon_0 \oint \kappa \mathbf{E} \cdot d\mathbf{A} = (1)\epsilon_0 E_0 A = q$$

or

$$E_0 = \frac{q}{\epsilon_0 A} = \frac{7.02 \times 10^{-10} \text{ C}}{(8.85 \times 10^{-12} \text{ F/m})(115 \times 10^{-4} \text{ m}^2)}$$
$$= 6900 \text{ V/m} = 6.90 \text{ kV/m}. \qquad \text{(Answer)}$$

Note that we put $\kappa = 1$ in this equation because the Gaussian surface over which Gauss' law was integrated does not pass through any dielectric. Note too that the value of E_0 remains unchanged as the slab is introduced. It depends only on the free charge on the plates.

Table 3   Sample Problem 9: A Summary of Results

Quantity		No Slab	Partial Slab	Full Slab
C	pF	8.21	13.4	21.4
q	pC	702	702	702
q'	pC	—	433	433
V	V	85.5	52.3	32.8
E_0	kV/m	6.90	6.90	6.90[a]
E	kV/m	—	2.64	2.64

[a] Assumes that a very narrow gap is present.

(d) Calculate the electric field E in the dielectric slab.

Again we apply Eq. 34, this time to the lower Gaussian surface in Fig. 16. We find

$$\epsilon_0 \oint \kappa \mathbf{E} \cdot d\mathbf{A} = \kappa \epsilon_0 E A = q$$

or

$$E = \frac{q}{\kappa \epsilon_0 A} = \frac{E_0}{\kappa} = \frac{6.90 \text{ kV/m}}{2.61}$$
$$= 2.64 \text{ kV/m}. \qquad \text{(Answer)}$$

(e) What is the potential difference between the plates after the slab has been introduced?

Equation 6 yields

$$V = \int_{+}^{-} E \, ds = E_0 (d - b) + E b$$
$$= (6900 \text{ V/m})(0.0124 \text{ m} - 0.0078 \text{ m})$$
$$\quad + (2640 \text{ V/m})(0.0078 \text{ m})$$
$$= 52.3 \text{ V}. \qquad \text{(Answer)}$$

This contrasts with the original applied potential difference of 85.5 V.

(f) What is the capacitance with the slab in place? From Eq. 1,

$$C = \frac{q}{V} = \frac{7.02 \times 10^{-10} \text{ C}}{52.3 \text{ V}}$$
$$= 1.34 \times 10^{-11} \text{ F} = 13.4 \text{ pF}. \qquad \text{(Answer)}$$

Table 3 summarizes the results of this sample problem and also includes the results that would have followed if the dielectric slab had completely filled the space between the plates.

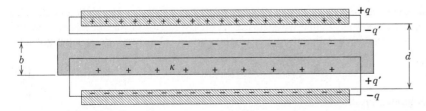

Figure 16   A parallel-plate capacitor containing a dielectric slab that only partially fills the space between the plates.

REVIEW AND SUMMARY

A Capacitor; Capacitance

A *capacitor* consists of two isolated conductors (plates) carrying equal and opposite charges $+q$ and $-q$. The capacitance C is defined from

$$q = CV \qquad [1]$$

where V is the potential difference between the plates. The SI unit of capacitance is the farad (1 farad = 1 F = 1 coulomb/volt). See Sample Problem 3.

Evaluating Capacitance

We generally evaluate the capacitance of a capacitor by (1) assuming charge q to have been placed on the plates, (2) finding the electric field $\mathbf{E}$ due to this charge, (3) evaluating the potential difference V, and (4) calculating C from Eq. 1. Section 27–3 shows several important examples in detail. The results are summarized here and in Table 1.

A *parallel-plate capacitor* with plane parallel plates of area A and spacing d has capacitance

Parallel-Plate Capacitor

$$C = \frac{\epsilon_0 A}{d}; \quad \text{(parallel-plate capacitor)}. \qquad [9]$$

See Sample Problem 1. A *cylindrical capacitor* consists of two long coaxial cylinders of length L. The inner and outer radii are a and b, and the capacitance is

Cylindrical Capacitor

$$C = 2\pi\epsilon_0 \frac{L}{\ln(b/a)} \quad \text{(cylindrical capacitor)}. \qquad [14]$$

See Sample Problem 2. A *spherical capacitor* with concentric spherical plates of inner and outer radii a and b has capacitance

Spherical Capacitor

$$C = 4\pi\epsilon_0 \frac{ab}{b-a} \quad \text{(spherical capacitor)}. \qquad [17]$$

If $b \to \infty$ and $a = R$, as for the earth in Sample Problem 4, we have the capacitance of an isolated sphere:

Isolated Sphere

$$C = 4\pi\epsilon_0 R \quad \text{(isolated sphere)}. \qquad [18]$$

The *equivalent capacitances* C_{eq} of combinations of individual capacitors arranged in *series* and in *parallel* are

Capacitors in Series and in Parallel

$$C_{eq} = \sum_n C_n \quad \text{(capacitors in parallel)} \qquad [19]$$

and

$$\frac{1}{C_{eq}} = \sum_n \frac{1}{C_n} \quad \text{(capacitors in series)}. \qquad [20]$$

Sample Problem 5 illustrates how these can be combined to calculate the capacitance of more complicated series-parallel combinations.

The *potential energy* U of a charged capacitor, given by

Potential Energy

$$U = \frac{q^2}{2C} = \frac{1}{2}CV^2 \quad \text{(potential energy)}, \qquad [21,22]$$

is the work required to charge it. (See Sample Problem 6.) This energy is conveniently thought of as stored in the electric field $\mathbf{E}$ associated with the capacitor. By extension we can associate stored energy with an electric field generally, no matter what its origin. The *energy density u* is given by

Energy Density

$$u = \frac{1}{2}\epsilon_0 E^2 \quad \text{(energy density)} \qquad [23]$$

in which it is assumed that the field $\mathbf{E}$ exists in a vacuum. See Sample Problem 7.

Capacitance with a
Dielectric

If the space between the plates of a capacitor is completely filled with a dielectric material, the capacitance C is increased by a factor κ, called the *dielectric constant,* which is characteristic of the material (see Table 2). In a region completely filled by a dielectric, all electrostatic equations containing ϵ_0 are to be modified by replacing ϵ_0 by $\kappa\epsilon_0$. Sample Problem 8 shows some examples.

The effects of adding a dielectric can be understood physically in terms of the action of an electric field on the permanent or induced electric dipoles in the dielectric slab. As Fig. 14 shows, the result is the formation of induced surface charges, which results in a weakening of the field within the body of the dielectric.

When a dielectric is present, Gauss' law may be generalized to

Gauss's Law with a
Dielectric

$$\epsilon_0 \oint \kappa \mathbf{E} \cdot d\mathbf{A} = q \quad \text{(Gauss' law with dielectric).} \qquad [18]$$

Here q includes only the free charge, the induced surface charge being accounted for by including κ inside the integral. Sample Problem 9 applies this relation in an important special case.

QUESTIONS

1. A capacitor is connected across a battery. (*a*) Why does each plate receive a charge of exactly the same magnitude? (*b*) Is this true even if the plates are of different sizes?

2. Can there be a potential difference between two adjacent conductors that carry the same amount of positive charge?

3. A sheet of aluminum foil of negligible thickness is placed between the plates of a capacitor as in Fig. 17. What effect has it on the capacitance if (*a*) the foil is electrically insulated and (*b*) the foil is connected to the upper plate?

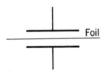

Figure 17 Question 3.

4. You are given two capacitors, C_1 and C_2, in which $C_1 > C_2$. How could things be arranged so that C_2 could hold more charge than C_1?

5. In Fig. 2 suppose that a and b are nonconductors, the charge being distributed arbitrarily over their surfaces. (*a*) Would Eq. 1 ($q = CV$) hold, with C independent of the charge arrangements? (*b*) How would you define V in this case?

6. You are given a parallel-plate capacitor with square plates of area A and separation d, in a vacuum. What is the qualitative effect of each of the following on its capacitance? (*a*) Reduce d. (*b*) Put a slab of copper between the plates, touching neither plate. (*c*) Double the area of both plates. (*d*) Double the area of one plate only. (*e*) Slide the plates parallel to each other so that the area of overlap is, say, 50%. (*f*) Double the potential difference between the plates. (*g*) Tilt one plate so that the separation remains d at one end but is $\frac{1}{2}d$ at the other.

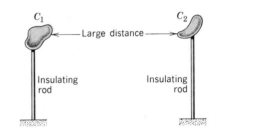

Figure 18 Question 7.

7. You have two isolated conductors, each of which has a certain capacitance; see Fig. 18. If you join these conductors by a fine wire, how do you calculate the capacitance of the combination? In joining them with the wire, have you connected them in parallel or in series?

8. Capacitors often are stored with a wire connected across their terminals. Why is this done?

9. If you were not to neglect the fringing of the electric field lines in a parallel-plate capacitor, would you calculate a higher or a lower capacitance?

10. Two circular copper disks are facing each other a certain distance apart. In what ways could you reduce the capacitance of this combination?

11. Would you expect the dielectric constant of a material to vary with temperature? If so, how? Does whether or not the molecules have permanent dipole moments matter here?

12. Discuss similarities and differences when (*a*) a dielectric slab and (*b*) a conducting slab are inserted between the plates of a parallel-plate capacitor. Assume the slab thicknesses to be one-half the plate separation.

13. An oil-filled, parallel-plate capacitor has been designed to have a capacitance C and to operate safely at or below a certain maximum potential difference V_m without arcing over. However, the designer did not do a good job and the capacitor occasionally arcs over. What can be done to redesign the capacitor, keeping C and V_m unchanged and using the same dielectric?

14. A dielectric object in a nonuniform electric field experiences a net force. Why is there no net force if the field is uniform?

15. A stream of tap water can be deflected if a charged rod is brought close to the stream. Explain carefully how this happens.

16. Water has a high dielectric constant. Why isn't it used ordinarily as a dielectric material in capacitors?

17. A dielectric slab is inserted in one end of a charged parallel-plate capacitor (the plates being horizontal and the charging battery having been disconnected) and then released. Describe what happens. Neglect friction.

18. A parallel-plate capacitor is charged by using a battery, which is then disconnected. A dielectric slab is then slipped between the plates. Describe qualitatively what happens to the charge, the capacitance, the potential difference, the electric field, and the stored energy.

19. While a parallel-plate capacitor remains connected to a battery, a dielectric slab is slipped between the plates. Describe qualitatively what happens to the charge, the capacitance, the potential difference, the electric field, and the stored energy. Is work required to insert the slab?

20. Two identical capacitors are connected as shown in Fig. 19. A dielectric slab is slipped between the plates of one capacitor, the battery remaining connected. Describe qualitatively what happens to the charge, the capacitance, the potential difference, the electric field, and the stored energy for each capacitor.

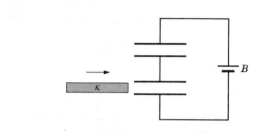

Figure 19 Question 20.

EXERCISES AND PROBLEMS

Section 27–2 Capacitance
1E. An electrometer is a device used to measure static charge. Unknown charge is placed on the plates of a capacitor and the potential difference is measured. What minimum charge can be measured by an electrometer with a capacitance of 50 pF and a voltage sensitivity of 0.15 V?

2E. The two metal objects in Fig. 20 have net charges of $+70$ pC and -70 pC, and this results in a 20-V potential difference between them. (a) What is the capacitance of the system? (b) If the charges are changed to $+200$ pC and -200 pC, what does the capacitance become? (c) What does the potential difference become?

Figure 20 Exercise 2.

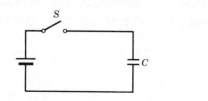

Figure 21 Exercise 3.

3E. The capacitor in Fig. 21 has a capacitance of 25 μF and is initially uncharged. The battery supplies 120 V. After switch S has been closed for a long time, how much charge will have passed through the battery?

Section 27–3 Calculating the Capacitance
4E. If we solve Eq. 9 for ϵ_0, we see that its SI units are farad/meter. Show that these units are equivalent to those obtained earlier for ϵ_0, namely coulomb²/newton·meter².

5E. A parallel-plate capacitor has circular plates of 8.2-cm radius and 1.3-mm separation. (a) Calculate the capacitance. (b) What charge will appear on the plates if a potential difference of 120 V is applied?

6E. You have two flat metal plates, each of area 1.0 m², with which to construct a parallel-plate capacitor. If its capacitance is to be 1.0 F, what must be the separation between the plates? Could this capacitor actually be constructed?

7E. The plate and cathode of a vacuum tube diode are in the form of two concentric cylinders with the cathode as the central cylinder. The cathode diameter is 1.6 mm and the plate diameter 18 mm with both elements having a length of 2.4 cm. Calculate the capacitance of the diode.

8E. The plates of a spherical capacitor have radii 38 mm and 40 mm. (a) Calculate the capacitance. (b) What must be the plate area of a parallel-plate capacitor with the same plate separation and capacitance?

9E. After you walk over a carpet on a dry day, your hand comes close to a metal doorknob and a 5-mm spark results. Such a spark means that there must have been a potential difference of possibly 15 kV between you and the doorknob. Assuming this potential difference, how much charge did you accumulate in walking over the carpet? For this extremely rough calculation, assume that your body can be represented by a uniformly charged conducting sphere 25 cm in radius and isolated from its surroundings.

10E. Two sheets of aluminum foil have a separation of 1.0 mm, a capacitance of 10 pF, and are charged to 12 V. (*a*) Calculate the plate area. The separation is now decreased by 0.10 mm with the charge held constant. (*b*) What is the new capacitance? (*c*) By how much does the potential difference change? Explain how a microphone might be constructed using this principle.

11P. Show that the plates of a parallel-plate capacitor attract each other with a force given by

$$F = \frac{q^2}{2\epsilon_0 A}.$$

Prove this by calculating the work necessary to increase the plate separation from x to $x + dx$, the charge q remaining constant.

12P. Using the result of Problem 11 show that the force per unit area (the *electrostatic stress*) acting on either capacitor plate is given by $\frac{1}{2}\epsilon_0 E^2$. Actually, this result is true in general, for a conductor of *any* shape with an electric field **E** at its surface.

13P. In Section 3 the capacitance of a cylindrical capacitor was calculated. Using the approximation (see Appendix G) that $\ln(1 + x) \approx x$ when $x \ll 1$, show that the capacitance approaches that of a parallel plate capacitor when the spacing between the two cylinders is small.

14P. Suppose that the two spherical shells of a spherical capacitor have their radii approximately equal. Under these conditions the device approximates a parallel-plate capacitor with $b - a = d$. Show that Eq. 17 does indeed reduce to Eq. 9 in this case.

15P. A spherical drop of mercury of radius R has a capacitance given by $C = 4\pi\epsilon_0 R$ (see Section 3). If two such drops combine to form a single larger drop, what is its capacitance?

16P. A capacitor is to be designed to operate, with constant capacitance, in an environment of fluctuating temperature. As shown in Fig. 22, the capacitor is a parallel-plate type with plastic "spacers" to keep the plates aligned. (*a*) Show that the rate of change of capacitance C with temperature T is given by

$$\frac{dC}{dT} = C\left(\frac{1}{A}\frac{dA}{dT} - \frac{1}{x}\frac{dx}{dT}\right),$$

where A is the plate area and x the plate separation. (*b*) If the plates are aluminum, what should be the coefficient of thermal

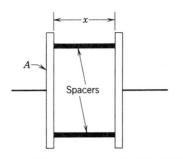

Figure 22 Problem 16.

expansion of the spacers in order that the capacitance not vary with temperature? (Ignore the effect of the spacers on the capacitance.)

17P*. A soap bubble of radius R_0 is slowly given a charge q. Because of mutual repulsion of the surface charges, the radius increases slightly to R. The air pressure inside the bubble drops, because of the expansion, to $p(V_0/V)$ where p is the atmospheric pressure, V_0 is the initial volume, and V the final volume. Show that

$$q^2 = 32\pi^2\epsilon_0 pR(R^3 - R_0^3).$$

(*Hint:* Consider the forces acting on a small area of the charged bubble. These are due to (*i*) gas pressure, (*ii*) atmospheric pressure, (*iii*) electrostatic stress; see Problem 12.)

Section 27–4 Capacitors in Series and Parallel

18E. How many 1.0-μF capacitors must be connected in parallel to store a charge of 1.0 C with a potential of 110 V across the capacitors?

19E. In Fig. 23 find the equivalent capacitance of the combination. Assume that $C_1 = 10\ \mu$F, $C_2 = 5.0\ \mu$F, and $C_3 = 4.0\ \mu$F.

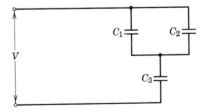

Figure 23 Exercise 19; Problems 27 and 48.

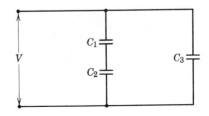

Figure 24 Exercise 20 and Problem 28.

20E. In Fig. 24 find the equivalent capacitance of the combination. Assume that $C_1 = 10 \mu F$, $C_2 = 5.0 \mu F$, and $C_3 = 4.0 \mu F$.

21E. Each of the uncharged capacitors in Fig. 25 has a capacitance of 25 μF. A potential difference of 4200 V is established when the switch is closed. How many coulombs of charge then pass through the meter A?

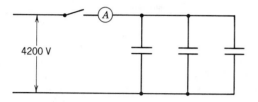

Figure 25 Exercise 21.

22E. A 6.0-μF capacitor is connected in series with a 4.0-μF capacitor; a potential difference of 200 V is applied across the pair. (a) Calculate the equivalent capacitance. (b) What is the charge on each capacitor? (c) What is the potential difference across each capacitor?

23E. Work the previous problem for the same two capacitors connected in parallel.

24P. (a) Three capacitors are connected in parallel. Each has plate area A and plate spacing d. What must be the spacing of a single capacitor of plate area A if its capacitance equals that of the parallel combination? (b) What must be the spacing if the three capacitors are connected in series?

25P. A potential difference of 300 V is applied to a 2.0-μF capacitor and an 8.0-μF capacitor connected in series. (a) What are the charge and the potential difference for each capacitor? (b) The charged capacitors are disconnected from each other and from the battery. They are then reconnected with their positive plates together and their negative plates together, no external voltage being applied. What are the charge and the potential difference for each? (c) The charged capacitors in (a) are reconnected with plates of *opposite* sign together. What are the steady-state charge and potential difference for each?

26P. In Fig. 26 a variable air capacitor of the type used in tuning radios is shown. Alternate plates are connected to-

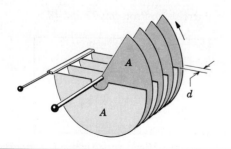

Figure 26 Problem 26.

gether, one group being fixed in position, the other group being capable of rotation. Consider a pile of n plates of alternate polarity, each having an area A and separated from adjacent plates by a distance d. Show that this capacitor has a maximum capacitance of

$$C = \frac{(n-1)\epsilon_0 A}{d}.$$

27P. In Fig. 23 suppose that capacitor C_3 breaks down electrically, becoming equivalent to a conducting path. What *changes* in (a) the charge and (b) the potential difference occur for capacitor C_1? Assume that $V = 100$ V.

28P. In Fig. 24 find (a) the charge, (b) the potential difference, and (c) the stored energy for each capacitor. Assume the numerical values of Problem 20, with $V = 100$ V.

29P. You have several 2.0-μF capacitors, each capable of withstanding 200 V without breakdown. How would you assemble a combination having an equivalent capacitance of (a) 0.40 μF or of (b) 1.2 μF, each capable of withstanding 1000 V?

30P. Figure 27 shows two capacitors in series, the rigid center section of length b being movable vertically. Show that the equivalent capacitance of the series combination is independent of the position of the center section and is given by

$$C = \frac{\epsilon_0 A}{a - b}.$$

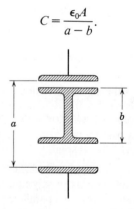

Figure 27 Problem 30.

31P. A 100-pF capacitor is charged to a potential difference of 50 V, the charging battery then being disconnected. The capacitor is then connected in parallel with a second (initially uncharged) capacitor. If the measured potential difference drops to 35 V, what is the capacitance of this second capacitor?

32P. In Figure 28, capacitors $C_1 = 1.0 \mu F$ and $C_2 = 3.0 \mu F$ are each charged to a potential $V = 100$ V but with opposite polarity, so that points a and c are on the side of the respective positive plates of C_1 and C_2, and points b and d are on the side of the respective negative plates. Switches S_1 and S_2 are now closed. (a) What is the potential difference between points e and f? (b) What is the charge on C_1? (c) What is the charge on C_2?

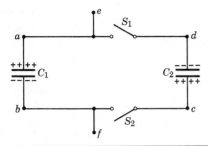

Figure 28 Problem 32.

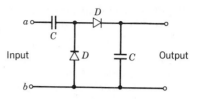

Figure 31 Problem 35.

33P. When switch S is thrown to the left in Fig. 29, the plates of the capacitor C_1 acquire a potential difference V_0. C_2 and C_3 are initially uncharged. The switch is now thrown to the right. What are the final charges q_1, q_2, q_3 on the corresponding capacitors?

erty that positive charge flows through it only in the direction of the arrow and negative charge flows through it only in the opposite direction.)

Section 27–5 Storing Energy in an Electric Field

36E. How much energy is stored in one cubic meter of air due to the "fair weather" electric field of strength 150 V/m?

37E. Attempts to build a controlled thermonuclear fusion reactor which, if successful, could provide the world with a vast supply of energy from heavy hydrogen in sea water, usually involve huge electric currents for short periods of time in magnetic field windings. For example, ZT-40 at Los Alamos Scientific Laboratory has rooms full of capacitors. One of the capacitor banks provides 61 mF at 10 kV. Calculate the stored energy (a) in joules, and (b) in kW·h.

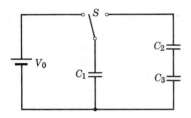

Figure 29 Problem 33.

38E. What capacitance is required to store an energy of 10 kW·h at a potential difference of 1000 V?

39E. A parallel-plate air capacitor has a capacitance of 130 pF. (a) What is the stored energy if the applied potential difference is 56 V? (b) Can you calculate the energy density for points between the plates?

34P. In Fig. 30 the battery B supplies 12 V. (a) Find the charge on each capacitor when switch S_1 is closed and (b) when (later) switch S_2 is also closed. Take $C_1 = 1.0\ \mu F$, $C_2 = 2.0\ \mu F$, $C_3 = 3.0\ \mu F$, and $C_4 = 4.0\ \mu F$.

40E. A parallel-plate air capacitor having area 40 cm^2 and spacing of 1.0 mm is charged to a potential difference of 600 V. Find (a) the capacitance, (b) the magnitude of the charge on each plate, (c) the stored energy, (d) the electric field between the plates, and (e) the energy density between the plates.

41E. Two capacitors, 2.0 μF and 4.0 μF, are connected in parallel across a 300-V potential difference. Calculate the total energy stored in the capacitors.

42E. Calculate the energy density of the electric field at distance r from the center of an electron at rest. If the electron is a point particle, what does this calculation yield for the energy density at points very close to the electron?

43E. A certain capacitor is charged to a potential V. If you wish to increase its stored energy by 10%, by what percentage should you increase V?

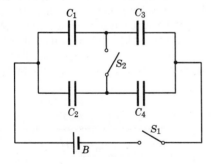

Figure 30 Problem 34.

44P. An isolated metal sphere whose diameter is 10 cm has a potential of 8000 V. Calculate the energy density in the electric field near the surface of the sphere.

35P. Figure 31 shows two identical capacitors C in a circuit with two (ideal) diodes D. A 100-V battery is connected to the input terminals, first with terminal a positive and later with terminal b positive. In each case, what is the potential difference across the output terminals? (An ideal diode has the prop-

45P. A parallel-connected bank of 2000 5.0-μF capacitors is used to store electric energy. What does it cost to charge this bank to 50,000 V, assuming a rate of 3.0¢/kW·h?

46P. For the capacitors of Problem 25, compute the energy stored for the three different connections of parts (a), (b), and (c). Compare your answers and explain any differences.

47P. One capacitor is charged until its stored energy is 4.0 J. A second uncharged capacitor is then connected to it in parallel. (a) If the charge distributes equally, what is now the total energy stored in the electric fields? (b) Where did the excess energy go?

48P. In Fig. 23 find (a) the charge, (b) the potential difference, and (c) the stored energy for each capacitor. Assume the numerical values of Exercise 19, with $V = 100$ V.

49P. A parallel-plate capacitor has plates of area A and separation d, and is charged to a potential difference V. The charging battery is then disconnected and the plates are pulled apart until their separation is $2d$. Derive expressions in terms of A, d, and V for (a) the new potential difference, (b) the initial and the final stored energy, and (c) the work required to separate the plates.

50P. A cylindrical capacitor has radii a and b as in Fig. 5. Show that half the stored electric potential energy lies within a cylinder whose radius is

$$r = \sqrt{ab}.$$

51P. Assume that the electron is not a point but a sphere of radius R over whose surface the electronic charge is uniformly distributed. (a) Determine the energy associated with the external electric field in vacuum of the electron as a function of R. (b) If you now associate this energy with the mass of the electron, you can, using $E = mc^2$, estimate the value of R. Evaluate this radius numerically; it is often called the *classical radius* of the electron.

Section 27–6 Capacitors with a Dielectric

52E. An air-filled parallel-plate capacitor has a capacitance of 1.3 pF. The separation of the plates is doubled and wax inserted between them. The new capacitance is 2.6 pF. Find the dielectric constant of the wax.

53E. Given a 7.4-pF air capacitor, you are asked to design a capacitor to store up to 7.4 μJ with a maximum potential difference of 652 V. What dielectric in Table 2 will you use to fill the gap in the air capacitor if you do not allow for a margin of error?

54E. For making a parallel-plate capacitor you have available two plates of copper, a sheet of mica (thickness = 0.10 mm, $\kappa = 5.4$), a sheet of glass (thickness = 2.0 mm, $\kappa = 7.0$), and a slab of paraffin (thickness = 1.0 cm, $\kappa = 2.0$). To obtain the largest capacitance, which sheet should you place between the copper plates?

55E. A parallel-plate air capacitor has a capacitance of 50 pF. (a) If its plates each have an area of 0.35 m², what is their separation? (b) If the region between the plates is now filled with material having a dielectric constant of 5.6, what is the capacitance?

56E. A coaxial cable used in a transmission line responds as a "distributed" capacitance to the circuit feeding it. Calculate the capacitance per meter for a cable having an inner radius of 0.10 mm and an outer radius of 0.60 mm. Assume that the space between the conductors is filled with polystyrene.

57P. A certain substance has a dielectric constant of 2.8 and a dielectric strength of 18 MV/m. If it is used as the dielectric material in a parallel-plate capacitor, what minimum area may the plates of the capacitor have in order that the capacitance be 7.0×10^{-2} μF and that the capacitor be able to withstand a potential difference of 4.0 kV?

58P. You are asked to construct a capacitor having a capacitance near 1 nF and a breakdown potential in excess of 10,000 V. You think of using the sides of a tall drinking glass (Pyrex), lining the inside and outside with aluminum foil (neglect the ends). What are (a) the capacitance and (b) breakdown potential? You use a glass 15 cm tall with an inner radius of 3.6 cm and an outer radius of 3.8 cm.

59P. You have been assigned to design a transportable capacitor that can store 250 kJ of energy. You select a parallel-plate type with dielectric. (a) What is the minimum capacitor volume achievable using a dielectric selected from those listed in Table 2 with values of dielectric strength? (b) Modern high-performance capacitors that can store 250 kJ have volumes of 0.087 m³. Assuming that the dielectric used has the same dielectric strength as in (a), what must be its dielectric constant?

60P. Two parallel-plate capacitors have the same plate area A and separation d, but the dielectric constants of the materials between their plates are $\kappa + \Delta\kappa$ and $\kappa - \Delta\kappa$, respectively. (a) Find the equivalent capacitance when they are connected in parallel. (b) If the total charge on the parallel combination is Q, what is the charge on the capacitor with the larger capacitance?

61P. A slab of copper of thickness b is thrust into a parallel-plate capacitor as shown in Fig. 32; it is exactly halfway between the plates. (a) What is the capacitance after the slab is introduced? (b) If a charge q is maintained on the plates, find the ratio of the stored energy before to that after the slab is inserted. (c) How much work is done on the slab as it is inserted? Is the slab sucked in or do you have to push it in?

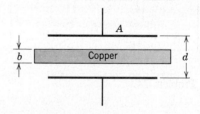

Figure 32 Problems 61 and 62.

62P. Reconsider Problem 61 assuming that the potential difference rather than the charge is held constant.

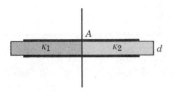

Figure 33 Problem 63.

63P. A parallel-plate capacitor is filled with two dielectrics as in Fig. 33. Show that the capacitance is given by

$$C = \frac{\epsilon_0 A}{d}\left(\frac{\kappa_1 + \kappa_2}{2}\right).$$

Check this formula for all the limiting cases that you can think of. (*Hint:* Can you justify regarding this arrangement as two capacitors in parallel?)

64P. A parallel-plate capacitor is filled with two dielectrics as in Fig. 34. Show that the capacitance is given by

$$C = \frac{2\epsilon_0 A}{d}\left(\frac{\kappa_1 \kappa_2}{\kappa_1 + \kappa_2}\right).$$

Check this formula for all the limiting cases that you can think of. (*Hint:* Can you justify regarding this arrangement as two capacitors in series?)

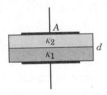

Figure 34 Problem 64.

65P. What is the capacitance of the capacitor in Fig. 35?

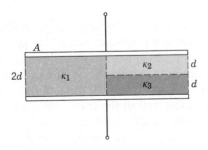

Figure 35 Problem 65.

Section 27–8 Dielectrics and Gauss' Law

66E. A parallel-plate capacitor has a capacitance of 100 pF, a plate area of 100 cm², and a mica dielectric ($\kappa = 5.4$). At 50-V potential difference, calculate (*a*) E in the mica, (*b*) the magnitude of the free charge on the plates, and (*c*) the magnitude of the induced surface charge.

67E. In Sample Problem 9, suppose that the battery remains connected during the time that the dielectric slab is being introduced. Calculate (*a*) the capacitance, (*b*) the charge on the capacitor plates, (*c*) the electric field in the gap, (*d*) the electric field in the slab, after the slab is introduced.

68P. Two parallel plates of area 100 cm² are each given equal but opposite charges of 8.9×10^{-7} C. The electric field within the dielectric material filling the space between the plates is 1.4×10^6 V/m. (*a*) Calculate the dielectric constant of the material. (*b*) Determine the magnitude of the charge induced on each dielectric surface.

69P. A parallel-plate capacitor has plates of area 0.12 m² and a separation of 1.2 cm. A battery charges the plates to a potential difference of 120 V and is then disconnected. A dielectric slab of thickness 40 mm and dielectric constant 4.8 is then placed symmetrically between the plates. (*a*) Find the capacitance before the slab is inserted. (*b*) What is the capacitance with the slab in place? (*c*) What is the free charge q before and after the slab is inserted? (*d*) Determine the electric field in the space between the plates and dielectric. (*e*) What is the electric field in the dielectric? (*f*) With the slab in place what is the potential difference across the plates? (*g*) How much external work is involved in the process of inserting the slab?

70P. In the capacitor of Sample Problem 9 (Fig. 16), (*a*) what fraction of the energy is stored in the air gaps? (*b*) What fraction is stored in the slab?

71P. A dielectric slab of thickness b is inserted between the plates of a parallel-plate capacitor of plate separation d. Show that the capacitance is given by

$$C = \frac{\kappa \epsilon_0 A}{\kappa d - b(\kappa - 1)}.$$

(*Hint:* Derive the formula following the pattern of Sample Problem 9.) Does this formula predict the correct numerical result of Sample Problem 9? Verify that the formula gives reasonable results for the special cases of $b = 0$, $\kappa = 1$, and $b = d$.

CHAPTER 28

CURRENT AND RESISTANCE

The 1870s saw the dawn of the electrical age as copper wires began to snake into our homes. The age of the semiconductor began in the 1950s. The 1980s may well herald the age of the superconductor. These materials—which have no electrical resistance at all—have many tantalizing and potentially useful properties. We see here a small magnet floating in air above a disk of superconducting material that is immersed in a bath of liquid nitrogen. . . . Can widespread use of high-speed levitated trains be far behind?*

28–1 Moving Charges and Electric Currents

The previous five chapters dealt with *electrostatics,* that is, with charges at rest. With this chapter we begin our study of *electric currents,* that is, of charges in motion.

Examples of electric currents abound, ranging from the large currents that constitute lightning strokes to the tiny nerve currents that regulate our muscular activity.

---

* See Essay 12, *Superconductivity,* by Peter Lindenfeld and Essay 13, *Magnetic Flight,* by Gerard K. O'Neill, for a fuller account of the magnetic properties and applications of superconductors.

The currents in household wiring, in light bulbs, and in electric appliances are familiar to all. Charge carriers of *both* signs flow in the ionized gases of fluorescent lamps, in the batteries of transistor radios, and in car batteries. Electric currents in semiconductors are to be found in pocket calculators and in the chips that control microwave ovens and electric dishwashers. A beam of electrons moves through an evacuated space in the picture tube of a TV set.

On a global scale, charged particles trapped in the Van Allen radiation belts surge back and forth above the atmosphere between the north and the south magnetic poles. We see the effects of occasional "spillover" from

---

these belts in the auroral displays of the northern and southern polar regions. On the scale of the solar system, enormous currents of protons, electrons, and ions fly radially outward from the sun as the *solar wind*. On the galactic scale, cosmic rays, which are largely energetic protons, stream through the galaxy.

Although an electric current is a stream of moving charges, not all moving charges constitute an electric current. If we are to say that an electric current passes through a given surface, there must be a net flow of charge through that surface. Two examples clarify our meaning.

First Example. The conduction electrons in an isolated length of copper wire are in random motion at speeds of the order of 10^6 m/s. If you pass a hypothetical plane through such a wire, conduction electrons pass through it *in both directions* at the rate of many billions per second. However, there is no *net* transport of charge and thus no current. However, if you connect the ends of the wire to a battery, you bias the flow — ever so slightly — in one direction so that there now is a net transport of charge and thus an electric current.

Second Example. The flow of water through a garden hose represents the directed flow of positive charge at a rate of perhaps several million coulombs per second. There is no net transport of charge, however, because there is a parallel flow of negative charge of exactly the same amount moving in exactly the same direction.

In this chapter we restrict ourselves largely to the study — within the framework of classical physics — of *steady* currents of *conduction electrons* moving through *metallic conductors* such as copper wires.

28-2 Electric Current

As Fig. 1*a* reminds us, an isolated conductor in electrostatic equilibrium — whether charged or not — is all at the same potential. No electric field can exist within it or parallel to its surface. Although conduction electrons are available, no net electric force acts on them and thus there is no current.

If, as in Fig. 1*b*, we insert a battery in the loop, the conducting loop is no longer an equipotential. Electric fields act in its interior, exerting forces on the conduction electrons and establishing a current. After a very short time, the electron flow reaches a steady-state condition. The situation is then completely analogous to the steady-state fluid flow that we studied in Chapter 16.

Figure 2 shows a section of a conductor, part of a

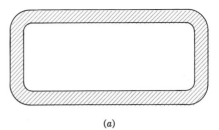

(a)

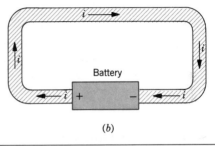

(b)

Figure 1 (*a*) A loop of copper in electrostatic equilibrium. The entire loop is at a single potential and the electric field for all internal points is zero. (*b*) Adding a battery imposes a potential difference. An electric field appears inside the conductor and causes charges to move around the loop, constituting a current.

conducting loop in which current has been established. The amount of charge dq that passes through a hypothetical plane, such as *xx*, is proportional to the length of time dt required for all the charge dq to pass through that plane. The proportionality constant is the current i; therefore,

$$\boxed{dq = i \, dt}$$ (definition of current). (1)

Using Eq. 1, we can find the charge that passes through plane *xx* in a time interval extending from 0 to t

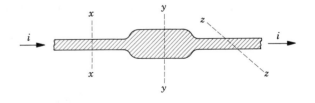

Figure 2 The current i through the conductor has the same value at planes *xx*, *yy*, and *zz*. The current density **J** at plane *yy* is smaller than at plane *xx* because the cross-sectional area of the conductor is greater at *yy*.

from

$$q = \int dq = \int_0^t i \, dt, \qquad (2)$$

in which the current i may — or may not — be a function of time.

Under steady-state conditions, the current is the same for planes yy and zz and indeed for all planes that pass completely through the conductor, no matter what their location or orientation. This follows from the fact that charge is *conserved*. Under the steady-state conditions that we have assumed, an electron must enter the conductor at one end for every electron that leaves at the other. In the same way, if we have a steady flow of water through a garden hose, a drop of water must leave the nozzle for every drop that enters the hose at the other end. The amount of water in the hose is a conserved quantity.

The SI unit for current is the coulomb per second or the ampere (abbr. A); that is,

1 ampere = 1 A = 1 coulomb/second = 1 C/s.

The ampere is an SI base unit; the coulomb is defined in terms of the ampere, as we discussed earlier in Chapter 23. We shall present the formal definition of the ampere in Chapter 31.

Current, defined by Eq. 1, is a scalar because both q and t in that equation are scalars. This may cause some difficulty because we often represent a current in a wire by an arrow to indicate the direction in which the charges are moving. Such arrows are not vectors, however, because they do not obey the laws of vector addition. Figure 3a shows a conductor splitting at a junction into two branches. Because charge is conserved, the magnitudes of the currents in the branches must add to yield the magnitude of the current in the original conductor, or

$$i_0 = i_1 + i_2. \qquad (3)$$

As Fig. 3b suggests, bending or reorienting the wires in space does not change the validity of Eq. 3. Current arrows are not vectors; they only show a direction (or sense) of flow along a conductor, not a direction in space.

The Directions of Currents. In Fig. 1b we drew the current arrows in the direction that a positive charge carrier — repelled by the positive battery terminal and attracted by the negative terminal — would circulate around the loop. Actually, the charge carriers in the copper loop of Fig. 1b are electrons, which carry a negative charge. They circulate in a direction opposite to that of the current arrows. Recall also that in a fluorescent lamp,

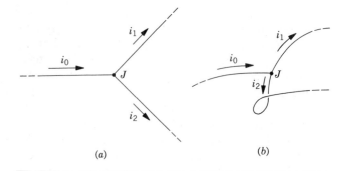

Figure 3 The relation $i_0 = i_1 + i_2$ is true at junction J no matter what the orientation in space of the three wires. Currents are scalars, not vectors.

charge carriers of *both* signs are present. Since positive and negative charge carriers move in opposite directions, we must choose which direction a current arrow represents.

In drawing the current arrows in Fig. 1b in a clockwise direction, we were following this convention:

The current arrow is drawn in the direction that positive carriers would move, even if the actual carriers are not positive.

It is only when we are interested in the detailed mechanism of charge transport that we need to pay attention to what the signs of the charge carriers actually are.

This convention is possible only because a positive charge carrier moving from left to right has the same external effect as a negative carrier moving from right to left.* In Fig. 4, for example, the two previously un-

---

* This is not entirely true. In Section 30–4 we discuss the *Hall effect*, which can be used in simple cases to establish the actual sign of the charge carriers.

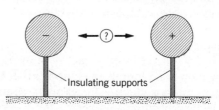

Figure 4 The charges on the two previously uncharged spheres arose by transferring charge from one sphere to the other. You cannot tell, after the event, whether positive charge was transferred from left to right or negative charge from right to left; the end result is the same.

charged spheres may have been charged by transporting positive charge from left to right or negative charge from right to left; both processes lead to the same end result.

Sample Problem 1 Water flows through a garden hose at a rate R of 450 cm³/s. To what current of negative charge does this correspond?

The current of negative charge is the rate at which molecules pass through any plane that cuts across the hose times the amount of negative charge carried by each molecule. If ρ is the density of water and M is its molecular weight, the rate (in moles per second) at which water is flowing through the plane is $R\rho/M$. If N_A is the Avogadro constant, the rate dN/dt at which molecules pass through the plane is

$$\frac{dN}{dt} = \frac{R\rho N_A}{M}$$

$$= \frac{\left(450 \times 10^{-6} \frac{m^3}{s}\right)\left(1000 \frac{kg}{m^3}\right)\left(6.02 \times 10^{23} \frac{molecules}{mol}\right)}{0.018 \text{ kg/mol}}$$

$$= 1.51 \times 10^{25} \text{ molecules/s}.$$

Each water molecule contains 10 electrons, 8 for the oxygen atom and 1 each for the two hydrogens. Each electron carries a charge of $-e$ so that the magnitude of the current corresponding to this movement of negative charge is

$$i = \frac{dq}{dt} = (1.51 \times 10^{25} \text{ molecules/s})(10 \text{ electrons/molecule})$$

$$\times (1.60 \times 10^{-19} \text{ C/electron})$$

$$= 2.4 \times 10^7 \text{ C/s} = 2.4 \times 10^7 \text{ A} = 24 \text{ MA}. \quad \text{(Answer)}$$

This current of negative charge is exactly compensated by a current of positive charge associated with the (positively charged) nuclei of the three atoms that make up the water molecule. Thus, there is no net flow of charge through the hose.

28-3 Current Density

Sometimes we are interested in the current i in a particular conductor. At other times we take a localized view

and are interested in the flow of charge at a particular point within a conductor. A (positive) charge carrier at a given point will flow in the direction of the electric field **E** at that point. To describe this flow, we introduce the *current density* **J**, a vector quantity that points in the direction of this field.

Figure 5a shows a simple case in which a current i is uniformly distributed over the cross section of a uniform conductor. The current density J in this case is constant for all points within the conductor and is related to the current i by

$$J = i/A, \tag{4}$$

in which A is the cross-sectional area of the conductor. The SI unit for current density is the ampere per square meter (abbr. A/m²). As comparison of Figs. 5a and 5b shows, the direction of the current density **J** is that of the electric field **E** regardless of the sign of the charge carriers.

For any surface—whether plane or not—through which there is a current i, the current density **J** for points on that surface is related to i by*

$$i = \oint \mathbf{J} \cdot d\mathbf{A}. \tag{5}$$

For the cross section xx in Fig. 2, **J** is constant in magnitude and direction and Eq. 5 readily reduces to Eq. 4.

In Section 16-9 we showed that, in fluid flow, the vector field represented by the array of velocity vectors of the fluid particles could be represented by a system of streamlines. The vector field represented by the array of current density vectors within a conductor can be represented in the same way. Figure 6, which is actually Fig. 19 of Chapter 16, can represent either the flow of a fluid

---

* Equation 5 shows that current is the flux of the current density through a surface. You may wish to review Section 25-4, in which we discuss a different flux, namely, the flux of the electric field through a surface.

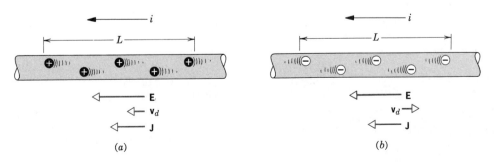

(a)   (b)

Figure 5 (a) Positive carriers drift in the direction of the applied electric field **E**. (b) Negative carriers drift in the opposite direction. By convention, the direction of the current density **J** and the sense of the current arrow are drawn *as if* the charge carriers were positive.

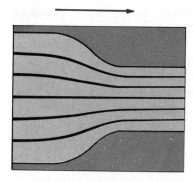

Figure 6 Streamlines representing the current density vectors in the flow of charge through a constricted conductor. Compare with Fig. 19 of Chapter 16, which shows the streamlines representing the velocity vectors in fluid flow.

through a constricted pipe or the flow of charge through a constricted conductor.

Calculating the Drift Speed. The conduction electrons in a copper conductor have *randomly directed* velocities of the order of 10^6 m/s. The *directed* flow of the conduction electrons for a typical current, that in household wiring, for example, is characterized by an average *drift speed* that is very much smaller, possibly of the order of 10^{-3} m/s.

For an apt analogy, consider a large crowd of people running around in random directions and jostling each other constantly. If the crowd is standing on a nearly level surface that slopes ever so slightly in a particular direction, the crowd will work its way slowly down the slope in that direction. A small but directed "drift speed" will be superimposed on the random jostling motion. For the conduction electrons, it is this drift speed that determines the current.

Now let us estimate the drift speed. Figure 5a shows the charge carriers in a wire moving to the left with an assumed constant drift speed v_d.* The number of carriers in a length L of the wire is nAL, where n is the number of carriers per unit volume and AL is the volume of a specified length of wire. A charge of magnitude

$$\Delta q = (nAL)e$$

passes out of this volume, through the left end of the conductor, in a time interval given by

$$\Delta t = L/v_d.$$

---

* Following convention, we assume that the charge carriers are positive, even though in fact they are (negative) conduction electrons.

The current in the wire is

$$i = \frac{\Delta q}{\Delta t} = \frac{nALe}{L/v_d} = nAev_d \qquad (6)$$

Solving for v_d and recalling Eq. 4 ($J = i/A$), we obtain

$$v_d = \frac{i}{nAe} = \frac{J}{ne}$$

or, extended to vector form,

$$\boxed{\mathbf{J} = (ne)\mathbf{v}_d.} \qquad (7)$$

Here the product ne, whose SI units are C/m^3, is the carrier charge density. For positive carriers, which we always assume, ne is positive and Eq. 7 predicts that $\mathbf{J}$ and $\mathbf{v}_d$ point in the same direction.

Sample Problem 2 One end of an aluminum wire whose diameter is 2.5 mm is welded to one end of a copper wire whose diameter is 1.8 mm. The composite wire carries a steady current i of 1.3 A. What is the current density in each wire?

We may take the current density as constant within each wire except for points near the junction. The current density is given by Eq. 4,

$$J = \frac{i}{A}.$$

The cross-sectional area A of the aluminum wire is

$$A_{Al} = \tfrac{1}{4}\pi d^2 = (\pi/4)(2.5 \times 10^{-3} \text{ m})^2 = 4.91 \times 10^{-6} \text{ m}^2$$

so that

$$J_{Al} = \frac{1.3 \text{ A}}{4.91 \times 10^{-6} \text{ m}^2}$$
$$= 2.6 \times 10^5 \text{ A/m}^2 = 26 \text{ A/cm}^2. \qquad \text{(Answer)}$$

As you can verify, the cross-sectional area of the copper wire proves to be 2.54×10^{-6} m^2 so that

$$J_{Cu} = \frac{1.3 \text{ A}}{2.54 \times 10^{-6} \text{ m}^2}$$
$$= 5.1 \times 10^5 \text{ A/m}^2 = 51 \text{ A/cm}^2. \qquad \text{(Answer)}$$

The fact that the wires are of different materials does not enter here.

Sample Problem 3 What is the drift speed of the conduction electrons in the copper wire of Sample Problem 2?

The drift speed is given by Eq. 7,

$$v_d = \frac{J}{ne}.$$

In copper, there is very nearly one conduction electron per atom on the average. The number n of electrons per unit volume is therefore the same as the number of atoms per unit volume and is given by

$$\frac{n}{N_A} = \frac{\rho_m}{M} \quad \text{or} \quad \left(\frac{\text{atoms/m}^3}{\text{atoms/mol}} = \frac{\text{mass/m}^3}{\text{mass/mol}}\right).$$

Here ρ_m is the (mass) density of copper, N_A is the Avogadro constant, and M is the atomic weight of copper.* Thus,

$$n = \frac{N_A \rho_m}{M}$$

$$= \frac{(6.02 \times 10^{23} \text{ mol}^{-1})(9.0 \times 10^3 \text{ kg/m}^3)}{64 \times 10^{-3} \text{ kg/mol}}$$

$$= 8.47 \times 10^{28} \text{ electrons/m}^3.$$

We then have

$$v_d = \frac{5.1 \times 10^5 \text{ A/m}^2}{(8.47 \times 10^{28} \text{ electrons/m}^3)(1.60 \times 10^{-19} \text{ C/electron})}$$

$$= 3.8 \times 10^{-5} \text{ m/s} = 14 \text{ cm/h}. \quad \text{(Answer)}$$

You may well ask: "If the electrons drift so slowly, why do the room lights turn on so quickly after I throw the switch?" Confusion on this point results from not distinguishing between the drift speed of the electrons and the speed at which *changes* in the electric field configuration travel along wires. This latter speed approaches that of light; electrons everywhere in the wire begin drifting almost at once. Similarly, when you turn the valve on your garden hose, with the hose full of water, a pressure wave travels along the hose at the speed of sound in water. The speed at which the water moves through the hose—measured perhaps with a dye marker—is much lower.

Sample Problem 4 A strip of silicon has a width w of 3.2 mm, a thickness t of 250 μm, and carries a current i of 5.2 mA. The silicon is an *n-type semiconductor*, having being "doped" with a controlled phosphorus impurity. The doping has the effect of greatly increasing n, the number of charge carriers per unit volume, as compared with the value for pure silicon. In this case, $n = 1.5 \times 10^{23}$ m^{-3}. (a) What is the current density in the strip?

From Eq. 4,

$$J = \frac{i}{wt} = \frac{5.2 \times 10^{-3} \text{ A}}{(3.2 \times 10^{-3} \text{ m})(250 \times 10^{-6} \text{ m})}$$

$$= 6500 \text{ A/m}^2. \quad \text{(Answer)}$$

(b) What is the drift speed?

From Eq. 7,

$$v_d = \frac{J}{ne} = \frac{6500 \text{ A/m}^2}{(1.5 \times 10^{23} \text{ m}^{-3})(1.60 \times 10^{-19} \text{ C})}$$

$$= 0.27 \text{ m/s} = 27 \text{ cm/s}. \quad \text{(Answer)}$$

The drift speed (0.27 m/s) of the electrons in this silicon semiconductor is much greater than the drift speed (3.8×10^{-5} m/s) of the conduction electrons in the metallic copper conductor of Sample Problem 3. This is because the density of charge carriers (1.5×10^{23} m^{-3}) in the silicon semiconductor is much smaller than the density of charge carriers (8.47×10^{28} m^{-3}) in the copper conductor; the fewer charge carriers that are available, the faster they must drift to transport the same amount of charge per unit time.

28-4 Resistance and Resistivity

If we apply the same potential difference between the ends of geometrically similar rods of copper and of glass, very different currents result. The characteristic of the conductor that enters here is its *resistance*. We determine the resistance of a conductor between any two points by applying a potential difference V between those points and measuring the current i that results. The resistance R is then

$$\boxed{R = V/i} \quad \text{(definition of } R\text{).} \quad \text{(8)}$$

The SI unit for resistance that follows from Eq. 8 is the volt/ampere. This combination occurs so often that we give it a special name, the *ohm* (symbol Ω). That is,

$$1 \text{ ohm} = 1 \ \Omega = 1 \text{ volt/ampere} = 1 \text{ V/A.} \quad \text{(9)}$$

A conductor whose function in a circuit is to provide a specified resistance is called a *resistor;* see Fig. 7. We represent a resistor in a circuit diagram by the symbol -\/\/\/-.

The resistance of a conductor depends on the manner in which the potential difference is applied to it. Figure 8, for example, shows a given potential difference applied in two different ways to the same conductor. As the current density streamlines suggest, the currents in the two cases—and hence the measured resistances—will be quite different.

The flow of charge through a resistor is often compared with the flow of water through a pipe when there is a pressure difference between the ends of the pipe. The pump that establishes the pressure difference between

---

* We use the subscript m to make it clear that the density referred to here is a mass density (kg/m^3), not a charge density (C/m^3).

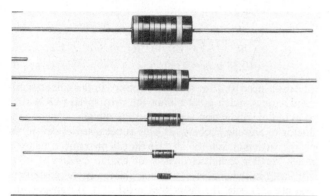

Figure 7 An assortment of resistors. The circular bands are color coding marks that identify the value of the resistance and the maximum allowable power dissipation.

the ends of the pipe can be compared with the battery that establishes the potential difference between the ends of the resistor. The flow of water (m³/s) can be compared with the flow of charge (C/s or A). The rate of flow of water for a given pressure difference depends on the characteristics of the pipe. Is it long or short? Is it narrow or wide? Is it empty or filled with sand or gravel? These characteristics of the pipe are analogous to those that determine the resistance of the resistor.

As we have done several times in other connections, we often wish to take a local view and deal not with a particular object but with a substance. We do this by focusing not on the potential difference V across a particular resistor but on the electric field **E** at a point in a resistive material. Instead of dealing with the current i through the resistor, we deal with the current density **J** at the point in question. Instead of the resistance R we define a new quantity, the *resistivity* ρ, from

$$\rho = E/J \qquad \text{(definition of } \rho\text{)}. \qquad (10)$$

The SI units of E are V/m and those of J are A/m². The SI units of ρ can then be seen to be Ω·m, pronounced "ohm·meter."*

$$\left(\frac{V/m}{A/m^2}\right)\left(\frac{1\,\Omega}{1\,V/A}\right) = \Omega \cdot m.$$

The conversion factor in the second parentheses is based on Eq. 8, the definition of the ohm.

We can write Eq. 10 in vector form as

$$\mathbf{E} = \rho \mathbf{J}. \qquad (11)$$

Equations 10 and 11 hold only for isotropic materials—materials in which the electrical properties are the same in all directions.

We often speak of the *conductivity* σ of a material. This is simply the reciprocal of its resistivity, or†

$$\sigma = 1/\rho \qquad \text{(definition of } \sigma\text{)}. \qquad (12)$$

The SI units of σ are (Ω·m)⁻¹, pronounced "reciprocal ohm·meters."‡ Table 1 lists the resistivities of some materials.

Calculating the Resistance. If we know the resistivity of a substance such as copper, we should be able to calculate the resistance of a length of wire of given diameter made of that substance. Let A (see Fig. 9) be the cross-sectional area of the wire, let L be its length, and let a

---

* An instrument used to measure resistance is called an *ohmmeter;* it is quite different from the *ohm·meter* (Ω·m), a unit of resistivity. In the same spirit, the *micrometer* (mi·crom′·e·ter) is an instrument used by machinists; it is quite different from the *micrometer* (mi′·cro·me·ter), a unit of length.

† We have used ρ earlier to represent both mass density and charge density. Also, we have used σ earlier to represent surface charge density. We use these symbols here with entirely different meanings; there are just not enough good symbols to go around.

‡ The usage mhos/m for conductivity is also not uncommon.

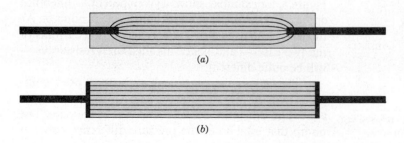

Figure 8 Two ways of applying a potential difference to a conducting rod. The heavy black conductors are assumed to have negligible resistance. Which rod will have the smaller measured resistance?

Table 1 Resistivities of Some Materials at Room Temperature (20°C)

Material	Resistivity ρ $\Omega \cdot m$	Temperature coefficient of resistivity $\alpha \, K^{-1}$
Typical Metals		
Silver	1.62×10^{-8}	4.1×10^{-3}
Copper	1.69×10^{-8}	4.3×10^{-3}
Aluminum	2.75×10^{-8}	4.4×10^{-3}
Tungsten	5.25×10^{-8}	4.5×10^{-3}
Iron	9.68×10^{-8}	6.5×10^{-3}
Platinum	$10.6 \ \times 10^{-8}$	3.9×10^{-3}
Manganin[a]	$48.2 \ \times 10^{-8}$	0.002×10^{-3}
Typical Semiconductors		
Silicon pure	2.5×10^{3}	-70×10^{-3}
Silicon n-type[b]	8.7×10^{-4}	
Silicon p-type[c]	2.8×10^{-3}	
Typical Insulators		
Glass	$10^{10} - 10^{14}$	
Fused quartz	$\sim 10^{16}$	

[a] An alloy specifically designed to have a small value of α.
[b] Pure silicon "doped" with phosphorus impurities to a charge carrier density of $10^{23} \, m^{-3}$
[c] Pure silicon "doped" with aluminum impurities to a charge carrier density of $10^{23} \, m^{-3}$.

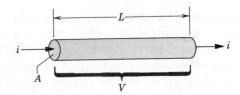

Figure 9 A potential difference V is applied between the ends of a wire of length L and cross section A, establishing a current i.

est interest when we are making electrical measurements on specific conductors. They are the quantities that we read directly on meters. We turn to the microscopic quantities E, J, and ρ when we are interested in the fundamental electrical behavior of matter, as we are in the research area of solid-state physics.

Variation with Temperature. The values of most physical properties vary with temperature and resistivity is no exception. Figure 10, for example, shows the variation of this property for copper over a wide range. The relation for copper—and for metals in general—is fairly linear over a rather broad range of temperatures. We can write, as an empirical approximation that is good enough for essentially all engineering purposes,

$$\rho - \rho_0 = \rho_0 \alpha (T - T_0). \tag{16}$$

potential difference V exist between its ends. If the streamlines representing the current density are uniform throughout the wire, the electric field and the current density will be constant for all points within the wire and will have the values

$$E = V/L \quad \text{and} \quad J = i/A. \tag{13}$$

We can then combine Eqs. 10 and 13 to write

$$\rho = \frac{E}{J} = \frac{V/L}{i/A}. \tag{14}$$

But V/i is the resistance R, which allows us to recast Eq. 14 as

$$R = \rho \frac{L}{A}. \tag{15}$$

We stress that Eq. 15 can only be applied to a homogeneous isotropic conductor of uniform cross section and provided the potential difference is applied as in Fig. 8b.

The macroscopic quantities V, i, and R are of great-

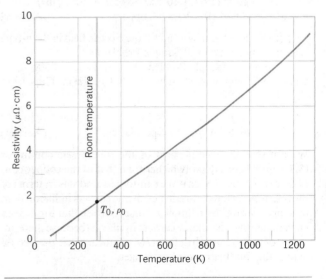

Figure 10 The resistivity of copper as a function of temperature. The dot on the curve marks a convenient room-temperature reference point ($T_0 = 293$ K, $\rho_0 = 1.69$ $\mu\Omega \cdot cm$).

Here T_0 is a selected reference temperature and ρ_0 is the resistivity at that temperature. Often we choose $T_0 = 293$ K (room temperature), for which $\rho_0 = 1.69\ \mu\Omega \cdot$ cm.

Because temperature enters Eq. 16 only as a difference, it does not matter whether you use the Celsius or the Kelvin scales in that equation because the sizes of the degree on these scales are identical. The quantity α in Eq. 16, called the *temperature coefficient of resistivity,* is chosen so that the equation gives the best agreement with experiment for temperatures in the chosen range. Some values of α for different metals are listed in Table 1.

The variation of resistivity with temperature is so reproducible that a *platinum resistance thermometer* has been adopted as a secondary thermometric standard for measuring temperatures in the range 14–900 K on the International Practical Temperature Scale; see Section 19–5. In using this device to measure temperatures, terms proportional to $(T - T_0)^2$ and $(T - T_0)^3$ are added to the right side of Eq. 16, yielding an equation of improved precision.

Sample Problem 5   (a) What is the strength of the electric field present in the copper conductor of Sample Problem 2?

In Sample Problem 2 we found the current density J to be 5.1×10^5 A/m²; from Table 1 we see that the resistivity ρ for copper is $1.69 \times 10^{-8}\ \Omega \cdot$m. Thus, from Eq. 11

$$E = \rho J = (1.69 \times 10^{-8}\ \Omega \cdot \text{m})(5.1 \times 10^5\ \text{A/m}^2)$$
$$= 8.6 \times 10^{-3}\ \text{V/m} \quad \text{(copper).} \qquad \text{(Answer)}$$

(b) What is the magnitude of the electric field in the *n*-type silicon semiconductor of Sample Problem 4?

From that Sample Problem we find that $J = 6500$ A/m² and from Table 1 we see that $\rho = 8.7 \times 10^{-4}\ \Omega \cdot$m. Thus, from Eq. 11

$$E = \rho J = (8.7 \times 10^{-4}\ \Omega \cdot \text{m})(6500\ \text{A/m}^2)$$
$$= 5.7\ \text{V/m} \quad (\text{n-type silicon).} \qquad \text{(Answer)}$$

We see that the electric field in the silicon semiconductor (5.7 V/m) is considerably higher than that in the copper conductor (8.7×10^{-3} V/m). We can understand this in terms of the much lower concentration of charge carriers in silicon than in copper. From the relation $J = nev_d$, we see that for a given current density, the charge carriers in silicon (because there are so few of them) must drift faster, which means that the electric field acting on them must be stronger.

Sample Problem 6   A rectangular block of iron has dimensions $1.2 \times 1.2 \times 15$ cm. (a) What is the resistance of the block measured between the two square ends? The resistivity of iron at room temperature is $9.68 \times 10^{-8}\ \Omega \cdot$m.

The area of a square end is $(1.2 \times 10^{-2}$ m$)^2$ or 1.44×10^{-4} m². From Eq. 15,

$$R = \frac{\rho L}{A} = \frac{(9.68 \times 10^{-8}\ \Omega \cdot \text{m})(0.15\ \text{m})}{1.44 \times 10^{-4}\ \text{m}^2}$$
$$= 1.0 \times 10^{-4}\ \Omega = 100\ \mu\Omega. \qquad \text{(Answer)}$$

(b) What is the resistance between two opposing rectangular faces?

The area of a rectangular face is $(1.2 \times 10^{-2}$ m$)(0.15$ m$)$ or 1.80×10^{-3} m². From Eq. 15,

$$R = \frac{\rho L}{A} = \frac{(9.68 \times 10^{-8}\ \Omega \cdot \text{m})(1.2 \times 10^{-2}\ \text{m})}{1.80 \times 10^{-3}\ \text{m}^2}$$
$$= 6.5 \times 10^{-7}\ \Omega = 0.65\ \mu\Omega. \qquad \text{(Answer)}$$

We assume in each case that the potential difference is applied to the block in such a way that the surfaces between which the resistance is desired are equipotentials. Otherwise, Eq. 15 would not be valid.

28–5   Ohm's Law

Figure 11*a* shows a "black box" with a potential difference V applied between its terminals. Figure 11*b* shows the current i that results for various values and polarities of V. The plot is a straight line passing through the origin, which means that the resistance V/i of the device in the box is a constant, independent of the potential difference used to measure it.

Figure 11*c* is a plot for a repetition of the experiment with a different object in the box. The V–i curve in the forward direction in this case is not a straight line and there is a total lack of symmetry when the polarity of the applied potential difference is reversed.

We say that the object in the box of Fig. 11*b*— which turns out to be a 1000-Ω resistor—obeys *Ohm's law.* The object in the box of Fig. 11*c*—which turns out to be a so-called *pn* junction diode—does not.

> *A conducting device obeys Ohm's law if its resistance between any two points is independent of the magnitude and polarity of the potential difference applied between those points.*

Modern microelectronics—and therefore much of the character of our present technological civilization—depends almost totally on devices that do *not* obey Ohm's law. Your pocket calculator, for example, is full of them.

It is a common error to say that Eq. 8 ($V = Ri$) is a statement of Ohm's law. Not true! This equation is sim-

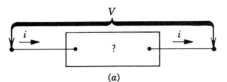

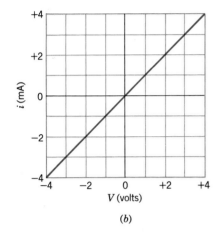

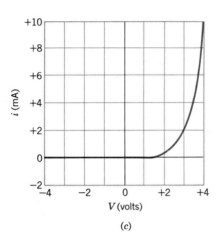

Figure 11 (*a*) A "black box" to whose terminals a potential difference *V* is applied, establishing a current *i*. (*b*) A *V*–*i* plot when the object in the box is a 1000-Ω resistor. (*c*) A *V*–*i* plot for a semiconducting *pn* junction diode.

ply the defining equation for resistance and applies to all conducting devices whether or not they obey Ohm's law. The essence of Ohm's law is that the *V*–*i* curve is linear; that is, the value of *R* is independent of the value of *V*.

We can also express Ohm's law from the local point of view, in which we focus our interest on conducting materials rather than on conducting devices. The relevant relation is Eq. 11 ($\mathbf{E} = \rho\mathbf{J}$), which is the local analog of Eq. 8 ($V = Ri$).

> *A conducting material obeys Ohm's law if its resistivity is independent of the magnitude and direction of the applied electric field.*

All homogeneous materials, be they conductors like copper or semiconductors like silicon (doped or pure), obey Ohm's law for some range of values of the electric field. If the field is too strong, however, there are departures from Ohm's law in all cases.

28–6 Ohm's Law: A Microscopic View

To find out *why* a given material such as copper or silicon obeys Ohm's law, we must look into the details of the conduction process at the atomic level. Here we consider only conduction in metals, such as copper.* We base our analysis on the *free-electron model,* in which we assume that the conduction electrons in the metal are free to move throughout the volume of the sample, like the molecules of a gas in a closed container.†

According to classical physics, the electrons should have a Maxwellian speed distribution somewhat like that of the molecules in a gas. In such a distribution, as we have seen, the average electron speed would be proportional to the square root of the absolute temperature; see Section 21–7. The motions of the electrons, however, are not governed by the laws of classical physics but by those of quantum physics. As it turns out, an assumption that is much closer to the quantum reality is that the electrons move with a single effective speed v_{eff}. For copper, $v_{\text{eff}} \approx 1.6 \times 10^6$ m/s, essentially independent of the temperature.

When we apply an electric field to the metal specimen, the electrons modify their random motions slightly and drift very slowly—in a direction opposite to that of the field—with an average drift speed v_d. As we saw in

---

* In Section 28–8 we consider the conduction mechanism in semiconductors.
† We assume that the electrons do not interact with each other but only with the atoms at the lattice sites. We assume further that after such a collision the electron emerges with a random velocity, with no memory—so to speak—of its initial velocity.

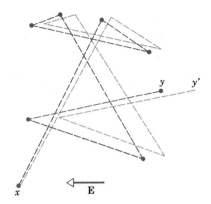

Figure 12 The colored lines show an electron moving from x to y, making six collisions en route. The gray lines show what its path *might* have been in the presence of an applied electric field **E**. Note the steady drift in the direction of $-$**E**. (Actually, the gray lines should be slightly curved, to represent the parabolic paths followed by the electrons between collisions under the influence of the electric field.)

Sample Problem 3, the drift speed ($\sim 4 \times 10^{-5}$ m/s) in a typical metallic conductor is less than the effective speed (1.6×10^6 m/s) by many orders of magnitude. Figure 12 suggests the relation between these two speeds. The colored lines show a possible random path for an electron in the absence of an applied field; the electron proceeds from x to y, making six collisions along the way. The light gray lines show how the same events *might* have occurred if an electric field **E** had been applied. We see that the electron drifts steadily to the right, ending at y' rather than at y. In preparing Fig. 12, we assumed that $v_d \approx 0.02\ v_{\text{eff}}$; actually it is more like $v_d \approx 10^{-9}\ v_{\text{eff}}$, so that the drift displayed in the figure is greatly exaggerated.

If an electron of mass m is placed in an electric field E, it will experience an acceleration given by Newton's second law:

$$a = \frac{F}{m} = \frac{eE}{m}. \qquad (17)$$

The nature of the collisions experienced by the electrons is such that, after a typical collision the electron will — so to speak — completely lose its memory of its accumulated drift velocity. The electron will then start off afresh after every encounter, moving off in a random direction. In the average time interval τ until the next collision, an average electron will change its velocity by an amount $a\tau$. As the Special Note that appears just before Sample Problem 7 makes clear, we identify this with the drift

speed v_d and we write, using Eq. 17,

$$v_d = a\tau = \frac{eE\tau}{m}. \qquad (18)$$

Combining this result with Eq. 7 yields

$$v_d = \frac{J}{ne} = \frac{eE\tau}{m},$$

which we can write as

$$E = \left(\frac{m}{e^2 n\tau} \right) J.$$

Comparing this with Eq. 11 ($E = \rho J$) leads to

$$\rho = \frac{m}{e^2 n\tau}. \qquad (19)$$

Equation 19 may be taken as a statement that metals obey Ohm's law if we can show that ρ is a constant, independent of the strength of the applied electric field E. Because n, m, and e are constant, this reduces to convincing ourselves that τ, the mean free time between collisions, is a constant, independent of the strength of the applied electric field.

Figure 12 reminds us that the speed distribution of the conduction electrons is only minimally affected by even a rather strong electric field. Speaking loosely, a given electron in its random motion is scarcely aware whether the electric field is ON or OFF: its "collision experience" remains essentially unchanged. We may be confident that, whatever the value of τ for a metal such as copper in the absence of a field, its value remains essentially unchanged when a field is applied.

A Special Note.* There is a natural inclination to write Eq. 18 as $v_d = \frac{1}{2}a\tau$, reasoning that $a\tau$ is the electron's *final velocity* and that its *average velocity* is just half that value. Indeed, this assumption was made by the distinguished physicist Paul Drude (1863–1906) who first proposed the free-electron model of conduction. Equation 18, however, is correct as it stands, without the factor of $\frac{1}{2}$.

This factor would be appropriate if we followed a typical electron, averaged its velocity over its mean time τ between collisions, and called that the drift speed.

---

* We are grateful to Professor Philip A. Casabella, who suggested the substance of this argument to us. For a fuller account, see *Electricity and Magnetism*, 2nd ed., by Edward Purcell, McGraw-Hill, New York, 1985 (Section 4.4). Factors such as 2, $\sqrt{2}$, and 2π that require special understanding are not uncommon in physics.

However, the drift speed ($=J/ne$) is proportional to the current density and must be the velocity averaged over *all* the electrons *at one instant of time.*

When we take an average over all electrons, the random, thermal velocities clearly average to zero and make no contribution to the drift velocity. Thus, we need only concern ourselves with the effect of the electric field on the velocity of the electrons.

At any given instant, each electron has a velocity component at, produced by the electric field, where t is the time since the *last* collision experienced by that electron. The average velocity for all the electrons at our given instant—which is the drift speed—is the average value of at. Since the acceleration $a (= eE/m)$ is the same for all electrons, the average value of at is $a\tau$, where τ is the average time since the last collision. It may take some thought to convince yourself that—since you are free to start your stopwatch for each electron at the time it makes a collision—the average time since the last collision, the average time to the next collision, and the average time between collisions, all have the same value τ.

Sample Problem 7   (a) What is the mean free time τ between collisions for the conduction electrons in copper?

From Eq. 19 we have

$$\tau = \frac{m}{ne^2\rho}$$

$$= \frac{9.1 \times 10^{-31}\ \text{kg}}{(8.47 \times 10^{28}\ \text{m}^{-3})(1.60 \times 10^{-19}\ \text{C})^2(1.69 \times 10^{-8}\ \Omega\cdot\text{m})}$$

$$= 2.5 \times 10^{-14}\ \text{s.} \qquad \text{(Answer)}$$

We picked up the value of n, the number of conduction electrons per unit volume in copper, from Sample Problem 3; the value of ρ comes from Table 1.

(b) What is the mean free path λ for these collisions? Assume an effective speed v_{eff} of 1.6×10^6 m/s.

As in Section 21-6, we define the mean free path from

$$\lambda = \tau v_{\text{eff}} = (2.5 \times 10^{-14}\ \text{s})(1.6 \times 10^6\ \text{m/s})$$
$$= 4.0 \times 10^{-8}\ \text{m} = 40\ \text{nm.} \qquad \text{(Answer)}$$

This is about 150 times the distance between nearest-neighbor ions in a copper lattice. A full quantum physics treatment reveals that we cannot view a "collision" as a direct interaction between an electron and an ion. Rather, it is an interaction between an electron and the thermal vibrations of the lattice, lattice imperfections, or lattice impurity atoms. An electron can pass very freely through an "ideal" lattice, that is, a geometrically "perfect" lattice close to the absolute zero of temperature. Mean free paths as large as 10 cm have been observed

under such conditions. Charge carriers in superconductors experience no collisions at all.

28-7   Energy and Power in Electric Circuits

Figure 13 shows a circuit consisting of a battery B connected to a "black box." A steady current i exists in the connecting wires and a steady potential difference V_{ab} exists between the terminals a and b. The box might contain a resistor, a storage battery, or a motor, among many other possibilities.

If a charge element dq moves through the box from terminal a to terminal b, its potential energy will be reduced by $dq\ V_{ab}$. The conservation of energy principle tells us that this energy must appear elsewhere in some form or other. What that form will be depends on what is in the box. In a time interval dt, the energy transferred within the box is then

$$dU = dq\ V_{ab} = i\ dt\ V_{ab}.$$

The *rate P* of energy transfer is dU/dt, or

$$\boxed{P = iV_{ab}} \qquad \text{(rate of electrical energy transfer).}$$
$$(20)$$

If the device in the box is an ideal motor connected to a mechanical load, the energy appears as mechanical work. If the device is a storage battery that is being charged, the energy appears as stored chemical energy in this second battery. If the device is a resistor, the energy

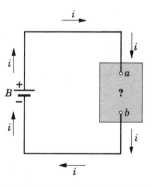

Figure 13   A battery B sets up a current i in a circuit containing a "black box," that is, a box whose contents are unknown to us.

appears as internal thermal energy, revealing itself as a temperature rise of the resistor.

The unit of power that follows from Eq. 20 is the volt · ampere. We can write it as

$$1 \text{ V}\cdot\text{A} = 1 \text{ V}\cdot\text{A} \left(\frac{1 \text{ J}}{1 \text{ V}\cdot\text{C}} \right) \left(\frac{1 \text{ C}}{1 \text{ A}\cdot\text{s}} \right) = 1 \text{ J/s} = 1 \text{ W}.$$

The conversion factor in the first set of parentheses comes from the definition of the volt; that in the second set of parentheses comes from the definition of the coulomb. Recall that we introduced the watt as a unit of mechanical work in Section 7–6.

The course of an electron moving through a resistor at constant drift speed is much like that of a stone falling through water at constant terminal speed. The average kinetic energy of the electron as it moves remains constant so that its lost electric potential energy must appear as thermal energy in the resistor. On a microscopic scale we can understand this since collisions between the electrons and the lattice increase the amplitude of the thermal lattice vibrations; this reveals itself as an increase in the temperature of the lattice.

For a resistor we can combine Eqs. 8 ($R = V/i$) and 20 to obtain, for the rate of electric energy dissipation in a resistor, either

$$\boxed{P = i^2 R} \qquad \text{(resistive dissipation)} \qquad (21)$$

or

$$\boxed{P = \frac{V^2}{R}} \qquad \text{(resistive dissipation).} \qquad (22)$$

Although Eq. 20 applies to electric energy transfers of all kinds, Eqs. 21 and 22 apply only to the transfer of electric potential energy to thermal energy in a resistor.

Sample Problem 8 You are given a length of heating wire made of a nickel–chromium–iron alloy called Nichrome; it has a resistance R of 72 Ω. Can you obtain more heat by winding the wire into a single coil or by cutting the wire in two and winding two separate coils? In each case the coils are to be connected individually across a 120-V line.

The power P for a single coil is, from Eq. 22,

$$P = \frac{V^2}{R} = \frac{(120 \text{ V})^2}{72 \ \Omega} = 200 \text{ W}. \qquad \text{(Answer)}$$

The power for a coil of half length (and thus half resistance) is

$$P' = \frac{V^2}{\frac{1}{2}R} = \frac{(120 \text{ V})^2}{36 \ \Omega} = 400 \text{ W}. \qquad \text{(Answer)}$$

There are two half-coils so that the power obtained from both of them is 800 W, or four times that for a single coil. This would seem to suggest that you could buy a heating coil, cut it in half, and reconnect it to obtain four times the heat output. Why is this not such a good idea?

Sample Problem 9 A wire whose length L is 2.35 m and whose diameter d is 1.63 mm carries a current i of 1.24 A. The wire dissipates thermal energy at the rate P of 48.5 mW. Of what is the wire made?

We can identify the material by its resistivity. From Eqs. 15 and 21 we have

$$P = i^2 R = \frac{i^2 \rho L}{A} = \frac{4 i^2 \rho L}{\pi d^2},$$

in which A $(= \frac{1}{4}\pi d^2)$ is the cross-sectional area of the wire. Solving for ρ, the resistivity of the material of which the wire is made, yields

$$\rho = \frac{\pi P d^2}{4 i^2 L} = \frac{(\pi)(48.5 \times 10^{-3} \text{ W})(1.63 \times 10^{-3} \text{ m})^2}{(4)(1.24 \text{ A})^2(2.35 \text{ m})}$$

$$= 2.80 \times 10^{-8} \ \Omega \cdot \text{m}. \qquad \text{(Answer)}$$

Inspection of Table 1 reveals the material to be aluminum.

28–8 Semiconductors (Optional)

Semiconducting devices are at the heart of the microelectronic revolution that has so influenced our lives. Table 2 compares the properties of silicon—a typical semiconductor—with those of copper—a typical metallic conductor. We see that, compared with copper, silicon (1) has many fewer charge carriers, (2) has a much higher resistivity, and (3) has a temperature coefficient of resistivity that is both large and negative. That is, although the resistivity of copper increases with temperature, that of pure silicon decreases.

The resistivity of pure silicon is so high that it is virtually an insulator and is thus of not much direct use in microelectronic circuits. The property that makes it useful is that—as Table 1 shows—its resistivity can be reduced in a controlled way by adding minute amounts of specific foreign "impurity" atoms, a process called *doping*.

We may fairly conclude that, because their electrical properties are so different, the fundamental conduction process in silicon must be quite different from that for copper. We explore these differences in some detail in Chapter 46 of the extended version of this book, restricting ourselves here to a broad outline.

Table 2   Some Electric Properties of Two Elements[a]

Property	Unit	Copper	Silicon
Type of material	—	Metal	Semiconductor
Density of charge carriers	m^{-3}	9×10^{28}	1×10^{16}
Resistivity	$\Omega \cdot m$	2×10^{-8}	3×10^{3}
Temperature coefficient of resistivity	K^{-1}	$+4 \times 10^{-3}$	-70×10^{-3}

[a] Data rounded off to one significant
figure for easy comparison.

We saw in Section 8–10 (see Fig. 17 of that chapter) that electrons in isolated atoms occupy quantized energy levels, each level containing a single electron. Electrons in solids also occupy quantized levels, as Fig. 14 shows. These levels—whose number is very great—are tightly compressed into allowed *bands* of closely spaced levels. The bands are separated by *gaps,* which represent ranges of energy that electrons may not possess.

In a metallic conductor such as copper (see Fig. 14a), the highest band that contains any electrons—called the *valence band*—is only partially filled. If an applied electric field is to establish a current, it must be possible for the conduction electrons to increase their energies. In a metal such as copper, this poses no problem because many vacant energy levels are readily at hand within the valence band.

In an insulator (Fig. 14b), the valence band is completely filled. The next higher available vacant levels lie in an empty band (called the *conduction band*) separated from the valence band by a considerable energy gap. If an electric field is applied, no current can occur because there is no mechanism by which an electron can increase its energy; the energy jump to the nearest vacant energy level is simply too great.

A semiconductor (Fig. 14c) is like an insulator except that the energy gap between the conduction band and the valence band is small enough so that the probability that electrons might "jump the gap" by thermal agitation is not vanishingly small. More important is the fact that controlled impurities—deliberately added—can contribute charge carriers to the conduction band.* Most semiconducting devices, such as transistors and junction diodes, are fabricated by the selective doping of different regions of the silicon matrix with different kinds of impurity atoms.

Let us now look again at Eq. 19, the expression for the resistivity of a conductor, with the band-gap picture in mind:

$$\rho = \frac{m}{ne^2\,\tau}. \tag{23}$$

Consider how the variables n and τ change as the temperature is increased, n being the number of charge carriers per unit volume and τ the mean time between collisions of the charge carriers.

In a conductor, n is large but very closely constant; that is, its value does not change appreciably with temperature. The increase of resistivity with temperature for metals is caused by an increase in the collision rate of the charge carriers, which shows up in Eq. 23 as a decrease in τ, the mean time between collisions.

In a semiconductor, n is small but increases very rapidly with temperature as the increased thermal agitation makes more charge carriers available. This causes the decrease of resistivity with temperature displayed in Table 2. The same increase in collision rate that we noted

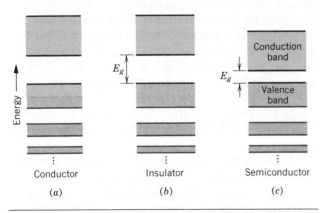

Figure 14   The allowed energy levels for the electrons in a solid form a pattern of allowed bands and forbidden gaps. (a) In a metallic conductor, the valence band is only partially filled. (b) In an insulator, the valence band is completely filled and the gap between the valence band and the conduction band is relatively large. (c) A semiconductor resembles an insulator except that the gap between bands is relatively small.

_____

* Vacancies (called *holes*) in the valence band can also serve as charge carriers. Details are given in Chapter 46.

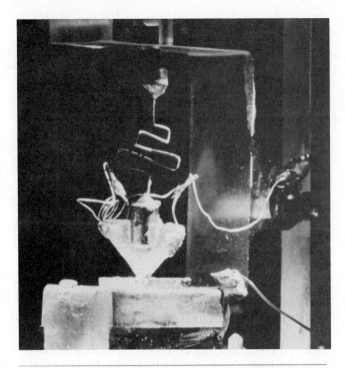

Figure 15 A model of the first transistor. Today, many thousands of such devices can be placed on a thin wafer a few millimeters on edge.

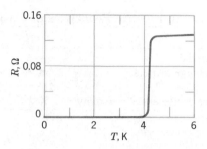

Figure 16 The resistivity of mercury drops to zero at a temperature of about 4 K. Mercury is solid at this low temperature.

for metals also occurs for semiconductors but its effect is swamped by the rapid increase in the number of charge carriers.

We begin to see how the band-gap picture—which is based solidly on quantum physics—can account for the properties of semiconductors. It is no accident that the transistor (see Fig. 15) was discovered by three physicists (William Shockley, John Bardeen, and Walter Brattain) as a specific application of quantum physics to solid materials. These physicists earned the 1956 Nobel Prize for their work.

28-9 Superconductors (Optional)

In 1911, the Dutch physicist Kammerlingh Onnes discovered that the resistivity of mercury absolutely disappears at temperatures below about 4 K; see Fig. 16. This phenomenon of *superconductivity* is of vast potential importance in technology because it means that charges can flow through a conductor without thermal losses. Currents induced in a superconducting ring, for example, have persisted for several years without diminution,

no battery of any kind being present in the circuit. A large superconducting ring is now being used in Tacoma, Washington to store electrical energy. It takes in up to five megawatts during peaks of supply and releases the energy during peaks in demand.

The problem with the technological development of superconductivity has always been the low temperatures that were necessary to maintain it. The magnets in the large Fermilab accelerator, for example, are energized by currents in superconducting coils, which must be maintained at ~4 K, the temperature of liquid helium.

In 1986, however, new ceramic materials were discovered that become superconducting at considerably higher temperatures. As we write this, temperatures as high as 160 K have been reported. As you read this, considerably higher temperatures will almost certainly have been reported, with room-temperature superconductivity a distinct possibility. The normal boiling temperature of liquid nitrogen is 77 K so that this inexpensive coolant—which is cheaper than bottled water—can be used in place of the much more expensive liquid helium. Conjectured applications run the gamut from magnetically levitated trains to desk-top main-frame computers to powerful motors the size of a walnut.

Superconductivity must not be thought of as simply a dramatic improvement in the normal conductivity process that we described in Section 28-6. The two processes are completely different. In fact, the best normal conductors, such as silver or copper, do not become superconducting; on the other hand, the recently discovered "supersuperconductors" are ceramic materials, which—as far as normal conduction is concerned—are insulators.

The mechanism of superconductivity remained unexplained for some 60 years after the discovery of the phenomenon. Then John Bardeen,* Leon Cooper, and Robert Schrieffer advanced a theoretical explanation, for which they were jointly awarded the 1972 Nobel Prize. The heart of the BCS theory — as it is called, after the initials of its developers — is the assumption that the charge carriers are not single electrons but pairs of electrons. These *Cooper pairs* behave like single particles, with properties dramatically different from those of single electrons.

---

* Yes, this is the same John Bardeen who shared the 1956 Nobel Prize for discovering the transistor; see Section 28–8. Professor Bardeen is the only person who has earned two Nobel Prizes in the same field.

Electrons normally repel each other so that some special mechanism is needed to induce them to form a pair. A semiclassical picture that helps in understanding this quantum BCS phenomenon is as follows: An electron plows through the lattice, distorting it slightly and thus leaving in its wake a very short-lived concentration of enhanced positive charge. If a second electron is nearby at the right moment, it may well be attracted to this region by the positive charge, thus forming a pair with the first electron. It is known that the newly discovered superconductors operate by means of Cooper pairs but, as of 1988, there is no universal agreement as to the mechanism by which these pairs are formed.

REVIEW AND SUMMARY

Current i

An *electric current i* in a conductor is defined by

$$dq = i \, dt. \tag{1}$$

Here dq is the amount of (positive) charge that passes in time dt through a hypothetical surface that cuts across the conductor; see Sample Problem 1. The direction of electric current is the direction in which positive charge carriers would move. The SI unit of electric current is the *ampere* ($= 1$ C/s; abbr. A).

Current (a scalar) is related to *current density* $\mathbf{J}$ (a vector) by

Current Density $\mathbf{J}$

$$i = \int \mathbf{J} \cdot d\mathbf{A} \tag{5}$$

where $d\mathbf{S}$ is an element of area and the integral is taken over any surface cutting across the conductor. The direction of $\mathbf{J}$ at any point is that in which a positive charge carrier would move if placed at that point; see Fig. 5 and Sample Problem 2.

When an electric field $\mathbf{E}$ is established in a conductor the charge carriers (assumed positive) acquire a *mean drift speed* $\mathbf{v}_d$ in the direction of $\mathbf{E}$, related to the current density by

The Mean Drift Speed of the Charge Carriers

$$\mathbf{J} = (ne)\mathbf{v}_d \tag{7}$$

where (ne) is the charge density; see Sample Problems 3 and 4.

The *resistance R* between any two equipotential surfaces of a conductor is defined from

The Resistance of a Conductor

$$R = V/i \quad \text{(definition of } R) \tag{8}$$

where V is the potential difference between those surfaces and i is the current. The SI unit of R is the *ohm* ($= 1$ V/A; abbr. Ω). Similar equations define the *resistivity ρ* and *conductivity σ* of a material:

Resistivity and Conductivity

$$\rho = \frac{1}{\sigma} = \frac{E}{j} \quad \text{(definitions of } \rho \text{ and } \sigma) \tag{10,12}$$

where E is the applied electric field. The SI unit of resistivity is the ohm-meter (abbr. $\Omega \cdot$ m); see Table 1 and Sample Problem 5. Eq. 10 corresponds to the vector equation

$$\mathbf{E} = \rho \mathbf{J}. \tag{11}$$

The resistance R for a cylindrical conductor of any cross-sectional shape is

$$R = \rho \frac{L}{A};$$ [15]

see Sample Problem 6.

The resistivity ρ for most materials changes with temperaure. For many materials, including metals, the empirical linear relationship is

The Change of ρ with Temperature

$$\rho - \rho_0 = \rho_0 \alpha (T - T_0)].$$ [16]

Here T_0 is a reference temperature, ρ_0 is the resistivity at T_0, and α is a mean temperature coefficient of resistivity; see Table 1.

Ohm's Law

A given *conductor* obeys *Ohm's law* if its resistance R, defined by Eq. 8, is independent of the applied potential difference V; compare Figs. 11b and 11c. A given *material* obeys Ohm's law if its resistivity, defined by Eq. 10, is independent of the magnitude and direction of the applied electric field **E**.

By treating the conduction electrons in a metal like the molecules of a gas it is possible to derive for the resistivity of a metal

Resistivity of a Metal

$$\rho = \frac{m}{e^2 n \tau}.$$ [19]

Here n is the number of electrons per unit volume and τ is the mean time between the collisions of an electron with the ion cores of the lattice. The discussion based on Fig. 12 shows that τ is independent of E and thus accounts for the fact that metals obey Ohm's law; see Sample Problem 7.

The power P or rate of energy transfer in an electric device across which a potential difference V_{ab} is maintained is

Power

$$P = i V_{ab} \quad \text{(rate of electrical energy transfer).}$$ [20]

If the device is a resistor we can write this as

Resistive Dissipation

$$P = i^2 R = \frac{V^2}{R} \quad \text{(resistive dissipation);}$$ [21,22]

see Sample Problems 8 and 9. In a resistor electrical potential energy is transferred to the lattice by the drifting charge carriers, appearing as internal thermal energy.

Semiconductors

Semiconductors are materials with few conduction electrons but with available conduction-level states close, in energy, to their valence bands. These then become conductors either by thermal agitation of electrons or, more importantly, by *doping* the material with other atoms which contribute electrons or holes to the conduction band.

Superconductors

Superconductors lose all electrical resistance at temperatures below some critical value. Recent research has discovered materials with increasingly higher critical temperatures, leading to the possibility of room temperature (or, at worst, liquid nitrogen temperature) superconducting devices.

QUESTIONS

1. What conclusions can you draw by applying Eq. 5 to a closed surface through which a number of wires pass in random directions, carrying steady currents of different magnitudes?

2. In our convention for the direction of current arrows (*a*) would it have been more convenient, or even possible, to have assumed all charge carriers to be negative? (*b*) Would it have been more convenient, or even possible, to have labeled the electron as positive, the proton as negative, etc?

3. List in tabular form similarities and differences between the flow of charge along a conductor, the flow of water through a horizontal pipe, and the conduction of heat through a slab. Consider such ideas as what causes the flow, what opposes it, what particles (if any) participate, and the units in which the flow may be measured.

4. Explain in your own words why we can have $\mathbf{E} \neq 0$ inside a conductor in this chapter whereas we took $\mathbf{E} = 0$ for granted in Section 7 of Chapter 25.

5. Let a battery be connected to a copper cube at two corners defining a body diagonal. Pass a hypothetical plane completely through the cube, tilted at an arbitrary angle. (a) Is the current i through the plane independent of the position and orientation of the plane? (b) Is there any position and orientation of the plane for which $\mathbf{J}$ is a constant in magnitude, direction, or both? (c) Does Eq. 5 hold for all orientations of the plane? (d) Does Eq. 5 hold for a closed surface of arbitrary shape, which may or may not lie entirely within the cube?

6. A potential difference V is applied to a copper wire of diameter d and length L. What is the effect on the electron drift speed of (a) doubling V, (b) doubling L, and (c) doubling d?

7. Why is it not possible to measure the drift speed for electrons by timing their travel along a conductor?

8. A potential difference V is applied to a circular cylinder of carbon by clamping it between circular copper electrodes, as in

Copper

Carbon

Copper

Figure 17 Question 8.

Fig. 17. Discuss the difficulty of calculating the resistance of the carbon cylinder using the relation $R = \rho L/A$.

9. How would you measure the resistance of a pretzel-shaped metal block? Give specific details to clarify the concept.

10. Sliding across the seat of an automobile can generate potentials of several thousand volts. Why isn't the slider electrocuted?

11. Discuss the difficulties of testing whether the filament of a light bulb obeys Ohm's law.

12. How does the relation $V = iR$ apply to resistors that do *not* obey Ohm's law?

13. A cow and a man are standing in a meadow when lightning strikes the ground nearby. Why is the cow more likely to be killed than the man? The responsible phenomenon is called "step voltage."

14. The gray lines in Fig. 12 should be curved slightly. Why?

15. A fuse in an electrical circuit is a wire that is designed to melt, and thereby open the circuit, if the current exceeds a predetermined value. What are some characteristics of an ideal fuse wire?

16. Why does an incandescent light bulb grow dimmer with use?

17. The character and quality of our daily lives is influenced greatly by devices that do not obey Ohm's law. What can you say in support of this claim?

18. From a student's paper: "The relationship $R = V/i$ tells us that the resistance of a conductor is directly proportional to the potential difference applied to it." What do you think of this proposition?

19. Carbon has a negative temperature coefficient of resistivity. This means that its resistivity drops as its temperature increases. Would its resistivity disappear entirely at some high enough temperature?

20. What special characteristics must heating wire have?

21. Equation 21 ($P = i^2R$) seems to suggest that the rate of increase of thermal energy in a resistor is reduced if the resistance is made less; Eq. 22 ($P = V^2/R$) seems to suggest just the opposite. How do you reconcile this apparent paradox?

22. Why do electric power companies reduce voltage during times of heavy demand? What is being saved?

23. Is the filament resistance lower or higher in a 500-W light bulb than in a 100-W bulb? Both bulbs are designed to operate on 120 V.

24. Five wires of the same length and diameter are connected in turn between two points maintained at constant potential difference. Will thermal energy be developed at the faster rate in the wire of (a) the smallest or (b) the largest resistance?

25. Why is it better to send 10,000 kW of electric power long distances at 10,000 volts rather than at 220 volts?

EXERCISES AND PROBLEMS

Section 28–2   Electric Current

1E. A current of 5.0 A exists in a 10-Ω resistor for 4.0 min. (a) How many coulombs and (b) how many electrons pass through any cross section of the resistor in this time?

2E. The current in the electron beam of a typical video dis-

play terminal is 200 μA. How many electrons strike the screen each second?

3P. You are given an isolated conducting sphere of 10-cm radius. One wire carries a current of 1.0000020 A into it. Another wire carries a current of 1.0000000 A out of it. How long would it take for the sphere to increase in potential by 1000 V?

4P. The belt of an electrostatic generator is 50 cm wide and travels at 30 m/s. The belt carries charge into the sphere at a rate corresponding to 100 μA. Compute the surface charge density on the belt. See Section 11 of Chapter 26.

Section 28–3 Current Density

5E. We have 2.0×10^8 doubly charged positive ions per cubic centimeter, all moving north with a speed of 1.0×10^5 m/s. (a) What is the current density **J**, in magnitude and direction? (b) Can you calculate the total current i in this ion beam? If not, what additional information is needed?

6E. A small but measurable current of 1.2×10^{-10} A exists in a copper wire whose diameter is 2.5 mm. Calculate (a) the current density and (b) the electron drift speed. See Sample Problem 3.

7E. A fuse in an electrical circuit is a wire that is designed to melt, and thereby open the circuit, if the current exceeds a predetermined value. Suppose that the material composing the fuse melts once the current density rises to 440 A/cm². What diameter of cylindrical wire should be used to limit the current to 0.50 A?

8E. The (United States) National Electric Code, which sets maximum safe currents for rubber-insulated copper wires of various diameters, is given (in part) below. Plot the safe current density as a function of diameter. Which wire gauge has the maximum safe current density?

Gauge[a]	4	6	8	10	12	14	16	18
Diameter (mils)[b]	204	162	129	102	81	64	51	40
Safe current (A)	70	50	35	25	20	15	6	3

[a] A way of identifying the wire diameter.
[b] 1 mil = 10^{-3} in.

9E. A current is established in a gas discharge tube when a sufficiently high potential difference is applied across the two electrodes in the tube. The gas ionizes; electrons move toward the positive terminal and singly charged positive ions toward the negative terminal. What are the magnitude and direction of the current in a hydrogen discharge tube in which 3.1×10^{18} electrons and 1.1×10^{18} protons move past a cross-sectional area of the tube each second?

10E. A p-n junction is formed from two different semiconducting materials in the form of identical cylinders with radius 0.165 mm, as depicted in Fig. 18. In one application 3.50×10^{15} electrons per second flow across the junction from the n to the p side while 2.25×10^{15} holes per second flow from the p to the n side. (A hole acts like a particle with charge $+1.6 \times 10^{-19}$ C.) What are (a) the total current and (b) the current density?

11P. Near the earth, the density of protons in the solar wind (see Section 1) is 8.7 cm⁻³ and their speed is 470 km/s. (a) Find the current density of these protons. (b) If the earth's magnetic

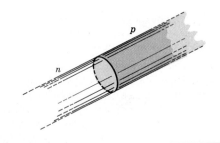

Figure 18   Exercise 10.

field did not deflect them, the protons would strike the earth. What total current would the earth receive?

12P. In a hypothetical fusion research lab, high temperature helium gas is completely ionized, each helium atom being separated into two free electrons and the remaining positively charged nucleus (alpha particle). An applied electric field causes the alpha particles to drift to the east at 25 m/s while the electrons drift to the west at 88 m/s. The alpha particle density is 2.8×10^{15} cm⁻³. Calculate the net current density; specify the current direction.

13P. How long does it take electrons to get from a car battery to the starting motor? Assume the current is 300 A and the electrons travel through a copper wire with cross-sectional area 0.21 cm² and length 0.85 m. See Sample Problem 3.

14P. A steady beam of alpha particles ($q = 2e$) traveling with constant kinetic energy 20 MeV carries a current 0.25 μA. (a) If the beam is directed perpendicular to a plane surface, how many alpha particles strike the surface in 3.0 s? (b) At any instant, how many alpha particles are there in a given 20-cm length of the beam? (c) Through what potential difference was it necessary to accelerate each alpha particle from rest to bring it to an energy of 20 MeV?

15P. (a) The current density across a cylindrical conductor of radius R varies according to the equation

$$J = J_0(1 - r/R),$$

where r = distance from the axis. Thus, the current density is a maximum J_0 at the axis $r = 0$ and decreases linearly to zero at the surface $r = R$. Calculate the current in terms of J_0 and the conductor's cross-sectional area $A = \pi R^2$. (b) Suppose that, instead, the current density is a maximum J_0 at the surface and decreases linearly to zero at the axis, so that

$$J = J_0 r/R.$$

Calculate the current. Why is the result different from (a)?

Section 28–4 Resistance and Resistivity

16E. A steel trolley-car rail has a cross-sectional area of 56 cm². What is the resistance of 10 km of rail? The resistivity of the steel is 3.0×10^{-7} $\Omega \cdot$m.

17E. A conducting wire has a 1.0-mm diameter, a 2.0-m

length, and a 50-mΩ resistance. What is the resistivity of the material?

18E. A human being can be electrocuted if a current as small as 50 mA passes near the heart. An electrician working with sweaty hands makes good contact with the two conductors he is holding. If his resistance is 2000 Ω, what might the fatal voltage be?

19E. A coil is formed by winding 250 turns of insulated gauge 16 copper wire (diameter = 1.3 mm) in a single layer on a cylindrical form whose radius is 12 cm. What is the resistance of the coil? Neglect the thickness of the insulation. See Table 1.

20E. A wire 4.0 m long and 6.0 mm in diameter has a resistance of 15 mΩ. If a potential difference of 23 V is applied between the ends, (a) what is the current in the wire? (b) What is the current density? (c) Calculate the resistivity of the wire material. Can you identify the material? See Table 1.

21E. A wire of Nichrome (a nickel-chromium–iron alloy commonly used in heating elements) is 1.0 m long and 1.0 mm² in cross-sectional area. It carries a current of 4.0 A when a 2.0-V potential difference is applied between its ends. Calculate the conductivity σ of Nichrome.

22E. (a) At what temperature would the resistance of a copper conductor be double its resistance at 20°C? (Use 20°C as the reference point in Eq. 16; compare your answer with Fig. 10.) (b) Does this same temperature hold for all copper conductors, regardless of shape or size?

23E. The copper windings of a motor have a resistance of 50 Ω at 20°C when the motor is idle. After running for several hours the resistance rises to 58 Ω. What is the temperature of the windings? Ignore changes in the dimensions of the windings. See Table 1.

24E. Using data from Fig. 11c, plot the resistance of the *pn* junction as a function of applied potential difference.

25E. A 4.0-cm-long caterpillar crawls in the direction of electron drift along a 5.2-mm-diameter bare copper wire that carries a current of 12 A. (a) What is the potential difference between the two ends of the caterpillar? (b) Is its tail positive or negative compared to its head? (c) How much time could it take the caterpillar to crawl 1.0 cm and still keep up with the drifting electrons in the wire?

26E. A cylindrical copper rod of length L and cross-sectional area A is reformed to twice its original length with no change in volume. (a) Find the new cross-sectional area. (b) If the resistance between its ends was R before the change, what is it after the change?

27E. A wire with a resistance of 6.0 Ω is drawn out through a die so that its new length is three times its original length. Find the resistance of the longer wire, assuming that the resistivity and density of the material are not changed during the drawing process.

28E. A certain wire has a resistance R. What is the resistance of a second wire, made of the same material, that is half as long and has half the diameter?

29P. What must be the diameter of an iron wire if it is to have the same resistance as a copper wire 1.2-mm in diameter, both wires being the same length?

30P. Two conductors are made of the same material and have the same length. Conductor A is a solid wire of diameter 1.0 mm. Conductor B is a hollow tube of outside diameter 2.0 mm and inside diameter 1.0 mm. What is the resistance ratio, R_A/R_B, measured between their ends?

31P. A copper wire and an iron wire of the same length have the same potential difference applied to them. (a) What must be the ratio of their radii if the current is to be the same? (b) Can the current density be made the same by suitable choices of the radii?

32P. A square aluminum rod is 1.3 m long and 5.2 mm on edge. (a) What is the resistance between its ends? (b) What must be the diameter of a circular 1.3-m copper rod if its resistance is to be the same?

33P. A potential difference V is applied to a wire of cross section A, length L, and resistivity ρ. You want to change the applied potential difference and draw out the wire so the power dissipated is increased by a factor of 30 and the current is increased by a factor of 4. What should be the new values of L and A?

34P. A rod of a certain metal is 1.6 m long and 5.5 mm in diameter. The resistance between its ends (at 20°C) is 1.09×10^{-3} Ω. A round disk is formed of this same material, 2.00 cm in diameter and 1.00 mm thick. (a) What is the material? (b) What is the resistance between the opposing round faces, assuming equipotential surfaces?

35P. An electrical cable consists of 125 strands of fine wire, each having 2.65-μΩ resistance. The same potential difference is applied between the ends of each strand and results in a total current of 0.75 A. (a) What is the current in each strand? (b) What is the applied potential difference? (c) What is the resistance of the cable?

36P. A common flashlight bulb is rated at 0.30 A and 2.9 V, the values of the current and voltage under operating conditions. If the resistance of the bulb filament when cold is 1.1 Ω, calculate the temperature of the filament when the bulb is on. The filament is made of tungsten.

37P. When 115 V is applied across a 0.30-mm radius, 10-m-long wire, the current density is 1.4×10^4 A/m². Find the resistivity of the wire.

38P. A block in the shape of a rectangular solid has a cross-sectional area of 3.50 cm², a length of 15.8 cm, and a resistance of 935 Ω. The material of which the block is made has 5.33×10^{22} conduction electrons/m³. A potential difference of 35.8 V is maintained between its ends. (a) What is the current in the block? (b) If the current density is uniform, what is its value? (*c*) What is the drift velocity of the conduction electrons? (*d*) What is the electric field in the block?

39P. Copper and aluminum are being considered for a high-voltage transmission line that must carry a current of 60 A. The resistance per unit length is to be 0.15 Ω/km. Compute for each choice of cable material (*a*) the current density and (*b*) the mass per meter of the cable. The densities of copper and aluminum are 8960 and 2700 kg/m^3, respectively.

40P. In the lower atmosphere of the earth there are negative and positive ions, created by radioactive elements in the soil and cosmic rays from space. In a certain region, the atmospheric electric field strength is 120 V/m, directed vertically down. Due to this field, singly charged positive ions, 620 per cm^3, drift downward and singly charged negative ions, 550 per cm^3, drift upward; see Fig. 19. The measured conductivity is $2.7 \times 10^{-14}/\Omega \cdot$m. (*a*) Calculate the ion drift speed, assumed the same for positive and negative ions, and (*b*) the current density.

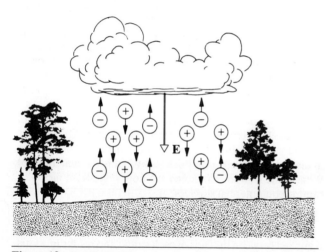

Figure 19 Problem 40.

41P. If the gauge number of a wire is increased by 6, the diameter is halved; if a gauge number is increased by 1, the diameter decreases by the factor $2^{1/6}$ (see the table in Exercise 8). Knowing this, and also knowing that 1000 ft of #10 copper wire has a resistance of approximately 1.00 Ω, estimate the resistance of 25 ft of #22 copper wire.

42P. When a metal rod is heated, not only its resistance but also its length and its cross-sectional area change. The relation $R = \rho L/A$ suggests that all three factors should be taken into account in measuring ρ at various temperatures. (*a*) If the temperature changes by 1.0°C, what percentage changes in R, L, and A occur for a copper conductor? (*b*) What conclusion do you draw? The coefficient of linear expansion is 1.7×10^{-5}/°C.

43P. A resistor is in the shape of a truncated right circular cone (Fig. 20). The end radii are a and b, the altitude is L. If the taper is small, we may assume that the current density is uniform across any cross section. (*a*) Calculate the resistance of this object. (*b*) Show that your answer reduces to $\rho(L/A)$ for the special case of zero taper ($a = b$).

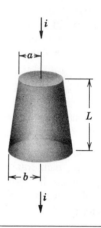

Figure 20 Problem 43.

Section 28–6 Ohm's Law—A Microscopic View

44P. Show that, according to the free-electron model of electrical conduction in metals and classical physics, the resistivity of metals should be proportional to $\sqrt{T}$, where T is absolute temperature. (See Eq. 21 of Chapter 21.)

Section 28–7 Energy and Power in Electric Circuits

45E. A student kept his 9.0-V, 7.0-W portable radio turned on from 9:00 p.m. until 2:00 a.m. How much charge went through it?

46E. An x-ray tube takes a current of 7.0 mA and operates at a potential difference of 80 kV. What power in watts is dissipated?

47E. Thermal energy is developed in a resistor at a rate of 100 W when the current is 3.0 A. What is the resistance?

48E. The headlights of a moving car draw about 10 A from the 12-V alternator, which is driven by the engine. Assume the alternator is 80 percent efficient and calculate the horsepower the engine must supply to run the lights.

49E. A space heater, operating from a 120-V line, has a hot resistance of 14 Ω. (*a*) At what rate is electrical energy transfered into heat? (*b*) At 5.0¢/kW·h, what does it cost to operate the device for 5.0 h?

50E. An unknown resistor is connected between the terminals of a 3.0-V battery. The power dissipated in the resistor is

0.54 W. The same resistor is then connected between the terminals of a 1.5-V battery. What power is dissipated in this case?

51E. A 500-W space heater operates from a 120-V line. (*a*) What is its (hot) resistance? (*b*) At what rate do electrons flow through any cross section of the filament?

52E. The National Board of Fire Underwriters has fixed safe current-carrying capacities for various sizes and types of wire. For #10 rubber-coated copper wire (diameter = 0.10 in.) the maximum safe current is 25 A. At this current, find (*a*) the current density, (*b*) the electric field, (*c*) the potential difference for 1000 ft of wire, and (*d*) the rate at which thermal energy is developed for 1000 ft of wire.

53E. A potential difference of 1.2 V is applied to a 33-m length of #18 copper wire (diameter = 0.040 in.). Calculate (*a*) the current, (*b*) the current density, (*c*) the electric field, and (*d*) the rate at which thermal energy is developed in the wire.

54P. A cylindrical resistor of radius 5.0 mm and length 2.0 cm is made of material that has a resistivity of 3.5×10^{-5} $\Omega \cdot$m. What are (*a*) the current density and (*b*) the potential difference when the power dissipation is 1.0 W?

55P. A heating element is made by maintaining a potential difference of 75 V along the length of a Nichrome wire with a 2.6×10^{-6} m² cross-section and a resistivity of 5.0×10^{-7} $\Omega \cdot$m. (*a*) If the element dissipates 5000 W, what is its length? (*b*) If a potential difference of 100 V is used to obtain the same power output, what should the length be?

56P. A 1250-W radiant heater is constructed to operate at 115 V. (*a*) What will be the current in the heater? (*b*) What is the resistance of the heating coil? (*c*) How much thermal energy is generated in one hour by the heater?

57P. A 100-W light bulb is plugged into a standard 120-V outlet. (*a*) How much does it cost per month to leave the light turned on? Assume electric energy costs 6¢/kW·h. (*b*) What is the resistance of the bulb? (*c*) What is the current in the bulb? (*d*) Is the resistance different when the bulb is turned off?

58P. A Nichrome heater dissipates 500 W when the applied potential difference is 110 V and the wire temperature is 800°C. How much power would it dissipate if the wire temperature were held at 200°C by immersion in a bath of cooling oil? The applied potential difference remains the same; α for Nichrome at 800°C is 4.0×10^{-4}/°C.

59P. A beam of 16-MeV deuterons from a cyclotron falls on a copper block. The beam is equivalent to a current of 15 μA. (*a*) At what rate do deuterons strike the block? (*b*) At what rate is thermal energy produced in the block?

60P. An electron linear accelerator produces a pulsed beam of electrons. The pulse current is 0.50 A and the pulse duration 0.10 μs. (*a*) How many electrons are accelerated per pulse? (*b*) What is the average current for a machine operating at 500 pulses/s? (*c*) If the electrons are accelerated to an energy of 50 MeV, what are the average and peak power outputs of the accelerator?

61P. A coil of current-carrying Nichrome wire is immersed in a liquid contained in a calorimeter. When the potential difference across the coil is 12 V and the current through the coil is 5.2 A, the liquid boils at a steady rate, evaporating at the rate of 21 mg/s. Calculate the heat of vaporization of the liquid, in cal/g.

62P. A resistance coil, wired to an external battery, is placed inside an adiabatic cylinder fitted with a frictionless piston and containing an ideal gas. A current i = 240 mA flows through the coil, which has a resistance R = 550 Ω. At what speed v must the piston, mass m = 12 kg, move upward in order that the temperature of the gas remains unchanged? See Fig. 21.

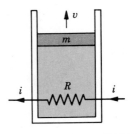

Figure 21 Problem 62.

63P. A 500-W heating unit is designed to operate from a 115-V line. (*a*) By what percentage will its heat output drop if the line voltage drops to 110 V? Assume no change in resistance. (*b*) Taking the variation of resistance with temperature into account, would the actual heat output drop be larger or smaller than that calculated in (*a*)?

64P. An electric immersion heater normally takes 100 min to bring cold water in a well-insulated container to a certain temperature, after which a thermostat switches the heater off. One day the line voltage is reduced by 6.0 percent because of a laboratory overload. How long will it now take to heat the water? Assume that the resistance of the heating element is the same for each of these two modes of operation.

65P. A 400-W immersion heater is placed in a pot containing 2.0 liters of water at 20°C. (*a*) How long will it take to bring the water to boiling temperature, assuming that 80% of the available energy is absorbed by the water? (*b*) How much longer will it take to boil half the water away?

66P. A 30-μF capacitor is connected across a programmed power supply. During the interval from $t = 0$ to $t = 3$ s the output voltage of the supply is given by $V(t) = 6 + 4t - 2t^2$ volts. At $t = 0.5$ s find (*a*) the charge on the capacitor, (*b*) the current into the capacitor, and (*c*) the power output from the power supply.

CHAPTER 29

ELECTROMOTIVE FORCE AND CIRCUITS

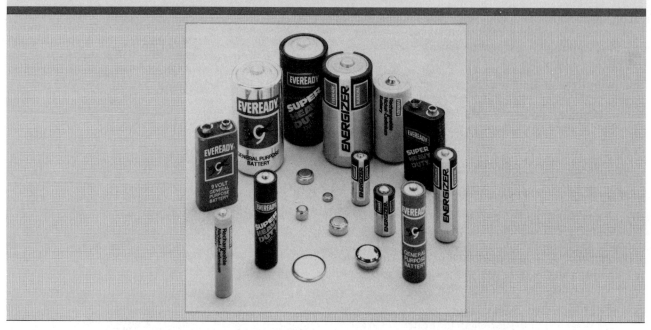

Batteries, whose bewildering variety is indicated by this small sample, free us from the tyranny of the extension cord and make possible activities that we could not otherwise imagine. They bring light and sound to remote places and make it possible to do calculations on a mountain top. Did you know that every Polaroid film pack contains a battery?

29–1 "Pumping" Charges

If you want to make charge carriers flow through a resistor, you must establish a potential difference between its ends. One way to do this is to connect each end of the resistor to a conducting sphere, one sphere being charged negatively and the other positively, as in Fig. 1. The trouble with this scheme is that the flow of charge acts to discharge the spheres, bringing them quickly to the same potential. When that happens, the flow of charge stops.

To maintain a steady flow of coolant in the cooling system of your car, you need a *water pump*, a device that—by doing work on the fluid—maintains a pressure difference between its input and its output ends. In the electrical case, we need a *charge pump*, a device that

—by doing work on the charge carriers—maintains a potential difference between its terminals. We call such a device a seat of *electromotive force* (symbol $\mathscr{E}$; abbr. emf).

A common seat of emf is the *battery*, used to power devices from wristwatches to submarines. The seat of

Figure 1 A steady current cannot exist because there is no mechanism to maintain a steady potential difference across the resistor. When the energy stored in the electric fields of the charged spheres has all been transferred to thermal energy in the resistor, the current must stop.

emf that most influences our daily lives, however, is the *electric generator,* whose output potential difference is led into our homes and workplaces from (usually) a remote generating plant. *Solar cells,* long familiar as the winglike panels on spacecraft, also dot the countryside for domestic applications. Less familiar seats of emf are the *fuel cells* that power the Space Shuttles and the *thermopiles* that provide onboard electric power for some spacecraft and for remote stations in Antarctica and elsewhere. Another example is the *electrostatic generator,* in which the potential difference is maintained by the mechanical movement of charge on an insulating belt. Living systems, ranging from electric eels and human beings to plants, are also seats of emf.

Although the devices that we have listed differ widely in their modes of operation, they all perform the basic function of a seat of emf: They are able to do work on charge carriers and thus maintain a potential difference between their output terminals.

29-2   Work, Energy, and Electromotive Force

Figure 2*a* shows a seat of emf (which, for concreteness, you may think of as a battery) as part of a simple circuit. The seat maintains its upper terminal positive and its lower terminal negative, as shown by the + and − signs. We represent its emf by an arrow placed next to the seat and pointing in the direction in which the seat, acting alone, would cause a positive charge carrier to move in the external circuit. In Fig. 2*a*, this direction is clockwise. We draw a small circle on the tail of an emf arrow so that we do not confuse it with a current arrow.

Within the seat of emf, the circulating positive charge carriers must move from a region of low potential (the negative terminal) to a region of high potential (the positive terminal). This is just opposite to the direction in which the electric field between the terminals would compel them to move.

We conclude that there must be some source of energy within the seat, enabling it to do work on the charges and thus forcing them to move as they do. The energy source may be chemical, as in a battery or a fuel cell. It may involve mechanical forces, as in a conventional generator or an electrostatic generator. Temperature differences may supply the motive power, as in a thermopile; or solar energy may supply it, as in a solar cell.

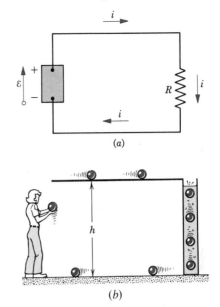

Figure 2   (*a*) A simple electric circuit, in which the emf $\mathscr{E}$ does work on the charge carriers and maintains a steady current through the resistor. (*b*) Its gravitational analog. Work done by the person maintains a steady flow of bowling balls through the viscous medium.

Let us analyze the circuit of Fig. 2*a* from the point of view of work and energy transfers. In any time interval dt, a charge dq passes through any cross section of this circuit. In particular, this amount of charge must enter the seat of emf at its low-potential end and must leave at its high-potential end. The seat must do an amount of work dW on the charge element dq to force it to move in this way. We define the emf of the seat from

$$\mathscr{E} = \frac{dW}{dq} \qquad \text{(definition of } \mathscr{E}\text{)}. \qquad (1)$$

The SI unit for emf that follows from Eq. 1 is the joule per coulomb, which we have met before (Eq. 5 of Chapter 26) and have called the *volt* (abbr. V). An ideal battery with an emf of 2.1 V would maintain a 2.1-V potential difference between its terminals.* The electromotive force, incidentally, is not a force; we measure its magnitude in volts and not in newtons. The name—like

---

* A *real* battery—in contrast to an ideal battery—only does so on open circuits, that is, if no current is passing through the battery. We clarify this point in the following section.

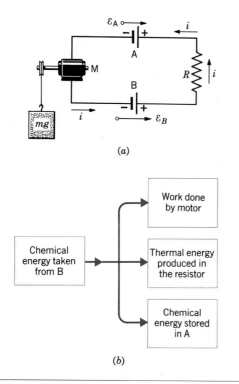

(a)

(b)

Figure 3 (a) $\mathcal{E}_B > \mathcal{E}_A$ so that battery **B** determines the direction of the current in this single-loop circuit. (b) Energy transfers in this circuit.

many other names in physics—is involved with the early history of the subject.

Figure 2b shows a gravitational analog to the circuit of Fig. 2a. In Fig. 2a, the seat of emf—which we take to be a battery—does work on the charge carriers, depleting its store of chemical energy. This energy appears as thermal energy in the resistor. In Fig. 2b, the person, in lifting the bowling balls from the floor to the shelf, does work on these "mass carriers." The balls roll slowly along the shelf, dropping from the right end into a cylinder of viscous oil. They sink to the bottom at an essentially constant terminal speed, are removed by a trapdoor mechanism not shown, and roll back along the floor to their starting position. The work done by the person, at the expense of her store of internal biochemical energy, appears as thermal energy in the viscous fluid, whose temperature rises slightly.

The circulation of charges in Fig. 2a will eventually stop if the battery does not replenish its store of chemical energy by being recharged. The circulation of bowling balls in the circuit of Fig. 2b will also eventually stop if

the person does not replenish her store of biochemical energy by eating.

Figure 3a shows a circuit containing two storage batteries, A and B, a resistor R, and an (ideal) electric motor used to lift a weight. The batteries are connected so that they tend to send charges around the circuit in opposite directions. The actual direction of current is determined by battery B, which has the larger emf. Figure 3b shows the energy transfers in this circuit. The chemical energy in B is steadily depleted, the energy appearing in the three forms shown on the right. Battery A is being charged while battery B is being discharged. That is, the energy of charges traversing battery A decreases as they pass through that battery; therefore, the charges do work on battery A, which is stored internally as chemical energy.

29-3 Calculating the Current

We present here two equivalent ways to calculate the current in the simple circuit of Fig. 4, one method based on considerations of energy conservation and the other on the concept of potential.

Energy Method.   Joule's law ($P = i^2R$; see Eq. 21 of Chapter 28) tells us that in a time interval dt an amount of energy given by $i^2R\,dt$ will appear in the resistor of Fig. 4 as thermal energy. During this same interval, a charge $dq\ (= i\,dt)$ will have moved through the seat of emf and the seat will have done work on this charge, according to Eq. 1:

$$dW = \mathcal{E}\,dq = \mathcal{E}i\,dt.$$

From the principle of conservation of energy, the work done by the seat of emf must equal the thermal energy that appears in the resistor, or

$$\mathcal{E}i\,dt = i^2R\,dt.$$

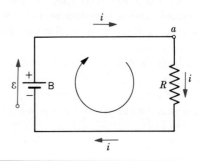

Figure 4   A single-loop circuit, in which $i = \mathcal{E}/R$.

Solving for i, we obtain

$$i = \frac{\mathcal{E}}{R}.\qquad(2)$$

Potential Method. If electric potential is to have any meaning, a given point in a circuit can have only a single value of the potential at a given time. Let us start at any point in the circuit of Fig. 4 and, in imagination, go around the circuit in either direction, adding algebraically the potential differences that we encounter. When we arrive at our starting point we must have returned to our starting potential. We formalize this in a statement that holds not only for single-loop circuits such as that of Fig. 4 but for any complete loop in a multiloop circuit:

Loop Rule.* The algebraic sum of the changes in potential encountered in a complete traversal of any closed circuit must be zero.

In a gravitational analog, this is no more than saying that any point on the side of a mountain must have a unique value of the gravitational potential, that is, a unique elevation above sea level. If you start from any point and return to it after walking around on the mountain, the algebraic sum of the changes in elevation that you encounter must be zero.

In Fig. 4, let us start at point a, whose potential is V_a, and traverse the circuit clockwise. In going through the resistor, there is a change in potential of $-iR$. The minus sign shows that the top of the resistor is at a higher potential than the bottom. As we traverse the battery from bottom to top, there is an increase in potential of $+\mathcal{E}$. Adding the potential changes algebraically to the initial potential must yield the initial potential. Thus,

$$V_a - iR + \mathcal{E} = V_a.$$

Solving for i leads directly to Eq. 2.

These two methods for finding the current are completely equivalent because, as we learned in Section 26–2, potential differences are defined in terms of work and energy.

To prepare for the study of circuits more complex

than that of Fig. 4, let us set down two "rules" for finding potential differences:

Resistor Rule. If you traverse a resistor in the direction of the current, the change in potential is $-iR$: in the opposite direction it is $+iR$. In a gravitational analog: if you walk downstream in a brook, your elevation decreases; if you walk upstream, it increases.

Emf Rule. If you traverse a seat of emf in the direction of the emf, the change in potential is $+\mathcal{E}$; in the opposite direction it is $-\mathcal{E}$.

These rules, which follow from our previous discussion, are not meant to be memorized. You should understand them so thoroughly that it becomes trivial to rederive them every time you use them.

29–4 Other Single-Loop Circuits

In this section we extend the simple circuit of Fig. 4 in two ways.

Internal Resistance. Figure 5a shows a circuit that emphasizes that all real seats of emf have an intrinsic internal resistance r. This resistance cannot be removed — although we usually wish it could be — because it is an inherent part of the device. Although we represent the emf and the internal resistance by separate symbols in Fig. 5a, they physically occupy the same region of space.

If we apply the loop rule, starting at b and going around clockwise, we obtain

$$V_b + \mathcal{E} - ir - iR = V_b$$

or

$$+\mathcal{E} - ir - iR = 0.\qquad(3)$$

Verify that we have properly applied the rules for signs given at the end of the preceding section. Compare Eq. 3 with Fig. 5b, which shows the changes in potential graphically. It is helpful to imagine Fig. 5b folded into a cylinder, with labeled points b connected, to suggest the continuity of the closed loop.

If we solve Eq. 3 for the current, we find

$$i = \frac{\mathcal{E}}{R + r}.\qquad(4)$$

As it must, this equation reduces to Eq. 2 for the case of $r = 0$.

---

* The loop rule is sometimes referred to as *Kirchhoff's loop rule*, after Gustav Robert Kirchhoff (pronounced Keerk-hoff; 1824–1887).

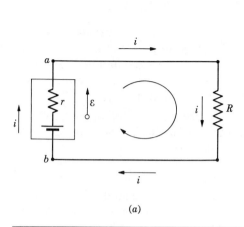

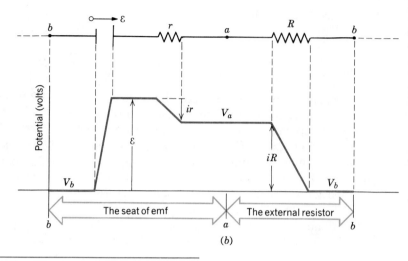

Figure 5 (*a*) A single-loop circuit, containing a seat of emf having an internal resistance *r*. (*b*) The circuit is shown spread out at the top. The potentials encountered in traversing the circuit clockwise from *b* are shown at the bottom.

Resistances in Series. Figure 6 shows three resistors connected *in series,* a battery being connected across the output terminals of the combination.

> *Resistors are said to be connected in series if the sum of the potential differences across each is equal to the potential difference applied to the combination.*

We seek the single resistance R_{eq} that is equivalent to this series combination.

Recall (see Section 27–4) that the definition of a series connection given above also applies to capacitors.* The definition requires that capacitors in series each have the same charge *q*; it requires that resistors in series each have the same current *i*. The equivalent resistance that we seek is the single resistance R_{eq} that, substituted for the series combination between the terminals *a* and *b*, will leave the current *i* unchanged.

Let us apply the loop rule, starting from terminal *a* and going clockwise around the circuit of Fig. 6. We find

$$-iR_1 - iR_2 - iR_3 + \mathcal{E} = 0,$$

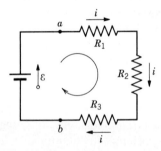

Figure 6 Three resistors are connected in series between points *a* and *b*. What is their equivalent resistance?

or

$$i = \frac{\mathcal{E}}{R_1 + R_2 + R_3}. \tag{5}$$

For the equivalent resistance we have

$$i = \frac{\mathcal{E}}{R_{eq}}. \tag{6}$$

Comparison of Eqs. 5 and 6 shows that

$$R_{eq} = R_1 + R_2 + R_3.$$

The extension to *n* resistors is straightforward and is

$$\boxed{R_{eq} = \sum_n R_n} \quad \text{(resistances in series).} \tag{7}$$

---

* For both resistors and capacitors in series, there is the implicit assumption that there are no side branches between the terminal points of the series combination.

Comparison with Eq. 19 of Chapter 27 shows that resistors in series follow the same rule as capacitors in parallel; to find the equivalent value of either capacitance or resistance for these arrangements you simply add the individual values.

29-5 Potential Differences

We often want to find the potential difference between two points in a circuit. In Fig. 5a, for example, what will a voltmeter read if we touch the positive probe to point a and the negative probe to point b? To find out, let us start from point a and traverse the circuit clockwise to point b, passing through resistor R. If V_a and V_b are the potentials at a and b, respectively, we have

$$V_a - iR = V_b$$

because (according to our resistor rule) we experience a decrease in potential in going through a resistor in the direction of the current. We rewrite this as

$$V_a - V_b = +iR, \tag{8}$$

which tells us that point a is more positive in potential than point b. Combining Eq. 8 with Eq. 4, we have

$$V_a - V_b = \mathcal{E}\frac{R}{R+r}. \tag{9}$$

> *To find the potential difference between any two points in a circuit, start at one point and traverse the circuit to the other, following any path, and add algebraically the changes in potential that you encounter.*

Let us again calculate $V_a - V_b$, starting again from point a but this time proceeding counterclockwise to b through the seat of emf. We have

$$V_a + ir - \mathcal{E} = V_b$$

or

$$V_a - V_b = \mathcal{E} - ir.$$

Again, combining this relation with Eq. 4 leads to Eq. 9.

Note that $V_a - V_b$ in Fig. 5 is the potential difference of the battery across the battery terminals. We see from Eq. 9 that $V_a - V_b$ is only equal to the emf $\mathcal{E}$ if the battery has no internal resistance ($r = 0$) or if the battery is an open circuit ($R \rightarrow \infty$).

Sample Problem 1 What is the current in the circuit of Fig. 7a? The emfs and the resistors have the following values:

$$\mathcal{E}_1 = 2.1 \text{ V}, \quad \mathcal{E}_2 = 4.4 \text{ V},$$
$$r_1 = 1.8 \ \Omega, \quad r_2 = 2.3 \ \Omega, \quad R = 5.5 \ \Omega.$$

The two emfs are connected so that they oppose each other but $\mathcal{E}_2$, because it is larger than $\mathcal{E}_1$, controls the direction of the current in the circuit, which is counterclockwise. The loop rule, applied clockwise from point a, yields

$$-\mathcal{E}_2 + ir_2 + iR + ir_1 + \mathcal{E}_1 = 0. \tag{10}$$

Check that this same equation results by going around counterclockwise or by starting at some point other than a. Also, compare this equation term by term with Fig. 7b, which shows the potential changes graphically.

Solving Eq. 10 for the current i, we obtain

$$\begin{aligned}
i &= \frac{\mathcal{E}_2 - \mathcal{E}_1}{R + r_1 + r_2} \\
&= \frac{4.4 \text{ V} - 2.1 \text{ V}}{5.5 \ \Omega + 1.8 \ \Omega + 2.3 \ \Omega} \\
&= 0.2396 \text{ A} \approx 240 \text{ mA}. \quad \text{(Answer)}
\end{aligned}$$

It is not necessary to know the direction of the current in advance. To show this, let us assume that the current in Fig. 7a is clockwise, that is, reverse the direction of the current arrow in Fig. 7a. The loop rule would then yield (going clockwise from a)

$$-\mathcal{E}_2 - ir_2 - iR - ir_1 + \mathcal{E}_1 = 0$$

or

$$i = -\frac{\mathcal{E}_2 - \mathcal{E}_1}{R + r_1 + r_2}.$$

Substituting numerical values (see above) yields $i = -240$ mA for the current. The minus sign is a signal that the current is in the opposite direction from that which we have assumed.

In more complex circuits involving many loops and branches, it is often impossible to know in advance the actual directions for the currents in all parts of the circuit. The procedure is to choose initially the current directions for each branch arbitrarily. If you get an answer with a positive sign for a particular current, you have chosen its direction correctly; if you get a negative sign, the current is opposite in direction to that chosen. In all cases, the numerical value will be correct.

Sample Problem 2 (a) What is the potential difference between points a and b in Fig. 7a?

This potential difference is the terminal potential difference of the emf $\mathcal{E}_2$. Let us start at point b and traverse the circuit counterclockwise to point a, passing directly through

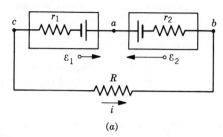

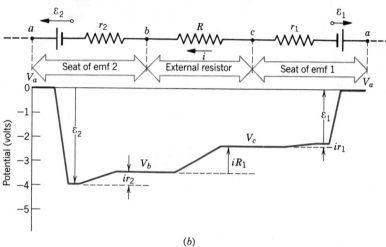

(a)

(b)

Figure 7   Sample Problems 1 and 2. *(a)* A single-loop circuit containing two seats of emf. *(b)* The potentials encountered in traversing this circuit clockwise from point *a*.

the seat of emf. We find

$$V_b - ir_2 + \mathcal{E}_2 = V_a$$

or

$$
\begin{aligned}
V_a - V_b &= -ir_2 + \mathcal{E}_2 \\
&= -(0.2396 \text{ A})(2.3 \text{ }\Omega) + 4.4 \text{ V} \\
&= +3.85 \text{ V}. \quad\quad\text{(Answer)}
\end{aligned}
$$

Thus, *a* is more positive than *b* and the potential difference between them (3.85 V) is *smaller* than the emf (4.4 V); see Fig. 7*b*.

We can verify this result by starting at point *b* in Fig. 7*a* and traversing the circuit clockwise to point *a*. For this different path we find

$$V_b + iR + ir_1 + \mathcal{E}_1 = V_a$$

or

$$
\begin{aligned}
V_a - V_b &= iR + ir_1 + \mathcal{E}_1 \\
&= (0.2396 \text{ A})(5.5 \text{ }\Omega + 1.8 \text{ }\Omega) + 2.1 \text{ V} \\
&= +3.85 \text{ V}, \quad\quad\text{(Answer)}
\end{aligned}
$$

exactly as before. The potential difference between two points has the same value for all paths connecting those points.

(b) What is the potential difference between points *a* and *c* in Fig. 7*a*?

Note that this potential difference is the terminal potential difference of the emf $\mathcal{E}_1$. Let us start at *c* and traverse the circuit clockwise to point *a*. We find

$$V_c + ir_1 + \mathcal{E}_1 = V_a$$

or

$$
\begin{aligned}
V_a - V_c &= ir_1 + \mathcal{E}_1 \\
&= (0.2396 \text{ A})(1.8 \text{ }\Omega) + 2.1 \text{ V} \\
&= +2.5 \text{ V}. \quad\quad\text{(Answer)}
\end{aligned}
$$

This tells us that *a* is at a higher potential than *c*. The terminal potential difference (2.5 V) is in this case *larger* than the emf (2.1 V); see Fig. 7*b*. Charge is being forced through $\mathcal{E}_1$ in a direction opposite to the one in which it would send charge if it were acting by itself; if $\mathcal{E}_1$ were a storage battery it would be charging at the expense of $\mathcal{E}_2$.

29–6   Multiloop Circuits

Figure 8 shows a circuit containing two loops. For simplicity, we have assumed ideal batteries, that is, batteries with no internal resistance. There are two *junctions* in this circuit, at *b* and *d*, and there are three *branches*

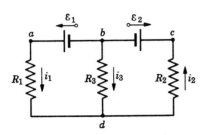

Figure 8 A simple multiloop circuit. What are the currents in the three branches?

connecting these junctions. The branches are the left branch *(bad)*, the right branch *(bcd)*, and the central branch *(bd)*. What are the currents in the three branches?

We label the currents i_1, i_2, and i_3, as shown. Current i_1 has the same value for any cross section of branch *bad*. Similarly, i_2 has the same value everywhere in the right branch and i_3 in the central branch. You can choose the directions of the currents arbitrarily. By studying the signs of the emfs in Fig. 8, you should be able to convince yourself that the current in the central branch must point up, not down as we have drawn it. We have deliberately chosen it pointing in the wrong direction so that we may see how the algebra will automatically correct such wrong guesses.

Consider junction *d*. The three currents carry charge either toward this junction or away from it. In a steady-state condition, charge does not pile up at any junction, nor does it drain away from it. Thus, the rate at which charge is delivered to the junction must equal the rate at which it is taken away, or

$$i_1 + i_3 = i_2. \qquad (11)$$

You can easily check that applying this theorem to junction *a* leads to exactly this same equation.* Equation 11 suggests a general principle:

Junction rule.† The sum of the currents approaching any junction must be equal to the sum of the currents leaving that junction.

This rule is simply a statement of the conservation of

charge. Thus, our basic tools for solving complex circuits are the *loop rule,* which is based on the conservation of energy, and the *junction rule,* which is based on the conservation of charge.

Equation 11 will give us any one of the currents if we know the other two. To solve the problem completely, we need more information; we can find it by applying the loop rule. If we traverse the left loop of Fig. 8 in a counterclockwise direction, this rule gives

$$\mathcal{E}_1 - i_1 R_1 + i_3 R_3 = 0. \qquad (12)$$

The right loop yields

$$-i_3 R_3 - i_2 R_2 - \mathcal{E}_2 = 0. \qquad (13)$$

Equations 11, 12, and 13 are three simultaneous equations involving the three currents as variables. Solving for the three unknowns we find, after a little algebra,

$$i_1 = \frac{\mathcal{E}_1(R_2 + R_3) - \mathcal{E}_2 R_3}{R_1 R_2 + R_2 R_3 + R_1 R_3} \quad \text{(left branch),} \qquad (14)$$

$$i_2 = \frac{\mathcal{E}_1 R_3 - \mathcal{E}_2(R_1 + R_3)}{R_1 R_2 + R_2 R_3 + R_1 R_3} \quad \text{(central branch),} \qquad (15)$$

and

$$i_3 = -\frac{\mathcal{E}_1 R_2 + \mathcal{E}_2 R_1}{R_1 R_2 + R_2 R_3 + R_1 R_3} \quad \text{(right branch).} \qquad (16)$$

Be sure to supply the missing steps.

Equation 16 shows that no matter what the numerical values of the resistances and the emfs, the current i_3 will have a negative sign. Thus — as we knew all along — the current is opposite in direction to that shown in Fig. 8. The currents i_1 and i_2 may be in either direction, depending on the numerical values of the resistances and the emfs.

When we have a complex array of equations such as Eqs. 14, 15, and 16, where there is always a chance that we have made a mistake in algebra, it is a good idea to check the equations to make sure that they give expected results in simple special cases. One such case arises if we put $R_3 = \infty$, which corresponds to clipping the central resistor out of the circuit with a pair of cutters. The circuit then becomes a single-loop circuit and Eqs. 14, 15, and 16 predict that

$$i_1 = i_2 = \frac{\mathcal{E}_1 - \mathcal{E}_2}{R_1 + R_2} \quad \text{and} \quad i_3 = 0.$$

These results are what we expect. How many other simple special cases can you identify?

It might occur to you that you can apply the loop

---

* It can be shown that, for a circuit with *n* junctions, there are only *n* − 1 independent junction equations. In this case, *n* = 2 so that there is just one independent equation.

† The junction rule is sometimes referred to as *Kirchhoff's junction rule.*

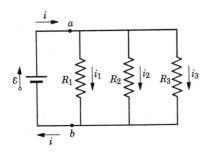

Figure 9 Three resistors are connected in parallel across points a and b. What is their equivalent resistance?

rule to a large loop, consisting of the entire circuit $abcda$ in Fig. 8. The rule yields for this loop

$$\mathscr{E}_1 - \mathscr{E}_2 - i_2 R_2 - i_1 R_1 = 0,$$

which is nothing more than the sum of Eqs. 12 and 13. Thus, the large loop does not yield another independent equation. In solving multiloop circuits, you will never find more independent equations than there are variables, no matter how many times you apply the loop rule and the junction rule. When you have written down as many independent equations as you need, stop.

Resistors in Parallel. Figure 9 shows three resistors connected across the same seat of emf. Resistors across which the same potential difference is applied are said to be *in parallel.* What is the equivalent resistance R_{eq} of this parallel combination? By *equivalent resistance,* we mean the single resistance that, substituted for the parallel combination between terminals a and b, would leave the current i unchanged.

The currents in the three branches of Fig. 9 are

$$i_1 = \frac{V}{R_1}, \quad i_2 = \frac{V}{R_2}, \quad \text{and} \quad i_3 = \frac{V}{R_3},$$

where V is the potential difference between a and b. If we apply the junction rule at point a, we find

$$i = i_1 + i_2 + i_3 = V\left(\frac{1}{R_1} + \frac{1}{R_2} + \frac{1}{R_3}\right). \quad (17)$$

If we replace the parallel combination by the equivalent resistance, we have

$$i = \frac{V}{R_{eq}}. \quad (18)$$

Comparing Eqs. 17 and 18 leads to

$$\frac{1}{R_{eq}} = \frac{1}{R_1} + \frac{1}{R_2} + \frac{1}{R_3}. \quad (19)$$

Table 1   Resistors and Capacitors in Series and in Parallel[a]

	Series	*Parallel*
Resistors	$R_{eq} = \sum_n R_n$ (29–7) Same current	$\dfrac{1}{R_{eq}} = \sum_n \dfrac{1}{R_n}$ (29–20) Same potential difference
Capacitors	$\dfrac{1}{C_{eq}} = \sum_n \dfrac{1}{C_n}$ (27–20) Same charge	$C_{eq} = \sum_n C_n$ (27–19) Same potential difference

[a] The chapter and equation numbers are displayed below the formulas.

Extending this result to the case of n resistors, we find

$$\boxed{\frac{1}{R_{eq}} = \sum_n \frac{1}{R_n}} \quad \text{(resistors in parallel).} \quad (20)$$

For the case of two resistors, the equivalent resistance is their product divided by their sum.* That is,

$$R_{eq} = \frac{R_1 R_2}{R_1 + R_2} = \left(\frac{R_2}{R_1 + R_2}\right) R_1 = \left(\frac{R_1}{R_1 + R_2}\right) R_2.$$

Note that, because the fractions in the parentheses above are each less than unity, the equivalent resistance is smaller than either of the two combining resistors. A little thought will convince you that this remains true for any number of resistors connected in parallel. Also note that the formula for resistors in parallel is identical in form to the formula for capacitors in series. Table 1 summarizes the relations for resistors and capacitors in series and in parallel.

Sample Problem 3 Figure 10 shows a circuit whose elements have the following values:

$$\mathscr{E}_1 = 2.1 \text{ V}, \quad \mathscr{E}_2 = 6.3 \text{ V},$$
$$R_1 = 1.7 \ \Omega, \quad R_2 = 3.5 \ \Omega.$$

(a) Find the currents in the three branches of the circuit.

Let us draw and label the currents as shown in the figure, choosing the current directions arbitrarily. Applying the junc-

---

* If you accidentally took the equivalent resistance to be the sum divided by the product, you would notice at once that this result would be dimensionally incorrect.

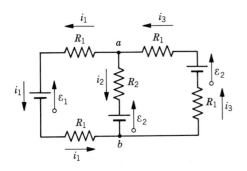

Figure 10   Sample Problem 3. A multiloop circuit. What are the currents in the three branches? What is the potential difference between points *a* and *b*?

tion rule at *a*, we find

$$i_3 = i_1 + i_2. \tag{21}$$

Now let us start at point *a* and traverse the left-hand loop in a counterclockwise direction. We find

$$-i_1 R_1 - \mathcal{E}_1 - i_1 R_1 + \mathcal{E}_2 + i_2 R_2 = 0$$

or

$$2i_1 R_1 - i_2 R_2 = \mathcal{E}_2 - \mathcal{E}_1. \tag{22}$$

If we traverse the right-hand loop in a clockwise direction from point *a*, we find

$$+i_3 R_1 - \mathcal{E}_2 + i_3 R_1 + \mathcal{E}_2 + i_2 R_2 = 0$$

or

$$i_2 R_2 + 2i_3 R_1 = 0. \tag{23}$$

Equations 21, 22, and 23 are three independent simultaneous equations involving the three variables i_1, i_2, and i_3. We can solve these equations for these variables, obtaining, after a little algebra,

$$i_1 = \frac{(\mathcal{E}_2 - \mathcal{E}_1)(2R_1 + R_2)}{4R_1(R_1 + R_2)},$$

$$= \frac{(6.3 \text{ V} - 2.1 \text{ V})(2 \times 1.7 \ \Omega + 3.5 \ \Omega)}{(4)(1.7 \ \Omega)(1.7 \ \Omega + 3.5 \ \Omega)}$$

$$= 0.82 \text{ A}, \qquad \text{(Answer)}$$

$$i_2 = -\frac{\mathcal{E}_2 - \mathcal{E}_1}{2(R_1 + R_2)},$$

$$= -\frac{6.3 \text{ V} - 2.1 \text{ V}}{(2)(1.7 \ \Omega + 3.5 \ \Omega)} = -0.40 \text{ A}, \qquad \text{(Answer)}$$

and

$$i_3 = \frac{(\mathcal{E}_2 - \mathcal{E}_1)(R_2)}{4R_1(R_1 + R_2)}$$

$$= \frac{(6.3 \text{ V} - 2.1 \text{ V})(3.5 \ \Omega)}{(4)(1.7 \ \Omega)(1.7 \ \Omega + 3.5 \ \Omega)} = 0.42 \text{ A}. \qquad \text{(Answer)}$$

The signs of the currents tell us that we have guessed correctly about the directions of i_1 and i_3 but that we are wrong about the direction of i_2; it should point up — and not down — in the central branch of the circuit of Fig. 10.

(b) What is the potential difference between points *a* and *b* in the circuit of Fig. 10?

We have, assuming the current directions shown in the figure,

$$V_a - i_2 R_2 - \mathcal{E}_2 = V_b,$$

or

$$V_a - V_b = \mathcal{E}_2 + i_2 R_2$$
$$= 6.3 \text{ V} + (-0.40 \text{ A})(3.5 \ \Omega) = +4.9 \text{ V}. \quad \text{(Answer)}$$

The positive sign of this result tells us that *a* is more positive in potential than *b*. From study of the circuit, this is what we would expect because all three batteries have their positive terminals on the top side of the figure.

Sample Problem 4   Figure 11*a* shows a cube made of 12 resistors, each of resistance *R*. Find R_{12}, the equivalent resistance of a cube edge.

Although this problem can be attacked by "brute force" methods, using the loop and junction rules, the symmetry of the connections suggests that there must be a neater method. The key is the realization that, from considerations of symmetry alone, points 3 and 6 must be at the same potential. So must points 4 and 5.

If two points in a circuit have the same potential, the currents in the circuit do not change if you connect these points by a wire. There will be no current in the wire because there is no potential difference between its ends. Electrically, then, points 3 and 6 are a single point; so are points 4 and 5.

This allows us to redraw the cube as in Fig. 11*b*. From this point, it is simply a matter of reducing the circuit between the input terminals to a single resistor, using the rules for resistors in series and in parallel. In Fig. 11*c*, we make a start by replacing five parallel resistor combinations by their equivalents, each of resistance $\frac{1}{2}R$.

In Fig. 11*d*, we have added the three resistors that are in series in the right-hand loop, obtaining a single equivalent resistance of 2*R*. In Fig. 11*e*, we have replaced the two resistors that now form the right-hand loop by a single equivalent resistor $\frac{2}{3}R$. In so doing, it is useful to recall that the equivalent resistance of two resistors in parallel is equal to their product divided by their sum.

In Fig. 11*f*, we have added the three series resistors of Fig. 11*e*, obtaining $\frac{7}{6}R$ and in Fig. 11*g* we have reduced this parallel combination to the single equivalent resistance that we seek, namely,

$$R_{12} = \tfrac{7}{12}R. \qquad \text{(Answer)}$$

You can also use these methods to find R_{13}, the equivalent resistance of a cube across a face diagonal and R_{17}, the equivalent resistance across a body diagonal.

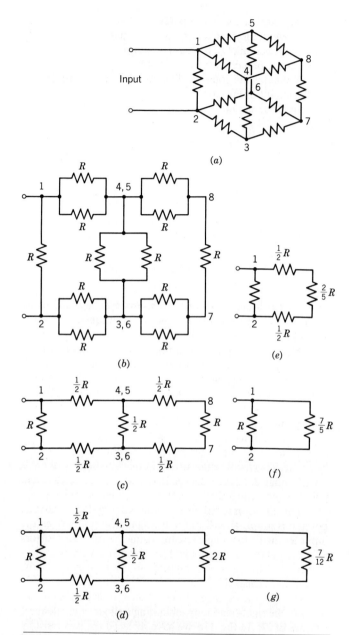

Figure 11 Sample Problem 4. (a) A cube formed of identical resistors. The remaining figures show how to reduce these 12 resistors to a single equivalent resistance.

29–7 Measuring Instruments

Several electric measuring instruments involve circuits that can be analyzed by the methods of this chapter. We discuss three of them.

1. The Ammeter. An instrument used to measure

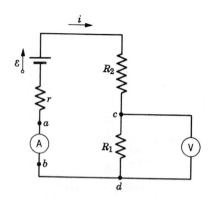

Figure 12 A single-loop circuit, showing how to connect an ammeter (A) and a voltmeter (V).

currents is called an *ammeter.* To measure the current in a wire, you usually have to break or cut the wire and insert the ammeter so that the current to be measured passes through the meter; see Fig. 12.

It is essential that the resistance R_A of the ammeter be very small compared to other resistances in the circuit. Otherwise, the very presence of the meter will change the current to be measured. In the circuit of Fig. 12, the required condition, assuming that the voltmeter is not connected, is

$$R_A \ll (r + R_1 + R_2).$$

2. The Voltmeter. A meter to measure potential differences is called a *voltmeter.* To find the potential difference between any two points in the circuit, the voltmeter terminals are connected between those points, without breaking the circuit; see Fig. 12.

It is essential that the resistance R_V of a voltmeter be very large compared to any circuit element across which the voltmeter is connected. Otherwise, the meter itself becomes an important circuit element and will alter the potential difference that is to be measured. In Fig. 12, the required condition is

$$R_V \gg R_1.$$

Often a single unit is packaged so that, by external switching, it can serve as either an ammeter or a voltmeter—and usually also as an *ohmmeter,* designed to measure the resistance of any element connected between its terminals. Such a versatile unit is called a *multimeter;* see Fig. 13. Its output readings may take the form of a pointer moving over a scale or of a digital display.

3. The Potentiometer. A *potentiometer* is a device

Figure 13 A typical multimeter, used to measure currents, potential differences, and resistances.

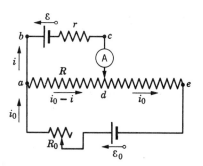

Figure 14 The rudiments of a potentiometer, used to compare emfs.

for measuring an unknown emf $\mathcal{E}_x$ by comparing it with a known standard emf $\mathcal{E}_s$.

Figure 14 shows its rudiments. The resistor that extends from a to e is a carefully made precision resistor with a sliding contact shown positioned at d. The resistance R in the figure is the resistance between points a and d.

When using the instrument, $\mathcal{E}_s$ is first placed in the position $\mathcal{E}$ and the sliding contact is adjusted until the current i is zero as noted on the sensitive ammeter A. The potentiometer is then said to be *balanced,* the value of R at balance being R_s. In this balance condition we have, considering the loop $abcda$,

$$\mathcal{E}_s = i_0 R_s. \qquad (24)$$

Because $i = 0$ in branch $abcd,$ the internal resistance r of the standard source of emf does not enter.

The process is now repeated with $\mathcal{E}_x$ substituted for $\mathcal{E}_s$, the potentiometer being balanced once more. The current i_0 remains unchanged and the new balance condition is

$$\mathcal{E}_x = i_0 R_x. \qquad (25)$$

From Eqs. 24 and 25 we then have

$$\mathcal{E}_x = \mathcal{E}_s \frac{R_x}{R_s}. \qquad (26)$$

Thus, the unknown emf can be found in terms of the known emf by making two adjustments of the precision resistor. In practice, potentiometers are conveniently

packaged units, containing a built-in *standard cell* that, after calibration at the National Bureau of Standards or elsewhere, serves as a convenient reference standard $\mathcal{E}_s$. Switching arrangements for interchanging the standard and the unknown emfs are also incorporated. The sliding contact usually moves over a scale on which the value of the unknown emf can be read directly, without having to perform the calculation required by Eq. 26.

29-8 *RC Circuits*

The preceding sections dealt with circuits in which the circuit elements were resistors and in which the currents did not vary with time. Here we introduce the capacitor as a circuit element, which will lead us to the study of time-varying currents.

Charging a Capacitor. In the circuit of Fig. 15, let us throw switch S from the indicated position to position a, thus introducing an emf $\mathcal{E}$ into the circuit and *charging* the capacitor C through the resistor R. How will the charging current i vary with time?

Let us apply the loop rule to the circuit of Fig. 15,

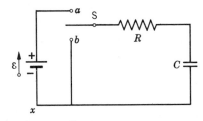

Figure 15 When switch S is closed on a, the capacitor C is *charged* through the resistor R. When the switch is afterward closed on b, the capacitor *discharges* through R.

going around in a clockwise direction. We have

$$\mathcal{E} - iR - \frac{q}{C} = 0,$$

in which q/C is the potential difference between the capacitor plates. We rearrange this equation as

$$iR + \frac{q}{C} = \mathcal{E}. \qquad (27)$$

We cannot immediately solve Eq. 27 because it contains two variables, the current i and the charge q. However, these variables are not independent but are related by

$$i = \frac{dq}{dt}.$$

Substituting for i in Eq. 27, we find

$$\boxed{R \frac{dq}{dt} + \frac{q}{C} = \mathcal{E}} \quad \text{(charging equation).} \quad (28)$$

This is the differential equation that describes the variation with time of the charge q in the circuit of Fig. 13. Our task is to find the function $q(t)$ that satisfies this equation and also satisfies the requirement that the capacitor be initially uncharged. This requirement, namely that $q = 0$ at $t = 0$, is typical of the *initial conditions* that we impose — in general — on differential equations.

Although Eq. 28 is not hard to solve, we choose to avoid mathematical complexity by simply presenting the solution, which is (we claim)

$$\boxed{q = C\mathcal{E}(1 - e^{-t/RC})} \quad \text{(charge).} \quad (29)$$

Note that Eq. 29 does indeed satisfy our required initial condition, in that $q = 0$ at $t = 0$. The derivative of $q(t)$ is

$$\boxed{i = \frac{dq}{dt} = \left(\frac{\mathcal{E}}{R}\right) e^{-t/RC}} \quad \text{(current).} \quad (30)$$

If we substitute q from Eq. 29 and i from Eq. 30 into Eq. 28, the differential equation does indeed reduce to an identity, as you should be certain to verify. Thus, Eq. 29 is indeed a solution of Eq. 28.

We can measure $q(t)$ experimentally by measuring a quantity proportional to it, namely, V_C, the potential difference across the capacitor. From Eq. 29 we have

$$V_C = \frac{q}{C} = \mathcal{E}(1 - e^{-t/RC}) \qquad (31)$$

Similarly, we can measure $i(t)$ by measuring V_R, the potential difference across the resistor. From Eq. 30 we have

$$V_R = iR = \mathcal{E}e^{-t/RC}. \qquad (32)$$

Figure 16 shows plots of V_C and V_R. We note that V_C and V_R add up at each instant to $\mathcal{E}$, as Eq. 27 requires.

The Time Constant. The product RC that appears in Eqs. 29 and 30 has the dimensions of time (because the exponent in those equations must be dimensionless). RC is called the *capacitive time constant* of the circuit and is represented by the symbol τ. It is the time at which the charge on the capacitor has increased to a factor of $(1 - e^{-1})$ or about 63% of its equilibrium value. To show this, let us put $t = RC$ in Eq. 29. We find

$$q = C\mathcal{E}(1 - e^{-1}) = 0.63 \, C\mathcal{E}.$$

Because $C\mathcal{E}$ is the equilibrium charge on the capacitor, corresponding to $t \to \infty$ in Eq. 29, we have proved our point.

The Charging Process. It is easy enough to derive equations such as Eqs. 29 and 30 without any real physical understanding of what is going on. Let us therefore look at the charging process in an RC circuit qualitatively and physically.

When switch S in Fig. 15 is first closed on a, there is no charge on the capacitor and therefore no potential difference between its plates. Thus, the full potential difference $\mathcal{E}$ appears across the resistor, setting up an initial current $\mathcal{E}/R$. Once the current starts, charges

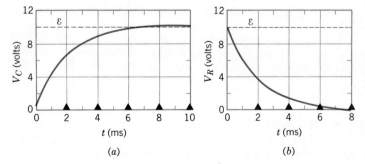

(a)

(b)

Figure 16 (*a*) A plot of Eq. 31, which traces the buildup of charge on the capacitor of Fig. 15. (*b*) A plot of Eq. 32, which traces the decline of the charging current in the circuit of Fig. 15. The curves are plotted for $R = 2000 \, \Omega$, $C = 1 \, \mu$F, and $\mathcal{E} = 10$ V; the filled triangles represent successive time constants.

begin to appear on the capacitor plates and a potential difference q/C builds up between those plates. This in turn means that the potential difference across the resistor must decrease by this same amount. This decrease in the potential difference across R means that the charging current is reduced. Thus, the charge on the capacitor builds up and the charging current through the resistor falls off until the capacitor is fully charged. In this condition, the full emf $\mathcal{E}$ is being applied to the capacitor, there being no potential drop across the resistor. This is precisely the reverse of the initial state, in which the full emf $\mathcal{E}$ appeared across the resistor. Review the derivations of Eqs. 29 and 30 and study Fig. 16 with the arguments of this paragraph in mind.

Discharging a Capacitor.   Assume now that the capacitor of Fig. 15 is fully charged to the voltage of the emf and that the switch S is then thrown from a to b so that capacitor C can *discharge* through resistor R. How does the discharge current in this single-loop circuit vary with time?

Equation 28 continues to hold except that now there is no longer an emf in the circuit. Putting $\mathcal{E} = 0$ in this equation, we find

$$\boxed{R\frac{dq}{dt} + \frac{q}{C} = 0}\quad \text{(discharging equation).}\quad (33)$$

The solution to this differential equation is

$$\boxed{q = q_0 e^{-t/RC}}\quad \text{(charge),}\quad (34)$$

in which $q_0 (= C\mathcal{E}$ in our case) is the initial charge on the capacitor. You can verify by substitution that Eq. 34 is indeed a solution of Eq. 33.

The capacitive time constant RC governs the discharging process as well as the charging process. We see that at $t = RC$ the capacitor charge is reduced to $C\mathcal{E}e^{-1}$, which is about 37% of its initial charge.

The current during discharge follows from differentiating Eq. 34 and is

$$\boxed{i = \frac{dq}{dt} = -\frac{q_0}{RC}e^{-t/RC} = -i_0 e^{-t/RC}}\quad \text{(current).}\quad (35)$$

Here $\mathcal{E}/R$ is the magnitude of the initial current, i_0, corresponding to $t = 0$. This is reasonable because the potential difference across the fully charged capacitor is $\mathcal{E}$. The negative sign shows that the discharge current is in the opposite direction to the charging current, as we

Table 2   RC Charging and Discharging[a]

	Charging	Discharging
Differential equation	$R\dfrac{dq}{dt} + \dfrac{q}{C} = \mathcal{E}$	$R\dfrac{dq}{dt} + \dfrac{q}{C} = 0$
	(29-28)	(29-33)
Charge	$q = C\mathcal{E}(1 - e^{-t/RC})$	$q = q_0 e^{-t/RC}$
	(29-29)	(29-34)
Current	$i = (\mathcal{E}/R)e^{-t/RC}$	$i = -(q_0/RC)e^{-t/RC}$
	(29-30)	(29-35)

[a] The chapter and equation numbers are displayed below the formulas.

expect. Table 2 summarizes the equations for the charging and discharging of a capacitor.

Sample Problem 5   A capacitor C is discharging through a resistor R. (a) After how many time constants will its charge fall to one-half of its initial value?

The charge on the capacitor varies according to Eq. 34,

$$q = q_0 e^{-t/RC},$$

in which q_0 is the initial charge. We are asked to find the time t at which $q = \frac{1}{2}q_0$, or

$$\tfrac{1}{2}q_0 = q_0 e^{-t/RC}.$$

Canceling q_0 and then taking logarithms of each side, we find

$$0 - \ln 2 = -\frac{t}{RC}$$

or

$$t = (\ln 2)\,RC = 0.69\,RC.\quad \text{(Answer)}$$

Thus, the charge will drop to half its initial value after 0.69 time constants.

(b) After how many time constants will the stored energy drop to half of its initial value?

The energy of the capacitor is

$$U = \frac{q^2}{2C} = \frac{q_0^2}{2C}\,e^{-2t/RC} = U_0 e^{-2t/RC},$$

in which U_0 is the initial stored energy. We are asked to find the time at which $U = \frac{1}{2}U_0$, or

$$\tfrac{1}{2}U_0 = U_0 e^{-2t/RC}.$$

Canceling U_0 and then taking logarithms of each side, we obtain

$$-\ln 2 = -2t/RC$$

or

$$t = RC\,\frac{\ln 2}{2} = 0.35\,RC.\quad \text{(Answer)}$$

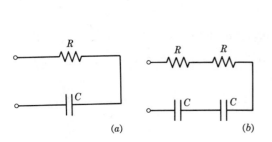

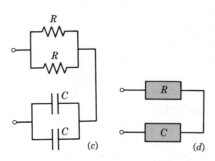

Figure 17 Sample Problem 6. The capacitive time constant of all these circuits is RC.

Thus, the stored energy will drop to half its initial value after 0.35 time constants have elapsed. This remains true no matter what the initial stored energy may be. It takes longer (0.69 RC) for the *charge* to fall to half its initial value than for the *energy* (0.35 RC) to fall to half its initial value. Does this result surprise you?

Sample Problem 6 The time constant τ of the circuit of Fig. 17a is clearly RC. (a) What is the time constant of the circuit of Fig. 17b?

The equivalent resistance is $2R$ and the equivalent capacitance is $\frac{1}{2}C$ so that

$$\tau = (2R)(\tfrac{1}{2}C) = RC, \qquad \text{(Answer)}$$

just as for the circuit of Fig. 17a.

(b) What is the time constant of the circuit of Fig. 17c?

The equivalent resistance is $\frac{1}{2}R$ and the equivalent capacitance is $2C$ so that, once again,

$$\tau = (\tfrac{1}{2}R)(2C) = RC. \qquad \text{(Answer)}$$

These results are evidence of the reciprocal nature of the formulas for combining resistors and capacitors in series and in parallel; see Table 1. The time constant of the generalized circuit of Fig. 17d is also RC, provided that the box marked R is any configuration whatever of identical resistors and the box marked C is the same configuration of identical capacitors.*

---

* This generalization was pointed out to us by Professor Andrew L. Gardner.

REVIEW AND SUMMARY

A *seat of electromotive force* is a device (a battery or a generator, say) that supplies energy (chemical or mechanical, say) to maintain a potential difference between its output terminals. If dW is the energy the seat provides to force positive charge dq through the seat from the negative to the positive terminal then

Electromotive Force

$$\mathcal{E} = \frac{dW}{dq} \quad \text{(definition of } \mathcal{E}\text{)}. \qquad [1]$$

The volt is the SI unit of emf as well as of potential difference. Study Fig. 2.

Several *rules* or *methods* are stated here in sufficient generality that they can be used, not only for the circuits of this chapter, but for more complex situations to be encountered later:

Energy Method

Energy Method *The net total energy delivered by seats of emf must be balanced by energy dissipated by or stored in the circuit; see Fig. 3.*

Potential Method

Potential Method *The potential difference between any two points is the algebraic sum of the changes in potential encountered in traversing the circuit from one point to the other along any path.*

The change in potential in traversing a *resistor* in the direction of the current is $-iR$; in the opposite direction it is $+iR$. The change in potential in traversing a *seat of emf* in the direction of the emf is $+\mathcal{E}$; in the opposite direction it is $-\mathcal{E}$. See Sample Problem 2.

Loop Rule

Loop Rule *The algebraic sum of the changes in potential encountered in a complete traversal of any closed circuit must be zero;* see Figs. 5 and 6 and Sample Problem 1.

Junction Rule

Junction Rule *The sum of the currents entering any junction must be equal to the sum of the currents leaving that junction.* See Fig. 8 and the analysis of Section 29–6.

The current in a single-loop circuit containing single resistor R and an emf $\mathcal{E}$ with internal resistance r (see Fig. 5) is

$$i = \frac{\mathcal{E}}{R + r}, \qquad [4]$$

which reduces to Eq. 2 when $r = 0$.

Series Elements

Resistors, and other circuit elements, are in *series* if the sum of the individual potential differences is the potential difference of the combination; see Fig. 6. The equivalent resistance of resistors in series is

$$R_{eq} = \sum_n R_n \quad \text{(resistors in series).} \qquad [7]$$

Parallel Elements

Resistors, and other circuit elements, are in *parallel* if they share a common potential difference; see Fig. 7. The equivalent resistance of resistors in parallel is

$$\frac{1}{R_{eq}} = \sum_n \frac{1}{R_n} \quad \text{(resistors in parallel).} \qquad [20]$$

Sample Problem 3 illustrates the use of the circuit rules and Sample Problem 4 shows an interesting application of the analysis of a circuit using equivalent resistances of series and parallel combinations. Section 29–7 uses the circuit rules to analyze the operation of conventional ammeters, voltmeters, and potentiometers.

RC Circuits

When an emf $\mathcal{E}$ is applied to a resistor R and capacitor C in series, as in Fig. 15, the charge on the capacitor increases according to

$$q = C\mathcal{E}(1 - e^{-t/RC}) \quad \text{(charging capacitor),} \qquad [29]$$

in which $C\mathcal{E} = q_0$ is the equilibrium charge and $RC = \tau$ is the *capacitive time constant* of the circuit. When the emf is replaced by a short circuit, the charge on the capacitor decays according to

$$q = q_0 e^{-t/RC} \quad \text{(discharging capacitor).} \qquad [34]$$

See Sample Problem 5.

QUESTIONS

1. Does the direction of the emf provided by a battery depend on the direction of current flow through the battery?

2. In Fig. 3 discuss what changes would occur if we increased the mass m by such an amount that the "motor" reversed direction and became a "generator," that is, a seat of emf.

3. Discuss in detail the statement that the energy method and the loop theorem method for solving circuits are perfectly equivalent.

4. Devise a method for measuring the emf and the internal resistance of a battery.

5. How could you calculate V_{ab} in Fig. 5a by following a path from a to b that does not lie in the conducting circuit?

6. A 25-W, 120-V bulb glows at normal brightness when connected across a bank of batteries. A 500-W, 120-V bulb glows only dimly when connected across the same bank. How could this happen?

7. Under what circumstances can the terminal potential difference of a battery exceed its emf?

8. Automobiles generally use a 12-V electrical system. Years ago a 6-volt system was used. Why the change? Why not 24 V?

9. The loop theorem is based on the conservation of energy principle and the junction theorem on the conservation of charge principle. Explain just how these theorems are based on these principles.

10. Under what circumstances would you want to connect batteries in parallel? In series?

11. Under what circumstances would you want to connect resistors in parallel? In series?

12. What is the difference between an emf and a potential difference?

13. Referring to Fig. 8 use a qualitative physical argument to convince yourself that i_3 is drawn in the wrong direction.

14. Explain in your own words why the resistance of an ammeter should be very small whereas that of a voltmeter should be very large.

15. Do the junction and loop theorems apply in a circuit containing a capacitor?

16. Show that the product RC in Eqs. 29 and 30 has the dimensions of time, that is, that 1 second = 1 ohm × 1 farad.

17. A capacitor, resistor, and battery are connected in series. The charge that the capacitor stores is unaffected by the resistance of the resistor. What purpose then is served by the resistor?

18. Explain why, in Sample Problem 5, the energy falls to half its initial value more rapidly than does the charge.

19. The light flash in a camera is produced by the discharge of a capacitor across the lamp. Why don't we just connect the photoflash lamp directly across the power supply used to charge the capacitor?

20. Does the time required for the charge on a capacitor in an RC circuit to build up to a given fraction of its equilibrium value depend on the value of the applied emf? Does the time required for the charge to change by a given amount depend on the applied emf?

21. A capacitor is connected across the terminals of a battery. Does the charge that eventually appears on the capacitor plates depend on the value of the internal resistance of the battery?

22. Devise a method whereby an RC circuit can be used to measure very high resistances.

23. In Fig. 15 suppose that switch S is closed on a. Explain why, in view of the fact that the negative terminal of the battery is not connected to resistance R, the initial current in R should be $\mathcal{E}/R$, as Eq. 30 predicts.

EXERCISES AND PROBLEMS

Section 29–2 Work, Energy, and Electromotive Force

1E. (a) How much work does a 12-V seat of emf do on an electron as it passes through from the positive to the negative terminal? (b) If 3.4×10^{18} electrons pass through each second, what is the power output of the seat?

2E. A 5.0-A current is set up in an external circuit by a 6.0-V storage battery for 6.0 min. By how much is the chemical energy of the battery reduced?

3E. A standard flashlight battery can deliver about 2.0 W·h of energy before it runs down. (a) If a battery costs 80 cents, what is the cost of operating a 100-W lamp for 8.0 h using batteries? (b) What is the cost if power provided by an electric utility company, at 12 cents per kW·h, is used?

4P. A certain 12-V car battery carries an initial charge of 120 A·h. Assuming that the potential across the terminals stays constant until the battery is completely discharged, for how many hours can it deliver energy at the rate of 100 W?

Section 29–5 Potential Differences

5E. In Fig. 18 $\mathcal{E}_1 = 12$ V and $\mathcal{E}_2 = 8$ V. What is the direction of the current in the resistor? Which emf is doing positive work? Which point, A or B, is at the higher potential?

6E. A wire of resistance 5.0 Ω is connected to a battery whose emf $\mathcal{E}$ is 2.0 V and whose internal resistance is 1.0 Ω. In 2.0 min (a) how much energy is transferred from chemical to electrical form? (b) How much energy appears in the wire as thermal energy? (c) Account for the difference between (a) and (b).

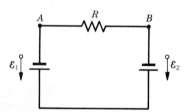

Figure 18 Exercise 5.

7E. In Fig. 5a put $\mathcal{E} = 2.0$ V and $r = 100$ Ω. Plot (a) the current and (b) the potential difference across R, as functions of R over the range 0 to 500 Ω. Make both plots on the same graph. (c) Make a third plot by multiplying together, for each value of R, the two curves plotted. What is the physical significance of this plot?

8E. Assume that the batteries in Fig. 19 have negligible internal resistance. Find: (a) the current in the circuit, (b) the power dissipated in each resistor, and (c) the power supplied or absorbed by each emf.

9E. A 12-V emf car battery with an internal resistance of 0.04 Ω is being charged with a current of 50 A. (a) What is the potential difference across its terminals? (b) At what rate is heat being dissipated in the battery? (c) At what rate is electric energy being converted to chemical energy? (d) What are the answers to (a) and (b) when the battery is used to supply 50 A to the starter motor?

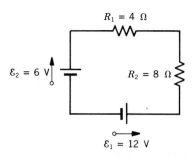

Figure 19 Exercise 8.

10E. In Fig. 20 the potential at point P is 100 V. What is the potential at point Q?

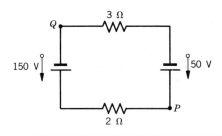

Figure 20 Exercise 10.

11E. The section of circuit AB (see Fig. 21) absorbs 50 W of power when a current $i = 1.0$ A passes through it in the indicated direction. (a) What is the potential difference between A and B? (b) If the element C does not have internal resistance, what is its emf? (c) What is its polarity?

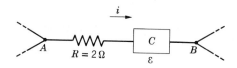

Figure 21 Exercise 11.

12E. In Fig. 6 calculate the potential difference across R_2, using two different paths. Assume $\mathscr{E} = 12$ V, $R_1 = 3.0$ Ω, $R_2 = 4.0$ Ω, and $R_3 = 5.0$ Ω.

13E. In Fig. 7a calculate the potential difference between a and c by considering a path that contains R and $\mathscr{E}_2$. (See Sample Problem 2.)

14E. A gasoline gauge for an automobile is shown schematically in Fig. 22. The indicator (on the dashboard) has a resistance of 10 Ω. The tank unit is simply a float connected to a resistor that has a resistance of 140 Ω when the tank is empty,

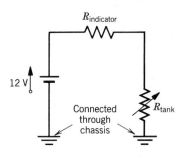

Figure 22 Exercise 14.

20 Ω when it is full, and varies linearly with the volume of gasoline. Find the current in the circuit when the tank is (a) empty; (b) half full; (c) full.

15P. (a) In Fig. 23 what value must R have if the current in the circuit is to be 1.0 mA? Take $\mathscr{E}_1 = 2.0$ V, $\mathscr{E}_2 = 3.0$ V, and $r_1 = r_2 = 3.0$ Ω. (b) What is the rate at which thermal energy appears in R?

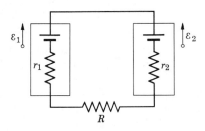

Figure 23 Problem 15.

16P. Thermal energy is to be generated in a 0.10-Ω resistor at the rate of 10 W by connecting it to a battery whose emf is 1.5 V. (a) What is the internal resistance of the battery? (b) What potential difference exists across the resistor?

17P. The current in a single-loop circuit is 5.0 A. When an additional resistance of 2.0 Ω is inserted in series, the current drops to 4.0 A. What was the resistance in the original circuit?

18P. Power is supplied by an emf $\mathscr{E}$ to a transmission line with resistance R. Find the ratio of the power dissipated in the line for $\mathscr{E} = 110,000$ V to that dissipated for $\mathscr{E} = 110$ V, assuming the power supplied is the same for the two cases.

19P. The starting motor of an automobile is turning slowly and the mechanic has to decide whether to replace the motor, the cable, or the battery. The manufacturer's manual says that the 12-V battery can have no more than 0.020 Ω internal resistance, the motor no more than 0.200 Ω resistance, and the cable no more than 0.040 Ω resistance. The mechanic turns on the motor and measures 11.4 V across the battery, 3.0 V across the cable, and a current of 50 A. Which part is defective?

20P. Two batteries having the same emf $\mathscr{E}$ but different internal resistances r_1 and r_2 $(r_1 > r_2)$ are connected in series to an external resistance R. (a) Find the value of R that makes the potential difference zero between the terminals of one battery. (b) Which battery is it?

21P. A solar cell generates a potential difference of 0.10 V when a 500-Ω resistor is connected across it and a potential difference of 0.15 V when a 1000-Ω resistor is substituted. What are (a) the internal resistance and (b) the emf of the solar cell? (c) The area of the cell is 5.0 cm^2 and the intensity of light striking it is 2.0 mW/cm^2. What is the efficiency of the cell for converting light energy to heat in the 1000-Ω external resistor?

22P. (a) In the circuit of Fig. 5a show that the power delivered to R as thermal energy is a maximum when R is equal to the internal resistance r of the battery. (b) Show that this maximum power is $P = \mathscr{E}^2/4r$.

23P. Conductors A and B, having equal lengths of 40 m and a common diameter of 2.6 mm, are connected in series. A potential difference of 60 V is applied between the ends of the composite wire. The resistances of the wires are 0.127 and 0.729 Ω, respectively. Determine: (a) the current density in each wire; (b) the potential difference across each wire. (c) Identify the wire materials. See Table 1 in Chapter 28.

24P. A battery of emf $\mathscr{E} = 2.0$ V and internal resistance $r = 0.50$ Ω is driving a motor. The motor is lifting a 2.0-N mass at constant speed $v = 0.50$ m/s. Assuming no power losses, find (a) the current i in the circuit and (b) the potential difference V across the terminals of the motor. (c) Discuss the fact that there are two solutions to this problem.

25P. A temperature-stable resistor is made by connecting a resistor made of silicon in series with one made of iron. If the required total resistance is 1000 Ω at 20°C, what should be the resistances of the two resistors? See Table 1 in Chapter 28.

Section 29-6 Multiloop Circuits

26E. Four 18-Ω resistors are connected in parallel across a 25-V battery. What is the current through the battery?

27E. A total resistance of 3.0 Ω is to be produced by connecting an unknown resistance to a 12 Ω resistance. What must be the value of the unknown resistance and how should it be connected?

28E. By using only two resistors—singly, in series, or in parallel—you are able to obtain resistances of 3, 4, 12, and 16 Ω. What are the separate resistances of the resistors?

29E. In Fig. 24 find the current in each resistor and the potential difference between a and b. Put $\mathscr{E}_1 = 6.0$ V, $\mathscr{E}_2 = 5.0$ V, $\mathscr{E}_3 = 4.0$ V, $R_1 = 100$ Ω, and $R_2 = 50$ Ω.

30E. Figure 25 shows a circuit containing an ammeter and three switches, labeled S_1, S_2, and S_3. Find the readings of the ammeter for all possible combinations of switch settings. Put $\mathscr{E} = 120$ V, $R_1 = 20$ Ω, and $R_2 = 10$ Ω. Assume that the ammeter and battery have no resistance.

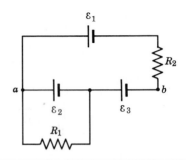

Figure 24 Exercise 29.

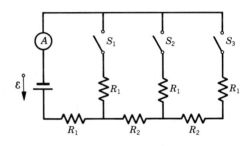

Figure 25 Exercise 30.

31E. In Fig. 26, find the equivalent resistance between points (a) A and B, (b) A and C, and (c) B and C.

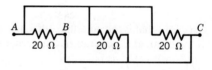

Figure 26 Exercise 31.

32E. In Fig. 27, find the equivalent resistance between points D and E.

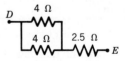

Figure 27 Exercise 32.

33E. Two light bulbs, one of resistance R_1 and the other of resistance R_2 $(<R_1)$ are connected (a) in parallel and (b) in series. Which bulb is brighter in each case?

34E. In Fig. 8 calculate the potential difference between points c and d by as many paths as possible. Assume that $\mathscr{E}_1 = 4.0$ V, $\mathscr{E}_2 = 1.0$ V, $R_1 = R_2 = 10$ Ω, and $R_3 = 5.0$ Ω.

35E. Nine copper wires of length l and diameter d are connected in parallel to form a single composite conductor of resistance R. What must be the diameter D of a single copper wire of length l if it is to have the same resistance?

36E. A 120-V power line is protected by a 15-A fuse. What is the maximum number of 500-W lamps that can be simultaneously operated in parallel on this line?

37E. A circuit containing five resistors connected to a 12-V battery is shown in Fig. 28. What is the voltage drop across the 5.0-Ω resistor?

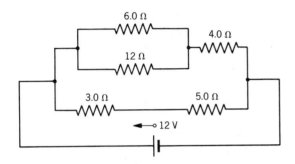

Figure 28 Exercise 37.

38E. For manual control of the current in a circuit, you can use a parallel combination of variable resistors of the sliding contact type, as in Fig. 29, with $R_1 = 20R_2$. (a) What procedure is used to adjust the current to the desired value? (b) Why is the parallel combination better than a single variable resistor?

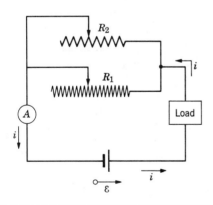

Figure 29 Exercise 38.

39P. Two resistors R_1 and R_2 may be connected either in series or parallel across a (resistanceless) battery with emf $\mathcal{E}$. We desire the thermal energy transfer rate for the parallel combination to be five times that for the series combination. If $R_1 = 100$ Ω, what is R_2?

40P. You are given a number of 10-Ω resistors, each capable of dissipating only 1.0 W. What is the minimum number of such resistors that you need to combine in series or parallel

combinations to make a 10-Ω resistor capable of dissipating at least 5.0 W?

41P. Two batteries of emf $\mathcal{E}$ and internal resistance r are connected in parallel across a resistor R, as in Fig. 30a. (a) For what value of R is the thermal energy delivered to the resistor a maximum? (b) What is the maximum energy dissipation rate?

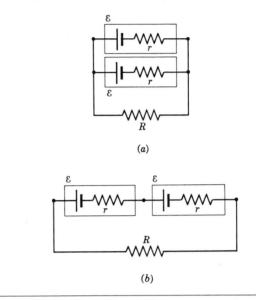

Figure 30 Problems 41 and 43.

42P. In Fig. 31 imagine an ammeter inserted in the branch containing R_3. (a) What will it read, assuming $\mathcal{E} = 5.0$ V, $R_1 = 2.0$ Ω, $R_2 = 4.0$ Ω, and $R_3 = 6.0$ Ω? (b) The ammeter and the source of emf are now physically interchanged. Show that the ammeter reading remains unchanged.

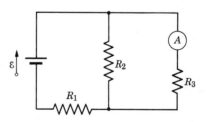

Figure 31 Problem 42.

43P. You are given two batteries of emf $\mathcal{E}$ and internal resistance r. They may be connected either in series or in parallel and are used to establish a current in a resistor R, as in Fig. 30. (a) Derive expressions for the current in R for both methods of connection. (b) Which connection yields the larger current if $R > r$ and if $R < r$?

44P. *N* identical batteries of emf $\mathcal{E}$ and internal resistance *r* may be connected all in series or all in parallel. Show that each arrangement will give the same current in an external resistor *R*, if *R* = *r*.

45P. (*a*) Calculate the current through each source of emf in Fig. 32. (*b*) Calculate V_{ab}. Assume that $R_1 = 1.0\ \Omega$, $R_2 = 2.0\ \Omega$, $\mathcal{E}_1 = 2.0$ V, and $\mathcal{E}_2 = \mathcal{E}_3 = 4.0$ V.

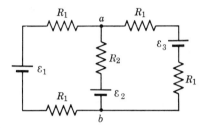

Figure 32 Problem 45.

46P. A three-way 120-V lamp bulb, rated for 100-200-300 W burns out a filament. Afterwards, the bulb operates at the same intensity on its lowest and its highest switch positions but does not operate at all on the middle position. (*a*) How are the filaments wired up inside the bulb and how is the switch constructed? (*b*) Calculate the resistances of the filaments.

47P. What current, in terms of $\mathcal{E}$ and *R*, does the ammeter *A* in Fig. 33 read? Assume that *A* has zero resistance.

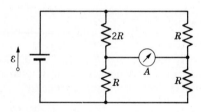

Figure 33 Problem 47.

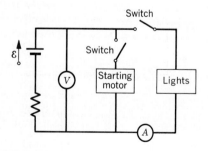

Figure 34 Problem 48.

48P. When the lights of an automobile are switched on, an ammeter in series with them reads 10 A and a voltmeter con-

nected across them reads 12 V. See Fig. 34. When the electric starting motor is turned on, the ammeter reading drops to 8.0 A and the lights dim somewhat. If the internal resistance of the battery is 0.050 Ω and that of the ammeter is negligible, what are (*a*) the emf of the battery and (*b*) the current through the starting motor when the lights are on?

49P. (*a*) In Fig. 35 what is the equivalent resistance of the network shown? (*b*) What are the currents in each resistor? Put $R_1 = 100\ \Omega$, $R_2 = R_3 = 50\ \Omega$, $R_4 = 75\ \Omega$, and $\mathcal{E} = 6.0$ V.

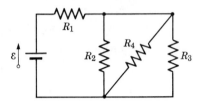

Figure 35 Problem 49.

50P. (*a*) In Fig. 36 what power appears as thermal energy in R_1? In R_2? In R_3? (*b*) What power is supplied by $\mathcal{E}_1$? By $\mathcal{E}_2$? (*c*) Discuss the energy balance in this circuit. Assume that $\mathcal{E}_1 = 3.0$ V, $\mathcal{E}_2 = 1.0$ V, $R_1 = 5.0\ \Omega$, $R_2 = 2.0\ \Omega$, and $R_3 = 4.0\ \Omega$.

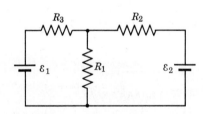

Figure 36 Problem 50.

51P. (*a*) In the circuit of Fig. 37, for what value of *R* will the battery deliver energy to the circuit at a rate of 60 W? (*b*) What is the maximum power the battery can deliver to the circuit? (*c*) What is the minimum power the battery can deliver?

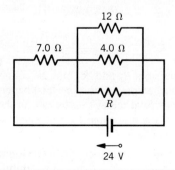

Figure 37 Problem 51.

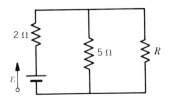

Figure 38 Problem 52.

52P. In the circuit of Fig. 38, $\mathscr{E}$ has a constant value but R can be varied. Find the value of R that results in the maximum heating in that resistor.

53P. In Fig. 39, find the equivalent resistance between points (a) F and H and (b) F and G.

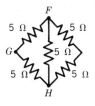

Figure 39 Problem 53.

54P. A copper wire of radius $a = 0.25$ mm has an aluminum jacket of outside radius $b = 0.38$ mm. (a) There is a current $i = 2.0$ A in the composite wire. Using Table 1 in Chapter 28, calculate the current in each material. (b) What is the wire length if a potential difference $V = 12$ V maintains the current?

55P. Figure 40 shows a battery connected across a uniform resistor R_0. A sliding contact can move across the resistor from $x = 0$ at the left to $x = 10$ cm at the right. Find an expression for the power dissipated in the resistor R as a function of x. Plot the function for $\mathscr{E} = 50$ V, $R = 2000$ Ω, and $R_0 = 100$ Ω.

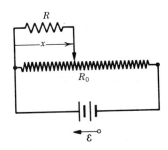

Figure 40 Problem 55.

56P. Twelve resistors, each of resistance R ohms, form a cube (see Fig. 11a). (a) Find R_{13}, the equivalent resistance of a face diagonal. (b) Find R_{18}, the equivalent resistance of a body diagonal. See Sample Problem 4.

Section 29-7 Measuring Instruments

57E. A simple ohmmeter is made by connecting a 1.5-V flashlight battery in series with a resistor R and a 1.0-mA ammeter, as shown in Fig. 41. R is adjusted so that when the circuit terminals are shorted together the meter deflects to its full-scale value of 1.0 mA. What external resistance across the terminals results in a deflection of (a) 10%, (b) 50%, and (c) 90% of full scale? (d) If the ammeter has a resistance of 20 Ω and the internal resistance of the battery is negligible, what is the value of R?

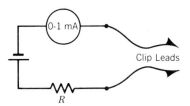

Figure 41 Exercise 57.

58P. In Fig. 12 assume that $\mathscr{E} = 3.0$ V, $r = 100$ Ω, $R_1 = 250$ Ω, and $R_2 = 300$ Ω. If $R_V = 5.0$ kΩ, what percent error is made in reading the potential difference across R_1? Ignore the presence of the ammeter.

59P. In Fig. 12 assume that $\mathscr{E} = 5.0$ V, $r = 2.0$ Ω, $R_1 = 5.0$ Ω, and $R_2 = 4.0$ Ω. If $R_A = 0.10$ Ω, what percent error is made in reading the current? Assume that the voltmeter is not present.

60P. *Resistance measurement.* A voltmeter (resistance R_V) and an ammeter (resistance R_A) are connected to measure a resistance R, as in Fig. 42a. The resistance is given by $R = V/i$, where V is the voltmeter reading and i is the current in the resistor R. Some of the current registered by the ammeter (i') goes through the voltmeter so that the ratio of the meter readings ($= V/i'$) gives only an *apparent* resistance reading R'.

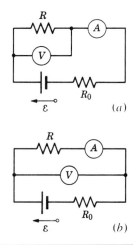

Figure 42 Problems 60, 61, and 62.

Show that R and R' are related by

$$\frac{1}{R} = \frac{1}{R'} - \frac{1}{R_V}.$$

Note that as $R_V \to \infty$, $R' \to R$.

61P. *Resistance measurement.* If meters are used to measure resistance, they may also be connected as they are in Fig. 42b. Again the ratio of the meter readings gives only an apparent resistance R'. Show that R' is related to R by

$$R = R' - R_A,$$

in which R_A is the ammeter resistance. Note that as $R_A \to 0$, $R' \to R$.

62P. In Fig. 42 the ammeter and voltmeter resistances are 3.00 Ω and 300 Ω, respectively. If $R = 85$ Ω, (a) What will the meters read for the two different connections? (b) What apparent resistance R' will be computed in each case? Take $\mathcal{E} = 12$ V and $R_0 = 100$ Ω.

63P. *The Wheatstone bridge.* In Fig. 43 R_s is to be adjusted in value until points a and b are brought to exactly the same potential. (One tests for this condition by momentarily connecting a sensitive ammeter between a and b; if these points are at the same potential, the ammeter will not deflect.) Show that when this adjustment is made, the following relation holds:

$$R_x = R_s(R_2/R_1).$$

An unknown resistor (R_x) can be measured in terms of a standard (R_s) using this device, which is called a Wheatstone bridge.

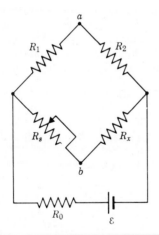

Figure 43 Problems 63 and 64.

64P. If points a and b in Fig. 43 are connected by a wire of resistance r, show that the current in the wire is

$$i = \frac{\mathcal{E}(R_s - R_x)}{(R + 2r)(R_s + R_x) + 2R_sR_x}.$$

where $\mathcal{E}$ is the emf of the battery. Assume that R_1 and R_2 are equal ($R_1 = R_2 = R$) and that R_0 equals zero. Is this formula consistent with the result of Problem 63?

Section 29–8 RC Circuits

65E. In an RC series circuit $\mathcal{E} = 12$ V, $R = 1.4$ MΩ, and $C = 1.8$ μF. (a) Calculate the time constant. (b) Find the maximum charge that will appear on the capacitor during charging. (c) How long does it take for the charge to build up to 16 μC?

66E. How many time constants must elapse before a capacitor in an RC circuit is charged to within 1.0 percent of its equilibrium charge?

67E. From Fig. 16, deduce the value of the time constant of the RC circuit. Can the values of R and C be found?

68E. A 15-kΩ resistor and a capacitor are connected in series and a 12-V potential is suddenly applied. The potential across the capacitor rises to 5.0 V in 1.3 μs. (a) Calculate the time constant. (b) Find the capacitance of the capacitor.

69P. An RC circuit is discharged by closing a switch at time $t = 0$. The initial potential difference across the capacitor is 100 V. If the potential difference has decreased to 1.0 V after 10 s, (a) calculate the time constant of the circuit. (b) What will be the potential difference at $t = 17$ s?

70P. Figure 44 shows the circuit of a flashing lamp, like those attached to barrels at highway construction sites. The fluorescent lamp L is connected in parallel across the capacitor C of an RC circuit. Current passes through the lamp only when the potential across it reaches the breakdown voltage V_L; in this event, the capacitor discharges through the lamp and it flashes for a very short time. Suppose that two flashes per second are needed. Using a lamp with breakdown voltage $V_L = 72$ V, a 95-V battery, and a 0.15-μF capacitor, what should be the resistance R of the resistor?

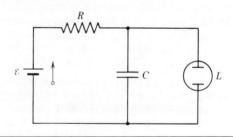

Figure 44 Problem 70.

71P. A 1.0-μF capacitor with an initial stored energy of 0.50 J is discharged through a 1.0-MΩ resistor. (a) What is the initial charge on the capacitor? (b) What is the current through the resistor when the discharge starts? (c) Determine V_C, the voltage across the capacitor, and V_R, the voltage across the resistor, as functions of time. (d) Express the rate of generation of thermal energy in the resistor as a function of time.

72P. A 3.0-MΩ resistor and a 1.0-μF capacitor are connected in a single-loop circuit with a seat of emf with $\mathcal{E} = 4.0$ V. At 1.0 s after the connection is made, what are the rates at which (a) the charge of the capacitor is increasing, (b) energy is being stored in the capacitor, (c) thermal energy is appearing in the resistor, and (d) energy is being delivered by the seat of emf?

73P. The potential difference between the plates of a leaky 2.0-μF capacitor drops to one-fourth its initial value in 2.0 s. What is the equivalent resistance between the capacitor plates?

74P. Prove that when switch S in Fig. 15 is thrown from a to b all the energy stored in the capacitor is transformed into thermal energy in the resistor. Assume that the capacitor is fully charged before the switch is thrown.

75P. An initially uncharged capacitor C is fully charged by a constant emf $\mathcal{E}$ in series with a resistor R. (a) Show that the final energy stored in the capacitor is half the energy supplied by the emf. (b) By direct integration of i^2R over the charging time, show that the thermal energy dissipated by the resistor is also half the energy supplied by the emf.

76P. A controller on an electronic arcade game consists of a variable resistor connected across the plates of a 0.22-μF capacitor. The capacitor is charged to 5.0 V, then discharged through the resistor. The time for the potential difference across the plates to decrease to 0.80 V is measured by an internal clock. If the range of discharge times that can be handled effectively is from 10 μs to 6.0 ms, what should be the range of the resistor?

77P. The circuit of Fig. 45 shows a capacitor C, two batteries, two resistors, and a switch S. Initially S has been open for a long time. It is then closed for a long time. By how much does the

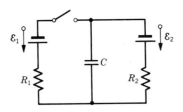

Figure 45 Problem 77.

charge on the capacitor change over this period? Assume $C = 10$ μF, $\mathcal{E}_1 = 1.0$ V, $\mathcal{E}_2 = 3.0$ V, $R_1 = 0.20$ Ω, and $R_2 = 0.40$ Ω.

78P. In the circuit of Fig. 46, $\mathcal{E} = 1.2$ kV, $C = 6.5$ μF, $R_1 = R_2 = R_3 = 0.73$ MΩ. With C completely uncharged, the switch S is suddenly closed ($t = 0$). (a) Determine the currents through each resistor for $t = 0$ and $t = \infty$. (b) Draw qualitatively a graph of the potential drop V_2 across R_2 from $t = 0$ to $t = \infty$. (c) What are the numerical values of V_2 at $t = 0$ and $t = \infty$? (d) Give the physical meaning of "$t = \infty$" in this case.

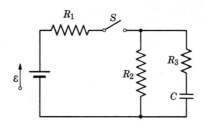

Figure 46 Problem 78.

ESSAY 11
EXPONENTIAL GROWTH

ALBERT A. BARTLETT

UNIVERSITY OF
COLORADO, BOULDER

The greatest shortcoming of the human race is our inability to understand the exponential function.

The importance of the exponential function arises, not so much from its appearance in physics (as in the RC circuit), but because it describes steady growth. To see this, let us assume a quantity N is growing steadily a fixed fraction (or percent) per unit time. We would describe this by writing

$$\frac{dN}{Ndt} = k \qquad (1)$$

where (dN/N) is the fractional change in N in a time dt. A quantity growing 4% per year would have $k = 0.04$ year^{-1}.

Let us rearrange Eq. 1:

$$\frac{dN}{dt} = kN. \qquad (2)$$

This equation reflects the fact that the rate of change of N is proportional to N. Equation 2 may be integrated to give

$$N = N_0 e^{kt} \qquad (3)$$

where N_0 is the size of N at the time $t = 0$. The quantity e is the base of natural logarithms (2.718 . . .). For example, suppose the quantity N is growing 5% per year. Then $(dN/N) = 0.05$ when $dt = 1$ year, and $k = 0.05$ year^{-1}. Equation 2 becomes

$$\frac{dN}{dt} = 0.05\, N \qquad (2^1)$$

and Eq. 3 becomes

$$N = N_0 e^{0.05t} \qquad (3^1)$$

Table 1 shows the increase in size of a quantity that is growing 5% per unit time. We can interpret the table as follows. If a quantity is growing steadily at 5% per year, then in 10 years it will have increased in size by 64.9%. If a quantity is growing 5% per day, then in 100 days it will be 148.4 times as large as it was initially. It is easy to calculate that in 1095 days the quantity growing 5% per day would increase in size by a factor equal to Avogadro's number! We note from Table 1 that as time goes on the size of N becomes very large, ultimately approaching infinity. This leads to a profound conclusion. It is impossible for steady growth of any real thing to continue forever —or even for a long period of time.

Our society worships growth, yet growth for long periods of time is impossible. This is the conflict that is recognized in the opening statement.

Table 1 Steady Growth of 5% per Unit Time. The Time Units Could Be Seconds, Days, Years, or Any Unit of Time.

Time Units	Value of N, as a Multiple of N_0.
0	1.000
1	1.051
2	1.105
5	1.284
10	1.649
20	2.718
50	12.18
100	148.4
200	2.20×10^4
500	7.20×10^{10}
1000	5.18×10^{21}

Table 2 The Size of the Growing Quantity in Units of the Doubling Time T_2

Time in Units of T_2	Size in Units of N_0
0	$2^0 = 1$
1	$2^1 = 2$
2	$2^2 = 4$
3	$2^3 = 8$
4	$2^4 = 16$
5	$2^5 = 32$
10	$2^{10} = 1024$
n	2^n

The Doubling Time, T_2

Let us now calculate the period of time T_2 for the growing quantity to double in size ($N = 2N_0$):

$$2N_0 = N_0 e^{kT_2}$$
$$\ln(2) = kT_2$$
$$T_2 = \frac{\ln 2}{k} = \frac{0.693}{k} = \frac{69.3}{100\,k} \approx \frac{70}{P}. \tag{4}$$

Here P is the percent growth per unit time, which is $100k$. Thus one finds the doubling time by dividing 70 by the percentage growth per unit time. For example, if something is growing at a rate of 5% per year, the doubling time is $(70/5) = 14$ years. If a quantity is growing steadily at a rate of 5% per day, then the quantity will double in size in $(70/5) = 14$ days! This is the origin of the "Rule of 70" (sometimes referred to as the "Rule of 72") that is used by bankers and investment counselors.

It should be clear that the time T_3 for the growing quantity to triple in size will be given by

$$T_3 = \frac{\ln 3}{k} \approx \frac{110}{P}. \tag{5}$$

We can express Eq. 3 in terms of powers of 2 instead of powers of e:

$$N = N_0(2)^{t/T_2}. \tag{6}$$

Table 2 shows the way the growth develops when Eq. 6 is used. In estimating the effects of steady growth it is convenient to remember that

$$2^{10} \approx 10^3 \tag{7}$$

Example

When compound interest is calculated annually it provides an example of growth that is approximately exponential. A sum of money ($N_0 = \$1$) is placed in a savings account that adds 7% interest at the end of each year. The interest in the first year is $0.07 \times \$1$, so that at the end of the year the account contains $1.07.

If the interest is calculated monthly at a rate of ($7/12 = 0.5833\%$ per month), the account at the end of the year will contain $1.072, which is slightly larger than the amount when the interest is calculated only once at the end of the year. If one calculates the interest continuously (as many banks do) the account grows in accord with Eq. 3. Thus in 1 year of continuous compounding of the interest the value of the account is

$$N = 1 \times e^{0.07 \times 1} = \$1.0725 \ldots.$$

In 10 years, the value is

$$N = 1 \times e^{0.07 \times 10} = 2.014.$$

In 100 years we have

$$N = 1 \times e^{0.07 \times 100} = \$1096.63$$

and in 1000 years we have

$$N = 1 \times e^{0.07 \times 1000} = \$2.52 \times 10^{30}.$$

The growth of the account is ultimately so rapid that the banking laws prohibit one from leaving sums of money at interest untouched for long periods of time.

This problem can also be used to illustrate the ease with which exponential arithmetic can be worked out in one's head. For $k = 0.07$ ($P = 7\%$) the doubling time is $70/7 = 10$ years. Thus in 200 years we are dealing with 20 doublings in which time $1 becomes 2^{20}. Equation 7 allows us to estimate the value of 2^{20}:

$$2^{20} = 2^{10} \times 2^{10} \approx 10^3 \times 10^3 = 10^6.$$

Thus $1 at an annual interest rate of 7% compounded continuously would become approximately one million dollars in 200 years. The actual value is 1.20×10^6.

Problem

In 1985 the assessed valuation of New York City was 39.4×10^9. In the year 1626 Manhattan Island was purchased for $24. If the $24 had been placed in a savings account, what interest rate (compounded continuously) would be required so that the account in 1985 would contain a sum of money equal to the assessed valuation of New York City in 1985? In this example, $t = 1985 - 1626 = 359$ years and

$$39.4 \times 10^9 = 24 \, e^{k \times 359}.$$

Solving, we find that k has the value 0.0591, which corresponds to an annual interest rate of 5.91 percent compounded continuously.

Example

On July 7, 1986, reports appeared in the press saying that the world population had reached the number 5×10^9 people. The growth rate is reported to be 1.7% per year, ($T_2 = 41$ years). If this growth rate continued unchanged (a) How long a time is required for the world population to increase by one million people? Since $dN \ll N$

we can use Eq. 2:

$$dN = kN \, dt$$
$$10^6 = 0.017 \times 5.0 \times 10^9 \, dt$$
$$dt = 1.2 \times 10^{-2} \text{ years} = 4.3 \text{ days!}$$

(b) How long a time would be required for the mass of people to equal the mass of the earth? (Assume the mass of a person is 65 kg.)

$$5.98 \times 10^{24} = (5.0 \times 10^9 \times 65)e^{0.017t}$$

Solving for t gives 1.8×10^3 years! The impossibility of steady growth of populations for long periods of time is clear from this example.

Illustration, The Power of Powers of Two

Legend has it that the game of chess was invented by a mathematician who worked for a king. The king was so pleased that he asked the mathematician to name the reward he wanted. The mathematician said, "Take my new chess board and on the first square place one grain of wheat; on the next square double the one and put 2 grains. On the third square double the 2 and put 4 grains . . . Just keep doubling this way until you have doubled for every square. That will be an adequate payment." Table 3 indicates the details of the filling the board. This total number of grains is approximately 500 times the 1976 worldwide harvest of wheat! How did we get such a large number? We started with *one* grain but we let the number grow steadily until it had doubled 63 times. Table 3 illustrates a further important aspect of steady growth: the number of grains required on any square is one grain more than the total of all the grains on *all* of the preceding squares!

Example

In the mid-1970s the American electric power industries placed ads in magazines calling attention to the fact that for a century their industry had been growing steadily at a rate of about 7% per year and they expected this growth to continue steadily in the future. As we saw with the chessboard and Table 3, this would require that the electric generating capacity constructed in the United States in each future decade be equal to the total generating capacity in place at the start of that decade! It would also require that the consumption of electrical energy in the United States in any coming decade be equal to the total of all the electrical energy that had been consumed in the entire previous history of the electrical generating industry in the United States prior to the start of that decade! (See Table 3.) The financial trauma of the electrical

Table 3

Square Number	Grains on Square	Total Grains on Board
1	1	1
2	2	3
3	4	7
4	8	15
5	16	31
64	2^{63}	$2^{64} - 1 = 1.8 \times 10^{19}$

generating industry in the United States in the late 1970s and early 1980s was due in part to the failure to recognize the simple facts of the arithmetic of steady growth. The industry was not prepared for the inevitable reductions in their rate of growth.

Steady Growth in a Finite Environment

Fossil fuels are finite resources. What happens if we plan to try to maintain steady growth in the rate of consumption of a fossil fuel? We can illustrate this with the story of bacteria in a bottle.

Bacteria grow by doubling; one bacterium divides to become two; the two divide to become four; the four divide to become eight, etc. Suppose we had bacteria that doubled in number this way every minute. Suppose we put one bacterium in an empty bottle at 11:00 A.M. and later we observed that the bottle was full at 12:00 noon.

Here we have our example of ordinary steady growth with a doubling time of one minute in the finite environment of one bottle. Let's examine three questions.

1. At what time was the bottle half full?
    Answer: at 11:59 — because the bacteria double in number every minute!

2. If you were an average bacterium in that bottle, at what time would you first realize that you were running out of space?

Table 4 shows the last minutes in the bottle and may help you to decide on your answer to this question.

Table 4

Time	Fraction Full (used)	Fraction Empty (available)
11:55	$\frac{1}{32}$	$\frac{31}{32}$
11:56	$\frac{1}{16}$	$\frac{15}{16}$
11:57	$\frac{1}{8}$	$\frac{7}{8}$
11:58	$\frac{1}{4}$	$\frac{3}{4}$
11:59	$\frac{1}{2}$	$\frac{1}{2}$
12:00	1	Zero

3. Suppose that at 11:58 some of the bacteria realize that they are running out of space so they launch a great search for new bottles. Their search yields three new bottles, which is three times as much resource as they had known previously. Surely this will make them self-sufficient in space! How long can the growth continue because of this tremendous discovery? The answer is given in Table 5.

If we plan to have continued steady growth in our rates of consumption of finite resources of fossil fuels, then this example shows that enormous discoveries of new reserves will permit the growth to continue for only very short periods of time.

Table 5

Time	Bottles Filled	Bottles Unfilled
12:00	1	3
12:01	2	2
12:02	4	Zero

CHAPTER 30

THE MAGNETIC FIELD

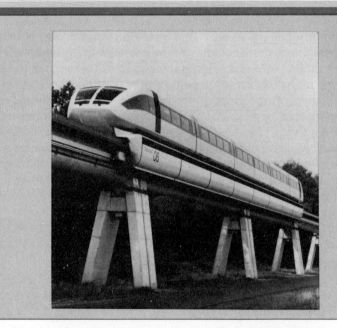

The TRANSRAPID 06, a 200-passenger intercity transit vehicle undergoing tests in West Germany. It has no wheels. Supported and propelled by electromagnetic forces, it has achieved a speed of 300 km/h. New developments in superconducting materials make the prospects for such trains considerably brighter.

30-1 The Magnetic Field

As you fasten a note to the refrigerator door with a small magnet, you become aware with your fingertips that the space around the magnet has special properties. The space near a rubbed plastic rod also has special properties. In that case, we learned to understand them by postulating that there is an *electric field* **E** at all points near the rod. By analogy, it seems logical to postulate that there is a *magnetic field,* which we represent by the symbol **B**, at all points near a magnet.

Figure 1 suggests, by means of iron filings, the magnetic field in the space surrounding a small *permanent magnet.* In another familiar type of magnet, a coil is wound around an iron core, the strength of the external magnetic field being determined by the current in the coil. In industry, such *electromagnets* are used for sorting scrap iron (Fig. 2). Figure 3 shows another type of electromagnet, found in research laboratories the world over.

In electrostatics we represented the relation between electric field **E** and electric charge by

$$\text{electric charge} \rightleftarrows \mathbf{E} \rightleftarrows \text{electric charge.} \quad (1)$$

That is, electric charges set up an electric field and this field in turn exerts an (electric) force on another charge that may be placed in it.

Symmetry—a powerful tool that we have used many times before—suggests that we set up a similar relation for magnetism, namely,

$$\text{``magnetic charge''} \rightleftarrows \mathbf{B} \rightleftarrows \text{``magnetic charge,''} \quad (2)$$

Petroleum is perhaps the most widely used fossil fuel. It has no serious competition as *the* fuel for transportation. Petroleum consumption worldwide grew at a steady rate of about 7% per year for a century before leveling off (zero growth) in the late 1970s. In the United States we equate growth of the rate of consumption with prosperity. The less-developed nations aspire to have growth in their rates of consumption of petroleum in order to improve the quality of life for their people (whose numbers are growing steadily). The world and U.S. situations in regard to petroleum, were summarized in *Science*.[2] "In a recent U.S. Geological Survey report, government researchers specializing in the estimation of oil resources issued a strong warning—the amount of oil left to slake the world's thirst for energy is smaller than some optimists would like to think and may be even smaller than conservative estimates. . . . This new oil [yet to be found] plus reserves known to exist in the ground would last the world about 60 years at present rates of consumption [zero growth]. The estimated U.S. supply from undiscovered resources and demonstrated reserves is 36 years at present rates of production [zero growth] or 19 years in the absence of imports."

In 1977 James Schlesinger, the U.S. Secretary of Energy, noted that in the energy crisis, "we have a classic case of exponential growth against a finite source."[3] The economic trauma of the period of the energy crisis in the late 1970s was followed by a period of declining prices and apparent abundance of fossil fuels. During this time the arithmetic seems to have been forgotten.

Summary

We can summarize a few of the facts of steady growth as they are revealed by this review of the exponential function.

1. Steady growth occurs whenever the rate of change of a quantity is proportional to that quantity.
2. Steady growth is characterized by a constant doubling time.
3. In a modest number of doubling times, steady growth can produce very large numbers.
4. In steady growth, the increase in one doubling time is equal to the sum of all preceding growth.
5. When one has steady growth in a finite environment, the end comes frighteningly fast.
6. Growth is always a short-term transient phenomenon.

The greatest shortcoming of the human race is our inability to understand the exponential function.

BIBLIOGRAPHY   (1) A. A. Bartlett, *Civil Engineering*, Dec. 1969, p. 71.   (2) *Science*, Jan. 27, 1984, p. 382.   (3) *Time*, April 25, 1977, p. 27. Extended discussions of this material are available in the magazine, *The Physics Teacher, 14*, 1976, p. 393; *14*, 1976, p. 485; *15*, 1977, p. 37; *15*, 1977, p. 98; *15*, 1977, p. 225; *16*, 1978, p. 23; *16*, 1978, p. 92; *16*, 1978, p. 158; *17*, 1979, p. 23; and in the *American Journal of Physics, 46*, 1978, p. 876; *54*, 1986, p. 398.

Figure 1   Iron filings sprinkled on a sheet of paper tell us that there is a bar magnet underneath.

Figure 2   An industrial electromagnet, used to load scrap iron.

in which **B** is the magnetic field. The only problem with this idea is that there seem to be no magnetic charges. That is, there are no isolated point objects from which magnetic field lines emerge. Certain current theories predict that such *magnetic monopoles* should exist and many physicists have looked for them, but so far their existence has not been confirmed.

From where then does the magnetic field come? Experiment shows that it comes from *moving* electric charges. A charge sets up an *electric* field whether the charge is at rest or is moving. A charge sets up a *magnetic* field, however, *only* if it is moving. Where are these moving electric charges? In the permanent magnet of Fig. 1, they are the spinning and circulating electrons in the iron atoms that make up the magnet. In the electromagnets of Figs. 2 and 3, they are the electrons drifting through the coils of wire that surround these magnets.

In magnetism then we think in terms of

$$\text{moving electric charge} \rightleftharpoons \mathbf{B}$$
$$\rightleftharpoons \text{moving electric charge.} \quad (3)$$

Because an electric current in a wire is a stream of moving charges, we can also write Eq. 3 as

$$\text{electric current} \rightleftharpoons \mathbf{B} \rightleftharpoons \text{electric current.} \quad (4)$$

Equations 3 and 4 tell us (1) a moving charge or a current sets up a magnetic field and also (2) if we place a moving charge, or a wire carrying a current, in a magnetic field, a magnetic force will act on it. It was the Danish physicist Hans Christian Oersted who (in 1820) first linked the

Figure 3   A research-type electromagnet, showing iron frame F, pole faces P, and coils C. The pole faces are 30 cm in diameter.

then separate sciences of current electricity and magnetism by showing that an (electric) current in a wire could deflect a (magnetic) compass needle.

An electric motor illustrates Eq. 4 very well. In most motors (1) a current in a wire (the field coils) sets up a magnetic field and (2) the magnetic field in turn exerts a force on a second current-carrying wire (the armature windings), causing the shaft to rotate.

In this chapter, we deal with only half of the story summarized by Eqs. 3 and 4. That is, we assume that a magnetic field exists—perhaps in the space between the pole faces of an electromagnet such as that of Fig. 3—and we ask what force this field exerts on charges that move through it. In the next chapter, we deal with the other half of the story, namely, where does the magnetic field come from in the first place?

30-2 The Definition of B

We defined the electric field **E** at a point by putting a test charge q at rest at that point and measuring the (electric) force $\mathbf{F}_E$ that acted on this charge. We then defined **E** from

$$\mathbf{F}_E = q\mathbf{E}. \tag{5}$$

If we had a magnetic monopole available, we could define **B** in a similar way. Because such particles have not been found in nature, we must define **B** in another way, in terms of the magnetic force exerted on a moving electric charge, as Eq. 3 suggests.

How do we find the magnitude and direction of **B** at any given point?* In principle, we fire a test charge q through the point, in various directions and with various speeds. For every such trial, we record the force $\mathbf{F}_B$ (if any) that acts on the moving charge. If we analyze the results of many such trials, we find that the vector $\mathbf{F}_B$ is related to the velocity vector **v** by

$$\mathbf{F}_B = q\,\mathbf{v} \times \mathbf{B} \tag{6}$$

in which **B** is the magnetic field vector. This equation is our defining equation for **B**.† As we shall see later, the direction of **B** found from analyzing the force on a moving charge is the same as the direction in which a compass needle would point.

Here are some of the things that you can verify from Eq. 6, and from Fig. 4 which represents it.

1. The magnetic force $\mathbf{F}_B$ always acts sideways, that is, at right angles to the velocity vector. This means that a (steady) magnetic field can neither speed up nor slow down a moving charged particle. It can deflect the particle sideways but it cannot change the particle's ki-

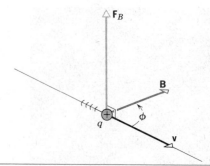

Figure 4 A particle with a positive charge q moving with velocity **v** through a magnetic field **B** experiences a magnetic deflecting force $\mathbf{F}_B$. See Eq. 6.

netic energy. Figure 5 shows a beam of electrons in a cathode ray tube being deflected sideways by a magnetic field that has a direction at right angles to the plane of the figure.

2. A magnetic field exerts no force on a charge that moves parallel (or antiparallel) to it. From Eq. 6 we see that the magnitude of the magnetic deflecting force is given by

$$F_B = qvB \sin \phi. \tag{7}$$

This tells us that when **v** is parallel or antiparallel to **B**—which corresponds to $\phi = 0°$ or $180°$—the magnetic deflecting force is indeed zero.

3. The *maximum* value of the deflecting force

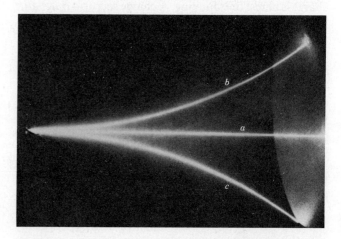

Figure 5 A triple exposure of an electron beam in a cathode ray tube. In (*a*) there is no external magnetic field. In (*b*) a magnetic field points out of the plane of the figure. In (*c*) a magnetic field points into the plane of the figure.

---

* As a preliminary, we had better make sure that there is no complicating *electric* field acting at that point. We can do this by putting a *stationary* charge there and verifying that no force acts on it.

† You may wish to review Section 3–7, which deals with vector products.

($=qvB$) occurs when the test charge is moving at right angles to the magnetic field ($\phi = 90°$).

4. The magnitude of the deflecting force is directly proportional to q and to v. The greater the charge of the test body and the faster it is moving, the greater the magnetic deflecting force.

5. The direction of the magnetic deflecting force depends on the sign of q, positive and negative test charges with equal velocities being deflected in opposite directions.

It is impressive that all this information can be compressed into an equation as modest in structure as Eq. 6. This economy of expression is one of the reasons that we find vectors useful in physics.

To develop a feeling for Eq. 6, consider Fig. 6, which shows some tracks left by fast charged particles moving through a *bubble chamber* at the Lawrence Berkeley Laboratory. The chamber, which is filled with liquid hydrogen, is immersed in a strong uniform magnetic field that points out of the plane of the figure.

At point P in Fig. 6 an incoming gamma ray— which leaves no track because it is uncharged—triggers an event in the chamber that imparts energy to two electrons (tracks marked e^-) and to one positive electron (track marked e^+). Check from Eq. 6 that the three tracks bend in the directions that you expect.

The SI unit for **B** that follows from Eqs. 6 and 7 is the newton/(coulomb · meter/second). For convenience, this is called the *tesla* (abbr. T). Recalling that a coulomb/second is an ampere, we have

$$1 \text{ tesla} = 1 \text{ T} = 1 \text{ N/A} \cdot \text{m}. \qquad (8)$$

Table 1 Some Magnetic Fields^a

At the surface of a neutron star (calculated)	10^8 T
The magnet of Fig. 3	1.5 T
Near a small bar magnet	10^{-2} T
At the surface of the earth	10^{-4} T
In interstellar space	10^{-10} T
Smallest value in a magnetically shielded room	10^{-14} T

^{a} Approximate values.

An earlier (non-SI) unit for **B**, still in common use, is the *gauss;* the relation is

$$1 \text{ tesla} = 10^4 \text{ gauss.} \qquad (9)$$

Table 1 shows the magnetic fields that occur in a few situations. Note that the earth's magnetic field near the earth's surface is about 10^{-4} T ($= 100 \ \mu$T or 1 gauss).

We can represent magnetic fields by field lines, just as we did for the electric field. The same rules apply. That is, (1) the direction of the tangent to a magnet field line at any point gives the direction of **B** at that point, and (2) the spacing of the lines is a measure of the magnitude of **B**. Therefore, the magnetic field is strong where the lines are close together and conversely.

Figure 7 (compare Fig. 1) shows how the magnetic field near a bar magnet can be represented by magnetic field lines. Note that the lines pass right through the magnet, forming closed loops. The external magnetic effects of a bar magnet are strongest near its ends. The end from which the field lines emerge is called a *north pole,* the other end being the *south pole.*

Opposite magnetic poles attract each other. From this fact and from the fact that the north pole of a com-

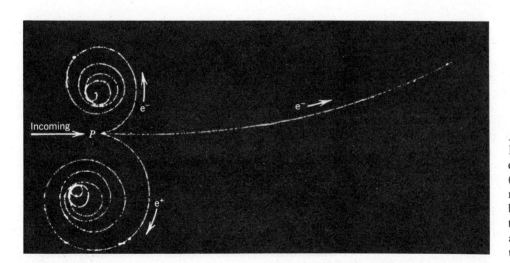

Figure 6 The tracks of two electrons (e^-) and a positron (e^+) in a bubble chamber. A magnetic field fills the chamber, pointing out of the plane of the figure. What can you tell about the particles by examining their tracks?

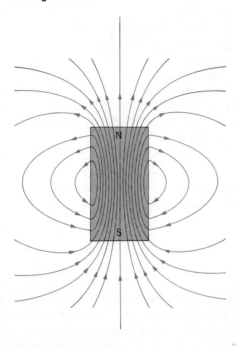

Figure 7 The magnetic field lines for a bar magnet. The lines form closed loops, leaving the magnet at its north pole and entering it at its south pole.

pass needle points north, we conclude that the earth's magnetic pole in the northern hemisphere is a south magnetic pole. That is, in the Arctic regions, the lines of the earth's magnetic field point generally down, into the earth's surface. The earth's magnetic pole in Antarctica is a north magnetic pole. That is, lines of the earth's magnetic field in this region point generally up, out of the earth's surface.

Sample Problem 1 A uniform magnetic field **B**, with magnitude 1.2 mT, points vertically upward throughout the volume of the room in which you are sitting. A 5.3-MeV proton moves horizontally from south to north through a certain point in the room. What magnetic deflecting force acts on the proton as it passes through this point? The proton mass is 1.67×10^{-27} kg.

The magnetic deflecting force depends on the speed of the proton, which we can find from $K = \frac{1}{2}mv^2$. Solving for v, we find

$$v = \sqrt{\frac{2K}{m}} = \sqrt{\frac{(2)(5.3 \text{ MeV})(1.60 \times 10^{-13} \text{ J/MeV})}{1.67 \times 10^{-27} \text{ kg}}}$$

$$= 3.2 \times 10^7 \text{ m/s}.$$

Equation 7 then yields

$$\begin{aligned} F_B &= qvB \sin \phi \\ &= (1.60 \times 10^{-19} \text{ C})(3.2 \times 10^7 \text{ m/s}) \\ &\quad \times (1.2 \times 10^{-3} \text{ T})(\sin 90°) \\ &= 6.1 \times 10^{-15} \text{ N}. \end{aligned}$$ (Answer)

This may seem like a small force but it acts on a particle of small mass, producing a large acceleration, namely,

$$a = \frac{F_B}{m} = \frac{6.1 \times 10^{-15} \text{ N}}{1.67 \times 10^{-27} \text{ kg}} = 3.7 \times 10^{12} \text{ m/s}^2.$$

It remains to find the direction of $\mathbf{F}_B$. In Eq. 6, **v** points horizontally from south to north and **B** points vertically up. The rule for the direction of vector products (see Section 3–7) shows us that the deflecting force $\mathbf{F}_B$ must point horizontally from west to east, as Fig. 8 shows.

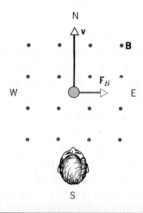

Figure 8 Sample Problem 1. A view from above of a student sitting in a room filled with a magnetic field and watching a moving proton deflect toward the east. The magnetic field points vertically upward in the room.

If the charge of the particle had been negative, the magnetic deflecting force would have pointed in the opposite direction, that is, horizontally from east to west. This is predicted automatically by Eq. 6, if we substitute $-e$ for q.

In this calculation, we used the (approximate) classical expression ($K = \frac{1}{2}mv^2$) for the kinetic energy of the proton rather than the (exact) relativistic expression (see Eq. 27 of Chapter 7). The criterion for when the classical expression may safely be used is that $K \ll mc^2$, where mc^2 is the rest energy of the particle. In this case, $K = 5.3$ MeV and the rest energy of a proton is 938 MeV. This proton passes the test and we were justified in treating it as "slow," that is, in using the classical $K = \frac{1}{2}mv^2$ formula for the kinetic energy. In dealing with energetic particles, we must always be alert to this point.

30-3 Discovering the Electron

A beam of electrons can be deflected by a magnetic field. You are perhaps more familiar with this fact than you realize. Such deflections paint the images on our television screens and trace out the letters on the screens of our word processors. It was not always so.

At the end of the last century, the cathode ray tube, far from being found in every home, was the last word in advanced laboratory research instrumentation. In 1897, J. J. Thomson at Cambridge University showed that the "rays" that cause the glass walls of such tubes to glow were streams of negatively charged particles, which he called *corpuscles.* We now call them *electrons.*

What Thomson did was measure the ratio of the mass m to the charge q of the cathode ray particles. In Fig. 9, which is a modern version of his apparatus, electrons are emitted from hot filament F and accelerated by an applied potential difference V. Passing through a slit in screen C, they then enter a region in which they move at right angles to an electric field **E** and a magnetic field **B**, these two fields being at right angles to each other. The beam, striking fluorescent screen S, is visible as a spot of light.

Study of Fig. 9 shows that, no matter what the sign of the charge carried by the particle, the electric field and the magnetic field will deflect it in opposite directions. In particular, if the particle is negatively charged, the electric field will deflect the particle up and the magnetic field will deflect it down.

Suppose the two fields in Fig. 9 are adjusted so that the two deflecting forces just cancel. From Eqs. 5 and 6 we then have

$$qE = qvB$$

or

$$v = \frac{E}{B}. \tag{10}$$

The arrangement of crossed electric and magnetic fields thus acts as a *velocity filter* and allows us to measure the speed of the particles that pass through it.

Thomson's procedure was equivalent to the following: (1) Set $E = 0$ and $B = 0$ and note the position of the undeflected beam spot. (2) Apply the electric field **E**, measuring on the fluorescent screen the beam deflection that it causes. (3) Leaving **E** in place, apply a magnetic field **B** and adjust its value until the beam deflection is restored to zero.

The deflection of an electron in a purely electric field (step 2 above), measured at the far edge of the deflecting plates, is analyzed in Sample Problem 8 of Chapter 24. The deflection is shown there to be where v is the

$$y = \frac{qEL^2}{2mv^2}, \tag{11}$$

electron speed and L is the length of the plates. The deflection y cannot be measured directly but it can be calculated from the measured displacement of the spot on the screen. The *direction* of the deflection allows one to determine the sign of the charge carried by the particle.

Eliminating v between Eqs. 10 and 11 leads to

$$\frac{m}{q} = \frac{B^2L^2}{2yE}, \tag{12}$$

in which all quantities on the right can be measured.

Thomson put forward the daring and important claim—which turned out to be correct—that his corpuscles are a constituent of all matter. He further con-

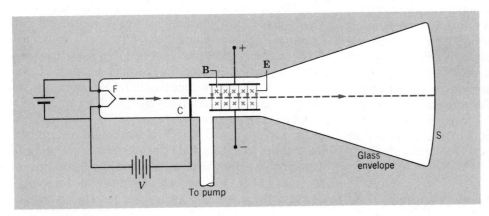

Figure 9 A modern version of J. J. Thomson's apparatus for measuring the ratio of mass to charge for the electron. The electric field **E** is established by connecting a battery across the plate terminals. The magnetic field **B** is set up by means of a current in a system of coils (not shown).

cluded that his corpuscles were lighter than the lightest known atom (hydrogen) by a factor of more than 1000.* It was his m/q measurement, coupled with the boldness and accuracy of these two claims, that constitutes the "discovery of the electron," with which he is generally credited. Direct measurement of the electronic charge soon followed and, within a few years, acceptance of the electron as a particle of nature was firmly established. When you hear our times referred to as the *electronic* age, think of J. J. Thomson and his corpuscles.

30–4 The Hall Effect

A beam of electrons in a vacuum can be deflected sideways by a magnetic field. Do you suppose that it is possible to deflect the drifting conduction electrons in a copper wire sideways by means of a magnetic field? In 1879, Edwin H. Hall, then a 24-year old graduate student of Henry A. Rowland at the Johns Hopkins University, showed that it is. This *Hall effect* allows us to find out whether the charge carriers in a conductor carry a positive or a negative charge. Beyond that, we can measure the number of such carriers per unit volume of the conductor.

Figure 10*a* shows a copper strip of width d, carrying a current i whose conventional direction is from top to bottom. The charge carriers are electrons and, as we know, they drift (with drift speed v_d) in the opposite direction. At the instant shown in Fig. 10*a*, an external magnetic field **B**, pointing into the plane of the figure, has just been turned on. From Eq. 6 we see that a magnetic deflecting force will act on each drifting electron, pushing it toward the right edge of the strip.

As time goes on, electrons will move to the right, mostly piling up on the right edge of the strip, leaving uncompensated positive charges in their fixed positions near the left edge. Thus, a constant transverse electric field **E** will build up at all points within the strip, pointing from left to right as shown in Fig. 10*b*. This field will exert an electric force on each electron, tending to push it to the left.

An equilibrium quickly develops in which the electric force on each electron builds up until it just cancels the magnetic force. When this happens, as Fig. 10*b* shows, the drifting electrons are in balance as far as their sideways motions are concerned. They drift along the

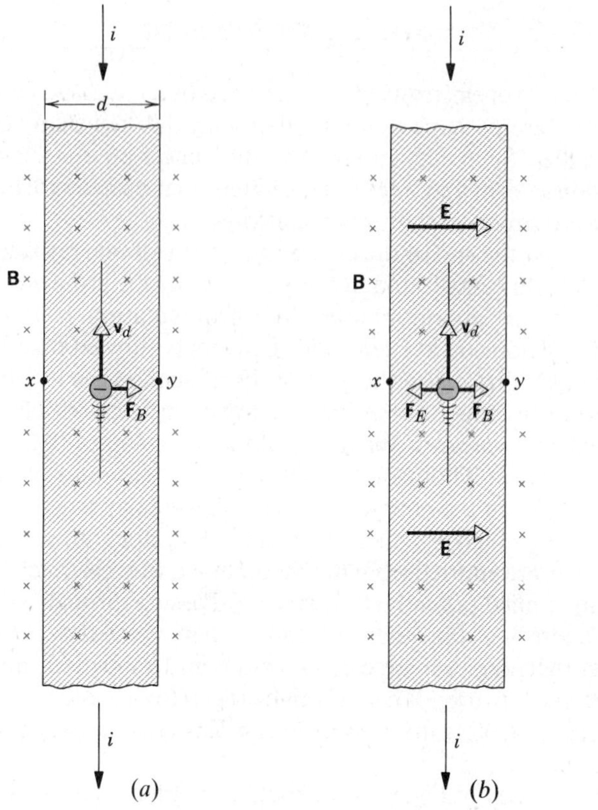

Figure 10 A strip of copper immersed in a magnetic field **B** carries a current i: (*a*) the situation immediately after the magnetic field is turned on and (*b*) the situation at equilibrium, which quickly follows. Note that negative charges pile up on the right side of the strip, leaving uncompensated positive charges on the left. Point x is at a higher potential than point y.

strip toward the top of the figure with no net wandering either to the right or to the left.

The electric field **E** that builds up is associated with a *Hall potential difference V*, where $E = V/d$. We can measure V by connecting the terminals of a voltmeter between points x and y in Fig. 10*b*. From its polarity, we can find the sign of the charge carriers. Let us see how.

In Fig. 10*b* we assumed that the charge carriers were electrons and thus negatively charged. Check carefully that, had the charge carriers been positive, the directions of the vectors v_d and **E** in Fig. 10*b* would reverse but those of the vectors F_E and F_B would remain unchanged. Thus, drifting positive charges would also be pushed to the right in Fig. 10*b*, leaving uncompensated negative charges on the left. The polarity of the Hall potential difference V would be opposite to that for negative charge carriers.

In Chapter 28, which deals with direct currents, we often assumed the charge carriers to be positive when, in fact, we knew that they were negative. We justified this on the basis that it made no difference to our measurements of current and potential difference. The Hall effect, however, is one case in which it *does* make a difference.

Now for the quantitative part. At equilibrium (Fig. 10*b*) the electric and magnetic forces are in balance, or, from Eqs. 5 and 7,

$$(-e)E = (-e)v_d B. \tag{13}$$

The drift speed v_d is given from Eq. 7 of Chapter 28 as

$$v_d = \frac{J}{ne} = \frac{i}{neA}, \tag{14}$$

in which $J(= i/A)$ is the current density in the strip, A being the cross-sectional area of the strip.

Substituting V/d for E in Eq. 13 and eliminating v_d by means of Eq. 14, we obtain

$$\boxed{n = \frac{Bi}{Vte}} \tag{15}$$

in which $t(= A/d)$ is the thickness of the strip. Thus, we can find n, the density of charge carriers, in terms of quantities that we can measure.

It is also possible to use the Hall effect to measure directly the drift speed v_d of the charge carriers, which you may recall is of the order of centimeters per hour.* In this clever experiment, the metal strip is moved mechanically through the magnetic field in a direction opposite to that of the drift velocity of the charge carriers. The speed of the moving specimen is then adjusted until the Hall potential difference vanishes. At this condition, the velocity of the moving specimen is equal in magnitude but opposite in direction to the velocity of the charge carriers; the velocity of the charge carriers *with respect to the magnetic field* is thus zero and there is no Hall effect.

The Hall effect has been, and continues to be, tremendously useful in helping us to understand electrical

---

* "In Memoriam J. Jaumann: A Direct Demonstration of the Drift Velocity in Metals," by W. Klein, *American Journal of Physics,* January 1987.

conduction in metals and semiconductors. To interpret it fully, however, we must replace the classical derivation that we have just given by one based on quantum physics. The 1985 Nobel Prize for physics was awarded for a fundamental discovery about the quantized nature of resistance, based on Hall effect measurements.

Sample Problem 2 A strip of copper 150 μm thick is placed in magnetic field $\mathbf{B} = 0.65$ T and a current $i = 23$ A is set up in the strip. What Hall potential difference V will appear across the width of the strip?

In Sample Problem 3 of Chapter 28, we calculated the number of charge carriers per unit volume for copper, finding

$$n = 8.47 \times 10^{28} \text{ electrons/m}^3.$$

From Eq. 15 then,

$$V = \frac{Bi}{net} = \frac{(0.65 \text{ T})(23 \text{ A})}{(8.47 \times 10^{28} \text{ m}^{-3})(1.60 \times 10^{-19} \text{ C})(150 \times 10^{-6} \text{ m})}$$
$$= 7.4 \times 10^{-6} \text{ V} = 7.4 \ \mu\text{V}. \tag{Answer}$$

This potential difference, though small, is readily measurable.

30-5   A Circulating Charge

If a particle moves in a circle at constant speed, we can be sure that the net force acting on the particle is constant in magnitude and points toward the center of the circle, at right angles to the particle's velocity. Think of a stone tied to a string and whirled in a circle on a smooth horizontal surface, or of a satellite moving in a circular orbit around the earth. In the first case, the tension in the string provides the necessary sideways force. In the second case, it is the earth's gravitational attraction that does so.

Figure 11 shows another example. Here a beam of electrons is projected in the plane of the figure with speed v from an *electron gun* G. The electrons find themselves moving in a region filled with a uniform magnetic field $\mathbf{B}$, whose direction is out of the plane of the figure. A sideways magnetic deflecting force $\mathbf{F} (= q\mathbf{v} \times \mathbf{B})$ acts on them so that they follow a circular path. The path is visible because atoms of the residual gas emit light when some of the circulating electrons collide with them.

If we apply Newton's second law to a particle of charge q moving at right angles to a uniform magnetic

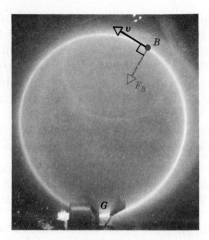

Figure 11 Electrons circulating in a chamber containing gas at low pressure. A uniform magnetic field **B**, pointing out of the plane of the figure at right angles to it, fills the chamber. Note the radially-directed magnetic deflecting force **F**$_B$.

field **B**, we find

$$qvB = \frac{mv^2}{r}$$

or

$$r = \frac{mv}{qB} \quad \text{(radius)}, \tag{16}$$

which gives the radius of the path.

The angular frequency ω, the frequency v, and the period T are given by

$$\omega = \frac{v}{r} = \frac{qB}{m} \quad \text{(angular frequency)}, \tag{17}$$

$$v = \frac{\omega}{2\pi} = \frac{qB}{2\pi m} \quad \text{(frequency)}, \tag{18}$$

and

$$T = \frac{1}{v} = \frac{2\pi m}{qB} \quad \text{(period)}. \tag{19}$$

Note that none of these three related quantities depends on the speed of the particle.* Fast particles move in large

---

* This is true only to the extent that the speed of the particle is much less than the speed of light.

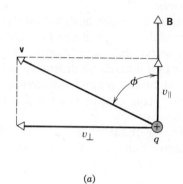

(a)

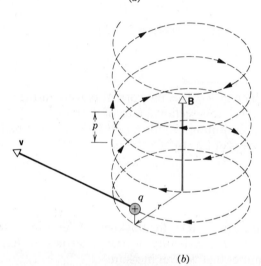

(b)

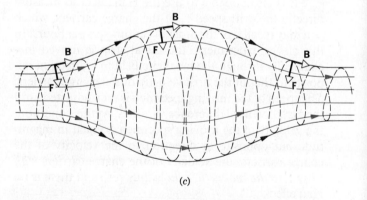

(c)

Figure 12 (a) An electron moves in a magnetic field, its velocity making an angle ϕ with the field direction. (b) The electron follows a helical path, of radius r and pitch p. (c) A charged particle spiraling in an inhomogeneous magnetic field. The particle can become trapped, spiraling back and forth between the strong field regions at either end. Note that the magnetic force vectors at each end of the *magnetic bottle* have a component pointing toward the center of the bottle.

circles and slow ones in small circles but all particles with the same value of the ratio q/m take the same time T (the period) to complete one round trip.

If the velocity of the particle has a component parallel to the (uniform) magnetic field, the particle will spiral about the direction of the field in a helical path. Figure 12*a*, for example, shows the velocity vector **v** resolved into two components, one parallel to **B** and one at right angles to it. Thus,

$$v_{\parallel} = v \cos \phi \quad \text{and} \quad v_{\perp} = v \sin \phi. \quad (20)$$

The electron moves in an upward spiraling helix, as Fig. 12*b* shows.

Figure 12*c* shows a charged particle spiraling in a highly *nonuniform* magnetic field. The spiraling particle, reflected from the high-field regions at either end of this *magnetic bottle,* can become trapped for substantial periods of time. Figure 13 shows electrons and protons trapped in this way by the earth's magnetic field, forming the Van Allen radiation belts. These belts—naturally occurring magnetic bottles—were discovered in 1958 when a particle counter aboard an early satellite *(Explorer 1)* stopped functioning above a certain altitude. Scientists learned later that it had been jammed by the unexpectedly high counting rate. Electrons and protons

Figure 14   This photo, transmitted from the satellite *Dynamics Explorer I,* clearly shows the auroral oval and also the dark and light hemispheres of the earth.

spiral back and forth in these belts, bouncing between the strong-field regions near the earth's magnetic poles in a few seconds.

Electrons and protons, streaming out from the sun as a *solar wind,* greatly distort the earth's magnetic field, producing electric currents that stretch it out into a *magnetotail* some 1000 earth radii long. Magnetic substorms in the magnetotail occasionally inject charged particles into the earth's upper atmosphere at high latitudes, forming the familiar aurora. Figure 14, transmitted from the *Dynamics Explorer I* satellite, shows that the aurora is concentrated in an oval around the earth's magnetic pole.

Sample Problem 3   The electrons shown circulating in Fig. 11 have a kinetic energy of 22.5 eV. A uniform magnetic field emerges from the plane of the figure and has a magnitude of 4.55×10^{-4} T. (a) What is the orbit radius?

Calculating the speed from the kinetic energy (just as we did in Sample Problem 1) yields $v = 2.81 \times 10^6$ m/s.* From

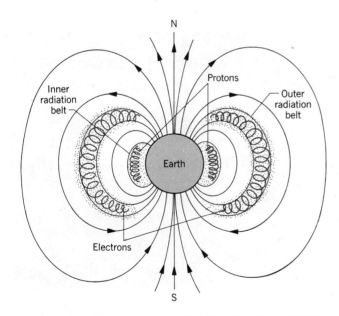

Figure 13   The earth's magnetic field, showing protons and electrons trapped in the Van Allen radiation belts. The magnetic north and south poles of the earth are shown. The figure is a figure of revolution about this N–S axis.

---

* The kinetic energy of the electron (= 22.5 eV) is such a tiny fraction of its rest energy (= 0.511 MeV or 5.11×10^5 eV) that we are easily justified in using the classical (that is, the nonrelativistic) formula for the kinetic energy.

Eq. 16 then,

$$r = \frac{mv}{qB} = \frac{(9.11 \times 10^{-31} \text{ kg})(2.81 \times 10^6 \text{ m/s})}{(1.60 \times 10^{-19} \text{ C})(4.55 \times 10^{-4} \text{ T})}$$

$$= 3.52 \text{ cm}. \qquad \text{(Answer)}$$

(b) What is the frequency v of the circulating electrons? From Eq. 18,

$$v = \frac{qB}{2\pi m} = \frac{(1.60 \times 10^{-19} \text{ C})(4.55 \times 10^{-4} \text{ T})}{2\pi(9.11 \times 10^{-31} \text{ kg})}$$

$$= 1.27 \times 10^7 \text{ Hz} = 12.7 \text{ MHz}. \qquad \text{(Answer)}$$

Note that this result does not depend on the speed of the particle but only on the nature of the particle (that is, on the ratio q/m) and on the strength B of the magnetic field.

(c) What is the period of revolution T?

$$T = \frac{1}{v} = \frac{1}{1.27 \times 10^7 \text{ Hz}} = 7.86 \times 10^{-8} \text{ s}$$

$$= 78.6 \text{ ns}. \qquad \text{(Answer)}$$

Verify, using the right-hand rule for vector products, that Eq. 6 gives the correct direction of rotation for the circulating electrons in Fig. 11, that is, counterclockwise.

Sample Problem 4 Suppose that the velocity vector of the electron in Sample Problem 3 makes an angle ϕ of 65.5° with the direction of the magnetic field, as Fig. 12a shows. (a) What is the radius of its helical path?

From Eq. 16, the radius of the helix is

$$r = \frac{mv_\perp}{qB} = \frac{m(v \sin \phi)}{qB}$$

$$= \frac{(9.11 \times 10^{-31} \text{ kg})(2.81 \times 10^6 \text{ m/s})(\sin 65.5°)}{(1.60 \times 10^{-19} \text{ C})(4.55 \times 10^{-4} \text{ T})}$$

$$= 3.20 \text{ cm}. \qquad \text{(Answer)}$$

Note that this is less than the value (= 3.52 cm) calculated in the previous problem because we used only a component of $\mathbf{v}$ rather than v in our calculation.

(b) What is the pitch p of the helix (that is, the distance between corresponding points on adjacent loops)?

We note first that the period T of rotation of the particle, being independent of its speed, is the same as in the previous Sample Problem. From Fig. 12b then,

$$p = v_\parallel T = (v \cos \phi)T$$

$$= (2.81 \times 10^6 \text{ m/s})(\cos 65.5°)(7.86 \times 10^{-8} \text{ s})$$

$$= 9.16 \times 10^{-2} \text{ m} = 9.16 \text{ cm}. \qquad \text{(Answer)}$$

30–6 Cyclotrons and Synchrotrons

What is the ultimate structure of matter? This question has always been at the cutting edge of physics. One way

of getting at this problem is to allow an energetic charged particle (a proton, for example) to slam into a solid target. Better yet, allow two such energetic protons to collide head-on. Analyzing the debris from such collisions is the most useful way to learn about the nature of the subatomic particles of matter. The Nobel Prizes in Physics for 1976 and 1984 were awarded for just such studies.

How then can a proton acquire kinetic energy? The direct approach is to allow the proton to "fall" through a potential difference V, thereby increasing its kinetic energy by eV. As we want higher and higher energies, however, it becomes more and more difficult to establish the necessary potential difference.

A better way is to arrange for the proton to circulate in a magnetic field and to give it a modest electrical "kick" once per revolution. For example, if a proton circulates 100 times in a magnetic field and if it receives an energy boost of 100 keV every time it completes an orbit, it will end up with a kinetic energy of (100)(100 keV) or 10 MeV. We describe two devices based on this principle.

The Cyclotron. Figure 15 shows a beam of accelerated particles emerging from an early cyclotron through a thin metal foil that separates the vacuum of the cyclotron chamber from the air of the room.

Figure 16 is a top view of the region of a cyclotron in which the particles (protons, say) circulate. The two hol-

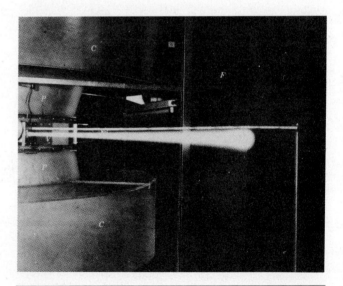

Figure 15 An early cyclotron at the University of Pittsburgh, showing an accelerated beam of deuterons emerging into the air of the laboratory. Note vacuum chamber V, magnet frame F, magnet pole faces P, and magnet coils C. The rule is 6 ft long.

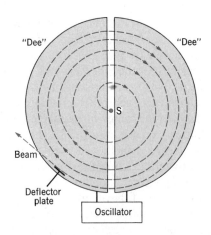

Figure 16   The elements of a cyclotron, showing the ion source S and the dees. A uniform magnetic field emerges from the plane of the figure. The circulating protons spiral outward within the hollow dees, gaining energy every time they cross the gap between the dees.

low D-shaped objects (open on their straight edges) are made of copper sheet. These *dees,* as they are called, form part of an electrical oscillator, which establishes an alternating potential difference across the gap between them. The dees are immersed in a magnetic field ($B = 1.5$ T) whose direction is out of the plane of the figure. This field is set up by a large electromagnet, not shown in Fig. 16 but apparent in Fig. 15.

Suppose that a proton, injected at the center of the cyclotron in Fig. 16, finds the dee that it faces to be negative. It will accelerate toward this dee and will enter it. Once inside, it is screened from electric fields by the copper walls of the dee. The magnetic field, however, is not screened by the (nonmagnetic) copper dees so that the proton bends in a circular path whose radius, which depends on its speed, is given by Eq. 16, or $r = mv/qB$.

Let us assume that at the instant the proton emerges from the dee the accelerating potential difference has changed sign. Thus, the proton *again* faces a negatively charged dee and is *again* accelerated. This process continues, the circulating proton always being in step with the oscillations of the dee potential, until the proton spirals out to the edge of the dee system.

The key to the operation of the cyclotron is that the frequency v at which the proton circulates in the field must be equal to the fixed frequency v_{osc} of the electrical oscillator, or

$$v = v_{osc} \qquad \text{(resonance condition).} \qquad (21)$$

This *resonance condition* says that, if the energy of the circulating proton is to increase, energy must be fed to it at a frequency v_{osc} that is equal to the natural frequency v at which the proton circulates in the magnetic field.

Combining Eqs. 18 and 21 allows us to write the resonance condition as

$$qB = 2\pi m v_{osc}. \qquad (22)$$

For the proton, q and m are fixed. The oscillator (we assume) is designed to work at a single fixed frequency v_{osc}. We then "tune" the cyclotron by varying B until Eq. 22 is satisfied and a beam of energetic protons appears.

The Proton Synchrotron.   At proton energies above 50 MeV, the conventional cyclotron begins to fail because one of its assumptions—that the frequency of revolution of a charged particle circulating in a magnetic field is independent of its speed—is true only for speeds that are much less than the speed of light. As the proton speed increases, we must treat the problem by relativistic rules.

We know from relativity theory that as the speed of the circulating proton approaches that of light it takes a longer and longer time to make the trip around its orbit. This means that the frequency of revolution of the circulating proton decreases steadily. Thus, the protons get out of step with the oscillator—whose frequency remains fixed at v_{osc}—and eventually the energy of the circulating proton stops increasing.

There is another problem. For a 500-GeV proton in a magnetic field of 1.5 T, the radius of curvature is 1.1 km. A magnet of the conventional cyclotron type of this size would be impossibly expensive, the area of its pole faces being about 1000 acres.

The *proton synchrotron* is designed to meet these two difficulties. The magnetic field B and the oscillator frequency v_{osc}, instead of having fixed values as in the conventional cyclotron, are made to vary with time during the accelerating cycle. If this is done properly, (1) the frequency of the circulating protons remains in step with the oscillator at all times and (2) the protons follow a circular—not a spiral—path. Thus, the magnet need only be ring-shaped, rather than solid. The ring, however, must be large if high energies are to be achieved. Indeed, the area enclosed by the ring of the proton synchrotron at the Fermi National Accelerator Laboratory (Fermilab) is large enough to be used as a test area for the rejuvenation of the primeval midwestern prairie.

Figure 17 is a view inside the accelerator tunnel at Fermilab. Energetic protons, traveling inside a highly evacuated pipe about 2 inches in diameter, curve gently

Figure 17 A view along the tunnel of the proton synchrotron at Fermilab. The tunnel circumference is 6.3 km.

around the 4-mile circumference of the magnet ring. The protons make about 400,000 round trips to reach their full energy of 1 TeV (= 10^{12} eV).

The accelerator, called the *Tevatron,* is operated in a mode such that 1-TeV protons and 1-TeV antiprotons collide head-on with each other, making an energy of 2 TeV available in the center-of-mass reference frame of the two colliding particles. As of 1988, this was the highest collision energy ever achieved by an accelerator. Figure 18 shows an aerial view of the ring and the associated laboratory buildings.

The cry goes up for still more energetic protons. Figure 19 shows (smallest circle) the Fermilab ring and (next largest circle) the accelerator ring at the European Center for Particle Physics (CERN). The largest circle represents the Superconducting Super Collider (SSC), which—now in advanced design stage—is planned to generate proton–antiproton collision energies of 20 TeV. The ring—some 52 miles in circumference and planned to house 10,000 superconducting magnets—is displayed against a satellite photo of Washington, DC to show the scale. The SSC ring will be about the size of the beltway system of highways that surrounds this city.

Figure 18 An aerial view of the Fermi National Accelerator Laboratory at Batavia, Illinois.

Figure 19   The largest circle shows the planned Superconducting Super Collider (SSC), superimposed (for scale purposes only) on a satellite photo of the Washington, DC area. The intermediate circle is the European accelerator at CERN in Switzerland and the smallest circle is the Fermilab accelerator. All circles are drawn to correspond to the same assumed magnetic field.

Sample Problem 5   The cyclotron shown in Fig. 15 operated at an oscillator frequency of 12 MHz and had a dee radius $R = 53$ cm. (a) What value of the magnetic field B was needed to allow deuterons to be accelerated?

A deuteron has the same charge as a proton but approximately twice the mass ($m = 3.34 \times 10^{-27}$ kg). From Eq. 22,

$$B = \frac{2\pi m v_{osc}}{q} = \frac{(2\pi)(3.34 \times 10^{-27}\text{ kg})(12 \times 10^6\text{ s}^{-1})}{1.60 \times 10^{-19}\text{ C}}$$

$$= 1.6\text{ T.} \qquad\qquad\qquad\text{(Answer)}$$

Note that, to allow protons to be accelerated, the magnitude of the magnetic field would have to be reduced by a factor of 2, assuming that the oscillator frequency remained fixed at 12 MHz.

(b) What deuteron energy results?

From Eq. 16, the speed of a deuteron circulating with a radius equal to the dee radius R is given by

$$v = \frac{RqB}{m} = \frac{(0.53\text{ m})(1.60 \times 10^{-19}\text{ C})(1.6\text{ T})}{3.34 \times 10^{-27}\text{ kg}}$$

$$= 4.06 \times 10^7\text{ m/s.}$$

This speed corresponds to a kinetic energy of

$$K = \tfrac{1}{2}mv^2$$
$$= \tfrac{1}{2}(3.34 \times 10^{-27}\text{ kg})(4.06 \times 10^7\text{ m/s})^2$$
$$\times (1\text{ MeV}/1.60 \times 10^{-13}\text{ J})$$
$$= 17\text{ MeV.} \qquad\qquad\text{(Answer)}$$

As Fig. 15 shows, deuterons of this energy, as they emerge from the near vacuum of the cyclotron chamber, have a range in air of about 1.5 m.

30-7   The Magnetic Force on a Current

We have already seen (in connection with the Hall effect) that a magnetic field exerts a sideways force on the conduction electrons in a wire. This force must be transmitted bodily to the wire itself, because the conduction electrons cannot escape sideways out of the wire.

In Fig. 20a, a wire, carrying no current, hangs vertically between the pole faces of a permanent magnet. The magnetic field points outward from the page. In Fig. 20b, a current is set up in the wire, pointing upward; the wire deflects to the right. In Fig. 20c, we reverse the direction of the current and the wire deflects to the left. Note that the wire deflects sideways to the direction of the current, just as we expect.

Let us look at the magnetic deflection of a wire more closely, tying it to the magnetic forces that act on the individual charge carriers. Figure 21 shows an enlarged view of a wire carrying a current i, the axis of the wire being at right angles to a magnetic field **B**. We see one of

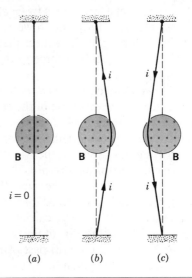

Figure 20   A flexible wire passes between the pole faces of an electromagnet. (a) There is no current in the wire. (b) A current is established. (c) The same as (b) except that the direction of the current is reversed. The connections for getting the current into the wire at one end and out of it at the other end are not shown.

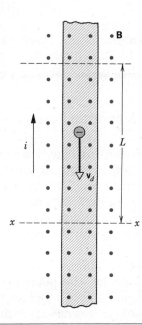

Figure 21 A close-up view of a length L of the wire of Fig. 20b. The current direction is upward, which means that electrons drift downward. A magnetic field emerges from the plane of the figure so that the wire is deflected to the right.

the conduction electrons, drifting downward with an assumed drift speed v_d. Equation 7, in which we must put $\phi = 90°$, tells us that a force given by $(-e)v_d B$ must act on each such electron. From Eq. 6 we see that this force must point to the right. We expect then that the wire as a whole will experience a force to the right, in agreement with Fig. 20b.

If, in Fig. 21, we were to reverse *either* the direction of the magnetic field *or* the direction of the current, the force on the wire would reverse, pointing now to the left. Note too that it doesn't matter whether we consider negative charges drifting downward in the wire (the actual case) or positive charges drifting upward. The direction of the deflecting force on the wire is the same. We are safe then in dealing with the conventional direction of current, which assumes positive charge carriers.

Consider a length L of the wire of Fig. 21. The electrons in this section of wire will drift past plane xx in a time L/v_d, carrying a charge given by

$$q = i\left(\frac{L}{v_d}\right)$$

through that plane. Substituting this into Eq. 7 yields

$$F_B = qv_d B \sin \phi$$
$$= (iL/v_d)(v_d)B \sin 90°$$

or

$$F_B = iLB. \tag{23}$$

This equation gives the force that acts on a segment of a straight wire of length L, immersed in a magnetic field that is at right angles to the wire.

If the magnetic field is *not* at right angles to the wire, as in Fig. 22, the magnetic force is given by a generalization of Eq. 23, or

$$\boxed{\mathbf{F}_B = i\mathbf{L} \times \mathbf{B}} \quad \text{(force on a current).} \tag{24}$$

Here $\mathbf{L}$ is a vector that points along the wire segment in the direction of the (conventional) current.

Equation 24 is equivalent to Eq. 6 in that either can be taken as the defining equation for $\mathbf{B}$. In practice, we define $\mathbf{B}$ from Eq. 24. It is much easier to measure the magnetic force acting on a wire than on a single moving charge.

If a wire is not straight, we can imagine it broken up into small straight segments and apply Eq. 24 to each such segment. The force on the wire as a whole is then the vector sum of all the forces on the segments that make it up. In the differential limit, we can write

$$d\mathbf{F}_B = i\,d\mathbf{L} \times \mathbf{B}, \tag{25}$$

and we can find the resultant force on any given structure of currents by integrating Eq. 25 over that structure.

In using Eq. 25, bear in mind that there is no such thing as an isolated, current-carrying wire segment of length dL. There must always be a way to introduce the current into the segment at one end and to take it out at the other.

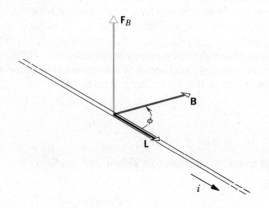

Figure 22 A wire segment of length L makes an angle ϕ with a magnetic field. Compare carefully with Fig. 4.

Sample Problem 6 A straight, horizontal stretch of copper wire carries a current $i = 28$ A. What are the magnitude and direction of the magnetic field needed to "float" the wire, that is, to balance its weight? Its linear density is 46.6 g/m.

Figure 23 shows the arrangement. For a length L of wire we have (see Eq. 23)

$$mg = LiB,$$

or

$$B = \frac{(m/L)g}{i} = \frac{(46.6 \times 10^{-3} \text{ kg/m})(9.8 \text{ m/s}^2)}{28 \text{ A}}$$

$$= 1.6 \times 10^{-2} \text{ T } (= 160 \text{ gauss}). \qquad \text{(Answer)}$$

This is about 160 times the strength of the earth's magnetic field.

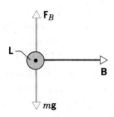

Figure 23 Sample Problem 6. A wire (shown in cross section) can be made to "float" in a magnetic field. The current in the wire emerges from the figure and the magnetic field points to the right.

Sample Problem 7 Figure 24 shows a wire segment, placed in a uniform magnetic field **B** that points out of the plane of the figure. If the segment carries a current i, what resultant magnetic force **F** acts on it?

The force that acts on each straight section, from Eq. 23, has the magnitude

$$F_1 = F_3 = iLB$$

and points down, as shown by the arrows in the figure.

A segment of the arc of length dL has a force dF acting on it, whose magnitude is given by

$$dF = iB\, dL = iB(R\, d\theta)$$

and whose direction is radially toward O, the center of the arc. Note that only the downward component of this force element is effective. The horizontal component is canceled by an oppositely directed horizontal component associated with a symmetrically located segment on the opposite side of the arc.

Thus, the total force on the central arc points down and is given by

$$F_2 = \int_0^\pi dF \sin\theta = \int_0^\pi (iBR\, d\theta)\sin\theta$$

$$= iBR \int_0^\pi \sin\theta\, d\theta = 2iBR.$$

The resultant force on the entire wire is then

$$F = F_1 + F_2 + F_3 = iLB + 2iBR + iLB$$
$$= 2iB(L + R). \qquad \text{(Answer)}$$

Note that this force is just the same as the force that would act on a straight wire of length $2(L + R)$. This would be true no matter what the shape of the central segment, shown as a semicircle in Fig. 24. Can you convince yourself that this is so?

30-8 Torque on a Current Loop

Much of the world's work is done by electric motors. The forces that do this work are the magnetic forces that we studied in the previous section; that is, they are the forces that a magnetic field exerts on a wire that carries a current.

Figure 25 shows a simple motor, in which the two

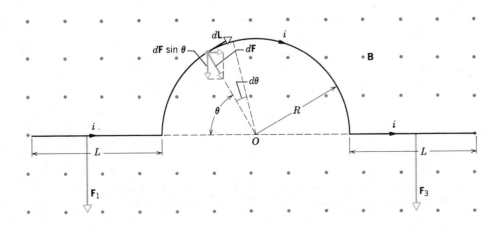

Figure 24 Sample Problem 7. A wire segment carrying a current i is immersed in a magnetic field. The resultant force on the wire is directed downward.

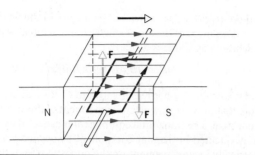

Figure 25 The rudiments of an electric motor. A rectangular coil, carrying a current and free to rotate about a fixed axis, is placed in a magnetic field. A commutator (not shown) reverses the direction of the current every half revolution so that the magnetic torque always acts in the same direction.

magnetic forces combine to exert a torque on a current loop, tending to rotate it about the central axis. Although we omit many essential details of how an electric motor works, it seems clear that the action of a magnetic field in exerting a torque on a current loop is at the heart of it.

Figure 26 shows two views of a rectangular loop of sides a and b, carrying a current i and immersed in a uniform magnetic field **B**. We place it in the field so that its long sides, labeled 1 and 3, are at right angles to the field direction. Wires to lead the current into and out of the loop are needed but, for simplicity, we do not show them.

Imagine now that the loop is held at rest, so that **n**, a vector normal to its plane, makes an angle θ with the field direction, as Fig. 26b shows. Note that we have

defined the direction of **n** by a right-hand rule. That is, (1) curl the fingers of your right hand around the coil in the direction of the current; (2) your extended thumb then points in the direction of **n**. What net force and what net torque act on the loop in this position?

The net force is the vector sum of the forces acting on each of the four sides of the loop. For side 2 the vector **L** in Eq. 24 points in the direction of the current and has the magnitude b. The angle between **L** and **B** for side 2 (see Fig. 26b) is $90° - \theta$. Thus, the magnitude of the force acting on this side is

$$F_2 = ibB \sin(90° - \theta) = ibB \cos \theta. \quad (26)$$

You can show that the force **F**$_4$ acting on side 4 has the same magnitude as **F**$_2$ but points in the opposite direction. Thus, **F**$_2$ and **F**$_4$, taken together, cancel out exactly. Their net force is zero and, because they have the same line of action, so is their net torque.

The situation is different for sides 1 and 3. Here the common magnitude of **F**$_1$ and **F**$_3$ is iaB and they point in opposite directions so that they do not tend to move the coil bodily. However, as Fig. 26b shows, these two forces do *not* share the same line of action so they *do* tend to turn the coil. There is a net torque.

The magnitude τ' of the torque due to forces **F**$_1$ and **F**$_3$ is (see Fig. 26b)

$$\tau' = (iaB)(b/2)(\sin \theta) + (iaB)(b/2)(\sin \theta) = iaB \sin \theta.$$

This torque acts on every turn of the coil. If there are N turns, the total torque is

$$\tau = N\tau' = NiabB \sin \theta = (NiA)B \sin \theta. \quad (27)$$

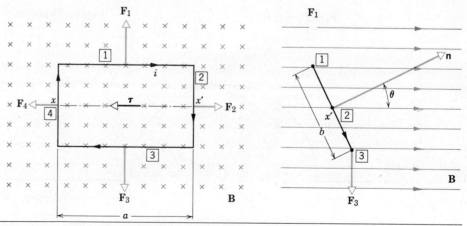

Figure 26 A rectangular coil carrying a current i is placed in a uniform magnetic field. A torque acts to align the normal vector **n** with the direction of the field.

in which $A (= ab)$ is the area of the coil. The quantities in parentheses ($= NiA$) are grouped together because they are all properties of the coil: its number of turns, its area, and the current it carries. We can show (see Problem 57) that this equation holds for all plane loops, no matter what their shape.

Instead of watching the motion of the coil, it is simpler to watch the vector **n**, which is normal to its plane. Equation 27 tells us that a current-carrying coil placed in a magnetic field will tend to rotate so that this normal vector points in the field direction. This is just what a compass needle does.

Sample Problem 8   Analog voltmeters and ammeters, in which the reading is displayed by the deflection of a pointer over a scale, work by measuring the torque exerted by a magnetic field on a current loop. Figure 27 shows the rudiments of a *galvanometer,* on which both analog ammeters and analog voltmeters are based. The coil is 2.1 cm high and 1.2 cm wide; it has 250 turns and is mounted so that it can rotate about a vertical axis in a uniform radial magnetic field with $B = 0.23$ T. A spring Sp provides a countertorque that balances the magnetic torque, resulting in a steady angular deflection ϕ corresponding to a given steady current i in the coil. If a current of 100 μA produces an angular deflection of 28°, what must be the torsional constant of the spring (see Eq. 14-20)?

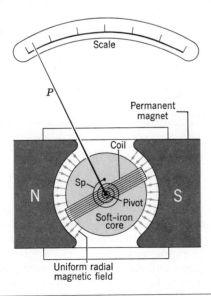

Figure 27   Sample Problem 8. The rudiments of a galvanometer. Depending on the external circuit, this device can be wired up as either a voltmeter or an ammeter.

Setting the magnetic torque equal to the spring torque (see Eq. 27) yields

$$\tau = NiAB \sin \theta = \kappa \phi, \qquad (28)$$

in which ϕ is the angular deflection of the pointer and A ($= 2.52 \times 10^{-4}$ m²) is the area of the coil. Note that the normal to the plane of the coil (that is, the pointer) is always at right angles to the (radial) magnetic field so that $\theta = 90°$ for all pointer positions.

Solving Eq. 28 for κ, we find

$$\kappa = \frac{NiAB \sin \theta}{\phi}$$

$$= \frac{(250)(100 \times 10^{-6} \text{ A})(2.52 \times 10^{-4} \text{ m}^2)(0.23 \text{ T})(\sin 90°)}{28°}$$

$$= 5.2 \times 10^{-8} \text{ N} \cdot \text{m/degree.} \qquad \text{(Answer)}$$

Many modern ammeters and voltmeters are of the digital, direct-reading type and operate in a way that does not involve a moving coil.

30-9   A Magnetic Dipole

In physics, we like to identify the main features of a problem, ignoring details that do not matter. In this spirit we describe the current loop of the preceding section by a single vector **μ**, its *magnetic dipole moment*. We take the direction of **μ** to be that of the normal vector **n** to the plane of the loop, as in Fig. 26*b*. We take as the magnitude of **μ** the quantity NiA. Thus, we can rewrite Eq. 27 as

$$\tau = \mu B \sin \theta, \qquad (29)$$

in which θ is the angle between the vectors **μ** and **B**.

We can generalize this to the vector relation

$$\tau = \mu \times \mathbf{B}, \qquad (30)$$

which reminds us very much of the corresponding equation for the torque exerted by an *electric* field on an *electric* dipole, namely (see Eq. 25 of Chapter 24),

$$\tau = \mathbf{p} \times \mathbf{E}.$$

In each case the torque exerted by the external field — be it magnetic or electric — is equal to the vector product of the corresponding dipole moment and the field vector.

If a magnetic field exerts a torque on a magnetic dipole, then work must be done to change the orientation of the dipole. The magnetic dipole must have a *magnetic potential energy* that depends on its orientation in the external field. For electric dipoles we have shown (see Eq. 28 of Chapter 24) that

$$U(\theta) = -\mathbf{p} \cdot \mathbf{E}.$$

In strict analogy, we can write for the magnetic case

$$U(\theta) = -\boldsymbol{\mu} \cdot \mathbf{B}. \tag{31}$$

Thus, a magnetic dipole has its lowest energy ($= -\mu B \cos 0° = -\mu B$) when it is lined up with the magnetic field. It has its greatest energy ($= -\mu B \cos 180° = +\mu B$) when it points in a direction opposite to the field. The difference in energy between these two positions is

$$\begin{aligned}\Delta U &= U(180°) - U(0°) \\ &= (-\mu B \cos 180°) - (-\mu B \cos 0°) \\ &= (+\mu B) - (-\mu B) = 2\mu B.\end{aligned} \tag{32}$$

This much work must be done by an external agent to turn a magnetic dipole through 180°, starting from its lined-up position.

So far, we have identified a magnetic dipole as a current loop. However, a simple bar magnet is also a magnetic dipole. So is a rotating sphere of charge. The earth itself is a magnetic dipole. Finally, most subatomic particles, including the electron, the proton, and the neutron, have magnetic dipole moments. As we shall see, all these quantities can be viewed — in some sense or other — as current loops. Here, for comparison, are some approximate magnetic dipole moments:

A proton	1.4×10^{-26} J/T
An electron	9.3×10^{-24} J/T
The coil of Sample Problem 9	5.4×10^{-6} J/T
A small bar magnet	5 J/T
The earth	8.0×10^{22} J/T

Sample Problem 9 (a) What is the magnetic dipole moment of the coil of Sample Problem 8, assuming that it carries a current of 85 μA?

The *magnitude* of the magnetic dipole moment of the coil, whose area A is 2.52×10^{-4} m², is

$$\begin{aligned}\mu &= NiA \\ &= (250)(85 \times 10^{-6}\ \text{A})(2.52 \times 10^{-4}\ \text{m}^2) \\ &= 5.36 \times 10^{-6}\ \text{A} \cdot \text{m}^2 = 5.36 \times 10^{-6}\ \text{J/T}. \text{(Answer)}\end{aligned}$$

Show that these two sets of units are identical. Note that the second set of units follows logically from Eq. 31.

The *direction* of $\boldsymbol{\mu}$, as inspection of Fig. 27 shows, is that of the pointer. You can verify this by showing that, if we assume $\boldsymbol{\mu}$ to be in the pointer direction, the torque predicted by Eq. 30 is such that it would indeed move the pointer clockwise across the scale.

(b) The magnetic dipole moment of the coil is lined up with an external magnetic field whose strength is 0.85 T. How much work is required to turn the coil end for end?

The work is equal to the increase in potential energy, which is

$$\begin{aligned}W &= \Delta U = 2\mu B = 2(5.36 \times 10^{-6}\ \text{J/T})(0.85\ \text{T}) \\ &= 9.1 \times 10^{-6}\ \text{J} = 9.1\ \mu\text{J}. \tag{Answer}\end{aligned}$$

This is about equal to the work needed to lift an aspirin tablet through a vertical height of about 3 mm.

REVIEW AND SUMMARY

Magnetic Field B

The magnetic field $\mathbf{B}$ is defined in terms of the sideways force $\mathbf{F}_B$ acting on a test particle with charge q and moving with velocity $\mathbf{v}$,

$$\mathbf{F}_B = q\mathbf{v} \times \mathbf{B} \tag{6}$$

study Figs. 4 through 7 and Sample Problem 1. The SI unit for $\mathbf{B}$ is the *tesla* (abbr. T) where $1\ \text{T} = 1\ \text{N/(A} \cdot \text{m)} = 10^4$ gauss. Some representative magnetic fields are listed in Table 1.

e/m for the Electron

J. J. Thomson, in his 1897 discovery of the electron, used both magnetic and electric fields (see Section 30–3 and Fig. 9) to determine its charge-to-mass ratio.

The Hall Effect

When a conducting strip of thickness t carrying a current i is placed in a magnetic field $\mathbf{B}$, some charge carriers (with charge e) build up on the sides of the conductor, as illustrated in Fig. 10. A potential difference V builds up across the strip. The polarity of V gives the sign of the charge carriers; the density of charge carriers may be calculated from

$$n = \frac{Bi}{Vte}. \tag{15}$$

Sample Problem 2 illustrates the magnitudes of the quantities involved.

A charged particle with mass m and charge q moving with velocity $\mathbf{v}$ perpendicular to a magnetic field $\mathbf{B}$ will travel in a circle of radius

A Charged Particle Circulating in a Magnetic Field

$$r = \frac{mv}{qB} \quad \text{(radius)}. \qquad [16]$$

Its frequency of revolution in the field (the *cyclotron frequency*) is

$$v = \frac{\omega}{2\pi} = \frac{1}{T} = \frac{qB}{2\pi m} \quad \text{(frequency, period)}. \qquad [18,19]$$

See Figs. 11 through 14 and Sample Problems 3 and 4.

Cyclotrons and Synchrotrons

A cyclotron is a particle accelerator that uses a magnetic field to hold a charged particle in a circular orbit so that a modest accelerating potential may act on it repeatedly, resulting in high energies. Because the moving particle gets out of step with the oscillator as its speed approaches that of light, there is an upper limit to the energy attainable with the cyclotron. A synchrotron avoids this difficulty. Here both B and the oscillator frequency v_{osc} are programmed to change cyclically so that the particle can not only go to high energies but can do so at a constant orbital radius; this allows the use of a ring magnet rather than a solid magnet, at great saving in cost. See Section 30–6, especially Figs. 15 through 19 and Sample Problem 5, for an introduction to some exciting applications in modern physics.

Magnetic Force on a Current

A straight wire carrying a current i in a uniform magnetic field experiences a sideways force

$$\mathbf{F}_B = i\mathbf{L} \times \mathbf{B}. \qquad [24]$$

The sideways force acting on a current element $id\mathbf{L}$ in a magnetic field is

$$d\mathbf{F} = id\mathbf{L} \times \mathbf{B}. \qquad [25]$$

The direction of the length element $d\mathbf{L}$ is that of the current density vector $\mathbf{J}$. See Figs. 20 through 22 and Sample Problems 6 and 7.

Torque on a Current Loop

A current loop (area A, current i with N turns) in a uniform magnetic field $\mathbf{B}$ will experience a torque τ given by

$$\boldsymbol{\tau} = \boldsymbol{\mu} \times \mathbf{B}. \qquad [30]$$

Here $\boldsymbol{\mu}$ is the *magnetic dipole moment* with magnitude $\mu = NiA$ and a direction perpendicular to the plane of the loop in the right-hand-rule direction. This torque is the operating principle in electric motors and analog voltmeters and ammeters. See Figs. 25 through 27 and Sample Problems 8 and 9. Bar magnets, molecules, atoms, basic particles (electrons, protons, neutrons, etc.) all have magnetic dipole properties.

Orientation Energy of a Magnetic Dipole

The potential energy of orientation of a magnetic dipole in a magnetic field is

$$U(\theta) = -\boldsymbol{\mu} \cdot \mathbf{B}. \qquad [31]$$

See Sample Problem 9.

QUESTIONS

1. Of the three vectors in the equation $\mathbf{F}_B = q\mathbf{v} \times \mathbf{B}$, which pairs are always at right angles? Which may have any angle between them?

2. Why do we not simply define the direction of the magnetic field $\mathbf{B}$ to be the direction of the magnetic force that acts on a moving charge?

3. Imagine that you are sitting in a room with your back to one wall and that an electron beam, traveling horizontally from the back wall toward the front wall, is deflected to your right. What is the direction of the uniform magnetic field that exists in the room?

4. How could we rule out that the forces between two magnets are electrostatic forces?

5. If an electron is not deflected in passing through a certain region of space, can we be sure that there is no magnetic field in that region?

6. If a moving electron is deflected sideways in passing through a certain region of space, can we be sure that a magnetic field exists in that region?

7. A beam of electrons can be deflected either by an electric field or by a magnetic field. Is one method better than the other? . . . in any sense easier?

8. A charged particle passes through a magnetic field and is deflected. This means that a force acted on it and changed its momentum. Where there is a force, there must be a reaction force. On what object does it act?

9. Imagine the room in which you are seated to be filled with a uniform magnetic field with **B** pointing vertically downward. At the center of the room two electrons are suddenly projected horizontally with the same initial speed but in opposite directions. (a) Describe their motions. (b) Describe their motions if one particle is an electron and one a positron, that is, a positively charged electron. (The electrons will gradually slow down as they collide with molecules of the air in the room.)

10. In Fig. 6 why are the low-energy electron tracks spirals? That is, why does the radius of curvature change in the constant magnetic field in which the chamber is immersed?

11. What are the primary functions of (a) the electric field and (b) the magnetic field in the cyclotron?

12. What central fact makes the operation of a conventional cyclotron possible? Ignore relativistic considerations.

13. A bare copper wire emerges from one wall of a room, crosses the room, and disappears into the opposite wall. You are told that there is a steady current in the wire. How can you find its direction? Describe as many ways as you can think of. You may use any reasonable piece of equipment, but you may not cut the wire.

14. In Section 7 we state that a magnetic field **B** exerts a sideways force on the conduction electrons in, say, a copper wire carrying a current i. We have tacitly assumed that this same force acts on the conductor itself. Are there some missing steps in this argument? If so, supply them.

15. A current in a magnetic field experiences a force. Therefore, it should be possible to pump conducting liquids by sending a current through the liquid (in an appropriate direction) and letting it pass through a magnetic field. Design such a pump. This principle is used to pump liquid sodium (a conductor, but highly corrosive) in some nuclear reactors, where it is used as a coolant. What advantages would such a pump have?

16. An airplane is flying west in level flight over Massachusetts, where the earth's magnetic field is directed downward below the horizontal in a northerly direction. As a result of the magnetic force on the free electrons in its wings, one of its wingtips will have more electrons than the other. Which one (right or left) is it? Will the answer be different if the plane is flying east?

17. A conductor, even though it is carrying a current, has zero net charge. Why, then, does a magnetic field exert a force on it?

18. You wish to modify a galvanometer (see Sample Problem 8) to make it into (a) an ammeter and (b) a voltmeter. What do you need to do in each case?

19. A rectangular current loop is in an arbitrary orientation in an external magnetic field. How much work is required to rotate the loop about an axis perpendicular to its plane?

20. Equation 30 ($\tau = \mu \times \mathbf{B}$) shows that there is no torque on a current loop in an external magnetic field if the angle between the axis of the loop and the field is (a) 0° or (b) 180°. Discuss the nature of the equilibrium (that is, is it stable, neutral, or unstable?) for these two positions.

21. In Sample Problem 9 we showed that the work required to turn a current loop end-for-end in an external magnetic field is $2\mu B$. Does this result hold no matter what the original orientation of the loop was?

22. Imagine that the room in which you are seated is filled with a uniform magnetic field with **B** pointing vertically upward. A circular loop of wire has its plane horizontal. For what direction of current in the loop, as viewed from above, will the loop be in stable equilibrium with respect to forces and torques of magnetic origin?

23. The torque exerted by a magnetic field on a magnetic dipole can be used to measure the strength of that magnetic field. For an accurate measurement, does it matter whether the dipole moment is small or not? Recall that, in the case of measurement of an electric field, the test charge was to be as small as possible so as not to disturb the source of the field.

24. You are given a smooth sphere the size of a ping-pong ball and told that it contains a magnetic dipole. What experiments would you carry out to find the magnitude and the direction of its magnetic dipole moment?

EXERCISES AND PROBLEMS

Section 30–2 The Definition of B

1E. Express magnetic field B and magnetic flux Φ_B in terms of the dimensions $M, L, T,$ and Q (mass, length, time, and charge).

2E. Four particles follow the paths shown in Fig. 28 as they pass through the magnetic field there. What can one conclude about the charge of each particle?

3E. An electron in a TV camera tube is moving at 7.2 ×

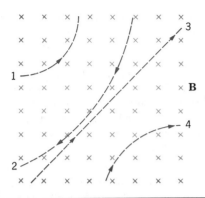

Figure 28 Exercise 2.

10^6 m/s in a magnetic field of strength 83 mT. (*a*) Without knowing the direction of the field, what could be the greatest and least magnitudes of the force the electron could feel due to the field? (*b*) At one point the acceleration of the electron is 4.9×10^{14} m/s^2. What is the angle between the electron's velocity and the magnetic field?

4E. A proton traveling at 23° with respect to a magnetic field of strength 2.6 mT experiences a magnetic force of 6.5×10^{-17} N. Calculate (*a*) the speed and (*b*) the kinetic energy in eV of the proton.

5P. An electron has a velocity given in m/s by $\mathbf{v} = 2.0 \times 10^6\mathbf{i} + 3.0 \times 10^6\mathbf{j}$. It enters a magnetic field given in T by $\mathbf{B} = 0.030\mathbf{i} - 0.15\mathbf{j}$. (*a*) Find the magnitude and direction of the force on the electron. (*b*) Repeat your calculation for a proton having the same velocity.

6P. An electron in a uniform magnetic field has a velocity $\mathbf{v} = 40\mathbf{i} + 35\mathbf{j}$ km/s. It experiences a force $\mathbf{F} = -4.2\mathbf{i} + 4.8\mathbf{j}$ fN. If $B_x = 0$, calculate the magnetic field.

7P. The electrons in the beam of a television tube have an energy of 12 keV. The tube is oriented so that the electrons move horizontally from magnetic south to magnetic north. The vertical component of the earth's magnetic field points down and has a magnitude of 55 μT. (*a*) In what direction will the beam deflect? (*b*) What is the acceleration of a given electron due to the magnetic field? (*c*) How far will the beam deflect in moving 20 cm through the television tube?

8P*. An electron has an initial velocity $12\mathbf{j} + 15\mathbf{k}$ km/s and a constant acceleration of $\mathbf{i}(2.0 \times 10^{12}$ m/s$^2)$ in a region in which uniform electric and magnetic fields are present. If $\mathbf{B} = 400\mathbf{i}$ μT, find the electric field $\mathbf{E}$.

Section 30–3   Discovering the Electron
9E. A typical cathode-ray oscilloscope employs a cathode-ray tube in which electric fields are used for both horizontal and vertical deflections of the electron beam, but that is otherwise similar to the tube shown in Fig. 30–12. Figure 29 shows the face of such a tube. The solid straight line results when the

electron beam is repeatedly swept left to right by a time-varying electric field. If a uniform magnetic field is applied perpendicularly inward through the face of the tube, one might expect the horizontal line to be shifted or tilted. Which of the four dashed lines in the figure will be the line that will result?

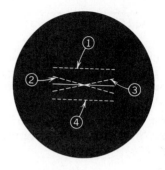

Figure 29 Exercise 9.

10E. A 10-keV electron moving horizontally enters a region of space in which there is a downward-directed electric field of magnitude 10 kV/m. (*a*) What are the magnitude and direction of the (smallest) magnetic field that will allow the electron to continue to move horizontally? Ignore the gravitational force, which is rather small. (*b*) Is it possible for a proton to pass through this combination of fields undeflected? If so, under what circumstances?

11E. An electric field of 1.5 kV/m and a magnetic field of 0.40 T act on a moving electron to produce no force. (*a*) Calculate the minimum electron speed v. (*b*) Draw the vectors $\mathbf{E}$, $\mathbf{B}$, and $\mathbf{v}$.

12P. An electron is accelerated through a potential difference of 1.0 kV and directed into a region between two parallel plates separated by 20 mm with a potential difference of 100 V between them. If the electron enters moving perpendicular to the electric field between the plates, what magnetic field is necessary perpendicular to both the electron path and the electric field so that the electron travels in a straight line?

13P. An ion source is producing ions of ^{6}Li (mass = 6.0 u) each carrying a single positive elementary charge ($+e$). The ions are accelerated by a potential difference of 10 kV and pass horizontally into a region in which there is a vertical magnetic field $B = 1.2$ T. Calculate the strength of the smallest electric field to be set up over the same region that will allow the ^{6}Li ions to pass through undeflected.

Section 30–4   The Hall Effect
14E. Figure 30 shows the cross section of a sample carrying a current directed out of the page. (*a*) Which pair of the four terminals (a, b, c, d) should be used to measure the Hall voltage if the magnetic field is in the $+x$ direction and the charge

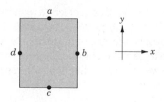

Figure 30 Exercise 14.

carriers are negative? What is the expected polarity of this voltage? (b) Repeat if the magnetic field is in the $-y$ direction and the charge carriers are positive. (c) Discuss the situation if the magnetic field is in the $+z$ direction.

15E. Show that, in terms of the Hall electric field E and the current density J, the number of charge carriers per unit volume is given by

$$n = \frac{JB}{eE}.$$

16P. In a Hall-effect experiment, a current of 3.0 A lengthwise in a conductor 1.0 cm wide, 4.0 cm long, and 10 μm thick produced a tranverse Hall voltage (across the width) of 10 μV when a magnetic field of 1.5 T is passed perpendicularly through the thin conductor. From these data, find (a) the drift velocity of the charge carriers and (b) the number density of charge carriers. (c) Show on a diagram the polarity of the Hall voltage with a given current and magnetic field direction, assuming the charge carriers are (negative) electrons.

17P. (a) Show that the ratio of the Hall electric field E to the electric field E_C responsible for the current is

$$\frac{E}{E_C} = \frac{B}{ne\rho},$$

where ρ is the resistivity of the material. (b) Compute the ratio numerically for Sample Problem 2. See Table 1 in Chapter 28.

18P. A metal strip 6.5 cm long, 0.85 cm wide, and 0.76 mm thick moves with constant velocity v through a magnetic field $B = 1.2$ mT perpendicular to the strip, as shown in Fig. 31. A

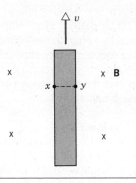

Figure 31 Problem 18.

potential difference of 3.9 μV is measured between points x and y across the strip. Calculate the speed v.

Section 30–5   A Circulating Charge

19E. Magnetic fields are often used to bend a beam of electrons in physics experiments. What uniform magnetic field, applied perpendicular to a beam of electrons moving at 1.3×10^6 m/s, is required to make the electrons travel in a circular arc of radius 0.35 m?

20E. (a) In a magnetic field with $B = 0.50$ T, for what path radius will an electron circulate at 0.10 the speed of light? (b) What will be its kinetic energy in eV? Ignore the small relativistic effects.

21E. What uniform magnetic field must be set up in space to permit a proton of speed 1.0×10^7 m/s to move in a circle the size of the earth's equator?

22E. A 1.2-keV electron is circulating in a plane at right angles to a uniform magnetic field. The orbit radius is 25 cm. Calculate (a) the speed of the electron, (b) the magnetic field, (c) the frequency of revolution, and (d) the period of the motion.

23E. An electron is accelerated from rest by a potential difference of 350 V. It then enters a uniform magnetic field of magnitude 200 mT, its velocity being at right angles to this field. Calculate (a) the speed of the electron and (b) the radius of its path in the magnetic field.

24E. *Time-of-flight spectrometer.* S. A. Goudsmit devised a method for measuring accurately the masses of heavy ions by timing their period of revolution in a known magnetic field. A singly charged ion of iodine makes 7.00 rev in a field of 45.0 mT in 1.29 ms. Calculate its mass, in atomic mass units. Actually, the mass measurements are carried out to much greater accuracy than these approximate data suggest.

25E. An alpha particle ($q = +2e$, $m = 4.0$ u) travels in a circular path of radius 4.5 cm in a magnetic field with $B = 1.2$ T. Calculate (a) its speed, (b) its period of revolution, (c) its kinetic energy in eV, and (d) the potential difference through which it would have to be accelerated to achieve this energy.

26E. (a) Find the frequency of revolution of an electron with an energy of 100 eV in the earth's magnetic field of 35 μT. (b) Calculate the radius of the path of this electron if its velocity is perpendicular to the magnetic field.

27E. A beam of electrons whose kinetic energy is K emerges from a thin-foil "window" at the end of an accelerator tube. There is a metal plate a distance d from this window and at right angles to the direction of the emerging beam. See Fig. 32. Show that we can prevent the beam from hitting the plate if we apply a magnetic field B such that

$$B \geq \sqrt{\frac{2mK}{e^2 d^2}},$$

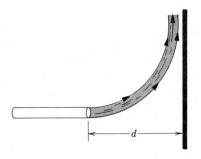

Figure 32 Exercise 27.

in which m and e are the electron mass and charge. How should **B** be oriented?

28P. In a nuclear experiment a 1.0-MeV proton moves in a uniform magnetic field in a circular path. What energy must (a) an alpha particle and (b) a deuteron have if they are to circulate in the same orbit? (Recall that for an alpha particle $q = +2e$, $m = 4.0$ u.)

29P. A proton, a deuteron, and an alpha particle, accelerated through the same potential difference, enter a region of uniform magnetic field, moving at right angles to **B**. (a) Compare their kinetic energies. If the radius of the proton's circular path is 10 cm, what are the radii of (b) the deuteron and (c) the alpha-particle paths?

30P. A proton, a deuteron, and an alpha particle with the same kinetic energies enter a region of uniform magnetic field, moving at right angles to **B**. Compare the radii of their circular paths.

31P. *Mass spectrometer.* Figure 33 shows an arrangement

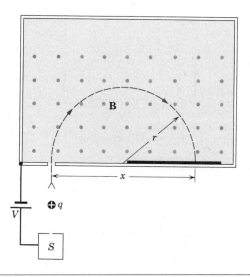

Figure 33 Problems 31, 32 and 33.

used to measure the masses of ions. An ion of mass m and charge $+q$ is produced essentially at rest in source S, a chamber in which a gas discharge is taking place. The ion is accelerated by potential difference V and allowed to enter a magnetic field **B**. In the field it moves in a semicircle, striking a photographic plate at distance x from the entry slit. Show that the ion mass m is given by

$$m = \frac{B^2 q}{8V} x^2.$$

32P. Two types of singly-ionized atoms having the same charge q and mass differing by a small amount Δm are introduced into the mass spectrometer described in Problem 31. (a) Calculate the difference in mass in terms of V, q, m (of either), B, and the distance Δx between the spots on the photographic plate. (b) Calculate Δx for a beam of singly ionized chlorine atoms of masses 35 and 37 u if $V = 7.3$ kV and $B = 0.50$ T.

33P. In a mass spectrometer (see Problem 31) used for commercial purposes, uranium ions of mass 3.92×10^{-25} kg and charge 3.2×10^{-19} C are separated from related species. The ions are first accelerated through a potential difference of 100 kV and then pass into a magnetic field, where they are bent in a path of radius 1.0 m. After traveling through 180°, they are collected in a cup after passing through a slit of width 1.0 mm and a height of 1.0 cm. (a) What is the magnitude of the (perpendicular) magnetic field in the separator? If the machine is designed to separate out 100 mg of material per hour, calculate (b) the current of the desired ions in the machine and (c) the thermal energy dissipated in the cup in one hour.

34P. Bainbridge's mass spectrometer, shown in Fig. 34, separates ions having the same velocity. The ions, after entering through slits S_1 and S_2, pass through a velocity selector composed of an electric field produced by the charged plates P and P', and a magnetic field **B** perpendicular to the electric field and the ion path. Those ions that pass undeviated through the crossed **E** and **B** fields enter into a region where a second magnetic field **B'** exists, and are bent into circular paths. A photographic plate registers their arrival. Show that $q/m = E/(rBB')$, where r is the radius of the circular orbit.

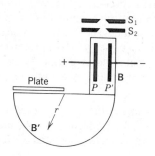

Figure 34 Problem 34.

35P. A 2.0-keV positron is projected into a uniform magnetic field **B** of 0.10 T with its velocity vector making an angle of 89° with **B**. Find (a) the period, (b) the pitch p, and (c) the radius r of the helical path. See Fig. 12b.

36P. A neutral particle is at rest in a uniform magnetic field of magnitude B. At time $t = 0$ it decays into two charged particles each of mass m. (a) If the charge of one of the particles is $+q$, what is the charge of the other? (b) The two particles move off in separate paths both of which lie in the plane perpendicular to **B**. At a later time the particles collide. Express the time from decay until collision in terms of m, B, and q.

37P. (a) What speed would a proton need to circle the earth at the equator, if the earth's magnetic field is everywhere horizontal there and directed along longitudinal lines? Relativistic effects must be taken into account. Take the magnitude of the earth's magnetic field to be 41 μT at the equator. (Hint: Replace the momentum mv in Eq. 16 with the relativistic momentum given in Eq. 20 of Chapter 9.) (b) Draw the velocity and magnetic field vectors corresponding to this situation.

Section 30–6 Cyclotrons and Synchrotrons

38E. In a certain cyclotron a proton moves in a circle of radius 0.50 m. The magnitude of the magnetic field is 1.2 T. (a) What is the cyclotron frequency? (b) What is the kinetic energy of the proton, in eV?

39E. A physicist is designing a cyclotron to accelerate protons to one-tenth the speed of light. The magnet used will produce a field of 1.4 T. Calculate (a) the radius of the cyclotron and (b) the corresponding oscillator frequency. Relativity considerations are not significant.

40P. The cyclotron of Sample Problem 5 was normally adjusted to accelerate deuterons. (a) What energy of protons could it produce, using the same oscillator frequency as that used for deuterons? (b) What magnetic field would be required? (c) What energy of protons could be produced if the magnetic field was left at the value used for deuterons? (d) What oscillator frequency would then be required? (e) Answer the same questions for alpha particles, instead of protons. (For an alpha particle, $q = +2e$, $m = 4.0$ u.)

41P. A deuteron in a cyclotron is moving in a magnetic field with $B = 1.5$ T and an orbit radius of 50 cm. Because of a grazing collision with a target, the deuteron breaks up, with a negligible loss of kinetic energy, into a proton and a neutron. Discuss the subsequent motions of each. Assume that the deuteron energy is shared equally by the proton and neutron at breakup.

42P. Estimate the total path length traversed by a deuteron in the cyclotron of Sample Problem 5 during the acceleration process. Assume an accelerating potential between the dees of 80 kV.

Section 30–7 The Magnetic Force on a Current

43E. Figure 35 shows a magnet and a straight wire in which electrons are flowing out of the page at right angles to it. In which case will there be a force on the wire that points toward the top of the page?

Figure 35 Exercise 43.

44E. A horizontal conductor in a power line carries a current of 5000 A from south to north. The earth's magnetic field (60 μT) is directed toward the north and is inclined downward at 70° to the horizontal. Find the magnitude and direction of the magnetic force on 100 m of the conductor due to earth's field.

45E. A wire 1.8 m long carries a current of 13 A and makes an angle of 35° with a uniform magnetic field $B = 1.5$ T. Calculate the magnetic force on the wire.

46P. A wire of 62 cm length and mass 13 g is suspended by a pair of flexible leads in a magnetic field of 0.44 T. What are the magnitude and direction of the current required to remove the tension in the supporting leads? See Fig. 36.

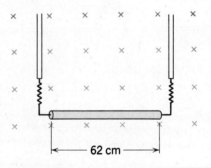

Figure 36 Problem 46.

47P. A wire 50 cm long lying along the x axis carries a current of 0.50 A in the positive x direction. A magnetic field is present that is given in T by $\mathbf{B} = 0.0030\mathbf{j} + 0.010\mathbf{k}$. Find the components of the force on the wire.

48P. A metal wire of mass m slides without friction on two

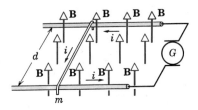

Figure 37 Problem 48.

horizontal rails spaced a distance d apart, as in Fig. 37. The track lies in a vertical uniform magnetic field **B**. A constant current i flows from generator G along one rail, across the wire, and back down the other rail. Find the velocity (speed and direction) of the wire as a function of time, assuming it to be at rest at $t = 0$.

49P. Figure 38 shows a wire of arbitrary shape carrying a current i between points a and b. The wire lies in a plane at right angles to a uniform magnetic field **B**. Prove that the force on the wire is the same as that on a straight wire carrying a current i directly from a to b. (*Hint:* Replace the wire by a series of "steps" parallel and perpendicular to the straight line joining a and b.)

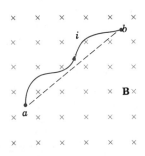

Figure 38 Problem 49.

50P. A long, rigid conductor, lying along the x axis, carries a current of 5.0 A in the $-x$ direction. A magnetic field **B** is present, given by $\mathbf{B} = 3\mathbf{i} + 8x^2\mathbf{j}$, with x in meters and **B** in mT. Calculate the force on the 2.0-m segment of the conductor that lies between $x = 1.0$ m and $x = 3.0$ m.

51P. Consider the possibility of a new design for an electric train. The engine is driven by the force due to the vertical component of the earth's magnetic field on a conducting axle. Current is passed down one rail, into a conducting wheel, through the axle, through another conducting wheel, and then back to the source via the other rail. (*a*) What current is needed to provide a modest 10-kN force? Take the vertical component of the earth's field to be 10 μT and the length of the axle to be 3.0 m. (*b*) How much power would be lost for each ohm of

resistance in the rails? (*c*) Is such a train totally unrealistic or just marginally unrealistic?

52P. A 1.0-kg copper rod rests on two horizontal rails 1.0 m apart and carries a current of 50 A from one rail to the other. The coefficient of static friction is 0.60. What is the smallest magnetic field (not necessarily vertical) that would cause the bar to slide?

Section 30–8 Torque on a Current Loop

53E. A single-turn current loop, carrying a current of 4.0 A, is in the shape of a right triangle with sides 50 cm, 120 cm, and 130 cm. The loop is in a uniform magnetic field of magnitude 75 mT whose direction is parallel to the current in the 130-cm side of the loop. (*a*) Find the magnetic force on each of the three sides of the loop. (*b*) Show that the total magnetic force on the loop is zero.

54E. Figure 39 shows a rectangular, 20-turn loop of wire, 10 cm by 5.0 cm. It carries a current of 0.10 A and is hinged at one side. It is mounted with its plane at an angle of 30° to the direction of a uniform magnetic field of 0.50 T. Calculate the torque about the hinge line acting on the loop.

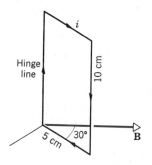

Figure 39 Exercise 54.

55E. A stationary, circular wall clock has a face with a radius of 15 cm. Six turns of wire are wound around its perimeter; the wire carries a current 2.0 A in the clockwise direction. The clock is located where there is a constant, uniform external magnetic field of 70 mT (but the clock still keeps perfect time). At exactly 1:00 p.m., the hour hand of the clock points in the direction of the external magnetic field. (*a*) After how many minutes will the minute hand point in the direction of the torque on the winding due to the magnetic field? (*b*) What is the magnitude of this torque?

56P. A length L of wire carries a current i. Show that if the wire is formed into a circular coil, the maximum torque in a given magnetic field is developed when the coil has one turn only and the maximum torque has the magnitude

$$\tau = \frac{1}{4\pi}\, L^2 iB.$$

57P. Prove that the relation $\tau = NiAB \sin \theta$ holds for closed loops of arbitrary shape and not only for rectangular loops as in Fig. 26. (*Hint:* Replace the loop of arbitrary shape by an assembly of adjacent long, thin, approximately rectangular, loops that are nearly equivalent to it as far as the distribution of current is concerned.)

58P. A closed loop of wire carries a current i. The loop is in a uniform magnetic field **B**. Show that the total magnetic force on the loop is zero. Does your proof also hold for a nonuniform magnetic field?

59P. Figure 40 shows a wire ring of radius a at right angles to the general direction of a radially-symmetric diverging magnetic field. The magnetic field at the ring is everywhere of the same magnitude B, and its direction at the ring is everywhere at an angle θ with a normal to the plane of the ring. The twisted lead wires have no effect on the problem. Find the magnitude and direction of the force the field exerts on the ring if the ring carries a current i as shown in the figure.

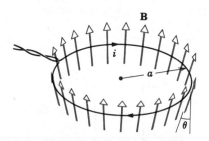

Figure 40 Problem 59.

60P. A certain galvanometer has a resistance of 75.3 Ω; its needle experiences a full-scale deflection when a current of 1.62 mA passes through its coil. (*a*) Determine the value of the auxiliary resistance required to convert the galvanometer into a voltmeter that reads 1.00 V at full-scale deflection. How is it to be connected? (*b*) Determine the value of the auxiliary resistance required to convert the galvanometer into an ammeter that reads 50.0 mA at full-scale deflection. How is it to be connected?

61P. Figure 41 shows a wooden cylinder with a mass $m = 0.25$ kg and a length $L = 0.10$ m, with $N = 10$ turns of wire wrapped around it longitudinally, so that the plane of the wire loop contains the axis of the cylinder. What is the least current through the loop that will prevent the cylinder from rolling down a plane inclined at an angle θ to the horizontal, in the presence of a vertical, uniform magnetic field of 0.50 T, if the plane of the windings is parallel to the inclined plane?

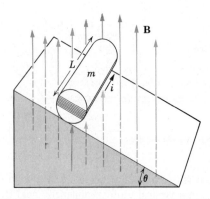

Figure 41 Problem 61.

Section 30–9 A Magnetic Dipole

62E. A circular coil of 160 turns has a radius of 1.9 cm. (*a*) Calculate the current that results in a magnetic moment of 2.3 A·m². (*b*) Find the maximum torque that the coil, carrying this current, can experience in a uniform 35-mT magnetic field.

63E. The magnetic dipole moment of the earth is 8.0×10^{22} J/T. Assume that this is produced by charges flowing in the molten outer core of the earth. If the radius of the circular path is 3500 km, calculate the required current.

64E. A circular wire loop whose radius is 15 cm carries a current of 2.6 A. It is placed so that the normal to its plane makes an angle of 41° with a uniform magnetic field of 12 T. (*a*) Calculate the magnetic dipole moment of the loop. (*b*) What torque acts on the loop?

65E. A single-turn current loop, carrying a current of 5.0 A, is in the shape of a right triangle with sides 30, 40, and 50 cm. The loop is in a uniform magnetic field of magnitude 80 mT whose direction is parallel to the current in the 50-cm side of the loop. Find the magnitude of (*a*) the magnetic dipole moment of the loop and (*b*) the torque on the loop.

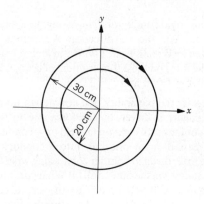

Figure 42 Exercise 66.

66E. Two concentric circular loops (radii 20 cm and 30 cm) in the xy plane each carry a clockwise current of 7.0 A, as shown in Fig. 42. (*a*) Find the net magnetic moment of this system. (*b*) Repeat if the current in the inner loop is reversed.

67P. A circular loop of wire having a radius of 8.0 cm carries a current of 0.20 A. A unit vector parallel to the dipole moment μ of the loop is given by $0.60\mathbf{i} - 0.80\mathbf{j}$. If the loop is located in a magnetic field given in T by $\mathbf{B} = 0.25\mathbf{i} + 0.30\mathbf{k}$, find (*a*) the magnitude and direction of the torque on the loop and (*b*) the magnetic potential energy of the loop.

68P. Figure 43 shows a current loop *ABCDEFA* carrying a current $i = 5.0$ A. The sides of the loop are parallel to the coordinate axes, with $AB = 20$ cm, $BC = 30$ cm, and $FA = 10$ cm. Calculate the magnitude and direction of the magnetic dipole moment of this loop. (*Hint:* Imagine equal and opposite

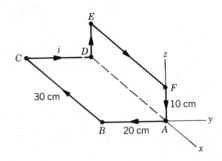

Figure 43 Problem 68.

currents i in the line segment *AD*, then treat the two rectangular loops *ABCDA* and *ADEFA*.)

CHAPTER 31
AMPERE'S LAW

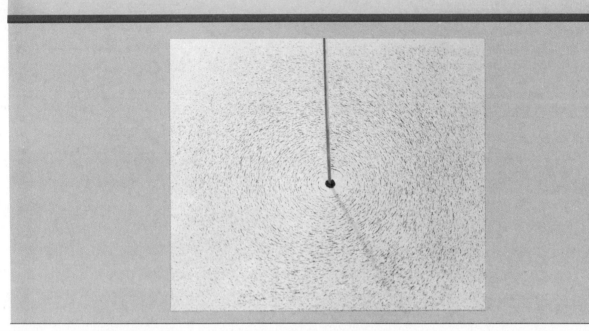

In 1820, the Danish physicist Hans Christian Oersted discovered that a current in a wire causes a nearby compass needle to deflect. In modern language, we say that a current in a wire sets up a magnetic field *in the surrounding space, as the iron filings in this photo attest. Oersted's discovery was a real breakthrough in a deliberate search for a connection between the then separate sciences of electricity and magnetism.*

31–1 Current and the Magnetic Field

A basic fact of *electrostatics* is that two charges exert forces on each other. Earlier, we wrote

$$\text{charge} \rightleftharpoons \text{electric field} \rightleftharpoons \text{charge} \qquad (1)$$

in which we introduced the electric field $\mathbf{E}$ as an intermediary in this interaction. Equation 1 suggests that (1) charges generate electric fields and (2) electric fields exert forces on charges.

A basic fact of *magnetism* is that two parallel wires carrying currents also exert forces on each other. By analogy with Eq. 1, we can write

$$\text{current} \rightleftharpoons \text{magnetic field} \rightleftharpoons \text{current} \qquad (2)$$

in which we introduce the magnetic field $\mathbf{B}$ as an intermediary. Equation 2 suggests that (1) currents generate magnetic fields and (2) magnetic fields exert forces on currents. We dealt with the second part of this interaction in Chapter 30. We deal with the first part in this chapter.

31–2 Calculating the Magnetic Field

The central question of this chapter is

How can you calculate the magnetic field that a given distribution of currents sets up in the surrounding space?

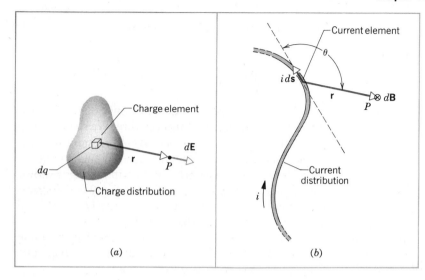

Figure 1 (*a*) A charge element *dq* establishes a differential electric field element *d*E at point *P*. (*b*) A current element *i d*s establishes a differential magnetic field *d*B at point *P*. The symbol ⊗ (the tail of an arrow) shows that the element *d*B points *into* the page.

Let us recall the equivalent central question in electrostatics:

How can you calculate the electric field that a given distribution of charges sets up in the surrounding space?

We learned to do this in Chapter 24, for static charge distributions such as a uniform sphere, line, ring, or disk. Our approach was to divide the charge distribution into charge elements *dq* as in Fig. 1*a*. We then calculated the field *d*E set up by a given charge element at an arbitrary field point *P*. Finally, we calculated **E** at point *P* by integrating *d*E over the entire charge distribution.

The *magnitude* of *d*E in such calculations is given by

$$dE = \left(\frac{1}{4\pi\epsilon_0}\right)\frac{dq}{r^2} \qquad (3)$$

in which *r* is the distance from the charge element to point *P*. Equation 3 is essentially Coulomb's law, the quantity in the parentheses being a familiar constant. For a positive element of charge, the *direction* of *d*E is that of **r**, where **r** is the vector pointing *from* the charge element *dq to* the field point *P*.

We can express both the magnitude and the direction of *d*E by writing Eq. 3 in vector form, as

$$d\mathbf{E} = \left(\frac{1}{4\pi\epsilon_0}\right)\frac{dq}{r^3}\mathbf{r} \qquad \text{(Coulomb's law)}, \quad (4)$$

which shows formally that, for a positive element of charge, the direction of *d*E is the direction of **r**. It may seem that Coulomb's law has suddenly become an inverse cube law — rather than an inverse square law — but that is not the case. The exponent 3 in the denominator is needed because we have added a factor of magnitude *r* in the numerator; Eq. 4 is still an inverse square law.

In the magnetic case, we proceed by analogy. Figure 1*b* shows a wire of arbitrary shape carrying a current *i*. What is the magnetic field **B** at an arbitrary field point *P* near this wire? We first break up the wire into differential current elements *i d*s, corresponding to the charge elements *dq* of Fig. 1*a*. Here the vector *d*s is a differential element of length, pointing tangent to the wire in the direction of the current. We note at once a complexity in our analogy to the electrostatic case: The differential charge element *dq* is a *scalar* but the differential current element *i d*s is a *vector*.

The *magnitude* of the magnetic field contribution set up at point *P* by a given current element is

$$dB = \frac{\mu_0}{4\pi}\frac{i\,ds\,\sin\theta}{r^2}. \qquad (5)$$

Here μ_0 is a constant, called the *permeability constant*, whose value is exact, by definition,

$$\mu_0 = 4\pi \times 10^{-7}\ \text{T·m/A} \approx 1.26 \times 10^{-6}\ \text{T·m/A}. \qquad (6)$$

This constant plays a role in magnetic problems much

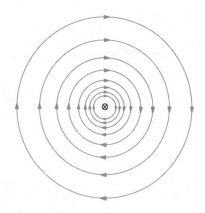

Figure 2 The lines of the magnetic field for a current i in a long straight wire are concentric circles. Their direction is given by a right-hand rule. Note that the current is directed into the page.

like the role that the permittivity constant ϵ_0 plays in electrostatic problems.

The *direction* of $d\mathbf{B}$ in Fig. 1b is that of the vector $d\mathbf{s} \times \mathbf{r}$, where $\mathbf{r}$ is a vector that points *from* the current element *to* the point P at which you wish to know the field. The symbol $\otimes$ in Fig. 1b (representing the tail of an arrow) shows that $d\mathbf{B}$ at point P is directed into the plane of the page at right angles. We can write the expression for $d\mathbf{B}$ in vector form as

$$d\mathbf{B} = \left(\frac{\mu_0}{4\pi}\right)\frac{i\, d\mathbf{s} \times \mathbf{r}}{r^3} \qquad \text{(Biot–Savart law).} \qquad (7)$$

Here again, Eq. 7 remains an inverse square law because a factor of magnitude r in the numerator cancels one of the factors of r in the denominator. Equation 7, which we can call Coulomb's law for magnetism, is more often called *the law of Biot and Savart*.* It contains within its structure information about both the magnitude and the direction of the differential field element $d\mathbf{B}$.

Equation 4 is our basic tool for calculating the electric field set up by a given distribution of charge. In the same way, Eq. 7 is our basic tool for calculating the magnetic field set up at various points by a given distribution of current.

A Long Straight Wire. The simplest problem in magnetism is that of a long straight wire carrying a current i. Below, we shall use the law of Biot and Savart to prove that the magnitude of the magnetic field at a perpendicu-

---

* Rhymes with "Leo and bazaar."

lar distance r from such a wire is given by

$$B(r) = \frac{\mu_0 i}{2\pi r} \qquad \text{(long straight wire).} \qquad (8)$$

The magnitude of $\mathbf{B}$ depends only on the current and on radial distance r from the wire. We shall show in our derivation that the lines of $\mathbf{B}$—consistent with the picture shown on the opening page of this chapter—form concentric circles around the wire, as Fig. 2 shows. The increase in spacing of the lines in Fig. 2 with increasing distance from the wire represents the $1/r$ decrease in the magnitude of $\mathbf{B}$ predicted by Eq. 8.

An All-Purpose Right-Hand Rule. Here is a simple right-hand rule for finding the direction of the magnetic field set up by a current in a long wire:

Grasp the wire in your right hand with your extended thumb pointing in the direction of the current. Your fingers will then naturally curl around in the direction of the magnetic field lines.

This rule is a specific application of a more general right-hand rule that we shall find useful. In magnetism we shall find many situations with cylindrical symmetry in which there is what we can call a "curly element" (in this case, the circular magnetic field lines) and a "straight element" (in this case, the current in the wire). Your right hand also has a "curly element" (your fingers) and a "straight element" (your extended thumb.) By matching your right hand to the physical situation, you can find the directions in which various physical quantities point. We shall clarify this general rule by applying it to new situations as they arise.

Proof of Equation 8. Figure 3, which is just like Fig. 1b except that the wire is straight, illustrates the problem. The differential magnetic field set up at point P by the current element $i\, d\mathbf{s}$ is given in magnitude by Eq. 5, or

$$dB = \frac{\mu_0}{4\pi}\frac{i\, ds \sin\theta}{r^2}. \qquad (9)$$

The direction of $d\mathbf{B}$ in Fig. 3 is that of the vector $d\mathbf{s} \times \mathbf{r}$, namely, into the plane of the figure at right angles.

Note that $d\mathbf{B}$ at point P has this same direction for every current element into which the wire can be divided. Thus, to find the total magnetic field $\mathbf{B}$ at point P, we integrate Eq. 9, obtaining

$$B = \int dB = \frac{\mu_0 i}{4\pi}\int_{s=-\infty}^{s=+\infty}\frac{\sin\theta\, ds}{r^2}. \qquad (10)$$

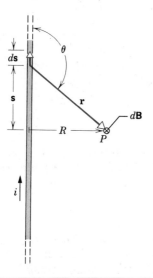

Figure 3 Calculating the magnetic field set up by a current i in a long straight wire. The field $d\mathbf{B}$ associated with the current element $i \, d\mathbf{s}$ points into the page, as shown.

The variables θ, s, and r in this equation are not independent, being related by

$$\sin \theta = \sin (\pi - \theta) = \frac{R}{\sqrt{s^2 + R^2}}$$

and

$$r = \sqrt{s^2 + R^2}.$$

With these substitutions, Eq. 10 becomes

$$B = \frac{\mu_0 i}{4\pi} \int_{-\infty}^{+\infty} \frac{R}{(s^2 + R^2)^{3/2}} \, ds = \frac{\mu_0 i}{4\pi R} \left| \frac{s}{(s^2 + R^2)^{1/2}} \right|_{-\infty}^{+\infty}$$

$$= \frac{\mu_0}{2\pi} \frac{i}{R}.$$

With a small change in notation, we have Eq. 8, the relation we set out to prove.

Sample Problem 1   The magnetic field a distance $r = 2.3$ cm from the axis of a long straight wire is 13 mT. What is the current in the wire? The earth's magnetic field has a strength of only about 0.1 mT so that we may ignore its influence in this problem.

Solving Eq. 8 for i, we obtain

$$i = \frac{2\pi B r}{\mu_0} = \frac{(2\pi)(13 \times 10^{-3} \text{ T})(2.3 \times 10^{-2} \text{ m})}{4\pi \times 10^{-7} \text{ T} \cdot \text{m/A}}$$

$$= 1500 \text{ A.} \qquad \text{(Answer)}$$

31-3   The Magnetic Force on a Wire Carrying a Current

We recall from Section 30-7 that a section of a long straight wire of length L, carrying a current i and placed in a uniform magnetic field, experiences a sideways deflecting force given by

$$\mathbf{F} = i \, \mathbf{L} \times \mathbf{B}_{\text{ext}}. \qquad (11)$$

We introduce the subscript in $\mathbf{B}_{\text{ext}}$ to remind ourselves that the magnetic field in this force equation must be the field set up by some external agent, an electromagnet of the type shown in Fig. 3 of Chapter 30 being one possibility.

In particular, the *external* field that appears in Eq. 11 must be distinguished carefully from the *intrinsic* field $\mathbf{B}_{\text{intr}}$ that is set up by the current in the wire itself. In Fig. 2, for example, the field shown is the intrinsic field resulting from the current in the wire. There is no external field in that figure so that, according to Eq. 11, no magnetic deflecting force acts on the wire. In the corresponding electrostatic case, the electric field set up by a point charge (which we could have called $\mathbf{E}_{\text{intr}}$) exerted no electric force on that charge.

Figure 4 shows the lines of the *resultant* magnetic field $\mathbf{B}$ associated with a current in a wire that is oriented at right angles to a uniform external magnetic field $\mathbf{B}_{\text{ext}}$. At any point, the resultant field $\mathbf{B}$ will be the vector sum

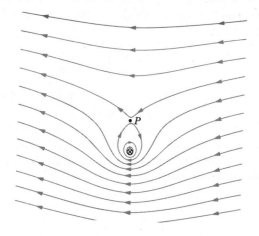

Figure 4   A long straight wire carrying a current i into the page is immersed in a uniform external magnetic field $\mathbf{B}_{\text{ext}}$. The field lines shown represent the resultant field formed by combining vectorially at each point the uniform external field and the intrinsic field associated with the current in the wire.

of $\mathbf{B}_{ext}$ and $\mathbf{B}_{intr}$, or

$$\mathbf{B} = \mathbf{B}_{ext} + \mathbf{B}_{intr}. \qquad (12)$$

These two fields tend to cancel each other above the wire and to reinforce each other below it. At point P in Fig. 4, $\mathbf{B}_{ext}$ and $\mathbf{B}_{intr}$ cancel exactly. Very near the wire, $\mathbf{B}_{intr}$ dominates and the field lines are closely represented by concentric circles, like those of Fig. 2. Far from the wire, $\mathbf{B}_{ext}$ dominates and the field lines are represented closely by uniformly spaced parallel lines.

Michael Faraday, who originated the concept of lines of force, endowed them with more reality than we currently give them. He imagined that, like stretched rubber bands, they represent the site of mechanical forces. Using Faraday's analogy, can you not readily believe that the wire in Fig. 4 will be deflected upward? Verify that it will, using Eq. 11.

31–4 Two Parallel Conductors

Two long parallel wires carrying currents exert forces on each other. Figure 5 shows two such wires, separated by a distance d and carrying currents i_a and i_b. Let us analyze the forces that these wires exert on each other, in terms of Eq. 2:

$$\text{current} \rightleftharpoons \text{magnetic field} \rightleftharpoons \text{current}.$$

Wire a in Fig. 5 produces a magnetic field $\mathbf{B}_a$ at all points. The magnitude of $\mathbf{B}_a$ at the site of wire b is, from Eq. 8,

$$B_a = \frac{\mu_0 i_a}{2\pi d}. \qquad (13)$$

The right-hand rule tells us that the direction of $\mathbf{B}_a$ at wire b is down, as the figure shows.

Wire b, which carries a current i_b, finds itself immersed in an external magnetic field $\mathbf{B}_a$. A length L of this wire will experience a sideways magnetic force given by Eq. 11, whose magnitude is

$$F_{ba} = i_b L B_a = \frac{\mu_0 L i_b i_a}{2\pi d}. \qquad (14)$$

The rule for vector products tells us that $\mathbf{F}_{ba}$ lies in the plane of the wires and points to the left in Fig. 5.

We could equally well have computed the force on wire a by determining the magnetic field that wire b produces at the site of wire a. For parallel currents, this force would point to the right in Fig. 5, which means that

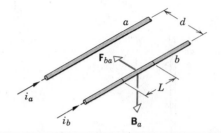

Figure 5 Two parallel wires carrying currents in the same direction attract each other. The magnetic field at wire b set up by the current in wire a is shown.

the two wires would attract each other. Note that (1) the forces on the two wires form an action–reaction pair and (2) the *external* field in which either wire finds itself is the *intrinsic* field of the other wire.

You should be able to show that, for antiparallel currents, the two wires repel each other. The rule is:

Parallel currents attract and antiparallel currents repel.

This rule is opposite to the rule for the forces between charges in this sense: Although like (parallel) currents attract each other, like (same sign) charges repel each other.

The force acting between currents in parallel wires is the basis for the definition of the ampere, which is one of the seven SI base units.* The definition, adopted in 1946, is:

The ampere is that constant current which, if maintained in two straight parallel conductors of infinite length, of negligible circular cross section, and placed 1 meter apart in vacuum, would produce on each of these conductors a force equal to 2×10^{-7} newtons per meter of length.

In practice, multiturn coils of carefully controlled geometries are substituted for the "conductors of infinite length" of the definition.

Sample Problem 2 Show that Eq. 14 is consistent with the definition of the ampere given above.

Let us put $i_a = i_b = 1$ A and $d = 1$ m in that equation. We

---

* See Appendix A for a list of the SI base units, along with their definitions.

find

$$\frac{F}{L} = \frac{\mu_0 i_a i_b}{2\pi d} = \frac{(4\pi \times 10^{-7} \text{ T·m/A})(1 \text{ A})(1 \text{ A})}{(2\pi)(1 \text{ m})}$$

$$= 2 \times 10^{-7} \text{ T·A} = 2 \times 10^{-7} \text{ N/m}. \qquad \text{(Answer)}$$

The unit transformation is helped by a study of the units in Eq. 14, in which we see that

$$1 \text{ N} = 1 \text{ T·A·m},$$

or $1 \text{ T·A} = 1 \text{ N/m}$.

Sample Problem 3 Two long parallel wires a distance $2d$ apart carry equal currents i in opposite directions, as shown in Fig. 6a. Derive an expression for $B(x)$, the magnitude of the resultant magnetic field for points at a distance x from the point midway between the wires, as shown.

Study of the figure and the use of the right-hand rule show that the fields set up by the currents in the individual wires point in the same direction for all points between the wires. From Eq. 8 we then have

$$B(x) = B_a(x) + B_b(x) = \frac{\mu_0 i}{2\pi(d+x)} + \frac{\mu_0 i}{2\pi(d-x)}$$

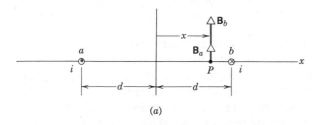

(a)

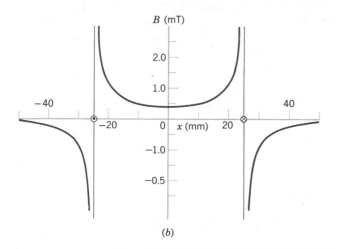

(b)

$$= \frac{\mu_0 i d}{\pi(d^2 - x^2)}. \qquad \text{(Answer)} \quad \text{(15)}$$

Inspection of this relation shows that (1) $B(x)$ is symmetrical about the midpoint ($x = 0$); (2) $B(x)$ has its minimum value ($= \mu_0 i/\pi d$) at this point; and (3) $B(x) \to \infty$ as $x \to \pm d$. At these locations, the point P in Fig. 6a is within the wires on their axes. Our derivation of Eq. 8, however, is valid only for points outside the wires so that Eq. 15 above holds only up to the surface of the wires.*

Figure 6b shows a plot of Eq. 15 for $i = 25$ A and $2d = 50$ mm. We leave it as an exercise to show that Eq. 15 holds also for points beyond the wires, that is, for points with $|x| > d$.

Sample Problem 4 Figure 7a shows two long parallel wires carrying currents i_1 and i_2 in the directions shown. What are the magnitude and the direction of the resultant magnetic field at point P? Assume the following values: $i_1 = 15$ A, $i_2 = 32$ A, and $d = 5.3$ cm.

Figure 7b shows the individual magnetic fields $\mathbf{B}_1$ and $\mathbf{B}_2$ set up by the currents i_1 and i_2, respectively. Verify that their

---

* Actually, for points on the axes of the long straight wires of Fig. 6, symmetry considerations lead us to conclude that the intrinsic magnetic field must be zero. This conclusion follows because there is no preferred direction for the field at such points; see Fig. 11.

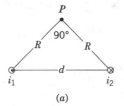

(a)

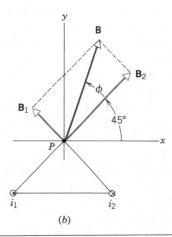

(b)

Figure 6 Sample Problem 3. (a) Two parallel wires carry currents of the same magnitude in opposite directions. At points between the wires, such as P, the magnetic fields for the separate currents point in the same direction. (b) A plot of $B(x)$ for $i = 25$ A and a wire separation of 50 mm.

Figure 7 Sample Problem 4. (a) Two wires carry currents in opposite directions. What is the magnetic field at P? (b) The separate fields combine vectorially to yield the resultant field $\mathbf{B}$.

directions are correctly given by the right-hand rule. The magnitudes of these fields are given by Eq. 8 as

$$B_1 = \frac{\mu_0 i_1}{2\pi R} = \frac{\mu_0 i_1}{2\pi(d/\sqrt{2})} = \frac{\sqrt{2}\mu_0}{2\pi d} i_1$$

and

$$B_2 = \frac{\mu_0 i_1}{2\pi R} = \frac{\mu_0 i_2}{2\pi(d/\sqrt{2})} = \frac{\sqrt{2}\mu_0}{2\pi d} i_2,$$

in which we have replaced R by its equal, $d/\sqrt{2}$.

The magnitude of the resultant magnetic field $\mathbf{B}$ is

$$B = \sqrt{B_1^2 + B_2^2} = \frac{\sqrt{2}\mu_0}{2\pi d} \sqrt{i_1^2 + i_2^2}$$

$$= \frac{(\sqrt{2})(4\pi \times 10^{-7} \text{ T} \cdot \text{m/A}) \sqrt{(15 \text{ A})^2 + (32 \text{ A})^2}}{(2\pi)(5.3 \times 10^{-2} \text{ m})}$$

$$= 1.89 \times 10^{-4} \text{ T} \approx 190 \ \mu\text{T}. \qquad \text{(Answer)}$$

The angle ϕ between $\mathbf{B}$ and $\mathbf{B}_2$ in Fig. 7b follows from

$$\phi = \tan^{-1} \frac{B_1}{B_2} = \tan^{-1} \frac{i_1}{i_2}$$

$$= \tan^{-1} \frac{15 \text{ A}}{32 \text{ A}} = 25°.$$

The angle between $\mathbf{B}$ and the x axis is then

$$\phi + 45° = 25° + 45° = 70°. \qquad \text{(Answer)}$$

31-5 Ampere's Law

In electrostatics, we can use Coulomb's law—that workhorse of electrostatics—to calculate the electric field caused by any charge distribution. For complex distributions, we may have to resort to a computer but we can always get a numerical answer to any accuracy we wish. However, when we draw together the laws of electromagnetism in Table 2 of Chapter 37 (Maxwell's equations), we do not represent the field of electrostatics by Coulomb's law but rather by Gauss' law. In electrostatics, where the charges are stationary or only slowly moving, these two laws are equivalent. However, Gauss' law is more compatible in form with the other equations of electromagnetism than is Coulomb's law and allows us to solve electric field problems of appropriately high symmetry with ease and elegance.*

The situation in magnetism is similar. We can cal-

culate the magnetic field caused by any current distribution, using the law of Biot and Savart—the magnetic equivalent of Coulomb's law. Again, in difficult cases, we may have to resort to a numerical calculation, using a computer. However, if we turn to Table 2 in Chapter 37 and examine the collected equations of electromagnetism (Maxwell's equations), we do not find the law of Biot and Savart among them. In its place we find *Ampere's law*, first advanced by Andre Marie Ampère (1775–1836) for whom the SI unit of current is named. Both Ampere's law and the law of Biot and Savart are relations between a current distribution and the magnetic field that it generates. Ampere's law, however, has a simplicity and form that makes it more compatible with the other equations of electromagnetism and—in the spirit of Gauss' law—allows us to solve magnetic field problems of appropriately high symmetry with ease and elegance.

Our plan is to display Ampere's law:

$$\oint \mathbf{B} \cdot d\mathbf{s} = \mu_0 i \qquad \text{(Ampere's law)} \qquad (16)$$

and then to become familiar with it by using it. Ampere's law is applied to an arbitrary closed loop, called an *Amperian loop;* the circle on the integral sign indicates that the quantity $\mathbf{B} \cdot d\mathbf{s}$ is to be integrated around that (closed) loop. The current i in Eq. 16 is the net current enclosed by the loop. Speaking loosely, Ampere's law relates the distribution of the magnetic field at points on the loop to the current that passes through the loop.

Let us examine Ampere's law by seeing how to apply it in the situation of Fig. 8. The figure shows the cross sections of three long straight wires that pierce the plane of the page at right angles to it. The wires carry currents i_1, i_2, and i_3 in the directions shown. The arbi-

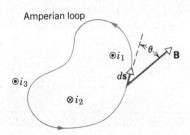

Figure 8 Ampere's law is applied to an arbitrary Amperian loop that encloses two long straight wires but excludes a third wire. Note the directions of the currents.

---

* Gauss' law is also more general than Coulomb's law in situations involving electric fields set up by rapidly moving charges.

trary Amperian loop to which we intend to apply Ampere's law lies entirely in the plane of the figure and threads its way among the wires, enclosing two of them but excluding the third.

We divide the Amperian loop of Fig. 8 into differential line segments of length $d\mathbf{s}$, one of which we show. At this line element, the magnetic field will have a particular value $\mathbf{B}$, also shown in the figure. Because of symmetry, $\mathbf{B}$ must lie in the plane of the figure, making an angle θ with the extended direction of the line element $d\mathbf{s}$.

The quantity $\mathbf{B} \cdot d\mathbf{s}$ on the left side of Eq. 16 is a scalar product and has the value $B \cos \theta \, ds$. The integral on the left side of Eq. 16 then becomes

$$\oint \mathbf{B} \cdot d\mathbf{s} = \oint B \cos \theta \, ds.$$

This *line integral,* as integrals of this type are called, instructs us to go around the Amperian loop of Fig. 8, adding (that is, integrating) the quantity $B \cos \theta \, ds$ as we go. We choose, arbitrarily, to traverse the loop in a counterclockwise sense.

So much for the left side of Eq. 16. The term i on the right side is the *net* current encircled by the loop, the currents being added algebraically. For the special case of Fig. 8, we have

$$i = i_1 - i_2.$$

For a counterclockwise traversal of the loop, currents pointing out of the loop (out of the page in Fig. 8) are taken as positive, those pointing inward being negative.* Note that i_3 in Fig. 8 is not included in calculating i; it lies outside the loop and is not encircled by it.

Applying Ampere's law (Eq. 16) to the situation of Fig. 8 then gives us

$$\oint B \cos \theta \, ds = \mu_0 \, (i_1 - i_2).$$

This result, even though we can push it no further, demonstrates the power and elegance of Ampere's law. Because the symmetry is not high enough, we cannot explicitly evaluate the closed line integral on the left side of this equation for the arbitrary Amperian loop shown.

However, from the right hand side of the equation we know what that value must be. It depends *only* on the net current passing through the surface having the Amperian loop as a boundary. Note the similarity to Gauss's law. There the integral of $\mathbf{E}$ over any closed surface depends *only* on the net charge enclosed by that surface. These two laws are part of Nature's table of integrals.

We turn now to the simpler and familiar case of a single long straight wire carrying a current i, which *does* have enough symmetry so that we can use Ampere's law to find the magnetic field, $\mathbf{B}$. As Fig. 9 shows, we take our Amperian loop to be a concentric circle of radius r. This choice permits us to take full advantage of the cylindrical symmetry of the problem. Because of this symmetry, we conclude that $\mathbf{B}$ has the same magnitude, B, at every point on the circular Amperian loop. It remains to decide whether $\mathbf{B}$ is everywhere *tangent* to that loop or *perpendicular* to it, these two possibilities being equally symmetrical. We can rely on a simple compass experiment to show that the former possibility is the case; that is, $\mathbf{B}$ is tangent to the loop. (Can you show that radial lines of $\mathbf{B}$ are not consistent with Ampere's law?)

We assume further that the direction of $\mathbf{B}$ is that given by the right-hand rule of Section 31–2. Thus, $\mathbf{B}$ and $d\mathbf{s}$ point in the same direction, the angle θ between them being zero. The left-hand side of Ampere's law then becomes

$$\oint \mathbf{B} \cdot d\mathbf{s} = \oint B \cos \theta \, ds = B \oint ds = B(2\pi r).$$

Note that $\oint ds$ above is simply the circumference of the circular loop, which is $2\pi r$. The right side of Ampere's law is simply $\mu_0 i$ so that

$$B(2\pi r) = \mu_0 i$$

or

$$B = \frac{\mu_0 i}{2\pi r}. \tag{17}$$

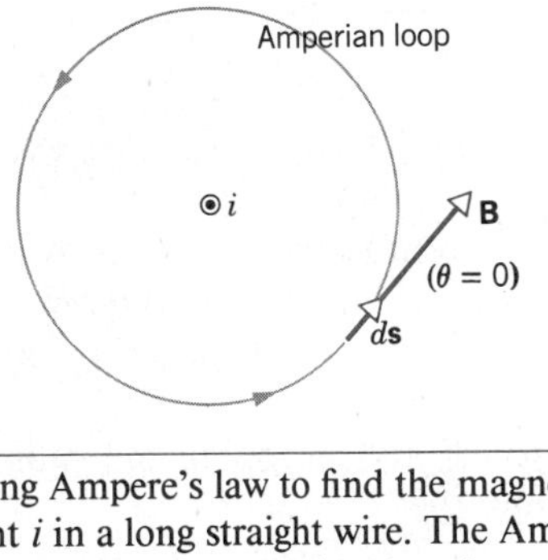

Figure 9 Using Ampere's law to find the magnetic field set up by a current i in a long straight wire. The Amperian loop is a concentric circle that lies outside the wire.

---

* Our right-hand rule applies here, in this form: If the fingers of your right hand (the curly element) represent the direction of traversal around the Amperian loop, then your extended right thumb (the straight element) represents the positive direction for currents encircled by the loop.

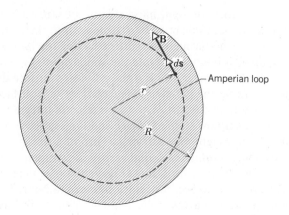

Figure 10 Using Ampere's law to find the magnetic field set up by a current i in a long straight wire of circular cross section. The Amperian loop is drawn inside the wire. The current is uniformly distributed over the cross section of the wire and emerges from the page.

This is precisely Eq. 8, which we derived earlier — with considerably more effort — using the law of Biot and Savart.

Figure 10 shows another situation in which we can usefully apply Ampere's law. It shows a cross section of a long straight wire of radius R, carrying a current i_0 uniformly distributed over the cross section of the wire. What magnetic field does this wire set up, both for points outside the wire and for points inside the wire?

For outside points, for which $r > R$, the answer is given by Eq. 8. Figure 10 shows an Amperian loop of radius r that is suitable for considering inside points, that is, points for which $r < R$. Symmetry suggests that **B** is tangent to the loop, as shown. Ampere's law

$$\oint \mathbf{B} \cdot d\mathbf{s} = \mu_0 i$$

becomes

$$(B)(2\pi r) = \mu_0 i = \mu_0 i_0 \left(\frac{\pi r^2}{\pi R^2} \right).$$

Note that the current i that appears in Ampere's law is not the total current i_0 in the wire but only that fraction of the total current that is enclosed by the Amperian loop.

Solving for B and dropping the subscript on the current, we find

$$B = \left(\frac{\mu_0 i}{2\pi R^2} \right) r, \qquad (18)$$

which shows that, within the wire, B is proportional to r, starting from a value of zero at the center of the wire.

At the surface of the wire ($r = R$), Eq. 18 reduces to the same expression found by putting $r = R$ in Eq. 8 ($B = \mu_0 i / 2\pi R$). That is, the expressions for the magnetic field outside the wire and inside the wire yield the same result at the surface of the wire.

Sample Problem 5 A long straight wire of radius $R = 1.5$ mm carries a steady current i_0 of 32 A. (a) What is the magnetic field at the surface of the wire?

Equations 8 and 18 each apply. From the former, we have

$$B = \frac{\mu_0 i}{2\pi r} = \frac{(4\pi \times 10^{-7} \text{ T} \cdot \text{m/A})(32 \text{ A})}{(2\pi)(1.5 \times 10^{-3} \text{ m})}$$
$$= 4.27 \times 10^{-3} \text{ T} \approx 4.3 \text{ mT}. \qquad \text{(Answer)}$$

(b) What is the magnetic field at $r = 1.2$ mm?

Such points lie inside the wire so that Eq. 18 applies. We have

$$B = \frac{\mu_0 i r}{2\pi R^2} = \frac{(4\pi \times 10^{-7} \text{ T} \cdot \text{m/A})(32 \text{ A})(1.2 \times 10^{-3} \text{ m})}{(2\pi)(1.5 \times 10^{-3} \text{ m})^2}$$
$$= 3.41 \times 10^{-3} \text{ T} \approx 3.4 \text{ mT}. \qquad \text{(Answer)}$$

Figure 11 is a plot of the magnetic field, both inside and outside the wire. Note that it reaches its maximum value at the surface of the wire.

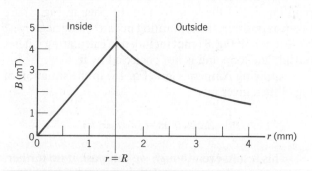

Figure 11 The magnetic field for the conductor of Fig. 10 and Sample Problem 5, both inside and outside the wire. The maximum field occurs at the surface of the wire.

31-6 Solenoids and Toroids

We now turn our attention to another problem of high symmetry in which Ampere's law will prove useful. It is the magnetic field set up by the current in a long tightly

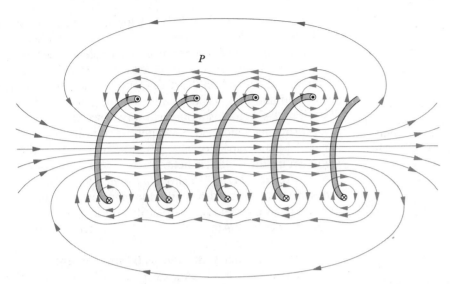

Figure 12 A section of a "stretched-out" solenoid, showing the magnetic field lines. Only the back portions of the separated windings are shown.

wound helical coil. Such a coil is called a *solenoid.* We assume here that its length is much greater than its diameter.

Figure 12 shows a section of a "stretched-out" solenoid. The solenoid field is the vector sum of the fields set up by the individual turns. For points close to these turns, the observer is not aware that the wire is bent into an arc. The wire behaves magnetically almost like a long straight wire and the lines of **B** associated with each single turn are almost concentric circles. Figure 12 suggests that the field tends to cancel between the turns. It also suggests that, at points inside the solenoid and reasonably far from the wires, **B** is parallel to the solenoid axis. In the limiting case of tightly packed square wires, the solenoid becomes essentially a cylindrical current sheet and the requirements of symmetry make the statement just given rigorously true.

For points such as *P* in Fig. 12, the field set up by the upper part of the solenoid turns (marked ⊙) points to the left and tends to cancel the field set up by the lower part of the turns (marked ⊗), which points to the right. As the solenoid approaches the configuration of an infinitely long cylindrical current sheet, the magnetic field **B** outside the solenoid approaches zero. Taking the external field to be zero is an excellent assumption for a real solenoid if its length is much greater than its diameter and if we consider points near the central region of the solenoid, far from its ends. The direction of the magnetic field along the solenoid axis follows from our all-purpose right-hand rule, interpreted this way: Grasp the solenoid with your right hand so that your fingers (the curly element) follow the direction of the current in the windings.

Your extended right thumb (the straight element) will then point in the direction of the axial magnetic field.

Figure 13 shows the lines of **B** for a real solenoid. The spacing of the lines of **B** in the central plane shows that the internal field is fairly strong and uniform over its

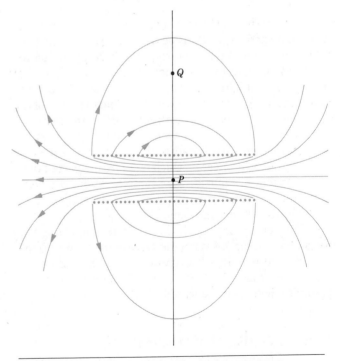

Figure 13 Magnetic field lines for a solenoid of finite length. Note that the field is strong and uniform at internal points such as *P* but is relatively weak for external points such as *Q*.

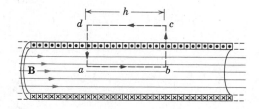

Figure 14 An application of Ampere's law to a section of a long idealized solenoid. The Amperian loop is the rectangle *abcd*.

cross section and that the external field is relatively weak.

Let us apply Ampere's law

$$\oint \mathbf{B} \cdot d\mathbf{s} = \mu_0 i \qquad (19)$$

to the rectangular Amperian loop *abcd* in the ideal (infinite) solenoid of Fig. 14. We write the integral $\oint \mathbf{B} \cdot d\mathbf{s}$ as the sum of four integrals, one for each path segment:

$$\oint \mathbf{B} \cdot d\mathbf{s} = \int_a^b \mathbf{B} \cdot d\mathbf{s} + \int_b^c \mathbf{B} \cdot d\mathbf{s} + \int_c^d \mathbf{B} \cdot d\mathbf{s} + \int_d^a \mathbf{B} \cdot d\mathbf{s}. \qquad (20)$$

The first integral on the right of Eq. 20 is Bh, where B is the magnitude of $\mathbf{B}$ inside the solenoid and h is the arbitrary length of the path from a to b. Note that path ab, though parallel to the solenoid axis, is some distance r from it; it will turn out that B inside the solenoid is constant over its cross section and thus independent of r.

The second and the fourth integrals on the right of Eq. 20 are zero because for every element of these paths $\mathbf{B}$ is at right angles to the path (or is zero) and thus $\mathbf{B} \cdot d\mathbf{s}$ is zero. The third integral, which includes the part of the rectangle that lies outside the solenoid, is zero because $B = 0$ for all external points. Thus, $\oint \mathbf{B} \cdot d\mathbf{s}$ for the entire rectangular path has the value Bh.

The net current i enclosed by the rectangular Amperian loop in Fig. 14 is not the same as the current i_0 in the solenoid windings because the windings pass more than once through this loop. Let n be the number of turns per unit length of the solenoid; then

$$i = i_0(nh).$$

Ampere's law (Eq. 19) then becomes

$$Bh = \mu_0 i_0 nh$$

or

$$\boxed{B = \mu_0 i_0 n} \qquad \text{(ideal solenoid).} \qquad (21)$$

Although we derived Eq. 21 for an infinitely long solenoid, it holds quite well for actual solenoids if we apply it only at internal points near the solenoid center. Equation 21 is consistent with the experimental fact that B does not depend on the diameter or the length of the solenoid and that B is constant over the solenoid cross section. A solenoid is a practical way to set up a known uniform magnetic field for experimentation, just as a parallel-plate capacitor is a practical way to set up a known uniform electric field.

A Toroid. Figure 15 shows a *toroid*, which we may describe as a solenoid bent into the shape of a doughnut. What magnetic field is set up at its interior points? We can find out from Ampere's law and from certain considerations of symmetry.

From symmetry, the lines of $\mathbf{B}$ form concentric circles inside the toroid, as shown in the figure. Let us choose a concentric circle of radius r as an Amperian loop and traverse it in the clockwise direction. Ampere's law (see Eq. 19) yields

$$(B)(2\pi r) = \mu_0 i_0 N,$$

where i_0 is the current in the toroid windings (and is positive) and N is the total number of turns. This gives

$$\boxed{B = \frac{\mu_0 i_0 N}{2\pi} \frac{1}{r}} \qquad \text{(toroid).} \qquad (22)$$

In contrast to the solenoid, B is not constant over the cross section of a toroid. It is easy to show, from Ampere's law, that $B = 0$ for points outside an ideal toroid.

Close inspection of Eq. 22 will justify our earlier statement that "a toroid is a solenoid bent into the shape

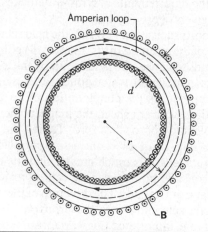

Figure 15 A toroid. The internal magnetic field can be found by applying Ampere's law to the Amperian loop shown.

of a doughnut." The denominator in Eq. 22, which is $2\pi r$, is the central circumference of the toroid and $N/2\pi r$ is just n, the number of turns per unit length. With this substitution, Eq. 22 reduces to $B = \mu_0 i_0 n$, the equation for the magnetic field in the central region of a solenoid.

The direction of the magnetic field within a toroid follows from our all-purpose right-hand rule: Grasp the toroid with the fingers of your right hand curling in the direction of the current in the windings; your extended right thumb points in the direction of the magnetic field.

Toroids form the central feature of the *tokamak,* a device showing promise as the basis for a fusion power reactor. We discuss its mode of operation in Chapter 48 of the extended version of this book.

Sample Problem 6 A solenoid has a length $L = 1.23$ m and an inner diameter $d = 3.55$ cm. It has five layers of windings of 850 turns each and carries a current $i_0 = 5.57$ A. What is B at its center?

From Eq. 21

$$B = \mu_0 i_0 n = (4\pi \times 10^{-7} \text{ T·m/A})(5.57 \text{ A})\left(\frac{5 \times 850 \text{ turns}}{1.23 \text{ m}}\right)$$

$$= 2.42 \times 10^{-2} \text{ T} = 24.2 \text{ mT}. \quad \text{(Answer)}$$

Note that Eq. 21 applies even if the solenoid has more than one layer of windings because the diameter of the windings does not enter into the equation.

31-7   A Current Loop as a Magnetic Dipole

So far we have studied the magnetic field set up by a long straight wire, a solenoid, and a toroid. We turn our attention here to the field set up by a single current loop. We learned in Section 30-9 that such a loop behaves like a magnetic dipole in that, if we place it in an external magnetic field, a torque τ given by

$$\tau = \mu \times \mathbf{B} \quad (23)$$

acts on it. Here μ is the magnetic dipole moment of the loop, given in magnitude by NiA where N is the number of turns, i is the current in the loop, and A is the area enclosed by the loop.

The direction of μ is given by our all-purpose right-hand rule: Grasp the loop so that the fingers of your right hand curl around the loop in the direction of the current; your extended thumb then points in the direction of the dipole moment μ, which happens also to be the direction of the magnetic field on the axis of the loop.

We turn now to the other aspect of the current loop as a magnetic dipole: What magnetic field does it set up in the surrounding space? The problem does not have enough symmetry to make Ampere's law useful so that we must turn to the law of Biot and Savart. We consider only points on the axis of the loop, which we take to be a z axis. We shall show below that the solution is

$$B(z) = \frac{\mu_0 i R^2}{2(R^2 + z^2)^{3/2}} \quad (24)$$

in which R is the radius of the circular loop and z is the distance of the point in question from the center of the loop.

For axial points far from the loop, we have $z \gg R$ in Eq. 24. With that approximation, this equation reduces to

$$B(z) \approx \frac{\mu_0 i R^2}{2z^3}.$$

Recalling that πR^2 is the area A of the loop and extending our result to include a loop of N turns, we can write this equation as

$$B(z) = \frac{\mu_0}{2\pi} \frac{NiA}{z^3}$$

or, since $\mathbf{B}$ and μ have the same direction,

$$\mathbf{B}(z) = \frac{\mu_0}{2\pi} \frac{\mu}{z^3} \quad \text{(current loop),} \quad (25)$$

where μ is the magnetic dipole moment of the current loop.

Thus, we have two ways in which we can regard a current loop as a magnetic dipole: It experiences a torque when we place it in an external magnetic field; it generates its own intrinsic magnetic field given, for distant points along the axis, by Eq. 25.

Equation 25 reminds us of the result of Eq. 11 of Chapter 24 for the *electric* field on the axis of an *electric* dipole; namely,

$$\mathbf{E}(z) = \frac{1}{2\pi\epsilon_0} \frac{\mathbf{p}}{z^3}$$

in which $\mathbf{p}$ is the electric dipole moment. Table 1 is a summary of the properties of electric and magnetic dipoles as we have developed them so far. The symmetry between the two sets of equations is striking.

Proof of Equation 24. Figure 16 shows a circular loop of radius R carrying a current i. Consider a point P on the axis of the loop, a distance z from its plane. Let us apply

Table 1 Some Dipole Equations

Property	Dipole Type	Relation	Equation Number
Torque in an external field	Electric	$\boldsymbol{\tau} = \mathbf{p} \times \mathbf{E}$	24–25
	Magnetic	$\boldsymbol{\tau} = \boldsymbol{\mu} \times \mathbf{B}$	30–30
Energy in an external field	Electric	$U = -\mathbf{p} \cdot \mathbf{E}$	24–28
	Magnetic	$U = -\boldsymbol{\mu} \cdot \mathbf{B}$	30–31
Field at distant axial points	Electric	$\mathbf{E}(z) = \dfrac{1}{2\pi\epsilon_0} \dfrac{\mathbf{p}}{z^3}$	24–11
	Magnetic	$\mathbf{B}(z) = \dfrac{\mu_0}{2\pi} \dfrac{\boldsymbol{\mu}}{z^3}$	31–25

the law of Biot and Savart to a current element located at the left side of the loop. The vector $d\mathbf{s}$ for this element points out of the page at right angles. The angle θ between $d\mathbf{s}$ and the vector $\mathbf{r}$ in Fig. 16 is 90° and the plane formed by these two vectors is at right angles to the plane of the figure. From the law of Biot and Savart, the differential field $d\mathbf{B}$ set up by this current element is at right angles to this plane and thus lies in the plane of the figure and at right angles to $\mathbf{r}$, as Fig. 16 shows.

Let us resolve $d\mathbf{B}$ into two components, one, $d B_\parallel$,

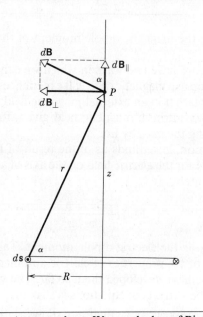

Figure 16 A current loop. We use the law of Biot and Savart to find the magnetic field at axial points.

along the axis of the loop and another, $d B_\perp$, at right angles to this axis. Only $d B_\parallel$ contributes to the total magnetic field B at point P. This follows because, from symmetry, the axial direction is the only preferred direction, so that the vector sum of all magnetic field components at right angles to the axis must add up to zero. This leaves only the axial components and we have

$$B = \int dB_\parallel,$$

where the integral is a simple scalar integration.

For the element shown in Fig. 16, the law of Biot and Savart (Eq. 5) gives

$$dB = \left(\frac{\mu_0}{4\pi}\right) \frac{i\, ds \sin 90°}{r^2}.$$

We also have

$$dB_\parallel = dB \cos \alpha.$$

Combining these two relations, we have

$$dB_\parallel = \frac{\mu_0 i \cos \alpha \, ds}{4\pi r^2}. \tag{26}$$

Figure 16 shows that r and α are not independent but are related to each other. Let us express each in terms of the variable z, the distance of point P from the center of the loop. The relations are

$$\cos \alpha = \frac{R}{r} = \frac{R}{\sqrt{R^2 + z^2}} \tag{27}$$

and

$$r = \sqrt{R^2 + z^2}. \tag{28}$$

Substituting Eqs. 27 and 28 into Eq. 26, we find

$$dB = \frac{\mu_0 i R}{4\pi (R^2 + z^2)^{3/2}} \, ds.$$

Note that i, R, and z have the same values for all current elements. Integrating this equation and noting that $\int ds$ is simply the circumference of the loop, we find

$$B = \int dB = \frac{\mu_0 i R}{4\pi (R^2 + z^2)^{3/2}} \int ds$$

or

$$B(x) = \frac{\mu_0 i R^2}{2(R^2 + z^2)^{3/2}},$$

which is Eq. 24, the relation we sought to prove.

REVIEW AND SUMMARY

The Biot-Savart Law

The magnetic field set up by a current-carrying conductor can be found from the *Biot-Savart law*. This asserts that the contribution $d\mathbf{B}$ to the field set up by a current element $i\,d\mathbf{s}$ at a point distant $\mathbf{r}$ from the current element has a magnitude given by

$$d\mathbf{B} = \left(\frac{\mu_0}{4\pi}\right) \frac{i\,d\mathbf{s} \times \mathbf{r}}{r^3} \quad \text{(Biot-Savart law)}. \tag{7}$$

μ_0

Study Fig. 3 carefully. The quantity μ_0, called the permeability constant, has the value $4\pi \times 10^{-7}$ T·m/A $\approx 1.26 \times 10^{-6}$ T·m/A.

A Long, Straight Wire

For a *long straight wire* carrying a current i the Biot-Savart law gives, for the magnetic field at a distance r from the wire,

$$B(r) = \frac{\mu_0 i}{2\pi r} \quad \text{(long straight wire)}; \tag{8}$$

see Sample Problem 1. Figure 2 shows the concentric field lines. The right-hand rule described in Section 31–2 is helpful in relating the directions of $\mathbf{B}$ and $i\,d\mathbf{s}$ in this and other magnetic problems.

$\mathbf{B}_{ext}$, $\mathbf{B}_{intr}$, and Their Resultant

If a wire carrying a current is placed in an external magnetic field $\mathbf{B}_{ext}$, a force given by Eq. 30–24 will act on it. $\mathbf{B}_{ext}$ must not be confused with $\mathbf{B}_{intr}$, the intrinsic magnetic field set up by the current in the wire itself. Figure 4 shows field lines representing the vector sum of $\mathbf{B}_{ext}$ and $\mathbf{B}_{intr}$ for a particular case.

Parallel wires carrying currents in the same (opposite) direction attract (repel) each other. For a length L of the wire the magnitude of the force on either wire is

The Force Between Parallel Wires

$$F_{ba} = i_b L B_a = \frac{\mu_0 L i_a i_b}{2\pi d}, \tag{14}$$

where d is the wire separation. The ampere is defined from this relation. See Sample Problems 2–4.

For current distributions of high symmetry *Ampere's law,*

Ampere's Law

$$\oint \mathbf{B} \cdot d\mathbf{s} = \mu_0 i \quad \text{(Ampere's law)} \tag{16}$$

can be used in placed of the Biot-Savart law to calculate the magnetic field. One first chooses a closed *Amperian loop* around which to apply the law. Study the fairly general case of Fig. 8 to learn how to evaluate the line integral $\oint \mathbf{B} \cdot d\mathbf{s}$ and to find i. See Sample Problem 5.

Using Ampere's law we show that B inside a *long solenoid,* carrying current i_0, at points near its center is given by

A Solenoid

$$B = \mu_0 i_0 n \quad \text{(ideal solenoid)} \tag{21}$$

where n is the number of turns per unit length; see Sample Problem 6. Also, B inside a *toroid* is

A Toroid

$$B = \frac{\mu_0 i_0 N}{2\pi} \frac{1}{r} \quad \text{(toroid)}; \tag{22}$$

See Fig. 15 and Sample Problem 6.

The magnetic field produced by a *current loop* (a *magnetic dipole*) at a distance z along its axis is parallel to the axis and has magnitude

Field of a Magnetic Dipole

$$\mathbf{B}(z) = \frac{\mu_0}{2\pi} \frac{\mu}{z^3} \quad \text{(current loop)}, \tag{25}$$

μ being the dipole moment of the loop. Table 1 summarizes and compares the impressively symmetric properties of electric and magnetic dipoles.

QUESTIONS

1. A beam of 20-MeV protons emerges from a cyclotron. Do these particles cause a magnetic field?

2. Discuss analogies and differences between Coulomb's law and the Biot-Savart law.

3. Consider a magnetic field line. Is the magnitude of **B** constant or variable along such a line? Can you give an example of each case?

4. In electronics, wires that carry equal but opposite currents are often twisted together to reduce their magnetic effect at distant points. Why is this effective?

5. Drifting electrons constitute the current in a wire and a magnetic field is associated with this current. What current and magnetic field would be measured by an observer moving along with the drifting electrons?

6. Consider two charges, first (*a*) of the same sign and then (*b*) of opposite signs, that are moving along separated parallel paths with the same velocity. Compare the directions of the mutual electric and magnetic forces in each case.

7. Is there any way to set up a magnetic field other than by causing charges to move?

8. Like currents attract and unlike currents repel. Like stationary charges repel and unlike stationary charges attract. Is this a paradox?

9. Figure 17 shows four vertical wires carrying equal currents in the same direction. What is the direction of the force on the left-hand wire, caused by the currents in the other three wires?

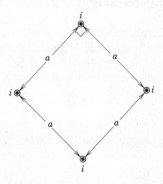

Figure 17   Question 9.

10. Two long parallel conductors carry equal currents *i* in the same direction. Sketch roughly the resultant lines of **B** due to the action of both currents. Does your figure suggest an attraction between the wires?

11. A current is sent through a vertical spring from whose lower end a weight is hanging; what will happen?

12. Two long straight wires pass near one another at right angles. If the wires are free to move, describe what happens when currents are sent through both of them.

13. Two fixed wires cross each other perpendicularly so that they do not actually touch but are close to each other, as shown in Fig. 18. Equal currents *i* exist in each wire in the directions indicated. In what region(s) will there be some points of zero net magnetic field?

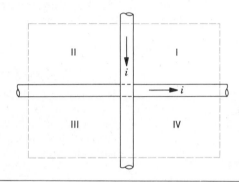

Figure 18   Question 13.

14. A messy loop of limp wire is placed on a smooth table and anchored at points *a* and *b* as shown in Fig. 19. If a current *i* is now passed through the wire, will it try to form a circular loop or will it try to bunch up further?

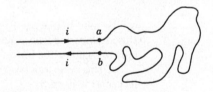

Figure 19   Question 14.

15. Can the path of integration around which we apply Ampere's law pass through a conductor?

16. Suppose we set up a path of integration around a cable that contains 12 wires with different currents (some in opposite directions) in each wire. How do we calculate *i* in Ampere's law in such a case?

17. Apply Ampere's law qualitatively to the three paths shown in Fig. 20.

18. Discuss analogies and differences between Gauss's law and Ampere's law.

19. A steady longitudinal uniform current is set up in a long

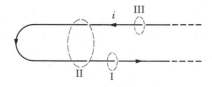

Figure 20 Question 17.

copper tube. Is there a magnetic field (a) inside and/or (b) outside the tube?

20. A long straight wire of radius R carries a steady current i. How does the magnetic field generated by this current depend on R? Consider points both outside and inside the wire.

21. A long straight wire carries a constant current i. What does Ampere's law require for (a) a loop that encloses the wire but is not circular, (b) a loop that does not enclose the wire, and (c) a loop that encloses the wire but does not all lie in one plane?

22. Two long solenoids are nested on the same axis, as Fig. 21 shows. They carry identical currents but in opposite directions. If there is no magnetic field inside the inner solenoid, what can you say about n, the number of turns per unit length, for the two solenoids? Which one, if either, has the larger value?

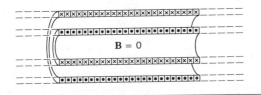

Figure 21 Question 22.

23. If the windings on a solenoid are considered to be helical, rather than being equivalent to a cylindrical current sheet, then the magnetic field outside is not zero strictly but is a circling field, just the same as the field of a straight wire. Explain.

24. The magnetic field at the center of a circular current loop has the value $\mu_0 i/2R$; see Eq. 24. However, the *electric* field at the center of a ring of charge is *zero*. Why this difference?

25. A steady current is set up in a cubical network of resistive wires, as in Fig. 22. Use symmetry arguments to show that the magnetic field at the center of the cube is zero.

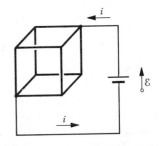

Figure 22 Question 25.

26. Does Eq. 21 ($B = \mu_0 i_0 n$) hold for a solenoid of square cross section?

27. Give details of three ways in which you can measure the magnetic field **B** at a point P, a perpendicular distance r from a long straight wire carrying a constant current i. Base them on: (a) projecting a particle of charge q through point P with velocity **v**, parallel to the wire; (b) measuring the force per unit length exerted on a second wire, parallel to the first wire and carrying a current i'; (c) measuring the torque exerted on a small magnetic dipole located a perpendicular distance r from the wire.

28. How might you measure the magnetic dipole moment of a compass needle?

29. A circular loop of wire lies on the floor of the room in which you are sitting. It carries a constant current i in a clockwise direction, as viewed from above. What is the direction of the magnetic dipole moment of this current loop?

30. Is **B** uniform for all points within a circular loop of wire carrying a current? Explain.

EXERCISES AND PROBLEMS

Section 31–2 Calculating the Magnetic Field

1E. A #10 bare copper wire (2.6 mm in diameter) can carry a current of 50 A without overheating. For this current, what is the magnetic field at the surface of the wire?

2E. The magnitude of the magnetic field 88 cm from the axis of a long straight wire is 7.3 μT. Calculate the current in the wire.

3E. A surveyor is using a magnetic compass 20 ft below a power line in which there is a steady current of 100 A. Will this interfere seriously with the compass reading? The horizontal component of the earth's magnetic field at the site is 20 μT (= 0.20 gauss).

4E. The 25-kV electron gun in a TV tube fires an electron beam 0.22 mm in diameter at the screen, 5.6×10^{14} electrons arriving each second. Calculate the magnetic field produced by the beam at a point 1.5 mm from the axis of the beam.

5E. Figure 23 shows a 3.0-cm segment of wire, centered at the origin, carrying a current of 2.0 A in the $+y$ direction. (Of course this segment must be part of some complete circuit.) To calculate the **B** field at a point P several meters from the origin

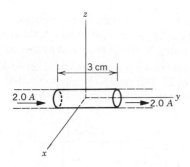

Figure 23 Exercise 5.

one may use the Biot-Savart law in the form $B = (\mu_0/4\pi)i \Delta s \sin \theta/r^2$ with $\Delta s = 3.0$ cm. This is because r and θ are essentially constant over the segment of wire. Calculate **B** (magnitude and direction) at the following (x, y, z) locations: (a) (0, 0, 5 m), (b) (0, 6 m, 0), (c) (7 m, 7 m, 0), (d) (−3 m, −4 m, 0).

6E. A long wire carrying a current of 100 A is placed in a uniform external magnetic field of 5.0 mT (= 50 gauss). The wire is at right angles to this magnetic field. Locate the points at which the resultant magnetic field is zero.

7E. At a position in the Philippines the earth's magnetic field of 39 μT is horizontal and due north. The net field is zero exactly 8.0 cm above a long straight horizontal wire that carries a constant current. (a) What is the magnitude of the current and (b) what is the direction of the current?

8E. A positive point charge of magnitude q is a distance d from a long straight wire carrying a current i and is traveling with speed v perpendicular to the wire. What are the direction and magnitude of the force acting on it if the charge is moving (a) toward, or (b) away from the wire?

9E. A long straight wire carries a current of 50 A. An electron, traveling at 1.0×10^7 m/s, is 5.0 cm from the wire. What force acts on the electron if the electron velocity is directed (a) toward the wire, (b) parallel to the wire, and (c) at right angles to the directions defined by (a) and (b)?

10E. A straight conductor carrying a current i is split into identical semicircular turns as shown in Fig. 24. What is the magnetic field at the center C of the circular loop so formed?

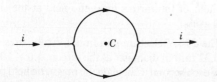

Figure 24 Exercise 10.

11E. Two infinitely long wires carry equal currents, i. Each follows a 90° arc on the circumference of the same circle of

radius R in the configuration shown in Fig. 25. Show, without doing a detailed calculation, that **B** at the center of the circle is the same as the **B** field a distance R below an infinite straight wire carrying a current i to the left.

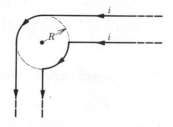

Figure 25 Exercise 11.

12P. Use the Biot-Savart law to calculate the magnetic field **B** at C, the common center of the semicircular arcs AD and HJ, of radii R_2 and R_1, respectively, forming part of the circuit $ADJHA$ carrying current i, as shown in Fig. 26.

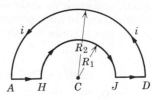

Figure 26 Problem 12.

13P. A long hairpin is formed by bending a piece of wire as shown in Fig. 27. If the wire carries a 10-A current, (a) what are the direction and magnitude of **B** at point a? (b) At point b? Take $R = 5.0$ mm.

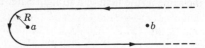

Figure 27 Problem 13.

14P. A wire carrying current i has the configuration shown in Fig. 28. Two semi-infinite straight sections, each tangent to the

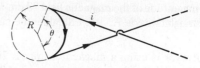

Figure 28 Problem 14.

same circle, are connected by a circular arc, of angle θ, along the circumference of the circle, with all sections lying in the same plane. What must θ be in order for B to be zero at the center of the circle?

15P. Consider the circuit of Fig. 29. The curved segments are arcs of circles of radii a and b. The straight segments are along the radii. Find the magnetic field $\mathbf{B}$ at P, assuming a current i in the circuit.

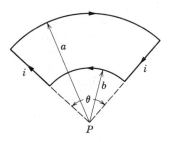

Figure 29 Problem 15.

16P. The wire shown in Fig. 30 carries a current i. What is the magnetic field $\mathbf{B}$ at the center C of the semicircle arising from (a) each straight segment of length L, (b) the semicircular segment of radius R, and (c) the entire wire?

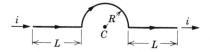

Figure 30 Problem 16.

17P. A straight wire segment of length L carries a current i. Show that the magnetic field $\mathbf{B}$ associated with this segment, at a distance R from the segment along a perpendicular bisector (see Fig. 31), is given in magnitude by

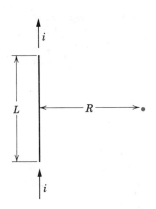

Figure 31 Problem 17.

$$B = \frac{\mu_0 i}{2\pi R} \frac{L}{(L^2 + 4R^2)^{1/2}}.$$

Show that this expression reduces to an expected result as $L \to \infty$.

18P. A square loop of wire of edge a carries a current i. Show that the value of B at the center is given by

$$B = \frac{2\sqrt{2}\mu_0 i}{\pi a}.$$

(*Hint:* See Problem 17.)

19P. Show that B at the center of a rectangular loop of wire of length L and width W, carrying a current i, is given by

$$B = \frac{2\mu_0 i}{\pi} \frac{(L^2 + W^2)^{1/2}}{LW}.$$

Show that this reduces to a result consistent with Sample Problem 3 for $L \gg W$.

20P. A square loop of wire of edge a carries a current i. Show that B for a point on the axis of the loop and a distance x from its center is given by

$$B(x) = \frac{4\mu_0 i a^2}{\pi(4x^2 + a^2)(4x^2 + 2a^2)^{1/2}}.$$

Show that the result is consistent with the result of Problem 18.

21P. You are given a length L of wire in which a current i may be established. The wire may be formed into a circle or a square. Show that the square yields the greater value for B at the central point.

22P. A straight section of wire of length L carries a current i. Show that the magnetic field associated with this segment at P a perpendicular distance D from one end of the wire (see Fig. 32) is given by

$$B = \frac{\mu_0 i}{4\pi D} \frac{L}{(L^2 + D^2)^{1/2}}.$$

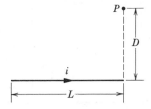

Figure 32 Problem 22.

23P. A current i flows in a straight wire segment of length a, as in Fig. 33. Show that the magnetic field at point Q in that figure is zero and that the field at P is given in magnitude by

$$B = \frac{\sqrt{2}\mu_0 i}{8\pi a}.$$

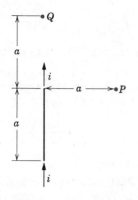

Figure 33 Problem 23.

24P. Find the magnetic field **B** at point P in Fig. 34. See Problem 23.

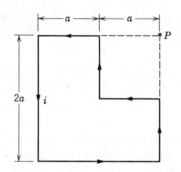

Figure 34 Problem 24.

25P. Calculate **B** at point P in Fig. 35. Assume that $i = 10$ A and $a = 8.0$ cm.

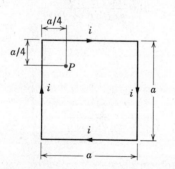

Figure 35 Problem 25.

26P. Figure 36 shows a cross section of a long, thin ribbon of width w that is carrying a uniformly-distributed total current i into the page. Calculate the magnitude and direction of the magnetic field, **B**, at a point P in the plane of the ribbon at a distance d from its edge. (*Hint:* Imagine the ribbon to be constructed from many long, thin, parallel wires.)

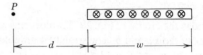

Figure 36 Problem 26.

Section 31–4 Two Parallel Conductors

27E. Two long parallel wires are 8.0 cm apart. What equal currents must flow in the wires if the magnetic field halfway between them is to have a magnitude of 300 μT? Consider both (a) parallel and (b) antiparallel currents.

28E. Two long straight parallel wires, separated by 0.75 cm, are perpendicular to the plane of the page as shown in Fig. 37. Wire W_1 carries a current of 6.5 A into the page. What must be the current (magnitude and direction) in wire W_2 for the resultant magnetic field at point P to be zero?

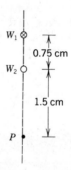

Figure 37 Exercise 28.

29E. Two long parallel wires a distance d apart carry currents of i and $3i$ in the same direction. Locate the point or points at which their magnetic fields cancel.

30E. Figure 38 shows five long parallel wires in the xy plane. Each wire carries a current $i = 3.0$ A in the positive x direction. The separation between adjacent wires is $d = 8.0$ cm. Find the magnetic force per meter exerted on each of these five wires.

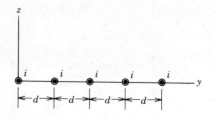

Figure 38 Exercise 30.

31E. For the wires in Sample Problem 3, show that Eq. 15 holds for points beyond the wires, that is, for points with $|x| > d$.

32E. Two long straight parallel wires 10 cm apart each carry a current of 100 A. Figure 39 shows a cross section, with the wires running perpendicular to the page and point P lying on the perpendicular bisector of d. Find the magnitude and direction of the magnetic field at P when the current in the left-hand wire is out of the page and the current in the right-hand wire is (a) out of the page and (b) into the page.

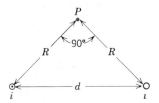

Figure 39 Exercise 32.

33P. In Fig. 6a assume that both currents are in the same direction, out of the plane of the figure. Show that the magnetic field in the plane defined by the wires is

$$B(x) = \frac{\mu_0 i x}{\pi(x^2 - d^2)}.$$

Assume $i = 10$ A and $d = 2.0$ cm in Fig. 6a and plot $B(x)$ for the range -2 cm $< x < +2$ cm. Assume that the wire diameters are negligible.

34P. Four long copper wires are parallel to each other, their cross section forming a square 20 cm on edge. A 20-A current is set up in each wire in the direction shown in Fig. 40. What are the magnitude and direction of **B** at the center of the square?

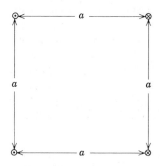

Figure 40 Problems 34, 35, and 36.

35P. Suppose, in Fig. 40, that the currents are all out of the page. What is the force per unit length (magnitude and direction) on any one wire? In the case of parallel motion of charged particles in a plasma, this is known as the pinch effect.

36P. In Fig. 40 what is the force per unit length acting on the lower left wire, in magnitude and direction?

37P. Two long wires a distance d apart carry equal antiparallel currents i, as in Fig. 41. (a) Show that **B** at point P, which is equidistant from the wires, is given by

$$B = \frac{2\mu_0 i d}{\pi(4R^2 + d^2)}.$$

(b) In what direction does **B** point?

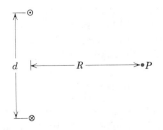

Figure 41 Problem 37.

38P. Figure 42 shows a long wire carrying a current of 30 A. The rectangular loop carries a current of 20 A. Calculate the resultant force acting on the loop. Assume that $a = 1.0$ cm, $b = 8.0$ cm, and $L = 30$ cm.

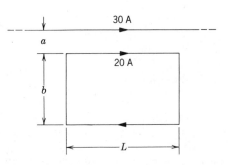

Figure 42 Problem 38.

39P. Figure 43 shows an idealized schematic of an "electromagnetic rail gun," designed to fire projectiles at speeds up to 10 km/s. (The feasibility of these devices as defenses against ballistic missiles is being studied.) The projectile P sits between and in contact with two parallel rails along which it can slide. A generator G provides a current that flows up one rail, across the projectile, and back down the other rail. (a) Let $w =$ distance between the rails, r the radius of the rails (presumed circular) and i the current. Show that the force on the projectile is to the right and given approximately by

$$F = \frac{1}{2}(i^2\mu_0/\pi)\ln\left(\frac{w + r}{r}\right).$$

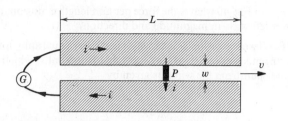

Figure 43 Problem 39.

(*b*) If the projectile (in this case a test slug) starts from the left end of the rail at rest, find the speed *v* at which it is expelled at the right. Assume that $i = 450$ kA, $w = 12$ mm, $r = 6.7$ cm, $L = 4.0$ m, and that the mass of the slug is $m = 10$ g.

Section 31–5 Ampere's Law

40E. Each of the indicated eight conductors in Fig. 44 carries 2.0 A of current into or out of the page. Two paths are indicated for the line integral $\oint \mathbf{B} \cdot d\mathbf{s}$. What is the value of the integral for (*a*) the dotted path? (*b*) the dashed path?

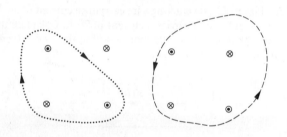

Figure 44 Exercise 40.

41E. Eight wires cut the page perpendicularly at the points shown in Fig. 45. A wire labeled with the integer *k* ($k = 1$, $2, \ldots, 8$) bears the current ki_0. For those with odd *k*, the current is out of the page; for those with even *k* it is into the page. Evaluate $\oint \mathbf{B} \cdot d\mathbf{s}$ along the closed path shown in the direction shown.

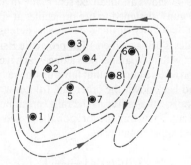

Figure 45 Exercise 41.

42E. Figure 46 shows a cross section of a long cylindrical conductor of radius *a*, carrying a uniformly distributed current *i*. Assume $a = 2.0$ cm and $i = 100$ A and plot $B(r)$ over the range $0 < r < 6.0$ cm.

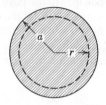

Figure 46 Exercise 42.

43E. In a certain region there is a uniform current density of 15 A/m² in the positive *z* direction. What is the value of $\oint \mathbf{B} \cdot d\mathbf{s}$ when the line integral is taken along the four straight-line segments from $(4d, 0, 0)$ to $(4d, 3d, 0)$ to $(0, 0, 0)$ to $(4d, 0, 0)$, where $d = 20$ cm?

44P. Two square conducting loops carry currents of 5.0 A and 3.0 A as shown in Fig. 47. What is the value of the line integral $\oint \mathbf{B} \cdot d\mathbf{s}$ for each of the two closed paths shown?

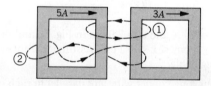

Figure 47 Problem 44.

45P. Show that a uniform magnetic field **B** cannot drop abruptly to zero as one moves at right angles to it, as suggested by the horizontal arrow through point *a* in Fig. 48. (*Hint:* Apply Ampere's law to the rectangular path shown by the dashed lines.) In actual magnets "fringing" of the lines of **B** always occurs, which means that **B** approaches zero in a grad-

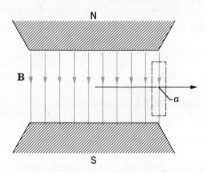

Figure 48 Problem 45.

ual manner. Modify the **B** lines in the figure to indicate a more realistic situation.

46P. Figure 49 shows a cross section of a hollow cylindrical conductor of radii a and b, carrying a uniformly distributed current i. (a) Show that $B(r)$ for the range $b < r < a$ is given by

$$B(r) = \frac{\mu_0 i}{2\pi(a^2 - b^2)}\left(\frac{r^2 - b^2}{r}\right).$$

(b) Test this formula for the special cases of $r = a$, $r = b$, and $b = 0$. (c) Assume $a = 2.0$ cm, $b = 1.8$ cm, and $i = 100$ A and plot $B(r)$ for the range $0 < r < 6$ cm.

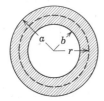

Figure 49   Problem 46.

47P. Figure 50 shows a cross section of a long conductor of a type called a coaxial cable. Its radii (a, b, c) are shown in the figure. Equal but opposite currents i exist in the two conductors. Derive expressions for $B(r)$ in the ranges (a) $r < c$, (b) $c < r < b$, (c) $b < r < a$, and (d) $r > a$. (e) Test these expressions for all the special cases that occur to you. (f) Assume $a = 2.0$ cm, $b = 1.8$ cm, $c = 0.40$ cm, and $i = 120$ A and plot $B(r)$ over the range $0 < r < 3$ cm.

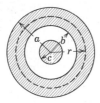

Figure 50   Problem 47.

48P. The current density inside a long, solid, cylindrical wire of radius a is in the direction of the axis and varies linearly with radial distance r from the axis according to $J = J_0 r/a$. Find the magnetic field inside the wire.

49P. A long circular pipe, with an outside radius of R, carries a (uniformly distributed) current of i_0 (into the paper as shown in Fig. 51). A wire runs parallel to the pipe at a distance of $3R$ from center to center. Calculate the magnitude and direction of the current in the wire that would cause the resultant magnetic field at the point P to have the same magnitude, but the

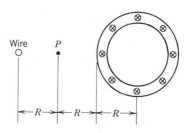

Figure 51   Problem 49.

opposite direction, as the resultant field at the center of the pipe.

50P. Figure 52 shows a cross section of a long cylindrical conductor of radius a containing a long cylindrical hole of radius b. The axes of the two cylinders are parallel and are a distance d apart. A current i is uniformly distributed over the cross-hatched area in the figure. (a) Use superposition ideas to show that the magnetic field at the center of the hole is

$$B = \frac{\mu_0 i d}{2\pi(a^2 - b^2)}.$$

(b) Discuss the two special cases $b = 0$ and $d = 0$. (c) Can you use Ampere's law to show that the magnetic field in the hole is uniform? (*Hint:* Regard the cylindrical hole as filled with two equal currents moving in opposite directions, thus canceling each other. Each of these currents must have the same current density as that in the actual conductor. Thus we superimpose the fields due to two complete cylinders of current, of radii a and b, each cylinder having the same current density.)

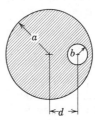

Figure 52   Problem 50.

51P. Figure 53 shows a cross section of an infinite conducting sheet with a current per unit x-length λ emerging from the page at right angles. (a) Use the right-hand rule and symmetry arguments to convince yourself that the magnetic field **B** is constant for all points P above the sheet (and for all points P' below it) and is directed as shown. (b) Use Ampere's law to prove that $B = \frac{1}{2}\mu_0\lambda$.

52P*. In a certain region there is a magnetic field given in mT by **B** = 3**i** + 8(x^2/d^2)**j** where x is the x-coordinate distance in

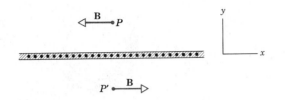

Figure 53 Problem 51.

meters and d is a constant, with units of length. Some current must be flowing in the region to cause the specified **B** field. (a) Evaluate the integral $\oint \mathbf{B} \cdot d\mathbf{s}$ along the straight path from $(d, 0, 0)$ to $(d, d, 0)$. (b) Let $d = 0.5$ m in the expression for **B** and apply Ampere's law to determine what current flows through a square of side length $d (= 0.5$ m$)$ that lies in the first quadrant of the xy plane with one corner at the origin. (c) Is this current in the **k** or $-$**k** direction?

Section 31-6 Solenoids and Toroids

53E. A solenoid 95 cm long has a radius of 2.0 cm, a winding of 1200 turns and carries a current of 3.6 A. Calculate the strength of the magnetic field inside the solenoid.

54E. A 200-turn solenoid having a length of 25 cm and diameter of 10 cm carries a current of 0.30 A. Calculate the magnitude of the magnetic field **B** near the center of the solenoid.

55E. A solenoid 1.3 m long and 2.6 cm in diameter carries a current of 18 A. The magnetic field inside the solenoid is 23 mT. Find the length of the wire forming the solenoid.

56E. A toroid having a square cross section, 5.0 cm on edge, and an inner radius of 15 cm has 500 turns and carries a current of 0.80 A. What is the magnetic field inside the toroid at (a) the inner radius and (b) the outer radius of the toroid?

57E. Show that if the thickness of a toroid is very small compared to its radius of curvature (very skinny toroid), then Eq. 21 for the field inside a toroid reduces to Eq. 20 for the field inside a solenoid. Explain why this result is expected.

58P. Treat a solenoid as a continuous cylindrical current sheet, whose current per unit length, measured parallel to the cylinder axis, is λ. B inside a solenoid is given by Eq. 21 ($B = \mu_0 i_0 n$). Show that this can also be written as $B = \mu_0 \lambda$. This is the value of the *change* in **B** that you encounter as you move from inside the solenoid to outside, through the solenoid wall. Show that this same change occurs as you move through an infinite plane current sheet such as that of Fig. 53 (see Problem 51). Does this equality surprise you?

59P. In Section 6 we showed that the magnetic field at any radius r inside a toroid is given by

$$B = \frac{\mu_0 i_0 N}{2\pi r}.$$

Show that as you move from a point just inside a toroid to a

point just outside the magnitude of the *change* in **B** that you encounter—at any radius r—is just $\mu_0 \lambda$. Here λ is the current per unit length for any circumference ($= 2\pi r$) of the toroid. Compare the similar result found in Problem 58. Is the equality surprising?

60P. A long solenoid with 10 turns/cm and a radius of 7.0 cm carries a current of 20 mA. A current of 6.0 A flows in a straight conductor along the axis of the solenoid. (a) At what radial distance from the axis will the direction of the resulting **B** field be at 45° from the axial direction? (b) What is the magnitude of the magnetic field?

61P. A long solenoid has 100 turns per centimeter and carries current i. An electron moves within the solenoid in a circle of radius 2.30 cm perpendicular to the solenoid axis. The speed of the electron is $0.046c$ (c = speed of light). Find the current i in the solenoid.

62P. An interesting (and frustrating) effect occurs when one attempts to confine a collection of electrons and positive ions (a plasma) in the magnetic field of a toroid. Particles whose motion is perpendicular to the **B** field will not execute circular paths because the field strength varies with radial distance from the axis of the toroid. This effect, which is shown (exaggerated) in Fig. 54, causes particles of opposite sign to drift in opposite directions parallel to the axis of the toroid. (a) What is the sign of the charge on the particle whose path is sketched in the figure? (b) If the particle path has a radius of curvature of 11 cm when its radial distance from the axis of the toroid is 125 cm, what will be the radius of curvature when the particle is 110 cm from the axis?

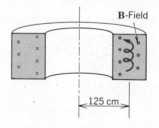

Figure 54 Problem 62.

Section 31-7 A Current Loop as a Magnetic Dipole

63E. What is the magnetic dipole moment μ of the solenoid described in Exercise 54?

64E. Figure 55a shows a length of wire carrying a current i and bent into a circular coil of one turn. In Fig. 55b the same length of wire has been bent more sharply, to give a double loop of smaller radius. (a) If B_a and B_b are magnitudes of the magnetic fields at the centers of the two loops, what is the ratio B_b/B_a? (b) What is the ratio of their dipole moments, μ_b/μ_a?

65E. Figure 56 shows an arrangement known as a Helmholtz coil: It consists of two circular coaxial coils each of N turns and

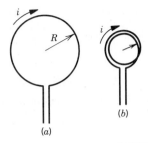

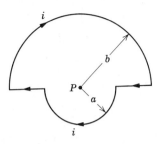

Figure 55 Exercise 64.

Figure 57 Problem 68.

radius R, separated by a distance R. They carry equal currents i in the same direction. Find the magnetic field at P, midway between the coils.

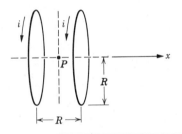

Figure 56 Exercise 65; Problems 69 and 70.

66E. A student makes an electromagnet by winding 300 turns of wire around a wooden cylinder of diameter 5.0 cm. The coil is connected to a battery producing a current of 4.0 A in the wire. (*a*) What is the magnetic moment of this device? (*b*) At what axial distance $R \gg d$ will the magnetic field of this dipole be 5.0 μT (approximately one-tenth that of the earth's magnetic field)?

67E. The magnetic field $B(x)$ for various points on the axis of a square current loop of side a is given in Problem 20. (*a*) Show that the axial field for this loop for $x \gg a$ is that of a magnetic dipole (see Eq. 25). (*b*) What is the magnetic dipole moment of this loop?

68P. You are given a closed circuit with radii a and b, as shown in Fig. 57, carrying a current i. (*a*) What are the magnitude and direction of **B** at point P? (*b*) Find the magnetic dipole moment of the circuit.

69P. Two 300-turn coils each carry a current i. They are arranged a distance apart equal to their radius, as in Fig. 56. For $R = 5.0$ cm and $i = 50$ A, plot B as a function of distance x along the common axis over the range $x = -5$ cm to $x = +5$ cm, taking $x = 0$ at the midpoint P. (Such coils provide an especially uniform field B near point P.) (*Hint:* See Eq. 24.)

70P. In Exercise 65 (Fig. 56) let the separation of the coils be a variable s (not necessarily equal to the coil radius R). Show that the first derivative of the magnetic field (dB/dx) vanishes at the midpoint P regardless of the value of s. Why would you expect this to be true from symmetry? (*b*) Show that the second derivative of the magnetic field (d^2B/dx^2) also vanishes at P provided $s = R$. This accounts for the uniformity of B near P for this particular coil separation.

71P. A circular loop of radius 12 cm carries a current of 15 A. A second loop of radius 0.82 cm, having 50 turns and a current of 1.3 A is at the center of the first loop. (*a*) What magnetic field **B** does the large loop set up at its center? (*b*) What torque acts on the small loop? Assume that the planes of the two loops are at right angles and that the magnetic field due to the large loop is essentially uniform throughout the volume occupied by the small loop.

72P. A conductor carries a current of 6.0 A along the closed path *abcdefgha* involving 8 of the 12 edges of a cube of side 10 cm as shown in Fig. 58. (*a*) Why can one regard this as the superposition of three square loops: *bcfgb*, *abgha*, and *cdefc*? (*b*) Find the magnetic dipole moment $\boldsymbol{\mu}$ (magnitude and direction) of this current. (*c*) Calculate **B** at the points $(x, y, z) = (0, 5$ m, $0)$ and $(5$ m, $0, 0)$.

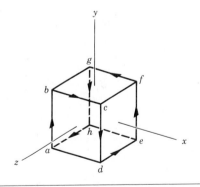

Figure 58 Problem 72.

73P. (*a*) A long wire is bent into the shape shown in Fig. 59, without cross contact at P. The radius of the circular section is

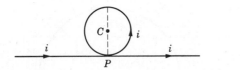

Figure 59 Problem 73.

R. Determine the magnitude and direction of **B** at the center *C* of the circular portion when the current *i* is as indicated. (*b*) The circular part of the wire is rotated without distortion about its (dashed) diameter perpendicular to the straight portion of the wire. The magnetic moment associated with the circular loop is now in the direction of the current in the straight part of the wire. Determine **B** at *C* in this case.

74P*. A thin plastic disk of radius *R* has a charge *q* uniformly distributed over its surface. If the disk is rotated at an angular frequency ω about its axis, show that (*a*) the magnetic field at the center of the disk is

$$B = \frac{\mu_0 \omega q}{2\pi R}$$

and (*b*) the magnetic dipole moment of the disk is

$$\mu = \frac{\omega q R^2}{4}.$$

(*Hint:* The rotating disk is equivalent to an array of current loops.)

75P*. Derive the solenoid equation (Eq. 21) starting from the expression for the field on the axis of a circular loop (Eq. 24). (*Hint:* Subdivide the solenoid into a series of current loops of infinitesimal thickness and integrate. See Fig. 16.)

CHAPTER 32

FARADAY'S LAW OF INDUCTION

Faraday's law of induction is closer than you may think. Behind these slots are copper wires that may snake out of your house and over the countryside for hundreds of miles. If you could follow them, you would almost certainly find yourself at an electric generator, in which a potential difference is established between the wires by Faraday's law. On your way from the wall plug to the generator, you would have to portage around several transformers, in which Faraday's law is again at work, stepping up or down the potential difference between the wires.

32-1 Two Symmetries

If you put a closed conducting loop in an external magnetic field and send a current through it, a torque will act on the loop tending to cause it to rotate.* We summarize this as follows:

$$\text{current} \rightarrow \text{torque} \quad \text{(loop in magnetic field)}. \quad (1)$$

Equation 1 is the principle of the electric motor.

Symmetry—on which we lean so much in physics

* See Section 30-8.

—compels us to ask: "What if we try it the other way around? Suppose that we put a closed conducting loop in an external magnetic field and rotate it by exerting a torque on it from some external source. Will an electric current appear in the loop?" That is, can we write

$$\text{torque} \rightarrow \text{current} \quad \text{(loop in magnetic field)}? \quad (2)$$

This does indeed happen! It is the principle of the electric generator. The law that governs the appearance of this current is called *Faraday's law of induction*. Physicists have learned to be guided in thinking about nature by such pleasing symmetries as those displayed by Eqs. 1

and 2. Symmetry is beauty, in physics as well as in art.

There is another symmetry — and a curious one — at the human level associated with the law of induction. This law was discovered in 1831 by Michael Faraday and also, independently and at about the same time, by the American physicist Joseph Henry. The self-educated Faraday was apprenticed at age 14 to a London bookbinder. He wrote: "There were plenty of books there and I read them." Henry was apprenticed at age 13 to a watchmaker in Albany, New York.

In later years, Faraday was appointed Director of the Royal Institution in London, whose founding was due in large part to an American, Benjamin Thomson (Count Rumford). Henry, on the other hand, became Secretary (that is, Director) of the Smithsonian Institution in Washington, DC, which was founded by an endownment from an Englishman, James Smithson.

Even though Faraday published his results first, which gave him priority of discovery, the SI unit of inductance (see Chapter 33) is called the *henry* (abbr. H). On the other hand, the SI unit of capacitance, as we have seen, is called the *farad* (abbr. F). In Chapter 35 we shall study the *electromagnetic oscillator,* in which energy is shuttled back and forth between inductive and capacitive forms, just as it is between kinetic and potential forms in an oscillating pendulum. It is pleasant to see the names of these two gifted scientists so closely interwoven in the operation of the electromagnetic oscillator, a device made possible by their joint discovery of electromagnetic induction.

32–2 Two Experiments

Two simple tabletop experiments will guide us to an understanding of Faraday's law of induction.

First Experiment. Figure 1 shows the terminals of a wire loop connected to a sensitive galvanometer G that can detect the presence of a current in the loop. Normally, we would not expect this meter to deflect because there is no battery in the circuit. However, if you push a bar magnet toward the loop a curious thing happens. While the magnet is moving (and *only* while it is moving) the meter deflects, showing that a current has been set up in the coil. Furthermore, the faster you move the magnet, the greater the deflection. When you stop moving the magnet, the deflection stops and the needle of the meter returns to zero. If you move the magnet *away* from the loop, the meter again deflects while the magnet

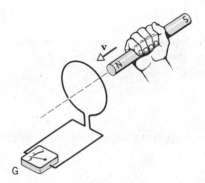

Figure 1 Galvanometer G deflects when the magnet is moving with respect to the coil.

is moving, but in the opposite direction, which tells us that the current in the loop is in the opposite direction.

If you reverse the magnet end for end, so that the south pole (rather than the north pole) faces the loop, the experiments work just as before except that the directions of the deflections are reversed. Further tests would convince you that what matters is the relative motion of the magnet and the loop. It makes no difference whether you move the loop toward the magnet or the magnet toward the loop.

The current that appears in the loop in this experiment is called an *induced current* and we say that it is set up by an *induced electromotive force.* Such induced electromotive forces play an important part in our daily lives. The chances are good that the lights in the room in which you are reading this book are operated by an induced electromotive force produced in a commercial electric generator.

Second Experiment. Here we use the apparatus of Fig. 2. The loops are close to each other but remain at rest and have no direct electrical contact. If you close switch

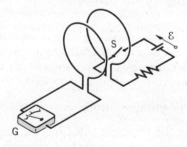

Figure 2 Galvanometer G deflects momentarily when switch S is closed or opened. No physical motion of the coils is involved.

S, thus setting up a current in the right-hand loop, the meter in the left-hand loop deflects momentarily and then returns to zero. If you than open the switch, thus interrupting this current, the meter again deflects but in the opposite direction. None of the apparatus is physically moving in this experiment.

Only when the current in the right-hand loop is rising or falling does an induced electromotive force appear in the left-hand loop. When there is a steady current in the right-hand loop, there is no induced electromotive force, no matter how large that steady current may be.

In thinking about these two experiments, we are led to conclude the following:

An induced electromotive force appears only when something is changing. In a static situation, in which no physical objects are moving and the currents are steady, there is no induced electromotive force. The key word is change.

32-3 Faraday's Law of Induction

Faraday had the insight to see the common feature of the two experiments described in the preceding section. In language that he might have used:

An induced emf appears in the left-hand loop of Figs. 1 and 2 only when the number of magnetic field lines that pass through that loop is changing.

The actual *number* of lines of the magnetic field that passes through the left-hand loop at any moment is of no concern; it is the *rate at which this number is changing* that determines the induced electromotive force.

In the experiment of Fig. 1, the lines originate in the bar magnet and the number of lines passing through the left-hand loop increased when you brought the magnet closer and decreased when you pulled the magnet away.

In the experiment of Fig. 2, the magnetic lines are associated with the current in the right-hand loop. The number of lines through the left-hand loop increased (from zero) when you closed switch S and decreased (back to zero) when you then opened this switch.

A Quantitative Treatment.   Consider a surface— which may or may not be plane—bounded by a closed loop. We represent the number of magnetic lines that pass through that surface by the *magnetic flux Φ_B* for

that surface, where Φ_B is defined from*

$$\boxed{\Phi_B = \oint \mathbf{B} \cdot d\mathbf{A}} \quad \text{(magnetic flux defined).} \quad (3)$$

Here $d\mathbf{A}$ is a differential element of surface area and the integration is to be carried out over the entire surface.

If the magnetic field has a constant magnitude B and is everywhere at right angles to a plane surface of area A, Eq. 3 reduces to

$$\boxed{\Phi_B = BA} \quad \text{(special case of Eq. 3),} \quad (4)$$

in which Φ_B is the absolute value of the flux.

From Eqs. 3 and 4 we see that the SI unit for the magnetic flux is the tesla·meter^2, to which we give the name *weber* (abbr. Wb). That is,

$$1 \text{ weber} = 1 \text{ Wb} = 1 \text{ T·m}^2. \quad (5)$$

Having established the definition of magnetic flux, we are now ready to state Faraday's law of induction in a quantitative way:

The induced emf in a circuit is equal (except for a change in sign) to the rate at which the magnetic flux through that circuit is changing with time.

In equation form this law becomes

$$\boxed{\mathcal{E} = -\frac{d\Phi_B}{dt}} \quad \text{(Faraday's law).} \quad (6)$$

If the rate of change of flux is in webers per second, the induced emf will be in volts. The minus sign has to do with the direction of the induced emf, as we explain at the end of this section.

If we change the magnetic flux through a coil of N turns, an induced emf appears in every turn and these emfs—like those of batteries connected in series—are to be added. If the coil is so tightly wound that each turn can be said to occupy the same region of space, the flux

---

* This definition of the flux of the magnetic field **B** is just like the definition of the flux of the electric field **E** given in Section 25-4 (which you may wish to review) or the flux of the current density **J** given by Eq. 5 of Chapter 28. Incidentally, we may add magnetic flux to our growing list of physical quantities that are quantized when examined on a fine enough scale; see Essay 12, *Superconductivity,* by Peter Lindenfeld for some details.

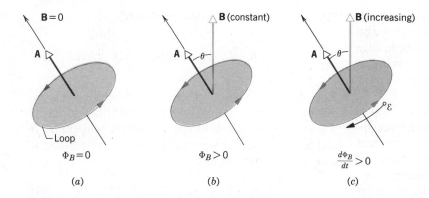

Figure 3 (*a*) A flat surface of area A bounded by a conducting loop. The positive direction of circulation around the loop is related to the vector **A** by the right-hand rule. (*b*) A steady magnetic field now pierces the loop; the magnetic flux through the loop is positive. (*c*) The magnitude of the magnetic field increases with time. The induced emf is in the negative direction of circulation, as shown.

Φ_B through each turn will be the same. Such coils are described as *close packed*. The flux through each turn is also the same for (ideal) solenoids and toroids (see Section 31–6). The induced electromotive force in all such arrangements is then

$$\mathcal{E} = -N\frac{d\Phi_B}{dt} \quad \text{(Faraday's law).} \quad (7)$$

A Word About Signs (Optional). The minus signs in Eqs. 6 and 7 guide us in finding the *direction* of the induced emf in a particular situation. Although we shall rely on Lenz's law (see Section 32–4) to give us this information, a well-trained physicist should also be able to deduce this direction in a formal manner directly from Faraday's law. Let us do so now.

We start with Eq. 3, the definition of the magnetic flux. Flux, which is a scalar, can be either positive or negative, a matter we have not yet addressed. In defining the flux of the *electric field* in connection with Gauss' law (see Section 25–4), we dealt only with closed surfaces and defined the direction of the area element $d\mathbf{A}$ uniquely, as the outward direction. In defining the flux of the magnetic field, however, we deal with a surface that is not closed and we must make an arbitrary decision as to which direction from that surface we take as the positive direction for the area element vector $d\mathbf{A}$.

In Fig. 3*a*, for example, we consider a plane surface of area A bounded by a loop and we arbitrarily take the direction shown in the figure as the direction of the area vector **A**. Our all-purpose right-hand rule then tells us that the direction around the loop indicated by the arrows will be our positive reference direction.

In Fig. 3*b*, we show a (constant) magnetic field **B** enclosed by the loop, the direction of **B** making an angle θ with the area vector **A**, where $\theta < 90°$. From Eq. 3 then, the flux Φ_B enclosed by the loop is positive; that is, $\Phi_B > 0$. If now we allow the strength of the magnetic

field to *increase* with time, as in Fig. 3*c*, the flux through the loop will also increase with time. That is, $d\Phi_B/dt$ is also a *positive* quantity. From Faraday's law (Eq. 6), the minus sign requires that the induced emf be *negative*, which corresponds to the direction shown for the emf arrow in Fig. 3*c*. For practice, you might work out the direction of the induced emf assuming that **B** is decreasing in magnitude or that **B** points in the opposite direction to that shown in Figs. 3*b* and 3*c* and is either increasing or decreasing in magnitude.

Sample Problem 1 The long solenoid S of Fig. 4 has 220 turns/cm and carries a current $i = 1.5$ A; its diameter d is 3.2 cm. At its center we place a 130-turn close-packed coil C of diameter 2.1 cm. The current in the solenoid is reduced to zero and then increased to 1.5 A in the other direction at a steady rate over a period of 50 ms. What is the absolute value (that is, the magnitude without regard for sign) of the induced emf that appears in the central coil while the current in the solenoid is being changed?

The induced emf follows from Faraday's law (Eq. 7), in which we ignore the minus sign because we are concerned only with the absolute value of the emf. Thus,

$$\mathcal{E} = \frac{N\,\Delta\Phi_B}{\Delta t}$$

in which N is the number of turns in the inner coil C.

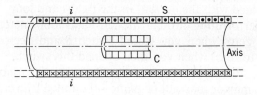

Figure 4 Sample Problem 1. A coil C is located inside a solenoid S. When the current in the solenoid is changed, an induced emf appears in the coil.

The absolute value of the initial flux through each turn of this coil is given from Eq. 4 as

$$\Phi_B = BA.$$

The magnetic field B at the center of the solenoid is given by Eq. 21 of Chapter 31 or

$$B = \mu_0 in$$

$$= (4\pi \times 10^{-7} \text{ T} \cdot \text{m/A})(1.5 \text{ A})(220 \text{ turns/cm})(100 \text{ cm/m})$$

$$= 4.15 \times 10^{-2} \text{ T}.$$

The area of the central coil (not of the solenoid) is given by $\frac{1}{4}\pi d_C^2$, where d_C is the diameter of coil C, and works out to be $3.46 \times 10^{-4} \text{ m}^2$. The absolute value of the initial flux through each turn of the coil is then

$$\Phi_B = (4.15 \times 10^{-2} \text{ T})(3.46 \times 10^{-4} \text{ m}^2)$$

$$= 1.44 \times 10^{-5} \text{ Wb} = 14.4 \ \mu\text{Wb}.$$

The flux changes sign as the current is changed so that the absolute value of the *change* in flux $\Delta\Phi_B$ for each turn of the central coil is thus $2 \times 14.4 \ \mu\text{Wb}$ or $28.8 \ \mu\text{Wb}$. This change occurs in 50 ms, giving for the magnitude of the induced emf,

$$\mathscr{E} = \frac{N \Delta\Phi_B}{\Delta t} = \frac{(130 \text{ turns})(28.8 \times 10^{-6} \text{ Wb})}{50 \times 10^{-3} \text{ s}}$$

$$= 7.5 \times 10^{-2} \text{ V} = 75 \text{ mV}. \qquad \text{(Answer)}$$

We shall explain in the next section how to use Lenz's law to find the *direction* of the induced emf.

32-4   Lenz's Law

If you drop a hammer from the top of a ladder, you do not need a rule for signs to tell you whether the hammer will move toward the center of the earth or in the opposite direction. If asked how you know that the hammer will fall, your best answer is: "It always has in the past." If pressed for a more formal answer, you might say: "If it falls, gravitational potential energy decreases and kinetic energy increases, which is acceptable. If the hammer were to rise, however, its potential energy and its kinetic energy would *both* increase, a violation of the conservation of energy principle." In this section, we want to apply this kind of reasoning to find the direction in which an induced emf will act. That is, it will act in a direction that is consistent with the conservation of energy principle.

In 1834, just 3 years after Faraday put forward his law of induction, Heinrich Friedrich Lenz gave us this rule (known as *Lenz's law*) for determining the direction

of an induced current in a closed conducting loop:

An induced current in a closed conducting loop will appear in such a direction that it opposes the change that produced it.

The minus sign in Faraday's law carries with it this symbolic notion of opposition.

Lenz's law refers to induced *currents* and not to induced emfs, which means that we can apply it directly only to closed conducting loops. If the loop is not closed, however, we can usually think in terms of what would happen if it were closed and in this way find the direction of the induced emf.

To understand Lenz's law, let us apply it to a specific case, namely, the first of Faraday's experiments, shown in Fig. 1. We interpret Lenz's law as applied to this experiment in two different but equivalent ways.

First Interpretation.   A current loop, like a bar magnet, has a north pole and a south pole. In each case the north pole is the region *from which* the magnetic lines emerge. If the loop is to oppose the motion of the magnet toward it, the face of the loop toward the magnet must become a north pole; see Fig. 5. The two north poles — one of the current loop and one of the magnet — will then repel each other. The right-hand rule applied to the wire of the loop shows that to produce a north pole on the face of the loop toward the magnet, the current must be as shown. Specifically, the current will be counterclockwise as we sight along the magnet toward the loop.

In the language of Lenz's law, the pushing of the magnet is the "change" that produces the induced current and that current acts to oppose the "push." If you pull the magnet away from the coil, the induced current will oppose the "pull" by creating a south pole on the face of the loop toward the magnet. This would require that the direction of the induced current be reversed.

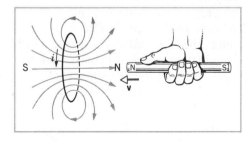

Figure 5   Lenz's law at work. If you push the magnet toward the loop, the induced current points as shown, setting up a magnetic field that opposes the motion of the magnet.

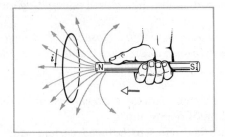

Figure 6 Lenz's law at work. If you push the magnet toward the loop, you increase the magnetic flux through the loop. The induced current in the loop sets up a magnetic field that opposes this increase in flux.

Whether you push or pull the magnet, the motion will always be opposed.

Second Interpretation. Let us now apply Lenz's law to the experiment of Fig. 1 in a different way. Figure 6 shows the lines of **B** for the bar magnet.* From this point of view the "change" referred to in Lenz's law is the increase in Φ_B through the loop caused by bringing the magnet closer. The induced current opposes this change by setting up a field of its own that opposes this increase. Thus, the magnetic field caused by the induced current must point from left to right through the plane of the loop of Fig. 6, in agreement with our earlier conclusion.

The induced magnetic field does not intrinsically oppose the *magnetic field* of the magnet; it opposes the *change* in this field, which in this case is the increase in magnetic flux through the loop. If you withdraw the magnet, you reduce Φ_B through the loop. The induced magnetic field will now oppose this decrease in Φ_B (that is, the change) by *reinforcing* the magnetic field. In each case the induced field opposes the change that gives rise to it.

Lenz's Law and Energy Conservation. Think what would happen if Lenz's law were turned the other way around, that is, if the induced current acted to *aid* the change that produced it. That would mean, for example, that a *south* pole would appear on the face of the loop in Fig. 5 as you pushed the *north* pole of a magnet toward it.

You would then only need to push a resting magnet slightly to get it moving and the action would be self-perpetuating. The magnet would accelerate toward the

loop, gaining kinetic energy as it did so. At the same time thermal energy would appear in the loop. This would indeed be a something-for-nothing situation! Needless to say, it does not happen. Lenz's law is no more than a statement of the principle of conservation of energy in a form suitable for use in circuits in which there are induced currents.

Whether you push the magnet toward the loop in Fig. 1 or pull it away from the loop, you will always experience a resisting force and will thus have to do work. From the conservation of energy principle, this work must be exactly equal to the thermal energy that appears in the coil because these are the only two energy transfers that take place in this isolated system. The faster you move the magnet, the more rapidly you do work, and the greater the rate of production of thermal energy in the coil. If you cut the loop and then do the experiment, there will be no induced current, no thermal energy, no resisting force on the magnet, and no work required to move it. There will still be an emf in the loop but, like a battery in an open circuit, it will not set up a current.

32-5 Induction: A Quantitative Study

Figure 7 shows a rectangular loop of wire of width L, one end of which is in a uniform external magnetic field that is directed into the plane of the loop at right angles. This field may be produced, for example, by a large electromagnet. The dashed lines in Fig. 7 show the assumed limits of the magnetic field, the fringing of the field at its edges being neglected. You are asked to pull this loop to the right at a constant speed v.

The setup of Fig. 7 does not differ in any essential way from the setup of Fig. 6. In each case a magnet and a conducting loop are in relative motion; in each case the flux of the field through the loop is changing with time. It is true that in Fig. 6 the flux is changing because **B** is changing and in Fig. 7 the flux is changing because the effective loop area is changing but that difference is not essential. The important difference between the two arrangements for our purposes is that the arrangement of Fig. 7 is easier to calculate. Let us then calculate the rate at which you do mechanical work as you pull steadily on the loop in Fig. 7.

The Mechanical Work. If we know both the force F that you must exert and the speed v with which you pull the loop, we can calculate the rate at which you must do

---

* There are two magnetic fields in this experiment—one associated with the current in the loop and one associated with the bar magnet. You must always be certain with which one you are dealing.

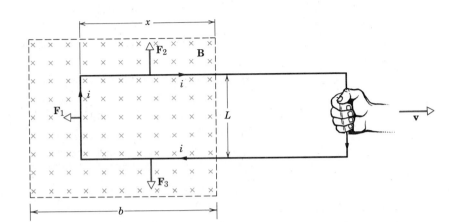

Figure 7 You pull a closed conducting loop out of a magnetic field at constant speed. While the loop is moving, a clockwise induced current appears in the loop. Thermal energy appears in the loop at a rate equal to the rate at which mechanical work is done on the loop.

work. The rate is

$$P = Fv. \qquad (8)$$

Because the current-carrying loop is immersed in an external magnetic field, forces $\mathbf{F}_1$, $\mathbf{F}_2$, and $\mathbf{F}_3$ will act on three of the conductors that make up the loop, as Fig. 7 shows. Their magnitudes and directions follow from Eq. 24 of Chapter 30, or

$$\mathbf{F} = i\,\mathbf{L} \times \mathbf{B}.$$

Because $\mathbf{F}_2$ and $\mathbf{F}_3$ are equal and opposite, they cancel. $\mathbf{F}_1$ is the force that opposes your effort to move the loop. F is then given in magnitude by

$$F = F_1 = iLB \sin 90° = iLB. \qquad (9)$$

The induced current in the loop is given by

$$i = \frac{\mathcal{E}}{R}, \qquad (10)$$

where R is the loop resistance.

From Lenz's law, this current (and thus also $\mathcal{E}$) must be clockwise in Fig. 7. By circulating with this sense, it sets up a magnetic field within the loop that opposes the "change," that is, the decrease in Φ_B caused by pulling the loop out of the field.

We can now use Faraday's law to find the induced emf in the loop; from Eq. 7 (neglecting the minus sign because we are only interested in the absolute value of the emf) we have

$$\mathcal{E} = \frac{d\Phi_B}{dt}.$$

The absolute value of the flux ϕ_B enclosed by the loop in Fig. 7 is

$$\Phi_B = BLx,$$

where Lx is the area of that part of the loop in which the magnetic field is not zero. The emf is then

$$\mathcal{E} = \frac{d}{dt} BLx = BL \frac{dx}{dt} = BLv,$$

in which we have replaced dx/dt by v, the speed at which you are pulling the loop out of the field. Note that if v is constant—which we assume—the induced emf will also be constant.

From Eq. 10, the induced current in the loop is

$$i = \frac{\mathcal{E}}{R} = \frac{BLv}{R}. \qquad (11)$$

The force that you would have to exert is

$$F = iLB = \frac{B^2 L^2 v}{R} \qquad (12)$$

and the rate that you would do work is then, from Eq. 8,

$$P = Fv = \frac{B^2 L^2 v^2}{R} \quad \text{(rate of doing work).} \qquad (13)$$

The Thermal Energy. Now let us find the rate at which thermal energy appears in the loop as you pull it along at constant speed. We calculate it from Eq. 21 of Chapter 28 or

$$P = i^2 R. \qquad (14)$$

Substituting for i from Eq. 11, we find

$$P = \left(\frac{BLv}{R}\right)^2 R = \frac{B^2 L^2 v^2}{R} \quad \text{(thermal energy rate),} \qquad (15)$$

which is exactly equal to the rate at which you are doing work on the loop; see Eq. 13. Thus, the work that you do in pulling the loop through the magnetic field appears as

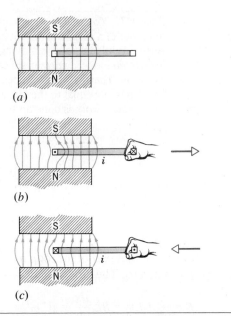

(a)

(b)

(c)

Figure 8 The pattern of the lines of force strongly suggests that any attempt to move the closed conducting loop, in either direction, will give rise to an opposing force. (Recall that field lines act somewhat like stretched rubber bands.)

thermal energy in the loop, manifesting itself as a small increase in the temperature of the loop.

Figure 8 shows a side view of the loop in the magnetic field. In Fig. 8a the loop is stationary; in Fig. 8b you are pulling it to the right; in Fig. 8c you are pushing it to the left. The magnetic field lines in these figures represent the *resultant* magnetic field produced by adding vectorially the external field of the electromagnet and the self-field — if any — set up by the induced current in the loop. The patterns of these lines suggest convincingly that, no matter which way you move the loop, you experience a resisting force.

Sample Problem 2 Suppose that the "loop" in Fig. 7 is actually a tightly wound coil of 85 turns, made of copper wire. Suppose further that $L = 13$ cm, $B = 1.5$ T, $R = 6.2$ Ω, and $v = 18$ cm/s. (a) What induced emf appears in the coil?

The induced emf appears in every turn of the coil so that, from Eq. 11, its absolute value is

$$\mathcal{E} = NBLv = (85 \text{ turns})(1.5 \text{ T})(0.13 \text{ m})(0.18 \text{ m/s})$$
$$= 2.98 \text{ V} \approx 3.0 \text{ V}. \qquad \text{(Answer)}$$

(b) What is the induced current?

We have

$$i = \frac{\mathcal{E}}{R} = \frac{2.98 \text{ V}}{6.2 \text{ Ω}} = 0.48 \text{ A}. \qquad \text{(Answer)}$$

(c) What force must you exert on the coil to pull it along? The force is given from Eq. 12 as

$$F = NiLB = (85 \text{ turns})(0.48 \text{ A})(0.13 \text{ m})(1.5 \text{ T})$$
$$= 8.0 \text{ N} \ (= 1.8 \text{ lb}). \qquad \text{(Answer)}$$

(d) At what rate must you do work to pull the coil along? The power you must exert follows from

$$P = Fv = (8.0 \text{ N})(0.18 \text{ m/s})$$
$$= 1.4 \text{ W}. \qquad \text{(Answer)}$$

Thermal energy appears in the loop at this same rate.

Sample Problem 3 Figure 9a shows a rectangular loop of resistance R, width L, and length b being pulled at constant speed v through a region of depth d in which a uniform magnetic field **B** is set up by an electromagnet. (a) Plot the flux Φ_B through the loop as a function of the position x of the right side of the loop. Assume that $L = 40$ mm, $b = 10$ cm, $d = 15$ cm, $R = 1.6$ Ω, $B = 2.0$ T, and $v = 1.0$ m/s.

The flux is zero when the loop is not in the field; it is BLb $(= 8$ mWb$)$ when the loop is entirely in the field; it is BLx when the loop is entering the field and $BL[b - (x - d)]$ when the loop is leaving the field. These conclusions, which you should verify, are plotted in Fig. 9b.

(b) Plot the induced emf as a function of the position of the loop.

The induced emf (see Eq. 6) is given by $-d\Phi_B/dt$, which we can write as

$$\mathcal{E} = -\frac{d\Phi_B}{dt} = -\frac{d\Phi_B}{dx}\frac{dx}{dt} = -\frac{d\Phi_B}{dx}v,$$

where $d\phi_B/dx$ is the slope of the curve of Fig. 9b. The emf is plotted as a function of x in Fig. 9c.

Lenz's law, from the arguments that we have given earlier, shows that when the loop is entering the field, the emf acts counterclockwise in Fig. 9a; when the loop is leaving the field, the emf is clockwise in that figure. There is no emf when the loop is either entirely out of the field or entirely in it because, in these two situations, the flux through the loop is not changing with time.

(c) Plot the rate of production of thermal energy in the loop as a function of the position of the loop.

This is given by $P = \mathcal{E}^2/R$. It may be calculated by squaring the ordinate of the curve of Fig. 9c and dividing by R. The result is plotted in Fig. 9d.

In practice, the external magnetic field B cannot drop sharply to zero at its boundary but must approach zero

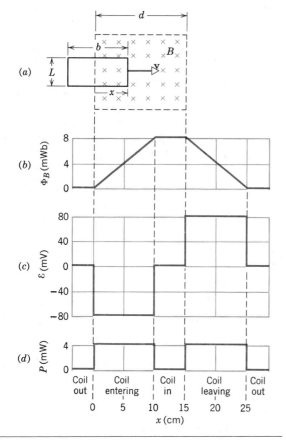

Figure 9 Sample Problem 3. (*a*) A closed conducting loop is pulled at constant speed completely through a magnetic field. (*b*) The flux through the loop for various positions of the right side of the loop. (*c*) The induced emf for various positions. (*d*) The rate at which thermal energy appears in the loop for various positions.

smoothly. The result will be a rounding of the corners of the curves plotted in Fig. 9. What changes would occur in these curves if the loop were cut, so that it no longer formed a closed conducting path?

32-6 Induced Electric Fields

Let us place a copper ring of radius r in a uniform external magnetic field, as in Fig. 10a. The field—neglecting fringing—fills a cylindrical volume of radius R. Suppose that you increase the strength of this field at a steady rate, perhaps by increasing—in an appropriate way—the current in the windings of the electromagnet that sets up

the field. The magnetic flux through the ring will then change at a steady rate and—by Faraday's law—an induced emf and thus an induced current will appear in the ring. From Lenz's law you can deduce that the direction of the induced current is counterclockwise in Fig. 10a.

If there is a current in the copper ring, an electric field must be present at various points within the ring, set up by the changing magnetic flux. This *induced electric field* **E** is just as real as an electric field set up by static charges; each field, no matter what its source, will exert a force $q_0\mathbf{E}$ on a test charge. By this line of reasoning, we are lead to a useful and informative restatement of Faraday's law of induction:

A changing magnetic field produces an electric field.

The striking feature of this statement is that the electric fields are induced even if there is no copper ring.

To fix these ideas, consider Fig. 10b, which is just like Fig. 10a except the copper ring has been replaced by a hypothetical circular path of radius r. We assume as before that the magnetic field **B** is increasing in magnitude at the same constant rate dB/dt.

The circular path in Fig. 10b encloses a magnetic flux Φ_B. Because this flux is changing with time, an induced emf given by $\mathcal{E} = -d\Phi_B/dt$ will appear around the path. The electric fields induced at various points around the circular path must—from symmetry—be tangent to the circle, as Fig. 10b shows.* Thus, the electric lines of force set up by the changing magnetic field are in this case a set of concentric circles, as in Fig. 10c.

As long as the magnetic field is increasing with time, the electric field represented by the circular lines of force in Fig. 10c will be present. If the magnetic field remains constant with time, there will be no induced electric field and thus no lines of force. If the magnetic field is *decreasing* with time (at the same rate), the lines of force will be as they are in Fig. 10c but will point in the opposite direction. All this is what we have in mind when we say: "A changing magnetic field produces an electric field."

A Reformulation of Faraday's Law. Consider a test charge q_0 moving around the circular path of Fig. 10b.

---

* Arguments of symmetry would also permit the lines of **E** around the circular path to be *radial*, rather than tangential. However, such radial lines would imply that there are free charges distributed symmetrically about the axis of symmetry; there are no such charges.

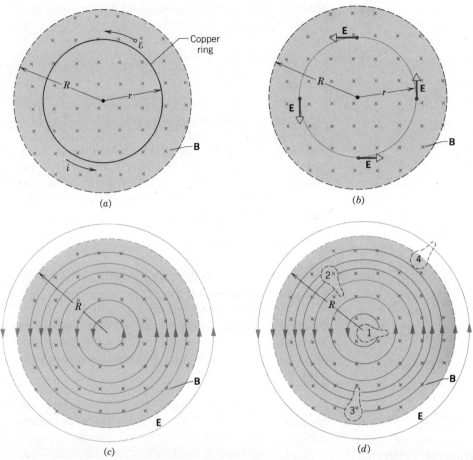

Figure 10 (*a*) If the magnetic field increases at a steady rate, a constant induced current appears, as shown, in the copper ring of radius *r*. (*b*) Induced electric fields appear at various points even when the ring is removed. (*c*) The complete picture of the induced electric fields, displayed as lines of force. (*d*) Four similar closed paths, around which an induced emf may—or may not—appear.

The work W done on it in one revolution by the electric field is, in terms of our definition of an emf, simply $\mathscr{E}q_0$. From another point of view, the work is $(q_0E)(2\pi r)$, where q_0E is the magnitude of the force acting on the test charge and $2\pi r$ is the distance over which that force acts. Setting these two expressions for W equal to each other and canceling q_0, we find

$$\mathscr{E} = E\, 2\pi r. \qquad (16)$$

In a more general case than that of Fig. 10*b*, we can write

$$\mathscr{E} = \oint \mathbf{E} \cdot d\mathbf{s}. \qquad (17)$$

This integral reduces at once to Eq. 16 if we evaluate it for the special case of Fig. 10*b*.

If we combine Eq. 17 with Eq. 6 ($\mathscr{E} = -\,d\Phi_B/dt$), we can write Faraday's law of induction as

$$\boxed{\oint \mathbf{E} \cdot d\mathbf{s} = -\frac{d\Phi_B}{dt}} \quad \text{(Faraday's law).} \quad (18)$$

This is the form in which Faraday's law is displayed in Table 2 of Chapter 37, this table being our summary of Maxwell's equations of electromagnetism. Equation 18 is again what we have in mind when we say: "A changing magnetic field produces an electric field." The changing magnetic field appears on the right side of this equation, the electric field on the left.

Faraday's law in the form of Eq. 18 can be applied to *any* closed path that can be drawn in a changing magnetic field. Figure 10*d*, for example, shows four such paths, all having the same shape and area but located in different positions in the changing field. For paths 1 and 2, the induced emf is the same because these paths lie entirely in the magnetic field and thus have the same value of $d\Phi_B/dt$. Note that, even though the emf $\mathscr{E}$ ($= \oint \mathbf{E} \cdot d\mathbf{s}$) is the same for these two paths, the distribution of the electric field vectors around these paths is different, as indicated by the pattern of electric lines of force. For path 3 the induced emf is smaller because Φ_B (and hence $d\Phi_B/dt$) is smaller, and for path 4 the induced emf is zero, even though the electric field is not zero at any point on the path.

A New Look at Electric Potential.   The induced electric fields that are set up by the induction process are not associated with static charges but with a changing magnetic flux. Although electric fields set up in either way exert forces on test charges, there is an important difference between them. The simplest evidence of this difference is that the lines of force associated with induced electric fields form closed loops, as in Fig. 10c. Lines associated with static charges never do so but must start on positive charges and end on negative charges.

In a more formal sense, we can state the difference between electric fields set up by induction and those set up by static charges in these words:

> *Electric potential has meaning only for electric fields set up by static charges; it has no meaning for electric fields set up by induction.*

We can understand this qualitatively by considering what happens to a test charge that makes a single journey around the circular path in Fig. 10b. It starts at a certain point and, after it returns to that same point, has experienced an emf $\mathcal{E}$ of, let us say, 5 V. Its potential should have increased by this amount. This is impossible, however, because otherwise the same point in space would have two different values of potential. We can only conclude that potential has no meaning for electric fields that are set up by changing magnetic fields.

We can take a more formal look by recalling Eq. 8 of Chapter 26, which defines the potential difference between two points i and f:

$$V_f - V_i = \frac{W_{if}}{q_0} = -\int_i^f \mathbf{E} \cdot d\mathbf{s}. \qquad (19)$$

In Chapter 26 we had not yet encountered Faraday's law of induction so that the electric fields involved in the derivation of Eq. 19 were those due to static charges. If i and f in Eq. 19 are the same point, the path connecting them is a closed loop, V_i and V_f are identical, and Eq. 19 reduces to

$$\oint \mathbf{E} \cdot d\mathbf{s} = 0. \qquad (20)$$

However, when a changing magnetic flux is present, this integral is *not* zero but is $-d\Phi_B/dt$, as Eq. 18 asserts. Again, we conclude that electric potential has no meaning for electric fields associated with induction.

Sample Problem 4   In Fig. 10b, assume that $R = 8.5$ cm and that $dB/dt = 0.13$ T/s. (a) What is the magnitude of the electric field $\mathbf{E}$ for $r = 5.2$ cm?

From Faraday's law (Eq. 18) we have

$$(E)(2\pi r) = -\frac{d\Phi_B}{dt}.$$

We note that $r < R$. The flux Φ_B through a closed path of radius r is then

$$\Phi_B = B(\pi r^2).$$

so that

$$(E)(2\pi r) = -(\pi r^2)\frac{dB}{dt}.$$

Solving for E and dropping the minus sign, we find

$$E = \tfrac{1}{2}(dB/dt)\,r. \qquad (21)$$

Note that the induced electric field E depends on dB/dt but not on B. For $r = 5.2$ cm, we have, for the magnitude of $\mathbf{E}$,

$$\begin{aligned} E &= \tfrac{1}{2}(dB/dt)r = (\tfrac{1}{2})(0.13 \text{ T/s})(5.2 \times 10^{-2} \text{ m}) \\ &= 0.0034 \text{ V/m} = 3.4 \text{ mV/m}. \qquad \text{(Answer)} \end{aligned}$$

(b) What is the magnitude of the induced electric field for $r = 12.5$ cm?

In this case we have $r > R$ so that the entire flux of the magnet passes through the circular path. Thus,

$$\Phi_B = B(\pi R^2).$$

From Faraday's law (Eq. 18) we then find

$$(E)(2\pi r) = \frac{d\Phi_B}{dt} = -(\pi R^2)\frac{dB}{dt}.$$

Solving for E and again dropping the minus sign, we find

$$E = \tfrac{1}{2}(dB/dt)R^2\,\frac{1}{r}. \qquad (22)$$

Interestingly, an electric field is induced in this case even at points that are well outside the (changing) magnetic field, an

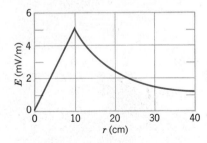

Figure 11   A plot of the induced electric field $E(r)$ for the conditions of Sample Problem 4.

important result that makes transformers possible. For $r = 12.5$ cm, Eq. 22 gives

$$E = \frac{(\frac{1}{2})(0.13 \text{ T/s})(8.5 \times 10^{-2} \text{ m})^2}{12.5 \times 10^{-2} \text{ m}}$$

$$= 3.8 \times 10^{-3} \text{ V/m} = 3.8 \text{ mV/m}. \quad \text{(Answer)}$$

Equations 21 and 22 yield the same result, as they must, for $r = R$. Figure 11 shows a plot of $E(r)$ based on these two equations.

32–7 The Betatron

The betatron is a device used to accelerate electrons to high energies by allowing them to be acted on by induced electric fields. Although the betatron is not widely used today, we describe it because it is a perfect example of the reality of these induced fields.

Figure 12 shows a cross section of a betatron in a plane that contains its vertical symmetry axis. The (time-varying) magnetic field that is shown there has several functions: (1) It guides the electrons in a circular path. (2) The changing magnetic flux generates an electric field that accelerates the electrons in this path. (3) It keeps the radius of the electron orbit essentially constant during the acceleration process. (4) It injects the electrons into the orbit initially and extracts them from the orbit after they have reached their full energy. (5) It provides a restoring force that resists any tendency for the electrons to stray away from their orbit, either vertically

or radially. It is remarkable that it is possible to do all these things by proper shaping and control of the magnetic field. Although the concept of the betatron had been proposed by others, it was Don W. Kerst, at the University of Illinois, who first succeeded in 1941 in providing a magnetic field that performed all these functions in a working betatron.

The object marked D in Fig. 12 is an evacuated ceramic "doughnut," inside of which the electrons circulate and are accelerated. Their orbit is a circle of constant radius R, its plane being at right angles to the page. In the figure, the electrons are circulating counterclockwise as viewed from above. Thus, we see them emerging from the figure on the left $\odot$ and entering it on the right $\otimes$.

The alternating magnetic field shown in Fig. 12 is produced by means of an alternating current in coils (not shown) around the iron pole pieces. The field B_{orb} at the orbit position serves to guide the electrons in their orbit. The field in the area enclosed by the orbit, whose average value is $B_{\text{av}}(= 2B_{\text{orb}})$, contributes to the central flux Φ_B. It is the time variation of this central flux that induces the electric field that acts on the electrons and accelerates them.

Sample Problem 5 In a 100-MeV betatron built at the General Electric Company, the orbit radius R is 84 cm. The magnetic field in the region enclosed by the orbit rises periodically (60 times per second) from zero to a maximum average value

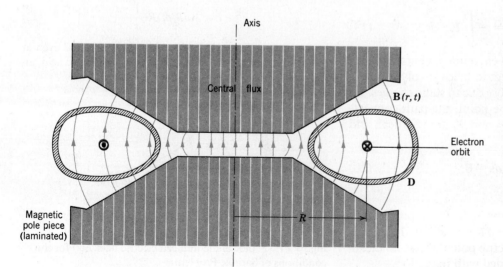

Figure 12 A cross section of a betatron, showing the orbit of the accelerating electrons and a "snapshot" of the time-varying magnetic field at a certain moment during the acceleration cycle.

$B_{av,m} = 0.80$ T in an accelerating interval of one fourth of a period, or 4.2 ms. (a) How much energy does the electron gain in one average trip around its orbit in this changing flux?

The central flux rises during the accelerating interval from zero to a maximum of

$$\Phi_m = (B_{av,m})(\pi R^2)$$
$$= (0.80 \text{ T})(\pi)(0.84 \text{ m})^2 = 1.8 \text{ Wb}.$$

The average value of $d\Phi_B/dt$ during the accelerating interval is then

$$\left(\frac{d\Phi_B}{dt}\right)_{av} = \frac{1.8 \text{ Wb}}{4.2 \times 10^{-3} \text{ s}} = 430 \text{ Wb/s}.$$

From Faraday's law (Eq. 6) this is also the average emf in volts. Thus, the electron increases its energy by an average of 430 eV per revolution in this changing flux. To achieve its full final energy of 100 MeV, it has to make about 230,000 revolutions in its orbit, a total path length of about 1200 km.

(b) What is the *average* speed of an electron during its acceleration cycle?

The length of the acceleration cycle is given as 4.2 ms and the path length is calculated above to be 1200 km. The average speed is then

$$\bar{v} = \frac{1200 \times 10^3 \text{ m}}{4.2 \times 10^{-3} \text{ s}} = 2.86 \times 10^8 \text{ m/s}. \quad \text{(Answer)}$$

This is 95% of the speed of light. The actual speed of the fully accelerated electron, when it has reached its final energy of 100 MeV, can be shown to be 99.9987% of the speed of light.

32-8   Physics: A Human Enterprise (Optional)

The world honors some of its citizens by naming things after them. On the endless list are countries (Bolivia), states (Pennsylvania), cities (Seattle), universities (Harvard), mountains (Pike's Peak), automobiles (Chevrolet), birds (Bachman's warbler), plants (poinsettia), minerals (dolomite), and food containers (Mason jars and Tupperware).

It is much the same in physics. Among the ways in which we honor some of those who have contributed to our discipline is attaching their name to something. Because of his wide range of interests and the remarkable standard of his achievements, Michael Faraday has been especially honored in this way:

1. The Farad.   As we have seen, the unit of capacitance is named in honor of his early discoveries in this field.

2. Faraday's Law of Induction.   This entire chapter deals with this important law, of which Faraday was the co-discoverer.

3. The Faraday.   This physical constant, named to recognize Faraday's extensive work in electrochemistry, is the charge possessed by 1 mole of single-ionized atoms. Its value is 96,580.35 C.

4. The Faraday Effect.   This effect, which we do not discuss in this book, deals with the rotation of the plane of polarization of polarized light (see Section 38-7) in a solution of organic materials such as sugar.

Most SI units are named after an individual closely associated with the branch of physics with which the unit deals. There are—among others—the newton (force; Isaac Newton), the ampere (current; Andre Marie Ampère), the coulomb (charge; Charles August Coulomb), the joule (energy; James Prescott Joule), the hertz (frequency; Heinrich Hertz), the watt (power; James Watt), the kelvin (temperature; Lord Kelvin), and the henry (inductance; Joseph Henry).

Units are only one of many things that are named after physicists. Among items mentioned in this book we can list—proceeding alphabetically—Archimedes' principle, Bernoulli's equation, the Carnot cycle, the Doppler shift, the Eötvös experiment, the Foucault pendulum, Gauss' law, the Hall effect, the Ives-Stillwell experiment, Joule's law, the Kelvin scale, Lenz's law, the Mach number, Newton's rings, Ohm's law, the Planck constant, Rayleigh's criterion, Schrödinger's equation, Torricelli's law, the Van de Graaff accelerator, Wien's law, Young's modulus, and the Zeeman effect.

This practice is a constant reminder that physics is a human enterprise; that behind every discovery and every new insight, there are the minds and hands of people not unlike ourselves. It is of course true that many physicists —equally meritorious—do not happen to be recognized in this particular way.

You may wish to read about the lives and works of physicists for whom readable and authoritative biographies exist. Table 1 lists some of these, starting (arbitrarily) with Michael Faraday.

It has often been said that physics is a young person's game and that, if one has not made some contribution of merit by age 30, there is little hope for the future! Although this is far from true, it *is* true that many physicists did make major contributions in the full flush of youth, Newton at 24 and Einstein at 26 being examples that stand out. As a more useful chronological guide than birth and death date, we list in Table 1 the year in which the physicist in question became 30.

Table 1 Biographical Information About Some Physicists[a]

Person	Year Turned 30	Reference
Michael Faraday	1821	*Michael Faraday,* by L. Pearce Williams, Basic Books, New York, 1964
Marie Curie	1897	*Madame Curie,* by Eve Curie, Doubleday, Doran & Company, New York, 1937
Albert Einstein	1909	*Subtle is the Lord . . . ,* by Abraham Pais, Clarendon Press, New York, 1982 *Einstein,* by Jeremy Bernstein, The Viking Press, New York, 1973.
H. G. J. Moseley	—[b]	*H. G. J. Moseley,* by J. L. Heilbron, University of California Press, Berkeley, 1974
Niels Bohr	1915	*Niels Bohr,* by Ruth Moore, Alfred A. Knopf, New York, 1966
I. I. Rabi	1928	*Rabi: American Physicist,* by John S. Rigden, Basic Books, New York, 1987
Enrico Fermi	1931	*Atoms in the Family,* by Laura Fermi, The University of Chicago Press, Chicago, 1954
Ernest Lawrence	1931	*Lawrence and Oppenheimer,* by Nuel Pharr Davis,
Robert Oppenheimer	1934	Simon and Schuster, New York, 1968
Hans A. Bethe	1936	*Hans Bethe: Prophet of Energy,* by Jeremy Bernstein, Basic Books, New York, 1980
Luis W. Alvarez	1941	*Alvarez: Adventures of a Physicist,* by Luis W. Alvarez, Basic Books, New York, 1987
Richard Feynman	1948	*Surely You're Joking, Mr. Feynman,* by Richard P. Feynman, Bantam Books, New York, 1986
Jeremy Bernstein	1959	*The Life it Brings: One Physicist's Beginnings,* by Jeremy Bernstein, Ticknor & Fields, New York, 1987

[a] For fascinating accounts of the life and work of many prominent physicists, read:

The Physicists — The History of a Scientific Community in Modern America, by Daniel J. Kevles, Vintage Books, New York, 1979
The Second Creation — Makers of the Revolution in Twentieth-Century Physics, by Robert P. Crease and Charles C. Mann, Macmillan Publishing, New York, 1986
From Falling Bodies to Radio Waves — Classical Physicists and Their Discoveries, by Emilio Segre, W. H. Freeman, New York, 1984
From X-Rays to Quarks — Modern Physicists and Their Discoveries, by Emilio Segre, W. H. Freeman, New York, 1984

[b] Moseley, who laid the experimental basis for the concept of atomic number, never reached 30. He was killed by a sniper's bullet in 1915, during World War I.

REVIEW AND SUMMARY

The flux Φ_B for a given surface immersed in a magnetic field **B** is defined from

Definition of Magnetic Flux

$$\Phi_B = \int \mathbf{B} \cdot d\mathbf{A} \qquad [3]$$

where the integral is taken over the surface. Review Section 25-4, which deals with the corresponding flux Φ_E for a surface immersed in an electric field **E**. The SI unit of magnetic flux is the weber, where 1 Wb = 1 T·m².

Faraday's law of induction states that if Φ_B for a surface bounded by a closed loop changes with time an emf given by

Faraday's Law of Induction

$$\mathcal{E} = -N\frac{d\Phi_B}{dt} \quad \text{(Faraday's law)} \qquad [7]$$

appears in the loop. If the loop is a conductor, a current will be set up. The two experiments of Section 32–2 and Sample Problem 1 illustrate this law.

Lenz's Law

Lenz's law specifies the direction of the current induced in a closed conducting loop by a changing magnetic flux. The law states: *An induced current in a closed conducting loop will appear in such a direction that it opposes the change that produced it.* Lenz's law is a consequence of the

Conservation of Energy

conservation of energy principle. Section 32–5, for example, shows that work is needed to pull a closed conducting loop out of a magnetic field and that this energy is accounted for as thermal energy of the loop material. If the direction of the induced current were opposite to that predicted by Lenz's law, work would be done *by* the loop and energy would not be conserved.

An induced emf will be present even if the loop through which the magnetic flux is changing is not a physical conductor but an imaginary line. The changing flux induces an electric field **E** at every point of such a loop, the emf being related to **E** by

Emf and the Induced Electric Field

$$\mathcal{E} = \oint \mathbf{E} \cdot d\mathbf{s}. \qquad [17]$$

The integral is taken around the loop. Combining Eqs. 7 and 17 lets us write Faraday's law in its most general form, or

Faraday's Law of Induction

$$\oint \mathbf{E} \cdot d\mathbf{s} = -\frac{d\Phi_B}{dt} \quad \text{(Faraday's law)}. \qquad [18]$$

The essence of this law is that *a changing magnetic field* ($d\Phi_B/dt$) *induces an electric field* (**E**). Sample Problem 4 illustrates the law in this form.

Induced Electric Fields

Electric fields induced by changing magnetic fields differ from those associated with static charges in that their lines of force form closed loops, as in Fig. 10. Electric potential has no meaning for such induced electric fields; the forces associated with them are nonconservative.

The Betatron

The betatron (Fig. 11) uses induced electric fields to accelerate electrons to high energy. Sample Problem 5 shows how to use Faraday's law of induction to calculate the average energy gain per revolution of the orbit for a particular machine.

QUESTIONS

1. Are induced emfs and currents different in any way from emfs and currents provided by a battery connected to a conducting loop?

2. Is the size of the voltage induced in a coil through which a magnet moves affected by the strength of the magnet? If so, explain how.

3. Explain in your own words the difference between a magnetic field **B** and the flux of a magnetic field Φ_B. Are they vectors or scalars? In what units may each be expressed? How are these units related? Are either or both (or neither) properties of a given point in space?

4. Can a charged particle at rest be set in motion by the action of a magnetic field? If not, why not? If so, how?

5. You drop a bar magnet along the axis of a long copper tube. Describe the motion of the magnet and the energy interchanges involved. Neglect air resistance.

6. You are playing with a metal loop, moving it back and forth in a magnetic field, as in Fig. 8. How can you tell, without detailed inspection, whether or not the loop has a narrow saw cut across it, rendering it nonconducting?

7. Figure 13 shows an inclined wooden track that passes, for part of its length, through a strong magnetic field. You roll a copper disk down the track. Describe the motion of the disk as it rolls from the top of the track to the bottom.

8. Figure 14 shows a copper ring, hung from a ceiling by two

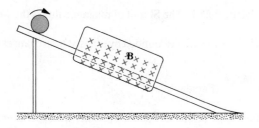

Figure 13 Question 7.

Figure 14 Question 8.

threads. Describe in detail how you might most effectively use a bar magnet to get this ring to swing back and forth.

9. Is an emf induced in a long solenoid by a bar magnet that moves inside it along the solenoid axis? Explain your answer.

10. Two conducting loops face each other a distance d apart (Fig. 15). An observer sights along their common axis from left to right. If a clockwise current i is suddenly established in the larger loop, by a battery not shown, (a) what is the direction of the induced current in the smaller loop? (b) What is the direction of the force (if any) that acts on the smaller loop?

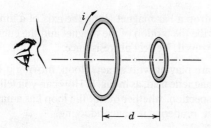

Figure 15 Question 10.

11. What is the direction of the induced emf in coil Y of Fig. 16 (a) when coil Y is moved toward coil X? (b) When the

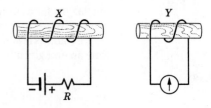

Figure 16 Question 11.

current in coil X is decreased, without any change in the relative positions of the coils?

12. The north pole of the magnet is moved away from a copper ring, as in Fig. 17. In the part of the ring farthest from the reader, which way does the current point?

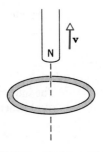

Figure 17 Question 12.

13. A circular loop moves with constant velocity through regions where uniform magnetic fields of the same magnitude are directed into or out of the plane of the page, as indicated in Fig. 18. At which of the seven indicated positions will the emf be (a) clockwise? (b) counterclockwise? (c) zero?

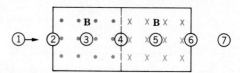

Figure 18 Question 13.

14. A short solenoid carrying a steady current is moving toward a conducting loop as in Fig. 19. What is the direction of the induced current in the loop as one sights toward it as shown?

15. The resistance R in the left-hand circuit of Fig. 20 is being increased at a steady rate. What is the direction of the induced current in the right-hand circuit?

16. What is the direction of the induced current through resistor R in Fig. 21 (a) immediately after switch S is closed, (b)

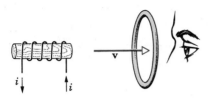

Figure 19 Question 14.

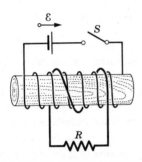

Figure 20 Question 15.

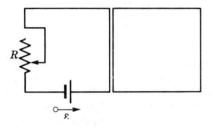

Figure 21 Question 16.

some time after switch S was closed, and (c) immediately after switch S is opened? (d) When switch S is held closed, from which end of the longer coil do field lines emerge? This is the effective north pole of the coil. (e) How do the conduction electrons in the coil containing R know about the flux within the long coil? What really gets them moving?

17. In Faraday's law of induction, does the induced emf depend on the resistance in the circuit? If so, how?

18. Suppose that the direction of induced emfs was governed by what we can call the Antilenz law: The induced current will appear in such a direction that it aids the change that produced it. Design a machine based on this law that would make a lot of money for you. (Alas, the Antilenz law is *false.*)

19. The loop of wire shown in Fig. 22 rotates with constant angular speed about the x axis. A uniform magnetic field **B**, whose direction is that of the positive y axis is present. For what portions of the rotation is the induced current in the loop (a) from P to Q, (b) from Q to P, (c) zero? (d) Repeat if the direction of rotation is reversed from that shown in the figure.

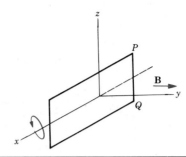

Figure 22 Question 19.

20. In Fig. 23 the straight movable wire segment is moving to the right with a constant velocity **v**. An induced current appears in the direction shown. What is the direction of the uniform magnetic field (assumed constant and perpendicular to the page) in region A?

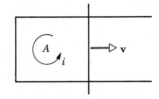

Figure 23 Question 20.

21. A conducting loop, shown in Fig. 24, is removed from the permanent magnet by pulling it vertically upward. (a) What is the direction of the induced current? (b) Is a force required to remove the loop? (c) Does the total amount of thermal energy produced in removing the loop depend on the time taken to remove it?

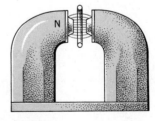

Figure 24 Question 21.

22. A plane closed loop is placed in a uniform magnetic field. In what ways can the loop be moved without inducing an emf? Consider motions of both translation and rotation.

23. *Eddy currents.* A sheet of copper is placed in a magnetic field as shown in Fig. 25. If we attempt to pull it out of the field

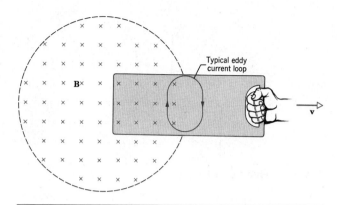

Figure 25 Question 23.

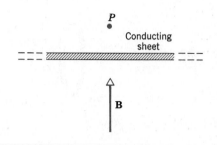

Figure 27 Question 25.

or push it further in, a resisting force automatically appears. Explain its origin. (*Hint:* Currents, called eddy currents, are induced in the sheet in such a way as to oppose the motion.)

24. *Magnetic damping.* A strip of copper is mounted as a pendulum about *O* in Fig. 26. It is free to swing through a magnetic field normal to the page. If the strip has slots cut in it as shown, it can swing freely through the field. If a strip without slots is substituted, the vibratory motion is strongly damped. Explain the observations. (*Hint:* Use Lenz's law; consider the paths that the charge carriers in the strip must follow if they are to oppose the motion.)

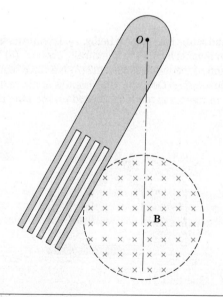

Figure 26 Question 24.

25. *Electromagnetic shielding.* Consider a conducting sheet lying in a plane perpendicular to a magnetic field **B**, as shown in Fig. 27. (*a*) If **B** suddenly changes, the full change in **B** is not immediately detected at points near *P*. Explain. (*b*) If the resistivity of the sheet is zero, the change is not ever detected at *P*. Explain. (*c*) If **B** changes periodically at high frequency and the conductor is made of a material of low resistivity, the region near *P* is almost completely shielded from the changes in flux. Explain. (*d*) Why is such a conductor not useful as a shield from static magnetic fields?

26. (*a*) In Fig. 10*b*, need the circle of radius *r* be a conducting loop in order that **E** and $\mathscr{E}$ be present? (*b*) If the circle of radius *r* were not concentric (moved slightly to the left, say), would $\mathscr{E}$ change? Would the configuration of **E** around the circle change? (*c*) For a concentric circle of radius *r*, with $r > R$, does an emf exist? Do electric fields exist?

27. A copper ring and a wooden ring of the same dimensions are placed so that there is the same changing magnetic flux through each. Compare the induced electric fields in the two rings.

28. An airliner is cruising in level flight over Alaska, where the earth's magnetic field has a large downward component. Which of its wingtips (right or left) has more electrons than the other?

29. In Fig. 10*d* how can the induced emfs around paths 1 and 2 be identical? The induced electric fields are much weaker near path 1 than near path 2, as the spacing of the lines of force shows. See also Fig. 11.

30. Show that, in the betatron of Fig. 12, the directions of the lines of **B** are correctly drawn to be consistent with the direction of circulation shown for the electrons.

31. In the betatron of Fig. 12 you want to increase the orbit radius by suddenly imposing an additional central flux $\Delta\Phi_B$ (set up by suddenly establishing a current in an auxilliary coil not shown). Should the lines of **B** associated with this flux increment be in the same direction as the lines shown in the figure or in the opposite direction? Assume that the magnetic field at the orbit position remains relatively unchanged by this flux increment.

32. In the betatron of Fig. 12, why is the iron core of the magnet made of laminated sheets rather than of solid metal as for the cyclotron of Section 6 of Chapter 30? (*Hint:* Consider the implications of Questions 24 and 25.)

EXERCISES AND PROBLEMS

Section 32-3 Faraday's Law of Induction

1E. At a certain location in the northern hemisphere, the earth's magnetic field has a magnitude of 42 μT and points downward at 57° to the vertical. Calculate the flux through a horizontal surface of area 2.5 m²; see Fig. 28.

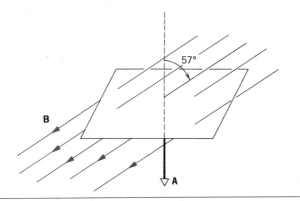

Figure 28 Exercise 1.

2E. A small loop of area A is inside of, and has its axis in the same direction as, a long solenoid of n turns per unit length and current i. If $i = i_0 \sin \omega t$, find the emf in the loop.

3E. A circular UHF television antenna has a diameter of 11 cm. The magnetic field of a TV signal is normal to the plane of the loop and, at one instant of time, its magnitude is changing at the rate 0.16 T/s. The field is uniform. What is the emf in the antenna?

4E. A uniform magnetic field **B** is perpendicular to the plane of a circular wire loop of radius r. The magnitude of the field varies with time according to $B = B_0 e^{-t/\tau}$ where B_0 and τ are constants. Find the emf in the loop as a function of time.

5E. In Fig. 29 the magnetic flux through the loop shown increases according to the relation

$$\Phi_B = 6t^2 + 7t,$$

where Φ_B is in milliwebers and t is in seconds. (a) What is the magnitude of the emf induced in the loop when $t = 2.0$ s? (b) What is the direction of the current through R?

6E. The magnetic field through a one-turn loop of wire 12 cm in radius and 8.5 Ω in resistance changes with time as shown in Fig. 30. Calculate the emf in the loop as a function of time. Consider the time intervals (a) $t = 0$ to $t = 2$ s; (b) $t = 2$ s to $t = 4$ s; (c) $t = 4$ s to $t = 6$ s. The (uniform) magnetic field is at right angles to the plane of the loop.

7E. A loop antenna of area A and resistance R is perpendicular to a uniform magnetic field **B**. The field drops linearly to

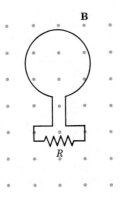

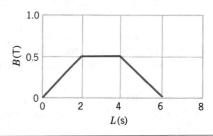

Figure 29 Exercise 5 and Problem 17.

Figure 30 Exercise 6.

zero in a time interval Δt. Find an expression for the total thermal energy dissipated in the loop.

8E. A uniform magnetic field is normal to the plane of a circular loop 10 cm in diameter made of copper wire (diameter = 2.5 mm). (a) Calculate the resistance of the wire. (See Table 1 in Chapter 28.) (b) At what rate must the magnetic field change with time if an induced current of 10 A is to appear in the loop?

9P. The current in the solenoid of Sample Problem 1 changes, not as in that Sample Problem, but according to $i = 3.0t + 1.0t^2$, where i is in amperes and t is given in seconds. (a) Plot the induced emf in the coil from $t = 0$ to $t = 4$ s. (b) The resistance of the coil is 0.15 Ω. What is the current in the coil at $t = 2.0$ s?

10P. In Fig. 31 a 120-turn coil of radius 1.8 cm and resistance 5.3 Ω is placed outside a solenoid like that of Sample Problem 1. If the current in the solenoid is changed as in that Sample Problem, (a) what current appears in the loop while the solenoid current is being changed? (b) How do the conduction electrons in the loop "get the message" from the solenoid that they should move to establish a current? After all, the magnetic flux is entirely confined to the interior of the solenoid.

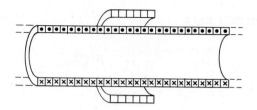

Figure 31 Problem 10.

11P. A long solenoid with a radius of 25 mm has 100 turns/cm. A single loop of wire of radius 5.0 cm is placed around the solenoid, the axis of the loop and the solenoid coinciding. The current in the solenoid is reduced from 1.0 A to 0.50 A at a uniform rate over a time interval of 10 ms. What emf appears in the loop?

12P. Derive an expression for the flux through a toroid of N turns carrying a current i. Assume that the windings have a rectangular cross section of inner radius a, outer radius b, and height h.

13P. A toroid having a 5.0-cm square cross section and an inside radius of 15 cm has 500 turns of wire and carries a current of 0.80 A. What is the magnetic flux through the cross section?

14P. You are given 50 cm of copper wire (diameter = 1.0 mm). It is formed into a circular loop and placed at right angles to a uniform magnetic field that is increasing with time at the constant rate of 10 mT/s. At what rate is thermal energy generated in the loop?

15P. A closed loop of wire consists of a pair of equal semicircles, radius 3.7 cm, lying in mutually perpendicular planes. The loop was formed by folding a circular loop along a diameter until the two halves became perpendicular. A uniform magnetic field **B** of magnitude 76 mT is directed perpendicular to the fold diameter and makes equal angles (= 45°) with the planes of the semicircles as shown in Fig. 32. The magnetic field is reduced at a uniform rate to zero during a time interval 4.5 ms. Determine the magnitude of the induced emf and the direction of the induced current in the loop during this interval.

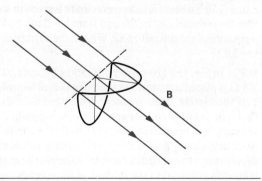

Figure 32 Problem 15.

16P. Figure 33 shows two parallel loops of wire having a common axis. The smaller loop (radius r) is above the larger loop (radius R), by a distance $x \gg R$. Consequently the magnetic field, due to the current i in the larger loop, is nearly constant throughout the smaller loop. Suppose that x is increasing at the constant rate $dx/dt = v$. (a) Determine the magnetic flux across the area bounded by the smaller loop as a function of x. (b) Compute the emf generated in the smaller loop. (c) Determine the direction of the induced current flowing in the smaller loop. (*Hint:* See Equation 25 in Chapter 31.)

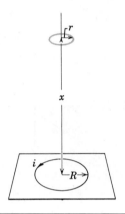

Figure 33 Problem 16.

17P. In Fig. 29 let the flux for the loop be $\Phi_B(0)$ at time $t = 0$. Then let the magnetic field **B** vary in a continuous but unspecified way, in both magnitude and direction, so that at time t the flux is represented by $\Phi_B(t)$. (a) Show that the net charge $q(t)$ that has passed through resistor R in time t is

$$q(t) = \frac{1}{R}[\Phi_B(0) - \Phi_B(t)],$$

independent of the way **B** has changed. (b) If $\Phi_B(t) = \Phi_B(0)$ in a particular case we have $q(t) = 0$. Is the induced current necessarily zero throughout the interval $0 \rightarrow t$?

18P. A hundred turns of insulated copper wire are wrapped around a wooden cylindrical core of cross-sectional area 1.2×10^{-3} m². The two terminals are connected to a resistor. The total resistance in the circuit is 13 Ω. If an externally applied uniform longitudinal magnetic field in the core changes from 1.6 T in one direction to 1.6 T in the opposite direction, how much charge flows through the circuit? (*Hint:* See Problem 17.)

19P. A square wire loop with 2.0-m sides is perpendicular to a uniform magnetic field, with half the area of the loop in the field, as shown in Fig. 34. The loop contains a 20-V battery with negligible internal resistance. If the magnitude of the field varies with time according to $B = 0.042 - 0.87t$, with B in tesla and t in seconds, (a) what is the total emf in the circuit? (b) What is the direction of the current through the battery?

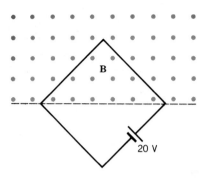

Figure 34 Problem 19.

20P. A wire is bent into three circular segments of radius $r = 10$ cm as shown in Fig. 35. Each segment is a quadrant of a circle, ab lying in the xy plane, bc lying in the yz plane, and ca lying in the zx plane. (a) If a uniform magnetic field **B** points in the positive x direction, what is the magnitude of the emf developed in the wire when B increases at the rate of 3.0 mT/s? (b) What is the direction of the current in the segment bc?

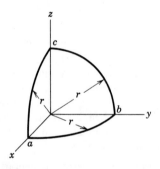

Figure 35 Problem 20.

21P. Two long, parallel copper wires (diameter = 2.5 mm) carry currents of 10 A in opposite directions. (a) If their centers are 20 mm apart, calculate the flux per meter of wire that exists in the space between the axes of the wires. (b) What fraction of this flux lies inside the wires? (c) Repeat the calculation of (a) for parallel currents.

Section 32–5 Induction: A Quantitative Study
22E. An automobile having a radio antenna 1.1 m long travels at 90 km/h in a region where the earth's magnetic field is 55 μT. Is an emf induced in the antenna? If so, what is its maximum possible value?

23E. A spaceship 12 m wide moves at a speed of 2.4×10^7 m/s through a weak interstellar magnetic field of 0.36 nT. Regard the ship as a metal rod 12 m long and assume it moves perpendicular to the field. Calculate the emf generated across the width of the ship.

24E. A circular loop of wire 10 cm in diameter is placed with

its normal making an angle of 30° with the direction of a uniform 0.50-T magnetic field. The loop is "wobbled" so that its normal rotates in a cone about the field direction at the constant rate of 100 rev/min; the angle between the normal and the field direction (= 30°) remains unchanged during the process. What emf appears in the loop?

25E. A metal rod moves with constant velocity along two parallel metal rails, connected with a strip of metal at one end, as shown in Fig. 36. A magnetic field $B = 0.35$ T points out of the page. (a) If the rails are separated by 25 cm and the speed of the rod is 55 cm/s, what emf is generated? (b) If the rod has a resistance of 18 Ω and the rails have negligible resistance, what is the current in the rod?

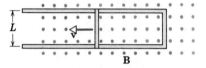

Figure 36 Exercises 25 and 26.

26E. Figure 36 shows a conducting rod of length L being pulled along horizontal, frictionless, conducting rails at a constant velocity **v**. A uniform vertical magnetic field **B** fills the region in which the rod moves. Assume that $L = 10$ cm, $v = 5.0$ m/s, and $B = 1.2$ T. (a) What is the induced emf in the rod? (b) What is the current in the conducting loop? Assume that the resistance of the rod is 0.40 Ω and that the resistance of the rails is negligibly small. (c) At what rate is thermal energy being generated in the rod? (d) What force must be applied to the rod by an external agent to maintain its motion? (e) At what rate does this external agent do work on the rod? Compare this answer with the answer to (c).

27E. In Fig. 37 a conducting rod of mass m and length L slides without friction on two long horizontal rails. A uniform vertical magnetic field **B** fills the region in which the rod is free to move. The generator G supplies a constant current i that flows down one rail, across the rod, and back to the generator along the other rail. Find the velocity of the rod as a function of time, assuming it to be at rest at $t = 0$.

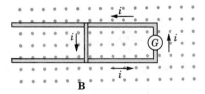

Figure 37 Exercise 27 and Problem 34.

28P. A circular loop made of a stretched conducting elastic material has a 12 cm radius. It is placed with its plane at right angles to a uniform 0.80-T magnetic field. When released, the radius of the loop starts to shrink at an instantaneous rate of 75 cm/s. What emf is induced in the loop at that instant?

29P. Two straight conducting rails form a right angle where their ends are joined. A conducting bar in contact with the rails starts at the vertex at time $t = 0$ and moves with a constant velocity of 5.2 m/s to the right, as shown in Fig. 38. A 0.35 T magnetic field points out of the page. Calculate (a) the flux through the triangle formed by the rails and bar at $t = 3.0$ s and (b) the emf around the triangle at that time. (c) In what manner does the emf around the triangle vary with time?

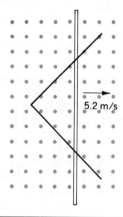

Figure 38 Problem 29.

30P. A stiff wire bent into a semicircle of radius a is rotated with a frequency v in a uniform magnetic field, as suggested in Fig. 39. What are (a) the frequency and (b) the amplitude of the emf induced in the loop?

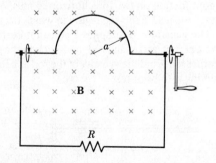

Figure 39 Problem 30.

31P. *Alternating current generator.* A rectangular loop of N turns and of length a and width b is rotated at a frequency v in a uniform magnetic field **B**, as in Fig. 40. (a) Show that an in-

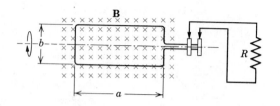

Figure 40 Problem 31.

duced emf given by

$$\mathscr{E} = 2\pi v N a b B \sin 2\pi v t = \mathscr{E}_0 \sin 2\pi v t$$

appears in the loop. This is the principle of the commercial alternating-current generator. (b) Design a loop that will produce an emf with $\mathscr{E}_0 = 150$ V when rotated at 60 rev/s in a magnetic field of 0.50 T.

32P. A generator consists of 100 turns of wire formed into a rectangular loop 50 cm by 30 cm, placed entirely in a uniform magnetic field with magnitude $B = 3.5$ T. What is the maximum value of the emf produced when the loop is spun at 1000 revolutions per minute about an axis perpendicular to **B**?

33P. Calculate the average power supplied by the generator of Problem 31b if it is connected to a circuit of 42-Ω resistance.

34P. In Exercise 27 (see Fig. 37) the constant-current generator G is replaced by a battery that supplies a constant emf $\mathscr{E}$. (a) Show that the velocity of the rod now approaches a constant terminal value **v** and give its magnitude and direction. (b) What is the current in the rod when this terminal velocity is reached? (c) Analyze both this situation and that of Exercise 27 from the point of view of energy transfers.

35P. At a certain place, the earth's magnetic field has magnitude $B = 0.59$ gauss and is inclined downwards at an angle of 70° to the horizontal. A flat horizontal circular coil of wire with a radius of 10 cm has 1000 turns and a total resistance of 85 Ω. It is connected to a galvanometer with 140 Ω resistance. The coil is flipped through a half revolution about a diameter, so it is again horizontal. How much charge flows through the galvanometer during the flip? (*Hint:* See Problem 17.)

36P. Figure 41 shows a rod of length L caused to move at constant speed v along horizontal conducting rails. In this case the magnetic field in which the rod moves is not uniform but is provided by a current i in a long parallel wire. Assume that $v = 5.0$ m/s, $a = 10$ mm, $L = 10$ cm, and $i = 100$ A. (a) Calculate the emf induced in the rod. (b) What is the current in the conducting loop? Assume that the resistance of the rod is 0.40 Ω and that the resistance of the rails is negligible. (c) At what rate is thermal energy being generated in the rod? (d) What force must be applied to the rod by an external agent to maintain its motion? (e) At what rate does this external agent do work on the rod? Compare this answer to (c).

37P. For the situation shown in Fig. 42, $a = 12$ cm, $b =$

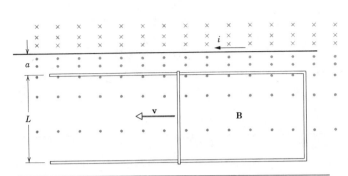

Figure 41   Problem 36.

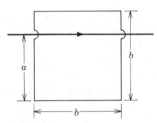

Figure 42   Problem 37.

16 cm. The current in the long straight wire is given by $i = 4.5t^2 - 10t$, where i is in amperes and t is in seconds. (*a*) Find the emf in the square loop at $t = 3.0$ s. (*b*) What is the direction of the induced current in the loop?

38P.   In Fig. 43, the square has sides of length 2.0 cm. A magnetic field points out of the page; its magnitude is given by $B = 4t^2y$, where B is in tesla, t is in seconds, and y is in meters. Determine the emf around the square at $t = 2.5$ s and give its direction.

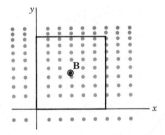

Figure 43   Problem 38.

39P.   A rectangular loop of wire with length a, width b, and resistance R is placed near an infinitely long wire carrying current i, as shown in Fig. 44. The distance from the long wire to the center of the loop is r. Find (*a*) the magnitude of the

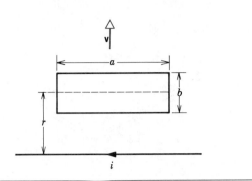

Figure 44   Problem 39.

magnetic flux through the loop and (*b*) the current in the loop as it moves away from the long wire with speed v.

40P.   Figure 45 shows a "homopolar generator," a device with a solid conducting disk as rotor. This machine can produce a greater emf than wire loop rotors, since they can spin at a much higher angular speed before centrifugal forces disrupt the rotor. (*a*) Show that the emf produced is given by

$$\mathcal{E} = \pi v B R^2$$

where v is the spin frequency, R the rotor radius, and B the uniform magnetic field perpendicular to the rotor. (*b*) What torque must be provided by the motor spinning the rotor when the output current is i?

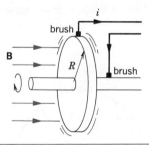

Figure 45   Problem 40.

41P.   A rod with length l, mass m, and resistance R slides without friction down parallel conducting rails of negligible resistance, as in Fig. 46. The rails are connected together at the bottom as shown, forming a conducting loop with the rod as the top member. The plane of the rails makes an angle θ with the horizontal and a uniform vertical magnetic field **B** exists throughout the region. (*a*) Show that the rod acquires a steady-state terminal velocity whose magnitude is

$$v = \frac{mgR}{B^2l^2}\frac{\sin\theta}{\cos^2\theta}.$$

(*b*) Show that the rate at which thermal energy is being generated in the rod is equal to the rate at which the rod is losing

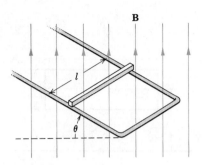

Figure 46 Problem 41.

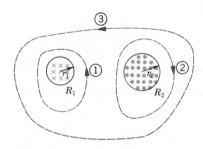

Figure 48 Exercise 44.

gravitational potential energy. (c) Discuss the situation if **B** were directed down instead of up.

42P*. A wire whose cross-sectional area is 1.2 mm² and whose resistivity is 1.7×10^{-8} $\Omega \cdot$m is bent into a circular arc of radius $r = 24$ cm as shown in Fig. 47. An additional straight length of this wire, OP, is free to pivot about O and makes sliding contact with the arc at P. Finally, another straight length of this wire, OQ, completes the circuit. The entire arrangement is located in a magnetic field $B = 0.15$ T directed out of the plane of the figure. The straight wire OP starts from rest with $\theta = 0$ and has a constant angular acceleration of 12 rad/s². (a) Find the resistance of the loop OPQO as a function of θ. (b) Find the magnetic flux through the loop as a function of θ. (c) For what value of θ is the induced current in the loop a maximum? (d) What is the maximum value of the induced current in the loop?

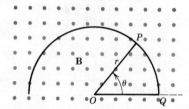

Figure 47 Problem 42.

Section 32–6 Induced Electric Fields

43E. A long solenoid has a diameter of 12 cm. When a current i is passed through its windings, a uniform magnetic field $B = 30$ mT is produced in its interior. By decreasing i, the field is caused to decrease at the rate 6.5 mT/s. Calculate the magnitude of the induced electric field (a) 2.2 cm and (b) 8.2 cm from the axis of the solenoid.

44E. Figure 48 shows two circular regions R_1 and R_2 with radii $r_1 = 20$ cm and $r_2 = 30$ cm, respectively. In R_1 there is a uniform magnetic field $B_1 = 50$ mT into the page and in R_2 there is a uniform magnetic field $B_2 = 75$ mT out of the page

(ignore any fringing of these fields). Both fields are decreasing at the rate 8.5 mT/s. Calculate the integral $\oint \mathbf{E} \cdot d\mathbf{s}$ for each of the three indicated paths.

45P. Early in 1981 the Francis Bitter National Magnet Laboratory at M.I.T. commenced operation of a 3.3-cm diameter cylindrical magnet, which produces a 30 T field, then the world's largest steady-state field. The field can be varied sinusoidally between the limits of 29.6 T and 30 T at a frequency of 15 Hz. When this is done, what is the maximum value of the induced electric field at a radial distance of 1.6 cm from the axis? This magnet is described in *Physics Today*, August 1984. (*Hint:* See Sample Problem 4.)

46P. Figure 49 shows a uniform magnetic field **B** confined to a cylindrical volume of radius R. **B** is decreasing in magnitude at a constant rate of 10 mT/s. What is the instantaneous acceleration (direction and magnitude) experienced by an electron placed at a, at b, and at c? Assume $r = 5.0$ cm. (The necessary fringing of the field beyond R will not change your answer as long as there is axial symmetry about the perpendicular axis through b.)

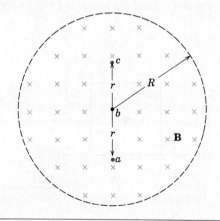

Figure 49 Problem 46.

47P. Prove that the electric field **E** in a charged parallel-plate capacitor cannot drop abruptly to zero as one moves at right angles to it, as suggested by the arrow in Fig. 50 (see point a). In

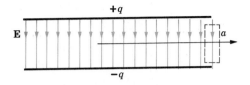

Figure 50   Problem 47.

actual capacitors fringing of the lines of force always occurs, which means that **E** approaches zero in a continuous and gradual way; compare with Problem 45, Chapter 31. (*Hint:* Apply Faraday's law to the rectangular path shown by the dashed lines.)

Section 32–7   The Betatron

48E.   Figure 51*a* shows a top view of the electron orbit in a betatron. Electrons are accelerated in a circular orbit in the *xy* plane and then withdrawn to strike the target *T*. The magnetic field **B** is along the *z* axis (the positive *z* axis is out of the page). The magnetic field B_z along this axis varies sinusoidally as shown in Fig. 51*b*. Recall that the magnetic field must (*i*) guide the electrons in their circular path, and (*ii*) generate the electric field that accelerates the electrons. Which quarter cycle(s) in Fig. 51*b* are suitable (*a*) according to (*i*), (*b*) according to (*ii*), (*c*) for operation of the betatron?

49E.   In a certain betatron the radius of the electron orbit is $r = 32$ cm and the magnetic field at this radius is given by $B_{orb} = (0.28) \sin 120\pi t$, where *t* in seconds gives B_{orb} in tesla. (*a*) Calculate the induced electric field felt by the electrons at $t = 0$. (*b*) Find the acceleration of the electrons at this instant. Ignore relativistic effects.

50P.   Some measurements of the maximum magnetic field as a function of radius for a betatron are:

r, cm	*B*, tesla	*r*, cm	*B*, tesla
0	0.400	81.2	0.409
10.2	0.950	83.7	0.400
68.2	0.950	88.9	0.381
73.2	0.528	91.4	0.372
75.2	0.451	93.5	0.360
77.3	0.428	95.5	0.340

Show by graphical analysis that the relation $\bar{B} = 2B_{orb}$ mentioned in Section 32–7 as essential to betatron operation is satisfied at the orbit radius, $R = 84$ cm. (*Hint:* Note that

$$\bar{B} = \frac{1}{\pi R^2} \int_0^R B(r) 2\pi r \, dr$$

and evaluate the integral graphically.)

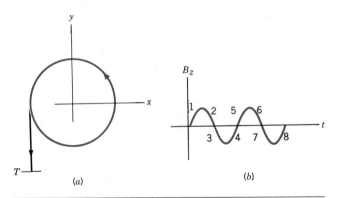

(a)   (b)

Figure 51   Problem 48.

ESSAY 12
SUPERCONDUCTIVITY

PETER LINDENFELD

RUTGERS, THE STATE
UNIVERSITY OF NEW
JERSEY

Introduction

In 1911, when H. Kamerlingh Onnes and his assistant Gilles Holst at the University of Leiden in the Netherlands cooled some mercury to a temperature lower than anyone ever had before, they could not believe that it had no electrical resistance at all (Figs. 1 and 2). It seemed that some accidental short circuit was preventing a proper measurement. Only after they tried again and actually saw the resistance come and go on heating and cooling through the "transition temperature" did they realize that something totally unexpected was happening. This was the first in a string of surprises that superconductivity has provided for us since that time.

It turned out that the loss of resistance is just part of what makes superconductors interesting. They also have magnetic properties unlike those of any other materials. On the one hand they can generate extremely high magnetic fields, on the other hand they allow the measurement, control, and utilization of weaker magnetic fields than

Figure 1 Kamerlingh Onnes was the first person to liquify helium. This achievement made it possible to do experiments a few degrees above absolute zero, and led to the discovery of superconductivity.

Figure 2 This graph was used to describe the discovery of superconductivity. The horizontal axis is the temperature of a sample of mercury in kelvins. The vertical axis is the electrical resistance of the sample in arbitrary units.

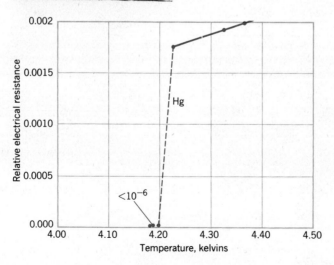

Figure 3 (*a*) Superconducting quantum interference devices immersed in liquid helium detect the magnetic fields generated by brain activity. (*b*) The graph shows the magnetic field generated when a person hears a 600-Hz tone. Adjacent magnetic field lines differ by 10 fT or 10^{-14} T. This experiment makes it possible to determine which part of the brain participates in hearing.

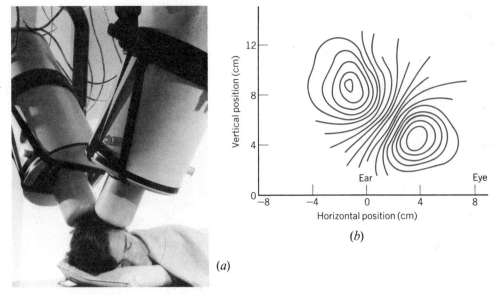

(*a*)

(*b*)

had been thought possible. The reason that their marvellous properties were not more fully exploited is, of course, that until 1986 they were known to occur only at temperatures below about 20 K, or 20 Celsius degrees above absolute zero.

In spite of the considerable cost and complication of the refrigeration, both the high-field and the low-field applications of superconductivity were being developed more and more even before superconductivity was found to exist also at much higher temperatures. At one end of the scale are the superconducting magnets of the accelerator at the Fermi National Laboratory, which accelerates elementary particles to the highest energies, and at the other end the "SQUIDS" that record the magnetic fields generated by the human brain (Fig. 3). In early 1987 the science of superconductivity was revolutionized by the discovery of materials that become superconducting above 90 K.

Magnets and Large-Scale Applications

An electromagnet is a device with zero efficiency. It may be expensive to build and to operate, it may be an essential part of a generator or accelerator, but if, as is usually the case, its function is to produce a constant magnetic field, all the energy supplied to it is dissipated as heat energy. This fact is most evident when we consider permanent magnets, which generate magnetic fields without any energy input. Each electron is a little permanent magnet, and the only reason that we are not more often aware of the magnetic properties of matter is that the magnetic fields of different electrons usually cancel because of their different orientations.

In iron and other "magnetic" materials the electrons can be lined up, but even at best they produce fields of only about 2 T (tesla). For larger fields current-carrying coils are invariably used. Ordinarily a significant amount of energy must be supplied and is wasted as heat at the rate of I^2R watts (where I is the current and R the resistance). More or less elaborate systems must be used to carry the heat away. To make R zero by using superconducting coils is clearly a huge advantage.

Kamerlingh Onnes was immediately aware of the possibilities of generating large fields without useless heat, but saw his hopes dashed with the first experiments. In the

Figure 4 Each line represents the "critical field curve" of an element. Superconductivity can exist only at the combinations of temperature and magnetic field below the line. Alloys discovered in the 1950s extended the range of critical temperatures by a factor of about three and the range of critical fields by more than a hundred.

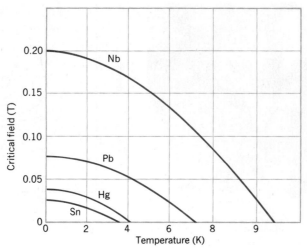

superconductors known at that time superconductivity is destroyed by magnetic fields less than $\frac{1}{10}$ T (Fig. 4).

Today we know that magnetism and superconductivity are natural enemies. Macroscopic magnetic properties depend on electrons that are lined up parallel to one another, while superconductivity requires pairs of electrons with their spins in opposite directions so that their magnetic fields cancel. A magnetic field causes a torque on an electron that tends to line it up with the field. It therefore acts to break up the superconducting pairs and so to destroy superconductivity.

Figure 5 A record speed of 321 miles per hour was set in 1979 by a magnetically levitated train in Japan. This train floats without touching the ground as a result of the repulsion between its superconducting magnets and the magnetic field that they induce in the tracks.

With hindsight we can see that there were times when the discovery of high-field superconductivity was tantalizingly close, but we cannot be too surprised that it was not until the 1950s, about 40 years after the discovery of superconductivity, that materials and processes were developed that allowed the first superconducting high-field magnets to be made.

Today almost every large physics laboratory and many small ones have superconducting magnets. They are used in the most powerful particle accelerators, and the realization of magnetic fusion devices for energy generation is expected to depend on them. But there are also applications that were not even imagined before the advent of superconductivity. One of these is the development of magnetically levitated trains or "electromagnetic flight" as it is sometimes more imaginatively called. The flight here is only a very small distance above the ground, just enough to keep car and rails from touching, so that friction between the two is eliminated, and with it a major obstacle to the achievement of higher speeds (Fig. 5).

Forces on Magnets and Superconductors

Put a bar magnet, an electron, or a current-carrying coil in a uniform magnetic field. Each of these objects has a north pole and a south pole. The north pole experiences a force in the direction of the magnetic field, the south pole a force of equal magnitude in the opposite direction. There is no net force. Unless the object is already lined up with the field there will, however, be a torque tending to make the north–south axis parallel to the field. (Although we are using the language of poles, the description in terms of forces on currents is entirely equivalent and leads to the same result.)

Figure 6 The forces in a magnetic field on permanent magnets (grey) and on superconductors (colored).

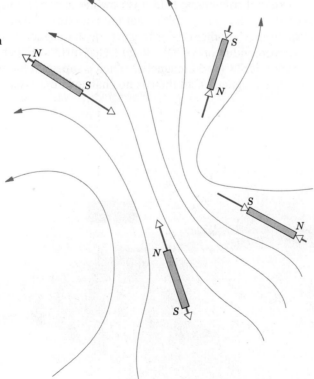

The situation is quite different if the field is not uniform. Suppose, for example, that the field lines converge, indicating that the field becomes stronger where the lines are closer together. There will be a torque as before, but when the north pole points along the field it will now be in a stronger field than the south pole. The net result is that the whole object (magnet, coil, or electron) is pulled toward the strongest part of the field. If the field lines diverge it is the south pole that is in the stronger field and the object will be pulled back, again toward the stronger field. A small magnet has a field that is strongest close to it, and two such magnets will attract (Fig. 6).

In a superconductor the magnetization is in the direction opposite to that of the external magnetic field. This is called diamagnetism. We can see how it arises by considering a cylinder or a closed coil made of perfectly conducting wire, but without a current source, and, at least initially, without any current. We do not have a magnetized object to start with, but currents, and hence magnetization, may be induced in accordance with Faraday's law.

If we take the coil to a region where there is a magnetic field, the increase in magnetic flux through the coil will produce an induced emf, and with it an induced current and an induced magnetic field. In accord with Lenz's law the induced field will be in a direction opposite to the change in the flux through the coil, in this case opposite to the direction of the increasing external field through the superconducting coil.

The induced emf disappears as soon as the coil comes to rest and the field through it stops increasing. In a coil made of normal wire the current and its magnetic field will then also cease. In a perfectly conducting coil, however, or on the surface of a cylinder made of perfectly conducting material, the current will continue to flow even without an emf, since the resistance is zero and no energy input is required.

In an external converging field it is now the opposite pole, the south pole, which is in the stronger field. The result is that the net force is toward the weakest part of the field. In the field of another coil or magnet the diamagnet is repelled!

For a superconductor we have to go a step further. Experiment shows that a superconductor is always diamagnetic. When a superconductor is cooled from above to below T_c in a magnetic field there is no change in the external flux, so that Faraday's

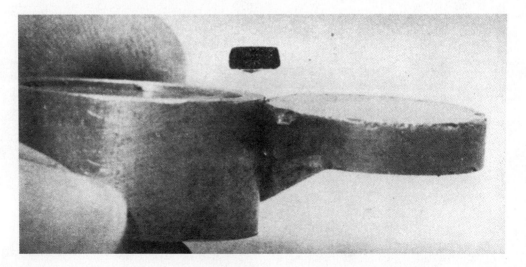

Figure 7   A sample of superconducting yttrium barium copper oxide ($T_c = 95$ K) floating above a permanent magnet.

law would not predict induced magnetization. Yet the superconductor becomes diamagnetic. This is called the Meissner effect, first demonstrated by Meissner and Ochsenfeld in 1933. It shows that the properties of a superconductor cannot simply be described by saying that it is a perfect conductor. The diamagnetism is illustrated in Figure 7, which shows the repulsion of a superconductor by a permanent magnet.

Flux Quantization and Small-Scale Applications

The use of superconductors to detect extremely small magnetic fields depends on two phenomena, flux quantization and "Josephson tunneling," named after Brian Josephson, who predicted the effect while he was still a student at Cambridge University in England.

Just as electric charge is quantized and occurs only in multiples of the electronic charge e (1.6×10^{-19} C) so is the magnetic flux through a superconducting loop. The flux quantum is equal to $h/2e$ (where h is Planck's constant), equal to about 2×10^{-15} T·m². This tiny amount of flux and even small fractions of it can be detected by means of the Josephson effect.

What Josephson showed is that the pairs of electrons in a superconducting current can move ("tunnel") through a thin insulating barrier. A loop of superconducting material with such a barrier or "tunnel junction" can still be superconducting. Suppose now that we try to increase the magnetic flux through the loop. Since the flux is quantized it cannot increase continuously. Instead the current through the loop will change so as to keep the flux constant. The junction, however, cannot support more than a very small current. When this amount is reached, the junction ceases momentarily to be superconducting and allows the flux through the loop to change discontinuously. Under the right conditions the change can be made equal to just one flux quantum in one step. By counting the number of steps, the flux through the coil, and therefore also the magnetic field itself, can be determined with great precision.

A loop that contains one or more Josephson junctions for flux detection or measurement is called a "superconducting quantum interference device" or SQUID (Fig. 8).

The discontinuous, stepwise change of flux is also the basis for the use of Josephson junction devices as memory and processing elements in digital computers. Because of their superconductivity and resulting lack of heat dissipation, they can be packed very closely together.

Figure 8 (a) Schematic diagram of a two-junction SQUID. (b) Data from a circuit containing a two-junction SQUID. The graph shows the variation of the current as a function of magnetic field. Each cycle represents a change in flux by one flux quantum.

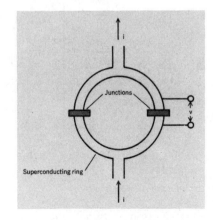

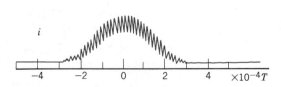

Materials

After the discovery in 1911 of superconductivity in mercury, other elements in the same region of the periodic table of elements were also found to be superconductors. Tin, indium, lead, and thallium led the way, with transition temperatures (T_c) ranging from 2.4 K for thallium to 7.2 K for lead. All were discovered in Kamerlingh Onnes' laboratory, which was then the only place where liquid helium, and with it these low temperatures, were available. In 1923 Toronto joined the exclusive club, and Berlin in 1925, where Walther Meissner and his co-workers soon discovered superconductivity in a different part of the periodic table, among the "transition elements," including niobium, which remains, with its T_c of 9.2 K, the element with the highest transition temperature. Further progress was made when it was realized that metallic compounds could have even higher values. By 1940 the record holder was NbC with a T_c of 10.1 K, in 1954 it was Nb_3Sn with T_c equal to 18 K. In 1973 Nb_3Ge, with a T_c of 23.2 K was in first place, and remained there for 14 years until the dramatic discoveries of 1986 and 1987 of superconductivity in ceramic oxides, first demonstrated by Müller and Bednorz at about 35K in lanthanum-barium-copper oxide, and later by Chu and his collaborators near 91 K in yttrium-barium-copper oxide (Fig. 9).

Together with the advances in T_c came increases in the magnetic fields where superconductivity can persist, and which can therefore be generated by superconducting magnets. For about 40 years the compounds of niobium held undisputed first place in both areas. In 1987 the ceramic oxides became the focus of intense research toward further improvements and new applications.

Figure 9 The highest superconducting transition temperatures (in kelvins) from 1911 to 1987.

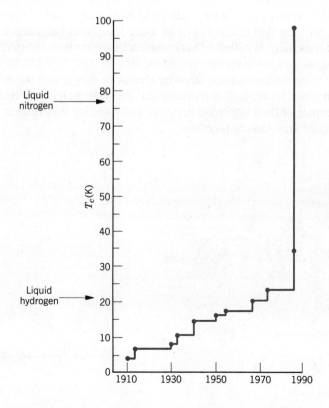

Theory

Until 1957 superconductivity eluded almost all fundamental understanding. Much had been learned empirically but there was hardly even the beginning of a theory based on atomic and electronic properties. The frustration among physicists is illustrated by the statement of Felix Bloch, a pioneer in the theory of conductivity, when he said "superconductivity is impossible."

The reason for the difficulty is easy enough to understand. It was necessary to find how electrons can cooperate more fully than they normally do, in spite of their mutual repulsion as described by Coulomb's law. The subtle mechanism was finally discovered by Bardeen, Cooper, and Schrieffer ("BCS"). Their theory describes the behavior of superconductors so well and in so much detail that it was widely thought that except for following up some loose ends there would be little further interest in the subject among physicists. It was not the first time that a field of inquiry had been declared to be in terminal decline when in fact it was entering a time of rebirth full of new and unexpected discoveries.

The BCS mechanism depends on the fact that the negatively charged electrons are moving through a lattice of positively-charged ions. The ions are attracted toward the path of an electron. Another electron that comes along before the ions return to their equilibrium positions finds itself in an environment distorted by the passage of the first electron and in this way, indirectly, the two electrons interact.

The BCS theory allows us to understand the delicate balance between opposing tendencies that, under the right conditions, can lead to superconductivity. Its quantitative details describe the properties of the superconductors known at that time very fully, including the parameters that affect the transition temperature. Yet in the 30 years following its development there were only minor discoveries of new superconductors, and almost no increase in their transition temperatures. It began to be suspected that there might be physical laws that prevented higher values of T_c, and papers were written to show why improvements were likely to be impossible.

This era ended abruptly in 1986 with the discovery of a whole new class of materials with much higher transition temperatures. Once again nature showed that she holds unsuspected secrets, which she is, however, willing, sparingly, to uncover to diligent and imaginative search.

CHAPTER 33

INDUCTANCE

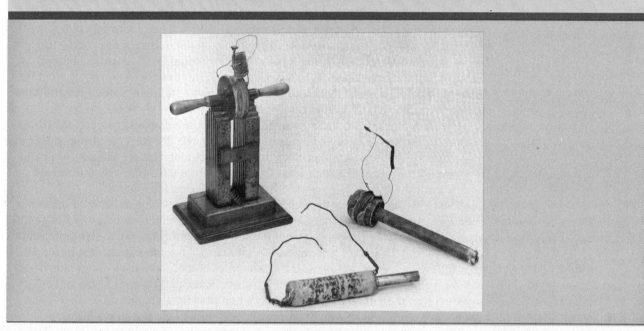

The figure shows the modest kind of apparatus with which Michael Faraday discovered the law of induction. It must be remembered, however, that in those early days such amenities as insulated wire were not commercially available. It is said that Faraday insulated his wires by wrapping them with strips from one of his wife's petticoats. There are not many fields of physics today in which significant new results can be obtained with such simple apparatus.

33-1 Capacitors and Inductors

We found in Chapter 27 that a *capacitor* (symbol ⊣⊢; see Fig. 1a) is an arrangement that we can conveniently use to set up a known *electric* field in a given region of space. We took the parallel plate arrangement as a convenient prototype.

Symmetrically, we can define an *inductor* (symbol ⎓⎓⎓⎓; see Fig. 1b) as an arrangement that we can conveniently use to set up a known *magnetic* field in a specified region. We take a long solenoid (more specifically, a short length near the center of a long solenoid) as a convenient prototype.

We can express the connection between capacitors and inductors in symbolic form:

inductor : magnetic field : : capacitor : electric field.

Physicists delight in such symmetries, parallelisms and equivalencies. Apart from their inherent aesthetic appeal, as a practical matter we are able to learn new things by leaning on old knowledge. For example, we shall learn that energy can be stored in the magnetic field of an inductor just as it can in the electric field of a capacitor.

Furthermore, we have seen that, if we connect a battery to a capacitor and a resistor in series, the circuit does not come to its final equilibrium state at once but

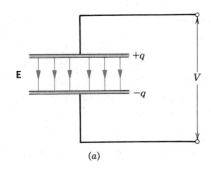

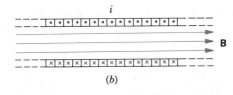

Figure 1 (a) A *capacitor,* of the parallel-plate geometry, displaying its associated *electric* field. (b) An *inductor,* (the central portion of a long solenoid), displaying its associated *magnetic* field.

approaches it exponentially. We shall learn in this chapter that the same thing is true if we connect a battery to an inductor and a resistor in series.

Let us start by defining the *inductance* of an inductor.

33-2 Inductance

A *capacitor* is an arrangement such that, if you place a charge q on its plates, a potential difference V appears across them, the *capacitance C* being given by

$$C = \frac{q}{V} \quad \text{(capacitance defined)}. \quad (1)$$

The SI unit of capacitance is, as we have seen, the *farad,* named after Michael Faraday.

An *inductor* is an arrangement such that, if you establish a current i in its windings, a magnetic flux Φ links each winding, the *inductance L* being given by

$$\boxed{L = \frac{N\Phi}{i}} \quad \text{(inductance defined)}, \quad (2)$$

in which N is the number of turns. The product $N\Phi$ is called the number of *flux linkages.*

The SI unit of magnetic flux is the tesla · meter² so that the SI unit of inductance is the $\text{T} \cdot \text{m}^2/\text{A}$. We call this the *henry* (abbr. H), after the American physicist Joseph Henry, the co-discoverer of the law of induction and a contemporary of Faraday. Thus

$$1 \text{ henry} = 1 \text{ H} = 1 \text{ T} \cdot \text{m}^2/\text{A}. \quad (3)$$

Throughout this chapter we assume that all inductors, no matter what their geometrical arrangement, have no magnetic materials such as iron in their vicinity.

The Inductance of a Solenoid. Consider a long solenoid of cross-sectional area A. What is the inductance per unit length near its center?

To use the defining equation for inductance (Eq. 2) we must calculate the number of flux linkages set up by a given current in the solenoid windings. Consider a length l near the center of this solenoid. The number of flux linkages for this section of the solenoid is

$$N\Phi = (nl)(BA)$$

in which n is the number of turns per unit length of the solenoid and B is the magnetic field within the solenoid.

B is given by Eq. 21 of Chapter 31

$$B = \mu_0 ni$$

so that, from Eq. 2,

$$L = \frac{N\Phi}{i} = \frac{(nl)(B)(A)}{i} = \frac{(nl)(\mu_0 ni)(A)}{i} = \mu_0 n^2 lA. \quad (4)$$

Thus the inductance per unit length for a long solenoid near its center is then

$$\boxed{L/l = \mu_0 n^2 A} \quad \text{(solenoid)}. \quad (5)$$

The inductance—like the capacitance—depends only on geometrical factors. The dependence on the square of the number of turns per unit length is to be expected. If you triple n you not only triple the number of turns (N) but you also triple the flux (Φ) through each turn, a factor of nine for the flux linkages and hence (see Eq. 2) for the inductance.

If the solenoid is very much longer than its radius, then Eq. 4 gives its inductance to a good approximation. This approximation neglects the spreading of the magnetic field lines near the ends of the solenoid, just as the parallel-plate capacitor formula ($C = \varepsilon_0 A/d$) neglects the fringing of the electric field lines near the edges of the capacitor plates.

Inductance of a Toroid. Figure 2 shows a toroid of N turns and of rectangular cross section with the dimensions indicated. What is its inductance?

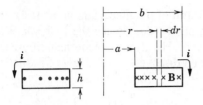

Figure 2 A cross section of a toroid, showing the current in the windings and the associated magnetic field.

Once more, to use the defining equation for inductance (Eq. 2), we must calculate the number of flux linkages that are set up by a given current. To do so, we must know how the magnetic field within the toroid depends on the current in its windings. We solved this problem in Chapter 31, where we learned that the magnetic field, which is not uniform over the cross section of a toroid, is given by Eq. 22 of that chapter, or

$$B = \frac{\mu_0 i N}{2\pi r}, \qquad (6)$$

in which i is the current in the toroid windings.*

The flux Φ over the toroid cross section must be found by integration. If $h\, dr$ is the area of the elementary strip shown between the dashed lines in Fig. 2, we have, using Eq. 6

$$\Phi = \int \mathbf{B} \cdot d\mathbf{A} = \int_a^b (B)(h\, dr) = \int_a^b \frac{\mu_0 i N}{2\pi r} h\, dr$$
$$= \frac{\mu_0 i N h}{2\pi} \int_a^b \frac{dr}{r} = \frac{\mu_0 i N h}{2\pi} \ln \frac{b}{a}.$$

The inductance then follows from Eq. 2, its defining equation,

$$L = \frac{N\Phi}{i} = \frac{\mu_0 i N^2 h}{2\pi i} \ln \frac{b}{a},$$

or

$$\boxed{L = \frac{\mu_0 N^2 h}{2\pi} \ln \frac{b}{a}} \qquad \text{(toroid).} \qquad (7)$$

Note again that the inductance depends only on geometrical factors and that the number of turns enters as a square.

---

* Equation 6 holds no matter what the shape or dimensions of the toroid cross section.

Recall (see Section 27–3) that a capacitance can always be written as the *permittivity constant* ϵ_0 times a quantity with the dimensions of a length; thus ϵ_0 can be expressed in farads per meter. We see from Eq. 7 that an inductance can be written as the *permeability constant* μ_0 times a quantity with the dimensions of a length. This means that the permeability constant μ_0 can be expressed in henrys per meter, or

$$\mu_0 = 4\pi \times 10^{-7} \text{ T} \cdot \text{m/A} = 4\pi \times 10^{-7} \text{ H/m.} \qquad (8)$$

Sample Problem 1 The toroid of Fig. 2 has $N = 1250$ turns, $a = 52$ mm, $b = 95$ mm, and $h = 13$ mm. What is its inductance?

From Eq. 7

$$L = \frac{\mu_0 N^2 h}{2\pi} \ln \frac{b}{a}.$$
$$= \frac{(4\pi \times 10^{-7} \text{ H/m})(1250)^2(13 \times 10^{-3} \text{ m})}{2\pi} \ln \frac{95 \text{ mm}}{52 \text{ mm}}$$
$$= 2.45 \times 10^{-3} \text{ H} = 2.45 \text{ mH.} \qquad \text{(Answer)}$$

33–3 Self-Induction

If two coils—which we can now call inductors—are near each other, a current i in one coil will set up a magnetic flux Φ through the second coil. We learned in Chapter 32 that, if we change this flux by changing the current, an induced emf will appear in the second coil according to Faraday's law; see Fig. 2 of Chapter 32.

An induced emf also appears in a coil if we change the current in that same coil; see Fig. 3.

This is called *self-induction* and the electromotive force produced is called a *self-induced emf.* It obeys Faraday's law of induction just as other induced emfs do.

For any inductor, Eq. 2 tells us that

$$N\Phi = Li. \qquad (9)$$

Faraday's law tells us

$$\mathcal{E} = -\frac{d(N\Phi)}{dt}. \qquad (10)$$

By combining Eqs. 9 and 10 we can write, for the self-in-

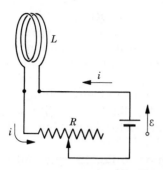

Figure 3 If the current in the coil L is changed, by varying the contact position on resistor R, a self-induced emf will appear in the coil.

duced emf

$$\boxed{\mathcal{E} = -L\frac{di}{dt}} \quad \text{(self-induced emf).} \quad (11)$$

Thus in any inductor (such as a coil, a solenoid, or a toroid) a self-induced emf appears whenever the current changes with time. The magnitude of the current itself has no influence on the induced emf; only the rate of change of the current counts.

You can find the *direction* of a self-induced emf from Lenz's law. The minus sign in Eq. 11 represents the fact that — as the law states — the self-induced emf acts to oppose the change that brings it about.

Suppose that, as in Fig. 4a, you set up a current i in a coil and arrange to have it increase with time at a rate di/dt. In the language of Lenz's law this increase in the current is the "change" that the self-induction must oppose. To do so a self-induced emf must appear in the coil,

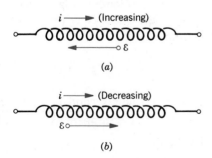

Figure 4 (a) If the current i is increasing, the induced emf appears in a direction such that it opposes the increase. (b) If the current i is decreasing, the induced emf appears in a direction such that it opposes the decrease.

pointing — as the figure shows — so as to oppose the increase in the current. If you cause the current to decrease with time, as in Fig. 4b, the self-induced emf must point in a direction that tends to oppose the decrease in the current, as the figure shows.

33-4 An LR Circuit

In Section 29-8 we saw that if you suddenly introduce an emf $\mathcal{E}$ into a single-loop circuit containing a resistor R and a capacitor C, the charge does not build up immediately to its final equilibrium value $C\mathcal{E}$ but approaches it in an exponential fashion described by Eq. 29 of Chapter 29, or

$$q = C\mathcal{E}(1 - e^{-t/\tau_c}). \quad (12)$$

The rate at which the charge builds up is determined by the capacitive time constant τ_C, defined from

$$\tau_C = RC. \quad (13)$$

If you suddenly remove the emf from this same circuit, the charge does not immediately fall to zero but approaches zero in an exponential fashion, described by Eq. 34 of Chapter 29, or

$$q = q_0 e^{-t/\tau_c}. \quad (14)$$

The same time constant τ_C describes the fall of the charge as well as its rise.

An analogous slowing of the rise (or fall) of the current occurs if we introduce an emf $\mathcal{E}$ into (or remove it from) a single-loop circuit containing a resistor R and an inductor L. With the switch S in Fig. 5 closed on a, for example, the current in the resistor starts to rise. If the inductor were not present, the current would rise rapidly to a steady value $\mathcal{E}/R$. Because of the inductor, however, a self-induced emf, which we label $\mathcal{E}_L$, appears in the circuit and, from Lenz's law, this emf opposes the rise of current, which means that it opposes the battery emf $\mathcal{E}$ in

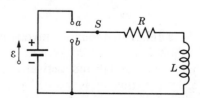

Figure 5 An LR circuit. When switch S is closed on a, the current rises and approaches a limiting value of $\mathcal{E}/R$.

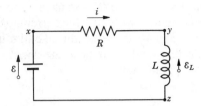

Figure 6 The circuit of Fig. 5 with the switch closed on *a*. Apply the loop theorem clockwise, starting at *x*.

polarity. Thus the resistor responds to the difference between two emfs, a constant one $\mathcal{E}$ due to the battery and a variable one $\mathcal{E}_L$ $(= -L\,di/dt)$ due to self-induction. As long as this second emf is present, the current in the resistor will be less than $\mathcal{E}/R$.

As time goes on, the rate at which the current increases becomes less rapid and the magnitude of the self-induced emf, which is proportional to di/dt, becomes smaller. Thus the current in the circuit approaches $\mathcal{E}/R$ asymptotically.

Now let us analyze the situation quantitatively. With the switch *S* in Fig. 5 thrown to *a*, the circuit is equivalent to that of Fig. 6. Let us apply the loop theorem, starting at *x* in this figure and going clockwise around the loop. For the direction of current shown, *x* will be higher in potential than *y*, which means that we encounter a drop in potential of $-iR$ as we traverse the resistor. Point *y* is higher in potential than point *z* because, for an increasing current, the induced emf will oppose the rise of the current by pointing as shown. Thus, as we traverse the inductor from *y* to *z*, we encounter a drop in potential of $-L(di/dt)$. We encounter a rise in potential of $+\mathcal{E}$ in traversing the battery from *z* to *x*. The loop theorem thus gives

$$-iR - L\frac{di}{dt} + \mathcal{E} = 0$$

or

$$\boxed{iR + L\frac{di}{dt} = \mathcal{E}} \qquad \text{(LR circuit).} \quad (15)$$

Equation 15 is a differential equation involving the variable *i* and its first derivative di/dt. We seek the function $i(t)$ such that when it and its first derivative are substituted in Eq. 15, the equation is satisfied and the initial condition $i(0) = 0$ is also satisfied.

Although there are formal rules for solving various classes of differential equations (and Eq. 15 can, in fact,

be easily solved by direct integration, after rearrangement), we often find it simpler to guess at the solution, guided by physical reasoning and by previous experience. We can test any proposed solution by substituting it in the differential equation and seeing whether this equation reduces to an identity. In this case, we will be guided by the fact that our solution should be closely analogous to Eq. 12 for the build-up of charge in an *RC* circuit.

The solution to Eq. 15 which satisfies the initial condition is, we then claim,

$$i = \frac{\mathcal{E}}{R}(1 - e^{-Rt/L}). \quad (16)$$

To test this solution by substitution, we find the derivative di/dt, which is

$$\frac{di}{dt} = \frac{\mathcal{E}}{L}e^{-Rt/L}. \quad (17)$$

Substituting *i* and di/dt into Eq. 15 leads to an identity, as you can easily check. Thus Eq. 16 is indeed a solution of Eq. 15.

We can rewrite Eq. 16 as

$$\boxed{i = \frac{\mathcal{E}}{R}(1 - e^{-t/\tau_L}),} \qquad \text{(rise of current)} \quad (18)$$

in which τ_L, the *inductive time constant,* is given by

$$\boxed{\tau_L = L/R} \qquad \text{(time constant).} \quad (19)$$

Figure 7 shows how the potential difference V_R across the resistor $(= iR)$ and V_L across the inductor $(= L\,di/dt)$ vary with time for particular values of $\mathcal{E}$, *L*, and *R*. Compare this figure carefully with the corresponding figure for an *RC* circuit (Fig. 16 of Chapter 29).

To show that the quantity τ_L $(= L/R)$ has the dimensions of time we put

$$\frac{1\text{ henry}}{\text{ohm}}$$

$$= \frac{1\text{ henry}}{\text{ohm}}\left(\frac{1\text{ volt}\cdot\text{second}}{\text{henry}\cdot\text{ampere}}\right)\left(\frac{1\text{ ohm}\cdot\text{ampere}}{\text{volt}}\right)$$

$$= 1\text{ second.}$$

The first quantity in parentheses is a conversion factor based on Eq. 11. The second conversion factor is based on the relation $V = iR$.

The physical significance of the time constant fol-

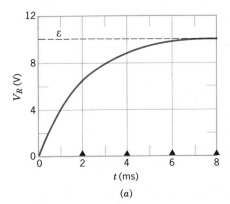

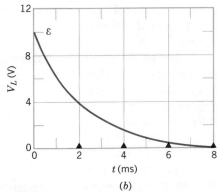

(a)

(b)

Figure 7   The variation with time of (a) V_R, the potential difference across the resistor in the circuit of Fig. 6 and (b) V_L, the potential difference across the inductor in that circuit. The triangles represent successive inductive time constants ($t = \tau_L, 2\tau_L, 3\tau_L, \ldots$). The figure is plotted for $R = 2000\,\Omega$, $L = 4.0\,\text{H}$, and $\mathscr{E} = 10\,\text{V}$.

lows from Eq. 18. If we put $t = \tau_L = L/R$ in this equation, it reduces to

$$i = \frac{\mathscr{E}}{R}(1 - e^{-1}) = 0.63\,\frac{\mathscr{E}}{R}.$$

Thus the time constant τ_L is that time at which the current in the circuit will reach within $1/e$ (about 37%) of its final equilibrium value (see Fig. 7).

If the switch S in Fig. 5, having been left in position a long enough time for the equilibrium current $\mathscr{E}/R$ to be established, is thrown to b, the effect is to remove the battery from the circuit.* The differential equation that governs the subsequent decay of the current in the circuit can be found by putting $\mathscr{E} = 0$ in Eq. 15, or

$$L\frac{di}{dt} + iR = 0.$$

You can show by the test of substitution that the solution of this differential equation which satisfies the initial condition $i(0) = i_0 = \dfrac{\mathscr{E}}{R}$ is

$$\boxed{i = \frac{\mathscr{E}}{R}\,e^{-t/\tau_L} = i_0 e^{-t/\tau_L}} \qquad \text{(decay of current).} \quad (20)$$

We see that both the rise in current (Eq. 18) and the decay of the current (Eq. 20) in an LR circuit are governed by the same inductive time constant, τ_L.

---

* The connection to b must be made before the connection to a is broken. A switch that does this is called a *make-before-break* switch.

We have used i_0 in Eq. 20 to represent the current at time $t = 0$, which in our case happened to be $\mathscr{E}/R$ but could have been any initial value. The time dependence of the decaying current is identical to that of the decaying potential difference, plotted in Fig. 7b.

Sample Problem 2   A solenoid has an inductance of 53 mH and a resistance of 0.37 Ω. If it is connected to a battery, how long will it take for the current to reach one-half its final equilibrium value?

The equilibrium value of the current is reached as $t \to \infty$; from Eq. 16 it is $\mathscr{E}/R$. If the current has half this value at a particular time t_0, this equation becomes

$$\frac{1}{2}\frac{\mathscr{E}}{R} = \frac{\mathscr{E}}{R}(1 - e^{-t_0/\tau_L}).$$

Solving for t_0 by rearranging and taking the (natural) logarithm of each side, we find

$$t_0 = \tau_L \ln 2$$
$$= \ln 2\,\frac{L}{R} = \ln 2\,\frac{53 \times 10^{-3}\,\text{H}}{0.37\,\Omega}$$
$$= 0.10\,\text{s} = 100\,\text{ms.} \qquad \text{(Answer)}$$

33-5   Energy and the Magnetic Field

When we lift a stone we do work, which we can get back by lowering the stone. It is convenient to think of the energy being stored temporarily in the gravitational field of the earth and the lifted stone and being withdrawn from this field when we lower the stone.

When we pull two unlike charges apart we like to say that the resulting electric potential energy is stored in the electric field of the charges. We can get it back from the field by letting the charges move closer together again.

In the same way we can consider energy to be stored in a magnetic field. For example, two long, rigid, parallel wires carrying current in the same direction attract each other, and we must do work to pull them apart. We can get this stored energy back at any time by letting the wires move back to their original positions.

To derive a quantitative expression for the energy stored in the magnetic field, consider Fig. 6, which shows a source of emf $\mathcal{E}$ connected to a resistor R and an inductor L. Equation 15

$$\mathcal{E} = iR + L\frac{di}{dt}, \tag{21}$$

is the differential equation that describes the growth of current in this circuit. We stress that this equation follows immediately from the loop theorem and that the loop theorem in turn is an expression of the principle of conservation of energy for single-loop circuits. If we multiply each side of Eq. 21 by i, we obtain

$$\mathcal{E}i = i^2R + Li\frac{di}{dt}, \tag{22}$$

which has the following physical interpretation in terms of work and energy:

1. If a charge dq passes through the seat of emf $\mathcal{E}$ in Fig. 6 in time dt, the seat does work on it in amount $\mathcal{E}\,dq$. The rate of doing work is $(\mathcal{E}\,dq)/dt$, or $\mathcal{E}i$. Thus the left term in Eq. 22 is the rate at which the seat of emf delivers energy to the circuit.

2. The second term in Eq. 22 is the rate at which energy appears as thermal energy in the resistor.

3. Energy that does not appear as thermal energy must, by our hypothesis, be stored in the magnetic field. Since Eq. 22 represents a statement of the conservation of energy for LR circuits, the last term must represent the rate dU_B/dt at which energy is stored in the magnetic field, or

$$\frac{dU_B}{dt} = Li\frac{di}{dt}. \tag{23}$$

We can write this as

$$dU_B = Li\,di.$$

Integrating yields

$$U_B = \int dU_B = \int_0^i Li\,di$$

or

$$\boxed{U_B = \tfrac{1}{2}Li^2} \quad \text{(magnetic energy),} \tag{24}$$

which represents the total energy stored by an inductor L carrying a current i.

We can compare this relation with the expression for the energy associated with a capacitor C carrying a charge q, namely,

$$U_E = \frac{q^2}{2C}. \tag{25}$$

Here the energy is stored in an electric field. In each case the expression for the stored energy was derived by setting it equal to the work that must be done to set up the field.

Sample Problem 3 A coil has an inductance of 53 mH and a resistance of 0.35 Ω. (a) If a 12-V emf is applied, how much energy is stored in the magnetic field after the current has built up to its equilibrium value?

The stored energy is given by Eq. 24

$$U_B = \tfrac{1}{2}Li^2.$$

To find the equilibrium stored energy, we must substitute the equilibrium current in this expression. From Eq. 16 the equilibrium current is

$$i_\infty = \frac{\mathcal{E}}{R} = \frac{12\text{ V}}{0.35\ \Omega} = 34.3\text{ A}.$$

The substitution yields

$$U_{B\infty} = \tfrac{1}{2}Li_\infty^2 = (\tfrac{1}{2})(53 \times 10^{-3}\text{ H})(34.3\text{ A})^2$$
$$= 31\text{ J}. \tag{Answer}$$

(b) After how many time constants will half of this equilibrium energy be stored in the magnetic field?

We are asked: At what time will the relation

$$U_B = \tfrac{1}{2}U_{B\infty}$$

be satisfied? Equation 24 allows us to rewrite this as

$$\tfrac{1}{2}Li^2 = (\tfrac{1}{2})\tfrac{1}{2}Li_\infty^2$$

or

$$i = (1/\sqrt{2})i_\infty.$$

But i is given by Eq. 16 and i_∞ (see above) is $\mathscr{E}/R$, so that

$$\frac{\mathscr{E}}{R}(1 - e^{-t/\tau_L}) = \frac{\mathscr{E}}{\sqrt{2}R}.$$

This can be written as

$$e^{-t/\tau_L} = 1 - 1/\sqrt{2} = 0.293,$$

which yields

$$\frac{t}{\tau_L} = -\ln 0.293 = 1.23$$

or

$$t = 1.23\,\tau_L. \qquad \text{(Answer)}$$

Thus the stored energy will reach half of its equilibrium value after 1.23 time constants.

Sample Problem 4 A 3.56-H inductor is placed in series with a 12.8-Ω resistor, an emf of 3.24 V being suddenly applied to the combination. At 0.278 s (which is one inductive time constant) after the contact is made, (a) what is the rate P at which energy is being delivered by the battery?

The rate, P, is available from $P = \mathscr{E}i$. The current is given by Eq. 18, or

$$i = \frac{\mathscr{E}}{R}(1 - e^{-t/\tau_L}),$$

which, after one time constant, becomes

$$i = \frac{3.24\text{ V}}{12.8\ \Omega}(1 - e^{-1}) = 0.1600\text{ A}.$$

The rate at which the battery delivers energy is then

$$P = \mathscr{E}i = (3.24\text{ V})(0.1600\text{ A})$$
$$= 0.5184\text{ W} \approx 518\text{ mW}. \qquad \text{(Answer)}$$

(b) At what rate P_R does energy appear as thermal energy in the resistor?

This is given by

$$P_R = i^2R = (0.1600\text{ A})^2(12.8\ \Omega)$$
$$= 0.3277\text{ W} \approx 328\text{ mW}. \qquad \text{(Answer)}$$

(c) At what rate P_B is energy being stored in the magnetic field?

This is given by Eq. 23, which requires that we know di/dt. Differentiating Eq. 18 yields

$$\frac{di}{dt} = \frac{\mathscr{E}}{R}\frac{R}{L}(e^{-t/\tau_L}) = \frac{\mathscr{E}}{L}e^{-t/\tau_L}$$

After one time constant we have

$$\frac{di}{dt} = \frac{3.24\text{ V}}{3.56\text{ H}}e^{-1} = 0.3348\text{ A/s}.$$

From Eq. 23 the desired rate is then

$$P_B = \frac{dU_B}{dt} = Li\frac{di}{dt}$$
$$= (3.56\text{ H})(0.1600\text{ A})(0.3348\text{ A/s})$$
$$= 0.1907\text{ W} \approx 191\text{ mW}. \qquad \text{(Answer)}$$

Note that, as required by energy conservation,

$$P = P_R + P_B,$$

or

$$P = 0.3277\text{ W} + 0.1907\text{ W} = 0.5184\text{ W} \approx 518\text{ mW}.$$

33-6 Energy Density and the Magnetic Field

So far we have dealt with the energy U stored in the magnetic field of a specific current-carrying inductor. Here we turn our attention to the magnetic field itself—regardless of its source—and seek an expression for the *energy density* u_B, the energy per unit volume stored at any point in the field.

Consider a length l near the center of a long solenoid; Al is the volume associated with this length. The stored energy must lie entirely within this volume because the magnetic field outside such a solenoid is essentially zero. Moreover, the stored energy must be uniformly distributed throughout the volume of the solenoid because the magnetic field is uniform everywhere inside.

Thus, we can write

$$u_B = \frac{U_B}{Al}$$

or, since

$$U_B = \tfrac{1}{2}Li^2,$$

we have

$$u_B = \frac{(L/l)i^2}{2A}.$$

We can substitute for L/l for the solenoid from Eq. 5, obtaining

$$u_B = \tfrac{1}{2}\mu_0 n^2 i^2.$$

From Eq. 21 of Chapter 31 ($B = \mu_0 in$) we can write this finally as

$$\boxed{u_B = \frac{B^2}{2\mu_0}} \qquad \text{(magnetic energy density).} \quad (26)$$

Table 1 Some Corresponding Electrical and Magnetic Quantities

Definition	$C = q/V$	$L = N\Phi/i$
Dimensions	$C = \epsilon_0 \times$ a length	$L = \mu_0 \times$ a length
Constants	$\epsilon_0 = 8.85$ pF/m	$\mu_0 = 1.26$ μH/m
Energy storage	$U_C = \frac{1}{2} CV^2 = \frac{q^2}{2C}$	$U_L = \frac{1}{2} Li^2 = \frac{(N\Phi)^2}{2L}$
Energy density	$u_E = (\epsilon_0/2)E^2$	$u_B = (1/2\mu_0)B^2$
Time constant	$\tau_C = RC$	$\tau_L = L/R$

This equation gives the energy density stored at any point where the magnetic field is B. Even though we derived Eq. 26 by considering a special case, the solenoid, it remains true for all magnetic field configurations, no matter how they are generated. Equation 26 is to be compared with Eq. 23 of Chapter 27, namely

$$u_E = \tfrac{1}{2}\epsilon_0 E^2, \tag{27}$$

which gives the energy density (in a vacuum) at any point in an electric field. Note that both u_B and u_E are proportional to the square of the appropriate field quantity, B or E.

The solenoid plays a role in relation to magnetic fields similar to the role the parallel-plate capacitor plays with respect to electric fields. In each case we have a simple device that can be used for setting up a uniform field throughout a well-defined region of space and for deducing, in a simple way, some properties of these fields. Table 1 compares some corresponding electrical and magnetic quantities.

Sample Problem 5 A long coaxial cable (Fig. 8) consists of two concentric cylinders with radii a and b. Its central conductor carries a steady current i, the outer conductor providing the return path. (a) Calculate the energy stored in the magnetic field between the conductors for a length l of such a cable.

Consider a volume dV consisting of a cylindrical shell whose radii are r and $r + dr$ and whose length is l. The energy dU contained within this shell is

$$dU = u_B \, dV.$$

The energy density, from Eq. 26, is

$$u_B = \frac{B^2}{2\mu_0}.$$

In the space between the two conductors Ampere's law,

$$\oint \mathbf{B} \cdot d\mathbf{s} = \mu_0 i,$$

leads to

$$(B)(2\pi r) = \mu_0 i,$$

or

$$B = \frac{\mu_0 i}{2\pi r}.$$

The energy density is then

$$u_B = \frac{1}{2\mu_0} \left(\frac{\mu_0 i}{2\pi r} \right)^2 = \frac{\mu_0 i^2}{8\pi^2 r^2}.$$

The volume, dV, of the shell is $(2\pi r l)(dr)$ so the energy dU contained within the shell is

$$dU = \frac{\mu_0 i^2}{8\pi^2 r^2} (2\pi r l)(dr) = \frac{\mu_0 i^2 l}{4\pi} \frac{dr}{r}.$$

The total energy follows by integrating this expression over the volume between the cylinders.

$$U = \int dU = \frac{\mu_0 i^2 l}{4\pi} \int_a^b \frac{dr}{r} = \frac{\mu_0 i^2 l}{4\pi} \ln \frac{b}{a}. \tag{28}$$

No energy is stored outside the outer conductor or inside the hollow inner conductor because the magnetic field is zero there, as you can show from Ampere's law. Note that we could also have arrived at this result by calculating the inductance of a length l of the cable, using Eq. 2, then used Eq. 24 to find the stored energy.

(b) What is the stored energy per unit length for $a = 1.2$ mm, $b = 3.5$ mm, and $i = 2.7$ A?

From Eq. 28 we have

$$U/l = \frac{\mu_0 i^2}{4\pi} \ln \frac{b}{a} = \frac{(4\pi \times 10^{-7} \text{ H/m})(2.7 \text{ A})^2}{4\pi} \ln \frac{3.5 \text{ mm}}{1.2 \text{ mm}}$$
$$= 7.8 \times 10^{-7} \text{ J/m} = 780 \text{ nJ/m}. \tag{Answer}$$

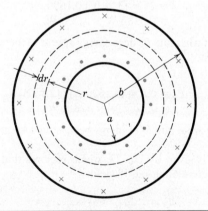

Figure 8 Sample Problem 5. A cross section of a long coaxial cable, of inner radius a and outer radius b. The dots and crosses represent equal but opposite currents in the two thin concentric conductors.

Sample Problem 6 Compare the energy required to set up, in a cube 10 cm on edge (a) a uniform electric field of 100 kV/m and (b) a uniform magnetic field of 1.0 T. Both these fields would be judged reasonably large but readily available in the laboratory.

(a) In the electric case we have, where V_0 is the volume of the cube,

$$\begin{aligned}
U_E = u_E V_0 &= \tfrac{1}{2}\epsilon_0 E^2 V_0 \\
&= (\tfrac{1}{2})(8.85 \times 10^{-12}\ \text{F/m})(10^5\ \text{V/m})^2(0.1\ \text{m})^3 \\
&= 4.4 \times 10^{-5}\ \text{J} = 44\ \mu\text{J}. \qquad \text{(Answer)}
\end{aligned}$$

(b) In the magnetic case, we have

$$U_B = u_B V_0 = \frac{B^2}{2\mu_0} V_0 = \frac{(1.0\ \text{T})^2(0.1\ \text{m})^3}{(2)(4\pi \times 10^{-7}\ \text{T}\cdot\text{m/A})}$$
$$= 398\ \text{J} \approx 400\ \text{J}. \qquad \text{(Answer)}$$

In terms of fields normally available in the laboratory, much larger amounts of energy can be stored in a magnetic field than in an electric one, the ratio being about 10^7 in this example. Conversely, much more energy is required to set up a magnetic field of "reasonable" laboratory magnitude than is required to set up an electric field of similarly reasonable magnitude.

33–7 Mutual Induction (Optional)

In this section we return to the case of two interacting coils, which we first discussed in Section 32–2, and we treat it in a somewhat more formal manner. In Fig. 2 of that chapter, we saw that if two coils are close together, a steady current i in one coil will set up a magnetic flux Φ linking the other coil. If we change i with time, an emf $\mathcal{E}$ given by Faraday's law appears in the second coil; we called this process *induction*. We could better have called it *mutual induction*, to suggest the mutual interaction of the two coils and to distinguish it from *self-induction*, in which only one coil is involved.

Let us look a little more quantitatively at mutual induction. Figure 9a shows two circular close-packed coils near each other and sharing a common axis.* There is a steady current i_1 in coil 1, set up by the battery in the external circuit. This current produces a magnetic field suggested by the lines of B_1 in the figure. Coil 2 is connected to a sensitive galvanometer G but contains no battery; a magnetic flux Φ_{21} (the flux through coil 2

---

* For ease of representation, the two coils in Fig. 9 are not actually drawn as close-packed, which requires that every turn of a given coil enclose the same flux.

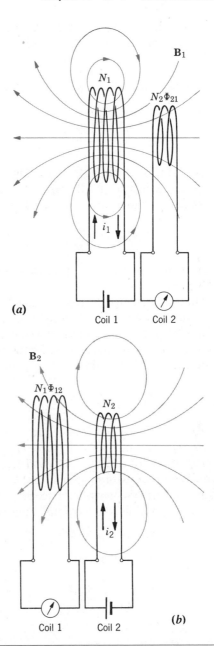

Figure 9 Mutual inductance. (a) If the current in coil 1 changes, an emf will be induced in coil 2. (b) If the current in coil 2 changes, an emf will be induced in coil 1.

associated with the current in coil 1) is linked by its N_2 turns.

We define the mutual inductance M_{21} of coil 2 with respect to coil 1 as

$$M_{21} = \frac{N_2 \Phi_{21}}{i_1}. \qquad (29)$$

Compare this with Eq. 2 ($L = N\Phi/i$), the definition of (self) inductance. We can recast Eq. 29 as

$$M_{21}i_1 = N_2\Phi_{21}.$$

If, by external means, we cause i_1 to vary with time, we have

$$M_{21}\frac{di_1}{dt} = N_2\frac{d\Phi_{21}}{dt}.$$

The right side of this equation, from Faraday's law, is, apart from a change in sign, just the emf $\mathcal{E}_2$ appearing in coil 2 due to the changing current in coil 1, or

$$\mathcal{E}_2 = -M_{21}\frac{di_1}{dt}, \qquad (30)$$

which you should compare with Eq. 11 ($\mathcal{E} = -L\,di/dt$) for self-inductance.

Let us now interchange the roles of coils 1 and 2, as in Fig. 9b. That is, we set up a current i_2 in coil 2, by means of a battery, and this produces a magnetic flux Φ_{12} that links coil 1. If we change i_2 with time, we have, by the same argument given above,

$$\mathcal{E}_1 = -M_{12}\frac{di_2}{dt}. \qquad (31)$$

Thus we see that the emf induced in either coil is proportional to the rate of change of current in the other coil. The proportionality constants M_{21} and M_{12} seem to be different. We assert, without proof, that they are in fact the same so that no subscripts are needed. This conclusion is true but is in no way obvious. Thus, we have

$$M_{21} = M_{12} = M \qquad (32)$$

and we can rewrite Eqs. 30 and 31 as

$$\boxed{\mathcal{E}_2 = -M\,di_1/dt} \qquad (33)$$

and

$$\boxed{\mathcal{E}_1 = -M\,di_2/dt.} \qquad (34)$$

The induction is indeed mutual. The SI unit for M (as for L) is the henry.

Sample Problem 7 Figure 10 shows two circular close-packed coils, the smaller (radius R_2) being coaxial with the larger (radius R_1) and in the same plane. (a) Derive an expres-

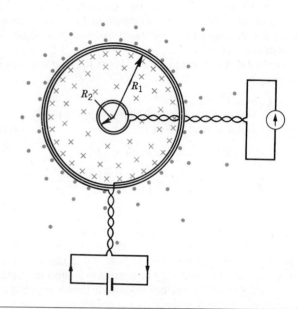

Figure 10 Sample Problem 7. A small coil is placed at the center of a large coil. What is their mutual inductance?

sion for the coefficient of mutual inductance M for this arrangement of these two coils, assuming that $R_1 \gg R_2$.

As the figure suggests we imagine that we establish a current i_1 in the larger coil and we note the magnetic field B_1 that it sets up. The value of B_1 at the center of this coil is (see Eq. 24 of Chapter 31, with $z = 0$)

$$B_1 = \frac{\mu_0 i_1 N_1}{2R_1}.$$

Because we have assumed that $R_1 \gg R_2$, we may take B_1 to be the magnetic field at all points within the boundary of the smaller coil. The flux linkages for the smaller coil are then

$$N_2\Phi_{21} = N_2(B_1)(\pi R_2^2) = \frac{\pi\mu_0 N_1 N_2 R_2^2 i_1}{2R_1}.$$

From Eq. 29 we then have

$$M = \frac{N_2\Phi_{21}}{i_1} = \frac{\pi\mu_0 N_1 N_2 R_2^2}{2R_1}. \qquad \text{(Answer)}$$

(b) What is the value of M for $N_1 = N_2 = 1200$ turns, $R_2 = 1.1$ cm and $R_1 = 15$ cm.

The above equation yields

$$M = \frac{(\pi)(4\pi \times 10^{-7}\ \text{H/m})(1200)(1200)(0.011\ \text{m})^2}{(2)(0.15\ \text{m})}$$

$$= 2.29 \times 10^{-3}\ \text{H} \approx 2.3\ \text{mH}. \qquad \text{(Answer)}$$

Consider the situation if we reverse the roles of the two coils in Fig. 10, that is, if we set up a current i_2 in the smaller coil and try to calculate M from Eq. 29

$$M = \frac{N_1 \Phi_{12}}{i_2}.$$

The calculation of Φ_{12} (the flux of the smaller coil's magnetic field encompassed by the larger coil) is not simple. If we were to calculate it numerically using a computer, we would find exactly 2.3 mH, as above! This should make us appreciate the fact that Eq. 32 ($M_{12} = M_{21} = M$) is not obvious.

REVIEW AND SUMMARY

An Inductor

An *inductor* is an arrangement that can be used to set up a known magnetic field in a specified region. If a current i is established through each of the N windings of an inductor, a magnetic flux Φ links those windings. The *inductance* L of the inductor is

Inductance

$$L = \frac{N\Phi}{i} \quad \text{(inductance defined).} \tag{2}$$

The SI unit of inductance is the *henry* (abbr. H), with

$$1 \text{ henry} = 1 \text{ H} = 1 \text{ T} \cdot \text{m}^2/\text{A}. \tag{3}$$

The inductance per unit length of a long solenoid near its center is

Solenoid Inductance

$$L/l = \mu_0 n^2 A \quad \text{(solenoid)} \tag{5}$$

and the inductance of a rectangular toroid (height h, inner and outer radii a and b, N total turns) is

Toroid Inductance

$$L = \frac{\mu_0 N^2 h}{2\pi} \ln \frac{b}{a} \quad \text{(toroid);} \tag{7}$$

see Sample Problem 1.

If a current i in a coil changes with time, an emf described by Faraday's law is induced in the coil itself. This self-induced emf is given by

Self-induction

$$\mathcal{E} = -L \frac{di}{dt} \quad \text{(self-induced emf).} \tag{11}$$

The direction of $\mathcal{E}$ is found from Lenz's law; the self-induced emf acts to oppose the change that brings it about.

If a constant emf $\mathcal{E}$ is introduced into a single-loop circuit containing a resistor R and an inductor L (see Fig. 6), the current rises to an equilibrium value of $\mathcal{E}/R$ according to

Series *LR* Circuits

$$i = \frac{\mathcal{E}}{R} (1 - e^{-t/\tau_L}) \quad \text{(rise of current).} \tag{18}$$

Here $\tau_L \, (= L/R)$ governs the rate of rise of the current and is called the inductive time constant of the circuit. Figure 7, which shows the potential differences across the resistor and the inductor while the current is rising, deserves careful study; see also Sample Problem 2. The decay of the current when the source of constant emf is removed is given by

$$i = i_0 e^{-t/\tau_L} \quad \text{(decay of current).} \tag{20}$$

By applying the conservation-of-energy principle to the rise of current in the *LR* circuit of Fig. 6 we deduce that, if an inductor L carries a current i, an energy given by

Storage of Energy by an Inductor

$$U_B = \tfrac{1}{2} L i^2 \quad \text{(magnetic energy)} \tag{24}$$

can be viewed as stored in its magnetic field. Sample Problems 3 and 4 analyze the energy transfers in an *LR* circuit.

Applying Eq. 24 to a section of a long solenoid leads us to deduce the general result that, if B is the magnetic field at any point, the density of stored magnetic energy at that point is

Magnetic Field Energy

$$u_B = \frac{B^2}{2\mu_0} \quad \text{(magnetic energy density).} \qquad [26]$$

Sample Problems 5 and 6 deal with magnetic energy. Table 1 displays some interesting and useful analogies between electrical and magnetic quantities.

If two coils or similar conductors are near each other (see Fig. 9) a changing current in either coil can induce an emf in the other. This mutual induction phenomenon is described by

Mutual Induction

$$\mathscr{E}_2 = -M \, di_1/dt \quad \text{and} \quad \mathscr{E}_1 = -M \, di_2/dt. \qquad [33,34]$$

where M (measured in henries) is the coefficient of mutual inductance for the coil arrangement; see Sample Problem 7.

QUESTIONS

1. Explain how a long straight wire can show self-induction effects. How would you go about looking for them?

2. If the flux passing through each turn of a coil is the same, the inductance of the coil may be computed from $L = N\Phi_B/i$ (Eq. 2). How might one compute L for a coil for which this assumption is not valid?

3. Show that the dimensions of the two expressions for L, $N\Phi_B/i$ (Eq. 2) and $\mathscr{E}/(di/dt)$ (Eq. 11), are the same.

4. You want to wind a coil so that it has resistance but essentially no inductance. How would you do it?

5. Is the inductance per unit length for a solenoid near its center the same as, less than, or greater than the inductance per unit length near its ends? Justify your answer.

6. Explain why the self-inductance of a coaxial cable is expected to increase when the radius of the outer conductor is increased, the radius of the inner conductor remaining fixed.

7. A steady current is set up in a coil with a very large inductive time constant. When the current is interrupted with a switch, a heavy arc tends to appear at the switch blades. Explain why. (*Note:* Interrupting currents in highly inductive circuits can be destructive and dangerous.)

8. Suppose that you connect an ideal (that is, essentially resistanceless) coil across an ideal (again, essentially resistanceless) battery. You might think that, because there is no resistance in the circuit, the current would jump at once to a very large value. On the other hand, you might think that, because the inductive time constant ($= L/R$) is extremely large, the current would rise very slowly, if at all. What actually happens?

9. In an LR circuit like that of Fig. 6, can the self-induced emf ever be larger than the battery emf?

10. In an LR circuit like that of Fig. 6, is the current in the resistor always the same as the current in the inductor?

11. In the circuit of Fig. 5 the self-induced emf is a maximum at the instant the switch is closed on a. How can this be — there is no current in the inductor at this instant?

12. The switch in Fig. 5, having been closed on a for a "long" time, is thrown to b. What happens to the energy that is stored in the inductor?

13. A coil has a (measured) inductance L and a (measured) resistance R. Is its inductive time constant necessarily given by $\tau_L = L/R$? Bear in mind that we derived that equation (see Fig. 5) for a situation in which the inductive and resistive elements are physically separated. Discuss.

14. Figure 7a in this chapter and Figure 16b in Chapter 29, are plots of $V_R(t)$ for, respectively, an LR circuit and an RC circuit. Why are these two curves so different? Account for each in terms of physical processes going on in the appropriate circuit.

15. Two solenoids, A and B, have the same diameter and length and contain only one layer of copper windings, with adjacent turns touching, insulation thickness being negligible. Solenoid A contains many turns of fine wire and solenoid B contains fewer turns of heavier wire. (a) Which solenoid has the larger self-inductance? (b) Which solenoid has the larger inductive time constant? Justify your answers.

16. Can you make an argument based on the manipulation of bar magnets to suggest that energy may be stored in a magnetic field?

17. Draw all the formal analogies that you can think of between a parallel-plate capacitor (for electric fields) and a long solenoid (for magnetic fields).

18. In each of the following operations energy is expended. Some of this energy is returnable (can be reconverted) into electrical energy that can be made to do useful work and some becomes unavailable for useful work or is wasted in other ways. In which case will there be the *least* percentage of returnable electrical energy? (a) Charging a capacitor; (b) charging a storage battery; (c) sending a current through a resistor; (d) setting up a magnetic field; (e) moving a conductor in a magnetic field.

19. The current in a solenoid is reversed. What changes does this make in the magnetic field **B** and the energy density u at various points along the solenoid axis?

20. Commercial devices such as motors and generators that are involved in the transformation of energy between electrical and mechanical forms involve magnetic rather than electrostatic fields. Why should this be so?

21. A heavy current is passed, clockwise, through both coils of a standard mutual inductance shown in Fig. 11. Q is the horizontal midpoint of the large coil whose ends are P and S. The midpoint of the small coil starts a distance x from R. Describe its subsequent motion.

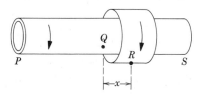

Figure 11   Question 21.

22. In a case of mutual induction, such as in Fig. 9, is self-induction also present? Discuss.

23. You are given two similar flat circular coils of N turns each. The centers of the coils are maintained at a fixed distance apart. For what orientation will their mutual inductance M be the greatest? For what orientation will it be the least?

24. A flat circular coil is placed completely outside a long solenoid, near its center, the axes of the coil and the solenoid being parallel. Describe the mutual induction, if any, of this combination.

25. A circular coil of N turns surrounds a long solenoid. Is the mutual inductance greater when the coil is near the center of the solenoid or when it is near one end? Justify your answer.

26. A coil, connected to an ac household power outlet, carries an alternating current. Explain why the amplitude of the current in the coil might change if you place a conducting metal object near it. Design a metal detector based on this principle.

27. A long cylinder is wound from left to right with one layer of wire, giving it n turns per unit length with a self-inductance of L_1, as in Fig. 12a. If the winding is now continued, in the same *sense* but returning from right to left, as in Fig. 12b, so as to give a second layer also of n turns per unit length, then what is the value of the self-inductance? Explain.

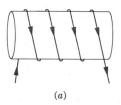

(a)

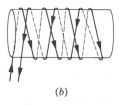

(b)

Figure 12   Question 27.

EXERCISES AND PROBLEMS

Section 33–2   Inductance

1E. The inductance of a close-packed coil of 400 turns is 8.0 mH. Calculate the magnetic flux through the coil when the current is 5.0 mA.

2E. A circular coil has a 10-cm radius and consists of 30 closely-wound turns of wire. An externally-produced magnetic field of 2.6 mT is perpendicular to the coil. (a) If no current is in the coil, what is the flux linkage? (b) When the current in the coil is 3.8 A in a certain direction, the net flux through the coil is found to vanish. What is the inductance of the coil?

3E. A solenoid is wound with a single layer of insulated copper wire (diameter, 2.5 mm). It is 4.0 cm in diameter and 2.0 m long. (a) How many windings are on the solenoid? (b) What is the inductance per meter for the solenoid near its center? Assume that adjacent wires touch and that insulation thickness is negligible.

4P. A long thin solenoid can be bent into a ring to form a toroid. Show that if the solenoid is long and thin enough, the equation for the inductance of a toroid (Eq. 7) is equivalent to that for a solenoid of the appropriate length (Eq. 5).

5P. *Inductors in series.* Two inductors L_1 and L_2 are connected in series and are separated by a large distance. (a) Show that the equivalent inductance is given by

$$L_{eq} = L_1 + L_2.$$

(b) Why must their separation be large for this relationship to hold? (c) What is the generalization of (a) for N inductors in series?

6P. *Inductors in parallel.* Two inductors L_1 and L_2 are connected in parallel and separated by a large distance. (a) Show that the equivalent inductance is given by

$$\frac{1}{L_{eq}} = \frac{1}{L_1} + \frac{1}{L_2}.$$

(b) Why must their separation be large for this relationship to hold? (c) What is the generalization of (a) for N inductors in parallel?

7P. A wide copper strip of width W is bent into a piece of slender tubing of radius R with two plane extensions, as shown in Fig. 13. A current i flows through the strip, distributed uniformly over its width. In this way a "one-turn solenoid" has been formed. (a) Derive an expression for the magnitude of the magnetic field **B** in the tubular part (far away from the edges). (Hint: Assume that the field outside this one-turn solenoid is negligibly small.) (b) Find also the inductance of this one-turn solenoid, neglecting the two plane extensions.

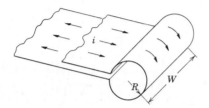

Figure 13 Problem 7.

8P. Two long parallel wires, each of radius a, whose centers are a distance d apart carry equal currents in opposite directions. Show that, neglecting the flux within the wires themselves, the inductance of a length l of such a pair of wires is given by

$$L = \frac{\mu_0 l}{\pi} \ln \frac{d-a}{a}.$$

See Sample Problem 3, Chapter 31. (Hint: Calculate the flux through a rectangle of which the wires form two opposite sides.)

Section 33–3   Self-Induction

9E. At a given instant the current and the induced emf in an inductor are as indicated in Fig. 14. (a) Is the current increasing or decreasing? (b) The emf is 17 V and the rate of change of the current is 25 kA/s; what is the value of the inductance?

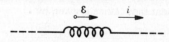

Figure 14 Exercise 9.

10E. A 12-H inductor carries a steady current of 2.0 A. How can a 60-V self-induced emf be made to appear in the inductor?

11E. A long cylindrical solenoid with 100 turns per cm has a radius of 1.6 cm. Assume the magnetic field it produces is parallel to its axis and is uniform in its interior. (a) What is its

inductance per meter of length? (b) If the current changes at the rate 13 A/s, what emf is induced per meter?

12E. The inductance of a closely wound N-turn coil is such that an emf of 3.0 mV is induced when the current changes at the rate 5.0 A/s. A steady current of 8.0 A produces a magnetic flux of 40 μWb through each turn. (a) Calculate the inductance of the coil. (b) How many turns does the coil have?

13P. The current i through a 4.6-H inductor varies with time t as shown on the graph of Fig. 15. The inductor has a resistance of 12 Ω. Calculate the induced emf during the time intervals (a) $t = 0$ to $t = 2$ ms, (b) $t = 2$ ms to $t = 5$ ms, (c) $t = 5$ ms to $t = 6$ ms. (Ignore the behavior at the ends of the intervals.)

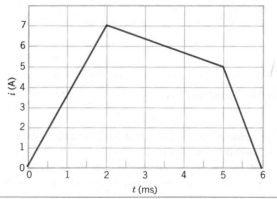

Figure 15 Problem 13.

Section 33–4   An LR Circuit

14E. The current in an LR circuit builds up to one-third of its steady-state value in 5.0 s. Calculate the inductive time constant.

15E. How many "time constants" must we wait for the current in an LR circuit to build up to within 0.10% of its equilibrium value?

16E. The current in an LR circuit drops from 1.0 A to 10 mA in the first second following removal of the battery from the circuit. If L is 10 H, find the resistance R in the circuit.

17E. How long would it take, following the removal of the battery, for the potential difference across the resistor in an LR circuit ($L = 2.0$ H, $R = 3.0$ Ω) to decay to 10% of its initial value?

18E. (a) Consider the LR circuit of Fig. 5. In terms of the battery emf $\mathcal{E}$, what is the self-induced emf $\mathcal{E}_L$ when the switch has just been closed on a? (b) What is $\mathcal{E}_L$ after two time constants? (c) After how many time constants will $\mathcal{E}_L$ be just one half of the battery emf $\mathcal{E}$?

19E. A solenoid having an inductance of 6.3 μH is connected in series with a 1.2-kΩ resistor. (a) If a 14-V battery is switched across the pair, how long will it take for the current through the resistor to reach 80% of its final value? (b) What is the current through the resistor after one time constant?

20E. The flux linkage through a certain coil of 0.75-Ω resistance is 26 mWb when there is a current of 5.5 A in it. (a) Calculate the inductance of the coil. (b) If a 6.0-V battery is suddenly connected across the coil, how long will it take for the current to rise from 0 to 2.5 A?

21P. Suppose the emf of the battery in the circuit of Fig. 6 varies with time t so the current is given by $i(t) = 3.0 + 5.0t$, where i is in amperes and t is in seconds. Take $R = 4.0\ \Omega$, $L = 6.0$ H, and find an expression for the battery emf as a function of time. (*Hint:* Apply the loop theorem.)

22P. At $t = 0$ a battery is connected to an inductor and resistor connected in series. The table below gives the measured potential difference, in volts, across the inductor as a function of time, in ms, following the connection of the battery. Deduce (a) the emf of the battery and (b) the time constant of the circuit.

t (ms)	V_L (V)	t (ms)	V_L (V)
1.0	18.2	5.0	5.98
2.0	13.8	6.0	4.53
3.0	10.4	7.0	3.43
4.0	7.90	8.0	2.60

23P. A 45-V potential difference is suddenly applied to a coil with $L = 50$ mH and $R = 180\ \Omega$. At what rate is the current increasing after 1.2 ms?

24P. A wooden toroidal core with a square cross section has an inner radius of 10 cm and an outer radius of 12 cm. It is wound with one layer of wire (diameter, 1.0 mm; 'resistance' 0.02 Ω/m). What are (a) the inductance and (b) the inductive time constant? Ignore the thickness of the insulation.

25P. In Fig. 16, $\mathscr{E} = 100$ V, $R_1 = 10\ \Omega$, $R_2 = 20\ \Omega$, $R_3 = 30\ \Omega$, and $L = 2.0$ H. Find the values of i_1 and i_2 (a) immediately after switch S is closed; (b) a long time later; (c) immediately after switch S is opened again; (d) a long time later.

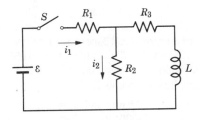

Figure 16 Problem 25.

26P. In the circuit shown in Fig. 17, $\mathscr{E} = 10$ V, $R_1 = 5.0\ \Omega$, $R_2 = 10\ \Omega$, and $L = 5.0$ H. For the two separate conditions (I)

switch S just closed and (II) switch S closed for a long time, calculate (a) the current i_1 through R_1, (b) the current i_2 through R_2, (c) the current i through the switch, (d) the potential difference across R_2, (e) the potential difference across L, and (f) di_2/dt.

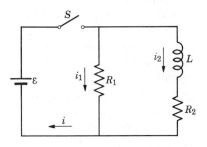

Figure 17 Problem 26.

27P. In Fig. 18 the component in the upper branch is an ideal 3-A fuse. It has zero resistance as long as the current through it remains less than 3 A. If the current reaches 3 A, it "blows" and thereafter it has infinite resistance. Switch S is closed at time $t = 0$. (a) When does the fuse blow? (b) Sketch a graph of the current i through the inductor as a function of time. Mark the time at which the fuse blows.

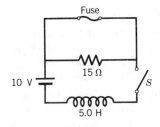

Figure 18 Problem 27.

28P*. For the circuit shown in Fig. 19, switch S is closed at time $t = 0$. Thereafter the constant current source, by varying its emf, maintains a constant current i out of its upper terminal. (a) Derive an expression for the current through the inductor as a function of time. (b) Show that the current through the

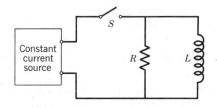

Figure 19 Problem 28.

resistor equals the current through the inductor at time $t = (L/R) \ln 2$.

Section 33–5   Energy and the Magnetic Field

29E. The magnetic energy stored in a certain inductor is 25 mJ when the current is 60 mA. (*a*) Calculate the inductance. (*b*) What current is required for the magnetic energy to be four times as much?

30E. A 90-mH toroidal inductor encloses a volume of 0.020 m³. If the average energy density in the toroid is 70 J/m³, what is the current?

31E. Consider the circuit of Fig. 6. In terms of the time constant, at what instant after the battery is connected will the energy stored in the magnetic field of the inductor be half its steady-state value?

32E. A coil with an inductance of 2.0 H and a resistance of 10 Ω is suddenly connected to a resistanceless battery with $\mathcal{E} = 100$ V. (*a*) What is the equilibrium current? (*b*) How much energy is stored in the magnetic field when this current exists in the coil?

33E. A coil with an inductance of 2.0 H and a resistance of 10 Ω is suddenly connected to a resistanceless battery with $\mathcal{E} = 100$ V. At 0.10 s after the connection is made, what are the rates at which (*a*) energy is being stored in the magnetic field, (*b*) thermal energy is appearing, and (*c*) energy is being delivered by the battery?

34P. Suppose that the inductive time constant for the circuit of Fig. 6 is 37 ms and the current in the circuit is zero at time $t = 0$. At what time does the rate at which energy is dissipated in the resistor equal the rate at which energy is being stored in the inductor?

35P. A coil is connected in series with a 10-kΩ resistor. When a 50-V battery is applied to the two, the current reaches a value of 2.0 mA after 5.0 ms. (*a*) Find the inductance of the coil. (*b*) How much energy is stored in the coil at this same moment?

36P. For the circuit of Fig. 6, assume that $\mathcal{E} = 10$ V, $R = 6.7$ Ω, and $L = 5.5$ H. The battery is connected at time $t = 0$. (*a*) How much energy is delivered by the battery during the first 2.0 s? (*b*) How much of this energy is stored in the magnetic field of the inductor? (*c*) How much has been dissipated in the resistor?

37P. (*a*) Find an expression for the energy density as a function of the radial distance for the toroid of Sample Problem 1. (*b*) Integrating the energy density over the volume of the toroid, calculate the total energy stored in the field of the toroid; assume $i = 0.50$ A. (*c*) Using Eq. 18 in this chapter evaluate the energy stored in the toroid directly from the inductance and compare with (*b*).

38P. A solenoid, with length 80 cm and radius 5.0 cm, consists of 3000 turns distributed uniformly over its length. Its total resistance is 10 Ω. At the instant 5.0 ms after it is con-

nected to a 12-V battery, (*a*) how much energy is stored in its magnetic field and (*b*) how much energy has been supplied by the battery up to that time? (Neglect end effects.)

39P. Prove that, after switch *S* in Fig. 5 is thrown from *a* to *b*, all the energy stored in the inductor ultimately appears as thermal energy in the resistor.

Section 33–6   Energy Density and the Magnetic Field

40E. A solenoid 85 cm long has a cross sectional area of 17 cm². There are 950 turns of wire carrying a current of 6.6 A. (*a*) Calculate the magnetic field energy density inside the solenoid. (*b*) Find the total energy stored in the magnetic field inside the solenoid. (Neglect end effects)

41E. What must be the magnitude of a uniform electric field if it is to have the same energy density as that possessed by a 0.50-T magnetic field?

42E. The magnetic field in the interstellar space of our galaxy has a magnitude of about 10^{-10} T. How much energy is stored in this field in a cube 10 light-years on edge? (For scale, note that the nearest star is 4.3 light-years distant and the radius of our galaxy is about 8×10^4 light-years.)

43E. Use the result of Sample Problem 5 to obtain an expression for the inductance of a length *l* of the coaxial cable.

44E. A circular loop of wire 50 mm in radius carries a current of 100 A. (*a*) Find the magnetic field strength at the center of the loop. (*b*) Calculate the energy density at the center of the loop.

45P. A length of copper wire carries a current of 10 A, uniformly distributed. Calculate (*a*) the magnetic energy density and (*b*) the electric energy density at the surface of the wire. The wire diameter is 2.5 mm and its resistance per unit length is 3.3 Ω/km.

46P. (*a*) What is the magnetic energy density of the earth's magnetic field of 50 μT? (*b*) Assuming this to be relatively constant over distances small compared with the earth's radius and neglecting the variations near the magnetic poles, how much energy would be stored in a shell between the earth's surface and 16 km above the surface?

Section 33–7   Mutual Induction

47E. Two coils are at fixed locations. When coil 1 has no current and the current in coil 2 increases at the rate 15 A/s, the emf in coil 1 is 25 mV. (*a*) What is their mutual inductance? (*b*) When coil 2 has no current and coil 1 has a current of 3.6 A, what is the flux linkage in coil 2?

48E. Coil 1 has $L_1 = 25$ mH and $N_1 = 100$ turns. Coil 2 has $L_2 = 40$ mH and $N_2 = 200$ turns. The coils are rigidly positioned with respect to each other, their coefficient of mutual inductance M being 3.0 mH. A 6.0-mA current in coil 1 is changing at the rate of 4.0 A/s. (*a*) What flux Φ_{12} links coil 1 and what self-induced emf appears there? (*b*) What flux Φ_{21} links coil 2 and what mutually induced emf appears there?

49P. In Fig. 20 two coils are connected as shown. The coils separately have inductances L_1 and L_2. The coefficient of mutual inductance is M. (*a*) Show that this combination can be replaced by a single coil of equivalent inductance given by

$$L_{eq} = L_1 + L_2 + 2M.$$

(*b*) How could the coils in Fig. 20 be reconnected to yield an equivalent inductance of

$$L_{eq} = L_1 + L_2 - 2M?$$

Note that this problem is an extension of Problem 5, in which the requirement that the coils be far apart is removed.

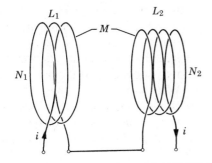

Figure 20 Problem 49.

50P. A coil C of N turns is placed around a long solenoid S of radius R and n turns per unit length, as in Fig. 21. Show that the coefficient of mutual inductance for the coil-solenoid combination is given by

$$M = \mu_0 \pi R^2 n N.$$

Explain why M does not depend on the shape, size, or possible lack of close-packing of the coil.

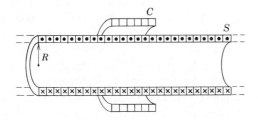

Figure 21 Problem 50.

51P. Figure 22 shows a coil of N_2 turns linked as shown to a toroid of N_1 turns. The inner toroid radius is a, its outer radius is b, and its height is h, as in Fig. 22. Show that the coefficient of mutual inductance M for the toroid-coil combination is

$$M = \frac{\mu_0 N_1 N_2 h}{2\pi} \ln \frac{b}{a}.$$

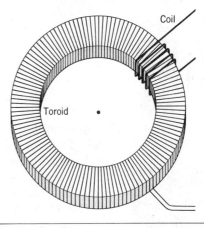

Figure 22 Problem 51.

52P. Figure 23 shows, in cross section, two coaxial solenoids. Show that the coefficient of mutual inductance M for a length l of this solenoid-solenoid combination is given by

$$M = \pi R_1^2 l \mu_0 n_1 n_2,$$

in which n_1 and n_2 are the respective numbers of turns per unit length and R_1 is the radius of the inner solenoid. Why does M depend on R_1 and not on R_2?

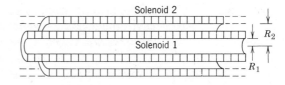

Figure 23 Problem 52.

53P. A rectangular loop of N close-packed turns is positioned near a long straight wire as in Fig. 24. (*a*) What is the coefficient of mutual inductance M for the loop-wire combination? (*b*) Evaluate M for $N = 100$, $a = 1.0$ cm, $b = 8.0$ cm, and $l = 30$ cm.

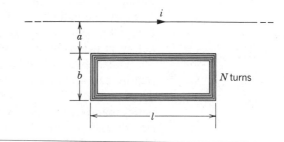

Figure 24 Problem 53.

ESSAY 13
MAGNETIC FLIGHT

GERARD K. O'NEILL

O'NEILL
COMMUNICATIONS, INC.

Imagine traveling from New York to Orlando, Florida, in half an hour, at a cruising speed of a thousand meters per second (2300 miles per hour)! And imagine making that journey in the greatest comfort, in a seat with plenty of legroom, with unlimited carry-on baggage as close as you like it. If that's hard enough to believe, imagine further that your vehicle makes not a sound as it goes, is unseen by anyone outside it, and uses less energy in getting you there than the chemical energy in a liter of gasoline. Your imagination can take a break now, because all of those things are within the range of engineering based on the physics of magnetic flight.

As we know all too well by observing any large highway, or any railway or airport, all of our present transport systems move people only at the price of noise pollution, and by using a great deal of energy. What would be the characteristics of an ideal inter-city transportation system? Most of us would list high speed, economy, and comfort. But we'd also list attributes to make the system easy on our environment: silence, avoiding using valuable land area, and freedom from chemical pollution. Magnetic flight can achieve all of these.

By "magnetic flight" we mean the support of a vehicle not by wheels on a road or track, nor by the aerodynamic forces of flight through the air, but by the strong lift provided by the interaction of an electric current with a magnetic field. We provide our lifting force in that manner, with no physical contact and no need for an aerodynamic medium such as air. Using that method we can see at once that the ideal environment for a magnetic-flight vehicle is a vacuum. Why move through the air when we don't need it, and when it only causes drag? And if we don't need lifting wings, why not make the vehicle long and slim, and let it fly in vacuum in a small-diameter tunnel?

We've become quite used to the idea that high speed can be achieved only in aircraft, and that to reach that speed at a price we can afford we must resign ourselves to traveling with only a tiny amount of personal space, hunched in minimal seats while we stare at our knees. In magnetic flight all such constraints vanish. Traveling in vacuum, a vehicle indeed should be of small diameter, so that its enclosing tunnel need not be too large. But there is no aerodynamic limit on the vehicle's length. Before looking into the physics of magnetic flight, but simply knowing that it exists, we can draw up a first design for our "magnetic floater" vehicle. See Figs. 1 and 2.

As we check the physics of magnetic flight, I'll mention just one more of its unique characteristics: it has no particular speed limit, such as the speed of sound. Your magnetic-flight trip from New York to Orlando, or a similar trip from Boston to San Francisco in just over an hour, would cost no more in energy if carried out at half or at twice the speed of 1000 m/s chosen for our examples.

Statements such as these suggest ultra-high technology, and would seem to require new scientific breakthroughs. Not at all. The basic facts of magnetic flight depend only on Maxwell's four equations, the fundamental equations of electromagnetics that govern everything from the operation of an electric toaster to the transmission of starlight. And Maxwell's equations were completed by 1865, more than a century ago. Indeed, recent breakthroughs in high-temperature superconductors will make "floater" systems a little easier to engineer, and will make them a little cheaper

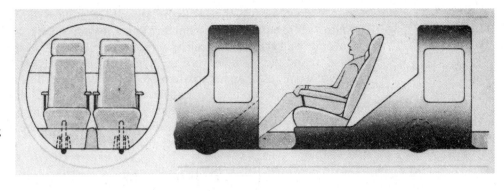

Figure 1 Side and front views of the modular carriage of a floater car. Each module has reclining seats for two passengers, and has a baggage compartment.

Figure 2 Boarding area for a floater high-speed underground transport system. The passenger-seating "carriage" of the floater vehicle rolls out of the air-lock at the far end into the central fenced area. The fences then sink into the floor to permit passengers to step on or off the carriage conveniently. When all passengers have boarded, the carriage returns through the air-lock into the enclosing shell of the vehicle, which maintains normal air pressure during flight.

to operate—but without them all of the foregoing statements could still come true within a few years.

We can measure the lifting force on which magnetic flight depends by an experiment in a basic geometry: imagine two circles of current-carrying wire, of equal diameter, spaced apart by a short distance. Each current produces a magnetic field at the other. It is the interaction of the upper current's magnetic field with the lower current that lifts the lower coil.

Let's do the experiment. We use two circular wire loops of diameter one meter. Each wire turns out to weigh 4.5 newtons (about one pound). When spaced so that the wire centers are 0.065 meter apart, we find that the lower wire can be supported when we pass a current of 788 amperes through each wire. Notice that the currents have to circulate in the same direction; if we connect the wires for currents in opposite directions, the wires repel rather than attract each other. By the way, it makes no difference whether we use single wires carrying 788 amperes, or coils of 788 turns, carrying one ampere in each individual turn. Only the product of amperes and turns of wire matters.

It's always well to check theory against experiment, and vice versa. In this case our theory comes from Maxwell's equations. In one of its simplest applications, it says that when we space two very long straight wires a distance s apart (in meters), and pass currents through the wires, the force between the wires, per meter of length, will be

$$\text{force/meter} = \left(\frac{\mu_0}{2\pi}\right)\left(\frac{1}{s}\right)(\text{current in first wire})(\text{current in second wire}).$$

In metric units of amperes, meters, and newtons, that is

$$\text{force/meter} = \left(\frac{2 \times 10^{-7}}{s}\right)(\text{current in first wire})(\text{current in second wire}).$$

That's a clue to the force between currents in other geometries: it tells us that the forces between currents are stronger when the currents are close to each other. With that clue, rather than using a more complicated formula for the exact force between two circular currents, let's approximate our geometry by observing that most of the force between the two circular coils must be coming from portions of the currents that are close to each other. In the circular coils, a current section in the lower coil must be affected most by the current section just above it, and the distant part of the current in the upper coil must not be contributing much to the force on that current section.

The geometry of two long straight wires is plain enough that a physicist or engineer could calculate the formula above, even without using a reference book. The exact formula for two circles of wire, though, is another story: to calculate that requires more advanced mathematics and a lot more work. Knowing that, let's be bold and bend those long straight wires into circles, and persist in using the formula above to find the force between them. Of course our answer won't be precisely right, because we've made an approximation by our boldness. But if we come anywhere near right, it will confirm our understanding and assure us that we haven't lost a factor of 10 or $\frac{1}{10}$ in our work. Let's try it. Bending the long wires into circles, the total length of each circle is pi times the diameter, or 3.14 meters. The long-wire formula gives us a force per meter of

$$\left(\frac{2 \times 10^{-7}}{.065}\right)(788)^2 = 1.9 \text{ newtons/meter of length.}$$

We multiply that force per meter by the 3.14 meters of total length, and find

$$(1.9 \text{ newtons/meter})(3.14 \text{ meters}) = 6.0 \text{ newtons.}$$

Not bad! Recall that the weight of the lower wire was 4.5 newton, which also had to equal the magnetic force on the wire when it lifted free. We got within 33% of the experimental result, while having made a pretty outrageous approximation to arrive at our estimate. That process of making a "quick and dirty" check calculation as a first step before working out a problem in complete and accurate detail is exactly the way that experienced physicists and engineers do their work. The quick check confirms basic understanding and makes sure that careless mistakes in factors of 10 haven't been made. Only when that's done—and it doesn't take long—does it make sense to take the time to work out detailed formulas and arrive at precise answers.

Just as in the case of an iron nail attracted to a permanent magnet, the attraction of the two currents in our example is unstable. Once it moves, the nail flies to and sticks to the magnet. In the same way the lower wire of Fig. 3 jumps upward to the fixed wire when the current reaches 788 amperes. But our experiment lacks vital elements of a practical magnetic flight system: a sensing mechanism to monitor the height of the wire, and control circuits to reduce the upper current when the lower wire is too high, or to increase the upper current when the lower wire is too low. With those additional elements our experiment can be made stable. The same principles are used in modern aircraft. Those aircraft are made stable by the addition of elements to sense roll, pitch, and yaw, directing motors to drive the aircraft control surfaces.

In West Germany, attractive magnetic lift has been carried to a late stage of experimental development, with the support of the West German government. Electronic systems for sensing and control of the currents have been made so reliable that they have been certified for full-size systems carrying passengers. Over many trips during a West German transport fair held several years ago, more than 40,000 people were carried in an attractive lift vehicle. I rode in a similar vehicle built by JAL, Japan Air Lines, at a site near Tokyo in the early 1980s. Attractive magnetic lift has strong points in its favor: it is so efficient that lifting a mass of several tons requires no more power than one can draw from a car battery. The corresponding currents are low enough that the system can be built out of ordinary copper, aluminum, and steel, rather than needing superconductors. Because the control of position is electronic,

Figure 3 Experiment on magnetic lift. Each wire is a circle of diameter one meter.

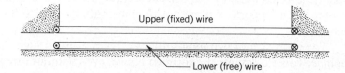

Figure 4 Essentials of a floater vehicle. The pressure shell encloses the passenger carriage of Fig. 1. The current loops function like the free wire of Fig. 3, providing lift.

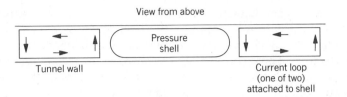

View from above

Pressure shell

Tunnel wall

Current loop (one of two) attached to shell

control circuits using the memory of a computer can compensate for irregularities in the track. The quality of the ride can therefore be quite smooth.

But there is an alternative, which can be demonstrated by inverting our experiment, placing the fixed wire below and the floating wire above it. In that case we must also reverse one of the currents. Then, even without control circuits, the experimental setup is stable in vertical motion. That alternative, called "repulsive magnetic lift," has been developed by the Japanese National Railways to the point of a full-size two-car train, running on a test track several kilometers long. Many passengers have taken demonstration rides on that train.

Armed with the knowledge of the fundamental formulas of magnetic lift, we can move further toward the full design of our "floater" vehicle. We'll need a passenger compartment, a long slim cylinder, enclosed to hold normal atmospheric pressure inside. That's much like the fuselage of an airliner. Instead of wings, we'll have long rectangular loops of wire ahead of and behind the passenger compartment. They're the equivalents of the lower wire in our experiment. Because the vehicle will fly in vacuum, it doesn't need an aerodynamic shape. Our floater can be as angular and lumpy as a spacecraft that never needs to enter an atmosphere (Fig. 4).

We have two basic choices in providing the stationary current—the equivalent of the current in the upper wire—to make the magnetic field that will lift the lower current. One way is by induction: the rapid motion of the floater's currents could induce currents in a passive sheet of metal, a "guideway," below the vehicle, and the magnetic field of those induced currents would repel and so lift the moving current. That is repulsive magnetic lift, which requires wider spacings and strong fixed super-currents in the vehicle.

Alternatively we can use attractive magnetic lift. That way, our "guideway" can consist of a series of pairs of short, nearly rectangular loops of wire, so arranged that in each section of the tunnel we can choose to divide the supporting current unequally between the left and right loops. The division will be determined by a central computer, based on the exact measured trajectories of the vehicles. The system is adaptive, maintaining the path of the vehicles straight even though the tunnel itself may have moved a little out of line over the years. To save power, we need only turn on the supporting currents in a particular section of the tunnel when the vehicle is approaching. As soon as it's past, we can turn off the currents again (Fig. 5).

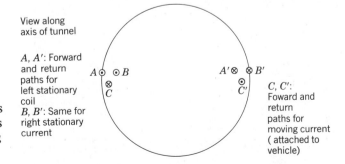

View along axis of tunnel

A, A': Forward and return paths for left stationary coil
B, B': Same for right stationary current

C, C': Foward and return paths for moving current (attached to vehicle)

Figure 5 View along axis of tunnel, of current paths for stationary and moving coils.

There is another alternative that also uses attractive magnetic lift. We can replace the fixed currents by a simple, passive steel guideway, as is done in the West German system. The vehicle can have ordinary currents in copper or aluminum (not superconductors). Those currents can be controlled by a computer to rise or fall as needed to keep the vehicle on a smooth trajectory, even though the guideway is irregular because of movement of the ground since its construction.

Now for the matter of acceleration and deceleration. We accomplish them in the same way as in our experiment: our vehicle can have circular current-loops centered on the tunnel, and we can install controllable current-loops in the tunnel walls (Fig. 6). We can pull on our vehicle as it approaches one of the loops, and push on it, with a current in the opposite direction, when it is past. We reverse the currents to provide deceleration as the vehicle approaches its destination. Notice how simple our vehicle is now. Its currents are low enough that they can be sustained by batteries, and the vehicle has almost no moving parts: only those of an air-conditioning system to keep the air fresh and at the right temperature. The trip times are so short that meal services aren't required, and as the entire floater system is controlled by a computer, monitored at a single control center, there doesn't have to be a flight crew. There can be entertainment systems, probably in the form of flat-screen television sets at the front of each compartment, and stereo programs as we are used to on airplane flights.

When the additional physics is worked out, one finds that a floater system uses very little energy. In cruising flight, the drag force is about $\frac{1}{100}$ of the weight of the vehicle. (By contrast, the figure for the best jet aircraft is about $\frac{1}{14}$.) Energy is fed into kinetic form as the floater is accelerated to its high speed, but most of that energy is recovered and returned to the floater system electrical power grid during the deceleration process. The system runs as a motor during acceleration, and as a generator during deceleration.

Two other kinds of power loss must be considered in the design of working magnetic flight systems: power for the ordinary currents in copper or aluminum, and power to maintain the moderate vacuum needed in the tunnels. As the numbers for the coil sizes work out, the power lost to the coil currents can be made small compared to the drag losses. The power for pumps to maintain the vacuum is also small, because the tunnels can be made relatively leak-tight. As an illustration, it is common in pipelines the size of the magnetic flight tunnels to have no leaks over hundreds of kilometers length that are large enough for measurable amounts of oil or natural gas to leak out.

With all of the system design elements in place, we can now calculate trip times and the amount of energy required for a journey. We've chosen a cruising speed of 1000 m/s so that the curves in the underground tunnel don't have to be excessively long. From the formula for acceleration in circular motion,

$$a = \frac{v^2}{r}$$

Figure 6 Side view of tunnel, with current loops providing forward acceleration.

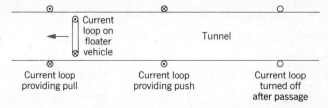

with a the acceleration in meters per second squared, v the velocity in meters per second, and r the radius of the circle in meters, one finds that limiting the transverse acceleration to two tenths of the free-fall acceleration g requires curves with radii of at least 500 kilometers (just over 300 miles). Commercial passengers in airliners are often subjected to much more than that, but two tenths of g is a standard figure for railroads. Military pilots flying supersonic aircraft at 1000 m/s experience accelerations according to the same formula. But they are willing to take accelerations 50 times higher than those of railway passengers (10 g) in order to turn in radii of about 10 kilometers.

To be gentle to our passengers, we'll limit the forward acceleration and deceleration to 3 m/s², about 0.3 g. To reach our cruising speed of 1000 m/s from a standing start will require, using the formula

$$\text{velocity} = (\text{acceleration})(\text{time}),$$

a time just under six minutes. The deceleration time will be the same. During the acceleration and deceleration phases the average speed of the vehicle will be half of its cruising speed, or 500 m/s. The distance covered during acceleration will be

$$\begin{aligned}(\text{distance}) &= (\text{average velocity})(\text{time})\\ &= (500 \text{ m/s})(333 \text{ s}) = 1.7 \times 10^5 \text{ m}\end{aligned}$$

For the New York to Orlando trip, the distance is 1507 km. Subtracting 170 km at each end for acceleration and deceleration, we're left with 1170 km of cruising flight. At our cruising speed of 1000 m/s that will take 1200 s, or 20 min. The entire trip will therefore require just about 30 min.

The energy per passenger required for the journey can be calculated from the fact that the drag on the vehicle will be about $\frac{1}{100}$ of its weight. The number of seats in the vehicle doesn't matter; it could be as few as 25 or as many as 300. Suppose each passenger has an average mass of 77 kilograms (170 pounds), and that the average passenger brings 23 kilograms of baggage (50 pounds). That brings the total mass for each passenger to a round number of 100 kilograms. Assuming that the vehicle mass is 100 kilograms per passenger, then we have 200 kilograms total, for a weight of

$$\begin{aligned}F = ma &= (200 \text{ kg})(9.8 \text{ m/s}^2)\\ &= 1960 \text{ newtons per passenger.}\end{aligned}$$

Over a total distance of 1507 km, the drag force will cost us, in energy,

$$\begin{aligned}\text{energy} = (\text{force})(\text{distance}) &= (\text{weight}/100) \times \text{distance}\\ &= (1960 \text{ N})(1{,}507{,}000 \text{ m})(\tfrac{1}{100})\\ &= 29.5 \text{ million joules per passenger.}\end{aligned}$$

That is the energy released by burning about one liter of gasoline. The kinetic energy of motion is

$$\begin{aligned}\text{kinetic energy} = (\tfrac{1}{2})(\text{mass})(\text{velocity squared}) &= \tfrac{1}{2}mv^2\\ &= (\tfrac{1}{2})(200 \text{ kilogram})(1000 \text{ m/s})^2\\ &= 106 \text{ million joules per passenger.}\end{aligned}$$

Interestingly enough, that's about 3.6 times higher than the energy per passenger lost by the vehicle to drag during the entire journey. It tells us that if we accelerated to our full cruising speed in the first 170 km out of the New York terminal, as assumed above, we could coast all the way to Orlando!

Given the advantages of "floater" magnetic flight systems, it will be only a question of time before such systems are built.

CHAPTER 34

MAGNETISM AND MATTER

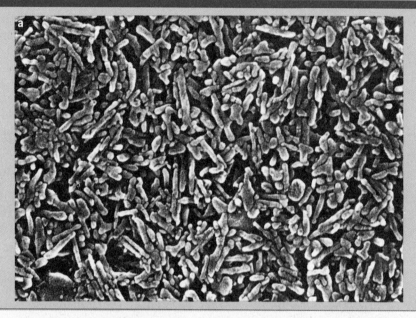

The figure shows an electron micrograph of the needle-like magnetic particles in a floppy disk. Current hard disk drives can store information magnetically at densities exceeding one million bits per cm². If scaled up to aircraft size, the problem of controlling the position of the read-record head of a such a drive is equivalent to that of flying a jumbo jet at 500 mi/h at 0.1 in. above the ground. The market for devices in which information is stored magnetically is estimated at 35 billion dollars per year.

34-1 Magnets

The most familiar magnets may well be the small decorative devices used to fasten notes to the refrigerator door. Magnets play a much larger part in our daily lives than that, however. Even if we restrict ourselves solely to the household environment, we can find magnets or magnetic materials in the motors of our electric appliances, in TVs and VCRs; in doorbells and thermostats; in relays, circuit breakers, and the electric utility meter; in loudspeakers, Walkman headsets, floppy disks, and aquarium pumps. Magnets or magnetic materials are in the ink on checks and on dollar bills; in answering machines, telephones, tape decks, credit cards, audio cassettes, cupboard doors, and portable chess sets. Can you think of any household application that we have left out? We have ignored entirely the many uses of magnets and of magnetic materials in industry and in scientific research.

We start our study of magnetic materials by looking at Fig. 1, which shows iron filings sprinkled on a sheet of cardboard under which there is a short bar magnet. The pattern of the filings suggests that the magnet has two

Figure 1   A bar magnet is a magnetic dipole. The iron filings suggest the magnetic lines of force.

poles, similar to the positive and the negative charges of an electric dipole. We conventionally label the magnetic poles as *north* and *south*. The north magnetic pole is the one from which the magnetic field lines emerge from the magnet into the surrounding space; these field lines then reenter the magnet at its south magnetic pole.

All attempts to isolate these poles fail. If we break the magnet, as in Fig. 2, we end up with three smaller magnets, each with its north and south pole. We can push this breakup of the magnet as far as its constituent atoms and electrons and we still fail to find anything that we can call an isolated magnetic pole, or a *magnetic monopole,* as we have come to call it. We conclude:

The simplest magnetic structure that can exist in nature is the magnetic dipole. *There are no magnetic monopoles, that is, there are no magnetic structures analogous to isolated electric charges.*

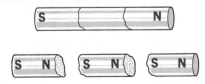

Figure 2   If you break a magnet, each fragment becomes a separate magnet, with its own north and south pole.

The fundamental magnetic dipole in nature—the one responsible for the magnetic properties of bulk matter—is that associated with the electron.

34-2   Magnetism and the Electron

Electrons can generate magnetism in three ways.

1.   The Magnetism of Moving Charges.   By streaming through an evacuated space or drifting through a conducting wire, electrons—like other charged particles—can set up an external magnetic field. We have studied fields produced in this way in earlier chapters and will not pursue this aspect of magnetism further here.

2.   Magnetism and Spin.   An isolated electron can be viewed classically as a tiny spinning negative charge, with an intrinsic *spin angular momentum S.* Associated with this spin angular momentum is an intrinsic *spin magnetic moment* μ_s. The magnitude of the spin angular momentum, as predicted by quantum theory and as measured in the laboratory, is

$$S = \frac{h}{4\pi} = 5.2729 \times 10^{-35} \text{ J} \cdot \text{s}.$$

Here h is the Planck constant, the central constant of quantum physics.

When we deal with magnetism at the level of electrons and atoms, we find it convenient to adopt a non-SI unit for measuring magnetic moments. It is called the *Bohr magneton* (symbol μ_B) and is defined in terms of three fundamental constants of nature as

$$\boxed{1 \ \mu_B = \frac{eh}{4\pi m} = 9.27 \times 10^{-24} \text{ J/T}}$$

(the Bohr magneton),   (1)

in which e is the elementary charge, and m is the electron mass. Expressed in these units, the magnitude of the spin magnetic moment of the electron is*

$$\mu_s = 1 \ \mu_B. \qquad (2)$$

Figure 3 shows the associated electric and magnetic

---

*The quantum theory of the electron (called *quantum electrodynamics,* or QED) predicts that the 1 in Eq. 2 should be replaced by 1.001 159 652 193. This remarkable prediction has been verified by equally remarkable experiments. However, for our purposes, 1 will do nicely.

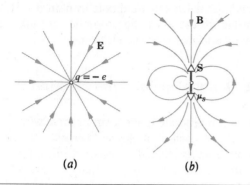

Figure 3 The lines of (a) the electric field and (b) the magnetic field for an isolated electron.

field lines for an isolated resting electron. The lines of the electric field point radially inward, as we expect for a particle with a negative charge. The magnetic lines, however, are those of a magnetic dipole. Note that the direction of the intrinsic spin magnetic moment vector for the electron (labeled μ_s) is opposite to that for the intrinsic spin angular momentum vector (labeled **S**). This is just what we would expect classically — as you should be sure to verify — for a spinning *negative* charge. Table 1 summarizes the properties of the free electron.

3. Magnetism and Orbital Motion. Electrons attached to atoms exist in states that have an intrinsic *orbital angular momentum* L_{orb}, corresponding classically to the motion of the electron in an orbit around the nucleus of the atom. These orbiting electrons are equivalent to tiny current loops and have an *orbital magnetic moment* μ_{orb} associated with them. Like essentially all physical properties examined at the atomic level, the orbital magnetic moment of the electron is quantized, being restricted to integral multiples of the Bohr magneton.

With so many possibilities for magnetism, you may well ask, "Every solid contains electrons; why isn't *everything* magnetic? Why can only magnetized iron and a few other substances pick up nails?" There are two answers.

Our first answer is that, in most cases, the magnetic

Table 1   Some Properties of the Electron[a]

Mass	m	9.1094×10^{-31} kg
Charge	$-e$	-1.6022×10^{-19} C
Angular momentum	S	5.2729×10^{-35} J·s
Magnetic moment	μ_s	9.2848×10^{-24} J/T

[a] Although these quantities are known to considerably higher precision, we display only 5 significant figures here.

moments of the electrons in a solid combine so as to cancel each other in their external effects. It is only when atoms contain unpaired electrons and special circumstances permit the large-scale alignment of their dipole moments that the familiar external effects are possible.

Our second answer is that, in a sense, everything *is* magnetic. When we speak popularly of magnetism, we almost always mean *ferromagnetism,* the familiar strong magnetism of the bar magnet or the compass needle. As we shall see, however, there are other kinds of magnetism — whose magnetic forces are too feeble to detect with our fingertips — that occur rather generally in matter.

Sample Problem 1 Devise a method for measuring the magnetic dipole moment μ for a bar magnet.

(a) Place the magnet in a uniform external magnetic field **B**, with μ making an angle θ with **B**. A torque τ will act on the magnet, given by Eq. 30 of Chapter 30, or

$$\tau = \mu \times \mathbf{B}. \qquad (3)$$

The magnitude of this torque is

$$\tau = \mu B \sin \theta. \qquad (4)$$

Clearly we can find μ if we measure τ, B, and θ.

(b) A second technique is to suspend the magnet from its center of mass and to allow it to oscillate about its stable equilibrium position in the external field B. For small oscillations, $\sin \theta$ can be replaced by θ and Eq. 4 becomes

$$\tau = -(\mu B)\theta = -\kappa\theta, \qquad (5)$$

where κ is a constant. We have inserted the minus sign to show that τ is a *restoring torque,* acting always in the opposite direction to the angular displacement θ.

Since τ is proportional to θ, the condition for simple angular harmonic motion is met. The period of oscillation T is given by Eq. 21 of Chapter 14.

$$T = 2\pi \sqrt{\frac{I}{\kappa}} = 2\pi \sqrt{\frac{I}{\mu B}}. \qquad (6)$$

in which I is the rotational inertia. With this equation we can find μ from the measured quantities T, B, and I.

34-3   Orbital Angular Momentum and Magnetism

Here we derive the relation between the orbital **magnetic** moment of an electron and its orbital angular **momentum.**

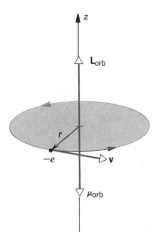

Figure 4   A classical representation of an electron circulating in an orbit of radius r with constant speed v. The orbital angular momentum vector and the orbital magnetic moment vector are both shown. Note that—because the electron carries a negative charge—these vectors point in opposite directions.

Figure 4 shows an electron moving in a circular orbit of radius r with speed v. The circulating electron is equivalent to a single-turn current loop. The magnetic moment of a current loop (see Eq. 27 of Chapter 30) is

$$\mu_{orb} = iA. \tag{7}$$

The current, i, is the amount of charge that passes any point in the orbit divided by the time required. The charge involved is e and the time, T, for one revolution is

$$T = \frac{2\pi r}{v} \tag{8}$$

so that

$$i = \frac{e}{(2\pi r/v)}.$$

The area of the single-turn loop (orbit) is $A = \pi r^2$, giving

$$\mu_{orb} = \frac{e}{(2\pi r/v)}\,\pi r^2 = \tfrac{1}{2}evr. \tag{9}$$

The angular momentum, on the other hand, is

$$L_{orb} = mvr. \tag{10}$$

Combining Eqs. 9 and 10 and generalizing to a vector formulation, we find

$$\mu_{orb} = -\frac{e}{2m}\,\mathbf{L}_{orb} \quad \text{(orbital motion),} \tag{11}$$

as the relation between the orbital angular momentum $\mathbf{L}_{orb}$ and orbital magnetic moment μ_{orb}. The minus sign arises because the orbiting electron carries a negative charge. Convince yourself that this requires that the orbital angular momentum vector and the orbital magnetic moment vector point in opposite directions.

Quantum physics tells us that the orbital angular momentum of an electron is quantized and that its smallest (non-zero) value is $h/2\pi$. Substituting this for L_{orb} in Eq. 11 yields, considering magnitudes only,

$$\mu_{orb} = \frac{e}{2m}\frac{h}{2\pi} = \frac{eh}{4\pi m}. \tag{12}$$

We recognize this quantity (see Eq. 1) as the *Bohr magneton*, the unit in which magnetic moments are expressed at the level of atoms and electrons. Physically then:

A Bohr magneton is equal to the orbital magnetic moment of an electron circulating in an orbit with the smallest allowed (non-zero) value of orbital angular momentum.

Sample Problem 2   What value for the Bohr magneton may be found from Eq. 12, its defining equation?

We have, from this equation,

$$\mu_B = \frac{eh}{4\pi m} = \frac{(1.60 \times 10^{-19}\ \text{C})(6.63 \times 10^{-34}\ \text{J}\cdot\text{s})}{(4\pi)(9.11 \times 10^{-31}\ \text{kg})}$$

$$= 9.27 \times 10^{-24}\ \text{C}\cdot\text{J}\cdot\text{s/kg}$$
$$= 9.27 \times 10^{-24}\ \text{J/T}. \quad \text{(Answer)}$$

The unit transformation follows from the relation $F = iLB$, which shows that $1\ \text{T} = 1\ \text{N/A}\cdot\text{m} = 1\ \text{kg/C}\cdot\text{s}$.

34-4   Gauss' Law for Magnetism

Gauss' law for magnetism, which is one of the basic equations of electromagnetism (see Table 2 of Chapter 37), is a formal way of stating that conclusion that has been forced on us by the facts of magnetism, namely, that isolated magnetic poles do not exist. This equation asserts that the magnetic flux Φ_B through any closed Gaussian surface must be zero, or

$$\boxed{\Phi_B = \oint \mathbf{B}\cdot d\mathbf{A} = 0}$$

(Gauss' law for magnetism).   (13)

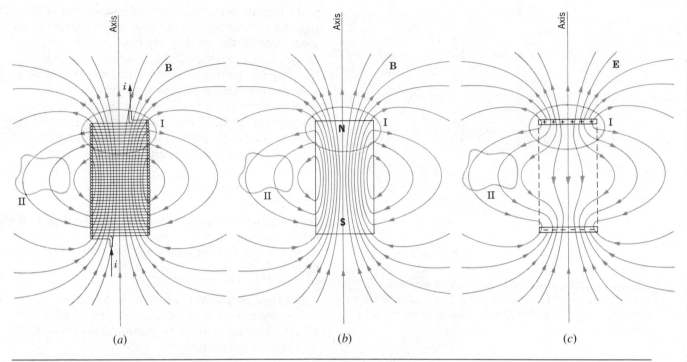

Figure 5 The lines of the magnetic field for (a) a short solenoid and (b) a short bar magnet. In each case the top end of the structure is a north magnetic pole. (c) The lines of the electric field for two charged disks. At large distances, all three fields resemble those for a dipole. The curves labeled I or II represent closed Gaussian surfaces.

We contrast this equation with Gauss' law for electricity, which is

$$\Phi_E = \epsilon_0 \oint \mathbf{E} \cdot d\mathbf{A} = q$$

(Gauss' law for electricity). (14)

In both of these laws, the integral is to be taken over the entire closed Gaussian surface. The fact that a zero appears at the right of Eq. 13, but not at the right of Eq. 14, means that in magnetism there is to be no counterpart to the free charge q in electricity.

Figure 5a suggests a Gaussian surface, marked I, enclosing one end of a short solenoid. Such a solenoid, as we have seen, sets up a magnetic dipole field at large distances; the end of the solenoid enclosed by surface I behaves, for such distant points, like a north magnetic pole. Note that the lines of the magnetic field **B** enter the Gaussian surface inside the solenoid and leave it outside the solenoid. No lines originate or terminate inside this surface; in other words, there are no *sources* or *sinks* of **B**; in still other words, there are no free magnetic poles. Thus the total flux Φ_B for surface I in Fig. 5a is zero, as Gauss' law for magnetism (Eq. 13) requires.

We also have $\Phi_B = 0$ for surface II in Fig. 5a and indeed for any closed surface that can be drawn in this figure. The situation is just the same if we replace the short solenoid by a short bar magnet, as in Fig. 5b. Here too $\Phi_B = 0$ for any closed surface that we can draw.

Figure 5c shows a close electrostatic analog to the two magnetic dipoles that we have been discussing. It consists of two oppositely charged circular disks facing each other as shown. The electric field **E** set up at distant points by this arrangement is also that of a dipole. In this case there is a net (outward) flux of the field lines for the Gaussian surface marked I; there *is* a source of the field, namely, the positive charges enclosed by the surface. (The negative charges of the other disk constitute a *sink* of the electric field.) Of course, for a Gaussian surface such as that marked II in Fig. 5c, we have $\Phi_E = 0$ because this surface happens to enclose no charge.

34–5 The Magnetism of the Earth

William Gilbert, physician in residence to Queen Elizabeth I and author of *De Magnete,* the first systematic survey of magnetic phenomena, knew in 1600 that the

Figure 6 Sir William Gilbert showing Queen Elizabeth I a spherical magnet whose magnetic field represents that of the earth. In those early days of bold seafaring exploration and uncertain navigation, the discovery that the earth was a huge magnet was a practical discovery of prime importance.

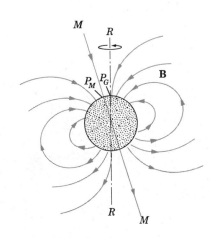

Figure 7 The earth's magnetic field represented as a dipole field. Its axis MM makes an angle of 11.5° with the earth's rotational axis RR. P_G and P_M are, respectively, the earth's geographic north pole and its magnetic north pole. The latter is actually the south pole of the earth's internal magnetic dipole.

earth was a huge magnet; see Fig. 6. As Fig. 7 suggests, the earth's magnetic field for points near its surface can be represented by a magnetic dipole located near the center of the earth.* To a good approximation, the earth's magnetic dipole moment has a magnitude of 8.0×10^{22} J/T; the dipole axis, shown as MM in Fig. 7, makes an angle of 11.5° with the earth's rotation axis, shown as RR in that figure. MM intersects the earth's surface at two points, defining what cartographers call the *north magnetic pole* (in northwest Greenland) and the *south magnetic pole* (in Antarctica). In general, lines of **B** for the earth's field emerge from the earth's surface in the southern hemisphere and reenter it in the northern hemisphere; thus the magnetic pole of the earth's dipole, deep inside the earth in the nothern hemisphere, is actually a south magnetic pole, just like the bottom end of the bar magnet of Fig. 5*b*.

Because of its practical applications in navigation, communication, and prospecting, the earth's magnetic field at its surface has been studied extensively for many years. The quantities of interest, as for any vector field, are the magnitude and direction of the field at different locations on the earth's surface and in the surrounding space. Field directions near the earth's surface are conveniently specified, with reference to the earth itself, in terms of the field declination (the angle between true

geographic north and the horizontal component of the field) and the inclination (the angle between a horizontal plane and the field direction). There are a variety of commercially-available magnetometers used to measure these quantities with high precision, but rough measurements can be made with a compass and a dip meter. The compass is just a magnet mounted horizontally, so that it can rotate freely about a vertical axis. The angle between its direction and true geographic north is the field declination. A dip meter is a similar magnet mounted for free rotation in a vertical plane. When aligned with its plane of rotation parallel to the compass direction, the angle between the needle and the horizontal is the inclination. The strength of the field can be estimated by measuring the frequency of the oscillation of either instrument after being displaced from its equilibrium position. The calculations are the same as we described earlier in Sample Problem 1.

As an example, the north pole of a compass needle in Tucson, Arizona, pointed about 13° east of geographic north (the declination) in 1964. The magnitude of the horizontal component of the earth's field was 26 μT (= 0.26 gauss), and the north end of a dip needle pointed downward making an angle of about 59° (the inclination) with a horizontal plane. As we expect from Fig. 7, lines of B are *entering* the earth's surface at this point.

At any point on the earth's surface the observed magnetic field may differ appreciably from the idealized,

---

* The magnetic field shown in Fig. 7 is idealized, suggesting what the field would be like for an isolated dipole, uninfluenced by the effects of the solar wind; see Fig. 9.

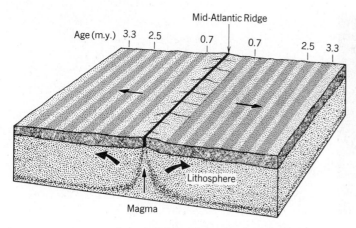

Figure 8 A magnetic profile of the sea floor on either side of the Mid-Atlantic Ridge. Points at which the earth's magnetic field has reversed its polarity are indicated. The sea floor, extruded from the ridge and spreading out as part of the continental drift system, displays a record of the past magnetic history of the earth's core.

best-fit, dipole field, in both magnitude and direction. At any location this observed field varies with time, by measurable amounts over a period of a few years and by substantial amounts over, say, 100 years. Thus between 1580 and 1820 the direction of the compass needle at London changed by 35°.

In spite of these local variations, the best-fit dipole field changes only slowly over such time periods, which are after all very short compared to the age of the earth. The variation of the earth's field over much longer intervals can be studied by measuring the weak intrinsic magnetism of the ocean floor on either side of the Mid-Atlantic Ridge. These deposits are formed by molten magma that oozes from the ridge. The magma solidifies and spreads out laterally at the rate of a few centimeters per year. The weak magnetism of the solidified magma preserves a "frozen-in" record of the earth's magnetic field at the time of solidification and thus allows us to study the direction and magnitude of the earth's magnetic field in the distant past; see Fig. 8. Such studies tell us, among other things, that the earth's field completely changes its direction (reverses its polarity) every million years or so.

There is at present no satisfactory detailed explanation for the origin of the earth's magnetic field. It seems certain that it arises in some way from current loops induced in the liquid and highly conducting outer region of the earth's core. The precise mode of action of this internal geomagnetic "dynamo" and the source of energy needed to keep it operating remain matters of continuing research interest.

Our moon, having no molten core, has no magnetic field. Most of the other planets in our solar system, Mercury and Jupiter among them, have magnetic fields. So do the sun and many other stars. In particular, neutron stars are thought to have magnetic fields of many millions of teslas in strength. There is also a magnetic field associated with our home galaxy. This field is weak (≈ 2 pT) but also important because of the immense volume it occupies.

At distances above the earth of the order of a few radii the earth's dipole field becomes modified in a major way by the action of the *solar wind*, which consists of streams of charged particles that constantly pour out of the sun. Figure 9 shows the resultant magnetic field of the earth at such large distances. A long *magnetotail* stretches outward away from the sun, extending for many thousands of earth diameters. The study of the

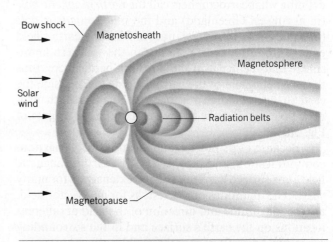

Figure 9 The earth's magnetic field, a composite of the earth's intrinsic dipole field as severely modified by the solar wind. The long *magnetotail* stretches out for several thousand earth diameters in a direction downstream from the solar wind.

magnetic field configurations of the earth and of our sister planets is high on the priority list of the space exploration programs of several countries.

Sample Problem 3   From data given earlier in this section find (a) the vertical component B_v of the earth's magnetic field and (b) the magnitude of the resultant magnetic field B at Tucson.

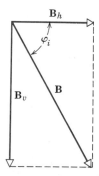

Figure 10   Sample Problem 3. The earth's magnetic field, along with its horizontal and vertical components, at Tucson, Arizona.

Figure 10 shows the situation. We have

(a)
$$B_v = B_h \tan \phi_i$$
$$= (26 \ \mu T)(\tan 59°)$$
$$= 43 \ \mu T = 0.43 \text{ gauss.} \qquad \text{(Answer)}$$

and

(b)
$$B = \sqrt{B_h^2 + B_v^2}$$
$$= \sqrt{(26 \ \mu T)^2 + (43 \ \mu T)^2}$$
$$= 50 \ \mu T = 0.50 \text{ gauss.} \qquad \text{(Answer)}$$

Note that the magnetic declination at Tucson plays no role in this problem.

34-6   Paramagnetism

Magnetism as we know it in our daily experience is an important but special branch of the subject called *ferromagnetism;* we discuss this in Section 34–8. Here we discuss a weaker and thus less familiar form of magnetism called *paramagnetism.*

For most atoms and ions, the magnetic effects of the electrons, including both their spins and orbital motions,

exactly cancel so that the atom or ion is not magnetic. This is true for the rare gases such as neon and for ions such as Cu^+, which make up ordinary copper.* For other atoms or ions the magnetic effects of the electrons do not cancel, so that the atom as a whole has a magnetic dipole moment μ. Examples are found among the transition elements, such as Mn^{++}; the rare earths, such as Gd^{+++}, and the actinide elements, such as U^{++++}.

If we place a sample of N atoms, each of which has a magnetic dipole moment μ, in a magnetic field, the elementary atomic dipoles tend to line up with the field. This tendency to align is called *paramagnetism.* For perfect alignment, the sample as a whole would have a magnetic dipole moment of $N\mu$. However, the aligning process is seriously disturbed by thermal agitation. The importance of thermal agitation may be measured by comparing two energies: One ($= \frac{3}{2} kT$) is the mean translational kinetic energy of an atom at temperature T. The other ($= 2 \ \mu B$) is the difference in energy between an atom whose dipole moment is lined up with the magnetic field and one pointing in the opposite direction. As Sample Problem 4 shows, the effect of the collisions at ordinary temperatures and fields is very important. The sample acquires a magnetic moment when placed in an external magnetic field, but this moment is usually very much smaller than the maximum possible moment $N\mu$.

We can express the extent to which a given specimen of a material is magnetized by dividing its measured magnetic moment by its volume. This vector quantity, the magnetic moment per unit volume, is called the *magnetization* **M** of the sample.

In 1895 Pierre Curie discovered experimentally that the magnetization **M** of a paramagnetic specimen is directly proportional to **B**, the effective magnetic field in which the specimen is placed, and inversely proportional to the kelvin temperature T. In equation form

$$\boxed{M = C \left(\frac{B}{T} \right)} \qquad \text{(Curie's law),} \qquad (15)$$

in which C is a constant. This equation is known as *Curie's law* and the constant C is called *the Curie constant.* The law is physically reasonable in that increasing B tends to align the elementary dipoles in the specimen,

---

* Cu^+ stands for a neutral copper atom from which one electron has been removed, leaving a net charge for the copper ion of $+ e$. Similarly, Mn^{++} stands for a neutral manganese atom from which two electrons have been removed, etc.

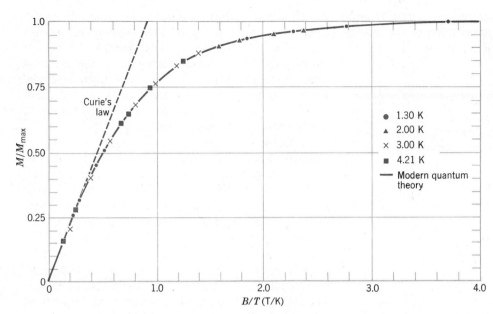

Figure 11 The ratio M to M_{max} for a paramagnetic salt, measured as a function of magnetic field for various low temperatures. After W. E. Henry.

that is, to increase M, whereas increasing T tends to interfere with this alignment, that is, to decrease M. Curie's law is well verified experimentally, provided that the ratio B/T does not become too large.

M cannot increase without limit, as Curie's law implies, but must approach a value M_{max} $(= \mu N/V)$ corresponding to the complete alignment of the N dipoles contained in the volume V of the specimen. Figure 11 shows this saturation effect for a sample of potassium chromium sulfate. The chromium ions are responsible for all the paramagnetism of this salt, all its other elements being paramagnetically inert.

It is not easy to achieve anything like total alignment in a paramagnetic sample. Even at a temperature as low as 1.3 K, the magnetic field required to achieve 99.5% saturation in the material to which Fig. 11 refers, is about 5 T (= 50,000 gauss). The curve that passes through the experimental points in Fig. 11 is calculated from quantum physics; we see that it is in excellent agreement with experiment.

Sample Problem 4 A paramagnetic gas, whose atoms have a magnetic dipole moment of 1.0 Bohr magnetons, is placed in an external magnetic field of magnitude 1.5 T. At room temperature ($T = 300$ K), calculate and compare U_T, the mean kinetic energy of translation $(= \frac{3}{2} kT)$, and U_B the magnetic energy $(= 2 \mu B)$.

We have

$$U_T = (\tfrac{3}{2}) kT = (\tfrac{3}{2})(1.38 \times 10^{-23} \text{ J/K})(300 \text{ K})$$
$$= 6.2 \times 10^{-21} \text{ J} = 0.039 \text{ eV}. \qquad \text{(Answer)}$$

Using Eq. 1,

$$U_B = 2 \mu B = (2)(9.27 \times 10^{-24} \text{ J/T})(1.5 \text{ T})$$
$$= 2.8 \times 10^{-23} \text{ J} = 0.00017 \text{ eV}. \qquad \text{(Answer)}$$

Because U_T equals about 230 U_B, we see that energy exchanges in collisions can interfere seriously with the alignment of the dipoles with the external field.

34-7 Diamagnetism (Optional)

Perhaps the earliest observation in electrostatics is that uncharged bits of paper will be *attracted* to a charged rod if placed in the nonuniform *electric* field near the tip of the rod; see Fig. 14 of Chapter 27. The molecules of the paper do not have intrinsic electric dipole moments and we account for the attraction in terms of dipole moments that are induced in the paper by the action of the external electric field.

A parallel effect occurs in magnetism. Some materials, called *diamagnetic,* do not have intrinsic magnetic dipoles (that is, they are not paramagnetic), but dipole moments may be induced in them by the action of an external magnetic field. If a sample of such a material is placed in the nonuniform magnetic field near one pole of a strong magnet, a (very weak) force will act on the sample. In contrast to the electric case, however, the sample will not be attracted toward the pole of the magnet but *repelled.*

We trace this difference in behavior between the electric and the magnetic cases to the fact that induced electric dipoles point in the same direction as the external electric field but induced magnetic dipoles point in the *opposite* direction to the external magnetic field.

Diamagnetism is a manifestation of Faraday's law of induction acting on the atomic electrons, whose motions—on a classical picture—are equivalent to tiny current loops.* The fact that the induced magnetic moment is *opposite* to the direction of the inducing magnetic field may be viewed as a consequence of Lenz's law, acting on the atomic scale.†

Diamagnetism is present in all atoms, but if these atoms happen to have intrinsic magnetic dipole moments, the diamagnetic effects are masked by the far stronger paramagnetic or ferromagnetic behavior.

34-8   Ferromagnetism

When we speak of magnetism in everyday conversation we almost certainly have in mind an image of a bar magnet picking up nails or a magnetic disk clinging to a refrigerator. We almost certainly do not have in mind the relatively weak paramagnetic or diamagnetic effects discussed in the two preceding sections.

It turns out that for iron and several other elements (most notably cobalt, nickel, gadolinium and dysprosium) and for many alloys of these and other elements a special interaction, called *exchange coupling,* serves to align the atomic dipoles in rigid parallelism, in spite of the randomizing tendency of the thermal motions of the atoms. This phenomenon, called *ferromagnetism,* is a purely quantum effect and cannot be explained in terms of classical physics.

If the temperature is raised above a certain critical value, called the *Curie temperature,* the exchange coupling ceases to be effective and most such materials become simply paramagnetic. For iron the Curie temperature is 1043 K (= 770°C). Ferromagnetism is evidently a property not only of the individual atom or ion but also of the interaction of each atom or ion with its neighbors in the crystal lattice of the solid.

To study the magnetization of a ferromagnetic material such as iron, it is convenient to form it into a toroidal or doughnut-shaped ring, as in Fig. 12. Primary coil P, containing n turns per unit length, is wrapped around it. This coil is essentially a long solenoid bent into a circle; if it carries a current i_p, we can find the magnetic field within the toroidal space from Eq. 21 of Chapter 31:

$$B_0 = \mu_0 n i_p. \qquad (16)$$

Note carefully that B_0 is the field that would be present within the toroid if the iron core were not in place. The arrangement of Fig. 12 is called a *Rowland ring,* after the physicist H. A. Rowland (1848–1901), who devised it.

The actual magnetic field B in the toroidal space of the Rowland ring of Fig. 12, with the iron core in place, is greater than B_0, commonly by a large factor. We can write

$$B = B_0 + B_M, \qquad (17)$$

where B_M is the contribution of the iron core to the total magnetic field B. B_M is associated with the alignment of the elementary atomic dipoles in the iron and is proportional to the magnetization M of the iron, that is, to its

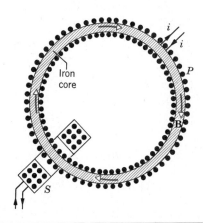

Figure 12   A specimen of iron whose ferromagnetic properties are being studied is formed into a Rowland ring. Primary coil P is used to magnetize the ring. Secondary coil S is used to measure the total magnetic field within the specimen.

---

* Interestingly, diamagnetism was discovered (and so named) by Michael Faraday, another of his many accomplishments.

† We do not push the semiclassical explanation of this very weak effect further. For a more complete treatment, see *Electricity and Magnetism,* by Edward M. Purcell, McGraw-Hill Book Company, New York 1985, 2nd ed., Section 11.5.

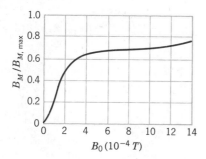

Figure 13 A *magnetization curve* for the core material of the Rowland ring of Fig. 12. On the vertical axis, 1.0 corresponds to complete alignment of the atomic dipoles within the specimen.

magnetic moment per unit volume. B_M has a certain maximum value, $B_{M,\text{max}}$, corresponding to complete alignment of the atomic dipoles.

It is instructive to plot, for any ferromagnetic specimen that can be formed into a Rowland ring, the ratio of B_M to $B_{M,\text{max}}$ as a function of B_0. Such a *magnetization curve,* as it is called, is displayed in Fig. 13. It is similar to the magnetization curve of Fig. 11 for a paramagnetic substance. Both are measures of the extent to which an applied magnetic field can succeed in aligning the elementary dipoles that make up the material in question.

For the ferromagnetic core material to which Fig. 13 refers, the alignment of the dipoles is about 75% complete for $B_0 \approx 1 \times 10^{-3}$ T. If B_0 were increased to 1 T, the fractional saturation of the specimen would increase to about 99.7%. Such a large value of B_0, if plotted in Fig. 13, would be about 95 ft to the right of the origin on the horizontal axis; complete saturation of the dipoles is approached only with considerable difficulty.

The use of iron in transformers, electromagnets, and other devices greatly increases the strength of the magnetic field that can be generated by a given current in a given set of windings. That is, very often, $B_M \gg B_0$ in Eq. 17. However, when B_0 is very large the presence of iron offers no advantage because of the saturation effect suggested in Fig. 13. When generating magnetic fields greater than this saturation limit, it is often simpler to abandon the use of iron and to rely on the "brute force" application of very large currents.

The magnetization curve for paramagnetism (Fig. 11) is explained in terms of the mutually opposing tendencies of alignment with the external field and of randomization because of the temperature motions. In ferromagnetism, however, we have assumed that adjacent atomic dipoles are locked in rigid parallelism. Why, then, does the magnetic moment of the specimen not reach its saturation value for very low—even zero—values of B_0? Why is not every iron nail a strong permanent magnet?

To understand this consider first a specimen of a ferromagnetic material such as iron that is in the form of a single crystal. That is, the arrangement of atoms that make it up—its crystal lattice—extends with unbroken regularity throughout the volume of the specimen. Such a crystal will, in its normal unmagnetized state, be made up of a number of *magnetic domains.* These are regions of the crystal throughout which the alignment of the atomic dipoles is essentially perfect. For the crystal as a whole, however, the domains are so oriented that they largely cancel each other as far as their external magnetic effects are concerned.

Figure 14 is a magnified photograph of such an assembly of domains in a single crystal of nickel. It was made by sprinkling a colloidal suspension of finely powdered iron oxide on a properly etched surface of the crystal. The domain boundaries, which are thin regions in which the alignment of the elementary dipoles changes from a certain orientation in one domain to a quite different orientation in the other, are the sites of intense, but highly localized and nonuniform magnetic

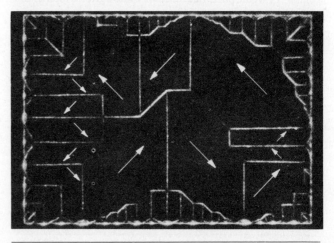

Figure 14 Domain patterns for a single crystal of nickel. The white lines show the boundaries of the domains. The arrows show the orientation of the magnetic dipoles within the domains.

fields. The suspended colloidal particles are attracted to these regions. Although the atomic dipoles in the individual domains are completely aligned, the crystal as a whole may have a very small resultant magnetic moment.

Actually a piece of iron as we ordinarily find it — an iron nail, say — is not a single crystal but is an assembly of many tiny crystals, randomly arranged; we call it a *polycrystalline solid*. Each tiny crystal, however, has its array of domains, just as in Fig. 14. Let us magnetize such a specimen by placing it in an external magnetic field of gradually increasing strength. Two effects take place, both contributing to the magnetization curve of Fig. 14. One is a growth in size of the domains that are favorably oriented at the expense of those that are not. Second, the orientation of the dipoles within a domain may swing around as a unit, becoming closer to the field direction.

Measuring B_M. It is possible to find B_M for a ferromagnetic specimen by measuring B, calculating B_0 from Eq. 16, and then subtracting, using Eq. 17:

$$B_M = B - B_0. \qquad (18)$$

B can be measured by setting up an induced current in secondary coil S of Fig. 12. With the iron core initially unmagnetized, set up a steady current i_p in the primary coil P. The magnetic flux linking secondary coil S (of resistance R_s and N_s turns) will rise from zero to a value BA in a time Δt, A being the cross-sectional area of the iron core.

From Faraday's law of induction an emf whose average value is

$$\mathcal{E}_s = -N_s \frac{d\Phi}{dt} = -\frac{N_s BA}{\Delta t}$$

will appear in the secondary coil. The average current in this coil during the time that the flux is changing will then be

$$i_s = \frac{\mathcal{E}_s}{R_s} = \frac{N_s BA}{R_s \Delta t}.$$

We have then for B

$$B = \frac{(i_s \Delta t)R_s}{N_s A} = \frac{\Delta q \, R_s}{N_s A},$$

in which $\Delta q \, (= i_s \, \Delta t)$ is the charge that passes through the secondary coil during the time that the magnetic field

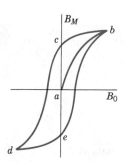

Figure 15 A magnetization curve (*ab*) for a ferromagnetic specimen and an associated hysteresis loop (*bcdeb*).

in the iron core is changing.* Thus a measurement of Δq (using an instrument called a ballistic galvanometer or in some other way) yields a measurement of the magnetic field B within the iron core.

Hysteresis. Magnetization curves for ferromagnetic materials do not retrace themselves as we increase and then decrease the external magnetic field B_0. Figure 15 shows the following operations with a Rowland ring: (1) starting with the iron unmagnetized (point *a*), increase the toroid current until $B_0 \,(= \mu_0 n i)$ has the value corresponding to point *b*; (2) reduce the current in the toroid winding back to zero (point *c*); (3) reverse the toroid current and increase it in magnitude until point *d* is reached; (4) reduce the current to zero again (point *e*); (5) reverse the current once more until point *b* is reached again.

The lack of retraceability shown in Fig. 15 is called *hysteresis*. Note that at points *c* and *e* the iron core is magnetized, even though there is no current in the toroid windings; this is the familiar phenomenon of permanent magnetism.

Hysteresis can be understood on the basis of the magnetic domain concept. Evidently the motions of the domain boundaries and the reorientations of the domain directions are not totally reversible. When the applied magnetic field B_0 is increased and then decreased back to its initial value the domains do not return completely to their original configuration but retain some "memory" of the initial increase. The "memory" of magnetic materials is essential for the magnetic storage of information, as on cassette tapes or computer disks.

---

* Although the rate of increase of the flux will not be strictly linear, and the time Δt will not be well-defined, the charge Δq is a quantity that can be precisely measured.

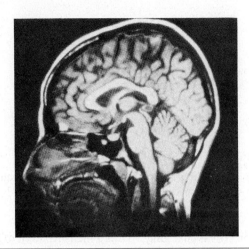

Figure 16 A cross section of a human head, taken by magnetic resonance imaging (MRI) techniques. It shows detail not visible on x-ray images and involves no radiation health risk to the patient.

34-9 Nuclear Magnetism—An Aside

The nuclei of many atoms are also magnetic dipoles. Their magnitudes tend to be about a thousand times smaller than those associated with the atomic electrons so that nuclear magnetism does not contribute in a measurable way to the gross magnetic properties of solids. It is nevertheless a subject of vital interest, both for what it can tell us about the internal structures of atoms and nuclei and for its practical applications. Figure 16, for example, shows a cross section of a human head taken by recently developed magnetic resonance imaging (MRI) techniques. For another example, much information about the structure of our galaxy and our universe comes to us from observations made with radio telescopes. Radiation at a wavelength of 21 cm, emitted from hydrogen atoms when the relative orientation of their nuclear and their electronic magnetic dipole moments changes, has proven particularly useful.

REVIEW AND SUMMARY

Magnetic Poles and Dipoles

There is no convincing evidence that magnetic monopoles (the magnetic equivalent of free electric charges) exist; see Fig. 2. The simplest sources of the magnetic field are magnetic dipoles, associated either with the orbital motions of electrons in atoms (see Fig. 4) or with the intrinsic spin motions (see Fig. 3b) of electrons, protons, and many other such particles. The magnetic dipole moments associated with electron motion are measured in terms of the *Bohr magneton* μ_B, where

The Bohr Magneton

$$1 \, \mu_B = \frac{eh}{4\pi m} = 9.27 \times 10^{-24} \text{ J/T} \quad \text{(the Bohr magneton).} \qquad [1]$$

The dipole moment associated with intrinsic electron spin is almost exactly $-1 \, \mu_B$, the minus sign indicating that the magnetic dipole moment vector is opposite the direction of the spin angular momentum vector. The dipole moment associated with electron orbital motion is $\mu_{orb} = -\frac{e}{2m} \mathbf{L}_{orb}$; see Sample Problem 2.

Gauss' law for magnetism,

Gauss' Law for Magnetism

$$\Phi_B = \oint \mathbf{B} \cdot d\mathbf{A} = 0 \quad \text{(Gauss' law for magnetism),} \qquad [13]$$

states that the magnetic flux through any closed Gaussian surface must be zero. See Fig. 5 for illustrations and for a comparison with Gauss' law for electrostatics. Equation 13 is simply a formal statement of the observation that there are no magnetic monopoles.

The Earth's Magnetic Field

The earth's magnetic field is approximately that of a dipole, with a dipole moment of 8.0×10^{22} J/T, near the earth's center. The dipole moment makes an angle of 11.5° with the earth's rotation axis, lines of **B** emerging from the earth's southern hemisphere. The field is thought to originate in current loops induced in the earth's liquid outer core by a mechanism not yet fully understood.

The direction of the local magnetic field at any point is given by its angle of *declination* (in a horizontal plane) from true north and its angle of *inclination* (in a vertical plane) from the horizontal; see Sample Problem 3.

Paramagnetism

  In the nature of their response to an external magnetic field materials may be broadly grouped as diamagnetic, paramagnetic, or ferromagnetic. Paramagnetic materials, which are (weakly) attracted by a magnetic pole, have intrinsic magnetic dipole moments which tend to line up with an external magnetic field, thus enhancing the field. This tendency is interfered with (see Sample Problem 4) by thermal agitation. The *magnetization M* of a specimen, which is its magnetic moment per unit volume, is given approximately by Curie's law, or

$$M = C\frac{B}{T} \quad \text{(Curie's law).} \qquad [15]$$

For strong enough fields or low enough temperatures this law breaks down as the atomic dipoles approach complete alignment and produce a maximum magnetization $M_{\text{max}} = \mu N/V$; see the *magnetization curve* of Fig. 11.

Diamagnetism

  Diamagnetic materials are (weakly) repelled by the pole of a strong magnet. The atoms of such materials do not have intrinsic magnetic dipole moments. A dipole moment may be induced, however, by an external magnetic field, its direction being opposite to that of the field.

Ferromagnetism

  In ferromagnetic materials such as iron a quantum interaction between neighboring atoms locks the atomic dipoles in rigid parallelism in spite of the disordering tendency of thermal agitation. This interaction abruptly disappears at a well-defined Curie temperature, above which the material becomes simply paramagnetic.

Magnetization Curves

  Figure 12 shows how a known external magnetic field, B_0 can be applied to a ring-shaped ferromagnetic specimen. The resultant magnetic field can be measured by an induction method and a magnetization curve such as that of Fig. 13 can be plotted. The course of this curve as B_0 increases corresponds to the growth of favorably oriented magnetic domains in the material; see Fig. 14 and the related text.

Hysteresis

  As Fig. 15 shows, ferromagnetic magnetization curves do not retrace themselves, a phenomenon called hysteresis. Some alignment of dipoles remains even when the external magnetic field is completely removed; the result is the familiar "permanent" magnet.

Nuclear Magnetism

  The nuclei of many atoms are magnetic dipoles, a fact of some importance in studying nuclear structure. This property also allows us to develop ingenious strategies for studying the location and density of particular atoms; magnetic resonance imaging (MRI) in medicine and mapping the density of hydrogen in the universe are important examples.

QUESTIONS

1. Two iron bars are identical in appearance. One is a magnet and one is not. How can you tell them apart? You are not permitted to suspend either bar as a compass needle or to use any other apparatus.

2. Two iron bars always attract, no matter the combination in which their ends are brought near each other. Can you conclude that one of the bars must be unmagnetized?

3. How can you determine the polarity of an unlabeled magnet?

4. The neutron, which has no charge, has a magnetic dipole moment. Is this possible on the basis of classical electromagnetism, or does this evidence alone indicate that classical electromagnetism has broken down?

5. Must all permanent magnets have identifiable north and south poles? Consider geometries other than the bar or horseshoe magnet.

6. A certain short iron rod is found, by test, to have a north pole at each end. You sprinkle iron filings over the rod. Where (in the simplest case) will they cling? Make a rough sketch of what the lines of **B** must look like, both inside and outside the rod.

7. Starting with *A* and *B* in the positions and orientations shown in Fig. 17, with *A* fixed but *B* free to rotate, what happens (*a*) if *A* is an electric dipole and *B* is a magnetic dipole; (*b*) if *A* and *B* are both magnetic dipoles; (*c*) if *A* and *B* are both electric dipoles? Answer the same questions if *B* is fixed and *A* is free to rotate.

8. Cosmic rays are charged particles that strike our atmo-

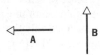

Figure 17 Question 7.

sphere from some external source. We find that more low-energy cosmic rays reach the earth near the north and south magnetic poles than at the (magnetic) equator. Why is this so?

9. How might the magnetic dipole moment of the earth be measured?

10. Give three reasons for believing that the flux Φ_B of the earth's magnetic field is greater through the boundaries of Alaska than through those of Texas.

11. You are a manufacturer of compasses. (*a*) Describe ways in which you might magnetize the needles. (*b*) The end of the needle that points north is usually painted a characteristic color. Without suspending the needle in the earth's field, how might you find out which end of the needle to paint? (*c*) Is the painted end a north or a south magnetic pole?

12. Would you expect the magnetization at saturation for a paramagnetic substance to be very much different from that for a saturated ferromagnetic substance of about the same size? Why or why not?

13. The magnetization induced in a given diamagnetic sphere by a given external magnetic field does not vary with temperature, in sharp contrast to the situation in paramagnetism. Explain this behavior in terms of the description that we have given of the origin of diamagnetism.

14. Explain why a magnet attracts an unmagnetized iron object such as a nail.

15. Does any net force or torque act on (*a*) an unmagnetized iron bar or (*b*) a permanent bar magnet when placed in a uniform magnetic field?

16. A nail is placed at rest on a smooth tabletop near a strong magnet. It is released and attracted to the magnet. What is the source of the kinetic energy it has just before it strikes the magnet?

17. Superconductors are said to be perfectly diamagnetic. Explain.

18. Compare the magnetization curves for a paramagnetic substance (see Fig. 11) and for a ferromagnetic substance (see Fig. 13). What would a similar curve for a diamagnetic substance look like?

19. Why do iron filings line up with a magnetic field, as in Fig. 1? After all, they are not intrinsically magnetized.

20. The earth's magnetic field can be represented closely by that of a magnetic dipole located at or near the center of the earth. The earth's magnetic poles can be thought of as (*a*) the points where the axis of this dipole passes through the earth's surface or as (*b*) the points on the earth's surface where a dip needle would point vertically. Are these necessarily the same points?

21. A "friend" borrows your favorite compass and paints the entire needle red. When you discover this you are lost in a cave and have with you two flashlights, a few meters of wire, and (of course) this book. How might you discover which end of your compass needle is the north-seeking end?

22. A Rowland ring is supplied with a constant current. What happens to the magnetic induction in the ring if a small slot is cut out of the ring, leaving an air gap?

23. How can you magnetize an iron bar if the earth is the only magnet around?

24. How would you go about shielding a certain volume of space from constant external magnetic fields? If you think it can't be done, explain why.

EXERCISES AND PROBLEMS

Section 34–2 Magnetism and the Electron

1E. Using the values of (spin) angular momentum S and (spin) magnetic moment μ_s given in Table 1 for the free electron, show that

$$\mu_s = \frac{e}{m} S.$$

Verify that the units and dimensions are consistent. This result is a prediction of a relativistic theory of the electron advanced by P. A. M. Dirac in 1928.

2P. Figure 18 shows four arrangements of pairs of small compass needles, set up in a space in which there is no external magnetic field. Identify the equilibrium in each case as stable or unstable. For each pair consider only the torque acting on one needle due to the magnetic field set up by the other. Explain your answers.

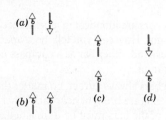

Figure 18 Problem 2.

3P. A simple bar magnet hangs from a string as in Fig. 19. A uniform magnetic field **B** directed horizontally is then established. Sketch the resulting orientation of the string and the magnet.

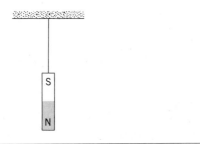

Figure 19 Problem 3.

Section 34–3 Orbital Angular Momentum and Magnetism

4E. In the lowest energy state of the hydrogen atom the most probable distance between the single orbiting electron and the central proton is 5.2×10^{-11} m. Calculate (a) the electric field and (b) the magnetic field set up by the proton at this distance, measured along the proton's axis of spin. The charge and magnetic moment of the proton are $+1.6 \times 10^{-19}$ C and 1.4×10^{-26} J/T, respectively.

5P. A charge q is distributed uniformly around a thin ring of radius r. The ring is rotating about an axis through its center and perpendicular to its plane at an angular speed ω. (a) Show that the magnetic moment due to the rotating charge is

$$\mu = \tfrac{1}{2} q \omega r^2.$$

(b) What is the direction of this magnetic moment if the charge is positive?

Section 34–4 Gauss' Law for Magnetism

6E. Imagine rolling a sheet of paper into a cylinder and placing a bar magnet near its end as shown in Fig. 20. (a) Sketch the **B** field lines as they cross the paper cylinder. (b) What can you say about the sign of $\mathbf{B} \cdot d\mathbf{A}$ for every $d\mathbf{A}$ of this paper cylinder? (c) Does this contradict Gauss' law for magnetism? Explain.

Figure 20 Exercise 6.

7E. The magnetic flux through each of five faces of a die (singular of "dice") is given by $\Phi_B = \pm N$ Wb, where N ($= 1$ to 5) is the number of spots on the face. The flux is positive (outward) for N even and negative (inward) for N odd. What is the flux through the sixth face of the die?

8P. A Gaussian surface in the shape of a right circular cylinder has a radius of 12 cm and a length of 80 cm. Through one end there is an inward magnetic flux of 25 μWb. At the other end there is a uniform magnetic field of 1.6 mT, normal to the surface and directed outward. What is the net magnetic flux through the curved surface?

9P*. Two wires, parallel to the z axis and a distance $4r$ apart, carry equal currents i in opposite directions, as shown in Fig. 21. A circular cylinder of radius r and length L has its axis on the z axis, midway between the wires. Use Gauss' law for magnetism to calculate the net outward magnetic flux through the half of the cylindrical surface above the x axis. (*Hint:* Find the flux through that portion of the xz plane that is within the cylinder.)

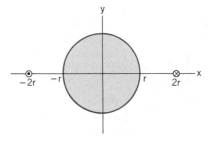

Figure 21 Problem 9.

Section 34–5 The Magnetism of the Earth

10E. In New Hampshire the average horizontal component of the earth's magnetic field in 1912 was 16 μT and the average inclination or "dip" was 73°. What was the corresponding magnitude of the earth's magnetic field?

11E. In Sample Problem 3 the vertical component of the earth's magnetic field in Tucson, Arizona, was found to be 43 μT. Assume this is the average value for all of Arizona, which has an area of 295,000 square kilometers, and calculate the net magnetic flux through the rest of the earth's surface (the entire surface excluding Arizona). Is the flux outward or inward?

12E. The earth has a magnetic dipole moment of 8.0×10^{22} J/T. (a) What current would have to be set up in a single turn of wire going around the earth at its magnetic equator if we wished to set up such a dipole? (b) Could such an arrangement be used to cancel out the earth's magnetism at points in space well above the earth's surface? (c) On the earth's surface?

13P. The magnetic field of the earth can be approximated as a dipole magnetic field, with horizontal and vertical components, at a point a distance r from the earth's center, given by

$$B_h = \frac{\mu_0 \mu}{4\pi r^3} \cos \lambda_m, \quad B_v = \frac{\mu_0 \mu}{2\pi r^3} \sin \lambda_m,$$

where λ_m is the *magnetic latitude* (latitude measured from the magnetic equator toward the north or south magnetic pole). The magnetic dipole moment $\mu = 8.0 \times 10^{22}$ A·m². (a) Show that the strength at latitude λ_m is given by

$$B = \frac{\mu_0 \mu}{4\pi r^3} \sqrt{1 + 3 \sin^2 \lambda_m}.$$

(b) Show that the inclination ϕ_i of the magnetic field is related to the magnetic latitude λ_m by

$$\tan \phi_i = 2 \tan \lambda_m.$$

14P. Use the results displayed in Problem 13 to predict the value of the earth's magnetic field, magnitude, and inclination at (a) the magnetic equator; (b) a point at magnetic latitude $60°$; (c) the north magnetic pole.

15P. Find the altitude above the earth's surface where the earth's magnetic field has a magnitude one-half the surface value at the same magnetic latitude. (Use the dipole field approximation given in Problem 13.)

16P. Using the dipole field approximation to the earth's magnetic field (see Problem 13), calculate the maximum strength of the magnetic field at the core-mantle boundary, which is 2900 km below the earth's surface.

17P. Use the results displayed in Problem 13 to calculate the magnitude and inclination angle of the earth's magnetic field at the north geographic pole. (*Hint:* The angle between the magnetic axis and the rotational axis of the earth is $11.5°$.) Why do the calculated values probably not agree with the measured values?

Section 34–6 Paramagnetism

18E. A 0.50-T magnetic field is applied to a paramagnetic gas whose atoms have an intrinsic magnetic dipole moment of 1.0×10^{-23} J/T. At what temperature will the mean kinetic energy of translation of the gas atoms be equal to the energy required to reverse such a dipole end for end in this magnetic field?

19E. A cylindrical rod magnet has a length of 5.0 cm and a diameter of 1.0 cm. It has a uniform magnetization of 5.3×10^3 A/m. What is its magnetic dipole moment?

20E. A paramagnetic substance is (weakly) attracted to a pole of a magnet. Figure 22 shows a model of this phenomenon. The "paramagnetic substance" is a current loop L, which is placed on the axis of a bar magnet nearer to its north pole than its south pole. Because of the torque $\tau = \mu \times B$ exerted on the loop by the B field of the bar magnet, the magnetic dipole moment μ of the loop will align itself to be parallel to B. (a) Make a sketch showing the B field lines due to the bar magnet. (b) Show the direction of the current i in the loop. (c) Using

$d\mathbf{F} = i\,d\mathbf{s} \times \mathbf{B}$ show from (a) and (b) that the net force on L is toward the north pole of the bar magnet.

21P. The paramagnetic salt to which the magnetization curve of Fig. 11 applies is to be tested to see whether it obeys Curie's law. The sample is placed in a 0.50-T magnetic field that remains constant throughout the experiment. The magnetization M is then measured at temperatures ranging from 10 to 300 K. Would it be found that Curie's law is valid under these conditions?

22P. A sample of the paramagnetic salt to which the magnetization curve of Fig. 11 applies is held at room temperature (300 K). At what applied magnetic field would the degree of magnetic saturation of the sample be (a) 50 percent? (b) 90 percent? (c) Are these fields attainable in the laboratory?

23P. A sample of the paramagnetic salt to which the magnetization curve of Fig. 11 applies is immersed in a magnetic field of 2.0 T. At what temperature would the degree of magnetic saturation of the sample be (a) 50 percent? (b) 90 percent?

24P. An electron with kinetic energy K_e travels in a circular path that is perpendicular to a uniform magnetic field, subject only to the force of the field. (a) Show that the magnetic dipole moment due to its orbital motion has magnitude $\mu = K_e/B$ and that it is in the direction opposite to that of $\mathbf{B}$. (b) What is the magnitude and direction of the magnetic dipole moment of a positive ion with kinetic energy K_i under the same circumstances? (c) An ionized gas consists of 5.3×10^{21} electrons/m³ and the same number of ions/m³. Take the average electron kinetic energy to be 6.2×10^{-20} J and the average ion kinetic energy to be 7.6×10^{-21} J. Calculate the magnetization of the gas for a magnetic field of 1.2 T.

25P. Consider a solid containing N atoms per unit volume, each atom having a magnetic dipole moment μ. Suppose the direction of μ can be only parallel or antiparallel to an externally applied magnetic field $\mathbf{B}$ (this will be the case if μ is due to the spin of a single electron). According to statistical mechanics, it can be shown that the probability of an atom being in a state with energy U is proportional to $e^{-U/(kT)}$ where T is the temperature and k is Boltzmann's constant. Thus, since $U = -\mu \cdot \mathbf{B}$, the fraction of atoms whose dipole moment is parallel to $\mathbf{B}$ is proportional to $e^{(\mu B)/(kT)}$ and the fraction of atoms whose dipole moment is antiparallel to $\mathbf{B}$ is proportional to $e^{-(\mu B)/(kT)}$. (a) Show that the magnetization of this solid is $M = N\mu \tanh(\mu B/kT)$. Here tanh is the hyperbolic tangent function: $\tanh(x) = (e^x - e^{-x})/(e^x + e^{-x})$. (b) Show that (a) reduces to $M = N\mu^2 B/kT$ for $\mu B \ll kT$. (c) Show that (a) reduces to $M = N\mu$ for $\mu B \gg kT$. (d) Show that (b) and (c) agree qualitatively with Fig. 11.

Section 34–7 Diamagnetism

26E. A diamagnetic substance is (weakly) repelled by a pole of a magnet. Figure 22 shows a model of this phenomenon. The "diamagnetic substance" is a current loop L that is placed on the axis of a bar magnet nearer to its north pole than its

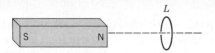

Figure 22 Exercises 20 and 26.

south pole. Because the substance is diamagnetic the magnetic moment μ of the loop will align itself to be antiparallel to the **B** field of the bar magnet. (a) Make a sketch showing the **B** field lines due to the bar magnet. (b) Show the direction of the current i in the loop. (c) Using $d\mathbf{F} = i\,d\mathbf{s} \times \mathbf{B}$, show from (a) and (b) that the net force on L is away from the north pole of the bar magnet.

27P. Analyze qualitatively the appearance of induced magnetic dipole moments in diamagnetism from the point of view of Faraday's law of induction. [*Hint:* See Fig. 10b, Chapter 32. Also, note that for orbiting electrons the inductive effects (any change in speed) persist after the magnetic field has stopped changing; they vanish only when the field is removed.]

Section 34–8 Ferromagnetism

28E. Measurements in mines and boreholes indicate that the temperature in the earth increases with depth at the average rate of 30°C/km. Assuming a surface temperature of 10°C, at what depth does iron cease to be ferromagnetic? (The Curie temperature of iron varies very little with pressure.)

29E. The exchange coupling mentioned in Section 8 as being responsible for ferromagnetism is *not* the mutual magnetic interaction energy between two elementary magnetic dipoles. To show this calculate (a) the magnetic field a distance of 10 nm away along the dipole axis from an atom with magnetic dipole moment 1.5×10^{-23} J/T (cobalt), and (b) the minimum energy required to turn a second identical dipole end for end in this field. Compare with the results of Sample Problem 4. What do you conclude?

30E. The saturation magnetization of the ferromagnetic metal nickel is 4.7×10^5 A/m. Calculate the magnetic moment of a single nickel atom. (The density of nickel is 8.90 g/cm^3 and its atomic mass is 58.71.)

31E. The dipole moment associated with an atom of iron in an iron bar is 2.1×10^{-23} J/T. Assume that all the atoms in the bar, which is 5.0 cm long and has a cross-sectional area of 1.0 cm^2, have their dipole moments aligned. (a) What is the dipole moment of the bar? (b) What torque must be exerted to hold this magnet at right angles to an external field of 1.5 T? The density of iron is 7.9 g/cm^3.

32P. The magnetic dipole moment of the earth is 8.0×10^{22} J/T. (a) If the origin of this magnetism were a magnetized iron sphere at the center of the earth, what would be its radius? (b) What fraction of the volume of the earth would such a sphere occupy? Assume complete alignment of the dipoles. The density of the earth's inner core is 14 g/cm^3. The magnetic dipole moment of an iron atom is 2.1×10^{-23} J/T. (*Note:* The earth's inner core is in fact thought to be in both liquid and solid form and partly iron, but a permanent magnet as the source of the earth's magnetism has been ruled out by several considerations. For one, the temperature is certainly above the Curie point.)

33P. Figure 23 shows the apparatus used in a lecture demonstration of para- and diamagnetism. A sample of the magnetic material is suspended by a string ($L = 2$ m) in a region ($d = 2$ cm) between the two poles of a powerful electromagnet. Pole P_1 is sharply pointed and pole P_2 is rounded as indicated. Any deflection of the string from the vertical is visible to the audience by means of an optical projection system (not shown). (a) First a bismuth (highly diamagnetic) sample is used. When the electromagnet is turned on, the sample is observed to deflect slightly (about 1 mm) toward one of the poles. What is the direction of this deflection? (b) Next an aluminum (paramagnetic, conducting) sample is used. When the electromagnet is turned on, the sample is observed to deflect strongly (about 1 cm) toward one pole for about a second and then deflect moderately (a few mm) toward the other pole. Explain and indicate the direction of these deflections. (*Hint:* Note that the sample is a conductor.) (c) What would happen if a ferromagnetic sample were used?

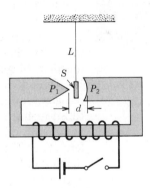

Figure 23 Problem 33.

34P. A Rowland ring is formed of ferromagnetic material. It is circular in cross section, with an inner radius of 5.0 cm and an outer radius of 6.0 cm and is wound with 400 turns of wire. (a) What current must be set up in the windings to attain a toroidal field $B_0 = 0.20$ mT? (b) A secondary coil wound around the toroid has 50 turns and has a resistance of 8.0 Ω. If, for this value of B_0, we have $B_M = 800B_0$, how much charge moves through the secondary coil when the current in the toroid windings is turned on?

ESSAY 14
MAGNETISM AND LIFE

CHARLES P. BEAN

RENSSELAER
POLYTECHNIC
INSTITUTE

Introduction

Owing to the apparent mystery of magnetic forces, people of the past, and even today, have looked for effects of the magnetic field on human and other animal life. In many cases they have believed they have found such effects but never in a way that could be replicated. Recently, however, a young scientist has discovered a completely replicable effect of the earth's magnetic field on a class of living organisms—the magnetotactic bacteria. This essay tells you about that discovery and some of its consequences. Before telling that story, it is necessary to give some historical and scientific perspective.

We said that magnetism was mysterious. Take two permanent magnets and have them approach one another. In one orientation, they attract one another through empty space and in another repel. We have all felt the strangeness of this effect. For the five-year-old Einstein, the observation of the deflection of a compass needle caused him first to think about fields of force and what they might be. Our present understanding, due largely to Einstein's thoughts, is that there exists only one field, the electromagnetic field, and that our perception depends on our motion with respect to that field. For instance, what we call a magnetic field is caused by the motion of electric charges with respect to us. One consequence is that, since the relative velocities of motion are usually much less than the speed of light, magnetic forces are usually much less than electrical forces. Electrical forces, for instance, hold atoms and solids together. Magnetic forces provide usually only a small fraction of the total binding.

In common terms, when we speak of a magnet we mean the type that sticks on refrigerator doors—the permanent magnet. In nature, it principally exists in one form called magnetite with the chemical formula, Fe_3O_4. More informatively, it can be written as $FeO \cdot Fe_2O_3$ to show that each molecule has one ferrous (Fe^{++}) ion and two ferric (Fe^{+++} ions). It is an unremarkable-looking black stone (lodestone), but one that led to the great age of geographic discovery in the twelfth to sixteenth centuries as well as to many aspects of modern science.

The details of the discovery of the properties of magnetite are lost in antiquity. The fact that magnetite can attract iron was known to the Chinese over 2000 years ago and knowledge of that interaction either spread to the Mediterranean or was independently discovered there in classical times. The crucial discovery that magnetite tends to align itself in the earth's field was, apparently, a discovery of the Chinese. (Curiously enough, they called it a "south pointer.") Further it was found by experiment that needles of iron could be magnetized by stroking them with lodestone or, alternatively, by heating the needles red hot and allowing them to cool in the earth's magnetic field. A magnetized needle suspended in water or on a string was found to point roughly north and south. With this discovery, ships could safely sail the open ocean even if the stars were invisible. For the first time a ship could follow a charted course of exploration and return to its home port.

This essay shows that a large class of bacteria, some billions of years ago, developed a magnetic guidance system that guided them in their movements. Thus one of man's great discoveries is now known to have been antedated by the simplest organism on earth.

Magnetic Navigation by Bacteria

Bacteria are single-cell organisms that live everywhere. They flourish both inside and on the surface of our bodies. They can be found in hot springs at 85°C and at the bottom of the ocean. Typically, they are a few micrometers in size and so can only be seen with a microscope. Owing to the resolution limit imposed by the wavelength of visible light (about 0.4 μm in water), their detailed interior structure cannot be seen with the optical microscope. An electron microscope is required to see the finer points of their structure.

Despite their small size, bacteria show great powers of adaptation to local conditions and a wide range of behavior. For instance, a large class of bacteria can function best in conditions under which oxygen is not present. It is thought that these arose early in the development of the earth before the evolution of plants and the consequent release of oxygen to the atmosphere. (Oxygen is a waste product of photosynthesis just as carbon dioxide is of our metabolism.) Now these so-called anaerobic bacteria can be found today in many aquatic environments with low oxygen levels. Decaying animal and vegetable matter provide these conditions. Typically the waste product of anaerobic metabolism is methane—also known as marsh gas. Most such bacteria can swim using one or more appendages—called flagellae when they are long and few and pili when mutiple and short. In addition they have receptors that can sense chemicals and dissolved gases in the water and so can swim toward food and away from poisons. For the anaerobic bacterium oxygen is such a poison.

In 1975, Richard Blakemore made a remarkable discovery. At the time he was a graduate student in microbiology at the University of Massachusetts working at the Woods Hole Oceanographic Institute on Cape Cod. His area of research concerned the role of anaerobic bacteria in the ecology of muds and swamps. He took some mud from a nearby saltwater pond, mixed it with a little seawater, put a drop on a microscope slide and observed it at high power. Doubtless this had been done tens of thousands of times before, dating back to the days of Pasteur in the 1850s. But Blakemore noticed what no one had recorded ever noticing earlier. In some drops the bacteria swam to one side of the drop. Were they swimming away or toward the room light? He covered the microscope and its internal light with a box. He turned the microscope around. He moved the microscope to another room. In each case the bacteria continued to swim in the same geographic direction—north. Biological and chemical laboratories are well equipped with small permanent magnets coated with plastic. In conjunction with a larger external rotating magnet they are used to stir and mix solutions. Blakemore picked up such a stirrer and brought it near the drop. In one orientation, the stirrer bar did not affect the motion, in the opposite orientation the direction of motion of the bacteria was reversed! This observation, when the stirrer magnet was tested with a compass, showed that almost all the bacteria swam in the direction of the north-seeking end of the magnetic compass.

Here was a discovery that was unique in the history of magnetism and biology. It showed a direct and repeatable influence of the earth's magnetic field on a living organism. Blakemore and co-workers quickly showed that life was not essential to the orientation of bacteria. Killed bacteria would rotate to follow an imposed magnetic field but, since they were dead, would not migrate in that field.

The Mechanism of Orientation

Each discovery in science leads to more questions. In this case, they were: what is the physical mechanism of the effect and what are its biological implications? For the first

Figure 1 A magnetotactic bacterium as seen in a transmission electron micrograph. The unusual feature of the internal structure of this type of bacterium is a chain of electron dense particles roughly 50 nm in diameter. These are composed of magnetite (Fe_3O_4) and are each a fully magnetized permanent magnet. The chain functions as a compass that orients the bacterium in the same direction as the lines of force of the earth's magnetic field. The species shown here has a flagellum at each end. It is capable of swimming forward or backward. Many species have flagellae only at one end. (R. P. Blakemore and R. B. Frankel, *Scientific American,* December 1981.)

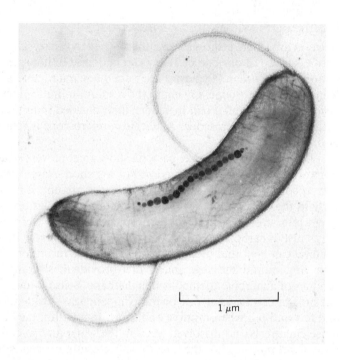

1 μm

question, Blakemore reasoned that the simplest mechanism was to have each bacterium be, in itself, a small compass that would rotate in a magnetic field. An electron micrograph of a typical bacterium (Fig. 1) showed a novel alignment of electron-dense particles within a typical bacterium that responded to a magnetic field. Could they be the compass? In discussing this conjecture with Edward M. Purcell of Harvard, the latter suggested that these were probably single-domain magnetic particles of some iron-containing substance and, if so, one consequence would be that a short pulse of a high oppositely-directed magnetic field would reverse the magnetization of each bacterium before it could rotate to accommodate itself to the new field. An experiment, in conjunction with Adrianus J. Kalmijn of the Woods Hole Oceanographic Institute, showed just such an effect. After such a pulse, the bacteria were not south-seeking.

Using the physics of magnetism developed in Chapter 34, one can estimate how much material would be necessary to orient a bacterium in the earth's magnetic field. Sample Problem 4 shows that an atom with a magnetic moment of one Bohr magneton is not significantly aligned at room temperature in a field of 1.5 T. To have significant alignment the magnetic energy μB must be comparable or greater than the energy associated with thermal disorder, kT. (The factors of 2 and $\frac{2}{3}$ used in Sample Problem 4 are not necessary in this order of magnitude calculation.) This requirement says that

$$\mu B > kT \quad \text{or} \quad \mu > kT/B$$

where μ is the total magnetic moment of the bacterium, k is Boltzmann's constant (1.38×10^{-23} J/K), T is the absolute temperature (300 K), and B is the earth's magnetic field ($\sim 5 \times 10^{-5}$ T). Using these figures

$$\mu > (1.38 \times 10^{-23})(3 \times 10^2)/(5 \times 10^{-5}) = 8.3 \times 10^{-17} \text{ J/T}.$$

If magnetite is fully magnetized, it has a magnetic moment per unit volume of 5×10^5 J/Tm^3. (This value corresponds to 4 Bohr magnetons per molecule of Fe_3O_4, the moment of Fe^{+2}. The magnetic moments of the Fe^{+3} ions exactly cancel one another.) Since the total magnetic moment, μ, is given by the product of the magnetic moment per unit volume and the volume, V, we derive that the volume, V, of fully magnetized material needed for significant alignment is

$$V > (8.3 \times 10^{-17})/(5 \times 10^5) = 1.7 \times 10^{-22} \text{ m}^3$$

These lower limits can be compared with the volume estimated from the electron micrograph shown in Fig. 1. We see that the chain of particles has 20 members of diameter ~ 50 nm. Assuming them to be spheres, their total volume is $20\pi(50 \times 10^{-9})^3/6 = 13 \times 10^{-22}$ m^3 or about eight times the minimum calculated above. This implies, of course, a magnetic moment about eight times that for which $\mu B = kT$. Consequently, the bacteria are well aligned in the earth's magnetic field.

Another method of calculating the magnetic moment of the material comes from an ingenious experiment by Adrianus Kalmijn. He observes, using a microscope, a single live bacterium swimming along a field direction. For low fields the bacterium is buffeted by the thermal motions of the water molecules and its average forward velocity is low. For higher fields the motion of the bacterium closely follows the field lines. Kalmijn notes the time it takes the bacterium to move between two fixed lines in the optical field of the microscope, then reverses the magnetic field. The bacterium makes a U-turn and goes in the opposite direction. He has watched hundreds of round trips and averaged times to obtain an average velocity. A specimen result is shown in Fig. 2. The interpretation of this result is that the bacterium swims always at $117 \ \mu$m/s but forward motion only results from the component of velocity parallel to the field direction. At low fields the bacterium swims in many directions with only slow progress along the field direction. At high fields the bacterium swims only parallel to the field. Conceptually this is identical to the problem of calculating the average moment of a paramagnet such as the chromium salt shown in Fig. 12 of Chapter 34. The experimental points can be fitted with two parameters, the swimming velocity when fully aligned and the magnetic moment of the bacterium. The solid curve of the figure is a theoretical curve with a maximum speed of $117 \ \mu$m/s and a

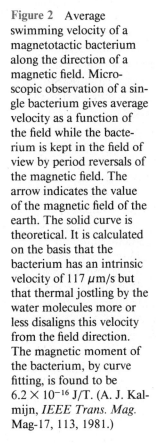

Figure 2 Average swimming velocity of a magnetotactic bacterium along the direction of a magnetic field. Microscopic observation of a single bacterium gives average velocity as a function of the field while the bacterium is kept in the field of view by period reversals of the magnetic field. The arrow indicates the value of the magnetic field of the earth. The solid curve is theoretical. It is calculated on the basis that the bacterium has an intrinsic velocity of $117 \ \mu$m/s but that thermal jostling by the water molecules more or less disaligns this velocity from the field direction. The magnetic moment of the bacterium, by curve fitting, is found to be 6.2×10^{-16} J/T. (A. J. Kalmijn, *IEEE Trans. Mag.* Mag-17, 113, 1981.)

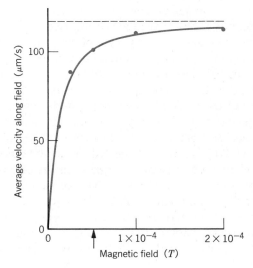

magnetic moment of 6.2×10^{-16} J/T. For the saturation magnetization of 5×10^5 J/T·m³ given earlier, this corresponds to about 12×10^{-22} m³ of Fe_3O_4. This value is very close to that derived from examination of the photograph in Fig. 1 although it is not the same bacterium. Thus two lines of evidence are consistent with the hypothesis that the mechanism of alignment is simply that of a compass.

The Consequences of Orientation in the Earth's Magnetic Field

In biology, one often looks for an advantage gained by an organism with a special sense. For instance, bats can both emit and detect ultrasound. By use of this faculty they can both fly in the dark and locate prey, such as moths. In the case of the magnetotactic bacteria, one can identify a clear advantage. Figure 8 of Chapter 34 shows the earth's magnetic field represented as a dipole. In the Northern hemisphere, there is a downward component. Bacteria swimming along the field line would go down. Thus if anaerobic bacteria were stirred from their usual environment, they would be aided in their return to the mud. One consequence of this concept is that bacteria in the Southern hemisphere would have south-seeking magnetic moments. An expedition by Blakemore and collaborators to New Zealand showed exactly that. At the equator, one finds both polarities but the advantage is not completely clear since there the bacteria are constrained to move in horizontal lines.

In each population of bacteria there may be a few bacteria per thousand of the "wrong" polarity. Investigators have taken a sample of mud and water and put it in a field whose vertical component is reversed. The vast majority of the bacteria swim to the surface and the consequent oxygen-rich environment. In that environment their metabolism and reproductive capacity are lowered. The former aberrant bacteria who now go to the mud are favored. After a matter of eight weeks almost all bacteria have reversed polarity. We know no mechanism whereby bacteria could rotate their internal magnetic particles. Most probably the new population of bacteria are descendents of the few aberrant bacteria. If so, the experiment shows, in microcosm, an evolutionary adaptation to environmental change. (Can you think of an experiment to prove whether the new population truly descends from the aberrant bacteria?)

Do Other Organisms Respond to the Earth's Magnetic Field?

Just as mariners use a magnetic compass to guide them, migratory birds and nectar-seeking bees could use a magnetic sense. Over the years many investigators have explored this possibility. They have attached magnets and dummy magnets to birds and claimed an altered behavior for the case of magnets. The suggestion has been made that pigeons not only detect direction but can detect a change in magnetic field strength of 2 parts in 10^4. Bees are thought by some to use a magnetic map and convey, by dances, directions to prospective foragers. (J. L. Gould in *American Scientist,* May–June 1980 gives a good review of the understanding of possible magnetic sensitivity of birds and bees.) Contrary to the case of Blakemore's analysis of magnetotactic bacteria, no one has found a mechanism of these possible effects. If it were a local compass, it would have to be connected to the nervous system of these higher organisms rather than act as a passive torque as in the case of the bacteria. No such connection has been found. Indeed no one has been able to condition a bird or bee to be attracted or repelled by a magnetic field. Consequently, most biophysicists do not believe that the case for magnetic sensitivity of birds and bees has been proven. But the question is still open. Another Blakemore may yet make a discovery that revolutionizes this field as thoroughly as did Blakemore's original observation of magnetotactic bacteria.

CHAPTER 35

ELECTROMAGNETIC OSCILLATIONS

The radar beams that make modern air travel possible are generated in the electromagnetic oscillator of a ground-based radar transmitter. When the beam strikes the aircraft it is detected by a receiver containing another electromagnetic oscillator tuned to the same frequency. The detected radar beam triggers a transponder beacon (still another electromagnetic oscillator) which transmits coded identity and altitude information, all of which is detected by a tuned receiver at the ground antenna and displayed on the air traffic controller's video terminal. Radio communication with the aircraft involves two other electromagnetic oscillators, one in the transmitter and another in the receiver.

35–1 New Physics—Old Mathematics

In this chapter we describe how electric charge q varies with time in a circuit made up of an inductor L, a capacitor C, and a resistor R. From another point of view, we are going to describe how energy shuttles back and forth between the magnetic field of the inductor and the electric field of the capacitor, being gradually dissipated—as the oscillations proceed—as thermal energy in the resistor.

We have studied all these things before, in another context. In Chapter 14 we saw how displacement x varies with time in a mechanical oscillating system made up of a block of mass m, a spring k, and a viscous or frictional element such as a dash pot; see Fig. 17 of Chapter 14. We also studied how energy shuttles back and forth between the kinetic energy of the oscillating mass and the potential energy of the spring, being gradually dissipated—as the oscillations proceed—as thermal energy.

The parallel between the two idealized systems is exact and the controlling differential equations are identical. Thus, we need to learn no new mathematics; we can simply change the symbols and give our full attention to the physics of the situation.

35–2 *LC* Oscillations: Qualitative

Of the three circuit elements, resistance R, capacitance C, and inductance L, we have so far studied the combinations RC (Section 29–8) and RL (Section 33–4). In these two kinds of circuits we found that the charge, current, and potential difference both grow and decay exponentially. The time scale of the growth or decay is given by the *time constant* τ, which is either capacitive or inductive as the case may be.

We now examine the remaining two-element circuit, *LC*. We shall see that in this case the charge, current, and potential difference vary not exponentially (with time constant τ) but *sinusoidally* (with angular frequency ω). In other words, the circuit *oscillates*. Let us find out what is going on in such a circuit, from a physical point of view.

Assume that initially the capacitor C in Fig. 1a car-

ries a charge q and the current i in the inductor is zero.* At this instant the energy stored in the electric field of the capacitor is given by Eq. 21 of Chapter 27:

$$U_E = \frac{q^2}{2C}.$$ (1)

The energy stored in the magnetic field of the inductor, given by

$$U_B = \tfrac{1}{2}Li^2,$$ (2)

is zero because the current is zero.

The capacitor now starts to discharge through the inductor, positive charge carriers moving counterclockwise, as shown in Fig. 1b. This means that a current i, given by dq/dt and pointing down in the inductor, is

---

* When we deal with sinusoidally oscillating electrical quantities such as charge, current, or potential difference, we represent their instantaneous values by small letters (q, i, and v) and their oscillation amplitudes by capital letters (Q, I, and V).

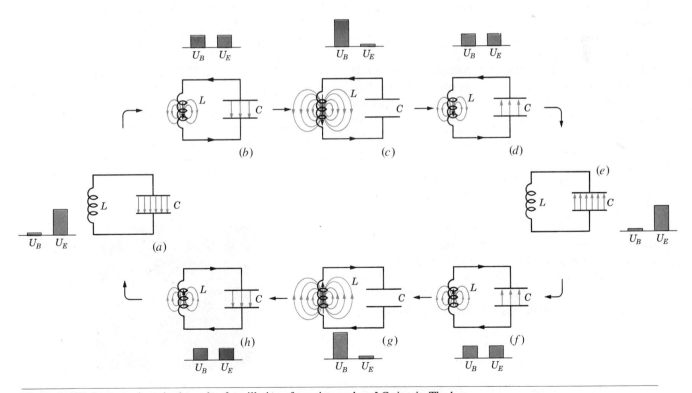

Figure 1 Eight stages in a single cycle of oscillation of a resistanceless *LC* circuit. The bar graphs by each figure show the stored magnetic and electric energies. The vertical arrows on the inductor axis show the current.

established. As q decreases, the energy stored in the electric field in the capacitor also decreases. This energy is transferred to the magnetic field that appears around the inductor because of the current i that is building up there. Thus, the electric field decreases, the magnetic field builds up, and energy is transferred from the former to the latter.

At a time corresponding to Fig. 1c, all the charge on the capacitor will have disappeared. The electric field in the capacitor will be zero, the energy stored there having been transferred entirely to the magnetic field of the inductor. According to Eq. 2, there must then be a current—and indeed one of maximum value—in the inductor. Note that even though q equals zero, the current (which is the rate at which q is changing with time) is not zero at this time.

The large current in the inductor in Fig. 1c continues to transport positive charge from the top plate of the capacitor to the bottom plate, as shown in Fig. 1d; energy now flows from the inductor back to the capacitor as the electric field builds up again. Eventually, the energy will have been transferred completely back to the capacitor, as in Fig. 1e. The situation of Fig. 1e is like the initial situation, except that the capacitor is charged oppositely.

The capacitor will start to discharge again, the current now being clockwise, as in Fig. 1f. Reasoning as before, we see that the circuit eventually returns to its initial situation and that the process continues at a definite frequency ν to which corresponds a definite angular frequency ω $(= 2\pi\nu)$. Once started, such LC oscillations (in the ideal case that we are describing, in which the circuit contains no resistance) continue indefinitely, energy being shuttled back and forth between the electric field in the capacitor and the magnetic field in the inductor. Any configuration in Fig. 1 can be set up as an initial condition. The oscillations will then continue from that point, proceeding clockwise around the figure. Compare these oscillations carefully with those of the block–spring system described in Fig. 6 of Chapter 8.

To find the charge q as a function of time, we can use a voltmeter to measure the time-varying potential difference v_C that exists across capacitor C. The relation

$$v_C = \left(\frac{1}{C}\right) q$$

shows that v_C is proportional to q. To measure the current, we can insert a small resistance R in the circuit and measure the time-varying potential difference v_R across

it. This is proportional to i through the relation

$$v_R = Ri.$$

We assume here that R is so small that its effect on the behavior of the circuit is negligible. Both q and i, or more correctly v_C and v_R which are proportional to them, can be displayed on the screen of a cathode ray oscilloscope, as suggested by Fig. 2.

In an actual LC circuit, the oscillations will not continue indefinitely because there is always some resistance present that will drain energy from the electric and magnetic fields and dissipate it as thermal energy; the circuit will become warmer. The oscillations, once started, will die away as Fig. 3 suggests. Compare this figure with Fig. 18 of Chapter 14, which shows the decay of the mechanical oscillations of a block–spring system caused by frictional damping.

It is possible to sustain the electromagnetic oscillations if you arrange to supply, automatically and periodi-

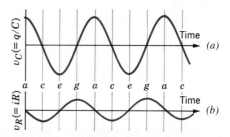

Figure 2 (a) The potential difference across the capacitor of the circuit of Fig. 1 as a function of time. This quantity is proportional to the charge on the capacitor. (b) A quantity proportional to the current in the circuit of Fig. 1. The letters indicate the corresponding oscillation stages marked in Fig. 1.

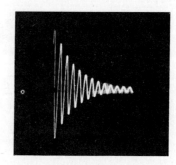

Figure 3 A photograph of an oscilloscope trace showing how the oscillations in an LCR circuit die away because energy is dissipated in the resistor in thermal form.

cally (once a cycle, say), enough energy from an outside source to compensate for that dissipated as thermal energy. We are reminded of a clock escapement, which is a device for feeding energy from a spring or a falling weight into an oscillating pendulum, thus compensating for frictional losses that would otherwise cause the oscillations to die away. *LC* oscillators, whose frequency may be varied between specified limits, are commercially available as packaged units over a wide range of frequencies, extending from low audiofrequencies (lower than 10 Hz) to microwave frequencies (higher than 10 GHz).

Sample Problem 1   A 1.5-μF capacitor is charged to 57 V. The charging battery is then disconnected and a 12-mH coil is connected across the capacitor, so that *LC* oscillations occur. What is the maximum current in the coil? Assume that the circuit contains no resistance.

From the conservation of energy principle, the maximum stored energy in the capacitor must equal the maximum stored energy in the inductor. This leads, from Eqs. 1 and 2, to

$$\frac{Q^2}{2C} = \tfrac{1}{2}LI^2,$$

where I is the maximum current and Q is the maximum charge. Note that the maximum current and the maximum charge do not occur at the same time but one-fourth of a cycle apart; see Figs. 1 and 2. Solving for I and substituting CV for Q, we find

$$I = V\sqrt{\frac{C}{L}} = (57 \text{ V})\sqrt{\frac{1.5 \times 10^{-6} \text{ F}}{12 \times 10^{-3} \text{ H}}}$$

$$= 0.637 \text{ A} \approx 640 \text{ mA}. \qquad \text{(Answer)}$$

Sample Problem 2   A 1.5-mH inductor in an *LC* circuit stores a maximum energy of 17 μJ. What is the peak current I?

When the current has its maximum value, the energy is all stored in the inductor, none being at that instant in the capacitor. If we solve Eq. 2 ($U_B = \tfrac{1}{2}LI^2$) for I we find

$$I = \sqrt{\frac{2U_B}{L}} = \sqrt{\frac{(2)(17 \times 10^{-6} \text{ J})}{1.5 \times 10^{-3} \text{ H}}}$$

$$= 0.151 \text{ A} \approx 150 \text{ mA}. \qquad \text{(Answer)}$$

35-3   Developing the Mechanical Analogy

Let us look a little closer at the analogy between the *LC* system of Fig. 1 and the block–spring system of Fig. 6 of

Table 1   The Energy in Two Oscillating Systems Compared

Mechanical System (Figure 8-6)		Electromagnetic System (Figure 1)	
Element	Energy	Element	Energy
Spring	Potential, $\tfrac{1}{2}kx^2$	Capacitor	Electric, $\tfrac{1}{2}(1/C)q^2$
Block	Kinetic, $\tfrac{1}{2}mv^2$	Inductor	Magnetic, $\tfrac{1}{2}Li^2$
$v = dx/dt$		$i = dq/dt$	

Chapter 8. In the oscillating block–spring system, two kinds of energy occur. One is potential energy of the compressed or extended spring; the other is kinetic energy of the moving block. These are given by the familiar formulas in the first column of Table 1. The table suggests that a capacitor is in some mathematical way like a spring and an inductor is like a mass and that certain electromagnetic quantities "correspond" to certain mechanical ones, namely,

q corresponds to x,
i corresponds to v,
C corresponds to $1/k$,
L corresponds to m.

Comparison of Fig. 1 with Fig. 6 of Chapter 8 shows how close the correspondence is. Note how v and i correspond in the two figures; also x and q. Note too how in each case the energy alternates between two forms, magnetic and electric for the *LC* system, and kinetic and potential for the block–spring system.

In Section 14-3 we saw that the natural angular frequency of oscillation of a (frictionless) block–spring system is

$$\boxed{\omega = \sqrt{\frac{k}{m}}} \qquad \text{(block–spring system).} \qquad (3)$$

The method of correspondences suggests that to find the natural frequency for a (resistanceless) *LC* circuit, k should be replaced by $1/C$ and m by L, obtaining

$$\boxed{\omega = \frac{1}{\sqrt{LC}}} \qquad \text{(\textit{LC} system).} \qquad (4)$$

This result is indeed correct, as we show in the next section.

35–4 *LC* Oscillations: Quantitative

Here we want to show explicitly that Eq. 4 for the angular frequency of the *LC* oscillations is correct. At the same time, we want to examine even more closely the analogy between the *LC* oscillations and the block–spring oscillations. We start by extending somewhat our earlier treatment of the mechanical block–spring oscillator.

The Block–Spring Oscillator. We analyzed block–spring oscillations in Chapter 14 in terms of energy transfers and did not—at that early stage—derive the fundamental differential equation that governs that oscillatory motion. We do so now.

We can write, for the mechanical energy of a block–spring oscillator at any instant,*

$$U = U_m + U_k = \tfrac{1}{2}mv^2 + \tfrac{1}{2}kx^2, \qquad (5)$$

where U_m and U_k are, respectively, the kinetic energy of the moving block and the potential energy of the stretched or compressed spring. If there is no friction—which we assume—the total energy U remains constant with time, even through v and x vary. In more formal language, $dU/dt = 0$. This leads to

$$\frac{dU}{dt} = \frac{d}{dt}(\tfrac{1}{2}mv^2 + \tfrac{1}{2}kx^2) = mv\frac{dv}{dt} + kx\frac{dx}{dt} = 0. \qquad (6)$$

But $dx/dt = v$ and $dv/dt = d^2x/dt^2$. With these substitutions, Eq. 6 becomes

$$\boxed{m\frac{d^2x}{dt^2} + kx = 0} \qquad \text{(block–spring oscillations).} \qquad (7)$$

Equation 7 is the fundamental *differential equation* that governs the frictionless block–spring oscillations. It involves the displacement x and its second derivation with respect to time.

The general solution to Eq. 7, that is, the function $x(t)$ that describes the block–spring oscillations, is (as we have seen)

$$\boxed{x = X\cos(\omega t + \phi)} \qquad \text{(displacement),} \qquad (8)$$

in which X is the amplitude of the mechanical oscillations.

---

* Earlier, we used E to represent the mechanical energy; we make small changes in notation here to conform to the usage of the parallel electrical situations.

The *LC* Oscillator. Now let us analyze the oscillations of a (resistanceless) *LC* circuit, proceeding in exactly the same way as we have just done for the block–spring oscillator. The total energy U present at any instant in an oscillating *LC* circuit is given by

$$U = U_B + U_E = \tfrac{1}{2}Li^2 + \frac{q^2}{2C},$$

which expresses the fact that at any arbitrary time the energy is stored partly in the magnetic field in the inductor and partly in the electric field in the capacitor. Since we have assumed the circuit resistance to be zero, there is no energy transfer to thermal energy and U remains constant with time, even though i and q vary. In more formal language, dU/dt must be zero. This leads to

$$\frac{dU}{dt} = \frac{d}{dt}\left(\tfrac{1}{2}Li^2 + \frac{q^2}{2C}\right) = Li\frac{di}{dt} + \frac{q}{C}\frac{dq}{dt} = 0. \qquad (9)$$

Now, $dq/dt = i$ and $di/dt = d^2q/dt^2$. With these substitutions, Eq. 9 becomes

$$\boxed{L\frac{d^2q}{dt^2} + \frac{1}{C}q = 0} \qquad \text{(\textit{LC} oscillations).} \qquad (10)$$

This is the *differential equation* that describes the oscillations of a (resistanceless) *LC* circuit. A careful comparison of Eq. 10 with Eq. 7 shows that the two equations are exactly of the same mathematical form, differing only in the symbols used.

Since the differential equations are mathematically identical, their solutions must also be mathematically identical. Because q corresponds to x, we can write the general solution of Eq. 10, giving q as a function of time, by analogy to Eq. 8 as

$$\boxed{q = Q\cos(\omega t + \phi)} \qquad \text{(charge),} \qquad (11)$$

where Q is the amplitude of the charge variations and ω is the angular frequency of the electromagnetic oscillations.

We can test whether Eq. 11 is indeed a solution of Eq. 10 by substituting it and its second derivative in that equation. To find the second derivative, we write

$$\frac{dq}{dt} = i = -\omega Q\sin(\omega t + \phi)$$

and

$$\frac{d^2q}{dt^2} = -\omega^2 Q\cos(\omega t + \phi).$$

Substituting q and d^2q/dt^2 into Eq. 10, we have

$$-L\omega^2Q\cos(\omega t + \phi) + \frac{1}{C}Q\cos(\omega t + \phi) = 0.$$

Canceling $Q\cos(\omega t + \phi)$ and rearranging lead to

$$\omega = \frac{1}{\sqrt{LC}}. \qquad (12)$$

Thus, if ω is given the constant value $1/\sqrt{LC}$, Eq. 11 is indeed a solution of Eq. 10. Equation 12 is exactly the same expression for ω given by Eq. 4, which we arrived at by the method of correspondences.

The phase constant ϕ in Eq. 10 is determined by the conditions that prevail at $t = 0$. If we put $\phi = 0$, for example, Eq. 11 requires that, at $t = 0$, $q = Q$ and i ($= dq/dt) = 0$; these are the initial conditions represented by Fig. 1a.

The stored electric energy in the LC circuit, using Eq. 1, is

$$U_E = \frac{q^2}{2C} = \frac{Q^2}{2C}\cos^2(\omega t + \phi) \qquad (13)$$

and the magnetic energy, using Eq. 2, is

$$U_B = \tfrac{1}{2}Li^2 = \tfrac{1}{2}L\omega^2Q^2\sin^2(\omega t + \phi).$$

Substituting the expression for ω (see Eq. 4) into this last equation, we have

$$U_B = \frac{Q^2}{2C}\sin^2(\omega t + \phi). \qquad (14)$$

Figure 4 shows plots of $U_E(t)$ and $U_B(t)$ for the case of $\phi = 0$. Note that (1) the maximum values of U_E and U_B are the same ($= Q^2/2C$); (2) at any instant the sum of U_E and U_B is a constant ($= Q^2/2C$); and (3) when U_E has its maximum value, U_B is zero and conversely. Compare

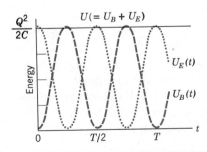

Figure 4 The stored magnetic energy (---) and the stored electric energy ($\cdots$) in the circuit of Fig. 1 as a function of time. Note that their sum remains constant. T is the period of oscillation.

this discussion with that given in Section 14-4 for the energy transfers in a block–spring system.

Sample Problem 3 (a) In an oscillating LC circuit, what value of charge, expressed in terms of the maximum charge Q, is present on the capacitor when the energy is shared equally between the electric and the magnetic field? Assume that $L = 12$ mH and $C = 1.7 \mu$F.

The problem requires that $U_E = \tfrac{1}{2}U_{E,m}$. The instantaneous and maximum stored energy in the capacitor are, respectively,

$$U_E = \frac{q^2}{2C} \quad \text{and} \quad U_{E,m} = \frac{Q^2}{2C},$$

so that

$$\frac{q^2}{2C} = \frac{1}{2}\frac{Q^2}{2C}$$

or

$$q = \frac{1}{\sqrt{2}}Q = 0.707\, Q. \qquad \text{(Answer)}$$

(b) How much time is required for this condition to arise, assuming the capacitor to be fully charged initially?

We write, putting $\phi = 0$ in Eq. 11 and using the result just found,

$$\frac{q}{Q} = \frac{Q\cos \omega t}{Q} = \frac{1}{\sqrt{2}} \quad \text{or} \quad \omega t = 45° = \frac{\pi}{4}\text{ rad},$$

which corresponds to $\tfrac{1}{8}$ of one period of oscillation. The angular frequency ω is found from Eq. 4, or

$$\omega = \frac{1}{\sqrt{LC}} = \frac{1}{\sqrt{(12 \times 10^{-3}\text{ H})(1.7 \times 10^{-6}\text{ F})}}$$
$$= 7.00 \times 10^3\text{ rad/s}.$$

The time t is then

$$t = \frac{\pi/4\text{ rad}}{\omega} = \frac{\pi}{(4)(7.00 \times 10^3\text{ rad/s})}$$
$$= 1.12 \times 10^{-4}\text{ s} \approx 110\ \mu\text{s}. \qquad \text{(Answer)}$$

Convince yourself that the frequency ν and the period T of the oscillation are ~ 1.1 kHz and $\sim 900\ \mu$s, respectively.

35-5 Damped LC Oscillations

If resistance R is present in an LC circuit, the total electromagetic energy U is no longer constant but decreases with time as it is transformed steadily to thermal energy in the resistor. As we shall see, the analogy with the damped block–spring oscillator of Section 14–8 is exact. As before, we have

$$U = U_B + U_E = \tfrac{1}{2}Li^2 + \frac{q^2}{2C}. \tag{15}$$

U is no longer constant but rather

$$\frac{dU}{dt} = -i^2R, \tag{16}$$

the minus sign signifying that the stored energy U decreases with time, being converted to thermal energy at the rate i^2R. Differentiating Eq. 15 and combining the result with Eq. 16, we have

$$Li\frac{di}{dt} + \frac{q}{C}\frac{dq}{dt} = -i^2R.$$

Substituting dq/dt for i and d^2q/dt^2 for di/dt and dividing by i, we obtain

$$\boxed{L\frac{d^2q}{dt^2} + R\frac{dq}{dt} + \frac{1}{C}q = 0} \quad \text{(LCR circuit),} \tag{17}$$

which is the differential equation that describes the damped LC oscillations. If we put $R = 0$, this equation reduces—as it must—to Eq. 10, which is the differential equation that described the undamped LC oscillations.

We state without proof that the general solution of Eq. 17 can be written in the form

$$\boxed{q = Qe^{-Rt/2L} \cos(\omega' t + \phi),} \tag{18}$$

in which

$$\omega' = \sqrt{\omega^2 - (R/2L)^2}.$$

Equation 18 is the exact equivalent of Eq. 31 of Chapter 14, the equation for the displacement as a function of time in damped simple harmonic motion.

Equation 18, which can be described as a cosine function with an amplitude that decreases exponentially with time, is the equation of the decay curve of Fig. 3. The frequency ω' is strictly less than the frequency ω ($= 1/\sqrt{LC}$) of the undamped oscillations but here we shall consider only those cases in which the resistance R is so small that we can put $\omega' = \omega$ with negligible error. Recall that we made a similar assumption of small damping of block–spring oscillations in Section 14–8.

Sample Problem 4 A circuit has $L = 12$ mH, $C = 1.6 \ \mu$F, and $R = 1.5 \ \Omega$. (a) After what time t will the amplitude of the charge oscillations drop to one-half of its initial value?

This will occur when the amplitude factor $e^{-Rt/2L}$ in Eq. 18 has the value $\frac{1}{2}$, or

$$e^{-Rt/2L} = \tfrac{1}{2}.$$

Taking the logarithm of each side gives us

$$(-Rt/2L)(\ln e) = \ln 1 - \ln 2.$$

But $\ln e = 1$ and $\ln 1 = 0$, so that

$$t = \frac{2L}{R}\ln 2 = \frac{(2)(12 \times 10^{-3} \text{ H})(\ln 2)}{1.5 \ \Omega}$$

$$= 0.0111 \text{ s} \approx 11 \text{ ms.} \qquad \text{(Answer)}$$

(b) To how many periods of oscillation does this correspond?

The number of oscillations is the elapsed time divided by the period, which is related to the angular frequency ω by $T = 2\pi/\omega$. The angular frequency is

$$\omega = \frac{1}{\sqrt{LC}} = \frac{1}{\sqrt{(12 \times 10^{-3} \text{ H})(1.6 \times 10^{-6} \text{ F})}}$$

$$= 7216 \text{ rad/s} \approx 7200 \text{ rad/s}.$$

The period of oscillation is then

$$T = \frac{2\pi}{\omega} = \frac{2\pi}{7216 \text{ rad/s}} = 8.707 \times 10^{-4} \text{ s}$$

The elapsed time, expressed in terms of the period of oscillation, is then

$$\frac{t}{T} = \frac{0.0111 \text{ s}}{8.707 \times 10^{-4} \text{ s}} \approx 13 \qquad \text{(Answer)}$$

Thus, the amplitude drops to one-half after about 13 cycles of oscillation. By comparison, the damping in this example is less severe than that shown in Fig. 3, where the amplitude drops to one-half in about three cycles.

35–6 Forced Oscillations and Resonance

We have discussed the free oscillations of an LC circuit and also the damped oscillations, in which a resistive element R is present. If the damping is small enough— and we have assumed that it is—both kinds of oscillation have an angular frequency given by

$$\boxed{\omega_0 = \frac{1}{\sqrt{LC}}} \quad \text{(natural frequency),} \tag{19}$$

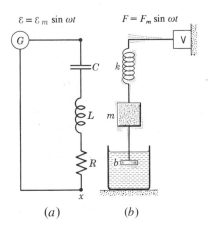

$\mathcal{E} = \mathcal{E}_m \sin \omega t$ $F = F_m \sin \omega t$

(a)   (b)

Figure 5 Forced oscillations at angular frequency ω in (a) an electromagnetic oscillating system and (b) a corresponding mechanical oscillating systems. Corresponding elements in the two systems are drawn opposite each other.

which we call the *natural frequency* of the oscillating system. Although earlier we identified this simply as ω, we relabel it here as ω_0.

Suppose now that we impress a time-varying emf given by

$$\mathcal{E} = \mathcal{E}_m \sin \omega t \qquad (20)$$

on the system by an external generator. Here ω, which can be varied at will, is the frequency imposed by this external source. We describe such oscillations as *forced*. When the emf described by Eq. 20 is first applied, there will be certain time-varying transient currents in the circuit. Our interest, however, is in the sinusoidal currents that exist in the circuit after these start-up transients have died away.

Whatever the (constant) natural frequency ω_0 may be, the oscillations of charge, current, or potential difference in the circuit must occur at the external or driving frequency ω.

Figure 5 compares the electromagnetic oscillating system with a corresponding mechanical system. A vibrator V, which imposes an external alternating force, corresponds to generator G, which imposes an external alternating emf. Other quantities "correspond" as before. Note incidentally that, although we use the same pictorial symbol for a spring and an inductor, they are not corresponding elements. In the appropriate differen-

tial equations, a spring is described mathematically like a capacitor and an inductor like a massive body.

We shall solve the problem of the forced oscillations of an LCR circuit exactly in Chapter 36, which deals with alternating currents. Here we content ourselves with giving the solution and examining some graphical results.

The electrical variable of most interest in the circuit of Fig. 5a is the current and we assert that this may be written

$$i = I \sin(\omega t - \phi). \qquad (21)$$

The current amplitude I in Eq. 21 is a measure of the response of the circuit of Fig. 5a to the impressed emf. It is reasonable to suppose, from experience (in pushing swings, for example), that I will be larger the closer the driving frequency ω is to the natural frequency ω_0 of the system. In other words, we expect that a plot of I versus ω will exhibit a maximum when

$$\boxed{\omega = \omega_0} \quad \text{(resonance)}, \qquad (22)$$

which we call the *resonance condition*.

Figure 6 shows three plots of I as a function of the ratio ω/ω_0, each plot corresponding to a different value of the resistance R. We see that each of these resonance peaks does indeed have a maximum value when the resonance condition of Eq. 22 is satisfied. Note that as R is decreased, the resonance peak becomes sharper, as shown by the three horizontal arrows drawn at the half-maximum level of each curve.

Figure 6 suggests the common experience of tuning a radio set. In turning the tuning knob, we are adjusting the natural frequency ω_0 of an internal LC circuit to match the frequency ω of the signal transmitted by the antenna of the broadcasting station; we are looking for resonance. In a metropolitan area, where there are many signals whose frequencies are often close together, sharpness of tuning becomes important.

Figure 6 finds a counterpart in Fig. 20 of Chapter 14, which shows resonance peaks for the forced oscillations of a mechanical oscillator such as that of Fig. 5b. In this case also, the maximum response occurs when $\omega = \omega_0$ and the resonance peaks become sharper as the damping factor (the coefficient b) is reduced. A careful observer will note that the curves of Fig. 6 and of Fig. 20 of Chapter 14 do not exactly "correspond." The former is a plot of current amplitude, the mechanical quantity that corresponds to the current being the velocity. The latter figure, however, is not a plot of velocity amplitude

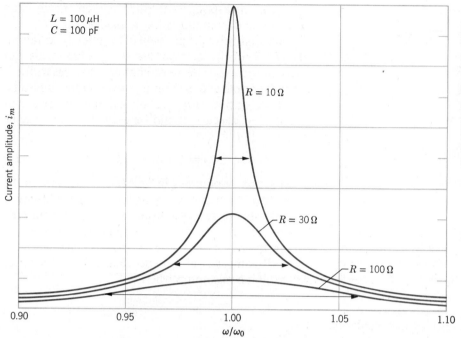

$L = 100\,\mu\text{H}$
$C = 100\,\text{pF}$

$R = 10\,\Omega$

$R = 30\,\Omega$

$R = 100\,\Omega$

Current amplitude, i_m

0.90   0.95   1.00   1.05   1.10

ω/ω_0

Figure 6 Resonance curves for forced oscillations in the circuit of Fig. 5a. The values of L and C are the same for all three curves but the values of R are different, as marked on the curves. The horizontal arrows on each curve measure its width at the half-maximum level, a measure of the sharpness of the resonance. Note that the current amplitude is a maximum in each case at resonance ($\omega/\omega_0 = 1$).

versus frequency but of displacement amplitude versus frequency. Nevertheless, both sets of curves illustrate the resonance phenomenon.

35–7 Other Oscillators: A Taste of Electronics (Optional)

We must not leave the impression that all oscillators are based on LC circuits like that of Fig. 1.

Crystal Oscillators. When you see the word quartz on your wristwatch or wall clock it means that the device is based on a *quartz crystal oscillator*. Quartz has the interesting property that, if you cut it into a thin wafer and squeeze it, equal but opposite charges appear on the opposing surfaces of the wafer; conversely, if you apply a potential difference between opposing faces of the wafer, the dimensions of the wafer change.

This property of *piezoelectricity** gives a convenient coupling between the mechanical oscillations of the crystal, which occur at a very sharply defined frequency, and the electrical properties of the circuit of which the

crystal is a part. Quartz crystal oscillators are used in situations—such as watches and radio transmitters—where stability of the oscillator frequency is a fundamental concern. Piezoelectric crystals are used in other applications, in which a mechanical movement must be converted into an electrical signal—as in a phonograph pickup arm—or an electrical signal into a mechanical movement—as in a sonar transmitter.

Feedback Oscillators. When you push the buttons on your telephone handset, the tones that you hear are generated by another kind of oscillator, based on resistive and capacitive elements alone, with no inductance involved. Such oscillators are useful in microelectronics because—although it is easy enough to embody a resistor or a capacitor in a microelectronic chip—it is more difficult to embody an inductor.

Figure 7 suggests schematically how such oscillators work. Figure 7a shows a simple amplifier, which has the property that it amplifies an input signal v_i by a factor g to generate an output signal v_o or

$$g = \frac{v_o}{v_i} \quad \text{(Fig. 7a)}.$$

The dimensionless quantity g, called the *voltage gain* of the amplifier, is greater than unity.

Figure 7b shows how this basic amplifier can be turned into an oscillator by the mechanism of *positive*

---

* The word *piezo-* comes from the Greek "to push." Piezoelectricity was discovered in 1880 by the brothers Pierre and Jacques Curie when they were, respectively, 21 and 24 years old.

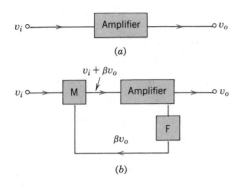

(a)

$v_i + \beta v_o$

(b)

Figure 7 (a) A simple amplifier, showing the input and the output signals. (b) The amplifier turned into an oscillator by positive feedback. The feedback circuit F sends a fraction β of the output signal back to the input, where it is added to the input signal in mixer M.

feedback. A fraction βv_o of the output signal (where $\beta < 1$) is sampled by the feedback circuitry (box F) and fed back to the input, where it is mixed (in box M) with the original input signal in such a way as to reinforce that signal. If it is desired that the circuit oscillate at a specific frequency, the feedback circuit must be so designed that the fed-back signal joins the input signal after a delay corresponding to one period of oscillation at that frequency. In that way, the input signal will be reinforced in the proper phase and energy will be added to the input circuit to compensate for resistive losses.

In Fig. 7b, we can write the gain of the *amplifier alone* as

$$g = \frac{v_o}{v_i + \beta v_o} \quad \text{(amplifier alone).} \quad (23)$$

However, the effective gain g' of the *overall circuit* of Fig. 7b is

$$g' = \frac{v_o}{v_i} \quad \text{(overall circuit).} \quad (24)$$

Combining Eqs. 23 and 24 leads to

$$g' = \frac{g}{1 - \beta g}, \quad (25)$$

which shows that g' can be much greater than g. In fact, g' can, in principle, be made infinitely great if the feedback loop is designed so that $\beta g = 1$. In practice,* this means that the oscillator will not require an input signal to cause it to oscillate; the slightest random electrical disturbance will start it off. It will then oscillate at the resonant frequency determined by the design of the frequency-dependent feedback circuit.

---

* In practice, an infinite gain cannot be achieved because, in all amplifiers, the intrinsic gain g decreases as the magnitude of the input signal increases.

REVIEW AND SUMMARY

In a (resistanceless) oscillating LC circuit, energy may be stored in the capacitor or in the inductor, according to

LC Energy Transfers

$$U_E = \frac{Q^2}{2C} \quad \text{and} \quad U_B = \frac{1}{2} Li^2. \quad [1,2]$$

The total energy $E \, (= U_E + U_B)$ remains constant as energy oscillates back and forth between these elements, as in Fig. 1; see Sample Problems 1 and 2.

A Mechanical Analogy

Energy transfers in an oscillating circuit are analogous to those in an oscillating mass-spring system. The correspondences outlined in Table 1 allow us to predict that the angular frequency of the LC oscillations will be given by

$$\omega = \frac{1}{\sqrt{LC}} \quad (LC \text{ circuit}). \quad [4]$$

The conservation of energy principle leads to

A Quantitative Solution

$$L \frac{d^2q}{dt^2} + \frac{q}{C} = 0 \quad (LC \text{ oscillations}) \quad [10]$$

as the differential equation of free resistanceless LC oscillations. The solution is

$$q = Q \cos(\omega t + \phi) \quad \text{(charge)} \tag{17}$$

with ω given by Eq. 4. The charge amplitude Q and the phase constant ϕ are fixed by the initial conditions of the system. See Sample Problem 3.

$U_E(t)$ and $U_B(t)$

Given $q(t)$, and its derivative $i(t)$, we can find the energies $U_E(t)$ and $U_B(t)$ of Fig. 1. See Eqs. 13 and 14 and Fig. 4.

Damped Oscillations

For damped LC oscillations a dissipative element R is present and the conservation of energy principle shows the differential equation to be

$$L \frac{d^2q}{dt^2} + R \frac{dq}{dt} + \frac{1}{C} q = 0 \quad \text{(LCR circuit).} \tag{17}$$

Its solution is

$$q = Q e^{-Rt/2L} \cos(\omega' t + \phi). \tag{18}$$

We consider only low damping situations, in which ω' may be set equal to the ω of Eq. 4. See Sample Problem 4.

Forced Oscillations

An LCR circuit such as that of Fig. 5a may be set into *forced oscillation* at angular frequency ω by an impressed emf such as

$$\mathcal{E} = \mathcal{E}_m \sin \omega t. \tag{20}$$

Here we relabel the (constant) *natural frequency* $(= 1/\sqrt{LC})$ of the resonant system as ω_0, reserving ω for the (variable) frequency of the impressed emf. The current in the circuit is

$$i = I \sin(\omega t - \phi). \tag{21}$$

Resonance

The current amplitude I has a maximum value when $\omega = \omega_0$, a condition called *resonance*. Figure 6 shows that resonance peaks becomes sharper as the circuit resistance R is decreased. Review the analogy to forced mechanical oscillations (see Fig. 5b).

Quartz Oscillators

Quartz crystal oscillators depend on the *piezoelectric* property of quartz; stresses resulting from mechanical oscillation cause alternating electric potentials that may then be used as part of electrical timing circuits of various kinds. *Feedback oscillators* are amplifier-feedback combinations with overall gains near $+1$. The circuit oscillates at a resonant frequency determined by the properties of the feedback circuit.

Feedback Oscillators

QUESTIONS

1. Why doesn't the LC circuit of Fig. 1 simply stop oscillating when the capacitor has been completely discharged?

2. How might you start an LC circuit into oscillation with its initial condition being represented by Fig. 1c? Devise a switching scheme to bring this about.

3. The lower curve (b) in Figure 2 is proportional to the derivative of the upper curve (a). Explain why.

4. In an oscillating LC circuit, assumed resistanceless, what determines (a) the frequency and (b) the amplitude of the oscillations?

5. In connection with Figs. 1c and g, explain how there can be a current in the inductor even though there is no charge on the capacitor.

6. In Fig. 1, what changes are required if the oscillations are to proceed counterclockwise around the figure?

7. In Fig. 1, what phase constants ϕ in Eq. 11 would permit the eight circuit situations shown to serve in turn as initial conditions?

8. What constructional difficulties would you encounter if you tried to build an LC circuit of the type shown in Fig. 1 to oscillate (a) at 0.01 Hz, or (b) at 10^{10} Hz?

9. Two inductors L_1 and L_2 and two capacitors C_1 and C_2 can be connected in series according to the arrangement in Fig. 8a or Fig. 8b. Are the frequencies of the two oscillating circuits equal? Consider the cases (i) $C_1 = C_2$, $L_1 = L_2$; (ii) $C_1 \neq C_2$, $L_1 \neq L_2$.

10. In the mechanical analogy to the oscillating LC circuit, what mechanical quantity corresponds to potential difference?

11. Discuss the assertion that the resonance curves of Fig. 6 and Chapter 14, Fig. 20 cannot truly be compared because the

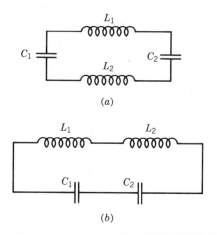

(a)

(b)

Figure 8 Question 9.

former is a plot of the current amplitude (I) and the latter of the displacement amplitude (x_m). Are these "corresponding" quantities? Does it make any difference if our purpose is only to exhibit the resonance phenomenon?

12. In comparing an electromagnetic oscillating system to a mechanical oscillating system, to what mechanical properties are the following electromagnetic properties analogous: capacitance, resistance, charge, electric field energy, magnetic field energy, inductance, current?

13. Two identical springs are joined and connected to an object with mass m, the arrangement being free to oscillate on a horizontal frictionless surface as in Fig. 9. Sketch the electromagnetic analog of this mechanical oscillating system.

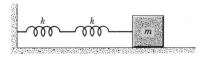

Figure 9 Question 13.

14. Explain why it is not possible to have (a) a real LC circuit without resistance, (b) a real inductor without inherent capacitance, or (c) a real capacitor without inherent inductance. Discuss the practical validity of the LC circuit of Fig. 1, in which each of the above realities is ignored.

15. All practical LC circuits have to contain some resistance. However, one can buy a packaged audio oscillator in which the output maintains a constant amplitude indefinitely and does not decay, as it does in Fig. 3. How can this happen?

16. What would a resonance curve for $R = 0$ look like if plotted in Fig. 6?

17. What is the difference between free, damped, and forced oscillating circuits?

EXERCISES AND PROBLEMS

Section 35–2 *LC* Oscillations—Qualitative

1E. What is the capacitance of an LC circuit if the maximum charge on the capacitor is 1.6 μC and the total energy is 140 μJ?

2E. 1.5-mH inductor in an LC circuit stores a maximum energy of 10 μJ. What is the peak current?

3E. In an oscillating LC circuit $L = 1.1$ mH and $C = 4.0$ μF. The maximum charge on C is 3.0 μC. Find the maximum current.

4E. An LC circuit consists of a 75-mH inductor and a 3.6-μF capacitor. If the maximum charge on the capacitor is 2.9 μC, (a) what is the total energy in the circuit and (b) what is the maximum current?

5E. For a certain LC circuit the total energy is converted from electrical energy in the capacitor to magnetic energy in the inductor in 1.5 μs. (a) What is the period of oscillation? (b) What is the frequency of oscillation? (c) How long after the magnetic energy is a maximum will it be a maximum again?

Section 35–3 Developing the Mechanical Analogy

6E. A 0.50-kg body oscillates on a spring that, when extended 2.0 mm from equilibrium, has a restoring force of 8.0 N. (a) Calculate the angular frequency of oscillation. (b) What is its period of oscillation? (c) What is the capacitance of the analogous LC system if L is chosen to be 5.0 H?

7P. The energy in an LC circuit containing a 1.25-H inductor is 5.70 μJ. The maximum charge on the capacitor is 175 μC. Find (a) the mass, (b) the spring constant, (c) the maximum displacement, and (d) the maximum speed for the analogous mechanical system.

Section 35–4 *LC* Oscillations—Quantitative

8E. LC oscillators have been used in circuits connected to loudspeakers to create some of the sounds of "electronic music." What inductance must be used with a 6.7-μF capacitor to produce a frequency of 10 kHz, about the middle of the audible range of frequencies?

9E. What capacitance would you connect across a 1.3-mH inductor to make the resulting oscillator resonate at 3.5 kHz?

10E. In an LC circuit with $L = 50$ mH and $C = 4.0$ μF, the current is initially a maximum. How long will it take before the capacitor is fully charged for the first time?

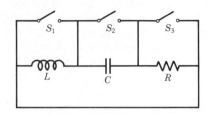

Figure 10   Exercise 11.

11E. Consider the circuit shown in Fig. 10. With switch S_1 closed and the other two switches open, the circuit has a time constant τ_C (Section 8, Chapter 29). With switch S_2 closed and the other two switches open, the circuit has a time constant τ_L (Section 4, Chapter 33). With switch S_3 closed and the other two switches open, the circuit oscillates with a period T. Show that $T = 2\pi\sqrt{\tau_C\tau_L}$.

12E. Derive the differential equation for an LC circuit (Eq. 10) using the loop theorem.

13E. A single loop consists of several inductors $(L_1, L_2, \ldots)$, several capacitors $(C_1, C_2, \ldots)$, and several resistors $(R_1, R_2, \ldots)$ connected in series as shown, for example, in Fig. 11a. Show that, regardless of the sequence of these circuit elements in the loop, the behavior of this circuit is identical to that of a simple damped LC circuit shown in Fig. 11b. (*Hint:* Consider the loop equation.)

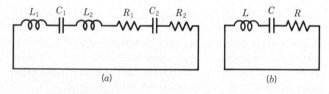

(a)   (b)

Figure 11   Exercise 13.

14P. An oscillating LC circuit consisting of a 1.0-nF capacitor and a 3.0-mH coil has a peak voltage of 3.0 V. (a) What is the maximum charge on the capacitor? (b) What is the peak current through the circuit? (c) What is the maximum energy stored in the magnetic field of the coil?

15P. An LC circuit has an inductance of 3.0 mH and a capacitance of 10 μF. (a) Calculate the angular frequency and (b) the period of the oscillation. (c) At time $t = 0$ the capacitor is charged to 200 μC, and the current is zero. Sketch roughly the charge on the capacitor as a function of time.

16P. In an LC circuit in which $C = 4.0$ μF the maximum potential difference across the capacitor during the oscillations is 1.5 V and the maximum current through the inductor is 50 mA. (a) What is the inductance L? (b) What is the frequency of the oscillations? (c) How much time does it take for the charge on the capacitor to rise from zero to its maximum value?

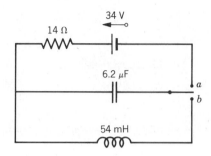

Figure 12   Problem 17.

17P. In the circuit shown in Fig. 12 the switch has been in position a for a long time. It is now thrown to b. (a) Calculate the frequency of the resulting oscillating current. (b) What will be the amplitude of the current oscillations?

18P. You are given a 10-mH inductor and two capacitors, of 5.0-μF and 2.0-μF capacitance. List the resonant frequencies that can be generated by connecting these elements in various combinations.

19P. An LC circuit oscillates at 10.4 kHz. (a) If the capacitance is 340 μF, what is the inductance? (b) If the maximum current is 7.20 mA, what is the total energy in the circuit? (c) Calculate the maximum charge on the capacitor.

20P. (a) In an oscillating LC circuit, in terms of the maximum charge on the capacitor, what value of charge is present when the energy in the electric field is one-half that in the magnetic field? (b) What fraction of a period must elapse following the time the capacitor is fully charged for this condition to arise?

21P. At some instant in an oscillating LC circuit, three-fourths of the total energy is stored in the magnetic field of the inductor. (a) In terms of the maximum charge on the capacitor, what is the charge on the capacitor at this instant? (b) In terms of the maximum current in the inductor, what is the current in the inductor at this instant?

22P. An inductor is connected across a capacitor whose capacitance can be varied by turning a knob. We wish to make the frequency of the LC oscillations vary linearly with the angle of rotation of the knob, going from 2×10^5 Hz to 4×10^5 Hz as the knob turns through 180°. If $L = 1.0$ mH, plot C as a function of angle for the 180° rotation.

23P. A variable capacitor with a range from 10 to 365 pF is used with a coil to form a variable-frequency LC circuit to tune the input to a radio. (a) What ratio of maximum to minimum frequencies may be tuned with such a capacitor? (b) If this capacitor is to tune from 0.54 to 1.60 MHz, the ratio computed in (a) is too large. By adding a capacitor in parallel to the variable capacitor this range may be adjusted. How large should this capacitor be and what inductance should be chosen in order to tune the desired range of frequencies?

24P. In an LC circuit $L = 25$ mH and $C = 7.8$ μF. At time $t = 0$ the current is 9.2 mA, the charge on the capacitor is 3.8 μC, and the capacitor is charging. (a) What is the total energy in the circuit? (b) What is the maximum charge on the capacitor? (c) What is the maximum current? (d) If the charge on the capacitor is given by $q = Q \cos(\omega t + \phi)$, what is the phase angle ϕ? (e) Suppose the data are the same, except that the capacitor is discharging at $t = 0$. What then is the phase angle ϕ?

25P. In an oscillating LC circuit $L = 3.0$ mH and $C = 2.7$ μF. At $t = 0$ the charge on the capacitor is zero and the current is 2.0 A. (a) What is the maximum charge that will appear on the capacitor? (b) In terms of the period T of oscillation, how much time will elapse after $t = 0$ until the energy stored in the capacitor will be increasing at its greatest rate? (c) What is this greatest rate at which energy flows into the capacitor?

26P. In an LC circuit with $C = 64$ μF the current as a function of time is given by $i = (1.6)\sin(2500t + 0.68)$, where t is in seconds, i in amperes, and the phase angle in radians. (a) How soon, after $t = 0$, will the current reach its maximum value? (b) Calculate the inductance. (c) Find the total energy in the circuit.

27P. The resonant frequency of a series circuit containing inductance L_1 and capacitance C_1 is ω_0. A second series circuit, containing inductance L_2 and capacitance C_2, has the same resonant frequency. In terms of ω_0, what is the resonant frequency of a series circuit containing all four of these elements? Neglect resistance. (*Hint:* Use the formulas for equivalent capacitance and equivalent inductance.)

28P. Three identical inductors L and two identical capacitors C are connected in a two-loop circuit as shown in Fig. 13. (a) Suppose the currents are as shown in Fig. 13a. What is the current in the middle inductor? Write down the loop equations

and show that they are satisfied provided that the current oscillates with angular frequency $\omega = 1/\sqrt{LC}$. (b) Now suppose the currents are as shown in Fig. 13b. What is the current in the middle inductor? Write down the loop equations and show that they are satisfied provided the current oscillates with angular frequency $\omega = 1/\sqrt{3LC}$. (c) In view of the fact that the circuit can oscillate at two different frequencies, show that it is not possible to replace this two-loop circuit by an equivalent single-loop LC circuit.

29P*. In Fig. 14 the 900-μF capacitor is initially charged to 100 V and the 100-μF capacitor is uncharged. Describe in detail how one might charge the 100-μF capacitor to 300 V by manipulating switches S_1 and S_2.

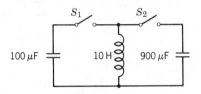

Figure 14 Problem 29.

Section 35–5 Damped LC Oscillations

30E. What resistance R should be connected to an inductor $L = 220$ mH and capacitor $C = 12$ μF in series in order that the maximum charge on the capacitor decay to 99% of its initial value in 50 cycles?

31E. Consider a damped LC circuit. (a) Show that the damping term $e^{-Rt/2L}$ (which involves L but not C) can be rewritten in the more symmetric manner (involving both L and C) as $e^{-\pi R\sqrt{C/L}(t/T)}$. Here T is the period of oscillation (neglecting resistance). (b) Using (a), show that the SI unit of $\sqrt{L/C}$ is "ohm." (c) Using (a), show that the condition that the fractional energy loss per cycle be small is $R \ll \sqrt{L/C}$.

32P. In a damped LC circuit, find the time required for the maximum energy present in the capacitor during one oscillation to fall to one-half of its initial value. Assume $q = Q$ at $t = 0$.

33P. A single loop circuit consists of a 7.2-Ω resistor, a 12-H inductor, and a 3.2-μF capacitor. Initially the capacitor has a charge of 6.2 μC and the current is zero. Calculate the charge on the capacitor N complete cycles later for $N = 5, 10,$ and 100.

34P. (a) By direct substitution of Eq. 18 into Eq. 17, show that $\omega' = \sqrt{(1/LC) - (R/2L)^2}$. (b) By what fraction does the frequency of oscillation shift when the resistance is increased from 0 to 100 Ω in a circuit with $L = 4.4$ H and $C = 7.3$ μF?

35P*. In a damped LC circuit show that the fraction of the energy lost per cycle of oscillation, $\Delta U/U$, is given to a close approximation by $2\pi R/\omega L$. The quantity $\omega L/R$ is often called the Q of the circuit (for "quality"). A "high-Q" circuit has low resistance and a low fractional energy loss per cycle ($= 2\pi/Q$).

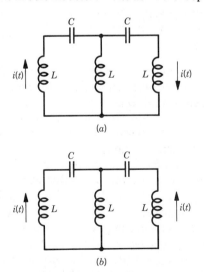

Figure 13 Problem 28.

CHAPTER 36

ALTERNATING CURRENTS

It is one thing to invent the electric light bulb and quite another to devise an electric power distribution system so that the benefits of electric light can be enjoyed on a city-wide basis. Thomas Edison, who did the former, did not succeed so well in the latter because of his commitment to direct-current distribution systems. Charles Proteus Steinmetz, a gifted engineer and mathematician, favored alternating current systems in the fierce debate on this subject that raged at the end of the last century. His side won.

36-1 Why Alternating Current?

Most homes and offices in this country are wired for *alternating current,* that is, for currents whose value varies with time in a sinusoidal fashion, changing direction (most commonly) 120 times per second. At first sight this may seem a strange arrangement. We have seen that the drift speed of the conduction electrons in household wiring may typically be 4×10^{-5} m/s.* If we now reverse their direction every $\frac{1}{120}$ s, such electrons could move only about 3×10^{-7} m in one half cycle. At this rate, a typical electron could drift past no more than

---
* See Sample Problem 3, Chapter 28.

about ten atoms in the copper lattice before it was required to reverse itself. How, you may wonder, can the electron ever get anywhere?

Although this question may be worrisome, it is a needless concern. The conduction electrons do not have to "get anywhere." When we say that the current in a wire is one ampere we mean that charge carriers pass through any plane cutting across that wire at the rate of one coulomb per second. The speed at which the carriers cross that plane does not enter directly; one ampere may correspond to a lot of charge carriers moving very slowly or to a few moving very rapidly. Furthermore, the signal to the electrons to reverse directions—which originates in the alternating emf provided by the generator—is

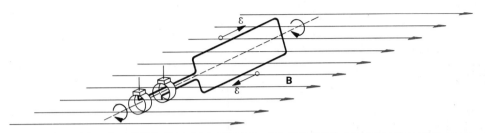

Figure 1   The basic principle of an alternating-current generator is a conducting loop rotated in an external magnetic field. In practice, the alternating emf induced in the loop of many turns of wire is made accessible by means of slip rings attached to the rotating shaft and each connected to one end of the wire.

propagated along the conductor at a speed close to that of light. All electrons, no matter where they are located, get their reversal instructions at about the same instant. Finally, we note that many devices, such as electric light bulbs or toasters, do not care in what direction the electrons are moving as long as they are indeed moving and thus delivering energy to the conductor in thermal form.

The basic advantage of alternating currents is this: *As the current alternates, so does the magnetic field that surrounds the conductor.* This makes possible the use of Faraday's law of induction which, among other things, means that we can step up or step down the magnitude of an alternating potential difference at will, using a transformer, as we shall show later in this chapter. Finally, alternating currents are more readily adaptable to rotating machinery such as generators or motors than are direct currents. If, for example, we rotate a coil in an external magnetic field, as in Fig. 1, the emf induced in the coil is an alternating emf. To extract an alternating potential difference from the coil and then to transform it into a potential difference of constant magnitude and polarity to supply a direct current power distribution system is a challenging engineering problem.

Alternating emfs and the alternating currents generated by them are central—not only to power distribution systems as we have discussed above—but also to radio, television, satellite communications systems, computer systems, and much else that helps to fashion our modern lifestyle.

36-2   Our Plan for this Chapter

In Section 35-6 we saw that if an alternating emf given by

$$\mathcal{E} = \mathcal{E}_m \sin \omega t \qquad (1)$$

is applied to a circuit such as that of Fig. 2, an alternating current given by

$$i = I \sin(\omega t - \phi) \qquad (2)$$

is established in the circuit.* Just as in direct current circuits, the alternating current i in the circuit of Fig. 2 has the same value at any given instant in all parts of the (single-loop) circuit. Furthermore, the angular frequency ω of the current in Eq. 2 is necessarily the same as the angular frequency of the generator that appears in Eq. 1.

The basic characteristics of the alternating emf supplied by the generator are its amplitude $\mathcal{E}_m$ and its angular frequency ω. The basic characteristics of the circuit of Fig. 2 are the resistance R, the capacitance C, and the inductance L. The basic characteristics of the alternating current given by Eq. 2 are its amplitude I and its phase constant ϕ. Our aim in this Chapter can be expressed as follows:

Given	Find
Generator: $\mathcal{E}_m$ and ω	
	I and ϕ
Circuit: R, C, and L	

Rather than attempt to find I and ϕ by solving the differential equation that applies to the circuit of Fig. 2 we will use a geometrical method, introducing the concept of *phasors*.

---

* Throughout this chapter, small letters, such as i, will represent instantaneous, time-varying quantities and capital letters, such as I, will represent the corresponding amplitudes.

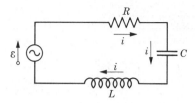

Figure 2 A single-loop circuit containing a resistor, a capacitor, and an inductor. A generator sets up an alternating emf that establishes an alternating current.

36–3 Three Simple Circuits

Let us first simplify the problem suggested by Fig. 2 by considering three simpler circuits, each containing the alternating current generator and only one other element, R, C, or L. We start with R.

A Resistive Circuit. Figure 3a shows a circuit containing a resistive element only, acted on by the alternating emf of Eq. 1. By the loop theorem the alternating potential difference across the resistor is

$$v_R = V_R \sin \omega t. \tag{3}$$

From the definition of resistance we can also write

$$i_R = \frac{v_R}{R} = \frac{V_R}{R} \sin \omega t = I_R \sin \omega t. \tag{4}$$

By comparison with Eq. 2 we see that, in this case of a purely resistive load, the phase constant $\phi = 0°$. From Eq. 4 we also see that the voltage amplitude and the current amplitude are related by

$$\boxed{V_R = I_R R} \quad \text{(resistor)}. \tag{5}$$

Although we developed this relation for the circuit of Fig. 3a, it applies to an individual resistor in any alternating current circuit whatever, no matter how complex.

Comparison of Eqs. 3 and 4 shows that the time-varying quantities v_R and i_R are in phase, which means that their corresponding maxima occur at the same time. Figure 3b, a plot of $v_R(t)$ and $i_R(t)$, illustrates this.

Figure 3c shows a useful geometrical way of looking at the same situation, the method of *phasors*. The two open arrows—the phasors—rotate counterclockwise about the origin with an angular frequency ω. The *length* of a phasor is proportional to the *amplitude* of the alternating quantity involved, that is, to V_R and to I_R. The *projection* of a phasor on the *vertical* axis is proportional to the *instantaneous value* of this alternating quantity,

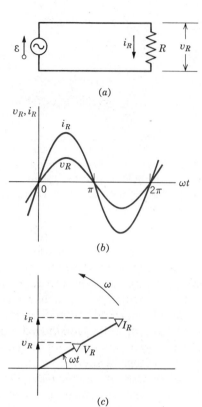

Figure 3 (a) A resistor is connected across an alternating-current generator. (b) The current and the potential difference across the resistor are in phase. (c) A phasor diagram shows the same thing.

that is, to v_R and to i_R. That v_R and i_R are in phase follows from the fact that their phasors lie along the same line in Fig. 3c. Follow the rotation of the phasors in this figure and convince yourself that it completely and correctly describes Eqs. 3 and 4.

A Capacitive Circuit. Figure 4a shows a circuit containing a capacitor only, acted on by the alternating emf of Eq. 1. From the loop theorem the potential difference across the capacitor is

$$v_C = V_C \sin \omega t. \tag{6}$$

From the definition of capacitance we can also write

$$q_C = C v_C = (C V_C) \sin \omega t. \tag{7}$$

Our concern, however, is with the current rather than with the charge. Thus we differentiate Eq. 7 to find

$$i_C = \frac{dq_C}{dt} = \omega C V_C \cos \omega t. \tag{8}$$

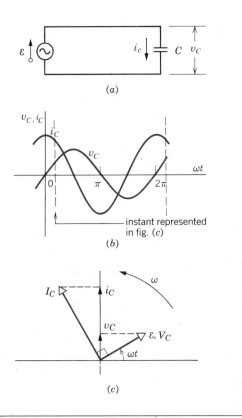

Figure 4   (*a*) A capacitor is connected across an alternating-current generator. (*b*) The potential difference across the capacitor lags the current by 90°. (*c*) A phasor diagram shows the same thing.

We now recast Eq. 8 in two ways. First, for reasons of symmetry of notation, we introduce the quantity X_C, called the *capacitive reactance* of the capacitor at the frequency in question. It is defined from

$$X_C = \frac{1}{\omega C} \quad \text{(capacitive reactance),} \quad (9)$$

and we see that its value depends not only on the value of the capacitance but also on the frequency at which the capacitor is used. We know from the definition of the capacitive time constant ($\tau = RC$) that the SI unit for C can be expressed as second/ohm. Applying this knowledge to Eq. 9 shows that the SI unit of X_C is the *ohm*, just as for resistance R.

As our second modification of Eq. 8, we replace cos ωt by a phase-shifted sine, namely

$$\cos \omega t = \sin(\omega t + 90°).$$

You can easily verify this identity by expanding the right-hand side above according to the formula for $\sin(\alpha + \beta)$ listed in Appendix G.

With these two modifications, Eq. 8 becomes

$$i_C = \left(\frac{V_C}{X_C}\right) \sin(\omega t + 90°) = I_C \sin(\omega t + 90°). \quad (10)$$

A comparison with Eq. 2 shows that, in this case of a purely capacitive load, the phase constant $\phi = -90°$. We see also from Eq. 10 that the voltage amplitude and the current amplitude are related by

$$V_C = I_C X_C \quad \text{(capacitor).} \quad (11)$$

Although we developed this relation for the specific circuit of Fig. 4*a*, it holds for an individual capacitor in any alternating current circuit, no matter how complex.

Comparison of Eqs. 6 and 10, or inspection of Fig. 4*b*, shows that the quantities v_C and i_C are 90°, or one-quarter cycle, out of phase. Furthermore, we see that i_C *leads* v_C, which means that, if you monitored the current i_C and the potential difference v_C in the circuit of Fig. 4*a*, you would find that i_C reached its maximum *before* v_C did, by one-quarter cycle.

This relation between i_C and v_C is illustrated with equal clarity from the phasor diagram of Fig. 4*c*. As the phasors representing these two quantities rotate counterclockwise, we see that the phasor labeled I_C does indeed lead that labeled V_C, and by an angle of 90°. That is, the phasor I_C coincides with the vertical axis one-quarter cycle before the phasor V_C does. Be sure to convince yourself that the phasor diagram of Fig. 4*c* is consistent with Eqs. 6 and 10.

An Inductive Circuit.   Figure 5*a* shows a circuit containing an inductive element only, acted on by the alternating emf of Eq. 1. From the loop theorem, we can write

$$v_L = V_L \sin \omega t. \quad (12)$$

From the definition of inductance we can also write

$$v_L = L \frac{di_L}{dt}. \quad (13)$$

If we combine these two equations we have

$$\frac{di_L}{dt} = \left(\frac{V_L}{L}\right) \sin \omega t. \quad (14)$$

Our concern, however, is with the current rather than with its time derivative. We find it by integrating Eq. 14,

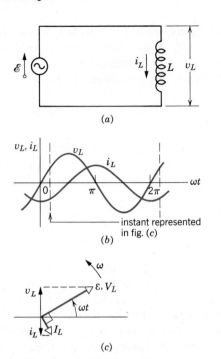

(a)

(b)

instant represented
in fig. (c)

(c)

Figure 5 (a) An inductor is connected across an alternating-current generator. (b) The potential difference across the inductor leads the current by 90°. (c) A phasor diagram shows the same thing.

or

$$i_L = \int di_L = \frac{V_L}{L} \int_0^t \sin \omega t \, dt$$

$$= -\left(\frac{V_L}{\omega L}\right) \cos \omega t. \qquad (15)$$

Again, we recast this equation in two ways. First, once more for reasons of symmetry of notation, we introduce the quantity X_L, called the *inductive reactance* of the inductor at the frequency in question. It is defined from

$$\boxed{X_L = \omega L} \quad \text{(inductive reactance).} \qquad (16)$$

Analysis shows that the SI units of X_L is the *ohm*, just as it is for X_C and for R. Second, we replace $-\cos \omega t$ in Eq. 15 by a phase-shifted sine, namely

$$-\cos \omega t = \sin(\omega t - 90°).$$

You can verify this identity by expanding its right-hand side according to the formula for $\sin(\alpha - \beta)$ given in Appendix G.

With these two changes, Eq. 15 becomes

$$i_L = \left(\frac{V_L}{X_L}\right) \sin(\omega t - 90°) = I_L \sin(\omega t - 90°). \qquad (17)$$

By comparison with Eq. 2 we see that, in this case of a purely inductive load, the phase constant $\phi = +90°$. From Eq. 17 we also see that the current amplitude and the voltage amplitude are related by

$$\boxed{V_L = I_L X_L} \quad \text{(inductor).} \qquad (18)$$

Although we developed Eq. 18 for the specific circuit of Fig. 5a, it holds for an individual inductor in any alternating current circuit, no matter how complex.

Comparison of Eqs. 12 and 17, or inspection of Fig. 5b, shows that the quantities i_L and v_L are 90° out of phase. In this case, however, i_L *lags* v_L. That is, if you monitored the current i_L and the potential difference v_L in the circuit of Fig. 5a, you would find that i_L reached its maximum value *after* v_L did, by one-quarter cycle.

The phasor diagram of Fig. 5c also contains this information. As the phasors rotate in the figure, we see that the phasor labeled I_L does indeed lag that labeled V_L, and by an angle of 90°. Be sure to convince yourself that Fig. 5c does indeed represent Eqs. 12 and 17.

Table 1 summarizes what we have learned about the relations between the current i and the voltage v for each of the three kinds of circuit elements.

Sample Problem 1 In Fig. 4a let $C = 15 \ \mu F$, $\nu = 60$ Hz, and $\mathscr{E}_m = V_C = 36$ V. (a) Find the capacitive reactance X_C.

From Eq. 9 we have

$$X_C = \frac{1}{\omega C} = \frac{1}{2\pi \nu C}$$

$$= \frac{1}{(2\pi)(60 \text{ Hz})(15 \times 10^{-6} \text{ F})} = 177 \ \Omega. \qquad \text{(Answer)}$$

Understand that a capacitive reactance, although measured in ohms, is not a resistance. It is what the definition of Eq. 9 says it is.

(b) Find the current amplitude I_C in this circuit.

From Eq. 11 (see also Table 1) we have

$$I_C = \frac{V_C}{X_C} = \frac{36 \text{ V}}{177 \ \Omega} = 0.203 \text{ A} \approx 200 \text{ mA}. \qquad \text{(Answer)}$$

We see that, although a reactance is not a resistance, the capacitive reactance plays the same role for a capacitor that the resistance does for a resistor. Note also that, if you doubled the

Table 1   Phase and Amplitude Relations for Alternating Currents and Voltages

Circuit Element	Symbol	Impedance*	Phase of the Current	Phase Angle ϕ	Amplitude Relation
Resistor	R	R	In phase with v_R	$0°$	$V_R = I R$
Capacitor	C	X_C	Leads v_C by 90°	$-90°$	$V_C = I X_C$
Inductor	L	X_L	Lags v_L by 90°	$+90°$	$V_L = I X_L$

Many students have remembered the phase relations from:

"ELI the ICE man"

Here L and C stand for inductance and capacitance; E stands for voltage and I for current. Thus in an inductive circuit (ELI) the current (I) lags the voltage (E).

* As we shall see, *impedance* is a general term that includes both resistance and reactance.

frequency, the capacitive reactance would drop to half its value and the current amplitude would double. We can understand this physically: To get the same value of V_C you must deliver the same charge to the capacitor ($V_C = q/C$). If the frequency doubles, then you have only half the time to deliver this charge so that the maximum current must double. To sum up: For capacitors, the higher the frequency, the lower the reactance.

Sample Problem 2   In Fig. 5a let L = 230 mH, v = 60 Hz, and $\mathcal{E}_m = V_L$ = 36 V. (a) Find the inductive reactance X_L.

(a) From Eq. 16

$$X_L = \omega L = 2\pi v L = (2\pi)(60 \text{ Hz})(230 \times 10^{-3} \text{ H})$$
$$= 87 \ \Omega. \qquad \text{(Answer)}$$

(b) Find the current amplitude I_L in the circuit. From Eq. 18

$$I_L = \frac{V_L}{X_L} = \frac{36 \text{ V}}{87 \ \Omega} = 0.414 \text{ A} \approx 410 \text{ mA.} \quad \text{(Answer)}$$

Note that, if you doubled the frequency, the inductive reactance would double and the current amplitude would be cut in half. We can also understand this physically: To get the same value of V_L, you must change the current at the same rate ($V_L = L \ di/dt$). If the frequency doubles, you cut the time of change in half so that the maximum current is also cut in half. To sum up: For inductors, the higher the frequency, the higher the reactance.

36-4   The Series *LCR* Circuit

We are now ready to consider the full problem posed by Fig. 2, in which the applied alternating emf is (see Eq. 1)

$$\mathcal{E} = \mathcal{E}_m \sin \omega t \quad \text{(applied emf)} \qquad (19)$$

and the resulting alternating current is (see Eq. 2)

$$i = I \sin(\omega t - \phi) \quad \text{(the alternating current).} \quad (20)$$

Our task, we recall, is to find the current amplitude I and the phase constant ϕ. As a starting point we apply the loop theorem to the circuit of Fig. 2, obtaining

$$\mathcal{E} = v_R + v_C + v_L. \qquad (21)$$

This relation among these four time-varying quantities remains true at all instants of time.

Consider now the phasor diagram of Fig. 6a. It shows, at an arbitrary instant of time, the as yet unknown alternating current. Its maximum value I, its phase ($\omega t - \phi$), and its instantaneous value i are all shown. Recall that, although the potential differences across the circuit elements in Fig. 2 all vary in time with different phases, the current i is common to all elements; there is only one current.

From the phase information summarized in Table 1 we can then draw Fig. 6b, in which we show also the three phasors representing the voltages across the three circuit elements R, C, and L at that same instant. As we have learned, the current is in phase with v_R, it leads v_C by 90°, and it lags behind v_L by 90°.

Note that the algebraic sum of the projections of the phasors V_R, V_C, and V_L on the vertical axis is just the right-hand side of Eq. 21. This sum of projections must then equal the left-hand side of that equation, which is $\mathcal{E}$, the projection of the phasor $\mathcal{E}_m$.

In vector operations the (algebraic) sum of the pro-

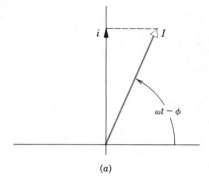

(a)

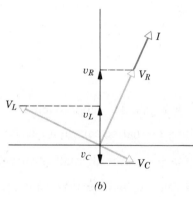

(b)

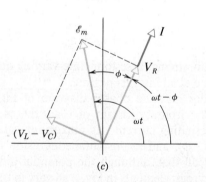

(c)

Figure 6 (*a*) A phasor representing the alternating current in the *LCR* circuit of Fig. 2. The amplitude *I*, the instantaneous value *i*, and the phase ($\omega t - \phi$) are all shown. (*b*) Phasors representing the alternating potential differences across the resistor, the capacitor, and the inductor are added. Note their phase differences with respect to the alternating current. (*c*) A phasor representing the alternating emf is added.

jections of a set of vectors on a given axis is equal to the projection on that axis of the (vector) sum of those vectors. It follows that the phasor $\mathcal{E}_m$ is equal to the (vector) sum of the phasors V_R, V_C, and V_L, as is shown in Fig. 6*c*. The figure also shows the angle ϕ, the phase difference

between the current and applied emf that appears in Eqs. 1 and 2 (and also in Eqs. 19 and 20).

In Fig. 6*c* we have formed the phasor difference $V_L - V_C$ and we note that it is at right angles to V_R. We see from the figure that

$$\mathcal{E}_m^2 = V_R^2 + (V_L - V_C)^2.$$

From the amplitude information displayed in Table 1 we can write this as

$$\mathcal{E}_m^2 = (IR)^2 + (IX_L - IX_C)^2,$$

which we can rearrange in the form

$$I = \frac{\mathcal{E}_m}{\sqrt{R^2 + (X_L - X_C)^2}}. \qquad (22)$$

The denominator in Eq. 21 is called the *impedance* Z of the circuit for the frequency in question; that is

$$\boxed{Z = \sqrt{R^2 + (X_L - X_C)^2}} \quad \text{(impedance defined)}. \quad (23)$$

We can then write Eq. 22 as

$$I = \frac{\mathcal{E}_m}{Z}. \qquad (24)$$

If we substitute for X_C and X_L from Eqs. 9 and 16 we can write Eq. 22 more explicitly as

$$\boxed{I = \frac{\mathcal{E}_m}{\sqrt{R^2 + (\omega L - 1/\omega C)^2}}}$$

(current amplitude). (25)

This relation is half of the solution of the problem we set for ourselves in Section 36-2. It is an expression for the current amplitude I in Eq. 2 in terms of $\mathcal{E}_m$, ω, R, C, and L.

Equation 25 is essentially the equation of the resonance curves of Fig. 6 of Chapter 35. Inspection of Eq. 25 shows that the maximum value of I occurs when

$$\frac{1}{\omega C} = \omega L \quad \text{or} \quad \omega = \frac{1}{\sqrt{LC}}. \quad \text{(resonance)}$$

This is precisely the resonance condition that we discussed in connection with Fig. 6 of Chapter 35. The value of I at resonance is, as we see from Eq. 25, just $\mathcal{E}_m/R$. Again this agrees with Fig. 6 of Chapter 35, in which we see that the resonance maximum increases as the circuit resistance decreases.

The Phase Constant. It remains to find an equivalent expression for the phase constant ϕ in Eq. 2. From Fig. 6c and Table 1 we can write

$$\tan \phi = \frac{V_L - V_C}{V_R} = \frac{IX_L - IX_C}{IR}$$

or

$$\boxed{\tan \phi = \frac{X_L - X_C}{R}} \quad \text{(phase constant).} \quad (26)$$

Thus, we have now solved the second half of our problem; we have expressed ϕ in terms of ω, R, C, and L. Note that $\mathcal{E}_m$ is not involved in this case.

We drew Fig. 6c arbitrarily with $X_L > X_C$, that is, we assumed the circuit of Fig. 1 to be more inductive than capacitive. Study of Eq. 25 shows that—as far as the amplitude of the current is concerned—the quantity $X_L - X_C$ is squared so that it does not matter whether $X_L > X_C$ or $X_L < X_C$. However, in using Eq. 26 to compute the phase of the current with respect to the applied emf, it does matter. In Fig. 6c, the phase constant is positive, as befits an inductive load (see Table 1).

Two Limiting Cases. In one limiting case we can put $R = X_L = 0$ in Eqs. 25 and 26, leading to $I = \mathcal{E}_m/X_C$ and $\tan \phi = -\infty$. The physical interpretation of this is that the circuit is now purely capacitive, as in Fig. 4a, and the phase constant ϕ is $-90°$, as in Fig. 4b,c.

In a second limiting case we can put $R = X_C = 0$ in Eqs. 25 and 26, leading to $I = \mathcal{E}_m/X_L$ and $\tan \phi = +\infty$. This corresponds to a purely inductive circuit, as in Fig. 5a and the phase constant ϕ is $+90°$, as in Figs. 5b,c.

We note that the currents described by Eqs. 20, 22, and 26 is the *steady-state* current that occurs after the alternating emf has been applied for some time. When the emf is first applied to a circuit, there is also a *transient current* whose duration is determined by the effective circuit time constants $\tau_L = L/R$ and $\tau_C = RC$. This transient current can be quite substantial and can, for example, destroy a motor on start-up if it is not properly allowed for in the circuit design.

Sample Problem 3 In Fig. 2 let $R = 160$ ohm, $C = 15$ μF, $L = 230$ mH, $v = 60$ Hz, and $\mathcal{E}_m = 36$ V. (a) Find the capacitive reactance X_C.

$$X_C = 177 \ \Omega, \quad \text{as in Sample Problem 1.} \quad \text{(Answer)}$$

(b) Find the inductive reactance X_L.

$$X_L = 87 \ \Omega, \quad \text{as in Sample Problem 2.} \quad \text{(Answer)}$$

Note that $X_C > X_L$ so that the circuit is more capacitive than inductive.

(c) Find the impedance Z for the circuit.
From Eq. 23,

$$Z = \sqrt{R^2 + (X_L - X_C)^2}$$
$$= \sqrt{(160 \ \Omega)^2 + (87 \ \Omega - 177 \ \Omega)^2} = 184 \ \Omega. \quad \text{(Answer)}$$

(d) Find the current amplitude I.
From Eq. 24,

$$I = \frac{\mathcal{E}_m}{Z} = \frac{36 \text{ V}}{184 \ \Omega} = 0.196 \text{ A} \approx 200 \text{ mA}. \quad \text{(Answer)}$$

(e) Find the phase constant ϕ in Eq. 2 (or Eq. 20)
From Eq. 26 we have

$$\tan \phi = \frac{X_L - X_C}{R} = \frac{87 \ \Omega - 177 \ \Omega}{160 \ \Omega} = -0.563$$

Thus we have

$$\phi = \tan^{-1}(-0.563) = -29.4°. \quad \text{(Answer)}$$

As Table 1 suggests, a negative phase constant is appropriate for a capacitive load.

36-5 Power in Alternating-Current Circuits

In the LCR circuit of Fig. 2 the source of energy is the alternating-current generator. Of the energy that it provides some is stored in the electric field in the capacitor, some in the magnetic field in the inductor, and some is dissipated as thermal energy in the resistor. In steady-state operation—which we assume—the average energy stored in the capacitor and in the inductor remains constant. The net flow of energy is thus from the generator to the resistor, as in Fig. 7, where it is transformed from electromagnetic to thermal form.

The instantaneous rate at which energy is transformed in the resistor can be written, with the help of Eq. 2, as

$$P = i^2 R = [I \sin(\omega t - \phi)]^2 R$$
$$= I^2 R[\sin(\omega t - \phi)]^2. \quad (27)$$

The *average* rate at which power is transferred to the resistor, however, which is our real concern, is the average of Eq. 27 over time. Figure 8b shows that the average value of $\sin^2 \theta$ over one complete cycle, where θ is any angular variable, is just one-half. Note in Fig. 8b how the (shaded) parts of the curve that lie above the horizontal

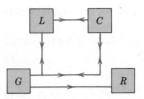

Figure 7 The flow of energy in the *LCR* circuit of Fig. 2. All energy comes from generator *G*. During each half cycle the capacitor *C* and the inductor *L* receive energy and give out as much as they have received. Some of this energy (and, at the resonant frequency, all of it) is shuttled back and forth between the capacitor and the inductor. Energy continues to flow from the generator to resistor *R*, where it appears in thermal form.

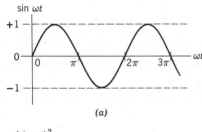

(a)

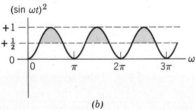

(b)

Figure 8 (a) A plot of sin θ versus θ. Its average value over one cycle is zero. (b) A plot of sin² θ versus θ. Its average value over one cycle is one-half. The angle θ represents any angular quantity.

line marked $\frac{1}{2}$ just fill in the empty spaces below that line. Thus we can write Eq. 27 as

$$P_{av} = \tfrac{1}{2}I^2 R = (I/\sqrt{2})^2 R. \tag{28}$$

We choose to call $I/(\sqrt{2})$ the *root-mean-square,* or *rms,* value of the current i and we write, in place of Eq. 28,

$$\boxed{P_{av} = I_{rms}{}^2 R} \quad \text{(average power).} \tag{29}$$

The rms notation is appropriate. First we *square* the current I, obtaining I^2. Then we average it (that is, we find its *mean* value), obtaining $\frac{1}{2}I^2$, as in Fig. 8b. Finally we take its *square root,* obtaining $I/\sqrt{2}$, which we relabel as I_{rms}.

Equation 29 looks much like Eq. 21 of Chapter 28; the message is that if we use rms quantities for *I*, for *V*, and for $\mathcal{E}$, the average power dissipation, which is all that usually matters, will be the same for alternating-current circuits as for direct-current circuits with a constant emf.

Alternating-current instruments, such as ammeters and voltmeters, are usually calibrated to read I_{rms}, V_{rms}, and $\mathcal{E}_{rms}$. Thus, if you plug an alternating-current voltmeter into a household electric outlet and it reads 120 V, that is an rms voltage. The *maximum* value of the potential difference at the outlet is $\sqrt{2} \times (120\text{ V})$ or 170 V. The sole reason for using rms values in alternating-current circuits is to let us use the familiar direct-current power relationships of Section 28–7. We summarize the relations between the maximum and the rms values for the three variables of interest:

$$I_{rms} = I/\sqrt{2}, \quad V_{rms} = V/\sqrt{2}, \quad \text{and}$$
$$\mathcal{E}_{rms} = \mathcal{E}_m/\sqrt{2}. \tag{30}$$

Because the proportionality factor in Eq. 30 ($= 1/\sqrt{2}$) is the same for all three variables, we can write the important Eqs. 24 and 25 as

$$I_{rms} = \frac{\mathcal{E}_{rms}}{Z} = \frac{\mathcal{E}_{rms}}{\sqrt{R^2 + (X_L - X_c)^2}}, \tag{31}$$

and, indeed, this is the form that we almost always use.

We can recast Eq. 29 in a useful equivalent way by combining it with the relationship $I_{rms} = \mathcal{E}_{rms}/Z$ (see Eq. 31) or

$$P_{av} = \frac{\mathcal{E}_{rms}}{Z} I_{rms} R = \mathcal{E}_{rms} I_{rms} (R/Z). \tag{32}$$

From Fig. 6b and Table 1, however, we see that R/Z is just the cosine of the phase constant ϕ. Thus

$$\cos \phi = \frac{V_R}{\mathcal{E}_m} = \frac{IR}{IZ} = \frac{R}{Z}.$$

Equation 32 then becomes

$$\boxed{P_{av} = \mathcal{E}_{rms} I_{rms} \cos \phi} \quad \text{(average power),} \tag{33}$$

in which cos ϕ is called the *power factor.* Since cos $\phi = \cos(-\phi)$, Eq. 33 is independent of whether the phase constant ϕ is positive or negative.

If we wish to deliver maximum power to a resistor in an *LCR* circuit, we should keep the power factor cos ϕ as close to unity as possible. This is equivalent to keeping the phase constant ϕ in Eq. 2 as close to zero as possible. If, for example, a load is highly inductive, it can be made

Table 2   The Average Power Transferred from the Generator for Three Special Cases

Circuit Element	Impedance Z	Phase Constant ϕ	Power Factor $\cos \phi$	Average Power P_{av}
R	R	Zero	1	$\mathcal{E}_{rms}I_{rms}$
C	X_C	$-90°$	Zero	Zero
L	X_L	$+90°$	Zero	Zero

less so by adding more capacitance to the circuit, thus reducing the phase constant and increasing the power factor in Eq. 33. Power utility companies place capacitors throughout their transmission system to bring this about.

Table 2 shows that Eq. 33 gives reasonable results in the three special cases in which the single circuit element present is a resistor (as in Fig. 3a), a capacitor (as in Fig. 4a), or an inductor (as in Fig. 5a).

Sample Problem 4   Consider again the circuit of Fig. 2, using the same parameters that we used in Sample Problem 3, namely: $R = 160 \ \Omega$, $C = 15 \ \mu F$, $L = 230$ mH, $\nu = 60$ Hz, and $\mathcal{E}_m = 36$ V. (a) Find the rms electromotive force, $\mathcal{E}_{rms}$.

$$\mathcal{E}_{rms} = \mathcal{E}_m/\sqrt{2} = 36 \text{ V}/\sqrt{2} = 25.5 \text{ V} \approx 26 \text{ V}. \quad \text{(Answer)}$$

(b) Find the rms current, I_{rms}.

In Sample Problem 3 we see that $I = 0.196$ A. We then have

$$I_{rms} = I/\sqrt{2} = 0.196 \text{ A}/\sqrt{2} = 0.139 \text{ A} \approx 0.14 \text{ A}. \quad \text{(Answer)}$$

(c) Find the power factor, $\cos \phi$.

In Sample Problem 3 we found that the phase constant ϕ was $-29.4°$. Thus

$$\text{Power factor} = \cos (-29.4°) = 0.871 \approx 0.87. \quad \text{(Answer)}$$

(d) Find the average power P_{av} dissipated in the resistor. From Eq. 29 we have

$$P_{av} = I_{rms}^2 R = (0.139 \text{ A})^2 (160 \ \Omega) = 3.1 \text{ W}. \quad \text{(Answer)}$$

Alternatively, Eq. 33 yields

$$
\begin{aligned}
P_{av} &= \mathcal{E}_{rms}I_{rms} \cos \phi \\
&= (25.5 \text{ V})(0.139 \text{ A})(0.871) = 3.1 \text{ W}, \quad \text{(Answer)}
\end{aligned}
$$

in full agreement. Note that, to get agreement of these results to two significant figures, we had to use three significant figures for the currents and voltages. These precautions against numerical rounding errors should not detract from the fact the Eqs. 29 and 33 are identically equal.

36-6   The Transformer

For alternating-current circuits, the average power dissipation in a resistive load is given by Eq. 33*

$$P_{av} = IV. \quad (34)$$

This means that, for a given power requirement, we have a range of choices, from a relatively large current I and a relatively small potential difference V or just the reverse, provided only that their product remains constant.

In electric power distribution systems it is desirable, both for reasons of safety and for efficient design of equipment, to deal with relatively low voltages at both the generating end (the electric power plant) and the receiving end (the home or factory). Nobody wants an electric toaster or a child's electric train to operate at, say, 10 kV.

On the other hand, in the transmission of electric energy from the generating plant to the consumer, we want the lowest practical current (and thus the largest practical potential difference) so as to minimize the I^2R ohmic losses in the transmission line. $V = 500$ kV is not uncommon. Thus, there is a fundamental mismatch between the requirements for efficient transmission on the one hand and efficient and safe generation and consumption on the other.

We need a device that can, as design considerations require, raise or lower the potential difference in a circuit, keeping the product current $\times$ voltage essentially constant. The *transformer* of Fig. 9 is such a device. It has no moving parts, operates by Faraday's law of induction, and has no direct-current counterpart of equivalent simplicity.

In Fig. 9 two coils are shown wound around a soft iron core. The primary winding, of N_p turns, is connected to an alternating-current generator whose emf $\mathcal{E}$ is given by

$$\mathcal{E} = \mathcal{E}_m \sin \omega t. \quad (35)$$

The secondary winding, of N_s turns, is an open circuit as long as switch S is open, which we assume for the present. Thus there is no secondary current. We assume further that the resistances of the primary and secondary

---

* In this Section, we assume a purely resistive load, so that $\phi = 0$ (and thus $\cos \phi$, which is the power factor, is equal to unity) in Eq. 33. Furthermore, we follow conventional practice and drop the subscripts identifying rms quantities. Practicing engineers and scientists assume that all time-varying current and voltages are reported as their rms values; that is what the meters read.

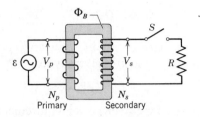

Figure 9 An ideal transformer, showing two coils wound on a soft iron core.

windings and also the magnetic hysteresis losses in the iron core are negligible. Well-designed, high-capacity transformers can have energy losses as low as one percent so that our assumption of an ideal transformer is not unreasonable.

For the above conditions the primary winding is a pure inductance; compare Fig. 5a. Thus the (very small) primary current, called the magnetizing current I_{mag}, lags the primary potential difference V_p by 90°; the power factor ($= \cos \phi$ in Eq. 33) is zero and thus no power is delivered from the generator to the transformer.

However, the small alternating primary current I_{mag} induces an alternating magnetic flux Φ_B in the iron core and we assume that all this flux links the turns of the secondary windings. From Faraday's law of induction the induced emf per turn $\mathscr{E}_{turn}$ is the same for both the primary and secondary windings. Thus, assuming that the symbols represent rms values, we can write

$$\mathscr{E}_{turn} = \frac{d\Phi_B}{dt} = \frac{V_p}{N_p} = \frac{V_s}{N_s}$$

or

$$V_s = V_p(N_s/N_p) \quad \text{(transformation of voltage).} \quad (36)$$

If $N_s > N_p$, we speak of a *step-up transformer;* if $N_s < N_p$, we speak of a *step-down transformer.*

In all of the above we have assumed an open-circuit secondary so that no power is transmitted through the transformer. Now let us close switch S in Fig. 9, thus connecting the secondary winding to a resistive load R. In general the load would also contain inductive and capacitive elements, but we confine ourselves to this special case.

Several things happen when we close switch S. (1) An alternating current I_s appears in the secondary circuit, with a corresponding power dissipation $I_s^2 R$ ($= V_s^2/R$) in the resistive load. (2) This current induces its own alternating magnetic flux in the iron core and this

flux induces (from Faraday's law and Lenz's law) an opposing emf in the primary windings.*(3) V_p, however, cannot change in response to this opposing emf because it must always equal the emf that is provided by the generator; closing switch S cannot change this fact. (4) To ensure this, a new alternating current I_p must appear in the primary circuit, its magnitude and phase constant being just that needed to cancel the opposing emf generated in the primary windings by I_s.

Rather than analyze the above rather complex process in detail, we take advantage of the overall view provided by the conservation of energy principle. For an ideal transformer with a resistive load this tells us that

$$I_p V_p = I_s V_s.$$

Because Eq. 36 holds whether or not the secondary circuit of Fig. 9 is closed, we then have

$$I_s = I_p(N_p/N_s) \quad \text{(transformation of currents)} \quad (37)$$

as the transformation relation for currents.

Finally, knowing that $I_s = V_s/R$, we can use Eqs. 36 and 37 to obtain

$$I_p = \frac{V_p}{(N_p/N_s)^2 R},$$

which tells us that, from the point of view of the primary circuit, the equivalent resistance of the load is not R but

$$R_{eq} = (N_p/N_s)^2 R \quad \text{(transformation of resistances).} \quad (38)$$

Equation 38 suggests still another function for the transformer. We have seen that, for maximum transfer of energy from a seat of emf to a resistive load, the resistance of the generator and the resistance of the load must be equal.† The same relation holds for ac circuits except that the *impedance* (rather than the resistance) of the generator must be matched to that of the load. It often happens—as when we wish to connect a speaker to an amplifier—that this condition is far from met, the amplifier being of high impedance and the speaker of low impedance. We can match the impedances of the two devices by coupling them through a transformer with a suitable turns ratio.

---

* In Chapter 33 we neglected the magnetic effects resulting from induced currents. Here, however, the magnetic effects of current induced in the secondary winding is not only not small, it is essential to the operation of the transformer!

† See Problem 22 of Chapter 29.

Sample Problem 5 A transformer on a utility pole operates at $V_p = 8.5$ kV on the primary side and supplies electric energy to a number of nearby houses at $V_s = 120$ V, both quantities being rms values. Assume an ideal transformer, a resistive load, and a power factor of unity. (a) What is the turns ratio N_p/N_s of this step-down transformer?

From Eq. 36 we have

$$\frac{N_p}{N_s} = \frac{V_p}{V_s} = \frac{8.5 \times 10^3 \text{ V}}{120 \text{ V}} = 70.8 \approx 71. \quad \text{(Answer)}$$

(b) The rate of average energy consumption in the houses served by the transformer at a given time is 78 kW. What are the rms currents in the primary and secondary windings of the transformer?

From Eq. 33 we have (with $\cos \phi = 1$)

$$I_p = \frac{P_{av}}{V_p} = \frac{78 \times 10^3 \text{ W}}{8.5 \times 10^3 \text{ V}} = 9.18 \text{ A} \approx 9.2 \text{ A} \quad \text{(Answer)}$$

and

$$I_s = \frac{P_{av}}{V_s} = \frac{78 \times 10^3 \text{ W}}{120 \text{ V}} = 650 \text{ A}. \quad \text{(Answer)}$$

(c) What is the equivalent resistive load in the secondary circuit?

Here we have

$$R_s = \frac{V_s}{I_s} = \frac{120 \text{ V}}{650 \text{ A}} = 0.185 \ \Omega \approx 0.19 \ \Omega. \quad \text{(Answer)}$$

(d) What is the equivalent resistive load in the primary circuit?

Here we have

$$R_p = \frac{V_p}{I_p} = \frac{8.5 \times 10^3 \text{ V}}{9.18 \text{ A}} = 926 \ \Omega \approx 930 \ \Omega. \quad \text{(Answer)}$$

We can verify this from Eq. 38, which we write as

$$R_p = (N_p/N_s)^2 R_s = (70.8)^2 (0.185 \ \Omega)$$
$$= 925 \ \Omega \approx 930 \ \Omega. \quad \text{(Answer)}$$

Except for rounding errors, the two results are in full agreement.

REVIEW AND SUMMARY

Current Amplitude and Phase

The basic problem in alternating-current analysis is to find expressions for the current amplitude I and the phase angle ϕ in

$$i = I \sin(\omega t - \phi) \qquad [2]$$

when an emf given by $\mathcal{E} = \mathcal{E}_m \sin \omega t$ is applied to a circuit; we use the series LCR circuit of Fig. 2 as an example. The *phase constant* ϕ is an angle that indicates the extent to which the current lags the applied emf.

Single Circuit Elements

The alternating potential difference across a resistor has amplitude $V = IR$; the current is in phase with the potential difference. For a *capacitor,* $V = IX_C$, $X_C = 1/\omega C$ being the *capacitive reactance;* current leads potential difference by 90°. For an *inductor,* $V = IX_L$, $X_L = \omega L$ being the *inductive reactance;* current lags potential difference by 90°. These results are summarized in Table 1.

Phasors

Phasors provide a powerful mathematical tool for representing alternating currents and voltages (and other phase-related quantities). We represent the maximum value (amplitude) of any alternating quantity by an open arrow rotating counterclockwise about the origin at the angular frequency ω. The projection of this arrow on the vertical axis gives the instantaneous value. The voltage and currents of simple R, C, and L elements are represented by phasors in Figs. 3, 4, and 5; these should be studied carefully. See also Sample Problems 1 and 2.

Series LCR Circuit

The phasor diagrams Fig. 6b, drawn with the help of Table 1, shows the relationship of the potential drops to the currents for the three elements in the series LCR circuit of Fig. 2. The loop theorem (see Eq. 21) allows construction of the circuit emf phasor, as in Fig. 6c. Derivations based on that figure yield

Current Amplitude

$$I = \frac{\mathcal{E}_m}{\sqrt{R^2 + (X_L - X_C)^2}} = \frac{\mathcal{E}_m}{\sqrt{R^2 + (\omega L - 1/\omega C)^2}} \quad \text{(current amplitude)} \qquad [22,25]$$

and

Phase

$$\tan \phi = \frac{X_L - X_C}{R} \quad \text{(phase constant).} \tag{26}$$

Introducing the impedance,

Impedance

$$Z = \sqrt{R^2 + (X_L - X_C)^2} \quad \text{(impedance),} \tag{23}$$

allows us to write Eq. 22 as $I = \mathcal{E}_m/Z$. See Sample Problem 3 for a summary of this analysis.

Resonance

Equation 25 is the equation of the *resonance curves* of Fig. 6 in Chapter 35. Peak current amplitude occurs when $X_C = X_L$ (the resonance condition). It has the value $\mathcal{E}_m/R$; the phase angle ϕ is zero at resonance.

In the series *LCR* circuit of Fig. 2, the *average power* output P_{av} of the generator is delivered to the resistor, where it appears as thermal energy:

Power

$$P_{av} = (I_{rms})^2 R = \mathcal{E}_{rms} I_{rms} \cos \phi \quad \text{(average power).} \tag{29,33}$$

Rms Notation

Here "rms" stands for *root-mean-square;* rms quantities are related to maximum quantities by $I_{rms} = I/\sqrt{2}$ and $\mathcal{E}_{rms} = \mathcal{E}_m/\sqrt{2}$. Alternating-current voltmeters and ammeters have their scales adjusted to read rms values. The term $\cos \phi$ above is called the *power factor.* Table 2 examines P_{av} for the three special cases of Figs. 3, 4, and 5. See Sample Problem 4.

Transformers

A *transformer* (assumed "ideal"; see Fig. 9) is a soft-iron yoke on which are wound a primary coil of N_p turns and a secondary coil of N_s turns. If the primary is connected to an alternating current generator, the primary and secondary voltages are related by

$$V_s = V_p(N_s/N_p) \quad \text{(transformation of voltage).} \tag{36}$$

The currents are related by

$$I_s = I_p(N_p/N_s) \quad \text{(transformation of currents)} \tag{37}$$

and the effective resistance of the circuit is

$$R_{eq} = (N_p/N_s)^2 \quad \text{(transformation of resistances).} \tag{38}$$

See Sample Problem 5.

QUESTIONS

1. In the relation $\omega = 2\pi\nu$ when using SI units we measure ω in radians per second and ν in hertz or cycles per second. The radian is a measure of angle. What connection do angles have with alternating currents?

2. If the output of an ac generator such as that in Fig. 1 is connected to an *LCR* circuit such as that of Fig. 2, what is the ultimate source of the power dissipated in the resistor?

3. Why would power distribution systems be less effective without alternating emfs?

4. In the circuit of Fig. 2, why is it safe to assume that (*a*) the alternating current of Eq. 2 has the same angular frequency ω as the alternating emf of Eq. 1, and (*b*) that the phase angle ϕ in Eq. 2 does not vary with time? What would happen if either of these (true) statements were false?

5. How does a phasor differ from a vector? We know, for example, that emfs, potential differences, and currents are not vectors. How then can we justify constructions such as Fig. 6?

6. Would any of the discussion of Section 3 be invalid if the phasor diagrams were to rotate in the clockwise direction, rather than the counterclockwise direction that we assumed?

7. Suppose that, in a series *LCR* circuit, the frequency of the applied voltage is changed continuously from a very low value to a very high value. How does the phase constant change?

8. Does it seem intuitively reasonable that the capacitive reactance ($= 1/\omega C$) should vary inversely with the angular frequency, whereas the inductive reactance ($= \omega L$) varies directly with this quantity?

9. During World War II, at a large research laboratory in this country, an alternating current generator was located a mile or so from the laboratory building it served. A technician increased the speed of the generator to compensate for what he called "the loss of frequency along the transmission line" connecting the generator with the laboratory building. Comment on this procedure.

10. Discuss in your own words what it means to say that an alternating current "leads" or "lags" an alternating emf.

11. If, as we stated in Section 4, a given circuit is "more inductive than capacitive," that is, that $X_L > X_C$, (a) does this mean, for a fixed angular frequency, that L is relatively "large" and C is relatively "small," or L and C are both relatively "large"? (b) For fixed values of L and C does this mean that ω is relatively "large" or relatively "small"?

12. How could you determine, in a series LCR circuit, whether the circuit frequency is above or below resonance?

13. What is wrong with this statement: "If $X_L > X_C$, then we must have $L > 1/C$"?

14. How, if at all, must Kirchoff's rules (the loop and junction theorems) for direct current circuits be modified when applied to alternating current circuits?

15. Do the loop theorem and the junction theorem apply to multiloop ac circuits as well as to multiloop dc circuits?

16. In Sample Problem 4 what would be the effect on P_{av} if you increased (a) R, (b) C, and (c) L? How would ϕ in Eq. 33 change in these three cases?

17. Do commercial power station engineers like to have a low power factor or a high one, or does it make any difference to them? Between what values can the power factor range? What determines the power factor; is it characteristic of the generator, of the transmission line, of the circuit to which the transmission line is connected, or of some combination of these?

18. Can the instantaneous power delivered by a source of alternating current ever be negative? Can the power factor ever be negative? If so, explain the meaning of these negative values.

19. In a series LCR circuit the emf is leading the current for a particular frequency of operation. You now lower the frequency slightly. Does the total impedance of the circuit increase, decrease, or stay the same?

20. If you know the power factor ($= \cos \phi$ in Eq. 33) for a given LCR circuit, can you tell whether or not the applied alternating emf is leading or lagging the current? If so, how? If not, why not?

21. What is the permissible range of values of the phase angle ϕ in Eq. 2? Of the power factor in Eq. 33?

22. Why is it useful to use the rms notation for alternating currents and voltages?

23. You want to reduce your electric bill. Do you hope for a small or a large power factor or does it make any difference? If it does, is there anything that you can do about it? Discuss.

24. In Eq. 33 is ϕ the phase angle between $\mathscr{E}(t)$ and $i(t)$ or between $\mathscr{E}_{rms}$ and i_{rms}? Explain.

25. A doorbell transformer is designed for a primary rms input of 120 V and a secondary rms output of 6 V. What would happen if the primary and secondary connections were accidentally interchanged during installation? Would you have to wait for someone to push the doorbell to find out? Discuss.

26. You are given a transformer enclosed in a wooden box, its primary and secondary terminals being available at two opposite faces of the box. How could you find its turns ratio without opening the box?

27. In the transformer of Fig. 9, with the secondary on open circuit, what is the phase relationship between (a) the impressed emf and the primary current, (b) the impressed emf and the magnetic field in the transformer core, and (c) the primary current and the magnetic field in the tranformer core?

28. What are some applications of a step-up transformer? a step-down transformer?

29. What determines which winding of a transformer is the primary and which the secondary? Can a transformer have a single primary and two secondaries? a single secondary and two primaries?

30. Instead of the 120-V, 60-Hz current typical of the United States, Europeans use 240-V, 50 Hz alternating currents. While on vacation in Europe, you would like to use some of your American appliances, such as a clock, an electric razor, and a hair dryer. Can you do so simply by plugging in a 2:1 step-up transformer? Explain why this apparently simple step may or may not suffice.

EXERCISES AND PROBLEMS

Section 36–3 Three Simple Circuits

1E. Let Eq. 1 describe the effective emf available at an ordinary 60-Hz ac outlet. What angular frequency ω does this correspond to? How does the utility company establish this frequency?

2E. A 1.5-μF capacitor is connected as in Fig. 4a to an ac generator with $\mathscr{E}_m = 30$ V. What is the amplitude of the resulting alternating current if the frequency of the emf is (a) 1.0 kHz, (b) 8.0 kHz?

3E. A 50-mH inductor is connected as in Fig. 5a to an ac generator with $\mathscr{E}_m = 30$ V. What is the amplitude of the resulting alternating current if the frequency of the emf is (a) 1.0 kHz, (b) 8.0 kHz?

4E. A 50-Ω resistor is connected as in Fig. 3a to an ac generator with $\mathscr{E}_m = 30$ V. What is the amplitude of the resulting alternating current if the frequency of the emf is (a) 1.0 kHz, (b) 8.0 kHz?

5E. A 45-mH inductor has a reactance of 1.3 kΩ. (a) What is the frequency? (*b*) What is the capacitance of a capacitor with the same reactance at that frequency? (*c*) If the frequency is doubled, what are the reactances of the inductor and capacitor?

6E. A 1.5-μF capacitor has a capacitive reactance of 12 Ω. (*a*) What must be the frequency? (*b*) What will be the capacitive reactance if the frequency is doubled?

7E. (*a*) At what frequency would a 6.0-mH inductor and a 10-μF capacitor have the same reactance? (*b*) What would this reactance be? (*c*) Show that this frequency would be equal to the natural frequency of free *LC* oscillations.

8P. The output of an ac generator is $\mathcal{E} = \mathcal{E}_m \sin \omega t$, with $\mathcal{E}_m = 25$ V and $\omega = 377$ rad/s. It is connected to a 12.7-H inductor. (*a*) What is the maximum value of the current? (*b*) When the current is a maximum, what is the emf of the generator? (*c*) When the emf of the generator is -12.5 V and increasing in magnitude, what is the current? (*d*) For the conditions of part (*c*), is the generator supplying energy to or taking energy from the rest of the circuit?

9P. The ac generator of Problem 8 is connected to a 4.15-μF capacitor. (*a*) What is the maximum value of the current? (*b*) When the current is a maximum, what is the emf of the generator? (*c*) When the emf of the generator is -12.5 V and increasing in magnitude, what is the current? (*d*) For the conditions of part (*c*), is the generator supplying energy to or taking energy from the rest of the circuit?

10P. The output of an ac generator is given by $\mathcal{E} = \mathcal{E}_m \sin(\omega t - \pi/4)$, where $\mathcal{E}_m = 30$ V and $\omega = 350$ rad/s. The current is given by $i(t) = I \sin(\omega t - 3\pi/4)$, where $I = 620$ mA. (*a*) At what time, after $t = 0$, does the generator emf first reach a maximum? (*b*) At what time, after $t = 0$, does the current first reach a maximum? (*c*) The circuit contains a single element other than the generator. Is it a capacitor, and inductor, or a resistor? Justify your answer. (*d*) What is the value of the capacitance, inductance, or resistance, as the case may be?

11P. The output of an ac generator is given by $\mathcal{E} = \mathcal{E}_m \sin(\omega t - \pi/4)$, where $\mathcal{E}_m = 30$ V and $\omega = 350$ rad/s. The current is given by $i(t) = I \sin(\omega t + \pi/4)$, where $I = 620$ mA. (*a*) At what time, after $t = 0$, does the generator emf first reach a maximum? (*b*) At what time, after $t = 0$, does the current first reach a maximum? (*c*) The circuit contains a single element other than the generator. Is it a capacitor, an inductor, or a resistor? Justify your answer. (*d*) What is the value of the capacitance, inductance, or resistance, as the case may be?

12P. *Three-phase power transmission.* A three-phase generator *G* produces electrical power that is transmitted by means of three wires as shown in Fig. 10. The potentials (relative to a common reference level) of these wires are $V_1 = A \sin(\omega t)$, $V_2 = A \sin(\omega t - 120°)$, and $V_3 = A \sin(\omega t - 240°)$. Some industrial equipment (e.g., motors) has three terminals and is designed to be connected directly to these three wires. To use a more conventional two-terminal device (e.g., a light bulb), one connects it to any two of the three wires. Show that the poten-

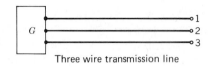

Three wire transmission line

Figure 10 Problem 12.

tial difference between *any two* of the wires (*i*) oscillates sinusoidally with angular frequency ω, and (*ii*) has amplitude $A\sqrt{3}$.

Section 36–4 The Series *LCR* Circuit

13E. (*a*) Recalculate all the quantities asked for in Sample Problem 3 if the capacitor is removed from the circuit, all other parameters in that sample problem remaining unchanged. (*b*) Draw to scale a phasor diagram like that of Fig. 6*c* for this new situation.

14E. (*a*) Recalculate all the quantities asked for in Sample Problem 3 if the inductor is removed from the circuit, all other parameters in that sample problem remaining unchanged. (*b*) Draw to scale a phasor diagram like that of Fig. 6*c* for this new situation.

15E. (*a*) Recalculate all the quantities asked for in Sample Problem 3 for $C = 70$ μF, the other parameters in that sample problem remaining unchanged. (*b*) Draw to scale a phasor diagram like that of Fig. 6*c* for this new situation and compare the two diagrams closely.

16E. Consider the resonance curves of Fig. 6, Chapter 35. (*a*) Show that for frequencies above resonance the circuit is predominantly inductive and for frequencies below resonance it is predominantly capacitive. (*b*) How does the circuit behave at resonance? (*c*) Sketch a phasor diagram like that of Fig. 6*c* for conditions at a frequency higher than resonance, at resonance, and lower than resonance.

17P. Verify mathematically that the following geometrical construction correctly gives both the impedance Z and the phase constant ϕ. Referring to Fig. 11, (*i*) draw an arrow in the $+y$ direction of magnitude X_C, (*ii*) draw an arrow in the $-y$ direction of magnitude X_L, (*iii*) draw an arrow of magnitude R in the $+x$ direction. Then the magnitude of the "resultant" of these arrows is Z and the angle (measured below the $+x$ axis) of this resultant is ϕ.

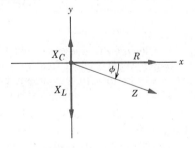

Figure 11 Problem 17.

18P. Can the amplitude of the voltage across an inductor be greater than the amplitude of the generator emf in an *LCR* circuit? Consider an *LCR* circuit with $\mathcal{E}_m = 10$ V, $R = 10$ Ω, $L = 1.0$ H, and $C = 1.0$ μF. Find the amplitude of the voltage across the inductor at resonance.

19P. A coil of self-inductance 88 mH and unknown resistance and a 0.94-μF capacitor are connected in series with an oscillator of frequency 930 Hz. If the phase angle between the applied voltage and current is 75°, what is the resistance of the coil?

20P. When the generator emf in Sample Problem 3 is a maximum, what is the voltage across (a) the generator, (b) the resistor, (c) the capacitor, and (d) the inductor? (e) By summing these with appropriate signs, verify that the loop rule is satisfied.

21P. An *LCR* circuit such as that of Fig. 2 has $R = 5.0$ Ω, $C = 20$ μF, $L = 1.0$ H, and $\mathcal{E}_m = 30$ V. (a) At what angular frequency ω_0 will the current have its maximum value, as in the resonance curves of Fig. 6, Chapter 35? (b) What is this maximum value? (c) At what two angular frequencies ω_1 and ω_2 will the current amplitude have one-half of this maximum value? (d) What is the fractional half-width $[= (\omega_1 - \omega_2)/\omega_0]$ of the resonance curve?

22P. For a certain *LCR* circuit the maximum generator emf is 125 V and the maximum current is 3.20 A. If the current leads the generator emf by 0.982 rad, (a) what is the impedance and (b) what is the resistance of the circuit? (c) Is the circuit predominantly capacitive or inductive?

23P. In a certain *LCR* circuit, operating at 60 Hz, the maximum voltage across the inductor is twice the maximum voltage across the resistor, while the maximum voltage across the capacitor is the same as the maximum voltage across the resistor. (a) By what phase angle does the current lag the generator emf? (b) If the maximum generator emf is 30 V, what should be the resistance of the circuit to obtain a maximum current of 300 mA?

24P. The circuit of Sample Problem 3, to which the phasor diagram of Fig. 6c corresponds, is not in resonance. (a) How can you tell? (b) What capacitor would you combine in parallel with the capacitor already in the circuit to bring resonance about? (c) What would the current amplitude then be?

25P. A resistor-inductor-capacitor combination, R_1, L_1, C_1 has a resonant frequency that is just the same as that of a different combination, R_2, L_2, C_2. You now connect the two combinations in series. Show that this new circuit also has the same resonant frequency as the separate individual circuits.

26P. A high-impedance ac voltmeter is connected in turn across the inductor, the capacitor, and the resistor in a series circuit having an ac source of 100 V (rms) and gives the same reading in volts in each case. What is this reading?

27P. Show that the fractional half-width of a resonance curve is given to a close approximation by

$$\frac{\Delta\omega}{\omega} = \sqrt{\frac{3C}{L}}\, R,$$

in which ω is the angular frequency at resonance and $\Delta\omega$ is the width of the resonance curve at half-amplitude. Note that $\Delta\omega/\omega$ decreases with R, as Fig. 6, Chapter 35, shows. Use this formula to check the answer to part (d) of Problem 21.

28P*. The ac generator in Fig. 12 supplies 120 V (rms) at 60 Hz. With the switch open as in the diagram, the resulting current leads the generator emf by 20°. With the switch in position 1 the current lags the generator emf by 10°. When the switch is in position 2 the rms current is 2.0 A. Find the values of R, L, and C.

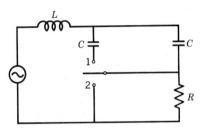

Figure 12 Problem 28.

Section 36–5 Power in Alternating Current Circuits

29E. What is the maximum value of an ac voltage whose rms value is 100 volts?

30E. What direct current will produce the same amount of heat, in a particular resistor, as an alternating current that has a maximum value of 2.6 A?

31E. Calculate the average power dissipated in the circuits of Exercises 3, 4, 13, and 14.

32E. Show that the average power delivered to the circuit of Fig. 2 can also be written as

$$P_{av} = \mathcal{E}_{rms}^2\, R/Z^2.$$

Show that this expression gives reasonable results for a purely resistive circuit, for an *LCR* circuit at resonance, for a purely capacitive circuit, and for a purely inductive circuit.

33E. An electric motor connected to a 120-V, 60-Hz power outlet does mechanical work at the rate of 0.10 hp (1 hp = 746 W). If it draws an rms current of 0.65 A, what is its effective resistance, in terms of power transfer? Would this be the same as the resistance of its coils, as measured with an ohmmeter with the motor disconnected from the power outlet?

34E. An air conditioner connected to a 120-V rms ac line is equivalent to a 12-Ω resistance and a 1.3-Ω inductive reactance in series. (*a*) Calculate the impedance of the air conditioner. (*b*) Find the average power supplied to the appliance.

35E. An electric motor has an effective resistance of 32 Ω and an inductive reactance of 45 Ω when working under load. The rms voltage across the alternating source is 420 V. Calculate the rms current.

36P. Show mathematically, rather than graphically as in Fig. 8*b*, that the average value of $\sin^2(\omega t - \phi)$ over an integral number of quarter-cycles is one-half.

37P. For an *LCR* circuit show that in one cycle with period T (*a*) the energy stored in the capacitor does not change; (*b*) the energy stored in the inductor does not change; (*c*) the generator supplies energy $(\frac{1}{2}T)\mathcal{E}_m I \cos \phi$; and (*d*) the resistor dissipates energy $(\frac{1}{2}T)RI^2$. (*e*) Show that the quantities found in (*c*) and (*d*) are equal.

38P. In an *LCR* circuit, $R = 16.0$ Ω, $C = 31.2$ μF, $L = 9.20$ mH, and $\mathcal{E} = \mathcal{E}_m \sin \omega t$, with $\mathcal{E}_m = 45$ V, and $\omega = 3000$ rad/s. For time $t = 0.442$ ms find (*a*) the rate at which energy is being supplied by the generator, (*b*) the rate at which energy is being stored in the capacitor, (*c*) the rate at which energy is being stored in the inductor, and (*d*) the rate at which energy is being dissipated in the resistor. (*e*) What is the meaning of a negative result for any of parts (*a*), (*b*), and (*c*)? (*f*) Show that the results of parts (*b*), (*c*), and (*d*) sum to the result of part (*a*).

39P. In Fig. 13 show that the power dissipated in the resistor R is a maximum when $R = r$, in which r is the internal resistance of the ac generator. In the text we have tacitly assumed, up to this point, that $r = 0$.

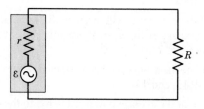

Figure 13 Problems 39 and 48.

40P. Fig. 14 shows an ac generator connected to a "black box" through a pair of terminals. The box contains an *LCR* circuit, possibly even a multiloop circuit, whose elements and arrangements we do not know. Measurements outside the box reveal that

$$\mathcal{E}(t) = (75 \ V) \sin \omega t$$

and

$$i(t) = (1.2 \ A) \sin(\omega t + 42°).$$

(*a*) What is the power factor? (*b*) Does the current lead or lag

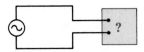

Figure 14 Problem 40.

the emf? (*c*) Is the circuit in the box largely inductive or largely capacitive in nature? (*d*) Is the circuit in the box in resonance? (*e*) Must there be a capacitor in the box? an inductor? a resistor? (*f*) What power is delivered to the box by the generator? (*g*) Why don't you need to know the angular frequency ω to answer all these questions?

41P. In an *LCR* circuit such as that of Fig. 2 assume that $R = 5.0$ Ω, $L = 60$ mH, $v = 60$ Hz, and $\mathcal{E}_m = 30$ V. For what values of the capacitance would the average power dissipated in the resistor be (*a*) a maximum and (*b*) a minimum? (*c*) What are these maximum and minimum powers? (*d*) The corresponding phase angles? (*e*) The corresponding power factors?

42P. A typical "light dimmer" used to dim the stage lights in a theater consists of a variable inductor L connected in series with the light bulb B as shown in Fig. 15. The power supply is 120 V (rms) at 60 Hz; the light bulb is marked "120 V, 1000 W." (*a*) What maximum inductance L is required if the power in the light bulb is to be varied by a factor of five? Assume that the resistance of the light bulb is independent of its temperature. (*b*) Could one use a variable resistor instead of an inductor? If so, what maximum resistance is required? Why isn't this done?

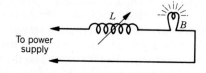

Figure 15 Problem 42.

43P. In Fig. 16, $R = 15$ Ω, $C = 4.7$ μF, and $L = 25$ mH. The generator provides a sinusoidal voltage of 75 V (rms) and frequency $v = 550$ Hz. (*a*) Calculate the rms current amplitude. (*b*) Find the rms voltages V_{ab}, V_{bc}, V_{cd}, V_{bd}, V_{ad}. (*c*) What average power is dissipated by each of the three circuit elements?

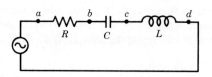

Figure 16 Problem 43.

Section 36–6 The Transformer

44E. A generator supplies 100 V to the primary coil of a transformer of 50 turns. If the secondary coil has 500 turns, what is the secondary voltage?

45E. A transformer has 500 primary turns and 10 secondary turns. (*a*) If V_P for the primary is 120 V (rms), what is V_S for the secondary, assumed on open circuit? (*b*) If the secondary is now connected to a resistive load of 15 Ω, what are the currents in the primary and secondary windings?

46E. Figure 17 shows an "autotransformer." It consists of a single coil (with an iron core). Three "taps" are provided. Between taps T_1 and T_2 there are 200 turns and between taps T_2 and T_3 there are 800 turns. Any two taps can be considered the "primary terminals" and any two taps can be considered the "secondary terminals." List all the ratios by which the primary voltage may be changed to a secondary voltage.

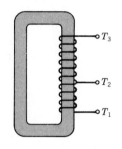

Figure 17 Exercise 46.

47P. An ac generator delivers power to a resistive load in a remote factory over a two-cable transmission line. At the factory a step-down transformer reduces the voltage from its (rms) transmisson value V_t to a much lower value, safe and convenient for use in the factory. The transmission line resistance is 0.30 Ω/cable and the power delivered to the factory is 250 kW. Calculate the voltage drop along the transmission line and the power dissipated in the line as thermal energy. Assume (*a*) $V_t = 80$ kV, (*b*) $V_t = 8.0$ kV, and (*c*) $V_t = 0.80$ kV and comment on the acceptability of each choice.

48P. *Impedance matching.* In Fig. 13 let the rectangular box on the left represent the (high-impedance) output of an audio amplifier, with $r = 1000$ Ω. Let $R = 10$ Ω represent the (low-impedance) coil of a loudspeaker. In Problem 39 we learned that for maximum transfer of power to the load R we must have $R = r$ and that is not true in this case. However, we also learned, in Section 6, that a transformer can be used to "transform" resistances, making them behave electrically as if they were larger or smaller than they actually are. Sketch the primary and secondary coils of a transformer to be introduced between the "amplifier" and the "speaker" in Fig. 13 to "match the impedances." What must be the turns ratio?

CHAPTER 37

MAXWELL'S EQUATIONS

In days of old, knights wore heraldic symbols on their shields to give information about themselves to the outside world. Today, that role has been taken over by messages on T-shirts and bumper stickers. The message here is: "If you know that these are Maxwell's equations, we share certain insights." The form in which the equations are displayed in this case (differential form) is not the same as the form (integral form) in which we display them in this book. Nevertheless, the content of the equations is identical.

37-1 Pulling Things Together

Fourteen chapters ago, when we began our venture into electromagnetism, we promised that, at some stage, we would assemble what we had learned into a single set of equations, called Maxwell's equations. The time has come to do so. Our plan is to assemble these equations in a preliminary way, at which time we shall discover—using arguments of symmetry—that there is an important term missing in one of them. We then develop this term and, finally, display the equations in their complete form.

All equations of physics that serve, as these do, to correlate experiments in a vast area and to predict new results have a certain beauty about them that can be appreciated, by those who understand them, on an aesthetic level. This is true for Newton's laws of motion, for the laws of thermodynamics, for the theory of relativity, and for the theories of quantum physics.

As for Maxwell's equations, the physicist Ludwig Boltzmann (quoting a line from Goethe) wrote: "Was it a God who wrote these lines. . . ." In more recent times J. R. Pierce, in a book chapter entitled "Maxwell's Wonderful Equations" writes: "To anyone who is motivated by anything beyond the most narrowly practical, it is worthwhile to understand Maxwell's equations simply for the good of his soul." The scope of these equations has been well summarized by the remark that Maxwell's equations account for the facts that a compass needle points north, that light bends when it enters water, and that your car starts when you turn the ignition key. These equations are the basis for the operation of all such electromagnetic and optical devices as electric motors, telescopes, cyclotrons, eyeglasses, television transmitters and receivers, telephones, electromagnets, radar, and microwave ovens.

James Clerk Maxwell, who was born in the same year that Faraday discovered the law of induction, died at age 48 in 1879, the year that Einstein was born.* One is reminded that Newton was born in the year that Galileo, his illustrious predecessor, died. Maxwell spent much of his short but highly productive life providing a theoretical basis for the experimental discoveries of Faraday. It is fair to say that Einstein was led to his theory of relativity by means of his close scrutiny of Maxwell's equations. Einstein, a great admirer of Maxwell, once wrote of him, "Imagine his feelings when the differential equations he had formulated proved to him that electromagnetic fields spread in the form of polarized waves and with the speed of light!"

37–2 Maxwell's Equations: A Tentative Listing

When we studied classical mechanics and thermodynamics, our aim was to identify the smallest, most compact set of equations or laws that would define the subject as completely as possible. In mechanics we found this in Newton's three laws of motion and in the associated force laws, such as Newton's law of gravitation. In thermodynamics we found it in the three laws that we have described in Chapters 19, 20, and 22.

---

* See "James Clerk Maxwell," by James R. Newman, *Scientific American,* June 1955.

We have now reached the point in our studies of electromagnetism at which we can begin to assemble its basic equations. Table 1 shows a tentative set of them, pulled together from earlier sections of this book. As you examine this short list, it may occur to you that you have encountered many more than four equations during the course of the last 14 chapters! That is certainly true but most of those equations—the expression for the electric field strength on the axis of an electric dipole, for example—apply to special situations and are not basic, in the sense that they are derived from more fundamental equations. You may still ask: "What about Coulomb's law and the law of Biot and Savart? We certainly treated them as fundamental equations." However, as we pointed out earlier, these laws may be viewed as fundamental only for stationary or slowly moving charges. We generalize them in Table 1 by Eq. I (for Coulomb's law) and Eq. IV (for the law of Biot and Savart.) Equations I and IV hold for rapidly time varying as well as for static or near-static situations.

The missing term to which we referred above will prove to be no trifling correction but will round out the complete description of electromagnetism and, beyond this, will establish optics as an integral part of electromagnetism. In particular, it will allow us to prove—as we shall do in the following chapter—that the speed of light c in free space is related to purely electric and magnetic quantities by

$$c = \frac{1}{\sqrt{\epsilon_0 \mu_0}} \qquad \text{(the speed of light).} \qquad (1)$$

It will also lead us to the concept of the electromagnetic spectrum, which lies behind the experimental discovery of radio waves.

We have seen how the principle of symmetry permeates physics and how it has often led to new insights or discoveries. For example: If body A attracts body B with a force $\mathbf{F}$, then perhaps body B attracts body A with a

Table 1   The Basic Equations of Electromagnetism: A Tentative List

Number	Name	Equation	Reference
I	Gauss' law for electricity	$\oint \mathbf{E} \cdot d\mathbf{A} = q/\epsilon_0$	25–9
II	Gauss' law for magnetism	$\oint \mathbf{B} \cdot d\mathbf{A} = 0$	34–13
III	Faraday's law of induction	$\oint \mathbf{E} \cdot d\mathbf{s} = -d\Phi_B/dt$	32–18
IV	Ampere's law	$\oint \mathbf{B} \cdot d\mathbf{s} = \mu_0 i$	31–16

force $-\mathbf{F}$ (it does). For another example: If there is a negative electron, there may well be a positive electron (there is).

Let us examine Table 1 from this point of view. First, we say that when we are dealing with symmetry considerations alone (that is, not making quantitative calculations) we can ignore the quantities ϵ_0 and μ_0. These constants result from our choice of unit systems and play no role in arguments of symmetry. In fact, the equations displayed on the T-shirt in the photo at the beginning of this chapter are expressed in a unit system in which $\epsilon_0 = \mu_0 = 1$.

With this in mind, we see that the left sides of the equations in Table 1 are completely symmetrical, in pairs. Equations I and II are surface integrals of $\mathbf{E}$ and $\mathbf{B}$, respectively, over closed surfaces. Equations III and IV are line integrals of $\mathbf{E}$ and $\mathbf{B}$, respectively, around closed loops. (Note in passing that, if we had substituted Coulomb's law for Eq. I and the law of Biot and Savart for Eq. IV, these appealing symmetries would be entirely missing.)

The right sides of these equations, however, do not seem symmetrical at all. We identify two kinds of asymmetry, which we discuss separately.

The First Asymmetry. This asymmetry deals with the apparent fact that although there are isolated centers of charge (electrons and protons, say) isolated centers of magnetism (magnetic monopoles, see Section 34–1) do not seem to exist in nature. That is how we interpret the fact that a q occurs on the right side of Eq. I but no corresponding magnetic quantity appears on the right side of Eq. II. In the same way, the term $\mu_0 i \,(= \mu_0 \, dq/dt)$ appears on the right of Eq. IV but no similar term (a current of magnetic monopoles) appears on the right of Eq. III.

These considerations of symmetry—coupled with the detailed predictions of certain preliminary theories of the nature of elementary particles and forces—have motivated physicists to search for the magnetic monopole in great earnest and in many ways; none has been found. It may yet be found. It is as though nature were hinting and guiding physicists in their explorations.

The Second Asymmetry. This one sticks out like a sore thumb. On the right side of Eq. III (Faraday's law of induction), we find the term $-d\Phi_B/dt$ and we interpret this law loosely by saying:

If you change a magnetic field ($d\Phi_B/dt$), you produce an electric field ($\oint \mathbf{E} \cdot d\mathbf{s}$).

We learned this in Section 32–2 where we showed that if you shove a bar magnet through a closed conducting loop, you do indeed induce an electric field, and thus a current, in that loop.

From the principle of symmetry, we are entitled to suspect that the symmetrical relation holds, that is:

If you change an electric field ($d\Phi_E/dt$), you produce a magnetic field ($\oint \mathbf{B} \cdot d\mathbf{s}$).

This conclusion from symmetry arguments proves to be correct when submitted to the test of laboratory experiment. This supposition supplies us with the important "missing" term in Eq. IV in Table 1; we develop this idea fully in the following section.

37–3 Induced Magnetic Fields

Here we discuss in detail the evidence for the supposition of the previous section, namely: "A changing electric field induces a magnetic field." Although we shall be guided by considerations of symmetry alone, we shall also point to direct experimental verification.

Figure 1a shows a uniform electric field $\mathbf{E}$ filling a cylindrical region of space. It might be produced by a circular parallel-plate capacitor, as suggested in Fig. 1b. We assume that E is increasing at a steady rate dE/dt, which means that charge must be supplied to the capacitor plates at a steady rate; to supply this charge requires a steady current i into the positive plate and an equal steady current i out of the negative plate.

Direct experiment shows that a magnetic field is set up by this changing electric field both inside the changing electric field and in the space outside that field. Figure 1a shows $\mathbf{B}$ for four selected points. This figure suggests a beautiful example of the symmetry of nature. A changing magnetic field induces an electric field (Faraday's law); now we see that a changing electric field induces a magnetic field.

To describe this new effect quantitatively, we are guided by analogy with Faraday's law of induction,

$$\oint \mathbf{E} \cdot d\mathbf{s} = -\frac{d\Phi_B}{dt}$$

(Faraday's law of induction),   (2)

which asserts that an electric field (left side) is produced

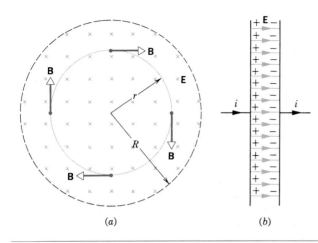

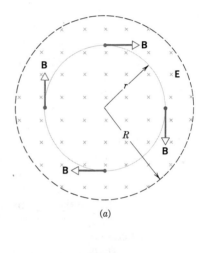

of Chapter 32, displays these electromagnetic symmetries more clearly. Figure 2a shows a magnetic field produced by a changing electric field; Fig. 2b shows the converse. In each figure the appropriate flux, Φ_E or Φ_B, is increasing. However, experiment requires that the lines of **B** in Fig. 2a be clockwise whereas those of **E** in Fig. 2b must be counterclockwise. This is why Eqs. 2 and 4 differ by a minus sign.

In Section 31-2 we saw that a magnetic field can also be set up by a current in a wire. We described this quantitatively by Ampere's law, which we now recognize

Figure 1   (a) A uniform electric field **E**, which is increasing in magnitude, fills a cylindrical region of space. The magnetic fields induced by this changing electric field are shown for four points on a circle of arbitrary radius r. (b) Such a changing electric field might be produced by charging a parallel-plate capacitor, as shown in this side view.

by a changing magnetic field (right side). For the symmetrical counterpart we might well venture to write

$$\oint \mathbf{B} \cdot d\mathbf{s} = -\frac{d\Phi_E}{dt} \quad \text{(not correct)}. \qquad (3)$$

This is certainly symmetrical with Eq. 2, but there are two things wrong with it. The first is that experiment requires that we replace the minus sign by a plus sign. This in itself is a kind of symmetry and, in any case, nature requires it.

The second difficulty with Eq. 3 is a formal one, based on the fact that we are committed to SI units. In this unit system, Eq. 3 is not dimensionally correct; to make it correct, we must insert a factor $\mu_0\epsilon_0$ on the right-hand side. Thus, the (correct) symmetrical counterpart of Eq. 2, which we may well call Maxwell's law of induction, is

$$\boxed{\oint \mathbf{B} \cdot d\mathbf{s} = +\mu_0\epsilon_0 \frac{d\Phi_E}{dt}}$$

(Maxwell's law of induction).   (4)

Note that in Eq. 2 the "induction" refers to the induced electric field; in Eq. 4 it refers to the induced magnetic field.

Figure 2, in which we compare Fig. 1a and Fig. 10b

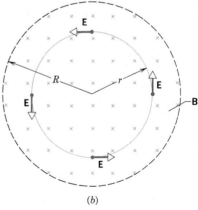

Figure 2   (a) The same as Fig. 1a; a changing electric field induces a magnetic field. (b) The same as Fig. 10b of Chapter 32; a changing magnetic field induces an electric field. In each case the fields marked by the crosses are increasing in magnitude. Note that the induced fields in these two cases point in opposite directions.

as being incomplete in form. Thus,

$$\oint \mathbf{B} \cdot d\mathbf{s} = \mu_0 i \quad \text{(Ampere's law—incomplete)}, \quad (5)$$

in which i is the conduction current passing through the Amperian loop around which the line integral is taken.

Thus, there are at least two ways of setting up a magnetic field: (1) by a changing electric field (Eq. 4) and (2) by a current (Eq. 5). In general, we must allow for both possibilities. By combining Eqs. 4 and 5, we arrive at the law in its complete form,

$$\oint \mathbf{B} \cdot d\mathbf{s} = +\mu_0 \epsilon_0 \frac{d\Phi_E}{dt} + \mu_0 i$$

$$\text{(Ampere–Maxwell law)}. \quad (6)$$

Maxwell is responsible for this important generalization of Ampere's law. It is a central and vital contribution, as we have pointed out earlier.

In Chapter 31 we assumed that no changing electric fields were present so that the term $d\Phi_E/dt$ in Eq. 6 was zero. In the discussion leading to Eq. 3, we assumed that there were no conduction currents in the space containing the electric field. Thus, the term i in Eq. 6 was zero. We see now that each of these situations is a special case.

We cannot claim to have derived Eq. 6 from deeper principles. Although our symmetry arguments should have made this equation at least reasonable, it basically must stand or fall on whether or not its predictions agree with experiment. As we shall see in Chapter 38, this agreement is complete and impressive.

Sample Problem 1 A parallel-plate capacitor with circular plates is being charged as in Fig. 1b. (a) Derive an expression for the induced magnetic field at various radii r for the case of $r < R$.

There are no conduction currents between the plates so that $i = 0$ in Eq. 6, leaving

$$\oint \mathbf{B} \cdot d\mathbf{s} = \mu_0 \epsilon_0 \frac{d\Phi_E}{dt}. \quad (7)$$

We can write this equation, for the case of $r < R$ in Fig. 1a, as

$$(B)(2\pi r) = \mu_0 \epsilon_0 \frac{d}{dt} [(E)(\pi r^2)] = \mu_0 \epsilon \pi r^2 \frac{dE}{dt}.$$

Solving for B, we find

$$B = \tfrac{1}{2}\mu_0 \epsilon_0 r \frac{dE}{dt} \quad (r < R). \quad \text{(Answer)}$$

We see that $B = 0$ at the center of the capacitor, where $r = 0$, and that B increases linearly with r out to the edge of the circular capacitor plates.

(b) Evaluate B for $r = R = 55$ mm and for $dE/dt = 1.5 \times 10^{12}$ V/m·s.

From the expression just derived, we have

$$B = (\tfrac{1}{2})(4\pi \times 10^{-7} \text{ T·m/A})(8.85 \times 10^{-12} \text{ C}^2/\text{N·m}^2)$$
$$\times (55 \times 10^{-3} \text{ m})(1.5 \times 10^{12} \text{ V/m·s})$$
$$= 4.59 \times 10^{-7} \text{ T} \approx 460 \text{ nT}. \quad \text{(Answer)}$$

Be sure to check the unit cancellations.

(c) Derive an expression for the induced magnetic field in this example for the case of $r > R$.

For points with $r > R$, the electric field E is equal to zero, so that Eq. 7 becomes

$$(B)(2\pi r) = \mu_0 \epsilon_0 \frac{d}{dt} [(E)(\pi R^2)] = \mu_0 \epsilon_0 \pi R^2 \frac{dE}{dt}.$$

Solving for B, we find

$$B = \frac{\mu_0 \epsilon_0 R^2}{2r} \frac{dE}{dt} \quad (r > R). \quad \text{(Answer)}$$

Note that the two expressions for B that we have derived yield the same result—as we expect—for $r = R$. Furthermore, the value of B at $r = R$, which we calculated in (b) above, is the maximum value that occurs for any value of r.

The induced magnetic field calculated in (b) above is so small that it can scarcely be measured with simple apparatus. This is in sharp contrast to induced electric fields (Faraday's law), which can be demonstrated easily. This experimental difference is in part due to the fact that induced emfs can easily be multiplied by using a coil of many turns. No technique of comparable simplicity exists for induced magnetic fields. In experiments involving oscillations at very high frequencies, dE/dt above can be very large, resulting in significantly larger values of the induced magnetic field. In any case, the experiment of this Sample Problem has been done and the presence of the induced magnetic fields verified quantitatively.

37-4 Displacement Current

If you look closely at the right side of Eq. 6, you will see that the term $\epsilon_0 \, d\Phi_E/dt$ must have the dimensions of a current. Even though no motion of charge is involved, there are advantages in giving this term the name *displacement current* and representing it by the symbol i_d.*

---

* The word *displacement* was introduced for historical reasons that need not concern us here.

That is,

$$i_d = \epsilon_0 \frac{d\Phi_E}{dt} \quad \text{(displacement current).} \quad (8)$$

Thus, we can say that a magnetic field can be set up either by a *conduction current i* or by a *displacement current i_d* and we can rewrite Eq. 6 as

$$\oint \mathbf{B} \cdot d\mathbf{s} = \mu_0(i_d + i)$$

(Ampere–Maxwell law). (9)

By generalizing the definition of current in this way, we can hold on to the notion that current is continuous, a principle established for steady conduction current in Section 28-2. In Fig. 1b, for example, a (conduction) current i enters the positive plate and leaves the negative plate. The conduction current is not continuous across the capacitor gap because no charge is transported across this gap. However, the displacement current i_d in the gap will prove to be exactly i, thus retaining the concept of the continuity of current.

To calculate the displacement current, recall that E in the gap of the capacitor of Fig. 1b is given by Eq. 4 of Chapter 27, or

$$E = \frac{q}{\epsilon_0 A},$$

in which q is the charge on the positive capacitor plate and A is the plate area. Differentiation gives

$$\frac{dE}{dt} = \frac{1}{\epsilon_0 A} \frac{dq}{dt} = \frac{1}{\epsilon_0 A} i,$$

so that we can write for the conduction current

$$i = \epsilon_0 A \frac{dE}{dt} \quad \text{(conduction current).} \quad (10)$$

The displacement current, defined by Eq. 8, is

$$i_d = \epsilon_0 \frac{d\Phi_E}{dt} = \epsilon_0 \frac{d(EA)}{dt}$$

or

$$i_d = \epsilon_0 A \frac{dE}{dt} \quad \text{(displacement current).} \quad (11)$$

As Eqs. 10 and 11 show, the conduction current in the connecting wires of the capacitor of Fig. 1b and the displacement current in the gap between the plates have the same value. When the capacitor is fully charged, up to the value of the applied emf, the current in the connecting wires drops to zero. The electric field between the plates assumes a steady value so that $dE/dt = 0$, which means that the displacement current also drops to zero.

The displacement current i_d, given by Eq. 11, has a direction as well as a magnitude. The direction of the conduction current i is that of the conduction current density vector $\mathbf{J}$. Similarly, the direction of the displacement current i_d is that of the displacement current density vector $\mathbf{J}_d$ which—as we see from Eq. 11—is just $\epsilon_0(d\mathbf{E}/dt)$. The right-hand rule applied to $\mathbf{J}_d$ gives the direction of the associated magnetic field, just as it does for the conduction current density $\mathbf{J}$.

Sample Problem 2 What is the displacement current for the situation of Sample Problem 1?

From Eq. 8, the definition of displacement current,

$$i_d = \epsilon_0 \frac{d\Phi_E}{dt} = \epsilon_0 \frac{d}{dt}[(E)(\pi R^2)] = \epsilon_0 \pi R^2 \frac{dE}{dt}$$

$$= (8.85 \times 10^{-12} \text{ C}^2/\text{N} \cdot \text{m}^2)(\pi)(55 \times 10^{-3} \text{ m})^2$$
$$\times (1.5 \times 10^{12} \text{ V/m} \cdot \text{s})$$
$$= 0.126 \text{ A} \approx 130 \text{ mA.} \quad \text{(Answer)}$$

You may be asking yourself: "This is a reasonably large current. Yet in Sample Problem 1(b) we saw that it produced a magnetic field of only 460 nT at the edge of the capacitor plates. Why such a very small value of the magnetic field?" It is true that a conduction current of 130 mA in a thin wire would produce a much larger magnetic field at the surface of the wire, easily detectable by a compass needle.

The difference is *not* caused by the fact that one current is a conduction current and the other is a displacement current. Under the same conditions, both kinds of current are equally effective in generating a magnetic field. The difference arises because the conduction current, in this case, is confined to a thin wire but the displacement current is spread out over an area equal to the surface area of the capacitor plates. Thus, the capacitor behaves like a "fat wire" of radius 55 mm, carrying a (displacement) current of 130 mA. Its largest magnetic effect, which occurs at the capacitor edge, is much smaller than would be the case at the surface of a thin wire.

37-5 Maxwell's Equations

Equation 6 completes our presentation of the basic equations of electromagnetism, called Maxwell's equa-

Table 2   Maxwell's Equations[a]

Number	Name	Equation	Describes	Reference
I	Gauss' law for electricity	$\oint \mathbf{E} \cdot d\mathbf{A} = q/\epsilon_0$	Charge and the electric field	Chapter 25
II	Gauss' law for magnetism	$\oint \mathbf{B} \cdot d\mathbf{A} = 0$	The magnetic field	Chapter 34
III	Faraday's law	$\oint \mathbf{E} \cdot d\mathbf{s} = -\dfrac{d\Phi_B}{dt}$	An electric field produced by a changing magnetic field	Chapter 32
IV	Ampere–Maxwell law	$\oint \mathbf{B} \cdot d\mathbf{s} = \mu_0\epsilon_0\dfrac{d\Phi_E}{dt} + \mu_0 i$	A magnetic field produced by a changing electric field or by a current or both	Chapters 31 and 37

[a] Written on the assumption that no dielectric or magnetic materials are present.

tions. We display them in Table 2, which rounds out the preliminary listing of Table 1 by supplying the missing term in Eq. IV of that table. These equations are, of course, not purely theoretical speculations but were developed to explain certain crucial laboratory experiments and observations. We list some of these in Table 3.

Maxwell described his theory of electromagnetism in a lengthy *Treatise on Electricity and Magnetism,* published in 1873, just 6 years before his death. The *Treatise* makes difficult reading and, as a matter of fact, does not contain Maxwell's equations in the form in which we have presented them (nor in the form shown in the photo at the head of this chapter.) It fell to Oliver Heaviside (1850–1925), described as "an unemployed, largely self-educated former telegrapher" who "set out in the 1870's to master electromagnetic theory" to cast Maxwell's theory into the form of the four equations that we know today.

Maxwell's ideas were slow to be adopted by the generation of practical men, trained in rule-of-thumb methods, who were his contemporaries.* Figure 3, first published in 1888, suggests the nature of the controversy. It shows W. H. Preece ("Experience"), the head of

the British telegraph service and a leading "practical man," in triumph over Oliver Lodge ("Experiment"), a prominent supporter of Maxwellian ideas. The triumph

Figure 3   An 1888 cartoon suggesting the opposition of those trained in rule-of-thumb methods to the introduction of Maxwellian ideas. The triumph of the "practical men" (as they called themselves) slowly declined as it became more and more apparent that it was Maxwell's equations that were "practical."

* See "Practice vs. Theory—The British Electrical Debate, 1888–1891," by Bruce J. Hung, *Isis,* September 1983.

Table 3 Some Crucial Electromagnetic Experiments

1. Like charges repel and unlike charges attract, with a force that varies as the inverse square of their separation.
2. A charge placed on an insulated conductor moves entirely to its outer surface.
3. It has not been possible to verify the existence of a magnetic monopole.
4. A bar magnet, thrust through a closed coil, will set up a current in that coil.
5. If the current in a coil changes, a current will be set up in a second closed coil, placed nearby.
6. A current in a wire can influence a compass needle.
7. Parallel wires carrying currents in the same direction attract each other.
8. The speed of light can be calculated from purely electric and magnetic measurements.

was short-lived; the year in which the cartoon appeared was the year in which Heinrich Hertz, guided by Maxwell's theory, discovered radio waves.

We suggested in Section 37–2 that Maxwell's equations (as they appear in Table 2) bear the same relation to electromagnetism that Newton's laws of motion do to mechanics. There is, however, an important difference. Einstein presented his special theory of relativity in 1905, roughly 200 years after Newton's laws appeared and about 40 years after Maxwell's equations. As it turns out, Newton's laws had to be drastically modified in cases in which the relative speeds approached that of light. However, no changes whatever were required in Maxwell's equations; they are totally consistent with the special theory of relativity. In fact, Einstein's theory grew out of his deep and careful thinking about the electromagnetic equations of James Clerk Maxwell.

REVIEW AND SUMMARY

In Table 1 we summarize the basic equations of electromagnetism as they have been presented in earlier chapters. In studying them for symmetry we come to see that, to make Ampere's law symmetrical with Faraday's law, we must write it as

Maxwell's Extension of Ampere's Law

$$\oint \mathbf{B} \cdot d\mathbf{s} = \mu_0 \epsilon_0 \frac{d\Phi_E}{dt} + \mu_0 i \quad \text{(Ampere-Maxwell law)}. \qquad [6]$$

The new first term on the right states that *a changing electric field ($d\Phi_E/dt$) generates a magnetic field ($\oint \mathbf{B} \cdot d\mathbf{s}$)*. It is the symmetrical counterpart of Faraday's law: *A changing magnetic field ($d\Phi_B/dt$) generates an electric field ($\oint \mathbf{E} \cdot d\mathbf{s}$)*. See Sample Problem 1.

We define displacement current as

Displacement Current

$$i_d = \epsilon_0 \frac{d\Phi_E}{dt} \quad \text{(displacement current)}. \qquad [8]$$

Equation 6 then becomes

$$\oint \mathbf{B} \cdot d\mathbf{s} = \mu_0(i_d + i) \quad \text{(Ampere–Maxwell law)}. \qquad [9]$$

We thus retain the notion of continuity of current (conduction current + displacement current). Sample Problem 2 illustrates this for the charging capacitor of Fig. 1 and Sample Problem 1. Displacement current involves a changing electric field and *not* a transfer of charge.

Maxwell's Equations

Maxwell's equations, displayed in Table 2, summarize all of electromagnetism and form its foundation. They will be used later and warrant careful study.

QUESTIONS

1. In your own words explain why Faraday's law of induction (see Table 2) can be interpreted by saying, "A changing magnetic field generates an electric field."

2. If a uniform flux Φ_E through a plane circular ring decreases with time, is the induced magnetic field (as viewed along the direction of **E**) clockwise or counterclockwise?

3. Why is it so easy to show that "a changing magnetic field produces an electric field" but so hard to show in a simple way that "a changing electric field produces a magnetic field"?

4. In Fig. 1*a* consider a circle with $r > R$. How can a magnetic field be induced around this circle, as Sample Problem 1 shows? After all, there is no electric field at the location of this circle and $dE/dt = 0$ here.

5. In Fig. 1*a*, **E** is into the figure and is increasing in magnitude. Find the direction of **B** if, instead, (*a*) **E** is into the figure and decreasing, (*b*) **E** is out of the figure and increasing, (*c*) **E** is out of the figure and decreasing, and (*d*) **E** remains constant.

6. In Fig. 1*c*, Chapter 35, a displacement current is needed to maintain continuity of current in the capacitor. How can one exist, considering that there is no charge on the capacitor?

7. In Figs. 1*a,b* what is the direction of the displacement current i_d? In this same figure, can you find a rule relating the direction (*a*) of **B** and **E**? (*b*) of **B** and dE/dt?

8. What advantages are there in calling the term $\epsilon_0 d\Phi_E/dt$ in Eq. IV, Table 2 a displacement current?

9. Can a displacement current be measured with an ammeter? Explain.

10. Why are the magnetic effects of conduction currents in wires so easy to detect but the magnetic effects of displacement currents in capacitors so hard to detect?

11. In Table 2 there are three kinds of apparent lack of symmetry in Maxwell's equations. (*a*) The quantities ϵ_0 and/or μ_0 appear in I and IV but not in II and III. (*b*) There is a minus sign in III but no minus sign in IV. (*c*) There are missing "magnetic pole terms" in II and III. Which of these represent genuine lack of symmetry? If magnetic monopoles were discovered, how would you rewrite these equations to include them? (*Hint:* Let *p* be the magnetic pole strength.)

EXERCISES AND PROBLEMS

Section 37–2 Maxwell's Equations — A Tentative Listing
1E. Verify the numerical value of the speed of light from Eq. 1 and show that the equation is dimensionally correct. (See Appendix B.)

2E. (*a*) Show that $\sqrt{\mu_0/\epsilon_0} = 377 \, \Omega$ (called the "impedance of free space"). (*b*) Show that the angular frequency of ordinary 60 Hz ac is 377 rad/s. (*c*) Compare (*a*) with (*b*). Do you think that this coincidence is the reason that 60 Hz was originally chosen as the frequency for ac generators? Recall that, in Europe, 50 Hz is used.

Section 37–3 Induced Magnetic Fields
3E. For the situation of Sample Problem 1, where is the induced magnetic field equal to one-half of its maximum value?

4P. Suppose that a circular-plate capacitor has a radius R of 30 mm and a plate separation of 5.0 mm. A sinusoidal potential difference with a maximum value of 150 V and a frequency of 60 Hz is applied between the plates. Find $B_m(R)$, the maximum value of the induced magnetic field at $r = R$.

5P. For the conditions of Problem 4, plot $B_m(r)$ for the range $0 < r < 10$ cm.

Section 37–4 Displacement Current
6E. Prove that the displacement current in a parallel-plate capacitor can be written as

$$i_d = C \frac{dV}{dt}.$$

7E. You are given a 1.0-μF parallel-plate capacitor. How would you establish an (instantaneous) displacement current of 1.0 A in the space between its plates?

8E. In Sample Problem 1 show that the *displacement current density* J_d is given, for $r < R$, by

$$J_d = \epsilon_0 \frac{dE}{dt}.$$

9E. Fig. 4 shows the plates P_1 and P_2 of a circular parallel-plate capacitor of radius R. They are connected as shown to long straight wires in which a constant conduction current i exists. A_1, A_2, and A_3 are hypothetical circles of radius r, two of them outside the capacitor and one between the plates. Show that the magnetic field at the circumference of each of these circles is given by

$$B = \frac{\mu_0 i}{2\pi r}.$$

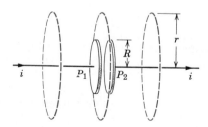

Figure 4 Exercise 9.

10P. In Sample Problem 1 show that the expressions derived for $B(r)$ can be written as

$$B(r) = \frac{\mu_0 i_d}{2\pi r} \quad (r \geq R)$$

and

$$B(r) = \frac{\mu_0 i_d r}{2\pi R^2} \quad (r \le R).$$

Note that these expressions are of just the same form as those derived in Chapter 31 except that the conduction current i has been replaced by the displacement current i_d.

11P. A parallel-plate capacitor with circular plates 20 cm in diameter is being charged as in Fig. 1b. The displacement current density throughout the region is uniform, into the paper in the diagram, and has a value of 20 A/m². (a) Calculate the magnetic field B at a distance $r = 50$ mm from the axis of symmetry of the region. (b) Calculate dE/dt in this region.

12P. A uniform electric field collapses to zero from an initial strength of 6.0×10^5 N/C in a time of 15 μs in the manner shown in Fig. 5. Calculate the displacement current, through a 1.6 m² region perpendicular to the field, during each of the time intervals (a), (b), and (c) shown on the graph. (Ignore the behavior at the ends of the intervals.)

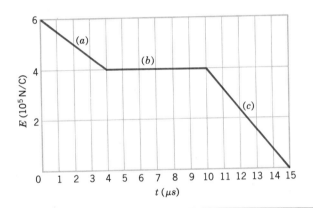

Figure 5 Problem 12.

13P. A parallel-plate capacitor has square plates 1.0 m on a side as in Fig. 6. There is a charging current of 2.0 A flowing into (and out of) the capacitor. (a) What is the displacement current through the region between the plates? (b) What is dE/dt in this region? (c) What is the displacement current through the square dashed path between the plates? (d) What is $\oint \mathbf{B} \cdot d\mathbf{s}$ around this square dashed path?

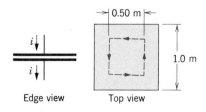

Figure 6 Problem 13.

14P. In 1929 M. R. Van Cauwenberghe succeeded in measuring directly, for the first time, the displacement current i_d between the plates of a parallel-plate capacitor to which an alternating potential difference was applied, as suggested by Fig. 1. He used circular plates whose effective radius was 40 cm and whose capacitance was 100 pF. The applied potential difference had a maximum value V_m of 174 kV at a frequency of 50 Hz. (a) What maximum displacement current was present between the plates? (b) Why was the applied potential difference chosen to be as high as it is? (The delicacy of these measurements is such that they were only performed in a direct manner more than 60 years after Maxwell enunciated the concept of displacement current! The experiment is described in *Journal de Physique*, No. 8, 1929.)

15P. The capacitor in Fig. 7 consisting of two circular plates with radius $R = 18$ cm is connected to a source of emf $\mathcal{E} = \mathcal{E}_m \sin \omega t$, where $\mathcal{E}_m = 220$ V and $\omega = 130$ rad/s. The maximum value of the displacement current is $i_d = 7.6$ μA. Neglect fringing of the electric field at the edges of the plates. (a) What is the maximum value of the current i? (b) What is the maximum value of $d\Phi_E/dt$, where Φ_E is the electric flux through the region between the plates? (c) What is the separation d between the plates? (d) Find the maximum value of the magnitude of **B** between the plates at a distance $r = 11$ cm from the center.

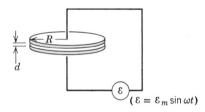

Figure 7 Problem 15.

Section 37-5 Maxwell's Equations

16E. Which of Maxwell's equations (Table 2) is most closely associated with each of the crucial experiments listed in Table 3?

17P. *A self-consistency property of two of the Maxwell equations* (numbers III and IV in Table 2). Two adjacent closed paths *abefa* and *bcdeb* share the common edge *bc* as shown in Fig. 8. (a) We may apply $\oint \mathbf{E} \cdot d\mathbf{s} = -d\Phi_B/dt$ (III) to each of

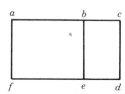

Figure 8 Problem 17.

these two closed paths separately. Show that, from this alone, Eq. III is *automatically* satisfied for the composite closed path *abcdefa*. (*b*) Repeat using Eq. IV.

18P. *A self-consistency property of two of the Maxwell equations* (numbers I and II in Table 2). Two adjacent closed parallelepipeds share a common face as shown in Fig. 9. (*a*) We may apply $\oint \mathbf{E} \cdot d\mathbf{A} = q/\epsilon_0$ (I) to each of these two closed surfaces separately. Show that, from this alone, Eq. I is *automatically* satisfied for the composite closed surface. (*b*) Repeat using Eq. II.

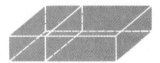

Figure 9 Problem 18.

19P. Maxwell's equations as displayed in Table 2 are written on the assumption that no dielectric materials are present.

How should the equations be written if this restriction is removed?

20P. A long cylindrical conducting rod with radius R is centered on the x axis as shown in Fig. 10. A narrow saw cut is made in the rod at $x = b$. A conduction current i, increasing with time and given by $i = \alpha t$, flows toward the right in the rod; α is a (positive) proportionality constant. At $t = 0$ there is no charge on the cut faces near $x = b$. (*a*) Find the magnitude of the charge on these faces, as a function of time. (*b*) Use Eq. I in Table 2 to find E in the gap as a function of time. (*c*) Sketch the lines of **B** for $r < R$, where r is the distance from the x axis. (*d*) Use Eq. IV in Table 2 to find $B(r)$ in the gap for $r < R$. (*e*) Compare the above answer with $B(r)$ in the rod for $r < R$.

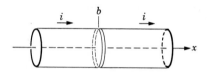

Figure 10 Problem 20.

CHAPTER 38

ELECTROMAGNETIC WAVES

A rainbow over Woolsthorpe Manor, Newton's birthplace. (Newton's analysis of the rainbow in his book Opticks *begins, "This Bow never appears, but where it rains in the Sunshine . . . ") Maxwell and his wondrous equations have taught us to view a beam of sunlight as a configuration of electric and magnetic fields, traveling through space. The sunlight is here spread out for us in the rainbow by wavelength, displaying a tiny segment of the vast electromagnetic spectrum.*

38-1 "Maxwell's Rainbow"

Maxwell's crowning achievement was to show that optics, the study of visible light, is a branch of electromagnetism and that a beam of light is a traveling configuration of electric and magnetic fields. This chapter, the last in our study of strictly electric and magnetic phenomena, is intended to round out that study and to form a bridge to the subject of optics.

In Maxwell's day, visible light and the adjoining infrared and ultraviolet radiations were the only electro-magnetic radiations known. Spurred on by Maxwell's predictions, however, Heinrich Hertz discovered what we now call radio waves and verified that they move through the laboratory at the same speed as visible light. It is for this achievement that Hertz is commemorated by using his name as the SI unit of frequency.

As Fig. 1 shows, we now know an entire spectrum of electromagnetic waves, referred to by one imaginative writer as "Maxwell's rainbow." Think of the extent to which we are bathed in electromagnetic radiation from various regions of the spectrum. The sun, whose radia-

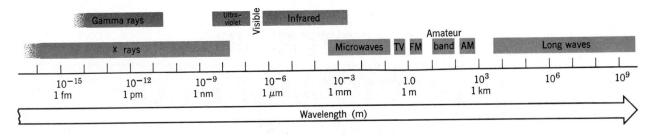

Figure 1 The electromagnetic spectrum. See "The Allocation of the Radio Spectrum," by Charles Lee Jackson, *Scientific American*, February 1980, for a description of the complex process of allocating frequencies in the range from ~ 10 kHz to ~ 300 GHz.

tions define the environment to which we as a species have evolved and adapted, is the dominant source. We are also crisscrossed by radio and TV signals. Microwaves from radar systems and from telephone relay systems may reach us. There are electromagnetic waves from light bulbs, from the heated engine blocks of automobiles, from x-ray machines, from lightning flashes, and from radioactive materials buried in the earth. Beyond this, radiation reaches us from stars and other objects in our galaxy and from other galaxies. We are even exposed, however weakly, to radiation (wavelength ≈ 2 mm) from the primeval fireball, thought by many to be associated with the creation of our universe. Electromagnetic waves also travel in the other direction. Television signals, transmitted from Earth since about 1950, have now taken news about us to whatever technically sophisticated inhabitants there may be on whatever planets may encircle the nearest 400 or so stars.

The wavelength scale in Fig. 1 (and similarly the corresponding frequency scale) is drawn so that each scale marker represents a change in wavelength (and correspondingly in frequency) by a factor of 10. The scale is open-ended, the wavelengths of electromagnetic waves having no inherent upper or lower bounds.

Certain regions of the electromagnetic spectrum in Fig. 1 are identified by familiar labels, *x rays* and *microwaves* being examples. These labels denote roughly defined wavelength ranges within which certain kinds of sources and detectors of the radiations in question are in common use. Other regions of Fig. 1, of which those labeled *TV* and *AM* are examples, represent specific wavelength bands assigned by law for certain commercial or other purposes. Only the most familiar of such allocated bands are shown.

There are no gaps in the electromagnetic spectrum. For example, we can produce electromagnetic radiation with wavelengths in the millimeter range either by mi-

crowave techniques (microwave oscillators) or by infrared techniques (heated sources). We stress that all electromagnetic waves, no matter where they lie in the spectrum, travel through free space with the same speed *c*.

The visible region of the spectrum is of course of particular interest to us. Figure 2 shows the relative sensitivity of the eye of an assumed standard human observer to radiations of various wavelength. The center of the visible region is about 555 nm; light of this wavelength produces the sensation that we call yellow-green.

The limits of the visible spectrum are not well defined because the eye sensitivity curve approaches the axis asymptotically at both long and short wavelengths. If we take the limits, arbitrarily, as the wavelengths at which the eye sensitivity has dropped to 1% of its maximum value, these limits are about 430 and 690 nm, less than a factor of two in wavelength. The eye can detect radiation beyond these limits if it is intense enough. In many experiments in physics we use photographic plates

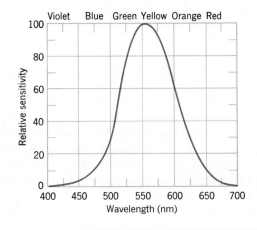

Figure 2 The relative sensitivity of the human eye at different wavelengths.

or light-sensitive electronic detectors in place of the human eye.

38-2   Generating an Electromagnetic Wave

Let us see how electromagnetic waves are generated. Some radiations such as x rays, gamma rays, and visible light come from sources that are of atomic or nuclear size, where quantum physics rules. To simplify matters, we restrict ourselves here to that region of the spectrum ($\lambda \approx 1$ m) in which the source of the radiation (a short-wave radio antenna, say) is both macroscopic and of manageable dimensions.

Figure 3 shows, in broad outline, a generator of such waves. At its heart is an *LC oscillator,* which establishes an angular frequency ω ($= 1/\sqrt{LC}$). Charges and currents in this circuit vary sinusoidally at this frequency, as

depicted in Fig. 1 of Chapter 35. An external source — possibly a battery — supplies the energy needed to compensate both for thermal losses in the circuit and for energy carried away by the radiated electromagnetic wave.

The *LC* oscillator of Fig. 3 is transformer-coupled to a *transmission line,* which might be a coaxial cable among other possibilities, that feeds the *antenna.* The two branches of the antenna alternate sinusoidally in potential at the angular frequency ω set by the oscillator, causing charges to surge back and forth along the antenna axis. The effect is that of an electric dipole whose electric dipole moment varies sinusoidally with time.

Figure 4 shows the electromagnetic wave generated by the accelerating charges in the oscillating electric dipole antenna. The electric and magnetic field lines, which form a figure of revolution about the dipole axis, travel away from the antenna with speed c. The intensity

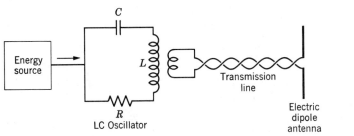

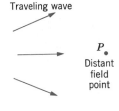

Figure 3   An arrangement for generating a traveling electromagnetic wave in the shortwave radio region of the spectrum. P is a distant point at which an observer can monitor the wave.

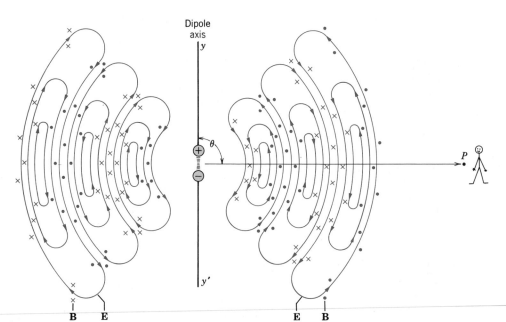

Figure 4   A close-up view of the oscillating dipole antenna of Fig. 3, showing the electric and magnetic field lines associated with the radiated electromagnetic wave. The dots and crosses, as usual, represent field lines emerging from and entering into the plane of the figure. The field pattern close to the antenna (the *near field*) is more complicated and is not shown.

of the traveling wave in any direction is proportional to $\sin^2 \theta$, being zero in the direction of the axis of the dipole ($\theta = 0$ or $180°$) and a maximum in the equatorial plane of the dipole ($\theta = 90°$). The electromagnetic field pattern close to the antenna (the *near field*) is complicated and is not shown in Fig. 4. The field pattern that *is* shown in that figure (the *radiation field*) holds for all radial distances from the antenna such that $r \gg \lambda$.

38-3 The Traveling Electromagnetic Wave — Qualitative

Consider now an observer stationed at some distant fixed point P, far enough from the antenna of Fig. 3 so that the wave fronts sweeping past him would be essentially plane. How would such an observer describe the varying electric and magnetic field patterns that constitute the wave?

Figure 5 suggests how the electric field **E** and the magnetic field **B** change with time as the wave sweeps past our stationary observer. Note that **E** and **B** are perpendicular to each other and to the direction of propagation of the wave and that they are in phase. That is, both wave components reach their maxima at the same time.

We postulate that our observer would quantify his observations by writing down, for the magnitudes of the electric and magnetic field vectors,

$$E = E_m \sin (kx - \omega t)$$
(electric component)   (1)

and

$$B = B_m \sin (kx - \omega t)$$
(magnetic component).   (2)

Figure 5 The electromagnetic wave moves directly toward an observer stationed at a remote field point, such as P in Fig. 4. The "patches" show the electric and magnetic field patterns that the observer, facing the oncoming wave, would measure during one period of the wave as the wave sweeps past his position.

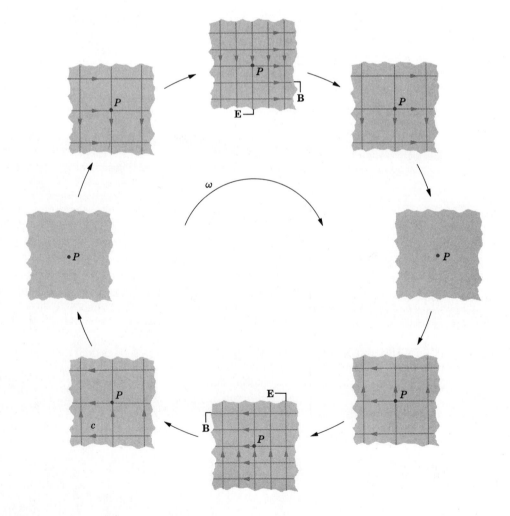

Here x is the distance (measured from any convenient origin) in the direction in which the wave is traveling. The speed of the wave, which we represent by the symbol c, is ω/k; see Eq. 14 of Chapter 17.

Equation 1 describes a time-varying electric field; from Maxwell's law of induction, it should be associated with a magnetic field. On the other hand, Eq. 2 describes a time-varying magnetic field; from Faraday's law of induction, it should be associated with an electric field. As a matter of fact, the magnetic field associated with Eq. 1 is precisely the field described by Eq. 2. Furthermore, the electric field associated with Eq. 2 is precisely the field described by Eq. 1.

What a neat arrangement! The two wave components—the electric and the magnetic—feed on each other. The spatial variation of each is associated with the time variation of the other, the entire pattern forming a consistent whole, traveling through space with speed c and completely described by Maxwell's equations.

Let us write down these equations as they apply to an electromagnetic wave traveling through free space. In free space there are no charges ($q = 0$) and no conduction currents ($i = 0$). Equations identified as III and IV in Table 2 of Chapter 37 then become

$$\oint \mathbf{E} \cdot d\mathbf{s} = -\frac{d\Phi_\mathbf{B}}{dt}$$

(Faraday's law of induction)   (3)

and

$$\oint \mathbf{B} \cdot d\mathbf{s} = \mu_0 \epsilon_0 \frac{d\Phi_\mathbf{E}}{dt}$$

(Maxwell's law of induction).   (4)

Our question now is:

Is our description of the electromagnetic wave, summarized by Eqs. 1 and 2, consistent with Maxwell's equations as represented by Eqs. 3 and 4?

The answer, which we shall prove in the next section is: Yes, provided that

$$\frac{E_m}{B_m} = c$$

(magnitude ratio),   (5)

where c ($= \omega/k$) is the speed of the waves, given by

$$c = \frac{1}{\sqrt{\mu_0 \epsilon_0}}$$

(the wave speed).   (6)

The requirement expressed by Eq. 5 is not surprising. If the electric and the magnetic components of the wave are as intricately intertwined as we have described, their wave amplitudes E_m and B_m cannot be independent of each other.

Equation 6 is the basis for Maxwell's claim that his electromagnetic equations apply to light and that optics is a branch of electromagnetism. When the electromagnetic quantities μ_0 and ϵ_0 are substituted into Eq. 6, the speed that results is indeed the speed of light.

A Special Note on Eq. 6 (Optional).   The quantities μ_0 and ϵ_0 in Eq. 6 are characteristics of the SI unit system. In Maxwell's day that unit system did not yet exist and Eq. 6 would have appeared in a different but totally equivalent form that need not concern us here. What is important is that Maxwell predicted that certain experiments —purely electrical and magnetic in character and in which light beams played no essential role—would yield a speed equal to the speed of light. Maxwell's own words convey his excitement:

> "The velocity of transverse undulations in our hypothetical medium, calculated from the electromagnetic experiments of MM Kohlrausch and Weber, agrees so exactly with the velocity of light calculated from the optical experiments of M Fizeau, that we can scarcely avoid the inference that *light consists in the transverse undulations of the same medium which is the source of the electric and magnetic phenomena.*"

The language reflects the insights and the vocabulary of an earlier day, but it is still clear that the speed of light has been measured by strictly electrical and magnetic experiments. The emphasis is Maxwell's.

After the SI unit system was introduced, Maxwell's prediction was cast in the form of Eq. 6. From the beginning, μ_0 in that equation was given an arbitrarily assigned value, exact by definition:

$$\mu_0 = 4\pi \times 10^{-7} \text{ H/m} \quad \text{(exact by definition).} \quad (7)$$

The quantity ϵ_0 can be measured in the laboratory, the most common method being to measure the capacitance of a parallel-plate capacitor of known plate area A and plate separation d and to compute ϵ_0 from the relation

$C = \epsilon_0 A/d$. One can then calculate the speed of light c from Eq. 6 and compare it with the measured value.

Since 1983, however, as part of the redefinition of the meter, the speed of light has been given an assigned value, exact and by definition, of

$$c = 299{,}792{,}458 \text{ m/s} \quad \text{(exact by definition).} \quad (8)$$

Today, our confidence in Maxwell's theory is so great that we now *assume* Eq. 6 to be correct and we use it to assign an exact value to ϵ_0:

$$\epsilon_0 = 1/c^2 \mu_0 \quad \text{(exact by definition).} \quad (9)$$

Thus, since 1983, because of our redefinition of the meter and of our confidence in Maxwell's prediction, measuring the speed of light and verifying Eq. 6 have lost their significance as possible experiments.

38-4 The Traveling Electromagnetic Wave — Quantitative (Optional)

In this Section, we show that Eqs. 5 and 6 do indeed follow from Maxwell's equations. Figure 6 shows a three-dimensional "snapshot" of a plane wave traveling in the x direction and displaying the instantaneous value of **E** and **B** whose magnitudes are given by Eqs. 1 and 2. The lines of **E** are chosen to be parallel to the y axis and those of **B** to the z axis.

Figure 7 shows two sections through the three-dimensonal diagram of Fig. 6. In Fig. 7a the plane of the page is the xy plane and in Fig. 7b it is the xz plane. Note that, consistent with Eqs. 1 and 2, **E** and **B** are in phase, that is, at any point through which the wave is moving they reach their maximum values at the same time. We divide our proof into two parts, dealing separately with the induced electric field and the induced magnetic field.

The Induced Electric Field. The shaded rectangle of dimensions dx and h in Fig. 7a is fixed at a particular point P on the x axis. As the wave passes over it, the magnetic flux Φ_B through the rectangle will change and — according to Faraday's law of induction — induced electric fields should appear around the rectangle. These induced electric fields are, in fact, simply the electric component of the traveling electromagnetic wave.

Let us apply Lenz's law. The flux Φ_B for the shaded rectangle of Fig. 7a is decreasing with time because the wave is moving through the rectangle to the right and a region of weaker magnetic fields is moving into the rectangle. The induced electric field will act to oppose this change, which means that if we imagine the boundary of the rectangle to be a conducting loop, a counterclockwise induced current would appear in it. This current would produce a field of **B** that, within the rectangle, would point out of the page, thus opposing the decrease in Φ_B. There is, of course, no conducting loop, but the net induced electric field does indeed act counterclockwise around the rectangle because $E + dE$, the field magnitude at the right edge of the rectangle, is greater than E, the field magnitude at the left edge. Thus, the electric field configuration is entirely consistent with the concept that it is induced by the changing magnetic field.

For a more detailed analysis let us apply Eq. 3, Faraday's law of induction,

$$\oint \mathbf{E} \cdot d\mathbf{s} = -\frac{d\Phi_B}{dt} \quad \text{(Faraday's law of induction),} \quad (10)$$

going counterclockwise around the shaded rectangle of Fig. 7a. There is no contribution to the integral from the top or bottom of the rectangle because **E** and $d\mathbf{s}$ are at right angles here. The integral then becomes

$$\oint \mathbf{E} \cdot d\mathbf{s} = [(E + dE)(h) - [(E)(h)] = h \, dE. \quad (11)$$

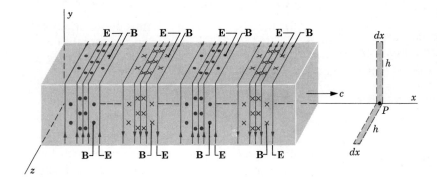

Figure 6 Another view of the plane electromagnetic wave traveling directly toward an observer stationed at point P in Fig. 4. The lines, dots, and crosses represent the electric and magnetic components of the wave. The shaded rectangles at P refer to Fig. 7.

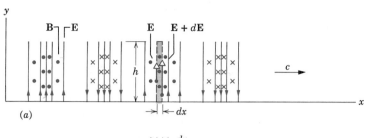

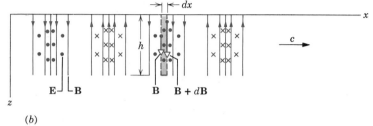

Figure 7 (*a*) The wave of Fig. 6 viewed in the *xy* plane. As the wave sweeps past, the magnetic flux through the shaded rectangle changes, inducing an electric field. (*b*) The wave of Fig. 6 viewed in the *xz* plane. As the wave sweeps past, the electric flux through the shaded rectangle changes, inducing a magnetic field.

The flux Φ_B through this rectangle is

$$\Phi_B = (B)(h\ dx), \tag{12}$$

where B is the (average) magnitude of B at the rectangular strip and $h\ dx$ is the area of the strip. Differentiating Eq. 12 gives

$$\frac{d\Phi_B}{dt} = h\ dx\ \frac{dB}{dt}. \tag{13}$$

If we substitute Eqs. 11 and 13 into Faraday's law (Eq. 10), we find

$$h\ dE = -h\ dx\ \frac{dB}{dt}$$

or

$$\frac{dE}{dx} = -\frac{dB}{dt}. \tag{14}$$

Actually, both B and E are functions of *two* variables, x and t. In evaluating dE/dx we assume that t is constant because Fig. 7*a* is an "instantaneous snapshot." Also, in evaluating dB/dt we assume that x is constant since what we want is the time rate of change of B at a particular place, the point P in Fig. 7*a*. The derivatives under these circumstances are called *partial derivatives,* and a special notation is used for them. In this notation Eq. 14 becomes

$$\frac{\partial E}{\partial x} = -\frac{\partial B}{\partial t}. \tag{15}$$

The minus sign in his equation is appropriate and neces-

sary because, although E is increasing with x at the site of the shaded rectangle in Fig. 7*a*, B is decreasing with t. Since $E(x, t)$ and $B(x, t)$ are known (see Eqs. 1 and 2), Eq. 15 reduces to

$$kE_m \cos(kx - \omega t) = \omega B_m \cos(kx - \omega t).$$

In Section 17–5 we saw that the ratio ω/k for a traveling wave is just its speed, which we choose here to call c. The above equation then becomes

$$\frac{E_m}{B_m} = c \quad \text{(magnitude ratio)}, \tag{16}$$

which is just Eq. 5. We have accomplished the first part of our task.

The Induced Magnetic Field. We now turn to Fig. 7*b*, in which the flux Φ_E for the shaded rectangle is decreasing with time as the wave moves through it. Let us apply Eq. 4, Maxwell's law of induction,

$$\boxed{\oint \mathbf{B} \cdot d\mathbf{s} = \epsilon_0 \mu_0 \frac{d\Phi_E}{dt}}$$

(Maxwell's law of induction). (17)

The changing flux Φ_E will induce a magnetic field at points around the periphery of the rectangle. This induced magnetic field is simply the magnetic component of the electromagnetic wave. Thus, the electric and the magnetic components of the wave are intimately connected, each depending on the time rate of change of the other.

The integral in Eq. 17, evaluated by proceeding

counterclockwise around the shaded rectangle of Fig. 7b, is

$$\oint \mathbf{B} \cdot d\mathbf{s} = [-(B + dB)(h) + [(B)(h)] = -h\, dB, \quad (18)$$

where B is the magnitude of B at the left edge of the strip and $B + dB$ is its magnitude at the right edge.

The flux Φ_E through the rectangle of Fig. 7b is

$$\Phi_E = (E)(h\, dx). \quad (19)$$

Differentiating Eq. 19 gives

$$\frac{d\Phi_E}{dt} = h\, dx\, \frac{dE}{dt}. \quad (20)$$

If we substitute Eqs. 18 and 20 into Maxwell's law of induction (Eq. 17), we find

$$-h\, dB = \epsilon_0\, \mu_0 \left(h\, dx\, \frac{dE}{dt} \right)$$

or, changing to partial derivatives as we did before,

$$-\frac{\partial B}{\partial x} = \epsilon_0\, \mu_0\, \frac{\partial E}{\partial t}. \quad (21)$$

Again, the minus sign in this equaton is necessary because, although B is increasing with x at point P in the shaded rectangle in Fig. 7b, E is decreasing with t.

Evaluating Eq. 21 by using Eqs. 1 and 2 leads to

$$-kB_m \cos(kx - \omega t) = -\epsilon_0\, \mu_0\, \omega E_m \cos(kx - \omega t),$$

which we can write as

$$\frac{E_m}{B_m} = \frac{1}{\epsilon_0\, \mu_0\, (\omega/k)} = \frac{1}{\epsilon_0\, \mu_0\, c}. \quad (22)$$

Combining Eqs. 16 and 22 leads at once to

$$c = \frac{1}{\sqrt{\epsilon_0\, \mu_0}} \quad \text{(the wave speed)}, \quad (23)$$

which is exactly Eq. 6. We have completed our proof that Eqs. 1 and 2, describing a traveling electromagnetic wave, are consistent with Maxwell's equations.

38-5 Energy Transport and the Poynting Vector

All sunbathers know that an electromagnetic wave can transport energy and can deliver it to a body on which it falls. The rate of energy transport per unit area (that is to say, the intensity) in such a wave is described by a vector

S, called the *Poynting vector* after John Henry Poynting (1852–1914), who first discussed its properties. S is defined from

$$\boxed{\mathbf{S} = \frac{1}{\mu_0}\, \mathbf{E} \times \mathbf{B}} \quad \text{(Poynting vector)} \quad (24)$$

its SI units being watts/meter^2. The direction of $\mathbf{S}$ at any point gives the direction of energy transport at that point.

Although we shall restrict our studies to energy transported by a plane electromagnetic wave, Eq. 24 holds in any situation in which related electric and magnetic fields are present. By applying Eq. 24, you can use the Poynting vector to trace the direction of energy flow from an alternating current generator, through a transformer, along a transmission line, through another transformer, and into your electric toaster.[*] Along the transmission line, the vector $\mathbf{S}$ has a large component parallel to the line to represent the energy flow from the generator to your house. It will also have a small component pointing directly into the conductors forming the line, to account for the i^2R resistive losses in the line.

Let us test that Eq. 24 gives a correct direction for $\mathbf{S}$ by applying it to the plane electromagnetic wave shown in Fig. 6. Note that at places where the lines of $\mathbf{E}$ point in the direction of increasing y, the lines of $\mathbf{B}$ point in the direction of increasing z. At other places, where the lines of $\mathbf{E}$ point in the direction of decreasing y, the lines of $\mathbf{B}$ point in the direction of decreasing z. In both cases, however, the product $\mathbf{E} \times \mathbf{B}$ points in the direction of increasing x, that is, the direction in which the wave is traveling. We conclude that—as far as predicting the correct direction of energy transport in the plane wave of Fig. 6 is concerned—Eq. 24 gives the right answer.

The magnitude of S for the traveling electromagnetic wave of Fig. 6 is given by[†]

$$S = \frac{1}{\mu_0}\, EB, \quad (25)$$

in which S, E, and B are instantaneous values. E and B are so closely coupled to each other that we only need to deal with one of them; we choose E, largely because most instruments for detecting electromagnetic waves deal

---

[*] See, for example, "The Poynting Vector Field and Energy Flow Within a Transformer," by F. Herrmann and Bruno Schmid, *American Journal of Physics,* June 1986.

[†] $\mathbf{E}$ and $\mathbf{B}$ are at right angles to each other, so that $\mathbf{E} \times \mathbf{B} = EB \sin 90° = EB$.

with the electric rather than the magnetic component of the wave.

Equation 5 ($E_m/B_m = c$) expresses the relation between the amplitudes of these two wave components. From Eqs. 1 and 2, we see that this same relation must hold between the instantaneous values, that is

$$\frac{E_m}{B_m} = \frac{E}{B} = c. \qquad (26)$$

Thus we can replace B in Eq. 25 by E/c, obtaining

$$\boxed{S = \frac{1}{c\mu_0} E^2} \quad \text{(energy flow; plane wave)} \quad (27)$$

as the Poynting equation for the special case of a plane electromagnetic wave. We shall show below that this equation can also be derived from information we already have about energy storage in electric and magnetic fields.

In practice, we are more interested in the average value of the time-varying quantity S, which is the *intensity I* of the wave. From Eq. 27 and Eq. 1 we can then write

$$I = \overline{S} = \frac{1}{c\mu_0} \overline{E^2} = \frac{1}{c\mu_0} E_m^2 \overline{\sin^2(kx - \omega t)}. \quad (28)$$

We have many times used the fact that the average value of the square of the sine of any angle (over an integral number of half-waves) is one-half. Furthermore, $E_m = \sqrt{2}\, E_{\text{rms}}$. With these two substitutions, Eq. 28 becomes

$$\boxed{I = \overline{S} = \frac{1}{c\mu_0} E_{\text{rms}}^2} \quad \text{(energy flow; plane wave).} \quad (29)$$

This useful equation follows directly from Eq. 27, whose proof we now present.

Proof of Eq. 27   (Optional).   Figure 8 shows a travel-ing plane wave, along with a thin "box" of thickness dx and face area A. The box, a mathematical construction, is fixed with respect to the axes and the wave moves through it. At any instant the energy stored in the box is

$$dU = dU_E + dU_B = (u_E + u_B)(A\, dx)$$

where u_E and u_B are, respectively, the electric and the magnetic energy densities and $A\, dx$ is the volume of the box. From Eq. 23 of Chapter 27 and Eq. 26 of Chapter 33, we can write this as

$$dU = \left(\frac{1}{2}\epsilon_0 E^2 + \frac{1}{2\mu_0} B^2\right)(A\, dx). \quad (30)$$

We can use Eq. 6 to eliminate ϵ_0 and can (again) replace B by E/c. The result is

$$dU = \left(\frac{E^2}{c^2\mu_0}\right)(A\, dx). \quad (31)$$

Although we here write dU in terms of E, we must remember that the energy stored in an electromagnetic wave is evenly distributed between the electric and the magnetic field components of the wave.

This energy dU will pass through the right face of the box of Fig. 8 in a time dt equal to dx/c. Thus, the energy per unit area per unit time, which is S, is given by

$$S = \frac{dU}{A\, dt} = \frac{(E^2/c^2\mu_0)(A\, dx)}{A\,(dx/c)} = \frac{1}{c\mu_0} E^2, \quad (32)$$

which (see Eq. 27) is just what we set out to prove.

Sample Problem 1 An observer is 1.8 m from a point light source whose power output P is 250 W. Calculate the rms values of the electric and the magnetic fields at the position of the observer. Assume that the source radiates uniformly in all directions.

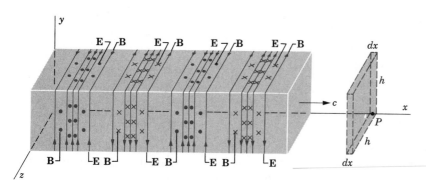

Figure 8 The wave of Fig. 6 once more. It transports energy through a hypothetical rectangular box held fixed at position P. Note that, at all points in the wave, the vector **E × B** points in the direction in which the wave is moving.

The intensity of the light at a distance r from the source is given by

$$I = \frac{P}{4\pi r^2},$$

where $4\pi r^2$ is the area of a sphere of radius r centered on the source. The intensity is also given by Eq. 29, so that

$$I = \frac{P}{4\pi r^2} = \frac{1}{c\mu_0} E_{rms}^2.$$

The rms electric field is then

$$\begin{aligned}
E_{rms} &= \sqrt{\frac{Pc\mu_0}{4\pi r^2}} \\
&= \sqrt{\frac{(250 \text{ W})(3.00 \times 10^8 \text{ m/s})(4\pi \times 10^{-7} \text{ H/m})}{(4\pi)(1.8 \text{ m})^2}} \\
&= 48.1 \text{ V/m} \approx 48 \text{ V/m}. \qquad \text{(Answer)}
\end{aligned}$$

The rms value of the magnetic field follows from Eq. 5 and is

$$\begin{aligned}
B_{rms} &= \frac{E_{rms}}{c} = \frac{48.1 \text{ V/m}}{3.00 \times 10^8 \text{ m/s}} \\
&= 1.6 \times 10^{-7} \text{ T}. \qquad \text{(Answer)}
\end{aligned}$$

Note that E_{rms} ($= 48$ V/m) is appreciable as judged by ordinary laboratory standards but that B_{rms} ($= 0.0016$ gauss) is quite small. This helps to explain why most instruments used for the detection and measurement of electromagnetic waves respond to the electric component of the wave. It is wrong, however, to say that the electric component of an electromagnetic wave is "stronger" than the magnetic component. You cannot compare quantities that are measured in different units. As we have seen, the electric and the magnetic components are on an absolutely equal basis as far as the propagation of the wave is concerned, their average energies—which *can* be compared—being exactly equal.

38-6 Radiation Pressure

We all know that electromagnetic waves transport energy. Perhaps you also know that such waves can transport linear momentum. That is, it is possible to exert a pressure (a *radiation pressure*) on an object by shining a light on it. Such forces must be small in relation to forces of our daily experience because, for example, we do not feel a recoil force when we turn on a flashlight.

Let a parallel beam of radiation, light for example, fall on an object for a time Δt and be entirely absorbed by the object. Maxwell showed that, if an energy U is ab-

sorbed during this interval, the magnitude of the momentum p delivered to the object is given by

$$\boxed{p = \frac{U}{c}} \qquad \text{(total absorption),} \qquad (33)$$

where c is the speed of light. The direction of **p** is the direction of the incident beam. If the energy U is entirely reflected, the magnitude of the momentum delivered at normal incidence will be twice that given above, or

$$\boxed{p = \frac{2U}{c}} \qquad \text{(total reflection).} \qquad (34)$$

In the same way, twice as much momentum is delivered to an object when a perfectly elastic tennis ball is bounced from it as when it is struck by a perfectly inelastic ball (a lump of putty, say) of the same mass and velocity. If the light energy U is partly reflected and partly absorbed, the delivered momentum will lie between U/c and $2U/c$.

The first measurement of radiation pressure was made in 1901–1903 by Nichols and Hull at Dartmouth College and by Lebedev in Russia, about 30 years after the existence of such effects had been predicted theoretically by Maxwell. Nichols and Hull measured radiation pressures and verified Eqs. 33 and 34, using a torsion balance technique. They allowed light to fall on mirror M as in Fig. 9; the radiation pressure caused the balance arm to turn through a measured angle θ, twisting the torsion fiber F. Assuming a suitable calibration for their torsion fiber, the experimenters could calculate a nu-

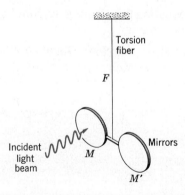

Figure 9 The arrangement of Nichols and Hull for measuring radiation pressure. Many details, essential to the success of this delicate experiment, are omitted.

merical value for this pressure. Nichols and Hull measured the intensity of their light beam by allowing it to fall on a blackened metal disk of known absorptivity and by measuring the temperature rise of this disk. In a particular run these experimenters measured a radiation pressure of 7.01 $\mu N/m^2$; the predicted value was 7.05 $\mu N/m^2$, in excellent agreement. Assuming a mirror area of 1 cm^2, this represents a force on the mirror of only 7×10^{-10} N, a remarkably small force.

The development of laser technology has permitted the achievement of radiation pressures much higher than those discussed so far in this section. This comes about because a beam of laser light — unlike a beam of light from a small lamp filament — can be focused to a tiny spot only a few wavelengths in diameter. This permits the delivery of very large energy fluxes to small objects placed at the focal spot.*

Sample Problem 2   A beam of light with an energy flux S of 12 W/cm^2 falls perpendicularly on a perfectly reflecting plane mirror of 1.5-cm^2 area. What force acts on the mirror?

From Newton's second law, the average force on the mirror is given by

$$F = \frac{\Delta p}{\Delta t},$$

where Δp is the momentum transferred to the mirror in time Δt. From Eq. 34 we have

$$\Delta p = \frac{2 \, \Delta U}{c} = \frac{2SA \, \Delta t}{c}.$$

We have then for the force

$$F = \left(\frac{2S}{c}\right) A = \frac{(2)(12 \times 10^4 \text{ W/m}^2)}{3.00 \times 10^8 \text{ m/s}} (1.5 \times 10^{-4} \text{ m}^2)$$
$$= 1.2 \times 10^{-7} \text{ N}. \qquad \text{(Answer)}$$

This is a very small force, about equal to the weight of a very small grain of table salt. Note that the quantity $2S/c$, set in the large parentheses above, is just the radiation pressure.

38-7   Polarization

Sharp-eyed travelers will notice that most TV antennas on British rooftops are vertical but in the United States

---

* See "The Pressure of Laser Light," by Arthur Ashkin, *Scientific American*, February 1972.

they are horizontal. The difference comes about because the British elect to transmit their TV signals so that the electric field vectors oscillate in a vertical plane. In this country — with equal arbitrariness — we have chosen to have them oscillate in a horizontal plane.

The wave characteristic that enters here is its *polarization*. The transverse electromagnetic wave of Fig. 6 is said to be *polarized* (more specifically, *plane polarized*) in the y direction, which means that the alternating electric field vectors are parallel to this direction for all points in the wave. (The magnetic field vectors are parallel to the z direction, but in dealing with polarization questions we focus our attention on the electric field, to which most detectors of electromagnetic radiation are sensitive.)

Figure 10, which is simply another representation of the traveling electromagnetic wave of Fig. 6, is drawn to stress the polarization feature. The wave in this figure is traveling in the x direction and is polarized in the y direction. The plane defined by the direction of propagation (the x axis) and the direction of polarization (the y axis) is called the *plane of vibration*.

Electromagnetic waves in the radio and microwave range readily exhibit polarization. Such a wave, generated by the surging of charge up and down in the dipole that forms the transmitting antenna of Fig. 11, has (for points along the horizontal axis) an electric field vector **E** that is parallel to the dipole axis. The plane of vibration of the transmitted wave is the plane of the figure. See also Fig. 12a, which shows a polarized beam emerging at right angles from the plane of the figure.

When the polarized wave of Fig. 11 falls on a second dipole connected to a microwave receiver, the alternating electric component of the wave will cause electrons to surge back and forth in the receiving antenna, producing a reading on the detector. If we turn the receiving antenna through 90° about the direction of propagation, the detector reading drops to zero. In this orientation the electric field vector is not able to cause charge to move along the dipole axis because it points at right angles to this axis. Check this out with your rabbit-ears TV antenna.

In radio and microwave sources the elementary radiators, which are the electrons surging back and forth in the transmitting antenna, act in unison; we say that they form a *coherent* source. In common light sources, however, such as the sun or a fluorescent lamp, the elementary radiators, which are the atoms that make up the source, act independently. Because of this difference the light propagated from such sources in a given direction

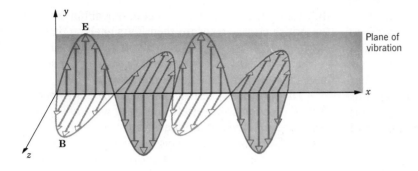

Figure 10 Another representation of the traveling electromagnetic wave of Fig. 6, showing the instantaneous electric and magnetic field vectors at various points along the *x* axis. The wave is *polarized* in the *y* direction, the *plane of vibration* being the *xy* plane.

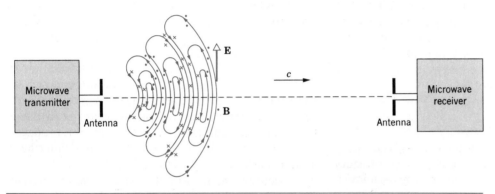

Figure 11 The electromagnetic wave generated by the transmitter is polarized, the plane of vibration being the plane of the page. The receiver can detect this wave with maximum effectiveness if (as shown) its receiving antenna lies in this plane. If the receiving antenna were to be turned at right angles to this plane, no signal would be detected.

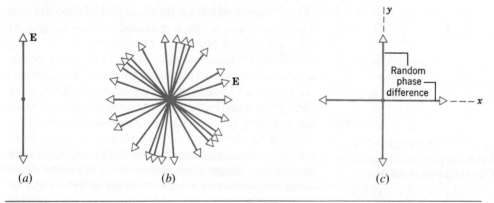

(*a*)   (*b*)   (*c*)

Figure 12 (*a*) A polarized wave moving out of the plane of the page, showing only the electric field vector. (*b*) An unpolarized wave, such as might be emitted by a light bulb. It is made up of a random array of polarized wave trains. (*c*) A second way of looking at an unpolarized wave. It can be viewed as the superposition of two plane polarized waves at right angles to each other but with a random phase difference between them.

consists of many independent wavetrains whose planes of vibration are randomly oriented about the direction of propagation, as in Fig. 12*b*. Such light is *unpolarized.* Furthermore, the random orientation of the planes of vibration conceals the transverse nature of the waves. To study this transverse nature, we must find a way to unscramble the different planes of vibration.

Figure 13 shows unpolarized light falling on a sheet

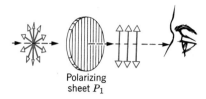

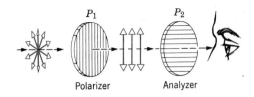

Polarizing
sheet P_1

Polarizer   Analyzer

Figure 13 Unpolarized light (see Fig. 12b) is polarized by the action of a single polarizing sheet.

Figure 15 Unpolarized light is not transmitted by crossed polarizing sheets.

of commercial polarizing material called *Polaroid.* There exists in the sheet a certain characteristic polarizing direction, shown by the parallel lines.

The sheet will transmit only those wavetrain components whose electric vectors vibrate parallel to the polarizing direction and will absorb those that vibrate at right angles to this direction.

The light emerging from a polarizing sheet will be polarized. The polarizing direction of the sheet is established during the manufacturing process by embedding certain long-chain molecules in a flexible plastic sheet and then stretching the sheet so that the molecules are aligned parallel to each other. Such a sheet absorbs radiation polarized in a direction parallel to the long molecules; radiation perpendicular to them passes through.

. In Fig. 14 the polarizing sheet or *polarizer* lies in the plane of the page and the direction of propagation is into the page. The arrow **E** shows the plane of vibration of a randomly selected wavetrain falling on the sheet. It can be broken down into two vector components, E_x (of

magnitude $E \sin \theta$) and E_y (of magnitude $E \cos \theta$). Only E_y will be transmitted; E_x will be absorbed within the sheet.

Let us place a second polarizing sheet P_2 (usually called, when so used, an *analyzer*) as in Fig. 15. If we rotate P_2 about the direction of propagation, there are two positions, 180° apart, at which the transmitted light intensity is almost zero; these are the positions in which the polarizing directions of P_1 and P_2 are at right angles.

If the amplitude of the polarized light falling on P_2 is E_m, the amplitude of the light that emerges is $E_m \cos \theta$, where θ is the angle between the polarizing directions of P_1 and P_2. Recalling that the intensity of the light beam is proportional to the square of the amplitude, we see that the transmitted intensity I varies with θ according to

$$\boxed{I = I_m \cos^2 \theta} \quad \text{(law of Malus)} \quad (35)$$

in which I_m is the maximum value of the transmitted intensity. This maximum occurs when the polarizing directions of P_1 and P_2 are parallel, that is, when $\theta = 0$ or 180°. Figure 16a, in which two overlapping polarizing sheets are in the parallel position ($\theta = 0$ or 180° in Eq. 35), shows that the light transmitted through the region of overlap has its maximum value. In Fig. 16b one or the other of the sheets has been rotated through 90° so that θ in Eq. 35 has the value 90° or 270°; the light transmitted through the region of overlap is now a minimum.

We shall see in Chapter 39 that light can also be polarized by being reflected at a certain critical angle from a sheet of glass or other dielectric material. This fact was discovered by chance in 1809 by Etienne Louis Malus, as he was gazing at the reflection of the setting sun in the Luxembourg Palace in Paris. Malus was not gazing through a polarizing sheet—such sheets were not invented until many years later—but through a crystal of calcite, which (like many other naturally-occurring crystals) also has polarizing properties. Equation 35, developed by Malus as a follow-up to his chance observation, is called the *law of Malus.*

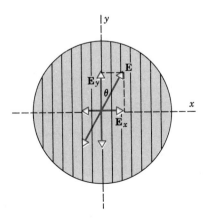

Figure 14 Another view of Fig. 13, showing a wavetrain represented by **E** falling on the polarizing sheet. Only the component E_y will be transmitted, the component E_x being absorbed within the sheet.

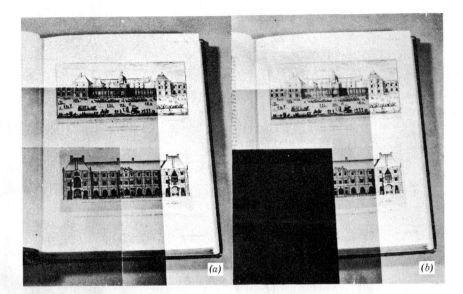

Figure 16 Two sheets of Polaroid are laid over a drawing of the Luxembourg Palace in Paris. In (*a*) the axes of the sheets are parallel, so that light is transmitted through both sheets. In (*b*) one of the sheets has been turned through 90° so that the condition of Fig. 15 prevails and no light passes through. (Malus discovered the phenomenon of polarization by reflection while looking through a polarizing crystal at sunlight reflected from the windows of this building.)

We can find out a lot about an object by examining the polarization state of the light emitted by or scattered from that object. For example, from studies of the polarization of light reflected from grains of cosmic dust present in our home galaxy, we can tell that they have been oriented in the weak galactic magnetic field ($\sim 10^{-8}$ T) so that their long dimension is parallel to this field. Polarization studies have suggested that Saturn's rings consist of ice crystals. We can find the size and shape of virus particles by the polarization of ultraviolet light scattered from them. We can learn a lot about the structure of atoms and nuclei from polarization studies of their emitted radiations. Thus, we have a useful research technique for structures ranging in size from a galaxy ($\sim 10^{+20}$ m) to a nucleus ($\sim 10^{-14}$ m).

Sample Problem 3 Two polarizing sheets have their polarizing directions parallel so that the intensity I_m of the transmitted light is a maximum. Through what angle must either sheet be turned if the intensity is to drop by one-half?

From Eq. 35, since $I = \frac{1}{2}I_m$, we have

$$\tfrac{1}{2}I_m = I_m \cos^2 \theta$$

or

$$\theta = \cos^{-1} \pm \frac{1}{\sqrt{2}} = \pm 45° \text{ and } \pm 135°. \quad \text{(Answer)}$$

The same effect is obtained no matter which sheet is rotated or in which direction.

38–8 The Speed of Electromagnetic Waves

The speed of electromagnetic radiation in free space—which we shall call the speed of light—is central not only to Maxwell's theory of electromagnetism but also to Einstein's theory of relativity. Its value has been measured so often and so precisely that, in a sense, we can say that it has been "measured to death." We have in mind the fact that the speed of light has been assigned an exact value by definition as part of the 1983 redefinition of the meter, namely

$$c = 299{,}792{,}458 \text{ m/s} \quad \text{(exactly)}.$$

Thus, the long history of measuring the speed of light is over. Today, if you send a light beam from one point to another (in free space) and measure its transit time, you are not measuring the speed of light; you are measuring the distance between the two points.

An Historical Perspective. Galileo was perhaps the first to try to measure the speed of light. In his chief work, *Two New Sciences,* published in 1638, is a conversation among three fictitious persons called Salviati, Simplicio, and Sagredo (who evidently represents Galileo himself). Here is part of what they say about the speed of light:

Simplicio: Everyday experience shows that the propagation of light is instantaneous; for when we see a piece of artillery fired, at a great distance, the flash reaches our eyes without lapse of time; but the sound reaches the ear only after a noticeable interval.

Sagredo: Well, Simplicio, the only thing I am able to infer from this familiar bit of experience is that sound, in reaching our ear, travels more slowly than light; it does not inform me whether the coming of the light is instantaneous or whether, although extremely rapid, it still occupies time.

Sagredo then describes a possible method for measuring the speed of light. He and an assistant stand facing each other some distance apart, at night. Each carries a lantern, which can be covered or uncovered at will. Galileo started the experiment by uncovering his lantern. When the light reached the assistant he uncovered his own lantern, whose light was then seen by Galileo. Galileo tried to measure the time between the instant at which he uncovered his own lantern and the instant at which the light from his assistant's lantern reached him. The method fails because human reaction times are far too slow.

Then followed various astronomical methods and then methods involving the timing of light pulses produced by shining light through a toothed wheel or reflecting it from a rotating mirror. The most recent (and perhaps the last) measurement of the speed of light involved measuring the frequency of light trapped as standing waves in a laser cavity of known length and computing c from the relation $c = \lambda v$. Table 3 of Chapter 23 points out some landmarks in the history of the measurement of the speed of light.

Einstein and the Speed of Light. When we say that the speed of sound in dry air at $0°C$ is 331.7 m/s, we imply a reference frame fixed with respect to the air mass through which the sound wave is moving. When we say that the speed of electromagnetic waves in free space is 299,792,458 m/s, what reference frame are we talking about? It cannot be the medium through which the light wave travels because, in contrast to sound, no medium is required.

Physicists of the nineteenth century, influenced as they then were by an analogy between light waves and sound waves or other purely mechanical disturbances, did not accept the idea of a wave requiring no medium. They postulated the existence of an ether, which was a tenuous substance that filled all space and served as a medium of transmission for light.

Although it proved useful for many years, the ether concept did not survive the test of experiment. In particular, careful attempts to measure the speed of the earth through the ether always gave the result of zero. Physicists were not willing to believe that the earth was permanently at rest in the ether and that all other bodies in the universe were in motion through it.

In 1905 Einstein resolved the dilemma by making a bold postulate to which we have already referred in Section 17–7 which you may wish to reread at this time. If a number of observers are moving (at uniform velocity) with respect to each other and to a light source and if each observer measures the speed of the light emerging from the source, they will all measure the same value. In formal language:

The speed of light in free space has the same value c in all directions and in all inertial reference frames.

This is a fundamental assumption of Einstein's theory of relativity, which we develop in detail in Chapter 42. The postulate does away with the need for an ether by asserting that the speed of light is the same in all reference frames; none is singled out as fundamental. The theory of relativity, derived from this postulate, has been tested many times, and agreement with the predictions of theory has always emerged. These agreements, extending over half a century, lend strong support to Einstein's basic postulate about light propagation.

REVIEW AND SUMMARY

The Electromagnetic Spectrum

Maxwell's equations predict the existence of a spectrum of electromagnetic waves (see Fig. 1) that travel through free space with a common speed c. Figure 4 shows the spacial distribution of fields produced by the dipole shortwave-radio antenna of Fig. 3. The wave is propagated because the electric and magnetic fields are generated by the alternating magnetic and electric fields, respectively, in accord with Faraday's law and the Ampere–Maxwell law.

The electric and magnetic fields of the wave have the forms

The Fields

$$E = E_m \sin(kx - \omega t) \quad \text{and} \quad B = B_m \sin(kx - \omega t). \qquad [1,2]$$

By applying Maxwell's equations we can show that

Magnitude Ratio and Wave Speed

$$\frac{E_m}{B_m} = c = \frac{1}{\sqrt{\epsilon_0 \mu_0}}.$$ [16,23]

The Poynting vector, defined from

Energy Flow

$$\mathbf{S} = \frac{1}{\mu_0} \mathbf{E} \times \mathbf{B} \quad \text{(Poynting vector)},$$ [24]

gives the energy flux (W/m^2) in an electromagnetic wave. The intensity of the wave (the average of S) is

Intensity

$$I = \overline{S} = \frac{1}{c\mu_0} E_{rms}^2 \quad \text{(energy flow; plane wave)}.$$ [29]

See Sample Problem 1.

Radiation Pressure
Electromagnetic radiation falling on a surface exerts a radiation pressure on it. Given the energy flux S falling normally on a plane mirror, Sample Problem 2 shows how to calculate the radiation pressure $p = 2S/c$.

Polarization
An electromagnetic wave from an antenna like that of Fig. 3 is polarized; its electric field vectors are parallel. The direction of the electric field $\mathbf{E}$ is called the direction of polarization; the plane containing $\mathbf{E}$ and the direction of propagation is called the plane of vibration. See Figs. 10, 11, and 12a. Light from an "ordinary" source such as the sun is unpolarized; its energy is emitted in wavetrains of finite length whose orientation about the propagation direction is random; see Fig. 12b.

Unpolarized Light

Commercial Polarizing Sheets
A sheet of Polaroid only transmits light with electric field components parallel to a given direction, strongly absorbing the others; the emerging light is polarized. The intensity of polarized light passing through a polarizing sheet is reduced from I_m to I, where

Law of Malus

$$I = I_m \cos^2 \theta \quad \text{(law of Malus)}.$$ [35]

Here θ (see Fig. 14) is the angle between the plane of vibration and the polarizing direction of the sheet; see Figs. 15 and 16 and Sample Problem 3.

Einstein's Postulate
Einstein postulated that no medium is required for the propagation of light and that the speed of light in free space has the same value c in all directions and in all inertial reference frames. This experimentally-verified prediction leads directly to the theory of relativity.

QUESTIONS

1. Electromagnetic waves reach us from the farthest depths of space. From the information they carry, can we tell what the universe is like at the present moment? At any selected time in the past?

2. If you are asked on an examination what fraction of the electromagnetic spectrum lies in the visible range, what would you reply?

3. Why are danger signals in red when the eye is most sensitive to yellow-green?

4. Comment on this definition of the limits of the spectrum of visible light, given by a physiologist: "The limits of the visible spectrum occur when the eye is no better adapted than any other organ of the body to serve as a detector."

5. How might an eye-sensitivity curve like that of Fig. 2 be measured?

6. In connection with Fig. 2, (a) do you think it possible that the wavelength of maximum sensitivity could vary if the intensity of the light is changed? (b) What might the curve of Fig. 2 look like for a color-blind person who could not, for example, distinguish red from green?

7. What feature of light corresponds to loudness in sound?

8. List several ways in which radio waves differ from visible light waves. In what ways are they the same?

9. Suppose that human eyes were insensitive to visible light but were very sensitive to infrared light. What environmental changes would be needed if you were to (a) walk down a long corridor and (b) drive a car? Could the phenomenon of color exist? How would traffic lights have to be modified?

10. Speaking loosely we can say that the electric and the magnetic components of a traveling electromagnetic wave "feed on each other." What does this mean?

11. "Displacement currents are present in a traveling electromagnetic wave and we may associate the magnetic field component of the wave with these currents." Is this statement true? Discuss it in detail.

12. H. G. Wells, in his novel *The Invisible Man,* described a concoction developed by a "mad scientist" that would render the person who drank it invisible. Give arguments to prove that a truly invisible man would be blind.

13. Can an electromagnetic wave be deflected by a magnetic field? By an electric field?

14. How does a microwave oven cook food? You can boil water in a plastic bag in such an oven. How can this happen?

15. Why is Maxwell's modification of Ampere's law (that is, the term $\epsilon_0 d\Phi_E/dt$ in Table 2, Chapter 37) needed to understand the propagation of electromagnetic waves?

16. If you were to calculate the Poynting vector for various points in and around a transformer, what would you expect the field pattern of these vectors to look like? Assume that an alternating potential difference has been applied to the primary windings and that a resistive load is connected across the secondary.

17. Can an object absorb light energy without having linear momentum transferred to it? If so, give an example. If not, explain why.

18. When you turn on a flashlight does it experience any force associated with the emission of the light?

19. We associated energy and linear momentum with electromagnetic waves. Is angular momentum present also?

20. What is the relation, if any, between the intensity I of an electromagnetic wave and the magnitude S of its Poynting vector?

21. As you recline in a beach chair in the sun, why are you so conscious of the thermal energy delivered to you but totally unresponsive to the linear momentum delivered from the same source? Is it true that when you catch a hard-pitched baseball you are conscious of the energy delivered but not of the momentum?

22. As we normally experience them, radio waves are almost always polarized and visible light is almost always unpolarized. Why should this be so?

23. Sample Problem 3 shows that, when the angle between the two polarizing directions is turned from 0° to 45°, the intensity of the transmitted beam drops to one half its initial value. What happens to this "missing" energy?

24. You are given a number of polarizing sheets. Explain how you would use them to rotate the plane of polarization of a plane polarized wave through any given angle. How could you do it with the least energy loss?

25. What determines the desirable length and orientation of the rabbit ears antenna on a portable TV set?

26. Why do sunglasses made of polarizing materials have a marked advantage over those that simply depend on absorption effects? What disadvantages might they have?

27. Why aren't sound waves polarized?

28. Unpolarized light falls on two polarizing sheets so oriented that no light is transmitted. If a third polarizing sheet is placed between them, can light be transmitted? If so, explain how.

29. Find a way to identify the polarizing direction of a sheet of Polaroid. No marks appear on the sheet.

30. When observing a clear sky through a polarizing sheet, you find that the intensity varies on rotating the sheet. This does not happen when viewing a cloud through the sheet. Why?

31. How could Galileo have tested experimentally that reaction times were an overwhelming source of error in his attempt to measure the speed of light, described in Section 8?

32. Can you think of any "everyday" observation (that is, without experimental apparatus) to show that the speed of light is not infinite? Think of lightning flashes, possible discrepancies between the predicted time of sunrise and the observed time, radio communications between earth and astronauts in orbiting space ships, and so on.

33. Atoms are mostly empty space. However, the speed of light passing through a transparent solid made up of such atoms is often considerably less than the speed of light in free space. How can this be?

34. In a vacuum, does the speed of light depend on (*a*) the wavelength, (*b*) the frequency, (*c*) the intensity, (*d*) the state of polarization, (*e*) the speed of the source, or (*f*) the speed of the observer?

EXERCISES AND PROBLEMS

Section 38–1 "Maxwell's Rainbow"

1E. Project Seafarer was an ambitious program to construct an enormous antenna, buried underground on a site about 4000 square miles in area. Its purpose was to transmit signals to submarines while they were deeply submerged. If the effective wavelength was 10^4 earth radii, what would be (*a*) the frequency and (*b*) the period of the radiations emitted? Ordinarily electromagnetic radiations do not penetrate very far into conductors such as sea water. Can you think of any reason why such ELF (extremely low frequency) radiations should penetrate more effectively? Think of the limiting case of zero frequency. (Why not transmit signals at zero frequency?)

2E. (*a*) How long does it take a radio signal to travel 150 km from a transmitter to a receiving antenna? (*b*) We see a full moon by reflected sunlight. How much earlier did the light that enters our eye leave the sun? The earth–moon and the earth–

sun distances are 3.8×10^5 km and 1.5×10^8 km. (*c*) What is the round-trip travel time for light between Earth and a spaceship orbiting Saturn, 1.3×10^9 km distant? (*d*) The Crab nebula, which is about 6500 light-years distant, is thought to be the result of a supernova explosion recorded by Chinese astronomers in A.D. 1054. In approximately what year did the explosion actually occur?

3E. (*a*) The wavelength of the most energetic x rays produced when electrons accelerated to 18 GeV in the Stanford Linear Accelerator slam into a solid target is 0.067 fm. What is the frequency of these x rays? (*b*) A VLF (very low frequency) radio wave has a frequency of only 30 Hz. What is its wavelength?

4E. (*a*) At what wavelengths does the eye of a standard observer have half its maximum sensitivity? (*b*) What are the wavelength, the frequency, and the period of the light for which the eye is the most sensitive?

5E. In the electromagnetic spectrum displayed in Fig. 1, the wavelength scale markers represent successive powers of 10 and are evenly spaced. Show that the frequency markers must also be evenly spaced, with the same interval.

6P. The radiation from a certain HeNe laser, although centered on 632.8 nm, has a finite "linewidth" of 0.010 nm. Calculate the linewidth in frequency units.

7P. One method for measuring the speed of light, based on observations by Roemer in 1676, consisted in observing the apparent times of revolution of one of the moons of Jupiter.

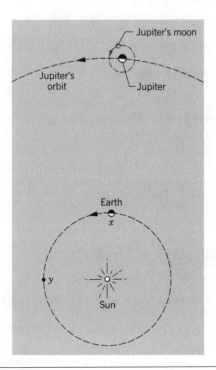

Figure 17   Problem 7.

The true period of revolution is 42.5 h. (*a*) Taking into account the finite speed of light, how would you expect the apparent time of revolution to alter as the earth moves in its orbit from point *x* to point *y* in Fig. 17? (*b*) What observations would be needed to compute the speed of light? Neglect the motion of Jupiter in its orbit. Fig. 17 is not drawn to scale.

Section 38–2 Generating an Electromagnetic Wave

8E. Find the angle from the axis of an oscillating dipole at which the intensity of the radiation field is one-half its value in the equatorial plane.

9E. What inductance is required with a 17-pF capacitor in order to construct an oscillator capable of generating 550-nm (i.e., visible) electromagnetic waves? Comment on your answer.

10P. Figure 18 shows an *LC* oscillator connected by a transmission line to an antenna of a so-called *magnetic* dipole type. Compare with Fig. 3, which shows a similar arrangement but with an *electric* dipole type of antenna. (*a*) What is the basis for the names of these two antenna types? (*b*) Draw figures corresponding to Figs. 4 and 5 to describe the electromagnetic wave that sweeps past the observer at point *P* in Fig. 18.

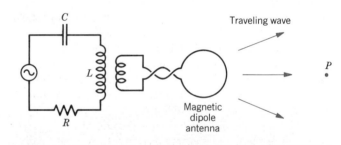

Figure 18   Problem 10.

Section 38–4 The Traveling Electromagnetic Wave— Quantitative

11E. A plane electromagnetic wave has a maximum electric field of 3.2×10^{-4} V/m. Find the maximum magnetic field.

12E. The electric field associated with a plane electromagnetic wave is given by $E_x = 0$; $E_y = 0$; $E_z = 2 \cos[\pi \times 10^{15}(t - x/c)]$, with $c = 3.0 \times 10^8$ m/s and all quantities in SI units. The wave is propagating in the $+x$ direction. Write expressions for the components of the magnetic field of the wave.

13P. Start from Eqs. 15 and 21 and show that $E(x, t)$ and $B(x, t)$, the electric and magnetic field components of a plane traveling electromagnetic wave, must satisfy the "wave equations"

$$\frac{\partial^2 E}{\partial t^2} = c^2 \frac{\partial^2 E}{\partial x^2}$$

and

$$\frac{\partial^2 B}{\partial t^2} = c^2 \frac{\partial^2 B}{\partial x^2}.$$

14P. (*a*) Show that Eqs. 1 and 2 satisfy the wave equations displayed in Problem 13. (*b*) Show that any expressions of the form

$$E = E_m f(kx \pm \omega t)$$

and

$$B = B_m f(kx \pm \omega t),$$

where $f(kx \pm \omega t)$ denotes an arbitrary function, also satisfy these wave equations.

Section 38–5 Energy Transport and the Poynting Vector

15E. Show, by finding the direction of the Poynting vector **S**, that the directions of the electric and magnetic fields at all points in Figs. 4, 5, 6, and 7 are consistent at all times with the assumed directions of propagation.

16E. Currently-operating Nd:glass lasers can provide 100 TW of power in 1.0-ns pulses at a wavelength of 0.26 μm. How much energy is contained in a single pulse?

17E. Our closest stellar neighbor, α-Centauri, is 4.3 light-years away. It has been suggested that TV programs from our planet have reached this star and may have been viewed by the hypothetical inhabitants of a hypothetical planet orbiting this star. A TV station on earth has a power output of 1.0 MW. What is the intensity of its signal at α-Centauri?

18E. An electromagnetic wave is traveling in the negative y direction. At a particular position and time, the electric field is along the positive z axis and has a magnitude of 100 V/m. What are the direction and magnitude of the magnetic field at that position and at that time?

19E. The earth's mean radius is 6.37×10^6 m and the mean earth–sun distance is 1.5×10^8 km. What fraction of the radiation emitted by the sun is intercepted by the disc of the earth?

20E. The radiation emitted by a laser is not exactly a parallel beam; rather, the beam spreads out in the form of a cone with circular cross section. The angle θ of the cone, see Fig. 19, is called the *full-angle beam divergence*. An argon laser, radiating at 514.5 nm, is aimed at the moon in a ranging experiment. If the beam has a full-angle beam divergence of 0.88 μrad, what area on the moon's surface is illuminated by the laser?

Figure 19 Exercise 20.

21E. The intensity of direct solar radiation that was unabsorbed by the atmosphere on a particular summer day is 100 W/m². How close would you have to stand to a 1.0-kW electric heater to feel the same intensity? Assume that the heater radiates uniformly in all directions.

22E. Show that in a plane traveling electromagnetic wave the average intensity, that is, the average rate of energy transport per unit area, is given by

$$\bar{S} = \frac{E_m^2}{2\mu_0 c} = \frac{cB_m^2}{2\mu_0}.$$

23E. What is the average intensity of a plane traveling electromagnetic wave if B_m, the maximum value of its magnetic field component, is 1.0×10^{-4} T (= 1.0 gauss)? (*Hint:* See Exercise 22.)

24E. In a plane radio wave the maximum value of the electric field component is 5.0 V/m. Calculate (*a*) the maximum value of the magnetic field component and (*b*) the wave intensity. (*Hint:* See Exercise 22.)

25P. You walk 150 m directly toward a street lamp and find that the intensity increases to 1.5 times the intensity at your original position. How far from the lamp were you first standing? (The lamp radiates uniformly in all directions.)

26P. Prove that, for any point in an electromagnetic wave such as that of Fig. 6, the time-averaged density of the energy stored in the electric field equals that of the energy stored in the magnetic field.

27P. Sunlight strikes the earth, outside its atmosphere, with an intensity of 1.4 kW/m². Calculate E_m and B_m for sunlight, assuming it to be a plane wave.

28P. The maximum electric field at a distance of 10 m from a point light source is 2.0 V/m. Calculate (*a*) the maximum value of the magnetic field, (*b*) the average intensity, and (*c*) the power output of the source.

29P. A cube of edge a has its edges parallel to the x, y, and z axes of a rectangular coordinate system. A uniform electric field **E** is parallel to the y axis and a uniform magnetic field **B** is parallel to the x axis. Calculate (*a*) the rate at which, according to the Poynting vector point of view, energy may be said to pass through each face of the cube and (*b*) the net rate at which the energy stored in the cube may be said to change.

30P. Frank D. Drake, an active investigator in the SETI (Search for Extra-Terrestrial Intelligence) program, has said that the large radio telescope in Arecibo, Puerto Rico, ". . . can detect a signal which lays down on the entire surface of the earth a power of only one picowatt." See Fig. 20. (*a*) What is the power actually received by the Arecibo antenna for such a signal? The antenna diameter is 1000 ft. (*b*) What would be the power output of a source at the center of our galaxy that could provide such a signal? The galactic center is 2.2×10^4 ly away. Take the source as radiating uniformly in all directions.

Figure 20 Problem 30.

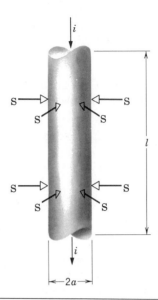

Figure 21 Problem 35.

31P. A HeNe laser, radiating at 632.8 nm, has a power output of 3.0 mW and a full-angle beam divergence (see Exercise 20) of 0.17 mrad. (*a*) What is the intensity of the beam 40 m from the laser? (*b*) What would be the power output of an isotropic source that provides this same intensity at the same distance?

32P. An airplane flying at a distance of 10 km from a radio transmitter receives a signal of power 10 μW/m². Calculate (*a*) the amplitude of the electric field at the airplane due to this signal; (*b*) the amplitude of the magnetic field at the airplane; (*c*) the total power radiated by the transmitter, assuming the transmitter to radiate uniformly in all directions.

33P. During a test, a NATO surveillance radar system, operating at 12 GHz with 180 kW of output power, attempts to detect an incoming "enemy" aircraft at 90 km. The target aircraft is designed to have a very small effective area for reflection of radar waves of 0.22 m². Assume that the radar beam spreads out isotropically into the forward hemisphere both upon transmission and reflection and ignore absorption in the atmosphere. For the reflected beam as received back at the radar site, calculate (*a*) the intensity, (*b*) maximum value of the electric field vector, and (*c*) the rms value of the magnetic field.

34P. A copper wire (diameter, 2.5 mm; resistance, 1.0 Ω per 300 m) carries a current of 25 A. Calculate (*a*) E, (*b*) B, and (*c*) S for a point on the surface of the wire.

35P. Figure 21 shows a cylindrical resistor of length *l*, radius *a*, and resistivity ρ, carrying a current *i*. (*a*) Show that the Poynting vector S at the surface of the resistor is everywhere directed normal to the surface, as shown. (*b*) Show that the rate *P* at which energy flows into the resistor through its cylindrical surface, calculated by integrating the Poynting vector over this surface, is equal to the rate at which thermal energy is produced; that is,

$$\int \mathbf{S} \cdot d\mathbf{A} = i^2 R,$$

where *d*A is an element of area of the cylindrical surface. This suggests that, according to the Poynting vector point of view, the energy that appears in a resistor as thermal energy does not enter it through the connecting wires but through the space around the wires and the resistor.

Section 38–6 Radiation Pressure

36E. Show (*a*) that the force *F* exerted by a laser beam of intensity *I* on a perfectly reflecting object of area *A* normal to the beam is given by $F = 2IA/c$, and (*b*) that the pressure $P = 2I/c$.

37E. High-power lasers are used to compress gas plasmas by radiation pressure. The reflectivity of a plasma is unity if the electron density is high enough. A laser generating pulses of radiation of peak power 1.5×10^3 MW is focused onto 1.0 mm² of high-electron-density plasma. Find the pressure exerted on the plasma.

38E. The average intensity of the solar radiation that falls normally on a surface just outside the earth's atmosphere is 1.4 kW/m². (*a*) What radiation pressure is exerted on this surface, assuming complete absorption? (*b*) How does this pressure compare with the earth's sea-level atmospheric pressure, which is 1.0×10^5 N/m²?

39E. Radiation from the sun striking the earth has an intensity of 1.4 kW/m². (*a*) Assuming that the earth behaves like a flat disk at right angles to the sun's rays and that all the incident energy is absorbed, calculate the force on the earth due to radiation pressure. (*b*) Compare it with the force due to the sun's gravitational attraction.

40E. What is the radiation pressure 1.5 m away from a

500-W light bulb? Assume that the surface on which the pressure is exerted faces the bulb and is perfectly absorbing and that the bulb radiates uniformly in all directions.

41P. A plane electromagnetic wave, with wavelength 3.0 m, travels in free space in the $+x$ direction with its electric vector **E**, of amplitude 300 V/m, directed along the y axis. (*a*) What is the frequency v of the wave? (*b*) What is the direction and amplitude of the magnetic field associated with the wave? (*c*) If $E = E_m \sin(kx - \omega t)$, what are the values of k and ω? (*d*) What is the time-averaged rate of energy flow in W/m² associated with this wave? (*e*) If the wave falls upon a perfectly absorbing sheet of area 2.0 m², at what rate would momentum be delivered to the sheet and what is the radiation pressure exerted on the sheet?

42P. A helium-neon laser of the type often found in physics laboratories has a beam power output of 5.0 mW at a wavelength of 633 nm. The beam is focused by a lens to a circular spot whose effective diameter may be taken to be 2.0 wavelengths. Calculate (*a*) the intensity of the focused beam, (*b*) the radiation pressure exerted on a tiny perfectly-absorbing sphere whose diameter is that of the focal spot, (*c*) the force exerted on this sphere, and (*d*) the acceleration imparted to it. Assume a sphere density of 5.0×10^3 kg/m³.

43P. A laser has a power output of 4.6 W and a beam diameter of 2.6 mm. If it is aimed vertically upward, what is the height H of a perfectly reflecting cylinder that can be made to "hover" by the radiation pressure exerted by the beam? Assume that the density of the cylinder is 1.2 g/cm³. See Fig. 22.

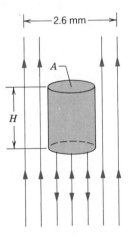

Figure 22 Problem 43.

44P. Radiation of intensity I is normally incident on an object that absorbs a fraction f of it and reflects the rest. What is the radiation pressure?

45P. Prove, for a plane wave at normal incidence on a plane surface, that the radiation pressure on the surface is equal to the energy density in the beam outside the surface. This relation holds no matter what fraction of the incident energy is reflected.

46P. Prove, for a stream of bullets striking a plane surface at normal incidence, that the "pressure" is twice the kinetic energy density in the stream above the surface; assume that the bullets are completely absorbed by the surface. Constrast this with the behavior of light; see Problem 45.

47P. A small spaceship whose mass, with occupant, is 1.5×10^3 kg is drifting in outer space, where no gravitational field exists. If the astronaut turns on a 10-kW laser beam, what speed would the ship attain in one day because of the reaction force associated with the momentum carried away by the beam?

48P. It has been proposed that a spaceship might be propelled in the solar system by radiation pressure, using a large sail made of foil. How large must the sail be if the radiation force is to be equal in magnitude to the sun's gravitational attraction? Assume that the mass of the ship + sail is 1500 kg, that the sail is perfectly reflecting, and that the sail is oriented at right angles to the sun's rays. See Appendix C for needed data.

49P. A particle in the solar system is under the combined influence of the sun's gravitational attraction and the radiation force due to the sun's rays. Assume that the particle is a sphere of density 1.0×10^3 kg/m³ and that all of the incident light is absorbed. (*a*) Show that all particles with radius less than some critical radius, R_0, will be blown out of the solar system. (*b*) Calculate R_0. Note that R_0 does not depend on the distance from the particle to the sun.

Section 38–7 Polarization

50E. The magnetic field equations for an electromagnetic wave in free space are $B_x = B \sin(ky + \omega t)$, $B_y = B_z = 0$. (*a*) What is the direction of propagation? (*b*) Write the electric field equations. (*c*) Is the wave polarized? If so, in what direction?

51E. A beam of unpolarized light of intensity 0.01 W/m² falls at normal incidence upon a polarizing sheet. (*a*) Find the maximum value of the electric field of the transmitted beam. (*b*) What is the radiation pressure exerted on the polarizing sheet?

52E. Unpolarized light falls on two polarizing sheets placed one on top of the other. What must be the angle between the characteristic directions of the sheets if the intensity of the transmitted light is one-third the intensity of the incident beam? Assume that each polarizing sheet is ideal, that is, that it reduces the intensity of unpolarized light by exactly 50%.

53E. Three polarizing plates are stacked. The first and third are crossed; the one between has its axis at 45° to the axes of the other two. What fraction of the intensity of an incident unpolarized beam is transmitted by the stack?

54P. An unpolarized beam of light is incident on a group of four polarizing sheets that are lined up so that the characteristic direction of each is rotated by 30° clockwise with respect to the preceding sheet. What fraction of the incident intensity is transmitted?

55P. A beam of polarized light strikes two polarizing sheets. The characteristic direction of the second is 90° with respect to the incident light. The characteristic direction of the first is at angle θ with respect to the incident light. Find angle θ for a transmitted beam intensity that is 0.10 times the incident beam intensity.

56P. A beam of light is plane polarized in the vertical direction. The beam falls at normal incidence on a polarizing sheet with its polarizing direction at 70° to the vertical. The transmitted beam falls, also at normal incidence, on a second polarizing sheet with its polarizing direction horizontal. If the intensity of the original beam is 43 W/m², what is the intensity of the beam transmitted by the second sheet?

57P. Suppose that in Problem 56 the incident beam was unpolarized. What now is the intensity of the beam transmitted by the second sheet?

58P. A beam of light is a mixture of polarized light and unpolarized light. When it is sent through a Polaroid sheet, we find that the transmitted intensity can be varied by a factor of five depending on the orientation of the Polaroid. Find the relative intensities of these two components of the incident beam.

59P. It is desired to rotate the plane of vibration of a beam of polarized light by 90°. (a) How might this be done using only Polaroid sheets? (b) How many sheets are required in order for the total intensity loss to be less than 40%?

60P. At a beach the light is generally partially polarized. At a particular beach on a particular day near sundown the horizontal component of the electric field vector is 2.3 times the vertical component. A standing sunbather puts on polaroid sunglasses; the glasses suppress the horizontal field component. (a) What fraction of the light energy received before the glasses were put on now reaches the eyes? (b) The sunbather, still wearing the glasses, lies on his side. What fraction of the light energy received before the glasses were put on reaches the eyes now?

ESSAY 15
PHYSICS AND TOYS

RAYMOND C. TURNER
CLEMSON UNIVERSITY

Figure 1 Photograph of the flicker light.

The basic principles of physics can often be demonstrated with ordinary toys. By understanding how these toys work, you can better understand the world around you.

Toys are often used to illustrate various physical principles of mechanics. You are probably already aware that the motion of *weebles* that wobble but don't fall down and other balancing toys can be explained using the concepts of center of mass, torque, and equilibrium.[1] Similarly, a *water rocket* can be used to demonstrate conservation of momentum, while a *hot wheels* race car and track can be analyzed by using the law of conservation of energy.[2] This article will be limited, however, to discussing several toys that deal with magnetism and light.

You have probably seen a *flicker light* in a toy store or a gift shop. What makes its filament vibrate? An example of such a lamp is shown in Fig. 1. Its operation can be analyzed by considering the Lorentz magnetic force on a current element, $d\mathbf{F} = i\,d\mathbf{s} \times \mathbf{B}$. This type of lamp has a small permanent magnet mounted near the lamp's filament. When there is a current in the filament, the magnetic field of the magnet exerts a force on the filament, pushing it in a direction perpendicular to both the field and the current. Since the lamp is operated from the usual household 60 Hz ac line, the current changes direction 120 times each second, and this changes the direction of the force at the same rate. This causes the filament to vibrate back and forth, and the light to appear to flicker. But how can this explanation be verified? Usually the magnet in a flicker light is mounted inside the glass bulb, but the particular lamp shown in the figure has its magnet attached to the ring on the outside of the bulb. This lets you move the magnet relative to the filament, thus changing the magnetic field. The farther you move the magnet from the lamp, the smaller the magnetic field at the filament, and the smaller the force on the filament. You can further verify this explanation by operating the lamp from a dc voltage source. In this case there is a force in only one direction, and the filament is simply pushed one way instead of vibrating.

A very clever magnetic toy is the *seal and ball,* which is shown in Fig. 2.[3] When the toy is placed on a flat table and the seal is pushed near the ball, the ball begins to rotate. As long as the seal is kept near the ball, the ball continues to rotate. What is clever about the toy is that this is done with just two permanent magnets, one in the seal and one in the ball. You can understand the operation of the toy by considering the forces and torques that the magnets exert on each other. The magnet in the ball is cylindrical in shape with its magnetic moment μ along the cylinder axis. The magnet is positioned below the center of the ball, so that the ball's center of mass is on its vertical axis, but below its geometric center, and its magnetic moment is vertical. The magnet in the seal has its magnetic moment at an angle of about 45° to the vertical. By separately considering the horizontal and vertical components of the seal's magnetic moment, you can understand the motion of the ball. The horizontal component

Figure 2 Photograph of the seal and ball.

sets up a horizontal magnetic field **B** at the ball. This will exert a torque $\tau = \mu \times \mathbf{B}$ on the ball, which will tip the ball slightly as indicated in Fig. 3. This tipping will move the center of mass of the ball slightly off the vertical axis (exaggerated in the figure). The vertical component of the seal's magnetic moment is essentially parallel to the ball's magnetic moment. This is the same as having two magnets side by side with their north poles together and their south poles together, which gives a repulsive force **F** between the magnets. This means that the ball will be pushed in a direction away from the seal. This magnetic force will exert a torque about the vertical axis, which, coupled with the frictional force of the table on the ball, will cause the ball to rotate. In short, the horizontal component of the seal's magnetic moment exerts a magnetic torque on the ball that causes the ball to tip slightly about a horizontal axis, and the magnetic repulsive force then exerts a torque causing the ball to rotate about the vertical axis. You can now also understand the source of energy that is necessary if the ball is to gain kinetic energy. The seal exerts a repulsive force on the ball, so the ball must in turn exert a repulsive force on the seal. In order for you to push the seal toward the ball you must do work against this repulsive force. This work done by you then increases the kinetic energy of the ball. By understanding magnetic forces and torques you can understand the operation of this clever toy.

A group of toys often seen in gift shops are shown in Fig. 4. These are the *rolling dolphin,* the *space circle* and the *mystery top.*[4] All of these are dynamic toys with one feature in common; they all keep moving. The dolphins roll, the circle swings, and the top spins. How do they work? Why do they keep moving? They all operate on the

Figure 3 View of the ball facing the seal. The rectangle is the side view of the flat cylindrical magnet. (*a*) The horizontal component **B** of the magnetic field from the seal exerts a clockwise torque on the magnetic moment **μ**, causing the ball to tip about point *P*. (*b*) The component of the magnetic field of the seal that is parallel to **μ** will exert a repulsive force **F**. This will cause the ball to rotate about the vertical axis.

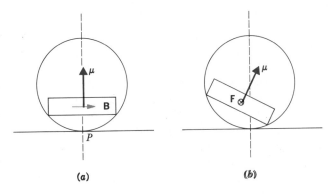

(*a*)   (*b*)

Figure 4 Photograph of the space circle, mystery top, and rolling dolphin.

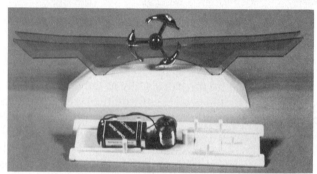

Figure 5 Photograph of the rolling dolphin open.

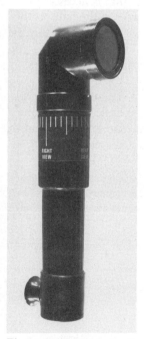

Figure 6 Photograph of the rotating periscope.

same principle, and you can get a clue to this mechanism by opening the base of one and looking inside. The *rolling dolphin* with its base open is shown in Fig. 5. You find that there is a coil of wire (actually, two concentric coils), a transistor, and a battery. In addition, you find that each dolphin has a small magnet in it. The battery provides the energy to keep the dolphins rolling, but you must understand Faraday's law and Lenz's law in order to understand the operation of the toy. As the magnet in one of the dolphins passes near the coil, it induces a voltage in the coil as predicted by Faraday's law. This voltage is applied to a transistor circuit in such a way as to turn on a second circuit consisting of the other coil and the battery. This second coil is wound in the opposite direction from the first. According to Lenz's law, the voltage induced in the first coil is such as to oppose the motion of the dolphin. The current in the second coil is then in a direction to enhance the motion. The dolphin is given a small pull if it is coming in toward the coil, or a small push if it is moving away; and it doesn't matter whether the dolphin is coming in from the right or the left. While the other two toys shown in the figure look quite different from the *rolling dolphin,* they operate in exactly the same way. Each toy contains a moving magnet, which, as predicted by Faraday's law, induces a voltage in a coil, and this voltage turns on a second circuit. The current in this second circuit is such as to increase the speed of the moving magnet.

A toy periscope can be used to demonstrate reflection of light and image formation with flat mirrors. A particularly interesting version is a *rotating periscope* shown in Fig. 6. The top of the periscope can be rotated so that you can use it to look behind you. But when you do this, you find that the image is inverted. Why is this if the image is normal when you are looking straight ahead? Examination of the periscope shows that it contains two mirrors each at a 45° angle to the vertical. Light from an object will reflect down from the top mirror, and then out from the bottom mirror. An

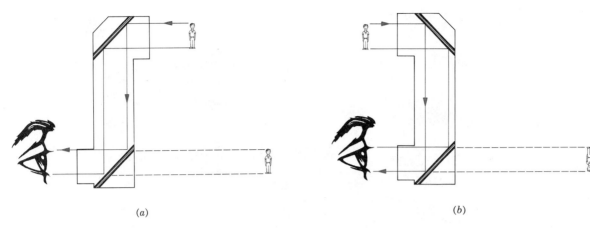

(a)   (b)

Figure 7 Image formation with the periscope. (*a*) The image is upright when the periscope is in its normal position. (*b*) The image is inverted when one periscope mirror is rotated 180°.

upright image will normally be observed, as sketched in Fig. 7*a*. If the top mirror is reversed, however, then the reflected light rays form an inverted image as shown in Fig. 7*b*. By using the law of reflection and tracing light rays, you can readily see why the image is upright in a normal periscope, but is inverted when one mirror is turned 180°. You can ask another question. What would you observe if the top mirror is rotated 90° so that you are looking to the side? You decide the answer to that question.

The *radiometer* shown in Fig. 8 has been sold in novelty shops for many years, but its operation is probably a mystery to most people.[5] It consists of four vanes, black on one side and white on the other, that are free to rotate about a vertical axis inside a glass housing. If the radiometer is placed in a beam of sunlight, or near a reasonably bright light bulb, the vanes begin to rotate rapidly. A possible cause of the rotation is radiation pressure from the light. Assume that the light is incident normal to the surface of the vanes as sketched in Fig. 9. The light striking the black side of the vane would be absorbed, transferring the light's total momentum to the vane; light incident on the white side of the vane would be reflected, reversing the light's momentum and transferring twice as much momentum to the vane. The net momentum transfer would cause the vanes to rotate, with the black side of the vanes leading and the white side following. But if you observe the rotation, you will see that the white side of the vanes is always leading. The problem is that the bulb does not have a good

Figure 8 Photograph of the radiometer.

Figure 9 Drawing of light incident on black and white radiometer vanes.

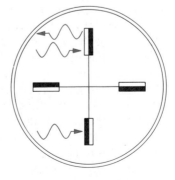

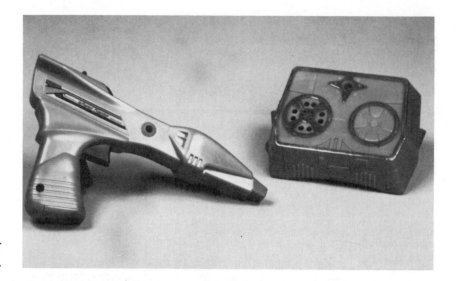

Figure 10 Photograph of the "laser gun" and target.

vacuum in it, and the air in the bulb plays an important role in the motion. Since the black side of the vane absorbs the light's energy, while the white side reflects it, the black side of the vane becomes hotter than the white side. This temperature difference causes the vanes to move due both to a circulation of the air in the bulb and to a momentum transfer. Air molecules that come in contact with the vane will leave the vane with an energy that is dependent on the vane's temperature. Molecules leaving the hotter black vane will on the average have a larger energy and thus a larger momentum than those leaving the cooler white vane. The recoil momentum of the vane will be larger for the hotter black side than for the cooler white side. As a result, the vane will gain a net momentum, rotating in such a way that the white side is always leading. This is still a momentum transfer that helps cause the vanes to rotate, but it is not the momentum of the light that does it.

A relatively new toy that is popular now is sometimes called *laser tag.* As shown in Fig. 10, it consists of a "laser gun" that emits an invisible beam of electromagnetic radiation and a sensor that detects the beam and indicates if the wearer has been "tagged." What are the properties of this electromagnetic radiation? There are any number of questions that you might ask about the radiation, and then answer by experimentation. Does the beam travel in straight lines, or will it bend around corners? Can the beam be reflected? If it can, what materials are reflecting, and does it obey the law of reflection? Can it be refracted and if it can, can it be focused with a lens? What is the maximum distance at which the beam can be detected, and can this range be extended by using a lens? Are there materials that will transmit this radiation, but not visible light; that is, can the beam go through some kinds of walls? Are there materials that transmit visible light, but not this radiation, so you can build a transparent shield? You can undoubtedly think of many more questions that can be asked, and then can be investigated experimentally. You may be able to guess the answers to some of the above questions by knowing that the beam of electromagnetic radiation from the "laser gun" is, in fact, infrared light.

These are only a few of the many toys that can be used to illustrate basic physical principles. Only your imagination and ingenuity limit you in your application of the fundamental laws of physics to ordinary objects, even toys. Science can be fun!

REFERENCES [1] "Toys in Physics Teaching: Balancing Man," by Raymond C. Turner, *American Journal of Physics,* January 1987. [2] "Hot Wheels Physics," by Stanley J. Briggs, *The Physics Teacher,* May 1970. [3] "A Physics Toy: Magnetic Seal and Ball," by Raymond C. Turner, *The Physics Teacher,* December 1987. [4] "How Things Work: A Spinning Top," by H. Richard Crane, *The Physics Teacher,* February 1984. [5] "Running Crooke's Radiometer Backwards," by Frank S. Crawford, *American Journal of Physics,* November 1985.

CHAPTER 39

GEOMETRICAL OPTICS

We are so accustomed to mirrors that we forget what a marvelous thing it is to be able to form an image of ourselves in such a simple way. This three-dimensional phantom—rigidly coupled to our own movements—looks back at us from the solid space behind the mirror where we know that only darkness exists. The Hall of Mirrors at Glacier Park in Lucerne recaptures some of the magic.

39-1 Geometrical Optics

If you are attending an outdoor concert and somebody stands up in front of you, you can still hear the music but you can no longer see the stage. Why this difference in behavior between sound waves and light waves? We trace it to the fact that the wavelength of sound (about 1 m) is about the same size as the "obstacle" but the wavelength of light (about 500 nm or 5×10^{-7} m) is very much smaller.

This experience illustrates that, under some circumstance, waves behave to a good approximation as if they traveled in straight lines, being blocked by barriers and casting sharp shadows. We need only to agree not to place in the path of the wave any obstacle or aperture, such as a mirror, lens, slit, or baffle, unless its dimensions are much larger than the wavelength. For light waves this useful special case of wave behavior is called *geometrical optics;* it is the subject of this chapter.

39-2 Reflection and Refraction

Figure 1*a* shows a beam of light falling on a plane glass surface. Part of the light is reflected from the surface and part of it is transmitted into the glass. Note that the

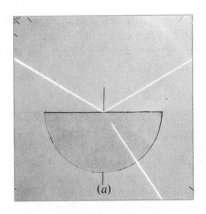

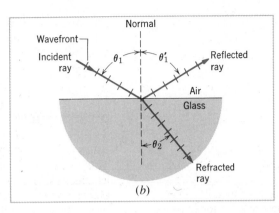

(a)

(b)

Figure 1 (*a*) A photograph showing the reflection and the refraction of an incident light beam as it falls on a plane glass surface. (*b*) A representation using rays. The angles of incidence (θ_1), of reflection (θ_1'), and of refraction (θ_2) are marked. Note that each angle is measured from the surface normal to the appropriate ray.

transmitted beam is bent or *refracted* as it passes through the surface.*

Let us define some useful quantities. In Fig. 1*b* we represent the incident beam and also the reflected and the refracted beams by *rays,* which are lines drawn at right angles to the wavefronts. The *angle of incidence* θ_1, the *angle of reflection* θ_1', and the *angle of refraction* θ_2 are also shown in the figure. Note that each of these angles is measured between the normal to the surface and the appropriate ray. The plane that contains both the incident ray and a line normal to the surface is called the *plane of incidence.* (In Fig. 1*b*, the plane of incidence is the plane of the printed page.)

Experiment shows that reflection and refraction are governed by these laws:

The law of reflection: The reflected ray lies in the plane of incidence and

$$\theta_1' = \theta_1 \quad \text{(reflection).} \tag{1}$$

The law of refraction: The refracted ray lies in the plane of incidence and

$$n_1 \sin \theta_1 = n_2 \sin \theta_2 \quad \text{(refraction).} \tag{2}$$

Here n_1 is a dimensionless constant called the *index of refraction* of medium 1 and n_2 is the index of refraction of medium 2. Eq. 2 is called Snell's law. As we shall see in Section 40-2, the index of refraction of a substance is

---

* *Refracted* comes from the Latin for broken, as in fractured leg. If you dip a pencil on a slant part way into a fishbowl, the pencil seems to be "broken."

c/v, where c is the speed of light in free space and v is its speed in the substance in question.

Table 1 shows the indices of refraction for some common substances; note that $n \approx 1$ for air. It is the high index of refraction of diamond ($n = 2.42$, versus ~ 1.5 for a typical glass) that accounts for its sparkle. As Fig. 2 shows, the index of refraction for a given substance varies with wavelength.

We give here two examples of clever applications of the laws of reflection and refraction, chosen from a very long list.

Reflection: A Neat Application.   A plane mirror will reflect a beam of light back toward its source if the mirror is positioned exactly right. A *corner reflector,* which is three mirrors fastened together as an inside corner of a cube, has the property that it will reflect a beam directly backward *no matter how the reflector is oriented.*

Luminous "cat's eyes," imbedded in the roadway, work on this principle. In the Physics Building at the University of Washington, three mirrors are mounted in a corner of the ceiling in a stairwell so that you can see

Table 1   Some Indices of Refraction[a]

Medium	Index	Medium	Index
Vacuum (exactly)	1.00000	Typical crown glass	1.52
Air (STP)	1.00029	Sodium chloride	1.54
Water (20°C)	1.33	Polystyrene	1.55
Acetone	1.36	Carbon disulfide	1.63
Ethyl alcohol	1.36	Heavy flint glass	1.65
Sugar solution (30%)	1.38	Sapphire	1.77
Fused quartz	1.46	Heaviest flint glass	1.89
Sugar solution (80%)	1.49	Diamond	2.42

[a] For a wavelength of 589 nm (yellow sodium light).

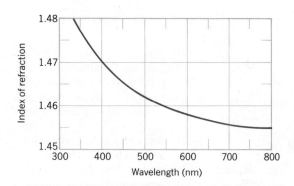

Figure 2 The index of refraction as a function of wavelength for fused quartz with respect to air. Shorter wavelengths, corresponding to higher indices of refraction, are bent through larger angles on entering quartz.

your own image from every step. Figure 3 shows a laser geodynamic satellite (LAGEOS) studded with 426 corner reflectors. Range measurements made by bouncing laser signals from it permit measurement of continental drift to within two centimeters. Finally, there is a corner reflector on the moon, from which we can bounce signals at will.

Refraction: A Neat Application. Investigators often want to trace glass fragments found at the scene of an accident so as to identify a hit-and-run motorist. The index of refraction of such fragments can be found by immersing a fragment in a transparent liquid whose index of refraction is particularly temperature-sensitive. The temperature of the liquid is then adjusted until the

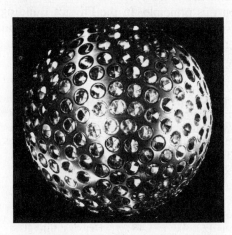

Figure 3 A LAser GEOdynamic Satellite (LAGEOS), launched into a 3600-mi high orbit in 1976. Corner reflectors permit precise location of its position by means of laser-pulse ranging. The satellite is 24 in. in diameter.

glass fragment disappears visually, a signal that there is no longer any refraction at the glass–liquid interface. The index of refraction of the glass is the same as the known index of the liquid at that temperature.

Sample Problem 1 An incident beam falls on the plane polished surface of a block of fused quartz, making an angle of 31.25° with the normal. This beam contains light of two wavelengths, 404.7 nm and 508.6 nm. The indices of refraction for quartz at these wavelengths are 1.4697 and 1.4619, respectively; the index of refraction for air may be taken as 1.0003 at both wavelengths. What is the angle between the two refracted rays?

From Eq. 2 we have, for the 404.7-nm ray,

$$n_1 \sin \theta_1 = n_2 \sin \theta_2$$

or

$$(1.0003) \sin 31.25° = (1.4697) \sin \theta_2,$$

which leads to

$$\theta_2 = \sin^{-1}\left(\sin 31.25° \, \frac{1.0003}{1.4697}\right) = 20.6761°.$$

In the same way we have, for the 508.6-nm ray,

$$\theta_2' = \sin^{-1}\left(\sin 31.25° \, \frac{1.0003}{1.4619}\right) = 20.7915°.$$

The angle $\Delta\theta$ between the rays is

$$\Delta\theta = 20.7915° - 20.6761°$$
$$= 0.1154° \approx 6.9 \text{ min of arc.} \quad \text{(Answer)}$$

The component with the shorter wavelength, having the larger index of refraction, has the smaller angle of refraction and is thus bent through the larger angle.

39–3 Total Internal Reflection

Figure 4 shows rays from a point source in glass falling on a glass–air interface. As the angle of incidence θ is increased, we reach a situation (see ray e) at which the refracted ray points along the surface, the angle of refraction being 90°. For angles of incidence larger than this *critical angle θ_c*, there is no refracted ray, and we speak of *total internal reflection*.

We find the critical angle by putting $\theta_2 = 90°$ in the law of refraction (see Eq. 2):

$$n_1 \sin \theta_c = n_2 \sin 90°,$$

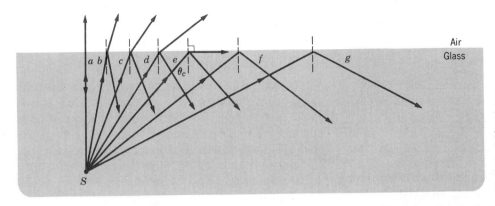

Figure 4 Total internal reflection of light from a point source S occurs for all angles of incidence greater than the critical angle θ_c. At the critical angle, the refracted ray points along the air–glass interface.

or

$$\theta_c = \sin^{-1} \frac{n_2}{n_1} \qquad \text{(critical angle)}. \qquad (3)$$

The sine of an angle cannot exceed unity so that we must have $n_2 < n_1$. This tells us that total internal reflection cannot occur when the incident light is in the medium of lower index of refraction. The word *total* means just that; the reflection occurs with no loss of intensity. In ordinary reflection from a mirror — by way of contrast — there is an intensity loss of about 4%.

Total internal reflection makes possible the *colonofiberscope* and other fiber optical devices by means of which physicians can visually inspect many internal body sites; see Fig. 5. In these devices, a bundle of fibers

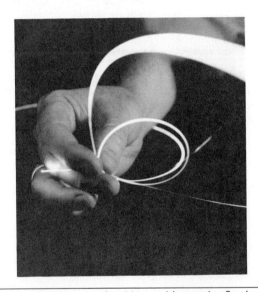

Figure 6 Light is transmitted by total internal reflection through an optical fiber of the kind used in lightwave communication systems.

transmits an image that can be inspected visually outside the body.

One rapidly growing application of total internal reflection is lightwave communication over an optical fiber network that now spans the country and will soon span the Atlantic Ocean; see Fig. 6. As Fig. 7 shows, the fiber consists of a central core that is graded smoothly into an outer cladding layer of a material of lower index of refraction. Only those rays that are internally reflected can be propagated along the cable. To reduce attenuation of the signal as it passes along the cable, materials of extreme purity have been developed. If sea water were as transparent as the glass from which optical fibers are made, it would be possible to see the sea bottom by reflected sunlight at a depth of several miles.

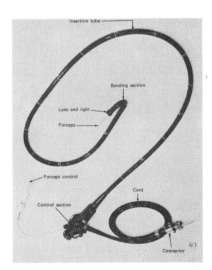

Figure 5 A *colonofiberscope,* used to examine the intestinal tract.

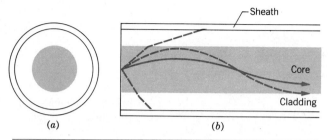

Figure 7 (a) An optical fiber shown in cross section. The fiber diameter is about that of a human hair. (b) A transverse view, showing propagation by total internal reflection. The glass core, the glass-graded cladding (of lower index than the core) and the protective sheath are shown.

Sample Problem 2 Figure 8 shows a triangular prism of glass in air, a ray incident normal to one face being totally reflected. If θ_1 is 45°, what can you say about the index of refraction of the glass?

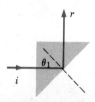

Figure 8 Sample Problem 2. The incident ray is totally internally reflected at the glass–air interface.

The angle θ_1 must be equal to or greater than the critical angle θ_c, where θ_c is given by Eq. 3:

$$\theta_c = \sin^{-1} \frac{n_2}{n_1} = \sin^{-1} \frac{1}{n}$$

in which, for all practical purposes, the index of refraction of air ($= n_2$) is set equal to unity. Since total internal reflection occurs, then θ_c must be less than 45° and

$$n > \frac{1}{\sin 45°} = 1.41. \qquad \text{(Answer)}$$

If the index of refraction of the glass is less than 1.41, total internal reflection will not occur.

39-4 Polarization by Reflection

If you rotate a polarizing sunglass lens in front of one eye, you can reduce or eliminate the glare from sunlight re-

flected from water or from any glossy surface. Such reflected light is fully or partially polarized by the process of reflection from the surface.

Figure 9 shows an unpolarized beam falling on a glass surface. The electric field vector for each wavetrain in the beam can be resolved into a *perpendicular component* (perpendicular to the plane of incidence), represented by dots in Fig. 9, and a *parallel component* (lying in the plane of incidence), represented by arrows. For the unpolarized incident light, these two components are of equal amplitude.

Experimentally, for glass or other dielectric materials, there is a particular angle of incidence, called the *Brewster angle* θ_B, at which the reflection for the parallel component is zero. This means that the beam reflected from the glass for this angle of incidence is completely polarized, with its plane of vibration at right angles to the plane of Fig. 9.

Since the parallel component of a ray incident at the Brewster angle is not reflected, it must be fully transmitted. Just as total internal reflection permits lossless reflection, the phenomenon of Fig. 9 permits lossless transmission (of the parallel component). In a gas laser, light must be reflected back and forth through the laser cavity by external mirrors, perhaps several hundred times. If the ends of the laser cavity are sealed by *Brewster windows* (that is, windows tilted at the Brewster angle) beam losses by reflection on successive passage through the windows can be prevented.

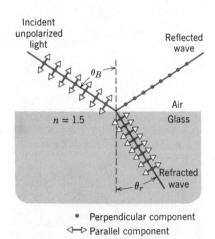

Figure 9 For a particular angle of incidence, called the *Brewster angle* θ_B, the parallel component of the incident ray is transmitted without loss. As a consequence, the reflected ray contains no parallel component and is thus completely polarized. The transmitted ray is partially polarized.

Brewster's Law. At the Brewster angle we find experimentally that the reflected and the refracted beams are at right angles, or, equivalently (Fig. 9),

$$\theta_B + \theta_r = 90°.$$

From the law of refraction we have

$$n_1 \sin \theta_B = n_2 \sin \theta_r.$$

Combining these equations leads to

$$n_1 \sin \theta_B = n_2 \sin (90° - \theta_B) = n_2 \cos \theta_B,$$

or

$$\theta_B = \tan^{-1} \frac{n_2}{n_1},$$

where the incident ray is in medium one and the refracted ray in medium two. We can write this as

$$\boxed{\theta_B = \tan^{-1} n} \qquad \text{(Brewster's law),} \qquad (4)$$

where $n(= n_2/n_1)$ is the index of refraction of medium two with respect to medium one. Equation 4 is known as *Brewster's law* after Sir David Brewster, who deduced it empirically in 1812.*

Sample Problem 3 We wish to use a plate of glass ($n = 1.57$) as a polarizer. (a) What is the Brewster angle? That is, at what angle of incidence will the reflected beam be fully polarized?
From Eq. 4

$$\theta_B = \tan^{-1} n = \tan^{-1} 1.57 = 57.5°. \qquad \text{(Answer)}$$

(b) What angle of refraction corresponds to this angle of incidence?
Since $\theta_B + \theta_r = 90°$ we have

$$\theta_r = 90° - \theta_B$$
$$= 90° - 57.5° = 32.5°. \qquad \text{(Answer)}$$

39-5   A Plane Mirror

Perhaps our simplest optical experience is looking in a mirror. Figure 10 shows a point source of light O, which we will call the *object,* placed at a distance o in front of a plane mirror. The light that falls on the mirror is represented by rays emanating from O. At the point in Fig. 10

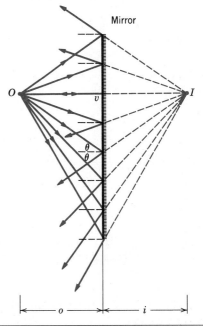

Figure 10 A point object forms a virtual image in a plane mirror. The rays *appear* to diverge from I but actually no light is present at this point.

at which each ray from O strikes the mirror, we construct a reflected ray. If we extended the reflected rays backward, they intersect in a point I that we call the *image* of the object O. The image is as far behind the mirror as the object is in front of it. This fact can easily be brought home if you photograph yourself in a mirror with a sonar autofocus camera; the sound waves will focus the camera on the mirror surface and your image, which is well behind that surface, will be out of focus.

When we look into the mirror, the reflected rays really seem to be coming from the image point, although we know that they do not. We call such an image a *virtual* image, meaning that light does not actually pass though it. We know from daily experience how "real" such a virtual image appears to be and how definite is its location in the space behind the mirror, even though this space may, in fact, be occupied by a brick wall.

Figure 11 shows two rays selected from the bundle of rays in Fig. 10. One strikes the mirror at v, along a perpendicular line. The other strikes it at an arbitrary point a, making an angle of incidence θ with the normal at that point. Elementary geometry shows that the right triangles $aOva$ and $aIva$ are congruent and thus

$$\boxed{i = -o} \qquad \text{(plane mirror).} \qquad (5)$$

---

* Brewster was also the inventor of the *kaleidoscope,* a word from the Greek meaning: "seeing beautiful shapes." See Essay 16.

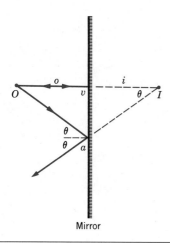

Figure 11 Two rays from Fig. 10. Ray *Oa* makes an arbitrary angle θ with the normal to the mirror surface.

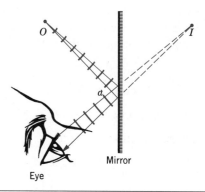

Figure 12 A pencil of rays from *O* enters the eye after reflection at the mirror. Only a small portion of the mirror near *a* is effective. The small arcs represent portions of spherical wavefronts. The eye "thinks" that the light is coming from *I*.

We introduce the minus sign to signal that the image is virtual, a point we will expand upon below.

Only rays that lie fairly close together can enter the eye after reflection at a mirror. For the eye position shown in Fig. 12, only a small patch of the mirror near point *a* (a patch smaller than the pupil of our eye) is used in forming the image. You might experiment with a mirror, closing one eye and looking at the image of a small object such as the tip of a pencil. Then move your fingertip over the mirror surface until you cannot see the image. Only that small portion of the mirror under your fingertip was used to bring the image to you.

Image Reversal. As Fig. 13*a* shows, the image of a left hand is a right hand, and it is often said that a mirror reverses right and left. Perceptive students then often ask: Why then does the mirror not also reverse up and down?

Figure 13*b* sheds some light on the matter. Of the three arrows that form the object, the images of the two that lie in a plane parallel to the mirror are just like their objects. The image of the arrow that points toward the mirror, however, is reversed front to back. Thus, it is more accurate to say that a mirror reverses front to back than that it reverses right to left. The transformation of a left hand into a right hand is accomplished, in a sense, by transferring the front of the hand to the back of the hand directly through the hand.

Sample Problem 4 Kareem Abdul-Jabbar is 7'2" (= 218 cm) tall. How tall must a vertical mirror be if he is to be able to see his entire length?

In Fig. 14, the top of Kareem's head (*h*), his eyes (*e*), and

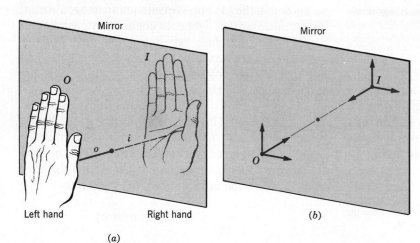

(*a*)

Figure 13 (*a*) The object *O* is a left hand; its image *I* is a right hand. (*b*) Study of a reflected three-arrow object shows that a mirror interchanges front and back, rather than right and left.

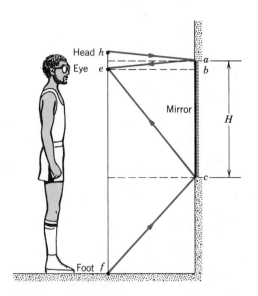

Figure 14   Sample Problem 4. A "full-length mirror" need only be half your height.

the bottoms of his feet (f) are marked by dots. The figure shows the paths followed by rays that leave his head and his feet and enter his eyes, reflecting from the mirror at points a and c, respectively. The mirror need occupy only the vertical distance H between those points.

We see that

$$ab = \tfrac{1}{2}he \quad \text{and} \quad bc = \tfrac{1}{2}ef.$$

Thus,

$$H = ac = ab + bc = \tfrac{1}{2}(he + ef)$$
$$= (\tfrac{1}{2})(218 \text{ cm}) = 109 \text{ cm}. \qquad \text{(Answer)}$$

The mirror need be no taller than half the height of the person. If you have a full-length mirror available, you might experiment by taping newspaper over those portions of the mirror that do not contribute to your image. You will find that what you have left is just half of your height. Mirrors that extend below point c just allow you to look at an image of the floor. Your horizontal distance from the mirror makes no difference.

39-6   Spherical Mirrors

In the last section we discussed image formation in a plane mirror. Now we will see what happens to the image if the surface of the mirror is curved. In particular, we will consider a spherical mirror, which is simply a small section of the surface of a sphere. A plane mirror is in fact a spherical mirror with an infinitely large radius of curvature.

Let us start with a plane mirror (Fig. 15a) and make it *concave* ("caved in") toward the observer but with r, its *radius of curvature,* still very large, as in Fig. 15b. Note that C, its *center of curvature,* is on the left in Fig. 15b. Compared with a plane mirror, two things happen. First, the image moves farther behind the mirror (that is, i takes on a larger negative value). Second, the size of the image increases. For a plane mirror, the image is exactly the same size as the object. For this curved mirror, the image is *larger* than the object. This is the principle of the makeup mirror or the shaving mirror, which are slightly concave and magnify the face. The relatively narrow angle of divergence of the reflected rays in Fig. 15b suggests that such a mirror possesses a narrower field of view than does a plane mirror; the image of your face is bigger but you cannot see as much of it.

If we curve the plane mirror to make it *convex* toward the observer, as in Fig. 15c, the image moves closer to the mirror and shrinks. Such convex mirrors are used as right-hand side-view mirrors in automobiles and as surveillance mirrors in buses and supermarkets. The wide angle of divergence of the reflected rays in Fig. 15c suggests that such a mirror possesses a wider field of view than does a plane mirror; the driver can see the whole interior of the bus. Note that the images are virtual for all three cases shown in Fig. 15.

As Fig. 16 shows, we can also use a spherical mirror to bring incident light to a focus. For a concave mirror (Fig. 16a) the focus is *real;* you can ignite a piece of paper by focusing sunlight on it in this way. Reflecting telescopes also use this principle to bring distant starlight to a focus. For a convex mirror (Fig. 16b) the focus is *virtual.* The point F in each of these figures is called the *focal point* and the distance f is the *focal length.* As we shall prove in Section 39-11, there is a simple relation between the distance o of the object from the mirror, the distance i of the image from the mirror, and the focal length f of the mirror. It is

$$\boxed{\frac{1}{o} + \frac{1}{i} = \frac{1}{f}} \qquad \text{(spherical mirror),} \qquad (6)$$

in which f is related to r, the radius of curvature of the mirror, by

$$\boxed{f = \tfrac{1}{2}r} \qquad \text{(spherical mirror).} \qquad (7)$$

We can combine these relations and write the mirror equation also in the form

$$\boxed{\frac{1}{o} + \frac{1}{i} = \frac{2}{r}} \qquad \text{(spherical mirror).} \qquad (8)$$

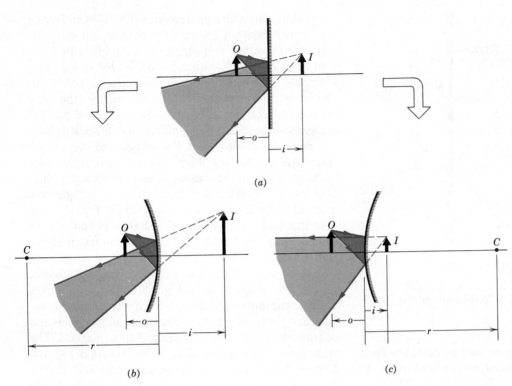

(a)

(b)

(c)

Figure 15 (a) An object forms a virtual image in a plane mirror. (b) If the mirror is bent so that it becomes *concave*, the image moves farther away and becomes larger. (c) If the plane mirror is bent so that it becomes *convex*, the image moves closer and becomes smaller.

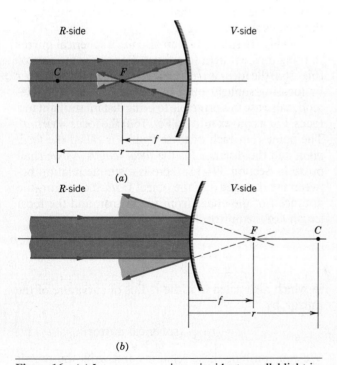

(a)

(b)

Figure 16 (a) In a concave mirror, incident parallel light is brought to a real focus at F, on the R-side of the mirror. (b) In a convex mirror, incident parallel light is made to seem to diverge from a virtual focus at F, on the V-side of the mirror.

We can test Eq. 8 by letting r become infinitely large, thus describing a plane mirror. If we do so, Eq. 8 reduces to $i = -o$, which is exactly the relation (see Eq. 5) that we gave earlier for such a mirror. We can test Eq. 6 by letting o become infinitely large, thus describing parallel incident light from an infinitely distant object as in Fig. 16. Equation 6 then reduces to $i = f$, showing that the image does indeed appear at the focal point.

In Fig. 17 the three figures show what happens as the object is moved from a point close to a concave mirror (Fig. 17a) to the focal point (Fig. 17b), and finally to a position beyond the focal point (Fig. 17c). Note that the image is *virtual* in the first figure but *real* in the other two. The image is *upright* in the first two figures but *inverted* in the third. In studying this figure it is well to follow the image position as the object moves in a continuous fashion from a point close to the mirror out to infinity. (Where is the image when the object is at the center of curvature?)

We shall show in Section 39–7 that the *lateral magnification m* of a mirror is given by

$$m = -\frac{i}{o}$$ (lateral magnification). (9)

For a plane mirror, for which $i = -o$ (see Eq. 5), we have $m = +1$. The magnification of 1 means that the image is the same size as the object. The plus sign, as we shall see, means that the image is *upright;* a minus sign would signify an *inverted* image.

Equations 6, 7, 8, and 9 hold for all mirrors, whether plane, concave, or convex. In using these equations, however, we must be careful about the signs of the quantities $o, i, r, f,$ and m that enter into them.

If we set it up properly, our rule for signs will apply not only to mirrors but also to spherical refracting surfaces and to lenses. We start by calling the front of the mirror—where only real images can be formed—the

R-side (*R* for real) of the mirror. Similarly, we call the back of the mirror—where only virtual images can be formed—the *V*-side (*V* for virtual) of the mirror. Our rule for signs is: Associate *positive* with *Real, R-side,* and up*Right;* associate *negative* with *Virtual, V-side,* and in*Verted.* In particular:

o is positive if the object is real,*
i is positive if the image is real (that is, if I is on the *R*-side),
r is positive if C is on the *R*-side,
f is positive if the focus is real (that is, if F is on the *R*-side),
m is positive if the image is up*Right*.

We will clarify the use of these rules in the sample problems.†

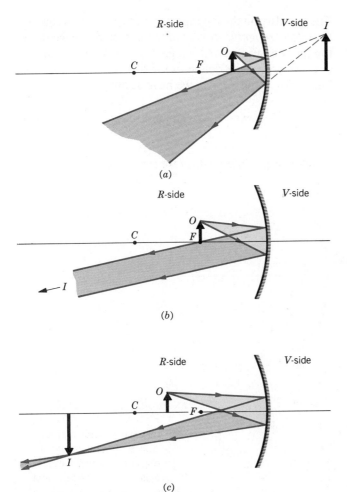

(a)

(b)

(c)

Figure 17 An object is moved at a constant slow speed from a position near the surface of a concave mirror out to infinity. How does its image move? This figure shows three stages of the motion.

Sample Problem 5 A convex mirror (see Fig. 15c) has a radius of curvature of 22 cm. (a) If an object is placed 14 cm away from the mirror, where is its image?

As Fig. 15c shows, the center of curvature for this mirror is on the *V*-side of the mirror. According to our rule for signs, the radius of curvature r is negative, as it is for all convex mirrors. We have then, from Eq. 8

$$\frac{1}{o} + \frac{1}{i} = \frac{2}{r}$$

or

$$\frac{1}{+14 \text{ cm}} + \frac{1}{i} = \frac{2}{-22 \text{ cm}},$$

which yields

$$i = -6.2 \text{ cm}, \qquad \text{(Answer)}$$

in agreement with the graphical prediction. The negative sign for i reminds us that the image is on the *V*-side of the mirror and thus is virtual.

(b) What is the magnification in this case?
From Eq. 9 we have

$$m = -\frac{i}{o} = -\frac{-6.2 \text{ cm}}{+14 \text{ cm}} = +0.44. \qquad \text{(Answer)}$$

---

* Although objects, like images, can be virtual, we consider here only the familiar case of real objects.
† You may wish to use Eq. 9 to check on the following limerick, composed by Robert G. Greenler of the Optical Society of America
> To a politician who would reveal
> An Image with public appeal
> Said his optical friend
> "In my view I contend
> If it's Upright it's Virtual, not Real!"

Thus the image is smaller than the object and, according to our rule for signs, it is upright. This is in full qualitative agreement with Fig. 15c.

39–7 Ray Tracing

Figures 18a and 18b show an object O in front of a concave mirror. We can locate an image of any off-axis point graphically by tracing any two of four special rays. Thus:

1. A ray that strikes the mirror parallel to its axis passes through the focal point (ray 1 in Fig. 18a).

2. A ray that strikes the mirror after passing through the focal point emerges parallel to the axis (ray 2 in Fig. 18a).

3. A ray that strikes the mirror after passing through the center of curvature C returns along itself (ray 3 in Fig. 18b).

4. A ray that strikes the vertex of the mirror will be reflected at an equal angle with the axis of the mirror (ray 4 in Fig. 18b).

The four rays are displayed on two separated plots to minimize clutter. Our descriptions of the rays need to be modified only slightly to apply to convex mirrors, as in Figs. 18c and 18d.

The Magnification. Here we wish to derive Eq. 9 ($m = -i/o$), the expression for the magnification of an object reflected in a mirror. Consider ray 4 in Fig. 18b. It is drawn to be reflected at the vertex v, making equal angles with the axis of the mirror at that point.

For the two similar right triangles in the figure we can write

$$\frac{Ib}{Oa} = \frac{vb}{va}.$$

The quantity on the left (apart from a question of sign) is the *lateral magnification m* of the mirror. Since we want to represent an *inverted* image by a *negative* magnification, we arbitrarily define m for this case as $-(Ib/Oa)$. Since $vb = i$ and $va = o$, we have at once

$$m = -\frac{i}{o} \quad \text{(magnification)}, \qquad (10)$$

which is Eq. 9, the relation we set out to prove.

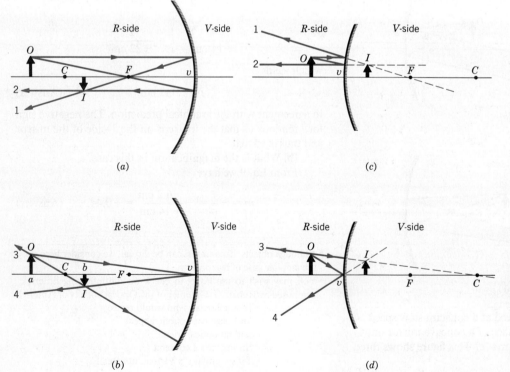

(a)

(c)

(b)

(d)

Figure 18 (a,b) Four rays that may be drawn graphically to find the image of an object in a concave mirror. Note that the image is real, inverted, and smaller than the object. (c,d) Four similar rays for the case of a convex mirror. Note that the image is virtual, upright, and smaller than the object.

39-8   Spherical Refracting Surfaces

In Fig. 19a light rays from a point object O fall on a convex spherical refracting surface of radius of curvature r. The surface separates two media, the index of refraction of the medium containing the incident light being n_1 and that on the other side of the surface being n_2. After refraction at the surface, the rays join to form a real image I.

As we shall prove in Section 39-11, the image distance i is related to the object distance o, the radius of curvature r, and the two indices of refraction, by

$$\boxed{\frac{n_1}{o} + \frac{n_2}{i} = \frac{n_2 - n_1}{r}}$$ (single surface).   (11)

This equation is quite general and holds whether the refracting surface is convex (Fig. 19a) or concave (Fig. 19b) and also for the case in which $n_2 < n_1$ (Fig. 19c).

The rule for signs for a single spherical refracting surface is the same as for spherical mirrors. There is a difference, however; for mirrors the R-side (where real images are formed) is the side toward which the incident light is *reflected*. For refracting surfaces, the R-side is the side toward which the incident light is *transmitted*. Figure 20 clarifies this distinction.

Sample Problem 6   Locate the image for the geometry shown in Fig. 19a, assuming the radius of curvature r to be 11 cm, $n_1 = 1$, and n_2 to be 1.9. Let the object be 19 cm to the

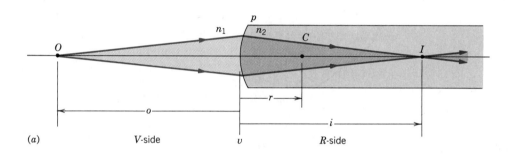

(a)   V-side   R-side

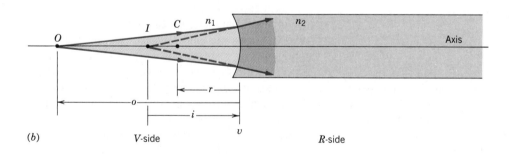

(b)   V-side   R-side

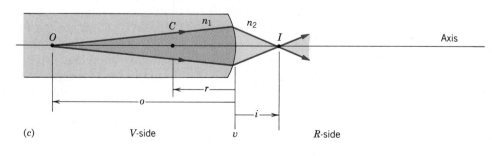

(c)   V-side   R-side

Figure 19   (a) A real image is formed by refraction at a convex spherical boundary between two media; in this case $n_2 > n_1$. (b) A virtual image is formed by refraction at a concave spherical boundary between two media; as in (a), $n_2 > n_1$. (c) The same as (b) except that $n_2 < n_1$.

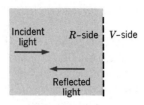

(spherical mirror)

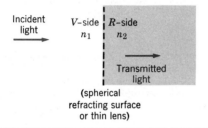

(spherical
refracting surface
or thin lens)

Figure 20 Real images are formed on the same side as the incident light for mirrors but on the opposite side for refracting surfaces and for lenses.

left of the vertex v.

From Eq. 11,

$$\frac{n_1}{o} + \frac{n_2}{i} = \frac{n_2 - n_1}{r},$$

we have

$$\frac{1}{+19 \text{ cm}} + \frac{1.9}{i} = \frac{1.9 - 1}{+11 \text{ cm}}.$$

Notice that r is positive because the center of curvature C of the surface in Fig. 19a lies on the R-side. If we solve the above equation for i, we find

$$i = 65 \text{ cm.} \qquad \text{(Answer)}$$

This result agrees with Fig. 19a and is consistent with the sign conventions. The light actually passes through the image point I so the image is real, as indicated by the positive sign that we found for i. Remember also that n_1 ($= 1$ in this case) always refers to the medium on the side of the surface from which the light comes.

39–9 Thin Lenses

In most refraction situations there is more than one refracting surface. This is true even for a spectacle lens, the light passing from air into glass and then from glass into air. We consider here only the special case of a *thin lens,* that is, a lens in which the thickness of the lens is small

compared to the object distance, the image distance, or either of the two radii of curvature. For such a lens—as we shall prove in Section 39–11—these quantities are related by

$$\frac{1}{o} + \frac{1}{i} = \frac{1}{f} \qquad \text{(thin lens)} \qquad (12)$$

in which the focal length f of the lens is given by

$$\frac{1}{f} = (n - 1)\left(\frac{1}{r_1} - \frac{1}{r_2}\right) \qquad \text{(thin lens)}. \qquad (13)$$

Note that Eq. 12 is the same equation that we used for spherical mirrors. Equation 13 is often called the *lens makers equation;* it relates the focal length of the lens to the index of refraction n of the lens material and the radii of curvature of the two surfaces.

In Eq. 13, r_1 is the radius of curvature of the lens surface on which the light first falls and r_2 is that of the second surface. If the lens is immersed in a medium for which the index of refraction is not unity, Eq. 13 still holds; simply replace n in that formula by $n_{\text{lens}}/n_{\text{medium}}$. The same rules for signs apply to lenses as to mirrors and spherical refracting surfaces.

A thin lens has *two* focal points, symmetrically placed on either side of the lens; we must be careful to distinguish between them.

Figure 21a shows the formation of a real inverted image in a converging lens, that is, in a lens that would cause incident parallel light to converge to a real focus. C_1 and C_2 are the centers of curvature of the first and the second surfaces, respectively. C_1 is on the R-side so that r_1 is positive but C_2 is on the V-side so that r_2 is negative. For such a (converging) lens you can show (using Eq. 13) that the focal length f is positive. A converging lens is thicker at the center than at the edges.

Figure 21b shows the formation of a virtual, upright image in a *diverging* lens, that is, in a lens that would cause incident parallel light to diverge from a virtual focus. C_1 and C_2 are the centers of curvature of the first and the second surfaces, respectively. C_1 is now on the V-side so that r_1 is negative and C_2 is now on the R-side so that r_2 is positive. For such a (diverging) lens you can show (using Eq. 13) that the focal length f is negative. A diverging lens is thinner at the center than at the edges.

Graded Index Lenses. All rays that pass through a lens from a real point object to a real point image must contain the same number of wavelengths, even though

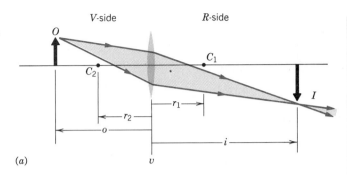

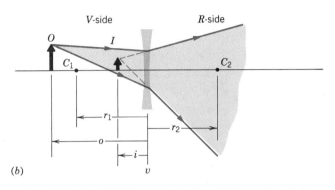

Figure 21 (a) A real, inverted image is formed by a converging lens. Such a lens has a positive focal length and is thicker in the center than at the edges. (b) A virtual, erect image is formed by a diverging lens. Such a lens has a negative focal length and is thinner at the center than at the edges.

their geometrical path lengths are necessarily different. In lens of the type we have been discussing, this is accomplished by causing the thickness of the lens to vary with radial distance from the central axis.

The same effect can be accomplished with a flat lens, if the index of refraction is caused to vary appropriately with axial distance. Figure 22 shows such a *graded index lens,* designed for use in a photo-copying machine.

Ray Tracing. We can locate the image of an off-axis point graphically as we show in Fig. 23 for a converging lens. With small changes, the construction can easily be adapted to diverging lenses. Thus:

1. A ray parallel to the axis and falling on the lens passes through the focal point F_2 (ray x in Fig. 23a).

2. A ray falling on a lens after passing through the focal point F_1 will emerge from the lens parallel to the axis (ray y).

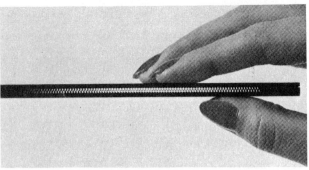

Figure 22 A side-view of a graded index lens. The lens is flat and focuses the light because its index of refraction (not its thickness) decreases in a controlled way with radial distance from the lens axis.

3. A ray falling on the lens at its center will pass through undeflected. There is no deflection because the lens, near its center, behaves like a thin piece of glass with parallel sides. The direction of the light rays is not changed and the sideways displacement can be neglected because the lens thickness has been assumed to be negligible (ray z).

Sample Problem 7 (a) The lens of Fig. 21a has radii of curvature of magnitude 42 cm and is made of glass with $n = 1.65$. Compute its focal length.

Since C_1 lies on the R-side of the lens in Fig. 21a, r_1 is positive ($= +42$ cm). Since C_2 lies on the V-side, r_2 is negative ($= -42$ cm). Substituting in Eq. 13 yields

$$\frac{1}{f} = (n-1)\left(\frac{1}{r_1} - \frac{1}{r_2}\right) = (1.65 - 1)\left(\frac{1}{+42\ \text{cm}} - \frac{1}{-42\ \text{cm}}\right)$$

or

$$f = +32\ \text{cm}. \qquad \text{(Answer)}$$

A positive focal length indicates that, in agreement with what we have said above, parallel incident light converges after refraction to form a real focus.

(b) Find the focal length for the lens of Fig. 21b, again assuming 42-cm radii and $n = 1.65$.

In Fig. 21b, C_1 lies on the V-side of the lens so that r_1 is negative ($= -42$ cm). Since r_2 is positive ($= +42$ cm), Eq. 13 yields

$$\frac{1}{f} = (n-1)\left(\frac{1}{r_1} - \frac{1}{r_2}\right) = (1.65 - 1)\left(\frac{1}{-42\ \text{cm}} - \frac{1}{+42\ \text{cm}}\right)$$

or

$$f = -32\ \text{cm}. \qquad \text{(Answer)}$$

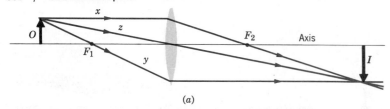

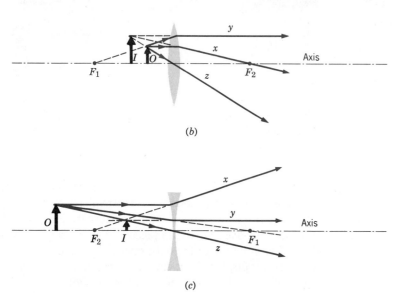

(a)

(b)

(c)

Figure 23 Three special rays allow us to locate an image formed by a thin lens.

A negative focal length indicates that, in agreement with what we have said above, incident light diverges after refraction to form a virtual image.

Sample Problem 8 A converging thin lens like that of Fig. 21a has a focal length of $+24$ cm. An object is placed 9.5 cm from the lens. Note that this is between the focal point and the lens. (a) Where is the image?

From Eq. 12

$$\frac{1}{o} + \frac{1}{i} = \frac{1}{f}$$

we have

$$\frac{1}{+9.5 \text{ cm}} + \frac{1}{i} = \frac{1}{+24 \text{ cm}},$$

which yields

$$i = -16 \text{ cm}. \qquad \text{(Answer)}$$

The minus sign means that the image is on the V-side of the lens and is thus virtual.

(b) What is the lateral magnification of the lens for this object?

The lateral magnification is given by Eq. 9, which holds for thin lenses as well as for mirrors:

$$m = -\frac{i}{o} = -\frac{-16 \text{ cm}}{+9.5 \text{ cm}} = +1.7. \qquad \text{(Answer)}$$

The plus sign signifies an upright image.

39-10 Optical Instruments

The human eye is a remarkably effective organ, but its range can be extended in many ways by optical instruments such as spectacles, simple magnifiers, motion picture projectors, cameras (including TV cameras), microscopes, and telescopes. In many cases these devices extend the scope of our vision beyond the visible range; satellite-borne infrared cameras and x-ray microscopes are examples.

In almost all cases of modern sophisticated optical instruments, the mirror and thin lens formulas hold only as approximations. In typical laboratory microscopes the lens can by no means be considered "thin." In most optical instruments lenses are compound, that is, they are made of several components, the interfaces rarely being exactly spherical. Figure 24, for example, shows the components of a typical zoom lens, commonly used in TV cameras to provide a 20-to-one range in focal lengths.

In what follows we describe three optical instruments, assuming, for simplicity of illustration only, that the thin lens formula applies.

Simple Magnifier. The normal human eye can focus a sharp image of an object on the retina if the object O is located anywhere from infinity to a certain point called

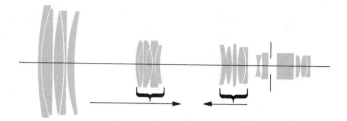

Figure 24   The components of a zoom lens in a TV camera. The central sections of the lens system move as shown. None of these lenses is "thin" and the paraxial approximation is not imposed. This is the real world of high-performance geometrical optics.

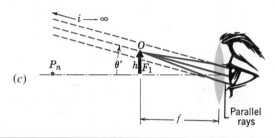

Figure 25   (*a*) An object of height *h* is placed at the near point of a human eye. (*b*) The object is moved closer but now the observer cannot bring it into focus. (*c*) A converging lens is placed close to the eye so that the rays from the object appear to come from an infinite distance and the eye can easily focus on them. A "magnifier" simply permits you to bring the object closer to your eye. Not to scale.

the *near point* P_n. If you move the object closer than the near point, the perceived retinal image becomes fuzzy. The location of the near point normally varies with age. We have all heard about people who claim not to need glasses but who read their newspapers at arm's length; their near points are receding! Find your own near point by moving this page closer to your eyes, considered separately, until you reach a position at which the image begins to become indistinct. In what follows, we take the near point to be 15 cm from the eye, a typical value for 20-year-olds.

Figure 25*a* shows an object *O* placed at the near point P_n. The size of the perceived image on the retina is determined by the angle θ. One way to make the object seem larger is to move it closer to your eye, as in Fig. 25*b*. The image on your retina is now larger, but the object is now so close that the eye cannot bring it into focus. We can give the eye some help by inserting a converging lens (of focal length f) just in front of the eye as in Fig. 25*c*. The eye now perceives an image at infinity, rays from which are comfortably focused by the eye.

The angle of the image rays is now θ', where $\theta' > \theta$. The *angular magnification* m_θ (not to be confused with the lateral magnification m) can be found from

$$m_\theta = \theta'/\theta$$

This is just the ratio of the sizes of the two images on the eye's retina. From Fig. 25,

$$\theta \approx h/15 \text{ cm} \quad \text{and} \quad \theta' \approx h/f,$$

so that

$$m_\theta \approx \frac{15 \text{ cm}}{f} \quad \text{(simple magnifier).} \quad (14)$$

Lens aberrations limit the angular magnifications for a single lens to something less than 10. This is enough, however, for stamp collectors and for actors portraying Sherlock Holmes.

Compound Microscope.   Figure 26 shows a thin lens version of a compound microscope, used for viewing small objects that are very close to the objective lens of the instrument. The object *O*, of height *h*, is placed just outside the first focal point F_1 of the objective lens, whose focal length is f_{ob}. A real, inverted image *I* of height h' is formed by the objective, the lateral magnification being given by Eq. 9, or

$$m = -\frac{h'}{h} = -\frac{s \tan \theta}{f_{ob} \tan \theta} = -\frac{s}{f_{ob}}. \quad (15)$$

As usual, the minus sign indicates an inverted image.

The distance *s* (called the *tube length*) is so chosen that the image *I* falls on the first focal point F_1' of the eyepiece, which than acts as a simple magnifier as described above. Parallel rays enter the eye and a final image I' forms at infinity. The final magnification *M* is given by the product of the linear magnification *m* for

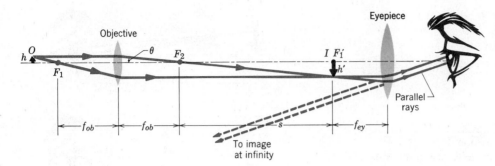

Figure 26 A thin-lens version of a compound microscope. Not to scale.

the objective lens, given by Eq. 15, and the angular magnification of the eyepiece, given by Eq. 14, or

$$M = m \times m_\theta = -\frac{s}{f_{ob}} \frac{15 \text{ cm}}{f_{ey}} \quad \text{(microscope)}. \quad (16)$$

Refracting Telescope. Like microscopes, telescopes come in a large variety of forms. The form we describe here is the simple refracting telescope that consists of an objective lens and an eyepiece, both represented in Fig. 27 by thin lenses, although in practice, as for microscopes, they will each be compound lens systems.

At first glance it may seem that the lens arrangements for telescopes and for microscopes are similar. However, telescopes are designed to view large objects, such as galaxies, stars, and planets, at large distances, whereas microscopes are designed for just the opposite purpose. Note also that in Fig. 27 the second focal point of the objective F_2 coincides with the first focal point of the eyepiece F_1', but in Fig. 26 these points are separated by the tube length s.

In Fig. 27 parallel rays from a distant object strike the objective lens, making an angle θ with the telescope axis and forming a real, inverted image at the common focal point F_2, F_1'. This image acts as an object for the eyepiece and a (still inverted) virtual image is formed at

infinity. The rays defining the image make an angle θ_{ey} with the telescope axis.

The angular magnification m_θ of the telescope is θ_{ey}/θ_{ob}. For paraxial rays (rays close to the axis) we can write $\theta_{ob} = h'/f_{ob}$ and $\theta_{ey} = -h'/f_{ey}$ or

$$m_\theta = -\frac{f_{ob}}{f_{ey}} \quad \text{(telescope)}. \quad (17)$$

Magnification is only one of the design factors of an astronomical telescope and is indeed easily achieved (How?). A good telescope needs *light-gathering power,* which determines how bright the image is. This is important when viewing faint objects such as distant galaxies and is accomplished by making the objective lens diameter as large as possible. *Field of view* is another important parameter. An instrument designed for galactic observation (narrow field of view) must be quite different from one designed for the observation of meteors (wide field of view). The telescope designer must also take account of lens and mirror aberrations including *spherical aberration* (that is, lenses and mirrors with truly spherical surfaces do not form sharp images) and *chromatic aberration* (that is, for simple lenses the focal length varies with wavelength so that fuzzy images are formed, displaying unnatural colors). There is also *resolving power,* which describes the ability of any optical

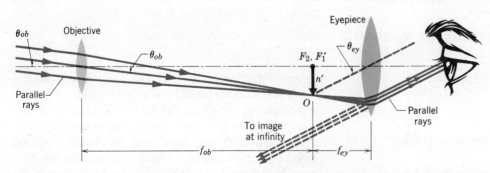

Figure 27 A thin-lens version of a refracting telescope. Not to scale.

instrument to distinguish between two objects (stars, say) whose angular separation is small. This by no means exhausts the design parameters of astronomical telescopes. We could also make a similar listing for any high-performance optical instrument.

39-11 Three Proofs (Optional)

The Spherical Mirror Formula (Eq. 8). Figure 28 shows a point object O placed on the axis of a concave spherical mirror beyond its center of curvature C. A ray from O that makes an angle α with the axis intersects the axis at I after reflection from the mirror at a. A ray that leaves O along the axis will be reflected back along itself at v and will also pass through I. Thus, I is the image of O; it is a *real* image because light actually passes through it. Let us find the image distance i in Fig. 28.

A useful theorem is that the exterior angle of a triangle is equal to the sum of the two opposite interior angles. Applying this to triangles $OaCO$ and $OaIO$ in Fig. 28 yields

$$\beta = \alpha + \theta \quad \text{and} \quad \gamma = \alpha + 2\theta.$$

If we eliminate θ between these two equations, we find

$$\alpha + \gamma = 2\beta. \tag{18}$$

We can write angles α, β, and γ, in radian measure, as

$$\alpha \approx \frac{av}{v0} = \frac{av}{o},$$

$$\beta = \frac{av}{vC} = \frac{av}{r}, \tag{19}$$

$$\gamma \approx \frac{av}{vI} = \frac{av}{i}.$$

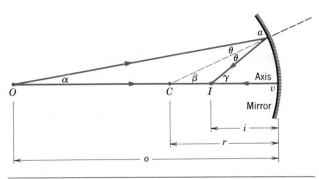

Figure 28 A point object O forms a real point image I after reflection from a concave spherical mirror.

Only the equation for β is exact, because the center of curvature of arc av is at C and not at O or I. However, the equations for α and γ are approximately correct if these angles are small enough. Substituting Eqs. 19 into Eq. 18, using Eq. 7 to replace r by $2f$ and canceling av leads exactly to Eq. 8, the relation that we set out to prove.

The Refracting Surface Formula (Eq. 11). The incident ray in Fig. 29 that falls on point a is refracted there according to

$$n_1 \sin \theta_1 = n_2 \sin \theta_2.$$

If α is small, θ_1 and θ_2 will also be small and we can replace the sines of these angles by the angles themselves. Thus, the above equation becomes

$$n_1 \theta_1 \approx n_2 \theta_2. \tag{20}$$

We again use the theorem that the exterior angle of a triangle is equal to the sum of the two opposite interior angles. Applying this to triangles $COaC$ and $ICaI$ yields

$$\theta_1 = \alpha + \beta \quad \text{and} \quad \beta = \theta_2 + \gamma. \tag{21}$$

If we eliminate θ_1 and θ_2 from Eqs. 20 and 21, we find

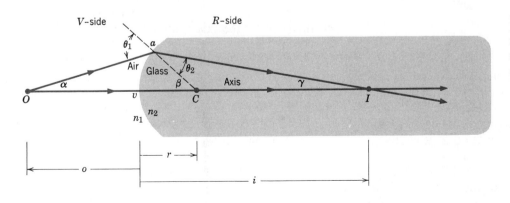

Figure 29 A point object O forms a real point image I after refraction at a spherical convex surface between two media.

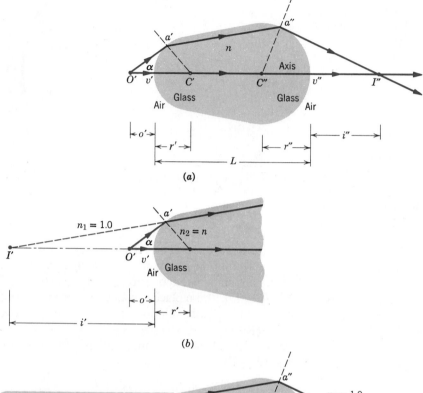

Figure 30 (*a*) Two rays from O' form a real image at I'' after being refracted at two spherical surfaces, the first surface being converging and the second diverging. (*b*) The first surface and (*c*) the second surface, shown separately. The vertical scale has been exaggerated for clarity.

$$n_1\, \alpha + n_2\, \gamma = (n_2 - n_1)\beta. \qquad (22)$$

In radian measure the angles α, β, and γ are

$$\alpha \approx \frac{av}{o}; \quad \beta = \frac{av}{r}; \quad \gamma \approx \frac{av}{i}. \qquad (23)$$

Only the second of these equations is exact. The other two are approximate because I and O are not the centers of circles of which av is an arc. However, for α small enough we can make the inaccuracies in Eqs. 23 as small as we wish. Substituting Eqs. 23 into Eq. 22 leads directly to Eq. 11, the relation we set out to prove.

The Thin Lens Formulas (Eqs. 12 and 13). Our plan is to consider each lens surface separately, using the image formed by the first surface as an object for the second.

Figure 30*a* shows such a thick glass "lens" of length L whose surfaces are ground to radii r' and r''. A point object O' is placed near the left surface as shown. A ray leaving O' along the axis is not deflected on entering or leaving the lens.

A second ray leaving O' at an angle α with the axis strikes the surface at point a', is refracted, and strikes the second surface at point a''. The ray is again refracted and crosses the axis at I'', which, being the intersection of two rays from O'', is the image of point O', formed after refraction at two surfaces.

Figure 30*b* shows the first surface, which forms a virtual image of O' at I'. To locate I', we use Eq. 11,

$$\frac{n_1}{o} + \frac{n_2}{i} = \frac{n_2 - n_1}{r}.$$

Putting $n_1 = 1$ and $n_2 = n$ and bearing in mind that the image distance is negative (that is, $i = -i'$ in Fig. 30b), we obtain

$$\frac{1}{o'} - \frac{n}{i'} = \frac{n-1}{r'}. \tag{24}$$

In this equation i' will be a positive number because we have arbitrarily introduced the minus sign appropriate to a virtual image.

Figure 30c shows the second surface. Unless an observer at point a'' were aware of the existence of the first surface, he would think that the light striking that point originated at point I' in Fig. 30b and that the region to the left of the surface was filled with glass. Thus, the (virtual) image I' formed by the first surface serves as a real object O'' for the second surface. The distance of this object from the second surface is

$$o = i' + L. \tag{25}$$

In applying Eq. 11 to the second surface, we insert $n_1 = n$ and $n_2 = 1$ because the object behaves as if it were imbedded in glass. If we use Eq. 25, Eq. 11 becomes

$$\frac{n}{i' + L} + \frac{1}{i''} = \frac{1 - n}{r'}. \tag{26}$$

Let us now assume that the thickness L of the "lens" in Fig. 30a is so small that we can neglect it in comparison with other linear quantities in this figure (such as o', i', o'', i'', r', and r''). In all that follows we make this *thin-lens approximation*. Putting $L = 0$ in Eq. 26 leads to

$$\frac{n}{i'} + \frac{1}{i''} = -\frac{n-1}{r''}. \tag{27}$$

Adding Eqs. 24 and 27 leads to

$$\frac{1}{o'} + \frac{1}{i''} = (n-1)\left(\frac{1}{r'} - \frac{1}{r''}\right).$$

Finally, calling the original object distance simply o and the final image distance simply i leads to

$$\frac{1}{o} + \frac{1}{i} = (n-1)\left(\frac{1}{r'} - \frac{1}{r''}\right) \tag{28}$$

which, with a small change in notation, are Eqs. 12 and 13, the relations we set out to prove.

REVIEW AND SUMMARY

Geometrical Optics

Light is most accurately described as an electromagnetic wave, whose speed and other properties are derivable from Maxwell's equations. *Geometrical optics* is an approximate treatment in which the waves can be represented by straight-line rays; it is valid if the waves do not encounter obstacles comparable in size to the wavelength of the radiation.

Reflection and Refraction

When a light ray falls on a boundary between two transparent media a *reflected* and a *refracted* ray generally appear. Both rays remain in the plane of incidence. The *angle of reflection* is equal to the angle of incidence and the *angle of refraction* is related to the angle of incidence by

$$n_1 \sin \theta_1 = n_2 \sin \theta_2 \quad \text{(refraction)}. \tag{1}$$

See Fig. 1 and Sample Problem 1.

A wave encountering a boundary for which a transmitted wave would have a higher speed will experience *total internal reflection* (see Fig. 4 and Sample Problem 2) if its angle of incidence is at least equal to θ_c, where

Total Internal Reflection

$$\theta_c = \sin^{-1} \frac{n_2}{n_1} \quad \text{(critical angle)}. \tag{3}$$

A reflected wave will be *polarized*, with its **E** vector perpendicular to the plane of incidence, if it strikes a boundary at the *polarizing angle* θ_B, where

Polarization by Reflection

$$\theta_B = \tan^{-1}(n_2/n_1) \quad \text{(Brewster's law)}. \tag{4}$$

See Fig. 9 and Sample Problem 3.

Rays diverging from a point object O can recombine to form an (approximately) point image I if they encounter a spherical mirror, a spherical refracting surface, or a thin lens. For rays sufficiently close to the axis we have the following (in which o is the *object distance* and i is the *image distance*):

1. *Spherical mirror:*

Mirrors

$$\frac{1}{o} + \frac{1}{i} = \frac{1}{f} = \frac{2}{r} \quad \text{(spherical mirror).}$$ [6,8]

See Figs. 15–17 and Sample Problem 6. A *plane mirror* is a special case for which $r \to \infty$, yielding $o = -i$; see Figs. 10–14 and Sample Problem 4.

2. *Spherical refracting surface:*

Refracting Surface

$$\frac{n_1}{o} + \frac{n_2}{i} = \frac{n_2 - n_1}{r} \quad \text{(single surface).}$$ [11]

See Fig. 19 and Sample Problem 6.

3. *Thin lens:*

Thin Lens

$$\frac{1}{o} + \frac{1}{i} = \frac{1}{f} = (n - 1) \left(\frac{1}{r_1} - \frac{1}{r_2} \right) \quad \text{(thin lenses).}$$ [12,13]

See Fig. 21 and Sample Problems 7 and 8.

The Sign Conventions

The rule for signs is: Associate *positive* with *Real, R-side,* and *upRight;* associate *negative* with *Virtual, V-side,* and *inVerted.* The R-side for mirrors is the side toward which the incident light is *reflected;* for refraction it is the side toward which the incident light is *transmitted.* See Fig. 20.

Images of extended objects may be found graphically by ray tracing; see Fig. 18 for mirrors and Fig. 23 for thin lenses. The *lateral magnification* in these two cases is given by

Lateral Magnification

$$m = -i/o,$$ [9]

a positive value of m corresponding to an erect image.

Optical Instruments

Three simplified treatments of optical instruments are given:

1. The *simple magnifier* (Fig. 25). The angular magnification is given by

$$m_\theta = \frac{(15 \text{ cm})}{f} \quad \text{(simple magnifier).}$$ [14]

2. The *compound microscope* (Fig. 26). The angular magnification is given by

$$M = m \times m_\theta = \frac{s}{f_{ob}} \frac{(15 \text{ cm})}{f_{ey}} \quad \text{(microscope).}$$ [16]

3. The *refracting telescope* (Fig. 27). The overall angular magnification is given by

$$m_\theta = -\frac{f_{ob}}{f_{ey}} \quad \text{(telescope).}$$ [17]

QUESTIONS

1. Describe what your immediate environment would be like if all objects were totally absorbing. Sitting in a chair in a room, could you see anything? If a cat entered the room could you see it?

2. Can you think of a simple test or observation to prove that the law of reflection is the same for all wavelengths, under conditions in which geometrical optics prevail?

3. A street light, viewed by reflection across a body of water in which there are ripples, appears very elongated. Explain.

4. Shortwave broadcasts from Europe are heard in the United States even though the path is not a straight line. Explain how.

5. The travel time of signals from satellites to receiving stations on the earth varies with the frequency of the signal. Why?

6. By what percent does the speed of blue light in fused quartz differ from that of red light?

7. Can (a) reflection phenomena or (b) refraction phenomena be used to determine the wavelength of light?

8. Describe and explain what a fish sees as it looks in various directions above its "horizon."

9. What is a plausible explanation for the observation that a street appears darker when wet than when dry?

10. Light, traveling through vacuum from a distant stationary source, strikes your eye. If the source starts to move rapidly toward you, how does this affect (*a*) the wavelength, (*b*) the frequency, and (*c*) the speed of the light?

11. Design a periscope, taking advantage of total internal reflection. What are the advantages compared with silvered mirrors?

12. For a plane mirror, what is the focal length? The magnification?

13. At night, in a lighted room, you blow a smoke ring toward a window pane. If you focus your eyes on the ring as it approaches the pane it will seem to go right through the glass into the darkness beyond. What is the explanation of this illusion?

14. How do your eyes adjust for seeing objects at different distances from you?

15. A machinist whose eyesight is failing finds that he can read his micrometer scale more easily if he squints at it through a tiny aperture formed by coiling his index finger into the base of his thumb. Although less bright, the image is sharper than when he looks at the scale directly. What is the explanation?

16. In driving a car you sometimes see vehicles such as ambulances with letters printed on them in such a way that they read in the normal fashion when you look at them through the rear-view mirror. Print your name so that it may be so read.

17. Brewster's law determines the polarizing angle on reflection from a material such as glass; see Fig. 9. A plausible interpretation for zero reflection of the parallel component at that angle is that the charges in the glass are caused to oscillate parallel to the reflected ray by this component and produce no radiation in this direction. Explain this and comment on the plausibility.

18. Explain how polarization by reflection could occur if the light is incident on the interface from the side with the higher index of refraction (glass to air, for example).

19. We all know that when we look into a mirror right and left are reversed. Our right hand will seem to be a left hand; if we part our hair on the left it will seem to be parted on the right, etc. Can you think of a system of mirrors that would let us see ourselves as others see us? If so draw it and prove your point by sketching some typical rays.

20. Devise a system of plane mirrors that will let you see the back of your head. Trace the rays to prove your point.

21. Can a virtual image be photographed by exposing a film at the location of the image? Explain.

22. If converging rays fall on a plane mirror, is the image virtual?

23. It is a bright sunny day and you want to create a rainbow in your back yard, using your garden hose. Exactly how do you go about it? Incidentally, why can't you walk under, or go to the end of, a rainbow?

24. Is it possible, by using one or more prisms, to recombine into white light the color spectrum formed when white light passes through a single prism? If yes, explain how.

25. How does atmospheric refraction affect the apparent time of sunset?

26. Stars twinkle but planets don't. Why?

27. In many city buses a convex mirror is suspended over the door, in full view of the driver. Why not a plane or a concave mirror?

28. What approximations were made in deriving the mirror equation (Eq. 8):

$$\frac{1}{o} + \frac{1}{i} = \frac{2}{r}?$$

29. Under what conditions will a spherical mirror, which may be concave or convex, form (*a*) a real image, (*b*) an inverted image, and (*c*) an image smaller than the object?

30. Can a virtual image be projected onto a screen? Photographed? If you put a piece of paper at the site of a virtual image (assuming a high-intensity light beam) will it ignite after sufficient exposure?

31. You are looking at a dog through a glass window pane. Where is the image of the dog? Is it real or virtual? Is it upright or inverted? What is the magnification? (*Hint:* Think of the window pane as the limiting case of a thin lens in which the radii of curvature have been allowed to become infinitely large.)

32. In some cars the right side mirror bears the notation: "Objects in the mirror are closer than they appear." What feature of the mirror requires this warning? What advantage does the mirror have to compensate for this disadvantage? Do cars viewed in this mirror seem to be moving faster or slower than they would be if viewed in a plane mirror?

33. We have all seen TV pictures of a baseball game shot from a camera located somewhere behind second base. The pitcher and the batter are about 60 ft apart but they look much closer on the TV screen. Why are images viewed through a telephoto lens foreshortened in this way?

34. An unsymmetrical thin lens forms an image of a point object on its axis. Is the image location changed if the lens is reversed?

35. Why has a lens two focal points and a mirror only one?

36. Under what conditions will a thin lens, which may be converging or diverging, form (*a*) a real image, (*b*) an inverted image, and (*c*) an image smaller than the object?

37. A diver wants to use an air-filled plastic bag as a converging lens for underwater visibility. Sketch a suitable cross section for the bag.

38. What approximations were made in deriving the thin

lens equation (Eq. 12):

$$\frac{1}{o} + \frac{1}{i} = \frac{1}{f}?$$

39. A concave mirror and a converging lens have the same focal length in air. Do they have the same focal length when immersed in water? If not, which has the greater focal length?

40. Under what conditions will a thin lens have a lateral magnification (a) of -1 and (b) of $+1$?

41. How does the focal length of a thin glass lens for blue light compare with that for red light, assuming the lens is (a) diverging and (b) converging?

42. Does the focal length of a lens depend on the medium in which the lens is immersed? Is it possible for a given lens to act as a converging lens in one medium and a diverging lens in another medium?

43. Are the following statements true for a glass lens in air? (a) A lens that is thicker at the center than at the edges is a converging lens for parallel light. (b) A lens that is thicker at the edges than at the center is a diverging lens for parallel light.

44. Under what conditions would the lateral magnification ($m = -i/o$) for lenses and mirrors become infinite? Is there any practical significance to such a condition?

45. Is the focal length of a spherical mirror affected by the medium in which it is immersed? of a thin lens? Why the difference, if any?

46. Why is the magnification of a simple magnifier (see the derivation leading to Eq. 14) defined in terms of angles rather than image/object size?

47. Ordinary spectacles do not magnify but a simple magnifier does. What, then, is the function of spectacles?

48. The *"f-number"* of a camera lens (see Problem 88) is its focal length divided by its aperture (effective diameter). Why is this useful to know in photography? How can the *f*-number of the lens be changed? How is exposure time related to *f*-number?

49. Does it matter whether (a) an astronomical telescope, (b) a compound microscope, (c) a simple magnifier, (d) a camera, including a TV camera, or (e) a projector, including a slide projector and a motion picture projector, produces upright or inverted images? What about real or virtual images?

50. The unaided human eye produces a real but inverted image on the retina. (a) Why then don't we perceive objects such as people and trees as upside down? (b) We don't, of course, but suppose that we wore special glasses so that we did. If you then turned this book upside down, could you read this question with the same facility that you do now?

51. Which of the following: a converging lens, a diverging lens, a concave mirror, a convex mirror, a plane mirror, is used (a) as a magnifying glass? (b) as the reflector in the lamphouse of a slide projector? (c) as the objective of a reflecting telescope? (d) in a kaleidoscope? (e) as the eyepiece of opera glasses? (f) to obtain a wider angle rear view from the driver's seat in a car?

52. In William Golding's *Lord of the Flies* the character Piggy uses his glasses to focus the sun's rays and kindle a fire. Later, the boys abuse Piggy and break his glasses. He is unable to identify them at close range because he is nearsighted. Find the flaw in this narrative. (*Boston Globe,* December 17, 1985, Letters)

53. Explain the function of the objective lens of a microscope; why use an objective lens at all? Why not just use a very powerful simple magnifier?

54. Why do astronomers use optical telescopes in looking at the sky? After all, the stars are so far away that they still appear to be points of light, without any detail discernible.

55. A watchmaker uses diverging eyeglasses for driving, no glasses for reading, and converging glasses in his occupational work. Is he nearsighted or farsighted? Explain. (See Problem 87.)

EXERCISES AND PROBLEMS

Section 39–2   Reflection and Refraction

1E. In Fig. 31 find the angles (a) θ_1 and (b) θ_2.

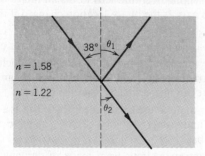

Figure 31 Exercise 1.

2E. Light in vacuum is incident on the surface of a glass slab. In the vacuum the beam makes an angle of 32° with the normal to the surface, while in the glass it makes an angle of 21° with the normal. What is the index of refraction of the glass?

3E. The speed of yellow sodium light in a certain liquid is measured to be 1.92×10^8 m/s. What is the index of refraction of this liquid, with respect to air, for sodium light?

4E. What is the speed in fused quartz of light of wavelength 550 nm? (See Fig. 2.)

5E. *Cerenkov radiation.* When an electron moves through a medium at a speed exceeding the speed of light in that medium, it radiates electromagnetic energy (the Cerenkov effect). What minimum speed must an electron have in a liquid of refractive index 1.54 in order to radiate?

6E. A laser beam travels along the axis of a straight section of pipeline, one mile long. The pipe normally contains air at standard temperature and pressure, but it may also be evacuated. In which case would the travel time for the beam be greater and by how much?

7E. When the rectangular metal tank in Fig. 32 is filled to the top with an unknown liquid, an observer with eyes level with the top of the tank can just see the corner E. Find the index of refraction of the liquid.

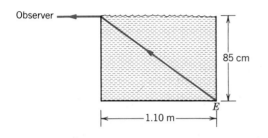

Figure 32 Exercise 7.

8E. Claudius Ptolemy (c. A.D. 150) gave the following measured values for the angle of incidence θ_1 and the angle of refraction θ_2 for a light beam passing from air to water:

θ_1	θ_2	θ_1	θ_2
10°	8°	50°	35°
20°	15°30′	60°	40°30′
30°	22°30′	70°	45°30′
40°	29°	80°	50°

Are these data consistent with the law of refraction? If so, what index of refraction results? These data are interesting as perhaps the oldest recorded physical measurements.

9P. A bottom-weighted 2.0-m-long vertical pole extends from the bottom of a swimming pool to a point 0.5 m above the water. Sunlight is incident at 55° above the horizon. What is the length of the shadow of the pole on the level bottom of the pool?

10P. Prove that a ray of light incident on the surface of a sheet of plate glass of thickness t emerges from the opposite face parallel to its initial direction but displaced sideways, as in Fig. 33. Show that, for small angles of incidence θ, this displacement is given by

$$x = t\theta \frac{n-1}{n}$$

where n is the index of refraction and θ is measured in radians.

11P. Ocean waves moving at a speed of 4.0 m/s are approaching a beach at an angle of 30° to the normal, as shown in Fig. 34. Suppose the water depth changes abruptly and the

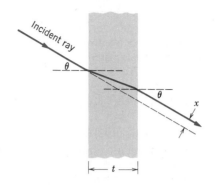

Figure 33 Problem 10.

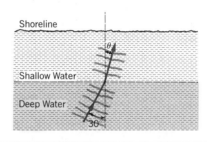

Figure 34 Problem 11.

wave speed drops to 3.0 m/s. Close to the beach, what is the angle θ between the direction of wave motion and the normal? (Assume the same law of refraction as for light.) Explain why most waves come in normal to a shore even though at large distances they approach at a variety of angles.

12P. A 60° prism is made of fused quartz. A ray of light falls on one face, making an angle of 35° with the normal. Trace the ray through the prism graphically with some care, showing the paths traversed by rays representing (a) blue light, (b) yellow-green light, and (c) red light. (See Fig. 2.)

13P. A penny lies at the bottom of a pool with depth d and index of refraction n, as shown in Fig. 35. Show that light rays that are close to the normal appear to come from a point $d_a = d/n$ below the surface. This distance is the apparent depth of the pool.

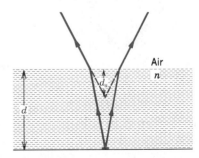

Figure 35 Problem 13.

14P. As an example of the importance of the paraxial ray assumption, consider this problem. You place a coin at the bottom of a swimming pool filled with water ($n = 1.33$) to a depth of 2.4 m. What is the apparent depth of the coin below the surface when viewed (a) at near normal incidence (that is, by paraxial rays) and (b) by rays that leave the coin making an angle of 30° with the normal (that is, definitely not paraxial rays)?

15P. Figure 36 shows a small light bulb suspended 250 cm above the surface of the water in a swimming pool. The water is 200 cm deep and the bottom of the pool is a large mirror. Where is the image of the light bulb? Consider only paraxial rays near normal incidence.

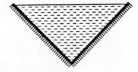

Figure 36 Problem 15.

16P. A layer of water ($n = 1.33$) 20 mm thick floats on carbon tetrachloride ($n = 1.46$) 40 mm thick. How far below the water surface, viewed at normal incidence, does the bottom of the tank seem to be?

17P. Two perpendicular mirrors form the sides of a vessel filled with water, as shown in Fig. 37. A light ray is incident from above, normal to the water surface. (a) Show that the emerging ray is parallel to the incident ray. Assume that there are two reflections at the mirror surfaces. (b) Repeat the analysis for the case of oblique incidence, the ray lying in the plane of the figure.

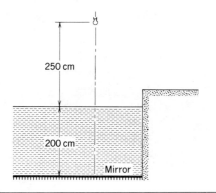

Figure 37 Problem 17.

18P. The index of refraction of the earth's atmosphere decreases monotonically with height from its surface value (about 1.00029) to the value in space (about 1.00000) at the top of the atmosphere. This continuous (or graded) variation can

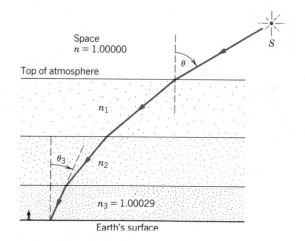

Figure 38 Problem 18.

be approximated by considering the atmosphere to be composed of three (or more) plane parallel layers in each of which the index of refraction is constant. Thus, in Fig. 38, $n_3 > n_2 > n_1 > 1.00000$. Consider a ray of light from a star S that strikes the top of the atmosphere at an angle θ with the vertical. (a) Show that the apparent direction θ_3 of the star with the vertical as seen by an observer at the earth's surface is given from

$$\sin \theta_3 = \left(\frac{1}{n_3}\right) \sin \theta.$$

(*Hint:* Apply the law of refraction to successive pairs of layers of the atmosphere; ignore the curvature of the earth.) (b) Calculate the shift in position of a star observed to be 20° from the vertical. (The very small effects due to atmospheric refraction can be most important; for example, they must be taken into account in using navigation satellites to obtain accurate fixes of position on the earth.)

19P. An incident ray falls on one face of a glass prism in air. The angle of incidence θ is chosen so that the emerging ray also makes an angle θ with the normal to the other face, the ray passing symmetrically through the prism; see Fig. 39. Show

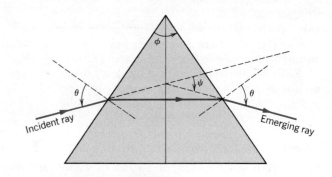

Figure 39 Problem 19.

that the index of refraction n of the glass prism is given by

$$n = \frac{\sin \frac{1}{2}(\psi + \phi)}{\sin \frac{1}{2}\phi},$$

where ϕ is the vertex angle of the prism between the two faces and ψ is the deviation angle through which the light beam has been turned in passing through the prism. (Under these conditions the deviation angle ψ has the minimum value, with respect to rotation of the prism, and is called the *angle of minimum deviation*.)

20P. A ray of light goes through an equilateral prism in the position of minimum deviation. The total deviation is 30°. What is the index of refraction of the prism? See Problem 19.

Section 39–3 Total Internal Reflection

21E. The refractive index of benzene is 1.8. What is the critical angle of incidence for a light ray traveling in benzene toward a plane layer of air above it?

22E. A light ray falls on a square glass slab as in Fig. 40. What must be the minimum index of refraction of the glass if total internal reflection occurs at the vertical face?

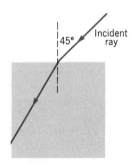

Figure 40   Exercise 22.

23E. A ray of light is incident normally on the face ab of a glass prism ($n = 1.52$), as shown in Fig. 41. (a) Assuming that the prism is immersed in air, find the largest value for the angle ϕ so that the ray is totally reflected at face ac. (b) Find ϕ if the prism is immersed in water.

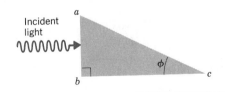

Figure 41   Exercise 23.

24E. A fish is 2.00 m below the surface of a smooth lake. At what angle above the horizontal must it look to see the light from a small fire burning at the water's edge 100 m away? Take the index of refraction for water to be 1.33.

25E. A point source is 80 cm below the surface of a body of water. Find the diameter of the largest circle at the surface through which light can emerge from the water.

26P. A glass cube has a small spot at its center. (a) What parts of the cube face must be covered to prevent the spot from being seen, no matter what the direction of viewing? (b) What fraction of the cube surface must be so covered? Assume a cube edge of 10 mm and an index of refraction of 1.5. (Neglect the subsequent behavior of an internally reflected ray.)

27P. A plane wave of white light traveling in fused quartz strikes a plane surface of the quartz, making an angle of incidence θ. Is it possible for the internally reflected beam to appear (a) bluish or (b) reddish? (c) Roughly what value of θ must be used? (*Hint:* White light will appear bluish if wavelengths corresponding to red are removed from the spectrum.)

28P. A given monochromatic light ray, initially in air, strikes a 90° prism at P (see Fig. 42) and is refracted there and at Q to such an extent that it just grazes the right-hand prism surface at Q. (a) Determine the index of refraction of the prism for this wavelength in terms of the angle of incidence θ_1 that gives rise to this situation. (b) Give a numerical upper bound for the index of refraction of the prism. Show, by ray diagrams, what happens if the angle of incidence at P is (c) slightly greater than θ_1 or (d) is slightly less than θ_1.

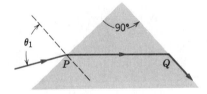

Figure 42   Problem 28.

29P. A glass prism with an apex angle of 60° has $n = 1.60$. (a) What is the smallest angle of incidence for which a ray can enter one face of the prism and emerge from the other? (b) What angle of incidence would be required for the ray to pass through the prism symmetrically? See Problem 19.

30P. A point source of light is placed a distance h below the surface of a large deep lake. (a) Show that the fraction f of the light energy that escapes directly from the water surface is independent of h and is given by

$$f = \frac{1}{2}(1 - \sqrt{1 - 1/n^2})$$

where n is the index of refraction of water. (*Note:* Absorption within the water and reflection at the surface (except where it is total) have been neglected.) (b) Evaluate this fraction for $n = 1.33$.

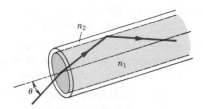

Figure 43 Problem 31.

31P. An optical fiber consists of a glass core (index of refraction n_1) surrounded by a coating (index of refraction $n_2 < n_1$). Suppose a beam of light enters the fiber from air at an angle θ with the fiber axis as shown in Fig. 43. (*a*) Show that the greatest possible value of θ for which a ray can be propagated down the fiber is given by $\theta = \sin^{-1} \sqrt{n_1^2 - n_2^2}$. (*b*) Assume the glass and coating indices of refraction are 1.58 and 1.53, respectively, and calculate the value of this angle.

32P. In an optical fiber (see preceding problem), different rays travel different paths along the fiber, leading to different travel times. This causes a light pulse to spread out as it travels along the fiber, resulting in information loss. The delay time should be minimized in designing a fiber. Consider a ray that travels a distance L along a fiber axis and another that is reflected, at the critical angle, as it travels to the same destination as the first. (*a*) Show that the difference Δt in the times of arrival is given by

$$\Delta t = \frac{L}{c} \frac{n_1}{n_2} (n_1 - n_2),$$

where n_1 is the index of refraction of the glass core and n_2 is the index of refraction of the fiber coating. (*b*) Evaluate Δt for the fiber of Problem 31, with $L = 300$ m.

33P*. Sound waves generated in the earth by the detonation of a small amount of explosive obey the same laws of reflection, refraction, and total internal reflection as do light rays. Detectors, set up on the surface in a straight line from the detonation point S (see Fig. 44) detect the arrival of the sound waves. Suppose that a layer of soil in which the sound speed is v_1 covers solid bedrock in which the sound speed is v_2; suppose also, as is often the case, that $v_2 > v_1$. Waves arrive at a detector by two routes: (*i*) a direct surface wave; (*ii*) a wave striking the interface of soil and bedrock at the critical angle for total internal reflection; this wave travels along the boundary at speed v_2, generating waves that return to the surface, leaving the interface at the same angle as that of incidence. (Waves simply reflected from the interface are not considered.) (*a*) Show that the travel time of these critically reflected waves is given by

$$t_c = \frac{2D \sqrt{v_2^2 - v_1^2}}{v_1 v_2} + \frac{x}{v_2},$$

where D is the thickness of the upper layer and x the distance from detonation S to the detector. (*b*) Show that beyond a certain distance x^*, the critically reflected wave arrives before

the direct wave, and that

$$x^* = 2D \sqrt{\frac{1 + n}{1 - n}}, \quad n = v_1/v_2$$

and, therefore, by determining x^* the thickness D of the upper layer is determined. This method is widely employed in determining the suitability of land areas for construction purposes, to trace subsurface water-bearing zones, etc., and is called *seismic surveying*.

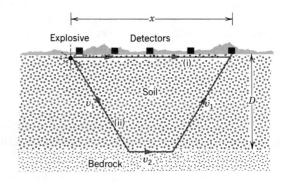

Figure 44 Problem 33.

34P*. A seismic surveying team is attempting to determine the depth to a horizontal bedrock layer. Detectors are placed 200 m apart. The data, shown graphically in Fig. 45, shows the time from the explosion to the arrival of the first sound waves at each detector plotted against the distance to the detector. Apply the theory of Problem 33 and find (*a*) the sound speeds v_1 and v_2, and (*b*) the depth to the bedrock layer.

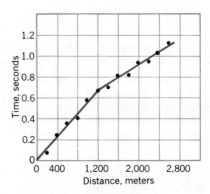

Figure 45 Problem 34.

Section 39-4 Polarization by Reflection
35E. (*a*) At what angle of incidence will the light reflected from water be completely polarized? (*b*) Does this angle depend on the wavelength of the light?

36E. Light traveling in water of refractive index 1.33 is incident upon a plate of glass of refractive index 1.53. At what angle of incidence is the reflected light completely plane polarized?

37E. Calculate the range of polarizing angles for white light incident on fused quartz. Assume that the wavelength limits are 400 and 700 nm and use the dispersion curve of Fig. 2.

38P. When red light in vacuum is incident at the polarizing angle on a certain glass slab, the angle of refraction is 32°. What are (*a*) the index of refraction of the glass and (*b*) the polarizing angle?

Section 39–5   A Plane Mirror

39E. A small object is 10 cm in front of a plane mirror. If you stand behind the object, 30 cm from the mirror, and look at its image, for what distance must you focus your eyes?

40E. Suppose you wished to photograph an object seen in a plane mirror. If the object is 5.0 m to your right and 1.0 m closer to the plane of the mirror than you, for what distance must you focus the lens of your camera, which you are holding 4.3 m from the mirror?

41E. You are standing in front of a large plane mirror, contemplating your image. If you move toward the mirror at speed *v*, at what speed does your image move toward you? Report this speed both (*a*) in your own reference frame and (*b*) in the reference frame of the room in which the mirror is at rest.

42E. Figure 46 shows an incident ray striking plane mirrors *MM'* and *M'M''*. Find the angle between the incoming ray *i* and the outgoing ray *r*. The two mirrors are at right angles.

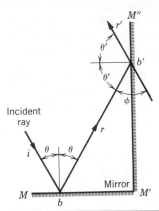

Figure 46   Problem 42.

43P. Figure 47 shows (top view) that Bernie *B* is walking directly toward the center of a vertical mirror. How close to the mirror will he be when Sarah *S* is just able to see him? Take *d* = 3.0 m.

44P. Prove that if a plane mirror is rotated through an angle α, the reflected beam is rotated through an angle 2α. Show that this result is reasonable for α = 45°.

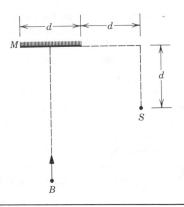

Figure 47   Problem 43.

45P. In Fig. 12 you rotate the mirror 30° counterclockwise about its bottom edge, leaving the point object *O* in place. Is the image point displaced? If so, where is it? Can the eye still see the image without being moved? Sketch a figure showing the new situation.

46P. A small object *O* is placed one-third of the way between two parallel plane mirrors as in Fig. 48. Trace appropriate bundles of rays for viewing the four images that lie closest to the object.

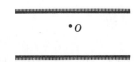

Figure 48   Problem 46.

47P. Two plane mirrors make an angle of 90° with each other. What is the largest number of images of an object placed between them that can be seen by a properly placed eye? The object need not lie on the mirror bisector.

48P. A point object is 10 cm away from a plane mirror while the eye of an observer (pupil diameter 5.0 mm) is 20 cm away. Assuming both the eye and the point to be on the same line perpendicular to the surface, find the area of the mirror used in observing the reflection of the point.

49P. You put a point source of light *S* a distance *d* in front of

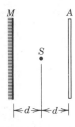

Figure 49   Problem 49.

a screen A. How is the intensity at the center of the screen changed if you put a mirror M a distance d behind the source, as in Fig. 49? (*Hint:* Recall from Chapter 38 the variation of intensity with distance from a point source of light.)

50P. Figure 50 shows an idealized submarine periscope (submarine not shown). The periscope consists of two parallel plane mirrors set at 45° to the periscope axis. An object (arrow) is sighted at a distance D from the scope, as shown. Due to the action of the mirrors, describe the image as seen by the submarine officer peering into the scope. Specifically, is the image (*a*) real or virtual; (*b*) upright or inverted; (*c*) magnified; if so, by how much? (*d*) Find the distance of the image from the bottom mirror.

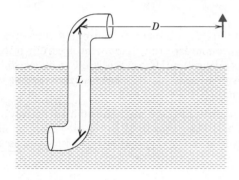

Figure 50 Problem 50.

51P. Solve Problem 47 if the angle between the mirrors is (*a*) 45°, (*b*) 60°, (*c*) 120°, the object always being placed on the bisector of the mirrors.

52P*. A *corner reflector,* much used in optical, microwave, and other applications, consists of three plane mirrors fastened together as the corner of a cube. It has the property that an incident ray is returned, after three reflections, with its direction exactly reversed. Prove this result.

Section 39-6 Spherical Mirrors

53E. A concave shaving mirror has a radius of curvature of 35 cm. It is positioned so that the image of a man's face is 2.5 times the size of his face. How far is the mirror from the man's face?

54E. For clarity, the rays in figures like Fig. 15*b* are not drawn paraxial enough for Eq. 8 to hold with great accuracy. With a ruler, measure r and o in this figure and calculate, from Eq. 8, the predicted value of i. Compare this with the measured value of i.

55P. Fill in the table below, each column of which refers to a spherical mirror and a real object. Check your results by ray-tracing. Distances are in centimeters; if a number has no plus or minus sign in front of it, find the correct sign.

56P. A short linear object of length L lies on the axis of a spherical mirror, a distance o from the mirror. (*a*) Show that its image will have a length L' where

$$L' = L\left(\frac{f}{o-f}\right)^2.$$

(*b*) Show that the *longitudinal magnification* m' ($= L'/L$) is equal to m^2 where m is the lateral magnification.

57P. (*a*) A luminous point is moving at speed v_0 toward a spherical mirror, along its axis. Show that the speed at which the image of this point object is moving is given by

$$v_I = -\left(\frac{r}{2o-r}\right)^2 v_0.$$

(*Hint:* Start from Eq. 8.) (*b*) Assume that the mirror is concave, with $r = 15$ cm (and thus $f = 7.5$ cm) and that $v_0 = 5.0$ cm/s. Find the speed of the image if the object is far outside the focal point ($o = 30$ cm). (*c*) If it is close to the focal point ($o = 8.0$ cm). (*d*) If it is very close to the mirror ($o = 0.1$ cm).

Section 39-8 Spherical Refracting Surfaces

58P. A parallel beam of light from a laser falls on a solid transparent sphere of index of refraction n, as shown in Fig. 51.

Table for Problem 55

	a	*b*	*c*	*d*	*e*	*f*	*g*	*h*
Type	Concave						Convex	
f(cm)	20		+20			20		
r (cm)					−40		40	
i (cm)					−10		4	
o (cm)	+10	+10	+30	+60				+24
m		+1		−0.5		+0.10		0.50
Real image?		no						
Upright image?								no

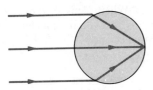

Figure 51 Problem 58.

(a) Show that the beam cannot be brought to a focus at the back of the sphere unless the beam width is small compared with the radius of the sphere. (b) If the condition in (a) is satisfied, what is the index of refraction of the sphere? (c) What index of refraction, if any, will focus the beam at the center of the sphere?

59P. Fill in the table below, each column of which refers to a spherical surface separating two media with different indices of refraction. Distances are measured in centimeters. The object is real in all cases. Draw a figure for each situation and construct the appropriate rays graphically. Assume a point object.

60P. A narrow parallel incident beam falls on a solid glass sphere at normal incidence. Locate the image in terms of the index of refraction n and the sphere radius r.

Section 39–9 Thin Lenses

61E. An object is 20 cm to the left of a thin diverging lens having a 30 cm focal length. Where is the image formed? Obtain the image position both by calculation and also from a ray diagram.

62E. Two converging lenses, with focal lengths f_1 and f_2, are positioned a distance $f_1 + f_2$ apart, as shown in Fig. 52. Ar-

rangements like this are called *beam expanders* and are often used to increase the diameters of light beams from lasers. (a) If W_1 is the incident beam width, show that the width of the emerging beam is $W_2 = (f_2/f_1)W_1$. (b) Show how a combination of one diverging and one converging lens can also be arranged as a beam expander. Incident rays parallel to the axis should exit parallel to the axis.

63E. Calculate the ratio of the intensity of the beam emerging from the beam expander of Exercise 62 to the intensity of the laser beam.

64E. A double-convex lens is to be made of glass with an index of refraction of 1.5. One surface is to have twice the radius of curvature of the other and the focal length is to be 60 mm. What are the radii?

65E. You focus an image of the sun on a screen, using a thin lens whose focal length is 20 cm. What is the diameter of the image? (See Appendix C for needed data on the sun.)

66E. A lens is made of glass having an index of refraction of 1.5. One side of the lens is flat and the other convex with a radius of curvature of 20 cm. (a) Find the focal length of the lens. (b) If an object is placed 40 cm to the left of the lens, where will the image be located?

67E. Using the lensmaker's formula (Eq. 13), decide which of the thin lenses in Fig. 53 are converging and which diverging for parallel incident light.

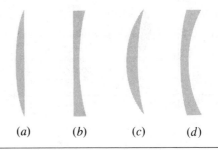

Figure 53 Exercise 67.

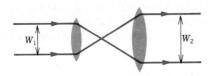

Figure 52 Exercise 62.

Table for Problem 59

	a	b	c	d	e	f	g	h
n_1	1.0	1.0	1.0	1.0	1.5	1.5	1.5	1.5
n_2	1.5	1.5	1.5		1.0	1.0	1.0	
o (cm)	+10	+10		+20	+10		+70	+100
i (cm)		−13	+600	−20	−6	−7.5		+600
r (cm)	+30		+30	−20		−30	+30	−30
Real image?								

68E. Show that the focal length f for a thin lens whose index of refraction is n and which is immersed in a fluid whose index of refraction is n' is given by

$$\frac{1}{f} = \frac{n - n'}{n'} \left(\frac{1}{r'} - \frac{1}{r''} \right).$$

69E. A movie camera with a lens of focal length 75 mm takes a picture of a 180-cm-high person standing 27 m away. What is the height of the image of the person on the film?

70P. You have a supply of flat glass disks ($n = 1.5$) and a lens-grinding machine that can be set to grind radii of curvature of either 40 cm or 60 cm. You are asked to prepare a set of six lenses like those shown in Fig. 54. What will be the focal

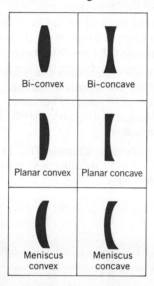

Bi-convex	Bi-concave
Planar convex	Planar concave
Meniscus convex	Meniscus concave

Figure 54 Problem 70.

length of each lens? Will the lens form a real or a virtual image of the sun? (*Note:* Where you have a choice of radii of curvature, select the smaller one.)

71P. The formula

$$\frac{1}{o} + \frac{1}{i} = \frac{1}{f}$$

is called the *Gaussian* form of the thin lens formula. Another form of this formula, the *Newtonian* form, is obtained by considering the distance x from the object to the first focal point and the distance x' from the second focal point to the image. Show that

$$xx' = f^2.$$

72P. To the extent possible, fill in the table below, each column of which refers to a thin lens. Distances are in centimeters; if a number (except in row n) has no plus sign or minus sign in front of it, find the correct sign. Draw a figure for each situation and construct the appropriate rays graphically. The object is real in all cases.

73P. A converging lens with a focal length of $+20$ cm is located 10 cm to the left of a diverging lens having a focal length of -15 cm. If a real object is located 40 cm to the left of the first lens, locate and describe completely the image formed.

74P. An object is placed 1.0 m in front of a converging lens, of focal length 0.50 m, which is 2.0 m in front of a plane mirror. (*a*) Where is the final image, measured from the lens, that would be seen by an eye looking toward the mirror through the lens? (*b*) Is the final image real or virtual? (*c*) Is the final image upright or inverted? (*d*) What is the lateral magnification?

75P. An upright object is placed a distance in front of a converging lens equal to twice the focal length f_1 of the lens. On the other side of the lens is a converging mirror of focal length f_2

Table for Problem 72

	a	b	c	d	e	f	g	h	i
Type	converging								
f (cm)	10	+10	10	10					
r_1 (cm)					+30	−30	−30		
r_2 (cm)					−30	+30	−60		
i (cm)									
o (cm)	+20	+5	+5	+5	+10	+10	+10	+10	+10
n					1.5	1.5	1.5		
m			>1	<1				0.5	0.5
Real image?									yes
Upright image?							yes		

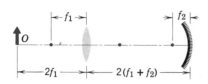

Figure 55 Problem 75.

separated from the lens by a distance $2(f_1 + f_2)$; see Fig. 55. (a) Find the location, nature, and relative size of the final image, as seen by an eye looking toward the mirror through the lens. (b) Draw the appropriate ray diagram.

76P. An illuminated arrow forms a real inverted image of itself at a distance $d = 40$ cm, measured along the optic axis of a lens; see Fig. 56. The image is just half the size of the object. (a) What kind of lens must be used to produce this image? (b) How far from the object must the lens be placed? (c) What is the focal length of the lens?

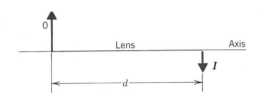

Figure 56 Problem 76.

77P. An object is 20 cm to the left of a lens with a focal length of $+10$ cm. A second lens of focal length $+12.5$ cm is 30 cm to the right of the first lens. (a) Using the image formed by the first lens as the object for the second, find the location and relative size of the final image. (b) Verify your conclusions by drawing the lens system to scale and constructing a ray diagram. (c) Describe the final image.

78P. Two thin lenses of focal lengths f_1 and f_2 are in contact. Show that they are equivalent to a single thin lens with a focal length given by

$$f = \frac{f_1 f_2}{f_1 + f_2}.$$

79P. The *power P* of a lens is defined by $P = 1/f$, where f is the focal length. The unit of power is the *diopter*, where 1 diopter = 1/meter. (a) Why is this a reasonable definition to use for lens power? (b) Show that the net power of two lenses in contact is given by $P = P_1 + P_2$, where P_1 and P_2 are the powers of the separate lenses. (Hint: See Problem 78.)

80P. An illuminated slide is mounted 44 cm from a screen. How far from the slide must a lens of focal length 11 cm be placed in order to focus an image on the screen?

81P. Show that the distance between a real object and its real

image formed by a thin converging lens is always greater than or equal to four times the focal length of the lens.

82P. A luminous object and a screen are a fixed distance D apart. (a) Show that a converging lens of focal length f will form a real image on the screen for two positions that are separated by

$$d = \sqrt{D(D - 4f)}.$$

(b) Show that the ratio of the two image sizes for these two positions is

$$\left(\frac{D - d}{D + d} \right)^2.$$

Section 39–10 Optical Instruments

83E. A microscope of the type shown in Fig. 26 has a focal length for the objective lens of 4.0 cm and for the eyepiece lens of 8.0 cm. The distance between the lenses is 25 cm. (a) What is the distance s in Fig. 26? (b) To reproduce the conditions of Fig. 26 how far beyond F_1 in that figure should the object be placed? (c) What is the lateral magnification m of the objective? (d) What is the angular magnification m_θ of the eyepiece? (e) What is the overall magnification M of the microscope?

84E. The magnifying power of an astronomical telescope in normal adjustment is 36, and the diameter of the objective lens is 75 mm. What is the minimum diameter of the eyepiece required to collect all the light entering the objective from a distant point source on the axis of the instrument?

85P. In connection with Fig. 25c, (a) show that if the object O is moved from the first focal point F_1 toward the eye, the image moves in from infinity and the angle θ' (and thus the angular magnification m_θ) is increased. (b) If you continue this process, at what image location will m_θ have its maximum usable value? (c) Show that the maximum usable value of m_θ is $1 + (15 \text{ cm})/f$. (d) Show that in this situation the angular magnification is equal to the linear magnification.

86P. *The eye—the basic optical instrument:* Fig. 57a suggests a normal human eye. Parallel rays entering a relaxed eye gazing at infinity produce a real, inverted image on the retina. The eye thus acts as a converging lens. Most of the refraction occurs at the outer surface of the eye, the *cornea*. Assume a focal length f for the eye of 2.50 cm. In Fig. 57b the object is moved in to a distance $o = 40$ cm from the eye. To form an image on the retina the effective focal length of the eye must be reduced to f'. This is done by the action of the ciliary muscles that change the shape of the lens and thus the effective focal length of the eye. (a) Find f' from the above data. (b) Would the effective radii of curvature of the lens become larger or smaller in the transition from Fig. 57a to 57b? (In the figure the structure of the eye is only roughly suggested and Fig. 57b is not to scale.)

87P. In an eye that is *farsighted* the eye focuses parallel rays so that the image would form behind the retina, as in Fig. 58a. In an eye that is *nearsighted* the image is formed in front of the

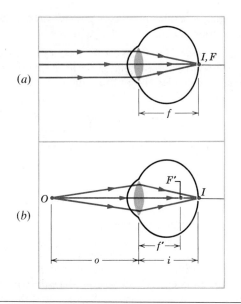

Figure 57 Problem 86.

retina, as in Fig. 58b. (a) How would you design a corrective lens for each eye defect? Make a ray diagram for each case. (b) If you need spectacles only for reading, are you nearsighted or farsighted? (c) What is the function of bifocal spectacles, in which the upper parts and lower parts have different focal lengths?

88P. *The camera:* Figure 59 shows an idealized camera focused on an object at infinity. A real, inverted image I is

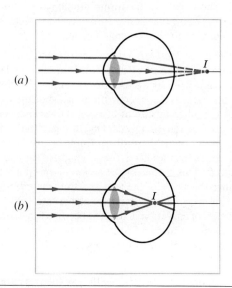

Figure 58 Problem 87.

formed on the film, the image distance i being equal to the (fixed) focal length f ($= 5.0$ cm, say) of the lens system. In Fig. 59b the object O is closer to the camera, the object distance o being, say, 100 cm. To focus an image I on the film, we must extend the lens away from the camera (why?). (a) Find i' in Fig. 59b. (b) By how much must the lens be moved? Note that the camera differs from the eye (see Problem 86) in this respect. In the camera, f remains constant and the image distance i must be adjusted by moving the lens. For the eye the image distance i remains constant and the focal length f is adjusted by distorting the lens. Compare Fig. 57 and Fig. 59 carefully.

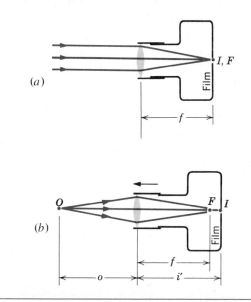

Figure 59 Problem 88.

89P. *The reflecting telescope:* Isaac Newton, having convinced himself (erroneously as it turned out) that chromatic aberration was an inherent property of refracting telescopes, invented the reflecting telescope, shown schematically in Fig. 60. He presented his second model of this telescope, which has a magnifying power of 38, to the Royal Society, which still has it. In Fig. 60 incident light falls, closely parallel to the telescope

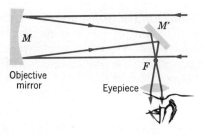

Figure 60 Problem 89.

axis, on the objective mirror M. After reflection from small mirror M' (the figure is not to scale), the rays form a real, inverted image in the focal plane through F. This image is then viewed through an eyepiece. (*a*) Show that the angular magnification m_θ is also given by Eq. 17, or

$$m_\theta = -f_{ob}/f_{ey}$$

where f_{ob} is the focal length of the objective mirror and f_{ey} that of the eyepiece. (*b*) The 200-in. mirror in the reflecting telescope at Mt. Palomar in California has a focal length of 16.8 m. Estimate the size of the image formed in the focal plane of this mirror when the object is a meter stick 2.0 km away. Assume parallel incident rays. (*c*) The mirror of a different reflecting astronomical telescope has an effective radius of curvature ("effective" because such mirrors are ground to a parabolic rather than a spherical shape, to eliminate spherical aberration defects) of 10 m. To give an angular magnification of 200, what must be the focal length of the eyepiece?

90P. In a compound microscope, the object is 10 mm from the objective lens. The lenses are 300 mm apart and the intermediate image is 50 mm from the eyepiece. What magnification is produced?

ESSAY 16
KALEIDOSCOPES

JEARL WALKER
CLEVELAND STATE
UNIVERSITY

Modern kaleidoscopes offer brightly colored displays with subtle symmetries. I do not mean the inexpensive toys that yield a few murky images. Instead, I refer to the kaleidoscopes that are now an art form. They contain not only fine-quality mirrors that yield hundreds of clear images but also lenses, filters, and other devices that alter and color the images, sometimes creating vivid illusions. What makes a kaleidoscope display pleasing images and how are the images created? Were it a matter of only simple reflections, kaleidoscopes would be largely identical. Such is not the case, for modern kaleidoscopes come in wide variety.

Before describing the novel kaleidoscopes, I shall explore how simple arrangements of mirrors give multiple images of objects. Mount a vertical mirror on a table and place a small object, for example, a coin, in front of it. When you peer down into the mirror you see the coin and also its virtual image that seems to lie on the tabletop within the mirror, as distant from the lower edge of the mirror as the coin is. One mirror yields one virtual image of the coin.

Next, add another identical mirror, with its vertical edge running alongside a vertical edge of the first mirror. Adjust the angle between the mirror planes to be 60°. When you look down into the mirrors you again see the coin directly but now you also see five virtual images of it in pie-slice sectors that are clustered around the vertex at which the mirrors meet on the table. The composite display has a sixfold symmetry.

One way to locate the images is by means of "virtual mirrors." Note that the lower edges of the mirrors show up in the display as four virtual images. The images can be considered to be mirrors that reflect images just as real mirrors do. Figure 1 shows where the various images seem to lie. To eliminate the complexity of perspective, the plan is an overhead view, although such a viewing direction would scarcely allow you to see any of the reflected images.

The direct view a is reflected by mirror A to form sector b. The reflection amounts mathematically to a simple rotation. Imagine rotating sector a around a hinge lying

Figure 1 A sixfold cluster around a 60° vertex.

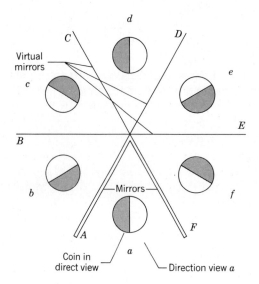

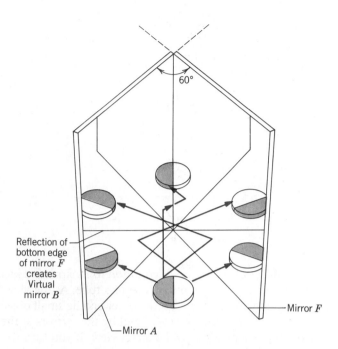

60°

Reflection of
bottom edge
of mirror F
creates
Virtual
mirror B

Mirror F

Mirror A

Figure 2 Reflections from
the mirrors.

along A. The sector leaves the plane of the sketch and then returns to it, forming
sector b.

The image of mirror F in sector b is labeled B. It is a virtual mirror in the sense
that it too can serve as a hinge. This time mentally rotate sector b around a hinge
along B. The sector leaves the plane of the sketch and then returns to it to form sector
c. Continue the procedure. Rotate sector c around the virtual mirror labeled C so that
it forms the rearmost sector d. The procedure works just as well in the counterclockwise direction, with sector a generating sector f, followed by f generating e, and finally
e generating d. Note that sector d is the same whether you consider a clockwise or
counterclockwise generation of the sectors. Such an arrangement is said to be unambiguous. If you change your perspective into the mirrors, the contents of sector d (as
well as the other sectors) are unchanged.

The images may be produced mathematically by mental rotations of the sectors,
but they are physically produced by reflections of light rays from the real mirrors as
shown in Fig. 2. The image of the coin in sector b results from a reflection of rays by
mirror A: A ray leaves the coin and then reflects from mirror A to you. You mentally
extrapolate the ray backward, concluding that the light originated where you perceive
the coin to lie in sector b. The image in sector f also depends on a single reflection of
light but the images in sectors c and e require an additional reflection. For example,
the image in sector c involves light that leaves the coin, reflects from mirror F and
then from mirror A.

Sector d is more complex, because it depends on three reflections of the light. For
example, light might leave the coin, reflect from A, then F, and finally A again.
Instead, the sequence might be F, A, and then F. Which sequence is important depends on your perspective into the mirror system. If you face toward mirror A, you
intercept light undergoing the reflection sequence of A, F, and A. Note the pattern in
the number of reflections in the cluster. The farther from the direct view a sector is,
the greater the number of reflections must be.

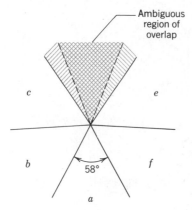

Ambiguous region of overlap

Figure 3 Effect of decreasing the angle between the mirrors.

This cluster of six sectors is common to inexpensive kaleidoscopes having two mirrors angled at 60°. The mirrors lie in a tube. At the viewing end you peer through a small opening at the center. At the other end lie small objects, often colored bits of plastic. By looking at the vertex formed by the mirrors at the far end, you see a cluster of six sectors, one of which is the direct view. If you turn the kaleidoscope, the colored bits of plastic fall into a new arrangement. Part of the kaleidoscope's charm is that the sixfold symmetry remains even when the bits of plastic fall chaotically into place.

An inexpensive kaleidoscope usually has mirrors that consist of a glass layer with a reflecting metallic coating on the rear surface. With such mirrors, light weakly reflects from the front surface of the glass and then more strongly reflects from the metal coating. The resulting double image has fuzzy edges. If the light undergoes several reflections from the mirrors, the image becomes even fuzzier. Better-quality kaleidoscopes come with "front surface mirrors" that have the metallic coating on the front of the glass, thus eliminating the double reflection and maintaining sharp edges on the images.

Return to the tabletop experiment. If you decrease the angle between the mirrors, the sectors narrow in angle while sector *d* begins to separate into two sectors that overlap at the rear of the display as shown in Fig. 3. The overlap is ambiguous: Its contents depend on which mirror you look into. If you face primarily toward mirror *A*, you see one type of scene at the rear of the display, whereas if you face toward mirror *F*, you see quite a different scene there. Moreover, the overlap of sectors at the rear leaves at least one of them smaller than the other sectors. The ambiguity and the unequal angular width of the sectors at the rear spoil the symmetry.

Are there any other angles between the mirrors that yield unambiguous sectors and offer unmarred symmetry? Yes, there are several, each an even divisor of 360°. For example, an angle of 45° gives an eightfold symmetry, with the rearmost sector requiring four reflections from the mirrors. The largest angle that gives unambiguous images is 180°, which is the same as saying that a single mirror can be mounted in the kaleidoscope tube. However, the twofold symmetry that results is dull. More interesting is the fourfold symmetry that is created when the angle is 90°.

Return to the tabletop arrangement with the mirror angle at 60°. Add a third identical mirror so that the edges at either end of the mirror system form an equilateral triangle. Look down into the assembly with a coin lying between the mirrors. The addition of the third mirror dramatically increases the number of images, with images extending in all directions in the plane of the tabletop as far as you can see. In princi-

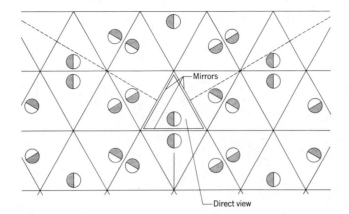

Figure 4 A portion of the image field produced by three mirrors.

ple, the number of images is infinite. However, the quality of the mirrors limits how far from the direct view you can see distinct images.

Part of the image plane is indicated in Fig. 4. Note that each vertex of the equilateral triangle of the direct view has a cluster of sectors with sixfold symmetry. The plane is completely covered with such triangles and clusters. This display is typical of what you see in a kaleidoscope where the ends of the mirrors form an equilateral triangle. The display is unambiguous. When you change your angle of view into the mirror system, the contents of the sectors are unchanged. If the direct view contains a variety of brightly colored objects, the image field can be dazzling. You see a repetition of patterns while also seeing the sixfold symmetry at each vertex.

The triangular pieces of the image field are created by reflections of light from the mirrors. At the vertices of the direct view, the number of reflections required is the same as when only two mirrors are in place. Sectors more distant from the direct view require more reflections. Figure 5 indicates the number of reflections required for some of the sectors.

Most three-mirror kaleidoscopes are based on the equilateral triangle design. What happens when the angles between the mirrors are changed? Suppose, for example, the ends of the mirrors form an isosceles triangle with angles of 50°, 65°, and 65°. Since the angles are not even divisors of 360°, many of the sectors are ambiguous, changing in content and angular size as you vary your angle of view into the kaleidoscope.

The most expensive kaleidoscope I own, costing well over $1000, contains

Figure 5 The number in an image triangle is the number of reflections involved in producing the image that appears to be in that triangle. The triangles continue out to infinity in all directions in the image plane.

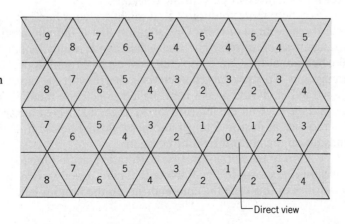

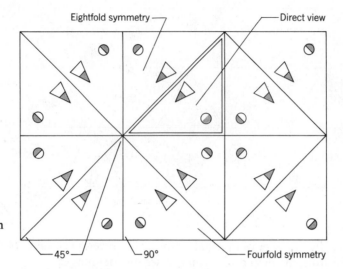

Figure 6 A mirror system giving two types of symmetry.

mirrors that form an isosceles triangle with angles of 22.5°, 78.75°, and 78.75°. At first glance, the image field produced by the instrument is enchanting because the 22.5° angle generates a 16-fold symmetry around one corner of the direct view. However, when I look at the rest of the image field, I see incomplete reflections that ruin any sense of symmetry.

Are there any mirror arrangements other than an equilateral triangle that give completely unambiguous image fields and full symmetry? Yes, there are, one arrangement being long known. Four mirrors forming a rectangle give unambiguous images with fourfold symmetry. I found two more solutions that were even more delightful because they simultaneously create more than one type of symmetry. One solution is a right triangle with two 45° angles (see Fig. 6). This mirror system gives eightfold symmetry around the 45° vertices and fourfold symmetry around the 90° vertices throughout the image field.

The other solution is the best, because it offers three types of symmetry. It is a right triangle with angles of 60° and 30° (see Fig. 7). The 90° vertices have fourfold symmetry, the 60° vertices have sixfold symmetry, and the 30° vertices have 12-fold symmetry. Soon after I published this design, two kaleidoscope makers incorporated it in their instruments. Even though I had calculated and mapped the image field, I was still stunned by the beauty of the images when I finally had the opportunity to peer into a kaleidoscope with the design.

Are there more solutions? You might try your hand at designing kaleidoscopes with three or more mirrors. One solution that is initially promising is a hexagonal arrangement of mirrors. The image field can certainly be filled with hexagons much like hexagonal tiles can completely cover a bathroom floor. However, the images are ambiguous. To show this characteristic, begin with the direct view that is hexagonal as shown in Fig. 8. Add a coin near the left side. Then mentally rotate the direct view around the edge on the left (mirror *A*) to get a reflected hexagon *b*. Then rotate *b* around virtual mirror *B* to get another reflected hexagon *c*. The image of the coin ends up on the bottom right of hexagon *c*.

Return to the initial hexagon *a*. Rotate it around mirror *C* to get another reflected hexagon *c* as shown in Fig. 9. Note that *c* fits exactly where it did with the clockwise rotations but this time the coin ends up on the bottom left. Thus, the hexagonal

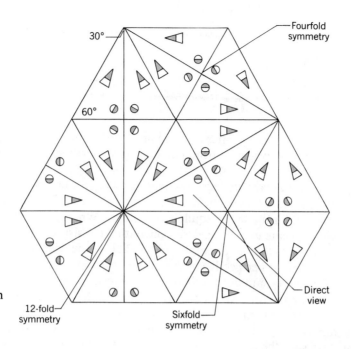

Figure 7 A mirror system giving three types of symmetry.

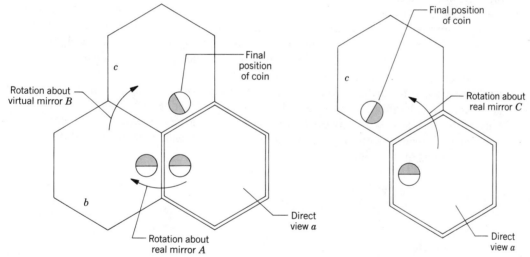

Figure 8 Clockwise reflection of a hexagon.

Figure 9 Counterclockwise reflection of a hexagon.

system is ambiguous. When you look into a kaleidoscope with a hexagonal mirror system, your angle of view determines what you see in *c*. For some angles you might even see two coins in *c*.

Many kaleidoscopes are open at the far end. When you hold them toward a colorful or moving scene, you see a mosaic of slightly different images. If the far end is equipped with a convex lens that compresses the external scene, the mosaic is even more striking. The pieces of the mosaic differ because they depend on the angle at which light rays enter the far end of the tube. The pieces that are near the direct view involve rays that strike a mirror with an incident angle that is close to 90° and then reflect once or twice before reaching you as shown in Fig. 10. These pieces are similar to the direct scene. The pieces far from the direct scene require rays that strike a

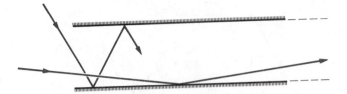

Figure 10 Rays entering an open kaleidoscope.

mirror with a much smaller incident angle and then reflect many times before reaching you. These rays originate from parts of the scenery well away from what you see in the direct view.

One of my favorite kaleidoscopes has slanted mirrors. At the far end of the instrument, the edges of the mirrors form a small equilateral triangle. The edges at the viewing end form a larger equilateral triangle. When you look into the instrument you see what appears to be a geodesic sphere that consists of many equilateral triangles and that seems to float in empty space. Each of the triangular pieces contains a slightly different perspective of the scenery toward which the kaleidoscope is pointed.

You can best understand how the geodesic sphere is created by looking down into a single mirror held on a table. Begin with the mirror vertical. The image of the table you see in the mirror is a flat continuation of the table you see directly. Now tilt the mirror away from you. The image of the table tilts down, rotating around its juncture with the table you see directly. In the kaleidoscope, the slanted mirrors tilt the images so that they seem to be glued to the side of a sphere. The images that are farther from the direct view require more reflections from the mirrors, which increases the tilt of the image and creates the illusion of there being a sphere in front of you.

A giant version of this type of kaleidoscope was once planned for Disney World by Stephen Hines, an optical engineer and designer. Spectators were to ride through the kaleidoscope toward the floating sphere. The sight would have been spectacular. However, considerations of expense reduced the kaleidoscope to one that is viewed only at the end. Still, with mirrors that are nearly 3 m long and with a triangular opening at the viewing end that is over a meter long on each edge, the scaled-down kaleidoscope is quite a sight.

CHAPTER 40
INTERFERENCE

The pyramid-shaped PAVE PAWS radar on Cape Cod sweeps each of its two fan-shaped microwave beams back and forth through an angle of ±60° every microsecond, searching for submarine-launched missiles. The two beams are generated by the constructive interference of microwave signals from the thousands of antenna elements — shown above — that cover two faces of the pyramid. The pyramid is as immobile as its Egyptian counterparts, the sweeping being done by periodically and electronically varying the phases of the microwave signals fed to these elements.*

40-1 Interference

Sunlight, as the rainbow shows us, is a composite of all the colors of the visible spectrum. The colors reveal themselves in the rainbow because the incident wavelengths are bent through different angles as they pass through raindrops that form the bow. However, the striking colors of soap bubbles, oil slicks, peacock feathers, and the throats of hummingbirds are not produced by *refraction* but by constructive and destructive *interference* of light reflected from the upper and lower surfaces of a thin film. The interfering waves combine either to enhance or to suppress certain colors in the spectrum of the incident sunlight.

This selective enhancement or suppression of selected wavelengths has many applications. When light falls on an ordinary glass surface, for example, about 4% of the incident energy is reflected, thus weakening the transmitted beam by that amount. This unwanted loss of light can be a real problem in optical systems with many

---

* The coverage is ± 120° in azimuth and a range of 3000 miles; see "Phased-Array Radars," by Eli Brookner, *Scientific American*, February 1985.

components. A thin transparent film, deposited on the optical surface, can largely suppress the reflected light (and thus enhance the transmitted light) by destructive interference. The bluish cast of your camera lens shows the presence of such a coating.

Sometimes we wish to enhance—rather than reduce—the reflectivity of a surface and this too can be done with interference coatings. In fact, an interference stack of a number of films, with differing thicknesses and indices of refraction, can be designed to give almost any desired wavelength profile for the reflected or the transmitted light.* For example, windows can be provided with coatings that have high reflectivity in the infrared, thus admitting the visible component of sunlight but reflecting its infrared or heating component.

The interference of light is not restricted to light reflected from the two surfaces of a thin film. Such phenomena can occur in any situation in which light from a single source is divided into two sub-beams that recombine after following paths of different lengths. To understand interference we must go beyond the restrictions of geometric optics and employ the full power of wave optics. In fact, the existence of interference phenomena—as we shall see—is perhaps our most convincing evidence that light is a wave.

40-2   Light as a Wave

The first person to advance a convincing wave theory for light was the Dutch physicist Christian Huygens, in 1678. While much less comprehensive than the later electromagnetic theory of Maxwell, it was simpler mathematically and remains useful today. Its great merit is that it can account for the laws of reflection and refraction in terms of a wave picture and that it gives physical meaning to the index of refraction.

Huygens' wave theory is based on a geometrical construction that allows us to tell where a given wavefront will be at any time in the future if we know its present position. This construction is based on *Huygens' principle,* which is:

> *All points on a wave front serve as point sources of spherical secondary wavelets. After a time t, the new position of the wave front will be the surface of tangency to these secondary wavelets.*

* See "Optical Interference Coatings," by Philip Baumeister and Gerald Pincus, *Scientific American,* December 1970.

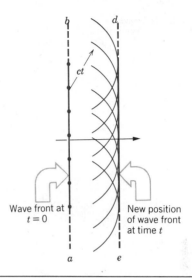

Figure 1   The propagation of a plane wave in free space, as portrayed by Huygens' principle.

Here is a simple example: Given a wave front, *ab* in Fig. 1, in a plane wave in free space, where will the wave front be a time *t* later? We let several points on this plane (see dots) serve as centers for secondary spherical wavelets. In a time *t*, the radius of these spherical waves is *ct*, where *c* is the speed of light in free space. We represent the plane of tangency to these spheres at time *t* by *de*. As we expect, it is parallel to plane *ab* and a perpendicular distance *ct* from it. Thus, plane wave fronts are propagated as planes and with speed *c*. Note that the Huygens method involves a three-dimensional construction and that Fig. 1 is the intersection of this construction with the plane of the page.

The Law of Refraction.   We now use Huygens' principle to derive the law of refraction. Figure 2 shows four stages in the refraction of three wave fronts at a plane interface between air (medium 1) and glass (medium 2). We choose the wave fronts in the incident beam to be separated, arbitrarily, by λ_1, the wavelength in medium 1. Let the speed of light in air be v_1 and that in glass be v_2. We assume that $v_2 < v_1$, which happens to be true.

The angle θ_1 in Fig. 2a is the angle between the wave front and the surface; this is the same as the angle between the *normal* to the wave front (that is, the incident ray) and the *normal* to the surface; thus, θ_1 is the angle of incidence. The time ($= \lambda_1/v_1$) for a Huygens wavelet to expand from point *e* in Fig. 2b to include point *c* will equal the time ($= \lambda_2/v_2$) for a wavelet in the glass to expand at the reduced speed v_2 from *h* to include *e'*.

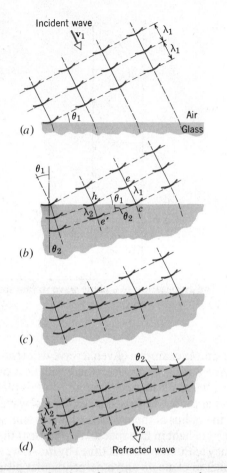

Figure 2 The refraction of a plane wave at a plane surface, as portrayed by Huygens' principle. Note that the wavelength in glass is smaller than that in air. For simplicity, the reflected wave is not shown.

Equating these times, we have

$$\frac{\lambda_1}{\lambda_2} = \frac{v_1}{v_2}, \tag{1}$$

which shows that the wavelength is proportional to the speed in the medium.

The refracted wave front must be tangent to an arc of radius λ_2 centered on h. Since c lies on the new wave front, the tangent must pass through this point also. Note that θ_2, the angle between the refracted wave front and the surface, is equal to the angle of refraction.

For the right triangles hce and hce' we may write

$$\sin \theta_1 = \frac{\lambda_1}{hc} \quad \text{(for } hce)$$

and

$$\sin \theta_2 = \frac{\lambda_2}{hc} \quad \text{(for } hce').$$

Dividing these two equations and using Eq. 1, we find

$$\frac{\sin \theta_1}{\sin \theta_2} = \frac{\lambda_1}{\lambda_2} = \frac{v_1}{v_2}. \tag{2}$$

We can define an *index of refraction* for each medium as the ratio of the speed of light in free space to the speed of light in the medium. Thus,

$$\boxed{n = \frac{c}{v}.} \tag{3}$$

In particular, for our two media, we have

$$n_1 = \frac{c}{v_1} \quad \text{and} \quad n_2 = \frac{c}{v_2}. \tag{4}$$

If we combine Eqs. 2 and 4 we find

$$\frac{\sin \theta_1}{\sin \theta_2} = \frac{c/v_2}{c/v_1} = \frac{n_2}{n_1} \tag{5}$$

or

$$n_1 \sin \theta_1 = n_2 \sin \theta_2 \quad \text{(law of refraction),} \tag{6}$$

which is the law of refraction.

Counting Waves. When two waves from the same source rejoin after following separate paths, we are less concerned with the separate geometric lengths of those paths than with the number of waves each path contains. It is the difference in this number of waves between the two paths that determines the phase difference between the waves and thus the intensity of the light. If one or both paths are through media with different indices of refraction, we must take into account the fact that the wavelength of light depends on the index of refraction.

From Eq. 1 we can write, for the wavelength λ_n in a given medium in terms of the wavelength λ in a vacuum,

$$\lambda_n = \lambda \frac{v}{c} \tag{7}$$

or

$$\boxed{\lambda_n = \frac{\lambda}{n},} \tag{8}$$

where n is the index of refraction of the medium with

respect to a vacuum. Thus, as Fig. 2 makes clear, the higher the index of refraction, the shorter the wavelength of the light.

If rays diverging from a point source form a real image after passing through a converging lens, the rays travel over paths of different *geometric* length but all paths contain the same number of waves; otherwise the rays would not be in phase when they recombine at the image point. It is, in fact, the function of the lens — taking advantage of the fact that the wavelength in glass is less than that in air — to cause the rays to follow paths that contain the same number of waves. For a converging lens a short path, directly along the lens axis, passes through the largest thickness of glass. A path that passes through the edge of the lens passes through a shorter thickness of glass, which just compensates for its greater geometric length.

40-3   Diffraction

To understand the interference of two combining waves, we must first understand the central features of the diffraction of waves, a subject that we explore much more fully in Chapter 41. If a wave falls on a barrier that has an opening of dimensions similar to the wavelength, the wave will flair out into the region beyond. This phenomenon, called *diffraction,* is in the spirit of the spreading out of the wavelets in the Huygens construction of Fig. 1. Figure 3 shows the diffraction of water waves in a shallow

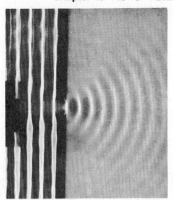

Figure 3   The diffraction of water waves in a ripple tank. Waves moving from left to right flare out through an opening in a barrier into the space beyond. Note that the opening is about the same size as the wavelength.

ripple tank. These waves were produced by tapping the water surface periodically with a flat stick.

Figure 4*a* shows the situation schematically for an incident plane wave of wavelength λ falling on a slit of width $a = 6.0 \lambda$. The light clearly flares out on the far side of the slit. Figures 4*b* ($a = 3.0 \lambda$) and 4*c* ($a = 1.5 \lambda$) illustrate the main feature of diffraction: The narrower the slit, the greater the diffraction.

We see here the limitations of geometric optics, whose central feature is a ray following a linear path. If we try to form such a ray physically by allowing light to fall on a narrow slit, or a series of narrow slits, we are

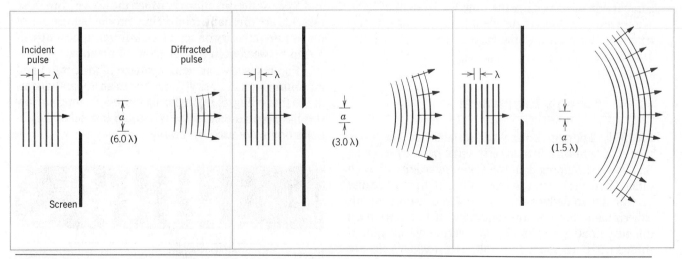

Figure 4   Diffraction represented schematically. For a given wavelength λ, the diffraction is more pronounced the smaller the slit width a. The figures show the cases for $a = 6.0 \lambda$, $a = 3.0 \lambda$, and $a = 1.5 \lambda$.

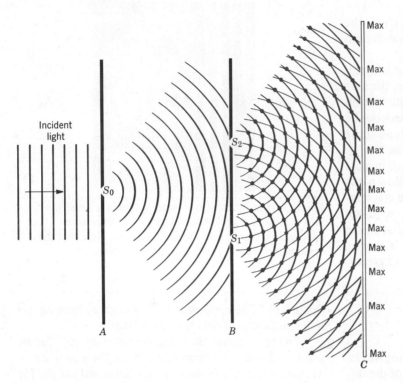

Max
Max
Max
Max
Max
Max
Max
Max
Max
Max
Max
Max
Max

S_2
S_1
S_0

Incident light

A B C

Figure 5 In Thomas Young's interference experiment, light diffracted from pinhole S_0 falls on pinholes S_1 and S_2 in screen B. Light diffracted from these two pinholes overlaps in the region between screens B and C, producing an interference pattern on screen C.

foiled at every turn by diffraction. Indeed, the more we try to define the ray physically (that is, the narrower we make the slits), the greater is the spreading out by diffraction. We said in Chapter 39 that geometric optics only holds if we place no barriers, slits, or other apertures in the path of the light beam if their dimensions are comparable to or smaller than the wavelength of light. We see now that this is equivalent to saying that geometric optics holds only to the extent that we can neglect diffraction.

40–4 Young's Experiment

In 1801, Thomas Young first established the wave theory of light on a firm experimental basis by showing that two overlapping light waves can *interfere* with each other. His experiment was especially convincing because he was able to deduce the wavelength of light from his experiments, the first measurement of this important quantity. Young's value for the average wavelength of sunlight, 570 nm in modern units, is remarkably close to the modern value of 555 nm.

Young allowed sunlight to fall on a pinhole S_0 punched in a screen A. As represented in Fig. 5, the

emerging light spreads out by diffraction and falls on pinholes S_1 and S_2 punched into screen B. Diffraction occurs again at the two pinholes and two overlapping spherical waves expanded into the space to the right of screen B, where they can interfere with each other. The lines of dots extending radially outward from the slits S_1 and S_2 show the directions in which the waves from these two slits are mutually reinforcing, producing regions of maximum intensity on screen C. Minima appear on this screen between each pair of adjacent maxima.

Figure 6 shows an actual pattern of maximum and minimum light intensity, generated in an apparatus similar to that of Fig. 5. In preparing this figure, long narrow slits have been used in screens A and B instead of the pinholes originally used by Young.

Figure 6 An interference pattern produced by the arrangement shown in Fig. 5, with narrow slits substituted for the pinholes. The alternating maxima and minima are called *interference fringes*.

Figure 7 shows an interference pattern set up by overlapping water waves in a ripple tank. The waves are generated by two spheres connected to the same mechanical vibrator and oscillating up and down through the water surface. The two spheres of Fig. 7 are to be compared to the pinholes S_1 and S_2 of Fig. 5.

Let us analyze Young's experiment qualitatively, assuming that the incident light consists of a single wavelength only. In Fig. 8, P is an arbitrary point on screen C, which is a distance D from the screen containing the two narrow slits, S_1 and S_2. Let θ be the angle between cP and the horizontal line cO. Let b be a point on the line PS_1 such that the distances PS_2 and Pb are equal. The waves emitted at the slits, S_1 and S_2, are initially in phase with each other. We can see in the figure though that the wave emitted at the slits, S_1 and S_2 are initially in phase with wave emitted at S_2. The difference in the path lengths of these two waves is simply S_1b. The nature of the interference of the two waves at P depends entirely on the length of the path difference. For example, if the path difference S_1b contains an integral number of wavelengths ($S_1b = m\lambda$, where m is an integer), the two waves will be in phase at P and will interfere constructively, producing a maximum in the intensity of light on the screen at that point.

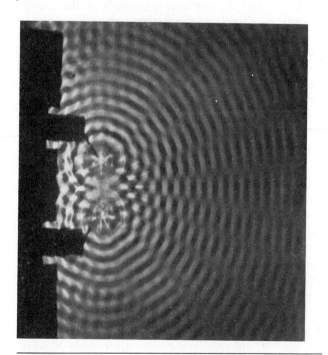

Figure 7 An interference pattern produced by water waves in a ripple tank. Compare with the region between screens B and C in Fig. 5.

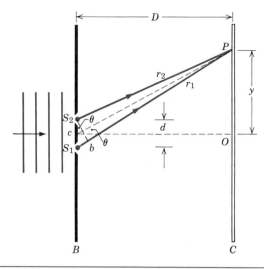

Figure 8 Waves from slits S_1 and S_2 combine at P, an arbitrary point on the screen. The angle θ serves as a convenient locator for P. Actually, $D \gg d$, the figure being distorted for clarity.

Now for the quantitative part. Let us assume that the slit spacing d is very much smaller than the distance D between the two screens (the ratio d/D in Fig. 8 has been greatly exaggerated for clarity). This means that the lines S_1P and S_2P are nearly parallel to each other.* The triangle S_1S_2b is then a right triangle, and the angle between S_1S_2 and bS_2 is equal to θ. Thus, the length of the side S_1b in this triangle is equal to $d \sin \theta$. This is the path difference of the two waves so that the conditions for a maximum intensity at P is

$$d \sin \theta = m\lambda, \qquad m = 0, 1, 2, \ldots$$

(maxima). (9)

For each maximum above O in Fig. 8 there is a symmetrically located maximum below O. There is a central maximum, described by $m = 0$.

For a minimum at P, $S_1b (= d \sin \theta)$ must contain a half-integral number of wavelengths, or

$$d \sin \theta = (m + \tfrac{1}{2})\lambda, \qquad m = 0, 1, 2, \ldots$$

(minima). (10)

---

* We often put a lens behind the two slits, the screen being in the focal plane of the lens. Under these conditions the rays focussed at P are strictly parallel on striking the lens and we can drop the requirement that $d \ll D$. The lens may in practice be the lens and cornea of the eye, the screen being the retina.

Sample Problem 1 What is the linear distance on screen C in Fig. 8 between adjacent maxima? The wavelength λ is 546 nm; the slit separation d is 0.12 mm, and the slit–screen separation D is 55 cm.

We assume from the start that the angle θ in Fig. 8 will be small enough to permit us to use the approximations

$$\sin \theta \approx \tan \theta \approx \theta,$$

in which θ is to be expressed in radian measure. From Fig. 8 we see that

$$\tan \theta \approx \theta = \frac{y}{D}.$$

From Eq. 9 we have

$$\sin \theta \approx \theta = \frac{m\lambda}{d}.$$

If we equate these two expressions for θ and solve for y, we find

$$y_m = \frac{m\lambda D}{d}. \tag{11}$$

For the adjacent fringe, we have

$$y_{m+1} = \frac{(m+1)\lambda D}{d}. \tag{12}$$

We find the fringe separation by subtracting Eq. 11 from Eq. 12, or

$$\begin{aligned}
\Delta y = y_{m+1} - y_m &= \frac{\lambda D}{d} \\
&= \frac{(546 \times 10^{-9} \text{ m})(55 \times 10^{-2} \text{ m})}{0.12 \times 10^{-3} \text{ m}} \\
&= 2.50 \times 10^{-3} \text{ m} \approx 2.5 \text{ mm}. \quad \text{(Answer)}
\end{aligned}$$

As long as d and θ in Fig. 8 are small, the separation of the interference fringes is independent of m; that is, the fringes are evenly spaced. If the incident light contains more than one wavelength, the separate interference patterns, which will have different fringe spacings, will be superimposed.

40–5 Coherence

If we are to have well-defined interference fringes on screen C in Fig. 5, the light waves that travel from S_1 and S_2 to any point on this screen must have a well-defined phase difference ϕ that remains constant with time. That will be the case for the experiment of Fig. 5 because the two slits S_1 and S_2 are illuminated by light diffracted from the same source slit S_0. The two beams emerging

from slit S_1 and S_2 are said to be *completely coherent*, because their phase difference does not change with time. Coherent beams in the radiofrequency region of the electromagnetic spectrum can be formed by connecting two separate antennas to the same oscillator.

Dark fringes on screen C in Fig. 5 correspond to places at which the phase difference ϕ of the waves arriving from the two slits is π, 3π, 5π, $\ldots$. Bright fringes are places at which this phase difference is 0, 2π, 4π, 6π, $\ldots$.

Imagine, however, that the slits S_1 and S_2 in Fig. 5 are replaced by two completely independent light sources, such as two fine incandescent wires placed side by side. No interference fringes would appear on screen C but only a relatively uniform illumination. We can understand this if we assume that for completely independent light sources the phase difference between the two beams arriving at any point on screen C in Fig. 5 will vary with time in a random way. At a certain instant conditions may be right for cancellation and a short time later they may be right for reinforcement. The eye, however, cannot follow these rapid variations and sees only a uniform illumination. The intensity at any point on screen C is equal to the sum of the intensities that each source S_1 and S_2 produces separately at that point. Under these conditions the two beams emerging from S_1 and S_2 are said to be *completely incoherent*. Incoherent beams in the radiofrequency part of the electromagnetic spectrum are formed by connecting two antennas to completely independent oscillators.

For completely coherent light beams we (1) combine the amplitudes vectorially, taking the (constant) phase difference properly into account, and then (2) square this resultant amplitude to obtain a quantity proportional to the resultant intensity. For completely incoherent light beams, on the other hand, we (1) square the individual amplitudes to obtain quantities proportional to the individual intensities and then (2) add the individual intensities to obtain the resultant intensity. This procedure is in agreement with the experimental fact that for completely independent light sources the resultant intensity at every point is always greater than the intensity produced at that point by either light source acting alone.

In common sources of visible light, such as incandescent wires or an electric discharge passing through a gas, the fundamental light emission processes occur in individual atoms and these atoms do not act together in a cooperative (that is, coherent) way. The act of light emission by a single atom in a typical case takes about 10^{-8} s

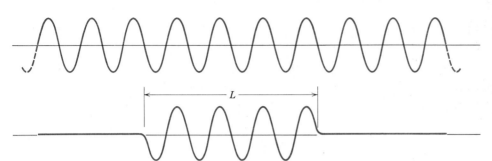

Figure 9 *(a)* A section of a pure sine wave, with no beginning and no ending. *(b)* A *wavetrain,* of finite length L. Light from ordinary incandescent sources is emitted as wavetrains, originating with individual atoms.

and the emitted light is properly described as a *wavetrain* (Fig. 9*b*) rather than as a wave (Fig. 9*a*). For emission times such as these, the wavetrains are a few meters long.

Interference effects from ordinary light sources may be produced by putting a narrow slit (S_0 in Fig. 5) directly in front of the source. This ensures that, by diffraction, the same families of wavetrains fall on slits S_1 and S_2. The diffracted beams emerging from S_1 and S_2 are thus coherent with respect to each other. When the phase of the light emitted from S_0 changes, this change is transmitted simultaneously to S_1 and S_2. Thus, at any point on screen C, a constant phase difference is maintained between the beams from these two slits and a stationary interference pattern occurs.

The lack of coherence of the light from common sources such as the sun or incandescent lamp filaments is due to the fact that the emitting atoms in such sources act independently rather than cooperatively. Since 1960 it has been possible to make light sources in which the atoms do act cooperatively, the emitted light being highly coherent. Such devices are the familiar *lasers,* a coined word derived from: *l*ight *a*mplification through the *s*timulated *e*mission of *r*adiation. Laser light (which we shall describe more fully in Chapter 45 of the extended version of this book) is not only highly coherent but also extremely monochromatic, highly directional, and focusable to a spot whose dimensions are of the order of magnitude of one wavelength. This "light fantastic," as it has been called, has a bewildering variety of applications, among which are telephone communication over optical fibers, spotwelding detached retinas, hydrogen fusion research, measuring continental drift, and holography, a technique in which three-dimensional images can be produced from information stored in a two-dimensional matrix.

40-6 Intensity in Double-Slit Interference

Equations 9 and 10 tell us how to locate the maxima and the minima of the double-slit interference fringes on screen C of Fig. 8 as a function of the angle θ in that figure. Note that θ is our position locator on the screen, every screen point P being associated with a definite value of θ. Here we wish to derive an expression for the intensity I of the fringes as a function of θ.

Let us assume that the electric field components of the light waves arriving at point P in Fig. 8 from the two slits vary with time as

$$E_1 = E_0 \sin \omega t \qquad (13)$$

and

$$E_2 = E_0 \sin(\omega t + \phi), \qquad (14)$$

where ω is the angular frequency of the waves and ϕ is the phase difference between them. We shall show below that these two waves will combine at P to produce an intensity of illumination I given by

$$\boxed{I = 4I_0 \cos^2 \tfrac{1}{2}\phi,} \qquad (15)$$

where

$$\boxed{\phi = \frac{2\pi d}{\lambda} \sin \theta.} \qquad (16)$$

In Eq. 15, I_0 is the intensity on the screen associated with light from one of the two slits, the other slit being temporarily covered. We assume that the slits are so narrow in comparison to the wavelength that this single-slit intensity is essentially uniform over the region of the screen in which we wish to examine the fringes.

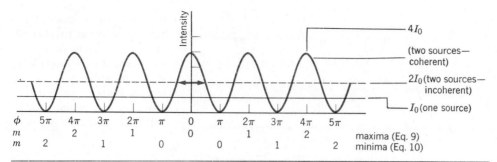

Figure 10 A plot of Eq. 15, showing the intensity of a double-slit interference pattern as a function of the phase difference between the waves from the two slits. I_0 is the (uniform) intensity that would appear on the screen if one slit were covered. The average intensity of the fringe pattern is $2I_0$ and the *maximum* intensity is $4I_0$.

Equations 15 and 16, which together tell us how the intensity I of the fringe pattern varies with the angle θ in Fig. 8, necessarily contain information about the location of the maxima and the minima. Let us see if we can extract it.

Study of Eq. 15 shows that intensity maxima will occur when

$$\tfrac{1}{2}\phi = m\pi, \qquad m = 0, 1, 2, \ldots \qquad (17)$$

If we put this result into Eq. 16 we find

$$2m\pi = \frac{2d\pi}{\lambda}\sin\theta$$

or

$$d\sin\theta = m\lambda, \qquad m = 0, 1, 2, \ldots \text{ (maxima)}, \quad (18)$$

which is exactly Eq. 9, the expression that we derived earlier for the location of the maxima.

The minima in the fringe pattern occur when

$$\tfrac{1}{2}\phi = (m + \tfrac{1}{2})\pi, \qquad m = 0, 1, 2, \ldots.$$

If we combine this relation with Eq. 16, we are led at once to

$$d\sin\theta = (m + \tfrac{1}{2})\lambda, \qquad m = 0, 1, 2, \ldots \text{ (minima)}, \quad (19)$$

which is just Eq. 10, the expression derived earlier for the location of the fringe minima.

Figure 10, which is a plot of Eq. 15, shows the intensity pattern for double-slit interference as a function of the phase angle ϕ. The horizontal solid line is I_0, the (uniform) intensity on the screen if one of the slits is covered up. Note from Eq. 15 that the intensity (which is always positive) varies from zero at the fringe minima to $4I_0$ at the fringe maxima.

If the two sources (slits) that illuminate the screen were *incoherent,* so that no enduring phase relation existed between them, there would be no fringe pattern and the intensity would have the uniform value $2I_0$ for all points on the screen; see the horizontal dashed line in Fig. 10.

Interference cannot create or destroy energy but merely redistributes it over the screen. Thus, the *average* intensity on the screen must be the same, whether the sources are coherent or not. This follows at once from Eq. 15; if we substitute one-half for the average value of the cosine-squared term, this equation reduces to $\bar{I} = 2I_0$. We have many times used the fact that the average value of the square of a sine or a cosine term over one or more half-cycles is one-half.

Proof of Eqs. 15 and 16. (Optional). We choose to combine E_1 and E_2, given by Eqs. 13 and 14, respectively, by the method of *phasors,* which we encountered in our studies of alternating currents in Chapter 36. The method will be especially useful later, when we want to combine a large number of wave disturbances with differing phases.

In Fig. 11a, we represent E_1 of Eq. 13 by a rotating phasor of amplitude E_0. In that figure, the alternating wave disturbance E_1 is simply the projection of the phasor E_0 on the vertical axis.

A second wave disturbance E_2, which has the same amplitude E_0 but a phase difference ϕ with respect to E_1 (see Eq. 14), can be represented (see Fig. 11b) as the projection on the vertical axis of a second phasor of magnitude E_0, which makes a fixed angle ϕ with the first phasor. As this figure shows, the sum of E_1 and E_2, which is the instantaneous amplitude of the resultant wave, is the sum of the projections of the two phasors on the vertical axis. This is revealed more clearly if we redraw the phasors, as in Fig. 11c, placing the foot of one arrow

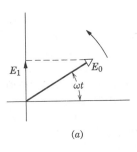

(a)

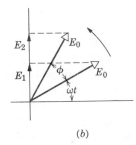

(b)

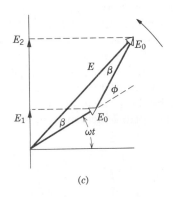

(c)

Figure 11 (*a*) A time-varying wave disturbance E_1 is represented as the projection of a rotating *phasor*. (*b*) Two phasors, with a constant phase difference ϕ between them. (*c*) Another way of drawing (*b*).

at the head of the other, maintaining the proper phase difference, and letting the whole assembly rotate counterclockwise about the origin.

In Fig. 11*c*, this sum can also be regarded as the projection on a vertical axis of a phasor of amplitude E, which makes a phase angle β with respect to the phasor that generates E_1. Note that the (algebraic) sum of the projections of the two phasors is equal to the projection of the (vector) sum of the two phasors.

In most problems in optics we are concerned only with the amplitude E of the resultant wave and not with its time variation. This is because the eye and most measuring instruments respond to the intensity of the light

(that is, to the square of the amplitude) and cannot detect the rapid time variations. For light at the center of the visible spectrum ($\lambda \approx 550$ nm), the frequency $\nu (= c/\lambda)$ is about 5×10^{14} Hz. Usually, we need not consider the rotation of the phasors and can confine our attention to finding the amplitude of the resultant phasor.

Let us find the amplitude E in Fig. 11*c*. From the theorem that an exterior angle (ϕ) is equal to the sum of the two opposite interior angles ($\beta + \beta$) we see from Fig. 11*c* that $\beta = \frac{1}{2}\phi$. Thus, we have

$$E = 2(E_0 \cos \beta) = 2E_0 \cos \tfrac{1}{2}\phi. \qquad (20)$$

If we square each side of this relation we obtain

$$E^2 = 4E_0^2 \cos^2 \tfrac{1}{2}\phi. \qquad (21)$$

We learned in Section 38-5 that the intensity of a wave is proportional to the square of its amplitude. Thus, we can write for Eq. 21

$$I = 4I_6 \cos^2 \tfrac{1}{2}\phi,$$

which is Eq. 15, one of the equations that we set out to prove.

It remains to prove Eq. 16, which relates the phase difference ϕ between the waves arriving at any point P on the screen of Fig. 8 to the angle θ that serves as a locator of that point.

The phase difference ϕ in Eq. 14 is associated with a path difference $S_1 b$ in Fig. 8. If $S_1 b$ is $\frac{1}{2}\lambda$, ϕ will be π; if $S_1 b$ is λ, ϕ will be 2π, and so on. This suggests

$$\text{phase difference} = \frac{2\pi}{\lambda} \, (\text{path difference}). \qquad (22)$$

The path difference $S_1 b$ in Fig. 8 is just $d \sin \theta$ so that Eq. 22 becomes

$$\phi = \frac{2\pi d}{\lambda} \sin \theta,$$

which is just Eq. 16, the other equation that we set out to prove.

In a more general case we might want to find the resultant of a number ($n > 2$) of sinusoidally varying wave disturbances. The general procedure is as follows:

1. Construct a series of phasors representing the functions to be added. Draw them end to end, maintaining the proper phase relations between adjacent phasors.

2. Construct the vector sum of his array. Its length gives the amplitude of the resultant. The angle between it and the first phasor is the phase of the resultant with

respect to this first phasor. The projection of this phasor on the vertical axis gives the time variation of the resultant wave disturbance.

Sample Problem 2 Find graphically, using phasor methods, the resultant $E(t)$ of the following wave disturbances:

$$E_1 = E_0 \sin \omega t$$
$$E_2 = E_0 \sin(\omega t + 15°)$$
$$E_3 = E_0 \sin(\omega t + 30°)$$
$$E_4 = E_0 \sin(\omega t + 45°).$$

Figure 12 shows the assembly of four phasors that represent these functions. We find by graphical measurement with a ruler and a protractor that the amplitude E_R is 3.8 times as long as E_0 and that the resultant wave makes a phase angle ϕ_0 of 22.5° with respect to E_1. In other words,

$$E(t) = E_1 + E_2 + E_3 + E_4$$
$$= 3.8 \sin(\omega t + 22.5°). \qquad \text{(Answer)}$$

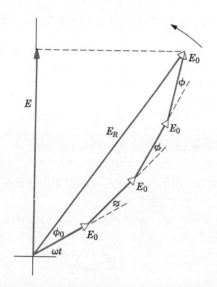

Figure 12 Sample Problem 2. Four wave disturbances added graphically, using the method of phasors. E is the amplitude of the resultant wave.

Check this result by direct trigonometric calculation or by geometric calculation from the phasor diagram of Fig. 12.

40–7 Interference from Thin Films

The colors that we see when sunlight falls on a soap bubble, an oil slick, or a ruby-throated hummingbird are caused by the interference of light waves reflected from the front and back surfaces of a thin transparent film. The film thickness is typically of the order of magnitude of the wavelength of the light involved. Thin-film technology, including the deposition of multiple-layered films, is highly developed and is widely used for the control of the reflection and/or transmission of light or radiant heat at optical surfaces.

Figure 13 shows a transparent film of uniform thickness d illuminated by monochromatic light of wavelength λ from a point source S. The eye is so positioned that a particular incident ray i from the source enters the eye, as ray r_1, having been reflected from the front surface of the film at a. The incident ray also enters the film at a as a refracted ray and is reflected from the back surface of the film at b; it then reemerges from the front surface of the film at c and also enters the eye, as ray r_2. The geometry of Fig. 13 is such that rays r_1 and r_2 are parallel. Having originated in the same point source, they are also coherent and are thus in a position to interfere. Because these two rays have traveled over paths of different lengths, have traversed different media, and — as we shall see — have suffered different kinds of reflections at a and at b, there will be a phase difference between them. The intensity perceived by the eye, as the parallel rays from the region a,c of the film enter it, is determined by this phase difference.

For near-normal incidence ($\theta_i \approx 0°$ in Fig. 13) the

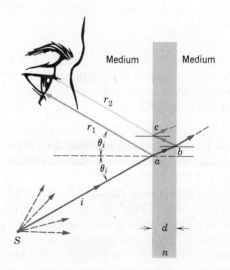

Figure 13 A thin film is viewed by light reflected from source S. Waves reflected from the front surface and the back surface enter the eye as shown, the intensity of the resultant wave being determined by the phase difference between the combining waves. The medium is assumed to be air.

geometric path difference for the two rays from S will be close to $2d$. We might expect the resultant wave reflected from the film to be an interference maximum if the distance $2d$ is an integral number of wavelengths. This statement must be modified for two reasons, which we discuss in turn.

First Reason. In counting the number of wavelengths in the path difference $2d$, we must be sure to use the wavelength in the medium λ_n and not the wavelength in air λ. These two quantities are related by (see Eq. 8)

$$\lambda_n = \frac{\lambda}{n} \qquad (23)$$

in which n is the index of refraction of the film material with respect to air.

Second Reason. To bring out the second point, let us assume that the film is so thin that $2d$ is very much less than one wavelength. We might assume the phase difference between the two waves would be close to zero and we would expect such a film to appear bright on reflection. However, it appears dark. This is clear from Fig. 14, which shows a thin vertical film of soapy water, viewed by reflected monochromatic light. The action of gravity produces a wedge-shaped film, extremely thin at its top edge. To explain this and many similar phenomena, we assume that one or the other of the two rays of Fig. 13 suffers an abrupt phase change of 180° when it is reflected from the air–film interface.

Figure 14 A soapy water film on a wire loop, viewed by reflected light. The black segment at the top is not a torn film. It arises because the film, by drainage, is so thin there that there is destructive interference between waves reflected from its two surfaces.

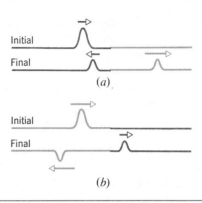

Figure 15 Phase changes on reflection at a junction between two stretched composite strings of different linear densities. The wave speed is greater in the lighter string. (*a*) The incident pulse is in the heavier string. (*b*) The incident pulse is in the lighter string.

In Section 17–12 we discussed phase changes on reflection for transverse waves in strings. To extend these ideas, consider the composite string of Fig. 15, which consists of two parts with different linear densities, stretched to the same tension. Recall from Eq. 26 of Chapter 17 that the wave speed is greater in the lighter string.

In Fig. 15*a* the incident pulse is in the heavier string and in Fig. 15*b* it is in the lighter string. In each case, the transmitted pulse suffers no change in sign as it passes through the boundary.

The behavior of the reflected pulses is different, however. When the incident pulse approaches a junction beyond which its speed will be greater (Fig. 15*a*), the reflected pulse suffers no phase change. However, when the incident pulse approaches a junction beyond which its speed is less (Fig. 15*b*), the reflected pulse is reversed in sign. That is, it undergoes a phase change of 180°.

Figure 15*a* suggests a light wave in glass approaching a surface beyond which there is a medium of lower index of refraction (which implies a larger wave speed) such as air. Figure 15*b* suggests a light wave in air approaching glass. The optical situation can be summed up as follows:

When reflection occurs from an interface beyond which the medium has a lower index of refraction, the reflected wave undergoes no phase change; when the medium beyond the interface has a higher index, there is a phase change of 180°. The transmitted wave undergoes no phase change in either case.

We are now able to take into account both factors that determine the nature of the interference, namely, the dependence of the wavelength on the index of refraction and phase changes on reflection. For the two rays of Fig. 13 to combine to give a maximum intensity, assuming normal incidence, we must have

$$2d = (m + \tfrac{1}{2})\lambda_n, \qquad m = 0, 1, 2, \dots$$

The term $\tfrac{1}{2}\lambda_n$ is introduced because of the phase change on reflection, a phase change of $180°$ being equivalent to half a wavelength. Substituting λ/n for λ_n yields finally

$$\boxed{2nd = (m + \tfrac{1}{2})\lambda,} \quad m = 0, 1, 2 \dots$$
$$\text{(maxima).} \quad (24)$$

The condition for a minimum intensity is

$$\boxed{2nd = m\lambda,} \quad m = 0, 1, 2, \dots \quad \text{(minima).} \quad (25)$$

These equations hold when the index of refraction of the film is either greater or less than the indices of the media on each side of the film. In these cases, there will be a relative phase change of $180°$ for reflections at the two surfaces since there will be a phase reversal at one reflection but not at the other. A water film in air and an air film in the space between two glass plates provide examples of cases to which Eqs. 24 and 25 apply.

If the film thickness is not uniform, as in Fig. 14, where the film is wedge shaped because of the action of gravity, constructive interference will occur in certain parts of the film and destructive interference will occur in others. Lines of maximum and of minimum intensity will appear, called *fringes of constant thickness*. Each fringe is the locus of points for which the film thickness d is constant. If the film is illuminated with white light rather than monochromatic light, the light reflected from various parts of the film will be modified by the various constructive or destructive interferences that occur. This accounts for the brilliant colors of soap bubbles and oil slicks.

Sample Problem 3 A water film ($n = 1.33$) in air is 320 nm thick. If it is illuminated with white light at normal incidence, what color will it appear to be in reflected light?

If we solve Eqs. 24 and 25 for λ, we find

$$\lambda = \frac{2nd}{m + \tfrac{1}{2}} = \frac{(2)(1.33)(320 \text{ nm})}{m + \tfrac{1}{2}}$$

$$= \frac{851 \text{ nm}}{m + \tfrac{1}{2}} \quad \text{(maxima)}$$

and

$$\lambda = \frac{851 \text{ nm}}{m} \quad \text{(minima).}$$

Maxima and minima occur for the wavelengths shown in the following table, in which IR and UV stand for infrared and ultraviolet, respectively:

m	0 (max)	1 (min)	1 (max)	2 (min)	2 (max)
λ (nm)	1702	851	567	426	340
Color	IR	IR	Yellow-green	Deep purple	UV

Only the maximum corresponding to $m = 1$ lies in the visible region (see Fig. 2 of Chapter 38): Light of this wavelength appears yellow-green. If white light is used to illuminate the film, the yellow-green component will be enhanced when viewed by reflection.

Sample Problem 4 Lenses are often coated with thin films of transparent substances like magnesium fluoride (MgF_2; $n = 1.38$) to reduce the reflection from the glass surface. How thick a coating is needed to produce a minimum reflection at the center of the visible spectrum ($\lambda = 550$ nm)?

We assume that the light strikes the lens at near-normal incidence, θ being exaggerated for clarity in Fig. 16. We seek the film thickness that will bring about destructive interference between rays r_1 and r_2.

Equation 25 does *not* apply in this case because a phase change of $180°$ is now associated with *each* ray and not with

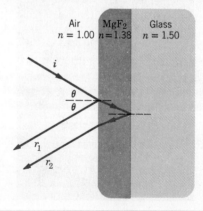

Figure 16 Sample Problem 4. Unwanted reflections from glass can be suppressed (at a chosen wavelength) by coating the glass with a thin transparent film of magnesium fluoride of properly chosen thickness.

only one ray as in Fig. 13. At both the front and rear surfaces of the MgF$_2$ film in Fig. 16 the reflection is from a medium of greater index of refraction so that there is no *net* change in phase produced by the two reflections.

This means that, to find the thickness of the thinnest possible film, the path difference between the two rays (which is $2d$) must equal one-half of a wavelength of light in the medium (= $\frac{1}{2}\lambda_n = \lambda/2n$). Thus,

$$2d = \frac{\lambda}{2n}$$

or

$$d = \frac{\lambda}{4n} = \frac{550 \text{ nm}}{(4)(1.38)} = 100 \text{ nm}. \qquad \text{(Answer)}$$

Note that this answer is independent of the index of refraction of the glass (or of the air, for that matter) as long as $n_{\text{glass}} > 1.38$.

40-8   Michelson's Interferometer

An *interferometer* is a device that can be used to measure lengths or changes in length with great accuracy by means of interference fringes. We describe the form originally devised and built by A. A. Michelson in 1881.* Consider light that leaves point P on extended source S (Fig. 17) and falls on a *beam splitter M*. This is a mirror with the property that it transmits half the incident light, reflecting the rest; in the figure we have assumed, for convenience, that this mirror possesses negligible thickness. At M the light divides into two waves. One proceeds by transmision toward mirror M_1; the other proceeds by reflection toward M_2. The waves are reflected at each of these mirrors and are sent back along their directions of incidence, each wave eventually entering the eye. Since the waves are coherent, being derived from the same point on the source, they will interfere.

The path difference for the two waves when they recombine is $2d_2 - 2d_1$ and anything that changes this path difference will cause a change in the relative phase of the two waves as they enter the eye. As an example, if mirror M_2 is moved by a distance $\frac{1}{2}\lambda$, the path difference changes by λ and the observer will see the fringe pattern shift by one fringe. If the mirrors M_1 and M_2 are exactly

* See "Michelson: America's First Nobel Prize Winner in Science," by R. S. Shankland, *The Physics Teacher*, January 1977.

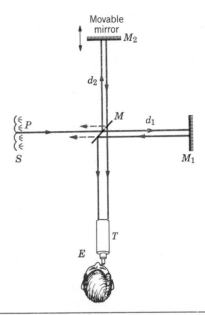

Figure 17   Michelson's interterometer, showing the path of a ray originating at point P of an extended source S. The ray splits, the two subrays following different paths and then recombining. The fringe pattern can be changed by changing the length of one path by moving mirror M_2.

perpendicular to each other, the effect is that of light from an extended source S falling on a uniformly thick slab of air, between glass, whose thickness is equal to $d_2 - d_1$.

By such techniques the lengths of objects can be expressed in terms of the wavelength of light. In Michelson's day, the standard of length—the meter—was chosen by international agreement to be the distance between two fine scratches on a certain metal bar preserved at Sèvres, near Paris. Michelson was able to show, using his interferometer, that the standard meter was equivalent to 1,553,163.5 wavelengths of a certain monochromatic red light emitted from a light source containing cadmium. For this careful measurement, Michelson received the 1907 Nobel prize in physics. His work laid the foundation for the eventual abandonment (in 1961) of the meter bar as a standard of length and for the redefinition of the meter in terms of the wavelength of light. In 1983, as we have seen, even this wavelength standard was not precise enough to meet the growing requirements of science and technology and was replaced by a new standard based on a defined value for the speed of light.

REVIEW AND SUMMARY

Huygens' Principle

The three-dimensional transmission of waves, including light, may often be predicted by *Huygens' principle,* which states: All points on a wave front can serve as point sources of spherical secondary wavelets; see Fig. 1.

The *law of refraction* can be derived from Huygens' principle by presuming that the index of refraction of any medium is $n = c/v$, v being the speed of light in the medium; see Fig. 2.

Geometrical Optics and Diffraction

Attempts to isolate a ray by forcing light through a narrow slit fail because of *diffraction,* the flaring out of the light into the geometrical shadow of the slit; see Figs. 3 and 4. If such slits are present, the approximations of geometrical optics (Chapter 39) fail and the full treatment of wave optics must be used.

Interference

In *Young's double-slit interference experiment* light from a slit in screen *A* of Fig. 5 flares out (by diffraction) and falls on the two slits in screen *B*. The light from these slits also flares out in the region beyond *B* and interference occurs between the two overlapping wavelets. A fringe pattern like that of Fig. 6 is formed on screen *C*. Figure 7 shows the same effect in water waves.

The light intensity at any point on screen *C* of Fig. 5 depends in part on the path difference between two rays drawn to that point from the two slits. If this difference is an integral number of wavelengths, the waves will interfere constructively and an intensity maximum results. If it is a half-integral multiple, there is destructive interference and an intensity minimum. Analysis shows that the conditions for maximum and minimum intensity are

Intensity in Two-slit Interference

$$d \sin \theta = m\lambda \qquad m = 0, 1, 2 \ldots \quad \text{(maxima)} \qquad [9]$$

$$d \sin \theta = (m + \tfrac{1}{2})\lambda \quad m = 0, 1, 2 \ldots \quad \text{(minima).} \qquad [10]$$

See Sample Problem 1.

Coherence and Incoherence

If two overlapping light waves are to interfere at a given point, the phase difference between them must remain constant with time; the waves must be *coherent.* Very long wavetrains must be involved; see Fig. 9. When two coherent waves overlap, the resulting intensity may be found by the phasor method, as in Fig. 11. In this method the amplitude E_θ of the electric field vector of the resultant wave is calculated, taking the phase difference between the two combining waves properly into account. The intensity I_θ of the resultant wave is then taken to be proportional to E_θ^2. As applied to Young's double-slit experiment we have, for the interference of two beams with intensity I_0:

Intensity in Young's Experiment

$$I_\theta = 4I_0 \cos^2(\tfrac{1}{2}\phi), \quad \text{where} \quad \phi = \left(\frac{2\pi d}{\lambda}\right) \sin \theta. \qquad [15,16]$$

Equations 9 and 10, which identify the positions of the fringe maxima and minima, are contained within this relation. Figure 11 should be studied closely. Sample Problem 2 shows how to apply the phasor strategy to analyze four-slit interference.

When light falls on a *thin transparent film,* the light waves reflected from the front and the rear surfaces, as in Fig. 13, interfere. For near-normal incidence the conditions for a maximum or a minimum intensity of the light reflected from the film are

Thin-film Interference

$$2nd = (m + \tfrac{1}{2})\lambda \quad m = 0, 1, 2 \ldots \quad \text{(maxima)} \qquad [24]$$

$$2nd = m\lambda \qquad m = 0, 1, 2 \ldots \quad \text{(minima).} \qquad [25]$$

Phase Change on Reflection

Here the medium in which the film is immersed is assumed to be the same on each side, n is the index of the film material with respect to this medium, and λ is the wavelength in vacuum. The formula also assumes that one of the interfering waves has undergone a "hard" reflection (that is, from a higher index medium; compare Fig. 15b) and the other a "soft" reflection (that is, from a lower index medium; compare Fig. 15a). Hard reflections involve a phase change of 180°; soft reflections do not. Sample Problems 3 and 4 discuss cases in which some assumptions of the above formula do not hold and the problem must be solved from first principles.

The Michelson
Interferometer

In *Michelson's interferometer* (Fig. 17) a light beam is split into two sub-beams which, after traversing paths of different lengths, are recombined so that they interfere and form a fringe pattern. By varying the path length of one of the sub-beams distances can be accurately expressed in terms of wavelengths of light. Thus, in Fig. 17, as mirror M_2 is moved through the distance in question, one simply counts the number of fringes (wavelengths) that pass through the field of view of the observer.

QUESTIONS

1. Light has (*a*) a wavelength, (*b*) a frequency, and (*c*) a speed. Which, if any, of these quantities remains unchanged when light passes from a vacuum into a slab of glass?

2. Why is it plausible that the wavelength of light should change when the light passes from air into glass but that its frequency should not?

3. The speed and wavelength of, say, red light that we see in air is reduced when the light passes into water. Would that light then appear to be another color—blue, perhaps—if you viewed it from under the water surface?

4. Would you expect sound waves to obey the laws of reflection and of refraction obeyed by light waves? Does Huygens' principle apply to sound waves in air? If Huygens' principle predicts the laws of reflection and refraction, why is it necessary or desirable to view light as an electromagnetic wave, with all its attendant complexity?

5. In Young's double-slit interference experiment, using a monochromatic laboratory light source, why is screen *A* in Fig. 5 necessary?

6. What changes occur in the pattern of interference fringes if the apparatus of Fig. 8 is placed under water?

7. Do interference effects occur for sound waves? Recall that sound is a longitudinal wave and that light is a transverse wave.

8. If interference between light waves of different frequencies is possible, one should observe light beats, just as one obtains sound beats from two sources of sound with slightly different frequencies. Discuss how one might experimentally look for this possibility.

9. Why are parallel slits preferable to the pinholes that Young used in demonstrating interference?

10. Is coherence important in reflection and refraction?

11. If your source of light is a laser beam, you do not need the equivalent of screen *A* in Fig. 5. Why?

12. Describe the pattern of light intensity on screen *C* in Fig. 8 if one slit is covered with a red filter and the other with a blue filter, the incident light being white.

13. If one slit in Fig. 8 is covered, what change would occur in the intensity of light in the center of the screen?

14. We are all bathed continuously in electromagnetic radiation, from the sun, from radio and TV signals, from the stars and other celestial objects. Why don't these waves interfere with each other?

15. Is polarization or interference a better test for identifying waves? Do they give the same information?

16. What causes the fluttering of a TV picture when an airplane flys overhead?

17. Is it possible to have coherence between light sources emitting light of different wavelengths?

18. Each slit in Fig. 8 is covered with a sheet of Polaroid, the polarizing directions of the two slits being at right angles. What is the pattern of light intensity on screen *C*? (The incident light is unpolarized.)

19. Suppose that the film coating in Fig. 16 had a refractive index greater than that of the glass. Could it still be nonreflecting? If so, what difference would it make?

20. What are the requirements for maximum intensity when viewing a thin film by *transmitted* light?

21. Why does a film (soap bubble, oil slick, etc.) have to be "thin" to display interference effects? Or does it? How thin is "thin"?

22. Why do coated lenses (see Sample Problem 4) look purple by reflected light?

23. Ordinary store windows or home windows reflect light from both their interior and exterior plane surfaces. Why then don't we see interference effects?

24. A person wets his eyeglasses to clean them. As the water evaporates he notices that for a short time the glasses become markedly less reflecting. Explain why.

25. A lens is coated to reduce reflection, as in Sample Problem 4. What happens to the energy that had previously been reflected? Is it absorbed by the coating?

26. Consider the following objects that produce colors when exposed to sunlight: (1) soap bubbles, (2) rose petals, (3) the inner surface of an oyster shell, (4) thin oil slicks, (5) nonreflecting coatings on camera lenses, and (6) peacock tail feathers. The colors displayed by all but one of these are purely interference phenomena, no pigments being involved. Which one is the exception? Why do the others seem to be "colored"?

27. An automobile directs its headlights onto the side of a barn. Why are interference fringes not produced in the region in which light from the two beams overlaps?

28. A soap film on a wire loop held in air appears black at its thinnest portion when viewed by reflected light. On the other hand, a thin oil film floating on water appears bright at its thinnest portion when similarly viewed from the air above. Explain these phenomena.

29. If the pathlength to the movable mirror in Michelson's interferometer (see Fig. 17) greatly exceeds that to the fixed mirror (say by more than a meter) the fringes begin to disappear. Explain why. Lasers greatly extend this range. Why?

30. How would you construct an acoustical Michelson interferometer to measure sound wavelengths? Discuss differences from the optical interferometer.

EXERCISES AND PROBLEMS

Section 40-2   Light as a Wave

1E. The wavelength of yellow sodium light in air is 589 nm. (a) What is its frequency? (b) What is its wavelength in glass whose index of refraction is 1.52? (c) From the results of (a) and (b) find its speed in this glass.

2E. How much faster does light travel in sapphire than in diamond? See Table 1, Chapter 39.

3E. Derive the law of reflection using Huygens' principle.

4P. Light of free-space wavelength 600 nm travels 1.6 μm in a medium of index of refraction 1.5. Find (a) the wavelength in the medium, and (b) the phase difference after moving that distance, with respect to light traveling the same distance in free space.

5P. One end of a stick is dragged through water at a speed v, which is greater than the speed u of water waves. Applying Huygens' construction to the water waves, show that a conical wavefront is set up and that its half-angle α is given by

$$\sin \alpha = u/v.$$

This is familiar as the bow wave of a ship and the shock wave caused by an object moving through air with a speed exceeding that of sound, as in Fig. 17, Chapter 18.

Section 40-4   Young's Experiment

6E. Monochromatic green light, wavelength = 550 nm, illuminates two parallel narrow slits 7.7 μm apart. Calculate the angular deviation of the third order, $m = 3$, bright fringe (a) in radians, and (b) in degrees.

7E. What is the phase difference between the waves from the two slits arriving at the mth dark fringe in a Young's double-slit experiment?

8E. In an experiment using Young's arrangement to demonstrate the interference of light, the separation d of the two narrow slits is doubled. In order to maintain the same spacing of the fringes, how must the distance D of the screen from the slits be altered? (The wavelength of the light remains unchanged.)

9E. Young's experiment is performed with blue-green light of wavelength 500 nm. The slits are 1.2 mm apart and the screen is 5.4 m from the slits. How far apart are the bright fringes?

10E. Find the slit separation of a double-slit arrangement that will produce interference fringes 0.018 rad apart on a distant screen. Assume sodium light ($\lambda = 589$ nm).

11E. A double-slit arrangement produces interference fringes for sodium light ($\lambda = 589$ nm) that are 0.0035 rad apart. For what wavelength would the angular separation be 10% greater?

12E. In a double-slit arrangement the slits are separated by a distance equal to 100 times the wavelength of the light passing through the slits. (a) What is the angular separation in radians between the central maximum and an adjacent maximum? (b) What is the linear distance between these maxima if the screen is at a distance of 50 cm from the slits?

13E. In a double-slit experiment, $\lambda = 546$ nm, $d = 0.10$ mm, and $D = 20$ cm. What is the linear distance between the fifth maximum and seventh minimum from the central maximum?

14E. A double-slit arrangement produces interference fringes for sodium light ($\lambda = 589$ nm) that are 0.20° apart. What is the angular fringe separation if the entire arrangement is immersed in water ($n = 1.33$)?

15P. In a double-slit experiment the distance between slits is 5.0 mm and the slits are 1.0 m from the screen. Two interference patterns can be seen on the screen, one due to light with wavelength 480 nm and the other due to light with wavelength 600 nm. What is the separation on the screen between the third-order interference fringes of the two different patterns?

16P. In Young's interference experiment in a large ripple tank (see Fig. 7) the coherent vibrating sources are placed 120 mm apart. The distance between maxima 2.00 m away is 180 mm. If the speed of ripples is 25 cm/s, calculate the frequency of the vibrators.

17P. If the distance between the first and tenth minima of a double-slit pattern is 18 mm and the slits are separated by 0.15 mm with the screen 50 cm from the slits, what is the wavelength of the light used?

18P. As shown in Fig. 18, A and B are two identical radiators of waves that are in phase and of the same wavelength λ. The radiators are separated by distance 3.0λ. Find the largest distance from A, along the line Ax, for which destructive interference occurs. Express this in terms of λ.

Figure 18 Problem 18.

19P. A thin flake of mica ($n = 1.58$) is used to cover one slit of a double-slit arrangement. The central point on the screen is occupied by what used to be the seventh bright fringe. If $\lambda = 550$ nm, what is the thickness of the mica?

20P. Sketch the interference pattern expected from using two pinholes, rather than narrow slits.

21P. In Fig. 19, the source emits monochromatic light of wavelength λ. S is a narrow slit in an otherwise opaque screen I. A plane mirror, whose surface includes the axis of the lens shown, is located a distance h below S. Screen II is at the focal plane of the lens. (a) Find the condition for maximum and minimum brightness of fringes on screen II in terms of the usual angle θ, the wavelength, and the distance h. (b) Show that fringes appear only in region A (above the axis of the lens), but not in region B (below the axis of the lens) or in both regions A and B. (*Hint:* Consider the image of S formed by the mirror.) This arrangement for obtaining a double-slit interference pattern from a single slit is called *Lloyd's mirror*.

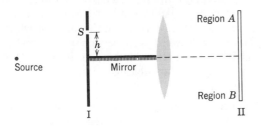

Figure 19 Problem 21.

22P. Two coherent radio point sources separated by 2.0 m are radiating in phase with $\lambda = 0.50$ m. A detector moved in a circular path around the two sources in a plane containing them will show how many maxima?

23P. Interference fringes are produced using white light with a double-slit arrangement. A piece of parallel-sided mica of refractive index 1.6 is placed in front of one of the slits, as a result of which the center of the fringe system moves to the left a distance subsequently shown to accommodate 30 dark bands when light of wavelength 480 nm is used. What is the thickness of the mica?

24P. In the front of a lecture hall, a coherent beam of mono-

chromatic light from a helium–neon laser ($\lambda = 632.8$ nm) illuminates a double slit. From there it travels a distance $d = 20$ m to a mirror at the back of the hall, and returns the same distance to a screen. (a) In order that the distance between interference maxima be 10 cm what should be the distance between the two slits? (b) State what you will see if the lecturer slips a thin sheet of cellophane over one of the slits. The path through the cellophane contains 2.5 more waves than a path through air of the same geometric thickness.

25P. One slit of a double-slit arrangement is covered by a thin glass plate of refractive index 1.4, and the other by a thin glass plate of refractive index 1.7. The point on the screen where the central maximum fell before the glass plates were inserted is now occupied by what had been the $m = 5$ bright fringe before. Assume $\lambda = 480$ nm and that the plates have the same thickness t and find the value of t.

26P. Sodium light ($\lambda = 589$ nm) falls on a double slit of separation $d = 2.0$ mm. The slit-screen distance $D = 40$ mm. What percent error is made by using Eq. 9 to locate the tenth bright fringe on the screen?

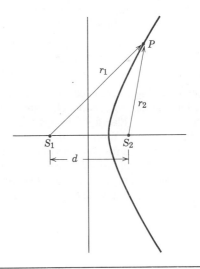

Figure 20 Problem 27.

27P. Two point sources, S_1 and S_2 in Fig. 20, emit coherent waves. Show that curves, such as that given, over which the phase difference for rays r_1 and r_2 is a constant, are hyperbolas. (*Hint:* A constant phase difference implies a constant difference in length between r_1 and r_2.) The OMEGA system of sea navigation relies on this principle. S_1 and S_2 are phase-locked transmitters. The ship's navigator notes the received phase difference on an oscilloscope and locates the ship on a hyperbola. Reception of signals from a third transmitter is needed to determine the position on that hyperbola.

Section 40-5 Coherence

28E. The *coherence length* of a wavetrain is the distance over which the phase constant is the same. (*a*) If an individual atom emits coherent light for 1×10^{-8} s, what is the coherence length of the wavetrain? (*b*) Suppose this wavetrain is separated into two parts with a partially reflecting mirror and later reunited after one beam travels 5 m and the other 10 m. Do the waves produce interference fringes observable by a human eye?

Section 40-6 Intensity in Double-slit Interference

29E. Add the following quantities graphically, using the phasor method (see Sample Problem 2):

$$y_1 = 10 \sin \omega t,$$
$$y_2 = 15 \sin(\omega t + 30°),$$
$$y_3 = 5 \sin(\omega t - 45°).$$

30E. Source A of long-range radio waves leads source B by $90°$. The distance r_A to a detector is greater than the distance r_B by 100 m. What is the phase difference at the detector? Both sources have a wavelength of 400 m.

31E. Light of wavelength 600 nm is incident normally on two parallel narrow slits separated by 0.60 mm. Sketch the intensity pattern observed on a distant screen as a function of angle θ for the range of values $0 \le \theta \le 0.0040$ radians.

32P. Two waves of the same frequency have amplitudes 1 and 2, respectively. They interfere at a point where their phase difference is $60°$. What is the resultant amplitude?

33P. S_1 and S_2 in Fig. 21 are effective point sources of radiation, excited by the same oscillator. They are coherent and in phase with each other. Placed 4.0 m apart, they emit equal amounts of power in the form of 1.0-m wavelength electromagnetic waves. (*a*) Find the positions of the first (that is, the nearest), the second, and the third maxima of the received signal, as the detector is moved out along Ox. (*b*) Is the intensity at the nearest minimum equal to zero? Justify your answer.

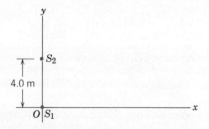

Figure 21 Problem 33.

34P. Find the sum of the following quantities (*a*) graphically, using phasors, and (*b*) using trigonometry:

$$y_1 = 10 \sin \omega t.$$
$$y_2 = 8 \sin(\omega t + 30°).$$

35P. Show that the half-width $\Delta\theta$ of the double-slit interference fringes (see horizontal arrow in Fig. 10) is given by

$$\Delta\theta = \frac{\lambda}{2d}$$

if θ is small enough so that $\sin\theta \approx \theta$.

36P*. One of the slits of a double-slit system is wider than the other, so that the amplitude of the light reaching the central part of the screen from one slit, acting alone, is twice that from the other slit, acting alone. Derive an expression for the intensity I in terms of θ, corresponding to Eqs. 15 and 16.

Section 40-7 Interference from Thin Films

37E. A thin coating of refractive index 1.25 is applied to a glass camera lens to minimize the intensity of the light reflected from the lens. In terms of λ, the wavelength in air of the incident light, what is the smallest thickness of the coating that is needed?

38E. We wish to coat a flat slab of glass ($n = 1.50$) with a transparent material ($n = 1.25$) so that light of wavelength 600 nm (in vacuum) incident normally is not reflected. What minimum thickness could the coating have?

39E. A thin film in air is 0.41 μm thick and is illuminated by white light normal to its surface. Its index of refraction is 1.5. What wavelengths within the visible spectrum will be intensified in the reflected beam?

40E. A disabled tanker leaks kerosene ($n = 1.2$) into the Persian Gulf, creating a large slick on top of the water ($n = 1.3$). (*a*) If you are looking straight down from an airplane onto a region of the slick where its thickness is 460 nm, for which wavelength(s) of visible light is the reflection the greatest? (*b*) If you are scuba-diving directly under this same region of the slick, for which wavelength(s) of visible light is the transmitted intensity the strongest?

41E. Light of wavelength 585 nm is incident normally on a thin soapy film ($n = 1.33$) suspended in air. If the film is 0.00121 mm thick, determine whether it appears bright or dark when observed from a point near the light source.

42E. A lens is coated with a thin transparent film to minimize reflection of the red component of white light. The index of refraction of the film is 1.30 and that of the lens is 1.65. What minimum thickness of film is needed? The wavelength of the light in air is 680 nm.

43E. In costume jewelry, rhinestones (made of glass with $n = 1.50$) are often coated with silicon monoxide ($n = 2.0$) to make them more reflective. How thick should the coating be to achieve strong reflection for 560-nm light, incident normally?

44E. A soap film ($n = 1.33$) is illuminated by light of wavelength 624 nm in air, incident normally. What is (*a*) the smallest thickness and (*b*) the second smallest thickness that results in no reflection?

45P. A plane wave of monochromatic light falls normally on a uniformly thin film of oil that covers a glass plate. The wavelength of the source can be varied continuously. Complete destructive interference of the reflected light is observed for wavelengths of 500 and 700 nm and for no wavelengths between them. If the index of refraction of the oil is 1.3 and that of the glass is 1.5, find the thickness of the oil film.

46P. White light reflected at perpendicular incidence from a soap film has, in the visible spectrum, an interference maximum at 600 nm and a minimum at 450 nm with no minimum in between. If $n = 1.33$ for the film, what is the film thickness, assumed uniform?

47P. A sheet of glass having an index of refraction of 1.40 is to be coated with a film of material having a refractive index of 1.55 such that green light (wavelength = 525 nm) is preferentially transmitted. (a) What is the minimum thickness of the film that will achieve the result? (b) Why are other parts of the visible spectrum not also preferentially transmitted? (c) Will the transmission of any colors be sharply reduced?

48P. A plane monochromatic light wave in air falls at normal incidence on a thin film of oil that covers a glass plate. The wavelength of the source may be varied continuously. Complete destructive interference in the reflected beam is observed for wavelengths of 500 and 700 nm and for no wavelength in between. The index of refraction of glass is 1.5. Show that the index of refraction of the oil must be less than 1.5.

49P. A thin film of acetone (index of refraction = 1.25) is coating a thick glass plate (index of refraction = 1.50). Plane light waves of variable wavelengths are incident normal to the film. When one views the reflected wave, it is noted that complete destructive interference occurs at 600 nm and constructive interference at 700 nm. Calculate the thickness of the acetone film.

50P. An oil drop ($n = 1.20$) floats on a water ($n = 1.33$) surface and is observed from above by reflected light (see Fig. 22). (a) Will the outer (thinnest) regions of the drop correspond to a bright or a dark region? (b) Approximately how thick is the oil film where one observes the third blue region from the outside of the drop? (c) Why do the colors gradually disappear as the oil thickness becomes larger?

51P. From a medium of index of refraction n_1, monochromatic light of wavelength λ falls normally on a thin film of uniform thickness and index of refraction n_2. The transmitted light travels in a medium with index of refraction n_3. Find expressions for the minimum film thickness (in terms of λ and the indices of refraction) for the following cases: (a) $n_1 < n_2 > n_3$—minimum reflected light; (b) $n_1 < n_2 > n_3$—maximum transmitted light; (c) $n_1 < n_2 < n_3$—minimum reflected light; (d) $n_1 < n_2 < n_3$—maximum transmitted light; and (e) $n_1 < n_2 < n_3$—maximum reflected light.

52P. In Sample Problem 4 assume that there is zero reflection for light of wavelength 550 nm at normal incidence. Calculate the factor by which the reflection is diminished by the coating at 450 and at 650 nm.

53P. A broad source of light ($\lambda = 680$ nm) illuminates normally two glass plates 120 mm long that touch at one end and are separated by a wire 0.048 mm in diameter at the other end (Fig. 23). How many bright fringes appear over the 120 mm distance?

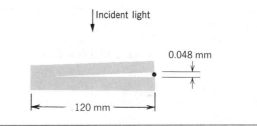

Figure 23 Problems 53 and 54.

54P. In Fig. 23 white light is incident from above. (a) Observed from above, why does the region near the edge, where the two glass plates touch, appear black? (b) For what part of the visible spectrum does destructive interference next occur? (c) What color does an observer see where this destructive interference occurs?

55P. A perfectly flat piece of glass ($n = 1.5$) is placed over a perfectly flat piece of black plastic ($n = 1.2$) as shown in Fig. 24a. They touch at A. Light of wavelength 600 nm is incident

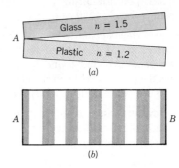

Figure 24 Problem 55.

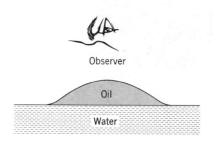

Figure 22 Problem 50.

normally from above. The location of the dark fringes in the reflected light is shown on the sketch of Fig. 24b. (a) How thick is the space between the glass and the plastic at B? (b) Water (n = 1.33) seeps into the region between the glass and plastic. How many dark fringes are seen when all the air has been displaced by water? (The straightness and equal spacing of the fringes is an accurate test of the flatness of the glass.)

56P. A transparent liquid of refractive index 4/3 is allowed to displace the air from an air wedge that is formed from two glass plates touching each other along one edge. What happens, as a result, to the spacing of the dark lines caused by interference of the monochromatic light reflected back?

57P. Light of wavelength 630 nm is incident normally on a thin wedge-shaped film with index of refraction 1.5. There are 10 bright and nine dark fringes over the length of film. By how much does the film thickness change over this length?

58P. Two pieces of plate glass are held together in such a way that the air space between them forms a very thin wedge. Light of wavelength 480 nm strikes the upper surface perpendicularly and is reflected from the lower surface of the top glass and the upper surface of the bottom glass, thereby producing a series of interference fringes. How much thicker is the air wedge at the sixteenth fringe than it is at the sixth?

59P. In an air wedge formed by two plane glass plates, touching each other along one edge, there are 4001 dark lines observed when viewed by reflected monochromatic light. When the air between the plates is evacuated, only 4000 such lines are observed. Calculate the index of refraction of the air from this data.

60P. *Newton's rings.* Figure 25 shows a lens with radius of curvature R resting on an accurately plane glass plate and illuminated from above by light with wavelength λ. Figure 26 shows that circular interference fringes (Newton's rings) appear, associated with the variable thickness d of the air film between the lens and the plate. Find the radii r of the circular interference maxima assuming that $r/R \ll 1$.

61P. In a Newton's rings (see Problem 60) experiment the radius of curvature R of the lens is 5.0 m and its diameter is 20 mm. (a) How many rings are produced? (b) How many rings would be seen if the arrangement were immersed in water (n = 1.33)? Assume that λ = 589 nm.

62P. A Newton's rings apparatus is used to determine the radius of a lens (see Fig. 25). The radii of the nth and (n + 20)th bright rings are measured and found to be 0.162 cm and 0.368 cm, respectively, in light of wavelength 546 nm. Calculate the radius of curvature of the lower surface of the lens.

63P. In the Newton's rings experiment, use the result of Problem 60 to show that the difference in radius between adjacent rings (maxima) is given by

$$\Delta r = r_{m+1} - r_m \approx \tfrac{1}{2}\sqrt{\lambda R/m},$$

assuming $m \gg 1$.

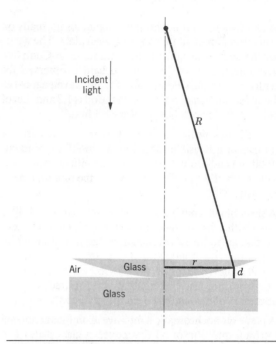

Figure 25 Problems 60–64.

Figure 26 Problems 60–64.

64P. In the Newton's rings experiment, use the result of Problem 63 to show that the *area* between adjacent rings (maxima) is given by

$$A = \pi\lambda R,$$

assuming $m \gg 1$. Note that this area is independent of m.

Section 40–8   Michelson's Interferometer

65E. If mirror M_2 in Michelson's interferometer is moved through 0.233 mm, 792 fringes are counted with a light meter. What is the wavelength of the light?

66E. A thin film with $n = 1.40$ for light of wavelength 589 nm is placed in one arm of a Michelson interferometer. If a shift of 7.0 fringes occurs, what is the film thickness?

67P. An airtight chamber 5.0 cm long with glass windows is placed in one arm of a Michelson interferometer as indicated in Fig. 27. Light of wavelength $\lambda = 500$ nm is used. The air is slowly evacuated from the chamber using a vacuum pump. While the air is being removed, 60 fringes are observed to pass through the view. From these data, find the index of refraction of air at atmospheric pressure.

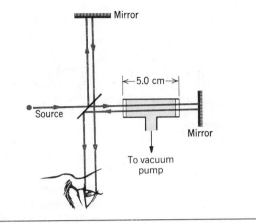

Figure 27   Problem 67.

68P. Write an expression for the intensity observed in Michelson's interferometer (Fig. 17) as a function of the position of the movable mirror. Measure the position of the mirror from the point at which $d_1 = d_2$.

ESSAY 17
LIGHTWAVE COMMUNICATIONS USING OPTICAL FIBERS

SUZANNE R. NAGEL

AT&T BELL LABORATORIES

A revolutionary new technology—lightwave communications—is transforming the communications networks of the world. Huge quantities of information—voice signals, video and digital data—can be rapidly and efficiently transmitted from one place to another by using an ever expanding grid of optical fibers. These hair-thin strands of glass carry information over very long distances in the form of pulses of light. Why is communicating with light so significant and how do these optical fiber "lightguides" work? Let's briefly explore the answer to these questions.

A basic communications system consists of a transmitter (signal source), in which information is encoded, a transmission medium (signal carrier) and a receiver (signal detector) that decodes or reconstructs the original information. Most modern communication systems are "digital" because of the excellent transmission quality that can be achieved. In a simple digital communication system, information is encoded into binary digits consisting of zeros or ones. Before we examine lightwave communications in particular, let's explore how digital encoding is done for one simple voice signal. When we talk, our voice has a maximum frequency of about 4000 hertz (cycles/sec). If this voice signal is sampled as a function of time, it can be accurately represented using a digital code of zeros and ones. Figure 1 is a simple representation of how this is done using a technique called pulse code modulation (PCM). A very small portion of a voice signal is represented in this figure. The voice signal is sampled as shown by the dots. At any sampling time, a binary amplitude code is assigned (right hand axis) based on the relative amplitude of the signal which ranges from 0 to 7 (left-hand axis). Thus, in this example, a relative amplitude of 1 is represented by 001, 7 by 111, etc. The resulting digital-pulse-coded signal is shown in the lower portion of the figure. Your voice signal can thus be represented by a sequence of 0's

Figure 1 Pulse code modulation technique. The relative amplitide of an analogue signal such as a voice is sampled at a specific sampling rate, and assigned a binary amplitude code. A binary-pulse-coded signal results which can be used for generating a transmission signal.

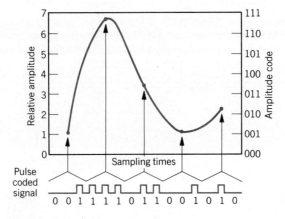

Figure 2 Schematic representation of a lightwave telecommunication system. All information is encoded in a binary data stream of zero's and one's and joined together through a multiplexer. The signal from the multiplexer is used to turn a laser or light emitting diode (LED) on and off at a given transmission rate. The light generated in this manner (represented by *hv*) is transmitted over an optical fiber where the output signal falls on a photodetector. The electrical signal generated at the photodetector is fed into a demultiplexer which separates the various signals and routes them to their final destination.

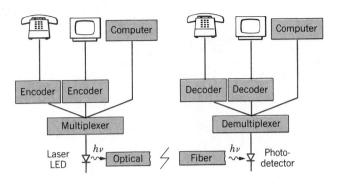

Lightwave Telecommunication System

and 1's. To represent accurately a voice signal in digital form, roughly 64,000 pulse coded signal bits per second are needed. In other words, your voice, as you speak, is accurately represented by a sequence of 64,000 zeros and ones per second! Each zero or one is a "bit" of information.

Why can a lightwave communication system transmit so much digital information relative to conventional communication systems? Because the rate at which information can be transmitted is directly related to the signal frequency. Light has a frequency in the range of $10^{14} - 10^{15}$ hertz, compared to radio frequencies of $\sim 10^6$ hertz and microwave frequencies of $10^8 - 10^{10}$ hertz. Therefore, a transmission system that operates at the frequency of light can theoretically transmit information at a higher rate than systems that operate at radio or microwave frequencies. The digital transmission rate is defined as the number of bits/second that are transmitted.

A simple lightwave telecommunication system is shown in Figure 2. The information that is transmitted can be telephone voice signals, video signals, or digital data from a computer. Voice and video signals are encoded into a binary sequence of zeros and ones. All of these signals are blended together into a single very high-data-rate stream in the multiplexer unit. For simplicity let us only consider blending or multiplexing voice signals together for transmission. Each voice signal requires 6.4×10^4 bits/sec. If the data rate of the system is 1 Gbit/sec (1×10^9 bits/sec), the number of voice channels that can be multiplexed together is approximately 15,000! (1×10^9 divided by 6.4×10^4.) How is this actually done? In the lightwave transmitter, each "one" corresponds to an electrical pulse and each "zero" corresponds to the absence of an electrical pulse. These electrical pulses are used to turn a light source on and off very rapidly, much like turning a light switch on and off. The light source can be a laser or a light emitting diode (LED). Thus, in the transmitter in a lightwave communication system, information is blended together into a very high data rate sequence of electrical pulses, which in turn are used to turn a light source on and off very rapidly: all binary encoded information is thus transformed into a timed sequence of flashes of light for transmission.

The next important part of the lightwave communication system is the transmission medium. Although in principle these flashes of light could be transmitted through the open atmosphere, much the same as radio signals are, it would be very difficult to build a practical telecommunications system in this way. Instead, glass fiber "lightguides" are used to carry the light from the transmitter to the third part of the lightwave system, the receiver. In the receiver, each pulse of light is detected by a photodetector. As each pulse of light arrives at the photodetector, a pulse of electrical

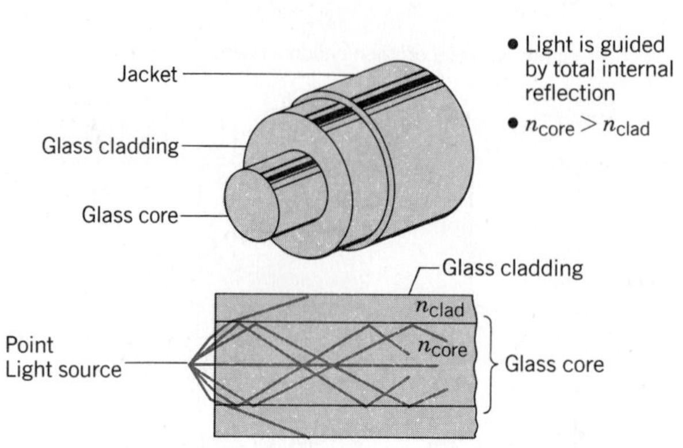

Figure 3 Simple light-guiding structure for an optical transmission medium. The core and cladding index of refraction are represented by n_{core} and n_{clad}, respectively.

current is produced. In this way, the optical pulses are converted back to electrical pulses. The receiver also has a "demultiplexer" which separates the signals and converts them back into voice, video and computer data.

Although this is only a very basic description of how a lightwave communication system operates, it shows the basic approach: information is converted into pulses of light that are transmitted over some distance through an optical fiber, then converted back into information.

Now we can examine in a little more detail how an optical fiber is used to transmit information in the form of pulses of light. The first important property of a fiber is that it be able to guide light from one place to another. The basic principle that an optical fiber uses is "total internal reflection," shown for a lightguide structure in Figure 3. A fiber structure consists of a central core of material that has a higher refractive index than the surrounding material, called the "cladding." Remember, the refractive index is the ratio of the speed of light in a perfect vacuum relative to its speed through the material. A fiber made from glass will usually also have a plastic jacket on the outside to protect the glass from mechanical abrasion and other environmental effects. A light source such as a laser or LED is placed close to the fiber core. The light source gives off or emits a "cone" of light that is coupled into the core of the fiber. For this light to be guided, it must meet the requirement for total internal reflection. As is shown in the diagram, some angles of light are guided in "zig-zag" paths in the fiber while others are not. If the angle of light from the light source is too large, the light gets refracted into the cladding instead of reflected within the core. Although the details of light guidance are far more complex than this simple description, the basic principle of total internal reflection is useful for understanding how light is guided within the core of the optical fiber.

While this simple type of fiber is very useful for guiding light over short distances (meters), more specialized fiber structures are required to guide light over the long distances used in telecommunication systems. Let us examine what determines how closely spaced in time individual pulses can be coupled into fibers. Remember, we want to carry as many pulses per second as possible through the fiber. Figure 4 shows three different types of fiber. First, let us look at the simple fiber shown at the top. It is called a "multimode" fiber because many angles of light are guided and it is said to have a step index because it has a core of constant refractive index surrounded by a cladding of lower refractive index. The refractive index profile shown depicts the relative value of the index as a function of position across the fiber cross section. A pulse of light from the source is coupled into the fiber. Many beams, or angles of

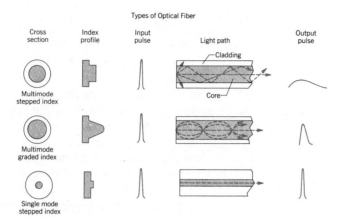

Figure 4 Types of optical fiber.

light, all traveling at the same speed, are guided in the core. But the beam that travels down the center has a shorter distance to travel than those beams that have to zig-zag back and forth. As a result, the "narrow" pulse of light that was initially coupled into the fiber is considerably broadened after traveling many kilometers through the fiber, due to the different relative distances each beam has to travel. It is this effect that limits how closely together the input pulses can be spaced and can be detected without overlapping at the output end.

Two types of fibers are used in lightwave systems to overcome the limits of this pulse broadening effect. One type of fiber is shown in the center of Figure 4 and is called a multimode graded index fiber. Note that the core of the fiber has a gradually changing refractive index profile. This type of profile results in the guided angles of light being bent in gentle periodic paths as they travel through the fiber. How does this help? The beam that travels down the center of the fiber still has the shortest distance to travel, but it travels the slowest since it is in the highest refractive index part of the fiber. A beam that travels along the periodic path spends most of its time in a lower index portion of the fiber, so it travels at a higher speed! Thus, the longer the distance a given beam has to travel, the faster on average it can be made to travel. By very carefully controlling the exact shape of the refractive index profile in the fiber, all angles of guided light can be made to travel through the fiber in roughly the same amount of time! This thus decreases the amount of spreading of the output pulse and extends the distance over which pulses of light can be transmitted before they start to overlap. The third type of fiber depicted on the bottom of Figure 4 completely eliminates the pulse spreading due to different angles of light being guided. This type of fiber is called a single mode fiber, and only the on-axis beam of light is guided. This is achieved by using very small refractive index differences between the core and cladding, and very small cores. This type of fiber can carry the data at the very highest bit rates.

Although the details of light guidance and pulse broadening are far more complicated than this simple description, you can begin to understand how the fiber is used to transmit data at high rates. Lasers or LED's are turned on and off very rapidly in a specific sequence corresponding to ones and zeros; each pulse is individually transmitted through the fiber; and at the output end of the fiber the arriving pulse of light is detected. The broadening of the pulse of light in the fiber as it travels plays an important role in determining the maximum rate at which information can be transmitted, as well as the distance over which information can be transmitted.

A second important property of the fiber is its optical attenuation, since this also determines how far the signal can be transmitted and can still be intense enough to be detected. When light travels through a material, it is not perfectly conducted, and the intensity of the signal decreases with distance. The light traveling through a fiber is diminished by both absorption and scattering. Very simply, absorption results in a fraction of the intensity of the light being transferred to the material itself instead of being transmitted. In contrast, scattering results in the light being diffused or deflected in many different directions. Therefore, some of the light is scattered out from the core rather than being transmitted, causing a decrease in the light signal intensity. In glass fibers made from very high purity, low scattering silica glasses, the amount of scattering and absorption is extremely low at wavelengths of light in the near-infrared. These wavelengths are just a little longer than the wavelengths of the visible spectrum. Remember, visible light ranges from 380 nm (violet) to 700 nm (red). The most common transmission wavelength is 1300 nm (approximately 2×10^{14} hertz). In silica glass fibers, more than 95 percent of the light is transmitted over a distance of one kilometer. This very high transparency allows light to be transmitted over distances of 20 to 200 kilometers and still be intense enough for the signal to be detected by a photodetector device in the receiver.

While this discussion of lightwave communication and the optical fiber transmission medium has been simple and qualitative, the interested student will find a wealth of reference material available to explore this emerging area of technology in greater depth. A few advanced references are listed at the end of this discussion.

At this time, the quest to utilize the frequency potential of communication systems based on light continues, and many advances are still being made. In the years 1987 and 1988, telecommunication systems are being installed that operate at 1.7 Gbits/sec (1.7×10^9 bits/sec), corresponding to $\sim 25,000$ voices being carried over a fiber that is roughly the size of a human hair! Impressive as this may seem, it is still several orders magnitude below the theoretical capacity. In the coming years, these fiber-based lightwave communications systems will increasingly be used to carry voice, video, and computer data across the country and the world as well as to and from homes and businesses.

REFERENCES FOR FURTHER READINGS *Optical Fiber Telecommunications,* S. E. Miller and A. Chynoweth (eds.), Academic Press, 1979, 705 pp. J. E. Midwinter, *Optical Fibers for Transmission,* John Wiley & Sons, Inc. 1979, 410 pp. A. W. Snyder and J. D. Love, *Optical Waveguide Theory,* Chapman and Hall, 1983, 734 pp. *Optical Fiber Communications,* Vol. 1, Fiber Fabrication, T. Li, (ed.) Academic Press, 1985, 363 pp. D. J. Morris, *Pulse Code Formats for Fiber Optical Data Communication,* Marcel Decker, Inc., 1983, 217 pp.

CHAPTER 41
DIFFRACTION

About 160,000 years ago, a star in a nearby galaxy blew up. In February of 1987 the light from this supernova reached Earth and was first identified by Ian Shelton, a Canadian astronomer working at the Las Campanas observatory in Chile. The cross that is such a feature of this photo has nothing to do with the explosion but is formed by the diffraction of light by the internal supporting structure of the telescope. We shall see in this chapter that the ability of all optical instruments to form sharp images of point sources is limited by the diffraction of light.

41–1 Diffraction and the Wave Theory of Light

In Chapter 40 we defined diffraction rather loosely as the flaring out of light as it emerges from a narrow slit. Its main feature is: for a given wavelength, the narrower the slit the more the flaring out. Figure 1 shows a narrow slit and the diffraction pattern formed when monochromatic light from a distant source falls on it. The incident light certainly "flares out" into the geometric shadow of the slit, but there is more to it than that. There is a broad and intense central maximum—which was our principal concern in Chapter 40—but there are also secondary maxima and minima. Figure 2 shows the diffraction pattern formed when light from a distant point source falls on a razor blade.

If geometrical optics prevailed, light would form a sharp shadow of the slit edges, as in Fig. 1b. Figure 1c is what we actually see. The assumption of geometrical optics—that light travels in straight lines—though useful in many situations, is only an approximation.

Today, diffraction finds a ready explanation in terms of the wave theory of light. This theory, advanced by Huygens and used by Young to explain double-slit interference, was very slow in being adopted, largely because it ran counter to the theory of Newton, who held

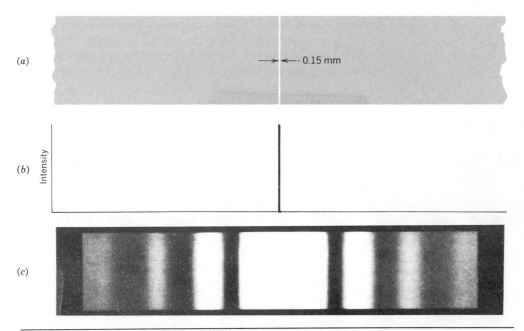

(a)

(b)

(c)

→ ← 0.15 mm

Intensity

Figure 1   (a) A slit in an otherwise opaque screen. The slit width (0.15 mm) has been somewhat exaggerated for clarity. (b) The pattern you would expect on a diffusing screen if light *truly* traveled in straight lines. (c) The pattern that actually occurs. The light flares out substantially beyond the slit. Note the broad central diffraction maximum, the regions of zero intensity, and the secondary diffraction maxima.

the view that light was a stream of particles.

Support for Newton's views prevailed in French scientific circles. Enter Augustin Fresnel,* a young military engineer who followed his passion for optics largely in the spare time that he could wrench from his military duties. Fresnel believed in the wave theory of light and submitted memoirs to the French Academy of Sciences describing his experiments and his wave-theory explanations of them.

In 1819, the Academy, dominated by the supporters of Newton† and thinking to challenge the wave point of view, organized a prize competition for an essay on the subject of diffraction. Fresnel won. The Newtonians, however, were neither converted nor silenced. One of them, Poisson, pointed out the "strange result" that, if Fresnel's theories were correct, a bright spot should appear in the center of the shadow of a sphere. The prize

* Pronounced Fra-nel′.

† In 1819, Newton's great work *Opticks* had been in print for 115 years. For an appreciation of Newton's views on light, from today's perspective, see I. Bernard Cohen's preface to the 1952 reprinting of this work (Dover Publications).

Figure 2   The diffraction pattern of a razor blade in monochromatic light. Note the fringes of alternating maximum and minimum intensity.

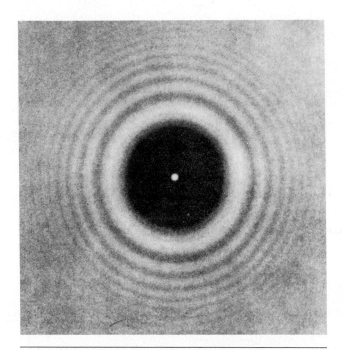

Figure 3 The diffraction pattern of a disk. Note the concentric diffraction rings and—especially—Fresnel's bright spot in the center of the pattern. If geometrical optics prevailed, the center of the pattern (being the most effectively screened) is the last place you would expect to find a bright spot.

committee arranged a test of this prediction and discovered (see Fig. 3) that the predicted *Fresnel bright spot,* as we call it today, was indeed there! Nothing builds confidence in a theory so much as having one of its unexpected and counterintuitive predictions verified by experiment.

41–2 Diffraction from a Single Slit— Locating the Minima

Let us consider how light is diffracted by a long narrow slit of width a. Figure 4 shows a plane wave falling at normal incidence on such a slit. Let us focus our attention on the central point P_0 of screen C. A set of horizontal, parallel rays (not shown in the figure) emerging from the slit will be focused at P_0. These rays all contain the same number of wavelengths and, since they are in phase at the plane of the slit, they will still be in phase at P_0 and the central point of the diffraction pattern that appears on screen C has a maximum intensity.

We now consider another point on the screen. Light rays that reach P_1 in Fig. 4 leave the slit at an angle θ as shown. The ray xP_1 determines θ because it passes undeflected through the center of the lens. Ray r_1 originates at the top of the slit and ray r_2 at its center. If θ is chosen so that the distance bb' in the figure is one-half a wave-

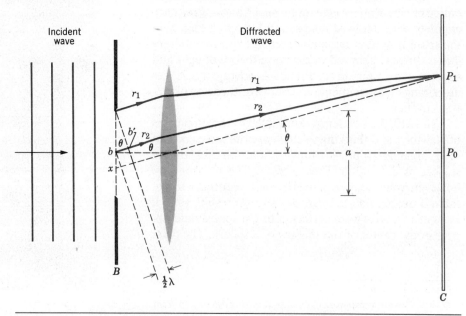

Figure 4 Conditions at the first minimum of the diffraction pattern. Angle θ is such that the distance bb' is one-half of a wavelength.

Figure 5   A marine radar antenna. You can tell from its dimensions that its transmitted beam is fan-shaped, being broad in the vertical plane and narrow in the horizontal plane.

length, r_1 and r_2 will be out of phase and will produce no effect at P_1. In fact, every ray from the upper half of the slit will be canceled by a ray from the lower half, originating at a point $a/2$ below the first ray. The point P_1, the first minimum of the diffraction pattern, will have zero intensity.

The condition shown in Fig. 4 is

$$\frac{a}{2} \sin \theta = \frac{\lambda}{2},$$

or

$$a \sin \theta = \lambda \quad \text{(first minimum).} \quad (1)$$

Equation 1 shows that the central maximum becomes *wider* as the slit becomes *narrower*. That is, for a fixed wavelength, θ increases as a decreases, just as Fig. 4 of Chapter 40 suggests. For $a = \lambda$ we have $\theta = 90°$, so that the central maximum fills the entire forward hemisphere.

Equation 1 helps us to understand why the antennas of marine radars (see Fig. 5) are much longer in the horizontal dimension than in the vertical dimension. The microwave beam from such radars is fan shaped, being very narrow in the horizontal plane (hence the long horizontal dimension) and wide in the vertical plane (hence the short vertical dimension).

In Fig. 6 the slit is divided into four equal zones, with a ray shown leaving the top of each zone. Let θ be chosen so that the distance bb' is one-half a wavelength. Rays r_1 and r_2 will then cancel at P_2. Rays r_3 and r_4 will also be half a wavelength out of phase and will also cancel. Consider four other rays, emerging from the slit a given distance below the four rays above. The two rays

below r_1 and r_2 will exactly cancel, as will the two rays below r_3 and r_4. We can proceed across the entire slit and conclude again that no light reaches P_2; we have located a second point of zero intensity.

The condition described (see Fig. 6) requires that

$$\frac{a}{4} \sin \theta = \frac{\lambda}{2},$$

or

$$a \sin \theta = 2\lambda \quad \text{(second minimum).} \quad (2)$$

By extension of Eqs. 1 and 2, the general formula for the minima in the diffraction pattern on screen C can be seen to be

$$\boxed{a \sin \theta = m\lambda \quad m = 1, 2, 3, \ldots} \quad \text{(minima).} \quad (3)$$

There is a maximum approximately halfway between each adjacent pair of minima. There are also minima and maxima on the bottom half of the screen, the pattern being symmetrical about P_0 in Fig. 6.

Sample Problem 1   A slit of width a is illuminated by white light. For what value of a will the first minimum for red light ($\lambda = 650$ nm) fall at $\theta = 15°$?

At the first minimum, $m = 1$ in Eq. 3. Solving for a, we then find

$$a = \frac{m\lambda}{\sin \theta} = \frac{(1)(650 \text{ nm})}{\sin 15°}$$

$$= 2511 \text{ nm} \approx 2.5 \ \mu\text{m}. \qquad \text{(Answer)}$$

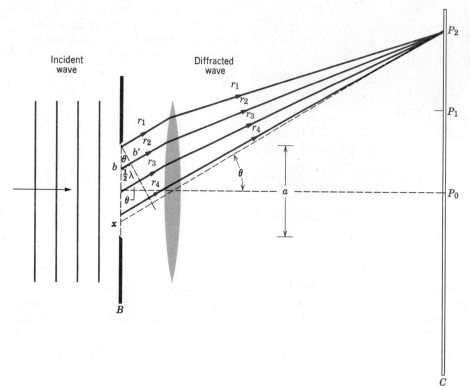

Figure 6 Conditions at the second minimum of the diffraction pattern of a single slit. The path difference bb' is one-half a wavelength.

For the incident light to flare out that much ($\pm 15°$) the slit must be very fine indeed, amounting to about 4 times the wavelength. Note that a fine human hair may only be about 100 μm in diameter.

Sample Problem 2 In Sample Problem 1, what is the wavelength λ' of the light whose first diffraction maximum (not counting the central maximum) falls at 15°, thus coinciding with the first minimum for red light?

This maximum is about halfway between the first and second minima. We can find it without too much error by putting $m = 1.5$ in Eq. 3, or*

$$a \sin \theta = 1.5 \lambda'.$$

From Sample Problem 1, however,

$$a \sin \theta = \lambda.$$

Dividing gives

$$\lambda' = \frac{\lambda}{1.5} = \frac{650 \text{ nm}}{1.5} = 433 \text{ nm.} \qquad \text{(Answer)}$$

Light of this wavelength is violet. The second maximum for light of wavelength 430 nm will always coincide with the first

minimum for light of wavelength 650 nm, no matter what the slit width. If the slit is relatively narrow, the angle θ at which this overlap occurs will be relatively large, and conversely.

41-3 Diffraction: A Closer Look

Figure 7 shows the generalized diffraction situation. The curved surfaces on the left represent wavefronts of the incident light falling on the diffracting object B which, in the case we examined in Section 41-2, was a narrow slit. In Fig. 7 we show the diffracting object as an opaque screen containing an aperture of arbitrary shape. C in Fig. 7 is a screen or photographic film that receives the light that passes through or around the diffracting object.

We can calculate the pattern of light intensity on screen C by subdividing the wavefront into elementary areas dA, each of which becomes a source of an expanding Huygens wavelet. The light intensity at an arbitrary point P is found by superimposing the wave disturbances (that is, the E-vectors) caused by the wavelets reaching P from all these elementary sources. Instead of a screen with a hole in it, the diffracting object could equally well

---

* $m = 1.43$ turns out to be a much better value but $m = 1.5$ will do for our purpose.

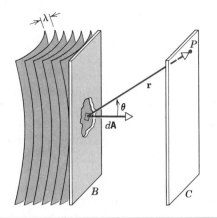

Figure 7 The elements of diffraction. Coherent wavefronts of light fall on screen *B*, which contains an aperture of arbitrary shape. The diffraction pattern is formed on diffusing screen *C*.

be a wire mesh, a screen with a single slit (Fig. 1), or a razor blade (Fig. 2).

The wave disturbances reaching *P* differ in amplitude and in phase because the elementary radiators are at varying distances from *P*. Diffraction calculations— simple in principle—may become difficult in practice. The calculation must be repeated for every point on screen *C* at which we wish to know the light intensity.

Figure 8*a* shows light from source *S* falling on a slit in screen *B* and forming a diffraction pattern on screen *C*. The rays falling on the slit and emerging from it are not parallel, which makes the calculation of the light intensity for various points on screen *C* more difficult.

A simplification results if source *S* and screen *C* are

Figure 8 (*a*) Light from point source *S* illuminates a slit in screen *B*. The intensity at point *P* on screen *C* depends on the relative phases of the light received from the various parts of the slit. (*b*) If source *S* and screen *C* are moved to a large distance from the slit, both the incident and the emergent light at *B* are closely parallel and the diffraction calculations are much simplified. (*c*) Rather than move the source and the screen, you can put each in the focal plane of a lens, as shown; the light entering and emerging from the slit is parallel, as in (*b*).

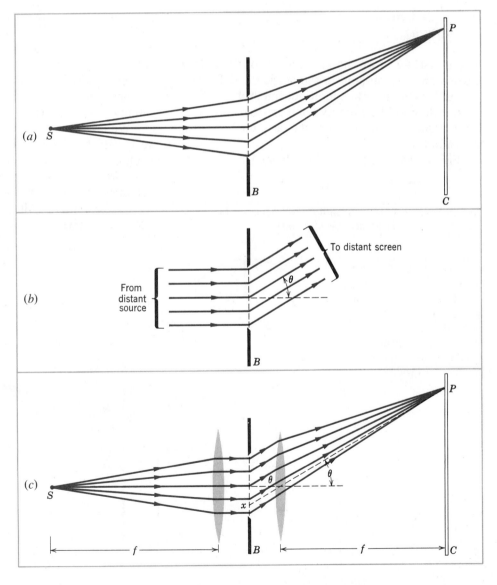

moved to a large distance from the diffraction aperture, as in Fig. 8b. The wavefronts arriving at the diffracting aperture from the distant source S are now planes, and the rays associated with these wavefronts are parallel to each other. This condition can be established in the laboratory by using two converging lenses, as in Fig. 8c. The first of these converts the diverging wave from the source into a plane wave. The second lens causes plane waves leaving the diffracting aperture to converge to point P. All rays that illuminate P will leave the diffracting aperture parallel to the dashed line Px drawn from P through the center of this second (thin) lens. In this chapter, we shall assume that the conditions described in Figs. 8b and 8c prevail.*

41–4   Diffraction from a Single Slit — Qualitative

In Section 41–2 we saw how to find the positions of the maxima and the minima in the single-slit diffraction pattern. Now we turn to a more general problem: Find an expression for the intensity I of the pattern as a function of position on the screen, that is, as a function of the angle θ in Fig. 6.

Figure 9 shows a slit of width a divided into N paral-

---

* The general case of Fig. 8a is called *Fresnel diffraction;* the special case of Fig. 8c is called *Fraunhofer diffraction.*

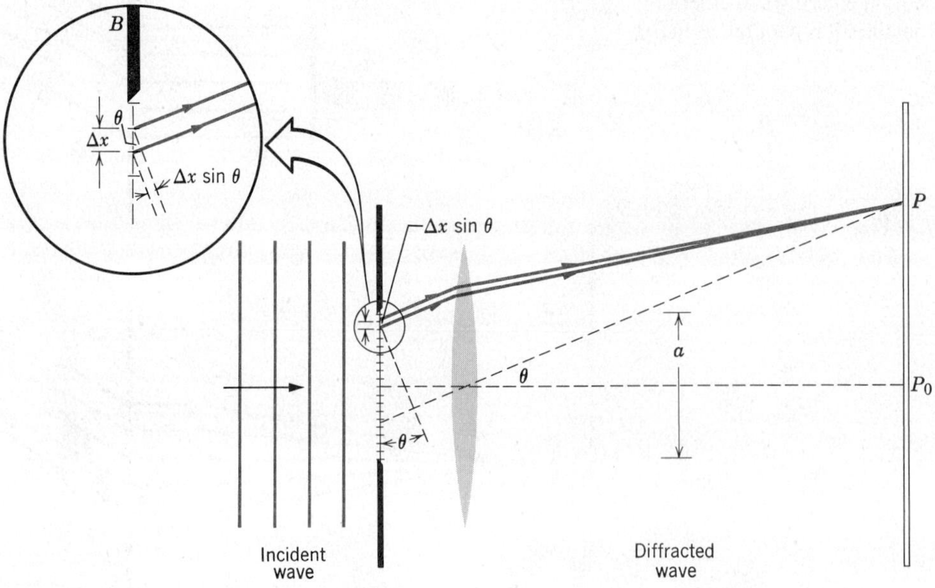

lel strips of width Δx. Each strip acts as a radiator of Huygens' wavelets and produces a characteristic wave disturbance at point P. We may take the amplitudes ΔE_0 of the wave disturbances at P from the various strips as equal if θ in Fig. 9 is not too large.

The wave disturbances from adjacent strips have a constant phase difference $\Delta\phi$ between them at P given by

$$\text{phase difference} = \left(\frac{2\pi}{\lambda}\right)(\text{path difference})$$

or

$$\Delta\phi = \left(\frac{2\pi}{\lambda}\right)(\Delta x \sin\theta), \tag{4}$$

where $\Delta x \sin\theta$, as the figure insert shows, is the path difference for rays originating at the top edges of adjacent strips. Thus at point P, N vectors with the same amplitude ΔE_0, the same wavelength λ, and the same phase difference $\Delta\phi$ between adjacent members, combine to produce a resultant disturbance. We ask: What is the amplitude E_θ of the resultant wave disturbance?

We find the answer by representing the individual wave disturbances ΔE_0 by phasors and calculating the resultant phasor amplitude, just as we did in Section 40–6. We do so qualitatively, reserving a quantitative treatment for the following section.

At the center of the diffraction pattern, θ equals zero, and the phase shift between adjacent strips (see Eq.

Figure 9   A slit of width a is divided into N strips of width Δx. The insert shows conditions at the second strip more clearly. In the differential limit, the slit is divided into an infinite number of strips of differential width dx.

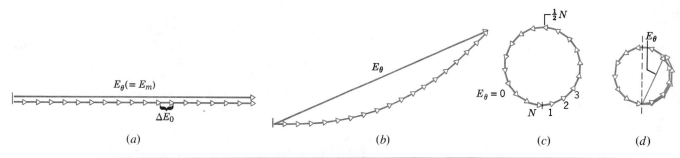

Figure 10   Phasors in single-slit diffractions. Conditions at (*a*) the central maximum, (*b*) a direction θ slightly removed from the central maximum, (*c*) the first minimum, and (*d*) the first maximum beyond the central maximum. The figure corresponds to $N = 18$ in Fig. 9.

4) is also zero. As Fig. 10*a* shows, the phasor arrows in this case are laid end to end and the amplitude of the resultant has its maximum value E_m. This corresponds to the center of the central maximum.

As we move to a value of θ other than zero, $\Delta\phi$ assumes a definite nonzero value (again see Eq. 4), and the array of arrows is now as shown in Fig. 10*b*. The resultant amplitude E_θ is less than before. Note that the length of the "arc" of small arrows is the same for both figures and indeed for all figures in this series. As θ increases further, a situation is reached (Fig. 10*c*) in which the chain of arrows curls around through 360°, the tip of the last arrow touching the foot of the first arrow. This corresponds to $E_\theta = 0$, that is, to the first minimum. For this condition the ray from the top of the slit (marked 1 in Fig. 10*c*) is 180° out of phase with the ray from the center of the slit (marked $\frac{1}{2}N$ in Fig. 10*c*). These phase relations are consistent with Fig. 4, which also represents the first minimum.

As θ increases further, the phase shift continues to increase, and the chain of arrows coils around through an angular distance greater than 360°, as in Fig. 10*d*, which corresponds to the first maximum beyond the central maximum. The intensity of this maximum is much smaller than that of the central maximum. In making this comparison, recall that the arrows marked E_θ in Fig. 10*a* correspond to the *amplitudes* of the wave disturbance and not to the *intensities*. The amplitudes must be squared to obtain the corresponding relative intensities.

41–5   Diffraction from a Single Slit— Quantitative

Equation 3 tells us how to locate the minima of the single-slit diffraction pattern on screen *C* of Fig. 9 as a

function of the angle θ in that figure. Note that θ is our position locator, every screen point *P* being associated with a definite value of θ. Here we wish to derive an expression for the intensity *I* of the pattern as a function of θ.

We state, and shall prove below, that the intensity is given by

$$I = I_m \left(\frac{\sin \alpha}{\alpha} \right)^2 \tag{5}$$

where

$$\alpha = \tfrac{1}{2}\phi = \left(\frac{\pi a}{\lambda} \right) \sin \theta. \tag{6}$$

We can here view α as a convenient transfer angle, serving as a connection between θ of Fig. 9 and *I* of Eq. 5. In Eq. 5, I_m is the maximum value of the intensity, which occurs at the center of the diffraction pattern, corresponding to $\theta = 0$. Figure 11 shows plots of the intensity of a single-slit diffraction pattern, calculated from Eqs. 5 and 6. Note that, as the slit width is decreased from 10λ (in Fig. 11*c*) to 5λ (in Fig. 11*b*) to λ (in Fig. 11*a*), the central maximum becomes correspondingly broader.

Equations 5 and 6, which together tell us how the intensity of the diffraction pattern varies with the angle θ in Fig. 11, necessarily contain information about the location of the intensity minima. Let us see if we can extract it.

Study of Eq. 5 shows that intensity minima will occur when

$$\alpha = m\pi \quad m = 1, 2, 3, \ldots \tag{7}$$

If we put this result into Eq. 6 we find

$$m\pi = \frac{\pi a}{\lambda} \sin \theta,$$

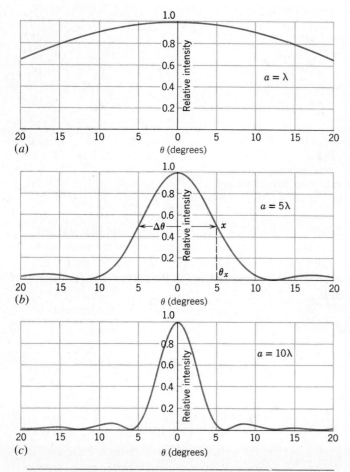

Figure 11 The relative intensity in single-slit diffraction for three different values of the ratio a/λ. The wider the slit, the narrower is the central diffraction peak.

or

$$a \sin \theta = m\lambda \quad m = 1, 2, 3, \ldots \quad \text{(minima)} \quad (8)$$

which is exactly Eq. 3, the expression that we derived earlier for the location of the minima.

Proof of Eq. 5. The arc of small arrows in Fig. 12 shows the phasors representing, in amplitude and phase, the wave disturbances that reach an arbitrary point P on the screen of Fig. 9, corresponding to a particular angle θ. The resultant amplitude at P is E_θ. If we divide the slit of Fig. 9 into infinitesimal strips of width dx, the arc of arrows in Fig. 12 approaches the arc of a circle, its radius R being indicated in that figure. The length of the arc is E_m, the amplitude at the center of the diffraction pattern, for at the center of the pattern the wave disturbances are all in phase and this "arc" becomes a straight line as in Fig. 10a.

The angle ϕ in the lower part of Fig. 12 is the difference in phase between the infinitesimal vectors at the left and the right ends of the arc E_m. This means that ϕ is the phase difference between rays from the top and the bottom of the slit of Fig. 6. From geometry we see that ϕ is also the angle between the two radii marked R in Fig. 12. From this figure we can write

$$E_\theta = 2R \sin \tfrac{1}{2}\phi.$$

In radian measure ϕ (from the figure) is

$$\phi = \frac{E_m}{R}.$$

Combining yields

$$E_\theta = \left(\frac{E_m}{\tfrac{1}{2}\phi}\right) \sin \frac{1}{2}\phi.$$

We have several times made use of the fact that the intensity of a wave is proportional to the square of its amplitude. If we do so again with the above equation, substituting α for $\tfrac{1}{2}\phi$, we are led to an expression for the intensity as a function of θ that we can write as

$$I = I_m \left(\frac{\sin \alpha}{\alpha}\right)^2.$$

This is exactly Eq. 5, one of the two equations we set out to prove.

It remains to relate the angle α to the angle θ. The phase difference ϕ between rays from the top and the

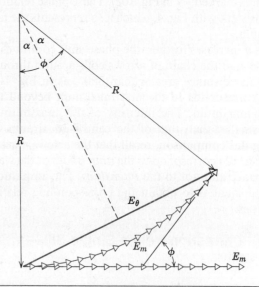

Figure 12 A construction used to calculate the intensity in single-slit diffraction. The situation corresponds to that of Fig. 10b.

bottom of the slit in Fig. 9 is related to a path difference for those rays by Eq. 4, or

$$\phi = \left(\frac{2\pi}{\lambda}\right)(a \sin \theta).$$

But $\phi = 2\alpha$, so that this equation readily reduces to Eq. 6.

Sample Problem 3 Find the intensities of the secondary maxima in the single-slit diffraction pattern of Fig. 1, measured relative to the intensity of the central maximum.

The secondary maxima lie approximately halfway between the minima, which are given by Eq. 7 ($\alpha = m\pi$). The maxima are then given (approximately) by

$$\alpha = (m + \tfrac{1}{2})\pi \quad m = 1, 2, 3, \ldots$$

If we substitute this result into Eq. 5 we obtain

$$\frac{I}{I_m} = \left(\frac{\sin \alpha}{\alpha}\right)^2 = \left(\frac{\sin(m + \tfrac{1}{2})\pi}{(m + \tfrac{1}{2})\pi}\right)^2 \quad m = 1, 2, 3, \ldots$$

The first secondary maximum occurs for $m = 1$, its relative intensity being

$$\frac{I}{I_m} = \left(\frac{\sin(1 + \tfrac{1}{2})\pi}{(1 + \tfrac{1}{2})\pi}\right)^2 = \left(\frac{\sin 1.5\pi}{1.5\pi}\right)^2 = \left(\frac{\sin 270°}{1.5\pi}\right)^2$$

$$= 4.50 \times 10^{-2} \approx 4.5\%. \quad \text{(Answer)}$$

For $m = 2$ and $m = 3$ we find that $I/I_m = 1.6\%$ and 0.83%, respectively.

The successive maxima decrease rapidly in intensity. The pattern of Fig. 1 has been deliberately overexposed to reveal these faint secondary maxima.

41-6 Diffraction from a Circular Aperture

Here we consider diffraction by a circular aperture of diameter d, the aperture constituting the boundary of a converging lens.

Figure 13 shows the image of a distant point source of light (a star, for instance) formed on a photographic film placed in the focal plane of a converging lens. It is not a point, as the (approximate) geometrical optics treatment suggests, but a circular disk surrounded by several progressively fainter secondary rings. Comparison with Fig. 1 leaves little doubt that we are dealing with a diffraction phenomenon in which, however, the aperture is a circle rather than a rectangular slit. The ratio d/λ, where d is the diameter of the lens (or of a circular

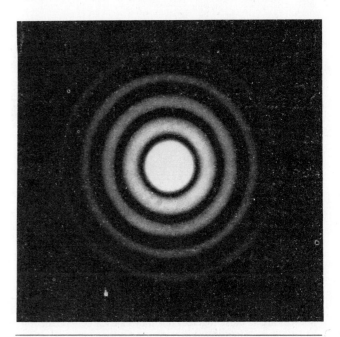

Figure 13 The diffraction pattern of a circular aperture. Note the central maximum and the circular secondary maxima. The figure has been deliberately overexposed to bring out these secondary maxima, which are much less intense than the central disk.

aperture placed in front of the lens), determines the scale of the diffraction pattern, just as the ratio a/λ does for a slit.

Analysis shows that the first minimum for the diffraction pattern of a circular aperture of diameter d is given by

$$\sin \theta = 1.22 \frac{\lambda}{d}$$

$$\text{(1st minimum; circular aperture).} \quad (9)$$

This is to be compared with Eq. 1, or

$$\sin \theta = \frac{\lambda}{a} \quad \text{(1st minimum; single slit),} \quad (10)$$

which locates the first minimum for a long narrow slit of width a. The factor 1.22 emerges from the mathematical analysis when we integrate over the elementary radiators into which the circular aperture may be divided.

Resolving Power. The fact that lens images are diffraction patterns is important when we wish to distinguish two distant point objects whose angular separation is small. Figure 14 shows the visual appearances and the corresponding intensity patterns for two distant point objects (stars, say) with small angular separations. In

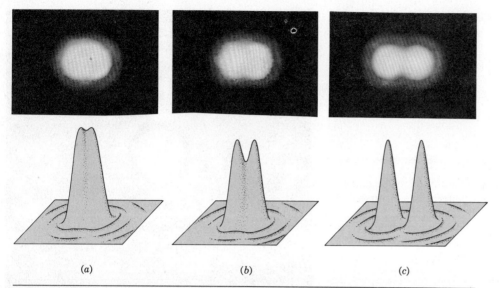

(a)   (b)   (c)

Figure 14 The images of two point sources (stars) are formed by a converging lens. In (a) the sources are so close together that they can scarcely be distinguished. In (b) they can be marginally distinguished and in (c) they are clearly distinguished. The criterion in (b) (Rayleigh's criterion) is that the central maximum of one diffraction pattern coincide with the first minimum of the other. Computer-generated profiles of the intensities are shown below the images.

Figure 14a the objects are not resolved; that is, they cannot be distinguished from a single point object. In Fig. 14b they are barely resolved and in Fig. 14c they are fully resolved.

In Fig. 14b the angular separation of the two point sources is such that the maximum of the diffraction pattern of one source falls on the first minimum of the diffraction pattern of the other, a condition called *Rayleigh's criterion* for resolvability. From Eq. 9, two objects that are barely resolvable by this criterion must have an angular separation θ_R of

$$\theta_R = \sin^{-1} \frac{1.22\lambda}{d}.$$

Since the angles involved are small, we can replace $\sin \theta_R$ by θ_R expressed in radians:

$$\boxed{\theta_R = 1.22 \frac{\lambda}{d}} \qquad \text{(Rayleigh's criterion).} \quad (11)$$

If the angular separation θ between the objects is greater than θ_R, we can resolve the two objects; if it is significantly less, we cannot. The objects must be of comparable brightness for Rayleigh's criterion to be useful.

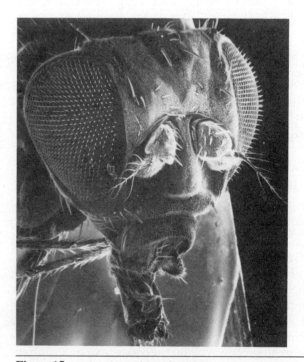

Figure 15 A head-on view of a midge, suggesting the fine resolution that can be obtained with the small wavelengths that are possible in an electron microscope.

When we wish to use a lens to resolve objects of small angular separation, it is desirable to make the central disk of the diffraction pattern as small as possible. This can be done (see Eq. 11) by increasing the lens diameter or by using a shorter wavelength.

To reduce diffraction effects in microscopes we often use ultraviolet light, which, because of its shorter wavelength, permits finer detail to be examined than would be possible for the same microscope operated with visible light. In Chapter 44 of the extended version of this text, we show that beams of electrons behave like waves under some circumstances. In the electron microscope such beams may have an effective wavelength as much as 10^5 times shorter than the wavelength of visible light. This permits the detailed examination of tiny structures that would be hopelessly blurred by diffraction if viewed with an optical microscope; see Fig. 15.

Sample Problem 4 A converging lens 32 mm in diameter has a focal length f of 24 cm. (a) What angular separation must two distant point objects have to satisfy Rayleigh's criterion? Assume that $\lambda = 550$ nm.

From Eq. 11

$$\theta_R = 1.22\,\frac{\lambda}{d} = \frac{(1.22)(550 \times 10^{-9}\ \text{m})}{32 \times 10^{-3}\ \text{m}}$$
$$= 2.10 \times 10^{-5}\ \text{rad} = 4.3\ \text{arc seconds.}\quad \text{(Answer)}$$

(b) How far apart are the centers of the diffraction patterns in the focal plane of the lens?

The linear separation is

$$\Delta x = f\theta = (0.24\ \text{m})(2.10 \times 10^{-5}\ \text{rad})$$
$$= 5.0\ \mu\text{m.}\quad \text{(Answer)}$$

This is about 9 wavelengths of the light employed.

41-7 Diffraction from a Double Slit (Optional)

In Young's double-slit experiment (Section 40-4) we assumed that the slits were arbitrarily narrow, that is, that $a \ll \lambda$. For such narrow slits, the central part of the screen on which the light falls is uniformly illuminated by the diffracted waves from each slit. When such waves interfere, they produce interference fringes of uniform intensity.

In practice, for visible light, the condition $a \ll \lambda$ is

usually not met. For such relatively wide slits, the intensity of the interference fringes formed on the screen will *not* be uniform. Instead, the intensity of the fringes will be governed by an intensity envelope that is the diffraction pattern of a single slit.

Figure 16a, for example, suggests the double-slit fringe pattern that would occur if the slits were infinitely narrow. Figure 16b shows the diffraction pattern of an actual slit; the broad central maximum and one weaker secondary maximum (at $\pm 17°$) are shown. Figure 16c, which is found by multiplying Figs. 16a and 16b, shows the resulting interference pattern. We see that the positions of the fringes remain unchanged but their intensities are indeed governed by the single-slit diffraction pattern.

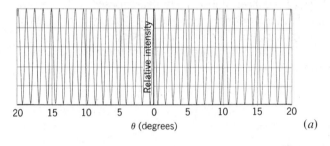

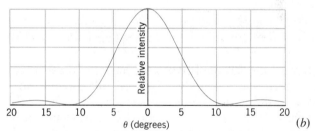

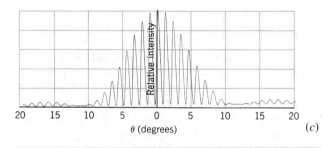

Figure 16 (a) The uniform fringe pattern to be expected from a double slit with vanishingly narrow slits. (b) The diffraction pattern of a typical slit of width a (not vanishingly narrow). (c) The fringe pattern formed by two of these slits. The pattern is equivalent to the pattern in (a) multiplied point by point by the pattern in (b).

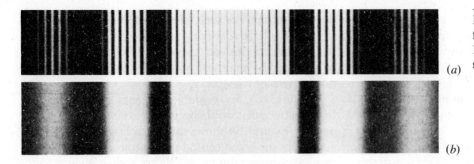

(a)

(b)

Figure 17 (a) Interference fringes for a double-slit system; compare Fig. 16c. (b) The diffraction pattern of a single slit; compare Fig. 16(b).

Figure 17a shows an actual double-slit interference–diffraction pattern. If one slit is covered, the single-slit diffraction pattern of Fig. 17b results. Note that Fig. 17a corresponds to Fig. 16c and Fig. 17b to Fig. 16b. In making these comparisons, bear in mind that Fig. 17 has been deliberately overexposed to bring out the faint secondary maxima and that two secondary maxima (rather than one) are shown.

The intensity of double-slit interference pattern is given by

$$I = I_m(\cos \beta)^2 \left(\frac{\sin \alpha}{\alpha}\right)^2 \quad \text{(double slit)} \quad (12)$$

in which

$$\beta = \left(\frac{\pi d}{\lambda}\right) \sin \theta \quad \text{and} \quad \alpha = \left(\frac{\pi a}{\lambda}\right) \sin \theta. \quad (13)$$

Here d is the distance between the centers of the slits and a is the slit width. A study of Eq. 12 should convince you that it is the product of the interference pattern for a pair of narrow slits with slit separation d (see Eqs. 15 and 16 of Chapter 40) and the diffraction pattern for a single slit of width a (see Eqs. 5 and 6).

Let us look at Eqs. 12 and 13 more closely. If we put $a = 0$ in Eq. 13, for example, then $\alpha = 0$ and $\sin \alpha/\alpha = 1$. Equation 12 then reduces, as it must, to an equation describing the interference pattern for a pair of vanishingly narrow slits with slit separation d. Similarly, putting $d = 0$ is equivalent physically to causing the two slits to merge into a single slit of width a. Putting $d = 0$ in Eq. 13 yields $\beta = 0$ and $\cos^2 \beta = 1$. In this case Eq. 12 reduces, as it must, to an equation describing the diffraction pattern for a single slit of width a.

The double-slit pattern described by Eqs. 12 and 13 and displayed in Fig. 17a combines interference and diffraction in an intimate way. Both are superposition effects and depend on adding wave disturbances at a given point, taking phase differences properly into account. If the waves to be combined originate from a finite (and usually small) number of elementary coherent sources—as in Young's double-slit experiment—we call the process *interference*. If the waves to be combined originate by subdividing a wavefront into infinitesimal coherent sources of differential size—as we did in Fig. 6—we call the process *diffraction*. This distinction between interference and diffraction (which is somewhat arbitrary and not always adhered to) is convenient, but we should not lose sight of the fact that both are superposition effects and that often both are present simultaneously, as in Fig. 17a.

Sample Problem 5 In a double-slit experiment, the distance D of the screen from the slits is 52 cm, the wavelength λ is 480 nm, the slit separation d is 0.12 mm, and the slit width a is 0.025 mm. (a) What is the spacing between adjacent fringes?

The intensity pattern is given by Eq. 12, the fringe spacing being determined by the interference factor $\cos^2 \beta$. From Sample Problem 1, Chapter 40, we have

$$\Delta y = \frac{\lambda D}{d}.$$

Substituting yields

$$\Delta y = \frac{(480 \times 10^{-9} \text{ m})(52 \times 10^{-2} \text{ m})}{0.12 \times 10^{-3} \text{ m}}$$

$$= 2.080 \times 10^{-3} \text{ m} \approx 2.1 \text{ mm}. \quad \text{(Answer)}$$

(b) What is the distance from the central maximum to the first minimum of the fringe envelope?

The angular position of the first minimum follows from Eq. 1, or

$$\sin \theta = \frac{\lambda}{a} = \frac{480 \times 10^{-9} \text{ m}}{25 \times 10^{-6} \text{ m}} = 0.0192.$$

This is so small that, with little error, we can put $\sin \theta \approx$

$\overline{\tan} \theta \approx \theta$. Thus

$$y = D \tan \theta \approx D\theta = (52 \times 10^{-2} \text{ m})(0.0192)$$
$$= 9.98 \times 10^{-3} \text{ m} \approx 10 \text{ mm.} \qquad \text{(Answer)}$$

You can show that there are about 9 fringes in the central peak of the diffraction envelope.

Sample Problem 6 What requirements must be met for the central maximum of the envelope of the double-slit interference pattern to contain exactly 11 fringes?

The required condition will be met if the sixth minimum of the interference factor ($\cos^2 \beta$ in Eq. 12) coincides with the first minimum of the diffraction factor ($\sin \alpha/\alpha)^2$ in Eq. 12.

The sixth minimum of the interference factor occurs when

$$\beta = (11/2)\pi$$

in Eq. 12. The first minimum in the diffraction term occurs for

$$\alpha = \pi$$

in Eq. 12. Dividing (see Eq. 13) yields

$$\frac{\beta}{\alpha} = \frac{d}{a} = \frac{11}{2}.$$

This condition depends only on the ratio of the slit separation d to the slit width a and not at all on the wavelength. For long waves the pattern will be broader than for short waves, but there will always be 11 fringes in the central peak of the envelope.

41-8 Multiple Slits

A logical extension of Young's double-slit interference experiment is to increase the number of slits from two to a larger number N. An arrangement like that of Fig. 18, usually involving many more slits—as many as 10^3/mm

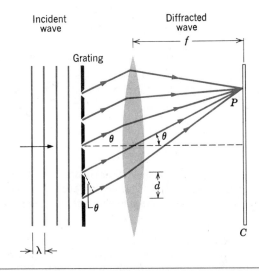

Figure 18 An idealized diffraction grating containing 5 slits. The figure is distorted for clarity.

is not uncommon—is called a *diffraction grating*. As for a double slit, the intensity pattern that results when monochromatic light falls on a grating consists of a series of interference fringes.

Figure 19 compares the intensity patterns for $N = 2$ and for $N = 5$, showing only the central maximum of the single-slit diffraction envelope. We see two important changes that occur when we increase the number of slits from two to five. (1) The fringes become narrower. (2) Faint secondary maxima (three in this case) appear between the fringes. As N increases, perhaps to 10^4 for a useful grating, the fringes become very sharp indeed and the secondary maxima, while increasing in number, become so reduced in intensity as to be negligible in their effects; we shall ignore them in what follows.

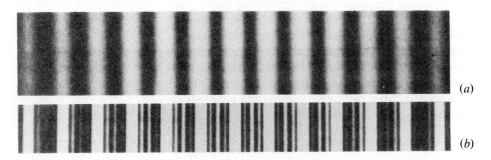

(a)

(b)

Figure 19 (a) A diffraction pattern for a two-slit diffraction "grating." (b) The same for a five-slit grating. Note that the fringes are sharper and that secondary maxima of low intensity make their appearance. The pattern for the five-slit grating was deliberately overexposed to bring out these faint secondary maxima.

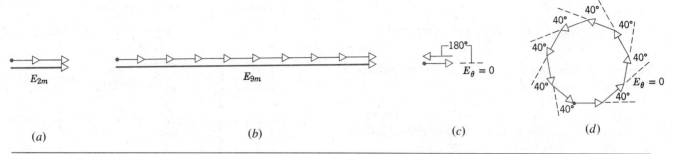

Figure 20 Drawings (*a*) and (*b*) show conditions at the central maximum for a two-slit and a nine-slit grating, respectively. Drawings (*c*) and (*d*) show conditions at the minimum of zero intensity that lies on either side of this central maximum.

Positions of the Fringes. A sharply defined fringe (or *line*) will occur when $d \sin \theta$, which is the path difference between rays from adjacent slits in Fig. 18, is equal to an integral number of wavelengths, or

$$\boxed{d \sin \theta = m\lambda} \quad m = 0, 1, 2, \ldots \text{ (maxima).} \quad (14)$$

Here m is called the *order number* of the line in question, $m = 0$ corresponding to the central line. This equation is identical with Eq. 9 of Chapter 40, which locates the intensity maxima for a double slit. The locations of the diffracted lines are thus determined only by the ratio λ/d and are independent of N.

Width of the Fringes. Here we seek to understand how the fringes narrow into sharp lines as N is increased. We use a graphical argument, based on phasors. Figures 20*a* and 20*b* show conditions at the central maximum for a two-slit and a nine-slit grating. The small arrows represent the amplitudes of the wave disturbances arriving at the center of the screen.

Consider the angle $\Delta\theta$, corresponding to the position of zero intensity that lies on either side of the central maximum. Figures 20*c* and 20*d* show the phasors at this point. The phase difference between waves from adjacent slits, which is zero at the central maximum, must increase by an amount $\Delta\phi$ chosen so that the array of the phasors just closes on itself, yielding zero resultant intensity. For $N = 2$, $\Delta\phi = 2\pi/2$ (= 180°); for $N = 9$, $\Delta\phi = 2\pi/9$ (= 40°). In the general case of N slits, the phase difference at the first intensity minimum is given by

$$\Delta\phi = \frac{2\pi}{N}. \quad (15)$$

We see that, as N increases, the phase difference between adjacent slits that corresponds to the first intensity minimum becomes smaller and smaller. This phase difference for adjacent waves corresponds to a path difference ΔL given by Eq. 22 of Chapter 40, or

$$\text{path difference} = \left(\frac{\lambda}{2\pi}\right) \text{(phase difference)}$$

or, from Eq. 15,

$$\Delta L = \left(\frac{\lambda}{2\pi}\right)(\Delta\phi) = \left(\frac{\lambda}{2\pi}\right)\left(\frac{2\pi}{N}\right) = \frac{\lambda}{N}. \quad (16)$$

From Fig. 18, however, the path difference ΔL at the first minimum is also given by $d \sin \Delta\theta$, so that we can write

$$d \sin \Delta\theta = \frac{\lambda}{N},$$

or (because $\Delta\theta$ is very small for a sharp maximum)

$$\Delta\theta = \frac{\lambda}{Nd} \quad \text{(central maximum).} \quad (17)$$

We interpret $\Delta\theta$ as the *line width* of the central maximum. Equation 17 shows specifically that, if λ and d are fixed, the line width $\Delta\theta$ of the central maximum will decrease as we increase N. That is, the central maximum becomes sharper as we add more slits to the grating. (Comparison of Eq. 17 with Eq. 3 suggests that, as far as the location of its first diffraction minimum is concerned, the grating behaves like a single slit of width Nd.)

Equation 17 gives the line width for the central line only. We state without proof that the general expression for the width of lines of other orders, that occur at an angle θ, is

$$\boxed{\Delta\theta = \frac{\lambda}{Nd \cos \theta}} \quad \text{(line width)} \quad (18)$$

in which θ defines the direction in which the line in question occurs. (Note that, in this case, the grating behaves like a single slit whose width in the direction defined by θ is $Nd \cos \theta$, the *projected* width of the grating.)

41-9   Diffraction Gratings

The grating spacing d for a typical grating that contains 12,000 "slits" distributed over a 1-in. width is 25.4 mm/12,000 or 2100 nm. Gratings are widely used to measure wavelengths and to study the structure and intensity of spectral lines.

Gratings are made by ruling equally spaced parallel grooves on a glass or a metal plate, using a diamond cutting point whose motion is automatically controlled by an elaborate ruling engine. Gratings ruled on metal are called *reflection gratings* because the interference effects are viewed in reflected rather than in transmitted light. Once such a master grating has been prepared, replicas can be formed by pouring a liquid plastic on the grating, allowing it to harden, and stripping it off. The stripped plastic, fastened to a flat piece of glass or other backing, forms a good grating.

Figure 21 shows a simple grating spectroscope, used for viewing the spectrum of a light source, assumed to emit a number of discrete wavelengths, or spectral lines.

The light from source S is focused by lens L_1 on a slit S_1 placed in the focal plane of lens L_2. The parallel light emerging from collimator C falls on grating G. Parallel rays associated with a particular maximum occurring at angle θ fall on lens L_3, being brought to a focus in a plane $F - F'$. The image formed in this plane is examined, using a magnifying lens arrangement E, called an eyepiece. A symmetrical interference pattern is formed on the other side of the central position, as shown by the dotted lines. The entire spectrum can be viewed by rotating telescope T through various angles. Instruments used for scientific research or in industry are more complex than the simple arrangement of Fig. 21. They invariably employ photographic or photoelectric recording and are called *spectrographs.*

Grating instruments can be used to make absolute measurements of wavelength, since the grating spacing d in Eq. 14 can be measured accurately with a microscope. Several spectra are normally produced in such instruments, corresponding to $m = 1, 2, 3, \ldots$ in Eq. 14; see Fig. 22. This may cause some confusion if the spectra overlap. Further, this multiplicity of spectra reduces the recorded intensity of any given spectrum line because the available energy is divided among a number of spectra. However, by controlling the shape of the grating rulings, a large fraction of the energy can be concentrated in a particular order; this is called *blazing.*

Figure 21   A simple type of grating spectroscope used to analyze the wavelengths of the light emitted by source S.

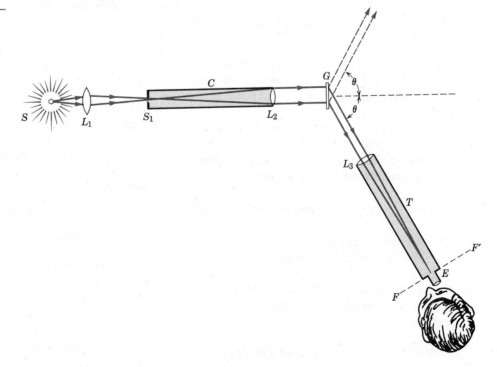

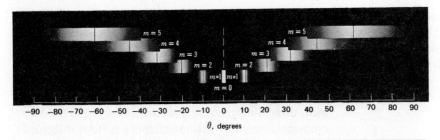

Figure 22 The spectrum of white light as viewed in an instrument like that of Fig. 21. The different orders, identified by the order number m, are shown separated vertically for clarity. As actually viewed, they overlap. The central line in each order corresponds to $\lambda = 550$ nm, the center of the visible spectrum.

41-10 Gratings: Dispersion and Resolving Power (Optional)

Dispersion. It is desirable in a grating that wavelengths that are close together be as widely separated in angle as possible. That is, we wish a grating to have a large *dispersion D*, defined from

$$D = \frac{\Delta\theta}{\Delta\lambda} \quad \text{(dispersion defined)}. \quad (19)$$

Here $\Delta\theta$ is the angular separation of two lines whose wavelengths differ by $\Delta\lambda$. We show below that the dispersion of a grating is given by

$$\boxed{D = \frac{m}{d\cos\theta}} \quad \text{(dispersion of a grating)}. \quad (20)$$

Thus, to achieve high dispersion, use a grating of small grating spacing (small d) and work in high orders (large m). Note that the dispersion does not depend on the number of rulings.

Resolving Power. To separate lines whose wavelengths are close together, the line widths should be as small as possible. Expressed otherwise, the grating should have a large *resolving power R*, defined from

$$R = \frac{\lambda}{\Delta\lambda} \quad \text{(resolving power defined)}. \quad (21)$$

Here λ is the mean wavelength of two spectrum lines that can barely be recognized as separate and $\Delta\lambda$ is the wavelength difference between them. The smaller $\Delta\lambda$ is, the closer the lines can be and still be resolved. We shall show below that the resolving power of a grating is given by the simple expression

$$\boxed{R = Nm} \quad \text{(resolving power of a grating)}. \quad (22)$$

It is to achieve high *dispersion* that grating rulings are closely spaced (d small; see Eq. 20). It is to achieve high *resolving power* that many rulings are used (N large; see Eq. 22).

Proof of Eq. 20. Let us start with Eq. 14, the expression for the angular positions of the lines in the diffraction pattern of a grating:

$$d\sin\theta = m\lambda.$$

Let us regard θ and λ as variables and take differentials of this expression. We find

$$d\cos\theta\,d\theta = m\,d\lambda.$$

For small enough angles, we can write these differentials as small differences, thus

$$d\cos\theta\,\Delta\theta = m\,\Delta\lambda \quad (23)$$

or

$$\frac{\Delta\theta}{\Delta\lambda} = \frac{m}{d\cos\theta}.$$

The ratio on the left is simply D (see Eq. 19) so that we have indeed derived Eq. 20.

Proof of Eq. 22. We start with Eq. 23, which was derived from Eq. 14, the expression for the angular position of the lines in the diffraction pattern formed by a grating. Thus

$$d\cos\theta\,\Delta\theta = m\,\Delta\lambda. \quad (24)$$

Here $\Delta\lambda$ is the small wavelength difference between the two waves and $\Delta\theta$ is the angular separation between them. If $\Delta\theta$ is to be the largest angle that will permit the two lines to be resolved, it must (by Rayleigh's criterion) be equal to the width of the line, which is given by Eq. 18, or

$$\Delta\theta = \frac{\lambda}{Nd\cos\theta} \quad \text{(line width)}. \quad (25)$$

Table 1   Three Gratings[a]

Grating	N	d nm	θ	D °/μm	R
A	10,000	2540	13.3°	23.2	10,000
B	20,000	2540	13.3°	23.2	20,000
C	10,000	1370	15.5°	43.4	10,000

[a] For $\lambda = 589$ nm and $m = 1$.

If we substitute Eq. 25 into Eq. 24, we find

$$d \cos \theta = \frac{\lambda}{Nd \cos \theta} = m \, \Delta\lambda,$$

from which it readily follows that

$$R = \frac{\lambda}{\Delta\lambda} = Nm.$$

This is Eq. 22, which we set out to derive.

Dispersion and Resolving Power Compared.   The resolving power of a grating must not be confused with its dispersion. Table 1 shows the characteristics of three gratings, each illuminated with light of $\lambda = 589$ nm, the diffracted light being viewed in the first order ($m = 1$ in Eq. 14). You should verify that the values of D and R are given in the table can be calculated from Eqs. 20 and 22, respectively.

For the conditions noted in Table 1, gratings A and B have the same *dispersion* and A and C have the same *resolving power.*

Figure 23 shows the intensity patterns (line shapes) that would be produced by these gratings for two lines of wavelength λ_1 and λ_2, in the vicinity of $\lambda = 589$ nm. Grating B, which has high resolving power, has narrow lines and is inherently capable of distinguishing lines that are much closer together in wavelength than those in the figure. Grating C, which was high dispersion, produces twice the angular separation between lines λ_1 and λ_2 as does grating B.

Sample Problem 7   A diffraction grating has 1.26×10^4 rulings uniformly spaced over 25.4 mm. It is illuminated at normal incidence by yellow light from a sodium vapor lamp. This light contains two closely spaced lines (the well-known sodium doublet) of wavelengths 589.00 nm and 589.59 nm. (a) At what angle will the first-order maximum occur for the first of these wavelengths?

The grating spacing d is given by

$$d = \frac{L}{N} = \frac{25.4 \times 10^{-3} \text{ m}}{1.26 \times 10^4}$$

$$= 2.016 \times 10^{-6} \text{ m} = 2016 \text{ nm}.$$

The first-order maximum corresponds to $m = 1$ in Eq. 14. We thus have

$$\theta = \sin^{-1} \frac{m\lambda}{d} = \sin^{-1} \left(\frac{(1)(589.00 \text{ nm})}{2016 \text{ nm}} \right)$$

$$= 16.99° \approx 17.0°. \qquad \text{(Answer)}$$

(b) What is the angular separation between these two lines (in first order)?

Here the *dispersion* of the grating comes into play. From Eq. 20, the dispersion is

$$D = \frac{m}{d \cos \theta} = \frac{1}{(2016 \text{ nm})(\cos 16.99°)}$$

$$= 5.187 \times 10^{-4} \text{ rad/nm}.$$

From Eq. 19, the defining equation for dispersion, we have

$$\Delta\theta = D \, \Delta\lambda$$
$$= (5.187 \times 10^{-4} \text{ rad/nm})(589.59 \text{ nm} - 589.00 \text{ nm})$$
$$= 3.06 \times 10^{-4} \text{ rad} = 0.0175° = 1.05 \text{ arc min}. \quad \text{(Answer)}$$

As long as the grating spacing d remains fixed, this result holds no matter how many lines there are in the grating.

(c) How close in wavelength can two lines be (in first order) and still be resolved by this grating?

Here the *resolving power* of the grating comes into play. From Eq. 22, the resolving power is

$$R = Nm = (1.26 \times 10^4)(1) = 1.26 \times 10^4.$$

From Eq. 21, the defining equation for resolving power, we have,

$$\Delta\lambda = \frac{\lambda}{R} = \frac{589 \text{ nm}}{1.26 \times 10^4} = 0.047 \text{ nm}. \qquad \text{(Answer)}$$

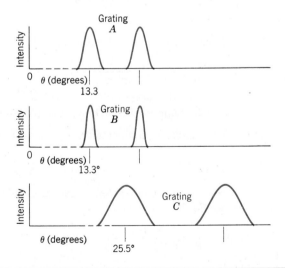

Figure 23   The intensity patterns for light of wavelengths λ_1 and λ_2 falling on the gratings of Table 1. Grating B has the highest resolving power and grating C the highest dispersion.

Thus, this grating can easily resolve the two sodium lines, which have a wavelength separation of 0.59 nm. Note that this result depends only on the number of grating rulings and is independent of d, the spacing between adjacent rulings.

(d) How many rulings can a grating have and just resolve the sodium doublet lines?

From Eq. 21, the defining equation for R, the grating must have a resolving power of

$$R = \frac{\lambda}{\Delta\lambda} = \frac{589 \text{ nm}}{0.59 \text{ nm}} = 998.$$

From Eq. 23, the number of rulings needed to achieve this resolving power (in first order) is

$$N = \frac{R}{m} = \frac{998}{1} = 998 \text{ rulings}. \qquad \text{(Answer)}$$

Since the grating has about 13 times as many rulings as this, it can easily resolve the sodium doublet lines, as we have already shown in (c) above.

Sample Problem 8 A grating has 8200 lines uniformly spaced over 25.4 mm and is illuminated by light from a mercury vapor discharge. (a) What is the expected dispersion, in the third order, in the vicinity of the intense green line ($\lambda =$ 546 nm)?

The grating spacing is given by
$$d = \frac{L}{N} = \frac{25.4 \times 10^{-3} \text{ m}}{8200}$$
$$= 3.098 \times 10^{-6} \text{ m} \approx 3100 \text{ nm}.$$

We must find the angle θ at which the line in question occurs. From Eq. 14, we have

$$\theta = \sin^{-1}\frac{m\lambda}{d} = \sin^{-1}\frac{(3)(546 \text{ nm})}{3100 \text{ nm}}$$
$$= 31.9°.$$

We can now calculate the dispersion. From Eq. 20

$$D = \frac{m}{d\cos\theta} = \frac{3}{(3100 \text{ nm})(\cos 31.9°)}$$
$$= 1.14 \times 10^{-3} \text{ rad/nm}$$
$$= 0.0653°/\text{nm} = 3.92 \text{ arc min/nm}. \quad \text{(Answer)}$$

(b) What is the resolving power of this grating in the fifth order? From Eq. 22

$$R = Nm = (8200)(5) = 4.10 \times 10^4. \qquad \text{(Answer)}$$

Thus, near $\lambda = 546$ nm and in fifth order, a wavelength difference given by (see Eq. 21)

$$\Delta\lambda = \frac{\lambda}{R} = \frac{546 \text{ nm}}{4.10 \times 10^4} = 0.013 \text{ nm}$$

can be resolved.

41–11 X-ray Diffraction

X rays are electromagnetic radiation whose wavelengths are of the order of 1 Å (= 10^{-10} m).* Compare this with 550 nm (= 5.5×10^{-7} m) for the center of the visible spectrum. Figure 24 shows how the x rays are produced when electrons from a heated filament F are accelerated by a potential difference V and strike a metal target T.

For such small wavelengths a standard optical diffraction grating, as normally employed, cannot be used to discriminate between different wavelengths. For $\lambda = 1$ Å (= 0.1 nm) and $d = 3000$ nm, for example, Eq. 14 shows that the first-order maximum occurs at

$$\theta = \sin^{-1}\frac{m\lambda}{d} = \sin^{-1}\frac{(1)(0.1 \text{ nm})}{3000 \text{ nm}} = 0.0019°.$$

This is too close to the central maximum to be practical. A grating with $d \approx \lambda$ is desirable, but, since x-ray wavelengths are about equal to atomic diameters, such gratings cannot be constructed mechanically.

In 1912 it occurred to the German physicist Max von Laue that a crystalline solid, consisting as it does of a regular array of atoms, might form a natural three-dimensional "diffraction grating" for x rays. The idea is that in a crystal, such as sodium chloride (NaCl), there is a basic unit of atoms (called the *unit cell*) that repeats itself throughout the array. In NaCl four sodium ions

---

* The *ångström* (1 Å = 10^{-10} m) is "a unit to be used with SI for a limited time." Its usefulness stems from the fact that atoms are about this same size.

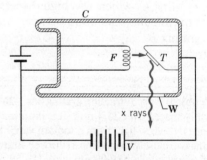

Figure 24 X rays are generated when electrons from heated filament F, accelerated through a potential difference V, strike a metal target T. W is a "window"—transparent to x rays—in the evacuated chamber C.

and four chlorine ions are associated with each unit cell. Figure 25 represents a section through a crystal of NaCl and identifies this basic unit. This crystal has a cubic structure and the unit cell is thus itself a cube, measuring a_0 on the side.

Each unit cell in the crystal acts as a diffracting center, just as a ruled groove acts as a linear diffracting center in a conventional grating. For both the conventional grating and the crystal grating, the intensity at any point outside the array is determined by the phase difference and intensity of radiation diffracted from each center to the point in question.

Figure 26a is a copy of Fig. 25b with a set of dashed lines added. These lines represent one of a large number of arbitrary families of planes that can be drawn through the crystal, each such family having a characteristic *interplanar spacing d*. Figure 26b shows how a strong diffracted beam can be "reflected" from such a family of

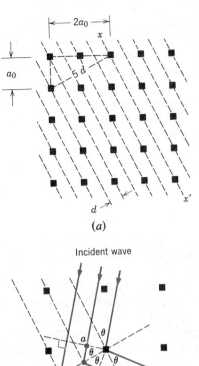

(a)

(b)

Figure 26 (a) A section through the NaCl cell lattice of Fig. 25b. The dashed sloping lines represent an arbitrary family of planes, with interplanar spacing d. (b) An incident beam falls on the family of planes shown in (a). A strong diffracted beam will be formed if Bragg's law is satisfied.

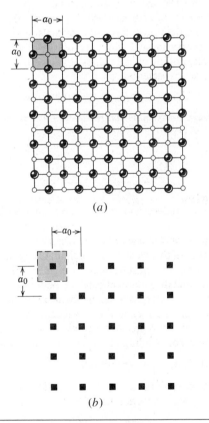

(a)

(b)

Figure 25 (a) A plane through a crystal of sodium chloride, showing the sodium and the chlorine ions. (b) The corresponding unit cells in this section, each cell being represented by a small black square.

planes if the incident beam falls on this family in a direction that is proper for its wavelength.

If we consider only one of the planes that make up the family shown in Fig. 26b, mirrorlike "reflection" occurs for *any* incident angle. To have a constructive interference in the beam diffracted in the direction θ from the *entire family* of planes, the rays from the separate planes must reinforce each other. This means that the path difference for rays from adjacent planes (*abc* in Fig. 26b) must be an integral number of wavelengths, or

$$2d \sin \theta = m\lambda \qquad m = 1, 2, 3, \ldots$$

(Bragg's law).   (26)

Each of the many families of planes has its own characteristic interplanar spacing d. Note in Fig. 26b that, contrary to the practice in optics, the angle θ is measured from the surface of the crystal, not from the normal to that surface. Bragg's law is named after the British physicist W. L. Bragg, who first derived it. He and his father shared the 1915 Nobel Prize for their use of x rays to study the structures of crystals.

For the planes of Fig. 26a you can show that the interplanar spacing d is related to the unit cell dimension a_0 by

$$d = \frac{a_0}{\sqrt{5}}. \tag{27}$$

This relation suggests how the dimensions of the unit cell can be found if the interplanar spacing is measured by x-ray diffraction methods.

X-ray diffraction is a powerful tool for studying both x-ray spectra and the arrangement of atoms in crystals. To study spectra, a particular set of crystal planes, having a known spacing d, is chosen. Diffraction from these planes diffracts different wavelengths at different angles. A detector, then, that can discriminate one angle from another can be used to determine the wavelength of radiation reaching it. On the other hand, we can study the crystal itself, using a monochromatic x-ray beam, determining not only the spacings of various crystal planes but also the structure of the unit cell. The DNA

molecule, and many other equally complex structures, have been mapped by x-ray diffraction methods.

Sample Problem 9 At what angles must an x-ray beam with $\lambda = 1.10$ Å fall on the family of planes represented in Fig. 26b if a diffracted beam is to exist? Assume the material to be sodium chloride ($a_0 = 5.63$ Å).

The interplanar spacing d for these planes is given by Eq. 27 or

$$d = \frac{a_0}{\sqrt{5}} = \frac{5.63 \text{ Å}}{\sqrt{5}} = 2.518 \text{ Å}.$$

Equation 26 gives

$$\theta = \sin^{-1} \frac{m\lambda}{2d} = \sin^{-1}\left(\frac{(m)(1.10 \text{ Å})}{(2)(2.518 \text{ Å})}\right)$$
$$= \sin^{-1}(0.2184 \, m).$$

Diffracted beams are possible for $\theta = 12.6°$ ($m = 1$), $\theta = 25.9°$ ($m = 2$), $\theta = 40.9°$ ($m = 3$), and $\theta = 60.9°$ ($m = 4$). Higher order beams cannot exist because they require that $\sin \theta > 1$.

Actually, the unit cell in cubic crystals such as NaCl has diffraction properties such that the intensity of diffracted x-ray beams corresponding to odd values of m is zero. Thus the only beams that are expected are

$$\theta = 25.9° \quad (m = 2) \quad \text{and} \quad \theta = 60.9°(m = 4). \quad \text{(Answer)}$$

REVIEW AND SUMMARY

Diffraction

Diffraction, which occurs when a wave encounters an obstacle or hole whose size is comparable to the wavelength of the wave, is convincing evidence of the wave theory of light. Using the Huygens' construction, the wave is divided at the obstruction into infinitesimal wavelets that then interfere with each other as they proceed (see Fig. 7). We treat diffraction effects that occur a large distance from the obstruction, often studied experimentally by interposing a lens so that patterns that would otherwise occur at infinite distance are focused onto a screen at the lens focal plane.

Single-slit Diffraction

Waves passing through a long narrow slit of width a produce a *single-slit diffraction pattern* with a central maximum together with minima corresponding to diffraction angles θ that satisfy

$$a \sin \theta = m\lambda \quad m = 1, 2, 3, \ldots \text{ (minima)}. \tag{3}$$

See Sample Problems 1 and 2. Using *phasor diagrams* to add the Huygens' wavelets, we show in Section 41–5 that the diffracted intensity for a given diffraction angle θ is

$$I = I_m \left(\frac{\sin \alpha}{\alpha}\right)^2 \quad \text{where} \quad \alpha = \frac{\pi a}{\lambda} \sin \theta. \tag{6}$$

See Sample Problem 3 for a calculation of the intensity of the secondary maxima.

Diffraction by a *circular* aperture or lens with diameter d also produces a central maximum

Circular Diffraction

with a first minimum at a diffraction angle θ given by

$$\sin \theta = 1.22 \frac{\lambda}{d} \quad \text{(1st minimum, circular aperature)}. \qquad [9]$$

See Fig. 13.

Rayleigh's criterion suggests that two objects viewed through a telescope or microscope are on the verge of resolvability if the central diffraction maximum of one is at the first minimum of the other. Their angular separation must be at least

Rayleigh's Criterion

$$\theta_R = 1.22 \frac{\lambda}{d} \quad \text{(Rayleigh's criterion)}. \qquad [11]$$

in which d is the diameter of the objective lens; Sample Problem 4.

Waves passing through two slits, each of width a, whose centers are distance d apart, display diffraction patterns whose intensity I at various diffraction angles θ is given by

Double-slit Diffraction

$$I = I_m \cos^2 \beta \left(\frac{\sin \alpha}{\alpha} \right)^2 \quad \text{(double slit)}, \qquad [12]$$

with $\beta = (\pi d/\lambda) \sin \theta$ and α the same as for single-slit diffraction. See Sample Problems 5 and 6.

Diffraction by N *multiple slits* results in principal maxima whenever

Multiple-slit Diffraction

$$d \sin \theta = m\lambda \quad m = 0, 1, 2 \ldots \quad \text{(maxima)} \qquad [14]$$

with the angular width of the maxima given by

$$\Delta\theta = \frac{\lambda}{Nd \cos \theta} \quad \text{(line width)}. \qquad [18]$$

Diffraction Gratings

A *diffraction grating* is a series of "slits" used to separate an incident wave into different wavelength components whose principal diffraction maxima are directionally dispersed by the grating. A grating is characterized by two parameters, the dispersion D and the resolving power R.

$$D = \frac{\Delta\theta}{\Delta\lambda} = \frac{m}{d \cos \theta} \quad \text{and} \quad R = \frac{\lambda}{\Delta\lambda} = Nm. \qquad [19-22]$$

See Fig. 23 and Sample Problems 7 and 8.

X-ray Diffraction

The regular array of atoms in a crystal is a three-dimensional diffraction grating for short-wavelength waves such as x rays. The atoms can be visualized as being arranged in planes with characteristic interplanar spacing d. Diffraction maxima (constructive interference) occur if the incident direction of the wave, measured from the surface of a plane of atoms, and the wavelength λ of the radiation satisfy *Bragg's law:*

Bragg's Law

$$2d \sin \theta = m\lambda \quad m = 1, 2, 3 \ldots \quad \text{(Bragg's law)}. \qquad [26]$$

See Sample Problem 9.

QUESTIONS

1. Why is the diffraction of sound waves more evident in daily experience than that of light waves?

2. Sound waves can be diffracted. About what width of a single slit should you use if you wish to broaden the distribution of an incident plane sound wave of frequency 1 kHz?

3. Why aren't sound waves polarized?

4. Why do radio waves diffract around buildings, although light waves do not?

5. A loud-speaker horn, used at a rock concert, has a rectangular aperture 1 m high and 30 cm wide. Will the pattern of sound intensity be broader in the horizontal plane or in the vertical?

6. For what kind of waves could a long picket fence be considered a useful diffraction grating?

7. A particular radar antenna is designed to give accurate measurements of the altitude of an aircraft but less accurate measurements of its direction in a horizontal plane. Must the height-to-width ratio of the radar antenna be less than, equal to, or greater than unity?

8. In single-slit diffraction, what is the effect of increasing (a) the wavelength and (b) the slit width?

9. Why are the colors in the spectrum of a light source called "lines"?

10. While listening to the car radio, you may have noticed that the AM signal fades—but the FM signal doesn't—when you drive under a bridge. Could diffraction have anything to do with this?

11. What will the single-slit diffraction pattern look like if $\lambda > a$?

12. What would the pattern on a screen formed by a double slit look like if the slits did not have the same width? Would the locations of the fringes be changed?

13. The shadow of a vertical flagpole cast by the sun has clearly defined edges near its base, but less-well-defined edges near its top end. Why?

14. A crossed diffraction grating has lines ruled in two directions, at right angles to each other. Predict the pattern of light intensity on the screen if light is sent through such a grating.

15. Sunlight falls on a single slit of width 1 μm. Describe qualitatively what the resulting diffraction pattern looks like.

16. In Fig. 6, rays r_1 and r_3 are in phase; so are r_2 and r_4. Why isn't there a maximum intensity at P_2 rather than a minimum?

17. When the atmosphere is not quite clear, one may sometimes see colored circles concentric with the sun or moon, generally not more than four or five times the diameter of the sun or moon and invariably having the inner edge blue. What is the explanation of this phenomenon?

18. When we speak of diffraction by a single slit we imply that the width of the slit must be much less than its length. Suppose that, in fact, the length was equal to twice the width. Make a rough guess at what the diffraction pattern would look like.

19. In connection with Fig. 4 we stated, correctly, that the optical pathlengths from the slit to point P_0 are all the same. Why?

20. Distinguish clearly between θ, α, and ϕ in Eqs. 5 and 6.

21. If we were to redo our analysis of the properties of lenses in Chapter 39 by the methods of geometric optics but without restricting our considerations to paraxial rays and to "thin" lenses, would diffraction phenomena emerge from the analysis? Discuss.

22. The double-slit pattern of Fig. 27a seen with a monochromatic light source is somehow changed to the pattern of Fig. 27b. Consider the following possible changes in conditions: (1) The wavelength of the light was decreased. (2) The wavelength of the light was increased. (3) The width of each slit was increased. (4) The separation of the slits was increased. (5) The separation of the slits was decreased. (6) The width of each slit

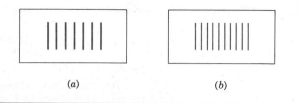

Figure 27 Question 22.

was decreased. Which selection(s) of the above changes would explain the alteration of the pattern?

23. We have seen that diffraction limits the resolving power of optical telescopes (see Fig. 14). Does it also do so for large radio telescopes?

24. In double-slit interference patterns such as that of Fig. 17a we said that the interference fringes were modulated in intensity by the diffraction pattern of a single slit. Could we reverse this statement and say that the diffraction pattern of a single slit is intensity-modulated by the interference fringes? Discuss.

25. Assume that the limits of the visible spectrum are 430 and 680 nm. How would you design a grating, assuming that the incident light falls normally on it, such that the first-order spectrum barely overlaps the second-order spectrum?

26.. A glass prism can form a spectrum. Explain how. How many "orders" of spectra will a prism produce?

27. For the simple spectroscope of Fig. 21 show (a) that θ increases with λ for a grating and (b) that θ decreases with λ for a prism.

28. Explain in your own words why increasing the number of slits, N, in a diffraction grating sharpens the maxima. Why does decreasing the wavelength do so? Why does increasing the slit spacing, d, do so?

29. How much information can you discover about the structure of a diffraction grating by analyzing the spectrum it forms of a monochromatic light source? Let $\lambda = 589$ nm, for an example.

30. (a) Why does a diffraction grating have closely spaced rulings? (b) Why does it have a large number of rulings?

31. Two nearly-equal wavelengths are incident on a grating of N slits and are not quite resolvable. However, they become resolved if the number of slits is increased. Formulas aside, is the explanation of this that: (a) more light can get through the grating? (b) the principal maxima become more intense and hence resolvable? (c) the diffraction pattern is spread more and hence the wavelengths become resolved? (d) there are a larger number of orders? or (e) the principal maxima become narrower and hence resolvable?

32. How can the resolving power of a lens be increased?

33. The relation $R = Nm$ suggests that the resolving power of a given grating can be made as large as desired by choosing an arbitrarily high order of diffraction. Discuss this possibility.

34. Show that at a given wavelength and a given angle of diffraction the resolving power of a grating depends only on its width $W (= Nd)$.

35. How would you experimentally measure (*a*) the dispersion D and (*b*) the resolving power R of a grating spectrograph?

36. For a given family of planes in a crystal, can the wavelength of incident x rays be (*a*) too large or (*b*) too small to form a diffracted beam?

37. If a parallel beam of x rays of wavelength λ is allowed to fall on a randomly-oriented crystal of any material, generally no intense diffracted beams will occur. Such beams appear if (*a*) the x-ray beam consists of a continuous distribution of wavelengths rather than a single wavelength or (*b*) the specimen is not a single crystal but a finely divided powder. Explain each case.

38. Does an x-ray beam undergo refraction as it enters and leaves a crystal? Explain your answer.

EXERCISES AND PROBLEMS

Section 41–2   Diffraction from a Single Slit—Locating the Minima

1E. When monochromatic light is incident on a slit 0.022 mm wide, the first diffraction minimum is observed at an angle of 1.8° from the direction of the direct beam. What is the wavelength of the incident light?

2E. Monochromatic light of wavelength 441 nm falls on a narrow slit. On a screen 2.0 m away, the distance between the second minimum and the central maximum is 1.5 cm. (*a*) Calculate the angle of diffraction θ of the second minimum. (*b*) Find the width of the slit.

3E. Light of wavelength 633 nm is incident on a narrow slit. The angle between the first minimum on one side of the central maximum and the first minimum on the other side is 1.2°. What is the width of the slit?

4E. A single slit is illuminated by light whose wavelengths are λ_a and λ_b, so chosen that the first diffraction minimum of the λ_a component coincides with the second minimum of the λ_b component. (*a*) What relationship exists between the two wavelengths? (*b*) Do any other minima in the two patterns coincide?

5E. The distance between the first and fifth minima of a single-slit pattern is 0.35 mm with the screen 40 cm away from the slit, using light having a wavelength of 550 nm. (*a*) Calculate the diffraction angle θ of the first minimum. (*b*) Find the width of the slit.

6E. For a single slit the first minimum occurs at $\theta = 90°$, thus filling the forward hemisphere beyond the slit with light. What must be the ratio of the slit width to the wavelength for this to take place?

7E. A plane wave, wavelength = 590 nm, falls on a slit with $a = 0.40$ mm. A thin converging lens, focal length = +70 cm, is placed behind the slit and focuses the light on a screen. (*a*) How far is the screen behind the lens? (*b*) What is the linear distance on the screen from the center of the pattern to the first minimum?

8P. A slit 1.0 mm wide is illuminated by light of wavelength 589 nm. We see a diffraction pattern on a screen 3.0 m away. What is the distance between the first two diffraction minima on either side of the central diffraction maximum?

9P. Sound waves with frequency 3000 Hz diffract out of a speaker cabinet with a 0.30-m diameter opening into a large auditorium. Where does a listener standing against a wall 100 m from the speaker have the most difficulty hearing? Assume a sound speed of 343 m/s.

10P. Manufacturers of wire (and other objects of small dimensions) sometimes use a laser to continually monitor the thickness of the product. The wire intercepts the laser beam, producing a diffraction pattern like that of a single slit of the same width as the wire diameter; see Fig. 28. Suppose a He–Ne laser, wavelength 632.8 nm, illuminates a wire, the diffraction pattern being projected onto a screen 2.6 m away. If the desired wire diameter is 1.37 mm, what would be the observed distance between the two tenth order minima on each side of the central maximum?

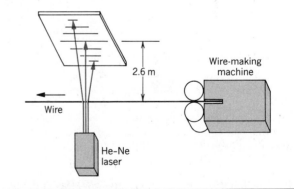

Figure 28   Problem 10.

Section 41–5   Diffraction from a Single Slit—Quantitative

11E. A 0.10-mm-wide slit is illuminated by light with a wavelength 589 nm. Consider rays for which $\theta = 30°$ in Fig. 8*b* and calculate the phase difference at the screen of Huygens' wavelets from the top and midpoint of the slit. (*Hint:* See Eq. 4.)

12E. Monochromatic light with wavelength 538 nm falls on a slit with width 0.025 mm. The distance from the slit to a screen is 3.5 m. Consider a point on the screen 1.1 cm from the central maximum. (a) Calculate θ. (b) Calculate α. (c) Calculate the ratio of the intensity at this point to the intensity at the central maximum.

13P. If you double the width of a single slit, the intensity of the central maximum of the diffraction pattern increases by a factor of four, even though the energy passing through the slit only doubles. Explain this quantitatively.

14P. *Babinet's principle:* A monochromatic beam of parallel light is incident on a "collimating" hole of diameter $x \gg \lambda$. Point P lies in the geometrical shadow region on a distant screen, as shown in Fig. 29a. Two obstacles, shown in Fig. 28b, are placed in turn over the collimating hole. A is an opaque circle with a hole in it and B is the "photographic negative" of A. Using superposition concepts, show that the intensity at P is identical for each of the two diffracting objects A and B. In this connection, it can be shown that the diffraction pattern of a wire is that of a slit of equal width. See "Measuring the Diameter of a Hair by Diffraction" by S.M. Curry and A.L. Schawlow, *American Journal of Physics,* May 1974.

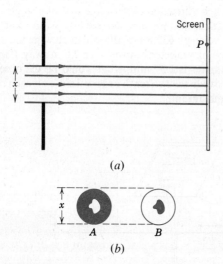

(a)

(b)

Figure 29 Problem 14.

15P. The full width at half maximum (FWHM) of the central diffraction maximum is defined as the angle between the two points in the pattern where the intensity is one-half that at the center of the pattern. (See Fig. 11b.) (a) Show that the intensity drops to one-half of the maximum value when $\sin^2 \alpha = \alpha^2/2$. (b) Verify that $\alpha = 1.39$ radians (about 80°) is a solution to the transcendental equation of part (a). (c) Show

that the FWHM is $\Delta\theta = 2 \sin^{-1}(0.443\lambda/a)$. (d) Calculate the FWHM of the central maximum for slits whose widths are 1, 5, and 10 wavelengths.

16P. (a) Show that the values of α at which intensity maxima for single-slit diffraction occur can be found exactly by differentiating Eq. 5 with respect to α and equating to zero, obtaining the condition

$$\tan \alpha = \alpha.$$

(b) Find the values of α satisfying this relation by plotting graphically the curve $y = \tan \alpha$ and the straight line $y = \alpha$ and finding their intersections or by using a pocket calculator to find an appropriate value of α by trial and error. (c) Find the (nonintegral) values of m corresponding to successive maxima in the single-slit pattern. Note that the secondary maxima do not lie exactly halfway between minima.

17P*. Derive this expression for the intensity pattern for a three-slit "grating":

$$I = \tfrac{1}{9}I_m(1 + 4 \cos \phi + 4 \cos^2 \phi),$$

where

$$\phi = \frac{2\pi d \sin \theta}{\lambda}.$$

Assume that $a \ll \lambda$ and be guided by the derivation of the corresponding double-slit formula (Eq. 15 of Chapter 40).

Section 41–6 Diffraction from a Circular Aperture

18E. The two headlights of an approaching automobile are 1.4 m apart. At what (a) angular separation and (b) maximum distance will the eye resolve them? Assume a pupil diameter of 5.0 mm and a wavelength of 550 nm. Also assume that diffraction effects alone limit the resolution.

19E. An astronaut in a satellite claims he can just barely resolve two point sources on the earth, 160 km below him. Calculate their (a) angular and (b) linear separation, assuming ideal conditions. Take $\lambda = 540$ nm, and the pupil diameter of the astronaut's eye to be 5.0 mm.

20E. Find the separation of two points on the moon's surface that can just be resolved by the 200-in. (= 5.1-m) telescope at Mount Palomar, assuming that this distance is determined by diffraction effects. The distance from the earth to the moon is 3.8×10^5 km. Assume a wavelength of 550 nm.

21E. The wall of a large room is covered with acoustic tile in which small holes are drilled 5.0 mm from center to center. How far can a person be from such a tile and still distinguish the individual holes, assuming ideal conditions? Assume the diameter of the pupil of the observer's eye to be 4.0 mm and the wavelength to be 550 nm.

22E. The pupil of a person's eye has a diameter of 5.0 mm. What distance apart must two small objects be if, when 250 mm from the eye, their images are just resolved when they are illuminated with light of wavelength 500 nm?

23E. Under ideal conditions, estimate the linear separation of two objects on the planet Mars that can just be resolved by an observer on earth (*a*) using the naked eye, and (*b*) using the 200-in. (= 5.1 m) Mount Palomar telescope. Use the following data: distance to Mars = 8.0×10^7 km; diameter of pupil = 5.0 mm; wavelength of light = 550 nm.

24E. If Superman really had x-ray vision at 0.10 nm wavelength and a 4.0 mm pupil diameter, at what maximum altitude could he distinguish villains from heroes assuming the minimum detail required was 5.0 cm?

25E. A navy cruiser employs radar with a wavelength of 1.6 cm. The circular antenna has a diameter of 2.3 m. At a range of 6.2 km, what is the smallest distance that two speedboats can be from each other and still be resolved as two separate objects by the radar system?

26P. Nuclear-pumped x-ray lasers are seen as a possible weapon to destroy ICBM booster rockets at ranges up to 2000 km. One limitation on such a device is the spreading of the beam due to diffraction, with resulting dilution of beam intensity. Consider such a laser operating at a wavelength of 1.4 nm. The lasing element is a wire with diameter 0.20 mm. (*a*) Calculate the diameter of the central beam at the target 2000 km away. (*b*) By what factor is the beam intensity reduced in transit to target? (The laser is fired from space, so that atmospheric absorption can be ignored.)

27P. The paintings of Georges Seurat consist of closely spaced small dots ($\approx$ 2 mm in diameter) of pure pigment, as indicated in Fig. 30. The illusion of color mixing occurs because the pupils of the observer's eyes diffract light entering them. Calculate the minimum distance an observer must stand from such a painting to achieve the desired blending of color. Take the wavelength of the light to be 550 nm and the diameter of the pupil to be 1.5 mm.

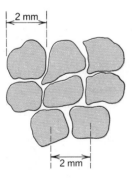

Figure 30 Problem 27.

28P. (*a*) A circular diaphragm 60 cm in diameter oscillates at a frequency of 25 kHz in an underwater source of sound used for submarine detection. Far from the source the sound intensity is distributed as a diffraction pattern for a circular hole whose diameter equals that of the diaphragm. Take the speed of sound in water to be 1450 m/s and find the angle between the normal to the diaphragm and the direction of the first minimum. (*b*) Repeat for a source having an (audible) frequency of 1.0 kHz.

29P. In June 1985 a laser beam was fired from the Air Force Optical Station on Maui, Hawaii, and reflected back from the shuttle *Discovery* as it sped by, 220 miles overhead. The diameter of the central maximum of the beam at the shuttle position was said to be 30 ft and the beam wavelength was 500 nm. What is the effective diameter of the laser aperture at the Maui ground station? (*Hint:* A laser beam spreads because of diffraction; assume a circular exit aperture.)

30P. A "spy in the sky" satellite orbiting at 160 km above the earth's surface has a lens with a focal length of 3.6 m. Its resolving power for objects on the ground is 30 cm; it could easily measure the size of an aircraft's air intake. What is the effective lens diameter, determined by diffraction consideration alone? Assume $\lambda = 550$ nm. Far more effective satellites are reported to be in operation today. See "Reconnaissance and Arms Control" by Ted Greenwood, *Scientific American,* February 1973.

31P. Millimeter-wave radar generates a narrower beam than conventional microwave radar. This makes them less vulnerable to antiradar missiles. (*a*) Calculate the angular width, from first minimum to first minimum, of the central "lobe" produced by a 220-GHz radar beam emitted by a 0.55-m diameter circular antenna. (The frequency is chosen to coincide with a low-absorption atmospheric "window.") (*b*) Calculate the same quantity for the ship's radar described in Exercise 25.

32P. (*a*) How small is the angular separation of two stars if their images are barely resolved by the Thaw refracting telescope at the Allegheny Observatory in Pittsburgh? The lens diameter is 76 cm and its focal length is 14 m. Assume $\lambda = 550$ nm. (*b*) Find the distance between these barely resolved stars if each of them is 10 light years distant from the earth. (*c*) For the image of a single star in this telescope, find the diameter of the first dark ring in the diffraction pattern, as measured on a photographic plate placed at the focal plane. Assume that the star image structure is associated entirely with diffraction at the lens aperture and not with (small) lens "errors," etc.

33P. It can be shown that, except for $\theta = 0$, a circular obstacle produces the same diffraction pattern as a circular hole of the same diameter. Furthermore, if there are many such obstacles, such as water droplets located randomly, then the interference effects vanish leaving only the diffraction associated with a single obstacle. (*a*) Explain why one sees a "ring" around the moon on a foggy night. The ring is usually reddish in color; explain why. (*b*) Calculate the size of the water droplets in the air if the ring around the moon appears to have a diameter 1.5 times that of the moon. The angular diameter of the moon in the sky is 0.5°. (*c*) At what distance from the moon might a bluish ring be seen? Sometimes the rings are white; why? (*d*)

The color arrangement is opposite to that in a rainbow: why should this be so?

34P. In a Soviet–French experiment to monitor the moon's surface with a light beam, pulsed radiation from a ruby laser ($\lambda = 0.69 \, \mu$m) was directed to the moon through a reflecting telescope with a mirror radius of 1.3 m. A reflector on the moon behaved like a circular plane mirror with radius 10 cm, reflecting the light directly back toward the telescope on earth. The reflected light was then detected by a photometer after being brought to a focus by this telescope. What fraction of the original light energy was picked up by the detector? Assume that for each direction of travel all the energy is in the central diffraction circle.

Section 41–7 Diffraction from a Double Slit

35E. Suppose that, as in Sample Problem 6, the envelope of the central peak contains 11 fringes. How many fringes lie between the first and second minima of the envelope?

36E. For $d = 2a$ in Fig. 31, how many interference fringes lie in the central diffraction envelope?

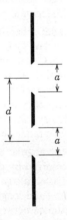

Figure 31 Exercise 36.

37P. If we put $d = a$ in Fig. 31, the two slits coalesce into a single slit of width $2a$. Show that Eq. 12 reduces to the diffraction pattern for such a slit.

38P. (a) Design a double-slit system in which the fourth fringe, not counting the central maximum, is missing. (b) What other fringes, if any, are also missing?

39P. Two slits of width a and separation d are illuminated by a coherent beam of light of wavelength λ. What is the linear separation of the interference fringes observed on a screen that is at a distance D away?

40P. (a) How many complete fringes appear between the first minima of the fringe envelope to either side of the central maximum for a double-slit pattern if $\lambda = 550$ nm, $d = 0.15$ mm, and $a = 0.030$ mm? (b) What is the ratio of the in-

tensity of the third fringe to the side of the center to that of the central fringe?

41P. Light of wavelength 440 nm passes through a double slit, yielding the diffraction pattern of intensity I versus deflection angle θ shown in Fig. 32. Calculate (a) the slit width, and (b) the slit separation. (c) Verify the displayed intensities of the $m = 1$ and $m = 2$ interference fringes.

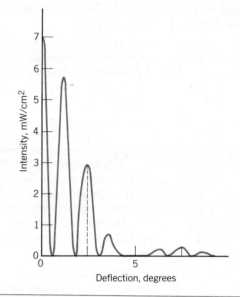

Figure 32 Problem 41.

42P. An acoustic double-slit system (slit separation d, slit width a) is driven by two loudspeakers as shown in Fig. 33. By use of a variable delay line, the phase of one of the speakers may be varied. Describe in detail what changes occur in the inten-

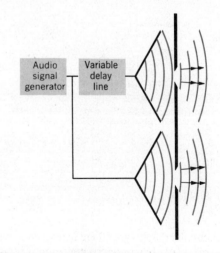

Figure 33 Problem 42.

sity pattern at large distances as this phase difference is varied from zero to 2π. Take both interference and diffraction effects into account.

Section 41-9 Diffraction Gratings

43E. A diffraction grating 20 mm wide has 6000 rulings. (*a*) Calculate the distance *d* between adjacent rulings. (*b*) At what angles will maximum-intensity beams occur if the incident radiation has a wavelength of 589 nm?

44E. A diffraction grating has 200 rulings/mm, and a strong diffracted beam is noted at $\theta = 30°$. (*a*) What are the possible wavelengths of the incident light? (*b*) What colors are they?

45E. A grating has 315 rulings/mm. For what wavelengths in the visible spectrum can fifth-order diffraction be observed?

46E. Given a grating with 400 lines/mm, how many orders of the entire visible spectrum (400–700 nm) can be produced?

47E. A diffraction grating 3.0 cm wide produces a deviation of 33° in the second order with light of wavelength 600 nm. What is the total number of lines on the grating?

48E. A diffraction grating exactly one centimeter wide has 10,000 parallel slits. Monochromatic light that is incident normally is deviated through 30° in first order. What is the wavelength of the light?

49P. Light of wavelength 600 nm is incident normally on a diffraction grating. Two adjacent principal maxima occur at $\sin \theta = 0.2$ and $\sin \theta = 0.3$, respectively. The fourth order is missing. (*a*) What is the separation between adjacent slits? (*b*) What is the smallest possible individual slit width? (*c*) Name all orders actually appearing on the screen with the values derived in (*a*) and (*b*).

50P. A diffraction grating is made up of slits of width 300 nm with a 900-nm separation between centers. The grating is illuminated by monochromatic plane waves, $\lambda = 600$ nm, the angle of incidence being zero. (*a*) How many diffraction maxima are there? (*b*) What is the width of the spectral lines observed in first order if the grating has 1000 slits? See Eq. 18.

51P. Assume that the limits of the visible spectrum are arbitrarily chosen as 430 and 680 nm. Calculate the number of rulings per mm of a grating that will spread the first-order spectrum through an angular range of 20°.

52P. With light from a gaseous discharge tube incident normally on a grating with a distance 1.73 μm between adjacent slit centers, a green line appears with sharp maxima at measured transmission angles $\theta = \pm 17.6°, 37.3°, -37.1°, 65.2°$, and $-65.0°$. Compute the wavelength of the green line that best fits the data.

53P. Assume that light is incident on a grating at an angle ψ as shown in Fig. 34. Show that the condition for a diffraction maximum is

$$d(\sin \psi + \sin \theta) = m\lambda, \quad m = 0, 1, 2, \ldots$$

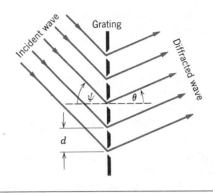

Figure 34

Only the special case $\psi = 0$ has been treated in this chapter (compare with Eq. 14).

54P. A transmission grating with $d = 1.50$ μm is illuminated at various angles of incidence by light of wavelength 600 nm. Plot as a function of angle of incidence (0 to 90°) the angular deviation of the first-order diffracted beam from the incident direction. See Problem 53.

55P. Two spectral lines have wavelengths λ and $\lambda + \Delta\lambda$, respectively, where $\Delta\lambda \ll \lambda$. Show that their angular separation $\Delta\theta$ in a grating spectrometer is given approximately by

$$\Delta\theta = \frac{\Delta\lambda}{\sqrt{(d/m)^2 - \lambda^2}},$$

where *d* is the separation of adjacent slit centers and *m* is the order at which the lines are observed. Notice that the angular separation is greater in the higher orders than in lower orders.

56P. White light (400 nm $< \lambda <$ 700 nm) is incident on a grating. Show that, no matter what the value of the grating spacing *d*, the second- and third-order spectra overlap.

57P. Show that in a grating with alternately transparent and opaque strips of equal width, all the even orders (except $m = 0$) are absent.

58P. A grating has 350 rulings/mm and is illuminated at normal incidence by white light. A spectrum is formed on a screen 30 cm from the grating. If a 10-mm square hole is cut in the screen, its inner edge being 50 mm from the central maximum and parallel to it, what range of wavelengths passes through the hole?

59P. Derive Eq. 18, that is, the expression for the width of lines other than the central maximum.

Section 41-10 Gratings: Dispersion and Resolving Power

60E. The *D* line in the spectrum of sodium is a doublet with wavelengths 589.0 nm and 589.6 nm. Calculate the minimum number of lines in a grating needed to resolve this doublet in the second order spectrum. See Sample Problem 7.

61E. A grating has 600 rulings/mm and is 5.0 mm wide. (*a*) What is the smallest wavelength interval that can be resolved in the third order at $\lambda = 500$ nm? (*b*) How many higher orders can be seen?

62E. A source containing a mixture of hydrogen and deuterium atoms emits light containing two closely spaced red colors at $\lambda = 656.3$ nm whose separation is 0.18 nm. Find the minimum number of lines needed in a diffraction grating that can resolve these lines in the first order.

63E. (*a*) How many rulings must a 4.0-cm-wide diffraction grating have to resolve the wavelengths 415.496 nm and 415.487 nm in the second order? (*b*) At what angle are the maxima found?

64E. In a particular grating the sodium doublet (see Sample Problem 7) is viewed in third order at 10° to the normal and is barely resolved. Find (*a*) the grating spacing and (*b*) the total width of the rulings.

65E. Show that the dispersion of a grating can be written as

$$D = \frac{\tan \theta}{\lambda}.$$

66E. A grating has 40,000 rulings spread over 76 mm. (*a*) What is its expected dispersion D for sodium light ($\lambda = 589$ nm) in the first three orders? (*b*) What is its resolving power in these orders?

67P. Light containing a mixture of two wavelengths, 500 nm and 600 nm, is incident normally on a diffraction grating. It is desired (1) that the first and second principal maxima for each wavelength appear at $\theta \leq 30°$, (2) that the dispersion be as high as possible, and (3) that the third order for 600 nm be a missing order. (*a*) What should be the separation between adjacent slits? (*b*) What is the smallest possible individual slit width? (*c*) Name all orders for 600 nm that actually appear on the screen with the values derived in (*a*) and (*b*).

68P. In Problem 50, calculate the product of line width and resolving power of the grating in first order.

69P. A diffraction grating has a resolving power $R = \lambda/\Delta\lambda = Nm$. (*a*) Show that the corresponding frequency range $\Delta\nu$ that can just be resolved is given by $\Delta\nu = c/Nm\lambda$. (*b*) From Fig. 18, show that the "times of flight" of the two extreme rays differ by an amount $\Delta t = (Nd/c) \sin \theta$. (*c*) Show that $(\Delta\nu)(\Delta t) = 1$, this relation being independent of the various grating parameters. Assume $N \gg 1$.

Section 41–11 X-ray Diffraction
70E. The x-ray wavelength 0.12 nm is found to reflect in the second order from the face of a lithium fluoride crystal at a Bragg angle of 28°. Calculate the distance between adjacent crystal planes.

71E. A beam of x rays of wavelength 30 pm is incident on a calcite crystal of lattice spacing 0.30 nm. What is the smallest angle between the crystal planes and the x-ray beam that will result in constructive reflection of the x rays?

72E. Monochromatic high-energy x rays are incident on a crystal. If first-order reflection is observed at Bragg angle 3.4°, at what angle would second-order reflection be expected?

73E. An x-ray beam, containing radiation of two distinct wavelengths, is scattered from a crystal, yielding the intensity spectrum shown in Fig. 35. The interplanar spacing of the scattering planes is 0.94 nm. Determine the wavelengths of the x rays present in the beam.

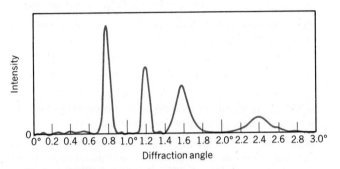

Figure 35 Problem 73.

74E. In comparing the wavelengths of two monochromatic x-ray lines, it is noted that line *A* gives a first-order reflection maximum at a glancing angle of 23° to the smooth face of a crystal. Line *B*, known to have a wavelength of 97 pm, gives a third-order reflection maximum at an angle of 60° from the same face of the same crystal. (*a*) Calculate the interplanar spacing. (*b*) Find the wavelength of line *A*.

75E. Monochromatic x rays are incident on a set of NaCl crystal planes whose interplanar spacing is 39.8 pm. When the beam is rotated 60° from the normal, first-order Bragg reflection is observed. What is the wavelength of the x rays?

76P. Prove that it is not possible to determine both wavelength of radiation and spacing of Bragg reflecting planes in a crystal by measuring the angles for Bragg reflection in several orders.

77P. Assume that the incident x-ray beam in Fig. 36 is not monochromatic but contains wavelengths in a band from 95 to 130 pm. Will diffracted beams, associated with the planes shown, occur? If so, what wavelength is diffracted? Assume $d = 0.275$ nm.

78P. First-order Bragg scattering from a certain crystal occurs at an angle of incidence of 63.8°; see Fig. 37. The wavelength of the x rays is 0.26 nm. Assuming that the scattering is from the dashed planes shown, find the unit cell size a_0.

79P. Consider an infinite two-dimensional square lattice as in Fig. 25*b*. One interplanar spacing is obviously a_0 itself. (*a*)

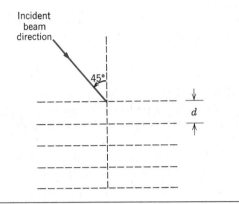

Figure 36   Problems 77 and 80.

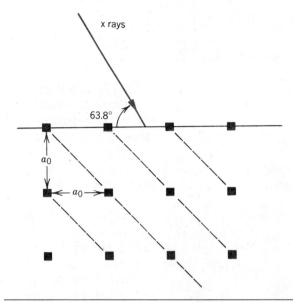

Figure 37   Problem 78.

Calculate the next five smaller interplanar spacings by sketching figures similar to Fig. 26a. (b) Show that your answers obey the general formula

$$d = a_0/\sqrt{h^2 + k^2}$$

where h and k are both relatively prime integers that have no common factor other than unity.

80P.   Monochromatic x rays ($\lambda = 0.125$ nm) fall on a crystal of sodium chloride, making an angle of 45° with the reference line shown in Fig. 36. The planes shown are those of Fig. 26a,

for which $d = 0.252$ nm. Through what angles must the crystal be turned to give a diffracted beam associated with the planes shown? Assume that the crystal is turned about an axis that is perpendicular to the plane of the page.

ESSAY 18
HOLOGRAPHY

TUNG H. JEONG

LAKE FOREST COLLEGE

What Is a Hologram and How Does It "Work"?

A basic understanding of *holography*[1,2,3] requires a review of many major principles in physical and geometric optics found in preceding chapters. They include *interference* and *diffraction,* as applied to visible light.

A *hologram* is a recording on a light-sensitive medium, such as a photographic plate, of *interference patterns* formed between two or more beams of light derived from the same laser.

In making a hologram, part of the output from a *laser* is spread out by a *lens* or curved *mirror* and directed onto the plate. This is called the *reference beam (R)*. The remainder of the light illuminates a three-dimensional object being recorded. The light scattered by the object toward the plate is called the *object beam (O)*. Because all the light is from the same laser, the two beams are mutually *coherent* and form distinct interference patterns.

When a hologram is illuminated by *R*, it behaves as a complex *grating* and diffracts the light. The diffraction pattern precisely recreates the *wavefronts* that emanated from the original object.

To understand this process better, first consider the simplest of all objects, a point in space. Figure 1*a* shows two beams, situated far from the plate and interfering at 90° with respect to each other. The interference pattern is precisely the same as that from a *Young's double slit* with very wide separation. The exposed and developed hologram is a *diffraction grating* consisting of $d = \lambda$, where λ is the wavelength of the laser.

Figure 1*b* shows how to reconstruct the *wavefront* of *O*. Laser light from *R* illuminates the hologram. The diffracted light, according to the equation $m\lambda = d \sin \theta$, yields $\theta = 90°$, for $m = 1$. There is no room for higher orders. Thus all the diffracted light forms a *virtual image* of *O*. If *R* were directed backwards (called a *conjugate beam R′*) as shown in Figure 1*c*, *O* is reconstructed backward also. A screen placed in the location of *O′* will show the *real image*.

If we replace the point *O* with a three-dimensional object illuminated by laser light, the new object beam *O* consists of a large collection of point sources representing the *scattering* centers on the object. The recording on the plate now consists of a *superposition* of gratings. When illuminated by *R* (or *R′*), a virtual (or real) image of the object can be observed.

The above recording is called a "laser transmission" hologram.[2] It has the remarkable property that any small area on it is capable of recreating a complete picture of the object.

Figure 2 represents the general interference pattern between *R* and *O* on a plane containing the sources.[4] The lines represent the locations of *maxima;* halfway between them are the *minima.* An analog using water waves can be produced in a *ripple tank.* The perpendicular bisector of *RO* is the *zeroth order* interference, the loci of points that have the same *optical path* to *R* and *O*.

In three-dimensional space, the pattern is a figure of revolution with *RO* as the axis. It is a family of hyperboloids with foci *R* and *O*.

In region *A* sufficiently far from *R* and *O*, the pattern is precisely that of the Young's double-slit interference as discussed before. Region *B* consists of waves moving in opposite directions, forming *standing waves.* The *antinodes* along the line joining *R* and *O* are separated by $(\tfrac{1}{2})\lambda$. Region *C* represents *fringes* of a *Michelson interferometer.*

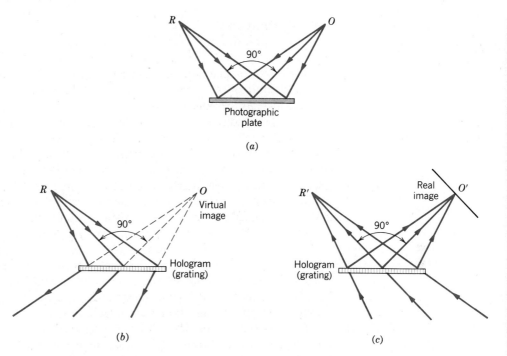

(a)

Figure 1 (*a*) Light from two widely separated coherent point sources R and O produces fine interference fringes on the photoplate, which becomes a diffraction grating upon chemical processing. (*b*) When R alone is directed at the grating, the diffracted light precisely recreates the wavefronts of 0. (*c*) If R were directed in a reverse direction, the real image of 0 can be projected on a screen.

(b)　　　　(c)

Generally, the distance *RO* is many thousand wavelengths of light, and the patterns are microscopic, beyond visual resolution. Figure 2 is a special case in which *R* and *O* are only a few wavelengths apart for simplicity.

If a hologram is made by placing the plate in region *B*, parallel to the zeroth order of Fig. 2, it records the standing wave pattern. The photographic emulsion on the plate is generally $10 \, \lambda$ thick, thus it records up to 20 hyperboloidal planes. This remarkable "white light reflection" hologram[3] can be viewed with a point source of *incandescent* light from *R*. Because it is a "volume" hologram, it performs *Bragg diffraction* and selects the same wavelength λ from the white light and reconstruct the wavefront of *O*.

Figure 2 The general interference pattern between two point sources. It is a family of hyperbolas

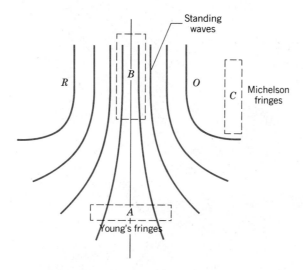

Making Holograms

Figure 3 show a simple system for making laser transmission holograms.[5] The output from a 1- to 5-milliwatt HeNe laser is spread by a front-surfaced curved mirror. Some light arrives at the plate directly and serves as the reference beam, where R is the *focal point* of the mirror. Another part of the light illuminates the object and is scattered onto the plate as the object beam.

The plate and object are supported by a steel plate on top of an inflated rubber tube, which absorbs mechanical vibrations from beneath. All components are held down by magnets or glue. This is necessary because during the exposure, which may be several seconds long, any relative movement between the object and the plate will smear the microscopic interference patterns being recorded, resulting in failure.

In general, much more complicated optical arrangements are necessary in order to illuminate large scenes in more artistic or useful ways.

Using a ruby laser, which can emit more than one joule of light energy in less than 20 nanoseconds, a hologram can be made of deep transient scenes. Although the subjects are generally in motion, the brief exposure time allows the recording of the necessary patterns without smearing. For human and animal subjects, eye-safety rules must be followed. These include closing the eyes, wearing goggles or opaque contact lenses, or using ingenious illumination techniques.

Figure 4 is a photograph of the reconstructed wavefronts from such a hologram. The subject is the author in a laboratory. One can look around through this small hologram as if it were a window on the wall of the laboratory and see everything in full dimension. If an undiverged laser beam were directed at any small area of the hologram in a reversed direction of R, a real image could be projected.

Figure 5 shows a setup for recording a white-light reflection hologram. Notice that the object is in contact with the plate and on the opposite site of the mirror.[5]

Because it is a volume hologram, the separation between *Bragg planes* can be shifted by chemically swelling or shrinking the emulsion. This allows one to reconstruct the image in any color desired. Artists can produce beautiful work in multiple

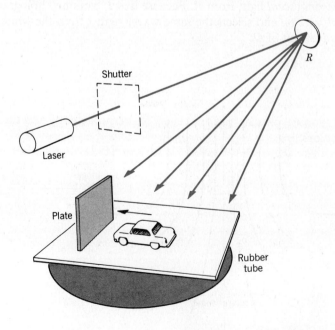

Figure 3 Simple configuration for making a transmission hologram viewable with a laser light.

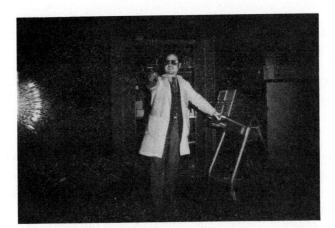

Figure 4 Deep scene image of the author from a transmission hologram made with a pulsed ruby laser.

Figure 5 System for making the simplest hologram viewable with incandescent light.

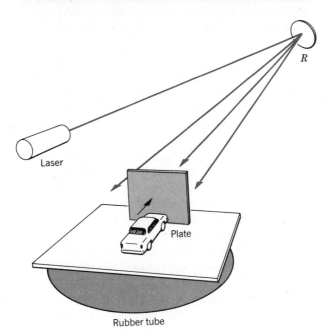

color by making multiple exposures, and swelling the emulsion between exposures. Upon developing, each exposure shrinks differently and the image appears in many colors.

Other Types of Holograms

Many variations from the above two basic types are now available for artistic as well as technical applications. The main hybrid holograms are the following.

Rainbow hologram[6,7]: A white-light viewable transmission hologram that offers three-dimensionality only in the horizontal direction. The color observed depends on the vertical viewing angle. When such holograms are aluminized in the back, they can be attached to credit cards or other surfaces and serves as pseudo-reflection holograms.

Integral stereogram[8]: A hologram synthesized from a series of motion picture frames, allowing live subjects as well as outdoor scenes to be recorded without initial laser illumination. Computer-generated images can also be synthesized.

Focused image hologram[9]: A hologram made using the real image of an object such as *O*, located through the plane of the plate. The real image is usually from another hologram.

Holographic cinema[10]: By making a projection screen in the form of a large hologram that behaves as a multiple elliptical mirror, real images from a series of holograms can be projected to each viewer in an auditorium.

There are numerous other variations the reader can find in the open literature.

Other Applications

Although most holograms encountered are made for visual observations, much more work in the field is aimed at technical applications. Some major fields include "holographic interferometry," "HOE" (holographic optical elements), and "phase conjugation," an important and exotic application that will play a major role in the new field of "optronics" (optical electronics). A brief discussion of each follows.

Interferometry[11]: When two separate exposures are made, and the object undergoes minute mechanical change in between, two slightly different images will appear at reconstruction. The precise change is revealed visually through the mutual interference of the two reconstructed wavefronts. Figure 6 shows a "real-time" observation of a growing mushroom.[12] The virtual image from a hologram made a few seconds before is interfering with light from the actual mushroom, resulting in "live" fringes that allows precise interpretation. Fringes caused by the swaying motion are separated from those caused by the net growth by using additional techniques.

Holographic optical elements (HOE): Diffraction gratings can now be made holographically. Indeed, a transmission hologram of a point source nearby behaves as a *concave grating.* HOEs are used for laser scanning at store checkout counters. HUDs (head up displays) are transparent holograms on windshields of aircrafts (in the future, cars) that relay visual information to the pilot looking straight ahead. HOEs are generally more compact and less costly than their classical counterparts, and can perform optical functions otherwise impossible.

Phase conjugation[13]: When *R'* is directed through a hologram, *O'* is directed backward (time-reversed beam) toward the object. Consider a material such as barium titanate, which is capable of recording a hologram, without any chemical processing, as soon as sufficient energy from *O* and *R* fall on it. If an intense conjugate beam (the "pump" beam) is present simultaneously, *O'* further illuminates the actual object and

Figure 6 "Real time" observation of a live specimen through a hologram made moments before.

causes it to become brighter with time. This is a form of light amplification. The process has innumerable other novel applications and has already become a field of research by itself.

REFERENCES [1]D. Gabor, Proc. Roy. Soc. (London) **A197,** 454–487 (1949). E. N. Leith and J. Upatnieks, [2]J. Opt. Soc. Am. **54,** 1295 (1964). [3]Y. N. Denisyuk, Opt. Spectgry. (USSR) **18** (2), 152 (1965). [4]T. Jeong, Am. J. Phys. August, 714–717 (1975). [5]T. Jeong, *Laser Holography,* published by Thomas A. Edison Foundation, 21000 W. Ten Mile Road, Southfield, MI 48075. [6]S. Benton, Proc. of the Int'l. Sym. on Display Holography, Vol. I, 5–14 (1982) (Published by Holography Workshops, Lake Forest College, Lake Forest, IL 60045). [7]S. St. Cyr, Proc. of the Int'l. Sym. on Display Holography, Vol. II, 191–221 (1985). [8]W. Molteni, Proc. of the Int'l. Sym. on Display Holography, Vol. I, 15–26 (1982). [9]H. Bjelkhagen, Proc. of the Int'l. Sym. on Display Holography, Vol. I, 45–54 (1982). [10]T. Jeong, *Holography,* pub. by Integraf, Box 586, Lake Forest, IL 60045. [11]N. Abramson, *Holograms and Their Evaluations,* Academic Press, (1981). [12]T. Jeong, Pro. of S. P. I. E. Vol. 746, 16–19 (1987). [13]R. A. Fisher, *Optical Phase Conjugation,* Academic Press (1983).

CHAPTER 42
RELATIVITY*

"Everything should be as simple as possible—but not simpler."
Albert Einstein (1879–1955)

This statue of Albert Einstein, by the sculptor Robert Berks, is located on Constitution Avenue in Washington, DC, in front of the National Academy of Sciences building. It is a favorite of children, many of whom like to sit on his lap.

42-1 What Is Relativity All About?

In 1905, the 26-year-old Albert Einstein (see Fig. 1) put forward his *special theory of relativity.*† At that time, Einstein was Technical Expert (Third Class) in the Swiss Patent Office, working on physics in his spare time and in what has been termed "splendid isolation" from phys-

---

† Einstein also put forward a *general theory* of relativity, in 1917. It deals with the interpretation of gravity—not as a force but as a curvature in space and time. Although we do not deal with the general theory in this chapter, we gave a preview of its central principle, the principle of equivalence, in Section 15-10. In this chapter, the word *relativity* will always refer to the *special theory,* in which gravity plays no role.

---

icists in the academic community. During that same year, he published a second paper on relativity and two other world-class papers on entirely different subjects, for one of which he was later awarded the Nobel prize.

Relativity, an aesthetically appealing theory, is about the nature of space and time. It has survived every one of the many searching experimental tests to which it has been subjected during the last eight decades. Its status today is such that, if an experimental result is proposed that is inconsistent with relativity, physicists

---

* A fuller treatment of relativity can be found in *Basic Concepts in Relativity and Early Quantum Theory,* 2nd ed., by Robert Resnick and David Halliday, John Wiley & Sons, New York, 1985.

Figure 1 Einstein in the early 1900s, at his desk in the Bern Patent Office.

Table 1   Earlier Sections on Relativity

Section	Chapter Title	Section Title
4–9	Motion in a Plane	Relative Motion at High Speeds
7–7	Work and Energy	Kinetic Energy at High Speeds
8–9	The Conservation of Energy	Mass and Energy
10–6	Collisions	Reactions and Decay Processes
15–10	Gravity	A Closer Look at Gravity
17–7	Waves—I	The Speed of Light
38–8	Electromagnetic Waves	The Speed of Electromagnetic Waves

everywhere would conclude that there must be something wrong with the experiment. Asked about the influence of relativity on the development of physics, one physicist replied, "Well, relativity is simply *there*."

Relativity has a reputation, among those who have not studied it, as a difficult subject. It is not mathematical complexity that stands in the way of understanding; if you can solve a quadratic equation, you are overqualified. The difficulty lies entirely with the fact that relativity forces us to reexamine critically our ideas of space and time.

Our life experiences are restricted in that we have no direct experience with tangible objects moving faster than a tiny fraction of the speed of light. It is no wonder that our ideas of space and time, molded by this restricted experience, are also restricted. In much the same way a bacterium, spending its life in a fluid environment dominated by viscous forces, knows nothing of gravity. Our advice: Be receptive to new ideas and keep an open mind.

42-2   Our Plan

Relativity rests on two postulates, which we shall present in the next section. We ask you to accept them provisionally but uncritically and not to say to yourself, "Well, this postulate can't be true because. . . ." It is an admirable trait to question all statements in physics but

in this case we propose the following:

1.   Accept the postulates provisionally.
2.   Examine, with us, the consequences that flow from these postulates.
3.   Examine, again with us, the universal agreement of these consequences with experiment.*

If you follow this course, you can master the basic ideas of relativity. You will come to see how natural it is, how it simply extends the classical view, and how much common sense it contains.

Relativity is so important for physics that we did not feel that we could postpone discussing it until so late in the book. Therefore, to provide some foretaste of the subject, we inserted a number of optional relativity-related sections at appropriate places in earlier chapters; they are listed in Table 1. Throughout this chapter, we suggest at appropriate points that you go back and read (or reread) this earlier material.

42-3   The Postulates

1. The Postulate of Relativity.   *The laws of physics are the same for observers in all inertial*

---

* See "Modern Tests of Special Relativity," by Mark P. Haugan and Clifford M. Will, *Physics Today,* May 1987, for a summary of the scope and precision of these experimental tests and of their impressive confirmation of the predictions of relativity.

reference frames. No frame is singled out as preferred.

It was assumed by Galileo that the laws of *mechanics* were the same in all inertial reference frames. (Newton's first law of motion is one important consequence.) Einstein extended that idea to include *all* the laws of physics, including especially electromagnetism and optics. This postulate does *not* say that the measured values of all physical quantities are the same for all inertial observers; most are not. It is the *laws of physics* that relate these measurements to each other that are the same. Stated another way, one of the tests of any proposed law of physics is that it must satisfy the relativity postulate.

2. The Postulate of the Speed of Light. *The speed of light in free space has the same value c in all directions and in all inertial reference frames.*

We can also phrase this postulate to say that — as Fig. 2 whimsically suggests — there is in nature an *ultimate speed c,* the same in all directions and in all inertial reference frames. Light happens to travel at this ultimate speed, as do massless particles such as neutrinos. The ultimate speed *c* also sets a limit to which any material particle such as an electron can be accelerated.*

We first introduced this postulate in Section 17–7. We suggest that you reread that section, study Fig. 9 of that chapter carefully, and become thoroughly familiar with exactly what the second postulate says.

The Ultimate Speed.   The reality of the existence of an ultimate speed for accelerated electrons is shown in a 1964 experiment of W. Bertozzi. He accelerated electrons to various measured speeds and — by an independent calorimetric method — also measured their kinetic energies. Figure 3 shows that as the force that acts on a very fast electron is increased, its measured kinetic energy increases toward very large values but its speed does not increase appreciably. Electrons have been accelerated to at least 0.999 999 999 95 times the speed of light but — close though it may be — that speed is still less than the ultimate speed *c*.

Testing the Speed of Light Postulate.   If the speed of light is the same in all inertial reference frames, then the speed of light emitted by a moving source should be the

---

* Relativity can be developed — with exactly the same results — without reference to light. See "Relativity Without Light," by N. David Mermin, *American Journal of Physics,* February 1984.

Figure 2   A whimsical illustration for an article "The Ultimate Speed Limit," by Isaac Asimov (*Saturday Review,* July 1972).

same as the speed of light emitted by a source that is at rest in the laboratory. This claim has been tested directly, in an experiment of high precision. The "light source" was the *neutral pion* (symbol π^0), an unstable, short-lived particle that may be produced by collisions in a particle

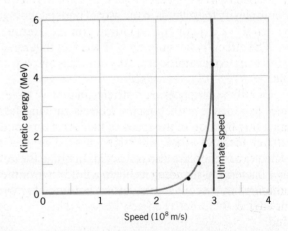

Figure 3   The dots show the measured values of the kinetic energy of an electron plotted against its measured speed. No matter how much energy you impart to an electron (or to any other particle, for that matter) you can never get its speed to equal or to exceed the ultimate limiting speed *c*. This "brick wall" can be approached as closely as you like but never reached. The curve is the prediction of relativity theory (see Eq. 37).

accelerator. It decays into two gamma rays by the process

$$\pi^0 \rightarrow \gamma + \gamma. \qquad (1)$$

Gamma rays are part of the electromagnetic spectrum and obey the postulate of the speed of light, just as visible light does.

In a 1964 experiment, physicists at CERN, the European particle-physics laboratory near Geneva, generated a beam of pions moving at a speed of $0.99975c$ with respect to the laboratory. The experimenters then measured the speed of the gamma rays emitted from these very rapidly-moving sources. Their results for the speed of light were

From the moving pions: 2.9977×10^8 m/s;
From a resting source (accepted value):
                                   2.9979×10^8 m/s.

There seems little doubt that the speed of light emitted by these pions—which were racing along at almost the speed of light—is the same as we would measure if the pions had been at rest in the laboratory.

Sample Problem 1   A 20-GeV electron, such as might be generated in the Stanford Linear Accelerator, can be shown to have a speed $v = 0.999\,999\,999\,67c$. If such an electron raced a light pulse to the nearest star outside the solar system (Proxima Centauri, 4.3 light years or 4.0×10^{16} m distant), by how much time would the light pulse win the race?

If L is the distance to the star, the difference in travel times is

$$\Delta t = \frac{L}{v} - \frac{L}{c} = L\,\frac{c-v}{vc}.$$

Now v is so close to c that we can put $v = c$ in the denominator of this expression (but not in the numerator!). If we do so, we find

$$\Delta t = \frac{L}{c}\left(1 - \frac{v}{c}\right) = \frac{(4.0 \times 10^{16}\text{ m})(1 - 0.999\,999\,999\,67)}{3.00 \times 10^8\text{ m/s}}$$
$$= 0.044\text{ s} = 44\text{ ms.} \qquad \text{(Answer)}$$

The 20-GeV electron certainly comes close to the ultimate speed c but it does not equal or surpass it.

42-4   Measuring an Event

An *event* is something that happens to which an observer can assign three space coordinates and one time coordi-

nate. Among many possible events are (1) the turning on or off of a tiny light bulb, (2) the collision of two particles, (3) the passage of a pulse of light through a specified point in space, or (4) the coincidence of the hand of a clock with a marker on the rim of the clock. An observer, fixed in an inertial reference frame, may assign to event A (the turning on of a light bulb, say) the following spacetime coordinates:*

Record of Event A	
Coordinate	*Value*
x	3.58 m
y	-1.29 m
z	2.77 m
t	34.5 s

A given event may be recorded by any number of observers, each in their own inertial reference frame. In general, all such observers will assign different spacetime coordinates to the same event. Note that an event does not, in any sense, "belong" to a particular inertial reference frame. An event is just something that happens and anyone may look at it and assign spacetime coordinates to it.

We need to understand in some detail how a single observer, fixed in an inertial reference frame, assigns space and time coordinates to a single event. Many of the procedures that we outline will seem totally impractical. However, they are *thought procedures,* outlining *in principle* how the measurements are to be carried out.

The Space Coordinates.   We imagine the observer's coordinate system fitted with a close-packed, three-dimensional array of measuring rods, one set of rods parallel to each of the three coordinate axes. Thus, if the event is the turning on of a small light bulb, the observer need only read the three space coordinates at the location of the bulb.

The Time Coordinate.   For the time coordinate, we imagine that every point of intersection of the array of measuring rods has a tiny clock, which the observer can read by the light generated by the event. Figure 4 suggests the "jungle gym" of meter rods and clocks that we have described.

The array of clocks must be synchronized properly. You may think that it is enough to assemble a set of

---

* Space and time are so closely linked in relativity that we describe them collectively as *spacetime,* without a hyphen.

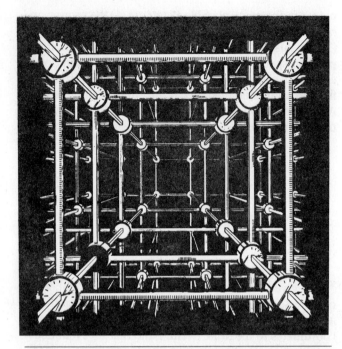

Figure 4 Suggesting the "jungle gym" of rods and clocks used by an observer in an inertial reference frame to assign spacetime coordinates to an event.

identical clocks, set them all to the same time, and then move them to their assigned positions. However, we are committed to question everything. How do we know, for example, that moving the clocks does not change their rates? (Actually, it does.) We must put the clocks in place and *then* synchronize them.

If we had a method of transmitting signals at infinite speed, synchronization would be a simple matter. However, no known signal has this property. We choose light (interpreted broadly to include the entire electromagnetic spectrum) to send out our synchronizing signals because, in free space, light travels at the highest possible speed, the limiting speed c.

Here is one of many ways that we might synchronize an array of clocks with the help of light signals: The observer enlists the help of a large number of temporary helpers, one for each clock. The observer then stands at a point selected as the origin and sends out a pulse of light when the origin clock reads $t = 0$. When the light pulse reaches each helper, that helper sets his or her clock to read $t = r/c$, where r is the distance of the helper from the origin.

All these procedures refer to a *single* observer in a *single* inertial reference frame. Other observers in other inertial reference frames must have a similar array of rods and clocks in order to assign spacetime coordinates to the events that they obseve.

42–5 Simultaneous Events

Suppose that one observer (Sam) notes that two independent events (event Red and event Blue) occur at the same time. Suppose also that another observer (Sally), who is moving at a constant velocity **v** with respect to Sam, also records these same two events. Will Sally also find that they occurred at the same time?

The answer is that in general she will not. Let us be clear about what we are saying:

> *If two observers are in relative motion, they will not, in general, agree as to whether two events are simultaneous. If one observer finds them to be simultaneous, the other will not, and conversely.*

We cannot say that one observer is right and the other wrong. The situation is completely symmetrical and there is no reason to choose one observer over the other. We conclude the following:

> *Simultaneity is not an absolute concept but a relative one, depending on the state of motion of the observer.*

Of course, if the relative speed of the observers is very much less than the speed of light, the measured departures from simultaneity become so small that they are not noticeable. Such is the case for all our experiences of daily living; this is why the relativity of simultaneity is unfamiliar.

42–6 Simultaneity: A Closer Look

Let us clarify the relativity of simultaneity by a specific example. We base our analysis directly on the postulates of relativity, no clocks or measuring rods being directly involved.

Figure 5 shows two long spaceships (the *SS Fitzgerald* and the *SS Lorentz*), which can serve as inertial reference frames for observer Sam and observer Sally. The two observers are stationed at the midpoints of their ships. The ships are separating along a common x axis, the relative velocity of *Fitzgerald* with respect to *Lorentz*

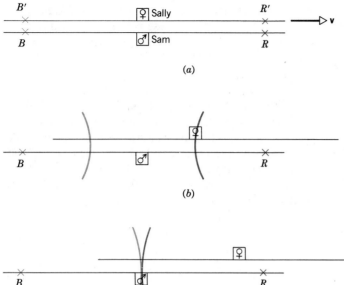

(a)

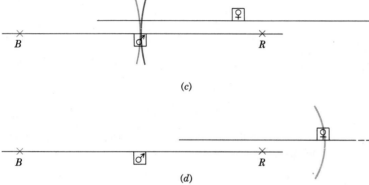

(b)

(c)

(d)

Figure 5   The two spaceships are represented by simple straight lines. The figure shows the situation from Sam's point of view, in which Sally's spaceship is moving to the right with speed v. (a) Light waves leave the site of the Red event (RR') and the Blue event (BB'). Successive drawings correspond to the assumption that Sam will perceive events Red and Blue to be simultaneous. (b) The Red wave front reaches Sally. (c) Both wave fronts reach Sam. (d) The Blue wave front reaches Sally; this last figure is not to the scale of the previous three.

being **v**. Figure 5a shows the ships with the two observer stations momentarily aligned opposite each other.

Two meteorites strike the ships, one setting off a red flare (event Red) and the other a blue flare (event Blue). Each event leaves a permanent mark on each ship, at positions R,R' and B,B'.

Let us suppose that the expanding wave fronts from the two events happen to reach Sam at the same time, as Fig. 5c shows. Let us further suppose that, after the episode, Sam finds, by measurement, that he was stationed exactly halfway between the markers B and R on his ship. He will say:

Sam: Light from event Red and event Blue reached me at the same time. From the marks on my spaceship, I find that I was standing halfway between the two sources when the light from them reached me. Therefore, event Red and event Blue are simultaneous events.

As study of Fig. 5 shows, however, the expanding wave front from event Red will reach Sally *before* the expand-

ing wave front from event Blue does. She will say:

Sally: Light from event Red reached me before light from event Blue did. From the marks on my spaceship, I found that I too was standing halfway between the two sources. Therefore, the events were *not* simultaneous; event Red occurred first, followed by event Blue.

Their reports do not agree. Nevertheless, both observers are correct. That is what the relativity of simultaneity is all about.

It is essential to understand that there is only one wave front expanding from the site of each event and that *this wave front travels with the same speed c in both reference frames,* exactly as the speed of light postulate requires.

It *might* have happened that the meteorites struck the ships in such a way that they appeared simultaneous to Sally. Sam would then declare them not to be simultaneous. The experiences of the two observers are exactly symmetrical. (Note that we have avoided saying anything like: The meteorites struck the ships simultaneously. That would raise the question: Simultaneous in which reference frame? We always said: The meteorites struck the ships *in such a way that.* . . .)

42-7 The Relativity of Time

The relativity of simultaneity is closely related to the relativity of time. That is, if different observers measure the time interval between a given pair of events, they will in general not agree as to how long that interval is. We describe a simple case, again basing our analysis directly on the postulates of relativity.

In Fig. 6, Sally is in a train that is moving with uniform velocity **v** with respect to the station. She has an electronic clock, which she uses to measure the time Δt_0 between two events:

Event 1. The turning on of a flash bulb B.

Event 2. The arrival of the light back at its source after reflection from a mirror on the ceiling.

For the time interval between these two events, Sally finds

$$\Delta t_0 = \frac{2D}{c} \quad \text{(Sally)}, \qquad (2)$$

where D is the distance between the source and the mirror. Note that for Sally these two events occur *at the same place* and she can time the interval between them with *a single clock C located at that place.* A time interval measured with a single resting clock is called a *proper time interval,* identified by the subscript zero.

Consider now how these same two events look to Sam, who is standing on the station platform as the train goes by. Because of the postulate of the speed of light, the light travels at the same speed c for Sam as for Sally. It travels a larger distance for Sam, however, namely, $2L$. The time interval measured by Sam between these two events is

$$\Delta t = \frac{2L}{c} \quad \text{(Sam)}, \qquad (3)$$

where

$$L = \sqrt{(\tfrac{1}{2}v\,\Delta t)^2 + D^2}. \qquad (4)$$

From Eq. 2, we can write this as

$$L = \sqrt{(\tfrac{1}{2}v\,\Delta t)^2 + (\tfrac{1}{2}c\,\Delta t_0)^2}. \qquad (5)$$

If we eliminate L between Eqs. 3 and 5 and solve for Δt, we find

$$\Delta t = \frac{\Delta t_0}{\sqrt{1 - (v/c)^2}}. \qquad (6)$$

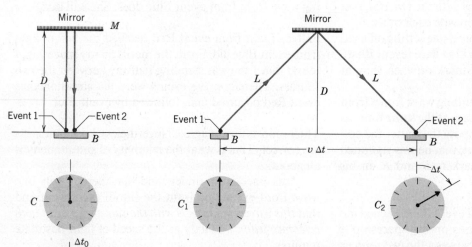

Figure 6 (*a*) Sally times the light pulse excursion using a single resting clock C, reporting a proper time Δt_0. (*b*) Sam, watching the moving train, requires two synchronized clocks, C_1 and C_2, to measure the elapsed time Δt.

Note that for Sam the two events occur *in different places* and, to measure Δt, he must use *two synchronized clocks, C_1 and C_2, located at different places in his reference frame.* The interval that he measures is *not* a proper time interval because he does not measure it with a single resting clock. For this reason, the situations for Sam and Sally (unlike their situations in Fig. 5) are *not* symmetrical.*

We can rewrite Eq. 6 as

$$\Delta t = \frac{\Delta t_0}{\sqrt{1 - \beta^2}} \quad \text{(time dilation),} \quad (7)$$

in which we have replaced the dimensionless ratio v/c by the symbol β, which we call the *speed parameter.*

Because $\beta < 1$, we always have $\Delta t > \Delta t_0$. Sam, standing on the station platform, might say:

Sam: Sally and I have identical clocks and we each measure the time interval between the same two events. Sally's reference frame is special because, for her, the events occur at the same place so that she can use a single clock to measure the time interval. For me, the events occur at different places and I must use *two* synchronized clocks, one at the location of each event. I find that — no matter how fast (or in what direction) the train is moving — my measured value for the time interval is always greater than Sally's.

This *time dilation effect†* is very real and has nothing to do with any mechanical change that takes place in a clock because of its motion. It is simply the nature of time.

The Lorentz Factor. We can also write Eq. 7 in the form

$$\Delta t = \gamma \, \Delta t_0 \quad \text{(time dilation)} \quad (8)$$

in which the dimensionless quantity γ, called the *Lorentz factor,* is given by

$$\gamma = \frac{1}{\sqrt{1 - \beta^2}} \quad \text{(Lorentz factor).} \quad (9)$$

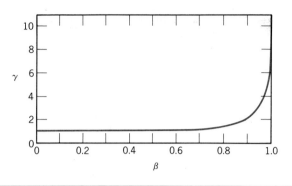

Figure 7 A plot of the Lorentz factor γ as a function of the speed parameter $\beta \, (= v/c)$.

This combination of variables occurs often in relativity. Figure 7 shows how γ varies with the speed parameter β. As Table 2 shows, the Lorentz factor γ, which is always greater than 1, is our guide to the importance of relativity to the problem at hand.

A Test of Time Dilation: Microscopic Clocks. Subatomic particles called *muons* are unstable and, when at rest in the laboratory, decay with an average lifetime of 2.200 μs. This average lifetime, measured for resting muons with a single resting laboratory clock, is thus a *proper time interval* and we can label it Δt_0.

In a 1968 experiment at CERN, a beam of muons, circulating in a storage ring of 2.5-m radius, was accelerated to a speed of $0.9966c$.* The average lifetime of these muons was then measured while they were in flight at this very high speed.

The accelerated muons can serve as tiny moving clocks, to which Eq. 8, $\Delta t = \gamma \, \Delta t_0$, applies. We first calculate the Lorentz factor γ for the moving muons from Eq. 9,

$$\gamma = \frac{1}{\sqrt{1 - (v/c)^2}} = \frac{1}{\sqrt{1 - (0.9966)^2}}$$
$$= 12.14,$$

which is substantially greater than one. Equation 8 yields

$$\Delta t = \gamma \, \Delta t_0 = (12.14)(2.200 \, \mu s) = 26.7 \, \mu s.$$

The measured value was 26.2 μs, in excellent agreement

---

* To make them symmetrical, you would have to move Sally's flash–mirror arrangement to the platform and give it to Sam. Then Sam would measure a proper time interval and Sally would measure a dilated interval.

† To dilate is to expand or stretch.

---

* You may object that these muons, in uniform circular motion with a centripetal acceleration of $10^{15}g$, are not in an inertial reference frame so that relativity does not apply. It can be shown, however, that it does apply in this case. Indeed, this experiment can be taken as a verification of that prediction.

Table 2   The Speed Parameter and the Lorentz Factor

Particle	Speed Parameter β	Lorentz Factor γ	Can I use Newtonian Mechanics?[a]
Fastest aircraft (Mach 6.72)	0.0000068	1.000000. . .	Certainly
Earth's orbital speed	0.000099	1.000000005	Yes
1-keV electron	0.063	1.0020	Yes (?)
Around the world in one second	0.13	1.009	Maybe
1-MeV electron	0.94	2.9	No
1-GeV electron	0.999 999 88	2000	Never
Electron in Sample Problem 1	0.999 999 999 67	40,000	Don't even think of it

[a] Relativity applies at *all* speeds, Newtonian mechanics only at speeds much less than the speed of light.

within the experimental error. The time dilation factor is universally accepted and indeed turned to advantage in the design of certain high-energy particle experiments. One physicist has written: "We frequently transport beams of unstable particles over long distances such that no particles would be left in the beam without the help of Einstein's factor."

A Test of Time Dilation: Macroscopic Clocks.   In October 1977, Joseph Hafele and Richard Keating carried out what must have been a grueling experiment. They flew four portable atomic clocks twice around the world on commercial airlines, once in each direction. Their purpose was ". . . to test Einstein's theory of relativity with macroscopic clocks." As we have just seen, the time dilation predictions of Einstein's theory had already been confirmed on a microscopic scale but there is great comfort in seeing a confirmation made with an actual clock. Such measurements became possible only because of the very high precision of modern atomic clocks. Hafele and Keating verified the predictions of the theory to within 5–10%.*

A few years later, physicists at the University of Maryland carried out a similar experiment with improved precison. They flew an atomic clock round and round over Chesapeake Bay for flights of 15-h duration and succeeded in checking the time dilation prediction to better than 1%. Today, when atomic clocks are transported from one place to another for calibration or other purposes, the effect of time dilation caused by their motion must always be taken into account.

The Twin Paradox.   Time dilation applies not only to clocks but to all naturally occurring time intervals including pulse rates and average lifetimes. This gives rise to the twin paradox, which is no paradox at all to those who understand relativity and accept its predictons.

One twin sister embarks on a round-trip journey to a star in a very fast spaceship while the other twin remains on earth. As measured by the earth-bound twin, her sister's heart beat, respiration rate, and indeed her aging processes slow down because of the time dilation effect. The traveling twin, however, measuring her pulse rate with her wristwatch (a co-traveling clock!) is totally unaware of any such changes and judges herself to be normal in all such respects. Nevertheless, when the spaceship returns, the traveling twin will be found to be younger than the stay-at-home twin. The stay-at-home twin may in fact have died and be represented only by her aging greatgrandchildren!

Some will say that the traveling twin could argue: Can I not regard myself as stationary and my earth-bound sister as traveling? If so, she should be younger than I when we meet and not the other way around. Clearly, A cannot be younger than B and also B younger than A; that is the paradox.

The *answer to the paradox* is that the twins are *not* in symmetrical situations and the traveling twin *cannot* make the argument outlined above. If the traveling twin is to return, she must at some point turn around, which

* Einstein's *general* theory of relativity, which predicts that the rate of a clock is influenced by gravity, also plays a role in this experiment.

involves a very necessary and readily measurable acceleration. Put another way, the traveling twin (in the simplest case) switches inertial reference frames, from an outward bound frame to an inward bound frame. The stay-at-home twin undergoes no acceleration and remains in a single reference frame throughout. It is the traveling twin—and not the stay-at-home twin—who will be the younger when they meet again.*

Although this experiment has not (yet) been carried out with actual twins, the round-the-world experiment of Hafele and Keating, mentioned above, amounts to the same thing. Instead of twin people, they used twin clocks, the stay-at-home clock remaining at the Bureau of Standards and the traveling clock going round the world, and necessarily accelerating in the process. When the two clocks "met" at the end of the trip, the traveling clock (in terms of accumulated time) was indeed "younger" than the stay-at-home clock, by 273 ns for the westward circuit.

Sample Problem 2   At what relative speed would a moving clock appear, to a stationary observer, to run at half the rate observed by a person moving with the clock?

The person moving with the clock records a proper time Δt_0, since the clock is at rest relative to him. The person who is watching the moving clock records a dilated time Δt for that clock. That is, if $\Delta t_0 = 1$ h then $\Delta t = 2$ h and, in general, $\Delta t = 2\Delta t_0$. From Eq. 9 then,

$$\gamma = 2 = \frac{1}{\sqrt{1 - \beta^2}}.$$

Squaring both sides yields

$$(4)(1 - \beta^2) = 1$$

or

$$\beta = \sqrt{3/4} = 0.866. \qquad \text{(Answer)}$$

Thus, a clock must be traveling at about 87% of the speed of light for the time dilation factor to amount to a factor of 2. Such a speed corresponds to circling the globe at the equator 6.7 times per second.

---

*For a detailed account, with the arguments carefully spelled out, see *Basic Concepts,* 2nd ed., Supplementary Topic B (The Twin Paradox), by Robert Resnick and David Halliday, John Wiley & Sons, New York, 1985.

42-8   The Relativity of Length

If you want to measure the length of a rod that is at rest with respect to you, you can—at your leisure—note the positions of its end points on a long stationary scale and subtract the two readings. If the rod is moving, however, you must note the positions of the end points *simultaneously* (in your reference frame) or your measurement cannot be called a length. Figure 8 suggests the difficulty of trying to measure the length of a swimming goldfish by locating its ends at different times. Because simultaneity is relative and it enters into length measurements, you expect that length is also a relative quantity.

If the length of an object at rest in your reference frame—called its *proper length*—is L_0, the length that you would measure if the rod is moving past you (parallel to itself) at speed $v(= \beta c)$. is given by

$$L = L_0\sqrt{1 - \beta^2} = \frac{L_0}{\gamma} \qquad \text{(length contraction).} \quad (10)$$

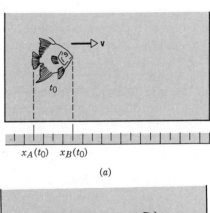

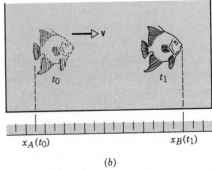

Figure 8   If you want to measure the length of your pet goldfish while it is swimming, you must mark the positions of its head and its tail simultaneously (in your reference frame), as in (a), rather than at arbitrary times, as in (b).

Because the Lorentz factor γ is always greater than unity, the length of a moving rod is always measured to be smaller than the length of the rod when it is at rest. Like the time dilation, the length contract is very real. We can summarize both phenomena as follows:

If two events occur at the same place in an inertial reference frame, the time interval Δt_0 between them, measured by a single resting clock, is called a proper time interval. All other inertial observers will measure a larger value for this interval.

The length L_0 of a rod, measured in an inertial reference frame in which the rod is at rest, is called its proper length. All other inertial observers will measure a shorter length.

The questions, "Does the rod *really* shrink?" and "Do the atoms in the rod *really* get pushed closer together?" are not proper questions within the framework of relativity. The length of a rod is what you measure it to be and motion affects measurements.

Proof of Eq. 10.   The length contraction is a direct consequence of time dilation. Consider once more our two observers. Sally is seated on a train moving through a station and Sam is again on the station platform. They both want to measure the length of the platform. Sam, using a tape measure, finds the length to be L_0, a *proper length* because the platform is at rest with respect to him. Sam also notes that a marker on the train covers this length in a time $\Delta t = L_0/v$, where v is the speed of the train. That is,

$$L_0 = v\,\Delta t \quad \text{(Sam)}. \tag{11}$$

This time interval Δt is not a proper time interval because the two events that define it (marker passes back of platform and marker passes front of platform) occur at two different places and Sam must use two synchronized clocks to measure the time interval Δt.

For Sally, however, the platform is moving. She sees it approach and then recede (at speed v) and finds that the two events measured by Sam occur *at the same place* in her reference frame. She can time them with a single resting clock so that the interval Δt_0 that she measures is a proper time interval. To her, the length L of the platform is given by

$$L = v\,\Delta t_0 \quad \text{(Sally)}. \tag{12}$$

If we divide Eq. 12 by Eq. 11 and use Eq. 8, the time

dilation equation, we have

$$\frac{L}{L_0} = \frac{v\,\Delta t_0}{v\,\Delta t} = \frac{1}{\gamma},$$

or

$$L = \frac{L_0}{\gamma}, \tag{13}$$

which is exactly Eq. 10, the length contraction equation.

Sample Problem 3   Two spaceships, each of proper length $L_0 = 230$ m, pass each other as in Fig. 9. Sally, located at point A on one of the spaceships, measures a time interval of 3.57 μs for the second ship to pass her. What is the relative speed parameter β of the two ships? Let AB be the coincidence of points A and B and AC the coincidence of points A and C.

The time interval between events AB and AC, measured by Sally using a single clock at A, is a *proper time interval* $\Delta t_0 = 3.57$ μs. The length L that Sally measures for the other ship is

$$L = v\,\Delta t_0 = \beta c\,\Delta t_0.$$

However, Sally knows that the *proper* length of the other ship is $L_0 = 230$ m, where (see Eq. 10)

$$L = \frac{L_0}{\gamma} = L_0\sqrt{1 - \beta^2}.$$

Setting these two values for L equal to each other, we find

$$\beta c\,\Delta t_0 = L_0\sqrt{1 - \beta^2}.$$

If we square each side of this equation and solve for β, we find, after a little algebra,

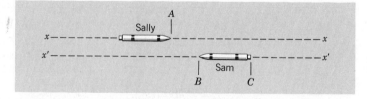

Figure 9   Sample Problem 3. Sally measures the length of Sam's spaceship as it drifts past.

$$\beta = \frac{L_0}{\sqrt{(c\,\Delta t_0)^2 + L_0^2}}$$

$$= \frac{230\ \text{m}}{\sqrt{(3.00 \times 10^8\ \text{m/s})^2 (3.57 \times 10^{-6}\ \text{s})^2 + (230\ \text{m})^2}}$$

$$= 0.210. \tag{Answer}$$

Thus, the ships are separating at about 21% of the speed of light.

42-9   The Transformation Equations

As Fig. 10 shows, observer S sees observer S' moving with speed v in the positive direction of their common xx' axis. Observer S reports spacetime coordinates x, y, z, t for an event and observer S' reports x', y', z', t' for the same event. How are these sets of numbers related?

We claim at once (although it requires proof) that the y and z coordinates, which are at right angles to the motion, are not affected by the motion. That is, we will always have $y = y'$ and $z = z'$. Our interest then reduces to the relation between x and x' and between t and t'.

The Galilean Transformation Equations. In prerelativity days, the relations we sought would be given by

$$x' = x - vt$$
$$t' = t.$$

(Galilean transformation equations; valid at low speeds only).   (14)

The first of these equations seems to follow from Fig. 10, coupled with the assumption that each observer chooses $t = t' = 0$ to represent the instant that their origins coincided.

The second of these equations would have seemed so obvious to prerelativity physicists that it would seem strange to even write it down. After all, they might have said: "Time is absolute and the same for everybody," or perhaps more likely: "Time is time!"

The fact that Eqs. 14 seem almost obviously true is a reflection of the fact that all our experience with space and time coordinates is limited to the very special case of $v \ll c$. In fact, at speeds comparable to the speed of light, each of the Galilean transformation equations fails to agree with experiment.

The Lorentz Transformation Equations.* We state without proof that the correct transformation equations, which remain valid for all speeds up to the speed of light, can be derived from the postulates of relativity. The results are

$$x' = \gamma(x - vt)$$
$$t' = \gamma(t - vx/c^2).$$

(Lorentz transformation equations; valid at all speeds).   (15)

---

*You may wonder why we do not call these the *Einstein transformation equations* (and why not the *Einstein factor* for γ). The great Dutch physicist H. A. Lorentz actually derived these equations before Einstein did but (as Lorentz graciously conceded) he did not take the further bold step of interpreting these equations as describing the true nature of space and time. It is this interpretation that is at the heart of relativity.

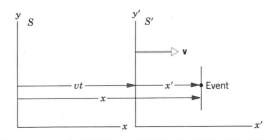

Figure 10 Two inertial reference frames share a common xx' axis. Frame S' is receding from frame S with speed v.

Note how, in the second Lorentz equation, the variable x is bound up with the determination of t'. Time and space are closely intertwined in relativity.

It is a formal requirement of relativistic equations that they should reduce to familiar classical equations if we let c approach infinity. After all, if the speed of light were infinitely great, *all* finite speeds would be "low" and classical equations would never fail. If we let $c \to \infty$ in Eqs. 15, $\gamma \to 1$ and these equations reduce—as we expect—to the Galilean equations (Eqs. 14).

Equations 15 are written in a form that is useful if we are given x and t and wish to find x' and t'. We may wish to go the other way, however. In that case we simply solve Eqs. 15 for x and t, obtaining

$$x = \gamma(x' + vt')$$
$$t = \gamma(t' + vx'/c^2).$$

(16)

Comparison shows that, starting from either Eqs. 15 or Eqs. 16, you can find the other set by interchanging primed and unprimed quantities and reversing the sign of the relative velocity v.

Equations 15 and 16 relate the coordinates of a single event as seen by two observers. Sometimes we want to know not the coordinates of a single event but the differences between coordinates for a pair of events. That is, if we label our events 1 and 2, we may want to know

$$\Delta x = x_2 - x_1 \quad \text{and} \quad \Delta t = t_2 - t_1,$$

as seen by observer S, and

$$\Delta x' = x_2' - x_1' \quad \text{and} \quad \Delta t' = t_2' - t_1',$$

as observed by S'.

Table 3 displays the Lorentz equations in difference form, suitable for analyzing pairs of events. The equations in the table were derived by simply taking differences between the four variables displayed in Eqs. 15 and 16.

Table 3 The Lorentz Transformation Equations[a]

1. $\Delta x = \gamma(\Delta x' + v\,\Delta t')$	1′. $\Delta x' = \gamma(\Delta x - v\,\Delta t)$
2. $\Delta t = \gamma(\Delta t' + v\,\Delta x'/c^2)$	2′. $\Delta t' = \gamma(\Delta t - v\,\Delta x/c^2)$

$$\gamma = \frac{1}{\sqrt{1 - (v/c)^2}} = \frac{1}{\sqrt{1 - \beta^2}}$$

[a] Written for pairs of events, as difference equations.

42–10 Some Consequences of the Lorentz Equations

Here we use the transformation equations of Table 3 to affirm some of the conclusions that we reached earlier by arguments based directly on the postulates.

Simultaneity. Consider Eq. 2 of Table 3,

$$\Delta t = \gamma(\Delta t' + v\,\Delta x'/c^2) \quad \text{(a Lorentz equation).} \quad (17)$$

If two events occur at different places in S', then $\Delta x'$ in this equation is not zero. It follows that even if the events are simultaneous in S' ($\Delta t' = 0$), they will not be simultaneous in S. The time interval in S will be $\Delta t = \gamma v\,\Delta x'/c^2$. This is in accord with the conclusion we reached in Section 42–6.

Time Dilation. Suppose now that two events occur at the same place in S' ($\Delta x' = 0$) but at different times ($\Delta t' \neq 0$). Equation 17 then reduces to

$$\Delta t = \gamma\,\Delta t' \quad \text{(events in same place in } S'\text{).} \quad (18)$$

This confirms time dilation. Because the two events occur at the same place in S', the time interval $\Delta t'$ between them can be measured with a single clock, located at that place. Under these conditions, the measured interval is a *proper time interval*, and we can label it Δt_0. Thus, Eq. 18 becomes

$$\Delta t = \gamma\,\Delta t_0 \quad \text{(time dilation),}$$

which is exactly Eq. 8, the time dilation equation.

Length Contraction. Consider Eq. 1′ of Table 3,

$$\Delta x' = \gamma(\Delta x - v\,\Delta t) \quad \text{(a Lorentz equation).} \quad (19)$$

If a rod lies parallel to the xx' axis and is at rest in reference frame S', an observer can measure its length at leisure. The value $\Delta x'$ that is obtained by subtracting the coordinates of the end points of the rod will be its *proper length L_0*.

The rod is moving in frame S. Thus, Δx can only be identified as the length L of the rod if the coordinates of the end points are measured *simultaneously*, that is, if

$\Delta t = 0$. If we put $\Delta x' = L_0$, $\Delta x = L$, and $\Delta t = 0$ in Eq. 19, we find

$$L = \frac{L_0}{\gamma} \quad \text{(length contraction),} \quad (20)$$

which is exactly Eq. 10, the length contraction formula.

Sample Problem 4 In inertial frame S, a blue light flashes, followed after 5.35 μs by a red flash. The separation of the two flashes is $\Delta x = 2.45$ km, with the red flash occurring at the larger value of x. S' is moving in the direction of increasing x with a speed parameter $\beta = 0.855$. What is the distance between the two events and the time interval between them as measured in S'?

Equations 1′ and 2′ of Table 3, with v replaced by βc, are

$$\Delta x' = \gamma(\Delta x - \beta c\,\Delta t) \quad (21)$$

and

$$\Delta t' = \gamma(\Delta t - \beta\,\Delta x/c). \quad (22)$$

We are told that

$$\Delta x = x_R - x_B = 2.45 \text{ km} = 2450 \text{ m}$$

and

$$\Delta t = t_R - t_B = 5.35 \ \mu\text{s} = 5.35 \times 10^{-6} \text{ s}$$

and that

$$\gamma = \frac{1}{\sqrt{1 - \beta^2}} = \frac{1}{\sqrt{1 - (0.855)^2}} = 1.928.$$

Thus, we have, from Eq. 21,

$$\begin{aligned}
\Delta x' &= (1.928)[2450 \text{ m} - (0.855) \\
&\quad \times (3.00 \times 10^8 \text{ m/s})(5.35 \times 10^{-6} \text{ s})] \\
&= 2078 \text{ m} \approx 2.08 \text{ km} \quad \text{(Answer)}
\end{aligned}$$

and from Eq. 22,

$$\begin{aligned}
\Delta t' &= (1.928)\left(5.35 \times 10^{-6} \text{ s} - \frac{(0.855)(2450 \text{ m})}{3.00 \times 10^8 \text{ m/s}}\right) \\
&= -3.147 \times 10^{-6} \text{ s} \approx -3.15 \ \mu\text{s}. \quad \text{(Answer)}
\end{aligned}$$

We conclude that in S' the red flash is also more distant but that distance is 2.08 km (rather than 2.45 km). The negative sign in the last result tells us that in S'—contrary to what is observed in S—the red flash occurs first. The time between the flashes in S' is 3.15 μs (not 5.35 μs).

Sample Problem 5 A plane flying at a speed u travels from Seattle to Atlanta. In inertial frame S, which is fixed with respect to the ground, the plane's takeoff from Seattle and its landing in Atlanta are separate events, separated in space by a distance Δx and in time by an interval Δt. Assume that an observer S', moving with respect to observer S, measures these same two events. Is it possible for these events to be recorded reversed in sequence, that is, can an observer in S' see the plane land in Atlanta before it takes off from Seattle?

Consider Eq. 2' of Table 3,

$$\Delta t' = \gamma(\Delta t - \beta \, \Delta x/c) \quad \text{(a Lorentz equation).}$$

The quantities Δx (the distance from Seattle to Atlanta) and Δt (the flight time) are both positive quantities. Let us find the value of β such that $\Delta t'$ would be negative. Study of the Lorentz equation above shows that we must have

$$\frac{\beta \, \Delta x}{c} > \Delta t.$$

But $\Delta x/\Delta t$ is just the speed u of the plane in S. We then have the requirement that

$$\beta > \frac{c}{u}.$$

But, since the plane cannot exceed the speed of light, $c/u > 1$ so that the requirement is

$$\beta > 1. \quad \text{(Answer)}$$

This requirement is impossible to fulfill because it would require observer S' to be traveling at a speed greater than the speed of light. A plane cannot land before it takes off, even in special relativity! The takeoff and landing of the plane are not *independent* events because one must necessarily occur before the other, in *all* reference frames. It is never possible to reverse the sequence of events that are causally related. If A causes B, then all observers will agree that A precedes B; you cannot be born before your mother is born!

42-11   The Transformation of Velocities

Here we wish to use the Lorentz equations to compare the velocities that two observers in different inertial reference frames S and S' would measure for the same moving particle.

Suppose that the particle, moving with constant speed parallel to the xx' axis, sends out two signals as it moves. Each observer measures the spatial interval and the time interval between these two events. These four measurements are related by Eqs. 1 and 2 of Table 3, or

$$\Delta x = \gamma(\Delta x' + u \, \Delta t') \quad \text{and} \quad \Delta t = \gamma(\Delta t' + u \, \Delta x'/c^2),$$

in which we now represent by u the velocity of S' with respect to S. If we divide the first of these equations by

the second, we find

$$\frac{\Delta x}{\Delta t} = \frac{\Delta x' + u \, \Delta t'}{\Delta t' + u \, \Delta x'/c^2}.$$

Dividing the numerator and the denominator of the right side by $\Delta t'$, we find

$$\frac{\Delta x}{\Delta t} = \frac{\Delta x'/\Delta t' + u}{1 + u(\Delta x'/\Delta t')/c^2}.$$

But, in the differential limit, $\Delta x/\Delta t$ is v, the velocity measured in S and $\Delta x'/\Delta t'$ is v', the velocity measured in S'. We have finally that

$$\boxed{v = \frac{v' + u}{1 + uv'/c^2}} \quad \text{(relativistic velocity law)} \quad (23)$$

as the relativistic velocity transformation law. We discussed this law earlier, using a slightly different notation, in optional Section 4-9. You may wish to reread that section and, in particular, to study Sample Problems 10 and 11 of Chapter 4. Equation 23 reduces to the classical, or Galilean, velocity transformation law,

$$v = v' + u \quad \text{(classical velocity law),} \quad (24)$$

when we apply the formal test of letting $c \to \infty$.

42-12   The Doppler Effect

We discussed the Doppler effect for sound waves in air in Section 18-7. In that case, there were two Doppler formulas because the velocities of the source and of the detector with respect to the transmitting medium (air) must be considered separately. In prerelativity days, physicists believed that a medium (called the *luminiferous ether*) was needed to support the propagation of light. Thus, the two Doppler formulas for sound were taken over for light by simply substituting the speed of light c for the speed of sound.

With the advent of relativity, Einstein declared that the propagation of light requires no medium and that the only relevant velocity is the relative velocity of the source and the detector. Thus, there should be just one Doppler formula for light, not two, and it must be derived from relativity theory.

The classical and the relativistic Doppler formulas for the shifted frequency are

Prerelativity theory:
source fixed, $\qquad v = v_0(1 - \beta) \qquad (25)$
detector receding

Relativity theory:
source and detector separating

$$v = v_0 \sqrt{\frac{1-\beta}{1+\beta}}. \qquad (26)$$

Prerelativity theory:
source receding, detector fixed

$$v = v_0 \frac{1}{1+\beta}. \qquad (27)$$

Here $\beta = v/c$, where v is the relative speed of separation of source and detector and c is the speed of light. The quantity v_0 is the frequency (a *proper* frequency) that would be measured by an observer for whom the source is at rest. The frequency v is that which would be measured by an observer in relative motion with respect to the source. All three formulas apply to the case in which source and detector are moving farther apart. If they are moving closer together, it is only necessary to change the sign of β in the above three equations.

To test the theory of relativity, we must be able to show by experiment that Doppler shifts for light (or other electromagnetic radiation) obey Eq. 26 and not Eqs. 25 or 27. At first sight this does not look too difficult because the equations look so different. The three equations, however, are deceptively similar at low speeds, which is the speed region in which we must conduct our tests.

If $\beta \ll 1$ and if we expand Eqs. 25, 26, and 27 as power series in β, keeping no terms higher than β^2, we find that all three equations reduce to

$$v \approx v_0 (1 - \beta + C\beta^2) \qquad (28)$$

The equations differ only in the numerical constant C, the coefficient of the term in β^2. The predicted values for C are

Equation	Value of C
25	0
26	$\frac{1}{2}$
27	1

For a source moving at low speed, β will be small and β^2 correspondingly smaller so that clever experiments are needed. In 1938, H. E. Ives and G. R. Stilwell of the Bell Telephone Laboratories devised just such an experiment. They sent the light beam back and forth through the apparatus in such a way that the term containing β in Eq. 28 effectively canceled out, leaving only the crucial term in β^2. These workers verified the prediction of relativity theory to a precision of a few percent. A repetition of the experiment in 1985 with modern techniques improved the agreement with the prediction of relativity to 1 part in 25,000 or 0.004%.

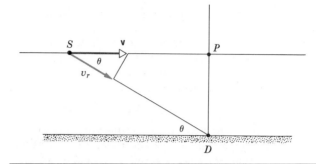

Figure 11 A light source S travels with velocity **v** past a detector at D. As the source passes through point P, the radial component of its velocity is zero. According to relativity, there should be a transverse Doppler effect at this point; classical theory predicts that there should not be.

Transverse Doppler Effect. Our entire discussion of the Doppler effect, both here and in Chapter 18, is for the case in which the source and the detector are moving directly toward each other or directly away from each other. Suppose, however, that a light source S is moving past a detector D as shown in Fig. 11. When the moving source passes through point P in Fig. 11, the radial component of its velocity is zero. The classical and the relativistic predictions for the Doppler effect in his case are

$$\text{Prerelativity theory} \quad v = v_0 \qquad (29)$$

and

$$\text{Relativity theory} \quad v = v_0 \sqrt{1 - \beta^2}. \qquad (30)$$

That is, there is (Eq. 30) or there is not (Eq. 29) a *transverse Doppler effect.*

The transverse Doppler effect is really another test of time dilation. If we write Eq. 30 in terms of the period T of oscillation of the emitted rays instead of the frequency, we have, recalling that $T = 1/v$,

$$T = \frac{T_0}{\sqrt{1 - \beta^2}} = \gamma T_0, \qquad (31)$$

in which $T_0 (= 1/v_0)$ is the *proper period* of the source. As comparison with Eq. 5 shows, Eq. 31 is simply the time dilation formula. We can understand this physically if we think of the moving source as a moving clock, beating out electromagnetic oscillations at a proper frequency v_0 and a corresponding proper period T_0. We expect to measure a longer period (that is, a smaller frequency) for such a moving clock.

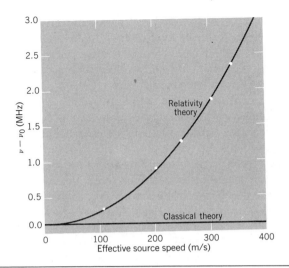

Figure 12   The results of Kundig on the transverse Doppler effect. The experiments fall very nicely on the curve predicted by relativity theory and not at all on the curve predicted by classical theory.

A Transverse Doppler Experiment.   In 1963, W. Kundig obtained excellent quantitative data confirming Eq. 30 to within an experimental error of about 1%. In his experiment, a radioactive source emitting gamma rays was located on the rotor of a centrifuge. By sensitive techniques, Kundig was able to measure the shift in frequency detected by a detector placed on the rim of the centrifuge. Figure 12 shows his results, which confirm the prediction of the relativity theory beyond any question.

42-13   A New Look at Momentum

Suppose that a number of observers, each in his or her own inertial reference frame, watch an isolated collision between two particles. In classical mechanics, we have seen that—even though the observers measure different velocities for the colliding particles—they all find that the law of conservation of momentum holds. That is, they find that the momentum of the system of particles after the collision is the same as it was before the collision.

How is this situation affected by relativity? We find that, if we continue to define the momentum $\mathbf{p}$ of a particle as $m\mathbf{v}$, the product of its mass and its velocity, momentum is *not* conserved for all inertial observers.

We have two choices: (1) Give up the law of conservation of momentum. (2) See if we can redefine the momentum of a particle in some new way so that the law of conservation of momentum still holds. We choose the second of these alternatives.

Consider a particle moving with constant speed v in the x direction. Classically, its momentum is

$$p = mv = m\frac{\Delta x}{\Delta t} \quad \text{(momentum—classical)}, \quad (32)$$

in which Δx is the distance covered in time Δt. As we turn our thoughts to relativity, this classical definition is suspect from the beginning because it sets an upper limit ($= mc$) on the momentum that a particle may attain. If momentum is to be useful in relativity, it must be able to attain high values, without any discernible upper limit.

It seems clear that we must generalize the classical definition of momentum. There is no way to derive a generalized definition rigorously from a restricted one. We can only make an educated guess and then test it by seeing whether, using this new definition, momentum is conserved in collisions between particles.

The definition we propose is

$$p = m\frac{\Delta x}{\Delta t_0}.$$

Here, as before, Δx is the distance covered by a moving particle as viewed by an observer watching that particle. However, Δt_0 is the time to cover that distance, measured not by the observer watching the moving particle but by an observer moving with the particle. The particle is at rest with respect to this second observer so that the time that observer measures is a proper time Δt_0.

Using the time dilation formula (Eq. 8), we can then write

$$p = m\frac{\Delta x}{\Delta t_0} = m\frac{\Delta x}{\Delta t}\frac{\Delta t}{\Delta t_0} = m\frac{\Delta x}{\Delta t}\gamma.$$

But $\Delta x/\Delta t$ is just the particle velocity v so that

$$\boxed{p = \gamma mv} \quad \text{(momentum—relativistic)}. \quad (33)$$

Note that this differs from the classical definition of Eq. 32 only by the Lorentz factor γ. Unlike the classical definition, this relativistic definition permits the momentum p to approach infinitely large values as the particle speed v approaches the speed of light as a limiting value.

We can generalize the definition of Eq. 33 to vector

form as

$$\boxed{\mathbf{p} = \gamma m\mathbf{v}} \quad \text{(momentum—relativistic).} \quad (34)$$

We introduced this definition without elaboration in Section 9–4 as a foretaste of things to come; see Eq. 20 of Chapter 9. We state without further proof that, if we adopt the definition of momentum presented in Eq. 34, we can continue to use the conservation-of-momentum principle up to the very highest particle speeds.

Relativistic Mass. We can also write Eq. 33 in the form

$$p = m'v, \quad (35)$$

in which m', called the *relativistic mass* of the particle, is given by

$$\boxed{m' = \gamma m} \quad \text{(relativistic mass).} \quad (36)$$

To distinguish between the two masses, we now refer to m as the *rest mass* of the particle. As its name implies, the rest mass of a particle is the mass measured in a reference frame in which the particle is at rest. As Eq. 35 shows, we can carry over the classical definition of momentum by simply substituting the relativistic mass for the rest mass. The relativistic mass m' increases with velocity (because the Lorentz factor γ in Eq. 36 does so), approaching an infinite value as the speed of the particle approaches the speed of light.*

42–14 A New Look at Energy

In Section 7–7 we presented, without elaboration, the following relativistic expression for the kinetic energy of a particle:

$$K = mc^2 \left(\frac{1}{\sqrt{1 - (v/c)^2}} - 1 \right),$$

which we can now write as

$$\boxed{K = mc^2(\gamma - 1)} \quad \text{(relativistic kinetic energy).} \quad (37)$$

We showed in Section 7–7 that—unlikely as it may seem—this expression reduces to the familiar classical $K = \frac{1}{2}mv^2$ at low speeds. The derivation of Eq. 37 follows exactly the same path that we followed in deriving the classical kinetic energy expression ($K = \frac{1}{2}mv^2$) in Section 7–5 except that we use the relativistic, rather than the classical, expression for the momentum. In both the classical case and the relativistic case, we set the kinetic energy K equal to the work that must be done to accelerate the particle from rest to its observed speed.*

Let us point to some of the consequences of Eq. 37. We start by defining the *total energy E* of a particle as γmc^2. With the help of Eq. 37 we can then write

$$\boxed{\begin{aligned} E &= \gamma mc^2 \\ &= mc^2 + K \end{aligned}} \quad \begin{aligned} &\text{(total energy;} \\ &\text{single particle).} \end{aligned} \quad (38)$$

We interpret the total energy E of the moving particle as made up of mc^2, which we call the *rest energy* of the particle, and K, its kinetic energy. Table 1 of Chapter 8 lists the rest energies of a few particles and other objects. The rest energy of an electron, for example, is 0.511 MeV and for a proton it is 938.3 MeV.

The total energy of a system of particles is

$$\begin{aligned} E &= \sum E_i = \sum (\gamma mc^2) \\ &= \sum mc^2 + \sum K \end{aligned} \quad \begin{aligned} &\text{(total energy;} \\ &\text{system of particles).} \end{aligned} \quad (39)$$

In relativity, the *conservation of energy principle* is stated as follows:

> *For an isolated system of particles, the total energy E of the system, defined by Eq. 39, remains constant, no matter what interactions may occur among the particles.*

Thus, in any isolated reaction or decay process involving two or more particles, the total energy of the system after the interaction must remain equal to the total energy after the reaction. During the reaction, the total rest energy of the interacting particles may change but the total kinetic energy must then also change by an equal amount in the opposite direction to compensate.

We note from Eqs. 36 and 39 that the total energy E of a system of particles can also be written as $\sum(m'c^2)$, where m' is the relativistic mass. Thus, the conservation

---

* In treatments in which relativistic mass plays a central role, it is usually represented by m, the rest mass being then represented by m_0. In this book, m will always represent the rest mass.

* See the footnote reference on the opening page of this chapter for details of the proof.

Figure 13   The fireball of a nuclear explosion, a striking example of the conversion of (rest) mass into energy.

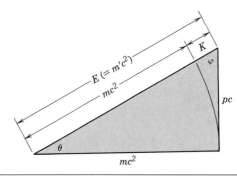

Figure 14   A useful mnemonic device for remembering the relativistic relations among the total energy E, the rest energy mc^2, the kinetic energy K, and the momentum p.

of total energy is equivalent to the conservation of relativistic mass. Therefore, although *rest* mass may not be conserved for an isolated system, *relativistic* mass is always conserved.

Considerations of this sort are at the root of Einstein's well-known $E = mc^2$ relation, which asserts that rest energy is freely convertible into other forms. All reactions — whether chemical or nuclear — in which energy is released or absorbed involve a corresponding change in the rest energy of the reactants; Fig. 13 is a particularly striking example. We discussed the $E = mc^2$ relation in detail in Section 8–9; see, in particular, Sample Problems 7, 8, and 9 in that Section.

Momentum and Kinetic Energy.   In classical mechanics, the momentum p of a particle is mv and its kinetic energy K is $\frac{1}{2}mv^2$. If we eliminate v between these two expressions, we find a direct relation between the momentum and the kinetic energy:

$$p^2 = 2Km \quad \text{(classical)}. \tag{40}$$

We can find a similar connection in relativity by eliminating v between the relativistic definition of momentum (Eq. 33) and the relativistic definition of kinetic energy (Eq. 37). Doing so leads, after some algebra, to

$$\boxed{(pc)^2 = K^2 + 2Kmc^2} \quad \text{(relativistic)}. \tag{41}$$

With the aid of Eq. 38, we can transform Eq. 41 into a relation between the momentum p and the total energy E of a particle, or

$$\boxed{E^2 = (pc)^2 + (mc^2)^2} \quad \text{(relativistic)}, \tag{42}$$

The right triangle of Fig. 14 helps to keep these useful relations in mind. You can also show that, in that triangle,

$$\sin \theta = \beta \quad \text{and} \quad \sin \phi = 1/\gamma. \tag{43}$$

Sample Problem 6   An electron has a kinetic energy K of 2.53 MeV. (a) What is its total energy E?

From Eq. 38 we have

$$E = mc^2 + K.$$

Earlier, we listed mc^2 for an electron as 0.5110 MeV so that

$$E = 0.511 \text{ MeV} + 2.53 \text{ MeV} = 3.04 \text{ MeV}. \quad \text{(Answer)}$$

(b) What is its momentum p?
From Eq. 42,

$$E^2 = (pc)^2 + (mc^2)^2,$$

we can write

$$(3.04 \text{ MeV})^2 = (pc)^2 + (0.511 \text{ MeV})^2$$

or

$$pc = \sqrt{(3.04 \text{ MeV})^2 - (0.511 \text{ MeV})^2}$$
$$= 3.00 \text{ MeV}.$$

It is customary to report momentum in particle physics in units of energy divided by c. Thus,

$$p = 3.00 \text{ MeV}/c. \quad \text{(Answer)}$$

(c) What is the electron's relativistic mass?
From Eqs. 36 and 38 we see that

$$E = \gamma mc^2 = m'c^2,$$

so that

$$m' = \frac{E}{c^2} = \frac{E}{mc^2} m$$

$$= \frac{3.04 \text{ MeV}}{0.511 \text{ MeV}} m = 5.95m. \qquad \text{(Answer)}$$

Thus, the electron's relativistic mass is 5.95 times its rest mass. From Eq. 36 we see that this is just the Lorentz factor ($\gamma = 5.95$).

Sample Problem 7 A proton is accelerated to 1.08 TeV in the Fermilab accelerator. (a) What is its speed parameter β?

Let us write Eq. 37 in the form

$$K = \gamma mc^2 - mc^2.$$

If we solve for γ we find

$$\gamma = \frac{K + mc^2}{mc^2} = \frac{K}{mc^2} + 1$$

$$= \frac{1.08 \times 10^{12} \text{ eV}}{938.3 \times 10^6 \text{ eV}} + 1 = 1151.0 + 1 = 1152.$$

This Lorentz factor γ ($= 1152$) is much greater than unity, indicating that relativity is absolutely vital in this problem. This is borne out by our observation that the kinetic energy of the moving proton (1.08×10^{12} eV) is more than 1000 times greater than its rest energy (938.3×10^6 eV).

Knowing γ, it would seem a straightforward matter to solve Eq. 9 for the speed parameter β. It simplifies our work greatly, however, to realize that β will be very close indeed to unity and that we would do well to solve not for β itself but for $1 - \beta$, its departure from unity.

Equation 9 then becomes

$$\gamma = \frac{1}{\sqrt{1 - \beta^2}} = \frac{1}{\sqrt{(1 - \beta)(1 + \beta)}} \approx \frac{1}{\sqrt{(2)(1 - \beta)}}.$$

We have here taken advantage of the fact that β is so close to unity that $1 + \beta$ is very close to 2.

If we solve the above equation for $1 - \beta$, we find

$$1 - \beta = \frac{1}{2\gamma^2} = \frac{1}{(2)(1152)^2} = 3.77 \times 10^{-7}.$$

Finally, we find for β

$$\beta = 1 - (1 - \beta) = 1 - 3.77 \times 10^{-7}$$

$$= 0.999\ 999\ 623. \qquad \text{(Answer)}$$

If you think that proceeding by approximation did not simplify the work, try solving the problem without making any approximation. You will overload your calculator.

If you have any lingering doubts about relativity theory, try solving this problem using the classical kinetic energy formula, $K = \frac{1}{2}mv^2$. You will find a speed 48 times the speed of light!

(b) How much slower than the speed of light is this high-energy proton moving?

We can write

$$\Delta v = c - v = c(1 - v/c) = c(1 - \beta).$$

Using data from (a), we find

$$\Delta v = c(1 - \beta) = (3.00 \times 10^8 \text{ m/s})(3.77 \times 10^{-7})$$

$$= 113 \text{ m/s}. \qquad \text{(Answer)}$$

42–15 The Common Sense of Relativity

We have come to a point from which we can look back and think about the common sense of relativity. We must start by agreeing that relativity permeates physics to its roots and that the theory has passed all experimental tests without the emergence of the slightest flaw. It is a theory that is aesthetically pleasing, inherently simple, comprehensively consistent, of great predictive value, and highly practical. If, for example, engineers were to design a high-energy particle accelerator ignoring relativity, the accelerator can be guaranteed not to work.

Are not the two postulates of the theory reasonable? Most of us willingly embrace the first postulate, the postulate of relativity, which gives a larger scope to a concept already familiar to Galileo. Einstein's second postulate, which asserts that there is in nature an ultimate limiting speed at which signals and energy can be transmitted from point to point, seems more reasonable the longer we think about it. If it were not true, it would then be possible to transmit signals instantaneously to all parts of the universe. It is really this assumption of instantaneous action-at-a-distance (which underlies classical physics) that is unreasonable. If there *is* a limiting speed, then from the principle of relativity, should it not be the same for all observers, regardless of their state of motion?

The relativity of simultaneity, the shrinking of moving rods, and the slowing down of moving clocks may disturb the rigidly classical mind; but are not all these phenomena based on measurement, following procedures derived in a reasonable way from the postulates? If the limiting speed c were only a lot smaller (perhaps 1000 mi/h), would not all these phenomena seem to be common sense of the highest order? Is not the relativity of the effects just what we would expect? That is, if A's clock seems to B to run slow, then do we not find it comfortably consistent that B's clock will seem to A to run slow? After all, who are A and B that we should be forced to choose between them?

Relativity broadens immeasurably our view of the world around us. In classical physics we have the separate laws of the conservation of mass and of energy. Is it not a great step forward to replace these by the single conservation law of total energy (or of relativistic mass)? There are other examples that we have not been able to explore fully. For example, in relativity space and time are linked together as spacetime and the electric field and the magnetic field are linked as aspects of a single electromagnetic field. The thoughtful student must conclude with us that, in relativity, we have the very model of a theory. We owe it all to Albert Einstein, who described well the nature of his contemplation of the world around him when he remarked: "I want to know God's thoughts . . . the rest are details." Figure 15 showed the tribute paid to Einstein by the cartoonist Herblock at the time of Einstein's death in 1955.

Figure 15 A cartoon by Herblock, published at Einstein's death in 1955.

REVIEW AND SUMMARY

The Postulates

Einstein's *special theory of relativity* is based on two postulates:

1. The law of physics are the same for observers in all inertial reference frames. No frame is singled out as preferred.
2. The speed of light in free space has the same value c in all directions and in all inertial reference frames.

The free-space speed of light c is an ultimate speed that cannot be exceeded by any entity carrying energy or information. See Sample Problem 1.

Coordinates of an Event

Three space coordinates and one time coordinate specify an *event*. One problem of special relativity is to relate these coordinates as assigned by two observers in uniform motion with respect to each other. Section 42–4 describes thought experiments by which a single observer might assign such coordinates.

Simultaneous Events

If two observers are in relative motion, they will not, in general, agree as to whether two events are simultaneous or not. If one observer finds two events at different locations to be simultaneous, the other will not, and conversely. Simultaneity is *not* an absolute concept but a relative one, depending on the motion of the observer. In Section 42–6, we show that this relativity of simultaneity is a direct consequence of the finite ultimate speed c.

Relativity of Time

If events occur at the same place in an inertial reference frame, the time interval Δt_0 is the *proper time* of the events. *All other observers will measure a larger value for this interval.* For an observer moving with speed v with respect to the original inertial frame, the measured time interval is

Time Dilation

$$\Delta t = \frac{\Delta t_0}{\sqrt{1 - (v/c)^2}} = \frac{\Delta t_0}{\sqrt{1 - \beta^2}} = \gamma \, \Delta t_0 \quad \text{(time dilation).} \qquad [7,8]$$

Here $\beta = v/c$ is the *speed parameter* and $\gamma = 1/\sqrt{1 - \beta^2}$ is the *Lorentz factor;* see Table 2. An

Relativity of Length

important consequence is that moving clocks run slow as measured by an observer at rest. See Sample Problem 2.

The length L_0 of a rod, measured in an inertial reference frame in which the rod is at rest, is called its *proper length. All other inertial observers will measure a shorter length.* For an observer moving with speed v with respect to the original inertial frame, the measured length is

Length Contraction

$$L = L_0\sqrt{1 - \beta^2} = \frac{L_0}{\gamma} \quad \text{(length contraction)}. \qquad [10]$$

See Sample Problem 3.

The Transformation Equations

The *Lorentz transformation equations* relate the coordinates of a single event as seen by two observers, S and S', with S' moving with respect to S with a velocity v in the positive x direction. The two perpendicular coordinates (y and z) are not affected. The x and t coordinates are related by

The Lorentz Transformation

$$
\begin{aligned}
x' &= \gamma(x - vt) & \text{(Lorentz transformation equations;} \\
t' &= \gamma(t - vx/c^2) & \text{valid at all speeds).}
\end{aligned}
\qquad [15]
$$

See Sample Problems 4 and 5.

Transformation of Velocities

A particle moving with speed v' in the positive x' direction in an inertial frame of reference S' that itself is moving with speed u parallel to the x direction of a second interial frame S will be measured in S as having speed

$$v = \frac{v' + u}{1 + uv'/c^2} \quad \text{(relativistic velocity law)}. \qquad [23]$$

If a source with frequency v_0 moves directly away from a detector with relative velocity $\mathbf{v}$, the frequency v measured by the detector is

Longitudinal Doppler Effect

$$v = v_0 \sqrt{\frac{1 - \beta}{1 + \beta}}. \qquad [26]$$

If the relative motion of the source is transverse, the Doppler formula is

Transverse Doppler Effect

$$v = v_0\sqrt{1 - \beta^2}. \qquad [30]$$

This *transverse Doppler effect* is entirely a consequence of time dilation.

Momentum and Energy

Presuming the validity of generalized laws of conservation of momentum and energy, the expressions for linear momentum $\mathbf{p}$, kinetic energy K, and total energy E that must be used for particles in relativistic dynamics are

$$\mathbf{p} = \gamma m\mathbf{v} = m'\mathbf{v} \quad \text{(momentum—relativistic)}, \qquad [34]$$

$$K = mc^2(\gamma - 1) \quad \text{(relativistic kinetic energy)}, \qquad [37]$$

and

$$E = \gamma mc^2 = mc^2 + K = m'c^2 \quad \text{(total energy, single particle)}. \qquad [38]$$

Here m is the *rest mass* of the particle and $m' = \gamma m$ is its *relativistic mass.* With these definitions, conservation of total energy for a system of particles takes the form

$$E = \sum (\gamma mc^2) = \sum mc^2 + \sum K \quad \text{(total energy, system of particles)}. \qquad [39]$$

Two additional relationships, derivable from the definitions, are often useful:

$$(pc)^2 = K^2 + 2Kmc^2 \quad \text{(relativistic)} \qquad [41]$$

and

$$E^2 = (pc)^2 + (mc^2)^2 \quad \text{(relativistic)}. \qquad [42]$$

See Sample Problems 6 and 7 for representative calculations.

QUESTIONS

1. How would you test a proposed reference frame to find out whether or not it is an inertial frame?

2. Give examples in which effects associated with the earth's rotation are significant enough in practice to rule out a laboratory frame as being a good enough approximation to an inertial frame.

3. The speed of light in a vacuum is a true constant of nature, independent of the wavelength of the light or the choice of an (inertial) reference frame. Is there any sense, then, in which Einstein's second postulate can be viewed as contained within the scope of his first postulate?

4. Discuss the problem that young Einstein grappled with; that is, what would be the appearance of an electromagnetic wave to a person running along with it at speed c?

5. Quasars (*quasi-stellar objects*) are the most intrinsically luminous objects in the universe. Many of them fluctuate in brightness, often on a time scale of a day or so. How can the rapidity of these brightness changes be used to estimate an upper limit to the size of these objects? (*Hint:* Separated points cannot change in a coordinated way unless information is sent from one to the other.)

6. The sweep rate of the tail of a comet can exceed the speed of light. Explain this phenomenon and show that there is no contradiction with relativity.

7. Borrowing two phrases from Herman Bondi, we can catch the spirit of Einstein's two postulates by labeling them: (1) the principle of "the irrelevance of velocity" and (2) the principle of "the uniqueness of light." In what senses are velocity irrelevant and light unique in these two statements?

8. A beam from a laser falls at right angles on a plane mirror and rebounds from it. What is the speed of the reflected beam if the mirror is (*a*) fixed in the laboratory? (*b*) Moving directly toward the laser with speed v?

9. Give an example from classical physics in which the motion of a clock affects its rate, that is, the way it runs. (The magnitude of the effect may depend on the detailed nature of the clock.)

10. Although in relativity (where motion is relative and not absolute) we find that "moving clocks run slow," this effect has nothing to do with the motion altering the way a clock works. What does it have to do with?

11. We have seen that if several observers watch two events, labeled A and B, one of them may say that event A occurred first but another may claim that it was event B that did so. What would you say to a friend who asked you which event *really did* occur first?

12. Two events occur at the same place and at the same time for one observer. Will they be simultaneous for all other observers? Will they also occur at the same place for all other observers?

13. Two observers, one at rest in S and one at rest in S', each carry a meter stick oriented parallel to their relative motion. *Each* observer finds upon measurement that the *other* observer's meter stick is shorter than his own meter stick. Does this seem like a paradox to you? Explain. (*Hint:* Compare the following situation. Harry waves goodbye to Walter who is in the rear of a station wagon driving away from Harry. Harry says that Walter gets smaller. Walter says that Harry gets smaller. Are they measuring the same thing?)

14. How does the concept of simultaneity enter into the measurement of the length of a body?

15. In relativity the time and space coordinates are intertwined and treated on a more or less equivalent basis. Are time and space fundamentally of the same nature, or is there some essential difference between them that is preserved even in relativity?

16. In the "twin paradox," explain (in terms of heartbeats, physical and mental activities, and so on) why the younger returning twin has not lived any longer than his own proper time even though his stay-at-home brother may say that he has. Hence, explain the remark: "You age according to your own proper time."

17. Can we simply substitute m' for m in classical equations to obtain the correct relativistic equations? Give examples.

18. If zero-mass particles have a speed c in one reference frame, can they be found at rest in any other frame? Can such particles have any speed other than c?

19. How many relativistic expressions can you think of in which the Lorentz factor γ enters as a simple multiplier?

20. In a given magnetic field, would a proton or an electron, traveling at the same speed, have the greater frequency of revolution? See Problem 57.

21. Is the rest mass of a stable, composite particle (a gold nucleus, for example) greater than, equal to, or less than the sum of the rest masses of its constituents? Explain.

22. "The rest mass of the electron is 0.511 MeV." Exactly what does this statement mean?

23. "The relation $E = mc^2$ is essential to the operation of a power plant based on nuclear fission but has only a negligible relevance for a fossil-fuel plant." Is this a true statement? Explain why or why not.

24. A hydroelectric plant generates electricity because water falls under gravity through a turbine, thereby turning the shaft of a generator. According to the mass–energy concept, must the appearance of energy (the electricity) be identified with a mass decrease somewhere? If so, where?

25. Exactly why is it that, kilogram for kilogram, nuclear explosions release so much more energy than do TNT explosions?

26. A hot metallic sphere cools off as it rests on the pan of a scale. If the scale were sensitive enough, would it indicate a change in rest mass? If so, woudl it be an increase or a decrease?

27. Some say that relativity complicates things. Give examples to the contrary, wherein relativity simplifies matters.

EXERCISES AND PROBLEMS

Section 42-3 The Postulates

1E. What fraction of the speed of light does each of the following speeds represent; that is, what is the speed parameter β? (a) A typical rate of continental drift (1 in. per year). (b) A typical drift speed for electrons in a current-carrying conductor (0.5 mm/s). (c) A highway speed limit of 55 mi/h. (d) The root-mean-square speed of a hydrogen molecule at room temperature. (e) A supersonic plane flying at Mach 2.5 (1200 km/h). (f) The escape speed of a projectile from the surface of the earth. (g) The speed of the earth in its orbit around the sun. (h) A typical recession speed of a distant quasar (3.0×10^4 km/s).

2E. Quite apart from effects due to the earth's rotational and orbital motions, a laboratory frame is not strictly an inertial frame because a particle placed at rest there will not, in general, remain at rest; it will fall under gravity. Often, however, events happen so quickly that we can ignore free fall and treat the frame as inertial. Consider, for example, a 1.0-MeV electron (for which $v = 0.992c$) projected horizontally into a laboratory test chamber and moving through a distance of 20 cm. (a) How long would it take, and (b) how far would the electron fall during this interval? What can you conclude about the suitability of the laboratory as an inertial frame in this case?

3P. Find the speed of a particle that takes 2 years longer than light to travel a distance of 6.0 ly.

Section 42-7 The Relativity of Time

4E. What must be the speed parameter β if the Lorentz factor γ is to be (a) 1.01? (b) 10? (c) 100? (d) 1000?

5E. The mean lifetime of muons stopped in a lead block in the laboratory is measured to be 2.2 μs. The mean lifetime of high-speed muons in a burst of cosmic rays observed from the earth is measured to be 16 μs. Find the speed of these cosmic ray muons.

6P. An unstable high-energy particle enters a detector and leaves a track 1.05 mm long before it decays. Its speed relative to the detector was $0.992c$. What is its proper lifetime? That is, how long would it have lasted before decay had it been at rest with respect to the detector?

7P. A pion is created in the higher reaches of the earth's atmosphere when an incoming high-energy cosmic-ray particle collides with an atomic nucleus. A pion so formed descends toward earth with a speed of $0.99c$. In a reference frame in which they are at rest, pions decay with a mean life of 26 ns. As

measured in a frame fixed with respect to the earth, how far (on the average) will such a typical pion move through the atmosphere before it decays?

8P. You wish to make a round trip from earth in a spaceship, traveling at constant speed in a straight line for six months and then returning at the same constant speed. You wish further, on your return, to find the earth as it will be a thousand years in the future. (a) How fast must you travel? (b) Does it matter whether or not you travel in a straight line on your journey? If, for example, you traveled in a circle for one year, would you still find that a thousand years had elapsed by earth clocks when you returned?

Section 42-8 The Relativity of Length

9E. A rod lies parallel to the x axis of reference frame S, moving along this axis at a speed of $0.63c$. Its rest length is 1.7 m. What will be its measured length in frame S?

10E. The length of a spaceship is measured to be exactly half its rest length. (a) What is the speed of the spaceship relative to the observer's frame? (b) By what factor do the spaceship's clocks run slow, compared to clocks in the observer's frame?

11E. A 100-MeV electron, for which $\beta = 0.999987$, moves along the axis of an evacuated tube that has a length of 3.0 m as measured by a laboratory observer S with respect to whom the tube is at rest. An observer S' moving with the electron, however, would see this tube moving past with speed $v (= \beta c)$. What length would this observer measure for the tube?

12E. The rest radius of the earth is 6370 km and its orbital speed about the sun is 30 km/s. By how much would the earth's diameter appear to be shortened to an observer stationed so as to be able to watch the earth move past him at this speed?

13E. A spaceship of rest length 130 m drifts past a timing station at a speed of $0.74c$. (a) What is the length of the spaceship as measured by the timing station? (b) What time interval between the passage of the front and back end of the ship will the station monitor record?

14P. A space traveler takes off from earth and moves at speed $0.99c$ toward the star Vega, which is 26 ly distant. How much time will have elapsed by earth clocks (a) when the traveler reaches Vega? (b) when the earth observers receive word from him that he has arrived? (c) How much older will the earth observers calculate the traveler to be when he reaches Vega than he was when he started the trip?

15P. An airplane whose rest length is 40 m is moving at a

uniform velocity with respect to the earth at a speed of 630 m/s. (*a*) By what fraction of its rest length will it appear to be shortened to an observer on earth? (*b*) How long would it take by earth clocks for the airplane's clock to fall behind by 1 μs? (Assume that only special relativity applies.)

16P. (*a*) Can a person, in principle, travel from earth to the galactic center (which is about 23,000 ly distant) in a normal lifetime? Explain, using either time-dilation or length-contraction arguments. (*b*) What constant speed would be needed to make the trip in 30 y (proper time)?

Section 42–10 Some Consequences of the Lorentz Equations

17E. Observer *S* assigns the following spacetime coordinates to an event:

$$x = 100 \text{ km}, \quad t = 200 \text{ } \mu s.$$

What are the coordinates of this event in frame *S'*, which moves in the direction of increasing *x* with speed 0.95*c*? Assume *x* = *x'* at *t* = *t'* = 0.

18E. Observer *S* reports that an event occurred on his *x* axis at *x* = 3.0 × 10⁸ m at a time *t* = 2.50 s. (*a*) Observer *S'* is moving in the direction of increasing *x* at a speed of 0.40*c*. What coordinates would he report for the event? (*b*) What coordinates would he report if he were moving in the direction of *decreasing x* at this same speed?

19E. Inertial frame *S'* moves at a speed of 0.60*c* with respect to frame *S*. Two events are recorded. In frame *S*, event 1 occurs at the origin at *t* = 0 and event 2 occurs on the *x* axis at *x* = 3.0 km and at *t* = 4.0 μs. What times of occurrence does observer *S'* record for these same events? Explain the difference in the time order.

20E. An experimenter arranges to trigger two flashbulbs simultaneously, a blue flash located at the origin of his reference frame and a red flash at *x* = 30 km. A second observer, moving at a speed of 0.25*c* in the direction of increasing *x*, also views the flashes. (*a*) What time interval between them does she find? (*b*) Which flash does she say occurs first?

21E. In Table 3 the Lorentz transformation equations in the right-hand column can be derived from those in the left-hand column simply by (1) exchanging primed and unprimed quantities and (2) changing the sign of *v*. Verify this procedure by deriving one set of equations directly from the other by algebraic manipulation.

22P. A clock moves along the *x* axis at a speed of 0.60*c* and reads zero as it passes the origin. (*a*) Calculate the Lorentz factor. (*b*) What time does the clock read as it passes *x* = 180 m?

23P. An observer *S* sees a flash of red light 1200 m from his position and a flash of blue light 720 m closer to him and on the same straight line. He measures the time interval between the occurrence of the flashes to be 5.00 μs, the red flash occurring

first. (*a*) What is the relative velocity **v** (magnitude and direction) of a second observer *S'* who would record these flashes as occurring at the same place? (*b*) From the point of view of *S'*, which flash occurs first? (*c*) What time interval between them would *S'* measure?

24P. In Problem 23, observer *S* sees the two flashes in the same positions, but they now occur closer together in time. How close together in time can they be and still have it possible to find a frame *S'* in which they occur at the same place?

Section 42–11 The Transformation of Velocities

25E. A particle moves along the *x'* axis of frame *S'* with a speed of 0.40*c*. Frame *S'* moves with a speed of 0.60*c* with respect to frame *S*. What is the measured speed of the particle in frame *S*?

26E. Frame *S'* moves relative to frame *S* at 0.62*c* in the direction of increasing *x*. In frame *S'* a particle is measured to have a velocity of 0.47*c* in the direction of increasing *x'*. (*a*) What is the velocity of the particle with respect to frame *S*? (*b*) What would be the velocity of the particle with respect to *S* if it moved (at 0.47*c*) in the direction of *decreasing x'* in the *S'* frame? In each case, compare your answers with the predictions of the classical velocity transformation equation.

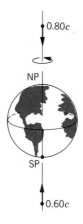

Figure 16 Exercise 27.

27E. One cosmic-ray particle approaches the earth along its axis with a velocity of 0.80*c* toward the North Pole and another, with a velocity 0.60*c*, toward the South Pole. See Fig. 16. What is the relative speed of approach of one particle with respect to the other? (*Hint:* It is useful to consider the earth and one of the particles as the two inertial reference frames.)

28E. Galaxy A is reported to be receding from us with a speed of 0.35*c*. Galaxy B, located in precisely the opposite direction, is also found to be receding from us at this same speed. What recessional speed would an observer on Galaxy A find (*a*) for our galaxy? (*b*) for Galaxy B?

29E. It is concluded from measurements of the red shift of the emitted light that quasar Q_1 is moving away from us at a speed of $0.80c$. Quasar Q_2, which lies in the same direction in space but is closer to us, is moving away from us at speed $0.40c$. What velocity for Q_2 would be measured by an observer on Q_1?

30P. A spaceship whose rest length is 350 m has a speed of $0.82c$ with respect to a certain reference frame. A micrometeorite, also with a speed of $0.82c$ in this frame, passes the spaceship on an antiparallel track. How long does it take this object to pass the spaceship?

31P. To circle the earth in low orbit a satellite must have a speed of about 17,000 mi/h. Suppose that two such satellites orbit the earth in opposite directions. (*a*) What is their relative speed as they pass? Evaluate using the classical Galilean velocity transformation equation. (*b*) What fractional error was made because the (correct) relativistic transformation equation was not used?

32P. A spaceship, at rest in a certain reference frame S, is given a speed increment of $0.50c$. It is then given a further $0.50c$ increment in this new frame, and this process is continued until its speed with respect to its original frame S exceeds $0.999c$. How many increments does it require?

Section 42–12   The Doppler Effect

33E. A spaceship, moving away from the earth at a speed of $0.90c$, reports back by transmitting on a frequency (measured in the spaceship frame) of 100 MHz. To what frequency must earth receivers be tuned to receive these signals?

34E. In the spectrum of quasar 3C9, some of the familiar hydrogen lines appear but they are shifted so far toward the red that their wavelengths are observed to be three times as large as that observed in the light from hydrogen atoms at rest in the laboratory. (*a*) Show that the classical Doppler equation gives a velocity of recession greater than c. (*b*) Assuming that the relative motion of 3C9 and the earth is entirely one of recession, find the recession speed predicted by the relativistic Doppler equation.

35E. Give the Doppler wavelength shift $\lambda - \lambda_0$, if any, for the sodium D_2 line (589.00 nm) emitted from a source moving in a circle with constant speed ($= 0.10c$) as measured by an observer fixed at the center of the circle.

36P. A spaceship is receding from the earth at a speed of $0.20c$. A light on the rear of the ship appears blue ($\lambda = 450$ nm) to passengers on the ship. What color would it appear to an observer on earth?

37P. A radar transmitter T is fixed to a reference frame S' that is moving to the right with speed v relative to reference frame S (see Fig. 17). A mechanical timer (essentially a clock) in frame S', having a period τ_0 (measured in S') causes transmitter T to emit radar pulses, which travel at the speed of light and are received by R, a receiver fixed in frame S. (*a*) What would be the period τ of the timer relative to observer A, who is

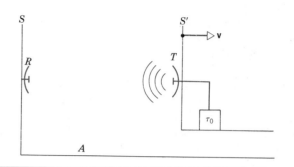

Figure 17   Problem 37.

fixed in frame S? (*b*) Show that the receiver R would observe the time interval between pulses arriving from T, not as τ or as τ_0, but as

$$\tau_R = \tau_0 \sqrt{\frac{c + v}{c - v}}.$$

(*c*) Explain why the observer at R measures a different period for the transmitter than does observer A, who is in the same reference frame. (*Hint:* A clock and a radar pulse are not the same.)

Section 42–14   A New Look at Energy

38E. How much work must be done to increase the speed of an electron from rest (*a*) to $0.50c$? (*b*) to $0.990c$? (*c*) to $0.9990c$?

39E. An electron is moving at a speed such that it could circumnavigate the earth at the equator in one second. (*a*) What is its speed, in terms of the speed of light? (*b*) Its kinetic energy K? (*c*) What percent error do you make if you use the classical formula to calculate K?

40E. Find the speed parameter β and the Lorentz factor γ for an electron whose kinetic energy is (*a*) 1.0 keV; (*b*) 1.0 MeV; (*c*) 1.0 GeV.

41E. Find the speed parameter β and the Lorentz factor γ for a particle whose kinetic energy is 10 MeV if the particle is (*a*) an electron; (*b*) a proton; (*c*) an alpha particle.

42E. Verify the statement in Sample Problem 5, Chapter 32, that a 100-MeV electron travels at 99.9987% of the speed of light.

43E. A particle has a speed of $0.99c$ in a laboratory reference frame. What are its kinetic energy, its total energy, and its momentum if the particle is (*a*) a proton or (*b*) an electron?

44E. The United States consumed about 2.2×10^{12} kWh of electrical energy in 1979. How much matter would have to vanish to account for the generation of this energy? Does it make any difference to your answer if this energy is generated in oil-burning, nuclear, or hydroelectric plants?

45E. Quasars are thought to be the nuclei of active galaxies in the early stages of their formation. A typical quasar radiates energy at the rate of 10^{41} W. At what rate is the mass of this quasar being reduced to supply this energy? Express your answer in solar mass units per year, where one solar mass unit (smu = 2×10^{30} kg) is the mass of our sun.

46P. How much work must be done to increase the speed of an electron from (a) $0.18c$ to $0.19c$? (b) $0.98c$ to $0.99c$? Note that the speed increase (= $0.01c$) is the same in each case.

47P. What is the speed of a particle (a) whose kinetic energy is equal to twice its rest energy? (b) whose total energy is equal to twice its rest energy?

48P. (a) What potential difference would accelerate an electron to the speed of light, according to classical physics? (b) With this potential difference, what speed would the electron actually attain?

49P. A particle has a momentum equal to mc. (a) What is its speed? (b) Its relativistic mass? (c) Its kinetic energy?

50P. What must be the momentum of a particle with rest mass m in order that its total energy be three times its rest energy?

51P. Consider the following, all moving in free space: a 2.0-eV photon, a 0.40-MeV electron, and a 10-MeV proton. (a) Which is moving the fastest? (b) The slowest? (c) Which has the greatest momentum? (d) The least? (*Note:* A photon is a light-particle of zero rest mass.)

52P. A 5-grain aspirin tablet has a mass of 320 mg. For how many miles would the energy equivalent of this mass, in the form of gasoline, power an automobile? Assume 30 mi/gal and a heat of combustion of 1.3×10^8 J/gal for the gasoline.

53P. (a) If the kinetic energy K and the momentum p of a particle can be measured, it should be possible to find its rest mass m and thus identify the particle. Show that

$$m = \frac{(pc)^2 - K^2}{2Kc^2}.$$

(b) Show that this expression reduces to an expected result as $u/c \rightarrow 0$, in which u is the speed of the particle. (c) Find the rest mass of a particle whose kinetic energy is 55.0 MeV and whose momentum is 121 MeV/c; express your answer in terms of the rest mass of the electron.

54P. In a high-energy collision of a primary cosmic-ray particle near the top of the earth's atmosphere, 120 km above sea level, a pion is created with a total energy E of 1.35×10^5 MeV, traveling vertically downward. In its proper frame this pion decays 35 ns after its creation. At what altitude above sea level does the decay occur? The rest energy of a pion is 139.6 MeV.

55P. The average lifetime of muons at rest is 2.2 μs. A laboratory measurement on the decay in flight of the muons in a beam emerging from a particle accelerator yields an average lifetime of 6.9 μs. what is: (a) The speed of these muons in the laboratory? (b) The relativistic mass (in terms of m_e, the rest

mass of the electron)? (c) The kinetic energy? (d) The momentum? The rest mass of a muon is 207 times that of an electron.

56P. (a) How much energy is released in the explosion of a fission bomb containing 3.0 kg of fissionable material? Assume that 0.10 percent of the rest mass is converted to released energy. (b) What mass of TNT would have to explode to provide the same energy release? Assume that each mole of TNT liberates 3.4 MJ of energy on exploding. The molecular weight of TNT is 0.227 kg/mol. (c) For the same mass of explosive, how much more effective are nuclear explosions than TNT explosions? That is, compare the fractions of the rest mass that are converted to energy in each case.

57P. In Section 5 of Chapter 30 we showed that a particle of charge q and mass m moving with speed v perpendicular to a uniform magnetic field B moves in a circle of radius r given by (see Eq. 16, Chapter 30)

$$r = \frac{mv}{qB}.$$

Also, it was demonstrated that the period T of the circular motion is independent of the speed of the particle. Now, these results hold only if $v \ll c$. For particles moving faster, the radius of the circular path can be shown to be

$$r = \frac{p}{qB} = \frac{m(\gamma v)}{qB} = \frac{mv}{qB\sqrt{1 - \beta^2}}.$$

This equation is valid at all speeds. Compute the radius of the path of a 10-MeV electron moving perpendicular to a uniform 2.2-T magnetic field. Use both the (a) "classical" and (b) "relativistic" formulas. (c) Calculate the true period of the circular motion. Is the result independent of the speed of the electron?

58P. Ionization measurements show that a particular nuclear particle carries a double charge (= $2e$) and is moving with a speed of $0.71c$. Its measured radius of curvature in a magnetic field of 1.00 T is 6.28 m. Find the rest mass of the particle and identify it. (*Hint:* Light nuclear particles are made up of neutrons [which carry no charge] and protons [charge = $+e$], in roughly equal numbers. Take the rest mass of either of these particles to be 1.00 u. See Problem 57.)

59P. A 10-GeV proton in the cosmic radiation approaches the earth in the plane of its geomagnetic equator, in a region over which the earth's average magnetic field is 55 μT. What is the radius of its curved path in that region?

60P. A 2.50-MeV electron moves at right angles to a magnetic field in a path whose radius of curvature is 3.0 cm. What is the magnetic field B?

61P. The proton synchrotron at Fermilab accelerates protons to a kinetic energy of 500 GeV. At this energy, calculate (a) the Lorentz factor, (b) the speed parameter, and (c) the magnetic field at the proton orbit that has a radius of curvature of 750 m. (The proton has a rest energy of 938.3 MeV.)

APPENDIX A
THE INTERNATIONAL SYSTEM OF UNITS (SI)*

1. The SI Base Units

Quantity	Name	Symbol	Definition
length	meter	m	". . . the length of the path traveled by light in vacuum in 1/299,792,458 of a second." (1983)
mass	kilogram	kg	". . . this prototype [a certain platinum-iridium cylinder] shall henceforth be considered to be the unit of mass." (1889)
time	second	s	". . . the duration of 9,192,631,770 periods of the radiation corresponding to the transition between the two hyperfine levels of the ground state of the cesium-133 atom." (1967)
electric current	ampere	A	". . . that constant current which, if maintained in two straight parallel conductors of infinite length, of negligible circular cross section, and placed 1 meter apart in vacuum, would produce between these conductors a force equal to 2×10^{-7} newton per meter of length." (1946)
thermodynamic temperature	kelvin	K	". . . the fraction 1/273.16 of the thermodynamic temperature of the triple point of water." (1967)
amount of substance	mole	mol	". . . the amount of substance of a system which contains as many elementary entities as there are atoms in 0.012 kilogram of carbon 12." (1971)
luminous intensity	candela	cd	". . . the luminous intensity, in the perpendicular direction, of a surface of 1/600,000 square meter of a blackbody at the temperature of freezing platinum under a pressure of 101.325 newton per square meter." (1967)

---

* Adapted from " The International System of Units (SI)," National Bureau of Standards Special Publication 330, 1972 edition. The definitions above were adopted by the General Conference of Weights and Measures, an international body, on the dates shown. In this book we do not use the candela.

2. Some SI Derived Units

Quantity	Name of Unit	Symbol	
area	square meter	m^2	
volume	cubic meter	m^3	
frequency	hertz	Hz	s^{-1}
mass density (density)	kilogram per cubic meter	kg/m^3	
speed, velocity	meter per second	m/s	
angular velocity	radian per second	rad/s	
acceleration	meter per second squared	m/s^2	
angular acceleration	radian per second squared	rad/s^2	
force	newton	N	$kg \cdot m/s^2$
pressure	pascal	Pa	N/m^2
work, energy, quantity of heat	joule	J	$N \cdot m$
power	watt	W	J/s
quantity of electricity	coulomb	C	$A \cdot s$
potential difference, electromotive force	volt	V	W/A
electric field strength	volt per meter	V/m	
electric resistance	ohm	Ω	V/A
capacitance	farad	F	$A \cdot s/V$
magnetic flux	weber	Wb	$V \cdot s$
inductance	henry	H	$V \cdot s/A$
magnetic flux density	tesla	T	Wb/m^2
magnetic field strength	ampere per meter	A/m	
entropy	joule per kelvin	J/K	
specific heat capacity	joule per kilogram kelvin	$J/(kg \cdot K)$	
thermal conductivity	watt per meter kelvin	$W/(m \cdot K)$	
radiant intensity	watt per steradian	W/sr	

3. The SI Supplementary Units

Quantity	Name of Unit	Symbol
plane angle	radian	rad
solid angle	steradian	sr

APPENDIX B

SOME FUNDAMENTAL CONSTANTS OF PHYSICS

Constant	Symbol	Computational Value	Best (1986) value Value[a]	Best (1986) value Uncertainty[b]
Speed of light in a vacuum	c	3.00×10^8 m/s	2.99792458	exact
Elementary charge	e	1.60×10^{-19} C	1.60217738	0.30
Electron rest mass	m_e	9.11×10^{-31} kg	9.1093897	0.59
Permittivity constant	ϵ_0	8.85×10^{-12} F/m	8.85418781762	exact
Permeability constant	μ_0	1.26×10^{-6} H/m	1.25663706143	exact
Electron rest mass[c]	m_e	5.49×10^{-4} u	5.48579902	0.023
Neutron rest mass[c]	m_n	1.0087 u	1.008664704	0.014
Hydrogen atom rest mass[c]	m_{1_H}	1.0078 u	1.007825035	0.011
Deuterium atom rest mass[c]	m_{2_H}	2.0141 u	2.0141019	0.053
Helium atom rest mass[c]	$m_{4_{He}}$	4.0026 u	4.0026032	0.067
Electron charge to mass ratio	e/m_e	1.76×10^{11} C/kg	1.75881961	0.30
Proton rest mass	m_p	1.67×10^{-27} kg	1.6726230	0.59
Ratio of proton mass to electron mass	m_p/m_e	1840	1836.152701	0.020
Neutron rest mass	m_n	1.68×10^{-27} kg	1.6749286	0.59
Muon rest mass	m_μ	1.88×10^{-28} kg	1.8835326	0.61
Planck constant	h	6.63×10^{-34} J·s	6.6260754	0.60
Electron Compton wavelength	λ_c	2.43×10^{-12} m	2.42631058	0.089
Universal gas constant	R	8.31 J/mol·K	8.314510	8.4

Constant	Symbol	Computational Value	Best (1986) value	
			Value[a]	Uncertainty[b]
Avogadro constant	N_A	6.02×10^{23} mol^{-1}	6.0221367	0.59
Boltzmann constant	k	1.38×10^{-23} J/K	1.380657	11
Molar volume of ideal gas at STP[d]	V_m	2.24×10^{-2} m^3/mol	2.241409	8.4
Faraday constant	F	9.65×10^4 C/mol	9.6485309	0.30
Stefan–Boltzmann constant	σ	5.67×10^{-8} W/m$^2 \cdot$K^4	5.67050	34
Rydberg constant	R	1.10×10^7 m^{-1}	1.0973731534	0.0012
Gravitational constant	G	6.67×10^{-11} m^3/s$^2 \cdot$kg	6.67260	100
Bohr radius	r_B	5.29×10^{-11} m	5.29177249	0.045
Electron magnetic moment	μ_e	9.28×10^{-24} J/T	9.2847700	0.34
Proton magnetic moment	μ_P	1.41×10^{-26} J/T	1.41060761	0.34
Bohr magneton	μ_B	9.27×10^{-24} J/T	9.2740154	0.34
Nuclear magneton	μ_N	5.05×10^{-27} J/T	5.0507865	0.34

[a] Same unit and power of ten as the computational value.

[b] Parts per million.

[c] Mass given in unified atomic mass units, where 1 u $= 1.6605402 \times 10^{-27}$ kg.

[d] STP—standard temperature and pressure $= 0°$C and 1.0 bar.

* The values in this table were largely selected from a longer list in *Symbols, Units and Nomenclature in Physics* (IUPAP), prepared by E. Richard Cohen and Pierre Giacomo, 1986.

APPENDIX C

SOME ASTRONOMICAL DATA

Some Distances from the Earth

To the moon*	3.82×10^8 m
To the sun*	1.50×10^{11} m
To the nearest star (Proxima Centauri)	4.04×10^{16} m
To the center of our galaxy	2.2×10^{20} m
To the Andromeda Galaxy	2.1×10^{22} m
To the edge of the observable universe	$\sim 10^{26}$ m

* Mean distance.

The Sun, the Earth and the Moon

Property	Unit	Sun[a]	Earth	Moon
Mass	kg	1.99×10^{30}	5.98×10^{24}	7.36×10^{22}
Mean radius	m	6.96×10^8	6.37×10^6	1.74×10^6
Mean density	kg/m³	1410	5520	3340
Surface gravity	m/s²	274	9.81	1.67
Escape velocity	km/s	618	11.2	2.38
Period of rotation[c]	—	37 d—poles[b] 26 d—equator	23 h 56 min	27.3 d

[a] The sun radiates energy at the rate of 3.90×10^{26} W; just outside the earth's atmosphere solar energy is received, assuming normal incidence, at the rate of 1340 W/m².
[b] The sun—a ball of gas—does not rotate as a rigid body.
[c] Measured with respect to the distant stars.

Some Properties of the Planets

	Mercury	Venus	Earth	Mars	Jupiter	Saturn	Uranus	Neptune	Pluto
Mean distance from sun, 10^6 km	57.9	108	150	228	778	1,430	2,870	4,500	5,900
Period of revolution, y	0.241	0.615	1.00	1.88	11.9	29.5	84.0	165	248
Period of rotation[a], d	58.7	−243[b]	0.997	1.03	0.409	0.426	−0.451[b]	0.658	6.39
Orbital speed, km/s	47.9	35.0	29.8	24.1	13.1	9.64	6.81	5.43	4.74
Inclination of axis to orbit	<28°	~3°	23.5°	24.0°	3.08°	26.7°	82.1°	28.8°	?
Inclination of orbit to earth's orbit	7.00°	3.39°	—	1.85°	1.30°	2.49°	0.77°	1.77°	17.2°
Eccentricity of orbit	0.206	0.0068	0.0167	0.0934	0.0485	0.0556	0.0472	0.0086	0.250
Equatorial diameter, km	4,880	12,100	12,800	6,790	143,000	120,000	51,800	49,500	3,000(?)
Mass (earth = 1)	0.0558	0.815	1.000	0.107	318	95.1	14.5	17.2	0.01(?)
Density (water = 1)	5.60	5.20	5.52	3.95	1.31	0.704	1.21	1.67	?
Surface gravity[c], m/s^2	3.78	8.60	9.78	3.72	22.9	9.05	7.77	11.0	0.3(?)
Escape velocity[c], km/s	4.3	10.3	11.2	5.0	59.5	35.6	21.2	23.6	0.9(?)
Known satellites	0	0	1	2	16 + ring	17 + rings	15 + rings	2 + rings(?)	1

[a] Measured with respect to the distant stars.
[b] Tne sense of rotation is opposite to that of the orbital motion
[c] Measured at the planet's equator.

APPENDIX D

PROPERTIES OF THE ELEMENTS

Element	Symbol	Atomic number, Z	Atomic mass, g/mol	Density, g/cm³ at 20°C	Melting point, °C	Boiling point, °C	Specific heat, J/(g·C°) at 25°C
Actinium	Ac	89	(227)	—	1323	(3473)	0.092
Aluminum	Al	13	26.9815	2.699	660	2450	0.900
Americium	Am	95	(243)	11.7	1541	—	—
Antimony	Sb	51	121.75	6.62	630.5	1380	0.205
Argon	Ar	18	39.948	1.6626×10^{-3}	−189.4	−185.8	0.523
Arsenic	As	33	74.9216	5.72	817 (28 at.)	613	0.331
Astatine	At	85	(210)	—	(302)	—	—
Barium	Ba	56	137.34	3.5	729	1640	0.205
Berkelium	Bk	97	(247)	—	—	—	—
Beryllium	Be	4	9.0122	1.848	1287	2770	1.83
Bismuth	Bi	83	208.980	9.80	271.37	1560	0.122
Boron	B	5	10.811	2.34	2030	—	1.11
Bromine	Br	35	79.909	3.12 (liquid)	−7.2	58	0.293
Cadmium	Cd	48	112.40	8.65	321.03	765	0.226
Calcium	Ca	20	40.08	1.55	838	1440	0.624
Californium	Cf	98	(251)	—	—	—	—
Carbon	C	6	12.01115	2.25	3727	4830	0.691
Cerium	Ce	58	140.12	6.768	804	3470	0.188
Cesium	Cs	55	132.905	1.9	28.40	690	0.243
Chlorine	Cl	17	35.453	3.214×10^{-3} (0°C)	−101	−34.7	0.486
Chromium	Cr	24	51.996	7.19	1857	2665	0.448
Cobalt	Co	27	58.9332	8.85	1495	2900	0.423
Copper	Cu	29	63.54	8.96	1083.40	2595	0.385
Curium	Cm	96	(247)	—	—	—	—
Dysprosium	Dy	66	162.50	8.55	1409	2330	0.172
Einsteinium	Es	99	(254)	—	—	—	—
Erbium	Er	68	167.26	9.15	1522	2630	0.167
Europium	Eu	63	151.96	5.245	817	1490	0.163
Fermium	Fm	100	(257)	—	—	—	—
Fluorine	F	9	18.9984	1.696×10^{-3} (0°C)	−219.6	−188.2	0.753
Francium	Fr	87	(223)	—	(27)	—	—
Gadolinium	Gd	64	157.25	7.86	1312	2730	0.234
Gallium	Ga	31	69.72	5.907	29.75	2237	0.377
Germanium	Ge	32	72.59	5.323	937.25	2830	0.322
Gold	Au	79	196.967	19.32	1064.43	2970	0.131

Element	Symbol	Atomic number, Z	Atomic mass, g/mol	Density, g/cm³ at 20°C	Melting point, °C	Boiling point, °C	Specific heat, J/(g·C°) at 25°C
Hafnium	Hf	72	178.49	13.09	2227	5400	0.144
Helium	He	2	4.0026	0.1664×10^{-3}	-269.7	-268.9	5.23
Holmium	Ho	67	164.930	8.79	1470	2330	0.165
Hydrogen	H	1	1.00797	0.08375×10^{-3}	-259.19	-252.7	14.4
Indium	In	49	114.82	7.31	156.634	2000	0.233
Iodine	I	53	126.9044	4.94	113.7	183	0.218
Iridium	Ir	77	192.2	22.5	2447	(5300)	0.130
Iron	Fe	26	55.847	7.87	1536.5	3000	0.447
Krypton	Kr	36	83.80	3.488×10^{-3}	-157.37	-152	0.247
Lanthanum	La	57	138.91	6.189	920	3470	0.195
Lawrencium	Lw	103	(257)	—	—	—	—
Lead	Pb	82	207.19	11.36	327.45	1725	0.129
Lithium	Li	3	6.939	0.534	180.55	1300	3.58
Lutetium	Lu	71	174.97	9.849	1663	1930	0.155
Magnesium	Mg	12	24.312	1.74	650	1107	1.03
Manganese	Mn	25	54.9380	7.43	1244	2150	0.481
Mendelevium	Md	101	(256)	—	—	—	—
Mercury	Hg	80	200.59	13.55	-38.87	357	0.138
Molybdenum	Mo	42	95.94	10.22	2617	5560	0.251
Neodymium	Nd	60	144.24	7.00	1016	3180	0.188
Neon	Ne	10	20.183	0.8387×10^{-3}	-248.597	-246.0	1.03
Neptunium	Np	93	(237)	19.5	637	—	1.26
Nickel	Ni	28	58.71	8.902	1453	2730	0.444
Niobium	Nb	41	92.906	8.57	2468	4927	0.264
Nitrogen	N	7	14.0067	1.1649×10^{-3}	-210	-195.8	1.03
Nobelium	No	102	(255)	—	—	—	—
Osmium	Os	76	190.2	22.57	3027	5500	0.130
Oxygen	O	8	15.9994	1.3318×10^{-3}	-218.80	-183.0	0.913
Palladium	Pd	46	106.4	12.02	1552	3980	0.243
Phosphorus	P	15	30.9738	1.83	44.25	280	0.741
Platinum	Pt	78	195.09	21.45	1769	4530	0.134
Plutonium	Pu	94	(244)	—	640	3235	0.130
Polonium	Po	84	(210)	9.24	254	—	—
Potassium	K	19	39.102	0.86	63.20	760	0.758
Praseodymium	Pr	59	140.907	6.769	931	3020	0.197
Promethium	Pm	61	(145)	—	(1027)	—	—
Protactinium	Pa	91	(231)	—	(1230)	—	—
Radium	Ra	88	(226)	5.0	700	—	—
Radon	Rn	86	(222)	9.96×10^{-3} (0°C)	(-71)	-61.8	0.092
Rhenium	Re	75	186.2	21.04	3180	5900	0.134
Rhodium	Rh	45	102.905	12.44	1963	4500	0.243
Rubidium	Rb	37	85.47	1.53	39.49	688	0.364
Ruthenium	Ru	44	101.107	12.2	2250	4900	0.239
Samarium	Sm	62	150.35	7.49	1072	1630	0.197
Scandium	Sc	21	44.956	2.99	1539	2730	0.569
Selenium	Se	34	78.96	4.79	221	685	0.318
Silicon	Si	14	28.086	2.33	1412	2680	0.712
Silver	Ag	47	107.870	10.49	960.8	2210	0.234
Sodium	Na	11	22.9898	0.9712	97.85	892	1.23
Strontium	Sr	38	87.62	2.60	768	1380	0.737

Element	Symbol	Atomic number, Z	Atomic mass, g/mol	Density, g/cm³ at 20°C	Melting point, °C	Boiling point, °C	Specific heat, J/(g·C°) at 25°C
Sulfur	S	16	32.064	2.07	119.0	444.6	0.707
Tantalum	Ta	73	180.948	16.6	3014	5425	0.138
Technetium	Tc	43	(99)	11.46	2200	—	0.209
Tellurium	Te	52	127.60	6.24	449.5	990	0.201
Terbium	Tb	65	158.924	8.25	1357	2530	0.180
Thallium	Tl	81	204.37	11.85	304	1457	0.130
Thorium	Th	90	(232)	11.66	1755	(3850)	0.117
Thulium	Tm	69	168.934	9.31	1545	1720	0.159
Tin	Sn	50	118.69	7.2984	231.868	2270	0.226
Titanium	Ti	22	47.90	4.507	1670	3260	0.523
Tungsten	W	74	183.85	19.3	3380	5930	0.134
Uranium	U	92	(238)	19.07	1132	3818	0.117
Vanadium	V	23	50.942	6.1	1902	3400	0.490
Xenon	Xe	54	131.30	5.495×10^{-3}	−111.79	−108	0.159
Ytterbium	Yb	70	173.04	6.959	824	1530	0.155
Yttrium	Y	39	88.905	4.472	1526	3030	0.297
Zinc	Zn	30	65.37	7.133	419.58	906	0.389
Zirconium	Zr	40	91.22	6.489	1852	3580	0.276

The values in parentheses in the column of atomic masses are the mass numbers of the longest-lived isotopes of those elements that are radioactive. Melting points and boiling points in parentheses are uncertain.

All the physical properties are given for a pressure of one atmosphere except where otherwise specified.

The data for gases are valid only when these are in their usual molecular state, such as H_2, He, O_2, Ne, etc. The specific heats of the gases are the values at constant pressure.

Source: Adapted from Wehr, Richards, Adair, *Physics of the Atom,* 4th ed., Addison-Wesley, Reading, MA, 1984.

APPENDIX E

PERIODIC TABLE OF THE ELEMENTS

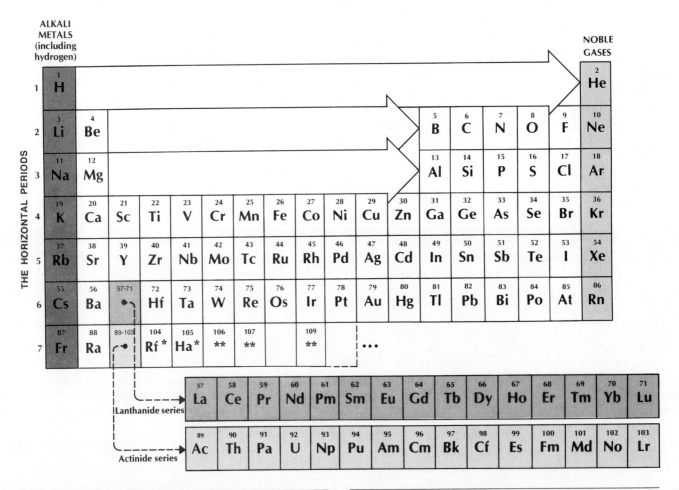

* The names of these elements (Rutherfordium and Hahnium) have not been accepted because of conflicting claims of discovery. A group in the USSR has proposed the names Kurchatovium and Neilsbohrium.

** Discovery of these three elements has been reported but names for them have not yet been proposed.

APPENDIX F

CONVERSION FACTORS

Conversion factors may be read off directly from the tables. For example, 1 degree = 2.778×10^{-3} revolutions, so $16.7° = 16.7 \times 2.778 \times 10^{-3}$ rev. The SI quantities are capitalized.

Adapted in part from G. Shortley and D. Williams, *Elements of Physics,* Prentice-Hall, Englewood Cliffs, NJ, 1971.

Plane Angle

	°	′	″	RADIAN	rev
1 degree =	1	60	3600	1.745×10^{-2}	2.778×10^{-3}
1 minute =	1.667×10^{-2}	1	60	2.909×10^{-4}	4.630×10^{-5}
1 second =	2.778×10^{-4}	1.667×10^{-2}	1	4.848×10^{-6}	7.716×10^{-7}
1 RADIAN =	57.30	3438	2.063×10^5	1	0.1592
1 revolution =	360	2.16×10^4	1.296×10^6	6.283	1

Solid Angle

1 sphere = 4π steradians = 12.57 steradians

Length

	cm	METER	km	in.	ft	mi
1 centimeter =	1	10^{-2}	10^{-5}	0.3937	3.281×10^{-2}	6.214×10^{-6}
1 METER =	100	1	10^{-3}	39.37	3.281	6.214×10^{-4}
1 kilometer =	10^5	1000	1	3.937×10^4	3281	0.6214
1 inch =	2.540	2.540×10^{-2}	2.540×10^{-5}	1	8.333×10^{-2}	1.578×10^{-5}
1 foot =	30.48	0.3048	3.048×10^{-4}	12	1	1.894×10^{-4}
1 mile =	1.609×10^5	1609	1.609	6.336×10^4	5280	1

1 angström = 10^{-10} m
1 nautical mile = 1852 m
  = 1.151 miles = 6076 ft
1 fermi = 10^{-15} m

1 light-year = 9.460×10^{12} km
1 parsec = 3.084×10^{13} km
1 fathom = 6 ft
1 Bohr radius = 5.292×10^{-11} m

1 yard = 3 ft
1 rod = 16.5 ft
1 mil = 10^{-3} in.
1 nm = 10^{-9} m

Area

	METER²	cm²	ft²	in.²
1 SQUARE METER =	1	10^4	10.76	1550
1 square centimeter =	10^{-4}	1	1.076×10^{-3}	0.1550
1 square foot =	9.290×10^{-2}	929.0	1	144
1 square inch =	6.452×10^{-4}	6.452	6.944×10^{-3}	1

1 square mile = 2.788×10^7 ft² = 640 acres   1 acre = 43,560 ft²
1 barn = 10^{-28} m²   1 hectare = 10^4 m² = 2.471 acre

Volume

	METER³	cm³	L	ft³	in.³
1 CUBIC METER =	1	10^6	1000	35.31	6.102×10^4
1 cubic centimeter =	10^{-6}	1	1.000×10^{-3}	3.531×10^{-5}	6.102×10^{-2}
1 liter =	1.000×10^{-3}	1000	1	3.531×10^{-2}	61.02
1 cubic foot =	2.832×10^{-2}	2.832×10^4	28.32	1	1728
1 cubic inch =	1.639×10^{-5}	16.39	1.639×10^{-2}	5.787×10^{-4}	1

1 U.S. fluid gallon = 4 U.S. fluid quarts = 8 U.S. pints = 128 U.S. fluid ounces = 231 in.³
1 British imperial gallon = 277.4 in³ = 1.201 U.S. fluid gallons

Mass

Quantities in the colored areas are not mass units but are often used as such. When we write, for example, 1 kg "=" 2.205 lb this means that a kilogram is a *mass* that *weighs* 2.205 pounds under standard condition of gravity (g = 9.80665 m/s²).

	g	KILOGRAM	slug	u	oz	lb	ton
1 gram =	1	0.001	6.852×10^{-5}	6.022×10^{23}	3.527×10^{-2}	2.205×10^{-3}	1.102×10^{-6}
1 KILOGRAM =	1000	1	6.852×10^{-2}	6.022×10^{26}	35.27	2.205	1.102×10^{-3}
1 slug =	1.459×10^4	14.59	1	8.786×10^{27}	514.8	32.17	1.609×10^{-2}
1 u =	1.661×10^{-24}	1.661×10^{-27}	1.138×10^{-28}	1	5.857×10^{-26}	3.662×10^{-27}	1.830×10^{-30}
1 ounce =	28.35	2.835×10^{-2}	1.943×10^{-3}	1.718×10^{25}	1	6.250×10^{-2}	3.125×10^{-5}
1 pound =	453.6	0.4536	3.108×10^{-2}	2.732×10^{26}	16	1	0.0005
1 ton =	9.072×10^5	907.2	62.16	5.463×10^{29}	3.2×10^4	2000	1

1 metric ton = 1000 kg

Density

Quantities in the colored areas are weight densities and, as such, are dimensionally different from mass densities. See note for mass table.

	slug/ft³	KILOGRAM/ METER³	g/cm³	lb/ft³	lb/in.³
1 slug per ft³ =	1	515.4	0.5154	32.17	1.862×10^{-2}
1 KILOGRAM per METER³ =	1.940×10^{-3}	1	0.001	6.243×10^{-2}	3.613×10^{-5}
1 gram per cm³ =	1.940	1000	1	62.43	3.613×10^{-2}
1 pound per ft³ =	3.108×10^{-2}	16.02	1.602×10^{-2}	1	5.787×10^{-4}
1 pound per in.³ =	53.71	2.768×10^4	27.68	1728	1

Time

	y	d	h	min	SECOND
1 year =	1	365.25	8.766×10^3	5.259×10^5	3.156×10^7
1 day =	2.738×10^{-3}	1	24	1440	8.640×10^4
1 hour =	1.141×10^{-4}	4.167×10^{-2}	1	60	3600
1 minute =	1.901×10^{-6}	6.944×10^{-4}	$1.66i \times 10^{-2}$	1	60
1 SECOND =	3.169×10^{-8}	1.157×10^{-5}	2.778×10^{-4}	1.667×10^{-2}	1

Speed

	ft/s	km/h	METER/ SECOND	mi/h	cm/s
1 foot per second =	1	1.097	0.3048	0.6818	30.48
1 kilometer per hour =	0.9113	1	0.2778	0.6214	27.78
1 METER per SECOND =	3.281	3.6	1	2.237	100
1 mile per hour =	1.467	1.609	0.4470	1	44.70
1 centimeter per second =	3.281×10^{-2}	3.6×10^{-2}	0.01	2.237×10^{-2}	1

1 knot = 1 nautical mi/h = 1.688 ft/s   1 mi/min = 88.00 ft/s = 60.00 mi/h

Force

Quantities in the colored areas are not force units but are often used as such. For instance, if we write 1 gram-force "=" 980.7 dynes, we mean that a gram-mass experiences a force of 980.7 dynes under standard conditions of gravity ($g = 9.80665$ m/s²)

	dyne	NEWTON	lb	pdl	gf	kgf
1 dyne =	1	10^{-5}	2.248×10^{-6}	7.233×10^{-5}	1.020×10^{-3}	1.020×10^{-6}
1 NEWTON =	10^5	1	0.2248	7.233	102.0	0.1020
1 pound =	4.448×10^5	4.448	1	32.17	453.6	0.4536
1 poundal =	1.383×10^4	0.1383	3.108×10^{-2}	1	14.10	1.410×10^{-2}
1 gram-force =	980.7	9.807×10^{-3}	2.205×10^{-3}	7.093×10^{-2}	1	0.001
1 kilogram-force =	9.807×10^5	9.807	2.205	70.93	1000	1

Pressure

	atm	dyne/cm²	inch of water	cm Hg	PASCAL	lb/in.²	lb/ft²
1 atmosphere =	1	1.013×10^6	406.8	76	1.013×10^5	14.70	2116
1 dyne per cm² =	9.869×10^{-7}	1	4.015×10^{-4}	7.501×10^{-5}	0.1	1.405×10^{-5}	2.089×10^{-3}
1 inch of water[a] at 4°C =	2.458×10^{-3}	2491	1	0.1868	249.1	3.613×10^{-2}	5.202
1 centimeter of mercury[a] at 0°C =	1.316×10^{-2}	1.333×10^4	5.353	1	1333	0.1934	27.85
1 PASCAL =	9.869×10^{-6}	10	4.015×10^{-3}	7.501×10^{-4}	1	1.450×10^{-4}	2.089×10^{-2}
1 pound per in.² =	6.805×10^{-2}	6.895×10^4	27.68	5.171	6.895×10^3	1	144
1 pound per ft² =	4.725×10^{-4}	478.8	0.1922	3.591×10^{-2}	47.88	6.944×10^{-3}	1

[a] Where the acceleration of gravity has the standard value 9.80665 m/s².

1 bar = 10^6 dyne/cm² = 0.1 MPa   1 millibar = 10^3 dyne/cm² = 10^2 Pa   1 torr = 1 millimeter of mercury

Energy, Work, Heat

Quantities in the colored areas are not properly energy units but are included for convenience. They arise from the relativistic mass-energy equivalence formula $E = mc^2$ and represent the energy released if a kilogram or unified atomic mass unit (u) is completely converted to energy.

	Btu	erg	ft·lb	hp·h	JOULE	cal	kW·h	eV	MeV	kg	u
1 British thermal unit =	1	1.055×10^{10}	777.9	3.929×10^{-4}	1055	252.0	2.930×10^{-4}	6.585×10^{21}	6.585×10^{15}	1.174×10^{-14}	7.070×10^{12}
1 erg =	9.481×10^{-11}	1	7.376×10^{-8}	3.725×10^{-14}	10^{-7}	2.389×10^{-8}	2.778×10^{-14}	6.242×10^{11}	6.242×10^5	1.113×10^{-24}	670.2
1 foot-pound =	1.285×10^{-3}	1.356×10^7	1	5.051×10^{-7}	1.356	0.3238	3.766×10^{-7}	8.464×10^{18}	8.464×10^{12}	1.509×10^{-17}	9.037×10^9
1 horsepower-hour =	2545	2.685×10^{13}	1.980×10^6	1	2.685×10^6	6.413×10^5	0.7457	1.676×10^{25}	1.676×10^{19}	2.988×10^{-11}	1.799×10^{16}
1 JOULE =	9.481×10^{-4}	10^7	0.7376	3.725×10^{-7}	1	0.2389	2.778×10^{-7}	6.242×10^{18}	6.242×10^{12}	1.113×10^{-17}	6.702×10^9
1 calorie =	3.969×10^{-3}	4.186×10^7	3.088	1.560×10^{-6}	4.186	1	1.163×10^{-6}	2.613×10^{19}	2.613×10^{13}	4.660×10^{-17}	2.806×10^{10}
1 kilowatt-hour =	3413	3.6×10^{13}	2.655×10^6	1.341	3.6×10^6	8.600×10^5	1	2.247×10^{25}	2.247×10^{19}	4.007×10^{-11}	2.413×10^{16}
1 electron volt =	1.519×10^{-22}	1.602×10^{-12}	1.182×10^{-19}	5.967×10^{-26}	1.602×10^{-19}	3.827×10^{-20}	4.450×10^{-26}	1	10^{-6}	1.783×10^{-36}	1.074×10^{-9}
1 million electron volts =	1.519×10^{-16}	1.602×10^{-6}	1.182×10^{-13}	5.967×10^{-20}	1.602×10^{-13}	3.827×10^{-14}	4.450×10^{-20}	10^6	1	1.783×10^{-30}	1.074×10^{-3}
1 kilogram =	8.521×10^{13}	8.987×10^{23}	6.629×10^{16}	3.348×10^{10}	8.987×10^{16}	2.146×10^{16}	2.497×10^{10}	5.610×10^{35}	5.610×10^{29}	1	6.022×10^{26}
1 unified atomic mass unit =	1.415×10^{-13}	1.492×10^{-3}	1.101×10^{-10}	5.559×10^{-17}	1.492×10^{-10}	3.564×10^{-11}	4.146×10^{-17}	9.32×10^8	932.0	1.661×10^{-27}	1

Power

	Btu/h	ft·lb/s	hp	cal/s	kW	WATT
1 British thermal unit per hour =	1	0.2161	3.929×10^{-4}	6.998×10^{-2}	2.930×10^{-4}	0.2930
1 foot-pound per second =	4.628	1	1.818×10^{-3}	0.3239	1.356×10^{-3}	1.356
1 horsepower =	2545	550	1	178.1	0.7457	745.7
1 calorie per second =	14.29	3.088	5.615×10^{-3}	1	4.186×10^{-3}	4.186
1 kilowatt =	3413	737.6	1.341	238.9	1	1000
1 WATT =	3.413	0.7376	1.341×10^{-3}	0.2389	0.001	1

Magnetic Flux

	maxwell	WEBER
1 maxwell =	1	10^{-8}
1 WEBER =	10^{8}	1

Magnetic Field

	gauss	TESLA	milligauss
1 gauss =	1	10^{-4}	1000
1 TESLA =	10^{4}	1	10^{7}
1 milligauss =	0.001	10^{-7}	1

$1 \text{ tesla} = 1 \text{ weber/meter}^2$

APPENDIX G
MATHEMATICAL FORMULAS

Geometry

Circle of radius r: circumference $= 2\pi r$; area $= \pi r^2$.

Sphere of radius r: area $= 4\pi r^2$; volume $= \frac{4}{3}\pi r^3$.

Right circular cylinder of radius r and height h: area $= 2\pi r^2 + 2\pi rh$; volume $= \pi r^2 h$.

Triangle of base a and altitude h: area $= \frac{1}{2}ah$.

Quadratic Formula

If $ax^2 + bx + c = 0$, then $x = \dfrac{-b \pm \sqrt{b^2 - 4ac}}{2a}$.

Trigonometric Functions of Angle θ

$\sin\theta = \dfrac{y}{r}$ $\quad \cos\theta = \dfrac{x}{r}$

$\tan\theta = \dfrac{y}{x}$ $\quad \cot\theta = \dfrac{x}{y}$

$\sec\theta = \dfrac{r}{x}$ $\quad \csc\theta = \dfrac{r}{y}$

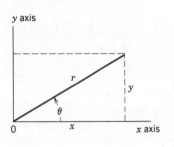

Pythagorean Theorem

$a^2 + b^2 = c^2$

Triangles

Angles A, B, C

Opposite sides a, b, c

$A + B + C = 180°$

$\dfrac{\sin A}{a} = \dfrac{\sin B}{b} = \dfrac{\sin C}{c}$

$c^2 = a^2 + b^2 - 2ab \cos C$

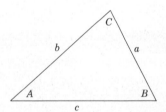

Mathematical Signs and Symbols

$=$ equals
$\approx$ equals approximately
$\neq$ is not equal to
$\equiv$ is identical to, is defined as
$>$ is greater than ($\gg$ is much greater than)
$<$ is less than ($\ll$ is much less than)
$\geq$ is greater than or equal to (or, is no less than)
$\leq$ is less than or equal to (or, is no more than)
$\pm$ plus or minus ($\sqrt{4} = \pm 2$)
$\propto$ is proportional to
Σ the sum of
$\bar{x}$ the average value of x

Trigonometric Identities

$\sin(90° - \theta) = \cos \theta$
$\cos(90° - \theta) = \sin \theta$
$\sin \theta / \cos \theta = \tan \theta$
$\sin^2 \theta + \cos^2 \theta = 1 \quad \sec^2 \theta - \tan^2 \theta = 1 \quad \csc^2 \theta - \cot^2 \theta = 1$
$\sin 2\theta = 2 \sin \theta \cos \theta$
$\cos 2\theta = \cos^2 \theta - \sin^2 \theta = 2 \cos^2 \theta - 1 = 1 - 2 \sin^2 \theta$
$\sin(\alpha \pm \beta) = \sin \alpha \cos \beta \pm \cos \alpha \sin \beta$
$\cos(\alpha \pm \beta) = \cos \alpha \cos \beta \mp \sin \alpha \sin \beta$

$$\tan(\alpha \pm \beta) = \frac{\tan \alpha \pm \tan \beta}{1 \mp \tan \alpha \tan \beta}$$

$\sin \alpha \pm \sin \beta = 2 \sin \tfrac{1}{2}(\alpha \pm \beta) \cos \tfrac{1}{2}(\alpha \mp \beta)$

Binomial Theorem

$$(1 \pm x)^n = 1 \pm \frac{nx}{1!} + \frac{n(n-1)}{2!}x^2 + \cdots \quad (x^2 < 1)$$

$$(1 \pm x)^{-n} = 1 \mp \frac{nx}{1!} + \frac{n(n+1)x^2}{2!} + \cdots \quad (x^2 < 1)$$

Exponential Expansion

$$e^x = 1 + x + \frac{x^2}{2!} + \frac{x^3}{3!} + \cdots$$

Logarithmic Expansion

$$\ln(1 + x) = x - \tfrac{1}{2}x^2 + \tfrac{1}{3}x^3 - \cdots \quad (|x| < 1)$$

Trigonometric Expansions (θ in radians)

$$\sin \theta = \theta - \frac{\theta^3}{3!} + \frac{\theta^5}{5!} - \cdots$$

$$\cos \theta = 1 - \frac{\theta^2}{2!} + \frac{\theta^4}{4!} - \cdots$$

$$\tan \theta = \theta + \frac{\theta^3}{3} + \frac{2\theta^5}{15} + \cdots$$

Products of Vectors

Let $\mathbf{i}$, $\mathbf{j}$, $\mathbf{k}$ be unit vectors in the x, y, z directions. Then

$$\mathbf{i} \cdot \mathbf{i} = \mathbf{j} \cdot \mathbf{j} = \mathbf{k} \cdot \mathbf{k} = 1, \quad \mathbf{i} \cdot \mathbf{j} = \mathbf{j} \cdot \mathbf{k} = \mathbf{k} \cdot \mathbf{i} = 0,$$

$$\mathbf{i} \times \mathbf{i} = \mathbf{j} \times \mathbf{j} = \mathbf{k} \times \mathbf{k} = 0,$$

$$\mathbf{i} \times \mathbf{j} = \mathbf{k}, \quad \mathbf{j} \times \mathbf{k} = \mathbf{i}, \quad \mathbf{k} \times \mathbf{i} = \mathbf{j}.$$

Any vector $\mathbf{a}$ with components a_x, a_y, a_z along the x, y, z axes can be written

$$\mathbf{a} = a_x\mathbf{i} + a_y\mathbf{j} + a_z\mathbf{k}.$$

Let $\mathbf{a}$, $\mathbf{b}$, $\mathbf{c}$ be arbitrary vectors with magnitudes a, b, c. Then

$$\mathbf{a} \times (\mathbf{b} + \mathbf{c}) = (\mathbf{a} \times \mathbf{b}) + (\mathbf{a} \times \mathbf{c})$$

$$(s\mathbf{a}) \times \mathbf{b} = \mathbf{a} \times (s\mathbf{b}) = s(\mathbf{a} \times \mathbf{b}) \quad (s = \text{a scalar}).$$

Let θ be the smaller of the two angles between $\mathbf{a}$ and $\mathbf{b}$. Then

$$\mathbf{a} \cdot \mathbf{b} = \mathbf{b} \cdot \mathbf{a} = a_x b_x + a_y b_y + a_z b_z = ab \cos \theta$$

$$\mathbf{a} \times \mathbf{b} = -\mathbf{b} \times \mathbf{a} = \begin{vmatrix} \mathbf{i} & \mathbf{j} & \mathbf{k} \\ a_x & a_y & a_z \\ b_x & b_y & b_z \end{vmatrix} = (a_y b_z - b_y a_z)\mathbf{i}$$

$$+ (a_z b_x - b_z a_x)\mathbf{j} + (a_x b_y - b_x a_y)\mathbf{k}$$

$$|\mathbf{a} \times \mathbf{b}| = ab \sin \theta$$

$$\mathbf{a} \cdot (\mathbf{b} \times \mathbf{c}) = \mathbf{b} \cdot (\mathbf{c} \times \mathbf{a}) = \mathbf{c} \cdot (\mathbf{a} \times \mathbf{b})$$

$$\mathbf{a} \times (\mathbf{b} \times \mathbf{c}) = (\mathbf{a} \cdot \mathbf{c})\mathbf{b} - (\mathbf{a} \cdot \mathbf{b})\mathbf{c}$$

Derivatives and Integrals

In what follows, the letters u and v stand for any functions of x, and a and m are constants. To each of the indefinite integrals should be added an arbitrary constant of integration. The *Handbook of Chemistry and Physics* (CRC Press Inc.) gives a more extensive tabulation.

1. $\dfrac{dx}{dx} = 1$

1. $\displaystyle\int dx = x$

2. $\dfrac{d}{dx}(au) = a\dfrac{du}{dx}$

2. $\displaystyle\int au\, dx = a\int u\, dx$

3. $\dfrac{d}{dx}(u + v) = \dfrac{du}{dx} + \dfrac{dv}{dx}$

3. $\displaystyle\int (u + v)\, dx = \int u\, dx + \int v\, dx$

4. $\dfrac{d}{dx}x^m = mx^{m-1}$

4. $\displaystyle\int x^m\, dx = \dfrac{x^{m+1}}{m+1}\quad (m \neq -1)$

5. $\dfrac{d}{dx}\ln x = \dfrac{1}{x}$

5. $\displaystyle\int \dfrac{dx}{x} = \ln |x|$

6. $\dfrac{d}{dx}(uv) = u\dfrac{dv}{dx} + v\dfrac{du}{dx}$

6. $\displaystyle\int u\dfrac{dv}{dx}\, dx = uv - \int v\dfrac{du}{dx}\, dx$

7. $\dfrac{d}{dx}e^x = e^x$

7. $\displaystyle\int e^x\, dx = e^x$

8. $\dfrac{d}{dx}\sin x = \cos x$

8. $\displaystyle\int \sin x\, dx = -\cos x$

9. $\dfrac{d}{dx}\cos x = -\sin x$

9. $\displaystyle\int \cos x\, dx = \sin x$

10. $\dfrac{d}{dx}\tan x = \sec^2 x$

10. $\displaystyle\int \tan x\, dx = \ln|\sec x|$

11. $\dfrac{d}{dx}\cot x = -\csc^2 x$

11. $\displaystyle\int \sin^2 x\, dx = \tfrac{1}{2}x - \tfrac{1}{4}\sin 2x$

12. $\dfrac{d}{dx}\sec x = \tan x \sec x$

12. $\displaystyle\int e^{-ax}\, dx = -\dfrac{1}{a}e^{-ax}$

13. $\dfrac{d}{dx}\csc x = -\cot x \csc x$

13. $\displaystyle\int xe^{-ax}\, dx = -\dfrac{1}{a^2}(ax + 1)e^{-ax}$

14. $\dfrac{d}{dx}e^u = e^u\dfrac{du}{dx}$

14. $\displaystyle\int x^2 e^{-ax}\, dx = -\dfrac{1}{a^3}(a^2x^2 + 2ax + 2)e^{-ax}$

15. $\dfrac{d}{dx}\sin u = \cos u\dfrac{du}{dx}$

15. $\displaystyle\int_0^\infty x^n e^{-ax}\, dx = \dfrac{n!}{a^{n+1}}$

16. $\dfrac{d}{dx}\cos u = -\sin u\dfrac{du}{dx}$

16. $\displaystyle\int_0^\infty x^{2n} e^{-ax^2}\, dx = \dfrac{1\cdot 3\cdot 5\,\cdots\,(2n-1)}{2^{n+1}a^n}\sqrt{\dfrac{\pi}{a}}$

APPENDIX H

WINNERS OF THE NOBEL PRIZE IN PHYSICS*

1901	Wilhelm Konrad Röntgen	1845–1923	for the discovery of x-rays
1902	Hendrik Antoon Lorentz	1853–1928	for their researches into the influence of magnetism upon radiation
	Pieter Zeeman	1865–1943	phenomena
1903	Antoine Henri Becquerel	1852–1908	for his discovery of spontaneous radioactivity
	Pierre Curie	1859–1906	for their joint researches on the radiation phenomena discovered
	Marie Sklowdowska-Curie	1867–1934	by Professor Henri Becquerel
1904	Lord Rayleigh (John William Strutt)	1842–1919	for his investigations of the densities of the most important gases and for his discovery of argon
1905	Philipp Eduard Anton von Lenard	1862–1947	for his work on cathode rays
1906	Joseph John Thomson	1856–1940	for his theoretical and experimental investigations on the conduction of electricity by gases
1907	Albert Abraham Michelson	1852–1931	for his optical precision instruments and metrological investigations carried out with their aid
1908	Gabriel Lippmann	1845–1921	for his method of reproducing colors photographically based on the phenomena of interference
1909	Guglielmo Marconi	1874–1937	for their contributions to the development
	Carl Ferdinand Braun	1850–1918	of wireless telegraphy
1910	Johannes Diderik van der Waals	1837–1932	for his work on the equation of state for gases and liquids
1911	Wilhelm Wien	1864–1928	for his discoveries regarding the laws governing the radiation of heat
1912	Nils Gustaf Dalén	1869–1937	for his invention of automatic regulators for use in conjunction with gas accumulators for illuminating lighthouses and buoys
1913	Heike Kamerlingh Onnes	1853–1926	for his investigations of the properties of matter at low temperatures which led, *inter alia,* to the production of liquid helium
1914	Max von Laue	1879–1960	for his discovery of the diffraction of Röntgen rays by crystals
1915	William Henry Bragg	1862–1942	for their services in the analysis of crystal structure by means of
	William Lawrence Bragg	1890–1971	x-rays
1917	Charles Glover Barkla	1877–1944	for his discovery of the characteristic x-rays of the elements
1918	Max Planck	1858–1947	for his discovery of energy quanta
1919	Johannes Stark	1874–1957	for his discovery of the Doppler effect in canal rays and the splitting of spectral lines in electric fields
1920	Charles-Édouard Guillaume	1861–1938	for the service he has rendered to precision measurements in Physics by his discovery of anomalies in nickel steel alloys
1921	Albert Einstein	1879–1955	for his services to Theoretical Physics, and especially for his

* See *Nobel Lectures, Physics,* 1901–1970, Elsevier Publishing Company for biographies of the awardees and for lectures given by them on receiving the prize.

			discovery of the law of the photoelectric effect
1922	Niels Bohr	1885–1962	for the investigation of the structure of atoms, and of the radiation emanating from them
1923	Robert Andrews Millikan	1868–1953	for his work on the elementary charge of electricity and on the photoelectric effect
1924	Karl Manne Georg Siegbahn	1888–1979	for his discoveries and research in the field of x-ray spectroscopy
1925	James Franck	1882–1964	for their discovery of the laws governing the impact of an electron upon an atom
	Gustav Hertz	1887–1975	
1926	Jean Baptiste Perrin	1870–1942	for his work on the discontinuous structure of matter, and especially for his discovery of sedimentation equilibrium
1927	Arthur Holly Compton	1892–1962	for his discovery of the effect named after him
	Charles Thomson Rees Wilson	1869–1959	for his method of making the paths of electrically charged particles visible by condensation of vapor
1928	Owen Willans Richardson	1879–1959	for his work on the thermionic phenomenon and especially for the discovery of the law named after him
1929	Prince Louis-Victor de Broglie	1892–1987	for his discovery of the wave nature of electrons
1930	Sir Chandrasekhara Venkata Raman	1888–1970	for his work on the scattering of light and for the discovery of the effect named after him
1932	Werner Heisenberg	1901–1976	for the creation of quantum mechanics, the application of which has, among other things, led to the discovery of the allotropic forms of hydrogen
1933	Erwin Schrödinger	1887–1961	for the discovery of new productive forms of atomic theory
	Paul Adrien Maurice Dirac	1902–1984	
1935	James Chadwick	1891–1974	for his discovery of the neutron
1936	Victor Franz Hess	1883–1964	for the discovery of cosmic radiation
	Carl David Anderson	1905–1984	for his discovery of the positron
1937	Clinton Joseph Davisson	1881–1958	for their experimental discovery of the diffraction of electrons by crystals
	George Paget Thomson	1892–1975	
1938	Enrico Fermi	1901–1954	for his demonstrations of the existence of new radioactive elements produced by neutron irradiation, and for his related discovery of nuclear reactions brought about by slow neutrons
1939	Ernest Orlando Lawrence	1901–1958	for the invention and development of the cyclotron and for results obtained with it, especially for artificial radioactive elements
1943	Otto Stern	1888–1969	for his contribution to the development of the molecular ray method and his discovery of the magnetic moment of the proton
1944	Isidor Isaac Rabi	1898–1988	for his resonance method for recording the magnetic properties of atomic nuclei
1945	Wolfgang Pauli	1900–1958	for the discovery of the Exclusion Principle (Pauli Principle)
1946	Percy Williams Bridgman	1882–1961	for the invention of an apparatus to produce extremely high pressures, and for the discoveries he made therewith in the field of high-pressure physics
1947	Sir Edward Victor Appleton	1892–1965	for his investigations of the physics of the upper atmosphere, especially for the discovery of the so-called Appleton layer
1948	Patrick Maynard Stuart Blackett	1897–1974	for his development of the Wilson cloud chamber method, and his discoveries therewith in nuclear physics and cosmic radiation
1949	Hideki Yukawa	1907–1981	for his prediction of the existence of mesons on the basis of theoretical work on nuclear forces
1950	Cecil Frank Powell	1903–1969	for his development of the photographic method of studying nuclear processes and his discoveries regarding mesons made with this method
1951	Sir John Douglas Cockcroft	1897–1967	for their pioneer work on the transmutation of atomic nuclei by artificially accelerated atomic particles
	Ernest Thomas Sinton Walton	1903–	
1952	Felix Bloch	1905–1983	for their development of new methods for nuclear magnetic precision methods and discoveries in connection therewith
	Edward Mills Purcell	1912–	

1953	Frits Zernike	1888–1966	for his demonstration of the phase-contrast method, especially for his invention of the phase-contrast microscope
1954	Max Born	1882–1970	for his fundamental research in quantum mechanics, especially for his statistical interpretation of the wave function
	Walther Bothe	1891–1957	for the coincidence method and his discoveries made therewith
1955	Willis Eugene Lamb	1913–	for his discoveries concerning the fine structure of the hydrogen spectrum
	Polykarp Kusch	1911–	for his precision determination of the magnetic moment of the electron
1956	William Shockley	1910–	for their researches on semiconductors and their discovery of the transistor effect
	John Bardeen	1908–	
	Walter Houser Brattain	1902–1987	
1957	Chen Ning Yang	1922–	for their penetrating investigation of the parity laws which has led to important discoveries regarding the elementary particles
	Tsung Dao Lee	1926–	
1958	Pavel Aleksejevič Čerenkov	1904–	for the discovery and the interpretation of the Cerenkov effect
	Il' ja Michajlovič Frank	1908–	
	Igor' Evgen' evič Tamm	1895–1971	
1959	Emilio Gino Segrè	1905–	for their discovery of the antiproton
	Owen Chamberlain	1920–	
1960	Donald Arthur Glaser	1926–	for the invention of the bubble chamber
1961	Robert Hofstadter	1915–	for his pioneering studies of electron scattering in atomic nuclei and for his thereby achieved discoveries concerning the structure of the nucleons
	Rudolf Ludwig Mössbauer	1929–	for his researches concerning the resonance absorption of γ-rays and his discovery in this connection of the effect which bears his name
1962	Lev Davidovič Landau	1908–1968	for his pioneering theories of condensed matter, especially liquid helium
1963	Eugene P. Wigner	1902–	for his contributions to the theory of the atomic nucleus and the elementary particles, particularly through the discovery and application of fundamental symmetry principles
	Maria Goeppert Mayer	1906–1972	for their discoveries concerning nuclear shell structure
	J. Hans D. Jensen	1907–1973	
1964	Charles H. Townes	1915–	for fundamental work in the field of quantum electronics which has led to the construction of oscillators and amplifiers based on the maser-laser principle
	Nikolai G. Basov	1922–	
	Alexander M. Prochorov	1916–	
1965	Sin-itiro Tomonaga	1906–1979	for their fundamental work in quantum electrodynamics, with deep-ploughing consequences for the physics of elementary particles
	Julian Schwinger	1918–	
	Richard P. Feynman	1918–1988	
1966	Alfred Kastler	1902–1984	for the discovery and development of optical methods for studying Hertzian resonance in atoms
1967	Hans Albrecht Bethe	1906–	for his contributions to the theory of nuclear reactions, especially his discoveries concerning the energy production in stars
1968	Luis W. Alvarez	1911–1988	for his decisive contribution to elementary particle physics, in particular the discovery of a large number of resonance states, made possible through his development of the technique of using hydrogen bubble chamber and data analysis
1969	Murray Gell-Mann	1929–	for his contributions and discoveries concerning the classification of elementary particles and their interactions
1970	Hannes Alvén	1908–	for fundamental work and discoveries in magneto-hydrodynamics with fruitful applications in different parts of plasma physics
	Louis Néel	1904–	for fundamental work and discoveries concerning antiferromagnetism and ferrimagnetism which have led to important applications in solid state physics
1971	Dennis Gabor	1900–1979	for his discovery of the principles of holography

1972	John Bardeen	1908–	for their development of a theory of superconductivity
	Leon N. Cooper	1930–	
	J. Robert Schrieffer	1931–	
1973	Leo Esaki	1925–	for his discovery of tunneling in semiconductors
	Ivar Giaever	1929–	for his discovery of tunneling in superconductors
	Brian D. Josephson	1940–	for his theoretical prediction of the properties of a super-current through a tunnel barrier
1974	Antony Hewish	1924–	for the discovery of pulsars
	Sir Martin Ryle	1918–1984	for his pioneering work in radioastronomy
1975	Aage Bohr	1922–	for the discovery of the connection between collective motion and particle motion and the development of the theory of the structure of the atomic nucleus based on this connection
	Ben Mottelson	1926–	
	James Rainwater	1917–	
1976	Burton Richter	1931–	for their (independent) discovery of an important fundamental particle.
	Samuel Chao Chung Ting	1936–	
1977	Philip Warren Anderson	1923–	for their fundamental theoretical investigations of the electronic structure of magnetic and disordered systems
	Nevill Francis Mott	1905–	
	John Hasbrouck Van Vleck	1899–1980	
1978	Peter L. Kapitza	1894–1984	for his basic inventions and discoveries in low-temperature physics
	Arno A. Penzias	1926–	for their discovery of cosmic microwave
	Robert Woodrow Wilson	1936–	background radiation
1979	Sheldon Lee Glashow	1932–	for their unified model of the action of the weak and electromagnetic forces and for their prediction of the existence of neutral currents
	Abdus Salam	1926–	
	Steven Weinberg	1933–	
1980	James W. Cronin	1931–	for the discovery of violations of fundamental symmetry principles in the decay of neutral K mesons
	Val L. Fitch	1923–	
1981	Nicolaas Bloembergen	1920–	for their contribution to the development of laser spectroscopy
	Arthur Leonard Schawlow	1921–	
	Kai M. Siegbahn	1918–	for his contribution of high-resolution electron spectroscopy
1982	Kenneth Geddes Wilson	1936–	for his method of analyzing the critical phenomena inherent in the changes of matter under the influence of pressure and temperature
1983	Subrehmanyan Chandrasekhar	1910–	for his theoretical studies of the structure and evolution of stars
	William A. Fowler	1911–	for his studies of the formation of the chemical elements in the universe
1984	Carlo Rubbia	1934–	for their decisive contributions to the large project, which led to the discovery of the field particles W and Z, communicators of the weak interaction
	Simon van der Meer	1925–	
1985	Klaus von Klitzing	1943–	for his discovery of the quantized Hall resistance
1986	Ernst Ruska	1906–	for his invention of the electron microscope;
	Gerd Binnig	1947–	for their invention of the scanning-tunneling
	Heinrich Rohrer	1933–	electron microscope
1987	Karl Alex Müller	1927–	for their discovery of a new
	J. George Bednorz	1950–	class of superconductors.

ANSWERS TO ODD-NUMBERED EXERCISES AND PROBLEMS

CHAPTER 1

3. (a) 186 mi. (b) 3.0×10^8 mm. **5.** (a) 10^9. (b) 10^{-4}. (c) 9.14×10^5. **7.** 32.2 km. **9.** 0.0195 km^3 **11.** (a) 251 ft^2. (b) 23.3 m^2. (c) 3060 ft^3. (d) 86.6 m^3. **13.** 844 km. **15.** (a) 11.3 m^2/L. (b) 1.13×10^4 m^{-1}. (c) 2.17×10^{-3} gal/ft^2. **17.** (a) $d_{sun}/d_{moon} = 400$. (b) $V_{sun}/V_{moon} = 6.4 \times 10^7$. (c) 3.5×10^3 km. **19.** (a) 31 m. (b) 21 m. (c) Lake Ontario. **21.** 52.6 min; 5.2% **23.** 720 days. **25.** (a) Yes. (b) 8.6 s. **27.** 0.12 AU/min. **29.** 3.3 ft. **31.** 2.03 h. **33.** 5.9×10^{26}. **35.** 9.01×10^{49}. **37.** 1.32×10^9 kg. **39.** 3.80 mg/s. **41.** (a) 0.282 nm. (b) 0.416 nm.

CHAPTER 2

1. 3 μm. **3.** 0.41 s. **5.** 2 cm/y. **7.** 6.71×10^8 mi/h, 9.84×10^8 ft/s, 1.00 ly/y. **9.** (a) 5.7 ft/s. (b) 7.0 ft/s. **11.** (a) 0, -2, 0, 12 m. (b) $+12$ m. (c) $+7$ m/s. **13.** 48 km/h. **15.** 2.56 m/s^2. **19.** (a) The signs of v and a are: OA: $+, -$; AB: 0, 0; BC: $+, +$; CD: $+, 0$. (b) No. **21.** (e) Situations a, b, and d. **23.** (a) 80 m/s. (b) 110 m/s. (c) 20 m/s^2. **25.** 100 m. **27.** (a) 28.5 cm/s. (b) 18.0 cm/s. (c) 40.5 cm/s. (d) 28.1 m/s. (e) 30.4 cm/s. **29.** (a) L/T^2, ft/s^2; L/T^3, ft/s^3. (b) 2.0 s. (c) 24 ft. (d) -16 ft. (e) 3, 0, -9, -24 ft/s. (f) 0, -6, -12, -18 ft/s^2. **31.** 2.78 m/s^2 (9.4 ft/s^2). **33.** 0.10 m. **35.** (a) 1.6 m/s. (b) 17.6 m/s. **37.** 20.7g. **39.** (a) 25g. (b) 400 m. **41.** Both accelerations are equal to 0.278 m/s^2. **43.** (a) 5.0 m/s^2. (b) 4.0 s. (c) 6.0 s. (d) 90 m. **45.** (a) 5.00 m/s. (b) 1.67 m/s^2. (c) 7.50 m. **47.** (a) 0.75 s. (b) -20 ft/s^2. **49.** (a) 73.5 mi/h. (b) 49.0 s. (c) 26.1 mi/h. **51.** (a) 34.7 ft. (b) 41.6 s. **53.** Yes. **55.** (a) 29.4 m. (b) 2.45 s. **57.** 183 m/s. **61.** (a) 27.4 m/s. (b) 5.35 m/s. (c) 1.50 m. **63.** 3.96 m/s. **67.** 858 m/s^2, up. **69.** 1.23, 4.90, 19.6, 30.6 cm. **71.** (a) 8.85 m/s. (b) 1.00 m. **73.** 22.2 cm and 88.9 cm below the nozzle. **75.** (a) 1.54 s. (b) 27.1 m/s. **77.** 96g. **79.** (a) 17 s. (b) 290m. **81.** (a) 76 m above the ground. (b) 4.1 s. **83.** (a) 5.45 s. (b) 41.4 m/s. **85.** 0.75 in. (1.91 cm).

CHAPTER 3

1. The displacements should be (a) parallel; (b) antiparallel; (c) perpendicular. **3.** (a) 370 m, 35.8° north of east. (b) Displacement magnitude = 370 m; distance walked = 425 m. **5.** 81 km, 40° north of east. **7.** (a) 38 units at 320°. (b) 130 units at 1.2°. (c) 62 units at 130°. **9.** $a_x = -2.51$; $a_y = -6.86$. **11.** $r_x = 13.0$ m; $r_y = 7.50$ m. **13.** (a) 14.1

cm, 45° left of straight down. (b) 20.0 cm, vertically up. (c) Zero. **15.** 4.74 km. **17.** 168 cm, 32.5° above the floor. **19.** (a) 21.0 ft. (b) No; yes; yes. (c) $10\mathbf{i} + 12\mathbf{j} + 14\mathbf{k}$ if the x, y, z axes are parallel to the respective dimensions. (d) 26.1 ft. **21.** $r_x = 11.8$, $r_y = -5.8$, $r_z = -2.8$. **23.** (a) $8\mathbf{i} + 2\mathbf{j}$, 8.2, 14°. (b) $2\mathbf{i} - 6\mathbf{j}$, 6.32, 288°. **25.** 5.00, 323°; 10.0, 53.1°; 11.2, 26.6°; 11.2, 79.7°. (e) 11.2, 260°. The angles are with the $+x$ axis. **27.** (a) $r_x = 1.59$, $r_y = 12.1$. (b) 12.2. (c) 82.5°. **29.** 3390 ft, horizontally. **31.** (a) -2.83, -2.83; $+5.00$, 0; 3.00, 5.20; m. (b) 5.17, 2.37; m. (c) 5.69 m, 24.6° north of east. (d) 5.69 m, 24.6° south of west. **35.** (b) 11,200 km. **37.** (a) 10.0 m, north. (b) 7.50 m, south. **43.** (a) 30.0. (b) A vector of magnitude 52.0, pointing as given by the right-hand rule. **45.** (a) 0. (b) -16. (c) -9. **49.** (a) $11\mathbf{i} + 5\mathbf{j} - 7\mathbf{k}$; 14.0. (b) 120°. **53.** (a) 2.97. (b) $1.51\mathbf{i} + 2.67\mathbf{j} - 1.36\mathbf{k}$. (c) 48.5°. **55.** 70.5°.

CHAPTER 4

1. (a) 671 mi, 63.4° south of east. (b) 298 mi/h, 63.4° south of east. (c) 400 mi/h. **3.** (a) 6.79 km/h. (b) 6.96°. **5.** (a) $6\mathbf{i} - 106\mathbf{j}$, m. (b) $19\mathbf{i} - 224\mathbf{j}$, m/s. (c) $24\mathbf{i} - 336\mathbf{j}$, m/s^2. **7.** 60.0°. **9.** (a) $-18\mathbf{i}$, m/s^2. (b) 0.750 s. (c) Never. (d) 2.19 s. **11.** 5.44×10^{-15} m. **13.** (a) 62.5 ms. (b) 1600 ft/s. **15.** (a) 3.03 s. (b) 758 m. (c) 29.7 m/s. **17.** (a) 16.3 m/s, 23.0° above the horizontal. (b) 27.5 m/s, 56.9° below the horizontal. **19.** (a) 32.4 m. (b) 37.7 m. **23.** (a) 87.5 ft. (b) 93.8 ft/s. (c) 169 ft. **25.** 33 cm. **31.** His record would have been longer by about 1 cm. **33.** (a) 256 m/s. (b) 44.9 s. **35.** 78.3 ft/s (24.0 m/s) at 64.8° (64.9°) above the horizontal. **37.** (b) 6.29°; 83.7°. **39.** 30 m. **41.** The third. **43.** (a) 202 m/s. (b) 806 m. (c) 161 m/s; 171 m/s. **45.** 78.5°. **47.** 24.9 m. **49.** Between the angles 31° and 63° above the horizontal. **51.** (a) 309 s. (b) 105 km. (c) 139 km. **53.** 4.00 m/s^2. **55.** (a) 22.3 m. (b) 15.2 s. **57.** (a) 4.6×10^{12} m. (b) 2.8 d. **59.** (a) 7.35 km. (b) 79.7 km/h. **61.** (a) 18.5 m/s. (b) 35.6. **63.** 2.58 cm/s^2. **65.** (a) 6.46 m. (b) 5.44×10^{-9} m/s^2. **67.** (a) 92. (b) 9.6. (c) $92 = (9.6)^2$. **69.** (a) 5 km/h, upstream. (b) 1 km/h, downstream. **71.** Wind blows from the west at 55.1 mi/h. **73.** $0.830c$. **75.** The motorist is approaching the police car along the line of sight at 100 km/h. **77.** 0.964 m/s, vertically up. **79.** 80.1 m/s. **81.** 185 km/h, 22° south of west. **83.** 92.6° from the direction of motion of the car. **85.** (a) Head the boat 25.3° upstream. (b) 12.6 min.

CHAPTER 5

1. Both speeds are about 10 m/s. **3.** (a) 10 lb. (b) 10 lb. **5.** (a) 247 m/s^2. (b) 1.24×10^5 N. **7.** 16.4 N. **9.** (a) 2.2×10^{-3} N. (b) 3.7×10^{-3} N. **11.** (a) 0.619 m/s^2. (b) 0.130 m/s^2. (c) 2.60 m. **13.** (a) 1.10 N. **15.** (a) 22 N, 2.2 kg. (b) 1100 N, 110 kg. (c) 1.6×10^4 N, 1.6×10^3 kg. **17.** (a) 10 N; 2.04 kg. (b) Zero; 2.04 kg. **19.** 1.47 N. **21.** 8.0 cm/s^2. **23.** (a) 13.1 N. (b) 3.27 m/s^2. **25.** (a) 41.7 N. (b) 72.1 N. (c) 4.90 m/s^2. **27.** (a) 0.022 m/s^2. (b) 82,100 km. (c) 1900 m/s. **29.** (a) 1.09×10^{-15} N. (b) 8.93×10^{-30} N. **31.** (a) 1300 lb (5400 N). (b) 5.5 s. (c) 50 ft (15 m). (d) 2.7 s. **33.** 10.2 m/s^2. **35.** (a) 110 lb. (b) 110 lb. **37.** (a) $30\mathbf{i} + 30\mathbf{j}$, m/s. (b) $450\mathbf{i} + 225\mathbf{j}$, m. **39.** (a) 65 N. (b) 49.1 N. **41.** (a) 218 kN. (b) 50.4 kN. **43.** (a) 0.970 m/s^2. (b) $T_1 = 11.6$ N, $T_2 = 34.9$ N. **45.** (a) 620 N. (b) 580 N. **47.** 94 tons **49.** (a) 3260 N. (b) 2720 kg. (c) 1.20 m/s^2. **51.** (a) 1.23; 2.46; 3.69; 4.92, N. (b) 6.15 N. (c) 0.25 N. **53.** Lower the object with an acceleration $\geq$ 4.2 ft/s^2. **55.** (a) 0.50 slugs (7.3 kg). (b) 20 lb (89 N). **57.** (a) 4.90 m/s^2. (b) 1.96 m/s^2. (c) 118 N. **59.** (a) 1.39×10^8 N, 6.94×10^6 N. (b) 4.17 y, 4.25 y. **61.** (a) 2.18 m/s^2. (b) 116 N. (c) 21.0 m/s^2. **63.** (a) 4.55 m/s^2. (b) 2.59 m/s^2. **65.** (a) 9.40 km. (b) 60.8 km. **67.** (a) 466 N. (b) 527 N.

CHAPTER 6

1. (a) 198 N. (b) 123 N. **3.** (b) 72.7 N. (c) 26.9 N. **5.** 50.4 N. **7.** 8820 N. **9.** 9.31 m/s^2. **11.** (a) 20.5 lb (90.2 N). (b) 16.0 lb (70.4 N). (c) 2.88 ft/s^2 (0.882 m/s^2). **13.** (a) No. (b) A 12-lb force to the left and a 5.0-lb force up. **15.** 7.51 m/s^2. **17.** $\mu_s = 0.577$. $\mu_k = 0.541$. **19.** (a) 0.11 m/s^2, 0.23 m/s^2. (b) 0.041, 0.028. **21.** 36 m. **23.** (a) 68 lb. (b) 4.2 ft/s^2. **25.** (a) 66 N. (b) 2.29 m/s^2. **27.** (a) $\mu_k mg(\sin\theta - \mu_k \cos\theta)$. (b) $\theta_0 = \tan^{-1}\mu_s$. **29.** (a) 3030 N (681 lb). (b) 914 N (206 lb). **31.** (a) 8.57 N. (b) 46.2 N. (c) 38.6 N. **33.** (a) Zero. (b) 13.7 ft/s^2 down the incline. (c) 5.10 ft/s^2 down the incline. **35.** (a) 13.1 N. (b) 1.63 m/s^2. **37.** (a) 1.05 N, in tension. (b) 3.62 m/s^2. (c) The rod is under compression. **39.** (a) 6.1 m/s^2. (b) 0.98 m/s^2. **41.** $g(\sin\theta - \sqrt{2}\mu_k \cos\theta)$. **43.** 9.90 s. **45.** 3.75. **47.** 12.1 cm. **49.** 9.68. **51.** (a) 726 lb (3210 N). (b) Yes. **53.** 0.080. **55.** (a) 3.14 ft/s (0.963 m/s). (b) 0.0205 (0.0206). **57.** (a) 2.20×10^6 m/s. (b) 9.13×10^{22} m/s^2, toward the nucleus. (c) 8.32×10^{-8} N. **59.** 178 km/h. **61.** (a) 30 cm/s. (b) 180 cm/s^2, radially inward. (c) 3.6×10^{-3} N. (d) 0.37. **63.** (a) 275 N. (b) 877 N. **65.** (a) 175 lb. (b) 50 lb. **67.** (a) At the bottom of the circle. (b) 30.9 ft/s. **69.** (a) 9.46 m/s. (b) 19.6 m. **71.** 13°. **73.** (b) 8.74 N. (c) 37.9 N. (d) 6.45 m/s.

CHAPTER 7

1. (a) 564 J. (b) -525 J. (c) Zero. (d) Zero. (e) 39 J. **3.** (a) 200 N; 625 m; -1.25×10^5 J. (b) 500 N; 250 m; -1.25×10^5 J.

5. (a) $-3Mgd/4$. (b) Mgd. **7.** 411 ft·lb (550 J). **9.** (a) 30.1 J. (b) -30.1 J. (c) Zero. (d) 0.225. **11.** $+25$ J.
15. (a) 0.0433 J. (b) 0.130 J. **17.** 1.21×10^6 m/s. **19.** (a) 2.9×10^7 m/s. (b) 1.3 MeV. **21.** -20.2 ft·lb (-24.4 J).
23. $AB : +$; $BC : 0$; $CD : -$; $DE : +$. **25.** 7.94 J. **27.** 102 ft (31.2 m); no. **29.** Man, 2.4 m/s; boy, 4.8 m/s.
31. -11.7 J. **33.** (a) 8800 ft·lb (1.164×10^4 J). (b) -8000 ft·lb (-1.058×10^4 J). (c) 800 ft·lb (1060 J). (d) 17.9 ft/s (5.43
m/s). **35.** (a) 53.8 m/s. (b) 52.4 m/s. (c) 75.7 m, below. **37.** 0.191. **39.** (a) 7.11×10^{10} J. (b) 1185 MW. **41.** 693 W
(0.935 hp). **43.** 227 hp (1.69×10^5 W.) **45.** 5.5×10^6 N. **47.** 24.2 W. **49.** (a) 3.35 kW. (b) 1.80 kW. **51.** 880 MW.
53. (a) 487 W. (b) -487 W. **55.** (a) 2.2×10^5 ft·lb (3.0×10^5 J). (b) 13 hp (9.7 kW). (c) 26 hp (19 kW). **57.** (a) 0.83,
2.5, 4.2 J. (b) 5.0 W. **59.** (a) 0.771 mi. (b) 71.3 kW. **61.** 0.24 hp. **63.** (a) 2.1×10^6 kg. (b) $\sqrt{100 + 1.46t}$ m/s.
(c) $(1.5 \times 10^6)/\sqrt{100 + 1.46t}$ N. (d) 6.69 km. **65.** (a) 112 rev/min. (b) 18.6 W. **67.** (a) 79.1 keV. (b) 3.11 MeV. (c) 10.9
MeV.

CHAPTER 8

1. 88.9 N/cm. **3.** (a) 7.94×10^4 J. (b) 1.84 W. **5.** (a) 2695 MJ. (b) 2695 MW. (c) 236 M\$. **7.** (a) 196 J. (b) 167 J.
(c) 12.9 m/s. **9.** (a) 4.00×10^4 J. (b) 4.00×10^4 J. **11.** (a) v_0. (b) $\sqrt{v_0^2 + gh}$. (c) $\sqrt{v_0^2 + 2gh}$. **13.** 56.0 m/s.
15. (a) 19.3 J, assuming $U(0) = 0$. (b) 22.5 m/s. **17.** (a) 7.84 N/cm. (b) 62.7 J. (b) 80.0 cm. **19.** (a) mgL (b) $\sqrt{2gL}$.
21. (a) 2.82 m/s. (b) 2.71 m/s. **23.** 6.51 m. **25.** (a) 34.7 cm. (b) 169 cm/s. **27.** (a) 39.2 ft/s. (b) 4.24 in. **29.** (a) 4.98
m/s. (b) 79.4°. (c) 64.0 J. **31.** $mgL/32$. **33.** 10 cm. **35.** 1.25 cm. **37.** It comes close to breaking, but does not break.
39. (a) $2\sqrt{gL}$. (b) $5mg$. (c) 71°. **41.** (a) 39.5 cm. (b) 3.65 cm. **45.** (c) -1.2×10^{-19} J. (d) 2.2×10^{-19} J. (e) $\simeq 1 \times 10^{-9}$ N.
47. 39.3 kW. **49.** 738 m. **51.** (a) 1.53 MJ. (b) -5.10 MJ. (c) 1.02 MJ. (d) 62.6 m/s. **53.** (a) -0.900 J. (b) -0.459 J.
(c) 1.04 m/s. **55.** (a) 17.8 ft/s. (b) 17.7 ft. **57.** 4.3 m. **59.** (a) 31.0 J. (b) 5.33 m/s. (c) Conservative. **63.** In the
center of the flat part. **65.** (a) 24 ft/s. (b) 3.0 ft. (c) 9.0 ft. (d) 48.8 ft. **67.** (a) 9.00×10^{15} J. (b) 285,000 years.
69. (a) 5.01×10^{16} J. (b) 557 g. **71.** 1.10 kg. **73.** 266 times the equatorial circumference of the earth. **75.** 191.
77. (a) 2.46×10^{15} s^{-1}. (b) Emitted.

CHAPTER 9

1. 4640 km. **3.** $x_{cm} = 1.07$ m; $y_{cm} = 1.33$ m. **5.** $x_{cm} = -0.25$ m; $y_{cm} = 0$. **7.** Within the iron 2.68 cm from the
iron-aluminum boundary and 1.40 cm below the top surface. **9.** $x_{cm} = y_{cm} = 20$ cm; $z_{cm} = 16$ cm. **11.** 6.19 m.
13. (a) Down; $mv/(m + M)$. (b) Balloon again stationary. **15.** (a) L. (b) Zero. **17.** 58 kg. **19.** (a) Midway between
them. (b) It moves 1.0 mm toward the heavier body. (c) $0.0016g$, down. **21.** 7560 slug·ft/s, in the direction of motion.
23. (a) 3.80 kg·m/s. (b) 1810 J. (c) 8.44 mm/s. **25.** 1.92×10^{-21} kg·m/s. **27.** (a) 4.00×10^4 kg·m/s. (b) West.
29. (a) 6.94×10^4 J. (b) 3.56×10^4 kg·m/s, 38.7° south of east. **31.** $0.707c$. **33.** 0.57 m/s ($= 1.9$ ft/s), toward the center
of mass. **35.** It increases by 4.42 m/s. **37.** (a) $10\mathbf{i} + 15\mathbf{j}$, m/s. (b) 500 J is lost. **39.** (a) 5.89 ms. (b) 935 N. **41.** 14
m/s, 135° from either other piece. **43.** (a) 0.476 m/s. (b) 0.452 m/s. **45.** (a) 721 m/s. (b) 937 m/s. **47.** 216. **49.** One
chunk comes to rest. The other moves ahead with a speed of 4.0 m/s. **51.** (a) 0.482 g. (b) 7230 N. **53.** (a) $+216$ J.
(b) 1180 N. (c) $+432$ J. **55.** 103 m/s. **57.** (a) 2.72. (b) 7.39. **59.** 0.0022. **61.** 6.1 s. **63.** Fast barge: 46 N more; slow
barge: no change. **65.** (a) 328 ft (100 m). (b) 0.503. **67.** (a) 862 N. (b) 2.42 m/s. **69.** (a) 1600 J. (b) 12.9 N. (c) 0.509
m/s^2. (d) Braking force $= 12.9$ N.

CHAPTER 10

1. 399 N·s. **3.** 2.5 m/s. **5.** $2\mu u$. **7.** 53.5 m/s. **9.** (a) 2.30 N·s. (b) 2.30 N·s. (c) 2300 N. (d) 57.5 J. **11.** 38 km/s.
13. 1.5 N. **15.** (a) 3.68 m/s. (b) 1.33 N·s. (c) 175 N. **17.** (a) 1.95×10^5 kg·m/s for each direction of thrust. (b) Backward:
$+66.1$ MJ; forward: -50.9 MJ; sideways: $+7.6$ MJ. **19.** 41.7 cm/s. **21.** (a) 1.90 m/s, to the right. (b) Yes. **23.** 4.2
m/s. **25.** 7.8% **27.** (a) 74.4 m/s. (b) 81.5 m/s; 84.1 m/s. **29.** (a) 2.47 m/s. (b) 1.23 m/s. **33.** $m_1/3$. **35.** (a) 23.3 ft
above top of shaft. (b) 23.3 ft below top of shaft. **37.** 3.0 m/s. **39.** 310 m/s. **41.** (a) 2.66 m/s. (b) 1420 m/s. **43.** 185
tons. **45.** $\frac{1}{6}mv^2$. **47.** 12.9 tons. **49.** 25 cm. **51.** $\sqrt{2E\frac{M+m}{mM}}$. **53.** 4.15×10^5 m/s. **55.** (a) $(-1.00\mathbf{i} + 0.668\mathbf{j}) \times 10^{-19}$
kg·m/s. (b) 1.19×10^{-12} J. **57.** 38.3 mi/h; 63.4° south of west. **59.** (a) 1010 m/s; 9.46° clockwise from the $+x$ axis.
(b) 3.23 MJ. **63.** $d\left(\frac{m_1}{m_1 + m_2}\right)^2$. **65.** $\mathbf{v}_2$ and $\mathbf{v}_3$ will be at 30° to $\mathbf{v}_0$ and will have a magnitude of 6.9 m/s. $\mathbf{v}_1$ will be in
the opposite direction to $\mathbf{v}_0$ and will have magnitude 2.0 m/s. **67.** 8.12 MeV. **69.** (a) -1.21 MeV. (b) 2.05 MeV.

CHAPTER 11

1. (a) 1.50 rad. (b) 85.9°. (c) 1.49 m. (d) 37.8 m. **3.** (a) $a + 3bt^2 - 4ct^3$. (b) $6bt - 12ct^2$. **5.** (a) $\omega(2) = 4.00$
rad/s; $\omega(4) = 28.0$ rad/s. (b) 12.0 rad/s^2. (c) $\alpha(2) = 6.00$ rad/s^2; $\alpha(4) = 18.0$ rad/s^2. **7.** (a) $\omega_0 + at^4 - bt^3$.

(b) $\omega_0 t + at^5/5 - bt^4/4$. **9**. 14.4 **11**. (a) 4.00 m/s. (b) No. **13**. (a) -156 rev/min^2. (b) 19.5. **15**. 200 rev/min.
17. (a) 2.00 rad/s^2. (b) 5.00 rad/s. (c) 10.0 rad/s. (d) 75.0 rad. **19**. (a) 2.0 rev/s. (b) 3.8 s. **21**. (a) 13.5 s. (b) 27.0 rad/s. **23**. (a) 1.04 rev/s^2. (b) 4.8 s. (c) 9.6 s. (d) 48. **25**. (b) -2.30×10^{-9} rad/s^2. (c) $\simeq 4610$. (d) 24 ms. **27**. (a) 3.49 rad/s. (b) 20.6 in./s. (c) 10.0 in./s. **29**. (a) 20.9 rad/s. (b) 12.5 m/s. (c) 800 rev/min^2 (d) 600. **31**. (a) 2.0×10^{-7} rad/s. (b) 30 km/s. (c) 6.0 mm/s^2, toward the sun. **33**. (a) 2.50×10^{-3} rad/s. (2.50×10^{-3} rad/s). (b) 66.0 ft/s^2 (20.2 m/s^2). (c) Zero (Zero). **35**. (a) 7.27×10^{-5} rad/s. (b) 355 m/s. (c) 7.27×10^{-5} rad/s, 463 m/s. **37**. (a) 40.2 cm/s^2. (b) 2.36×10^3 m/s^2. (c) 83.2 m. **39**. (a) 3.8×10^3 rad/s. (b) 190 m/s. **41**. 16.4 s. **43**. (a) 22.4 rad/s. (b) 5.38 km. (c) 1.15 h. **45**. 12.4 kg·m^2. **47**. First cylinder: 1079 J; second cylinder: 9708 J. **49**. (a) 1305 g·cm^2. (b) 545 g·cm^2. (c) 1850 g·cm^2. **51**. (a) $5ml^2 + \frac{8}{3}Ml^2$. (b) $(\frac{5}{2}m + \frac{4}{3}M)l^2\omega^2$. **53**. (a) 9.71×10^{37} kg·m^2. (b) 2.57×10^{29} J. (c) 1.94×10^9 y. **55**. $M(a^2 + b^2)/3$. **57**. 0.0971 kg·m^2. **59**. 4.59 N·m. **61**. 137 N·m. **63**. (a) $r_2 F_2 \sin\theta_2 - r_1 F_1 \sin\theta_1$. (b) 3.85 N·m, into page. **65**. (a) 28.2 rad/s^2. (b) 338 N·m. **67**. (a) 0.791 kg·m^2. (b) 1.79×10^{-2} N·m. **69**. (a) 420 rad/s^2. (b) 495 rad/s. **71**. Small sphere: (a) 0.689 N·m. (b) 3.05 N. Large sphere: (a) 9.84 N·m. (b) 11.5 N. **73**. 1.73 m/s^2; 6.92 m/s^2. **75**. (a) 1.88×10^{12} J/s. (b) -2.67×10^{-22} rad/s^2. (c) 4.06×10^9 N. **77**. 140 cm/s. **79**. (a) 19.8 kJ. (b) 1.32 kW.
81. 5.42 m/s. **83**. (a) 8.3×10^{28} N·m. (b) 2.6×10^{29} J. (c) 3.0×10^{21} kW. **85**. $\sqrt{\frac{2gh}{1+2M/3m+I/Mr^2}}$. **87**. (a) 42.1 km/h. (b) 3.09 rad/s^2. (c) 7.57 kW.

CHAPTER 12

1. 1.00 **3**. -3.15 J. **5**. 1/50. **7**. (a) 44.8 ft·lb. (b) 11.2 ft; no. **9**. (a) $mR^2/2$. (b) A solid circular cylinder. **11**. (a) $mg(R - r)$; 2/7. (b) $17mg/7$. **13**. (a) 62.6 rad/s. (b) 4.01 m. **15**. (a) 12.5 cm/s^2. (b) 4.38 s. (c) 27.2 rev/s. **19**. 1.25×10^8 kg·m^2/s. **21**. (a) 12 kg·m^2/s, out of page. (b) 3.0 N·m, out of page. **23**. (a) Zero. (b) 8.0i + 8.0k, N·m. **27**. (a) 600k, kg·m^2/s. (b) 720k, kg·m^2/s. **29**. (a) 0.528 kg·m^2/s. (b) 4200 rev/min. **31**. (a) -4.11 m/s^2. (b) -16.4 rad/s^2. (c) -2.54 N·m. **33**. (a) $\frac{1}{2}mgt^2 v_0 \cos\theta_0$. (b) $mgt v_0 \cos\theta_0$. (c) $mgt v_0 \cos\theta_0$. **35**. $\omega_0 \left(\frac{R_1 I_2}{R_2 I_1} + \frac{R_2}{R_1}\right)^{-1}$. **37**. (a) $14ml^2$. (b) $4ml^2\omega$. (c) $14ml^2\omega$. **39**. (a) 1/3. (b) 1/9. **41**. (a) 3.6 rev/s. (b) 3.0. **43**. (a) 267 rev/min. (b) 2/3. **45**. 3.0 min. **47**. 12.7 rad/s, clockwise viewed from above. **49**. (a) $7ML^2/12$. (b) $7ML^2\omega_0/12$; down. (c) $14\omega_0/5$. (d) $21ML^2\omega_0^2/40$. **51**. (a) $(I\omega_0 - mRv)/(I + mR^2)$. (b) No. **53**. The day would be longer by about 0.5 s. **55**. (a) 0.148 rad/s. (b) 0.0123. (c) 181°. **57**. 0.429 rev/min.

CHAPTER 13

1. (a) Two. (b) Seven. **3**. 8.67 N **5**. (a) $-27i + 2j$. (b) 176° counterclockwise from $+x$ axis. **7**. 7920 N. **9**. 0.29. **11**. (a) 625 lb (2780 N). (b) 875 lb (3890 N). **13**. Left pedestal: 1160 N (tension); right pedestal: 1740 N (compression). **15**. Three-fourths the length of the beam from the man at the end. **17**. $F_{muscle} = 1920$ N, up, three times the weight; $F_{bone} = 2560$ N, down, four times the weight. **19**. (a) Bottom: $2W$; sides: W. (b) $\sqrt{2}W$. **21**. (a) 29°. (b) $T_1 = 49$ lb, $T_2 = 28$ lb, $T_3 = 57$ lb. **25**. (a) 408 N. (b) $F_h = 245$ N (right); $F_v = 163$ N (up). **27**. (a) 5.0 N. (b) 30 N. (c) 1.33 m. **31**. 30 lb (130 N) vertical; 18 lb (80 N) horizontal, oppositely directed. **33**. 2.20 m. **35**. (a) 1.50 m. (b) $F_h = 433$ N; $F_v = 250$ N. **37**. (a) $a_1 = L/2$, $a_2 = L/4$, $a_3 = L/6$, $a_4 = L/8$; $h = 25L/24$. (b) $b_1 = L/2$, $b_2 = 5L/8$; $h = 9L/8$. (c) $c_1 = 2L/3$, $c_2 = L/2$; $h = 7L/6$. **39**. (a) 47 lb. (b) $F_A = 120$ lb, $F_E = 72$ lb. **41**. (i) $\mu < L/2h$. (ii) $\mu > L/2h$. **43**. (a) Slides at 31°. (b) Tips at 34°. **45**. 7.5×10^{10} N/m^2. **47**. (a) 0.803 mm. (b) 2.42 cm. **49**. (a) 6.50×10^6 N/m^2. (b) 1.15 mm. **51**. (a) 1.43×10^9 N. (b) 75.

CHAPTER 14

1. (a) 0.500 s. (b) 2.00 Hz. (c) 18.0 cm. **3**. (a) 245 N/m. (b) 0.284 s. **5**. 708 N/m. **7**. $\nu > 498$ Hz. **9**. (a) 100 N/m. (b) 0.449 s. **11**. (a) 6.28×10^5 rad/s. (b) 1.59 mm. **13**. (a) 1.00 mm. (b) 0.754 m/s. (c) 568 m/s^2. **15**. (a) 1.29×10^5 N/m (8880 lb/ft). (b) 2.68 Hz (2.68 Hz). **17**. (a) 4.0 s. (b) $\pi/2$ rad/s. (c) 0.37 cm. (d) $(0.37$ cm)$\cos\frac{\pi}{2}t$. (e) $(-0.58$ cm/s$)\sin\frac{\pi}{2}t$. (f) 0.58 cm/s. (g) 0.91 cm/s^2. (h) Zero. (i) 0.58 cm/s. **19**. %16 m/s (23.6 ft/s). **21**. 1.60 kg. **23**. (a) 1.59 Hz. (b) 100 cm/s, 0. (c) 1000 cm/s^2, 10 cm. (d) $-(10$ N/m$)x$. **25**. 21.6 cm. **27**. (a) 25 cm. (b) 2.2 Hz. **29**. (a) 0.500 m. (b) -0.251 m; 3.06 m/s. **31**. (a) $0.183A$. (b) Same direction. **37**. $k_1 = (n+1)k/n$; $k_2 = (n+1)k$. **39**. 3.74×10^{-2} J. **41**. (a) 133 N/m. (b) 0.621 s. (c) 1.61 Hz. (d) 0.050 m. (e) 0.166 J. (f) 0.505 m/s. **43**. 7.25×10^6 N/m. (b) 49,400. **45**. (a) $mv/(m+M)$. (b) $mv/\sqrt{k(m+M)}$. **47**. (a) 80 sin 2000t, N, with t in s. (b) 3.14 ms. (c) 4.00 m/s. (d) 0.0800 J. **51**. (a) 1/16 J. (b) 1/32 J. **53**. 0.0792 kg·m^2. **55**. 12.0 s. **57**. 8.28 s. **59**. 9.47 m/s^2. **61**. 8.77 s. **63**. $2\pi\sqrt{(l^2 + 12d^2)/12gd}$. **65**. $2\pi\sqrt{(R^2 + 2d^2)/2gd}$. **67**. (a) 0.205 kg·m^2. (b) 47.7 cm. (c) 1.50 s.

71. (a) $\pi\sqrt{(L^2 + 12x^2)/3gx}$. **73.** (a) 0.35 Hz. (b) 0.39 Hz. (c) Zero. **77.** $(r/R)\sqrt{k/m}$. **81.** 0.386. **83.** (a) 0.102 kg/s. (b) 0.137 J. **85.** $k = 490$ N/cm; $b = 1090$ kg/s. **87.** 1.88 in.

CHAPTER 15

1. Only 0.20 revolutions **3.** (a) 3×10^{-8} N. (b) 2×10^{-6} N. (c) 1×10^{-6} N. (d) No, the force due to the planet is 30-70 times greater. **5.** 2.60×10^5 km. **7.** 0.10%. **9.** (a) $G(M_1 + M_2)m/a^2$. (b) GM_1m/b^2. (c) Zero. **13.** (a) 9.83 m/s^2. (b) 9.84 m/s^2. (c) 9.79 m/s^2. **15.** $\frac{GmM}{d^2}\left[1 - \frac{1}{8(1-R/2d)^2}\right]$. **19.** (a) 1.62 m/s^2. (b) 4.93 s. **21.** 0.245 oz. **23.** (a) 3.4×10^{-2} m/s^2. (b) 5.9×10^{-3} m/s^2. (c) 1.39×10^{-10} m/s^2. **25.** 9.78 m/s^2. **27.** (b) 1.9 h. **31.** (b) 250 m; 50 m. (c) 293 m; 7.0 m. **35.** (a) 3.30×10^{-7} N, toward the 60-kg sphere. (b) -2.22×10^{-7} J. **37.** 220 km/s. **41.** $2 \times 10^{-6}R_s$. **43.** $-Gm(M_e/R + M_m/r)$. **45.** (a) 1700 m/s. (b) 250 km. (c) 1400 m/s. **47.** (a) 2.24×10^{-7} rad/s. (b) 89.5 km/s. **49.** (a) $GMmx/(x^2 + R^2)^{3/2}$. (b) $v^2 = 2GM\left(\frac{1}{R} - \frac{1}{\sqrt{R^2+x^2}}\right)$. **53.** 1.87 y. **55.** 5.93×10^{24} kg. **57.** 0.354 lunar months. **59.** 3.5 y. **61.** 5.01×10^7 m or 7.2 solar radii. **67.** 16.6 km/s. **69.** 81.3°. **71.** (a) 2.8 y. (b) 1.0×10^{-4}. **73.** (a) 1/2. (b) 1/2. (c) B, by 8.4×10^7 ft·lb. **75.** (a) 54.1 km/s. (b) 1.00 km/s. **77.** (a) No. (b) Same. (c) Yes. **79.** (a) $r^{3/2}$. (b) r^{-1}. (c) $r^{1/2}$. (d) $r^{-1/2}$. **81.** (a) 7.54 km/s. (b) 97.3 min. (c) 405 km; 7.68 km/s; 92.3 min. (d) 3.18×10^{-3} N. (e) If the earth and satellite are considered isolated, yes. **83.** 1.64 cm/s, to the west.

CHAPTER 16

1. 1000 kg/m^3. **3.** 1.1×10^5 Pa or 1.1 atm. **5.** 2.88×10^4 N. **7.** 6.0 lb/in.^{2}. **9.** (b) 6000 lb. **11.** 230 lb/in.^{2} (1.6×10^6 Pa). **13.** 132 km. **15.** (a) 600; 30; 80 tons. (b) No: even though the previous answers are changed to 3100, 280, 760 tons, atmospheric pressure acts on each side of the concrete walls and cancels out. **17.** 7.23×10^5 N. **19.** $\rho gA(h_2 - h_1)^2/4$. **23.** 1.74 km. **25.** (a) 2.22. (b) 3.55. **27.** 10.3 m. **29.** (a) fA/a. (b) 20 lb. **31.** 1070 g. **33.** 1.47 g/cm^3. **35.** 600 kg/m^3. **37.** (a) 670 kg/m^3. (b) 740 kg/m^3. **39.** 395 kg. **41.** (a) 1.22 kg. (b) 1340 kg/m^3. **43.** 0.126 m^3. **45.** Five. **47.** (a) 1.80 m^3. (b) 4.75 m^3. **49.** 2.79 g/cm^3. **53.** (a) 15 gal/min. (b) 95.2%. **55.** 65.8 W. **57.** (a) 2.5 m/s. (b) 2.57×10^5 Pa. **59.** 254 lb/in.^{2}. **61.** 1.41×10^5 N·m **63.** (a) 5.5×10^{-2} ft^3/s. (b) 3.0 ft. **67.** 862 kg, assuming $\rho_{air} = 1.2$ kg/m^3. **69.** 63.3 m/s. **71.** (a) 560 Pa. (b) 5040 N. **73.** 39.9 m/s. **75.** 31 m/s. **77.** (b) $H - h$. (c) $H/2$. **79.** (a) $\sqrt{2g(h_2 + d)}$. (b) $p_{atm} - \rho g(h_2 + d + h_1)$. (c) 10.3 m. **81.** 2.4 ft^3/s. **83.** 5 min 42 s.

CHAPTER 17

1. (a) 75 Hz. (b) 13.3 ms. **3.** (a) 1.67 s. (b) 2.00 m/s. (c) 3.33 m. (d) 15 cm. **5.** (a) 0.68 s. (b) 1.47 Hz. (c) 2.06 m/s. **7.** (c) 200 cm/s; $-x$ direction. **9.** (d) 5 cm/s. **11.** (a) 2.0 mm; 95.5 Hz; 30 mm/s; 31.4 cm. (b) 1.2 m/s. **13.** (a) 6.0 cm. (b) 100 cm. (c) 2.0 Hz. (d) 200 cm/s. (e) $-x$ direction. (f) 75 cm/s. (g) -2.03 cm. **15.** (a) 6.00 m/s. (b) $y = 0.30\sin(\pi x/12 + 50\pi t)$, with x, y in cm and t in s. **19.** 129 m/s (430 ft/s). **21.** 135 N. **25.** (a) 15 m/s. (b) 0.036 N. **27.** $y = (0.12 \text{ mm})\cos[(141 \text{ m}^{-1})x - (628 \text{ s}^{-1})t]$. **29.** (a) 0.64 Hz. (b) 130 cm. (c) $y = 5\sin(0.1x - 4.0t)$, with x, y in cm and t in s. (d) 6400 dyne. **31.** 2.63 m from the end of the wire from which the later disturbance originates. **35.** 44.9 Hz. **37.** $1.41y_m$. **39.** $\lambda = 2\sqrt{d^2 + 4(H + h)^2} - 2\sqrt{d^2 + 4H^2}$. **43.** (a) 2.0 Hz; 200 cm; 400 cm/s. (b) $x = 50$ cm, 150 cm, 250 cm, etc. (c) $x = 0$, 100 cm, 200 cm, etc. **47.** (a) 833 Hz. (b) 0.418 m. **49.** (a) 66.1 m/s. (b) 26.4 Hz. **51.** (a) 144 m/s. (b) 3.0 m; 1.5 m. (c) 48.0 Hz; 96.0 Hz. **53.** (a) 0.25 cm; 120 cm/s. (b) 3.0 cm. (c) zero. **55.** (a) 105 Hz. (b) 157.5 m/s. **59.** (a) 50 Hz. (b) $y = 0.50\sin\pi[(x \pm 100t)]$, with x in m, y in cm, t in s. **61.** (a) 8.1 lb (36 N). **63.** (a) 323 Hz. (b) six.

CHAPTER 18

1. 104 kHz. **3.** (a) 2.29, 0.229, 22.9 kHz. (b) 1.14, 0.114, 11.4 kHz. **5.** (a) ~ 3%. **7.** The radio listener; the travel time factor is 52.5. **9.** 7.87×10^{10} Pa. **11.** If only the length is uncertain, it must be known to within 10^{-4} cm. If only the time is imprecise, the uncertainty must be no more than one part in 10^5. **13.** 43.6 m. **15.** 40.7 m. **17.** 236°. **19.** (a) $343(1+2n)$ Hz, with n being an integer from 0 to 28. (b) $686n$ Hz, with n being an integer from 1 to 29. **21.** (a) 0.0796 W/m^2. (b) 0.0127 W/m^2. **23.** 35.7 nm. **27.** (a) 39.7 μW/m^2. (b) 166 nm. (c) 0.923 Pa. **29.** (a) 57.8. (b) 2.99×10^{-4}. **31.** 17.5 cm. **33.** (a) $I \propto r^{-1}$. (b) $A \propto r^{-1/2}$. **35.** 5000; 71; 71. **37.** 2 cm. **39.** (a) 77.8 dB. (b) 4.48 nW. **41.** At ± 0, 0.572, 1.14, 1.72, 2.29 m from the midpoint. **43.** 64.7 Hz, 129 Hz. **45.** 346 m. **49.** (a) 405 m/s. (b) 596 N. **51.** 5.00 cm from one end. **53.** Water filled to a height of 7/8, 5/8, 3/8, 1/8 meter. **55.** (a) 1130, 1500, and

1880 Hz. **57**. 12 ft. **59**. (a) Node. (c) 22 s. **61**. 387 Hz. **63**. 505, 507, 508 Hz or 501, 503, 508 Hz. **65**. (a) Ten. (b) Four. **67**. Zero. **69**. 17.5 kHz. **71**. (a) 526 Hz. (b) 555 Hz. **73**. 1100 m/s. **75**. 30°. **77**. 33.0 km. **81**. 155 Hz. **83**. (a) 467 Hz. (b) 494 Hz. **85**. (a) 0.90 ft. (b) 1580. (c) 1080 ft/s. (d) 0.68 ft. **87**. 41 kHz. **89**. (a) 2.0 kHz. (b) 2.0 kHz. **91**. (a) 485.8 Hz. (b) 500.0 Hz. (c) 486.2 Hz. (d) 500.0 Hz. **93**. 1.2×10^6 m/s, receding. **95**. No.

CHAPTER 19

1. 2.71 K. **3**. 31.5 **5**. 291.1 K. **7**. 348 K. **9**. (a) $10.000\,\Omega$, $4.124 \times 10^{-3}/°C$, $-1.779 \times 10^{-6}/°C^2$. **11**. 1/2. **13**. 7 Celsius. **15**. No; 310 K = 98.6°F. **17**. (a) $-40°$. (b) $575°$. (c) Celsius and Kelvin cannot give the same reading. **19**. 1.1 cm. **21**. 2.731 cm. **23**. (a) $12.8 \times 10^{-6}/°F$. (b) 0.169 in. **25**. $31 \times 10^{-6}/°C$. **27**. 0.32 cm^2 (0.050 in^2). **29**. 29 cm^3. **31**. 0.432 cm^3. **33**. $-157°C$. **41**. 9.07 s, the clock running slow. **43**. (a) $+1.06 \times 10^{-5}$. (b) -1.54×10^{-5}. **45**. 7.5 cm. **47**. Increases by 0.10 mm. **49**. 2.64×10^8 Pa. **51**. 66.4°C. **53**. 0.266 mm.

CHAPTER 20

1. 333 J. **3**. 35.7 m^3. **5**. 6.66×10^{12} J. **7**. 42.7 kJ. **9**. 250 g. **11**. 706 ft/s. **13**. 1.17°C. **15**. 2 min 41 s. **17**. (a) 20300 cal. (b) 1110 cal. (c) 873°C. **19**. 45.4°C. **21**. (a) 18,700. (b) 10.4 h. **23**. 2.8 days. **25**. 81.8 cal. **27**. 13.5°C. **29**. (a) 5.25°C, no ice remaining. (b) 0°C, 59.7 g of ice left. **31**. 8.72 g. **33**. $C \propto T^{-2/3}$. **35**. A: 120 J; B: 75 J; C: 30 J. **37**. (a) -200 J. (b) -293 J. (c) -93 J. **39**. -1.19 cal. **41**. 7.95 kcal. **43**. (a) 0.360 mg/s. (b) 0.814 J/s. (c) -0.694 J/s. **45**. (a) 1.2 W/m·K; 0.70 Btu/ft·°F·h. (b) 0.0297 ft^2·°F·h/Btu. **47**. 1660 J/s. **51**. Arrangement (b). **53**. (a) 1.95 MW. (b) 0.218 kW. **55**. 2.02×10^7 J. **57**. 0.394 cm/h. **59**. Cu-Al, 84.3°C; Al-Brass, 57.6°C.

CHAPTER 21

1. (a) 0.0127. (b) 7.65×10^{21}. **3**. 6560. **5**. Number of molecules in the ink $\approx 3 \times 10^{16}$. Number of people $\approx 5 \times 10^{20}$. Statement is wrong, by a factor of about 20,000. **7**. (a) 5.47×10^{-8} mol. (b) 3.29×10^{16}. **9**. (a) 106. (b) 0.892 m^3. **11**. 27.0 lb/in^2. **13**. (a) 2.50×10^{25}. (b) 1.20 kg. **15**. 4240 ft·lb (5700 N·m). **17**. 1/5. **19**. 103 cm^3. **21**. 198°F. **23**. 2.0×10^5 Pa. **25**. 183 m/s. **27**. 9.53×10^6 m/s. **29**. 307°C. **31**. 1.89×10^4 dyne/cm^2. **33**. (a) 0.0353 eV; 0.0482 eV. (b) 3400 J; 4650 J. **35**. 9.1×10^{-6}. **37**. (a) 6.75×10^{-20} J. (b) 10.7. **39**. 0.323 nm. **41**. 15 cm. **43**. (a) 3.26×10^{10}. (b) 173 m. **45**. (a) 22.5 L. (b) 2.25. (c) 8.4×10^{-5} cm. (d) Same as (c). **49**. (a) 3.18 cm/s. (b) 3.37 cm/s. (c) 4.00 cm/s. **51**. (a) $\bar{v}$, v_{rms}, v_p. (b) v_p, $\bar{v}$, v_{rms}. **53**. (a) 1.0×10^4 K; 1.6×10^5 K. (b) 440 K; 7000 K. **55**. 4.71 **57**. (a) $2g/3v_0$. (b) $N/3$. (c) $1.22v_0$. (d) $1.31v_0$. **59**. $RT \ln(V_f/V_i)$. **61**. (a) 15.9 J. (b) 34.4 J/mol·K. (c) 26.2 J/mol·K. **63**. $(n_1C_1 + n_2C_2 + n_3C_3)/(n_1 + n_2 + n_3)$ **65**. 7940 J. **67**. (a) 6980 J. (b) 4990 J. (c) 1990 J. (d) 2990 J. **71**. (a) 2.5 atm, 340 K. (b) 0.40 L. **73**. 1500 N·m$^{2.2}$. **75**. 1.40. **79**. (a) $p_0/3$. (b) Polyatomic (ideal). (c) $K_f/K_i = 1.44$. **81**. Q, W and ΔU are all greatest for (a) and least for (c). **83**. (a) In joules, in the order Q, ΔU, W: 1→2: 3740, 3740, 0; 2→3: 0, -1810, 1810; 3→1: -3220, -1930, -1290; Cycle: 520, 0, 520. (b) $V_2 = 0.0246$ m^3; $p_2 = 2$ atm; $V_3 = 0.0373$ m^3; $p_3 = 1$ atm.

CHAPTER 22

1. (a) 30.8 %. (b) 16 kcal. **3**. 25.4%. **5**. (a) 7200 J. (b) 960 J. (c) 13%. **7**. (a) 49.4 kcal. (b) 31.0 kJ. **9**. (a) 0.071 J. (b) 0.50 J. (c) 2.0 J. (d) 5.0 J. **11**. 99.999947%. **13**. 75. **17**. (a) 94 J. (b) 230 J. **19**. 13 J. **21**. (a) 1.115 kcal/s. (b) 0.995 kcal/s. **23**. $e = 1/(K + 1)$. **27**. $[1 - (T_2/T_1)]/[1 - (T_4/T_3)]$. **29**. (a) 78%. (b) 81.5 kg/s. **31**. 0.139. **33**. (a) $+0.602$ cal/K. (b) -0.602 cal/K. **35**. -0.30 cal/g·K. **37**. 8.785×10^{-3} cal/K. **39**. 4450 cal. **41**. (a) 1.95 J/K. (b) 0.650 J/K. (c) 0.217 J/K. (d) 0.072 J/K. **43**. (a) 57°C. (b) -5.27 cal/K. (c) $+5.95$ cal/K. (d) $+0.68$ cal/K. **45**. $+0.153$ cal/K. **47**. (a) 320 K. (b) Zero. (c) $+0.41$ cal/K. **49**. (a) $p_1/3$; $p_1/3^{1.4}$; $T_1/3^{0.4}$. (b) In the order W, Q, ΔU, ΔS: 1 → 2: $1.10RT_1$, $1.10RT_1$, 0, $1.10R$; 2 → 3: 0, $-0.889RT_1$, $-0.889RT_1$, $-1.10R$; 3 → 1: $-0.889RT_1$, 0, $0.889RT_1$, 0. **51**. (a) -225 cal/K. (b) $+225$ cal/K. **53**. (a) $3p_0V$. (b) $6RT_0$; $(3/2)R \ln 2$. (c) Both are zero.

CHAPTER 23

1. (a) 8.99×10^9 N. (b) 8990 N. **3**. 1.39 m. **5**. (a) 4.9×10^{-7} kg. (b) 7.1×10^{-11} C. **7**. $3F/8$. **9**. (a) $q_1 = 9q_2$. (b) $q_1 = -25q_2$. **11**. 1.2×10^{-5} C and 3.8×10^{-5} C. **13**. 14 cm from the positive charge, 24 cm from the negative charge. **15**. (a) A charge $-4q/9$ must be located on the line segment joining the two positive charges, a distance $L/3$ from the $+q$ charge. **17**. (a) $Q = -2\sqrt{2}q$. (b) No. **19**. (b) $\pm 2.4 \times 10^{-8}$ C. **21**. (a) $\frac{L}{2}\left(1 + \frac{1}{4\pi\epsilon_0}\frac{qQ}{Wh^2}\right)$. (b) $\sqrt{\frac{3}{4\pi\epsilon_0}\frac{qQ}{W}}$ **23**. 3.78 N.

25. 1.93 MC.   **27**. (a) 8.99×10^{-19} N. (b) 625.   **29**. 11.9 cm.   **31**. 1.34 days.   **33**. 1.3×10^7 C.   **35**. 1.52×10^8 N.
37. (a) Positron. (b) Electron.   **39**. (a) 510 N. (b) 7.7×10^{28} m/s^2.   **41**. (a) $(Gh/2\pi c^3)^{1/2}$. (b) 1.61×10^{-35} m.

CHAPTER 24

1. 3.51×10^{15} m/s^2.   **3**. (a) 4.8×10^{-13} N. (b) 4.8×10^{-13} N.   **5**. (a) 1.5×10^3 N/C. (b) 2.4×10^{-16} N, up. (c) 1.6×10^{-26} N.
(d) 1.5×10^{10}.   **7**. (a) 6.4×10^{-18} N. (b) ~ 20 N/C.   **9**. To the right in the figure.   **15**. $+1.00$ μC.   **17**. (a) The larger
charge produces a field of 1.3×10^5 N/C at the site of the smaller; the smaller produces a field of 5.3×10^4 N/C at the site
of the larger. (b) 1.1×10^{-2} N.   **19**. 2.86×10^{21} V/m; radially outward.   **21**. (a) $0.172a$ to the right of the $+2q$ charge.
23. 50 cm from q_1 and 100 cm from q_2.   **25**. (a) 4.8×10^{-8} N/C. (b) 7.7×10^{-27} N.   **27**. 1.02×10^5 N/C, upward.
29. 6.63×10^{-15} N.   **31**. $\frac{1}{4\pi\epsilon_0}\frac{p}{r^3}$; antiparallel to p.   **41**. (a) 0.104 μC. (b) 1.31×10^{17}. (c) 4.96×10^{-6}.   **43**. (a) 2.46×10^{17}
m/s^2. (b) 0.122 ns. (c) 1.83 mm.   **45**. (a) 7.12 cm. (b) 28.5 ns. (c) 11.2%.   **47**. $5e$.   **49**. (a) 0.245 N, 11.3° clockwise
from the +x axis. (b) $x = 108$ m; $y = -21.6$ m.   **51**. (a) $-\mathbf{j}(2.1 \times 10^{15}$ m/s$^2)$. (b) $\mathbf{i}(1.5 \times 10^5$ m/s$) - \mathbf{j}(2.8 \times 10^8$ m/s$)$.
53. (a) $2\pi\sqrt{\frac{l}{|g-qE/m|}}$. (b) $2\pi\sqrt{\frac{l}{g+qE/m}}$.   **55**. (a) 9.30×10^{-15} C·m. (b) 2.05×10^{-11} J.   **57**. $2pE\cos\theta_0$.

CHAPTER 25

1. (a) 693 kg/s. (b) 693 kg/s. (c) 347 kg/s. (d) 347 kg/s. (e) 575 kg/s.   **3**. (a) Zero. (b) -3.92 N·m^2/C. (c) Zero. (d) Zero
for each field.   **7**. 2.03×10^5 N·m^2/C.   **9**. $q/6\epsilon_0$.   **11**. 3.54 μC.   **13**. Through each of the three faces meeting at q: zero;
through each of the three other faces: $q/24\epsilon_0$.   **15**. (a) 36.6 μC. (b) 4.14×10^6 N·m^2/C.   **17**. (a) $-Q$. (b) $-Q$. (c) $-(Q+q)$.
(d) Yes.   **21**. (a) 0.317 μC. (b) 0.143 μC.   **23**. (a) $E = \lambda/2\pi\epsilon_0 r$. (b) Zero.   **25**. 3.61 nC.   **27**. (a) 2.3×10^6 N/C,
radially out. (b) 4.5×10^5 N/C, radially in.   **29**. (b) $\rho R^2/2\epsilon_0 r$.   **31**. (a) 5.30×10^7 N/C. (b) 59.9 N/C.   **33**. 5.0 nC/m^2.
35. (a) Zero. (b) $E = \sigma/\epsilon_0$, to the left. (c) Zero.   **37**. 4.94×10^{-22} C/m^2.   **39**. -7.51nC.   **41**. (a) Zero. (b) 2.9×10^4
N/C. (c) 200 N/C.   **43**. (a) Zero. (b) $q_a/4\pi\epsilon_0 r^2$. (c) $(q_a + q_b)/4\pi\epsilon_0 r^2$.   **45**. (a) $E = q/4\pi\epsilon_0 r^2$, radially outward. (b) Same
as (a). (c) No. (d) Yes, charges are induced on the surfaces. (e) yes. (f) No. (g) No.   **47**. (a) $E = (q/4\pi\epsilon_0 a^3)r$. (b) $E =$
$q/4\pi\epsilon_0 r^2$. (c) Zero. (d) Zero. (e) Inner, $-q$; Outer, zero.   **49**. (a) 4.0×10^6 N/C. (b) Zero.

CHAPTER 26

1. 1.2 GeV.   **3**. (a) 3.00×10^{10} J. (b) 7.75 km/s. (c) 9.0×10^4 kg (99 tons).   **5**. (a) 2.46 V. (b) 2.46 V. (c) Zero.
7. 2.90 kV.   **9**. 8.8 mm.   **11**. (a) $-qr^2/(8\pi\epsilon_0 R^3)$. (b) $q/(8\pi\epsilon_0 R)$.   **15**. (a) 4.5 m. (b) No.   **17**. -1.1 nC.   **21**. No.
23. 637 mV.   **25**. (a) 0.54 mm. (b) 790 V.   **27**. (a) 38.0 s. (b) 280 days.   **29**. (a) $Q/4\pi\epsilon_0 r$. (b) $\frac{\rho}{3\epsilon_0}\left(\frac{3}{2}r_2^2 - \frac{1}{2}r^2 - \frac{r_1^3}{r}\right)$;
$\rho = Q/\frac{4\pi}{3}(r_2^3 - r_1^3)$. (c) $\frac{\rho}{2\epsilon_0}(r_2^2 - r_1^2)$, with ρ as in (b).   **33**. 667 N/C.   **35**. $\frac{1}{2\pi\epsilon_0}\frac{p\cos\theta}{r^3}$.   **37**. 39.0 V/m, toward $x = 0$.
39. (a) $\frac{\lambda}{4\pi\epsilon_0}\ln\left(\frac{L+y}{y}\right)$. (b) $\frac{\lambda}{4\pi\epsilon_0}\frac{L}{y(L+y)}$. (c) Zero.   **41**. (a) $qd/2\pi\epsilon_0 a(a + d)$.   **43**. (a) -6.8MV. (b) -1.9 J.   **45**. (a) 0.484
eV. (b) Zero.   **47**. -1.24×10^{-6} J.   **49**. (a) -7.8×10^5 V; $+0.6 \times 10^5$ V. (b) 2.5 J. (c) Work is converted into potential
energy.   **51**. (a) 27 V. (b) -27eV. (c) 13.6 eV. (d) 13.6 eV.   **53**. 2.48 km/s.   **55**. 5.85×10^7 m/s.   **57**. $qQ/4\pi\epsilon_0 K$.
59. 0.32 km/s.   **61**. 1.6×10^{-9} m.   **65**. (a) $V_1 = V_2$. (b) $q_1 = q/3$, $q_2 = 2q/3$.   **67**. (a) -0.12 V. (b) 1.8×10^{-8} N/C,
radially inward.   **69**. (a) 12,000 N/C. (b) 1800 V. (c) 5.77 cm.   **71**. (a) 0.11 mC; 1.1 μC.   **73**. (a) 3.2×10^{-13} J or 2.0
MeV. (b) 1.6×10^{-13} J or 1.0 MeV. (c) The proton.   **75**. (a) $r > 9.0$ cm. (b) 2.7 kW. (c) 20 μC/m^2.

CHAPTER 27

1. 7.5 pC.   **3**. 3.0 mC.   **5**. (a) 144 pF. (b) 17.3 nC.   **7**. 0.551 pF.   **9**. 4.2×10^{-7} C.   **15**. $5.05\pi\epsilon_0 R$.   **19**. 3.16 μF.
21. 315 mC.   **23**. (a) 10.0 μF. (b) $q_4 = 0.800$ mC; $q_6 = 1.20$ mC. (c) 200 V.   **25**. (a) $q_2 = q_8 = 0.48$ mC; $V_2 = 240$ V;
$V_8 = 60$ V. (b) $q_2 = 0.19$ mC; $q_8 = 0.77$ mC; $V_2 = V_8 = 96$ V. (c) $q_2 = q_8 = 0$; $V_2 = V_8 = 0$.   **27**. (a) $+7.9 \times 10^{-4}$
C. (b) $+79$ V.   **29**. (a) Five in series. (b) Three arrays as in (a) in parallel. There are other possibilities.   **31**. 43 pF.
33. $q_1 = \frac{C_1 C_2 + C_1 C_3}{C_1 C_2 + C_1 C_3 + C_2 C_3}C_1 V_0$; $q_2 = q_3 = \frac{C_2 C_3}{C_1 C_2 + C_1 C_3 + C_2 C_3}C_1 V_0$.   **35**. First case: 50 V; second case: zero.   **37**. (a)
3.05 MJ. (b) 0.847 kW·h.   **39**. (a) 0.204 μJ. (b) No.   **41**. 0.27 J.   **43**. 4.88%.   **45**. 27.8¢.   **47**. (a) 2.0 J.   **49**. (a) $2V$.
(b) $U_i = \epsilon_0 A V^2/2d$; $U_f = 2U_i$. (c) $\epsilon_0 A V^2/2d$.   **51**. (a) $e^2/8\pi\epsilon_0 R$. (b) 1.41 fm.   **53**. Pyrex.   **55**. (a) 6.20 cm. (b) 280
pF.   **57**. 0.63 m^2.   **59**. (a) 2.85 m^3. (b) 10.1×10^3.   **61**. (a) $\epsilon_0 A/(d - b)$. (b) $d/(d - b)$. (c) $-q^2 b/2\epsilon_0 A$; sucked in.
65. $\frac{\epsilon_0 A}{4d}\left(\kappa_1 + \frac{2\kappa_2\kappa_3}{\kappa_2+\kappa_3}\right)$.   **67**. (a) 13.4 pF. (b) 1.15 nC. (c) 1.13×10^4 N/C. (d) 4.33×10^3 N/C.   **69**. (a) 89 pF. (b) 120 pF.
(c) 11 nC; 11 nC. (d) 10 KV/m. (e) 2.1 kV/m. (f) 88 V. (g) 0.17 μJ.

CHAPTER 28

1. (a) 1200 C. (b) 7.5×10^{21}. **3.** 5.6 ms. **5.** (a) 6.4 A/m^2, north. **7.** 0.380 mm. **9.** 0.67 A, toward the negative terminal. **11.** (a) 0.654 μA/m^2. (b) 83.4 MA. **13.** 13.4 min. **15.** (a) $J_0A/3$. (b) $2J_0A/3$. **17.** 1.96×10^{-8} $\Omega\cdot$m. **19.** 2.40 Ω. **21.** 2.0×10^6 $(\Omega\cdot$m$)^{-1}$. **23.** 57.2°C. **25.** (a) 0.384 mV. (b) Negative. (c) 3 min

58 s. **27.** 54 Ω. **29.** 2.87 mm. **31.** (a) 2.39, iron being larger. (b) No. **33.** New length = $1.369L$; new area = $0.730A$. **35.** (a) 6.00 mA. (b) 1.59×10^{-8} V. (c) 21.2 pΩ. **37.** 8.21×10^{-4} $\Omega\cdot$m. **39.** (a) 5.32×10^5 A/m^2 for copper; 3.27×10^5 A/m^2 for aluminum. (b) 1.01 kg/m for copper; 0.495 kg/m for aluminum. **41.** 0.40 Ω. **43.** (a) $R = \rho L/\pi ab$. **45.** 14.0 kC. **47.** 11.1 Ω. **49.** (a) 1.0 kW. (b) 25 cents. **51.** (a) 28.8 Ω. (b) 2.60×10^{19} s^{-1}. **53.** (a) 1.74 A. (b) 2.15 MA/m^2. (c) 36.3 mV/m. (d) 2.09 W. **55.** (a) 5.85 m. (b) 10.4 m. **57.** (a) $4.46 for a 31-day month. (b) 144 Ω. (c) 0.833 A. **59.** (a) 9.4×10^{13} s^{-1}. (b) 240 W. **61.** 710 cal/g. **63.** (a) 8.6%. (b) Smaller. **65.** (a) 28 min. (b) 1.6 h.

CHAPTER 29

1. (a) 1.92×10^{-18} J (= 12 eV). (b) 6.53 W. **3.** (a) $320. (b) 9.6¢. **5.** Counterclockwise; #1; B. **9.** (a) 14 V. (b) 100 W. (c) 600 W. (d) 10 V; 100 W. **11.** (a) 50 V. (b) 48 V. (c) B is the negative terminal. **13.** 2.5 V. **15.** (a) 990 Ω. (b) 9.4×10^{-4} W. **17.** 8.0 Ω. **19.** The cable. **21.** (a) 1000 Ω. (b) 300 mV. (c) 0.225%. **23.** (a) 1.32×10^7 A/m^2 in each. (b) $V_A = 8.90$ V; $V_B = 51.1$ V. (c) A: Copper; B: Iron. **25.** Silicon: 85.0 Ω; iron: 915 Ω. **27.** 4.00 Ω. **29.** $i_1 = 50$ mA; $i_2 = 60$ mA; $V_{ab} = 9.0$ V. **31.** (a) 6.67 Ω. (b) 6.67 Ω. (c) Zero. **33.** (a) R_2. (b) R_1. **35.** $3d$. **37.** 7.50 V. **39.** 38 Ω or 260 Ω. **41.** (a) $R = r/2$. (b) $P_{max} = \mathcal{E}^2/2r$. **43.** (a) $2\mathcal{E}/(2r + R)$, series; $2\mathcal{E}/(r + 2R)$, parallel. (b) Series if $r < R$; parallel if $R < r$. **45.** (a) Left branch, 0.67 A, down; center branch, 0.33 A, up; right branch, 0.33 A, up. (b) 3.3 V. **47.** $\mathcal{E}/7R$. **49.** (a) 120 Ω. (b) $i_1 = 50$ mA; $i_2 = i_3 = 20$ mA; $i_4 = 10$ mA. **51.** (a) 19.5 Ω. (b) 82.3 W. (c) 57.6 W. **53.** (a) 2.50 Ω. (b) 3.125 Ω. **55.** (50 kW) $\left(\frac{x}{2000+10x-x^2}\right)^2$, x in cm. **57.** (a) 13.5 kΩ. (b) 1500 Ω. (c) 167 Ω. (d) 1480 Ω. **59.** 0.9% **65.** (a) 2.52 s. (b) 21.6 μC. (c) 3.40 s. **67.** 2.2 ms; no. **69.** (a) 2.17 s. (b) 39.6 mV. **71.** (a) 10^{-3} C. (b) 10^{-3} A. (c) $V_C = 10^3 e^{-t}$, $V_R = 10^3 e^{-t}$, volts. **73.** 0.72 MΩ. **77.** Decreases by 13.3 μC.

CHAPTER 30

1. M/QT; ML^2/QT. **3.** (a) 9.56×10^{-14} N; zero. (b) 27.8°. **5.** (a) 6.2×10^{-14}**k**, N. (b) -6.2×10^{-14}**k**, N. **7.** (a) East. (b) 6.28×10^{14} m/s^2. (c) 2.98 mm. **9.** 2 **11.** (a) 3.75 km/s. **13.** (b) 680 kV/m. **17.** (b) 2.84×10^{-3}. **21.** 1.6×10^{-8} T. **23.** (a) 1.11×10^7 m/s. (b) 0.316 mm. **25.** (a) 2.60×10^6 m/s. (b) 0.217 μs. (c) 0.140 MeV. (d) 70 kV. **29.** (a) $K_p = K_d = K_\alpha/2$. (b) $R_d = R_\alpha = 14$ cm. **33.** (a) 495 mT. (b) 22.7 mA. (c) 8.17 MJ. **35.** (a) 0.357 ns. (b) 0.17 mm. (c) 1.5 mm. **37.** (a) 2.9998×10^8 m/s. **41.** Neutron moves tangent to original path; proton moves in a circular orbit of radius 25 cm. **43.** (b). **45.** 20.1 N. **47.** $(-2.5\mathbf{j} + 0.75\mathbf{k}) \times 10^{-3}$ N. **51.** (a) 3.3×10^9 A. (b) 1.0×10^{17} W. (c) Totally unrealistic. **53.** (a) 0; 1.38 mN; 1.38 mN. **55.** (a) 20 min. (b) 5.94×10^{-2} N·m. **59.** $2\pi aiB\sin\theta$, normal to the plane of the loop (up). **61.** 2.45 A. **63.** 2.08 GA. **67.** (a)(14 N·m)$\left(-\frac{2}{3}\mathbf{i} - \frac{1}{2}\mathbf{j} + \frac{5}{9}\mathbf{k}\right)$. (b) 6.0×10^{-4} J.

CHAPTER 31

1. 7.69 mT. **3.** Yes. **5.** (a) 0.24**i**, nT. (b) Zero. (c) 43.3**k**, pT. (d) -0.144**k**, nT. **7.** (a) 15.6 A. (b) West to east. **9.** (a) 3.2×10^{-16} N, parallel to the current. (b) 3.2×10^{-16} N, radially outward if **v** is parallel to the current. (c) Zero. **13.** (a) 1.03 mT, out of figure. (b) 0.80 mT, out of figure. **15.** $\frac{\mu_0 i\theta}{4\pi}\left(\frac{1}{b} - \frac{1}{a}\right)$, out of page. **25.** 200 μT, into the page. **27.** (a) It is impossible to have other than $B = 0$ midway between them. (b) 30.0 A. **29.** At all points between the wires, on a line parallel to them, at a distance $d/4$ from the wire carrying current i. **35.** $0.338\mu_0 i^2/a$, towards the center of the square. **37.** (b) To the right. **39.** (b) 2.31 km/s. **41.** $+5\mu_0 i_0$. **43.** 4.52×10^{-6} T·m. **47.** (a) $\mu_0 ir/2\pi c^2$. (b) $\mu_0 i/2\pi r$. (c) $\frac{\mu_0 i}{2\pi(a^2-b^2)}\left(\frac{a^2-r^2}{r}\right)$. (d) Zero. **49.** $3i_0/8$, into the page. **53.** 5.71 mT. **55.** 108 m. **61.** 2.72 kA. **63.** 0.471 A·m^2. **65.** $8\mu_0 Ni/5\sqrt{5}R$. **67.** (b) ia^2. **71.** (a) 78.5 μT. (b) 1.08×10^{-6} N·m. **73.** (a) $(\mu_0 i/2R)(1 + 1/\pi)$, out of page. (b) $(\mu_0 i/2\pi R)\sqrt{1 + \pi^2}$, 18° out of page.

CHAPTER 32

1. 57.2 μWb. **3.** 1.52 mV. **5.** (a) 31 mV. (b) Right to left. **7.** $A^2 B^2/R\Delta t$. **9.** (b) 58.0 mA. **11.** 1.2 mV. **13.** 1.15 μWb. **15.** 51.4 mV; clockwise when viewed along the direction of **B**. **17.** (b) No. **19.** (a) 21.74 V. (b) Counterclockwise. **21.** (a) 13 μWb/m. (b) 17%. **23.** 104 mV. **25.** (a) 48.1 mV. (b) 2.67 mA. **27.** $BiLt/m$, away from G. **29.** (a)

85.2 T·m². (b) 56.8 V. (c) Linearly. **31.** Design it so that $Nab = 5m^2/2\pi$. **33.** 268 W. **35.** 1.55×10^{-5} C. **37.** (a) 0.598 μV. (b) Counterclockwise. **39.** (a) $\frac{\mu_0 ia}{2\pi}\ln\left(\frac{2r+b}{2r-b}\right)$. (b) $2\mu_0 iabv/\pi R(4r^2 - b^2)$. **43.** (a) 71.5 μV/m. (b) 143 μV/m. **45.** 0.151 V/m. **49.** (a) 33.8 V/m. (b) 5.94×10^{12} m/s².

CHAPTER 33

1. 0.100 μWb. **3.** (a) 800. (b) 2.53×10^{-4} H. **7.** (a) $\mu_0 i/W$. (b) $\pi\mu_0 R^2/W$. **9.** (a) Decreasing. (b) 0.68 mH. **11.** (a) 0.101 H. (b) 1.31 V. **13.** (a) 1.34 kA. (b) 256 A. (c) 1.92 kA. **15.** 6.91. **17.** 1.54 s. **19.** (a) 8.45 ns. (b) 7.77 mA. **21.** $42 + 20t$, V. **23.** 12.0 A/s. **25.** (a) $i_1 = i_2 = 3.33$ A. (b) $i_1 = 4.55$ A; $i_2 = 2.73$ A. (c) $i_1 = 0$; $i_2 = 1.82$ A. (d) $i_1 = i_2 = 0$. **27.** (a) 1.50 s. **29.** (a) 13.9 H. (b) 120 mA. **31.** $1.23\tau_L$. **33.** (a) 240 W. (b) 150 W. (c) 390 W. **35.** (a) 97.9 H. (b) 0.196 mJ. **37.** (a) $\mu_0 i^2 N^2/8\pi^2 r^2$. (b) 0.306 mJ. (c) 0.306 mJ. **41.** 1.5×10^8 V/m. **43.** $(\mu_0 l/2\pi)\ln(b/a)$. **45.** (a) 1.0 J/m³. (b) 4.8×10^{-15} J/m³. **47.** (a) 1.67 mH. (b) 6.01 mWb. **53.** (a) $\frac{\mu_0 Nl}{2\pi}\ln\left(1 + \frac{b}{a}\right)$. (b) 13.2 μH.

CHAPTER 34

5. (b) In the direction of $\mathbf{r} \times \mathbf{v}$, $\mathbf{r}$ and $\mathbf{v}$ being the position and velocity of any point on the ring. **7.** +3 Wb. **9.** $(\mu_0 iL/\pi)\ln 3$. **11.** 12.7 MWb, outward. **15.** 1660 km. **17.** 61.1 μT; 84.2°. **19.** 20.8 mJ/T. **21.** Yes. **23.** (a) 3.7 K. (b) 1.3 K. **29.** (a) 3.00 μT. (b) 5.63×10^{-10} eV. **31.** (a) 8.9 A·m². (b) 13 N·m.

CHAPTER 35

1. 9.14 nF. **3.** 45.2 mA. **5.** (a) 6.00 μs. (b) 167 kHz. (c) 3.00 μs. **7.** (a) 1.25 kg. (b) 3.72×10^4 N/m. (c) 1.75×10^{-4} m. (d) 3.02×10^{-2} m/s. **9.** 1.59 μF. **15.** (a) 5770 rad/s. (b) 1.09 ms. **17.** (a) 275 Hz. (b) 364 mA. **19.** (a) 0.689 μH. (b) 17.9 pJ. (c) 0.110 μC. **21.** $Q/2$. (b) $0.866I$. **23.** (a) 6.0:1. (b) 36 pF; 0.22 mH. **25.** (a) 0.18 mC. (b) $T/8$. (c) 66.7 W. **27.** ω_0. **29.** Let $T_2 = 0.596$ s be the period of the inductor and 900 μF capacitor and $T_1 = 0.199$ s the period of inductor and 100 μF capacitor. Close S_2, wait $T_2/4$; quickly close S_1, then open S_2; wait $T_1/4$ and then open S_1. **33.** 5.85 μC; 5.52 μC; 1.93 μC.

CHAPTER 36

1. 377 rad/s. **3.** (a) 0.955 A. (b) 0.119 A. **5.** (a) 4.60 kHz. (b) 26.6 pF. (c) $X_L = 2.6$ kΩ; $X_C = 0.65$ kΩ. **7.** (a) 0.65 kHz. (b) 24 Ω. **9.** (a) 39.1 mA. (b) Zero. (c) 33.8 mA. (d) Supplying energy. **11.** (a) 6.73 ms. (b) 2.24 ms. (c) Capacitor. (d) 59.0 μF. **13.** (a) $X_C = 0$; $X_L = 87$ Ω; $Z = 182$ Ω; $I = 198$ mA; $\phi = 28.5°$. **15.** (a) $X_C = 37.9$ Ω; $X_L = 87$ Ω; $Z = 167$ Ω; $I = 216$ mA; $\phi = 17.1°$. **19.** 89.0 Ω. **21.** (a) 224 rad/s. (b) 6.00 A. (c) 228 rad/s; 219 rad/s. (d) 0.039. **23.** (a) 45°. (b) 70.7 Ω. **29.** 141 V. **31.** Zero; Zero; 3.14 W; 1.82 W. **33.** 177 Ω. **35.** 7.61 A. **41.** (a) 117 F. (b) Zero. (c) 90 W; zero. (d) 0°; 90°. (e) 1; 0. **43.** (a) 2.59 A. (b) 38.8 V; 159 V, 224 V, 64.2 V; 75 V. (c) 100 W for R; zero for L and C. **45.** (a) 2.4 V. (b) 3.2 mA; 0.16 A. **47.** (a) 1.9 V; 5.8 W. (b) 19 V; 0.58 kW. (c) 0.19 kV; 58 kW.

CHAPTER 37

3. At $r = 27.5$ mm and $r = 110$ mm. **7.** Change the potential difference between the plates at the rate of 1.0 MV/s. **11.** (a) 0.63 μT. (b) 2.3×10^{12} V/m·s. **13.** (a) 2.0 A. (b) 2.3×10^{11} V/m·s. (c) 0.50 A. (d) 0.63 μT·m. **15.** (a) 7.60 μA. (b) 859 kV·m/s. (c) 3.39 mm. (d) 5.16 pT.

CHAPTER 38

1. (a) 4.71×10^{-3} Hz. (b) 3 min 32 s. **3.** (a) 4.5×10^{24} Hz. (b) 1.0×10^4 km, or 1.6 earth radii. **7.** (a) It would steadily increase. (b) The summed discrepancies between the apparent times of eclipse and those observed from x; the radius of the earth's orbit. **9.** 5.0×10^{-21} H. **11.** 1.07 pT. **17.** 4.8×10^{-29} W/m². **19.** 4.51×10^{-10}. **21.** 89.2 cm. **23.** 1.2 MW/m². **25.** 817 m. **27.** (a) 1.03 kV/m; 3.43 μT. **29.** (a) $\pm EBa^2/\mu_0$ for faces parallel to the xy plane; zero through each of the other four faces. (b) Zero. **31.** (a) 82.6 W/m². (b) 1.66 MW. **33.** (a) 1.53×10^{-17} W/m². (b) 107 nV/m. (c) 0.252 fT. **37.** 10 MPa. **39.** (a) 6.0×10^8 N. (b) $F_{grav} = 3.6 \times 10^{22}$ N. **41.** (a) 100 MHz. (b) 1.0 μT along the z axis. (c) 2.1 m^{-1}; 6.3×10^8 rad/s. (d) 120 W/m². (e) 8.0×10^{-7} N; 4.0×10^{-7} N/m². **43.** 491 nm. **47.** 1.92 mm/s. **49.** (b) 600 nm. **51.** (a) 1.94 V/m. (b) 1.67×10^{-11} N/m². **53.** 1/4. **55.** 20° or 70°. **57.** 19.0 W/m². **59.** (b) 5 sheets.

CHAPTER 39

1. (a) 38.0°. (b) 52.9°. **3.** 1.56. **5.** 1.9×10^8 m/s. **7.** 1.26. **9.** 1.07 m. **11.** 22.0°. **15.** 401 cm beneath the mirror surface. **21.** 33.7°. **23.** (a) 49°. (b) 28°. **25.** 182 cm. **27.** (a) Yes. (b) No. (c) 43°. **29.** (a) 35.6°. (b) 53.1°. **31.** (b) 23.2°. **35.** (a) 53°. (b) Yes. **37.** 55.50° to 55.77°. **39.** 40 cm. **41.** (a) $2v$. (b) v. **43.** 1.50 m. **47.** Three. **49.** New illumination is 10/9 of the old. **51.** (a) 7. (b) 5. (c) 2. **53.** 10.5 cm. **55.** For alternate vertical columns. (a) +, +40, −20, +2, no, yes. (c) Concave, +40, +60, −2, yes, no. (e) Convex, −20, +20, +0.5, no, yes. (g) −20, −, −, +5, +0.8, no, yes. **57.** (b) 0.556 cm/s. (c) 11.25 m/s. (d) 5.14 cm/s. **59.** For alternate vertical columns: (a) −18, no. (c) +71, yes. (e) +30, no. (g) −26, no. **61.** $i = -12.0$ cm. **63.** f_1^2/f_2^2. **65.** 1.85 mm. **67.** (a) Converging. (b) Diverging. (c) Converging. (d) Diverging. **69.** 5.14 cm. **72.** Alternate vertical columns (an X means that the quantity cannot be found from the data given): (a) +, X, X, +20, X, −1, yes, no. (c) Converging, +, X, X, −10, X, no, yes. (e) Converging, +30, −15, +1.5, no, yes. (g) Diverging, −120, −9.2, +0.92, no, yes. (i) Converging, +3.3, X, X, +5, X, −, no. **73.** Upright, virtual, 30 cm to the left of the second lens. **75.** (a) The final image coincides in location with the object. It is real, inverted and $m = -1.0$. **77.** (a) Coincides in location with the original object and is enlarged 5.0 times. (c) Virtual and inverted. **83.** (a) 13.0 cm. (b) 1.23 cm. (c) −3.25. (d) 3.13. (e) −10.2. **89.** (b) 8.4 mm. (c) 2.5 cm.

CHAPTER 40

1. (a) 5.1×10^{14} Hz. (b) 388 nm. (c) 1.98×10^8 m/s. **7.** $(2m+1)\pi$. **9.** 2.25 mm. **11.** 648 nm. **13.** 1.6 mm. **15.** 0.072 mm. **17.** 600 nm. **19.** 6.6 μm. **21.** (a) $2h \sin\theta = m\lambda$ (minimum); $2h \sin\theta = (m + \frac{1}{2})\lambda$ (maximum). **23.** 9.0 μm. **25.** 8.0 μm. **29.** $y = 27 \sin(\omega t + 8.5°)$. **33.** (a) 1.17; 3.00; 7.50 m. (b) No. **37.** $\lambda/5$. **39.** 492 nm. **41.** Bright. **43.** 70.0 nm. **45.** 673 nm. **47.** (a) 169 nm. (c) Blue-violet will be sharply reduced. **49.** 840 nm. **51.** (a) $\lambda/2n_2$. (b) $\lambda/2n_2$. (c) $\lambda/4n_2$. (d) $\lambda/4n_2$. (e) $\lambda/2n_2$. **53.** 141. **55.** (a) 1800 nm. (b) 8. **57.** 1.89 μm. **59.** 1.00025. **61.** (a) 34. (b) 46. **65.** 588 nm. **67.** 1.003.

CHAPTER 41

1. 691 nm. **3.** 60.4 μm. **5.** (a) 2.19×10^{-4} rad. (b) 2.51 mm. **7.** (a) 70.0 cm. (b) 1.03 mm. **9.** 41.2 m from perpendicular to speaker. **11.** 160°. **15.** (d) 53°; 10°; 5.1°. **19.** (a) 1.32×10^{-4} rad. (b) 21.1 m. **21.** 30 m. **23.** (a) 1.1×10^4 km. (b) 11 km. **25.** 52.6 m. **27.** 4.5 m. **29.** 4.73 cm. **31.** (a) 0.347°. (b) 0.973°. **33.** (b) 0.07 mm. (c) Three times the lunar diameter; if water droplets of various sizes are present, a number of rings of different colors are present, giving a whitish appearance. (d) These halos are a diffraction effect, a rainbow being formed by refraction. **35.** 5. **39.** $\lambda D/d$. **41.** (a) 5.05 μm. (b) 20.2 μm. **43.** (a) 3330 nm. (b) ±10.2°; 20.7°; 32.0°; 45.0°; 62.2°. **45.** All wavelengths shorter than 635 nm. **47.** 13,600. **49.** (a) 6.0 μm. (b) 1.5 μm. (c) $m = 0, 1, 2, 3, 5, 6, 7, 9$. **51.** 1100. **61.** (a) 55.6 pm. (b) None. **63.** (a) 23,100. (b) 28.7°. **67.** (a) 2400 nm. (b) 800 nm. (c) $m = 0, 1, 2$. **71.** 2.87°. **73.** 26.2 pm; 39.4 pm. **75.** 39.8 pm. **77.** Yes; $m = 3$ for $\lambda = 0.124$ nm; $m = 4$ for $\lambda = 0.097$ nm. **79.** (a) $a_0/\sqrt{2}$; $a_0/\sqrt{5}$; $a_0/\sqrt{10}$; $a_0/\sqrt{13}$; $a_0/\sqrt{17}$.

CHAPTER 42

1. (a) 3×10^{-18}. (b) 2×10^{-12}. (c) 8.2×10^{-8}. (d) 6.4×10^{-6}. (e) 1.1×10^{-6}. (f) 3.7×10^{-5}. (g) 9.9×10^{-5}. (h) 0.10. **3.** $0.750c$. **5.** $0.991c$. **7.** 54.7 m. **9.** 1.32 m. **11.** 1.53 cm. **13.** (a) 87.4 m. (b) 394 ns. **15.** (a) 2.21×10^{-12}. (b) 5.25 d. **17.** $x' = 138$ km; $t' = -374\,\mu$s. **19.** $t_1' = 0$; $t_2' = -250\,\mu$s. **23.** (a) S' must move towards S, along their common axis, at a speed of $0.480c$. (b) The 'red' flash (suitably Doppler shifted). (c) 4.39 μs. **25.** $0.806c$. **27.** $0.946c$. **29.** $0.588c$, recession. **31.** (a) 34,000 mi/h. (b) 6.4×10^{-10}. **33.** 22.9 MHz. **35.** +2.97 nm. **39.** (a) $0.134c$. (b) 4.65 keV. (c) 1.94%. **41.** (a) 0.9988; 20.6. (b) 0.145; 1.01. (c) 0.073; 1.0027. **43.** (a) 5.71 GeV; 6.65 GeV; 6.58 GeV/c. (b) 3.11 MeV; 3.62 MeV; 3.59 MeV/c. **45.** 18 smu/y. **47.** (a) $0.943c$. (b) $0.866c$. **49.** (a) $0.707c$. (b) $1.414m$. (c) $0.414mc^2$. **51.** (a) The photon. (b) The proton. (c) The proton. (d) The photon. **53.** (a) $207m_e$; the particle is a muon. **55.** (a) $0.948c$. (b) $649m_e$. (c) 226 MeV. (d) 316 MeV/c. **57.** (a) 4.85 mm. (b) 16.0 mm. (c) 0.335 ns; no. **59.** 660 km. **61.** (a) 534. (b) 0.99999825. (c) 2.23 T.

PHOTO CREDITS

40: Courtesy Goodyear Tire and Rubber. **Essay 6** Page E6-1: Courtesy Peter J. Brancazio. **Chapter 17** Opener: Scott Ransom/Taurus Photos. Fig. 14: From *Atlas of Optical Phenomena,* M. Cagnet, et al., Springer-Verlag, Prentice-Hall, 1962. Fig. 18: From PSSC, *Physics,* D. C. Heath, Lexington, Mass., 1960, with permission. Fig. 19: Anthony French, Department of Physics, MIT, Cambridge, Mass. Fig. 20: Courtesy Dr. T. D. Rossing, Northern Illinois University. **Essay 7** Page E7-1: Courtesy John S. Rigden. Fig 1: copyright © Lincoln Center for the Performing Arts/Photo by Ezra Stoller/Esto. Fig. 2: Courtesy Symphony Hall, Boston. Figs. 3 and 4: Courtesy Lincoln Center for the Performing Arts/Photo by Suzanne Faulkner Stevens copyright © 1976. **Chapter 18** Opener: copyright © Eric Kroll/Taurus Photos. Fig. 1:(*a*) Wesmar, Marine Systems Division (*b*) Courtesy Dr. John C. Birnholz and the Rush–Presbyterian-St. Luke's Medical Center, Chicago. Reprinted with permission of *American Scientist.* Copyright © Courtesy John S. Foster, Stanford University. Photo by C. F. Quate. Fig. 8: From "Acoustics of Violins" by G. M. Hutchins, revised for *Scientific American* article, Oct. 1981. Courtesy Carleen Hutchins, Catgut Acoustical Society. Photo by Dr. Raul A. Stetson. Fig. 18: Courtesy U. S. Army Ballistic Research Laboratory. Fig. 20: Courtesy John S. Foster, Stanford University. Photo by C. F. Quate. Fig. 32: Courtesy Mount Wilson and Palomar Observatories. **Chapter 19** Opener: Courtesy NASA. Fig. 8: AP-Wide World Photos. Fig. 23: Palmer/Monkmeyer. **Essay 9** Page E9-1: copyright © Richard Howard. Photo supplied by Bill Sweet, American Institute of Physics. **Chapter 20** Opener: Courtesy Smithsonian Institution. Fig. 9: Courtesy A. A. Bartlett, University of Colorado. Fig. 11: Courtesy Soehngen. **Essay 10** Page E10-1: Courtesy Dr. Jearl Walker. Fig. 2: Courtesy R. H. Kindig, Denver, Colorado. Fig. 5: Jeff Werner. **Chapter 21** Opener: Courtesy Dr. Harold E. Edgerton, MIT. Fig. 4: Courtesy U. S. Department of Energy, Oak Ridge, Tenn. Fig. 16: Courtesy NASA. **Chapter 22** Opener: Atomic Industrial Forum, New York. Fig. 3: "Power from the Sea" by Terry R. Penny and Desikan Bharathan. Copyright © 1987 by *Scientific American, Inc.* All rights reserved. Fig. 9: Courtesy The Bryant Day & Night and Payne Brands of Carrier Corp. **Chapter 23** Opener: The Seattle Times. Fig. 2: Courtesy Xerox Corp. Fig. 3: AT&T Bell Laboratories. Fig. 7: From *Introduction to the Dectection of Nuclear Particles in a Bubble Chamber,* Ealing Press, 1969. Courtesy Lawrence Berkeley Radiation Laboratories, University of California at Berkeley. **Chapter 24** Opener: Courtesy Research-Cottrell. Figs. 13 and 14: *Scientific American,* copyright © 1979 by IBM Corp. Reprinted by permission. **Chapter 25** Opener: The Art Museum, Princeton University. The John B. Putnam, Jr., Memorial Collection. **Chapter 26** Opener: Copyright © Michael Philip Manheim. Fig. 17: Courtesy E. Philip Krider, Institute of Atmospheric Physics, University of Arizona. Fig. 18: Courtesy Ford Motor Co., Technical Photographic Services. Fig. 20: Courtesy Purdue University. Fig. 27: Courtesy NASA. **Chapter 27** Opener: Roger Ressmeyer/Wheeler Pictures. Fig. 1: Courtesy Sprague Electric Co. Fig. 10: Courtesy of the Director of The Royal Institution. **Chapter 28** Opener: Courtesy IBM Corp. Fig. 7: Courtesy Allen-Bradley Co. Fig. 15: Courtesy AT&T. **Chapter 29** Opener: Courtesy Union Carbide. Fig. 13: Courtesy Simpson Electric Co. **Essay 11** Page E11-1: Courtesy A. A. Bartlett, University of Colorado. **Chapter 30** Opener: Courtesy Dornier-System, GmBH, Friedrichschafen. Fig. 1: D. C. Heath and Co., with Education Development Center. Fig. 2: Hugh Rogers/Monkmeyer. Fig. 3: Courtesy Varian Associates. Fig. 5: Courtesy Professor J. le P. Webb, University of Sussex, Brighton, England. Fig. 6: Courtesy Lawrence Berkeley Radiation Laboratory, University of California. Fig. 11: Courtesy Professor J. le P. Webb, University of Sussex, Brighton, England. Fig. 14: Courtesy Dr. L. A. Frank, University of Iowa. Fig. 15: Courtesy A. J. Allen. Figs. 17 and 18: Courtesy Fermi National Accelerator Laboratory. Fig. 19: Courtesy NASA. **Chapter 31** Opener: D. C. Heath and Co. with Education Development Center. **Chapter 32** Opener: Courtesy Alice Halliday. **Essay 12** Page E12-1: Courtesy Brian Holton. Fig. 1: Drawing by nephew of H. Kamerlingh Onnes, from *Superconductivity* by D. Shoenberg, Ph. D, Cambridge University Press, 1952. Fig. 3: From *Physics Today,* March 1986, p. 37. Courtesy New York University Medical Center and Biomagnetic Technologies, Inc. Photo by Hank Morgan. Fig. 5: From *SCC,* 1987, p. 37. Courtesy Japanese National Railways, and Railways Graphics Co., Ltd. Fig. 7: From *Physics Bulletin,* Vol. 38, No. 6, June 1987. Courtesy Birmingham Superconductivity Consortium. Fig. 8: From *Physics Today,* August 1971, p. 32. **Chapter 33** Opener: Courtesy the Director of The Royal Institution. **Essay 13** Page E13-1: Courtesy Dr. Gerard O'Neill. Figs. 1 and 2: *2081: A Hopeful View of the Human Future* by Gerard K. O'Neill , pp. 126 and 127. Simon & Schuster, 1981. **Chapter 34** Opener: Courtesy Dr. James U. Lemke, Recording Physics, Inc., San Diego. Fig. 6: Courtesy Colchester and Essex Museum. Fig. 14: Courtesy R. W. DeBlois. Fig. 16: Courtesy General Electric Medical Systems, Inc. **Essay 14** E14-1: Courtesy Charles Bean. Fig. 1: Courtesy R. Blakemore and N. Blakemore. **Chapter 35** Opener: Courtesy Federal Aviation Administration. **Chapter 36** Opener: Courtesy Con Edison. **Chapter 37** Opener: Anne Manning. Fig. 3: "Ajax Defying the Lightning." From the cover of Electrical Plant, December 1888. By permission, British Post Office. **Chapter 38** Opener: Courtesy American Institute of Physics, Niels Bohr Library. Photo by Roy L. Bishop. Fig. 20: Courtesy NASA. **Essay 15** Page E15-1 and Figs. 1, 2, 4–6, 8, and 10: Courtesy Raymond C. Turner. **Chapter 39** Opener: Piergiorgio Scharandis/Black Star. Fig. 1: PSSC *Physics,* 2nd Ed., copyright © 1965, D. C. Heath & Co. with Education Development Center, Newton, Mass. Fig. 5: Courtesy Olympus Corp. Fig. 6: Courtesy AT&T Bell Labs. Fig. 22: Courtesy Minolta Corp. **Essay 16** Page E16-1: Courtesy Dr. Jearl Walker. **Chapter 40** Opener: Courtesy Raytheon Co. Photo by Eli Brookner. Fig. 6: From *Atlas of Optical Phenomena* by Cagnet, et al., Springer-Verlag, Prentice-Hall, 1962. Fig. 7: Education Development Center, Newton, Mass. **Essay 17** Page E17-1: Courtesy Suzanne R. Nagel. **Chapter 41** Opener: Courtesy European Southern Observatory. Fig. 1: Courtesy *Atlas of Optical Phenomena* by Cagnet, et al., Springer-Verlag. Fig. 2: From Sears, Zemansky and Young, *University Physics,* 5th Ed., copyright © 1976, Addison-Wesley, Reading, Mass. Fig. 3: *Atlas of Optical Phenomena* by Cagnet, et al., Springer-Verlag, 1962. Fig. 7: Courtesy Viking Yacht Co. Inc. Fig. 13: *Atlas of Optical Phenomena* by Cagnet, et al., Springer-Verlag, 1962. Fig. 15: Courtesy Professor L. M. Beidleu, Florida State University, Tallahassee. Figs. 17 and 19: *Atlas of Optical Phenomena* by Cagnet, et al., Springer-Verlag, 1962. **Essay 18** Page E18-1: Courtesy Richard Smith. Fig. 4: Courtesy Edward Wesly. Fig. 6: Tong H. Jeong. **Chapter 42** Opener: Courtesy Alice Halliday. Fig. 1: With permission of The Hebrew University of Jerusalem, Israel. Fig. 4: From *Spacetime Physics* by Edwin Taylor and John Archibald Wheeler. Copyright © 1963, 1966, W. H. Freeman and Company. Reprinted with permission. Fig. 13: Courtesy Los Alamos National Laboratory. Fig. 15: The Washington Post.

INDEX

SELECTED TABLES*

* Selected from the approximately 90 tables in the complete book.

SOME SIGN CONVENTIONS

Description	Convention
Doppler effect Section 18–7 Section 42–12	Associate *toward* with *frequency increase* and choose signs to make this happen; consider the effect of each motion separately.
Thermodynamics Section 20–5	*Heat:* Positive when *added to* the system. *Work:* Positive when *done by* the system.
Potential differences for circuit elements Section 29–3	*Resistor:* Positive when traversed in a direction opposite to the current *emf $\mathcal{E}$:* Positive when traversed through the source from the negative to the positive terminal.
Alternating currents Chapter 36	Phase relations among current i, emf $\mathcal{E}$, capacitance C and inductance L. "ELI the ICE man" $\mathcal{E}$ leads i in inductive circuits. i leads $\mathcal{E}$ in capacitive circuits.
Geometrical optics Chapter 39	For mirrors, single surfaces, and lenses, associate *positive* with *Real, R-side,* and *upRight*. Thus: o Real object f Real focus (Focal point on R-side) i Real image (on R-side) m upRight image r Center of curvature on R-side